# Encyclopedia of Chemical Physics and Physical Chemistry

Volume II: Methods

Online version at www.ecppc.iop.org

# Encyclopedia of Chemical Physics and Physical Chemistry

## Volume II: Methods

Edited by

## John H Moore

*University of Maryland*

and

## Nicholas D Spencer

*ETH-Zürich*

Institute of Physics Publishing
Bristol and Philadelphia

*British Library Cataloguing-in-Publication Data*

A catalogue record for this book is available from the British Library.

ISBN 0 7503 0798 6 (Vol. I)
    0 7503 0799 4 (Vol. II)
    0 7503 0800 1 (Vol. III)
    0 7503 0313 1 (3 Vol. set)

*Library of Congress Cataloging-in-Publication Data are available*

**Online version of encyclopedia at www.ecppc.iop.org**

**Jacket illustration.** Atomic force microscope image showing the spherulitic morphology of a biodegradable polyesterurethane, spin-cast from 1 wt. % solution in 1-methyl-2-pyrrolidone onto a silicon wafer at 2000 rpm. Diameter is 100 microns. (Kirill Feldman, ETH-Zürich.)

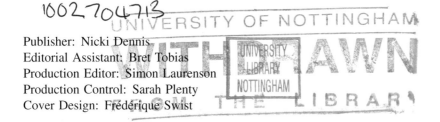

Publisher: Nicki Dennis
Editorial Assistant: Bret Tobias
Production Editor: Simon Laurenson
Production Control: Sarah Plenty
Cover Design: Frédérique Swist

Published by Institute of Physics Publishing, wholly owned by The Institute of Physics, London

Institute of Physics Publishing, Dirac House, Temple Back, Bristol BS1 6BE, UK

US Office: Institute of Physics Publishing, The Public Ledger Building, Suite 1035, 150 South Independence Mall West, Philadelphia, PA 19106, USA

Typeset in TeX using the IOP Bookmaker Macros
Printed in the UK by MPG Books Ltd, Bodmin, Cornwall

# Contents

**VOLUME I: FUNDAMENTALS**

**PART A1**        **MICROSCOPICS**                                                                     **1**

    A1.1  The quantum mechanics of atoms and molecules    3
        *J F Stanton*
    A1.2  Internal molecular motions    49
        *M E Kellman*
    A1.3  Quantum mechanics of condensed phases    79
        *J R Chelikowsky*
    A1.4  The symmetry of molecules    123
        *P Jensen and P R Bunker*
    A1.5  Intermolecular interactions    161
        *A J Thakkar*
    A1.6  Interaction of light with matter: a coherent perspective    187
        *D J Tannor*
    A1.7  Surfaces and interfaces    245
        *J A Yarmoff*

**PART A2**        **THERMODYNAMICS AND STATISTICAL MECHANICS**                                         **281**

    A2.1  Classical thermodynamics    283
        *R L Scott*
    A2.2  Statistical mechanics of weakly interacting systems    331
        *R C Desai*
    A2.3  Statistical mechanics of strongly interacting systems: liquids and solids    379
        *J C Rasaiah*
    A2.4  Fundamentals of electrochemistry    477
        *A Hamnett*
    A2.5  Phase transitions and critical phenomena    523
        *R L Scott*

**PART A3**        **DYNAMICAL PROCESSES**                                                               **571**

    A3.1  Kinetic theory: transport and fluctuations    573
        *J R Dorfman*
    A3.2  Non-equilibrium thermodynamics    597
        *R F Fox*

A3.3    Dynamics in condensed phase (including nucleation)                       617
        *R C Desai*

A3.4    Gas-phase kinetics                                                       653
        *D Luckhaus and M Quack*

A3.5    Ion chemistry                                                            683
        *A A Viggiano and T M Miller*

A3.6    Chemical kinetics in condensed phases                                    711
        *J Schroeder*

A3.7    Molecular reaction dynamics in the gas phase                             745
        *D M Neumark*

A3.8    Molecular reaction dynamics in condensed phases                          759
        *G A Voth*

A3.9    Molecular reaction dynamics: surfaces                                    775
        *G R Darling, S Holloway and C Rettner*

A3.10   Reactions on surfaces: corrosion, growth, etching and catalysis          795
        *T P St Clair and D W Goodman*

A3.11   Quantum mechanics of interacting systems: scattering theory              827
        *G C Schatz*

A3.12   Statistical mechanical description of chemical kinetics: RRKM            865
        *W L Hase*

A3.13   Energy redistribution in reacting systems                                897
        *R Marquardt and M Quack*

A3.14   Nonlinear reactions, feedback and self-organizing reactions              937
        *S K Scott*

## VOLUME II: METHODS

## PART B1            DETERMINING MATERIALS AND MOLECULAR PROPERTIES            961

B1.1    Electronic spectroscopy                                                  963
        *S J Strickler*

B1.2    Vibrational spectroscopy                                                 991
        *C Schmuttenmaer*

B1.3    Raman spectroscopy                                                      1017
        *D J Ulness, J C Kirkwood and A C Albrecht*

B1.4    Microwave and terahertz spectroscopy                                    1063
        *G A Blake*

B1.5    Nonlinear optical spectroscopy of surfaces and interfaces               1089
        *J I Dadap and T F Heinz*

B1.6    Electron-impact spectroscopy                                            1127
        *J H Moore*

B1.7    Mass spectrometry                                                       1147
        *P M Mayer*

B1.8    Diffraction: x-ray, neutron and electron                                1175
        *E Prince*

B1.9    Scattering: light, neutrons, x-rays                                     1197
        *B S Hsiao and B Chu*

B1.10    Coincidence techniques                                                          1227
         *M A Coplan*

B1.11    NMR of liquids                                                                  1245
         *O W Howarth*

B1.12    NMR of solids                                                                   1273
         *R Dupree and M E Smith*

B1.13    NMR relaxation rates                                                            1307
         *J Kowalewski*

B1.14    NMR imaging (diffusion and flow)                                                1325
         *U Goerke, P J McDonald and R Kimmich*

B1.15    EPR methods                                                                     1351
         *S Weber*

B1.16    Chemically-induced nuclear and electron polarization
         (CIDNP and CIDEP)                                                               1389
         *E J Harbron and M D E Forbes*

B1.17    Microscopy: electron (SEM and TEM)                                              1419
         *R R Schröder and M Müller*

B1.18    Microscopy: light                                                               1447
         *H Kiess*

B1.19    Scanning probe microscopies                                                     1467
         *N D Spencer and S P Jarvis*

B1.20    The surface forces apparatus                                                    1517
         *M Heuberger*

B1.21    Surface structural determination: diffraction methods                           1537
         *M A Van Hove*

B1.22    Surface characterization and structural determination: optical methods          1563
         *F Zaera*

B1.23    Surface structural determination: particle scattering methods                   1583
         *J W Rabalais*

B1.24    Rutherford backscattering, resonance scattering, PIXE and
         forward (recoil) scattering                                                     1611
         *C C Theron, V M Prozesky and J W Mayer*

B1.25    Surface chemical characterization                                               1633
         *L Coulier and J W Niemantsverdriet*

B1.26    Surface physical characterization                                               1649
         *W T Tysoe and G Wu*

B1.27    Calorimetry                                                                     1679
         *K N Marsh*

B1.28    Electrochemical methods                                                         1701
         *A W E Hodgson*

B1.29    High-pressure studies                                                           1731
         *M F Nicol*

**PART B2        DYNAMIC MEASUREMENTS**                                                  **1741**

B2.1     Ultrafast spectroscopy                                                          1743
         *W F Beck*

| B2.2 | Electron, ion and atom scattering | 1773 |
| | *M R Flannery* | |
| B2.3 | Reactive scattering | 1819 |
| | *P J Dagdigian* | |
| B2.4 | NMR methods for studying exchanging systems | 1847 |
| | *A D Bain* | |
| B2.5 | Gas-phase kinetics studies | 1871 |
| | *D Luckhaus and M Quack* | |

**PART B3**    **TECHNIQUES FOR APPLYING THEORY**    **1905**

| B3.1 | Quantum structural methods for atoms and molecules | 1907 |
| | *J Simons* | |
| B3.2 | Quantum structural methods for the solid state and surfaces | 1947 |
| | *F Starrost and E A Carter* | |
| B3.3 | Statistical mechanical simulations | 1983 |
| | *M P Allen* | |
| B3.4 | Quantum dynamics and spectroscopy | 2027 |
| | *S M Anderson, R G Sadygov and D Neuhauser* | |
| B3.5 | Optimization and reaction path algorithms | 2061 |
| | *P Pulay and J Baker* | |
| B3.6 | Mesoscopic and continuum models | 2087 |
| | *M Müller* | |

**VOLUME III: APPLICATIONS**

**PART C1**    **MICROSCOPIC SYSTEMS**    **2111**

| C1.1 | Clusters | 2113 |
| | *L-S Wang* | |
| C1.2 | Fullerenes | 2131 |
| | *D M Guldi* | |
| C1.3 | Van der Waals molecules | 2157 |
| | *J M Hutson* | |
| C1.4 | Atom traps and studies of ultracold systems | 2175 |
| | *J Weiner* | |
| C1.5 | Single molecule spectroscopy | 2199 |
| | *A Myers Kelley* | |

**PART C2**    **EXTENDED AND MACROSCOPIC SYSTEMS**    **2229**

| C2.1 | Polymers | 2231 |
| | *P Robyr* | |
| C2.2 | Liquid crystals | 2259 |
| | *I W Hamley* | |
| C2.3 | Micelles | 2285 |
| | *J Texter* | |

C2.4 Organic films (Langmuir–Blodgett films and self-assembled monolayers) 2317
*G Hähner*

C2.5 Introducing protein folding using simple models 2345
*D Thirumalai and D K Klimov*

C2.6 Colloids 2367
*J S van Duijneveldt*

C2.7 Catalysis 2395
*B C Gates*

C2.8 Corrosion 2415
*P Schmuki and M J Graham*

C2.9 Tribology 2437
*A J Gellman*

C2.10 Surface electrochemistry 2445
*H-H Strehblow and D Lützenkirchen-Hecht*

C2.11 Ceramic processing 2457
*K G Ewsuk*

C2.12 Zeolites 2473
*A Kogelbauer and R Prins*

C2.13 Plasma chemistry 2491
*M Schmidt and K Becker*

C2.14 Biophysical chemistry 2509
*J J Ramsden*

C2.15 Optoelectronics 2545
*W L Wilson*

C2.16 Semiconductors 2567
*H Temkin and S K Estreicher*

C2.17 Nanocrystals 2589
*V L Colvin and D M Mittleman*

C2.18 Etching and deposition 2613
*H P Gillis*

**PART C3** **CHEMICAL KINETICS AND DYNAMICS** **2631**

C3.1 Transient kinetic studies 2633
*R A Goldbeck and D S Kliger*

C3.2 Electron transfer reactions 2657
*G C Walker and D N Beratan*

C3.3 Energy transfer in gases 2681
*G W Flynn*

C3.4 Electronic energy transfer in condensed phases 2701
*A A Demidov and D L Andrews*

C3.5 Vibrational energy transfer in condensed phases 2717
*L K Iwaki and D D Dlott*

C3.6 Chaos and complexity in chemical systems 2737
*R Kapral and S J Fraser*

Index 2759

# PART B1

---

# DETERMINING MATERIALS AND MOLECULAR PROPERTIES

# B1.1
# Electronic spectroscopy

*S J Strickler*

### B1.1.1  Introduction

Optical spectroscopy is the study of the absorption and emission of light by atoms, molecules, or larger assemblies. Electronic spectroscopy is the branch of the field in which the change produced by the absorption or emission is a rearrangement of the electrons in the system. These changes are interpreted in terms of the quantum theory of electronic structure. To a first approximation, the rearrangements usually correspond to an electron being transferred from one orbital to another, and a transition will be described in terms of those orbitals. The wavelengths or frequencies of transitions help identify atoms and molecules and give information about their energy levels and hence their electronic structure and bonding. Intensities of absorption or emission give information about the nature of the electronic states and help to determine concentrations of species. In the case of molecules, along with the rearrangements of electrons there are usually changes in nuclear motions, and the vibrational and rotational structure of electronic bands give valuable insights into molecular structure and properties. For all these reasons, electronic spectroscopy is one of the most useful tools in chemistry and physics.

Most electronic transitions of interest fall into the visible and near-ultraviolet regions of the spectrum. This range of photon energies commonly corresponds to electrons being moved among valence orbitals. These orbitals are important to an understanding of bonding and structure, so are of particular interest in physical chemistry and chemical physics. For this reason, most of this chapter will concentrate on visible and near-UV spectroscopy, roughly the region between 200 and 700 nm, but there are no definite boundaries to the wavelengths of interest. Some of the valence orbitals will be so close in energy as to give spectra in the near-infrared region. Conversely, some valence transitions will be at high enough energy to lie in the vacuum ultraviolet, below about 200 nm, where air absorbs strongly and instrumentation must be evacuated to allow light to pass. In this region are also transitions of electrons to states of higher principal quantum number, known as Rydberg states. At still higher energies, in the x-ray region, are transitions of inner-shell electrons, and their spectroscopy has become an extremely useful tool, especially for studying solids and their surfaces. However, these other regions will not be covered in detail here.

Section B1.1.2 provides a brief summary of experimental methods and instrumentation, including definitions of some of the standard measured spectroscopic quantities. Section B1.1.3 reviews some of the theory of spectroscopic transitions, especially the relationships between transition moments calculated from wavefunctions and integrated absorption intensities or radiative rate constants. Because units can be so confusing, numerical factors with their units are included in some of the equations to make them easier to use. Vibrational effects, the Franck–Condon principle and selection rules are also discussed briefly. In the final section, B1.1.4, a few applications are mentioned to particular aspects of electronic spectroscopy.

## B1.1.2    Experimental methods

### *B1.1.2.1    Standard instrumentation*

There are two fundamental types of spectroscopic studies: absorption and emission. In absorption spectroscopy an atom or molecule in a low-lying electronic state, usually the ground state, absorbs a photon to go to a higher state. In emission spectroscopy the atom or molecule is produced in a higher electronic state by some excitation process, and emits a photon in going to a lower state. In this section we will consider the traditional instrumentation for studying the resulting spectra. They define the quantities measured and set the standard for experimental data to be considered.

### *(a) Emission spectroscopy*

Historically, emission spectroscopy was the first technique to be extensively developed. An electrical discharge will break up most substances into atoms and small molecules and their ions. It also excites these species into nearly all possible stable states, and the higher states emit light by undergoing transitions to lower electronic states. In typical classical instruments light from the sample enters through a slit, is collimated by a lens or mirror, is dispersed by a prism or grating so that different colours or wavelengths are travelling in different directions and is focused on a detector. Gratings are generally favoured over prisms. A concave grating may do the collimation, dispersion and focusing all in one step. The angle at which the light is reflected by a plane grating is simply related to the spacing, $d$, of the grooves ruled on the grating and the wavelength, $\lambda$, of the light:

$$\pm n\lambda = d(\sin\alpha + \sin\beta)$$

where $\alpha$ and $\beta$ are the angles of incidence and reflection measured from the normal, and $n$ is the order of the reflection, i.e. the number of cycles by which wave fronts from successive grooves differ for constructive interference. (The angles may be positive or negative depending on the experimental arrangement.) The first determinations of the absolute wavelengths of light and most of our knowledge of energies of excited states of atoms and molecules came from measuring these angles. In everyday use now the wavelength scale of an instrument can be calibrated using the wavelengths of known atomic lines. Grating instruments do have the disadvantage in comparison with prism spectrometers that different orders of light may overlap and need to be sorted out.

The earliest detector was the human eye observing the different colours. More versatile is a photographic plate, where each wavelength of light emitted shows up as a dark line (an image of the entrance slit) on the plate. It has the advantage that many wavelengths are measured simultaneously. Quantitative measurements are easier with a photomultiplier tube placed behind an exit slit. The spectrum is obtained as the different wavelengths are scanned across the slit by rotating the grating. The disadvantage is that measurements are made one wavelength at a time. With the advent of solid-state electronics, array detectors have become available that will measure many wavelengths at a time much like a photographic plate, but which can be read out quickly and quantitatively into a computer or other data system.

The design and use of spectrographs or spectrometers involves a compromise between resolution—how close in wavelength two lines can be and still be seen as separate—and sensitivity—how weak a light can be observed or how long it takes to make a measurement. Books have been written about the design of such instruments [1], and the subject cannot be pursued in this work.

Larger molecules generally cannot be studied in quite the same way, as an electric discharge merely breaks them up into smaller molecules or atoms. In such a case excited states are usually produced by optical excitation using light of the same or higher energy. Many modern fluorimeters are made with two monochromators, one to select an excitation wavelength from the spectrum of a suitable lamp, and the other to observe the emission as discussed above. Most studies of large molecules are done on solutions because vapour pressures are too low to allow gaseous spectra.

The fundamental measurements made in emission spectroscopy are the wavelengths or frequencies and the intensities of emission lines or bands. The problem with intensity measurements is that the efficiency of a dispersing and detecting system varies with wavelength. Relative intensities at a single wavelength are usually quite easily measured and can be used as a measure of concentration or excitation efficiency. Relative intensities of lines at nearly the same wavelength, say different rotational lines in a given band, can usually be obtained by assuming that the efficiency is the same for all. But the absolute intensity of a band or relative intensities of well separated bands require a calibration of the sensitivity of the instrument as a function of wavelength. In favourable cases this may be done by recording the spectrum of a standard lamp that has in turn been calibrated to give a known spectrum at a defined lamp current. For critical cases it may be necessary to distinguish between intensities measured in energy flow per unit time or in photons per unit time.

## (b) Absorption spectroscopy

Absorption spectroscopy is a common and well developed technique for studying electronic transitions between the ground state and excited states of atoms or molecules. A beam of light passes through a sample, and the amount of light that is absorbed during the passage is measured as a function of the wavelength or frequency of the light. The absorption is measured by comparing the intensity, $I$, of light leaving the sample with the intensity, $I_0$, entering the sample. The transmittance, $T$, is defined as the ratio

$$T = I/I_0.$$

It is often quoted as a percentage. In measuring the spectra of gases or solutions contained in cells, $I_0$ is usually taken to be the light intensity passing through an empty cell or a cell of pure solvent. This corrects well for reflection at the surfaces, absorption by the solvent or light scattering, which are not usually the quantities of interest.

It is usually convenient to work with the decadic absorbance, $A$, defined by

$$A = \log(I_0/I) = -\log T.$$

The unmodified term absorbance usually means this quantity, though some authors use the Napierian absorbance $B = -\ln T$. The absorbance is so useful because it normally increases linearly with path length, $l$, through the sample and with the concentration, $c$, of the absorbing species within the sample. The relationship is usually called Beer's law:

$$A = \varepsilon c l.$$

The quantity $\varepsilon$ is called the absorption coefficient or extinction coefficient, more completely the molar decadic absorption coefficient; it is a characteristic of the substance and the wavelength and to a lesser extent the solvent and temperature. It is common to take path length in centimetres and concentration in moles per litre, so $\varepsilon$ has units of $l\ mol^{-1}\ cm^{-1}$. The electronic absorption spectrum of a compound is usually shown as a plot of $\varepsilon$ versus wavelength or frequency.

Another useful quantity related to extinction coefficient is the cross section, $\sigma$, defined for a single atom or molecule. It may be thought of as the effective area blocking the beam at a given wavelength, and the value may be compared with the size of the molecule. The relationship is

$$\sigma = (\ln 10)\varepsilon/N_A$$

where $N_A$ is Avogadro's number. If $\varepsilon$ is in $l\ mol^{-1}\ cm^{-1}$ and $\sigma$ is desired in $cm^2$ the relationship may be written

$$\sigma = (3.8235 \times 10^{-21}\ cm^3\ mol\ l^{-1})\varepsilon.$$

The standard instrument for measuring an absorption spectrum is a double-beam spectrophotometer. A typical instrument uses a lamp with a continuous spectrum to supply the light, usually a tungsten lamp for the visible, near-infrared, and near-ultraviolet regions and a discharge lamp filled with hydrogen or deuterium for farther in the ultraviolet. Light from the source passes through a monochromator to give a narrow band of wavelengths, and is then split into two beams. One beam passes through a cell containing the sample, the other through an identical reference cell filled with solvent. These beams define $I$ and $I_0$, respectively. The beams are monitored by a detection system, usually a photomultiplier tube. An electronic circuit measures the ratio of the two intensities and displays the transmittance or absorbance. The wavelength is varied by scanning the monochromator, and the spectrum may be plotted on a chart recorder.

A different design of instrument called a diode array spectrometer has become popular in recent years. In this instrument the light from the lamp passes through the sample, then into a spectrometer to be dispersed, and then is focused onto an array of solid-state detectors arranged so that each detector element measures intensity in a narrow band of wavelengths—say one detector for each nanometre of the visible and ultraviolet regions. The output is digitized and the spectrum displayed on a screen, and it can be read out in digital form and processed with a computer. The complete spectrum can be recorded in a few seconds. This is not formally a double-beam instrument, but because a spectrum is taken so quickly and handled so easily, one can record the spectrum of a reference cell and the sample cell and then compare them in the computer, so it serves the same purpose. The available instruments do not give quite the resolution or versatility of the standard spectrophotometers, but they are far quicker and easier to use.

### B1.1.2.2    Some modern techniques

The traditional instruments for measuring emission and absorption spectra described above set the standard for the types of information which can be obtained and used by spectroscopists. In the more recent past, several new techniques have become available which have extended the range of spectroscopic measurements to higher resolution, lower concentrations of species, weaker transitions, shorter time scales, etc. Many studies in electronic spectroscopy as a branch of physical chemistry or chemical physics are now done using these new techniques. The purpose of this section is to discuss some of them.

### (a) Lasers

The foremost of the modern techniques is the use of lasers as spectroscopic tools. Lasers are extremely versatile light sources. They can be designed with many useful properties (not all in the same instrument) such as high intensity, narrow frequency bandwidth with high-frequency stability, tunability over reasonable frequency ranges, low-divergence beams which can be focused into very small spots, or pulsed beams with very short time durations. There are nearly as many different experimental arrangements as there are experimenters, and only a few examples will be mentioned here.

While a laser beam can be used for traditional absorption spectroscopy by measuring $I$ and $I_0$, the strength of laser spectroscopy lies in more specialized experiments which often do not lend themselves to such measurements. Other techniques are commonly used to detect the absorption of light from the laser beam. A common one is to observe fluorescence excited by the laser. The total fluorescence produced is normally proportional to the amount of light absorbed. It can be used as a measurement of concentration to detect species present in extremely small amounts. Or a measurement of the fluorescence intensity as the laser frequency is scanned can give an absorption spectrum. This may allow much higher resolution than is easily obtained with a traditional absorption spectrometer. In other experiments the fluorescence may be dispersed and its spectrum determined with a traditional spectrometer. In suitable cases this could be the emission from a single electronic–vibrational–rotational level of a molecule and the experimenter can study how the spectrum varies with level.

Other methods may also be useful for detecting the absorption of laser radiation. For example, the heat generated when radiation is absorbed can be detected in several ways. One way observes the defocusing of the laser beam when the medium is heated and its refractive index changes. Another way, called photoacoustic spectroscopy, detects sound waves or pressure pulses when light is absorbed from a pulsed laser. Still another method useful with high-intensity pulsed lasers is to measure light absorption by the excited states produced. This is often useful for studying the kinetics of the excited species as they decay or undergo reactions.

Another example of a technique for detecting absorption of laser radiation in gaseous samples is to use multiphoton ionization with intense pulses of light. Once a molecule has been electronically excited, the excited state may absorb one or more additional photons until it is ionized. The electrons can be measured as a current generated across the cell, or can be counted individually by an electron multiplier; this can be a very sensitive technique for detecting a small number of molecules excited.

## (b) Excited-state lifetimes

Measurements of the decay rates of excited states are important, both for the fundamental spectroscopic information they can give, and for studies of other processes such as energy transfer or photochemistry. The techniques used vary greatly depending on the time scale of the processes being studied. For rather long time scales, say of the order of a millisecond or longer, it is rather simple to excite the molecules optically, cut off the exciting light, and watch the decay of emission or some other measurement of excited-state concentration.

For fluorescent compounds and for times in the range of a tenth of a nanosecond to a hundred microseconds, two very successful techniques have been used. One is the phase-shift technique. In this method the fluorescence is excited by light whose intensity is modulated sinusoidally at a frequency $f$, chosen so its period is not too different from the expected lifetime. The fluorescent light is then also modulated at the same frequency but with a time delay. If the fluorescence decays exponentially, its phase is shifted by an angle $\Delta\phi$ which is related to the mean life, $\tau$, of the excited state. The relationship is

$$\tan \Delta\phi = 2\pi f \tau.$$

The phase shift is measured by comparing the phase of the fluorescence with the phase of light scattered by a cloudy but non-fluorescent solution.

The other common way of measuring nanosecond lifetimes is the time-correlated single-photon counting technique [2]. In this method the sample is excited by a weak, rapidly repeating pulsed light source, which could be a flashlamp or a mode-locked laser with its intensity reduced. The fluorescence is monitored by a photomultiplier tube set up so that current pulses from individual photons can be counted. It is usually arranged so that at most one fluorescence photon is counted for each flash of the excitation source. A time-to-amplitude converter and a multichannel analyser (equipment developed for nuclear physics) are used to determine, for each photon, the time between the lamp flash and the photon pulse. A decay curve is built up by measuring thousands of photons and sorting them by time delay. The statistics of such counting experiments are well understood and very accurate lifetimes and their uncertainties can be determined by fitting the resulting decay curves.

One advantage of the photon counting technique over the phase-shift method is that any non-exponential decay is readily seen and studied. It is possible to detect non-exponential decay in the phase-shift method too by making measurements as a function of the modulation frequency, but it is more cumbersome.

At still shorter time scales other techniques can be used to determine excited-state lifetimes, but perhaps not as precisely. Streak cameras can be used to measure faster changes in light intensity. Probably the most useful techniques are pump–probe methods where one intense laser pulse is used to excite a sample and a weaker pulse, delayed by a known amount of time, is used to probe changes in absorption or other properties caused by the excitation. At short time scales the delay is readily adjusted by varying the path length travelled by the beams, letting the speed of light set the delay.

*(c) Photoelectron spectroscopy*

Only brief mention will be made here of photoelectron spectroscopy. This technique makes use of a beam of light whose energy is greater than the ionization energy of the species being studied. Transitions then occur in which one of the electrons of the molecule is ejected. Rather than an optical measurement, the kinetic energy of the ejected electron is determined. Some of the technology is described in section B1.6 on electron energy-loss spectroscopy. The ionization energy of the molecule is determined from the difference between the photon energy and the kinetic energy of the ejected electron.

A useful light source is the helium resonance lamp which produces light of wavelength 58.4 nm or a photon energy of 21.2 eV, enough to ionize any neutral molecule. Often several peaks can be observed in the photoelectron spectrum corresponding to the removal of electrons from different orbitals. The energies of the peaks give approximations to the orbital energies in the molecule. They are useful for comparison with theoretical calculations.

An interesting variation on the method is the use of a laser to photodetach an electron from a negative ion produced in a beam of ions. Since it is much easier to remove an electron from a negative ion than from a neutral molecule, this can be done with a visible or near-ultraviolet laser. The difference between the photon energy and the electron energy, in this case, gives the electron affinity of the neutral molecule remaining after the photodetachment, and may give useful energy levels of molecules not easily studied by traditional spectroscopy [3].

*(d) Other techniques*

Some other extremely useful spectroscopic techniques will only be mentioned here. Probably the most important one is spectroscopy in free jet expansions. Small molecules have often been studied by gas-phase spectroscopy where sharp rotational and vibrational structure gives detailed information about molecular states and geometries. The traditional techniques will often not work for large molecules because they must be heated to high temperatures to vaporize them and then the spectra become so broad and congested that detailed analysis is difficult or impossible. In jet spectroscopy the gaseous molecules are mixed with an inert gas and allowed to expand into a vacuum. The nearly adiabatic expansion may cool them rapidly to temperatures of a few degrees Kelvin while leaving them in the gas phase. The drastic simplification of the spectrum often allows much more information to be extracted from the spectrum.

Fourier-transform instruments can also be used for visible and ultraviolet spectroscopy. In this technique, instead of dispersing the light with a grating or prism, a wide region of the spectrum is detected simultaneously by splitting the beam into two components, one reflected from a stationary mirror and one from a movable mirror, and then recombining the two beams before they enter the detector. The detected intensity is measured as a function of the position of the movable mirror. Because of interference between the two beams, the resulting function is the Fourier transform of the normal spectrum as a function of wavelength. This offers some advantages in sensitivity and perhaps resolution because the whole spectrum is measured at once rather than one wavelength at a time. The technique is not too common in electronic spectroscopy, but is very widely used for the infrared. It is described more fully in the chapter on vibrational spectroscopy, section B1.2. While the light sources and detectors are different for the visible and ultraviolet region, the principles of operation are the same.

### B1.1.3   Theory

The theory of absorption or emission of light of modest intensity has traditionally been treated by time-dependent perturbation theory [4]. Most commonly, the theory treats the effect of the oscillating electric field of the light wave acting on the electron cloud of the atom or molecule. The instantaneous electric field is assumed to be uniform over the molecule but it oscillates in magnitude and direction with a frequency $v$. The energy of a system of charges in a uniform electric field, **E**, depends on its dipole moment according to

$$E = -\mu \cdot \mathbf{E}$$

where the dipole moment is defined by

$$\mu = \sum_i q_i r_i$$

and the $q_i$ and $r_i$ are the charges and positions of the particles, i.e. the electrons and nuclei. The result of the time-dependent perturbation theory is that the transition probability for a transition between one quantum state $i$ and another state $j$ is proportional to the absolute value squared of the matrix element of the electric dipole operator between the two states

$$\mu_{ij} = \int \Psi_i^* \mu \Psi_j \, d\tau$$

$$\text{transition probability} \propto |\mu_{ij}|^2. \tag{B1.1.1}$$

The transition occurs with significant probability only if the frequency of the light is very close to the familiar resonance condition, namely $hv = \Delta E$, where $h$ is Planck's constant and $\Delta E$ is the difference in energy of the two states. However, transitions always occur over a range of frequencies because of various broadening effects; if nothing else, as required by the uncertainty principle, the states will not have precisely defined energies if they have finite lifetimes.

### B1.1.3.1  *Absorption spectroscopy*

*(a) Integrated absorption intensity*

The relationship between the theoretical quantity $\mu_{ij}$ and the experimental parameter $\varepsilon$ of absorption spectroscopy involves, not the value of $\varepsilon$ at any one wavelength, but its integral over the absorption band. The relationship is

$$\int \varepsilon \, d\tilde{v} = \frac{2\pi^2 N_A \tilde{v}}{3hc\varepsilon_0 \ln 10} |\mu_{ij}|^2 = (2.512 \times 10^{19} \, \text{l mol}^{-1} \, \text{cm}^{-3}) \frac{\tilde{v}}{e^2} |\mu_{ij}|^2. \tag{B1.1.2}$$

We will quote a numerical constant in some of these equations to help with actual calculations. The units can be very confusing because it is conventional to use non-SI units for several quantities. The wavenumber value, $\tilde{v}$, is usually taken to be in $\text{cm}^{-1}$. The extinction coefficient is conveniently taken in units of $\text{l mol}^{-1} \, \text{cm}^{-1}$. We have inserted the factor $e^2$ into the equation because values of $\mu_{ij}$ are usually calculated with the charges measured in units of the electron charge. For the sake of consistency, we have quoted the numerical factor appropriate for $\mu/e$ taken in centimetres, but the values are easily converted to use other length units such as Ångstroms or atomic units. The value of $\tilde{v}$ in the right-hand side of the equation is to be interpreted as a suitable average frequency for the transition. This causes no difficulty unless the band is very broad. Some of the difficulties with different definitions of intensity terms have been discussed by Hilborn [5].

*(b) Oscillator strength*

A related measure of the intensity often used for electronic spectroscopy is the oscillator strength, $f$. This is a dimensionless ratio of the transition intensity to that expected for an electron bound by Hooke's law forces so as to be an isotropic harmonic oscillator. It can be related either to the experimental integrated intensity or to the theoretical transition moment integral:

$$f = \frac{4\varepsilon_0 m_e c^2 \ln 10}{e^2 N_A} \int \varepsilon \, d\tilde{\nu} = (4.319 \times 10^{-9} \text{ mol l}^{-1} \text{ cm}^2) \int \varepsilon \, d\tilde{\nu} \tag{B1.1.3}$$

or

$$f = \frac{8\pi^2 m_e c \tilde{\nu}}{3he^2} |\boldsymbol{\mu}_{ij}|^2 = (1.085 \times 10^{11} \text{ cm}^{-1}) \frac{\tilde{\nu}}{e^2} |\boldsymbol{\mu}_{ij}|^2. \tag{B1.1.4}$$

The harmonically bound electron is, in a sense, an ideal absorber since its harmonic motion can maintain a perfect phase relationship with the oscillating electric field of the light wave. Strong electronic transitions have oscillator strengths of the order of unity, but this is not, as sometimes stated, an upper limit to $f$. For example, some polyacetylenes have bands with oscillator strengths as high as 5 [6]. There is a theorem, the Kuhn–Thomas sum rule, stating that the sum of the oscillator strengths of all electronic transitions must be equal to the number of electrons in an atom or molecule [7].

In the above discussion we have used the electric dipole operator $\boldsymbol{\mu}$. It is also sometimes possible to observe electronic transitions occurring due to interaction with the magnetic field of the light wave. These are called magnetic dipole transitions. They are expected to be weaker than electric dipole transitions by several orders of magnitude. If account is taken of a variation of the field of the light wave over the size of the molecule it is possible to treat quadrupole or even higher multipole transitions. These are expected to be even weaker than typical magnetic dipole transitions. We will concentrate on the more commonly observed electric dipole transitions.

Equation (B1.1.1) for the transition moment integral is rather simply interpreted in the case of an atom. The wavefunctions are simply functions of the electron positions relative to the nucleus, and the integration is over the electronic coordinates. The situation for molecules is more complicated and deserves discussion in some detail.

### (c) Transition moments for molecules

Electronic spectra are almost always treated within the framework of the Born–Oppenheimer approximation [8] which states that the total wavefunction of a molecule can be expressed as a product of electronic, vibrational, and rotational wavefunctions (plus, of course, the translation of the centre of mass which can always be treated separately from the internal coordinates). The physical reason for the separation is that the nuclei are much heavier than the electrons and move much more slowly, so the electron cloud normally follows the instantaneous position of the nuclei quite well. The integral of equation (B1.1.1) is over all internal coordinates, both electronic and nuclear. Integration over the rotational wavefunctions gives rotational selection rules which determine the fine structure and band shapes of electronic transitions in gaseous molecules. Rotational selection rules will be discussed below. For molecules in condensed phases the rotational motion is suppressed and replaced by oscillatory and diffusional motions.

In this section we concentrate on the electronic and vibrational parts of the wavefunctions. It is convenient to treat the nuclear configuration in terms of normal coordinates describing the displacements from the equilibrium position. We call these nuclear normal coordinates $Q_j$ and use the symbol $Q$ without a subscript to designate the whole set. Similarly, the symbol $x_i$ designates the coordinates of the $i$th electron and $x$ the whole set of electronic coordinates. We also use subscripts l and u to designate the lower and upper electronic states of a transition, and subscripts $a$ and $b$ to number the vibrational states in the respective electronic states. The total wavefunction $\Psi$ can be written

$$\Psi_{la}(x, Q) = \psi_l(x, Q)\phi_{la}(Q) \qquad \Psi_{ub}(x, Q) = \psi_u(x, Q)\phi_{ub}(Q).$$

Here each $\phi(Q)$ is a vibrational wavefunction, a function of the nuclear coordinates $Q$, in first approximation usually a product of harmonic oscillator wavefunctions for the various normal coordinates. Each $\psi(x, Q)$ is the electronic wavefunction describing how the electrons are distributed in the molecule. However, it has the nuclear coordinates within it as parameters because the electrons are always distributed around the nuclei and follow those nuclei whatever their position during a vibration. The integration of equation (B1.1.1) can be carried out in two steps—first an integration over the electronic coordinates $x$, and then integration over the nuclear coordinates $Q$. We define an electronic transition moment integral which is a function of nuclear position:

$$\boldsymbol{\mu}_{\mathrm{lu}}(Q) = \int \psi_{\mathrm{l}}^*(x, Q)\boldsymbol{\mu}\psi_{\mathrm{u}}(x, Q)\,\mathrm{d}x. \tag{B1.1.5}$$

We then integrate this over the vibrational wavefunctions and coordinates:

$$\boldsymbol{\mu}_{\mathrm{l}a,\mathrm{u}b} = \int \Psi_{\mathrm{l}a}^*\boldsymbol{\mu}\Psi_{\mathrm{u}b}\,\mathrm{d}\tau = \int \phi_{\mathrm{l}a}^*(Q)\boldsymbol{\mu}_{\mathrm{lu}}(Q)\phi_{\mathrm{u}b}(Q)\,\mathrm{d}Q. \tag{B1.1.6}$$

This last transition moment integral, if plugged into equation (B1.1.2), will give the integrated intensity of a vibronic band, i.e. of a transition starting from vibrational state $a$ of electronic state l and ending on vibrational level $b$ of electronic state u.

*(d) The Franck–Condon principle*

The electronic transition moment of equation (B1.1.5) is related to the intensity that the transition would have if the nuclei were fixed in configuration $Q$, but its value may vary with that configuration. It is often useful to expand $\boldsymbol{\mu}_{\mathrm{lu}}(Q)$ as a power series in the normal coordinates, $Q_i$:

$$\boldsymbol{\mu}_{\mathrm{lu}}(Q) = \boldsymbol{\mu}_{\mathrm{lu}}(0) + \sum_i \left(\frac{\partial\boldsymbol{\mu}_{\mathrm{lu}}}{\partial Q_i}\right)_0 Q_i + \cdots. \tag{B1.1.7}$$

Here $\boldsymbol{\mu}_{\mathrm{lu}}(0)$ is the value at the equilibrium position of the initial electronic state.

In many cases the variation is not very strong for reasonable displacements from equilibrium, and it is sufficient to use only the zero-order term in the expansion. If this is inserted into equation (B1.1.6) we get

$$\boldsymbol{\mu}_{\mathrm{l}a,\mathrm{u}b} = \boldsymbol{\mu}_{\mathrm{lu}}(0) \int \phi_{\mathrm{l}a}^*(Q)\phi_{\mathrm{u}b}(Q)\,\mathrm{d}Q$$

and using this in equation (B1.1.2) for the integrated intensity of a vibronic band we get the relationship

$$\int \varepsilon\,\mathrm{d}\tilde{\nu} = \frac{2\pi^2 N_\mathrm{A}\tilde{\nu}_{\mathrm{l}a\to\mathrm{u}b}}{3hc\varepsilon_0 \ln 10}|\boldsymbol{\mu}_{\mathrm{lu}}(0)|^2 \left|\int \phi_{\mathrm{l}a}^*(Q)\phi_{\mathrm{u}b}(Q)\,\mathrm{d}Q\right|^2. \tag{B1.1.8}$$

The last factor, the square of the overlap integral between the initial and final vibrational wavefunctions, is called the Franck–Condon factor for this transition.

The Franck–Condon principle says that the intensities of the various vibrational bands of an electronic transition are proportional to these Franck–Condon factors. (Of course, the frequency factor must be included for accurate treatments.) The idea was first derived qualitatively by Franck through the picture that the rearrangement of the light electrons in the electronic transition would occur quickly relative to the period of motion of the heavy nuclei, so the position and momentum of the nuclei would not change much during the transition [9]. The quantum mechanical picture was given shortly afterwards by Condon, more or less as outlined above [10].

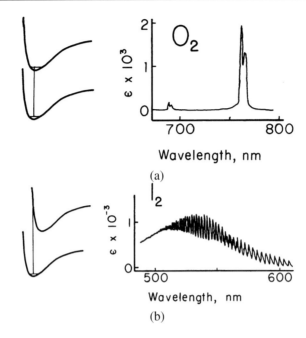

**Figure B1.1.1.** (a) Potential curves for two states with little or no difference in the equilibrium position of the upper and lower states. A transition of $O_2$, with displacement only 0.02 Å, is shown as an example. Data taken from [11]. Most of the intensity is in the 0–0 vibrational band with a small intensity in the 1–0 band. (b) Potential curves for two states with a large difference in the equilibrium position of the two states. A transition in $I_2$, with a displacement of 0.36 Å, is shown as an example. Many vibrational peaks are observed.

   The effects of the principle are most easily visualized for diatomic molecules for which the vibrational potential can be represented by a potential energy curve. A typical absorption starts from the lowest vibrational level of the ground state (actually a thermal distribution of low-lying levels). A useful qualitative statement of the Franck–Condon principle is that vertical transitions should be favoured. Figure B1.1.1(a) illustrates the case where the potential curve for the excited state lies nearly directly above that for the ground state. Then by far the largest overlap of excited state wavefunctions with the lowest level of the ground state will be for the $v = 0$ level, and we expect most intensity to be in the so-called 0–0 band, i.e. from $v = 0$ in the lower state to $v = 0$ in the upper state. A case in point is the transition of the $O_2$ molecule at about 750 nm in the near-infrared. (This is actually a magnetic dipole transition rather than electric dipole, so it is very weak, but the vibrational effects are the same.) Both ground and excited state have a $(\pi^*)^2$ electron configuration and nearly the same equilibrium bond length, only 0.02 Å different. The spectrum shows most of the intensity in the 0–0 band with less than one tenth as much in the 1–0 band [11].

   Figure B1.1.1(b) shows a contrasting case, where the potential curve for the excited state is displaced considerably relative to the ground-state curve. Then a vertical transition would go to a part of the excited-state curve well displaced from the bottom, and the maximum overlap and greatest intensity should occur for high-lying levels. There results a long progression of bands to various vibrational levels. The spectrum of $I_2$ is shown as an illustration; here the displacement between the two minima is about 0.36 Å. Many vibronic transitions are seen. One can observe the excited-state levels getting closer together and converging as the dissociation limit is approached, and part of the absorption goes to continuum states above the dissociation energy. (The long-wavelength part of the spectrum is complicated by transitions starting from thermally excited vibrational levels of the ground state.)

*(e) Beyond Franck–Condon*

There are cases where the variation of the electronic transition moment with nuclear configuration cannot be neglected. Then it is necessary to work with equation (B1.1.6) keeping the dependence of $\mu_{\mathrm{lu}}$ on $Q$ and integrating it over the vibrational wavefunctions. In most such cases it is adequate to use only the terms up to first-order in equation (B1.1.7). This results in 'modified Franck–Condon factors' for the vibrational intensities [12].

*(f) Total intensity of an electronic transition*

Equation (B1.1.8) gives the intensity of one vibronic band in an absorption spectrum. It is also of interest to consider the total integrated intensity of a whole electronic transition, i.e. the sum of all the vibronic bands corresponding to the one electronic change. In the most common absorption spectroscopy experiment we can assume that all transitions originate in the lowest vibrational level of the ground electronic state, which we can designate as level l0. The transitions can go to various levels u$b$ of the upper electronic state. The total integrated intensity is then obtained by summing over the index $b$ which numbers the excited state vibrational levels.

$$\int \varepsilon \, \mathrm{d}\tilde{\nu} = \frac{2\pi^2 N_{\mathrm{A}}}{3hc\varepsilon_0 \ln 10} \sum_b \tilde{\nu}_{\mathrm{l0} \to \mathrm{u}b} \left| \int \phi_{\mathrm{l0}}^*(Q) \mu_{\mathrm{lu}}(Q) \phi_{\mathrm{u}b}(Q) \, \mathrm{d}Q \right|^2. \tag{B1.1.9}$$

This equation can be simplified if the frequency term $\tilde{\nu}_{\mathrm{l0} \to \mathrm{u}b}$ is removed from the summation. One way to do this is to incorporate it into the integral on the left-hand side by writing $\int \varepsilon \, \mathrm{d} \ln \tilde{\nu}$. The alternative is to use an appropriate average $\tilde{\nu}$ outside the sum, choosing the proper average by making the expressions equal. Often it is enough to pick an average by eye, but if high accuracy is important the value to use is given by

$$\langle \tilde{\nu} \rangle = \frac{\int \varepsilon \, \mathrm{d}\tilde{\nu}}{\int \varepsilon \, \mathrm{d} \ln \tilde{\nu}}.$$

With the frequency removed from the sum, (B1.1.9) has just a sum over vibrational integrals. Because all the vibrational wavefunctions for a given potential surface will form a complete set, it is possible to apply a sum rule to simplify the resulting expression:

$$\sum_b \left| \int \phi_{\mathrm{l0}}^*(Q) \mu_{\mathrm{lu}}(Q) \phi_{\mathrm{u}b}(Q) \, \mathrm{d}Q \right|^2 = \int \phi_{\mathrm{l0}}^*(Q) |\mu_{\mathrm{lu}}(Q)|^2 \phi_{\mathrm{l0}}(Q) \, \mathrm{d}Q$$

i.e. the sum is just the mean value of $|\mu_{\mathrm{lu}}(Q)|^2$ in the initial vibrational state. Then the total integrated intensity of the electronic band is given by

$$\int \varepsilon \, \mathrm{d}\tilde{\nu} = \frac{2\pi^2 N_{\mathrm{A}} \langle \tilde{\nu} \rangle}{3hc\varepsilon_0 \ln 10} \int \phi_{\mathrm{l0}}^*(Q) |\mu_{\mathrm{lu}}(Q)|^2 \phi_{\mathrm{l0}}(Q) \, \mathrm{d}Q. \tag{B1.1.10}$$

If we can get by with using only the zero-order term of (B1.1.7), we can take $\mu_{\mathrm{lu}}$ out of the integral and use the fact that $\phi_{\mathrm{l0}}$ is normalized. The last equation then simplifies further to

$$\int \varepsilon \, \mathrm{d}\tilde{\nu} = \frac{2\pi^2 N_{\mathrm{A}} \langle \tilde{\nu} \rangle}{3hc\varepsilon_0 \ln 10} |\mu_{\mathrm{lu}}(0)|^2 = (2.512 \times 10^{19} \, \mathrm{l \, mol}^{-1} \, \mathrm{cm}^{-3}) \frac{\langle \tilde{\nu} \rangle}{e^2} |\mu_{\mathrm{lu}}(0)|^2. \tag{B1.1.11}$$

Equations (B1.1.10) and (B1.1.11) are the critical ones for comparing observed intensities of electronic transitions with theoretical calculations using the electronic wavefunctions. The transition moment integral $\mu_{\mathrm{lu}}$ calculated from electronic wavefunctions is related to the absorption intensity integrated over the whole electronic transition. It is found that simple forms of electronic wavefunctions often do not give very good intensities, and high-quality wavefunctions are required for close agreement with experiment.

*B1.1.3.2   Emission spectroscopy*

The interpretation of emission spectra is somewhat different but similar to that of absorption spectra. The intensity observed in a typical emission spectrum is a complicated function of the excitation conditions which determine the number of excited states produced, quenching processes which compete with emission, and the efficiency of the detection system. The quantities of theoretical interest which replace the integrated intensity of absorption spectroscopy are the rate constant for spontaneous emission and the related excited-state lifetime.

*(a) Emission rate constant*

Einstein derived the relationship between spontaneous emission rate and the absorption intensity or stimulated emission rate in 1917 using a thermodynamic argument [13]. Both absorption intensity and emission rate depend on the transition moment integral of equation (B1.1.1), so that gives us a way to relate them. The symbol $A$ is often used for the rate constant for emission; it is sometimes called the Einstein $A$ coefficient. For emission in the gas phase from a state $i$ to a lower state $j$ we can write

$$A_{i \to j} = \frac{16\pi^3}{3h\varepsilon_0} \tilde{\nu}_{i \to j}^3 |\boldsymbol{\mu}_{ij}|^2 = (7.235 \times 10^{10} \text{ cm s}^{-1}) \frac{\tilde{\nu}_{i \to j}^3}{e^2} |\boldsymbol{\mu}_{ij}|^2. \tag{B1.1.12}$$

*(b) Molecular emission and the Franck–Condon principle*

For molecules we can use Born–Oppenheimer wavefunctions and talk about emission from one vibronic level to another. Equations (B1.1.5), (B1.1.6) and (B1.1.7) can be used just as they were for absorption. If we have an emission from vibronic state u$b$ to the lower state l$a$, the rate constant for emission would be given by

$$A_{ub \to la} = \frac{16\pi^3}{3h\varepsilon_0} \tilde{\nu}_{ub \to la}^3 \left| \int \phi_{ub}^*(Q) \boldsymbol{\mu}_{ul}(Q) \phi_{la}(Q) \, \mathrm{d}Q \right|^2. \tag{B1.1.13}$$

If we can use only the zero-order term in equation (B1.1.7) we can remove the transition moment from the integral and recover an equation involving a Franck–Condon factor:

$$A_{ub \to la} = \frac{16\pi^3}{3h\varepsilon_0} |\boldsymbol{\mu}_{ul}(0)|^2 \tilde{\nu}_{ub \to la}^3 \left| \int \phi_{ub}^*(Q) \phi_{la}(Q) \, \mathrm{d}Q \right|^2. \tag{B1.1.14}$$

Now the spectrum will show various transitions originating with state u$b$ and ending on the various vibrational levels l$a$ of the lower electronic state. Equation (B1.1.14) (or (B1.1.13) if we have to worry about variation of transition moment) gives us a way of comparing the intensities of the bands. The intensities will be proportional to the $A_{ub \to la}$ provided that we measure the intensity in photons per unit time rather than the more conventional units of energy per unit time. The first part of the expression in (B1.1.14) is the same for all the transitions. The part that varies between bands is the Franck–Condon factor multiplied by the cube of the frequency. Equation (B1.1.8) for absorption intensity also had a frequency factor, but the variation in frequency has more effect in emission spectroscopy because it appears to a higher power. Equation (B1.1.14) embodies the Franck–Condon principle for emission spectroscopy.

*(c) Excited-state lifetime*

We now discuss the lifetime of an excited electronic state of a molecule. To simplify the discussion we will consider a molecule in a high-pressure gas or in solution where vibrational relaxation occurs rapidly, we will assume that the molecule is in the lowest vibrational level of the upper electronic state, level u0, and we will further assume that we need only consider the zero-order term of equation (B1.1.7). A number of radiative transitions are possible, ending on the various vibrational levels $la$ of the lower state, usually the ground state. The total rate constant for radiative decay, which we will call $A_{u0 \to 1}$, is the sum of the rate constants, $A_{u0 \to la}$. By summing the terms in equation (B1.1.14) we can get an expression relating the radiative lifetime to the theoretical transition moment $\mu_{ul}$. Further, by relating the transition moment to integrated absorption intensity we can get an expression for radiative rate constant involving only experimental quantities and not dependent on the quality of the electronic wavefunctions:

$$A_{u0 \to 1} = \frac{16\pi^3 n^2}{3h\varepsilon_0} \langle \tilde{\nu}_f^{-3} \rangle_{Av}^{-1} \frac{g_1}{g_u} |\mu_{ul}(0)|^2$$

$$= (7.235 \times 10^{10} \text{ cm s}^{-1}) n^2 \frac{g_1 \langle \tilde{\nu}_f^{-3} \rangle_{Av}^{-1}}{g_u e^2} |\mu_{ul}(0)|^2 \tag{B1.1.15}$$

or

$$A_{u0 \to 1} = \frac{8\pi c n^2 \ln 10}{N_A} \langle \tilde{\nu}_f^{-3} \rangle_{Av}^{-1} \frac{g_1}{g_u} \int \varepsilon \, d \ln \tilde{\nu}$$

$$= (2.881 \times 10^{-9} \text{ s}^{-1} \text{ l}^{-1} \text{ mol cm}^4) \langle \tilde{\nu}_f^{-3} \rangle_{Av}^{-1} n^2 \frac{g_1}{g_u} \int \varepsilon \, d \ln \tilde{\nu}. \tag{B1.1.16}$$

These equations contain the peculiar average fluorescence frequency $\langle \tilde{\nu}_f^{-3} \rangle_{Av}^{-1}$, the reciprocal of the average value of $\tilde{\nu}^{-3}$ in the fluorescence spectrum. It arises because the fluorescence intensity measured in photons per unit time has a $\nu^3$ dependence. For completeness we have added a term $n^2$, the square of the refractive index, to be used for molecules in solution, and the term $g_1/g_u$, the ratio of the degeneracies of the lower and upper electronic states, to allow for degenerate cases [14]. It is also possible to correct for a variation of transition moment with nuclear configuration if that should be necessary [15].

$A_{u0 \to 1}$ is the first-order rate constant for radiative decay by the molecule. It is the reciprocal of the intrinsic mean life of the excited state, $\tau_0$:

$$1/\tau_0 = A_{u0 \to 1}.$$

If there are no competing processes the experimental lifetime $\tau$ should equal $\tau_0$. Most commonly, other processes such as non-radiative decay to lower electronic states, quenching, photochemical reactions or energy transfer may compete with fluorescence. They will reduce the actual lifetime. As long as all the processes are first-order in the concentration of excited molecules, the decay will remain exponential and the mean life $\tau$ will be reduced by a factor of the fluorescence quantum yield, $\Phi_f$, the fraction of the excited molecules which emit:

$$\tau = \Phi_f \tau_0. \tag{B1.1.17}$$

### B1.1.3.3  Selection rules

Transition intensities are determined by the wavefunctions of the initial and final states as described in the last sections. In many systems there are some pairs of states for which the transition moment integral vanishes while for other pairs it does not vanish. The term 'selection rule' refers to a summary of the conditions for non-vanishing transition moment integrals—hence observable transitions—or vanishing integrals so no observable transitions. We discuss some of these rules briefly in this section. Again, we concentrate on electric dipole transitions.

### (a) Atoms

The simplest case arises when the electronic motion can be considered in terms of just one electron: for example, in hydrogen or alkali metal atoms. That electron will have various values of orbital angular momentum described by a quantum number $l$. It also has a spin angular momentum described by a spin quantum number $s$ of $\frac{1}{2}$, and a total angular momentum which is the vector sum of orbital and spin parts with quantum number $j$. In the presence of a magnetic field the component of the angular momentum in the field direction becomes important and is described by a quantum number $m$. The selection rules can be summarized as

$$\Delta l = \pm 1 \qquad \Delta j = 0, \pm 1 \qquad \Delta m = 0, \pm 1.$$

This means that one can see the electron undergo transitions from an s orbital to a p orbital, from a p orbital to s or d, from a d orbital to p or f, etc, but not s to s, p to p, s to d, or such. In terms of state designations, one can have transitions from $^2S_{1/2}$ to $^2P_{1/2}$ or to $^2P_{3/2}$, etc.

In more complex atoms there may be a strong coupling between the motion of different electrons. The states are usually described in terms of the total orbital angular momentum $L$ and the total spin angular momentum $S$. These are coupled to each other by an interaction called spin–orbital coupling, which is quite weak in light atoms but gets rapidly stronger as the nuclear charge increases. The resultant angular momentum is given the symbol $J$. States are named using capital letters S, P, D, F, G, . . . to designate $L$ values of 0, 1, 2, 3, 4, . . .. A left superscript gives the multiplicity ($2S + 1$), and a right subscript gives the value of $J$, for example $^1S_0$, $^3P_1$ or $^2D_{5/2}$.

There is a strict selection rule for $J$:

$$\Delta J = 0, \pm 1 \qquad \text{with the restriction that } J = 0 \text{ to } J = 0 \text{ is forbidden.}$$

There are approximate selection rules for $L$ and $S$, namely

$$\Delta L = 0, \pm 1 \qquad \text{and} \qquad \Delta S = 0.$$

These hold quite well for light atoms but become less dependable with greater nuclear charge. The term 'intercombination bands' is used for spectra where the spin quantum number $S$ changes: for example, singlet–triplet transitions. They are very weak in light atoms but quite easily observed in heavy ones.

### (b) Electronic selection rules for molecules

Atoms have complete spherical symmetry, and the angular momentum states can be considered as different symmetry classes of that spherical symmetry. The nuclear framework of a molecule has a much lower symmetry. Symmetry operations for the molecule are transformations such as rotations about an axis, reflection in a plane, or inversion through a point at the centre of the molecule, which leave the molecule in an equivalent configuration. Every molecule has one such operation, the identity operation, which just leaves the molecule alone. Many molecules have one or more additional operations. The set of operations for a molecule form a mathematical group, and the methods of group theory provide a way to classify electronic and vibrational states according to whatever symmetry does exist. That classification leads to selection rules for transitions between those states. A complete discussion of the methods is beyond the scope of this chapter, but we will consider a few illustrative examples. Additional details will also be found in section A1.4 on molecular symmetry.

In the case of linear molecules there is still complete rotational symmetry about the internuclear axis. This leads to the conservation and quantization of the component of angular momentum in that direction. The quantum number for the component of orbital angular momentum along the axis (the analogue of $L$ for an atom) is called $\Lambda$. States which have $\Lambda = 0, 1, 2, \ldots$ are called $\Sigma$, $\Pi$, $\Delta$, . . . (analogous to S, P, D, . . . of atoms). $\Sigma$ states are non-degenerate while $\Pi$, $\Delta$, and higher angular momentum states are always doubly degenerate because the angular momentum can be in either direction about the axis. $\Sigma$ states need an additional symmetry designation. They are called $\Sigma^+$ or $\Sigma^-$ according to whether the electronic wavefunction is symmetric or antisymmetric to the symmetry operation of a reflection in a plane containing the internuclear axis. If the molecule has a centre of symmetry like $N_2$ or $CO_2$, there is an additional symmetry classification, g or u, depending on whether the wavefunction is symmetric or antisymmetric with respect to inversion through that centre. Symmetries of states are designated by symbols such as $\Pi_g$, $\Pi_u$, $\Sigma_g^+$, etc. Finally, the electronic wavefunctions will have a spin multiplicity $(2S+1)$, referred to as singlet, doublet, triplet, etc. The conventional nomenclature for electronic states is as follows. The state is designated by its symmetry with a left superscript giving its multiplicity. An uppercase letter is placed before the symmetry symbol to indicate where it stands in order of energy: the ground state is designated X, higher states of the same multiplicity are designated A, B, C, ... in order of increasing energy, and states of different multiplicity are designated a, b, c, ... in order of increasing energy. (Sometimes, after a classification of states as A, B, C, etc has become well established, new states will be discovered lying between, say, the B and C states. Then the new states may be designated B', B'' and so on rather than renaming all the states.) For example, the $C_2$ molecule has a singlet ground state designated as X $^1\Sigma_g^+$, a triplet state designated a $^3\Pi_u$ lying only 700 cm$^{-1}$ above the ground state, another triplet 5700 cm$^{-1}$ higher in energy designated b $^3\Sigma_g^-$, a singlet state designated A $^1\Pi_u$ lying 8400 cm$^{-1}$ above the ground state and many other known states [16]. A transition between the ground state and the A state would be designated as A $^1\Pi_u$–X $^1\Sigma_g^+$. The convention is to list the upper state first regardless of whether the transition is being studied in absorption or emission.

The electronic selection rules for linear molecules are as follows. $\Delta\Lambda = 0, \pm 1$. $\Delta S = 0$. Again, these are really valid only in the absence of spin–orbital coupling and are modified in heavy molecules. For transitions between $\Sigma$ states there is an additional rule that $\Sigma^+$ combines only with $\Sigma^+$ and $\Sigma^-$ combines only with $\Sigma^-$, so that transitions between $\Sigma^+$ states and $\Sigma^-$ states are forbidden. If the molecule has a centre of symmetry then there is an additional rule requiring that g↔u, while transitions between two g states or between two u states are forbidden.

We now turn to electronic selection rules for symmetrical nonlinear molecules. The procedure here is to examine the structure of a molecule to determine what symmetry operations exist which will leave the molecular framework in an equivalent configuration. Then one looks at the various possible point groups to see what group would consist of those particular operations. The character table for that group will then permit one to classify electronic states by symmetry and to work out the selection rules. Character tables for all relevant groups can be found in many books on spectroscopy or group theory. Here we will only pick one very simple point group called $C_{2v}$ and look at some simple examples to illustrate the method.

The $C_{2v}$ group consists of four symmetry operations: an identity operation designated $E$, a rotation by one-half of a full rotation, i.e. by 180°, called a $C_2$ operation and two planes of reflection passing through the $C_2$ axis and called $\sigma_v$ operations. Examples of molecules belonging to this point group are water, $H_2O$; formaldehyde, $H_2CO$; or pyridine, $C_5H_5N$. It is conventional to choose a molecule-fixed axis system with the $z$ axis coinciding with the $C_2$ axis. If the molecule is planar, it is conventionally chosen to lie in the $yz$ plane with the $x$ axis perpendicular to the plane of the molecule [17]. For example, in $H_2CO$ the $C_2$ or $z$ axis lies along the C–O bond. One of the $\sigma_v$ planes would be the plane of the molecule, the $yz$ plane. The other reflection plane is the $xz$ plane, perpendicular to the molecular plane.

Table B1.1.1 gives the character table for the $C_{2v}$ point group as it is usually used in spectroscopy. Because each symmetry operation leaves the molecular framework and hence the potential energy unchanged, it should not change the electron density or nuclear position density: i.e. the square of an electronic or vibrational wavefunction should remain unchanged. In a group with no degeneracies like this one, that means that a wavefunction itself should either be unchanged or should change sign under each of the four symmetry operations. The result of group theory applied to such functions is that there are only four possibilities for how they change under the operations, and they correspond to the four irreducible representations designated as $A_1$, $A_2$, $B_1$ and $B_2$. The characters may be taken to describe what happens in each symmetry. For example, a function classified as $B_1$ would be unchanged by the identity operation or by $\sigma_v(xz)$ but would be changed in sign by the $C_2$ or $\sigma_v(yz)$ operations. Every molecular orbital and every stationary state described by a

**Table B1.1.1.** Character table for the $C_{2v}$ point group.

| $C_{2v}$ | $E$ | $C_2$ | $\sigma_v(xz)$ | $\sigma_v(yz)$ | | |
|---|---|---|---|---|---|---|
| $A_1$ | 1 | 1 | 1 | 1 | $z$ | $x^2, y^2, z^2$ |
| $A_2$ | 1 | 1 | $-1$ | $-1$ | $R_z$ | $xy$ |
| $B_1$ | 1 | $-1$ | 1 | $-1$ | $x, R_y$ | $xz$ |
| $B_2$ | 1 | $-1$ | $-1$ | 1 | $y, R_x$ | $yz$ |

many-electron wavefunction can be taken to belong to one of these symmetry classes. The same applies to vibrations and vibrational states.

The last two columns of the character table give the transformation properties of translations along the $x$, $y$, and $z$ directions, rotations about the three axes represented by $R_x$, etc, and products of two coordinates or two translations represented by $x^2$, $xy$, etc. The information in these columns is very useful for working out selection rules.

Whenever a function can be written as a product of two or more functions, each of which belongs to one of the symmetry classes, the symmetry of the product function is the direct product of the symmetries of its constituents. This direct product is obtained in non-degenerate cases by taking the product of the characters for each symmetry operation. For example, the function $xy$ will have a symmetry given by the direct product of the symmetries of $x$ and of $y$; this direct product is obtained by taking the product of the characters for each symmetry operation. In this example it may be seen that, for each operation, the product of the characters for $B_1$ and $B_2$ irreducible representations gives the character of the $A_2$ representation, so $xy$ transforms as $A_2$.

The applications to selection rules work as follows. Intensities depend on the values of the transition moment integral of equation (B1.1.1):

$$\mu_{ij} = \int \Psi_i^* \mu \Psi_j \, d\tau.$$

An integral like this must vanish by symmetry if the integrand is antisymmetric under any symmetry operation, i.e. it vanishes unless the integrand is totally symmetric. For $C_{2v}$ molecules that means the integrand must have symmetry $A_1$. The symmetry of the integrand is the direct product of the symmetries of the three components in the integral. The transition moment operator is a vector with three components, $\mu_x$, $\mu_y$ and $\mu_z$, which transform like $x$, $y$ and $z$, respectively. To see if a transition between state $i$ and state $j$ is allowed, one determines the symmetries of the three products containing the three components of $\mu$, i.e. $\Psi_i^* \mu_x \Psi_j$, $\Psi_i^* \mu_y \Psi_j$ and $\Psi_i^* \mu_z \Psi_j$. If any one of them is totally symmetrical, the transition is formally allowed. If none of the three is totally symmetrical the transition is forbidden. It should be noted that being allowed does not mean that a transition will be strong. The actual intensity depends on the matrix element $\mu_{ij}$, whose value will depend on the details of the wavefunctions.

There is a further item of information in this procedure. If one of the three component integrals is non-zero, the molecule will absorb light polarized along the corresponding axis. For example, if $\mu_{y,.ij} = \int \Psi_i^* \mu_y \Psi_j \, d\tau$ is non-zero, the transition will absorb light polarized along the $y$ axis. One may be able to observe this polarization directly in the spectrum of a crystal containing the molecules in known orientations. Alternatively, in the gas phase one may be able to tell the direction of polarization in the molecular framework by looking at the intensity distribution among the rotational lines in a high-resolution spectrum.

Analogous considerations can be used for magnetic dipole and electric quadrupole selection rules. The magnetic dipole operator is a vector with three components that transform like $R_x$, $R_y$ and $R_z$. The electric quadrupole operator is a tensor with components that transform like $x^2$, $y^2$, $z^2$, $xy$, $yz$ and $xz$. These latter symmetries are also used to get selection rules for Raman spectroscopy. Character tables for spectroscopic use usually show these symmetries to facilitate such calculations.

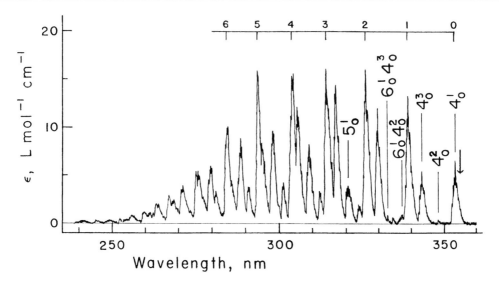

**Figure B1.1.2.** Spectrum of formaldehyde with vibrational resolution. Several vibronic origins are marked. One progression in $\nu_2$ starting from the $4_0^1$ origin is indicated on the line along the top. A similar progression is built on each vibronic origin. Reprinted with permission from [20]. Copyright 1982, American Chemical Society.

When spectroscopists speak of electronic selection rules, they generally mean consideration of the integral over only the electronic coordinates for wavefunctions calculated at the equilibrium nuclear configuration of the initial state, $Q = 0$,

$$\mu_{lu}(0) = \int \psi_l^*(x, 0)\mu\psi_u(x, 0)\, dx. \tag{B1.1.18}$$

If one of the components of this electronic transition moment is non-zero, the electronic transition is said to be allowed; if all components are zero it is said to be forbidden. In the case of diatomic molecules, if the transition is forbidden it is usually not observed unless as a very weak band occurring by magnetic dipole or electric quadrupole interactions. In polyatomic molecules forbidden electronic transitions are still often observed, but they are usually weak in comparison with allowed transitions.

The reason they appear is that symmetric polyatomic molecules always have some non-totally-symmetric vibrations. When the nuclear framework is displaced from the equilibrium position along such a vibrational coordinate, its symmetry is reduced. It can then be thought of as belonging, for the instant, to a different point group of lower symmetry, and it is likely in that group that the transition will be formally allowed. Even though $\mu_{lu}(0)$ from equation (B1.1.18) is zero, $\mu_{lu}(Q)$ from equation (B1.1.5) will be non-zero for some configurations $Q$ involving distortion along antisymmetric normal coordinates. The total integrated intensity of an electronic band is given by (B1.1.10). It involves the square of the electronic transition moment averaged over the initial vibrational state, including the configurations in which the transition moment is not zero. In suitable cases it may be possible to calculate the first-order terms of equation (B1.1.7) from electronic wavefunctions and use them in equation (B1.1.5) to calculate an integrated absorption intensity to compare with the observed integrated intensity or oscillator strength [18].

The spin selection rule for polyatomic molecules is again $\Delta S = 0$, no change in spin in the absence of spin–orbital coupling. This rule becomes less valid when heavy atoms are included in the molecule. Spin-changing transitions can be observed by suitable techniques even in hydrocarbons, but they are quite weak. When spin–orbital coupling is important it is possible to use the symmetries of the spin wavefunctions, assign symmetries to total orbital-plus-spin wavefunctions and use group theory as above to get the selection rules.

Most stable polyatomic molecules whose absorption intensities are easily studied have filled-shell, totally symmetric, singlet ground states. For absorption spectra starting from the ground state the electronic selection rules become simple: transitions are allowed to excited singlet states having symmetries the same as one of the coordinate axes, $x$, $y$ or $z$. Other transitions should be relatively weak.

*(c) Vibronic selection rules*

Often it is possible to resolve vibrational structure of electronic transitions. In this section we will briefly review the symmetry selection rules and other factors controlling the intensity of individual vibronic bands.

In the Born–Oppenheimer approximation the vibronic wavefunction is a product of an electronic wave-function and a vibrational wavefunction, and its symmetry is the direct product of the symmetries of the two components. We have just discussed the symmetries of the electronic states. We now consider the symmetry of a vibrational state. In the harmonic approximation vibrations are described as independent motions along normal modes $Q_i$ and the total vibrational wavefunction is a product of functions, one wavefunction for each normal mode:

$$\phi(Q) = \varphi_{v_1}(Q_1)\varphi_{v_2}(Q_2)\varphi_{v_3}(Q_3)\ldots. \tag{B1.1.19}$$

Each such normal mode can be assigned a symmetry in the point group of the molecule. The wavefunc-tions for non-degenerate modes have the following simple symmetry properties: the wavefunctions with an odd vibrational quantum number $v_i$ have the same symmetry as their normal mode $Q_i$; the ones with an even $v_i$ are totally symmetric. The symmetry of the total vibrational wavefunction $\phi(Q)$ is then the direct product of the symmetries of its constituent normal coordinate functions $\varphi_{v_i}(Q_i)$. In particular, the lowest vibrational state, with all $v_i = 0$, will be totally symmetric. The states with one quantum of excitation in one vibration and zero in all others will have the symmetry of that one vibration. Once the symmetry of the vibrational wave-function has been established, the symmetry of the vibronic state is readily obtained from the direct product of the symmetries of the electronic state and the vibrational state. This procedure gives the correct vibronic symmetry even if the harmonic approximation or the Born–Oppenheimer approximation are not quite valid.

The selection rule for vibronic states is then straightforward. It is obtained by exactly the same procedure as described above for the electronic selection rules. In particular, the lowest vibrational level of the ground electronic state of most stable polyatomic molecules will be totally symmetric. Transitions originating in that vibronic level must go to an excited state vibronic level whose symmetry is the same as one of the coordinates, $x$, $y$, or $z$.

One of the consequences of this selection rule concerns forbidden electronic transitions. They cannot occur unless accompanied by a change in vibrational quantum number for some antisymmetric vibration. Forbidden electronic transitions are not observed in diatomic molecules (unless by magnetic dipole or other interactions) because their only vibration is totally symmetric; they have no antisymmetric vibrations to make the transitions allowed.

The symmetry selection rules discussed above tell us whether a particular vibronic transition is allowed or forbidden, but they give no information about the intensity of allowed bands. That is determined by equation (B1.1.9) for absorption or (B1.1.13) for emission. That usually means by the Franck–Condon principle if only the zero-order term in equation (B1.1.7) is needed. So we take note of some general principles for Franck–Condon factors (FCFs). Usually the normal coordinates of the upper and lower states are quite similar. (When they are not it is called a Duschinsky effect [19] and the treatment becomes more complicated.) Because of the product form of the vibrational wavefunctions of equation (B1.1.19) the FCF is itself a product of FCFs for individual normal modes. If there is little or no change in the geometry of a given normal mode, the FCF for that mode will be large only if its vibrational quantum number does not change. But for modes for which there is a significant change in geometry, the FCFs may be large for a number of vibrational levels in the final state. The spectrum then shows a series of vibronic peaks differing in energy by the frequency of that vibration. Such a series is referred to as a progression in that mode.

Most commonly, the symmetry point group of the lower and upper states will be the same. Then only totally symmetric vibrations can change equilibrium positions—a change in a non-totally-symmetric mode would mean that the states have configurations belonging to different point groups. So one may expect to see progressions in one or more of the totally symmetric vibrations but not in antisymmetric vibrations. In symmetry-forbidden electronic transitions, however, one will see changes of one quantum (or possibly other odd numbers of quanta) in antisymmetric vibrations as required to let the transition appear.

An example of a single-absorption spectrum illustrating many of the effects discussed in this section is the spectrum of formaldehyde, $H_2CO$, shown in figure B1.1.2 [20]. This shows the region of the lowest singlet–singlet transition, the $A\,^1A_2$—$X\,^1A_1$ transition. This is called an $n \to \pi^*$ transition; the electronic change is the promotion of an electron from a non-bonding orbital (symmetry $B_2$) mostly localized on the oxygen atom into an antibonding $\pi^*$ orbital (symmetry $B_1$) on the C–O bond. By the electronic selection rules, a transition from the totally symmetric ground state to a $^1A_2$ state is symmetry forbidden, so the transition is quite weak with an oscillator strength of $2.4 \times 10^{-4}$. The transition is appearing with easily measured intensity due to coupling with antisymmetric vibrations. Most of the intensity is induced by distortion along the out-of-plane coordinate, $Q_4$. This means that in equation (B1.1.7) the most significant derivative of $\mu_{1u}$ is the one with respect to $Q_4$. The first peak seen in figure B1.1.2, at 335 nm, has one quantum of vibration $\nu_4$ excited in the upper state. The band is designated as $4_0^1$ which means that vibration number 4 has 1 quantum of excitation in the upper state and 0 quanta in the lower state. The symmetry of $Q_4$ is $B_1$, and combined with the $A_2$ symmetry of the electronic state it gives an upper state of vibronic symmetry $B_2$, the direct product $A_2 \times B_1$. It absorbs light with its electric vector polarized in the $y$ direction, i.e. in plane and perpendicular to the C–O bond.

If the 0–0 band were observable in this spectrum it would be called the origin of the transition. The $4_0^1$ band is referred to as a vibronic origin. Starting from it there is a progression in $\nu_2$, the C–O stretching mode, which gets significantly longer in the upper state because the presence of the antibonding $\pi$ electron weakens the bond. Several of the peaks in the progression are marked along the line at the top of the figure.

Several other vibronic origins are also marked in this spectrum. The second major peak is the $4_0^3$ band, with three quanta of $\nu_4$ in the upper state. This upper state has the same symmetry as the state with one quantum. Normally, one would not expect much intensity in this peak, but it is quite strong because the excited state actually has a non-planar equilibrium geometry, i.e. it is distorted along $Q_4$. Every vibronic origin including this one has a progression in $\nu_2$ built on it. The intensity distribution in a progression is determined by the Franck–Condon principle and, as far as can be determined, all progressions in this spectrum are the same.

At 321 nm there is a vibronic origin marked $5_0^1$. This has one quantum of $\nu_5$, the antisymmetric C–H stretching mode, in the upper state. Its intensity is induced by a distortion along $Q_5$. This state has $B_2$ vibrational symmetry. The direct product of $B_2$ and $A_2$ is $B_1$, so it has $B_1$ vibronic symmetry and absorbs x-polarized light. One can also see a $4_0^2 6_0^1$ vibronic origin which has the same symmetry and intensity induced by distortion along $Q_6$.

A very weak peak at 348 nm is the $4_0^2$ origin. Since the upper state here has two quanta of $\nu_4$, its vibrational symmetry is $A_1$ and the vibronic symmetry is $A_2$, so it is forbidden by electric dipole selection rules. It is actually observed here due to a magnetic dipole transition [21]. By magnetic dipole selection rules the $^1A_2$–$^1A_1$ electronic transition is allowed for light with its magnetic field polarized in the $z$ direction. It is seen here as having about 1% of the intensity of the symmetry-forbidden electric dipole transition made allowed by vibronic coupling, or an oscillator strength around $10^{-6}$. This illustrates the weakness of magnetic dipole transitions.

### (d) Rotational selection rules

If the experimental technique has sufficient resolution, and if the molecule is fairly light, the vibronic bands discussed above will be found to have a fine structure due to transitions among rotational levels in the two states. Even when the individual rotational lines cannot be resolved, the overall shape of the vibronic band will be related to the rotational structure and its analysis may help in identifying the vibronic symmetry. The analysis of the band appearance depends on calculation of the rotational energy levels and on the selection rules and relative intensity of different rotational transitions. These both come from the form of the rotational wavefunctions and are treated by angular momentum theory. It is not possible to do more than mention a simple example here.

The simplest case is a $^1\Sigma{-}^1\Sigma$ transition in a linear molecule. In this case there is no orbital or spin angular momentum. The total angular momentum, represented by the quantum number $J$, is entirely rotational angular momentum. The rotational energy levels of each state approximately fit a simple formula:

$$E_J = BJ(J + 1) - DJ^2(J + 1)^2.$$

The second term is used to allow for centrifugal stretching and is usually small but is needed for accurate work. The quantity $B$ is called the rotation constant for the state. In a rigid rotator picture it would have the value

$$B = \frac{h^2}{8\pi^2 I}$$

and is usually quoted in reciprocal centimetres. $I$ is the moment of inertia. In an actual molecule which is vibrating, the formula for $B$ must be averaged over the vibrational state, i.e. one must use an average value of $1/I$. As a result $B$ varies somewhat with vibrational level. The values of $B$ and the moments of inertia obtained from the spectra are used to get structural information about the molecule. The bonding and hence the structure will be different in the two states, so the $B$ values will generally differ significantly. They are called $B'$ and $B''$. The convention is to designate quantities for the upper state with a single prime and quantities for the lower state with a double prime.

The rotational selection rule for a $^1\Sigma{-}^1\Sigma$ transition is $\Delta J = \pm 1$. Lines which have $J' = J'' - 1$ are called P lines and the set of them is called the P branch of the band. Lines for which $J' = J'' + 1$ are called R lines and the set of them the R branch. (Although not seen in a $^1\Sigma{-}^1\Sigma$ transition, a branch with $J' = J''$ would be called a Q branch, one with $J' = J'' - 2$ would be an O branch, or one with $J' = J'' + 2$ would be an S branch, etc.) Individual lines may be labeled by giving $J''$ in parentheses like R(1), P(2), etc. For lines with low values of $J$, the R lines get higher in energy as $J$ increases while the P lines get lower in energy with increasing $J$. If $B''$ and $B'$ are sufficiently different, which is the usual case in electronic spectra, the lines in one of the two branches will get closer together as $J$ increases until they pile up on top of each other and then turn around and start to move in the other direction as $J$ continues to increase. The point at which the lines pile up is called a band head. It is often the most prominent feature of the band. If $B'' > B'$, this will happen in the R branch and the band head will mark the high-energy or short-wavelength limit of each vibronic band. Such a band is said to be shaded to the red because absorption or emission intensity has a sharp onset on the high-energy side and then falls off gradually on the low-energy or red side. This is the most common situation where the lower state is bound more tightly and has a smaller moment of inertia than the upper state. But the opposite can occur as well. If $B'' < B'$ the band head will be formed in the P branch on the low-energy side of the vibronic band, and the band will be said to be shaded to the violet or shaded to the blue. Note that the terms red for the low-energy direction and violet or blue for the high-energy direction are used even for spectra in the ultraviolet or infrared regions where the actual visible red and blue colours would both be in the same direction.

The analysis of rotational structure and selection rules for transitions involving $\Pi$ or $\Delta$ states becomes considerably more complicated. In general, Q branches will be allowed as well as P and R branches. The coupling between different types of angular momenta—orbital angular momentum of the electrons, spin angular momentum for states of higher multiplicity, the rotational angular momentum and even nuclear spin terms—is a complex subject which cannot be covered here: the reader is referred to the more specialized literature.

*B1.1.3.4  Perturbations*

Spectroscopists working with high-resolution spectra of small molecules commonly fit series of rotational lines to formulae involving the rotational constants, angular momentum coupling terms, etc. However, occasionally they find that some lines in the spectrum are displaced from their expected positions in a systematic way. Of course, a displacement of a line from its expected position means that the energy levels of one of the states are displaced from their expected energies. Typically, as $J$ increases some lines will be seen to be displaced in one direction by increasing amounts up to a maximum at some particular $J$, then for the next $J$ the line will be displaced in the opposite direction, and then as $J$ increases further the lines will gradually approach their expected positions. These displacements of lines and of state energies are called perturbations [22].

They are caused by interactions between states, usually between two different electronic states. One hard and fast selection rule for perturbations is that, because angular momentum must be conserved, the two interacting states must have the same $J$. The interaction between two states may be treated by second-order perturbation theory which says that the displacement of a state is given by

$$\Delta E_1 = \frac{|H'_{12}|^2}{E_1^\circ - E_2^\circ}$$

where $H'_{12}$ is the matrix element between the two states of some small term $H'$ in the Hamiltonian which is unimportant in determining the expected energies of the states. This interaction always has the effect of pushing the two states apart in energy by equal and opposite displacements inversely proportional to the zero-order separation of the two states. The perturbation is observed when the vibronic level of the state with the larger $B$ value lies slightly lower in energy than a vibronic level of the other state. Then with increasing $J$ the energy of the rotating level of the first state gets closer and closer to the energy of the second state and finally passes it and then gets farther away again. The maximum displacements occur at the $J$ values where the two energies are the closest.

The spectral perturbations are observed in a transition involving one of the interacting states. Sometimes it is possible also to see an electronic transition involving the other of the interacting states, and then one should see equal but opposite displacements of rotational levels with the same $J$.

An interesting example occurs in the spectrum of the $C_2$ molecule. The usual rule of absorption spectroscopy is that the transitions originate in the ground electronic state because only it has sufficient population. However, in $C_2$ transitions were observed starting both from a $^3\Pi_u$ state and from a $^1\Sigma_g^+$ state, so it was not clear which was the ground state. The puzzle was solved by Ballik and Ramsay [23] who observed perturbations in a $^3\Sigma_g^- - ^3\Pi_u$ transition due to displacements of levels in the $^3\Sigma_g^-$ state. They then reinvestigated a $^1\Pi_u - ^1\Sigma_g^+$ transition known as the Phillips system, and they observed equal and opposite displacements of levels in the $^1\Sigma_g^+$ state, thus establishing that the $^1\Sigma_g^+$ and $^3\Sigma_g^-$ states were perturbing each other. For example, in the $v = 4$ vibrational level of the $^1\Sigma_g^+$ state, the $J = 40$ rotational level was displaced to lower energy by $0.26~cm^{-1}$; correspondingly, in the $v = 1$ vibrational level of the $^3\Sigma_g^-$ state, the $J = 40$ level was displaced upwards by $0.25~cm^{-1}$. The values have an uncertainty of $0.02~cm^{-1}$, so the displacements of the levels with the same $J$ are equal and opposite within experimental error. Similarly, the $J = 42$ level in the $^1\Sigma_g^+$ state was displaced upwards by $0.17~cm^{-1}$ and the $J = 42$ level of the $^3\Sigma_g^-$ state displaced downwards by $0.15~cm^{-1}$. These observations established that these particular levels were very close to each other in energy and the authors were able to prove that the $^1\Sigma_g^+$ was lower by about $600~cm^{-1}$ than the $^3\Pi_u$ state and was the ground state. Absorption spectra of $C_2$ are typically observed in the vapour over graphite at high temperatures. At 2500 K the value of $kT$ is about $1700~cm^{-1}$, much greater than the energy separation of the two states. Since the $^1\Sigma_g^+$ state is non-degenerate and the $^3\Pi_u$ state has a sixfold degeneracy, most of the molecules are actually in the upper state. This accounts for the observation of absorptions starting from both states.

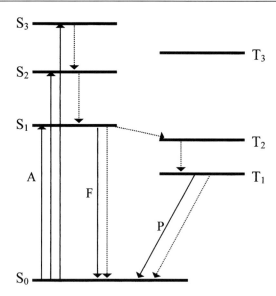

**Figure B1.1.3.** State energy diagram for a typical organic molecule. Solid arrows show radiative transitions; A: absorption, F: fluorescence, P: phosphorescence. Dotted arrows: non-radiative transitions.

The perturbations in this case are between a singlet and a triplet state. The perturbation Hamiltonian, $H'$, of the second-order perturbation theory is spin–orbital coupling, which has the effect of mixing singlet and triplet states. The magnitude of the perturbations can be calculated fairly quantitatively from high-quality electronic wavefunctions including configuration interaction [24].

## B1.1.4   Examples

### B1.1.4.1   Photophysics of molecules in solution

To understand emission spectroscopy of molecules and/or their photochemistry it is essential to have a picture of the radiative and non-radiative processes among the electronic states. Most stable molecules other than transition metal complexes have all their electrons paired in the ground electronic state, making it a singlet state. Figure B1.1.3 gives a simple state energy diagram for reference. Singlet states are designated by the letter S and numbered in order of increasing energy. The ground state is called $S_0$. Excited singlet states have configurations in which an electron has been promoted from one of the filled orbitals to one of the empty orbitals of the molecule. Such configurations with two singly occupied molecular orbitals will give rise to triplet states as well as singlet states. A triplet state results when one of the electrons changes its spin so that the two electrons have parallel spin. Each excited singlet state will have its corresponding triplet state. Because the electron–electron repulsion is less effective in the triplet state, it will normally be lower in energy than the corresponding singlet state.

Spectroscopists observed that molecules dissolved in rigid matrices gave both short-lived and long-lived emissions which were called fluorescence and phosphorescence, respectively. In 1944, Lewis and Kasha [25] proposed that molecular phosphorescence came from a triplet state and was long-lived because of the well known spin selection rule $\Delta S = 0$, i.e. interactions with a light wave or with the surroundings do not readily change the spin of the electrons. Typical singlet lifetimes are measured in nanoseconds while triplet lifetimes of organic molecules in rigid solutions are usually measured in milliseconds or even seconds. In liquid media where diffusion is rapid the triplet states are usually quenched, often by the nearly ubiqitous molecular oxygen.

Because of that, phosphorescence is seldom observed in liquid solutions. In the spectroscopy of molecules the term fluorescence is now usually used to refer to emission from an excited singlet state and phosphorescence to emission from a triplet state, regardless of the actual lifetimes.

If a light beam is used to excite one of the higher singlet states, say $S_2$, a very rapid relaxation occurs to $S_1$, the lowest excited singlet state. This non-radiative process just converts the difference in energy into heat in the surroundings. A radiationless transition between states of the same multiplicity is called internal conversion. Relaxation between states of the same multiplicity and not too far apart in energy is usually much faster than radiative decay, so fluorescence is seen only from the $S_1$ state. These radiationless processes in large molecules are the analogue of the perturbations observed in small molecules. They are caused by small terms in the Hamiltonian such as spin–orbital coupling or Born–Oppenheimer breakdown, which mix electronic states. The density of vibrational levels of large molecules can be very high and that makes these interactions into irreversible transitions to lower states.

Once the excited molecule reaches the $S_1$ state it can decay by emitting fluorescence or it can undergo a further radiationless transition to a triplet state. A radiationless transition between states of different multiplicity is called intersystem crossing. This is a spin-forbidden process. It is not as fast as internal conversion and often has a rate comparable to the radiative rate, so some $S_1$ molecules fluoresce and others produce triplet states. There may also be further internal conversion from $S_1$ to the ground state, though it is not easy to determine the extent to which that occurs. Photochemical reactions or energy transfer may also occur from $S_1$.

Molecules which reach a triplet state will generally relax quickly to state $T_1$. From there they can emit phosphorescence or decay by intersystem crossing back to the ground state. Both processes are spin forbidden and again often have comparable rates. The $T_1$ state is often also important for photochemistry because its lifetime is relatively long. Both phosphorescence and intersystem crossing are dependent on spin–orbital coupling and are enhanced by heavy atoms bound to the molecule or in the environment. They are also enhanced by the presence of species with unpaired electrons such as $O_2$ because electron exchange can effect such transitions without actually requiring the spin of an electron to be reversed. $O_2$ is found to quench both fluorescence and phosphorescence, and it is often necessary to remove oxygen from solutions for precise emission measurements.

## B1.1.4.2  *Widths and shapes of spectral lines*

High-resolution spectroscopy used to observe hyperfine structure in the spectra of atoms or rotational structure in electronic spectra of gaseous molecules commonly must contend with the widths of the spectral lines and how that compares with the separations between lines. Three contributions to the linewidth will be mentioned here: the natural line width due to the finite lifetime of the excited state, collisional broadening of lines, and the Doppler effect.

The most fundamental limitation on sharpness of spectral lines is the so-called natural linewidth. Because an excited state has a finite lifetime, the intensity of light it emits falls off exponentially as a function of time. A beam of light whose intensity varies with time cannot have a single frequency. Its spectral distribution is the Fourier transform of its temporal shape. For an exponential decay the spectral distribution will have the form

$$I(\nu) = I(\nu_0)\frac{(1/4\pi\tau)^2}{(\nu - \nu_0)^2 + (1/4\pi\tau)^2} \tag{B1.1.20}$$

where $\nu_0$ is the frequency of the centre of the band and $\tau$ is the mean life of the excited state. The same formula applies to lines in the absorption spectrum. This shape is called a Lorentzian shape. Its full width at half maximum (FWHM) is $1/(2\pi\tau)$. The shorter the lifetime, the broader the line. Another way to think about the width is to say that the energy of a state has an uncertainty related to its lifetime by the uncertainty principle. If the transition is coupling two states, both of which have finite lifetimes and widths, it is necessary to combine the effects.

Spectral lines are further broadened by collisions. To a first approximation, collisions can be thought of as just reducing the lifetime of the excited state. For example, collisions of molecules will commonly change the rotational state. That will reduce the lifetime of a given state. Even if the state is not changed, the collision will cause a phase shift in the light wave being absorbed or emitted and that will have a similar effect. The line shapes of collisionally broadened lines are similar to the natural line shape of equation (B1.1.20) with a lifetime related to the mean time between collisions. The details will depend on the nature of the intermolecular forces. We will not pursue the subject further here.

A third source of spectral broadening is the Doppler effect. Molecules moving toward the observer will emit or absorb light of a slightly higher frequency than the frequency for a stationary molecule; those moving away will emit or absorb a slightly lower frequency. The magnitude of the effect depends on the speed of the molecules. To first order the frequency shift is given by

$$\Delta \nu = \nu_0 (\nu_x / c)$$

where $\nu_x$ is the component of velocity in the direction of the observer and $c$ is the speed of light.

For a sample at thermal equilibrium there is a distribution of speeds which depends on the mass of the molecules and on the temperature according to the Boltzmann distribution. This results in a line shape of the form

$$I(\nu) = I(\nu_0) \exp \left[ \frac{-Mc^2}{2\nu_0^2 RT} (\nu - \nu_0)^2 \right]$$

where $M$ is the atomic or molecular mass and $R$ the gas constant. This is a Gaussian line shape with a width given by

$$\text{FWHM} = \frac{2\nu_0}{c} \left( \frac{2RT \ln 2}{M} \right)^{1/2}.$$

The actual line shape in a spectrum is a convolution of the natural Lorentzian shape with the Doppler shape. It must be calculated for a given case as there is no simple formula for it. It is quite typical in electronic spectroscopy to have the FWHM determined mainly by the Doppler width. However, the two shapes are quite different and the Lorentzian shape does not fall off as rapidly at large $(\nu - \nu_0)$. It is likely that the intensity in the wings of the line will be determined by the natural line shape.

Collisional broadening is reduced by the obvious remedy of working at low pressures. Of course, this reduces the absorption and may require long path lengths for absorption spectroscopy. Doppler widths can be reduced by cooling the sample. For many samples this is not practical because the molecules will have too low a vapour pressure. Molecular beam methods and the newer technique of jet spectroscopy can be very effective by restricting the motion of the molecules to be at right angles to the light beam. Some other techniques for sub-Doppler spectroscopy have also been demonstrated using counter-propagating laser beams to compensate for motion along the direction of the beam. The natural linewidth, however, always remains when the other sources of broadening are removed.

### B1.1.4.3   Rydberg spectra

The energies of transitions of a hydrogen atom starting from the ground state fit exactly the equation

$$E_n = E_I - R/n^2$$

where $R$ is the Rydberg constant, $E_I$ is the ionization energy of the atom (which in hydrogen is equal to the Rydberg constant) and $n$ is the principal quantum number of the electron in the upper state. The spectrum shows a series of lines of increasing $n$ which converge to a limit at the ionization energy.

Other atoms and molecules also show similar series of lines, often in the vacuum ultraviolet region, which fit approximately a similar formula:

$$E_n = E_{\mathrm{I}} - \frac{R}{(n-\delta)^2}.$$

Such a series of lines is called a Rydberg series [26]. These lines also converge to the ionization energy of the atom or molecule, and fitting the lines to this formula can give a very accurate value for the ionization energy. In the case of molecules there may be resolvable vibrational and rotational structure on the lines as well.

The excited states of a Rydberg series have an electron in an orbital of higher principal quantum number, $n$, in which it spends most of its time far from the molecular framework. The idea is that the electron then feels mainly a Coulomb field due to the positive ion remaining behind at the centre, so its behaviour is much like that of the electron in the hydrogen atom. The constant $\delta$ is called the quantum defect and is a measure of the extent to which the electron interacts with the molecular framework. It has less influence on the energy levels as $n$ gets larger, i.e. as the electron gets farther from the central ion. The size of $\delta$ will also depend on the angular momentum of the electron. States of lower angular momentum have more probability of penetrating the charge cloud of the central ion and so may have larger values of $\delta$. Actual energy levels of Rydberg atoms and molecules can be subject to theoretical calculations [27]. Sometimes the higher states have orbitals so large that other molecules may fall within their volume, causing interesting effects [28].

### B1.1.4.4  Multiphoton spectroscopy

All the previous discussion in this chapter has been concerned with absorption or emission of a single photon. However, it is possible for an atom or molecule to absorb two or more photons simultaneously from a light beam to produce an excited state whose energy is the sum of the energies of the photons absorbed. This can happen even when there is no intermediate stationary state of the system at the energy of one of the photons. The possibility was first demonstrated theoretically by Maria Göppert-Mayer in 1931 [29], but experimental observations had to await the development of the laser. Multiphoton spectroscopy is now a useful technique [30, 31].

The transition probability for absorption of two photons can be described in terms of a two-photon cross section $\delta$ by

$$-\mathrm{d}I = \delta I^2 N \,\mathrm{d}l$$

where $I$ is a photon flux in photons cm$^{-2}$ s$^{-1}$, $N$ is the number of molecules per cubic centimetre, and $l$ is distance through the sample. Measured values of $\delta$ are of the order of $10^{-50}$ cm$^4$ s photon$^{-1}$ molecule$^{-1}$ [32]. For molecules exposed to the intensity of sunlight at the earth's surface this would suggest that the molecule might be excited once in the age of the universe. However, the probability is proportional to the square of the light intensity. For a molecule exposed to a pulsed laser focused to a small spot, the probability of being excited by one pulse may be easily observable by fluorescence excitation or multiphoton ionization techniques.

One very important aspect of two-photon absorption is that the selection rules for atoms or symmetrical molecules are different from one-photon selection rules. In particular, for molecules with a centre of symmetry, two-photon absorption is allowed only for g↔g or u↔u transitions, while one-photon absorption requires g↔u transitions. Therefore, a whole different set of electronic states becomes allowed for two-photon spectroscopy. The group-theoretical selection rules for two-photon spectra are obtained from the symmetries of the $x^2$, $xy$, etc. terms in the character table. This is completely analogous to the selection rules for Raman spectroscopy, a different two-photon process.

A good example is the spectrum of naphthalene. The two lowest excited states have $B_{2u}$ and $B_{1u}$ symmetries and are allowed for one-photon transitions. A weak transition to one of these is observable in the two-photon spectrum [33], presumably made allowed by vibronic effects. Much stronger two-photon transitions are observable at somewhat higher energies to a $B_{3g}$ and an $A_g$ state lying quite close to the energies predicted by theory many years earlier [34].

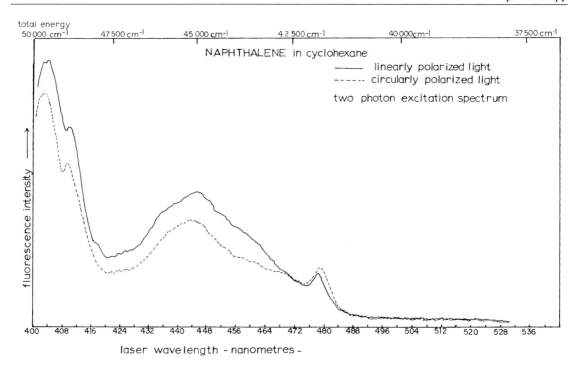

**Figure B1.1.4.** Two-photon fluorescence excitation spectrum of naphthalene. Reprinted from [35]. Courtesy, Tata McGraw-Hill Publishing Company Ltd, 7 West Patel Nagar, New Dehli, 110008, India.

An interesting aspect of two-photon spectroscopy is that some polarization information is obtainable even for randomly oriented molecules in solution by studying the effect of the relative polarization of the two photons. This is readily done by comparing linearly and circularly polarized light. Transitions to $A_g$ states will absorb linearly polarized light more strongly than circularly polarized light. The reverse is true of transitions to $B_{1g}$, $B_{2g}$, or $B_{3g}$ states. The physical picture is that the first photon induces an oscillating u-type polarization of the molecule in one direction. To get to a totally symmetric $A_g$ state the second photon must reverse that polarization, so is favoured for a photon of the same polarization. However, to get to, say, a $B_{3g}$ state, the second photon needs to act at right angles to the first, and in circularly polarized light that perpendicular polarization is always strong. Figure B1.1.4 shows the two-photon fluorescence excitation spectrum of naphthalene in the region of the g states [35]. One peak shows stronger absorption for circularly polarized, one for linearly polarized light. That confirms the identification as $B_{3g}$ and $A_g$ states respectively.

Three-photon absorption has also been observed by multiphoton ionization, giving Rydberg states of atoms or molecules [36]. Such states usually require vacuum ultraviolet techniques for one-photon spectra, but can be done with a visible or near-ultraviolet laser by three-photon absorption.

*B1.1.4.5   Other examples*

Many of the most interesting current developments in electronic spectroscopy are addressed in special chapters of their own in this encyclopedia. The reader is referred especially to sections B2.1 on ultrafast spectroscopy, C1.5 on single molecule spectroscopy, C3.2 on electron transfer, and C3.3 on energy transfer. Additional topics on electronic spectroscopy will also be found in many other chapters.

# References

[1] Sawyer R A 1951 *Experimental Spectroscopy* 2nd edn (Englewood Cliffs, NJ: Prentice-Hall)

[2] O'Conner D V and Phillips D 1984 *Time-Correlated Single Photon Counting* (London: Academic)

[3] Wenthold P and Lineberger W C 1999 Negative ion photoelectron spectroscopy studies of organic reactive intermediates *Accts. Chem. Res.* **32** 597–604

[4] Pauling L and Wilson E B 1935 *Introduction to Quantum Mechanics* (New York: McGraw-Hill) pp 294–314

[5] Hilborn H 1982 Einstein coefficients, cross sections, $f$ values, dipole moments and all that *Am. J. Phys.* **50** 982–6

[6] Kuhn H 1958 Oscillator strength of absorption bands in dye molecules *J. Chem. Phys.* **29** 958–9

[7] Kauzmann W 1957 *Quantum Chemistry* (New York: Academic) pp 651–3

[8] Born M and Oppenheimer R 1927 Concerning the quantum theory of molecules *Ann. Phys., Lpz* **84** 457–84

[9] Franck J 1925 Elementary processes of photochemical reactions *Trans. Faraday Soc.* **21** 536

[10] Condon E U 1928 Nuclear motion associated with electron transitions in diatomic molecules *Phys. Rev.* **32** 858–72
Condon E U 1947 The Franck–Condon principle and related topics *Am. J. Phys.* **15** 365–79

[11] Greenblatt G D, Orlando J J, Burkholder J B and Ravishankara A R 1990 Absorption measurements of oxygen between 330 and 1140 nm *J. Geophys. Res.* **95** 18 577–82

[12] Strickler S J and Vikesland J P 1974 $^3B_1$–$^1A_1$ transition of $SO_2$ gas. I. Franck–Condon treatment and transition moments *J. Chem. Phys.* **60** 660–3

[13] Einstein A 1917 On the quantum theory of radiation *Phys. Z.* **18** 121–8

[14] Strickler S J and Berg R A 1962 Relationship between absorption intensity and fluorescence lifetime of molecules *J. Chem. Phys.* **37** 814–22

[15] Strickler S J, Vikesland J P and Bier H D 1974 $^3B_1$–$^1A_1$ transition of $SO_2$ gas. II. Radiative lifetime and radiationless processes *J. Chem. Phys.* **60** 664–7

[16] Herzberg G, Lagerquist A and Malmberg C 1969 New electronic transitions of the $C_2$ molecule in absorption in the vacuum ultraviolet region *Can. J. Phys.* **47** 2735–43

[17] Mulliken R S 1955 Report on notation for the spectra of polyatomic molecules *J. Chem. Phys.* **23** 1997–2011

[18] Robey M J, Ross I G, Southwood-Jones R V and Strickler S J 1977 *A priori* calculations on vibronic coupling in the $^1B_{2u}$–$^1A_g$ (3200 Å) and higher transitions of naphthalene *Chem. Phys.* **23** 207–16

[19] Duschinsky F 1937 On the interpretation of electronic spectra of polyatomic molecules. I. Concerning the Franck–Condon Principle *Acta Physicochimica URSS* **7** 551

[20] Strickler S J and Barnhart R J 1982 Absolute vibronic intensities in the $^1A_2 \leftarrow {}^1A_1$ absorption spectrum of formaldehyde *J. Phys. Chem.* **86** 448–55

[21] Callomom J H and Innes K K 1963 Magnetic dipole transition in the electronic spectrum of formaldehyde *J. Mol. Spectrosc.* **10** 166–81

[22] Lefebvre-Brion H and Field R W 1986 *Perturbations in the Spectra of Diatomic Molecules* (Orlando: Academic)

[23] Ballik E A and Ramsay D A 1963 The $A'^3\Sigma_g^-$–$X'^3\Pi_u$ band system of the $C_2$ molecule *Astrophys. J.* **137** 61–83

[24] Langhoff S R, Sink M L, Pritchard R H, Kern C W, Strickler S J and Boyd M J 1977 *Ab initio* study of perturbations between the $X^1\Sigma_g^+$ and $b^3\Sigma_g^-$ states of the $C_2$ molecule *J. Chem. Phys.* **67** 1051–60

[25] Lewis G N and Kasha M 1944 Phosphorescence and the triplet state *J. Am. Chem. Soc.* **66** 2100–16

[26] Duncan A B F 1971 *Rydberg Series in Atoms and Molecules* (New York: Academic)

[27] Sandorfy C (ed) 1999 *The Role of Rydberg States in Spectroscopy and Photochemistry* (London: Kluwer Academic)

[28] Merkt F 1997 Molecules in high Rydberg states *Ann. Rev. Phys. Chem.* **48** 675–709

[29] Göppert-Mayer M 1931 Concerning elementary processes with two quanta *Ann. Phys.* **9** 273–94

[30] Ashfold M N R and Howe J D 1994 Multiphoton spectroscopy of molecular species *Ann. Rev. Phys. Chem.* **45** 57–82

[31] Callis P R 1997 Two-photon induced fluorescence *Ann. Rev. Phys. Chem.* **48** 271–97

[32] Monson P R and McClain W M 1970 Polarization dependence of the two-photon absorption of tumbling molecules with application to liquid 1-chloronaphthalene and benzene *J. Chem. Phys.* **53** 29–37

[33] Mikami N and Ito M 1975 Two-photon excitation spectra of naphthalene and naphthalene-$d_8$ *Chem. Phys. Lett.* **31** 472–8

[34] Pariser R 1956 Theory of the electronic spectra and structure of the polyacenes and of alternant hydrocarbons *J. Chem. Phys.* **24** 250–68

[35] Strickler S J, Gilbert J V and McClanaham J E 1984 Two-photon absorption spectroscopy of molecules *Lasers and Applications* eds H D Bist and J S Goela (New Delhi: Tata McGraw-Hill) pp 351–61

[36] Johnson P M 1976 The multiphoton ionization spectrum of benzene *J. Chem. Phys.* **64** 4143–8

# Further Reading

The pre-eminent reference works in spectroscopy are the set of books by G Herzberg.

Herzberg G 1937 *Atomic Spectra and Atomic Structure* (New York: Prentice-Hall)

Herzberg G 1950 *Molecular Spectra and Molecular Structure I. Spectra of Diatomic Molecules* 2nd edn (Princeton, NJ: Van Nostrand)

Herzberg G 1945 *Molecular Spectra and Molecular Structure II. Infrared and Raman Spectra of Polyatomic Molecules* (Princeton, NJ: Van Nostrand)

Herzberg G 1966 *Molecular Spectra and Molecular Structure III. Electronic Spectra and Electronic Structure of Polyatomic Molecules* (Princeton, NJ: Van Nostrand)

Huber K P and Herzberg G 1979 *Molecular Spectra and Molecular Structure IV. Constants of Diatomic Molecules* (New York: Van Nostrand-Reinhold)

Herzberg G 1971 *The Spectra and Structures of Simple Free Radicals: An Introduction to Molecular Spectroscopy* (Ithaca, NY: Cornell University Press)

# B1.2
# Vibrational spectroscopy

*Charles Schmuttenmaer*

## B1.2.1   Introduction

### B1.2.1.1   Overview

Vibrational spectroscopy provides detailed information on both structure and dynamics of molecular species. Infrared (IR) and Raman spectroscopy are the most commonly used methods, and will be covered in detail in this chapter. There exist other methods to obtain vibrational spectra, but those are somewhat more specialized and used less often. They are discussed in other chapters, and include: inelastic neutron scattering (INS), helium atom scattering, electron energy loss spectroscopy (EELS), photoelectron spectroscopy, among others.

Vibrational spectra span the frequency range 10–4000 cm$^{-1}$ (10 cm$^{-1}$ = 1.2 meV = 0.03 kcal mol$^{-1}$ = 0.11 kJ mol$^{-1}$, and 4000 cm$^{-1}$ = 496 meV = 10.3 kcal mol$^{-1}$ = 42.9 kJ mol$^{-1}$), depending on the strength of the bond and the reduced mass of the vibrational mode. Very weakly bound species, such as clusters bound only by van der Waals forces, or condensed phase modes with very large effective reduced masses, such as intermolecular modes in liquids or solids, or large amplitude motions in proteins or polymers, have very low frequency vibrations of 10–100 cm$^{-1}$. Modes involving many atoms in moderately large molecules absorb in the range 300–1200 cm$^{-1}$. The region 600–1200 cm$^{-1}$ is referred to as the fingerprint region because while many organic molecules will all have bands due to vibrations of C–H, C=O, O–H and so on, there will be low frequency bands unique to each molecule that involve complicated motions of many atoms. The region 1200–3500 cm$^{-1}$ is where most functional groups are found to absorb. Thus, the presence or absence of absorption at a characteristic frequency helps to determine the identity of a compound. The frequency of the absorption can be justified in terms of the masses of the atoms participating, the type of motion involved (stretch versus bend) and the bond strengths. The $H_2$ molecule has a reasonably large force constant and the smallest reduced mass, which results in the highest vibrational frequency at 4400 cm$^{-1}$. The width and intensity of an absorption feature provide information in addition to the absorption frequency. In favourable cases in the gas phase the width is determined by the vibrational lifetime or even the lifetime of a molecule if the vibrational energy is greater than the bond strength. The intensity yields information on the nature of the vibrational motion, and can also be used to determine the temperature of the sample.

Infrared and Raman spectroscopy each probe vibrational motion, but respond to a different manifestation of it. Infrared spectroscopy is sensitive to a change in the dipole moment as a function of the vibrational motion, whereas Raman spectroscopy probes the change in polarizability as the molecule undergoes vibrations. Resonance Raman spectroscopy also couples to excited electronic states, and can yield further information regarding the identity of the vibration. Raman and IR spectroscopy are often complementary, both in the type of systems that can be studied, as well as the information obtained.

Vibrational spectroscopy is an enormously large subject area spanning many scientific disciplines. The methodology, both experimental and theoretical, was developed primarily by physical chemists and has branched far and wide over the last 50 years. This chapter will mainly focus on its importance with regard to physical chemistry.

### B1.2.1.2   Infrared spectroscopy

For many chemists, the most familiar IR spectrometer is the dual beam instrument that covers the region 900–3400 $cm^{-1}$; it is used for routine analysis and compound identification. Typically, each of the functional groups of a molecule have unique frequencies, and different molecules have different combinations of functional groups. Thus, every molecule has a unique absorption spectrum. Of course, there can be situations where two molecules are similar enough that their spectra are indistinguishable on a system with moderate signal-to-noise ratio, or where there are strong background absorptions due to a solvent or matrix that obscures the molecular vibrations, so that it is not possible to distinguish all compounds under all circumstances; but it is usually quite reliable, particularly if one is comparing the spectrum of an unknown to reference spectra of a wide variety of compounds. Ease of implementation and reasonably unambiguous spectra have led to the widespread use of IR spectroscopy outside of physical chemistry.

Within physical chemistry, the long-lasting interest in IR spectroscopy lies in structural and dynamical characterization. High resolution vibration–rotation spectroscopy in the gas phase reveals bond lengths, bond angles, molecular symmetry and force constants. Time-resolved IR spectroscopy characterizes reaction kinetics, vibrational lifetimes and relaxation processes.

### B1.2.1.3   Raman spectroscopy

Raman spectrometers are not as widespread as their IR counterparts. This is partially due to the more stringent requirements on light source (laser) and monochromator. As is the case with IR spectroscopy, every molecule has a unique Raman spectrum. It is also true that there can be ambiguity because of molecular similarities or impurities in the sample. Resonance Raman spectroscopy allows interfering bands to be eliminated by selectively exciting only specific species by virtue of their electronic absorption, or coupling to a nearby chromophore. This is particularly helpful in discriminating against strong solvent bands. For example, the first excited electronic state of water is at about 7 eV ($\sim$175 nm excitation wavelength), whereas many larger molecules have electronic transitions at much lower photon energy. By using the resonant enhancement of the Raman signal from the excited electronic state, it is possible to obtain a factor of $10^6$ enhancement of the dissolved molecule.

One of the well known advantages of resonance Raman spectroscopy is that samples dissolved in water can be studied since water is transparent in the visible region. Furthermore, many molecules of biophysical interest assume their native state in water. For this reason, resonance Raman spectroscopy has been particularly strongly embraced in the biophysical community.

## B1.2.2   Theory

### B1.2.2.1   Classical description

Both infrared and Raman spectroscopy provide information on the vibrational motion of molecules. The techniques employed differ, but the underlying molecular motion is the same. A qualitative description of IR and Raman spectroscopies is first presented. Then a slightly more rigorous development will be described. For both IR and Raman spectroscopy, the fundamental interaction is between a dipole moment and an electromagnetic field. Ultimately, the two can only couple with each other if they have the same frequency, otherwise the time average of their interaction energy is zero.

The most important consideration for a vibration to be IR active is that its dipole moment *changes* upon vibration. That is to say, its dipole derivative must be nonzero. The time-dependence of the dipole moment

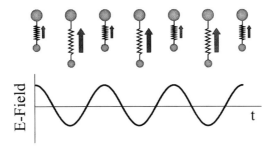

**Figure B1.2.1.** Schematic representation of the dependence of the dipole moment on the vibrational coordinate for a heteronuclear diatomic molecule. It can couple with electromagnetic radiation of the same frequency as the vibration, but at other frequencies the interaction will average to zero.

**Figure B1.2.2.** Schematic representation of the polarizability of a diatomic molecule as a function of vibrational coordinate. Because the polarizability *changes* during vibration, Raman scatter will occur in addition to Rayleigh scattering.

for a heteronuclear diatomic is shown in figure B1.2.1. Classically, an oscillating dipole radiates energy at the oscillation frequency. In a sense, this is true for a vibrating molecule in that when it is in an excited vibrational state it can emit a photon and lose energy. However, there are two fundamental differences. First, it does not continuously radiate energy, as a classical oscillator would. Rather, it vibrates for long periods without radiating energy, and then there is an instantaneous jump to a lower energy level accompanied by the emission of a photon. It should be noted that vibrational lifetimes in the absence of external perturbations are quite long, on the millisecond timescale. The second difference from a classical oscillator is that when a molecule is in its ground vibrational state it cannot emit a photon, but is still oscillating. Thus, the dipole can oscillate for an indefinitely long period without radiating any energy. Therefore, a quantum mechanical description of vibration must be invoked to describe molecular vibrations at the most fundamental level.

The qualitative description of Raman scattering is closely related. In this case, the primary criterion for a Raman active mode is that the polarizability of the molecule changes as a function of vibrational coordinate. An atom or molecule placed in an electric field will acquire a dipole moment that is proportional to the size of the applied field and the ease at which the charge distribution redistributes, that is, the polarizability. If the polarizability changes as a function of vibration, then there will be an induced dipole whose magnitude changes as the molecule vibrates, as depicted in figure B1.2.2, and this can couple to the EM field. Of course, the applied field is oscillatory, not static, but as we will see below there will still be scattered radiation that is related to the vibrational frequency. In fact, the frequency of the scattered light will be the sum and difference of the laser frequency and vibrational frequency.

Before presenting the quantum mechanical description of a harmonic oscillator and selection rules, it is worthwhile presenting the energy level expressions that the reader is probably already familiar with. A vibrational mode $v$, with an equilibrium frequency of $\tilde{v}_e$ (in wavenumbers) has energy levels (also in wavenumbers) given by $E_v = \tilde{v}_e(v + 1/2)$, where $v$ is the vibrational quantum number, and $v \geq 0$. The notation can become a bit confusing, so note that: $v$ (Greek letter nu) identifies the vibration, $\tilde{v}_e$ is the

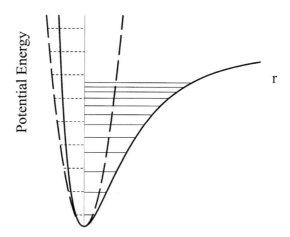

**Figure B1.2.3.** Comparison of the harmonic oscillator potential energy curve and energy levels (dashed lines) with those for an anharmonic oscillator. The harmonic oscillator is a fair representation of the true potential energy curve at the bottom of the well. Note that the energy levels become closer together with increasing vibrational energy for the anharmonic oscillator. The anharmonicity has been greatly exaggerated.

vibrational frequency (in wavenumbers), and $v$ (italic letter 'v') is the vibrational quantum number. It is trivial to extend this expression to a molecule with $n$ uncoupled harmonic modes,

$$E_{v_1,v_2,v_3,\ldots,v_n} = \sum_{i=1}^{n} \tilde{\nu}_i (v_i + 1/2) \qquad (B1.2.1)$$

where $\tilde{\nu}_i$ is the equilibrium vibrational frequency of the $i$th mode.

Of course, real molecules are not harmonic oscillators, and the energy level expression can be expanded in powers of $(v + 1/2)$. For a single mode we have

$$E_v = \tilde{\nu}_e(v + 1/2) - \tilde{x}_e \tilde{\nu}_e(v + 1/2)^2 + \cdots$$

where $\tilde{x}_e$ is the anharmonicity constant. This allows the spacing of the energy levels to decrease as a function of vibrational quantum number. Usually the expansion converges rapidly, and only the first two terms are needed. Harmonic oscillator and anharmonic oscillator potential energy curves with their respective energy levels are compared in figure B1.2.3.

There usually is rotational motion accompanying the vibrational motion, and for a diatomic, the energy as a function of the rotational quantum number, $J$, is

$$E_J = B_e J(J + 1) - D_e[J(J + 1)]^2$$

where $B_e$ and $D_e$ are the equilibrium rotational constant and centrifugal distortion constant respectively. The rotational constant is related to the moment of inertia through $B_e = h/8\pi^2 I_e c$, where $h$ is Planck's constant, $I_e$ is the equilibrium moment of inertia, and $c$ is the speed of light in vacuum. As the molecule rotates faster, it elongates due to centrifugal distortion. This increases the moment of inertia and causes the energy levels to become closer together. This is accounted for by including the second term with a negative sign. Overall, the vibration–rotation term energy is given by

$$E_{v,J} = \tilde{\nu}_e(v + 1/2) - \tilde{x}_e \tilde{\nu}_e(v + 1/2)^2 + B_e J(J + 1) - D_e[J(J + 1)]^2 - \alpha_e(v + 1/2)J(J + 1). \qquad (B1.2.2)$$

The only term in this expression that we have not already seen is $\alpha_e$, the vibration–rotation coupling constant. It accounts for the fact that as the molecule vibrates, its bond length changes which in turn changes the moment of inertia. Equation (B1.2.2) can be simplified by combining the vibration–rotation constant with the rotational constant, yielding a vibrational-level-dependent rotational constant,

$$B_v = B_e - \alpha_e(v + 1/2)$$

so the vibration–rotation term energy becomes

$$E_{v,J} = \tilde{v}_e(v + 1/2) - \tilde{x}_e\tilde{v}_e(v + 1/2)^2 + B_v J(J + 1) - D_e[J(J + 1)]^2.$$

### B1.2.2.2  Quantum mechanical description

The quantum mechanical treatment of a harmonic oscillator is well known. Real vibrations are not harmonic, but the lowest few vibrational levels are often very well approximated as being harmonic, so that is a good place to start. The following description is similar to that found in many textbooks, such as McQuarrie (1983) [2]. The one-dimensional Schrödinger equation is

$$-\frac{\hbar^2}{2\mu}\frac{d^2\psi}{dx^2} + U(x)\psi(x) = E\psi(x) \tag{B1.2.3}$$

where $\mu$ is the reduced mass, $U(x)$ is the potential energy, $\psi(x)$ is the wavefunction and $E$ is the energy. The harmonic oscillator wavefunctions which solve this equation yielding the energies as given in equation (B1.2.1) are orthonormal, and are given by

$$\psi_v(x) = N_v H_v(\alpha^{1/2}x)\, e^{-\alpha x^2/2}$$

where $\alpha = (k\mu/\hbar^2)$, $N_v$ are normalization constants given by

$$N_v = \frac{1}{(2^v v!)^{1/2}}\left(\frac{\alpha}{\pi}\right)^{1/4}$$

and $H_v$ are the Hermite polynomials. The Hermite polynomials are defined as [1]

$$H_v(x) = (-1)^v\, e^{x^2}\frac{d^v}{dx^v}\, e^{-x^2}.$$

Upon inspection, the first three are seen to be

$$H_0(x) = 1 \qquad H_1(x) = 2x \qquad H_2(x) = 4x^2 - 2$$

and higher degree polynomials are obtained using the recursion relation

$$H_{v+1}(x) = 2x H_v(x) - 2v H_{v-1}(x). \tag{B1.2.4}$$

The first few harmonic oscillator wavefunctions are plotted in figure B1.2.4.

Upon solving the Schrödinger equation, the energy levels are $E_v = \tilde{v}_e(v + 1/2)$, where $\tilde{v}_e$ is related to the force constant $k$ and reduced mass $\mu$ through

$$\tilde{v}_e = \frac{1}{2\pi c}\left(\frac{k}{\mu}\right)^{1/2}.$$

Since the reduced mass depends only on the masses of the atoms ($1/\mu = 1/m_1 + 1/m_2$ for a diatomic), which are known, measurement of the vibrational frequency allows the force constant to be determined.

$\psi$                                          $|\psi|^2$

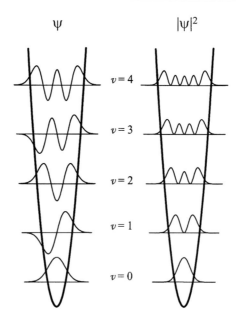

**Figure B1.2.4.** Lowest five harmonic oscillator wavefunctions $\psi$ and probability densities $|\psi|^2$.

The electric dipole selection rule for a harmonic oscillator is $\Delta v = \pm 1$. Because real molecules are not harmonic, transitions with $|\Delta v| > 1$ are weakly allowed, with $\Delta v = \pm 2$ being more allowed than $\Delta v = \pm 3$ and so on. There are other selection rules for quadrupole and magnetic dipole transitions, but those transitions are six to eight orders of magnitude weaker than electric dipole transitions, and we will therefore not concern ourselves with them.

The selection rules are derived through time-dependent perturbation theory [1, 2]. Two points will be made in the following material. First, the Bohr frequency condition states that the photon energy of absorption or emission is equal to the energy level separation of the two states. Second, the importance of the transition dipole moment is shown and, furthermore, it is also shown that the transition dipole moment for a vibrational mode is in fact the change in dipole as a function of vibration, that is, the dipole derivative. The time-dependent Schrödinger equation is

$$\hat{H}\Psi = i\hbar \frac{\partial \Psi}{\partial t}. \tag{B1.2.5}$$

The time- and coordinate-dependent wavefunction for any given state is

$$\Psi_v(x, t) = \psi_v(x)\, e^{-iE_v t/\hbar}$$

where $\psi_v(x)$ is a stationary state wavefunction obtained by solving the time-independent Schrödinger equation (B1.2.3). The Hamiltonian is broken down into two parts, $\hat{H} = \hat{H}_0 + \hat{H}^{(1)}$, where $\hat{H}_0$ is the Hamiltonian for the isolated molecule, and $\hat{H}^{(1)}$ describes the interaction Hamiltonian of the molecule with the electromagnetic field. In particular, the interaction energy between a dipole and a monochromatic field is

$$\hat{H}^{(1)} = -\mu E = -\mu E_0 \cos 2\pi \nu t.$$

Consider a two-state system, where

$$\Psi_1(x, t) = \psi_1(x)\, e^{-iE_1 t/\hbar} \qquad \text{and} \qquad \Psi_2(x, t) = \psi_2(x)\, e^{-iE_2 t/\hbar}.$$

These wavefunctions are orthogonal. Assume that the system is initially in state 1, and that the interaction begins at $t = 0$. Since there are only two states, at any later time the wavefunction for the system is

$$\Psi(t) = a_1(t)\Psi_1(t) + a_2(t)\Psi_2(t)$$

where the coefficients $a_1(t)$ and $a_2(t)$ are to be determined. We know that $a_1(0) = 1$ and $a_2(0) = 0$ from the initial conditions. By substitution into equation (B1.2.5), we have

$$a_1(t)\hat{H}^{(1)}\Psi_1 + a_2(t)\hat{H}^{(1)}\Psi_2 = i\hbar\Psi_1\frac{da_1}{dt} + i\hbar\Psi_2\frac{da_2}{dt}.$$

We can find the time-dependent coefficient for being in state 2 by multiplying from the left by $\psi_2^*$, and integrating over spatial coordinates:

$$a_1(t)\int\psi_2^*\hat{H}^{(1)}\Psi_1\,d\tau + a_2(t)\int\psi_2^*\hat{H}^{(1)}\Psi_2\,d\tau = i\hbar\frac{da_1}{dt}\int\psi_2^*\Psi_1\,d\tau + i\hbar\frac{da_2}{dt}\int\psi_2^*\Psi_2\,d\tau.$$

This expression can be simplified considerably since $\psi_2$ and $\psi_1$ are orthogonal which implies that $\psi_2$ and $\Psi_1$ are orthogonal as well. Furthermore, $a_1(t) \approx a_1(0) = 1$ and $a_2(t) \approx a_2(0) = 0$ since $\hat{H}^{(1)}$ is a small perturbation:

$$i\hbar\frac{da_2}{dt} = \exp\left[\frac{-i(E_1 - E_2)t}{\hbar}\right]\langle\psi_2|\hat{H}^{(1)}|\psi_1\rangle \tag{B1.2.6}$$

where we have used Dirac bracket notation for the integral, i.e.

$$\langle\psi_2|\hat{H}^{(1)}|\psi_1\rangle \equiv \int\psi_2^*\hat{H}^{(1)}\psi_1\,d\tau.$$

In order to evaluate equation (B1.2.6), we will consider the electric field to be in the $z$-direction, and express the interaction Hamiltonian as

$$\hat{H}^{(1)} = -\mu_z E_{0z}\cos 2\pi\nu t = -\frac{\mu_z E_{0z}}{2}(e^{i2\pi\nu t} + e^{-i2\pi\nu t}).$$

Before substituting everything back into equation (B1.2.6), we define the *transition dipole moment* between states 1 and 2 to be the integral

$$(\mu_z)_{21} \equiv \langle\psi_2|\mu_z|\psi_1\rangle. \tag{B1.2.7}$$

Now, we substitute it into equation (B1.2.6) to get

$$\frac{da_2}{dt} \propto (\mu_z)_{21}E_{0z}\left\{\exp\left[\frac{i(E_2 - E_1 + h\nu)t}{\hbar}\right] + \exp\left[\frac{i(E_2 - E_1 - h\nu)t}{\hbar}\right]\right\}$$

and then integrate from 0 to $t$ to obtain

$$a_2(t) \propto (\mu_z)_{21}E_{0z}\left\{\frac{1 - \exp[i(E_2 - E_1 + h\nu)t/\hbar]}{E_2 - E_1 + h\nu} + \frac{1 + \exp[i(E_2 - E_1 - h\nu)t/\hbar]}{E_2 - E_1 - h\nu}\right\}.$$

There are two important features of this result. The energy difference between states 1 and 2 is $\Delta E = E_2 - E_1$. When $\Delta E \approx h\nu$, the denominator of the second term becomes very small, and this term dominates. This is the well known Bohr frequency condition. The second important feature is that $(\mu_z)_{21}$ must be nonzero for an allowed transition, which is how the selection rules are determined.

The molecular dipole moment (not the transition dipole moment) is given as a Taylor series expansion about the equilibrium position

$$\mu = \mu_0 + \left(\frac{d\mu}{dx}\right)x + \left(\frac{d^2\mu}{dx^2}\right)x^2 + \cdots = \mu_0 + \mu_1 x + \mu_2 x^2 + \cdots$$

where $x$ is the displacement from equilibrium, $\mu_0$ is the permanent dipole moment and $\mu_1$ is the dipole derivative and so on. It is usually fine to truncate the expansion after the second term. Now we need to evaluate $(\mu_z)_{21}$ for the harmonic oscillator wavefunctions. Using equation (B1.2.7), we have

$$(\mu_z)_{j,i} = N_j N_i \int_{-\infty}^{\infty} H_{xj}(\alpha^{1/2}x)\, e^{-\alpha x^2/2} \mu H_{xi}(\alpha^{1/2}x)\, e^{-\alpha x^2/2}\, dx$$

which can be expanded by substituting $\mu = \mu_0 + \mu_1 x$:

$$(\mu_z)_{j,i} = N_j N_i \mu_0 \int_{-\infty}^{\infty} H_{xj}(\alpha^{1/2}x)\, e^{-\alpha x^2/2} H_{xi}(\alpha^{1/2}x)\, e^{-\alpha x^2/2}\, dx$$

$$+ N_j N_i \mu_1 \int_{-\infty}^{\infty} H_{xj}(\alpha^{1/2}x)\, e^{-\alpha x^2/2} x H_{xi}(\alpha^{1/2}x)\, e^{-\alpha x^2/2}\, dx. \qquad \text{(B1.2.8)}$$

The first term is zero if $i \neq j$ due to the orthogonality of the Hermite polynomials. The recursion relation in equation (B1.2.4) is rearranged

$$x H_v(x) = v H_{v-1}(x) + \frac{1}{2} H_{v+1}(x)$$

and substituted into the second term in equation (B1.2.8),

$$(\mu_z)_{j,i} = \frac{N_j N_i}{\alpha} \mu_1 \int_{-\infty}^{\infty} H_{xj}(\xi) \left[ x i H_{xi-1}(\xi) + \frac{1}{2} H_{xi+1}(\xi) \right] e^{-\xi^2}\, d\xi$$

where we have let $\xi = \alpha^{1/2}x$. Clearly, this integral will be nonzero only when $j = (i \pm 1)$, yielding the familiar $\Delta v = \pm 1$ harmonic oscillator selection rule. Furthermore, the overtone intensities for an anharmonic oscillator are obtained in a straightforward manner by determining the eigenfunctions of the energy levels in a harmonic oscillator basis set, and then summing the weighted contributions from the harmonic oscillator integrals.

### B1.2.2.3   Raman spectroscopy

*Normal Raman spectroscopy*

Raman scattering has been discussed by many authors. As in the case of IR vibrational spectroscopy, the interaction is between the electromagnetic field and a dipole moment, however in this case the dipole moment is induced by the field itself. The induced dipole is $\mu_{\text{ind}} = \alpha E$, where $\alpha$ is the polarizability. It can be expressed in a Taylor series expansion in coordinate displacement

$$\alpha = \alpha_0 + \left(\frac{d\alpha}{dx}\right)x + \left(\frac{d^2\alpha}{dx^2}\right)x^2 + \cdots = \alpha_0 + \alpha' x + \alpha'' x^2 + \cdots.$$

Here, $\alpha_0$ is the static polarizability, $\alpha'$ is the change in polarizability as a function of the vibrational coordinate, $\alpha''$ is the second derivative of the polarizability with respect to vibration and so on. As is usually the case, it is possible to truncate this series after the second term. As before, the electric field is $E = E_0 \cos 2\pi \nu_0 t$, where $\nu_0$ is the frequency of the light field. Thus we have

$$\mu_{\text{ind}} = (\alpha_0 + \alpha' x) E_0 \cos 2\pi \nu_0 t. \tag{B1.2.9}$$

The time dependence of the displacement coordinate for a mode undergoing harmonic oscillation is given by $x = x_m \cos 2\pi \nu_v t$, where $x_m$ is the amplitude of vibration and $\nu_v$ is the vibrational frequency. Substitution into equation (B1.2.9) with use of Euler's half-angle formula yields

$$\mu_{\text{ind}} = \alpha_0 E_0 \cos 2\pi \nu_0 t + \alpha' (x_m \cos 2\pi \nu_v t) E_0 \cos 2\pi \nu_0 t$$
$$= \alpha_0 E_0 \cos 2\pi \nu_0 t + \frac{E_0 \alpha' x_m}{2} [\cos 2\pi (\nu_0 - \nu_v) t + \cos 2\pi (\nu_0 + \nu_v) t].$$

The first term results in Rayleigh scattering which is at the same frequency as the exciting radiation. The second term describes Raman scattering. There will be scattered light at $(\nu_0 - \nu_v)$ and $(\nu_0 + \nu_v)$, that is at sum and difference frequencies of the excitation field and the vibrational frequency. Since $\alpha' x_m$ is about a factor of $10^6$ smaller than $\alpha_0$, it is necessary to have a very efficient method for dispersing the scattered light.

The bands on the low frequency side of the excitation frequency $(\nu_0 - \nu_v)$ are referred to as the Stokes lines, consistent with the terminology used in fluorescence, whereas those on the high frequency side $(\nu_0 + \nu_v)$ are the anti-Stokes lines. It is a bit unfortunate that this terminology was chosen, since the Raman process is fundamentally different from fluorescence. In particular, fluorescence is the result of a molecule absorbing light, undergoing vibrational relaxation in the upper electronic state, and re-emitting a photon at a lower frequency. The timescale for fluorescence is typically of the order of nanoseconds. The Raman process, on the other hand, is an instantaneous scattering process that occurs on a femtosecond timescale. The photon is never absorbed by the molecule. It is usually clear whether fluorescence or Raman scattering is being observed, but there are situations where it is ambiguous. We shall not pursue the issue any further here, however.

It is well known that the intensity of scattered light varies as the fourth power of the frequency, and based on this alone one would predict the Stokes lines to be less intense than the anti-Stokes by a factor of

$$\frac{I_{\text{Stokes}}}{I_{\text{anti-Stokes}}} = \frac{(\nu_0 - \nu_v)^4}{(\nu_0 + \nu_v)^4}$$

which is 0.68 for a 1000 cm$^{-1}$ vibration excited at 488 nm (20 492 cm$^{-1}$). In reality, the Stokes lines are *more* intense, typically by a factor of 2 to 10 000, as the vibrational frequency varies from 200 to 2000 cm$^{-1}$. This is easily justified when the Boltzmann population of initial states is taken into account. As seen in figure B1.2.5, the Stokes transitions correspond to the molecule being initially in a low energy state, usually $v = 0$, whereas it must be in an excited vibrational state if it is going to undergo an anti-Stokes process. The ratio of populations of two energy levels is given by

$$P_{12} = e^{-\Delta E_{12}/kT}$$

where $\Delta E_{12}$ is the energy difference between the two levels, $k$ is Boltzmann's constant and $T$ is the temperature in Kelvin. Since the Stokes lines are more intense than anti-Stokes, only the Stokes lines for a Raman spectrum are typically presented, and the abscissa is labelled with the frequency offset $(\nu_0 - \nu_v)$ rather than the actual frequency of scattered light.

Additional information about the vibration can be obtained through the depolarization ratio. This is the ratio of the intensity of scattered light that is polarized in a plane perpendicular to the incident radiation relative to that the scattered light that is polarized parallel to the incident polarization, $\rho = I_\perp / I_\parallel$. For totally

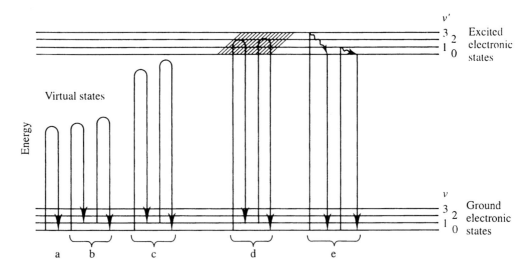

**Figure B1.2.5.** Comparison of several light scattering processes. (a) Rayleigh scattering, (b) Stokes and anti-Stokes Raman scattering, (c) pre-resonance Raman scattering, (d) resonance Raman scattering and (e) fluorescence where, unlike resonance Raman scattering, vibrational relaxation in the excited state takes place. From [3], used with permission.

symmetric modes, $\rho = 0$, while $0 < \rho < 3/4$ for non-totally symmetric modes [1, 3]. The polarization ratio can actually be greater than $3/4$ for a resonantly enhanced Raman band [3].

Consistent with the notion that Raman scattering is due to a change in polarizability as a function of vibration, some of the general features of Raman spectroscopy [3] are:

(1)   it is more sensitive to stretching modes than bending modes, especially totally symmetric modes;

(2)   the intensity increases with bond order (i.e. double bond vibrations are more intense than single bond vibrations);

(3)   for modes involving only one bond, the intensity increases with increasing atomic number of the atoms;

(4)   for cyclic molecules, breathing modes are strongest.

*Resonance Raman spectroscopy*

Resonance Raman spectroscopy has been discussed by many authors [4–15]. If the excitation frequency is resonant with an excited electronic state, it is possible to dramatically increase the Raman cross-section. It is still a scattering process without absorption, but the light and polarizability can now couple much more efficiently. Resonant enhancements of $10^4$ to $10^6$ are achievable. A second important fact is that the vibrational modes that are coupled to the electronic transition are selectively enhanced, which greatly aids in structural determination. Another implication is that much lower concentrations can be used. A typical neat liquid might have a concentration of roughly 5 to 50 mol $l^{-1}$. It is possible for resonant enhanced Raman bands of a molecule to be as strong or stronger than the solvent bands even at a concentration of $10^{-6}$ M. This is useful because it is often desirable to maintain low enough concentrations such that the solute molecules do not interact with each other or aggregate. Finally, excited state vibrational dephasing rates can be determined.

The sum-over-states method for calculating the resonant enhancement begins with an expression for the resonance Raman intensity, $I_{i,f}$, for the transition from initial state $i$ to final state $f$ in the ground electronic state, and is given by [14]

$$I_{i,f} = \frac{2^7 \pi^5}{3^2 c^4} I_0 \sum_{\rho,\sigma} |(\alpha_{\rho\sigma})_{i,f}|^2 (\nu_L \pm \nu_v)^4$$

where $I_0$ is the incident intensity, $\nu_L$ is the laser frequency, $\nu_v$ is the vibrational frequency and $(\alpha_{\rho\sigma})_{i,f}$ is the $\rho\sigma$th element of the Raman scattering tensor,

$$(\alpha_{\rho\sigma})_{i,f} = \sum_r \frac{(M_\rho)_{i,r}(M_\sigma)_{r,f}}{E_{r,i} - E_L + i\Gamma_r} + \frac{(M_\sigma)_{i,r}(M_\rho)_{r,f}}{E_{r,f} + E_L + i\Gamma_r}. \qquad (B1.2.10)$$

Here, $(M_\rho)_{a,b}$ is the $\rho$th component of the electronic transition moment from state $a$ to $b$, $E_{r,i}$ and $E_{r,f}$ are the energy differences between the resonant excited state and the initial and final states, respectively, $E_L - h\nu_L$ is the photon energy of the laser radiation, and $\Gamma_r$ is the excited state vibrational dephasing rate. The Raman scattering tensor has two contributions, the so-called $A$ and $B$ terms:

$$(\alpha_{\rho\sigma})_{i,f} = A + B.$$

The $B$-type enhancement is smaller than the $A$-type, and is usually, though not always, negligible. Therefore, we will concentrate on the $A$ term [14],

$$A = \frac{1}{h}[\langle g|\rho|e\rangle\langle e|\sigma|g\rangle] \sum_r \frac{\langle i \mid r\rangle\langle r \mid f\rangle}{E_{i,r} - E_L + i\Gamma_r}$$

where $\rho$ and $\sigma$ are pure electronic transition moments, and $\langle i \mid r\rangle$ and $\langle r \mid f\rangle$ are vibrational overlap integrals. In many cases, a single diagonal component of the scattering tensor dominates, which leads to a simplified expression,

$$\alpha_{zz} = \frac{1}{h}|\langle e|M_z|g\rangle|^2 \sum_r \frac{\langle i \mid r\rangle\langle r \mid f\rangle}{E_{i,r} - E_L + i\Gamma_r} \qquad (B1.2.11)$$

where $\langle e|M_z|g\rangle$ is the transition dipole moment for the electronic transition between ground state $g$ and electronically excited state $e$, and $r$ are the vibrational levels in the excited electronic state with vibrational energy $E_{i,r}$ relative to the initial vibrational level in the ground electronic state. From the form of this equation, we see that the enhancement depends on the overlap of the ground and excited state wavefunctions (Franck–Condon overlap), and is strongest when the laser frequency equals the energy level separation. Furthermore, systems with very large dephasing rates will not experience as much enhancement, all other factors being equal.

If there is only significant overlap with one excited vibrational state, equation (B1.2.11) simplifies further. In fact, if the initial vibrational state is $v_i = 0$, which is usually the case, and there is not significant distortion of the molecule in the excited electronic state, which may or may not hold true, then the intensity is given by

$$I \propto \frac{1}{(E_{00} - E_L)^2 + \Gamma^2}.$$

It is also possible to determine the resonant Raman intensities via a time-dependent method [16]. It has the advantages that the vibrational eigenstates of the excited electronic state need not be known, and that it provides a physical picture of the dynamics in the excited state. Assuming only one ground vibrational state is occupied, the Raman cross section is [16, 17]

$$\sigma_R(\omega_0) = \frac{8\pi \omega_0 \omega_s^3 (M_{ge}^0)^4}{9\hbar^2 c^4} \left| \int_0^\infty \langle f \mid i(t)\rangle \exp[i(\omega_L + \omega_i)t] \exp[-g(t)] \, dt \right|^2$$

where $\omega_L$ is the laser frequency, $\omega_s$ is the frequency of the scattered light, $(M_{ge}^0)$ is the electronic transition dipole, $\hbar\omega_i$ is the zero point vibrational energy of the initial state, $\langle f \mid i(t) \rangle$ is the overlap of the final state with the time-evolving state on the upper electronic state potential energy surface and the $\exp[-g(t)]$ term accounts for solvent-induced electronic dephasing. Unlike the sum-over-states method, the only excited state information needed is the potential energy surface in the immediate vicinity of where the initial state is projected onto the excited state.

### B1.2.3    Spectrometers

*B1.2.3.1    Infrared spectrometers*

In the most general terms, an infrared spectrometer consists of a light source, a dispersing element, a sample compartment and a detector. Of course, there is tremendous variability depending on the application.

*Light sources*

Light sources can either be broadband, such as a Globar, a Nernst glower, an incandescent wire or mercury arc lamp; or they can be tunable, such as a laser or optical parametric oscillator (OPO). In the former case, a monochromator is needed to achieve spectral resolution. In the case of a tunable light source, the spectral resolution is determined by the linewidth of the source itself. In either case, the spectral coverage of the light source imposes limits on the vibrational frequencies that can be measured. Of course, limitations on the dispersing element and detector also affect the overall spectral response of the spectrometer.

Desirable characteristics of a broadband light source are stability, brightness and uniform intensity over as large a frequency range as possible. Desirable characteristics of a tunable light source are similar to those of a broadband light source. Furthermore, its wavelength should be as jitter-free as possible. Having a linewidth of 0.001 cm$^{-1}$ is meaningless if the frequency fluctuates by 0.01 cm$^{-1}$ on the timescale of scanning over an absorption feature.

The region 4500–2850 cm$^{-1}$ is covered nicely by f-centre lasers, and 2800–1000 cm$^{-1}$ by diode lasers (but there are gaps in coverage). Difference frequency crystals allow two visible beams whose frequencies differ by the wavelengths of interest to be mixed together. Using different laser combinations, coverage from greater than 5000–1000 cm$^{-1}$ has been demonstrated. The spectral range is limited only by the characteristics of the difference frequency crystal. The ultimate resolution of a laser spectrometer is dictated by the linewidth of the tunable light source. If the linewidth of the light source is broader than the absorption linewidth, then sensitivity is diminished, and the transition will appear broader than it actually is.

When extremely high resolution is not required, an attractive alternative is found in OPOs. An OPO is based on a nonlinear crystal that converts an input photon into two output photons whose energies add up to the input photon's energy. They can be rapidly tuned over a relatively large range, 5000–2200 cm$^{-1}$, depending on the nonlinear crystal. In the IR, commonly used crystals are $LiNbO_3$ and KDP ($KH_2PO_4$). The wavelengths of the two output beams are determined by the angle of the nonlinear crystal. That is, it will only function when the index of refraction of the crystal in the direction of propagation is identical for all three beams. If narrow linewidths are required, it is necessary to seed the OPO with a weak beam from a diode laser. Thus, the parametric oscillation does not have to build up out of the noise, and is therefore more stable.

*Dispersing elements*

Dispersing elements must be used when broadband IR light sources are employed; either diffraction gratings or prisms can be used. Gratings are made by depositing a metal film, usually gold, on a ruled substrate. They can be used over a broad spectrum because the light never penetrates into the substrate. However, the reflectivity of the coating can be wavelength dependent. The useable range is determined by the pitch of the rulings. A grating suitable for use from 1000 to 3000 cm$^{-1}$ would have 100–200 lines mm$^{-1}$. If there are too many lines per millimetre, then it acts like a mirror rather than a grating. If it has too few, then the efficiency is greatly reduced.

Prisms can also be used to disperse the light. They are much easier to make than gratings, and are therefore less expensive, but that is their only real advantage. Since the light has to pass through them, it is necessary to find materials that do not absorb. Greater dispersion is obtained by using materials with as high an index of refraction as possible. Unfortunately, materials with high index are also close to an absorption, which leads to a large nonlinear variation in index as a function of frequency.

When dispersing elements are used, the resolution of the spectrometer is determined by the entrance slit width, the exit slit width, the focal length and the dispersing element itself. Resolving power is defined as

$$R = \lambda/\Delta\lambda = \nu/\Delta\nu$$

that is, the central wavelength (or frequency) of the light exiting the spectrometer relative to its linewidth (or bandwidth). Higher resolving powers allow closely spaced lines to be distinguished.

*Detectors*

The detector chosen is just as important as the light source. If the sample is absorbing light, but the detector is not responding at that frequency, then changes in absorption will not be recorded. In fact, one of the primary limitations faced by spectroscopists working in the far-IR region of the spectrum (10–300 cm$^{-1}$) has been lack of highly sensitive detectors. The type of detector used depends on the frequency of the light. Of course, at any frequency, a bolometer can be used, assuming the detection element absorbs the wavelength of interest. A bolometer operates on the principle of a temperature-dependent resistance in the detector element. If the detector is held at a very low temperature, 2–4 K, and has a small heat capacity, then incoming energy will heat it, causing the resistance to change. While these detectors are general, they are susceptible to missing weak signals because they are swamped by thermal blackbody radiation, and are not as sensitive as detectors that have a response tied to the photon energy. At frequencies between 4000 and 1900 cm$^{-1}$, InSb photodiodes are used, HgCdTe detectors are favoured for frequencies between 2000 and 700 cm$^{-1}$, and copper-doped germanium (Cu:Ge) photoconductors are used for frequencies between 1000 and 300 cm$^{-1}$ [18]. More recently, HgCdTe array detectors have become available that respond in the range 3500–900 cm$^{-1}$ [19].

### B1.2.3.2    *Raman spectrometers*

While Raman spectroscopy was first described in a paper by C V Raman and K S Krishnan in 1928, it has only come into widespread use in the last three decades owing to the ready availability of intense monochromatic laser light sources. Oddly enough, prior to 1945, Raman spectroscopy using Hg arc lamps was the method of choice for obtaining vibrational spectra since there was not yet widespread availability of IR spectrometers [5]. Just as with IR spectrometers, a Raman spectrometer consists of a light source, a dispersing element, a sample compartment and a detector. Again, there is tremendous variability depending on the application.

*Light sources*

The light source must be highly monochromatic so that the Raman scattering occurs at a well-defined frequency. An inhomogeneously broadened vibrational linewidth might be of the order of 20 cm$^{-1}$, therefore, if the excitation source has a wavelength of 500 nm, its linewidth must be less than 0.5 nm to ensure that it does not further broaden the line and decrease its intensity. This type of linewidth is trivial to achieve with a laser source, but more difficult with an arc lamp source.

Fixed frequency laser sources are most commonly used. For example, the $Ar^+$ laser has lines at 514, 496, 488, 476 and 458 nm. Sometimes a helium–neon laser is used (628 nm), or a doubled or tripled YAG (532 or 355 nm, respectively). Other wavelengths are generated by employing a Raman shifter with a variety of different gases. It is also desirable to have tunability, in order to carry out resonance Raman studies wherein one selectively measures the vibrations most strongly coupled to the electronic state being excited. Tunable lasers can be either line-tunable or continuously tunable. Thanks to the high sensitivity of photomultiplier tubes, the light source need only provide moderate power levels, $\sim$10–100 mW for example. In fact, one must be careful not to use too much power and damage the sample.

*Dispersing elements*

Due to the rather stringent requirements placed on the monochromator, a double or triple monochromator is typically employed. Because the vibrational frequencies are only several hundred to several thousand $cm^{-1}$, and the linewidths are only tens of $cm^{-1}$, it is necessary to use a monochromator with reasonably high resolution. In addition to linewidth issues, it is necessary to suppress the very intense Rayleigh scattering. If a high resolution spectrum is not needed, however, then it is possible to use narrow-band interference filters to block the excitation line, and a low resolution monochromator to collect the spectrum. In fact, this is the approach taken with Fourier transform Raman spectrometers.

*Detectors*

Because the scattered light being detected is in the visible to near UV region, photomultiplier tubes (PMTs) are the detector of choice. By using a cooled PMT, background counts can be reduced to a level of only a few per second. For studies with higher signal levels, array detectors such as optical multichannel analysers (OMAs), or CCD arrays can be used. This allows a complete spectrum to be obtained without having to scan the monochromator.

*Resolution*

The resolution of the Raman spectrum is determined by the monochromator. Furthermore, since the light being measured is in the visible region, usually around 20 000 $cm^{-1}$, the resolution of the monochromator must be significantly better than that of its IR counterpart because the resolving power is described by $\Delta \nu / \nu$. That is, for a 2000 $cm^{-1}$ vibration a resolving power of 20 000 is needed to get the same resolution as that obtained in an IR spectrometer with a resolution of only 2000.

*B1.2.3.3   Fourier transform techniques*

Fourier transform techniques do not change the underlying absorption or scattering mechanisms that couple light with matter. The data acquisition and processing are significantly different, however. In short, the difference is that the data are collected interferometrically in the time domain and then Fourier transformed into the frequency domain for subsequent analysis. These types of instruments are often used in experiments where broad spectral coverage with moderate sensitivity and frequency resolution is needed. This is often encountered when other aspects of the experiment are more difficult, such as surface studies. There is, however, ongoing research directed towards time-resolved FTIR with nanosecond time resolution [20, 21]. The basic requirements are a broadband light source, a beamsplitter, two delay lines (one fixed and one variable), a detector, and a computer to run the show and Fourier-transform the data.

The underlying concept behind an FTIR spectrometer can be understood when considering what happens when a beam of monochromatic light with wavelength $\lambda$ is sent through a Michelson interferometer, shown schematically in figure B1.2.6 [22]. If the pathlength difference between the two beams is zero, then they

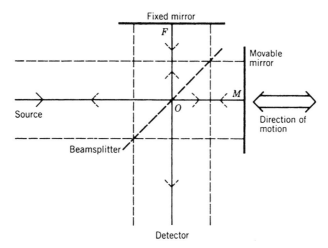

**Figure B1.2.6.** Schematic representation of a Michelson interferometer. From Griffiths P R and de Haseth J A 1986 *Fourier transform infrared spectroscopy Chemical Analysis* ed P J Elving and J D Winefordner (New York: Wiley). Reprinted by permission of John Wiley and Sons Inc.

will constructively interfere, yielding an intensity at the detector of $I_0$. Now consider what happens if the movable mirror is displaced by a distance $d = \lambda/4$. This will cause the optical path of that beam to change by an amount $\delta = \lambda/2$, and lead to destructive interference at the detector, with no power being measured. If the mirror is displaced by another $\lambda/4$, then the two beams will once again constructively interfere, leading to full intensity at the detector. The intensity as a function of mirror position is shown in figure B1.2.7(a). The intensity as a function of optical delay, $\delta$, is described by

$$I(\delta) = \frac{I(\tilde{\nu})}{2}[1 + \cos(2\pi\tilde{\nu}\delta)]$$

where $\tilde{\nu}$ is the the frequency of interest and $I(\tilde{\nu})$ is its intensity. Similarly, figure B1.2.7(b) shows the intensity as a function of mirror position when two frequencies with equal amplitudes are present, and $f_1 = 1.2f_2$. Figure B1.2.7(c) depicts these same two frequencies, but with the amplitude of the lower frequency twice as large as that of the higher frequency. Finally, figure B1.2.7(d) shows the result for a Gaussian distribution of frequencies. For a discrete distribution of frequencies, the intensity as a function of optical delay is

$$I(\delta) = \frac{1}{2}\sum_{i=1}^{n} I(\tilde{\nu}_i)[1 + \cos(2\pi\tilde{\nu}_i\delta)]$$

and if there is a continuous distribution of frequencies, it is

$$I(\delta) = \frac{1}{2}\int_0^\infty I(\tilde{\nu}_i)[1 + \cos(2\pi\tilde{\nu}\delta)]\,d\tilde{\nu}.$$

While the data are collected in the time domain by scanning a delay line, they are most easily interpreted in the frequency domain. It is straightforward to connect the time and frequency domains through a Fourier transform

$$I(\tilde{\nu}) = 4\int_0^\infty [I(\delta) - I(0)/2]\cos(2\pi\tilde{\nu}\delta)\,d\delta.$$

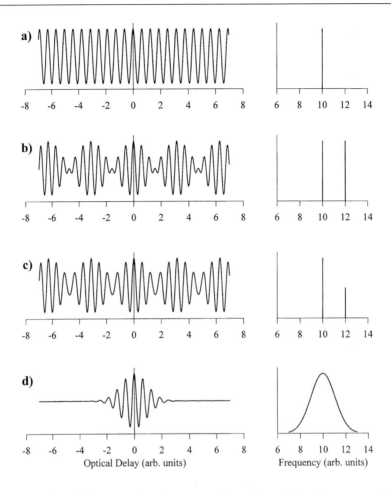

**Figure B1.2.7.** Time domain and frequency domain representations of several interferograms. (a) Single frequency, (b) two frequencies, one of which is 1.2 times greater than the other, (c) same as (b), except the high frequency component has only half the amplitude and (d) Gaussian distribution of frequencies.

Two scans are required to obtain an absorption spectrum. First, a blank reference scan is taken that characterizes the broadband light source. Then a scan with the sample in place is recorded. The ratio of the sample power spectrum to the reference power spectrum is the transmission spectrum. If the source has stable output, then a single reference scan can be used with many sample scans.

The interferogram is obtained by continuously scanning the movable mirror and collecting the intensity on the detector at regular intervals. Thus, each point corresponds to a different mirror position, which in turn corresponds to a different optical delay. Several scans can be averaged together because the scanning mechanism is highly reproducible.

The spectral resolution of the Fourier transformed data is given by the wavelength corresponding to the maximum optical delay. We have $\lambda_{max} = \delta_{max}$, and therefore $\Delta\tilde{\nu} = 1/\delta_{max}$. Therefore, if 0.1 cm$^{-1}$ spectral resolution is desired, the optical delay must cover a distance of 10 cm. The high frequency limit for the transform is the Nyquist frequency which is determined by the wavelength that corresponds to two time steps: $\lambda_{min} = 2\Delta\delta$, which leads to $\tilde{\nu}_{max} = 1/(2\Delta\delta)$. The factor of two arises because a minimum of two data points per period are needed to sample a sinusoidal waveform. Naturally, the broadband light source will determine the actual content of the spectrum, but it is important that the step size be small enough to accommodate the highest frequency components of the source, otherwise they will be folded into lower frequencies, which is known as 'aliasing'. The step size for an FT–Raman instrument must be roughly ten times smaller than that in the IR, which is one of the reasons that FT–Raman studies are usually done with near-IR excitation.

There are several advantages of FT techniques [23]. One is the Jaquinot (or throughput) advantage. Since there are fewer optical elements and no slits, more power reaches the detector. Assuming that the noise source is detector noise, which is true in the IR, but not necessarily for visible and UV, the signal-to-noise (S/N) ratio will increase. A second advantage of FT techniques is that they benefit from the Fellget (or multiplex) advantage. That is, when collecting the data, absorptions at all frequencies simultaneously contribute to the interferogram. This is contrasted with a grating spectrometer where the absorption is measured only at a single frequency at any given time. Theoretically, this results in an increase in the S/N ratio by a factor of

$$\sqrt{\frac{\tilde{\nu}_{max}}{\Delta \tilde{\nu}}} = \sqrt{\frac{\delta_{max}}{2\Delta \delta}}.$$

This assumes that both spectra have the same resolution, and that it takes the same amount of time to collect the whole interferogram as is required to obtain one wavelength on the dispersive instrument (which is usually a reasonable assumption). Thus, $\tilde{\nu}_{max}/\Delta \tilde{\nu}$ interferograms can be obtained and averaged together in the same amount of time it takes to scan the spectrum with the dispersive instrument. Since the S/N ratio scales with the square root of the number of scans averaged, the square root of this number is the actual increase in S/N ratio.

## B1.2.4   Typical examples

There are thousands of scientists whose work can be classified as vibrational spectroscopy. The following examples are meant to show the breadth of the field, but cannot be expected to constitute a complete representation of all the fields where vibrational spectroscopy is important.

### B1.2.4.1   Laser IR

For the highest resolution and sensitivity, laser-based spectrometers must be used. These have the advantage that the resolution depends on the linewidth of the laser, rather than the monochromator. Furthermore, at any given moment, all of the power is at the frequency of interest, rather than being spread out over the whole IR spectrum. Due to the fact that the emission from any given laser typically has only 100–1500 cm$^{-1}$ of tunability, and there can be difficulty maintaining narrow linewidths, the spectral coverage can be limited.

High resolution spectroscopic measurements in the gas phase yield the most detailed structural information possible. For example, measurements of weakly-bound complexes in the far-IR [24, 25] and IR [26, 27] have provided the most exact information on their structure and steady-state dynamics. Of course, a much higher level of theory must be used than was presented in section B1.2.2.1. Quite often the modes are so strongly coupled that all vibrational degrees of freedom must be treated simultaneously. Coriolis coupling and symmetry-allowed interactions among bands, i.e. Fermi resonances, are also quite significant, and must be treated explicitly. Direct measurement of the low frequency van der Waals modes in weakly-bound complexes has been discussed in section B1.4, *Microwave and Terahertz Spectroscopy*, and will not be repeated here.

A recent review of high-resolution, direct IR laser absorption spectroscopy in supersonic slit jets provides a prototypical example [26]. Figure B1.2.8 displays the experimental set-up. There are three different IR sources employed. The first utilizes difference frequency mixing of a single mode tunable ring dye laser with a single mode Ar$^+$ laser in a LiNbO$_3$ crystal to obtain light in the 2–4 $\mu$m region (5000–2500 cm$^{-1}$) with a frequency stability of 2 MHz ($6.7 \times 10^{-5}$ cm$^{-1}$). The second uses cryogenically cooled, single mode, lead-salt diodes to cover 4–10 $\mu$m (2500–1000 cm$^{-1}$) with a frequency resolution of 15 MHz ($5 \times 10^{-4}$ cm$^{-1}$).

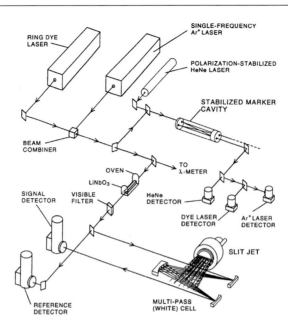

**Figure B1.2.8.** Schematic diagram of a slit jet spectrometer for high resolution IR studies of weakly-bound species. From [26], used with permission.

The third is a difference frequency scheme between a single mode dye laser and a single mode Nd:YAG laser which accesses wavelengths below 2 $\mu$m (frequencies greater than 5000 cm$^{-1}$). A long pathlength is achieved by combining a slit jet expansion with a multipass cell. This enhances the sensitivity by over two orders of magnitude.

One of the systems studied is the Ar–HF van der Waals (vdW) dimer. This rare gas–molecule complex has provided a wealth of information regarding intermolecular interactions. The vdW complex can be thought of in terms of the monomer subunit being perturbed through its interaction with the rare gas. The rotational motion is hindered, and for certain modes can be thought of as extremely large amplitude bending motion. Furthermore, there is low frequency intermolecular stretching motion. In a weakly-bound complex, the bends, stretches and internal rotations are all coupled with each other. The large spectral coverage has allowed for the measurement of spectra of Ar–HF with $v_{HF} = 1$ and 2. By combining these measurements with those made in the far-IR [28], intermolecular potential energy surfaces for $v_{HF} = 0$, 1 and 2 have been determined. Quite small amounts of internal energy ($\sim$150 cm$^{-1}$) allow the complex to fully sample the angular degree of freedom. Interestingly, the barrier to internal rotation is about the same for $v_{HF} = 0$ or 1, but significantly larger when $v_{HF} = 2$.

In addition to the dependence of the intermolecular potential energy surface on monomer vibrational level, the red-shifting of the monomer absorption as a function of the number of rare gas atoms in the cluster has been studied. The band origin for the $v_{HF} = 1 \leftarrow 0$ vibration in a series of clusters Ar$_n$–HF, with $0 < n < 5$, was measured and compared to the HF vibrational frequency in an Ar matrix ($n = \infty$). The monomer vibrational frequency $v_{HF}$ red shifts monotonically, but highly nonlinearly, towards the matrix value as sequential Ar atoms are added. Indeed, roughly 50% of the shift is already accounted for by $n = 3$.

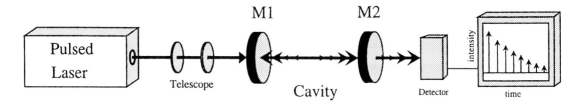

**Figure B1.2.9.** Schematic representation of the method used in cavity ringdown laser absorption spectroscopy. From [33], used with permission.

*Cavity ringdown spectroscopy*

The relatively new technique of cavity ringdown laser absorption spectroscopy, or CRLAS [29–32], has proven to be exceptionally sensitive in the visible region of the spectrum. Recently, it has been used in the IR to measure O–H and O–D vibrations in weakly-bound water and methanol clusters [33]. The concept of cavity ringdown is quite straightforward. If a pulse of light is injected into a cavity composed of two very highly reflective mirrors, only a very small fraction will escape upon reflection from either mirror. If mirrors with 99.996% reflectivity are used, then the photons can complete up to 15 000 round trip passes, depending on other loss factors in the cavity. This makes the effective pathlength up to 30 000 longer than the physical pathlength of the sample. Currently, dielectric coatings for IR wavelengths are not as efficient as those for the visible, and the highest reflectivity available is about 99.9–99.99%, which leads to sensitivity enhancements of several hundred to several thousand. It is best to use pulses with a coherence length that is less than the cavity dimensions in order to avoid destructive interference and cancellation of certain frequencies.

The light leaking out of the cavity will decay exponentially, with a time constant that reflects the round-trip losses. When an absorbing sample is placed in the cavity, there are additional losses and the exponential time constant will become shorter. More highly absorbing samples will affect the time constant to a larger extent, and the absolute absorption is determined. The experiment is shown schematically in figure B1.2.9. One of the most important attributes of CRLAS is that it is relatively insensitive to laser pulse intensity fluctuations since the ringdown time constant, not the transmitted intensity, is measured.

To date, the IR-CRLAS studies have concentrated on water clusters (both $H_2O$ and $D_2O$), and methanol clusters. Most importantly, these studies have shown that it is in fact possible to carry out CRLAS in the IR. In one study, water cluster concentrations in the molecular beam source under a variety of expansion conditions were characterized [34]. In a second study OD stretching bands in $(D_2O)_n$ clusters were measured [35]. These bands occur between 2300 and 2800 cm$^{-1}$, a spectral region that has been largely inaccessible with other techniques. Cooperative effects in the hydrogen-bonded network were measured, as manifested in red shifts of OD stretches for clusters up to $(D_2O)_8$. These data for additional isotopes of water are necessary to fully characterize the intermolecular interactions.

For methanol clusters [36], it was found that the dimer is linear, while clusters of 3 and 4 molecules exist as monocyclic ring structures. There also is evidence that there are two cyclic ring trimer conformers in the molecular beam.

*B1.2.4.2   Resonance Raman spectroscopy*

The advantages of resonance Raman spectroscopy have already been discussed in section B1.2.2.3. For these reasons it is rapidly becoming the method of choice for studying large molecules in solution. Here we will present one study that exemplifies its attributes. There are two complementary methods for studying proteins. First, it is possible to excite a chromophore corresponding to the active site, and determine which modes interact with it. Second, by using UV excitation, the amino acids with phenyl rings (tryptophan and tyrosine,

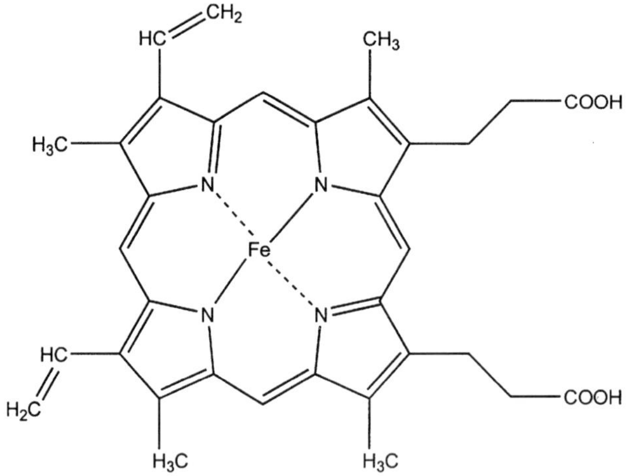

**Figure B1.2.10.** Structure of the protohaeme unit found in haemoglobin and myoglobin.

and a small contribution from phenylalanine) can be selectively excited [4]. The frequency shifts in the resonance Raman spectrum associated with them provide information on their environment.

There has been extensive work done on myoglobin, haemoglobin, Cytochrome-$c$, rhodopsin and bacteriorhodopsin. In fact, there are literally hundreds of articles on each of the above subjects. Here we will consider haemoglobin [12]. The first three of these examples are based on the protohaeme unit, shown in figure B1.2.10.

Haemoglobin is made up of four protohaeme subunits, two are designated $\alpha$ and two are designated $\beta$. The protohaeme unit with surrounding protein is shown in figure B1.2.11. The binding curve of $O_2$ suggests a two state model, where haemoglobin binds $O_2$ more strongly when the concentration is higher. Thus, it will strongly bind $O_2$ in the lungs where the concentration is high, and then release it in tissues where the $O_2$ concentration is low. The high affinity and low affinity states are referred to as the R state and T state, respectively [37]. The R and T states each have a distinct structure, and have been characterized crystallographically. Therefore, there are three primary issues:

(1)   What is the behaviour of the haeme macrocycles?

(2)   What is the interaction between the iron and the nearby histidines?

(3)   What are the structural dynamics of the tetramer as a whole? The haemoglobin bound to $O_2$ (HbO$_2$) is not photoactive, so the CO adduct, HbCO, is used instead.

Information about the haeme macrocycle modes is obtained by comparing the resonance Raman spectra of deoxyHb with HbCO. The d–d transitions of the metal are too weak to produce large enhancement, so the Soret band of the macrocycle, a $\pi$ to $\pi^*$ transition, is excited instead. It has been found that the vinyl groups do not participate in the conjugated system [4]. This is based on the fact that the vinyl C=C stretch does not exhibit resonant enhancement when the $\pi$ to $\pi^*$ transition is excited. On the other hand, it is found that both totally symmetric and nontotally symmetric modes of the macrocycle are all at a lower frequency in the HbCO photoproduct spectra relative to deoxyHb. This is interpreted to mean that the photoproduct has a slightly expanded core relative to the deoxy structure [12, 13]. Given that there is a structural change in the haeme centre, it is expected that the interaction of the iron with the proximal histidine should also be affected.

The Fe–N$^{His}$ mode is at 222 cm$^{-1}$ in the R state and 207 cm$^{-1}$ in the T state for the $\alpha$ subunits, but only shifted to 218 cm$^{-1}$ T state for the $\beta$ subunits. This is consistent with the interpretation that the Fe-imidazole interactions are weakened more in the T state of the $\alpha$ subunits than $\beta$ subunits. Time-resolved resonance

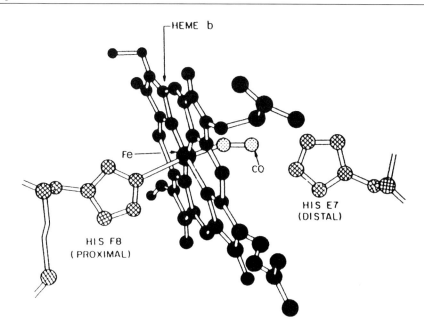

**Figure B1.2.11.** Biologically active centre in myoglobin or one of the subunits of haemoglobin. The bound CO molecule as well as the proximal and distal histidines are shown in addition to the protohaeme unit. From Rousseau D L and Friedman J M 1988 *Biological Applications of Raman Spectroscopy* vol 3, ed T G Spiro (New York: Wiley). Reprinted by permission of John Wiley and Sons Inc.

Raman studies have shown that the R $\rightarrow$ T switch is complete on a 10 $\mu$s timescale [38]. Finally, UV excitation of the aromatic protein side chains yields frequency shifts that indicate a change in the quaternary structure, and this occurs on the same timescale as the frequency shifts in the Fe–N$^{His}$ modes.

### B1.2.4.3   *Time-resolved spectroscopy*

Time-resolved spectroscopy has become an important field from x-rays to the far-IR. Both IR and Raman spectroscopies have been adapted to time-resolved studies. There have been a large number of studies using time-resolved Raman [39], time-resolved resonance Raman [7] and higher order two-dimensional Raman spectroscopy (which can provide coupling information analogous to two-dimensional NMR studies) [40]. Time-resolved IR has probed neutrals and ions in solution [41, 42], gas phase kinetics [43] and vibrational dynamics of molecules chemisorbed and physisorbed to surfaces [44]. Since vibrational frequencies are very sensitive to the chemical environment, pump-probe studies with IR probe pulses allow structural changes to be monitored as a reaction proceeds.

As an illustrative example, consider the vibrational energy relaxation of the cyanide ion in water [45]. The mechanisms for relaxation are particularly difficult to assess when the solute is strongly coupled to the solvent, and the solvent itself is an associating liquid. Therefore, precise experimental measurements are extremely useful. By using a diatomic solute molecule, this system is free from complications due to coupling of vibrational modes in polyatomics. Furthermore, the relatively low frequency stretch of roughly 2000 cm$^{-1}$ couples strongly to the internal modes of the solvent.

Infrared pulses of 200 fs duration with 150 cm$^{-1}$ of bandwidth centred at 2000 cm$^{-1}$ were used in this study. They were generated in a two-step procedure [46]. First, a $\beta$-BaB$_2$O$_4$ (BBO) OPO was used to convert the 800 nm photons from the Ti:sapphire amplifier system into signal and idler beams at 1379 and 1905 nm, respectively. These two pulses were sent through a difference frequency crystal ($AgGaS_2$) to yield pulses centred at 2000 cm$^{-1}$. A 32-element array detector was used to simultaneously detect the entire bandwidth of the pulse [45].

Two isotopic combinations of CN$^-$ ($^{13}C^{15}N^-$ which has a stretching frequency of 2004 cm$^{-1}$ and $^{12}C^{14}N^-$ with a stretching frequency of 2079 cm$^{-1}$) in both $H_2O$ and $D_2O$ yield a range of relaxation times. In particular, it is found that the vibrational relaxation time decreases from 120 to 71 ps in $D_2O$ as the vibrational frequency increases from 2004 to 2079 cm$^{-1}$. However, in $H_2O$, the relaxation time is roughly 30 ps for both isotopomers. The vibrational relaxation rate is highly correlated to the IR absorption cross section of the solvent at the frequency of solute vibration, which indicates that Coulombic interactions have a dominant role in the vibrational relaxation of CN$^-$.

### B1.2.4.4 Action spectroscopy

The term 'action spectroscopy' refers to those techniques that do not directly measure the absorption, but rather the consequence of photoabsorption. That is, there is some measurable change associated with the absorption process. There are several well known examples, such as photoionization spectroscopy [47], multi-photon ionization spectroscopy [48], photoacoustic spectroscopy [49], photoelectron spectroscopy [50, 51], vibrational predissociation spectroscopy [52] and optothermal spectroscopy [53, 54]. These techniques have all been applied to vibrational spectroscopy, but only the last one will be discussed here.

Optothermal spectroscopy is a bolometric method that monitors the energy in a stream of molecules rather than in the light beam. A well collimated molecular beam is directed toward a liquid helium cooled bolometer. There will be energy deposited in the bolometer from the translational kinetic energy of the molecules as well as any internal energy they may have. A narrow linewidth ($\sim$2 MHz) infrared laser illuminates the molecular beam, typically in a multipass geometry. As the laser frequency is scanned the molecules will absorb energy when the frequency corresponds to a transition frequency. At that point, the energy deposited in the bolometer will change.

The optothermal spectrum will faithfully represent the absorption spectrum provided that the molecules do not fluoresce prior to arrival at the detector. Fluorescence is not a problem because infrared fluorescence lifetimes are of the order of milliseconds. The transit time from the laser-molecular beam interaction region to the detector is tens of $\mu$s. Furthermore, it is possible to determine if the absorbing species is a stable monomer, or weakly bound cluster. When a stable monomer absorbs a photon, the amount of energy measured by the bolometer increases. When a weakly-bound species absorbs a photon greater than the dissociation energy, vibrational predissociation will take place and the dissociating fragments will not hit the bolometer element, leading to a decrease in energy measured by the bolometer. It is also possible to place the bolometer off-axis from the collimated molecular beam so that only dissociating molecules will register a signal.

A nice example of this technique is the determination of vibrational predissociation lifetimes of (HF)$_2$ [55]. The HF dimer has a nonlinear hydrogen bonded structure, with nonequivalent HF subunits. There is one 'free' HF stretch ($\nu_1$), and one 'bound' HF stretch ($\nu_2$), which rapidly interconvert. The vibrational predissociation lifetime was measured to be 24 ns when exciting the free HF stretch, but only 1 ns when exciting the bound HF stretch. This makes sense, as one would expect the bound HF vibration to be most strongly coupled to the weak intermolecular bond.

### B1.2.4.5 Microscopy

It is possible to incorporate a Raman or IR spectrometer within a confocal microscope. This allows the spatial resolution of the microscope and compound identification of vibrational spectroscopy to be realized simultaneously. One of the reasons that this is a relatively new development is because of the tremendous

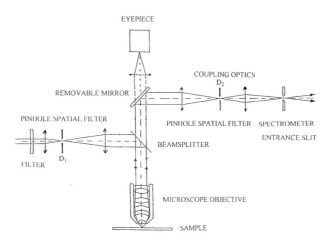

**Figure B1.2.12.** Schematic diagram of apparatus for confocal Raman microscopy. From [3], used with permission.

volume of data generated. For example, if a Raman microscope has roughly 1 $\mu$m spatial resolution, and an area of 100 $\mu$m $\times$ 100 $\mu$m is to be imaged, and the frequency region from 800 cm$^{-1}$–3400 cm$^{-1}$ is covered with 4 cm$^{-1}$ spectral resolution, then the data set has 6 million elements. Assuming each value is represented with a 4 byte number, the image would require 24 MBytes of storage space. While this is not a problem for current computers, the capacity of a typical hard drive on a PC from around 1985 (IBM 8088) was only 20 MByte. Also, rapid data transfer is needed to archive and retrieve images. Furthermore, in order to obtain the spectrum at any spatial position, array detectors (or FT methods) are required. A representative experimental set-up is shown in figure B1.2.12 [3].

Raman microscopy is more developed than its IR counterpart. There are several reasons for this. First, the diffraction limit for focusing a visible beam is about 10 times smaller than an IR beam. Second, Raman spectroscopy can be done in a backscattering geometry, whereas IR is best done in transmission. A microscope is most easily adapted to a backscattering geometry, but it is possible to do it in transmission.

Raman microscopy is particularly adept at providing information on heterogeneous samples, where a conventional spectrometer would average over many domains. It has found applications in materials science and catalysis; earth, planetary and environmental sciences; biological and medical sciences; and even in art history and forensic science [3]. For example, consider a hypothetical situation where someone owned what they believed to be a mediæval manuscript from the 12th century. One could easily identify a small region of green colour to non-destructively analyse using Raman microscopy. If it happens that the dye is identified as phthalocyanine green, which was not discovered until 1938, then one can be certain that the manuscript is not authentic, or that it has undergone restoration relatively recently.

## B1.2.5   Conclusions and future prospects

Vibrational spectroscopy has been, and will continue to be, one of the most important techniques in physical chemistry. In fact, the vibrational absorption of a *single* acetylene molecule on a Cu(100) surface was recently reported [56]. Its endurance is due to the fact that it provides detailed information on structure, dynamics and environment. It is employed in a wide variety of circumstances, from routine analytical applications, to identifying novel (often transient) species, to providing some of the most important data for advancing the understanding of intramolecular and intermolecular interactions.

## References

[1]  Califano S 1976 *Vibrational States* (New York: Wiley)
[2]  McQuarrie D A 1983 *Quantum Chemistry* (Mill Valley, CA: University Science Books)
[3]  Turrell G and Corset J (eds) 1996 *Raman Microscopy: Developments and Applications* (New York: Academic)
[4]  Asher S A 1993 UV resonance Raman-spectroscopy for analytical, physical and biophysical chemistry 2 *Anal. Chem.* **65** A201–10
[5]  Asher S A 1993 UV resonance Raman-spectroscopy for analytical, physical and biophysical chemistry 1 *Anal. Chem.* **65** A59–66
[6]  Asher S A and Chi Z H 1998 UV resonance Raman studies of protein folding in myoglobin and other proteins *Biophys. J.* **74** A29
[7]  Bell S E J 1996 Time-resolved resonance Raman spectroscopy *Analyst* **121** R107–20
[8]  Biswas N and Umapathy S 1998 Resonance Raman spectroscopy and ultrafast chemical dynamics *Curr. Sci.* **74** 328–40
[9]  Chi Z H, Chen X G, Holtz J S W and Asher S A 1998 UV resonance Raman-selective amide vibrational enhancement: quantitative methodology for determining protein secondary structure *Biochemistry* **37** 2854–64
[10] Hoskins L C 1984 Resonance Raman-spectroscopy of beta-carotene and lycopene—a physical-chemistry experiment *J. Chem. Educ.* **61** 460–2
[11] Johnson B R, Kittrell C, Kelly P B and Kinsey J L 1996 Resonance Raman spectroscopy of dissociative polyatomic molecules *J. Phys. Chem.* **100** 7743–64
[12] Kincaid J R 1995 Structure and dynamics of transient species using time-resolved resonance Raman-spectroscopy *Biochemical Spectroscopy Methods Enzymol.* vol 246, ed K Sauer (San Diego, CA: Academic) pp 460–501
[13] Spiro T G and Czernuszewicz R S 1995 Resonance Raman-spectroscopy of metalloproteins *Biochemical Spectroscopy Methods Enzymol.* vol 246, ed K Sauer (San Diego, CA: Academic) pp 416–60
[14] Strommen D P and Nakamoto K 1977 Resonance Raman-spectroscopy *J. Chem. Educ.* **54** 474–8
[15] Mathies R A 1995 Biomolecular vibrational spectroscopy *Biochemical Spectroscopy Methods Enzymol.* vol 246, ed K Sauer (San Diego, CA: Academic) pp 377–89
[16] Heller E J, Sundberg R L and Tannor D 1982 Simple aspects of Raman scattering *J. Phys. Chem.* **86** 1822–33
[17] Zhong Y and McHale J L 1997 Resonance Raman study of solvent dynamics in electron transfer. II. Betaine-30 in $CH_3OH$ and $CD_3OD$ *J. Chem. Phys.* **107** 2920–9
[18] Gruebele M H W 1988 *Infrared Laser Spectroscopy of Molecular Ions and Clusters* (Berkeley: University of California)
[19] Kidder L H, Levin I W, Lewis E N, Kleiman V D and Heilweil E J 1997 Mercury cadmium telluride focal-plane array detection for mid-infrared Fourier-transform spectroscopic imaging *Opt. Lett.* **22** 742–4
[20] Pibel C D, Sirota E, Brenner J and Dai H L 1998 Nanosecond time-resolved FTIR emission spectroscopy: monitoring the energy distribution of highly vibrationally excited molecules during collisional deactivation *J. Chem. Phys.* **108** 1297–300
[21] Leone S R 1989 Time-resolved FTIR emission studies of molecular photofragmentation *Accounts Chem. Res.* **22** 139–44
[22] Elving P J and Winefordner J D (eds) 1986 *Fourier Transform Infrared Spectroscopy* (New York: Wiley)
[23] Skoog D A, Holler F J and Nieman T A 1998 *Principles of Instrumental Analysis* 5th edn (Philadelphia: Harcourt Brace)
[24] Saykally R J and Blake G A 1993 Molecular-interactions and hydrogen-bond tunneling dynamics—some new perspectives *Science* **259** 1570–5
[25] Liu K, Brown M G and Saykally R J 1997 Terahertz laser vibration rotation tunneling spectroscopy and dipole moment of a cage form of the water hexamer *J. Phys. Chem.* A **101** 8995–9010
[26] Nesbitt D J 1994 High-resolution, direct infrared-laser absorption-spectroscopy in slit supersonic jets—intermolecular forces and unimolecular vibrational dynamics in clusters *Ann. Rev. Phys. Chem.* **45** 367–99
[27] Bacic Z and Miller R E 1996 Molecular clusters: structure and dynamics of weakly bound systems *J. Phys. Chem.* **100** 12 945–59
[28] Dvorak M A, Reeve S W, Burns W A, Grushow A and Leopold K R 1991 Observation of three intermolecular vibrational-states of Ar-HF *Chem. Phys. Lett.* **185** 399–402
[29] O'Keefe A and Deacon D A G 1988 Cavity ring-down optical spectrometer for absorption-measurements using pulsed laser sources *Rev. Sci. Instrum.* **59** 2544–51
[30] Scherer J J, Paul J B, O'Keefe A and Saykally R J 1997 Cavity ringdown laser absorption spectroscopy: history, development, and application to pulsed molecular beams *Chem. Rev.* **97** 25–51
[31] Zalicki P and Zare R N 1995 Cavity ring-down spectroscopy for quantitative absorption-measurements *J. Chem. Phys.* **102** 2708–17
[32] Lehmann K K and Romanini D 1996 The superposition principle and cavity ring-down spectroscopy *J. Chem. Phys.* **105** 10 263–77
[33] Scherer J J *et al* 1995 Infrared cavity ringdown laser-absorption spectroscopy (IR-CRLAS) *Chem. Phys. Lett.* **245** 273–80
[34] Paul J B, Collier C P, Saykally R J, Scherer J J and O'Keefe A 1997 Direct measurement of water cluster concentrations by infrared cavity ringdown laser absorption spectroscopy *J. Phys. Chem.* A **101** 5211–14
[35] Paul J B, Provencal R A and Saykally R J 1998 Characterization of the $(D_2O)(2)$ hydrogen-bond-acceptor antisymmetric stretch by IR cavity ringdown laser absorption spectroscopy *J. Phys. C: Solid State Phys.* A **102** 3279–83
[36] Provencal R A *et al* 1999 Infrared cavity ringdown spectroscopy of methanol clusters: single donor hydrogen bonding *J. Chem. Phys.* **110** 4258–67
[37] Monod J, Wyman J and Changeux J P 1965 On the nature of allosteric transitions: a plausible model *J. Mol. Biol.* **12** 88–118
[38] Scott T W and Friedman J M 1984 Tertiary-structure relaxation in haemoglobin—a transient Raman-study *J. Am. Chem. Soc.* **106** 5677–87
[39] Shreve A P and Mathies R A 1995 Thermal effects in resonance Raman-scattering—analysis of the Raman intensities of rhodopsin and of the time-resolved Raman-scattering of bacteriorhodopsin *J. Phys. Chem.* **99** 7285–99

[40] Tokmakoff A, Lang M J, Larsen D S, Fleming G R, Chernyak V and Mukamel S 1997 Two-dimensional Raman spectroscopy of vibrational interactions in liquids *Phys. Rev. Lett.* **79** 2702–5

[41] Owrutsky J C, Raftery D and Hochstrasser R M 1994 Vibrational-relaxation dynamics in solutions *Ann. Rev. Phys. Chem.* **45** 519–55

[42] Hamm P, Lim M, DeGrado W F and Hochstrasser R M 1999 The two-dimensional IR nonlinear spectroscopy of a cyclic penta-peptide in relation to its three-dimensional structure *Proc. Natl Acad. Sci. USA* **96** 2036–41

[43] Zheng Y F, Wang W H, Lin J G, She Y B and Fu K J 1992 Time-resolved infrared studies of gas-phase coordinatively unsaturated photofragments (Eta-5-C5h5)Mn(Co)X (X = 2 and 1) *J. Phys. Chem.* **96** 7650–6

[44] Cavanagh R R, Heilweil E J and Stephenson J C 1994 Time-resolved measurements of energy-transfer at surfaces *Surf. Sci.* **300** 643–55

[45] Hamm P, Lim M and Hochstrasser R M 1997 Vibrational energy relaxation of the cyanide ion in water *J. Chem. Phys.* **107** 10 523–31

[46] Seifert F, Petrov V and Woerner M 1994 Solid-state laser system for the generation of midinfrared femtosecond pulses tunable from 3.3-Mu-M to 10-Mu-M *Opt. Lett.* **19** 2009–11

[47] Wight C A and Armentrout P B 1993 Laser photoionization probes of ligand-binding effects in multiphoton dissociation of gas-phase transition-metal complexes *ACS Symposium Series* **530** 61–74

[48] Belbruno J J 1995 Multiphoton ionization and chemical-dynamics *Int. Rev. Phys. Chem.* **14** 67–84

[49] Crippa P R, Vecli A and Viappiani C 1994 Time-resolved photoacoustic-spectroscopy—new developments of an old idea *J. Photochem. Photobiol. B-Biol.* **24** 3–15

[50] Muller-Dethlefs K and Schlag E W 1998 Chemical applications of zero kinetic energy (ZEKE) photoelectron spectroscopy *Angew. Chem.-Int. Edit.* **37** 1346–74

[51] Bailey C G, Dessent C E H, Johnson M A and Bowen K H 1996 Vibronic effects in the photon energy-dependent photoelectron spectra of the $CH_3CN^-$ dipole-bound anion *J. Chem. Phys.* **104** 6976–83

[52] Ayotte P, Bailey C G, Weddle G H and Johnson M A 1998 Vibrational spectroscopy of small $Br \cdot (H_2O)_n$ and $I \cdot (H_2O)_n$ clusters: infrared characterization of the ionic hydrogen bond *J. Phys. Chem. A* **102** 3067–71

[53] Miller R E 1990 Vibrationally induced dynamics in hydrogen-bonded complexes *Accounts Chem. Res.* **23** 10–16

[54] Lehmann K K, Scoles G and Pate B H 1994 Intramolecular dynamics from eigenstate-resolved infrared-spectra *Ann. Rev. Phys. Chem.* **45** 241–74

[55] Huang Z S, Jucks K W and Miller R E 1986 The vibrational predissociation lifetime of the HF dimer upon exciting the free-H stretching vibration *J. Chem. Phys.* **85** 3338–41

[56] Stipe B C, Rezaei M A and Ho W 1998 Single-molecule vibrational spectroscopy and microscopy *Science* **280** 1732–5

## Further Reading

Wilson E B Jr, Decius J C and Cross P C 1955 *Molecular Vibrations: The Theory of Infrared and Raman Vibrational Spectra* (New York: Dover)

A classic text on molecular vibrations.

Herzberg G 1945 *Molecular Spectra and Molecular Structure II. Infrared and Raman Spectra of Polyatomic Molecules* (New York: Van Nostrand Reinhold)

Comprehensive treatment of vibrational spectroscopy, including data for a wide variety of molecules.

Califano S 1976 *Vibrational States* (New York: Wiley)

Similar to Wilson, Decius and Cross, but somewhat more modern. First three chapters provide a nice overview.

McQuarrie D A 1983 *Quantum Chemistry* (Mill Valley, CA: University Science Books)

The sections on vibrations and spectroscopy are somewhat more accessible mathematically than the previous three books.

Turrell G and Corset J (eds) 1996 *Raman Microscopy: Developments and Applications* (New York: Academic Press)

In addition to covering Raman microscopy, this book has a wealth of information on Raman instrumentation in general.

Elving P J and Winefordner J D (eds) 1986 *Fourier Transform Infrared Spectroscopy* (New York: Wiley)

Comprehensive coverage of all fundamental aspects of Fourier transform infrared spectroscopy.

# B1.3
# Raman spectroscopy

*Darin J Ulness, Jason C Kirkwood and A C Albrecht*

### B1.3.1   Introduction

Light, made monochromatic and incident upon a sample, is scattered and transmitted at the incident frequency—a process known as Rayleigh scattering. Early in 1928 C V Raman and K S Krishnan reported [1–3] the visual discovery of a new form of secondary radiation. Sunlight, passed through a blue–violet filter, was used as the light source incident upon many different organic liquids and even vapours. A green filter was placed between the sample and the viewer. The filters were sufficiently complementary to suppress the strong Rayleigh scattering and leave at longer wavelengths a feeble new kind of scattered radiation. Impurity fluorescence was discounted, for the signal was robust under purification and it also was strongly polarized. It was suggested that the incident photons had undergone inelastic scattering with the material—much as the Compton scattering of x-rays by electrons. Almost simultaneously, G Landsberg and L Mandelstam [4] reported a new kind of secondary radiation from crystalline quartz illuminated by the lines of the mercury vapour lamp. Spectrograms revealed a feeble satellite line to the red of each of the Rayleigh scattered mercury lines. In each case the displacement came close to a characteristic vibrational frequency of quartz at $\sim 480$ cm$^{-1}$. They wondered whether their new secondary radiation might not be of the same type as that seen by Krishnan and Raman. Such inelastic scattering of photons soon came to be called (spontaneous) Raman scattering or, given its quantized energy displacements, simply (spontaneous) Raman spectroscopy [5].

The 70 years since these first observations have witnessed dramatic developments in Raman spectroscopy, particularly with the advent of lasers. By now, a large variety of Raman spectroscopies have appeared, each with its own acronym. They all share the common trait of using high energy ('optical') light to probe small energy level spacings in matter.

This chapter is aimed at any scientist who wishes to become acquainted with this broad and interesting field. At the start we outline and present the principles and modern theoretical structure that unifies the many versions of Raman spectroscopies currently encountered. Then we sketch, briefly, individual examples from the contemporary literature of many of the Raman spectroscopies, indicating various applications. Though the theoretical structure is intended to stand on its own when discussing any one subject, there is no pretence of completeness, nor depth—this is not a review. But it is hoped that through the selected citations the interested reader is easily led into the broader literature on any given topic—including investigations yet to appear.

The study of small energy gaps in matter using the 'optical' spectral region (say the near-IR, visible and UV) offers many advantages over direct one-photon spectroscopies in the IR, far IR or even the microwave. First, it is instrumentally convenient. Second, one can readily avoid the problem of absorbing solvents when studying dissolved solutes. Finally, in the optical region, additional strong resonances (usually involving electronic transitions) are available which can enhance the Raman scattered intensity by many orders of magnitude. This has created the very lively field of resonance Raman spectroscopy—arguably one of the most popular current Raman based spectroscopies. Resonance Raman spectroscopy not only provides greatly amplified signals from specific Raman resonances in the ground state, but it also exposes useful properties of the upper electronic potential energy hypersurface that is reached in the optical resonance.

In general, the coupling of light with matter has as its leading term the electric field component of the electromagnetic (EM) radiation—extending from the microwave into the vacuum ultraviolet. In the optical region (or near optical), where the Raman spectroscopies are found, light fields oscillate in the 'peta-hertz' range (of the order of $10^{15}$ cycles per second). So for technical reasons the signals are detected as photons ('quadrature in the field')—not as the oscillating field itself. As in any spectroscopy, Raman spectro-scopies measure eigenvalue differences, dephasing times (through the bandwidths, or through time-resolved measurements) and quantitative details concerning the strength of the light/matter interaction—through the scattering cross-sections obtained from absolute Raman intensities. The small energy gap states of matter that are explored in Raman spectroscopy include phonon-like lattice modes (where Brillouin scattering can be the Raman scattering equivalent), molecular rotations (rotational Raman scattering), internal vibrations of molecules (vibrational Raman scattering) and even low lying electronic states (electronic Raman scattering). Also spin states may be probed through the magnetic component of the EM field. With the introduction of lasers, the Raman spectroscopies have been brought to a new level of sensitivity as powerful analytic tools for probing samples from the microscopic level (microprobes) to remote sensing. Not only have lasers inspired the development of new techniques and instrumentation, but they also have spawned more than 25 new kinds of Raman spectroscopy. As we shall see, this growth in experimental diversity has been accomplished by increasingly comprehensive theoretical understanding of all the spectroscopies.

The many Raman spectroscopies are found as well defined subgroups among the 'electric field' spectro-scopies in general. (In this chapter, the magnetic field spectroscopies are mentioned only in passing.) First, except for very high intensities, the energy of interaction of light with matter is sufficiently weak to regard modern spectroscopies as classifiable according to perturbative orders in the electric field of the light. Thus any given spectroscopy is regarded as being *linear* or *nonlinear* with respect to the incident light fields. In another major classification, a given spectroscopy (linear or nonlinear) is said to be *active* or *passive* [6, 7]. The *active* spectroscopies are those in which the principal event is a change of state population in the material. In order to conserve energy this must be accompanied by an appropriate change of photon numbers in the light field. Thus net energy is transferred between light and matter in a manner that survives averaging over many cycles of the perturbing light waves. We call these the Class I spectroscopies. They constitute all of the well known absorption and emission spectroscopies—whether they are one-photon or multi-photon. The *passive* spectroscopies, called Class II, arise from the momentary exchange of energy between light and matter that induces a macroscopically coherent, oscillating, electrical polarization (an oscillating electric dipole density wave) in the material. As long as this coherence is sustained, such polarization can serve as a source term in the wave equation for the electric field. A new (EM) field (the signal field) is produced at the frequency of the oscillating polarization in the sample. Provided the polarization wave retains some coherence, and matches the signal field in direction and wavelength (a condition called 'phase matching'), the new EM field can build up and escape the sample and ultimately be measured 'in quadrature' (as photons). In their extreme form, when no material resonances are operative, the Class II events ('spectroscopies' is a misnomer in the absence of resonances!) will alter only the states of the EM radiation and none of the material. In this case, the material acts *passively* while 'catalysing' alterations in the radiation. When resonances are present, Class II events become spectroscopies; some net energy may be transferred between light and matter, even as one focuses experimentally not on population changes in the material, but on alterations of the radiation. Class II events include all of the resonant and nonresonant, linear and nonlinear *dispersions*. Examples of Class II spectroscopies are classical diffraction and reflection (strongest at the linear level) and a whole array of light-scattering phenomena such as frequency summing (harmonic generation), frequency differencing, free induction decay and optical echoes.

The Class II (passive) spectroscopies

(i)   may or may not contain resonances,

(ii)  can appear at all orders of the incident field (but only at odd order for isotropic media (gases, liquids, amorphous solids)),

(iii) have a cross-section that is quadratic in concentration when the signal wave is homodyne detected and

(iv)  require phase-matching through experimental design, because the signal field must be allowed to build up from the induced polarization over macroscopic distances.

In tables B1.3.1 and B1.3.2, we assemble the more than twenty-five Raman spectroscopies and order them according to their degree of nonlinearity and by their class (table B1.3.1 for Class I, table B1.3.2 for Class II).

A diagrammatic approach that can unify the theory underlying these many spectroscopies is presented. The most complete theoretical treatment is achieved by applying statistical quantum mechanics in the form of the time evolution of the light/matter *density operator*. (It is recommended that anyone interested in advanced study of this topic should familiarize themselves with density operator formalism [8–12]. Most books on nonlinear optics [13–17] and nonlinear optical spectroscopy [18, 19] treat this in much detail.) Once the density operator is known at any time and position within a material, its matrix in the eigenstate basis set of the constituents (usually molecules) can be determined. The ensemble averaged electrical polarization, $P$, is then obtained—the centrepiece of all spectroscopies based on the electric component of the EM field.

Following the section on theory, the chapter goes on to present examples of most of the Raman spectroscopies that are organized in tables B1.3.1 and B1.3.2.

The Class I (active) spectroscopies, both linear and nonlinear [6, 7],

(i)   always require resonances,

(ii)  appear only at odd order in the incident fields, which are acting 'maximally in quadrature'. (That is, in polarizing the medium all but one of the fields act in pairs of Fourier components—also known as 'in quadrature' or as 'conjugate pairs'. When this happens, the electrical polarization must carry the same frequency, wavelength, and wave vector as the odd, unpaired field. Since this odd order polarization acts conjugately with the odd incident field, the 'photon' picture of the spectroscopy survives.),

(iii) have a cross-section that is linear in concentration of the resonant species and

(iv)  do not require phase matching by experimental design, for it is automatic.

## B1.3.2   Theory[1]

The oscillating electric dipole density, $P$ (the polarization), that is induced by the total incident electric field, $E$ is the principal property that generates all of the spectroscopies, both linear and nonlinear. The energy contained in $P$ may be used in part (or altogether) to shift the population of energy states of the material, or it may in part (or altogether) reappear in the form of a new EM field oscillating at the same frequency. When the population changes, or energy loss or gain in the light is detected, one is engaged in a Class I spectroscopy. On the other hand, when properties of the new field are being measured—such as its frequency, direction (wavevector), state of polarization and amplitude (or intensity)—one has a Class II spectroscopy.

[1]   A version of this material appears in a special issue of the *Journal of Physical Chemistry* dedicated to the Proceedings of the International Conference on Time-Resolved Vibrational Spectroscopy (TRVS IX), May 16–22 1999, Tucson, Arizona. See: Kirkwood J C, Ulness D J and Albrecht A C 2000 On the classification of the electric field spectroscopies: applications to Raman scattering *J. Phys. Chem.* A **104** 4167–73.

**Table B1.3.1.** Several Raman spectroscopies classified according to their spectroscopic class[a] and their wave-mixing energy level (WMEL) diagram(s)[b].

| Technique | Acronym | Representative Diagram(s) |
|---|---|---|
| CLASS I—THIRD ORDER | | |
| Spontaneous Raman | SR | A,B |
| Stimulated Raman gain spectroscopy | SRS | A,B,G,H |
| Stimulated Raman loss spectroscopy | SRL | Figure B1.3.3(b) |
| Surface enhanced Raman scattering | SERS | A,B |
| Fourier transform surface enhanced Raman scattering | FT-SERS | A,B |
| Fourier transform spontaneous Raman | FT-SR | A,B |
| Hardamard transform Raman scattering | HTRS | A,B |
| Resonance Raman scattering | RRS | A,B ($D_R$ only), figure B1.3.5 |
| Dissociative resonance Raman scattering | DRRS | A,B ($D_R$ only), figure B1.3.5 |
| Transient resonance Raman scattering | TRRS | A,B ($D_R$ only), figure B1.3.5 |
| Time resolved resonance Raman scattering | TRRR | A,B ($D_R$ only), figure B1.3.5 |
| Photoacoustic spontaneous Raman spectroscopy | PARS | A,B |
| Optoacoustic Raman spectroscopy | OARS | A,B |
| CLASS I—FIFTH ORDER | | |
| Hyper-Raman scattering | HRS | Not shown |
| Surface-enhanced hyper-Raman scattering | SEHRS | Not shown |
| Surface-enhanced resonance hyper-Raman scattering | SERHRS | Not shown |

[a] The class of a particular spectroscopy determines the general quantities that can be measured by a given technique—although the choice of a given spectroscopy depends greatly on experimental considerations and the expertise and/or resources of the experimenter. That being said, Class I or full quadrature spectroscopies have the potential to measure Raman frequencies, differential cross-sections, depolarization ratios and Raman lineshapes, each of which are discussed in the text. Class II spectroscopies, in addition to potentially measuring the same quantities as that in Class I, have the added features of greater control over the signal (highly directional) and measurement of the resonant–nonresonant ratio for the electric susceptibility.

[b] The wave-mixing energy level (WMEL) diagrams represent a specific Liouville pathway for the evolution of the density operator. The lettered diagrams listed here refer to the 'window' stage ($W$) of a particular WMEL diagram, as discussed in the text, since the 'doorway' stages ($D$) are common to all spectroscopies (unless otherwise noted). The diagrams for these doorway and window stages are given in figure B1.3.2.

Normally the amplitude of the total incident field (or intensity of the incident light) is such that the light/matter coupling energies are sufficiently weak not to compete seriously with the 'dark' matter Hamiltonian. As already noted, when this is the case, the induced polarization, $P$ is treated perturbatively in orders of the total electric field. Thus one writes

$$P = \overbrace{X^{(1)}E}^{P^{(1)}} + \overbrace{X^{(2)}EE}^{P^{(2)}} + \overbrace{X^{(3)}EEE}^{P^{(3)}} + \cdots \overbrace{X^{(s)}E\ldots E}^{P^{(s)}} + \cdots \tag{B1.3.1}$$

where the successive terms clearly appear with increasing order of nonlinearity in the total field, $E$. At this point, all properties appearing in equation (B1.3.1) are mathematically pure real. The response function of the material to the electric field acting at $s$th order is the electrical susceptibility, $X^{(s)}$ (it is an $s + 1$ rank tensor). Each element of this tensor will carry $s + 1$ subscripts, a notation that is used, understandably, only when necessary. Furthermore, the events at, say the $s$th order, are sometimes referred to as '$s + 1$-wave-mixing'—the additional field being the new EM field derived from $P^{(s)}$.

**Table B1.3.2.** Several Raman spectroscopies classified according to their spectroscopic class[a] and their wave-mixing energy level (WMEL) diagram(s)[b].

| Technique | Acronym | Representative Diagram(s) |
|---|---|---|
| CLASS II—THIRD ORDER | | |
| Coherent Raman scattering | CRS | C,D,E,F |
| Interferometric coherent Raman scattering | $I^{(2)}$CRS | C,D,E,F |
| Coherent anti-Stokes Raman scattering | CARS | E,F |
| Interferometric coherent anti-Stokes Raman scattering | $I^{(2)}$CARS | E,F |
| Coherent Stokes Raman scattering | CSRS | C,D |
| Interferometric coherent Stokes Raman scattering | $I^{(2)}$CSRS | C,D |
| Coherent Raman ellipsometry | CRE | C,D,E,F |
| Raman induced Kerr effect spectroscopy | RIKES | B,G |
| Impulsive stimulated Raman scattering | ISRS | C,D,E,F |
| CLASS II—FOURTH ORDER | | |
| BioCARS | BioCARS | Not shown |
| CLASS II—FIFTH ORDER | | |
| Raman quasi-echo | QE | Figure B1.3.8 |
| Coherent higher order Raman excitation spectroscopy | CHORES | Not shown |
| Coherent hyper-Raman scattering | CHRS | Not shown |
| CLASS II—SEVENTH ORDER | | |
| Raman echo | RE | Figure B1.3.7 |

[a,b] See table B1.3.1.

All *nonlinear* (electric field) spectroscopies are to be found in all terms of equation (B1.3.1) except for the first. The latter exclusively accounts for the standard linear spectroscopies—one-photon absorption and emission (Class I) and linear dispersion (Class II). For example, the term at third order contains by far the majority of the modern Raman spectroscopies (tables B1.3.1 and B1.3.2).

It is useful to recognize that in the laboratory one normally configures an experiment to concentrate on one particular kind of spectroscopy, even while all possible light/matter events must be occurring in the sample. How can one isolate from equation (B1.3.1) a spectroscopy of interest? In particular, how does one uncover the Raman spectroscopies? We shall see how passage to the complex mathematical representation of the various properties is invaluable in this process. It is useful to start by addressing the issue of distinguishing Class I and Class II spectroscopies at any given order of nonlinearity.

*B1.3.2.1 Class I and Class II spectroscopies and the complex susceptibilities*

All Class I spectroscopies at any order can be exposed by considering the long-term exchange of energy between light and matter as judged by the nonvanishing of the induced power density over many cycles of the field. The instantaneous power density at $s$th order is given by $\{E \cdot \frac{\partial P^{(s)}}{\partial t}\}$. For this product to survive over time, we ask that its normalized integral over a time $T$, which is much longer than the optical period of the field, should not vanish. This is called the cycle averaged, $s$th order power density. It is expressed as

$$W^{(s)} = \frac{1}{T} \int_0^T dt \left\{ E \cdot \frac{\partial P^{(s)}}{\partial t} \right\}. \tag{B1.3.2}$$

For $W^{(s)} > 0$, one has absorption; for $W^{(s)} < 0$, emission. Multiphoton absorption and emission fall into this class. The Class I Raman spectroscopies clearly exhibit a net absorption of energy in Stokes scattering and a net emission of energy in anti-Stokes scattering. Though $P^{(s)}$ involve $s$ actions of the total field, the light/matter energy exchange is always in the language of photons. A net energy in the form of photons is destroyed (absorbed) as the quantum state population of the material moves upward in energy; a net energy in the form of photons is created (emitted) as the population moves downward in energy. To survive the integration, the instantaneous power density should not oscillate rapidly (if at all), certainly not at optical frequencies. Since $E$ is intended to consist entirely of fields that oscillate at optical frequencies, the power density can have a non-oscillating term only when the field appears altogether an *even* number of times. Since it appears $s$ times in $P^{(s)}$ (equation (B1.3.1)) and once in $E$, all Class I spectroscopies exist only when $s + 1$ is even, or $s$ is odd (see equation (B1.3.2)).

Furthermore, the non-oscillating component of the integrand can best be sorted out by going to the complex representation of the total field, the polarization, and the susceptibility. The mathematically pure real quantities in equation (B1.3.2) can be written in their complex representation as follows:

$$E = \tfrac{1}{2}(\varepsilon + \varepsilon^*) \tag{B1.3.3}$$

$$P^{(s)} = \tfrac{1}{2}(p^{(s)} + p^{(s)*}) \tag{B1.3.4}$$

and

$$X^{(s)} = \tfrac{1}{2}(\chi^{(s)} + \chi^{(s)*}) \tag{B1.3.5}$$

in which $\varepsilon$, $p^{(s)}$ and $\chi^{(s)}$ are, in general, complex quantities whose real parts are given by $E$, $P^{(s)}$ and $X^{(s)}$, respectively. Introducing equations (B1.3.4)–(B1.3.5) into equation (B1.3.2), and applying the cycle average theorem for the integral [20], one finds that for all spectroscopies at $s$th order involving long-term light/matter energy exchange (Class I, in particular), the signal measured as a net energy exchange, $S_{\mathrm{I}}^{(s)}$, is proportional to the cycle averaged power density, or $S_{\mathrm{I}}^{(s)} \propto W^{(s)} \propto \mathrm{Im}\,\chi^{(s)}$. If the complex susceptibility, $\chi^{(s)}$, is pure real, there can be no long term energy exchange between light and matter. There can be no Class I spectroscopies based on a susceptibility component for which $\mathrm{Im}\,\chi^{(s)} = 0$.

Consider an ensemble composed of $N$ constituents (such as molecules) per unit volume. The (complex) density operator for this system is developed perturbatively in orders of the applied field, and at $s$th order is given by $\rho^{(s)}$. The (complex) $s$th order contribution to the ensemble averaged polarization is given by the trace over the eigenstate basis of the constituents of the product of the dipole operator, $N\mu$ and $\rho^{(s)}$: $p^{(s)} = \mathrm{Tr}\{N\mu\rho^{(s)}\}$. In turn, an expression for $\chi^{(s)}$ is obtained, which, in the frequency domain, consists of a numerator containing a product of $(s + 1)$ transition moment matrix elements and a denominator of $s$ complex energy factors. These complex energies express light/matter resonances and allow $\chi^{(s)}$ to become a complex quantity, its $\mathrm{Im}\,\chi^{(s)}$ part (pure real) being responsible for Class I spectroscopies. The light/matter resonances introduce the imaginary component to $\chi^{(s)}$ and permit a Class I spectroscopy to exist.

As noted, the Class II spectroscopies are based on detecting the new EM field that is derived from the induced polarization, $P^{(s)}$, at $s$th order. Here $P^{(s)}$, oscillating at optical frequencies, acts as the source term in Maxwell's equation to create the new optical field, $E_{\mathrm{new}}$, at the same frequency. Again, we recognize $\varepsilon_{\mathrm{new}} \propto p^{(s)} \propto \chi^{(s)}$.

Since optical fields oscillate too quickly for direct detection, they are measured 'in quadrature'—as photons (see below). There are two ways to achieve quadrature. One is *homodyne* detection in which the new field is measured at its quadrature, $\varepsilon_{\mathrm{new}}\varepsilon_{\mathrm{new}}^* = |\varepsilon_{\mathrm{new}}|^2$. These signals must be proportional to $|\chi^{(s)}|^2$. Thus $S_{\mathrm{II}}^{(s)}$ (homodyne) $\propto |\chi^{(s)}|^2$ and all phase information in $\chi^{(s)}$ is lost. Such is the case for almost all of the Class II spectroscopies, especially the Raman events at third order.

The second way to achieve quadrature is to introduce another field, $E_{\mathrm{lo}}$, (called a local oscillator) designed in frequency and wavevector to conjugate (go into quadrature) in its complex representation with the new field of interest. Thus in the heterodyne case, the signal photons are derived from $\varepsilon_{\mathrm{new}}\varepsilon_{\mathrm{lo}}^*$, or $S_{\mathrm{II}}^{(s)}$ (heterodyne) $\propto \chi^{(s)}$. In heterodyne detected $s + 1$ wave mixing, phase information is retained and one can take a full measure of the complex susceptibility, including its phase. The phase of the complex induced polarization, $p^{(s)}$, determines how its energy will partition between Class I (absorbed or emitted) and Class II (a new EM wave is launched) spectroscopies.

Consider all of the spectroscopies at third order ($s = 3$). To be as general as possible, suppose the total incident field consists of the combination of three experimentally distinct fields ($j = 1, 2, 3$). These can differ in any combination of their frequency, polarization and direction of incidence (wavevector). Thus the total field is written as

$$E = \sum_{j=1}^{3} E_j. \tag{B1.3.6}$$

In using the complex representation (equation (B1.3.4)), the $j$th electric field is given as

$$E_j = \tfrac{1}{2}(\varepsilon_j + \varepsilon_j^*) \tag{B1.3.7}$$

where (using Euler's identity)

$$\varepsilon_j = E_j^0 \, e^{-i(\mathbf{k}_j \cdot \mathbf{r} - \omega_j t)} \tag{B1.3.8}$$

$$\varepsilon_j^* = (E_j^0)^* \, e^{i(\mathbf{k}_j^* \cdot \mathbf{r} - \omega_j^* t)}. \tag{B1.3.9}$$

Here $E_j^0$ is the amplitude of the $j$th field and the real part of $\omega_j$ is its (circular) frequency or 'colour'. The real part of $\mathbf{k}_j$ is the product of the unit vector of incidence inside the sample, $\mathbf{e}_k$, and its amplitude, $\frac{2\pi n_j}{\lambda_j}$. Here $\frac{\lambda_j}{n_j}$ is the wavelength of the $j$th field inside the sample—$\lambda_j$ being the wavelength inside a vacuum and $n_j$ being the (real) index of refraction of the sample at $\omega_j$. As implied, all three properties may be complex: the amplitude because of an added phase to the field and/or a field that is elliptically (or circularly) polarized; the frequency because the field may be growing or decaying in time and the wavevector because the field may be decaying and growing according to its location within the sample.

The total field (equation (B1.3.6) with equation (B1.3.7)) is now

$$E = \frac{1}{2} \sum_{j=1}^{3} (\varepsilon_j + \varepsilon_j^*). \tag{B1.3.10}$$

It is only a matter of inserting this 'hexanomial', equation (B1.3.10), into equation (B1.3.1) to organize all possible three-beam spectroscopies that might appear at any given order.

### B1.3.2.2 The 'generators' for all third order spectroscopies from the complex representation of the field

In order to develop the theoretical structure that underlies each of the many Raman spectroscopies at third order, we use the above complex representation of the incident fields to produce the 'generators' of all possible electric field spectroscopies at third order. After this exercise, it is a simple matter to isolate the subset that constitutes the entire family of Raman spectroscopies.

At third order, one must expand $\frac{1}{8} (\sum_{i=1}^{3}(\varepsilon_i + \varepsilon_i^*) \sum_{j=1}^{3}(\varepsilon_j + \varepsilon_j^*) \sum_{k=1}^{3}(\varepsilon_k + \varepsilon_k^*))$ to enumerate the 'generators' of all possible third order spectroscopies. In this case, any given generator consists of an ordered list of these complex fields, such as $\varepsilon_i \varepsilon_j \varepsilon_k^*$. The ordering of the fields in each generator represents a time ordering of the actions of the applied fields. This can be of physical significance. Clearly this expansion must give 216 terms ($6^3$). These 216 terms or generators can be arranged into 108 pairs of mutually conjugate generators, since the total electric field is itself a quantity that is pure real. Of these 108 paired terms, exactly 27 are in the category of what is termed *nondegenerate* four wave mixing (ND4WM), where the signal frequency must be very far from any of the incident optical frequencies. These 27 pairs can generate only Class II spectroscopies and (it turns out) none of the Raman spectroscopies. The generic pair for these ND4WM processes is $\varepsilon_i \varepsilon_j \varepsilon_k$, $+\varepsilon_i^* \varepsilon_j^* \varepsilon_k^*$, where each of $i$, $j$ or $k$ can be fields 1, 2 or 3. (Henceforth the factor of $\frac{1}{8}$ shall be suppressed since it is common to all 216 generators.) The simple algebra of exponents is applied to such a product using equations (B1.3.9) and (B1.3.10). Thus, one sees how the polarization wave generated from such a term must oscillate at a frequency much larger than any of the incident colours, namely at the (real part of) $\omega_p = \omega_i + \omega_j + \omega_k$. The polarization wave must have a wavevector given by

$$\text{Re}(\mathbf{k}_p) = \text{Re}(\mathbf{k}_i + \mathbf{k}_j + \mathbf{k}_k) = 2\pi \left( \left(\frac{n_i}{\lambda_i}\right) e_{\mathbf{k}_i} + \left(\frac{n_j}{\lambda_j}\right) e_{\mathbf{k}_j} + \left(\frac{n_k}{\lambda_k}\right) e_{\mathbf{k}_k} \right).$$

The appropriate (complex) susceptibility tensor for this generator is $\chi^{(3)}(\omega_p = \omega_i + \omega_j + \omega_k)$.

When resonances, or near resonances, are present in the 4WM process, the ordering of the field actions in the perturbative treatment (equation (B1.3.1)), can be highly significant. Though the three-colour generators $(\varepsilon_i \varepsilon_j \varepsilon_k, \varepsilon_j \varepsilon_k \varepsilon_i, \varepsilon_k \varepsilon_i \varepsilon_j, \ldots)$ have identical frequency and wavevector algebra, their associated susceptibility functions $(\chi^{(3)}(\omega_p = \omega_i + \omega_j + \omega_k), \chi^{(3)}(\omega_p = \omega_j + \omega_k + \omega_i), \chi^{(3)}(\omega_p = \omega_k + \omega_i + \omega_j), \ldots)$ are, in general, different. As a result of the different colour ordering, two of their three energy denominator factors must differ. For this reason, the field ordering in each generator, together with its own response function, must be regarded individually.

The fourth electromagnetic wave, $E_{new}$, shall henceforth be called the signal wave, $E_s = \frac{1}{2}(\varepsilon_s + \varepsilon_s^*)$. It always must carry the same frequency as that of the polarization wave ($\omega_p = \omega_s$). It is launched by the collapse of this wave, provided that the polarization extends coherently over at least a few wavelengths of the incident light and that $\mathbf{k}_p = \mathbf{k}_s$.

The latter condition corresponds to the phase matching requirement already mentioned—the wavelength and direction of the material polarization wave must match those of the new EM wave as closely as possible. However, for all Class I spectroscopies, this condition is automatically achieved because of quadrature. In fact, this is true for all 'quadrature' spectroscopies—the Class I spectroscopies being the principal such, but, as noted, it is a nontrivial requirement in the 'nonquadrature' Class II spectroscopies, particularly in optically dispersive media.

In the complex mathematical representation, 'quadrature' means that, at the $(s + 1)$ wave mixing level, the product of $s$ input fields constituting the $s$th order generator and the signal field can be organized as a product of $(s + 1)/2$ conjugately paired fields. Such a pair for field $i$ is given by $\varepsilon_i \varepsilon_i^* = |\varepsilon_i|^2$. One sees that the exponent algebra for such a pair removes all dependence on the $\text{Re}(\omega_i)$ and $\text{Re}(\mathbf{k}_i)$, thus automatically removing all oscillations as well as satisfying the phase-matching requirement. A necessary (but not sufficient) requirement for full quadrature is for $s$ to be odd and also that half of the $s + 1$ fields be found to act conjugately with respect to the other half. Thus, for $s + 1$ fields, whenever the number of nonconjugate fields differs from the number of conjugate fields, one can only have 'nonquadrature'. Phase matching then must become an issue. This must always be the case for odd-wave mixing ($s$ is even), and it is also true for the above set of 27 frequency summing generators (for ND4WM). Thus for the currently considered generator, $\varepsilon_i \varepsilon_j \varepsilon_k$, all three fields ($i$, $j$ and $k$) act nonconjugately, so quadrature at the four-wave level simply is not possible.

### B1.3.2.3    The field generators for all third order Raman spectroscopies

We are now prepared to uncover all of the third order Raman spectroscopies (tables B1.3.1 and B1.3.2) and, in doing so, indicate how one might proceed to reveal the Raman events at higher order as well. Of the 108 pairs of conjugate generators, the remaining 81, unlike the above 27, are characterized by having one of the three fields acting conjugately with respect to the remaining two. Nine of these 81 terms generate fully degenerate 4WM. They constitute the 'one-colour' terms: $\varepsilon_i^* \varepsilon_i \varepsilon_i + \text{CC}, \varepsilon_i \varepsilon_i^* \varepsilon_i + \text{CC}$ and $\varepsilon_i \varepsilon_i \varepsilon_i^* + \text{CC}$ ($i = 1, 2,$ or 3) with susceptibility $\chi^{(3)}(\omega_s = \omega_i)$, since $\omega_s = -\omega_i + \omega_i + \omega_i = \omega_i$ with wavevector $\mathbf{k}_s = -\mathbf{k}_i + \mathbf{k}_i + \mathbf{k}_i = \mathbf{k}_i$.

Their full degeneracy is evident since all four fields carry the same frequency (apart from sign). Resonances appear in the electric susceptibilities when, by choice of incident colours and their signs, one or more of their energy denominators ($s$ in number at $s$th order) approaches a very small value because the appropriate algebraic colour combination matches material energy gaps. All Raman spectroscopies must, by definition, contain at least one low frequency resonance. When using only optical frequencies, this can only be achieved by having two fields acting conjugately and possessing a difference frequency that matches the material resonance. Further, they must act in the first two steps along the path to the third order polarization of the sample. These first two steps together prepare the Raman resonant material coherence and can be referred to as the 'doorway' stage of the Raman 4WM event.

Suppose the incident colours are such that the required Raman resonance ($\omega_R$) appears at $\omega_1 - \omega_2 \approx \omega_R$ (with $\omega_1 > \omega_2$). Thus the appropriate generators for the two-step doorway stage, common to all of the Raman spectroscopies, must be $\varepsilon_1 \varepsilon_2^*$ and $\varepsilon_2^* \varepsilon_1$, (and their conjugates, $\varepsilon_1^* \varepsilon_2$ and $\varepsilon_2 \varepsilon_1^*$). These differ only in the permutation of the ordering of the actions of fields 1 and 2. The usual algebra of exponents tells us that this doorway stage produces an intermediate polarization that oscillates at $\omega_1 - \omega_2 \approx \omega_R$. The resonantly produced Raman coherence having been established, it remains only to probe this intermediate polarization, that is, to convert it into an optical polarization from which the signal at optical frequencies is prepared. This is accomplished by the action of the third field which converts the low frequency Raman coherence into an optical one. This in turn, leads to the signal. This last two-step event can be referred to as the 'window' stage of the 4WM process. Obviously there are at most only six possible choices for this third field: $\varepsilon_j$ and $\varepsilon_j^*$ with $j = 1, 2, 3$. In this way we have isolated from the original 108 4WM generators the 12 that are responsible for all of the Raman spectroscopies. The choice of the probe field determines the frequency of the signal as well as its wavevector. While all types of Raman signal must be present at some level of intensity in any given experiment, it is a matter of experimental design—the detected $\omega_s$ and the aperture selected $\mathbf{k}_s$—as to which one Raman spectroscopy is actually being studied.

This classification by field generators in the complex field representation goes a long way towards organizing the nonlinear spectroscopies that carry at least one resonance. However it must be remembered that, in emphasizing a field ordering that locates the essential Raman resonance, we have neglected any other possible resonant and all nonresonant contributions to the third order polarization. While any additional resonance(s) are important to both Class I and Class II spectroscopies, the nonresonant contributions play no role in Class I spectroscopies, but must not be ignored in Class II studies. If one starts with the generator $\varepsilon_1 \varepsilon_2^* \varepsilon_3$ and then permutes the ordering of the three distinct complex field amplitudes, one arrives at 6 (3!) generators, only two of which induce the Raman resonance at $\omega_1 - \omega_2 \approx \omega_R$ ($\varepsilon_1 \varepsilon_2^* \varepsilon_3$ and $\varepsilon_2^* \varepsilon_1 \varepsilon_3$). The remaining four can only polarize the material without this Raman resonance ($\omega_3 \neq \omega_1, \omega_2$), but, if otherwise resonant, it can interfere with the Raman lineshape. Also when the issue of time resolved spectroscopies arises, the time ordering of field actions is under experimental control, and the active field generators are limited not only by requirements of resonance but by their actual time ordering in the laboratory.

In any case, the polarizing action upon the material by a given generator must be followed in more detail. In density matrix evolution, each specified field action transforms either the 'ket' or the 'bra' side of the density matrix. Thus for any specified $s$th order generator, there can be $2^s$ detailed paths of evolution. In addition, evolution for each of the $s!$ generators corresponding to all possible field orderings must be considered. One then has altogether $2^s s!$ paths of evolution. For $s = 3$ there are 48—eight for each of six generators.

*B1.3.2.4   Time evolution of the third order polarization by wave mixing energy level (WMEL) diagrams. The Raman spectroscopies classified*

The general task is to trace the evolution of the third order polarization of the material created by each of the above 12 Raman field operators. For brevity, we choose to select only the subset of eight that is based on two colours only—a situation that is common to almost all of the Raman spectroscopies. Three-colour Raman

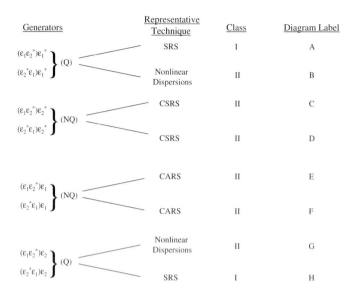

**Figure B1.3.1.** The eight third order, two-colour, Raman generators. Their full quadrature (Q) or nonquadrature (NQ) is indicated. The acronym of one of the related Raman spectroscopies is given, and each is labelled by spectroscopic class. Finally each generator is identified according to its window stage WMEL diagram (found alphabetically labelled in the second column of figures B1.3.2(a) and (b)).

studies are rather rare, but are most interesting, as demonstrated at both third and fifth order by the work in Wright's laboratory [21–24]. That work anticipates variations that include infrared resonances and the birth of doubly resonant vibrational spectroscopy (DOVE) and its two-dimensional Fourier transform representations analogous to 2D NMR [25].

Interestingly, three-colour spectroscopies at third order can only be of Class II, since the generators cannot possibly contain any quadrature. Maximal quadrature is necessary for Class I.

For the two-colour spectroscopies, maximal quadrature is possible and both Class I and Class II events are accounted for. Thus we trace diagrammatically the evolution to produce a signal field caused by the eight generators: (i) the four $(\varepsilon_1 \varepsilon_2^*)\varepsilon_1^*$, $(\varepsilon_1 \varepsilon_2^*)\varepsilon_2^*$, $(\varepsilon_1 \varepsilon_2^*)\varepsilon_1$ and $(\varepsilon_1 \varepsilon_2^*)\varepsilon_2$ and (ii) the four with the first two fields permuted, $(\varepsilon_2^* \varepsilon_1)\varepsilon_1^*$, $(\varepsilon_2^* \varepsilon_1)\varepsilon_2^*$, $(\varepsilon_2^* \varepsilon_1)\varepsilon_1$ and $(\varepsilon_2^* \varepsilon_1)\varepsilon_2$.

In each case we have indicated the doorway stage using parentheses. We note how in each of the two groups ((i) and (ii)) of the four generators, the 'outer' two are 'maximally in quadrature' and may reach full quadrature depending on the conjugation of the signal fields. Also in each group of four, the inner two generators contain no quadrature among the three fields, so these can only lead to nonquadrature 4WM and therefore only to Class II spectroscopies, regardless of the conjugation of the signal field. As already noted, the generators that are fully conjugate to these eight need not be considered, for in the end we can always take the real part of the complex polarization produced by a given generator.

In figure B1.3.1, these eight generators are organized for both the $(\varepsilon_1 \varepsilon_2^*)$ and the $(\varepsilon_2^* \varepsilon_1)$ doorway stages. Their potentially full quadrature (Q) or their certain nonquadrature (NQ) is indicated. Further, their spectroscopic class is identified—all of the NQ generators necessarily are assigned to Class II. The two potentially quadrature type generators may be assigned either to Class I or to Class II. The Raman spectroscopic examples of each are given. Finally, each is identified with its separate, alphabetically labelled, window stage 'wave mixing energy level' (WMEL) diagram. These WMEL diagrams are discussed next.

The density operator is a quantum statistical way of treating time evolution of a system that is described partly quantum mechanically (the constituent entities such as the spectroscopically active molecules or 'chromophores'), and partly statistical (the 'bath' of the surrounding micro-environment that dynamically perturbs the quantum levels of the chromophore). This matter–bath interaction is an important, usually dominant, source of damping of the macroscopic coherence. The radiative blackbody (bb) background is another ever-present bath that imposes radiative damping of excited states, also destroying coherences.

For $s = 3$, the time evolution of the system is tracked by following the stepwise changes in the bra state, $\langle j|$, or the ket state, $|k\rangle$, of the system caused by each of the three successive field interventions. This perturbative evolution of the density operator, or of the density matrix, is conveniently depicted diagrammatically using double sided Feynman diagrams or, equivalently, the WMEL diagrams. The latter are preferred since light/matter resonances are explicitly exposed. In WMEL diagrams, the energy levels of the constituents of the matter are laid out as solid horizontal lines to indicate the states (called 'real') that are active in a resonance, and as dashed horizontal lines (or no lines) when they serve as nonresonant ('virtual') states. The perturbative evolution of the density matrix is depicted using vertically oriented arrows for each of the field actions that appears in a given generator. These arrows are placed from left to right in the diagram in the same order as the corresponding field action in the generator. The arrow length is scaled to the frequency of the acting field. Solid arrows indicate evolution from the old ket (tail of arrow) to the new ket (head of arrow); dashed arrows indicate evolution from the old bra (tail of arrow) to the new bra (head of arrow). For a field acting nonconjugately, like $\varepsilon_i$, the frequency is positively signed, $\omega_i$, and the arrow for a ket change points up and that for a bra change points down. When the field acts conjugately, $\varepsilon_i^*$, the frequency is negatively signed, $-\omega_i$, and a ket changing arrow points down, while a bra changing arrow points up. These rules allow one to depict diagrammatically any and all density matrix evolutions at any order. Given the option of a bra or a ket change at each field action, one sees how a given $s$th order generator leads to $2^s$ diagrams, or paths of evolution. Normally only some (if any) encounter resonances. A recipe has been published [6] that allows one to translate any WMEL diagram into the analytic expression for its corresponding electrical susceptibility. After $s$ arrows have appeared (for an $s$th order evolution), the $(s + 1)$th field is indicated for any WMEL diagram of the nonquadrature class by a vertical wavy *line segment* whose vertical length scales to the signal frequency. For the WMEL diagrams of the full quadrature sort, the $(s + 1)$th field must be conjugate to one of the incident fields, so the wavy segment becomes a wavy arrow; either solid (ket-side action) or dashed (bra-side action).

Of the four possible WMEL diagrams for each the $(\varepsilon_1\varepsilon_2^*)$ and $(\varepsilon_2^*\varepsilon_1)$ doorway generators, only one encounters the Raman resonance in each case. We start with two parallel horizontal solid lines, together representing the energy gap of a Raman resonance. For ket evolution using $(\varepsilon_1\varepsilon_2^*)$, we start on the left at the lowest solid line (the ground state, $g$) and draw a long solid arrow pointing up $(+\omega_1)$, followed just to the right by a shorter solid arrow pointing down $(-\omega_2)$ to reach the upper solid horizontal line, $f$. The head of the first arrow brings the ket to a virtual state, from which the second arrow carries the ket to the upper of the two levels of the Raman transition. Since the bra is until now unchanged, it remains in $g$ ($\langle g|$); this doorway event leaves the density matrix at second order off-diagonal in which $\rho_{fg}^{(2)}$ is not zero. Thus a Raman coherence has been established. Analogously, the $(\varepsilon_2^*\varepsilon_1)$ doorway action on the ket side must be short solid arrow down $(-\omega_2)$ from $g$ to a virtual ket state, then long arrow up $(+\omega_1)$ to $f$ from the virtual state. This evolution also produces $\rho_{fg}^{(2)}$. Both doorway actions contain the same Raman resonance denominator, but differ in the denominator appearing at the first step; the downward $(\varepsilon_2^*\varepsilon_1)$ action is inherently anti-resonant ('N' for nonresonant) in the first step, the upward $(\varepsilon_1\varepsilon_2^*)$ action is potentially resonant ('R' for resonant) in the first step and is therefore stronger. Accordingly, we distinguish these two doorway events by labels $D_N$ and $D_R$, respectively (see figure B1.3.2). In resonance Raman spectroscopy, this first step in $D_R$ is fully resonant and overwhelms $D_N$. (The neglect of $D_N$ is known as the rotating wave approximation.) It is unnecessary to explore the bra-side version of these doorway actions, for they would appear in the fully conjugate version of these doorway events. Each of the doorway steps, $D_R$ and $D_N$, may be followed by any one of eight window events. The WMEL diagrams for the window events consist of the arrow for the last step of the third order polarization and the wavy segment for the signal wave. There are eight such window diagrams since each

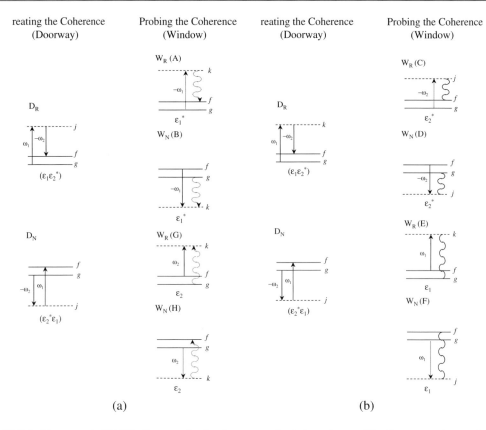

**Figure B1.3.2.** The separate WMEL diagrams for the doorway and window stages of the Raman spectroscopies. Solid and dashed vertical arrows correspond to ket- and bra-side light/matter interactions, respectively. The signal field is denoted by the vertical wavy line (arrow). The ground and final molecular levels (solid horizontal lines) are labelled $g$ and $f$, while the virtual levels (dashed horizontal lines) are labelled $j$ and $k$. The associated generators are given below each diagram. The doorway/window stages are classified as potentially resonant ($D_R/W_R$) or certainly nonresonant ($D_N/W_N$). In addition, the window stages are labelled alphabetically in order to distinguish the Raman techniques by their window stage WMEL diagram(s) (as in tables B1.3.1 and B1.3.2 and figure B1.3.1). (a) The doorway and window stage WMEL diagrams for SR, SRS and RRS. (b) The doorway and window stage WMEL diagrams for CARS and CSRS.

of the two steps can involve two colours and either bra- or ket-side evolution. These eight window WMEL diagrams are shown in figures B1.3.2(a) and (b) and are identified alphabetically. These also carry potentially resonant and anti-resonant properties in the third energy denominator (the first window step) and accordingly are labelled $W_R$ and $W_N$, where, as before, $W_R > W_N$. If the third step is completely resonant, $W_R \gg W_N$, and $W_N$ may be completely neglected (as with $D_R \gg D_N$).

It is now possible to label every one of the Raman spectroscopies listed in tables B1.3.1 and B1.3.2 according to its essential doorway/window WMEL diagram. This is shown in the third column of those tables. Again, the analytic form of the associated susceptibilities is obtained by recipe from the diagrams. When additional resonances are present, other WMEL diagrams must be included for both Class I and Class II spectroscopies. For the Class II spectroscopies, all of the nonresonant WMEL diagrams must be included as well.

*B1.3.2.5   The microscopic hyperpolarizability tensor, orientational averaging, the Kramers–Heisenberg expression and depolarization ratios*

As implied by the trace expression for the macroscopic optical polarization, the macroscopic electrical susceptibility tensor at any order can be written in terms of an ensemble average over the microscopic nonlinear polarizability tensors of the individual constituents.

*(a) Microscopic hyperpolarizability and orientational averaging*

Consider an isotropic medium that consists of independent and identical microscopic chromophores (molecules) at number density $N$. At $s$th order, each element of the macroscopic susceptibility tensor, given in laboratory Cartesian coordinates $A, B, C, D$, must carry $s + 1$ (laboratory) Cartesian indices ($X, Y$ or $Z$) and therefore number altogether $3^{(s+1)}$. Thus the third order susceptibility tensor contains 81 elements. Each tensor element of the macroscopic susceptibility is directly proportional to the sum over all elements of the corresponding microscopic, or molecular, hyperpolarizability tensor. The latter are expressed in terms of the four local (molecule based) Cartesian coordinates, $a, b, c, d$ (each can be $x, y,$ or $z$—accounting for all 81 elements of the *microscopic* tensor). To account for the contribution to the macroscopic susceptibility from each molecule, one must sum over all molecules in a unit volume, explicitly treating all possible orientations, since the projection of a microscopic induced dipole onto the $X, Y$ and $Z$ laboratory coordinates depends very much on the molecular orientation. This is accomplished by averaging the microscopic hyperpolarizability contribution to the susceptibility over a normalized distribution of orientations, and then simply multiplying by the number density, $N$. Let the orientational averaging be denoted by $\langle \ldots \rangle$. For any macroscopic tensor element, $\chi^{(3)}_{ABCD}$, one finds

$$\chi^{(3)}_{ABCD} = \frac{L^3}{\hbar^3 \epsilon_0} N \sum_{a,b,c,d} \langle (A, a)(B, b)(C, c)(D, d) \rangle \gamma_{abcd} \qquad (\text{B1.3.11})$$

where $(A, a)$ etc are the direction cosines linking the specified local Cartesian axes with specified laboratory axes and $L$ is a 'local' field factor. (The field experienced by the molecule is the incident field altered by polarization effects.) Often the $\hbar^{-s}$ factor is absorbed into the definition of the $s$th order hyperpolarizability.

The microscopic polarization involves four molecular based transition moment vectors—the induced dipole is along $a$, the first index. The transition moments along $b, c$ and $d$ are coupled to the laboratory axes, $B, C$ and $D$, respectively, along which the successive incident (or black-body) fields are polarized. The four transition moment unit vectors have been extracted and projected onto the laboratory axes: $A$—the direction of the induced macroscopic polarization, $B$—the polarization of the first acting field, $C$—the polarization of the second acting field and $D$—the polarization of the third acting field. The product of the four direction cosines is subjected to the orientational averaging process, as indicated. Each such average belongs to its corresponding scalar tensor element $\gamma_{abcd}$. It is important to sum over all local Cartesian indices, since all elements of $\gamma$ can contribute to each tensor element of $\chi^{(3)}$. At third order, the averaging over the projection of microscopic unit vectors to macroscopic, such as those in equation (B1.3.11), is identical to that found in two-photon spectroscopy (another Class I spectroscopy at third order). This has been treated in a general way (including circularly polarized light) according to molecular symmetry considerations by Monson and McClain [26].

For an isotropic material, all orientations are equally probable and all such products that have an odd number of 'like' direction cosines will vanish upon averaging[2]. This restricts the nonvanishing tensor elements to those such as $\chi^{(3)}_{AAAA}$, $\chi^{(3)}_{ABBA}$ etc. Similarly for the elements $\gamma_{abcd}$. Such orientational averaging is crucial in dictating how the signal field in any spectroscopy is polarized. In turn, polarization measurements can lead to important quantitative information about the elements of the macroscopic and microscopic tensors.

---

[2]  In fact averaging over an odd number of direction cosines need not always vanish for an isotropic system. This is the case for solutions containing chiral centres which may exhibit even order signals such as 'BioCARS' in table B1.3.2.

The passage from microscopic to macroscopic (equation (B1.3.11)) clearly exposes the additivity of the microscopic hyperpolarizabilities. Significantly, it is seen immediately why $\chi^{(3)}$ is linear in concentration, $N$. This brings out one of the major distinctions between the Class I and Class II spectroscopies (see item (iii) in section B1.3.1). The signals from Class I, being proportional to Im $\chi^{(3)}$, are *linear* in concentration. Those (homodyne signals) from Class II are proportional to $|\chi^{(3)}|^2$ and therefore must be *quadratic* in concentration. (However, Class II spectroscopies that are heterodyne detected are proportional to $\chi^{(3)}$ and are linear in $N$.)

*(b) The microscopic hyperpolarizability in terms of the linear polarizability: the Kramers–Heisenberg equation and Placzek linear polarizability theory of the Raman effect*

The original Placzek theory of Raman scattering [30] was in terms of the linear, or first order microscopic polarizability, $\alpha$ (a second rank tensor), not the third order hyperpolarizability, $\gamma$ (a fourth rank tensor). The Dirac and Kramers–Heisenberg quantum theory for linear dispersion did account for Raman scattering. It turns out that this link of properties at third order to those at first order works well for the electronically nonresonant Raman processes, but it cannot hold rigorously for the fully (triply) resonant Raman spectroscopies. However, provided one discards the important line shaping phenomenon called 'pure dephasing', one can show how the third order susceptibility does reduce to the treatment based on the (linear) polarizability tensor [6, 27].

What is the phenomenon 'pure dephasing' that one cannot formally encounter in the linear polarizability theories of Raman spectroscopies? It arises when theory is obliged to treat the environment of the spectroscopically active entity as a 'bath' that statistically modulates its states. In simple terms, there are two mechanisms for the irreversible decay of a coherent state such as a macroscopically coherent polarization wave. One involves the destruction by the bath of the local induced dipoles that make up the wave (a lifetime effect); the other involves the bath induced randomization of these induced dipoles (without their destruction). This latter mechanism is called pure dephasing. Together, their action is responsible for dephasing of a pure coherence. In addition, if the system is inherently inhomogeneous in the distribution of the two-level energy gap of the coherence, the local coherences will oscillate at slightly different frequencies causing, as these walk off, the macroscopic coherence, and its signal, to decay—even while the individual local coherences might not. This is especially important for the Class II spectroscopies. Unlike the first two dephasing mechanisms, this third kind can be reversed by attending to signals generated by the appropriate Fourier components of the subsequent field actions. The original macroscopic coherence will be reassembled (at least partially) in due course, to produce a renewed signal, called an 'echo'. For an electronic coherence made by a single optical field, this happens at the four-wave mixing level. However for the Raman spectroscopies, *two* (conjugately acting) optical fields are needed to create the vibrational coherence, and hence the true Raman echo appears at the eight-wave mixing level where $\chi^{(7)}$ is the important susceptibility. (We shall see that a *quasi* Raman echo can be exploited at the $\chi^{(5)}$ level.)

The important and frequently ignored fact is that Raman theory based on the polarizability tensor cannot contain the randomization mechanism for dephasing (pure dephasing). This mechanism is especially important in electronically resonant Raman spectroscopy in the condensed phase. The absence of pure dephasing in linear polarizability theory arises simply because the perturbative treatment upon which it is based involves the *independent* evolution of the bra and the ket states of the system. Conversely, the third order susceptibility approach, based on the perturbative development of the density operator, links together the evolution of the bra and ket states and easily incorporates pure dephasing. Furthermore, in resonance Raman spectroscopy, it is the pure dephasing mechanism that governs the interesting competition between resonance fluorescence and resonance Raman scattering [6]. In the linear polarization theory these are fixed in their relation, and the true resonance fluorescence component becomes an indistinguishable part of the Raman line shape. (However, interestingly, if the exciting light itself is incoherent, or the exciting light consists of sufficiently short pulses, even the linear polarizability theory tells how the resonance fluorescence-like component becomes distinguishable from the Raman-like signal [28].)

At the linear level, the microscopic induced dipole vector on a single molecule in the local Cartesian coordinate system is simply written as $\mu^{(1)} = \alpha E$ where $E$ is the applied field also expressed in the local Cartesian system. In full matrix language, in which the local second rank polarizability tensor is exposed, we can write:

$$\begin{bmatrix} \mu_x \\ \mu_y \\ \mu_z \end{bmatrix} = \begin{bmatrix} \alpha_{xx} & \alpha_{xy} & \alpha_{xz} \\ \alpha_{yx} & \alpha_{yy} & \alpha_{yz} \\ \alpha_{zx} & \alpha_{zy} & \alpha_{zz} \end{bmatrix} \begin{bmatrix} E_x \\ E_y \\ E_z \end{bmatrix}.$$

If we neglect pure dephasing, the general tensor element of the third order hyperpolarizability relates to those of the first order polarizability tensor according to

$$\gamma_{abcd}(\omega_s = \omega_1 - \omega_2 - \omega_3) = \frac{\alpha_{bc}(\omega_1, \omega_2)\alpha_{ad}^*(\omega_s, \omega_3)}{\omega_1 - \omega_2 - \omega_R + i\gamma_R}. \tag{B1.3.12}$$

Here, the linear polarizability, $\alpha_{bc}(\omega_1, \omega_2)$, corresponds to the doorway stage of the 4WM process while $\alpha_{ad}^*(\omega_s, \omega_3)$ to the window stage. We also see the (complex) Raman resonant energy denominator exposed. Of the three energy denominator factors required at third order, the remaining two appear, one each, in the two linear polarizability tensor elements.

In fact, each linear polarizability itself consists of a sum of two terms, one potentially *resonant* and the other *anti-resonant*, corresponding to the two doorway events, $D_R$ and $D_N$, and the window events, $W_R$ and $W_N$, described above. The hyperpolarizability chosen in equation (B1.3.12) happens to belong to the $\varepsilon_1\varepsilon_2^*\varepsilon_3^*$ generator. As noted, such three-colour generators cannot produce Class I spectroscopies (full quadrature with three colours is not possible). Only the two-colour generators are able to create the Class I Raman spectroscopies and, in any case, only two colours are normally used for the Class II Raman spectroscopies as well.

For linear polarizability elements that are pure real, we see that (from equation (B1.3.12))

$$\text{Im}[\gamma_{abcd}] = \frac{-\gamma_R}{(\omega_1 - \omega_2 - \omega_R)^2 + \gamma_R^2}\alpha_{bc}\alpha_{ad}.$$

When $\omega_1 = \omega_3$ (two colours), this is relevant to the Class I Raman spectroscopies (see section B1.3.2.1). In this case we expose a Lorentzian Raman lineshape with an HWHM of $\gamma_R$. At this point, the notation for the elements of the polarizability tensor suppresses the identity of the Raman transition, so it is now necessary to be more specific.

Consider Raman transitions between thermalized *molecular* eigenstate $g$ (ground) and *molecular* eigenstate $f$ (final). The quantum mechanical expression for $\alpha_{bc}$ responding to colours $i$ and $j$ is the famous (thermalized) *Kramers–Heisenberg equation* [29]

$$(\alpha_{bc}(\omega_i, \omega_j))_{gf} = \sum_n \left( \frac{\langle g|\mu_b|n\rangle\langle n|\mu_c|f\rangle}{\omega_{ng} - \omega_i} + \frac{\langle g|\mu_c|n\rangle\langle n|\mu_b|f\rangle}{\omega_{ng} + \omega_j} \right) \tag{B1.3.13}$$

where the notation on the left-hand side recognizes the Raman transition between molecular eigenstates $g$ and $f$. The sum on $n$ is over all molecular eigenstates. One should be reminded that the 'ground state' should actually be a thermal distribution over Boltzmann weighted states. Thus at the hyperpolarizability level, one would write $\sum_g W_g(\gamma_{abcd})_{gf}$, where $W_g$ is the appropriate Boltzmann weighting factor, $\exp[-\hbar\omega_g/kT]$. This detail is suppressed in what follows.

To branch into electronic, vibrational and rotational Raman spectroscopy, the Born–Oppenheimer (B–O) approximation must be introduced, as needed, to replace the molecular eigenstates as rovibronic products. For example, consider vibrational Raman scattering within the ground electronic state (or, analogously, within any other electronic state). For scattering between vibrational levels $v$ and $v'$ in the ground electronic state, we expand the molecular eigenstate notation to $|g\rangle \equiv |g\rangle|v\rangle$ and $|f\rangle \equiv |g\rangle|v'\rangle$ (the intermediate states, $|n\rangle$, may be left as molecular eigenstates). The curved bracket refers to the electronic eigenstate and the straight bracket to the vibrational states (where until now it referred to the molecular eigenstate). Now equation (B1.3.13) becomes

$$(\alpha_{bc}(\omega_i, \omega_j))_{vv'} = \langle v| \left[ \sum_n^\infty \left( \frac{(g|\mu_b|n)\langle n|\mu_c|g)}{\omega_{ng} - \omega_i} + \frac{(g|\mu_b|n)\langle n|\mu_c|g)}{\omega_{ng} + \omega_j} \right) \right] |v'\rangle.$$

We note that the expression in brackets is just the $b\,c$ tensor element of the *electronic* polarizability in the ground electronic state, $\alpha_{bc}^{el}(\omega_i, \omega_j)$. Thus

$$(\alpha_{bc}(\omega_i, \omega_j))_{vv'} = \langle v|\alpha_{bc}^{el}(\omega_i, \omega_j)|v'\rangle. \tag{B1.3.14}$$

Since the vibrational eigenstates of the ground electronic state constitute an orthonormal basis set, the off-diagonal matrix elements in equation (B1.3.14) will vanish unless the ground state electronic polarizability depends on nuclear coordinates. (This is the Raman analogue of the requirement in infrared spectroscopy that, to observe a transition, the electronic dipole moment in the ground electronic state must properly vary with nuclear displacements from equilibrium.) Indeed such electronic properties do depend on nuclear coordinates, for in the B–O approximation electronic eigenstates are parametrized by the positional coordinates of the constituent nuclei. For matrix elements in vibrational space, these coordinates become variables and the electronic polarizability (or the electronic dipole moment) is expanded in a Taylor series in nuclear displacements. Usually the normal mode approximation is introduced into the vibrational space problem (though it need not be) and the expansion is in terms of the normal displacement coordinates, $\{\triangle Q_i\}$, of the molecule. Thus for the electronic polarizability we have:

$$\alpha_{bc}^{el} = (\alpha_{bc}^{el})_0 + \sum_{i=1} \left( \frac{\partial \alpha_{bc}^{el}}{\partial Q_i} \right)_0 \triangle Q_i + \frac{1}{2} \sum_{i=1} \sum_{j=1} \left( \frac{\partial^2 \alpha_{bc}^{el}}{\partial Q_i \partial Q_j} \right)_0 \triangle Q_i \triangle Q_j + \cdots \tag{B1.3.15}$$

the leading term of which is unable to promote a vibrational transition because $\langle v|(\alpha_{bc}^{el})_0|v'\rangle = (\alpha_{bc}^{el})_0\langle v|v'\rangle = (\alpha_{bc}^{el})_0\delta_{vv'}$. (The $v = v'$ situation corresponds to Rayleigh scattering for which this leading term is the principal contributor.) The $\triangle Q_i$ in the second term is able to promote the scattering of fundamentals, since $\langle v|\triangle Q_i|v+1\rangle$ need not vanish. The $\triangle Q_i \triangle Q_j$ in the third set of terms can cause scattering of the first overtones ($i = j$) and combination states ($i \neq j$), etc, for the subsequent terms. As usual in spectroscopy, point group theory governs the selection rules for such matrix elements. As already noted these are identical to the two-photon selection rules [26], though here in vibrational space.

This linear polarizability theory of Raman scattering [30] forms the basis for bond polarizability theory of the Raman effect. Here the polarizability derivative is discussed in terms of its projection onto bonds of a molecule and the concept of additivity and the transferability of such bond specific polarizability derivatives can be discussed, and even semiquantitatively supported. Further, the vibronic (vibrational–electronic) theory of Raman scattering appears at this level. It introduces the Herzberg–Teller development for the nuclear coordinate dependence of electronic states, therefore that of the electronic transition moments and hence that of the electronic polarizability. This leads to the so-called 'A', 'B' and 'C' terms for Raman scattering, each having a different analytical form for the dispersive behaviour of the Raman cross-section as the exciting light moves from a nonresonant region towards an electronically resonant situation. An early review of these subjects can be found in [31].

For excitation at a wavenumber $\bar{v}_1$, ($\bar{v}_1 = \omega_1/2\pi c$), and the Raman wavenumber at $\bar{v}_R$, the total Raman cross-section for scattering in isotropic media[3,4] onto a spherical surface of $4\pi$ radians, for all analysing polarizations, for excitation with linearly polarized or unpolarized light, and integration over the Raman line,

---

[3] Here, we have averaged over all possible orientations of the molecules. (See [26].)
[4] Raman cross-sections, based on the linear polarizability, are now routinely subject to quantum chemical calculations. These may be found as options in commercial packages such as 'Gaussian 98' (Gaussian Inc., Pittsburgh, PA).

we have in terms of rotational invariants of the linear polarizability:

$$\sigma_R(\bar{v}_1) = \frac{32\pi^3}{9}\left(\frac{e^2}{4\pi\varepsilon_0\hbar c}\right)^2 \bar{v}_1(\bar{v}_1 \pm \bar{v}_R)^3 \{\Sigma^0 + \Sigma^1 + \Sigma^2\}. \tag{B1.3.16}$$

The upper sign is for anti-Stokes scattering, the lower for Stokes scattering. The factor in the parentheses is just the fine-structure constant and $\Sigma^0$, $\Sigma^1$, $\Sigma^2$ are the three rotationally invariant tensor elements of the hyperpolarizability (or the linear polarizability when pure dephasing is ignored), which are given by:

$$\Sigma^0 = \frac{1}{3}\sum_{\rho,\sigma}\gamma_{\sigma\rho\rho\sigma} \approx \frac{1}{3\Omega}\left|\sum_\rho \alpha_{\rho\rho}\right|^2 \tag{B1.3.17}$$

$$\Sigma^1 = \frac{1}{2}\sum_{\rho,\sigma}(\gamma_{\rho\rho\sigma\sigma} - \gamma_{\rho\sigma\rho\sigma}) \approx \frac{1}{4\Omega}\sum_{\rho,\sigma\neq\rho}|\alpha_{\rho\sigma} - \alpha_{\sigma\rho}|^2 \tag{B1.3.18}$$

and

$$\Sigma^2 = \frac{1}{2}\sum_{\rho,\sigma}(\gamma_{\rho\rho\sigma\sigma} + \gamma_{\rho\sigma\rho\sigma}) - \frac{1}{3}\sum_{\rho,\sigma}\gamma_{\rho\sigma\sigma\rho} \approx \frac{1}{4\Omega}\sum_{\rho,\sigma\neq\rho}|\alpha_{\rho\sigma} + \alpha_{\sigma\rho}|^2 + \frac{1}{6\Omega}\sum_{\rho,\sigma}|\alpha_{\rho\rho} - \alpha_{\sigma\sigma}|^2 \tag{B1.3.19}$$

where $\Omega$ is the Raman resonant energy denominator, $\omega_1 - \omega_2 - \omega_R + i\gamma_R$. With appropriate algebra, one finds that their sum is given by: $\Sigma^0 + \Sigma^1 + \Sigma^2 = \sum_{\rho,\sigma}\gamma_{\rho\rho\sigma\sigma}$, or in terms of the linear polarizability tensor elements: $\Sigma^0 + \Sigma^1 + \Sigma^2 = \frac{1}{\Omega}\sum_{\rho,\sigma}|\alpha_{\rho\sigma}|^2$.

Experimentally, it is these invariants (equations (B1.3.17)–(B1.3.19)) that can be obtained by scattering intensity measurements, though clearly not by measuring the total cross-section only.

Measurement of the total Raman cross-section is an experimental challenge. More common are reports of the differential Raman cross-section, $d\sigma_R/d\Omega$, which is proportional to the intensity of the scattered radiation that falls within the element of solid angle $d\Omega$ when viewing along a direction that is to be specified [15]. Its value depends on the design of the Raman scattering experiment.

In the appendix, we present the differential Raman scattering cross-section for viewing along any wavevector in the scattering sphere for both linearly and circularly polarized excitation. The more conventional geometries used for exciting and analysing Raman scattering are discussed next.

Suppose the exciting beam travels along $X$, and is linearly (l) polarized along $Z$. A popular experimental geometry is to view the scattered light along $Y$ (at $\pi/2$ radians to the plane defined by the wavevector, and the polarization unit vector, $e_Z$, of the exciting light). One analyses the $Z$ polarized component of the scattered light, called $I_\parallel$ ($I_Z$), and the $X$ polarized component, called $I_\perp$ ($I_X$). (Careful work must properly correct for the finite solid angle of detection.) The two intensities are directly proportional to the differential cross-section given by

$$\left(\frac{d\sigma}{d\Omega}\right)_\parallel = 4\pi^2\left(\frac{e^2}{4\pi\varepsilon_0\hbar c}\right)^2 \bar{v}_1(\bar{v}_1 \pm \bar{v}_R)^3\left(\frac{5\Sigma^0 + 2\Sigma^2}{15}\right) \tag{B1.3.20}$$

and

$$\left(\frac{d\sigma}{d\Omega}\right)_\perp = 4\pi^2\left(\frac{e^2}{4\pi\varepsilon_0\hbar c}\right)^2 \bar{v}_1(\bar{v}_1 \pm \bar{v}_R)^3\left(\frac{5\Sigma^1 + 3\Sigma^2}{30}\right). \tag{B1.3.21}$$

The depolarization ratio is defined as

$$\rho_l = \frac{I_\perp}{I_\parallel} \tag{B1.3.22}$$

which for the present case of an orientationally averaged isotropic assembly of Raman scatterers reduces to:

$$\rho_l = \frac{5\Sigma^1 + 3\Sigma^2}{10\Sigma^0 + 4\Sigma^2}. \tag{B1.3.23}$$

This result is general, for it includes the case where the tensor elements are complex, regardless of whether or not the hyperpolarizability tensor is built of the linear polarizability.

Another frequent experimental configuration uses naturally (n) polarized incident light, with the same viewing geometry and polarization analysis. Such light may be regarded as polarized equally along $Z$ (as before) and along the viewing axis, $Y$. Given the $I_\perp$ and $I_\parallel$ as defined in the linearly polarized experiment, one can reason that now with naturally polarized excitation $I_Z \propto I_\parallel + I_\perp$ (where the additional $I_\perp$ term along $Z$ originates from the $Y$ polarized excitation). Similarly we expect that $I_X \propto 2I_\perp$, one $I_\perp$ from each of the two excitation polarizations. The differential Raman cross-sections for naturally polarized excitation are defined as

$$\left(\frac{d\sigma}{d\Omega}\right)_X = 4\pi^2 \left(\frac{e^2}{4\pi\varepsilon_0\hbar c}\right)^2 \bar{v}_1(\bar{v}_1 \pm \bar{v}_R)^3 \left(\frac{5\Sigma^1 + 3\Sigma^2}{30}\right)$$

and

$$\left(\frac{d\sigma}{d\Omega}\right)_Z = 4\pi^2 \left(\frac{e^2}{4\pi\varepsilon_0\hbar c}\right)^2 \bar{v}_1(\bar{v}_1 \pm \bar{v}_R)^3 \left(\frac{10\Sigma^0 + 5\Sigma^1 + 7\Sigma^2}{60}\right).$$

Thus one predicts that the depolarization ratio for excitation with natural light should be

$$\rho_n = \frac{I_X}{I_Z} = \frac{2I_\perp}{I_\parallel + I_\perp} = \frac{2\rho_l}{1 + \rho_l} = \frac{10\Sigma^1 + 6\Sigma^2}{10\Sigma^0 + 5\Sigma^1 + 7\Sigma^2} \tag{B1.3.24}$$

or

$$\rho_l = \frac{\rho_n}{2 - \rho_n}$$

two well known relations. Of course, if one does not measure the polarization of the scattered light for either experiment, the detector collects signal from the sum of the differential cross sections, $(\frac{d\sigma}{d\Omega})_{\parallel(Z)} + (\frac{d\sigma}{d\Omega})_{\perp(X)}$.

Similar reasoning shows that were one to view along the $X$, $Y$ and $Z$ axes and polarization analyse the signal each time, whether excited by linearly or by naturally polarized light, the total intensity should be given by $I_{\text{total}} \propto I_\parallel + 2I_\perp$. Given equation (B1.3.23), if we add its denominator to twice the numerator we find that $I_{\text{total}} \propto \{\Sigma^0 + \Sigma^1 + \Sigma^2\}$, a reassuring result.

Knowledge of the depolarization ratios allows one to classify easily the Raman modes of a molecule into symmetric and asymmetric vibrations. If a molecule is undergoing a totally symmetric vibration, the depolarization ratio, $\rho_l$ ($\rho_n$) will be less than $3/4$ ($6/7$) and we say that the vibration is polarized (p). On the other hand, for asymmetric vibrations, the depolarization ratio will have a value close to $3/4$ ($6/7$) and we say these vibrations are depolarized (dp) [34]. It should be stated that these values for $\rho$ are only valid when the scattered radiation is collected at right angles to the direction of the incident light. If different geometry is used, $\rho_l$ and $\rho_n$ are accordingly changed (see the appendix).

An interesting phenomenon called the 'noncoincidence effect' appears in the Raman spectroscopies. This is seen when a given Raman band shows a peak position and a bandwidth that differs (slightly) with the polarization. It can be attributed to varying sensitivity of the different tensor elements to interchromophore interactions.

## B1.3.3  Raman spectroscopy in modern physics and chemistry

Raman spectroscopy is pervasive and ever changing in modern physics and chemistry. In this section of the chapter, sources of up-to-date information are given followed by brief discussions of a number of currently employed Raman based techniques. It is impractical to discuss every possible technique and impossible to predict the many future novel uses of Raman scattering that are sure to come, but it is hoped that this section will provide a firm launching point into the modern uses of Raman spectroscopy for present and future readers.

### B1.3.3.1  Sources of up-to-date information

There are three very important sources of up-to-date information on all aspects of Raman spectroscopy. Although papers dealing with Raman spectroscopy have appeared and will continue to appear in nearly every major chemical physics–physical chemistry based serial, *The Journal of Raman Spectroscopy* [35] is solely devoted to all aspects, both theoretical and experimental, of Raman spectroscopy. It originated in 1973 and continues to be a constant source of information on modern applications of Raman spectroscopy.

*Advances in Infrared and Raman Spectroscopy* [36] provides review articles, both fundamental and applied, in the fields of both infrared and Raman spectroscopy. This series aims to review the progress in all areas of science and engineering in which application of these techniques has a significant impact. Thus it provides an up-to-date account of both the theory and practice of these two complementary spectroscopic techniques.

The third important source for information on modern Raman spectroscopy are the books cataloguing the proceedings of the *International Conference on Raman Spectroscopy (ICORS)* [37]. *ICORS* is held every two years at various international locations and features hundreds of contributions from leading research groups covering all areas of Raman spectroscopy. Although the published presentations are quite limited in length, they each contain references to the more substantial works and collectively provide an excellent overview of current trends in Raman spectroscopy. A 'snapshot' or brief summary of the 1998 conference appears at the end of this chapter.

Through these three serials, a researcher new to the field, or one working in a specialized area of Raman spectroscopy, can quickly gain access to its current status.

### B1.3.3.2  Survey of techniques

With the theoretical background presented in the previous sections, it is now possible to examine specific Raman techniques. Of the list in table B1.3.1, we briefly discuss and provide references to additional information for the Class I spectroscopies—spontaneous Raman scattering (SR), Fourier transform Raman scattering (FTRS), resonance Raman scattering (RRS), stimulated Raman scattering (SRS), and surface enhanced Raman scattering (SERS)—and in table B1.3.2, the Class II spectroscopies—coherent Raman scattering (CRS), Raman induced Kerr-effect spectroscopy (RIKES), Raman scattering with noisy light, time resolved coherent Raman scattering (TRCRS), impulsive stimulated Raman scattering (ISRS) and higher order and higher dimensional Raman scattering.

First we discuss some Class I spectroscopies.

### (a) Spontaneous Raman scattering (SR)

Conventional spontaneous Raman scattering is the oldest and most widely used of the Raman based spectroscopic methods. It has served as a standard technique for the study of molecular vibrational and rotational levels in gases, and for both intra- and inter-molecular excitations in liquids and solids. (For example, a high resolution study of the vibrons and phonons at low temperatures in crystalline benzene has just appeared [38].)

In this earliest of Raman spectroscopies, there is only one incident field (originally sunlight or lines of the mercury lamp; today a single laser source). This is field 1 in the above language and it appears in quadrature in the two generators, $(\varepsilon_1 \varepsilon_2^*)\varepsilon_1^*$ and $(\varepsilon_2^* \varepsilon_1)\varepsilon_1^*$, relevant to SR. Figure B1.3.2(a) shows that these generators lead to four WMEL diagrams: $D_R W_R(A)$, $D_R W_N(B)$, $D_N W_R(A)$, $D_N W_N(B)$. The first is the strongest contributor (it is potentially resonant, R, in both the first and last steps). The last term is the weakest (being nonresonant, N, in both the first and last steps). We note that at the 4WM level, in all four terms, not only is field 1 in quadrature, but field 2 is likewise in quadrature (since for window events A and B, we have

$\varepsilon_s = \varepsilon_2$, namely the signal field is conjugate to the action of field 2). Now, since quadrature means photons, the Raman scattering event has destroyed a photon at $\omega_1$, while it has created a new photon at $\omega_2$.

The unique feature in spontaneous Raman spectroscopy (SR) is that field 2 is not an incident field but (at room temperature and at optical frequencies) it is resonantly drawn into action from the zero-point field of the ubiquitous blackbody (bb) radiation. Its active frequency is spontaneously selected (from the infinite colours available in the blackbody) by the resonance with the Raman transition at $\omega_1 - \omega_2 = \omega_R$ in the material. The effective bb field intensity may be obtained from its energy density per unit circular frequency, $\frac{\hbar \omega_2^3}{\pi^2 c^3}$, the Einstein $A$ coefficient at $\omega_2$. When the polarization field at frequency $\omega_s$, $p^{(3)}(\omega_s = \omega_1 - \omega_2 - \omega_1)$, produces an electromagnetic field which acts conjugately with this selected blackbody field (at $\omega_s = \omega_2$), the scattered Raman photon is created. Thus, one simply has growth of the blackbody radiation field at $\omega_2$, since full quadrature removes all oscillatory behaviour in time and all wavelike properties in space.

Unlike the typical laser source, the zero-point blackbody field is spectrally 'white', providing all colours, $\omega_2$, that seek out all $\omega_1 - \omega_2 = \omega_R$ resonances available in a given sample. Thus all possible Raman lines can be seen with a single incident source at $\omega_1$. Such multiplex capability is now found in the Class II spectroscopies where broadband excitation is obtained either by using modeless lasers, or a femtosecond pulse, which on first principles must be spectrally broad [32]. Another distinction between a coherent laser source and the blackbody radiation is that the zero-point field is spatially isotropic. By performing the simple wavevector algebra for SR, we find that the scattered radiation is isotropic as well. This concept of spatial incoherence will be used to explain a certain 'stimulated' Raman scattering event in a subsequent section.

For SR, a Class I spectroscopy, there must be a net transfer of energy between light and matter which survives averaging over many cycles of the optical field. Thus, the material must undergo a state population change such that the overall energy (light and matter) may be conserved. In Stokes vibrational Raman scattering (figure B1.3.3(a)), the chromophore is assumed to be in the ground vibrational state $|g\rangle$. The launching of the Stokes signal field creates a population shift from the ground state $|g\rangle$ to an excited vibrational state $|f\rangle$. Conversely, in anti-Stokes vibrational Raman scattering (figure B1.3.3(b)), the chromophore is assumed to be initially in an excited vibrational state, $|f\rangle$. Thus, the launching of the anti-Stokes field leaves the chromophore in the ground vibrational state, $|g\rangle$. This process is typically weaker than the Stokes process since it requires that an excited vibrational population exist (usually $W_f \ll W_g$). In thermal equilibrium, the intensity of the anti-Stokes frequencies compared to the Stokes frequencies clearly is reduced by the Boltzmann factor, $W_f/W_g = \exp[-\hbar \omega_{fg}/kT]$ [17]. Let us now discuss the apparatus used for the production and detection of the Raman scattered radiation.

### (b) Raman instrumentation

Dramatic advances in Raman instrumentation have occurred since 1928. At the beginning, various elemental lamps served as the incident light source and photographic plates were used to detect the dispersed scattered light. Mercury arc lamps were primarily used since they had strong emission lines in the blue region of the visible spectrum (see equation (B1.3.16)). As in all spectroscopies, detection devices have moved from the photographic to the photoelectric.

Even while Raman spectrometers today incorporate modern technology, the fundamental components remain unchanged. Commercially, one still has an excitation source, sample illuminating optics, a scattered light collection system, a dispersive element and a detection system. Each is now briefly discussed.

Continuous wave (CW) lasers such as $Ar^+$ and He–Ne are employed in commonplace Raman spectrometers. However laser sources for Raman spectroscopy now extend from the edge of the vacuum UV to the near infrared. Lasers serve as an energetic source which at the same time can be highly monochromatic, thus effectively supplying the single excitation frequency, $\bar{\nu}_1$. The beams have a small diameter which may be focused quite easily into the sample and are convenient for remote probing. Finally, almost all lasers

a)

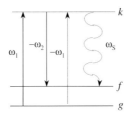

(b)

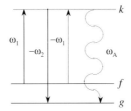

**Figure B1.3.3.** The full WMEL diagram (of the $D_R W_R$ sort) for spontaneous (or stimulated) (a) Stokes Raman scattering and (b) anti-Stokes Raman scattering. In Stokes scattering, the chromophore is initially in the ground vibrational state, $g$, and $\omega_1 > \omega_2$. In spontaneous anti-Stokes scattering, the chromophore must be initially in an excited vibrational state, $f$. Also note that in (b), $\omega_2$ is (arbitrarily) defined as being greater than $\omega_1$.

are linearly polarized, which makes measurements of the depolarization ratio, equation (B1.3.22), relatively straightforward.

The laser beam is typically focused onto the sample with a simple optical lens system (though microscopes are also used). The resultant scattered radiation is often collected at 90° from the incident beam propagation and focused onto the entrance of a monochromator. The monochromator removes extraneous light from the Raman radiation by use of a series of mirrors and optical gratings. Two features of the monochromator for maximum resolution are the slit width and, for monochromatic detection, the scanning speed. The resolution of the Raman spectrum is optimized by adjusting the slit width and scanning rate, while still maintaining a strong signal intensity.

After passing through the monochromator, the signal is focused onto the detection device. These devices range from photomultiplier tubes (PMTs) in which the signal is recorded at each frequency and the spectrum is obtained by scanning over a selected frequency range, to multichannel devices, such as arrays of photodiodes and charge coupling devices, which simultaneously detect the signal over a full frequency range. One may choose a specific detection device based on the particulars of an experiment. Sensitive detection of excited vibrational states that are produced in the Class I Raman spectroscopies is an alternative that can include acoustic detection of the heat released and resonance enhanced multiphoton ionization (REMPI). For special applications microspectroscopic techniques and fibre optic probes ('optodes') are used.

This basic instrumentation, here described within the context of spontaneous Raman scattering, may be generalized to most of the other Raman processes that are discussed. Specific details can be found in the citations.

*(c) Fourier transform Raman scattering (FTRS)*

Normal spontaneous Raman scattering suffers from lack of frequency precision and thus good spectral subtractions are not possible. Another limitation to this technique is that high resolution experiments are often difficult
to perform [39]. These shortcomings have been circumvented by the development of Fourier transform (FT)
Raman spectroscopy [40]. FT Raman spectroscopy employs a long wavelength laser to achieve viable interferometry, typically a Nd:YAG operating at 1.064 $\mu$m. The laser radiation is focused into the sample where
spontaneous Raman scattering occurs. The scattered light is filtered to remove backscatter and the Raman
light is sent into a Michelson interferometer. An interferogram is collected and detected on a near-infrared
detector (typically an $N_2$ cooled Ge photoresist). The detected signal is then digitized and Fourier transformed
to obtain a spectrum [41]. This technique offers many advantages over conventional Raman spectroscopy.
The 1.064 $\mu$m wavelength of the incident laser is normally far from electronic transitions and reduces the
likelihood of fluorescence interference. Also near-IR radiation decreases sample heating, thus higher powers
can be tolerated [42]. Since the signal is obtained by interferometry, the FT instrument records the intensity
of many wavelengths. Such simultaneous detection of a multitude of wavelengths is known as the multiplex
advantage, and leads to improved resolution, spectral acquisition time and signal to noise ratio over ordinary
spontaneous Raman scattering [39]. The improved wavelength precision gained by the use of an interferometer permits spectral subtraction, which is effective in removing background features [42]. The interferogram
is then converted to a spectrum by Fourier transform techniques using computer programs. By interfacing
the Raman setup with an FT-IR apparatus, one has both IR and Raman capabilities with one instrument.

Other than the obvious advantages of reduced fluorescence and high resolution, FT Raman is fast, safe
and requires minimal skill, making it a popular analytic tool for the characterization of organic compounds,
polymers, inorganic materials and surfaces and has been employed in many biological applications [41].

It should be noted that this technique is not without some disadvantages. The blackbody emission
background in the near IR limits the upper temperature of the sample to about 200 °C [43]. Then there is
the $\bar{v}^4$ dependence of the Raman cross-section (equations (B1.3.16) and (B1.3.20)–(B1.3.21)) which calls for
an order of magnitude greater excitation intensity when exciting in the near-IR rather than in the visible to
produce the same signal intensity [39].

*(d) Resonance Raman scattering (RRS)*

As the incident frequencies in any Raman spectroscopy approach an electronic transition in the material, the
$D_R W_R$ term in Raman scattering is greatly enhanced. One then encounters an extremely fruitful and versatile
branch of spectroscopy called resonance Raman scattering (RRS). In fact, it is fair to say that in recent years
RRS (Class I or Class II) has become the most popular form of Raman based spectroscopy. On the one hand,
it offers a diagnostic approach that is specific to those subsystems (even minority components) that exhibit the
resonance, even the very electronic transition to which the experiment is tuned. On the other hand, it offers a
powerful tool for exploring potential-energy hypersurfaces in polyatomic systems. It forms the basis for many
time resolved resonant Raman spectroscopies (TRRRSs) that exploit the non-zero vibronic memory implied
by an electronic resonance. It has inspired the time-domain theoretical picture of RRS which is formally the
appropriate transform of the frequency domain picture [44, 45]. Here the physically appealing picture arises
in which the two-step doorway event ($D_R$) prepares (vertically upward) a vibrational wave-packet that moves
(propagates) on the upper electronic state potential energy hypersurface. In the window stage ($W_R$) of the 4WM
event, this packet projects (vertically downward), accordingly to its lingering time on the upper surface, back
onto the ground state to complete the third order induced optical polarization that leads to the new fourth wave.

RRS has also introduced the concept of a 'Raman excitation profile' ($REP_j$ for the $j$th mode) [46–51].
An $REP_j$ is obtained by measuring the resonance Raman scattering strength of the $j$th mode as a function
of the excitation frequency [52, 53]. How does the scattering intensity for a given (the $j$th) Raman active

vibration vary with excitation frequency within an electronic absorption band? In turn, this has led to transform theories that try to predict the $REP_j$ from the ordinary absorption band (ABS), or the reverse. Thus one has the so-called forward transform, ABS $\rightarrow REP_j$, and the inverse transform, $REP_j \rightarrow$ ABS [54–56]. The inverse transform is a formal method that transforms an observed $REP_j$ into the electronic absorption band that is responsible for resonantly scattering mode $j$. This inverse transform raises theoretical issues concerning the frequently encountered problem of phase recovery of a complex function (in this case the complex Raman susceptibility), knowing only its amplitude [57].

One group has successfully obtained information about potential energy surfaces without measuring REPs. Instead, easily measured second derivative absorption profiles are obtained and linked to the full RRS spectrum taken at a single incident frequency. In this way, the painstaking task of measuring a REP is replaced by carefully recording the second derivative of the electronic absorption spectrum of the resonant transition [58, 59].

The fitting parameters in the transform method are properties related to the two potential energy surfaces that define the electronic resonance. These curves are obtained when the two hypersurfaces are cut along the $j$th normal mode coordinate. In order of increasing theoretical sophistication these properties are: (i) the relative position of their minima (often called the displacement parameters), (ii) the force constant of the vibration (its frequency), (iii) nuclear coordinate dependence of the electronic transition moment and (iv) the issue of mode mixing upon excitation—known as the Duschinsky effect—requiring a multidimensional approach.

We have seen how, by definition, all Raman spectroscopies must feature the difference frequency resonance that appears following the two-step doorway stage of the 4WM process. Basically, RRS takes advantage of achieving additional resonances available in the two remaining energy denominator factors found at third order (and, of course, still more at higher order). The two remaining energy factors necessarily involve an algebraic sum of an odd number of optical frequencies (one for the first step in the doorway stage, and three for the initial step in the window event). Since the algebraic sum of an odd number of optical frequencies must itself be optical, these additional resonances must be at optical frequencies. Namely, they must correspond to electronic transitions, including (in molecules) their dense rotational– (or librational–) vibrational substructure. The literature is filled with a great many interesting RRS applications, extending from resonances in the near-IR (dyes and photosynthetic pigments for example) to the deep UV (where backbone electronic resonances in proteins and nucleic acids are studied). These increasingly include TRRRS in order to follow the folding/unfolding dynamics of substructures (through the chromophore specificity of RRS) in biologically important molecules [60–62].

The reader must turn to the literature to amplify upon any of these topics. Here we return to the two-colour generator/WMEL scheme to see how it easily can be adapted to the RRS problem.

Let us consider RRS that contains both of the available additional resonances, as is normally the case (though careful choices of colours and their time sequence can isolate one or the other of these). First, we seek out the doorway events that contain not only the usual Raman resonance after the two fields, 1 and 2, have acted conjugately, but also the new resonance that appears after the first field has acted. The appropriate doorway generators remain $\varepsilon_1 \varepsilon_2^*$ and $\varepsilon_2^* \varepsilon_1$, in order to retain the Raman resonance. There are now two fully resonant doorway WMEL diagrams, which we shall call $D_{RR}(A_{12^*})$ and $D_{RR}(B_{12^*})$. These diagrams are shown in figure B1.3.4, in which the full manifold of sublevels for each electronic state is intimated. In doorway channel A, vibrational coherences are produced in the ground electronic state $g$, as usual, but now they are enhanced by the electronic resonance in step 1 ($\varepsilon_1$). However, in doorway channel B, vibrational coherences are produced in the excited electronic state, $e$, as well. Interestingly, since A is a ket–ket event and B is a ket–bra event (which differ by an overall sign), these two coherences must differ in phase by $\pi$ (180°). This may be important in any 4WM experiment that is phase sensitive (heterodyned Class II) and in which the window event does not reverse this phase difference.

Frequently, femtosecond pulses are used in such electronically resonant spectroscopy. Such pulses usually have near-transform limited bandwidths and can spectrally embrace a fair range of vibrational coherences.

a)

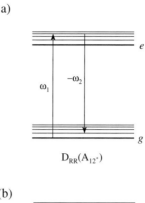

$$D_{RR}(A_{12^*})$$

(b)

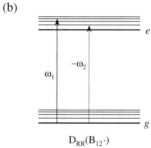

$$D_{RR}(B_{12^*})$$

**Figure B1.3.4.** The two fully resonant doorway stages for resonance Raman scattering (RRS), in which the manifold of vibrational sublevels for each electronic state is indicated. (a) Doorway stage $D_{RR}(A_{12^*})$, in which a vibrational coherence is produced in the ground electronic state, $g$. It is a ket–ket evolution. (b) Doorway stage $D_{RR}(B_{12^*})$, where a vibrational coherence is created in the excited electronic state, $e$. It is a ket–bra evolution. The coherences in both doorway stages are enhanced by the electronic resonance in their identical step 1 (generated by $\varepsilon_1$).

Thus, even when a single central colour is chosen to define the femtosecond exciting pulse, actually a broad band of colours is available to provide a range for both $\omega_1$ and $\omega_2$. Instead of preparing single well-defined vibrational coherences using sharply defined colours, now vibrational wavepackets are made in both the ground (channel A) and excited states (channel B). These must evolve in time, for they are not eigenstates in either potential energy hypersurface. The nature of the 4WM signal is sensitive to the location of the wavepacket in either hypersurface at the time the window event takes place. Such wavepackets, prepared from a spectrum of colours contained coherently within a femtosecond pulse, are termed 'impulsively' prepared (see impulsive Raman scattering for further details).

Whenever coherences in the upper manifold are particularly short lived, doorway channel A will dominate the evolution of the polarization at least at later times. In that case, the fringes seen in the 4WM signal as the time between the doorway and window stages is altered (with a delay line) reflect those Raman frequencies in the ground state that can be spectrally embraced by the femtosecond pulse. Then the Fourier transform of the fringes leads to the conventional spontaneous RRS of the ground state. Indeed, in the absence of electronic resonance, channel B reverts to a purely nonresonant doorway event ($D_N$) and only channel A reveals Raman resonances—those in the electronic ground state [62].

Whatever the detection technique, the window stage of the 4WM event must convert these evolved vibrational wavepackets into the third order polarization field that oscillates at an ensemble distribution of optical frequencies. One must be alert to the possibility that the window event after doorway channel B may involve resonances from electronic state manifold $e$ to some higher manifold, say $r$. Thus channel B followed by an $\varepsilon_3$ (ket) or a $\varepsilon_3^*$ (bra) event might be enhanced by an $e$-to-$r$ resonance. However, it is normal to confine the window event to the $e$-to-$g$ resonances, but this is often simply for lack of substantive $e$-to-$r$ information. Given that the third field action can be of any available colour, and considering only $g$-to-$e$ resonances, one has, for any colour $\omega_3$, only two possibilities following each doorway channel. Channel A should be followed by an $\varepsilon_3$ (*ket up*) or an $\varepsilon_3^*$ (*bra up*) event; channel B should be followed by a $\varepsilon_3$ (*bra down*) or a $\varepsilon_3^*$ (*ket down*) event.

Before looking more closely at these, it is important to recognize another category of pump–probe Raman experiments. These are often referred to as 'transient Raman' pump–probe studies. In these, a given system is 'pumped' into a transient condition such as an excited vibronic state, or a photochemical event such as dissociation or radical formation [63–65]. Such pumping can be achieved by any means—even by high energy radiation [66–68]—though normally laser pumping is used. The product(s) formed by the pump step is then studied by a Raman probe (often simply spontaneous Raman, sometimes CARS). Since the transient state is normally at low concentrations, the Raman probing seeks out resonant enhancement, as we are describing, and also means must be taken to stay away from the luminescence background that is invariably caused by the pump event. Often, time gated Fourier transform Raman in the near-IR is employed (that spectral region being relatively free of interfering fluorescence) and yet upper $e$-to-$r$ type resonances may still be available for RRS. Since transient systems are 'hot' by their very nature, both anti-Stokes as well as Stokes spontaneous Raman scattering can be followed to time the vibrational relaxation in transient excited states (see [69, 70]).

Returning to the original pump–probe RRS, it is a simple matter to complete the 4WM WMEL diagrams for any proposed RRS. Usually RRS experiments are of the full quadrature sort, both spontaneous RRS as well as homodyne detected femtosecond RRS. The latter fit most pump–probe configurations.

Let us, for example, present the full WMEL diagrams for full quadrature RRS with two colours, 1 and 2. (Recall that three colours cannot lead to full $Q$ at the 4WM level.) Given the $\varepsilon_1 \varepsilon_2^*$ doorway generator for channels A and B, the generator for the first step of the window event must either be $\varepsilon_1^*$ or $\varepsilon_2$, and the corresponding signal must conjugate either with $\varepsilon_2^*$ ($\varepsilon_s = \varepsilon_2$) or with $\varepsilon_1$ ($\varepsilon_s^* = \varepsilon_1^*$), respectively. In the former case, field 1 will have acted twice (and conjugately) to help produce the third order polarization and signal field directed along $k_2$. In the latter case, field 2 has acted twice (and conjugately) to help produce a third order polarization along $-k_1$. Of the two fields, the most intense clearly is the candidate for the twice acting field. The weaker field, then, is examined for the signal.

In addition, we have asked that a third resonance exist in the window stage. For channel A this requires that the window event begin either with *ket up* using $\varepsilon_2$, or with *bra up* using $\varepsilon_1^*$, while the window event following channel B should begin either with *bra down* using $\varepsilon_2$ or with *ket down* using $\varepsilon_1^*$. The corresponding four WMEL diagrams are shown in figure B1.3.5.

In these WMEL diagrams, the outcome of the collapse of the polarization, the second step of the window stage (the curvy line segment), depends on the phase difference between the induced $s$th order polarization and the new $(s + 1)$th electromagnetic field [7]. If this difference is $0°$, then the energy contained in the polarization is fully converted into population change in the medium (pure Class I spectroscopies). At third order, channel A populates vibrational levels in the ground electronic state; channel B populates vibrational levels in the excited electronic state. However if this phase difference is $90°$, the energy is converted fully into the new $(s + 1)$th electromagnetic field and the material is unchanged (pure Class II spectroscopies). If the phase difference lies between 0 and $90°$, as is almost always the case, then both outcomes occur to some extent and the experimentalist may perform either a Class I or a Class II spectroscopy.

As we have already seen, in such a full quadrature situation, phase matching is automatic (the signal is collinear with either of the incident fields), so the experiment then measures changes in one or more properties of one of the incident fields—either the first appearing light pulse, or the second appearing light pulse. These are distinguished both in order of appearance and by their wavevector of incidence. At full quadrature, the obvious property to measure is simply the intensity (Class I) as one (or more) of the time parameters is changed. With this example, it should be a simple matter to explore WMEL diagrams for any other RRS spectroscopies, in particular the nonquadrature ones.

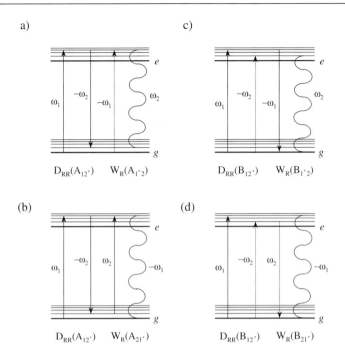

**Figure B1.3.5.** Four WMEL diagrams for fully resonant Raman scattering (RRS). Diagrams (a) and (b) both have doorway stage $D_{RR}(A_{12^*})$ (Figure B1.3.4(a)), in which a vibrational coherence is created in the ground electronic state, $g$. For the window event in (a), field 1 promotes the bra from the ground electronic state, $g$, to an excited electronic bra state, $e$. In this window stage $W_R(A_{1^*2})$, the third field action helps produce the third order polarization, which in turn gives rise to a signal field with frequency $\omega_2$. For the window event $W_R(A_{21^*})$ in (b), field 2 acts to promote the ket in the ground electronic state into one in the excited electronic state. Now a signal field with frequency $-\omega_1$ is created. Both diagrams (c) and (d) have doorway stage $D_{RR}(B_{12^*})$ (figure B1.3.4(b)), in which a vibrational coherence is created in the excited electronic state, $e$. At the window stage $W_R(B_{1^*2})$ (c) field 1 demotes the ket from one in electronic state $e$ to one in the ground electronic state $g$. A consequent third order polarization leads to a signal field at frequency $\omega_2$. At the window stage $W_R(B_{21^*})$ (d) the bra is demoted by field 2 to the ground electronic state to produce a signal field at frequency $-\omega_1$. Full quadrature is achieved in all four diagrams.

*(e) Stimulated Raman scattering (SRS)*

Since the inception of the laser, the phenomenon called stimulated Raman scattering (SRS) [71–74] has been observed while performing spontaneous Raman experiments. Stimulated Raman scattering is inelastic light scattering by molecules caused by presence of a light field that is stronger than the zero-point field of the blackbody. Thus SRS overtakes the spontaneous counterpart. SRS, like SR, is an active Class I spectroscopy. For the case of Stokes scattering, this light field may be a second laser beam at frequency $\omega_2$, or it may be from the polarization field that has built up from the spontaneous scattering event. Typically in an SRS experiment one also observes frequencies other than the Stokes frequency: those at $\omega_1 - 2\omega_R$, $\omega_1 - 3\omega_R$, $\omega_1 - 4\omega_R$ etc which are referred to as higher order Stokes stimulated Raman scattering and at $\omega_1 + \omega_R$ (anti-Stokes) and $\omega_1 + 2\omega_R$, $\omega_1 + 3\omega_R$, $\omega_1 + 4\omega_R$ (higher order anti-Stokes stimulated Raman scattering) [19]. This type of scattering may arise only after an appreciable amount of Stokes scattering is produced (unless the system is not at thermal equilibrium).

First order stimulated Stokes scattering experiences an exponential gain in intensity as the fields propagate through the scattering medium. This is given by the expression [75]

$$I_S(z) = I_S(0) \, e^{g_S I_L z}$$

where $z$ is the path length, $I_L$ is the intensity of the incident radiation and $I_S(0)$ is the intensity of the quantum noise associated with the vacuum state [15], related to the zero-point bb field. $g_S$ is known as the stimulated Raman gain coefficient, whose maximum occurs at exact resonance, $\omega_L - \omega_S = \omega_R$. For a Lorentzian lineshape, the maximum gain coefficient is given by

$$g_{S_{\text{max}}} = \frac{8\pi^2 c^2 N}{\hbar \omega_L \omega_S^2 n_S n_L} \frac{d\sigma_R}{d\Omega} \frac{1}{\gamma_R}$$

where $\omega_S$ is the frequency of the Stokes radiation and $\gamma_R$ is the HWHM of the Raman line (in units of circular frequency) [75]. This gain coefficient is seen to be proportional to the spontaneous differential Raman cross section $(d\sigma_R/d\Omega)$, (the exact nature of which depends on experimental design (see equations (B1.3.20) and (B1.3.21)) and also to the number density of scatterers, $N$. This frequency dependent gain coefficient may also be written in terms of the third order nonlinear Raman susceptibility, $\chi^{(3)}(\omega_S)$. As with SR, only the imaginary part of $\chi^{(3)}(\omega_S)$ contributes to Stokes amplification (the real part accounts for intensity dependent (nonlinear) refractive indices). For the definition of $\chi^{(3)}$ used here (equations (B1.3.11) and (B1.3.12)), their relation is given by

$$g_S(\omega_S) = \frac{-2\omega_S}{\varepsilon_0^2 n_S n_L c^2} \, \text{Im} \, \chi_{\text{AAAA}}^{(3)}(\omega_S)$$

where the magnitude of $\text{Im} \, \chi^{(3)}(\omega_S)$ is negative, thus leading to a positive gain coefficient [33]. For this expression, only the component of the scattered radiation parallel to the incident light is analysed.

Once an appreciable amount of Stokes radiation is generated, enough scatterers are left in an excited vibrational state for the generation of anti-Stokes radiation. Also, the Stokes radiation produced may now act as incident radiation in further stimulated Raman processes to generate the higher order Stokes fields. Although the Stokes field is spatially isotropic, scattered radiation in the forward and backward directions with respect to the incident light traverses the longest interaction length and thus experiences a significantly larger gain (typically several orders of magnitude larger than in the other directions [33]). Thus the first and higher order Stokes frequencies lie along the direction of the incident beam.

This is not the case for stimulated anti-Stokes radiation. There are two sources of polarization for anti-Stokes radiation [17]. The first is analogous to that in figure B1.3.3(b) where the action of the blackbody $(-\omega_2)$ is replaced by the action of a previously produced anti-Stokes wave, with frequency $\omega_A$. This radiation actually experiences an attenuation since the value of $\text{Im} \, \chi^{(3)}(\omega_A)$ is positive (leading to a negative 'gain' coefficient). This is known as the stimulated Raman loss (SRL) spectroscopy [76]. However the second source of anti-Stokes polarization relies on the presence of Stokes radiation [17]. This anti-Stokes radiation will emerge from the sample in a direction given by the wavevector algebra: $\mathbf{k}_A = 2\mathbf{k}_1 - \mathbf{k}_S$. Since the Stokes radiation is isotropic $(-\mathbf{k}_S)$, the anti-Stokes radiation (and subsequent higher order radiation) is emitted in the form of concentric rings.

## (f) Surface enhanced Raman scattering (SERS)

We have seen that the strength of Raman scattered radiation is directly related to the Raman scattering cross-section $(\sigma_R)$. The fact that this cross-section for Raman scattering is typically much weaker than that for absorption $(\sigma_R \ll \sigma_{\text{abs}})$ limits conventional SR as a sensitive analytical tool compared to (linear) absorption techniques. The complication of fluorescence in the usual Raman techniques of course tends to decrease the signal-to-noise ratio.

It was first reported in 1974 that the Raman spectrum of pyridine is enhanced by many orders of magnitude when the pyridine is adsorbed on silver metal [77]. This dramatic increase in the apparent Raman cross-section in the pyridine/silver system was subsequently studied in more detail [78, 79]. The enhancement of the Raman scattering intensity from molecules adsorbed to surfaces (usually, though not exclusively, noble metals) has come to be called the surface enhanced Raman spectroscopy (SERS). Since these early discoveries, SERS has been intensively studied for a wide variety of adsorbate/substrate systems [80–83]. The pyridine/silver system, while already thoroughly studied, still remains a popular choice among investigators both for elucidating enhancement mechanisms and for analytical purposes. The fluorescence of the adsorbed molecules does not experience similarly strong enhancement, and often is actually quenched. So, since the signal is dominated by the adsorbate molecules, fluorescence contamination is relatively suppressed.

The metal substrate evidently affords a huge ($\sim 10^{10}$ and even as high as $10^{14}$ [84, 85]) increase in the cross-section for Raman scattering of the adsorbate. There are two broad classes of mechanisms which are said to contribute to this enhancement [86–88]. The first is based on electromagnetic effects and the second on 'chemical' effects. Of these two classes the former is better understood and, for the most part, the specific mechanisms are agreed upon; the latter is more complicated and is less well understood. SERS enhancement can take place in either physisorbed or chemisorbed situations, with the chemisorbed case typically characterized by larger Raman frequency shifts from the bulk phase.

The substrate is, of course, a necessary component of any SERS experiment. A wide variety of substrate surfaces have been prepared for SERS studies by an equally wide range of techniques [87]. Two important substrates are electrochemically prepared electrodes and colloidal surfaces (either deposited or in solution).

Aside from the presence of a substrate, the SERS experiment is fundamentally similar to the standard conventional Raman scattering experiment. Often a continuous wave laser, such as an argon ion laser, is used as the excitation source, but pulsed lasers can also be used to achieve time resolved SERS. Also, as in conventional Raman scattering, one can utilize pre-resonant or resonant conditions to perform resonant SERS (often denoted SERRS for surface enhanced resonant Raman scattering). SERRS combines the cross-section enhancement of SERS with the electronic resonance enhancement of resonance Raman scattering. In fact, through SERRS, one can achieve extraordinary sensitivity, with reports appearing of near-single-molecule-based signals [84, 85, 89].

We now move on to some Class II spectroscopies.

### (g) Coherent Raman scattering (CRS)

The major Class II Raman spectroscopy is coherent Raman scattering (CRS) [90–93]. It is an extremely important class of nearly degenerate four-wave mixing spectroscopies in which the fourth wave (or signal field) is a result of the coherent stimulated Raman scattering. There are two important kinds of CRS distinguished by whether the signal is anti-Stokes shifted (to the blue) or Stokes shifted (to the red). The former is called CARS (coherent anti-Stokes Raman scattering) and the latter is called CSRS (coherent Stokes Raman scattering). Both CARS and CSRS involve the use of two distinct incident laser frequencies, $\omega_1$ and $\omega_2$ ($\omega_1 > \omega_2$). In the typical experiment $\omega_1$ is held fixed while $\omega_2$ is scanned. When $\omega_1 - \omega_2$ matches a Raman frequency of the sample a resonant condition results and there is a strong gain in the CARS or CSRS signal intensity. The complete scan of $\omega_2$ then traces out the CARS or CSRS spectrum of the sample. (Figure B1.3.2(b) shows representative WMEL diagrams for the CARS and CSRS processes.) There are, in actuality, 48 WMEL diagrams (including the nonresonant contributions) that one must consider for either of these two processes. These have been displayed in the literature ([98] (CARS) and [99] (CSRS)). For both processes, a pair of field–matter interactions produces a vibrational coherence between states $|g\rangle$ and $|f\rangle$ (see $D_R$ and $D_N$ of figure B1.3.2(b)). For the CARS process, the third field, having frequency $\omega_1$, acts in phase (the same Fourier component) with the first action of $\omega_1$ to produce a polarization that is anti-Stokes shifted from $\omega_1$ (see $W_R$(E) and $W_N$(F) of figure B1.3.2(b)). For the case of CSRS the third field action has frequency $\omega_2$ and acts in phase with the earlier action of $\omega_2$ ($W_R(C)$ and $W_N(D)$ of figure B1.3.2(b)). Unlike the Class I spectroscopies, no fields in CARS or CSRS (or any homodyne detected Class II spectroscopies) are in quadrature at the polarization level.

Since homodyne detected CRS is governed by the modulus square of $\chi^{(3)}$ ($I_{CRS} \propto |\chi^{(3)}|^2$), its lineshape is not a symmetric lineshape like those in the Class I spectroscopies, but it depends on both the resonant and nonresonant components of $\chi^{(3)}$, $\chi_R^{(3)}$ and $\chi_N^{(3)}$, respectively. Thus

$$|\chi^{(3)}|^2 = |\chi_R^{(3)} + \chi_N^{(3)}|^2 = |\chi_R^{(3)}|^2 + \chi_N^{(3)}(\chi_R^{(3)} + \chi_R^{*(3)}) + (\chi_N^{(3)})^2$$

and one is faced with both an absorptive component $|\chi_R^{(3)}|^2$ and a dispersive component, $\chi_N^{(3)}(\chi_R^{(3)} + \chi_R^{*(3)})$. ($\chi_N^{(3)}$ can, to a very good approximation, be taken to be a (pure real) constant over the width of the Raman line.) As a result, the CRS lineshape is asymmetric and more complicated due to this nonresonant background interference.

The primary advantages of CARS and CSRS include an inherently stronger signal than spontaneous Raman scattering (the incident fields are stronger than the zero-point blackbody fields) and one that is directional (phase matched). These characteristics combine to give the technique a much lower vulnerability to sample fluorescence and also an advantage in remote sensing. For CARS, fluorescence is especially avoided since the signal emerges to the blue of the incident laser frequencies and fluorescence must be absent (unless it were biphotonically induced). The primary disadvantage of CARS and CSRS is the interference of the nonresonant part of $\chi^{(3)}$ in the form of the dispersive cross-term. A class of techniques called polarization CRS utilizes the control over the polarization of the input beams to suppress the nonresonant background interference [94, 95]. On the other hand, the background interference necessarily carries information about the nonresonant component of the electric susceptibility which is sometimes a sought after quantity.

*(h) Raman induced Kerr effect spectroscopy (RIKES)*

The nonresonant background prevalent in CARS experiments (discussed above), although much weaker than the signals due to strong Raman modes, can often obscure weaker modes. Another technique which can suppress the nonresonant background signal is Raman induced Kerr-effect spectroscopy or RIKES [96, 97].

A RIKES experiment is essentially identical to that of CW CARS, except the probe laser need not be tunable. The probe beam is linearly polarized at $0°$ ($\rightarrow$), while the polarization of the tunable pump beam is controlled by a linear polarizer and a quarter waveplate. The pump and probe beams, whose frequency difference must match the Raman frequency, are overlapped in the sample (just as in CARS). The strong pump beam propagating through a nonlinear medium induces an anisotropic change in the refractive indices seen by the weaker probe wave, which alters the polarization of a probe beam [96]. The signal field is polarized orthogonally to the probe laser and any altered polarization may be detected as an increase in intensity transmitted through a crossed polarizer. When the pump beam is linearly polarized at $45°$ ($\nearrow$), contributions from the nonlinear background susceptibility exist ($\chi_{eff}^{(3)} = 3[\chi_{AABB}^{(3)} + \chi_{ABAB}^{(3)}]$). If the quarter-wave plate is adjusted to give circularly polarized light ($\circlearrowleft$), the nonresonant background will disappear ($\chi_{eff}^{(3)} = 3i[\chi_{AABB}^{(3)} - \chi_{ABAB}^{(3)}]$), provided $\chi_{AABB}^{NR} = \chi_{ABAB}^{NR}$ [19].

A unique feature of this Class II spectroscopy is that it occurs in full quadrature, and thus the phase-matching condition is automatically fulfilled for every propagation direction and frequency combination (for isotropic media). Characteristic WMEL diagrams for this process are given by diagrams B and G of figure B1.3.2(a). From these diagrams, one may notice that the frequency of the signal field is identical to that of one of the incident fields, thus one must carefully align the crossed polarizer to eliminate contamination by the probe beam.

A common technique used to enhance the signal-to-noise ratio for weak modes is to inject a local oscillator field polarized parallel to the RIKE field at the detector. This local oscillator field is derived from

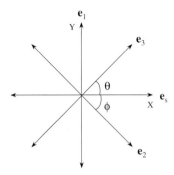

**Figure B1.3.6.** The configuration of the unit polarization vectors $\mathbf{e}_1$, $\mathbf{e}_2$, $\mathbf{e}_3$ and $\mathbf{e}_s$ in the laboratory Cartesian basis as found in the ASTERISK technique.

the probe laser and will add coherently to the RIKE field [96]. The relative phase of the local oscillator and the RIKE field is an important parameter in describing the optical heterodyne detected (OHD)-RIKES spectrum. If the local oscillator at the detector is in phase with the probe wave, the heterodyne intensity is proportional to $\mathrm{Re}\{\chi_{\mathrm{eff}}^{(3)}\}$. If the local oscillator is in phase quadrature with the probe field, the heterodyne intensity becomes proportional to $\mathrm{Im}\{\chi_{\mathrm{eff}}^{(3)}\}$. Thus, in addition in to signal-to-noise improvements, OHD-RIKES, being a heterodyne method, demonstrates a phase sensitivity not possible with more conventional homodyne techniques.

Still another spectroscopic technique used to suppress the nonresonant background is ASTERISK. The setup is identical to a conventional CARS experiment, except three independent input fields of frequencies, $\omega_1$, $\omega_2$ and $\omega_3$ are used. The relative polarization configuration (not wavevectors) for the *three* incident fields and the analyser ($\mathbf{e}_s$) is shown in figure B1.3.6, where the signal generated at $\omega_s$ will be polarized in the $x$-direction. (Both $\theta$ and $\phi$ are defined to be positive angles, as denoted in figure B1.3.6.) The recorded spectra will be relatively free of the nonresonant background (for $\theta = \phi \simeq 45°$, the detected intensity will be proportional to $|\chi_{\mathrm{ABAB}}^{(3)} - \chi_{\mathrm{ABBA}}^{(3)}|^2$); however one must satisfy the phase matching condition: $\Delta\mathbf{k} = \mathbf{k}_1 - \mathbf{k}_2 + \mathbf{k}_3 - \mathbf{k}_s$. In transparent materials, a greater than three-orders-of-magnitude reduction of the nonresonant background occurs as compared to its 25–100-fold suppression by RIKES [96].

*(i) Raman scattering with noisy light*

In the early 1980s, it was shown that noisy light offers a time resolution of the order of the noise correlation time, $\tau_c$, of the light (typically tens to several hundreds of femtoseconds)—many orders of magnitude faster than the temporal profile of the light (which is often several nanoseconds, but in principle can be CW) [100–102]. A critical review of many applications of noisy light (including CRS) is given by Kobayashi [103]. A more recent review by Kummrow and Lau [104] contains an extensive listing of references.

A typical noisy light based CRS experiment involves the splitting of a noisy beam (short autocorrelation time, broadband) into identical twin beams, B and B′, through the use of a Michelson interferometer. One arm of the interferometer is computer controlled to introduce a relative delay, $\tau$, between B and B′. The twin beams exit the interferometer and are joined by a narrowband field, M, to produce the CRS-type third order polarization in the sample ($\bar{\omega}_B - \omega_M \approx \omega_R$). The delay between B and B′ is then scanned and the frequency-resolved signal of interest is detected as a function of $\tau$ to produce an *interferogram*. As an interferometric spectroscopy, it has come to be called $\mathrm{I}^{(2)}\mathrm{CRS}$ ($\mathrm{I}^{(2)}\mathrm{CARS}$ and $\mathrm{I}^{(2)}\mathrm{CSRS}$), in which the '$\mathrm{I}^{(2)}$' refers to the two, twin incoherent beams that are interferometrically treated [105]. The theory of $\mathrm{I}^{(2)}\mathrm{CRS}$ [106, 107] predicts that the so-called radiation difference oscillations (RDOs) should appear in the 'monochromatically' detected

$I^{(2)}$CRS interferogram with a frequency of $\Delta \equiv \omega_M + 2\omega_R - \omega_D$, where $\omega_D$ is the detected frequency, $\omega_M$ is the narrowband frequency and $\omega_R$ the Raman (vibrational) frequency. Since $\omega_D$ and $\omega_M$ are known, $\omega_R$ may be extracted from the experimentally measured RDOs. Furthermore, the dephasing rate constant, $\gamma_R$, is determined from the observed decay rate constant, $\gamma$, of the $I^{(2)}$CRS interferogram. Typically for the $I^{(2)}$CRS signal $\omega_D \approx \omega_M + 2\omega_R$ and thus $\Delta \approx 0$. That is, the RDOs represent strongly *down-converted* (even to zero frequency) Raman frequencies. This down-conversion is one of the chief advantages of the $I^{(2)}$CRS technique, because it allows for the characterization of vibrations using optical fields but with a much smaller interferometric sampling rate than is needed in FT-Raman or in FT-IR. More explicitly, the Nyquist sampling rate criterion for the RDOs is much smaller than that for the vibration itself, not to mention that for the near-IR FT-Raman technique already discussed. This is particularly striking for high energy modes such as the C–H vibrations [108]. Modern applications of $I^{(2)}$CRS now utilize a 'two-dimensional' time–frequency detection scheme which involves the use of a CCD camera to detect an entire $I^{(2)}$CARS spectrum at every delay time [109]. These are called Raman spectrograms and allow for a greatly enhanced level of precision in the extraction of the Raman parameters—a precision that considerably exceeds the instrumental uncertainties.

The understanding of the underlying physical processes behind $I^{(2)}$CRS (and noisy light spectroscopies in general) has been aided by the recent development of a diagrammatic technique called factorized time correlation (FTC) diagram analysis for properly averaging over the noise components in the incident light [110–112] in any noisy light based spectroscopy (linear or nonlinear).

### (j) Time resolved coherent Raman scattering (TRCRS)

With the advent of short pulsed lasers, investigators were able to perform time resolved coherent Raman scattering. In contrast to using femtosecond pulses whose spectral width provides the two colours needed to produce Raman coherences, discussed above, here we consider pulses having two distinct centre frequencies whose difference drives the coherence. Since the 1970s, picosecond lasers have been employed for this purpose [113, 114], and since the late 1980s femtosecond pulses have also been used [115]. Here we shall briefly focus on the two-colour femtosecond pulsed experiments since they and the picosecond experiments are very similar in concept.

The TR-CRS experiment requires a femtosecond scale light source (originally a rhodamine 6G ring dye laser [115, 116]) and a second longer pulsed (typically several picoseconds) laser operating at a different frequency. The femtosecond source at one colour is split into two pulses having a relative and controllable delay, $\tau$, between them. Each of these two pulses acts once and with the same Fourier component, one in the doorway stage, the other in the window stage. The third, longer pulsed field at the second colour and in a conjugate manner participates with one of the femtosecond pulses in the doorway event to produce the Raman coherence. This polarization then launches the TR-CRS signal field which can be either homodyne or heterodyne detected. This signal must decay with increasing $\tau$ as the Raman coherence is given time to decay before the window event takes place.

For homodyne detection, the TR-CRS intensity (for Lorentzian Raman lines) is of the form [115]

$$I_{TR-CRS} \propto \left| \sum_j e^{-2\gamma_j \tau} e^{-i\omega_j \tau + i\phi_j} \right|^2 \tag{B1.3.25}$$

where $j$ runs over all Raman active modes contained within the bandwidth of the femtosecond scale pulse. The parameters $\gamma_j$, $\omega_j$ and $\phi_j$ are the dephasing rate constant, the Raman frequency and phase for the $j$th mode. One can see from equation (B1.3.25) that for a single mode $I_{TR-CRS}$ is a simple exponential decay, but when more modes are involved $I_{TR-CRS}$ will reveal a more complicated beat pattern due to the cross-terms.

TR-CRS has been used to study many molecules from benzene [115–118] to betacarotene [119].

*(k) Impulsive stimulated Raman scattering (ISRS)*

In discussing RRS above, mention is made of the 'impulsive' preparation of wavepackets in both the excited electronic potential surface and the ground state surface. In the absence of electronic resonance only the latter channel is operative and ground state wavepackets can be prepared in transparent materials using the spectral width of femtosecond light to provide the necessary colours. Such impulsive stimulated Raman scattering (ISRS) was first performed by Nelson *et al* on a variety of systems including acoustic modes in glasses [120] and librational and intramolecular vibrations in liquids [121].

To date, there are two types of configuration employed in ISRS: three-pulse [121] and two-pulse [121]. In both cases, (an) excitation pulse(s) provide(s) the necessary frequencies to create a vibrational wavepacket which proceeds to move within the potential surface. After a delay time $\tau$, a probe pulse having the same central frequency enters the sample and converts the wavepacket into an optical polarization and the coherent fourth wave is detected. For the case of two-pulse ISRS, the transmitted intensity along the probe pulse is followed. In three-pulse ISRS (defined by three wavevectors), the coherently scattered radiation is detected along its unique wavevector. The intensity of the scattered (transmitted) pulse as a function of $\tau$ shows damped oscillations at the frequency of the Raman mode (roughly the reciprocal of the recurrence time as the packet oscillates between the two walls of the potential curve). If the pulse durations are longer than the vibrational period (the spectral width is too small to embrace the resonance), no such oscillations can occur. Since in ISRS the spectral width of each pulse is comparable to the Raman frequency, each pulse contains spectral components that produce Stokes and anti-Stokes scattering. These oscillations occur due to the interference between the Stokes and anti-Stokes scattering processes [122]. These processes differ in phase by 180° (the WMEL rules can show this). This expected phase difference has been demonstrated when heterodyne detection is used (the optical Kerr effect probed by an $E_{lo}$) and the signal is frequency resolved [123].

As already mentioned, electronically resonant, two-pulse impulsive Raman scattering (RISRS) has recently been performed on a number of dyes [124]. The main difference between resonant and nonresonant ISRS is that the beats occur in the absorption of the probe rather than the spectral redistribution of the probe pulse energy [124]. These beats are $\frac{\pi}{2}$ out of phase with respect to the beats that occur in nonresonant ISRS (cosine-like rather than sinelike). RISRS has also been shown to have the phase of oscillation depend on the detuning from electronic resonance and it has been shown to be sensitive to the vibrational dynamics in both the ground and excited electronic states [122, 124].

*(l) Higher order and higher dimensional time resolved techniques*

Of great interest to physical chemists and chemical physicists are the broadening mechanisms of Raman lines in the condensed phase. Characterization of these mechanisms provides information about the microscopic dynamical behaviour of material. The line broadening is due to the interaction between the Raman active chromophore and its environment.

It has been shown that spontaneous or even coherent Raman scattering cannot be used to distinguish between the fast homogeneous and the slow inhomogeneous broadening mechanisms in vibrational transitions [18, 125]. One must use higher order (at least fifth order) techniques if one wishes to resolve the nature of the broadening mechanism. The ability of these higher order techniques to make this distinction is based on the echo phenomena very well known for NMR and mentioned above for D4WM with electronic resonances. The true Raman echo experiment is a time resolved seventh order technique which has recently been reported by Berg *et al* [126–129]. It is thus an 8WM process in which two fields are needed for each step in the normal 4WM echo. A Raman echo WMEL diagram is shown in figure B1.3.7. It is seen that, as in CRS, the first two pulsed field actions create a vibrational coherence. This dephases until the second pair of field actions creates a vibrational population. This is followed by two field actions which again create a vibrational coherence

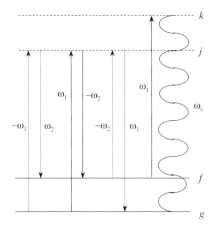

**Figure B1.3.7.** A WMEL diagram for the seventh order Raman echo. The first two field actions create the usual Raman vibrational coherence which dephases and, to the extent that inhomogeneity is present, also weakens as the coherence from different chromophores 'walks off'. Then such dephasing is stopped when a second pair of field actions converts this coherence into a population of the excited vibrational state $f$. This is followed by yet another pair of field actions which reconvert the population into a vibrational coherence, but now one with phase opposite to the first. Now, with time, the 'walked-off' component of the original coherence can reassemble into a polarization peak that produces the Raman echo at frequency $\omega_s = 2\omega_1 - \omega_2 = \omega_1 + \omega_{fg} = \omega_2 + 2\omega_{fg}$.

but, now, with opposite phase to the first coherence. Hence one obtains a partial rephasing, or echo, of the macroscopic polarization. The final field action creates the seventh order optical polarization which launches the signal field (the eighth field). Just as for the spin echo in NMR or the electronic echo in 4WM, the degree of rephasing (the magnitude of the echo) is determined by the amount of slow time scale (inhomogeneous) broadening of the two-level system that is present. Spectral diffusion (the exploration of the inhomogeneity by each chromophore) destroys the echo.

An alternative fifth order Raman quasi-echo experiment can also be performed [130–134]. Unlike the true Raman echo which involves only two vibrational levels, this process requires the presence of three very nearly evenly spaced levels. A WMEL diagram for the Raman quasi-echo process is shown in figure B1.3.8. Here again the first two field actions create a vibrational coherence which is allowed to dephase. This is followed by a second pair of field actions, which, instead of creating a population, creates a different vibrational coherence which is of opposite phase and roughly the same order of magnitude as the initial coherence. This serves to allow a rephasing of sorts, the quasi-echo, provided the levels are in part inhomogeneously spread. The final field action creates the fifth order optical polarization that launches the signal field (the sixth field in this overall six-wave mixing process).

As one goes to higher orders, there are many other processes that can and do occur. Some are true fifth or seventh order processes and others are 'cascaded' events arising from the sequential actions of lower order process [135]. Many of these cascaded sources of polarization interfere with the echo and quasi-echo signal and must be handled theoretically and experimentally.

The key for optimally extracting information from these higher order Raman experiments is to use two time dimensions. This is completely analogous to standard two-dimensional NMR [136] or two-dimensional 4WM echoes. As in NMR, the extra dimension gives information on coherence transfer and the coupling between Raman modes (as opposed to spins in NMR).

With the wealth of information contained in such two-dimensional data sets and with the continued improvements in technology, the Raman echo and quasi-echo techniques will be the basis for much activity

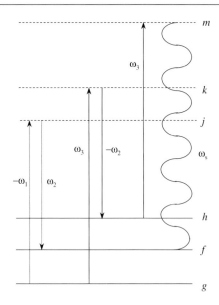

**Figure B1.3.8.** A WMEL diagram for the three-colour fifth order 'quasi-Raman echo'. As usual, the first pair of field actions creates the $fg$ Raman coherence which is allowed both to dephase and 'walk off' with time. This is followed by a second pair of field actions, which creates a different but oppositely phased Raman coherence (now $hf$) to the first. Its frequency is at $\omega_3 - \omega_1 = \omega_{hf}$. Provided that $\omega_{hf}$ frequencies are identified with an inhomogeneous distribution that is similar to those of the $\omega_{fg}$ frequencies, then a quasi-rephasing is possible. The fifth field action converts the newly rephased Raman polarization into the quasi-echo at $\omega_s = 2\omega_3 - \omega_1 = \omega_3 + \omega_{hf}$.

and will undoubtedly provide very exciting new insights into condensed phase dynamics in simple molecular materials to systems of biological interest.

*(m) Some other techniques*

This survey of Raman spectroscopies, with direct or implicit use of WMEL diagrams, has by no means touched upon all of the Raman methods (tables B1.3.1 and B1.3.2). We conclude by mentioning a few additional methods without detailed discussion but with citation. The first is Raman microscopy/imaging [137]. This technique combines microscopy and imaging with Raman spectroscopy to add an extra dimension to the optical imaging process and hence to provide additional insight into sample surface morphology. (Microscopy is the subject of another chapter in this encyclopedia.) The second is Raman optical activity [138, 139]. This technique discriminates small differences in Raman scattering intensity between left and right circularly polarized light. Such a technique is useful for the study of chiral molecules. The third technique is Raman-SPM [140]. Here Raman spectroscopy is combined with scanning probe microscopy (SPM) (a subject of another chapter in this encyclopedia) to form a complementary and powerful tool for studying surfaces and interfaces. The fourth is photoacoustic Raman spectroscopy (PARS) which combines CRS with photoacoustic absorption spectroscopy. Class I Raman scattering produces excited vibrational or rotational states in a gas whose energies are converted to bulk translational heating. A pressure wave is produced that is detected as an acoustic signal [141, 142]. The fifth is a novel 5WM process ($\chi^{(4)}$) which can occur in noncentrosymmetric isotropic solutions of chiral macromolecules [143]. This technique has been given the acronym BioCARS since it has the potential to selectively record background free vibrational spectra of biological molecules [143, 144]. It could be generalized to BioCRS (to include both BioCSRS and BioCARS). The signal is quite weak and can be enhanced with electronic resonance.

Finally, we touch upon the general class of spectroscopies known as hyper-Raman spectroscopies. The spontaneous version is called simply hyper-Raman scattering, HRS, a Class I spectroscopy, but there is also the coherent Class II version called CHRS (CAHRS and CSHRS) [145–147]. These 6WM spectroscopies depend on $\chi^{(5)}$ (Im $\chi^{(5)}$ for HRS) and obey the three-photon selection rules. Their signals are always to the blue of the incident beam(s), thus avoiding fluorescence problems. The selection rules allow one to probe, with optical frequencies, the usual IR spectrum (one photon), not the conventional Raman active vibrations (two photon), but also new vibrations that are symmetry forbidden in both IR and conventional Raman methods.

Although the fifth order hyper- (H-) Raman analogues exist for most of the Raman spectroscopies at third order (HRS [148], RHRS [149, 150], CHRS [145–147], SEHRS [151–153], SERHRS [154]), let us illustrate the hyper-Raman effect with HRS as the example. In this three-photon process, the scattered radiation contains the frequencies $2\omega_1 \pm \omega_R$ (Stokes and anti-Stokes hyper-Raman scattering), at almost twice the frequency of the incident light. The WMEL diagrams are identical to those of SR, except each single action of the incident laser field ($\omega_1$) must be replaced by two simultaneous actions of the laser field. (The WMEL diagrams for any other 'hyper' process may be obtained in a similar manner.) Experimentally, this phenomenon is difficult to observe ($I_{HRS}/I_{SR} \sim 10^{-5}$); however again electronic resonance enhancement is seen to greatly increase the signal intensity [148].

### B1.3.4   Applications

To emphasize the versatility of Raman spectroscopy we discuss just a few selected applications of Raman based spectroscopy to problems in chemical physics and physical chemistry.

#### B1.3.4.1   *Applications in surface physics*

In addition to the many applications of SERS, Raman spectroscopy is, in general, a useful analytical tool having many applications in surface science. One interesting example is that of carbon surfaces which do not support SERS. Raman spectroscopy of carbon surfaces provides insight into two important aspects. First, Raman spectral features correlate with the electrochemical reactivity of carbon surfaces; this allows one to study surface oxidation [155]. Second, Raman spectroscopy can probe species at carbon surfaces which may account for the highly variable behaviour of carbon materials [155]. Another application to surfaces is the use of Raman microscopy in the nondestructive assessment of the quality of ceramic coatings [156]. Finally, an interesting type of surface which does allow for SERS are Mellfs (metal liquid-like films) [157]. Mellfs form at organic–aqueous interfaces when colloids of organic and aqueous metal sols are made. Comparisons with resonance Raman spectra of the bulk solution can give insight into the molecule–surface interaction and adsorption [157].

#### B1.3.4.2   *Applications in combustion chemistry*

Laser Raman diagnostic techniques offer remote, nonintrusive, nonperturbing measurements with high spatial and temporal resolution [158]. This is particularly advantageous in the area of combustion chemistry. Physical probes for temperature and concentration measurements can be debatable in many combustion systems, such as furnaces, internal combustors etc., since they may disturb the medium or, even worse, not withstand the hostile environments [159]. Laser Raman techniques are employed since two of the dominant molecules associated with air-fed combustion are $O_2$ and $N_2$. Homonuclear diatomic molecules unable to have a nuclear coordinate-dependent dipole moment cannot be diagnosed by infrared spectroscopy. Other combustion species include $CH_4$, $CO_2$, $H_2O$ and $H_2$ [160]. These molecules are probed by Raman spectroscopy to determine the temperature profile and species concentration in various combustion processes.

For most practical applications involving turbulent flames and combustion engines, CRS is employed. Temperatures are derived from the spectral distribution of the CRS radiation. This may either be determined by scanning the Stokes frequency through the spectral region of interest or by exciting the transition in a single laser shot with a broadband Stokes beam, thus accessing all Raman resonances in a broad spectral region (multiplexing) [161]. The spectrum may then be observed by a broadband detector such as an optical multichannel analyser [162]. This broadband approach leads to weaker signal intensities, but the entire CRS spectrum is generated with each pulse, permitting instantaneous measurements [163]. Concentration measurements can be carried out in certain ranges (0.5–30% [161]) by using the nonresonant susceptibility as an *in situ* reference standard [158]. Thus fractional concentration measurements are obtained from the spectral profile.

### B1.3.4.3  Water

Arguably the single most important molecule is water. It is widely accepted that water is a prerequisite for life. In spite of its relatively simple molecular structure, water is one of the most curious and complicated liquids known. The source of this complexity and hence of the majority of water's unusual properties is hydrogen bonding where the intermolecular force leads to a variety of water 'oligomers'. Raman spectroscopy has proven to be of great use in probing not only the *intramolecular* vibrational modes of water, but also the *intermolecular* librational and translational modes as well [164–167]. The librational and translational Raman bands range from tens to nearly one thousand wavenumbers and give insight into the microscopic structure and dynamics of water and, in particular, into hydrogen bond stretching and bending.

### B1.3.4.4  Art, archeology and biomedicine

In principle, Raman spectroscopy can be a powerful diagnostic tool in a wide variety of fields that include art, archeology and biomedicine. It is a noninvasive, nondestructive method and, with its characteristically narrow vibrational spectra, it provides sensitive information that discriminates at the molecular level. In such applications, the general challenge is to suppress fluorescence interference and also to reject the strong Rayleigh signal from materials that are often fluorescent and highly scattering. Further, miniaturization and portability of a microspectroscopic Raman probe should give it a great advantage in diagnosis and imaging.

The spread of Raman spectroscopy to probe highly fluorescent and scattering materials began about 15 years ago, when near-infrared Fourier transform Raman spectroscopy (NIR-FTRS) became established (see Fourier transform Raman scattering). While NIR-FTRS, usually using fibre optical microprobes, has so far been the principal method for diagnostic studies, the use of SERS (and FT-SERS) is also finding its place. Recall how in SERS contact of the chromophore with a metal (mainly Ag, Au or Cu) enhances the Raman cross-section by orders of magnitude while at the same time fluorescence is quenched. The use of metal colloid hydrosols to activate SERS in solutions of chromophores has greatly extended the technique.

One successful alternative to the NIR-FT Raman systems, which usually excite with the 1064 nm fundamental wavelength of the Nd:YAG, is a portable miniaturized apparatus that excites with a powerful NIR diode laser (tuned to around 785 nm to match the notch filter—for Rayleigh rejection) [168]: this system, using optical fibre probing, is based on a state-of-the-art spectrograph and is designed for *in vivo* clinical use for early detection and treatment of disease. At the excitation wavelength, it is able to penetrate pigmented tissue to a few millimetres and nonpigmented tissue to a centimetre. Called an *in vivo* Raman spectroscopic system (IVRS), it is not FT, but its performance is compared with advantage to NIR-FTRS. Clearly such apparatus can be equally useful for molecular level diagnostics in other fields such as art and archeology. For further details and entrance into the literature, the reader is directed to journal volumes dedicated to the topic [169–173].

In all applications, good Raman data banks for reference materials are needed. Then, once a spectrum is acquired, powerful computer based techniques must be used to achieve spectral comparisons of the full Raman spectrum of sampled materials with those of the reference data. One of these [171] employs neural network analysis to compare the Raman spectrum of neoplastic tissue (brain, genitourinary tract, breast and colon) to that of the normal tissue with impressive success.

The diagnostic themes in art and archeology are to identify pigments and materials that are used in a given sample, to date them (and determine changes with time) and to authenticate the origins of a given specimen. Diagnosis for forensic purposes may also be considered. In medicine, the principal aim is to provide fast (tens of seconds or faster for *in vivo* studies), noninvasive, nondestructive Raman optical methods for the earliest detection of disease, malignancies being those of immediate interest. In a recent review [173], reference is made to the large variety of subjects for Raman probing of diseased tissues and cells. These include plaques in human arteries and malignancies in breasts, lungs, brain, skin and the intestine. Hair and nails have been studied for signs of disease—metabolic or toxic. Teeth, as well as implants and prostheses, have been examined. Foreign inclusions and chemical migration from surgical implants have been studied.

Another application to biomedicine is to use Raman probing to study DNA biotargets and to identify sequence genes. In fact SERS has been applied to such problems [174].

### B1.3.4.5  *Analytical/industrial applications*

Examples that use Raman spectroscopy in the quantitative analysis of materials are enormous. Technology that takes Raman based techniques outside the basic research laboratory has made these spectroscopies also available to industry and engineering. It is not possible here to recite even a small portion of applications. Instead we simply sketch one specific example.

Undeniably, one of the most important technological achievements in the last half of this century is the microelectronics industry, the computer being one of its outstanding products. Essential to current and future advances is the quality of the semiconductor materials used to construct vital electronic components. For example, ultra-clean silicon wafers are needed. Raman spectroscopy contributes to this task as a monitor, in real time, of the composition of the standard SC-1 cleaning solution (a mixture of water, $H_2O_2$ and $NH_4OH$) [175] that is essential to preparing the ultra-clean wafers.

### B1.3.5  **A snapshot of Raman activity in 1998**

In conclusion, we attempt to provide a 'snapshot' of current research in Raman spectroscopy. Since any choice of topics must be necessarily incomplete, and certainly would reflect our own scientific bias, we choose, instead, an arbitrary approach (at least one not that is not biased by our own specialization). Thus an abbreviated summary of the topics just presented in the keynote/plenary lectures at *ICORS XVI* in Cape Town, South Africa, is presented. Each of the 22 lectures appears in the *Proceedings (and Supplement) of the 16th International Conference on Raman Spectroscopy (1998)*, edited by A M Heyns (Chichester: Wiley) in a four-page format, almost all containing a short list of references. Rather than ourselves searching for seminal citations, we instead give the e-mail address of the principal author, when available. Though the intent in this procedure is to expose the wide scope of current Raman activity, it is hoped that the reader who is looking for more details will not hesitate to seek out the author in this fashion, and that the authors will not feel put upon by this manner of directing people to their work. To relate to tables B1.3.1 and B1.3.2, the acronym is given for the principal Raman spectroscopy that is used for each entry.

*Keynote lecture.* T G Spiro, e-mail address: `spiro@princeton.edu` (RRS and TRRRS). Review of protein dynamics followed by TRRRS selective to specific structural and prosthetic elements.

*Plenary 1.* D A Long, tel.: +44-1943 608 472. Historical review of the first 70 years of Raman spectroscopy

*Plenary 2.* S A Asher *et al*, e-mail address: `asher@vms.cis.pitt.edu/asher+` (RRS, TRRRS). UV RRS is used to probe methodically the secondary structure of proteins and to follow unfolding dynamics. Developing a library based approach to generalize the method to any protein.

*Plenary 3.* Ronald E Hester *et al*, e-mail address: `reh@york.ac.uk` (SERS). Use of dioxane envelope to bring water insoluble chromophores (chlorophylls) into contact with aqueous silver colloids for SERS enhancement. PSERRS—'protected surface-enhanced resonance Raman spectroscopy'.

*Plenary 4.* George J Thomas Jr *et al*, e-mail address: `thomasgj@cctr.umkc.edu` (RS). Protein folding and assembly into superstructures. (Slow) time resolved RS probing of virus construction via protein assembly into an icosahedral (capsid) shell.

*Plenary 5.* Manuel Cardona, e-mail address: `cardona@cardix.mpi-stuttgart.de` (RS). Studies of high $T_c$ superconductors. These offer all possible Raman transitions—phonons, magnons, free carrier excitations, pair breaking excitations and mixed modes.

*Plenary 6.* Shu-Lin Zhang *et al*, e-mail address: `slzhang@pku.edu.cn` (RS). Studies of phonon modes of nanoscale one-dimensional materials. Confinement and defect induced Raman transitions.

*Plenary 7.* S Lefrante, e-mail address: `lefrant@cnrs-imn.fr` (RRS and SERS). Raman studies of electronic organic materials from conjugated polymers to carbon nanotubes. New insight into chain length distribution, charge transfer and diameter distribution in carbon nanotubes offered by Raman probing.

*Plenary 8.* J Greve *et al*, e-mail address: `J.Greve@tn.utwente.nl` (RS). Confocal direct imaging Raman microscope (CDIRM) for probing of the human eye lens. High spatial resolution of the distribution of water and cholesterol in lenses.

*Plenary 9.* J W Nibler *et al*, e-mail address: `niblerj@chem.orst.edu` (CARS and SRS). High resolution studies of high lying vibration–rotational transitions in molecules excited in electrical discharges and low density monomers and clusters in free jet expansions. Ionization detected (REMPI) SRS or IDSRS. Detect Raman lines having an FWHM of 30 MHz ($10^{-3}$ cm$^{-1}$), possibly the sharpest lines yet recorded in RS. Line broadening due to saturation and the ac Stark effect is demonstrated.

*Plenary 10.* Hiro-o Hamaguchi, e-mail address: `hhama@chem.s.u-tokyo.ac.jp` (time and polarization resolved multiplex 2D-CARS). Two-dimensional (time and frequency) CARS using broadband dye source and streak camera timing. Studies dynamic behaviour of excited (pumped) electronic states. Follows energy flow within excited molecules. Polarization control of phase of signal (NR background suppression).

*Plenary 11.* W Kiefer *et al*, e-mail address: `wolfgang.kiefer@mail.uni-wue.de` (TR CARS). Ultrafast impulsive preparation of ground state and excited state wavepackets by impulsive CARS with REMPI detection in potassium and iodine dimers.

*Plenary 12.* Soo-Y Lee, e-mail address: `scileesy@nus.edu.sg` (RRS). Addresses fundamental theoretical questions in the phase recovery problem in the inverse transform (REP to ABS). See above.

*Plenary 13.* Andreas Otto, e-mail address: `otto@rz.uni-duesseldorf.de` (SERS). A survey of problems and models that underlie the SERS effect, now two decades old. Understanding the role of surface roughness in the enhancement.

*Plenary 14.* A K Ramdas *et al*, e-mail address: `akr@physics.purdue.edu` (RS). Electronic RS studies of doped diamond as potential semiconducting materials. A Raman active 1s ($p_{3/2}$)–1s ($p_{1/2}$) transition of a hole trapped on a boron impurity both in natural and $^{13}$C diamond. A striking sensitivity of the transition energy to the isotopic composition of the host lattice.

*Plenary 15.* B Schrader *et al*, e-mail address: `bernhard.schrader@uni-essen.de` (NIR-FTRS). A review of the use of Raman spectroscopy in medical diagnostics. Its possibilities, limitations and expectations. Emphasizes the need for a library of reference spectra and the applications of advanced analysis (chemometry) for comparing patient/library spectra.

*Plenary 16.* N I Koroteev *et al*, e-mail address: `Koroteev@nik.phys.msu.su` (CARS/CSRS, CAHRS, BioCARS). A survey of the many applications of what we call the Class II spectroscopies from third order and beyond. 2D and 3D Raman imaging. Coherence as stored information, quantum information (the 'qubit'). Uses terms CARS/CSRS regardless of order. BioCARS is fourth order in optically active solutions.

*Plenary 17*. P M Champion *et al*, e-mail address: champ@neu.edu (TRRRS). Femtosecond impulsive preparation and timing of ground and excited state Raman coherences in heme proteins. Discovery of coherence transfer along a de-ligation coordinate. See above for further comment.

*Plenary 18*. Robin J H Clark, e-mail address: r.j.h.clark@ucl.ac.uk (RS). Reports on recent diagnostic probing of art works ranging from illuminated manuscripts, paintings and pottery to papyri and icons. Nondestructive NIR microscopic RS is now realistic using CCD detection. Optimistic about new developments.

*Plenary 19*. H G M Edwards, e-mail address: h.g.m.edwards@bradford.ac.uk (NIR-FTRS). A review of recent applications of RS to archeology—characterizing ancient pigments, human skin, bone, ivories, teeth, resins, waxes and gums. Aging effects and dating possibilities. Emphasizes use of microscopic Raman.

*Plenary 20*. M Grimsditch, e-mail address: marcos_grimsditch@qmgate.anl.gov (magnetic field based RS). Low frequency Raman scattering from acoustic phonons is known as Brillouin scattering (BS). However any kind of small quantum Raman scattering is likewise called BS. Ferromagnetic materials offer spin that precesses coherently in the presence of an applied field. Such spin waves or magnons can undergo quantum jumps by the inelastic scattering of light. Experiments (and energy level spacing theory) involving surface magnons in very thin multilayer slabs (such as Fe/Cr/Fe) are discussed. The energy spacing (the Brillouin spectrum) depends on the applied magnetic field, and the RS (or BS) theory is driven by the magnetic component of the electromagnetic field, not the electric (as discussed exclusively in the present chapter).

*Plenary 21A*. Alian Wang *et al*, e-mail address: alianw@levee.wustl.edu (RS). (Unable to attend ICORS, but abstract is available in proceedings.) With technological advances, Raman spectroscopy now has become a field tool for geologists. Mineral characterization for terrestrial field work is feasible and a Raman instrument is being designed for the next rover to Mars, scheduled for 2003.

*Plenary 21B*. A C Albrecht *et al*, e-mail address: aca7@cornell.edu ($I^{(2)}$CRS) (substituting for plenary 21A). Discusses four new applications using a 'third' approach to the Class II spectroscopies (see above). Raman spectrograms from $I^{(2)}$CARS and $I^{(3)}$CARS are seen to (i) decisively discriminate between proposed mechanisms for dephasing of the ring breathing mode in liquid benzene, (ii) detect the presence of memory in the Brownian oscillator model of dephasing, (iii) determine with very high accuracy Raman frequency shifts and bandwidths with changing composition in binary mixtures. Moreover these are successfully related to the partial pressures as they change with composition and (iv) to provide a new, definitive, way to discriminate between two competing processes at fifth order (6WM)—cascaded third order or true fifth order.

Clearly the broad survey of current activity in Raman spectroscopy revealed by this simple snapshot promises an exciting future that is likely to find surprising new applications, even as present methods and applications become refined.

## Appendix

Here we examine the viewing angle dependence of the differential Raman cross-section for the cases of linearly polarized and circularly polarized incident light[5]. The angles used in such experiments are shown in figure B1.3.A.9. Experiments involving circularly polarized light are entirely defined in terms of the scattering angle, $\zeta$, the angle between the wavevectors of the incident ($\mathbf{k}_i$) and scattered ($\mathbf{k}_s$) light. For experiment involving linearly polarized light, two angles are needed: $\xi$, the angle between $\mathbf{k}_s$ and the unit vector along the direction of polarization of the incident light ($\mathbf{e}_i$) and $\eta$, the polar angle of $\mathbf{k}_s$ in the plane

---

[5]  This treatment is essentially that given in [176].

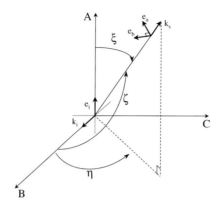

**Figure B1.3.A.9.** Diagram depicting the angles used in scattering experiments employing linearly and circularly polarized light. The subscripts i and s refer to the incident and scattered beam respectively.

perpendicular to $\mathbf{e}_i$. The polarization of the scattered light is analysed along the two axes $\mathbf{e}_a$ and $\mathbf{e}_b$, where $\mathbf{e}_b$ is chosen to be perpendicular to $\mathbf{e}_i$. The differential Raman cross-sections for the two analysing directions are [176]

$$\left(\frac{d\sigma}{d\Omega}\right)_a = 4\pi^2\left(\frac{e^2}{4\pi\varepsilon_0\hbar c}\right)^2 \bar{v}_1(\bar{v}_1 \pm \bar{v}_R)^3 \left(\begin{matrix} \frac{\cos^2\xi(5\Sigma^1+3\Sigma^2)}{30} \\ +\frac{\sin^2\xi(10\Sigma^0+4\Sigma^2)}{30} \end{matrix}\right) \qquad (B1.3.A1)$$

and

$$\left(\frac{d\sigma}{d\Omega}\right)_b = 4\pi^2\left(\frac{e^2}{4\pi\varepsilon_0\hbar c}\right)^2 \bar{v}_1(\bar{v}_1 \pm \bar{v}_R)^3 \left(\frac{5\Sigma^1 + 3\Sigma^2}{30}\right). \qquad (B1.3.A2)$$

The total differential cross-section (equation (B1.3.A1) + equation (B1.3.A2)) is then

$$\left(\frac{d\sigma}{d\Omega}\right)_{\xi,\eta} = \left(\frac{d\sigma}{d\Omega}\right)\left(1 - \frac{1-\rho_l}{1+\rho_l}\cos^2\xi\right) \qquad (B1.3.A3)$$

and the depolarization ratio

$$\rho(\xi,\eta) = \frac{\rho_l}{1-(1-\rho_l)\cos^2\xi} \qquad (B1.3.A4)$$

where $\rho_l$ is given by equation (B1.3.23) and $(d\sigma/d\Omega)$ is the total differential cross section at $90°$, $(d\sigma/d\Omega)_\parallel + (d\sigma/d\Omega)_\perp$, (equations (B1.3.20) and (B1.3.21)). From expressions (B1.3.A3) and (B1.3.A4), one can see that no new information is gained from a linearly polarized light scattering experiment performed at more than one angle, since the two measurables at an angle $(\xi,\eta)$ are given in terms of the corresponding quantities at $90°$.

For a circularly polarized light experiment, one can measure the cross sections for either right (r) or left (l) polarized scattered light. Suppose that right polarized light is made incident on a Raman active sample. The general expressions for the Raman cross sections are [176]

$$\left(\frac{d\sigma}{d\Omega}\right)_r = 4\pi^2\left(\frac{e^2}{4\pi\varepsilon_0\hbar c}\right)^2 \bar{v}_1(\bar{v}_1 \pm \bar{v}_R)^3 \left(\begin{matrix} \frac{1}{12}(1+\cos\zeta)^2\Sigma^0 \\ +\frac{1}{24}[(1+\cos\zeta)^2 + 2\sin^2\zeta]\Sigma^1 \\ +\frac{1}{120}[(1+\cos\zeta)^2 + 12(1-\cos\zeta)]\Sigma^2 \end{matrix}\right)$$

and

$$\left(\frac{d\sigma}{d\Omega}\right)_l = 4\pi^2\left(\frac{e^2}{4\pi\varepsilon_0\hbar c}\right)^2 \bar{v}_1(\bar{v}_1 \pm \bar{v}_R)^3 \left(\begin{matrix} \frac{1}{12}(-1+\cos\zeta)^2\Sigma^0 \\ +\frac{1}{24}[(-1+\cos\zeta)^2 + 2\sin^2\zeta]\Sigma^1 \\ +\frac{1}{120}[(-1+\cos\zeta)^2 + 12(1+\cos\zeta)]\Sigma^2 \end{matrix}\right).$$

In analogy with the depolarization ratio for linearly polarized light, the ratio of the two above quantities is known as the reversal coefficient, $R(\zeta)$, given by

$$R(\zeta) \equiv \frac{\left(\frac{d\sigma}{d\Omega}\right)_l}{\left(\frac{d\sigma}{d\Omega}\right)_r} = \frac{1 - \frac{1-\rho_l}{2(1+\rho_l)} \sin^2 \zeta - \frac{1-R(0)}{1+R(0)} \cos \zeta}{1 - \frac{1-\rho_l}{2(1+\rho_l)} \sin^2 \zeta + \frac{1-R(0)}{1+R(0)} \cos \zeta} \qquad (B1.3.A5)$$

where the zero angle reversal coefficient, $R(0)$, is $6\Sigma^2/(10\Sigma^0 + 5\Sigma^1 + \Sigma^2)$. Measurement of the reversal coefficient (equation (B1.3.A5)) at two appropriate scattering angles permits one to determine both $\rho_l$ and $R(0)$. Thus, only with circularly polarized light is one able to quantify all three rotational tensor invariants [176].

## Acknowledgments

One of us, ACA, also wishes to thank Dr David Klug for his provision of space, time, and a fine scientific atmosphere during a two month sabbatical leave at Imperial College, which turned out to involve considerable work on this material. The most helpful editing efforts of Professor R J H Clark have been very much appreciated. Finally, gratitude goes to Dr Gia Maisuradze for a careful reading of the manuscript and to Ms Kelly Case for her most attentive and effective reading of the proofs.

## References

[1] Raman C V and Krishnan K S 1928 A new type of secondary radiation *Nature* **121** 501–2
[2] Raman C V 1928 A new radiation *Indian J. Phys.* **2** 387–98
[3] Raman C V and Krishnan K S 1928 A new class of spectra due to secondary radiation. Part I *Indian J. Phys.* **2** 399–419
[4] Landsberg G and Mandelstam L 1928 Eine neue Erscheinung bei der Lichtzerstreuung in Krystallen *Naturwissenschatten* **16** 557–8
[5] Long D A 1988 Early history of the Raman effect *Int. Rev. Phys. Chem.* **7** 314–49
[6] Lee D and Albrecht A C 1985 A unified view of Raman, resonance Raman, and fluorescence spectroscopy (and their analogues in two-photon absorption) *Advances in Infrared and Raman Spectroscopy* vol 12, ed R J H Clark and R E Hester (New York: Wiley) pp 179–213
[7] Lee D and Albrecht A C 1993 On global energy conservation in nonlinear light matter interaction: the nonlinear spectroscopies, active and passive *Adv. Phys. Chem.* **83** 43–87
[8] Fano U 1957 Description of states in quantum mechanics by density matrix and operator techniques *Rev. Mod. Phys.* **29** 74–93
[9] ter Haar D 1961 Theory and applications of the density matrix *Rep. Prog. Phys.* **24** 304–62
[10] Cohen-Tannoudji C, Diu B and Laloë F 1977 *Quantum Mechanics* (New York: Wiley)
[11] Sakurai J J 1994 *Modern Quantum Mechanics* revised edn (Reading, MA: Addison-Wesley)
[12] Louisell W H 1973 *Quantum Statistical Properties of Radiation* (New York: Wiley)
[13] Shen Y R 1984 *The Principles of Nonlinear Optics* (New York: Wiley)
[14] Butcher P N and Cotter D 1990 *The Elements of Nonlinear Optics* (Cambridge: Cambridge University Press)
[15] Boyd R W 1992 *Nonlinear Optics* (San Diego, CA: Academic)
[16] Schubert M and Wilhiemi B 1986 *Nonlinear Optics and Quantum Electronics* (New York: Wiley)
[17] Yariv A 1989 *Quantum Electronics* (New York: Wiley)
[18] Mukamel S 1995 *Principles of Nonlinear Optical Spectroscopy* (New York: Oxford University Press)
[19] Levenson M D 1982 *Introduction to Nonlinear Laser Spectroscopy* (New York: Academic)
[20] Loudon R 1983 *The Quantum Theory of Light* (New York: Oxford University Press)
[21] Wright J C, Labuda M J, Zilian A, Chen P C and Hamilton J P 1997 New selective nonlinear vibrational spectroscopies *J. Luminesc.* **72–74** 799–801
[22] Labuda M J and Wright J C 1997 Measurement of vibrationally resonant $\chi^{(3)}$ and the feasibility of new vibrational spectroscopies *Phys. Rev. Lett.* **79** 2446–9
[23] Wright J C, Chen P C, Hamilton J P, Zilian A and Labuda M J 1997 Theoretical foundations for a new family of infrared four wave mixing spectroscopies *Appl. Spectrosc.* **51** 949–58
[24] Chen P C, Hamilton J P, Zilian A, Labuda M J and Wright J C 1998 Experimental studies for a new family of infrared four wave mixing spectroscopies *Appl. Spectrosc.* **52** 380–92
[25] Tokmakoff A, Lang M J, Larson D S and Fleming G R 1997 Intrinsic optical heterodyne detection of a two-dimensional fifth order Raman response *Chem. Phys. Lett.* **272** 48–54

[26] Monson P R and McClain W 1970 Polarization dependence of the two-photon absorption of tumbling molecules with application to liquid 1-chloronapthalene and benzene *J. Chem. Phys.* **53** 29–37

[27] Lee S Y 1983 Placzek-type polarizability tensors for Raman and resonance Raman scattering *J. Chem. Phys.* **78** 723–34

[28] Melinger J S and Albrecht A C 1986 Theory of time- and frequency-resolved resonance secondary radiation from a three-level system *J. Chem. Phys.* **84** 1247–58

[29] Kramers H A and Heisenberg W 1925 Über die Streuung von Strahlung durch Atome *Z. Phys.* **31** 681–708

[30] Placzek G 1934 The Rayleigh and Raman scattering *Handbuch der Radiologie* ed E Marx (Leipzig: Akademische) (Engl. transl. UCRL 526(L) 1962 Clearinghouse USAEC transl. A Werbin)

[31] Tang J and Albrecht A C 1970 Developments in the theories of vibrational Raman intensities *Raman Spectroscopy: Theory and Practice* vol 2, ed H A Szymanski (New York: Plenum) pp 33–68

[32] Toleutaev B N, Tahara T and Hamaguchi H 1994 Broadband ($1000$ cm$^{-1}$) multiplex CARS spectroscopy: application to polarization sensitive and time-resolved measurements *Appl. Phys.* **59** 369–75

[33] Laubereau A 1982 Stimulated Raman scattering *Non-Linear Raman Spectroscopy and its Chemical Applications* ed W Kiefer and D A Long (Dordrecht: Reidel)

[34] Woodward L A 1967 General introduction *Raman Spectroscopy: Theory and Practice* vol 1, ed H A Szymanski (New York: Plenum)

[35] Long D A (ed) *J. Raman Spectrosc.* (Chichester: Wiley)

[36] Clark R J H and Hester R E (eds) *Adv. Infrared Raman Spectros.* (London: Heyden)

[37] *ICORS '96: XVth Int. Conf. on Raman Spectroscopy* (New York: Wiley)

[38] Pinan J P, Ouillon R, Ranson P, Becucci M and Califano S 1998 High resolution Raman study of phonon and vibron bandwidths in isotropically pure and natural benzene crystal *J. Chem. Phys.* **109** 1–12

[39] Ferraro J R and Nakamoto K 1994 *Introductory Raman Spectroscopy* (San Diego: Academic)

[40] Hirschfeld T and Chase B 1986 FT-Raman spectroscopy: development and justification *Appl. Spectrosc.* **40** 133–9

[41] McCreery R L 1996 Instrumentation for dispersive Raman spectroscopy *Modern Techniques in Raman Spectroscopy* ed J J Laserna (New York: Wiley)

[42] Hendra P J 1996 Fourier transform Raman spectroscopy *Modern Techniques in Raman Spectroscopy* ed J J Laserna (New York: Wiley)

[43] Hendra P J, Jones C and Warnes G 1991 *Fourier Transform Raman Spectroscopy: Instrumentation and Chemical Applications* (New York: Ellis Horwood)

[44] Heller E J 1981 The semiclassical way to molecular spectroscopy *Accounts Chem. Res.* **14** 368–75

[45] Myers A B 1997 'Time dependent' resonance Raman theory *J. Raman Spectrosc.* **28** 389–401

[46] Hizhnyakov V and Tehver I 1988 Transform method in resonance Raman scattering with quadratic Franck–Condon and Herzberg–Teller interactions *J. Raman Spectrosc.* **19** 383–8

[47] Page J B and Tonks D L 1981 On the separation of resonance Raman scattering into orders in the time correlator theory *J. Chem. Phys.* **75** 5694–708

[48] Champion P M and Albrecht A C 1981 On the modeling of absorption band shapes and resonance Raman excitation profiles *Chem. Phys. Lett.* **82** 410–13

[49] Cable J R and Albrecht A C 1986 The inverse transform in resonance Raman scattering *Conf. sponsored by the University of Oregon* ed W L Peticolas and B Hudson

[50] Remacle F and Levine R D 1993 Time domain information from resonant Raman excitation profiles: a direct inversion by maximum entropy *J. Chem. Phys.* **99** 4908–25

[51] Albrecht A C, Clark R J H, Oprescu D, Owens S J R and Svensen C 1994 Overtone resonance Raman scattering beyond the Condon approximation: transform theory and vibronic properties *J. Chem. Phys.* **101** 1890–903

[52] Weber A (ed) 1979 *Raman Spectroscopy of Gases and Liquids* (New York: Springer)

[53] Page J B 1991 Many-body problem to the theory of resonance Raman scattering by vibronic systems *Top. Appl. Phys.* **116** 17–72

[54] Cable J R and Albrecht A C 1986 A direct inverse transform for resonance Raman scattering *J. Chem. Phys.* **84** 4745–54

[55] Joo T and Albrecht A C 1993 Inverse transform in resonance Raman scattering: an iterative approach *J. Phys. Chem.* **97** 1262–4

[56] Lee S-Y 1998 Forward and inverse transforms between the absorption lineshape and Raman excitation profiles *XVIth Int. Conf. on Raman Spectroscopy* ed A M Heyns (New York: Wiley) pp 48–51

[57] Lee S-Y and Feng Z W 1996 Reply to the comment on the inversion of Raman excitation profiles *Chem. Phys. Lett.* **260** 511–13

[58] Marzocchi M P, Mantini A R, Casu M and Smulevich G 1997 Intramolecular hydrogen bonding and excited state proton transfer in hydroxyanthraquinones as studied by electronic spectra, resonance Raman scattering, and transform analysis *J. Chem. Phys.* **108** 1–16

[59] Mantini A R, Marzocchi M P and Smulevich G 1989 Raman excitation profiles and second-derivative absorption spectra of beta-carotene *J. Chem. Phys.* **91** 85–91

[60] Asher S A, Chi Z, Holtz J S W, Lednev I K, Karnoup A S and Sparrow M C 1998 UV resonance Raman studies of protein structure and dynamics *XVIth Int. Conf. on Raman Spectroscopy* ed A M Heyns (New York: Wiley) pp 11–14

[61] Zhu I, Widom A and Champion P M 1997 A multidimensional Landau–Zener description of chemical reaction dynamics and vibrational coherence *J. Chem. Phys.* **107** 2859–71

[62] Champion P M, Rosca F, Chang W, Kumar A, Christian J and Demidov A 1998 Femtosecond coherence spectroscopy of heme proteins *XVIth Int. Conf. on Raman Spectroscopy* ed A M Heyns (New York: Wiley) pp 73–6

[63] Johnson B R, Kittrell C, Kelly P B and Kinsey J L 1996 Resonance Raman spectroscopy of dissociative polyatomic molecules *J. Phys. Chem.* **100** 7743–64

[64] Kung C Y, Chang B-Y, Kittrell C, Johnson B R and Kinsey J L 1993 Continuously scanned resonant Raman excitation profiles for iodobenzene excited in the B continuum *J. Phys. Chem.* **97** 2228–35

[65] Galica G E, Johnson B R, Kinsey J L and Hale M O 1991 Incident frequency dependence and polarization properties of the $CH_3I$ Raman spectrum *J. Phys. Chem.* **95** 7994–8004

[66] Tripathi G N R and Schuler R H 1984 The resonance Raman spectrum of phenoxyl radical *J. Chem. Phys.* **81** 113–21

[67] Tripathi G N R and Schuler R H 1982 Time-resolved resonance Raman scattering of transient radicals: the p-aminophenoxyl radical *J. Chem. Phys.* **76** 4289–90

[68] Qin L, Tripathi G N R and Schuler R H 1987 Radiolytic oxidation of 1,2,4-benzenetriol: an application of time-resolved resonance Raman spectroscopy to kinetic studies of reaction intermediates *J. Chem. Phys.* **91** 1905–10

[69] Okamoto H, Nakabayashi T and Tasumi M 1997 Analysis of anti-Stokes RRS excitation profiles as a method for studying vibrationally excited molecules *J. Phys. Chem.* **101** 3488–93

[70] Sakamoto A, Okamoto H and Tasumi M 1998 Observation of picosecond transient Raman spectra by asynchronous Fourier transform Raman spectroscopy 1998 *Appl. Spectrosc.* **52** 76–81

[71] Bloembergen N 1967 The stimulated Raman effect *Am. J. Phys.* **35** 989–1023

[72] Wang C-S 1975 The stimulated Raman process *Quantum Electronics* vol 1A, ed H Rabin and C L Tang (New York: Academic) pp 447–72

[73] Kaiser W and Maier M 1972 Stimulated Rayleigh, Brillouin and Raman spectroscopy *Laser Handbook* vol 2, ed F T Arecchi and E O Schult-Dubois (Amsterdam: North-Holland) pp 1077–150

[74] Wang C-S 1969 Theory of stimulated Raman scattering *Phys. Rev.* **182** 482–94

[75] Maier M, Kaiser W and Giordmaine J A 1969 Backward stimulated Raman scattering *Phys. Rev.* **177** 580–99

[76] Jones W T and Stoicheff B P 1964 Inverse Raman spectra: induced absorption at optical frequencies *Phys. Rev. Lett.* **13** 657–9

[77] Fleischmann M, Hendra P J and McQuillan A J 1974 Raman spectra of pryridine adsorbed at a silver electrode *Chem. Phys. Lett.* **26** 163–6

[78] Albrecht M G and Creighton J A 1977 Anomalously intense Raman spectra of pyridine at a silver electrode *J. Am. Chem. Soc.* **99** 5215–17

[79] Jeanmaire D L and Van Duyne R P 1977 Part I: heterocyclic, aromatic and aliphatic amines adsorbed on the anodized silver electrode *J. Electroanal. Chem.* **84** 1–20

[80] See articles in the special issue on Raman spectroscopy and surface phenomena 1991 *J. Raman Spectrosc.* **22** 727–839

[81] See articles in the special issue on Raman spectroscopy in surface science 1985 *Surf. Sci.* **158** 1–693

[82] Freunscht P, Van Duyne R P and Schneider S 1997 Surface-enhanced Raman spectroscopy of trans-stilbene adsorbed on platinum- or self-assembled monolayer-modified silver film over nanosphere surfaces *Chem. Phys. Lett.* **281** 372–8

[83] Yang W H, Hulteen J C, Schatz G C and Van Duyne R P 1996 A surface-enhanced hyper-Raman and surface-enhanced Raman scattering study of trans-1,2-bis(4-pyridyl)ethylene adsorbed onto silver film over nanosphere electrodes. Vibrational assignments: experiments and theory *J. Chem. Phys.* **104** 4313–26

[84] Nie S and Emory S R 1997 Probing single molecules and single nanoparticles by surface-enhanced Raman scattering *Science* **275** 1102–6

[85] Nie S and Emory S R 1997 Near-field surface-enhanced Raman spectroscopy on single silver nanoparticles *Anal. Chem.* **69** 2631–5

[86] Furtak T E and Reyes J 1980 A critical analysis of the theoretical models for the giant Raman effect from adsorbed molecules *Surf. Sci.* **93** 351–82

[87] Rupérez A and Laserna J J 1996 Surface-enhanced Raman spectroscopy *Modern Techniques in Raman Spectroscopy* ed J J Laserna (New York: Wiley) pp 227–64

[88] Otto A 1991 Surface-enhanced Raman scattering of adsorbates *J. Raman Spectrosc.* **22** 743–52

[89] Kneipp K, Wang Y, Kneipp H, Itzkan I, Dasari R R and Feld M S 1996 Approach to single molecule detection using surface-enhanced Raman scattering *ICORS '98: XVth Int. Conf. on Raman Spectroscopy* ed S A Asher and P B Stein (New York: Wiley) pp 636–7

[90] Eesley G L 1981 *Coherent Raman Spectroscopy* (Oxford: Pergamon)

[91] Marowsky G and Smirnov V N (eds) 1992 *Coherent Raman Spectroscopy: Recent Advances* (Berlin: Springer)

[92] Gomez J S 1996 Coherent Raman spectroscopy *Modern Techniques in Raman Spectroscopy* ed J J Laserna (New York: Wiley) pp 305–42

[93] Castellucci E M, Righini R and Foggi P (eds) 1993 *Coherent Raman Spectroscopy* (Singapore: World Scientific)

[94] Oudar J-L, Smith R W and Shen Y R 1979 Polarization-sensitive coherent anti-Stokes Raman spectroscopy *Appl. Phys. Lett.* **34** 758–60

[95] Schaertel S A, Lee D and Albrecht A C 1995 Study of polarization CRS and polarization $I^{(2)}$CRS with applications *J. Raman Spectrosc.* **26** 889–99

[96] Eesley G L 1978 Coherent Raman spectroscopy *J. Quant. Spectrosc. Radiat. Transfer* **22** 507–76

[97]  Heiman D, Hellwarth R W, Levenson M D and Martin G 1976 Raman-induced Kerr effect *Phys. Rev. Lett.* **36** 189–92
[98]  Schaertel S A, Albrecht A C, Lau A and Kummrow A 1994 Interferometric coherent Raman spectroscopy with incoherent light: some applications *Appl. Phys.* B **59** 377–87
[99]  Schaertel S A and Albrecht A C 1994 Interferometric coherent Raman spectroscopy: resonant and non-resonant contributions *J. Raman Spectrosc.* **25** 545–55
[100]  Morita N and Yajima T 1984 Ultrafast-time-resolved coherent transient spectroscopy with incoherent light *Phys. Rev.* A **30** 2525–36
[101]  Asaka S, Nakatsuka H, Fujiwara M and Matsuoka M 1984 Accumulated photon echoes with incoherent light in $Nd^{3+}$-doped silicate glass *Phys. Rev.* A **29** 2286–9
[102]  Beech R and Hartmann S R 1984 Incoherent photon echoes *Phys. Rev. Lett.* **53** 663–6
[103]  Kobayashi T 1994 Measurement of femtosecond dynamics of nonlinear optical responses *Modern Nonlinear Optics* part 3, ed M Evans and S Kielich *Adv. Chem. Phys.* **85** 55–104
[104]  Kummrow A and Lau A 1996 Dynamics in condensed molecular systems studied by incoherent light *Appl. Phys.* B **63** 209–23
[105]  Albrecht A C, Smith S P, Tan D, Schaertel S A and DeMott D 1995 Ultrasharp spectra and ultrafast timing from noisy coherence in four wave mixing *Laser Phys.* **5** 667–75
[106]  Dugan M A, Melinger J S and Albrecht A C 1988 Terahertz oscillations from molecular liquids in CSRS/CARS spectroscopy with incoherent light *Chem. Phys. Lett.* **147** 411–19
[107]  Dugan M A and Albrecht A C 1991 Radiation–matter oscillations and spectral line narrowing in field-correlated four-wave mixing I: theory *Phys. Rev.* A **43** 3877–921
[108]  Ulness D J, Stimson M J, Kirkwood J C and Albrecht A C 1997 Interferometric downconversion of high frequency molecular vibrations with time–frequency-resolved coherent Raman scattering using quasi-cw noisy laser light: C–H stretching modes of chloroform and benzene *J. Phys. Chem.* A **101** 4587–91
[109]  Stimson M J, Ulness D J and Albrecht A C 1996 Frequency and time resolved coherent Raman scattering in $CS_2$ using incoherent light *Chem. Phys. Lett.* **263** 185–90
[110]  Ulness D J and Albrecht A C 1996 Four-wave mixing in a Bloch two-level system with incoherent laser light having a Lorentzian spectral density: analytic solution and a diagrammatic approach *Phys. Rev.* A **53** 1081–95
[111]  Ulness D J and Albrecht A C 1997 A theory of time resolved coherent Raman scattering with spectrally tailored noisy light *J. Raman Spectrosc.* **28** 571–8
[112]  Ulness D J, Kirkwood J C, Stimson M J and Albrecht A C 1997 Theory of coherent Raman scattering with quasi-cw noisy light for a general lineshape function *J. Chem. Phys.* **107** 7127–37
[113]  Laubereau A and Kaiser W 1978 Vibrational dynamics of liquids and solids investigated by picosecond light pulses *Rev. Mod. Phys.* **50** 607–65
[114]  Laubereau A and Kaiser W 1978 Coherent picosecond interactions *Coherent Nonlinear Optics* ed M S Feld and V S Letokov (Berlin: Springer) pp 271–92
[115]  Leonhardt R, Holzapfel W, Zinth W and Kaiser W 1987 Terahertz quantum beats in molecular liquids *Chem. Phys. Lett.* **133** 373–7
[116]  Okamoto H and Yoshihara K 1990 Femtosecond time-resolved coherent Raman scattering under various polarization and resonance conditions *J. Opt. Soc.* B **7** 1702–8
[117]  Joo T, Dugan M A and Albrecht A C 1991 Time resolved coherent Stokes Raman spectroscopy (CSRS) of benzene *Chem. Phys. Lett.* **177** 4–10
[118]  Joo T and Albrecht A C 1993 Femtosecond time-resolved coherent anti-Stokes Raman spectroscopy of liquid benzene: a Kubo relaxation function analysis *J. Chem. Phys.* **99** 3244–51
[119]  Okamoto H and Yoshihara K 1991 Femtosecond time-resolved coherent Raman scattering from $\beta$-carotene in solution. Ultrahigh frequency (11 THz) beating phenomenon and sub-picosecond vibrational relaxation *Chem. Phys. Lett.* **177** 568–71
[120]  Yan Y X, Gamble E B and Nelson K A 1985 Impulsive stimulated Raman scattering: general importance in femtosecond laser pulse interactions with matter, and spectroscopic applications *J. Chem. Phys.* **83** 5391–9
[121]  Ruhman S, Joly A G and Nelson K A 1987 Time-resolved observations of coherent molecular vibrational motion and the general occurrence of impulsive stimulated scattering *J. Chem. Phys.* **86** 6563–5
[122]  Walsh A M and Loring R F 1989 Theory of resonant and nonresonant impulsive stimulated Raman scattering *Chem. Phys. Lett.* **160** 299–304
[123]  Constantine S, Zhou Y, Morais J and Ziegler L D 1997 Dispersed optical heterodyne birefringence and dichroism of transparent liquids *J. Phys. Chem.* A **101** 5456–62
[124]  Walmsley I A, Wise F W and Tang C L 1989 On the difference between quantum beats in impulsive stimulated Raman scattering and resonance Raman scattering *Chem. Phys. Lett.* **154** 315–20
[125]  Loring R F and Mukamel S 1985 Selectivity in coherent transient Raman measurements of vibrational dephasing in liquids *J. Chem. Phys.* **83** 2116–28
[126]  Vanden Bout D and Berg M 1995 Ultrafast Raman echo experiments in liquids *J. Raman Spectrosc.* **26** 503–11
[127]  Muller L J, Vanden Bout D and Berg M 1993 Broadening of vibrational lines by attractive forces: ultrafast Raman echo experiments in a $CH_3I:CDCl_3$ mixture *J. Chem. Phys.* **99** 810–19

[128] Vanden Bout D, Fretas J E and Berg M 1994 Rapid, homogeneous vibrational dephasing in ethanol at low temperatures determined by Raman echo measurements *Chem. Phys. Lett.* **229** 87–92

[129] Muller L J, Vanden Bout D and Berg M 1991 Ultrafast Raman echoes in liquid acetonitrile *Phys. Rev. Lett.* **67** 3700–3

[130] Tokmakoff A and Fleming G R 1997 Two-dimensional Raman spectroscopy of the intermolecular modes of liquid $CS_2$ *J. Chem. Phys.* **106** 2569–82

[131] Tokmakoff A, Lang M J, Larson D S, Fleming G R, Chernyak V and Mukamel S 1997 Two-dimensional Raman spectroscopy of vibrational interactions in liquids *Phys. Rev. Lett.* **79** 2702–5

[132] Steffen T and Duppen K 1996 Time resolved four- and six-wave mixing in liquids I. Theory *J. Chem. Phys.* **105** 7364–82

[133] Steffen T and Duppen K 1996 Time resolved four- and six-wave mixing in liquids II. Experiment *J. Chem. Phys.* **106** 3854–64

[134] Khidekel V and Mukamel S 1995 High-order echoes in vibrational spectroscopy of liquids *Chem. Phys. Lett.* **240** 304–14

[135] Ivanecky J E III and Wright J C 1993 An investigation of the origins and efficiencies of higher order nonlinear spectroscopic processes *Chem. Phys. Lett.* **206** 437–44

[136] Ernst R R, Bodenhausen G and Wokaun A 1987 *Principles of Nuclear Magnetic Resonance in One and Two Dimensions* (Oxford: Clarendon)

[137] Turrell G and Dhamelincourt P 1996 Micro-Raman spectroscopy *Modern Techniques in Raman Spectroscopy* ed J J Laserna (New York: Wiley) pp 109–42

[138] Barron L D, Hecht L, Bell A F and Wilson G 1996 Raman optical activity: an incisive probe of chirality and biomolecular structure and dynamics *ICORS '96: XVth Int. Conf. on Raman Spectroscopy* ed S A Asher and P B Stein (New York: Wiley) pp 1212–15

[139] Hecht L and Barron L D 1996 Raman optical activity *Modern Techniques in Raman Spectroscopy* ed J J Laserna (New York: Wiley) pp 265–342

[140] Ren B, Li W H, Mao B W, Gao J S and Tian Z Q 1996 Optical fiber Raman spectroscopy combined with scattering tunneling microscopy for simultaneous measurements *ICORS '96: XVth Int. Conf. on Raman Spectroscopy* ed S A Asher and P B Stein (New York: Wiley) pp 1220–1

[141] Barrett J J and Berry M J 1979 Photoacoustic Raman spectroscopy (PARS) using cw laser sources *Appl. Phys. Lett.* **34** 144–6

[142] Siebert D R, West G A and Barrett J J 1980 Gaseous trace analysis using pulsed photoacoustic Raman spectroscopy *Appl. Opt.* **19** 53–60

[143] Koroteev N I 1995 BioCARS—a novel nonlinear optical technique to study vibrational spectra of chiral biological molecules in solution *Biospectroscopy* **1** 341–50

[144] Koroteev N I 1996 Optical rectification, circular photogalvanic effect and five-wave mixing in optically active solutions *Proc. SPIE* **2796** 227–38

[145] Akhmanov S A and Koroteev N I 1981 *Methods of Nonlinear Optics in Light Scattering Spectroscopy* (Moscow: Nauka) (in Russian)

[146] Cho M 1997 Off-resonant coherent hyper-Raman scattering spectroscopy *J. Chem. Phys.* **106** 7550–7

[147] Yang M, Kim J, Jung Y and Cho M 1998 Six-wave mixing spectroscopy: resonant coherent hyper-Raman scattering *J. Chem. Phys.* **108** 4013–20

[148] Ziegler L D 1990 Hyper-Raman spectroscopy *J. Raman Spectrosc.* **71** 769–79

[149] Ziegler L D and Roebber J L 1987 Resonance hyper-Raman scattering of ammonia *Chem. Phys. Lett.* **136** 377–82

[150] Chung Y C and Ziegler L D 1988 The vibronic theory of resonance hyper-Raman scattering *J. Chem. Phys.* **88** 7287–94

[151] Golab J T, Sprague J R, Carron K T, Schatz G C and Van Duyne R P 1988 A surface enhanced hyper-Raman scattering study of pyridine adsorbed onto silver: experiment and theory *J. Chem. Phys.* **88** 7942–51

[152] Kneipp K, Kneipp H and Seifert F 1994 Near-infrared excitation profile study of surface-enhanced hyper-Raman scattering and surface-enhanced Raman scattering by means of tunable mode-locked Ti:sapphire laser excitation *Chem. Phys. Lett.* **233** 519–24

[153] Yu N-T, Nie S and Lipscomb L 1990 Surface-enhanced hyper-Raman spectrosocpy with a picosecond laser. New vibrational information for non-centrosymmetric carbocyanine molecules adsorbed on colloidal silver *J. Raman Spectrosc.* **21** 797–802

[154] Baranov A V, Bobovich Y S and Petrov V I 1993 Surface-enhanced resonance hyper-Raman (SERHR) spectroscopy of photochromic molecules *J. Raman Spectrosc.* **24** 695–7

[155] McCreery R L, Liu Y-C, Kagen M, Chen P and Fryling M 1996 Resonance and normal Raman spectroscopy of carbon surfaces: relationships of surface structure and reactivity *ICORS '96: XVth Int. Conf. on Raman Spectroscopy* ed S A Asher and P B Stein (New York: Wiley) pp 566–7

[156] Evans R, Smith I, Münz W D, Williams K J P and Yarwood J 1996 Raman microscopic studies of ceramic coatings based on titanium aluminum nitride *ICORS '96: XVth Int. Conf. on Raman Spectroscopy* ed S A Asher and P B Stein (New York: Wiley) pp 596–7

[157] Al-Obaidi A H R, Rigby S J, Hegarty J N M, Bell S E J and McGarvey J J 1996 Direct formation of silver and gold metal liquid-like films (MELLFS) from thiols and sols without organic solvents: SERS and AFM studies *ICORS '96: XVth Int. Conf. on Raman Spectroscopy* ed S A Asher and P B Stein (New York: Wiley) pp 590–1

[158] Hall R J and Boedeker L R 1984 CARS thermometry in fuel-rich combustion zones *Appl. Opt.* **23** 1340–6

[159] Stenhouse I A, Williams D R, Cole J B and Swords M D 1979 CARS measurements in an internal combustion engine *Appl. Opt.* **18** 3819–25

[160] Bechtel J H and Chraplyvy A R 1982 Laser diagnostics of flame, combustion products and sprays *Proc. IEEE* **70** 658–77
[161] Eckbreth A C, Dobbs G M, Stufflebeam J H and Tellex P A 1984 CARS temperature and species measurements in augmented jet engine exhausts *Appl. Opt.* **23** 1328–39
[162] Gross L P, Trump D D, MacDonald B G and Switzer G L 1983 10-Hz coherent anti-Stokes Raman spectroscopy apparatus for turbulent combustion studies *Rev. Sci. Instrum.* **54** 563–71
[163] Eckbreth A C 1988 Nonlinear Raman spectroscopy for combustion diagnostics *J. Quant. Spectrosc. Radiat. Transfer* **40** 369–83
[164] Walrafen G E 1967 Raman spectral studies of the effects of temperature on water structure *J. Chem. Phys.* **47** 114–26
[165] De Santis A, Sampoli M, Mazzacurati V and Ricci M A 1987 Raman spectra of water in the translational region *Chem. Phys. Lett.* **133** 381–4
[166] Nardone M, Ricci M A and Benassi P 1992 Brillouin and Raman scattering from liquid water *J. Mol. Struct.* **270** 287–99
[167] Carey D M and Korenowski G M 1998 Measurement of the Raman spectrum of liquid water *J. Chem. Phys.* **108** 2669–75
[168] Shim M G and Wilson B C 1997 Development of an in vivo Raman spectroscopic system for diagnostic applications *J. Raman Spectrosc.* **28** 131–42
[169] Spiro T G 1988 *Biological Applications of Raman Spectroscopy* (New York: Wiley)
[170] Puppels G J and Greve J 1996 Whole cell studies and tissue characterization by Raman spectroscopy *Biomedical Applications of Spectroscopy* vol 25, ed R J H Clark and R E Hester (Chichester: Wiley) pp 1–47
[171] Sockalingum G D, Chourpa I, Millot J M, Nabiev I, Sharonov S and Manfai M 1996 Optical microspectroscopy and imaging analysis: probing drugs and metal ions in living cells *Biomedical Applications of Spectroscopy* vol 25, ed R J H Clark and R E Hester (Chichester: Wiley) pp 48–87
[172] Yu N T, Li X Y and Kuck J F R 1996 Biomedical applications of Raman spectroscopy: Eye Lens Research and Cardiovascular Diseases *Biomedical Applications of Spectroscopy* vol 25, ed R J H Clark and R E Hester (Chichester: Wiley) pp 143–84
[173] Lawson E E, Barry B W, Williams A C and Edwards H G M 1997 Biomedical applications of Raman spectroscopy *J. Raman Spectrosc.* **28** 77–197
[174] Vo-Dihn T, Houck K and Stokes D L 1994 Surface-enhanced Raman gene probes *Anal. Chem.* **66** 3379–83
[175] Pelletier M J 1996 Raman spectroscopy: exploring the world outside the laboratory *ICORS '96: XVth Int. Conf. on Raman Spectroscopy* ed S A Asher and P B Stein (New York: Wiley) pp 1080–1
[176] Mortensen O S and Hassing S 1980 Polarization and interference phenomena in resonance Raman scattering *Advances in Infrared and Raman Spectroscopy* vol 6, ed R J H Clark and R E Hester (London: Heyden) pp 1–48

## Further Reading

Szymanski H A (ed) 1967 and 1970 *Raman Spectroscopy: Theory and Practice* vols 1 and 2 (New York: Plenum)

A valuable reference to anyone involved in this field.

Laserna J J (ed) 1996 *Modern Techniques in Raman Spectroscopy* (New York: Wiley)

Discusses modern techniques with emphasis on practical applications. Suitable for graduate students and researchers.

Ferraro J R and Nakamoto K 1994 *Introductory Raman Spectroscopy* (San Diego: Academic)

An introductory overview of Raman scattering. Suitable for someone beginning in this field.

Morresis A, Mariani L, Distefano M R and Giorgine M G 1995 Vibrational relaxation processes in isotropic molecular liquids. A critical comparison *J. Raman Spectrosc.* **26** 179–216

Provides an extensive literature survey of theoretical and experimental papers in the area of vibrational relaxation.

# B1.4
# Microwave and terahertz spectroscopy

*Geoffrey A Blake*

## B1.4.1 Introduction

Spectroscopy, or the study of the interaction of light with matter, has become one of the major tools of the natural and physical sciences during this century. As the wavelength of the radiation is varied across the electromagnetic spectrum, characteristic properties of atoms, molecules, liquids and solids are probed. In the optical and ultraviolet regions ($\lambda \sim 1$ $\mu$m up to 100 nm) it is the electronic structure of the material that is investigated, while at infrared wavelengths ($\sim$1–30 $\mu$m) the vibrational degrees of freedom dominate.

Microwave spectroscopy began in 1934 with the observation of the $\sim$20 GHz absorption spectrum of ammonia by Cleeton and Williams. Here we will consider the microwave region of the electromagnetic spectrum to cover the 1 to $100 \times 10^9$ Hz, or 1 to 100 GHz ($\lambda \sim 30$ cm down to 3 mm), range. While the ammonia microwave spectrum probes the inversion motion of this unique pyramidal molecule, more typically microwave spectroscopy is associated with the pure rotational motion of gas phase species.

The section of the electromagnetic spectrum extending roughly from 0.1 to $10 \times 10^{12}$ Hz (0.1–10 THz, 3–300 cm$^{-1}$) is commonly known as the far-infrared (FIR), submillimetre or terahertz (THz) region, and therefore lies between the microwave and infrared windows. Accordingly, THz spectroscopy shares both scientific and technological characteristics with its longer- and shorter-wavelength neighbours. While rich in scientific information, the FIR or THz region of the spectrum has, until recently, been notoriously lacking in good radiation sources—earning the dubious nickname 'the gap in the electromagnetic spectrum'. At its high-frequency boundary, most coherent photonic devices (e.g. diode lasers) cease to radiate due to the long lifetimes associated with spontaneous emission at these wavelengths, while at its low-frequency boundary parasitic losses reduce the oscillatory output from most electronic devices to insignificant levels. As a result, existing coherent sources suffer from a number of limitations. This situation is unfortunate since many scientific disciplines—including chemical physics, astrophysics, cosmochemistry and planetary/atmospheric science to name but a few—rely on *high-resolution* THz spectroscopy (both in a spectral and temporal sense). In addition, technological applications such as ultrafast signal processing and massive data transmission would derive tremendous enhancements in rate and volume throughput from frequency-agile THz synthesizers.

In general, THz frequencies are suitable for probing low-energy light–matter interactions, such as rotational transitions in molecules, phonons in solids, plasma dynamics, electronic fine structure in atoms, thermal imaging of cold sources and vibrational–rotation-tunnelling behaviour in weakly bound clusters. Within the laboratory, THz spectroscopy of a variety of molecules, clusters and condensed phases provides results that are critical to a proper interpretation of the data acquired on natural sources, and also leads to a better understanding of important materials—particularly hydrogen-bonded liquids, solids and polymers that participate in a variety of essential (bio)chemical processes.

For remote sensing, spectroscopy at THz frequencies holds the key to our ability to remotely sense environments as diverse as primaeval galaxies, star and planet-forming molecular cloud cores, comets and planetary atmospheres. In the dense interstellar medium characteristic of sites of star formation, for example, scattering of visible/UV light by sub-micron-sized dust grains makes molecular clouds optically opaque and lowers their internal temperature to only a few tens of Kelvin. The thermal radiation from such objects therefore peaks in the FIR and only becomes optically thin at even longer wavelengths. Rotational motions of small molecules and rovibrational transitions of larger species and clusters thus provide, in many cases, the only or the most powerful probes of the dense, cold gas and dust of the interstellar medium.

Since the major drivers of THz technology have been scientists, particularly physicists and astrophysicists seeking to carry out fundamental research, and not commercial interests, a strong coupling of technology development efforts for remote sensing with laboratory studies has long characterized spectroscopy at microwave and THz frequencies. In many respects the field is still in its infancy, and so this chapter will present both an overview of the fundamentals of microwave and THz spectroscopy as well as an assessment of the current technological state of the art and the potential for the future. We will begin with a brief overview of the general characteristics of THz spectrometers and the role of incoherent sources and detection strategies in the THz region, before turning to a more detailed description of the various coherent THz sources developed over the past decade and their applications to both remote sensing and laboratory studies.

## B1.4.2   Incoherent THz sources and broadband spectroscopy

### B1.4.2.1   Principles and instrumentation

Like most other fields of spectroscopy, research at THz frequencies in the first half of the twentieth century was carried out with either dispersive (i.e. grating-based) or Fourier transform spectrometers. The much higher throughput of Fourier transform spectrometers compared to those based on diffraction gratings has made THz Fourier transform spectroscopy, or THz FTS, the most popular incoherent technique for acquiring data over large regions of the THz spectrum. This is especially true for molecular line work where THz FTS resolutions of order 50–100 MHz or better have been obtained [1]. With large-format detector arrays, such as those available at optical through near- to mid-infrared wavelengths, grating- or Fabry–Perot-based instruments can provide superior sensitivity [2], but have not yet been widely utilized at THz frequencies due to the great difficulty of fabricating arrays of THz detectors.

The components of THz spectrometers can be grouped into three main categories: sources (e.g. lasers, Gunn oscillators, mercury-discharge lamps), propagating components (e.g. lenses, sample cells, filters) and detectors (e.g. bolometers, pyroelectric detectors, photoacoustic cells). Propagating components in the FIR are well established (see [3] for an excellent overview of technical information). In the area of detectors, recent progress has placed them ahead of source technology. For example, spider-web Si bolometers developed by Bock et al [4] have an electrical NEP (noise-equivalent power) of $4 \times 10^{-17}$ W Hz$^{-1/2}$ when cooled to 300 mK. For those who desire less exotic cryogenic options, commercially available Si-composite bolometers offer an electrical NEP of $1 \times 10^{-13}$ W Hz$^{-1/2}$, operating at liquid helium temperature (4.2 K), and an electrical NEP of $3 \times 10^{-15}$ W Hz$^{-1/2}$, operating at 1.2 K (pumped L$_{He}$). Combining these into large-format arrays remains a considerable technological challenge, although arrays of several tens of pixels on a side are now beginning to make their way into various telescopes such as the Caltech Submillimeter Observatory and the James Clerk Maxwell Telescope [5, 6]. In addition, their electrical bandwidths are typically only between a few hundred hertz and 1 kHz. Fast-modulation schemes cannot therefore be used, and careful attention must be paid to $1/f$ noise in experiments with Si bolometers. Hot-electron bolometers based on InSb offer electrical bandwidths of $\lesssim$1 MHz, but without cyclotron-assisted resonance, InSb THz bolometers cannot be used above $\nu = 25$–30 cm$^{-1}$ [7]. Similarly, photoconductors based on Ga:Ge offer high electrical speed and good quantum efficiency, but due to the bandgap of the material are unusable below 50–60 cm$^{-1}$ [8].

In practice, the NEP of a room-temperature THz spectrometer is usually limited by fluctuations (shot-noise) in the ambient blackbody radiation. Using an optical bandwidth $\Delta\nu = 3$ THz (limited by, for example,

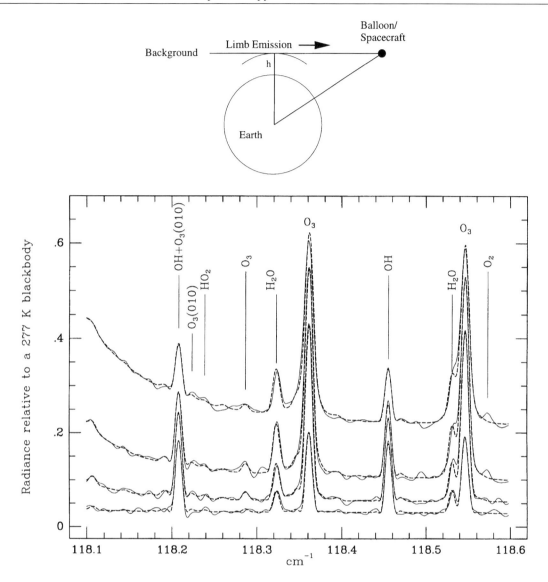

**Figure B1.4.1.** Top: schematic illustration of the observing geometry used for limb sounding of the Earth's atmosphere. Bottom: illustrative stratospheric OH emission spectra acquired by the SAO FIRS-2 far-infrared balloon-borne FTS in autumn 1989. The spectra are from a range of tangent heights ($h =$ tangent height in the drawing), increasing toward the bottom, where the data are represented by solid curves; nonlinear least-square fits to the measurements, based on a combination of laboratory data, the physical structure of the stratosphere and a detailed radiative transfer calculation, are included as dashed curves. The OH lines are $F_1$ ($^2\Pi_{3/2}$), $7/2^- \rightarrow 5/2^+$ and $7/2^+ \rightarrow 5/2^-$ (the hyperfine structure is unresolved in these measurements). Other major contributing lines are also identified [10].

a polyethylene/diamond dust window), a field of view (at normal incidence) $\theta = 9°$ and a detecting diameter (using a so-called Winston cone, which condenses the incident radiation onto the detecting element) $d =$ 1.1 cm, values that are typical for many laboratory applications, the background-limited NEP of a bolometer is given by

$$\mathrm{NEP} = \frac{2kT\pi d \sin(\theta/2)(\Delta f \Delta \nu)^{1/2}}{\lambda} = 0.2 \text{ pW} \qquad (\text{B1.4.1})$$

where $k$ is the Boltzmann constant, $T = 300$ K, $\Delta f$ (the electronic amplified bandwidth) $= 1$ Hz, and $\lambda$ (the band-centre wavelength) $\approx 200$ $\mu$m. The equation above uses the Rayleigh–Jeans law, which is valid for $\nu \le kT_{300 \text{ K}}\, h^{-1} = 17$ THz. Therefore, for laboratory absorption experiments, a typical FIR detector provides an estimated detection limit (NEP/source power) of $10^{-4}$ with a source output of 20 nW. In general, high-sensitivity bolometers saturate at an incident-power level of $\approx 1$ $\mu$W or less, resulting in an ultimate detection limit of $10^{-7}$. For yet higher dynamic range, a filter element (e.g. cold grating, prism, or etalon) must be placed before the detector to reduce background noise, or the background temperature must be lowered. Note the $(\Delta \nu)^{1/2}$ dependence in equation B1.4.1, which means that the optical bandwidth must be reduced to $\sim$30 GHz to drop the NEP by a factor of ten. *Thus, unlike shorter wavelength regions of the electromagnetic spectrum, due to the high background luminosity in the THz, spectroscopic sensitivity in the laboratory is limited by the source power, in comparison to the background power, incident on the detector—not by shot-noise of the available spectroscopic light sources.* As higher-power light sources are developed at THz frequencies, lower-NEP detectors can be utilized that are less prone to saturation, and shot-noise will become the limiting factor, as it is in other regions of the electromagnetic spectrum.

For broadband THz FTS instruments this large background actually leads to a 'multiplex disadvantage' in that the room-temperature laboratory background can easily saturate the sensitive THz detectors that are needed to detect the feeble output of incoherent THz blackbody sources, which drop rapidly as the wavelength increases. The resulting sensitivity is such that signal-to-noise ratios in excess of 100 are difficult to generate at the highest feasible resolutions of 50–100 MHz [1], which is still quite large compared to the 1–2 MHz Doppler-limited line widths at low pressure. For low-resolution work on condensed phases, or for the acquisition of survey spectra, however, THz FTS remains a popular technique. Beyond wavelengths of $\sim$1 mm, the sensitivity of FTS is so low that the technique is no longer competitive with the coherent approaches described below.

*B1.4.2.2    THz FTS studies of planetary atmospheres*

Rather different circumstances are encountered when considering THz remote sensing of extraterrestrial sources. The major source of THz opacity in the Earth's atmosphere is water vapour, and from either high, dry mountain sites or from space there are windows in which the background becomes very small. Incoherent instruments which detect the faint emission from astronomical sources can therefore be considerably more sensitive than their laboratory counterparts. Again, grating- or etalon-based and FTS implementations can be considered, with the former being preferred if somewhat coarse spectral resolution is desired or if large-format detector arrays are available.

In planetary atmospheres, a distinct advantage of THz studies over those at optical and infrared wavelengths is the ability to carry out spectroscopy without a background or input source such as the sun. Global maps of a wide variety of species can therefore be obtained at any time of day or night and, when taken at high enough spectral resolution, the shapes of the spectral lines themselves also contain additional information about vertical abundance variations and can be used to estimate the atmospheric temperature profile. The 'limb sounding' geometry, in which microwave and THz emission from the limb of a planetary atmosphere is imaged by an orbiting spacecraft or a balloon, is particularly powerful in this regard, and excellent reviews are available on this subject [9]. The observing geometry is illustrated in figure B1.4.1, which also presents a portion of the Earth's stratospheric emission spectrum near 118 cm$^{-1}$ obtained by the balloon-borne Smithsonian Astrophysical Observatory limb-sounding THz FTS [10].

### B1.4.3 Coherent THz sources and heterodyne spectroscopy

*B1.4.3.1 Principles and instrumentation*

The narrow cores of atmospheric transitions shown in figure B1.4.1 can be used, among other things, to trace the wind patterns of the upper atmosphere. For such work, or for astronomical remote-sensing efforts, resolutions of the order of 30–300 m s$^{-1}$ are needed to obtain pressure broadening or kinematic information, which correspond to spectral resolutions of $(\nu/\Delta \nu) \sim 1$–$10 \times 10^6$. Neither FTS nor grating-based spectrometers can provide resolution at this level and so other techniques based on coherent radiation sources must be used. The most important of these is called heterodyne spectroscopy. Heterodyne spectroscopy uses nonlinear detectors called mixers, in order to downconvert the high-frequency THz radiation into radiofrequency or microwave signals that can be processed using commercial instrumentation. An outline of a heterodyne receiver is presented in figure B1.4.2. In such a receiver, an antenna (dish) collects radiation from space, and this radiation is focused onto a detector operating in heterodyne mode, which means that the incoming signal is mixed with the output of a coherent source (called the local oscillator, or LO). Now, if a device can be constructed that responds quadratically, rather than linearly, to the two input beams, the output, $S(t)$, of such a device is given by

$$S(t) \propto e(t)^2$$
$$S(t) \propto 1/2(E_1^2 + E_1^2) + 1/2[E_1^2 \cos(2\omega_1 t) + E_2^2 \cos(2\omega_2 t)]$$
$$+ E_1 E_2 \cos[(\omega_1 t + \omega_2 t) + (\phi_1 + \phi_2)]$$
$$+ \underbrace{E_1 E_2 \cos[(\omega_1 - \omega_2)t + (\phi_1 - \phi_2)]}_{\text{DIFFERENCE-FREQUENCY FIELD}} \qquad \text{(B1.4.2)}$$

where the identities $2 \cos \alpha \cos \beta = \cos(\alpha + \beta) + \cos(\alpha - \beta)$ and $\cos^2 \alpha = 1/2(1 + \cos 2\alpha)$ are invoked. The last term is a field oscillating at a frequency equal to the difference between the two incidental fields—representing the beat-note between $\nu_1$ and $\nu_2$.

If $\nu_1 = \nu_{LO}$ and $\nu_2 = \nu_{SKY} \approx \nu_{LO}$, the sum frequency lies at THz frequencies, while the difference lies at radio or microwave frequencies, and is called the intermediate frequency, or IF. The IF can be amplified and recorded (for example, on a spectrum analyser or by a digital correlator or filterbank) across a range of frequencies simultaneously. The fact that the IF power is proportional to the product of the remote-signal power and the LO power results in two main advantages: (a) the signal-to-noise ratio is enhanced by using an LO power that is much higher than that of the remote signal and (b) the spectral resolution is set by the linewidth of the LO, which can be as narrow as desired.

Among the first THz mixers to be constructed were those based on room-temperature Schottky diodes [11]. Over the past decade, new mixers based on superconducting tunnel junctions have been developed that have effective noise levels only a few times the quantum limit of $T_{\text{mixer}} = h\nu/k$ [12]. However, certain conditions must be met in order to exploit these advantages. For example, while the LO power should be strong compared to the remote signal and to the noise, it must not be so strong as to saturate the detecting element. Roughly speaking, the optical coupled LO power level is about 1 $\mu$W for SIS (superconductor–insulator–superconductor) mixers and 100 nW for superconductive Nb HEBs (hot-electron bolometers, or transition edge bolometers, TEBs). In addition, since the overlap between the remote signal and the LO is important, the spatial distribution of the LO output must be well coupled to the receiver's antenna mode. Although this requirement imposes experimental complexity, it also provides excellent rejection of ambient background radiation.

As noted above, at THz frequencies the Rayleigh–Jeans approximation is a good one, and it is typical to report line intensities and detector sensitivities in terms of the Rayleigh–Jeans equivalent temperatures. In frequencies range where the atmospheric transmission is good, or from airborne or space-borne platforms,

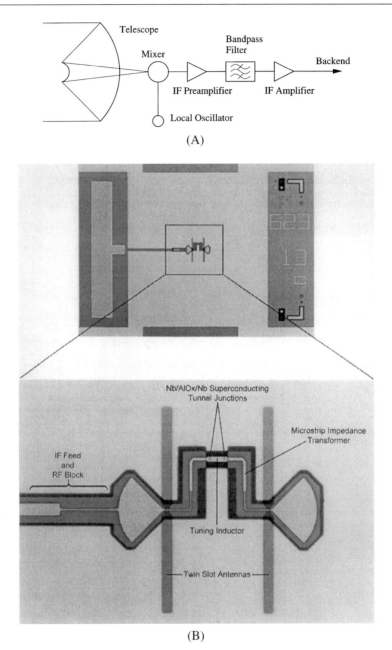

**Figure B1.4.2.** (A) Basic components of an astronomical heterodyne receiver. The photomicrograph in (B) presents the heart of a quasi-optical SIS mixer and its associated superconducting tuning circuits, while the image in (C) shows the fully assembled mixer, as it would be incorporated into a low-temperature cryostat (J Zmuidzinas, private communication).

the effective background temperature is only a few tens of Kelvin. Under such conditions, SIS mixers based on Nb, a particular implementation of which is pictured in figure B1.4.2, can now perform up to 1.0 THz with $T_{n,\text{mixer}} = 130$ K [13]. The earliest SIS microwave and millimetre-wave receivers utilized waveguide

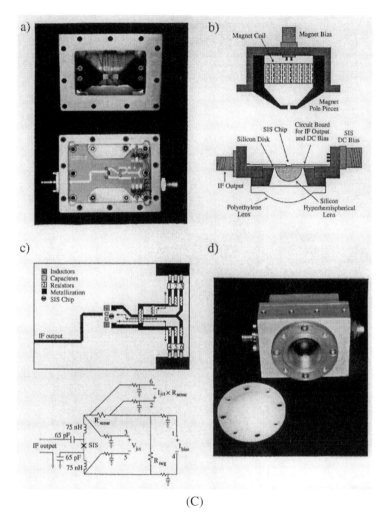

(C)

**Figure B1.4.2.** (Continued)

components, but as the operating frequencies have been pushed into the THz region, quasi-optical designs such as those shown in figure B1.4.2 become attractive. Such designs may also be easier to incorporate into THz receiver arrays. Recently, alternative superconducting Nb hot-electron mixers that rely on a diffusion-based relaxation mechanism have been demonstrated with $T_{mixer} = 750$ K between 1 and 2.5 THz [14]. These devices are expected to operate up to at least several tens of THz—*if coherent sources are available as LOs.* Thus, THz source technology plays a key role in setting the spectroscopic sensitivity for both laboratory and remote-sensing experiments.

### B1.4.3.2   *THz remote sensing with heterodyne receivers*

Heterodyne spectroscopy has been particularly critical to the study of the Earth's stratosphere, where the improved resolution and sensitivity compared to the FTS spectra shown in figure B1.4.1 have led to the collection of global maps of species important to ozone chemistry and atmospheric dynamics ($O_3$, $ClO$, $SO_2$,

$H_2O$, $O_2$, etc: see [9] for an extended overview), and of the dense interstellar medium. Although human beings have been systematically observing astronomical objects for thousands of years, until the advent of radioastronomy in the 1960s we possessed little knowledge of what, if anything, exists in the space between stars. Optical observations revealed only stars, galaxies and nebulae; if matter existed in the vast, dark interstellar medium, it was not detectable. However, the discovery of the first polyatomic microwave in the interstellar medium, water ($H_2O$), ammonia ($NH_3$) and formaldehyde ($H_2CO$), by microwave remote sensing (in 1968 [15]) set off an exciting era of discovery.

To date, researchers have identified more than 100 different molecules, composed of up to 13 atoms, in the interstellar medium [16]. Most were initially detected at microwave and (sub)millimetre frequencies, and the discoveries have reached far beyond the mere existence of molecules. Newly discovered entities such as diffuse interstellar clouds, dense (or dark) molecular clouds and giant molecular cloud complexes were characterized for the first time. Indeed, radioastronomy (which includes observations ranging from radio to submillimetre frequencies) has dramatically changed our perception of the composition of the universe. Radioastronomy has shown that most of the mass in the interstellar medium is contained in so-called dense molecular clouds, which have tremendous sizes of 1–100 light years, average gas densities of $10^2$–$10^3$ cm$^{-3}$, and temperatures in the range of 10–600 K. An overview of the THz emission from a cold, dense interstellar cloud is presented in figure B1.4.3 [17].

In addition to striking differences from one cloud to another, each dense molecular cloud is inhomogeneous, containing clumps, or cores, of higher-density material situated within envelopes of somewhat lower density. Many of these higher-density cores are active sites of star formation, with the youngest stars being detectable only in the IR or FIR. Star formation is of major interest in astrophysics, and it contains a wealth of interesting chemical reactions and physical phenomena (for excellent reviews, see [18–20]). Optical observations are unable to characterize interstellar clouds due to absorption and scattering, both of which have an inverse wavelength dependence, by the pervasive dust particles inside these clouds. Thus, microwave and THz spectroscopy is responsible for identifying most of the hundred or so interstellar molecules to data, and continues to dominate the fields of molecular astrophysics and interstellar chemistry. The power of heterodyne spectroscopy in examining the differences between dense clouds undergoing star (and presumably planet) formation is shown in the right panel of figure B1.4.3, which depicts the 345 GHz spectral line surveys of three regions of the W3 giant molecular cloud complex [21]. From such studies, which reveal dramatic differences in the THz spectrum of various objects, molecular astrophysicists hope to classify the evolutionary state of the cloud, just as optical spectra are used to classify stars.

High angular resolution studies with modern THz telescopes and interferometric arrays can even probe the material destined to become part of planetary systems. In the accretion discs around young stars, for example, a range of simple organic species have now been detected at (sub)millimetre wavelengths [22]. Such accretion disks are the assembly zones of planets, and the first steps in THz imaging their outer regions and in understanding the means by which they evolve have been taken [23]. The physical and chemical conditions in these objects can now be compared to that observed in primitive solar system objects such as comets and icy satellites. The recent apparitions of comets Hyakutake and Hale–Bopp, for example, have provided a wealth of new observations at IR, THz and microwave frequencies that have led to a much improved understanding of the origin and evolution of planetesimals in the outer solar system [24]. Future work in high-resolution THz imaging will be dramatically enhanced by the Atacama Large Millimeter Array (ALMA), which will operate over the 1 cm to 350 $\mu$m interval at an altitude of 16 000 ft in the Atacama desert of northern Chile [25].

Ultimately, studies from ground-based observatories are limited by absorption in the Earth's atmosphere. For example, no studies above $\approx$1 THz are possible even from mountain-top observatories. Two major instruments, SOFIA and FIRST, are poised to change this situation dramatically. SOFIA (for Stratospheric Observatory For Infrared Astronomy [26]) will carry a 2.7 m telescope in a 747SP aircraft to altitudes of 41 000–45 000 feet. At these altitudes, nearly 60–70% of the THz spectrum up to the mid-IR is accessible

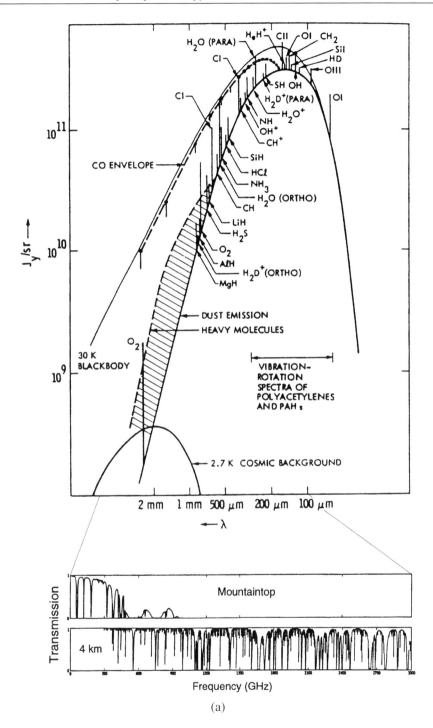

**Figure B1.4.3.** (a) A schematic illustration of the THz emission spectrum of a dense molecular cloud core at 30 K and the atmospheric transmission from ground and airborne altitudes (adapted, with permission, from [17]). (b) The results of 345 GHz molecular line surveys of three cores in the W3 molecular cloud; the graphics at left depict the evolutionary state of the dense cores inferred from the molecular line data [21].

CHEMICAL EVOLUTION IN THE W 3 MASSIVE STAR-FORMING REGION

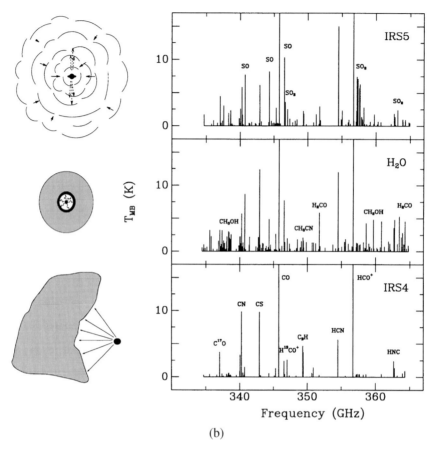

(b)

**Figure B1.4.3.** (Continued)

(figure B1.4.3), and both incoherent and heterodyne spectrometers are being constructed as part of the initial instrument suite. SOFIA will become operational in 2002. On somewhat longer timescales, FIRST (the Far-InfraRed Space Telescope [27]) will carry 0.4–1.2 THz SIS and 1.2–1.9/2.4–2.7 THz antenna coupled HEB receivers at the focal plane of a 4 m telescope into space. Like SOFIA, FIRST will also include incoherent spectrometers and imagers to take advantage of the low background flux and to survey larger regions of the THz spectrum and sky.

### B1.4.4   Spectroscopy with tunable microwave and THz sources

*B1.4.4.1   Principles and background*

Long before the invention of the laser, coherent radiation sources such as electron beam tubes (e.g. klystrons and backward wave oscillators, or BWOs) were in use by microwave spectroscopists to examine the direct absorption rotational spectroscopy of molecules. Being the interface between microwave spectroscopy (generally associated with low-energy rotational transitions of molecules) and mid-IR spectroscopy (generally associated with vibrational transitions of molecules), THz spectra such as those outlined in figure B1.4.3 probe the high-frequency rotations of molecules and certain large-amplitude vibrational motions. As the rotational energy levels of a molecule depend largely on its moment of inertia, which is determined by the molecular structure, high-resolution spectroscopy of gas phase molecules provides the most precise information available on molecular geometries.

Rotational transition frequencies acquired in the THz region expand upon and complement those acquired in the microwave. Two types of molecules undergo rotational transitions that fall in the FIR: molecules with rotation about an axis having a small moment of inertia, and molecules in high-$J$ states. FIR spectra of the first type of molecules are important for determining their equilibrium geometry, as many light molecules ($H_2O$, $NH_3$, HF, etc) only have transitions in the submillimetre and FIR regions. Due to the high rotational energy in the second type of species (high-$J$ molecules), interactions between the vibrational and rotational motions, namely, centrifugal distortion and Coriolis perturbations, become important. Given high enough spectral resolution and accuracy ($\Delta \nu / \nu \leq 10^{-5}$), shifts in rotational frequencies and changes in selection rules resulting from these interactions become significant. Thus, FIR spectroscopy of high-$J$ transitions enables detailed characterization of molecular Hamiltonians far beyond the rigid rotor approximation, giving more accurate zero-point rotational constants and rough estimates of the shapes of potential energy surfaces. Finally, for very large molecules or weakly bound clusters, the softest vibrational degrees of freedom can be proved at THz frequencies, as is outlined in greater detail below.

### B1.4.4.2    *Fourier transform microwave spectroscopy*

At microwave frequencies, direct absorption techniques become less sensitive than those in the THz region due to the steep dependence of the transition intensities with frequency. A variant of heterodyne spectroscopy, pioneered by Flygare and Balle [28], has proven to be much more sensitive. In this approach, molecules are seeded into or generated by a pulsed molecular beam which expands into a high-$Q$ microwave cavity. The adiabatic expansion cools the rotational and translation degrees of freedom to temperatures near 1–10 K, and thus greatly simplifies the rotational spectra of large molecules. In addition, the low-energy collisional environment of the jet can lead to the growth of clusters held together by weak intermolecular forces.

A microwave pulse from a tunable oscillator is injected into the cavity by an antenna, and creates a coherent superposition of rotational states. In the absence of collisions, this superposition emits a free-induction decay signal, which is detected with an antenna-coupled microwave mixer similar to those used in molecular astrophysics. The data are collected in the time domain and Fourier transformed to yield the spectrum whose bandwidth is determined by the quality factor of the cavity. Hence, such instruments are called Fourier transform microwave (FTMW) spectrometers (or Flygare–Balle spectrometers, after the inventors). FTMW instruments are extraordinarily sensitive, and can be used to examine a wide range of stable molecules as well as highly transient or reactive species such as hydrogen-bonded or refractory clusters [29, 30].

An outline of an FTMW instrument used in the study of large, polar, carbonaceous species is shown in figure B1.4.4. In this instrument, the FTMW cavity is mated to a pulsed electric discharge/supersonic expansion nozzle [31, 32]. Long-chain carbon species, up to that of $HC_{17}N$, as shown in figure B1.4.4, can be studied with this technique, as can a wide variety of other molecules and clusters. With the jet directed along the axis of the cavity, the resolution is highly sub-Doppler, with the slight complication that a Doppler doublet is formed by the difference between the laboratory and molecular beam reference frames. Studies of the rotational spectra of hydrogen-bonded clusters have also been carried out by several groups using FTMW instruments, a topic we shall return to later. FT-THz instruments can, in principle, be built using the highly sensitive THz SIS or HEB mixers outlined above, and would have extraordinary sensitivities. In order to saturate the rotational or rovibrational transitions, however, high-power THz oscillators are needed, but are not yet available.

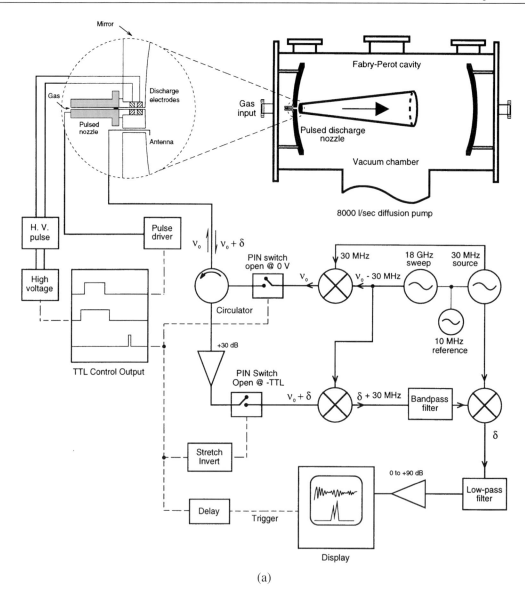

(a)

**Figure B1.4.4.** (a) An outline of the Harvard University electric discharge supersonic nozzle/Fourier transform microwave spectrometer. (b) The rotational states of $HC_{17}N$ observed with this apparatus [31].

### B1.4.4.3  CW THz sources and molecular spectroscopy

At frequencies up to ~150–200 GHz, solid-state sources such as YIG-tuned oscillators or Gunn diode oscillators are now available with power outputs of up to 100 mW. The harmonic generation of such millimetre-wave sources is relatively efficient for doubling and tripling ($\geq$10–15%), but for higher harmonics the power drops rapidly ($P_{\text{out}}(1 \text{ THz}) \leq 0.1$–10 $\mu$W). Nevertheless, harmonic generation was used as early as the 1950s to record the submillimetre wave spectra of stable molecules [33]. Harmonics from optimized solid-state millimetre-wave sources are now used to drive astronomical heterodyne receivers up to 900–1100 GHz [34], and the prospects for operation up to 2–3 THz are promising.

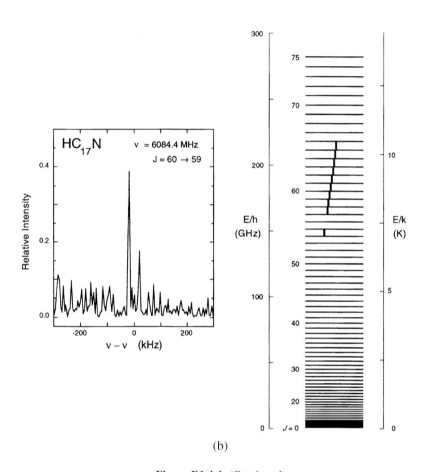

(b)

**Figure B1.4.4.** (Continued)

Even higher output power ($\sim$1–10 mW) is available from rapidly tunable BWOs up to 1–15 THz. BWOs are capable laboratory sources where they operate, and offer wide tunability and excellent spectral purity, especially when phase locked to the harmonics of lower-frequency microwave or millimetre-wave oscillators [35]. The high output power of BWOs and the relatively strong intrinsic strengths of pure rotational transitions of polar molecules gives BWO spectrometers very high sensitivity, and also enables them to utilize nonlinear methods such as Lamb dip spectroscopy. An example for the CO molecule is presented in figure B1.4.5 [36]. The resulting resolution is truly exceptional and leads to among the most precise molecular constants ever determined. Pioneered in the former Soviet Union, THz BWOs are finding increased applications in a number of laboratories.

BWOs are typically placed in highly overmoded waveguides, and the extension of this technology to higher frequencies will, by necessity, require a number of innovative solutions to the very small-scale structures that must be fabricated. Their size and weight also preclude them from space-based applications (e.g. FIRST). Thus, electronic oscillators are unlikely to cover all THz spectroscopy and/or remote sensing applications over the full range of 1–10 THz in the short term, and a number of alternative THz radiation sources are therefore under investigation. Among the most promising of these, over the long term, are engineered materials such as

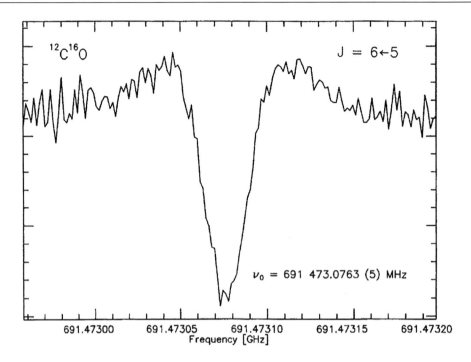

**Figure B1.4.5.** The Lamb dip spectrum of the CO 6–5 transition obtained with the Cologne THz BWO spectrometer. The dip is of order 30–40 kHz in width and the transition frequency is determined to 0.5 kHz [36].

quantum wells that either possess high nonlinearity for THz mixing experiments [37] or that can be tailored to provide direct emission in the FIR [38]. Tunable laser sources are the ultimate goal of such development programmes, but while THz spontaneous emission has been observed, laser action is still some time in the future.

A number of mixing experiments have therefore been used to generate both pulses and CW THz radiation. Among these, diode-based mixers used as upconvertors (that is, heterodyne spectroscopy 'in reverse') have been the workhorse FIR instruments. Two such techniques have produced the bulk of the spectroscopic results: (1) GaAs-based diode mixers that generate tunable sidebands on line-tunable FIR molecular gas lasers [39, 40], and (2) $CO_2$ laser-based THz difference frequency generation in ultrafast metal–insulator–metal (MIM) diodes [41, 42]. Both types of mixers have sufficient instantaneous bandwidth to place any desired millimetre-wave frequency on the carrier radiation, and respond well at THz frequencies. Having been used for many years in astronomical receiver applications, GaAs mixer technology is more mature than that for MIM diodes, and its conversion, noise and coupling mechanisms are better understood at present. In addition, their conversion efficiencies are good up to at least 4–5 THz, and several to several tens of microwatts are available from GaAs Schottky diode laser sideband generators. It is thus possible to construct THz spectrometers based on laser sideband generation that operate at or near the shot-noise limit, and this sensitivity has been used to investigate a wide range of interesting reactive and/or transient species, as is described in section B1.4.4.6.

While the conversion efficiency of MIM diodes is not as good as that of their GaAs counterparts, they are considerably faster, having also been used at IR and even visible wavelengths! Thus, MIM-based THz spectrometers work over wider frequency ranges than do GaAs FIR laser sideband generators, but with less output power. As described above, this translates directly into spectrometer sensitivity due to the high laboratory background in the THz region. Thus, where intense FIR gas laser lines are available, sideband generators are to be preferred, but at present only MIM-diode spectrometers can access the spectroscopically important region above 200 cm$^{-1}$ [42]. Also, because it is easy to block $CO_2$ laser radiation with a variety of reststrahlen solid-state filters, there is no fixed-frequency FIR gas laser carrier to reject in MIM spectrometers, and this simplifies the overall experimental design. MIM spectrometers that perform third-order mixing of two $CO_2$ laser lines and a tunable microwave source have also been constructed. This approach leads to very wide tunability and eliminates the need to scan the $CO_2$ lasers, and only decreases the output power by a small amount. Thus, it is possible to phase lock the $CO_2$ and microwave sources, leading to a direct synthesis approach to THz radiation to very high frequencies indeed. This is not feasible for the FIR laser sideband generators, and so the MIM approach has provided extremely accurate THz frequency standards for calibration gases such as CO, HF and HCl, which can then be used as secondary standards in a number of other techniques [41].

While extremely useful in the laboratory, the size and power requirements of both FIR laser sideband generators and MIM-based $CO_2$ laser spectrometers are excessive for space-borne applications. Research on other THz generation approaches by mixing has therefore continued. One particularly interesting approach from a technology and miniaturization point of view is optical heterodyne conversion, in which optical radiation is converted to THz light by semiconducting materials pumped above their band gaps. The use of optical or near-IR lasers to drive the process results in wide tunability and spectral coverage of the THz spectrum. Such approaches, described next, also have the considerable advantage of levering the rapid technological innovations in diode-pumped lasers and fast optoelectronic devices required for emerging industries such as optical telecommunications and optical computing. They therefore 'break the mould' of traditional THz LO development by small groups focused on scientific problems, and rapid developments can be expected with little or no investment by the THz community. Finally, as described next, both ultrafast time-resolved and CW high-resolution spectroscopies can be carried out using these approaches.

## B1.4.4.4   *Time domain THz spectroscopy*

Free-electron lasers have long enabled the generation of extremely intense, sub-picosecond THz pulses that have been used to characterize a wide variety of materials and ultrafast processes [43]. Due to their massive size and great expense, however, only a few research groups have been able to operate them. Other approaches to the generation of sub-picosecond THz pulses have therefore been sought, and one of the earliest and most successful involved semiconducting materials. In a photoconductive semiconductor, carriers (for n-type material, electrons) in the valence band absorb the incident radiative power (if $h\nu_0 \geq E_{bandgap}$) and are injected into the conduction band. Once they are in the conduction band, the electrons become mobile, and, if there is an applied bias, they begin to drift toward the photoconductor electrodes. Some of the electrons will reach the electrodes while some will encounter sites of ionic impurities. The latter electrons are trapped by the impurity sites and removed from the conduction band. As early as two decades ago, pulsed optical lasers were used by Auston and co-workers to generate and detect electrical pulses in DC biased voltage transmission lines [44]. For the earliest used materials such as GaAs and Si, the pulse widths were of order nanoseconds due to their long recombination times. The discovery of materials such as radiation-damaged silicon-on-sapphire and of low-temperature-grown (LTG) GaAs changed this situation dramatically.

Especially with LTG GaAs, materials became available that were nearly ideal for time-resolved THz spectroscopy. Due to the low growth temperature and the slight As excess incorporated, clusters are formed which act as recombination sites for the excited carriers, leading to lifetimes of $\leq 250$ fs [45]. With such recombination lifetimes, THz radiators such as dipole antennae or log-periodic spirals placed onto optoelectronic substrates and pumped with ultrafast lasers can be used to generate sub-picosecond pulses with optical bandwidths of 2–4 THz. Moreover, *coherent* sub-picosecond detection is possible, which enables both the real and imaginary refractive indices of materials to be measured. The overall sensitivity is $>10^4$, and a variety of solid-state and gas phase THz spectra have been acquired with such systems [46, 47], an excellent overview of which may be found in [48].

Recently, it has been shown that both the detection and generation of ultrafast THz pulses can be carried out using the electro-optic effect in thin films of materials such as ZnTe, GaAs and InP that are pumped in the near-IR [49]. The generation efficiency is similar to that of the photoconducting antenna approach, but the electro-optic scheme offers two extremely significant advantages. First, the detection bandwidth can be extremely large, up to 30–40 THz under optimum conditions [49]. Second, it is possible to directly *image* the THz field with such spectrometers. Such approaches therefore make possible the THz imaging of optically opaque materials with a compact, all solid-state, room-temperature system [50]!

The great sensitivity and bandwidth of electro-optic approaches to optical–THz conversion also enable a variety of new experiments in condensed matter physics and chemistry to be conducted, as is outlined in figure B1.4.6. The left-hand side of this figure outlines the experimental approach used to generate ultrafast optical and THz pulses with variable time delays between them [51]. A mode-locked Ti:sapphire laser is amplified to provide approximately 1 W of 100 fs near-IR pulses at a repetition rate of 1 kHz. The ~850 nm light is divided into three beams, two of which are used to generate and detect the THz pulses, and the third of which is used to optically excite the sample with a suitable temporal delay. The right-hand panel presents the measured relaxation of an optically excited TBNC molecule in liquid toluene. In such molecules, the charge distribution changes markedly in the ground and electronically excited states. In TBNC, for example, the excess negative charge on the central porphyrin ring becomes more delocalized in the excited state. The altered charge distribution must be accommodated by changes in the surrounding solvent. This so-called solvent reorganization could only be indirectly probed by Stokes shifts in previous optical–optical pump–probe experiments, but the optical–THz approach enables the solvent response to be *directly* investigated. In this case, at least three distinct temporal response patterns of the toluene solvent can be seen that span several temporal decades [51]. For solid-state spectroscopy, ultrafast THz studies have enabled the investigation of coherent oscillation dynamics in the collective (phonon) modes of a wide variety of materials for the first time [49].

### B1.4.4.5  *THz optical-heterodyne conversion in photoconductors*

For CW applications of optical-heterodyne conversion, two laser fields are applied to the optoelectronic material. The non-linear nature of the electro-optic effect strongly suppresses continuous emission relative to ultrashort pulse excitation, and so most of the CW research carried out to date has used photoconductive antennae. The CW mixing process is characterized by the average drift velocity $\bar{v}$ and carrier lifetime $\tau_0$ of the mixing material, typically LTG GaAs. If $\tau_0 \bar{v} \geq \delta$, the electrode spacing, then a significant amount of current will be generated by the photo-excitation. That is

$$i(t) = \frac{N_c(t) e \bar{v}}{\delta} \tag{B1.4.3}$$

where $N_c$ is the number of carriers. The rate equation for the photo-excitation-recombination process can be written as

$$\frac{dN_c(t)}{dt} = a P_0 - \frac{N_c(t)}{\tau_0} \tag{B1.4.4}$$

where $a$ is a proportionality constant and $P_0$ is the average power of the incident field over a few optical periods. The time-average is of interest here because $\tau_0$ in a semiconductor is always longer than an optical cycle and, therefore, the output current will not respond directly to the optical oscillations. Since the THz waves are generated with optical light, $\omega \equiv (\omega_1 - \omega_2) \ll \omega_1 \approx \omega_2$. Thus, integrating the right-hand side of the expression above over a few cycles of $\omega_1 \approx \omega_2$ yields

$$P_0 = E_1^2 + E_2^2 + 2 E_1 E_2 \cos(\omega t + \phi_1 - \phi_2). \tag{B1.4.5}$$

Substitution yields

$$\frac{dN_c}{dt} = a[E_1^2 + E_2^2 + 2E_1E_2\cos(\omega t + \phi_1 - \phi_2)] - \frac{N_c}{\tau_0}. \tag{B1.4.6}$$

By solving the differential equation above and using the single-pump case ($P_2 = 0$) to determine $a$, it can be shown that the photo-current is

$$i(t) = \frac{e\eta}{h\nu_0}\frac{\tau_0\bar{v}}{\delta}\left[P_1 + P_2 + 2\sqrt{\frac{P_1P_2}{1 + \omega^2\tau_0^2}}\cos(\omega t - \phi)\right] \tag{B1.4.7}$$

where $\nu_0 \equiv \nu_1 \approx \nu_2$, $P_1$ and $P_2$ are the incident optical powers, $\phi = \tan^{-1}(\omega\tau_0)$ and $\eta$, the quantum efficiency, is the number of carriers excited per incident photon. Recognizing that $\bar{v} = \mu E$ (where $\mu$ is the carrier mobility and $E$ is the electric field),

$$i(t) = \frac{e\eta}{h\nu_0}\frac{\tau_0\mu E}{\delta}\left[P_1 + P_2 + 2\sqrt{\frac{P_1P_2}{1 + \omega^2\tau_0^2}}\cos(\omega t - \phi)\right]. \tag{B1.4.8}$$

Separating the DC and oscillatory parts of the above equation gives

$$i_{dc} = \frac{e\eta}{h\nu_0}\frac{\tau_0\mu E}{\delta}(P_1 + P_2) \tag{B1.4.9}$$

$$i_{THz}(t) = \frac{e\eta}{h\nu_0}\frac{\tau_0\mu E}{\delta}2\sqrt{\frac{P_1P_2}{1 + \omega^2\tau_0^2}}\cos(\omega t - \phi). \tag{B1.4.10}$$

Thus, the beating of the two incident optical fields generates a modulation in the photo-excitation of the carriers, which in turn results in an oscillating electrical signal. Initial microwave experiments using an interdigitated electrode geometry by Brown *et al* [52] showed a flat frequency response up to 25 GHz with a conversion efficiency of 0.14%, in agreement with the signal level predicted by the theoretical analysis outlined above. At THz frequencies, the power decays rapidly from such structures due to the parasitic capacitance of the electrode structure and the finite carrier lifetime. Free-space radiation is generated by coupling the electrodes to a planar THz antenna. At 1 THz, the observed conversion efficiency is roughly $3 \times 10^{-6}$, and the damage threshold is of order 1 mW $\mu$m$^{-2}$. To alleviate these limitations, travelling-wave structures have now been developed that eliminate the capacitive roll-off and allow large-device active areas to be pumped. Powers in excess of 1 $\mu$W can now be achieved above 2 THz for input drive levels of 300–400 mW [53].

This power level is sufficient for laboratory spectroscopy or for use as a THz local oscillator, and such travelling-wave structures can be used over at least a decade of frequency (0.3–3 THz, for example) without moving parts. Further, compact, all solid-state spectrometers, such as that outlined in figure B1.4.7, can now be constructed using CW diode laser and optical tapered amplifier technology [54]. The major challenge in working with diode lasers is, in fact, their instantaneous line widths ($\geq 15$ MHz) and long-term frequency stability ($\sim 100$–200 MHz), both of which need considerable improvement to be useful as THz LOs. The main source of both instabilities is the notorious susceptibility of diode lasers to optical feedback. However, this susceptibility can be used to one's advantage by sending a small fraction of the laser output into a high-finesse ($F \geq 60$) optical cavity such that the diode laser 'sees' optical feedback only at cavity resonances. By locking the diode lasers to different longitudinal modes of an ultrastable reference cavity, it is possible to construct direct synthesis spectrometers that can be absolutely calibrated to $\Delta\nu/\nu < 10^{-8}$ or better. To demonstrate the continuous tunability and frequency stability of such an instrument, the lower panel of figure B1.4.7 presents a submillimetre spectrum of acetonitrile in which the transition frequencies are measured to better than 50 kHz. Future improvements to such systems should allow similar measurements on both stable and transient species up to at least 5–6 THz.

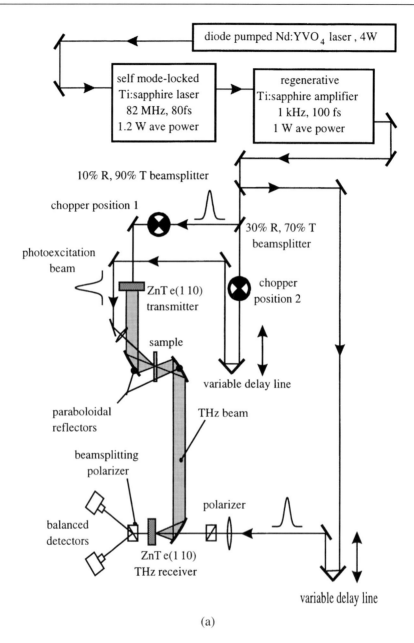

(a)

**Figure B1.4.6.** Left: an experimental optical THz pump–probe set-up using sub-picosecond THz pulse generation and detection by the electro-optic effect. Right: the application of such pulses to the relaxation of optically excited TBNC in toluene. The THz electric field used for these experiments is shown in the upper-right inset. Three exponential decay terms, of order 2, 50 and 700 ps, are required to fit the observed temporal relaxation of the solvent [51].

## B1.4.4.6  *THz spectroscopy of hydrogen-bonded and refractory clusters*

Among the most interesting transient species that can be studied at THz frequencies are those involving collections of molecules held together by van der Waals or hydrogen bonding forces. In no small measure this is true

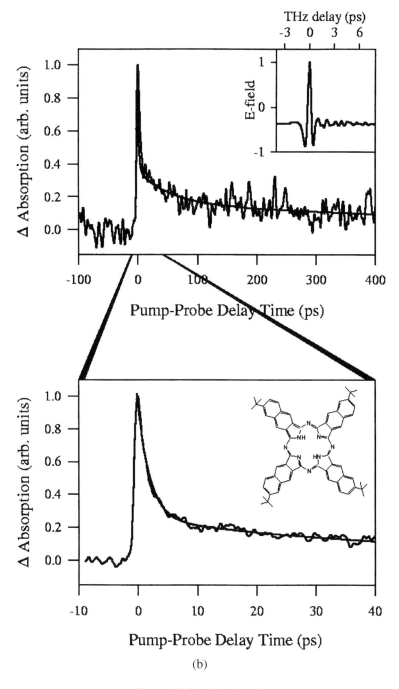

(b)

**Figure B1.4.6.** (Continued)

because hydrogen bonds are ubiquitous in nature. From the icy mantles covering interstellar dust to the nuclei of living cells, hydrogen bonds play crucial roles in the regulation and evolution of both inorganic and living systems. Accurate, fully anisotropic, descriptions of the intermolecular forces involved in these and other weak

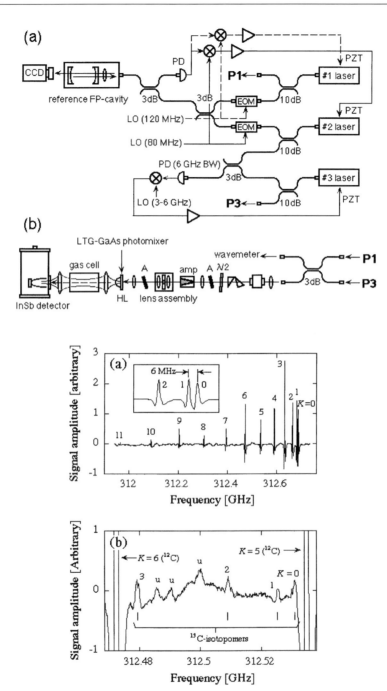

**Figure B1.4.7.** Top: THz generation by optical-heterodyne conversion in low-temperature GaAs. (a) The three DBR laser system that synthesizes a precise difference frequency for the THz photomixer spectrometer, (b) the MOPA system and the set-up for spectroscopy. Bottom: second-derivative absorption spectrum of the $CH_3CN$ $J_K = 16_K \rightarrow 17_K$ rotational transitions near 312 GHz. (a) The spectrum for ordinary $^{12}CH_3$ $^{12}CN$. The inset is an expanded view of the $K = 0–2$ lines. (b) The $K = 0–3$ lines of $CH_3$ $^{13}CN$ [54].

interactions are therefore assuming an increasingly pivotal role in modern molecular science, particularly in molecular biology [54]. Within chemical physics, the anisotropy of intermolecular forces plays a central role in understanding the dynamics associated with photoinitiated reactions in clusters [55], to name but one example.

Over the past century, much of the data underlying current descriptions of van der Waals and hydrogen bonds were obtained from measurements of second virial coefficients, pressure broadening, and other classical properties. Experimental advances during the past three decades have led to many techniques capable of interrogating intermolecular forces, most notably scattering experiments and the spectroscopic study of isolated clusters. Despite this long-standing interest, however, truly quantitative, microscopic models of these forces have only become available in recent years as advances in *ab initio* theory, high-resolution spectroscopy and eigenvalue generation on complex potential energy surfaces have converged, to enable the fitting of fully anisotropic force fields to experimental data for systems with two, three and four degrees of freedom [57].

The weak interactions in clusters are mathematically modelled by means of a multi-dimensional intermolecular potential energy surface, or IPS. Microwave spectroscopy, carried out primarily with the elegant, extraordinarily sensitive and sub-Doppler resolution FTMW technique outlined previously, proves the very lowest region of the ground-state surface; visible and IR spectroscopy probes states above the dissociation energy of the adduct (the latter of which are described elsewhere in this encyclopaedia). None are, in themselves, direct probes of the total ground-state IPS. Indeed, while microwave, IR and UV/Vis instruments have produced structural parameters and dynamical lifetimes for literally dozens of binary (and larger) weakly bound complexes (WBCs) over the past two decades [58–61], recent calculations which explicitly allow coupling between all the degrees of freedom present in the cluster reveal that structural parameters alone are not sufficient to accurately characterize the IPS [62].

Weak interactions are characterized by binding energies of at most a few kcal/mole and by IPSs with a very rich and complex topology connected by barriers of at most a few hundred $cm^{-1}$. Rotational, tunnelling and intermolecular vibrational states can therefore become quite strongly mixed, hence the general term of vibration–rotation–tunnelling (VRT) spectroscopy for the study of eigenvalues supported by an IPS [63, 64]. The VRT states in nearly all systems lie close to or above the tunnelling barriers, and therefore sample *large* regions of the potential surface. In addition, as they become spectroscopically observable, the number, spacings and intensities of the tunnelling splittings are intimately related to the nature of the tunnelling *paths* over the potential surface.

Thus, by measuring the intermolecular vibrations of a WBC, ultimately with resolution of the rotational, tunnelling and hyperfine structure, the most sensitive measure of the IPS is accessed directly. The difficulty of measuring these VRT spectra is the fact that they lie nearly exclusively at THz frequencies. As expected, the 'stiffer' the interaction, the higher in frequency these modes are found. In general, the total 0.3–30 THz interval must be accessed, although for the softest or heaviest species the modes rarely lie above 10–15 THz.

For WBCs composed of stable molecules, planar jet expansions produce sufficiently high concentrations that direct absorption THz studies can be pursued for clusters containing $\lesssim 6$ small molecules using FIR laser sidebands. Research on water clusters has been particularly productive, and has been used to investigate the structures and large-amplitude dynamics of the clusters outlined in the top panel of figure B1.4.8 [65–68]. In addition, as the bottom panel illustrates, not only are the VRT modes directly sampled by such work, but the available spectral resolution of $\lesssim 1$ MHz enables full rotational resolution along with a detailed investigation of the VRT and hyperfine splittings. The high resolution is also essential in untangling the often overlapping bands from the many different clusters formed in the supersonic expansion.

For clusters beyond the dimer, each of the monomers can both accept and donate hydrogen bonds, which leads to a rich suite of large-amplitude motions. Their spectroscopic manifestations are illustrated for the water trimer in figure B1.4.9. The most facile motion in this system is the 'flipping' of one of the non-bonded hydrogen atoms through the plane of the oxygen atoms. This motion is sufficiently fast that it produces symmetric top rovibrational spectra even though at any one instant the molecule is always asymmetric. Six of these flipping motions lead to the same structure, a process know as 'pseudorotation', and leads to the manifold

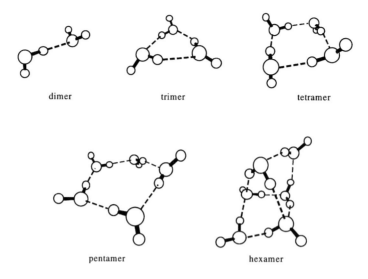

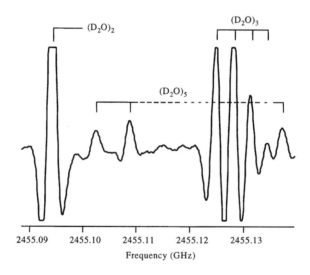

**Figure B1.4.8.** Top: the lowest-energy structures of water clusters, $(H_2O)_n$, from $n = 2$–6. Bottom: a sample $\sim$2.5 THz spectrum of such clusters formed in a pulsed planar supersonic expansion [65].

of states produced in the bottom panel of figure B1.4.9. The exchange of bound *versus* free hydrogen atoms in a monomer leads to the hyperfine splittings of the individual transitions, as is illustrated in the top panel of the same figure. From a comparison of such spectra with detailed calculations, a variety of IPS properties can be extracted to experimental precision.

Molecules like those presented in figure B1.4.4 form another interesting suite of targets from a THz perspective. Such chains can be treated as rigid rods, and as they get longer their lowest bending frequencies

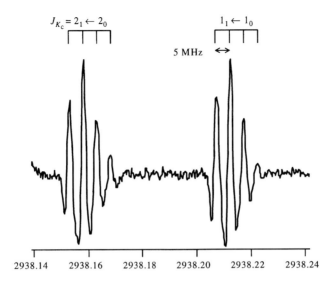

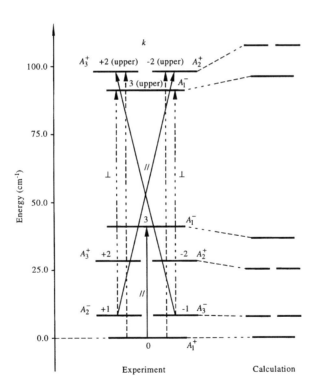

**Figure B1.4.9.** Top: rotation–tunnelling hyperfine structure in one of the 'flipping' modes of $(D_2O)_3$ near 3 THz. The small splittings seen in the $Q$-branch transitions are induced by the bound–free hydrogen atom tunnelling by the water monomers. Bottom: the low-frequency torsional mode structure of the water dimer spectrum, including a detailed comparison of theoretical calculations of the dynamics with those observed experimentally [65]. The symbols next to the arrows depict the parallel ($\Delta k = 0$) *versus* perpendicular ($\Delta K = \pm 1$) nature of the selection rules in the pseudorotation manifold.

move rapidly into the FIR. For example, the lowest frequencies of a variety of chains are as follows:

Cyanopolyynes, $HC_3N \rightarrow HC_{25}N$ $\quad$ $222 \rightarrow 8$ cm$^{-1}$
Polyacetylenes, $HC_4N \rightarrow HC_{20}H$ $\quad$ $220 \rightarrow 19$ cm$^{-1}$
Carbon clusters, $C_3 \rightarrow C_{20}$ $\quad$ $63 \rightarrow 17$ cm$^{-1}$
$C_nN$ radicals, $C_3N \rightarrow C_{19}N$ $\quad$ $144 \rightarrow 4$ cm$^{-1}$.

A real advantage of working in the FIR is that both polar and non-polar chains may be searched for. Indeed, the lowest bending frequency of $C_3$ has been studied in the laboratory [69], and tentatively detected toward the galactic centre source Sgr B2 [70]. Other large molecules such as polycyclic aromatic hydrocarbons (anthracene, pyrene, perylene, etc) or 'biomolecules' such as glycine or uracil also possess low-frequency FIR vibrations, and can be produced in sizable quantities in supersonic expansions through heated planar nozzles [71]. The study of such species is important cosmochemically, but is quite difficult at microwave frequencies where the rotational spectra are weak, and nearly impossible at IR or optical wavelengths due to the extinction present in dense molecular clouds and young stellar objects.

## B1.4.5  Outlook

Technology developments are revolutionizing the spectroscopic capabilities at THz frequencies. While no one technique is ideal for all applications, both CW and pulsed spectrometers operating at or near the fundamental limits imposed by quantum mechanics are now within reach. Compact, all-solid-state implementations will soon allow such spectrometers to move out of the laboratory and into a wealth of field and remote-sensing applications. From the study of the rotational motions of light molecules to the large-amplitude vibrations of clusters and the collective motions of condensed phases, microwave and THz spectroscopy opens up new windows to a wealth of scientifically and technologically important fields. Over the coming decade, truly user-friendly and extraordinarily capable instruments should become cost affordable and widely available, enabling this critical region of the electromagnetic spectrum to be fully exploited for the first time.

## References

[1]   Johns J W C 1985 High resolution far-infrared (20–350 cm$^{-1}$) spectra of several isotopic species of $H_2O$ *J. Opt. Soc. Am.* B **2** 1340–54
[2]   McLean I S 1995 Infrared array detectors—performance and prospects *IAU Symp.* **167** 69–78
[3]   Kimmitt M F 1970 *Far-Infrared Techniques* (London: Pion)
[4]   Bock J J, Chen D, Mauskopf P D and Lange A E 1995 A novel bolometer for infrared and millimeter-wave astrophysics *Space Sci. Rev.* **74** 229–35
[5]   Lis D C, Serabyn E, Keene J, Dowell C D, Benford D J, Phillips T G, Hunter T R and Wang N 1998 350 micro continuum imaging of the Orion A molecular cloud with the submillimeter high angular resolution camera (SHARC) *Astrophys. J.* **509** 299–308
[6]   Ivison R J, Smail I, Le Borgne J F, Blain A W, Kneib J P, Bezecourt J, Kerr T H and Davies J K 1997 A hyperluminous galaxy at $z = 2.8$ found in a deep submillimetre survey *Mon. Not. R. Astron. Soc.* **298** 583–93
[7]   Putley E H 1964 The ultimate sensitivity of sub-mm detectors *Infra. Phys.* **4** 1–35
[8]   Beeman J W and Haller E E 1994 Ga:Ge photoconductor arrays—design considerations and quantitative analysis of prototype single pixels *Infra. Phys. Technology* **35** 827–36
[9]   Waters J W 1993 Microwave limb sounding *Atmospheric Remote Sensing by Microwave Radiometry* ed M A Janssen (New York: Wiley) pp 383–496
[10]   Chance K, Traub W A, Johnson D G, Jucks K W, Ciarpallini P, Stachnik R A, Salawitch R J and Michelsen H A 1996 Simultaneous measurements of stratospheric $HO_x$, $NO_x$, and $Cl_{-x}$: comparison with a photochemical model *J. Geophys. Res.* **101** 9031–43
[11]   Winnewisser G 1994 Submillimeter and infrared astronomy: recent scientific and technical developments *Infra. Phys. Tech.* **35** 551–67
[12]   Carlstron J and Zmuidzinas J 1996 Millimeter and submillimeter techniques *Review of Radio Science 1993–1996* ed W Ross Stone (Oxford: Oxford University Press) pp 839–92
[13]   Bin M, Gaidis M C, Zmuidzinas J and Phillips T G 1997 Quasi-optical SIS mixers with normal tuning structures *IEEE Trans. Appl. Supercond.* **7** 3584–8

[14] Karasik B S, Gaidis M C, McGrath W R, Bumble B and LeDuc H G 1997 A low noise 2.5 THz superconducting NB hot-electron mixer *IEEE Trans. Appl. Supercond.* **7** 3580–3

[15] Barrett A H 1983 The beginnings of molecular radio astronomy *Serendipitous Discoveries in Radio Astronomy* ed K Kellerman and B Sheets (Green Bank, WV: NRAO)

[16] Ohishi M 1997 Observations of hot cores *IAU Symp.* **178** 61–74

[17] Phillips T G and Keene J 1992 Submillimeter astronomy *Proc. IEEE* **80** 1662–78

[18] Hartquist T W and Williams D A (eds) 1998 *The Molecular Astrophysics of Stars and Galaxies* (Oxford: Oxford University Press)

[19] Herbst E 1995 Chemistry in the interstellar medium *Ann. Rev. Phys. Chem.* **46** 27–53

[20] van Dishoeck E F and Blake G A 1998 Chemical evolution of star forming regions *Ann. Rev. Astron. Ap.* **36** 317–68

[21] van Dishoeck E F 1997 The importance of high resolution far-infrared spectroscopy of the interstellar medium *Proc. ESA Symp.* **SP-401** 81–90

[22] Dutrey A, Guilloteau S and Guelin M 1997 Chemistry of protosolar-like nebulae: the molecular content of the DM Tau and GG Tau disks *Astron. Astrophys.* **317** L55–8

[23] Sargent A I 1997 Protostellar and protoplanetary disks *IAU Symp.* **170** 151–8

[24] Biver N *et al* 1997 Evolution of the outgassing of comet Hale–Bopp (C/1995 O1) from radio observations *Science* **275** 1915–18

[25] For an overview of ALMA, see http://www.nrao.edu

[26] Erickson E F 1995 SOFIA—the next generation airborne observatory *Space Sci. Rev.* **74** 91–100

[27] Batchelor M, Adler D and Trogus W 1996 New plans for FIRST *Missions to the Moon & Exploring the Cold Universe* **18** 185–8

[28] Balle T J and Flygare W H 1981 A Fourier transform microwave spectrometer *Rev. Sci. Instrum.* **52** 33–45

[29] Munrow M R, Pringle W C and Novick S E 1999 Determination of the structure of the argon cyclobutanone van der Waals complex *J. Chem. Phys.* **103** 2256–61

[30] Lovas F J, Suenram R D, Ogata T and Yamamoto S 1992 Microwave spectra and electric dipole moments for low-J levels of interstellar radicals—SO, CCS, CCCS, c-H$_3$, CH$_2$CC and c-C$_3$H$_2$ *Astrophys. J.* **399** 325–9

[31] Thaddeus P, McCarthy M C, Travers M, Gottlieb C and Chen W 1998 New carbon chains in the laboratory and in interstellar space *Faraday Soc. Discuss.* **109** 121–36

[32] McCarthy M C, Travers M J, Kovacs A, Chen W, Novick S E, Gottlieb C A and Thaddeus P 1997 *Science* **275** 518–20

[33] Gordy W 1960 *Proc. Symp. MM Waves* (Brooklyn: Polytechnic Press)

[34] Rothermel H, Phillips T G and Keene J 1989 *Int. J. Infra. MM Waves* **10** 83–100

[35] Lewen F, Gendriesch R, Pak I, Paveliev D G, Hepp M, Schider R and Winnewisser G 1998 Phase locked backward wave oscillator pulsed beam spectrometer in the submillimeter wave range *Rev. Sci. Instrum.* **69** 32–9

[36] Winnewisser G, Belov S P, Klaus T and Schieder R 1997 Sub-Doppler measurements on the rotational transitions of carbon monoxide *J. Mol. Spectrosc.* **184** 468–72

[37] Maranowski K D, Gossard A C, Unterrainer K and Gornik E 1996 Far-infrared emission from parabolically graded quantum wells *Appl. Phys. Lett.* **69** 3522–4

[38] Xu B, Hu Q and Melloch M R 1997 Electrically pumped tunable terahertz emitter based on inter-subband transitions *Appl. Phys. Lett.* **71** 440–2

[39] Bicanic D D, Zuiberg B F J and Dymanus 1978 *Appl. Phys. Lett.* **32** 367–9

[40] Blake G A, Laughlin K B, Cohen R C, Busarow K L, Gwo D H, Schmuttenmaer C A, Steyert D W and Saykally R J 1991 Tunable far-infrared spectrometers *Rev. Sci. Instrum.* **62** 1693–700

[41] Varberg T D and Evenson K M 1992 Accurate far-infrared rotational frequencies of carbon monoxide *Astrophys. J.* **385** 763–5

[42] Odashima H, Zink L R and Evenson K M 1999 Tunable far-infrared spectroscopy extended to 9.1 THz *Opt. Lett.* **24** 406–7

[43] Kono J, Su M Y, Inoshita T, Noda T, Sherwin M S, Allen S J and Sakaki H 1997 Resonant THz optical sideband generation from confined magnetoexcitons *Phys. Rev. Lett.* **79** 1758–61

[44] For a historical review see Lee C H 1984 *Picosecond Optoelectronic Devices* (New York: Academic)

[45] Gupta S, Frankel M Y, Valdmanis J A, Whitaker J F, Mourou G A, Smith F W and Calawa A R 1991 Subpicosecond carrier lifetime in GaAs grown by MBE at low temperatures *Appl. Phys. Lett.* **59** 3276–8

[46] Jeon T I and Grischkowsky D 1998 Characterization of optically dense, doped semiconductors by reflection THz time domain spectroscopy *Appl. Phys. Lett.* **72** 3032–4

[47] Cheville R A and Grioschkowsky D 1999 Far-infrared foreign and self-broadened rotational linewidths of high temperature water vapor *J. Opt. Soc. Am. B* **16** 317–22

[48] Nuss M C and Orenstein J 1998 Terahertz time domain spectroscopy *Millimeter Submillimeter Wave Spectrosc. Solids* **74** 7–50

[49] Han P Y, Cho G C and Zhang X C 1999 Mid-IR THz beam sensors: exploration and application for phonon spectroscopy, ultrafast phenomena in semiconductors III *Proc. SPIE* **3624** 224–33

[50] Koch M, Hunsche S, Schuacher P, Nuss M C, Feldmann J and Fromm J 1998 THz-imaging: a new method for density mapping of wood *Wood Sci. Technol.* **32** 421–7

[51] Venables D S and Schmuttenmaer C A 1998 Far-infrared spectra and associated dynamics in acetonitrile-water mixtures measured with femtosecond THz pulse spectroscopy *J. Chem. Phys.* **108** 4935–44

[52] Brown E R, McIntosh K A, Smith F W, Manfra M J and Dennis C L 1993 Measurements of optical-heterodyne conversion in low-temperature grown GaAs *Appl. Phys. Lett.* **62** 1206–8

[53] Mastuura S, Blake G A, Wyss R, Pearson J, Kadow C, Jackson A and Gossard A C 1999 A travelling-wave photomixer based on angle-tuned phase matching *Appl. Phys. Lett.* **74** 2872–4

[54] Matsuura S, Chen P, Blake G A, Pearson J and Pickett H M 1999 A tunable, cavity-locked diode laser system for terahertz photomixing *IEEE Micro. Theory Technol.* **48** 380–7

[55] Stone A J 1996 *The Theory of Intermolecular Forces* (Oxford: Oxford University Press)

[56] Ionov S I, Ionov P I and Wittig C 1994 Time resolved studies of photoinitiated reactions in binary and larger $(N_2O)_m(HI)_n$ complexes *Discuss. Faraday Soc.* **97** 391–400

[57] Cohen R C and Saykally R J 1991 Multidimensional intermolecular potential surfaces from VRT spectra of van der Waals complexes *Ann. Rev. Phys. Chem.* **42** 369–92

[58] Novick S, Leopold K and Klemperer W 1990 *Atomic and Molecular Clusters* ed E Bernstein (New York: Elsevier) p 359–91 (the clusters listing presented here is now maintained electronically by S Novick)

[59] Miller R E 1988 *Science* **240** 447–52

[60] Nesbitt D J 1990 *Dynamics of Polyatomic van der Waals Complexes* ed N Halberstadt and K C Janda (New York: Plenum) pp 461–70

[61] Chuang C, Andrews P M and Lester M I 1995 Intermolecular vibrations and spin-orbit predissociation dynamics of NeOH *J. Chem. Phys.* **103** 3418–29

[62] Leforestier C, Braly L B, Liu K, Elrod M J and Saykally R J 1997 Fully coupled 6-dimensional calculations of the water dimer VRT states with a split Wigner pseudo-spectral approach *J. Chem. Phys.* **106** 8527–44

[63] Saykally R J and Blake G A 1993 Molecular interactions and hydrogen bond tunneling dynamics: some new perspectives *Science* **259** 1570–5

[64] Cotti G, Linnartz H, Meerts W L, van der Avoird A and Olthof E 1996 Stark effect and dipole moments of $(NH_3)_2$ in different vibration–rotation–tunneling states *J. Chem. Phys.* **104** 3898–906

[65] Liu K 1997 *PhD Thesis* University of California at Berkeley

[66] Viant M R, Cruzan J D, Lucas D D, Brown M G, Liu K and Saykally R J 1997 Pseudorotation in water trimer isotopomers using terahertz laser spectroscopy *J. Phys. Chem.* **101** 9032–41

[67] Cruzan J D, Viant M R, Brown M G, Lucas D D, Liu K and Saykally R J 1998 Terahertz laser vibration–rotation–tunneling spectrum of the water pentamer-d(10). Constraints on the bifurcation tunneling dynamics *Chem. Phys. Lett.* **292** 667–76

[68] Liu K, Brown M G and Saykally R J 1997 Terahertz laser vibration rotation tunneling spectroscopy and dipole moment of a cage form of the water hexamer *J. Phys. Chem.* **101** 8995–9010

[69] Schmuttenmaer C A, Cohen R C, Pugliano N, Heath J R, Cooksy A L, Busarow K L and Saykally R J 1990 Tunable far-IR laser spectroscopy of jet-cooled carbon clusters—the $v_2$ bending vibration of $C_3$ *Science* **249** 897–900

[70] Heath J R, van Orden A, Hwang H J, Kuo E W, Tanaka K and Saykally R J 1994 Toward the detection of pure carbon clusters in the ISM *Adv. Space Res.* **15** 25–33

[71] Liu K, Fellers R S, Viant M R, McLaughlin R P, Brown M G and Saykally R J 1996 A long pathlength pulsed slit valve appropriate for high temperature operation—infrared spectroscopy of jet-cooled large water clusters and nucleotide bases *Rev. Sci. Instrum.* **67** 410–16

## Further Reading

Kimmitt M F 1970 *Far-Infrared Techniques* (London: Pion)

An excellent overview of optical approaches to THz spectrometers.

Janssen M A (ed) 1993 *Atmospheric Remote Sensing by Microwave Radiometry* (New York: Wiley)

The most complete guide to microwave and THz atmospheric sensing.

Carlstrom J and Zmuidzinas J 1996 Millimeter and submillimeter techniques *Review of Radio Science 1993–1996* ed W Ross Stone (Oxford: Oxford University Press) pp 839–82

A quite readable summary of heterodyne detection strategies.

Tsen K T (ed) 1999 Ultrafast phenomena in semiconductors III *Proc. SPIE* **624** 298

An overview of recent progress in this explosive field.

Gordy W and Cook R J 1991 *Microwave Molecular Spectra* (New York: Wiley)

The most complete textbook available, suitable for graduate students and researchers.

Stone A J 1996 *The Theory of Intermolecular Forces* (Oxford: Oxford University Press)

A thorough, advanced tutorial on the nature of clusters held together by intermolecular forces, and the theories that can be used to analyse them.

# B1.5
# Nonlinear optical spectroscopy of surfaces and interfaces

*Jerry I Dadap and Tony F Heinz*

## B1.5.1 Introduction

### B1.5.1.1 Nonlinear optics and spectroscopy

Nonlinear optics is the study of the interaction between intense electromagnetic radiation and matter. It describes phenomena arising when the response of a medium to the electric field of light leaves the linear regime associated with the familiar and ubiquitous effects, such as reflection, refraction and absorption, comprising classical optics. In the presence of a sufficiently intense light source, the approximation of linearity breaks down. A new and much broader class of optical phenomena may be observed. Prototypical among these nonlinear optical effects is the production of light at new frequencies. Indeed, nonlinear optics is generally considered to have begun in 1961 when Franken and coworkers demonstrated optical second-harmonic generation (SHG) by insertion of a quartz crystal along the path of a laser beam [1]. In addition to the generation of new frequencies from excitation of a monochromatic source, nonlinear optical effects lead to the coupling between beams of identical and disparate frequencies, as well as to the action of a beam of light on itself.

Given the complexity of materials, it is perhaps surprising that a linear response to an applied optical field should be so common. This situation reflects the fact that the strength of electric fields for light encountered under conventional conditions is minute compared to that of the electric fields binding atoms and solids together. The latter may, for example, be estimated as $E_a \sim 1$ V $\text{Å}^{-1} = 10^8$ V cm$^{-1}$. Since the irradiance of a light beam required to reproduce this electric field strength is $\sim 10^{13}$ W cm$^{-2}$, we may understand why a linear approximation of the material response is adequate for conventional light sources. With the advent of the laser, with its capability for producing high optical power and a high degree of coherence, this situation has changed. Under laser radiation, nonlinear optical effects are readily observed and widely exploited.

Over the past decades, nonlinear optics has come to have a broad impact on science and technology, playing a role in areas as diverse as telecommunications, materials processing and medicine. Within the context of chemical science, nonlinear optics is significant in providing new sources of coherent radiation, in permitting chemical processes to be induced under intense electromagnetic fields and in allowing matter to be probed by many powerful spectroscopic techniques. In this chapter, we shall be concerned only with the spectroscopic implications of nonlinear optics. In particular, our attention will be restricted to the narrowed range of spectroscopic techniques and applications related to probing surfaces and interfaces. This subject is a significant one. Surfaces and interfaces have been, and remain, areas of enormous scientific and technological importance. Sensitive and flexible methods of interface characterization are consequently of great value. As the advances discussed in this chapter reveal, nonlinear optics offers unique capabilities to address surface and interface analysis.

*B1.5.1.2   Probing surfaces and interfaces*

The distinctive chemical and physical properties of surfaces and interfaces typically are dominated by the nature of one or two atomic or molecular layers [2, 3]. Consequently, useful surface probes require a very high degree of sensitivity. How can this sensitivity be achieved? For many of the valuable traditional probes of surfaces, the answer lies in the use of particles that have a short penetration depth through matter. These particles include electrons, atoms and ions, of appropriate energies. Some of the most familiar probes of solid surfaces, such as Auger electron spectroscopy (AES), low-energy electron diffraction (LEED), electron energy loss spectroscopy (EELS) and secondary ion mass spectroscopy (SIMS), exploit massive particles both approaching and leaving the surface. Other techniques, such as photoemission spectroscopy and inverse photoemission spectroscopy, rely on electrons for only half of the probing process, with photons serving for the other half. These approaches are complemented by those that directly involve the adsorbate of interest, such as molecular beam techniques and temperature programmed desorption (TPD). While these methods are extremely powerful, they are generally restricted to—or perform best for—probing materials under high vacuum conditions. This is a significant limitation, since many important systems are intrinsically incompatible with high vacuum (such as the surfaces of most liquids) or involve interfaces between two dense media. Scanning tunnelling microscopy (STM) is perhaps the electron-based probe best suited for investigations of a broader class of interfaces. In this approach, the physical proximity of the tip and the probe permits the method to be applied at certain interfaces between dense media.

Against this backdrop, the interest in purely optical probes of surfaces and interfaces can be easily understood. Since photons can penetrate an appreciable amount of material, photon-based methods are inherently appropriate to probing a very wide class of systems. In addition, photons, particularly those in the optical and infrared part of the spectrum, can be produced with exquisite control. Both the spatial and temporal properties of light beams can be tailored to the application. Particularly noteworthy is the possibility afforded by laser radiation of having either highly monochromatic radiation or radiation in the form of ultrafast (femtosecond) pulses. In addition, sources of high brightness are available, together with excellent detectors. As a consequence, within the optical spectral range several surface and interface probes have been developed. (Complementary approaches also exist within the x-ray part of the spectrum.) For optical techniques, one of the principal issues that must be addressed is the question of surface sensitivity. The ability of light to penetrate through condensed media was highlighted earlier as an attractive feature of optics. It also represents a potential problem in achieving the desired surface sensitivity, since one expects the bulk contribution to dominate that from the smaller region of the surface or interface.

Depending on the situation at hand, various approaches to achieving surface sensitivity may be appropriate for an optical probe. Some schemes rely on the presence of distinctive spectroscopic features in the surface region that may be distinguished from the response of the bulk media. This situation typically prevails in surface infrared spectroscopy [4] and surface-enhanced Raman scattering (SERS) [5–7]. These techniques provide very valuable information about surface vibrational spectra, although the range of materials is somewhat restricted in the latter case. Ellipsometry [8–10] permits a remarkably precise determination of the reflectivity of an interface through measurements of the polarization properties of light. It is a powerful tool for the analysis of thin films. Under appropriate conditions, it can be pushed to the limit of monolayer sensitivity. Since the method has, however, no inherent surface specificity, such applications generally require accurate knowledge and control of the relevant bulk media. A relatively recent addition to the set of optical probes is reflection difference absorption spectroscopy (RDAS) [10, 11]. In this scheme, the lowered symmetry of certain surfaces of crystalline materials is exploited in a differential measurement of reflectivity that cancels out the optical response of the bulk media.

In this chapter, we present a discussion of the nonlinear spectroscopic methods of second-harmonic generation (SHG) and sum-frequency generation (SFG) for probing surfaces and interfaces. While we have previously described the relative ease of observing nonlinear optical effects with laser techniques, it is still clear that linear optical methods will always be more straightforward to apply. Why then is nonlinear spectroscopy an attractive option for probing surfaces and interfaces? First, we may note that the method retains all of the advantages associated with optical methods. In addition, however, for a broad class of material systems the technique provides an *intrinsic* sensitivity to interfaces on the level of a single atomic layer. This is a very desirable feature that is lacking in linear optical probes. The relevant class of materials for which the nonlinear approach is inherently surface sensitive is quite broad. It consists of interfaces between all pairs of centrosymmetric materials. (Centrosymmetric materials, which include most liquids, gases, amorphous solids and elemental crystals, are those that remain unchanged when the position of every point is inverted through an appropriate origin.) The surface sensitivity of the SHG and SFG processes for these systems arises from a simple symmetry property: the second-order nonlinear optical processes of SHG and SFG are forbidden in centrosymmetric media. Thus, the bulk of the material does not exhibit a significant nonlinear optical response. On the other hand, the interfacial region, which necessarily breaks the inversion symmetry of the bulk, provides the desired nonlinear optical response.

Because of the generality of the symmetry principle that underlies the nonlinear optical spectroscopy of surfaces and interfaces, the approach has found application to a remarkably wide range of material systems. These include not only the conventional case of solid surfaces in ultrahigh vacuum, but also gas/solid, liquid/solid, gas/liquid and liquid/liquid interfaces. The information attainable from the measurements ranges from adsorbate coverage and orientation to interface vibrational and electronic spectroscopy to surface dynamics on the femtosecond time scale.

### B1.5.1.3  *Scope of the chapter*

In view of the diversity of material systems to which the SHG/SFG method has been applied and the range of the information that the method has yielded, we cannot give a comprehensive account of the technique in this chapter. For such accounts, we must refer the reader to the literature, particularly as summarized in various review articles [12–27] and monographs [10, 28, 29]. Our aim here is only to present an overview of the subject in which we attempt to describe the basic principles. The chapter is organized in the following fashion. We first outline basic theoretical considerations relevant to the technique, both in a brief general discussion of nonlinear optics and in a specific description of the nonlinear response of interfaces. After a few words about experimental techniques for surface SHG/SFG measurement, we devote the remainder of the chapter to describing the type of information that may be extracted from the nonlinear measurements. We have attempted at least to mention the different classes of information that have been obtained, such as adsorbate coverage or vibrational spectroscopy. In most cases, the corresponding approach has been widely and fruitfully applied in many experimental studies. Although we offer some representative examples, space does not permit us to discuss these diverse applications in any systematic way.

## B1.5.2  Theoretical considerations

### B1.5.2.1  *General background on nonlinear optics*

#### (a) Anharmonic oscillator model

In order to illustrate some of the basic aspects of the nonlinear optical response of materials, we first discuss the anharmonic oscillator model. This treatment may be viewed as the extension of the classical Lorentz model of the response of an atom or molecule to include nonlinear effects. In such models, the medium is treated as a collection of electrons bound about ion cores. Under the influence of the electric field associated with an optical wave, the ion cores move in the direction of the applied field, while the electrons are displaced in the opposite direction. These motions induce an oscillating dipole moment, which then couples back to the radiation fields. Since the ions are significantly more massive than the electrons, their motion is of secondary importance for optical frequencies and is neglected.

While the Lorentz model only allows for a restoring force that is linear in the displacement of an electron from its equilibrium position, the anharmonic oscillator model includes the more general case of a force that varies in a nonlinear fashion with displacement. This is relevant when the displacement of the electron becomes significant under strong driving fields, the regime of nonlinear optics. Treating this problem in one dimension, we may write an appropriate classical equation of motion for the displacement, $x$, of the electron from equilibrium as

$$m \left[ \frac{d^2 x}{dt^2} + 2\gamma \frac{dx}{dt} + \omega_0^2 x - (b^{(2)} x^2 + b^{(3)} x^3 + \cdots) \right] = -eE(t). \tag{B1.5.1}$$

Here $E(t)$ denotes the applied optical field, and $-e$ and $m$ represent, respectively, the electronic charge and mass. The (angular) frequency $\omega_0$ defines the resonance of the harmonic component of the response, and $\gamma$ represents a phenomenological damping rate for the oscillator. The nonlinear restoring force has been written in a Taylor expansion; the terms $m(b^{(2)} x^2 + b^{(3)} x^3 + \cdots)$ correspond to the corrections to the harmonic restoring force, with parameters $b^{(2)}$, $b^{(3)}$, ... taken as material-dependent constants. In this equation, we have recognized that the excursion of the electron is typically small compared to the optical wavelength and have omitted any dependence of the driving term $-eE$ on the position of the electron.

Here we consider the response of the system to a monochromatic pump beam at a frequency $\omega$,

$$E(t) = E(\omega) \exp(-i\omega t) + \text{c.c.} \tag{B1.5.2}$$

where the complex conjugate (c.c.) is included to yield an electric field $E(t)$ that is a real quantity. We use the symbol $E$ to represent the electric field in both the time and frequency domain; the different arguments should make the interpretation clear. Note also that $E(\omega)$ and analogous quantities introduced later may be complex. As a first approximation to the solution of equation (B1.5.1), we neglect the anharmonic terms to obtain the steady-state motion of the electron

$$x(t) = \frac{-eE(\omega)}{m} \frac{\exp(-i\omega t)}{\omega_0^2 - 2i\gamma\omega - \omega^2} + \text{c.c.} \tag{B1.5.3}$$

This solution is appropriate for the regime of a weak driving field $E(t)$. If we now treat the material as a collection of non-interacting oscillators, we may write the induced polarization as a sum of the individual dipole moments over a unit volume, i.e. $P(t) = -Nex(t)$, where $N$ denotes the density of dipoles and local-field effects have been omitted. Following equation (B1.5.2), we express the $P(t)$ in the frequency domain as

$$P(t) = P(\omega) \exp(-i\omega t) + \text{c.c.} \tag{B1.5.4}$$

We may then write the amplitude for the harmonically varying polarization as proportional to the corresponding quantity for the driving field, $E(\omega)$:

$$P(\omega) = \chi^{(1)}(\omega) E(\omega). \tag{B1.5.5}$$

The constant of proportionality,

$$\chi^{(1)}(\omega) = \frac{Ne^2}{m} \frac{1}{\omega_0^2 - 2i\gamma\omega - \omega^2} \tag{B1.5.6}$$

represents the linear susceptibility of the material. It is related to the dielectric constant $\varepsilon(\omega)$ by $\varepsilon(\omega) = 1 + 4\pi \chi^{(1)}(\omega)$.

Up to this point, we have calculated the linear response of the medium, a polarization oscillating at the frequency $\omega$ of the applied field. This polarization produces its own radiation field that interferes with the

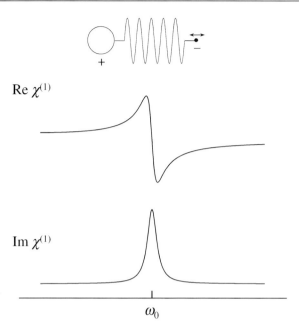

**Figure B1.5.1.** Anharmonic oscillator model: the real and imaginary parts of the linear susceptibility $\chi^{(1)}$ are plotted as a function of angular frequency $\omega$ in the vicinity of the resonant frequency $\omega_0$.

applied optical field. Two familiar effects result: a change in the speed of the light wave and its attenuation as it propagates. These properties may be related directly to the linear susceptibility $\chi^{(1)}(\omega)$. The index of refraction, $n = \mathrm{Re}[\sqrt{1 + 4\pi \chi^{(1)}}]$, is associated primarily with the real part of $\chi^{(1)}(\omega)$. It describes the reduced (phase) velocity, $c/n$, of the optical wave travelling through the medium compared to its speed, $c$, in vacuum. The imaginary part, $\mathrm{Im}[\chi^{(1)}(\omega)]$, on the other hand, gives rise to absorption of the radiation in the medium. The frequency dependence of the quantities $\mathrm{Re}[\chi^{(1)}(\omega)]$ and $\mathrm{Im}[\chi^{(1)}(\omega)]$ are illustrated in figure B1.5.1. They exhibit a resonant response for optical frequencies $\omega$ near $\omega_0$, and show the expected dispersive and absorptive lineshapes.

If we now include the anharmonic terms in equation (B1.5.1), an exact solution is no longer possible. Let us, however, consider a regime in which we do not drive the oscillator too strongly, and the anharmonic terms remain small compared to the harmonic ones. In this case, we may solve the problem perturbatively. For our discussion, let us assume that only the second-order term in the nonlinearity is significant, i.e. $b^{(2)} \neq 0$ and $b^{(i)} = 0$ for $i > 2$ in equation (B1.5.1). To develop a perturbational expansion formally, we replace $E(t)$ by $\lambda E(t)$, where $\lambda$ is the expansion parameter characterizing the strength of the field $E$. Thus, equation (B1.5.1) becomes

$$\ddot{x} + 2\gamma \dot{x} + \omega_0^2 x - b^{(2)} x^2 = -\lambda e E(t)/m. \tag{B1.5.7}$$

We then write the solution of equation (B1.5.7) as a power series expansion in terms of the strength $\lambda$ of the perturbation:

$$x = \lambda x^{(1)} + \lambda^2 x^{(2)} + \lambda^3 x^{(3)} + \cdots. \tag{B1.5.8}$$

If we substitute the expression for (B1.5.8) back into (B1.5.7) and require that the terms proportional to $\lambda$ and $\lambda^2$ on both sides of the resulting equation are equal, we obtain the equations

$$\ddot{x}^{(1)} + 2\gamma \dot{x}^{(1)} + \omega_0^2 x^{(1)} = -e E(t)/m \tag{B1.5.9}$$

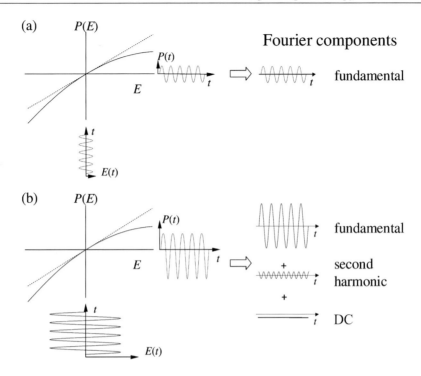

**Figure B1.5.2.** Nonlinear dependence of the polarization $P$ on the electric field $E$. (a) For small sinusoidal input fields, $P$ depends linearly on $E$; hence its harmonic content is mainly that of $E$. (b) For a stronger driving electric field $E$, the polarization waveform becomes distorted, giving rise to new harmonic components. The second-harmonic and DC components are shown.

$$\ddot{x}^{(2)} + 2\gamma \dot{x}^{(2)} + \omega_0^2 x^{(2)} - b^{(2)}(x^{(1)})^2 = 0. \tag{B1.5.10}$$

We immediately observe that the solution, $x^{(1)}$, to equation (B1.5.9) is simply that of the original harmonic oscillator problem given by equation (B1.5.3). Substituting this result for $x^{(1)}$ into the last term of equation (B1.5.10) and solving for the second-order term $x^{(2)}$, we obtain two solutions: one oscillating at twice the frequency of the applied field and a static part at a frequency of $\omega = 0$. This behaviour arises from the fact that the square of the term $x^{(1)}(t)$, which now acts as a source term in equation (B1.5.10), possesses frequency components at frequencies $2\omega$ and zero. The material response at the frequency $2\omega$ corresponds to SHG, while the response at frequency zero corresponds to the phenomenon of optical rectification. The effect of a linear and nonlinear relation between the driving field and the material response is illustrated in figure B1.5.2.

In analogy to equation (B1.5.3), we can write the steady-state solution to equation (B1.5.10) for the SHG process as

$$x^{(2)}(t) = x^{(2)}(2\omega) \exp(-2i\omega t) + \text{c.c.} \tag{B1.5.11}$$

The amplitude of the response, $x^{(2)}(2\omega)$, is given by the steady-state solution of equation (B1.5.10) as

$$x^{(2)}(2\omega) = \frac{(e/m)^2 b^{(2)} E(\omega)^2}{D(2\omega) D^2(\omega)} \tag{B1.5.12}$$

where the quantity $D(\omega)$ is defined as $D(\omega) \equiv \omega_0^2 - \omega^2 - 2i\gamma\omega$. Similarly, the amplitude of the response at

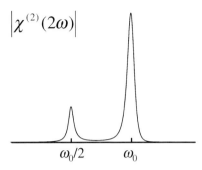

**Figure B1.5.3.** Magnitude of the second-order nonlinear susceptibility $\chi^{(2)}$ versus frequency $\omega$, obtained from the anharmonic oscillator model, in the vicinity of the single- and two-photon resonances at frequencies $\omega_0$ and $\omega_0/2$, respectively.

frequency zero for the optical rectification process is given by

$$x^{(2)}(0) = \frac{2(e/m)^2 b^{(2)} E(\omega) E(\omega)^*}{D(0) D(\omega) D(-\omega)}. \tag{B1.5.13}$$

Following the derivation of the linear susceptibility, we may now readily deduce the second-order susceptibility $\chi^{(2)}(2\omega = \omega + \omega)$ for SHG, as well as $\chi^{(2)}(0 = \omega - \omega)$ for the optical rectification process. Defining the second-order nonlinear susceptibility for SHG as the relation between the square of the relevant components of the driving fields and the nonlinear source polarization,

$$P(2\omega) = \chi^{(2)}(2\omega) E(\omega) E(\omega) \tag{B1.5.14}$$

we obtain

$$\chi^{(2)}(2\omega = \omega + \omega) = \frac{-(e^2/m^2) N b^{(2)}}{D(2\omega) D^2(\omega)}. \tag{B1.5.15}$$

As we shall discuss later in a detailed fashion, the nonlinear polarization associated with the nonlinear susceptibility of a medium acts as a source term for radiation at the second harmonic (SH) frequency $2\omega$. Since there is a definite phase relation between the fundamental pump radiation and the nonlinear source term, coherent SH radiation is emitted in well-defined directions. From the quadratic variation of $P(2\omega)$ with $E(\omega)$, we expect that the SH intensity $I_{2\omega}$ will also vary quadratically with the pump intensity $I_\omega$.

If we compare the nonlinear response of $\chi^{(2)}(2\omega = \omega + \omega)$ with the linear material response of $\chi^{(1)}(\omega)$, we find both similarities and differences. As is apparent from equation (B1.5.15) and shown pictorially in figure B1.5.3, $\chi^{(2)}(2\omega = \omega + \omega)$ exhibits a resonant enhancement for frequencies $\omega$ near $\omega_0$, just as in the case for $\chi^{(1)}(\omega)$. However, in addition to this so-called one-photon resonance, $\chi^{(2)}(2\omega = \omega + \omega)$ also displays a resonant response when $2\omega$ is near $\omega_0$ or $\omega \approx \omega_0/2$. This feature is termed a two-photon resonance and has no analogue in linear spectroscopy. Despite these differences between $\chi^{(1)}$ and $\chi^{(2)}$, one can see that both types of response provide spectroscopic information about the material system. A further important difference concerns symmetry characteristics. The linear response $\chi^{(1)}$ may be expected to be present in any material. The second-order nonlinear response $\chi^{(2)}$ requires the material to exhibit a non-centrosymmetric character, i.e. in the one-dimensional model, the $+x$ and $-x$ directions must be distinguishable. To understand this property, consider the potential energy associated with the restoring force on the electron. This potential may be written in the notation previously introduced as $V(x) = \frac{1}{2} m \omega_0^2 x^2 - \frac{1}{3} m b^{(2)} x^3 + \cdots$. The first term is allowed for all materials; the second term, however, cannot be present in centrosymmetric materials since it clearly differentiates between a displacement of $+x$ and $-x$. Thus $b^{(2)} = 0$ except in non-centrosymmetric materials. This symmetry distinction is the basis for the remarkable surface and interface sensitivity of SHG and SFG for bulk materials with inversion symmetry.

*(b) Maxwell's equations*

We now embark on a more formal description of nonlinear optical phenomena. A natural starting point for this discussion is the set of Maxwell equations, which are just as valid for nonlinear optics as for linear optics. In the absence of free charges and current densities, we have in cgs units:

$$\nabla \times H = \frac{1}{c}\frac{\partial D}{\partial t} \tag{B1.5.16}$$

$$\nabla \times E = -\frac{1}{c}\frac{\partial B}{\partial t} \tag{B1.5.17}$$

$$\nabla \cdot B = 0 \tag{B1.5.18}$$

$$\nabla \cdot D = 0 \tag{B1.5.19}$$

where $E$ and $H$ are the electric and magnetic field intensities, respectively; $D$ and $B$ are the electric displacement and magnetic induction, respectively. In the optical regime, we generally neglect the magnetic response of the material and take $B = H$. The material response is then incorporated into the Maxwell equations through the displacement vector $D$. This quantity is related to the electric field $E$ and the polarization $P$ (the electric dipole moment per unit volume) by

$$D = E + 4\pi P. \tag{B1.5.20}$$

The polarization $P$ is given in terms of $E$ by the constitutive relation of the material. For the present discussion, we assume that the polarization $P(r)$ depends only on the field $E$ evaluated at the same position $r$. This is the so-called dipole approximation. In later discussions, however, we will consider, in some specific cases, the contribution of a polarization that has a non-local spatial dependence on the optical field. Once we have augmented the system of equations (B1.5.16)–(B1.5.19) and equation (B1.5.20) with the constitutive relation for the dependence of $P$ on $E$, we may solve for the radiation fields. This relation is generally characterized through the use of linear and nonlinear susceptibility tensors, the subject to which we now turn.

*(c) Nonlinear optical susceptibilities*

If the polarization of a given point in space and time $(r, t)$ depends only on the driving electric field at the same coordinates, we may write the polarization as $P = P(E)$. In this case, we may develop the polarization in power series as $P = P_L + P_{NL} = P^{(1)} + P^{(2)} + P^{(3)} + \cdots$, where the linear term is $P_i^{(1)} = \sum_i \chi_{ij}^{(1)} E_j$ and the nonlinear terms include the second-order response $P_i^{(2)} = \sum_{jk} \chi_{ijk}^{(2)} E_j E_k$, the third-order response $P_i^{(3)} = \sum_{jkl} \chi_{ijkl}^{(3)} E_j E_k E_l$, and so forth. The coefficients $\chi_{ij}^{(1)}$, $\chi_{ijk}^{(2)}$, and $\chi_{ijkl}^{(3)}$ are, respectively, the linear, second-order nonlinear and the third-order nonlinear susceptibilities of the material. The quantity $\chi_{ijk...}^{(n)}$, it should be noted, is a tensor of rank $n + 1$, and $ijkl$ refer to indices of Cartesian coordinates.

The simple formulation just presented does not allow for the variation of the optical response with frequency, a behaviour of critical importance for spectroscopy. We now briefly discuss how to incorporate frequency-dependent behaviour into the polarization response. To treat the frequency response of the material, we consider an excitation electric field of the form of a superposition of monochromatic fields

$$E(t) = \sum_m E(\omega_m)\, e^{-i\omega_m t} \tag{B1.5.21}$$

where the summation extends over all positive and negative frequency components. Since $E(t)$ represents a physical field, it is constrained to be real and $E(-\omega_m) = E(\omega_m)^*$. In the same manner, we can write the polarization $P$ as

$$P(t) = \sum_n P(\omega_n) \, e^{-i\omega_n t}. \tag{B1.5.22}$$

Here the collection of frequencies in the summation may include new frequencies $\omega_n$ in addition to those in summation of equation (B1.5.21) for the applied field. The total polarization can be separated into linear, $P_L$, and nonlinear, $P_{NL}$, parts:

$$P(t) = P_L(t) + P_{NL}(t) = P^{(1)}(t) + P^{(2)}(t) + P^{(3)}(t) + \cdots \tag{B1.5.23}$$

where $P_L(t) = P^{(1)}(t)$ and $P_{NL}(t) = P^{(2)}(t) + P^{(3)}(t) + \cdots$, and the terms in $P$ correspond to an expansion in powers of the field $E$.

The linear susceptibility, $\chi_{ij}^{(1)}$, is the factor that relates the induced linear polarization to the applied field:

$$P_i^{(1)}(\omega_n) = \sum_i \chi_{ij}^{(1)}(\omega_n) E_j(\omega_n). \tag{B1.5.24}$$

$\chi_{ij}^{(1)}(\omega_n)$ in this formulation gives rise, as one expects, to a polarization oscillating as the applied frequency $\omega_n$, but may now incorporate a strength that varies with $\omega_n$, as illustrated earlier in the harmonic oscillator model. Similarly, we can define the corresponding frequency-dependent second-order, $\chi_{ijk}^{(2)}$, and third-order, $\chi_{ijkl}^{(3)}$, susceptibility tensors by

$$P_i^{(2)}(\omega_q + \omega_r) = p \sum_{jk} \chi_{ijk}^{(2)}(\omega_q + \omega_r, \omega_q, \omega_r) E_j(\omega_q) E_k(\omega_r) \tag{B1.5.25}$$

$$P_i^{(3)}(\omega_q + \omega_r + \omega_s) = p \sum_{jkl} \chi_{ijkl}^{(3)}(\omega_q + \omega_r + \omega_s, \omega_q, \omega_r, \omega_s) E_j(\omega_q) E_k(\omega_r) E_l(\omega_s) \tag{B1.5.26}$$

where the quantity $p$ is called the degeneracy factor and is equal to the number of distinct permutations of the applied frequencies $\{\omega_q, \omega_r\}$ and $\{\omega_q, \omega_r, \omega_s\}$ for the second- and third-order processes, respectively. The inclusion of the degeneracy factor $p$ ensures that the nonlinear susceptibility is not discontinuous when two of the fields become degenerate, e.g. the nonlinear susceptibility $\chi_{ijk}^{(2)}(\omega_1 + \omega_2, \omega_1, \omega_2)$ approaches the value $\chi_{ijk}^{(2)}(2\omega_1, \omega_1, \omega_1)$ as $\omega_2$ approaches the value $\omega_1$ [32]. As can be seen from equations (B1.5.25)–(B1.5.26), the first frequency argument of the nonlinear susceptibility is equal to the sum of the rest of its frequency arguments. Note also that these frequencies are not constrained to positive values and that the complete material response involves both the positive and negative frequency components.

We now consider some of the processes described by the nonlinear susceptibilities. For the case of the second-order nonlinear optical effects (equation (B1.5.25)), three processes can occur when the frequencies $\omega_1$ and $\omega_2$ are distinct. This can be seen by expanding equation (B1.5.25):

$$P_i^{(2)}(\omega_3) = 2 \sum_{jk} \chi_{ijk}^{(2)}(\omega_3, \omega_1, \omega_2) E_j(\omega_1) E_k(\omega_2) \tag{B1.5.27}$$

$$P_i^{(2)}(2\omega_\alpha) = \sum_{jk} \chi_{ijk}^{(2)}(2\omega_\alpha, \omega_\alpha, \omega_\alpha) E_j(\omega_\alpha) E_k(\omega_\alpha) \qquad \alpha = 1, 2 \tag{B1.5.28}$$

$$P_i^{(2)}(0) = 2 \left[ \sum_{jk} \chi_{ijk}^{(2)}(0, \omega_1, -\omega_1) E_j(\omega_1) E_k^*(\omega_1) + \sum_{jk} \chi_{ijk}^{(2)}(0, \omega_2, -\omega_2) E_j(\omega_2) E_k^*(\omega_2) \right]. \tag{B1.5.29}$$

These effects correspond, respectively, to the processes of sum-frequency generation (SFG), SHG and optical rectification.

For the case of third-order nonlinear optical effects (equation (B1.5.26)), a wide variety of processes are described by the different possible combinations of applied frequencies. We shall not attempt to catalogue them here. The most intuitive case is that of third-harmonic generation ($3\omega = \omega + \omega + \omega$), corresponding to addition of three equal frequencies. When one of the frequencies is zero (i.e. a DC field) one obtains the so-called electric-field induced SHG (EFISH) process ($2\omega = \omega + \omega + 0$). Several third-order effects have found significant use in both frequency and time-domain spectroscopy. These include notably coherent anti-Stokes Raman scattering, stimulated Raman scattering, general and degenerate four-wave mixing and two-photon absorption [30, 31].

*(d) Symmetry properties*

The second-order nonlinear susceptibility tensor $\chi^{(2)}(\omega_3, \omega_2, \omega_1)$ introduced earlier will, in general, consist of 27 distinct elements, each displaying its own dependence on the frequencies $\omega_1$, $\omega_2$ and $\omega_3 (= \pm\omega_1 \pm \omega_2)$. There are, however, constraints associated with spatial and time-reversal symmetry that may reduce the complexity of $\chi^{(2)}$ for a given material [32–34]. Here we examine the role of spatial symmetry.

The most significant symmetry property for the second-order nonlinear optics is inversion symmetry. A material possessing inversion symmetry (or centrosymmetry) is one that, for an appropriate origin, remains unchanged when all spatial coordinates are inverted via $r \rightarrow -r$. For such materials, the second-order nonlinear response vanishes. This fact is of sufficient importance that we shall explain its origin briefly. For a centrosymmetric material, $\chi^{(2)}$ should remain unchanged under an inversion operation, since the material by hypothesis does not change. On the other hand, the nature of the physical quantities $E$ and $P$ implies that they must obey $E \rightarrow -E$ and $P \rightarrow -P$ under the inversion operation [35]. The relation $P_i^{(2)} = \sum_{jk} \chi_{ijk}^{(2)} E_j E_k$ then yields $\sum_{jk} \chi_{ijk}^{(2)} E_j E_k = -\sum_{jk} \chi_{ijk}^{(2)} E_j E_k$, whence $\chi_{ijk}^{(2)} = 0$. A further useful symmetry relation applies specifically to SHG process. Since the incident fields $E_j(\omega)$ and $E_k(\omega)$ are identical, we may take $\chi_{ijk}^{(2)} = \chi_{ikj}^{(2)}$ without loss of information.

For materials that exhibit other classes of spatial symmetry not including centrosymmetry, we expect that $\chi_{ijk}^{(2)}$ will be non-vanishing but will display a simplified form [37]. To see how this might work, consider a crystal in which the $x$ and $y$ directions are equivalent. The nonlinear response of the medium would consequently be equivalent for an applied optical field polarized along either the $x$ or $y$ direction. It then follows, for example, that the nonlinear susceptibilities $\chi_{zxx}^{(2)}$ and $\chi_{zyy}^{(2)}$ are equal, as are other elements in which the indices $x$ and $y$ are exchanged. This reduces the complexity of $\chi^{(2)}$ significantly. Table B1.5.1 presents the form of the second-order nonlinear susceptibility relevant for the classes of symmetry encountered at isotropic and crystalline surfaces and interfaces.

*(e) Quantum mechanical description*

Having now developed some of the basic notions for the macroscopic theory of nonlinear optics, we would like to discuss how the microscopic treatment of the nonlinear response of a material is handled. While the classical nonlinear oscillator model provides us with a qualitative feeling for the phenomenon, a quantitative theory must generally begin with a quantum mechanical description. As in the case of linear optics, quantum mechanical calculations in nonlinear optics are conveniently described by perturbation theory and the density matrix formalism [36, 37]. The heart of this microscopic description is the interaction Hamiltonian $H_{int} = -\mu \cdot E(t)$, which characterizes the interaction of the system with the radiation and is treated as a perturbation. Here $\mu = -er$ is the electric dipole operator and $E(t)$ is the applied optical field. Applying this formalism with first-order perturbation theory to calculate the induced dipole moment $\mu$, we obtain a linear susceptibility that is proportional to the product of two matrix elements of $\mu$, $(\mu_i)_{gn}(\mu_j)_{ng}$, where $(\mu_i)_{gn} \equiv \langle g|\mu_i|n\rangle$.

**Table B1.5.1.** Independent non-vanishing elements of the nonlinear susceptibility, $\chi_s^{(2)}$ for an interface in the $xy$-plane for various symmetry classes. When mirror planes are present, at least one of them is perpendicular to the $y$-axis. For SHG, elements related by the permutation of the last two elements are omitted. For SFG, these elements are generally distinct; any symmetry constraints are indicated in parentheses. The terms enclosed in parentheses are antisymmetric elements present only for SFG. (After [71].)

| Symmetry class | Independent non-vanishing elements |
|---|---|
| 1 | $xxx, xxy, xyy, xyz, xxz, xzz, yxx, yxy, yxz, yyy, yyz, yzz, zxx,$ $zxy, zxz, zyy, zyz, zzz$ |
| $m$ | $xxx, xyy, xzx, xzz, yxy, yyz, zxx, zxz, zyy, zzz$ |
| 2 | $xzx, xyz, yxz, yzy, zxx, zyy, zxy, zzz$ |
| $2mm$ | $xzx, yzy, zxx, zyy, zzz$ |
| 3 | $xxx = -xyy = -yyx(= -yxy)$ $yyy = -yxx = -xyx(= -xxy), yzy = xzx,$ $zxx = zyy, xyz = -yxz$ $zzz, (zxy = -zyx)$ |
| $3m$ | $xxx = -xyy = -yxy(= -yyx)$ $yzy = xzx, zxx = zyy, zzz$ |
| $4, 6, \infty$ | $xxz = yyz, zxx = zyy, xyz = -yxz, zzz, (zxy = -zyx)$ |
| $4mm, 6mm, \infty m$ | $xxz = yyz, zxx = zyy, zzz$ |

This product may be viewed as arising from a process in which a photon is first destroyed in a real or virtual transition from a populated energy eigenstate $|g\rangle$ to an empty state $|n\rangle$; a photon is then emitted in the transition back to the initial state $|g\rangle$.

    The second-order nonlinear optical processes of SHG and SFG are described correspondingly by second-order perturbation theory. In this case, two photons at the driving frequency or frequencies are destroyed and a photon at the SH or SF is created. This is accomplished through a succession of three real or virtual transitions, as shown in figure B1.5.4. These transitions start from an occupied initial energy eigenstate $|g\rangle$, pass through intermediate states $|n'\rangle$ and $|n\rangle$ and return to the initial state $|g\rangle$. A full calculation of the second-order response for the case of SFG yields [37]

$$\chi_{ijk}^{(2)}(\omega_3, \omega_2, \omega_1) = -\frac{N}{\hbar^2} \sum_{g,n,n'} \frac{(\mu_i)_{gn}(\mu_j)_{nn'}(\mu_k)_{n'g}}{(\omega_3 - \omega_{ng} + i\Gamma_{ng})(\omega_2 - \omega_{n'g} + i\Gamma_{n'g})} \rho_g^{(0)} + \text{ seven similar terms.} \quad (\text{B1.5.30})$$

As for the linear response, the transitions occur through the electric-dipole operator $\mu$ and are characterized by the matrix elements $(\mu_i)_{gn}$. In equation (B1.5.30), the energy denominators involve the energy differences $\hbar\omega_{ng} \equiv E_n - E_g$ and widths $\hbar\Gamma_{ng}$ for transitions between eigenstates $|n\rangle$ and $|g\rangle$. The formula includes a sum over different possible ground states weighted by the factor $\rho_g^{(0)}$ representing the probability that state $|g\rangle$ is occupied. It is assumed that the material can be treated as having localized electronic states of density $N$ per unit volume and that their interaction and local-field effects may be neglected. Corresponding expressions result for delocalized electrons in crystalline solids [36, 37].

    The frequency denominators in the eight terms of equation (B1.5.30) introduce a resonant enhancement in the nonlinearity when any of the three frequencies ($\omega_1, \omega_2, \omega_3$) coincides with a transition from the ground state $|g\rangle$ to one of the intermediate states $|n'\rangle$ and $|n\rangle$. The numerator of each term, which consists of the product of the three dipole matrix elements $(\mu_i)_{gn}(\mu_j)_{nn'}(\mu_k)_{n'g}$, reflects, through its tensor character, the structural properties of the material, as well as the details of the character of the relevant energy eigenstates.

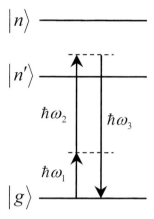

**Figure B1.5.4.** Quantum mechanical scheme for the SFG process with ground state $|g\rangle$ and excited states $|n'\rangle$ and $|n\rangle$.

### B1.5.2.2  Nonlinear optics of the interface

The focus of the present chapter is the application of second-order nonlinear optics to probe surfaces and interfaces. In this section, we outline the phenomenological or macroscopic theory of SHG and SFG at the interface of centrosymmetric media. This situation corresponds, as discussed previously, to one in which the relevant nonlinear response is forbidden in the bulk media, but allowed at the interface.

#### (a) Interfacial contribution

In order to describe the second-order nonlinear response from the interface of two centrosymmetric media, the material system may be divided into three regions: the interface and the two bulk media. The interface is defined to be the transitional zone where the material properties—such as the electronic structure or molecular orientation of adsorbates—or the electromagnetic fields differ appreciably from the two bulk media. For most systems, this region occurs over a length scale of only a few Ångströms. With respect to the optical radiation, we can thus treat the nonlinearity of the interface as localized to a sheet of polarization. Formally, we can describe this sheet by a nonlinear dipole moment per unit area, $P_s^{(2)}$, which is related to a second-order bulk polarization $P^{(2)}$ by $P^{(2)}(x, y, z, t) = P_s^{(2)}(x, y, t)\delta(z)$. Here $z$ is the surface normal direction, and the $x$ and $y$ axes represent the in-plane coordinates (figure B1.5.5).

The nonlinear response of the interface may then be characterized in terms of a surface (or interface) nonlinear susceptibility tensor $\chi_s^{(2)}$. This quantity relates the applied electromagnetic fields to the induced surface nonlinear polarization $P_s^{(2)}$:

$$P_s^{(2)}(\omega_3) = \chi_s^{(2)}(\omega_3 = \omega_1 + \omega_2) : E(\omega_1)E(\omega_2).  \tag{B1.5.31}$$

In this equation as well as in the succeeding discussions, we have suppressed, for notational simplicity, the permutation or degeneracy factor of two, required for SFG.

To define this model fully, we must specify the linear dielectric response in the vicinity of our surface nonlinear susceptibility. We do this in a general fashion by introducing a (frequency-dependent) linear dielectric response of the interfacial region $\varepsilon'$, which is bounded by the bulk media with dielectric constants $\varepsilon_1$ and $\varepsilon_2$. For simplicity, we consider all of these quantities to be scalar, corresponding to isotropic linear optical properties. The phenomenological model for second-order nonlinear optical effects is summarized in figure B1.5.5. (An alternative convention [38, 39] for defining the surface nonlinear susceptibility is one in which the fundamental fields $E(\omega_1)$ and $E(\omega_2)$ are taken as their value in the lower medium and the

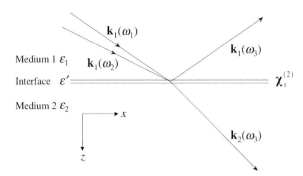

**Figure B1.5.5.** Schematic representation of the phenomenological model for second-order nonlinear optical effects at the interface between two centrosymmetric media. Input waves at frequencies $\omega_1$ and $\omega_2$, with corresponding wavevectors $k_1(\omega_1)$ and $k_1(\omega_2)$, are approaching the interface from medium 1. Nonlinear radiation at frequency $\omega_3$ is emitted in directions described by the wavevectors $k_1(\omega_3)$ (reflected in medium 1) and $k_2(\omega_3)$ (transmitted in medium 2). The linear dielectric constants of media 1, 2 and the interface are denoted by $\varepsilon_1$, $\varepsilon_2$, and $\varepsilon'$, respectively. The figure shows the $xz$-plane (the plane of incidence) with $z$ increasing from top to bottom and $z = 0$ defining the interface.

polarized sheet is treated as radiating in the upper medium. This convention corresponds in our model to the assignments of $\varepsilon'(\omega_1) = \varepsilon_2(\omega_1)$, $\varepsilon'(\omega_2) = \varepsilon_2(\omega_2)$, and $\varepsilon'(\omega_3) = \varepsilon_1(\omega_3)$.)

From the point of view of tensor properties, the surface nonlinear susceptibility $\chi^{(2)}_{s,ijk}$ is quite analogous to the bulk nonlinear response $\chi^{(2)}_{ijk}$ in a non-centrosymmetric medium. Consequently, in the absence of any symmetry constraints, $\chi_s^{(2)}$ will exhibit 27 independent elements for SFG and, because of symmetry for the last two indices, 18 independent elements for SHG. If the surface exhibits certain in-plane symmetry properties, then the form of $\chi_s^{(2)}$ will be simplified correspondingly. For the common situation of an isotropic surface, for example, the allowed elements of $\chi_s^{(2)}$ may be denoted as $\chi^{(2)}_{s,\perp\perp\perp}$, $\chi^{(2)}_{s,\perp\|\|}$, $\chi^{(2)}_{s,\|\perp\|}$ and $\chi^{(2)}_{s,\|\|\perp}$, where $\perp$ corresponds to the $z$ direction and $\|$ refers to either $x$ or $y$. For the case of SHG, where the fundamental frequencies are equal, $\omega_1 = \omega_2 \equiv \omega$, the tensor elements $\chi^{(2)}_{s,\|\perp\|}$ and $\chi^{(2)}_{s,\|\|\perp}$ are likewise equivalent. The non-vanishing elements of $\chi_s^{(2)}$ for other commonly encountered surface symmetries are summarized in table B1.5.1.

The linear and nonlinear optical responses for this problem are defined by $\varepsilon_1$, $\varepsilon_2$, $\varepsilon'$ and $\chi_s^{(2)}$, respectively, as indicated in figure B1.5.5. In order to determine the nonlinear radiation, we need to introduce appropriate pump radiation fields $E(\omega_1)$ and $E(\omega_2)$. If these pump beams are well-collimated, they will give rise to well-collimated radiation emitted through the surface nonlinear response. Because the nonlinear response is present only in a thin layer, phase matching [37] considerations are unimportant and nonlinear emission will be present in both transmitted and reflected directions. Here we model the pump beams associated with fields $E(\omega_1)$ and $E(\omega_2)$ as plane waves with wavevectors $k_1 = \hat{k}_1 \omega_1 \sqrt{\varepsilon_1(\omega_1)}/c$ and $k_2 = \hat{k}_2 \omega_2 \sqrt{\varepsilon_1(\omega_2)}/c$. The directions of the reflected and transmitted beams can then be obtained simply through conservation of the in-plane component of the wavevector, i.e. $k_{1x}(\omega_1) + k_{1x}(\omega_2) = k_{1x}(\omega_3) = k_{2x}(\omega_3)$. This is the nonlinear optical analogue of Snell's law. For the case of SHG, this equation may be reduced to $n(\omega) \sin \theta_\omega = n(2\omega) \sin \theta_{2\omega}$ for the angle of the incident pump radiation, $\theta_\omega$, and the angle of the emitted nonlinear beams, $\theta_{2\omega}$. The refractive indices in this equation correspond to those of the relevant bulk medium through which the beams propagate. For reflection in a non-dispersive medium, we obtain simply $\theta_\omega = \theta_{2\omega}$, as for the law of reflection. For the transmitted beam, the relation in the absence of dispersion reduces to the usual Snell's law for refraction.

A full solution of the nonlinear radiation follows from the Maxwell equations. The general case of radiation from a second-order nonlinear material of finite thickness was solved by Bloembergen and Pershan in 1962 [40]. That problem reduces to the present one if we let the interfacial thickness approach zero. Other equivalent solutions involved the application of the boundary conditions for a polarization sheet [14] or the use of a Green's function formalism for the surface [38, 39].

From such a treatment, we may derive explicit expressions for the nonlinear radiation in terms of the linear and nonlinear response and the excitation conditions. For the case of nonlinear reflection, we obtain an irradiance for the radiation emitted at the nonlinear frequency $\omega_3$ of

$$I(\omega_3) = \frac{8\pi^3 \omega_3^2 \sec^2\theta |e'(\omega_3) \cdot \chi_s^{(2)} : e'(\omega_1)e'(\omega_2)|^2 I(\omega_1)I(\omega_2)}{c^3[\varepsilon_1(\omega_3)\varepsilon_1(\omega_1)\varepsilon_1(\omega_2)]^{1/2}} \tag{B1.5.32}$$

where $I(\omega_1)$ and $I(\omega_2)$ denote the intensities of the pump beams at frequencies $\omega_1$ and $\omega_2$ incident from medium 1; $c$ is the speed of light in vacuum; $\theta$ is the angle of propagation direction of the nonlinear radiation relative to the surface normal. The vectors $e'(\omega_1)$, $e'(\omega_2)$ and $e'(\omega_3)$ represent the *unit* polarization vectors $\hat{e}_1(\omega_1)$, $\hat{e}_1(\omega_2)$, and $\hat{e}_1(\omega_3)$, respectively, adjusted to account for the linear optical propagation of the waves. More specifically, we may write

$$e'(\omega) = F_{1\to 2}\hat{e}_1(\omega). \tag{B1.5.33}$$

Here the 'Fresnel transformation' $F_{1\to 2}$ describes the relationship between the electric field $E\hat{e}_1$ in medium 1 (propagating towards medium 2) and the resulting field $Ee'$ at the interface. For light incident in the $x$–$z$ plane as shown in figure B1.5.5, $F_{1\to 2}$ is a diagonal matrix whose elements are

$$F_{1\to 2}^{xx} = 2\varepsilon_1 k_{2,z}/(\varepsilon_2 k_{1,z} + \varepsilon_1 k_{2,z}) \tag{B1.5.34}$$

$$F_{1\to 2}^{yy} = 2k_{1,z}/(k_{1,z} + k_{2,z}) \tag{B1.5.35}$$

$$F_{1\to 2}^{zz} = 2(\varepsilon_1\varepsilon_2/\varepsilon')k_{1,z}/(\varepsilon_2 k_{1,z} + \varepsilon_1 k_{2,z}) \tag{B1.5.36}$$

where the quantity $k_{i,z}$ denotes the magnitude of the $z$-component of the wavevector in medium $i$ at the relevant wavelength ($\omega_1$, $\omega_2$ or $\omega_3$).

The treatment of this section has been based on an assumed nonlinear surface response $\chi_s^{(2)}$ and has dealt entirely with electromagnetic considerations of excitation and radiation from the interface. A complete theoretical picture, however, includes developing a microscopic description of the surface nonlinear susceptibility. In the discussion in section B1.5.4, we will introduce some simplified models. In this context, an important first approximation for many systems of chemical interest may be obtained by treating the surface nonlinearity as arising from the composite of individual molecular contributions. The molecular response is typically assumed to be that of the isolated molecule, but in the summation for the surface nonlinear response, we take into account the orientational distribution appropriate for the surface or interface, as we discuss later. Local-field corrections may also be included [41, 42]. Such analyses may then draw on the large and well-developed literature concerning the second-order nonlinearity of molecules [43, 44]. If we are concerned with the response of the surface of a clean solid, we must typically adopt a different approach: one based on delocalized electrons. This is a challenging undertaking, as a proper treatment of the *linear* optical properties of surfaces of solids is already difficult [45]. Nonetheless, in recent years significant progress has been made in developing a fundamental theory of the nonlinear response of surfaces of both metals [46–51] and semiconductors [52–55].

*(b) Bulk contribution*

For centrosymmetric media the spatially local contribution to the second-order nonlinear response vanishes, as we have previously argued, providing the interface specificity of the method. This spatially local contribution, which arises in the quantum mechanical picture from the electric-dipole terms, represents the dominant response of the medium. However, if we consider the problem of probing interfaces closely, we recognize that we are comparing the nonlinear signal originating from an interfacial region of monolayer thickness with that of the bulk media. In the bulk media, the signal can build up over a thickness on the scale of the optical wavelength, as dictated by absorption and phase-matching considerations. Thus, a bulk nonlinear polarization that is much weaker than that of the dipole-allowed contribution present at the interface may still prove to be significant because of the larger volume contributing to the emission. Let us examine this point in a somewhat more quantitative fashion.

The higher-order bulk contribution to the nonlinear response arises, as just mentioned, from a spatially non-local response in which the induced nonlinear polarization does not depend solely on the value of the fundamental electric field at the same point. To leading order, we may represent these non-local terms as being proportional to a nonlinear response incorporating a first spatial derivative of the fundamental electric field. Such terms correspond in the microscopic theory to the inclusion of electric-quadrupole and magnetic-dipole contributions. The form of these bulk contributions may be derived on the basis of symmetry considerations. As an example of a frequently encountered situation, we indicate here the non-local polarization for SHG in a cubic material excited by a plane wave $E(\omega)$:

$$P_{b,i}^{(2)}(2\omega) = \gamma \nabla_i [E(\omega) \cdot E(\omega)] + \zeta E_i(\omega) \nabla_i E_i(\omega). \tag{B1.5.37}$$

The two coefficients $\gamma$ and $\zeta$ describe the material response and the Cartesian coordinate $i$ must be chosen as a principal axis of the material.

From consideration of the quantum mechanical expression of such a non-local response, one may argue that the dipole-forbidden bulk nonlinear polarization will have a strength reduced from that of the dipole-allowed response by a factor of the order of $(a/\lambda)$, with $a$ denoting a typical atomic dimension and $\lambda$ representing the wavelength of light. On the other hand, the relevant volume for the bulk contribution typically exceeds that of the interface by a factor of the order of $(\lambda/a)$. Consequently, one estimates that the net bulk and surface contributions to the nonlinear radiation may be roughly comparable in strength. In practice, the interfacial contribution often dominates that of the bulk. Nonetheless, one should not neglect *a priori* the possible role of the bulk nonlinear response. This situation of a possible bulk background signal comparable to that of the interface should be contrasted to the expected behaviour for a conventional optical probe lacking interface specificity. In the latter case, the bulk contribution would be expected to dominate that of the interface by several orders of magnitude.

*(c) Other sources*

As we have discussed earlier in the context of surfaces and interfaces, the breaking of the inversion symmetry strongly alters the SHG from a centrosymmetric medium. Surfaces and interfaces are not the only means of breaking the inversion symmetry of a centrosymmetric material. Another important perturbation is that induced by (static) electric fields. Such electric fields may be applied externally or may arise internally from a depletion layer at the interface of a semiconductor or from a double-charge layer at the interface of a liquid. Since the electric field is a polar vector, it acts to break the inversion symmetry and gives rise to dipole-allowed sources of nonlinear polarization in the bulk of a centrosymmetric medium. Assuming that the DC field, $E_{DC}$, is sufficiently weak to be treated in a leading-order perturbation expansion, the response may be written as

$$P_{DC}^{(3)}(2\omega) = \chi^{(3)} : E(\omega)E(\omega)E_{DC} \tag{B1.5.38}$$

where $\chi^{(3)}$ is the effective third-order response. This process is called electric-field-induced SHG or EFISH.

A different type of external perturbation is the application of a magnetic field. In contrast to the case of an electric field, an applied magnetic field does not lift the inversion symmetry of a centrosymmetric medium. Hence, it does not give rise to a dipole-allowed bulk polarization for SHG or SFG [23, 56]. A magnetic field can, however, modify the form and strength of the interfacial nonlinear response, as well as the

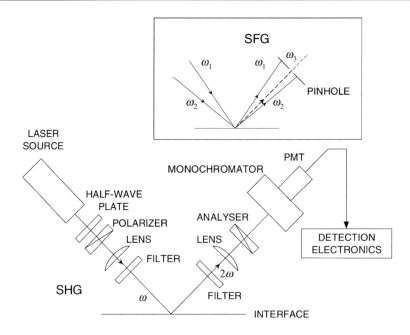

**Figure B1.5.6.** Experimental geometry for typical SHG and SFG measurements.

bulk quadrupole nonlinear susceptibilities. This process is termed magnetization-induced SHG or MSHG. Experiments exploiting both EFISH and MSHG phenomena are discussed in section B1.5.4.7.

### B1.5.3  Experimental considerations

In this section, we provide a brief overview of some experimental issues relevant in performing surface SHG and SFG measurements.

*B1.5.3.1  Experimental geometry*

The main panel of figure B1.5.6 portrays a typical setup for SHG. A laser source of frequency $\omega$ is directed to the sample, with several optical stages typically being introduced for additional control and filtering. The combination of a halfwave plate and a polarizer is used to specify the orientation of the polarization of the pump beam. It can also serve as a variable attenuator to adjust the intensity of the incoming beam. A lens then focuses the beam onto the sample. A low-pass filter is generally needed along the path of the fundamental radiation prior to the sample to remove any unwanted radiation at the frequency of the nonlinear radiation. This radiation may arise from previous optical components, including the laser source itself.

The reflected radiation consists of a strong beam at the fundamental frequency $\omega$ and a weak signal at the SH frequency. Consequently, a high-pass filter is introduced, which transmits the nonlinear radiation, but blocks the fundamental radiation. By inserting this filter immediately after the sample, we minimize the generation of other nonlinear optical signals from succeeding optical components by the fundamental light reflected from the sample. After this initial filtering stage, a lens typically recollimates the beam and an analyser is used to select the desired polarization. Although not always essential, a monochromator or bandpass filter is often desirable to ensure that only the SH signal is measured. Background signals near the SH frequency, but not associated with the SHG process, may arise from multiphoton fluorescence, hyper-Raman scattering and other nonlinear processes described by higher-order nonlinear susceptibilities.

Detection is usually accomplished through a photomultiplier tube. Depending on the nature of the laser source, various sensitive schemes for electronic detection of the photomultiplier may be employed, such as photon counting and gated integration. In conjunction with an optical chopper in the beam path, lock-in amplification techniques may also be advantageous.

One of the key factors for performing surface SHG/SFG measurements is to reduce all sources of background signals, since the desired nonlinear signal is always relatively weak. This goal is accomplished most effectively by exploiting the well-defined spectral, spatial and temporal characteristics of the nonlinear radiation. The first of these is achieved by spectral filtering, as previously discussed. The second may be achieved through the use of appropriate apertures; and the last, particularly for low-repetition rate systems, can be incorporated into the electronic detection scheme. When the excitation is derived from two non-collinear beams, the nonlinear emission will generally travel in a distinct direction (figure B1.5.6 inset). In this case, one can also exploit spatial filters to enhance spectral selectivity, since the reflected pump beams will travel in different directions. This property is particularly useful for SFG experiments in which the frequency of the visible beam and the SF signal are relatively similar.

For some experiments, it may be helpful to obtain a reference signal to correct for fluctuations and long-term drift in the pump laser. This correction is best accomplished by performing simultaneous measurements of the SHG or SFG from a medium that has a strong $\chi^{(2)}$ response in a separate detection arm. By this means, one may fully compensate for variations not only in pulse energy, but also in the temporal and spatial substructure of the laser pulses. Some experiments may require measurement of the phase of the nonlinear signal [57]. Such phase measurements rely on interference with radiation from a reference nonlinear source. The required interference can be achieved by placing a reference nonlinear crystal along the path of the laser beam immediately either before or after the sample [58]. For effective interference, we must control both the amplitude and polarization of the reference signal. This may be achieved by appropriate focusing conditions and crystal alignment. The phase of the reference signal must also be adjustable. Phase control may be obtained simply by translating the reference sample along the path of the laser, making use of the dispersion of the air (or other medium) through which the beams propagate.

### B1.5.3.2   Laser sources

In order to achieve a reasonable signal strength from the nonlinear response of approximately one atomic monolayer at an interface, a laser source with high peak power is generally required. Common sources include Q-switched ($\sim$10 ns pulsewidth) and mode-locked ($\sim$100 ps) Nd:YAG lasers, and mode-locked ($\sim$10 fs–1 ps) Ti:sapphire lasers. Broadly tunable sources have traditionally been based on dye lasers. More recently, optical parametric oscillator/amplifier (OPO/OPA) systems are coming into widespread use for tunable sources of both visible and infrared radiation.

In typical experiments, the laser fluence, or the energy per unit area, is limited to the sample's damage threshold. This generally lies in the range $\ll 1$ J cm$^{-2}$ and constrains our ability to increase signal strength by increasing the pump energy. Frequently, the use of femtosecond pulses is advantageous, as one may obtain a higher intensity (and, hence, higher nonlinear conversion efficiency) at lower fluence. In addition, such sources generally permit one to employ lower average intensity, which reduces average heating of the sample and other undesired effects [59]. Independently of these considerations, femtosecond lasers are, of course, also attractive for the possibilities that they offer for measurements of ultrafast dynamics.

### B1.5.3.3   Signal strengths

We now consider the signal strengths from surface SHG/SFG measurements. For this purpose, we may recast expression (B1.5.32) for the reflected nonlinear radiation in terms of the number of emitted photon per unit time as

$$S = \frac{8\pi^3 \omega_3}{\hbar c^3 [\varepsilon_1(\omega_3)\varepsilon_1(\omega_2)\varepsilon_1(\omega_1)]^{1/2}} \frac{\sec^2\theta \, P_{avg}(\omega_1) P_{avg}(\omega_2)}{t_p R_{rep} A} |e'(\omega_3) \cdot \chi_s^{(2)} : e'(\omega_1)e'(\omega_2)|^2. \qquad (B1.5.39)$$

The quantities in this formula are defined as in equation (B1.5.32), but with the laser parameters translated into more convenient terms: $P_{avg}$ is the average power at the indicated frequency; $t_p$ is the laser pulse duration; $R_{rep}$ is the pulse repetition rate; and $A$ is the irradiated area at the interface. The last three defined quantities are assumed to be equal for both excitation beams in an SFG measurement. If this is not the case, then $t_p$, $R_{rep}$, $A$, as well as the average power $P_{avg}(\omega_i)$, have to be replaced by the corresponding quantities within the window of spatial and temporal overlap.

From this expression, we may estimate typical signals for a surface SHG measurement. We assume the following as representative parameters: $\chi^{(2)} = 10^{-15}$ esu, $\varepsilon_1 = 1$, and $\sec^2\theta = 4$. For typical optical frequencies, we then obtain $S \approx 10^{-2}(P_{avg})^2/(t_p R_{rep} A)$, where $P_{avg}$, $A$, $R_{rep}$ and $t_p$ are expressed, respectively, in W, cm$^2$, Hz and s. Many recent SHG studies have been performed with mode-locked Ti:sapphire lasers. For typical laser parameters of $P_{avg} = 100$ mW, $A = 10^{-4}$ cm$^2$, $R_{rep} = 100$ MHz, and $t_p = 100$ fs, one then obtains $S \approx 10^5$ counts per second as a representative nonlinear signal.

### B1.5.3.4  Determination of nonlinear susceptibility elements

The basic physical quantities that define the material for SHG or SFG processes are the nonlinear susceptibility elements $\chi_{s,ijk}^{(2)}$. Here we consider how one may determine these quantities experimentally. For simplicity, we treat the case of SHG and assume that the surface is isotropic. From symmetry considerations, we know that $\chi_{s,ijk}^{(2)}$ has three independent and non-vanishing elements: $\chi_{s,\perp\perp\perp}^{(2)}$, $\chi_{s,\perp\|\|}^{(2)}$ and $\chi_{s,\perp\|\perp}^{(2)} = \chi_{s,\|\|\perp}^{(2)}$. The individual elements $\chi_{s,\perp\|\|}^{(2)}$ and $\chi_{s,\|\perp\|}^{(2)}$ can be extracted directly by an appropriate choice of input and output polarizations. Response from the $\chi_{s,\perp\|\|}^{(2)}$ element requires s-polarized pump radiation and produces p-polarized SH emission; excitation of the $\chi_{s,\|\perp\|}^{(2)}$ element requires a mixed-polarized pump radiation and can be isolated by detection of s-polarized radiation. The measurement of $\chi_{s,\perp\perp\perp}^{(2)}$ is bit more complicated: to isolate it would require a pump electric field aligned normal to the surface, thus implying a pump beam travelling parallel to the surface (the limit of grazing incidence).

An alternative scheme for extracting all three isotropic nonlinear susceptibilities can be formulated by examining equation (B1.5.39). By choosing an appropriate configuration and the orientation of the polarization of the SH radiation $e'(2\omega)$ such that the SHG signal vanishes, one obtains, assuming only surface contribution with real elements $\chi_{s,ijk}^{(2)}$,

$$e'(2\omega) \cdot \chi_s^{(2)} : e'(\omega)e'(\omega) = \sum_{ijk} e_i'(2\omega)\chi_{s,ijk}^{(2)}e_j'(\omega)e_k'(\omega) = 0.$$

Expanding this equation, we deduce that

$$e_\perp'(2\omega)[\chi_{s,\perp\perp\perp}^{(2)}e_\perp'(\omega)e_\perp'(\omega) + \chi_{s,\perp\|\|}^{(2)}e_\|'(\omega)e_\|'(\omega)] + 2e_\|'(2\omega)\chi_{s,\|\perp\|}^{(2)}e_\perp'(\omega)e_\|'(\omega) = 0. \qquad (B1.5.40)$$

The magnitudes of $e_i'$ ($i = \perp, \|$) contain the Fresnel factors from equations (B1.5.34)–(B1.5.36), which depend on the incident, reflected and polarization angles. Experimentally, one approach is to fix the input polarization and adjust the analyser to obtain a null in the SH signal [60]. By choosing distinct configurations such that the corresponding three equations from equation (B1.5.40) are linearly independent, the relative values of $\chi_{s,\perp\perp\perp}^{(2)}$, $\chi_{s,\perp\|\|}^{(2)}$ and $\chi_{s,\|\perp\|}^{(2)} = \chi_{s,\|\|\perp}^{(2)}$ may be inferred. This method has been implemented, for example, in determining the three susceptibility elements for the air/water interface [61]. The procedure just described is suitable for ascertaining the relative magnitudes of the allowed elements of $\chi_s^{(2)}$.

A determination of the absolute magnitude of the elements of $\chi_s^{(2)}$ may also be of value. In principle, this might be accomplished by careful measurements of signal strengths from equation (B1.5.32) or (B1.5.39). In practice, such an approach is very difficult, as it would require precise calibration of the parameters of the laser radiation and absolute detection sensitivity. The preferred method is consequently to compare the surface SHG or SFG response to that of a reference crystal with a known bulk nonlinearity inserted in place of the sample. Since the expected signal can be calculated for this reference material, we may then infer the absolute calibration of $\chi_s^{(2)}$ by comparison. It should be further noted that the phase of elements of $\chi_s^{(2)}$ may also be established by interference measurements as indicated in section B1.5.3.1.

## B1.5.4   Applications

The discussion of applications of the SHG and SFG methods in this section is directed towards an exposition of the different types of information that may be obtained from such measurements. The topics have been arranged accordingly into seven general categories that arise chiefly from the properties of the nonlinear susceptibility: surface symmetry and order, adsorbate coverage, molecular orientation, spectroscopy, dynamics, spatial resolution and perturbations induced by electric and magnetic fields. Although we have included some illustrative examples, a comprehensive description of the broad range of materials probed by these methods, and what has been learned about them, is clearly beyond the scope of this chapter.

### B1.5.4.1   Surface symmetry and order

Spatial symmetry is one of the basic properties of a surface or interface. If the symmetry of the surface is known *a priori*, then this knowledge may be used to simplify the form of the surface nonlinear susceptibility $\chi_s^{(2)}$, as discussed in section B1.5.2.2. Conversely, in the absence of knowledge of the surface symmetry, we may characterize the form of $\chi_s^{(2)}$ experimentally and then make inferences about the symmetry of the surface or interface. This provides some useful information about the material system under study, as we shall illustrate in this section. Before doing so, we should remark that the spatial properties being probed are averaged over a length scale that depends on the precise experimental geometry, but exceeds the scale of an optical wavelength. In the following paragraphs, we consider two of the interesting cases of surface symmetry: that corresponding to the surface of a crystalline material and that of a surface with chiral character.

### (a) Crystalline surfaces

All of the symmetry classes compatible with the long-range periodic arrangement of atoms comprising crystalline surfaces and interfaces have been enumerated in table B1.5.1. For each of these symmetries, we indicate the corresponding form of the surface nonlinear susceptibility $\chi_s^{(2)}$. With the exception of surfaces with four-fold or six-fold rotational symmetries, all of these symmetry classes give rise to a $\chi_s^{(2)}$ that may be distinguished from that of an isotropic surface with mirror symmetry, the highest possible surface symmetry.

An experimental analysis of the surface symmetry may be carried out in various ways. For a fixed crystal orientation, the surface symmetry may be probed by modifying the angle of incidence and polarization of the input and output beams. This approach is often employed for samples, such as those in ultrahigh vacuum, that are difficult to manipulate. An attractive alternative method is to probe the rotational anisotropy simply by recording the change in the nonlinear signal as the sample is rotated about its surface normal. The resulting data reflect directly the symmetry of the surface or interface. Thus, the rotational pattern for the (111) surface of a cubic centrosymmetric crystal with $3m$ symmetry will also have three-fold symmetry with mirror planes. A surface with four-fold symmetry, as in the case of the (001) surface of a cubic material, will give rise to a rotational anisotropy pattern that obeys four-fold symmetry. From table B1.5.1, we see that it will, however, do so in a trivial fashion by giving a response equivalent to that of an isotropic material, i.e. lacking any variation with rotation of the crystal surface.

As an illustration, we consider the case of SHG from the (111) surface of a cubic material ($3m$ symmetry). More general treatments of rotational anisotropy in centrosymmetric crystals may be found in the literature [62–64]. For the case at hand, we may determine the anisotropy of the radiated SH field from equation (B1.5.32) in conjunction with the form of $\chi_s^{(2)}$ from table B1.5.1. We find, for example, for the p-in/p-out and s-in/s-out polarization configurations:

$$E_{\text{p-in/p-out}}^{(2\omega)} = a_0 + a_3 \cos(3\psi) \tag{B1.5.41}$$

$$E_{\text{s-in/s-out}}^{(2\omega)} = b_3 \sin(3\psi) \tag{B1.5.42}$$

where $a_0$, $a_3$ and $b_3$ are constants. The angle $\psi$ corresponds to the rotation of the sample about its surface normal and is measured between the plane of incidence and the $[11\bar{2}]$ direction in the surface plane.

Figure B1.5.7 displays results of a measurement of the rotational anisotropy for an oxidized Si(111) surface [65]. For the case shown in the top panel, the results conform to the predictions of equation (B1.5.42) (with $I(2\omega) \propto |E^{(2\omega)}|^2$) for ideal $3m$ symmetry. The data clearly illustrate the strong influence of anisotropy on measured nonlinear signals. The lower panels of figure B1.5.7 are perturbed rotational anisotropy patterns. They correspond to data for vicinal Si(111) surfaces cut at $3°$ and $5°$, respectively, away from the true (111) orientation. The full lines fitting these data are obtained from an analysis in which the lowered symmetry of these surfaces is taken into account. The results show the sensitivity of this method to slight changes in the surface symmetry.

SH anisotropy measurements of this kind are of use in establishing the orientation of a crystal face of a material, as suggested by figure B1.5.7. The method is also of value for monitoring and study of crystal growth processes [66]. Consider, for example, the growth of Si on Si(111) surface. The crystalline surface exhibits strong rotational anisotropy, corresponding to the $3m$ symmetry of the surface. This will also be the case when a crystalline Si layer is grown on the sample. If, however, the overlayer is grown in an amorphous state, as would occur for Si deposition at room temperature, then the anisotropy will be reduced: the disordered overlayer will exhibit isotropic symmetry on the length scale of the optical wavelength. A further application of rotational anisotropy measurements has been found in the characterization of surface roughness [67].

### (b) Chiral interfaces

An important distinction among surfaces and interfaces is whether or not they exhibit mirror symmetry about a plane normal to the surface. This symmetry is particularly relevant for the case of isotropic surfaces ($\infty$-symmetry), i.e. ones that are equivalent in every azimuthal direction. Those surfaces that fail to exhibit mirror symmetry may be termed chiral surfaces. They would be expected, for example, at the boundary of a liquid comprised of chiral molecules. Magnetized surfaces of isotropic media may also exhibit this symmetry. (For a review of SHG studies of chiral interfaces, the reader is referred to [68].)

Given the interest and importance of chiral molecules, there has been considerable activity in investigating the corresponding chiral surfaces [68–70]. From the point of view of performing surface and interface spectroscopy with nonlinear optics, we must first examine the nonlinear response of the bulk liquid. Clearly, a chiral liquid lacks inversion symmetry. As such, it may be expected to have a strong (dipole-allowed) second-order nonlinear response. This is indeed true in the general case of SFG [71]. For SHG, however, the permutation symmetry for the last two indices of the nonlinear susceptibility tensor combined with the requirement of isotropic symmetry in three dimensions implies that $\chi^{(2)} = 0$. Thus, for the case of SHG, the surface/interface specificity of the technique is preserved even for chiral liquids.

A schematic diagram of the surface of a liquid of non-chiral (a) and chiral molecules (b) is shown in figure B1.5.8. Case (a) corresponds to $\infty m$-symmetry (isotropic with a mirror plane) and case (b) to $\infty$-symmetry (isotropic). For the $\infty m$-symmetry, the SH signal for the polarization configurations of s-in/s-out

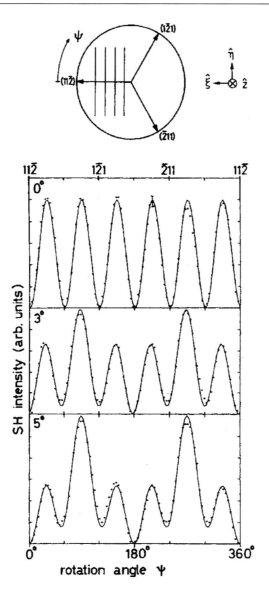

**Figure B1.5.7.** Rotational anisotropy of the SH intensity from oxidized Si(111) surfaces. The samples have either ideal orientation or small offset angles of 3° and 5° toward the [11$\bar{2}$] direction. Top panel illustrates the step structure. The points correspond to experimental data and the full lines to the prediction of a symmetry analysis. (From [65].)

and p-in/s-out vanish. From table B1.5.1, we find, however, that for the $\infty$-symmetry, an extra independent nonlinear susceptibility element, $\chi^{(2)}_{s,xyz} = -\chi^{(2)}_{s,yxz}$, is present for SHG. Because of this extra element, the SH signal for p-in/s-out configuration is no longer forbidden, and consequently, the SH polarization must no longer be strictly p-polarized. Figure B1.5.8(c) shows the SH signal passing through an analyser as a function of its orientation for a racemic mixture (squares) and for a non-racemic mixture (circle) of molecules in a Langmuir–Blodgett film [70]. For the racemic mixture (squares), which contains equal amounts of both enantiomers, the effective symmetry is $\infty m$. Hence, the p-in/s-out signal vanishes and the response curve

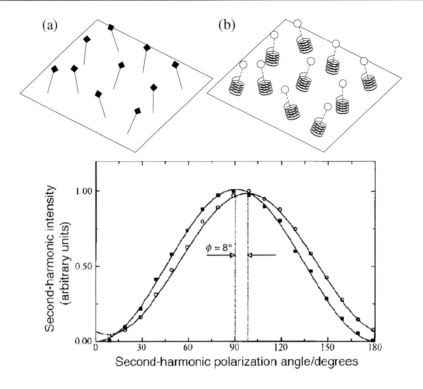

**Figure B1.5.8.** Random distribution of (a) non-chiral adsorbates that gives rise to a surface having effective ∞*m*-symmetry; (b) chiral molecules that gives rise to effective ∞-symmetry. (c) SH intensity versus the angle of an analyser for a racemic (squares) and a non-racemic (open circles) monolayer of chiral molecules. The pump beam was p-polarized; the SH polarization angles of 0° and 90° correspond to s- and p-polarization, respectively. (From [70].)

of figure B1.5.8 is centred at 90°. For the case of the non-racemic mixture, the effective symmetry is ∞. A p-in/s-out SH signal is present and leads, as shown in figure B1.5.8(c), to a displacement in the curve of the SH response versus analyser setting.

This effect of a change in the SH output polarization depending on the enantiomer or mixture of enantiomer is somewhat analogous to the linear optical phenomenon of optical rotary dispersion (ORD) in bulk chiral liquids. As such, the process for SH radiation is termed SHG-ORD [70]. In general, chiral surfaces will also exhibit distinct radiation characteristics for left- and right-polarized pump beams. Again, by analogy with the linear optical process of circular dichroism (CD), this effect has been termed SHG-CD [69].

### B1.5.4.2 Adsorbate coverage

A quantity of interest in many studies of surfaces and interfaces is the concentration of adsorbed atomic or molecular species. The SHG/SFG technique has been found to be a useful probe of adsorbate density for a wide range of interfaces. The surface sensitivity afforded by the method is illustrated by the results of figure B1.5.9 [72]. These data show the dramatic change in SH response from a clean surface of silicon upon adsorption of a fraction of a monolayer of atomic hydrogen.

We now consider how one extracts quantitative information about the surface or interface adsorbate coverage from such SHG data. In many circumstances, it is possible to adopt a purely phenomenological approach: one calibrates the nonlinear response as a function of surface coverage in a preliminary set of

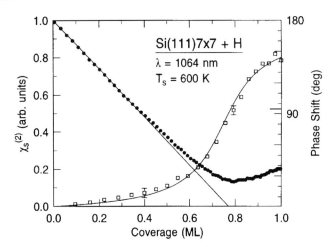

**Figure B1.5.9.** Dependence of the magnitude (full circles) and the phase (open squares) of the nonlinear susceptibility $\chi_s^{(2)}$ of Si(111)7 × 7 on the coverage of adsorbed atomic hydrogen for an excitation wavelength of 1064 nm. (From [72].)

experiments and then makes use of this calibration in subsequent investigations. Such an approach may, for example, be appropriate for studies of adsorption kinetics where the interest lies in the temporal evolution of the surface adsorbate density $N_s$.

For other purposes, obtaining a measure of the adsorbate surface density directly from the experiment is desirable. From this perspective, we introduce a simple model for the variation of the surface nonlinear susceptibility with adsorbate coverage. An approximation that has been found suitable for many systems is

$$\chi_s^{(2)}(N_s) = \chi_{s,0}^{(2)} + N_s \alpha^{(2)}. \tag{B1.5.43}$$

From a purely phenomenological perspective, this relationship describes a constant rate of change in the nonlinear susceptibility of the surface with increasing adsorbate surface density $N_s$. Within a picture of adsorbed molecules, $\alpha^{(2)}$ may be interpreted as the nonlinear polarizability of the adsorbed species. The quantity $\chi_{s,0}^{(2)}$ represents the nonlinear response in the absence of the adsorbed species.

If we consider the optical response of a molecular monolayer of increasing surface density, the form of equation (B1.5.43) is justified in the limit of relatively low density where local-field interactions between the adsorbed species may be neglected. It is difficult to produce any rule for the range of validity of this approximation, as it depends strongly on the system under study, as well as on the desired level of accuracy for the measurement. The relevant corrections, which may be viewed as analogous to the Clausius–Mossotti corrections in linear optics, have been the subject of some discussion in the literature [41, 42]. In addition to the local-field effects, the simple proportionality of variation in $\chi_s^{(2)}(N_s)$ with $N_s$ frequently breaks down for reasons related to the physical and chemical nature of the surface or interface. In particular, inhomogeneous surfaces in which differing binding sites fill in different proportions may give rise to a variation in $\chi_s^{(2)}(N_s)$ that would, to leading order, vary with relative populations of the different sites. Also, inter-adsorbate interactions that lead to shifts in energy levels or molecular orientation, as will be discussed later, will influence the nonlinear response in a manner beyond that captured in equation (B1.5.43).

Despite these caveats in the application of equation (B1.5.43), one finds that it provides reasonable accuracy in many experimental situations. The SH response for the H/Si system of figure B1.5.9, for example, is seen to obey the simple linear variation of $\chi_s^{(2)}(N_s)$ with $N_s$ of equation (B1.5.43) rather well up to an adsorbate coverage of about 0.5 monolayers. These data are also interesting because they show how destructive interference between the terms $\chi_{s,0}^{(2)}$ and $N_s \alpha^{(2)}$ can cause the SH signal to decrease with increasing $N_s$. In this particular example, the physical interpretation of this effect is based on the strong nonlinear response of the bare surface from the Si dangling bonds. With increasing hydrogen coverage, the concentration of dangling bonds is reduced and the surface nonlinearity decreases. For a system where $\chi_{s,0}^{(2)}$ is relatively small and the nonlinear response of the adsorbed species is significant, just the opposite trend for the variation of $\chi_s^{(2)}(N_s)$ with $N_s$ would occur.

The applications of this simple measure of surface adsorbate coverage have been quite widespread and diverse. It has been possible, for example, to measure adsorption isotherms in many systems. From these measurements, one may obtain important information such as the adsorption free energy, $\Delta G^\circ = -RT \ln(K_{eq})$ [21]. One can also monitor the kinetics of adsorption and desorption to obtain rates. In conjunction with temperature-dependent data, one may further infer activation energies and pre-exponential factors [73, 74]. Knowledge of such kinetic parameters is useful for technological applications, such as semiconductor growth and synthesis of chemical compounds [75]. Second-order nonlinear optics may also play a role in the investigation of physical kinetics, such as the rates and mechanisms of transport processes across interfaces [76].

Before leaving this topic, we would like to touch on two related points. The first concerns the possibility of an absolute determination of the surface adsorbate density. Equation (B1.5.43) would suggest that one might use knowledge, either experimental or theoretical, of $\alpha^{(2)}$ and an experimental determination of $\chi_s^{(2)}(N_s)$ and $\chi_{s,0}^{(2)}$ to infer $N_s$ in absolute terms. In practice, this is problematic. One experimental issue is that a correct measurement of $\chi_s^{(2)}$ in absolute terms may be difficult. However, through appropriate comparison with the response of a calibrated nonlinear reference material, we may usually accomplish this task. More problematic is obtaining knowledge of $\alpha^{(2)}$ for the adsorbed species. The determination of $\alpha^{(2)}$ in the gas or liquid phase is already difficult. In addition, the perturbation induced by the surface or interface is typically significant. Moreover, as discussed later, molecular orientation is a critical factor in determining the surface nonlinear response. For these reasons, absolute surface densities can generally be found from surface SHG/SFH measurements only if we can calibrate the surface nonlinear response at two or more coverages determined by other means. This situation, it should be noted, is not dissimilar to that encountered for many other common surface probes.

The second issue concerns molecular specificity. For a simple measurement of SHG at an arbitrary laser frequency, one cannot expect to extract information of the behaviour of a system with several possible adsorbed species. To make the technique appropriate for such cases, one needs to rely on spectroscopic information. In the simplest implementation, one chooses a frequency for which the nonlinear response of the species of interest is large or dominant. As will be discussed in section B1.5.4.4, this capability is significantly enhanced with SFG and the selection of a frequency corresponding to a vibrational resonance.

### B1.5.4.3  Molecular orientation

The nonlinear response of an individual molecule depends on the orientation of the molecule with respect to the polarization of the applied and detected electric fields. The same situation prevails for an ensemble of molecules at an interface. It follows that we may garner information about molecular orientation at surfaces and interfaces by appropriate measurements of the polarization dependence of the nonlinear response, taken together with a model for the nonlinear response of the relevant molecule in a standard orientation.

We now consider this issue in a more rigorous fashion. The inference of molecular orientation can be explained most readily from the following relation between the surface nonlinear susceptibility tensor $\chi_s^{(2)}$ and the molecular nonlinear polarizability $\alpha^{(2)}$:

$$\chi_{s,ijk}^{(2)} = N_s \sum_{i',j',k'} \langle T_{ii'} T_{jj'} T_{kk'} \rangle \alpha_{i'j'k'}^{(2)}. \tag{B1.5.44}$$

Here the $ijk$ coordinate system represents the laboratory reference frame; the primed coordinate system

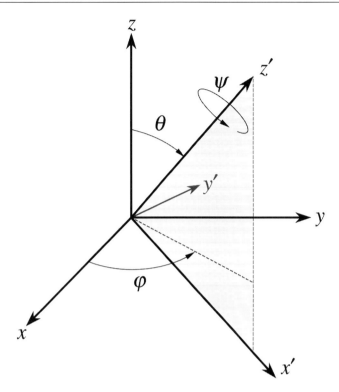

**Figure B1.5.10.** Euler angles and reference frames for the discussion of molecular orientation: laboratory frame $(x, y, z)$ and molecular frame $(x', y', z')$.

$i'j'k'$ corresponds to coordinates in the molecular system. The quantities $T_{ii'}$ are the matrices describing the coordinate transformation between the molecular and laboratory systems. In this relationship, we have neglected local-field effects and expressed the $\chi_s^{(2)}$ in a form equivalent to summing the molecular response over all the molecules in a unit surface area (with surface density $N_s$). (For simplicity, we have omitted any contribution to $\chi^{(2)}$ not attributable to the dipolar response of the molecules. In many cases, however, it is important to measure and account for the background nonlinear response not arising from the dipolar contributions from the molecules of interest.) In equation (B1.5.44), we allow for a distribution of molecular orientations and have denoted by $\langle\ \rangle$ the corresponding ensemble average:

$$\langle T_{ii'} T_{jj'} T_{kk'} \rangle = \int f(\theta, \varphi, \psi) T_{ii'} T_{jj'} T_{kk'}\, d\Omega. \qquad \text{(B1.5.45)}$$

Here $f(\theta, \varphi, \psi)$ is the probability distribution of finding a molecule oriented at $(\theta, \varphi, \psi)$ within an element $d\Omega$ of solid angle with the molecular orientation defined in terms of the usual Euler angles (figure B1.5.10).

Equation (B1.5.44) indicates that if we know $\chi_{s,ijk}^{(2)}$ and $\alpha_{i'j'k'}^{(2)}$, we may infer information about the third-order orientational moments $\langle T_{ii'} T_{jj'} T_{kk'} \rangle$. Since calibration of absolute magnitudes is difficult, we are generally concerned with a comparison of the relative magnitudes of the appropriate molecular ($\alpha^{(2)}$) and macroscopic ($\chi_s^{(2)}$) quantities. In practice, the complexity of the general relationship between $\alpha^{(2)}$ and $\chi_s^{(2)}$ means that progress requires the introduction of certain simplifying assumptions. These usually follow from symmetry considerations or, for the case of $\alpha^{(2)}$, from previous experimental or theoretical insight into the nature of the expected molecular response.

The approach may be illustrated for molecules with a nonlinear polarizability $\alpha^{(2)}$ dominated by a single axial component $\alpha^{(2)}_{z'z'z'}$, corresponding to a dominant nonlinear response from transitions along a particular molecular axis. Let us further assume that all in-plane direction of the surface or interface are equivalent. This would naturally be the case for a liquid or amorphous solid, but would not necessarily apply to the surface of a crystal. One then obtains from equation (B1.5.44), the following relations between the molecular quantities and surface nonlinear susceptibility:

$$\chi^{(2)}_{s,\perp\perp\perp} = \chi^{(2)}_{s,zzz} = N_s \langle \cos^3 \theta \rangle \alpha^{(2)}_{z'z'z'} \tag{B1.5.46}$$

$$\chi^{(2)}_{s,\perp\|\|} = \chi^{(2)}_{s,zxx} = \tfrac{1}{2} N_s \langle \cos^3 \theta \sin^2 \theta \rangle \alpha^{(2)}_{z'z'z'} \tag{B1.5.47}$$

$$\chi^{(2)}_{s,\|\perp\|} = \chi^{(2)}_{s,\|\|\perp} = \chi^{(2)}_{s,xzx} = \tfrac{1}{2} N_s \langle \cos^3 \theta \sin^2 \theta \rangle \alpha^{(2)}_{z'z'z'}. \tag{B1.5.48}$$

Notice that $\chi^{(2)}_{s,\perp\|\|} = \chi^{(2)}_{s,\|\perp\|} = \chi^{(2)}_{s,\|\|\perp}$, so that only two of the three nonlinear susceptibility tensor elements allowed for an isotropic surface are independent. From equations (B1.5.46)–(B1.5.48), we may form the ratio

$$\frac{2\chi^{(2)}_{s,\|\perp\|} + \chi^{(2)}_{s,\perp\perp\perp}}{\chi^{(2)}_{s,\perp\perp\perp}} = \frac{\langle \cos \theta \rangle}{\langle \cos^3 \theta \rangle}. \tag{B1.5.49}$$

Thus, a well-defined measure of molecular orientation is inferred from the measurement of the macroscopic quantities $\chi^{(2)}_{s,ijk}$. For the case of a narrow and isotropic distribution, i.e. $f(\theta) = \delta(\theta - \theta_0)$, the left-hand side term of equation (B1.5.49) becomes $\langle \cos \theta \rangle / \langle \cos^3 \theta \rangle = \sec^2 \theta_0$, for which the mean orientation $\theta_0$ is directly obtained. For a broad distribution, one may extract the mean orientation from such an expression for an assumed functional form.

As an example, the model described earlier for a molecule having a dominant nonlinear polarizability element $\alpha^{(2)}_{z'z'z'}$ has been applied to the determination of the molecular inclination $\theta$ between the molecular axis of a surfactant molecule, sodium-dodecylnaphtalene-sulphonate (SDNS) and the surface normal at the air/water interface [77]. This tilt angle $\theta$, shown in figure B1.5.11, was determined according to equations (B1.5.46)–(B1.5.48) under the assumption of a narrow orientational distribution. As the figure shows, the mean molecular orientation changes with increasing surface pressure $\pi$ as the molecules are forced into a more nearly vertical orientation.

In the literature, the interested reader may find treatment of molecules with differing, and more complex, behaviour for $\alpha^{(2)}$ [14]. Also of importance is the role of SFG for orientational analysis. Surface SFG may provide more orientational information, since there are additional independent elements of the surface nonlinear susceptibility. For example, an isotropic surface is characterized by three independent elements for SHG, but is described by four independent elements for SFG. The principal advantage of SFG over SHG, however, lies in its molecular specificity. As will be discussed in section B1.5.4.4, we may enhance the response of a molecular species of interest by choosing the appropriate infrared frequency for the SFG process. This behaviour helps to eliminate background signal and is useful in more complex systems with two or more molecular species present. Equally important, the excitation of a given vibration helps to define the form, and reduce the generality, of the molecular response $\alpha^{(2)}$. Further, the method may be applied for different vibrational resonances to deduce the orientation of different moieties of larger molecules [78].

### B1.5.4.4   Spectroscopy

The second-order nonlinear susceptibility describing a surface or interface, as indicated by the microscopic form of equation (B1.5.30), is resonantly enhanced whenever an input or output photon energy matches a transition energy in the material system. Thus, by scanning the frequency or frequencies involved in the

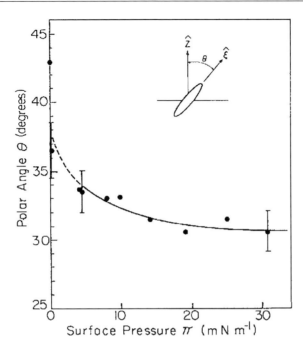

**Figure B1.5.11.** Tilt angle $\theta$ between the molecular axis of sodium-dodecylnaphtalene-sulphonate (SDNS) and the surface normal as a function of the surface pressure $\pi$ at the air/water interface. (From [77].)

surface nonlinear process, we may perform surface-specific spectroscopy. This method has been successfully applied to probe both electronic transitions and vibrational transitions at interfaces.

For studying electronic transitions at surfaces and interfaces, both SHG and SFG have been employed in a variety of systems. One particular example is that of the buried $CaF_2/Si(111)$ interface [79]. Figure B1.5.12(a) displays the experimental SH signal as a function of the photon energy of a tunable pump laser. An interface resonance for this system is found to occur for a photon energy near 2.4 eV. This value is markedly different from that of the energies of transitions in either of the bulk materials and clearly illustrates the capability of nonlinear spectroscopy to probe distinct electronic excitations of the interfacial region. The sharp feature appearing at 2.26 eV has been attributed to the formation of a two-dimensional exciton. It is important to point out that the measurement of the SHG signal alone does not directly show whether an observed resonance corresponds to a single- or a two-photon transition. To verify that the resonance enhancement does, in fact, correspond to a transition energy of 2.4 eV, a separate SF measurement (figure B1.5.12(b)) was performed. In this measurement, the tunable laser photon was mixed with another photon at a fixed photon energy (1.17 eV). By comparing the two sets of data, one finds that the resonance must indeed lie at the fundamental frequency of the tunable laser for this system.

The SHG/SFG technique is not restricted to interface spectroscopy of the delocalized electronic states of solids. It is also a powerful tool for spectroscopy of electronic transitions in molecules. Figure B1.5.13 presents such an example for a monolayer of the R-enantiomer of the molecule 2,2'-dihydroxyl-1,1'-binaphthyl, (R)-BN, at the air/water interface [80]. The spectra reveal two-photon resonance features near wavelengths of 332 and 340 nm that are assigned to the two lowest exciton-split transitions in the naphth-2-ol monomer of BN. An increase in signal at higher photon energies is also seen as a resonance as the $^1B_b$ state of the molecules is approached. The spectra in figure B1.5.13 have been obtained for differing polarization configurations. The arrangements of p-in/p-out and s-in/p-out will yield SH signals for any isotropic surface. In this case,

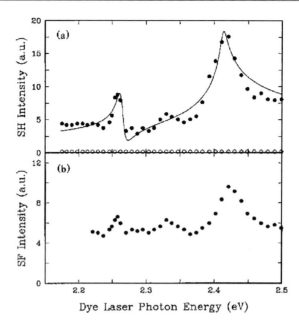

**Figure B1.5.12.** SH and SF spectra (full dots) for the $CaF_2/Si(111)$ interface: (a) SH intensity as a function of the photon energy of the tunable laser; (b) SF intensity obtained by mixing the tunable laser with radiation at a fixed photon energy of 1.17 eV. For comparison, the open circles in (a) are signals obtained for a native-oxide covered Si(111). The full line is a fit to the theory as discussed in [79].

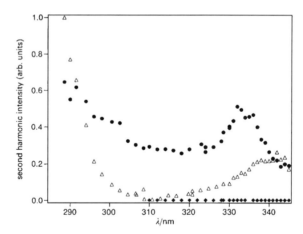

**Figure B1.5.13.** Spectra of the various non-chiral [p-in/p-out (filled circles) and s-in/p-out (filled diamonds)] and chiral [p-in/s-out (triangle)] SHG signals of (R)-BN molecules adsorbed at the air/water interface. (From [80].)

however, signal is also observed for the p-in/s-out configuration. This response arises from the $\chi^{(2)}_{s,xyz}$ element of the surface nonlinear response that is present because of the chiral character of the molecules under study, as previously discussed in section B1.5.4.1.

In addition to probing electronic transitions, second-order nonlinear optics can be used to probe vibrational resonances. This capability is of obvious importance and value for identifying chemical species at interfaces and probing their local environment. In contrast to conventional spectroscopy of vibrational transitions, which can also be applied to surface problems [4], nonlinear optics provides intrinsic surface specificity and is of particular utility in problems where the same or similar vibrational transitions occur at the interface as in the bulk media. In order to access the infrared region corresponding to vibrational transitions while maintaining an easily detectable signal, Shen and coworkers developed the technique of the infrared-visible sum-frequency generation [81]. In this scheme, a tunable IR source is mixed on the surface with visible light at a fixed frequency to produce readily detectable visible radiation. As the IR frequency is tuned through the frequency of a vibrational transition, the SF signal is resonantly enhanced and the surface vibration spectrum is recorded.

In order to examine the IR-visible SFG process more closely, let us consider an appropriate formula for the surface nonlinear susceptibility when the IR frequency $\omega_1 = \omega_{IR}$ is near a single vibrational resonance and the visible frequency $\omega_2 = \omega_{vis}$ is not resonant with an electronic transition. We may then write [81, 82]

$$\chi^{(2)}_{s,ijk}(\omega_{IR}) = \chi^{(2)NR}_{s,ijk} + \chi^{(2)R}_{s,ijk}(\omega_{IR}) = \chi^{(2)NR}_{s,ijk} + \sum_l \frac{A_{l,ijk}}{\omega_{IR} - \omega_l + i\Gamma_l} \tag{B1.5.50}$$

where $\chi^{(2)NR}_{s,ijk}$ and $\chi^{(2)R}_{s,ijk}$ are the non-resonant and resonant contributions to the signal, respectively; $A_{l,ijk}$, $\omega_l$, and $\Gamma_l$ are the strength, resonant frequency, linewidth of the $l$th vibrational mode. The quantity $A_l$ is proportional to the product of the first derivatives of the molecular dipole moment $\mu_i$ and of the electronic polarizability $\alpha_{jk}$ with respect to the $l$th normal coordinate $Q_l$:

$$A_{l,ijk} \propto \frac{\partial \mu_i}{\partial Q_l} \frac{\partial \alpha_{jk}}{\partial Q_l}. \tag{B1.5.51}$$

Consequently, in order for a vibrational mode to be observed in infrared-visible SFG, the molecule in its adsorbed state has to be both IR [$(d\mu_i/dQ_l) \neq 0$] and Raman [$(d\alpha_{jk}/dQ_l) \neq 0$] active.

The form of equation (B1.5.50) also allows us to make some remarks about the measured lineshapes in surface nonlinear spectroscopy. From this point of view, we may regard equation (B1.5.50) as being representative of the surface nonlinear response typically encountered near any resonance: It has a strongly varying resonant contribution together with a spectrally flat non-resonant background. The interesting aspect of this situation arises from the fact that we generally detect the *intensity* of the SH or SF signal, which is proportional to $|\chi^{(2)}_{s,ijk}|^2$. Consequently, interference between the resonant and non-resonant contributions is expected. Depending on the relative phase difference between these terms, one may observe various experimental spectra, as illustrated in figure B1.5.14. This type of behaviour, while potentially a source of confusion, is familiar for other types of nonlinear spectroscopy, such as CARS (coherent anti-Stokes Raman scattering) [30, 31] and can be readily incorporated into modelling of measured spectral features.

We now present one of the many examples of interfacial vibrational spectroscopy using SFG. Figure B1.5.15 shows the surface vibrational spectrum of the water/air interface at a temperature of 40 °C [83]. Notice that the spectrum exhibits peaks at 3680, 3400 and 3200 cm$^{-1}$. These features arise from the OH stretching mode of water molecules in different environments. The highest frequency peak is assigned to free OH groups, the next peak to water molecules with hydrogen-bonding to neighbours in a relatively disordered structure and the lowest frequency peak to water molecules in a well-ordered tetrahedrally bonded (ice-like) structure. In addition to the analogy of these assignments to water molecules in different bulk environments, the assignments are compatible with the measured temperature dependence of the spectra. The strong and narrow peak at 3680 cm$^{-1}$ provides interesting new information about the water surface. It indicates that a substantial fraction of the surface water molecules have unbonded OH groups protruding from the surface of the water into the vapour phase. This study exemplifies the unique capabilities of surface SFG, as there is no other technique that could probe the liquid/vapour interface in the presence of strong features of the vibrational modes of the bulk water molecules.

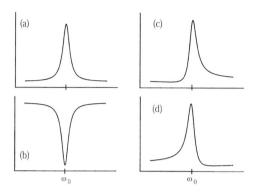

**Figure B1.5.14.** Possible lineshapes for an SFG resonance as a function of the infrared frequency $\omega_{IR}$. The measured SFG signal is proportional to $|\chi^{NR} + A/(\omega_{IR} - \omega_0 + i\Gamma)|^2$. Assuming both $\chi^{NR}$ and $\Gamma$ are real and positive, we obtain the lineshapes for various cases: (a) $\chi^{NR} \ll A/\Gamma$; (b) $A$ is purely imaginary and negative with $|\chi^{NR}\Gamma/A| > \frac{1}{2}$; (c) $A$ is real and positive; and (d) $A$ is real and negative. Note the apparent blue and red shifts of the peaks in cases (c) and (d), respectively.

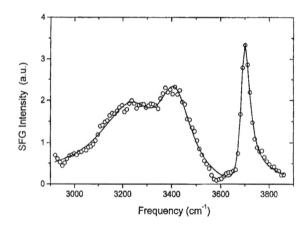

**Figure B1.5.15.** SFG spectrum for the water/air interface at 40 °C using the ssp polarization combination (s-, s- and p-polarized sum-frequency signal, visible input and infrared input beams, respectively). The peaks correspond to OH stretching modes. (After [83].)

An important consideration in spectroscopic measurements concerns the bandwidth of the laser sources. In order to resolve the vibrational resonances in a conventional approach, one needs, in the conventional scheme, a tunable source that has a narrow bandwidth compared to the resonance being studied. For typical resolutions, this requirement implies, by uncertainty principle, that IR pulses of picosecond or longer duration must be used longer. On the other hand, ultrafast pulsed IR sources with broad bandwidths are quite attractive from the experimental standpoint. In order to make use of these sources, two types of new experimental techniques have been introduced. One technique involves mixing the broadband IR source ($\sim 300$ cm$^{-1}$) with a narrowband visible input ($\sim 5$ cm$^{-1}$). By spectrally resolving the SF output, we may then obtain resolution of the IR spectrum limited only by the linewidth of the visible source [84, 85]. This result follows from the fact that $\omega_{IR} = \omega_{SF} - \omega_{vis}$ must be satisfied for the SFG process. The second new approach involves the application of a Fourier transform scheme [86]. This is accomplished by passing the IR pulses through an interferometer and then mixing these pairs of pulses with visible radiation at the surface.

### B1.5.4.5  Dynamics

Many of the fundamental physical and chemical processes at surfaces and interfaces occur on extremely fast time scales. For example, atomic and molecular motions take place on time scales as short as 100 fs, while surface electronic states may have lifetimes as short as 10 fs. With the dramatic recent advances in laser technology, however, such time scales have become increasingly accessible. Surface nonlinear optics provides an attractive approach to capture such events directly in the time domain. Some examples of application of the method include probing the dynamics of melting on the time scale of phonon vibrations [87], photoisomerization of molecules [88], molecular dynamics of adsorbates [89, 90], interfacial solvent dynamics [91], transient band-flattening in semiconductors [92] and laser-induced desorption [93]. A review article discussing such time-resolved studies in metals can be found in [94]. The SHG and SFG techniques are also suitable for studying dynamical processes occurring on slower time scales. Indeed, many valuable studies of adsorption, desorption, diffusion and other surface processes have been performed on time scales of milliseconds to seconds.

In a typical time-resolved SHG (SFG) experiment using femtosecond to picosecond laser systems, two (three) input laser beams are necessary. The pulse from one of the lasers, usually called the pump laser, induces the reaction or surface modification. This defines the starting point ($t = 0$). A second pulse (or a second set of synchronized pulses, for the case of SFG) delayed relative to the first pulse by a specified time $\Delta t$ is used to probe the reaction as it evolves. By varying this time delay $\Delta t$, the temporal evolution of the reaction can be followed. In order to preserve the inherent time resolution of an ultrafast laser, the relevant pulses are generally derived from a common source. For instance, in a basic time-resolved SHG experiment, where both the pump and probe pulses are of the same frequency, one simply divides the laser beam into two sets of pulses with a beam splitter. One of these pulses travels a fixed distance to the sample, while the other passes through a variable delay line to the sample. This approach provides a means of timing with sub-femtosecond accuracy, if desired. In some cases, at least one of the input beams has a different frequency from the others. Such pulses can be produced through processes such as harmonic generation or optical parametric generation from the main laser pulse.

As an example of this class of experiment, we consider an experimental study of the dynamics of molecular orientational relaxation at the air/water interface [90]. Such investigations are of interest as a gauge of the local environment at the surface of water. The measurements were performed with time-resolved SHG using Coumarin 314 dye molecules as the probe. In order to examine orientational motion, an anisotropic orientational distribution of molecules must first be produced. This is accomplished through a photoselection process in which the interface is irradiated by a linearly polarized laser pulse that is resonant with an electronic transition in the dye molecules. Those molecules that are oriented with their transition dipole moments parallel to the polarization of the pump beam are preferentially excited, producing an orientational anisotropy in the ground- and excited-state population. Subsequently, these anisotropic orientational distributions relax to the equilibrium configuration. The time evolution of the rotational anisotropy was followed by detecting the SH of a probe laser pulse as a function of the delay time, as shown in figure B1.5.16. Through a comparison of the results for different initial anisotropic distributions (produced by two orthogonal linearly-polarized pump beams, as shown in the figure, as well as by circularly-polarized pump radiation), one may deduce rates for both in-plane and out-of-plane orientational relaxation. The study yielded the interesting result that the orientational relaxation times at the liquid/vapour interface significantly exceeded those for the Coumarin molecules in the bulk of water. This finding was interpreted as reflecting the increased friction encountered in the surface region where the water molecules are more highly ordered than in the bulk liquid.

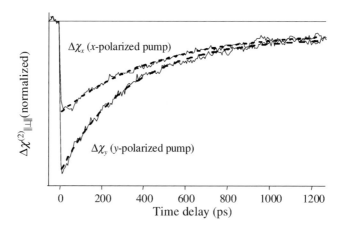

**Figure B1.5.16.** Rotational relaxation of Coumarin 314 molecules at the air/water interface. The change in the SH signal is recorded as a function of the time delay between the pump and probe pulses. Anisotropy in the orientational distribution is created by linearly polarized pump radiation in two orthogonal directions in the surface. (After [90].)

### B1.5.4.6 Spatial resolution

Another application of surface SHG or SFG involves the exploitation of the lateral resolution afforded by these optical processes. While the dimension of the optical wavelength obviously precludes direct access to the length scale of atoms and molecules, one can examine the micrometre and submicrometre length scale that is important in many surface and interface processes. Spatial resolution may be achieved simply by detecting the nonlinear response with a focused laser beam that is scanned across the surface [95]. Alternatively, one may illuminate a large area of the surface and image the emitted nonlinear radiation [96]. Applications of this imaging capability have included probing of magnetic domains [97] and spatially varying electric fields [98]. The application of near-field techniques may permit, as in linear optics, the attainment of spatial resolution below the diffraction limit. In a recent work [99], submicrometre spatial resolution was indeed reported by collecting the emitted SH radiation for excitation with a near-field fibre probe.

Diffraction measurements offer a complementary approach to the real-space imaging described earlier. In such schemes, periodically modulated surfaces are utilized to produce well-defined SH (or SF) radiation at discrete angles, as dictated by the conservation of the in-plane component of the wavevector. As an example of this approach, a grating in the surface adsorbate density may be produced through laser-induced desorption in the field of two interfering beams. This monolayer grating will readily produce diffracted SH beams in addition to the usual reflected beam. In addition to their intrinsic interest, such structures have permitted precise measurements of surface diffusion. One may accomplish this by observing the temporal evolution of SH diffraction efficiency, which falls as surface diffusion causes the modulation depth of the adsorbate grating to decrease. This technique has been applied to examine diffusion of adsorbates on the surface of metals [100] and semiconductors [101].

### B1.5.4.7 Electric and magnetic field perturbation

Probing electric and magnetic fields, and the effects induced by them, is of obvious interest in many areas of science and technology. We considered earlier the influence of such perturbations in a general fashion in section B1.5.2.2. Here we describe some related experimental measurements and applications. Electric fields act to break inversion and thus may yield bulk SHG and SFG signals from centrosymmetric media surrounding the interface, in addition to any field-dependent contribution of the interface itself. The high sensitivity of

SHG toward applied electric fields was first demonstrated in calcite by Terhune *et al* as early as 1962 [102]. Subsequent investigations of EFISH from centrosymmetric media have involved semiconductor/electrolyte and metal/electrolyte interfaces [103–105], as well as metal-oxide–semiconductor interfaces [106, 107].

The generality of the EFISH process has led to a variety of applications. These include probing the surface potential at interfaces involving liquids. Such measurements rely on the fact that the EFISH field is proportional to the voltage drop across the polarized layer provided, as is generally the case, that this region is thin compared to the scale of an optical wavelength [108]. This effect also serves as a basis for probing surface reactions involving changes of charge state, such as acid/base equilibria [109]. Another related set of applications involves probing electric fields in semiconductors, notably in centrosymmetric material silicon [98, 110, 111]. These studies have demonstrated the capability for spatial resolution, vector analysis of the electric field and, significantly, ultrafast (subpicosecond) time resolution. These capabilities of SHG complement other optical schemes, such as electro-optical and photoconductive sampling, for probing the dynamics of electric fields on very fast time scales.

The influence of an applied magnetic field, as introduced in section B1.5.2.2, is quite different from that of an applied electric field. A magnetic field may perturb the interfacial nonlinear response (and that of the weak bulk terms), but it does not lead to any dipole-allowed bulk nonlinear response. Thus, in the presence of magnetic fields and magnetization, SHG remains a probe that is highly specific to surfaces and interfaces. It may be viewed as the interface-sensitive analogue of linear magneto-optical effects. The first demonstration of the influence of magnetization on SHG was performed on an Fe(110) surface [112]. Subsequent applications have included examination of other materials for which both the bulk and surface exhibit magnetization. For these systems, surface specificity is of key importance. In addition, the technique has been applied to examine buried magnetic interfaces [113]. Excellent review articles on this subject matter are presented in [23] and [114].

### B1.5.4.8    *Recent developments*

Up to this point, our discussion of surface SHG and SFG has implicitly assumed that we are examining a smooth planar surface. This type of interface leads to well-defined and highly collimated transmitted and reflected beams. On the other hand, many material systems of interest in probing surfaces or interfaces are not planar in character. From the point of view of symmetry, the surface sensitivity for interfaces of centrosymmetric media should apply equally well for such non-planar interfaces, although the nature of the electromagnetic wave propagation may be modified to a significant degree. One case of particular interest concerns appropriately roughened surfaces of noble metals. These were shown as early as 1974 [115] to give rise to strong enhancements in Raman scattering of adsorbed species and led to extensive investigation of the phenomenon of surface-enhanced Raman scattering or SERS [5]. Significant enhancements in the SHG signals from such surfaces have also been found [116]. The resulting SH radiation is diffuse, but has been shown to preserve a high degree of surface sensitivity. Carrying this progression from planar surfaces one step further, researchers have recently demonstrated the possibility of probing the surfaces of small particles by SHG.

Experimental investigations of the model system of dye molecules adsorbed onto surfaces of polystyrene spheres have firmly established the sensitivity and surface specificity of the SHG method even for particles of micrometre size [117]. The surface sensitivity of the SHG process has been exploited for probing molecular transport across the bilayer in liposomes [118], for measurement of electrostatic potentials at the surface of small particles [119] and for imaging membranes in living cells [120]. The corresponding theoretical description of SHG from the surfaces of small spheres has been examined recently using the type of formalism presented earlier in this chapter [121]. Within this framework, the leading-order contributions to the SH radiation arise from the non-local excitation of the dipole and the local excitation of the quadrupole moments. This situation stands in contrast to linear optical (Rayleigh) scattering, which arises from the local excitation of the dipole moment.

## B1.5.5   Conclusion

In this brief chapter, we have attempted to describe some of the underlying principles of second-order nonlinear optics for the study of surfaces and interfaces. The fact that the technique relies on a basic symmetry consideration to obtain surface specificity gives the method a high degree of generality. As a consequence, our review of some of the applications of the method has necessarily been quite incomplete. Still, we hope that the reader will gain some appreciation for the flexibility and power of the method. Over the last few years, many noteworthy applications of the method have been demonstrated. Further advances may be anticipated from on-going development of the microscopic theory, as well as from adaptation of the macroscopic theory to new experimental conditions and geometries. At the same time, we see continual progress in the range and ease of use of the technique afforded by the impressive improvement of the performance and reliability of high-power laser sources.

## References

[1]   Franken P A, Hill A E, Peters C W and Weinreich G 1961 Generation of optical harmonics *Phys. Rev. Lett.* **7** 118
[2]   Somorjai G A 1981 *Chemistry in Two Dimensions* (Ithaca, NY: Cornell University Press)
[3]   Duke C B (ed) 1994 Surface science: the first thirty years *Surf. Sci.* **299/300** 1–1054
[4]   Dumas P, Weldon M K, Chabal Y J and Williams G P 1999 Molecules at surfaces and interfaces studied using vibrational spectroscopies and related techniques *Surf. Rev. Lett.* **6** 225–55
[5]   Chang R K and Furtak T E 1982 *Surface Enhanced Raman Scattering* (New York: Plenum)
[6]   Moskovits M 1985 Surface-enhanced spectroscopy *Rev. Mod. Phys.* **57** 783–826
[7]   Campion A and Kambhampati P 1998 Surface-enhanced Raman scattering *Chem. Soc. Rev.* **27** 241–50
[8]   Aspnes D E 1993 New developments in spectroellipsometry—the challenge of surfaces *Thin Solid films* **233** 1–8
[9]   Azzam R M A and Bashara N M 1977 *Ellipsometry and Polarized Light* (Amsterdam: North-Holland)
[10]  McGilp J F, Patterson C H and Weaire D L (ed) 1995 *Epioptics: Linear and Nonlinear Optical Spectroscopy of Surfaces and Interfaces* (Berlin: Springer)
[11]  Aspnes D E 1985 Above-bandgap optical anisotropies in cubic semiconductors: a visible–near ultraviolet probe of surfaces *J. Vac. Sci. Technol.* B **3** 1498–506
[12]  Richmond G L, Robinson J M and Shannon V L 1988 Second harmonic generation studies of interfacial structure and dynamics *Prog. Surf. Sci.* **28** 1–70
[13]  Shen Y R 1989 Surface-properties probed by second-harmonic and sum-frequency generation *Nature* **337** 519–25
[14]  Shen Y R 1989 Optical second harmonic-generation at interfaces *Ann. Rev. Phys. Chem.* **40** 327–50
[15]  Heinz T F 1991 Second-order nonlinear optical effects at surfaces and interfaces *Nonlinear Surface Electromagnetic Phenomena* ed H-E Ponath and G I Stegeman (Amsterdam: North-Holland) pp 353–416
[16]  Eisenthal K B 1992 Equilibrium and dynamic processes at interfaces by second harmonic and sum frequency generation *Ann. Rev. Phys. Chem.* **43** 627–61
[17]  Corn R M and Higgins D A 1994 Optical second-harmonic generation as a probe of surface-chemistry *Chem. Rev.* **94** 107–25
[18]  McGilp J F 1995 Optical characterisation of semiconductor surfaces and interfaces *Prog. Surf. Sci.* **49** 1–106
[19]  Reider G A and Heinz T F 1995 Second-order nonlinear optical effects at surfaces and interfaces: recent advances *Photonic Probes of Surfaces* ed P Halevi (Amsterdam: Elsevier) pp 413–78
[20]  Bain C D 1995 Sum-frequency vibrational spectroscopy of the solid–liquid interface *J. Chem. Soc. Faraday Trans.* **91** 1281–96
[21]  Eisenthal K B 1996 Liquid interfaces probed by second-harmonic and sum-frequency spectroscopy *Chem. Rev.* **96** 1343–60
[22]  Richmond G L 1997 Vibrational spectroscopy of molecules at liquid/liquid interfaces *Anal. Chem.* **69** A536–43
[23]  Rasing Th 1998 Nonlinear magneto-optical studies of ultrathin films and multilayers *Nonlinear Optics in Metals* ed K H Bennemann (Oxford: Clarendon) pp 132–218
[24]  Lüpke G 1999 Characterization of semiconductor interfaces by second-harmonic generation *Surf. Sci. Rep.* **35** 75–161
[25]  McGilp J F 1999 Second-harmonic generation at semiconductor and metal surfaces *Surf. Rev. Lett.* **6** 529–58
[26]  Miranda P B and Shen Y R 1999 Liquid interfaces: a study by sum-frequency vibrational spectroscopy *J. Phys. Chem.* B **103** 3292–307
[27]  Shultz M J, Schnitzer C, Simonelli D and Baldelli S 2000 Sum-frequency generation spectroscopy of the aqueous interface: ionic and soluble molecular solutions *Int. Rev. Phys. Chem.* **19** 123–53
[28]  Brevet P F 1997 *Surface Second Harmonic Generation* (Lausanne: Presses Polytechniques et Universitaires Romandes)
[29]  Bennemann K H (ed) 1998 *Nonlinear Optics in Metals* (Oxford: Clarendon)
[30]  Levenson M D and Kano S S 1988 *Introduction to Nonlinear Laser Spectroscopy* (Boston: Academic)
[31]  Mukamel S 1995 *Principles of Nonlinear Optical Spectroscopy* (Oxford: Oxford University Press)
[32]  Flytzanis C 1975 Theory of nonlinear optical susceptibilities *Quantum Electronics* vol 1A (New York: Academic)

[33]  Butcher P N and Cotter D 1990 *The Elements of Nonlinear Optics* (Cambridge: Cambridge University Press)

[34]  Boyd R W 1992 *Nonlinear Optics* (New York: Academic)

[35]  Jackson J D 1975 *Classical Electrodynamics* (New York: Wiley)

[36]  Bloembergen N 1965 *Nonlinear Optics* (New York: Benjamin)

[37]  Shen Y R 1984 *The Principles of Nonlinear Optics* (New York: Wiley)

[38]  Sipe J E 1987 New Green-function formalism for surface optics *J. Opt. Soc. Am.* B **4** 481–9

[39]  Mizrahi V and Sipe J E 1988 Phenomenological treatment of surface second-harmonic generation *J. Opt. Soc. Am.* B **5** 660–7

[40]  Bloembergen N and Pershan P S 1962 Light waves at the boundary of nonlinear media *Phys. Rev.* **128** 606–22

[41]  Ye P and Shen Y R 1983 Local-field effect on linear and nonlinear optical properties of adsorbed molecules *Phys. Rev.* B **28** 4288–94

[42]  Hayden L M 1988 Local-field effects in Langmuir–Blodgett films of hemicyanine and behenic acid mixtures *Phys. Rev.* B **38** 3718–21

[43]  Zyss J (ed) 1994 *Molecular Nonlinear Optics: Materials, Physics, and Devices* (Boston: Academic)

[44]  Prasad P N and Williams D J 1991 *Introduction to Nonlinear Optical Effects in Molecules and Polymers* (New York: Wiley)

[45]  Feibelman P J 1982 Surface electromagnetic fields *Prog. Surf. Sci.* **12** 287–407

[46]  Rudnick J and Stern E A 1971 Second-harmonic radiation from metal surfaces *Phys. Rev.* B **4** 4274–90

[47]  Keller O 1986 Random-phase-approximation study of the response function describing optical second-harmonic generation from a metal selvedge *Phys. Rev.* B **33** 990–1009

[48]  Liebsch A 1997 *Electronic Excitations at Metal Surfaces* (New York: Plenum)

[49]  Petukhov A V 1995 Sum-frequency generation on isotropic surfaces: general phenomenology and microscopic theory for jellium surfaces *Phys. Rev.* B **52** 16 901–11

[50]  Luce T A and Bennemann K H 1998 Nonlinear optical response of noble metals determined from first-principles electronic structures and wave functions: calculation of transition matrix elements *Phys. Rev.* B **58** 15 821–6

[51]  Schaich W L 2000 Calculations of second-harmonic generation for a jellium metal surface *Phys. Rev.* B **61** 10 478–83

[52]  Ghahramani E, Moss D J and Sipe J E 1990 Second-harmonic generation in odd-period, strained, $(Si)_n/(Ge)_n/Si$ superlattices and at Si/Ge interfaces *Phys. Rev. Lett.* **64** 2815–18

[53]  Gavrilenko V I and Rebentrost F 1995 Nonlinear optical susceptibility of the surfaces of silicon and diamond *Surf. Sci.* B **331–3** 1355–60

[54]  Mendoza B S, Gaggiotti A and Del Sole R 1998 Microscopic theory of second harmonic generation at Si(100) surfaces *Phys. Rev. Lett.* **81** 3781–4

[55]  Lim D, Downer M C, Ekerdt J G, Arzate N, Mendoza B S, Gavrilenko V I and Wu R Q 2000 Optical second harmonic spectroscopy of boron-reconstructed Si(001) *Phys. Rev. Lett.* **84** 3406–9

[56]  Pan R P, Wei H D and Shen Y R 1989 Optical second-harmonic generation from magnetized surfaces *Phys. Rev.* B **39** 1229–34

[57]  Stolle R, Marowsky G, Schwarzberg E and Berkovic G 1996 Phase measurements in nonlinear optics *Appl. Phys.* B **63** 491–8

[58]  Kemnitz K, Bhattacharyya K, Hicks J M, Pinto G R, Eisenthal K B and Heinz T F 1986 The phase of second-harmonic light generated at an interface and its relation to absolute molecular orientation *Chem. Phys. Lett.* **131** 285–90

[59]  Dadap J I, Hu X F, Russell N M, Ekerdt J G, Lowell J K and Downer M C 1995 Analysis of second-harmonic generation by unamplified, high-repetition-rate, ultrashort laser pulses at Si(001) interfaces *IEEE J. Selected Topics Quantum Electron* **1** 1145–55

[60]  Heinz T F, Tom H W K and Shen Y R 1983 Determination of molecular orientation of monolayer adsorbates by optical second-harmonic generation *Phys. Rev.* A **28** 1883–5

[61]  Goh M C, Hicks J M, Kemnitz K, Pinto G R, Bhattacharyya K, Heinz T F and Eisenthal K B 1988 Absolute orientation of water-molecules at the neat water-surface *J. Phys. Chem.* **92** 5074–5

[62]  Tom H W K, Heinz T F and Shen Y R 1983 Second-harmonic reflection from silicon surfaces and its relation to structural symmetry *Phys. Rev. Lett.* **51** 1983

[63]  Aktsipetrov O A, Baranova I M and Il'inskii Y A 1986 Surface contribution to the generation of reflected second-harmonic light for centrosymmetric semiconductors *Zh. Eksp. Teor. Fiz.* **91** 287–97 (Engl. transl. 1986 *Sov. Phys. JETP* **64** 167–73)

[64]  Sipe J E, Moss D J and van Driel H M 1987 Phenomenological theory of optical second- and third-harmonic generation from cubic centrosymmetric crystals *Phys. Rev.* B **35** 1129–41

[65]  van Hasselt C W, Verheijen M A and Rasing Th 1990 Vicinal Si(111) surfaces studied by optical second-harmonic generation: step-induced anisotropy and surface-bulk discrimination *Phys. Rev.* B **42** 9263–6

[66]  Heinz T F, Loy M M T and Iyer S S 1987 Nonlinear optical study of Si epitaxy *Mater. Res. Soc. Symp. Proc.* **55** 697

[67]  Dadap J I, Doris B, Deng Q, Downer M C, Lowell J K and Diebold A C 1994 Randomly oriented Ångstrom-scale microroughness at the $Si(100)/SiO_2$ interface probed by optical second harmonic generation *Appl. Phys. Lett.* **64** 2139–41

[68]  Verbiest T, Kauranen M and Persoons A 1999 Second-order nonlinear optical properties of chiral thin films *J. Mater. Chem.* **9** 2005–12

[69]  Petralli-Mallow T, Wong T M, Byers J D, Yee H I and Hicks J M 1993 Circular dichroism spectroscopy at interfaces—a surface second harmonic-generation study *J. Phys. Chem.* **97** 1383–8

[70]  Byers J D, Yee H I and Hicks J M 1994 A second harmonic generation analog of optical rotary dispersion for the study of chiral monolayers *J. Chem. Phys.* **101** 6233–41

[71]    Giordmaine J A 1965 Nonlinear optical properties of liquids *Phys. Rev.* A **138** 1599

[72]    Höfer U 1996 Nonlinear optical investigations of the dynamics of hydrogen interaction with silicon surfaces *Appl. Phys.* A **63** 533–47

[73]    Reider G A, Höfer U and Heinz T F 1991 Desorption-kinetics of hydrogen from the Si(111)7 × 7 surface *J. Chem. Phys.* **94** 4080–3

[74]    Höfer U, Li L P and Heinz T F 1992 Desorption of hydrogen from Si(100)2 × 1 at low coverages—the influence of $\pi$-bonded dimers on the kinetics *Phys. Rev.* B **45** 9485–8

[75]    Dadap J I, Xu Z, Hu X F, Downer M C, Russell N M, Ekerdt J G and Aktsiperov O A 1997 Second-harmonic spectroscopy of a Si(001) surface during calibrated variations in temperature and hydrogen coverage *Phys. Rev.* B **56** 13 367–79

[76]    Crawford M J, Frey J G, VanderNoot T J and Zhao Y G 1996 Investigation of transport across an immiscible liquid/liquid interface—electrochemical and second harmonic generation studies *J. Chem. Soc. Faraday Trans.* **92** 1369–73

[77]    Rasing Th, Shen Y R, Kim M W, Valint P Jr and Bock J 1985 Orientation of surfactant molecules at a liquid-air interface measured by optical second-harmonic generation *Phys. Rev.* A **31** 537–9

[78]    Zhuang X, Miranda P B, Kim D and Shen Y R 1999 Mapping molecular orientation and conformation at interfaces by surface nonlinear optics *Phys. Rev.* B **59** 12 632–40

[79]    Heinz T F, Himpsel F J, Palange E and Burstein E 1989 Electronic transitions at the $CaF_2$/Si(111) interface probed by resonant three-wave-mixing spectroscopy *Phys. Rev. Lett.* **63** 644–7

[80]    Hicks J M, Petralli-Mallow T and Byers J D 1994 Consequences of chirality in second-order nonlinear spectroscopy at interfaces *Faraday Disc.* **99** 341–57

[81]    Zhu X D, Suhr H and Shen Y R 1987 Surface vibrational spectroscopy by infrared-visible sum frequency generation *Phys. Rev.* B **35** 3047–59

[82]    Lin S H and Villaeys A A 1994 Theoretical description of steady-state sum-frequency generation in molecular absorbates *Phys. Rev.* A **50** 5134–44

[83]    Du Q, Superfine R, Freysz E and Shen Y R 1993 Vibrational spectroscopy of water at the vapor–water interface *Phys. Rev. Lett.* **70** 2313–16

[84]    Richter L T, Petralli-Mallow T P and Stephenson J C 1998 Vibrationally resolved sum-frequency generation with broad-bandwidth infrared pulses *Opt. Lett.* **23** 1594–6

[85]    van der Ham E W M, Vrehen Q H F and Eliel E R 1996 Self-dispersive sum-frequency generation at interfaces *Opt. Lett.* **21** 1448–50

[86]    McGuire J A, Beck W, Wei X and Shen Y R 1999 Fourier-transform sum-frequency surface vibrational spectroscopy with femtosecond pulses *Opt. Lett.* **24** 1877–9

[87]    Shank C V, Yen R and Hirlimann C 1983 Femtosecond-time-resolved surface structural dynamics of optically excited silicon *Phys. Rev. Lett.* **51** 900–2

[88]    Sitzmann E V and Eisenthal K B 1988 Picosecond dynamics of a chemical-reaction at the air–water interface studied by surface second-harmonic generation *J. Phys. Chem.* **92** 4579–80

[89]    Castro A, Sitzmann E V, Zhang D and Eisenthal K B 1991 Rotational relaxation at the air–water interface by time-resolved second-harmonic generation *J. Phys. Chem.* **95** 6752–3

[90]    Zimdars D, Dadap J I, Eisenthal K B and Heinz T F 1999 Anisotropic orientational motion of molecular adsorbates at the air–water interface *J. Chem. Phys.* **103** 3425–33

[91]    Zimdars D, Dadap J I, Eisenthal K B and Heinz T F 1999 Femtosecond dynamics of solvation at the air/water interface *Chem. Phys. Lett.* **301** 112–20

[92]    Lantz J M and Corn R M 1994 Time-resolved optical second harmonic generation measurements of picosecond band flattening processes at single crystal $TiO_2$ electrodes *J. Phys. Chem.* **98** 9387–90

[93]    Prybyla J A, Tom H W K and Aumiller G D 1992 Femtosecond time-resolved surface reaction: desorption of CO from Cu(111) in <325 fs *Phys. Rev. Lett.* **68** 503–6

[94]    Hohlfeld J, Conrad U, Müller Wellershoff S S and Matthias E 1998 Femtosecond time-resolved linear and second-order reflectivity of metals *Nonlinear Optics in Metals* ed K H Bennemann (Oxford: Clarendon) pp 219–67

[95]    Boyd G T, Shen Y R and Hansch T W 1986 Continuous-wave second-harmonic generation as a surface microprobe *Opt. Lett.* **11** 97–9

[96]    Schultz K A and Seebauer E G 1992 Surface diffusion of Sb on Ge(111) monitored quantitatively with optical second harmonic microscopy *J. Chem. Phys.* **97** 6958–67

[97]    Kirilyuk V, Kirilyuk A and Rasing Th 1997 A combined nonlinear and linear magneto-optical microscopy *Appl. Phys. Lett.* **70** 2306–8

[98]    Dadap J I, Shan J, Weling A S, Misewich J A, Nahata A and Heinz T F 1999 Measurement of the vector character of electric fields by optical second-harmonic generation *Opt. Lett.* **24** 1059–61

[99]    Smolyaninov I P, Zayats A V and Davis C C 1997 Near-field second-harmonic imaging of ferromagnetic and ferroelectric materials *Opt. Lett.* **22** 1592–4

[100]   Zhu X D, Rasing T H and Shen Y R 1988 Surface diffusion of CO on Ni(111) studied by diffraction of optical second-harmonic generation off a monolayer grating *Phys. Rev. Lett.* **61** 2883–5

[101]   Reider G A, Höfer U and Heinz T F 1991 Surface diffusion of hydrogen on Si(111)7*7 *Phys. Rev. Lett.* **66** 1994–7

[102]  Terhune R W, Maker P D and Savage C M 1962 Optical harmonic generation in calcite *Phys. Rev. Lett.* **8** 404

[103]  Lee C H, Chang R K and Bloembergen N 1967 Nonlinear electroreflectance in silicon and silver *Phys. Rev. Lett.* **18** 167–70

[104]  Aktsipetrov O A and Mishina E D 1984 Nonlinear optical electroreflection in germanium and silicon *Dokl. Akad. Nauk SSSR* **274** 62–5

[105]  Fischer P R, Daschbach J L and Richmond G L 1994 Surface second harmonic studies of Si(111)/electroltye and Si(111)/SiO$_2$/electrolyte interfaces *Chem. Phys. Lett.* **218** 200–5

[106]  Aktsipetrov O A, Fedyanin A A, Golovkina V N and Murzina T V 1994 Optical second-harmonic generation induced by a DC electric field at the Si–SiO$_2$ interface *Opt. Lett.* **19** 1450–2

[107]  Dadap J I, Hu X F, Anderson M H, Downer M C, Lowell J K and Aktsiperov O A 1996 Optical second-harmonic electroreflectance spectroscopy of a Si(0001) metal–oxide–semiconductor structure *Phys. Rev.* B **53** R7607–9

[108]  Zhao X L, Ong S W and Eisenthal K B 1993 Polarization of water-molecules at a charged interface. Second harmonic studies of charged monolayers at the air/water interface *Chem. Phys. Lett.* **202** 513–20

[109]  Ong S W, Zhao X L and Eisenthal K B 1992 Polarization of water-molecules at a charged interface: second harmonic studies of the silica water interface *Chem. Phys. Lett.* **191** 327–35

[110]  Nahata A, Heinz T F and Misewich J A 1996 High-speed electrical sampling using optical second-harmonic generation *Appl. Phys. Lett.* **69** 746–8

[111]  Ohlhoff C, Lupke G, Meyer C and Kurz H 1997 Static and high-frequency electric fields in silicon MOS and MS structures probed by optical second-harmonic generation *Phys. Rev.* B **55** 4596–606

[112]  Reif J, Zink J C, Schneider C-M and Kirschner J 1991 Effects of surface magnetism on optical second harmonic generation *Phys. Rev. Lett.* **67** 2878–81

[113]  Spierings G, Koutsos V, Wierenga H A, Prins M W J, Abraham D and Rasing Th 1993 Optical second harmonic generation study of interface magnetism *Surf. Sci.* **287–8** 747–9

Spierings G, Koutsos V, Wierenga H A, Prins M W J, Abraham D and Rasing Th 1993 Interface magnetism studied by optical second harmonic generation *J. Magn. Magn. Mater.* **121** 109–11

[114]  Vollmer R 1998 Magnetization-induced second harmonic generation from surfaces and ultrathin films *Nonlinear Optics in Metals* ed K H Bennemann (Oxford: Clarendon) pp 42–131

[115]  Fleischmann M, Hendra P J and McQuillan A J 1974 Raman-spectra of pyridine adsorbed at a silver electrode *Chem. Phys. Lett.* **26** 163–6

[116]  Chen C K, de Castro A R B and Shen Y R 1981 Surface enhanced second-harmonic generation *Phys. Rev. Lett.* **46** 145–8

[117]  Wang H, Yan E C Y, Borguet E and Eisenthal K B 1996 Second harmonic generation from the surface of centrosymmetric particles in bulk solution *Chem. Phys. Lett.* **259** 15–20

[118]  Srivastava A and Eisenthal K B 1998 Kinetics of molecular transport across a liposome bilayer *Chem. Phys. Lett.* **292** 345–51

[119]  Yan E C Y, Liu Y and Eisenthal K B 1998 New method for determination of surface potential of microscopic particles by second harmonic generation *J. Phys. Chem.* B **102** 6331–6

[120]  Campagnola P J, Wei M D, Lewis A and Loew L M 1999 High-resolution nonlinear optical imaging of live cells by second harmonic generation *Biophys. J.* **77** 3341–9

[121]  Dadap J I, Shan J, Eisenthal K B and Heinz T F 1999 Second-harmonic Rayleigh scattering from a sphere of centrosymmetric material *Phys. Rev. Lett.* **83** 4045–8

## Further Reading

General texts on nonlinear optics

Boyd R W 1992 Nonlinear Optics (New York: Academic)

Butcher P N and Cotter D 1990 *The Elements of Nonlinear Optics* (Cambridge: Cambridge University Press)

Shen Y R 1984 *The Principles of Nonlinear Optics* (New York: Wiley)

Yariv A 1989 *Quantum Electronics* 3rd edn (New York: Wiley)

General texts on nonlinear optical spectroscopy

Demtröder W 1996 *Laser Spectroscopy: Basic Concepts and Instrumentation* 2nd edn (Berlin: Springer)

Levenson M D and Kano S S 1988 *Introduction to Nonlinear Laser Spectroscopy* (Boston, MA: Academic)

Mukamel S 1995 *Principles of Nonlinear Optical Spectroscopy* (Oxford: Oxford University Press)

General texts on surface nonlinear optics

Brevet P F 1997 *Surface Second Harmonic Generation* (Lausanne: Presses Polytechniques et Universitaires Romandes)

Bennemann K H (ed) 1998 *Nonlinear Optics in Metals* (Oxford: Clarendon)

McGilp J F, Patterson C H and Weaire D L (eds) 1995 *Epioptics: Linear and Nonlinear Optical Spectroscopy of Surfaces and Interfaces* (Berlin: Springer)

# B1.6
# Electron-impact spectroscopy

*John H Moore*

### B1.6.0 Introduction

When one speaks of a 'spectrum', the dispersed array of colours from a luminous body comes to mind; however, in the most general sense, a spectrum is a record of the energy and probability of transitions between states of a substance. In electron spectroscopy the 'spectrum' takes the form of the energy distribution of electrons emanating from a sample. Electron spectroscopies are classified according to the phenomena giving rise to these electrons; historically, each technique has acquired an acronym until today one finds a veritable alphabet soup of electron spectroscopies in the scientific literature. For example, PES refers to photoelectron spectroscopy, a technique in which the detected electrons are emitted after the absorption of a photon induces transitions into the continuum beyond the first ionization potential of the sample. Electron-impact spectroscopies, the subject of this entry, entail the excitation of a transition by an electron impinging upon a sample with the subsequent measurement of the energy of the scattered electron. The spectrum is the scattered-electron intensity as a function of the difference between the incident- and scattered-electron energies—the *energy loss*.

The technologies of the various electron spectroscopies are similar in many ways. The techniques for measuring electron energies and the devices used to detect electrons are the same. All electron spectrometers must be housed in an evacuated container at pressures less than about $10^{-6}$ mbar since electrons cannot be transported through the atmosphere. Stray fields that perturb electron trajectories are a potential problem. To function correctly, an electron spectrometer must be shielded from the earth's magnetic field.

Electron-impact energy-loss spectroscopy (EELS) differs from other electron spectroscopies in that it is possible to observe transitions to states below the first ionization edge; electronic transitions to excited states of the neutral, vibrational and even rotational transitions can be observed. This is a consequence of the detected electrons not originating in the sample. Conversely, there is a problem when electron impact induces an ionizing transition. For each such event there are two outgoing electrons. To precisely account for the energy deposited in the target, the two electrons must be measured in coincidence.

In comparison to optical spectroscopy, electron-impact spectroscopy offers a number of advantages. Some of these are purely technological while others are a result of physical differences in the excitation mechanism. The energy of an electron can be varied simply and smoothly by scanning the voltage applied to, and hence the potential difference between, electrodes in the spectrometer. The same technology is applicable to electrons with energies, and energy losses, in the millielectronvolt (meV) range as in the kiloelectronvolt (keV) range. At least in principle, measurements analogous to IR spectroscopy can be carried out in the same instrument as measurements akin to x-ray spectroscopy. Unlike an optical instrument, the source intensity and the transmission of an electron spectrometer are nearly independent of energy, making the electron instrument more suitable for absolute intensity measurements; however, an electron spectrometer cannot always provide resolution comparable to that of an optical instrument. Studies of rotational and vibrational excitation, particularly in surface adsorbates, are routinely carried out by electron spectroscopy with a resolution of 2 to 5 meV (16 to 40 cm$^{-1}$). A resolution of 5 to 30 meV is typically obtained for electron-impact excitation of valence-electron transitions in atoms and molecules in the gas phase. Analogous studies by vis-UV spectroscopy easily provide 1 cm$^{-1}$ resolution. For inner-shell-electron excitation, electron spectroscopy provides resolution comparable or superior to x-ray spectroscopy with discrete-line-source x-ray tubes. X-ray synchrotron sources now becoming available will provide better resolution and intensity than can be achieved with an electron spectrometer, but it must be borne in mind that an electron spectrometer is a relatively inexpensive, table-top device, whereas a synchrotron is a remote, multimillion-dollar facility. In many applications, electron-scattering experiments are more sensitive than optical experiments. This is in part due to the superior sensitivity of electron detectors. An electron multiplier has essentially unit efficiency (100%); a photomultiplier or photodiode may have an efficiency of only a few per cent. For surface analysis, electron spectroscopies have a special advantage over optical techniques owing to the short range of electrons in solids.

The mechanism by which a transition is induced by electron impact depends on the nature of the coupling between the projectile electron and the target; this in turn is influenced by the velocity and closeness of approach of the projectile to the target. There is a wide range of possibilities. A high-energy projectile electron may pass quickly by, delivering only a photon-like electric-field pulse to the target at the instant of closest approach. Less probable are hard, billiard-ball-like collisions between the projectile and one target electron. At low energies, slower, more intimate collisions are characterized by many-electron interactions. Depending upon the mechanism, the momentum transferred from projectile to target can vary from the minimum necessary to account for the transition energy to many times more. The interaction influences the type of transition that can be induced and the way in which the projectile is scattered. It is even possible for the projectile electron to be exchanged for a target electron, thus allowing for electron-spin-changing transitions. This state of affairs is a contrast to optical excitation where the momentum transfer is a constant and only 'dipole-allowed' transitions occur with significant probability.

## B1.6.1   Technology

### B1.6.1.1   Cross section and signal intensity

The quantities to be measured in electron-impact spectroscopy are the probability of an electron impact's inducing a transition and the corresponding transition energy. The energy for the transition is taken from the kinetic energy of the projectile electron. Unlike the situation in optical spectroscopy, the exciting particle is not annihilated, but is scattered from the target at some angle to its initial direction. The scattering angle is a measure of the momentum transferred to the target, and, as such, is also an important variable.

The probability of a collision between an electron and a target depends upon the *impact parameter*, $b$, which is the perpendicular distance between the line of travel of the electron and the centre of force exerted by the target on the electron. The impact parameter is equivalent to the distance of closest approach if no potential is present between the electron and the target. For a hard-sphere collision between an infinitesimally small projectile and a target of radius $r$, the impact parameter must be less than $r$ for a collision to occur, and, from simple geometry, the scattering angle $\theta = 2 \arccos(b/r)$. The scattering angle is large for small impact parameters, while for 'grazing' collisions, where $b$ approaches $r$, the scattering angle is small. The probability of a collision is proportional to the cross sectional area of the target, $\pi r^2$. Real collisions between an electron and an atom involve at least central-force-field potentials and, frequently, higher-multipole potentials, but the billiard-ball scattering model is so pervasive that collision probabilities are almost always expressed as cross sections, often denoted by the symbol $\sigma$ with units of the order of the cross section of an atom, such as $10^{-16}$ cm$^2$ or square Ångströms (Å$^2$). The atomic unit of cross section is the Bohr radius squared ($r_0^2 = 0.28 \times 10^{-16}$ cm$^2$).

In a very simple form of electron spectroscopy, known as *electron transmission spectroscopy*, the attenuation of an essentially monoenergetic beam of electrons is measured after passage through a sample. If the target is very thin or of such low density that most electrons pass through unscattered, the attenuation is small and the transmitted current, I (in units of electrons per unit time, $s^{-1}$), compared to the incident current, $I_0$ ($s^{-1}$), is given by

$$\frac{I}{I_0} = e^{-\sigma n \ell}$$

where $n$ is the number density of particles in the target (typically in units of $cm^{-3}$), $\ell$ (cm) is the thickness of the target and $\sigma$ is the *total electron scattering cross section*. The cross section depends upon the energy, $E_0$, of the incident electrons: $\sigma = \sigma(E_0)$. The total electron scattering spectrum presented as $I/I_0$ as a function of $E_0$ bears an inverse relation to the cross section, the transmitted current decreasing as the cross section increases.

An electron-energy-loss spectrometer consists of an 'electron gun' that directs a collimated beam of electrons upon a sample, and an 'analyser' that collects electrons scattered in a particular direction (specified by $\theta$ and $\phi$ in spherical coordinates) and transmits to a detector those electrons with energy $E$. The electron-energy-loss spectrum is a plot of the scattered-electron current, $I_s$, arriving at the detector as a function of the energy loss, $(E_0 - E)$. The cross section for the inelastic scattering process giving rise to the observed signal depends upon the scattering angle: it is a *differential cross section*, $d\sigma/d\Omega$, where $d\Omega = \sin\theta \, d\theta \, d\phi$ in spherical coordinates. The magnitude of the cross section depends upon the incident electron energy: there will be a threshold below which the cross section is zero when the incident electron has insufficient energy to excite the transition, and there will be an incident electron energy for which the coupling between projectile and target is greatest and the cross section passes through a maximum. The instrument parameters, as well as the cross section, determine the actual signal level:

$$I_s = I_0 n \ell \left(\frac{d\sigma}{d\Omega}\right) \Delta\Omega$$

where $\Delta\Omega$ is the solid-angle field of view of the scattered-electron analyser.

The foregoing description of an electron energy-loss spectrometer assumes a monoenergetic incident electron beam and the excitation of a transition of negligible energy width. It is often the case that the transition intensity spans a range in energy, and, in addition, the incident beam has some energy spread and the analyser a finite bandpass. One must consider a cross section differential in both angle and energy, $d^2\sigma/d\Omega\,dE$. The signal intensity is then

$$I_s = I_0 n \ell \left(\frac{d^2\sigma}{d\Omega \, dE}\right) \Delta\Omega\Delta E$$

where $\Delta E$ represents a convolution of the energy spread of the source and the passband of the analyser.

### B1.6.1.2  *Electron optics*

The two essential elements of an electron spectrometer are the electrodes that accelerate electrons and focus them into a beam and the dispersive elements that sort electrons according to their energies. These serve the functions of lenses and prisms in an optical spectrometer. The same parameters are used to describe these elements in an electron spectrometer as in an optical spectrometer; the technology is referred to as *electron optics*.

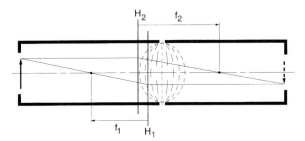

**Figure B1.6.1.** Equipotential surfaces have the shape of lenses in the field between two cylinders biased at different voltages. The focusing properties of the electron optical lens are specified by focal points located at focal lengths $f_1$ and $f_2$, measured relative to the principal planes, $H_1$ and $H_2$. The two principal rays emanating from an object on the left and focused to an image on the right are shown.

*(a) Electron lenses*

The typical electron-optical lens consists of a closely spaced pair of coaxial cylindrical tubes biased at different electrical potentials. The equipotential surfaces in the gap between the tubes assume shapes similar to those of optical lenses as illustrated in figure B1.6.1. An electron passing across these surfaces will be accelerated or decelerated, and its path will be curved to produce focusing. The main difference between the electron lens and an optical lens is that the quantity analogous to the refractive index, namely the electron velocity, varies continuously across an electrostatic lens, whereas a discontinuous change of refractive index occurs at the surface of an optical lens. Electron lenses are 'thick' lenses, meaning that their axial dimensions are comparable to their focal lengths. An important consequence is that the principal planes employed in ray tracing are separated from the midplane of the lens and lie to the low-velocity side of the lens gap, as shown in the figure. The design of these lenses is facilitated by tables of electron lens optical properties [1], and by computer programs that calculate the potential array for an arbitrary arrangement of electrodes, and trace electron trajectories through the resultant field [2]. In addition to cylindrical electrodes, electron lenses are sometimes created by closely spaced planar electrodes with circular apertures or slits. Shaped magnetic fields are also used to focus electrons, especially for electron energies much in excess of 10 keV where electrostatic focusing requires inconveniently high voltages.

*(b) Electron analysers*

An electron 'prism', known as an *analyser* or *monochromator*, is created by the field between the plates of a capacitor. The plates may be planar, simple curved, spherical, or toroidal as shown in figure B1.6.2. The trajectory of an electron entering the gap between the plates is curved as the electron is attracted to the positively biased (inner) plate and repelled by the negatively biased (outer) plate. The curvature of the trajectory is a function of the electron's kinetic energy so that the electrons in a beam projected between the plates are dispersed in energy. These devices are not only dispersive, but focusing; electrons of the same energy originating from a point source are brought to a point on a focal plane at the output side of the analyser. The energy passband, or *resolution*, of an electrostatic analyser is the range of energies, $\Delta E$, of electrons which, entering through a slit, are transmitted through the analyser to an exit slit. This quantity depends upon the width of the slits as well as the physical dimensions of the analyser. Fixing the slit widths and analyser dimension fixes the relative resolution, $\Delta E/E$, where $E$ is the nominal energy of transmitted electrons. The *resolving power* of an analyser is specified as $E/\Delta E$, the inverse of the relative resolution. For each type of analyser there is a simple relation between analyser dimensions and resolution; for example, the analyser

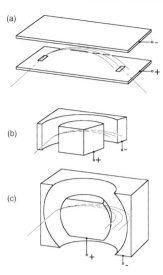

**Figure B1.6.2.** Electron analysers consisting of a pair of capacitor plates of various configurations: (a) the parallel-plate analyser, (b) the 127° cylindrical analyser and (c) the 180° spherical analyser. Trajectories for electrons of different energies are shown.

with hemispherical plates has relative resolution $\Delta E/E = w/2R$, where $w$ is the diameter of the entrance and exit apertures and $R$ is the mean radius of the plates.

For an analyser of fixed dimensions, the absolute resolution, $\Delta E$, can be improved, that is, made smaller, by reducing the pass energy, $E$. This is accomplished by decelerating the electrons to be analysed with a decelerating lens system at the input to the analyser. By this means, an absolute resolution as small as 2 meV has been achieved. For most practical analysers, $\Delta E/E$ is of the order of 0.01, and the pass energy in the highest-resolution spectrometers is of the order of 1 eV. The transmission of electrons of energy less than about 1 eV is generally not practical since unavoidable stray electric and magnetic fields produce unpredictable deflection of electrons of lower energies; even spectrometers of modest resolution require magnetic shielding to reduce the magnetic field of the earth by two to three orders of magnitude.

Magnetic fields are employed in several electron-energy analysers and filters (figure B1.6.3). For very-low-energy electrons (0 to 10 eV), the 'trochoidal analyser' has proven quite useful. This device employs a magnetic field aligned to the direction of the incident electrons and an electric field perpendicular to this direction. The trajectory of an electron injected into this analyser describes a spiral and the guiding centre of the spiral drifts in the remaining perpendicular direction. The drift rate depends upon the electron energy so that a beam of electrons entering the device is dispersed in energy at the exit. The projection of the trajectory on a plane perpendicular to the electric field direction is a troichoid, hence the name troichoidal analyser. The Wien filter is similar in that it uses crossed electric and magnetic fields; however, the fields are perpendicular to one another and both are perpendicular to the injected electron beam direction. The Coulomb force induced by the electric field, $E$, deflects electrons in one direction and the Lorentz force associated with the magnetic field, $B$, tends to deflect electrons in the opposite direction. The forces balance for one velocity, $v = |E|/|B|$, and electrons of this velocity are transmitted straight through the filter to the exit aperture. A magnetic field alone, perpendicular to the direction of an electron beam, will disperse electrons in energy. Sector magnets such as those used in mass spectrometers are used in electron spectrometers for very-high-energy electrons, the advantage over electrostatic deflectors being that large electrical potentials are not required. Another advantage is that deflection is in the direction parallel to the magnet pole face making it possible to view the

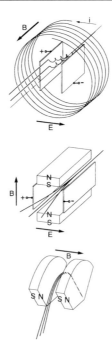

**Figure B1.6.3.** Electron energy analysers that use magnetic fields: (a) the trochoidal analyser employing an electromagnet, (b) the Wien filter and (c) the sector magnet analyser. Trajectories for electrons of different energies are shown.

entire dispersed spectrum at one time. By contrast, energy dispersion of an electron beam in an electrostatic device results in a significant portion of the dispersed electrons striking one or the other electrode.

*(c) Electron guns*

Thermionic emission from an electrically heated filament is the usual source of electrons in an electron spectrometer. Field emission induced by a large electric-field gradient at a sharply pointed electrode may be used when fine spatial resolution is required. For very special applications, laser-induced photoemission has been used to produce nearly monoenergetic electrons and electrons with spin polarization. The filament in a thermionic source is a wire or ribbon of tungsten or some other refractory metal, sometimes coated or impregnated with thoria to reduce the work function. The passage of an electrical current of a few amps heats the filament to 1500 to 2500 °C. As shown in figure B1.6.4 the filament is typically mounted in a diode arrangement, protruding through a cathode and a small distance from an anode with an aperture through which electrons are extracted. The cathode is biased 30 to 50 V negative with respect to the source and the anode 50 to 100 V positive. An approximately Maxwellian energy distribution is produced. Depending on the filament temperature and the potential drop across the hot portion of the filament, this distribution is 0.3 to 0.7 eV wide. For most applications, the electron beam extracted from a thermionic source is passed through a monochromator to select a narrower band of energies from that emanating from the source.

*(d) Electron spectrometers*

A typical electron energy-loss spectrometer is shown in figure B1.6.5. The major components are an electron source, a premonochromator, a target, an analyser and an electron detector. For gaseous samples, the target

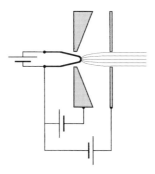

**Figure B1.6.4.** Diode electron source.

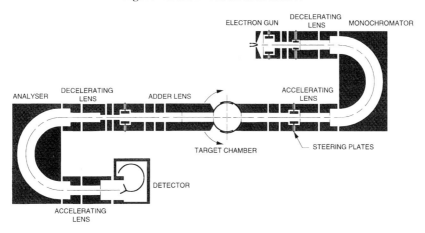

**Figure B1.6.5.** Typical electron energy-loss spectrometer.

may be a gas jet or the target may be a gas confined in a cell with small apertures for the incident beam and for the scattered electrons. The target may be a thin film to be viewed in transmission or a solid surface to be viewed in reflection. The analyser may be rotatable about the scattering centre so the angularly differential scattering cross section can be measured. Most often the detector is an electron multiplier that permits scattered electrons to be counted and facilitates digital processing of the scattered-electron spectrum. In low-resolution instruments, the scattered-electron intensity may be sufficient to be measured with a sensitive electrometer as an electron current captured in a 'Faraday cup'.

Electron lens systems between each component serve a number of functions. A lens following the source focuses electrons on the entrance aperture of the premonochromator and decelerates these electrons to the pass energy required to obtain the desired resolution. The lens following focuses electrons on the target and provides a variable amount of acceleration to permit experiments with different incident energies. The lens system at the input to the analyser again decelerates electrons and focuses on the entrance aperture of the analyser; however, this lens system has an additional function. The scattered-electron energy spectrum must be scanned. This is accomplished with an 'adder lens' that progressively adds back energy to the scattered electrons. The analyser is set to transmit electrons which have lost no energy; the energy-loss spectrum is then a plot of the detector signal as a function of the energy added by the adder lens. The alternative method of scanning is to vary the pass energy of the analyser; this has the disadvantage of changing the analyser resolution and transmission as the spectrum is scanned. The lens following the analyser accelerates transmitted electrons to an energy at which the detector is most sensitive, typically several hundred eV.

## B1.6.2  Theory

Inelastic electron collisions can be roughly divided into two regimes: those in which the kinetic energy of the projectile electron greatly exceeds the energy of the target atom or molecule's electrons excited by the collision, and those in which the projectile and target electron energies are comparable. In the higher-energy region the target electrons are little disturbed by the approach and departure of the projectile; the excitation occurs suddenly when the projectile is very close to the target. In the lower-energy region, the interaction proceeds on a time scale comparable to the orbital period of the target electrons; both projectile and target electrons make significant adjustments to one another's presence. In some such cases, it may even make sense to consider the electron–target complex as a transient negative ion.

### B1.6.2.1  Bethe–Born theory for high-energy electron scattering

Bethe provided the theoretical basis for understanding the scattering of fast electrons by atoms and molecules [3, 4]. We give below an outline of the quantum-mechanical approach to calculating the scattering cross section.

The Schrödinger equation for the projectile–target system is

$$\left[ -\frac{\hbar^2}{2m} \nabla_r^2 - \frac{\hbar^2}{2m} \sum_j \nabla_{r_j}^2 + V(r, r_j, r_N) \right] \phi(r, r_j) = \left[ \varepsilon_0 + \frac{\hbar^2 k_0^2}{2m} \right] \phi(r, r_j) \qquad (B1.6.1)$$

where $r$ gives the position of the projectile electron and the $r_j$ are the coordinates of the electrons in the target and the $r_N$ are the coordinates of the nuclei; $V(r, r_j, r_N)$ is the potential energy of interaction between the projectile and the particles (electrons and nuclei) that make up the target, as well as interactions between particles in the target; $\varepsilon_0$ is the energy of the target in its ground state; and $\hbar^2 k_0^2 / 2m$ is the initial kinetic energy of the projectile with wavevector $k_0$ and momentum $\hbar k_0$. The wave equation (B1.6.1) is inseparable because of terms in the potential-energy operator that go as $|r - r_j|$. It is thus impossible to obtain an analytic solution for the wave function of the scattered electron. An approximate solution can be obtained by expanding the wave function, $\phi(r, r_j)$, in the complete set of eigenfunctions of the target, $\chi_m(r_j)$, and of the projectile, $\Psi(r)$. This separates the wave equation into a set of coupled differential equations each of which manifests a discrete interaction coupling two states ($n$ and $m$) of the target. The interaction is described by a matrix element:

$$V_{mn} = \int \chi_n^*(r_j) V \chi_m(r_j) \, dr_j.$$

Approximate methods may be employed in solving this set of equations for the $\Psi(r)$; however, the asymptotic form of the solutions are obvious. For the case of elastic scattering

$$\Psi_0(r) \to e^{ik_0 z} + \frac{e^{ik_0 \cdot r}}{r} f_0(\theta, \phi)$$

the first term representing an incident plane wave moving in the $z$-direction in a spherical coordinate system and the second term an outgoing spherical wave modulated by a *scattering amplitude*, $f(\theta, \phi)$. For inelastic scattering, the solutions describe an outgoing wave with momentum $\hbar k$,

$$\Psi(r) \to \frac{e^{ik \cdot r}}{r} f(\theta, \phi).$$

In this case the projectile has imparted energy $\hbar^2 (k_0^2 - k^2)/2m$ to the target. Assuming the target is initially in its ground state ($m = 0$), the collision has excited the target to a state of energy $E_n = \hbar^2 (k_0^2 - k^2)/2m$.

The cross section for scattering into the differential solid angle $d\Omega$ centred in the direction $(\theta, \phi)$, is proportional to the square of the scattering amplitude:

$$\frac{d\sigma}{d\Omega} = \frac{k}{k_0}|f(\theta, \phi)|^2$$

where the ratio $k/k_0 = v/v_0$ accounts for the fact that, all other things being equal, the incident and scattered flux differ owing to the difference in velocity, $v_0$, of the incident electron compared to the velocity, $v$, of the scattered electron. As a consequence of the expansion of the total wave function, the scattering amplitude can also be decomposed into terms each of which refers to an interaction coupling specific states of the target:

$$f_{mn}(\theta, \phi) = \frac{m}{2\pi\hbar^2}\int e^{-i k \cdot r} V_{mn}\, e^{i k_0 \cdot r}\, dr = \frac{m}{2\pi\hbar^2}\int V_{mn}\, e^{i K \cdot r}\, dr \qquad (B1.6.2)$$

where $\hbar K = \hbar(k_0 - k)$ is the momentum transfer in the collision.

### B1.6.2.2  The Born approximation

In the high-energy regime it is appropriate to employ the Born approximation. There are three assumptions: (i) The incident wave is undistorted by the target. For a target in its ground state (specified by m = 0) this is equivalent to setting $V_{00} = 0$. (ii) There is no interaction between the outgoing electron and the excited target. For inelastic scattering with excitation of the final state n, this is equivalent to setting $V_{nn} = 0$. (iii) The excitation is a direct process with no involvement of intermediate states, thus $V_{mn} = 0$ unless $m = 0$. The scattering amplitude (equation (B1.6.2)) thus contains but one term:

$$f_{0n}(\theta, \phi) = \frac{m}{2\pi\hbar^2}\int e^{i K \cdot r}\chi_n^*(r_j)V\chi_0(r_j)\, dr_j\, dr. \qquad (B1.6.3)$$

When the potential consists of electron–electron and electron–nucleus Coulombic interactions,

$$V = \sum_j \frac{e^2}{|r - r_j|} - \sum_N \frac{Z_N e^2}{|r - r_N|}$$

substitution in (B1.6.3) yields

$$f_{0n}(\theta, \phi) = \frac{me^2}{2\pi\hbar^2}\int \chi_n^*(r_j)\sum_j \frac{e^{i K \cdot r}}{|r - r_j|}\chi_0(r_j)\, dr_j\, dr$$

the electron–nucleus terms having been lost owing to the orthonormality of the target wave functions. The important physical implication of the Born approximation becomes clear if one first performs the integration with respect to $r$, taking advantage of the transformation

$$\int \frac{e^{i K \cdot r}}{|r - r_j|}\, dr = \frac{4\pi}{K^2}e^{i K \cdot r_j}.$$

The differential cross section for inelastic collisions exciting the $n$th state of the target then takes the form

$$\left(\frac{d\sigma}{d\Omega}\right)_{0n} = \frac{4e^2 m^2}{\hbar^4 K^4}\frac{k}{k_0}\left|\int \chi_n^*(r_j)\sum_j e^{i K \cdot r_j}\chi_0(r_j)\, dr_j\right|^2 = \frac{4e^4 m^2}{\hbar^4 K^4}\frac{k}{k_0}|\varepsilon_n(K)|^2. \qquad (B1.6.4)$$

In this expression, factors that describe the incident and scattered projectile are separated from the square modulus of an integral that describes the role of the target in determining the differential cross section. The term preceding the integral, $4e^4m^2/\hbar^4K^4$, with units of area, is the Rutherford cross section for electron–electron scattering. The integral, represented by the quantity $\varepsilon_n(K)$, is known as the *inelastic scattering form factor*.

In the discussion above, scattering from molecules is treated as a superposition of noninteracting electron waves scattered from each atomic centre. In fact, there is a weak but observable interference between these waves giving rise to phase shifts associated with the different positions of the atoms in a molecule. This diffraction phenomenon produces oscillations in the differential cross section from which molecular structure information can be derived.

The interaction of the target with the incoming and outgoing electron wave must be considered at lower impact energies. This is achieved in the *distorted-wave approximation* by including $V_{00}$ and $V_{nn}$ in the calculation of the scattering amplitude. Higher-level calculations must also account for electron spin since spin exchange becomes important as the collision energy decreases.

### B1.6.2.3   *The generalized oscillator strength*

The Born approximation for the differential cross section provides the basis for the interpretation of many experimental observations. The discussion is often couched in terms of the *generalized oscillator strength*,

$$f_n(K) = \frac{2m}{\hbar^2}\frac{E_n}{K^2}\left|\int \chi_n^*(r_j)\sum_j e^{iK\cdot r_j}\chi_0(r_j)\,dr_j\right|^2 = \frac{2m}{\hbar^2}\frac{E_n}{K^2}|\varepsilon_n(K)|^2. \qquad (B1.6.5)$$

Assuming the validity of the Born approximation, an 'effective' generalized oscillator strength can be derived in terms of experimentally accessible quantities:

$$f_n(K) = \frac{\hbar^2}{2e^4m}\frac{k_0}{k}E_nK^2\left(\frac{d\sigma}{d\Omega}\right)_{0n}. \qquad (B1.6.6)$$

All the quantities on the right can be measured ($k_0$, $k$ and $K$ calculated from measurements of the incident energy, the energy-loss, and the scattering angle). For inelastic collisions resulting in transitions into the continuum beyond the first ionization potential, the cross section is measured per unit energy loss and the generalized oscillator strength *density* is determined:

$$\frac{df(K)}{dE} = \frac{\hbar^2}{2e^4m}\frac{k_0}{k}E_nK^2\frac{d^2\sigma}{d\Omega\,dE}. \qquad (B1.6.7)$$

The generalized oscillator strength provides the basis of comparison between electron energy-loss spectra and optical spectra; however, there is a problem with the determination of the absolute value of the generalized oscillator strength since measurements of the differential cross section can rarely be made on an absolute basis owing to the difficulty of accurately determining the target dimension and density. The problem is overcome by a normalization of experimentally determined generalized oscillator strengths according to the *Bethe sum rule* which requires that the sum of the $f_n(K)$ (equation B1.6.6) for all discrete transitions plus the integral of $df(K)/dE$ (equation B1.6.7) over the continuum adds up to the number of electrons in the target.

A particularly important property of the generalized oscillator strength is that, for high-energy, small-angle scattering, the generalized oscillator strength is approximately equal to the *optical oscillator strength*, $f_n^{opt}$, for electric-dipole transitions induced by photon absorption. That this is so can be seen from a power-series expansion of the form factor that appears in the expression for the generalized oscillator strength (equation B1.6.5):

$$\varepsilon_n(K) = \sum_{k=1}^{\infty}\int \chi_n^*(r_j)\sum_j\frac{(iK\cdot_j)^k}{k!}\chi_0(r_j)\,dr_j.$$

The operator in the first term goes as $r_j$, and is thus proportional to the optical *dipole transition moment*

$$M_{0n} = \int \chi_n^*(r_j) \sum_j er_j \chi_0(r_j) \, dr_j.$$

The second term is proportional to the optical quadrupole transition moment, and so on. For small values of momentum transfer, only the first term is significant, thus

$$\lim_{k \to 0} f_n(K) = f_n^{opt}.$$

The result is that the small-angle scattering intensity as a function of energy loss (the energy-loss spectrum) looks like the optical absorption spectrum. In fact, oscillator strengths are frequently more accurately and conveniently measured from electron-impact energy-loss spectra than from optical spectra, especially for higher-energy transitions, since the source intensity, transmission and detector sensitivity for an electron scattering spectrometer are much more nearly constant than in an optical spectrometer.

The proportionality of the generalized oscillator strength and the optical dipole oscillator strength appears to be valid even for incident-electron energies as low as perhaps 200 eV, but it is strictly limited to small-angle, forward scattering that minimizes momentum transfer in a collision [5]. Of course $K = 0$ is inaccessible in an inelastic collision; there must be at least sufficient momentum transferred to account for the kinetic energy lost by the projectile in exciting the target. For very-high-energy, small-angle scattering, the minimum momentum transfer is relatively small and can be ignored. At lower energies, an extrapolation technique has been employed in very accurate work.

On the other hand, there are unique advantages to electron-scattering measurements in the lower-energy, larger-scattering-angle regime in which the momentum transfer is larger than the minimum required to induce a transition. In this case, higher-order multipoles in the transition moment become significant, with the result that the cross section for the excitation of optically forbidden transitions increases relative to that for dipole-allowed transitions. The ability to vary the momentum transfer in electron-energy-loss spectroscopy yields a spectrum much richer than the optical spectrum.

### B1.6.2.4    *The Bethe surface: binary versus dipole collisions*

An important feature of electron-impact spectroscopy in comparison to photoabsorption is that the momentum transfer can be varied. As the scattering angle increases and the incident energy decreases, higher-order terms in the expansion of $\varepsilon_n(K)$ become relatively more important. Large-angle, high-momentum-transfer scattering results from small impact parameters. In this case the target experiences a nonuniform electric field as the electron passes by. Significant amplitude of higher-order multipoles in a nonuniform field permits the field to couple with higher-order multipoles of the target. Thus, for example, optically forbidden electric-quadrupole transitions are a significant feature of the low-energy, large-scattering-angle electron energy-loss spectrum. Selection rules for electronic excitation by electron impact have been treated in detail in the review by Hall and Read [6].

From the discussion above it appears that small-angle scattering events might better not be thought of as collisions at all. The excitation, which is photon-like, appears to be a consequence of the high-frequency electric pulse delivered to the target as the projectile electron passes rapidly by. These energetic, but glancing collisions are referred to as *dipole collisions*, in contradistinction to the larger-angle scattering regime of *binary collisions* where the projectile electron appears to undergo a hard collision with one of the target electrons.

A succinct picture of the nature of high-energy electron scattering is provided by the *Bethe surface* [4], a three-dimensional plot of the generalized oscillator strength as a function of the logarithm of the square of the momentum transfer, $\ln(K^2)$ and the energy-loss, $E_n$. To see how this works, consider the form of the

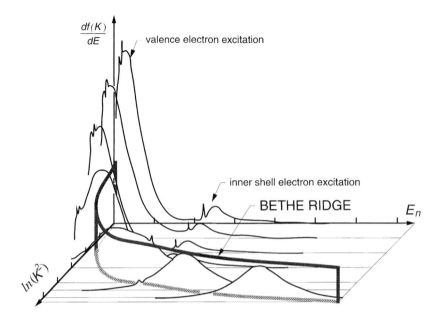

**Figure B1.6.6.** The Bethe surface. The sharp ridge corresponds to scattering from a single stationary target electron; the broadened ridge to scattering from the electrons in an atom or molecule.

Bethe surface for a 3D billiards game with a 'projectile' cue ball incident on a stationary billiard ball. This is a two-body problem so the energy-loss is uniquely determined by the momentum transfer: $E_n |\hbar K|^2/2m$. For each value of $K$, all the oscillator strength appears at a single value of $E_n$. The Bethe surface displays a sharp ridge extending from low values of $K^2$ and $E_n$ (corresponding to large-impact-parameter, glancing collisions) to high values (corresponding to near 'head on' collisions).

A schematic of the Bethe surface for electron scattering from an atom or molecule is shown in figure B1.6.6 superimposed on the surface for the two-body problem with a stationary target. The sharp ridge is broadened in the region of large momentum transfer and energy loss. These are hard collisions in which the projectile ejects an electron from the target. This is at least a three-body problem: because the recoil momentum of the ionic core is not accounted for, or, alternatively, since the target electron is not stationary, $K$ and $E_n$ are not uniquely related. The breadth and the shape of the ridge in the Bethe surface in the high-momentum region are a reflection of the momentum distribution of the target electron. Electron Compton scattering experiments [7, 8] and (e, 2e) experiments [9] are carried out in the high-momentum-transfer, large-energy-loss, region for the purpose of investigating the electron momentum distribution in atoms and molecules.

The dipole region is approached as one proceeds to the low-momentum-transfer portion of the Bethe surface. For sufficiently small momentum transfer, where $2m|\hbar K|^2$ approaches the binding energy of valence electrons in the target, a section through the Bethe surface has the appearance of the optical absorption spectrum. For values of the energy loss less than the first ionization energy of the target, sharp structure appears corresponding to the excitation of discrete, dipole-allowed transitions in the target. The ionization continuum extends from the first ionization potential, but, as in the photoabsorption spectrum, one typically sees sharp 'edges' as successive ionization channels become energetically possible. Also, as in the photoabsorption spectrum, resonance structures may appear corresponding to the excitation of metastable states imbedded in the continuum.

### B1.6.2.5   Low-energy electron scattering

Theoretically, the asymptotic form of the solution for the electron wave function is the same for low-energy projectiles as it is at high energy; however, one must account for the protracted period of interaction between projectile and target at the intermediate stages of the process. The usual procedure is to separate the incident-electron wave function into *partial waves*

$$e^{i k_0 \cdot r} = \sum_{l=0}^{\infty} (2l + 1) i^l P_l(\cos\theta) j_l(k_0 r)$$

($j_l$ a spherical Bessel function; $P_l$ a Legendre polynomial) and expand the wave function of the target in some complete basis set, typically the complete set of eigenfunctions for the unperturbed target. This approach allows for any distortion of the incoming and scattered electron waves, as well as any perturbation of the target caused by the approaching charge of the projectile. Each term in the expansion for the incident wave represents an angular momentum component: $l = 0$, the s-wave component; $l = 1$, the p-wave component; and so on. The scattering of the incident wave is treated component by component, the interaction with each giving rise to a phase shift in that component of the outgoing wave.

To see how this works, consider elastic scattering in a situation where the electron–target interaction can be described by a simple central-force-field potential, $V(r)$, that does not fall off faster than $r^{-1}$ at large $r$. In this case the wave equation for the projectile electron can be separated from the Schrödinger equation for the total electron–target system given above (equation (B1.6.1)). The wave equation for the projectile is

$$\left[ -\frac{\hbar^2}{2m} \nabla_r^2 + V(r) \right] \Psi(r) = \frac{\hbar^2 k_0^2}{2m} \Psi(r).$$

Since the potential depends only upon the scalar $r$, this equation, in spherical coordinates, can be separated into two equations, one depending only on $r$ and one depending on $\theta$ and $\phi$. The wave equation for the $r$-dependent part of the solution, $R(r)$, is

$$\left[ -\frac{\hbar^2}{2m} \frac{1}{r} \frac{\partial^2}{\partial r^2} r + \frac{\hbar^2}{2m} \frac{l(l+1)}{r^2} + V(r) \right] R(r) = \frac{\hbar^2 k_0^2}{2m} R(r) \tag{B1.6.8}$$

where $\hbar \sqrt{l(l+1)}/r$ is the orbital angular momentum associated with the $l$th partial wave. The solutions have the asymptotic form

$$R(r) \to \frac{1}{k_0 r} \sin\left( k_0 r - \frac{l\pi}{2} + \eta_l \right) \tag{B1.6.9}$$

and the calculation of the cross section is reduced to calculating the phase shift, $\eta_l$, for each partial wave. The phase shift is a measure of the strength of the interaction of a partial wave in the field of the target, as well as a measure of the time period of the interaction. High-angular-momentum components correspond to large impact parameters for which the interaction can generally be expected to be relatively weak. The exceptions are for the cases of long-range potentials, as when treating scattering from highly polarizable targets or from molecules with large dipole moments. In any event, only a limited number of partial waves need be considered in calculating the cross section—sometimes only one or two.

### B1.6.2.6   Resonances

The partial wave decomposition of the incident-electron wave provides the basis of an especially appealing picture of strong, low-energy resonant scattering wherein the projectile electron spends a sufficient period of time in the vicinity of the target that the electron–target complex is describable as a temporary negative ion.

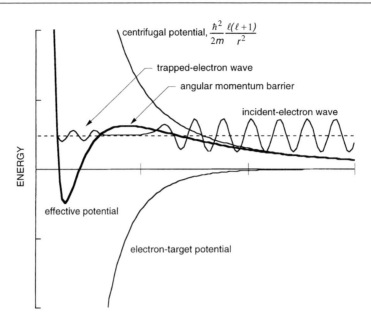

**Figure B1.6.7.** An angular momentum barrier created by the addition of the centrifugal potential to the electron–atom potential.

With the radial wave equation (B1.6.8) for the projectile in a central-force field as a starting point, define a fictitious potential

$$V'(r) = \frac{\hbar^2}{2m} \frac{l(l+1)}{r^2} + V(r).$$

This is a fictitious potential because it includes not only the true potential, $V(r)$, that contains the screened Coulomb potential and the polarization potential of the target, but also the term $(\hbar^2/2m)l(l+1)/r^2$ that arises from the centrifugal force acting on the projectile, a fictitious force associated with curvilinear motion. As shown in figure B1.6.7, this repulsive term may give rise to an *angular momentum barrier*. Some part of the incident-electron-wave amplitude may tunnel through the barrier to impinge upon the repulsive part of the true potential from which it is reflected to tunnel back out to join the incident wave. The superposition of these two waves produces the phase shift in the scattered wave (see equation (B1.6.9)). More interestingly, as shown in the figure, there are special incident electron energies for which the width of the well behind the barrier is equal to some integral multiple of the electron wavelength. A standing wave then persists, corresponding to an electron being temporarily trapped in the field of the target. This model describes the resonant formation of a metastable negative ion, or, more simply, a 'resonance'.

Finally, it must be recognized that all the above discussion assumes an isolated atomic or molecular target. To describe electron scattering in a complex target such as a solid, one must consider the extended nature of the valence-electron density that constitutes a kind of electron gas enveloping the array of positively charged atomic cores. Most calculations employ a so-called 'jellium' model in which the mobile electrons move in the field of a positive charge smeared out into a homogeneous neutralizing density.

### B1.6.3  Applications

Electron energy-loss spectroscopy is used for obtaining spectroscopic data as a convenient substitute for optical spectroscopy, and, taking advantage of differences in selection rules, as an adjunct to optical spectroscopy.

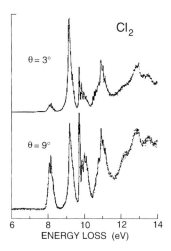

**Figure B1.6.8.** Energy-loss spectra of 200 eV electrons scattered from chlorine at scattering angles of 3° and 9° [10]. Optically forbidden transitions are responsible for the intensity in the 9° spectrum that does not appear in the 3° spectrum.

In addition, electron spectroscopy has many applications to chemical and structural analysis of samples in the gas phase, in the solid phase, and at the solid–gas interface.

### B1.6.3.1    *Valence-shell-electron spectroscopy*

Electronic transitions within the valence shell of atoms and molecules appear in the energy-loss spectrum from a few electron volts up to, and somewhat beyond, the first ionization energy. Valence-shell electron spectroscopy employs incident electron energies from the threshold required for excitation up to many kiloelectron volts. The energy resolution is usually sufficient to observe vibrational structure within the Franck–Condon envelope of an electronic transition. The sample in valence-shell electron energy-loss spectroscopy is most often in the gas phase at a sufficiently low pressure to avoid multiple scattering of the projectile electrons, typically about $10^{-3}$ mbar. Recently, electronic excitation in surface adsorbates has been observed in the energy-loss spectrum of electrons reflected from metallic substrates. When the measurements are carried out with relatively high incident-electron energies (many times the excitation and ionization energies of the target electrons) and with the scattered electrons detected in the forward direction (0° scattering angle), the energy-loss spectrum is essentially identical to the optical spectrum. As described above, this is the arrangement employed to determine oscillator strengths since forward scattering corresponds to collisions with the lowest momentum transfer. If the incident energy is reduced to about 100 eV (just a few times the target electron energies), symmetry-forbidden transitions can be uncovered and distinguished from optically allowed transitions by measuring the energy-loss spectrum at different scattering angles. An example is shown in figure B1.6.8. As the incident energy approaches threshold, it becomes possible to detect electron-spin-changing transitions.

### B1.6.3.2    *Inner-shell-electron energy-loss spectroscopy*

Inner-shell-electron energy-loss spectroscopy (ISEELS) refers to measurements of the energy lost by projectile electrons that have promoted inner-shell electrons into unfilled valence orbitals or into the ionization continuum beyond the valence shell. Inner-shell excitation and ionization energies fall in the region between 100 eV and several kiloelectron volts. The corresponding features in the energy-loss spectrum tend to be broad and diffuse. Inner-shell-hole states of neutral atoms and molecules are very short lived; the transition energies

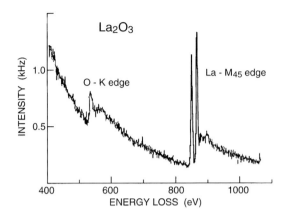

**Figure B1.6.9.** Energy-loss spectrum of $La_2O_3$ showing O K and La $M_{45}$ ionization edges with prominent 'white line' resonances at the La edge [11].

to these states are correspondingly uncertain. The energy of a transition into the continuum is completely variable. Inner-shell-electron ionization energies cannot be uniquely determined from a measurement of the energy lost by the scattered electron since an unknown amount of energy is carried away by the undetected ejected electron. To be more precise, the electron-impact ionization cross section depends upon the energy $(E_s)$ and angle $(\Omega_s)$ of the scattered electron as well as the energy $(E_e)$ and angle $(\Omega_e)$ of the ejected electron; it is a fourfold differential cross section: $\mathrm{d}^4\sigma/\mathrm{d}\Omega_s \,\mathrm{d}E_s \,\mathrm{d}\Omega_e \,\mathrm{d}E_e$. The intensity in the inner-shell energy-loss spectrum is proportional to this cross section integrated over $E_e$ and $\Omega_e$. For collisions in which the scattered electron is detected in the forward direction, the momentum transfer to the target is small and it is highly probable that the ejected-electron energy is very small. As a consequence, the ionization cross section for forward scattering is largest at the energy-loss threshold for ionization of each inner-shell electron and decreases monotonically, approximately as the inverse square of the energy loss. The basic appearance of each feature in the inner-shell electron energy-loss spectrum is that of a sawtooth that rises sharply at the low-energy 'edge' and falls slowly over many tens of electron volts. In addition, sharp structures may appear near the edge (figure B1.6.9).

In contrast to the broadly distributed valence-shell electron density in molecules and solids, inner-shell electron density is localized on a single atom. Molecular configuration and solid-state crystal structure have little effect upon inner-shell ionization energies. Each ionization edge in the energy-loss spectrum is characteristic of a particular type of atom; consequently, ISEELS has become an important technique for qualitative and quantitative elemental analysis. An especially useful elaboration of this technique is carried out in the transmission electron microscope where energy-loss analysis of electrons passing through a thin sample makes elemental analysis possible with the spatial resolution of the microscope [12]. A precision of the order of 100 ppm with nanometre spatial resolution has been achieved; this is close to single-atom sensitivity. A difficulty with ISEELS for analytical purposes is that each ionization edge is superimposed on the continuum associated with lower-energy ionizations of all the atoms in a sample. This continuum comprises an intense background signal which must be subtracted if the intensity at a characteristic edge is to be used as a quantitative measure of the concentration of a particular element in a sample.

Electrons arising from near-threshold inner-shell ionization have very little kinetic energy as they pass through the valence shell and may become trapped behind an angular momentum barrier in the exterior atomic potential, much as do low-energy incident electrons. This phenomenon, a wave-mechanical resonance as described above (section B1.6.2.6), gives rise to structure in the vicinity of an ionization edge in the energy-loss spectrum. For isolated molecular targets in the gas phase, the energies of near-edge resonances (relative to

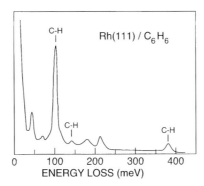

**Figure B1.6.10.** Energy-loss spectrum of 3.5 eV electrons specularly reflected from benzene absorbed on the rhenium(111) surface [15]. Excitation of C–H vibrational modes appears at 100, 140 and 372 meV. Only modes with a changing electric dipole perpendicular to the surface are allowed for excitation in specular reflection. The great intensity of the out-of-plane C–H bending mode at 100 meV confirms that the plane of the molecule is parallel to the metal surface. Transitions at 43, 68 and 176 meV are associated with Rh–C and C–C vibrations.

threshold) can be correlated with the eigenenergies of low-lying, unoccupied molecular orbitals (relative to the first ionization energy). Resonances in solid-state targets are especially prominent for fourth- and fifth-period elements where sharp threshold peaks known as 'white lines' are associated with electrons being trapped in vacant d and f bands (see figure B1.6.9). Resonance peaks have the effect of concentrating transition intensity into a narrow band of energies, thereby increasing the analytical sensitivity for these elements. Near-edge structure, being essentially a valence-shell or valence-band phenomenon, can provide important spectroscopic information about the chemical environment of the atoms in a sample.

### B1.6.3.3   *Reflected-electron energy-loss spectroscopy*

Vibrational spectroscopy of atoms and molecules near or on the surface of a solid has become an essential tool for the microscopic description of surface processes such as catalysis and corrosion. The effect of a surface on bonding is sensitively reflected by the frequencies of vibrational motions. Furthermore, since vibrational selection rules are determined by molecular symmetry that in turn is profoundly modified by the presence of a surface, it is frequently possible to describe with great accuracy the orientation of molecular adsorbates and the symmetry of absorption sites from a comparison of spectral intensities for surface-bound molecules to those for free molecules [13]. Electron spectroscopy has an advantage over optical methods for studying surfaces since electrons with energies up to several hundred electron volts penetrate only one or two atomic layers in a solid before being reflected, while the dimension probed by photons is of the order of the wavelength. Reflected-electron energy-loss spectroscopy (REELS) applied to the study of vibrational motion on surfaces represents the most highly developed technology of electron spectroscopy [14]. Incident electron energies are typically between 1 and 10 eV and sensitivity as low as a few per cent of a monolayer is routinely achieved (figure B1.6.10).

### B1.6.3.4   *Electron transmission spectroscopy*

An important feature of low-energy electron scattering is the formation of temporary negative ions by the resonant capture of incident electrons (see B1.6.2.6, above). These processes lead to sharp enhancements of the elastic-scattering cross section and often dominate the behaviour of the cross section for inelastic processes with thresholds lying close to the energy of a resonance [16]. Elastic-electron-scattering resonances are

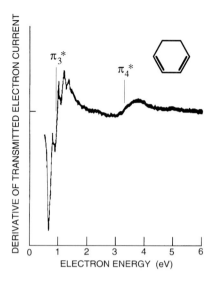

**Figure B1.6.11.** Electron transmission spectrum of 1, 3-cyclohexadiene presented as the derivative of transmitted electron current as a function of the incident electron energy [17]. The prominent resonances correspond to electron capture into the two unoccupied, antibonding $\pi^*$-orbitals. The $\pi_3^*$ negative ion state is sufficiently long lived that discrete vibronic components can be resolved.

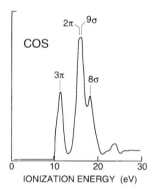

**Figure B1.6.12.** Ionization-energy spectrum of carbonyl sulphide obtained by dipole (e, 2e) spectroscopy [18]. The incident-electron energy was 3.5 keV, the scattered incident electron was detected in the forward direction and the ejected (ionized) electron detected in coincidence at 54.7° (angular anisotropies cancel at this 'magic angle'). The energy of the two outgoing electrons was scanned keeping the net energy loss fixed at 40 eV so that the spectrum is essentially identical to the 40 eV photoabsorption spectrum. Peaks are identified with ionization of valence electrons from the indicated molecular orbitals.

observed by electron transmission spectroscopy. Two types of resonance are distinguished: shape resonances and Feshbach resonances. *Shape resonances* arise when an electron is temporarily trapped in a well created in the 'shape' of the electron–target potential by a centrifugal barrier (figure B1.6.7). *Feshbach resonances* involve the simultaneous trapping of the projectile electron and excitation of a target electron. The resulting states owe their metastability to the absence of two-electron relaxation mechanisms. Shape resonances in the elastic-scattering cross section result in distinct peaks in the 0 to 10 eV electron transmission spectra of many molecules. The energy width of these features, in the range of 0.05 to 1 eV, is determined by the lifetime of the temporary negative ion. Negative ion resonance states in molecules are interpreted as the result of incident electrons being captured into low-lying unfilled molecular orbitals. The energy of resonances in electron transmission spectra can be correlated with the eigenenergies of the corresponding orbitals, much as peaks in the photoemission spectra of molecules are correlated with the eigenenergies of occupied orbitals. In this sense the two spectroscopies are complementary and yield a rather complete picture of the 'frontier' molecular orbitals. Electron transmission spectroscopy has focused largely on molecules with $\pi$-electron systems. In order to emphasize the abrupt change in cross section characteristic of a resonance, the spectra are presented as the first derivative of the transmitted current as a function of incident-electron energy. This is accomplished by modulating the incident-electron energy and detecting the modulated component of the transmitted current. An example is shown in figure B1.6.11. Feshbach resonances fall in the 0 to about 30 eV range. They are found in electron scattering from atoms, but rarely for molecules. The study of these resonances has contributed to the understanding of optically inaccessible excited states of atoms and ions.

### B1.6.3.5   Dipole (e, 2e) spectroscopy

The information from energy-loss measurements of transitions into the continuum, that is, ionizing excitations, is significantly diminished because the energy of the ionized electron is not known. The problem can be overcome by measuring simultaneously the energies of the scattered and ejected electrons. This is known as the (e, 2e) technique—the nomenclature is borrowed from nuclear physics to refer to a reaction with one free electron in the initial state and two in the final state. For spectroscopic purposes the experiment is carried out in the dipole scattering regime (see section B1.6.2.4). Two analyser/detector systems are used: one in the forward direction detects fast scattered electrons and the second detects slow electrons ejected at a large angle to the incident-electron direction (typically at the 'magic angle' of $54.7°$). In order to ensure that pairs of electrons originate from the same ionizing collision, the electronics are arranged to record only those events in which a scattered and ejected electron are detected in coincidence (see B.1.11, 'coincidence techniques'). The ionization energy, or binding energy, is unambiguously given by the difference between the incident electron energy and the sum of the energies of the scattered and ejected electrons detected in coincidence. Dipole (e, 2e) spectra (figure B1.6.12) are analogous to photoabsorption or photoelectron spectra obtained with tunable UV or x-ray sources.

## References

[1]  Harting E and Read F H 1976 *Electrostatic Lenses* (Amsterdam: Elsevier)
[2]  MacSimion C, McGilvery D C and Morrison R J S Montech Pty. Ltd., Monash University, Clayton, Victoria 3168, Australia
      Simion 3D version 6, Dahl D A, ms 2208, Idaho National Engineering Laboratory, PO Box 1625, Idaho Falls, ID 83415, USA
      Simion 3D version 6.0 for Windows, Princeton Electronic Systems, Inc, PO Box 8627, Princeton, NJ 08543, USA
      CPO-3D, RB Consultants Ltd, c/o Integrated Sensors Ltd, PO Box 88, Sackville Street, Manchester M60 1QD, UK. Fax: (UK)-61-200-4781.
[3]  Bethe H 1930 *Ann. Phys., Lpz.* **5** 325
[4]  Inokuti M 1971 *Rev. Mod. Phys.* **43** 297
[5]  Lassettre E N and Skerbele A and Dillon M A 1969 *J. Chem. Phys.* **50** 1829 and references therein to other work of Lassettre
[6]  Hall R I and Read F H 1984 *Electron–Molecule Collisions* ed I Shimamura and K Takayanagi (New York: Plenum)
[7]  Williams B (ed) 1977 *Compton Scattering* (New York: McGraw-Hill)
[8]  Bonham R A and Fink M 1974 *High Energy Electron Scattering (ACS Monograph 169)* (New York: Van Nostrand Reinhold) ch 5
[9]  Coplan M A, Moore J H and Doering J P 1994 *Rev. Mod. Phys.* **66** 985
[10] Spence D, Huebner R H, Tanaka H, Dillon M A and Wang R-G 1984 *J. Chem. Phys.* **80** 2989
[11] Manoubi T, Colliex C and Rez P 1990 *J. Electron. Spectros. Relat. Phenom.* **50** 1
[12] Leapman R D and Newbury D E 1993 *Anal. Chem.* **65** 2409
[13] Richardson N V and Bradshaw A M 1981 *Electron Spectroscopy: Theory, Techniques and Applications* vol 4, ed C R Brundle and A D Baker (London: Academic)
[14] Ibach H and Mills D L 1982 *Electron Energy Loss Spectroscopy and Surface Vibrations* (New York: Academic)
      Ibach H 1991 *Electron Energy Loss Spectrometers: the Technology of High Performance* (Berlin: Springer)

[15]  Koel B E and Somorjai G A 1983 *J. Electron. Spectrosc. Relat. Phenom.* **29** 287
[16]  Schulz G J 1973 *Rev. Mod. Phys.* **45** 378
      Schulz G J 1973 *Rev. Mod. Phys.* **45** 423
[17]  Giordan J C, McMillan M R, Moore J H and Staley S W 1980 *J. Am. Chem. Soc.* **102** 4870
[18]  Cook J P D, White M G, Brion C E, Schirmer J, Cederbaum L S and Von Niessen W 1981 *J. Electron Spectrosc. Relat. Phenom.* **22** 261

## Further Reading

Egerton R F 1986 *Electron Energy-Loss Spectroscopy in the Electron Microscope* (New York: Plenum)

This text covers quantitative analysis by electron energy-loss spectroscopy in the electron microscope along with instrumentation and applicable electron-scattering theory.

Joy D C 1986 The basic principles of EELS *Principles of Analytical Electron Microscopy* ed D C Joy, A D Romig Jr and J I Goldstein (New York: Plenum)

Good 20-page synopsis.

Moore J H, Davis C C and Coplan M A 1989 *Building Scientific Apparatus* 2nd edn (Redwood City, CA: Addison-Wesley) ch 5

The fundamentals of electron optical design.

McDaniel E W 1989 *Atomic Collisions: Electron and Photon Projectiles* (New York: Wiley)

This book provides one of the best descriptions of the physical basis and theory of electron scattering.

Bonham R A and Fink M 1974 *High Energy Electron Scattering (ACS Monograph 169)* (New York: Van Nostrand Reinhold)

This text deals with the theory and instrumentation for electron scattering at energies above about 10 keV, including an excellent discussion of the determination of atomic electron momentum from electron-scattering measurements.

Celotta R J and Huebner R H 1979 Electron impact spectroscopy: an overview of the low-energy aspects *Electron Spectroscopy: Theory, Techniques and Applications* vol 3, ed C R Brundle and A D Baker (London: Academic) ch 3

Although this review is a bit old it is still one of the best reviews of the theory and results of electron scattering from atoms and small molecules in the gas phase at energies below 1000 eV.

Bonham R A 1979 High energy electron impact spectroscopy *Electron Spectroscopy: Theory, Techniques and Applications* vol 3, ed C R Brundle and A D Baker (London: Academic)

This is a review of much of the material covered in the text by Bonham and Fink.

Kuyatt C E and Simpson J A 1967 Electron monochromator design *Rev. Sci. Instrum.* **38** 103

This is one of the first complete descriptions of the design of the modern electron spectrometer. All the important principles are presented and described.

Trajmar S and Cartwright D C 1984 Excitation of molecules by electron impact *Electron–Molecule Interactions and their Applications* vol 1, ed L G Christophorou (New York: Academic)

# B1.7
# Mass spectrometry

*Paul M Mayer*

## B1.7.1   Introduction

Mass spectrometry is one of the most versatile methods discussed in this encyclopedia. Ask a chemist involved in synthesis about mass spectrometry and they will answer that it is one of their most useful tools for identifying reaction products. An analytical chemist will indicate that mass spectrometry is one of the most sensitive detectors available for quantitative and qualitative analysis and is especially powerful when coupled to a separation technique such as gas chromatography. A physicist may note that high resolution mass spectrometry has been responsible for the accurate determination of the atomic masses listed in the periodic table. Biologists use mass spectrometry to identify high molecular weight proteins and nucleic acids and even for sequencing peptides. Materials scientists use mass spectrometry for characterizing the composition and properties of polymers and metal surfaces.

The mass spectrometer tends to be a passive instrument in these applications, used to record mass spectra. In chemical physics and physical chemistry, however, the mass spectrometer takes on a dynamic function as a tool for the investigation of the physico-chemical properties of atoms, molecules and ions. It is this latter application that is the subject of this chapter, and it is hoped that it will bring the reader to a new understanding of the utility of mass spectrometry in their research.

The chapter is divided into sections, one for each general class of mass spectrometer: magnetic sector, quadrupole, time-of-flight and ion cyclotron resonance. The experiments performed by each are quite often unique and so have been discussed separately under each heading.

## B1.7.2   Ion sources

A common feature of all mass spectrometers is the need to generate ions. Over the years a variety of ion sources have been developed. The physical chemistry and chemical physics communities have generally worked on gaseous and/or relatively volatile samples and thus have relied extensively on the two traditional ionization methods, electron ionization (EI) and photoionization (PI). Other ionization sources, developed principally for analytical work, have recently started to be used in physical chemistry research. These include fast-atom bombardment (FAB), matrix-assisted laser desorption ionization (MALDI) and electrospray ionization (ES).

### B1.7.2.1   Electron ionization (EI)

A schematic diagram of an electron ionization (EI) ion source is shown in figure B1.7.1. A typical source will consist of a block, filament, trap electrode, repeller electrode, acceleration region and a focusing lens. Sample vapour, introduced into the ion source (held at the operating potential of the instrument) through a variable leak valve or capillary interface, is ionized by electrons that have been accelerated towards the block by a

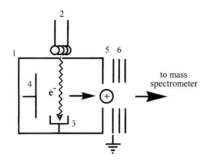

**Figure B1.7.1.** Schematic diagram of an electron ionization ion source: source block (1); filament (2); trap electrode (3); repeller electrode (4); acceleration region (5); focusing lens (6).

potential gradient and collected at the trap electrode. A repeller electrode nudges the newly formed ions out of the source through an exit slit, and they are accelerated to the operating kinetic energy of the instrument. A series of ion lenses is used to focus the ion beam onto the entrance aperture of the mass spectrometer.

Ionization with energetic electrons does not deposit a fixed amount of energy into a molecule. Rather, a 10 eV beam of electrons can deposit anywhere from 0 to 10 eV of energy. For this reason, most instruments relying on electron ionization (often referred to as 'electron impact' ionization, a misleading expression as no actual collision between a molecule and an electron takes place) use electron beams of fairly high energy. Most analytical instruments employ electron energies of $\sim$70 eV as it has been found that a maximum in the ion yield for most organic molecules occurs around this value. The resulting ion internal energies can be described by the Wannier threshold law [1], but at an electron energy of 70 eV this corresponds almost exactly to the photoelectron spectrum. Superimposed on this is the internal energy distribution of the neutral molecules prior to ionization. Since most ion sources operate at very low pressure ($\sim$10$^{-7}$ to 10$^{-6}$ Torr), the resulting ion population has a non-Boltzmann distribution of internal energies and thus it is difficult to discuss the resulting ion chemistry in terms of a thermodynamic temperature.

It is fairly difficult to obtain energy-selected beams of electrons (see chapter B1.7). Thus, electron beams employed in most mass spectrometers have broad energy distributions. One advantage of EI over photoionization, though, is that it is relatively simple to produce high energy electrons. All that is required is the appropriate potential drop between the filament and the ion source block. This potential drop is also continuously adjustable and the resulting electron flux often independent of energy.

### B1.7.2.2    Photoionization (PI)

Photoionization with photons of selected energies can be a precise method for generating ions with known internal energies, but because it is bound to a continuum process (the ejected electron can take on any energy), ions are usually generated in a distribution of internal energy states, according to the equation $h\nu + E_{\text{therm}} = IE_{\text{AB}} + E_{\text{AB}^{+\cdot}} + E_{\text{e}}$, where $h\nu$ is the photon energy, $E_{\text{therm}}$ is the average thermal energy of the molecule AB, $IE$ is the ionization energy of the molecule, $E_{\text{AB}^{+\cdot}}$ is the ion internal energy and $E_{\text{e}}$ is the kinetic energy of the departing electron. The deposition of energy into the ion typically follows the photoelectron spectrum up to the photon energy. Superimposed on this is the internal energy distribution of the original neutral molecules.

There are three basic light sources used in mass spectrometry: the discharge lamp, the laser and the synchrotron light source. Since ionization of an organic molecule typically requires more than 9 or 10 eV, light sources for photoionization must generate photons in the vacuum-ultraviolet region of the electromagnetic spectrum. A common experimental difficulty with any of these methods is that there can be no optical windows or lenses, the light source being directly connected to the vacuum chamber holding the ion source and mass spectrometer. This produces a need for large capacity vacuum pumping to keep the mass spectrometer at operating pressures. Multiphoton ionization with laser light in the visible region of the spectrum overcomes this difficulty.

### B1.7.2.3 Chemical ionization

A third method for generating ions in mass spectrometers that has been used extensively in physical chemistry is chemical ionization (CI) [2]. Chemical ionization can involve the transfer of an electron (charge transfer), proton (or other positively charged ion) or hydride anion (or other anion).

$$R^{+\cdot} + M \rightarrow R + M^{+\cdot}$$

$$RH^+ + M \rightarrow R + MH^+$$

$$RH^- + M \rightarrow R + MH^-.$$

The above CI reactions will occur if they are exothermic. In order for these reactions to occur with high efficiency, the pressure in the ion source must be raised to the milliTorr level. Also, the reagent species are often introduced in large excess so that they are preferentially ionized by the electron beam.

### B1.7.2.4 Other ionization methods

One feature common to all of the above ionization methods is the need to thermally volatilize liquid and solid samples into the ion source. This presents a problem for large and/or involatile samples which may decompose upon heating. Ionization techniques that have been developed to get around this problem include fast-atom bombardment (FAB) [3], matrix-assisted laser desorption ionization (MALDI) [4] and electrospray ionization (ES) [5] (figure B1.7.2). FAB involves bombarding a sample that has been dissolved in a matrix such as glycerol with a high energy beam of atoms. Sample molecules that have been protonated by the glycerol matrix are sputtered off the probe tip, resulting in gas-phase ions. If high energy ions are used to desorb the sample, the technique is called SIMS (secondary ion mass spectrometry). MALDI involves ablating a sample with a laser. A matrix absorbs the laser light, resulting in a plume of ejected material, usually containing molecular ions or protonated molecules. In electrospray, ions are formed in solution by adding protons or other ions to molecules. The solution is sprayed through a fine capillary held at a high potential relative to ground (several keV are common). The sprayed solution consists of tiny droplets that evaporate, leaving gas-phase adduct ions which are then introduced into a mass spectrometer for analysis.

### B1.7.2.5 Molecular beam sources

Sample can be introduced into the ion source in the form of a molecular beam [6, 7] (figure B1.7.3). Molecular beams are most often coupled to time-of-flight instruments for reasons that are discussed in section B1.7.5. The important advantage that molecular beams have over the other methods discussed in this section is their ability to cool the internal degrees of freedom of the sample. Collisions between a carrier gas (such as helium or argon) and the sample molecule in the rapidly expanding gas mixture results in rotational and vibrational cooling. Using this approach, the effective internal 'temperature' of the sample can be significantly less than ambient. One example of the benefits of using molecular beams is in photoionization. The photon energy can be more readily equated to the ion internal energy if the initial internal energy distribution of the neutral molecule is close to 0 K. This cooling also allows weakly bound species such as neutral clusters to be generated and their resulting PI or EI mass spectrum obtained.

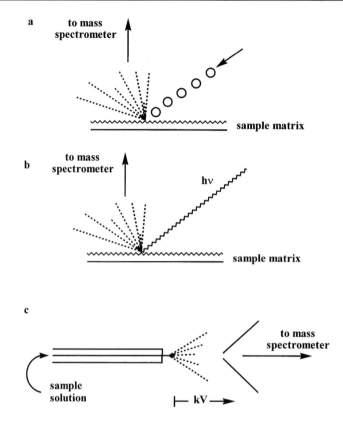

**Figure B1.7.2.** Schematic representations of alternative ionization methods to EI and PI: (a) fast-atom bombardment in which a beam of keV atoms desorbs solute from a matrix (b) matrix-assisted laser desorption ionization and (c) electrospray ionization.

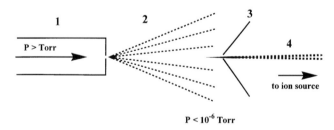

**Figure B1.7.3.** Schematic diagram of a molecular beam generator: nozzle (1); expansion region (2); skimmer (3) and molecular beam (4).

### B1.7.2.6   High pressure sources

There are two significant differences between a high pressure ion source and a conventional mass spectrometer ion source. High pressure sources typically have only two very small orifices (aside from the sample inlet), one to permit ionizing electrons into the source and one to permit ions to leave. Total ion source pressures of up to 5–10 Torr can be obtained allowing the sample vapour to reach thermal equilibrium with the walls of the source. Because the pressures in the source are large, a 70 eV electron beam is insufficient to effectively penetrate and ionize the sample. Rather, electron guns are used to generate electron translational energies of up to 1–2 keV.

### B1.7.3 Magnetic sector instruments

The first mass spectrometers to be widely used as both analytical and physical chemistry instruments were based on the deflection of a beam of ions by a magnetic field, a method first employed by J J Thomson in 1913 [8] for separating isotopes of noble gas ions. Modern magnetic sector mass spectrometers usually consist of both magnetic and electrostatic sectors, providing both momentum and kinetic energy selection. The term 'double-focusing' mass spectrometer refers to such a configuration and relates to the fact that the ion beam is focused at two places between the ion source of the instrument and the detector. It is also possible to add sectors to make three-, four-, five- and even six-sector instruments, though the larger of these are typically used for large molecule analysis. One of the staple instruments used in physical chemistry has been the reverse-geometry tandem sector mass spectrometer ('BE' configuration), which will be described below. The basic principles apply to any magnetic sector instrument configuration.

#### B1.7.3.1 Instrumentation

A schematic diagram of a reverse geometry mass spectrometer is shown in figure B1.7.4.

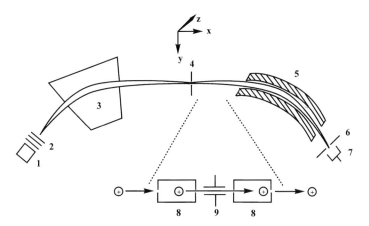

**Figure B1.7.4.** Schematic diagram of a reverse geometry (BE) magnetic sector mass spectrometer: ion source (1); focusing lens (2); magnetic sector (3); field-free region (4); beam resolving slits (5); electrostatic sector (6); electron multiplier detector (7). Second field-free region components: collision cells (8) and beam deflection electrodes (9).

#### (a) The magnetic sector

The magnetic sector consists of two parallel electromagnets surrounding an iron core. The ion beam travels through the flight tube perpendicular to the direction of the imposed magnetic field. The path of an ion travelling orthogonal to a magnetic field is described by a simple mathematical relationship:

$$r = \frac{mv}{Bze}. \tag{B1.7.1}$$

where $r$ is the radius of curvature of the path of the ion, $m$ is the ion's mass, $v$ is the velocity, $B$ is the magnetic field strength, $z$ is the number of charges on the ion and $e$ is the unit of elementary charge. Rearranged,

$$mv = rBze. \tag{B1.7.2}$$

Most instruments are configured with a fixed value for the radius of curvature, $r$, so changing the value of $B$ selectively passes ions of particular values of momentum, $mv$, through the magnetic sector. Thus, it is really the momentum that is selected by a magnetic sector, not mass. We can convert this expression to one involving the accelerating potential. Magnetic sector instruments typically operate with ion sources held at a potential of between 6 and 10 kV. This results in ions with keV translational kinetic energies. The ion kinetic energy can be written as $zeV = \frac{1}{2}mv^2$ and thus the ion velocity is given by the relationship

$$v = \left( \frac{2zeV}{m} \right)^{1/2}$$

and equation (B1.7.2) becomes

$$m/z = \frac{B^2 r^2 e}{2V}. \tag{B1.7.3}$$

In other words, ions with a particular mass-to-charge ratio, $m/z$, can be selectively passed through the magnetic sector by appropriate choice of a value of $V$ and $B$ (though normally $V$ is held constant and only $B$ is varied).

Magnetic sectors can be used on their own, or in conjunction with energy analysers to form a tandem mass spectrometer. The unique features of the reverse geometry instrument are presented from this point.

### (b) The field-free region

The momentum-selected ion beam passes through the field-free region (FFR) of the instrument on its way to the electrostatic sector. The FFR is the main experimental region of the magnetic sector mass spectrometer. Significant features of the FFR can be collision cells and ion beam deflection electrodes. One particular arrangement is shown in figure B1.7.4. A collision cell consists of a 2–3 cm long block of steel with a groove to pass the ion beam. A collision (target) gas can be introduced into the groove, prompting projectile–target gas collisions. The beam deflecting electrode assembly allows the ion beam to be deflected out of the beam path by the application of a potential difference across the assembly (see section B1.7.3.2).

### (c) The electrostatic sector (ESA)

The electrostatic sector consists of two curved parallel plates between which is applied a potential difference producing an electric field of strength $E$. Transmission of an ion through the sector is governed by the following relationship

$$\tfrac{1}{2}mv^2 = zeV = \tfrac{1}{2}zeEr.$$

This relationship gives an expression similar to equation (B1.7.1):

$$r = \frac{2V}{E}. \tag{B1.7.4}$$

Adjusting the potential across the ESA plates allows ions of selected translational kinetic energy to pass through and be focused, at which point a detector assembly is present to monitor the ion flux. Note that in equation (B1.7.4) neither the mass nor charge of the ion is present. So, isobaric ions with one, two, three etc. charges, accelerated by a potential drop, $V$, will pass through the ESA at a common value of $E$. An ion with +1 charge accelerated across a potential difference of 8000 V will have 8 keV translational kinetic energy and be transmitted through the ESA by a field $E_1$. An ion with +2 charges will have 16 keV translational energy but will experience a field strength equivalent to $2E_1$, and so forth.

*B1.7.3.2   Experiments using magnetic sector instruments*

A single magnetic sector can be used as a mass filter for other apparatus. However, much more information of the simple mass spectrum of a species can be obtained using the tandem mass spectrometer.

*(a) Mass-analysed ion kinetic energy spectrometry (MIKES)*

Ions accelerated out of the ion source with keV translational kinetic energies (and $m/z$ selected with the magnetic sector) will arrive in the FFR of the instrument in several microseconds. Ions dissociating on this timescale (with unimolecular decay rate constants between $10^2$ and $10^5$ s$^{-1}$, depending on the physical geometry of the instrument) have been given the name 'metastable ions' [9].

In the FFR of the sector mass spectrometer, the unimolecular decomposition fragments, A$^+$ and B, of the mass selected metastable ion AB$^+$ will, by the conservation of energy and momentum, have lower translational kinetic energy, $T$, than their precursor:

$$zT_{A^+} = \frac{1}{2}m_{A^+}v^2 \qquad zT_B = \frac{1}{2}m_{B\cdot}v^2 \qquad zT_{AB^{+\cdot}} = \frac{1}{2}m_{AB^+}v^2.$$

Thus we find

$$T_{A^+} = \frac{m_{A^+}}{m_{AB^{+\cdot}}}T_{AB^{+\cdot}}.$$

By scanning the ESA to pass ions with lower translational energies, the fragment ions will sequentially pass through to the detector (this is the so-called MS/MS, or MS$^2$, experiment). The final ion abundance kinetic energy spectrum (figure B1.7.5(a)) is converted to an ion abundance fragment $m/z$ spectrum by the above relationships. The MIKE spectrum is the end result of all low energy unimolecular processes of the selected ions, including isomerization. Thus, isomeric ions which interconvert on the $\mu$s timescale often have closely related, if not identical, MIKE spectra. There are several characteristic peak shapes expected in a MIKE spectrum that are summarized in figure B1.7.6.

*(b) Collision-induced dissociation (CID) mass spectrometry*

A collision-induced dissociation (CID) mass spectrum [10, 11] of mass selected ions is obtained by introducing a target gas into a collision cell in one of the field-free regions. The resulting high energy (keV) CID mass spectrum, obtained and analysed in the same way as a MIKE spectrum, contains peaks due to ions formed in virtually all possible unimolecular dissociation processes of the precursor ion (figure B1.7.5(b)). The timescale of the collision-induced fragmentation reactions is quite different from the MIKE experiment, ranging from the time of the collision event ($t \approx 10^{-15}$ s) to the time the ions exit the FFR. For this reason, isomerization reactions tend not to play a significant role in collision-induced reactions and thus the CID mass spectra are often characteristic of ion connectivity.

In collisional excitation, translational energy of the projectile ion is converted into internal energy. Since the excited states of the ions are quantized, so will the translational energy loss be. Under conditions of high energy resolution, it is possible to obtain a translational energy spectrum of the precursor ions exhibiting peaks that correspond to the formation for discrete excited states (translational energy spectroscopy [12]).

It is possible in a sector instrument to perform a variety of other experiments on the projectile ions. Many involve examining the products of charge exchange with the target gas, while others allow neutral species to be studied. Some of the more common experiments are summarized in figure B1.7.7. All of the experiments have been described for projectile cations, but anions can be studied in analogous manners.

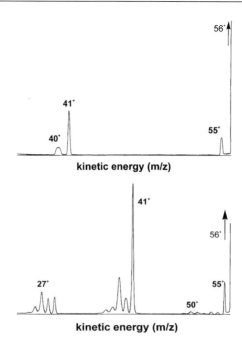

**Figure B1.7.5.** (a) MIKE spectrum of the unimolecular decomposition of 1-butene ions ($m/z$ 56). This spectrum was obtained in the second field-free region of a reverse geometry magnetic sector mass spectrometer (VG ZAB-2HF). (b) Collision-induced dissociation mass spectrum of 1-butene ions. Helium target gas was used to achieve 10% beam reduction (single collision conditions) in the second field-free region of a reverse geometry magnetic sector mass spectrometer (VG ZAB-2HF).

*(c) Kinetic energy release (KER) measurements*

In a unimolecular dissociation, excess product energy is typically distributed among the translational, rotational and vibrational modes in a statistical fashion. The experimentally observed phenomenon is the distribution of translational kinetic energies of the departing fragment ions (the kinetic energy release, KER) [9]. In magnetic sector instruments, the result of $x$-axial (i.e. along the beam path) KER is the observation of fragment ion peaks in the MIKE or CID spectra which have a broader kinetic energy spread than the precursor ion peak. In a CID, however, this spread is complicated by collisional scattering and so KER is most often discussed for peaks in MIKE spectra. If the mass spectrometer is operated under conditions of high resolution, obtained by narrowing the $y$-axis beam collimating slits throughout the instrument, the widths of the fragment ion peaks are indicative of this kinetic energy release. The measured value for the KER is typically expressed as the value at half-height of the fragment ion peak, $T_{0.5}$, and is calculated with the following equation

$$T_{0.5}(\text{meV}) = \frac{m_{AB^{+\cdot}}^2 (\Delta V_{0.5,A^+}^2 - \Delta V_{0.5,AB^{+\cdot}}^2)}{16(8)m_{A^+}m_B}$$

where $m$ is the mass of the various species, $\Delta V_{0.5}$ is the full width energy spread of the fragment and precursor ion peaks at half height and (8) represents the typically 8 keV translational kinetic energy of the precursor [9]. The resulting $T_{0.5}$ is in meV. Note that since knowledge of the internal energy distribution of the dissociating ions is lacking, the relationship between $T_{0.5}$ and the average KER is strictly qualitative, i.e. a large $T_{0.5}$ indicates a large average KER value. How statistical the distribution of product excess energies is will depend on the dynamics of the dissociation.

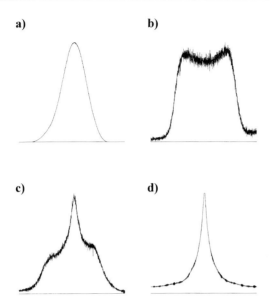

**Figure B1.7.6.** Fragment ion peak shapes expected in MIKE spectra: (a) typical Gaussian energy profile; (b) large average kinetic energy release causing $z$-axial discrimination of the fragment ions, resulting in a 'dished-top' peak; (c) competing fragmentation channels, each with its own distinct kinetic energy release, producing a 'composite' peak; (d) fragmentation occurring from a dissociative excited state.

## B1.7.4 Quadrupole mass filters, quadrupole ion traps and their applications

Another approach to mass analysis is based on stable ion trajectories in quadrupole fields. The two most prominent members of this family of mass spectrometers are the quadrupole mass filter and the quadrupole ion trap. Quadrupole mass filters are one of the most common mass spectrometers, being extensively used as detectors in analytical instruments, especially gas chromatographs. The quadrupole ion trap (which also goes by the name 'quadrupole ion store, QUISTOR', Paul trap, or just ion trap) is fairly new to the physical chemistry laboratory. Its early development was due to its use as an inexpensive alternative to tandem magnetic sector and quadrupole filter instruments for analytical analysis. It has, however, started to be used more in the chemical physics and physical chemistry domains, and so it will be described in some detail in this section.

The principles of operation of quadrupole mass spectrometers were first described in the late 1950s by Wolfgang Paul who shared the 1989 Nobel Prize in Physics for this development. The equations governing the motion of an ion in a quadrupole field are quite complex and it is not the scope of the present article to provide the reader with a complete treatment. Rather, the basic principles of operation will be described, the reader being referred to several excellent sources for more complete information [13–15].

### B1.7.4.1 The quadrupole mass filter

A schematic diagram of a quadrupole mass filter is shown in figure B1.7.8. In an ideal, three-dimensional, quadrupole field, the potential $\phi$ at any point $(x, y, z)$ within the field is described by equation (B1.7.5):

$$\phi = \frac{\phi_0}{r_0^2}(ax^2 + by^2 + cz^2) \tag{B1.7.5}$$

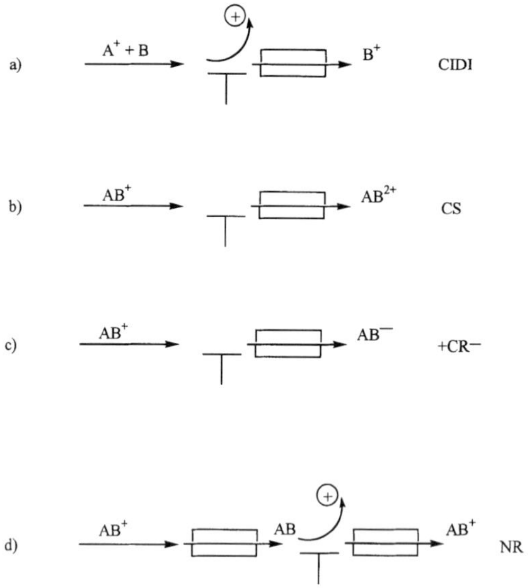

**Figure B1.7.7.** Summary of the other collision based experiments possible with magnetic sector instruments: (a) collision-induced dissociation ionization (CIDI) records the CID mass spectrum of the neutral fragments accompanying unimolecular dissociation; (b) charge stripping (CS) of the incident ion beam can be observed; (c) charge reversal (CR) requires the ESA polarity to be opposite that of the magnet; (d) neutralization–reionization (NR) probes the stability of transient neutrals formed when ions are neutralized by collisions in the first collision cell. Neutrals surviving to be collisionally reionized in the second cell are recorded as 'recovery' ions in the NR mass spectrum.

where $\phi_0$ is the applied potential, $r_0$ is half the distance between the hyperbolic rods and $a$, $b$ and $c$ are coefficients. The applied potential is a combination of a radio-frequency (RF) potential, $V \cos \omega t$, and direct current (DC) potential, $U$. The two can be expressed in the following relationship:

$$\phi_0 = U + V \cos \omega t$$

where $\omega$ is the angular frequency of the RF field (in rad s$^{-1}$) and is $2\pi$ times the frequency in Hertz. A potential applied across the rods (see figure B1.7.8) is flipped at radio-frequencies.

Ideally, the rods in a quadrupole mass filter should have a hyperbolic geometry, but more common is a set of four cylindrical rods separated by a distance $2r$, where $r \approx 1.16r_0$ (figure B1.7.8). This arrangement provides for an acceptable field geometry along the axis of the mass filter. Since there is no field along the axis of the instrument, equation (B1.7.5) simplifies to $\phi = (\phi_0/r_0^2)(ax^2 + by^2)$.

The values of $a$ and $b$ that satisfy this relationship are $a = 1$ and $b = -1$, so the potential inside the quadrupole mass filter takes on the form:

$$\phi = \frac{\phi_0}{r_0^2}(x^2 - y^2).$$

The equations of motion within such a field are given by:

$$\frac{d^2 x}{dt^2} + \frac{e}{mr_0^2}(U - V \cos \omega t)x = 0$$

$$\frac{d^2 y}{dt^2} + \frac{e}{mr_0^2}(U - V \cos \omega t)y = 0.$$

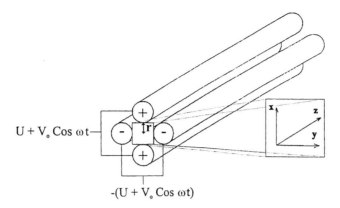

$U + V_o \cos \omega t$

$-(U + V_o \cos \omega t)$

**Figure B1.7.8.** Quadrupole mass filter consisting of four cylindrical rods, spaced by a radius $r$ (reproduced with permission of Professor R March, Trent University, Peterborough, ON, Canada).

Now, we can make three useful substitutions

$$a_x = -a_y = \frac{8eU}{m\omega^2 r_0^2} \qquad q_x = -q_y = \frac{8eV}{m\omega^2 r_0^2} \qquad \xi = \frac{\omega t}{2}$$

and the result is in the general form of the Mathieu equation [16]:

$$\frac{d^2 u}{d\xi^2} + (a_u - 2q_u \cos 2\xi)u = 0. \tag{B1.7.6}$$

Equation (B1.7.6) describes the ion trajectories in the quadrupole field (where $u$ can be either $x$ or $y$). The stable, bounded solutions to these equations represent conditions of stable, bounded trajectories in the quadrupole mass filter. A diagram representing the stable solutions to the equations for both the $x$- and $y$-axes (really the intersection of two sets of stability diagrams, one for the $x$-axis, one for the $y$-axis) is shown in figure B1.7.9(a). This figure represents the stability region closest to the axis of the instrument and is the most appropriate for the operation of the quadrupole. This stability region is also unique for a given $m/z$ ratio. Figure B1.7.9(b) represents the stability regions (transformed into axes of the applied DC and RF potential) for a series of ions with different $m/z$ values. The line running through the apex of each region is called the operating line and represents the conditions ($U$ and $V$) for the selective filtering of different mass ions through the instrument. The ratio of $U$ to $V$ is a constant along the operating line. It is apparent from figure B1.7.9(b) that the resolution of the quadrupole mass filter can be altered by changing the slope of the operating line. A greater slope means there is greater separation of the ions, but at the expense of sensitivity (fewer ions will have stable trajectories).

To be effective, it is necessary for the ions traversing the instrument to experience several RF cycles. Thus, unlike magnetic sector instruments, the ions formed in the ion source of a quadrupole mass filter apparatus are accelerated to only a few eV kinetic energy (typically 5–10 eV). The timescale of the experiment is therefore much longer than for magnetic sectors, ions having to survive milliseconds rather than microseconds to be detected.

### B1.7.4.2 Experiments using quadrupole mass filters

Aside from the single mass filter, the most common configuration for quadrupole mass spectrometers is the triple-quadrupole instrument. This is the simplest tandem mass spectrometer using quadrupole mass filters.

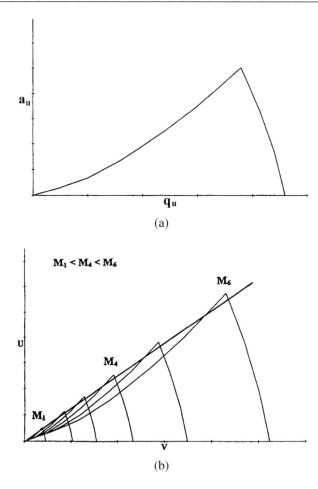

(a)

(b)

**Figure B1.7.9.** (a) Stability diagram for ions near the central axis of a quadrupole mass filter. Stable trajectories occur only if the $a_u$ and $q_u$ values lie beneath the curve. (b) Stability diagram (now as a function of $U$ and $V$) for six ions with different masses. The straight line running through the apex of each set of curves is the 'operating' line, and corresponds to values of $U/V$ that will produce mass resolution (reproduced with permission of Professor R March, Trent University, Peterborough, ON, Canada).

Typically, the first and last quadrupole are operated in mass selective mode as described above. The central quadrupole is usually an RF-only quadrupole. The lack of a DC voltage means that ions of all $m/z$ values will have stable trajectories through the filter.

*(a) Collision-induced dissociation*

Adding a collision gas to the RF-only quadrupole of a triple-quadrupole instrument permits collision-induced dissociation experiments to be performed. Unlike magnetic sector instruments, the low accelerating potential in quadrupole instruments means that low energy collisions occur. In addition, the time taken by the ions to traverse the RF only quadrupole results in many of these low energy collisions taking place. So, collisional excitation occurs in a multi-step process, rather than in a single, high-energy, process as in magnetic sector instruments.

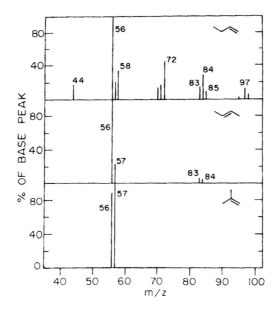

**Figure B1.7.10.** Three mass spectra showing the results of reactive collisions between a projectile ion $CH_3NH_2^{+\cdot}$ and three isomeric butenes. (Taken from Usypchuk L L, Harrison A G and Wang J 1992 Reactive collisions in quadrupole cells. Part 1. Reaction of $[CH_3NH_2]^{+\cdot}$ with isomeric butenes and pentenes *Org. Mass Spectrom.* **27** 777–82. Copyright John Wiley & Sons Limited. Reproduced with permission.)

### (b) Reactive collisions

Since ions analysed with a quadrupole instrument have low translational kinetic energies, it is possible for them to undergo bimolecular reactions with species inside an RF-only quadrupole. These bimolecular reactions are often useful for the structural characterization of isomeric species. An example of this is the work of Harrison and co-workers [17]. They probed the reactions of $CH_3NH_2^{+\cdot}$ ions with isomeric butenes and pentenes in the RF-only quadrupole collision cell of a hybrid BEqQ instrument. The mass spectra of the products of the ion–molecule reactions were distinct for the various isomers probed. Addition of the amine to the olefin followed by fragmentation produced characteristic iminium ions only for terminal olefins without substitution at the olefinic carbons (figure B1.7.10).

### (c) Equilibrium measurements

It is possible to determine the equilibrium constant, $K$, for the bimolecular reaction involving gas-phase ions and neutral molecules in the ion source of a mass spectrometer [18]. These measurements have generally focused on three properties, proton affinity (or gas-phase basicity) [19, 20], gas-phase acidity [21] and solvation enthalpies (and free energies) [22, 23]:

$$\text{proton affinity: } B_1H^+ + B_2 \rightleftharpoons B_1 + B_2H^+$$
$$\text{gas-phase acidity: } A_1H + A_2^- \rightleftharpoons A_1^- + A_2H$$
$$\text{solvation: } N + M^+ \rightleftharpoons MN^+.$$

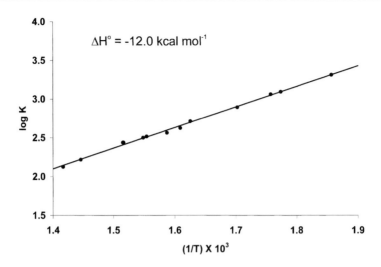

**Figure B1.7.11.** Van't Hoff plot for equilibrium data obtained for the reaction of isobutene with ammonia in a high pressure ion source (reproduced from data in [19]).

A common approach has been to measure the equilibrium constant, $K$, for these reactions as a function of temperature with the use of a variable temperature high pressure ion source (see section B1.7.2). The ion concentrations are approximated by their abundance in the mass spectrum, while the neutral concentrations are known from the sample inlet pressure. A van't Hoff plot of $\ln K$ versus $1/T$ should yield a straight line with slope equal to the reaction enthalpy (figure B1.7.11). Combining the PA with a value for $\Delta_{\mathrm{basicity}} G^0$ at one temperature yields a value for $\Delta S^0$ for the half-reaction involving addition of a proton to a species. While quadrupoles have been the instruments of choice for many of these studies, other mass spectrometers can act as suitable detectors [19, 20].

*(d) Selected ion flow tube*

Another example of the use of quadrupole filters in studying reactive collisions of gaseous ions is the selected ion flow tube (SIFT) [24]. This has been perhaps the most widely employed instrument for studying the kinetics of bimolecular reactions involving ions. Its development by N G Adams and D Smith sprang from the utility of the flowing afterglow (FA) technique (developed in the early 1960s by Ferguson, Fehsenfeld and Schmeltekopf [25, 26]) in the study of atmospheric reactions [27].

A schematic diagram of a SIFT apparatus is shown in figure B1.7.12. The instrument consists of five basic regions, the ion source, initial quadrupole mass filter, flow tube, second mass filter and finally the detector. The heart of the instrument is the flow tube, which is a steel tube approximately 1 m long and 10 cm in diameter. The pressure in the flow tube is kept of the order of 0.5 Torr, resulting in carrier gas flow rates of $\sim$100 m s$^{-1}$. Along the flow tube there are orifices that are used to introduce neutral reagents into the flow stream. Product ions arriving at the end of the flow tube are skimmed through a small orifice and mass analysed with a second quadrupole filter before being detected. The reactions occurring in the flow tube can be monitored as a function of carrier gas flow rate, and hence timescale. A detailed description of the extraction of rate constants from SIFT experiments is given by Smith and Adams [24]. Examples of the type of information obtained with the SIFT technique can be found in a recent series of articles by D Smith and co-workers [28–30].

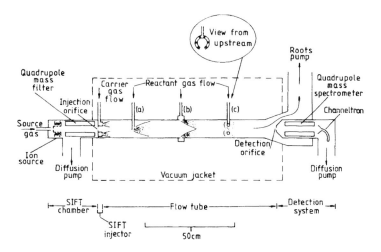

**Figure B1.7.12.** A schematic diagram of a typical selected-ion flow (SIFT) apparatus. (Smith D and Adams N G 1988 The selected ion flow tube (SIFT): studies of ion–neutral reactions *Advances in Atomic and Molecular Physics* vol 24, ed D Bates and B Bederson p 4. Copyright Academic Press, Inc. Reproduced with permission.)

*(e) Ion-guide instruments*

Another instrument used in physical chemistry research that employs quadrupole mass filters is the guided ion beam mass spectrometer [31]. A schematic diagram of an example of this type of instrument is shown in figure B1.7.13. A mass selected beam of ions is introduced into an ion guide, in which their translational energy can be precisely controlled down to a few fractions of an electron volt. Normally at these energies, divergence of the ion beam would preclude the observation of any reaction products. To overcome this, the ions are trapped in the beam path with either a quadrupole or octapole filter operated in RF-only mode (see above). The trapping characteristics (i.e., how efficiently the ions are 'guided' along the flight path) of the octapole filter are superior to the quadrupole filter and so octapoles are often employed in this type of apparatus. The operational principles of an octapole mass filter are analogous to those of the quadrupole mass filter described above.

Using a guided ion beam instrument the translational energy dependent reaction cross sections of endo-thermic fragmentation processes can be determined [32]. Modelling these cross sections ultimately yields their energy thresholds and a great deal of valuable thermochemical information has been derived with this technique. Precision of $\pm 0.2$ eV can be obtained for reaction thresholds. Bimolecular reactions can also be studied and reaction enthalpies derived from the analysis of the cross section data.

*B1.7.4.3 The quadrupole ion trap*

The quadrupole ion trap is the three dimensional equivalent to the quadrupole mass filter. A typical geom-etry consists of two hyperbolic endcap electrodes and a single ring electrode (figure B1.7.14). Unlike the quadrupole mass filter, however, the ion trap can be used both as a mass selective device or as an ion storage device. It is this latter ability that has led to the popularity and versatility of the ion trap. The theoretical treatment of ion trajectories inside the ion trap is similar to that presented above for the mass filter, except that now the field is no longer zero in the $z$-axis. It is convenient to use cylindrical coordinates rather than

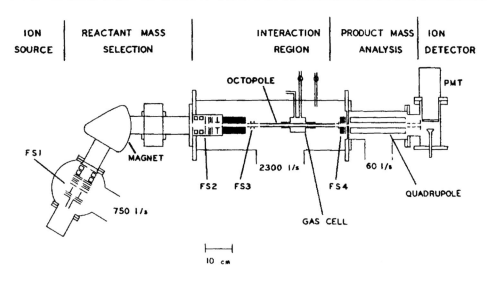

**Figure B1.7.13.** A schematic diagram of an ion-guide mass spectrometer. (Ervin K M and Armentrout P B 1985 Translational energy dependence of Ar⁺ + XY → ArX⁺ + Y from thermal to 30 eV c.m. *J. Chem. Phys.* **83** 166–89. Copyright American Institute of Physics Publishing. Reproduced with permission.)

Cartesian coordinates and the resulting relationship describing the potential inside the ion trap is given as:

$$\phi = \frac{\phi_0}{r_0^2}(r_0^2 - 2z_0^2)$$

where $r_0$ and $z_0$ are defined as in figure B1.7.14. The relationship $r_0^2 = 2z_0^2$ has usually governed the geometric arrangement of the electrodes. The equations of motion for an ion in the ion trap are analogous to those for the quadrupole mass filter:

$$\frac{d^2r}{dt^2} + \frac{2e}{mr_0^2}(U - V\cos\omega t)r = 0$$

$$\frac{d^2z}{dt^2} + \frac{4e}{mr_0^2}(U - V\cos\omega t)z = 0$$

and with the analogous substitutions,

$$a_z = -2a_r = \frac{-16eU}{m\omega^2(r_0^2 + 2z_0^2)}$$

$$q_z = -2q_r = \frac{-8eV}{m\omega^2(r_0^2 + 2z_0^2)}$$

$$\xi = \frac{\omega t}{2}.$$

the Mathieu equation is obtained.

   The Mathieu equation for the quadrupole ion trap again has stable, bounded solutions corresponding to stable, bounded trajectories inside the trap. The stability diagram for the ion trap is quite complex, but a subsection of the diagram, corresponding to stable trajectories near the physical centre of the trap, is shown in figure B1.7.15. The interpretation of the diagram is similar to that for the quadrupole mass filter.

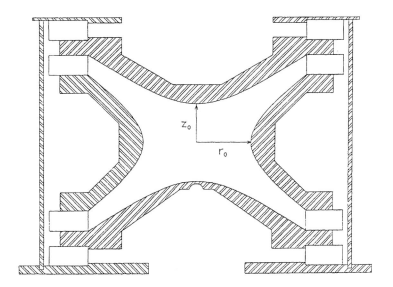

**Figure B1.7.14.** Schematic cross-sectional diagram of a quadrupole ion trap mass spectrometer. The distance between the two endcap electrodes is $2z_0$, while the radius of the ring electrode is $r_0$ (reproduced with permission of Professor R March, Trent University, Peterborough, ON, Canada).

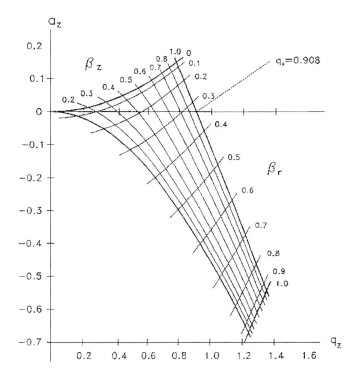

**Figure B1.7.15.** Stability diagram for ions near the centre of a quadrupole ion trap mass spectrometer. The enclosed area reflects values for $a_z$ and $q_z$ that result in stable trapping trajectories (reproduced with permission of Professor R March, Trent University, Peterborough, ON, Canada).

*B1.7.4.4    Experiments using quadrupole ion traps*

The ion trap has three basic modes in which it can be operated. The first is as a mass filter. By adjusting $U$ and $V$ to reside near the apex of the stability diagram in figure B1.7.15, only ions of a particular $m/z$ ratio will be selected by the trap. An operating line that intersects the stability diagram near the apex (as was done for the mass filter) with a fixed $U/V$ ratio describes the operation of the ion trap in a mass scanning mode. A second mode of operation is with the potential $\phi_0$ applied only to the ring electrode, the endcaps being grounded. This allows ions to be selectively stored. The application of an extraction pulse to the endcap electrodes ejects ions out of the trap for detection. A third mode is the addition of an endcap potential, $-U$. This mode permits mass selective storage in the trap, followed by storage of all ions. This latter method permits the ion trap to be used as a tandem mass spectrometer.

*(a) Ion isolation*

One of the principle uses of the ion trap is as a tandem-in-time mass spectrometer. Ions with a particular $m/z$ ratio formed in the ion trap, or injected into the trap from an external source, can be isolated by resonantly ejecting all other ions. This can be accomplished in a variety of ways. One method involves applying a broad-band noise field between the endcap electrodes to resonantly excite (in the axial direction) and eject all ions. The secular frequency of the ions to be stored is notched out of this noise field, leaving them in the trap. In another method, ions with lower and higher $m/z$ ratio can be ejected by adjusting the amplitude of the RF and DC potentials so that the $(a_z, q_z)$ value for the ion of interest lies just below the apex in figure B1.7.15.

*(b) Collision-induced dissociation*

A unique aspect of the ion trap is that the trapping efficiency is significantly improved by the presence of helium damping gas. Typical pressures of helium in the trap are $\sim 1$ milliTorr. Collisions between the trapped ions and helium gas effectively cause the ions to migrate towards the centre of the trap where the trapping field is most perfect. This causes significant improvements is sensitivity in analytical instruments. The presence of the damping gas also permits collision-induced dissociation to be performed. In the manner described above, ions with a particular $m/z$ ratio can be isolated in the trap. The axial component of the ion motion is then excited resonantly by applying a potential across the endcap electrodes. The amplitude of this potential is controlled to prevent resonant ejection of the ions, but otherwise this is a similar experiment to that described above for mass selective ejection. The potential (often called the 'tickle' potential), increases the kinetic energy of the selected ions which in turn increases the centre-of-mass collision energy with the helium damping gas. These collisions now excite the mass selected ions causing them to dissociate. After the 'tickle' period, the trap is returned to a state where all ions can be trapped and thus the mass spectrum of the fragmentation product ions can be obtained. The CID mass spectra that are obtained in this manner are similar to those obtained with triple quadrupole instruments, in that they are the result of many low energy collisions (figure B.1.7.16(a)). There have been attempts to measure unimolecular reaction kinetics by probing fragment ion intensities as a function of time after excitation [33].

With the right software controlling the instrument, it is possible for the above process to be repeated $n$ times, i.e., the ion trap is theoretically capable of $MS^n$ experiments, though the ion concentration in the trap is always the limiting factor in these experiments.

*(c) Bimolecular reactions*

The same procedure as outlined above can be used to study ion–molecule reactions [15, 34]. Mass-selected ions will react with neutral species inside the trap. The presence of the damping gas means that stable (thermodynamic and kinetic) complexes may be formed. Allowing the mass-selected ions to react in the trap, while storing all reaction products, allows the course of the reactions to be followed (figure B1.7.16(b)). Changing the storage time allows the relative abundance of reactant and product ions to be monitored as a function of time. This introduces the possibility of measuring ion–molecule reaction rate constants with the ion trap. Since the ion internal energy distribution in such an experiment is described by a temperature near the ambient temperature of the damping gas, ion–molecule reactions can be probed as a function of temperature by raising the trap temperature. Resonant excitation of the mass-selected ions (see CID section) effectively raises their internal energy, allowing endothermic reactions with neutral species to take place (there will be a limit on the endothermicity extending from the limit on internal excitation occurring in the resonant excitation process). The degree of axial resonant excitation can be controlled and there have been attempts to relate this to an effective 'temperature' [35].

## B1.7.5   Time-of-flight mass spectrometers

Probably the simplest mass spectrometer is the time-of-flight (TOF) instrument [36]. Aside from magnetic deflection instruments, these were among the first mass spectrometers developed. The mass range is theoretically infinite, though in practice there are upper limits that are governed by electronics and ion source considerations. In chemical physics and physical chemistry, TOF instruments often are operated at lower resolving power than analytical instruments. Because of their simplicity, they have been used in many spectroscopic apparatus as detectors for electrons and ions. Many of these techniques are included as chapters unto themselves in this book, and they will only be briefly described here.

### B1.7.5.1   Time-of-flight equations

The basic principle behind TOF mass spectrometry [36] is the equation for kinetic energy, $zeV = \frac{1}{2}mv^2$, where the translational kinetic energy of an ion accelerated out of the ion source by a potential drop, $V$ is $zeV$. If ions of mass $m$ are given $zeV$ kinetic energy, then the time, $t_d$, the ions take to travel a distance $d$ is given by:

$$t_d = d\sqrt{\frac{m}{2zeV}}. \tag{B1.7.7}$$

In the simplest form, $t_d$ reflects the time of flight of the ions from the ion source to the detector. This time is proportional to the square root of the mass, i.e., as the masses of the ions increase, they become closer together in flight time. This is a limiting parameter when considering the mass resolution of the TOF instrument.

The ion time of flight, as given by equation (B1.7.7), is oversimplified, however. There are a number of factors which change the final measured TOF. These are considered below.

### (a) Ion source residence time

A schematic diagram of a simple TOF instrument is shown in figure B1.7.17(a). Since the ion source region of any instrument has a finite size, the ions will spend a certain amount of time in the source while they are accelerating. If the initial velocity of the ions is $v_0$, the time spent in the source, $t_s$, is given by

$$t_s = \frac{v_0 m}{zeV/d_s}$$

where $v$ is the velocity of the ion, $d_s$ is the width of the ion source and $v/d_s$ represents the electric field strength inside the source. One obvious way to minimize this effect is to make the field strength as large as possible (by increasing $V$ or decreasing $d_s$).

Ions generated in the ion source region of the instrument may have initial velocities isotropically distributed in three dimensions (for gaseous samples, this initial velocity is the predicted Maxwell–Boltzmann

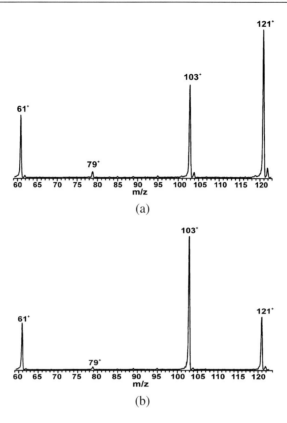

**Figure B1.7.16.** Mass spectra obtained with a Finnigan GCQ quadrupole ion trap mass spectrometer. (a) Collision-induced dissociation mass spectrum of the proton-bound dimer of isopropanol $[(CH_3)_2CHOH]_2H^+$. The $m/z$ 121 ions were first isolated in the trap, followed by resonant excitation of their trajectories to produce CID. Fragment ions include water loss ($m/z$ 103), loss of isopropanol ($m/z$ 61) and loss of 42 amu ($m/z$ 79). (b) Ion-molecule reactions in an ion trap. In this example the $m/z$ 103 ion was first isolated and then resonantly excited in the trap. Endothermic reaction with water inside the trap produces the proton-bound cluster at $m/z$ 121, while CID produces the fragment with $m/z$ 61.

distribution at the sample temperature). The time the ions spend in the source will now depend on the direction of their initial velocity. At one extreme, the ions may have a velocity $v_0$ in the direction of the extraction grid. The time spent in the source will be shorter than those with no component of initial velocity in this direction:

$$t_s = \frac{(v - v_0)m}{zeV/d_s}.$$

At the other extreme, ions with initial velocities in the direction opposite to the accelerating potential must first be turned around and brought back to their initial position. From this point their behaviour is the same as described above. The time taken to turn around in the ion source and return to the initial position ($t_r$) is given by:

$$t_r = \frac{2v_0m}{zeV/d_s}.$$

The final velocity of these two ions will be the same, but their final flight times will differ by the above turn-around time, $t_r$. This results in a broadening of the TOF distributions for each ion mass, and is another limiting factor when considering the mass (time) resolution of the instrument.

The final total ion time of flight in the TOF mass spectrometer with a single accelerating region can be written in a single equation, taking all of the above factors into account.

$$t_{TOF} = \frac{(2m)^{1/2}[(U_0 + zeV/d_s)^{1/2} \pm U_0^{1/2}]}{zeV/d_s} + \frac{(2m)^{1/2}d}{2(U_0 + zeV)^{1/2}}$$

where now the initial velocity has been replaced by the initial translation energy, $U_0$. This is the equation published in 1955 by Wiley and McLaren [37] in their seminal paper on TOF mass spectrometry.

### (b) Energy focusing and the reflectron TOF instrument

The resolution of the TOF instrument can be improved by applying energy focusing conditions that serve to overcome the above stated spread in initial translational energies of the generated ions. While there have been several methods developed, the most successful and the most commonly used method is the reflectron. The reflectron is an ion mirror positioned at the end of the drift tube that retards the ions and reverses their direction. Ions with a higher kinetic energy penetrate into the mirror to a greater extent than those with lower kinetic energies. The result is a focusing (in time) at the detector of ions having an initial spread of kinetic energies (figure B1.7.17(b)). The mirror also has the effect of increasing the drift length without increasing the physical length of the instrument.

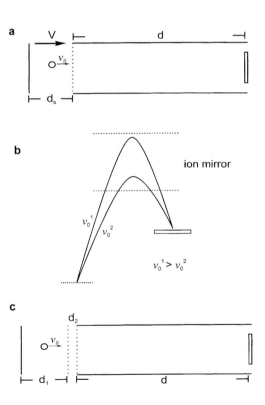

**Figure B1.7.17.** (a) Schematic diagram of a single acceleration zone time-of-flight mass spectrometer. (b) Schematic diagram showing the time focusing of ions with different initial velocities (and hence initial kinetic energies) onto the detector by the use of a reflecting ion mirror. (c) Wiley–McLaren type two stage acceleration zone time-of-flight mass spectrometer.

*(c) Spatial focusing*

Another consideration when gaseous samples are ionized is the variation in where the ions are formed in the source. The above arguments assumed that the ions were all formed at a common initial position, but in practice they may be formed anywhere in the acceleration zone. The result is an additional spread in the final TOF distributions, since ions made in different locations in the source experience different accelerating potentials and thus spend different times in the source and drift tube.

Spatial focusing in a single acceleration zone linear TOF instrument naturally occurs at a distance of $2d_s$ along the drift tube, which is seldom a practical distance for a detector. Wiley and McLaren described a two stage accelerating zone that allows spatial focusing to be moved to longer distances. An example of such an instrument is shown in figure B1.7.17(c). The main requirement is for the initial acceleration region to have a much weaker field than the second. The equation relating the instrumental parameters in figure B1.7.17(c) is:

$$d = d_1 k^{3/2} \left( 1 - \frac{1}{k + k^{1/2}} \frac{2d_2}{d_1} \right)$$

where $k = (\frac{1}{2} d_1 E_1 + d_2 E_2)/(\frac{1}{2} d_1 E_1)$, and $E_1$ and $E_2$ are the field strengths in regions 1 and 2. So, if the physical dimensions of the instrument $d_1$, $d_2$ and $d$ are fixed, a solution can be obtained for the relative field strengths necessary for spatial focusing.

*(d) Other ionization sources*

Other methods of sample introduction that are commonly coupled to TOF mass spectrometers are MALDI, SIMS/FAB and molecular beams (see section B1.7.2). In many ways, the ablation of sample from a surface simplifies the TOF mass spectrometer since all ions originate in a narrow space above the sample surface. This reduces many of the complications arising from the need for spatial focusing. Also, the initial velocity of ions generated are invariably in the TOF direction.

Molecular beam sample introduction (described in section B1.7.2), followed by the orthogonal extraction of ions, results in improved resolution in TOF instruments over effusive sources. The particles in the molecular beam typically have translational temperatures orthogonal to the beam path of only a few Kelvin. Thus, there is less concern with both the initial velocity of the ions once they are generated and with where in the ion source they are formed (since the particles are originally confined to the beam path).

### B1.7.5.2   *Experiments using TOF mass spectrometers*

Time-of-flight mass spectrometers have been used as detectors in a wider variety of experiments than any other mass spectrometer. This is especially true of spectroscopic applications, many of which are discussed in this encyclopedia. Unlike the other instruments described in this chapter, the TOF mass spectrometer is usually used for one purpose, to acquire the mass spectrum of a compound. They cannot generally be used for the kinds of ion–molecule chemistry discussed in this chapter, or structural characterization experiments such as collision-induced dissociation. However, they are easily used as detectors for spectroscopic applications such as multi-photoionization (for the spectroscopy of molecular excited states) [38], zero kinetic energy electron spectroscopy [39] (ZEKE, for the precise measurement of ionization energies) and coincidence measurements (such as photoelectron–photoion coincidence spectroscopy [40] for the measurement of ion fragmentation breakdown diagrams).

### B1.7.6   Fourier transform ion cyclotron resonance mass spectrometers

Fourier transform ion cyclotron resonance (FT-ICR) mass spectrometry is another in the class of trapping mass spectrometers and, as such is related to the quadrupole ion trap. The progenitor of FT-ICR, the ICR mass spectrometer, originated just after the Second World War when the cyclotron accelerator was developed into a means for selectively detecting ions other than protons. At the heart of ICR is the presence of a magnetic field that confines ions into orbital trajectories about their flight axis. Early ICR experiments mainly took advantage of this trapping and were focused on ion–molecule reactions. The addition of the three-dimensional trapping cell by McIver in 1970 [41, 42] led to improved storage of ions. In 1974 Comisarow and Marshall introduced the Fourier transform detection scheme that paved the way for FT-ICR [43, 44] which is now employed in virtually all areas in physical chemistry and chemical physics that use mass spectrometry.

### B1.7.6.1  Ion motion in magnetic and electric fields

Figure B1.7.18(a) shows a typical FT-ICR mass spectrometer cubic trapping cell. The principal axes are shown in the diagram, along with the direction of the imposed magnetic field. To understand the trajectory of an ion in such a field, the electrostatic and magnetic forces acting on the ion must be described [45]. If only a magnetic field is present, the field acts on the ions such that they take up circular orbits with a frequency defined by the ion mass:

$$\omega_c = \frac{zeB}{m}$$

where $ze$ is the charge on the ion, $B$ is the magnetic field strength (in tesla), $m$ is the ion mass and $\omega_c$ is the ion's cyclotron frequency (in rad s$^{-1}$). In modern FT-ICR instruments, magnetic fields of 6 T or more are common, with the latest being upwards of 20 T. These high field magnets are usually superconducting magnets, not unlike modern NMR instruments. Without trapping electrodes, the ions would describe a spiral trajectory and be lost from the trap.

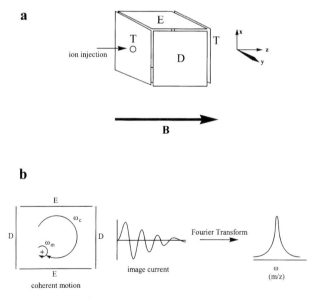

**Figure B1.7.18.** (a) Schematic diagram of the trapping cell in an ion cyclotron resonance mass spectrometer: excitation plates (E); detector plates (D); trapping plates (T). (b) The magnetron motion ($\omega_m$) due to the crossing of the magnetic and electric trapping fields is superimposed on the circular cyclotron motion ($\omega_c$) taken up by the ions in the magnetic field. Excitation of the cyclotron frequency results in an image current being detected by the detector electrodes which can be Fourier transformed into a secular frequency related to the $m/z$ ratio of the trapped ion(s).

The circular orbits described above are perturbed by an electrostatic field applied to the two endcap trapping electrodes (figure B1.7.18(b)). In addition to the trapping motion, the crossed electric and magnetic fields superimpose a magnetron motion, $\omega_m$, on the ions (figure B1.7.18(b)). An idealized trapping cell would produce a DC quadrupolar potential in three dimensions. The resulting ion motion is independent of the axial and radial position in the cell. In practice, however, the finite size of the trapping cell produces irregularities in the potential that affect ion motion.

*(a) Ion trapping*

The component of the DC quadrupolar potential in the $z$-axis direction is described by the following equation.

$$V(z) = \frac{V_T}{2} + \frac{kz^2}{2}$$

where $k$ is a constant and $V_T$ is the trapping potential applied to the endcap electrodes of the trapping cell. The derivative of this relationship yields a linear electric field along the $z$-axis.

$$E(z) = -kz.$$

From this relationship, an expression can be derived for the trapping frequency, $\omega_t$.

$$\omega_t = \left(\frac{kze}{m}\right)^{1/2}$$

which again is a function of ion charge and mass. In theory, this trapping frequency is harmonic and independent of the ion's position in the trapping cell, but in practice, the finite size of the trapping cell produces irregularities in $\omega_t$. The efficiency with which ions are trapped and stored in the FT-ICR cell diminishes as the pressure in the cell increases (as opposed to the quadrupole ion trap, which requires helium buffer gas for optimal trapping). For this reason, FT-ICR instruments are typically operated below $10^{-5}$ Torr (and usually closer to $10^{-8}$ Torr).

*(b) Ion detection*

In the other types of mass spectrometer discussed in this chapter, ions are detected by having them hit a detector such as an electron multiplier. In early ICR instruments, the same approach was taken, but FT-ICR uses a very different technique. If an RF potential is applied to the excitation plates of the trapping cell (figure B1.7.18(b)) equal to the cyclotron frequency of a particular ion $m/z$ ratio, resonant excitation of the ion trajectories takes place (without changing the cyclotron frequency). The result is ion trajectories of higher kinetic energy and larger radii inside the trapping cell. In addition, all of the ions with that particular $m/z$ ratio take up orbits that are coherent (whereas they were all out of phase prior to resonant excitation). This coherent motion induces an image current on the detector plates of the trapping cell that has the same frequency as the cyclotron frequency (figure B1.7.18(b)). This image current is acquired over a period of time as the ion packet decays back to incoherent motion. The digitized time-dependent signal can be Fourier transformed to produce a frequency spectrum with one component, a peak at the cyclotron frequency of the ions. It is possible to resonantly excite the trajectories of all ions in the trapping cell by the application of a broad-band RF excitation pulse to the excitation electrodes. The resulting time-dependent image current, once Fourier transformed, yields a frequency spectrum with peaks due to each ion $m/z$ in the trapping cell, and hence a mass spectrum. The intensities of the peaks are proportional to the concentrations of the ions in the cell.

*B1.7.6.2  Experiments using FT-ICR*

In many respects, the applications of FT-ICR are similar to those of the quadrupole ion trap, as they are both trapping instruments. The major difference is in the ion motion inside the trapping cell and the waveform detection. In recent years there have been attempts to use waveform detection methods with quadrupole ion traps [46].

*(a) Collision-induced dissociation and ion–molecule reactions*

As with the quadrupole ion trap, ions with a particular $m/z$ ratio can be selected and stored in the FT-ICR cell by the resonant ejection of all other ions. Once isolated, the ions can be stored for variable periods of time (even hours) and allowed to react with neutral reagents that are introduced into the trapping cell. In this manner, the products of bi-molecular reactions can be monitored and, if done as a function of trapping time, it is possible to derive rate constants for the reactions [47]. Collision-induced dissociation can also be performed in the FT-ICR cell by the isolation and subsequent excitation of the cyclotron frequency of the ions. The extra translational kinetic energy of the ion packet results in energetic collisions between the ions and background gas in the cell. Since the cell in FT-ICR is nominally held at very low pressures ($10^{-8}$ Torr), CID experiments using the background gas tend not to be very efficient. One common procedure is to pulse a target gas (such as Ar) into the trapping cell and then record the CID mass spectrum once the gas has been pumped away. CID mass spectra obtained in this way are similar to those obtained on triple quadrupole and ion trap instruments.

*(b) Kinetic studies*

Aside from the bimolecular reaction kinetics described above, it is possible to measure other types of kinetics with FT-ICR. Typically, for two species to come together in the gas phase to form a complex, the resulting complex will only be stable if a three body collision occurs. The third body is necessary to lower the internal energy of the complex below its dissociation threshold. Thus, complexes are generally made in high pressure ion sources. It is possible, however, for the complex to radiatively release excess internal energy. Dunbar and others [48, 49] have studied and modelled the kinetics of such 'radiative association' reactions in FT-ICR trapping cells because of the long time scales of the experiments (reaction progress is usually probed for many minutes). The rate constants for photon emission derived from the experimentally observed rate constants tend to be between 10 and 100 s$^{-1}$. It has also been found that ions can be dissociated by the absorption of blackbody radiation in the trapping cell (BIRD—blackbody infrared radiative dissociation) [50]. This technique, which is only feasible at the low pressures ($<10^{-8}$ Torr) and long trapping times inside the FT-ICR cell, allows the investigator to measure unimolecular decay rate constants of the order of $10^{-3}$ s$^{-1}$. Another approach to dissociation kinetics is time-resolved photodissociation [51]. Ions are photodissociated with laser light in the visible and near UV and the product ion intensity is monitored as a function of time. Rate constants from $10^{-3}$ s$^{-1}$ and higher can be measured with good precision using this technique.

*(c) Thermochemical studies: the bracketing method*

In an earlier section, measurements were described in which the equilibrium constant, $K$, for bimolecular reactions involving gas-phase ions and neutral molecules were determined. Another method for determining the proton or other affinity of a molecule is the bracketing method [52]. The principle of this approach is quite straightforward. Let us again take the case of a proton affinity determination as an example. In a reaction between a protonated base, $B_1H^+$ and a neutral molecule, $B_2$, proton transfer from $B_1$ to $B_2$ will presumably occur only if the reaction is exothermic (in other words, if the PA of $B_2$ is greater than that of $B_1$). So, by choosing a range of bases $B_1$ covering a range of PA values and reacting them with a molecule of unknown

PA, $B_2$, the reactions leading to $B_2H^+$ can be monitored by the presence of this latter ion in the mass spectrum. The PA of $B_2$ can quickly be narrowed down provided the reference values are well established. The nature of this experiment requires that a bimolecular reaction takes place between the reference base and unknown and thus these experiments are most commonly carried out in FT-ICR mass spectrometers, though quadrupole ion trap instruments have also been used.

## References

[1] Wannier G H 1953 The threshold law for single ionization of atoms or ions by electrons *Phys. Rev.* **90** 817–25
[2] Harrison A G 1992 *Chemical Ionization Mass Spectrometry* (Boca Raton, FL: Chemical Rubber Company)
[3] Barber N, Bordoli R S, Elliot G J, Sedgwick R D and Tyler A N 1982 Fast atom bombardment mass spectrometry *Anal. Chem.* **54** 645A–57A
[4] Karas M and Hillenkamp F 1988 Laser desorption ionization of proteins with molecular masses exceeding 10 000 Daltons *Anal. Chem.* **60** 2299–301
[5] Gaskell S J 1997 Electrospray: principles and practice *J. Mass Spectrom.* **32** 677–88
[6] DePaul S, Pullman D and Friedrich B 1993 A pocket model of seeded molecular beams *J. Phys. Chem.* **97** 2167–71
[7] Miller D R 1988 Free jet sources *Atomic and Molecular Beam Methods* ed G Scoles (New York: Oxford University Press)
[8] Thomson J J 1913 *Rays of Positive Electricity* (London: Longmans Green)
[9] Cooks R G, Beynon J H, Capriolo R M and Lester G R 1973 *Metastable Ions* (Amsterdam: Elsevier)
[10] Busch K L, Glish G L and McLuckey S A 1988 *Mass Spectrometry/Mass Spectrometry* (New York: VCH)
[11] Cooks R G (ed) 1978 *Collision Spectroscopy* (New York: Plenum)
[12] Hamdan M and Brenton A G 1991 High-resolution translational energy spectroscopy of molecular ions *Physics of Ion impact Phenomena* ed D Mathur (Berlin: Springer)
[13] Dawson P H 1976 *Quadrupole Mass Spectrometry and its Applications* (Amsterdam: Elsevier)
[14] March R E and Hughes R J 1989 *Quadrupole Storage Mass Spectrometry* (New York: Wiley–Interscience)
[15] March R E and Todd J F J 1995 *Practical Aspects of Ion Trap Mass Spectrometry* (Boca Raton, FL: Chemical Rubber Company)
[16] Mathieu E 1986 *J. Math. Pure Appl.* **13** 137
[17] Usypchuk L L, Harrison A G and Wang J 1992 Reactive collisions in quadrupole cells. Part I. Reaction of $[CH_3NH_2]^+$ with the isomeric butenes and pentenes *Org. Mass Spectrom.* **27** 777–82
[18] Kebarle P 1988 Pulsed electron high pressure mass spectrometer *Techniques for the Study of Ion-Molecule Reactions* ed J M Farrar and W H Saunders (New York: Wiley–Interscience)
[19] Szulejko J E and McMahon T B 1991 A pulsed electron beam, variable temperature, high pressure mass spectrometric re-evaluation of the proton affinity difference between 2-methylpropene and ammonia *Int. J. Mass Spectrom. Ion Proc.* **109** 279–94
[20] Szulejko J E and McMahon T B 1993 Progress toward an absolute gas-phase proton affinity scale *J. Am. Chem. Soc.* **115** 7839–48
[21] Berkowitz J, Ellison G B and Gutman D 1994 Three methods to measure RH bond energies *J. Phys. Chem.* **98** 2744–65
[22] Davidson W R, Sunner J and Kebarle P 1979 Hydrogen bonding of water to onium ions. Hydration of substituted pyridinium ions and related systems *J. Am. Chem. Soc.* **101** 1675–80
[23] Meot-Ner M 1984 Ionic hydrogen bond and ion solvation 2. Solvation of onium ions by 1–7 water molecules. Relations between monomolecular, specific and bulk hydration *J. Am. Chem. Soc.* **106** 1265–72
[24] Smith D and Adams N G 1988 The selected ion flow tube (SIFT): studies of ion–neutral reactions *Advances in Atomic and Molecular Physics* ed D Bates and B Bederson (Boston, MA: Academic)
[25] Ferguson E E, Fehsenfeld F C and Schmeltekopf A L 1969 *Adv. At. Mol. Phys.* **5** 1
[26] Ferguson E E, Fehsenfeld F C and Albritton D L 1979 Ion chemistry of the earth's atmosphere *Gas Phase Ion Chemistry* vol 1, ed M T Bowers (New York: Academic)
[27] Squires R R 1997 Atmospheric chemistry and the flowing afterglow technique *J. Mass Spectrum.* **32** 1271–72
[28] Spanel P and Smith D 1997 SIFT studies of the reactions of $H_3O^+$, $NO^-$ and $O_2^+$ with a series of alcohols *Int. J. Mass Spectrom. Ion Proc.* **167/168** 375–88
[29] Spanel P, Ji Y and Smith D 1997 SIFT study of the reactions of $H_3O^+$, $NO^-$ and $O_2^+$ with a series of aldehydes and ketones *Int. J. Mass Spectrom. Ion Proc.* **165/166** 25–37
[30] Spanel P and Smith D 1998 SIFT studies of the reactions of $H_3O^+$, $NO^-$ and $O_2^+$ with a series of volatile carboxylic acids and esters *Int. J. Mass Spectrom. Ion Proc.* **172** 137–47
[31] Ervin K M and Armentrout P B 1985 Translational energy dependence of $Ar^+ + XY \rightarrow ArX^+ + Y$ ($XY = D_2$, $D_2$, HD) from thermal to 30 eV c.m. *J. Chem. Phys.* **83** 166–89
[32] Rodgers M T, Ervin K M and Armentrout P B 1997 Statistical modeling of CID thresholds *J. Chem. Phys.* **106** 4499–508
[33] Asano K, Goeringer D and McLuckey S 1998 Dissociation kinetics in the quadrupole ion trap *Proc. 46th Conf. Am. Soc. Mass Spectrom.*
[34] Brodbelt J, Liou C-C and Donovan T 1991 Selective adduct formation by dimethyl ether chemical ionization is a quadrupole ion trap mass spectrometer and a conventional ion source *Anal. Chem.* **63** 1205–9
[35] Gronert S 1998 Estimation of effective ion temperatures in a quadrupole ion trap *J. Am. Soc. Mass Spectrom.* **9** 845–8

[36] Guilhaus M 1995 Principles and instrumentation in time-of-flight mass spectrometry: physical and instrumental concepts *J. Mass Spectrom.* **30** 1519–32

[37] Wiley W C and McLaren I H 1955 Time-of-flight mass spectrometer with improved resolution *Rev. Sci. Instrum.* **26** 1150–7

[38] Nesselrodt D R, Potts A R and Baer T 1995 Stereochemical analysis of methyl-substituted cyclohexanes using 2 + 1 resonance enhanced multiphoton ionization *Anal. Chem.* **67** 4322–9

[39] Muller-Dethlefs K and Schlag E W 1991 High resolution zero kinetic energy (ZEKE) photoelectron spectroscopy of molecular systems *Annu. Rev. Phys. Chem.* **42** 109–36

[40] Baer T 1986 *Adv. Chem. Phys.* **64** 111

[41] McIver R T 1970 A trapped ion analyzer cell for ion cyclotron resonance spectroscopy *Rev. Sci. Instrum.* **41** 555–8

[42] Vartanian V H, Anderson J S and Laude D A 1995 Advances in trapped ion cells for Fourier transform ion cyclotron resonance mass spectrometry *Mass Spec. Rev.* **41** 1–19

[43] Comisarow M B and Marshall A G 1996 Early development of Fourier transform ion cyclotron resonance (FT-ICR) spectroscopy *J. Mass Spectrom.* **31** 581–5

[44] Amster I J 1996 Fourier transform mass spectrometry *J. Mass Spectrom.* **31** 1325–37

[45] Freiser B S 1988 Fourier transform mass spectrometry *Techniques for the Study of Ion–Molecule Reactions* ed J M Farrar and W H Saunders (New York: Wiley–Interscience)

[46] Soni M, Frankevich V, Nappi M, Santini R E, Amy J W and Cooks R G 1996 Broad-band Fourier transform quadrupole ion trap mass spectrometry *Anal. Chem.* **68** 3341–20

[47] Grover R, Decouzon M, Maria P-C and Gal J-F 1996 Reliability of Fourier transform-ion cyclotron resonance determinations of rate constants for ion/molecule reactions *Eur. Mass Spectrom.* **2** 213–23

[48] Cheng Y-W and Dunbar R C 1995 Radiative association kinetics of methyl-substituted benzene ions *J. Phys. Chem.* **99** 10 802–7

[49] Fisher J J and McMahon T B 1990 Determination of rate constants for low pressure association reactions by Fourier transform-ion cyclotron resonance *Int. J. Mass Spectrom. Ion. Proc* **100** 707–17

[50] Dunbar R C and McMahon T B 1998 Activation of unimolecular reactions by ambient blackbody radiation *Science* **279** 194–7

[51] Lin C Y and Dunbar R C 1994 Time-resolved photodissociation rates and kinetic modeling for unimolecular dissociations of iodotoluene ions *J. Phys. Chem.* **98** 1369–75

[52] Born M, Ingemann S and Nobbering N M M 1994 Heats of formation of mono-halogen substituted carbenes. Stability and reactivity of CHX$^-$ (X = F, Cl, Br and I) radical anions *J. Am. Chem. Soc.* **116** 7210–17

## Further Reading

Cooks R G, Benynon J H, Capriolo R M and Lester G R 1973 *Metastable Ions* (Amsterdam: Elsevier)

This is the seminal book on metastable ions, their chemistry and experimental observation. It is a must for anyone starting out in gas-phase ion chemistry.

Busch K L, Glish G L and McLuckey S A 1988 *Mass Spectrometry/Mass Spectrometry* (New York: VCH)

This is one of the newer books covering tandem spectrometry and is a useful resource for the beginner and experienced mass spectrometrist.

Cooks R G (ed) 1978 *Collision Spectroscopy* (New York: Plenum)

This volume deals with the various physical methods for studying collisions between projectile and target species. Theories of collisional scattering, energy transfer and reactive interactions are presented.

Dawson P H 1967 *Quadrupole Mass Spectrometry and its Applications* (Amsterdam: Elsevier)

This is the standard reference volume for the theory and application of quadrupole mass spectrometry.

March R E and Rodd J F J 1995 *Practical Aspects of Ion Trap Mass Spectrometry* (Boca Raton, FL: Chemical Rubber Company)

This is a three volume set covering the physics and chemistry for quadrupole ion traps. It is a must for anyone using traps as part of their research.

Farrar J M and Saunders W H (eds) 1988 *Techniques for the Study of Ion–Molecule Reactions* (New York: Wiley–Interscience)

This volume contains excellent discussions of the various methods for studying ion–molecule reactions in the gas phase, including high pressure mass spectrometry, ion cyclotron resonance spectroscopy (and FT-ICR) and selected ion flow tube mass spectrometry.

# B1.8
# Diffraction: x-ray, neutron and electron

*Edward Prince*

## B1.8.1   Introduction

Diffraction is the deflection of beams of radiation due to interference of waves that interact with objects whose size is of the same order of magnitude as the wavelengths. Molecules and solids typically have interatomic distances in the neighbourhood of a few Ångströms (1 Å $= 10^{-10}$ m), comparable to the wavelengths of x-rays with energies of the order of 10 keV. Neutrons and electrons also have wave properties, with wavelengths given by the de Broglie relation, $\lambda = h/mv$, where $h$ is Planck's constant, $m$ is the mass of the particle and $v$ is its velocity. Neutron diffraction applications use neutrons with wavelengths in the range from 1 to 10 Å; most electron diffraction applications use wavelengths of the order of 0.05 Å, although low energy electron diffraction (LEED), used for studies of surfaces, employs electrons with wavelengths in the neighbourhood of 1 Å. All three techniques have extensive applications in physics, chemistry, materials science, mineralogy and molecular biology. Although perhaps the most familiar application is determination of the structures of crystalline solids, there are also applications to structural studies of amorphous solids, liquids and gases. Diffraction also plays an important role in imaging techniques such as electron microscopy.

## B1.8.2   Principles of diffraction

### B1.8.2.1   *The atomic scattering factor*

We shall first discuss the diffraction of x-rays from isolated atoms, because this case illustrates principles that can be generalized to most practical applications. Consider [1] an atom consisting of a nucleus surrounded by a spherically symmetric cloud of electrons that can be represented by a density function $\rho(r)$ and a plane wave that can be described by a vector $s_i$ normal to the wave front with magnitude $1/\lambda$, where $\lambda$ is the wavelength. According to Huygens's principle, each point within the electron cloud is the source of a spherical wavelet whose amplitude is proportional to $\rho(r)$. At a distance large compared with the dimensions of the atom this spherical wavelet will approximate a new plane wave and we are interested in the amplitude of the wave formed by the interference of the wavelets originating at all points within the electron cloud. Referring to figure B1.8.1, the difference in pathlength for a wave propagating in the direction of $s_f$ ($|s_f| = |s_i|$) and originating at point $r$, relative to the wave originating at the origin, is $\Delta l = \lambda r \cdot (s_f - s_i)$, and the difference in phase is therefore $\Delta \phi = 2\pi i r \cdot (s_f - s_i)$. The *atomic scattering factor*, $f(s_f)$, is the amplitude of the resultant wave in the direction parallel to $s_f$, which is the vector sum of all contributions. This is given by

$$f(s_f) = C \int \rho(r) \exp[2\pi i r \cdot (s_f - s_i)] d\tau \qquad (B1.8.1)$$

where the integral is over the volume of the atom and $C$ is a proportionality constant. $f(s_f)$ is therefore proportional to the Fourier transform of the electron density distribution, $\rho(r)$.

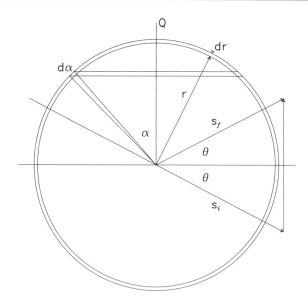

**Figure B1.8.1.** The atomic scattering factor from a spherically symmetric atom. The volume element is a ring subtending angle $\alpha$ with width $d\alpha$ at radius $r$ and thickness $dr$.

To evaluate this integral, first let $\boldsymbol{Q} = 2\pi(\boldsymbol{s}_f - \boldsymbol{s}_i)$, let $Q = |\boldsymbol{Q}|$, let $r = |\boldsymbol{r}|$ and let $\alpha$ be the angle between $\boldsymbol{r}$ and $\boldsymbol{Q}$. Now $|\boldsymbol{s}_f - \boldsymbol{s}_i| = 2|\boldsymbol{s}_i| \sin\theta = 2\sin\theta/\lambda$, so that

$$\boldsymbol{r} \cdot (\boldsymbol{s}_f - \boldsymbol{s}_i) = 2r|\boldsymbol{s}_i| \sin\theta \cos\alpha = 2r \sin\theta \cos\alpha/\lambda.$$

The area of a ring around $\boldsymbol{Q}$ with width $d\alpha$ at radius $r$ is $2\pi r^2 \sin\alpha \, d\alpha$, so

$$d\tau = 2\pi r^2 \sin\alpha \, d\alpha \, dr.$$

Because $\rho(\boldsymbol{r})$ is spherically symmetric, the number of electrons in this volume element is $\rho(r) \, d\tau$. Letting $x = Qr \cos\alpha$, $d\alpha = -dx/(Qr \sin\alpha)$. Then, making all substitutions,

$$f(Q) = C \int_0^\infty \frac{2\pi r^2}{Qr} \rho(r) \, dr \int_{-Qr}^{Qr} \exp(\mathrm{i}x) \, dx \tag{B1.8.2a}$$

$$= 4\pi C \int_0^\infty r^2 \rho(r) \frac{\sin(Qr)}{Qr} \, dr. \tag{B1.8.2b}$$

If $\theta = 0$, so that $Q = 0$, this reduces to

$$f(0) = 4\pi C \int_0^\infty r^2 \rho(r) \, dr \tag{B1.8.3}$$

so that the integral is the total charge in the electron cloud. The constant $C$ has the units of a length, and is conventionally chosen so that $f(Q)$ is a multiple of the 'classical electron radius', $2.818 \times 10^{-15}$ m.

The electron distribution, $\rho(r)$, has been computed by quantum mechanics for all neutral atoms and many ions and the values of $f(Q)$, as well as coefficients for a useful empirical approximation, are tabulated in the *International Tables for Crystallography* vol C [2]. In general, $f(Q)$ is a maximum equal to the nuclear charge, Z, for $Q = 0$ and decreases monotonically with increasing $Q$.

Because the neutron has a magnetic moment, it has a similar interaction with the clouds of unpaired d or f electrons in magnetic ions and this interaction is important in studies of magnetic materials. The magnetic analogue of the atomic scattering factor is also tabulated in the *International Tables* [3]. Neutrons also have direct interactions with atomic nuclei, whose mass is concentrated in a volume whose radius is of the order of $10^{-5}$ times the characteristic neutron wavelength. Thus $\rho(r)$ differs from zero only when $\sin(Qr)/Qr$ is effectively equal to one, so that $f(Q)$ is a constant independent of $Q$. Whereas the x-ray interaction depends on the total number of electrons in the cloud, and therefore on the nuclear charge, the neutron's interaction with a nucleus results from nuclear forces that vary in a haphazard manner from one isotope to another, lie within a rather narrow range and can even be negative, meaning that the Huygens wavelet from such a nucleus has a phase differing by $\pi$ from the phase of one from a nucleus whose scattering factor is positive. The neutron scattering factors, or *scattering lengths*, conventionally denoted by $b$, have magnitudes in the range $1$–$10 \times 10^{-15}$ m [4].

The atomic scattering factor for electrons is somewhat more complicated. It is again a Fourier transform of a density of scattering matter, but, because the electron is a charged particle, it interacts with the nucleus as well as with the electron cloud. Thus $\rho(r)$ in equation (B1.8.2b) is replaced by $\varphi(r)$, the electrostatic potential of an electron situated at radius $r$ from the nucleus. Under a range of conditions the electron scattering factor, $f_e(Q)$, can be represented in terms of the x-ray atomic scattering factor by the so-called Mott–Bethe formula,

$$f_e(Q) = 2\pi \frac{me^2}{h^2 \varepsilon_0} [Z - f_x(Q)]/Q^2 \qquad \text{(B1.8.4)}$$

where $m$ is the mass of the electron, $e$ is its charge and $\varepsilon_0$ is the permittivity of free space.

### B1.8.2.2  *Diffraction from clusters of atoms*

The derivation of equation (B1.8.1) makes no use of the assumption of spherical symmetry and it is, in fact, a very general result that the amplitude of a scattered wave is the Fourier transform of a density of scattering matter. Although there are examples of experimental observations of scattering from isolated atoms, one being the scattering of neutrons by a dilute solid solution of paramagnetic atoms in a diamagnetic matrix, diffraction from molecules, such as that of electrons in a gas, or from particles of contrasting density in a uniform medium are much more important. Examples of the latter are colloidal suspensions in a fluid and precipitates in an alloy. Note that the particles of contrasting density can also be voids: Babinet's principle requires that the amplitude of a scattered wave due to a negative difference be the complex conjugate of that due to a positive difference and, because the intensity of scattered radiation is proportional to the square of the modulus of the amplitude, the diffraction patterns are indistinguishable.

If individual scattering particles are far enough apart and their spatial distribution is such that the relative phases of their contributions to a scattered wave are random, the intensity distribution in the diffraction pattern will be the sum of contributions from all particles [5]. If the particles are identical (monodisperse) but have random orientations, or if they differ in size and shape (polydisperse), the resulting pattern will reflect an ensemble average over the sample. In either case there will be a spreading of the incident beam, so-called *small-angle scattering*. How small the angles are depends on the wavelength of the radiation and the size of the particles: long wavelengths give larger angles, but they also tend to be more strongly absorbed by the sample, so that there is a trade-off between resolution and intensity.

### B1.8.2.3  *Diffraction from crystalline solids*

#### (a) Bragg's law

The diffraction of x-rays was first observed in 1912 by Laue and coworkers [6]. A plausible, though un-documented, story says that the classic experiment was inspired by a seminar given by P P Ewald, whose

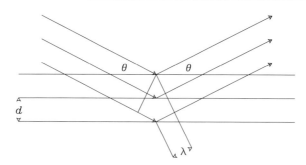

**Figure B1.8.2.** Bragg's law. When $\lambda = 2d \sin \theta$, there is strong, constructive interference.

doctoral thesis was a purely theoretical study of the interaction of electromagnetic waves with an array of dipoles located at the nodes of a three-dimensional lattice. At the time it was hypothesized that crystals were composed of parallelepipedal building blocks, *unit cells*, fitted together in three dimensions and that x-rays were short-wavelength, electromagnetic radiation, but neither hypothesis had been confirmed experimentally. The Laue experiment confirmed both, but the application of x-ray diffraction to the determination of crystal structure was introduced by the Braggs.

W L Bragg [7] observed that if a crystal was composed of copies of identical unit cells, it could then be divided in many ways into slabs with parallel, plane faces whose distributions of scattering matter were identical and that if the pathlengths travelled by waves reflected from successive, parallel planes differed by integral multiples of the wavelength there would be strong, constructive interference. Figure B1.8.2 shows a projection of parallel planes separated by a distance $d$ and a plane wave with wavelength $\lambda$ whose normal makes an angle $\theta$ with these reflecting planes. It is evident that the crests of waves reflected from successive planes will be in phase if $\lambda = 2d \sin \theta$, a relation that is known as *Bragg's law*. (It appears that W H Bragg played no role in the formulation of this relation, so it is correctly Bragg's law, not Braggs' law. In textbooks the relation is often stated in the form $n\lambda = 2d \sin \theta$, where $n$ is the 'order' of the reflection, but in crystallography the order is conventionally incorporated in the definition of $d$.)

*(b) The reciprocal lattice*

The vertices of the unit cells form an array of points in three-dimensional space, a *space lattice*. The edges of the parallelepiped can be defined by three non-coplanar vectors, $a$, $b$ and $c$, and then any lattice point can then be defined by a vector $r = ua + vb + wc$, where $u$, $v$ and $w$ are integers. Any point in the crystal can be specified by a vector $r + x$, where $x$ represents a vector within the unit cell. The periodicity of the crystal specifies that $\rho(r + x) = \rho(x)$. The families of parallel planes are specified by their *Miller indices*, conventionally denoted by $h$, $k$ and $l$. The three points that define the plane closest to the origin are $a/h$, $b/k$ and $c/l$, with the understanding that if any of the indices is equal to zero, the plane is parallel to the corresponding vector. To find solutions to the Bragg equation it is necessary to determine the value of $d$. If $a$, $b$ and $c$ are orthogonal, this is easy, but for most crystals they are not, and the computation is greatly simplified by use of the *reciprocal lattice*. The reciprocal lattice, which was introduced by J W Gibbs [8] and applied to the crystallographic problem by Ewald [9], is defined by three vectors, $a^* = (b \times c)/V$, $b^* = (c \times a)/V$ and $c^* = (a \times b)/V$, where $V = a \cdot b \times c$ is the volume of the unit cell. It can easily be shown [10] that the vector $d^* = ha^* + kb^* + lc^*$ is perpendicular to the planes defined by the Miller indices $h$, $k$ and $l$, and that $|d^*| = 1/d$. Bragg's law then becomes $\sin \theta = \lambda |d^*|/2$. Note that if $d^* = s_f - s_i$, as shown in figure B1.8.3, this condition will be satisfied on the surface of a sphere (the *Ewald sphere*) passing through the origin of the reciprocal lattice whose centre is at the point $-s_i$ in reciprocal space.

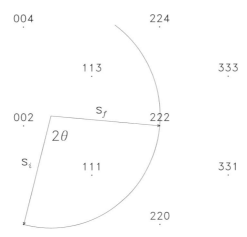

**Figure B1.8.3.** Ewald's reciprocal lattice construction for the solution of the Bragg equation. If $s_f - s_i$ is a vector of the reciprocal lattice, Bragg's law is satisfied for the corresponding planes. This occurs if a reciprocal lattice point lies on the surface of a sphere with radius $1/\lambda$ whose centre is at $-s_i$.

*(c) The structure amplitude*

The amplitude and therefore the intensity, of the scattered radiation is determined by extending the Fourier transform of equation (B1.8.1) over the entire crystal and Bragg's law expresses the fact that this transform has values significantly different from zero only at the nodes of the reciprocal lattice. The amplitude varies, however, from node to node, depending on the transform of the contents of the unit cell. This leads to an expression for the *structure amplitude*, denoted by $F(hkl)$, of the form

$$F(hkl) = C \int_0^c dz \int_0^b dy \int_0^a \rho(x, y, z) \exp\left[2\pi i\left(h\frac{x}{a} + k\frac{y}{b} + l\frac{z}{c}\right)\right] dx \qquad \text{(B1.8.5)}$$

where $a = |a|, b = |b|, c = |c|$ and $x$, $y$ and $z$ are the coordinates of a point in a (not necessarily orthogonal) Cartesian system defined by $a$, $b$ and $c$. Making use of the fact that the unit cell contents consist of atoms, each of which has its own atomic scattering factor, $f_j(d^*)$, this can be written

$$F(hkl) = C \sum_{j=1}^N f_j(d^*) \exp\left[2\pi i\left(h\frac{x}{a} + k\frac{y}{b} + l\frac{z}{c}\right)\right] \qquad \text{(B1.8.6)}$$

where $d^* = |d^*|$ and the sum is over $N$ atoms in the unit cell.

Equation (B1.8.6) assumes that all unit cells really *are* identical and that the atoms are fixed in their equilibrium positions. In real crystals at finite temperatures, however, atoms oscillate about their mean positions and also may be displaced from their average positions because of, for example, chemical inhomogeneity. The effect of this is, to a first approximation, to modify the atomic scattering factor by a convolution of $\rho(r)$ with a trivariate Gaussian density function, resulting in the multiplication of $f_j(d^*)$ by $\exp(-M)$, where

$$M = 8\pi^2 \overline{u}_j^2 \sin^2 \theta / \lambda^2 \qquad \text{(B1.8.7)}$$

and $\overline{u}_j^2$ is the mean square displacement of the centre of atom $j$ parallel to $d^*$. The factor $\exp(-M)$ is the *atomic displacement factor*, or, in older literature, the *temperature factor* or, for early workers in the field, the *Debye–Waller factor*.

*(d) Diffraction of neutrons from nonmagnetic and magnetic crystals*

Diffraction of neutrons [11] from nonmagnetic crystals is similar to that of x-rays, with the neutron atomic scattering factor substituted for the x-ray one. For magnetic crystals below their ordering temperatures, however, the neutron's magnetic moment interacts with an ordered array of electron spins, with the strength of the interaction being proportional to $\sin \alpha$, where $\alpha$ is the angle between $d^*$ and the electron spin axis, and the phases of the wavelets originating at different atoms depend on the relative orientations of their magnetic moments as well as on the path length. The electron spins may all point in the same direction, a *ferromagnet*, or those on different atoms may point in opposite directions, in equal numbers and in an ordered arrangement, an *antiferromagnet*, or they may be arranged in more complicated ways having a net magnetic moment, a so-called *ferrimagnet*. In the absence of an applied magnetic field the crystal tends to divide into domains in which the electron spins point along different, symmetry-equivalent directions and the diffracted intensities are averaged over the various possible values of the angle between the magnetic moment and $d^*$. Furthermore, the magnetic diffraction and the nuclear diffraction do not interfere with one another, and the nuclear and magnetic intensities simply add together, although in many cases the magnetic unit cell is larger than the nuclear unit cell, which produces additional diffraction peaks.

If a magnetic field is applied to the crystal, the domains become aligned and the nuclear and magnetic wavelets *do* interfere with one another. Then the amplitude of the diffracted wave depends on the orientation of the neutron spin. In special cases, $Co_{0.92}Fe_{0.08}$ is an example, the interference may be totally destructive for one neutron spin state and the diffracted beam becomes polarized. If a crystal of one of these materials is used as a monochromator, the diffraction of this polarized beam is a particularly sensitive probe for the study of magnetic structures.

*(e) Diffraction of electrons from crystals*

Diffraction of electrons from single crystals [12] differs from the diffraction of x-rays or neutrons because the interaction of electrons with matter is much stronger. Conventional electron diffraction is performed as an adjunct to electron microscopy. In fact, the same instrument, a transmission electron microscope, commonly serves for both, with the configuration of the electron optics determining whether a diffraction pattern is magnified, or the diffracted beams are recombined to form an image. Because of the strong interaction, specimens must be thin, and accelerating voltages must be large, of the order of 100 keV, so that the corresponding wavelength is much shorter than interatomic distances in a crystal, and the Ewald sphere is large compared with the spacing between points of the reciprocal lattice. As a result, when the direction of the incident beam is perpendicular to a reciprocal lattice plane, the small spread of the reciprocal lattice due to mosaic spread in the crystal produces a diffraction pattern that consists of spots with the structure of the reciprocal lattice plane.

Another mode of electron diffraction, low energy electron diffraction or LEED [13], uses incident beams of electrons with energies below about 100 eV, with corresponding wavelengths of the order of 1 Å. Because of the very strong interactions between the incident electrons and the atoms in the crystal, there is very little penetration of the electron waves into the crystal, so that the diffraction pattern is determined entirely by the arrangement of atoms close to the surface. Thus, in contrast to high energy diffraction, where the pattern is formed by transmission through a thin, crystalline film, and the diffracted beams make angles of only a fraction of a degree with the incident beam, the pattern in LEED is formed by reflection from a surface, and the diffracted beam may be in any direction away from the surface. Furthermore, because there is no significant interference with scattered wavelets coming from below the surface, the reciprocal lattice can be considered to consist not of points but of lines perpendicular to the surface that will always intersect the Ewald sphere, so that even with a monochromatic incident beam there will always be a pattern of spots on a photographic plate or a fluorescent screen.

Although the structure of the surface that produces the diffraction pattern must be periodic in two dimensions, it need not be the same substance as the bulk material. Thus LEED is a particularly sensitive tool for studying the structures and properties of thin layers adsorbed epitaxially on the surfaces of crystals.

### B1.8.2.4    Diffraction from noncrystalline solids

We have seen that the intensities of diffraction of x-rays or neutrons are proportional to the squared moduli of the Fourier transform of the scattering density of the diffracting object. This corresponds to the Fourier transform of a convolution, $P(s)$, of the form

$$P(s) = \int \rho(r)\rho(r+s)\, dr. \tag{B1.8.8}$$

The integrand in this expression will have a large value at a point $r$ if $\rho(r)$ and $\rho(r+s)$ are both large, and $P(s)$ will be large if this condition is satisfied systematically over all space. It is therefore a self- or autocorrelation function of $\rho(r)$. If $\rho(r)$ is periodic, as in a crystal, $P(s)$ will also be periodic, with a large peak when $s$ is a vector of the lattice and also will have a peak when $s$ is a vector between any two atomic positions. The function $P(s)$ is known as the *Patterson function*, after A L Patterson [14], who introduced its application to the problem of crystal structure determination.

### (a) Diffraction from glasses

There are two classes of solids that are not crystalline, that is, $\rho(r)$ is not periodic. The more familiar one is a glass, for which there are again two models, which may be called the random network and the random packing of hard spheres. An example of the first is silica glass or fused quartz. It consists of tetrahedral $SiO_4$ groups that are linked at their vertices by Si–O–Si bonds, but, unlike the various crystalline phases of $SiO_2$, there is no systematic relation between the orientations of neighbouring tetrahedra. In the random packing of spheres there is no regular arrangement of atoms even at short range and the coordination of any particular atom may have a wide variety of configurations. The two types of glass have similar diffraction properties, so we do not need to discuss them separately.

If the material is not periodic (but is isotropic), the integral in equation (B1.8.8) becomes spherically symmetric, and reduces for large values of $s$ ($=|s|$) to a constant equal to the average value of $\rho(r)^2$. In either the sphere-packing model or the random-network model, however, there is always a shortest interatomic distance and $\rho(r)$ falls to a small value between the atoms. The integrand will then have, on average, small values when $s$ is equal to an atomic radius and large values when $s$ is equal to a typical interatomic distance. The integrand and therefore $P(s)$, will have smaller ripples as $s$ increases through additional coordination shells. Because the diffracted intensity is proportional to the Fourier transform of $P(s)$, it will also have broad maxima and minima as $\sin\theta/\lambda$ increases.

### (b) Quasicrystals

The other type of noncrystalline solid was discovered in the 1980s in certain rapidly cooled alloy systems. D Shechtman and coworkers [15] observed electron diffraction patterns with sharp spots with fivefold rotational symmetry, a symmetry that had been, until that time, assumed to be impossible. It is easy to show that it is impossible to fill two- or three-dimensional space with identical objects that have rotational symmetries of orders other than two, three, four or six, and it had been assumed that the long-range periodicity necessary to produce a diffraction pattern with sharp spots could only exist in materials made by the stacking of identical unit cells. The materials that produced these diffraction patterns, but clearly could not be crystals, became known as *quasicrystals*.

Although details of quasicrystal structure remain uncertain, the circumstances under which diffraction patterns with 'impossible' symmetries can occur have become clear [16]. It is impossible to construct an object that has long-range periodicity using identical units with these symmetries, but it is not necessary for the object itself to have that symmetry. It is only necessary that its Patterson function be symmetric. The electron diffraction patterns observed by Shechtman actually have the symmetry of a regular icosahedron, and it is possible to build a structure with this symmetry using two rhombohedra, each having faces whose acute angle corners have an angle, $\alpha$, equal to $2\arctan[2/(1+\sqrt{5})] = 63.435°$. One of them has three acute angle corners meeting at a vertex, making a prolate rhombohedron, while the other has three obtuse angle corners meeting at a vertex, making an oblate rhombohedron. Large objects made from these two rhombohedra contain vectors parallel to all of the fivefold axes of the regular icosahedron, although different subsets of them appear in different, finite regions. More importantly, although there is no long range periodicity, departures from periodicity are bounded, which produces, as in a crystal, families of parallel planes with alternately higher and lower density. This in turn produces the observed, sharp diffraction spots.

## B1.8.3  Structure determination

We have thus far discussed the diffraction patterns produced by x-rays, neutrons and electrons incident on materials of various kinds. The experimentally interesting problem is, of course, the inverse one: given an observed diffraction pattern, what can we infer about the structure of the object that produced it? Diffraction patterns depend on the Fourier transform of a density distribution, but computing the inverse Fourier transform in order to determine the density distribution is difficult for two reasons. First, as can be seen from equation (B1.8.1), the Fourier transform is defined for all values of $s_f$, but it can be measured only for values of $|s_f - s_i|$ less than $2/\lambda$. For practical reasons $\lambda$ cannot be arbitrarily small, so that the Fourier transform can never be measured over its entire range. Second, the value of the Fourier transform is in general a complex number. Denoting $-h$, $-k$ and $-l$ by $\bar{h}$, $\bar{k}$ and $\bar{l}$, respectively, equation (B1.8.6) shows that, for a crystal, $F(\overline{hkl}) = F(hkl)^*$. Because the intensity is proportional to $|F(hkl)|^2 = F(hkl)F(hkl)^*$, it will be the same independent of the phase of $F(hkl)$. As a result, the structural information that can be determined is either restricted to averaged properties that do not depend on phase information, or the phase must be determined by methods other than the simple measurement of diffraction intensities.

### B1.8.3.1  Small-angle scattering

Materials have many properties that are important, scientifically and technologically, that do not depend on the details of long-range structure. For example, consider a solution of globular macromolecules in a solvent of contrasting scattering density. If the solution is not too highly concentrated, so that intermolecular interactions can be neglected, the diffraction pattern will be the sum of the diffraction patterns of all individual molecules. Under these conditions all diffracted radiation makes a small angle with the incident beam. Although all molecules are identical, they can have all possible orientations relative to the incident beam, so the diffraction will be that from a spherically averaged distribution. The intensity of diffraction is proportional to the squared modulus of the Fourier transform of the density distribution, which is the Fourier transform of its Patterson function. An expression for the intensity, $I(Q)$, can be derived by substituting $P(r)$ for $\rho(r)$ in equation (B1.8.2b), giving

$$I(Q) = C \int_0^\infty r^2 P(r) \frac{\sin(Qr)}{Qr} \, dr \qquad (B1.8.9a)$$

where $C$ is a scale factor dependent on the conditions of the experiment. This can be rewritten

$$QI(Q) = C \int_0^\infty r P(r) \sin(Qr) \, dr. \qquad (B1.8.9b)$$

The inverse of this Fourier sine transform is

$$r P(r) = C' \int_0^\infty Q I(Q) \sin(Qr) \, dQ. \tag{B1.8.10a}$$

It is conventional to express the structural information in terms of a *pair distance distribution function*, or PDDF [5], which is defined by $p(r) = r^2 P(r)$. Using this, equation (B1.8.10a) becomes

$$p(r) = C' \int_0^\infty I(Q) Q r \sin(Qr) \, dQ. \tag{B1.8.10b}$$

Equations (B1.8.9a) and (B1.8.10b) both involve integrals whose upper limits are infinite, but this does not present a serious problem, because $P(r)$ is zero for $r$ greater than the largest diameter of the molecule, and $I(Q)$ has a value significantly different from zero only for small angles. The functions $p(r)$ and $I(Q)$ both contain information about the sizes and shapes of the molecules. It is customary to plot the logarithm of $I(Q)$ as a function of $Q$ and such a plot for a spherical molecule has broad maxima with sharp minima between them, while less symmetric molecules produce curves with smaller ripples or a smooth falloff with increasing angle. A useful property of the molecule is the radius of gyration, $R_g$, which is a measure of the distribution of scattering density. This may be determined from the relation

$$R_g^2 = \frac{\int_0^\infty r^2 p(r) \, dr}{\int_0^\infty p(r) \, dr}. \tag{B1.8.11}$$

The volume of a uniform density molecule may be found from the relation

$$V = C'' \frac{\int_0^\infty p(r) \, dr}{\int_0^\infty Q^2 I(Q) \, dQ} \tag{B1.8.12}$$

where $C'' = 8\pi^3$ if the measurements are on an absolute scale. In practice, measurements are relative to a standard sample of known structure.

Although this discussion has been in terms of molecules in solution, the same principles apply to other cases, such as precipitates in an alloy or composites of ceramic particles dispersed in a polymer. The density, $\rho(r)$, is not relative to a vacuum, but is rather relative to a uniform medium. For x-rays this means electron densities, but for neutrons, because the atomic scattering factor is different from one isotope to another, an effect that is very large for the two stable isotopes of hydrogen, there can be wide variations in contrast depending on the isotopic compositions of the different components of the sample. This 'contrast variation' makes small-angle neutron scattering (SANS) a very versatile tool for the study of microstructure.

It has been shown that spherical particles with a distribution of sizes produce diffraction patterns that are indistinguishable from those produced by triaxial ellipsoids. It is therefore possible to assume a shape and determine a size distribution, or to assume a size distribution and determine a shape, but not both simultaneously.

### B1.8.3.2 *Pair distribution functions*

Another application in which useful information can be obtained in the absence of knowledge of the phase of the Fourier transform is the study of glasses and of crystals that contain short-range order but are disordered over long ranges. Here the objective is to determine a pair distribution function (PDF) [17], which is a generalization of the Patterson function that describes the probability of finding pairs of atoms separated by a vector $r_j - r_k$. For various reasons these studies are most easily done with neutrons and most of them have been done with glasses. In a glass the long-range structure may be assumed to be isotropic. Setting

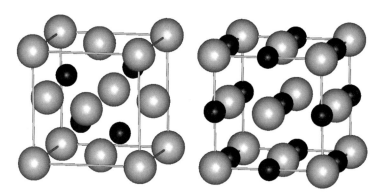

**Figure B1.8.4.** Two of the crystal structures first solved by W L Bragg. On the left is the structure of zincblende, ZnS. Each sulphur atom (large grey spheres) is surrounded by four zinc atoms (small black spheres) at the vertices of a regular tetrahedron, and each zinc atom is surrounded by four sulphur atoms. On the right is the structure of sodium chloride. Each chlorine atom (grey spheres) is surrounded by six sodium atoms (black spheres) at the vertices of a regular octahedron, and each sodium atom is surrounded by six chlorine atoms.

$r = |\boldsymbol{r}_j - \boldsymbol{r}_k|$, the particle density, $\rho(r)$, at distance $r$ from another particle, can be represented to a good approximation by

$$\rho(r) - \rho_0 = C \int_0^\infty Q[I_{obs}(Q) - I_{inc}] \sin(Qr) \, dQ \qquad (B1.8.13)$$

where $\rho_0$ is the mean overall density and $I_{inc}$ is an isotropic incoherent scattering that is the only source of scattering at sufficiently large $Q$.

### B1.8.3.3  Crystal structure determination

#### (a) Trial and error

Laue's original experiment established that x-rays were short-wavelength electromagnetic radiation and that crystals were composed of periodically repeated arrays of identical units, but it did not establish any scale for the wavelength or the sizes of the crystalline units. W L Bragg [18] observed that the positions of the spots in a diffraction photograph produced by zincblende, ZnS, could be explained by a model (see figure B1.8.4) in which the fundamental units were arranged on a face-centred cubic (f.c.c.) lattice. The same model explained the patterns of sodium chloride, potassium bromide and potassium iodide. (Interestingly, Bragg's initial model for potassium chloride was based on what is now called a primitive cubic lattice. This was an artifact resulting from the near identity of the atomic scattering factors of potassium and chlorine.) By observing the relative intensities of the diffraction spots and applying elementary principles of group theory, Bragg proposed models for the arrangements of atoms that turned out to be correct.

At about the same time R A Millikan [19] measured the charge on the electron. Dividing this into the electric charge required to electroplate a gram atomic weight of silver yielded a value, good to within 1%, for Avogadro's number, the number of formula units in a mole. Knowledge of the density and of Avogadro's number leads immediately to knowledge of the size of the unit cell, and thence to knowledge of the wavelength of the radiation producing the diffraction pattern. Bragg's models for the first crystal structures were deduced from Laue photographs, which use continuum radiation, but discovery of the characteristic spectral lines of the elements followed soon after, and H G J Moseley [20] used them to straighten out several anomalies in the periodic table of the elements and to predict the existence of several elements that had not previously been observed.

With the discovery of the x-ray line spectra it became possible to determine, at least relative to the slightly uncertain wavelengths, the sizes and also the symmetries, of the unit cells and the approximate sizes of the atoms. The theory of space groups, which had been worked out by mathematicians, principally A M Schönflies, in the 19th century, has always played a vital role in structural crystallography. With the structures of most of the solid elements and many of their binary compounds it was necessary only to calculate how many atoms would fit into the unit cell and to choose from a limited set of possible positions the ones that best accounted for the relative intensities of the diffraction spots. In sodium chloride, for example, the symmetry of the diffraction pattern shows that the unit cell is a cube and the fact that the indices, $h$, $k$ and $l$, are either all odd or all even shows that the cube is face centred. There is room for only four atoms each of sodium and chlorine and it is observed that those reflections with the indices all even are much stronger than those with the indices all odd. This is consistent with a model that has sodium atoms at the corners and at the centres of the faces of the cube and chlorine atoms at the centres of the edges and at the body centre.

Potassium chloride actually has the same structure as sodium chloride, but, because the atomic scattering factors of potassium and chlorine are almost equal, the reflections with the indices all odd are extremely weak, and could easily have been missed in the early experiments. The zincblende form of zinc sulphide, by contrast, has the same pattern of all odd and all even indices, but the pattern of intensities is different. This pattern is consistent with a model that again has zinc atoms at the corners and the face centres, but the sulphur positions are displaced by a quarter of the body diagonal from the zinc positions.

In all of these structures the atomic positions are fixed by the space group symmetry and it is only necessary to determine which of a small set of choices of positions best fits the data. According to the theory of space groups, all structures composed of identical unit cells repeated in three dimensions must conform to one of 230 groups that are formed by combining one of 14 distinct *Bravais lattices* with other symmetry operations. These include *rotation axes* of orders two, three, four and six and *mirror planes*. They also include *screw axes*, in which a rotation operation is combined with a translation parallel to the rotation axis in such a way that repeated application becomes a translation of the lattice, and *glide planes*, where a mirror reflection is combined with a translation parallel to the plane of half of a lattice translation. Each space group has a *general position* in which the three position coordinates, $x$, $y$ and $z$, are independent, and most also have *special positions*, in which one or more coordinates are either fixed or constrained to be linear functions of other coordinates. The properties of the space groups are tabulated in the *International Tables for Crystallography* vol A [21].

The first crystal structure to be determined that had an adjustable position parameter was that of pyrite, $FeS_2$. In this structure the iron atoms are at the corners and the face centres, but the sulphur atoms are further away than in zincblende along a different threefold symmetry axis for each of the four iron atoms, which makes the unit cell primitive.

Unfortunately for modern crystallographers, all of the crystal structures that could be solved by the choose-the-best-of-a-small-number-of-possibilities procedure had been solved by 1920. Bragg has been quoted as saying that the pyrite structure was 'very complicated', but he wrote, in about 1930, 'It must be realized, however, that (cases having one or two parameters) are still extremely simple. The more typical crystal may have ten, twenty, or forty parameters, to all of which values must be assigned before the analysis of the structure is complete.' This statement is read with amusement by a modern crystallographer, who routinely works with hundreds and frequently with thousands of parameters.

*(b) Patterson methods*

We have seen that the intensities of diffraction are proportional to the Fourier transform of the Patterson function, a self-convolution of the scattering matter and that, for a crystal, the Patterson function is periodic in three dimensions. Because the intensity is a positive, real number, the Patterson function is not dependent on phase and it can be computed directly from the data. The squared structure amplitude is

$$|F(hkl)|^2 = I(hkl)/Lp \tag{B1.8.14}$$

where $I(hkl)$ is the integrated intensity of the $hkl$ reflection, $L$ is the so-called *Lorentz factor*, which depends on the experimental geometry and $p$ is a polarization factor, which is equal to one for nuclear scattering of neutrons and depends on the scattering angle, $2\theta$, for x-rays. From this the Patterson function is

$$P(x, y, z) = \sum_{hkl} |F(hkl)|^2 \cos[2\pi(hx + ky + lz)] \tag{B1.8.15}$$

where the sum is over all values of $h$, $k$ and $l$. In practice $I(hkl)$ can be measured only over a finite range of $h$, $k$ and $l$ and the resulting truncation introduces ripples into the Patterson function.

The Patterson function has peaks corresponding to all interatomic vectors in the density function, the height of a peak being proportional to the product of the atomic scattering factors of the two atoms. Thus, although the Patterson function contains superpositions of the structure as if each atom is in turn placed at the origin and, therefore, has so many peaks that it is difficult to interpret except for very simple structures, there are several features that give important information about the underlying density function. If one or two of the atoms in the unit cell have much higher atomic numbers and, therefore, large values of the atomic scattering factors for x-rays, the peaks in the Patterson function that correspond to vectors between them will stand out from the rest. Peaks corresponding to vectors between the heavy atoms and lighter ones will also be higher than those corresponding to vectors between lighter atoms, which may reveal features of the environment of the heavy atom. If neutron diffraction is used to study crystals that contain atoms with negative scattering factors, especially hydrogen, but also manganese and titanium, the Patterson function will have negative regions corresponding to vectors between the negative scatterers and other atoms.

If the space group contains screw axes or glide planes, the Patterson function can be particularly revealing. Suppose, for example, that parallel to the $c$ axis of the crystal there is a $2_1$ screw axis, one that combines a $180°$ rotation with a translation of $c/2$. Then for an atom at position $(x, y, z)$ there will be another at $(-x, -y, z + \frac{1}{2})$. The section of the Patterson function at $z = \frac{1}{2}$ will therefore contain a peak at position $(2x, 2y)$ for every atom in the unique part of the cell, the *asymmetric unit*. Because this property was first applied to structure determination by D Harker, these special sections of Patterson functions are known as *Harker sections* [22]. If there is a single heavy atom in the asymmetric unit, the Harker section can completely determine the position of the atom. This plays a critical role in the method of *isomorphous replacement*, which we discuss below.

*(c) More trial and error*

With diffraction data alone, in the absence of phase information, it is always possible to put restrictions on the choice of space group and in many cases it is possible to determine the space group uniquely. Careful measurement of the positions of diffraction spots determines the dimensions of the unit cell and assigns it to one of seven symmetry systems, triclinic, monoclinic, orthorhombic, trigonal, tetragonal, hexagonal, and cubic. The 14 Bravais lattices divide into five basic types, designated primitive, single-face centred, all-face centred, body centred and rhombohedral, which can be distinguished by special patterns of observed and unobserved reflections. We have already discussed the all-face centred lattice, in which the indices are either all odd or all even. In a body centred cell the sum of the indices is always even, while in a primitive cell there are no restrictions.

If one or two of the indices are zeros, there may be additional restrictions. We have seen that a $2_1$ screw axis parallel to the $c$ axis of the unit cell produces pairs of atoms at $(x, y, z)$ and $(-x, -y, z + \frac{1}{2})$.

From equation (B1.8.6) we can write

$$F(00l) = C \sum_{j=1}^{N/2} f_j(d^*)\{\exp[2\pi i l z_j] + \exp[2\pi i l(\tfrac{1}{2} + z_j)]\} \tag{B1.8.16a}$$

which can be written

$$F(00l) = C \sum_{j=1}^{N/2} f_j(d^*)[1 + (-1)^l] \exp(2\pi i l z_j). \tag{B1.8.16b}$$

All terms in the sum vanish if $l$ is odd, so $(00l)$ reflections will be observed only if $l$ is even. Similar restrictions apply to classes of reflections with two indices equal to zero for other types of screw axis and to classes with one index equal to zero for glide planes. These *systematic absences*, which are tabulated in the *International Tables for Crystallography* vol A, may be used to identify the space group, or at least limit the choices.

The presence of a $2_1$ screw axis and a glide plane perpendicular to it implies also the existence of a centre of symmetry, so that, for an atom at $(x, y, z)$, there is another one at $(-x, -y, -z)$. Equation (B1.8.6) can then be written

$$F(hkl) = C \sum_{j=1}^{N/2} f_j(d^*)\{\exp[2\pi i(hx_j + ky_j + lz_j)] + \exp[-2\pi i(hx_j + ky_j + lz_j)]\}. \tag{B1.8.17}$$

The two exponential terms are complex conjugates of one another, so that all structure amplitudes must be real and their phases can therefore be only zero or $\pi$. (Nearly 40% of all known structures belong to monoclinic space group $P2_1/c$. The systematic absences of $(0k0)$ reflections when $k$ is odd and of $(h0l)$ reflections when $l$ is odd identify this space group and show that it is centrosymmetric.) Even in the absence of a definitive set of systematic absences it is still possible to infer the (probable) presence of a centre of symmetry. A J C Wilson [23] first observed that the probability distribution of the magnitudes of the structure amplitudes would be different if the amplitudes were constrained to be real from that if they could be complex. Wilson and co-workers established a procedure by which the frequencies of suitably scaled values of $|F|$ could be compared with the theoretical distributions for centrosymmetric and noncentrosymmetric structures. (Note that Wilson named the statistical distributions *centric* and *acentric*. These were not intended to be synonyms for centrosymmetric and noncentrosymmetric, but they have come to be used that way.)

The knowledge that a crystal structure is centrosymmetric reduces the phase problem to one of determining signs, but it is still a formidable one. An extended trial-and-error method uses all available information, including that derived from Patterson methods, numbers of special positions in the unit cell, known interatomic distances, likely group configurations etc, to guess a trial structure and compute from it a set of signs, which are then used to compute a density map, or, more likely, a difference map, in which the Fourier coefficients are the differences between the values of $F(hkl)$ computed from the trial structure and their observed values. Features of the difference map suggest modifications to the trial structure and a new set of signs is used to compute an updated map. With luck, this procedure will converge in a few iterations to a reasonable structure.

### (d) Direct methods

As the number of atoms in the asymmetric unit increases, the solution of a structure by any of these phase-independent methods becomes more difficult, and by 1950 a PhD thesis could be based on a single crystal structure. At about that time, however, several groups observed that the fact that the electron density must be non-negative everywhere could be exploited to place restrictions on possible phases. The first use of this fact was by D Harker and J S Kasper [24], but their relations were special cases of more general relations introduced by J Karle and H Hauptman [25]. Denoting by $h_i$ the set of indices $h_i, k_i, l_i$, the Karle–Hauptman condition states that all matrices of the form

$$
\begin{pmatrix}
F(\mathbf{0}) & F(h_1) & F(h_2) & \cdots & F(h_n) \\
F^*(h_1) & F(\mathbf{0}) & F(h_2 - h_1) & \cdots & F(h_n - h_1) \\
F^*(h_2) & F^*(h_2 - h_1) & F(\mathbf{0}) & \cdots & F(h_n - h_2) \\
\vdots & \vdots & \vdots & \ddots & \vdots \\
F^*(h_n) & F^*(h_n - h_1) & F^*(h_n - h_2) & \cdots & F(\mathbf{0})
\end{pmatrix}
$$

must be positive definite. Defining $U(h_i)$ by $U(h_i) = F(h_i)/F(\mathbf{0})$ and taking a $3 \times 3$ matrix for an example, this condition implies that the determinant

$$
D(h_1, h_2) = \begin{vmatrix}
1 & U(h_1) & U(h_2) \\
U^*(h_1) & 1 & U(h_2 - h_1) \\
U^*(h_2) & U^*(h_2 - h_1) & 1
\end{vmatrix} \geq 0. \tag{B1.8.18}
$$

From this some tedious but straightforward algebra leads to

$$
|[U(h_2 - h_1) - U(h_1)U(h_2)]|^2 < [1 - |U(h_1)|^2][1 - |U(h_2)|^2]. \tag{B1.8.19}
$$

The two factors on the right are both positive, real numbers less than one. If the magnitudes of $U(h_1)$ and $U(h_2)$ are both close to one, therefore, the magnitude of the difference between the terms within the brackets on the left (complex numbers in general) must be small.

Karle and Hauptman showed that the fact that the crystal is composed of discrete atoms implies that large enough determinants of the type in inequality (B1.8.18) must vanish, leading to exact relations among sets of phases. For structures with moderately large numbers of atoms in the asymmetric unit 'large enough' may be very large indeed, and relations such as inequality (B1.8.19) may not represent much of a restriction on the phase of $U(h_2 - h_1)$. Further development of these principles by Hauptman, Karle, I L Karle, M M Woolfson and many others [26] showed that, although no one of these relations would put a significant restriction on a phase, reasonable assumptions about probability distributions would lead to statistical tests that assigned high probabilities to sufficient numbers of phases so that the correct structure could be identified in a density map. These developments, together with the revolution in computing power, have made the solution of structures with up to several hundred atoms in the asymmetric unit a matter of routine.

*(e) Isomorphous replacement*

While direct methods have opened up structural chemistry with hundreds of atoms in the asymmetric unit, many of the most interesting studies are of biological macromolecules, particularly proteins, which may have thousands of atoms in the asymmetric unit. Furthermore, all biological molecules are chiral, which means that the space groups in which they crystallize can never possess centres of symmetry or mirror (or glide) planes. Although the phases of some sets of reflections may be restricted by symmetry, most structure amplitudes are complex and with large structures the statistical techniques do not supply sufficient information to be useful. The first successful method of determining phases in macromolecular structure studies was the method of *isomorphous replacement*, in which a crystal of a protein is treated chemically to incorporate a small number of heavy atoms into the crystal without disturbing very much the arrangement of the protein molecules. In favourable cases two or more heavy-atom derivatives can be prepared in which the arrangements of the heavy atoms are different. The contribution of the protein molecule to the structure amplitude is assumed to be the same in the derivatives as in the native protein and the interatomic vectors of the heavy atoms stand out sufficiently in a Patterson map to allow the heavy-atom positions to be determined.

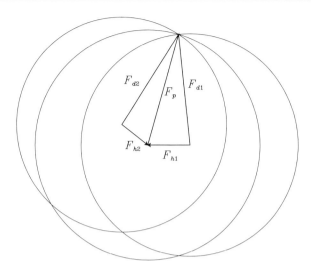

**Figure B1.8.5.** $F_p$, $F_{d1}$ and $F_{d2}$ are the measured structure amplitudes of a reflection from a native protein and from two heavy-atom derivatives. $F_{h1}$ and $F_{h2}$ are the heavy atom contributions. The point at which the three circles intersect is the complex value of $F_p$.

Referring to figure B1.8.5, the radii of the three circles are the magnitudes of the observed structure amplitudes of a reflection from the native protein, $F_p$ and of the same reflection from two heavy-atom derivatives, $F_{d1}$ and $F_{d2}$. We assume that we have been able to determine the heavy-atom positions in the derivatives and $F_{h1}$ and $F_{h2}$ are the calculated heavy-atom contributions to the structure amplitudes of the derivatives. The centres of the derivative circles are at points $-F_{h1}$ and $-F_{h2}$ in the complex plane, and the three circles intersect at one point, which is therefore the complex value of $F_p$. The phases for as many reflections as possible can then be used to compute a density map. The protein molecule is a chain of amino acid residues, whose sequence can be determined by biochemical means and each residue consists of a backbone portion, whose length is essentially the same for all amino acids and a side chain. With lots of both skill and luck, and sophisticated computer hardware and software, a model of the chain can be fitted into the density map to obtain a trial structure that can be refined.

### (f) Multiple-wavelength anomalous diffraction

A technique that employs principles similar to those of isomorphous replacement is *multiple-wavelength anomalous diffraction* (MAD) [27]. The expression for the atomic scattering factor in equation (B1.8.2b) is strictly accurate only if the x-ray wavelength is well away from any characteristic absorption edge of the element, in which case the atomic scattering factor is real and $F(\overline{hkl}) = F(hkl)^*$. Since the diffracted intensity is proportional to $|F(hkl)|^2$, the diffraction process in effect introduces a centre of symmetry into all data, a fact that is known as Friedel's law. If the wavelength is near an absorption edge, however, the atomic scattering factor becomes complex and the phases of the contributions of an atom to $F(hkl)$ and $F(\overline{hkl})$ differ. The increasingly widespread availability of synchrotron radiation has made it possible to collect diffraction data at several wavelengths, including some near an absorption edge of one or more of the elements in the crystal. The differences between the intensities of $hkl$ and $\overline{hkl}$ then serve in a role similar to that played by the differences between the intensities of native and derivative data in isomorphous replacement.

## B1.8.4  Experimental techniques

There are many experimental techniques for diffraction studies, depending on whether the materials producing the diffraction are crystalline or amorphous solids, liquids or gases. Crystalline materials are further subdivided according to whether the sample is a single crystal or a powder composed of many small crystals, frequently of more than one phase. All techniques include a source of radiation, a system for holding and manipulating the sample and a means of detecting the scattered radiation.

### B1.8.4.1  Sources of radiation

#### (a) X-rays

X-rays for diffraction are generated in two ways. The most common is to bombard a metallic anode in a vacuum tube with electrons emitted thermionically from a hot cathode, thereby exciting the characteristic radiation from the anode material, which is usually copper or molybdenum, although some other metals are used for special purposes. If the accelerating voltage in the tube is well above that required to eject a K shell electron from an atom of the anode material, most of the x-radiation emitted will be in the characteristic lines of the K series on top of a continuous, *Bremsstrahlung* spectrum. K$\beta$ and higher energy lines may be filtered out using a suitable metallic filter, or the characteristic line may be selected by reflection from a monochromator crystal.

The other type of x-ray source is an electron synchrotron, which produces an extremely intense, highly polarized and, in the direction perpendicular to the plane of polarization, highly collimated beam. The energy spectrum is continuous up to a maximum that depends on the energy of the accelerated electrons, so that x-rays for diffraction experiments must either be reflected from a monochromator crystal or used in the Laue mode. Whereas diffraction instruments using vacuum tubes as the source are available in many institutions worldwide, there are synchrotron x-ray facilities only in a few major research institutions. There are synchrotron facilities in the United States, the United Kingdom, France, Germany and Japan.

#### (b) Neutrons

Neutrons for diffraction experiments are also produced in two ways. Thermal neutrons from a nuclear reactor are reflected from a monochromator crystal and Bragg's law is satisfied for neutrons scattered from the sample by measuring the scattering angle, $2\theta$. In a spallation source short pulses of protons bombard a heavy metal target and high energy neutrons are produced by nuclear reactions. These neutrons interact with a moderator, giving a somewhat longer pulse of neutrons with a spectrum that extends down to thermal energies and therefore to wavelengths up to a few Ångströms. Diffraction from a sample a few metres away from the moderator is observed at a fixed angle and the relation between wavelength and velocity causes Bragg's law to be satisfied at some time after the initial pulse.

As with synchrotron x-rays, neutron diffraction facilities are available at only a few major research institutions. There are research reactors with diffraction facilities in many countries, but the major ones are in North America, Europe and Australia. The are fewer spallation sources, but there are major ones in the United States and the United Kingdom.

#### (c) Electrons

As noted earlier, most electron diffraction studies are performed in a mode of operation of a transmission electron microscope. The electrons are emitted thermionically from a hot cathode and accelerated by the electric field of a conventional electron gun. Because of the very strong interactions between electrons and matter, significant diffracted intensities can also be observed from the molecules of a gas. Again, the source of electrons is a conventional electron gun.

### B1.8.4.2 Detectors

Detectors for the three types of radiation are similar and may be classified in two categories, photographic and electronic. In addition to photographic films and plates, photographic detectors also include fluorescent screens and image plates, in which x-rays produce a latent image in a storage phosphor. In the dark the phosphor emits radiation very slowly, but exposure to light from a laser stimulates fluorescence, which then can be observed by a photomultiplier tube and converted to an electronic signal. Because neutrons interact weakly with most materials, image plates and fluorescent screens must contain one of the elements, such as gadolinium, that have isotopes with high absorption cross-sections. Photographic detection of neutrons usually uses a fluorescent screen to enhance the image.

There are many types of electronic detector. The original form of electronic detector was the Geiger counter, but it was replaced many years ago by the proportional counter, which allows selection of radiation of a particular type or energy. Proportional counters for x-rays are filled with a gas such as xenon, and those for neutrons are filled with a gas containing a neutron-absorbing isotope, usually $^3$He. Recently these gases have been used to construct position-sensitive area detectors for both x-rays and neutrons.

### B1.8.4.3 Single-crystal diffraction

Many different geometrical arrangements are commonly used for measurements of diffraction of x-rays and neutrons from single crystals. All have a mechanism for setting the crystal so that Bragg's law is satisfied for some set of crystal planes and for placing a detector in the proper position to observe the reflection. In the original x-ray diffraction experiments of Laue and co-workers the x-rays had a broad spectral distribution, so that for any angular position of a crystal and any interplanar spacing there were x-rays with the proper wavelength to satisfy Bragg's law. Laue photographs reveal the internal symmetry of the crystal and are therefore used to determine the symmetry and orientation of the crystal. For crystal structure determination it is necessary to measure accurate intensities and it is usual to use a monochromatic beam of x-rays or neutrons.

For diffraction studies with monochromatic radiation, the crystal is commonly mounted on an Eulerian cradle, which can rotate the crystal so that the normal to any set of planes bisects the angle between the incident and reflected beams, which is set for reflection from planes with a particular value of the interplanar spacing, $d$.

If the detection system is an electronic, area detector, the crystal may be mounted with a convenient crystal direction parallel to an axis about which it may be rotated under the control of a computer that also records the diffracted intensities. Because the orientation of the crystal is known at the time an x-ray photon or neutron is detected at a particular point on the detector, the indices of the crystal planes causing the diffraction are uniquely determined. If films are used, additional information is needed to index the pattern. The crystal is either oscillated through a narrow angular range, so that only a small number of planes can come into the reflecting position, or the film is moved, with a mask covering all but a small part of it, so that the exposed part of it is coordinated with the angular position of the crystal. There are two common moving-film methods, the Weissenberg method, in which a cylindrical film is moved parallel to the rotation axis of the crystal and the precession method, in which the normal to a reciprocal lattice plane is moved along a circular cone while a flat film and a circular slit are both moved in such a way that the positions of the spots on the film correspond to the points of the reciprocal lattice plane.

One form of electron diffraction is similar to the precession method, except that the 'single crystal' is a grain of a polycrystalline foil. Figure B1.8.6 shows an electron diffraction pattern produced when the beam is directed down a fivefold symmetry axis of a quasicrystal. Because of the very short wavelength the

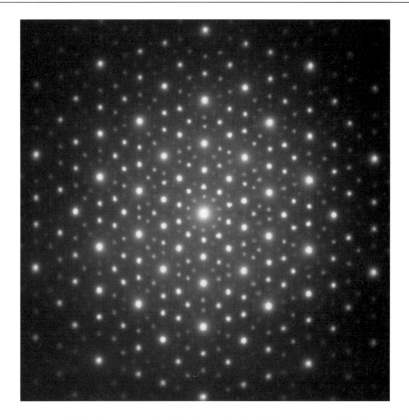

**Figure B1.8.6.** An electron diffraction pattern looking down the fivefold symmetry axis of a quasicrystal. Because Friedel's law introduces a centre of symmetry, the symmetry of the pattern is tenfold. (Courtesy of L Bendersky.)

cone angle is so small that it lies within the mosaic spread of the grain, and the resulting diffraction pattern, after magnification by the electron optics, closely resembles a precession pattern made with x-rays. In this technique the divergence of the electron beam is extremely small, and the diffraction spots correspond to lattice points in a plane of the reciprocal lattice that passes through the origin. The diffraction pattern therefore has a centre of symmetry. In convergent beam electron diffraction (CBED) [28] (see figure B1.8.7) the divergence of the electron beam is still only a few tenths of a degree, but the resultant smearing of the Ewald sphere allows it to intersect layers of the reciprocal lattice adjacent to the one passing through the origin, so that a region of broadened diffraction spots is surrounded by one or more rings of additional spots corresponding to points in these adjacent planes. Because Friedel's law does not apply in those planes, the pattern more closely reflects the true symmetry of the crystal.

An experimental technique that is useful for structure studies of biological macromolecules and other crystals with large unit cells uses neither the broad, 'white', spectrum characteristic of Laue methods nor a sharp, monochromatic spectrum, but rather a spectral band with $\Delta\lambda/\lambda \approx 20\%$. Because of its relation to the Laue method, this technique is called *quasi-Laue*. It was believed for many years that the Laue method was not useful for structure studies because reflections of different orders would be superposed on the same point of a film or an image plate. It was realized recently, however, that, if there is a definite minimum wavelength in the spectral band, more than 80% of all reflections would contain only a single order. Quasi-Laue methods are now used with both neutrons and x-rays, particularly x-rays from synchrotron sources, which give an intense, white spectrum.

**Figure B1.8.7.** A convergent beam diffraction pattern of the fivefold axis of a quasicrystal, as in figure B1.8.6. The diffraction rings show that the symmetry is fivefold, not tenfold. (Courtesy of L Bendersky.)

### B1.8.4.4  Powder diffraction

Many scientifically and technologically important substances cannot be prepared as single-crystals large enough to be studied by single crystal diffraction of x-rays and, especially, neutrons. If a sample composed of a very large number (of order $10^{10}$ or more) of very small ($10 \ \mu$m or smaller) crystals are irradiated by a monochromatic beam of x-rays or neutrons, there will be some crystals with the right orientation to reflect from all possible sets of crystal planes with interplanar spacings greater than $\lambda/2$. The resulting diffraction pattern contains intensity peaks that are characteristic for any crystalline compound. The pattern corresponds to a uniform distribution of reciprocal lattice points on the surface of a sphere and analysis of the pattern to determine the size and shape of the unit cell can be a difficult (but generally computationally tractable) problem. Nevertheless, in addition to structure determination, powder diffraction is an extremely powerful tool for phase identification and, because a mixture of crystalline phases will give the characteristic patterns of all phases present, quantitative phase analysis. Also if, because of mechanical deformation, for example, the powder sample is not spherically uniform, powder diffraction can reveal the nature of the preferred orientation. Because of the weak interaction and, therefore, high penetration of the neutron, neutron powder diffraction is particularly useful for preferred orientation (texture) studies of bulk materials.

X-ray powder diffraction studies are performed both with films and with counter diffractometers. The powder photograph was developed by P Debye and P Scherrer and, independently, by A W Hull. The Debye–Scherrer camera has a cylindrical specimen surrounded by a cylindrical film. In another commonly used powder camera, developed by A Guinier, a convergent beam from a curved, crystal monochromator passes through a thin, flat sample and is focused on the film. The common x-ray powder diffractometer uses so-called Bragg–Brentano (although it was apparently developed by W Parrish) focusing. A divergent beam from the line focus of an x-ray tube is reflected from a flat sample and comes to an approximate focus at a receiving slit.

Powder diffraction studies with neutrons are performed both at nuclear reactors and at spallation sources. In both cases a cylindrical sample is observed by multiple detectors or, in some cases, by a curved, position-sensitive detector. In a powder diffractometer at a reactor, collimators and detectors at many different $2\theta$ angles are scanned over small angular ranges to fill in the pattern. At a spallation source, pulses of neutrons of different wavelengths strike the sample at different times and detectors at different angles see the entire powder pattern, also at different times. These slightly displaced patterns are then 'time focused', either by electronic hardware or by software in the subsequent data analysis.

## B1.8.5 Frontiers

Starting from the truly heroic solution of the structure of penicillin by D C Hodgkin (née Crowfoot) and coworkers [29], x-ray diffraction has been the means of molecular structure determination (and the basis of many Nobel prizes in addition to Hodgkin's) of important compounds, including natural products, of which minute quantities were available for analysis, enabling chemical synthesis and further study. Knowledge of the molecular structure leads in turn to an understanding of reaction mechanisms and, in the case of biological molecules in particular, to an understanding of enzyme function and how drugs can be designed to promote desirable reactions and inhibit undesirable ones.

The development of neutron diffraction by C G Shull and coworkers [30] led to the determination of the existence, previously only a hypothesis, of antiferromagnetism and ferrimagnetism. More recently neutron diffraction, because of its sensitivity to light elements in the presence of heavy ones, played a crucial role in demonstrating the importance of oxygen content in high-temperature superconductors.

The development of synchrotron x-ray sources has resulted in a vast expansion of the capability of x-ray diffraction for determining macromolecular structure, but advances are still limited by the rarity and expense of synchrotron facilities. Correspondingly, the use of neutron diffraction has always been inhibited by the relatively low intensities available and the resulting need for large samples and long data collection times. With both synchrotrons and neutron sources observation time at existing facilities is chronically oversubscribed. Thus there is a need to develop both instruments and methodologies for maximum utilization of the sources.

## References

[1]  James R W 1965 *The Optical Principles of the Diffraction of X-Rays* (Cornell University Press) ch III
[2]  Maslen E N, Fox A G and O'Keefe M A 1999 X-ray scattering *International Tables for Crystallography* 2nd edn, vol C, ed A J C Wilson and E Prince (Dordrecht: Kluwer) section 6.1.1
[3]  Brown P J 1999 Magnetic form factors *International Tables for Crystallography* 2nd edn, vol C, ed A J C Wilson and E Prince (Dordrecht: Kluwer) section 4.4.5
[4]  Sears V F 1999 Scattering lengths for neutrons *International Tables for Crystallography* 2nd edn, vol C, ed A J C Wilson and E Prince (Dordrecht: Kluwer) section 4.4.4
[5]  Glatter O and May R 1999 Small angle techniques *International Tables for Crystallography* 2nd edn, vol C, ed A J C Wilson and E Prince (Dordrecht: Kluwer) chapter 2.6
[6]  Friedrich W, Knipping P and von Laue M 1912 Interferenz-Erscheinungen bei Röntgenstrahlen *Sitzungsberichte der Königlich Bayerischen Akademie der Wissenschaften zu München* pp 303–22
[7]  Bragg W L 1913 The diffraction of short electromagnetic waves by a crystal *Proc. Camb. Phil. Soc.* **17** 43–58
[8]  Gibbs J W 1928 *Elements of Vector Analysis (Collected Works of J. Willard Gibbs Vol II, Part 2)* (New York: Longmans Green)
[9]  Ewald P P 1921 Das reziproke Gitter in der Strukturtheorie *Z. Kristallogr.* **56** 129–56
[10] Prince E 1994 *Mathematical Techniques in Crystallography and Materials Science* 2nd edn (Heidelberg: Springer)
[11] Bacon G E 1962 *Neutron Diffraction* 2nd edn (Oxford: Clarendon)
[12] Thomas G and Goringe M J 1981 *Transmission Electron Microscopy of Materials* (New York: Wiley)

[13] van Hove M A, Weinberg W H and Chan C-H 1986 *Low Energy Electron Diffraction: Experiment, Theory, and Surface Structure Determination* (Berlin: Springer)

[14] Patterson A L 1934 A Fourier series method for for the determination of the components of interatomic distances in crystals *Phys. Rev.* **46** 372–6

[15] Shechtman D, Blech I, Gratias D and Cahn J W 1984 Metallic phase with long range orientational order and no translational symmetry *Phys. Rev. Lett.* **53** 1951–3

[16] Prince E 1987 Diffraction patterns from tilings with fivefold symmetry *Acta Crystallogr.* A **43** 393–400

[17] Toby B H and Egami T 1992 Accuracy of pair distribution function analysis applied to crystalline and noncrystalline materials *Acta Crystallogr.* A **48** 336–46

[18] Bragg W L 1913 The structure of some crystals as indicated by their diffraction of X-rays *Proc. R. Soc.* A **89** 248–60

[19] Millikan R A A new modification of the cloud method of determining the elementary electrical charge and the most probable value of that charge *Phil. Mag.* **19** 209–28

[20] Moseley H G J 1913 The high-frequency spectra of the elements *Phil. Mag.* **26** 1024–34

[21] Hahn Th (ed) 1992 *International Tables for Crystallography* vol A (Dordrecht: Kluwer)

[22] Harker D 1936 The application of the three-dimensional Patterson method and the crystal structures of proustite, $Ag_3AsS_3$, and pyrargyrite, $Ag_3SnS_3$ *J. Chem. Phys.* **4** 381–90

[23] Wilson A J C 1949 The probability distribution of X-ray intensities *Acta Crystallogr.* **2** 318–21

[24] Harker D and Kasper J S 1948 Phases of Fourier coefficients directly from crystal diffraction data *Acta Crystallogr.* **1** 70–5

[25] Karle J and Hauptman H 1950 The phases and magnitudes of the structure factors *Acta Crystallogr.* **3** 181–7

[26] Woolfson M M 1987 Direct methods—from birth to maturity *Acta Crystallogr.* A **43** 593–612

[27] Hendrickson W A 1991 Determination of macromolecular structures from anomalous diffraction of synchrotron radiation *Science* **254** 51–8

[28] Tanaka M and Terauchi M 1985 *Convergent-Beam Electron Diffraction* JEOL, Tokyo

[29] Crowfoot D, Bunn C W, Rogers-Low B W and Turner-Jones A 1949 The X-ray crystallographic investigation of the structure of penicillin *Chemistry of Penicillin* ed H T Clarke, J R Johnson and R Robinson (Princeton, NJ: Princeton University Press) pp 310–66

[30] Shull C G, Strauser W A and Wollan E O 1951 Neutron diffraction by paramagnetic and antiferromagnetic substances *Phys. Rev.* **83** 333–45

## Further Reading

Wilson A J C and Prince E (eds) 1999 *Mathematical, Physical and Chemical Tables (International Tables for Crystallography C)* 2nd edn (Dordrecht: Kluwer)

A comprehensive compilation of articles written by experts in their fields about all aspects of diffraction, with extensive lists of further references.

Zachariasen W H 1945 *Theory of X-ray Diffraction in Crystals* (New York: Dover)

The classic text on the diffraction of x-rays.

Bacon G E 1962 *Neutron Diffraction* 2nd ed (Oxford: Clarendon)

Likewise, the classic text on neutron diffraction.

Thomas G and Goringe M J 1981 *Transmission Electron Microscopy of Materials* (New York: Wiley)

A practical introduction to electron microscopy and diffraction.

# B1.9
# Scattering: light, neutrons, x-rays

*Benjamin S Hsiao and Benjamin Chu*

### B1.9.1    Introduction

Scattering techniques using light, neutrons and x-rays are extremely useful to study the structure, size and shape of large molecules in solids, liquids and solutions. The principles of the scattering techniques, which involve the interaction of radiation with matter, are the same. However, the data treatment for scattering from light, neutrons or x-rays can be quite different because the intrinsic property of each radiation and its interactions with matter are different. One major difference in the data treatment arises from the states of matter. Although the general equations describing the interaction between radiation and matter are valid for all classes of materials, unique analytical treatments have been made to suit the different states of matter.

In this chapter, the general principles of the scattering phenomenon and specific data treatments for the material (isotropic and anisotropic) in both solid and solution states are presented. These treatments are useful for the analysis of scattering data by light, neutrons or x-rays from different material systems such as crystalline polymers, complex fluids (including colloidal suspensions and solutions of biological species), multicomponent systems (including microemulsions and nanocomposites) and oriented polymers. For detailed theoretical derivations, the reader should refer to the many excellent textbooks and review articles that deal with the subjects of scattering from light [1–7], neutrons [7–11] and x-rays [7, 11–15]. We, however, will not discuss the detailed instrumentation for different scattering experiments as this topic has been well illustrated in some of the above references [3–7, 9–15]. Also absent will be the analysis for thin films and interfaces which warrants a separate chapter by itself. The main focus of this chapter is to provide a comprehensive overview to the field of scattering from materials (with emphasis on polymers) including an appropriate comparison between the different techniques. Selected example studies will also be included at the end of each section to illustrate the applications of some advanced scattering techniques.

### B1.9.2    Interaction of radiation and matter

In free space, electromagnetic radiation consists of simultaneous electric and magnetic fields, which vary periodically with position and time. These fields are perpendicular to each other and to the direction of wave propagation. The electromagnetic radiation consists of a wide range of wavelengths from $10^{-10}$ m (x-rays) to $10^4$ m (low frequency radio waves). Visible light has wavelengths from 400 nm (violet) to 700 nm (red), which constitutes a very small fraction of the electromagnetic spectrum.

As the electromagnetic radiation interacts with matter, the resultant radiation may follow several pathways depending on the wavelength and the material characteristics. Scattering without loss of energy is termed elastic. Elastic scattering from the periodic structure of matter emitting radiation of wavelength with the same magnitude leads to diffraction phenomena. The radiation may be slowed down by the refraction phenomenon. The radiation may also be absorbed by the material. The absorbed energy may be transferred to different

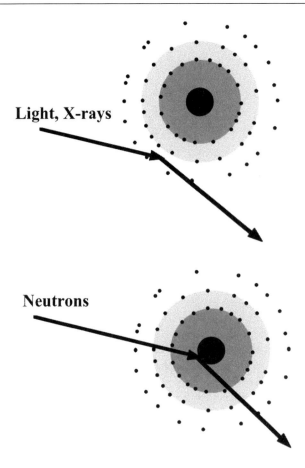

**Figure B1.9.1.** Diagrams showing that x-ray and light scattering involve extra-nuclear electrons, while neutron scattering depends on the nature of the atomic nucleus.

modes of motion, dissipated as heat or re-emitted as radiation at a different frequency. Energetic photons from x-rays or ultra-violet (UV) radiation can produce dissociation of chemical bonds leading to chemical reactions ejecting photoelectrons (known as x-ray photoelectron spectroscopy, XPS, or electron spectroscopy for chemical analysis, ESCA). Fluorescence occurs when the transfer of the residual energy to electronic modes takes place. Raman scattering occurs when the energy is transferred to or from rotational or vibrational modes. The characterization techniques based on these different phenomena are described in other chapters of section B.

Light scattering arises from fluctuations in refractive index or polarizability and x-ray scattering arises from fluctuations in electron density. Both are dependent on interactions of radiation with extra-nuclear electrons (figure B1.9.1) and will be discussed together. If we consider an incident beam having an electric field $E = E_0 \cos(2\pi r/\lambda - \omega t)$, where $E_0$ is the magnitude, $\lambda$ is the wavelength in vacuum, $r$ is the distance of the observer from the scatterer and $\omega$ is the angular frequency. From electromagnetic radiation incident upon an atom (with polarizability, $\alpha$), a dipole moment $m = \alpha E$ will be induced. The oscillating dipole will serve as a source of secondary radiation (this is scattering) with amplitude $E_s$ [16],

$$E_s = \frac{1}{c^2 r} \frac{\mathrm{d}^2 m}{\mathrm{d}t^2} \cos \varphi \qquad (B1.9.1)$$

where $c$ is the velocity of light and $\varphi$ is the angle between the plane of the polarization and the dipole moment. Thus we obtain

$$E_s = \frac{-\alpha E_0 \omega^2}{c^2 r} \cos \varphi \cos(\omega t - \phi) \tag{B1.9.2}$$

where $\phi$ is a phase angle which takes into account that the wave must travel a distance $(r = d)$ to reach the observer $(\phi = 2\pi d/\lambda)$. These equations presume that the electric field at the scattering position is not modified by the induced-dielectric environment (the Rayleigh–Gans approximation). Equation (B1.9.2) is thus termed Rayleigh scattering, which holds true for light scattering provided that the light frequency is small when compared with the resonance frequency of the electrons. For x-ray scattering, the frequency of electromagnetic radiation is higher than the resonance frequency of the electrons. In this case, Thomson scattering prevails and the scattering amplitude becomes

$$E_s = \frac{-e^2 \pi E_0}{m_0 c^2 r} \cos \varphi \cos(\omega t - \phi) \tag{B1.9.3}$$

where $e$ is the electron charge and $m_0$ is the electron mass. Thus, all electrons scatter x-rays equally and the x-ray scattering ability of an atom depends on the number of electrons, which is proportional to the atomic number, $Z$, in the atom. It should be noted that Rayleigh scattering is dependent on frequency and polarizability, but Thomson scattering is not. As a result, more polarizable molecules (larger, conjugated, more aromatic) are better Rayleigh scatterers than others. Neutron scattering depends upon nuclear properties being related to fluctuations in the neutron scattering cross section $\sigma$ between the scatterer and the surroundings. Hence, hydrogen can be a strong neutron scatterer in an isotope environment, but it is a weak electron scatterer.

The generalized scattering equation can be expressed by a complex exponential form

$$(E_s)_j = E_0 K_j \exp[i(\omega t - \phi_j)] \tag{B1.9.4}$$

where the subscript $j$ refers to the scattering from the $j$th element, $(E_s)_j$ represents the amplitude of the scattering of the $j$th scatterer and $K_j$ is proportional to the scattering power of the $j$th scatterer. For a collection of scattering elements, the total field strength (amplitude) of the scattered waves is

$$E_s = \sum_j (E_s)_j = E_0 \sum_j K_j \exp[i(\omega t - \phi_j)]. \tag{B1.9.5}$$

All scattering phenomena (light, x-rays and neutrons) can be interpreted in terms of this equation (B1.9.5). These techniques differ mainly in the structural entities that contribute to the $K_j$ term. For light, the refractive index or polarizability is the principal contributor; for x-rays, the electron density is the contributor; and for neutrons, the nature of the scattering nucleus is the contributor. Equation (B1.9.5) thus represents a starting point for the discussion of the interference problem presented below.

### B1.9.3 Light scattering

The intensity of light scattering, $I_s$, for an isolated atom or molecule is proportional to the mean squared amplitude

$$I_s = K \langle E_s^2 \rangle \tag{B1.9.6}$$

where the constant $K$ is equal to $c/4\pi$ for electromagnetic radiation and $\langle \rangle$ represents an average operation. Combining equations (B1.9.2) and (B1.9.6), we have

$$I_s = K \frac{\alpha^2 E_0^2 \omega^4}{c^4 r^2} \langle \cos^2 \varphi \cos^2(\omega t - \phi) \rangle. \tag{B1.9.7}$$

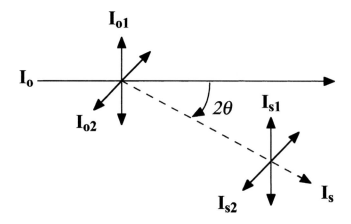

**Figure B1.9.2.** Resolution of a plane unpolarized incident beam into polarized scattering components.

As the average is over all values of time,

$$\langle \cos^2(\omega t - \phi_j) \rangle = \langle \cos^2 x \rangle = \left( \int_0^{2x} \cos^2 x \, dx \right) \left( \int_0^{2x} dx \right)^{-1} = \frac{1}{2} \tag{B1.9.8}$$

so equation (B1.9.7) can be simplified to

$$I_s = K \frac{\alpha^2 E_0^2 \omega^4}{2c^4 r^2} \cos^2 \varphi. \tag{B1.9.9}$$

The incident intensity, $I_0$, is given by

$$I_0 = K \langle E^2 \rangle = K E_0^2 \langle \cos^2(\omega t - \phi) \rangle = \tfrac{1}{2} K E_0^2. \tag{B1.9.10}$$

The ratio of the scattered to the incident intensity is given by

$$\frac{I_s}{I_0} = \frac{\alpha^2 \omega^4}{c^4 r^2} \cos^2 \varphi. \tag{B1.9.11}$$

Equation (B1.9.11) is valid only for plane polarized light. For unpolarized incident light, the beam can be resolved into two polarized components at right angles to each other. The scattered intensity can thus be expressed as (figure B1.9.2)

$$I_s = I_{s1} + I_{s2}. \tag{B1.9.12}$$

As a result, equation (B1.9.11) becomes

$$\frac{I_s}{I_0} = \frac{\alpha^2 \omega^4}{2c^4 r^2} (\cos^2 \varphi_1 + \cos^2 \varphi_2) = \frac{\alpha^2 \omega^4}{2c^4 r^2} (1 + \cos^2 2\theta) \tag{B1.9.13}$$

where we let $\varphi_1 = 0$ and $2\theta = \varphi_2$. Note that in light scattering, one often defines $\theta = \varphi_2$ with $\theta$ being the scattering angle. Herein, we define $2\theta$ as the scattering angle to be consistent with x-ray scattering described in B1.9.4. Since the frequency term may be converted to wavelength (in vacuum)

$$\frac{\omega^4}{c^4} = \frac{16\pi^4 \nu^4}{c^4} = \frac{16\pi^4}{\lambda^4} \tag{B1.9.14}$$

where $\nu$ is the frequency, equation (B1.9.13) becomes

$$\frac{I_s}{I_0} = \frac{8\pi^4\alpha^2}{\lambda^4 r^2}(1 + \cos^2 2\theta). \tag{B1.9.15}$$

This equation has the following implications.

(1) At $2\theta = 0°$, the scattering comprises both components of polarization of the incident beam; at $2\theta = 90°$, the scattering comprises only one half of the incident beam. Consequently, the scattered light at 90° will be plane polarized.

(2) Since $I_s \propto 1/\lambda^4$, this indicates that the shorter wavelengths (such as blue) scatter more than longer ones (such as red). Therefore, the light from a clear sky being blue is due to scattering by gas molecules in the atmosphere.

### B1.9.3.1 Scattering from a collection of objects

For random locations of $N$ scattering objects in volume $V$, the scattered intensity can be found by summing the scattering from each object:

$$\frac{I_s}{I_0} = \frac{8\pi^4 N\alpha^2}{\lambda^4 r^2}(1 + \cos^2 2\theta). \tag{B1.9.16}$$

A modified form of equation (B1.9.16) is usually used to express the scattering power of a system in terms of the 'Rayleigh ratio' defined as

$$R = \frac{(I_s/I_0)(r^2)}{V(1 + \cos^2 2\theta)} = \frac{8\pi^4\alpha^2}{\lambda^4}\left(\frac{N}{V}\right). \tag{B1.9.17}$$

In this case, the scattering serves as a means for counting the number of molecules (or particles, or objects) per unit volume ($N/V$). It is seen that the polarizability, $\alpha$, will be greater for larger molecules, which will scatter more. If we take the Clausius–Mosotti equation [16]:

$$(n^2 - 1)/(n^2 + 2) = \frac{4}{3}\pi\left(\frac{N}{V}\right)\alpha \tag{B1.9.18}$$

and consider $n \approx 1$ ($n$ is the refractive index), then

$$\alpha = \frac{n-1}{2\pi}\left(\frac{V}{N}\right). \tag{B1.9.19}$$

If the scattering particles are in a dielectric solvent medium with solvent refractive index $n_0$, we can define the excess polarizability ($\alpha_{ex} = \alpha(\text{solution}) - \alpha(\text{solvent})$) as

$$\alpha_{ex} = \frac{n^2 - n_0^2}{4\pi}\left(\frac{V}{N}\right). \tag{B1.9.20}$$

If the weight concentration $C$ is used, $C = MN/(N_a V)$, where $M$ is the molecular weight of the particle, $N_a$ is Avogadro's number and $n \approx n_o$, then the above relationship becomes

$$\alpha_{ex} = \frac{(n+n_0)(n-n_0)}{4\pi C}\frac{CV}{N} = \frac{n_0}{2\pi}\left(\frac{\partial n}{\partial C}\right)\left(\frac{M}{N_a}\right) \tag{B1.9.21}$$

where $\partial n/\partial C \approx (n - n_0)/C$ at constant temperature. The excess Rayleigh ratio $R_{ex} = R(\text{solution}) - R(\text{solvent})$ has the form

$$R_{ex} = \frac{2\pi^2 n_0^2}{\lambda^4 N_a} \left(\frac{\partial n}{\partial C}\right)^2 CM = HCM \tag{B1.9.22}$$

where $H$ is the optical constant for unpolarized incident light. For polarized light, the constant $H$ can be twice as large due to the factor $(1 + \cos^2 2\theta)$. It is noted that this factor of 2 depends on the definition of $R$, i.e., whether the $1 + \cos^2 2\theta$ term is absorbed by $R$.

$$R_{ex,p} = \frac{4\pi^2 n_0^2}{\lambda^4 N_a} \left(\frac{\partial n}{\partial C}\right)^2 CM = HCM \tag{B1.9.23}$$

where $R_{ex,p}$ is the excess Rayleigh ratio for polarized light. The above treatment is valid for molecules that are small compared to the wavelength of the incident beam.

### B1.9.3.2 Scattering from a solution of large molecules

For molecules having dimensions comparable with the wavelength, phase differences will occur between waves scattered from different regions of the molecule. These phase differences result in an angular dependence of the scattered intensity. The reduction may be expressed in terms of a particle interference factor $P(2\theta)$ such that

$$P(2\theta) = \frac{R_{ex}(\text{experimental})}{R_{ex}(\text{no interference})} \tag{B1.9.24}$$

where $2\theta$ is the scattering angle. We again caution the reader that the conventional symbol for the scattering angle by light is $\theta$. Herein we use the symbol $2\theta$ to be consistent with x-ray and neutron scattering described later. The interference factor also follows the expression

$$P(2\theta) = \left\langle \left[ \sum_j \alpha_j \cos(\omega t - \phi_j) \right]^2 \right\rangle \left\langle \left[ \sum_j \alpha_j \cos(\omega t) \right]^2 \right\rangle^{-1} \tag{B1.9.25}$$

where the summation is over all parts of a molecule.

The nature of $P(2\theta)$ may be qualitatively deduced from figure B1.9.3. For scattering in the forward direction $(\theta_1)$ the path difference between the rays from elements A and B of the molecule $(d_B - d_A)$ is less than that at the backward scattering angle $(2\theta_2)$ to observer $O_2$, $(d_B' - d_A')$. So, a greater phase difference occurs at $2\theta_2$. If the dimensions of the molecule (or particle) are less than the wavelength, destructive interference occurs and $P(2\theta)$ will decrease. If the molecular (or particle) dimensions are much greater than the probing wavelength, both destructive and constructive interference can occur leading to maxima and minima in $P(2\theta)$. For $2\theta = 0°$, no path difference exists and $P(2\theta) = 1$. Thus, the scattering technique can be used to estimate the size of the molecule (or particle).

Equation (B1.9.5) gives the total amplitudes of scattering from a collection of objects and is a good starting point for the derivation of interference phenomena associated with molecular size.

$$E_s = E_0 \sum_j K_j \exp[i\omega(t - d_j/c)] \tag{B1.9.26}$$

where $d_j$ is the distance between the scattering element and the observation point P (shown in figure B1.9.4). In figure B1.9.4, the following relationships can be approximated as

$$d_j = r_j \cdot s_0 + D - r_j \cdot s_1 = D + r_j \cdot (s_0 - s_1) = D + (r_j \cdot s) \tag{B1.9.27}$$

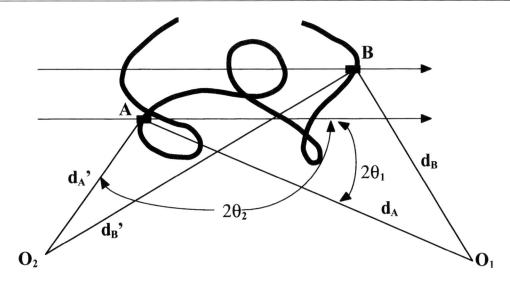

**Figure B1.9.3.** Variation of $P(2\theta)$ as a function of scattering angle.

where $D$ represents the distance between the observation point $P$ and the origin O, $s_0$ is the unit vector in the incident beam direction, $s_1$ is the unit vector in the scattered beam direction and $r_j$ is the vector to the $j$th scattering element. Equation (B1.9.26) thus becomes

$$E_s = E_0 \sum_j K_j\, e^{i\omega t}\, e^{-i\omega(r_j\cdot s)/c}\, e^{-i\omega D/c} = F\, e^{-ikD} \tag{B1.9.28}$$

where $F = \sum E_0 K_j\, e^{i\omega t}\, e^{-ik(r_j\cdot s)}$, which is defined as the structure or form factor of the object, and $k = \omega/c = 2\pi/\lambda$. For a system with a large number of scattering elements, the summation in equation (B1.9.28) may be replaced by an integral

$$F = E_0\, e^{i\omega t} \sum_j K_j \exp[-ik(r_j \cdot s)] = E_0\, e^{i\omega t} \int_r \rho(r) \exp[-ik(r_j \cdot s)]\, d^3r. \tag{B1.9.29}$$

The term $E_0\, e^{i\omega t}$ is related to the incident beam, which is often omitted in the theoretical derivation (as follows). The second term in (B1.9.29) represents the amplitude scattered by a three dimensional element with a volume element $d^3r$ and $\rho(r)$ being the density profile. From now on, we will simplify the symbol for the dot product of two vectors $r_j \cdot s$ as $r_j s$ and other similar products. If we consider the spherical polar coordinates (figure B1.9.5), this volume element becomes

$$d^3r = \int_{\phi=0}^{2\pi} \int_{\varphi=0}^{\pi} \int_{r=0}^{\infty} r^2 \sin\varphi\, dr\, d\varphi\, d\phi \tag{B1.9.30}$$

so that more generally,

$$F = \int_{\phi=0}^{2\pi} \int_{\varphi=0}^{\pi} \int_{r=0}^{\infty} \rho(r, \varphi, \phi)\, e^{-ik(r_j s)} r^2 \sin\varphi\, d\phi\, d\varphi\, dr. \tag{B1.9.31}$$

For spherically symmetric systems, we can derive the following expression from (B1.9.31) [17]:

$$F = 4\pi \int_{r=0}^{\infty} \rho(r)\frac{\sin qr}{qr} r^2\, dr \tag{B1.9.32}$$

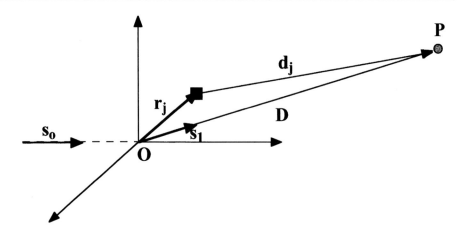

**Figure B1.9.4.** Geometrical relations between vectors associated with incident and scattered light.

where $q = 4\pi(\sin\theta)/\lambda$, with $2\theta$ being the scattering angle. The interference factor described previously (equation (B1.9.25)) may be expressed in terms of the intramolecular particle scattering factor as

$$P(2\theta) = \left( \int_{r=0}^{\infty} \rho(r) \frac{\sin qr}{qr} r^2 \, dr \right) \left( \int_{r=0}^{\infty} \rho(r) r^2 \, dr \right)^{-1}. \qquad (B1.9.33)$$

If we define the radius of gyration, $R_g$, by

$$R_g^2 = \left( \int_0^{\infty} \rho(r) r^2 \, dr \right) \left( \int_0^{\infty} \rho(r) \, dr \right)^{-1} \qquad (B1.9.34)$$

then equation (B1.9.33) can be expressed as [18]

$$P(2\theta) = 1 - \frac{q^2 R_g^2}{3} + \cdots = 1 - \frac{16\pi^2 R_g^2}{3\lambda^2} \sin^2(2\theta/2) + \cdots. \qquad (B1.9.35)$$

To measure the molecular weight of the molecule, we can modify equation (B1.9.23) to take into account the intramolecular interference in the dilute solution range,

$$\frac{HC}{R_{\text{ex}}} \cong \frac{1}{MP(2\theta)} + 2A_2C \qquad (B1.9.36)$$

where $A_2$ is the second virial coefficient. By combining (B1.9.35) with (B1.9.36), we obtain

$$\frac{HC}{R_{\text{ex}}} \cong \frac{1}{M} \left( 1 + \frac{q^2 R_g^2}{3} \right) + 2A_2C. \qquad (B1.9.37)$$

This is the basic equation for monodisperse particles in light scattering experiments. We can derive three relationships by extrapolation.

$$\lim_{C \to 0} \frac{HC}{R_{\text{ex}}} \cong \frac{1}{M} \left( 1 + \frac{q^2 R_g^2}{3} \right) \qquad (B1.9.38)$$

$$\lim_{q \to 0} \frac{HC}{R_{\text{ex}}} \cong \frac{1}{M} + 2A_2C \qquad (B1.9.39)$$

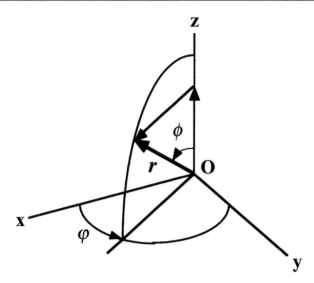

**Figure B1.9.5.** Geometrical relations between the Cartesian coordinates in real space, the spherical polar coordinates and the cylindrical polar coordinates.

$$\lim_{q \to 0, C \to 0} \frac{HC}{R_{\text{ex}}} \cong \frac{1}{M}. \tag{B1.9.40}$$

A graphical method, proposed by Zimm (thus termed the Zimm plot), can be used to perform this double extrapolation to determine the molecular weight, the radius of gyration and the second virial coefficient. An example of a Zimm plot is shown in figure B1.9.6, where the light scattering data from a solution of poly(tetrafluoroethylene) (PTFE) ($M_w = (2.9 \pm 0.2) \times 10^5$ g mol$^{-1}$; $A_2 = -(6.7 \pm 1.3) \times 10^{-5}$ mol cm$^3$ g$^{-2}$ and $R_g = 17.8 \pm 2.4$ nm) in oligomers of poly(chlorotrifluoroethylene) (as solvents) at 340 °C is shown [19]. The dashed lines represent the extrapolated values at $C = 0$ and $2\theta = 0$.

### B1.9.3.3 Calculate scattered intensity

There are two ways to calculate the scattered intensity. One is to first calculate the magnitude of the structure factor $F$ by summing the amplitude of scattering elements in the system and then multiply it by $F^*$ (the conjugate of $F$). This method is best for the calculation of the scattered intensity from discrete particles such as spheres or rods. Two examples are illustrated as follows.

(1)  The scattering from an *isolated sphere* may be calculated from equation (B1.9.32). This derivation assumes that the sphere is uniform, with its density profile $\rho(r) = \rho_0$ if $r < r_0$ and $\rho(r) = 0$ if $r > r_0$ (surrounded by a non-scattering material). With this assumption, equation (B1.9.32) becomes

$$F_{\text{sp}} = 4\pi \rho_0 \int_{r=0}^{r_0} \frac{\sin qr}{qr} r^2 \, \mathrm{d}r. \tag{B1.9.41}$$

By changing the variable, $x = qr$, we obtain

$$F_{\text{sp}} = \frac{4\pi \rho_0}{q^3} \int_{x=0}^{x=U} x \sin(x) \, \mathrm{d}x. \tag{B1.9.42}$$

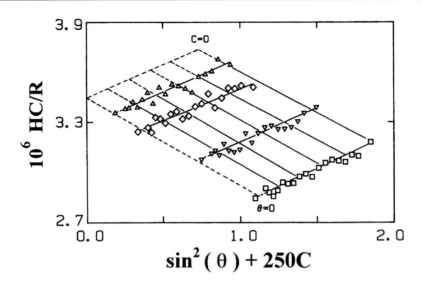

**Figure B1.9.6.** A typical Zimm plot; data obtained from a solution of poly(tetrafluoroethylene) (PTFE) ($M_w = (2.9 \pm 0.2) \times 10^5$ g mol$^{-1}$; $A_2 = -(6.7 \pm 1.3) \times 10^{-5}$ mol cm$^3$ g$^{-2}$ and $R_g = 17.8 \pm 2.4$ nm) in oligomers of poly(chlorotrifluoroethylene) (as solvents) at 340 °C. (Reprinted with permission from Chu *et al* [19].)

We can integrate the above equation by parts and derive

$$F_{sp} = \frac{4\pi\rho_0}{q^3}[\sin U - U\cos U] = V_{sp}\rho_0\Phi(U) \tag{B1.9.43}$$

where $V_{sp}$ is the sphere volume and

$$\Phi(U) = \text{the sphere scattering function} = \frac{3}{U^3}[\sin U - U\cos U] \tag{B1.9.44}$$

with the parameter $U = qr_0$. The intensity then is proportional to $F^2$ ($R = K_1 F F^*$, where $K_1$ is a calibration constant for light scattering).

(2) This treatment may be extended to *spheres containing a core–shell structure*. If the core density is $\rho_1$ in the range of 0 to $r_1$, the shell density is $\rho_2$ in the range of $r_1$ to $r_2$ and the density of the surrounding medium is $\rho_0$, then the magnitude of the structure factor becomes

$$F_{cs} = V_1(\rho_1 - \rho_2)\Phi(U_1) + V_2(\rho_2 - \rho_0)\Phi(U_2) \tag{B1.9.45}$$

where $V_i = (4/3)\pi r_i^3$ and $U_i = qr_i$.

The second method to calculate the scattered intensity (or $R$: the Rayleigh ratio) is to square the sum in $F$ directly.

$$R(q) = K_1 \sum_i \sum_j \rho_i \rho_j \exp[\mathrm{i}(qr_{ij})]. \tag{B1.9.46}$$

For a continuous system with assorted scattering elements, the sum can be replaced by integration and be expressed as:

$$R(q) = K_1 V \langle \eta^2 \rangle \int \gamma(r) \exp[\mathrm{i}(qr)]\,\mathrm{d}^3 r \tag{B1.9.47}$$

where $V$ is the scattering volume, $\eta$ is the heterogeneity of the fluctuation parameter defined as $\eta = \rho - \langle\rho\rangle$ and $\gamma(r)$ is termed the *correlation function* defined as

$$\gamma(r_{ij}) = \langle\eta_i\eta_j\rangle. \tag{B1.9.48}$$

Equation (B1.9.47) applies to the general scattering expression of any system. With spherical symmetry, the scattered intensity becomes [20]

$$R(q) = 4\pi K_1 V \langle\eta^2\rangle \int_0^\infty \gamma(r) r^2 \frac{\sin(qr)}{qr}\, dr. \tag{B1.9.49}$$

In the case of anisotropic systems with a cylindrical symmetry (such as rods or fibres), the scattered intensity can be expressed as (derivation to be made later in section B1.9.4):

$$R(q) = 4\pi K_1 V \langle\eta^2\rangle \int_0^\infty \gamma(r) J_0(q_r r) r\, dr \int_0^\infty \gamma(z) \cos(q_z z)\, dz \tag{B1.9.50}$$

where the subscript $r$ represents the operation along the radial direction, the subscript $z$ represents the operation along the cylinder axis, $J_0$ is the zero-order Bessel function and $q_r$ and $q_z$ are scattering vectors along the $r$ (equator) and $z$ (meridian) directions, respectively.

### B1.9.3.4 Estimate object size

One of the most important functions in the application of light scattering is the ability to estimate the object dimensions. As we have discussed earlier for dilute solutions containing large molecules, equation (B1.9.38) can be used to calculate the 'radius of gyration', $R_g$, which is defined as the mean square distance from the centre of gravity [12]. The combined use of equations (B1.9.38)–(B1.9.40) (the Zimm plot) will yield information on $R_g$, $A_2$ and molecular weight.

The above approximation, however, is valid only for dilute solutions and with assemblies of molecules of similar structure. In the event that concentration is high where intermolecular interactions are very strong, or the system contains a less defined morphology, a different data analysis approach must be taken. One such approach was derived by Debye *et al* [21]. They have shown that for a random two-phase system with sharp boundaries, the correlation function may carry an exponential form.

$$\gamma(r) = e^{-r/a_c} \tag{B1.9.51}$$

where $a_c$ is a correlation length describing the dimension of heterogeneities. Substitution of (B1.9.51) into (B1.9.47) gives rise to the expression

$$R(q) = \frac{8\pi K_1 \langle\eta^2\rangle a_c^3}{[1 + q^2 a_c^2]^2}. \tag{B1.9.52}$$

Based on this equation, one can make a 'Debye–Bueche' plot by plotting $[R(q)]^{-1/2}$ versus $q^2$ and determine the slope and the intercept of the curve. The correlation length thus can be calculated as [21]

$$a_c = \left(\frac{\text{slope}}{\text{intercept}}\right)^{1/2}. \tag{B1.9.53}$$

## B1.9.4 X-ray scattering

X-ray scattering arises from fluctuations in electron density. The general expression of the absolute scattered intensity $I_{abs}(q)$ (simplified as $I(q)$ from now on) from the three-dimensional objects immersed in a different density medium, similar to (B1.9.47), can be expressed as:

$$I(q) = K_x V \langle \eta^2 \rangle \int \gamma(r) e^{i(qr)} d^3 r \tag{B1.9.54}$$

where $V$ is the volume of the scatterer, $\langle \eta^2 \rangle$ is the square of the electron density fluctuations, $\gamma(r)$ is the correlation function and $K_x$ is a calibration constant depending on the incident beam intensity and the optical apparatus geometry (e.g. polarization factor) given by

$$K_x = I_0 \left( \frac{e^2}{m_0 c^2} \right)^2 \frac{1}{D_s^2} \frac{1 + \cos^2 2\theta}{2} \tag{B1.9.55}$$

where $e$, $m_0$ are the charge and mass of an electron, $c$ is the velocity of light, $D_s$ is the sample to detector distance and $2\theta$ is the scattering angle.

If the scattering system is isotropic, equation (B1.9.54) can be expressed in spherical polar coordinates (the derivation is similar to equation (B1.9.32)):

$$I(q) = 4\pi K_x V \langle \eta^2 \rangle \int_0^\infty \gamma(r) r^2 \frac{\sin(qr)}{qr} dr. \tag{B1.9.56}$$

This expression is very similar to (B1.9.49). If the scattering system is anisotropic, equation (B1.9.54) can then be expressed in cylindrical polar coordinates (see figure B1.9.5):

$$I(q) = I(q_r, \psi, q_z) = K_x V_s \langle \eta^2 \rangle \iiint \gamma(r, \varphi, z) e^{i(rq_r \cos(\varphi - \psi) + z q_z)} r \, dr \, d\varphi \, dz \tag{B1.9.57}$$

where $r$ represents the distance along the radial direction and $z$ represents the distance along the cylindrical axis (in real space), $q_r$ and $q_z$ are correspondent scattering momenta along the $r$ and $z$ directions in reciprocal space, $\varphi$ is the polar angle in real space and $\psi$ is the polar angle in reciprocal space. Equation (B1.9.57) is a general expression for cylinders without any assumptions. If we consider the scatterers having the geometry of a cylinder, it is reasonable to assume $\gamma(r, \varphi, z) = \gamma(r, \varphi)\gamma(z)$, which indicates that the two correlation functions along the radial (equatorial) direction and the cylindrical (meridional) direction are independent. In addition, the term $\gamma(\varphi) = 1$ can be applied, which represents the symmetry of a cylinder. Equation (B1.9.57) thus can be rewritten as

$$I(q) = K_x V \langle \eta^2 \rangle \int_0^\infty \gamma(r) \left( \int_0^{2\pi} e^{irq_r \cos(\varphi - \psi)} d\varphi \right) r \, dr \int_{-\infty}^\infty \gamma(z) e^{izq_z} dz. \tag{B1.9.58}$$

If we define $\delta = \varphi - \psi$ and $rq_r = u$, then

$$\int_0^{2\pi} e^{irq_r \cos(\varphi - \psi)} d\varphi = \int_0^{2\pi} e^{iu \cos \delta} d\delta = 2\pi J_0(u) \tag{B1.9.59}$$

where $J_0(u)$ is the zeroth order Bessel function of the first kind. Also, we have

$$\int_{-\infty}^\infty \gamma(z) e^{izq_z} dz = \int_{-\infty}^\infty \gamma(z)(\cos(zq_z) + i \sin(zq_z)) dz = \int_{-\infty}^\infty \gamma(z) \cos(zq_z) dz \tag{B1.9.60}$$

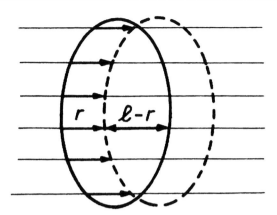

**Figure B1.9.7.** Diagram to illustrate the relationship between chord length ($\ell$) and $r$ in a particle.

because both $\gamma(z)$ and $\cos(zq_z)$ are even functions (i.e., $\gamma(z) = \gamma(-z)$ and $\cos(zq_z) = \cos(-zq_z)$) and $\sin(zq_z)$ is an odd function. The integral of the latter ($\gamma(z)\sin(zq_z)$, an odd function) is an even function integrated from $-\infty$ to $\infty$, which becomes zero. Combining equations (B1.9.59) and (B1.9.60) into equation (B1.9.58), we obtain

$$I(q) = K_x V \langle \eta^2 \rangle \int_0^\infty \gamma(r) 2\pi J_0(q_r r) r \, dr \int_0^\infty 2\gamma(z) \cos(q_z z) \, dz. \tag{B1.9.61}$$

This equation is the same as (B1.9.50) for light scattering.

### B1.9.4.1  Particle scattering

The general expression for particle scattering can best be described by the correlation function $\gamma(r)$. Using the definition in (B1.9.48), we have

$$\gamma(r) = (\Delta\rho)^2 \gamma_0(r) \tag{B1.9.62}$$

where $\Delta\rho$ is the electron density difference between the particle and the surrounding medium and is assumed to be constant, $\gamma(0) = 1$ and $\gamma(r) = 0$ for $r \geq D$ (the diameter of the particle). The function $\gamma_0(r)$ can be expressed by a distribution function $G(l)$ with $l$ being the *intersection length* or the *chord length* of the particle (figure B1.9.7) [12, 13]

$$\gamma_0(r) = \frac{1}{\bar{l}} \int_r^D (l - r) G(l) \, dl \tag{B1.9.63}$$

with

$$\bar{l} = \int_0^D l G(l) \, dl. \tag{B1.9.64}$$

By differentiation, we can derive

$$\frac{d\gamma_0(r)}{dr} = -\frac{1}{\bar{l}} \int_r^D G(l) \, dl$$

$$\frac{d^2\gamma_0(r)}{dr^2} = \frac{1}{\bar{l}} G(l). \tag{B1.9.65}$$

The distance distribution function $p(r)$ has a clear geometrical definition. It is defined as

$$p(r) = \gamma(r) r^2. \tag{B1.9.66}$$

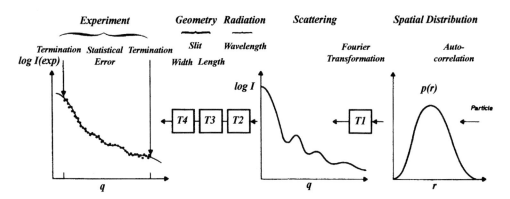

**Figure B1.9.8.** Schematic diagram of the relationship between a particle distribution and the measured experimental scattering data. This figure is duplicated from [14], with permission from Academic Press.

For homogeneous particles, it represents the number of distances within the particle. For inhomogeneous particles, it has to take into account the different electron density of the volume elements. Thus it represents the number of pairs of difference in electrons separated by the distance $r$. A qualitative description of shape and internal structure of the particle can be obtained directly from $p(r)$. In addition, several structural parameters can be determined quantitatively [22]. We can describe several analytical forms of the distance distribution function for different shapes of homogeneous particles as follows.

(1)  Globular particles

$$p(r) = 12x^2(2 - 3x + x^3) \qquad (B1.9.67)$$

   where $x = r/D$.

(2)  Rodlike particles

$$p(r) = \frac{1}{2\pi}\rho_c^2 A^2(L - r) \qquad (B1.9.68)$$

   where $\rho_c$ is the particle electron density, $A$ is the cross-section of the rod and $L$ is the length of the rod particle.

(3)  Flat particles, i.e. particles elongated in two dimensions (discs, flat prisms)

$$p(r) = \frac{16}{\pi}x(\arccos(x) - x\sqrt{(1 - x^2)}) \qquad (B1.9.69)$$

   where $x = r/D$.

   The radius of gyration of the whole particle, $R_g$, can be obtained from the distance distribution function $p(r)$ as

$$R_g^2 = \left(\int_0^\infty p(r)r^2\,dr\right)\left(\int_0^\infty p(r)\,dr\right)^{-1}. \qquad (B1.9.70)$$

This value can also be obtained from the innermost part of the scattering curve.

$$I(q) = I(0)\exp\left(-\frac{R_g^2 q^2}{3}\right) \qquad (B1.9.71)$$

where $I(0)$ is the scattered intensity at zero scattering angle. A plot of $\log I(q)$ versus $q^2$, which is known as the Guinier plot, should show a linear descent with a negative slope $(= -R_g^2/3)$ related to the radius of gyration.

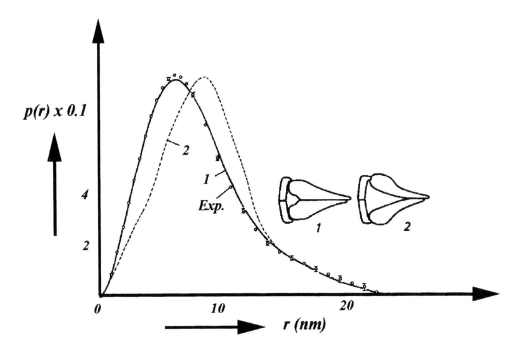

**Figure B1.9.9.** Comparison of the distance distribution function $p(r)$ of a RNA-polymerase core enzyme from the experimental data (open circle) and the simulation data (using two different models). This figure is duplicated from [27], with permission from Elsevier Science.

In practical data analysis, we are interested in extracting information about the size, shape and distribution of the scattering particles. The most widely used approach for this purpose is the indirect Fourier transformation method, pioneered by O Glatter [23–25]. This approach can be briefly illustrated as follows. In dilute solutions with spherical particles, where the interparticle scattering is negligible, the distance distribution function $p(r)$ can be directly calculated from the scattering data through Fourier transformation (combining equations (B1.9.56) and (B1.9.66))

$$I(q) = 4\pi K_x V \langle \eta^2 \rangle \int_0^\infty p(r) \frac{\sin(qr)}{qr} \, dr. \tag{B1.9.72}$$

This process is shown in figure B1.9.8, where $T_1$ represents the Fourier transformation. If there are additional instrumentation effects desmearing the data (such as the slit geometry, wavelength distribution etc), appropriate inverse mathematical transformations ($T_2$, $T_3$, $T_4$) can be used to calculate $p(r)$. If the particle has a certain shape, the $\sin(x)/x$ term in (B1.9.72) must be replaced by the form factor according to the shape assumed. If concentrations increase, the interparticle scattering (the so-called structure factor) should be considered, which will be discussed later in section B1.9.5. A unique example to illustrate the usefulness of the distance distribution function is the study of a DNA-dependent RNA polymerase core enzyme [26, 27] (figure B1.9.9). The RNA polymerase enzyme is known to have four subunits but in two possible configurations (model 1 and model 2). Model 1 has a configuration with the two larger subunits having a centre-to-centre distance of 5 nm, and model 2 has a more open configuration having a centre-to-centre distance of 7 nm. From the experimental data (open circle), it is clear that model 1 gives a better fit to the data.

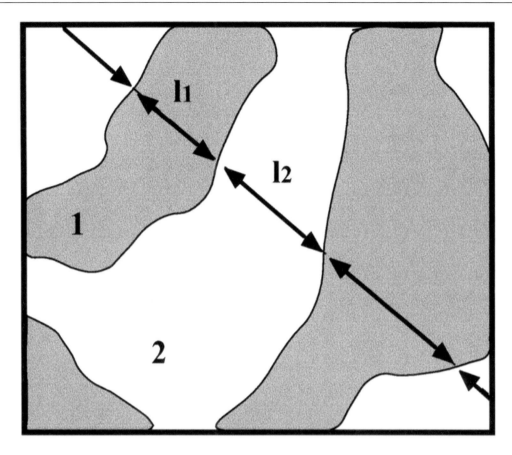

**Figure B1.9.10.** Non-particulate random two-phase system.

### B1.9.4.2   *Nonparticulate scattering*

If we consider the scattering from a general two-phase system (figure B1.9.10, distinguished by indices 1 and 2) containing constant electron density in each phase, we can define an average electron density $\overline{\rho}$ and a mean square density fluctuation as:

$$\overline{\rho} = \varphi_1 \rho_1 + \varphi_2 \rho_2 \tag{B1.9.73}$$

$$\overline{\eta^2} = (\rho_1 - \rho_2)^2 \varphi_1 \varphi_2 \tag{B1.9.74}$$

where $\varphi_1$ and $\varphi_2$ are the fractions of the total volume $V$. Equation (B1.9.74) is directly related to the invariant $Q$ of the system

$$Q = 2\pi^2 V \overline{\eta^2} = 2\pi^2 V (\rho_1 - \rho_2)^2 \varphi_1 \varphi_2. \tag{B1.9.75}$$

In this case, the correlation function $\gamma(r)$ becomes

$$\gamma(r) = (\rho_1 - \rho_2)^2 \varphi_1 \varphi_2 \gamma_0(r) \tag{B1.9.76}$$

where $\gamma_0(r)$ is the normalized correlation function, which is related to the geometry of the particle.

There are several geometrical variables that one can extract from the correlation function approach. First, a correlation volume $v_c$ can be defined, which is related to the extrapolated intensity at zero angle $I(0)$.

$$I(0) = V(\rho_1 - \rho_2)^2 \varphi_1 \varphi_2 v_c. \tag{B1.9.77}$$

With the use of invariant $Q$, we obtain

$$v_c = \frac{2\pi^2}{Q} I(0).$$

(B1.9.78)

If we consider the chord length distribution, we can express the alternating chords $l_1$ and $l_2$ as:

$$l_1 = 4\frac{V}{S}\varphi_1 \qquad l_2 = 4\frac{V}{S}\varphi_2$$

(B1.9.79)

where $S$ is the total surface area. The mean chord length thus becomes

$$\bar{l} = l_1\varphi_2 = l_2\varphi_1 = 4\frac{V}{S}\varphi_1\varphi_2.$$

(B1.9.80)

As we will discuss in the next section, the scattered intensity $I(q)$ at very large $q$ values will be proportional to the $q^{-4}$ term. This is the well known Porod approximation, which has the relationship

$$\lim_{q\to\infty} I(q) = V\varphi_1\varphi_2(\rho_1 - \rho_2)^2 \frac{8\pi}{\bar{l}} \frac{1}{q^4}.$$

(B1.9.81)

It is sometimes more convenient to normalize the absolute scattered intensity $I(q)$ and use the following expression:

$$\lim_{q\to\infty} I(q)q^4/Q = \frac{1}{\pi\varphi_1\varphi_2} \frac{S}{V}.$$

(B1.9.82)

Thus the ratio of $S/V$ can be determined by the limiting value of $I(q)q^4$. We will discuss the Porod approximation next.

### B1.9.4.3 Porod approximation

According to the Porod law [28], the intensity in the tail of a scattering curve from an *isotropic* two-phase structure having sharp phase boundaries can be given by equation (B1.9.81). In fact, this equation can also be derived from the general expression of scattering (B1.9.56). The derivation is as follows. If we assume $qr = u$ and use the Taylor expansion at large $q$, we can rewrite (B1.9.56) as

$$\begin{aligned}
I(q) &= \frac{4\pi K_x V \langle \eta^2 \rangle}{q^3} \int_0^\infty \gamma(r) u \sin(u)\, du \\
&= \frac{4\pi K_x V \langle \eta^2 \rangle}{q^3} \left( \int_0^\infty \gamma(0) u \sin(u)\, du + \int_0^\infty \frac{\gamma'(0)}{q} u^2 \sin(u)\, du + \int_0^\infty \frac{\gamma''(0)}{2q^2} u^3 \sin(u)\, du + \cdots \right)
\end{aligned}$$

(B1.9.83)

where $\gamma'$ and $\gamma''$ are the first and second derivatives of $\gamma$. The following expressions can be derived to simplify the above equation.

$$\int_0^\infty x \sin(x)\, dx = 0, \quad \int_0^\infty x^2 \sin(x)\, dx = -2, \quad \int_0^\infty x^3 \sin(x)\, dx = 0.$$

(B1.9.84)

Then

$$I(q) = -\frac{8\pi K_x V \langle \eta^2 \rangle}{q^4}\gamma'(0) + O(q^{-2n}, d^{2n-3}\gamma(0)/dr^{2n-3}) \qquad (n = 3, 4, 5\ldots).$$

(B1.9.85)

The second term in equation (B1.9.85) rapidly approaches zero in the large $q$ region, thus

$$\lim_{q \to \infty} I(q) = -\frac{8\pi K_x V \langle \eta^2 \rangle}{q^4} \gamma'(0). \qquad (B1.9.86)$$

This equation is the same as equation (B1.9.81), which is termed the Porod law. Thus, the scattered intensity $I(q)$ at very large $q$ values will be proportional to the $q^{-4}$ term, this relationship is valid only for sharp interfaces.

The least recognized forms of the Porod approximation are for the *anisotropic* system. If we consider the cylindrical scattering expression of equation (B1.9.61), there are two principal axes ($z$ and $r$ directions) to be discussed

$$I(q) = K_x V \langle \eta^2 \rangle \int_0^\infty \gamma(r) 2\pi J_0(q_r r) r \, dr \int_0^\infty 2\gamma(z) \cos(q_z z) \, dz \qquad (B1.9.61)$$

where $q_r$ and $q_z$ are the scattering vectors along the $r$ or $z$ directions, respectively.

(1) For the component of the scattered intensity along the equatorial direction (i.e., perpendicular to the cylinder direction, $q_z = 0$), equation (B1.9.61) can be simplified as

$$I(q_r) = K'_x V \langle \eta^2 \rangle \int_0^\infty \gamma(r) 2\pi J_0(q r) r \, dr \qquad (B1.9.87)$$

where $K'_x \, (=K_x \int_0^\infty 2\gamma(z) \, dz)$ is a new constant, because the integral of $\gamma(z)$ is a constant. In this equation, the correlation function $\gamma(r)$ can again be expanded by the Taylor series in the region of small $r$, which becomes

$$I(q_r) = \frac{2\pi K'_x V \langle \eta^2 \rangle}{q_r^2} \left( \int_0^\infty \gamma(0) u J_0(u) \, du + \int_0^\infty \frac{\gamma'(0)}{q_r} u^2 J_0(u) \, du + \int_0^\infty \frac{\gamma''(0)}{2q_r^2} u^3 J_0(u) \, du + \cdots \right). \qquad (B1.9.88)$$

The following relationships can be used to simplify the above equation

$$\int_0^\infty x J_0(x) \, dx = 0, \quad \int_0^\infty x^2 J_0(x) \, dx = -1, \quad \int_0^\infty x^3 J_0(x) \, dx = 0. \qquad (B1.9.89)$$

Thus, we obtain

$$I(q_r) = -\frac{2\pi K'_x V \langle \eta^2 \rangle}{q_r^3} \gamma'(0) + O(q_r^{-2n-1}, d^{2n-1} \gamma(0)/dr^{2n-1}) \qquad (n = 2, 3, 4 \ldots). \qquad (B1.9.90)$$

This expression holds true only in the large $q_r$ region in reciprocal space (or small $r$ in real space). Since the second term in equation (B1.9.88) rapidly approaches zero at large $q_r$, we have

$$\lim_{q_r \to \infty} I(q_r) = -\frac{2\pi K'_x V \langle \eta^2 \rangle}{q_r^3} \gamma'(0). \qquad (B1.9.91)$$

This equation is the Porod law for the large-angle tail of the scattering curve along the equatorial direction, which indicates that the equatorial scattered intensity $I(q_r)$ is proportional to $q_r^{-3}$ in the Porod region of an anisotropic system. Cohen and Thomas have derived the following relationships for the two-dimensional two-phase system (with sharp interfaces) such as fibres [29].

$$\langle \eta^2 \rangle = v_1(1 - v_1)(\rho_1 - \rho_2)^2 \qquad (B1.9.92)$$

$$\gamma'(0) = -\frac{L_I}{A} \frac{1}{\pi v_1(1 - v_1)} \qquad (B1.9.93)$$

where $\upsilon$ represents the area fraction of a phase, $L_l$ is the length of the interface between the two phases and $A$ is the total cross sectional area.

(2) For the component of the scattered intensity along the meridional direction (i.e., parallel to the cylinder direction, $q_r = 0$), equation (B1.9.61) can be rewritten as

$$I(q_z) = K_x'' V \langle \eta^2 \rangle \int_0^\infty 2\gamma(z) \cos(q_z z) \, dz \qquad (B1.9.94)$$

where $K_x'' (=K_x \int_0^\infty 2\pi \gamma(r) r \, dr)$ is a constant, because the integral of $\gamma(r) r$ is a constant. Again, we can expand the term $\gamma(z)$ using the Taylor series in the small $z$ region to derive the Porod law. Equation (B1.9.94) thus becomes ($q_z z = v$).

$$I(q_z) = \frac{2K_x'' V \langle \eta^2 \rangle}{q_z} \left( \int_0^\infty \gamma(0) \cos(v) \, dv + \int_0^\infty \frac{\gamma'(0)}{q_z} v \cos(v) \, dv + \int_0^\infty \frac{\gamma''(0)}{2q_z^2} v^2 \cos(v) \, dv + \cdots \right).$$
$$(B1.9.95)$$

The following expressions can be derived to simplify (B1.9.95)

$$\int_0^\infty \cos(x) \, dx = 0, \quad \int_0^\infty x \cos(x) \, dx = -1, \quad \int_0^\infty x^2 \cos(x) \, dx = 0. \qquad (B1.9.96)$$

Thus,

$$I(q_z) = -\frac{2K_x'' V_s \langle \eta^2 \rangle}{q_z^2} \gamma'(0) + O(q_z^{-2n}, d^{2n-1}\gamma(0)/dr^{2n-1}) \qquad (n = 2, 3, 4 \ldots). \qquad (B1.9.97)$$

Again, the second term in (B1.9.97) rapidly approaches zero at large $q_z$, thus we obtain

$$\lim_{q_z \to \infty} I(q_z) = -\frac{2K_x'' V \langle \eta^2 \rangle}{q_z^2} \gamma'(0). \qquad (B1.9.98)$$

If the scatterers are elongated along the fibre direction and the two phases and their interfaces have the same consistency in both radial and $z$ directions, a different expression of the Porod law can be derived.

$$\lim_{q_z \to \infty} I(q_z) = -\frac{2K_x'' V (\rho_1 - \rho_2)^2}{\pi q_z^2} \frac{C}{A} \qquad (B1.9.99)$$

This is the Porod law for the large angle tail of the scattering curve in the meridional direction. In this case, the scattered intensity is proportional to $q_z^{-2}$ at large scattering angles.

### B1.9.4.4 Scattering from semicrystalline polymers

Semicrystalline polymers are ideal objects to be studied by small-angle x-ray scattering (SAXS), because electron density variations of the semicrystalline morphology (with alternating crystalline and amorphous structures) have a correlation length of several hundred Ångströms, which falls in the resolution range of SAXS (1–100 nm). In addition, the semicrystalline structures can usually be described by assuming electron density variations to occur in one coordinate only. In this case, the scattered intensity $I(q)$ can be described by a one-dimensional correlation function $\gamma_1(r)$.

The scattered intensity measured from the isotropic three-dimensional object can be transformed to the one-dimensional intensity function $I_1(q)$ by means of the Lorentz correction [15]

$$I_1(q) = I(q) 4\pi q^2 \qquad (B1.9.100)$$

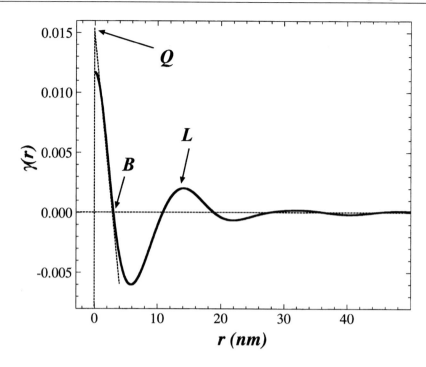

**Figure B1.9.11.** The analysis of correlation function using a lamellar model.

where the term $4\pi q^2$ represents the scattering volume correction in space. In this case, the correlation and interface distribution functions become

$$\gamma_1(r) = \left( \int_0^\infty I_1(q) \cos(qr) \, dq \right) \Big/ Q \tag{B1.9.101}$$

$$q_1(r) = \partial^2(\gamma_1(r))/\partial r^2 = \left( -\int_0^\infty I_1(q) q^2 \cos(qr) \, dq \right) \Big/ Q \tag{B1.9.102}$$

where $Q(= \int_0^\infty I_1(q) \, dq = 4\pi \int_0^\infty I(q) q^2 \, dq)$ is the invariant. The above two equations are valid only for lamellar structures.

Lamellar morphology variables in semicrystalline polymers can be estimated from the correlation and interface distribution functions using a two-phase model. The analysis of a correlation function by the two-phase model has been demonstrated in detail before [30, 31]. The thicknesses of the two constituent phases (crystal and amorphous) can be extracted by several approaches described by Strobl and Schneider [32]. For example, one approach is based on the following relationship:

$$x_1 x_2 = \frac{B}{L} \tag{B1.9.103}$$

where $x_1$ and $x_2$ are the linear fractions of the two phases within the lamellar morphology, $B$ is the value of the abscissa when the ordinate is first equal to zero in $\gamma_1(r)$ and $L$ represents the long period determined as the first maximum of $\gamma_1(r)$ (figure B1.9.11).

The analysis of the interface distribution function $g_1(r)$ is relatively straightforward [33]. The profile of $g_1(r)$ can be directly calculated from the Fourier transformation of the interference function or by taking

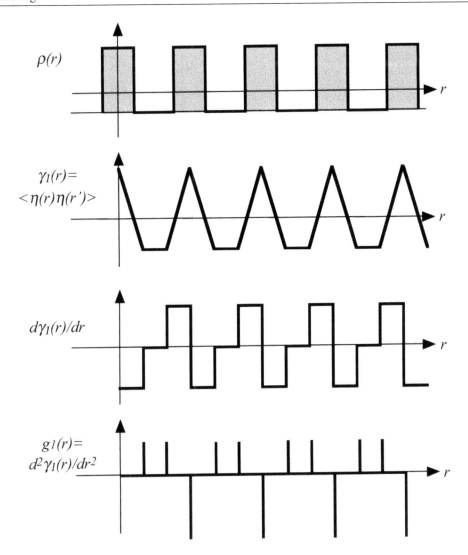

**Figure B1.9.12.** The schematic diagram of the relationships between the one-dimensional electron density profile, $\rho(r)$, correlation function $\gamma_1(r)$ and interface distribution function $g_1(r)$.

the second derivative of the correlation function (B1.9.102). In the physical sense, the interface distribution function represents the probability of finding an interface along the density profile. A positive value indicates an even number of interfaces within a real space distance with respect to the origin. A negative value indicates an odd number of interfaces within the corresponding distance. With lamellar morphology, odd numbers of interfaces correspond to integral numbers of long periods. The shape of the probability distribution with distance for a given interface manifests as the shape of the corresponding peak on the interface distribution function. These distributions can be deconvoluted to reveal more detailed morphological parameters [34]. The schematic diagram of the relationships between the one-dimensional electron density profile, $\rho(r)$, correlation function, $\gamma_1(r)$, and interface distribution function, $g_1(r)$, is shown in figure B1.9.12. In general, we find the values of the long period calculated from different methods, such as a conventional analysis by using Bragg's law, the correlation function and the interface distribution function, to be quite different.

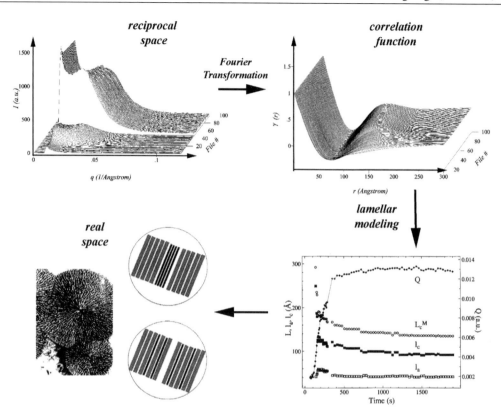

**Figure B1.9.13.** Time-resolved SAXS profiles during isothermal crystallization (230 °C) of PET (the first 48 scans were collected with 5 seconds scan time, the last 52 scans were collected with 30 seconds scan time); calculated correlation functions $\gamma(r)$ (normalized by the invariant $Q$) and lamellar morphological variables extracted from the correlation functions (invariant $Q$, long period $L_c^M$, crystal lamellar thickness $l_c$ and interlamellar amorphous thickness $l_a$).

However, their trends as functions of time and temperature are usually similar. The ordering of these long periods indicates the heterogeneity of the lamellar distributions in the morphology [35].

In oriented systems (fibres or stretched films), the scattered image often appears as a two-bar or a four-point pattern with the scattering maximum at or near the meridian (fibre axis). The one-dimensional scattered intensity along the meridian must be calculated by the projection method using the following formalism

$$I_1(q_z) = \int_0^\infty I(q_r, q_z) q_r \, \mathrm{d}q_r. \tag{B1.9.104}$$

This intensity can be used to calculate the correlation function (B1.9.101) and the interface distribution function (B1.9.102) and to yield the lamellar crystal and amorphous layer thicknesses along the fibre.

Recently, a unique approach for using the correlation function method has been demonstrated to extract morphological variables in crystalline polymers from time-resolved synchrotron SAXS data. The principle of the calculation is based on two alternative expressions of Porod's law using the form of interference function [33, 36]. This approach enables a continuous estimate of the Porod constant, corrections for liquid scattering and finite interface between the two phases, from the time-resolved data. Many detailed morphological variables such as lamellar long period, thicknesses of crystal and amorphous phases, interface thickness and scattering invariant can be estimated. An example analysis of isothermal crystallization in poly(ethyleneterephthalate) (PET) at 230 °C measured by synchrotron SAXS is illustrated here. Time-resolved synchrotron SAXS profiles after the removal of background scattering (air and windows), calculated correlation function profiles and morphological variables extracted by using the two-phase crystal lamellar model are shown in figure B1.9.13. Two distinguishable stages are seen in this figure (the first stage was collected at 5 seconds per scan; the later stage was collected at 30 seconds per scan). It is seen that the long period $L_c^M$ and crystal lamellar thickness $l_c$ decrease with time. This behaviour can be explained by the space filling of thinner secondary crystal lamellae after the initial formation of thicker primary lamellae during isothermal crystallization [36].

## B1.9.5   Neutron scattering

Neutron scattering depends upon nuclear properties, which are related to fluctuations in the neutron scattering cross section $\sigma$ between the scatterer and the surroundings. The scattered amplitude from a collection of scatterers can thus be written as (similar to (B1.9.29)):

$$F(q) = \sum b_j \exp(iqr_j) \tag{B1.9.105}$$

where $b_j$ is referred to as the 'scattering length' of the object $j$; its square is related to the scattering cross section ($\sigma_j = 4\pi b_j^2$) of the $j$th object. The value of $b_j$ depends on the property of the nucleus and is generally different for different isotopes of the same element. As neutron scattering by nature is a nuclear event and the wavelength used in neutron scattering is much larger than the nuclear dimensions, the intranuclear interference of waves due to neutron scattering need not be considered such that $b_j$ is normally independent of scattering angle.

In neutron scattering, the scattered intensity is often expressed in terms of the differential scattering cross section:

$$\frac{\partial \sigma}{\partial \Omega} = \sum_{i,j}^{N} \langle b_i b_j \exp(iq(r_j - r_i)) \rangle \tag{B1.9.106}$$

where $d\Omega$ is the small solid angle into which the scattered neutrons are accepted. We can mathematically divide the above equation into two parts:

$$\frac{\partial \sigma}{\partial \Omega} = \sum_{j=1}^{N} b_j^2 + \sum_{i \neq j}^{N} \langle b_i b_j \exp(iq(r_j - r_i)) \rangle = N\langle b^2 \rangle + \langle b \rangle^2 \sum_{i \neq j}^{N} \langle \exp(iq(r_j - r_i)) \rangle. \tag{B1.9.107}$$

If we define

$$\overline{\Delta b^2} = \langle b^2 \rangle - \langle b \rangle^2 \tag{B1.9.108}$$

we can obtain the following relationship:

$$\frac{\partial \sigma}{\partial \Omega} = N\overline{\Delta b^2} + \langle b \rangle^2 \sum_{i,j}^{N} \langle \exp(iq(r_j - r_i)) \rangle \tag{B1.9.109}$$

where $\langle b \rangle^2 = b^2$. The first term represents the incoherent scattering, which depends only on the fluctuations of the scattering length $b$ occupying the different positions. The second term is the coherent scattering, which depends on the positions of all the scattering centres and is responsible for the angular dependence of the scattered intensity. In this chapter, we shall only focus on the phenomenon of coherent scattering. Using the concept of correlation function (as in equation (B1.9.54)), we obtain

$$I(q) = \left( \frac{\partial \sigma}{\partial \Omega} \right)_{coh} = K_n b^2 \int \gamma(r) \, e^{-iqr} d^3 r \tag{B1.9.110}$$

where $K_n$ is a calibration constant depending on the incident flux and apparatus geometry. The expressions of scattered intensity for isotropic and anisotropic systems can be obtained similarly to equations (B1.9.49) and (B1.9.50).

### B1.9.5.1   Scattering from multicomponent systems

Equation (B1.9.110) is for a system containing only one type of scattering length. Let us consider the system containing more than one species, such as a two-species mixture with $N_1$ molecules of scattering length $b_1$ and $N_2$ of scattering length $b_2$. The coherent scattered intensity $I(q)$ becomes

$$I(q) = K_n \left\{ b_1^2 \sum_{i_1}^{N_1} \sum_{j_1}^{N_1} \langle \exp(-iqr_{ij}) \rangle + 2b_1 b_2 \sum_{i_1}^{N_1} \sum_{j_2}^{N_2} \langle \exp(-iqr_{ij}) \rangle + b_2^2 \sum_{i_2}^{N_2} \sum_{j_2}^{N_2} \langle \exp(-iqr_{ij}) \rangle \right\} \quad \text{(B1.9.111)}$$

or

$$I(q) = K_n \left\{ b_1^2 \int \gamma_{11}(r) \, e^{-iqr} \, d^3r + 2b_1 b_2 \int \gamma_{12}(r) \, e^{-iqr} \, d^3r + b_2^2 \int \gamma_{22}(r) \, e^{-iqr} \, d^3r \right\} \quad \text{(B1.9.112)}$$

where $\gamma_{ij}$ is the correlation function between the local density of constituents $i$ and $j$. We can define three partial structural factors, $S_{11}$, $S_{12}$ and $S_{22}$

$$S_{ij}(r) = K_n \int \gamma_{ij}(r) \, e^{-iqr} \, d^3r. \quad \text{(B1.9.113)}$$

Thus

$$I(q) = b_1^2 S_{11}(q) + 2b_1 b_2 S_{12}(q) + b_2^2 S_{22}(q). \quad \text{(B1.9.114)}$$

If the two constituting molecular volumes are identical in a two component system, we can obtain [37, 38]

$$S_{11}(q) = S_{22}(q) = -S_{12}(q) \quad \text{(B1.9.115)}$$

or

$$I(q) = (b_1 - b_2)^2 S_{11}(q) = (b_1 - b_2)^2 S_{22}(q) = -(b_1 - b_2)^2 S_{12}(q). \quad \text{(B1.9.116)}$$

Let us assume that we have $p + 1$ different species; equation (B1.9.111) can be generalized as

$$I(q) = \sum_{i=1}^{p} b_i^2 S_{ii}(q) + 2 \sum_{i<j}^{p} b_i b_j S_{ij}(q). \quad \text{(B1.9.117)}$$

### B1.9.5.2   Properties of S(q)

As we have introduced the structure factor $S(q)$ (B1.9.113), it is useful to separate this factor into two categories of interferences for a system containing $N$ scattering particles [9]:

$$S(q) = N[P(q) + NQ(q)]. \quad \text{(B1.9.118)}$$

The first term, $P(q)$, represents the interferences within particles and its contribution is proportional to the number of particle, $N$. The second term, $Q(q)$, involves interparticle interferences and is proportional to the number of pairs of particles, $N^2$.

Let us consider the scattered intensity from a binary incompressible mixture of two species (containing $N_1$ molecules of particle 1 and $N_2$ molecules of particle 2) as in (B1.9.112); we can rewrite the relationship as

$$I(q) = b_1^2 N_1 [P_1(q) + N_1 Q_{11}(q)] + b_2^2 N_2 [P_2(q) + N_2 Q_{22}(q)] + 2b_1 b_2 N_1 N_2 Q_{12}(q) \quad \text{(B1.9.119)}$$

where $P_i(q)$ is the intramolecular interference of species $i$ and $Q_{ij}(q)$ is the intermolecular interferences between species $i$ and $j$.

*B1.9.5.3   Deuterium labelling*

Two of the most important functions in the application of neutron scattering are the use of deuterium labelling for the study of molecular conformation in the bulk state and the use of deuterium solvent in polymer solutions. In the following, we will consider several different applications of the general formula to deuteration.

(1)   Let us first consider two identical polymers, one deuterated and the other not, in a melt or a glassy state. The two polymers (degree of polymerization $d$) differ from each other only by scattering lengths $b_H$ and $b_D$. If the total number of molecules is $N$, $x$ is the volume fraction of the deuterated species ($x = N_D/N$, with $N_D + N_H = N$). According to equation (B1.9.116), we obtain

$$I(q) = (b_D - b_H)^2 S_{DD}(q) \tag{B1.9.120}$$

or $S_{HH}$ or $-S_{HD}$. The coherent scattering factor can further be expressed in terms of $P(q)$ and $Q(q)$ as (see equation (B1.9.117))

$$S_{DD} = xNd^2 P(q) + x^2 N^2 d^2 Q(q) = (1-x)Nd^2 P(q) + (1-x)^2 d^2 N^2 Q(q) = -x(1-x)N^2 d^2 Q(q). \tag{B1.9.121}$$

Thus

$$NQ(q) = -P(q). \tag{B1.9.122}$$

By combining equations (B1.9.120) and (B1.9.122), we have

$$I(q) = (b_D - b_H)^2 x(1-x)Nd^2 P(q). \tag{B1.9.123}$$

(2)   Next, let us consider the case of a system made up of two polymers with different degrees of polymerization $d_D$ for the deuterated species and $d_H$ for the other. The generalized expression of B1.9.122 becomes:

$$-N_D d_D^2 P_D(q) = N_D^2 d_D^2 Q_D(q) + N_D N_H d_D d_H Q_{DH}(q). \tag{B1.9.124}$$

(3)   Let us take two polymers (one deuterated and one hydrogenated) and dissolve them in a solvent (or another polymer) having a scattering length $b_0$. The coherent scattered intensity can be derived from (B1.9.117), which gives

$$I(q) = (b_D - b_0)^2 S_{DD}(q) + (b_H - b_0)^2 S_{HH}(q) + 2(b_D - b_0)(b_H - b_0)S_{HD}(q) \tag{B1.9.125}$$

where

$$\begin{aligned} S_{DD} &= xNd^2 P(q) + x^2 N^2 d^2 Q(q) \\ S_{HH} &= (1-x)Nd^2 P(q) + (1-x)^2 N^2 d^2 Q(q) \\ S_{HD} &= -x(1-x)Nd^2 Q(q). \end{aligned} \tag{B1.9.126}$$

Thus, we have

$$I(q) = (b_D - b_H)^2 x(1-x)Nd^2 P(q) + (xb_D + (1-x)b_H - b_0)^2 Nd^2 [P(q) + dQ(q)]. \tag{B1.9.127}$$

The second term of the above equation gives an important adjustment to contrast variation between the solvent and the polymer. If one adjusts the scattering length $b_0$ of the solvent by using a mixture of deuterated and hydrogenated solvent such that

$$xb_D + (1-x)b_H - b_0 = 0 \tag{B1.9.128}$$

then we can obtain (B1.9.123). This experiment thus yields directly the form factor $P(q)$ of the polymer molecules in solution even at high polymer concentrations.

*B1.9.5.4   Analysis of molecular parameters*

There are many different data analysis schemes to estimate the structure and molecular parameters of polymers from the neutron scattering data. Herein, we will present several common methods for characterizing the scattering profiles, depending only on the applicable $q$ range. These methods, which were derived based on different assumptions, have different limitations. We caution the reader to check the limitations of each method before its application.

(1)  If we deal with a solution at very low concentrations, we can ignore the interactions between the particles and express the scattered intensity as

$$\log I(q) = \log I(0) - \tfrac{1}{3}q^2 R_g^2 + \cdots \tag{B1.9.129}$$

where $R_g$ is the radius of gyration. This equation is similar to (B1.9.35). If one plots $\log[I(q)]$ as a function of $q^2$, the initial part is a straight line with a negative slope proportional to $R_g$, which is called the Guinier plot. This approach is only suitable for scattering in the low $q R_g$ range and in dilute concentrations. A similar expression proposed by Zimm has a slightly different form:

$$\frac{Nd^2}{S(q)} = (1 + \tfrac{1}{3}q^2 R_g^2 + \cdots) \tag{B1.9.130}$$

where $N$ is the number of scattering objects and $d$ is the degree of polymerization. This equation is similar to equation (B1.9.35). Thus if one plots $1/S(q)$ as function $q^2$, the initial slope is $R_g^2/3$.

The above radius of gyration is for an isotropic system. If the system is anisotropic, the mean square radius of gyration is equal to

$$\langle R_g^2 \rangle = \langle R_x^2 \rangle + \langle R_y^2 \rangle + \langle R_z^2 \rangle \tag{B1.9.131}$$

where $R_x$, $R_y$ and $R_z$ are the components of the radius of gyration along the $x, y, z$ axes. For the isotropic system

$$\langle R_x^2 \rangle = \langle R_y^2 \rangle + \langle R_z^2 \rangle = \tfrac{1}{3}\langle R_g^2 \rangle. \tag{B1.9.132}$$

(2)  In the intermediate and high $q$ range, the analysis becomes quite different. The qualitative interpretation for the scattering profile at the high $q$ range may make the use of scaling argument proposed by de Gennes [39]. If we neglect the intermolecular interactions, we can write

$$S(q) = Nd^2 P(q R_g) = V \varphi d P(q R_g) \tag{B1.9.133}$$

where $V$ is the volume of the sample, $\varphi$ is the volume fraction occupied by the scattering units in polymer with a degree of polymerization $d$ and $q R_g$ is a dimensionless quantity which is associated with a characteristic dimension. Typically, the term $P(q R_g)$ can be approximated by $(q R_g)^{-\beta}$. Since the relationships between $R_g$ and $z$ are known as follows:

$$\begin{aligned}
\text{a Gaussian chain} \quad & R_g \approx d^{0.5} \\
\text{a chain with excluded volume (Flory)} \quad & R_g \approx d^{0.6} \\
\text{a rod} \quad & R_g \approx d^{1.0} \\
\text{general expression} \quad & R_g \approx d^a.
\end{aligned} \tag{B1.9.134}$$

This leads to the following equation:

$$S(q) = V \varphi d (q d^a)^{-\beta} = V \varphi q^{-\beta} d^{1-a\beta}. \tag{B1.9.135}$$

In order to have $S(q)$ independent of $d$, the power of $d$ must be zero giving $a = 1/\beta$. This gives rise to the following relationships:

$$
\begin{array}{rl}
\text{a Gaussian chain} & S(q) = q^{-2} \\
\text{a chain with excluded volume (Flory)} & S(q) = q^{-1.66} \\
\text{a rod} & S(q) = q^{-1}
\end{array}
\tag{B1.9.136}
$$

(3) The quantitative analysis of the scattering profile in the high $q$ range can be made by using the approach of Debye *et al* as in equation (B1.9.52). As we assume that the correlation function $\gamma(r)$ has a simple exponential form ($\gamma(r) = \exp(-r/a_c)$, where $a_c$ is the correlation length), the scattered intensity can be expressed as

$$
I(q) = \frac{8\pi a_c^3 b_v^2 \varphi (1-\varphi)}{(1+q^2 a_c^2)^2}
\tag{B1.9.137}
$$

where $\varphi$ is the volume fraction of the component (with scattering length $b_1$ and volume of the monomer of the polymer $v_1$), $(1-\varphi)$ is the volume fraction of the solvent (with scattering length $b_0$ and volume of the monomer of the other polymer $v_0$) and $b_v = b_1/v_1 - b_0/v_0$. Thus the correlation length $a_c$ can be calculated by plotting $I(q)^{-1/2}$ versus $q^2$, using equation (B1.9.53) (as light scattering).

A more general case of continuously varying density was treated by Ornstein and Zernicke for scattering of opalescence from a two-phase system [40]. They argued that

$$
\gamma(r) = \text{constant} \frac{\exp(-r/\xi)}{r}
\tag{B1.9.138}
$$

where $\xi$ is a characteristic length. This leads to

$$
S(q) = \frac{B}{1+q^2\xi^2}
\tag{B1.9.139}
$$

where $B$ is a constant.

### B1.9.5.5 Calculate thermodynamic parameter $\chi$

In polymer solutions or blends, one of the most important thermodynamic parameters that can be calculated from the (neutron) scattering data is the enthalpic interaction parameter $\chi$ between the components. Based on the Flory–Huggins theory [41, 42], the scattering intensity from a polymer in a solution can be expressed as

$$
\frac{1}{s(q)} = \frac{1}{\varphi z P(q)} + \frac{1}{\varphi_0} - 2\chi
\tag{B1.9.140}
$$

where $s(q) = S(q)/N_T$ corresponds to the scattered intensity from a volume equal to that of a molecule of solvent, $N_T$ is the ratio of the solution volume to the volume of one solvent molecule ($N_T = V/v_s$), $\varphi$ is the volume fraction occupied by the polymer and $\varphi_0$ is the volume fraction occupied by the solvent, $v_s$ is the solvent specific volume. For a binary polymer blend, equation (B1.9.140) can be generalized as:

$$
\frac{1}{s(q)} = \frac{1}{\varphi_1 z_1 P_1(q)} + \frac{1}{\varphi_2 z_2 P_2(q)} - 2\chi
\tag{B1.9.141}
$$

where the subscripts 1 and 2 represent a solution of polymer 1 in polymer 2. The generalized Flory–Huggins model in the de Gennes formalism with the random phase approximation has the form:

$$
\frac{V}{I(q)} = \frac{1}{b_v^2} \left\{ \frac{1}{(C_1 M_1/N_A) P_1(q)(v_1^s)^2} + \frac{1}{(C_2 M_2/N_A) P_2(q)(v_2^s)^2} - \frac{2\chi}{v_0} \right\}
\tag{B1.9.142}
$$

where $b_v$ is the contrast factor as described before, $N_A$ is Avogadro's number, $C$ is the concentration, $M$ is the molecular weight, $v_s$ is the specific volume ($C_1 v_1^s + C_2 v_2^s = 1$) and $v_0$ is an arbitrary reference volume.

In the case of low concentration and low $q$ expansion, equation (B1.9.142) can be expressed by replacing $P_i(q)$ with the Guinier approximation (B1.9.130)

$$\frac{V v_0 b_v^2}{I(q)} = \left\{ \frac{1}{\varphi_1 z_1} + \frac{1}{\varphi_2 z_2} - 2\chi \right\} + \frac{1}{3} \left\{ \frac{R_{g1}^2}{\varphi_1 z_1} + \frac{R_{g2}^2}{\varphi_2 z_2} \right\} q^2. \tag{B1.9.143}$$

Thus the slope of the $I(q)^{-1}$ *versus* $q^2$ plot is related to the values of two radii of gyration.

## B1.9.6  Concluding remarks

In this chapter, we have reviewed the general scattering principles from matter by light, neutrons and x-rays and the data treatments for the different states of matter. The interaction between radiation and matter has the same formalism for all three cases of scattering, but the difference arises from the intrinsic property of each radiation. The major difference in data treatments results from the different states of matter. Although we have provided a broad overview of the different treatments, the content is by no means complete. Our objective in this chapter is to provide the reader a general background for the applications of scattering techniques to materials using light, neutrons and x-rays.

## References

[1]   van de Hulst H C 1957 *Light Scattering by Small Particles* (New York: Wiley)
[2]   Kerker M 1969 *The Scattering of Light and Other Electromagnetic Radiation* (New York: Academic)
[3]   Berne B J and Pecora R 1976 *Dynamic Light Scattering* (New York: Wiley–Interscience)
[4]   Chen S H, Chu B and Nossal R (eds) 1981 *Scattering Techniques Applied to Supramolecular and Nonequilibrium Systems* (New York: Plenum)
[5]   Chu B 1991 *Laser Light Scattering, Basic Principles and Practice* 2nd edn (New York: Academic) (See also the first edition (published in 1974) that contains more mathematical derivations.)
[6]   Schmitz K S 1990 *An Introduction to Dynamic Light Scattering by Macromolecules* (New York: Academic)
[7]   Linden P and Zemb Th (eds) 1991 *Neutron, X-ray and Light Scattering: Introduction to an Investigation Tool for Colloidal and Polymeric Systems* (Amsterdam: North-Holland)
[8]   Lovesey S W 1984 *Theory of Neutron Scattering from Condensed Matter* vol 1 (Oxford: Oxford University Press)
[9]   Higgins J S and Benoˆit H C 1994 *Polymers and Neutron Scattering* (Oxford: Oxford Science)
[10]  Wignall G D 1987 *Encyclopedia of Polymer Science and Engineering* vol 12 (New York: Wiley) p 112
[11]  Wignall G D, Crist B, Russell T P and Thomas E L (eds) 1987 *Mater. Res. Soc. Symp. Proc.* (Pittsburgh, PA: Materials Research Society) vol 79
[12]  Guinier A and Fournet G 1955 *Small Angle Scattering of X-rays* (New York: Wiley)
[13]  Brumberger H (ed) 1967 *Small-Angle X-ray Scattering* (New York: Gordon and Breach)
[14]  Glatter O and Kratky O 1982 *Small Angle X-ray Scattering* (New York: Academic)
[15]  Baltá-Calleja F J and Vonk G G 1989 *X-ray Scattering of Synthetic Polymers* (New York: Elsevier)
[16]  Ditchburn R W 1953 *Light* (New York: Wiley–Interscience)
[17]  Debye P 1915 *Ann. Phys., Lpz.* **46** 809
[18]  Pusey P N and Tough R A 1985 *Dynamic Light Scattering* ed R Pecora (New York: Plenum) ch 4
[19]  Chu B, Wu C and Zuo J 1987 *Macromolecules* **20** 700
[20]  Debye P and Bueche A M 1949 *J. Appl. Phys.* **20** 518
[21]  Debye P, Anderson H R and Brumberger H 1957 *J. Appl. Phys.* **28** 679
[22]  Glatter O 1979 *J. Appl. Crystallogr.* **12** 166
[23]  Glatter O 1977 *J. Appl. Crystallogr.* **10** 415
[24]  Glatter O 1981 *J. Appl. Crystallogr.* **14** 101
[25]  Strey R, Glatter O, Schubert K V and Kaler E W 1996 *J. Chem. Phys.* **105** 1175
[26]  Meisenberger O, Pilz I and Heumann H 1980 *FEBS Lett.* **112** 39
[27]  Meisenberger O, Heumann H and Pilz I 1980 *FEBS Lett.* **122** 117
[28]  Porod G 1952 *Kolloid Z. Z. Polym.* **125** 51
      Porod G 1952 *Kolloid Z. Z. Polym.* **125** 108

[29] Cohen Y and Thomas E 1988 *Macromolecules* **21** 433
     Cohen Y and Thomas E 1988 *Macromolecules* **21** 436
[30] Vonk C G and Kortleve G 1967 *Colloid Polym. Sci.* **220** 19
[31] Vonk G G 1973 *J. Appl. Crystallogr.* **6** 81
[32] Strobl G R and Schneider M 1980 *J. Polym. Sci.* B **18** 1343
[33] Ruland W 1977 *Colloid. Polym. Sci.* **255** 417
[34] Stribeck N and Ruland W 1978 *J. Appl. Crystallogr.* **11** 535
[35] Santa Cruz C, Stribeck N, Zachmann H G and Balta-Calleja F J 1991 *Macromolecules* **24** 5980
[36] Hsiao B S and Verma R K 1998 *J. Synchrotron Radiat.* **5** 23
[37] Daoud M, Cotton J P, Farnoux B, Jannink G, Sarma S and Benoît H H 1975 *Macromolecules* **8** 804
[38] Akcasu A, Summerfield G C, Jahshan S N, Han C C, Kim C Y and Yu H 1980 *J. Polym. Sci. Polym. Phys.* **18** 863
[39] de Gennes P G 1979 *Scaling Concepts in Polymer Physics* (Ithaca, NY: Cornell University Press)
[40] Ornstein L S and Zernike F 1914 *Proc. Acad. Sci. Amsterdam* **17** 793
[41] Flory P J 1942 *J. Chem. Phys.* **10** 51
[42] Huggins M L 1942 *J. Phys. Chem.* **46** 151

# B1.10
# Coincidence techniques

## Michael A Coplan

### B1.10.1 Introduction

Time is a fundamental variable in almost all experimental measurements. For some experiments, time is measured directly, as in the determination of radioactive halflives and the lifetimes of excited states of atoms and molecules. Velocity measurements require the measurement of the time required for an object to travel a fixed distance. Time measurements are also used to identify events that bear some correlation to each other and as a means for reducing background noise in experiments that would otherwise be difficult or impossible to perform. Examples of time correlation measurements are so-called coincidence measurements where two or more separate events can be associated with a common originating event by virtue of their time correlation. Positron annihilation in which a positron and an electron combine to yield two gamma rays, is an example of this kind of coincidence measurement. Electron impact ionization of atoms in which an incident electron strikes an atom, ejects an electron and is simultaneously scattered is another example. In such an experiment the ejected and scattered electrons originate from the same event and their arrival at two separate detectors is correlated in time. In one of the very first coincidence experiments Bothe and Geiger [1] used the time correlation between the recoil electron and inelastically scattered x-ray photon, as recorded on a moving film strip, to identify Compton scattering of an incident x-ray and establish energy conservation on the microscopic level. An example of the use of time correlation to enhance signal-to-noise ratios can be found in experiments where there is a background signal that is uncorrelated with the signal of interest. The effect of penetrating high-energy charged particles from cosmic rays can be eliminated from a gamma ray detector by construction of an anti-coincidence shield. Because a signal from the shield will also be accompanied by a signal at the detector, these spurious signals can be eliminated.

Experiments in almost all subjects from biophysics to high-energy particle physics and cosmic ray physics use time correlation methods. There are a few general principles that govern all time correlation measurements and these will be discussed in sufficient detail to be useful in not only constructing experiments where time is a parameter, but also in evaluating the results of time measurements and optimizing the operating parameters. In a general way, all physical measurements either implicitly or explicitly have time as a variable. Recognizing this is essential in the design of experiments and analysis of the results. The rapid pace of improvements and innovation in electronic devices and computers have provided the experimenter with electronic solutions to experimental problems that in the past could only be solved with custom hardware.

### B1.10.2 Statistics

#### B1.10.2.1 Correlated and random events

Correlated events are related in time and this time relation can be measured either with respect to an external clock or to the events themselves. Random or uncorrelated events bear no fixed time relation to each other

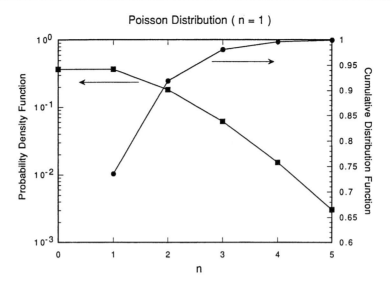

**Figure B1.10.1.** Poisson distribution for $\bar{n} = 1$ (left vertical axis). Cumulative Poisson distribution for $\bar{n} = 1$ (right vertical axis). The cumulative distribution is the sum of the values of the distribution from 0 to $n$, where $n = 1, 2, 3, 4, 5$ on this graph.

but, on the other hand, their very randomness allows them to be quantified. Consider the passing of cars on a busy street. It is possible to calculate the probability, $P_{\bar{n}}(n)$ that $n$ cars pass within a given time interval in terms of the average number of cars, $\bar{n}$, in that interval where $P_{\bar{n}}(n)$, the Poisson distribution is given by [2]

$$P_{\bar{n}}(n) = \frac{\bar{n}^n}{n!}\,\mathrm{e}^{-\bar{n}}.$$

For example, if one finds that 100 cars pass a fixed position on a highway in an hour, then the average number of cars per minute is 100/60. The probability that two cars pass in a minute is given by $\frac{1}{2}0.6^2\,\mathrm{e}^{-0.6} = 0.10$. The probability that three times the average number of cars pass per unit time is 0.02. $P_{\bar{n}}(n)$ and the integral of $P_{\bar{n}}(n)$ for $\bar{n} = 1$ are shown in figure B1.10.1. It is worthwhile to note that only a single parameter, $\bar{n}$, the average value, is sufficient to define the function; moreover, the function is not symmetric about $\bar{n}$.

This example of passing cars has implications for counting experiments. An arrangement for particle counting is shown in figure B1.10.2. It consists of the source of particles, a detector, preamplifier, amplifier/discriminator, counter, and a storage device for recording the results. The detector converts the energy of the particle to an electrical signal that is amplified by a low-noise preamplifier to a level sufficient to be amplified and shaped by the amplifier. The discriminator converts the signal from the amplifier to a standard electrical pulse of fixed height and width, provided that the amplitude of the signal from the amplifier exceeds a set threshold. The counter records the number of pulses from the discriminator for a set period of time. The factors that affect the measurement are counting rate, signal durations, and processing times.

In counting experiments, the instantaneous rate at which particles arrive at the detector can be significantly different from the average rate. In order to assess the rate at which the system can accept data, it is necessary to know how the signals from the detector, preamplifier, amplifier and discriminator each vary with time. For example, if signals are arriving at an average rate of 1 kHz at the detector, the average time between the start of each signal is $10^{-3}$ s. If we consider a time interval of $10^{-3}$ s, the average number of signals arriving in that interval is one; if as many as three pulses can be registered in the $10^{-3}$ s then 99% of all pulses will be counted. To accommodate the case of three pulses in $10^{-3}$ s it is necessary to recognize that the pulses

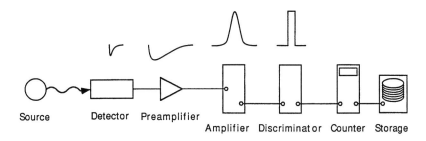

**Figure B1.10.2.** Schematic diagram of a counting experiment. The detector intercepts signals from the source. The output of the detector is amplified by a preamplifier and then shaped and amplified further by an amplifier. The discriminator has variable lower and upper level thresholds. If a signal from the amplifier exceeds the lower threshold while remaining below the upper threshold, a pulse is produced that can be registered by a preprogrammed counter. The contents of the counter can be periodically transferred to an on-line storage device for further processing and analysis. The pulse shapes produced by each of the devices are shown schematically above them.

will be randomly distributed in the $10^{-3}$ s interval. To register the three randomly distributed pulses in the $10^{-3}$ s interval requires approximately another factor of three in time resolution, with the result that the system must have the capability of registering events at a 10 kHz uniform rate in order to be able to register with 99% efficiency randomly arriving events arriving at a 1 kHz average rate. For 99.9% efficiency, the required system bandwidth increases to 100 kHz. Provided that the discriminator and counter are capable of handling the rates, it is then necessary to be sure that the duration of all of the electronic signals are consistent with the required time resolution. For 1 kHz, 10 kHz and 100 kHz bandwidths this means signal durations of 0.1, 0.01, and 0.001 ms, respectively. An excellent discussion of the application of statistics to physics experiments is given by Melissinos [3].

The processing times of the electronic units must also be taken into consideration. There are propagation delays associated with the active devices in the electronics circuits and delays associated with the actual registering of events by the counter. Processing delays are specified by the manufacturers of counters in terms of maximum count rate: a 10 MHz counter may only count at a 10 MHz rate if the input signals arriving at a uniform rate. For randomly arriving signals with a 10 MHz average rate, a system with a 100 MHz bandwidth is required to record 99% of the incoming events.

### B1.10.2.2  Statistical uncertainties

In counting experiments the result that is sought is a rate or the number of events registered per unit time. For convenience we divide the total time of the measurement, $T$, into $n$ equal time intervals, each of length $T/n$. If there are $N_i$ counts registered in interval $i$, then the mean or average rate is given by

$$\frac{1}{n} \sum_{i=1}^{n} \frac{N_i}{T/n} = \frac{N_N}{T}$$

where $N_N$ is the total number of counts registered during the course of the experiment. To assess the uncertainty in the overall measured rate, we assume that the individual rate measurements are statistically distributed about the average value, in other words, they arise from statistical fluctuations in the arrival times of the events and not from uncertainties in the measuring instruments. Here the assumption is that all events are registered with 100% efficiency and that there is negligible uncertainty in the time intervals over which the events are registered.

When the rate measurement is statistically distributed about the mean, the distribution of events can be described by the Poisson distribution, $P_{\bar{n}}(n)$, given by

$$P_{\bar{n}}(n) = \frac{\bar{n}^n}{n!}\, e^{-\bar{n}}$$

where $n$ is the number of events per unit time and the average number of events per unit time is $\bar{n}$. The uncertainty in the rate is given by the standard deviation and is equal to the square root of the average rate, $\sqrt{\bar{n}}$. Most significant is the relative uncertainty, $\sqrt{N/t_T}/N/t_T$. The relative uncertainty decreases as the number of counts per counting interval increases. For a fixed experimental arrangement the number of counts can be increased by increasing the time of the measurement. As can be seen from the formula, the relative uncertainty can be made arbitrarily small by increasing the measurement time; however, improvement in relative uncertainty is proportional to the reciprocal of the square root of the measurement time. To reduce the relative uncertainty by a factor of two, it is necessary to increase the measurement time by a factor of four. One soon reaches a point where the fractional improvement in relative uncertainty requires a prohibitively long measurement time.

In practice, the length of time for an experiment depends on the stability and reliability of the components. For some experiments, the solar neutrino flux and the rate of decay of the proton being extreme examples, the count rate is so small that observation times of months or even years are required to yield rates of sufficiently small relative uncertainty to be significant. For high count rate experiments, the limitation is the speed with which the electronics can process and record the incoming information.

In this section we have examined the issue of time with respect to the processing and recording of signals and also with regard to statistical uncertainty. These are considerations that are the basis for the optimization of more complex experiments where the time correlation between sets of events or among several different events are sought.

### B1.10.3  Time-of-flight experiments

Time-of-flight experiments are used to measure particle velocities and particle mass per charge. The typical experiment requires *start* and *stop* signals from detectors located at the beginning and end of the flight path, see figure B1.10.3. The *time-of-flight* is then the time difference between the signals from the stop and start detectors. The *start* signal is often generated by the opening of a shutter and *stop* by the arrival of the particle at a *stop* detector. Alternatively, the *start* signal may be generated by the particle passing through a *start* detector that registers the passing of the particle without altering its motion in a significant way. The result of accumulating thousands of time-of-flight signals is shown in figure B1.10.4, where the number of events with a given time-of-flight is plotted against time-of-flight specified as channel number. The data for the figure were acquired for a gas sample of hydrogen mixed with air. The different flight times reflect the fact that the ions all have the same kinetic energy.

#### B1.10.3.1   Sources of uncertainties

The measurement of velocity is given by $L/(t_2 - t_1)$, where $L$ is the effective length of the flight path and $t_2 - t_1$ is the difference in time between the arrival of the particle at the stop detector ($t_2$) and the start detector ($t_1$). The uncertainty in the velocity is

$$\frac{1}{\Delta t}\, \mathrm{d}L - \frac{L}{\Delta t^2}\, \mathrm{d}\Delta t$$

where $\Delta t = t_2 - t_1$.

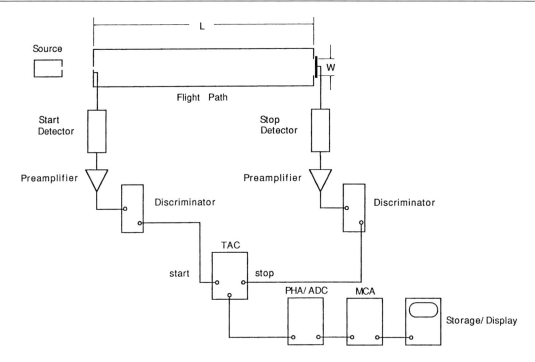

**Figure B1.10.3.** Time-of-flight experiment. Detectors at the beginning and end of the flight path sense the passage of a particle through the entrance and exit apertures. The width of the exit aperture, $W$, determines the amount of transverse velocity particles can have and still be detected at the end of the flight path. Transverse velocity contributes to the dispersion of flight times for identical particles with the same kinetic energies. Detector signals are amplified by a preamplifier; threshold discriminators produce standard pulses whenever the incoming signals exceed an established threshold level. Signals from the start and stop discriminators initiate the operation of the TAC (time-to-amplitude converter). At the end of the TAC cycle, a pulse is produced with amplitude proportional to the time between the start and stop signals. The TAC output pulse amplitude is converted to a binary number by a pulse height analyser (PHA) or analogue-to-digital converter (ADC). The binary number serves as an address for the multichannel analyser (MCA) that adds one to the number stored at the specified address. In this way, a collection of flight times, a time spectrum, is built up in the memory of the MCA. The contents of the MCA can be periodically transferred to a storage device for analysis and display.

The relative uncertainty is

$$\sqrt{\left(\frac{\mathrm{d}L}{L}\right)^2 + \left(\frac{\mathrm{d}t}{\Delta t}\right)^2}.$$

In a conventional time-of-flight spectrometer, the transverse velocities of the particles and the angular acceptance of the flight path and *stop* detector determine the dispersion in $L$. The dispersion in flight times is given by the time resolution of the *start* and *stop* detectors and the associated electronics. When a shutter is used there is also a time uncertainty associated with its opening and closing. A matter that cannot be overlooked in time-of-flight measurements is the rate at which measurements can be made. In the discussion the implicit assumption has been that only one particle is in the flight path at a time. If there is more than one, giving rise to multiple *start* and *stop* signals, it is not possible to associate a unique *start* and *stop* signal with each particle. It cannot be assumed that the first start signal and the first stop signals have been generated by the same particle, because the faster particles can overtake the slower ones in the flight path. When shutters are used, the opening time of the shutter must be sufficiently short to allow no more than one particle to enter

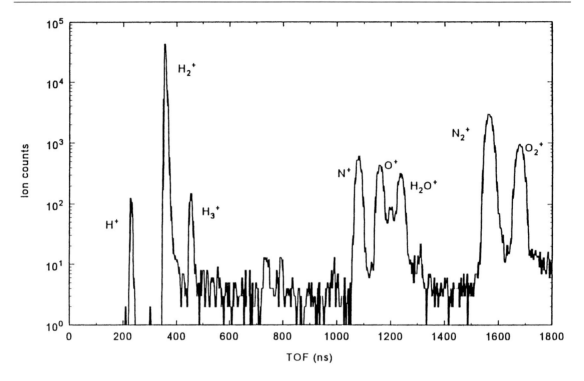

**Figure B1.10.4.** Time-of-flight histogram for ions resulting from the ionization of a sample of air with added hydrogen. The ions have all been accelerated to the same energy (2 keV) so that their time of flight is directly proportional to the reciprocal of the square root of their mass.

the flight path at a time. These considerations give rise to constraints on the rate which measurements can be made. If the longest time-of-flight is $T_{max}$, the particles must not be allowed to enter the flight path at a rate that exceeds $0.1/T_{max}$. Detector response time and the processing time of the electronics should be taken into consideration when calculating $T_{max}$. Time-of-flight experiments are inherently inefficient because the rate at which the shutter is opened is set by the time-of-flight of the slowest particle while the duration of the shutter opening is set by the total flux of particles incident on the shutter. The maximum uncertainty in the measured time-of-flight is, on the other hand, determined by $T_{min}$, the time for the fastest particle to traverse the flight path.

Many of the electronics in the time-of-flight system are similar to those in the counting experiment, with the exception of the time-to-amplitude converter (TAC) and analogue-to-digital converter (ADC). The TAC has *start* and *stop* inputs and an output. The output is a pulse of amplitude (height) proportional to the time difference between the *stop* and *start* pulses. Different scale settings allow adjustment of both the pulse height and time range. Typical units also include *true start* and *busy* outputs that allow monitoring of the input *start* rate and the interval during which the unit is busy processing input signal and therefore unavailable for accepting new signals.

The output of the TAC is normally connected to the input of a pulse height analyser (PHA), or ADC and a multichannel analyser (MCA). The PHA/ADC assigns a binary number to the height of the input pulse; 0 for a zero amplitude pulse and $2^n$ for the maximum amplitude pulse, where $n$ is an integer that can be selected according to the application. The binary number is used as the address for a multichannel analyser with $2^n$ address locations in the MCA. For each address from the PHA/ADC, a one is added to the contents of the address location. If, for example, a TAC has a range of 0 to 4 V output amplitude for *stop/start* time differences of 200 ns and the PHA/ADC assigns binary 0 to the 0 V amplitude signal and binary 255 to the 4 V amplitude signal, each of the 256 channels will correspond to 0.78 ns. With time, a histogram of flight times is built up in the memory of the MCA. Figure B1.10.4 is an illustration of a time-of-flight spectrum for a sample of air that has been ionized and accelerated to 2000 eV. The different flight times are a measure of the mass per charge of the ions.

An alternative to the TAC and PHA/ADC is the time-to-digital converter (TDC), a unit that combines the functions of the TAC and PHA/ADC. There are *start* and *stop* inputs and an output that provides a binary number directly proportional to the time difference between the stop and start signals. The TAC can be directly connected to a MCA or PC with the appropriate digital interface.

### B1.10.3.2  *Electronics limitations*

By storing the binary outputs of a PHA/ADC or a TDC in the form of a histogram in the memory of a MCA, computer analyses can be performed on the data that take full account of the limitations of the individual components of the measuring system. For the preamplifiers and discriminators, time resolution is the principal consideration. For the TAC, resolution and processing time can be critical. The PHA/ADC or a TDC provide the link between the analogue circuits of the preamplifiers and discriminators and the digital input of the MCA or PC. The conversion from analogue-to-digital by a PHA/ADC or TDC is not perfectly linear and the deviations from linearity are expressed in terms of differential and integral nonlinearities. Differential nonlinearity is the variation of input signal amplitude over the width of a single time channel. Integral nonlinearity is the maximum deviation of the measured time channel number from a least squares straight line fit to a plot of signal amplitude as a function of channel number.

### B1.10.4  Lifetime measurements

#### B1.10.4.1  *General considerations*

Lifetime measurements have elements in common with both counting and time-of-flight experiments [4, 5]. In a lifetime experiment there is an initiating event that produces the system that subsequently decays with the emission of radiation, particles or both. Decay is statistical in character; taking as an example nuclear decay, at any time $t$, each nucleus in a sample of $N$ nuclei has the same probability of decay in the interval $dt$. Those nuclei that remain then have the same probability of decaying in the next interval $dt$. The rate of decay is given by $dN/dt = -kN$. The constant $k$, with units 1/time, is called the *lifetime* and depends on the system and the nature of the decay. Integration of the first-order differential equation gives the exponential decay law, $N(t) = N_0 e^{-t/\tau}$, where $N_0$ is the number of systems (atoms, molecules, nuclei) initially created and $N(t)$ is the number that remain after a time $t$. The constant $\tau = 1/k$ can be obtained by measuring the time for the sample to decay to $1/e$ of the initial size.

A more direct method for lifetime measurements is the delayed coincidence technique [6] in which the time between an initiation event and the emission of a decay product is measured. A schematic diagram of an apparatus used for the measurement of atomic lifetimes is shown in figure B1.10.5. The slope of the graph of the natural log of the number of decay events as a function of time delay gives the lifetime directly. The precision with which the slope can be determined increases with the number of measurements. With $10^5$ separate time determinations, times to $10\tau$ can be sampled providing a range sufficient for a determination of $\tau$ to a few per cent. Enough time must be allowed between each individual measurement to allow for long decay times. This requires that the experimental conditions be adjusted so that on average one decay event is recorded for every 100 initiation events. The delayed coincidence method can routinely measure lifetimes from a few nanoseconds to microseconds. The lower limit is set by the excitation source and the time resolution of the detector and electronics. Lifetimes as short as 10 ps can be measured with picosecond pulsed laser excitation,

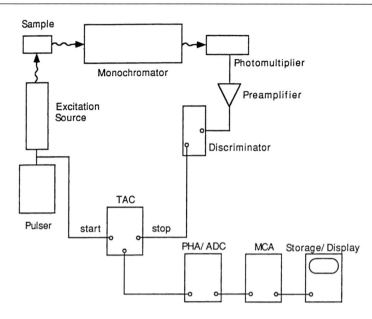

**Figure B1.10.5.** Lifetime experiment. A pulser triggers an excitation source that produces excited species in the sample. At the same time the pulser provides the *start* signal for a time-to-amplitude converter (TAC). Radiation from the decay of an excited species is detected with a photomultiplier at the output of a monochromator. The signal from the photomultiplier is amplified and sent to a discriminator. The output of the discriminator is the *stop* signal for the TAC. The TAC output pulse amplitude is converted to a binary number by a pulse height analyser (PHA) or analogue-to-digital converter (ADC). The binary number serves as an address for the multichannel analyser (MCA) that adds one to the number stored at the specified address. The pulser rate must accommodate decay times at least a factor of 10 longer than the lifetime of the specie under study. The excitation source strength and sample density are adjusted to have at most one detected decay event per pulse.

microchannel plate photomultipliers, a GHz preamplifier and a fast timing discriminator. Instrument stability and available time set the upper limit for lifetime measurements.

The delayed coincidence method has been applied to the fluorescent decay of laser excited states of biological molecules. By dispersing the emitted radiation from the decaying molecules with a polychromator and using an array of photodetectors for each wavelength region it is possible to measure fluorescent lifetimes as a function of the wavelength of the emitted radiation. The information is used to infer the conformation of the excited molecules [7].

### B1.10.4.2  *Multiple hit time-to-digital conversion*

Both lifetime and time-of-flight measurements have low duty cycles. In the case of the lifetime measurements sufficient time between initiation events must be allowed to accommodate the detection of long decays. Moreover, the signal rate has to be adjusted to allow for the detection of at most one decay event for every initiation event. In the case of time-of-flight measurements enough time must be allowed between measurements to allow for the slowest particle to traverse the flight path. As with the case of lifetime measurements, each initiation event must give rise to no more than one particle in the flight path.

In order to circumvent the signal limitations of lifetime and time-of-flight measurements multiple hit time-to-digital converters can be used. Typically, these instruments have from eight to sixteen channels and can be used in the 'common start' or 'common stop' mode. When used in the common start mode a single

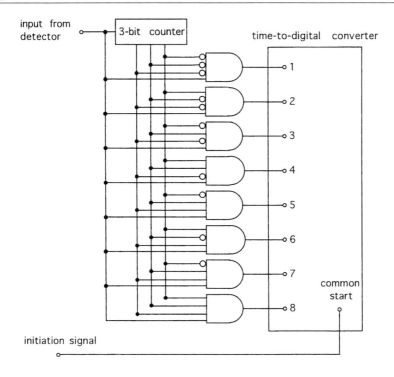

**Figure B1.10.6.** Multiple hit time-to-digital conversion scheme. Signals from the detector are routed to eight different TDC channels by a demultiplexer/data selector according to the digital output of the 3-bit counter that tracks the number of detector output signals following the initiation signal. The initiation signal is used as the common start. Not shown are the delays and signal conditioning components that are necessary to ensure the correct timing between the output of the counter and the arrival of the pulses on the common data line. Control logic to provide for the counter to reset after eight detector signals is also required.

start signal arms all of the channels with successive 'stop' signals directed to each of the channel inputs in sequence. In this way the signal rate can be increased by a factor equal to the number of channels in the TDC. A counter and demultiplexer/data selector [8] perform the function of routing the stop signals to the different channel inputs in sequence. A schematic diagram of the arrangement is shown in figure B1.10.6.

## B1.10.5 Coincidence experiments

Coincidence experiments explicitly require knowledge of the time correlation between two events. Consider the example of electron impact ionization of an atom, figure B1.10.7. A single incident electron strikes a target atom or molecule and ejects an electron from it. The incident electron is deflected by the collision and is identified as the scattered electron. Since the scattered and ejected electrons arise from the same event, there is a time correlation between their arrival times at the detectors.

Coincidence experiments have been common in nuclear physics since the 1930s. The widely used coincidence circuit of Rossi [9] allowed experimenters to determine, within the resolution time of the electronics of the day, whether two events were coincident in time. The early circuits were capable of submicrosecond resolution, but lacked the flexibility of today's equipment. The most important distinction between modern coincidence methods and those of the earlier days is the availability of semiconductor memories that allow

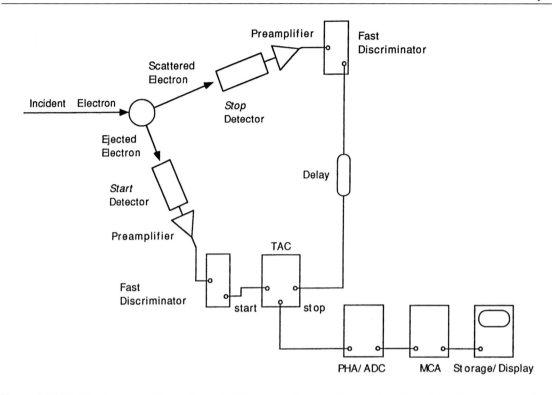

**Figure B1.10.7.** Electron impact ionization coincidence experiment. The experiment consists of a source of incident electrons, a target gas sample and two electron detectors, one for the scattered electron, the other for the ejected electron. The detectors are connected through preamplifiers to the inputs (start and stop) of a time-to-amplitude converter (TAC). The output of the TAC goes to a pulse-height-analyser (PHA) and then to a multichannel analyser (MCA) or computer.

one to now record precisely the time relations between all particles detected in an experiment. We shall see the importance of this in the evaluation of the statistical uncertainty of the results.

In a two detector coincidence experiment, of which figure B1.10.7 is an example, pulses from the two detectors are amplified and then sent to discriminators, the outputs of which are standard rectangular pulses of constant amplitude and duration. The outputs from the two discriminators are then sent to the *start* and *stop* inputs of a TAC or TDC. Even though a single event is responsible for ejected and scattered electrons, the two electrons will not arrive at the detectors at identical times because of differences in path lengths and electron velocities. Electronic propagation delays and cable delays also contribute to the *start* and *stop* signals not arriving at the inputs of the TAC or TDC at identical times; sometimes the *start* signal arrives first, sometimes the *stop* signal is first. If the signal to the *start* input arrives after the signal to the *stop*, that pair of events will not result in a TAC/TDC output. The result can be a 50% reduction in the number of recorded coincidences. To overcome this limitation a time delay is inserted in the *stop* line between the discriminator and the TAC/TDC. This delay can be a length of coaxial cable (typical delays are of the order of 1 ns/ft) or an electronic circuit. The purpose is to ensure that the stop signal always arrives at the *stop* input of the TAC/TDC after the *start* signal. A perfect coincidence will be recorded at a time difference approximately equal to the delay time. When a time window twice the duration of the delay time is used, perfect coincidence is at the centre of the time window and it is possible to make an accurate assessment of the background by considering the region to either side of the perfect coincidence region. An example of a time spectrum is shown in figure B1.10.8.

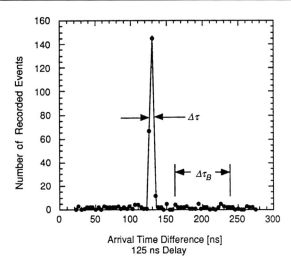

**Figure B1.10.8.** Time spectrum from a double coincidence experiment. Through the use of a delay in the lines of one of the detectors, signals that occur at the same instant in both detectors are shifted to the middle of the time spectrum. Note the uniform background upon which the true coincidence signal is superimposed. In order to decrease the statistical uncertainty in the determination of the true coincidence rate, the background is sampled over a time $\Delta\tau_B$ that is much larger than the width of the true coincidence signal, $\Delta\tau$.

### B1.10.5.1   Signal and background

Referring to figure B1.10.7, consider electrons from the event under study as well as from other events all arriving at the two detectors. The electrons from the event under study are correlated in time and result in a peak in the time spectrum centred approximately at the delay time. There is also a background level due to events that bear no fixed time relation to each other. If the average rate of the background events in each detector is $R_1$ and $R_2$, then the rate that two such events will be recorded within time $\Delta\tau$ is given by $R_B$, where

$$R_B = R_1 R_2 \Delta\tau.$$

Let the rate of the event under study be $R_A$. It will be proportional to the cross section for the process under study, $\sigma_A$, the incident electron current, $I_0$, the target density, $n$, the length of the target viewed by the detectors, $\ell$, the solid angles subtended by the detectors, $\Delta\omega_1$ and $\Delta\omega_2$ and the efficiency of the detectors, $\varepsilon_1$ and $\varepsilon_2$.

$$R_A = \sigma_A I_0 n \ell \Delta\omega_1 \Delta\omega_2 \varepsilon_1 \varepsilon_2.$$

The product $I_0 n \ell$, depends on the properties of the region from which the two electrons originate and is called the source function, $S$. The properties of the detectors are described by the product, $\Delta\omega_1 \Delta\omega_2 \varepsilon_1 \varepsilon_2$.

$$R_A = \sigma_A S \Delta\omega_1 \Delta\omega_2 \varepsilon_1 \varepsilon_2.$$

For the background, each of the rates, $R_1$ and $R_2$, will be proportional to the source function, the cross sections for single electron production and the properties of the individual detectors,

$$R_1 = S\sigma_1 \Delta\omega_1 \varepsilon_1 \qquad R_2 = S\sigma_2 \Delta\omega_2 \varepsilon_2.$$

Combining the two expressions for $R_1$ and $R_2$

$$R_B = S^2 \sigma_1 \sigma_2 \Delta\omega_1 \Delta\omega_2 \varepsilon_1 \varepsilon_2 \Delta\tau.$$

Comparing the expressions for the background rate and the signal rate one sees that the background increases as the square of the source function while the signal rate is proportional to the source function. The signal-to-background rate, $R_{AB}$, is then

$$R_{AB} = \frac{\sigma_A}{\sigma_1 \sigma_2} \frac{1}{S \Delta \tau}.$$

It is important to note that the signal is always accompanied by background. We now consider the signal and background after accumulating counts over a time $T$. For this it is informative to refer to figure B1.10.8 the time spectrum. The total number of counts within an arrival time difference $\Delta \tau$ is $N_T$ and this number is the sum of the signal counts, $N_A = R_A T$, and the background counts, $N_B = R_B T$,

$$N_T = N_A + N_B.$$

The determination of the background counts must come from an independent measurement, typically in a region of the time spectrum outside of the signal region, yet representative of the background within the signal region. The uncertainty in the determination of the signal counts is given by the square root of the uncertainties in the total counts and the background counts

$$\delta N_A = \sqrt{N_T + N_B}.$$

The essential quantity is the relative uncertainty in the signal counts, $\delta N_A / N_A$. This is given by

$$\delta N_A / N_A = \sqrt{N_T + N_B}/N_A = \sqrt{N_A + 2N_B}/N_A = \sqrt{R_A T + 2R_B T}/R_A T = \sqrt{\frac{1 + 2/R_{AB}}{R_A T}}.$$

Expressing $R_A$ in terms of $R_{AB}$ results in the following formula

$$\delta N_A / N_A = \sqrt{\frac{(R_{AB} + 2)\Delta \tau}{\frac{\sigma_A^2}{\sigma_1 \sigma_2} \Delta \omega_1 \Delta \omega_2 \varepsilon_1 \varepsilon_2 T}}.$$

There are a number of observations to be drawn from the above formula: the relative uncertainty can be reduced to an arbitrarily small value by increasing $T$, but because the relative uncertainty is proportional to $1/\sqrt{T}$, a reduction in relative uncertainty by a factor of two requires a factor of four increase in collection time. The relative uncertainty can also be reduced by reducing $\Delta \tau$. Here, it is understood that $\Delta \tau$ is the smallest time window that just includes all of the signal. $\Delta \tau$ can be decreased by using the fastest possible detectors, preamplifiers and discriminators and minimizing time dispersion in the section of the experiment ahead of the detectors.

The signal and background rates are not independent, but are coupled through the source function, $S$, as a consequence the relative uncertainty in the signal decreases with the signal-to-background rate, $R_{AB}$, a somewhat unanticipated result. Dividing $\delta N_A / N_A$ by its value at $R_{AB} = 1$ gives a reduced relative uncertainty $[\delta N_A / N_{AB}]_R$ equal to $\sqrt{(R_{AB} + 2)/3}$. A plot of $[\delta N_A / N_{AB}]_R$ as a function of $R_{AB}$ is shown in figure B1.10.9.

To illustrate this result, consider a case where there is one signal count and one background count in the time window $\Delta \tau$. The signal-to-background ratio is 1 and the relative uncertainty in the signal is $\sqrt{2 + 1}/1 = 1.7$. By increasing the source strength by a factor of 10 the signal will be increased by a factor of 10 and the background by a factor of 100. The signal-to-background ratio is now 0.1, but the relative uncertainty in the signal is $\sqrt{110 + 100}/10 = 1.45$, a clear improvement over the larger signal-to-background case.

Another method for reducing the relative uncertainty is to increase the precision of the measurement of background. The above formulae are based on an independent measurement of background over a time

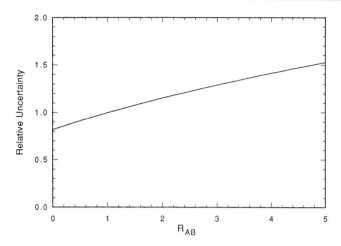

**Figure B1.10.9.** Plot of the reduced relative uncertainty of a double coincidence experiment as a function of the signal-to-background ratio. Note that the relative uncertainty decreases as the signal-to-background rate decreases.

window, $\Delta\tau$, that is equivalent to the time window within which the signal appears. If a larger time window for the background is used, the uncertainty in background determination can be correspondingly reduced. Let a time window of width $\Delta\tau_B$ be used for the determination of background, where $\Delta\tau/\Delta\tau_B = \rho < 1$. If the rate at which counts are accumulated in time window $\Delta\tau_B$ is $R_{BB}$, the background counts to be subtracted from the total counts in time window $\Delta\tau$ becomes $\rho R_{BB} T = \rho N_{BB}$. The uncertainty in the number of background counts to be subtracted from the total of signal plus background is $\sqrt{\rho^2 N_{BB}}$. The relative uncertainty of the signal is then

$$\sqrt{\frac{1 + 2\rho^2 N_{BB}/N_A}{N_A}} = \sqrt{\frac{1 + 2\rho N_B/N_A}{N_A}}.$$

The expression for the reduced relative uncertainty is then

$$[\delta N_A/N_{AB}]_R = \sqrt{\frac{(R_{AB} + 2\rho)}{3}}.$$

### B1.10.5.2   Detector alignment and source strength

For coincidence experiments where the detectors record time correlated events, maximum efficiency and sensitivity is attained when the detectors accept particles originating from the same source volume, in other words, the length, $\ell$, that defines the common volume of the source region seen by the detectors should be the same as the lengths and corresponding volumes for each of the detectors separately. If the two volumes are different, only the common volume seen by the detectors is used in the calculation of coincidence rate. Events that take place outside the common volume, but within the volume accepted by one of the detectors will only contribute to the background. This is shown schematically in figure B1.10.10. The situation is potentially complex if the source region is not uniform over the common volume viewed by the detectors and if the efficiencies of the detectors varies over the volume they subtend. Full knowledge of the geometric acceptances of the detectors and the source volume is necessary to accurately evaluate the size and strength of the source. Such an analysis is typically done numerically using empirical values of the view angles of the detectors and the spatial variation of target density and incident beam intensity.

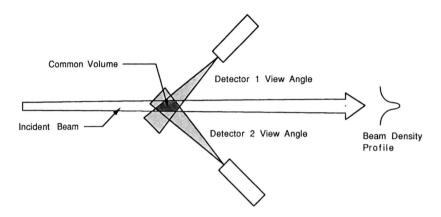

**Figure B1.10.10.** Schematic diagram of the effect of detector view angles on coincidence rate. The view angles of two detectors are shown along with the common view angle. Maximum signal collection efficiency is achieved when the individual view angles have the maximum overlap and when the overlap coincides with the maximum density of the incident beam.

### B1.10.5.3 Multiple parameter measurements

It is often the case that a time correlation measurement alone is not sufficient to identify a particular event of interest. For this reason, coincidence measurements are often accompanied by the measurement of another parameter such as energy, spin, polarization, wavelength, etc. For the case of the electron impact ionization experiment of figure B1.10.7, electrostatic energy analysers can be placed ahead of the detectors so that only electrons of a preselected energy arrive at the detectors. In this case the coincidence rate must take into account the energy bandpass of the analysers and the fact that in this example the energies of the scattered and ejected electrons are not independent, but are coupled through the relation that the energy of the incident electron equals the sum of the energies of the scattered and ejected electrons and the binding energy of the ejected electron

$$E_0 = E_1 + E_2 + BE,$$

where $E_0$ is the energy of the incident electron, $E_1$ is the energy of the scattered electron, $E_2$ is the energy of the ejected electron and $BE$ is the binding energy of the ejected electron. The coincidence rate now has an additional term, $\Delta E$, that is the overlap of the energy bandwidths of the two detectors, $\Delta E_1$ and $\Delta E_2$, subject to the energy conservation constraint. An often used approximation to $\Delta E$ is $\Delta E = \sqrt{\Delta E_1^2 + \Delta E_2^2}$. The background rates in the detectors depend only on the individual energy bandwidths, and are independent of any additional constraints. Maximum efficiency is achieved when the energy bandwidths are only just wide enough to accept the coincident events of interest.

### B1.10.5.4 Multiple detectors

The arrangement for a single coincidence measurement can be expanded to a multiple detector arrangement. If, for example, it is necessary to measure coincidence rates over an angular range, detectors can be placed at the angles of interest and coincidence rates measured between all pairs of detectors. This is a much more efficient way of collecting data than having two detectors that are moved to different angles after each coincidence determination. Depending on the signal and background rates, detector outputs can be multiplexed and a detector identification circuit used to identify those detectors responsible for each coincidence signal. If detectors are multiplexed, it is well to remember that the overall count rate is the sum of the rates for all of the detectors. This can be an important consideration when rates become comparable to the reciprocal of system dead time.

### B1.10.5.5   Preprocessing

It is often the case that signal rates are limited by the processing electronics. Consider a coincidence time spectrum of 100 channels covering 100 ns with a coincidence time window of 10 ns. Assume a signal rate of 1 Hz and a background rate of 1 Hz within the 10 ns window. This background rate implies a uncorrelated event rate of 10 kHz in each of the detectors. To register 99% of the incoming events, the dead time of the system can be no larger than 10 $\mu$s. The dead time limitation can be substantially reduced with a preprocessing circuit that only accepts events falling within 100 ns of each other (the width of the time spectrum). One way to accomplish this is with a circuit that incorporates delays and gates to only pass signals that fall within a 100 ns time window. With this circuit the number of background events that need to be processed each second is reduced from 10 000 to 10, and dead times as long as 10 ms can be accommodated while maintaining a collection efficiency of 99%. In a way this is an extension of the multiparameter processing in which the parameter is the time difference between processed events. A preprocessing circuit is discussed by Goruganthu *et al* [10].

### B1.10.5.6   Triple coincidence measurements

A logical extension of the coincidence measurements described above is the triple coincidence measurement shown schematically in figure B1.10.11. Taking as an example electron impact double ionization, the two ejected electrons and the scattered electron are correlated in time. On a three-dimensional graph with vertical axes representing the number of detected events and the horizontal axes representing the time differences between events at detectors 1 and 2 and 1 and 3, a time correlated signal is represented as a three-dimensional peak with a fixed base width in the horizontal plane. A triple coincidence time spectrum is shown in figure B1.10.12. Unlike the situation for double coincidence measurements the background has four sources, random rates in each of the three detectors, correlated events in detectors 1 and 2 with an uncorrelated signal in 3, correlated events in detectors 1 and 3 with an uncorrelated signal in 2 and correlated events in detectors 2 and 3 with an uncorrelated signal in detector 1. The first source gives a background that is uniform over the full horizontal plane. The three other sources produce two walls that are parallel to the two time axes and a third wall that lies at 45° to the time axes. These can be seen in figure B1.10.12.

For any region of the horizontal plane with dimensions $\Delta\tau_{12}\Delta\tau_{13}$, the uniform background rate is

$$R_{123} = R_1 R_2 R_3 \Delta\tau_{12}\Delta\tau_{13} = [S\sigma_1 \Delta\omega_1 \varepsilon_1][S\sigma_2 \Delta\omega_2 \varepsilon_2][S\sigma_3 \Delta\omega_3 \varepsilon_3]\Delta\tau_{12}\Delta\tau_{13}$$

where $R_1$, $R_2$ and $R_3$ are the random background rates in detectors 1, 2 and 3. The background due to two time correlated events and a single random event is $R_{12} + R_{13} + R_{23}$, where

$$R_{12} = [S\sigma_{12}\Delta\omega_1 \Delta\omega_2 \varepsilon_1 \varepsilon_2][S\sigma_3 \Delta\omega_3 \varepsilon_3]\Delta\tau_{13}$$

$$R_{13} = [S\sigma_{13}\Delta\omega_1 \Delta\omega_3 \varepsilon_1 \varepsilon_3][S\sigma_2 \Delta\omega_2 \varepsilon_2]\Delta\tau_{12}$$

$$R_{23} = [S\sigma_{23}\Delta\omega_2 \Delta\omega_3 \varepsilon_2 \varepsilon_3][S\sigma_1 \Delta\omega_1 \varepsilon_1]\Delta\tau_{23}.$$

The signal rate $R_A$, is

$$R_A = S\sigma_A \Delta\omega_1 \Delta\omega_2 \Delta\omega_3 \varepsilon_1 \varepsilon_2 \varepsilon_3$$

where the symbols have the same meaning as in the treatment of double coincidences.

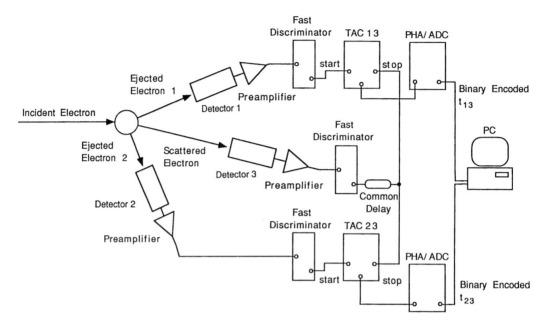

**Figure B1.10.11.** Electron impact double ionization triple coincidence experiment. Shown are the source of electrons, target gas, three electron detectors, one for the scattered electron and one for each of the ejected electrons. Two time differences, $t_{13}$ and $t_{23}$, are recorded for each triple coincidence. $t_{13}$ is the difference in arrival times of ejected electron 1 and the scattered electron; $t_{23}$ is the difference in arrival times of ejected electron 2 and the scattered electron. Two sets of time-to-amplitude converters (TACs) and pulse height analysers/analogue-to-digital converters (PHA/ADC) convert the times to binary encoded numbers that are stored in the memory of a computer. The data can be displayed in the form of a two-dimensional histogram (see figure B1.10.12).

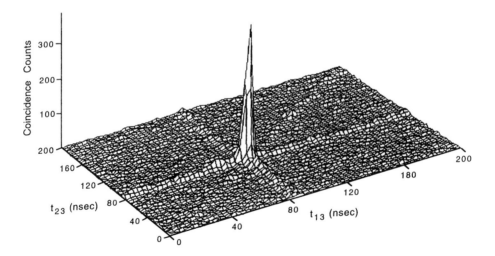

**Figure B1.10.12.** Schematic diagram of a two-dimensional histogram resulting from the triple coincidence experiment shown in figure B1.10.10. True triple coincidences are superimposed on a uniform background and three walls corresponding to two electron correlated events with a randomly occurring third electron.

If the signal falls within a two-dimensional time window $\Delta\tau_{12}\Delta\tau_{13}$, then the signal-to-background rate ratio is $R_A/[R_{123} + R_{12} + R_{13} + R_{23}] = \chi$, and the statistical uncertainty in the number of signal counts accumulated in time $T$ is

$$\delta N_A/N_A = \sqrt{\frac{1 + 2/\chi}{K_A S T}}$$

where $K_A = \sigma_A \Delta\omega_1 \Delta\omega_2 \Delta\omega_3 \varepsilon_1 \varepsilon_2 \varepsilon_3$. In contrast to simpler double coincidence experiments, $S$, the source function is not directly proportional to $1/\rho$. The full expression for $\delta N_A/N_A$ is

$$\delta N_A/N_A = \sqrt{\frac{1 + 2/\chi}{\dfrac{K_A}{2}\left[-\dfrac{K_1}{K_2} + \sqrt{\left(\dfrac{K_1}{K_2}\right)^2 + 4\left(\dfrac{K_A}{K_2}\right)\dfrac{1}{\chi}}\right] T}}$$

where

$$K_1 = \sigma_1 \Delta\omega_1 \varepsilon_1 \sigma_2 \Delta\omega_2 \varepsilon_2 \sigma_3 \Delta\omega_3 \varepsilon_3 \Delta\tau_{12}\Delta\tau_{13}$$

and

$$K_2 = \sigma_{12}\Delta\omega_1\Delta\omega_2\varepsilon_1\varepsilon_2\sigma_3\Delta\omega_3\varepsilon_3\Delta\tau_{13} + \sigma_{13}\Delta\omega_1\Delta\omega_3\varepsilon_1\varepsilon_3\sigma_2\Delta\omega_2\varepsilon_2\Delta\tau_{12} + \sigma_{23}\Delta\omega_2\Delta\omega_3\varepsilon_2\varepsilon_3\sigma_1\Delta\omega_1\varepsilon_1\Delta\tau_{23}.$$

As is the case for the double coincidence arrangement $\delta N_A/N_A$ is inversely proportional to the square root of the acquisition time, however $\delta N_A/N_A$ does not approach a minimum value in the limit of $\chi = 0$, but rather has a minimum in the region between $\chi = 0$ and $\chi = \infty$, the precise value of $\chi$ depending on the experimental conditions and the magnitudes of the cross sections for the different electron producing events. A detailed treatment of this topic with examples is given by Dupré *et al* [11].

## B1.10.6   Anti-coincidence

In high-energy physics experiments there can be many interfering events superimposed on the events of interest. An example is the detection of gamma rays in the presence of high-energy electrons and protons.

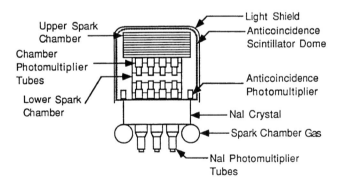

**Figure B1.10.13.** EGRET experiment showing spark chambers and photomultiplier tubes for the detection of gamma rays. The gamma ray detectors are surrounded by an anticoincidence scintillator dome that produces radiation when traversed by a high-energy charged particle but is transparent to gamma rays. Light pipes transmit radiation from the dome to photomultiplier tubes. A signal in a anticoincidence photomultiplier tube causes any corresponding signal in the gamma ray detectors to be ignored. The NaI crystal detector and photomultiplier tubes at the bottom of the unit provide high-resolution energy analysis of gamma rays passing through the spark chambers.

The electrons, protons and gamma rays all produce very similar signals in the solid state detectors that are used, and it is not possible to distinguish the gamma rays from the charged particles. A technique that is frequently used is to surround the gamma ray detectors with a plastic scintillation shield that produces a flash of light when traversed by a charged particle, but is transparent to the gamma rays. Light from the shield is coupled to photomultiplier detectors via light pipes. Signals that occur simultaneously in the photomultipliers and solid state detector are due to high-energy charged particles entering the instrument and are excluded from any analysis. An example of such an anti-coincidence circuit can be found in the energetic gamma ray experiment telescope (EGRET) on the gamma ray observatory (GRO) space craft that was launched in 1991 and continues to provide information on the energies and sources of gamma rays in the 20 MeV to 30 GeV energy range. Figure B1.10.13 is a schematic diagram of the EGRET experiment showing the anticoincidence shield.

## References

[1]   Bothe W and Geiger H 1925 Über das Wesen des Comptoneffekts; ein experimenteller Beitrag zur Theorie der Strahlung *Z. Phys.* **32** 639
[2]   Bevington P R 1969 *Data Reduction and Error Analysis for the Physical Sciences* (New York: McGraw Hill) pp 36–43
[3]   Melissinos A C 1966 *Experiments in Modern Physics* (New York: Academic) ch 10
[4]   Demas J N 1983 *Excited State Lifetime Measurements* (New York: Academic)
[5]   Lakowicz J R 1983 *Principles of Fluorescence Spectroscopy* (New York: Plenum) ch 3
[6]   Klose J Z 1966 Atomic lifetimes in neon I *Phys. Rev.* **141** 181
[7]   Knutson J R 1988 Time-resolved laser spectroscopy in biochemistry *SPIE* **909** 51–60
[8]   Millman J and Grabel A 1987 *Microelectronics* 2nd edn (New York: McGraw Hill) pp 279–84
[9]   Rossi B 1930 *Nature* **125** 636
[10]  Goruganthu R R, Coplan M A, Moore J H and Tossell J A 1988 (e,2e) momentum spectroscopic study of the interaction of $-CH_3$ and $-CF_3$ groups with the carbon–carbon triple bond *J. Chem. Phys.* **89** 25
[11]  Dupré C, Lahmam-Bennani A and Duguet A 1991 About some experimental aspects of double and triple coincidence techniques to study electron impact double ionizing processes *Meas. Sci. Technol.* **2** 327

# B1.11
# NMR of liquids

*Oliver W Howarth*

### B1.11.1  Introduction

Nuclear magnetic resonance (NMR) was discovered by Bloch, Purcell and Pound in 1945, as a development of the use of nuclear spins in low-temperature physics. Its initial use was for the accurate measurement of nuclear magnetic moments. However, increases in instrumental precision led to the detection of chemical shifts (B1.11.5) and then of spin–spin couplings (B1.11.6). This stimulated use by chemists. There have been spectacular improvements in sensitivity, resolution and computer control since then, so that NMR equipment is now essential in any laboratory for synthetic chemistry. Within moments or hours, it can determine the structure and, if desired, conformation of most medium-sized molecules in the solution phase. For this reason, a large pharmaceutical company will typically generate several hundred NMR spectra in one working day. NMR is also widely used in biochemistry for the much more challenging problem of determining the structures of smaller proteins and other biomolecules. The rates and extents of molecular motions can also be measured, through measuring the rates of energy transfer to, from and between nuclei (relaxation, B1.13). Outside of chemistry, it is used in a different mode for medical and other imaging, and for the detection of flow and diffusion in liquids (B1.14). It can also be used for clinical and *in vivo* studies, as the energies involved present no physiological dangers.

### B1.11.2  Nuclear spins

NMR depends on manipulating the collective motions of nuclear spins, held in a magnetic field. As with every rotatable body in nature, every nucleus has a spin quantum number $I$. If $I = 0$ (e.g. $^{12}C$, $^{16}O$) then the nucleus is magnetically inactive, and hence 'invisible'. If, as with most nuclei, $I > 0$, then the nucleus must possess a magnetic moment, because of its charge in combination with its angular momentum $\hbar\sqrt{I(I+1)}$. This makes it detectable by NMR. The most easily detected nuclei are those with $I = \frac{1}{2}$ and with large magnetic moments, e.g. $^{1}H$, $^{13}C$, $^{19}F$, $^{31}P$. These have two allowed states in a magnetic (induction) field $B_0$: $m_{1/2}$ and $m_{-1/2}$, with angular momentum components $\pm\hbar/2$. Thus these nuclear magnets lie at angles $\pm\cos^{-1}((\frac{1}{2})/(\frac{3}{4}))^{1/2} = 54.7°$ to $B_0$. They also precess rapidly, like all gyroscopes, at a rate $v_L$. This is called the Larmor frequency, after its discoverer. The $2I + 1$ permitted angles for other values of $I$ differ from the above angles, but they can never be zero or $180°$, because of the uncertainty principle.

The energy difference $\Delta E$ corresponding to the permitted transitions, $\Delta m = \pm 1$, is given by

$$\Delta E = \gamma h B_0/2\pi = h v_L.$$

Therefore, in NMR, one observes collective nuclear spin motions at the Larmor frequency. Thus the frequency of NMR detection is proportional to $B_0$. Nuclear magnetic moments are commonly measured either by their magnetogyric ratio $\gamma$, or simply by their Larmor frequency $v_L$ in a field where $^1$H resonates at 100 MHz (symbol $\Xi$).

### B1.11.2.1  Practicable isotopes

Almost all the stable elements have at least one isotope that can be observed by NMR. However, in some cases the available sensitivity may be inadequate, especially if the compounds are not very soluble. This may be because the relevant isotope has a low natural abundance, e.g. $^{17}$O, or because its resonances are extremely broad, e.g. $^{33}$S. Tables of such nuclear properties are readily available [1], and isotopic enrichment may be available as an expensive possibility. Broad resonances are common for nuclei with $I > 0$, because these nuclei are necessarily non-spherical, and thus have electric quadrupole moments that interact strongly with the electric field gradients present at the nuclei in most molecules. Linewidths may also be greatly increased by chemical exchange processes at appropriate rates (B2.7), by the presence of significant concentrations of paramagnetic species and by very slow molecular tumbling, as in polymers and colloids. For nuclei with $I = 1/2$, a major underlying cause of such broadening is the magnetic dipolar interaction of the nucleus under study with nearby spins. This and other interactions also lead to very large linewidths in the NMR spectra of solids and of near-solid samples, such as gels, pastes or the bead-attached molecules used in combinatorial chemistry. However, their effect may be reduced by specialized techniques (B1.13).

Fortunately, the worst broadening interactions are also removed naturally in most liquids and solutions, or at least greatly reduced in their effect, by the tumbling motions of the molecules, for many of the broadening interactions vary as $(3\cos^2\theta - 1)$ where $\theta$ is the angle between the H–H vector and $B_0$, and so they average to zero when $\theta$ covers the sphere isotropically. As a result, the NMR linewidths for the lighter spin-$\frac{1}{2}$ nuclei in smallish molecules will commonly be less than 1 Hz. Resolution of this order is necessary for the adequate resolution of the shifts and couplings described below. It requires expensive magnets and physically homogeneous samples. The presence of irregular interfaces degrades resolution by locally distorting the magnetic field, although the resulting spectra may still have enough resolution for some purposes, such as *in vivo* NMR.

It is occasionally desirable to retain a small proportion of molecular orientation, in order to quantitate the dipolar interactions present, whilst minimizing their contribution to the linewidth. Partial orientation may be achieved by using a nematic solvent. In large, magnetically anisotropic molecules it may occur naturally at the highest magnetic fields.

Figure B1.11.1 shows the range of radiofrequencies where resonances may be expected, between 650 and 140 MHz, when $B_0 = 14.1$ T, i.e. when the $^1$H resonance frequency is 600 MHz. There is one bar per stable isotope. Its width is the reported chemical shift range (B1.11.5) for that isotope, and its height corresponds to the log of the sensitivity at the natural abundance of the isotope, covering about six orders of magnitude. The radioactive nucleus $^3$H is also included, as it is detectable at safe concentrations and useful for chemical labelling. It is evident that very few ranges overlap. This, along with differences in linewidth, means that a spectrometer set to detect one nucleus is highly unlikely to detect any other in the same experiment.

### B1.11.2.2  Practicable samples

Once the above restrictions on isotope, solubility, chemical lability and paramagnetism are met, then a very wide range of samples can be investigated. Gases can be studied, especially at higher pressures. Solutions for $^1$H or $^{13}$C NMR are normally made in deuterated solvents. This minimizes interference from the solvent resonances and also permits the field to be locked to the frequency as outlined below. However, it is possible to operate with an unavoidably non-deuterated solvent, such as water in a biomedical sample, by using a range of solvent-suppression techniques. An external field–frequency lock may be necessary in these cases.

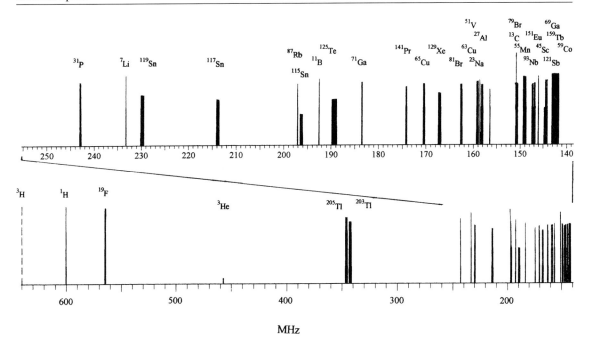

**Figure B1.11.1.** Resonance frequencies for different nuclei in a field of 14.1 T. Widths indicate the quoted range of shifts for each nucleus, and heights indicate relative sensitivities at the natural isotopic abundance, on a log scale covering approximately six orders of magnitude. Nuclei resonating below 140 MHz are not shown.

NMR spectra from a chosen nucleus will generally show all resonances from all such nuclei present in solution. Therefore, if at all possible, samples should be chemically pure, to reduce crowding in the spectra, and extraneous compounds such as buffers should ideally not contain the nuclei under study. When chromatographic separations are frequently and routinely essential, then on-line equipment that combines liquid chromatography with NMR is available [2]. Some separation into subspectra is also possible within a fixed sample, using specialized equipment, when the components have different diffusion rates. The technique is called diffusion ordered spectroscopy, or DOSY [3]. If the spectral crowding arises simply from the complexity of the molecule under study, as in proteins, then one can resort to selective isotopic labelling: for example, via gene manipulation. A wide range of experiments is available that are selective for chosen pairs of isotopes, and hence yield greatly simplified spectra.

The available sensitivity depends strongly on the equipment as well as the sample. $^1$H is the nucleus of choice for most experiments. 1 mg of a sample of a medium-sized molecule is adequate for almost all types of $^1$H-only spectra, and with specialized equipment one can work with nanogram quantities. At this lower level, the problem is not so much sensitivity as purity of sample and solvent. $^{13}$C NMR at the natural isotopic abundance of 1.1% typically requires 30 times this quantity of material, particularly if non-protonated carbon atoms are to be studied. Most other nuclei necessitate larger amounts of sample, although $^{31}$P [4] and $^{19}$F are useful exceptions. *In vivo* spectroscopy generally calls for custom-built probes suited for the organ or organism under study. Human *in vivo* spectra usually require a magnet having an especially wide bore, along with a moveable detection coil. They can reveal abnormal metabolism and internal tissue damage.

NMR can be carried out over a wide range of temperatures, although there is a time and often a resolution penalty in using temperatures other than ambient. An effective lower limit of $\sim -150\,^\circ$C is set by the lack of solvents that are liquid below this. Temperatures above $\sim 130\,^\circ$C require special thermal protection devices, although measurements have even been made on molten silicates.

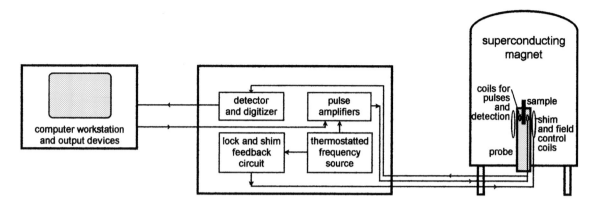

**Figure B1.11.2.** Simplified representation of an NMR spectrometer with pulsed RF and superconducting magnet. The main magnetic field $B_0$ is vertical and centred on the sample.

### B1.11.3    The NMR experiment

*B1.11.3.1    Equipment and resonance*

Figure B1.11.2 represents the essential components of a modern high-resolution NMR spectrometer, suitable for studies of dissolved samples. The magnet has a superconducting coil in a bath of liquid He, jacketed by liquid $N_2$. The resulting, persistent field $B_0$ ranges from 5.9 to 21.1 T, corresponding to $^1$H NMR at 250 to 900 MHz. This field can be adjusted to a homogeneity of better than 1 ppb over the volume of a typical sample. The sample is commonly introduced as 0.5 ml of solution held in a 5 mm OD precision glass tube. Microcells are available for very small samples, and special tubes are also available for, for example, pressurized or flow-through samples. The sample can be spun slowly in order to average the field inhomogeneity normal to the tube's axis, but this is not always necessary, for the reduction in linewidth may only be a fraction of a hertz. The field aligns the spins as described above, and a Boltzmann equilibrium then develops between the various $m_I$ Zeeman states. A typical population difference between the two $^1$H levels is 1 part in $10^5$. Thus $10^{-5}$ of the $^1$H spins in the sample are not paired. These are collectively called the bulk nuclear magnetization, and in many ways they behave in combination like a classical, magnetized gyroscope. The time required for the establishment of the Boltzmann equilibrium is approximately five times the spin–lattice relaxation time, $T_1$ (B1.14). Because the population difference between the two $^1$H levels is proportional to the field, the sensitivity of NMR also rises with the field.

The bulk magnetization is stimulated into precessional motion around $B_0$ by a radiofrequency (RF) pulse at $\nu_L$, applied through a solenoid-like coil whose axis is perpendicular to $B_0$. This motion amounts to a nuclear magnetic resonance. Typically, $10^{-5}$ s of RF power tilts the magnetization through 90° and thus constitutes a 90° pulse. The coil is then computer-switched from being a transmitter to becoming a receiver, with the free precessional motion of the magnetization generating a small RF voltage in the same coil, at or near the frequency $\nu_L$. The decay time of this oscillating voltage is of the order of $T_1$, unless speeded by contributions from exchange or field inhomogeneity. After appropriate amplification and heterodyne detection it is reduced to an audiofrequency and digitized. These raw data are commonly called the free induction decay, or FID, although the term applies more properly to the original bulk precession.

One further consequence of the use of a pulse, as against the older method of applying a single RF frequency, is that its underlying frequency is broadened over a range of the order of the reciprocal of the pulse length. Indeed, the RF power is in many cases virtually uniform or 'white' across a sufficient range to stimulate all the nuclei of a given isotope at once, even though they are spread out in frequency by differences of chemical shift. Thus, the repeat time for the NMR measurement is approximately $T_1$, typically a few seconds, rather than the considerably longer time necessary for sweeping the frequency or the field. Most FIDs are built up by gathering data after repeated pulses, each usually less than 90° in order to permit more rapid repetition. This helps to average away noise and other imperfections, relative to the signal, and it also permits more elaborate experiments involving sequences of multiple pulses and variable interpulse delays.

However, it also necessitates a strictly constant ratio of field to frequency, over the duration of the experiment. Although the master frequency source can be held very constant by a thermostatted source, the field is always vulnerable to local movements of metal, and to any non-persistence of the magnet current. Therefore the field is locked to the frequency through a feedback loop that uses continuous, background monitoring of the $^2$H solvent resonance. The probe containing the sample and coil will also normally have at least one further frequency channel, for decoupling experiments (B1.11.6). The lock signal is also simultaneously employed to maximize the field homogeneity across the sample, either manually or automatically, via low-current field correction 'shim' coils. A feedback loop maximizes the height of the lock signal and, because the peak area must nevertheless remain constant, thereby minimizes the peak's width.

The digitized FID can now be handled by standard computer technology. Several 'spectrum massage' techniques are available for reducing imperfections and noise, and for improving resolution somewhat. The FID is then converted into a spectrum by a discrete Fourier transformation. Essentially, digital sine and cosine waves are generated for each frequency of interest, and the FID is multiplied by each of these in turn. If the product of one particular multiplication does not average to zero across the FID, then that frequency, with that phase, is also present in the FID. The resulting plot of intensity versus frequency is the spectrum. It is normally 'phased' by appropriate combination of the sine and cosine components, so as to contain only positive-going peaks. This permits the measurement of peak areas by digital integration, as well as giving the clearest separation of peaks.

The information within the spectrum can then be presented in many possible ways. In a few cases, it is possible to identify the sample by a fully automatic analysis: for example, by using comparisons with an extensive database. However, most analyses require the knowledge outlined in the following sections.

Many other pulsed NMR experiments are possible, and some are listed in the final sections. Most can be carried out using the standard equipment described above, but some require additions such as highly controllable, pulsed field gradients, shaped RF pulses for (for example) single-frequency irradiations, and the combined use of pulses at several different frequencies.

## B1.11.4   Quantitation

The simplest use of an NMR spectrum, as with many other branches of spectroscopy, is for quantitative analysis. Furthermore, in NMR all nuclei of a given type have the same transition probability, so that their resonances may be readily compared. The area underneath each isolated peak in an NMR spectrum is proportional to the number of nuclei giving rise to that peak alone. It may be measured to ~1% accuracy by digital integration of the NMR spectrum, followed by comparison with the area of a peak from an added standard.

The absolute measurement of areas is not usually useful, because the sensitivity of the spectrometer depends on factors such as temperature, pulse length, amplifier settings and the exact tuning of the coil used to detect resonance. Peak intensities are also less useful, because linewidths vary, and because the resonance from a given chemical type of atom will often be split into a pattern called a multiplet. However, the relative overall areas of the peaks or multiplets still obey the simple rule given above, if appropriate conditions are met. Most samples have several chemically distinct types of (for example) hydrogen atoms within the molecules under study, so that a simple inspection of the number of peaks/multiplets and of their relative areas can help to identify the molecules, even in cases where no useful information is available from shifts or couplings.

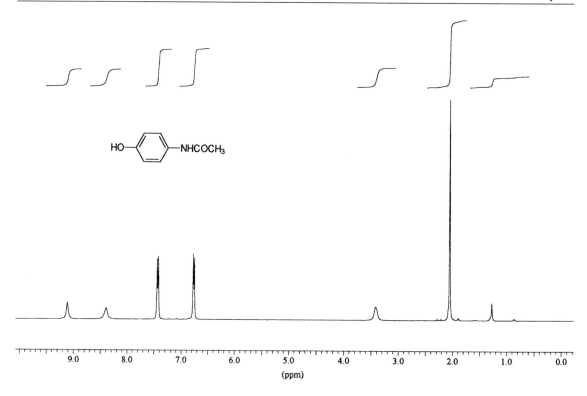

**Figure B1.11.3.** 400 MHz $^1$H NMR spectrum of paracetamol (structure shown) with added integrals for each singlet or multiplet arising from the paracetamol molecule.

This is illustrated in figure B1.11.3, the integrated $^1$H NMR spectrum of commercial paracetamol in deuteriodimethylsulfoxide solvent. The paracetamol itself gives five of the integrated peaks or multiplets. Two other integrated peaks at 3.4 and 1.3 ppm, plus the smaller peaks, arise from added substances. The five paracetamol peaks have area ratios (left to right) of 1:1:2:2:3. These tally with the paracetamol molecule (see diagram). The single H atoms are OH and NH respectively, the double ones are the two distinct pairs of hydrogens on the aromatic ring and the triple ones are the methyl group. Few other molecules will give these ratios, irrespective of peak position.

The other peaks demonstrate the power of NMR to identify and quantitate all the components of a sample. This is very important for the pharmaceutical industry. Most of the peaks, including a small one accidentally underlying the methyl resonance of paracetamol, arise from stearic acid, which is commonly added to paracetamol tablets to aid absorption. The integrals show that it is present in a molar proportion of about 2%. The broader peak at 3.4 ppm is from water, present because no attempt was made to dry the sample. Such peaks may be identified either by adding further amounts of the suspected substance, or by the more fundamental methods to be outlined below. If the sample were less concentrated, then it would also be possible to detect the residual hydrogen atoms in the solvent, i.e. from its deuteriodimethylsulfoxide-$d^5$ impurity, which resonates in this case at 2.5 ppm.

It is evident from the figure that impurities can complicate the use of NMR integrals for quantitation. Further complications arise if the relevant spins are not at Boltzmann equilibrium before the FID is acquired. This may occur either because the pulses are repeated too rapidly, or because some other energy input is present, such as decoupling. Both of these problems can be eliminated by careful timing of the energy inputs, if strictly accurate integrals are required.

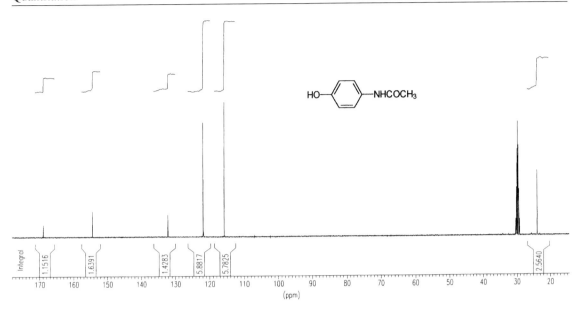

**Figure B1.11.4.** Hydrogen-decoupled 100.6 MHz $^{13}$C NMR spectrum of paracetamol. Both graphical and numerical peak integrals are shown.

Their effects are illustrated in figure B1.11.4, which is a $^1$H-decoupled $^{13}$C NMR spectrum of the same sample of paracetamol, obtained without such precautions. The main peak integrals are displayed both as steps and also as numbers below the peaks. The peaks from stearic acid are scarcely visible, but the dmso-d$^6$ solvent multiplet at 30 ppm is prominent, because $^{13}$C was also present at natural abundance in the solvent. The paracetamol integrals should ideally be in the ratios 1:1:1:2:2:1, corresponding to the carbonyl carbon, the four chemically distinct ring carbons and the methyl carbon to the right. However, the first three peaks correspond to carbons that have no attached H, and their integrals are reduced by factors of between 2 and 3. The methyl peak is also slightly reduced.

These reductions arose because the spectrum was obtained by the accumulation of a series of FIDs, generated by pulses spaced by less than the $5 \times T_1$ interval mentioned above, so that a full Boltzmann population difference was not maintained throughout. The shortfall was particularly acute for the unprotonated carbons, for these have long relaxation times, but it was also significant for the methyl group because of its rapid rotational motion. These losses of intensity are called 'saturation'. If they are intentionally introduced by selective irradiation just prior to the pulse, then they become the analogues of hole-bleaching in laser spectroscopy. They are useful for the selective reduction of large, unwanted peaks, as in solvent suppression. Also, because they are a property of the relevant nuclei rather than of their eventual chemical environment they can be used as transient labels in kinetic studies.

The integrals in figure B1.11.4 are, however, also distorted by a quite different mechanism. The spectrum was obtained in the standard way, with irradiation at all the relevant $^1$H frequencies so as to remove any couplings from $^1$H. This indirect input of energy partly feeds through to the $^{13}$C spins via dipolar coupling, to produce intensity gains at all peaks, but particularly at protonated carbons, of up to $\times 3$, and is an example of the nuclear Overhauser enhancement. The phenomenon is quite general wherever $T_1$ is dominated by dipolar interactions, and when any one set of spins is saturated, whether or not decoupling also takes place. It was discovered by A W Overhauser in the context of the dipolar interactions of electrons with nuclei [5]. The NOEs between $^1$H nuclei are often exploited to demonstrate their spatial proximity, as described

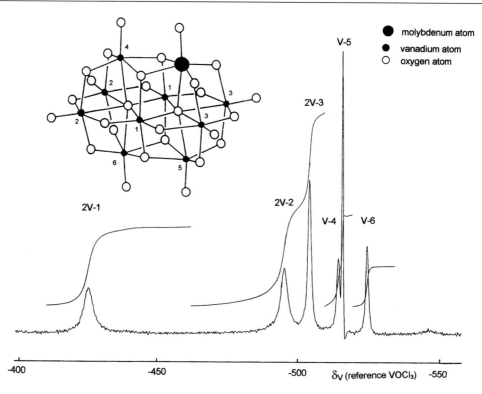

**Figure B1.11.5.** 105 MHz $^{51}$V NMR subtraction spectrum of the $[MoV_9O_{28}]^{5-}$ anion (structure shown). The integrals are sufficient to define the position of the Mo atom.

in the final section. It is possible to obtain decoupled $^{13}$C NMR spectra without the complications of the NOE, by confining the decoupling irradiation to the period of the FID alone, and then waiting for $10 \times T_1$ before repeating the process. However, the $\times 3$ enhancement factor is useful and also constant for almost all protonated carbons, so that such precautions are often superfluous. Both carbon and hydrogen integrals are particularly valuable for identifying molecular symmetry.

Figure B1.11.5 is an example of how relative integrals can determine structure even if the peak positions are not adequately understood. The decavanadate anion has the structure shown, where oxygens lie at each vertex and vanadiums at the centre of each octahedron. An aqueous solution of decavanadate was mixed with about 8 mol% of molybdate, and the three peaks from the remaining decavanadate were then computer-subtracted from the $^{51}$V NMR spectrum of the resulting equilibrium mixture [6]. The remaining six peaks arise from a single product and, although their linewidths vary widely, their integrals are close to being in a 2:2:2:1:1:1 ratio. This not only suggests that just one V has been replaced by Mo, but also identifies the site of substitution as being one of the four $MoO_6$ octahedra not lying in the plane made by vanadiums 1, 2 and 3. No other site or extent of substitution would give these integral ratios.

When $T_1$ is very short, which is almost always true with nuclei having $I > 1/2$, the dipolar contribution to relaxation will be negligible and, hence, there will be no contributions to the integral from either NOE or saturation. However, resonances more than about 1 kHz wide may lose intensity simply because part of the FID will be lost before it can be digitized, and resonances more than 10 kHz wide may be lost altogether. It is also hard to correct for minor baseline distortions when the peaks themselves are very broad.

## B1.11.5  Chemical shifts

Strictly speaking, the horizontal axis of any NMR spectrum is in hertz, as this axis arises from the different frequencies that make up the FID. However, whilst this will be important in the section that follows (on coupling), it is generally more useful to convert these frequency units into fractions of the total spectrometer frequency for the nucleus under study. The usual units are parts per million (ppm). Because frequency is strictly proportional to field in NMR, the same ppm units also describe the fractional shielding or deshielding of the main magnetic field, by the electron clouds around or near the nuclei under study. The likely range of such shieldings is illustrated in figures B1.11.1, B1.11.3 and B1.11.4.

### B1.11.5.1  Calibration

The ppm scale is always calibrated relative to the appropriate resonance of an agreed standard compound, because it is not possible to detect the NMR of bare nuclei, even though absolute shieldings can be calculated with fair accuracy for smaller atoms. Added tetramethylsilane serves as the standard for $^1$H, $^2$H, $^{13}$C and $^{29}$Si NMR, because it is a comparatively inert substance, relatively immune to changes of solvent, and its resonances fall conveniently towards one edge of most spectra. Some other standards may be unavoidably less inert. They are then contained in capillary tubes or in the annulus between an inner and an outer tube, where allowance must be made for the jump in bulk magnetic susceptibility between the two liquids. For dilute solutions where high accuracy is not needed, a resonance from the deuteriated solvent may be adequate as a secondary reference. The units of chemical shift are $\delta$. This unit automatically implies ppm from the standard, with the same sign as the frequency, and so measures deshielding. Hence $\delta = 10^6(\nu - \nu_{\text{ref}})/\nu_{\text{ref}}$.

### B1.11.5.2  Local contributions to the chemical shift

The shielding at a given nucleus arises from the virtually instantaneous response of the nearby electrons to the magnetic field. It therefore fluctuates rapidly as the molecule rotates, vibrates and interacts with solvent molecules. The changes of shift with rotation can be large, particularly when double bonds are present. For example, the $^{13}$C shift of a carbonyl group has an anisotropy comparable to the full spectrum width in figure B1.11.4. Fortunately, these variations are averaged in liquids, although they are important in the NMR of solids. This averaging process may be visualized by imagining the FID emitted by a single spin. If the emission frequency keeps jumping about by small amounts, then this FID will be made up from a series of short segments of sine waves. However, if the variations in frequency are small compared to the reciprocal of the segment lengths, then the composite wave will be close to a pure sine wave. Its frequency will be the weighted average of its hidden components and the rapid, hidden jumps will not add to the linewidth. This is described as 'fast exchange on the NMR timescale' and it is discussed more fully in B.2.7. The same principles apply to rapid chemical changes, such as aqueous protonation equilibria. In contrast, somewhat slower exchange processes increase linewidths, as seen for the OH and NH resonances in figure B1.11.3. In this sample, the individual molecules link transiently via hydrogen bonding, and are thus in exchange between different H-bonded oligomers.

    The two primary causes of shielding by electrons are diamagnetism and temperature-independent paramagnetism (TIP). Diamagnetism arises from the slight unpairing of electron orbits under the influence of the magnetic field. This always occurs so as to oppose the field and was first analysed by Lamb [7]. A simplified version of his formula, appropriate for a single orbital in isolated atoms, shows that the diamagnetic shielding contribution to $\delta$ is proportional to $-\rho_e\langle r^2\rangle$ where $\rho_e$ is the electron density and $\langle r^2\rangle$ is the average squared radius of the electron orbit [8]. Thus diamagnetic shielding is lowered and, hence, $\delta$ is increased, by the attachment of an electronegative group to the atom under study, for this decreases both $\rho_e$ and $\langle r^2\rangle$. Similarly, in conjugated systems, $\delta$ will often approximately map the distribution of electronic charge between the atoms.

**Table B1.11.1.** Effect of an electronegative substituent upon methyl shifts in X–$^{13}C^1H_3$.

| X | $\delta_H$ | $\delta_C$ | Electronegativity of X |
|---|---|---|---|
| Si(CH$_3$)$_3$ | 0.0 | 0.0 | 1.8 |
| H | 0.13 | −2.3 | 2.1 |
| CH$_3$ | 0.88 | 5.7 | 2.5 |
| I | 2.16 | −20.7[a] | 2.5 |
| NH$_2$ | 2.36 | 28.3 | 2.8 |
| Br | 2.68 | 10.0[a] | 2.8 |
| Cl | 3.05 | 25.1 | 3.0 |
| OH | 3.38 | 49.3 | 3.5 |
| F | 4.26 | 75.4 | 4.0 |

[a] Heavy-atom relativistic effects influence these shifts.

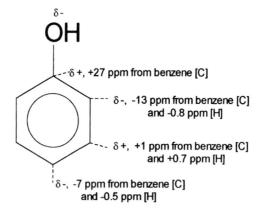

**Figure B1.11.6.** $^1$H and $^{13}$C chemical shifts in phenol, relative to benzene in each case. Note that δ (H or C) approximately follows δ (the partial charge at C).

The TIP contribution has the opposite effect on δ. It should not be confused with normal paramagnetism, for it does not require unpaired electrons. Instead, it arises from the physical distortion of the electron orbitals by the magnetic field. A highly simplified derivation of the effect of TIP upon δ shows it to be proportional to $+\langle r^{-3}\rangle/\Delta E$, with $r$ as above. $\Delta E$ is a composite weighting term representing the energy gaps between the occupied and the unoccupied orbitals. A small $\Delta E$ implies orbitals that are accessible and hence vulnerable to magnetic distortion. $\Delta E$ is particularly large for H, making the TIP contribution small in this case, but for all other atoms TIP dominates δ. The effect of an electronegative group on the TIP of an atom is to shrink the atom slightly and, thus, to increase δ, i.e. it influences chemical shift in the same direction as described above in the case of diamagnetism, but for a different reason.

These effects are clearly seen for saturated molecules in table B1.11.1, which correlates the $^1$H and $^{13}$C NMR chemical shifts of a series of compounds X–CH$_3$ with the approximate Pauling electronegativities of X. Both $\delta_H$ and $\delta_C$ increase with X as predicted, although the sequence is slightly disordered by additional, relativistic effects at the C atoms, when X is a heavy atom such as I. The correlations of shift with changes in the local electric charge and, hence, with orbital radius, are also seen for an aromatic system in figure B1.11.6. Here the hydrogen and carbon chemical shifts in phenol, quoted relative to benzene, correlate fairly well both with each other and also with the charge distribution deduced by other means.

**Table B1.11.2.** Fluorine-19 chemical shifts.

| Organic compounds | $\delta_F$ | Inorganic compounds | $\delta_F$ | Metal fluorides | $\delta_F$ |
|---|---|---|---|---|---|
| $CH_3F$ | $-272$ | HF | $-221$ | $MoF_6$ | $-278$ |
| $CH_3CH_2F$ | $-213$ | LiF | $-210$ | $SbF_5$ | $-108$ |
| $C_6F_6$ | $-163$ | $[BF_4]^-$ | $-163$ | $WF_6$ | $+166$ |
| $CH_2F_2$ | $-143$ | $BF_3$ | $-131$ | $ReF_7$ | $+345$ |
| $F_2C{=}CF_2$ | $-135$ | $ClF_3$ | $-4$ (1), $+116$ (2) | | |
| $H_2C{=}CF_2$ | $-81$ | $IF_7$ | $+170$ | | |
| $CF_3R$ | $-60$ to $-70$ | $IF_5$ | $+174$ (1), $+222$ (4) | | |
| $CF_2Cl_2$ | $-8$ | $ClF_5$ | $+247$ (1), $+412$ (4) | | |
| $CFCl_3$ | 0 (reference) | $XeF_2$ | $+258$ | | |
| $CF_2Br_2$ | $+7$ | $XeF_4$ | $+438$ | | |
| $CFBr_3$ | $+7$ | $XeF_6$ | $+550$ | | |
| | | $F_2$ | $+421.5$ | | |
| | | ClF | $+448.4$ | | |
| | | FOOF | $+865$ | | |

The effects of TIP also appear in figures B1.11.3 and B1.11.4. In the $^{13}C$ NMR spectrum, all the resonances of the $sp^2$ carbons lie above 100 ppm (a useful general rule of thumb) because $\Delta E$ is smaller for multiple bonds. The highest shifts are for the carbonyl C at 169 ppm and the ring C attached to oxygen at 155 ppm, because of the high electronegativity of O. In the $^1H$ spectrum, H atoms attached to $sp^2$ carbons also generally lie above 5 ppm and below 5 ppm if attached to $sp^3$ carbons. However, the NH and OH resonances have much less predictable shifts, largely governed by the average strength of the associated hydrogen bonds. When these are exceptionally strong, as in nucleic acids, $\delta_H$ can be as high as 30 ppm, whereas most normal H shifts lie in the range 0–10 ppm.

Table B1.11.2 gives a different example of the effects of TIP. In many fluorine compounds, $\Delta E$ is largely determined by the energy difference between the bonding and antibonding orbitals in the bond to F, and hence by the strength of this bond. The $^{19}F$ shifts correlate nicely in this way. For example, C–F bonds are notoriously strong and so give low values of $\delta_F$, whereas the remarkable reactivity of $F_2$ depends on the weakness of the F–F bond, which also gives $F_2$ a much higher chemical shift. The weakness of the bond to F is even more apparent in the chemical shifts of the explosive compounds $XeF_6$ and FOOF. In some compounds the spectra also reflect the presence of distinct types of fluorine within the molecule. For example, $ClF_3$ has a T-shaped structure with the Cl–F bond stronger for the stem of the T, giving this single F atom a lower shift.

The shifts for other nuclei are not usually so simple to interpret, because more orbitals are involved, and because these may have differing effects on $\langle r^{-3} \rangle$ and on $\Delta E$. Nevertheless, many useful correlations have been recorded [9]. A general rule is that the range of shifts increases markedly with increasing atomic number, mainly because of decreases in $\Delta E$ and increases in the number of outer-shell electrons. The shift range is particularly large when other orbital factors also lower and vary $\Delta E$, as in cobalt[III] complexes. However, the chemical shifts for isotopes having the same atomic number, such as $^1H$, $^2H$ and $^3H$, are almost identical. Each nucleus serves merely to report the behaviour of the same electron orbitals, except for very small effects of isotopic mass on these orbitals.

### B1.11.5.3  *Chemical shifts arising from more distant moieties*

Hydrogen shifts fall within a narrow range and, hence, are proportionally more susceptible to the influence of more distant atoms and groups. As an example, aromatic rings have a large diamagnetic anisotropy, because

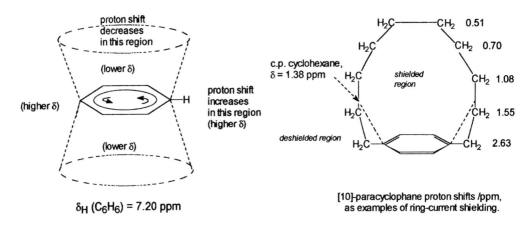

**Figure B1.11.7.** $^1$H chemical shifts in [10]-paracyclophane. They have values on either side of the 1.38 ppm found for large polymethylene rings and, thus, map the local shielding and deshielding near the aromatic moiety, as depicted in the upper part of the figure.

the effective radius of the electron orbit around the ring is large. This affects the chemical shifts of nearby atoms by up to 5 ppm, depending on their position relative to the ring. It also explains the relatively high shifts of the hydrogens directly attached to the ring. The diamagnetism of the ring is equivalent to a solenoid opposing $B_0$. Its lines of force necessarily loop backwards outside the ring and therefore deshield atoms in or near the ring plane. The effect is illustrated in figure B1.11.7 for a paracyclophane. The methylene shift values fall both above and below the value of 1.38, expected in the absence of the aromatic moiety, according to the average positioning of the methylene hydrogens relative to the ring plane. Similar but smaller shifts are observed with simple double and triple bonds, because these also generate a significant circulation of electrons. Polycyclic aromatic systems generate even larger shifts, in regions near to more than one ring.

Shifts are also affected by steric compression of any kind on the atom under study. The effect on a C atom can reduce $\delta_C$ by up to 10 ppm. For example, the $^{13}$C chemical shifts of the methyl carbons in but-2-ene are 5 ppm lower in the *cis* isomer than in the *trans*, because in the *cis* case the methyl carbons are only about 3 Å apart. Steric shifts are particularly important in the $^{13}$C NMR spectra of polymers, for they make the peak positions dependent on the local stereochemistry and, hence, on the tacticity of the polymer [10].

### B1.11.5.4   Shift reagents

Nearby paramagnetic molecules can also have a similar effect on shifts. Their paramagnetic centre will often possess a substantially anisotropic *g*-factor. If the paramagnetic molecule then becomes even loosely and temporarily attached to the molecule under study, this will almost always lead to significant shift changes, for the same reasons as with nearby multiple bonds. Furthermore, if the molecule under study and its paramagnetic attachment are both chiral, the small variations in the strength and direction of the attachment can effect a separation of the shifts of the two chiral forms and thus detect any enantiomeric excess. The paramagnetic species thus acts as a chiral shift reagent. In practice, the effect of the paramagnetic additive will be complicated, because there may be further shifts arising from direct leakage of free electron density between the paired molecules. There will also be broadening due to the large magnetic dipole of the unpaired electrons. These complications can be reduced by a judicious choice of shift reagent. It may also be possible to use a diamagnetic shift reagent such as 2,2′-trifluoromethyl-9-anthrylethanol and thus avoid the complications of paramagnetic broadening altogether, whilst hopefully retaining the induced chiral separation of shifts [11].

*B1.11.5.5 The prediction of chemical shifts*

Enormous numbers of chemical shifts have been recorded, particularly for $^1H$ and $^{13}C$. Many algorithms for the prediction of shifts have been extracted from these, so that the spectra of most organic compounds can be predicted at a useful level of accuracy, using data tables available in several convenient texts [12–15]. Alternatively, computer programs are available that store data from $10^4$–$10^5$ spectra and then use direct structure-comparison methods to predict the $^1H$ and $^{13}C$ chemical shifts of new species.

Shifts can also be predicted from basic theory, using higher levels of computation, if the molecular structure is precisely known [16]. The best calculations, on relatively small molecules, vary from observation by little more than the variations in shift caused by changes in solvent. In all cases, it is harder to predict the shifts of less common nuclei, because of the generally greater number of electrons in the atom, and also because fewer shift examples are available.

## B1.11.6 The detection of neighbouring atoms–couplings

The $^1H$ NMR spectrum in figure B1.11.2 shows two resonances in the region around 7 ppm, from the two types of ring hydrogens. Each appears as a pair of peaks rather than as a single peak. This splitting into a pair, or 'doublet', has no connection with chemical shift, because if the same sample had been studied at twice the magnetic field, then the splitting would have appeared to halve, on the chemical shift scale. However, the same splitting does remain strictly constant on the concealed, frequency scale mentioned previously, because it derives not from $B_0$, but instead from a fixed energy of interaction between the neighbouring hydrogen atoms on the ring. Each pair of peaks in the present example is called a doublet: a more general term for such peak clusters is a multiplet. The chemical shift of any symmetrical multiplet is the position of its centre, on the ppm scale, whether or not any actual peak arises at that point. The area of an entire multiplet, even if it is not fully symmetric, obeys the principles of section B1.11.4 above. It follows that the individual peaks in a multiplet may be considerably reduced in intensity compared with unsplit peaks, especially if many splittings are present.

The splittings are called $J$ couplings, scalar couplings or spin–spin splittings. They are also closely related to hyperfine splittings in ESR. Their great importance in NMR is that they reveal interactions between atoms that are chemically nearby and linked by a bonding pathway. The $J$ value of a simple coupling is the frequency difference in hertz between the two peaks produced by the coupling. It is often given by the symbol $^nJ$, where $n$ is the number of bonds that separate the coupled atoms. Thus, in combination with chemical shifts and integrals, they will often allow a chemical structure to be determined from a single $^1H$ NMR spectrum, even of a previously unknown molecule of some complexity. Their presence also permits a wide range of elegant experiments where spins are manipulated by carefully timed pulses. However, they also complicate spectra considerably if more than a few interacting atoms are present, particularly at low applied fields, when the chemical shift separations, converted back to hertz, are not large in comparison with the couplings. Fortunately some of the aforesaid elegant experiments can extract information on the connectivities of atoms, even when the multiplets are too complex or overlapped to permit a peak-by-peak analysis.

*B1.11.6.1 Coupling mechanisms*

The simplest mechanism for spin–spin couplings is described by the Fermi contact model. Consider two nuclei linked by a single electron-pair bond, such as $^1H$ and $^{19}F$ in the hydrogen fluoride molecule. The magnetic moment of $^1H$ can be either 'up' or 'down' relative to $B_0$, as described earlier, with either possibility being almost equally probable at normal temperatures. If the $^1H$ is 'up' then it will have a slight preferential attraction for the 'down' electron in the bond pair, i.e. the one whose magnetic moment lies antiparallel to it, for magnets tend to pair in this way. The effect will be to unpair the two electrons very slightly and, thus, to

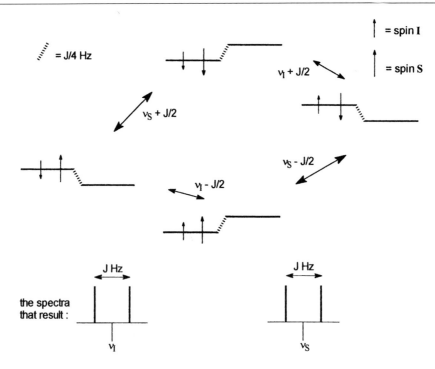

**Figure B1.11.8.** Combined energy states for two spins I and S (e.g. $^1$H and $^{13}$C) with an exaggerated representation of how their mutual alignment *via* coupling affects their combined energy.

make the $^{19}$F nucleus more likely to be close to the 'up' electron and thus slightly favoured in energy, if it is itself 'down'. The net result is that the $^1$H and $^{19}$F nuclei gain slightly in energy when they are mutually antiparallel. The favourable and unfavourable arrangements are summarized as follows:

energetically favourable:     H↑ e↓↑e ↓F     H↓ e↑↓e ↑F     energy gain $\frac{1}{4}J$

energetically unfavourable:   H↑ e↓↑e ↑F     H↓ e↑↓e ↓F     energy loss $\frac{1}{4}J$.

These energy differences then generate splittings as outlined in figure B1.11.8. If the energies of the two spins, here generalized to I and S, in the magnetic field are corrected by the above $\pm\frac{1}{4}J$ quantities, then the transitions will move from their original frequencies of $\nu_I$ and $\nu_S$ to $\nu_I \pm \frac{1}{2}J$ and $\nu_S \pm \frac{1}{2}J$. Thus, both the I and the S resonances will each be symmetrically split by $J$ Hz. In the case of $^1$H$^{19}$F, $^1J_{HF} = +530$ Hz. The positive sign applies when the antiparallel nuclear spin configuration is found, and the '1' superscript refers to the number of bonds separating the two spins.

Figure B1.11.8 does not apply accurately in the not uncommon cases when I and S have Larmor frequencies whose separation $\Delta\nu$ is not large compared with $J_{IS}$. In these cases the two spin states where I and S are antiparallel have very similar energies, so that the coupling interaction mixes their spin states, analogously to the non-crossing rule in UV–visible spectroscopy. As $\Delta\nu$ falls, the peak intensities in the multiplet alter so as to boost the component(s) nearest to the other multiplet, at the expense of the other components, whilst keeping the overall integral of the multiplet constant. This 'roofing' or 'tenting' is an example of a second-order effect on couplings. It is actually useful up to a point, in that it helps the analysis of couplings in a complex molecule, by indicating how they pair off. Some examples are evident in subsequent figures. Figure B1.11.10, for example, shows mutual roofing of the H-6a and H-6b multiplets, and also of H-3 with

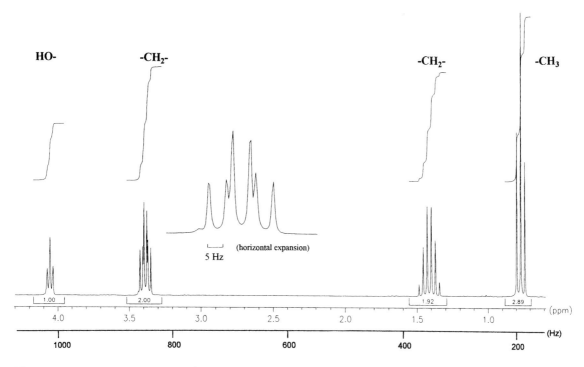

**Figure B1.11.9.** Integrated 250 MHz $^1$H NMR spectrum of dilute propan-1-ol in dimethylsulfoxide solvent. Here, the shift order parallels the chemical order. An expansion of the $H_2$-1 multiplet is included, as is the implicit frequency scale, also referenced here to TMS = 0.

H-2. Roofing is also seen, more weakly, in figure B1.11.9, where every multiplet is slightly tilted towards the nearest multiplet with which it shares a coupling. Indeed, it is unusual not to see second-order distortions in $^1$H NMR spectra, even when these are obtained at very high field.

As $\Delta\nu$ gets even smaller, the outer components of the multiplet become invisibly small. Further splittings may also appear in complex multiplets. In the extreme case that $\Delta\nu = 0$, the splittings disappear altogether, even though the physical coupling interaction is still operative. This is why the methyl resonance in figure B1.11.3 appears as a single peak. It is occasionally necessary to extract accurate coupling constants and chemical shifts from complex, second-order multiplets, and computer programs are available for this.

### B1.11.6.2  *Factors that determine coupling constants*

Because $J$ arises from the magnetic interactions of nuclei, the simplest factor affecting it is the product $\gamma_I\gamma_S$ of the two nuclear magnetogyric ratios involved. For example, $^1J_{DF}$ in $^2H^{19}F$ is 82 Hz, i.e. $^1J_{HF} \times \gamma_D/\gamma_H$. This totally predictable factor is sometimes discounted by quoting the reduced coupling constant $K_{IS} \equiv 4\pi^2 J_{IS}/h\gamma_I\gamma_S$.

A second determining factor in the Fermi contact mechanism is the requirement that the wavefunction of the bonding orbital has a significant density at each nucleus, in order for the nuclear and the electron magnets to interact. One consequence of this is that $K$ correlates with nuclear volume and therefore rises sharply for heavier nuclei. Thus the $^1K$ constants in the $XH_4$ series with X $=^{13}$C, $^{29}$Si, $^{73}$Ge, $^{119}$Sn and $^{207}$Pb are respectively +41.3, +84.9, +232, +430 and +938 N A$^{-2}$ m$^{-3}$. Here the average value of ($K$/nuclear mass number) = 3.5 $\pm$ 1.

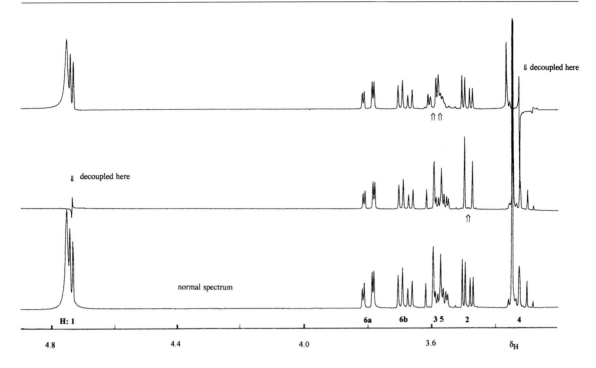

**Figure B1.11.10.** 400 MHz $^1$H NMR spectrum of methyl-$\alpha$-glucopyranose (structure as in figure B1.11.12) together with the results of decoupling at H-1 (centre trace) and at H-4 (upper trace).

The presence of a spin–spin splitting therefore means that the interacting atomic orbitals must each possess significant s character, because all orbitals other than s have zero density at the nucleus. The $^1J_{CH}$ values for $CH_4$, $H_2C=CH_2$ and $HC\equiv CH$ illustrate the dependence of $J$ upon s character. They are respectively +124.9, +156.2 and +249 Hz, and so they correlate unusually precisely with the s characters of the carbon orbitals bonding to H, these being respectively 25, 33.3 and 50%. A coupling also shows that the bond that carries the coupling cannot be purely ionic. A significant $J$ value demonstrates covalency, or at least orbital overlap of some kind.

The next large influence on $^nJ$ is the number of bonds $n$ in the bonding pathway linking the two spins and their stereochemistry. When $n > 1$, then there will be at least two electron pairs in the pathway, and $J$ will thus be attenuated by the extent to which one electron pair can influence the next. In simple cases, the insertion of a further electron pair will reverse the sign of $J$, because parallel electrons in different, interacting orbitals attract each other slightly, following Hund's rules. Thus, for a two-bond pathway, a favoured spin configuration will now be

energetically favourable:    $X\uparrow e \downarrow\uparrow e.e\uparrow\downarrow e \uparrow Y$.

Some other general rules have been extracted; they come particularly from $^nJ_{HH}$ data, but also have a wider validity in many cases.

- $n = 2$  $^2J$ is often low, because of competing pathways involving different orbitals. A typical value for chemically distinct H atoms in a methylene group is $-14$ Hz. However, it can rise to zero, or even to a positive value, in $C=CH_2$ moieties.
- $n = 3$  $^3J$ is almost always positive and its magnitude often exceeds that of $^2J$. It always depends in a predictable way on the dihedral angle $\phi$ between the outer two of the three bonds in the coupling pathway.

Karplus first showed theoretically that $^3J$ varies to a good approximation as $A \cos^2 \phi + B \cos \phi$, where $A$ and $B$ are constants, and also that $A \gg B$ [17]. His equation has received wide-ranging experimental verification. For a typical HCCH pathway, with not particularly electronegative substituents, $A = 13$ Hz and $B = -1$ Hz. This predicts $^3J = 14.0$ Hz for $\phi = 180°$ and 2.75 Hz for $\phi = 60°$. Thus, a typical HCCH$_3$ coupling, where the relevant dihedral angles are $60°$ (twice) and $180°$ (once) will average to $(2 \times 2.75 + 14)/3 = 6.5$ Hz, as the C–C bond rotates. Almost all other $^3J$ couplings follow a similar pattern, with $^3J$ being close to zero if $\phi = 90°$, even though the values of $A$ vary widely according to the atoms involved. The pattern may be pictured as a direct, hyperconjugative interaction between the outer bonding orbitals, including their rear lobes. It is only fully effective when these orbitals lie in the same plane. The Karplus equation offers a valuable way of estimating bond angles from couplings, especially in unsymmetrical molecular fragments such as CH–CH$_2$, where the presence of two $^3J$ couplings eliminates any ambiguities in the values of $\phi$.

- $n = 4$ $^4J$ couplings are often too small to resolve. However, they are important in cases where the relevant orbitals are aligned appropriately. If the molecular fragment HC–C=CH has the first CH bond approximately parallel to the C=C $\pi$ orbital, the resulting hyperconjugative interaction will give rise to an 'allylic' coupling of about $-2$ Hz. A similar coupling arises in a saturated HC–C–CH fragment if the bonds lie approximately in a W configuration. In such cases the rear lobes of the CH bonding orbitals touch, thus offering an extra electronic pathway for coupling. Indeed, all such contacts give rise to couplings, even if the formal bonding pathway is long.

Although the Fermi contact mechanism dominates most couplings, there are smaller contributions where a nuclear dipole physically distorts an orbital, not necessarily of s type [18]. There are many useful compilations of $J$ and $K$ values, especially for HH couplings (see [9], ch 4, 7–21 and [12–15]).

### B1.11.6.3  Multiple couplings

In principle, every nucleus in a molecule, with spin quantum number $I$, splits every other resonance in the molecule into $2I + 1$ equal peaks, i.e. one for each of its allowed values of $m_I$. This could make the NMR spectra of most molecules very complex indeed. Fortunately, many simplifications exist.

(i)    As described above, most couplings fall off sharply when the number of separating bonds increases.

(ii)   Nuclei with a low isotopic abundance will contribute correspondingly little to the overall spectra of the molecule. Thus, splittings from the 1.1% naturally abundant $^{13}$C isotope are only detectable at low level in both $^1$H and $^{13}$C NMR spectra, where they are called $^{13}$C sidebands.

(iii)  Nuclei with $I > 1/2$ often relax so rapidly that the couplings from them are no longer detected, for reasons analogous to shift averaging by chemical exchange. For example, couplings from Cl, Br and I atoms are invisible, even though these halogens all have significant magnetic moments.

(iv)   Chemical exchange will similarly average couplings to zero, if it takes place much faster than the value of the coupling in hertz.

(v)    Selected nuclei can also have their couplings removed deliberately (decoupled) by selective irradiation during the acquisition of the FID.

Despite these simplifications, a typical $^1$H or $^{19}$F NMR spectrum will normally show many couplings. Figure B1.11.9 is the $^1$H NMR spectrum of propan-1-ol in a dilute solution where the exchange of OH hydrogens between molecules is slow. The underlying frequency scale is included with the spectrum, in order to emphasize how the couplings are quantified. Conveniently, the shift order matches the chemical order of the atoms. The resonance frequencies of each of the 18 resolved peaks can be quantitatively explained by the four chemical shifts at the centre of each multiplet, plus just three values of $^3J$, two of these being in fact almost the same. If the hydrogen types are labelled in chemical order from the methyl hydrogens H$_3$-3 to the hydroxyl hydrogen then the three coupling types visibly present are $^3J_{23} = 2a$, $^3J_{12} = 2b$ and $^3J_{1-OH} = 2c$ Hz, with $a \approx b$ because of the strong chemical similarity of all the HCCH bonding pathways. Consider first the methyl resonance $H_3$-3, and its interaction with the two equivalent hydrogens H-2 and H-2′. Each of the three equivalent hydrogens of $H_3$-3 will first be split into a 1:1 doublet by H-2, with peak positions $\delta_3 \pm a$, i.e. $a$ Hz on either side of the true chemical shift $\delta_3$, reconverted here into frequency units. H-2′ will further split each component, giving three peaks with positions $\delta_3 \pm a \pm a$. When all the possible $\pm$ combinations are considered, they amount to peaks at $\delta_3$ (twice) plus $\delta_3 \pm 2a$ (once each) and are conveniently called a (1:2:1) triplet. The OH multiplet is similarly a (1:2:1) triplet with peaks at $\delta_{OH}$ (twice) plus $\delta_{OH} \pm 2c$ (once each).

Each H-1 hydrogen contributes to a more complex multiplet, with peaks at all combinations of the frequencies $\delta_1 \pm b \pm b \pm c$, thus making a triplet of doublets whose peak positions are $\delta_1 \pm c$ (twice each) and $\delta_1 \pm 2b \pm c$ (once each). Finally, each H-2 hydrogen contributes to the sextet at $\delta_2 = 1.42$ ppm, whose individual components appear at $\delta_2 \pm a \pm a \pm a \pm b \pm b$. If we take $b = a$, then the possible combinations of these frequencies amount to the following peak positions, with their relative heights given in brackets: $\delta_2 \pm 5a$ (1), $\delta_2 \pm 3a$ (5), $\delta_2 \pm a$ (10). Note that the relative intensities follow Pascal's triangle and also that the separation of the outermost line of a multiplet must always equal the sum of the underlying couplings, when only spin-1/2 nuclei are involved.

The same principles apply to couplings from spins with $I > 1/2$, where these are not seriously affected by relaxation. Figure B1.11.4 illustrates a common case. The solvent resonance at 30 ppm is a 1:3:6:7:6:3:1 multiplet arising from the $^{13}C^2H_3$ carbons in the solvent deuterioacetone. Each of the three deuterium nuclei splits the carbon resonance into a 1:1:1 triplet, corresponding to the three possible Zeeman states of an $I = 1$ nucleus. Thus the overall peak positions are the possible combinations of $\delta_C + (a$ or 0 or $- a) + (a$ or 0 or $- a) + (a$ or 0 or $- a)$, where the coupling constant $a$ is $\sim 20$ Hz.

With relatively simple spectra, it is usually possible to extract the individual coupling constants by inspection, and to pair them by size in order to discover what atoms they connect. However, the spectra of larger molecules present more of a challenge. The multiplets may overlap or be obscured by the presence of several unequal but similarly sized couplings. Also, if any chiral centres are present, then the two hydrogens in a methylene group may no longer have the same chemical shift, and in this case they will also show a mutual $^2J$ coupling. Fortunately, several powerful aids exist to meet this challenge: decoupling and a range of multidimensional spectra.

### B1.11.6.4  Decoupling

Several examples have already been given of resonances that merge because the underlying molecular or spin states interchange more rapidly than their frequency separation. Similar interchanges can also be imposed so as to remove a coupling in a controllable way. One irradiates a selected resonance so as to give its magnetization no preferred direction, within the timescale set by the coupling to be removed. This can be achieved selectively for any resonance removed from others by typically 20 Hz, although multiplets nearly as close as this will unavoidably be perturbed in a noticeable and predictable way by the irradiation process.

Figure B1.11.10 offers an example. It shows the 400 MHz $^1$H NMR spectrum of $\alpha$-1-methylglucopyranose, below two further spectra where $^1$H-decoupling has been applied at H-1 and H-4 respectively. The main results of the decouplings are arrowed. Irradiation at the chemical shift position of H-1 removes the smaller of the two couplings to H-2. This proves the saccharide to be in its $\alpha$ form, i.e. with $\phi \approx 60°$ rather than 180°, according to the Karplus relationship given previously. Note that both the H-1 resonance and the overlapping solvent peak are almost totally suppressed by the saturation, caused by the decoupling irradiation. The H-4 resonance is a near-triplet, created by the two large and nearly equal couplings to H-3 and H-5. In both these cases, $\phi \approx 180°$. The spectrum in this shift region is complicated by the methyl singlet and by some minor peaks from impurities. However, these do not affect the decoupling process, beyond being severely distorted by it. Genuine effects of decoupling are seen at the H-3 and the H-5 resonances only.

Single-frequency decoupling is easy and rapidly carried out. However, it may be limited by the closeness of different multiplets. Also, it will not normally be possible to apply more than one frequency of decoupling irradiation at a time. Fortunately, these disadvantages do not apply to the equivalent multidimensional methods.

It is also usually possible to remove all the couplings from a particular isotope, e.g. $^1$H, provided that one only wishes to observe the spectrum from another isotope, e.g. $^{13}$C. Either the decoupling frequency is noise-modulated to cover the relevant range of chemical shifts, or else the same decoupling is achieved more efficiently, and with less heating of the sample, by using a carefully designed, continuous sequence of composite pulses. Figure B1.11.4 is a $^1$H-decoupled $^{13}$C spectrum: this is sometimes abbreviated to $^{13}$C{$^1$H}. In the absence of decoupling, the resonances would each be split by at least four short- and long-range couplings from $^1$H atoms, and the signal-to-noise ratio would drop accordingly. The largest couplings would arise form the directly attached hydrogens. Thus, one might use a $^{13}$C NMR spectrum without decoupling to distinguish $CH_3$ (broadened 1:3:3:1 quartets) from $CH_2$ (1:2:1 triplets) from CH (1:1 doublets) from C (singlets). However, more efficient methods are available for this, such as the DEPT and INEPT pulse sequences outlined below. Two-dimensional methods are also available if one needs to detect the connectivities revealed by the various $^{13}$C–$^1$H couplings.

### B1.11.6.5  Polarization transfer

Couplings can also be exploited in a quite different way, that lies behind a wide range of valuable NMR techniques. If just one component peak in a multiplet X can be given a non-equilibrium intensity, e.g. by being selectively inverted, then this necessarily leads to large changes of intensity in the peaks of any other multiplet that is linked via couplings to X. Figure B1.11.11 attempts to explain this surprising phenomenon. Part A shows the four possible combined spin states of a $^{13}$C$^1$H molecular fragment, taken as an example. These are the same states as in figure B1.11.8, but attention is now drawn to the populations of the four spin states, each reduced by subtracting the 25% population that would exist at very low field, or alternatively at infinite temperature. The figures above each level are these relative differences, in convenient units. The intensity of any one transition, i.e. of the relevant peak in the doublet, is proportional to the difference of these differences, and is therefore proportionally relative to unity for any $^{13}$C transition at Boltzmann equilibrium, and 4 for any $^1$H transition.

The only alteration in part B is that the right hand $^1$H transition has been altered so as to interchange the relevant populations. In practice, this might be achieved either by the use of a highly selective soft pulse, or by a more elaborate sequence of two pulses, spaced in time so as the allow the Larmor precessions of the two doublet components to dephase by 180°. The result is to produce a $^{13}$C doublet whose relative peak intensities are 5: −3, in place of the original 1:1. Further, similar manipulations of the $^{13}$C spins can follow, so as to re-invert the −3 component. This results in a doublet, or alternatively a singlet after decoupling, that has four times the intensity it had originally. The underlying physical process is called polarization transfer. More generally, any nucleus with magnetogyric ratio $\gamma_A$, coupled to one with $\gamma_B$, will have its intensity altered by $\gamma_B/\gamma_A$. This gain is valuable if $\gamma_B \gg \gamma_A$. The technique is then called INEPT (insensitive nucleus enhancement by polarization transfer) [19]. It has an added advantage. The relaxation rate of a high-$\gamma$ nucleus is usually much less than that of a low-$\gamma$ nucleus, because it has a bigger magnetic moment to interact with its surroundings. In the INEPT experiment, the spin populations of the low-$\gamma$ nucleus are driven by the high-$\gamma$ nucleus, rather than by natural relaxation. Hence the repeat time for accumulating the FID is shortened to that appropriate for the high-$\gamma$ nucleus.

The same general methodology can also be applied to edit (for example) a decoupled $^{13}$C NMR spectrum into four subspectra, for the $CH_3$, $CH_2$, CH and C moieties separately. A common variant method called DEPT (distortionless enhancement by polarization transfer) uses non-standard pulse angles, and is a rapid and reliable way for assigning spectra of medium complexity [20].

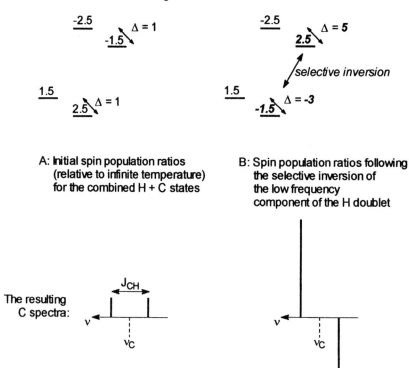

$\Delta$ measures the relative intensity of the C transitions.
Note the dramatic change in these after selective inversion.

A: Initial spin population ratios
(relative to infinite temperature)
for the combined H + C states

B: Spin population ratios following
the selective inversion of
the low frequency
component of the H doublet

The resulting
C spectra:

**Figure B1.11.11.** Polarization transfer from $^1H$ to $^{13}C$ (see the text). The inversion of one H transition also profoundly alters the C populations.

More generally, note that the application of almost any multiple pulse sequence, where at least two pulses are separated by a time comparable to the reciprocal of the coupling constants present, will lead to exchanges of intensity between multiplets. These exchanges are the physical method by which coupled spins are correlated in 2D NMR methods such as correlation spectroscopy (COSY) [21].

## B1.11.7   Two-dimensional methods

The remarkable stability and controllability of NMR spectrometers permits not only the precise accumulation of FIDs over several hours, but also the acquisition of long series of spectra differing only in some stepped variable such as an interpulse delay. A peak at any one chemical shift will typically vary in intensity as this series is traversed. All the sinusoidal components of this variation with time can then be extracted, by Fourier transformation of the variations. For example, suppose that the normal 1D NMR acquisition sequence (relaxation delay, 90° pulse, collect FID)$_n$ is replaced by the 2D sequence (relaxation delay, 90° pulse, delay $\tau$ – 90° pulse, collect FID)$_n$ and that $\tau$ is increased linearly from a low value to create the second dimension. The polarization transfer process outlined in the previous section will then cause the peaks of one multiplet to be modulated in intensity, at the frequencies of any other multiplet with which it shares a coupling.

The resulting data set constitutes a rectangular or square array of data points, having a time axis in both dimensions. This is converted via Fourier transformation in both dimensions, giving the corresponding array of points in a 2D spectrum, with each axis being in $\delta$ units for convenience. These are most conveniently plotted as a contour map. In the above example the experiment is COSY-90, '90' referring to the second pulse, and the map should be at least approximately symmetrical about its diagonal. Any off-diagonal or 'cross' peaks then indicate the presence of couplings. Thus, a cross-peak with coordinates ($\delta_A$, $\delta_B$) indicates a coupling that connects the multiplets A and B. The spectrum is normally simplified by eliminating superfluous, mirror-image peaks, either with phase-cycled pulses and appropriate subtractions or by the use of carefully controlled, linear pulsed field gradients. No special equipment is needed in a modern spectrometer, although the data sets are typically 1 Mbyte or larger. The time requirement is only about 16 times that for a 1D spectrum, in favourable cases, and may be less if pulsed field gradients are used.

### B1.11.7.1 Homonuclear COSY spectra

Figure B1.11.12 shows the 2D COSY-45 contour plot of the same $\alpha$-1-methylglucopyranose compound as in a previous figure. The corresponding 1D spectrum, plotted directly above, is in fact the projection of the 2D spectrum onto the horizontal shift axis. The hydrogen assignments are added underneath the multiplets. The analysis of such a spectrum begins by selecting one peak, identifiable either by its distinctive chemical shift or by mapping its coupling pattern onto the expected pattern of molecular connectivity. Here, the doublet at 4.73 ppm uniquely has the high shift expected when a CH group bears two O substituents. The coupling pattern can now be identified by noting the horizontal and vertical alignments of all the strongest diagonal and off-diagonal resonances. These alignments must be exact, within the limits of the digitization, as the COSY process does not cause any shifts. The H-1 to H-2 correlation exemplifies this precision, in that the cross-peak whose centre has coordinates at $\delta$ (3.47, 4.73) is precisely aligned with the centres of the H-2 and H-1 multiplets respectively. The same principle allows one to see that the cross-peak at $\delta$ (3.47, 3.59) arises from the (H-2 to H-3) coupling, whereas the more complex multiplet to the right of it must come from the overlap of the (H-3 to H-4) and the (H-5 to H-4) cross-peaks, respectively at $\delta$ (3.59, 3.33) and $\delta$ (3.56, 3.33). In this way it is possible to distinguish the H-3 and H-5 multiplets, respectively at 3.59 and 3.56 ppm, even though they overlap in the 1D spectrum. This illustrates the power of multidimensional NMR to separate overlapping resonances. In a similar way, the 2D spectrum makes it clear that the OCH$_3$ resonance at 3.35 ppm has no coupling connection with any other hydrogen in the saccharide, so that its shift overlap with H-4 is purely accidental.

Several other features of the figure merit attention. The two hydroxymethyl resonances, H-6a at 3.79 ppm and H-6b at 3.68 ppm, are separated in shift because of the chiral centres present, especially the nearby one at C-5. (One should note that a chiral centre will always break the symmetry of a molecule, just as an otherwise symmetrical coffee mug loses its symmetry whilst grasped by a right hand.) H-5 is thus distinctive in showing three coupling connections, and could be assigned by this alone, in the absence of other information. Also, the (H-6a to H-6b) cross-peak gives the general impression of a tilt parallel to the diagonal, whereas some of the other cross-peaks show the opposite tilt. This appearance of a tilt in the more complex cross-peaks is the deliberate consequence of completing the COSY pulse sequence with a 45° rather than a 90° pulse. It actually arises from selective changes of intensity in the component peaks of the off-diagonal multiplet. The 'parallel' pattern arises from negative values of the coupling constant responsible for the cross-peak, i.e. the 'active' coupling. It therefore usually shows that the active coupling is of $^2J$ or $^4J$ type, whereas an 'antiparallel' pattern usually arises from a $^3J$ active coupling.

Another advantage of a COSY spectrum is that it can yield cross-peaks even when the active coupling is not fully resolved. This can be useful for assigning methyl singlets, for example in steroids, and also for exploiting couplings comparable with the troublesomely large linewidths in protein spectra, for example, or the $^{11}B\{^1H\}$ spectra of boranes. However, it does also mean that longer-range couplings may appear unexpectedly, such as the weak (H-1 to H-3), (H-1 to H-6b) and (H-6a to H-4) couplings in figure B1.11.12. Their appearance can yield stereochemical information such as the existence of bonding pathways having appropriate conformations, and they are usually recognizable by their comparatively weak intensities.

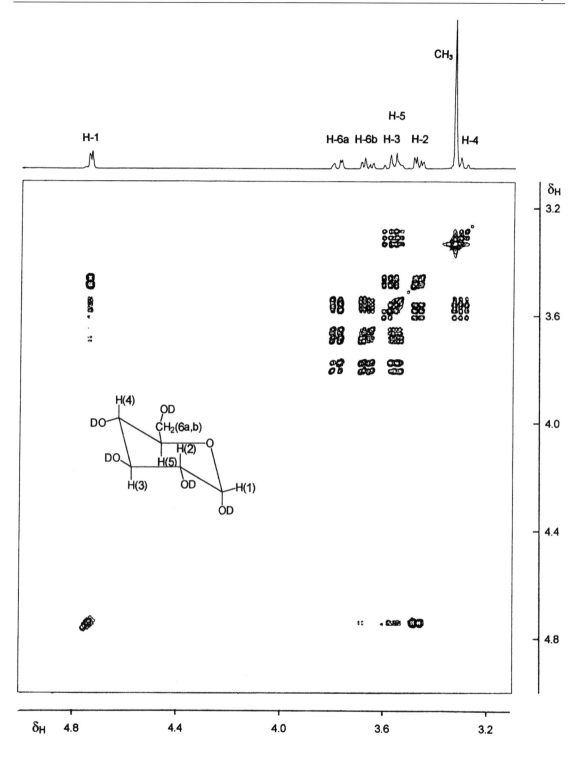

**Figure B1.11.12.** $^1$H–$^1$H COSY-45 2D NMR spectrum of methyl-$\alpha$-glucopyranose (structure shown). The coupling links and the approximate couplings can be deduced by inspection.

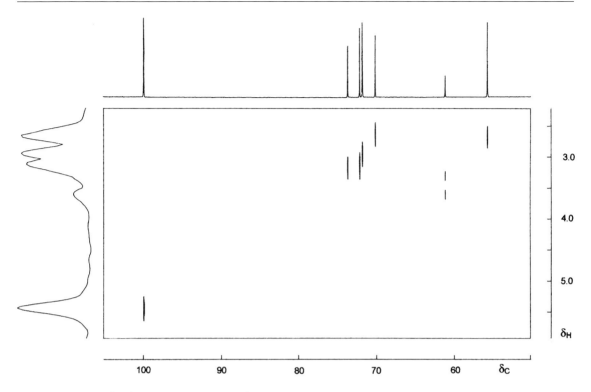

**Figure B1.11.13.** $^{13}$C–$^1$H shift correlation via $^1J_{CH}$. This spectrum of methyl-$\alpha$-glucopyranose (structure as in figure B1.11.12) permits unambiguous $^{13}$C assignments once the $^1$H assignments have been determined as in figure B1.11.12.

Many variations of the basic homonuclear COSY experiment have been devised to extend its range. A brief guide to some classes of experiment follows, along with a few of the common acronyms.

(i)   Experiments using pulsed field gradients [23]. These can be very rapid if concentration is not a limiting factor, but they require added equipment and software. Acronym: a prefix of 'g'.

(ii)  Multiple quantum methods. These employ more complex pulse sequences, with the aim of suppressing strong but uninteresting resonances such as methyl singlets. Acronyms include the letters MQ, DQ, etc. They are also useful in isotope-selective experiments, such as INADEQUATE [24]. For example, the $^{13}$C–$^{13}$C INADEQUATE experiment amounts to a homonuclear $^{13}$C COSY experiment, with suppression of the strong singlets from uncoupled carbons.

(iii) Extended or total correlation methods. These reveal linked couplings, involving up to every atom in a group, where at least one reasonably large coupling exists for any one member atom. They are valuable for identifying molecular moieties such as individual amino acids in a peptide, for they can remove ambiguities arising from peak overlaps. Acronym: TOCSY (total correlation spectroscopy).

(iv)  Correlations exploiting nuclear Overhauser enhancements (NOEs) in place of couplings, such as NOESY. These are discussed in section B1.11.8.

(v)   Correlations that emphasize smaller couplings. These are simply achieved by the addition of appropriate, fixed delays in the pulse sequence.

Homonuclear techniques such as $J$-resolved spectroscopy also exist for rotating all multiplets through $90°$, to resolve overlaps and also give a 1D spectrum from which all homonuclear couplings have been removed [26].

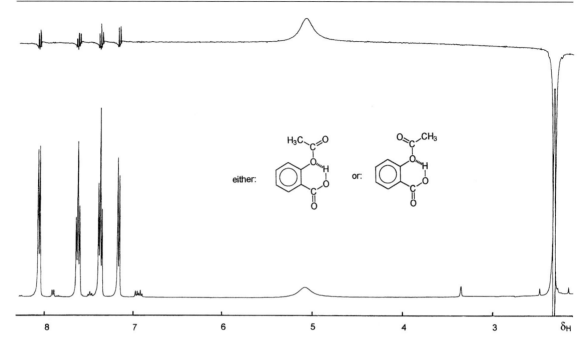

**Figure B1.11.14.** $^1$H NOE-difference spectrum (see the text) of aspirin, with pre-saturation at the methyl resonance, proving that the right-hand conformation is dominant.

The theory of these and other multidimensional NMR methods requires more than can be visualized using magnetization vectors. The spin populations must be expressed as the diagonal elements of a density matrix, which not only has off-diagonal elements indicating ordinary, single-quantum transitions such as those described earlier, but also other off-diagonal elements corresponding to multiple-quantum transitions or 'coherences'. The pulses and time developments must be treated as operators. The full theory also allows experiments to be designed so as to minimize artefacts [27–29].

### B1.11.7.2    Heteronuclear correlation spectra

Similar experiments exist to correlate the resonances of different types of nucleus, e.g. $^{13}$C with $^1$H, provided that some suitable couplings are present, such as $^1J_{CH}$. It is necessary to apply pulses at both the relevant frequencies and it is also desirable to be able to detect either nucleus, to resolve different peak clusters. Detection through the nucleus with the higher frequency is usually called reverse-mode detection and generally gives better sensitivity. The spectrum will have the two different chemical shift scales along its axes and therefore will not be symmetrical.

A $^1$H(detected)–$^{13}$C shift correlation spectrum (common acronym HMQC, for heteronuclear multiple quantum coherence, but sometimes also called COSY) is a rapid way to assign peaks from protonated carbons, once the hydrogen peaks are identified. With changes in pulse timings, this can also become the HMBC (heteronuclear multiple bond connectivity) experiment, where the correlations are made via the smaller $^2J_{CH}$ and $^3J_{CH}$ couplings. This helps to assign quaternary carbons and also to identify coupling and, hence, chemical links, where H–H couplings are not available. Similar experiments exist for almost any useful pairing of nuclei: those to $^{15}$N are particularly useful in the spectra of suitably labelled peptides. Figure B1.11.13 shows a simple $^{13}$C–$^1$H shift correlation spectrum of the same saccharide as in the previous two figures, made by exploiting the $^1J_{CH}$ couplings. The detection was via $^{13}$C and so the spectrum has good resolution in the $^{13}$C dimension, here plotted as the horizontal axis, but rather basic resolution in the vertical $^1$H dimension. The $^1$H resolution is nonetheless sufficient to show that C-6 is a single type of carbon attached to two distinct types of hydrogen. Inspection of the $^1$H axis shows that the approximate centre of each 2D multiplet matches the shifts in the previous spectra. this affords an unambiguous assignment of the carbons.

Even more complex sequences can be applied for the simultaneous correlation of, for example, $^{15}$N shifts with $^1$H–$^1$H COSY spectra. Because the correlation of three different resonances is highly specific, the chance of an accidental overlap of peaks is greatly reduced, so that very complex molecules can be assigned. Furthermore, the method once again links atoms where no simple H–H couplings are available, such as across a peptide bond. Such 3D NMR methods generate enormous data sets and, hence, very long accumulation times, but they allow the investigation of a wide range of labelled biomolecules.

## B1.11.8  Spatial correlations

$J$ couplings are not the only means in NMR for showing that two atoms lie close together. Pairs of atoms, particularly $^1$H, also affect each other through their dipolar couplings. Even though the dipolar splittings are averaged away when a molecule rotates isotropically, the underlying magnetic interactions are still the major contributors to the $^1$H $T_1$. Each pairwise interaction of a particular $H_A$ with any other $H_B$, separated by a distance $r_{AB}$ in a rigid molecule, will contribute to $1/T_{1A}$ in proportion to $(r_{AB})^{-6}$.

Although it is possible to detect this interaction by the careful measurement of all the $T_1$s in the molecule, it may be detected far more readily and selectively, *via* the mutual NOE. When a single multiplet in a $^1$H NMR spectrum is selectively saturated, the resulting input of energy leaks to other nuclei, through all the dipolar couplings that are present in the molecule [29]. This alters the intensities and likewise the integrals of their resonances. The A spin of an isolated pair of spins, A and B, in a molecule tumbling fairly rapidly, will have its intensity increased by the multiplicative factor $1 + \gamma_B/2\gamma_A$. This amounts to a gain of 50% when A and B are both $^1$H, although it will normally be less in practice, because of the competing interactions of other spins and other relaxation mechanisms. If B is $^1$H and A is $^{13}$C, the corresponding enhancement is 299%. In this case there need be no competition, for all the $^1$H resonances can be saturated simultaneously, using the techniques of broadband decoupling. Also, dipolar coupling usually dominates the relaxation of all carbons bearing hydrogens. The threefold intensity gain is a valuable bonus in $^{13}$C{$^1$H} NMR.

### B1.11.8.1  NOE-difference spectra

Even if the intensity changes are only of the order of 1%, they may nevertheless be reliably detected using difference spectra [30]. Figure B1.11.14 shows an NOE-difference $^1$H NMR spectrum of slightly impure aspirin, over a normal spectrum of the same sample. The difference spectrum was obtained by gently saturating the methyl singlet at 2.3 ppm, for a period of 2 s just before collecting the spectrum, and then by precisely subtracting the corresponding spectrum acquired without this pre-irradiation. As the pre-irradiation selectively whitewashes the methyl peak, the result of the subtraction is to create a downwards-going resultant peak. The other peaks subtract away to zero, unless the energy leakage from the pre-irradiation has altered their intensity. In the figure, the four aromatic H resonances to the left subtract to zero, apart from minor errors arising from slight instabilities in the spectrometer. In contrast, the broad OH resonance at 5 ppm in the upper spectrum proves that this hydrogen lies close to the CH$_3$ group. Thus, aspirin must largely possess the conformation shown on the right. If the left-hand conformation had been significant, then the aromatic H at 7.2 ppm would have received a significant enhancement.

NOE-difference spectroscopy is particularly valuable for distinguishing stereoisomers, for it relies solely on internuclear distances, and thus avoids any problems of ambiguity or absence associated with couplings. With smallish molecules, it is best carried out in the above 1D manner, because ~2 s are necessary for the transmission of the NOE. The transmission process becomes more efficient with large molecules and is almost optimal for proteins. However, problems can occur with molecules of intermediate size [31]. A 2D version of the NOE-difference experiment exists, called NOESY.

### B1.11.8.2    Protein structures

If multidimensional spectra of both COSY and NOESY types can be obtained for a protein, or any comparable structured macromolecule, and if a reasonably complete assignment of the resonances is achieved, then the NOESY data can be used to determine its structure [31–33]. Typically, several hundred approximate H–H distances will be found via the NOESY spectrum of a globular protein having mass around 20 000 Da. These can then be used as constraints in a molecular modelling calculation. The resulting structures can compare in quality with those from x-ray crystallography, but do not require the preparation of crystals. Related spectra can also elucidate the internal flexibility of proteins, their folding pathways and their modes of interaction with other molecules. Such information is vital to the pharmaceutical industry in the search for new drugs. It also underpins much biochemistry.

### References

[1] Harris R K 1996 Nuclear spin properties and notation *Encyclopedia of NMR* vol 5, ed D M Grant and R K Harris (Chichester: Wiley) pp 3301–14

[2] Dorn H C 1984 $^1$H NMR—a new detector for liquid chromatography *Anal. Chem.* **56** 747A–58A

[3] Morris K F and Johnson C S Jr 1993 Resolution of discrete and continuous molecular size distributions by means of diffusion-ordered 2D NMR spectroscopy *J. Am. Chem. Soc.* **115** 4291–9

[4] Quin L D and Verkade J G (eds) 1994 *Phosphorus-31 NMR Spectral Properties in Compound Characterization and Structural Analysis* (New York: VCH)

[5] Overhauser A W 1953 Polarization of nuclei in metals *Phys. Rev.* **92** 411–15

[6] Howarth O W, Pettersson L and Andersson I 1989 Monomolybdononavanadate and *cis*- and *trans*-dimolybdo-octavanadate *J. Chem. Soc. Dalton Trans.* 1915–23

[7] Lamb W 1941 Internal diamagnetic fields *Phys. Rev.* **60** 817–19

[8] Lynden-Bell R M and Harris R K 1969 *Nuclear Magnetic Resonance Spectroscopy* (London: Nelson) pp 81–3

[9] Mason J (ed) 1987 *Multinuclear NMR* (New York: Plenum) ch 7–21

[10] Tonelli A S 1996 *Polymer Spectroscopy* ed A Fawcett (Chichester: Wiley) ch 2

[11] Martin M M, Martin G J and Delpuech J-J (eds) 1980 Use of chemicals as NMR auxiliary reagents *Practical NMR Spectroscopy* (London: Heyden) ch 10

[12] Williams D H and Fleming I 1995 *Spectroscopic Methods in Organic Chemistry* (London: McGraw-Hill) ch 3

[13] Breitmaier E and Voelter W 1986 *Carbon-13 NMR Spectroscopy: High Resolution Methods and Applications in Organic Chemistry* (New York: VCH)

[14] Pretsch E, Clerc T, Seibl J and Simon W 1983 *Tables of Spectral Data for Structural Determination of Organic Compounds* Engl. edn. (Berlin: Springer)

[15] Silverstein R M, Bassler G C and Morrill T C 1981 *Spectrometric Identification of Organic Compounds* (New York: Wiley) ch 4 and 5

[16] Jameson C J and Mason J 1987 The chemical shift *Multinuclear NMR* ed J Mason (New York: Plenum) ch 3

[17] Karplus M 1959 Contact electron spin coupling of nuclear magnetic moments *J. Chem. Phys.* **30** 11–15

[18] Venanzi T J 1982 Nuclear magnetic resonance coupling constants and electronic structure in molecules *J. Chem. Educ.* **59** 144–8

[19] Morris G A and Freeman R 1979 Enhancement of nuclear magnetic resonance signals by polarization transfer *J. Am. Chem. Soc.* **101** 760–2

[20] Doddrell D M, Pegg D T and Bendall M R 1982 Distortionless enhancement of NMR signals by polarization transfer *J. Magn. Reson.* **48** 323–7

[21] Bax A and Freeman R 1981 Investigation of complex networks of spin–spin coupling by two-dimensional NMR *J. Magn. Reson.* **44** 542–61

[22] Hurd R E 1990 Gradient enhanced spectroscopy *J. Magn. Reson.* **87** 422–8

[23] Bax A, Freeman R and Kempsell S P 1980 Natural abundance $^{13}$C–$^{13}$C coupling observed via double quantum coherence *J. Am. Chem. Soc.* **102** 4849–51

[24] Braunschweiler L and Ernst R R 1983 Coherence transfer by isotropic mixing: application to proton correlation spectroscopy *J. Magn. Reson.* **53** 521–8

[25] Aue W P, Kharan J and Ernst R R 1976 Homonuclear broadband decoupling and two-dimensional J-resolved NMR spectroscopy *J. Chem. Phys.* **64** 4226–7

[26] Ernst R R, Bodenhausen G and Wokaun A 1987 *Principles of Nuclear Magnetic Resonance in One and Two Dimensions* (Oxford: Clarendon)
[27] Bax A 1982 *2-Dimensional Nuclear Magnetic Resonance in Liquids* (Delft: Delft University Press)
[28] Freeman R 1997 *Spin Choreography* (Oxford: Spektrum) ch 3
[29] Neuhaus D and Williamson M 1989 *The Nuclear Overhauser Effect in Structural and Conformational Analysis* (New York: VCH)
[30] Bax A and Davis D G 1985 Practical aspects of two-dimensional transverse NOE spectroscopy *J. Magn. Reson.* **63** 207–13
[31] Wüthrich K 1996 Biological macromolecules: structural determination in solution *Encyclopedia of NMR* vol 2, ed D M Grant and R K Harris (Chichester: Wiley) pp 932–9
[32] Markley J R and Opella S J (eds) 1997 *Biological NMR Spectroscopy* (Oxford: Oxford University Press)
[33] Oschkinat H, Müller T and Dieckmann T 1994 Protein structure determination with three- and four-dimensional spectroscopy *Angew. Chem. Int. Ed. Engl.* **33** 277–93

## Further Reading

Abraham R J, Fisher J and Loftus P 1988 *Introduction to NMR Spectroscopy* (Chichester: Wiley)

A first text that concentrates on $^1$H and $^{13}$C NMR. Level suitable for undergraduates, although complete beginners might need more help.

Williams D H and Fleming I 1995 *Spectroscopic Methods in Organic Chemistry* (London: McGraw-Hill)

Includes basic interpretation, 30 valuable tables of data, and a concise introduction to multidimensional NMR spectra from an interpretational point of view.

Mason J (ed) 1987 *Multinuclear NMR* (New York: Plenum)

The most recent comprehensive text concentrating on the entire Periodic Table. Individual elements are also covered from time to time in monographs and reviews, e.g. in *Progress in NMR Spectroscopy*.

Braun S, Kalinowski H-O and Berger S 1998 *150 and More Basic NMR Experiments* (Weinheim: VCH)

A fairly straightforward 'how to do it' book for the practising spectroscopist.

Sanders J K M and Hunter B J 1993 *Modern NMR Spectroscopy: a Guide for Chemists* (Oxford: Oxford University Press)

An informative second-level text with a practical flavour.

Martin G E and Zekster A S 1988 *Two-dimensional NMR Methods for Establishing Molecular Connectivity* (Weinheim: VCH)

Contains a wealth of practical experience.

Hoch J C and Stern A S 1996 *NMR Data Processing* (New York: Wiley-Liss)

Careful treatment of what happens to the data after they have been collected.

Ernst R R, Bodenhausen G and Wokaun A 1987 *Principles of Nuclear Magnetic Resonance in One and Two Dimensions* (Oxford: Clarendon)

Authoritative at a high theoretical level.

Freeman R 1996 *Spin Choreography* (Oxford: Spektrum)

Exceptionally elegant exposition of a wide range of advanced but much-used NMR concepts.

Grant D M and Harris R K (eds) 1996 *Encyclopedia of NMR* (Chichester: Wiley)

Very comprehensive eight-volume coverage.

# B1.12
# NMR of solids

## R Dupree and M E Smith

### B1.12.1   Introduction

Solid-state NMR has long been used by physicists to study a wide range of problems such as superconductivity, magnetism, the electronic properties of metals and semiconductors, ionic motion etc. The early experiments mostly used 'wide line' NMR where high resolution was not required but with the development of the technique, particularly the improvements in resolution and sensitivity brought about by magic angle spinning (B1.12.4.3), and decoupling and cross polarization (B1.12.4.4), solid-state NMR has become much more widely used throughout the physical and, most recently, biological sciences. Although organic polymers were the first major widespread application of high-resolution solid-state NMR, it has found application to many other types of materials, from inorganics such as aluminosilicate microporous materials, minerals and glasses to biomembranes. Solid-state NMR has become increasing multinuclear and the utility of the technique is evidenced by the steady and continued increase in papers that use the technique to characterize materials. There is no doubt that the solid-state NMR spectrometer has become a central piece of equipment in the modern materials physics and chemistry laboratories.

The principal difference from liquid-state NMR is that the interactions which are averaged by molecular motion on the NMR timescale in liquids lead, because of their anisotropic nature, to much wider lines in solids. Extra information is, in principle, available but is often masked by the lower resolution. Thus, many of the techniques developed for liquid-state NMR are not currently feasible in the solid state. Furthermore, the increased linewidth and the methods used to achieve high resolution put more demands on the spectrometer. Nevertheless, the field of solid-state NMR is advancing rapidly, with a steady stream of new experiments forthcoming.

This chapter summarizes the interactions that affect the spectrum, describes the type of equipment needed and the performance that is required for specific experiments. As well as describing the basic experiments used in solid-state NMR, and the more advanced techniques used for distance measurement and correlation, some emphasis is given to nuclei with spin $I > \frac{1}{2}$ since the study of these is most different from liquid-state NMR.

### B1.12.2   Fundamentals

#### B1.12.2.1   Interaction with external magnetic fields

NMR is accurately described as a coherent radiofrequency (RF) spectroscopy of the nuclear magnetic energy levels. The physical basis of the technique is the lifting of the degeneracy of the different $m_z$ nuclear spin states through interaction with an external magnetic field, creating a set of energy levels. The total energy separation between these levels is determined by a whole range of interactions. The nuclear spin Hamiltonian

**Table B1.12.1.** Summary of main interactions important to NMR.

| $H^m$ | Interaction | $T_{ij}$ | $A_i$ | Typical size (Hz) | Comments |
|---|---|---|---|---|---|
| $H^Z$ | Zeeman | Unitary | $B_0$ | $10^7$–$10^9$ | Interaction with main magnetic field |
| $H^{RF}$ | RF | Unitary | $B_1$ | $10^3$–$10^5$ | Interaction with RF field |
| $H^D$ | Dipolar | $D$ | $I$ | $10^3$–$10^4$ | Through space spin–spin interaction, axially symmetric traceless tensor |
| $H^{CS}$ | Chemical shielding | $\sigma$ | $B_0$ | $10^2$–$10^5$ | Alteration of the magnetic field by the electrons |
| $H^J$ | Indirect spin | $J$ | $I$ | $1$–$10^3$ | Spin–spin interaction mediated via the bonding electrons through the contact interaction |
| $H^P$ | Paramagnetic | | $S$ | $10^2$–$10^5$ | Interaction with isolated unpaired electrons |
| $H^K$ | Knight shift | $K$ | $S$ | $10^2$–$10^5$ | Interaction with conduction electrons via the contact interaction |
| $H^Q$ | Quadrupolar | $Eq$ | $I$ | $10^3$–$10^7$ | Interaction of nuclear quadrupolar moment with the electric field gradient ($q$) and $I$ twice to produce effectively an $I^2$ interaction |

has parts corresponding to the experimental conditions, which are termed external, and parts that result from the sample itself, which are called internal. The internal part provides information about the physical and electronic structure of the sample. The total interaction energy of the nucleus may be expressed as a sum of the individual Hamiltonians given in equation B1.12.1, (listed in table B1.12.1) and are discussed in detail in several excellent books [1–4].

$$H^{TOT} = H^Z + H^{RF} + H^D + H^{CS} + H^K + H^J + H^P + H^{Q(1)} + H^{Q(2)} + \cdots . \qquad (B1.12.1)$$

The large static applied magnetic field ($B_0$) produces the Zeeman interaction ($= -\hbar\omega_0 I_z$, where $I_z$ is the z-component of $I$ (the nuclear spin) with eigenvalues $m_z$ ($-I \leq m_z \leq I$), figure B1.12.1(a)) with the nuclear magnetic dipole moment $\mu$ ($= \gamma\hbar I$, where $\gamma$ is the gyromagnetic ratio of the nucleus). The $B_0$ field is taken to define the z-axis in the laboratory frame and gives an interaction energy of

$$H^Z = \mu B_0 = -\gamma\hbar B_0 m_z = -\hbar\omega_0 m_z \qquad (B1.12.2)$$

where $\omega_0$ is the Larmor frequency in angular frequency. In this chapter only the high field limit is considered whereby the nuclear spin states are well described by the Zeeman energy levels, and all the other interactions can be regarded as perturbations of these spin states. Any nucleus that possesses a magnetic moment is technically accessible to study by NMR, thus only argon and cerium of the stable elements are excluded as they possess only even–even isotopes.

For a spin-$\frac{1}{2}$ nucleus the two $m_z$ values $\pm\frac{1}{2}$ have energies of $\pm\gamma\hbar B_0/2$ giving an energy separation of $\gamma\hbar B_0$. Thermal equilibrium produces a Boltzmann distribution between these energy levels and produces the bulk nuclear magnetization of the sample through the excess population which for a sample containing a total of $N$ spins is $\sim N\gamma\hbar B_0/2kT$. For example for $^{29}$Si in an applied magnetic field of 8.45 T the excess in the populations at room temperature is only 1 in $10^5$ so that only a small number of the total number of spins contribute to the signal, and this is possibly the greatest weakness of NMR. As the magnetization is directly proportional to the applied magnetic field, sensitivity provides one of the drives to higher applied magnetic fields. The dependence on $\gamma$ explains why some nuclei are more favoured as the ease with which a nucleus can be observed depends upon the receptivity ($R_x$)

$$R_x = \gamma^3 C_x I_x (I_x + 1) \tag{B1.12.3}$$

where $C_x$ is the natural abundance of the isotope being considered [3].

In spectroscopy involving electromagnetic (em) radiation both spontaneous and stimulated events can occur. NMR is a relatively low-frequency em spectroscopy so that spontaneous events are very unlikely ($\sim 10^{-22}$ s$^{-1}$) and stimulated events therefore dominate. This means that NMR is a coherent spectroscopy so that the excess in population can be made to work in a concerted way. The NMR experiment involves measurement of the energy level separation by application of a time-varying magnetic field $B_1$ orthogonal to $B_0$. $B_1$ excites transitions (through $I_+$ and $I_-$, the conventional spin raising and lowering operators) when its frequency ($\omega$) is close to $\omega_0$, typically in the RF region of 10 MHz to 1 GHz. The Hamiltonian for this interaction is

$$H^{RF} = (-\gamma \hbar B_1 / 2)(I_+ e^{-i\omega t} + I_- e^{+i\omega t}). \tag{B1.12.4}$$

### B1.12.2.2  *Local interactions*

In diamagnetic insulating solids spin-$\frac{1}{2}$ nuclei experience a range of interactions that include magnetic dipolar ($H^D$) interaction through space with nearby nuclear magnetic moments, chemical shielding ($H^{CS}$), modification of the magnetic field at the nucleus due to the surrounding electrons and indirect spin–spin coupling ($H^J$)—interaction of magnetic moments mediated by intermediate electron spins. In materials that contain paramagnetic centres the unpaired electrons can interact strongly with the nuclei ($H^P$) and possibly cause very large shifts and severe broadening of the NMR signal. The fluctuating magnetic fields produced by the electron spins can produce very efficient relaxation. Hence, for solids where the nuclei are slowly relaxing and will dissolve paramagnetic ions, small amounts ($\sim 0.1$ mol%) are added to aid relaxation. In materials containing conduction electrons these can also interact strongly with the nuclear spin via a contact interaction $H^K$ that produces relaxation and a change in resonance position termed the Knight shift, both of which provide important information on the nature of the density of states at the Fermi surface. Nuclei with spin $I > \frac{1}{2}$ are also affected by the electric quadrupole interaction ($H^Q$), an interaction between the nuclear electric quadrupole moment and the gradient in the electric field at the nucleus. Although this is an electrical interaction it depends on the magnetic quantum number ($m_z$) so affects the NMR spectrum [2]. The background of the quadrupole interaction is given in the classic paper by Cohen and Reif [5].

All interactions associated with NMR can be expressed as tensors and may be represented by a general expression

$$H^m = k^m I_i T^m_{ij} A^m_j \tag{B1.12.5}$$

where $H^m$ is one of the component Hamiltonians in equation B1.12.1. For each interaction there is a constant $k$, a $3 \times 3$ second-rank tensor $T_{ij}$ and then another vector quantity that the spin ($I$) interacts with: this is either a field or a spin. Normally three numbers are needed to describe a $3 \times 3$ tensor relating two vectors and these numbers are usually the isotropic value, the anisotropy and the asymmetry [6]. Their exact definition can vary even though there are conventions that are normally used, so that any paper should be examined carefully to see how the quantities are being defined. Note that the chemical shielding is fundamentally described by the tensor $\sigma$ although in experiment data it is the chemical shift $\delta$ that is normally reported. The shielding and the shift are related by $\delta = 1 - \sigma$ [3, 6].

In a typical isotropic powder the random distribution of particle orientations means the principal axes systems (where the tensor only has diagonal elements) will be randomly distributed in space. In the presence of a large magnetic field, this random distribution gives rise to broadening of the NMR spectrum since the

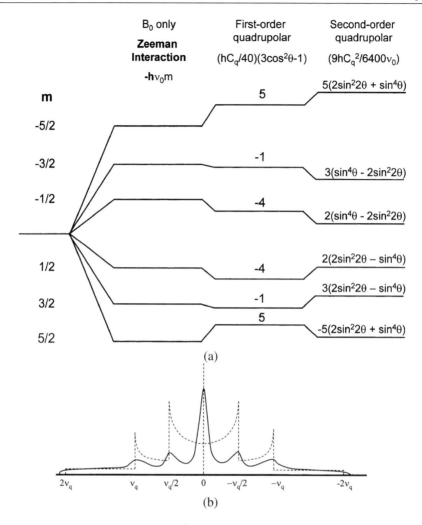

Figure B1.12.1. (a) Energy level diagram for an $I = \frac{5}{2}$ nucleus showing the effects of the Zeeman interaction and first- and second-order quadrupolar effect. The resulting spectra show static powder spectra for (b) first-order perturbation for all transitions and (c) second-order broadening of the central transition. (d) The MAS spectrum for the central transition $(A = (I(I+1) - 3/4)v_Q^2/v_0)$.

exact resonance frequency of each crystallite will depend on its orientation relative to the main magnetic field. Fortunately, to first order, all these interactions have similar angular dependences of $(3\cos^2\theta - 1 + \eta\sin^2\theta\cos^2\varphi)$ where $\eta$ is the asymmetry parameter of the interaction tensor ($\eta = 0$ for axial symmetry). Lineshapes can provide very important information constraining the local symmetry of the interaction that can often be related to some local structural symmetry.

Of the NMR-active nuclei around three-quarters have $I \geq 1$ so that the quadrupole interaction can affect their spectra. The quadrupole interaction can be significant relative to the Zeeman splitting. The splitting of the energy levels by the quadrupole interaction alone gives rise to pure nuclear quadrupole resonance (NQR) spectroscopy. This chapter will only deal with the case when the quadrupole interaction can be regarded as simply a perturbation of the Zeeman levels.

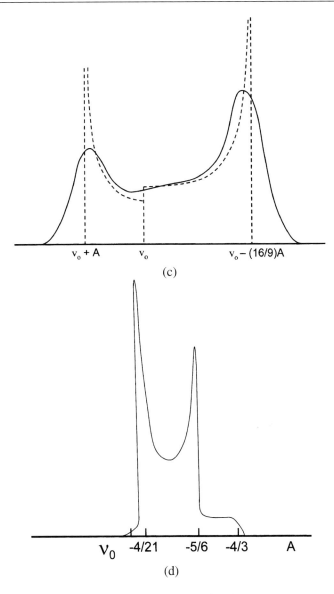

**Figure B1.12.1.** (Continued)

The electric field gradient is again a tensor interaction that, in its principal axis system (PAS), is described by the three components $V_{x'x'}$, $V_{y'y'}$ and $V_{z'z'}$, where $'$ indicates that the axes are not necessarily coincident with the laboratory axes defined by the magnetic field. Although the tensor is completely defined by these components it is conventional to recast these into 'the electric field gradient' $eq = V_{z'z'}$, the largest component, and the asymmetry parameter $\eta_Q = |V_{y'y'} - V_{x'x'}|/V_{z'z'}$. The electric field gradient is set up by the charge distribution outside the ion (e.g. $Al^{3+}$) but the initially spherical charge distribution of inner shells of electrons of an ion will themselves become polarized by the presence of the electrical field gradient to lower their energy in the electric field. This polarization produces an electric field gradient at the nucleus itself of $eq_n = eq(1 - \gamma_\infty)$ where $(1 - \gamma_\infty)$ is the Sternheimer antishielding factor which is a measure of the magnification of $eq$ due to this polarization of the core electron shells [7]. Full energy band structure calculations of electric field gradients show how important the contribution of the electrons on the ion itself are compared to the lattice. The quadrupole Hamiltonian (considering axial symmetry for simplicity) in the laboratory frame, with $\theta$ the angle between the $z'$-axis of the quadrupole PAS and $B_0$, is

$$H^Q = (hC_Q/(8I(2I-1)))[(3\cos^2\theta - 1)(3I_z - a) + 3\sin\theta\cos\theta(I_z(I_+ + I_-) + (I_+ + I_-)I_z) + 3\sin^2\theta/2(I_+^2 + I_-^2)]$$
(B1.12.6)

where $C_Q = (e^2 q Q/h)(1 - \gamma_\infty)$ and $a = I(I+1)$. In the limit $H^Z \gg H^Q$ a standard perturbation expansion using the eigenstates of $H^Z$ is applicable. The first-order term splits the spectrum into $2I$ components (figure B1.12.1(a)) of intensity $|\langle m-1|I_x|m\rangle|^2$ ($\propto a - m(m-1)$) at frequency

$$v_m^{(1)} = (3C_Q/4I(2I-1))(3\cos^2\theta - 1)(m_z - 1/2).$$
(B1.12.7)

This perturbation can cause the non-central transitions (i.e. $m_z \neq \frac{1}{2}$) to be shifted (figure B1.12.1(b)) sufficiently far from the Larmor frequency such that these transitions become difficult to observe with conventional pulse techniques. This is particularly important for spin-1 nuclei (of which the most important ones are $^2$H and $^{14}$N) as there is no central transition ($m_z = \frac{1}{2}$) and all transitions are broadened to first order. Fortunately, for non-integer quadrupolar nuclei for the central transition $v_m^{(1)} = 0$ and the dominant perturbation is second order only (equation B1.12.8) which gives a characteristic lineshape (figure B1.12.1(c) for axial symmetry):

$$v_m^{(2)} = (-9C_Q^2/64v_0 I^2(2I-1)^2)(a - 3/4)(1 - \cos^2\theta)(9\cos^2\theta - 1).$$
(B1.12.8)

This angular dependence is different from the first-order perturbations so that the conventional technique of removing linebroadening in solids, MAS (see below), cannot completely remove this interaction at the same time as removing the first-order broadening. Hence, the resolution of MAS spectra from quadrupolar nuclei is usually worse than for spin-$\frac{1}{2}$ nuclei and often characteristic lineshapes are observed. If this is the case, it is usually possible to deduce the NMR interactions $C_Q$, $\eta$ and $\delta_{iso}$ providing valuable information about the sample.

### B1.12.2.3  Basic experimental principles of FT NMR

The essence of NMR spectroscopy is to measure the separation of the magnetic energy levels of a nucleus. The original method employed was to scan either the frequency of the exciting oscillator or to scan the applied magnetic field until resonant absorption occurred. However, compared to simultaneous excitation of a wide range of frequencies by a short RF pulse, the scanned approach is a very time-inefficient way of recording the spectrum. Hence, with the advent of computers that could be dedicated to spectrometers and efficient Fourier transform (FT) algorithms, pulsed FT NMR became the normal mode of operation. Operating at constant field and frequency also produced big advantages in terms of reproducibility of results and stability of the applied magnetic field. In an FT NMR experiment a pulse of RF close to resonance, of duration $T_p$, is applied to a coil. If the pulse is applied exactly on resonance (i.e. the frequency of the applied em radiation exactly matches that required for a transition) it produces a resultant magnetic field orthogonal to $B_0$ in the rotating frame which causes a coherent oscillation of the magnetization. The magnetization is consequently tipped by an angle $\theta_p$ ($= \gamma B_1 T_p$) away from the direction defined by $B_0$. After a $\pi/2$-pulse all the magnetization is in a plane transverse to the direction to $B_0$ and, hence, is termed transverse magnetization. $B_0$ exerts a torque on the transverse magnetization which will consequently Larmor precess about $B_0$. The rotating magnetization is then providing an alternating flux linkage with the NMR coil that, through Faraday's law of electromagnetic induction, will produce a voltage in the NMR coil. The transverse magnetization decays through relaxation processes (see chapter B1.13). The observed signal is termed the free induction decay (FID). Adding coherently $n$ FIDs together improves $S/N$ by $n^{1/2}$ compared to a single FID.

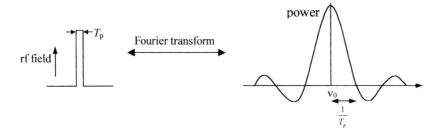

**Figure B1.12.2.** Power distribution for an RF pulse of duration $T_p$ applied at frequency $\nu_0$.

In the linear approximation there is a direct Fourier relationship between the FID and the spectrum and, in the great majority of experiments, the spectrum is produced by Fourier transformation of the FID. It is a tacit assumption that everything behaves in a linear fashion with, for example, uniform excitation (or effective RF field) across the spectrum. For many cases this situation is closely approximated but distortions may occur for some of the broad lines that may be encountered in solids. The power spectrum $P(\nu)$ of a pulse applied at $\nu_0$ is given by a $\text{sinc}^2$ function [8]

$$P(\nu) = [\sin^2 \pi T_p(\nu_0 - \nu)]/(\nu_0 - \nu)^2. \tag{B1.12.9}$$

The spectral frequency range covered by the central lobe of this $\text{sinc}^2$ function increases as the pulselength decreases. For a spectrum to be undistorted it should really be confined to the middle portion of this central lobe (figure B1.12.2). There are a number of examples in the literature of solid-state NMR where the resonances are in fact broader than the central lobe so that the 'spectrum' reported is only effectively providing information about the RF-irradiation envelope, not the shape of the signal from the sample itself. The $\text{sinc}^2$ function describes the best possible case, with often a much stronger frequency dependence of power output delivered at the probe-head. (It should be noted here that other excitation schemes are possible such as adiabatic passage [9] and stochastic excitation [10] but these are only infrequently applied.) The excitation/recording of the NMR signal is further complicated as the pulse is then fed into the probe circuit which itself has a frequency response. As a result, a broad line will not only experience non-uniform irradiation but also the intensity detected per spin at different frequency offsets will depend on this probe response, which depends on the quality factor ($Q$). The quality factor is a measure of the sharpness of the resonance of the probe circuit and one definition is the resonance frequency/halfwidth of the resonance response of the circuit (also $= \omega_0 L/R$ where $L$ is the inductance and $R$ is the probe resistance). Hence, the width of the frequency response decreases as $Q$ increases so that, typically, for a $Q$ of 100, the halfwidth of the frequency response at 100 MHz is about 1 MHz. Hence, direct FT-pulse observation of broad spectral lines becomes impractical with pulse techniques for linewidths greater than $\sim$200 kHz. For a great majority of NMR studies carried out on nuclei such as $^1$H, $^{13}$C and $^{29}$Si this does not really impose any limitation on their observation. Broader spectral lines can be reproduced by pulse techniques, provided that corrections are made for the RF-irradiation and probe responses but this requires careful calibration. Such corrections have been most extensively used for examining satellite transition spectra from quadrupolar nuclei [11].

Another problem in many NMR spectrometers is that the start of the FID is corrupted due to various instrumental deadtimes that lead to intensity problems in the spectrum. The spectrometer deadtime is made up of a number of sources that can be apportioned to either the probe or the electronics. The loss of the initial part of the FID is manifest in a spectrum as a rolling baseline and the preferential loss of broad components of the spectrum. In the best cases the deadtime is $\leq 2$ $\mu$s, but even this can still lead to severe distortion of broad spectral lines. Baseline correction can be achieved by use of either simple spline fits using spectrometer software (including back-prediction) or the use of analytical functions which effectively amount

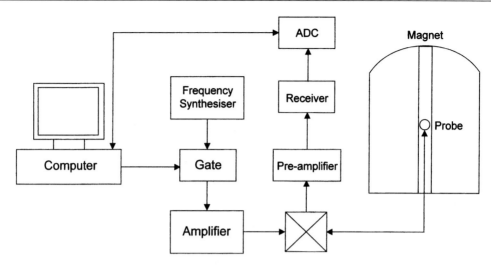

**Figure B1.12.3.** Schematic representation showing the components of a pulse FT NMR spectrometer.

to full-intensity reconstruction. Many spectrometer software packages now contain correction routines, but all such procedures should be used with extreme caution.

## B1.12.3   Instrumentation

### B1.12.3.1   Overview of a pulse FT NMR spectrometer

The basic components of a pulse FT NMR spectrometer are shown schematically in figure B1.12.3. It can be seen that, in concept, a NMR spectrometer is quite simple. There is a high-field magnet, which these days is nearly always a superconducting solenoid magnet that provides the basic Zeeman states on which to carry out the NMR experiment. The probe circuit containing the sample in the NMR coil is placed in the magnetic field. The probe is connected to the transmitter that is gated to form the pulses that produce the excitation. The probe is also connected to the receiver and it requires some careful design to ensure that the receiver that is sensitive to $\mu$V does not see any of the large-excitation voltages produced by the transmitter. The relatively simple concept of the experiment belies the extensive research and development effort that has gone into developing the components of NMR spectrometers. Although all components are important, emphasis is placed on three central parts of the spectrometer: namely the magnet, the probe and signal detection.

### B1.12.3.2   Magnets

Much effort goes into producing ever higher magnetic fields, and the highest currently commercially available for solid-state NMR is 18.8 T. Standard instruments are now considered to be 4.7–9.4 T. The drive for higher fields is based on the increased chemical shift dispersion (in hertz) and the increase in sensitivity via both the Boltzmann factor and higher frequency of operation. For solid-state NMR of half-integer spin quadrupolar nuclei there is the additional advantage that the second-order quadrupolar broadening of the central transition decreases inversely proportionally with $B_0$. Superconducting solenoids dominate based on Nb$_3$Sn or NbTi multifilament wire kept in liquid helium. However, fields and current densities now used are close to the critical limits of these materials demanding improved materials technology [12]. The principle of operation is very simple: a high current is passed through a long coil of wire, with typically 40–100 A of current circulating around several kilometres of wire. This means that the magnet stores significant amounts of energy in its field ($= 1/2LI^2$; $L$ is the solenoid's inductance and $I$ is the current flowing) of up to 10 MJ.

A superconducting magnet consists of a cryostat, main coil, superconducting shim set and a means for attaching the current supply to the main coil (figure B1.12.4). The cryostat consists of two vessels for the liquid cryogens, an inner one for helium and the outer one for nitrogen. Then, to insulate these, there are several vacuum jackets with a radiation shield. The aim is to reduce heat leakage to the inner chamber to conserve helium. Superconducting magnets in NMR are usually operated in persistent mode, which means that, after a current is introduced, the start and end of the main coil are effectively connected so that the current has a continuous path within the superconductor and the power supply can then be disconnected. To achieve this the circuits within the cryostat have a superconducting switch. The coil circuits are also designed to cope with a sudden, irreversible loss of superconductivity, termed a quench. There are resistors present (called dump resistors) to disperse the heating effect and prevent damage to the main coil when a quench occurs.

Higher magnetic fields exist than those used in NMR but the NMR experiment imposes constraints in addition to just magnitude. An NMR spectroscopy experiment also demands homogeneity and stability of the magnetic field. Long-term stability is aided by persistent-mode operation and the drift should be $<2 \times 10^{-7}$ a day. Homogeneity requirements for solid-state NMR experiments are typically up to $2 \times 10^{-9}$ over a volume of $\sim 1$ cm$^3$. The main coil alone is unable to produce this level of homogeneity so there is also a set of smaller superconducting coils called cryoshims. The number of these cryoshims depends on the design and also the purpose of the magnet (e.g. solid-state NMR, high-resolution NMR, imaging) but typically varies between three and eight. Although for many wide-line solid-state NMR experiments the homogeneity produced by the cryoshims is sufficient, most commercial spectrometers also have a room-temperature shim set which further improves the homogeneity of the magnetic field. A final consideration for the magnet is the accessible room-temperature bore of the magnet. A standard liquids magnet has a bore of 52 mm diameter. However most solid-state NMR spectroscopists prefer 89 mm as this gives much more room for the probe, allowing the use of larger and more robust electrical components for handling high powers and for accommodating some of the more specialist probe designs (e.g. double angle rotation, dynamic angle spinning (see section B1.12.4.5) etc).

## B1.12.3.3  *Probes*

The heart of an NMR spectrometer is the probe, which is essentially a tuned resonant circuit with the sample contained within the main inductance (the NMR coil) of that circuit. Usually a parallel tuned circuit is used with a resonant frequency of $\omega_0 = (LC)^{-1/2}$. The resonant frequency is obviously the most important probe parameter but the input impedance, which should be 50 $\Omega$, and $Q$ are also extremely important. Several designs of coil exist each having advantages in specific applications. The traditional coil design, particularly applicable at lower frequencies and for solids, is the conventional solenoid. For external access of large samples, Helmholtz or saddle coils are used, such as are in widespread use for high-resolution liquid-state NMR studies. For large samples, again with external access, birdcage coils are finding increasing uses especially in magnetic resonance imaging. There are competing requirements for the probe which can be characterized in terms of $Q$; pulselength ($\propto Q^{-1/2}$), deadtime ($\propto Q$), maximum voltage ($\propto Q$), bandwidth ($\propto Q^{-1}$) and sensitivity ($\propto Q^{1/2}$). A probe cannot be designed that will lead to all these being optimized simultaneously because of their differing $Q$-dependence so that compromise and focus on the most important aspects for a specific application is required. The probe needs to be constructed from robust electronic components, as they often have to withstand high voltages (many kilovolts).

Often linear circuit analysis is applied but in probes designed for solids, and therefore high-power operation, nonlinear effects can occur. Furthermore, in doubly and triply tuned probes used for decoupling, cross-polarization, and for some of the more sophisticated pulse sequences such as REDOR, TEDOR etc even small voltages generated at the second resonant frequency are unacceptable. They can swamp the NMR signal, given that the coil is part of more than one resonant circuit. Detailed consideration of the design criterion for double-tuned CP-MAS probes has been given [13]. These days, a number of sequences demand triply resonant circuits with two channels tunable over the lower frequency ranges and the third tunable to the high-$\gamma$ nuclei (e.g. $^1$H, $^{19}$F).

### B1.12.3.4  Signal detection

In addition to the deadtime mentioned earlier, magnetoacoustic ringing (up to 200 $\mu$s) can be very significant without careful probe design. Samples that exhibit piezoelectric behaviour can produce very long response times of up to 10 ms. The microvolt NMR signal generated in the coil needs amplification prior to detection and digitization. The first stage is about 30 dB amplification in the preamplifier with the most important characteristic being the noise figure (NF), essentially a measure of the noise added to the signal by the amplifier. Careful consideration is necessary for the production of a low noise figure and rapid recovery from saturation and, again, some compromise is required, with recovery of <2 $\mu$s. The preamp. should also have good linearity.

The amplified signal is passed to a double-balanced mixer configured as a phase-sensitive detector where the two inputs are the NMR signal ($\omega_0$) and the frequency of the synthesizer ($\omega_{\text{ref}}$) with the output proportional to $\cos((\omega_0 - \omega_{\text{ref}})t + \phi) + \cos((\omega_0 + \omega_{\text{ref}})t + \phi)$. The sum frequency is much larger than the total bandwidth of the spectrometer so it is lost, leaving only the difference frequency. Phase-sensitive detection is equivalent to examining the NMR signal in the rotating reference frame and if the frequencies $\omega_0$ and $\omega_{\text{ref}}$ are equal a constant output is obtained (neglecting relaxation effects). Most modern spectrometers employ two phase-sensitive detectors which have reference signals that differ in phase by $90°$, termed quadrature phase-sensitive detection [14]. This scheme can distinguish whether a signal is above or below the reference frequency and allows the transmitter frequency to be placed at the centre of the range of interest, improving pulse power efficiency and signal-to-noise. (This is not possible with a single detector, which can only provide the magnitude of the offset.) Imbalance in the two channels and non-$90°$ angles between them can give rise to quadrature images and should be minimized in the spectrometer set-up. Phase cycling includes application of different phase pulses and normally four phases $90°$ apart are available. However, much more sophisticated phase cycles are now required and variable phases can be generated by using a digital synthesizer. In particular, there is increasing demand for much smaller phase shifts than $90°$.

The signal is then digitized ready for storage in the computer memory. Digitizers are characterized by the number of bits (usually 12 or 16, determining the dynamic range), the rate of digitization (determining the dwell time) and the size of memory capable of storing the data points. Until very recently the ability to record narrow spectral objects over a broad range of frequencies has been limited, usually by the on-board computer memory, but commercial spectrometers have now addressed this problem.

## B1.12.4  Experimental techniques

### B1.12.4.1  Classification of nuclei

The sensitivity in an NMR experiment is directly proportional to the number of spins, making quantification of the amount of a particular element present straightforward, at least for spin $I = \frac{1}{2}$ nuclei. Furthermore, the large variation in the gyromagnetic ratio, $\gamma$, means that, except for pathological cases, the resonant frequencies are sufficiently different from one element to another that there is no possibility of confusing different elements. The ease of obtaining a spectrum, and to a certain extent the type of experiment undertaken, depends upon $\gamma$, the nuclear spin and the natural abundance of the isotope concerned. The large variation in $\gamma$ means that the sensitivity, which is proportional to $\gamma^3$ (see section B1.12.2.1), varies by $>10^4$ from (say) $^1$H and $^{19}$F which have the largest $\gamma$ to $^{109}$Ag which has one of the smallest. For spin $I = \frac{1}{2}$ nuclei the ability to obtain a useful signal is also dependent on the spin–lattice relaxation time, $T_1$, as well as the sensitivity. As the principal

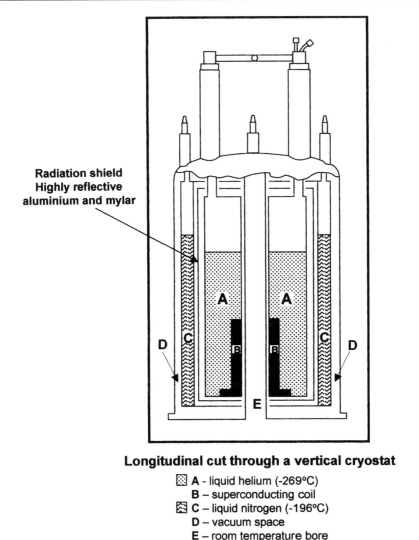

**Longitudinal cut through a vertical cryostat**

⊞ A - liquid helium (-269°C)
B – superconducting coil
▨ C – liquid nitrogen (-196°C)
D – vacuum space
E – room temperature bore

**Figure B1.12.4.** Construction of a high-field superconducting solenoid magnet.

cause of relaxation in spin-$\frac{1}{2}$ systems is usually dipolar and thus proportional to (at least) $\gamma^2$, relaxation times can be very long for low-gamma nuclei making experiments still more difficult.

Nuclei with spin $I > \frac{1}{2}$, about three-quarters of the periodic table, have a quadrupolar moment and as a consequence are affected by any electric field gradient present (see section B1.12.2.2). The lines can be very broad even when techniques such as MAS (see section B1.12.4.3) are used, thus it is convenient to divide nuclei into groups depending upon ease of observation and likely width of the line. This is done in table B1.12.2 for the most commonly studied elements. These have been divided into six categories: (a) spin $I = \frac{1}{2}$ nuclei which are readily observable, (b) spin $I = \frac{1}{2}$ nuclei which are observable with difficulty because either the isotopic abundance is low, in which case isotopic enrichment is sometimes used (e.g. $^{15}$N), or because the $\gamma$ is very small (e.g. $^{183}$W), (c) spin $I = 1$ nuclei where the quadrupolar interaction can lead to very broad lines (d) non-integer spin $I > \frac{1}{2}$ nuclei which are readily observable, (e) non-integer spin $I > \frac{1}{2}$ nuclei which are readily observable only in relatively symmetric environments because the quadrupolar interaction can strongly broaden the line and (f) those spin $I > \frac{1}{2}$ nuclei whose quadrupole moment is sufficiently large that they can only be observed in symmetric environments where the electric field gradient is small. Of course the boundaries between the categories are not 'rigid'; as technology improves and magnetic fields increase there is movement from (f) to (e) and (e) to (d). In each case the sensitivity at natural abundance relative to $^{29}$Si (for which resonance is readily observed) is given. More complete tables of nuclear properties relevant to NMR are given in several books on NMR [1, 2].

### B1.12.4.2   *Static broad-line experiments*

Static NMR powder patterns offer one way of characterizing a material, and if spectral features can be observed and the line simulated then accurate determination of the NMR interaction parameters is possible. The simplest experiment to carry out is one-pulse acquisition and despite the effects outlined above that corrupt the start of the time domain signal, the singularities are relatively narrow spectral features and their position can be recorded. The magnitude of the interaction can then be estimated. A common way to overcome deadtime problems is to form a signal with an effective time zero point outside the deadtime, i.e. an echo. There is a huge multiplicity of methods for forming such echoes. Most echo methods are two-pulse sequences, with the classic spin echo consisting of $90° - \tau - 180°$ which refocuses at $\tau$ after the second pulse. The echo decay shape is a good replica of the original FID and its observation can be used to obtain more reliable and quantitative information about solids than from the one-pulse experiment.

To accurately determine broad spectral lineshapes from echoes hard RF pulses are preferred for uniform excitation. Often an echo sequence with phase cycling first proposed by Kunwar *et al* [15] has been used, which combines phase cycling to remove quadrature effects and to cancel direct magnetization (the remaining FID) and ringing effects. The rotation produced by the second pulse in the two-pulse echo experiment is not critical. In practice, the best choice is to make the second pulse twice the length of the first, with the actual length a trade-off between sensitivity and uniformity of the irradiation. In recording echoes there is an important practical consideration in that the point of applying the echo is to move the effective $t = 0$ position for the FID to being outside the region where the signal is corrupted. However, in order that phasing problems do not re-emerge, a data sampling rate should be used that is sufficient to allow the effective $t = 0$ point to be accurately defined. If $T_2$ allows the whole echo (both before and after the maximum, i.e. $t = 0$) to be accurately recorded without an unacceptably large loss of intensity, there is then no need to accurately define the new $t = 0$ position. Fourier transformation of the whole echo (which effectively amounts to integration between $\pm\infty$) followed by magnitude calculation removes phasing errors producing a pure absorption lineshape with the signal-to-noise $\sqrt{2}$ larger than that obtained by transforming from the echo maximum.

Even if echoes are used, there are still difficulties in recording complete broad spectral lines with pulsed excitation. Several approaches have been adopted to overcome these difficulties based on the philosophy that although the line is broad it can be recorded using a series of narrow-banded experiments. One of these approaches is to carry out a spin-echo experiment using relatively weak RF pulses, recording only the intensity of the on-resonance magnetization and repeating the experiment at many frequencies to map out the lineshape. This approach has been successfully used in a series of studies. An example is $^{91}$Zr from the polymorphs of $ZrO_2$ (figure B1.12.5) where mapping out the lineshape clearly shows differences in NMR interaction [16]. This approach works but is extremely tedious because each frequency step requires accurate retuning of the probe. An alternative is to sweep the main magnetic field. There are several examples of sweeping the main magnetic field for solids dating from the earliest days of NMR but only a limited number reported using superconducting magnets, with a recent example being for $^{27}$Al in $\alpha$-$Al_2O_3$ [17]. It is now possible to have a single NMR spectrometer that is capable of both conventional high-resolution spectroscopy and also field sweep operation. As with the stepped-frequency experiment, relatively soft pulses are applied, and although strictly the on-resonance part of the magnetization should be used, it has been shown experimentally that

**Table B1.12.2.** Classification of nuclei according to spin and ease of observation.

| Isotope | Sensitivity at natural abundance relative to $^{29}$Si | Isotope | Sensitivity at natural abundance relative to $^{29}$Si |
|---|---|---|---|
| $I = \frac{1}{2}$ | | | |
| (a) Readily observable | | | |
| $^1$H | 2700 | $^{113}$Cd | 3.6 |
| $^{13}$C | 0.48 | $^{119}$Sn | 12 |
| $^{19}$F | 2200 | $^{125}$Te | 6.0 |
| $^{29}$Si | 1.00 | $^{195}$Pt | 9.1 |
| $^{31}$P | 180 | $^{199}$Hg | 2.6 |
| $^{77}$Se | 1.4 | $^{205}$Tl | 35 |
| $^{89}$Y | 0.32 | $^{207}$Pb | 5.6 |
| (b) Observable with difficulty or requiring isotopic enrichment | | | |
| $^{15}$N* | $1.0 \times 10^{-2}$ | $^{109}$Ag | 0.13 |
| $^{57}$Fe* | $1.0 \times 10^{-3}$ | $^{183}$W | $2.8 \times 10^{-2}$ |
| $^{109}$Rh | $8.5 \times 10^{-2}$ | | |
| (c) Integer $I = 1$ | | | |
| $^2$D* | $3.9 \times 10^{-3}$ | $^{14}$N | 2.7 |
| Non-integer $I > \frac{1}{2}$ | | | |
| (d) Readily observable | | | |
| $^7$Li | 730 | $^{23}$Na | 250 |
| $^9$Be | 37 | $^{27}$Al | 560 |
| $^{11}$B | 360 | $^{51}$V | 1030 |
| $^{17}$O* | $2.9 \times 10^{-2}$ | $^{133}$Cs | 130 |
| (e) Readily observable only in relatively symmetric environments | | | |
| $^{25}$Mg | 0.73 | $^{59}$Co | 750 |
| $^{33}$S* | $4.7 \times 10^{-2}$ | $^{65}$Cu | 96 |
| $^{37}$Cl | 1.7 | $^{67}$Zn | 0.32 |
| $^{39}$K | 1.3 | $^{71}$Ga | 150 |
| $^{43}$Ca* | $2.3 \times 10^{-2}$ | $^{81}$Br | 133 |
| $^{45}$Sc | 820 | $^{87}$Rb | 133 |
| $^{55}$Mn | 470 | $^{93}$Nb | 1300 |
| (f) Observable in very symmetric environment | | | |
| $^{49}$Ti | 0.44 | $^{105}$Pd | 0.68 |
| $^{53}$Cr | 0.23 | $^{115}$In | 910 |
| $^{61}$Ni | 0.11 | $^{121}$Sb | 250 |
| $^{73}$Ge | 0.30 | $^{127}$I | 260 |
| $^{75}$As | 69 | $^{135}$Ba | 0.89 |
| $^{87}$Sr | 0.51 | $^{139}$La | 160 |
| $^{91}$Zr | 2.9 | $^{209}$Bi | 380 |
| $^{95}$Mo | 1.4 | | |

* Isotopically enriched.

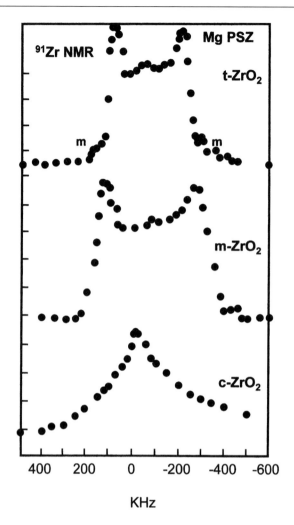

**Figure B1.12.5.** $^{91}$Zr static NMR lineshapes from $ZrO_2$ polymorphs using frequency-stepped spin echoes.

using the spin-echo intensity directly accurately reproduces the lineshape (figure B1.12.6) [18]. Recording the full distortion-free lineshape including the satellite transitions then allows accurate determination of the asymmetry parameter.

For quadrupolar nuclei, the dependence of the pulse response on $\nu_Q/\nu_1$ has led to the development of quadrupolar nutation, which is a two-dimensional (2D) NMR experiment. The principle of 2D experiments is that a series of FIDs are acquired as a function of a second time parameter (e.g. here the pulse length applied). A double Fourier transformation can then be carried out to give a 2D data set ($F1$, $F2$). For quadrupolar nuclei while the pulse is on the experiment is effectively being carried out at low field with the spin states determined by the quadrupolar interaction. In the limits $\nu_Q \ll \nu_1$ and $\nu_Q \gg \nu_1$ the pulse response lies at $\nu_1$ and $(I + \frac{1}{2})\nu_1$ respectively so is not very discriminatory. However, for $\nu_Q \sim \nu_1$ the pulse response is complex and allows $C_Q$ and $\eta$ to be determined by comparison with theoretical simulation. Nutation NMR of quadrupolar nuclei has largely been limited by the range of RF fields that puts $\nu_Q$ in the intermediate region. This approach has been extended by Kentgens by irradiating off-resonance, producing a larger effective nutation field, and matches $\nu_1$ to $\nu_Q$ [19].

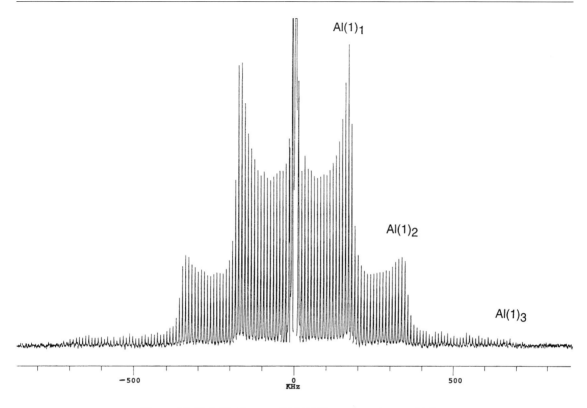

**Figure B1.12.6.** Field-swept $^{27}$Al NMR spectrum from $\alpha$-Al$_2$O$_3$.

### B1.12.4.3  Magic angle spinning

All of the anisotropic interactions described in section B1.12.2 are second-rank tensors so that, in polycrystalline samples, lines in solids can be very broad. In liquids, the rapid random molecular tumbling brought about by thermal motion averages the angular dependences to zero, leaving only the isotropic part of the interaction. The essence of the spinning technique is to impose a time dependence on these interactions externally so as to reduce the anisotropic part, which to first order has the same Legendre polynominal, $P_2(\cos\theta)$, dependence for all the interactions, in a similar but not identical manner as in liquids. This time dependence is imposed by rapidly rotating the whole sample container (termed the rotor) at the so-called 'magic angle' where $3\cos^2\theta = 1$, i.e. $\theta = 54°\ 44'$. 'Rapid' means that the spinning speed should be faster than the homogeneous linewidth. (A spectral line is considered homogeneous when all nuclei can be considered as contributing intensity to all parts of the line, so that the intrinsic linewidth associated with each spin is the same as the total linewidth.) Essentially, this is determined by the dipolar coupling: in proton-rich systems the proton–proton coupling can be >40 kHz, beyond the range of current technology where the maximum commercially available spinning speed is $\sim$35 kHz. Fortunately, in most other systems the line is inhomogeneously broadened, i.e. is made up of distinct contributions from individual spins in differently oriented crystallites which merge to give the composite line, the intrinsic width associated with each spin being considerably narrower than the total linewidth. To cause effective narrowing of these lines the spinning rate needs only to exceed the intrinsic linewidth and typically a few kilohertz is sufficient to narrow lines where the chemical shift is the dominant linebroadening mechanism. In figure B1.12.7 both the static and MAS $^{29}$Si spectrum of a sample of sodium disilicate (Na$_2$Si$_2$O$_5$) crystallized from a glass is shown as an example. Whilst the static spectrum

# $^{29}$Si Spectra of Sodium Disilicate Crystal

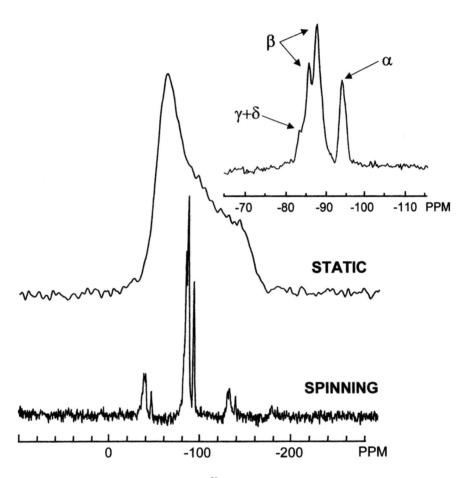

**Figure B1.12.7.** Static and MAS $^{29}$Si NMR spectra of crystalline $Na_2Si_2O_5$.

clearly indicates an axial chemical shift powder pattern, it gives no evidence of more than one silicon site. The MAS spectrum clearly shows four resolved lines from the different polymorphs present in the material whose widths are $\sim100$ times less than the chemical shift anisotropy. Note that if the spinning speed is less than the static width then a series of spinning sidebands are produced, separated from the isotropic line by the spinning speed (these are also visible in figure B1.12.7). For multisite systems, spinning sidebands can make interpretation of the spectrum more difficult and it may be necessary to do experiments at more than one spinning speed in order to determine the isotropic peak and to eliminate overlap between sidebands and main peaks. However, the presence of sidebands can be useful since from the amplitude of the spinning sideband envelope one can deduce the complete chemical tensor, giving additional information about the local site environment [20].

The interpretation of MAS experiments on nuclei with spin $I > \frac{1}{2}$ in non-cubic environments is more complex than for $I = \frac{1}{2}$ nuclei since the effect of the quadrupolar interaction is to spread the $\pm m \leftrightarrow \pm(m-1)$

**Table B1.12.3.** Spin-dependent factor $f(I)$ for the isotropic second-order quadrupole shift.

| | $I$ | | | |
|---|---|---|---|---|
| | $\frac{3}{2}$ | $\frac{5}{2}$ | $\frac{7}{2}$ | $\frac{9}{2}$ |
| $f(I)$ | $\frac{1}{3}$ | $\frac{2}{25}$ | $\frac{5}{147}$ | $\frac{1}{54}$ |

transition over a frequency range $(2m-1)\nu_Q$. This usually means that for non-integer nuclei only the $\frac{1}{2} \leftrightarrow -\frac{1}{2}$ transition is observed since, to first order in the quadrupolar interaction, it is unaffected. However, usually second-order effects are important and the angular dependence of the $\frac{1}{2} \leftrightarrow -\frac{1}{2}$ transition has both $P_2(\cos\theta)$ and $P_4(\cos\theta)$ terms, only the first of which is cancelled by MAS. As a result, the line is narrowed by only a factor of $\sim$3.6, and it is necessary to spin faster than the residual linewidth $\Delta\nu_Q$ where

$$\Delta\nu_Q = f(I)\frac{C_Q^2(6+\eta)^2}{224\nu_0} \tag{B1.12.10}$$

to narrow the line. The resulting lineshape depends upon both $C_Q$ and $\eta$ and is shown in figure B1.12.8 for several different asymmetry parameters. The centre of gravity of the line does not occur at the isotropic chemical shift (see figure B1.12.1(d)); there is a quadrupolar shift

$$\nu_{m,\mathrm{cg}}^{(2)} = \frac{3}{40}f(I)\frac{C_q^2}{\nu_0}\left(1+\frac{\eta^2}{3}\right). \tag{B1.12.11}$$

If the lineshape can be clearly distinguished, then $C_Q$, $\eta$ and $\delta_{\mathrm{iso}}$ can be readily determined even for overlapping lines, although it should be noted that most computer packages used to simulate quadrupolar lineshapes assume 'infinite' spinning speed. Significant differences between the experimental lineshape and simulation can occur if one is far from this limit and caution is required. A further problem for MAS of quadrupolar nuclei is that the angle must be set very accurately (which can be difficult and time consuming with some commercial probes) to obtain the true lineshape of broad lines. In many samples (e.g. glasses and other disordered systems) featureless lines are observed and the centre of gravity must then be used to estimate $\delta_{\mathrm{iso}}$ and the electric field gradient since

$$\delta_{\mathrm{iso}} = \delta_{\mathrm{cg}} - \frac{3}{40}f(I)\frac{C_q^2}{\nu_0^2}\left(1+\frac{\eta^2}{3}\right) \tag{B1.12.12}$$

and a plot of $\nu_{\mathrm{cg}}$ against $1/\nu_0^2$ will give $\delta_{\mathrm{iso}}$ and $C_Q^2(1+\eta^2/3)$. $f(I)$ is the spin-dependent factor $[I(I+1) - 3/4]/I^2(2I-1)^2$ and is given in table B1.12.3.

The second-order quadrupolar broadening of the $\frac{1}{2} \leftrightarrow -\frac{1}{2}$ transition can be further reduced by spinning at an angle other than 54.7° (VAS), the width being a minimum between 60–70°. The reduction is only $\sim$2 however, and dipolar and shift anisotropy broadening will be reintroduced, thus VAS has only found limited application.

*B1.12.4.4 Decoupling and cross-polarization*

In proton-containing systems such as organic materials the dipole–dipole interaction is usually sufficiently large that current spinning speeds are not sufficient to narrow the line. However, in a heteronuclear $(I, S)$ spin system if a large RF magnetic field is applied at the resonance frequency of the spin, $I$, one wishes to decouple, the magnetization will precess around the effective field in the rotating frame and the average $IS$

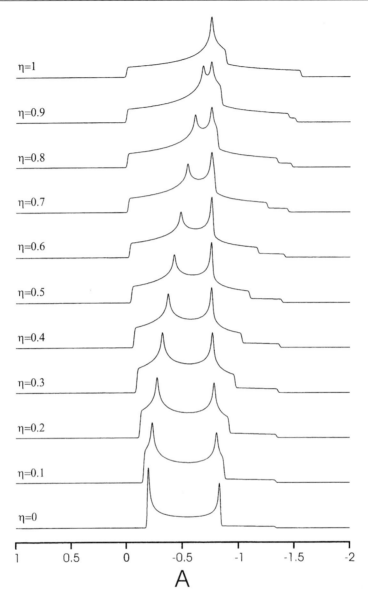

**Figure B1.12.8.** MAS NMR lineshapes from the central transition lineshape for non-integer quadrupole lineshapes with various $\eta_Q$ ($A = (I(I+1) - 3/4)\nu_Q^2/\nu_0$).

dipolar coupling will tend to zero. This technique is most commonly applied to remove the $^1$H–$^{13}$C dipolar coupling, but can be used for any system. Decoupling is usually combined with cross-polarization (CP) and the pulse sequence is shown in figure B1.12.9. A 90° pulse is applied to the $I$ spin system and the phase of the RF is then shifted by 90° to spin lock the magnetization in the rotating frame. The $S$ spin system RF is now turned on with an amplitude such that $\gamma_I B_{1I} = \gamma_S B_{1S}$ i.e. in the rotating frame both spin systems are precessing at the same rate (the Hartmann–Hahn condition) and thus magnetization can be transferred via the flip-flop term in the dipolar Hamiltonian. The length of time that the $S$ RF is on is called the contact time and

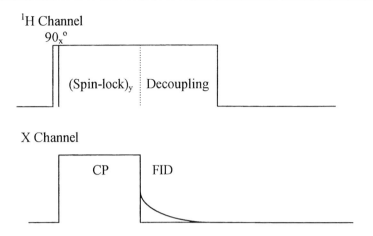

**Figure B1.12.9.** Pulse sequence used for CP between two spins $(I, S)$.

must be adjusted for optimum signal. Since magnetization transfer is via the dipole interaction, in general the signal from $S$ spin nuclei closest to the $I$ spins will appear first followed by the signal from $S$ spins further away. Thus one use of CP is in spectral editing. However, the main use is in signal enhancement since the $S$ spin magnetization will be increased by $(\gamma_I/\gamma_S)$ which for ${}^1\text{H}-{}^{13}\text{C}$ is $\sim4$ and for ${}^1\text{H}-{}^{15}\text{N}$ is $\sim9$. In addition, the $S$ spin relaxation time $T_1$ is usually much longer than that of the $I$ spin system (because $\gamma$ is smaller); as a consequence in CP one can repeat the experiment at a rate determined by $T_{1I}$ rather than $T_{1S}$ producing a further significant gain in signal to noise. For spin $I$, $S = \frac{1}{2}$ systems the equation describing the signal is given by

$$M_0(\text{CP}) = M_0\left(\frac{\gamma_I}{\gamma_S}\right)\left[\exp\left(-\frac{t}{T_{1\rho I}}\right) - \exp\left(-\left\{\frac{1}{T_{1\rho S}} + \frac{1}{T_{IS}}\right\}t\right)\right] \qquad (\text{B1.12.13})$$

where $T_{1\rho I}$ and $T_{1\rho S}$ are the relaxation times in the rotating frame of the $I$ and $S$ spin systems respectively and $T_{IS}$, which is dependent upon the strength of the dipole coupling between the $I$ and $S$ spin systems, is the characteristic time for the magnetization of the $S$ spin system to increase via contact with the $I$ spins. Generally, $T_{1\rho S}$ is very much longer than $T_{IS}$ and can be ignored and although $T_{1\rho I}$ is shorter it too is usually much longer than $T_{IS}$. In this case maximum signal will occur for a contact time $t \sim T_{IS}\ln(T_{IS}/T_{1\rho I})$. If one wishes to do quantitative measurements using CP it is necessary to plot signal amplitude versus contact time and then fit to equation B1.12.13. A typical plot is shown for glycine in figure B1.12.10. It can be seen from the inset that $T_{IS}$ for the $CH_2$ peak at $\sim40$ ppm is much shorter ($20~\mu s$) than that for the COOH peak at 176 ppm ($570~\mu s$). Carbon spectra can be very complex. Various pulse sequences have been used to simplify the spectra. The most common is dipolar dephasing, in which there is a delay after the contact time (during which no decoupling takes place) before the signal is acquired. The signal from strongly coupled carbon (e.g. $CH_2$) will decay rapidly during this time leaving only weakly coupled carbons visible in the spectrum.

Whilst MAS is effective (at least for spin $I = \frac{1}{2}$) in narrowing inhomogeneous lines, for abundant spin systems with a large magnetic moment such as protons (or fluorine) the homonuclear dipole coupling can be $>30$ kHz, too strong to be removed by MAS at currently available spinning speeds. Thus there have been many multiple pulse sequences designed to remove homonuclear coupling. When combined with MAS they are called CRAMPS (combined rotation and multiple pulse spectroscopy). The MREV-8 [6] sequence is one of the more commonly used and robust sequences; however, they all require a special probe because of the high powers needed, together with a very well tuned and set up spectrometer. Furthermore, to be effective the period of rotation needs to be much greater than the cycle time of the sequence, thus their use is somewhat restricted.

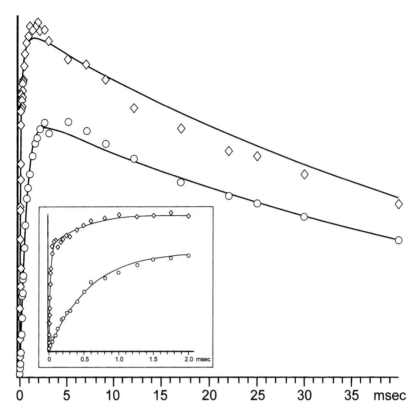

**Figure B1.12.10.** $^{13}$C CP data from the two carbons in glycine as a function of contact time. The signal for short contact times is shown in the inset where the effect of the different $T_{IS}$ values can be clearly seen.

### B1.12.4.5    High-resolution spectra from quadrupolar nuclei

Although MAS is very widely applied to non-integer spin quadrupolar nuclei to probe atomic-scale structure in solids, such as distinguishing $AlO_4$ and $AlO_6$ environments [21], simple MAS about a single axis cannot produce a completely averaged isotropic spectrum. As the second-order quadrupole interaction contains both second-rank ($\propto 3\cos^2\theta - 1$ ($P_2(\cos\theta)$)) and fourth-rank ($\propto 35\cos^4\theta - 30\cos^2\theta + 3$ ($P_4(\cos\theta)$)), the fourth-order Legendre polynomial)) terms it can be seen from figure B1.12.11 that spinning at 54.7° can only partially average $35\cos^4\theta - 30\cos^2\theta + 3$. If a characteristic well defined lineshape can be resolved the NMR interaction parameters can be deduced. Even if overlapping lines are observed, provided a sufficient number of features can be discerned, fitting, especially constrained by field variation, will allow the interactions to be accurately deduced. However, there are still many cases where better resolution from such nuclei would be extremely helpful, and over the last decade or so there have been many ingenious approaches to achieve this. Here, four of the main approaches will be briefly examined, namely satellite transition MAS spectroscopy, double-angle rotation (DOR), dynamic angle spinning (DAS) and multiple quantum (MQ) MAS. The latter three produce more complete averaging of the interactions by imposing more complex time dependences on the interactions. DOR and DAS do this directly on the spatial coordinates, whereas MQ also manipulates the spin part of the Hamiltonian. For each method the background to producing better-resolution solid-state NMR spectra will be given, and the pros and cons of each detailed. Extensive referencing to general reviews [22–24] and reviews of specific techniques is given where more details can be found. Then, by way of illustration, comparison will be made of these methods applied to $^{27}$Al NMR of kyanite.

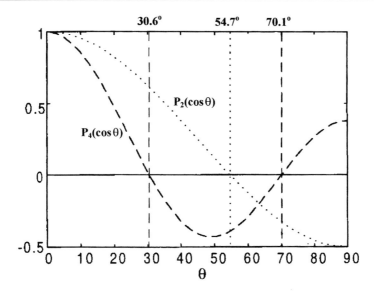

**Figure B1.12.11.** Angular variation of the second- and fourth-rank Legendre polynomials.

### (a) Satellite transition MAS

*Physical background.* MAS will narrow the inhomogeneously broadened satellite transitions to give a series of sharp sidebands whose intensity envelopes closely follow the static powder pattern so that the quadrupole interaction can be deduced. The work of Samoson [25] gave real impetus to satellite transition spectroscopy by showing that both the second-order quadrupolar linewidths and isotropic shifts are functions of $I$ and $m_z$. Some combinations of $I$ and $m_z$ produce smaller second-order quadrupolar effects on the satellite lines than for the central transition, thus offering better resolution and more accurate determination of $\delta_{iso, cs}$. The two cases where there are distinct advantages of this approach over using the central transition are $(\pm\frac{3}{2}, \pm\frac{1}{2})$ for $I = \frac{5}{2}$ and $(\pm\frac{5}{2}, \pm\frac{3}{2})$ for $I = \frac{9}{2}$. $^{27}$Al has been the focus of much of the satellite transition work and has been used for a range of compounds including ordered crystalline, atomically disordered and amorphous solids. The work of the group of Jaeger has greatly extended the practical implementation and application of this technique and a comprehensive review is given in [26].

*Advantages.* The experiment can be carried out with a conventional fast-spinning MAS probe so that it is straightforward to implement. For recording the satellite transition lineshapes it offers better signal-to-noise and is less susceptible to deadtime effects than static measurements. As the effects differ for each $m_z$ value, a single satellite transition experiment is effectively the same as carrying out multiple field experiments on the central transition.

*Disadvantages.* The magic angle must be extremely stable and accurately set. The spinning speed must show good stability over the duration of the experiment. The probe needs to be accurately tuned and careful correction for irradiation and detection variations with frequency, and baseline effects are necessary. The gain in resolution only applies to $I = \frac{5}{2}$ and $\frac{9}{2}$.

### (b) Double angle spinning

*Physical background.* DOR offers the most direct approach to averaging $P_2(\cos\theta)$ and $P_4(\cos\theta)$ terms simultaneously, by making the spinning axis a continually varying function of time [26]. To achieve this, DOR uses a spinning rotor (termed the inner rotor) which moves bodily by enclosing it in a spinning outer rotor, thereby forcing the axis of the inner rotor to describe a complicated but continuous trajectory as a function of time. The effect of this double rotation is to introduce modulation of the second-order quadrupolar frequency of the central transition in the laboratory frame of the form

$$H^{Q(2)}(t) = \frac{hC_Q^2}{8I(2I-1)\nu_0}[A_0(I, m, \eta_Q) + A_1(I, m, \eta_Q)P_2(\cos\beta_1)P_2(\cos\beta_2)F_1(\theta, \phi)$$
$$+ A_2(I, m, \eta_Q)P_4(\cos\beta_1)P_4(\cos\beta_2)F_2(\theta, \phi) + \text{terms} \propto \cos(\omega_1 t + \gamma_2)] \qquad \text{(B1.12.14)}$$

where $\beta_1$ and $\beta_2$ are the angles between the outer rotor and the magnetic field and the angle between the axes of the two rotors, respectively, and $\phi$ and $\theta$ describe the orientation of the principal axis system of a crystallite in the inner rotor. $\omega_1$ is the angular frequency of the outer rotor and $\gamma_2$ an angle describing the position of the outer rotor in the laboratory frame. More details of the functions $A$ and $F$ can be found by comparison with equations B1.12.10–B1.12.13 in [24]. $\beta_1$ and $\beta_2$ can be chosen so that $P_2(\cos\beta_1) = 0$ (54.74°) and $P_4(\cos\beta_2) = 0$ (30.56° or 70.15°).

Simulation of the complete DOR spectrum (centreband plus the spinning sidebands) will yield the NMR interaction parameters. However, it is most usual to perform the experiment to give improved resolution and simply quote the measured peak position, which appears at the sum of the isotropic chemical and second-order quadrupole shifts. DOR experiments at more than one applied magnetic field will allow these different isotropic contributions to be separated and hence provide an estimate of the quadrupole interaction. This approach is similar to that using the field variation of the centre of gravity of the MAS centreband (equation B1.12.12) but has the advantage that the narrower, more symmetric line makes determination of the correct position more precise. For experiments carried out at two magnetic fields where the Larmor frequencies are $\nu_{01}$ and $\nu_{02}$ for the measured DOR peak positions (in ppm) at the two magnetic fields at $\delta_{\text{dor1,2}}$ then

$$\delta_{\text{iso,cs}} = \frac{\nu_{01}^2\delta_{\text{dor1}} - \nu_{02}^2\delta_{\text{dor2}}}{\nu_{01}^2 - \nu_{02}^2} \qquad \text{(B1.12.15)}$$

and

$$\left[\frac{3(a-\frac{3}{4})}{40I^2(2I-1)^2}\right]C_Q^2\left(1 + \frac{\eta_Q^2}{3}\right) = \frac{\nu_{01}^2\nu_{02}^2(\delta_{\text{dor1}} - \delta_{\text{dor2}})}{\nu_{01}^2 - \nu_{02}^2}. \qquad \text{(B1.12.16)}$$

*Advantages.* DOR works well if the quadrupolar interaction is dominant and the sample is highly crystalline, with some extremely impressive gains in resolution. Provided that the correct RF-excitation conditions are employed the spectral information is directly quantitative.

*Disadvantages.* The technique requires investment in a specialized, complex probe-head that requires considerable experience to use effectively. The relatively slow rotation speed of the large outer rotor can lead to difficulties in averaging strong homogeneous interactions and produces many closely spaced spectral sidebands. In disordered solids where there is a distribution of isotropic chemical shifts, quite broad sidebands can result that may coalesce at the slow rotation rates used. Currently, the maximum actual spinning speed that can be routinely obtained in the latest system with active computer control of the gas pressures is ~1500 Hz. By the use of synchronous triggering [28] this effectively amounts to a spinning speed of 3 kHz. Undoubtedly the technology associated with the technique will continue to improve leading to increased spinning speeds and thus expanding the application of the technique. Also, the RF coil encloses the whole system, and the filling factor is consequently small leading to relatively low sensitivity. The large coil size also means that the RF generated is quite low and that double tuning for CP is difficult, although such an experiment has been performed.

*(c) Dynamic angle spinning*

*Physical background.* DAS is a 2D NMR experiment where the evolution of the magnetization is divided into two periods and the sample is spun about a different axis during each period [27, 29]. During the first

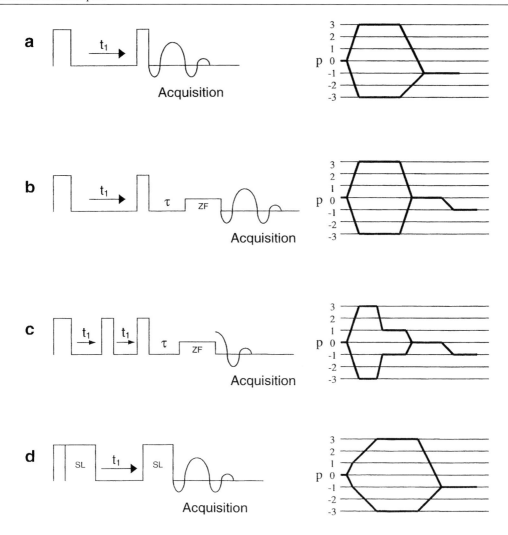

**Figure B1.12.12.** Pulse sequences used in multiple quantum MAS experiments and their coherence pathways for (a) two-pulse, (b) $z$-filter, (c) split-$t_1$ with $z$-filter and (d) RIACT (II).

evolution time $t_1$ the sample is spun at an angle of $\theta_1$. The magnetization is then stored along the $z$-axis and the angle of the spinning axis is changed to $\theta_2$. After the rotor is stabilized ($\sim$30 ms) the magnetization is brought into the $xy$-plane again and a signal is acquired. The second-order quadrupole frequency of an individual crystallite depends on the angle of the spinning axis. So during $t_1$ the quadrupole frequency will be $\nu_1$, and $\nu_2$ during $t_2$. If $\nu_2$ is of opposite sign to $\nu_1$ the magnetization from the crystallite will be at its starting position again at some time during $t_2$. One can choose both angles in such a way that the signals from each individual crystallite will be at the starting position at exactly the same time. In other words, an echo will form and the effect of the second-order quadrupolar broadening is removed at the point of echo formation. To achieve this cancellation to form an echo then

$$P_2(\cos\theta_1) = -kP_2(\cos\theta_2) \qquad \text{and} \qquad P_4(\cos\theta_1) = -kP_4(\cos\theta_2) \qquad \text{(B1.12.17)}$$

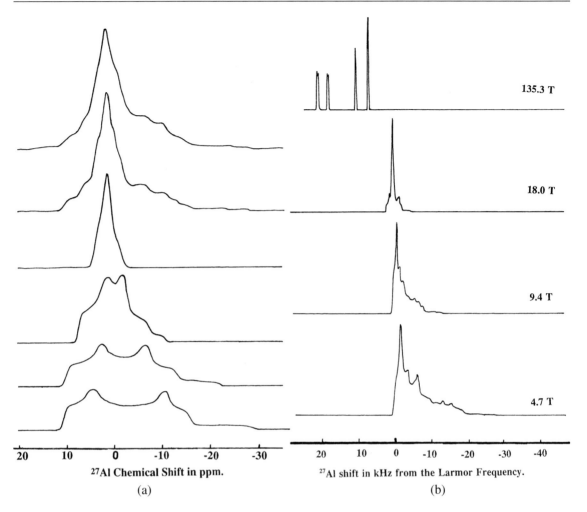

**27Al Chemical Shift in ppm.**

(a)

**27Al shift in kHz from the Larmor Frequency.**

(b)

**Figure B1.12.13.** $^{27}$Al MAS NMR spectra from kyanite (a) at 17.55 T along with the complete simulation and the individual components, (b) simulation of centreband lineshapes of kyanite as a function of applied magnetic field, and the satellite transitions showing (c) the complete spinning sideband manifold and (d) an expansion of individual sidebands and their simulation.

must both be satisfied where $k$ is the scaling factor. There is a continuous set of solutions for $\theta_1$ and $\theta_2$, the so-called DAS complementary angles, and each set has a different scaling factor. For these solutions, the second-order quadrupole powder pattern at $\theta_1$ is exactly the scaled mirror image of the pattern at $\theta_2$ and an echo will form at $t_2 = kt_1$. For the combination $\theta_1 = 30.56°$, $\theta_2 = 70.12°$ the $P_4(\cos\theta)$ terms are zero and the scaling factor $k = 1.87$. A practically favoured combination is $\theta_1 = 37.38°$, $\theta_2 = 79.19°$ as the scaling factor $k = 1$ and the spectra are exact mirror images so that an echo will form at $t_1 = t_2$. There are several ways in which the DAS spectra can be acquired; one can acquire the entire echo, which means that the resulting 2D spectrum will be sheared. Some additional processing is then required to obtain an isotropic spectrum in $F1$. The acquisition could also start at the position of the echo so that an isotropic spectrum in $F1$ is obtained directly. A third possibility is to carry the experiment out as a pseudo 1D experiment where only the top of the echo is acquired as a function of $t_1$. In this case the isotropic spectrum is acquired directly but there is no saving in the duration of the experiment.

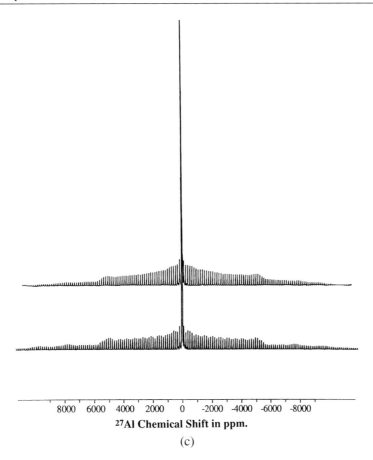

<sup></sup>**27Al Chemical Shift in ppm.**

(c)

**Figure B1.12.13.** (Continued)

*Advantages.* Compared to DOR, a small rotor can be used allowing relatively fast spinning speeds; high RF powers can be attained and if the coil is moved with the rotor a good filling factor can be obtained. In the isotropic dimension high-resolution spectra are produced and the second dimension retains the anisotropic information.

*Disadvantages.* Again, a specialized probe-head is necessary to allow the rotor axis to be changed, usually using a stepper motor and a pulley system. Angle switching needs to be as fast as possible and reproducible. For an RF coil that changes orientation with the rotor the RF field will vary with the angle and the RF leads need to be flexible and resistant to metal fatigue. A major limitation of the DAS technique is that it cannot be used on compounds with a short $T_1$ because of the time needed to reorient the spinning axis. If magnetization is lost through $T_1$ and/or spin diffusion effects the signal can become very weak. This factor has meant that DAS has been most useful on $^{17}O$ where the dipole–dipole interaction is weak but has not had the impact on NMR of nuclei such as $^{27}Al$ that it might otherwise have had.

*(d) MQMAS*

*Physical background.* Relatively recently (1995) a new experiment emerged that has already had a great impact on solid-state NMR spectroscopy of quadrupolar nuclei [30]. The 2D multiple quantum magic angle spinning (2D MQMAS) experiment greatly enhances resolution of the spectra of half-integer spin quadrupolar nuclei.

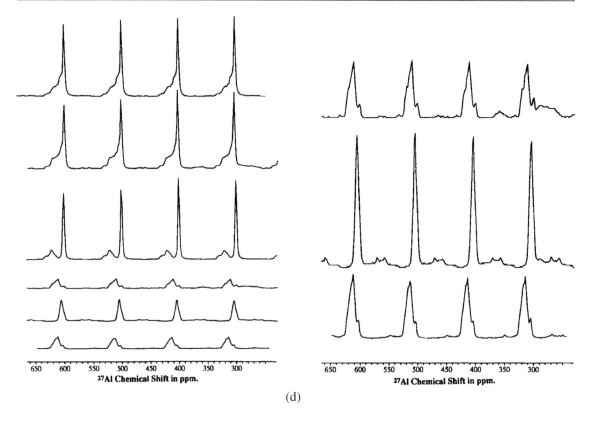

(d)

**Figure B1.12.13.** (Continued)

Basically, this experiment correlates the $(m, -m)$ multiple quantum transition to the $(\frac{1}{2}, -\frac{1}{2})$ transition. The resolution enhancement stems from the fact that the quadrupole frequencies for both transitions are correlated. At specific times the anisotropic parts of the quadrupole interaction are refocused and an echo forms. The frequency of an $(m, -m)$ transition is given by

$$\nu_p = C_0(p)\nu_0^Q - \frac{7}{18}C_4(p)\nu_4^Q(\theta, \phi) \tag{B1.12.18}$$

where $p = 2m$ is the order of the coherence, $p = 1$ for the $(\frac{1}{2}, -\frac{1}{2})$ transition, $p = 3$ for the $(\frac{3}{2}, -\frac{3}{2})$ transition, etc. The coefficients are defined as

$$C_0(p) = p\left(I(I+1) - \frac{3}{4}p^2\right)$$

$$C_4(p) = p\left(18I(I+1) - \frac{34}{4}p^2 - 5\right). \tag{B1.12.19}$$

The isotropic part $\nu_0^Q$ of the quadrupole frequency is

$$\nu_0^Q = -(\nu_q^2(3 + \eta_q^2))/90\nu_0 \qquad \text{[Hz]} \tag{B1.12.20}$$

and the anisotropic part $v_4^Q(\theta, \phi)$ is given by

$$v_4^Q(\theta, \phi) = (v_q^2/112v_0) \times [(7/18)(3-\eta_q \cos 2\phi)^2 \sin^4 \theta + (2\eta_q \cos 2\phi - 4 - (2/9)\eta_q^2) \sin^2 \theta + (2/45)\eta_q^2 + (4/5)].$$

(B1.12.21)

Numerous schemes exist that are used to obtain 2D MQMAS spectra [24]. The simplest form of the experiment is when the MQ transition is excited by a single, high-power RF pulse, after which the MQ coherence is allowed to evolve for a time $t_1$ (figure B1.12.12). After the evolution time, a second pulse is applied which converts the MQ coherence into a $p = -1$ coherence which is observed during $t_2$. The signal is then acquired immediately after the second pulse and the echo will form at a time $t_2 = |QA|t_1$. Both pulses are non-selective and will excite all coherences to a varying degree. Selection of the coherences of interest is achieved by cycling the pulse sequence through the appropriate phases. Phase cycles can be easily worked out by noting that an RF phase shift of $\phi$ degrees is seen as a phase shift of $p\phi$ degrees for a $p$-quantum coherence [31]. After a 2D Fourier transformation the resonances will show up as ridges lying along the quadrupole anisotropy (QA) axis. The isotropic spectrum can be obtained by projection of the entire 2D spectrum on a line through the origin ($v_1 = v_2 = 0$) perpendicular to the QA axis. Figure B1.12.12 shows some of the many different pulse sequences and their coherence pathways that can be used for the 2D MQMAS experiment.

The isotropic shift and the quadrupole-induced shift (QIS) can easily be obtained from the data. The QIS is different in both dimensions, however, and is given by

$$\delta_{qis}^p = \frac{C_0(p)v_0^Q}{pv_0} \times 10^6 \qquad \text{[in ppm]}.$$

(B1.12.22)

$\delta_{cg}^p$, the position of the centre of gravity with $p = 1$ for the single quantum dimension and $p = \Delta m$ for the multiple quantum dimension, is given by

$$\delta_{cg}^p = \delta_{iso} + \delta_{qis}^p.$$

(B1.12.23)

Hence the isotropic shift can be easily retrieved using

$$\delta_{iso} = \frac{C_0(p)\delta_{cg}^{p=1} - pC_0(1)\delta_{cg}^p}{C_0(p) - pC_0(1)} \qquad \text{[in ppm]}$$

(B1.12.24)

and the isotropic quadrupolar shift by

$$v_0^Q = \frac{(\delta_{cg}^{p=1} - \delta_{cg}^p)}{C_0(1) - C_0(p)/p} v_0 \qquad \text{[in Hz]}.$$

(B1.12.25)

Shearing of the data is performed to obtain isotropic spectra in the $F1$ dimension and to facilitate easy extraction of the 1D slices for different peaks. Shearing is a projection of points that lie on a line with a slope equal to the anisotropy axis onto a line that is parallel to the $F2$ axis [24]. Shearing essentially achieves the same as the split-$t_1$ experiment or delayed acquisition of the echo. Although sheared spectra may look more attractive, they do not add any extra information and they are certainly not necessary for the extraction of QIS and $\delta_{iso}$ values.

*Advantages.* The experiment can be readily carried out with a conventional probe-head, although the fastest spinning and highest RF powers available are useful. The pulse sequences are relatively easy to set up (compared to DAS and DOR) and the results are usually quite straightforward to interpret in terms of the number of sites and determination of the interactions.

*Disadvantages.* A researcher new to the subject will be confronted with a large number of schemes for collecting, processing and presenting the data which can be very confusing. The relationship between the measured peak positions and the NMR interaction parameters crucially depends on the processing and referencing conventions adopted. There is one clear distinction between the two main approaches: either the MQ evolution is regarded as having taken place only in the evolution time ($t_1$), or the period up to the echo is also regarded as being part of the evolution time which is then $(1 + QA)t_1$. A detailed critique of these two approaches and the consequences of adopting each has recently been given by Man [32]. Shearing data introduces an extra processing step, which may introduce artefacts. The key point in determining the quadrupole parameters is the accuracy of measuring the position of the centre of gravity of the resonance. Both the excitation efficiency as well as significant intensity in the spinning sidebands can adversely affect the accuracy of determining the centre of gravity. MQ spectra show a strong dependence on excitation efficiency that is strongly dependent on the value of $C_Q$ relative to the RF field strength. This means that often the spectra are non-quantitative and sites with a large $C_Q$ can be lost completely from the spectrum.

### (e) Application of high-resolution methods to $^{27}Al$ in Kyanite

Kyanite, a polymorph of $Al_2SiO_5$, has four distinct $AlO_6$ sites. The crystal structure and site geometries (e.g. Al–O bondlengths) are well characterized. The quadrupole interactions vary from 3.6 to 10.1 MHz, providing a good test of the different approaches to achieving high-resolution NMR spectra from quadrupolar nuclei. Both single-crystal studies [33] and NQR [34] have accurately determined the quadrupole interaction parameters. MAS studies of the central transition have been reported from 7 to 18.8 T [35–38]. An example of a 17.55 T spectrum is shown in figure B1.12.13(a) along with a simulation of the individual components and their sum. It can be clearly seen that at even this high field there is extensive overlap of the four components. However, many distinct spectral features exist that constrain the simulation and these can be followed as a function of applied magnetic field. As the field is increased the second-order quadrupole effects decrease. Comparing simulations at 4.7 and 9.4 T (figure B1.12.13(b)) the field variation simply scales the width of the spectrum, being a factor of two narrower (in hertz) at the higher field. By extending the simulation to higher fields, 18 T can be seen to be a poor choice as there is considerable overlap of the resonances with virtually no shift dispersion between the sites. 135.3 T would provide a completely resolved spectrum under simple MAS.

Satellite transition MAS NMR provides an alternative method for determining the interactions. The intensity envelope of the spinning sidebands are dominated by site A2 (using the crystal structure nomenclature) which has the smallest $C_Q$, resulting in the intensity for the transitions of this site being spread over the smallest range ($\propto C_Q$), and will have the narrowest sidebands ($\propto C_Q^2/\nu_0$) of all the sites (figure B1.12.13(c)) [37]. The simulation of this envelope provides additional constraints on the quadrupole interaction parameters for this site. Expanding the sidebands (shown for the range 650 to 250 ppm in figure B1.12.13(d)) reveals distinct second-order lineshapes with each of the four sites providing contributions from the $(\pm\frac{1}{2}, \pm\frac{3}{2})$ and $(\pm\frac{3}{2}, \pm\frac{5}{2})$ transitions. The improved resolution provided by the $(\pm\frac{1}{2}, \pm\frac{3}{2})$ transition compared to the central transition is clear. For the A2 site both of the contributing transitions are clearly seen while for the other three only the $(\pm\frac{1}{2}, \pm\frac{3}{2})$ contribution is really observable until the vertical scale is increased by six, which then shows the outer transition for all sites, especially A3. The simulation of all three transitions provides an internal check on the interaction parameters for each site.

DOR provides significantly higher resolution than MAS [37, 39]. At 11.7 T a series of relatively narrow resonances and accompanying sidebands are observed under DOR (figure B1.12.14(a)). The relatively slow spinning speed of the outer rotor results in numerous sidebands and the isotropic line is identified by collecting spectra at several different spinning speeds. If the isotropic position is then collected as a function of $B_0$ the quadrupole interaction parameters can then be deduced. MQ MAS provides an alternative approach for producing high resolution [38, 40], with the whole 2D data set shown at 9.4 T along with the isotropic projections at 11.7 and 18.8 T (figure B1.12.14(b)) [38]. All three isotropic triple quantum (3Q) projections show only three resolved lines as the NMR parameters from the two sites (A1, A4) with the largest $C_Q$ means that their resonances are superimposed at all fields. This is confirmed by the 9.4 T 3Q data where an RF field of 280 kHz was employed making the data more quantitative: the three resonances with isotropic shifts of 43.0, 21.1 and 8.0 ppm had intensities of 2.1:1.1:1.0 respectively. At 11.7 and 18.8 T the MQ MAS NMR

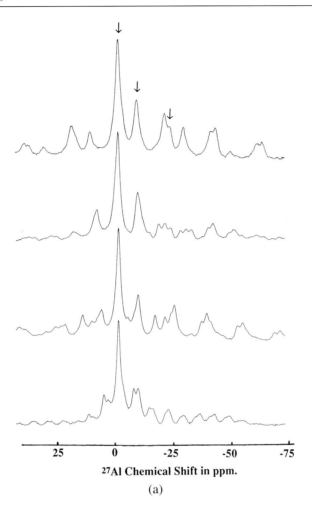

$$\downarrow$$

**27Al Chemical Shift in ppm.**

(a)

**Figure B1.12.14.** (a) $^{27}$Al DOR NMR spectrum of kyanite at 11.7 T at various spinning speeds and (b) the $^{27}$Al MQ MAS NMR spectrum of kyanite at 9.4 T (top) along with the isotropic projections at 11.7 and 18.8 T. (The MQ MAS NMR data are taken from [38] with the permission of Academic Press.)

spectra collected are not quantitative since the RF fields to excite the 3Q transitions were not strong enough. For $I = \frac{5}{2}$ the isotropic shifts are $-\frac{10}{17}$ of the value compared to direct MAS at the same field, so MQ data effectively produces results at a 'negative' applied magnetic field thereby more strongly constraining the NMR interaction parameters deduced from isotropic shift against $B_0^{-2}$ plots.

When $^{27}$Al MAS NMR spectra were first collected at moderate $B_0$ with relatively slow MAS rates (<4 kHz) there was much confusion about the quantitative integrity of such spectra. In fact, provided the correct excitations are employed (i.e. $v_1 \gg v_r$ and $2\pi v_1 T_p \ll 1$) all that is necessary to make MAS NMR spectra quantitative is to know what fraction of the different transitions are contributing to the centreband. In practice this usually amounts to estimating the fraction of the $(\frac{1}{2}, -\frac{1}{2})$ transition that contributes to the centreband, which depends on $v_Q^2/v_0 v_r$ [41]. At 17.55 T the intensity distribution between the sites in the centreband was 23%:27%:26%:24% (A1:A2:A3:A4). Once the correction factors are taken into account, within experimental error all sites are equally populated. DAS is not reported for kyanite because of its shortcomings for fast-relaxing nuclei such as $^{27}$Al. The data on kyanite demonstrates how accurately the

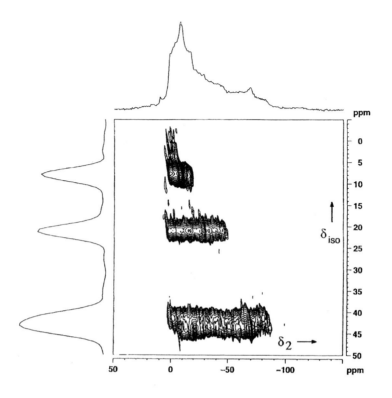

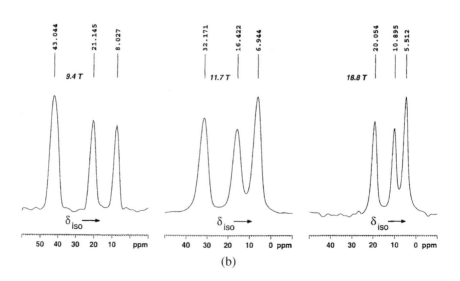

(b)

**Figure B1.12.14.** (Continued)

**(a)**

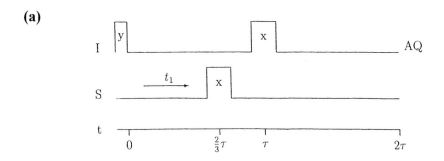

**(b)**

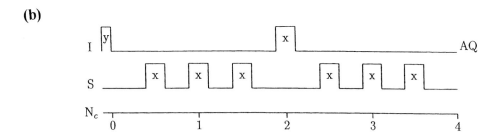

**(c)**

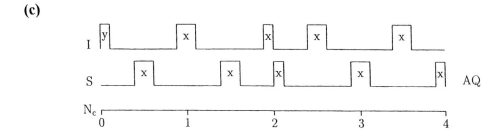

**(d)**

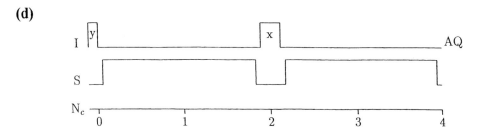

**Figure B1.12.15.** Some double-resonance pulse sequences for providing distance information in solids: (a) SEDOR, (b) REDOR, (c) TEDOR and (d) TRAPDOR. In all sequences the narrow pulses are 90° and the wide pulses 180°. For sequences that employ MAS the number of rotor cycles ($N_c$) is shown along the bottom.

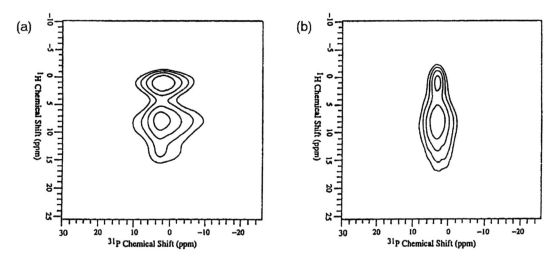

**Figure B1.12.16.** $^1$H–$^{31}$P HETCOR NMR spectra from (a) brushite and (b) bone. (Adapted from [48] with the permission of Academic Press.)

interaction parameters and intensities can be extracted for quadrupole nuclei by using a combination of advanced techniques.

### B1.12.4.6   Distance measurements, dipolar sequences and correlation experiments

The dipolar coupling between two nuclei $I$, $S$ is $(\mu_0/4\pi)(\hbar/2\pi)\gamma_I\gamma_S r_{IS}^{-3}$, thus in systems containing pairs of nuclei measurement of the dipolar coupling will give the distance between them. The simplest pulse sequence for doing this, SEDOR (spin echo double resonance) [2], is shown in figure B1.12.15(a). A normal 'spin echo' sequence is applied to spin $I$ which refocuses the heteronuclear dipole coupling and the chemical shift anisotropy producing an echo at $2\tau$. However, if a 180° pulse is applied to the $S$ spin system it will invert the sign of any $IS$ dipole coupling and reduce the echo intensity. By varying the 180° pulse position a set of difference signals can be used to determine the dipolar coupling and hence the internuclear distance for an isolated spin pair system. The SEDOR sequence is only useful for static samples and the range of distances accessible is restricted by $T_2$, however, it can give useful qualitative information in systems where the spins are not isolated pairs.

MAS averages the heteronuclear dipole interaction giving a much increased resolution and increased $T_2$. The REDOR (rotational echo double resonance) [42] sequence enables the heteronuclear dipole coupling to be measured for a spinning sample. There are several versions of this experiment; in one (figure B1.12.15(b)) a rotor-synchronized echo is applied to $I$ whilst a series of rotor-synchronized 180° pulses are applied to $S$. As with SEDOR, the attenuated signal is subtracted from the normal echo signal to determine the REDOR fraction. In suitably labelled systems dipolar couplings as small as 25 Hz have been measured corresponding to a distance of 6.7 Å [43].

TEDOR (transferred echo double resonance) [44, 45] is another experiment for measuring internuclear distances whilst spinning. In this (figure B1.12.15(c)) the $S$ spin is observed. There is first a REDOR sequence on the $I$ spin, the magnetization is then transferred by applying 90° pulses on both spins and this is followed by a REDOR sequence on the $S$ spins. Both REDOR and TEDOR require accurate setting of the 180° pulse and good long-term spectrometer stability to be effective. An advantage of the TEDOR sequence is that there is no background signal from uncoupled spins; however, the theoretical maximum efficiency is

50% giving a reduced signal. Whilst both REDOR and TEDOR can work well for spin $I$, $S = \frac{1}{2}$ pairs, for quadrupolar nuclei the situation is more complex [24]. The echo sequence must be applied to the quadrupolar nucleus if only one of the pair has spin $I > \frac{1}{2}$ since accurate inversion is rarely possible for spins $> \frac{1}{2}$. A sequence designed specifically for a quadrupolar nucleus is TRAPDOR (figure B1.12.15(d)) in which the dephasing spin is always quadrupolar. One cannot obtain accurate values of the dipolar coupling with this sequence but it can be used to give qualitative information about spatial proximity. There are many other pulse sequences designed to give distance information via the dipole–dipole interaction, most of limited applicability. One, designed specifically for homonuclei, that works well in labelled compounds is DRAMA [46] and a comparison of the more popular homonuclear sequences is given in [47].

In liquid-state NMR two-dimensional correlation of spectra has provided much useful information. However, in solids the linewidths ($T_2$), even under MAS, are usually such that experiments like COSY and INADEQUATE which rely on through-bond $J$ coupling cannot be used (although there are some notable exceptions). The most commonly used heteronuclear correlation (HETCOR) experiments in solids all rely on dipolar coupling and can thus be complicated by less local effects. Nevertheless, they are able to correlate specific sites in the MAS spectrum of one nucleus with sites of the second nucleus which are nearby. The simplest versions of the experiment are just a two-dimensional extension of CP in which the pulse that generates magnetization is separated from the matching pulses by a time which is incremented to give the second dimension. An example of the usefulness of the technique is shown in figure B1.12.16 where the $^1$H–$^{31}$P correlation of two inorganic hydrated phosphates, (a) brushite and (b) a bone, are shown [48]. In both cases two phosphorus sites that completely overlap in the 1D MAS spectrum are clearly visible in the 2D spectrum.

## References

[1] Abragam A 1983 *Principles of Nuclear Magnetism* (Oxford: Oxford University Press)
[2] Slichter C P 1990 *Principles of Magnetic Resonance* (Berlin: Springer)
[3] Harris R K 1984 *NMR Spectroscopy* (London: Pitman)
[4] Spiess H W and Schmidt-Rohr K *Multidimensional NMR of Polymers* (New York: Academic)
[5] Cohen M H and Reif F 1954 *Solid State Phys.* **5** 736
[6] Gerstein B C and Dybowski C 1985 *Transient Techniques in NMR of Solids* (New York: Academic)
[7] Sternheimer R M 1958 *Phys. Rev.* **37** 736
[8] Fukushima E and Roeder S B W 1981 *Experimental Pulse NMR* (Reading, MA: Addison-Wesley)
[9] Kentgens A P M 1991 *J. Magn. Reson.* **95** 619
[10] Liao M Y, Chew B G M and Zax D B 1995 *Chem. Phys. Lett.* **242** 89
[11] Kunath-Fandrei G 1998 *PhD Thesis* University of Jena (Aachen: Shaker)
[12] Laukien D D and Tschopp W H 1993 *Concepts in Magnetic Resonance* **6** 255
[13] Doty F D, Connick T J, Ni X Z and Clingan M N 1988 *J. Magn. Reson.* **77** 536
[14] Derome A E 1987 *Modern NMR Techniques for Chemistry Research* (Oxford: Pergamon)
[15] Kunwar A C, Turner G L and Oldfield E 1986 *J. Magn. Reson.* **69** 124
[16] Bastow T J and Smith M E 1994 *Solid State NMR* **3** 17
[17] Wu X, Juban E A and Butler L G 1994 *Chem. Phys. Lett.* **221** 65
[18] Poplett I J F and Smith M E 1998 *Solid State NMR* **11** 211
[19] Kentgens A P M 1998 *Prog. NMR Spectrosc.* **32** 141
[20] Herzfield J and Berger A E 1980 *J. Chem. Phys.* **73** 6021
[21] Smith M E 1993 *Appl. Magn. Reson.* **4** 1
[22] Freude D and Haase J 1993 *NMR Basic Principles and Progress* vol 29, ed P Diehl *et al* (Berlin: Springer) p 1
[23] Kentgens A P M 1997 *Geoderma* **80** 271
[24] Smith M E and van Eck E R H 1999 *Prog. NMR Spectrosc.* **34** 159
[25] Samoson A 1985 *Chem. Phys. Lett.* **119** 29
[26] Jaeger C 1994 *NMR Basic Principles and Progress* ed B Blumich and R Kosfeld (Berlin: Springer) vol 31, p 135
[27] Zwanziger J and Chmelka B F 1994 *NMR Basic Principles and Progress* ed B Blumich and R Kosfeld (Berlin: Springer) vol 31, p 202
[28] Samoson A and Lippmaa E 1989 *J. Magn. Reson.* **89** 410
[29] Mueller K T, Sun B Q, Chingas G C, Zwanziger J W, Terao T and Pines A 1990 *J. Magn. Reson.* **86** 470
[30] Medek A, Harwood J S and Frydman L 1995 *J. Am. Chem. Soc.* **117** 12 779
[31] Ernst R R, Bodenhausen G and Wokaun A 1987 *Principles of NMR in One and Two Dimensions* (Oxford: Clarendon)

[32] Man P P 1998 *Phys. Rev.* B **58** 2764
[33] Hafner S S and Raymond M 1967 *Am. Mineral.* **52** 1632
[34] Lee D and Bray P J 1991 *J. Magn. Reson.* **94** 51
[35] Lippmaa E, Samoson A and Magi M 1986 *J. Am. Chem. Soc.* **108** 1730
[36] Alemany L B, Massiot D, Sherriff B L, Smith M E and Taulelle F 1991 *Chem. Phys. Chem.* **117** 301
[37] Smith M E, Jaeger C, Schoenhofer R and Steuernagel S 1994 *Chem. Phys. Lett.* **219** 75
[38] Alemany L B, Steuernagel S, Amoureux J-P, Callender R L and Barron A R 1999 *Chem. Phys. Lett.* **14** 1
[39] Xu Z and Sherriff B L 1993 *Appl. Magn. Reson* **4** 203
[40] Baltisberger J H, Xu Z, Stebbins J F, Wang S H and Pines A 1996 *J. Am. Chem. Soc.* **118** 7209
[41] Massiot D, Bessada C, Coutures J P and Taulelle F 1990 *J. Magn. Reson.* **90** 231
[42] Gullion T and Schaefer J 1989 *Adv. Magn. Reson.* **13** 57
[43] Merritt M E, Goetz J, Whitney D, Chang C P P, Heux L, Halary J L and Schaefer J 1998 *Macromolecules* **31** 1214
[44] Van Eck E R H and Veeman W S 1993 *Solid State NMR* **2** 307
[45] Hing A W, Vega S and Schaefer J 1993 *J. Magn. Res.* A **103** 151
[46] Tycko R and Dabbagh G 1990 *Chem. Phys. Lett.* **173** 461
[47] Baldus M, Geurts D G and Meier B H 1998 *Solid State NMR* **11** 157
[48] Santos R A, Wind R A and Bronniman C E 1990 *J. Magn. Reson.* B **105** 183

## Further Reading

Stejskal E O and Memory J D 1994 *High Resolution NMR in the Solid State* (Oxford: Oxford University Press)

Introductory text, fairly mathematical, concentrates on spin $I = \frac{1}{2}$ systems, good references.

Slichter C P 1989 *Principles of Magnetic Resonance* 3rd edn (Berlin: Springer)

Comprehensive coverage from a physics viewpoint.

Engelhardt G and Michel D 1987 *High Resolution Solid State NMR of Silicates and Zeolites* (New York: Wiley)

Good coverage of NMR in these solids up to 1987.

Blümich B *et al* (eds) 1994 *Solid State NMR* vol 1 (Berlin: Springer)

NMR basic principles and progress; specialised monograph giving detailed descriptions of specific areas of solid state NMR.

Schmidt-Rohr K and Spiess H W 1994 *Multidimensional Solid State NMR and Polymers* (New York: Academic)

A comprehensive text which discusses advanced experiments with particular reference to polymers.

Fitzgerald J J (ed) 1999 *Solid State NMR Spectroscopy of Inorganic Materials* (Washington, DC: American Chemical Society)

Gives a range of current examples of solid state NMR applied to inorganic materials.

Grant D M and Harris R K (eds) 1996 *Encyclopaedia of NMR* (New York: Wiley)

Contains articles on all aspects of NMR.

# B1.13
# NMR relaxation rates

*Jozef Kowalewski*

## B1.13.1   Introduction

Nuclear magnetic resonance (NMR) spectroscopy deals with the interactions among nuclear magnetic moments (nuclear spins) and between the spins and their environment in the presence of a static magnetic field $B_0$, probed by radiofrequency (RF) fields. The couplings involving the spins are relatively weak which results, in particular in low-viscosity liquids, in narrow lines and spectra with a rich structure. The NMR experiments are commonly carried out in the time domain: the spins are manipulated by sequences of RF pulses and delays, creating various types of non-equilibrium spin state, and the NMR signal corresponding to magnetization components perpendicular to $B_0$ is detected as a function of time (the free induction decay, FID). In terms of the kinetics, the weakness of the interactions results in slow decays (typically milliseconds to seconds) of non-equilibrium states. The recovery processes taking the ensembles of nuclear spins back to equilibrium and some related phenomena, are called *nuclear spin relaxation*. The relaxation behaviour of nuclear spin systems is both an important source of information about the molecular structure and dynamics and a factor to consider in optimizing the design of experiments.

The concept of 'relaxation time' was introduced to the vocabulary of NMR in 1946 by Bloch in his famous equations of motion for nuclear magnetization vector $M$ [1]:

$$\frac{\mathrm{d}M_z}{\mathrm{d}t} = \frac{M_0 - M_z}{T_1}$$

$$\frac{\mathrm{d}M_{x,y}}{\mathrm{d}t} = -\frac{M_{x,y}}{T_2}. \tag{B1.13.1}$$

The phenomenological Bloch equations assume the magnetization component along $B_0$ (the longitudinal magnetization $M_z$) to relax exponentially to its equilibrium value, $M_0$. The time constant for the process is called the spin–lattice or longitudinal relaxation time, and is denoted $T_1$. The magnetization components perpendicular to $B_0$ (the transverse magnetization, $M_{x,y}$) are also assumed to relax in an exponential manner to their equilibrium value of zero. The time constant for this process is called the spin–spin or transverse relaxation time and is denoted $T_2$. The inverse of a relaxation time is called a relaxation rate.

The first microscopic theory for the phenomenon of nuclear spin relaxation was presented by Bloembergen, Purcell and Pound (BPP) in 1948 [2]. They related the spin–lattice relaxation rate to the transition probabilities between the nuclear spin energy levels. The BPP paper constitutes the foundation on which most of the subsequent theory has been built, but contains some faults which were corrected by Solomon in 1955 [3]. Solomon noted also that a correct description of even a very simple system containing two interacting spins, requires introducing the concept of cross-relaxation, or magnetization exchange between the spins.

The subsequent development has been rich and the goal of this entry is to provide a flavour of relaxation theory (section B1.13.2), experimental techniques (section B1.13.3) and applications (section B1.13.4).

The further reading list for this entry contains five monographs, a review volume and two extensive reviews from the early 1990s. The monographs cover the basic NMR theory and the theoretical aspects of NMR relaxation. The review volume covers many important aspects of modern NMR experiments in general and relaxation measurements in particular. The two reviews contain more than a thousand references to application papers, mainly from the eighties. The number of literature references provided in this entry is limited and, in particular in the theory and experiments sections, priority is given to reviews rather than to original articles.

## B1.13.2   Relaxation theory

We begin this section by looking at the Solomon equations, which are the simplest formulation of the essential aspects of relaxation as studied by NMR spectroscopy of today. A more general Redfield theory is introduced in the next section, followed by the discussion of the connections between the relaxation and molecular motions and of physical mechanisms behind the nuclear relaxation.

### B1.13.2.1   The Solomon theory for a two-spin system

Let us consider a liquid consisting of molecules containing two nuclei with the spin quantum number 1/2. We denote the two spins $I$ and $S$ and assume that they are distinguishable, i.e. either belong to different nuclear species or have different chemical shifts. The system of such two spins is characterized by four energy levels and by a set of transition probabilities between the levels, cf. figure B1.13.1. We assume at this stage that the two spins interact with each other only by the dipole–dipole (DD) interaction. The DD interaction depends on the orientation of the internuclear axis with respect to $B_0$ and is thus changed (and averaged to zero on a sufficiently long time scale) by molecular tumbling. Taking this motional variation into consideration, it is quite straightforward to use time-dependent perturbation theory to derive transition probabilities between the pairs of levels in figure B1.13.1. Briefly, the transition probabilities are related to spectral density functions (*vide infra*), which measure the intensity of the local magnetic fields fluctuating at the frequencies corresponding to the energy level differences in figure B1.13.1. Solomon [3] showed that the relaxation of the longitudinal magnetization components, proportional to the expectation values of $I_z$ and $S_z$ operators, was related to the populations of the four levels and could be described by a set of two coupled equations:

$$\frac{d\langle I_z \rangle}{dt} = -(W_0 + 2W_{1I} + W_2)(\langle I_z \rangle - I_z^0) - (W_2 - W_0)(\langle S_z \rangle - S_z^0)$$

$$\frac{d\langle S_z \rangle}{dt} = -(W_2 - W_0)(\langle I_z \rangle - I_z^0) - (W_0 + 2W_{1S} + W_2)(\langle S_z \rangle - S_z^0) \tag{B1.13.2}$$

or

$$\frac{d\langle I_z \rangle}{dt} = -\rho_I(\langle I_z \rangle - I_z^0) - \sigma_{IS}(\langle S_z \rangle - S_z^0)$$

$$\frac{d\langle S_z \rangle}{dt} = -\sigma_{IS}(\langle I_z \rangle - I_z^0) - \rho_S(\langle S_z \rangle - S_z^0). \tag{B1.13.3}$$

$I_z^0$ and $S_z^0$ are the equilibrium magnetization for the two spins and $\rho_I$ and $\rho_S$ are the corresponding decay rates (spin–lattice relaxation rates). The symbol $\sigma_{IS}$ denotes the cross-relaxation rate. The general solutions of equations (B1.13.2) or (B1.13.3) for $\langle I_z \rangle$ or $\langle S_z \rangle$ are sums of two exponentials, i.e. the longitudinal magnetizations in a two-spin system do not follow the Bloch equations. Solomon demonstrated also that the simple exponential relaxation behaviour of the longitudinal magnetization was recovered under certain limiting conditions.

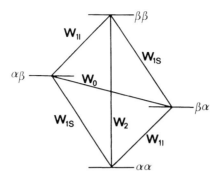

**Figure B1.13.1.** Energy levels and transition probabilities for an $IS$ spin system. (Reproduced by permission of Academic Press from Kowalewski J 1990 *Annu. Rep. NMR Spectrosc.* **22** 308–414.)

(i)  The two spins are identical (e.g. the two proton spins in a water molecule). We then have $W_{1I} = W_{1S} = W_1$ and the spin–lattice relaxation rate is given by:

$$T_1^{-1} = 2(W_1 + W_2).\qquad(B1.13.4)$$

(ii)  One of the spins, say $S$, is characterized by another, faster, relaxation mechanism. We can then say that the $S$ spin remains in thermal equilibrium on the time scale of the $I$-spin relaxation. This situation occurs in paramagnetic systems, where $S$ is an electron spin. The spin–lattice relaxation rate for the $I$ spin is then given by:

$$\rho_I = T_{1I}^{-1} = W_0 + 2W_1 + W_2.\qquad(B1.13.5)$$

(iii)  One of the spins, say $S$ again, is saturated by an intense RF field at its resonance frequency. These are the conditions applying for e.g. carbon-13 (treated as the $I$ spin) under broad-band decoupling of protons ($S$ spins). The relaxation rate is also in this case given by equation (B1.13.5). In addition, we then observe the phenomenon referred to as the (hetero)nuclear Overhauser enhancement (NOE), i.e. the steady-state solution for $\langle I_z \rangle$ is modified to

$$\langle I_z \rangle_{\text{steady state}} = I_z^0 + \frac{\sigma_{IS}}{\rho_I} S_z^0.\qquad(B1.13.6)$$

## B1.13.2.2  The Redfield theory

A more general formulation of relaxation theory, suitable for systems with scalar spin–spin couplings ($J$ couplings), is known as the Wangsness, Bloch and Redfield (WBR) theory or the Redfield theory [4]. In analogy with the Solomon theory, the Redfield theory is also based on the second-order perturbation theory, which in certain situations (unusual for nuclear spin systems in liquids) can be a limitation. Rather than dealing with the concepts of magnetizations or energy level populations in the Solomon formulation, the Redfield theory is given in terms of density operator (for a general review of the density operator formalism, see 'Further reading'). Briefly, the density operator describes the average behaviour of an ensemble of quantum mechanical systems. It is usually expressed by expansion in a suitable operator basis set. For the discussion of the $IS$ system above, the appropriate 16-dimensional basis set can e.g. consist of the unit operator, $E$, the operators corresponding to the Cartesian components of the two spins, $I_x$, $I_y$, $I_z$, $S_x$, $S_y$, $S_z$ and the products of the components of $I$ and the components of $S$. These 16 operators span the Liouville space for our two-spin system. If we concentrate on the longitudinal relaxation (the relaxation connected to the distribution of populations), the Redfield theory predicts the relaxation to follow a set of three coupled differential equations:

$$
\frac{\mathrm{d}}{\mathrm{d}t}
\begin{pmatrix}
\langle I_z \rangle \\
\langle S_z \rangle \\
\langle 2I_z S_z \rangle
\end{pmatrix}
= -
\begin{pmatrix}
\rho_I & \sigma_{IS} & \delta_{I,IS} \\
\sigma_{IS} & \rho_S & \delta_{S,IS} \\
\delta_{I,IS} & \delta_{S,IS} & \rho_{IS}
\end{pmatrix}
\begin{pmatrix}
\langle I_z \rangle - I_z^0 \\
\langle S_z \rangle - S_z^0 \\
\langle 2I_z S_z \rangle
\end{pmatrix}.
\tag{B1.13.7}
$$

The difference compared to equations (B1.13.2) or (B1.13.3) is the occurrence of the expectation value of the $2I_z S_z$ operator (the two-spin order), characterized by its own decay rate $\rho_{IS}$ and coupled to the one-spin longitudinal operators by the terms $\delta_{I,IS}$ and $\delta_{S,IS}$. We shall come back to the physical origin of these terms below.

The matrix on the rhs of equation (B1.13.7) is called the relaxation matrix. To be exact, it represents one block of a larger, block-diagonal relaxation matrix, defined in the Liouville space for the two-spin system. The remaining part of the large matrix (sometimes also called the relaxation supermatrix) describes the relaxation of coherences, which can be seen as generalizations of the transverse components of the magnetization vector. In systems without degeneracies, each of the coherences decays exponentially, with its own $T_2$. The Redfield theory can be used to obtain expressions for the relaxation matrix elements for arbitrary spin systems and for any type of relaxation mechanism. In analogy with the Solomon theory, also these more general relaxation rates are expressed in terms of various spectral density functions.

### B1.13.2.3  Molecular motions and spin relaxation

Nuclear spin relaxation is caused by fluctuating interactions involving nuclear spins. We write the corresponding Hamiltonians (which act as perturbations to the static or time-averaged Hamiltonian, determining the energy level structure) in terms of a scalar contraction of spherical tensors:

$$
H_1(t) = \sum_{q=-j}^{j} (-1)^q F^{(q)}(t) A^{(-q)}.
\tag{B1.13.8}
$$

$j$ is the rank of the tensor describing the relevant interactions, which can be 0, 1 or 2. $A^{(-q)}$ are spin operators and $F^{(q)}(t)$ represent classical functions related to the lattice, i.e. to the classically described environment of the spins. The functions $F^{(q)}(t)$ are, because of random molecular motions, stochastic functions of time. A fluctuating (stochastic) interaction can cause transitions between the energy levels of a spin system (and thus transfer the energy between the spin systems and its environment) if the power spectrum of the fluctuations contains Fourier components at frequencies corresponding to the relevant energy differences. In this sense, the transitions contributing to the spin relaxation and originating from randomly fluctuating interactions, are not really fundamentally different from the transitions caused by coherent interactions with electromagnetic radiation. According to theory of stochastic processes (the Wiener–Khinchin theorem) [5], the power available at a certain frequency, or the spectral density function $J(\omega)$ at that frequency, is obtained as a Fourier transform of a time correlation function (tcf) characterizing the stochastic process. A tcf $G(\tau)$ for a stochastic function of time $F^{(q)}(t)$ is defined:

$$
G(\tau) = \langle F^{(-q)}(t) F^{(q)}(t+\tau) \rangle
\tag{B1.13.9}
$$

with the corresponding spectral density:

$$
J(\omega) = \int_{-\infty}^{+\infty} G(\tau) \exp(i\omega\tau)\, \mathrm{d}\tau
\tag{B1.13.10}
$$

where the symbol $\langle\rangle$ denotes ensemble average. The function in the lhs of equation (B1.13.9) is called the auto-correlation function. The prefix 'auto' refers to the fact that we deal with an ensemble average of the product of a stochastic function taken at one point in time and the same function at another point in time,

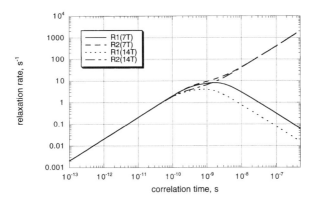

**Figure B1.13.2.** Spin–lattice and spin–spin relaxation rates ($R_1$ and $R_2$, respectively) for a carbon-13 spin directly bonded to a proton as a function of correlation time at the magnetic fields of 7 and 14 T.

the difference between the two time points being $\tau$. In certain situations in relaxation theory, we also need cross-correlation functions, where we average the product of two different stochastic functions, corresponding to different Hamiltonians of equation (B1.13.8).

We now come back to the important example of two spin 1/2 nuclei with the dipole–dipole interaction discussed above. In simple physical terms, we can say that one of the spins senses a fluctuating local magnetic field originating from the other one. In terms of the Hamiltonian of equation (B1.13.8), the stochastic function of time $F^{(q)}(t)$ is proportional to $Y_{2q}(\theta, \phi)/r_{IS}^3$, where $Y_{2q}$ is an $l = 2$ spherical harmonic and $r_{IS}$ is the internuclear distance. If the two nuclei are directly bonded, $r_{IS}$ can be considered constant and the random variation of $F^{(q)}(t)$ originates solely from the reorientation of the molecule-fixed internuclear axis with respect to the laboratory frame. The auto-tcf for $Y_{2q}$ can be derived, assuming that the stochastic time dependence can be described by the isotropic rotational diffusion equation. We obtain then:

$$G(\tau) = G(0) \exp(-\tau/\tau_c) \tag{B1.13.11}$$

with the corresponding spectral density:

$$J(\omega) = G(0)\frac{2\tau_c}{1 + \omega^2\tau_c^2}. \tag{B1.13.12}$$

We call $\tau_c$ the correlation time: it is equal to $1/6D_R$, where $D_R$ is the rotational diffusion coefficient. The correlation time increases with increasing molecular size and with increasing solvent viscosity. Equations (B1.13.11) and (B1.13.12) describe the rotational Brownian motion of a rigid sphere in a continuous and isotropic medium. With the Lorentzian spectral densities of equation (B1.13.12), it is simple to calculate the relevant transition probabilities. In this way, we can use e.g. equation (B1.13.5) to obtain $T_1^{-1}$ for a carbon-13 bonded to a proton as well as the corresponding $T_2^{-1}$, as a function of the correlation time for a given magnetic field, cf. figure B1.13.2. The two rates are equal in the region $\omega\tau_c \ll 1$, referred to as the 'extreme narrowing'. In the extreme narrowing regime, the relaxation rates are independent of the magnetic field and are simply given by a product of the square of an interaction strength constant (the dipole coupling constant, DCC) and the correlation time.

In order to obtain a more realistic description of reorientational motion of internuclear axes in real molecules in solution, many improvements of the tcf of equation (B1.13.11) have been proposed [6]. Some of these models are characterized in table B1.13.1. The entry 'number of terms' refers to the number of exponential functions in the relevant tcf or, correspondingly, the number of Lorentzian terms in the spectral density function.

**Table B1.13.1.** Selected dynamic models used to calculate spectral densities.

| Dynamic model | Parameters influencing the spectral densities | Number of terms | Comment | Ref. |
|---|---|---|---|---|
| Isotropic rotational diffusion | Rotational diffusion coefficient, $D_R = 1/6\tau_c$ | 1 | Useful as a first approximation | Further reading |
| Rotational diffusion, symmetric top | Two rotational diffusion coefficients, $D_{\parallel}$ and $D_{\perp}$, the angle $\theta$ between the symmetry axis and the internuclear axis | 3 | Rigid molecule, requires the knowledge of geometry | [7] |
| Rotational diffusion, asymmetric top | Three rotational diffusion coefficients, two polar angles $\theta$ and $\phi$ | 5 | Rigid molecule, rather complicated | [8] |
| Isotropic rotational diffusion with one internal degree of freedom | Rotational diffusion coefficient, $D_R$, internal motion rate parameter, angle between the internal rotation axis and the internuclear axis | 3 | Useful for e.g. methyl groups | [9] |
| 'Model-free' | Global and local correlation times, generalized order parameter, $S$ | 2 | Widely used for non-rigid molecules | [10] |
| Anisotropic model-free | $D_{\parallel}$ and $D_{\perp}$, $\theta$, local correlation time, generalized order parameter, $S$ | 4 | Allows the interpretation of data for non-spherical, non-rigid molecules | [11] |

### B1.13.2.4 Relaxation mechanisms

The DD interaction discussed above is just one—admittedly a very important one—among many possible sources of nuclear spin relaxation, collected in table B1.13.2 (not meant to be fully comprehensive). The interactions listed there are often used synonymously with 'relaxation mechanisms' and more detailed descriptions of various mechanisms can be found in 'Further reading'. We can note in table B1.13.2 that one particular type of motion—reorientation of an axis—can cause random variations of many $j = 2$ interactions: DD for different intramolecular axes, CSA, quadrupolar interaction. In fact, all the interactions with the same tensor rank can give rise to the interference or cross-terms with each other. In an important paper by Szymanski *et al* [20], the authors point out that the concept of a 'relaxation mechanism' is more appropriate to use referring to a pair of interactions, rather than to a single one. The role of interference terms has been reviewed by Werbelow [21]. They often contribute to relaxation matrices through coherence or polarization transfer and through higher forms of order or coherence. The relevant spectral densities are of cross-correlation (rather than auto-correlation) type. For example, the off-diagonal $\delta$-terms in the relaxation matrix of equation (B1.13.7), connecting the one- and two-spin longitudinal order, are caused by cross-correlation between the DD and CSA interactions, which explains why they are called 'cross-correlated relaxation rates'.

Before leaving the subsection on relaxation mechanisms, I wish to mention the connections between relaxation and chemical exchange (exchange of magnetic environments of spins by a chemical process). The chemical exchange and relaxation determine together the NMR lineshapes, the exchange affects the measured

**Table B1.13.2.** Interactions giving rise to nuclear spin relaxation.

| Interaction | Tensor rank | Process causing the random changes | Comment | Ref. |
|---|---|---|---|---|
| Intramolecular dipole-dipole (DD) | 2 | Reorientation of the inter-nuclear axis | Very common for $I = 1/2$ | Further reading |
| Intermolecular DD | 2 | Distance variation by translational diffusion | Less common | [12] |
| Chemical shift anisotropy (CSA) | 2 | Reorientation of the CSA principal axis | Increases with the square of the magnetic field | [13] |
| Intramolecular quadrupolar | 2 | Reorientation of the electric field gradient principal axis | Dominant for $I \geq 1$ (covalently bonded) | [14] |
| Intermolecular quadrupolar | 2 | Fluctuation of the electric field gradient, moving multipoles | Common for $I \geq 1$ in free ions in solution | [15] |
| Antisymmetric CSA | 1 | Reorientation of a pseudo-vector | Very uncommon | [13] |
| Spin-rotation | 1 | Reorientation and time dependence of angular momentum | Small molecules only | [16] |
| Scalar coupling | 0 | Relaxation of the coupled spin or exchange | Can be important for $T_2$ | Further reading |
| Hyperfine interaction (dipolar and scalar) | 2,0 | Electron relaxation, may be complicated | Paramagnetic systems and impurities | [17–19] |

relaxation rates and it can also act as a source of random variation of spin interactions. The relaxation effects of chemical exchange have been reviewed by Woessner [22].

## B1.13.3   Experimental methods

Relaxation experiments were among the earliest applications of time-domain high-resolution NMR spectroscopy, invented more than 30 years ago by Ernst and Anderson [23]. The progress of the experimental methodology has been enormous and only some basic ideas of the experiment design will be presented here. This section is divided into three subsections. The first one deals with Bloch equation-type experiments, measuring $T_1$ and $T_2$ when such quantities can be defined, i.e. when the relaxation is monoexponential. As a slightly oversimplified rule of thumb, we can say that this happens in the case of isolated spins. The two subsections to follow cover multiple-spin effects.

### B1.13.3.1   Spin–lattice and spin–spin relaxation rates

Measurements of spin–lattice relaxation time, $T_1$, are the simplest relaxation experiments. A straightforward method to measure $T_1$ is the inversion-recovery experiment, the principle of which is illustrated in figure B1.13.3. The equilibrium magnetization $M_0$ or $M_z(\infty)$ (cf. figure B1.13.3(A)), directed along $B_0$ (the $z$-axis), is first inverted by a $180°$ pulse (a $\pi$-pulse), a RF pulse with the duration $\tau_{180°}$ and the RF magnetic field $B_1$ (in the the direction perpendicular to $B_0$) chosen so that the magnetization is nutated by $180°$ around the $B_1$ direction. The magnetization immediately after the $180°$ pulse is directed along the $-z$ direction (cf. figure B1.13.3(B)) and starts to relax following equation (B1.13.1). After a variable delay $\tau$, when the

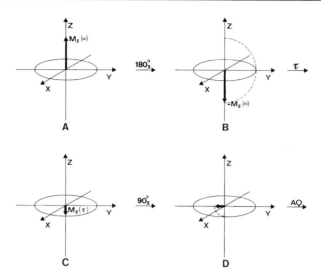

**Figure B1.13.3.** The inversion-recovery experiment. (Reproduced by permission of VCH from Banci L, Bertini I and Luchinat C 1991 *Nuclear and Electron Relaxation* (Weinheim: VCH).)

$M_z(\tau)$ has reached the stage depicted in figure B1.13.3(C), a 90° pulse is applied. This pulse nutates the magnetization along the $z$-axis to the $x$, $y$-plane (cf. figure B1.13.3(D)), where it can be detected in the form of a FID. If required, the experiment can be repeated a number of times to improve the signal-to-noise ratio, waiting for about $5T_1$ (recycle delay) between scans to allow for return to equilibrium. The subsequent Fourier transformation (FT) of the FID gives a spectrum. The experiment is repeated for different values of the delay $\tau$ and the measured line intensities are fitted to an exponential expression $S(\tau) = A + B \exp(-\tau/T_1)$. The inversion-recovery experiments are often performed for multiline spectra of low-natural abundance nuclei, such as $^{13}$C or $^{15}$N, under the conditions of broadband saturation (decoupling) of the abundant proton spins. The proton ($S$-spin) decoupling renders the relaxation of the $I$ spin of the dipolarly coupled $IS$-spin system monoexponential; we may say that the decoupling results in 'pseudo-isolated' $I$ spins. An example of a $^{13}$C inversion-recovery experiment for a trisaccharide, melezitose, is shown in figure B1.13.4.

NMR spectroscopy is always struggling for increased sensitivity and resolution, as well as more efficient use of the instrument time. To this end, numerous improvements of the simple inversion-recovery method have been proposed over the years. An early and important modification is the so-called fast inversion recovery (FIR) [25], where the recycle delay is made shorter than $5T_1$ and the experiment is carried out under the steady-state rather than equilibrium conditions. A still more time-saving variety, the super-fast inversion recovery (SUFIR) has also been proposed [26].

Several other improvements of the inversion-recovery scheme employ advanced tools of modern NMR spectroscopy: polarization transfer and two-dimensional spectroscopy (see further reading). The basic design of selected pulse sequences is compared with the simple inversion-recovery scheme in figure B1.13.5, taken from Kowalewski and Mäler [24], where references to original papers can be found. The figure B1.13.5(a), where thick rectangular boxes denote the 180° $I$-spin pulses and thin boxes the corresponding 90° pulses, is a representation of the inversion-recovery sequence with the continuous saturation of the protons. In figure B1.13.5(b), the inverting $I$-spin pulse is replaced by a series of pulses, separated by constant delays and applied at both the proton and the $I$-spin resonance frequencies, which creates a more strongly polarized initial $I$-spin state (the polarization transfer technique). In figure B1.13.5(c), a two-dimensional (2D) NMR technique is employed. This type of approach is particularly useful when the sample contains many heteronuclear $IS$ spin

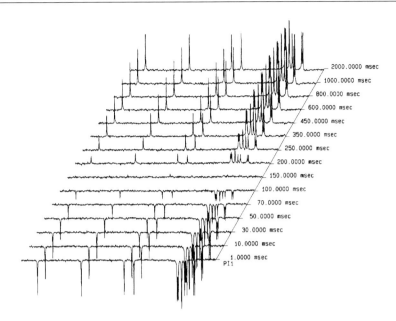

**Figure B1.13.4.** The inversion-recovery determination of the carbon-13 spin–lattice relaxation rates in melezitose. (Reproduced by permission of Elsevier from Kowalewski J and Mäler L 1997 *Methods for Structure Elucidation by High-Resolution NMR* ed Gy Batta, K E Kövér and Cs Szántay (Amsterdam: Elsevier) pp 325–47.)

pairs, with different $I$s and different $S$s characterized by slightly different resonance frequencies (chemical shifts), resulting in crowded spectra. In a generic 2D experiment, the NMR signal is sampled as a function of two time variables: $t_2$ is the running time during which the FID is acquired (different points in the FID have different $t_2$). In addition, the pulse sequence contains an evolution time $t_1$, which is systematically varied in the course of the experiment. The double Fourier transformation of the data matrix $S(t_1, t_2)$ yields a two-dimensional spectrum. In the example of figure B1.13.5(c), the polarization is transferred first from protons to the $I$ spins, in the same way as in figure B1.13.5(b). This is followed by the evolution time, during which the information on the various $I$-spin resonance frequencies is encoded. The next period is the analogue of the delay $\tau$ of the simple inversion-recovery experiment. The final part of the sequence contains an inverse polarization transfer, from $I$ spins to protons, followed by the proton detection. The resulting 2D spectrum, for a given delay $\tau$, has the proton chemical shifts on one axis and the shifts of the $J$-coupled $I$-spin on the other one. We can thus call the experiment the proton–$I$-spin correlation experiment. This greatly improves the spectral resolution. Spectra with several different $\tau$ delays are acquired and the $I$ spin $T_1$ is determined by fitting the intensity decay for a given peak in the 2D spectrum.

The discussion above is concerned with $T_1$ experiments under high resolution conditions at high magnetic field. In studies of complex liquids (polymer solutions and melts, liquid crystals), one is often interested to obtain information on rather slow (micro- or nanosecond time scale) motions measuring $T_1$ at low magnetic fields using the field-cycling technique [27]. The same technique is also invaluable in studies of paramagnetic solutions. Briefly, the non-equilibrium state is created by keeping the sample at a certain, moderately high polarizing field $B_0$ and then rapidly switching $B_0$ to another value, at which we wish to measure $T_1$. After the variable delay $\tau$, the field is switched again and the signal is detected.

The spin–spin relaxation time, $T_2$, defined in the Bloch equations, is simply related to the width $\Delta\nu_{1/2}$ of the Lorentzian line at the half-height: $\Delta\nu_{1/2} = 1/\pi T_2$. Thus, it is in principle possible to determine $T_2$ by measuring the linewidth. This simple approach is certainly useful for rapidly relaxing quadrupolar nuclei

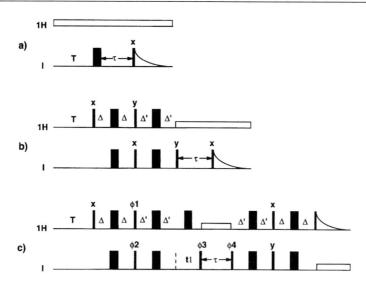

**Figure B1.13.5.** Some basic pulse sequences for $T_1$ measurements for carbon-13 and nitrogen-15. (Reproduced by permission of Elsevier from Kowalewski J and Mäler L 1997 *Methods for Structure Elucidation by High-Resolution NMR* ed Gy Batta, K E Kövér and Cs Szántay (Amsterdam: Elsevier) pp 325–47.)

and has also been demonstrated to work for $I = 1/2$ nuclei, provided the magnetic field homogeneity is ascertained. The more usual practice, however, is to suppress the inhomogenous broadening caused by the spread of the magnetic field values (and thus resonance frequencies) over the sample volume, by the spin-echo technique. Such experiments are in general more difficult to perform than the spin–lattice relaxation time measurements. The most common echo sequence, the $90°–\tau–180°–\tau$–echo, was originally proposed by Carr and Purcell [28] and modified by Meiboom and Gill [29]. After the initials of the four authors, the modified sequence is widely known as the CPMG method. The details of the behaviour of spins under the spin-echo sequence can be found in modern NMR monographs (see further reading) and will not be repeated here. We note, however, that complications can arise in the presence of scalar spin–spin couplings.

An alternative procedure for determining the transverse relaxation time is the so-called $T_{1\rho}$ experiment. The basic idea of this experiment is as follows. The initial $90°$ $I$-spin pulse is applied with $B_1$ in the $x$-direction, which turns the magnetization from the $z$-direction to the $y$-direction. Immediately after the initial pulse, the $B_1$ RF field is switched to the $y$-direction so that $M$ and $B_1$ become collinear. The notation $T_{1\rho}$ ($T_1$ in the rotating frame) alludes to the fact that the decay of $M$ along $B_1$ is similar to the relaxation of longitudinal magnetization along $B_0$. The $T_{1\rho}$ in liquids is practically identical to $T_2$. The measurements of $T_2$ or $T_{1\rho}$ can, in analogy with $T_1$ studies, also utilize the modern tools increasing the sensitivity and resolution, such as polarization transfer and 2D techniques.

### B1.13.3.2  *Cross-relaxation and nuclear Overhauser enhancement*

Besides measuring $T_1$ and $T_2$ for nuclei such as $^{13}$C or $^{15}$N, relaxation studies for these nuclei also include measurements of the NOE factor, cf. equation (B1.13.6). Knowing the $T_{1I}^{-1}$ ($\rho_I$) and the steady-state NOE (measured by comparing the signal intensities in the presence and in the absence of the saturating field), we can derive the cross-relaxation rate, $\sigma_{IS}$, which provides an additional combination of spectral densities, useful for e.g. molecular dynamics studies.

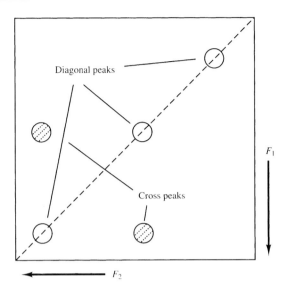

**Figure B1.13.6.** The basic elements of a NOESY spectrum. (Reproduced by permission of Wiley from Williamson M P 1996 *Encyclopedia of Nuclear Magnetic Resonance* ed D M Grant and R K Harris (Chichester: Wiley) pp 3262–71).

The most important cross-relaxation rate measurement are, however, performed in homonuclear networks of chemically shifted and dipolarly coupled proton spins. The subject has been discussed in two books [30, 31]. There is large variety of experimental procedures [32], of which I shall only mention a few. A simple method to measure the homonuclear NOE is the NOE-difference experiment, in which one measures spectral intensities using low power irradiation at selected narrow regions of the spectrum, before applying the observe pulse. This corresponds to different individual protons acting as the saturated $S$ spins. The difference spectrum is obtained by subtracting the spectrum obtained under identical conditions, but with the irradiation frequency applied in a region without any peaks to saturate. The NOE-difference experiments are most often applied in a semi-quantitative way in studies of small organic molecules.

For large molecules, such as proteins, the main method in use is a 2D technique, called NOESY (nuclear Overhauser effect spectroscopy). The basic experiment [33, 34] consists of three 90° pulses. The first pulse converts the longitudinal magnetizations for all protons, present at equilibrium, into transverse magnetizations which evolve during the subsequent evolution time $t_1$. In this way, the transverse magnetization components for different protons become labelled by their resonance frequencies. The second 90° pulse rotates the magnetizations to the $-z$-direction. The interval between the second and third pulse is called the mixing time, during which the spins evolve according to the multiple-spin version of equations (B1.13.2) and (B1.13.3) and the NOE builds up. The final pulse converts the longitudinal magnetizations, present at the end of the mixing time, into detectable transverse components. The detection of the FID is followed by a recycle delay, during which the equilibrium is recovered and by the next experiment, e.g. with another $t_1$. After acquiring the 2D data matrix $S(t_1, t_2)$ (for a given mixing time) and the double Fourier transformation, one obtains a 2D spectrum shown schematically in figure B1.13.6. The individual cross-relaxation rates for pairs of spins can be obtained by following the build-up of the cross-peak intensities as a function of the mixing time for short mixing times (the so-called initial rate approximation). For longer mixing times and large molecules, the cross-peaks show up in a large number of positions, because of multiple transfers called spin-diffusion. The analysis then becomes more complicated, but can be handled based on a generalization of the Solomon equation to many spins (the complete relaxation matrix treatment) [35].

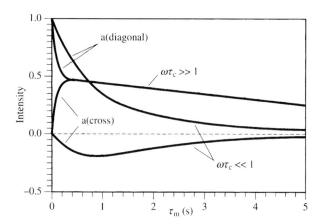

**Figure B1.13.7.** Simulated NOESY peak intensities in a homonuclear two-spin system as a function of the mixing time for two different motional regimes. (Reproduced by permission of Wiley from Neuhaus D 1996 *Encyclopedia of Nuclear Magnetic Resonance* ed D M Grant and R K Harris (Chichester: Wiley) pp 3290–301.)

As seen in equations (B1.13.2) and (B1.13.3), the cross-relaxation rate $\sigma_{IS}$ is given by $W_2 - W_0$, the difference between two transition probabilities. Assuming the simple isotropic rotational diffusion model, each of the transition probabilities is proportional to a Lorentzian spectral density (cf. equation (B1.13.12)), taken at the frequency of the corresponding transition. For the homonuclear case, $W_2$ corresponds to a transition at high frequency $(\omega_I + \omega_S) \approx 2\omega_I$, while $W_0$ is proportional to a Lorentzian at $(\omega_I - \omega_S) \approx 0$. When the product $\omega_I \tau_c$ is small, $W_2$ is larger than $W_0$ and the cross-relaxation rate is positive. When the product $\omega_I \tau_c$ is large, the Lorentzian function evaluated at $2\omega_I$ is much smaller than at zero-frequency and $\sigma_{IS}$ changes sign. The corresponding NOESY peak intensities in a two-spin system are shown as a function of the mixing time in figure B1.13.7. Clearly, the intensities of the cross-peaks for small molecules $(\omega^2 \tau_c^2 \ll 1)$ have one sign, while the opposite sign pertains for large molecules $(\omega^2 \tau_c^2 \gg 1)$. At a certain critical correlation time, we obtain no NOESY cross-peaks. In such a situation and as complement to the NOESY experiments, one can perform an experiment called ROESY (rotating frame Overhauser effect spectroscopy) [36]. The relation between NOESY and ROESY is similar to that between $T_1$ and $T_{1\rho}$.

### B1.13.3.3  *Cross-correlated relaxation*

Studies of cross-correlated relaxation have received increasing attention during the last decades. The general strategies for creating and detecting different types of spin-ordered state and for measuring the transfer rates have been discussed by Canet [37]. I shall concentrate here on measurements of the DD-CSA interference terms in two-spin systems, the $\delta_{I,IS}$ and $\delta_{S,IS}$ terms in equation (B1.13.7). Let us consider a system where the two spins have their resonances sufficiently far apart that we can construct pulses selective enough to manipulate one of them at a time (this is automatically fulfilled for a heteronuclear case; in the homonuclear this requires specially shaped low power RF pulses). One way to measure the longitudinal cross-correlated relaxation rates is to invert one of the spins by a 180° pulse and to detect the build-up of the two-spin order. The two-spin order $\langle 2I_z S_z \rangle$ is not detectable directly but, if one of the spins is exposed to a 90° pulse, the two-spin order becomes converted into a detectable signal in the form of an antiphase doublet, cf. figure B1.13.8 (the corresponding one-spin order subjected to a 90° pulse gives rise to an in-phase doublet). To separate the two types of order in a clean way, one can use an RF pulse trick called the double-quantum filter. There are many ways to optimize the above method as well as other schemes to measure the $\langle I_z \rangle$ to $\langle 2I_z S_z \rangle$ transfer rates. One such scheme uses a set of three selective NOESY experiments, where three 90° pulses strike the spins in the

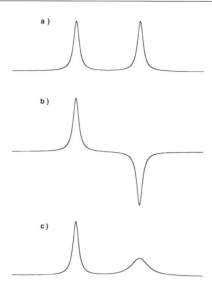

**Figure B1.13.8.** Schematic illustration of (a) an antiphase doublet, (b) an in-phase doublet and (c) a differentially broadened doublet. The splitting between the two lines is in each case equal to $J$, the indirect spin–spin coupling constant.

sequence $III$, $SSS$ and $IIS$ [38]. Another scheme uses an extended sequence of 180° pulses, followed by a detecting pulse [39].

The cross-correlation effects between the DD and CSA interactions also influence the transverse relaxation and lead to the phenomenon known as differential line broadening in a doublet [40], cf. figure B1.13.8. There is a recent experiment, designed for protein studies, that I wish to mention at the end of this section. It has been proposed by Pervushin *et al* [41], is called TROSY (transverse relaxation optimized spectroscopy) and employs the differential line-broadening in a sophisticated way. One works with the $^{15}$N–proton $J$-coupled spin system. When the system is subjected to a 2D nitrogen–proton correlation experiment, the spin–spin coupling gives rise to four lines (a doublet in each dimension). The differential line broadening results in one of the four lines being substantially narrower than the other ones. Pervushin *et al* [41] demonstrated that the broader lines can be suppressed, resulting in greatly improved resolution in the spectra of large proteins.

## B1.13.4    Applications

In this section, I present a few illustrative examples of applications of NMR relaxation studies within different branches of chemistry. The three subsections cover one 'story' each, in order of increasing molecular size and complexity of the questions asked.

### B1.13.4.1    Small molecules: the dual spin probe technique

Small molecules in low viscosity solutions have, typically, rotational correlation times of a few tens of picoseconds, which means that the extreme narrowing conditions usually prevail. As a consequence, the interpretation of certain relaxation parameters, such as carbon-13 $T_1$ and NOE for proton-bearing carbons, is very simple. Basically, the DCC for a directly bonded CH pair can be assumed to be known and the experiments yield a value of the correlation time, $\tau_c$. One interesting application of the measurement of $\tau_c$ is to follow its variation with the site in the molecule (motional anisotropy), with temperature (the correlation time increases often with decreasing temperature, following an Arrhenius-type equation) or with the composition of solution. The latter two types of measurement can provide information on intermolecular interactions.

Another application of the knowledge of $\tau_c$ is to employ it for the interpretation of another relaxation measurement in the same system, an approach referred to as the dual spin probe technique. A rather old, but illustrative, example is the case of *tris*(pentane-2,4-dionato)aluminium(III), Al(acac)$_3$. Dechter *et al* [42] reported measurements of carbon-13 spin–lattice relaxation for the methine carbon and of the aluminium-27 linewidth in Al(acac)$_3$ in toluene solution. $^{27}$Al is a quadrupolar nucleus ($I = 5/2$) and the linewidth gives directly the $T_2$, which depends on the rotational correlation time (which can be assumed the same as for the methine CH axis) and the strength of the quadrupolar interaction (the quadrupolar coupling constant, QCC). Thus, the combination of the carbon-13 and aluminium-27 measurements yields the QCC. The QCC in this particular case was also determined by NMR in a solid sample. The two measurements agree very well with each other (in fact, there is a small error in the paper [42]: the QCC from the linewidth in solution should be a factor of $2\pi$ larger than what is stated in the article). More recently, Champmartin and Rubini [43] studied carbon-13 and oxygen-17 (another $I = 5/2$ quadrupolar nucleus) relaxation in pentane-2,4-dione (Hacac) and Al(acac)$_3$ in solution. The carbon measurements were performed as a function of the magnetic field. The methine carbon relaxation showed, for both compounds, no field dependence, while the carbonyl carbon $T_1^{-1}$ increased linearly with $B_0^2$. This indicates the CSA mechanism and allows an estimate of its interaction strength, the anisotropy of the shielding tensor. Also this quantity could be compared with solid state measurements on Al(acac)$_3$ and, again, the agreement was good. From the oxygen-17 linewidth, the authors obtained also the oxygen-17 QCC. The chemically interesting piece of information is the observation that the QCC changes only slightly between the free acid and the trivalent metal complex.

### B1.13.4.2   *Oligosaccharides: how flexible are they?*

Oligosaccharides are a class of small and medium-sized organic molecules, subject to intense NMR work. I present here the story of two disaccharides, sucrose and $\alpha$-D-Man$p$-(1 $\rightarrow$ 3)-$\beta$-D-Glc$p$-OMe and a trisaccharide melezitose. McCain and Markley [44] published a carbon-13 $T_1$ and NOE investigation of sucrose in aqueous solution as a function of temperature and magnetic field. At low temperatures, a certain field dependence of the parameters could be observed, indicating that the extreme narrowing conditions might not be fulfilled under these circumstances. Taking the system out of extreme narrowing (which corresponds to correlation times of few hundred picoseconds) renders the relaxation rates field dependent and allows the investigators to ask more profound questions concerning the interaction strength and dynamics. Kovacs *et al* [45] followed up the McCain–Markley study by performing similar experiments on sucrose in a 7:3 molar D$_2$O/DMSO-d$_6$ solvent mixture. The solvent mixture has about four times higher viscosity than water and is a cryo-solvent. Thus, it was possible to obtain the motion of the solute molecule far from the extreme narrowing region and to make a quantitative determination of the effective DCC, in a way which is related to the Lipari–Szabo method [10]. Mäler *et al* [46] applied a similar experimental approach (but extended also to include the $T_2$ measurements) and the Lipari–Szabo analysis to study the molecular dynamics of melezitose in the same mixed solvent. Somewhat different dynamic behaviour of different sugar residues and the exocyclic hydroxymethyl groups was reported.

The question of the flexibility (or rigidity) of the glycosidic linkage connecting the different sugar units in oligosaccharides is a hot issue in carbohydrate chemistry. The findings of Mäler *et al* [46] could be interpreted as indicating a certain flexibility of the glycosidic linkages in melezitose. The question was posed more directly by Poppe and van Halbeek [47], who measured the intra- and interresidue proton–proton NOEs and rotating frame NOEs in sucrose in aqueous solution. They interpreted the results as proving the non-rigidity of the glycosidic linkage in sucrose. Mäler *et al* [48] investigated the disaccharide $\alpha$-D-Man$p$-(1 $\rightarrow$ 3)-$\beta$-D-Glc$p$-OMe by a combination of a carbon-13 spin–lattice relaxation study with measurements of the intra- and interresidue proton–proton cross-relaxation rates in a water–DMSO mixture. They interpreted their data in terms of the dynamics of the CH axes, the intra-ring HH axis and the trans-glycosidic HH axis being described by a single set of Lipari–Szabo parameters. This indicated that the inter-residue HH axis did not sense any additional mobility as compared to the other axes. We shall probably encounter a continuation of this story in the future.

### B1.13.4.3  *Human ubiquitin: a case history for a protein*

Human ubiquitin is a small (76 amino acid) and well characterized protein. I choose to illustrate the possibilities offered by NMR relaxation studies of proteins [49] through this example. The 2D NMR studies for the proton resonance assignment and a partial structure determination through NOESY measurements were reported independently by Di Stefano and Wand [50] and by Weber *et al* [51]. Schneider *et al* [52] studied a uniformly nitrogen-15 labelled species of human ubiquitin and reported nitrogen-15 $T_1$ (at two magnetic fields) and NOE values for a large majority of amide nitrogen sites in the molecule. The data were interpreted in terms of the Lipari–Szabo model, determining the generalized order parameter for the amide NH axes. It was thus possible to identify the more flexible and more rigid regions in the protein backbone. Tjandra *et al* [53] extended this work, both in terms of experiments (adding $T_2$ measurements) and in terms of interpretation (allowing for the anisotropy of the global reorientation, by means of the anisotropic Lipari–Szabo model [10]). This provided a more quantitative interpretation of the molecular dynamics. During the last two years, Bax and coworkers have, in addition, determined the CSA for the nitrogen-15 and proton for most of the amide sites, as well as for the $\alpha$-carbons, through measurements of the cross-correlated relaxation rates in the nitrogen-15–amide proton or the $C_\alpha$–$H_\alpha$ spin pairs [54–56]. The CSA values could, in turn, be correlated with the secondary structure, hydrogen bond length etc. It is not likely that the ubiquitin story is finished either.

### Acknowledgments

I am indebted to Dr Dan Bergman and Mr Tomas Nilsson for valuable comments on the manuscript.

### References

  [1]  Bloch F 1946 Nuclear induction *Phys. Rev.* **70** 460–74
  [2]  Bloembergen N, Purcell E M and Pound R V 1948 Relaxation effects in nuclear magnetic resonance absorption *Phys. Rev.* **73** 679–712
  [3]  Solomon I 1955 Relaxation processes in a system of two spins *Phys. Rev.* **99** 559–65
  [4]  Redfield A G 1996 Relaxation theory: density matrix formulation *Encyclopedia of Nuclear Magnetic Resonance* ed D M Grant and R K Harris (Chichester: Wiley) pp 4085–92
  [5]  Van Kampen N G 1981 *Stochastic Processes in Physics and Chemistry* (Amsterdam: North-Holland)
  [6]  Woessner D E 1996 Brownian motion and correlation times *Encyclopedia of Nuclear Magnetic Resonance* ed D M Grant and R K Harris (Chichester: Wiley) pp 1068–84
  [7]  Woessner D E 1962 Nuclear spin relaxation in ellipsoids undergoing rotational Brownian motion *J. Chem. Phys.* **37** 647–54
  [8]  Huntress W T 1970 The study of anisotropic rotation of molecules in liquids by NMR quadrupolar relaxation *Adv. Magn. Reson.* **4** 1–37
  [9]  Woessner D E 1962 Spin relaxation processes in a two-proton system undergoing anisotropic reorientation *J. Chem. Phys.* **36** 1–4
[10]  Lipari G and Szabo A 1982 Model-free approach to the interpretation of nuclear magnetic resonance relaxation in macromolecules 1. Theory and range of validity *J. Am. Chem. Soc.* **104** 4546–59
[11]  Barbato G, Ikura M, Kay L E, Pastor R W and Bax A 1992 Backbone dynamics of calmodulin studied by $^{15}$N relaxation using inverse detected two-dimensional NMR spectroscopy: the central helix is flexible *Biochemistry* **31** 5269–78
[12]  Hwang L-P and Freed J H 1975 Dynamic effects of pair correlation functions on spin relaxation by translational diffusion in liquids *J. Chem. Phys.* **63** 4017–25
[13]  Anet F A L and O'Leary D J 1992 The shielding tensor. Part II. Understanding its strange effects on relaxation *Concepts Magn. Reson.* **4** 35–52
[14]  Werbelow L G 1996 Relaxation theory for quadrupolar nuclei *Encyclopedia of Nuclear Magnetic Resonance* ed D M Grant and R K Harris (Chichester: Wiley) pp 4092–101
[15]  Roberts J E and Schnitker J 1993 Ionic quadrupolar relaxation in aqueous solution—dynamics of the hydration sphere *J. Phys. Chem.* **97** 5410–17

[16] McClung R E D 1996 Spin-rotation relaxation theory *Encyclopedia of Nuclear Magnetic Resonance* ed D M Grant and R K Harris (Chichester: Wiley) pp 4530–5

[17] Banci L, Bertini I and Luchinat C 1991 *Nuclear and Electron Relaxation* (Weinheim: VCH)

[18] Bertini I, Luchinat C and Aime S 1996 *NMR of Paramagnetic Substances* (Amsterdam: Elsevier)

[19] Kowalewski J 1996 Paramagnetic relaxation in solution *Encyclopedia of Nuclear Magnetic Resonance* ed D M Grant and R K Harris (Chichester: Wiley) pp 3456–62

[20] Szymanski S, Gryff-Keller A M and Binsch G A 1986 Liouville space formulation of Wangsness–Bloch–Redfield theory of nuclear spin relaxation suitable for machine computation. I. Fundamental aspects *J. Magn. Reson.* **68** 399–432

[21] Werbelow L G 1996 Relaxation processes: cross correlation and interference terms *Encyclopedia of Nuclear Magnetic Resonance* ed D M Grant and R K Harris (Chichester: Wiley) pp 4072–8

[22] Woessner D E 1996 Relaxation effects of chemical exchange *Encyclopedia of Nuclear Magnetic Resonance* ed D M Grant and R K Harris (Chichester: Wiley) pp 4018–28

[23] Ernst R R and Anderson W A 1986 Application of Fourier transform spectroscopy to magnetic resonance *Rev. Sci. Instrum.* **37** 93–102

[24] Kowalewski J and Mäler L 1997 Measurements of relaxation rates for low natural abundance $I = 1/2$ nuclei *Methods for Structure Elucidation by High-Resolution NMR* ed Gy Batta, K E Kövér and Cs Szántay (Amsterdam: Elsevier) pp 325–47

[25] Canet D, Levy G C and Peat I R 1975 Time saving in $^{13}$C spin-lattice relaxation measurements by inversion-recovery *J. Magn. Reson.* **18** 199–204

[26] Canet D, Mutzenhardt P and Robert J B 1997 The super fast inversion recovery (SUFIR) experiment *Methods for Structure Elucidation by High-Resolution NMR* ed Gy Batta, K E Kövér and Cs Szántay (Amsterdam: Elsevier) pp 317–23

[27] Koenig S H and Brown R D 1990 Field-cycling relaxometry of protein solutions and tissue—implications for MRI *Prog. Nucl. Magn. Reson. Spectrosc.* **22** 487–567

[28] Carr H Y and Purcell E M 1954 Effects of diffusion on free precession in nuclear magnetic resonance experiments *Phys. Rev.* **94** 630–8

[29] Meiboom S and Gill D 1958 Modified spin-echo method for measuring nuclear relaxation times *Rev. Sci. Instrum.* **29** 688–91

[30] Noggle J H and Schirmer R E 1971 *The Nuclear Overhauser Effect* (New York: Academic)

[31] Neuhaus D and Williamson M P 1989 *The Nuclear Overhauser Effect in Structural and Conformational Analysis* (New York: VCH)

[32] Neuhaus D 1998 Nuclear Overhauser effect *Encyclopedia of Nuclear Magnetic Resonance* ed D M Grant and R K Harris (Chichester: Wiley) pp 3290–301

[33] Jeener J, Meier B H, Bachmann P and Ernst R R 1979 Investigation of exchange processes by two-dimensional NMR spectroscopy *J. Chem. Phys.* **71** 4546–53

[34] Williamson M P 1996 NOESY *Encyclopedia of Nuclear Magnetic Resonance* ed D M Grant and R K Harris (Chichester: Wiley) pp 3262–71

[35] Borgias B A, Gochin M, Kerwood D J and James T L 1990 Relaxation matrix analysis of 2D NMR data *Prog. NMR Spectrosc.* **22** 83–100

[36] Bax A and Grzesiek S 1996 ROESY *Encyclopedia of Nuclear Magnetic Resonance* ed D M Grant and R K Harris (Chichester: Wiley) pp 4157–67

[37] Canet D 1989 Construction, evolution and detection of magnetization modes designed for treating longitudinal relaxation of weakly coupled spin 1/2 systems with magnetic equivalence *Prog. NMR Spectrosc.* **21** 237–91

[38] Di Bari L, Kowalewski J and Bodenhausen G 1990 Magnetization transfer modes in scalar-coupled spin systems investigated by selective 2-dimensional nuclear magnetic resonance exchange experiments *J. Chem. Phys.* **93** 7698–705

[39] Levitt M H and Di Bari L 1994 The homogeneous master equation and the manipulation of relaxation networks *Bull. Magn. Reson.* **16** 94–114

[40] Farrar T C and Stringfellow T C 1996 Relaxation of transverse magnetization for coupled spins *Encyclopedia of Nuclear Magnetic Resonance* ed D M Grant and R K Harris (Chichester: Wiley) pp 4101–7

[41] Pervushin K, Riek R, Wider G and Wüthrich K 1997 Attenuated $T_2$ relaxation by mutual cancellation of dipole–dipole coupling and chemical shift anisotropy indicates an avenue to NMR structures of very large biological macromolecules in solution *Proc. Natl Acad. Sci. USA* **94** 12 366–71

[42] Dechter J J, Henriksson U, Kowalewski J and Nilsson A-C 1982 Metal nucleus quadrupole coupling constants in aluminum, gallium and indium acetylacetonates *J. Magn. Reson.* **48** 503–11

[43] Champmartin D and Rubini P 1996 Determination of the O-17 quadrupolar coupling constant and of the C-13 chemical shielding tensor anisotropy of the CO groups of pentane-2,4-dione and beta-diketonate complexes in solution. NMR relaxation study *Inorg. Chem.* **35** 179–83

[44] McCain D C and Markley J L 1986 Rotational spectral density functions for aqueous sucrose: experimental determination using $^{13}$C NMR *J. Am. Chem. Soc.* **108** 4259–64

[45] Kovacs H, Bagley S and Kowalewski J 1989 Motional properties of two disaccharides in solutions as studied by carbon-13 relaxation and NOE outside of the extreme narrowing region *J. Magn. Reson.* **85** 530–41

[46] Mäler L, Lang J, Widmalm G and Kowalewski J 1995 Multiple-field carbon-13 NMR relaxation investigation on melezitose *Magn. Reson. Chem.* **33** 541–8

[47]  Poppe L and van Halbeek H 1992 The rigidity of sucrose—just an illusion? *J. Am. Chem. Soc.* **114** 1092–4

[48]  Mäler L, Widmalm G and Kowalewski J 1996 Dynamical behavior of carbohydrates as studied by carbon-13 and proton nuclear spin relaxation *J. Phys. Chem.* **100** 17 103–10

[49]  Dayie K T, Wagner G and Lefèvre J F 1996 Theory and practice of nuclear spin relaxation in proteins *Annu. Rev. Phys. Chem.* **47** 243–82

[50]  Di Stefano D L and Wand A J 1987 Two-dimensional [1]H NMR study of human ubiquitin: a main chain directed assignment and structure analysis *Biochemistry* **26** 7272–81

[51]  Weber P L, Brown S C and Mueller L 1987 Sequential [1]H NMR assignment and secondary structure identification of human ubiquitin *Biochemistry* **26** 7282–90

[52]  Schneider D M, Dellwo M J and Wand A J 1992 Fast internal main-chain dynamics of human ubiquitin *Biochemistry* **31** 3645–52

[53]  Tjandra N, Feller S E, Pastor R W and Bax A 1995 Rotational diffusion anisotropy of human ubiquitin from N-15 NMR relaxation *J. Am. Chem. Soc.* **117** 12 562–6

[54]  Tjandra N, Szabo A and Bax A 1996 Protein backbone dynamics and N-15 chemical shift anisotropy from quantitative measurement of relaxation interference effects *J. Am. Chem. Soc.* **118** 6986–91

[55]  Tjandra N and Bax A 1997 Solution NMR measurement of amide proton chemical shift anisotropy in N-15-enriched proteins. Correlation with hydrogen bond length *J. Am. Chem. Soc.* **119** 8076–82

[56]  Tjandra N and Bax A 1997 Large variations in C-13(alpha) chemical shift anisotropy in proteins correlate with secondary structure *J. Am. Chem. Soc.* **119** 9576–7

## Further Reading

Hennel J W and Klinowski J 1993 *Fundamentals of Nuclear Magnetic Resonance* (Harlow: Longman)

A good introductory textbook, includes a nice and detailed presentation of relaxation theory at the level of Solomon equations.

Goldman M 1988 *Quantum Description of High-Resolution NMR in Liquids* (Oxford: Clarendon)

A good introductory treatment of the density operator formalism and two-dimensional NMR spectroscopy, nice presentation of Redfield relaxation theory.

Canet D 1996 *Nuclear Magnetic Resonance Concepts and Methods* (Chicester: Wiley)

A good NMR textbook, with a nice chapter on relaxation.

Ernst R R, Bodenhausen G and Wokaun A 1987 *Principles of Nuclear Magnetic Resonance in One and Two Dimensions* (Oxford: Clarendon Press)

An extensive presentation of the modern NMR theory, standard text for 2D NMR.

Abragam A 1961 *Principles of Nuclear Magnetism* (Oxford: Clarendon)

An extensive presentation of the fundamentals of NMR. A very good chapter on relaxation, including general theory and mechanisms. A real classic 'still going strong'.

Batta Gy, Kövér K E and Szántay Cs (eds) 1997 *Methods for Structure Elucidation by High-Resolution NMR* (Amsterdam: Elsevier)

A nice collection of review articles dealing with various experimental aspects of modern NMR spectroscopy; several papers on relaxation.

Kowalewski J 1990 Nuclear spin relaxation in diamagnetic fluids Part 1 *Annu. Rep. NMR Spectrosc.* **22** 308–414; 1991 Part 2: *Annu. Rep. NMR Spectrosc.* **23** 289–374

An extensive review of the relaxation literature of the 1980s.

# B1.14
# NMR imaging (diffusion and flow)

## *Ute Goerke, Peter J McDonald and Rainer Kimmich*

### B1.14.1 Introduction

Nuclear magnetic resonance imaging and magnetic resonance determinations of flow and diffusion are increasingly coming to the fore as powerful means for characterizing dynamic processes in diverse areas of materials science covering physics, chemistry, biology and engineering. In this chapter, the basic concepts of the methods are introduced and a few selected examples presented to show the power of the techniques. Although flow and diffusion through bulk samples can be measured, they are primarily treated here as parameters to be mapped in an imaging experiment. To that end, imaging is dealt with first, followed by flow and diffusion, along with other contrast parameters such as spin-relaxation times and chemical shift.

The great diversity of applications of magnetic resonance imaging (MRI) has resulted in a plethora of techniques which at first sight can seem bemusing. However, at heart they are built on a series of common building blocks which the reader will progressively come to recognize. The discussion of imaging is focused very much on just three of these—slice selection, phase encoding and frequency encoding, which are brought together in perhaps the most common imaging experiment of all, the spin warp sequence [1–3]. This sequence is depicted in figure B1.14.1. Its building blocks are discussed in the following sections. Blocks for contrast enhancement and parameter selectivity can be added to the sequence.

### B1.14.2 Fundamentals of spatial encoding

The classical description of magnetic resonance suffices for understanding the most important concepts of magnetic resonance imaging. The description is based upon the Bloch equation, which, in the absence of relaxation, may be written as

$$\frac{dM(r, t)}{dt} = \gamma M(r, t) \times B(r, t).$$

The equation describes the manner in which the nuclear magnetization, $M$, at position $r$ and time $t$ precesses about the magnetic flux density, $B$, in which it is found. The constant $\gamma$ is the magnetogyric ratio of the nuclides under study. The precessional frequency, $\omega$, is given by the Larmor equation,

$$\omega = \gamma B. \tag{B1.14.1}$$

Magnetic resonance imaging, flow and diffusion all rely upon manipulating spatially varying magnetic fields in such a manner as to encode spatial information within the accumulated precession of the magnetization. In most MRI implementations, this is achieved with three orthogonal, constant, pulsed magnetic field gradients which are produced using purpose built current windings carrying switched currents. The gradient fields are superimposed on the normal static, applied field $B_0$. Although the current windings produce gradient fields

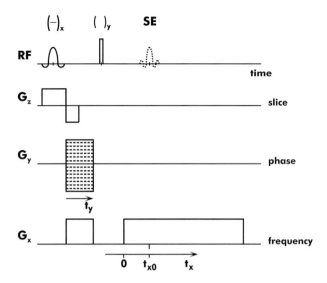

**Figure B1.14.1.** Spin warp spin-echo imaging pulse sequence. A spin echo is refocused by a non-selective 180° pulse. A slice is selected perpendicular to the $z$-direction. To frequency-encode the $x$-coordinate the echo $SE$ is acquired in the presence of the readout gradient. Phase-encoding of the $y$-dimension is achieved by incrementing the gradient pulse $G_y$.

with components in all directions, for sufficiently high applied flux densities and small enough gradients, it is sufficient to consider only the components in the direction of the applied field, conventionally assigned the $z$-direction. Accordingly, the gradients are referred to as $G_x = \frac{\partial B_z}{\partial x}$, $G_y = \frac{\partial B_z}{\partial y}$ and $G_z = \frac{\partial B_z}{\partial z}$. The local polarizing field at position $r$ is then given by

$$B(r) = (B_0 + G \cdot r)u_z.$$

The Bloch equation is simplified, and the experiment more readily understood, by transformation into a frame of reference rotating at the frequency $\omega_0 = \gamma B_0$ about the $z$-axis whereupon:

$$\frac{dM'(r, t)}{dt} = \gamma M'(r, t) \times (G \cdot r)u_z.$$

The transverse magnetization may be described in this frame by a complex variable, $m$, the real and imaginary parts of which represent the real and imaginary components of observable magnetization respectively:

$$m(r, t) = M'_x(r, t) + iM'_y(r, t).$$

The components of the Bloch equation are hence reduced to

$$\frac{dM'_z(r, t)}{dt} = 0$$

$$\frac{dm(r, t)}{dt} = -i\gamma(G \cdot r)m(r, t).$$

The $z$-component of the magnetization is constant. The evolution of the transverse magnetization is given by

$$m(r, t) = m(r, 0) \exp[-i\Omega(r)t] \qquad (B1.14.2)$$

where $\Omega(r) = \gamma G \cdot r$ is the offset frequency relative to the resonance frequency $\omega_0$. Spin position is encoded directly in the offset frequency $\Omega$. Measurement of the offset frequency forms the basis of the frequency encoding of spatial information, the first building block of MRI. It is discussed further in subsequent sections. In a given time $t = \tau$ the transverse magnetization accumulates a spatially varying phase shift, $\Omega(r)\tau$. Measurement of the phase shift forms the basis of another building block, the second encoding technique, phase encoding, which is also further discussed below. However, before proceeding with further discussion of either of these two, we turn our attention to the third key component of an imaging experiment—slice selection.

### B1.14.2.1  Slice selection

A slice is selected in NMR imaging by applying a radio frequency (RF) excitation pulse to the sample in the presence of a magnetic field gradient. This is in contrast to spatial encoding, where the magnetization following excitation is allowed to freely evolve in the presence of a gradient. A simple appreciation of how slice selection works is afforded by comparing the spread of resonance frequencies of the nuclei in the sample with the frequency bandwidth of the RF pulse. The resonance frequencies of nuclear spins in a sample placed centrally in a magnetic field, $B_0$, with a superimposed constant gradient $G$ vary linearly across the sample between $\omega_0 \pm \gamma G d_s / 2$ where $d_s$ is the dimension of the sample in the gradient direction. To a first approximation, the radio-frequency excitation pulse (of duration $t_w$ and carrier frequency $\omega_0$) contains frequency components in the range $\omega_0 \pm \pi/t_w$. If the pulse bandwidth is significantly less than the spread of frequencies within the sample, the condition for a so called *soft pulse*, then the pulse excites nuclei only in a central slice of the sample, perpendicular to the gradient, where the frequencies match. A slice at a position other than the centre of the sample can be chosen by offsetting the excitation carrier frequency, $\omega_0$.

A more detailed description of the action of an arbitrary pulse on a sample in a gradient can be obtained from a solution of the Bloch equations, either numerically or using advanced analytic techniques. In general this is complicated since the effective field in the rotating frame is composed not only of the spatially varying gradient field in the $z$-direction but also the transverse excitation field which, in general, varies with time. What follows is therefore an approximate treatment which nonetheless provides a surprisingly accurate description of many of the more commonly used slice selection pulses [4].

Suppose that a gradient, $G = G_z u_z$, is applied in the $z$-direction. In a local frame of reference rotating about the combined polarizing and gradient fields at the frequency $\omega = \omega_0 + \gamma(G \cdot r)$, an excitation pulse $B_1(t)$ applied at the central resonance frequency of $\omega_0$ and centred on $t = 0$ is seen to rotate at the offset frequency $\gamma(G \cdot r)$ and therefore to have components given by

$$B_1(t)\left\{ \cos\left[ \gamma(G \cdot r)\left(t + \frac{t_w}{2}\right)\right] u'_x + \sin\left[ \gamma(G \cdot r)\left(t + \frac{t_w}{2}\right)\right] u'_y \right\}.$$

The excitation field is the only field seen by the magnetization in the rotating frame. The magnetization precesses about it. Starting from equilibrium ($M_0 = M_0 u_z$), transverse components are created and develop according to

$$\frac{\mathrm{d}m(r,t)}{\mathrm{d}t} = i\gamma M'_z(r,t)B_1(t)\exp\left[ -i\gamma(G \cdot r)\left(t + \frac{t_w}{2}\right)\right].$$

The simplifying approximation of a linear response is now made, by which it is assumed that rotations about different axes may be decoupled. This is only strictly valid for small rotations, but is surprisingly good for larger rotations too. This means that $M'_z(r,t) \approx M_0(r) =$ constant. Accordingly, at the end of the pulse the transverse magnetization is given by

$$m\left(r, t = \frac{t_w}{2}\right) \approx i\gamma M_0(r)\exp\left[ -i\gamma(G \cdot r)\frac{t_w}{2}\right]\int_{-\frac{t_w}{2}}^{+\frac{t_w}{2}} B_1(t)\exp[-i\gamma(G \cdot r)t]\,\mathrm{d}t. \tag{B1.14.3}$$

The integral describes the spatial amplitude modulation of the excited magnetization. It represents the excitation or slice profile, $g(z)$, of the pulse in real space. As $B_1$ drops to zero for $t$ outside the pulse, the integration limits can be extended to infinity whereupon it is seen that the excitation profile is the Fourier transform of the pulse shape envelope:

$$g(z) = \gamma \int_{-\infty}^{+\infty} B_1(t) \exp(-i\gamma G_z t z) \, dt.$$

For a soft pulse with a rectangular envelope

$$B_1(t) = \begin{cases} B_1 & \text{for } -\frac{t_w}{2} \leq t \leq \frac{t_w}{2} \\ 0 & \text{otherwise} \end{cases}$$

and a carrier frequency resonant at the position $z = z_0$ the excitation profile is sinc shaped. In normalized form, it is:

$$\frac{g(z - z_0)}{g(0)} = \text{sinc}\left[\frac{1}{2\pi}(z - z_0)\gamma G_z t_w\right]$$

where

$$\text{sinc}(x) = \frac{\sin(\pi x)}{\pi x}.$$

A sinc-shape has side lobes which impair the excitation of a distinct slice. Other pulse envelopes are therefore more commonly used. Ideally, one would like a rectangular excitation profile which results from a sinc-shaped pulse with an infinite number of side lobes. In practice, a finite pulse duration is required and therefore the pulse has to be truncated, which causes oscillations in the excitation profile. Another frequently used pulse envelope is a Gaussian function:

$$B_1(t) = B_1(0) \, e^{-at^2}$$

which produces a Gaussian slice profile:

$$\frac{g(z - z_0)}{g(0)} = e^{-\Omega_{z_0}^2/(4a)}$$

where $\Omega_{z_0} = \gamma G_z(z - z_0)$ and $a$ is a pulse-width parameter. The profile is centred on $z_0$ at which position the transmitter frequency is resonant. The full width at half maximum of the slice, $\Delta z_{1/2}$, is determined by the constant:

$$\Delta z_{1/2} = \frac{4}{\gamma G_z}\sqrt{a \ln 2}.$$

For a given half width at half maximum in the time domain, $\Delta t_{1/2} = 2\sqrt{\frac{\ln 2}{a}}$, the slice width $\Delta z_{1/2}$ decreases with increasing gradient strength $G_z$.

Closer examination of equation (B1.14.3) reveals that, after the slice selection pulse, the spin isochromats at different positions in the gradient direction are not in phase. Rather they are rotated by $i \exp\left(-i\gamma G_z z \frac{t_w}{2}\right)$ and therefore have a phase $(\gamma G_z z t_w - \pi)/2$. Destructive interference of isochromats at different locations leads to cancellation of the total transverse magnetization.

The dephased isochromats may be refocused by applying one or more RF and trimming gradient pulses after the excitation. In the pulse sequence shown in figure B1.14.1 the magnetization is refocused by a negative trimming $z$-gradient pulse of the same amplitude, but half the duration as the original slice selection gradient pulse. The trimming gradient causes a precession of the magnetization at every location which exactly compensates that occurring during the initial excitation (save for the constant factor $\pi$). A similar effect may be achieved using a (non-selective) $180°$ pulse and a trimming gradient pulse of the same sign.

### B1.14.2.2 *Frequency encoding*

Once a slice has been selected and excited, it is necessary to encode the ensuing NMR signal with the coordinates of nuclei within the slice. For each coordinate ($x$ and $y$) this is achieved by one of two very closely related means, frequency encoding or phase encoding [1]. In this section we consider the former and in the next, the latter. In the section after that we show how the two are combined in the most common imaging experiment.

As before, we note that the resonance frequency of a nucleus at position $r$ is directly proportional to the combined applied static and gradient fields at that location. In a gradient $G = G_x u_x$, orthogonal to the slice selection gradient, the nuclei precess (in the usual frame rotating at $\omega_0$) at a frequency $\omega = \gamma G_x x$. The observed signal therefore contains a component at this frequency with an amplitude proportional to the local spin density. The total signal is of the form

$$S(t_x) = \int_{-\infty}^{+\infty} dx\, m(x)\, e^{-i\gamma G_x x t_x}$$

from which it is seen that the spin density in the $x$-direction is recovered by a Fourier transform of the signal with respect to time

$$m(x) \propto \int dt_x\, S(t_x) \exp(i\gamma G_x x t_x).$$

In practice, it is generally preferable to create and record the signal in the form of an echo well clear of pulse ring-down and other artifacts associated with defining the zero of time. This can be done either by first applying a negative gradient lobe followed by a positive gradient—a gradient echo—or by including a 180° inversion pulse between two positive gradients—a spin echo. Figure B1.14.1 demonstrates the spin echo. A trimming $x$-gradient of duration $t_y$ is placed before the 180° pulse which inverts the phase of the magnetization, so that, with reference to figure B1.14.1,

$$m(r, t_x) = m(r, 0)\, e^{i\gamma G_x x t_y}\, e^{-i\gamma G_x x t_x}.$$

The maximum signal appears at the echo centre when the exponent disappears for all $r$ at time $t_x = t_{x0}$. If the magnitude of the read gradient and the corresponding trimming gradient is the same, then $t_{x0} = t_y$. The magnetization profile is obtained by Fourier transforming the echo.

### B1.14.2.3 *Phase encoding*

If a gradient pulse is applied for a fixed evolution time $t_y$ the magnetization is dephased by an amount dependent on the gradient field. The signal phase immediately after the gradient varies linearly in the direction of the gradient. For a $y$-gradient $G = G_y u_y$ of duration $t_y$ as shown in figure B1.14.1 we have:

$$m(r, G_y) = m(r, 0)\, e^{-i\gamma G_y y t_y}$$

For phase encoding the phase twist is most commonly varied by incrementing $G_y$ in a series of subsequent transients as this results in a constant transverse relaxation attenuation of the signal at the measurement position. The signal intensity as a function of $G_y$ is

$$S(G_y) = \int_{-\infty}^{+\infty} dy\, m(y)\, e^{-i\gamma G_y y t_y}.$$

The magnetization profile in the $y$-direction is recovered by Fourier transformation with respect to $G_y$.

*B1.14.2.4   2D spin-echo FT imaging and k-space*

We now bring all the elements of the imaging experiment together within the typical spin-warp imaging sequence [1] previously depicted in figure B1.14.1. A soft $90°$ pulse combined with slice-selection gradient $G_z$ excites a slice in the $x/y$-plane. The spin isochromats are refocused by the negative $z$-gradient lobe. A subsequent spin echo, $SE$, is formed by a hard $180°$ pulse inserted after the slice selection pulse. The $G_x$-gradient either side of the $180°$ pulse first dephases and then rephases the magnetization so that an echo forms at twice the time separation, $\tau$, between the two pulses. The $x$-dimension is encoded by acquiring the echo in the presence of this gradient, often known as a readout gradient. The $y$-dimension is phase encoded using the gradient $G_y$ which is incremented in subsequently measured transients. The signal intensity $S$ of the echo is the superposition of all the transverse magnetization originating from the excited slice. The acquired data set is described by:

$$S(t_x, G_y) = \int_{-\infty}^{+\infty} \int_{-\infty}^{+\infty} dx\, dy\, m(x, y)\, e^{-i\gamma G_x x(t_x - t_{x0})}\, e^{i\gamma G_y y t_y}.$$

The magnetization density is recovered by a two-dimensional Fourier transform of the data with respect to $t_x$ and $G_y$.

From a more general point of view, components $k_j$, $j = x, y, z$ of a wave vector $k$ which describes the influence of all gradient pulses may be defined as follows: $k_j(t) = \int_0^t \gamma G_i(t')\, dt'$. For the 2D imaging pulse sequence discussed here, $k_x = \gamma G_x(t_x - t_{x0})$ and $k_y = -\gamma G_y t_y$ (the negative signs resulting from the inversion pulse). Spatial encoding is the sampling of $k$-space and the acquired data set is then:

$$S(k_x, k_y) = \int_{-\infty}^{+\infty} \int_{-\infty}^{+\infty} dx\, dy\, m(x, y)\, e^{-ik_x x}\, e^{-ik_y y}.$$

The image, i.e., the spatially resolved distribution of the magnetization $m(x, y)$, is reconstructed by two-dimensional Fourier transformation with respect to $k_x$ and $k_y$. In the absence of other interactions and encodings discussed below, it represents the spin density distribution of the sample.

There is of course no requirement to confine the slice selection to the $z$-gradient. The gradients may be used in any combination and an image plane selected in any orientation without recourse to rotating the sample.

Another frequently used imaging method is gradient-recalled spin-echo imaging (figure B1.14.2). In this method, the $180°$ pulse of the spin warp experiment is omitted and the first lobe of the $G_x$ gradient is instead inverted. Otherwise the experiment is the same. As the refocusing $180°$ pulse is omitted, the echo time $T_E$ can be adjusted to be shorter than in the spin-echo version. Therefore gradient-recalled echo and variants of this technique are used when samples with shorter $T_2$ are to be imaged. On the other hand, this method is more susceptible to off-resonance effects, e.g., due to chemical shift or magnetic field inhomogeneities.

The imaging methods just described require $n$ transients each containing $l$ data points to be acquired so as to construct a two-dimensional image of a matrix of $n \times l$ pixels. Since it is necessary to wait a time of the order of the spin–lattice relaxation time, $T_1$, for the spin system to recover between the collection of transients, the total imaging time is in excess of $nT_1$ for a single average. This may mean imaging times of the order of minutes. The $k$-space notation and description of imaging makes it easy to conceive of single transient imaging experiments in which, by judicious switching of the gradients so as to form multiple echoes, the whole of $k$-space can be sampled in a single transient.

Techniques of this kind go by the generic title of echo-planar imaging methods [5–8] and in the case of full three-dimensional imaging, echo-volumar imaging. A common echo-planar imaging pulse sequence— that for blipped echo-planar imaging—is shown in figure B1.14.3. Slice selection is as before. The alternate positive and negative pulses of $x$-gradient form repeated gradient echoes as $k$-space is repeatedly traversed

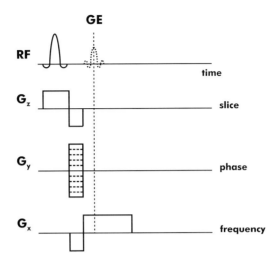

**Figure B1.14.2.** Gradient-recalled echo pulse sequence. The echo is generated by deliberately dephasing and refocusing transverse magnetization with the readout gradient. A slice is selected in the $z$-direction and $x$- and $y$-dimension are frequency and phase encoded, respectively.

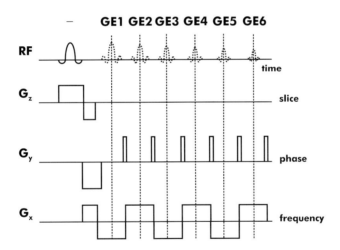

**Figure B1.14.3.** Echo-planar imaging (EPI) pulse sequence. In analogy to the pure gradient-echo recalled pulse sequence a series of echoes $GE1$–$GE6$ is refocused by alternatingly switching between positive and negative readout gradients. During the readout gradient switching a small phase-encoding gradient pulse (blip) is applied. The spatial phase encoding is hence stepped through the acquired echo train.

in the positive and negative $x$-directions. These echoes are used for frequency encoding. The initial negative pulse of $y$-gradient followed by the much smaller pulses of positive $y$-gradient ensure that each traverse in the $x$-direction is for a different value of $k_y$, starting from an extreme negative value.

In cases where it is not possible to rapidly switch the gradients, frequency-encoded profiles may be acquired in different directions by rotating the gradient and/or sample orientation between transients. The image is reconstructed using filtered back-projection algorithms [9, 10]. The two-dimensional raster of $k$-space for

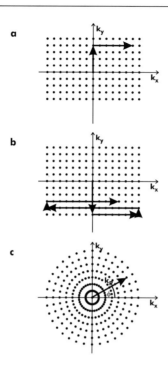

**Figure B1.14.4.** $k$-space representations of the (a) spin-echo imaging pulse sequence in figure B1.14.1, (b) echo-planar imaging sequence in figure B1.14.3 and (c) back-projection imaging. In (a) and (b) the components of the wave vectors $k_x$ and $k_y$ represent frequency and phase encoding, respectively. For back-projection (c) solely a readout gradient for frequency encoding is used. The direction of this gradient is changed stepwise from 0 to 180° in the $x/y$-plane (laboratory frame) by appropriate superposition of the $x$- and $y$-gradient. The related wave vector $k_R$ then rotates around the origin sampling the $k_x/k_y$-plane. As the sampled points are not equidistant in $k$-space, an image reconstruction algorithm different from the two-dimensional Fourier transform has to be used [9, 10].

the spin warp experiment shown in figure B1.14.1 is shown in figure B1.14.4(a) and that for the blipped echo-planar method in figure B1.14.4(b). The raster for back-projection methods is shown in figure B1.14.4(c).

*B1.14.2.5 Resolution*

The achievable spatial resolution is limited by several effects. The first is the maximum gradient strength and encoding times available. Bearing in mind equation (B1.14.1) and $t_x^{\mathrm{max}} = 1/\Delta f_{\mathrm{max}}$ the pixel size resulting from the Fourier transform of the frequency encoded echo is given by

$$\Delta x = \frac{2\pi}{\gamma G_x t_x^{\mathrm{max}}} \tag{B1.14.4}$$

where $t_x^{\mathrm{max}}$ is the total data acquisition time. The maximum useful encoding time is limited by the transverse relaxation time of the nuclei, $T_2$. Therefore the best resolution which can be achieved is of the order of [14]

$$\Delta x^{\mathrm{best}} = \frac{2\pi}{\gamma G_x T_2}. \tag{B1.14.5}$$

A similar result holds for phase encoding where the gradient strength appearing in the expression for the resolution is the maximum gradient strength available. According to these results the resolution can be improved by raising the gradient strength. However, for any spectrometer system, there is a maximum gradient strength which can be switched within a given rise time due to technical limitations. Although modern gradient coil sets are actively shielded to avoid eddy currents in the magnet, in reality systems with pulsed gradients in excess of 1 T m$^{-1}$ are rare. Since the $T_2$ of many commonly imaged, more mobile, samples is of the order of 10 ms, resolution limits in $^1$H imaging are generally in the range one to ten micrometres. At this resolution, considerable signal averaging is generally required in order to obtain a sufficient signal to noise ratio and the imaging time may extend to hours. Moreover, a slice thickness of typically 500 $\mu$m, which is significantly greater than the lateral resolution, is frequently used to improve signal to noise. In solids, $T_2$ is generally very much shorter than in soft matter and high resolution imaging is not possible without recourse either to sophisticated line narrowing techniques [11], to magic-echo refocusing variants [12] or to very high gradient methods such as stray field imaging [13].

Motion, and in particular diffusion, causes a further limit to resolution [14, 15]. First, there is a physical limitation caused by spins diffusing into adjacent voxels during the acquisition of a transient. For water containing samples at room temperature the optimal resolution on these grounds is about 5 $\mu$m. However, as will be seen in subsequent sections, diffusion of nuclei in a magnetic field gradient causes an additional attenuation of the signal in the time domain. In the presence of a steady gradient, it is $\exp(-\gamma^2 G^2 D t^3/3)$. Hence, the linewidth for spins diffusing in a gradient is of the order of

$$\Delta f_D = 0.6(\gamma^2 G^2 D/3)^{\frac{1}{3}}$$

where $D$ is the diffusion coefficient, so that the best resolution becomes in analogy to the derivation of the equations (B1.14.4) and (B1.14.5)

$$\Delta x^{\text{best}} \approx 2.6 \left( \frac{D}{\gamma G} \right)^{\frac{1}{3}}.$$

In practice, internal gradients inherent to the sample resulting from magnetic susceptibility changes at internal interfaces can dominate the applied gradients and lead to strong diffusive broadening just where image resolution is most required. Again the resolution limits tend to be on the ten micrometre scale.

## B1.14.3  Contrasts in MR imaging

Almost without exception, magnetic resonance 'images' are more than a simple reflection of nuclear spin density throughout the sample. They crudely visualize one or more magnetic resonance measurement parameters with which the NMR signal intensity is weighted. A common example of this kind is spin-relaxation-weighted image contrast. Mapping refers to encoding of the value of an NMR measurement parameter within each image pixel and thereby the creation of a map of this parameter. An example is a velocity map. The power of MRI compared to other imaging modalities is the large range of dynamic and microscopic structural contrast parameters which the method can encode, 'visualise' and map.

### B1.14.3.1  Relaxation

Transverse relaxation weighting is perhaps the most common form of contrast imposed on a magnetic resonance image. It provides a ready means of differentiating between more mobile components of the sample such as low viscous liquids which are generally characterized by long $T_2$ values of the order of seconds and less mobile components such as elastomers, fats and adsorbed liquids with shorter $T_2$ values of the order of tens of milliseconds [6]. Transverse relaxation contrast is, in fact, a natural consequence of the spin warp imaging technique described in the previous section. As already seen, data are recorded in the form of a spatially encoded spin echo. Only those nuclides in the sample with a $T_2$ of the order of, or greater than, the echo time, $T_E = 2\tau \approx 2t_y$, contribute significantly to the echo signal. Consequently, the image reflects the distribution of nuclides for which $T_2 \geq T_E$. Often, a crude distinction between two components in a sample, one more mobile than the other, is made on the basis of a single $T_2$-weighted image in which the echo time is chosen intermediate between their respective $T_2$ values. For quantitative $T_2$-mapping, images are recorded at a variety of echo times and subsequently analysed by fitting single- or multi-modal relaxation decays to the image intensity on a pixel by pixel basis. The fit parameters are then used to generate a true $T_2$-map in its own right.

An example of the application of $T_2$-weighted imaging is afforded by the imaging of the dynamics of chemical waves in the Belousov–Zhabotinsky reaction shown in figure B1.14.5 [16]. In these images, bright bands correspond to an excess of $Mn^{3+}$ ions with a long $T_2$ and dark bands to an excess of $Mn^{2+}$ ions with a short $T_2$.

Another powerful contrast parameter is spin–lattice, or $T_1$, relaxation. Spin–lattice relaxation contrast can again be used to differentiate different states of mobility within a sample. It can be encoded in several ways. The simplest is via the repetition time, $T_R$, between the different measurements used to collect the image data. If the repetition time is sufficiently long such that $T_R \gg T_1$ for all nuclei in the sample, then all nuclei will recover to thermal equilibrium between measurements and will contribute equally to the image intensity. However, if the repetition time is reduced, then those nuclei for which $T_R < T_1$ will not recover between measurements and so will not contribute to the subsequent measurement. A steady state rapidly builds up in which only those nuclei with $T_1 \ll T_R$ contribute in any significant manner. As with $T_2$-contrast, single images recorded with a carefully selected $T_R$ may be used to select crudely a short $T_1$ component of a sample.

The mathematical description of the echo intensity as a function of $T_2$ and $T_1$ for a repeated spin-echo measurement has been calculated on the basis that the signal before one measurement cycle is exactly that at the end of the previous cycle. Under steady state conditions of repeated cycles, this must therefore equal the signal at the end of the measurement cycle itself. For a spin-echo pulse sequence such as that depicted in figure B1.14.1, the echo magnetization is given by [17]

$$M_{\text{echo}} = M_0 \frac{[1 - 2\exp(-\frac{T_R - T_E/2}{T_1}) + \exp(-\frac{T_R}{T_1})]\exp(-\frac{T_E}{T_2})}{1 + \cos\theta \exp(-\frac{T_R}{T_1})} \sin\theta$$

where $M_0$ is the equilibrium magnetization and $\theta$ is the tip angle of the radio-frequency excitation pulse and where it is assumed that there is total dephasing of the magnetization between cycles. Other expressions applicable to other situations are to be found in the literature. In practice, some kind of relaxation-weighting of image contrast is always present and can hardly be avoided.

Other methods to encode $T_1$ contrast include saturation recovery [18, 19] and $T_1$ nulling techniques [20]. In the latter, a $180°$ pulse is applied some time $T_0$ before the image data acquisition. This pulse inverts the magnetization. In the interval $T_0$ the magnetization recovers according to $[1 - 2\exp(-t/T_1)]$ so that at the time of image data excitation it is $[1 - 2\exp(-T_0/T_1)]$. Nuclei for which $T_0 = 0.693T_1$ have zero magnetization at this time and so do not contribute to the signal intensity. This method may be used to suppress a large background component in an image, such as that due to bulk water. With saturation recovery, a train of radio-frequency pulses is applied to the sample some time $T_{SR}$ prior to the data acquisition sequence. The train of pulses, which are often applied with ever decreasing spacing between them, serves to saturate the equilibrium magnetization. Only those nuclides for which $T_{SR} < T_1$ recover to equilibrium prior to the image acquisition proper and so only these nuclei contribute to the image intensity. Multiple images recorded as a function of $T_{SR}$ may be used to build a true $T_1$-map by fitting a relaxation recovery curve to the data on a pixel by pixel basis. An example of brine in a sandstone core is depicted in figure B1.14.6 [21]. Both $M_0$- and $T_1$-maps, which were obtained from fitting with a stretched exponential, clearly show layers in the stone. The spin–lattice relaxation presumably correlates with spatially varying pore sizes and surface relaxivity.

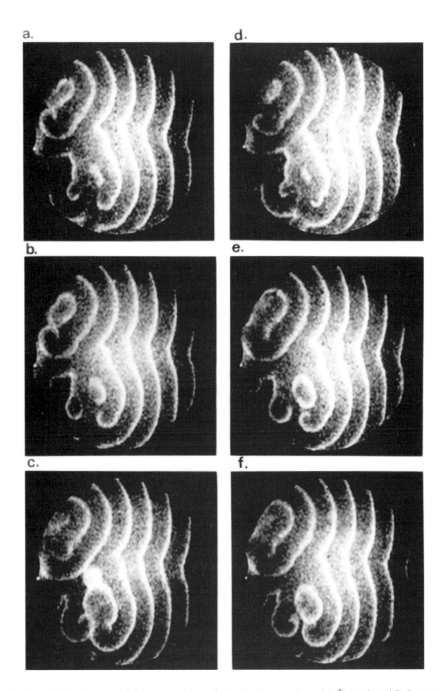

**Figure B1.14.5.** $T_2$-weighted images of the propagation of chemical waves in an $Mn^{2+}$ catalysed Belousov–Zhabotinsky reaction. The images were acquired in 40 s intervals (a) to (f) using a standard spin echo pulse sequence. The slice thickness is 2 mm. The diameter of the imaged pill box is 39 mm. The bright bands correspond to an excess of $Mn^{3+}$ ions with long $T_2$, dark bands to an excess of $Mn^{2+}$ with a short $T_2$. (From [16]).

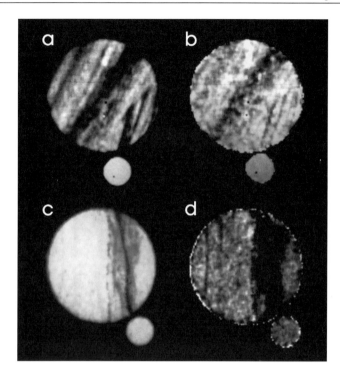

**Figure B1.14.6.** $T_1$-maps of a sandstone reservoir core which was soaked in brine. (a), (b) and (c), (d) represent two different positions in the core. For $T_1$-contrast a saturation pulse train was applied before a standard spin-echo imaging pulse sequence. A full $T_1$-relaxation recovery curve for each voxel was obtained by incrementing the delay between pulse train and imaging sequence. $M_0$- ((a) and (c)) and $T_1$-maps ((b) and (d)) were calculated from stretched exponentials which are fitted to the magnetization recovery curves. The maps show the layered structure of the sample. Presumably $T_1$-relaxation varies spatially due to inhomogeneous size distribution as well as surface relaxivity of the pores. (From [21].)

### B1.14.3.2  *Chemically resolved imaging*

In many instances, it is important that some form of chemical selectivity be applied in magnetic resonance imaging so as to distinguish nuclei in one or more specific molecular environment(s). There are many ways of doing this and we discuss here just three. The first option is to ensure that one of the excitation RF pulses is a narrow bandwidth, frequency selective pulse applied in the absence of any gradient [22]. Such a pulse can be made specific to one particular value of the chemical shift and thereby affects only nuclei with that chemical shift. In practice this can be a reasonable method for the specific selection of fat or oil or water in a mixed hydocarbon/water system.

A higher level of sophistication involves obtaining a full chemical shift spectrum within each image pixel. The chemical information can be encoded either before or after the image encoding. The important requirement is that the spin system is allowed to evolve without gradients and that data are recorded as a function of this chemical shift evolution time. The data can then be Fourier transformed with respect to the evolution time as well as the standard imaging variables so as to yield the spatially resolved chemical shift spectrum. In this respect chemical shift imaging is like a four-dimensional imaging experiment [23]. A post-encoding sequence suitable for this purpose is shown in figure B1.14.7. The chemical shift encoding part—a spin echo in the absence of gradients—comes after the image encoding part—a two-dimensional

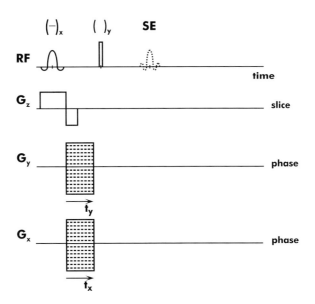

**Figure B1.14.7.** Chemical shift imaging sequence [23]. Both $x$- and $y$-dimensions are phase encoded. Since line-broadening due to acquiring the echo in the presence of a magnetic field gradient is avoided, chemical shift information is retained in the echo.

phase encoding experiment. The $180°$ pulse refocuses all chemical-shift-induced dephasing occurring during the spatial encoding.

The third alternative is a more robust, sensitive and specialized form of the first, in that only hydrogen nuclei indirectly spin–spin coupled to $^{13}$C in a specific molecular configuration are imaged. In achieving selectivity, the technique exploits the much wider chemical shift dispersion of $^{13}$C compared to $^1$H. The method involves cyclic transfer from selected $^1$H nuclei to indirectly spin–spin coupled $^{13}$C nuclei and back according to the sequence

$$^1\text{H} \rightarrow {}^{13}\text{C} \rightarrow {}^1\text{H}$$

Called CYCLCROP (cyclic cross polarization) [24], the method works by first exciting all $^1$H magnetization. Cross polarization pulses are then applied at the specific Larmor frequencies of the $^1$H–$^{13}$C pair of interest so as to transfer coherence from $^1$H to $^{13}$C. The transfer pulses must satisfy the Hartmann–Hahn condition

$$\gamma_C B_{1C} = \gamma_H B_{1H}$$

$B_{1H}$ and $B_{1C}$ are the excitation magnetic field strengths and must be applied for a time of the order of $1/J$ where $J$ is the spin–spin coupling constant. Following magnetization transfer, the $^{13}$C magnetization is stored along the $z$-axis and the remaining $^1$H magnetization from other molecular groupings is destroyed by a series of pulses and homospoil gradients. The stored magnetization specific to the coupled $^1$H–$^{13}$C pair is then returned to the $^1$H by a second pair of cross polarization pulses. CYCLCROP chemical selective excitation may replace the initial excitation in a standard 2DFT imaging experiment with the slice selection moved to the $180°$ pulse in order to yield a $^{13}$C edited image. A number of pulse schemes for the cross coupling are known, each with various advantages in terms of low radio-frequency power deposition, tolerance of pulse artifact, breadth of the spectral bandwidth etc.

In figure B1.14.8, CYCLCROP has been used to map $^{13}$C labelled sucrose in the stem of a castor bean seedling [25]. Its arrival and accumulation are visualized in a series of subsequently acquired $^{13}$C-selective images.

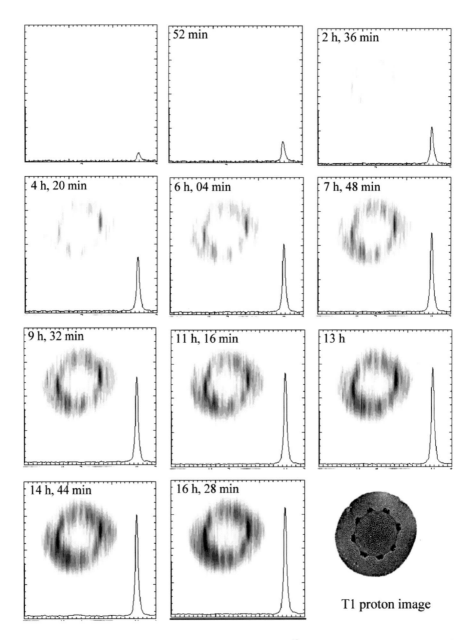

**Figure B1.14.8.** Time course study of the arrival and accumulation of $^{13}$C labelled sucrose in the stem of a castor bean seedling. The labelled tracer was chemically, selectively edited using CYCLCROP (cyclic cross polarization). The first image in the upper left corner was taken before the incubation of the seedling with enriched hexoses. The time given in each image represents the time elapsed between the start of the incubation and the acquisition. The spectrum in the lower right corner of each image shows the total intensity of $^{13}$C nuclei. At later times, enriched sucrose is visible in the periphery of the stem. Especially high intensities are detected in the vascular bundles. The last image represents a micrograph of the stem structure showing the position of the vascular bundles (dark features). (From [25]).

### B1.14.4   Flow and diffusion

NMR is an important technique for the study of flow and diffusion, since the measurement may be made highly sensitive to motion without in any way influencing the motion under study. In analogy to many non-NMR-methods, mass transport can be visualized by imaging the distribution of magnetic tracers as a function of time. Tracers may include paramagnetic contrast agents which, in particular, reduce the transverse relaxation time of neighbouring nuclei and therefore appear as $T_2$-contrast in an image. The $^{13}C$ cross polarization method with enriched compounds may also be used as a tracer experiment. More sophisticated tracer methods include so called 'tagging' experiments in which the excitation of nuclei is spatially selective. The spatial evolution of the selected nuclei is followed. This example is discussed in section B1.14.4.5.

Generally, however, the application of tracer methods remains a rarity compared to methods which directly exploit the motion sensitivity of the NMR signal. The detection of motion is based on the sensitivity of the signal phase to translational movements of nuclei in the presence of magnetic field gradients [26, 27]. Using the large magnetic field gradient in the stray field of superconducting magnets, displacements as small as 10 nm in slowly diffusing polymer melts can be detected. At the other extreme, velocities of the order of m s$^{-1}$, such as occur in blood-filled arteries, can be measured.

#### B1.14.4.1   Coherent and stationary flow

The displacement of a spin can be encoded in a manner very similar to that used for the phase encoding of spatial information [28–30]. Consider a spin $j$ with position $r(t)$ moving in a magnetic field gradient $G$. The accumulated phase, $\varphi_j$, of the spin at time $t$ is given by

$$\varphi_j(t) = -\gamma \int_0^t G(t') \cdot r_j(t') \, dt'. \tag{B1.14.6}$$

In order to encode displacement as opposed to average position, the gradient is applied in such a manner as to ensure that $\int_0^t G(t') \, dt' = 0$. Generally, this means applying gradient pulses in bipolar pairs or applying uni-modal gradient pairs either side of a 180° RF inversion pulse, with the advantage that the necessary careful balancing of the gradient amplitudes is more straightforward.

We first examine how this works for the case of coherent flow. A typical pulse sequence is shown in figure B1.14.9. This sequence creates a spin echo using two unipolar gradient pulses on either side of a 180° pulse. The duration of each gradient pulse of strength $G_v$ is $\delta$. The centres of the gradient pulses are separated by $\Delta$.

Under steady-state flow conditions (coherent motion), a Taylor series can be applied to describe the time-dependent position of the fluid molecules:

$$r(t) = r_0 + vt + \tfrac{1}{2}at^2 + \cdots.$$

If terms of higher order than linear in $t$ are neglected, the transverse magnetization evolves in the presence of the first bipolar gradient pulse according to (equations (B1.14.2) and (B1.14.6)):

$$m(t) = m(0)\,e^{-i\gamma G_v \cdot r_0 t - i\frac{1}{2}\gamma G_v \cdot v t^2} \qquad 0 \le t \le \delta.$$

The phase of the transverse magnetization is inverted by the 180° pulse and the magnetization after the second gradient pulse and therefore at the echo centre is:

$$m(\delta, \Delta) = m(0)\,e^{+\gamma G_v \cdot r_0 \delta + \gamma \frac{1}{2} G_v \cdot v \delta^2}\,e^{-\gamma G_v \cdot r_0[\delta+\Delta]-\gamma G_v \cdot v[(\delta+\Delta)^2-\Delta^2]}$$
$$= m(0)\,e^{-i\gamma G_v \cdot v \delta \Delta}.$$

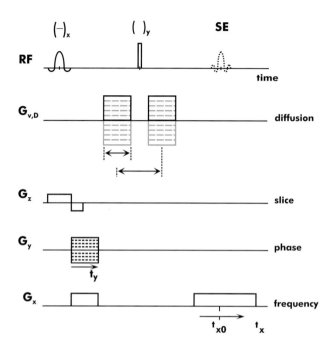

**Figure B1.14.9.** Imaging pulse sequence including flow and/or diffusion encoding. Gradient pulses $G_{v,D}$ before and after the inversion pulse are supplemented in any of the spatial dimensions of the standard spin-echo imaging sequence. Motion weighting is achieved by switching a strong gradient pulse pair $G_{v,D}$ (see solid black line). The steady-state distribution of flow (coherent motion) as well as diffusion (spatially incoherent motion) in a sample is encoded by incrementing $G_{v,D}$ (see dashed grey lines). The measured data set then consists of two spatial and a motion-encoded dimension. Velocity and/or diffusion maps can be rendered by three-dimensional Fourier transformation.

The echo phase does not depend on the initial position of the nuclei, only on their displacement, $v\Delta$, occurring in the interval between the gradient pulses. Analysis of the phase of the echo yields a measure of flow velocity in a bulk sample. Spatial resolution is easily obtained by the incorporation of additional imaging gradients. One way of doing this is illustrated in figure B1.14.9. The first part of the experiment is the same flow encoding experiment as just discussed. The velocity-encoded echo is the excitation for the subsequent 2DFT experiment which is as previously discussed. Where both stationary and moving spins are present, these superimpose in the image. A variety of methods exist for separating the two, including cycling the phase of the velocity encoding gradients or making measurements at two or more strengths of the velocity encoding gradient. In the latter, a wave number $k_v = \gamma G_v \cdot v$ can be defined adding an additional dimension to spatial $k$-space. Fourier transformation of this dimension directly produces the velocity spectrum of each voxel. As an example, velocity maps of flow through an extruder are shown in figure B1.14.10 [31].

### B1.14.4.2 Pulsating motion

Flow which fluctuates with time, such as pulsating flow in arteries, is more difficult to experimentally quantify than steady-state motion because phase encoding of spatial coordinate(s) and/or velocity requires the acquisition of a series of transients. Then a different velocity is detected in each transient. Hence the phase-twist caused by the motion in the presence of magnetic field gradients varies from transient to transient. However if the motion is periodic, e.g., $v(r, t) = v_0 \sin\{\omega_v t + \phi_0\}$ with a spatially varying amplitude $v_0 = v_0(r)$,

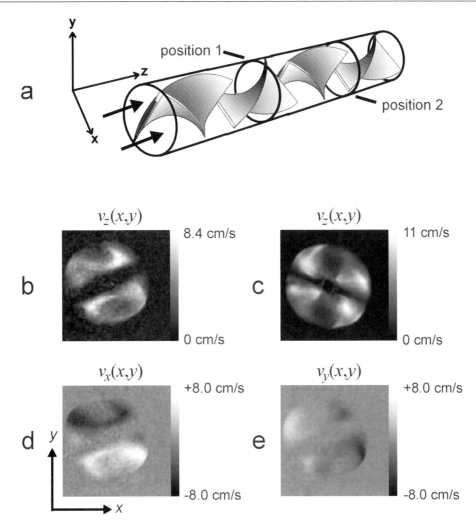

**Figure B1.14.10.** Flow through an KENICS mixer. (a) A schematic drawing of the KENICS mixer in which the slices selected for the experiment are marked. The arrows indicate the flow direction. Maps of the $z$-component of the velocity at position 1 and position 2 are displayed in (b) and (c), respectively. (d) and (e) Maps of the $x$- and the $y$-velocity component at position 1. The FOV (field of view) is 10 mm. (From [31].)

a pulsation frequency $\omega_v = \omega_v(r)$ and an arbitrary phase $\phi_0$, the phase modulation of the acquired data set is described as follows:

$$S(k_x, k_y, k_v) = \int_{-\infty}^{+\infty} \int_{\infty}^{+\infty} \int_{-\infty}^{+\infty} \mathrm{d}x\,\mathrm{d}y\,\mathrm{d}v\, m(x, y, v)\, \mathrm{e}^{-\mathrm{i}k_x x - \mathrm{i}k_y y + \mathrm{i}k_v v_0 \sin(\omega_v l_P T_R + \phi_0)}$$

where $t = l_P T_R$ is the time at which the $l_P$th transient is acquired. Since $k_y$ and $k_v$ are 'incremented' for phase encoding in subsequent transients, they are linked to the phase twist caused by $v(t)$ via the parameter $l_P$. The reconstruction of the dimensions $y$ and $v$ by Fourier transformation is therefore affected, making an interpretation difficult.

Nevertheless, averaging provides information about the motional parameters such as velocity amplitude and pulsation frequency. If the repetition time, the delay between subsequent transients, is not equal to a

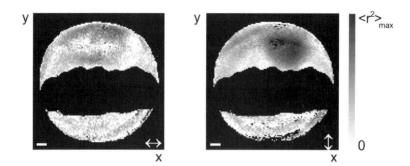

**Figure B1.14.11.** Amplitude-weighted images of (temporally) uncorrelated motions in a quail egg at an incubation period at about 140 h. A standard gradient echo sequence supplemented with strong bipolar gradient pulses for motion weighting was used. A high number of transients ($N_t = 490$) was acquired for each phase-encoding step to adequately average out temporal fluctuation of the motion. The intensity in the images shown corresponds to the signal ratio with and without motion weighting. Light grey shades hence represent no signal attenuation, darker shades strong signal attenuation due to uncorrelated motion. Pixels with signal below the noise level are set black as is the case in the egg yolk (black region in the middle of the egg) due to comparatively short $T_2$. The white double arrows indicate the probed velocity component. Both images show signal attenuation due to strong motion in the region above the egg yolk where the embryo presumably is located. Furthermore, the attenuation of the signal appears to be much stronger for the $y$-velocity component than for the $x$-component indicating strongly anisotropic motion. The white bar represents 2 mm. (From [32].)

multiple of the pulsation period the motion appears to be temporally uncorrelated. In this case, a temporal average over all velocity values is obtained by accumulating a sufficient number of transients. This causes a broad phase distribution resulting in signal attenuation similar to the one caused by diffusion, a spatially incoherent phenomenon. The images provide quantitative information about the distribution of motion and velocity amplitude. The temporal characteristics of the pulsation are detected by omitting phase encoding. Using a gradient pulse pair of constant magnitude for motion weighting the signal phase is then solely a function of the velocity. Since the $y$-dimension is no longer encoded, the observed velocity distribution is an average relative to this coordinate over spatially incoherent motion. The pulsation rate can be determined from the intensity modulation due to the motion in a series of transients.

Figure B1.14.11 shows the application of averaging techniques for the characterization of pulsating motions which start on about the fourth incubation day in quail eggs [32]. Dark areas in the incoherent motion-weighted images in figure B1.14.11 and represent strong motion, white no uncorrelated motion. They are localized at the suspected position of the embryo above the egg yolk which is the black region in the middle of the egg. The signal attenuation strongly depends on the probed velocity component indicating the anisotropic nature of the motion. In figure B1.14.12(a) and (b), a time series of profiles through the region of motion with spatially incoherent motion weighting were acquired before and after the start of pulsations. At the later incubation stage the modulation of the signal intensity due to temporally periodic motion is clearly visible. Fourier transformation (figure B1.14.12(d)) reveals a pulsation frequency of about 0.4 Hz.

### B1.14.4.3  Diffusion and pseudo-diffusion

If magnitude and/or direction of velocity vary on a length scale below spatial resolution, the detected motion is incoherent. (Self-) diffusion certainly falls into this category, but randomly oriented flow is also spatially incoherent. A well known example is the blood flow through brain capillaries, which are smaller than a voxel [33]. Incoherent flow is often referred to as pseudo-diffusion. An apparent diffusion coefficient which

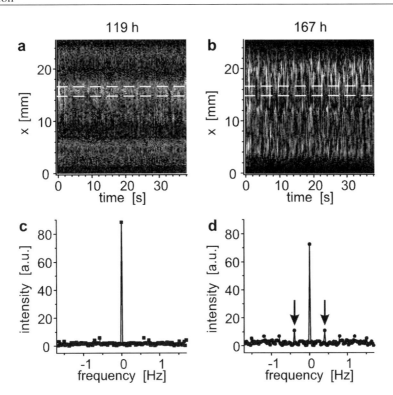

**Figure B1.14.12.** Study of the temporal fluctuation of motion in a quail egg at the incubation times 119 h and 167 h. Spatial phase encoding of the $y$-dimension was omitted to increase the rapidity of the imaging experiment. Profiles were obtained by 1D Fourier transformation of echoes which were acquired in the presence of a readout gradient for frequency encoding of the $x$-coordinate. A strong gradient pulse pair for (spatially) incoherent motion weighting was applied during the evolution period of the magnetization. A series of subsequent single-scan profiles were measured at the two different incubation times 119 h (a) and 167 h (c). Temporal fluctuations of the signal intensity which are not visible in (a) reveal themselves at the later incubation stage. The intensity modulation which was caused by temporally fluctuating motion was analysed by Fourier transformation. The spectra which were calculated from the integral intensities of the pixels between the two dashed lines in (a) and (c) are displayed in (b) and (d) for the two incubation times. The line at 0 Hz is due to the constant baseline offset. The arrows in (d) mark the peak representing a frequency of 0.4 Hz of the pulsating motion. (From [32].)

can be significantly bigger than the self-diffusion coefficient is then defined. Pulse sequences to measure coherent flow (figure B1.14.9) can also be used for (spatially) incoherent motion although the theory has to be reconsidered at this point [34–36].

The observed magnetization in the echo is the superposition of all the spins $j$ at different positions $r_j$:

$$m(T_E) = \langle m(0)\, e^{i\gamma G_D \cdot (r_j(\Delta) - r_j(0))\delta} \rangle_j$$
$$= \langle m(0)\, e^{i\varphi_j(t)} \rangle_j.$$

It is expanded into:

$$m(T_E) = m(0) \sum_j P(\varphi_j)\, e^{i\varphi_j(t)}$$

where $P(\varphi_j)$ is the probability of phase $\varphi_j$. The underlying motional process influences this phase distribution.

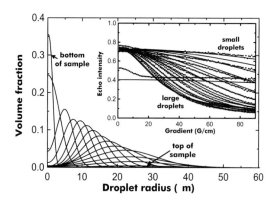

**Figure B1.14.13.** Derivation of the droplet size distribution in a cream layer of a decane/water emulsion from PGSE data. The inset shows the signal attenuation as a function of the gradient strength for diffusion weighting recorded at each position (top trace = bottom of cream). A Stokes-based velocity model (solid lines) was fitted to the experimental data (solid circles). The curious horizontal trace in the centre of the plot is due to partial volume filling at the water/cream interface. The droplet size distribution of the emulsion was calculated as a function of height from these NMR data. The most intense narrowest distribution occurs at the base of the cream and the curves proceed logically up through the cream in steps of 0.041 cm. It is concluded from these data that the biggest droplets are found at the top and the smallest at the bottom of the cream. (From [45].)

To demonstrate the principle, the simple case of normal isotropic diffusion will be discussed [27]. The solution of Fick's diffusion equation together with the central limit theorem implies that, for a constant gradient, a Gaussian phase distribution function with a mean squared phase twist $\overline{\varphi^2}(t)$ applies at any instant

$$P(\varphi) = \frac{1}{(2\pi\overline{\varphi^2})^{1/2}} \exp(-\varphi^2/\overline{\varphi^2})$$

leading to the transverse magnetization

$$m(t) = m(0)\,\mathrm{e}^{-\overline{\varphi^2(t)}/2}. \tag{B1.14.7}$$

The relationship between mean squared phase shift and mean squared displacement can be modelled in a simple way as follows: This motion is mediated by small, random jumps in position occurring with a mean interval $\tau_j$. If the jump size in the gradient direction is $\epsilon$, then after $n$ jumps at time $t = n\tau_j$, the displacement of a spin is

$$E(n\tau_j) = \sum_{i=1}^{n} \epsilon a_i$$

where $a_i$ is a randomly either $+1$ or $-1$. Hence, from the relation $\varphi = \gamma G_D \delta E(n\tau_j)$, the phase shift distribution imposed by the diffusion measurement gradients $G_D$ of duration $\delta$ is

$$\overline{\varphi^2} = \gamma^2 \delta^2 G_D^2 \epsilon^2 \overline{\left(\sum_{i=1}^{n} a_i\right)^2}.$$

The summation averages to $n$. Using the definition of the diffusion coefficient, $D = \epsilon^2/(2\tau_j)$, and the diffusion time, $\Delta = n\tau_j$, equation (B1.14.7) gives

$$m(t) = m(0)\exp(-\gamma^2\delta^2 G_D^2 D\Delta).$$

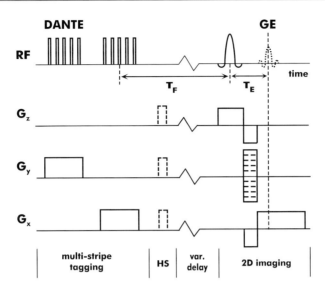

**Figure B1.14.14.** Pulse sequence for multi-plane tagging. A magnetization saturation grid is prepared in the multi-plane tagging section. After a certain time of flight $T_F$ this grid is imaged using a standard imaging pulse sequence. The motion of tagged spins is visualized by displacements of the grid lines.

This is the factor by which the echo magnetization is attenuated as a result of diffusion. More elaborate calculations, which account for phase displacements due to diffusion occurring during the application of the gradient pulses yield

$$m(t) = m(0) \exp[-\gamma^2 \delta^2 G_D^2 D(\Delta - \delta/3)].$$

This expression can be used for pulsed field gradient spin-echo experiments and also for spin-echo experiments in which the gradient is applied continuously.

A measure of the echo attenuation within each pixel of an image created using the pulse sequence of figure B1.14.9, perhaps by repeating the experiment with different values of $G_D$ and/or $\delta$, gives data from which a true diffusion map can be constructed [37, 38].

In principle, it is possible to measure both flow and diffusion in a single experiment: the echo is both phase shifted and attenuated. It is also possible to account for the presence of background gradients arising from the sample itself which can be significant. The exact form of the echo in these circumstances has been calculated and the results are to be found in the literature [39–42]. Furthermore, in systems in which $T_2$ is relatively short compared to $T_1$, the stimulated echo comprising three $90°$ pulses can be used instead of a $90°$–$t$–$180°$ spin-echo sequence [43]. In this case, the fact that the velocity encoding time, $\Delta$, can be made significantly longer outweighs the fact that only half of the magnetization is refocused after the third pulse.

### B1.14.4.4 Restricted diffusion pulsed field gradient microscopy

Diffusometry and spatially resolved magnetic resonance are usefully combined in an alternate technique to imaging which is increasingly coming to the fore. It has been dubbed both $q$-space microscopy [44] and restricted diffusion pulsed field gradient (PFG) microscopy. The method probes the microstructure of a sample on the micrometre scale—significantly smaller than conventional MRI permits—by measuring the effects of restricted diffusion of a translationally mobile species. The technique yields parameters characteristic of the average structure of the whole sample (the experiment done without spatial resolution) or of the sample within

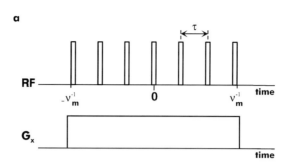

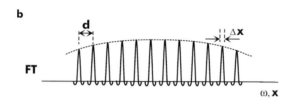

**Figure B1.14.15.** Preparation of a magnetization grid by means of DANTE pulse trains. The pulse sequence for one-dimensional tagging is shown in (a). As depicted in (b) the spectrum of a pulse train consists of a comb of peaks. To tag spins in certain regions of the sample the DANTE pulse train is applied while a magnetic field gradient $G_x$ is switched. The magnetization is then saturated in planes which are located at equidistant positions corresponding to the spectral peaks of the DANTE pulse train and which are orthogonal to the spatial coordinate $x = \omega/(\gamma G_x)$. A magnetization saturation grid is prepared by subsequently using the sequence (a) with magnetic field gradients in the two directions of the imaging plane coordinates. Saturated spins (bearing transverse magnetization) are visualized as black grid lines in an image. These lines of thickness $\Delta x = 2\pi v_m/(\gamma G_x)$ are separated by $d = 2\pi/(\gamma G_x \tau)$ where $\tau$ is the delay from one RF pulse to the next.

an image pixel (sub-millimetre scale) if the experiment is done in combination with conventional imaging. The method works because of a Fourier relationship which exists between the observed echo attenuation in a pulsed field gradient diffusometry experiment and the propagator, which describes the molecular motion $P(r; r', t)$. $P(r; r', t)$ is the conditional probability of finding a diffusing spin at location $r'$ at time $t$ given its initial location, $r$. This propagator depends intimately on the microstructure of the sample. Following on from the above analysis, the echo attenuation is

$$E(G_D, \delta, \Delta) = \int_r \rho(r) \int_{r'} P(r; r', \Delta) \exp[i\gamma\delta G_D \cdot (r - r')] \, dr' \, dr$$

where $\rho(r)$ is the spin density defined by the microstructure. In the long timescale limit, the diffusing spins sample the whole of the available space and $P(r; r', t)$ becomes independent of $r$ such that

$$P(r; r', \infty) = \rho(r') = \rho(r)$$

then

$$E(G_D, \delta, \infty) = \left| \int_r \rho(r) \exp[i\gamma\delta(G_D \cdot r)] \, dr \right|^2 = |S(q)|^2$$

where $q = (2\pi)^{-1}\gamma\delta G_D$ and $S(q)$ is a structure factor, defined by the above expression with direct analogies in optics and neutron scattering. Measurement of echo attenuation and hence $S(q)$ and calculations of microstructure have been reported for both model and real systems including porous media and emulsions.

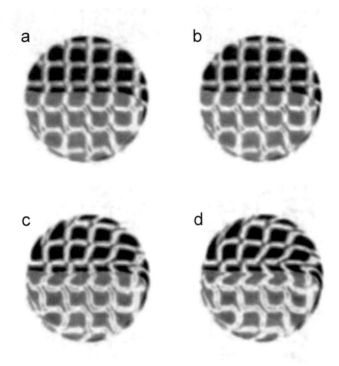

**Figure B1.14.16.** Multi-plane tagging experiment for a 1:1 water (bottom) and oil (top) mixture in the cylinder, which rotates clockwise. The rotation rate was 0.5 rev s$^{-1}$, and the tagging delay times $T_F$ are (a) 1 ms, (b) 25 ms, (c) 50 ms and (d) 100 ms. The interface between the fluids is clearly shown. The misalignment in the horizontal direction at the interface is caused by chemical shift between water and oil. Flow was mainly detected in a thin layer near the cylinder and a layer along the water–oil interface in the centre that flows to the right. (From [49].)

As an example figure B1.14.13 shows the droplet size distribution of oil drops in the cream layer of a decane-in-water emulsion as determined by PFG [45]. Each curve represents the distribution at a different height in the cream with large drops at the top of the cream. The inset shows the PFG echo decay trains as a function of height and the curves to which the data were fitted using a Stokes-velocity-based model of the creaming process.

*B1.14.4.5 Planar tagging*

A more qualitative means of visualising flows is by 'multi-plane tagging' [46, 47]. The principle of tagging experiments is to excite magnetic resonance only in stripes across the image plane and to observe the spatial evolution of these stripes with time (figure B1.14.14). The excitation can be achieved using a comb of radio-frequency pulses each of narrow flip angle applied to a sample in a magnetic field gradient (figure B1.14.15(a)) [48]. The Fourier transform of this excitation shows distinct maxima occurring at frequency intervals given by the reciprocal of the pulse spacing. The overall excitation bandwidth—which must be sufficient to cover the frequency spread of all nuclei in the sample—is determined by the individual pulse widths, and the sharpness of the maxima by the number of pulses. This frequency response is illustrated in figure B1.14.15(b) from which it is clear how the excitation is achieved. The action of the comb can be understood more qualitatively as follows. Each pulse tips all spins by a small angle. Between the pulses the spins precess by an amount dependent on their position in the gradient. Only at positions where the precessional phases between pulses is equal to $2n\pi$ do the pulses have a cumulative effect in tipping the magnetization through a large angle.

Following excitation of this kind, a standard spin warp imaging protocol can be used to create the image. By varying the delay between excitation and image acquisition, flow is visualized by different degrees of distortions of the saturated magnetization grid. An example of a rotating cylinder filled with water–oil mixture is shown in figure B1.14.16 [49].

## References

[1] Edelstein W A, Hutchison J M S, Johnson G and Redpath T W 1980 Spin warp NMR imaging and applications to human whole-body imaging *Phys. Med. Biol.* **25** 751–6

[2] Johnson G, Hutchison J M S, Redpath T W and Eastwood L M 1983 Improvements in performance time for simultaneous three-dimensional NMR imaging *J. Magn. Reson.* **54** 374–84

[3] Morris P G 1986 *Nuclear Magnetic Resonance Imaging in Medicine and Biology* (Oxford: Clarendon)

[4] Bailes D R and Bryant D J 1984 NMR Imaging *Contemp. Phys.* **25** 441–75

[5] Mansfield P 1977 Multi-planar image formation using NMR spin echoes *J. Phys. C: Solid State Phys.* **10** L55–L58

[6] Mansfield P and Morris P G 1982 NMR imaging in biomedicine (New York: Academic)

[7] Stehling M K, Turner R and Mansfield P 1991 Echo-planar imaging: magnetic-resonance-imaging in a fraction of a second *Science* **254** 43–50

[8] Mansfield P and Pykett I L 1978 Biological and medical imaging by NMR *J. Magn. Reson.* **29** 355–73

[9] Lauterbur P C 1973 Image formation by induced local interactions: examples employing nuclear magnetic resonance *Nature* **242** 190–1

[10] Lauterbur P C 1974 Magnetic resonance zeugmatography *Pure Appl. Chem.* **40** 149–57

[11] Smith M E and Strange J H 1996 NMR techniques in material physics: a review *Meas. Sci. Technol.* **7** 449–75

[12] Hafner S, Demco D E and Kimmich R 1996 Magic echoes and NMR imaging of solids *Solid State Nucl. Magn. Reson.* **6** 275–93

[13] McDonald P J 1997 Stray field magnetic resonance imaging *Prog. Nucl. Magn. Reson. Spectrosc.* **30** 69–99

[14] Callaghan P T 1993 *Principles of Nuclear Magnetic Resonance Microscopy* (Oxford: Clarendon)

[15] Ahn C B and Cho Z H 1989 A generalized formulation of diffusion effects in $\mu$m resolution nuclear magnetic-resonance imaging *Med. Phys.* **16** 22–8

[16] Tzalmona A, Armstrong R L, Menzinger M, Cross A and Lemaire C 1990 Detection of chemical waves by magnetic resonance imaging *Chem. Phys. Lett.* **174** 199–202

[17] Wehrli F W, MacFall J R, Shutts D, Breyer R and Herfkens R J 1984 Mechanisms of contrast in NMR imaging *J. Comput. Assist. Tomogr.* **8** 369–80

[18] Osment P A, Packer K J, Taylor M J, Attard J J, Carpenter T A, Hall L D, Harrod N J and Doran S J 1990 NMR imaging of fluids in porous solids *Phil. Trans. R. Soc.* A **333** 441–52

[19] Attard J J, Carpenter T A, Hall L D, Davies S, Taylor M J and Packer K J 1991 Spatially resolved $T_1$ relaxation measurements in reservoir cores *Magn. Reson. Imaging* **9** 815–19

[20] Balcom B J, Fischer A E, Carpenter T A and Hall L D 1993 Diffusion in aqueous gels—mutual diffusion-coefficients measured by one-dimensional nuclear-magnetic-resonance imaging *J. Am. Chem. Soc.* **115** 3300–305

[21] Attard J J, Doran S J, Herrod N J, Carpenter T A and Hall L D 1992 Quantitative NMR spin–lattice-relaxation imaging of brine in sandstone reservoir cores *J. Magn. Reson.* **96** 514–25

[22] Haase A and Frahm J 1985 Multiple chemical-shift-selective NMR imaging using stimulated echoes *J. Magn. Reson.* **64** 94–102

[23] Maudsley A A, Hilal S K, Perman W H and Simon H E 1983 Spatially resolved high-resolution spectroscopy by 4-dimensional NMR *J. Magn. Reson.* **51** 147–52

[24] Kunze C and Kimmich R 1994 Proton-detected $^{13}$C imaging using cyclic-J cross polarization *Magn. Reson. Imaging* **12** 805–10

[25] Heidenreich M, Spyros A, Köckenberger W, Chandrakumar N, Bowtell R and Kimmich R 1998 CYCLCROP mapping of $^{13}$C labelled compounds: perspectives in polymer sciences and plant physiology *Spatially Resolved Magnetic Resonance, Proc. 4th Int. Conf. on Magnetic Resonance Microscopy and Macroscopy* ed P Blümler, B Blümich, R E Botto and E Fukushima (Weinheim: Wiley–VCH) pp 21–52

[26] Hahn E L 1950 Spin Echoes *Phys. Rev.* **80** 580–94

[27] Carr H Y and Purcell E M 1954 Effects of diffusion on free precession in nuclear magnetic resonance experiments *Phys. Rev.* **94** 630–8

[28] Moran P R 1982 A flow velocity zeugmatographic interlace for NMR imaging in humans *Magn. Reson. Imaging* **1** 197–203

[29] Redpath T W, Norris D G, Jones R A and Hutchison J M S 1984 A new method of NMR flow imaging *Phys. Med. Biol.* **29** 891–5

[30] Bryant D J, Payne J A, Firmin D N and Longmore D B 1984 Measurement of flow with NMR imaging using a gradient pulse and phase difference technique *J. Comput. Assist. Tomogr.* **8** 588–93

[31] Rombach K, Laukemper-Ostendorf and Blümler P 1998 Applications of NMR flow imaging in materials science *Spatially Resolved Magnetic Resonance, Proc. 4th Int. Conf. on Magnetic Resonance Microscopy and Macroscopy* ed P Blümler, B Blümich, R E Botto and E Fukushima (Weinheim: Wiley–VCH) pp 517–29

[32] Goerke U and Kimmich R 1998 NMR-imaging techniques for quantitative characterization of periodic motions: 'incoherent averaging' and 'spectral side band analysis' *Spatially Resolved Magnetic Resonance, Proc. 4th Int. Conf. on Magnetic Resonance Microscopy and Macroscopy* ed P Blümler, B Blümich, R E Botto and E Fukushima (Weinheim: Wiley–VCH) pp 499–506

[33] Le Bihan D, Breton E, Lallemand D, Aubin M-L, Vignaud J and Laval-Jeantet M 1988 Separation of diffusion and perfusion in intravoxel incoherent motion MR imaging 1988 *Radiology* **168** 497–505

[34] Stepisnik J 1981 Analysis of NMR self-diffusion measurements by a density-matrix calculation *Physica* B/C **104** 350–64

[35] Stepisnik J 1985 Measuring and imaging of flow by NMR *Prog. Nucl. Magn. Reson. Spectrosc.* **17** 187–209

[36] Kärger J, Pfeifer H and Heink W 1988 Principles and application of self-diffusion measurements by nuclear magnetic resonance *Adv. Magn. Res.* **12** 1–89

[37] Taylor D G and Bushell M C 1985 The spatial-mapping of translational diffusion-coefficients by the NMR imaging technique *Phys. Med. Biol.* **30** 345–9

[38] Basser P J, Mattiello J and LeBihan D 1994 MR diffusion tensor spectroscopy and imaging *Biophys. J.* **66** 259–67

[39] Williams W D, Seymour E F W and Cotts R M 1978 A pulsed-gradient multiple-spin-echo NMR technique for measuring diffusion in the presence of background magnetic field gradients 1978 *J. Magn. Reson.* **31** 271–82

[40] Karlicek R F Jr and Lowe I J 1980 A modified pulsed gradient technique for measuring diffusion in the presence of large background gradients *J. Magn. Reson.* **37** 75–91

[41] Lucas A J, Gibbs S J, Jones E W G, Peyron M, Derbyshire J A and Hall L D 1993 Diffusion imaging in the presence of static magnetic-field gradients *J. Magn. Reson.* A **104** 273–82

[42] Lian J, Williams D S and Lowe I J 1994 Magnetic resonance imaging of diffusion in the presence of background gradients and imaging of background gradients *J. Magn. Reson.* A **106** 65–74

[43] Tanner J E 1970 Use of the stimulated echo in NMR diffusion studies *J. Chem. Phys.* **52** 2523–6

[44] Callaghan P T, Eccles C D and Xia Y 1988 NMR microscopy of dynamic displacements—*k*-space and *q*-space imaging *J. Phys. E: Sci. Instrum.* **21** 820–2

[45] McDonald P J, Ciampi E, Keddie J L, Heidenreich M and Kimmich R, Magnetic resonance determination of the spatial dependence of the droplet size distribution in the cream layer of oil-in-water emulsions: evidence for the effects of depletion flocculation *Phys. Rev.* E, submitted

[46] Axel L and Dougherty L 1989 Heart wall motion—improved method of spatial modulation of magnetization for MR imaging *Radiology* **172** 349–50

[47] Zerhouni E A, Parrish D M, Rodgers W J, Yang A and Shapiro E P 1988 Human-heart-tagging with MR imaging—a method for noninvasive assessment of myocardial motion *Radiology* **169** 59–63

[48] Morris G A and Freeman R 1978 Selective excitation in Fourier transform nuclear magnetic resonance *J. Magn. Reson.* **29** 433–62

[49] Jeong E K, Altobelli S A and Fukushima E 1994 NMR imaging studies of stratified flows in a horizontal rotating cylinder *Phys. Fluids* **6** 2901–6

## Further Reading

Blümich B and Kuhn W (eds) 1992 *Magnetic Resonance Microscopy Methods and Applications in Materials Science, Agriculture and Biomedicine* (Weinheim: Wiley–VCH)

Blümler P, Blümich B, Botto R E and Fukushima E (eds) 1998 *Spatially Resolved Magnetic Resonance, Proc. 4th Int. Conf. on Magnetic Resonance Microscopy and Macroscopy* (Weinheim: Wiley–VCH)

These two references give an excellent overview over the most recent examples in this research field.

Callaghan P T 1993 *Principles of Nuclear Magnetic Resonance Microscopy* (Oxford: Clarendon)

Kimmich R 1997 *NMR—Tomography, Diffusometry, Relaxometry* (Berlin: Springer)

Two standard text books which are recommended to researchers and graduate students who seek deeper insight into methodology and theory.

# B1.15
# EPR methods

*Stefan Weber*

### B1.15.1   Introduction

Systems containing unpaired electron spins, such as free radicals, biradicals, triplet states, most transition metal and rare-earth ions and some point defects in solids form the playground for electron paramagnetic resonance, EPR, also called electron spin resonance, ESR, or electron magnetic resonance, EMR. The fundamentals of EPR spectroscopy are very similar to the more familiar nuclear magnetic resonance (NMR) technique. Both deal with interactions of electromagnetic radiation with magnetic moments, which in the case of EPR arise from electrons rather than nuclei. With few exceptions, unpaired electrons lead to a non-vanishing spin of a particle that can be used as a spectroscopic probe. In EPR spectroscopy such molecules are studied by observing the magnetic fields at which they come into resonance with monochromatic electromagnetic radiation. Since species with unpaired electron spins are relatively rare compared to the multitude of species with magnetic nuclei, EPR is less widely applicable than NMR or even optical spectroscopy which has clear advantages with its ability to detect diamagnetic as well as paramagnetic states. What appears to be a drawback, however, can turn into an invaluable advantage, for instance, when selectively studying paramagnetic ions or molecules buried in a large protein environment. With its inherent specificity for those reactants, intermediates or products that carry unpaired electron spins, together with its high spectral resolution, EPR has excelled over many other techniques in, for example, unravelling the primary events of photosynthesis. Similarly, many key intermediates in this process have been identified by EPR. By appending a paramagnetic fragment—a so-called 'spin-label'—to a molecule of biological importance, in effect one has acquired a probe to supply data on the interactions and dynamics of biological molecules. Very many systems of biomedical interest have had their structure and function elucidated by application of modern EPR techniques. Also EPR has allowed chemists to probe into the details of reaction mechanisms by using the technique of spin trapping to identify reactive radical intermediates. As one last example of the many successes of EPR the identification of paramagnetic species in insulators and semiconductors is worth mentioning.

More than 50 years after its invention by the Russian physicist Zavoisky (for a review of the EPR history see [1]), advanced EPR techniques presently applied in the above mentioned areas of physics, chemistry and biology include time-resolved continuous wave (CW) and pulsed EPR (Fourier transform (FT) EPR and electron-spin echo (ESE) detected EPR) at various microwave (MW) frequencies and multiple-resonance EPR methods such as electron–nuclear double resonance (ENDOR) and electron–nuclear–nuclear TRIPLE resonance in the case of electron and nuclear transitions and electron–electron double resonance (ELDOR) in the case of different electron spin transitions. High-field/high-frequency EPR and ENDOR have left the developmental stage, and a wide range of significant applications continues to emerge. The range of multi-frequency EPR spectroscopy is now extending from radiofrequencies (RFs) in near-zero fields up to several hundred gigahertz in superconducting magnets or Bitter magnets.

In this article only the most important and frequently applied EPR methods will be introduced. For more extensive treatments of CW and pulsed EPR the reader is referred to some excellent review articles that will be specified in the respective sections of this article. A good starting point for further reading is provided by a number of outstanding textbooks which have been written on the various aspects of EPR in general [2–8]. Interested readers might also appreciate the numerous essays on various magnetic resonance topics that are published on a bimonthly basis in the educational journal 'Concepts in Magnetic Resonance'.

## B1.15.2   EPR background

### B1.15.2.1   Spins and magnetic moments

Historically, the recognition of electron spin can be traced back to the famous Stern–Gerlach experiment in the early 1920s. Stern and Gerlach observed that a beam of silver atoms was split into two components deflected in different directions when passing through an inhomogeneous magnetic field. The observation could only be explained with the concept of a half-integral angular momentum ascribed to an intrinsic spin of the electron. EPR spectroscopy relies on the behaviour of the electron angular momentum and its associated magnetic moment in an applied magnetic field.

If the angular momentum of a free electron is represented by a spin vector $\mathbf{S} = (S_x, S_y, S_z)$, the magnetic moment $\boldsymbol{\mu}_S$ is related to $\mathbf{S}$ by

$$\boldsymbol{\mu}_S = -g_e \beta_e \mathbf{S} \tag{B1.15.1}$$

where $g_e$ is a dimensionless number called the electron $g$-factor and $\beta_e = |e|\hbar/(2m_e) = 9.274\,0154 \times 10^{-24}$ J T$^{-1}$ is the Bohr magneton; $e$ is the electronic charge, $\hbar = h/(2\pi)$ is Planck's constant and $m_e$ is the electron mass. The negative sign in equation (B1.15.1) indicates that, because of the negative charge of the electron, the magnetic moment vector is antiparallel to the spin (since $g_e > 0$). In the quantum theory $\boldsymbol{\mu}_S$ and $\mathbf{S}$ are treated as (vector) operators. Suppose that the angular momentum operator $\mathbf{S}$ is defined in units of $\hbar$, then $\mathbf{S}^2$ has the eigenvalues $S(S+1)$, where $S$ is either integer or half integer. The magnitude of the angular momentum itself is given by the square root of the eigenvalue of $\mathbf{S}^2$, which is $\sqrt{S(S+1)}$. Any component of $\mathbf{S}$ (for example $S_z$) commutes with $\mathbf{S}^2$, so that simultaneously eigenvalues of both $\mathbf{S}^2$ and $S_z$ may be specified, which are $S(S+1)$ and $M_S$, respectively. $M_S$ has $(2S+1)$ allowed values running in integral steps from $-S$ to $+S$.

Classically, the interaction energy of a magnetic moment $\boldsymbol{\mu}_S$ in an applied magnetic field $\mathbf{B}$ is

$$E = -\boldsymbol{\mu}_S \cdot \mathbf{B}. \tag{B1.15.2}$$

For a quantum mechanical system $\boldsymbol{\mu}_S$ is replaced by the appropriate operator, equation (B1.15.1) to obtain the Hamiltonian for a free electron in a magnetic field,

$$\mathcal{H} = g_e \beta_e \mathbf{S} \cdot \mathbf{B}. \tag{B1.15.3}$$

If the magnetic field is $B_0$ in the $z$-direction, $\mathbf{B} = (0, 0, B_0)$, the scalar product simplifies and the Hamiltonian becomes

$$\mathcal{H} = g_e \beta_e S_z B_0. \tag{B1.15.4}$$

The eigenvalues of this Hamiltonian are simple, being only multiples $g_e \beta_e B_0$ of the eigenvalues of $S_z$. Therefore, the allowed energies are $E_{M_S} = g_e \beta_e M_S B_0$. For a simple system of one unpaired electron, $S = \frac{1}{2}$ and $M_S = \pm\frac{1}{2}$, which results in two energy states which are degenerate in zero field and whose energy separation increases linearly with $B_0$. This is summarized in figure B1.15.1 where the two states are also labelled with their eigenfunctions $|+\frac{1}{2}\rangle \equiv |\alpha\rangle$ and $|-\frac{1}{2}\rangle \equiv |\beta\rangle$ to indicate the $M_S = +\frac{1}{2}$ and $M_S = -\frac{1}{2}$ eigenstates for $S = \frac{1}{2}$, respectively. The lowest state has $M_S = -\frac{1}{2}$ (since $g_e > 0$), so that the projection of $\mathbf{S}$ along the

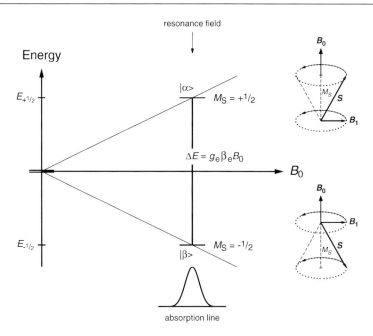

**Figure B1.15.1.** Energy levels for an electron ($S = \frac{1}{2}$) as a function of the applied magnetic field $B_0$. $E_{1/2} = +\frac{1}{2} g_e \beta_e B_0$ and $E_{-1/2} = -\frac{1}{2} g_e \beta_e B_0$ represent the energies of the $M_S = +\frac{1}{2}$ and $M_S = -\frac{1}{2}$ states, respectively.

$z$-axis, $S_z$, is antiparallel to the field, but in accordance with physical expectation the $z$-component of the magnetic moment is parallel to the field (see equation (B1.15.1)). The splitting of the electron spin energy levels by a magnetic field is referred to as the Zeeman effect.

Under the influence of the external magnetic field **B**, the spin **S** and the magnetic moment $\mu_S$ perform a precessional motion about the axis pointing along the direction of **B**. In the absence of additional magnetic fields, the angle between **S** and **B** does not change and the motion of the spin about the **B** axis may be illustrated as cones, as shown on the right-hand side of figure B1.15.1 for the two possible orientations of **S**. The $z$-component of the magnetic moment is sharply defined, while the $x$- and $y$-components are not, since they oscillate in the $xy$-plane. The precession frequency of $\mu_S$ and **S** about **B** is called the Larmor frequency, $\omega_0 = g_e \beta_e B_0 / \hbar$.

The transition between the two eigenstates can be induced by the application of microwave (MW) radiation with frequency $\omega$ and its magnetic field vector linearly polarized in the $xy$-plane. The oscillating $\mathbf{B}_1$-field of this radiation can be formally decomposed into two counter-rotating circularly polarized components, one of which is rotating in the same direction as the Larmor precession of the spins. The other component of the MW field does not interact significantly with the electron spins and can be neglected. If $\omega$ is different from $\omega_0$ the precessing magnetic moment will not seriously be affected by $\mathbf{B}_1$, for its component in the $xy$-plane will pass in and out of phase with $\mathbf{B}_1$ and there will be no resultant interaction. Transitions occur only near the resonance condition of

$$\omega = \omega_0 = g_e \beta_e B_0 / \hbar. \tag{B1.15.5}$$

On exact resonance, $\mu_S$ and $\mathbf{B}_1$ can remain in phase so the precessing magnetic moment experiences a constant field $\mathbf{B}_1$ in the $xy$-plane (see figure B1.15.1). It will respond to this by precessing about it with frequency $\omega_1 = g_e \beta_e B_1 / \hbar$.

In the language of quantum mechanics, the time-dependent $\mathbf{B}_1$-field provides a perturbation with a nonvanishing matrix element joining the stationary states $|\alpha\rangle$ and $|\beta\rangle$. If the rotating field is written in terms of an amplitude $B_1$ a perturbing term in the Hamiltonian is obtained

$$\mathcal{H}' = g_e \beta_e B_1 (S_x \cos(\omega t) + S_y \sin(\omega t)). \tag{B1.15.6}$$

The operators $S_x$ and $S_y$ have matrix elements between states $|\alpha\rangle$ and $|\beta\rangle$ or, in general, between states $|M_S\rangle$ and $|M_S \pm 1\rangle$ and consequently induce transitions between levels adjacent in energy.

If $B_1 \ll B_0$ first-order perturbation theory can be employed to calculate the transition rate for EPR (at resonance)

$$W_{\beta \leftarrow \alpha} = \pi \frac{\omega_1^2}{2} \rho(\omega). \tag{B1.15.7}$$

In equation (B1.15.7) $\rho(\omega)$ is the frequency distribution of the MW radiation. This result obtained with explicit evaluation of the transition matrix elements occurring for simple EPR is just a special case of a much more general result, Fermi's golden rule, which is the basis for the calculation of transition rates in general:

$$W_{M_S \leftarrow M_S+1} = 2\pi |\langle M_S|\mathcal{H}'|M_S + 1\rangle|^2 \rho(\omega). \tag{B1.15.8}$$

Using the selection rule for allowed transitions the relative intensity for the transition from the state $|M_S\rangle$ to $|M_S + 1\rangle$ is given by

$$W_{M_S \leftarrow M_S+1} \propto (S(S + 1) - M_S(M_S + 1)). \tag{B1.15.9}$$

The transition between levels coupled by the oscillating magnetic field $\mathbf{B}_1$ corresponds to the absorption of the energy required to reorient the electron magnetic moment in a magnetic field. EPR measurements are a study of the transitions between electronic Zeeman levels with $\Delta M_S = \pm 1$ (the selection rule for EPR).

The $g$-factor for a free electron, $g_e = 2.002\,319\,304\,386(20)$, is one of the most accurately known physical constants. For a magnetic field of 1 T ($= 10^4$ G) the resonance frequency $\omega/(2\pi)$ is 28.024 945 GHz, approximately three orders of magnitude larger than is required for any nuclear resonance (because $\beta_e/\beta_N \approx 1836$). This corresponds to a wavelength of about 10 mm and is in the microwave (MW) region of the electromagnetic spectrum.

### B1.15.2.2  Thermal equilibrium, magnetic relaxation and Lorentzian lineshape

Application of an oscillating magnetic field at the resonance frequency induces transitions in both directions between the two levels of the spin system. The rate of the induced transitions depends on the MW power which is proportional to the square of $\omega_1 = \gamma_e B_1$ (the amplitude of the oscillating magnetic field) (see equation (B1.15.7)) and also depends on the number of spins in each level. Since the probabilities of upward ($|\beta\rangle \to |\alpha\rangle$) and downward ($|\alpha\rangle \to |\beta\rangle$) transitions are equal, resonance absorption can only be detected when there is a population difference between the two spin levels. This is the case at thermal equilibrium where there is a slight excess of spins in the energetically lower $|\beta\rangle$-state. The relative population of the two-level system in thermal equilibrium is given by the Boltzmann distribution

$$\frac{N_\alpha}{N_\beta} = \exp\left(-\frac{\Delta E}{k_B T}\right) = \exp\left(-\frac{g_e \beta_e B_0}{k_B T}\right) \tag{B1.15.10}$$

where $N_\alpha$ and $N_\beta$ are the populations of the upper and lower spin states, respectively, $\Delta E$ is the energy difference separating the states, $k_B$ is the Boltzmann constant and $T$ is the temperature in Kelvin. The total number of spins is, of course, $N = N_\alpha + N_\beta$. Computation of the fractional excess of the lower level,

$$\frac{N_\beta - N_\alpha}{N} = \frac{1 - \exp(-g_e \beta_e B_0/(k_B T))}{1 + \exp(-g_e \beta_e B_0/(k_B T))} \tag{B1.15.11}$$

yields, for electrons in a magnetic field of 0.3 T at 300 K, a value of $7.6 \times 10^{-4}$, while for protons under the same conditions the value is only $1.2 \times 10^{-6}$. Thus, at thermal equilibrium, in EPR experiments one can virtually always take any nuclear spin state belonging to the same electron spin state to be equally populated. Because of the slightly larger number of spins occupying the lower energy level, there will be a net absorption of energy which results in an exponential decay of the initial population difference of the spin states. Eventually the levels would be equally populated (the spin system is then said to be saturated) if there were no radiationless processes that restored the thermal equilibrium distribution of the population by dissipating the energy absorbed by the spin system to other degrees of freedom. These nonradiative transitions between the two states $|\alpha\rangle$ and $|\beta\rangle$ are called spin–lattice relaxation. Spin–lattice relaxation is possible because the spin system is coupled to fluctuating magnetic fields driven by the thermal motions of the surroundings which are at thermal equilibrium. These fluctuations can stimulate spin flips and, therefore, this process leads to unequal probabilities of spontaneous transitions $|\alpha\rangle \rightarrow |\beta\rangle$ and $|\beta\rangle \rightarrow |\alpha\rangle$ and unequal populations at thermal equilibrium. In a magnetic resonance experiment one always has a competition between spin–lattice relaxation and the radiation field whose nature is to equalize the population of the levels. Qualitatively, $T_1$ is the time for the population difference to decay to $1/e$ of its equilibrium value after the perturbation (which in the case of magnetic resonance is the radiation field) is removed.

A second type of relaxation mechanism, the spin–spin relaxation, will cause a decay of the phase coherence of the spin motion introduced by the coherent excitation of the spins by the MW radiation. The mechanism involves slight perturbations of the Larmor frequency by stochastically fluctuating magnetic dipoles, for example those arising from nearby magnetic nuclei. Due to the randomization of spin directions and the concomitant loss of phase coherence, the spin system approaches a state of maximum entropy. The spin–spin relaxation disturbing the phase coherence is characterized by $T_2$.

A result of the relaxation processes is a shortened lifetime of the spin states giving rise to a broadening of the EPR line, which for most magnetic resonance lines dominated by homogeneous linewidth can be written as

$$f(\omega) = \frac{A\gamma M_0 T_2}{1 + \gamma^2 B_1^2 T_1 T_2 + T_2^2(\omega_0 - \omega)^2}. \tag{B1.15.12}$$

In equation (B1.15.12), $M_0$ is the $z$-component of the bulk magnetization vector, $\mathbf{M} = (1/V)\sum_i^N \boldsymbol{\mu}_i$ (unit $\mathrm{J\,T^{-1}\,m^{-3}}$), for an ensemble of $N$ spin magnetic moments at thermal equilibrium (in the absence of any resonant radiation), or in other words the net magnetic moment per unit volume, $\gamma = g_e \beta_e/\hbar$ is the gyromagnetic ratio and $A$ is a proportionality constant to include instrumental factors. The lineshape function $f(\omega)$ has a maximum at $\omega = \omega_0$ and it decreases for high power levels (i.e. for large $B_1$) and when the spin–lattice relaxation is not fast enough to maintain the population difference. This decrease is called saturation. If the saturation factor $s$ is defined by

$$s = \frac{1}{1 + \gamma^2 B_1^2 T_1 T_2} \tag{B1.15.13}$$

then $f(\omega)$ has the form

$$f(\omega) = \frac{As\gamma M_0 T_2}{1 + sT_2^2(\omega_0 - \omega)^2}. \tag{B1.15.14}$$

Well below saturation $s \approx 1$, and so the lineshape function becomes

$$f(\omega) = \frac{A\gamma M_0 T_2}{1 + T_2^2(\omega_0 - \omega)^2}. \tag{B1.15.15}$$

This is the famous Lorentzian function which is very often found for spectra of radicals in solution. In order to determine the relaxation times $T_1$ and $T_2$, a series of EPR spectra is recorded with the MW power varying from a condition of negligible saturation ($B_1^2 \gamma^2 T_1 T_2 \ll 1$; $s \approx 1$) to one of pronounced saturation ($B_1^2 \gamma^2 T_1 T_2 > 1$; $s < 1$). $T_2$ is then calculated from the linewidth below saturation by means of the expression

$$T_2 = \frac{2}{\Delta \omega_{1/2}^0} \qquad (B1.15.16)$$

where $\Delta \omega_{1/2}^0$ is the half width at half height of the magnetic resonance absorption line in the limit $B_1 \to 0$ ($s \to 1$). For $T_1$ one obtains

$$T_1 = \frac{\Delta \omega_{1/2}^0}{2} \left( \frac{1/s - 1}{\omega_1^2} \right). \qquad (B1.15.17)$$

One of the principal experimental advantages of this method of determining relaxation times is that it may be carried out with standard EPR spectrometers using CW-detected EPR lines [9, 10]. A discussion of more direct measurements of $T_1$ and $T_2$ using time-resolved EPR techniques is deferred to a later point (see sections B1.15.4 and B1.15.6.3(b)).

### B1.15.2.3  Spin Hamiltonian

To characterize and interpret EPR spectra one needs to obtain transition frequencies and transition probabilities between the $(2S+1)$ spin states. All interactions of the spins of electrons and nuclei with the applied magnetic field and with each other that lead to energy differences between states with different angular momenta have to be considered. The interactions are expressed in terms of operators representing the spins, with various coupling coefficients for the different interactions. The contributions of all these interactions make up the spin Hamiltonian that will be given in energy units throughout this text. Since, in principle, the spin Hamiltonian has no effect on the spatial part of the electronic wavefunction, the energy of the spin system in a certain state characterized by the quantum numbers $M_S$ and $M_I$ can be derived from the time-independent Schrödinger equation. The EPR spectrum is then interpreted as the allowed transitions between the eigenvalues of the spin Hamiltonian.

### (a) Electron Zeeman interaction

The first contribution considered here is the electron Zeeman interaction, i.e. the coupling of the magnetic dipole moment of the electron spin to the external magnetic field. For symmetry reasons the electron Zeeman interaction is isotropic for a free electron spin and is characterized by the Zeeman splitting constant $g_e$. The $g$-value of an unpaired electron in an atomic or molecular environment is very often different from $g_e$ and may also be anisotropic, i.e. dependent on the orientation of the system relative to the magnetic field $\mathbf{B}$. The deviation from the spin-only value of the $g$-factor and the anisotropy result from the contribution of the orbital angular momentum to the total angular momentum of the electron. This phenomenon is called spin–orbit coupling. It leads to an anisotropic electron Zeeman interaction (EZI) which is usually formulated as

$$\mathcal{H}_{\text{EZI}} = \beta_e \mathbf{B} \mathbf{g} \mathbf{S} \qquad (B1.15.18)$$

where the field and angular momentum vectors are coupled through a symmetric matrix $\mathbf{g}$ of dimension $3 \times 3$. In organic radicals orbital momenta are almost completely quenched by chemical bonding (with the exception of cases where the energies of the two orbitals are nearly degenerate), leading to only small deviations $\Delta g = |g_{ii} - g_e|$, $i = X, Y, Z$ of the principal values of $\mathbf{g}$ from the free-electron value $g_e$ (typically in a range from $10^{-5}$ to $5 \times 10^{-2}$). $g$-values very different from $g_e$ are expected for first-row transition metal ions and for rare-earth ions where spin–orbit coupling is more complete. The $\mathbf{g}$-matrix of organic radicals reflects certain features of the electronic wavefunction of the paramagnetic species. The spatial distribution of

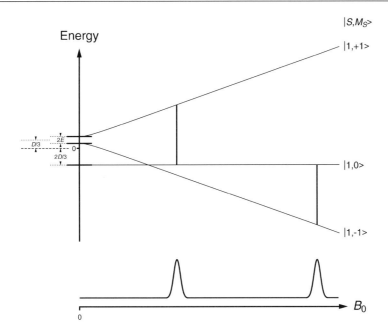

**Figure B1.15.2.** The state energies and corresponding eigenfunctions (high-field labels) as a function of the applied magnetic field $B_0$ for a system of spin $S = 1$ and $\mathbf{B} \parallel z$, shown for $D > 0$ and $E \neq 0$. The two primary transitions ($\Delta M_S = \pm 1$) are indicated for a constant frequency spectrum. Note that, because $E \neq 0$, the state energies vary nonlinearly with $B_0$ at low $B_0$.

the orbital carrying the unpaired spin can be influenced by interactions with other molecules, e.g. via hydrogen bonding. Therefore, a determination of the $g$-values and the orientation of the main axes of $\mathbf{g}$ with respect to the molecular axis frame can give highly specific information on the interaction of the molecule with its surrounding.

*(b) Electron spin–spin interaction*

An atom or a molecule with the total spin of the electrons $S = 1$ is said to be in a triplet state. The multiplicity of such a state is $(2S + 1) = 3$. Triplet systems occur in both excited and ground state molecules, in some compounds containing transition metal ions, in radical pair systems, and in some defects in solids.

For a system with $S = 1$, there are three sublevels characterized by $M_S = \pm 1$ and $M_S = 0$. In contrast to systems with $S = \frac{1}{2}$, these sublevels may not be degenerate in the absence of an external magnetic field (see figure B1.15.2). The lifting of degeneracy of the spin states at zero field is called zero-field splitting (ZFS) and it is common for systems with $S \geq 1$. For triplet states of organic molecules ($S = 1$) the ZFS arises from the dipolar interaction of the two magnetic moments of the electron spins with each other. The interaction is described by an additional term

$$\mathcal{H}_{ZFS} = \mathbf{SDS} \tag{B1.15.19}$$

that must be included in a spin Hamiltonian when $S \geq 1$. The spin–spin coupling (or ZFS) tensor $\mathbf{D}$ is a symmetric and traceless ($\sum_{i=X,Y,Z} D_{ii} = 0$) second-rank tensor. Therefore, $\mathbf{D}$ can be written in its principal axis frame with only two parameters

$$\mathcal{H}_{ZFS} = D(S_z^2 - \tfrac{1}{3}\mathbf{S}^2) + E(S_x^2 - S_y^2) \tag{B1.15.20}$$

where $D = \frac{3}{2}D_{ZZ}$ is the axial and $E = \frac{1}{2}(D_{XX} - D_{YY})$ is the rhombic zero-field parameter. One may define an asymmetry parameter $\eta_D = E/D$ of the **D**-tensor. The case of $\eta_D = 0$ (or $E = 0$) corresponds to an axially symmetric ZFS tensor ($D_{XX} = D_{YY}$) and two of the states will remain degenerate at zero magnetic field. The ZFS parameters can in general be determined from the EPR spectrum (for $\mathcal{H}_{ZFS} < \mathcal{H}_{EZI}$). In liquids the ZFS is averaged out to zero.

*(c) Electron spin exchange interaction*

For systems with two unpaired electrons, such as radical pairs, biradicals or triplets, there are four spin states that can be represented by the product functions $|M_{S1}\rangle \otimes |M_{S2}\rangle$, i.e. $|\alpha_1\alpha_2\rangle$, $|\alpha_1\beta_2\rangle$, $|\beta_1\alpha_2\rangle$ and $|\beta_1\beta_2\rangle$, where the subscripts indicate electron spin 1 and electron spin 2, respectively. In a paramagnetic centre of moderate size, however, it is more advantageous to combine these configurations into combination states because, in addition to the dipolar coupling between the spin magnetic moments, there is also an electrostatic interaction between the electron spins, the so-called exchange interaction, which gives rise to an energy separation between the singlet state, $|S\rangle = \frac{1}{\sqrt{2}}(|\alpha_1\beta_2\rangle - |\beta_1\alpha_2\rangle)$, and the triplet states ($|T_+\rangle = |\alpha_1\alpha_2\rangle$, $|T_0\rangle = \frac{1}{\sqrt{2}}(|\alpha_1\beta_2\rangle + |\beta_1\alpha_2\rangle)$, $|T_-\rangle = |\beta_1\beta_2\rangle$). The magnitude of the (isotropic) exchange interaction can be derived from the overlap of the wavefunctions and is described by the Hamiltonian

$$\mathcal{H}_{EX} = -2J\mathbf{S}_1 \cdot \mathbf{S}_2 \qquad (B1.15.21)$$

which denotes a scalar coupling between the spins $\mathbf{S}_1 = (S_{1x}, S_{1y}, S_{1z})$ and $\mathbf{S}_2 = (S_{2x}, S_{2y}, S_{2z})$. The energy separation between the $|S\rangle$ and $|T_0\rangle$ wavefunctions is determined by the exchange coupling constant $J$. For $J > 0$ the singlet state is higher in energy than the $|T_0\rangle$-state. The observed properties of the system depend on the magnitude of $J$. If it is zero the two spins behave completely independently and one would have a true biradical. At the other extreme, when $J$ is large the singlet lies far above the triplet and the magnetic resonance properties are solely determined by the interaction within the triplet manifold.

*(d) Hyperfine interaction*

The interaction of the electron spin's magnetic dipole moment with the magnetic dipole moments of nearby nuclear spins provides another contribution to the state energies and the number of energy levels, between which transitions may occur. This gives rise to the hyperfine structure in the EPR spectrum. The so-called hyperfine interaction (HFI) is described by the Hamiltonian

$$\mathcal{H}_{HFI} = \mathbf{SAI} \qquad (B1.15.22)$$

where **A** is the HFI matrix and $\mathbf{I} = (I_x, I_y, I_z)$ is the vector representation of the nuclear spin. The HFI consists of two parts and therefore, equation (B1.15.22) can be separated into the sum of two terms

$$\mathcal{H}_{HFI} = \mathbf{SA}^{dip}\mathbf{I} + a\mathbf{S} \cdot \mathbf{I} \qquad (B1.15.23)$$

where the first term describes the anisotropic dipolar coupling through space between the electron spin and the nuclear spin. $\mathbf{A}^{dip}$ is the symmetric and traceless dipolar HFI matrix. In the so-called point-dipole approximation, where both spins are assumed to be located, this part is given by

$$\mathcal{H} = -g\beta_e g_n \beta_n \left[ \frac{\mathbf{I} \cdot \mathbf{S}}{r^3} - \frac{3(\mathbf{Ir})(\mathbf{Sr})}{r^5} \right]. \qquad (B1.15.24)$$

In equation (B1.15.24), **r** is the vector connecting the electron spin with the nuclear spin, $r$ is the length of this vector and $g_n$ and $\beta_n$ are the $g$-factor and the Bohr magneton of the nucleus, respectively. The dipolar coupling is purely anisotropic, arising from the spin density of the unpaired electron in an orbital of non-spherical symmetry (i.e. in p, d or f-orbitals) with a vanishing electron density at the nucleus. Since $\mathbf{A}^{\mathrm{dip}}$ is traceless the dipolar interactions are averaged out in isotropic fluid solution and only the orientation-independent isotropic coupling represented by the second term in equation (B1.15.23) gives rise to the observed hyperfine coupling in the spectrum. This isotropic contribution is called the (Fermi) contact interaction arising from electrons in s orbitals (spherical symmetry) with a finite probability ($|\Psi(0)|^2$) of finding the electron at the nucleus. The general expression for the isotropic hyperfine coupling constant is

$$a = \frac{2\mu_0}{3} g\beta_e g_n \beta_n |\Psi(0)|^2. \tag{B1.15.25}$$

Hence, a measurement of hyperfine coupling constants provides information on spin densities at certain positions in the molecule and thus renders a map of the electronic wavefunction.

The simplest system exhibiting a nuclear hyperfine interaction is the hydrogen atom with a coupling constant of 1420 MHz. If different isotopes of the same element exhibit hyperfine couplings, their ratio is determined by the ratio of the nuclear $g$-values. Small deviations from this ratio may occur for the Fermi contact interaction, since the electron spin probes the inner structure of the nucleus if it is in an s orbital. However, this so-called hyperfine anomaly is usually smaller than 1%.

### (e) Nuclear Zeeman and nuclear quadrupole interaction

While all contributions to the spin Hamiltonian so far involve the electron spin and cause first-order energy shifts or splittings in the EPR spectrum, there are also terms that involve only nuclear spins. Aside from their importance for the calculation of ENDOR spectra, these terms may influence the EPR spectrum significantly in situations where the high-field approximation breaks down and second-order effects become important. The first of these interactions is the coupling of the nuclear spin to the external magnetic field, called the nuclear Zeeman interaction (NZI). Neglecting chemical shift anisotropies that are usually small and not resolved in ENDOR spectra it can be considered isotropic and written as

$$\mathcal{H}_{\mathrm{NZI}} = -g_n \beta_n \mathbf{B} \cdot \mathbf{I}. \tag{B1.15.26}$$

The negative sign in equation (B1.15.26) implies that, unlike the case for electron spins, states with larger magnetic quantum number have smaller energy for $g_n > 0$. In contrast to the $g$-value in EPR experiments, $g_n$ is an inherent property of the nucleus. NMR resonances are not easily detected in paramagnetic systems because of sensitivity problems and increased linewidths caused by the presence of unpaired electron spins.

Since atomic nuclei are not perfectly spherical their spin leads to an electric quadrupole moment if $I \geq 1$ which interacts with the gradient of the electric field due to all surrounding electrons. The Hamiltonian of the nuclear quadrupole interactions can be written as tensorial coupling of the nuclear spin with itself

$$\mathcal{H}_{\mathrm{NQI}} = \mathbf{IPI} \tag{B1.15.27}$$

where $\mathbf{P}$ is the quadrupole coupling tensor. Comparison with equation (B1.15.19) shows that the NQI can be formally treated in a way analogous to that for the ZFS. In liquids the NQI is averaged out to zero.

### (f) The complete spin Hamiltonian

The complete spin Hamiltonian for a description of EPR and ENDOR experiments is given by

$$\mathcal{H} = \mathcal{H}_{\mathrm{EZI}} + \mathcal{H}_{\mathrm{ZFS}} + \mathcal{H}_{\mathrm{EX}} + \mathcal{H}_{\mathrm{HFI}} + \mathcal{H}_{\mathrm{NZI}} + \mathcal{H}_{\mathrm{NQI}}. \tag{B1.15.28}$$

The approximate magnitudes of the terms in equation (B1.15.28) are shown in an overview in figure B1.15.3 (see also [2, 3, 11]).

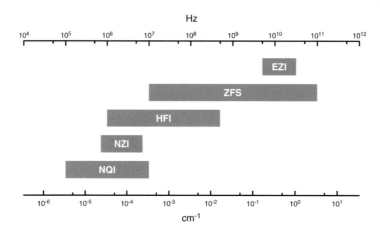

**Figure B1.15.3.** Typical magnitudes of interactions of electron and nuclear spins in the solid state (logarithmic scale).

### B1.15.3   EPR instrumentation

A typical CW-detected EPR spectrum is recorded by holding the frequency constant and varying the external magnetic field. In doing so one varies the separation between the energy levels to match it to the quantum of the radiation. Even though magnetic resonance could also be achieved by sweeping the frequency at a fixed magnetic field, high-frequency sources with a broad frequency range and simultaneously low enough noise characteristics have not yet been devised to be practical for frequency-swept EPR.

EPR absorption has been detected from zero magnetic field up to fields as high as 30 T corresponding to a MW frequency of $10^{12}$ Hz. There are various considerations that influence the choice of the radiation frequency. Higher frequencies, which require higher magnetic fields, give inherently greater sensitivity by virtue of a more favourable Boltzmann factor (see equation (B1.15.11)). However, several factors place limits on the frequency employed, so that frequencies in the MW region of the electromagnetic spectrum remain favoured. One limitation is the sample size; at frequencies around 40 GHz the dimensions of a typical resonant cavity are of the order of a few millimetres, thus restricting the sample volume to about 0.02 $\text{cm}^3$. The requirement to reduce the sample size roughly compensates for the sensitivity enhancement in going to higher fields and frequencies in EPR. However, the sensitivity advantage persists if only small quantities of the sample are available. This is often the case for biological samples, particularly when single-crystal studies are intended. Second, high frequencies require high magnetic fields that are homogeneous over the sample volume. Sufficiently homogeneous magnetic fields above 2.5 T are difficult to produce with electromagnets. Superconducting magnets are commercially available for magnetic fields up to 22 T, but they are expensive and provide only small room-temperature bores, thus limiting the space available for the resonator. Third, the small size of MW components for high frequencies makes their fabrication technically difficult and costly. These and other factors have resulted in a choice of the frequency region around 10 GHz (usually denominated the X-band region) as the resonance frequency of most commercially available spectrometers. The most common frequency bands for high-field/high-frequency EPR are Q-band (35 GHz) and W-band (95 GHz), where the wavelengths are 8 mm and 3 mm, respectively. In order to carry out EPR experiments in larger objects such as intact animals it appears necessary to use lower frequencies because of the large dielectric losses of aqueous samples. At L-band frequencies (1–2 GHz), with appropriate configurations of EPR resonators, whole animals the size of mice can be studied by insertion into the resonator. Table B1.15.1 lists typical frequencies and wavelengths, together with the resonant fields required for resonance of a free electron.

**Table B1.15.1.** Some frequencies and resonance fields (for $g = 2$) used in EPR.

| Typical EPR frequency $\nu$ (GHz) | Vacuum wavelength $\lambda$ (m) | Typical EPR field $B_0$ (T) | Band designation | Frequency (GHz) |
|---|---|---|---|---|
| 1.5 | $2 \times 10^{-1}$ | 0.054 | L | 0.39–1.55 |
| 3.0 | $1 \times 10^{-1}$ | 0.107 | S | 1.55–3.9 |
| 6.0 | $5 \times 10^{-2}$ | 0.214 | C | 3.9–6.2 |
| 9.5 | $3.2 \times 10^{-2}$ | 0.339 | X | 6.2–10.9 |
| 24 | $1.2 \times 10^{-2}$ | 0.856 | K | 10.9–36 |
| 36 | $8.3 \times 10^{-3}$ | 1.285 | Q | 36–46 |
| 50 | $6.0 \times 10^{-3}$ | 1.784 | V | 46–56 |
| 95 | $3.2 \times 10^{-3}$ | 3.390 | W | 56–100 |
| 140 | $2.1 \times 10^{-3}$ | 4.996 | D | |
| 250 | $1.2 \times 10^{-3}$ | 8.921 | — | |
| 360 | $8.3 \times 10^{-4}$ | 12.846 | — | |
| 604 | $5.0 \times 10^{-4}$ | 21.552 | — | |

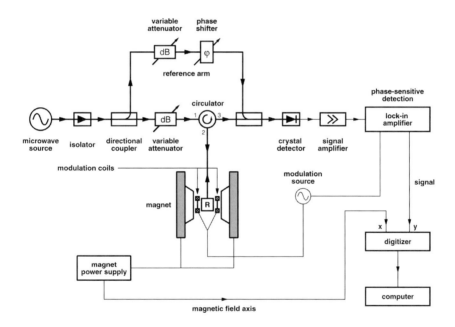

**Figure B1.15.4.** Block diagram of a typical EPR spectrometer operating at X-band frequencies.

The components of a typical EPR spectrometer operating at X-band frequencies [3, 4, 6] are shown in figure B1.15.4.

Up to the present the MW radiation has usually been provided by reflex klystrons, which essentially consist of a vacuum tube and a pair of electrodes to produce an electron beam that is velocity modulated by a radio-frequency (RF) electric field. The net effect is the formation of groups or bunches of electrons. Using a reflector the bunched electron beam is turned around and will debunch, giving up energy to the cavity, provided the RF frequency and the beam and reflector voltages are properly adjusted. This will set up one of several stable klystron modes at the cavity frequency $f_0$; the mode corresponding to the highest output of power is usually the one utilized. A typical klystron has a mechanical tuning range allowing the klystron cavity frequency to be tuned over a range of 5–50% around $f_0$. The adjustment of the reflector voltage allows one to vary the centre frequency of a given mode over a very limited range of 0.2–0.8%. Often a low-amplitude sine-wave reflector voltage modulation is employed as an integral part of an automatic frequency control (AFC). It is desirable that the klystron frequency be very stable; hence, fluctuations of the klystron temperature or of applied voltages must be minimized and mechanical vibrations suppressed. Klystrons are employed as generators of nearly monochromatic output radiation in the frequency range from 1 to 100 GHz. In commercial EPR spectrometers the klystron normally provides less than 1 W of continuous output power. Increasingly, however, solid-state devices, such as Gunn-effect oscillators and IMPATTs, are superseding klystron tubes.

Waveguides are commonly used to transmit microwaves from the source to the resonator and subsequently to the receiver. For not-too-high-frequency radiation ($\leq 10$ GHz) low-loss MW transmission can also be achieved using strip-lines and coaxial cables.

At the output of a klystron an isolator is often used to prevent back-reflected microwaves to perturb the on-resonant klystron mode. An isolator is a microwave-ferrite device that permits the transmission of microwaves in one direction and strongly attenuates their propagation in the other direction. The principle of this device involves the Faraday effect, that is, the rotation of the polarization planes of the microwaves.

The amount of MW power reaching the sample in the resonator is controlled by a variable attenuator. Like the isolator, the circulator is a non-reciprocal device that serves to direct the MW power to the resonator (port 1 $\rightarrow$ port 2) and simultaneously allows the signal reflected at resonance to go from the resonator directly to the receiver (port 2 $\rightarrow$ port 3).

Although achievement of resonance in an EPR experiment does not require the use of a resonant cavity, it is an integral part of almost all EPR spectrometers. A resonator dramatically increases the sensitivity of the spectrometer and greatly simplifies sample access. A resonant cavity is the MW analogue of a RF-tuned *RLC* circuit and many expressions derived for the latter may also be applied to MW resonators. A typical resonant cavity for microwaves is a box or a cylinder fabricated from high-conductivity metal and with dimensions comparable to the wavelength of the radiation. Each particular cavity size and shape can sustain oscillations in a number of different standing wave patterns, called resonator modes. Visual images and mathematical expressions of the distributions of the electric and magnetic field vectors within the cavity can be derived from Maxwell's equations with suitable boundary conditions. The locations of the maxima of the $\mathbf{B}_1$-field and the $\mathbf{E}$-field are different depending on the mode chosen for the EPR experiment. It is desirable to design the cavity in such a way that the $\mathbf{B}_1$ field is perpendicular to the external field $\mathbf{B}$, as required by the nature of the resonance condition. Ideally, the sample is located at a position of maximum $\mathbf{B}_1$, because below saturation the signal-to-noise ratio is proportional to $B_1$. Simultaneously, the sample should be placed at a position where the $\mathbf{E}$-field is a minimum in order to minimize dielectric power losses which have a detrimental effect on the signal-to-noise ratio.

The sharpness of the frequency response of a resonant system is commonly described by a factor of merit, called the quality factor, $Q = \nu/\Delta\nu$. It may be obtained from a measurement of the full width at half maximum $\Delta\nu$, of the resonator frequency response curve obtained from a frequency sweep covering the resonance. The sensitivity of a system (proportional to the inverse of the minimum detectable number of paramagnetic centres in an EPR cavity) critically depends on the quality factor

$$S \propto Q\eta \qquad (\text{B1.15.29})$$

where $\eta = \int_{\text{sample}} B_1^2 \mathrm{d}V / \int_{\text{cavity}} B_1^2 \mathrm{d}V$ is the filling factor. The cavity types most commonly employed in EPR are the rectangular–parallelepiped cavity and the cylindrical cavity. The rectangular cavity is typically operated in a transverse electric mode, TE$_{102}$, which permits the insertion of large samples with low dielectric constants. It is especially useful for liquid samples in flat cells, which may extend through the entire height of the cavity. In the cylindrical cavity a $TE_{011}$ mode is frequently used because of its fairly high $Q$-factor and the very strong $B_1$ along the sample axis.

Microwaves from the waveguide are coupled into the resonator by means of a small coupling hole in the cavity wall, called the iris. An adjustable dielectric screw (usually machined from Teflon) with a metal tip adjacent to the iris permits optimal impedance matching of the cavity to the waveguide for a variety of samples with different dielectric properties. With an appropriate iris setting the energy transmission into the cavity is a maximum and simultaneously reflections are minimized. The optimal adjustment of the iris screw depends on the nature of the sample and is found empirically.

Other frequently used resonators are dielectric cavities and loop-gap resonators (also called split-ring resonators) [12]. A dielectric cavity contains a diamagnetic material that serves as a dielectric to raise the effective filling factor by concentrating the $\mathbf{B}_1$ field over the volume of the sample. Hollow cylinders machined from fused quartz or sapphire that host the sample along the cylindrical axis are commonly used. Loop-gap resonators consist of one or a series of cylindrical loops interrupted by at least one or several gaps. Loops and gaps act as inductive and capacitive elements, respectively. With a suitable choice of loop and gap dimensions, resonators operating at different resonance frequencies over a wide range of the MW spectrum can be constructed. Loop-gap resonators typically have low $Q$-factors. Their broad-bandwidth frequency response, $\Delta\nu$, makes them particularly useful in EPR experiments where high time resolution, $\tau_{res} = 1/(2\pi\,\Delta\nu)$, because fast signal changes are required. Excellent filling factors, $\eta$, may be obtained with loop-gap devices; the high $\eta$ makes up for the typically low $Q$ to yield high sensitivity (see equation (B1.15.29)), valuable for small sample sizes and in pulsed EPR experiments. Coupling of microwaves into these cavities is most conveniently accomplished by a coupling loop that acts as an antenna. Typically, the distance between the antenna and the loop-gap resonator is varied in order to obtain optimal impedance matching.

When the applied magnetic field is swept to bring the sample into resonance, MW power is absorbed by the sample. This changes the matching of the cavity to the waveguide and some power is now reflected and passes via the circulator to the detector. This reflected radiation is thus the EPR signal.

The most commonly used detector in EPR is a semiconducting silicon crystal in contact with a tungsten wire, which acts as an MW rectifier. At microwatt powers, crystal detectors are typically non-linear and render a rectified current that is proportional to the MW power (i.e. proportional to $B_1^2$). In the milliwatt region, the rectified crystal current becomes proportional to the square root of the MW power (i.e. proportional to $B_1$), and the crystal behaves as a linear detector. In EPR spectroscopy it is preferred to operate the crystal rectifier in its linear regime. However, since the EPR signal is typically rather small, the diode needs to be biased to operate it at higher MW power levels. This can be done by slightly mismatching the cavity to the waveguide in order to increase the MW power back-reflected from the cavity, or by adding microwaves at a constant power level guided through the reference arm (often called the bypass arm) of the spectrometer. The reference arm takes microwaves from the waveguide ahead of the circulator and returns them with adjusted phase and power behind the circulator. When properly adjusted, the reference arm can also be used to detect the in-phase ($\chi'$) and out-of-phase ($\chi''$) components of the EPR signal with respect to the phase of the microwaves.

When sweeping the magnetic field through resonance, a crystal detector renders a slowly varying DC signal which is not readily processed and which is superimposed by low-frequency noise contributions. To overcome this, a phase-sensitive detection technique utilizing small-amplitude magnetic field modulation is employed in most EPR spectrometers. Modulation of the magnetic field is achieved by placing a pair of Helmholtz coils on each side of the cavity along the axis of the external magnetic field. An alternating current is fed through them and a small oscillating magnetic field is induced which is superimposed on the external magnetic field. The effect of the modulation is depicted in figure B1.15.5. Provided the amplitude of the modulation field is small compared to the linewidth of the absorption signal, $\Delta B_{1/2}$, the change in MW power at the detector will contain an oscillatory component at the modulation frequency whose amplitude will be proportional to the slope of the EPR line. A lock-in detector compares the modulated EPR signal from the crystal with a reference and only passes the components of the signal that have the proper frequency and phase.

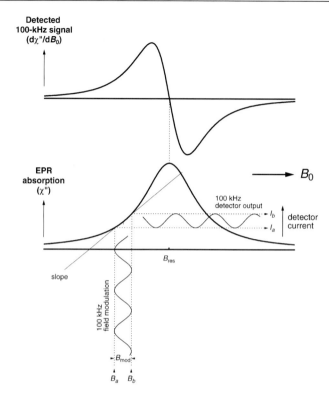

**Figure B1.15.5.** Effect of small-amplitude 100 kHz field modulation on the detector output current. The static magnetic field is modulated between the limits $B_a$ and $B_b$. The corresponding detector current varies between the limits $I_a$ and $I_b$. The upper diagram shows the recorded 100 kHz signal as a function of $B_0$. After [3].

The reference voltage comes from the same frequency generator that produces the field modulation voltage and this causes the EPR signal to pass through while most noise at frequencies other than the modulation frequency is suppressed. As a result of phase-sensitive detection using lock-in amplification one typically obtains the first derivative of the absorption line EPR signal. The application of field modulation, however, can cause severe lineshape distortion: to limit modulation-induced line broadening to below 1% of the undistorted linewidth, $\Delta B_{1/2}^\circ$ requires small modulation amplitudes ($B_{mod} < 0.15\,\Delta B_{1/2}^\circ$ for Lorentzian lineshapes and $B_{mod} < 0.3\,\Delta B_{1/2}^\circ$ for Gaussian lineshapes).

After the signal emerges from the lock-in amplifier it still contains a considerable amount of noise. Most of the noise contributions to the signal can be eliminated by passing the signal through a low-pass filter. The filter time constant is a measure of the cutoff frequency of the filter. If accurate linewidth and $g$-factor measurements are intended, one must be careful to employ a sufficiently short response time because lineshape distortions may occur as a result of too intense filtering.

Figure B1.15.6 presents a liquid-phase EPR spectrum of an organic radical measured using a conventional EPR spectrometer like the one depicted in figure B1.15.4. As is usual, the lines are presented as first derivatives $d\chi''/dB_0$ of the power absorbed by the spins. The spectrum shows a pronounced pattern of hyperfine lines arising from two different groups of protons (see also figure B1.15.9). The number, spacing and intensity of the lines provides information on the molecular and electronic structure of the molecule carrying the unpaired electron spin. The individual lines have a Lorentzian lineshape with a homogeneous linewidth determined by $T_2$. The most common case for inhomogeneously broadened lines giving rise to a Gaussian lineshape is

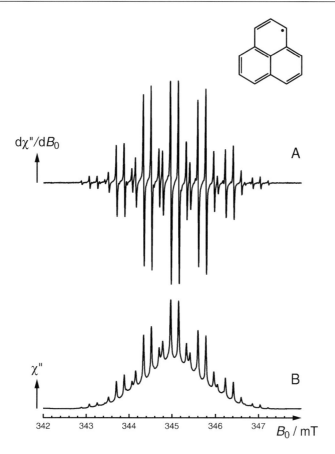

**Figure B1.15.6.** The EPR spectrum of the perinaphthenyl radical in mineral oil taken at room temperature. (A) First derivative of the EPR absorption $\chi''$ with respect to the external magnetic field, $B_0$. (B) Integrated EPR spectrum.

unresolved hyperfine interactions arising from a large number of nonequivalent nuclei and anisotropies of the hyperfine coupling which will persist when recording EPR spectra of radicals in solids.

EPR has been successfully applied to radicals in the solid, liquid and gaseous phase. Goniometer techniques have been adopted to measure anisotropic magnetic interactions in oriented (e.g. single-crystal) and partially oriented (e.g. film) samples as a function of the sample orientation with respect to the external field. Variable temperature studies can provide a great deal of information about a spin system and its interactions with its environment. Therefore, low-temperature as well as high-temperature EPR experiments can be conducted by either heating or cooling the entire cavity in a temperature-controlled cryostat or by heating or cooling the sample in a jacket inserted into the cavity. Specialized cavity designs have also been worked out to perform EPR studies under specific conditions (e.g. high pressures). Sample irradiation is facilitated through shielded openings in the cavity.

The practical goal of EPR is to measure a stationary or time-dependent EPR signal of the species under scrutiny and subsequently to determine magnetic interactions that govern the shape and dynamics of the EPR response of the spin system. The information obtained from a thorough analysis of the EPR signal, however, may comprise not only the parameters enlisted in the previous chapter but also a wide range of other physical parameters, for example reaction rates or orientation order parameters.

**B1.15.4    Time-resolved CW EPR methods**

Although EPR in general has the potential to follow the concentration changes of short-lived paramagnetic intermediates, standard CW EPR using field modulation for narrow-band phase-sensitive detection is geared for high sensitivity and correspondingly has only a mediocre time resolution. Nevertheless, transient free radicals in the course of (photo-)chemical processes can be studied by measuring the EPR line intensity of a spectral feature as a function of time at a fixed value of the external magnetic field. Typically, the optimum time response of a commercial spectrometer which uses a CW fixed-frequency lock-in detection is in the order of 20 $\mu$s. By use of field modulation frequencies higher than the 100 kHz usually employed in commercial instruments, the time resolution can be increased by about an order of magnitude, which makes this method well suited for the study of transient free radicals on a microsecond timescale.

*B1.15.4.1    Transient EPR spectroscopy*

The time resolution of CW EPR can be considerably improved by removing the magnetic field modulation completely. Rather, a suitably fast data acquisition system is employed to directly detect the transient EPR signal as a function of time at a fixed magnetic field. In transient EPR spectroscopy (TREPR) [13, 14] paramagnetic species (e.g. free radicals, radical pairs, triplets or higher multiplet states) are generated on a nanosecond timescale by a short laser flash or radiolysis pulse and the arising time-dependent signals are detected in the presence of a weak MW magnetic field. For this purpose the standard EPR spectrometer shown in figure B1.15.4 needs to be modified. The components for the field modulation may be removed, and the lock-in amplifier and digitizer are replaced by a fast transient recorder or a digital oscilloscope triggered by a photodiode in the light path of the laser. The response time of such a spectrometer is potentially controlled by the bandwidth of each individual unit. Provided that the MW components are adequately broadbanded and the laser flash is sufficiently short (typically a few nanoseconds), the time resolution can be pushed to the $10^{-8}$ s range (which is not far from the physical limit given by the inverse MW frequency) if a resonator with a low $Q$-value (and hence wide bandwidth) is used. Fortunately, the accompanying sensitivity loss (see equation (B1.15.29)) can be compensated to a large extent by using resonators with high filling factors [12]. Furthermore, excellent sensitivity is obtained in studies of photoprocesses where the light-generated paramagnetic species are typically produced in a state of high electron spin polarization (or, in other words, removed from the thermal equilibrium population of the electron spin states). Nevertheless, the EPR time profiles at a fixed magnetic field position are repeatedly measured in order to improve the signal-to-noise ratio by a factor $\sqrt{N}$, where $N$ is the number of time traces averaged.

Following the pioneering work by Kim and Weissman [15], it has been demonstrated that TREPR works for a broad range of resonance frequencies from 4 GHz (S-band) up to 95 GHz (W-band). As an example the time-dependent EPR signal of the photo-generated triplet state of pentacene in a *para*-terphenyl single crystal obtained by TREPR at X-band is shown in figure B1.15.7. Note that EPR signals taken in direct detection appear in the absorption (or dispersion) mode, not in the usual derivative form associated with field modulation and phase-sensitive detection. Therefore, positive signals indicate absorptive (A) and negative signals emissive (E) EPR transitions.

The following discussion of the time dependence of the EPR response in a TREPR experiment is based on the assumption that the transient paramagnetic species is long lived with respect to the spin relaxation parameters. When $\omega_1$ is large compared to the inverse relaxation times, i.e. $\omega_1 = \gamma B_1 \gg T_1^{-1}, T_2^{-1}$, or, in other words, for high MW powers, the signal exhibits oscillations with a frequency proportional to the MW magnetic field $B_1$. These so-called transient nutations are observed if resonance between an electron spin transition and a coherent radiation field is suddenly achieved. The phenomenon can be understood when viewing the motion of the magnetization vector, **M**, in a reference frame, $(x', y', z)$, rotating at the frequency $\omega$ of the MW field around the static external field (**B**$_0 \parallel z$). Hence, in the rotating frame $B_1$ is a stationary field

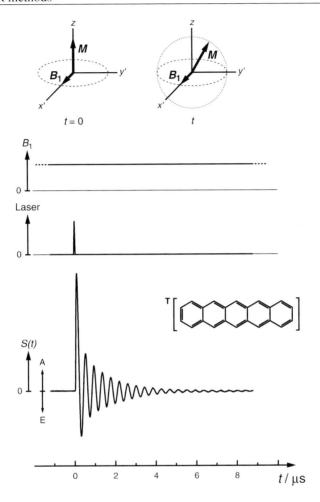

**Figure B1.15.7.** Transient EPR. Bottom: time-resolved EPR signal of the laser-flash-induced triplet state of pentacene in *p*-terphenyl. $B_1 = 0.085$ mT. Top: initially, the transient magnetization **M** is aligned along **B** $\parallel z$. In the presence of a MW magnetic field $\mathbf{B}_1$ the magnetization precesses about $\mathbf{B}_1 \parallel x'$ (rotating frame representation).

defined along the $x'$-axis as indicated in figure B1.15.7. Paramagnetic species are created during the laser flash at $t = 0$ with their spins aligned with the applied magnetic field, which causes an initial magnetization $M_z(0)$ in this direction. This is acted upon by the radiation field at or near resonance, which rotates it to produce an orthogonal component, $M_{y'}(t)$, to which the observed signal is proportional. The signal initially increases as $M_{y'}(t)$ grows under the rotation of **M** about $\mathbf{B}_1$ (at exact resonance) or $\mathbf{B}_0 + \mathbf{B}_1$ (near resonance), but, while the system approaches a new state via oscillations, **M** continually decreases under the influence of spin–spin relaxation which destroys the initial phase coherence of the spin motion within the $y'z$-plane. In solid-state TREPR, where large inhomogeneous EPR linewidths due to anisotropic magnetic interactions persist, the long-time behaviour of the spectrometer output, $S(t)$, is given by

$$S(t) \propto \omega_1 J_0(\omega_1 t) \, e^{-t/(2T_2)} \tag{B1.15.30}$$

where the oscillation of the transient magnetization is described by a Bessel function $J_0(\omega_1 t)$ of zeroth order, damped by the spin–spin relaxation time $T_2$. At low MW powers ($\omega_1^2 T_1 T_2 \ll 1$) an exponential decay of the

EPR signal is observed, governed by spin–lattice relaxation

$$S(t) \propto \omega_1 \, \mathrm{e}^{-t/T_1}. \tag{B1.15.31}$$

The rise time of the signals—independent of the chosen MW power—is proportional to the inverse inhomogeneous EPR linewidth. As can be seen from equations (B1.15.30) and (B1.15.31) a measurement of the $\omega_1$-dependence of the transient EPR signals provides a straightforward method to determine not only the relaxation parameters of the spin system but also the strength of the MW magnetic field $B_1$ at the sample. Spectral information can be obtained from a series of TREPR signals taken at equidistant magnetic field points covering the total spectral width. This yields a two-dimensional variation of the signal intensity with respect to both the magnetic field and the time axis. Transient spectra can be extracted from such a plot at any fixed time after the laser pulse as slices parallel to the magnetic field axis.

### B1.15.4.2  *MW-switched time integration method (MISTI)*

An alternative method to obtain accurate values of the spin–lattice relaxation time $T_1$ is provided by the TREPR technique with gated MW irradiation, also called the MW-switched time integration method (MISTI) [13, 14]. The principle is quite simple. The MW field is switched on with a variable delay $\tau$ after the laser flash. The amplitude of the transient signal plotted as a function of $\tau$ renders the decay of the spin-polarized initial magnetization towards its equilibrium value. This method is preferred over the TREPR technique at low MW power (see equation (B1.15.31)) since the spin system is allowed to relax in the absence of any resonant MW field in a true spin–lattice relaxation process. The experiment is carried out by adding a PIN diode MW switch between the MW source and the circulator (see figure B1.15.4), and set between a pair of isolators. Since only low levels of MW power are switched (typically less than 1 W), as opposed to those in ESE and FT EPR, the detector need not be protected against high incident power levels.

As a summary it may be of interest to point out why TREPR spectroscopy and related methods remain important in the EPR regime, even though pulsed EPR methods are becoming more and more widespread. (1) For the case of an inhomogeneously broadened EPR line the time resolution of TREPR compares favourably with pulsed techniques. (2) The low MW power levels commonly employed in TREPR spectroscopy do not require any precautions to avoid detector overload and, therefore, the full time development of the transient magnetization is obtained undiminished by any MW detection deadtime. (3) Standard CW EPR equipment can be used for TREPR requiring only moderate efforts to adapt the MW detection part of the spectrometer for the observation of the transient response to a pulsed light excitation with high time resolution. (4) TREPR spectroscopy proved to be a suitable technique for observing a variety of spin coherence phenomena, such as transient nutations [16], quantum beats [17] and nuclear modulations [18], that have been useful to interpret EPR data on light-induced spin-correlated radical pairs.

### B1.15.5  Multiple resonance techniques

In the previous chapters experiments have been discussed in which one frequency is applied to excite and detect an EPR transition. In multiple resonance experiments two or more radiation fields are used to induce different transitions simultaneously [19–23]. These experiments represent elaborations of standard CW and pulsed EPR spectroscopy, and are often carried out to complement conventional EPR studies, or to refine the information which can in principle be obtained from them.

### B1.15.5.1  *Electron–nuclear double resonance spectroscopy (ENDOR)*

It was noted earlier that EPR may at times be used to characterize the electronic structure of radicals through a measurement of hyperfine interactions arising from nuclei that are coupled to the unpaired electron spin.

In very large radicals with low symmetry, where the presence of many magnetic nuclei results in a complex hyperfine pattern, however, the spectral resolution of conventional EPR is very often not sufficient to resolve or assign all hyperfine couplings. It was as early as 1956 that George Feher demonstrated that by electron–nuclear double resonance (ENDOR) the spectral resolution can be greatly improved [24]. In ENDOR spectroscopy the electron spin transitions are still used as means of detection because the sensitivity of the electron resonance measurement is far greater than that of the nuclear resonance. In brief, an EPR transition is saturated, which leads to a collapse of the observed EPR signal as the corresponding state populations equalize. If one now simultaneously irradiates the spin system with an RF field in order to induce transitions between the nuclear sublevels, the condition of saturation in the EPR transition is lifted as the nuclear sublevel populations shift, and there is a partial recovery of the EPR signal.

ENDOR transitions can be easily understood in terms of a simple system consisting of a single unpaired electron spin ($S = \frac{1}{2}$) coupled to a single nuclear spin ($I = \frac{1}{2}$). The interactions responsible for the various splittings are summarized in the following static Hamiltonian:

$$\mathcal{H} = \mathcal{H}_{\text{EZI}} + \mathcal{H}_{\text{NZI}} + \mathcal{H}_{\text{HFI}} = \beta_e \mathbf{BgS} - g_n \beta_n \mathbf{B} \cdot \mathbf{I} + \mathbf{SAI}. \qquad (B1.15.32)$$

The coupling constants of the hyperfine and the electron Zeeman interactions are scalar as long as radicals in isotropic solution are considered, leading to the Hamiltonian

$$\mathcal{H} = g_e \beta_e \mathbf{B} \cdot \mathbf{S} - g_n \beta_n \mathbf{B} \cdot \mathbf{I} + a\mathbf{S} \cdot \mathbf{I}. \qquad (B1.15.33)$$

In the high-field approximation with $\mathbf{B} \parallel z$, the energy eigenvalues classified by the magnetic spin quantum numbers, $M_S$ and $M_I$, are given by

$$E_{M_S, M_I} = g_e \beta_e B_0 M_S - g_n \beta_n M_I B_0 + a M_S M_I \qquad (B1.15.34)$$

where $g_n$ and $a$ may be positive or negative, thus leading to a different ordering of the levels. The energy level diagram for the case $a < 0$ and $|a|/2 < g_n \beta_n B_0$ is shown in figure B1.15.8. Adopting the notation $\hbar \omega_e = g_e \beta_e B_0$ and $\hbar \omega_n = g_n \beta_n B_0$, two EPR transitions are obtained,

$$\omega_{\text{EPR}} = \omega_e \pm a/(2\hbar) \qquad (B1.15.35)$$

which obey the selection rule $\Delta M_S = \pm 1$ and $\Delta M_I = 0$. The two ENDOR transitions are

$$\omega_{\text{ENDOR}}^{\pm} = |\omega_n \pm a/(2\hbar)| \qquad (B1.15.36)$$

which satisfy the selection rule $\Delta M_S = 0$ and $\Delta M_I = \pm 1$. The absolute value is used in equation (B1.15.36) to take into account the two cases $|a|/(2\hbar) < |\omega_n|$ and $|a|/(2\hbar) > |\omega_n|$. The corresponding ENDOR spectra are shown schematically in figures B1.15.8(B) and (C). Irrespective of the EPR line monitored, two ENDOR lines, separated by $|a|/\hbar$ and centred at $|\omega_n|$, are observed. For $|a|/(2\hbar) > |\omega_n|$ the two ENDOR transitions are given by $|a|/(2\hbar) \pm |\omega_n|$: again, two lines are observed; however, separated by $2|\omega_n|$ and centred at $|a|/(2\hbar)$.

For the simple system discussed above the advantages of performing double resonance do not become so apparent: two lines are observed using either method, EPR or ENDOR. The situation dramatically changes when there are $i$ groups of nuclei present, each group consisting of $n_i$ magnetically equivalent nuclei with nuclear spin quantum number $I_i$, each one coupling to the unpaired electron spin with the hyperfine constant $a_i$. While the EPR spectrum will consist of $\Pi_i (2n_i I_i + 1)$ lines, for each group of equivalent nuclei, no matter how many nuclei there are or what their spin quantum number is, there will still be only two ENDOR lines separated by $|a_i|$ or $2\omega_{ni}$. Hence, with increasing number of groups of nuclei the number of ENDOR lines increases only in an additive way. Since the ENDOR spectral lines are comparable in width to EPR

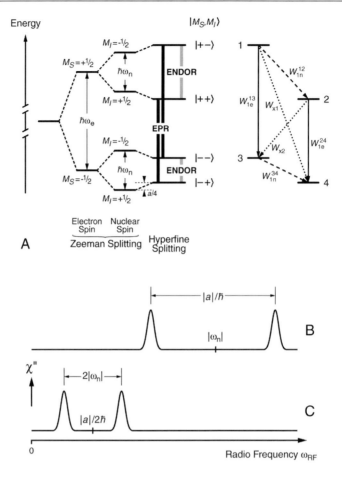

**Figure B1.15.8.** (A) Left side: energy levels for an electron spin coupled to one nuclear spin in a magnetic field, $S = I = \frac{1}{2}$, $g_n > 0$, $a < 0$, and $|a|/(2\hbar) < \omega_n$. Right side: schematic representation of the four energy levels with $|\pm \pm\rangle \equiv |M_S = \pm \frac{1}{2}, M_I = \pm \frac{1}{2}\rangle$. $|+-\rangle \equiv 1, |++\rangle \equiv 2, |--\rangle \equiv 3$ and $|-+\rangle \equiv 4$. The possible relaxation paths are characterized by the respective relaxation rates $W$. The energy levels are separated horizontally to distinguish between the two electron spin transitions. Bottom: ENDOR spectra shown when $|a|/(2\hbar) < |\omega_n|$ (B) and when $|\omega_n| < |a|/(2\hbar)$ (C).

lines, the reduced number of lines in the ENDOR spectrum results in a much greater effective resolution. Therefore, accurate values of the hyperfine couplings may be obtained from an ENDOR experiment even under conditions where the hyperfine pattern is not resolved in the EPR spectrum. In addition, ENDOR spectra become easier to interpret when there are nuclei with different magnetic moments involved. Their ENDOR lines normally appear in different frequency ranges and, from their Larmor frequencies, these nuclei can be immediately identified. ENDOR is also a well justified method when anisotropic hyperfine and nuclear quadrupole (for nuclei with $I \geq 1$) couplings in solids are to be measured. As an example, the ENDOR spectrum of the perinaphthenyl radical in liquid solution is depicted in figure B1.15.9 (see also figure B1.15.6 for a comparison with the CW EPR spectrum).

The ENDOR experiment is performed at a constant external magnetic field by applying MW and RF fields in a continuous fashion. This technique is called CW ENDOR spectroscopy. The design of an ENDOR spectrometer differs only slightly from a basic CW or pulsed EPR spectrometer. A coil located around the

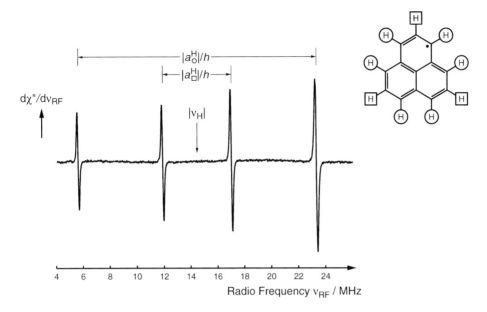

**Figure B1.15.9.** The ENDOR spectrum of the perinaphthenyl radical in mineral oil taken at room temperature. $|a_{\bigcirc}^{H}|/h = 17.68$ MHz and $|a_{\square}^{H}|/h = 5.12$ MHz are the hyperfine coupling constants for the protons in the position $\bigcirc$ and $\square$, respectively. $|a_{\bigcirc}^{H}|/(g\beta_e) = 0.631$ mT and $|a_{\square}^{H}|/(g\beta_e) = 0.183$ mT.

sample tube within the resonant EPR cavity is used as an element in a RF transmitter circuit and is the source of a RF field. The basic elements of the RF circuit include a low-power signal source or sweeper and a high-power amplifier to produce an RF output signal that can be scanned over a wide frequency range (e.g. for proton ENDOR at X-band from approximately 4 to 30 MHz) at a power level up to 1 kW. To carry out an ENDOR experiment, the magnetic field $B_0$ is set at the resonance of one of the observed EPR transitions. Then, the MW power is increased in order to partially saturate the EPR transition. The degree of saturation is provided by the saturation factor $s$ defined earlier (see equation (B1.15.13)). Finally, a strong RF oscillating field of varying frequency is applied to induce and saturate a transition within the nuclear sublevels. When the resonance condition for nuclear transitions is fulfilled, the saturated EPR transition can be desaturated by the ENDOR transition provided both transitions have energy levels in common. This desaturation of the EPR transition is detected as a change in the EPR absorption at characteristic frequencies $\omega_{\text{ENDOR}}$, and constitutes the ENDOR response.

Phenomenologically, the ENDOR experiment can be described as the creation of alternative relaxation paths for the electron spins, which are excited with microwaves. In the four-level diagram of the $S = I = \frac{1}{2}$ system described earlier (see figure B1.15.8) relaxation can occur via several mechanisms: $W_{1e}$ and $W_{1n}$ describe the relaxation rates of the electron spins and nuclear spins, respectively. $W_{x1}$ and $W_{x2}$ are cross-relaxation rates in which electron and nuclear spin flips occur simultaneously. Excitation, for example, of the EPR transition $|-+\rangle \leftrightarrow |++\rangle$ (i.e. $4 \leftrightarrow 2$) will equalize the population of both levels, 4 and 2, if the direct relaxation (characterized by the relaxation rate $W_{1e}^{24}$) cannot compete with the transitions induced by the resonant microwaves. Simultaneous application of an RF field at a frequency corresponding to the $|++\rangle \leftrightarrow |+-\rangle$ (i.e. $2 \leftrightarrow 1$) transition then opens a relaxation path via $W_{1e}^{13}$ and $W_{1n}^{34}$ or, more directly, via $W_{x1}$. The extent to which these relaxation bypasses can compete with the direct $W_{1e}^{24}$ route controls the degree of desaturation of the EPR line, and, therefore, determines the ENDOR signal intensity, which, consequently, does not generally reflect the number of contributing nuclei (in contrast to EPR and NMR). The signal intensity observed depends very critically on the balance between the various relaxation rates and the magnitude of the MW and RF fields, $B_1$ and $B_2$, respectively. Additional parameters to be varied in order to optimize the ENDOR signal-to-noise ratio are the radical concentration, the solvent viscosity and the temperature. The amplitude of ENDOR signals is furthermore influenced by the enhancement effect which occurs because the nucleus does not only experience the time-dependent magnetic field $B_2$ at the RF, but also an additional magnetic field component (the hyperfine field) due to the magnetic moment of the electron. Therefore, the effective field at the nucleus can be described as

$$B_2^{\text{eff}} = \kappa B_2 \qquad \text{(B1.15.37)}$$

where $\kappa$ is the hyperfine enhancement factor. For isotropic HFI $\kappa = |1 - M_S a/(\hbar\omega_n)|$. The hyperfine enhancement is one reason for the different intensities of the individual lines of an ENDOR line pair; at the same RF power the high-frequency line is usually more intense than the low-frequency one. Another reason for asymmetrical ENDOR line patterns is the effectiveness of the cross-relaxation paths: $W_{x1}$ is in general different from $W_{x2}$, thus leading to an asymmetrical relaxation network and, as a consequence, to unequal signal intensities. In spectra of single crystals, powders and noncrystalline solids, however, the enhancement factor is governed by different hyperfine tensor components. This often leads to unexpected intensity patterns within ENDOR line pairs.

Despite the increased resolution of ENDOR compared to EPR, some restrictions concerning the information contents of ENDOR spectra persist: (1) unlike EPR, the relative signal intensities of ENDOR line pairs belonging to different groups of nuclei do not give any indication of the relative number of nuclei belonging to the individual groups; (2) the ENDOR spectrum does not give the sign of the hyperfine couplings, that is, one does not know which ENDOR transition belongs to which electron spin state. Both problems are addressed in triple resonance, which can be seen as an extension to ENDOR spectroscopy. Therefore, triple resonance experiments are very often carried out in order to supplement ENDOR data.

### B1.15.5.2  *Electron–nuclear–nuclear triple resonance*

A refinement of the ENDOR experiment is electron–nuclear–nuclear triple resonance, now commonly denoted TRIPLE. In TRIPLE experiments one monitors the effect of a simultaneous excitation of two nuclear spin transitions on the level of the EPR absorption. Two versions, known as special TRIPLE (ST) and general TRIPLE (GT), are routinely performed on commercially available spectrometers.

#### (a) Special TRIPLE (ST)

The special TRIPLE technique [25, 26] is used for hyperfine couplings $|a|/(2\hbar) < |\omega_n|$. In a typical experiment, RF is generated at two frequencies: one is fixed at the free nuclear frequency $\omega_n$ appropriate to the sort of nuclei under scrutiny and the second is swept. These two frequencies are multiplied to obtain the sum and the difference frequencies, $\omega_n \pm \omega_{RF2}$, which are used to irradiate the sample. The experiment can be understood using the energy level and relaxation scheme of figure B1.15.8. Both ENDOR transitions, $\omega_{\text{ENDOR}}^{\pm}$ (i.e. transitions $3 \leftrightarrow 4$ and $1 \leftrightarrow 2$), associated with the same nucleus are simultaneously excited. In cases of vanishing cross-relaxation the second saturating RF field enhances the efficiency of the relaxation bypass, thus increasing the signal intensity, particularly in cases where $W_n$ is the rate-limiting step (because $W_n \ll W_e$). A second advantage of ST resonance over ENDOR is that when both RF fields are sufficiently strong to completely saturate nuclear transitions the EPR desaturation becomes independent of $W_n$. Consequently, the line intensities are no longer determined by the relaxation behaviour of the various nuclei, but rather reflect the number of nuclei involved in the transition. Finally, ST also has the advantage of higher resolution because the effective saturation of nuclear transitions results in smaller observed linewidths compared to ENDOR.

*(b) General TRIPLE (GT)*

In a general TRIPLE (GT) experiment one particular ENDOR transition is pumped with the first RF while the second RF is scanned over the whole range of nuclear resonances [27]. Therefore, nuclear transitions of different sets of nuclei of the same kind or of different kinds are saturated simultaneously and the effect on the ENDOR transitions for all the hyperfine couplings in the system is measured. Clearly, ST is included within GT. From the characteristic intensity changes of the high-frequency and low-frequency signals compared with those of the ENDOR signals the relative signs of the hyperfine coupling constants can easily be determined.

### B1.15.5.3 *Electron–electron double resonance (ELDOR)*

ELDOR is the acronym for electron–electron double resonance. In an ELDOR experiment [28] one observes a reduction in the EPR signal intensity of one hyperfine transition that results from the saturation of another EPR transition within the spin system. ELDOR measurements are still relatively rare but the experiment is firmly established in the EPR repertoire.

With help of the four-level diagram of the $S = I = \frac{1}{2}$ system (see figure B1.15.8) two common ways for recording ELDOR spectra will be illustrated. In frequency-swept ELDOR the magnetic field is set at a value that satisfies the resonance condition for one of the two EPR transitions, e.g. $4 \leftrightarrow 2$, at the fixed observe klystron frequency, $\omega_{obs}$. The pump klystron is then turned on and its frequency, $\omega_{pump}$, is swept. When the pump frequency passes through the value that satisfies the resonance condition of the $3 \leftrightarrow 1$ transition, there is a decrease in the signal at the frequency $\omega_{obs}$ which constitutes the ELDOR signal. In field-swept ELDOR the pumping and observing MW frequencies are held fixed at predetermined values and the magnetic field is swept through the region of resonance. ELDOR experiments are technically more difficult than ENDOR: simultaneous EPR in one magnetic field for two different transitions requires irradiation simultaneously at two MW frequencies. That is, one requires a resonator tunable to two MW frequencies separated by a multiple of the hyperfine coupling. The development of loop-gap and split-ring resonators has, because of their wide bandwidth and the feasibility of high filling factors, made ELDOR a truly practical technique.

To analyse ELDOR responses, the reduction of the observed EPR transition at $\omega_{obs}$ is expressed quantitatively in terms of the ELDOR reduction factor

$$R = \frac{I(\omega_{pump} = 0) - I(\omega_{pump})}{I(\omega_{pump} = 0)} \tag{B1.15.38}$$

where $I(\omega_{pump} = 0)$ is the observed EPR intensity with pump power off, and $I(\omega_{pump})$ is the intensity with pump power on. The ELDOR technique is very sensitive to the various relaxation mechanisms involved. For the $S = I = \frac{1}{2}$ system $R$ may be expressed in terms of the six relaxation rates between the four energy levels that are indicated in figure B1.15.8. With the assumption $W_e^{13} = W_e^{24} = W_e$ and $W_n^{12} = W_n^{34} = W_n$ the ELDOR reduction factor is given by

$$R = \frac{W_n^2 - W_{x1} W_{x2}}{W_e(2W_n + W_{x1} + W_{x2}) + (W_n + W_{x1})(W_n + W_{x2})} \tag{B1.15.39}$$

which shows that the ELDOR response will be a reduction if $W_n^2 > W_{x1}, W_{x2}$. If modulation of dipolar hyperfine couplings is the dominant relaxation mechanism, this condition can be fulfilled for dilute radical concentrations at low temperatures. At high concentrations or sufficiently high temperatures, Heisenberg spin exchange or chemical exchange, which tends to equalize the population of all spin levels, is the dominant ELDOR mechanism.

ELDOR has been employed to study a number of systems such as inorganic compounds, organic compounds, biologically important compounds and glasses. The potential of ELDOR for studying slow molecular motions has been recognized by Freed and coworkers [29, 30].

**B1.15.6   Pulsed EPR spectroscopy**

By far the greatest advantage of pulsed EPR [31, 32] lies in its ability to manipulate the spin system nearly at will and, thus, to measure properties that are not readily available from the CW EPR spectra. Nevertheless, EPR has long remained a domain of CW methods. In contrast to the rapid development of pulsed NMR spectroscopy, the utilization of the time domain in EPR took a much longer time, even though the underlying principles are essentially the same. There are several reasons for this slow development of pulsed EPR. (1) The large energies involved in electron spin interactions (see figure B1.15.3) can give rise to spectral widths of the order of 10–25% of the carrier frequency (at X-band) as opposed to the ppm scale which applies to NMR. Consequently, with the exception of some organic radicals in solution and a few defect centres in single crystals, it is technically impossible to excite the entire EPR spectrum by a pulse of electromagnetic radiation. (2) CW EPR records derivatives of absorption spectra by using magnetic field modulation in a range between 10 kHz and 100 kHz, a method that takes advantage of narrowband detection at the modulation frequency (see figure B1.15.5) and of better resolution of the derivative as compared to the absorption lineshape (see figure B1.15.6). Calculating the derivative from the absorption lineshape obtained with pulsed methods results in a decrease in the signal-to-noise ratio. For these reasons, CW EPR is typically more sensitive than pulsed EPR at a given resolution. (3) Finally, the fact that electron spin relaxation times are orders of magnitudes shorter than the nuclear spin relaxation times encountered in NMR makes the technology required to perform pulsed EPR experiments much more demanding.

In recent years, however, enormous progress has been made and with the availability of the appropriate MW equipment pulsed EPR has now emerged from its former shadowy existence. Fully developed pulse EPR instrumentation is nowadays commercially available [31, 33].

The practical goal for pulsed EPR is to devise and apply pulse sequences in order to isolate pieces of information about a spin system and to measure that information as precisely as possible. To achieve this goal it is necessary to understand how the basic instrumentation works and what happens to the spins during the measurement.

*B1.15.6.1   Pulses and their effects*

For an understanding of pulsed excitation of spin ensembles it is of fundamental importance to realize that radiation pulses actually contain ranges of frequencies: A burst of monochromatic microwaves at frequency $\omega_{MW}$ and of pulse duration $t_p$ translates into a frequency spectrum of the pulse that has field components at all frequencies. The amplitude of the field drops off as one moves away from the carrier frequency $\omega_{MW}$ according to $B_1(\omega) \propto B_1 \sin(\omega_{MW} t_p)/(\omega_{MW} t_p)$. The excitation bandwidth of a specific pulse depends only on the pulse duration, $t_p$.

The effect of an MW pulse on the macroscopic magnetization can be described most easily using a coordinate system $(x', y', z)$ which rotates with the frequency $\omega_{MW}$ about the $z$-axis defined by the applied field **B**. Initially, the net magnetic moment vector **M** is in its equilibrium position oriented parallel to the direction of the strong external field. In the rotating frame, $\mathbf{B}_1$ is a stationary field, which is assumed to be oriented parallel to the $x'$-axis of the rotating coordinate system. The result of applying a short intense MW pulse is to rotate the magnetization **M** about the axis defined by $\mathbf{B}_1$, i.e. the $x'$-axis, through the flip angle $\theta = \gamma_e B_1 t_p = \omega_1 t_p$, expressed in radians. When the duration of the MW radiation at a given MW power level is just long enough to flip **M** into the $x'y'$-plane, the pulse is defined as a $\pi/2$-pulse. Immediately after the cessation of the pulse, **M** has been rotated with its magnitude unaltered (if relaxation phenomena are negligible during the excitation pulse) to an orientation perpendicular to $\mathbf{B}_1$ at angle $\theta$ with respect to $B_0 \parallel z$. After this perturbation the system is then allowed to return to its equilibrium, or, after an appropriate delay, additional pulses with specific flip angles and phases are applied to further manipulate the spin system. With suitable apparatus, that is, the detection system aligned in the direction of the $y'$-axis of the rotating axis system, the temporal behaviour of the $y'$-component of the magnetization can be followed. The normalized FT of the function $M_{y'}(t)$ provides the lineshape which is analogous to that obtained from CW EPR experiments under nonsaturating conditions.

### B1.15.6.2 *Instrumentation*

The design of a pulsed EPR spectrometer depends heavily on the required pulse length and pulse power which in turn are mainly dictated by the relaxation times of the paramagnetic species to be studied, but also by the type of experiment performed. When pulses of the order of a few nanoseconds are required (either to compete with the relaxation times or to excite a broad spectral range) not only is high MW power needed to fulfill the condition $\gamma_e B_1 t_p = \pi/2$, but also the whole design of such a high power spectrometer becomes much more complex and the construction is more expensive. In most of today's pulsed EPR spectrometers the MW pulses are formed on a low-power level by fast switching diodes after the CW source. These low-power pulses are then fed into a pulsed high-power MW amplifier (typically a travelling-wave tube amplifier) capable of giving the requisite high power up to a few kW. The amplified pulses are then directed into a resonant cavity to excite the EPR transitions of the system. The MW power in the resonator grows as $[1 - \exp(-\omega_{MW}t/Q)]$ and decays as $\exp(-\omega_{MW}t/Q)$ in response to a square, resonant pulse. An important consideration in a pulsed EPR spectrometer is the detection deadtime, or how soon after a pulse the signal $M_{y'}(t)$ can be measured. Typically, the deadtime is taken to be the time when ringing from the resonator equals thermal noise. The choice of an appropriate resonator for pulsed EPR experiments is therefore always influenced by the conflicting demands for a short deadtime and good sensitivity: a low quality factor $Q$ for bandwidth coverage and fast instrumental response should be combined with concentrated MW fields for short pulses and high filling factor, the latter in partial compensation for the loss in sensitivity in favour of fast time resolution. The spin system responds to the exciting MW pulses by producing a signal at a later time when the incident pulses are off. A low-noise amplifier amplifies the signal to a level well above the noise floor of the detector. Standard Schottky barrier diodes can be used as detectors up to a bandwidth of 5 MHz. For broader bandwidths, multiplying mixers can be employed to downconvert the signal from the sample to a video signal centred at zero frequency. (The mixer output is the sum or the difference of two input frequencies and the signal amplitudes are proportional to the input amplitudes.) A quadrature mixer has two inputs, one for the signal and one for a reference from the master MW oscillator. Its two outputs are in quadrature with each other: one is out of phase by 90° and the other in phase with the pulse phase. Quadrature detection means that, unlike the case in other spectrometers, the reference arm needs no phase shifter since the phase of the recorded signal can be adjusted digitally by taking a linear combination of the two quadrature components. Typically, the low-noise MW amplifier and the mixer detector are very sensitive to high-power reflections from the cavity and, therefore, have to be protected during the excitation pulse. A gated PIN-diode switch in front of the amplifier strongly attenuates the input of the detection system during the excitation pulses and, therefore, avoids saturation or permanent damage.

### B1.15.6.3 *Pulse EPR methods*

#### (a) Fourier transform EPR (FT EPR)

All operating principles are the same as in FT NMR. A single short and intense MW pulse (typically a $\pi/2$-pulse along $x'$) is applied to flip the magnetizations into the $x'y'$-plane of the rotating frame (see figure B1.15.10(A)). The induced signal proportional to $M_{y'}$ will decay due to transverse relaxation or sample inhomogeneities. This process is called free induction decay (FID). The complete spectrum is obtained without the need of a field sweep via FT of the FID. Under most conditions, the FT EPR spectrum measured using a single excitation pulse corresponds exactly to the CW EPR spectrum. Todays state-of-the-art pulsed EPR spectrometers feature $\pi/2$-pulse lengths of $t_p \approx 5$ ns or less, corresponding to an excitation bandwidth of roughly 200 MHz.

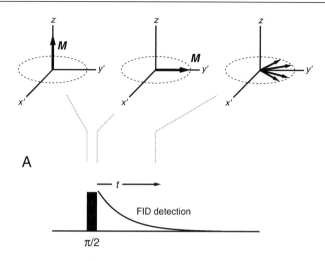

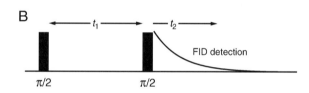

**Figure B1.15.10.** FT EPR. (A) Evolution of the magnetization during an FT EPR experiment (rotating frame representation). (B) The COSY FT EPR experiment.

Therefore, FT EPR is applicable to not too wide spectral patterns consisting of narrow lines (as typical for free radicals in solution) with long enough $T_2$ so that the FID does not die away before the deadtime has elapsed. In the case of inhomogeneously broadened EPR lines (as typical for free radicals in solids) the dephasing of the magnetizations of the individual spin packets (which all possess slightly different resonance frequencies) will be complete within the detection deadtime and, therefore, the FID signal will usually be undetectable.

In complete analogy to NMR, FT EPR has been extended into two dimensions. Two-dimensional correlation spectroscopy (COSY) is essentially subject to the same restrictions regarding excitation bandwidth and detection deadtime as was described for one-dimensional FT EPR. In 2D-COSY EPR a second time dimension is added to the FID collection time by a preparatory pulse in front of the FID detection pulse and by variation of the evolution time between them (see figure B1.15.10(B)). The FID is recorded during the detection period of duration $t_2$, which begins with the second $\pi/2$-pulse. For each $t_1$ the FID is collected, then the phase of the first pulse is advanced by 90°, and a second set of FIDs is collected. The two sets of FIDs, whose amplitudes oscillate as functions of $t_1$, then undergo a two-dimensional complex Fourier transformation, generating a spectrum over the two frequency variables $\omega_1$ and $\omega_2$. The peaks along the leading diagonal $\omega_1 = \omega_2$ correspond to the usual absorption spectrum, whereas the cross-peaks (peaks removed from the diagonal) provide evidence for cross-correlations.

*(b) Electron-spin echo (ESE) methods*

Under conditions where the rapid decay of the FID following a single excitation pulse is governed by inhomogeneous broadening the dephasing of the individual spin packets in the $x'y'$-plane can be reversed by the application of a second MW pulse in an ESE experiment (see figure B1.15.11) [34]. As before, the experiment begins with the net electron spin magnetization $\mathbf{M}$ aligned along the magnetic field direction $z$. At the end of the first $\pi/2$-pulse, which is applied at the Larmor precession frequency $\omega_0$, with the amplitude $B_1$ pointing along the $x'$-direction, the net magnetic moment is in the equatorial plane. Immediately, the magnetization starts to decay as different spin packets precess about $z$ at their individual Larmor frequencies $\omega_i \neq \omega_0$. In the rotating frame the contributions of different spin packets to the magnetization $M_{y'}$ appear to fan out as shown in figure B1.15.11(A): viewed in the laboratory frame some spins would appear to precess faster ($\omega_i > \omega_0$) and some slower ($\omega_i < \omega_0$) than the average. As a result, the FID decays rapidly and after a short time there is no detectable signal. From this FID an echo can be generated by means of a second MW pulse, applied at time $\tau$ after the first pulse. The second pulse is just long enough to turn the magnetization vectors through $180°$ about the $x'$-axis. The original precession frequencies and the directions of rotation of the individual components will remain unaltered and, therefore, the magnetizations will rotate toward each other in the $x'y'$-plane until they refocus after the same time $\tau$ into a macroscopic magnetic moment along the $-y'$-axis. At this point the spin alignment produces a microwave field $B_1$ in the cavity corresponding to an emission signal that is referred to as an echo [34].

As the spins precess in the equatorial plane, they also undergo random relaxation processes that disturb their movement and prevent them from coming together fully realigned. The longer the time $\tau$ between the pulses the more spins lose coherence and consequently the weaker the echo. The decay rate of the two-pulse echo amplitude is described by the phase memory time, $T_M$, which is the time span during which a spin can remember its position in the dephased pattern after the first MW pulse. $T_M$ is related to the homogeneous linewidth of the individual spin packets and is usually only a few microseconds, even at low temperatures.

The two-pulse sequence $\pi/2$-$\tau$-$\pi$-$\tau$ is not the only sequence which leads to the formation of an echo. A pulse sequence which has proven to have particular value consists of three $\pi/2$-pulses as depicted in figure B1.15.11(B). In this three-pulse sequence with pulse intervals $\tau$ and $T$ a so-called stimulated echo is formed after an interval $\tau$ following the third pulse. The mechanism of formation of the stimulated echo is a little more complicated than that of the primary echo and the reader is referred to some excellent review articles [32, 35, 36] for a comprehensive discussion of this topic. Here it is sufficient to mention that with the second $\pi/2$-pulse the $y'$-components of the dephased magnetization pattern are temporarily stored in the $x'z$-plane where they remain during the waiting time $T$. The third MW pulse brings the $M_z$ magnetizations back into the $x'y'$-plane, where they continue their time evolution and give rise to the stimulated echo at time $\tau$ after the third pulse. The characteristic time of the three-pulse echo decay as a function of the waiting time $T$ is much longer than the phase memory time $T_M$ (which governs the decay of a two-pulse echo as a function of $\tau$), since the phase information is stored along the $z$-axis where it can only decay via spin–lattice relaxation processes or via spin diffusion.

In electron-spin-echo-detected EPR spectroscopy, spectral information may, in principle, be obtained from a Fourier transformation of the second half of the echo shape, since it represents the FID of the refocused magnetizations, however, now recorded with much reduced deadtime problems. For the inhomogeneously broadened EPR lines considered here, however, the FID and therefore also the spin echo, show little structure. For this reason, the amplitude of the echo is used as the main source of information in ESE experiments. Recording the intensity of the two-pulse or three-pulse echo amplitude as a function of the external magnetic field defines electron-spin-echo- (ESE-) detected EPR spectroscopy. Such a field-swept ESE spectrum is similar to the conventional CW EPR spectrum except for the fact that the lines appear in absorption and not in the more familiar first derivative form.

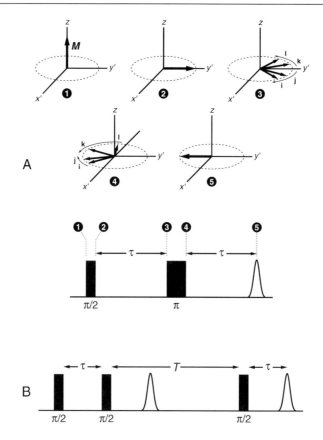

**Figure B1.15.11.** Formation of electron spin echoes. (A) Magnetization of spin packets $i$, $j$, $k$ and $l$ during a two-pulse experiment (rotating frame representation). (B) The pulse sequence used to produce a stimulated echo. In addition to this echo, which appears at $\tau$ after the third pulse, all possible pairs of the three pulses produce primary echoes. These occur at times $2\tau$, $2(\tau + T)$ and $(\tau + 2T)$.

ESE-detected EPR spectroscopy has been used advantageously for the separation of spectra arising from different paramagnetic species according to their different echo decay times. Furthermore, field-swept ESE spectroscopy is superior to conventional CW EPR when measuring very broad spectral features. This is because the field modulation amplitudes used in CW EPR to detect the first derivative of the signal are often too small compared to the width of the EPR line, so that the gradient of the absorption signal becomes very small. Such features are either invisible in CW EPR or obscured by baseline drifts, while they can be well distinguished in the absorption spectra with ESE-detected EPR.

In electron spin echo relaxation studies, the two-pulse echo amplitude, as a function of the pulse separation time $\tau$, gives a measure of the phase memory relaxation time $T_M$ from which $T_2$ can be extracted if $T_1$-effects are taken into consideration. Problems may arise from spectral diffusion due to incomplete excitation of the EPR spectrum. In this case some of the transverse magnetization may leak into adjacent parts of the spectrum that have not been excited by the MW pulses. Spectral diffusion effects can be suppressed by using the Carr–Purcell–Meiboom–Gill pulse sequence, which is also well known in NMR. The experiment involves using a sequence of $\pi$-pulses separated by $2\tau$ and can be denoted as $[\pi/2\text{-}(\tau\text{-}\pi\text{-}\tau\text{-}\text{echo})_n]$. A series of echoes separated by $2\tau$ is generated and the decay in their amplitudes is characterized by $T_M$.

The other important (spin–lattice) relaxation time $T_1$ is accessible with the help of an additional preparation, e.g. an inversion (i.e. a $\pi$-pulse) or saturation pulse (a single long pulse or a chain of short $\pi/2$-pulses) placed with a variable delay time $T$ in front of a two-pulse ESE sequence. The $T_1$-information is then extracted from the dependence of the echo amplitude on the interval $T$. Experiments of this type are generally called 'inversion recovery' or 'saturation recovery' experiments. In principle $T_1$ can also be estimated from the amplitude of the stimulated echo, as the $z$-magnetization relaxes towards thermal equilibrium during the variable pulse delay time $T$. Here, inaccuracies in measuring $T_1$ may again originate from spectral diffusion and the interaction between the electron spin and the nuclear spins which can affect the amplitude of the echo.

The electron-spin echo envelope modulation (ESEEM) phenomenon [37, 38] is of primary interest in pulsed EPR of solids, where anisotropic hyperfine and nuclear quadrupole interactions persist. The effect can be observed as modulations of the echo intensity in two-pulse and three-pulse experiments in which $\tau$ or $T$ is varied. In liquids the modulations are averaged to zero by rapid molecular tumbling. The physical origin of ESEEM can be understood in terms of the four-level spin energy diagram for the $S = I = \frac{1}{2}$ model system introduced earlier to describe ENDOR (see figure B1.15.8). So far, however, only isotropic hyperfine couplings have been considered, leading to an EPR spectrum of this system that comprises the two allowed transitions $1 \leftrightarrow 3$ and $2 \leftrightarrow 4$ with $\Delta M_S = \pm 1$ and $\Delta M_I = 0$. The situation is different for the case where the hyperfine couplings are anisotropic and, in particular, of the same order of magnitude as the nuclear Zeeman couplings. Because of the anisotropic nature of the interactions, the energy levels of the spin system are modified and the nuclear spin states are mixed. As a consequence, the transitions $1 \leftrightarrow 4$ and $2 \leftrightarrow 3$ involving a simultaneous nuclear spin transition (both forbidden for the isotropic case) are now also allowed to some extent. In figure B1.15.12(A) the evolution of this spin system in a two-pulse echo experiment (see figure B1.15.11(A)) is considered. Since there are four transitions, there are four components of the magnetization to keep track of. For simplicity, only the magnetizations of two transitions, $\omega_{24}$ and $\omega_{14}$ (labelled $a$ and $f$, respectively), originating from the same nuclear spin level in the lower electron spin manifold, are considered. By applying sufficiently short MW pulses both allowed and forbidden transitions are excited simultaneously. After the first $\pi/2$-pulse, the two sets of electrons are precessing at two different frequencies separated by the nuclear frequency. If spin packet $a$ is on resonance ($\omega_{24} = \omega_{MW}$), its component of the magnetization is fixed in the rotating frame, whereas the magnetization component $f$ precesses with frequency $\omega_{14} - \omega_{MW} = \omega_{12}$. After the time $\tau$, the $\pi$-pulse inverts the vector $a$ into the $-y'$-direction, and at the same time, because of the branching of transitions, gives rise to a new component $f'$. The effect of the $\pi$-pulse on the magnetization component $f$ is a rotation about $x'$ by $180°$ and the formation of a component $a'$ according to the ratio of the transition probabilities for allowed and forbidden transitions. $a$ and $a'$ will remain unaltered in the rotating frame (because they are on resonance), whereas the two vectors $f$ and $f'$ continue to precess with the off-resonance frequency $\omega_{12}$. At time $\tau$, $f$ will refocus with $a$ at the $-y'$-axis to form an echo, but the vectors $a'$ and $f'$ will not contribute, because they are no longer oriented along $-y'$. This results in a reduction of the echo intensity at time $\tau$. The component $f'$ will only contribute to the echo if, in time $\tau$, it precesses an integral number of times in the $x'y'$-plane. Therefore, the echo amplitude oscillates in proportion to $\cos(\omega_{14}\tau)$. The same holds for any combination of transitions with energy levels in common. Therefore, one expects the echo intensity to oscillate not only with the nuclear frequencies $\omega_{12}$ and $\omega_{34}$ but also with the sum and the difference of these frequencies. By FT of the echo envelope an ENDOR-like spectrum is obtained. The amplitudes of the modulation frequencies are determined by the depth parameter $k = 4I_a I_f = (B\omega_n/(\hbar\omega_{12}\omega_{34}))^2$, where $I_a$ and $I_f$ denote the intensities of the allowed and forbidden transitions, respectively. $B/\hbar$ is a measure of the anisotropy of the hyperfine coupling tensor. Large modulation amplitudes are expected for $\omega_{12}, \omega_{34} \to 0$. This is in contrast to ENDOR spectroscopy, where the enhancement factor and, therefore, the ENDOR line intensities, decrease for small nuclear transition frequencies. For small hyperfine coupling constants $\omega_{12} \approx \omega_{34} \approx \omega_n$ and $k \propto (B/(\hbar\omega_n))^2$. Again in contrast to ENDOR, the ESEEM modulation depth will increase for nuclei with smaller $\gamma_n$.

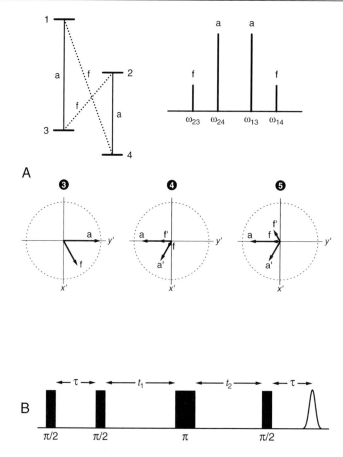

**Figure B1.15.12.** ESEEM spectroscopy. (A) Top: energy level diagram and the corresponding stick spectrum for the two allowed ($a$) and two forbidden ($f$) transitions. Bottom: time behaviour of the magnetization of an allowed ($a$) spin packet and a forbidden ($f$) spin packet during a two-pulse ESE sequence (see figure B1.15.11(A)). (B) The HYSCORE pulse sequence.

The stimulated (three-pulse) echo decay may also be modulated, but only by the nuclear frequencies $\omega_{12}$ and $\omega_{34}$ and not by their sum and difference frequencies. The qualitative reason for this is that the first pulse generates modulation at the nuclear frequencies; the second pulse additionally incorporates the sum and difference frequencies and the third pulse causes interference of the sum and difference frequencies to leave only the ENDOR frequencies. Apart from the depth parameter $k$ the modulation amplitudes of the ENDOR frequencies $\omega_{12}$ and $\omega_{34}$ are determined by $\sin^2(\omega_{34}\tau/2)$ and $\sin^2(\omega_{12}\tau/2)$, respectively. As a consequence, so-called blind spots can occur. For example, if $\frac{1}{2}\omega_{34}$ is an integral multiple of $\pi$, i.e. $\tau = 2\pi n/\omega_{34}$, then the modulation at $\omega_{12}$ is completely suppressed. Therefore, the dependence on the response on $\tau$ should also be examined.

The main advantage of the three-pulse ESEEM experiment as compared to the two-pulse approach lies in the slow decay of the stimulated echo intensity determined by $T_1$, which is usually much longer than the phase memory time $T_M$ that limits the observation of the two-pulse ESE.

More sophisticated pulse sequences have been developed to detect nuclear modulation effects. With a five-pulse sequence it is theoretically possible to obtain modulation amplitudes up to eight times greater than in a three-pulse experiment, while at the same time the unmodulated component of the echo is kept close to zero. A four-pulse ESEEM experiment has been devised to greatly improve the resolution of sum-peak spectra.

In the 2D three-pulse ESEEM technique both the time intervals $t_1 = \tau$ and $t_2 = T$ of a stimulated echo sequence are independently increased in steps (see figure B1.15.11(B)) [39]. If the spacing between the first two pulses, $\tau$, is varied over a sufficiently broad range, blind spots, which caused problems in one-dimensional spectra, do not arise in this 2D ESEEM method. Combination cross peaks arising from couplings with several inequivalent nuclei can be used to determine the relative signs of the hyperfine splittings. A disadvantage of the 2D three-pulse ESEEM technique is that the echo intensities decay at different rates along the two time axes, with $T_M$-relaxation along the $t_1$-axis and $T_1$-relaxation along the $t_2$-axis. As a result, the linewidths in the two frequency dimensions can differ by orders of magnitude.

An alternative 2D ESEEM experiment based on the four-pulse sequence depicted in figure B1.15.12(B) has been proposed by Mehring and coworkers [40]. In the hyperfine sublevel correlation (HYSCORE) experiment, the decay of the echo intensity as a function of $t_1$ is governed by $T_1$-relaxation, whereas the echo decay along the $t_2$-axis is determined by the $T_2$-relaxation of the nuclei. Since both relaxation processes are fairly slow, the resolution along both frequency dimensions is much increased compared to the 2D three-pulse ESEEM experiment. In the HYSCORE experiment, too, the positions of the cross-peaks can be used to determine the relative signs of the hyperfine coupling constants.

The ESEEM methods are best suited for the measurement of small hyperfine couplings, e.g. for the case of nuclear spins with small magnetic moment. Larger hyperfine interactions can be measured best by pulsed versions of ENDOR spectroscopy. These methods will be introduced as a final application of the pulsed excitation scheme introduced earlier. Pulsed ENDOR methods are double-resonance techniques wherein, at some particular time in an ESE pulse sequence, a RF pulse is applied that is swept in frequency to match resonance with the hyperfine-coupled nuclei. The typical pulse schemes for the most commonly used versions of pulsed ENDOR, termed Mims- [41] and Davies-type [42] ENDOR to acknowledge those who originally introduced them, are depicted in figure B1.15.13. In both experiments the ENDOR effect is manifested in a change of the ESE intensity when the RF field is on nuclear resonance. The ENDOR spectrum can thus be recorded by detecting the echo amplitude as a function of the frequency of the RF pulse.

The Davies-ENDOR technique is based on an inversion recovery sequence (see figure B1.15.13(A)). The experiment starts by interchanging the populations of levels 1 and 3 of one of the EPR transitions of the $S = I = \frac{1}{2}$ model spin system by means of a first selective MW $\pi$-pulse (with a strength $|\omega_1| \ll |A|/\hbar$). Neglecting relaxation during the time span $T$, a two-pulse ESE sequence performed after time $t = T$ produces an echo which is inverted with respect to a two-pulse ESE applied to the same spin system at thermal equilibrium. When, during the time span $T$, a selective RF $\pi$-pulse is applied, the two-pulse ESE will disappear as soon as the RF field is on resonance with one of the two transitions $1 \leftrightarrow 2$ or $3 \leftrightarrow 4$. This is because the populations of the nuclear sublevels are interchanged by the RF pulse, which simultaneously equalizes the populations of the on-resonant EPR transition. The ENDOR effect will not be observable if the preparation pulse (i.e. MW pulse 1) and/or the two-pulse ESE sequence are non-selective.

Mims ENDOR involves observation of the stimulated echo intensity as a function of the frequency of an RF $\pi$-pulse applied between the second and third MW pulse. In contrast to the Davies ENDOR experiment, the Mims-ENDOR sequence does not require selective MW pulses. For a detailed description of the polarization transfer in a Mims-type experiment the reader is referred to the literature [43]. Just as with three-pulse ESEEM, blind spots can occur in ENDOR spectra measured using Mims' method. To avoid the possibility of missing lines it is therefore essential to repeat the experiment with different values of the pulse spacing $\tau$. Detection of the echo intensity as a function of the RF frequency and $\tau$ yields a real two-dimensional experiment. An FT of the $\tau$-domain will yield cross-peaks in the 2D-FT-ENDOR spectrum which correlate different ENDOR transitions belonging to the same nucleus. One advantage of Mims ENDOR over Davies ENDOR is its larger echo intensity because more spins due to the nonselective excitation are involved in the formation of the echo.

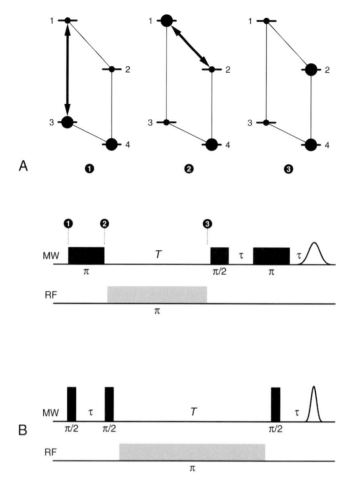

**Figure B1.15.13.** Pulsed ENDOR spectroscopy. (A) Top: energy level diagram of an $S = I = \frac{1}{2}$ spin system (see also figure B1.15.8(A)). The size of the filled circles represents the relative population of the four levels at different times during the (3 + 1) Davies ENDOR sequence (bottom). (B) The Mims ENDOR sequence.

Pulsed ENDOR offers several distinct advantages over conventional CW ENDOR spectroscopy. Since there is no MW power during the observation of the ESE, klystron noise is largely eliminated. Furthermore, there is an additional advantage in that, unlike the case in conventional CW ENDOR spectroscopy, the detection of ENDOR spin echoes does not depend on a critical balance of the RF and MW powers and the various relaxation times. Consequently, the temperature is not such a critical parameter in pulsed ENDOR spectroscopy. Additionally the pulsed technique permits a study of transient radicals.

More advanced pulsed techniques have also been developed. For a review of pulsed ENDOR techniques the reader is referred to [43–45].

### B1.15.7 High-field EPR spectroscopy

Since its discovery in 1944 by Zavoisky, EPR has typically been performed at frequencies below 40 GHz. This limitation was a technical one, but recent developments in millimetre and submillimetre wave frequency technology and magnetic field technology have enabled the exploration of ever higher EPR frequencies [46–48].

High-field/high-frequency EPR spectroscopy has a number of inherent advantages [47]. (1) The spectral resolution of $g$-factor differences and anisotropies greatly improves since the electron Zeeman interaction scales linearly with the magnetic field (see equation (B1.15.18)). If paramagnetic centres with different $g$-values or different magnetic sites of rather similar $g$-values are present, the difference in the spectral field positions of the resonances is proportional to the MW frequency $\omega$

$$\Delta B_0 = \frac{\hbar\omega}{\beta_e} \left( \frac{1}{g_1} - \frac{1}{g_2} \right). \tag{B1.15.40}$$

Even for a single radical the spectral resolution can be enhanced for disordered solid samples if the inhomogeneous linewidth is dominated by unresolved hyperfine interactions. Whereas the hyperfine line broadening is not field dependent, the anisotropic **g**-matrix contribution scales linearly with the external field. Thus, if the magnetic field is large enough, i.e. when the condition

$$\frac{\Delta g}{g_{iso}} B_0 > B_{1/2}^{HFI} \tag{B1.15.41}$$

is fulfilled, the powder spectrum is dominated by the anisotropic **g**-matrix. Equation (B1.15.41) may be considered as the high-resolution condition for solid-state EPR spectra, to be fulfilled only at high enough $B_0$. From figure B1.15.14 one sees that for a nitroxide spin label in a protein this is fulfilled almost completely at 95 GHz, but not at 10 GHz. In the case of well resolved $g$-anisotropy the extension to high-field ENDOR and ESEEM has the additional advantage of providing single-crystal-like hyperfine information when transitions are excited at field positions where only specific orientations of the **g**-matrix with respect to the external magnetic field contribute to the spectrum. (2) Relaxation times become longer for many systems at higher frequencies. (3) Particularly in studies of small samples, high-field/high-frequency EPR is typically more sensitive compared to EPR at X-band frequencies, by virtue of the increased Boltzmann factor (see equation (B1.15.10)). (4) For high-spin systems with zero-field splittings larger than the MW quantum it is impossible to observe all EPR transitions. Here, higher-frequency experiments are essential for recording the whole spectrum. (5) At high frequencies it becomes possible to violate the high-temperature approximation with standard cryogenic systems. This effect can be exploited to gain information on the absolute sign of parameters of the spin Hamiltonian.

Disadvantages of high-field/high-frequency EPR are mainly technical ones due to the limited availability of MW components operating at millimetre and submillimetre wavelengths and the high costs of spectrometer development. Furthermore, low-frequency EPR will not be completely superseded by high-field/high-frequency EPR because some experiments that rely on the violation of the high-field approximation no longer work when increasing the EPR frequency. The largest disadvantages occur in studies of proton interactions, where ESEEM is a convenient tool at X-band but cannot be used at W-band frequencies and above because of too small modulation depths for most systems. Nevertheless, these drawbacks are outweighed by the advantages to such an extent that high-field/high-frequency EPR methods will become more and more widespread as one overcomes the technical hurdles.

The early high-field/high-frequency EPR spectrometers were developed mostly on the basis of klystron MW generators. In the past few years, however, solid-state MW sources such as Gunn oscillators or IMPATT diodes were applied more frequently. To improve their mediocre frequency stability they need to be phase locked to a low-frequency stable reference source. If frequencies higher than approximately 150 GHz are required, the output frequency of the solid-state MW generator can be multiplied in a Schottky diode harmonic generator (multiplication factors between 2 and 5). To generate even higher frequencies up to 1 THz more exotic high-power pulsed and CW tube sources such as gyrotrons, extended interaction oscillators (EIOs), backward wave oscillators (BWOs) or magnetrons are available. Their spectral characteristics may be favourable; however, they typically require highly stabilized high-voltage power supplies. Still higher frequencies may be obtained using far-infrared gas lasers pumped for example by a $CO_2$ laser [49].

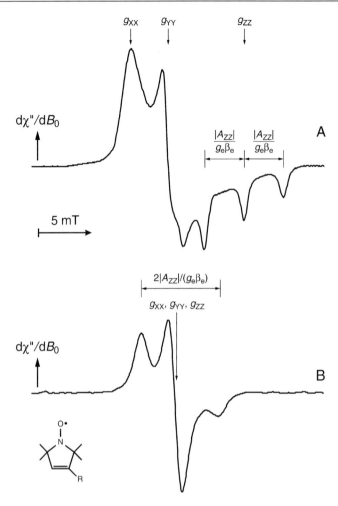

**Figure B1.15.14.** Comparison of 95.1 GHz (A) and 9.71 GHz (B) EPR spectra for a frozen solution of a nitroxide spin label attached to insulin measured at 170 K.

All the waveguide elements become very small at high frequencies as compared to the standard X-band or even Q-band. This produces high losses up to several decibels per metre due to waveguide imperfections. Therefore, if millimetre and submillimetre waves have to be transmitted over long (and straight) distances, oversized or corrugated waveguides are normally used because of their smaller ohmic losses. Corrugated waveguides have narrow grooves, each a quarter-wavelength deep, cut into the guide walls. The effect of the grooves is to destructively average the **E** field near the wall surface which cannot now have a non-zero component perpendicular to the surface. To couple to a resonant cavity, however, these waveguides need to be tapered back to the fundamental-mode waveguide. Very recently developed EPR spectrometers operating at 130 GHz [50], 250 GHz [51] and 360 GHz [52] (see figure B1.15.15) forgo waveguides for millimetre-wave transmission for the most part. Instead quasi-optic techniques are used [53]. Once millimetre-waves have been converted into Gaussian beams by means of corrugated feedhorns they can be transported and manipulated in free space using quasi-optical elements such as lenses (constructed from Teflon or high-density polyethylene) and off-axis mirrors. The losses in these elements are virtually negligible.

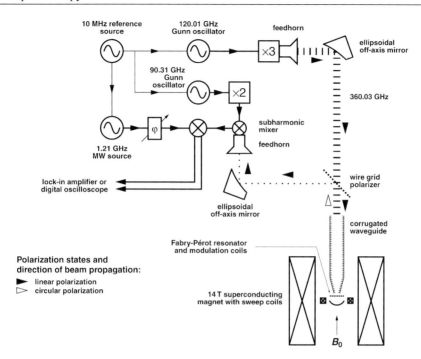

**Figure B1.15.15.** High-field/high-frequency EPR spectrometer operating at 360.03 GHz [52]. Microwaves at 360.03 GHz are produced by frequency multiplication of the output of a Gunn oscillator and are then passed into a corrugated feedhorn to set up a fundamental Gaussian beam. Refocusing and redirection of the beam is accomplished using off-axis mirrors. The microwaves then travel through a corrugated waveguide and are coupled into a Fabry–Pérot-type open cavity. Incident linearly polarized microwaves and circularly polarized EPR signals emitted from the resonator are separated by a wiregrid polarizer. Heterodyne phase-sensitive detection of the EPR signal is achieved in a subharmonic mixer using microwaves of 180.62 GHz to produce a 1.21 GHz intermediate frequency signal which is further down-converted in a quadrature mixer. All MW sources are phase locked to one common reference oscillator.

The mechanical specifications of cavities for millimetre waves are highly demanding regarding cavity dimensions, cavity surfaces and precision of coupling mechanisms due to the reduced dimensions at millimetre wavelengths. Furthermore, with increasing MW frequency resonators become more and more difficult to handle. Therefore, high-field/high-frequency EPR measurements are very often carried out without a cavity with the sample placed directly in a transmission waveguide. From the point of view of absolute sensitivity, however, small-volume cavities are evidently preferable. Typically used cavities for millimetre- and submillimetre-wave EPR are multi-mode Fabry–Pérot resonators [54] consisting of a confocal or semi-confocal arrangement of two mirrors placed at a particular distance apart. Fabry–Pérot resonators have been used successfully in the reflection and the transmission mode. A typical Fabry–Pérot resonator is extremely sensitive to displacements of the mirrors by as little as 0.1 $\mu$m. Also, the mechanical isolation of the cavity from the modulation coils has to be improved compared to lower frequency designs because of the larger modulation amplitudes and increased interaction forces with the larger static field.

Many different sorts of millimetre-wave detectors have been developed, each offering its own combination of advantages and drawbacks. For convenience, they may be divided into two general categories: bolometers and mixers (heterodyne detectors). A bolometer is a device which responds to a change in temperature produced when it absorbs incident radiation. The noise figure of an He-cooled bolometer is excellent; however, its small bandwidth limits its application to CW EPR experiments. Heterodyne detection systems

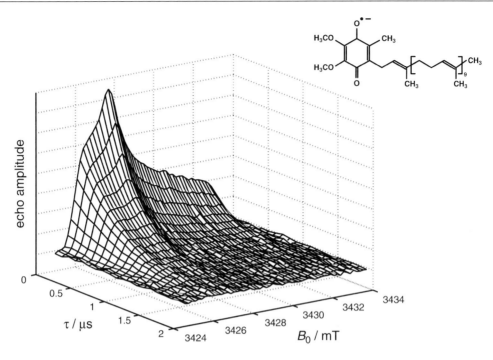

**Figure B1.15.16.** Two-pulse ESE signal intensity of the chemically reduced ubiquinone-10 cofactor in photosynthetic bacterial reaction centres at 115 K. MW frequency is 95.1 GHz. One dimension is the magnetic field value $B_0$; the other dimension is the pulse separation $\tau$. The echo decay function is anisotropic with respect to the spectral position.

transfer a signal band from a high frequency to a lower frequency where low-noise amplifiers are available. This is accomplished, for example, in a mixer by overlaying the signal with a monofrequent and stable local oscillator (LO) frequency to produce a (difference) intermediate frequency (IF) in the lower GHz range. With increasing frequency, however, the Schottky barrier diodes used in mixers become very sensitive to static electricity and mechanical stress, thus limiting their reliability.

Pulsed, or time-domain, EPR spectrometers have also been developed at higher frequencies up to 140 GHz [55, 56]. They are generally low-power units with characteristically long pulse lengths (typically 50 ns for a $\pi/2$-pulse) due to the limited MW powers available at millimetre wavelengths and the lack of fast-switching pulse-forming devices at these frequencies. In general, all the experiments outlined in the previous section can be performed, however, with the even more severe restriction to limited excitation bandwidths. Nevertheless, pulsed EPR performed at high frequencies has clear advantages when the poor orientation selection at low frequencies prevents the study of spectral anisotropies of the ESE decay, for example in relaxation measurements. As an example, figure B1.15.16 depicts the decay of a two-pulse ESE of a quinone radical as a function of the external magnetic field. Clearly, the echo decay governed by $T_2$ is different for selected field positions in the spectrum. This $T_2$ anisotropy can be analysed in terms of anisotropic motional processes of the radical in its molecular environment. Due to the low $g$-anisotropy (as is typical in biomolecules) this experiment would not have been successful at lower frequencies.

## References

[1] Eaton G R, Eaton S S and Salikhov K M 1998 *Foundations of Modern EPR* (Singapore: World Scientific)
[2] Atherton N M 1993 *Principles of Electron Spin Resonance* (Chichester: Ellis Horwood)

[3] Weil J A, Bolton J R and Wertz J E 1994 *Electron Paramagnetic Resonance* (New York: Wiley)

[4] Poole C P 1983 *Electron Spin Resonance: a Comprehensive Treatise on Experimental Techniques* 2nd edn (Mineola, NY: Dover)

[5] Gordy W 1980 *Theory and Applications of Electron Spin Resonance* (New York: Wiley)

[6] Alger R S 1968 *Electron Paramagnetic Resonance: Techniques and Applications* (New York: Wiley)

[7] Carrington A and McLachlan A D 1967 *Introduction to Magnetic Resonance* (New York: Harper and Row)

[8] Kurreck H, Kirste B and Lubitz W 1988 *Electron Nuclear Double Resonance Spectroscopy of Radicals in Solution: Application to Organic and Biological Chemistry* (Weinheim: VCH)

[9] Poole C P and Farach H A 1971 *Relaxation in Magnetic Resonance* (New York: Academic)

[10] Bertini I, Martini G and Luchinat C 1994 Relaxation, background, and theory *Handbook of Electron Spin Resonance* (ed C Poole and H Farach (New York: American Institute of Physics) ch 3, pp 51–77

[11] Slichter C P 1992 *Principles of Magnetic Resonance* (Berlin: Springer)

[12] Hyde J S and Froncisz W 1989 Loop gap resonators *Advanced EPR in Biology and Biochemistry* ed A J Hoff (Amsterdam: Elsevier) ch 7, pp 277–305

[13] Stehlik D, Bock C H and Thurnauer M 1989 Transient EPR-spectroscopy of photoinduced electronic spin states in rigid matrices *Advanced EPR in Biology and Biochemistry* ed A J Hoff (Amsterdam: Elsevier) ch 11, pp 371–403

[14] McLauchlan K A 1990 Continuous-wave transient electron spin resonance *Modern Pulsed and Continuous-Wave Electron Spin Resonance* ed L Kevan and M Bowman (New York: Wiley) ch 7, pp 285–363

[15] Kim S S and Weissman S I 1976 Detection of transient electron paramagnetic resonance *J. Magn. Reson.* **24** 167–9

[16] Furrer R, Fujara F, Lange C, Stehlik D, Vieth H and Vollmann W 1980 Transient ESR nutation signals in excited aromatic triplet states *Chem. Phys. Lett.* **75** 332–9

[17] Kothe G, Weber S, Bittl R, Ohmes E, Thurnauer M and Norris J 1991 Transient EPR of light-induced radical pairs in plant photosystem I: observation of quantum beats *Chem. Phys. Lett.* **186** 474–80

[18] Weber S, Ohmes E, Thurnauer M C, Norris J R and Kothe G 1995 Light-generated nuclear quantum beats: a signature of photosynthesis *Proc. Natl Acad. Sci. USA* **92** 7789–93

[19] Piekara-Sady L and Kispert L D 1994 ENDOR spectroscopy *Handbook of Electron Spin Resonance* ed C P Poole and H A Farach (New York: American Institute of Physics) ch 5, pp 311–57

[20] Möbius K, Plato M and Lubitz W 1982 Radicals in solution studied by ENDOR and TRIPLE resonance spectroscopy *Phys. Rep.* **87** 171–208

[21] Möbius K, Lubitz W and Plato M 1989 Liquid-state ENDOR and TRIPLE resonance *Advanced EPR in Biology and Biochemistry* ed A J Hoff (Amsterdam: Elsevier) ch 13, pp 441–99

[22] Möbius K and Biehl R 1979 Electron–nuclear–nuclear TRIPLE resonance of radicals in solution *Multiple Electron Resonance Spectroscopy* ed M M Dorio and J H Freed (New York: Plenum) ch 14, pp 475–507

[23] Leniart D S 1979 Instrumentation and experimental methods in double resonance *Multiple Electron Resonance Spectroscopy* ed M M Dorio and J H Freed (New York: Plenum) ch 2, pp 5–72

[24] Feher G 1956 Observation of nuclear magnetic resonances via the electron spin resonance line *Phys. Rev.* **103** 834–7

[25] Freed J H 1969 Theory of saturation and double resonance effects in ESR spectra. IV. Electron–nuclear triple resonance *J. Chem. Phys.* **50** 2271–2

[26] Dinse K P, Biehl R and Möbius K 1974 Electron nuclear triple resonance of free radicals in solution *J. Chem. Phys.* **61** 4335–41

[27] Biehl R, Plato M and Möbius K 1975 General TRIPLE resonance on free radicals in solution. Determination of relative signs of isotropic hyperfine coupling constants *J. Chem. Phys.* **63** 3515–22

[28] Hyde J S, Chien J C W and Freed J 1968 Electron–electron double resonance of free radicals in solution *J. Chem. Phys.* **48** 4211–26

[29] Bruno G and Freed J 1974 ESR lineshapes and saturation in the slow motional region: ELDOR *Chem. Phys. Lett.* pp 328–32

[30] Freed J 1979 Theory of multiple resonance and ESR saturation in liquids and related media *Multiple Electron Resonance Spectroscopy* ed M M Dorio and J H Freed (New York: Plenum) ch 3, pp 73–142

[31] Keijzers C P, Reijerse E and Schmidt J 1989 *Pulsed EPR: a New Field of Applications* (Amsterdam: North-Holland)

[32] Schweiger A 1991 Pulsed electron spin resonance spectroscopy: basic principles, techniques, and examples of applications *Angew. Chem. Int. Edn Engl.* **30** 265–92

[33] Bowman M K 1990 Fourier transform electron spin resonance *Modern Pulsed and Continuous-Wave Electron Spin Resonance* ed L Kevan and M Bowman (New York: Wiley) ch 1, pp 1–42

[34] Hahn E L 1950 Spin echoes *Phys. Rev.* **80** 580–94

[35] Ponti A and Schweiger A 1994 Echo phenomena in electron paramagnetic resonance spectroscopy *Appl. Magn. Reson.* **7** 363–403

[36] Schweiger A 1990 New trends in pulsed electron spin resonance methodology *Modern Pulsed and Continuous-Wave Electron Spin Resonance* ed L Kevan and M K Bowman (New York: Wiley) ch 2, pp 43–118

[37] Rowan L G, Hahn E L and Mims W B 1965 Electron-spin echo envelope modulation *Phys. Rev.* **137** A61–A71

[38] Mims W B 1972 Envelope modulation in spin-echo experiments *Phys. Rev.* B **5** 2409–19

[39] Merks R P J and de Beer R 1979 Two-dimensional Fourier transform of electron spin-echo envelope modulation. An alternative for ENDOR *J. Phys. Chem.* **83** 3319–22

[40] Höfer P, Grupp A, Nebenführ H and Mehring M 1986 Hyperfine sublevel correlation (HYSCORE) spectroscopy: a 2D ESR investigation of the squaric acid radical *Chem. Phys. Lett.* **132** 279–82

[41] Mims W B 1965 Pulsed ENDOR experiments *Proc. R. Soc.* **452** 452–7

[42]  Davies E R 1974 A new pulse ENDOR technique *Phys. Lett.* **47** 1–2

[43]  Gemperle C and Schweiger A 1991 Pulsed electron-nuclear double resonance methodology *Chem. Rev.* **91** 1481–505

[44]  Grupp A and Mehring M 1990 Pulsed ENDOR spectroscopy in solids *Modern Pulsed and Continuous-Wave Electron Spin Resonance* ed L Kevan and M K Bowman (New York: Wiley) ch 4, pp 195–229

[45]  Dinse K P 1989 Pulsed ENDOR *Advanced EPR in Biology and Biochemistry* ed A J Hoff (Amsterdam: Elsevier) ch 17, pp 615–30

[46]  Lebedev Y S 1990 High-frequency continuous-wave electron spin resonance *Modern Pulsed and Continuous-Wave Electron Spin Resonance* ed L Kevan and M K Bowman (New York: Wiley) ch 8, pp 365–404

[47]  Lebedev Y S 1994 Very-high-field EPR and its applications *Appl. Magn. Reson.* **7** 339–62

[48]  Budil D E, Earle K A, Lynch W B and Freed J H 1989 Electron paramagnetic resonance at 1 millimeter wavelengths *Advanced EPR in Biology and Biochemistry* ed A J Hoff (Amsterdam: Elsevier) ch 8, pp 307–40

[49]  Barra A L, Brunel L and Robert J 1990 EPR spectroscopy at very high field *Chem. Phys. Lett.* **165** 107–9

[50]  Reijerse E J, van Dam P J, Klaassen A A K, Hagen W, van Bentum P J M and Smith G 1998 Concepts in high-frequency EPR—applications to bio-inorganic systems *Appl. Magn. Reson.* **14** 153–67

[51]  Lynch W, Earle K and Freed J 1988 1-mm wave ESR spectrometer *Rev. Sci. Instrum.* **59** 1345–51

[52]  Fuchs M, Weber S, Möbius K, Rohrer M and Prisner T 1998 A submillimeter high-field EPR spectrometer using quasi-optical microwave bridge devices *Magnetic Resonance and Related Phenomena* ed D Ziessow, W Lubitz and F Lendzian (Berlin: Technische Universität Berlin)

[53]  Earle K, Budil D and Freed J 1996 Millimeter wave electron spin resonance using quasioptical techniques *Advances in Magnetic and Optical Resonance* vol 19, ed W Warren (San Diego: Academic) pp 253–323

[54]  Kogelnik H and Li T 1966 Laser beams and resonators *Proc. IEEE* **5** 88–105

[55]  Allgeier J, Disselhorst A, Weber R, Wenckebach W and Schmidt J 1990 High-frequency pulsed electron spin resonance *Modern Pulsed and Continuous-Wave Electron Spin Resonance* ed L Kevan and M K Bowman (New York: Wiley) ch 6, pp 267–83

[56]  Prisner T F 1997 Pulsed high-frequency/high-field EPR *Advances in Magnetic and Optical Resonance* vol 20, ed W Warren (San Diego: Academic) pp 245–99

# B1.16
# Chemically-induced nuclear and electron polarization (CIDNP and CIDEP)

## *Elizabeth J Harbron and Malcolm D E Forbes*

### B1.16.1　Introduction

Chemically-induced spin polarization was one of the last truly new physical phenomena in chemistry to be discovered and explained during this century. So unusual were the observations and so ground-breaking the theoretical descriptions that, over a very short time period, the chemist's way of thinking about free radical reactions and how to study them was fundamentally changed. After the earliest experimental reports of unusual phases of electron paramagnetic resonance (EPR) (1963) [1] and nuclear magnetic resonance (NMR) (1967) [2–4] transitions in thermal, photolytic and radiolytic reactions involving free radical intermediates, it took several years of theoretical development before the idea of the radical pair mechanism (RPM) was put forward to explain the results [5–9]. Gradually, the theory was tested and improved, and additional polarization mechanisms were discovered. The overall physical picture has stood the test of time and now both chemically-induced dynamic nuclear polarization (CIDNP) and its electron analogue (CIDEP) are well understood. The phenomena are exploited by many researchers who are trying to understand the kinetic and magnetic properties (and the links between them) of free radicals, biradicals and radical ion pairs in organic photochemistry, as well as photosynthetic reaction centres and other biologically relevant systems. The high structural resolution of NMR and EPR spectroscopies, combined with recent advances in fast data collection instrumentation and high powered pulsed lasers, has made time-resolved CIDNP and CIDEP experiments some of the most informative in the modern physical chemistry arsenal.

　　In spectroscopy it is common for transitions to be observed as absorptive lines because the Boltzmann distribution, at equilibrium, ensures a higher population of the lower state than the upper state. Examples where emission is observed, which are by definition non-equilibrium situations, are usually cases where excess population is created in the higher level by infusing energy into the system from an external source. For example, steady-state emission spectroscopy is used to measure fluorescence or phosphorescence from the excited states of organic molecules. The technique requires excitation to the upper energy levels first, then what is observed is a spontaneous emission. Another example is the laser, which is pumped with an external source such as a flash lamp or an electric arc to ensure a population inversion, and stimulated emission then occurs from the upper state upon absorption of another photon. What makes the non-Boltzmann NMR and EPR populations observed in CIDNP and CIDEP experiments so unusual is that *nuclear-spin dependent* chemical reactions (homolytic bond-breaking or forming) are responsible for the process. While it usually requires energy to break the bond, once it is broken the mixing of spin wavefunctions in the resulting radical pair, which will be described in detail in the following, is all that is necessary to make some NMR and EPR transitions appear with enhanced absorption (greater intensity than Boltzmann would predict) or even in emission (higher population in the excited state). The overall phase and magnitude of the polarization is

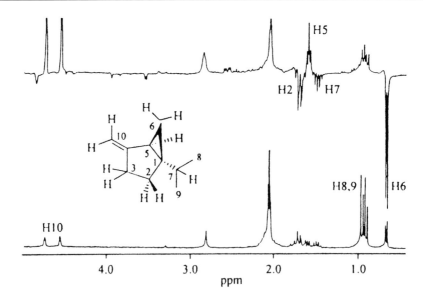

**Figure B1.16.1.** $^1$H CIDNP spectrum (250 MHz; top) observed during irradiation of chloranil with sabinene (**1**) in acetone-$d_6$ and dark spectrum (bottom). Assignments are based on the 2D $^1$H–$^1$H COSY spectrum. Reprinted from [10].

dependent on the nuclear spin projections of the nuclei (usually, but not always, protons) near the free radical site of the molecules in question. For this reason, it is easy to see why a suitable theoretical description of CIDNP took a long time to evolve. The idea that the nuclear spin-state energy level differences, which are much smaller than $kT$ at room temperature, could be responsible for different chemical reaction rates was a revolutionary and somewhat controversial one. As more and more experiments were performed to support this idea, it rapidly gained acceptance and, in fact, helped connect the solution dynamics of small molecules to spin quantum mechanics in a very natural and informative fashion.

We make one important note here regarding nomenclature. Early explanations of CIDNP invoked an Overhauser-type mechanism, implying a dynamic process similar to spin relaxation; hence the word 'dynamic' in the CIDNP acronym. This is now known to be incorrect, but the acronym has prevailed in its infant form.

The general phenomena of CIDNP and CIDEP are presented in figures B1.16.1 and B1.16.2. Figure B1.16.1 shows work by Roth *et al* [10] on radical cation structure in which the bottom trace is a 'dark' spectrum and the top trace is the CIDNP spectrum [10]. Figure B1.16.2 shows CIDEP spectra of radicals formed by decomposition of a fluorinated polymer initiator [11]. The NMR spectra in figure B1.16.1 and the EPR spectra in figure B1.16.2 can be recognized as spin-polarized by the presence of lines in emission and enhanced absorption. The origin of the CIDNP and CIDEP phenomena will be explored and explained in the following, and we will return to these examples for further analysis once the theory behind them is understood.

## B1.16.2   CIDNP

### B1.16.2.1   Spin Hamiltonian

CIDNP involves the observation of diamagnetic products formed from chemical reactions which have radical intermediates. We first define the geminate radical pair (RP) as the two molecules which are born in a radical reaction with a well defined phase relation (singlet or triplet) between their spins. Because the spin physics of the radical pair are a fundamental part of any description of the origins of CIDNP, it is instructive to begin

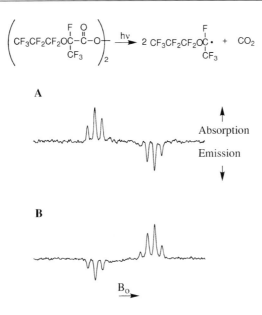

**Figure B1.16.2.** X-band TREPR spectra obtained at 0.1 $\mu$s after 308 nm photolysis of a fluorinated peroxide dimer in Freon 113 at room temperature. Part A is the A/E RPM spectrum obtained upon direct photolysis; part B is the E/A RPM spectrum obtained upon triplet sensitization of this reaction using benzophenone.

with a discussion of the radical-pair spin Hamiltonian. The Hamiltonian can be used in conjunction with an appropriate basis set to obtain the energetics and populations of the RP spin states. A suitable Hamiltonian for a radical pair consisting of radicals 1 and 2 is shown in equation (B1.16.1) below [12].

$$\hat{H}_{RP} = \beta_e \hbar^{-1} B_0 (g_1 \hat{S}_{1z} + g_2 \hat{S}_{2z}) + \hat{S}_1 \cdot \sum_j a_j \hat{I}_k + \hat{S}_2 \cdot \sum_k a_k \hat{I}_k - J(\tfrac{1}{2} + 2\hat{S}_1 \cdot \hat{S}_2) \qquad \text{(B1.16.1)}$$

The first term describes the electronic Zeeman energy, which is the interaction of the magnetic field with the two electrons of the radical pair with the magnetic field, $B_0$. The two electron spins are represented by spin operators $\hat{S}_{1z}$ and $\hat{S}_{2z}$. In this expression, $g$ is the $g$ factor, which is the chemical shift of the unpaired electrons. The other variables in the first term are constants: $\beta_e$ is the Bohr magneton, and $\hbar$ is Planck's constant divided by $2\pi$. The second term in the Hamiltonian describes the hyperfine interaction between each radical and the nuclei on that radical, where $\hat{S}$ is again the electron spin operator, $\hat{I}$ is the nuclear spin operator, and $a_j$ and $a_k$ are the hyperfine coupling constants. The hyperfine constants describe the coupling between electronic and nuclear spins; coupling between an electron and a proton is designated $a_H$. For most carbon-centred alkyl free radicals, $a_H$ for an electron and a proton on the same carbon is negative in sign while $a_H$ for a proton $\beta$ to an electron is positive. The sign of the hyperfine coupling constant will become an important issue in the analysis of CIDNP data below.

The final term in the radical pair Hamiltonian is the exchange interaction ($J$) between the unpaired electrons. This interaction is a scalar quantity that describes the coupling of the angular momenta of two radicals which are in close proximity. Its magnitude decreases exponentially with increasing inter-radical distance as shown by equation (B1.16.2), where $J_0$ is the exchange interaction at the point of closest contact, $r_0$; $r$ is the inter-radical distance; and $\lambda$ is a fall-off parameter generally accepted to be approximately 1 Å$^{-1}$ for isotropic solutions. The exchange interaction should not be confused with the quantum mechanical term

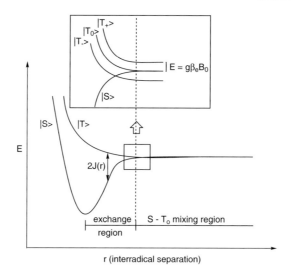

**Figure B1.16.3.** Energy levels versus inter-radical separation for a radical pair. The lower part of the figure shows the $S$ and $T$ levels in the absence of an external magnetic field while the inset shows the splitting of the triplet levels in the presence of a magnetic field, $B_0$.

exchange integral, although the two are related [13].

$$J = J_0\, e^{-\lambda(r-r_0)}. \tag{B1.16.2}$$

The exchange interaction results in an energy splitting between the singlet and triplet states of the RP as shown in figure B1.16.3, which shows a plot of the RP energy levels versus the inter-radical separation. The $S$ and $T$ levels shown in the lower part of the figure are in the absence of an external magnetic field. When an external field is applied, the triplet level is split into $T_+$, $T_0$ and $T_-$, as shown in the upper inset. At the high magnetic fields at which most CIDNP experiments are conducted, $T_+$ and $T_-$ are far away from $S$ and can be neglected in what is known as the high-field approximation.

When the inter-radical distance is very small, $J$ is large, and $S$ and $T_0$ are far apart and cannot mix. This is called the 'exchange region' as the electron spins are constrained by the exchange interaction to remain in their respective spin states. As the radicals diffuse apart, $J$ rapidly falls to zero, and the $S$ and $T_0$ levels become degenerate and are allowed to mix. If the radicals diffuse back to the region of large $J$, they must again be in either the $S$ or $T_0$ state, but some of the RPs which were formerly in one spin state will now be in the other. This phenomenon is known as intersystem crossing; it is a process critical to the understanding of CIDNP and will be explained in greater detail below.

### B1.16.2.2   Mechanism of intersystem crossing

Vector representations of the radical pair spin states, shown in figure B1.16.4A, help to explain how intersystem crossing occurs in RPs. The vector diagrams show the magnitude of the spin angular momentum vector and its $z$ component (parallel to $B_0$) for each of the two spins in the RP. Because the $x$ and $y$ components of spin are unspecified, each spin vector can be thought to precess around a cone. When considering only the influence of the interaction of the electron chemical shift with the magnetic field, the frequency of electron precession is given by the expression for the Larmor frequency, equation (B1.16.3). Additional interactions, such as electron–nuclear hyperfine couplings, must be added to or subtracted from the Larmor frequency in

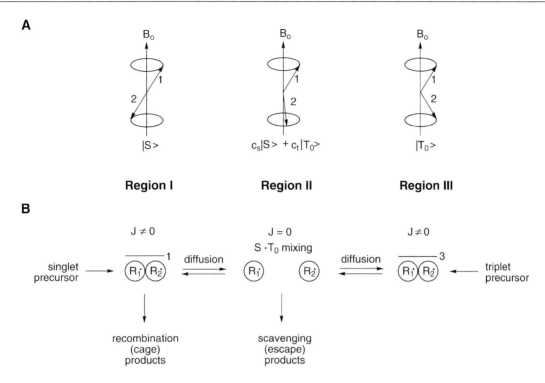

**Figure B1.16.4.** Part A is the vector representations of the $S$ state, an intermediate state, and the $T_0$ state of a radical pair. Part B is the radical reaction scheme for CIDNP.

order to determine the actual precessional frequency of a given electron;

$$\omega = g\beta_e\hbar^{-1}B_0. \tag{B1.16.3}$$

As mentioned previously, a geminate RP in close contact is constrained by the exchange interaction to remain in its initial spin state, $S$ or $T_0$, and vector representations of these states are shown in figure B1.16.4A. The exchange interaction prevents the two spins in the RP from precessing independently; so long as they precess at the same frequency, the two spins will remain in the same mutual orientation. Once the radicals have diffused to the region where $J$ is zero, however, they are free to precess independently of one another. At this point, differences in $g$ factor and/or hyperfine interaction will cause the radicals to precess at different frequencies. This difference in precessional frequencies is given by $Q$, as shown in equation (B1.16.4), where $m_{1i}$ and $m_{2j}$ are the nuclear magnetic quantum numbers for each member of the RP, respectively, and the other variables were defined previously;

$$\omega_1 - \omega_2 = 2Q = (g_1 - g_2)\beta_e\hbar^{-1}B_0 + \sum a_{1i}m_{1i} - \sum a_{2j}m_{2j}. \tag{B1.16.4}$$

Eventually, the two spins will fall out of step with one another and will oscillate between the $S$ state, intermediate states, and the $T_0$ state. An intermediate state is shown in region II of the vector diagram and can be described as a coherent superposition of the $S$ and $T_0$ states with the coefficients $c_S$ and $c_T$ delineating the amount of $S$ and $T_0$ 'character'. If the radicals diffuse back together to the region of large $J$, the RP is again required to be in either the $S$ or $T_0$ state. The squares of the coefficients $c_S$ and $c_T$ will determine the probability that the RP will jump to the $S$ or $T_0$ state at this inter-radical distance. Some RPs will now be in a different spin state than they were initially and are said to have undergone intersystem crossing, as mentioned above. While $S–T_0$ mixing occurs when the radicals in a RP are in the $J = 0$ region, the fact that intersystem crossing has occurred cannot be determined unless they diffuse back together and are forced by the exchange interaction to be in the $S$ or $T_0$ state.

By examining the expression for $Q$ (equation (B1.16.4)), it should now be clear that the nuclear spin state influences the difference in precessional frequencies and, ultimately, the likelihood of intersystem crossing, through the hyperfine term. It is this influence of *nuclear* spin states on *electronic* intersystem crossing which will eventually lead to non-equilibrium distributions of nuclear spin states, i.e. spin polarization, in the products of radical reactions, as we shall see below.

### B1.16.2.3    *Radical reaction scheme*

A general reaction scheme for CIDNP is shown in figure B1.16.4B, where the radical dynamics in each region correspond to the vector diagram for that region shown in figure B1.16.4A. A geminate RP is formed from a singlet or triplet precursor through bond cleavage or an electron transfer reaction. Thermal reactions proceed from the singlet state while photochemical reactions tend to occur from the triplet state; certain species, such as azo compounds, react from the singlet in photochemical reactions [14]. The RP is always formed in the same spin state as its precursor because the RP-forming reaction must conserve angular momentum. For the vast majority of reactions, recombination of singlet RPs is allowed while recombination of triplet RPs is forbidden. While there are exceptions [15], we will consider triplet RPs to be incapable of recombination throughout this explanation. It should also be noted that figure B1.16.4B shows radical-forming reaction pathways for both singlet and triplet RPs for the purpose of illustration, but it is to be understood that most radical-forming reactions occur from either a singlet (left side) or triplet (right side) precursor but not both.

The geminate RPs in figure B1.16.4B are indicated by a bar with the spin multiplicity. It is common to speak of geminate RPs as being in a 'cage', and this notion is central enough to our discussion of CIDNP to merit some discussion. The idea of the cage effect in radical chemistry stems from early work by Franck and Rabinowitch [16] in which they noted that radicals have an increased probability of recombination in solution as compared to the gas phase. While the term 'cage' may encourage one to picture a rigid ensemble of solvent molecules, the cage effect does not describe the influence of a static entity and is actually somewhat difficult to define.

While all agree that the cage is a concept critical to CIDNP, different researchers vary somewhat in their definition of it. Turro *et al* [17] conceive of a RP as guest and a solvent cage as host in an extremely short-lived 'collision complex' with a lifetime of about $10^{-11}$ s. Salikhov *et al* [12] define the cage as a region of effective recombination of two radicals in a RP. They note that radicals may diffuse to the second or third coordination sphere and still come back together to give products. Accordingly, the cage effect describes a twofold influence of condensed media: the two radicals in a RP are not only in close contact for a longer period of time than they would be in the gas phase but are also more likely to re-encounter one another after diffusing apart. Goez [18] describes the same effects in more abstract terms. He writes of the cage as a region of time and states that two radicals are in the cage so long as they have not lost each other for good. If two radicals in a RP diffuse apart but re-encounter, then they can be said to have been in the cage the entire time. If the same two radicals do not re-encounter, then they are said to have escaped the cage. These different perspectives on the cage are not meant to confuse the reader but are rather intended to present the general idea while conveying the complexity and importance of the concept in CIDNP.

Returning to figure B1.16.4, if the RP is initially in the singlet state, some of the geminate RPs will recombine in the cage in what are called *cage* or *recombination* products. In addition to the reformed radical precursor, recombination products can also include products from disproportionation reactions which occur in cage. A few singlet RPs may escape the cage instead of recombining; as mentioned before, triplet RPs cannot recombine, so they will also escape. Escaped radicals diffuse to region II, where $J$ is negligible, and may undergo $S$–$T_0$ mixing as described previously. From region II, the radicals may follow any of three different pathways.

(1)  The radicals may re-encounter one another following $S$–$T_0$ mixing. As they diffuse together into the region of large $J$, the radicals are again constrained to be in either the $S$ or the $T_0$ state. Some fraction of RPs will have undergone intersystem crossing.

(2)  An individual radical from the RP may encounter a radical from a different RP to form what are known as random RPs or F pairs. F pairs which happen to be in the singlet state have a high probability of recombining, so the remaining F pairs will be in the triplet state. Consequently, the initial condition for F pairs is the triplet state in nearly all cases.

(3)  An individual radical from the RP may be scavenged by a solvent or another chemical species to form diamagnetic products. Because the products are formed following escape from the cage, they are known as *escape* or *scavenging* products.

### B1.16.2.4  *Radical pair mechanism: net effect*

We have introduced the RP spin Hamiltonian, the mechanism of nuclear spin selective intersystem crossing, and the reaction scheme for RPs has been explained. We now possess all the tools we need to explain qualitatively how nuclear spin polarization arises and manifests itself in a CIDNP spectrum. The RPM may appear in a CIDNP spectrum as a *net* effect, a *multiplet* effect, or a combination of both. The net effect is observed when the g factor difference between radicals $R_1$ and $R_2$ ($\Delta g$) in a radical pair is large compared with the hyperfine interaction. The simplest example involves a RP with just one hyperfine interaction, as shown in figure B1.16.5. In this example we will set the following conditions: (1) the RP is initially a singlet; (2) the hyperfine coupling constant ($a_H$) is negative; and (3) the g factor for $R_1$ is greater than that for $R_2$. The recombination product is formed by in-cage recombination of $R_1$ and $R_2$, and a scavenging product is formed by the abstraction of, say, a halogen atom from the solvent following escape from the cage. The scavenging product is formed primarily from escaped triplet RPs although it should be noted that escaped singlets could also form the same product.

In this simple case, there are just two nuclear spin states, $\alpha$ and $\beta$. Equation (1.16.5) shows the calculation of the difference in electron precessional frequencies, $Q$, for nuclear spin states $\alpha$ (equation (B1.16.5a)) and $\beta$ (equation (B1.16.5b)).

$$2Q = \Delta g \beta_e \hbar^{-1} B_0 + (\tfrac{1}{2}) a_H \qquad\qquad (B1.16.5a)$$

$$2Q = \Delta g \beta_e \hbar^{-1} B_0 + (-\tfrac{1}{2}) a_H. \qquad\qquad (B1.16.5b)$$

Since $\Delta g$ is positive and $a_H$ is negative, $Q$ is larger for the $\beta$ state than for the $\alpha$ state. Radical pairs in the $\beta$ nuclear spin state will experience a faster intersystem crossing rate than those in the $\alpha$ state with the result that more RPs in the $\beta$ nuclear spin state will become triplets. The end result is that the scavenging product, which is formed primarily from triplet RPs, will have an excess of spins in the $\beta$ state while the recombination product, which is formed from singlet RPs, will have an excess of $\alpha$ nuclear spin states.

Relative populations for the $\alpha$ and $\beta$ states are indicated by the thickness of the lines in the diagrams at the bottom of figure B1.16.5, and the corresponding CIDNP spectrum is shown below each level diagram. The signal for the recombination product, with its excess of $\alpha$ spins, will be in enhanced absorption. This enhanced absorption is distinguished from a typical NMR absorptive signal by its abnormal intensity, which may be as much as 1000 times greater than a NMR signal from a Boltzmann population difference. The scavenging product, with its excess of $\beta$ spins, appears in emission in the CIDNP spectrum. Herein lies an extremely important feature of CIDNP: cage and escape products have opposite phases of polarization.

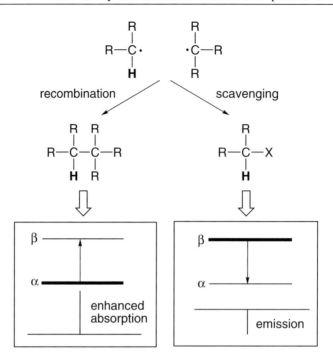

**Figure B1.16.5.** An example of the CIDNP net effect for a radical pair with one hyperfine interaction. Initial conditions: $g_1 > g_2$; $a_H$ negative; and the RP is initially singlet. Polarized nuclear spin states and schematic NMR spectra are shown for the recombination and scavenging products in the boxes.

It should be straightforward to see that changing the sign of $a_H$ in the previous example would change the values of $Q$ and would ultimately result in a flipping of the phase of the polarization. A rule for predicting the phase of the polarization for each product will be presented with the next example.

A slightly more complex system exhibiting the RPM net effect is presented in figure B1.16.6 [19]. In this case, radicals $R_1$ and $R_2$ each have one hyperfine coupling, so the two protons in the recombination product originate from different radicals. The four nuclear spin states and allowed transitions for the product are shown in figure B1.16.6A along with the NMR spectrum in the absence of spin polarization. Again, we must set some initial conditions for this CIDNP example: the RP is initially in the triplet state, both hyperfine coupling constants are positive and $g_1$ is greater than $g_2$. The values shown on each level in figure B1.16.6B are representative of the absolute values of $Q$ for each nuclear spin state and are proportional to the populations of those states. The phase of each transition can be determined by subtracting the $Q$ value of the upper level from that of the lower. For transitions 1 and 2, this value is $2a_1(\Delta g)$, which is positive and yields absorptive transitions. For transitions 3 and 4, the value $[-2a_2(\Delta g)]$ results in emission transitions. A stick plot of the CIDNP spectrum is shown in the figure. For the CIDNP net effect, each line within a multiplet will always have the same phase.

The phase of a transition in a CIDNP spectrum can be determined using rules developed by Kaptein [20]. The rule for the net effect is shown in equation (B1.16.6). For each term, the sign (+ or −) of that value is inserted, and the final sign determines the phase of the polarization: plus is absorptive and minus is emissive. The variables are defined in the caption to figure B1.16.7.

$$\Gamma_{net} = \mu\varepsilon\Delta g_i a_i. \tag{B1.16.6}$$

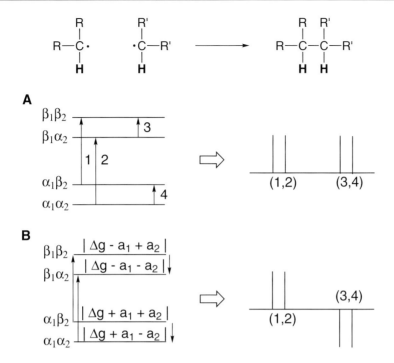

**Figure B1.16.6.** An example of CIDNP net effect for a radical pair with two hyperfine interactions. Part A shows the spin levels and schematic NMR spectrum for unpolarized product. Part B shows the spin levels and schematic NMR spectrum for polarized product. Populations are indicated on each level. Initial conditions: $g_1 > g_2$; $a_1 > 0$; $a_2 > 0$; spins on different radicals; the RP is initially triplet.

Kaptein's rule is applied below to each transition in the example in figure B1.16.6. It is important to choose $\Delta g$ correctly: $\Delta g$ is equal to $g_1 - g_2$ where $g_1$ describes the radical containing the nucleus of interest (often a proton) while $g_2$ is the other radical in the RP. The rule correctly predicts absorptive phase for NMR transitions 1 and 2 and emissive for NMR transitions 3 and 4.

$$\text{for 1 and 2: } \Gamma_{\text{net}} = (+)(+)(+)(+) = + = A$$
$$\text{for 3 and 4: } \Gamma_{\text{net}} = (+)(+)(-)(+) = - = E.$$

### B1.16.2.5  Radical pair mechanism: multiplet effect

The other RPM polarization pattern observed in CIDNP spectra is called the multiplet effect. In contrast to the net effect, the multiplet effect occurs when the hyperfine interactions are large compared with $\Delta g$. This is best explained by example, and the radical pair for a hypothetical case is shown in figure B1.16.8. We note that only the recombination product will be considered here. Both radicals are identical and have two protons with hyperfine coupling, $H_1$ and $H_2$. The initial conditions for this example are that $\Delta g$ is zero; the nuclear spin–spin coupling constant, $J_{12}$ is positive; $a_1$ is negative; $a_2$ is positive; the RP is initially a singlet; and the nuclear spins are both on the same radical. Values proportional to $Q$ are again shown on each nuclear spin level in figure B1.16.8. Because $\Delta g$ is zero, the Zeeman term in the equation for $Q$ is zero and, therefore, the value of $Q$ is proportional to the magnitude and the sign of the sum or difference of the hyperfine coupling

$$\boxed{\Gamma_{net} = \mu\varepsilon\Delta g_i\, a_i} \qquad\qquad \boxed{\Gamma_{mult} = \mu\varepsilon a_i\, a_j\, J_{ij}\, \sigma_{ij}}$$

$$+ = A \qquad\qquad\qquad\qquad + = E/A$$
$$- = E \qquad\qquad\qquad\qquad - = A/E$$

**Figure B1.16.7.** Kaptein's rules for net and multiplet RPM of CIDNP. The variables are defined as follows: $\mu = +$ for RP formed from triplet precursor or F pairs and $-$ for RP formed from singlet precursor. $\varepsilon = +$ for recombination (or disproportionation)/cage products and $-$ for scavenge/escape products. $\sigma_{ij} = +$ if nuclei $i$ and $j$ were on the same radical and $-$ if nuclei $i$ and $j$ were on different radicals. $\Delta g_i =$ sign of $(g_1 - g_2)$. $a =$ sign of hyperfine interaction. $J_{ij} =$ sign of exchange interaction.

constants, as shown in figure B1.16.8. Assuming that $a_1$ and $a_2$ are roughly equal in magnitude but opposite in sign, it should be clear that the $\alpha\alpha$ and $\beta\beta$ nuclear spin states will have very small $Q$ values while $\alpha\beta$ and $\beta\alpha$ will have larger $Q$ values. Accordingly, $\alpha\beta$ and $\beta\alpha$ will intersystem cross from singlet to triplet more quickly, and these levels will be depleted by escape from the cage relative to the $\alpha\alpha$ and $\beta\beta$ levels. The remaining populations are indicated by the width of the bars in the bottom of figure B1.16.8, and the transitions from more populated levels to less populated ones are shown. As shown in the stick plot CIDNP spectrum, the lines of each multiplet alternate in phase. Because the first line is emissive and the second line is absorptive, this pattern is called E/A for emissive/absorptive.

    Kaptein's rule for the multiplet effect is useful for predicting the phase of each transition, and it is similar to but has more variables than the rule for the net effect. The variables in equation (B1.16.7) are defined in figure B1.16.7. A final sign of plus predicts E/A phase while minus predicts A/E.

$$\Gamma_{mult} = \mu\varepsilon a_i a_j J_{ij}\sigma_{ij}. \tag{B1.16.7}$$

The application of Kaptein's rule to the example in figure B1.16.8 is shown below, and it correctly predicts E/A multiplets.

$$\Gamma_{mult} = (-)(+)(-)(+)(+)(+) = + = E/A.$$

One of the most attractive features of the CIDNP multiplet effect is that it allows determination of the sign of the $J$ coupling, which is often difficult to do by other methods.

### B1.16.2.6   *Examples of CIDNP*

While the stick plot examples already presented show net and multiplet effects as separate phenomena, the two can be observed in the same spectrum or even in the same NMR signal. The following examples from the literature will illustrate 'real life' uses of CIDNP and demonstrate the variety of structural, mechanistic, and spin physics questions which CIDNP can answer.

    Roth *et al* [10] have used CIDNP to study the structures of vinylcyclopropane radical cations formed from precursors such as sabinene (**1**).

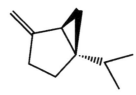

**1**

The radical cation of **1** (**1**$^{\cdot+}$) is produced by a photo-induced electron transfer reaction with an excited electron acceptor, chloranil. The major product observed in the CIDNP spectrum is the regenerated electron donor, **1**. The parameters for Kaptein's net effect rule in this case are that the RP is from a triplet precursor ($\mu$ is +), the recombination product is that which is under consideration ($\varepsilon$ is +) and $\Delta g$ is negative. This leaves the sign of the hyperfine coupling constant as the only unknown in the expression for the polarization phase. Roth *et al* [10] used the phase and intensity of each signal to determine the relative signs and magnitudes of the hyperfine coupling constant for each proton in **1**$^{\cdot+}$. Signals in enhanced absorption indicated negative hyperfine coupling constants while emissive signals indicated positive hyperfines.

The CIDNP spectrum is shown in figure B1.16.1 from the introduction, top trace, while a dark spectrum is shown for comparison in figure B1.16.1, bottom trace. Because the sign and magnitude of the hyperfine coupling constant can be a measure of the spin density on a carbon, Roth *et al* [10] were able to use the relative spin density of each carbon to determine that the structure of the radical cation **1**$^{\cdot+}$ is a delocalized one, shown below. This example demonstrates the use of CIDNP to determine the signs and relative magnitudes of the hyperfine coupling constants and to assign the structure of an intermediate.

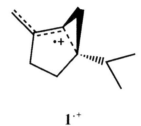

**1**$^{\cdot+}$

In a case involving both net and multiplet effects, Goez and Sartorius [21] studied the photoreaction of triethylamine with various triplet sensitizers containing carbonyl functionalities. In a two-step process, the amine (DH) first transfers an electron to the excited sensitizer and forms the aminium cation DH$^{\cdot+}$. The aminium cation is then deprotonated to form a neutral $\alpha$-aminoalkyl radical (D$^{\cdot}$), which can go on to form products. In this example, triethylamine (DH) was reacted with a variety of sensitizers, and N,N-diethylvinylamine was the polarized product which was studied. N,N-diethylvinylamine can be formed by two different pathways. If the deprotonation step occurs in cage, H$^{+}$ is transferred to the sensitizer, and polarization in the product arises from a neutral radical pair, AH$^{\cdot}$D$^{\cdot}$. If the deprotonation step occurs out of cage, then H$^{+}$ will be abstracted by free amine; in this case, polarization is formed from a radical ion pair, A$^{\cdot-}$–DH$^{\cdot+}$. The goal of this work was to determine the intermediates leading to N,N-diethylvinylamine; does the deprotonation step occur in cage or out of cage?

The radical cation and neutral radical derived from triethylamine are shown below.

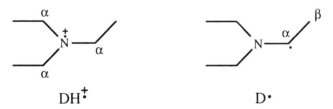

DH$^{\cdot+}$ has only one non-negligible hyperfine, $a_{H\alpha} = +19.0$ G while D$^{\cdot}$ has two significant hyperfines, $a_{H\alpha} = -13.96$ G and $a_{H\alpha} = +19.24$ G. Clearly, these two radicals will lead to very different polarizations in the CIDNP spectrum of both cage and escape products.

Figure B1.16.9 shows background-free, pseudo-steady-state CIDNP spectra of the photoreaction of triethylamine with (a) anthroquinone as sensitizer and (b) and (c) xanthone as sensitizer. Details of the pseudo-steady-state CIDNP method are given elsewhere [22]. In trace (a), no signals from the $\beta$ protons

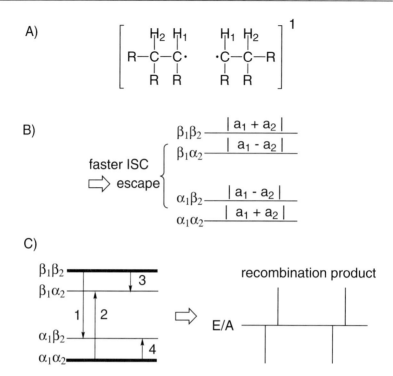

**Figure B1.16.8.** Example of CIDNP multiplet effect for a symmetric radical pair with two hyperfine interactions on each radical. Part A is the radical pair. Part B shows the spin levels with relative $Q$ values indicated on each level. Part C shows the spin levels with relative populations indicated by the thickness of each level and the schematic NMR spectrum of the recombination product.

of products 1 (recombination) or 2 (escape) are observed, indicating that the products observed result from the radical ion pair. Traces (b) and (c) illustrate a useful feature of pulsed CIDNP: net and multiplet effects may be separated on the basis of their radiofrequency (RF) pulse tip angle dependence [23]. Net effects are shown in trace (b) while multiplet effects can be seen in (c). Both traces show signals from the $\beta$ protons of products 1 and 2, indicating that these products were formed from a neutral radical pair intermediate. It was ultimately determined that the time scale of the deprotonation step relative to the lifetime of the radical ion pairs determined whether products were formed from radical ion pairs or from neutral radical pairs. The energetics of the system varied with the sensitizer, and results were compiled for a variety of sensitizers. This example illustrates the very common application of CIDNP as a mechanistic tool.

In an extension of traditional CIDNP methods, Closs and co-workers developed time-resolved CIDNP (TR-CIDNP) in the late 1970s [24–26]. The initial time-resolved experiments had a time resolution in the microsecond range [24], but a nanosecond method was later developed [27]. A typical pulse sequence for time-resolved CIDNP involves a series of saturation pulses to remove background signals from equilibrium polarization followed by a laser pulse to form the radical pairs. After a preset delay time, $\tau$, after the laser flash, a RF pulse is applied, and the FID of the product is acquired. Further details of this experiment are given in [26].

The first application of the time-resolved CIDNP method by Closs and co-workers involved the Norrish 1 cleavage of benzyl phenyl ketone [24, 25]. Geminate RPs may recombine to regenerate the starting material while escaped RPs may form the starting ketone (**12**), bibenzyl (**3**), or benzil (**4**), as shown below.

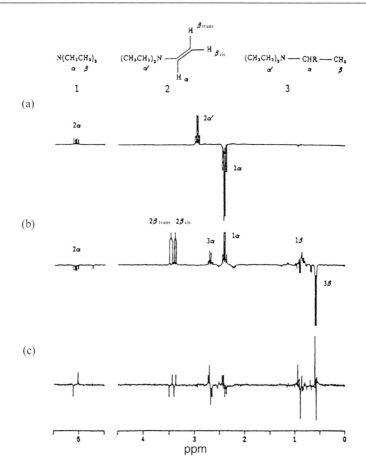

**Figure B1.16.9.** Background-free, pseudo-steady-state CIDNP spectra observed in the photoreaction of triethylamine with different sensitizers ((a), anthraquinone; (b), xanthone, CIDNP net effect; (c), xanthone, CIDNP multiplet effect, amplitudes multiplied by 1.75 relative to the centre trace) in acetonitrile-$d_3$. The structural formulae of the most important products bearing polarizations (1, regenerated starting material; 2, N,N-diethylvinylamine; 3, combination product of amine and sensitizer) are given at the top; R denotes the sensitizer moiety. The polarized resonances of these products are assigned in the spectra. Reprinted from [21].

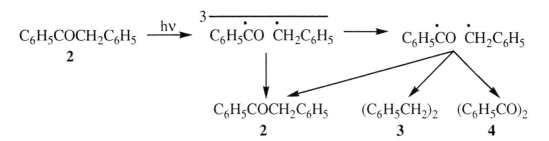

Closs *et al* [25] plotted the polarizations versus time of the starting ketone (**2**, trace B, emissive signal) and the bibenzyl escape product (**3**, trace A, absorptive signal), as shown in figure B1.16.10. Ketone **2** can be formed either by recombination of geminate pairs or the reaction of F pairs; in either case, the polarization will

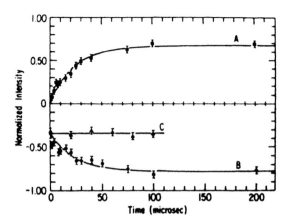

**Figure B1.16.10.** Intensity of CH$_2$ resonance as a function of delay time in A dibenzyl, B deoxybenzoin, and C deoxybenzoin in presence of thiol scavenger. Reprinted from [25].

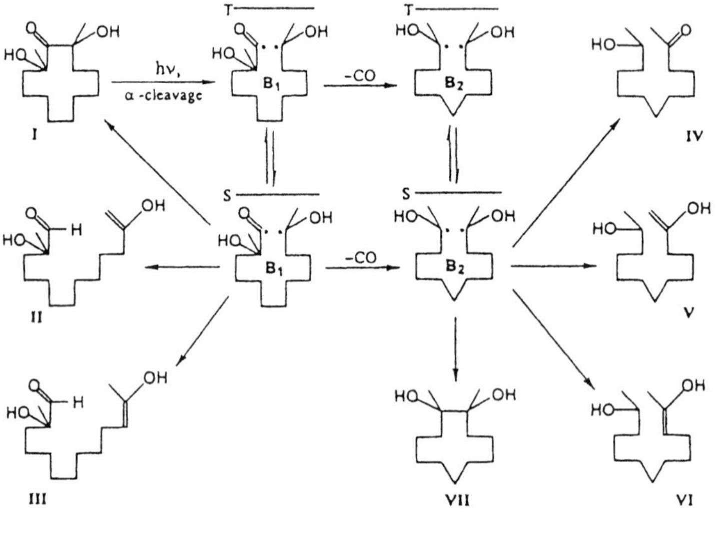

**Figure B1.16.11.** Biradical and product formation following photolysis of 2,12-dihydroxy-2,12-dimethylcyclododecanone. Reprinted from [28].

be emissive. At the earliest delay time (1 $\mu$s), the emissive signal of **2** is already present while no polarization from F pairs is apparent because there has not been sufficient time for the diffusion of radicals to occur. At later delay times, both absorptive and emissive polarizations grow until they reach a maximum. In order to demonstrate that much of the emissive polarization was due to production of **2** from F pairs, a thiol scavenger was added to trap the escaped benzoyl radicals. As shown in trace C, the emissive polarization is constant from the earliest delay time when the scavenger is present, indicating that much of the emissive polarization from geminate pairs is constant, while that from F pairs grows in with time. This was the first instance in which polarization from F pairs was shown to enhance the polarization of geminate products. In addition to establishing the utility of the time-resolved CIDNP method, this experiment was the first to demonstrate that polarization from cage and escape products could be separated based on the time scale of its production.

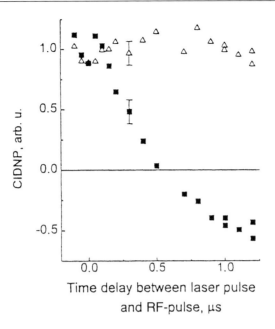

**Figure B1.16.12.** Experimental kinetics of the CIDNP net effect: ($\triangle$) for the aldehyde proton of the products II and III of primary biradical; ($\blacksquare$) for the CH$_3$CH(OH) protons of the products IV, V, and VI of secondary biradical. Reprinted from [28].

While the earliest TR-CIDNP work focused on radical pairs, biradicals soon became a focus of study. Biradicals are of interest because the exchange interaction between the unpaired electrons is present throughout the biradical lifetime and, consequently, the spin physics and chemical reactivity of biradicals are markedly different from radical pairs. Work by Morozova *et al* [28] on polymethylene biradicals is a further example of how this method can be used to separate net and multiplet effects based on time scale [28]. Figure B1.16.11 shows how the cyclic precursor, 2,12-dihydroxy-2,12-dimethylcyclododecanone, cleaves upon 308 nm irradiation to form an acyl–ketyl biradical, which will be referred to as the primary biradical since it is formed directly from the cyclic precursor. The acyl–ketyl primary biradical decarbonylates rapidly ($k_{CO} \geq 5 \times 10^7$ s$^{-1}$) to form a bis–ketyl biradical, which will be referred to as the secondary biradical. Both the primary and secondary biradicals can form a number of diamagnetic products, as shown in figure B1.16.11.

In the TR-CIDNP spectrum, the methyl protons of products IV, V, and VI, formed from the secondary biradical, show a combination of net and multiplet polarizations. Morozova *et al* [28] measured separately the time dependence of the net and multiplet polarizations for this group of protons, and the results are shown in figures B1.16.12 and B1.16.13, respectively. Clearly, the net and multiplet polarizations develop on different time scales; while the net polarization is constant after approximately 1 $\mu$s, the multiplet polarization takes much longer to evolve. It was determined that this difference arises because the net polarization in these products of the secondary biradical is actually inherited from the primary biradical, while the multiplet polarization is generated in the secondary biradical. If chemical transformation (decarbonylation in this case) is fast compared with the rate of intersystem crossing, then a secondary biradical or radical pair may inherit polarization from its precursor. This is known as the *memory effect* in CIDNP, and this work was the first report of the memory effect in biradicals. The polarization inherited from the primary biradical is net because $\Delta g > 0$ in the primary biradical; because the secondary biradical is symmetric, $\Delta g = 0$, and only multiplet polarization can be generated. It was also determined in this study that the kinetics of the net effect reflect the

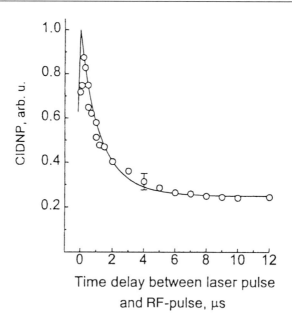

**Figure B1.16.13.** Kinetics of the CIDNP multiplet effect: (full curve) the calculated CIDNP kinetics for the product of disproportionation of bis–ketyl biradical; (O) experimental kinetics for the $CH_3CH(OH)$ protons of the products IV, V and VI of the secondary biradical. Reprinted from [28].

decay of the $T_0$ level while the multiplet effect corresponds to the decay of the $T_+$ and $T_-$ levels; the reasons for these observations are beyond the scope of this presentation, but the interested reader is directed to the references for additional details.

## B1.16.3   CIDEP

As the electron counterpart to CIDNP, CIDEP can provide different but complementary information on free radical systems. Whereas CIDNP involves the observation of diamagnetic products, the paramagnetic intermediates themselves are observed in CIDEP studies. Unlike CIDNP, CIDEP does not require chemical reaction for the formation of polarization, as we shall see below. In addition to the RPM, three other mechanisms may produce electron polarization: the triplet mechanism (TM), the radical-triplet pair mechanism (RTPM) and the spin-correlated radical pair (SCRP) mechanism. Some of these mechanisms provide information which can lead directly to the structural identification of radical intermediates while all of them supply data which may be used to elucidate mechanisms of radical reactions.

Experimentally, the observation of CIDEP is difficult but not impossible using a commercial steady-state EPR spectrometer with 100 kHz field modulation [29]. The success of this particular experiment requires large steady-state concentrations of radicals and rather slow spin relaxation, as the time response of the instrument is, at best, 10 $\mu$s. Therefore, the steady-state method works well for only a limited number of systems and generally requires a very strong CW lamp for irradiation. The time response can be shortened if the radicals are produced using a pulsed laser and the 100 kHz modulation is bypassed. The EPR signal is then taken directly from the preamplifier of the microwave bridge and passed to a boxcar signal averager [30] or transient digitizer [31]. At the standard X-band frequency the time response can be brought down to about 60 ns in this fashion. In this case the overall response becomes limited by the resonant microwave cavity quality factor [32]. At higher frequencies such as $Q$-band, the time response can be limited by laser pulse width or preamplifier rise time ($<10$ ns) [33]. Using the boxcar or digitizer method, CIDEP is almost always observable, and this so-called time-resolved (CW) EPR spectroscopy is the method of choice for many practitioners. The 'CW' in the name is used to indicate that the microwaves are always on during the experiment, even during the production of the radicals, as opposed to pulsed microwave methods such as electron-spin-echo or Fourier-transform (FT) EPR. Significant advantages in sensitivity with similar time response are available with FT-EPR, but there are also disadvantages in terms of the spectral width of the excitation that limit the application of this technique [34]. The TR (CW) method is the most facile and cost effective method for the observation of complete EPR spectra exhibiting CIDEP on the sub-microsecond time scale.

### B1.16.3.1    *Radical pair mechanism*

The RPM has already been introduced in our explanation of CIDNP. The only difference is that for the electron spins to become polarized, product formation from the geminate RP is not required. Rather, in a model first introduced by Adrian, so-called 'grazing encounters' of geminate radical pairs are all that is required [35]. Basically the RP must diffuse from a region where the exchange interaction is large to one where it is small, then back again. The spin wavefunction evolution that mixes the $S$ and $T_0$ electronic levels in the region of small $J$ leads to unequal populations of the $S$ and $T_0$ states in the region of large $J$. The magnitude of the RPM CIDEP is proportional to this population difference.

As for CIDNP, the polarization pattern is multiplet (E/A or A/E) for each radical if $\Delta g$ is smaller than the hyperfine coupling constants. In the case where $\Delta g$ is large compared with the hyperfines, net polarization (one radical A and the other E or *vice versa*) is observed. A set of rules similar to those for CIDNP have been developed for both multiplet and net RPM in CIDEP (equations (B1.16.8) and (B1.16.9)) [36]. In both expressions, $\mu$ is positive for triplet precursors and negative for singlet precursors. $J$ is always negative for neutral RPs, but there is evidence for positive $J$ values in radical ion reactions [37]. In equation (B1.16.8), $\Gamma_{\text{mult}} = +$ predicts E/A while $\Gamma_{\text{mult}} = -$ predicts A/E. For the net effect in equation (B1.16.9), $\Gamma_{\text{net}} = +$ predicts A while $\Gamma_{\text{net}}$ predicts E.

$$\Gamma_{\text{mult}} = -J\mu \tag{B1.16.8}$$

$$\Gamma_{\text{net}} = +J\Delta g\mu. \tag{B1.16.9}$$

Because the number of grazing encounters is a function of the diffusion coefficient, CIDEP by the RPM mechanism is a strong function of the viscosity of the solvent and, in general, the RPM becomes stronger with increasing viscosity. Pedersen and Freed [39] have developed analytical techniques for the functional form of the viscosity dependence of the RPM.

A typical example of RPM multiplet effects is shown in figure B1.16.2. Upon 308 nm laser irradiation, the O–O bond of a fluorinated peroxide dimer is cleaved to yield two radicals plus $CO_2$ as shown in figure B1.16.2. The radical signal is split into a doublet by the $\alpha$-fluorine atom, and each line in the doublet is split into a quartet by the adjacent $CF_3$ group (although not all lines are visible in these spectra). The A/E pattern of the spectrum in figure B1.16.2A indicates that the RP is formed from a singlet precursor. When benzophenone, a triplet sensitizer, is added to the system, the precursor becomes a triplet, and the polarization pattern is now E/A, as shown in figure B1.16.2B. These spectra demonstrate the utility of RPM CIDEP in determining the spin multiplicity of the precursor.

### B1.16.3.2    *Triplet mechanism*

A second mechanism of CIDEP is the triplet mechanism (TM) [40, 41]. As the name implies, this polarization is generated only when the RP precursor is a photoexcited triplet state. The polarization is produced during the intersystem crossing process from the first excited singlet state of the molecular precursor. It should be noted

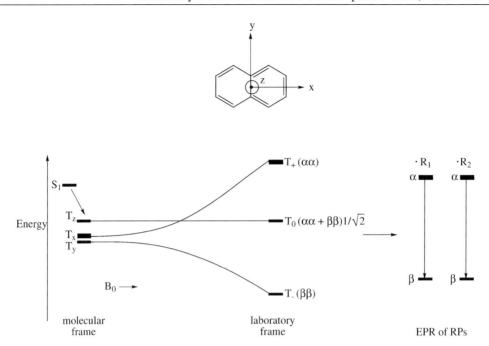

**Figure B1.16.14.** Top, the canonical axes for triplet naphthalene. The $z$-axis is directed out of the plane of the paper. Bottom, energy levels and relative populations during the CIDEP triplet mechanism process. See text for further details.

that this intersystem crossing, which will be explained in detail later, is to be distinguished from that described for RPs described above. Because the TM polarization is present before the triplet reacts to produce the RP, its phase is either net E or net A for both radicals. The origin of the polarization is as follows: in the intersystem crossing process, which is dominated by spin–orbit interactions, the most suitable basis set is one where the canonical orientations of the triplet state are represented. An example is shown in figure B1.16.14 where the directions of the triplet $T_x$, $T_y$ and $T_z$ basis functions for naphthalene are indicated in their usually defined orientations. These are also sometimes called the 'zero-field' basis functions. As the electrons undergo the intersystem crossing process, these are the orbital directions they 'see'. This is called the molecular frame of reference.

Because the spin–orbit interaction is anisotropic (there is a directional dependence of the 'view' each electron has of the relevant orbitals), the intersystem crossing rates from $S_1$ to each triplet level are different. Therefore, unequal populations of three triplet levels results. The $T_x$, $T_y$ and $T_z$ basis functions can be rewritten as linear combinations of the familiar $\alpha$ and $\beta$ spin $\frac{1}{2}$ functions and, consequently, they can also be rewritten as linear combinations of the high-field RP spin wavefunctions $T_+$, $T_0$ and $T_-$, which we have already described above. The net polarization generated in the zero-field basis is carried over to the high-field basis set and, consequently, the initial condition for the geminate RP is that the population of the triplet levels is not strictly equal ($\frac{1}{3}$ each). Exactly which triplet level is overpopulated depends on the sign of the zero-field splitting parameter $D$ in the precursor triplet state. A representative energy level diagram showing the flow of population throughout the intersystem crossing, RP and free radical stages is shown in figure B1.16.14.

The absolute magnitude of the TM polarization intensity is governed by the rate of rotation of the triplet state in the magnetic field. If the anisotropy of the zero-field states is very rapidly averaged (low viscosity), the TM is weak. If the experiment is carried out in a magnetic field where the Zeeman interaction is comparable with the $D$ value and the molecular tumbling rate is slow (high viscosity), the TM is maximized. Additional

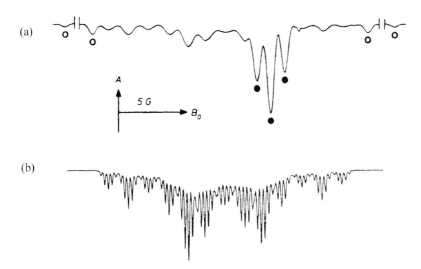

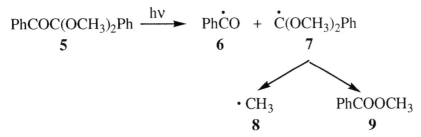

**Figure B1.16.15.** TREPR spectrum after laser flash photolysis of 0.005 M DMPA (5) in toluene. (a) 0.7 $\mu$s, 203 K, RF power 10 mW; O, lines ·CH$_3$ (8), spacing 22.8 G; •, benzoyl (6), remaining lines due to (7). (b) 2.54 $\mu$s, 298 K, RF power 2 mW to avoid nutations, lines of 7 only. Reprinted from [42].

requirements for a large TM polarization are: (1) the intersystem crossing rates from $S_1$ to $T_x$, $T_y$ and $T_z$ must be fast relative to the RP production step, and (2) the spin relaxation time in the excited triplet state should not be too short in order to ensure a large TM polarization.

An example of the triplet mechanism from work by Jent *et al* [42] is shown in figure B1.16.15. Upon laser flash photolysis, dimethoxyphenylacetophenone (DMPA, **5**) forms an excited singlet and undergoes fast intersystem crossing and subsequent photocleavage of the triplet to form radicals **6** and **7** as shown below. Radical **7** can subsequently fragment to form methyl radical (**8**) and methylbenzoate (**9**).

$$PhCOC(OCH_3)_2Ph \xrightarrow{\ h\nu\ } Ph\overset{\bullet}{C}O \ + \ \overset{\bullet}{C}(OCH_3)_2Ph$$
$$\quad\quad\ \ \textbf{5}\quad\quad\quad\quad\quad\quad\quad\quad\quad \textbf{6}\quad\quad\quad\quad\quad\textbf{7}$$

$$\cdot CH_3 \quad\quad\quad PhCOOCH_3$$
$$\textbf{8}\quad\quad\quad\quad\quad\quad \textbf{9}$$

Contradictory evidence regarding the reaction to form **8** and **9** from **7** led the researchers to use TREPR to investigate the photochemistry of DMPA. Figure B1.16.15A shows the TREPR spectrum of this system at 0.7 $\mu$s after the laser flash. Radicals **6**, **7** and **8** are all present. At 2.54 $\mu$s, only **7** can be seen, as shown in figure B1.16.15B. All radicals in this system exhibit an emissive triplet mechanism. After completing a laser flash intensity study, the researchers concluded that production of **8** from **7** occurs upon absorption of a second photon and not thermally as some had previously believed.

### B1.16.3.3    Radical-triplet pair mechanism

In the early 1990s, a new spin polarization mechanism was postulated by Paul and co-workers to explain how polarization can be developed in transient radicals in the presence of excited triplet state molecules (Blättler

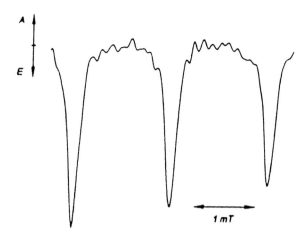

**Figure B1.16.16.** TREPR spectrum of TEMPO radicals in 1,2-epoxypropane solution with benzophenone, 1 $\mu$s after 308 nm laser flash. Reprinted from [45].

*et al* [43], Blättler and Paul [44], Goudsmit *et al* [45]). While the earliest examples of the radical-triplet pair mechanism (RTPM) involved emissive polarizations similar in appearance to triplet mechanism polarizations, cases have since been discovered in which absorptive and multiplet polarizations are also generated by RTPM.

Polarization obtained by RTPM is related to the RPM in that diffuse encounters are still required, but differs in that it involves the interaction of a photoexcited triplet state with a doublet state radical. When a doublet state (electron spin $\frac{1}{2}$) radical is present in high concentration upon production of a photoexcited triplet, the doublet and triplet interact to form new quartet and doublet states. When the two species find themselves in regions of effective exchange ($|J| > 0$), a fluctuating dipole–dipole interaction ($D$) induces transitions between states, leading to a population redistribution that is non-Boltzmann, i.e. CIDEP. This explanation of RTPM is only valid in regions of moderate viscosity. If the motion is too fast, the assumption of a static ensemble will break down. The resulting polarization is either net E or net A depending on the sign of $J$ (there is no dependence on the sign of $D$). This mechanism may also be observable in reactions where a doublet state radical produced by photolysis and unreacted triplets might collide. For this to happen, the triplet lifetimes, radical–triplet collision frequencies and triplet spin relaxation rates need to be of comparable time scales.

Figure B1.16.16 shows an example of RTPM in which the radical species is TEMPO (**10**), a stable nitroxide radical, while the triplet state is produced by photoexcitation of benzophenone (**11**) [45].

**10**                    **11**

The three-line spectrum with a 15.6 G hyperfine reflects the interaction of the TEMPO radical with the nitrogen nucleus ($I = 1$); the benzophenone triplet cannot be observed because of its short relaxation times. The spectrum shows strong net emission with weak E/A multiplet polarization. Quantitative analysis of the spectrum was shown to match a theoretical model which described the size of the polarizations and their dependence on diffusion.

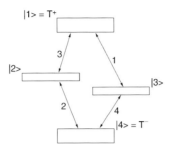

**Figure B1.16.17.** Level diagram showing the origin of SCRP polarization.

### B1.16.3.4  Spin correlated radical pair mechanism

The fourth and final CIDEP mechanism results from the observation of geminate radical pairs when they are still interacting, i.e. there is a measurable dipolar or exchange interaction between the components of the RP at the time of measurement. It is called the spin correlated radical pair (SCRP) mechanism and is found under conditions of restricted diffusion such as micelle-bound RPs [46, 47] or covalently linked biradicals [48] and also in solid state structures such as photosynthetic reaction centres [49] and model systems [50]. In this mechanism additional lines in the EPR spectrum are produced due to the interaction. If the $D$ or $J$ value is smaller than the hyperfines, then the spectrum is said to be first order, with each individual hyperfine line split by $2J$ or $2D$ into a doublet. The most unusual and immediately recognizable feature of the SCRP mechanism with small interactions is that each component of the doublet receives an opposite phase. For triplet precursors and negative $J$ values, which is the common situation, the doublets are E/A. The level diagram in figure B1.16.17 shows the origin of the SCRP polarization for such a system, considering only one hyperfine line.

When the $J$ or $D$ coupling exceeds the hyperfine couplings, the spectrum becomes second order and is much more complex. Lines are alternating with E or A phase, and, if $J$ or $D$ becomes even larger or becomes comparable with the Zeeman interaction, a net emission appears from $S–T_-$ mixing. In second-order spectra the $J$ coupling must be extracted from the spectrum by computer simulation. In terms of line positions and relative intensities, these spectra have a direct analogy to NMR spectroscopy: the first-order spectrum is the equivalent of the AX NMR problem (two sets of doublets), while the second-order spectrum is analogous to the AB ('roof' effect, intermediate $J$ values) or AA′ (convergence of lines, large $J$ values) nuclear spin system [51].

The SCRP polarization pattern can be complicated by other factors: chemical reaction can deplete the middle energy levels in Scheme Y as they have the most singlet character and are thus more likely to react upon encounter. Spin relaxation via either correlated (dipole–dipole) or uncorrelated ($g$ factor or hyperfine anisotropy) mechanisms can also redistribute populations on the TREPR time scale [52]. The most important relaxation mechanism is that due to modulation of the exchange interaction caused by conformational motion which changes the inter-radical distance on the EPR time scale. Here both the $T_1$ and $T_2$ relaxation processes are important [53]. Interestingly, $J$ modulation is also the process by which RPM is produced, and recently it has been demonstrated that at certain viscosities, both RPM and SCRP polarization patterns can be observed simultaneously in both micellar [54] and biradical-type RPs [55]. The presence of SCRP polarization in biradicals has enabled much information to be obtained regarding weak electronic couplings in flexible systems as a function of molecular structure, solvent, and temperature [56–59]. The spin polarization observed in EPR spectra of photosynthetic reaction centres has also proven informative in relating structure to function in those systems, especially in comparing structural parameters measured magnetically to those found by other methods such as x-ray crystallography [60].

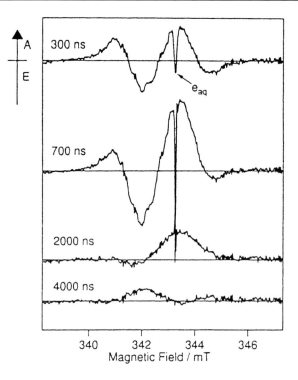

**Figure B1.16.18.** TREPR spectra observed after laser excitation of tetraphenylhydrazine in an SDS micelle at room temperature. Reprinted from [61].

Most observations of SCRP have been from triplet precursors, but Fukuju *et al* [61] have observed singlet-born SCRP upon photolysis of tetraphenylhydrazine in sodium dodecyl sulfate (SDS) micelles. The tetraphenylhydrazine (**12**) cleaves to form two diphenylamino (DPA) radicals, as shown below.

Figure B1.16.18 shows TREPR spectra of this system in SDS micelles at various delay times. The A/E/A/E pattern observed at early delay times is indicative of a singlet-born SCRP. Over time, a net absorptive component develops and, eventually, the system inverts to an E/A/E/A pattern at late delay times. The long lifetime of this SCRP indicates that the DPA radicals are hydrophobic enough that they prefer to remain in the micelle rather than escape. The time dependence of the spectra can be described by a kinetic model which considers the recombination process and the relaxation between all states of the RP. The reader is directed to the reference for further details.

### B1.16.3.5  *Further examples of CIDEP*

While each of the previous examples illustrated just one of the electron spin polarization mechanisms, the spectra of many systems involve polarizations from multiple mechanisms or a change in mechanism with delay time.

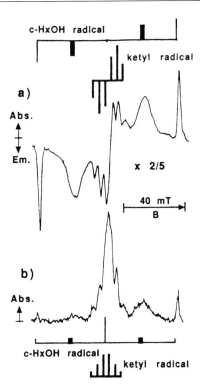

**Figure B1.16.19.** (a) CIDEP spectrum observed in the photolysis of xanthone ($1.0 \times 10^{-2}$ M) in cyclohexanol at room temperature. The stick spectra of the ketyl and cyclohexanol radicals with RPM polarization are presented. (b) CIDEP spectrum after the addition of hydrochloric acid (4.1 vol%; HCl 0.50 M) to the solution above. The stick spectra of the ketyl and cyclohexanol radicals with absorptive TM polarization are presented. The bold lines of the stick spectra of the cyclohexanol radical show the broadened lines due to ring motion of the radical. Reprinted from [62].

Work by Koga *et al* [62] demonstrates how the polarization mechanism can change upon alteration of the chemical environment. Upon laser flash photolysis, excited xanthone abstracts a proton from an alcohol solvent, cyclohexanol in this case. The xanthone ketyl radical (·XnH) and the alcoholic radical (·ROH) exhibit E/A RPM polarization with slight net emission, as shown in figure B1.16.19(a). Upon addition of HCl to the cyclohexanol solution, the same radicals are observed, but the polarization is now entirely an absorptive triplet mechanism, as shown in figure B1.16.19(b). It was determined that both $H^+$ and $Cl^-$ or HCl molecules must be present for this change in mechanism to occur, and the authors postulated that the formation of a charge transfer complex between xanthone and HCl in their ground state might be responsible for the observed change in polarization. This curious result demonstrates how a change in the CIDEP mechanism can yield information about chemical changes which may be occurring in the system.

Utilizing FT-EPR techniques, van Willigen and co-workers have studied the photoinduced electron transfer from zinc tetrakis(4-sulfonatophenyl)porphyrin (ZnTPPS) to duroquinone (DQ) to form $ZnTPPS^{3-}$ and $DQ^-$ in different micellar solutions [34, 63]. Spin-correlated radical pairs [$ZnTPPS^{3-} \ldots DQ^-$] are formed initially, and the SCRP lifetime depends upon the solution environment. The $ZnTPPS^{3-}$ is not observed due to its short $T_2$ relaxation time, but the spectra of $DQ^-$ allow for the determination of the location and stability of reactant and product species in the various micellar solutions. While DQ is always located within the micelle, the location of ZnTPPS and free $DQ^-$ depends upon the micellar environment.

ZnTPPS/DQ/Triton X-100

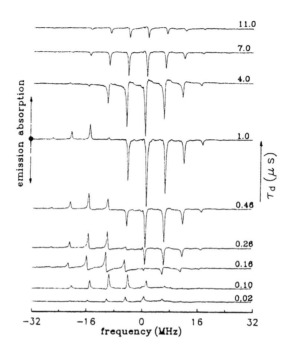

**Figure B1.16.20.** FTEPR spectra of photogenerated DQ⁻ in TX100 solution for delay times between laser excitation of ZnTPPS and microwave pulse ranging from 20 ns to 11 $\mu$s. The central hyperfine line ($M = 0$) is at $\approx -4.5$ MHz. Reprinted from [63].

Figure B1.16.20 shows spectra of DQ⁻ in a solution of TX100, a neutral surfactant, as a function of delay time. The spectra are qualitatively similar to those obtained in ethanol solution. At early delay times, the polarization is largely TM while RPM increases at later delay times. The early TM indicates that the reaction involves ZnTPPS triplets while the A/E RPM at later delay times is produced by triplet excited-state electron transfer. Calculation of relaxation times from spectral data indicates that in this case the ZnTPPS porphyrin molecules are in the micelle, although some may also be in the hydrophobic mantle of the micelle. Further, lineshape and polarization decay analyses indicate that free DQ⁻ radicals move from the micelle into the aqueous phase. The lack of observation of spin-correlated radical pairs indicates that they have dissociated prior to data acquisition. Small out-of-phase signal contributions at the earliest delay times show that the radical pair lifetime in TX100 solution is approximately 100 ns

FT-EPR spectra of the ZnTPPS/DQ system in a solution of cetyltrimethylammonium chloride (CTAC), a cationic surfactant, are shown in figure B1.16.21. As in the TX100 solution, both donor and acceptor are associated with the micelles in the CTAC solution. The spectra of DQ⁻ at delays after the laser flash of less than 5 $\mu$s clearly show polarization from the SCRP mechanism. While SCRPs were too short-lived to be observed in TX100 solution, they clearly have a long lifetime in this case. Van Willigen and co-workers determined that the anionic radicals ZnTPPS³⁻ and DQ⁻ remain trapped in the cationic micelles, i.e. an electrostatic interaction is responsible for the extremely long lifetime of the [ZnTPPS³⁻ ... DQ⁻] spin-correlated radical pair. The spectrum at delay times greater than 5 $\mu$s is again due to free DQ⁻. Linewidth analysis and relaxation time calculations indicate that the free DQ⁻ remains trapped in the cationic micelles. These results

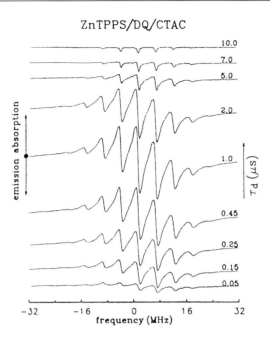

**Figure B1.16.21.** FTEPR spectra photogenerated DQ$^-$ in CTAC solution for delay times between laser excitation of ZnTPPS and microwave pulse ranging from 50 ns to 10 $\mu$s. The central hyperfine line ($M = 0$) is at $\approx$7 MHz. Reprinted from [63].

demonstrate the use of the CIDEP mechanisms in helping to characterize the physical environments of free radicals. In both cases shown here, spectral analysis allowed a determination of the lifetime of the initial SCRP and the location of the porphyrin and the free DQ$^-$.

Figure B1.16.22 shows a stick plot summary of the various CIDEP mechanisms and the expected polarization patterns for the specific cases detailed in the caption. Each mechanism clearly manifests itself in the spectrum in a different and easily observable fashion, and so qualitative deductions regarding the spin multiplicity of the precursor, the sign of $J$ in the RP and the presence or absence of SCRPs can immediately be made by examining the spectral shape. Several types of quantitative information are also available from the spectra. For example, if the molecular structure of one or both members of the RP is unknown, the hyperfine coupling constants and $g$-factors can be measured from the spectrum and used to characterize them, in a fashion similar to steady-state EPR. Sometimes there is a marked difference in spin relaxation times between two radicals, and this can be measured by collecting the time dependence of the CIDEP signal and fitting it to a kinetic model using modified Bloch equations [64].

If the rate of chemical decay of the RP is desired, the task is complex because the majority of the CIDEP signal decays via relaxation pathways on the 1–10 $\mu$s time scale, as opposed to chemical reaction rates which are nominally about an order of magnitude longer than this. There are two ways around this problem. The first is to use a transient digitizer or FT-EPR and signal average many times to improve the signal-to-noise ratio at long delay times where chemical reaction dominates the decay trace. The second is to return to the steady-state method described above and run what is called a 'kinetic EPR' experiment, where the light source is suddenly interrupted and the EPR signal decay is collected over a very long time scale. The beginning of the trace may contain both relaxation of CIDEP intensity as well as chemical decay; however, the tail end of this trace should be dominated by the chemical reaction rates. Much use has been made of kinetic EPR in measuring free radical addition rates in polymerization reactions [65, 66].

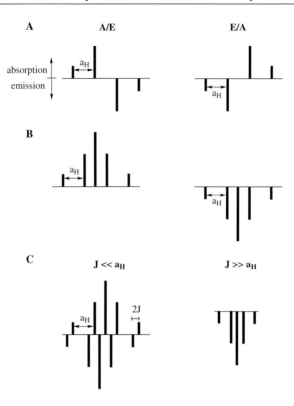

**Figure B1.16.22.** Schematic representations of CIDEP spectra for hypothetical radical pair $\cdot CH_3 + \cdot R$. Part A shows the A/E and E/A RPM. Part B shows the absorptive and emissive triplet mechanism. Part C shows the spin-correlated RPM for cases where $J \ll a_H$ and $J \gg a_H$.

From SCRP spectra one can always identify the sign of the exchange or dipolar interaction by direct examination of the phase of the polarization. Often it is possible to quantify the absolute magnitude of $D$ or $J$ by computer simulation. The shape of SCRP spectra are very sensitive to dynamics, so temperature and viscosity dependencies are informative when knowledge of relaxation rates of competition between RPM and SCRP mechanisms is desired. Much use of SCRP theory has been made in the field of photosynthesis, where structure/function relationships in reaction centres have been connected to their spin physics in considerable detail [67, 68].

## References

[1]  Fessenden R W and Schuler R H 1963 Electron spin resonance studies of transient alkyl radicals *J. Chem. Phys.* **39** 2147–95
[2]  Bargon J, Fischer H and Johnsen U 1967 Kernresonanz-Emissionslinien während rascher Radikalreaktionen *Z. Naturf.* a **20** 1551–5
[3]  Ward H R and Lawler R G 1967 Nuclear magnetic resonance emission and enhanced absorption in rapid organometallic reactions *J. Am. Chem. Soc.* **89** 5518–19
[4]  Lawler R G 1967 Chemically induced dynamic nuclear polarization *J. Am. Chem. Soc.* **89** 5519–21
[5]  Closs G L 1969 A mechanism explaining nuclear spin polarizations in radical combination reactions *J. Am. Chem. Soc.* **91** 4552–4
[6]  Closs G L and Trifunac A D 1969 Chemically induced nuclear spin polarization as a tool for determination of spin multiplicities of radical-pair precursors *J. Am. Chem. Soc.* **91** 4554–5
[7]  Closs G L and Trifunac A D 1970 Theory of chemically induced nuclear spin polarization. III. Effect of isotropic g shifts in the components of radical pairs with one hyperfine interaction *J. Am. Chem. Soc.* **92** 2183–4

[8] Kaptein R and Oosterhoff J L 1969 Chemically induced dynamic polarization II (relation with anomalous ESR spectra) *Chem. Phys. Lett.* **4** 195–7

[9] Kaptein R and Oosterhoff J L 1969 Chemically induced dynamic nuclear polarization III (anomalous multiplets of radical coupling and disproportionation products) *Chem. Phys. Lett.* **4** 214–16

[10] Roth H D, Weng H and Herbertz T 1997 CIDNP study and *ab initio* calculations of rigid vinylcyclopropane systems: evidence for delocalized 'ring-closed' radical cations *Tetrahedron* **53** 10 051–70

[11] Dukes K E 1996 *PhD Dissertation* University of North Carolina

[12] Salikhov K M, Molin Yu N, Sagdeev R Z and Buchachenko A L 1984 *Spin Polarization and Magnetic Effects in Radical Reactions* (Amsterdam: Elsevier)

[13] Michl J and Bonacic-Koutecky V 1990 *Electronic Aspects of Organic Photochemistry* (New York: Wiley)

[14] Porter N A, Marnett L J, Lochmüller C H, Closs G L and Shobtaki M 1972 Application of chemically induced dynamic nuclear polarization to a study of the decomposition of unsymmetric azo compounds *J. Am. Chem. Soc.* **94** 3664–5

[15] Closs G L and Czeropski M S 1977 Amendment of the CIDNP phase rules. Radical pairs leading to triplet states *J. Am. Chem. Soc.* **99** 6127–8

[16] Franck J and Rabinowitsch E 1934 Some remarks about free radicals and the photochemistry of solutions *Trans. Faraday Soc.* **30** 120–31

[17] Turro N J, Buchachenko A L and Tarasov V F 1995 How spin stereochemistry severely complicates the formation of a carbon–carbon bond between two reactive radicals in a supercage *Acc. Chem. Res.* **28** 69–80

[18] Goez M 1995 An introduction to chemically induced dynamic nuclear polarization *Concepts Magn. Reson.* **7** 69–86

[19] Pedersen J B 1979 *Theories of Chemically Induced Magnetic Polarization* (Odense: Odense University Press)

[20] Kaptein R 1971 Simple rules for chemically induced dynamic nuclear polarization *J. Chem. Soc. Chem. Commun.* 732–3

[21] Goez M and Sartorius I 1993 Photo-CIDNP investigation of the deprotonation of aminium cations *J. Am. Chem. Soc.* **115** 11 123–33

[22] Goez M 1995 Pulse techniques for CIDNP *Concepts Magn. Reson.* **7** 263–79

[23] Hany R, Vollenweider J-K and Fischer H 1988 Separation and analysis of CIDNP spin orders for a coupled multiproton system *Chem. Phys.* **120** 169–75

[24] Closs G L and Miller R J 1979 Laser flash photolysis with NMR detection. Microsecond time-resolved CIDNP: separation of geminate and random-phase polarization *J. Am. Chem. Soc.* **101** 1639–41

[25] Closs G L, Miller R J and Redwine O D 1985 Time-resolved CIDNP: applications to radical and biradical chemistry *Acc. Chem. Res.* **18** 196–202

[26] Miller R J and Closs G L 1981 Application of Fourier transform-NMR spectroscopy to submicrosecond time-resolved detection in laser flash photolysis experiments *Rev. Sci. Instrum.* **52** 1876–85

[27] Closs G L and Redwine O D 1985 Direct measurements of rate differences among nuclear spin sublevels in reactions of biradicals *J. Am. Chem. Soc.* **107** 6131–3

[28] Morozova O B, Tsentalovich Y P, Yurkovskaya A V and Sagdeev R Z 1998 Consecutive biradicals during the photolysis of 2,12-dihydroxy-2,12-dimethylcyclododecanone: low- and high-field chemically induced dynamic nuclear polarizations (CIDNP) study *J. Phys. Chem.* A **102** 3492–7

[29] Smaller B, Remko J R and Avery E C 1968 Electron paramagnetic resonance studies of transient free radicals produced by pulse radiolysis *J. Chem. Phys.* **48** 5174–81

[30] Trifunac A D, Thurnauer M C and Norris J R 1978 Submicrosecond time-resolved EPR in laser photolysis *Chem. Phys. Lett.* **57** 471–3

[31] Fessenden R W and Verma N C 1976 Time resolved electron spin resonance spectroscopy. III. Electron spin resonance emission from the hydrated electron. Possible evidence for reaction to the triplet state *J. Am. Chem. Soc.* **98** 243–4

[32] Forbes M D E, Peterson J and Breivogel C S 1991 Simple modification of Varian E-line microwave bridges for fast time-resolved EPR spectroscopy *Rev. Sci. Instrum.* **66** 2662–5

[33] Forbes M D E 1993 A fast 35 GHz time-resolved EPR apparatus *Rev. Sci. Instrum.* **64** 397–402

[34] van Willigen H, Levstein P R and Ebersole M H 1993 Application of Fourier transform electron paramagnetic resonance in the study of photochemical reactions *Chem. Rev.* **93** 173–97

[35] Adrian F J 1971 Theory of anomalous electron spin resonance spectra of free radicals in solution. Role of diffusion-controlled separation and reencounter of radical pairs *J. Chem. Phys.* **54** 3918–23

[36] Hore P J 1989 Analysis of polarized electron paramagnetic resonance spectra *Advanced EPR: Applications in Biology and Biochemistry* ed A J Hoff (Amsterdam: Elsevier) ch 12

[37] Sekiguchi S, Kobori Y, Akiyama K and Tero-Kubota S 1998 Marcus free energy dependence of the sign of exchange interactions in radical ion pairs generated by photoinduced electron transfer reactions *J. Am. Chem. Soc.* **120** 1325–6

[38] Pedersen J B and Freed J H 1973 Theory of chemically induced dynamic electron polarization. I *J. Chem. Phys.* **58** 2746–62

[39] Pedersen J B and Freed J H 1973 Theory of chemically induced dynamic electron polarization. II *J. Chem. Phys.* **59** 2869–85

[40] Wong S K, Hutchinson D A and Wan J K S 1973 Chemically induced dynamic electron polarization. II. A general theory for radicals produced by photochemical reactions of excited triplet carbonyl compounds *J. Chem. Phys.* **58** 985–9

[41] Atkins P W and Evans G T 1974 Electron spin polarization in a rotating triplet *Mol. Phys.* **27** 1633–44

[42] Jent F, Paul H and Fischer H 1988 Two-photon processes in ketone photochemistry observed by time-resolved ESR spectroscopy *Chem. Phys. Lett.* **146** 315–19

[43] Blättler C, Jent F and Paul H 1990 A novel radical-triplet pair mechanism for chemically induced electron polarization (CIDEP) of free radicals in solution *Chem. Phys. Lett.* **166** 375–80

[44] Blättler C and Paul H 1991 CIDEP after laser flash irradiation of benzil in 2-propanol. Electron spin polarization by the radical-triplet pair mechanism *Res. Chem. Intermed.* **16** 201–11

[45] Goudsmit G-H, Paul H and Shushin A I 1993 Electron spin polarization in radical-triplet pairs. Size and dependence on diffusion *J. Phys. Chem.* **97** 13 243–9

[46] Closs G L, Forbes M D E and Norris J R 1987 Spin-polarized electron paramagnetic resonance spectra of radical pairs in micelles. Observation of electron spin–spin interactions *J. Phys. Chem.* **91** 3592–9

[47] Buckley C D, Hunger D A, Hore P J and McLauchlan K A 1987 Electron spin resonance of spin-correlated radical pairs *Chem. Phys. Lett.* **135** 307–12

[48] Closs G L and Forbes M D E 1991 EPR spectroscopy of electron spin polarized biradicals in liquid solutions. Technique, spectral simulation, scope and limitations *J. Phys. Chem.* **95** 1924–33

[49] Norris J R, Morris A L, Thurnauer M C and Tang J 1990 A general model of electron spin polarization arising from the interactions within radical pairs *J. Chem. Phys.* **92** 4239–49

[50] Levanon H and Möbius K 1997 Advanced EPR spectroscopy on electron transfer processes in photosynthesis and biomimetic model systems *Ann. Rev. Biophys. Biomol. Struct.* **26** 495–540

[51] Friebolin H 1993 *Basic One- and Two-Dimensional NMR Spectroscopy* (New York: VCH)

[52] De Kanter F J J, den Hollander J A, Huizer A H and Kaptein R 1977 Biradical CIDNP and the dynamics of polymethylene chains *Mol. Phys.* **34** 857–74

[53] Avdievich N I and Forbes M D E 1995 Dynamic effects in spin-correlated radical pair theory: J modulation and a new look at the phenomenon of alternating line widths in the EPR spectra of flexible biradicals *J. Phys. Chem.* **99** 9660–7

[54] Forbes M D E, Schulz G R and Avdievich N I 1996 Unusual dynamics of micellized radical pairs generated from photochemically active amphiphiles *J. Am. Chem. Soc.* **118** 10 652–3

[55] Forbes M D E, Avdievich N I, Schulz G R and Ball J D 1996 Chain dynamics cause the disappearance of spin-correlated radical pair polarization in flexible biradicals *J. Phys. Chem.* **100** 13 887–91

[56] Maeda K, Terazima M, Azumi T and Tanimoto Y 1991 CIDNP and CIDEP studies on intramolecular hydrogen abstraction reaction of polymethylene-linked xanthone and xanthene. Determination of the exchange integral of the intermediate biradicals *J. Phys. Chem.* **95** 197–204

[57] Forbes M D E, Closs G L, Calle P and Gautam P 1993 The temperature dependence of the exchange coupling in polymethylene biradicals. Conclusions regarding the mechanism of the coupling *J. Phys. Chem.* **97** 3384–9

[58] Forbes M D E 1993 The effect of localized unsaturation on the scalar exchange coupling in flexible biradicals *J. Phys. Chem.* **97** 3390–5

[59] Forbes M D E 1993 The effect of $\pi$-system spacers on exchange couplings and end-to-end encounter rates in flexible biradicals *J. Phys. Chem.* **97** 3396–400

[60] Bittl R, van der Est A, Kamlowski A, Lubitz W and Stehlik D 1994 Time-resolved EPR of the radical pair $P_{865}^+Q_A^-$ in bacterial reaction centers. Observation of transient nutations, quantum beats and envelope modulation effects *Chem. Phys. Lett.* **226** 249–58

[61] Fukuju T, Yashiro H, Maeda K, Murai H and Azumi T 1997 Singlet-born SCRP observed in the photolysis of tetraphenylhydrazine in an SDS micelle: time dependence of the population of the spin states *J. Phys. Chem. A* **101** 7783–6

[62] Koga T, Ohara K, Kuwata K and Murai H 1997 Anomalous triplet mechanism spin polarization induced by the addition of hydrochloric acid in the photochemical system of xanthone in alcohol *J. Phys. Chem. A* **101** 8021–5

[63] Levstein P R and van Willigen H 1991 Photoinduced electron transfer from porphyrins to quinones in micellar systems: an FT-EPR study *Chem. Phys. Lett.* **187** 415–22

[64] Verma N C and Fessenden R W 1976 Time resolved ESR spectroscopy. IV. Detailed measurement and analysis of the ESR time profile *J. Chem. Phys.* **65** 2139–60

[65] Héberger K and Fischer H 1993 Rate constants for the addition of the 2-cyano-2-propyl radical to alkenes in solution *Int. J. Chem. Kin.* **25** 249–63

[66] Héberger K and Fischer H 1993 Rate constants for the addition of 2-hydroxy-2-propyl radical to alkenes in solution *Int. J. Chem. Kin.* **25** 913–20

[67] Thurnauer M C and Norris J R 1980 An electron spin echo phase shift observed in photosynthetic algae. Possible evidence for dynamic radical pair interactions *Chem. Phys. Lett.* **76** 557–61

[68] Prisner T F, van der Est A, Bittl R, Lubitz W, Stehlik D and Möbius K 1995 Time-resolved W-band (95 GHz) EPR spectroscopy of Zn-substituted reaction centers of *Rhodobacter sphaeroides* R-26 *Chem. Phys.* **194** 361–70

## Further Reading

Salikhov K M, Molin Yu N, Sagdeev R Z and Buchachenko A L 1984 *Spin Polarization and Magnetic Effects in Radical Reactions* (Amsterdam: Elsevier)

A detailed description of spin polarization theory.

Lepley A R and Closs G L (eds) 1973 *Chemically Induced Magnetic Polarization* (New York: Wiley)

An early summary of research in the field of CIDNP.

Muus L T (ed) 1977 *Chemically Induced Magnetic Polarization: Proc. NATO Advanced Study Institute (Sogesta, Urbino, Italy, April 17–30, 1977)* (Boston, MA: Reidel)

The proceedings of the first international meeting on spin polarization phenomena.

Carrington A and McLachlan A D 1979 *Introduction to Magnetic Resonance, with Applications to Chemistry and Chemical Physics* (New York: Wiley)

An excellent beginner's introduction to magnetic resonance, spin operators and their manipulation to predict and analyze spectra.

Weil J A, Bolton J R and Wertz J E 1994 *Electron Paramagnetic Resonance: Elementary Theory and Practical Applications* (New York: Wiley)

The standard monograph for those seeking an introduction to EPR spectroscopy.

Friebolin H 1993 *Basic One- and Two-Dimensional NMR Spectroscopy* (New York: VCH)

A basic introduction to NMR spectral analysis.

# B1.17
# Microscopy: electron (SEM and TEM)

*Rasmus R Schröder and Martin Müller*

## Abbreviations

| | |
|---|---|
| 2D | two-dimensional |
| 3D | three-dimensional |
| ssCCD | slow-scan charge coupled device |
| BSE | backscattered electrons |
| CTF | contrast transfer function |
| DQE | detection quantum efficiency |
| $E$ | electron energy |
| $E_0$ | electron rest energy |
| EDX | energy dispersive x-ray detection spectroscopy |
| EELS | electron energy loss spectroscopy |
| EFTEM | energy filtering transmission electron microscope |
| EM | electron microscope/microscopy |
| EPMA | electron probe micro analysis |
| ESEM | environmental scanning electron microscope |
| ESI | electron spectroscopic imaging |
| ESR | electron spin resonance |
| FEG | field emission gun |
| IP | imaging plate |
| LVSEM | low-voltage scanning electron microscope |
| MTF | modulation transfer function |
| NMR | nuclear magnetic resonance |
| PSF | point spread function |
| SE | secondary electron |
| SEM | scanning electron microscope |
| STEM | scanning transmission electron microscope |
| TEM | transmission electron microscope |

## B1.17.1 Introduction

The electron microscope (EM) was developed in the 1930s primarily as an imaging device which exceeded the resolution power of the light microscope by several orders of magnitude. With the evolution towards dedicated instruments designed to answer specific structural and analytical questions, electron microscopy (EM) has grown into a heterogeneous field of electron beam devices. These allow the study of the interaction

**Table B1.17.1.** Instrumental development of electron microscopy.

| Year | Event | Reference |
|------|-------|-----------|
| 1926 | Busch focuses an electron beam with a magnetic lens | |
| 1931 | Ruska and colleagues build the first TEM protoype | Knoll and Ruska (1932) [79] |
| 1935 | Knoll proves the concept of SEM | |
| 1938 | von Ardenne builds the first SEM prototype | |
| 1939 | Siemens builds the first commercial TEM | |
| 1965 | Cambridge Instruments builds the first commercial SEM | |
| 1968 | Crewe and colleagues introduce the FEG as electron beam source | Crewe *et al* (1968) [80] |
| 1968 | Crewe and colleagues build the first STEM prototype | Crewe *et al* (1968) [81] |
| 1995 | Zach proves the concept of a corrected LVSEM | Zach (1995) [10] |
| 1998 | Haider and colleagues prove the concept of the TEM spherical aberration corrector | Haider *et al* (1998) [55] |

of electrons with the sample, which can subsequently be interpreted as information about object structure or chemical composition. Therefore, EM must be compared to other high-resolution diffraction methods, such as x-ray or neutron scattering, or to spectroscopic techniques such as electron spin resonance spectroscopy (ESR) and nuclear magnetic resonance spectroscopy (NMR). More recent, non-diffractive techniques include scanning tunnelling microscopy (STM) and atomic force microscopy (AFM) (for a detailed discussion see chapter B1.19).

All these methods are used today to obtain structural and analytical information about the object (see also the specific chapters about these techniques). In the case of structural studies, x-ray crystallography is the method of choice if suitable three-dimensional (3D) crystals of the object are available. In fact, x-ray crystallography has provided a vast number of atomic structures of inorganic compounds or organic molecules. The main advantage of EM, however, is the possibility of directly imaging almost any sample, from large biological complexes consisting of many macromolecules down to columns of single atoms in a solid material. With modern instruments and applying specific preparation techniques, it is even possible to visualize beam-sensitive organic material at molecular resolution (of the order of 3–4 Å). Imaging at atomic resolution is almost routine in material science. Today, some challenges remain, including the combination of sub-Ångstrøm resolution with chemical analysis and the ability to routinely reconstruct the complete 3D spatial structure of a given sample.

The history of EM (for an overview see table B1.17.1) can be interpreted as the development of two concepts: the electron beam either illuminates a large area of the sample ('flood-beam illumination', as in the typical transmission electron microscope (TEM) imaging using a spread-out beam) or just one point, i.e. focused to the smallest spot possible, which is then scanned across the sample (scanning transmission electron microscopy (STEM) or scanning electron microscopy (SEM)). In both situations the electron beam is considered as a matter wave interacting with the sample and microscopy simply studies the interaction of the scattered electrons.

In principle, the same physico-chemical information about the sample can be obtained from both illumination principles. However, the difference of the achievable spatial resolutions of both illumination principles

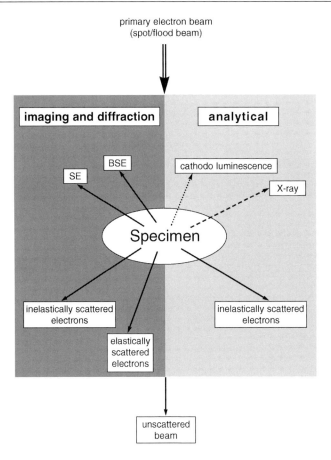

**Figure B1.17.1.** Schema of the electron specimen interactions and their potential use for structural and analytical studies.

illustrates the general difference of the two approaches. Spatial resolution in the case of flood-beam illumination depends on the point spread function (PSF) of the imaging lens and detector system and—for a typical TEM sample—on the interference effects representing phase shifts of the scattered electron wave. In general, resolution is described as a global phenomenon. For the scanned beam, all effects are local and confined to the illuminated spot. Thus, spatial resolution of any event is identical to the achievable spot diameter. Therefore, all steps in the further development of EM involve either the improvement of the PSF (either by improving sample preparation, imaging quality of the electron lenses, or improving spatial resolution of the electron detection devices) or the improvement of the spot size (by minimizing the size of the electron source using a field emission gun (FEG) and by improving the imaging quality of the electron lenses).

In general, EM using a focused beam provides higher spatial resolution if the spot size of the scanning beam is smaller than the delocalization of the event studied: consider, for example, inelastic scattering events which for signal-to-noise ratio (SNR) reasons can be imaged in energy filtering TEMs (EFTEM) at 1–2 nm resolution. In a dedicated STEM with an FEG electron source, a localization of the inelastic event comparable to the actual probe size of e.g. 2 Å can be expected. Moreover, this effect is enhanced by the electron detectors used in TEM and S(T)EM. For scanned-beam microscopy, detectors do not resolve spatial information, instead they are designed for highest detection quantum efficiency (DQE almost ideal). Such an ideal detection has only recently been reached for TEM by the use of slow-scan charge-coupled device

**Table B1.17.2.** Electron–specimen interactions.

|  | Elastic scattering | Inelastic scattering |
|---|---|---|
| Where<br>Scattering potential<br>Scattering angles<br>($E = 100$ keV) | Coulomb potential of nucleus<br>Localized<br>Large ($>10$ mrad) | At atomic shell electrons<br>Less localized<br>Smaller ($<10$ mrad) |
| Application | High-resolution signal<br>(TEM, STEM)<br>Back-scattering of electrons<br>(BSE signal in SEM) | Analytical signal<br>(TEM, STEM, SEM)<br>Emission of secondary electrons<br>(SE signal in SEM) |
| Used effects |  | Phonon excitation (20 meV–1 eV)<br>Plasmon and interband excitations (1–50 eV)<br>Inner-shell ionization ($\Delta E$ = ionization energy loss)<br>Emission of x-ray (continuous/characteristic, analytical EM) |

(ssCCD) cameras or imaging plates (IP). However, their spatial resolution is only moderate, compared to conventional, electron-sensitive photographic material.

For several reasons, such as ease of use, cost, and practicability, TEM today is the standard instrument for electron diffraction or the imaging of thin, electron-transparent objects. Especially for structural imaging at atomic level (spatial resolution of about 1 Å) the modern, aberration-corrected TEM seems to be the best instrument. SEM provides the alternative for imaging the surface of thick bulk specimens. Analytical microscopy can either be performed using a scanning electron probe in STEM and SEM (as for electron probe micro-analysis (EPMA), energy-dispersive x-ray spectroscopy (EDX) and electron energy loss spectroscopy (EELS)) or energy-selective flood-beam imaging in EFTEM (as for image-EELS and electron spectroscopic imaging (ESI)). The analytical EM is mainly limited by the achievable probe size and the detection limits of the analytical signal (number of inelastically scattered electrons or produced characteristic x-ray quanta). The rest of this chapter will concentrate on the structural aspects of EM. Analytical aspects are discussed in more detail in specialized chapters (see, for example, B1.6).

It is interesting to note the analogy of developments in light microscopy during the last few decades. The confocal microscope as a scanning beam microscope exceeds by far the normal fluorescence light microscope in resolution and detection level. Very recent advances in evanescent wave and interference microscopy seem to promise to provide even higher resolution (B1.18).

EM has been used in a wide variety of fields, from material sciences to cell and structural biology or medical research. In general, EM can be used for any high-resolution imaging of objects or their analytical probing. Modern instrumentation of STEM and SEM provides high-resolution instruments capable of probe sizes, in the case of TEM, of a few Ångstrøm or sub-Ångstrøm information limit. However, specimen properties and sample preparation limit the achievable resolution. Typical resolution obtained today range from atomic detail for solid materials, molecular detail with resolution in the order of 3–5 Å for crystalline biological samples, and about 1–2 nm for individual particles without a certain intrinsic symmetry. Recent publications on the different aspects of EM include Williams and Carter ([1], general text covering all the modern aspects of EM in materials science), the textbooks by Reimer ([2–4], detailed text about theory, instrumentation, and application), or—for the most complete discussion of all electron-optical and theoretical aspects—Hawkes and Kasper [5]. Additional research papers on specialized topics are referenced in text.

## B1.17.2    Interaction of electrons with matter and imaging of the scattering distribution

The interaction of electrons with the specimen is dominated by the Coulomb interaction of charged particles. For a summary of possible charge–charge interactions see figure B1.17.1. Elastic scattering by the Coulomb potential of the positively charged atomic nucleus is most important for image contrast and electron diffraction. This scattering potential is well localized, leads to large scattering angles and yields high-resolution structural information from the sample. In contrast, interactions with the atomic electrons lead to an energy loss of the incident electron by the excitation of different energy states of the sample, such as phonon excitation, plasmon excitation or inner-shell ionization (see table B1.17.2). Inelastic scattering processes are not as localized as the Coulomb potential of the nucleus, leading to smaller scattering angles (inelastic forward scattering), and are in general not used to obtain high-resolution structural information. Instead, inelastic scattering provides analytical information about the chemical composition and state of the sample.

The ratio of elastically to inelastically scattered electrons and, thus, their importance for imaging or analytical work, can be calculated from basic physical principles: consider the differential elastic scattering cross section

$$\frac{d\sigma_{el}}{d\Omega} = \frac{4Z^2 R^4 (1 + E/E_0)}{a_H^2} \frac{1}{[1 + (\theta/\theta_0)^2]} \tag{B1.17.1}$$

with the characteristic screening angle $\theta_0 = \lambda/(2\pi R)$ where $\theta$ denotes the scattering angle, $E$ is the electron energy, $E_0$ the electron rest energy, $+eZ$ is the charge of the nucleus, $R = a_H Z^{1/3}$, and $a_H = 0.0529$ nm is the Bohr radius. Compare this to the inelastic differential scattering cross section

$$\frac{d\sigma_{inel}}{d\Omega} = \frac{\lambda^4 (1 + E/E_0)^2}{4\pi^4 a_H^2} \frac{Z\left\{1 - \dfrac{1}{[1 + (\theta/\theta_0)^2]}\right\}}{(\theta^2 + \theta_E^2)^2} \tag{B1.17.2}$$

with the characteristic inelastic scattering angle $\theta_E = \Delta E/E \cdot (E + E_0)/(E + 2E_0)$ for a given energy loss $\Delta E$.

For large scattering angles, $\theta \gg \theta_0, \theta_E$ the ratio

$$\frac{d\sigma_{inel}/d\Omega}{d\sigma_{el}/d\Omega} = \frac{1}{Z} \tag{B1.17.3}$$

only depends on the atomic number $Z$, whereas for small angles $\theta$

$$d\sigma_{inel}/d\Omega > d\sigma_{el}/d\Omega \tag{B1.17.4}$$

for all $Z$. The total scattering cross sections are found by integrating the above equations [6]

$$\nu = \frac{\sigma_{inel}}{\sigma_{el}} \cong \frac{26}{Z} \tag{B1.17.5}$$

or experimentally [7]

$$\nu \cong \frac{18}{Z}. \tag{B1.17.6}$$

Equations (B1.17.1)–(B1.17.6) indicate that inelastic scattering is most important for light atoms, whereas elastic scattering dominates for large scattering angles and heavy atoms. Therefore, the high resolution image contrast—given by electrons scattered to large angles—for TEM and STEM is dominated by the elastic scattering process. Inelastically scattered electrons are treated either as background, or separated from the elastic image by energy-dispersive spectrometers.

In the case of SEM, both elastic and inelastic processes contribute to image contrast: elastic scattering to large angles (multiple elastic scattering resulting in a large scattering angle) produces backscattered electrons (BSE), which can be detected above the surface of the bulk specimen. Inelastic collisions can excite atomic electrons above the Fermi level, i.e. more energy is transferred to the electron than it would need to leave the sample. Such secondary electrons (SE) are also detected, and used to form SEM images. As will be discussed in B1.17.4 (specimen preparation), biological material with its light atom composition is often stained or coated with heavy metal atoms to increase either the elastic scattering contrast in TEM or the BSE signal in SEM. Unstained, native biological samples generally produce only little image contrast.

Inelastic scattering processes are not used for structural studies in TEM and STEM. Instead, the signal from inelastic scattering is used to probe the electron-chemical environment by interpreting the specific excitation of core electrons or valence electrons. Therefore, inelastic excitation spectra are exploited for analytical EM.

Next we will concentrate on structural imaging using only elastically scattered electrons. To obtain the structure of a scattering object in TEM it is sufficient to detect or image the scattering distribution. Consider the scattering of an incident plane wave $\psi_0(r) \equiv 1$ on an atomic potential $V(r)$. The scattered, outgoing wave will be described by the time-independent Schrödinger equation. Using Green's functions the wave function can be written as

$$\psi(r) = \exp(ikr) - \frac{m}{2\pi\hbar^2} \int \frac{\exp(ik|r - r'|)}{|r - r'|} V(r')\psi(r') \, dr' \tag{B1.17.7}$$

where $k$ denotes the initial wave vector, $m$ is equal to the electron mass. For large distances from the scattering centre $r \gg r'$. This can be approximated by

$$\psi(r) = \exp(ikr) + f(\theta)\frac{\exp(ikr)}{r} \tag{B1.17.8}$$

where $\theta$ denotes the scattering angle.

In this approximation, the wave function is identical to the incident wave (first term) plus an outgoing spherical wave multiplied by a complex scattering factor

$$f(\theta) = |f(\theta)| \exp(i\eta(\theta)) \tag{B1.17.9}$$

which can be calculated as

$$f(\theta) = -\frac{m}{2\pi\hbar^2} \int \exp(ik'r')V(r')\psi(r') \, dr' \tag{B1.17.10}$$

where $k' = kr'/r$. For a weak potential $V(r)$ it is possible to use the first Born approximation, i.e. $\psi(r')$ in equation (B1.17.10) can be replaced by the incident wave resulting in:

$$f(\theta) = -\frac{m}{2\pi\hbar^2} \int \exp(i(k - k')r')V(r') \, dr'. \tag{B1.17.11}$$

This equation describes the Fourier transform of the scattering potential $V(r)$. It should be noted that, in the Born approximation the scattering amplitude $f(\theta)$ is a real quantity and the additional phase shift $\eta(\theta)$ is zero. For atoms with high atomic number this is no longer true. For a rigorous discussion on the effects of the different approximations see [2] or [5].

In a diffraction experiment a quantity $|F(S)|^2$ can be measured which follows from equations (B1.17.8) and (B1.17.9) in Fourier space as

$$F(S) = \delta_0 + i|f(S)| \exp(i\eta(S)) \tag{B1.17.12}$$

where $S = k - k'$ denotes the scattering vector. Combining equations (B1.17.11) and (B1.17.12) leads to the conclusion that in a diffraction experiment the squared amplitude of the Fourier transform of the scattering potential $V(r)$ is measured. Similar formulae can be deduced for whole assemblies of atoms, e.g. macromolecules, resulting in the molecular transform instead of simple atomic scattering factors (for an introduction to the concept of molecular transforms see e.g. [8]). Such measurements are performed in x-ray crystallography, for example. To reconstruct the original scattering potential $V(r)$ it is necessary to determine the phases of the structure amplitudes to perform the reverse Fourier transform. However, if lenses are available for the particles used as incident beam—as in light and electron microscopy—a simple microscope can be built: the diffracted wave is focused by an objective lens into the back focal plane, where the scattered and unscattered parts of the wave are separated. Thus, the objective lens can simply be understood as a Fourier transform operator. In a subsequent imaging step by one additional lens, the scattered and unscattered waves are allowed to interfere again to form a direct image of the scattering potential. This can be understood as a second Fourier transform of the scattering factor (or molecular transform) recovering the spatial distribution of the scattering centres. The small angular scattering distribution of only 10–20 mrad results in a complication in the case of EM. The depth of focus is very large, i.e. it is not possible to recalculate the 3D distribution of the scattering potential but only its 2D projection along the incident beam. All scattering distributions, images, or diffraction patterns are always produced by the 2D projecting transmission function of the actual 3D object. Using a variety of tomographic data collections it is then possible to reconstruct the true 3D object (see below).

The above theory can also be applied to STEM, which records scattering distributions as a function of the scanning probe position. Images are then obtained by plotting the measured scattering intensities (i.e. in the case of the elastic scattering, the direct measurement of the scattering factor amplitudes) according to the probe position. Depending on the signal used, this leads to a conventional elastic dark-field image, or to STEM phase-contrast images [9].

In the case of the scanning electron microscope (SEM), images are formed by recording a specific signal resulting from the electron beam/specimen interaction as a function of the scanning probe position. Surface structures are generally described with the SE (secondary electron) signal. SEs are produced as a consequence of inelastic events. They have very low energies and, therefore, can leave the specimen and contribute to the imaging signal only when created very close to the specimen surface. The escape depth for the secondary electrons depends on the material. It is relatively large (tens of nanometres) for organic and biological material and small for heavy metals (1–3 nm). High-resolution topographic information (limited mainly by the diameter of the scanning electron beam) requires that the source of the signal is localized very close to the specimen surface. In the case of organic materials this localization can be achieved by a very thin metal coating (W, Cr, Pt; thickness = approximate SE escape depth).

The BSE signal is also frequently used for imaging purposes. BSE are electrons of the primary beam (scanning probe) that have been elastically or inelastically scattered in the sample. Their energy depends on their scattering history. When scattered from the surface, they may have lost no or very little energy and provide high topographic resolution. When multiply scattered inside the sample they may have lost several keV and transfer information from a large volume. This volume depends on the material as well as on the energy of the primary beam. BSE produce SE when passing through the SE escape zone. These SE are, however, not correlated with the position of the scanning probe and contribute a background noise which can obscure the high resolution topographic SE signal produced at the point of impact of the primary beam. The interpretation of high-resolution topographic images therefore depends on optimized handling of specimen properties, energy of the electron probe, metal coating and sufficient knowledge of the signal properties.

The discussion of electron–specimen interactions shows that, for a given incident electron dose, a certain quantity of resulting scattered electrons and secondary electrons or photons is produced. The majority of energy transfer into the specimen leads to beam damage and, finally, to the destruction of the sample structure.

Therefore it is desirable to simultaneously collect as much information from the interactions as possible. This concept could lead to an EM instrument based on the design of a STEM but including many different detectors for the elastic dark field, phase contrast, inelastically scattered electrons, BSE, SE, and EDX. The complexity of such an instrument would be enormous. Instead, specific instruments developed in the past can coarsely be categorized as TEM for structural studies on thin samples, STEM for analytical work on thin samples and SEM for analytical and surface topography studies.

### B1.17.3   Instrumentation

*B1.17.3.1   Electron beam instruments*

The general instrumentation of an EM very much resembles the way an ordinary, modern light microscope is built. It includes an electron beam forming source, an illumination forming condenser system and the objective lens as the main lens of the microscope. With such an instrumentation, one forms either the conventional bright field microscope with a large illuminated sample area or an illumination spot which can be scanned across the sample. Typical electron sources are conventional heated tungsten hairpin filaments, heated $LaB_6$-, or $CeB_6$-single-crystal electron emitters, or—as the most sophisticated source—FEGs. The latter sources lead to very coherent electron beams, which are necessary to obtain high-resolution imaging or very small electron probes.

Modern EMs use electromagnetic lenses, shift devices and spectrometers. However, electrostatic devices have always been used as electron beam accelerators and are increasingly being used for other tasks, e.g. as the objective lens (LVSEM, [10]).

EM instruments can be distinguished by the way the information, i.e. the interacting electrons, is detected. Figure B1.17.2 shows the typical situations for TEM, STEM, and SEM. For TEM the transmitted electron beam of the brightfield illumination is imaged simply as in an light microscope, using the objective and projective lenses as conventional imaging system. Combining such TEMs with energy-dispersive imaging elements (filters, spectrometers; see [14]) the modern generation of EFTEMs has been introduced in the last decade. In the case of STEM, the transmitted electrons are not again imaged by lenses, instead the scattered electrons are directly recorded by a variety of detectors. For SEM, the situation is similar to that of STEM. However, only the surface of a bulk specimen is scanned and the resulting backscattered or secondary electrons are recorded by dedicated detectors.

As a special development in recent years, SEMs have been designed which no longer necessitate high vacuum (environmental SEM, ESEM; variable pressure SEM, VPSEM). This development is important for the imaging of samples with a residual vapour pressure, such as aqueous biological or medical samples, but also samples in materials science (wet rock) or organic chemistry (polymers).

*B1.17.3.2   Electron detectors in TEM, STEM, and SEM*

Detectors in EM can be categorized according to their different spatial resolution or in relation to the time it takes to actually see and process the signal (real-time/on-line capability).

Historically, TEM—as an offspring of the electron oscillograph—uses a fluorescent viewing screen for the direct observation of impinging electrons by green fluorescent light. The spatial resolution of such screens is of the order of 30–50 $\mu$m. Coupled to a TV camera tube and computer frame-grabber cards, fluorescent screens are still used for the real-time recording of the image. Often the cameras are combined with silicon intensifier targets (SIT), which allows the detection of single electrons. This is of special importance for the study of beam-sensitive samples. In general, the spatial resolution of such combinations is very poor and dynamic range of the signal is limited to about 256 grey levels (8 bit).

The most common electron detector for TEM has been photographic emulsion. Its silver halide particles are sensitive to high-energy electrons and record images at very high resolution (grain size smaller than 10 $\mu$m)

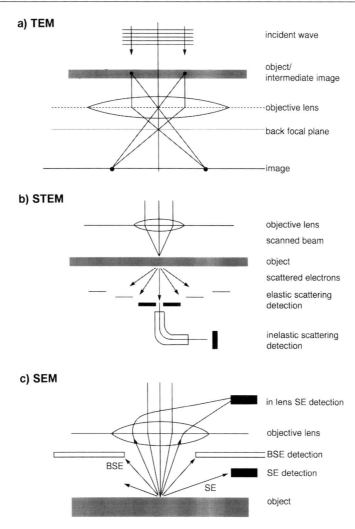

**Figure B1.17.2.** Typical electron beam path diagrams for TEM (a), STEM (b) and SEM (c). These schematic diagrams illustrate the way the different signals can be detected in the different instruments.

when exposed to the electron beam in the TEM. The resolution recorded on film depends on the scattering volume of the striking electron. For typical electron energies of 100–200 keV, its lateral spread is 10–20 $\mu$m. The dynamic range of photographic film is up to 4000 grey levels (12 bit). The most important advantage of film is its large detector size combined with high resolution: with a film size of $6 \times 9$ cm$^2$, up to $10^7$ image points can be recorded.

A recent development is the adaptation of IP to EM detection. IPs have been used for detecting x-rays, which generate multiple electron–electron hole pairs in a storage layer of BaFBr:Eu$^{2+}$. The pairs are trapped in Eu–F centres and can be stimulated by red light to recombine, thereby emitting a blue luminescent signal. Exposing IPs to high energetic electrons also produces electron–electron hole pairs. Scanning the IP with a red laser beam and detecting the blue signal via a photo-multiplier tube results in the readout of the latent image. The large detector size and their extremely high dynamic range of more than $10^7$ makes IPs the ideal detector for electron diffraction.

The only on-line detector for TEM with moderate-to-high spatial resolution is the slow-scan CCD camera. A light-sensitive CCD chip is coupled to a scintillator screen consisting of plastic, an yttrium–aluminium garnet (YAG) crystal, or phosphor powder. This scintillator layer deteriorates the original resolution of the CCD chip elements by scattering light into neighbouring pixels. Typical sizes of chips at present are $1024 \times 1024$ or $2048 \times 2048$ pixels of $(19\text{--}24 \ \mu m)^2$; the achievable dynamic range is about $10^5$ grey levels.

For all the detectors with spatially distinct signal recording, the numeric pixel size (such as scanning pixel size for photographic film and IP, or chip-element size for ssCCD) must be distinguished from the actual obtainable resolution. This resolution can be affected by the primary scattering process of electrons in the detecting medium, or by the scattering of a produced light signal or a scanning light spot in the detecting medium. Therefore, a point signal is delocalized, mathematically described by the PSF. The Fourier transform of the PSF is called the modulation transfer function (MTF), describing the spatial frequency response of the detector. Whereas the ideal detector has a MTF $= 1$ over the complete spatial frequency range, real detectors exhibit a moderate to strong fall-off of the MTF at the Nyquist frequency, i.e. their maximal detectable spatial resolution. In addition to spatial resolution, another important quantity characterizing a detector is the detection quantum efficiency (DQE). It is a measure of the detector noise and gives an assessment for the detection of single electrons.

For all TEM detectors the ssCCD has the best DQE. Depending on the scintillator, it is in the range of DQE $= 0.6\text{--}0.9$, comparable to IPs which show a DQE in the order of 0.5–0.8. The DQE of photographic emulsion is strongly dependent on electron dose and does not exceed DQE $= 0.2$. For a complete and up-to-date discussion on TEM electron detectors see the special issue of *Microscopy Research and Technique* (vol 49, 2000).

In SEM and STEM, all detectors record the electron current signal of the selected interacting electrons (elastic scattering, secondary electrons) in real time. Such detectors can be designed as simple metal-plate detectors, such as the elastic dark-field detector in STEM, or as electron-sensitive PMT. For a rigorous discussion of SEM detectors see [3].

Except for the phase-contrast detector in STEM [9], STEM and SEM detectors do not track the position of the recorded electron. The spatial information of an image is formed instead by assigning the measured electron current to the known position of the scanned incident electron beam. This information is then mapped into a 2D pixel array, which is depicted either on a TV screen or digitalized in a computer.

For the parallel recording of EEL spectra in STEM, linear arrays of semiconductor detectors are used. Such detectors convert the incident electrons into photons, using additional fluorescent coatings or scintillators in the very same way as the TEM detectors described above.

## B1.17.4  Specimen preparation

The necessity to have high vacuum in an electron beam instrument implies certain constraints on the specimen. In addition, the beam damage resulting from the interaction of electrons with the specimen (radiation damage) requires specific procedures to transfer the specimen into a state in which it can be analysed. During these procedures, which are more elaborate for organic than for inorganic solid-state materials, structural and compositional aspects of the specimen may be altered and consequently the corresponding information may be misleading or completely lost. It must be mentioned here that the EM is only the tool to extract information from the specimen. As well as having its own physical problems (CTF, beam damage, etc) it is—like any other microscopy—clearly not capable of restoring information that has been lost or altered during specimen preparation.

Specimens for (S)TEM have to be transparent to the electron beam. In order to get good contrast and resolution, they have to be thin enough to minimize inelastic scattering. The required thin sections of organic materials can be obtained by ultramicrotomy either after embedding into suitable resins (mostly epoxy- or methacrylate resins [11]) or directly at low temperatures by cryo-ultramicrotomy [12].

Ultramicrotomy is sometimes also used to produce thin samples of solid materials, such as metals [13] which are, however, preferentially prepared by chemical- or ion-etching (see [1]) and focused ion beam (FIB) techniques [14].

Bulk specimens for SEM also have to resist the impact of the electron beam instrument. While this is generally a minor problem for materials science specimens, organic and aqueous biological samples must be observed either completely dry or at low enough temperatures for the evaporation/sublimation of solvents and water to be negligible. Internal structures of aqueous biological samples can be visualized by cryosectioning or cryofracturing procedures [15, 16]. Similar procedures are used in the preparation of polymers and composites [17]. Fracturing and field-ion beam procedures are used to expose internal structures of semiconductors, ceramics and similar materials.

The preparation of biological specimens is particularly complex. The ultrastructure of living samples is related to numerous dynamic cellular events that occur in the range of microseconds to milliseconds [18]. Interpretable high-resolution structural information (e.g. preservation of dimensions, or correlation of the structural detail with a physiologically or biochemically controlled state) is therefore obtained exclusively from samples in which life has been stopped very quickly and with a sufficiently high time resolution for the cellular dynamics [19]. Modern concepts for specimen preparation therefore try to avoid traditional, chemical fixation as the life-stopping step because it is comparatively slow (diffusion limited) and cannot preserve all cellular components. Cryotechniques, often in combination with microscopy at low temperatures, are used instead. Very high cooling rates ($>10\,000$ K s$^{-1}$) are required to prevent the formation and growth of ice crystals, which would affect the structural integrity. Such high cooling rates, at the same time, result in a rapid arrest of the physiological events, i.e. produce a very high temporal resolution (microseconds to milliseconds) [20], in capturing dynamic processes in the cell [21, 22].

Despite the drawbacks of chemical-fixation based procedures [23, 24], most of our current knowledge on biological ultrastructure relies on this approach. In contrast to cryopreparative procedures, chemical fixation does not require special skills and instrumentation.

Cryoimmobilization procedures that lead to vitrification (immobilization of the specimen water in the amorphous state) are the sole methods of preserving the interactions of the cell constituents, because the liquid character of the specimen water is retained (reviewed in [25]).

Vitrification at ambient pressure requires very high cooling rates. It can be accomplished by the 'bare grid' approach for freezing thin ($<100$ nm) aqueous layers of suspensions containing isolated macromolecules, liposomes, viruses, etc. This technique was used to produce figure B1.17.6. It has developed into a powerful tool for structural biology, now providing subnanometre resolution of non-crystalline objects [26, 27]. The bare-grid technique permits imaging of macromolecules in functional states with sufficient resolution to allow the correlation with atomic data from x-ray diffraction of crystals [28, 29] (see also *Journal of Structural Biology*, vol 125, 1999).

High-pressure freezing is at present the only practicable way of cryoimmobilizing larger non-pretreated samples [30, 32]. At a pressure of 2100 bars, about the 10-fold greater thickness can be vitrified, as compared to vitrification at ambient pressure [33] and a very high yield of adequately frozen specimens (i.e. no detectable effects of ice crystal damage visible after freeze substitution) has been demonstrated by TEM of suspensions of micro-organisms, as well as for plant and animal tissue, provided that the thickness of the aqueous layer did not exceed 200 $\mu$m.

Biological material, immobilized chemically or by rapid freezing, must be transferred into an organic solvent that is compatible with the most frequently used hydrophobic resins. Chemically fixed materials are dehydrated in graded series of alcohol or acetone at room temperature. The ice of the frozen sample is dissolved at low temperature by a freeze-substitution process [34]. For TEM, the samples are embedded in resin [11], for SEM they are dried, most frequently by the critical-point drying technique, which avoids deleterious effects of the surface tension of the solvents. Dehydration and complete drying results in non-isotropic shrinkage of biological materials.

The information that can be extracted from inorganic samples depends mainly on the electron beam/specimen interaction and instrumental parameters [1], in contrast to organic and biological materials, where it depends strongly on specimen preparation.

For analytical SEM and non-destructive imaging (e.g. semiconductor, critical-dimension measurements (CD) and other quality control) adequate electron energies have to be selected in order to minimize charging up of the specimen. For high-resolution imaging of surfaces using the SE signal, the signal source often must be localized at the specimen surface by a thin metal coating layer [35].

## B1.17.5   Image formation and image contrast

Whereas electron optics, sample preparation and the interaction of electrons with the sample follow a common set of rules for all different kinds of EM (TEM, STEM, SEM), the image formation and image contrast in EM images is very technique-specific. In general, analytical imaging is distinguished from structural imaging, the latter being further classified either into the imaging of the projected 2D scattering potential for TEM and STEM or into the topographical imaging of a surface in SEM. The complete understanding of the electron–sample interaction and the increasingly better understanding of the sample preparation and reconstruction of the object from image contrast allows a quantitative interpretation of EM data. The electron microscope has evolved, over a long period, from a simple imaging microscope to a quantitative data collection device.

### B1.17.5.1   *Imaging of projected structure—the contrast transfer function (CTF) of TEM*

The discussion of the electron–specimen interaction has already provided the necessary physical principles leading to amplitude and phase changes of the scattered electron wave. Consider again the elastic scattering as described by equation (B1.17.1). In STEM the elastic scattering is measured by an angular detector, integrating over all electrons scattered to high angles (see figure B1.17.3(a)). For thin samples, the measured image contrast can be directly interpreted as the spatial distribution of different atomic composition. It corresponds to the pointwise measurement of the sample's scattering factor amplitude (see equation (B1.17.9))

$$f_{\text{int}} = \int_{\theta_{\text{det, min}}}^{\theta_{\text{det, max}}} \int_0^{2\pi} f(\theta) \, \mathrm{d}\varphi \, \mathrm{d}\theta \qquad (B1.17.13)$$

where $\theta$ denotes the scattering angle, constrained by the geometry of the angular detector; $\varphi$ denotes the azimuthal angle. According to the properties of the elastic scattering distribution (equation (B1.17.1)), the detected signal for a given detection angle interval depends strongly on the atomic number $Z$ of the scattering atom. This results in different contrast for different atomic composition ($Z$-contrast). With a different, position selective, detector it is also possible to measure the phase part of the scattering factor. The geometry of these detectors is illustrated in figure B1.17.3(b) [9].

It should be noted again that STEM provides lateral spatial discrimination of the 2D projected sample by the scanning of a point-like electron beam. Spatial resolution is thereby given by the focus of the incident beam, which is at present limited to a typical diameter of a few Ångstroms. Modern TEM, using the very coherent electron wave of a FEG and higher electron energies (see B1.17.3.1), delivers higher resolution, i.e. its information limit can be improved into the sub-Ångström regime. However, the correspondence of image contrast and scattering factor $f(\theta)$ (equation (B1.17.9)) is more complicated than in STEM. In TEM, image contrast can be understood either by interference effects between scattered and unscattered parts of the electron wave, or by simple removal of electrons scattered to higher angles (scattering contrast). The latter is important for imaging of strongly scattering objects consisting of heavy metal atoms. Electrons scattered to high angles are easily removed by a circular aperture (see figure B1.17.4(a)). Elastically scattered electrons not removed by such an aperture can form interference patterns with the unscattered part of the incident

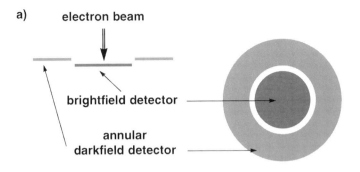

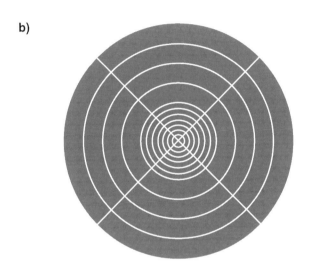

**Figure B1.17.3.** STEM detectors: (a) conventional bright and dark-field detectors, electrons are detected according to their different scattering angles, all other positional information is lost; (b) positional detector as developed by Haider and coworkers (Haider *et al* 1994).

electron wave (see figure B1.17.4(b)). Such interference patterns lead to either diffraction patterns or images of the sample, depending on the imaging conditions of the microscope. Therefore, any wave aberrations of the electron wave, both by the imaged sample (desired signal) and by lens aberrations or defocused imaging, result in a change of image contrast. Mathematically, this behaviour is described by the concept of contrast transfer in spatial frequency space (Fourier space of image), modelled by the CTF.

In the simplest case of bright-field imaging, the CTF can easily be deduced: the elastically scattered electron wave can be described using a generalized phase shift $\Phi_{\text{gen}}$ by

$$\psi_{\text{scattered}}(r) = \psi_0(r) \exp(i\Phi_{\text{gen}}(r)) \tag{B1.17.14}$$

where

$$\exp(i\Phi_{\text{gen}}(r)) = \exp(i\varphi_{\text{el}}(r) + \mu_{\text{el}}(r)) \tag{B1.17.15}$$

and $\varphi_{\text{el}}(r)$ and $\mu_{\text{el}}(r)$ denote the elastic phase and amplitude contrast potential. In the notation of equation (B1.17.15), $\varphi_{\text{el}}(r)$ denotes a positive phase potential whereas $\mu_{\text{el}}(r)$ denotes a negative absorption potential.

**a)    scattering contrast**

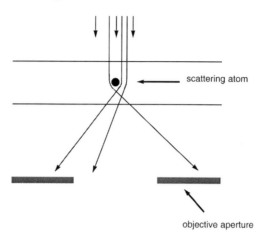

**b)    phase and amplitude contrast**

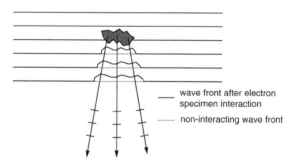

**Figure B1.17.4.** Visualization of image contrast formation methods: (a) scattering contrast and (b) interference contrast (weak phase/weak amplitude contrast).

Assuming a weak phase/weak amplitude object—compared to the unscattered part of the electron wave—the generalized phase shift can be reduced to

$$\exp(i\Phi_{gen}(\boldsymbol{r})) = 1 + i\varphi_{el}(\boldsymbol{r}) + \mu_{el}(\boldsymbol{r}). \tag{B1.17.16}$$

After propagation into the back focal plane of the objective lens, the scattered electron wave can be expressed in terms of the spatial frequency coordinates $\boldsymbol{k}$ as

$$\tilde{\psi}_{scattered}(\boldsymbol{k}) = (\delta(\boldsymbol{k}) + i\tilde{\varphi}_{el}(\mathrm{k}) + \tilde{\mu}_{el}(\boldsymbol{k})) \times \exp(-iW(\boldsymbol{k})). \tag{B1.17.17}$$

Here $W(\boldsymbol{k})$ denotes the wave aberration

$$W(\boldsymbol{k}) = \frac{\pi}{2}(C_s\lambda^3 k^4 - 2\Delta z\lambda k^2) \tag{B1.17.18}$$

with the objective lens spherical aberration $C_s$, the electron wave length $\lambda$ and the defocus $\Delta z$. It should be noted that, in the above formulae, the effect of inelastic scattering is neglected. For a rigorous discussion of image contrast, including inelastic scattering, see [36].

In the usual approximation of the object as a weak phase/weak amplitude object, this scattered wave can be used to calculate the intensity of the image transform as

$$\tilde{I}(k) = \tilde{\psi}_{\text{scattered}}(k) \otimes \tilde{\psi}^*_{\text{scattered}}(k). \tag{B1.17.19}$$

Calculating the convolution using equation (B1.17.17) and regrouping the terms yields the final equation for the image transform:

$$\tilde{i}(k) = \delta(k) - 2 \times (\tilde{\varphi}_{\text{el}}(k)) \sin(W(k)) + \tilde{\mu}_{\text{el}}(k) \cos(W(k)). \tag{B1.17.20}$$

The power spectrum of the image $PS(k)$ is then given by the expectation value $\langle \tilde{I}(k) \times \tilde{I}(k) \rangle$, normally calculated as the squared amplitude of the image transform. More detailed discussions of the above theory are found in [1, 2].

In the conventional theory of elastic image formation, it is now assumed that the elastic atomic amplitude scattering factor is proportional to the elastic atomic phase scattering factor, i.e.

$$\tilde{\mu}_{\text{el}}(k) = -A(k)\tilde{\varphi}(k) \equiv -A\tilde{\varphi}_{\text{el}}(k). \tag{B1.17.21}$$

The factor $A$ has been measured for a variety of samples, indicating that the approximation can be applied up to quasi-atomic resolution. In the case of biological specimens typical values of $A$ are of the order of 5–7%, as determined from images with a resolution of better than 10 Å [37, 38]. For an easy interpretation of image contrast and a retrieval of the object information from the contrast, such a combination of phase and amplitude information is necessary.

Figure B1.17.5 shows typical examples of the CTF for weak phase, weak amplitude or combined samples. The resulting effect on image contrast is illustrated in figure B1.17.6, which shows image averages of a protein complex embedded in a vitrified aqueous layer recorded at different defocus levels. The change in contrast and visible details is clear, but a direct interpretation of the image contrast in terms of object structure is not possible. To reconstruct the imaged complex, it is necessary to combine the information from the different images recorded with different defocus levels. This was first suggested by Schiske [39] and is normally applied to high-resolution images in materials science or biology. Such correction procedures are necessary to rectify the imaging aberrations from imperfect electron-optical systems, which result in a delocalized contrast of object points. In recent years, improvements of the electron-optical lenses have been made, and high-resolution imaging with localized object contrast will be possible in the future (see B1.17.5.3 and figure B1.17.9). It should be noted that any high-resolution interpretation of an EM image strongly depends on the correction of the CTF. In high-resolution TEM in materials science, sophisticated methods for this correction have been developed and are often combined with image simulations, assuming a certain atomic model structure of the sample. Three mainstream developments in this field are: (1) the use of focal series and subsequent image processing [40, 41], (2) electron holography [42–44] and (3) the development of corrected TEMs, which prohibit contrast delocalization (see B1.17.5.3).

### B1.17.5.2 Imaging of surface topology

SEMs are ideally suited to study highly corrugated surfaces, due to the large depth of focus. They are generally operated with lower beam energies (100 eV–30 keV) in order to efficiently control the volume in which the electron beam interacts with the sample to produce various specific signals (see figure B1.17.1) for imaging and compositional analysis. Modern SEM instruments (equipped with a field emission electron source) can scan the sample surface with a beam diameter of 1 nm or smaller, thus providing high-resolution structural information that can complement the information obtained from atomic force microscopy (AFM). In contrast to AFM, which directly provides accurate height information in a limited range, quantitative assessment of the surface topography by SEM is possible by measuring the parallax of stereo pairs [45].

**a) pure phase contrast**

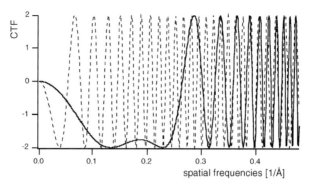

**b) pure amplitude contrast**

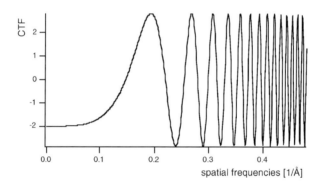

**c) real weak phase, weak amplitude object**

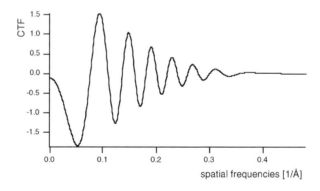

**Figure B1.17.5.** Examples of CTFs for a typical TEM (spherical aberration $C_s = 2.7$ mm, 120 keV electron energy). In (a) and (b) the idealistic case of no signal decreasing envelope functions [77] are shown. (a) Pure phase contrast object, i.e. no amplitude contrast; two different defocus values are shown (Scherzer focus of 120 nm underfocus (solid curve), 500 nm underfocus (dashed curve)); (b) pure amplitude object (Scherzer focus of 120 nm underfocus); (c) realistic case including envelope functions and a mixed weak amplitude/weak phase object (500 nm underfocus).

High-resolution topographic information is obtained by the secondary-electron signal (SE 1, see figure B1.17.7) produced at the point of impact of the primary beam (PE). The SE 1 signal alone is related to the

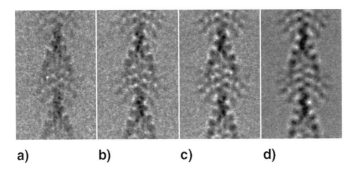

**Figure B1.17.6.** A protein complex (myosin S1 decorated filamentous actin) embedded in a vitrified ice layer. Shown is a defocus series at (a) 580 nm, (b) 1130 nm, (c) 1700 nm and (d) 2600 nm underfocus. The pictures result from averaging about 100 individual images from one electron micrograph; the decorated filament length shown is 76.8 nm.

position of the scanning beam. It depends on the distance the primary beam travels through the SE escape zone, where it releases secondary electrons that can leave the specimen surface, i.e. it depends on the angle of impact of the primary beam (see figure B1.17.7). The high-resolution topographic signal is obscured by other SE signals (SE 2, SE 3, figure B1.17.7) that are created by BSE (electrons of the primary beam, multiply scattered deeper inside the specimen) when they leave the specimen and pass through the SE escape zone (SE 2) and hit the pole pieces of the objective lens and/or the walls of the specimen chamber (SE 3). Additional background signal is produced by the primary beam striking the objective lens aperture.

High-resolution topographic imaging by secondary electrons therefore demands strategies (instrumentation, specimen preparation [35] and imaging conditions) that aim at enhancing the SE 1 signal and suppressing the background noise SE signals (e.g. [46]). Basically, the topographic resolution by SE 1 depends on the smallest spot size available and on the SE escape zone, which can be up to 100 nm for organic materials and down to 1–2 nm for metals.

Non-conductive bulk samples, in particular, are frequently rendered conductive by vacuum coating with metals using sputter or evaporation techniques. The metal coating should be of uniform thickness and significantly thinner than the smallest topographic details of interest. Metal coating provides the highest resolution images of surface details. It may, however, irreversibly destroy the specimen. An example of such a metal-coated sample is shown in figure B1.17.8.

SEM with low acceleration voltage (1–10 kV) (LVSEM) can be applied without metal coating of the sample, e.g. for quality control purposes in semiconductor industries, or to image ceramics, polymers or dry biological samples. The energy of the beam electrons (the acceleration voltage) should be selected so that charge neutrality is approached, i.e. the amount of energy that enters the sample also leaves the sample in the form of SE and BSE. Modern SEM instruments, equipped with FEGs provide an adequate spot size, although the spot size increases with decreasing acceleration voltage. The recent implementation of a cathode lens system [47] with very low aberration coefficients will allow the surfaces of non-metal coated samples at beam energies of only a few electronvolts to be imaged without sacrificing spot size. New contrast mechanisms and new experimental possibilities can be expected.

The fact that electron beam instruments work under high vacuum prohibits the analysis of aqueous systems, such as biological materials or suspensions, or emulsions without specimen preparation as outlined above. These preparation procedures are time consuming and are often not justified in view of the only moderate resolution required to solve a specific practical question (e.g. to analyse the grain size of powders, bacterial colonies on agar plates, to study the solidification of concrete, etc). Environmental SEM (ESEM) and 'high-pressure SEM' instruments are equipped with differentially pumped vacuum systems and

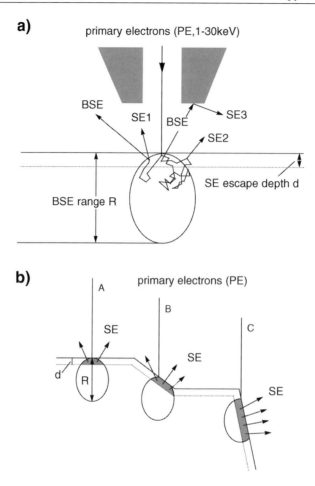

**Figure B1.17.7.** (a) Classification of the secondary electron signals. High-resolution topographic information is obtained by the SE 1 signal. It might be obscured by SE 2 and SE 3 signals that are created by the conversion of BSE. $d$: SE escape depth, $R$: range out of which BSE may leave the specimen. (b) SE signal intensity, $R > d$: the SE 1 signal depends on the angle of impact of the electron beam (PE). SE 1 can escape from a larger volume at tilted (B) surfaces and edges (C) than at orthogonal surfaces (A).

Peltier-cooled specimen stages, which allow wet samples to be observed at pressures up to 5000 Pa [48]. Evaporation of water from the specimen or condensation of water onto the specimen can thus be efficiently controlled. No metal coating or other preparative steps are needed to control charging of the specimen since the interaction of the electron beam with the gas molecules in the specimen chamber produces positive ions that can compensate surface charges. 'High-pressure SEM', therefore, can study insulators without applying a conductive coating. The high gas pressure in the vicinity of the specimen leads to a squirting of the electron beam. Thus the resolution-limiting spot size achievable on the specimen surface depends on the acceleration voltage, the gas pressure, the scattering cross section of the gas and the distance the electrons have to travel through the high gas pressure zone [49]. High-pressure SEM and ESEM is still under development and the scope of applications is expanding. Results to date consist mainly of analytical and low-resolution images (e.g. [50]).

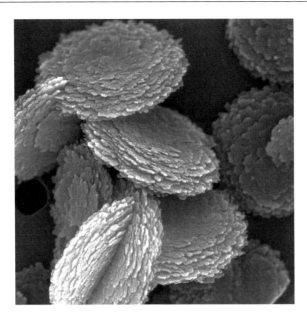

**Figure B1.17.8.** Iron oxide particles coated with 4 nm of Pt in an m-planar magnetron sputter coater (Hermann and Müller 1991). Micrographs were taken in a Hitachi S-900 'in-lens' field emission SEM at 30 000 primary magnification and an acceleration voltage of 30 kV. Image width is 2163 nm.

### B1.17.5.3   Modern developments of instruments affecting image contrast and resolution

As was discussed above, the image contrast is significantly affected by the aberrations of the electron-optical lenses. The discussion on the CTF showed that the broadening PSF of the TEM delocalizes information in an TEM image. This necessitates additional techniques to correct for the CTF, in order to obtain interpretable image information. Furthermore, it was discussed that the resolution of STEM and SEM depends on the size of the focused beam, which is also strongly dependent on lens aberrations. The leading aberrations in state-of-the-art microscopes are the spherical and chromatic aberrations in the objective lens. Correction of such aberrations was discussed as early as 1947, when Scherzer suggested the correction of electron optical lenses [51], but it was not until 1990 that a complete concept for a corrected TEM was proposed by Rose [52]. It was in the last decade that prototypes of such corrected microscopes were presented.

The first corrected electron-optical SEM was developed by Zach [10]. For low-voltage SEM (LVSEM, down to 500 eV electron energy instead of the conventional energies of up to 30 keV) the spot size is extremely large without aberration correction. Combining $C_s$ and $C_c$ correction and a electrostatic objective lens, Zach showed that a substantial improvement in spot size and resolution is possible. The achievable resolution in a LVSEM is now of the order of 1–2 nm. More recently, Krivanek and colleagues succeeded in building a $C_s$ corrected STEM [53, 54].

The construction of an aberration-corrected TEM proved to be technically more demanding: the point resolution of a conventional TEM today is of the order of 1–2 Å. Therefore, the aim of a corrected TEM must be to increase the resolution beyond the 1 Å barrier. This implies a great number of additional stability problems, which can only be solved by the most modern technologies. The first $C_s$ corrected TEM prototype was presented by Haider and coworkers [55]. Figure B1.17.9 shows the improvement in image quality and interpretability gained from the correction of the spherical aberration in the case of a materials science sample.

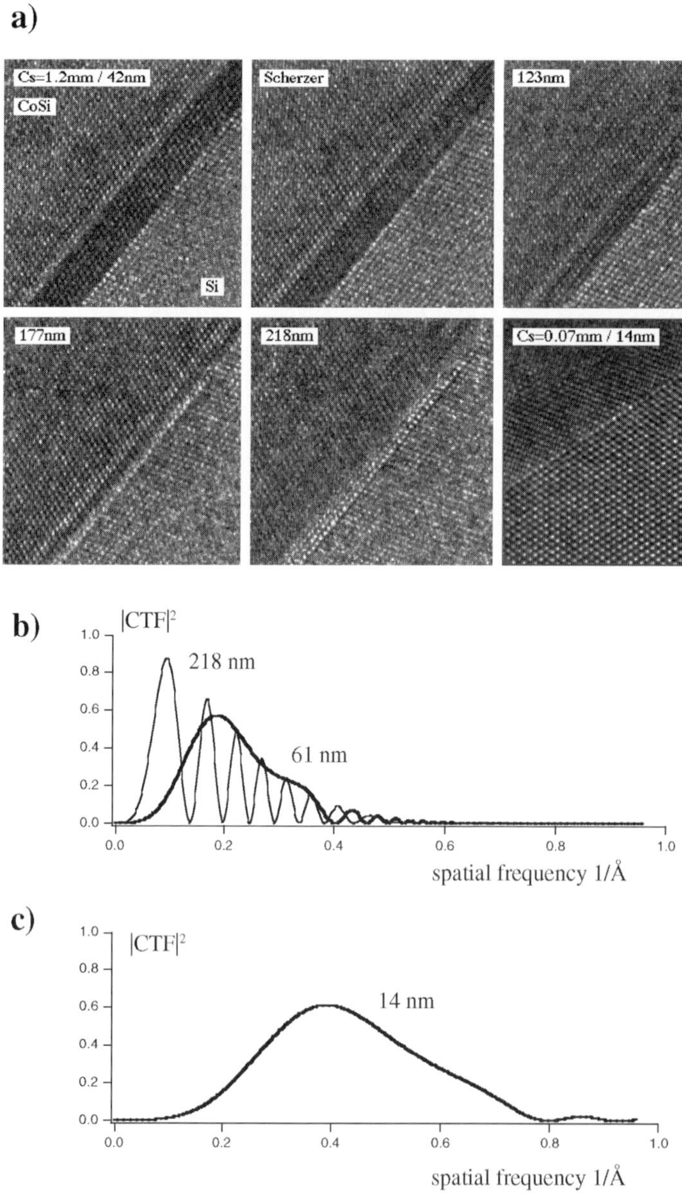

**a)**

**b)**

**c)**

**Figure B1.17.9.** A CoSi grain boundary as visualized in a spherical-aberration-corrected TEM (Haider *et al* 1998). (a) Individual images recorded at different defocus with and without correction of $C_s$; (b) CTFs in the case of the uncorrected TEM at higher defocus; (c) CTF for the corrected TEM at only 14 nm underfocus. Pictures by courtesy of M Haider and Elsevier.

The field of corrected microscopes has just begun with the instruments discussed above. The progress in this field is very rapid and proposals for a sub-Ångstrøm TEM (SATEM) or even the combination of this instrument with a corrected energy filter to form a sub-electronvolt/sub-Ångstrøm TEM (SESAM) are underway.

### B1.17.6 Analytical imaging, spectroscopy, and mass measurements

For a detailed discussion on the analytical techniques exploiting the amplitude contrast of inelastic images in ESI and image-EELS, see chapter B1.6 of this encyclopedia. One more recent but also very important aspect is the quantitative measurement of atomic concentrations in the sample. The work of Somlyo and colleagues [56], Leapman and coworkers [57, 58], and Door and Gängler [59] introduce techniques to convert measured intensities of inelastically scattered electrons directly into atomic concentrations.

For bio-medical or cell-biological samples, in particular, this provides a direct measurement of physiological ion concentrations. The main disadvantage of such methods is the almost unavoidable delocalization of free ions and the resulting change in concentration during sample preparation steps. The discussed rapid freezing and the direct observation of samples without any chemical treatment provides a very good compromise for organic samples. In materials science, delocalization does not seem to pose a major problem.

Another specialized application of EM image contrast is mass measurement. Using the elastic dark-field image in the STEM or the inelastic image in the EFTEM, a direct measurement of the scattering mass can be performed. For reviews on this technique see [60, 61].

### B1.17.7 3D object information

EM images are always either 2D projections of an interaction potential (see equations (B1.17.11) and (B1.17.12)) or a surface topology encoded in grey levels of individual image points (image contrast). The aim of EM image processing is to reconstruct the 3D object information from a limited number of such projections. This problem does not arise for all applications of EM. Very often in materials science the sample is prepared in such a way that one single projection image contains all the information necessary to answer a specified question. As an example, consider figure B1.17.9, which shows a Co–Si interface. The orientation of the sample is chosen to give a perfect alignment of atoms in the direction of the grain boundary. Imaging at atomic resolution allows the direct interpretation of the contrast as images of different atoms. One single exposure is, in principle, sufficient to collect all the information needed. It should be noted here again that this is only true for the spherical-aberration, $C_s$ corrected EM with its non-oscillating CTF (bottom right panel in figure B1.17.9(a)). As is obvious from the Co–Si interface (figure B1.17.9(a), finite $C_s$ imaging) and the defocus series of the biological sample (figure B1.17.6) more than one image has to be combined for conventional EMs. However, such a direct interpretation of one projected image to obtain the 3D structure information works only for samples that are ordered crystallographically at the atomic level. For other samples it is necessary to combine projections from different angles. Such samples are unique, non-crystallographic structures, e.g. structural defects in materials science, cellular compartments in cell biology, or macromolecules (protein complexes or single biological molecules) in high-resolution molecular imaging.

The generalized problem has been solved by tomography. In EM it is possible to tilt the sample along an axis which is perpendicular to the electron beam (single-axis tomography, see figure B1.17.10). If the sample can withstand a high cumulative electron dose, then an unlimited number of exposures at different defocus values and tilting angles can be recorded. If kinematical scattering theory can be applied (single elastic scattering, compare equation (B1.17.8)), then it is possible to correct all the effects of the CTF and each corrected projection image at one particular tilting angle corresponding to a section of the Fourier-transformed scattering potential (equations (B1.17.11) and (B1.17.12)). The combination of information from different tilting angles provides the determination of the structure factors in the complete 3D Fourier space. Finally, a simple mathematical inverse Fourier-transform produces a complete 3D reconstructed object. A geometric equivalent of the projection and reconstruction process is found in the sectioning of the Ewald-sphere with the 3D Fourier-transform of the scattering potential. For an introduction to the concepts of the Ewald-sphere and Fourier techniques in structure determination see [62].

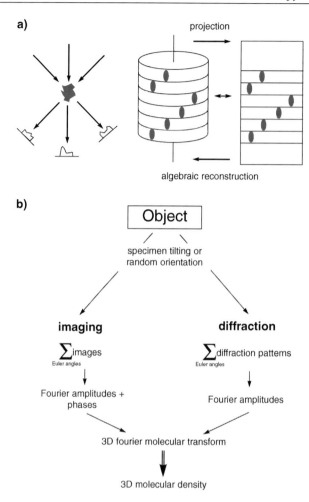

**Figure B1.17.10.** Principles of 3D reconstruction methods. (a) Principle of single axis tomography: a particle is projected from different angles to record corresponding images (left panel); this is most easily realized in the case of a helical complex (right panel). (b) Principle of data processing and data merging to obtain a complete 3D structure from a set of projections.

The basic reconstruction algorithms involved are mathematically well known and are well established [63]. A variety of concepts has evolved for single- and multi-axis tomography, combining projection information, calculating 3D object densities either with Fourier- or real-space algorithms (see figure B1.17.10 which shows some examples of the geometry used for the single- and multi-axis tomography). For a complete reference to the methods used see Frank [64] and the special issue of *Journal of Structural Biology* (vol 120, 1997).

The main disadvantage of the tomographic approach is the beam-induced destruction of the sample. In practice, one can record only a limited number of images. Therefore, it is not possible to correct the CTF completely or to obtain an infinite sampling of the projection angles from only one specimen. Two major approaches are used today: the single- and double-axis tomography of one individual object (e.g. cell organelles, see Baumeister and coworkers [65]) and, second, the imaging of many identical objects under different projection angles (see [64]; random conical tilt, angular reconstitution).

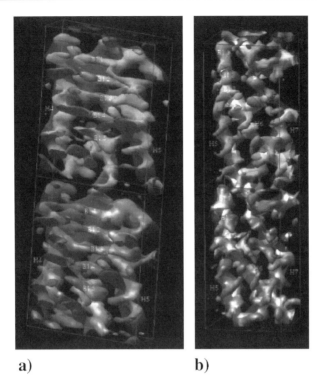

**a)**

**b)**

**Figure B1.17.11.** Reconstructed density of an $\alpha$, $\beta$-tubulin protein dimer as obtained from electron crystallography (Nogales *et al* 1997). Note the appearance of the $\beta$-sheets ((a), marked B) and the $\alpha$-helices ((b), marked H) in the density. In particular the right-handed $\alpha$-helix H6 is very clear. Pictures by courtesy of E Nogales and Academic Press.

For single-axis tomography, with its limited number of images and the subsequent coarse sampling in reciprocal Fourier space, only a moderate resolution can be expected. For chemically fixed samples with high image contrast from heavy atom staining it is possible to obtain a resolution of about 4 nm ([66], reconstruction of the centrosome). For native samples, true single-axis tomography without averaging over different samples results in even lower resolution. Today, sophisticated EM control software allows a fully automatic collection of tilted images [67], making single-axis tomography a perfect reconstruction tool for unique objects.

If many identical copies of the object under study are available, other procedures are superior. They rely on the fact that the individual molecules are oriented with respect to the incident electron beam. Such a situation is found mainly for native ice-embedded samples (compare the paragraph about preparation). In ice layers of sufficient thickness, no special orientation of the molecule is preferred. The obtained projection images from one, untilted image can then be classified and aligned in an angular reconstitution reconstruction process. By averaging large numbers of projection images, it is possible to correct for CTF effects [68] and to obtain an almost complete coverage of reciprocal Fourier space. If—for some reason—the object still shows a limited number of preferred orientations, an additional tilting of the sample again gives complete coverage of possible projection angles (random conical tilt method). Both methods have been successfully applied to many different biological samples (for an overview, see [64]).

An important point for all these studies is the possible variability of the single molecule or single particle studies. It is not possible, *a priori*, to exclude 'bad' particles from the averaging procedure. It is clear, however,

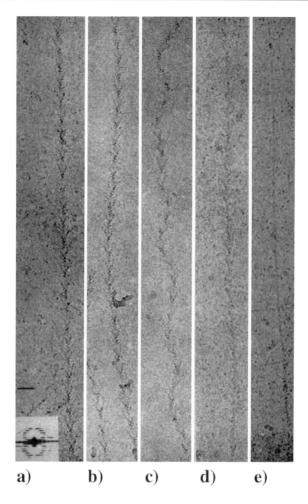

a)          b)          c)          d)          e)

**Figure B1.17.12.** Time-resolved visualization of the dissociation of myosin S1 from filamentous actin (see also figure B1.17.6). Shown are selected filament images before and after the release of a nucleotide analogue (AMPPNP) by photolysis: (a) before flashing, (b) 20 ms, (c) 30 ms, (d) 80 ms and (e) 2 s after flashing. Note the change in obvious order (as shown by the diffraction insert in (a)) and the total dissociation of the complex in (e). The scale bar represents 35.4 nm. Picture with the courtesy of Academic Press.

that high structural resolution can only be obtained from a very homogeneous ensemble. Various classification and analysis schemes are used to extract such homogeneous data, even from sets of mixed states [69]. In general, a typical resolution of the order of 1–3 nm is obtained today.

The highest resolutions of biological samples have been possible for crystalline samples (electron crystallography). 2D crystals of membrane proteins and one cytoskeletal protein complex have been solved at the 3–4 Å level combining imaging and electron diffraction, as pioneered by Henderson and coworkers [70–72], also see figure B1.17.11. Icosahedral virus particles are reconstructed from images to 8–9 Å resolution [26, 27], allowing the identification of alpha helices. Compared to single particles, these samples give much higher resolution, in part because much higher numbers of particles are averaged, but it is also possible that a crystallization process selects for the uniformity of the crystallizing object and leads to very homogeneous ensembles.

For electron crystallography, the methods to obtain the structure factors are comparable to those of conventional x-ray crystallography, except that direct imaging of the sample is possible. This means that both electron diffraction and imaging can be used, i.e. structure amplitudes are collected by diffraction, structure factor phases by imaging. For a general overview in the structure determination by electron crystallography, see [73]. The 3D structure of the sample is obtained by merging diffraction and imaging data of tilt series from different crystals. It is, therefore, a form of tomography adapted for a diffracting object.

Even though it is easy to get the phase information from imaging, in general, imaging at the desired high resolution (for structure determination work of the order of 3–4 Å) is very demanding. Specialized instrumentation (300 kV, FEG, liquid He sample temperature) have to be used to avoid multiple scattering, to allow better imaging (less imaging aberrations, less specimen charging which would affect the electron beam) and to reduce the effects of beam damage.

## B1.17.8 Time-resolved and *in situ* EM studies: visualization of dynamical events

As a result of the physical conditions in electron microscopes such as the high vacuum, the high energy load on the sample by inelastic scattering, or the artificial preparation of the sample by sectioning or thinning, it has become customary to think about samples as static objects, precluding the observation of their native structural changes during a reaction. In some studies, however, the dynamics of reactions have been studied for biological systems as well as in materials science. *In situ* microscopy was widely used in materials science in the 1960s and 1970s, when, for example, metal foils were studied, heated up in the EM, and reactions followed in a kind of time-lapse microscopy [74]. In recent years, similar experiments have been performed on semiconductors and ceramics, and a general new interest in *in situ* microscopy has developed.

Time-resolved EM in biological systems is a comparatively new and limited field. Simple time-lapse fixation of different samples of a reacting biological tissue has long been used, but the direct, temporal monitoring of a reaction was developed only with the invention of cryo-fixation techniques. Today, time-lapse cryo-fixation studies can be used in the case of systems with slow kinetics, i.e. reaction times of the order of minutes or slower. Here, samples of a reacting system are simply taken in certain time intervals and frozen immediately. For the study of very fast reactions, two approaches have been developed that couple the initiation of the reaction and the fixation of the system on a millisecond time scale. The reaction itself can be started either by a rapid mixing procedure [75] or by the release of a masked reaction partner photolysing caged compounds (see figure B1.17.12) [76]. For a review of time-resolved methods used in biological EM, see [19].

## References

[1] Williams D B and Carter C B 1996 *Transmission Electron Microscopy, A Textbook for Material Science* (New York: Plenum)
[2] Reimer L 1993 *Transmission Electron Microscopy* (Berlin: Springer)
[3] Reimer L 1998 *Scanning Electron Microscopy* (Berlin: Springer)
[4] Reimer L 1995 *Energy-Filtering Transmission Electron Microscopy* (Berlin: Springer)
[5] Hawkes P W and Kasper E 1989 *Principles of Electron Optics* vol 1 (London: Academic)
Hawkes P W and Kasper E 1989 *Principles of Electron Optics* vol 2 (London: Academic)
Hawkes P W and Kasper E 1994 *Principles of Electron Optics* vol 3 (London: Academic)
[6] Lenz F 1954 Zur Streuung mittelschneller Elektronen in kleinste Winkel *Z. Naturf.* a **9** 185–204
[7] Egerton R F 1976 Measurement of inelastic/elastic scattering ratio for fast electrons and its use in the study of radiation damage *Phys. Status Solidi* a **37** 663–8
[8] Cantor C R and Schimmel P R 1980 *Biophysical Chemistry, Part II: Techniques for the Study of Biological Structure and Function* (San Francisco: Freeman)
[9] Haider M, Epstein A, Jarron P and Boulin C 1994 A versatile, software configurable multichannel STEM detector for angle-resolved imaging *Ultramicroscopy* **54** 41–59
[10] Zach J 1989 Design of a high-resolution low-voltage scanning electron microscope *Optik* **83** 30–40
[11] Luft J H 1961 Improvements in epoxy resin embedding methods *J. Biophys. Biochem. Cytol.* **9** 409–14

[12]  Michel M, Gnägi H and Müller M 1992 Diamonds are a cryosectioner's best friend *J. Microsc.* **166** 43–56
[13]  Malis T F and Steele D 1990 Specimen preparation for TEM of materials II *Mat. Res. Soc. Symp. Proc.* **199** 3
[14]  Dravid V P 1998 *Hitachi Instrument News* 34th edn
[15]  Walther P, Hermann R, Wehrli E and Müller M 1995 Double layer coating for high resolution low temperature SEM *J. Microsc.* **179** 229–37
[16]  Walther P and Müller M 1999 Biological structure as revealed by high resolution cryo-SEM of block faces after cryo-sectioning *J. Microsc.* **196** 279–87
[17]  Roulin-Moloney A C 1989 Fractography and failure mechanisms of polymers and composites (London: Elsevier)
[18]  Plattner H 1989 *Electron Microscopy of Subcellular Dynamics* (London: CRC)
[19]  Knoll G 1995 Time resolved analysis of rapid events *Rapid Freezing, Freeze-fracture and Deep Etching* ed N Sievers and D Shotton (New York: Wiley–Lyss) p 105
[20]  Jones G J 1984 On estimating freezing times during tissue rapid freezing *J. Microsc.* **136** 349–60
[21]  van Harreveld A and Crowell J 1964 Electron microscopy after rapid freezing on a metal surface and substitution fixation *Anat. Rec.* **149** 381–6
[22]  Knoll G and Plattner H 1989 Ultrastructural analysis of biological membrane fusion and a tentative correlation with biochemical and biophysical aspects *Electron Microscopy of Subcellular Dynamics* ed H Plattner (London: CRC) pp 95–117
[23]  Hyatt M A 1981 Changes in specimen volume *Fixation for Electron Microscopy* ed M A Hyatt (New York: Academic) pp 299–306
[24]  Coetzee J and van der Merwe F 1984 Extraction of substances during glutaraldehyde fixation of plant cells *J. Microsc.* **135** 147–58
[25]  Dubochet J, Adrian M, Chang J, Homo J-C, Lepault J, McDowall A W and Schultz P 1988 Cryo-electron microscopy of vitrified specimens *Q. Rev. Biophys.* **21** 129–228
[26]  Böttcher B, Wynne S A and Crowther R A 1997 Determination of the fold of the core protein of hepatitis B virus by electron cryomicroscopy *Nature* **386** 88–91
[27]  Conway J F, Cheng N, Zlotnick A, Wingfield P T, Stahl S J and Steven A C 1997 Visualization of a 4-helix bundle in the hepatitis B virus capsid by cryo-electron microscopy *Nature* **386** 91–4
[28]  Rayment I, Holden H M, Whittaker M, Yohn C B, Lorenz M, Holmes K C and Milligan R A 1993 Structure of the actin–myosin complex and its implications for muscle contraction *Science* **261** 58–65
[29]  Schröder R R, Jahn W, Manstein D, Holmes K C and Spudich J A 1993 Three-dimensional atomic model of F-actin decorated with *Dictyostelium* myosin S1 *Nature* **364** 171–4
[30]  Riehle U and Höchli M 1973 The theory and technique of high pressure freezing *Freeze-Etching Technique and Applications* ed E L Benedetti and P Favard (Paris: Société Française de Microscopie Electronique) pp 31–61
[31]  Müller M and Moor H 1984 Cryofixation of thick specimens by high pressure freezing *The Science of Biological Specimen Preparation* ed J-P Revel, T Barnard and G H Haggis (O'Hare, IL: SEM, AMF 60666) pp 131–8
[32]  Moor H 1987 Theory and practice of high pressure freezing *Cryotechniques in Biological Electron Microscopy* ed R A Steinbrecht and K Zierold (Berlin: Springer) pp 175–91
[33]  Sartori N, Richter K and Dobochet J 1993 Vitrification depth can be increased more than 10-fold by high pressure freezing *J. Microsc.* **172** 55–61
[34]  Steinbrecht R A and Müller M 1987 Freeze-substitution and freeze-drying *Cryotechniques in Biological Electron Microscopy* ed R A Steinbrecht and K Zierold (Berlin: Springer) pp 149–72
[35]  Hermann R and Müller M 1991 High resolution biological scanning electron microscopy: a comparative study of low temperature metal coating techniques *J. Electron. Microsc. Tech.* **18** 440–9
[36]  Angert I, Majorovits E and Schröder R R 2000 Zero-loss image formation and modified contrast transfer theory of EFTEM *Ultramicroscopy* **81** 203–22
[37]  Toyoshima C, Yonekura K and Sasabe H 1993 Contrast transfer for frozen-hydrated specimens II. Amplitude contrast at very low frequencies. *Ultramicroscopy* **48** 165–76
[38]  Toyoshima C and Unwin P N T 1988 Contrast transfer for frozen-hydrated specimens: determination from pairs of defocus images *Ultramicroscopy* **25** 279–92
[39]  Schiske P 1968 Zur Frage der Bildrekonstruktion durch Fokusreihen *Proc. 14th Eur. Conf. on Electron Microscopy* p 145–6
[40]  Frank J and Penczek P 1995 On the correction of the contrast transfer function in biological electron microscopy *Optik* **98** 125–9
[41]  Thust A and Rosenfeld R 1998 State of the art of focal-series reconstruction in HRTEM *Electron Microscopy 1998: 14th Int. Conf. on Electron Microscopy (Cancun)* vol 1 (Bristol: Institute of Physics Publishing) pp 119–20
[42]  Tonomura A 1995 Recent developments in electron holography for phase microscopy *J. Electron Microsc.* **44** 425–35
[43]  Lichte H 1998 Gottfried Möllenstedt and his electron biprism: four decades of challenging and exciting electron physics *J. Electron Microsc.* **47** 387–94
[44]  Lehmann M, Lichte H, Geiger D, Lang G and Schweda E 1999 Electron holography at atomic dimensions—present state *Mater. Character.* **42** 249–63
[45]  Boyde A 1970 Practical problems and methods in the three-dimensional analysis of SEM images *Scanning Electron Microsc.* **105** 112
[46]  Peters K-R 1986 Working at higher magnifications in scanning electron microscopy with secondary and backscattered electrons on metal coated biological specimens and imaging macromolecular cell membrane structures *Science of Biological Specimen Preparation 1985* ed M Müller *et al* (O'Hare, IL: SEM, AMF 60666) pp 257–82

[47] Frank L, Müllerova I, Faulian K and Bauer E 1999 The scanning low-energy electron microscope: first attainment of diffraction contrast in the scanning electron microscope *Scanning* **21** 1–13

[48] Danilatos G D 1990 Design and construction of an environmental SEM *Scanning* **12** 23–7

[49] Adamaik B and Mathieu C 2000 The reduction of the beam gas interactions in the variable pressure scanning electron microscope with the use of helium gas *Scanning* **21** 178

[50] Manero J M, Masson D V, Marsal M and Planell J L 1999 Application of the technique of environmental scanning electron microscopy to the paper industry *Scanning* **21** 36–9

[51] Scherzer O 1947 Sphärische und chromatische Korrektur von Elektronen-Linsen *Optik* **2** 114–32

[52] Rose H 1990 Outline of a spherically corrected semiaplanatic medium-voltage transmission electron microscope *Optik* **85** 19–24

[53] Krivanek O L, Dellby N, Spence A J and Brown L M 1998 Spherical aberration correction in dedicated STEM *Electron Microscopy 1998: 14th Int. Conf. on Electron Microscopy (Cancun)* vol 1 (Bristol: Institute of Physics Publishing) pp 55–6

[54] Lupini A R and Krivanek O L 1998 Design of an objective lens for use in Cs-corrected STEM *Electron Microscopy 1998: 14th Int. Conf. on Electron Microscopy (Cancun)* vol 1 (Bristol: Institute of Physics Publishing) pp 59–60

[55] Haider M, Rose H, Uhlemann S, Schwab E, Kabius B and Urban K 1998 A spherical-aberration-corrected 200 keV transmission electron microscope *Ultramicroscopy* **75** 53–60

[56] Somlyo A V, Gonzalez-Serratos Y, Shuman H, McClellan G and Somlyo A P 1981 Calcium release and ionic changes in the sarcoplasmatic reticulum of tetanized muscle: an electron-probe study *J. Cell Biol.* **90** 577–94

[57] Leapman R D and Swyt C R 1988 Separation of overlapping core edges in electron energy loss spectra by multiple-least squares fitting *Ultramicroscopy* **26** 393–404

[58] Leapman R D, Hunt J A, Buchanan R A and Andrews S B 1993 Measurement of low calcium concentrations in cryosectioned cells by parallel-EELS mapping *Ultramicroscopy* **49** 225–34

[59] Door R and Gängler D 1995 Multiple least-squares fitting for quantitative electron energy-loss spectroscopy—an experimental investigation using standard specimens *Ultramicroscopy* **58** 197–210

[60] Engel A and Colliex C 1993 Application of scanning transmission electron microscopy to the study of biological structure *Curr. Opinion Biotechnol.* **4** 403–11

[61] Feja B and Aebi U 1999 Molecular mass determination by STEM and EFTEM: a critical comparison *Micron* **30** 299–307

[62] Holmes K C and Blow D M 1965 The use of x-ray diffraction in the study of protein and nucleic acid structure *Meth. Biochem. Anal.* **13** 113–239

[63] Klug A and Crowther R A 1972 Three-dimensional image reconstruction from the viewpoint of information theory *Nature* **238** 435–40

[64] Frank J 1996 *Three-Dimensional Electron Microscopy of Macromolecular Assemblies* (New York: Academic)

[65] Baumeister W, Grimm R and Walz T 1999 Electron tomography of molecules and cells *Trends Cell Biol.* **9** 81–5

[66] Ruiz T 1998 Conference talk *Gordon Conf. on Three-Dimensional Electron Microscopy*

[67] Koster A J, Grimm R, Typke D, Hegerl R, Stoschek A, Walz J and Baumeister W 1997 Perspectives of molecular and cellular electron tomography *J. Struct. Biol.* **120** 276–308

[68] Zhu J, Penczek P, Schröder R R and Frank J 1997 Three-dimensional reconstruction with contrast transfer function correction from energy-filtered cryoelectron micrographs: procedure and application to the 70S *Escherichia coli* ribosome *J. Struct. Biol.* **118** 197–219

[69] Agrawal R K, Heagle A B, Penczek P, Grassucci R A and Frank J 1999 EF-G-dependent GTP hydrolysis induces translocation accompanied by large conformational changes in the 70S ribosome *Nature Struct. Biol.* **6** 643–7

[70] Henderson R, Baldwin J M, Ceska T, Zemlin F, Beckmann E and Downing K 1990 Model for the structure of bacteriorhodopsin based on high-resolution electron cryo-microscopy *J. Mol. Biol.* **213** 899–929

[71] Kühlbrandt W, Wang D N and Fujiyoshi Y 1994 Atomic model of plant light-harvesting complex by electron crystallography *Nature* **367** 614–21

[72] Nogales E, Wolf S G and Downing K 1998 Structure of $\alpha,\beta$-tubulin dimer by electron crystallography *Nature* **391** 199–202

[73] Walz T and Grigorieff N 1998 Electron crystallography of two-dimensional crystals of membrane proteins *J. Struct. Biol.* **121** 142–61

[74] Butler E P and Hale K F 1981 *Practical Methods in Electron Microscopy* vol 9, ed A Glauert (New York: North-Holland)

[75] Berriman J and Unwin N 1994 Analysis of transient structures by cryo-electron microscopy combined with rapid mixing of spray droplets *Ultramicroscopy* **56** 241–52

[76] Menetret J-F, Hofmann W, Schröder R R, Rapp G and Goody R S 1991 Time-resolved cryo-electron microscopic study of the dissociation of actomyosin induced by photolysis of photolabile nucleotides *J. Mol. Biol.* **219** 139–43

[77] Frank J 1973 The envelope of electron microscopic transfer functions for partially coherent illumination *Optik* **38** 519–27

[78] Nogales E and Downing K C 1997 Visualizing the secondary structure of tubulin: three-dimensional map at 4Å *J. Struct. Biol.* **118** 119–27

[79] Knoll M and Ruska E 1932 Beitrag zur geometrischen Elektronenoptik I *Ann. Phys.* **12** 607–40

[80] Crewe A V, Eggenberger D N, Wall J and Welter L M 1968 Electron gun using a field emission source *Rev. Sci. Instrum.* **39** 576–86

[81] Crewe A V, Wall J and Welter L M 1968 A high resolution scanning transmission electron microscope *J. Appl. Phys.* **39** 5861–8

# B1.18
# Microscopy: light

*H Kiess*

## B1.18.1   Introduction

Light microscopy is of great importance for basic research, analysis in materials science and for the practical control of fabrication steps. When used conventionally it serves to reveal structures of objects which are otherwise invisible to the eye or magnifying glass, such as micrometre-sized structures of microelectronic devices on silicon wafers. The lateral resolution of the technique is determined by the wavelength of the light and the objective of the microscope. However, the quality of the microscopic image is not solely determined by resolution; noise and lack of contrast may also prevent images of high quality being obtained and the theoretical resolution being reached even if the optical components are ideal. The working range of the light microscope in comparison to other microscopic techniques is depicted schematically in table B1.18.1. Clearly, the light microscope has an operating range from about half a micrometer up to millimetres, although recent developments in improving resolution allow the lower limit to be pushed below half a micrometer.

Microscopes are also used as analytical tools for strain analysis in materials science, determination of refractive indices and for monitoring biological processes *in vivo* on a microscopic scale etc. In this case resolution is not necessarily the only important issue; rather it is the sensitivity allowing the physical quantity under investigation to be accurately determined.

**Table B1.18.1.** Overview of working ranges of various microscopic techniques (in $\mu$m).

| | |
|---|---|
| Light microscope | $0.5 \Leftrightarrow 1000$ |
| Scanning electron microscope | $0.05 \Leftrightarrow 1000$ |
| Transmission electron microscope | $0.001 \Leftrightarrow 10$ |
| Scanning probe microscope | $0.0001 \Leftrightarrow 100$ |

Light microscopy allows, in comparison to other microscopic methods, quick, contact-free and non-destructive access to the structures of materials, their surfaces and to dimensions and details of objects in the lateral size range down to about 0.2 $\mu$m. A variety of microscopes with different imaging and illumination systems has been constructed and is commercially available in order to satisfy special requirements. These include stereo, darkfield, polarization, phase contrast and fluorescence microscopes.

The more recent scanning light microscopes are operated in the conventional and/or in the confocal mode using transmitted, reflected light or fluorescence from the object. Operation in the confocal mode allows samples to be optically sectioned and 3D images of objects to be produced—an important aspect for imaging thick biological samples. The breakthrough for confocal microscopes was intimately connected with the advent of computers and data processing. The conventional microscope is then replaced by a microscopic *system* comprising the microscope, the scanning, illumination and light detection systems, the data processor and computer.

This overview will first deal with the optical aspects of conventional microscopes and the various means to improve contrast. Confocal microscopy, which in the last decade has become an important tool, especially for biology, is discussed in the final section.

## B1.18.2   Magnification, resolution and depth of focus

### B1.18.2.1   Magnification

Microscopes are imaging systems and, hence, the image quality is determined by lens errors, by structures in the image plane (e.g., picture elements of CCD cameras) and by diffraction. In addition, the visibility of objects with low contrast suffers from various noise sources such as noise in the illuminating system (shot noise), scattered light and by non-uniformities in the recording media. Interest often focuses on the achievable resolution, and discussions on limits to microscopy are then restricted to those imposed by diffraction (the so-called Abbe limit), assuming implicitly that lenses are free of errors and that the visual system or the image sensors are ideal. However, even under these conditions the Abbe limit of the resolution may not be reached if the contrast is insufficient and noise is high.

Before discussing the limits imposed by diffraction and the influence of contrast and noise on resolution, it is important to recall the basic principle of the light microscope: The objective lens provides a magnified real image of the object in the focal plane of the eyepiece. This image is then focused by the eyepiece onto the retina of the eye and is seen by the observer as a virtual image at about 25 cm distance, the normal distance for distinct vision (figure B1.18.1). The object is illuminated by the light of a lamp, either from below through the stage of the object holder if the object is transparent, or from the top if the object is non-transparent and reflecting. Organic objects containing fluorescent molecules are often investigated with an illuminating light beam that causes the sample to fluoresce. The exciting light is 'invisible' and the object is imaged and characterized by the emitted light.

The magnification $M_{\mathrm{microscope}}$ achievable by a microscope is the ratio of the scales of the virtual image and of the object. It can be easily seen from figure B1.18.1 that this ratio is given by

$$M_{\mathrm{microscope}} = M_{\mathrm{obj}} M_{\mathrm{eyepiece}}.$$

$M_{\mathrm{obj}}$ is the scale of magnification by the objective under the geometric conditions given by the microscope and $M_{\mathrm{eyepiece}} = \ell / f_{\mathrm{eyepiece}}$ is the magnification by the eyepiece, with the focal length $f_{\mathrm{eyepiece}}$ and $\ell = 25$ cm, the normal distance for distinct vision. The objectives are marked with the scale of magnification (e.g. 40:1) by the manufacturer and similarly, the eyepiece by its magnification under the given conditions (e.g. $5\times$). Multiplication of both numbers gives the magnification of the microscope. For practical reasons the magnification of the objective is not so high as to resolve all the details in the real image with the naked eye. The magnification is rather chosen to be about 500 $A_{\mathrm{obj}}$ to 1000 $A_{\mathrm{obj}}$, where $A_{\mathrm{obj}}$ is the numerical aperture of the objective (see the next section) The eyepiece is then necessary to magnify the real image so that it can conveniently be inspected.

### B1.18.2.2   Lateral resolution: diffraction limit

The performance of a microscope is determined by its objective. It is obvious that details of the object that are not contained in the real image (figure B1.18.1) cannot be made visible by the eyepiece or lens systems, whatever quality or magnification they may have. The performance is defined here as the size of the smallest lateral structures of the object that can be resolved and reproduced in the image. To fully assess resolution *and* image fidelity, the modulation transfer function of the imaging system has to be known (or for scanning

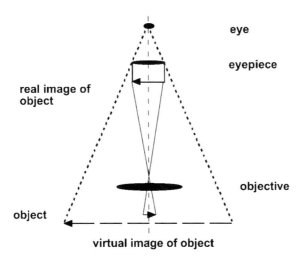

**Figure B1.18.1.** Light rays and imaging lenses in the microscope. The illumination system is not included. The real image is seen by the eye as a virtual image at 25 cm distance, the normal distance for distinct vision.

microscopes more conveniently the point spread function). The resolution is then given by the highest lateral frequency of an object which can just be transmitted by the optical system. Alternatively, one may consider the separation of two structure elements in the plane of the object, which are just discernible in the image [1–3]. Since an exact correlation exists between the pattern generated by the object in the exit pupil of the objective and the image, the limit on the resolution can be estimated simply. If the diffracted beams of zeroth and ±first order are collected by the lens, then an image of low fidelity of the structure, with the zeroth order only a grey area, is obtained. Hence, the limit to resolution is given whenever the zeroth- and first-order beams are collected (figure B1.18.2). If the diffracted light enters a medium of refractive index $n$, the minimal discernible separation $a_{min}$ of two structure elements is given by $n \sin \alpha = \lambda/a_{min}$. The expression $n \sin \alpha$ can be called the numerical aperture of the diffracted beam of first order which, for microscopes, is identical to the numerical aperture $A_{obj}$ of the objective lens. The numerical aperture of a lens is the product of the refractive index of the medium in front of the objective and of the sine of half of the angle whose vertex is located on the optical axis and being the starting point of a light cone of angle $\alpha$ which is just collected by the lens.

The smallest resolvable structure is thus $a_{min} = \lambda/A_{obj}$. If, in addition, the aperture of the illumination system is taken into account, one finds:

$$a_{min} = \lambda/(A_{obj} + A_{ill}).$$

The highest resolution is obtained if $A_{ill} = A_{obj}$. In this case $a_{min}$ equals about half the wavelength used for the illumination divided by the numerical aperture. Using blue or ultraviolet light for illumination, $a_{min}$ can reach values of 0.2 to 0.15 $\mu$m with a numerical aperture of the microscope of about 0.9.

The diffraction limit for resolution does not imply that objects of dimensions smaller than $a_{min}$ are not detectable. Single light-emitting molecules or scattering centres of atomic dimensions can be observed even though their size is below the resolution. For their detectability it is required that the separation of the centres is greater than the resolution and that the emitted signal is sufficiently high to become detectable by a light-sensitive device. Microscopy, for which these assumptions are fulfilled, has sometimes been called ultramicroscopy.

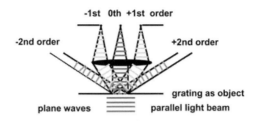

**Figure B1.18.2.** Diffraction figure of a grating: If only the zeroth-order beam were collected by the lens, only a bright area would be visible without any structure indicating the presence of the grating. If the zeroth- and ±first-order beams are collected, as indicated in the figure, the grating can be observed, albeit with incomplete object fidelity.

### B1.18.2.3    Contrast, noise and resolution

The resolution limited by diffraction assumes that illumination and contrast of the object are optimal. Here we discuss how noise affects the discernibility of small objects and of objects of weak contrast [4]. Noise is inherent in each light source due to the statistical emission process. It is, therefore, also a fundamental property by its very nature and limits image quality and resolution, just as diffraction is also responsible for the fundamental limit. Light passing through a test element in defined time slots $\Delta t$ will not contain the same number of photons in a series of successive runs. This is due to the stochastic emission process in the light source. Other sources of noise, such as inhomogeneities of recording media, will not be considered here.

It is assumed that the image can be divided up into a large number of picture elements, whose number will be of the order of $10^6$. If the contrast due to a structure in the object between two adjacent elements is smaller than the noise-induced fluctuations, the structure cannot be discerned, even if diffraction would allow this. Similarly, if the statistical excursion of the photon number in one or several of the elements is larger than or equal to the signal, then the noise fluctuations might be taken as true signals and lead to misinterpretations.

If viewed in transmission, the background brightness $B_b$ is higher than the light $B_o$ transmitted by an absorbing object. The contrast can then be defined as $C = (B_b - B_o)/B_b$ with $B_b \geq B_o \geq 0$ and $1 \geq C \geq 0$. It has been shown that density of photons $R$ (photons cm$^{-2}$) required to detect the contrast $C$ is

$$R = Nk^2/C^2.$$

Here, $N$ is the density of picture elements (cm$^{-2}$); if $N$ is high, the resolution is high, requiring, however, an increase of photon density over images with lower resolution. Also, low contrast requires greater photon density than high contrast in order to overcome false signals by noise fluctuations of adjacent picture elements. The factor $k$ reflects the random character of the photons (noise) and has to be chosen so as to protect against misinterpretations caused by noisy photon flux. $k$ depends somewhat on how well the image should be freed from noise-induced artefacts, a reasonable value being $k = 5$.

A summary of the diffraction- and noise-induced limitations of the resolution is qualitatively depicted in figure B1.18.3. With noise superimposed, the rectangular structure depicted in figure B1.18.3(a) becomes less defined with decreasing spacing and width of the rectangles. In figure B1.18.3(b), an assumed modulation transfer function of an objective is shown: that is, the light intensity in the image plane as a function of spatial frequency obtained by an object which sinusoidally modulates the transmitted light intensity. At low spatial frequencies, the amplitude is independent of frequency; at higher frequencies it drops linearly with increasing frequency. The root mean square (rms) noise, due to the statistical nature of the light, increases with spatial frequency. The intersection of the rms noise with the modulation transfer function gives the frequency at which noise becomes equal to ($k = 1$) or 1/25th ($k = 5$) of the signal. At high contrast, the decrease in image amplitude is usually determined by diffraction; at lower contrast, noise is predominant.

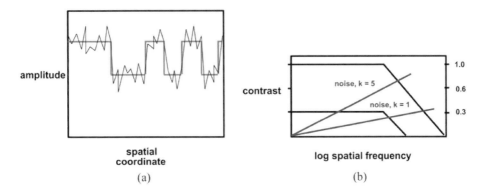

**Figure B1.18.3.** (a) Rectangular structure with noise superimposed: in the left half of the graph the rectangular structure is recognizable; in the right half, with the narrower spacing the rectangular structure would barely be recognizable without the guiding lines in the figure. (b) Modulation transfer function for a sinusoidal signal of constant amplitude as a function of frequency: at low frequency the amplitude of the transfer function is independent of the frequency. However, beyond a certain frequency the amplitude decreases with increasing frequency. This drop corresponds to a limitation of the resolution by diffraction. Noise increases with frequency and the crossings with the transfer function indicate where noise limits the resolution. At a contrast of 30% this cutoff frequency is exclusively determined by noise at point A; at 100% contrast the amplitude of the signal drops before the final signal is cut off at point B by noise. For $k = 5$, the noise at the crossing is $1/25$th of the signal; for $k = 1$, it is equal to the signal.

Under appropriate contrast and high light intensity, the resolution of planar object structures is diffraction limited. Noise in the microscopic system may also be important and may reduce resolution, if light levels and/or the contrasts are low. This implies that the illumination of the object has to be optimal and that the contrast of rather transparent or highly reflecting objects has to be enhanced. This can be achieved by an appropriate illumination system, phase- and interference-contrast methods and/or by data processing if electronic cameras (or light sensors) and processors are available. Last but not least, for low-light images, efforts can be made to reduce the noise either by averaging the data of a multitude of images or by subtracting the noise. Clearly, if the image is inspected by the eye, the number of photons, and hence the noise, are determined by the integration time of the eye of about 1/30 s; signal/noise can then only be improved, if at all possible, by increasing the light intensity. Hence, electronic data acquisition and processing can be used advantageously to improve image quality, since integration times can significantly be extended and noise suppressed.

### B1.18.2.4  Depth of focus

The depth of focus is defined as how far the object might be moved out of focus before the image starts to become blurred. It is determined (i) by the axial intensity distribution which an ideal object point suffers by imaging with the objective (point spread function), (ii) by geometrical optics and (iii) by the ability of the eye to adapt to different distances. In case (iii), the eye adapts and sees images in focus at various depths by successive scanning. Obviously, this mechanism is inoperative for image sensors and will not be considered here. The depth of focus caused by spreading the light intensity in axial direction (i) is given by

$$t_{PSF} = n\lambda/(A)^2$$

with $n$ the refractive index, $\lambda$ the wavelength of the light and $A$ the numerical aperture.

The focal depth in geometrical optics is based on the argument that 'points' of a diameter smaller than 0.15 mm in diameter cannot be distinguished. This leads to a focal depth of

$$t_{GO} = 0.15n/(AM_{microscope}).$$

The total depth of focus is the sum of both. It increases with the wavelength of the light, depends on the numerical aperture and the magnification of the microscope. For $\lambda = 550$ nm, a refractive index of 1 and a numerical aperture of 0.9, the depth of focus is in the region of 0.7 $\mu$m; with a numerical aperture of 0.4 it increases to about 5 $\mu$m. High-resolution objectives exclude the observation of details in the axial direction beyond their axial resolution. This is true for conventional microscopy, but not for scanning confocal microscopy, since optical sectioning allows successive layers in the bulk to be studied. Similarly, the field of view decreases with increasing resolution of the objective in conventional microscopy, whereas it is independent of resolution in scanning microscopy.

### B1.18.3   Contrast enhancement

In transmission microscopy, a transparent object yields low contrast. Molecular biological samples may be dyed in order to enhance contrast. However, this is in many cases neither possible nor desirable for various reasons, meaning that the object is only barely visible in outline and with practically no contrast. Similarly, if inorganic samples are to be investigated which are composites of materials of practically equal indices of refraction, the different components can only be distinguished in the microscope with great difficulty. This is all equally true for reflected light microscopy: the visibility and resolution of microscopic images suffer, if contrast is low. In order to cope with this, different illumination techniques are applied in order to enhance the contrast.

#### B1.18.3.1   Köhler's bright-field illumination system

Köhler's illumination system [5], which allows the field of view to be precisely illuminated, is schematically depicted in figure B1.18.4. The object is illuminated through a substage: the filament of a lamp is imaged by a collector lens into the focal plane of the condenser, where the condenser iris is located. Light from each point in the condenser iris passes through the object as a parallel beam inclined to the axis of the microscope at an angle depending on the position of the point in the iris. The parallel beams come to a focus at corresponding points in the focal plane of the objective. The collecter iris allows the area illuminated in the object plane to be varied. The condenser iris is the aperture of the illumination and its opening should be adjusted to the aperture of the objective lens and contrast properties of the object: if the apertures are equal, the highest resolution is achieved; if the illumination aperture is reduced, the contrast is enhanced. In practice, the aperture of the condenser is in most cases chosen to be smaller than the aperture of the objective lens in order to improve contrast.

#### B1.18.3.2   Enhanced contrast by dark-field illumination

Dark-field microscopy utilizes only those light beams from the condenser that have an aperture greater than that of the objective. This is in contrast to Köhler's illumination principle, where care is taken to adjust the aperture of the condenser by an iris to become equal to or smaller than that of the objective. A ring-type diaphragm is used to allow light beams to pass the condenser lens at an aperture greater than that of the objective lens. This is shown schematically in figure B1.18.5. In this arrangement, no direct light beams pass through the objective but only those which are diffracted or scattered by the object. If the direct light beam is blocked out, the background appears black instead of bright, thus increasing the contrast. Special condensers have been designed for dark-field illumination. Dark-field illumination has often been used in reflection.

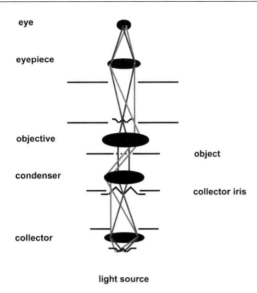

**Figure B1.18.4.** The most frequently used illumination system in bright-field microscopy.

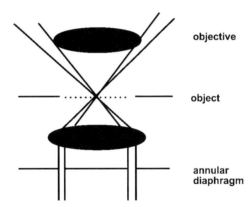

**Figure B1.18.5.** Dark-field illumination: the aperture of the objective is smaller than the aperture of the beams allowed by the annular diaphragm.

### B1.18.3.3    Zernike's phase contrast microscopy

Phase contrast microscopy [6, 7] is more sophisticated and universal than the dark-field method just described. In biology, in particular, microscopic objects are viewed by transmitted light and phase contrast is often used. Light passing through transparent objects has a different phase from light going through the embedding medium due to differences in the indices of refraction. The image is then a so-called phase image in contrast to an amplitude image of light absorbing objects. Since the eye and recording media in question respond to the intensity (amplitude) of the light and not to changes of the light phase, phase images are barely visible unless means are taken to modify the interference of the diffracted beams. The diffraction pattern of a phase grating is like that of an amplitude grating except that the zeroth-order beam is especially dominant in intensity. Zernike realized that modification of the zeroth-order beam will change the character of the image very effectively, by changing its phase and its intensity. For each object, depending on its character concerning

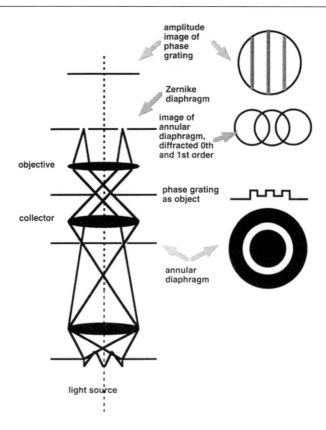

**Figure B1.18.6.** Schematic representation of Zernike's phase contrast method. The object is assumed to be a relief grating in a transparent material of constant index of refraction. Phase and amplitude are varied by the Zernike diaphragm, such that an amplitude image is obtained whose contrast is, in principle, adjustable.

the phase and amplitude, a 'Zernike diaphragm' (i.e. a diaphragm that affects the phase and the amplitude of the zeroth-order beam) can be constructed with an appropriate absorption and phase shift, which allows the weak-contrast image of the object to be transformed into an image of any desired contrast.

The principle of phase contrast microscopy is explained by figure B1.18.6. The object is assumed to be a linear phase grating. The diaphragm is annular, which means that only a small fraction of the diffracted light is covered by the Zernike diaphragm, as indicated in the figure for the first-order beams. In general, the Zernike diaphragm shifts the phase of the zeroth order by $\pi/2$ with respect to the diffracted beams. Since the intensity of the diffracted beams is much lower than that of the direct beam, the intensity of the zeroth-order beam is usually attenuated by adding an absorbing film. Clearly, all these measures indicate that images of high contrast cannot be combined with high intensity using this technique; a compromise between both has to be found depending on the requirements. The image fidelity of phase contrast imaging depends, therefore, on the width and light absorption of the Zernike diaphragm, in addition to the size and optical path difference created by the object under study and, finally, on the magnification.

At this point it is worth comparing the different techniques of contrast enhancements discussed so far. They represent spatial filtering techniques which mostly affect the zeroth order: dark field microscopy, which eliminates the zeroth order, the Schlieren method (not discussed here), which suppresses the zeroth order and one side band and, finally, phase contrast microscopy, where the phase of the zeroth order is shifted by $\pi/2$ and its intensity is attenuated.

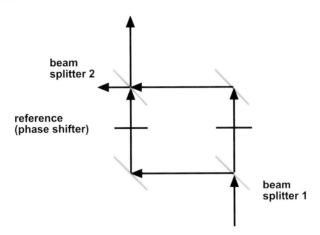

**Figure B1.18.7.** Principle for the realization of interference microscopy. The illuminating beam is split by beamsplitter 1 before passing the object so that the reference beam is not affected by the object. The separated beams interfere behind beamsplitter 2.

### B1.18.3.4    *Interference microscopy*

As already discussed, transparent specimens are generally only weakly visible by their outlines and flat areas cannot be distinguished from the surroundings due to lack of contrast. In addition to the phase contrast techniques, light interference can be used to obtain contrast [8, 9].

Transparent, but optically birefringent, objects can be made visible in the polarizing microscope if the two beams generated by the object traverse about the same path and are brought to interfere. In the case of optically isotropic bodies, the illuminating light beam has to be split into two beams: one that passes through the specimen and suffers phase shifts in the specimen which depend on the thickness and refraction index, and a second beam that passes through a reference object on a separate path (see figure B1.18.7). By superposing the two beams, phase objects appear in dark–bright contrast.

The differential interference contrast method utilizes the fact that, using a Wollaston prism, linearly polarized light can be split into two light beams of perpendicular polarization (figure B1.18.8). Since they are slightly parallel shifted, the two beams pass through the object at positions having different thickness and/or refractive index. The splitting of the beams is chosen to be sufficiently small not to affect the resolution. They are brought together again by a second Wollaston prism, and pass through the analyser. Since the beams are parallel and their waves planar in the object plane, the beams in the image plane are also parallel and the waves planar. Hence, the interference of the beams does not give rise to interference lines but to contrast, whose intensity depends on the phase difference caused by small differences of the refractive index. The image appears as a relief contrast which can be modified by changing the phase difference: for example, by moving the Wollaston prism perpendicularly to the optical axis.

The interpretation of such images requires some care, because the appearance of a relief structure may be misleading; it does not necessarily mean that the surface or thickness of the object is relief-like. Obviously, such a relief may also appear if samples are homogeneously thick, but composed of elements of different indices of refraction. Also, edges of the object may be missed if they are inappropriately oriented with respect to the polarization direction of the beams.

Interference microscopy is also possible in reflection. The surface structure of highly reflecting objects such as metals or metallized samples is frequently investigated in this way. Using multiple-beam interference [10], surface elevations as small as a few nanometres in height or depth can be measured. This is due to

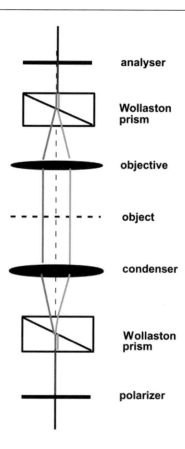

**Figure B1.18.8.** Differential interference contrast: the light beam is split into two beams by a Wollaston prism. The two beams pass the object at closely spaced positions and give, after interference, a contrast due to the phase difference.

the fact that the interference lines become very sharp if the monochromaticity of the light and the number of interfering beams are high.

### B1.18.3.5  Fluorescence microscopy

Fluorescence microscopy has been a very popular method of investigating biological specimens and obtaining contrast in otherwise transparent organic objects. The samples are stained with fluorescent dyes and illuminated with light capable of exciting the dye to fluoresce. The wavelength of the emitted light is Stokes-shifted to wavelengths longer than that of the primary beam. Since the quantum efficiency (the ratio of the numbers of emitted to exciting photons) of the dyes is often low and since the light is emitted in all directions, the image is of low intensity. Nevertheless, this technique allows images of high contrast and of high signal-to-noise ratio to be obtained. The principle of fluorescence microscopy is illustrated in figure B1.18.9 for the epifluorescence microscope. The primary excitation does not in principle directly enter the detector and thus provides the desired contrast between stained and unstained areas, which appear completely dark. It is obvious from the very nature of the preparation technique that, in addition to morphological structures, chemical and physicochemical features of the sample can be revealed if the dyes adsorb only at special chemical sites.

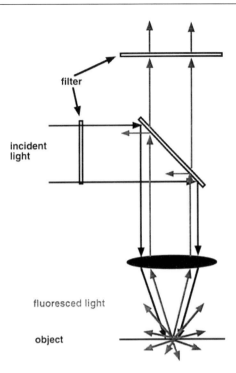

**Figure B1.18.9.** Epifluorescence microscope: the object is excited from the top and the fluorescent light is emitted in all directions, as indicated by the multitude of arrows in the object plane. The fluorescent light within the aperture of the objective gives rise to the image, showing that much of the fluorescent light is lost for imaging.

## B1.18.4   Scanning microscopy

In scanning microscopy, the object is successively scanned by a light spot, in contrast to conventional microscopy in which the entire object field is processed simultaneously. Thus, scanning represents a serial (and conventional microscopy a parallel) processing system. The requirements for the optical lenses are relaxed for the scanning microscope, because the whole field of view is no longer imaged at once, but the price that is paid is the need for reconstruction of the image from a set of data and the required precision for the scanning.

   A point light source is imaged onto the specimen by the objective and the transmitted light collected by the collector lens and detected by a broad-area detector; in the case of reflection microscopy, the objective lens also serves simultaneously as a collector (see figure B1.18.10). The resolution is solely determined by the objective lens, because the collector has no imaging function and only collects the transmitted light. The scanning is assumed in figure B1.18.10 to be based on the mechanical movement of the sample through the focal point of the objective. In this case, off-axis aberrations of the objective are avoided, the area to be imaged is not limited by the field of view of the objective and the image properties are identical and only determined by the specimen. The drawback of stage movement is the lower speed compared with beam scanning and the high mechanical precision required for the stage. Beam scanning allows the image to be reconstructed from the serially available light intensity data of the spots in real time [11, 12]. If a framestore is available, the image can be taken, stored, processed if desired and displayed. Processing of the electrical signal offers advantages. There is, for example, no need to increase contrast by stopping down the collector lens or by dark-field techniques which, in contrast to electronic processing, modify the resolution of the image.

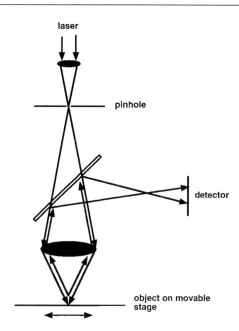

**Figure B1.18.10.** Scanning microscope in reflection: the laser beam is focused on a spot on the object. The reflected light is collected and received by a broad-area sensor. By moving the stage, the object can be scanned point by point and the corresponding reflection data used to construct the image. Instead of moving the stage, the illuminating laser beam can be used for scanning.

Scanning gives many degrees of freedom to the design of the optical system and the confocal arrangement is one of the most prominent, having revolutionized the method of microscopic studies, in particular of biological material. Since confocal microscopy has in recent years proved to be of great importance, it is discussed in some detail here.

## B1.18.5   Confocal scanning microscopy

### B1.18.5.1   *Principle and advantages of confocal microscopy*

The progress that has been achieved by confocal microscopy [13–16] is due to the rejection of object structures outside the focal point, rejection of scattered light and slightly improved resolution. These improvements are obtained by positioning pointlike diaphragms in optically conjugate positions (see figure B1.18.11). The rejection of structures outside the focal point allows an object to be optically sectioned and not only images of the surface are obtained by scanning but also of sections deep in a sample, so that three-dimensional microscopic images can be prepared as well as images of sections parallel to the optical axis. Therefore, internal structures in biological specimens can be made visible on a microsopic scale without major interference with the biological material by preparational procedures (fixation, dehydration etc) and without going through the painstaking procedure of mechanical sectioning. In addition, time-dependent studies of microscopic processes are possible. Obviously, there is a price to be paid: confocal microscopy requires serial data acquisition and processing and hence comprises a complete system whose cost exceeds that of a conventional microscope.

Figure B1.18.11 shows the basic arrangement of a confocal instrument. The important points are more easily presented for the reflection microscope, although everything also applies to transmission if modified appropriately. The broad-area detector is replaced by a point detector implemented by a pinhole placed in

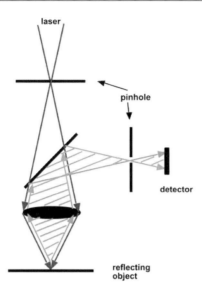

**Figure B1.18.11.** Confocal scanning microscope in reflection: the pinhole in front of the detector is in a conjugate position to the illumination pinhole. This arrangement allows the object to be optically sectioned. The lens is used to focus the light beam onto the sample and onto the pinhole. Thus, the resulting point spread function is sharpened and the resolution increased.

front of the detector at the conjugate position to the pinhole on the illumination side. This arrangement ensures that only light from the small illuminated volume is detected and light that stems from outside the focal point is strongly reduced in intensity. This is illustrated in figure B1.18.12, where a reflecting object is assumed to be below the focal plane: only a small fraction of the reflected light reaches the detector, since it is shielded by the pinhole. The intensity drops below detection threshold and no image can be formed.

### B1.18.5.2 Optical sectioning, smallest slice thickness and axial resolution

The fact that only points in the focal plane contribute to the image, whereas points above or below do not, allows optical sectioning. Thus, the object can be imaged layer by layer by moving them successively into the focal plane. For applications, it is important how thin a slice can be made by optical sectioning. A point-like object imaged by a microscope has a finite volume [13] which is sometimes called voxel, in analogy to pixel in two dimensions. Its extension in the axial direction determines the resolution in this direction ($z$) and the smallest thickness of a layer that can be obtained by optical sectioning. The intensity variation of the image of a point along the optical axis for the confocal arrangement is given by

$$I(u) = \{\sin(u/4)/u/4\}^4$$

with $u = (8\pi/\lambda)z \sin^2(\alpha/2)$. $\lambda$ is the wavelength of the light, and $n \sin \alpha$ is the numerical aperture of the lens. The function $I(u)$ is zero at $u = \pi, 2\pi \ldots$. If we take the spread of the function $I(u)$ between $u = \pm\pi$ as the smallest slice thickness $t$, one obtains

$$t = \lambda/(4 \sin^2 \alpha/2).$$

The slice thickness is proportional to the wavelength of the light and a function of the aperture angle. For $\lambda = 0.5\ \mu$m, the slice thickness is about $0.25\ \mu$m for $\alpha = \pi/2$. Obviously, the point spread function

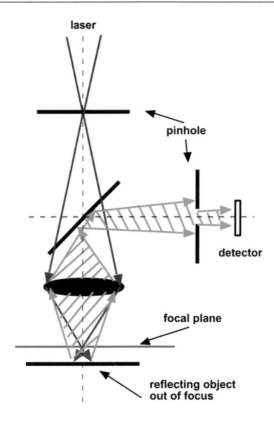

**Figure B1.18.12.** Illustration that only light reflected from object points in the focal plane contributes to the image. If the light is reflected from areas below the focal plane, only a small fraction can pass through the pinhole so that light from those areas does not contribute to the image. The pinhole in front of the detector is exaggerated in size for the sake of presentation.

serves also to determine the smallest separation that two points in the axial direction may have in order to be resolved. If the Rayleigh criterion is applied—intensity between the image points to be half of the intensity at maximum—then the resolution is in the range of 0.15–0.2 $\mu$m.

### B1.18.5.3   Range of depth for optical sectioning

The greatest depth at which a specimen can be optically sectioned is also of interest. This depth is limited by the working distance of the objective, which is usually smaller for objectives with greater numerical aperture. However, the depth imposed by the working distance of the objective is rarely reached, since other mechanisms provide constraints as well. These are light scattering and partial absorption of the exciting and emitted light, in the case of fluorescence microscopy. The exciting beam is partially absorbed by fluorophores until it reaches the focused volume. Hence, less light is emitted from a focused volume that is deep in the bulk of a sample. Thus, the intensity of the light reaching the detector decreases with increasing depth, so that for image formation laser power and/or integration time would have to be increased. Though technically possible, both cannot be increased beyond thresholds at which the samples, especially biological materials, are damaged.

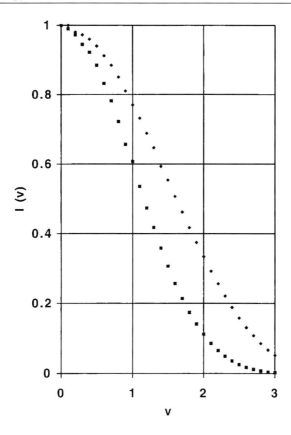

**Figure B1.18.13.** The point spread function in the confocal arrangement (full square) is sharpened in comparison with the conventional arrangement (full diamond). Therefore, the resolution is improved.

*B1.18.5.4 Lateral resolution*

The extension of the voxel in a radial direction gives information on the lateral resolution. Since the lateral resolution has so far not been discussed in terms of the point spread function for the conventional microscope, it will be dealt with here for both conventional and confocal arrangements [13]. The radial intensity distribution in the focal plane (perpendicular to the optical axis) in the case of a conventional microscope is given by

$$I_{\mathrm{m}}(v) = (2J_1(v)/v)^2$$

with $v = (2\pi/\lambda)rn \sin \alpha$, $\lambda$ is the wavelength of the light, $r$ is the radial coordinate, $n \sin \alpha$ is the numerical aperture, and $J_1(v)$ is the first-order Bessel function of the first kind. Zero intensity is at $v = 1.22\pi, 2.23\pi, 3.42\pi \ldots$.

For the confocal arrangement in transmission, the objective and the collector are used for imaging; in reflection the objective is used twice. Therefore, the radial intensity distribution in the image is the square of that of the conventional microscope:

$$I_{\mathrm{conf}}(v) = (2J_1(v)/v)^4.$$

$I_{\mathrm{conf}}(v)$ has the same zero points as $I_{\mathrm{m}}(v)$. However, in the confocal case the function is sharpened and the sidelobes are suppressed. The light intensity distributions for the conventional and the confocal case are depicted in figure B1.18.13. If the Rayleigh criterion for the definition of resolution is applied, one finds

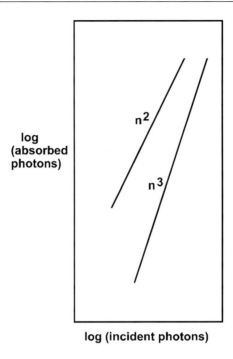

**log (incident photons)**

**Figure B1.18.14.** Schematic representation of the increase of absorption with photon density for two- and three-photon absorption.

that the lateral resolution in the confocal case is improved in comparison with conventional microscopy: obviously, the sharpened function in the confocal case allows two closely spaced points at smaller separation to be distinguished.

*B1.18.5.5   Contrast enhancement and practical limits to confocal one-photon-excitation fluorescence microscopy*

The methods to improve contrast described for conventional microscopy can also be applied to confocal microscopy [17]. However, because the images are obtained by scanning and data processing, the tools of image manipulation are also advantageously utilized to improve image quality. Nevertheless, all these methods have their limitations, as will be explained in the following example. Biological studies are often made with fluorescence. Usually the fluorophore is excited by one photon from the ground state to the excited state; the ratio of the number of photons emitted by the fluorophore to the number of exciting photons is, as a rule, significantly below one. Therefore, the number of photons collected per voxel is low, depending on the density of fluorophores, on the exciting light intensity and on the scan rate. The density of fluorophores is, in general, determined by the requirements of the experiment and cannot be significantly varied. In order to increase the signal, the scan rate would have to be lowered, and the number of scans and the exciting light intensity increased. However, extended exposure of the dyes leads to bleaching in the whole cone of illumination and hence to the number of layers to be sectioned. The number of layers is even more reduced if, in addition to bleaching the fluorophore, the excitation produces toxic products that modify or destroy the properties of living cells or tissues. Lasers could, in principle, supply higher light intensities, but saturation of emission of the dyes additionally limits the applicable power. In these circumstances, real improvement can only be reached if the voxel at the focal point could exclusively be excited by the incident light.

### B1.18.5.6 Confocal microscopy with multiphoton-excitation fluorescence

Usually a fluorophore is excited from its ground to its first excited state by a photon of an energy which corresponds to the energy difference between the two states. Photons of smaller energy are generally not absorbed. However, if their energy amounts to one-half or one-third (etc) of the energy difference, a small probability for simultaneous absorption of two or three (etc) photons exists since the energy condition for absorption is fulfilled. However, due to this small probability, the photon density has to be sufficiently high if two-, three- or $n$-photon absorption is to be observed [18]. In general, these densities can only be achieved by lasers of the corresponding power and with appropriate pulse width, since absorption by multiphoton processes increases with the $n$th power of the photon density (figure B1.18.14).

One-photon excitation has limitations due to the unwanted out-of-focus fluorophore absorption and bleaching, and light scattering. These drawbacks can be circumvented if multiphoton excitation of the fluorophore is used. Since it increases with the $n$th power of the photon density, significant absorption of the exciting light will only occur at the focal point of the objective where the required high photon density for absorption is reached. Consequently, fluorescent light will only be emitted from a volume element whose size will be determined by the intensity and power law dependence of the exciting radiation. Though confocal arrangement is not needed, it was shown that, in the confocal arrangement, the effective point spread function is less extended and, hence, the resolution improved [19, 20].

Thus, multiphoton excitation eliminates unwanted out-of-focus excitation, unnecessary phototoxity and bleaching. However, efficient power sources are required and, since the efficiency of multiphoton excitation is usually low, the times needed to generate images are increased.

### B1.18.5.7 The future: resolution beyond the diffraction limit in confocal fluorescence microscopy?

As light microscopy has many advantages over other microscopic techniques, the desire is to overcome the limit due to the extension of the point spread function or to reduce the emitting volume by multiphoton excitation. One proposal was made on the basis of fluorescence microscopy [21]. As discussed, in fluorescence microscopy, molecules are excited to emit light which is then used to form the microscopic image. If the exciting light is imaged onto a small volume of the sample, light emitted from this volume determines the spatial resolution (i.e. both in depth and the lateral direction). If the light-emitting volume can be reduced, resolution will be improved. This is achievable, in principle, by stimulated emission: if the stimulated emission rate is higher than the fluorescence decay and slower than the decay rate of intrastate vibrational relaxation, the emitting volume in the focal region shrinks. Estimates predict a resolution of 0.01–0.02 $\mu$m for continuous illumination of 1 mW and picosecond excitations of 10 MW cm$^{-2}$ at a rate of 200 kHz. If this idea can be reduced into practice, the diffraction limit would be overcome. The high resolution combined with the advantages of light microscopy over other microscopic methods would indeed represent a major breakthrough in this field.

## References

[1] Tappert J 1957 Bildentstehung im Mikroskop *Wissenschaft und Fortschritt* **7** 361
[2] Michel K 1981 *Die Grundzüge der Theorie des Mikroskops* 3rd edn (Stuttgart: Wissenschaftliche)
[3] Pluta M 1988 *Advanced Light Microscopy, vol 1: Principles and Basic Properties* (Amsterdam: Elsevier)
[4] Rose A 1973 *Vision, Human and Electronic* (New York: Plenum)
[5] Köhler A 1893 Ein neues Beleuchtungsverfahren für mikrophotographische Zwecke *Z. Wissensch. Mikr.* **10** 433
[6] Zernike F 1935 Das Phasenkontrastverfahren bei der mikroskopischen Beobachtung *Z. Phys.* **36** 848
[7] Zernike F 1942 Phase contrast, a new method for the microscopic observation of transparent objects I *Physica* **9** 686
    Zernike F 1942 Phase contrast, a new method for the microscopic observation of transparent objects II *Physica* **9** 974
[8] Pluta M 1989 *Advanced Light Microscopy, vol 2: Advanced Methods* (Amsterdam: Elsevier)
[9] Beyer H 1974 *Interferenzmikroskopie* (Leipzig: Akademic)
[10] Tolansky S 1960 *Surface Microtopography* (New York: Interscience)

[11] Shaw S L, Salmon E D and Quatrano R S 1995 Digital photography for the light microscope: results with a gated, video-rate CCD camera and NIH-image software *BioTechniques* **19** 946–55

[12] Kino G S and Xiao G Q 1990 Real time scanning optical microscope *Confocal Microscopy* ed T Wilson (New York: Academic)

[13] Wilson T (ed) 1990 Confocal microscopy *Confocal Microscopy* (New York: Academic)

[14] Hell S W and Stelzer E H K 1992 Fundamental improvement of resolution with a four Pi-confocal fluorescence microscope using two-photon excitation *Opt. Commun.* **93** 277–82

[15] Lindek St, Cremer Chr and Stelzer E H K 1996 Confocal theta fluorescence microscopy using two-photon absorption and annular apertures *Optik* **102** 131–4

[16] van Oijen A M, Kohler J, Schmidt J, Muller M and Brakenhoff G J 1998 3-Dimensional super-resolution by spectrally selective imaging *Chem. Phys. Lett.* **292** 183–7

[17] Török P, Sheppard C J R and Laczik Z 1996 Dark field and differential phase contrast imaging modes in confocal microscopy using a half aperture stop *Optik* **103** 101–6

[18] Wokosin D L, Centonze V, White J G, Armstrong D, Robertson G and Ferguson A I 1996 All-solid-state ultrafast lasers facilitate multiphoton excitation fluorescence imaging *IEEE J. Sel. Top. Quantum Electron.* **2** 1051–65

[19] Schrader M, Bahlmann K and Hell S W 1997 Three-photon-excitation microscopy: Theory, experiment, and applications *Optik* **104** 116–24

[20] Sako Y, Sekihata A, Yanagisawa Y, Yamamoto M, Shimada Y, Ozaki K and Kusumi A 1997 Comparison of two-photon excitation laser scanning microscopy with UV-confocal laser scanning microscopy in three-dimensional calcium imaging using the fluorescence indicator Indo-1 *J. Microsc.* **185** 9–20

[21] Hell S W and Kroug M 1995 Ground-state-depletion fluorescence microscopy: a concept for breaking the diffraction resolution limit *Appl. Phys.* B **60** 495–7

## Further Reading

*Books*

v Amelinck S, van Dyck D, van Landuyt J and van Trendelo G (eds) 1996 *Handbook of Microscopy, Application in Materials Science, Solid State Physics and Chemistry* 3 vols (Weinheim: VCH)

Pluta M 1988 *Advanced Microscopy* 3 vols (Amsterdam: Elsevier)

de Hoff R and Rhines F N 1991 *Quantitative Microscopy* (Lake Grove: Tech. Books)

Beyer H (ed) 1997 *Handbuch der Mikroskopie* (VEB-Verlag Technik)

Robenek H (ed) 1995 *Mikroskopie in Forschung und Technik* (GIT-Verlag)

Herman B and Jacobsen K 1990 *Optical Microscopy for Biology* (Wiley)

Brabury S and Everett B 1996 *Contrast Techniques in Light Microscopy, Microscopy Handbooks* 34 (Oxford: BIOS Scientific Publishers)

v Kriete (ed) 1992 *Visualization in Biomedical Microscopy, 3-d Imaging and Computer Visualization* (Weinheim: VCH)

Wilson T (ed) 1996 *Confocal Microscopy* (New York: Academic)

Cork T and Kino G S 1996 *Confocal Scanning Optical Microscopy and Related Imaging Systems* (New York: Academic)

Gu Min 1996 *Principles of Three Dimensional Imaging in Confocal Microscopes* (Singapore: World Scientific)

*Reviews*

Sheppard C J R 1987 Scanning optical microscopy *Adv. Opt. Electron Microscopy* **10** 1-98

Cooke P M 1996 Chemical microscopy *Anal. Chem.* **68** 333–78

Kapitza H G 1996 Confocal laser scanning microscopy for optical measurement of the microstructure of surfaces and layers *Tech. Mess.* **63** 136–41

Schroth D 1997 The confocal laser scanning microscopy. A new tool in materials testing *Materialpruefung* **39** 264

Chestnut M H 1997 Confocal microscopy of colloids *Curr. Opin. Colloid Interface Sci.* **2** 158–61

van Blaaderen A 1997 Quantitative real-space analysis of colloidal structures and dynamics with confocal scanning light microscopy *Prog. Colloid Polym. Sci.* **104** 59–65

Ribbe A E 1997 Laser scanning confocal microscopy in polymer science *Trends Polym. Sci.* **5** 333–7

Oliveira M J and Hemsley D A 1996 Optical microscopy of polymers *Sonderb. Prakt. Metallogr.* **27** 13–22

Nie Sh and Zare R N 1997 Optical detection of single molecules *Ann. Rev. Biophys. Biomol. Struct.* **26** 567–96

Masters B R 1994 Confocal redox imaging of cells *Adv. Mol. Cell Biol.* **8** 1–19

Ojcius D M, Niedergang F, Subtil A, Hellio R and Dautry-Varsat A 1996 Immunology and the confocal microscope *Res. Immunol.* **147** 175–88

Lemasters J J 1996 Confocal microscopy of single living cells *Chem. Anal., NY* **137** 157–77

Schrof W, Klingler J, Heckmann W and Horn D 1998 Confocal fluorescence and Raman microscopy in industrial research *Colloid Polym. Sci.* **276** 577–88

Sabri S, Richelme F, Pierres A, Benoliel A M and Bongrand P 1997 Interest of image processing in cell biology and immunology *J. Immunol. Methods* **208** 1–27

van Der Oord C J R, Jones G R, Shaw D A, Munro I H, Levine Y K and Gerritsen H C 1996 High-resolution confocal microscopy using synchrotron radiation *J. Microsc.* **182** 217–24

## *Applications*

Booker Gr, Laczik Z and Toeroek P 1995 *Applications of Scanning Infra-red Microscopy to Bulk Semiconductors (Inst. Phys. Conf. Ser., 146)* pp 681–92

Bhawalkar J D, Swiatkiewicz J, Pan S J, Samarabandu J K, Liou W S, He G S, Berezney R, Cheng P C and Prasad P N 1996 Three-dimensional laser scanning two-photon fluorescence confocal microscopy of polymer materials using a new, efficient upconverting fluorophore *Scanning* **18** 562–6

Ling X, Pritzker M D, Byerley J J and Burns C M 1998 Confocal scanning laser microscopy of polymer coatings *J. Appl. Polym. Sci.* **67** 149–58

Carlsson K and Liljeborg A 1997 Confocal fluorescence microscopy using spectral and lifetime information to simultaneously record four fluorophores with high channel separation *J. Microsc.* **185** 37–46

Wokosin D L and White J G 1997 Optimization of the design of a multiple-photon excitation laser scanning fluorescence imaging system *Proc. SPIE* **2984** 25–9

Fleury L, Gruber A, Draebenstedt A, Wrachtrup J and von Borczyskowski C 1997 Low-temperature confocal microscopy on individual molecules near a surface *J. Phys. Chem.* B **101** 7933–8

Peachey L D, Ishikawa H and Murakami T 1996 Correlated confocal and intermediate voltage electron microscopy imaging of the same cells using sequential fluorescence labeling fixation and critical point dehydration *Scanning Microsc. (Suppl)* **10** 237–47

Wolleschensky R, Feurer T, Sauerbrey R and Simon U 1998 Characterization and optimization of a laser-scanning microscope in the femtosecond regime *Appl. Phys.* B **67** 87–94

Llorca-Isern N and Espanol M 1997 Advanced microscopic techniques for surface characterization *Surf. Modif. Technol. XI (Proc. 11th Int. Conf.)* pp 722–35

Leonas K K 1998 Confocal scanning laser microscopy: a method to evaluate textile structures *Am. Dyest. Rep.* **87** 15–18

Wilson K R *et al* 1998 New ways to observe and control dynamics *Proc. SPIE* **3273** 214–18

Kim Ki H, So P T C, Kochevar I E, Masters B R and Gratton E 1998 Two-photon fluorescence and confocal reflected light imaging of thick tissue structures *Proc. SPIE* **3260** 46–57

# B1.19
# Scanning probe microscopies

*Nicholas D Spencer and Suzanne P Jarvis*

## B1.19.1   Introduction

The development of the scanning tunnelling microscope (STM) [1] was a revelation to the scientific community, enabling surface atomic features to be imaged in air with remarkably simple apparatus. The STM earned Binnig and Rohrer the Nobel prize for physics in 1986, and set the stage for a series of scanning probe microscopies (SPMs) based on a host of different physical principles, many of the techniques displaying nanometre resolution or better.

The methods have in turn launched the new fields of nanoscience and nanotechnology, in which the manipulation and characterization of nanometre-scale structures play a crucial role. STM and related methods have also been applied with considerable success in established areas, such as tribology [2], catalysis [3], cell biology [4] and protein chemistry [4], extending our knowledge of these fields into the nanometre world; they have, in addition, become a mainstay of surface analytical laboratories, in the worlds of both academia and industry.

Central to all SPMs (or 'local probe methods', or 'local proximal probes' as they are sometimes called) is the presence of a tip or sensor, typically of less than 100 nm radius, that is rastered in close proximity to—or in 'contact' with—the sample's surface. This set-up enables a particular physical property to be measured and imaged over the scanned area. Crucial to the development of this family of techniques were both the ready availability of piezoelements, with which the probe can be rastered with subnanometre precision, and the highly developed computers and stable electronics of the 1980s, without which the operation of SPMs as we know them would not have been possible.

A number of excellent books have been written on SPMs in general. These include the collections edited by Wiesendanger and Güntherodt [5] and Bonnell [6] as well as the monographs by Wiesendanger [7], DiNardo [8] and Colton [9].

## B1.19.2   Scanning tunnelling microscopy

### B1.19.2.1   *Principles and instrumentation*

Tunnelling is a phenomenon that involves particles moving from one state to another through an energy barrier. It occurs as a consequence of the quantum mechanical nature of particles such as electrons and has no explanation in classical physical terms. Tunnelling has been experimentally observed in many physical systems, including both semiconductors [10] and superconductors [11].

In STM, a sharp metal tip [12] is brought within less than a nanometre of a conducting sample surface, using a piezoelectric drive (figure B1.19.1). At these separations, there is overlap of the tip and sample wavefunctions at the gap, resulting in a tunnelling current of the order of nanoamps when a bias voltage

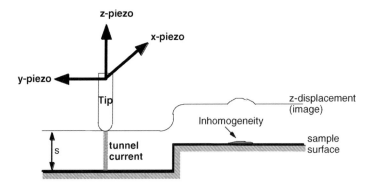

**Figure B1.19.1.** Principle of operation of a scanning tunnelling microscope. The $x$- and $y$-piezodrives scan the tip across the surface. In one possible mode of operation, the current from the tip is fed into a feedback loop that controls the voltage to the $z$-piezo, to maintain constant current. The line labelled $z$-displacement shows the tip reacting both to morphological and chemical (i.e. electronic) inhomogeneities. (Taken from [213].)

($\pm 10^{-3} - 4$ V) is applied to the tip [13]. The electrons flow from the occupied states of the tip to the unoccupied states in the sample, or *vice versa*, depending on the sign of the tip bias. The current is exponentially dependent on the tip–sample distance [14],

$$I = C \rho_t \rho_S \, e^{-s\sqrt{\phi}} \tag{B1.19.1}$$

where $s$ is the sample–tip separation, $\phi$ is a parameter related to the barrier between the sample and the tip, $\rho_t$ is the electron density of the tip, $\rho_S$ is the electron density of the sample and $C$, a constant, is a linear function of voltage. The exponential dependence on distance has several very important consequences. Firstly, it enables the local tip–sample spacing to be controlled very precisely ($< 10^{-2}$ Å) by means of a feedback loop connected to the $z$-piezo, using the tunnelling current as a control parameter. Secondly, it means that despite the fact that the tip may be many tens of nanometres in radius, the effective radius—through which most of the tunnelling takes place—is of atomic dimensions, yielding subnanometre spatial resolution. This tip may be rastered over an area that can range from hundredths of square nanometres to hundreds of square microns, and the surface topography—or more specifically the spatial distribution of particular electronic states—may thereby be imaged. Imaging (which may be done in air, in vacuum, or even under liquids) may be achieved either by monitoring the tunnelling current, in order to maintain a constant tip–sample separation and displaying the $z$-voltages as a function of $x$ and $y$ position, or by simply rastering the tip above the surface at a constant height, and plotting the tunnelling current on the $z$-axis. The former is known as *constant-current mode*, the latter as *constant-height mode* (figure B1.19.2). While constant-current mode is more stable for relatively rough surfaces, it is also somewhat slower than constant-height mode, because of its reliance on the feedback system, which sets a limit on the maximum scan speed.

The image obtained in a STM experiment is conventionally displayed on the computer screen as grey scales or false colour, with the lightest shades corresponding to peaks (or highest currents) and darkest shades corresponding to valleys (or lowest currents). With such graphic methods of data display, it is particularly tempting to interpret atomic-scale STM images as high-resolution topographs. However, it must be remembered that only electrons near the Fermi energy contribute to the tunnelling current, whereas all electrons contribute to the surface charge density. Since topography can reasonably be defined as a contour of constant surface charge density [15], STM images are intrinsically different from surface topographs.

In addition to its strong dependence on tip–sample separation, the tunnelling current is also dependent on the electron density of states (DOS) of both tip and sample (equation (B1.19.1)). This dependence can

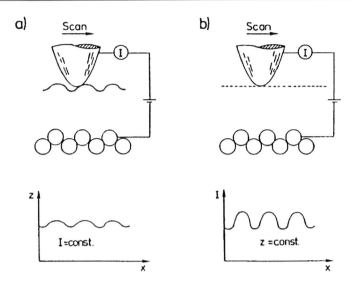

**Figure B1.19.2.** The two modes of operation for scanning tunnelling microscopes: (a) constant current and (b) constant height. (Taken from [214], figure 1.)

be exploited to produce a map of the local DOS under the tip by varying the applied voltage and measuring the tunnelling current. Both occupied and unoccupied electronic states can be probed by this method, which is known as scanning tunnelling spectroscopy (STS) [16]. The traditional method of mapping DOS is to use ultraviolet photoelectron spectroscopy (UPS) to measure occupied states and inverse photoemission spectroscopy (IPS) to measure empty states. However, it is important to remember that these data do not correspond exactly to those derived from STS measurements. Firstly, the STS spectrum is a convolution of tip and sample properties (a potential problem, should the tip become contaminated during the experiment). Secondly, since states near the upper edge of the energy range investigated see a lower barrier than those near the lower end, they contribute a greater tunnelling current, so that sensitivity to occupied states falls off with increasing energy below the Fermi level. Thirdly, STS is a much more surface- (and above-surface-) sensitive technique than UPS or IPS, meaning that surface electronic states contribute far more to the STS spectrum. This also means that the sensitivity to s, p, and d states is different in STS, due to the different degrees to which the electron density associated with these states extends out of the surface.

*B1.19.2.2    Applications of scanning tunnelling microscopy*

*(a) Semiconductors*

STM found one of its earliest applications as a tool for probing the atomic-level structure of semiconductors. In 1983, the $7 \times 7$ reconstructed surface of Si(111) was observed for the first time [17] in real space; all previous observations had been carried out using diffraction methods, the $7 \times 7$ structure having, in fact, only been hypothesized. By capitalizing on the spectroscopic capabilities of the technique it was also proven [18] that STM could be used to probe the electronic structure of this surface (figure B1.19.3).

A complete STS spectrum of the $7 \times 7$ reconstructed Si(111) surface displays remarkable correlation with the corresponding UPS and IPS spectra [19] (figure B1.19.4), showing the potential value of this approach. The high spatial resolution of the STS technique has also been demonstrated using a silicon surface containing impurity atoms [20] (figure B1.19.5), where the absence or presence of a band gap over *an individual atom* shows whether it belongs, respectively, to the silicon or to a metallic impurity.

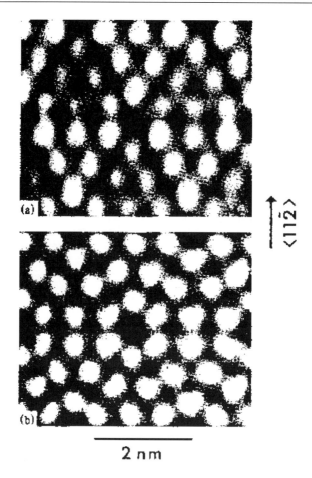

**Figure B1.19.3.** STM images of Si(111)-(7 × 7) measured with (a) −2 V and (b) +2 V applied to the sample. (Taken from [18], figure 1.)

Chemical reactions of ammonia with the silicon surface have also been clearly observed using STS [21], where the disappearance of the $\pi$ and $\pi^*$ states characteristic of the clean surface coincides with the formation of Si–H antibonding states corresponding to the dissociation of the ammonia on the Si surface.

Other semiconductors have also proved to be a fruitful ground for STM investigation. Zheng *et al* [22] have used the spatial resolution and electronic state sensitivity of STM to spatially display the electronic characteristics of single Zn impurity atoms in Zn-doped GaAs, both in filled and empty states, which show spherical and triangular symmetry, respectively. Upon imaging a number of Zn-induced features, a variety of different heights were recorded, corresponding to the depth of the impurity atoms within the sample. Thus STM was used to probe both the chemical nature and the 3D spatial location of the impurity atoms—an achievement that would have been inconceivable before the advent of STM.

STM has not as yet proved to be easily applicable to the area of ultrafast surface phenomena. Nevertheless, some success has been achieved in the direct observation of dynamic processes with a larger timescale. Kitamura *et al* [23], using a high-temperature STM to scan single lines repeatedly and to display the results as a time-*versus*-position pseudoimage, were able to follow the diffusion of atomic-scale vacancies on a heated Si(001) surface in real time. They were able to show that vacancy diffusion proceeds exclusively in one dimension, along the dimer row.

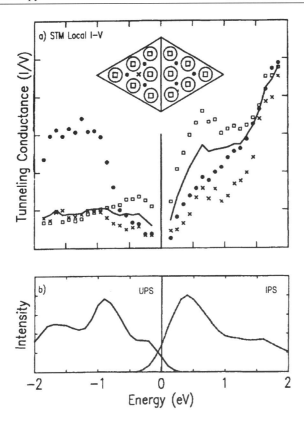

**Figure B1.19.4.** (a) Local conductance STS measurements at specific points within the Si(111)-(7 × 7) unit cell (symbols) and averaged over whole cell. (b) Equivalent data obtained by ultraviolet photoelectron spectroscopy (UPS) and inverse photoemission spectroscopy (IPS). (Taken from [19], figure 2.)

*(b) Metals*

STM has been applied with great success to the study of metals and adsorbate–metal systems [24]. This has naturally brought the technique into the mainstream of surface science, where structural information at the atomic level could previously only be obtained via diffraction methods such as low-energy electron diffraction (LEED) [25]. The STM can also provide a level of electronic information and visualization of the quantum mechanical behaviour of electrons that is unavailable from other methods: the images of copper and silver surfaces obtained by the groups of Eigler [26] and Avouris [27], showing standing waves produced by the defect-induced scattering of the 2D electron gas in surface states, bear eloquent testament to this (figure B1.19.6).

Surface reconstructions have been observed by STM in many systems, and the technique has, indeed, been used to confirm the 'missing row' structure in the 1 × 2 reconstruction of Au(110) [28]. As the temperature was increased within 10 K of the transition to the disordered 1 × 1 phase (700 K), a drastic reduction in domain size to ∼20–40 Å (i.e. less than the coherence width of LEED) was observed. In this way, the STM has been used to help explain and extend many observations previously made by diffraction methods.

STM studies of simple adsorbates on metal surfaces have proved challenging, partly due to the significant mobility of most small species on metals at room temperature, which therefore generally necessitates low-temperature operation. Additionally, since adsorbates can change the local density of states in the metal

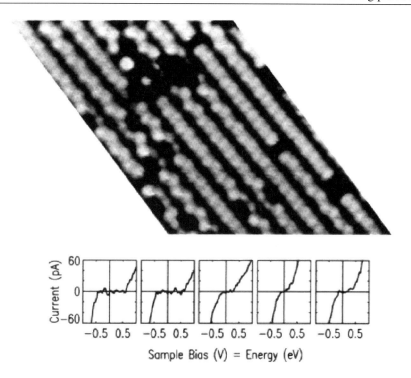

**Figure B1.19.5.** Tunnelling *I–V* curves acquired across a defect on Si(100). Away from the defect a bandgap can be seen. Over the defect itself, the bandgap disappears, suggesting that it possesses metallic character. This analysis was performed with spatial resolution of better than 1 nm. (Taken from [20], figures 2 and 6.)

surface, particular care must be taken not to interpret STM images of adsorbate–metal systems as simple topographs, but rather to capitalize on the technique's capability for observing unoccupied and occupied energy states. In this way, the bond between adsorbate and substrate can be investigated on a local level, subject to the restrictions on energy range mentioned above. By observing the electronic changes in the neighbourhood of an adsorption site, much can be learned about the range over which chemical bonds can act and influence each other.

Metal surfaces in motion have also been characterized by STM, one of the clearest examples being the surface diffusion of gold atoms on Au(111) [29] (figure B1.19.7). Surface diffusion of adsorbates on metals can be followed [30] provided that appropriate cooling systems are available, and STM has been successfully employed to follow the 2D dendritic growth of metals on metal surfaces [31].

*(c) Organic surfaces*

The operation of the STM depends on the conduction of electrons between tip and sample. This means, of course, that insulating samples are, in general, not accessible to STM investigations. Nevertheless, a large body of work [32] dealing with STM characterization of thin organic films on conducting substrates is now in the literature, and the technique provides local structural and electronic information that is essentially inaccessible by any other method.

STM of thin organic layers involves the tunnelling of current between the tip and the conducting substrate, underneath the organic layer. By choosing the tunnelling parameters appropriately [32] ($\approx$0.3–1 V, 0.05–1 nA for adsorbate, 0.1–0.3 V, 0.3–10 nA for substrate), the method can be used to image either the substrate or the

**Figure B1.19.6.** Constant current 50 nm × 50 nm image of a Cu(111) surface held at 4 K. Three monatomic steps and numerous point defects are visible. Spatial oscillations (electronic standing waves) with a periodicity of ~1.5 nm are evident. (Taken from [26], figure 1.)

adsorbate—or both simultaneously, if a suitable voltage programme is used—repeating each line scanned at both voltages. There is some evidence that the tip can damage the organic layer during the imaging process [33]. The precise mechanism by which insulating molecules are imaged remains a topic of much discussion. Although single organic molecules have been successfully imaged by STM [34], the majority of STM studies of organic species has concerned a single monolayer of molecules deposited by evaporation, or by self-assembly, or by Langmuir–Blodgett techniques [35]. Often these images corroborate what had already been deduced from painstaking LEED investigations: an example is the imaging of co-adsorbed arrays of benzene and mobile CO, as seen by Ohtani *et al* [36].

One class of large molecules that was investigated relatively early was liquid crystals [37, 38], and in particular the group 4-n-alkyl-4′-cyanbiphenyl (*m*CB). These molecules form a highly crystalline surface adlayer, and STM images clearly show the characteristic shape of the molecule (figure B1.19.8).

The self-assembly of alkanethiols on gold has been an important topic in surface chemistry over the last few years [39] and STM has contributed significantly to our understanding of these systems. In particular, the formation of etch pits on the surface of Au(111) following treatment with alkanethiols is a phenomenon that was first observed by STM [40]. The segregation of thiols of different molecular weight or functionality is proving to be a relevant issue in their application. Stranick *et al* [41] have used STM to show the segregation of thiols with only very slight molecular differences into domains of size 10–100 Å and their subsequent coalescence.

The STM study of biological macromolecules has also been an area of great activity, and the imaging of DNA has been one of the challenges of the STM technique [42] (figure B1.19.9). The elimination of artifacts has been a major issue in this story, and the work of Beebe *et al* [43] showing that 'DNA-like' structures were to be seen on the surface of clean graphite (HOPG) substrates was something of a milestone (figure B1.19.10).

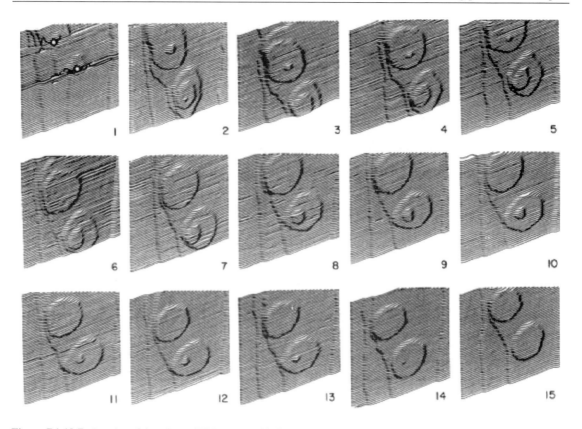

**Figure B1.19.7.** A series of time-lapse STM topographic images at room temperature showing a 40 nm × 40 nm area of Au(111). The time per frame is 8 min, and each took about 5 min to scan. The steps shown are one atomic unit in height. The second frame shows craters left after tip–sample contact, which are two and three atoms deep. During a 2 h period the small craters have filled completely with diffusing atoms, while the large craters continue to fill. (Taken from [29], figure 1.)

Other biomolecules imaged have included all DNA bases [44], polysaccharides [45] and proteins [46, 47]. In many cases there is strong evidence that the imaging process is facilitated by the presence of ultrathin (conducting) water films on the surface of the sample [48–50].

Lastly, STM has also been applied to the molecular-level imaging of polymer structures. In some cases these materials were deposited by Langmuir–Blodgett techniques [51], and in some cases by *in situ* polymerization [52]. Fujiwara *et al* [51] have used molecular dynamics simulations to interpret the images obtained from STM experiments. The combined use of these two techniques is proving to be a very powerful tool for understanding the conformation of polymer films on surfaces. They showed that the individual polyimide strands observed were aligned parallel to the deposition direction of the Langmuir–Blodgett film.

*(d) Electrochemistry*

The molecular-level observation of electrochemical processes is another unique application of STM [53, 54]. There are a number of experimental difficulties involved in performing electrochemistry with a STM tip and substrate, although many of these have been essentially overcome in the last few years.

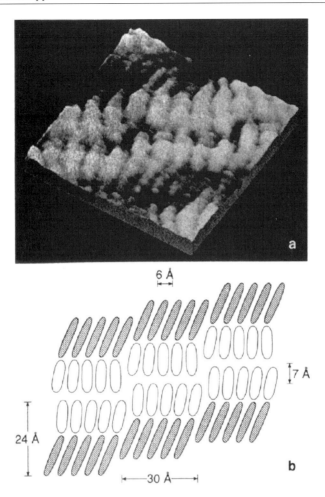

**Figure B1.19.8.** (a) STM image (5.7 nm × 5.7 nm) of 10-alkylcyanobiphenyl on graphite; (b) model showing the packing of the molecules. The shaded and unshaded segments represent the alkyl tails and the cyanobiphenyl head groups, respectively. (Taken from [38], figure 2.)

If the scanning tip is to be involved in electrochemical reactions, it is important to remember that at micrometre separations (i.e. when the tip is too far from the substrate for tunnelling to occur), the faradaic current is given by the equation [54]: $I_f \approx 4nFD_OC_Or$, where $D_O$ is the diffusion coefficient of a particular species, $F$ is Faraday's constant, $C_O$ is the concentration of the species in solution, $r$ is the radius of a disc of area equal to the effective exposed area of the tip and $n$ is the number of electrons involved in the reaction. The total tip current, $I$, when the separation is small enough for tunnelling to occur, is given by $I = I_f + I_t$, where $I_t$ is the tunnelling current, which is virtually independent of the total tip area exposed. In order to minimize $I_f$, so as to be able to perform meaningful STM experiments, the exposed tip must be made as small as possible, and a plethora of techniques has been developed [53] for insulating all but the very end of the tip.

Several designs for STM electrochemical cells have appeared in the literature [55]. In addition to an airtight liquid cell and the tip insulation mentioned above, other desirable features include the incorporation of a reference electrode (e.g. Ag/AgCl in saturated KCl) and a bipotentiostat arrangement, which allows the independent control of the two working electrodes (i.e. tip and substrate) [56] (figure B1.19.11).

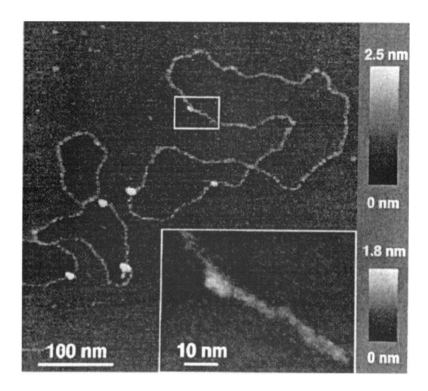

**Figure B1.19.9.** Plasmid DNA (pUC18) on mica imaged by STM at high resolution. The inset is a cut-out of a zoomed-in image taken immediately after the overview. (Taken from [42], figure 2.)

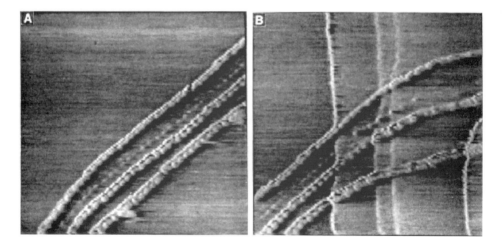

**Figure B1.19.10.** These images illustrate graphite (HOPG) features that closely resemble biological molecules. The surface features not only appear to possess periodicity (A), but also seem to meander across the HOPG steps (B). The average periodicity was $5.3 \pm 1.2$ nm. Both images measure 150 nm $\times$ 150 nm. (Taken from [43], figure 4.)

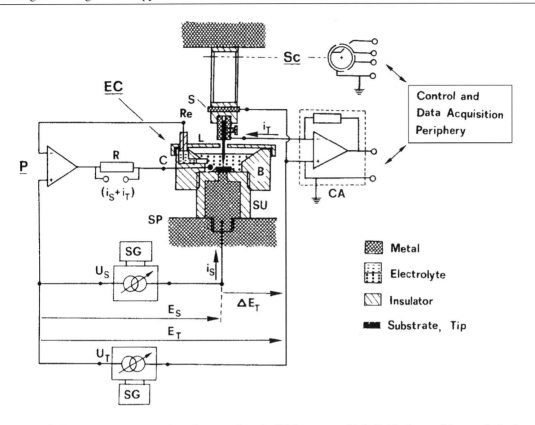

**Figure B1.19.11.** Schematic presentation of a potentiostatic STM system, with individual potential control of substrate and tip. Piezoelectric single-tube scanner Sc with titanium spacer plate S; electrochemical cell EC consisting of plexiglass beaker B with Pt counterelectrode C and Ag/AgCl reference electrode Re in 0.1 M NaCl; plexiglass lid L; PTFE support unit SU with epoxy-sealed substrate, mounted on support plate SP. Low-noise potentiostat P with low-impedance voltage units $U_S$ and $U_T$, both equipped with low-pass filter and signal generator SG; precision resistor R for measuring $(i_S + i_T)$; low-noise current amplifier CA for measuring $i_T$. (Taken from [56], figure 1.)

Examining electrodes and how they change under conditions of electrochemical reaction has been a major part of the electrochemical STM work performed until now. Many studies have revealed changes in surface reconstructions on silver and gold electrodes during electrochemical reactions [57], as well as increasing or decreasing surface roughness, depending on the conditions and electrolyte employed. Another field of activity has been the monitoring of metal deposition on electrodes [58], which is, of course, of tremendous practical importance. Since STM can image both periodic and non-periodic structures, it is of great utility, both in determining the geometric relationships between deposited metal and substrate, as well as in assessing the role of steps and defects in the deposition process [57].

Corrosion is another economically significant process that can be investigated on a molecular level, thanks to electrochemical STM. In addition to a number of academically interesting studies of systems such as the selective dissolution of copper from Cu–Au alloys [59], STM has also been used to investigate the properties of iron and steel under a variety of conditions designed to induce either passivation, corrosion, or electrochemical anodization [60, 61]. In the case of corrosion, STM has been used to monitor the growth of magnetite crystallites on the surface of the sample as it is taken through several successive cyclic voltammograms [61].

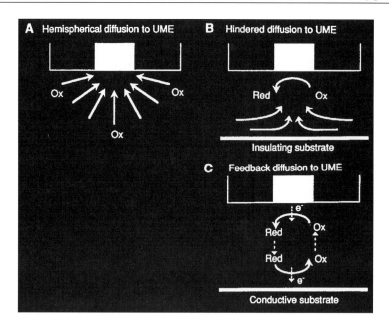

**Figure B1.19.12.** Basic principles of SECM. (a) With ultramicroelectrode (UME) far from substrate, diffusion leads to a steady-state current, $i_T, \infty$. (b) UME near an insulating substrate. Hindered diffusion leads to $i_T < i_T, \infty$. (c) UME near a conductive substrate. Positive feedback leads to $i_T > i_T, \infty$. (Taken from [62], figure 2.)

The technique of scanning electrochemical microscopy (SECM) [62] uses the same apparatus as in electrochemical STM, but instead of measuring tunnelling currents, the reaction $O + ne \rightarrow R$ (where O and R are oxidized and reduced species, respectively) is followed, by measuring the Faradaic current, $I_f$, at distances further from the substrate than those at which tunnelling will readily occur. The current, $I_f$, at distances far from the substrate surface, corresponds to the hemispherical diffusion of O to the tip surface (figure B1.19.12). As the tip nears the surface, this current is perturbed, either by hindered diffusion (lower current) or by reoxidation of R on the surface (higher current). The conductivity, potential, and electrochemical activity will therefore all influence $I_f$, which can thus be used to produce an electrochemical image of the surface—if plotted as a function of $x$ and $y$—as the tip is rastered over the surface. The technique has been used to image metals, polymers, biological materials and semiconductors.

*(e) Catalysis*

It has long been the goal of many catalytic scientists to be able to study catalysts on a molecular level under reaction conditions. Since the vast majority of catalytic reactions take place at elevated temperatures, the use of STM for such *in situ* catalyst investigations was predicated upon the development of a suitable STM reaction cell with a heating stage. This has now been done [3] by McIntyre *et al*, whose cell-equipped STM can image at temperatures up to 150 °C and in pressures ranging from ultrahigh vacuum up to several atmospheres. The set-up has been used for a number of interesting studies. In one mode of operation [63] (figure B1.19.13(a)), a Pt–Rh tip was first used to image clusters of carbonaceous species formed on a clean Pt(111) surface by heating a propylene adlayer to 550 K, and later to catalyze the rehydrogenation of the species (in a propylene/hydrogen atmosphere) at room temperature. The catalytic activity of the tip was induced by applying a voltage pulse, which presumably cleaned the surface of deactivating debris (figure B1.19.13(b)).

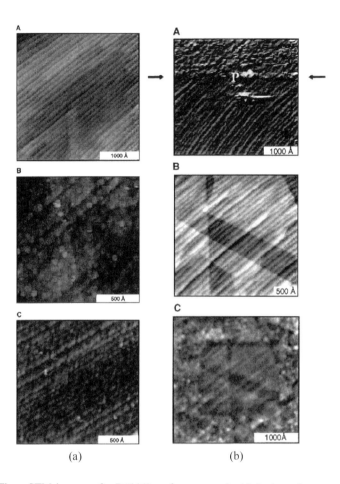

(a)                              (b)

**Figure B1.19.13.** (a) Three STM images of a Pt(111) surface covered with hydrocarbon species generated by exposure to propene. Images taken in constant-height mode. (A) after adsorption at room temperature. The propylidyne ($\equiv$C–CH$_2$–CH$_3$) species that formed was too mobile on the surface to be visible. The surface looks similar to that of the clean surface. Terraces ($\sim$10 nm wide) and monatomic steps are the only visible features. (B) After heating the adsorbed propylidyne to 550 K, clusters form by polymerization of the C$_x$H$_y$ fragments. The clusters are of approximately round shape with a diameter equal to the terrace width. They form rows covering the entire image in the direction of the step edges. (C) Rows of clusters formed after heating to 700 K. At this higher temperature, the carbonaceous clusters are more compact and slightly smaller in size, as they evolve to the graphitic form when H is lost completely. (b) The catalytic action of the STM Pt–Rh tip on a surface covered by carbonaceous clusters, as in figure B1.19.13(a). (B) Imaging was performed in 1 bar of a propene (10%) and hydrogen (90%) mixture at room temperature. (A) Carbon clusters were imaged in the top third of the image while the tip was inactive. A voltage pulse of 0.9 V was applied to the position marked P, leaving a mound of material 1.5 nm high. This process produced a chemically active Pt–Rh tip, which catalyzed the removal of all clusters in the remaining two-thirds of the image. Only the lines corresponding to the steps are visible. This image was illuminated from a near-incident angle to enhance the transition region where the tip was switched to its active state. (B) While the tip was in this catalytically active state, another area was imaged, and all of the clusters were again removed. (C) A slightly larger image of the area shown in (B) (centre square of this image), obtained after the tip was deactivated, presumably by contamination. The active-tip lifetime was of the order of minutes. (Taken from [63], figures 1 and 2.)

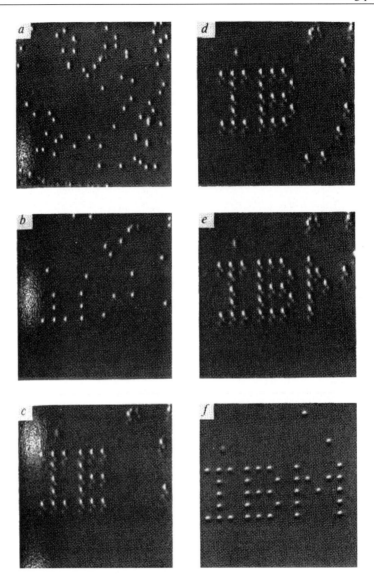

**Figure B1.19.14.** A sequence of STM images taken during the construction of a patterned array of xenon atoms on a Ni(100) surface. Grey scale is assigned according to the slope of the surface. The atomic structure of the nickel surface is not resolved. Each letter is 5 nm from top to bottom. (Taken from [65], figure 1.)

*(f) STM as a surface modification method*

Within a few years of the development of STM as an imaging tool, it became clear that the instrument could also find application in the manipulation of individual or groups of atoms on a surface [64]. Perhaps the most dramatic image originated from Eigler and Schweizer [65], who manipulated single physisorbed atoms of xenon on a Ni(110) surface, held at liquid helium temperature (figure B1.19.14). The tip–Xe distance was reduced (by raising the setpoint for the tunnelling current) until the tip–sample interaction became strong enough for the tip to be able to pick up the atom. After being moved to the desired location, the atom was

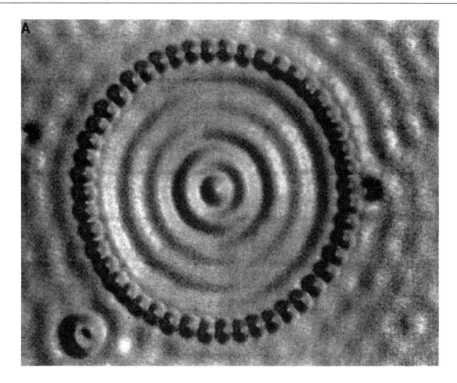

**Figure B1.19.15.** Spatial image of the eigenstates of a quantum corral. 48-atom Fe ring constructed on a Cu(111) surface. Average diameter of ring is 14.3 nm. The ring encloses a defect-free region of the surface. (Taken from [66], figure 2.)

removed by reversing the procedure. Using a similar experimental set-up, Crommie *et al* [66] have managed to shape the spatial distribution of electrons on an atomic scale, by building a ring of 48 iron adatoms (a 'quantum corral') on a Cu(111) surface, which confines the surface-state electrons of the copper by virtue of the scattering effect of the Fe atoms (figure B1.19.15). STS measurements of the local densities of states for the confined electrons correspond to the expected values for a 'particle-in-a-box', where the box is round and two-dimensional. In a similar way, Yokoyama *et al* [67] formed a pair of long straight chains of Al on the Si(001)-$c(4 \times 2)$ surface to create well defined 1D quantum wells. The electrons in the $\Pi^*$ surface states can propagate only in the dimer-row direction of Si(001)-$c(4 \times 2)$ because of nearly flat dispersion in the perpendicular direction. The STM/STS measurements of the standing-wave patterns and their discrete energy levels could be interpreted according to the '1D particle-in-a-box model'. This technique shows considerable promise for the further investigation of confined electrons and waveguides. There are numerous other means for moving atoms in surfaces, including voltage-pulsing techniques, which show promise as potential lithographic methods for silicon [68].

Finally, a technique that combines chemical vapour deposition (CVD) with STM has been devised by Kent *et al* [69]. The CVD gas used was iron pentacarbonyl, which is known to decompose under electron bombardment. Decomposition between tip and sample was found to occur at bias voltages above 5 V, forming iron clusters as small as 10 nm in diameter on the Si(111) substrate. Of particular practical interest is that arrays of 20 nm diameter dots have been shown to be magnetic, presenting a whole new range of possibilities for high-density data storage, as well as providing a convenient laboratory for nanometre-scale experiments in quantum magnetism.

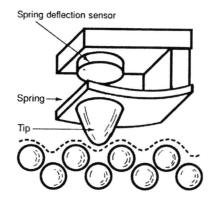

**Figure B1.19.16.** Schematic view of the force sensor for an AFM. The essential features are a tip, shown as a rounded cone, a spring, and some device to measure the deflection of the spring. (Taken from [74], figure 6.)

## B1.19.3   Force microscopy

*B1.19.3.1   Principles*

*(a) Background*

A major limitation of the scanning tunnelling microscope is its inability to analyse insulators, unless they are present as ultrathin films on conducting substrates. Soon after the development of the STM, work started on the development of an equivalent nanoscale microscope based on force instead of current as its imaging parameter [70]. Such an instrument would be equally adept at analysing both conducting and insulating samples. Moreover, the instrument already existed on a micro- and macro-scale as the stylus profilometer [71]; this is typically used to measure surface roughness in one dimension, although it had been extended into a three-dimensional imaging technique, with moderate resolution (0.1 $\mu$m lateral and 1 nm vertical), by Teague *et al* [72].

   The concept that Binnig and co-workers [73] developed, which they named the atomic force microscope (AFM, also known as the scanning force microscope, SFM), involved mounting a stylus on the end of a cantilever with a spring constant, $k$, which was lower than that of typical spring constants between atoms. This sample surface was then rastered below the tip, using a piezo system similar to that developed for the STM, and the position of the tip monitored [74]. The sample position ($z$-axis) was altered in an analogous way to STM, so as to maintain a constant displacement of the tip, and the $z$-piezo signal was displayed as a function of $x$ and $y$ coordinates (figure B1.19.16). The result is a force map, or image of the sample's surface [75], since displacements in the tip can be related to force by Hooke's Law, $F = -kz$, where $z$ is the cantilever displacement. In AFM, the displacement of the cantilever by the sample is very simply considered to be the result of long-range van der Waals forces and Born repulsion between tip and sample. However, in most practical implementations, meniscus forces and contaminants often dominate the interaction with interaction lengths frequently exceeding those predicted [76]. In addition, an entire family of force microscopies has been developed, where magnetic, electrostatic, and other forces have been measured using essentially the same instrument.

*(b) AFM instrumentation*

The first AFM used a diamond stylus, or 'tip' attached to a gold-foil cantilever, and much thought was given to the choice of an appropriate $k$-value [73]. While on the one hand a soft spring was necessary (in order

**Figure B1.19.17.** Commercially produced, microfabricated, V-shaped $Si_3N_4$ cantilever and tip for AFM (Taken from [215].)

to obtain the maximum deflection for a given force), it was desirable to have a spring with a high resonant frequency (10–100 kHz) in order to avoid sensitivity to ambient noise. The resonant frequency, $f_O$, is given by the equation

$$f_O = (1/2\pi)\sqrt{k/m_O}$$

where $m_O$ is the effective mass loading the spring. Thus, as $k$ is reduced to soften the spring, the mass of the cantilever must be reduced in order to keep the $k/m_O$ ratio as large as possible. Nowadays, cantilevers and integrated tips are routinely microfabricated out of silicon or silicon nitride. Typical dimensions of a cantilever are of the order of $1 \times 10 \times 100$ $\mu m^3$ [77], the exact dimensions depending on the intended use. Cantilevers designed to operate in contact with the surface, in a similar way to a surface profilometer, have low spring constants (usually less than 1 N m$^{-1}$) and correspondingly low resonant frequencies. Such levers are often fabricated in a V-shape configuration (figure B1.19.17), which makes for a greater stability towards lateral motion. If the tip is to come into hard contact with the surface, high-aspect-ratio tips are often desirable, with, typically, a radius of curvature of 10–30 nm. It is interesting to note that since experiments are generally carried out with contact forces on the order of nanonewtons, contact pressures in these experiments can be in the gigapascal range. In a stylus profilometer, the force exerted on the sample is some five orders of magnitude greater, but it is exerted over a larger contact area, leading to pressures in the tens of megapascals.

When the lever is intended for use with the tip separated from the surface, the lever stiffness is usually greater than 10 N m$^{-1}$ with a high resonant frequency. In this case, more care is taken to prepare tips with small radii of curvature—sometimes as low as 2 nm. However, in reality, most experiments are performed with tips of unknown radii or surface composition, apart from rare cases where the AFM has been combined with field ion microscopy [78] or a molecule or nanotube of known dimensions and composition has been attached to the tip [79]. It is likely that in most AFMs, microasperities and contaminants mediate the contact.

Detection of cantilever displacement is another important issue in force microscope design. The first AFM instrument used an STM to monitor the movement of the cantilever—an extremely sensitive method. STM detection suffers from the disadvantage, however, that tip or cantilever contamination can affect the instrument's sensitivity, and that the topography of the cantilever may be incorporated into the data. The most common methods in use today are optical, and are based either on the deflection of a laser beam [80], which has been bounced off the rear of the cantilever onto a position-sensitive detector (figure B1.19.18), or on an interferometric principle [81].

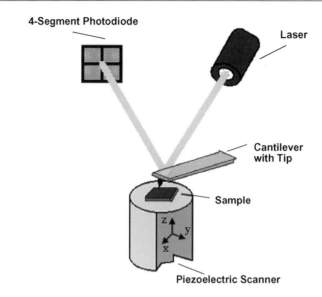

**Figure B1.19.18.** Schematic of an atomic force microscope showing the optical lever principle.

Lateral resolution in AFM is usually better than 10 nm and, by utilizing dynamic measurement techniques in ultrahigh vacuum, true atomic resolution can be obtained [82]. In hard contact with the surface, the atomic-scale structure may still appear to be present, but atomic-scale defects will no longer be visible, suggesting that the image is actually averaged over several unit cells. The precise way in which this happens is still the subject of debate, although the ease with which atomic periodicity can be observed with layered materials is probably due to the Moiré effect suggested by Pethica [83]. In this case a periodic image is formed by the sliding of planes directly under the tip caused by the lateral tip motion as the force varies in registry with unit lattice shear.

A further issue that should be considered when interpreting AFM images is that they are convolutions of the tip shape with the surface (figure B1.19.19). This effect becomes critical with samples containing 'hidden' morphology (or 'dead zones') on the one hand (such as deep holes into which the tip does not fit, or the underside of spherical features), or structure that is comparable in size to that of the tip on the other. While the hidden morphology cannot be regenerated, there have been several attempts to deconvolute tip shape and dimensions from AFM images ('morphological restoration') [84]. Some of these methods involve determining tip parameters by imaging a known sample, such as monodisperse nanospheres [85] or faceted surfaces [86]. Another approach is to analyse the AFM image as a whole, extracting a 'worst case' tip shape from common morphological features that appear in the image [87].

As with STM, the AFM can be operated in air, in vacuum or under liquids, providing a suitable cell is provided. Liquid cells (figure B1.19.20) are particularly useful for the examination of biological samples.

Analogously to STM, the image obtained in a force microscopy experiment is conventionally displayed on the computer screen as grey scales or false colour, with the lightest shades corresponding to peaks (or highest forces) and darkest shades corresponding to valleys (or lowest forces).

*(c) Forces in AFM*

Although imaging with force microscopy is usually achieved by means of rastering the sample in close proximity to the tip, much can be learned by switching off the *x*- and *y*-scanning piezos and following the

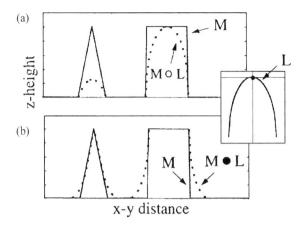

**Figure B1.19.19.** Examples of inaccessible features in AFM imaging. *L* corresponds to the AFM tip. The dotted curves show the image that is recorded in the case of (a) depressions on the underside of an object and (b) mounds on the top surface of an object. *M · L* and *M ∘ L* correspond to convolutions of the surface features with the tip shape. (Taken from [85], figure 2.)

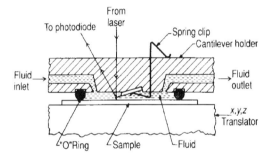

**Figure B1.19.20.** Cross section of an AFM fluid cell. (Taken from [216], figure 1.)

deflection of the cantilever as function of sample displacement, from large separations down to contact with the surface, and then back out to large separations. The deflection–displacement diagram is commonly known as a 'force curve' and this technique as 'force spectroscopy', although strictly speaking it is displacement that is actually measured, and further processing [88], reliant on an accurate spring constant, is necessary in order to convert the data into a true force–distance curve.

In order to obtain sensitive force curves, it is preferable to make dynamic force-gradient measurements with the force and energy being found by integration [89] or alternatively to measure frequency shift of the cantilever resonant frequency as a function of displacement. This method has become particularly popular in combination with non-contact mode AFM (see below). Although with this method it is not straightforward to relate frequency shifts to forces, it appears to be a promising technique for distinguishing between materials on the nanometre scale [90].

As the tip is brought towards the surface, there are several forces acting on it. Firstly, there is the spring force due to the cantilever, $F_S$, which is given by $F_S = -kz$. Secondly, there are the sample forces, which, in the case of AFM, may comprise any number of interactions including (generally attractive) van der Waals forces, chemical bonding interactions, meniscus forces or Born ('hard-sphere') repulsion forces. The total force gradient as the tip approaches the sample is the convolution of spring and sample force gradients, or

$$\frac{\partial F}{\partial D} = \frac{k(\partial^2 U/\partial D^2)}{k + \partial^2 U/\partial D^2}$$

where $U$ is the sample potential and $D$ the tip–sample separation. If the spring constant of the cantilever is comparable to the gradient of the tip–surface interaction, then at some point where $\partial^2 U/\partial D^2$ (negative for attraction) equals $k$, the total force gradient becomes instantaneously infinite, and the tip jumps towards the sample [91] (figure B1.19.21). This 'jump to contact' is analogous to the jump observed when two attracting magnets are brought together. The kinetic energy involved is often sufficient to damage the tip and sample, thus reducing the maximum possible resolution during subsequent imaging. Once a jump to contact occurs, the tip and sample move together (neglecting sample deformation for the time being) until the direction of sample travel is reversed. The behaviour is almost always hysteretic, in that the tip remains in contact with the sample due to adhesion forces, springing back to the equilibrium position when these have been exceeded by the spring force of the cantilever. The adhesion forces add to the total force exerted on the sample, and are often caused by tip contamination. It has been found that pretreating the tip in ozone and UV light, in order to remove organic contamination, reduces the adhesion observed, and improves image quality [92]. Image quality can also be enhanced by tailoring the imaging medium (i.e. in a liquid cell) to have a dielectric constant intermediate between those of the tip and the sample. This leads to a small, repulsive van der Waals force, which eliminates the jump to contact, and has been shown to improve resolution in a number of cases [93, 94], probably due to the fact that the tip is not damaged during the approach. It should be noted that the jump to contact may also be eliminated if a stiff cantilever is chosen, such that $k$, the force constant of the cantilever, is greater than $\partial^2 U/\partial D^2$ at all separations. This condition for jump-to-contact is insufficient if the stiffness of the cantilever is artificially enhanced using feedback [95] or if dynamic measurements are made [96, 97].

Since the AFM is commonly used under ambient conditions, it must be borne in mind that the sample is likely to be covered with multilayers of condensed water. Consequently, as the tip approaches the surface, a meniscus forms between tip and surface, introducing an additional attractive capillary force. Depending on the tip radius, the magnitude of this force can be equal to or greater than that of the van der Waals forces and is observed clearly in the approach curve [98]. In fact, this effect has been exploited for the characterization of thin liquid lubricant films on surfaces [95]. The capillary forces may be eliminated by operation in ultrahigh vacuum, provided both tip and sample are baked, or, most simply, by carrying out the experiment under a contamination-free liquid environment, using a liquid cell [99].

*(d) Non-contact AFM*

Non-contact AFM (NC-AFM) imaging is now a well established true-atomic-resolution technique which can image a range of metals, semiconductors and insulators. Recently, progress has also been made towards high-resolution imaging of other materials such as $C_{60}$, DNA and polypropylene. A good overview of recent progress is the proceedings from the First International Conference on NC-AFM [100].

Most NC-AFMs use a frequency modulation (FM) technique where the cantilever is mounted on a piezo and serves as the resonant element in an oscillator circuit [101, 102]. The frequency of the oscillator output is instantaneously modulated by variations in the force gradient acting between the cantilever tip and the sample. This technique typically employs oscillation amplitudes in excess of 20 nm peak to peak. Associated with this technique, two different imaging methods are currently in use: namely, fixed excitation and fixed amplitude. In the former, the excitation amplitude to the lever (via the piezo) is kept constant, thus, if the lever experiences a damping close to the surface the actual oscillation amplitude falls. The latter involves compensating the excitation amplitude to keep the oscillation amplitude of the lever constant. This mode also readily provides a measure of the dissipation during the measurement [100].

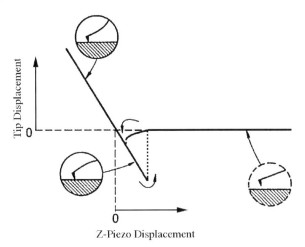

**Figure B1.19.21.** A plot of cantilever displacement as a function of tip sample separation during approach and retraction with an AFM. Note the adhesive forces upon retraction from the surface.

Although both methods have produced true-atomic-resolution images it has been very problematic to extract quantitative information regarding the tip–surface interaction as the tip is expected to move through the whole interaction potential during a small fraction of each oscillation cycle. For the same reason, it has been difficult to conclusively identify the imaging mechanism or the minimum tip–sample spacing at the turning point of the oscillation.

Many groups are now trying to fit frequency shift curves in order to understand the imaging mechanism, calculate the minimum tip–sample separation and obtain some chemical sensitivity (quantitative information on the tip–sample interaction). The most common methods appear to be perturbation theory for considering the lever dynamics [103], and quantum mechanical simulations to characterize the tip–surface interactions [104]. Results indicate that the interaction curve measured as a function of frequency shift does not correspond directly to the force gradient as first believed.

### (e) Intermittent contact AFM

A further variation is intermittent contact mode or 'TappingMode' [105] (TappingMode® is a trademark of Digital Instruments, Santa Barbara, CA.), where the tip is oscillated with large amplitudes (20–100 nm) near its resonant frequency, and the amplitude used to control the feedback loop. The system is adjusted so that the tip contacts the sample once within each vibrational cycle. Since the force on the sample in this mode is both small (<5 nN) and essentially normal to the surface, it is far less destructive than contact AFM, with its inherently large shear forces. This is of great importance when imaging biological materials; a further development of the intermittent-mode AFM, which allows it to be operated under *liquids* [106], extends the possibilities in this area even further.

### (f) Magnetic force microscopy

Magnetic forces may be exploited for the imaging of samples containing magnetic structure. Resolutions as high as 10 nm have been reported [107]. The central modification of the AFM needed to perform magnetic force microscopy (MFM) is the use of a magnetic tip, which often consists of an electrochemically etched ferromagnetic material, or a non-magnetic tip that has been coated with a magnetic thin film [108]. The experiment is run in non-contact mode, with the tip some 10 nm away from the surface. Detection at long

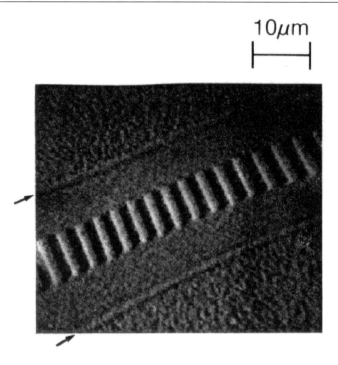

**Figure B1.19.22.** Magnetic force microscopy image of an 8 $\mu$m wide track on a magnetic disk. The bit transitions are spaced every 2 $\mu$m along the track. Arrows point to the edges of the DC-erased region. (Taken from [109], figure 7.)

range helps distinguish between magnetic and non-magnetic interactions. Greater sensitivity is obtained when the cantilever is oscillated and magnetic force gradients detected by changes in the resonant frequency as the tip approaches the magnetic surface. The method is unique in its ability to image magnetic structure in surfaces (figure B1.19.22) [109], which lies at the heart of magnetic data storage technology.

*(g) Lateral force microscopy*

Lateral force microscopy (LFM) has provided a new tool for the investigation of tribological (friction and wear) phenomena on a nanometre scale [110]. Alternatively known as friction force microscopy (FFM), this variant of AFM focuses on the lateral forces experienced by the tip as it traverses the sample surface, which correspond to the local coefficients of dynamic friction. LFM can therefore provide a frictional map of the surface with sub-nanometre resolution. It therefore has the potential to reveal chemical differences between regions of similar morphology, virtually down to the atomic scale.

The LFM method is an inherently contact-mode technique, and can be performed with an AFM, provided that there is some means of measuring the lateral tip displacement. Mate *et al* [111] were the first to modify their AFM in order to detect lateral forces and to observe frictional behaviour on the atomic scale. Their detection system was interferometric, and their cantilever and tip consisted of a shaped tungsten wire. Later developments using the laser beam-deflection method [112, 113] with two sets of position-sensing detectors (figure B1.19.23), enabled both lateral and normal forces to be measured simultaneously. Clearly, these two sets of forces are not entirely independent, since a lateral force will be felt as the tip is scanned over a step, for example, irrespective of the frictional coefficient at the step. However, by measuring the lateral force as the tip is scanned in both directions along the same line (producing a 'friction loop', figure B1.19.24), and subtracting one trace from the other, the frictional information can be separated from the purely morphological.

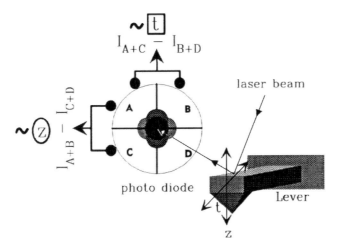

**Figure B1.19.23.** Principle of simultaneous measurement of normal and lateral (torsional) forces. The intensity difference of the upper and lower segments of the photodiode is proportional to the $z$-bending of the cantilever. The intensity difference between the right and left segments is proportional to the torsion, $t$, of the force sensor. (Taken from [110], figure 2.)

*(h) Mechanical imaging with force microscopy*

In AFM, the relative approach of sample and tip is normally stopped after 'contact' is reached. However, the instrument may also be used as a nanoindenter, measuring the penetration depth of the tip as it is pressed into the surface of the material under test. Information such as the elastic modulus at a given point on the surface may be obtained in this way [114], although producing enough points to synthesize an elastic modulus image is very time consuming.

Pulsed-force mode AFM (PFM-AFM) is a method introduced for fast mapping of local stiffness and adhesion with lower required data storage than recording force–distance curves at each point on the $x$–$y$ plane [115]. A sinusoidal or triangular modulation is applied between the tip and sample (either *via* lever or sample piezo) at a lower frequency than that of either the piezo or cantilever resonance frequency. Tip and sample then come into contact for part of each oscillation cycle. The deflection signal of the cantilever is put into sample-and-hold circuits. The peak displacement of the lever during the approach cycle is usually chosen for the feedback signal to the piezo, in order to maintain a constant force during scanning. Other sampling points can be chosen at any arbitrary timings within one cycle depending on the required property. For example, the local stiffness can be calculated from the slope obtained from subtracting the sample-and-hold signals at two different points on the linear part of the deflection–displacement curve. The adhesion force can be calculated from the difference between the largest negative deflection signal and the zero-deflection point.

Another method developed for imaging mechanical properties is ultrasonic force microscopy [116] (UFM). This technique involves carrying out contact AFM while oscillating the sample at high frequency: typically 200–700 kHz, chosen to be above the highest tip–sample resonance. If operating in the purely contact mode (i.e. if the tip does not leave the surface), the amplitude of the tip oscillations is determined by the elastic modulus of the tip–sample system, independent of the spring stiffness of the cantilever. The technique can be thought of as a fast-indentation system, and it samples a volume that has a radius some ten times the indentation depth. The overall spatial resolution is typically a few nanometres.

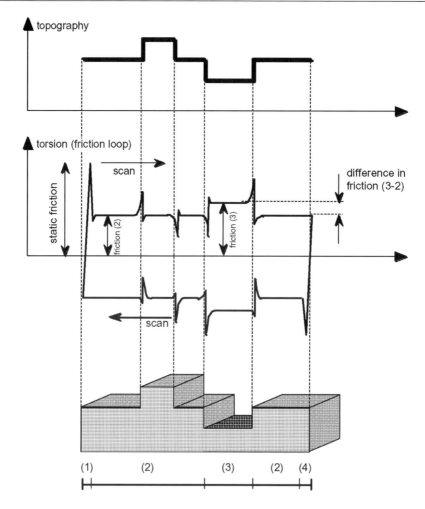

**Figure B1.19.24.** Friction loop and topography on a heterogeneous stepped surface. Terraces (2) and (3) are composed of different materials. In regions (1) and (4), the cantilever sticks to the sample surface because of static friction $F_{ST}$. The sliding friction is $t_1$ on part (2) and $t_3$ on part 3. In a torsional force image, the contrast difference is caused by the relative sliding friction, $\Delta F_{SL} = t_1 - t_3$. Morphological effects may be distinguished from frictional ones by their non-inverted behaviour upon scanning in the opposite direction. (Adapted from [110], figure 2.)

### B1.19.3.2 Applications of force microscopy

#### (a) Inorganic surfaces

True atomic resolution has been obtained on a wide range of inorganic surfaces including metals, semiconductors and insulators. Initially, imaging concentrated on Si (111) $7 \times 7$ as a means of demonstrating the true-atomic-resolution imaging capability of the technique [81]. Even with such a well understood surface, surprising results were obtained in the form of additional contrast revealed between different surface atoms. In the case of Erlandsson et al [117] their results showed that centre adatoms appeared to be 0.13 Å higher than the corner adatoms. They suggest that the additional contrast may be due to variation in chemical

**40 Å**

**Figure B1.19.25.** AFM image of Si(111)-(7 × 7) taken in the AC mode. Contrast can be observed between inequivalent adatoms. Image courtesy of R Erlandsson. (Taken from [217], figure 4.)

reactivity of the adatoms or to tip-induced, atomic relaxation effects reflecting the stiffness of the surface lattice (figure B1.19.25). Nakagiri *et al* [118] also saw additional contrast in their images of Si(111) 7 × 7. However, they observed the six atoms in one half of the unit cell to be brighter than in the other half. The two halves correspond to faulted and unfaulted halves of the unit cell according to the dimer–adatom stacking fault model [119]. At present they are not able to distinguish which atoms correspond to which half. The fact that this additional contrast varied depending on the precise experimental technique used indicates that different imaging mechanisms could be responsible as a result of the different tip material or height of the tip with respect to the surface.

Of particular interest are those surfaces where AFM has provided complementary information or revealed surface structure which could not be obtained by STM. One obvious application is the imaging of insulators such as NaCl(001) [120]. In this case it was possible to observe point defects and thermally activated atomic jump processes, although it was not possible to assign the observed maxima to anion or cation.

Another area where AFM has provided new information is the imaging of metal oxides such as $TiO_2$ [121, 122]. Although the surface of $TiO_2$(110) is observable with STM, only NC-AFM was able to image the bridging oxygen rows which are the outermost atoms on the surface (figure B1.19.26). True-atomic-resolution imaging is still a relatively recent development and the full power of the technique in imaging insulators such as $Al_2O_3$ or $SiO_2$ has yet to be demonstrated.

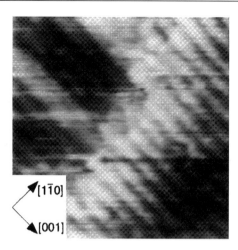

**Figure B1.19.26.** Highly resolved, non-contact AFM image of the $TiO_2(110)$-$(1 \times 1)$ surface ($8.5 \times 8.5$ nm$^2$) with a single step. The two-dimensional order of the bright spots ($0.65 \times 0.3$ nm$^2$) reproduces the alignment of the bridging oxygen atoms. (Taken from [121], figure 3.)

Even without atomic resolution, AFM has proved its worth as a technique for the local surface structural determination of a number of bio-inorganic materials, such as natural calcium carbonate in clam and sea-urchin shells [123], minerals such as mica [124] and molybdenite [125] as well as the surfaces of inorganic crystals, such as silver bromide [126] and sodium decatungstocerate [127]. This kind of information can prove invaluable in the understanding of phenomena such as biomineralization, the photographic process or catalysis, where the surface crystallography, especially the presence of defects and superstructures, can play an important role, but is difficult to determine by other methods. AFM has the considerable advantage that it can be used to examine powdered samples, either pressed into a pellet, if the contact mode is employed, or loosely dispersed on a surface, if intermittent or non-contact AFM is available.

AFM has also provided insights into the growth of metal clusters and films on mica. In the case of palladium [128], for example, it was found that clusters in the 50 nm range exhibited truncated triangular shapes. Epitaxial growth of silver on a mica surface [129] is seen to depend in a complex way on both substrate temperature and film thickness (figure B1.19.27), with island morphology giving way to channels, which become holes and then networks as the film thickness increases, the changes progressing more gradually as the substrate temperature is increased. The issue of surface roughness of silver films is central to the technique of surface-enhanced Raman spectroscopy, and AFM has been used to characterize films produced for this purpose [130]. AFM possesses a considerable advantage over electron microscopy in this kind of application, in that it can, if calibrated, automatically yield *quantitative* morphological and roughness data [131].

*(b) Organic surfaces*

AFM has been used to image several surfaces of organic crystals—such as tetracene [132] and pyrene [133]—and has produced images that can be compared to the unit cells expected from previous x-ray crystallographic studies. In the case of tetracene, the surface is merely a truncation of that expected from the bulk data. However, in the case of pyrene, where the bulk consists of dimer pairs, a surface reconstruction is evident in the AFM image, corresponding to the presence of monomer species. Reconstructions are common phenomena in metal surfaces, where LEED has frequently been used to detect them [25]. LEED has scarcely been used

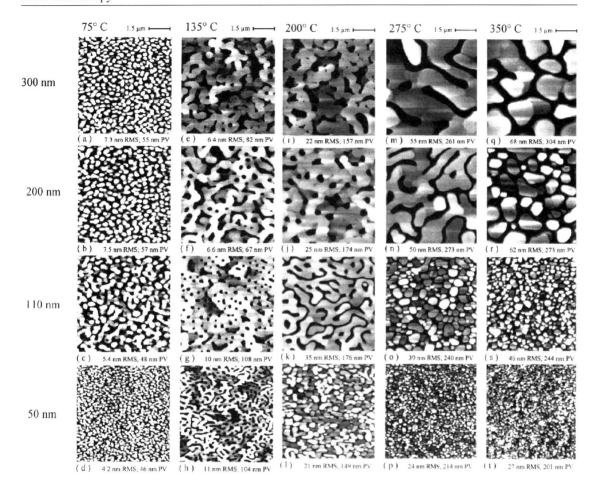

**Figure B1.19.27.** AFM topographic images ($7 \times 7 \ \mu m^2$) of 20 epitaxial Ag films on mica prepared at five substrate temperatures (75, 135, 200, 275, and 350 °C) and four film thicknesses (50, 110, 200, and 300 nm) using metal deposition rates of 0.1 to 0.2 nm s$^{-1}$. The vertical RMS and peak-to-valley roughness are indicated for each image. (Taken from [129], figure 1.)

to analyse organic crystal surfaces, however, due to problems associated with charging and/or degradation of the sample in an electron beam. AFM nicely fills this gap in the surface analytical arsenal.

The overwhelming majority of AFM studies on organic surfaces has concerned organic thin films on inorganic substrates and, in particular, those deposited *via* Langmuir–Blodgett or self-assembly processes [35]. These films have been an active research area for several years, frequently serving as models for complex systems, such as membranes. Thin organic films are also being developed for their nonlinear optical properties, as microlithographic resists and as sensor components [32].

Two Langmuir–Blodgett film systems that have been much studied by AFM are the calcium and barium arachidates, adsorbed as double layers on a silicon substrate, which has been pretreated so as to be hydrophobic [32]. These structures are formed by dipping the silicon into an arachidate film on a water trough and then removing the substrate again through the film. Repeating the process simply adds another double layer to the structure. As with organic crystals, more traditional surface-structure-determining approaches are often too destructive to allow analysis of systems such as these; possibly the most important and unique aspect

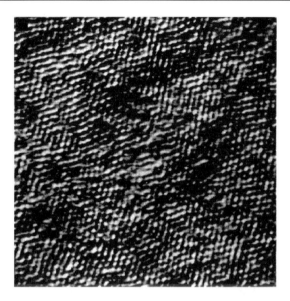

**Figure B1.19.28.** Molecular-scale image (2 nm × 20 nm) of a barium arachidate bilayer. Image was produced by averaging six images, but without filtering data. (Taken from [135], figure 1.)

of AFM in this context is that it is a *local probe*, and therefore capable of showing the enormous variety of defects in the films, both on a microscale (pores and islands in the films) and a nanoscale [134, 135] (twinning, defects and complexities in molecular packing) (figure B1.19.28). This topic has been dealt with at length by Frommer in her excellent review article [32].

Self-assembly of long-chain alkanethiols on the Au(111) surface has been studied by a number of different techniques including AFM, and it has been consistently shown that the molecules form a commensurate $(\sqrt{3} \times \sqrt{3})R30°$ structure. AFM studies can also provide additional information on the mechanical properties of the organic layers [136], which are interesting in that they serve as a model system for lubricants [137]. Above a critical applied load of 280 nN for the $C_{18}$ thiols, it has been found that the monolayers were disrupted, and that the subsequent image corresponded to that of the Au(111) substrate. However, on reducing the load to substantially below the critical value, the surface apparently healed, and the characteristic periodicity of the thiol overlayer returned. The exact way in which this phenomenon occurs is not completely understood [138]. Possibilities include displacement of the thiols by the tip, binding of the thiols to the tip, or desorption of the thiols into a liquid phase.

*(c) Polymer surfaces*

AFM is contributing significantly to our understanding of the surface structure of polymers, both on a microscale and on a molecular level. Segregation in the surface of block copolymers and polymer blends is often critical in determining technologically important properties, such as wettability or biocompatibility. It is also often difficult to image by optical or electron microscopy. AFM, on the other hand, offers a method for scrutinizing these materials down to the molecular level, without the need for surface preparation. AFM's ability to operate in a liquid environment makes it particularly useful in analysing polymers for medical applications, since these materials are designed to function surrounded by body fluids, which can influence the surface microstructure and nanostructure.

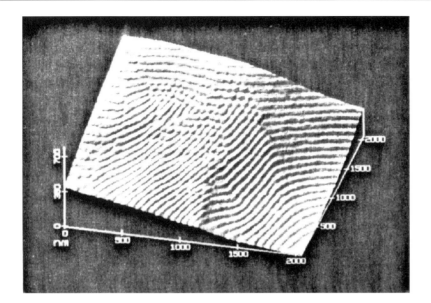

**Figure B1.19.29.** AFM image of polystyrene/polybutadiene copolymer, showing lamellar structure. (Taken from [140], figure 1.)

Several studies have concerned the microstructure of lamellae in materials such as the block copolymers polystyrene-*block*-poly-2-vinylpyridine [139] and polystyrene-*block*-polybutadiene [140], as well as single crystals of poly-para-xylylene [139], and reveal features (such as intersecting lamellae (figure B1.19.29)) that had not been previously observed.

LFM has also proved useful in the examination of polymer blends, since its ability to image effective frictional coefficients imparts a certain chemical sensitivity to the method [141]. A novel approach to discrimination between components of a polymer blend was adopted by Feldman *et al* [142], who used the low-refractive-index liquid, perfluorodecalin, as a medium for measurement of tip–polymer 'pull-off', thereby enhancing the London component of the Hamaker constant, improving the signal-to-noise ratio of the measurement. Using the same medium for lateral force (frictional) imaging, a striking contrast reversal was found between polystyrene and PMMA blend components, when the tip surface was changed from (hydrocarbon-covered) gold to silica (figure B1.19.30).

On the molecular level, spectacular AFM images have been obtained for a number of systems. In the case of isotactic polypropylene, for example, Snétivy and Vancso [143] have succeeded in imaging individual methyl groups on the polymer chain, and distinguishing between left- and right-handed helices in the crystalline i-polypropylene matrix (figure B1.19.31). The same group has also used AFM to image the phenylene groups in poly(*p*-phenyleneterephthalamide) fibres, and have used this data to show the existence of a new polymorphic form that had previously only been suggested by computer simulations.

*(d) Biological surfaces*

The AFM is now firmly established as a unique tool for the *in situ* investigation of biological surfaces [144], whether these be biomolecules, cell structures, or even viruses [145]. Often, a successful immobilization strategy has been key to successful imaging of the biological surface [146].

Lured by the promise of a new way to sequence the genetic code, and the prospect of nanomanipulation of nucleic acids, investigators have produced a plethora of papers in the area of AFM imaging of DNA. Notable

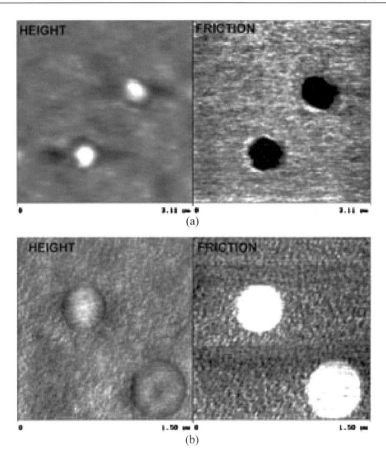

**Figure B1.19.30.** Height and friction images of a spin-cast polystyrene-poly(methyl methacrylate) blend obtained with (a) gold and (b) silica probes under perfluorodecalin. Note the reversal of frictional contrast and the high spatial resolution. (Taken from [142], figure 7.)

among these is the work of Bustamente *et al* [147], who developed one of the first reproducible methods for imaging nucleic acids. Their approach involved three important components: (1) the use of extremely sharp tips [148] (radius of curvature ≈10 nm), which are prepared by electron-beam deposition of a carbon whisker on the tip apex in an electron microscope [149], (2) the use of mica substrates that have been ion-exchanged with magnesium, in order to promote interaction with the phosphate groups on the DNA and (3) careful control of the relative humidity [150] (or operation under liquids), in order to prevent tip-induced sample movement. Using this approach, and imaging under 2-propanol, high-resolution images of plasmid DNA have been obtained [151] (figure B1.19.32), and the molecules dissected by momentarily increasing the AFM force [152]. Single- and double-stranded DNA [153] and RNA-polymerase-DNA complexes [154] have also been imaged using this approach, the helical pitch of the DNA deciphered [155], AFM used to determine local chirality of the DNA supercoiling [156] and even individual base pairs resolved [157].

Proteins have also been investigated extensively by AFM. Radmacher *et al* [158] monitored height changes in an adsorbed layer of lysozyme using intermittent-contact AFM, as the enzyme was exposed to a substrate molecule. The height changes were variously interpreted as due to conformational adaptations of the lysozyme, or to the different height of the enzyme–substrate complex. Many other proteins have

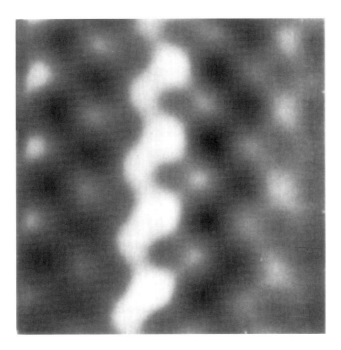

**Figure B1.19.31.** AFM image of a 2.7 nm × 2.7 nm area of a polypropylene surface, displaying methyl groups and right-and left-handed helices. (Taken from [143], figure 10.)

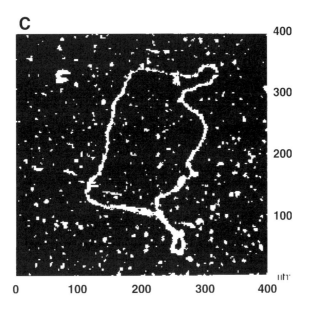

**Figure B1.19.32.** AFM image of Blue Script II plasmid (400 nm × 400 nm) in propanol, taken with 'super tip', prepared by carbon deposition on normal tip in SEM, followed by ion milling. (Taken from [152], figure 1.)

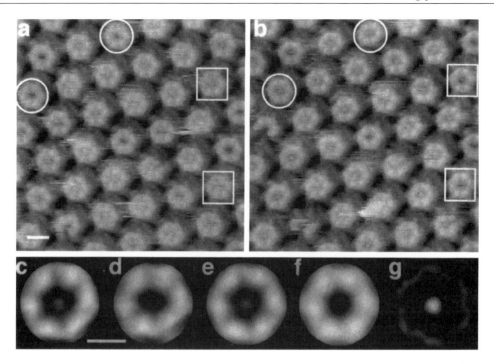

**Figure B1.19.33.** Conformational changes of the inner surface of an HPI layer. (a) The protruding cores are clearly visible, with some pores in an open conformation and others in an obstructed conformation. (b) Area shown in panel a imaged 5 min later. Some pores that were open earlier are now closed (circles), while closed ones have opened during this time interval (squares). Units were aligned and divided into two classes. The class averages exhibit a plugged (c) and an open hexamer (d). The difference map (g) represents the modulus of the height difference between the sixfold-symmetrized class averages ((e) and (f)). The full grey-level range corresponds to a vertical distance of 6 nm ((a) and (b)) and 3 nm ((c) to (f)). (Taken from [160], figure 3.)

been imaged, using various immobilization methods, and this area has been comprehensively reviewed in the literature [159]. Among the highest-resolution (<1 nm) examples have been those of Müller *et al* [160], who have studied the inner surface of the hexagonally packed intermediate (HPI) layers of cell envelope proteins, such as those in the bacterium *Deinococcus radiodurans*, where protein conformational changes can be observed (figure B1.19.33). AFM has also provided a new window into the channels present in the surfaces of living cells, and Lal *et al* [161] have imaged the channels formed when the responsible cell proteins (porins) are reconstituted as crystalline arrays. The resolution obtained was such that individual polar head groups of the lipid molecules could be discerned. Several mechanical studies on proteins using AFM have also been reported, notably the hysteretic unfolding of the giant muscle protein, titin, reported by Rief *et al* [162], where the importance of this protein as a 'strength reservoir' during muscle stretching was demonstrated for the first time.

In the area of cell and cell-structure imaging, AFM offers an advantage over optical and scanning electron microscopies in that it permits high-resolution imaging of living cells, and even the observation of dynamic phenomena. Henderson *et al* [163] observed the motion of filamentous actin in living glial cells (i.e., structures beneath the plasma membrane were imaged), and even performed nanosurgery on the cell with the AFM tip. Brandow *et al* [164] also cut into lipid membranes on a graphite surface using deliberately high force from the AFM tip, and found that the membranes healed themselves after sufficient time, but that the healing could be accelerated by rubbing the AFM tip perpendicular to the cut with a controlled force. The same group was also able to manipulate living glial cells [165] and even peel them away from a surface, if the normal force was appropriate. Thus, depending on the amount of force applied, the AFM could be used to cut, anneal, peel or image the sample.

Several groups have focused on the biochemical receptor–ligand interaction, measuring forces between individual molecular binding pairs, such as biotin and streptavidin [166, 167]. One approach involves coating the tip with either receptor or ligand, the sample with the complementary molecule, and then carefully monitoring tip–sample separation after contact. If the cantilever stiffness is appropriately chosen, the force–distance curve during separation appears to be 'quantized' in units corresponding to a single molecular-pair interaction. Lee *et al* [168] have also demonstrated and measured the interaction between single complementary strands of DNA base pairs using AFM. Several other examples of ligand–receptor interactions have been summarized in a review by Bongrand [169].

## (e) Colloids

Due to its sensitivity to small forces and its ability to operate in liquids, the AFM has opened up a new avenue of investigation into colloidal systems. A frequently used approach, developed by Ducker *et al* [170], involves the cementing of a colloidal-sized particle onto an AFM cantilever in place of the usual tip, and then monitoring interactions with some appropriate flat surface, or even another colloidal particle [171], under a variety of conditions. Larson *et al* [172] used this technique to investigate the interactions between a titania (rutile) particle ($\approx 9$ $\mu$m diameter) and a rutile single-crystal surface under various conditions of pH and ionic strength. From their experiments they were able to measure the van der Waals interaction between rutile surfaces directly, and to calculate the non-retarded Hamaker constant for the system. Biggs *et al* have applied a similar approach to the venerable subject of gold colloid stability [173], by immobilizing a $\sim 6$ $\mu$m gold sphere on a cantilever and measuring its interaction with a polished gold plate under solutions containing combinations of gold, citrate, chloride and a number of other ions of relevance to the colloid system. The authors were able to demonstrate the presence of a repulsive interaction between the gold surfaces due to adsorbed citrate or chloride; they showed that citrate adsorbed preferentially, and succeeded in measuring the surface potential of the gold as a function of anion concentrations.

## (f) Tribology

Force microscopy, and lateral (or frictional) force microscopy in particular, are having a tremendous impact in tribology—the science and technology of friction, wear, and lubrication. The interaction between moving surfaces (the central issue in tribology) is thought to consist of separate interactions between the many peaks in one surface with the many peaks in the countersurface. These peaks are known as asperities. It is this mode of interaction that leads to Amontons' empirical 'law' of friction (this law is generally attributed to Amontons, although initially observed by Leonardo da Vinci a century earlier), $F = \mu N$, where $F$ is the frictional force, $N$ is the normal force, and $\mu$ is the coefficient of friction. Notable by its absence in this equation is the *apparent* contact area between the two sliding bodies. In fact, as one might intuitively believe, the *actual* contact area, $A$, between the asperities is all-important, and the frictional force is proportional to $A$. However, $A$, in turn, increases in proportion to the normal force [174], (both due to 'flattening' of the asperities by the load and the creation of new load-bearing asperities, as the higher asperities are flattened) so that $A$ can be cancelled out of the resulting equation, leaving behind only the measurable quantities, $F$ and $N$. The importance of AFM in fundamental tribology research is that the AFM measurement can be thought of as a single asperity contact, i.e. the fundamental interaction in frictional behaviour.

Carpick *et al* [84] used AFM, with a Pt-coated tip on a mica substrate in ultrahigh vacuum, to show that if the deformation of the substrate and the tip–substrate adhesion are taken into account (the so-called

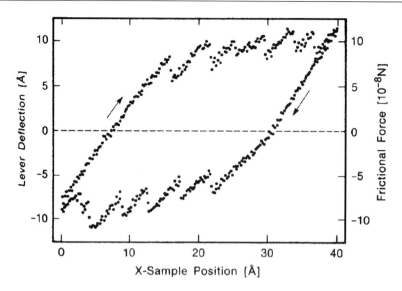

**Figure B1.19.34.** Cantilever deflection and corresponding frictional force in the $x$-direction as a function of sample position as a mica sample is scanned back and forth under a tungsten tip. (Taken from [124], figure 2.)

JKR model [175] of elastic adhesive contact), then the frictional force is indeed proportional to the contact area between tip and sample. However, under these single-asperity conditions, Amontons' law does not hold, since the 'statistical' effect of more asperities coming into play no longer occurs, and the contact area is not simply proportional to the applied load.

Mate *et al* [111], who pioneered LFM, used their instrument to show atomic-scale structure in the frictional force between a tungsten tip and surface graphite atoms. In fact, what these authors observed was stick–slip behaviour: an effect normally associated with macroscopic phenomena, such as the vibration induced in a violin string by the bow, or the squealing of an automobile's brakes. They also found that the frictional coefficient varied slightly with applied load, i.e. a deviation from Amontons' law. The same group went on to study mica surfaces [124], where a similar stick–slip behaviour was observed, the frictional coefficient varying with the unit cell periodicity of the mica cleavage plane (figure B1.19.34). Hu *et al* [176] examined mica surfaces by LFM using silicon nitride tips, and found that above normal loads of 10 nN, Amontons' law was obeyed, while at low loads, deviations from linearity were observed. They also showed that friction decreased substantially as a function of humidity, decreasing by an order of magnitude under liquid water, but was essentially invariant with scanning direction across the mica surface. Wear was also observed by these authors, who found that a threshold load value needed to be exceeded before layer-by-layer wear was initiated. Below this value, even multiple scans were not found to produce visible damage to the surface. A similar effect was observed for the system AgBr on NaCl(001) (SiO$_2$-coated tip) by Lüthi *et al* [177], where a wear onset was observed at around 14 nN. Interestingly, frictional coefficients were measured over a range of loads in this latter work and the $\mu$ measured on NaCl was found to be an order of magnitude lower than that measured on AgBr.

Numerous groups have applied LFM to address the issue of friction on thin organic films. These systems serve as useful models for the important macroscopic tribological issue of boundary lubrication. Overney *et al* [178] used LFM to examine mixed Langmuir–Blodgett films of perhydro arachidic acid and partially fluorinated carboxylic acid that had been transferred onto a silicon surface as a bilayer system, using poly(4-vinyl-N-methylpyridinium) as a countercation. The images clearly showed a difference in frictional coefficient between the two components, which segregated into submicron domains. Surprising, however, was the observation that the apparent frictional coefficient on the fluorinated component was a factor of four higher than that measured on the non-fluorinated one. The authors attributed this to the greater shear strength of fluorinated films, although more recent measurements using LB-deposited straight-chain acids of different lengths [179] suggest that the length of the molecule itself has a significant effect on mechanical properties, and therefore on the frictional coefficients measured.

Kim *et al* [180], using specially synthesized, end-functionalized alkanethiols, investigated mechanisms of friction by producing gold-supported monolayers containing varying quantities of various bulky endgroups. They found that the differences in friction were apparently due to differences in the size of the terminal groups, larger terminal groups (whether F-containing or not) giving rise to increasing interactions that provided pathways for energy dissipation, and therefore higher frictional losses.

The issues of the correlation of adhesion and of viscoelastic relaxation with friction are currently being investigated using AFM and LFM. Although friction does not correlate with the adhesion energy between two surfaces, there is increasing evidence that it is proportional to adhesion *hysteresis* [181]: i.e., the energy dissipated during the making and breaking of an adhesive bond. Marti *et al* [189] have shown that the adhesion hysteresis measured during force-curve measurements ($Si_3N_4$ tip) on silica and alumina surfaces under electrolytes is proportional to the LFM friction force measurements made under the same conditions. Haugstad *et al* [182] characterized fundamental aspects of friction on polymer surfaces by measuring frictional force on a gelatin film as a function of scanning frequency, i.e. velocity. The friction was found to increase at lower velocities, which can be explained by the onset of rubbery-state-type molecular relaxations, the viscoelastic dissipation correlating with frictional dissipation. Following intensive scanning at a particular site on the gelatin surface, a high-friction 'signature' could be observed in subsequent LFM measurements, and was explained by the perturbative effect of the previous frictional measurement. In this case, the LFM images were providing a map of molecular relaxation across the film.

### (g) Measurement of mechanical properties

The technological importance of thin films in such areas as semiconductor devices and sensors has led to a demand for mechanical property information for these systems. Measuring the elastic modulus for thin films is much harder than the corresponding measurement for bulk samples, since the results obtained by traditional indentation methods are strongly perturbed by the properties of the substrate material. Additionally, the behaviour of the film under conditions of low load, which is necessary for the measurement of thin-film properties, is strongly influenced by surface forces [75]. Since the force microscope is both sensitive to surface forces and has extremely high depth resolution, it shows considerable promise as a technique for the mechanical characterization of thin films.

A striking example of the use of the AFM as an indenter is provided by Burnham and Colton [184] (figure B1.19.35), where the differences (at <100 nm indentation) between the plastic behaviour of gold at 20 $\mu$N load and the elastic behaviour of graphite and elastomer at lower loads are clearly observable. The importance of surface forces in measuring mechanical properties at low load has been demonstrated by Salmeron *et al* [184], who showed that tip–sample adhesion can seriously perturb hardness measurements, not only for clean surfaces, but also for those covered by a layer of contamination. These authors also suggested that initial passivation of the surface (e.g., by sulfidation) might be an effective approach to overcoming these artefacts.

### (h) Chemical imaging

The imaging of surfaces according to chemical species is immensely useful in understanding the mechanisms of many complex technological systems on a fundamental level. Over the past two decades, Auger electron spectroscopy, x-ray photoelectron spectroscopy and secondary ion mass spectroscopy have been extended into

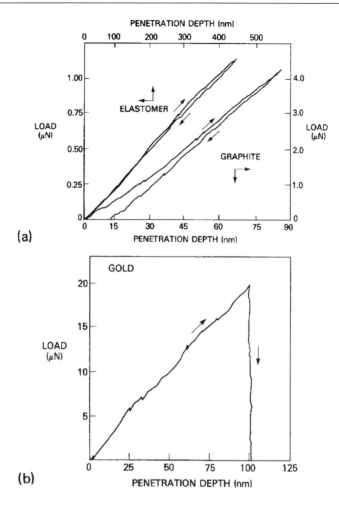

**Figure B1.19.35.** Experimental nanoindentation curves obtained with the AFM showing the loading and unloading behaviour of (a) an elastomer and highly oriented pyrolytic graphite and (b) a gold foil. (Taken from [183], figure 4.)

surface-sensitive chemical imaging methods [185]; nowadays the use of such techniques for troubleshooting has become virtually routine in the semiconductor industry, and has contributed significantly to our knowledge of catalyst systems, corrosion mechanisms and many other areas. However, these methods are not universally applicable since they are limited in spatial resolution (especially for insulating samples) and require the sample to be analysed in a vacuum. A chemically sensitive scanning force microscopy that could image the distribution of chemical species on an insulating surface with nanometre resolution under ambient conditions, or even under liquids, is therefore a highly desirable goal, and many research groups are active in this area.

A promising approach, still in the early stages of development, involves the functionalization of the AFM tip. In many cases this has been limited to the demonstration of the sensitivity of force curves to the chemical species present on the surface—a necessary first step for the development of an imaging methodology. Akari *et al* [186] have functionalized a gold-coated tip with COOH-terminated long-chain hydrocarbons, using a self-assembled monolayer procedure; they have shown that the contrast is considerably enhanced over that of the uncoated tip, when imaging a patterned mixed monolayer consisting of $CH_3$- and COOH-terminated molecules of similar length.

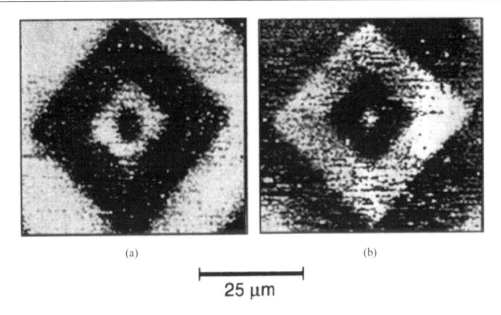

(a) (b)

25 μm

**Figure B1.19.36.** Image of the frictional force distribution of a pattern consisting of areas of CH$_3$-terminated and areas of COOH-terminated molecules attached to gold-coated silicon. The tip was also functionalized in (a) with CH$_3$ species and in (b) with COOH species. The bright regions correspond to the higher friction force, which in (a) is observed on the CH$_3$ areas and in (b) on the COOH areas. (Taken from [187], figure 3.)

LFM coupled with tip functionalization is a potentially important chemical imaging technique, since the tip functionality can be tailored so as to produce maximum contrast (i.e. maximum difference in frictional coefficient) between different chemical species on the surface. Frisbie *et al* [187] have examined a patterned surface of COOH and CH$_3$ groups, and have shown that the pattern could be readily imaged by a COOH- or CH$_3$-coated tip (figure B1.19.36) running in LFM mode, since the imaged frictional coefficients depended on the particular tip–surface species interaction. A potential pitfall with this technique is that both local chemistry and local mechanical properties, due to differences in molecular packing, contribute to the imaging contrast. This problem was elegantly sidestepped by McKendry *et al* who investigated chiral discrimination by chemical force microscopy [188]. In this case the interaction between the tip and surface can be changed without alteration of the mechanical properties or wetting energies of tip or surface. The chemical force microscope proved to be sufficiently sensitive to permit discrimination between enantiomers of simple chiral molecules in a friction image with more quantitative differences being obtained from adhesion force or 'pull-off' force histograms.

One potentially powerful approach to chemical imaging of oxides is to capitalize on the tip–surface interactions caused by the surface charge induced under electrolyte solutions [189]. The sign and the amount of the charge induced on, for example, an oxide surface under an aqueous solution is determined by the pH and ionic strength of the solution, as well as by the isoelectric point (IEP) of the sample. At pH values above the IEP, the charge is negative; below this value, the charge is positive. The same argument applies to a Si$_3$N$_4$ tip (normally an oxynitride at the surface), so that at every pH, either an attractive or a repulsive electrostatic tip–sample force will be superimposed upon the forces that are normally encountered in AFM (figure B1.19.37). By varying the pH and determining the value at which the charges switch sign, the IEP of the sample may be determined. Since this value is characteristic for a particular oxide, and the sample area probed is on the nanometre-squared scale, this appears to be a promising direction for the high-resolution chemical imaging of mixed oxide, oxide-covered alloy [190] or even protein [191] surfaces.

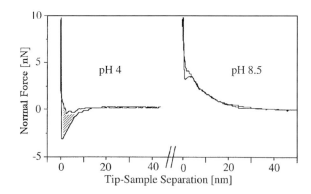

**Figure B1.19.37.** Normal force versus tip–sample distance curves for a $Si_3N_4$ tip on a $SiO_2$ surface under 1 mM NaCl solution at pH 4 and pH 9. (Taken from [189], figure 2.)

Finally, Berger *et al* [192] have developed a technique whereby an array of force curves is obtained over the sample surface ('force-curve mapping'), enabling a map of the tip–sample adhesion to be obtained. The authors have used this approach to image differently oriented phase domains of Langmuir–Blodgett-deposited lipid films.

### B1.19.4   Scanning near-field optical microscopy and other SPMs

Since the invention of the scanning tunnelling microscope, many other related techniques have been developed that combine the principles of piezo positioning with a feedback system, but rely on a surface interaction other than electron tunnelling or force sensing in order to produce images. The overwhelming majority of these scanning techniques are still in the very early stages of development and, in contrast to STM and AFM, have as yet revealed little that could not be better determined by other methods. Nevertheless, this is an area with tremendous promise, and a selection of these methods is therefore described below.

#### B1.19.4.1   *Scanning near-field optical microscopy*

Of the methods described in this section, scanning near-field optical microscopy (SNOM or NSOM) is the closest to being able to provide useful information that is unobtainable by other means. Indeed, this technique has already been made available as a commercial instrument. A detailed review of SNOM has been written by Pohl [193].

While the spatial resolution in classical microscopy is limited to approximately $\lambda/2$, where $\lambda$ is the optical wavelength (the so-called Abbé Limit [194], $\sim$0.2 $\mu$m with visible light), SNOM breaks through this barrier by monitoring the *evanescent* waves (of high spatial frequency) which arise following interaction with an object, rather than the *propagating* waves (of low spatial frequency), which are observed under far-field conditions. While the field intensity of propagating waves decays with the well known inverse-square dependence on distance from the source, evanescent waves decay much more rapidly, with an inverse-fourth power relationship. This means that the evanescent waves are almost completely damped out within a few nanometres of the source, and play no part in far-field (i.e. classical) optical measurements. Synge [195] was the first to discuss the possibility of exceeding the Abbé limit, as long ago as 1928, and suggested that by scanning a nanometre-sized aperture over the surface of the sample, a resolution higher than $\lambda/2$ could be obtained. This is analogous to the use of a stethoscope by a physician, where spatial information can be obtained with a resolution far greater than the acoustical wavelength [196]. This forms the basis of SNOM. In 1972, Ash *et al* [197] were able to show that this principle could indeed be demonstrated for microwave wavelengths (3 cm), but SNOM with visible light was not developed until the 1980s, when the technologies surrounding the STM became available.

The design of the SNOM in the first experiments consisted of a minute aperture, formed by a metallized glass fibre tip, which was rastered across a sample that was illuminated by a laser beam from behind (figure B1.19.38). The tip was maintained at a constant distance from the sample by using the tip–sample tunnelling current in a feedback loop. The resolution obtained was 25 nm ($\lambda/20$) [196, 198]. Subsequent variations to the experimental set-up have included the use of force interactions to maintain tip–sample separation [199] (enabling the imaging of insulators), as well as operation in transmission mode [200]. Recent applications have included the detection and spectroscopy of single molecules [201], where spectral differences corresponding to the different chemical environments of individual molecules can be discerned, and the orientation of each molecular dipole determined. In general, the combination of high lateral resolution in the optical near-field and spectroscopic information has been restricted to fluorescence and luminescence experiments. Recently, vibrational spectroscopy in the optical near-field has also been attempted [202–205] with surface-enhanced Raman scattering in particular, appearing to be a promising method for obtaining spectral, spatial and chemical information of molecular adsorbates with subwavelength lateral resolution. This has been implemented in illumination mode, where the incident light emerges from the fibre probe, to obtain surface-enhanced Raman spectra from single Ag colloid nanoparticles [206]. Raman chemical imaging on a scale of 100 nm has also been demonstrated by Deckert *et al* on dye-labelled DNA. On the nanometre scale a strong dependence of the enhanced Raman signal on substrate morphology is expected. It is therefore particularly useful to correlate near-field Raman spectra with topographic information on a nanoscale, as in the experiment of Zeisel *et al* [207] who investigate the near-field surface-enhanced Raman spectroscopy of dye molecules adsorbed on silver island films.

### B1.19.4.2   Scanning near-field acoustic microscopy (SNAM)

This corresponds to the physician's stethoscope case mentioned above, and has been realized [208] by bringing one leg of a resonating 33 kHz quartz tuning fork close to the surface of a sample, which is being rastered in the $x$–$y$ plane. As the fork-leg nears the sample, the fork's resonant frequency and therefore its amplitude is changed by interaction with the surface. Since the behaviour of the system appears to be dependent on the gas pressure, it may be assumed that the coupling is due to hydrodynamic interactions within the fork–air–sample gap. Since the fork tip–sample distance is approximately 200 $\mu$m ($\sim\lambda/20$), the technique is sensitive to the near-field component of the scattered acoustic signal. 1 $\mu$m lateral and 10 nm vertical resolutions have been obtained by the SNAM.

### B1.19.4.3   Scanning thermal profiler (STP)

This technique involves the scanning of a heated thermocouple tip above the surface of the sample (figure B1.19.39). Since the heat loss from the tip is highly dependent on the tip–sample spacing, the temperature of the tip can be used as a control parameter to monitor sample morphology and/or thermal properties [209]. The lateral resolution obtained with this method is of the order of 0.1 $\mu$m, with a vertical resolution of about 3 nm. The temperature sensitivity of the tip is $\sim$0.1 mK. An advantage of the technique is that morphology can be imaged at distances approximately equal to the desired lateral resolution.

An extension of this technique, known as scanning thermal microscopy [210] (SThM) combines the thermal profiling technique with modulated-temperature differential scanning calorimetry, by applying a sinusoidal modulation to the tip temperature. Using this technique, the spatial distribution of thermal properties of materials (such as conductivity and diffusivity) can be monitored, and subsurface features imaged (since the penetration depth of the evanescent temperature wave is frequency-dependent). Additionally, local calorimetric analysis can be used to probe thermally activated near-surface processes, such as glass transitions,

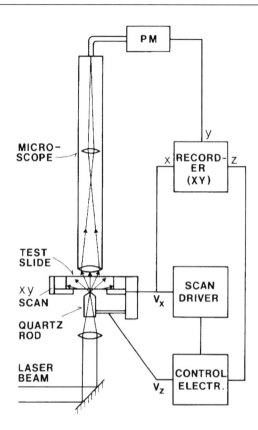

**Figure B1.19.38.** Schematic of a scanning near-field optical microscope (SNOM). (Taken from [196], figure 2.)

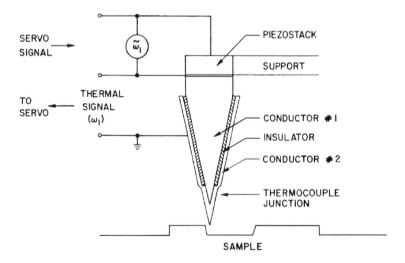

**Figure B1.19.39.** Schematic of the thermocouple probe in a scanning thermal profiler. The probe is supported on a piezoelectric element for modulation of tip–sample distance at frequency $\omega_1$ and to provide positioning. The AC thermal signal at $\omega_1$ is detected, rectified, and sent to the feedback loop, which supplies a voltage to the piezostack to maintain the average tip–sample spacing constant. (Taken from [209], figure 1.)

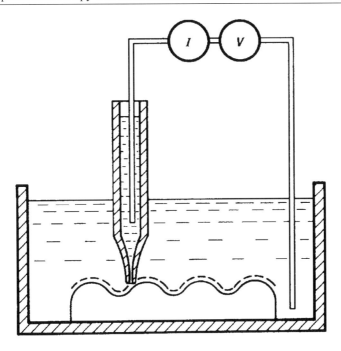

**Figure B1.19.40.** The scanning ion-conductance microscope (SICM) scans a micropipette over the contours of a surface, keeping the electrical conductance through the tip of the micropipette constant by adjusting the vertical height of the probe. (Taken from [211], figure 1.)

melting, crystallization or cure reactions. The technique has been used to great effect with polymer blends, where imaging contrast is caused by differences in the thermal properties of the individual components.

### B1.19.4.4 Scanning ion conductance microscopy (SICM)

This method relies on the simple principle that the flow of ions into an electrolyte-filled micropipette as it nears a surface is dependent on the distance between the sample and the mouth of the pipette [211] (figure B1.19.40). The probe height can then be used to maintain a constant current flow (of ions) into the micropipette, and the technique functions as a non-contact imaging method. Alternatively, the height can be held constant and the measured ion current used to generate the image. This latter approach has, for example, been used to probe ion flows through channels in membranes. The lateral resolution obtainable by this method depends on the diameter of the micropipette. Values of 200 nm have been reported.

### B1.19.4.5 Scanning micropipette molecule microscopy (SMMM)

The apparatus involved in this method is related to that of SICM, except that the micropipette is blocked by a permeable polymer plug, and connected to the inlet of a differentially pumped quadrupole mass spectrometer [212] (figure B1.19.38). Unlike most other scanning techniques, SMMM relies on a light microscope for positioning. Nevertheless, it is a unique spatially resolved sampling method for desorbing surface species under solution, and has numerous potential applications in biology and medicine. The diffusion of water through pores in a polymer membrane has been followed by using the set-up in figure B1.19.41, where diffusing HDO is converted to HD (with the unambiguous mass of 3 amu) prior to mass spectrometric detection by means of a uranium reduction furnace.

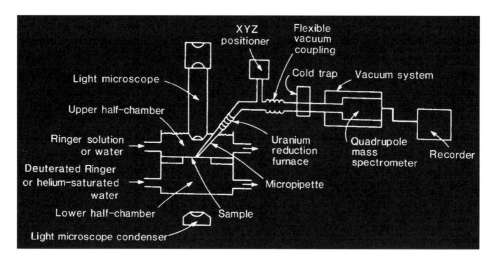

**Figure B1.19.41.** Schematic of the scanning micropipette molecule microscope. (Taken from [212], figure 1.)

## B1.19.5    Outlook

STM and its many related methods have opened new windows onto the nanometre-scale world. Within a short period of time, SPMs have become common fixtures, not only in surface science laboratories, but also in research groups working in areas as diverse as ceramics, polymers, cell biology, robotics, catalysis, and tribology. Clearly this trend will continue, as the concept of the proximal probe becomes combined with more and more of the macroscale analytical tools that we know today. It is also clear that the nanoscale chemical analysis of surfaces by means of SPM will become increasingly viable, with biological surfaces providing some of the most challenging and potentially fruitful analytical problems. Other suggested applications of SPMs are as high-density information storage systems, as selective molecular manipulators and as aids in microsurgery. With the field barely into its second decade, the technological and scientific possibilities of the SPM approach are immense.

## References

[1]   Binnig G, Rohrer H, Gerber Ch and Weibel E 1982 Tunnelling through a controllable vacuum gap *Appl. Phys. Lett.* **40** 178
[2]   Salmeron M B 1993 Use of the atomic force microscope to study mechanical properties of lubricant layers *MRS Bulletin* **XVIII-5** 20
       Overney R and Meyer E 1993 Tribological investigations using friction force microscopy *MRS Bulletin* **XVIII-5** 20
[3]   McIntyre B J, Salmeron M and Somorjai G A 1993 A variable pressure/temperature scanning tunnelling microscope for surface science and catalysis studies *Rev. Sci. Instrum.* **64** 687
[4]   Guckenberger R, Hartmann T, Wiegräbe W and Baumeister W 1995 The scanning tunnelling microscope in biology *Scanning Tunnelling Microscopy* vol II, ed R Wiesendanger and H-J Güntherodt (Berlin: Springer) ch 3
[5]   Wiesendanger R and Güntherodt H-J (eds) 1995 *Scanning Tunnelling Microscopy* (Berlin: Springer) vols I–III
[6]   Bonnell D A (ed) 1993 *Scanning Tunnelling Microscopy and Spectroscopy* (Weinheim: VCH)
[7]   Wiesendanger R 1994 *Scanning Probe Microscopy and Spectroscopy* (Cambridge: Cambridge University Press)
[8]   DiNardo N J 1994 *Nanoscale Characterization of Surfaces and Interfaces* (Weinheim: VCH)
[9]   Colton R J *et al* (eds) 1998 *Procedures in Scanning Probe Microscopies* (New York: Wiley)
[10]  Esaki L 1958 New phenomenon in narrow germanium p–n junction *Phys. Rev.* **109** 603
[11]  Josephson B D 1962 Possible new effects in superconductive tunnelling *Phys. Lett.* **1** 251
[12]  Rohrer G 1993 The preparation of tip and sample surfaces for STM experiments *Scanning Tunnelling Microscopy and Spectroscopy* ed D A Bonnell (Weinheim: VCH) ch 6
[13]  Wiesendanger R and Güntherodt H-J 1995 Introduction *Scanning Tunnelling Microscopy II* ed R Wiesendanger and H-J Güntherodt (Berlin: Springer) ch 1

[14] Bonnell D A 1993 Microscope design and operation *Scanning Tunnelling Microscopy and Spectroscopy* ed D A Bonnell (Weinheim: VCH) ch 2

[15] Tersoff J 1993 Theory of scanning tunnelling microscopy *Scanning Tunnelling Microscopy and Spectroscopy* ed D A Bonnell (Weinheim: VCH) ch 3

[16] Hamers R 1993 Methods of tunnelling spectroscopy with the STM *Scanning Tunnelling Microscopy and Spectroscopy* ed D A Bonnell (Weinheim: VCH) ch 4

[17] Binnig G, Rohrer H, Gerber Ch and Weibel E 1983 $7 \times 7$ reconstruction on Si(111) resolved in real space *Phys. Rev. Lett.* **50** 120

[18] Tromp R M, Hamers R J and Demuth J E 1986 Atomic and electronic contributions to Si(111)-($7 \times 7$) scanning-tunnelling-microscopy images *Phys. Rev.* B **34** 1388

[19] Hamers R J, Tromp R M and Demuth J E 1986 Surface electronic structure of Si(111)-($7 \times 7$) resolved in real space *Phys. Rev. Lett.* **56** 1972

[20] Hamers R J and Köhler U K 1989 Determination of the local electronic structure of atomic-sized defects on Si(001) by tunnelling spectroscopy *J. Vac. Sci. Technol.* A **7** 2854

[21] Hamers R, Avouris P and Boszo F 1987 Imaging of chemical-bond formation with the scanning tunnelling microscope: $NH_3$ dissociation on Si(001) *Phys. Rev. Lett.* **59** 2071

[22] Zheng Z F, Salmeron M B and Weber E R 1994 Empty state and filled state image of $Zn_{Ga}$ acceptor in GaAs studied by scanning tunnelling microscopy *Appl. Phys. Lett.* **64** 1836

[23] Kitamura N, Lagally M G and Webb M B 1993 Real-time observations of vacancy diffusion on Si(001)-($2 \times 1$) by scanning tunnelling microscopy *Phys. Rev. Lett.* **71** 2082

[24] Kuk Y 1994 STM on metals *Scanning Tunnelling Microscopy I* ed R Wiesendanger and H-J Güntherodt (Berlin: Springer) ch 3

[25] Somorjai G A 1994 *Introduction to Surface Chemistry and Catalysis* (New York: Wiley)

[26] Crommie M F, Lutz C P and Eigler D M 1993 Imaging standing waves in a two-dimensional electron gas *Nature* **363** 524

[27] Hasegawa Y and Avouris Ph 1993 Direct observation of standing wave formation at surface steps using scanning tunnelling spectroscopy *Phys. Rev. Lett.* **71** 1071

[28] Campuzano J C, Foster M S, Jennings G, Willis R F and Unertl W 1985 Au(110) ($1 \times 2$)-to-($1 \times 1$) phase transition: a physical realisation of the two-dimensional Ising model *Phys. Rev. Lett.* **54** 2684

[29] Jaklevic R C and Elie L 1988 Scanning-tunnelling-microscope observation of surface diffusion on an atomic scale: Au on Au(111) *Phys. Rev. Lett.* **60** 120

[30] Winterlin J and Behm R J 1994 Adsorbate covered metal surfaces and reactions on metal surfaces *Scanning Tunnelling Microscopy I* ed R Wiesendanger and H-J Güntherodt (Berlin: Springer) ch 4

[31] Hwang R Q, Schröder J, Günther C and Behm R J 1991 Fractal growth of two-dimensional islands: Au on Ru(0001) *Phys. Rev. Lett.* **67** 3279

[32] Frommer J E 1992 Rastertunnel- und Kraftmikroskopie in der Organischen Chemie *Angew. Chem.* **104** 1325

[33] Yackoboski K, Yeo Y H, McGonigal G C and Thomson D J 1992 Molecular position at the liquid/solid interface measured by voltage-dependent imaging with the STM *Ultramicroscopy* **42–44** 963

[34] Weiss P S and Eigler D M 1993 Site dependence of the apparent shape of a molecule in scanning tunnelling microscope images: benzene on Pt(111) *Phys. Rev. Lett.* **71** 3139

[35] Ulman A 1991 *Ultrathin Organic Films* (London: Academic)

[36] Ohtani H, Wilson R J, Chiang S and Mate C M 1988 Scanning tunnelling microscopy observations of benzene molecules on the Rh(111)-($3 \times 3$) ($C_6H_6$ + 2CO) surface *Phys. Rev. Lett.* **60** 2398

[37] Forster J S and Frommer J E 1988 Imaging of liquid crystals using a tunnelling microscope *Nature* **333** 542

[38] Smith D P E, Hörber H, Gerber Ch and Binnig G 1989 Smectic liquid crystal monolayers on graphite observed by scanning tunnelling microscopy *Science* **245** 43

[39] Nuzzo R G and Allara D L 1983 Adsorption of bifunctional organic disulfides on gold surfaces *J. Am. Chem. Soc.* **105** 4481

Nuzzo R G, Zegarski D R and Dubois L H 1987 Fundamental studies of the chemisorption of organosulfur compounds on Au(111). Implications for molecular self-assembly on gold surfaces *J. Am. Chem. Soc.* **109** 733

Bain C D, Troughton E B, Tao T, Evall J, Whitesides G M and Nuzzo R G 1989 Formation of monolayer films by the spontaneous assembly of organic thiols from solution onto gold *J. Am. Chem. Soc.* **111** 321

[40] Poirier G E and Tarlov M J 1994 The c($4 \times 2$) superlattice of n-alkanethiol monolayers self-assembled on Au(111) *Langmuir* **10** 2853

[41] Stranick S J, Parikh A N, Tao Y-T, Allara D L and Weiss P S 1994 Phase separation of mixed-composition self-assembled monolayers into nanometer scale molecular domains *J. Phys. Chem.* **98** 7636

[42] Guckenberger R, Heim M, Cevc G, Knapp H F, Wiegräbe W and Hillebrand A 1994 Scanning tunnelling microscopy of insulators and biological specimens based on lateral conductivity of ultrathin water films *Science* **266** 1538

[43] Clemmer C R and Beebe T P 1991 Graphite: a mimic for DNA and other biomolecules in scanning tunnelling microscope studies *Science* **25** 640

[44] Heckl W M, Smith D P E, Binnig G, Klagges H, Hänsch T W, and Maddocks J 1991 Two-dimensional ordering of the DNA base guanine observed by scanning tunnelling microscopy *Proc. Natl Acad. Sci., USA* **88** 8003

Heckl W M and Engel A 1995 *Visualisation of Nucleic Acids* ed G Morel (Boca Raton, FL: CRC Press)

Allen M J, Balooch M, Subbiah S, Tench R J, Balhorn R and Siekhaus W 1991 Scanning tunnelling microscope images of adenine and thymine at atomic resolution *Scanning Microsc.* **5** 625

Allen M J, Balooch M, Subbiah S, Tench R J, Balhorn R and Siekhaus W 1992 Analysis of adenine and thymine adsorbed on graphite by scanning tunnelling and atomic force microscopy *Ultramicrosc.* **42–44** 1049

[45] Rabe J P, Buchholz S and Ritcey A M 1990 Reactive graphite etch of an adsorbed organic monolayer—a scanning tunnelling microscopy study *J. Vac. Sci. Technol.* A **8** 679

Miles M J, Lee I and Atkins E D T 1991 Molecular resolution of polysaccharides by scanning tunnelling microscopy *J. Vac. Sci. Technol.* B **9** 1206

Tang S L, McGhie A J and Suna A 1993 Molecular-resolution imaging of insulating macromolecules with the scanning tunnelling microscope via a nontunnelling, electric-field-induced mechanism *Phys. Rev.* B **47** 3850

[46] Guckenberger R, Hartmann T and Knapp H F 1995 Recent developments *Scanning Tunnelling Microscopy II* ed R Wiesendanger and H-J Güntherodt (Berlin: Springer) ch 9

[47] Maaloum M, Chrétien D, Karsenti E and Hörber J K H 1994 Approaching microtubule structure with the scanning tunnelling microscope (STM) *J. Cell Sci.* **107** part II 3127

[48] Yuan J-Y, Shao Z and Gao C 1991 Alternative method of imaging surface topologies of nonconducting bulk specimens by scanning tunnelling microscopy *Phys. Rev. Lett.* **67** 863

[49] Guckenberger R, Heim M, Cevc G, Knapp H F, Wiegräbe W and Hillebrand A 1994 Scanning tunnelling microscopy of insulators and biological specimens based on lateral conductivity of ultrathin water films *Science* **266** 1538

[50] Fan F-R F and Bard A J 1995 STM on wet insulators: electrochemistry or tunnelling? *Science* **270** 1849

[51] Fujiwara I, Ishimoto C and Seto J 1991 Scanning tunnelling microscopy study of a polyimide Langmuir–Blodgett film *J. Vac. Sci. Technol.* B **9** 1148

Sotobayashi H, Schilling T and Tesche B 1990 Scanning tunnelling microscopy of polyimide monolayers prepared by the Langmuir–Blodgett technique *Langmuir* **6** 1246

[52] Grunze M, Unertl W N, Ganarajan S and French J 1988 Chemistry of adhesion at the polyimide–metal interface *Mater. Res. Soc. Symp. Proc.* **108** 189

[53] Siegenthaler H 1995 STM in electrochemistry *Scanning Tunnelling Microscopy II* ed R Wiesendanger and H-J Güntherodt (Berlin: Springer) ch 2

[54] Bard A J and Fan F F 1993 Applications in electrochemistry *Scanning Tunnelling Microscopy and Spectroscopy* ed D A Bonnell (Weinheim: VCH) ch 9

[55] Cataldi T R I, Blackham I G, Briggs G A D, Pethica J B and Hill H A O 1990 New insight for electrochemical electrode–surface investigations *J. Electroanal. Chem.* **290** 1

[56] Christoph R, Siegenthaler H, Rohrer H and Wiese H 1989 *In situ* scanning tunnelling microscopy at potential controlled Ag(100) substrates *Electrochim. Acta* **34** 1011

[57] Siegenthaler H 1995 Recent developments *Scanning Tunnelling Microscopy II* ed R Wiesendanger and H-J Güntherodt (Berlin: Springer) ch 9

[58] Magnussen O M, Hotlos J, Beitel G, Kolb D M and Behm R J 1991 Atomic structure of ordered copper adlayers on single-crystalline gold electrodes *J. Vac. Sci. Technol.* B **9** 969

[59] Chen S J, Sanz F, Ogletree D F, Hallmark V M, Devine T M and Salmeron M 1993 Selective dissolution of copper from Au-rich Cu–Au alloys: an electrochemical STM study *Surf. Sci.* **292** 289

[60] Ogura K, Tsujigo M, Sakurai K and Yano J 1993 Electrochemical coloration of stainless steel and the scanning tunnelling microscopic study *J. Electrochem. Soc.* **140** 1311

[61] Müller-Zülow B, Kipp S, Lacmann R and Schneeweiss M A 1994 Topological aspects of iron corrosion in alkaline solution by means of scanning force microscopy (SFM) *Surf. Sci.* **311** 153

[62] Bard A J, Fan F F, Pierce D T, Unwin P R, Wipf D O and Zhou F 1991 Chemical imaging of surfaces with the scanning electrochemical microscope *Science* **254** 68

[63] McIntyre B J, Salmeron M and Somorjai G A 1994 Nanocatalysis by the tip of a scanning tunnelling microscope operating inside a reactor cell *Science* **265** 1415

[64] Staufer U 1995 Surface modification with a scanning proximity probe microscope *Scanning Tunnelling Microscopy II* ed R Wiesendanger and H-J Güntherodt (Berlin: Springer) ch 8

[65] Eigler D M and Schweizer E K 1990 Positioning single atoms with a scanning tunnelling microscope *Nature* **344** 524

[66] Crommie M F, Lutz C P and Eigler D M 1993 Confinement of electrons to quantum corrals on a metal surface *Science* **262** 218

[67] Yokoyama T and Takayanagi K 1999 Size quantization of surface state electrons on the Si(001) surface *Phys. Rev.* B **59** 12 232

[68] Salling C T and Lagally M G 1994 Fabrication of atomic-scale structures on Si(001) surfaces *Science* **265** 502

[69] Kent A D, Shaw T M, Molnar S V and Awschalom D D 1993 Growth of high aspect ratio nanometer-scale magnets with chemical vapor deposition and scanning tunnelling microscopy *Science* **262** 1249

[70] Rugar D and Hansma P K 1990 Atomic force microscopy *Physics Today* **43**(10) 23

[71] Guenther K H, Wierer P G and Bennett J M 1984 Surface roughness measurements of low-scatter mirrors and roughness standards *Appl. Opt.* **23** 3820

[72] Teague E C, Scire F E, Backer S M and Jensen S W 1982 Three-dimensional stylus profilometry *Wear* **83** 1

[73] Binnig G, Quate C F and Gerber Ch 1986 Atomic force microscope *Phys. Rev. Lett.* **56** 930

[74] Hansma P K, Elings V B, Marti O and Bracker C E 1988 Scanning tunnelling microscopy and atomic force microscopy: application to biology and technology *Science* **242** 209

[75] Burnham N A and Colton R J 1993 Force microscopy *Scanning Tunnelling Microscopy and Spectroscopy* ed D A Bonnell (Weinheim: VCH) ch 7

[76] Hartmann U 1991 van der Waals interactions between sharp probes and flat sample surfaces *Phys. Rev.* B **43** 2404

[77] Wolter O, Bayer T and Greschner J 1991 Micromachined silicon sensors for scanning force microscopy *J. Vac. Sci. Technol.* B **9** 1353

[78] Cross G, Schirmeisen A, Stalder A, Grütter P, Tschudy M and Dürig U 1998 Adhesion interaction between atomically defined tip and sample *Phys. Rev. Lett.* **80** 4685

[79] Dai H, Hafner J H, Rinzler A G, Colbert D T and Smalley R E 1996 Nanotubes as nanoprobes in scanning probe microscopy *Nature* **384** 147

[80] Meyer G and Amer N M 1988 Novel optical approach to atomic force microscopy *Appl. Phys. Lett.* **53** 1045

[81] Sarid D, Pax P, Yi L, Howells S, Gallagher M, Chen T, Elings V and Bocek D 1992 Improved atomic force microscope using a laser diode interferometer *Rev. Sci. Instrum.* **63** 3905

Nonnenmacher M, Vaez-Iravani M and Wickramasinghe H K 1992 Attractive mode force microscopy using feedback-controlled fiber interferometer *Rev. Sci. Instrum.* **63** 5373

[82] Giessibl F J 1995 Atomic resolution of the silicon (111)-(7 × 7) surface by atomic force microscopy *Science* **260** 67

[83] Pethica J B 1986 Comment on interatomic forces in scanning tunnelling microscopy: giant corrugations of the graphite surface *Phys. Rev. Lett.* **57** 3235

[84] Carpick R W, Agraït N, Ogletree D F and Salmeron M 1996 Measurement of interfacial shear (friction) with an ultrahigh vacuum atomic force microscope *J. Vac. Sci. Technol.* B **14** 1289

[85] Wilson D L, Kump K S, Eppell S J and Marchant R E 1995 Morphological restoration of atomic force microscopy images *Langmuir* **11** 265

[86] Sheiko S S, Moller M, Reuvekamp E M C M and Zandbergen H W 1993 Calibration and evaluation of scanning-force-microscopy probes *Phys. Rev.* B **48** 5675

[87] Roberts C J, Williams P M, Davies J, Dawkes A C, Sefton J, Edwards J C, Haymes A G, Bestwick C, Davies M C and Tendler S J B 1995 Real-space differentiation of IgG and IgM antibodies deposited on microtiter wells by scanning force microscopy *Langmuir* **11** 1822

Shivji A P, Brown F, Davies M C, Jennings K H, Roberts C J, Tendler S J B, Wilkinson M J and Williams P M 1995 Scanning tunnelling microscopy studies of β-amyloid fibril structure and assembly *FEBS Lett.* **371** 25–8

[88] Ducker W A, Senden T J and Pashley R M 1992 Measurement of forces in liquids using a force microscope *Langmuir* **8** 1831

Ducker W A, Senden T J and Pashley R M 1991 Direct measurement of colloidal forces using an atomic force microscope *Nature* **353** 239

[89] Jarvis S P, Yamada H, Yamamoto S-I, Tokumoto H and Pethica J B 1996 Direct mechanical measurement of interatomic potentials *Nature* **384** 247

[90] Gotsmann B, Anczykowski B, Seidel C and Fuchs H 1999 Determination of tip–sample interaction forces from measured dynamic force spectroscopy curves *Appl. Surf. Sci.* **140** 314

[91] Tabor D and Winterton R H S 1969 The direct measurement of normal and retarded van der Waals forces *Proc. R. Soc.* A **312** 435

[92] Thundat T, Zheng X-Y, Chen G Y, Sharp S L, Warmack R J and Schowalter L J 1993 Characterization of atomic force microscope tips by adhesion force measurements *Appl. Phys. Lett.* **63** 2150

[93] Hutter J L and Bechhoefer J 1994 Measurement and manipulation of van der Waals forces in atomic-force microscopy *J. Vac. Sci. Technol.* B **12** 2251

[94] Hansma H G, Vesenka J, Siegerist C, Kelderman G, Morrett H, Sinsheimer R L, Bustamante C, Elings V and Hansma P K 1992 Reproducible imaging and dissection of plasmid DNA under liquid with the atomic force microscope *Science* **256** 1180

[95] Jarvis S P, Dürig U, Lantz M A, Yamada H and Tokumoto H 1998 Feedback stabilized force-sensors: a gateway to the direct measurement of interaction potentials *Appl. Phys.* A **66** S211

[96] Jarvis S P, Yamamoto S-I, Yamada H, Tokumoto H and Pethica J B 1997 Tip–surface interactions studied using a force controlled atomic force microscope in ultrahigh vacuum *Appl. Phys. Lett.* **70** 2238

[97] Giessibl F J 1997 Forces and frequency shifts in atomic-resolution dynamic-force microscopy *Phys. Rev.* B **56** 16 010

[98] Mate C M, Lorenz M R and Novotny V J 1989 Atomic force microscopy of polymeric liquid films *J. Chem. Phys.* **90** 7550

[99] Weisenhorn A L and Hansma P K 1989 Forces in atomic force microscopy in air and water *Appl. Phys. Lett.* **54** 2651

[100] Proceedings from the First International Conference on NC-AFM *Appl. Surf. Sci.* **140** 243

[101] Albrecht T R, Grütter P, Horne D and Rugar D 1991 Frequency modulation detection using high-Q cantilevers for enhanced force microscope sensitivity *J. Appl. Phys.* **69** 668

[102] Dürig U, Züger O and Stalder A 1992 Interaction force detection in scanning probe microscopy: methods and applications *J. Appl. Phys.* **72** 1778

[103] Sasaki N and Tsukada M 1999 Theory for the effect of the tip–surface interaction potential on atomic resolution in forced vibration system of noncontact AFM *Appl. Surf. Sci.* **140** 339

[104] Perez R, Payne M C, Stich I and Terakura K 1997 Role of covalent tip–surface interactions in noncontact atomic force microscopy on reactive surfaces *Phys. Rev. Lett.* **78** 678

[105] Zong Q, Inniss D, Kjoller K and Elings V B 1993 Fractured polymer/silica fiber surface studied by tapping mode atomic force microscopy *Surf. Sci. Lett.* **290** L688

[106] Hansma P K *et al* 1994 Tapping mode atomic force microscopy in liquids *Appl. Phys. Lett.* **64** 1738

[107] Hobbs P, Abraham D and Wickramasinghe H 1989 Magnetic force microscopy with 25 nm resolution *Appl. Phys. Lett.* **55** 2357

[108] Grütter P, Mamin H J and Rugar D 1995 Magnetic force microscopy (MFM) *Scanning Tunnelling Microscopy II* ed R Wiesendanger and H-J Güntherodt (Berlin: Springer) ch 5

[109] Rugar D, Mamin H J, Guenther P, Lambert S E, Stern J E, McFadyen I and Yogi T 1990 Magnetic force microscopy: general principles and application to longitudinal recording media *J. Appl. Phys.* **68** 1169

[110] Overney R and Meyer E 1993 Tribological investigations using friction force microscopy *MRS Bull.* **XVIII** 20

[111] Mate C M, Erlandsson R, McClelland G M and Chiang S 1987 Atomic-scale friction of a tungsten tip on a graphite surface *Phys. Rev. Lett.* **59** 1942

[112] Meyer G and Amer N M 1990 Simultaneous measurement of lateral and normal forces with an optical-beam-deflection atomic force microscope *Appl. Phys. Lett.* **57** 2089

[113] Marti O, Colchero J and Mlynek J 1990 Combined scanning force and friction microscopy of mica *Nanotechnology* **1** 141

[114] Burnham N A and Colton R J 1989 Measuring the nanomechanical properties and surface forces of materials using atomic force microscope *J. Vac. Sci. Technol.* A **7** 2906

[115] van der Werf K O, Putman C A J, Groth B G and Greve J 1994 Adhesion force imaging in air and liquid by adhesion mode atomic force microscopy *Appl. Phys. Lett.* **65** 1195

Miyatani T, Horii M, Rosa A, Fujihira M and Marti O 1997 Mapping of electric double-layer force between tip and sample surfaces in water with pulsed-force-mode atomic force microscopy *Appl. Phys. Lett.* **71** 2632

[116] Kolosov O and Yamanaka K 1993 Nonlinear detection of ultrasonic vibrations in an atomic force microscope *Japan. J. Appl. Phys.* **32** L1095

[117] Erlandsson R, Olsson L and Martensson P 1996 Inequivalent atoms and imaging mechanisms in AC-mode atomic force microscopy of Si(111) 7 × 7 *Phys. Rev.* B **54** 8309

[118] Nakagiri N, Suzuki M, Okiguchi K and Sugimura H 1997 Site discrimination of adatoms in Si(111)-7 × 7 by noncontact atomic force microscopy *Surf. Sci.* **375** L329

[119] Takayanagi K, Takashiro Y, Takahashi M and Takahashi S 1985 Structural analysis of Si(111)-7 × 7 by UHV-transmission electron diffraction and microscopy *J. Vac. Sci. Technol.* A **3** 1502

[120] Bammerlin M, Luthi R, Meyer E, Baratoff A, Lü J, Guggisberg M, Gerber Ch, Howald L and Gütherodt H-J 1997 True atomic resolution on the surface of an insulator via ultrahigh vacuum dynamic force microscopy *Probe Microsc.* **1** 3

[121] Fukui K, Onishi H and Iwasawa Y 1997 Atom-resolved image of the TiO$_2$(110) surface by noncontact atomic force microscopy *Phys. Rev. Lett.* **79** 4202

[122] Raza H, Pang C L, Haycock S A and Thornton G 1999 Non-contact atomic force microscopy imaging of TiO$_2$(100) surfaces *Appl. Surf. Sci.* **140** 271

[123] Friedbacher G, Hansma P K, Ramli E and Stucky G D 1991 Imaging powders with the atomic force microscope: from biominerals to commercial materials *Science* **253** 1261

[124] Erlandsson R, Hadzioannou G, Mate M, McClelland G and Chiang S 1988 Atomic scale friction between the muscovite mica cleavage plane and a tungsten tip *J. Chem. Phys.* **89** 5190

[125] Heckl W, Ohnesorge F, Binnig G, Specht M and Hashmi M 1991 Ring structures on natural molybden disulfide investigated by scanning tunnelling and scanning force microscopy *J. Vac. Sci. Technol.* B **9** 1072

[126] Hegenbart G and Müssig Th 1992 Atomic force microscopy studies of atomic structures on AgBr(111) surfaces *Surf. Sci. Lett.* **275** L655

[127] Keita B, Nadjo L and Kjoller K 1991 Surface characterization of a single crystal of sodium decatungstocerate (IV) by the atomic force microscope *Surf. Sci. Lett.* **256** L613

[128] Colchero J, Marti O, Mlynek J, Humbert A, Henry C R and Chapon C 1991 Palladium clusters on mica: a study by scanning force microscopy *J. Vac. Sci. Technol.* B **9** 794

[129] Baski A A and Fuchs H 1994 Epitaxial growth of silver on mica as studied by AFM and STM *Surf. Sci.* **313** 275

[130] van Duyne R P, Hulteen J C and Treichel D A 1993 Atomic force microscopy and surface-enhanced Raman spectroscopy. Ag island films and Ag film over polymer nanosphere surfaces supported on glass *J. Chem. Phys.* **99** 2101

[131] Roark S E and Rowlen K L 1993 Atomic force microscopy of thin Ag films. Relationship between morphology and optical properties *Chem. Phys. Lett.* **212** 50

[132] Overney R, Howald L, Frommer J, Meyer E and Güntherodt H 1991 Molecular surface structure of tetracene mapped by the atomic force microscope *J. Chem. Phys.* **94** 8441

[133] Overney R, Howald L, Frommer J, Meyer E, Brodbeck D and Güntherodt H 1992 Molecular surface structure of organic crystals observed by atomic force microscopy *Ultramicroscopy* **42–44** 983

[134] Schwartz D K, Viswanathan R and Zasadzinski J A N 1993 Commensurate defect superstructures in a Langmuir–Blodgett film *Phys. Rev. Lett.* **70** 1267

[135] Bourdieu L, Ronsin O and Chatenay D 1993 Molecular positional order in Langmuir–Blodgett films by atomic force microscopy *Science* **259** 798

[136] Liu G-Y and Salmeron M B 1994 Reversible displacements of chemisorbed n-alkanethiol molecules on Au(111) surface: an atomic force microscopy study *Langmuir* **10** 367

[137] Salmeron M, Neubauer G, Folch A, Tomitori M, Ogletree D F and Sautet P 1993 Viscoelastic and electrical properties of self-assembled monolayers on Au(111) films *Langmuir* **9** 3600

[138] Salmeron M, Liu G-Y and Ogletree D F 1995 Molecular arrangement and mechanical stability of self-assembled monolayers on Au(111) under applied load *Force in Scanning Probe Methods* ed H-J Güntherodt *et al* (Amsterdam: Kluwer)

[139] Grim P C M, Brouwer H J, Seyger R M, Oostergetel G T, Bergsma-Schutter W G, Arnberg A C, Güthner P, Dransfeld K and Hadziioannou G 1992 Investigation of polymer surfaces using scanning force microscopy (SFM) 'A new direct look on old polymer problems' *Makromol. Chem., Macromol. Symp.* **62** 141

[140] Annis B K, Noid D W, Sumpter B G, Reffner J R and Wunderlich B 1992 Application of atomic force microscopy (AFM) to a block copolymer and an extended chain polyethylene *Makromol. Chem., Rapid. Commun.* **13** 169

Annis B K, Schwark D W, Reffner J R, Thomas E L and Wunderlich B 1992 Determination of surface morphology of diblock copolymers of styrene and butadiene by atomic force microscopy *Makromol. Chem.* **193** 2589

[141] Krausch G, Hipp M, Böltau M, Mlynek J and Marti O 1995 High resolution imaging of polymer surfaces with chemical sensitivity *Macromolecules* **28** 260

[142] Feldman K, Tervoort T, Smith P and Spencer N D 1998 Toward a force spectroscopy of polymer surfaces *Langmuir* **14** 372

[143] Snétivy D and Vancso G J 1994 Atomic force microscopy of polymer crystals: 7. Chain packing, disorder and imaging of methyl groups in oriented isotactic polypropylene *Polymer* **35** 461

[144] Vansteenkiste S O, Davies M C, Roberts C J, Tendler S J B and Williams P M 1998 Scanning probe microscopy of biomedical interfaces *Prog. Surf. Sci.* **57** 95

[145] Ikai A, Yoshimura K, Arisaka F, Ritani A and Imai K 1993 Atomic force microscopy of bacteriophage T4 and its tube-baseplate complex *FEBS Lett.* **326** 39

[146] Wagner P 1998 Immobilization strategies for biological scanning probe microscopy *FEBS Lett.* **430** 112

[147] Bustamente C, Vesenka J, Tang C L, Rees W, Guthold M and Keller R 1992 Circular DNA molecules imaged in air by scanning force microscopy *Biochemistry* **31** 22

[148] Allen M J, Hud N V, Balooch M, Tench R J, Siekhaus W J and Balhorn R 1992 Tip-radius-induced artifacts in AFM images of protamine-complexed DNA fibers *Ultramicroscopy* **42–44** 1095

[149] Keller D and Chou C C 1992 Imaging steep, high structures by scanning force microscopy with electron beam deposited tips *Surf. Sci.* **268** 333

[150] Thundat T, Warmack R J, Allison D P, Bottomley L A, Lourenco A J and Ferrell T L 1992 Atomic force microscopy of deoxyribonucleic acid strands adsorbed on mica: the effect of humidity on apparent width and image contrast *J. Vac. Sci. Technol.* A **10** 630

[151] Bustamente C, Keller D and Yang G 1993 Scanning force microscopy of nucleic acids and nucleoprotein assemblies *Curr. Opin. Struct. Biol.* **3** 363

[152] Hansma H G, Vesenka J, Siegerist C, Kelderman G, Morrett H, Sinsheimer R L, Elings V, Bustamente C and Hansma P K 1992 Reproducible imaging and dissection of plasmid DNA under liquid with the atomic force microscope *Science* **256** 1180

[153] Hansma H G, Sinsheimer R L, Li M-Q and Hansma P K 1992 Atomic force microscopy of single- and double-stranded DNA *Nucleic Acids Res.* **20** 3585

[154] Rees W A, Keller R W, Vesenka J P, Yang G and Bustamente C 1993 Evidence of DNA bending in transcription complexes imaged by scanning force microscopy *Science* **260** 1646

[155] Hansma H G, Laney D E, Bezanilla M, Sinsheimer R L and Hansma P K 1995 Applications for atomic-force microscopy of DNA *Biophys. J.* **68** 1672

[156] Samori B, Siligardi G, Quagliariello C, Weisenhorn A L, Vesenka J and Bustamente C 1993 Chirality of DNA supercoiling assigned by scanning force microscopy *Proc. Natl Acad. Sci. USA* **90** 3598

[157] Hansma H G, Sinsheimer R L, Groppe J, Bruice T C, Elings V, Gurley G, Bezanilla M, Mastrangelo I A, Hough P V C and Hansma P K 1993 Recent advances in atomic force microscopy of DNA *Scanning* **15** 296

[158] Radmacher M, Fritz M, Hansma H G and Hansma P K 1994 Direct observation of enzyme activity with the atomic force microscope *Science* **265** 1577

[159] Bottomley L A, Coury J E and First P N 1996 Scanning probe microscopy *Anal. Chem.* **68** 185R

Shao Z and Yang J 1995 Progress in high-resolution atomic-force microscopy in biology *Qt. Rev. Biophys.* **28** 195

Shao Z, Mou J, Czajkowsky D M, Yang J and Yuan J 1996 Biological atomic force microscopy: what is achieved and what is needed *Adv. Phys.* **45** 1

Ikai A 1996 STM and AFM of bio/organic molecules and structures *Surf. Sci. Rep.* **26** 263

Lal R and John S A 1994 Biological applications of atomic-force microscopy *Am. J. Physiol.* **266** C1

Shao Z F, Yang J and Somlyo A P 1995 Biological atomic force microscopy: from microns to nanometers and beyond *Ann. Rev. Cell Dev. Biol.* **11** 241

Kasas S, Thompson N H, Smith B L, Hansma P K, Miklossy J and Hansma H G 1997 Biological applications of the AFM: from single molecules to organs *Int. J. Im. Syst. Technol.* **8** 151

[160] Müller D J, Baumeister W and Engel A 1996 Conformational change of the hexagonally packed intermediate layer of *Deinococcus radiodurans* monitored by atomic force microscopy *J. Bacteriol.* **178** 3025

Müller D W, Fotiadis D and Engel A 1998 Mapping flexible protein domains at subnanometre resolution with the atomic force microscope *FEBS Lett.* **430** 105

[161] Lal R, Kim H, Garavito R M and Arnsdorf M F 1993 Imaging of reconstituted biological channels at molecular resolution by atomic force microscopy *Am. J. Physiol.* **265** C851

[162] Rief M, Gautel M, Oesterhelt F, Fernandez J M and Gaub H E 1997 Reversible unfolding of individual titin immunoglobulin domains by AFM *Science* **276** 1109

[163] Henderson E, Haydon P G and Sakaguchi D S 1992 Actin filament dynamics in living glial cells imaged by atomic force microscopy *Science* **257** 1944

[164] Brandow S L, Turner D C, Ratna B R and Gaber B P 1993 Modification of supported lipid membranes by atomic force microscopy *Biophys. J.* **64** 898

[165] Parpura V, Haydon P G, Sakaguchi D S and Henderson E 1993 Atomic force microscopy and manipulation of living glial cells *J. Vac. Sci. Technol.* A **11** 773

[166] Lee G U, Kidwell D A and Colton R J 1994 Sensing discrete streptavidin–biotin interactions with atomic force microscopy *Langmuir* **10** 354

[167] Florin E-L, Moy V T and Gaub H E 1994 Adhesion forces between individual ligand-receptor pairs *Science* **264** 415

[168] Lee G U, Chrisey L A and Colton R J 1994 Direct measurement of the forces between complementary strands of DNA *Science* **266** 771

[169] Bongrand P 1999 Ligand–receptor interactions *Rep. Prog. Phys.* **62** 921

[170] Ducker W A, Senden T J and Pashley R M 1991 Direct measurement of colloidal forces using an atomic force microscope *Nature* **353** 239

[171] Li Y Q, Tao N J, Pan J, Garcia A A and Lindsay S M 1993 Direct measurement of interaction forces between colloidal particles using the scanning force microscope *Langmuir* **9** 637

[172] Larson I, Drummond C J, Chan D Y C and Grieser F 1993 Direct force measurements between $TiO_2$ surfaces *J. Am. Chem. Soc.* **115** 11 885

[173] Biggs S, Mulvaney P, Zukoski C F and Grieser F 1994 Study of anion adsorption at the gold-aqueous solution interface by atomic force microscopy *J. Am. Chem. Soc.* **116** 9150

[174] Greenwood J A 1967 On the area of contact between rough surfaces and flats *J. Lub. Tech. (ASME)* **1** 81

[175] Johnson K L, Kendall K and Roberts A D 1971 Surface energy and the contact of elastic solids *Proc. R. Soc.* A **324** 301
Sperling G 1964 *Dissertation* Karlsruhe Technische Hochschule

[176] Hu J, Xiao X-D, Ogletree D F and Salmeron M 1995 Atomic scale friction and wear of mica *Surf. Sci.* **327** 358

[177] Lüthi R, Meyer E, Haefke H, Howald L and Güntherodt H-J 1995 Nanotribology: an UHV-SFM study of thin films of AgBr(001) *Tribol. Lett.* **1** 23

[178] Overney R M, Meyer E, Frommer J, Brodbeck D, Lüthi R, Howald L, Güntherodt H-J, Fujihara M, Takano H and Gotoh Y 1992 Friction measurements of phase separated thin films with a modified atomic force microscope *Nature* **359** 133

[179] Brager W R, Koleske D D, Feldman K, Krueger D and Colton R J 1996 Small change-big effect: SPM studies of two-component fatty-acid monolayers *ACS Polymer Preprints* **37** 606

[180] Kim H I, Graupe M, Oloba O, Koini T, Imaduddin S, Lee T R and Perry S S 1999 Molecularly specific studies of the frictional properties of monolayer films: a systematic comparison of $CF_3$-, $(CH_3)_2CH$-, and $CH_3$-terminated films *Langmuir* **15** 3179

[181] Yoshizawa H, Chen Y-L and Israelachvili J N 1993 Fundamental mechanisms of interfacial friction. 1. Relation between adhesion and friction *J. Chem. Phys.* **97** 4128
Israelachvili J N, Chen Y-L and Yoshizawa H 1994 Relationship between adhesion and friction forces *J. Adhesion Sci. Technol.* **8** 1234

[182] Haugstad G, Gladfelter W L, Weberg E B, Weberg R T and Jones R R 1995 Friction force microscopy as a probe of molecular relaxation on polymer surfaces *Tribol. Lett.* **1** 253

[183] Burnham N A and Colton R J 1989 Measuring the nanomechanical properties and surface forces of materials using an atomic force microscope *J. Vac. Sci. Technol.* A **7** 2906

[184] Salmeron M, Folch A, Neubauer G, Tomitori M and Ogletree D F 1992 Nanometer scale mechanical properties of Au(111) thin films *Langmuir* **8** 2832

[185] Rivière J C 1990 *Practical Surface Analysis* 2nd edn, vol 1, ed D Briggs and M P Seah (Chichester: Wiley)
Briggs D 1990 *Practical Surface Analysis* 2nd edn, vol 2, ed D Briggs and M P Seah (Chichester: Wiley)

[186] Akari S, Horn D, Keller H and Schrepp W 1995 Chemical imaging by scanning force microscopy *Adv. Mater.* **7** 549

[187] Frisbie C D, Rozsnyai L F, Noy A, Wrighton M S and Lieber C M 1994 Functional group imaging by chemical force microscopy *Science* **265** 2071

[188] McKendry R, Theoclitou M-E, Rayment T and Abell C 1998 Chiral discrimination by chemical force microscopy *Nature* **391** 566

[189] Marti A, Hähner G and Spencer N D 1995 The sensitivity of frictional forces to pH on a nanometer scale: a lateral force microscopy study *Langmuir* **11** 4632
Hähner G, Marti A and Spencer N D 1997 The influence of pH on friction between oxide surfaces in electrolytes, studied with lateral force microscopy: application as a nanochemical imaging technique *Tribol. Lett.* **3** 359

[190] Sittig C, Hähner G, Marti A, Textor M, Spencer N D and Hauert R 1999 The implant material, Ti6Al7Nb: surface microstructure, composition, and properties *J. Mater. Sci.* **10** 191

[191] Bergasa F and Saenz J J 1992 Is it possible to observe biological macromolecules by electrostatic force microscopy? *Ultramicroscopy* **42–44** 1189

[192] Berger C E H, van der Werf K O, Kooyman R P H, de Grooth B G and Greve J 1995 Functional group imaging by adhesion AFM applied to lipid monolayers *Langmuir* **11** 4188

[193] Pohl D W 1991 Scanning near-field optical microscopy (SNOM) *Advances in Optical and Electron Microscopy* vol 12, ed R Barer and V E Cosslett (London: Academic)

[194] Abbé E 1873 *Archiv. Microskop. Anal.* **9** 413

[195] Synge E H 1928 Extending microscopic resolution into the ultra-microscopic region *Phil. Mag.* **6** 356

[196] Pohl D W, Denk W and Lanz M 1984 Optical stethoscopy: image recording with resolution $\lambda/20$ *Appl. Phys. Lett.* **44** 651

[197] Ash E A and Nichols G 1972 Super-resolution aperture scanning microscope *Nature* **237** 510

[198] Dürig U T, Pohl D W and Rohner F 1986 Near-field optical-scanning microscopy *J. Appl. Phys.* **59** 3318
    Pohl D W 1991 Scanning near-field optical microscopy (SNOM) *Adv. Opt. Electron. Microsc.* **12** 243

[199] Betzig E, Finn P L and Weiner J S 1992 Combined shear force and near-field scanning optical microscopy *Appl. Phys. Lett.* **60** 2484

[200] Fischer U Ch 1985 Optical characteristics of 0.1 $\mu$m circular apertures in a metal film as light sources for scanning ultramicroscopy *J. Vac. Sci. Technol.* B **3** 386
    Fischer U Ch, Dürig U T and Pohl D W 1988 Near-field scanning microscopy in reflection *Appl. Phys. Lett.* **52** 249
    Cline J A, Barshatzky H and Isaacson M 1991 Scanned-tip reflection-mode near-field scanning optical microscopy *Ultramicroscopy* **38** 299

[201] Betzig E and Chichester R J 1993 Single molecules observed by near-field scanning optical microscopy *Science* **262** 1422
    Trautman J K, Macklin J J, Brus L E and Betzig E 1994 Near-field spectroscopy of single molecules at room temperature *Nature* **369** 40

[202] Sharp S L, Warmack R J, Goudonnet J P, Lee I and Ferrell T L 1993 Spectroscopy and imaging using the photon scanning-tunnelling microscope *Acc. Chem. Res.* **26** 377

[203] Tsai D P, Othonos A, Moskovits M and Uttamchandani D 1994 Raman spectroscopy using a fibre optic probe with subwavelength aperture *Appl. Phys. Lett.* **64** 1768

[204] Smith D A, Webster S, Ayad M, Evans S D, Fogherty D and Batchelder D 1995 Development of a scanning near-field optical probe for localised Raman spectroscopy *Ultramicroscopy* **61** 247

[205] Takahashi S, Futamata M and Kojima I 1999 Spectroscopy with scanning near-field optical microscopy using photon tunnelling mode *J. Microscopy* **194** 519

[206] Emory S R and Nie S 1997 Near-field surface-enhanced Raman spectroscopy on single silver nanoparticles *Anal. Chem.* **69** 2631

[207] Zeisel D, Deckert V, Zenobi R and Vo-Dinh T 1998 Near-field surface-enhanced Raman spectroscopy of dye molecules adsorbed on silver island films *Chem. Phys. Lett.* **283** 381

[208] Guenther P, Fischer U Ch and Dransfeld K 1989 Scanning near-field acoustic microscopy *Appl. Phys.* B **48** 89

[209] Williams C C and Wickramasinghe H K 1986 Scanning thermal profiler *Appl. Phys. Lett.* **49** 1587

[210] Hammiche A, Hourston D J, Pollock H M, Reading M and Song M 1996 Scanning thermal microscopy: sub-surface imaging, thermal mapping of polymer blends, localised calorimetry *J. Vac. Sci. Technol.* B **14** 1486

[211] Hansma P K, Drake B, Marti O, Gould S A C and Prater C B 1989 The scanning ion-conductance microscope *Science* **243** 641

[212] Jarrell J A, King J G and Mills J W 1981 A scanning micropipette molecule microscope *Science* **211** 277

[213] Binnig *et al* 1982 Surface studies by scanning tunnelling microscopy *Phys. Rev. Lett.* **49** 57

[214] Hansma P K and Tersoff J 1987 Scanning tunneling microscopy *J. Appl. Phys.* **61** R1

[215] Weisenhorn A L 1991 *PhD Thesis* University of California, Santa Barbara

[216] Manne S, Hansma P K, Massie J, Elings V B and Gewirth A A 1991 Atomic-resolution electrochemistry with the atomic force microscope: copper deposition on gold *Science* **251** 183

[217] Jarvis S P and Tokumoto H 1997 Measurement and interpretation of forces in the atomic force microscope *Probe Microscopy* **1** 65

# B1.20
# The surface forces apparatus

*Manfred Heuberger*

## B1.20.1 Introduction

Compared with other direct force measurement techniques, a unique aspect of the surface forces apparatus (SFA) is to allow *quantitative measurement of surface forces and intermolecular potentials*. This is made possible by essentially three measures: (i) well defined contact geometry, (ii) high-resolution interferometric distance measurement and (iii) precise mechanics to control the separation between the surfaces.

It is remarkable that the roots of the SFA go back to the early 1960s [1]. Tabor and Winterton [2] and Israelachvili and Tabor [3] developed it to the current state of the art some 15 years before the invention of the more widely used atomic force microscope (AFM) (see chapter B1.19).

Although only a few dozen laboratories worldwide are actively practising the SFA technique, it has produced many notable findings and a fundamental understanding of surface forces [4]. These are applicable to various research fields such as polymer science [5–10], thin-film rheology [11–14], biology [15–19], liquid crystals [20–22], food sciences [12, 23, 24], molecular tribology [9, 14, 25–35] and automotive tribology [36]—to name just a few. Numerous experimental extensions of the SFA technique have been developed to take advantage of the unique experimental set-up.

One of the most important extensions is the measurement of lateral forces (friction). Friction measurements have accompanied the SFA technique since its early beginnings in the Cavendish laboratory in Cambridge [37] and a variety of different lateral force measurements are practised throughout the SFA community.

## B1.20.2 Principles

### B1.20.2.1 Direct force measurement

The measurement of surface forces calls for a rigid apparatus that exhibits a high force sensitivity as well as distance measurement and control on a subnanometre scale [38]. Most SFAs make use of an optical interference technique to measure distances and hence forces between surfaces. Alternative distance measurements have been developed in recent years—predominantly capacitive techniques, which allow for faster and simpler acquisition of an *averaged* distance [11, 39, 40] or even allow for simultaneous dielectric loss measurements at a confined interface.

The predominant method of measuring forces is to detect the proportional deflection of an elastic spring. The proportionality factor is commonly called the *spring constant* and, in the SFA, may range from some $10 \text{ N m}^{-1}$ to some $100 \text{ kN m}^{-1}$. It is essential that the apparatus frame and surface compliance be at least 1000 times stiffer than the force-measuring spring. In a typical SFA, one of the two surfaces is attached to the force-measuring spring, while the other surface is rigidly mounted to the apparatus frame. In this set-up,

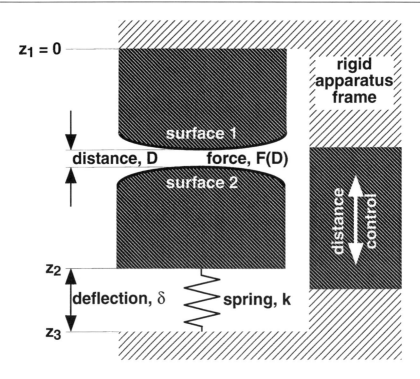

**Figure B1.20.1.** Direct force measurement via deflection of an elastic spring—essential design features of a direct force measurement apparatus.

shown schematically in figure B1.20.1, a set of at least three parameters must be controlled or measured simultaneously for a direct force measurement.

The accurate and absolute measurement of the distance, $D$, between the surfaces is central to the SFA technique. In a typical experiment, the SFA controls the base position, $z_3$, of the spring and simultaneously measures $D$, while the *spring constant*, $k$, is a known quantity. Ideally, the simple relationship $\Delta F(D) = k \Delta(D - z_3)$ applies. Since surface forces are of limited range, one can set $F(D = \infty) = 0$ to obtain an absolute scale for the force. Furthermore, $\delta F(D = \infty)/\delta D \approx 0$ so that one can readily obtain a calibration of the distance control at large distances relying on an accurate measurement of $D$. Therefore, $D$ and $F$ are obtained at high accuracy to yield $F(D)$, the so-called *force versus distance curve*.

Most interferometric SFAs allow one to measure the distance, $D$, as a function of a selected lateral dimension, $x'$, which can be used to obtain information about the entire contact geometry. Knowledge of the contact geometry, together with the assumption that the underlying intermolecular forces are additive, allows the *intermolecular potential* $W(D)$ to be deduced from the measured $F(D)$ using the so-called *Derjaguin approximation* [4, 41]. For the idealized geometry of an undeformed sphere of radius $R$ on a flat:

$$F(D) \approx 2\pi R W(D) \tag{B1.20.1}$$

where $R$ becomes an effective radius $R' = (R_1 R_2)^{0.5}$ for the case of two cylinders (SFA) with radii $R_1$ and $R_2$.

In accordance with equation (B1.20.1), one can plot the so-called *surface force parameter*, $P = F(D)/2\pi R$, versus $D$. This allows comparison of different direct force measurements in terms of intermolecular potentials $W(D)$, i.e. independent of a particular contact geometry. Figure B1.20.2 shows an example of the attractive van der Waals force measured between two curved mica surfaces of radius $R \approx 10$ mm.

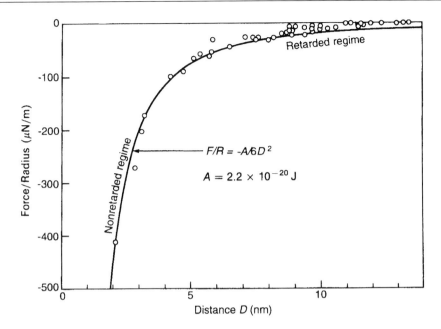

**Figure B1.20.2.** Attractive van der Waals potential between two curved mica surfaces measured with the SFA. (Reproduced with permission from [4], figure 11.6.)

An alternative way of measuring $F(D)$ is to control forces using magnetic fields instead of controlling distances [42, 43]. An electronically controlled current through one or more coils creates a magnetic gradient, which exerts a force on the force-measuring spring. The surfaces are allowed to equilibrate at a distance, $D$, which is then measured. This alternative set-up allows one to measure $D(F)$ instead of $F(D)$ allowing for a true equilibrium force measurement.

### B1.20.2.2   Sample preparation

Well defined contact geometry and absolute cleanliness are crucial factors for a successful SFA experiment. Therefore, two curved sheets of mica are brought into contact in crossed-cylinder geometry.

The sample preparation of these mica sheets is a delicate process that requires some experience and often takes 1–2 days prior to an SFA experiment. Through successive cleaving, one has to prepare 1–5 $\mu$m thick and uniform sheets of mica. Mica is a natural material that is available in different qualities [44].

Each newly cleaved mica surface is very clean. However, it is known that mica has a strong tendency to spontaneously adsorb particles [45] or organic contaminants [46], which may affect subsequent measurements. The mica sheets are cut into 10 mm × 10 mm sized samples using a hot platinum wire, then laid down onto a thick and clean 100 mm × 100 mm mica *backing sheet* for protection. On the backing sheet, the mica samples can be transferred into a vacuum chamber for thermal evaporation of typically 50–55 nm thick silver mirrors.

It was the idea of Winterton [2] to glue the otherwise fragile mica sheets onto polished *silica discs* to give them better mechanical stability, especially for friction experiments. The glue layer determines the final surface compliance of the silica/glue/mica stack which is typically around $4 \times 10^{10}$ Pa. The use of mica samples from the same original sheet guarantees that the interferometer will be perfectly symmetrical (see section B1.20.2.3).

The silica discs that now hold the back-silvered mica samples are finally mounted into the SFA so that the cylinder axes are crossed and the clean mica surfaces are facing each other.

### B1.20.2.3  Multiple beam interferometry

The absolute measurement of the distance, $D$, between the surfaces is central to the SFA technique. In *interferometric* SFAs, it is realized through an optical method called multiple beam interferometry (MBI), which has been described by Tolansky [47].

A 50 nm film of metal (silver) is deposited onto the atomically smooth mica sheets. White light with a coherence length of some 10 $\mu$m is directed normally through the mica sheets to illuminate the contact zone. The mica–silver interfaces have a reflectivity of typically 97% and form an optical resonator. A constructive interference occurs for light that has a wavelength equal to half the optical distance between the mirrors. This resonance is called *interference fringe of chromatic order $N = 1$*. The larger the optical distance, the more this resonance shifts towards the red end of the spectrum. In analogy to an organ pipe, one also observes harmonic fringes at higher frequencies which are identified with integer numbers $N = 2, 3, \ldots$. The emerging light from the silver/mica/mica/silver interferometer is focused onto the slit of an imaging spectrograph for further analysis. The light selected to enter the spectrograph entrance slit corresponds to a one-dimensional cut through the illuminated contact zone. The distance information along this (linear) dimension is maintained and the exit plane of the imaging spectrograph displays the interference pattern as a function of a position $x'$. Typically, one uses fringes in the visible spectrum ($15 < N < 35$) for distance measurements in the SFA, but other near-visible wavelengths can equally be employed. Generally, the distance resolution of MBI decreases linearly with increasing $N$ and longer wavelengths used.

Because the mica surfaces are curved, the optical distance between the mirrors is a function of the lateral dimension, $x'$, i.e. $T = T(x')$. Therefore, the wavelength of a given fringe becomes itself a function of $x'$. Since the chromatic order, $N$, is invariant within a given fringe, these fringes are commonly called fringes of equal chromatic order (FECO) (see figure B1.20.3).

The problem of calculating surface separations from a given FECO pattern is in general far more complex than it may seem at first sight. The main reason is the fact that the refractive index is different for each of the different layers in the interferometer. For the simplest case of two mica surfaces in contact, surrounded by a medium, one has to solve for a one-layer interferometer inside the contact area and for a three-layer interferometer outside the contact area—that is where light travels partially through the medium. The analytical treatment of a three-layer interferometer is considerably simplified if the two mica sheets have exactly the same thickness, i.e. when the interferometer is *symmetric*. Based on the work of Hunter *et al* [48], Israelachvili [49] has derived the first useful analytical expressions and methodology for the symmetrical three-layer interferometer in the SFA. Clarkson [50] extended MBI with a numerical analysis of asymmetric three-layer interferometers by applying the multilayer matrix method [51]. Later, Horn *et al* [52] derived useful analytical expressions for the asymmetric interferometer.

Nonetheless, the symmetric interferometer remains very useful, because there, the wavelengths of fringes with *even* chromatic order, $N$, strongly depend on the refractive index, $n_3$, of the central layer, whereas fringes with *odd* chromatic order are almost insensitive to $n_3$. This lucky combination allows one to measure the thickness as well as the refractive index of a layer between the mica surfaces independently and simultaneously [49].

To simplify FECO evaluation, it is common practice to experimentally filter out one of the components by the use of a linear polarizer after the interferometer. Mica birefringence can, however, be useful to study thin films of birefringent molecules [49] between the surfaces. Rabinowitz [53] has presented an eigenvalue analysis of birefringence in the multiple beam interferometer.

Partial reflections at the inner optical interfaces of the interferometer lead to so-called *secondary* and *tertiary* fringe patterns as can be seen from figure B1.20.4. These additional FECO patterns become clearly visible if the reflectivity of the silver mirrors is reduced. Methods for analysis of such secondary and tertiary FECO patterns were developed to extract information about the topography of non-uniform substrates [54].

In the *symmetric*, three-layer interferometer, only even-order fringes are sensitive to refractive index and it is possible to obtain spectral information of the confined film by comparison of the different intensities of

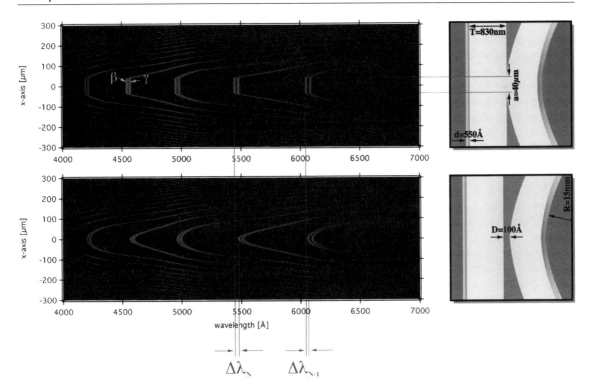

**Figure B1.20.3.** FECO allow optical distances in the SFA to be measured at subnanometre resolution. The FECOs depicted here belong to a symmetric, three-layer interferometer. The elastically deformed contact region appears as flattened fringes in the upper graph (vertical part). Once the surfaces are separated to a distance $D$ in a medium of index $n$, the wavelength shifts are measured to calculate $D$ and $n$. Since mica is birefringent, one can observe $\beta$ and $\gamma$ components for each fringe. To simplify data evaluation, birefringence can be suppressed using polarizing filters after the interferometer.

odd- and even-order fringes. The absorption spectrum of thin dye layers between mica was investigated by Müller and Mächtle [55, 56] using this method.

Instead of an absorbing dye layer between the mica, Levins *et al* [57] used thin metallic films and developed a method for FECO analysis using an extended spectral range.

The preparation of the reflecting silver layers for MBI deserves special attention, since it affects the optical properties of the mirrors. Another important issue is the optical phase change [58] at the mica/silver interface, which is responsible for a wavelength-dependent shift of all FECOs. The phase change is a function of silver layer thickness, $T$, especially for $T < 40$ nm [54]. The roughness of the silver layers can also have an effect on the resolution of the distance measurement [59, 60].

Another interesting extension of the FECO technique, using a capillary droplet of mercury as the second mirror, was developed by Horn *et al* [61]. The light from this special interferometer is analysed in reflection, which yields an inverted FECO pattern.

Every property of an interface that can be optically probed can, in principle, be measured with the SFA. This may include information obtainable from absorption spectroscopy [55], fluorescence, dichroism, birefringence, or nonlinear optics [43], some of which have already been realized.

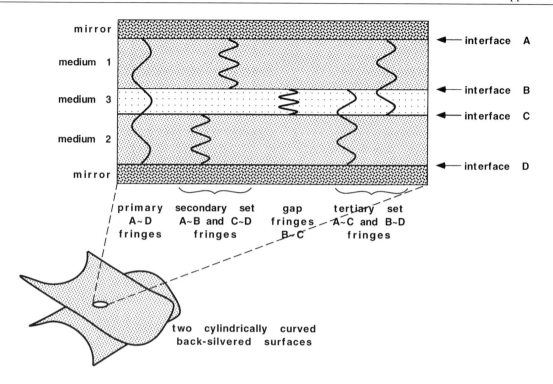

**Figure B1.20.4.** Cross-sectional sideview of a symmetric, three-layer interferometer illustrating the origin of primary, secondary, tertiary and gap FECOs. (Reproduced with permission from [54].)

*B1.20.2.4   Common designs and attachments*

Israelachvili and Adams [62] designed one of the first SFAs, known as the Mk I, in the mid-1970s. Later, Israelachvili developed the Mk II [63, 64], which is based on the Mk I but has improved mechanics—in particular, a double cantilever spring to avoid surface tilt upon deflection (force), as well as a number of new attachments for a variety of different experimental set-ups. The Mk I and Mk II designs served as basis for further versions. More recent and improved versions include the Mk III developed by Israelachvili [65], as well as the circular steel Mk IV developed by Parker *et al* [66] and the circular glass SFA designed by Klein [67]. Some of the most common designs are schematically reproduced in figure B1.20.5.

A considerable number of experimental extensions have been developed in recent years. Luckham *et al* [5] and Dan [68] review examples of dynamic measurements in the SFA. Studying the visco-elastic response of surfactant films [69] or adsorbed polymers [7, 9] promises to yield new insights into molecular mechanisms of frictional energy loss in boundary-lubricated systems [28, 70].

The measurement of lateral forces (friction and shear) in the SFA has recently been reviewed by Kumacheva [32]. To measure friction and shear response, one has to laterally drive one surface and simultaneously measure the response of a lateral spring mount. A variety of versions have been devised. Lateral drives are often based on piezoelectric or bimorph deflection [13, 71] or DC motor drives, whereas the response can be measured via strain gauges, bimorphs, capacitive or optical detection.

Another promising extension uses x-rays to probe the structure of confined molecules [72].

In summary, the SFA is a versatile instrument that represents a unique platform for many present and future implementations. Unlike any other experimental technique, the SFA yields quantitative insight into molecular dimensions, structures and dynamics under confinement.

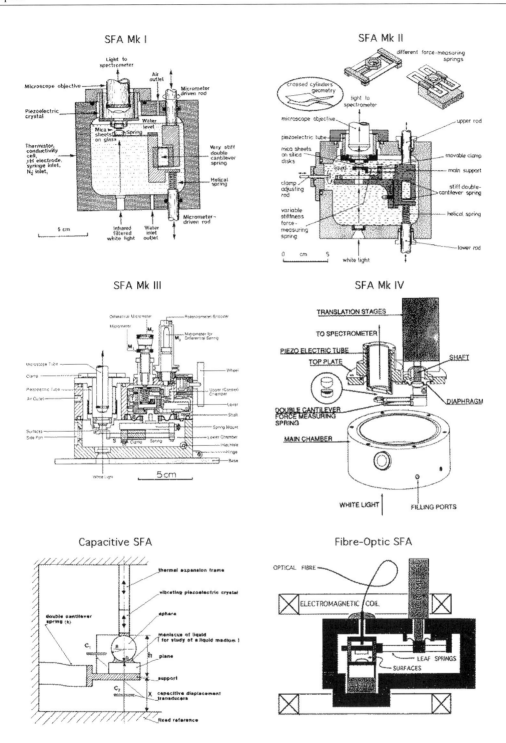

**Figure B1.20.5.** A selection of common SFA designs: SFA Mk I [62], SFA Mk II [64], SFA Mk III [65], SFA Mk IV [66], capacitive SFA [11], fibre-optic SFA [43]. Figures reproduced with permission from indicated references.

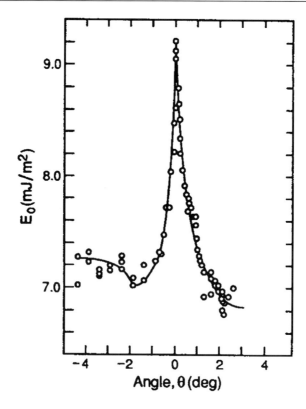

**Figure B1.20.6.** Short-range adhesion of a mica–mica contact as a function of the relative crystallographic orientation of the mica surfaces, measured in a dry nitrogen atmosphere. With permission from [94].

### B1.20.3    Applications

This section deals with some selected examples of typical SFA results, collected from various research areas. It is not meant to be a comprehensive review, rather a brief glance at the kind of questions that can be addressed with the SFA.

The earliest SFA experiments consisted of bringing the two mica sheets into contact in a controlled atmosphere (figure B1.20.6) or (confined) liquid medium [14, 27, 73–75]. Later, a variety of surfactant layers [76, 77], polymer surfaces [5, 9, 10, 13, 68, 78], polyelectrolytes [79], novel materials [80] or biologically relevant molecular layers [15, 19, 81–86] or model membranes [84, 87, 88] were prepared on the mica substrate. More recently, the SFA technique has also been extended to thick layers of other materials, such as silica [73, 89], polymer [10], as well as metals [59] and metal oxides [90, 91].

### B1.20.3.1    *Measuring short-range solvation and hydration forces*

The measurement of surface forces out-of-plane (normal to the surfaces) represents a central field of use of the SFA technique. Besides the ubiquitous van der Waals dispersion interaction between two (mica) surfaces in dry air (figures B1.20.2 and B1.20.6), there is a wealth of other surface forces arising when the surfaces are brought into contact in a liquid medium. Many of these forces result from the specific properties of the liquid medium and originate from a characteristic ordering or reorientation of atoms or molecules—processes which are often entropy driven.

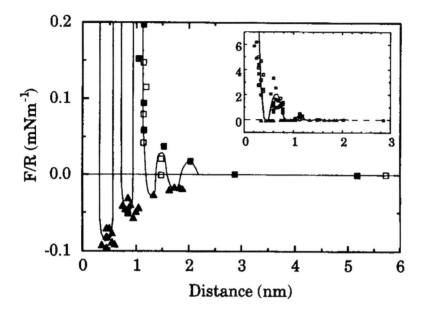

**Figure B1.20.7.** The solvation force of ethanol between mica surface. The inset shows the full scale of the experimental data. With permission from [75].

The well defined contact geometry and the ionic structure of the mica surface favours observation of *structural* and *solvation forces*. Besides a monotonic entropic repulsion one may observe superimposed periodic force modulations. It is commonly believed that these modulations are due to a metastable layering at surface separations below some 3–10 molecular diameters. These diffuse layers are very difficult to observe with other techniques [92]. The periodicity of these *oscillatory forces* is regularly found to correspond to the characteristic molecular diameter. Figure B1.20.7 shows a typical measurement of solvation forces in the case of ethanol between mica.

In the case of water, these forces are called *hydration* forces [93, 94]. The behaviour of water close to surfaces has attracted considerable attention due to its importance in the understanding of colloidal and biological interactions. Water seems to be a molecule of remarkable intermolecular interactions [95], mainly due to its capability to form hydrogen bonds. A number of aspects of water have been vigorously debated using results obtained with the SFA technique. These include the apparent viscosity in ultra-thin confined films [73] or the structure of water near surfaces, which is believed to give rise to *hydrophilic repulsion* or *hydrophobic attraction* [96, 97], or, indeed, the very origin of hydration forces and oscillatory forces [93].

### B1.20.3.2  *Measuring long-range DLVO-type interactions*

In a given aqueous electrolyte there is a certain population of hydrated ions and counterions. In addition, many surfaces, including mica, exhibit a net charge in aqueous solution. Counterions are known to form a diffuse screening layer near a charged molecule, particle or surface with roughly exponentially decreasing concentration into the solution. The characteristic decay length of this *double layer* is called the *Debye length* and decreases with increasing ionic strength. This double layer gives rise to an entropically driven, exponentially decreasing surface force, the so-called *double-layer repulsion*. Before the discovery of short-range forces (see above), it was commonly accepted that surface forces in liquid media could always be decomposed into an attractive dispersion component (van der Waals) and a repulsive double-layer force.

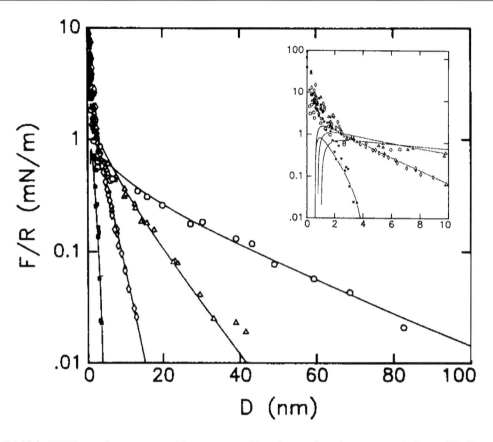

**Figure B1.20.8.** DLVO-type forces measured between two silica glass surfaces in aqueous solutions of NaCl at various concentrations. The inset shows the same data in the short-range regime up to $D = 10$ nm. The repulsive deviation at short range ($<2$ nm) is due to a monotonic solvation force, which seems not to depend on the salt concentration. Oscillatory surface forces are not observed. With permission from [73].

These two forces are combined in the well known Derjaguin–Landau, Verwey–Overbeck (DLVO) theory and a number of systems have been measured in the SFA to confirm the predictions of this theory. Figure B1.20.8 shows results obtained in aqueous solutions of NaCl on silica surfaces. The ranges of the observed long-range repulsion forces are in good agreement with the DLVO theory. The inset nicely demonstrates the effect of the monotonic short-range forces described in section B1.20.3.1. As a contrast to the results obtained on silica surfaces, figure B1.20.9 schematically displays the measured DLVO-type forces between mica surfaces in aqueous electrolytes. The main differences are that the monotonic hydration (solvation) force is dependent on the salt concentration and that there are oscillatory forces superimposed. On the mica surface, the monotonic hydration force hence seems to be mainly the result of the presence of hydrated ions in the double-layer, whereas the silica surface is strongly hydrophilic and hence 'intrinsically' hydrated [94].

### B1.20.3.3    Measuring biological interactions

Interactions between macromolecules (proteins, lipids, DNA, . . .) or biological structures (e.g. membranes) are considerably more complex than the interactions described in the two preceding paragraphs. The sum of all biological interactions at the molecular level is the basis of the complex mechanisms of *life*. In addition to

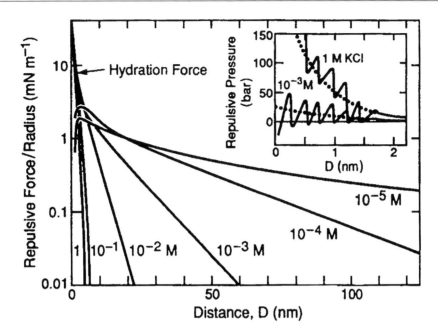

**Figure B1.20.9.** Schematic representation of DLVO-type forces measured between two mica surfaces in aqueous solutions of KNO$_3$ or KCl at various concentrations. The inset reveals the existence of oscillatory and monotonic structural forces, of which the latter clearly depend on the salt concentration. Reproduced with permission from [94].

computer simulations, direct force measurements [98], especially the surface forces apparatus, represent an invaluable tool to help understand the molecular interactions in biological systems.

Proteins can be physisorbed or covalently attached to mica. Another method is to immobilise and orient them by specific binding to receptor-functionalized planar lipid bilayers supported on the mica sheets [15]. These surfaces are then brought into contact in an aqueous electrolyte solution, while the pH and the ionic strength are varied. Corresponding variations in the force-versus-distance curve allow conclusions about protein conformation and interaction to be drawn [99]. The local electrostatic potential of protein-covered surfaces can hence be determined with an accuracy of ±5 mV.

A typical force curve showing the specific avidin–biotin interaction is depicted in figure B1.20.10. The SFA revealed the strong influence of hydration forces and membrane undulation forces on the specific binding of proteins to membrane-bound receptors [81].

Direct measurement of the interaction potential between tethered ligand (biotin) and receptor (strep-tavidin) have been reported by Wong *et al* [16] and demonstrate the possibility of controlling range and dynamics of specific biologic interactions *via* a flexible PEG-tether.

The adhesion and fusion mechanisms between bilayers have also been studied with the SFA [88, 100]. Kuhl *et al* [17] found that solutions of short-chained polymers (PEG) could produce a short-range depletion attraction between lipid bilayers, which clearly depends on the polymer concentration (figure B1.20.11). This *depletion attraction* was found to induce a membrane fusion within 10 minutes that was observed, in real-time, using FECO fringes. There has been considerable progress in the preparation of fluid membranes to mimic natural conditions in the SFA [87], which promises even more exciting discoveries in biologically relevant areas.

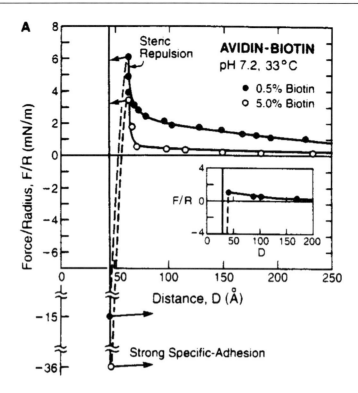

**Figure B1.20.10.** Typical force curve for a streptavidin surface interacting with a biotin surface in an aqueous electrolyte of controlled pH. This result demonstrates the power of specific protein interactions. Reproduced with permission from [81].

*B1.20.3.4   Measurements in molecular tribology*

Using friction attachments (see section B1.20.2.4), many remarkable discoveries related to thin-film and boundary lubrication have been made with the SFA. The dynamic aspect of confined molecules at a sliding interface has been extensively investigated and the SFA had laid the foundation for *molecular tribology* long before the AFM technique was available.

The often-cited Amontons' law [101, 102] describes friction in terms of a friction coefficient, which is, *a priori*, a material constant, independent of contact area or dynamic parameters, such as sliding velocity, temperature or load. We know today that all of these parameters can have a significant influence on the magnitude of the measured friction force, especially in thin-film and boundary-lubricated systems.

Using the SFA technique, it could be demonstrated that there is an intimate relationship between adhesion hysteresis and friction [28, 29, 68, 77]. Both processes dissipate energy through non-equilibrium mechanisms at the interface [30]. Friction can be represented as a sum of two terms, one adhesion related and the other load-related. It was recently shown with the SFA that, in the absence of adhesion, the load related portion linearly depends on the load and not necessarily on the *real contact area*, as commonly believed [25]. Figure B1.20.12 nicely illustrates this finding.

A traditional subject of discussion is the phenomenon of intermittent friction, also called *stick–slip* friction [35]. It is thought that interfacial molecules can switch between different dynamic states, namely between a solid-like and a liquid-like state [103]. It was also found that liquids become oriented [7] and solid-like, when confined in a narrow gap. A diffuse layering of the trapped molecules may occur (see also

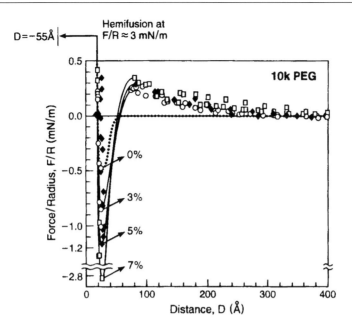

**Figure B1.20.11.** Force curves of DMPC/DPPE (dimyristoyl phosphatidylcholine and dipalmitoyl phosphatidylethanolamine) bilayers across a solution of PEG at different concentrations. Clearly visible is a concentration-dependent depletion attraction, with permission from [17].

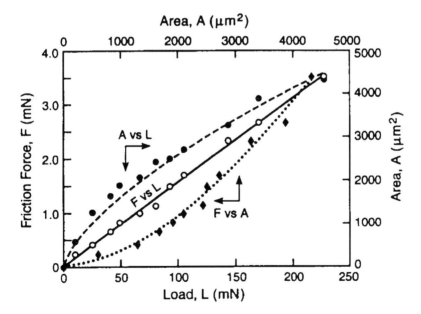

**Figure B1.20.12.** Measured friction force, $F$, and real contact area, $A$, against externally applied load, $L$, for two molecularly smooth mica surfaces sliding in 0.5 M KCl solution, i.e. in the absence of adhesion, with permission from [25].

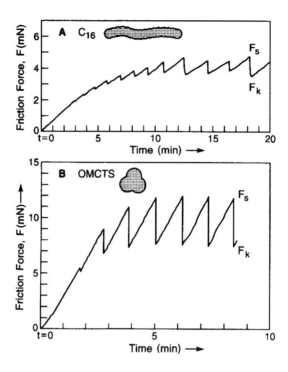

**Figure B1.20.13.** Temporal development of intermittent friction following commencement of sliding. The shape of the molecules has a great influence on history and time effects in the system. Reproduced with permission from [34].

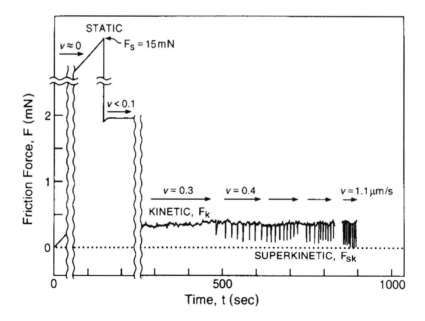

**Figure B1.20.14.** Dynamic friction at different velocities of DHDAA-coated mica (DHDAA = dihexadecyldimethylammonium acetate). Reproduced with permission from [70].

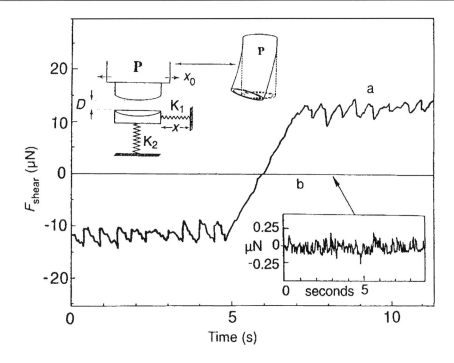

**Figure B1.20.15.** Shear force as a function of time for (a) bare mica in toluene and (b) polystyrene-covered mica in toluene. Reproduced with permission from [9].

section B1.20.3.1). The ordering mechanisms are particularly susceptible to the shape of the molecules and can spur substantial history and time effects, as illustrated in figure B1.20.13. Molecules in such ordered arrangements no longer behave as liquids and exhibit, for example, a finite yield stress [8]. When sliding above a critical speed, however, intermittent friction often disappears and the interface remains liquid-like at all times. This dynamic phenomenon, which occurs in the absence of hydrodynamic lubrication, is due to a time effect at the molecular level. Furthermore, on molecular layers of surfactants it was observed that, at even higher speeds, the system can enter a *superkinetic* regime with vanishingly small friction forces [70], as depicted in figure B1.20.14. In this high-speed regime, mechanisms of molecular entanglement are too slow to dissipate energy. New findings point out a more complex behaviour of such systems in terms of multiple relaxation mechanisms at the molecular interface [77]. It has also been shown experimentally [28] and by molecular dynamics simulation [104] that there is a potential to control interfacial dynamics with subnanometre out-of-plane excitations to achieve ultralow friction at arbitrarily slow sliding velocities.

Polymer-bearing surfaces have also attracted considerable attention in view of their particular friction properties and underlying mechanisms. Klein *et al* [9] have demonstrated the possibility of achieving ultra-low friction on surfaces covered with polymer brushes. Figure B1.20.15 displays the strong lubricating effect, which is due to a very fluid layer at the interface between polymer and solvent—a layer that remains fluid even under high compression. Accompanying molecular dynamics simulations [105] suggest that a dynamic disentanglement of the polymer chains is responsible for the observed reduction in friction.

## References

[1]   Bowden F P and Tabor D 1964 *Friction and Lubrication of Solids* Part II (Oxford: Oxford University Press)
[2]   Tabor D and Winterton R H S 1969 The direct measurement of normal and retarded van der Waals forces *Proc. R. Soc. London* A **312** 435–50
[3]   Israelachvili J N and Tabor D 1972 The measurement of van der Waals dispersion forces in the range 1.5 to 130 nm *Proc. R. Soc. London* A **331** 19–38
[4]   Israelachvili J N 1991 *Intermolecular and Surface Forces* 2nd edn (London: Academic)
[5]   Luckham P F and Manimaaran S 1997 Investigating adsorbed polymer layer behaviour using dynamic surface forces apparatuses— a review *Adv. Colloid Interface Sci.* **73** 1–46
[6]   Kelly T W *et al* 1998 Direct force measurements at polymer brush surfaces by atomic force microscopy *Macromolecules* **31** 4297–300
[7]   Dhinojwala A and Granick S 1997 Surface forces in the tapping mode: solvent permeability and hydrodynamic thickness of adsorbed polymer brushes *Macromolecules* **30** 1079–85
[8]   Luengo G, Israelachvili J N and Granick S 1996 Generalized effects in confined fluids: new friction map for boundary lubrication *Wear* **200** 328–35
[9]   Klein J *et al* 1994 Reduction of frictional forces between solid surfaces bearing polymer brushes *Nature* **370** 634–7
[10]  Mangipudi V S *et al* 1996 Measurement of interfacial adhesion between glassy polymers using the JKR method *Macromol. Symp.* **102** 131–43
[11]  Tonck A, Georges J M and Loubet J L 1988 Measurements of intermolecular forces and the rheology of dodecane between alumina surfaces *J. Colloid Interface Sci.* **126** 150–5
[12]  Borwankar R P and Case S E 1997 Rheology of emulsions, foams and gels *Curr. Opin. Colloid Interface Sci.* **2** 584–9
[13]  Luengo G *et al* 1997 Thin film rheology and tribology of confined polymer melts: contrasts with bulk properties *Macromolecules* **30** 2482–94
[14]  Demirel A L and Granick S 1996 Relaxations in molecularly thin liquid films *J. Phys.: Condens. Matter* **8** 9537–9
[15]  Leckband D 1995 The surface force apparatus—a tool for probing molecular protein interactions *Nature* **376** 617–18
[16]  Wong J Y *et al* 1997 Direct measurement of a tethered ligand–receptor interaction potential *Science* **275** 820–2
[17]  Kuhl T *et al* 1996 Direct measurement of polyethylene glycol induced depletion attraction between lipid bilayers *Langmuir* **12** 3003–14
[18]  Pincet F *et al* 1994 Long-range attraction between nucleosides with short-range specificity: direct measurements *Phys. Rev. Lett.* **73** 2780–3
[19]  Ionov R, De Coninck J and Angelova A 1996 On the origin of the long-range attraction between surface-confined DNA bases *Thin Solid Films* **284–285** 347–51
[20]  Richetti P *et al* 1996 Measurement of the interactions between two ordering surfaces under symmetric and asymmetric boundary conditions *Phys. Rev.* E **54** 1749–62
[21]  Idziak S H J *et al* 1996 Structure in a confined smectic liquid crystal with competing surface and sample elasticities *Phys. Rev. Lett.* **76** 1477–80
[22]  Idziak S H J *et al* 1996 Structure under confinement in a smectic-A and lyotropic surfactant hexagonal phase *Physica* B **221** 289–95
[23]  Giasson S, Israelachvili J N and Yoshizawa H 1997 Thin film morphology and tribology study of mayonnaise *J. Food Sci.* **62** 640–4
[24]  Luengo G *et al* 1997 Thin film rheology and tribology of chocolate *J. Food Sci.* **62** 767–72
[25]  Berman A, Drummond C and Israelachvili J N 1998 Amontons' law at the molecular level *Tribol. Lett.* **4** 95–101
[26]  Bhushan B, Israelachvili J N and Landman U 1995 Nanotribology: friction, wear and lubrication at the atomic scale *Nature* **374** 607–16
[27]  Granick S 1991 Motions and relaxations of confined liquids *Science* **253** 1374–9
[28]  Heuberger M, Drummond C and Israelachvili J N 1998 Coupling of normal and transverse motions during frictional sliding *J. Phys. Chem.* B **102** 5038–41
[29]  Israelachvili J N, Chen Y L and Yoshizawa H 1994 Relationship between adhesion and friction forces *J. Adhes. Sci. Technol.* **8** 1231–49
[30]  Israelachvili J N and Berman A 1995 Irreversibility, energy dissipation, and time effects in intermolecular and surface interactions *Israel J. Chem.* **35** 85–91
[31]  Krim J 1996 Friction at the atomic scale *Sci. Am.* **275** 74–80
[32]  Kumacheva E 1998 Interfacial friction measurement in surface force apparatus *Prog. Surf. Sci.* **58** 75–120
[33]  Reiter G, Demirel A L and Granick S 1994 From static to kinetic friction in confined liquid films *Science* **263** 1741–4
[34]  Yoshizawa H and Israelachvili J N 1993 Fundamental mechanisms of interfacial friction. 2. Stick–slip friction of spherical and chain molecules *J. Phys. Chem.* **97** 11 300–13
[35]  Yoshizawa H, McGuiggan P and Israelachvili J N 1993 Identification of a second dynamic state during stick-slip motion *Science* **259** 1305–8
[36]  Everson M P and Ohtani M 1998 New opportunities in automotive tribology *Tribol. Lett.* **5** 1–12

[37]  Tabor D 1992 *Fundamentals of Friction* ed I L Singer and H M Pollock (London: Kluwer)
[38]  Luesse C *et al* 1988 Drive mechanism for a surface force apparatus *Rev. Sci. Instrum.* **59** 811–2
[39]  Frantz P, Agrait N and Salmeron M 1996 Use of capacitance to measure surface forces. 1. Measuring distance of separation with enhanced spacial and time resolution *Langmuir* **12** 3289–94
[40]  Frantz P *et al* 1997 Use of capacitance to measure surface forces. 2. Application to the study of contact mechanics *Langmuir* **13** 5957–61
[41]  Derjaguin B V 1934 *Kolloid Zeitschrift* **69** 155–64
[42]  Stewart A M and Christenson H K 1990 Use of magnetic forces to control distance in a surface force apparatus *Meas. Sci. Technol.* **12** 1301–3
[43]  Frantz P *et al* 1997 Design of surface forces apparatus for tribology studies combined with nonlinear optical spectroscopy *Rev. Sci. Instrum.* **68** 2499–2504
[44]  Ribbe P H (ed) 1984 Micas *Reviews in Mineralogy* vol 13 (Chelsea, MI: BookCrafters)
[45]  Ohnishi S *et al* 1999 Presence of particles on melt-cut mica sheets *Langmuir* **15** 3312–6
[46]  Frantz P and Salmeron M 1998 Preparation of mica surfaces for enhanced resolution and cleanliness in the surface forces apparatus *Tribol. Lett.* **5** 151–3
[47]  Tolansky S 1948 *Multiple Beam Interferometry of Surfaces and Films* (Oxford: Oxford University Press)
[48]  Hunter S C and Nabarro F R N 1952 The origin of Glauert's superposition fringes *Phil. Mag.* **43** 538–46
[49]  Israelachvili J N 1973 Thin film studies using multiple beam interferometry *J. Colloid Interface Sci.* **44** 259–71
[50]  Clarkson M T 1989 Multiple-beam interferometry with thin metal films and unsymmetrical systems *J. Phys. D: Appl. Phys.* **22** 475–82
[51]  Born M and Wolf E 1980 *Principles of Optics* 6th edn (Oxford: Pergamon)
[52]  Horn R G and Smith D T 1991 Analytical solution for the three-layer multiple beam interferometer *Appl. Opt.* **30** 59–65
[53]  Rabinowitz P 1995 Eigenvalue analysis of the surface forces apparatus interferometer *J. Opt. Soc. Am.* A **12** 1593–601
[54]  Heuberger M, Luengo G and Israelachvili J N 1997 Topographic information from multiple beam interferometry in the surface forces apparatus *Langmuir* **13** 3839–48
[55]  Müller C, Mächtle P and Helm C A 1994 Enhanced absorption within a cavity. A study of thin dye layers with the surface forces apparatus *J. Phys. Chem.* **98** 11 119–25
[56]  Mächtle P, Müller C and Helm C A 1994 A thin absorbing layer at the center of a Fabry–Pérot interferometer *J. Physique* II **4** 481–500
[57]  Levins J M and Vanderlick T K 1994 Extended spectral analysis of multiple beam interferometry: a technique to study metallic films in the surface forces apparatus *Langmuir* **10** 2389–94
[58]  Farrell B, Bailey A I and Chapman D 1995 Experimental phase changes at the mica–silver interface illustrate the experimental accuracy of the central film thickness in a symmetrical three-layer interferometer *Appl. Opt.* **34** 2914–20
[59]  Levins J M and Vanderlick T K 1993 Impact of roughness of reflective films on the application of multiple beam interferometry *J. Colloid Interface Sci.* **158** 223–7
[60]  Levins J M and Vanderlick T K 1992 Reduction of the roughness of silver films by the controlled application of surface forces *J. Phys. Chem.* **96** 10 405–11
[61]  Horn R G *et al* 1996 The effect of surface and hydrodynamic forces on the shape of a fluid drop approaching a solid surface *J. Phys.: Condens. Matter* **8** 9483–90
[62]  Israelachvili J N and Adams G E 1976 Direct measurement of long range forces between two mica surfaces in aqueous KNO3 solutions *Nature* **262** 774
[63]  Israelachvili J N 1987 Direct measurements of forces between surfaces in liquids at the molecular level *Proc. Nat. Acad. Sci. USA* **84** 4722–5
[64]  Israelachvili J N 1989 Techniques for direct measurement of forces between surfaces in liquids at the atomic scale *Chemtracts—Anal. Phys. Chem.* **1** 1–12
[65]  Israelachvili J N and McGuiggan P M 1990 Adhesion and short-range forces between surfaces. I. New apparatus for surface force measurements *J. Mater. Res.* **5** 2223–31
[66]  Parker J L, Christenson H K and Ninham B W 1989 Device for measuring the force and separation between two surfaces down to molecular separations *Rev. Sci. Instrum.* **60**
[67]  Klein J 1983 *J. Chem. Soc. Faraday Trans.* I **79** 99
[68]  Dan N 1996 Time-dependent effects in surface forces *Current Opinion Colloid Interface Sci.* **1** 48–52
[69]  Kutzner H B, Luckham P F and Rennie J 1996 Measurement of the viscoelastic properties of thin surfactant films *Faraday Discuss.* **104** 9–16
[70]  Yoshizawa H, Chen Y L and Israelachvili J N 1993 Recent advances in molecular level understanding of adhesion, friction and lubrication *Wear* **168** 161–6
[71]  Peachey J, van Alsten J and Granick S 1991 Design of an apparatus to measure the shear response of ultrathin liquid films *Rev. Sci. Instrum.* **62** 463–73
[72]  Idziak S H J *et al* 1994 The x-ray surface forces apparatus: structure of a thin smectic liquid crystal film under confinement *Science* **264** 1915–8
[73]  Horn R, Smith D T and Haller W 1989 Surface forces and viscosity of water measured between silica sheets *Chem. Phys. Lett.* **162** 404–8

[74] Ruths M, Steinberg S and Israelachvili J N 1996 Effects of confinement and shear on the properties of thin films of thermotropic liquid crystal *Langmuir* **12** 6637–50

[75] Wanless E J and Christenson H K 1994 Interaction between surfaces in ethanol: adsorption, capillary condensation, and solvation forces *J. Chem. Phys.* **101** 4260–7

[76] Dedinaite A *et al* 1998 Interactions between modified mica surfaces in triglyceride media *Langmuir* **14** 5546–54

[77] Yamada S and Israelachvili J N 1998 Friction and adhesion hysteresis of fluorocarbon surfactant monolayer-coated surfaces measured with the surface forces apparatus *J. Phys. Chem.* B **102** 234–44

[78] Ruths M and Granick S 1998 Rate-dependent adhesion between opposed perfluoropoly (alkyl ether) layers: dependence on chain-end functionality and chain length *J. Phys. Chem.* B **102** 6056–63

[79] Lowack K and Helm C A 1998 Molecular mechanisms controlling the self-assembly process of polyelectrolyte multilayers *Macromolecules* **31** 823–33

[80] Luengo G *et al* 1997 Measurement of the adhesion and friction of smooth C-60 surfaces *Chem. Mater.* **9** 1166–71

[81] Leckband D *et al* 1994 Direct force measurements of specific and nonspecific protein interactions *Biochemistry* **33** 4611–23

[82] Chowdhury P B and Luckham P F 1995 Interaction forces between kappa-casein adsorbed on mica *Colloids Surfaces* B **4** 327–34

[83] Holmberg M *et al* 1997 Surface force studies of Langmuir–Blodgett cellulose films *J. Colloid Interface Sci.* **186** 369–81

[84] Kuhl T L *et al* 1994 Modulation of interaction forces between bilayers exposing short-chained ethylene oxide headgroups *Biophys. J.* **66** 1479–88

[85] Nylander T and Wahlgren N M 1997 Forces between adsorbed layers of beta-casein *Langmuir* **13** 6219–25

[86] Yu Z W, Calvert T L and Leckband D 1998 Molecular forces between membranes displaying neutral glycosphingolipids: evidence for carbohydrate attraction *Biochemistry* **37** 1540–50

[87] Seitz M *et al* 1998 Formation of tethered supported bilayers via membrane-inserting reactive lipids *Thin Solid Films* **327–9** 767–71

[88] Wolfe J *et al* 1991 The interaction and fusion of bilayers formed from unsaturated lipids *Eur. Biophys. J.* **19** 275–81

[89] Rutland M W and Parker J L 1994 Surface forces between silica surfaces in cationic surfactant solutions: adsorption and bilayer formation at normal and high pH *Langmuir* **10** 1110–21

[90] Xu Z H, Ducker W and Israelachvili J N 1996 Forces between crystalline alumina (sapphire) surfaces in aqueous sodium dodecyl sulfate surfactant solutions *Langmuir* **12** 2263–70

[91] Horn R G, Clarke D R and Clarkson M T 1988 Direct measurement of surface forces between sapphire crystals in aqueous solutions *J. Mater. Res.* **3** 413–6

[92] Cleveland J P, Schäffer T E and Hansma P K 1995 Probing oscillatory hydration potentials using thermal-mechanical noise in an atomic force microscope *Phys. Rev.* B **52** R8692–5

[93] Israelachvili J N and Pashley R M 1983 Molecular layering of water at surfaces and origin of repulsive hydration forces *Nature* **306** 249–50

[94] Israelachvili J N, McGuiggan P and Horn R 1992 Basic physics of interactions between surfaces in dry, humid, and aqueous environments *1st Int. Symp. on Semiconductor Wafer Bondings: Science, Technology and Applications* (Pennington, NJ: Electrochemical Society)

[95] Stanley H E 1999 Unsolved mysteries of water in its liquid and glass states *MRS Bull.* May 22–30

[96] Israelachvili J N 1996 Role of hydration and water structure in biological and colloidal interactions *Nature* **379** 219–25

[97] Müller H J 1998 Extraordinarily thick water films on hydrophilic solids: a result of hydrophobic repulsion? *Langmuir* **14** 6789–92

[98] Pierres A, Benoliel A M and Bongrand P 1996 Measuring bonds between surface-associated molecules *J. Immunol. Methods* **196** 105–20

[99] Leckband D *et al* 1993 Measurements of conformational changes during adhesion of lipid and protein (polylysine and S-layer) surfaces *Biotech. Bioeng.* **42** 167–77

[100] Helm C A, Israelachvili J N and McGuiggan P M 1992 Role of hydrophobic forces in bilayer adhesion and fusion *Biochemistry* **31** 1794–805

[101] Amontons G 1699 De la résistance causé dans les machines *Mémoires de l'Académie Royale* A 275–82

[102] Coulomb C A 1785 Théorie des machines simples *Mémoire de Mathématique et de Physique de l'Académie Royale* 161–342

[103] Gee M L *et al* 1990 Liquid to solidlike transitions of molecularly thin films under shear *J. Chem. Phys.* **93** 1895–905

[104] Gao J, Luedtke W and Landman U 1998 Friction control in thin-film lubrication *J. Phys. Chem.* B **102** 5033–7

[105] Grest G S 1996 Interfacial sliding of polymer brushes: a molecular dynamics simulation *Phys. Rev. Lett.* **76** 4979–82

## Further Reading

Derjaguin B V 1934 *Research in Surface Forces* (New York: Consultants Bureau)

    An old classic: four volumes.

Israelachvili J N 1991 *Intermolecular and Surface Forces* (London: Academic)

    The most often cited reference about surface forces and SFA.

Hutchings I M 1992 *Friction and Wear of Engineering Materials* (London: Arnold)

   A good introduction to tribology.

Bhushan B 1999 *Handbook of Micro/Nano Tribology* (Boca Raton, FL: CRC)

   A valuable reference to anyone involved with friction at small scales.

# B1.21
# Surface structural determination: diffraction methods

*Michel A Van Hove*

### B1.21.1  Introduction

Diffraction methods have provided the large majority of solved atomic-scale structures for both the bulk materials and their surfaces, mainly in the crystalline state. Crystallography by diffraction tends to filter out defects and focus on the periodic part of a structure. By adding contributions from very many unit cells, diffraction gives results that are, in effect, averaged over space and time. This is excellent for investigating stable states of solid matter as they occur in well crystallized samples; some forms of disorder can also be analysed reasonably well. Diffraction, however, is much less appropriate for examining inhomogeneous and time-dependent events such as transition states and pathways in chemical reactions.

For bulk structural determination (see chapter B1.9), the main technique used has been x-ray diffraction (XRD). Several other techniques are also available for more specialized applications, including: electron diffraction (ED) for thin film structures and gas-phase molecules; neutron diffraction (ND) and nuclear magnetic resonance (NMR) for magnetic studies (see chapters B1.12 and B1.13); x-ray absorption fine structure (XAFS) for local structures in small or unstable samples and other spectroscopies to examine local structures in molecules. Electron microscopy also plays an important role, primarily through imaging (see chapter B1.17).

At surfaces, the primary challenge is to obtain the desired surface sensitivity. Ideally, one wishes to gain structural information about those atomic layers which differ in their properties from the underlying bulk material. This means in practice extracting the structure of the first few monolayers, i.e. atoms within about 5–10 Å (0.5–1 nm) of the vacuum above the surface. The above-mentioned bulk methods, if applied unchanged, do not easily provide sensitivity to this very thin slice of matter. The challenge becomes even greater when dealing with an interface between two materials, including solid/liquid and solid/gas interfaces. A number of mechanisms are available to obtain surface sensitivity on the required depth scale. We shall describe some of them in the next section, with emphasis on the solid/vacuum interface.

However, it is necessary to first discuss the meaning of 'diffraction', because this concept can be interpreted in several ways. After these fundamental aspects are dealt with, we will take a statistical and historical view of the field. It will be seen that many different diffraction methods are available for surface structural determination.

It will also be useful to introduce concepts of two-dimensional ordering and the corresponding nomenclature used to characterize specific structures. We can then describe how the surface diffraction pattern relates to the ordering and, thus, provides important two-dimensional structural information.

We will, in the latter part of this discussion, focus only on those few methods that have been the most productive, with low-energy electron diffraction (LEED) receiving the most attention. Indeed, LEED has been the most successful surface structural method in two quite distinct ways. First, LEED has become an almost universal characterization technique for single-crystal surfaces: the diffraction pattern is easily imaged in real time and is very helpful in monitoring the state of the surface in terms of the ordering and, hence, also density, of adsorbed atoms and molecules. Second, LEED has been quite successful in determining the detailed atomic positions at a surface (e.g., interlayer distances, bond lengths and bond angles), especially for ordered structures. This relies primarily on simulating the intensity (current) of diffracted beams as a function of electron energy in order to fit assumed model structures to measured data. Because of multiple scattering, such simulation and fitting is a very different and much more difficult task than looking at a diffraction pattern.

We will close with a description of the state of the art and an outlook on the future of the field.

## B1.21.2    Fundamentals of surface diffraction methods

### B1.21.2.1    Diffraction

#### (a) Diffraction and structure

Diffraction is based on wave interference, whether the wave is an electromagnetic wave (optical, x-ray, etc), or a quantum mechanical wave associated with a particle (electron, neutron, atom, etc), or any other kind of wave. To obtain information about atomic positions, one exploits the interference between different scattering trajectories among atoms in a solid or at a surface, since this interference is very sensitive to differences in path lengths and hence to relative atomic positions (see chapter B1.9).

It is relatively straightforward to determine the *size and shape of the three- or two-dimensional unit cell* of a periodic bulk or surface structure, respectively. This information follows from the *exit directions* of diffracted beams relative to an incident beam, for a given crystal orientation: measuring those exit angles determines the unit cell quite easily. But no *relative positions* of atoms within the unit cell can be obtained in this manner. To achieve that, one must measure *intensities* of diffracted beams and then computationally analyse those intensities in terms of atomic positions.

With XRD applied to bulk materials, a detailed structural analysis of atomic positions is rather straight-forward and routine for structures that can be quite complex (see chapter B1.9): *direct methods* in many cases give good results in a single step, while the resulting atomic positions may be refined by iterative fitting procedures based on simulation of the diffraction process.

With ED, by contrast, the task is more complicated due to *multiple scattering* of the electrons from atom to atom (see chapter B1.17). Such multiple scattering is especially strong at the relatively low energies employed to study surfaces. This dramatically restricts the application of direct methods and strongly increases the computational cost of simulating the diffraction process. As a result, an iterative *trial-and-error fitting* is the method of choice with ED, even though it can be a slow process when many trial structures have to be tested.

Also, the result of any diffraction-based trial-and-error fitting is not necessarily unique: it is always possible that there exists another untried structure that would give a better fit to experiment. Hence, a multi-technique approach that provides independent clues to the structure is very fruitful and common in surface science: such clues include chemical composition, vibrational analysis and position restrictions implied by other structural methods. This can greatly restrict the number of trial structures which must be investigated.

#### (b) Non-periodic structures

Diffraction is not limited to periodic structures [1]. Non-periodic imperfections such as defects or vibrations, as well as sample-size or domain effects, are inevitable in practice but do not cause much difficulty or can be taken into account when studying the ordered part of a structure. Some other forms of disorder can also be handled quite well in their own right, such as *lattice-gas* disorder in which a given site in the unit cell is randomly occupied with less than 100% probability. At surfaces, lattice-gas disorder is very common when atoms or molecules are adsorbed on a substrate. The local adsorption structure in the given site can be studied in detail.

*(c) Non-planar initial waves*

More fundamental is the distinction between planar and spherical initial waves. In XRD, for instance, the incident x-rays are well described by plane waves; this is generally true of probes that are aimed at the sample from macroscopic distances, as is the case also in most forms of ED and ND. However, there are techniques in which a wave is generated locally within the sample, for instance through emission of an x-ray (by fluorescence) or an electron (by photoemission) from a sample atom. In such *point-source emission*, the wave which performs the useful diffraction initially has a spherical rather than planar character; it is centred on the nucleus of an atom, with a rapidly decaying amplitude as it travels away from the emitting site. (Depending on the excitation mechanism, this initial wave need not be spherically symmetrical, but may also have an angular variation, as given by spherical harmonics, for instance, or combinations thereof.)

This spherical outgoing wave can diffract only from atoms that are near to the emitting atom, mainly those atoms within a distance of a few atomic diameters. In these circumstances, the crystallinity of the sample is of less importance: the diffracting wave sees primarily the *local* atomic-scale neighbourhood of the emitting atoms. As long as the same local neighbourhood predominates everywhere in the sampled part of the surface, information about the structure of that neighbourhood can be extracted. It also helps very much if the local neighbourhood has a constant orientation, so that the experiment does not average over a multitude of orientations, since these tend to average out diffraction effects and thus wash away structural information.

*(d) Variety of diffraction methods*

From the above descriptions, it becomes apparent that one can include a wide variety of techniques under the label 'diffraction methods'. Table B1.21.1 lists many techniques used for surface structural determination, and specifies which can be considered diffraction methods due to their use of wave interference (table B1.21.1 also explains many technique acronyms commonly used in surface science). The diffraction methods range from the classic case of XRD and the analogous case of LEED to much more subtle cases like XAFS (listed as both SEXAFS (surface extended XAFS) and NEXAFS (near-edge XAFS) in the table).

XAFS is a good example of less obvious diffraction [2, 3]. In XAFS, an electron is emitted by an x-ray locally within the sample. It propagates away as a spherical wave, which is allowed to back-scatter from neighbouring atoms to the emitter atom. The back-scattered electron wave interferes at the emitting atom with the emitted wave, thereby modulating the probability of the emitting process itself when the energy (wavelength) is varied: as one cycles through constructive and destructive interferences, the emission probability oscillates with a period that reflects the interatomic distances. This emission probability is, however, measured through yet another process (e.g., absorption of the incident x-rays, or emission of other x-rays or other electrons), which oscillates in synchrony with the interference. Thus, the structure-determining diffraction is in such a case buried relatively deeply in the overall process, and does not closely resemble the classic plane-wave diffraction of XRD.

*B1.21.2.2   Surface sensitivity*

There are several approaches to gain the required surface sensitivity with diffraction methods. We review several of these here, emphasizing the case of solid/vacuum interfaces; some of these also apply to other interfaces.

*(a) Short mean free path*

One obvious method to obtain surface sensitivity is to choose probes and conditions that give shallow penetration. This can be achieved through a short *mean free path* $\lambda$, i.e. a short average distance until the probe (e.g.,

**Table B1.21.1.** Surface structural determination methods. The second column indicates whether a technique can be considered a diffraction method, in the sense of relying on wave interference. Also shown are statistics of surface structural determinations, extracted from the Surface Structure Database [14], up to 1997. Counted here are only 'detailed' and complete structural determinations, in which typically the experiment is simulated computationally and atomic positions are fitted to experiment. (Some structural determinations are performed by combining two or more methods: those are counted more than once in this table, so that the columns add up to more than the actual 1113 structural determinations included in the database.)

| Surface structural determination method | Diffraction method? | Number of structural determinations | Percentage of structural determinations |
|---|---|---|---|
| LEED | yes | 751 | 67.5 |
| IS (including LEIS, MEIS and HEIS for low-, medium- and high-energy ion scattering) | no | 102 | 9.2 |
| PD (covers a variety of other acronyms, like ARPEFS, ARXPD, ARXPS, ARUPS, NPD, OPD, PED) | yes | 88 | 7.9 |
| SEXAFS | yes | 67 | 6.0 |
| XSW | yes | 52 | 4.7 |
| XRD (also GIXS, GIXD) | yes | 40 | 3.6 |
| TOF-SARS (time-of-flight scattering and recoiling spectrometry) | no | 13 | 1.2 |
| NEXAFS (also called XANES) | yes | 11 | 1.0 |
| RHEED | yes | 10 | 0.9 |
| LEPD | yes | 5 | 0.4 |
| HREELS | yes | 4 | 0.4 |
| MEED | yes | 3 | 0.3 |
| AED | yes | 3 | 0.3 |
| SEELFS (surface extended energy loss fine structure) | yes | 2 | 0.2 |
| TED | yes | 1 | 0.1 |
| AD | yes | 1 | 0.1 |
| STM | no | 1 | 0.1 |

x-ray or electron) is absorbed by energy loss or is otherwise removed from the useful diffraction channels. For typical x-rays, $\lambda$ is of the order of micrometres in many materials, which is too large compared to the desired surface thickness [4].

But for electrons of low kinetic energies, i.e. $E \approx 10$–$1000$ eV, the mean free path $\lambda$ is of the order of 5–20 Å [5]. The mean free path has a minimum in the 100–200 eV range, with larger mean free paths existing both below and above this range.

Such ideal low mean free paths are the basis of LEED, the technique that has been used most for determining surface structures on the atomic scale. This is also the case of photoelectron diffraction (PD): here, the mean free path of the emitted electrons restricts sensitivity to a similar depth (actually double the depth of LEED, since the incident x-rays in PD are only weakly attenuated on this scale).

*(b) Grazing incidence and/or emergence*

Another approach to limit the penetration of the probe into the surface region is to use *grazing incidence* and/or *grazing emergence*; this works for those probes that already have a reasonably small mean free path $\lambda$. A grazing angle $\theta$ (measured from the surface normal, i.e., $\theta$ close to $90°$) then allows the probe to penetrate to a depth of only about $\lambda \cos(\theta)$. This approach is used primarily for higher-energy electrons above about 1000 eV in a technique called reflection high-energy electron diffraction (RHEED) [6].

With XRD, however, the mean free path is still too long to make this approach practical by itself [4]: as an example, to obtain even 100 Å penetration, one would typically need to use a grazing angle of about $0.05°$, which is technically extremely demanding. The penetration depth is proportional to the grazing angle of incidence at such small angles, so that a ten times smaller penetration depth requires a further tenfold reduction in grazing angle. In addition, such small grazing angles require samples with a flatness that is essentially impossible to achieve, in order that the x-rays see a flat surface rather than a set of ridges that shadow much of the surface.

*(c) Total external reflection*

In XRD, surface sensitivity can, however, be achieved through another phenomenon [4]: *total external reflection*. This also occurs at grazing angles of incidence, giving rise to the technique acronym of GIXS for grazing-incidence x-ray scattering. At angles within approximately $0.5°$ of $\theta = 90°$, x-rays cannot penetrate by refraction into materials: the laws of optics imply that the wave velocity of refracted waves in the material would have to be larger than the speed of light under those circumstances, which is impossible for propagating waves.

Instead, the incident wave is totally reflected. However, this is accompanied by a shallow penetration of waves that decay exponentially into the bulk while propagating parallel to the surface. Under such conditions, the decay length into the surface is of the order of 10–30 Å, as desired. This penetration depth depends on the material and not on the wavelength of the x-rays. Note that total external reflection does not require vacuum: it can occur at various kinds of interfaces, depending on the relative optical constants of the phases in contact.

*(d) High-surface area materials*

None of the above methods is sufficient for neutrons, however. Neutrons penetrate matter so easily that the only effective approach is to use materials with a very high surface-to-volume ratio. This can be accomplished with small particles and exfoliated graphite, for instance, but the technique has essentially been abandoned in surface studies [7, 8].

*(e) Superlattice diffraction*

One further method for obtaining surface sensitivity in diffraction relies on the presence of two-dimensional *superlattices* on the surface. As we shall see further below, these correspond to periodicities that are different from those present in the bulk material. As a result, additional diffracted beams occur (often called fractional-order beams), which are uniquely created by and therefore sensitive to this kind of surface structure. XRD, in particular, makes frequent use of this property [4]. Transmission electron diffraction (TED) also has used this property, in conjunction with ultrathin samples to minimize bulk contributions [9].

*(f) Hybrid methods*

As we have seen, the electron is the easiest probe to make surface sensitive. For that reason, a number of hybrid techniques have been designed that combine the virtues of electrons and of other probes. In particular, electrons and photons (x-rays) have been used together in techniques like PD [10] and SEXAFS (or EXAFS, which is the high-energy limit of XAFS) [2, 11]. Both of these rely on diffraction by electrons, which have been excited by photons. In the case of PD, the electrons themselves are detected after emission out of the surface, limiting the depth of 'sampling' to that given by the electron mean free path.

*(g) Elemental and chemical-state resolution*

With some techniques, another mechanism can give high surface sensitivity, namely *elemental resolution* through spectroscopic filtering of emitted electrons or x-rays. In this approach, one detects, by setting an energy window, only those electrons or x-rays that are emitted by a particular kind of atom, since each electronic level produces a line at a particular energy given by the level energy augmented by the excitation energy.

Thus, if a 'foreign' element is present only at the surface, one can detect a signal that only comes from that element and, therefore, only from the surface. Given sufficient energy resolution, one can even differentiate electrons coming from the same atoms in different bonding environments: e.g., in the case of a clean surface, atoms of the outermost layer *versus* bulk atoms [10]. This *chemical-state resolution* is due to the fact that electronic levels are shifted by bonding to other atoms, resulting in different emitted lines from atoms in different bonding situations.

Elemental and chemical-state resolution affords the possibility of detecting only a monolayer or even a fraction of a monolayer. This approach is prevalent in PD and in methods based on x-ray fluorescence.

It is also used in SEXAFS [11]: as we have seen, photoexcited electrons are back-reflected to the photoemitting atoms, thereby modulating the x-ray absorption cross-section through electron wave interference, after which a secondary electron or ion or fluorescent x-ray is ejected from the surface and finally detected. This latter ejection process provides surface sensitivity, through the electronic mean free path or the shallowness of ionic emission. However, elemental and chemical-state selection by energy filtering is essentially universal here, and again can give monolayer resolution with emission from foreign surface atoms different from the bulk atoms.

A similar device can be applied to a form of x-ray diffraction called the x-ray standing wave (XSW) method [12, 13], as detected by fluorescence. Here, x-ray waves reflected from bulk atomic planes form a standing wave pattern near the surface. The maxima and minima of this standing wave pattern can be arranged to fall at different locations on the atomic scale, by varying the energy and incidence angles. Thereby, the induced fluorescence varies with the location of those maxima and minima. Since the fluorescence is element specific, one can thus determine positions of foreign surface atoms relative to the extended bulk lattice (it remains difficult, however, to locate those substrate atoms that are close to the fluorescing surface atoms, because they are drowned by the bulk signal).

## B1.21.3    Statistics of full structural determinations

Many methods have been developed to determine surface structure: we have mentioned several in the previous section and there are many more. To get an idea of their relative usage and importance, we here examine historical statistics. We also review the kinds of surface structure that have been studied to date, which gives a feeling for the kinds of surface structures that current methods and technology can most easily solve. This will provide an overview of the range of surfaces for which detailed surface structures are known, and those for which very little is known.

As source of information we use the Surface Structure Database [14], a critical compilation of surface structures solved in detail, covering the period to the end of 1997. It contains 1113 structural determinations with, on average, two determinations for each structure: thus there are approximately 550 distinct solved structures available.

In terms of individual techniques, table B1.21.1 lists the breakdown totalled over time, counting from the inception of surface structural determination in the early 1970s. It is seen that LEED has contributed altogether about 67% of all structural determinations included in the database. The annual share of LEED was 100% until 1978, and has generally remained over 50% since then. In 1979 other methods started to produce structural determinations, especially PD, ion scattering (IS) and SEXAFS. XRD and then XSW started to contribute results in the period 1981–3.

As the table shows, a host of other techniques have contributed a dozen or fewer results each. It is seen that diffraction techniques have been very prominent in the field: the major diffraction methods have been LEED, PD, SEXAFS, XSW, XRD, while others have contributed less, such as NEXAFS, RHEED, low-energy position diffraction (LEPD), high-resolution electron energy loss spectroscopy (HREELS), medium-energy electron diffraction (MEED), Auger electron diffraction (AED), SEELFS, TED and atom diffraction (AD). The major non-diffraction method is IS, which is described in chapter B1.23.

The database provides interesting perspectives on the evolution of surface structural determination since its inception around 1970. Not surprisingly, there is a clear temporal trend toward more complex and more diverse materials, such as compound substrates, alloyed bimetallic surfaces, complex adsorbate-induced relaxations and reconstructions, epitaxial and pseudomorphic growth, alkali adsorption on semiconductor and transition metal substrates, and molecular adsorbates as well as co-adsorbates on metal surfaces. The complexity of some solved structures has grown to about 100 times that of the earliest structures. The range of structure types can also be gauged, for instance, from the list of substrate lattice categories included in the SSD database: bcc, $CdCl_2$, $CdI_2$, corundum, CsCl, CuAu I, $Cu_3Au$, diamond, fcc, fluorite, graphite, hcp, hexagonal, NaCl, perovskite, rutile, spinel, wurtzite, zincblende, $2H–MoS_2$, $2H–NbSe_2$ and $6H–SiC$.

Nonetheless, when counting all structures solved over time, one finds a strong predominance of studies in certain narrow categories, as exhibited by the following uneven statistics:

- fcc metals far outdistance any other substrate lattice type, with 60% of the total;
- the diamond lattice (C, Si and Ge) forms the next most numerous lattice category, about 10%, followed by the bcc (9%) and hcp (7%) lattices;
- elemental solids (with or without foreign adsorbates) form 85% of the substrates examined, the rest being metallic alloys (7%) or other compounds (8%);
- the surfaces of *non-reconstructed* elemental metal substrates (with or without adsorbates) constitute about 77% of the results; the remainder are *reconstructed*, i.e. have undergone a substantial structural change from the ideal termination of the bulk lattice, involving bond breaking and/or bond making;
- looking at electronic properties, metals again dominate heavily, with 81% of the total, followed by semiconductors (16%), insulators (3%) and semimetals (less than 1%);
- atomic overlayers comprise about 54% of all types of adsorption, as opposed to interstitial (1%) or substitutional (5%) underlayers, molecular overlayers (10%), multilayers (9%) or mixes of these adsorption modes.

There is much room for further study of various important categories of materials: one prominent example is oxides and other compounds (carbides, nitrides, . . .); another is all types of adsorption on oxides and other compounds.

However, recent advances in techniques will ensure further diversification and complexification of solved surface structures. The present maturity of techniques will thus increasingly allow the analysis of structures chosen for their practical interest rather than for their simplicity.

### B1.21.4  Two-dimensional ordering and nomenclature

In diffraction, the degree and kind of structural ordering is an important consideration, since the diffraction reflects those structural properties. As a result, diffraction methods are ideal for characterizing the degree and type of ordering that a surface exhibits. In particular, at surfaces, LEED has always been a favourite tool for 'fingerprinting' a particular state of ordering of a surface, enhancing experimental reproducibility. It is therefore useful to first briefly examine the forces that are responsible for the variety of ordering types that occur at surfaces. Then, we can introduce standard notation to succinctly describe specific forms of ordering that occur at surfaces.

### B1.21.4.1   Two-dimensional ordering

A large number of ordered surface structures can be produced experimentally on single-crystal surfaces, especially with adsorbates [15]. There are also many disordered surfaces. Ordering is driven by the interactions between atoms, ions or molecules in the surface region. These forces can be of various types: covalent, ionic, van der Waals, etc; and there can be a mix of such types of interaction, not only within a given bond, but also from bond to bond in the same surface. A surface could, for instance, consist of a bulk material with one type of internal bonding (say, ionic). It may be covered with an overlayer of molecules with a different type of intramolecular bonding (typically covalent); and the molecules may be held to the substrate by yet another form of bond (e.g., van der Waals).

Strong adsorbate–substrate forces lead to *chemisorption*, in which a chemical bond is formed. By contrast, weak forces result in *physisorption*, as one calls non-chemical 'physical' adsorption.

The balance between these different types of bonds has a strong bearing on the resulting ordering or disordering of the surface. For adsorbates, the relative strength of adsorbate–substrate and adsorbate–adsorbate interactions is particularly important. When adsorbate–substrate interactions dominate, well ordered overlayer structures are induced that are arranged in a *superlattice*, i.e. a periodicity which is closely related to that of the substrate lattice: one then speaks of *commensurate* overlayers. This results from the tendency for each adsorbate to seek out the same type of *adsorption site* on the surface, which means that all adsorbates attempt to bond in the same manner to substrate atoms.

An example of commensurate overlayers is provided by atomic sulfur chemisorbed on a Ni(100) surface: all S atoms tend to adsorb in the *fourfold coordinated hollow sites*, i.e. each S atom tries to bond to four Ni atoms. At typical high coverages and moderate temperatures, this results in an ordered array of S atoms on the Ni(100) surface. However, high temperatures will disorder such overlayers; also, this layer may be kinetically disordered during its formation, as a result of gradual addition of sulfur atoms before they manage to order. The same is often true of molecular adsorption. Although intramolecular bonding can be strong enough to keep an adsorbed molecular species intact despite its bonding to the substrate, there is usually only a relatively weak mutual interaction among adsorbed molecular species.

Relatively strong adsorbate–adsorbate interactions have a different effect: the adsorbates attempt to first optimize the bonding between them, before trying to satisfy their bonding to the substrate. This typically results in close-packed overlayers with an internal periodicity that it is not matched, or at least is poorly matched, to the substrate lattice. One thus finds well ordered overlayers whose periodicity is generally not closely related to the substrate lattice: this leads to so-called *incommensurate* overlayers. Such behaviour is best exemplified by very cohesive overlayers like graphite sheets or oxide thin films that adopt their own preferred lattice constant regardless of the substrate material on which they are adsorbed.

### B1.21.4.2   Coverage and monolayer definitions

It is useful to define the terms *coverage* and *monolayer* for adsorbed layers, since different conventions are used in the literature. The surface coverage measures the two-dimensional density of adsorbates. The most common definition of coverage sets it to be equal to one monolayer (1 ML) when each two-dimensional surface unit cell of the unreconstructed substrate is occupied by one adsorbate (the adsorbate may be an atom or a molecule). Thus, an overlayer with a coverage of 1 ML has as many atoms (or molecules) as does the outermost single atomic layer of the substrate.

However, many adsorbates cannot reach a coverage of 1 ML as defined in this way: this occurs most clearly when the adsorbate is too large to fit in one unit cell of the surface. For example, benzene molecules normally lie flat on a metal surface, but the size of the benzene molecule is much larger than typical unit cell areas on many metal surfaces. Thus, such an adsorbate will saturate the surface at a lower coverage than 1 ML; deposition beyond this coverage can only be achieved by starting the growth of a second layer on top of the first layer.

It is thus tempting to define the first saturated layer as being one monolayer, and this often done, causing some confusion. One therefore also often uses terms like *saturated monolayer* to indicate such a single adsorbate layer that has reached its maximal two-dimensional density. Sometimes, however, the word 'saturated' is omitted from this definition, resulting in a different notion of monolayer and coverage. One way to reduce possible confusion is to use, for contrast with the saturated monolayer, the term *fractional monolayer* for the term that refers to the substrate unit cell rather than the adsorbate size as the criterion for the monolayer density.

### B1.21.4.3  Two-dimensional crystallographic nomenclature

#### (a) Miller indices

Single-crystal surfaces are characterized by a set of Miller indices that indicate the particular crystallographic orientation of the surface plane relative to the bulk lattice [5]. Thus, surfaces are labelled in the same way that atomic planes are labelled in bulk x-ray crystallography. For example, a Ni(111) surface has a surface plane that is parallel to the (111) crystallographic plane of bulk nickel. Thus, the Ni(111) surface exposes a hexagonally close-packed layer of atoms, given that nickel has a face-centred close-packed (fcc) cubic bulk lattice, see figure B1.21.1(a). Some authors use the more correct notation {111} instead of (111), as is common in bulk crystallography to emphasize that the (111) plane is only one of several symmetrically equivalent plane orientations, like $(11\bar{1})$, $(\bar{1}11)$, etc. The {111} notation implicitly includes all such equivalent planes.

Figure B1.21.1 shows a number of other clean unreconstructed low-Miller-index surfaces. Most surfaces studied in surface science have low Miller indices, like (111), (110) and (100). These planes correspond to relatively close-packed surfaces that are atomically rather smooth. With fcc materials, the (111) surface is the densest and smoothest, followed by the (100) surface; the (110) surface is somewhat more 'open', in the sense that an additional atom with the same or smaller diameter can bond directly to an atom in the *second* substrate layer. For the hexagonal close-packed (hcp) materials, the (0001) surface is very similar to the fcc(111) surface: the difference only occurs deeper into the surface, namely in the fashion of stacking of the hexagonal close-packed monolayers onto each other (ABABAB... *versus* ABCABC..., in the convenient layer-stacking notation). The hcp($10\bar{1}0$) surface resembles the fcc(110) surface to some extent, in that it also presents open troughs between close-packed rows of atoms, exposing atoms in the second layer. With the body-centred cubic (bcc) materials, the (110) surface is the densest and smoothest, followed by the (100) surface; in this case, the (111) surface is rather more open and atomically 'rough'.

#### (b) High-Miller-index or stepped surfaces

The atomic structures of high-Miller-index surfaces are composed of *terraces*, separated by *steps*, which may have *kinks* in them [5]. Examples are shown in figure B1.21.2. Thus, the (755) surface of an fcc crystal consists of (111) terraces, six atoms deep (from one step to the next), separated by straight steps of (100) orientation and of single-atom height. The fcc(10,8,7) has 'kinks' in its step edges, i.e. the steps themselves are not straight. The steps and kinks provide a degree of roughness that can be very important as sites for chemical reactions or for nucleation of crystal growth.

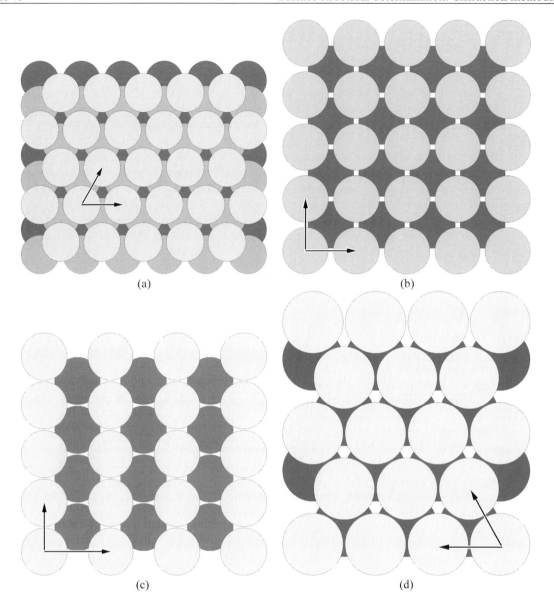

**Figure B1.21.1.** Atomic hard-ball models of low-Miller-index bulk-terminated surfaces of simple metals with face-centred close-packed (fcc), hexagonal close-packed (hcp) and body-centred cubic (bcc) lattices: (a) fcc(111)–(1 × 1); (b) fcc(100)–(1 × 1); (c) fcc(110)–(1 × 1); (d) hcp(0001)–(1 × 1); (e) hcp(10-10)–(1 × 1), usually written as hcp(10$\bar{1}$0)–(1 × 1); (f) bcc(110)–(1 × 1); (g) bcc(100)–(1 × 1) and (h) bcc(111)–(1 × 1). The atomic spheres are drawn with radii that are smaller than touching-sphere radii, in order to give better depth views. The arrows are unit cell vectors. These figures were produced by the software program BALSAC [35].

The step notation [5, 16] compacts the terrace/step information into the general form $w(h_t k_t l_t) \times (h_s k_s l_s)$. Here $(h_t k_t l_t)$ and $(h_s k_s l_s)$ are the Miller indices of the terrace plane and the step plane, respectively, while $w$ is the number of atoms that are counted in the width of the terrace, including the step-edge atom and the

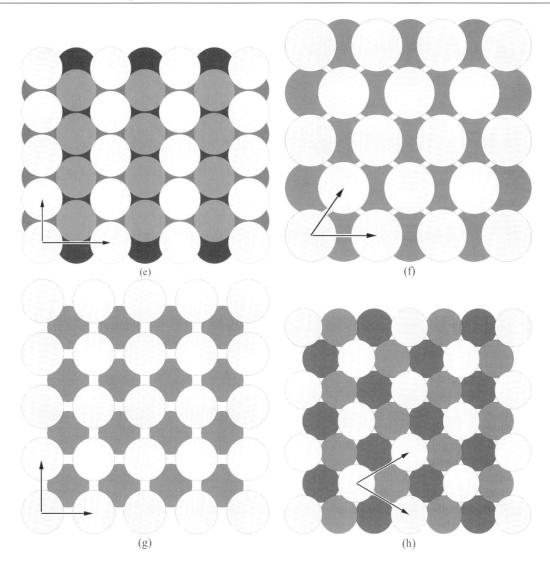

**Figure B1.21.1.** (Continued)

in-step atom. Thus, the fcc(755) surface can be denoted by $6(111) \times (100)$, since its terraces are six atoms in depth. A kinked surface, like fcc(10,8,7), can also be approximately expressed in this form: the step plane $(h_s k_s l_s)$ is a stepped surface itself, and thus has higher Miller indices than the terrace plane. However, the step notation does not exactly tell us the relative location of adjacent steps, and it is not entirely clear how the terrace width $w$ should be counted. A more complete microfacet notation is available to describe kinked surfaces generally [5].

*(c) Superlattices*

Many surfaces exhibit a different periodicity than expected from the bulk lattice, as is most readily seen in the diffraction patterns of LEED: often additional diffraction features appear which are indicative of a

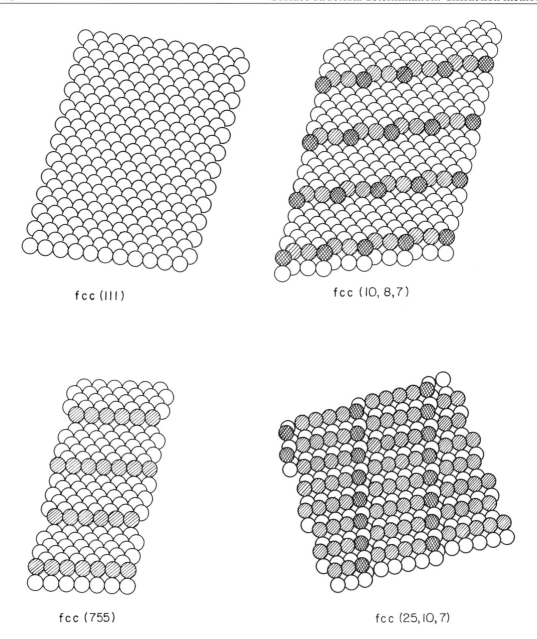

**Figure B1.21.2.** Atomic hard-ball models of 'stepped' and 'kinked' high-Miller-index bulk-terminated surfaces of simple metals with fcc lattices, compared with an fcc(111) surface: fcc(755) is stepped, while fcc(10,8,7) and fcc(25,10,7) are 'kinked'. Step-edge atoms are shown singly hatched, while kink atoms are shown cross-hatched.

*superlattice*. This corresponds to the formation of a new two-dimensional lattice on the surface, usually with some simple relationship to the expected 'ideal' lattice [5]. For instance, a layer of adsorbate atoms may occupy only every other equivalent adsorption site on the surface, in both surface dimensions. Such a lattice can be labelled $(2 \times 2)$: in each surface dimension the repeat distance is doubled relative to the ideal substrate. In this example, the unit cell of the original bulk-like surface is magnified by a factor of two in both directions, so that the new surface unit cell has dimensions $(2 \times 2)$ relative to the original unit cell. For instance, an oxygen overlayer on Pt(111), at a quarter-monolayer coverage, is observed to adopt an ordered $(2 \times 2)$ superlattice: this can be denoted as Pt(111)+$(2 \times 2)$–O, which provides a compact description of the main crystallographic characteristics of this surface. This particular notation is that of the Surface Structure Database [14]; other equivalent notations are also common in the literature, such as Pt(111)–$(2 \times 2)$–O or Pt(111)2 $\times$ 2–O.

This $(2 \times 2)$ notation can be generalized. First, it can take on the form $(m \times n)$, where the numbers $m$ and $n$ are two independent stretch factors for the two unit cell vectors. These numbers are often integers, but need not be. In addition, this new stretched unit cell can be rotated by any angle about the surface normal: this is denoted as $(m \times n)R\alpha°$, where $\alpha$ is the rotation angle in degrees [5, 17–19]; the suffix $R\alpha°$ is omitted when $\alpha = 0$, as is the case for Pt(111)+$(2 \times 2)$–O. This *Wood notation* [5, 19] allows the original unit cell to be stretched and rotated; however, it conserves the angle between the two unit cell vectors in the plane of the surface, therefore not allowing 'sheared' unit cells.

As a particular case, a surface may be given the Wood notation $(1 \times 1)$, as in Ni(111)–$(1 \times 1)$: this notation indicates that the two-dimensional unit cell of the surface has the same size as the two-dimensional unit cell of the bulk (111) layers. Thus, an ideally terminated bulk lattice without overlayers or reconstructions will carry the label $(1 \times 1)$.

The Wood notation can be generalized somewhat further, by adding either the prefix 'c' for centred, or the prefix 'p' for primitive. For instance, one may have a c$(2 \times 2)$ unit cell or a p$(2 \times 2)$ unit cell, the latter often abbreviated to $(2 \times 2)$ because it is identical to it. In a centred unit cell, the centre of the cell is an exact copy of the corners of the cell; this makes the cell non-primitive, i.e. it is no longer the smallest cell that, when repeated periodically across the surface, generates the entire surface structure. Nonetheless, the centred notation is often used because it can be quite convenient, as the next example will illustrate.

The c$(2 \times 2)$ unit cell can also be written as $(\sqrt{2} \times \sqrt{2})R45°$. Here, the original unit vectors of the $(1 \times 1)$ structure have both been stretched by factors $\sqrt{2}$ and then rotated by 45°. Thus, sulfur on Ni(100) forms an ordered half-monolayer structure that can be labelled as Ni(100)+c$(2 \times 2)$–S or, equivalently, Ni(100)+$(\sqrt{2} \times \sqrt{2})R45°$–S. The c$(2 \times 2)$ notation is clearly easier to write and also easier to convert into a geometrical model of the structure, and hence is the favoured designation.

A more general notation than Wood's is available for all kinds of unit cells, including those that are sheared, so that the superlattice unit cell can take on any shape, size and orientation. It is the *matrix notation*, defined as follows [5]. We connect the unit cell vectors $a'$ and $b'$ of the superlattice to the unit cell vectors $a$ and $b$ of the substrate by the general relations

$$a' = m_{11}a + m_{12}b$$

$$b' = m_{21}a + m_{22}b.$$

The coefficients $m_{11}$, $m_{12}$, $m_{21}$ and $m_{22}$ define the matrix $\mathbf{M} = \begin{pmatrix} m_{11} & m_{12} \\ m_{21} & m_{22} \end{pmatrix}$, which serves to denote the superlattice. The $(1 \times 1)$, $(2 \times 2)$ and c$(2 \times 2)$ lattices are then denoted respectively by the matrices $\mathbf{M} = \begin{pmatrix} 1 & 0 \\ 0 & 1 \end{pmatrix}$, $\mathbf{M} = \begin{pmatrix} 2 & 0 \\ 0 & 2 \end{pmatrix}$ and $\mathbf{M} = \begin{pmatrix} 1 & -1 \\ 1 & 1 \end{pmatrix}$. This allows the Ni(100)+c$(2 \times 2)$–S structure to be also written as Ni(100)+$\begin{pmatrix} 1 & -1 \\ 1 & 1 \end{pmatrix}$–S. Clearly, this notation is not as intuitive and compact as the c$(2 \times 2)$ Wood notation. However, when the Wood notation is not capable of a clear and compact notation, use of the matrix notation is necessary. Thus, a structure characterized by a matrix like $\mathbf{M} = \begin{pmatrix} 4 & -3 \\ 2 & 5 \end{pmatrix}$ could not be described in the Wood notation.

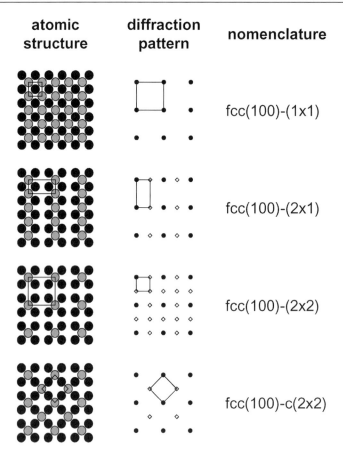

**Figure B1.21.3.** 'Direct' lattices (at left) and corresponding reciprocal lattices (at right) of a series of commonly occurring two-dimensional superlattices. Black circles correspond to the ideal $(1 \times 1)$ surface structure, while grey circles represent adatoms in the direct lattice (arbitrarily placed in 'hollow' positions) and open diamonds represent fractional-order beams in the reciprocal space. Unit cells in direct space and in reciprocal space are outlined.

In LEED experiments, the matrix **M** is determined by visual inspection of the diffraction pattern, thereby defining the periodicity of the surface structure: the relationship between surface lattice and diffraction pattern will be described in more detail in the next section.

A superlattice is termed *commensurate* when all matrix elements $m_{ij}$ are integers. If at least one matrix element $m_{ij}$ is an irrational number (not a ratio of integers), then the superlattice is termed *incommensurate*. A superlattice can be incommensurate in one surface dimension, while commensurate in the other surface dimension, or it could be incommensurate in both surface dimensions.

A superlattice can be caused by adsorbates adopting a different periodicity than the substrate surface, or also by a reconstruction of the clean surface. In figure B1.21.3 several superlattices that are commonly detected on low-Miller-index surfaces are shown with their Wood notation.

### B1.21.5   Surface diffraction pattern

The diffraction pattern observed in LEED is one of the most commonly used 'fingerprints' of a surface structure. With XRD or other non-electron diffraction methods, there is no convenient detector that images in

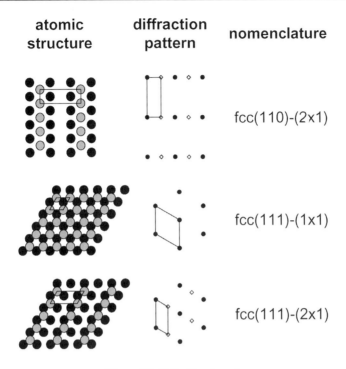

**atomic structure**   **diffraction pattern**   **nomenclature**

fcc(110)-(2×1)

fcc(111)-(1×1)

fcc(111)-(2×1)

**Figure B1.21.3.** (Continued)

real time the corresponding diffraction pattern. Point-source methods, like PD, do not produce a convenient spot pattern, but a diffuse diffraction pattern that does not simply reflect the long-range ordering.

So it is essential to relate the LEED pattern to the surface structure itself. As mentioned earlier, the diffraction pattern does not indicate relative atomic positions within the structural unit cell, but only the size and shape of that unit cell. However, since experiments are mostly performed on surfaces of materials with a known crystallographic bulk structure, it is often a good starting point to assume an ideally terminated bulk lattice; the actual surface structure will often be related to that ideal structure in a simple manner, e.g. through the creation of a superlattice that is directly related to the bulk lattice.

In this section, we concentrate on the relationship between diffraction pattern and surface lattice [5]. In direct analogy with the three-dimensional bulk case, the *surface lattice* is defined by two vectors $a$ and $b$ parallel to the surface (defined already above), subtended by an angle $\gamma$; $a$ and $b$ together specify one unit cell, as illustrated in figure B1.21.4. Within that unit cell atoms are arranged according to a *basis*, which is the list of atomic coordinates within that unit cell; we need not know these positions for the purposes of this discussion. Note that this unit cell can be viewed as being infinitely deep in the third dimension (perpendicular to the surface), so as to include all atoms below the surface to arbitrary depth.

There are several special shapes of the surface lattice, forming the five two-dimensional Bravais lattices shown in figure B1.21.4. The Bravais lattices form the complete list of possible lattices. They are characterized by unit cell vectors of equal length (in the case of the square and hexagonal lattices) and/or a subtended angle of 90° or 60° (for the square, rectangular and hexagonal lattices) or by completely general values (for the oblique lattice). The rectangular lattice comes in two varieties: primitive and centred. The centred lattice has the particularity that its atomic basis is duplicated: each atom is reproduced by displacement through the vector $1/2(a + b)$. The main value of the centred rectangular lattice is its convenience: it is easier to think in terms of the rectangle (with duplicated basis) than to think of the rhombus with arbitrary angle $\gamma$. One could

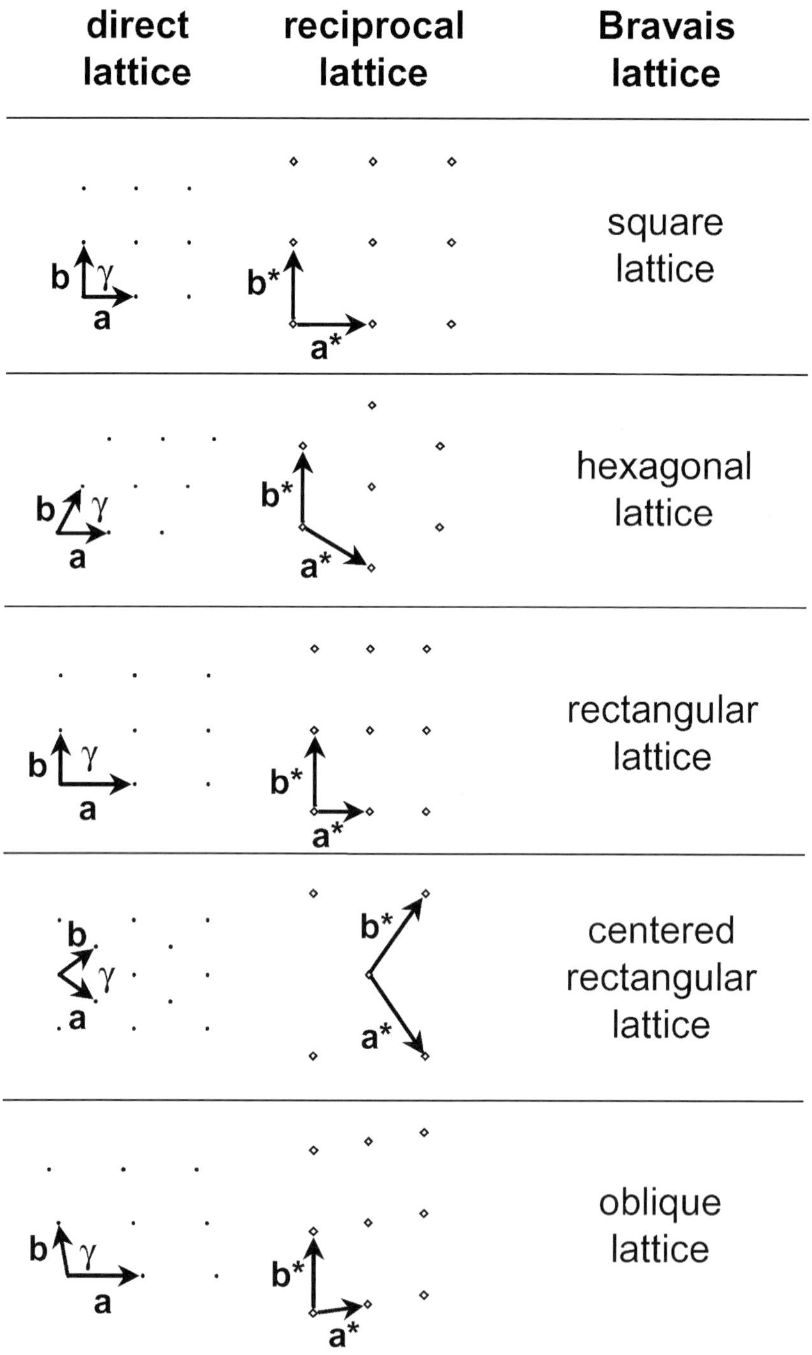

**Figure B1.21.4.** 'Direct' lattices (at left) and reciprocal lattices (middle) for the five two-dimensional Bravais lattices. The reciprocal lattice corresponds directly to the diffraction pattern observed on a standard LEED display. Note that other choices of unit cells are possible: e.g., for hexagonal lattices, one often chooses vectors *a* and *b* that are subtended by an angle $\gamma$ of 120° rather than 60°. Then the reciprocal unit cell vectors also change: in the hexagonal case, the angle between $a^*$ and $b^*$ becomes 60° rather than 120°.

also centre any of the other lattices, but one would only produce another instance of a square, rectangular or oblique lattice, i.e. nothing more convenient.

The diffraction of low-energy electrons (and any other particles, like x-rays and neutrons) is governed by the translational symmetry of the surface, i.e. the surface lattice. In particular, the directions of emergence of the diffracted beams are determined by conservation of the linear momentum parallel to the surface, $\hbar k_\parallel$. Here $k$ denotes the wavevector of the incident plane electron wave that represents the incoming electron beam. This conservation can occur in two ways. After the diffractive scattering, the parallel component of the momentum $\hbar k'_\parallel$ can be equal to that of the incident electron beam, i.e. $\hbar k'_\parallel = \hbar k_\parallel$; this corresponds to *specular* (mirror-like) reflection, with equal polar angles of incidence and emergence with respect to the surface normal, and with a simple reversal of the perpendicular momentum $\hbar k'_\perp = -\hbar k_\perp$.

Alternatively, the electron can exchange parallel momentum with the lattice, but only in well defined amounts given by vectors $\hbar g$ that belong to the *reciprocal lattice* of the surface. That is, the vector $g$ is a linear combination of two reciprocal lattice vectors $a^*$ and $b^*$, with integer coefficients. Thus, $g = ha^* + kb^*$, with arbitrary integers $h$ and $k$ (note that all the vectors $a$, $b$, $a^*$, $b^*$ and $g$ are parallel to the surface). The reciprocal lattice vectors $a^*$ and $b^*$ are related to the 'direct-space' lattice vectors $a$ and $b$ through the following non-transparent definitions, which also use a vector $n$ that is perpendicular to the surface plane, as well as vectorial dot and cross products:

$$a^* = 2\pi \left( \frac{b \times n}{a \cdot (b \times n)} \right) \qquad \text{and} \qquad b^* = 2\pi \left( \frac{n \times a}{b \cdot (n \times a)} \right).$$

These two equations are a special case of the corresponding three-dimensional definition, common in XRD, with the surface normal $n$ replacing the third lattice vector $c$.

Figure B1.21.4 illustrates the 'direct-space' and reciprocal-space lattices for the five two-dimensional Bravais lattices allowed at surfaces. It is useful to realize that the vector $a^*$ is always perpendicular to the vector $b$ and that $b^*$ is always perpendicular to $a$. It is also useful to notice that the length of $a^*$ is inversely proportional to the length of $a$, and likewise for $b^*$ and $b$. Thus, a large unit cell in direct space gives a small unit cell in reciprocal space, and a wide rectangular unit cell in direct space produces a tall rectangular unit cell in reciprocal space. Also, the hexagonal direct-space lattice gives rise to another hexagonal lattice in reciprocal space, but rotated by $90°$ with respect to the direct-space lattice.

The reciprocal lattices shown in figures B1.21.3 and B1.21.4 correspond directly to the diffraction patterns observed in LEED experiments: each reciprocal-lattice vector produces one and only one diffraction spot on the LEED display. It is very convenient that the hemispherical geometry of the typical LEED screen images the reciprocal lattice without distortion; for instance, for the square lattice one observes a simple square array of spots on the LEED display.

One of the spots in such a diffraction pattern represents the specularly reflected beam, usually labelled (00). Each other spot corresponds to another reciprocal-lattice vector $g = ha^* + kb^*$ and is thus labelled $(hk)$, with integer $h$ and $k$.

When a superlattice is present, additional spots arise in the diffraction pattern, as shown in figure B1.21.3 in terms of the reciprocal lattice: again, each reciprocal lattice point corresponds to a spot in a diffraction pattern. This can be easily understood from the fact that a larger unit cell in direct space imposes a smaller unit cell in reciprocal space. For instance, a $(2 \times 1)$ superlattice has a unit cell doubled in length in one surface direction relative to the $(1 \times 1)$ lattice, i.e. $a$ is replaced by $2a$. According to the above equations, this has no effect on $b^*$, but halves $a^*$. This is equivalent to allowing $h$ to be a half-integer in $g = ha^* + kb^*$, thus doubling the number of spots in the diffraction pattern. These additional spots are therefore often called half-order spots in the $(2 \times 1)$ case, or fractional-order spots in the general case.

With some practice, one can easily recognize specific superlattices from their LEED pattern. Otherwise, one can work through the above equations to connect particular superlattices to a given LEED pattern.

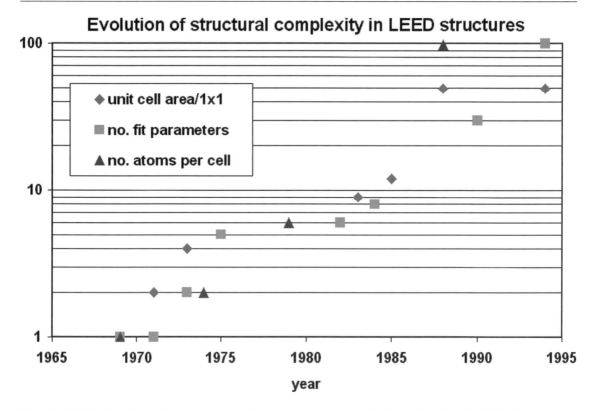

**Figure B1.21.5.** Evolution with time of the complexity of structural determination achievable with LEED. The unit cell area is measured relative to the unit cell area of the simple $(1 \times 1)$ structures studied in the early days: thus a $(n \times n)$ superstructure (due to reconstruction and/or adsorption) has a unit cell size of $n^2$. A $(7 \times 7)$ structure gives a complexity of 49 on this scale. The number of fit parameters measures the number of coordinates fitted to experiment in any given structure: a value of 1 was typical of many early determinations, when only one interlayer spacing was fitted to experiment. The Si(111)–$(7 \times 7)$ structure has over 100 fit parameters, if one allows only those structural changes in the top two double layers and the adatom layer that maintain the p3m1 symmetry of the substrate. The number of atoms per unit cell refers to so-called composite layers, which are groups of closely spaced layers: this number dramatically affects computation time in multiple-scattering methods. It has grown from 1 in the simplest structures to about 100 in the Si(111)–$(7 \times 7)$ structure.

A number of examples are given and discussed in some detail in [5]. A discussion can also be found there of the special case of stepped and kinked surfaces.

## B1.21.6   Diffraction pattern of disordered surfaces

Many forms of disorder in a surface structure can be recognized in the LEED pattern. The main manifestations of disorder are *broadening* and *streaking* of diffraction spots and *diffuse* intensity between spots [1].

Broadening of spots can result from thermal diffuse scattering and island formation, among other causes. The thermal effects arise from the disorder in atomic positions as they vibrate around their equilibrium sites; the sites themselves may be perfectly crystalline.

## Re(0001)-(2√3x2√3)R30°-6S

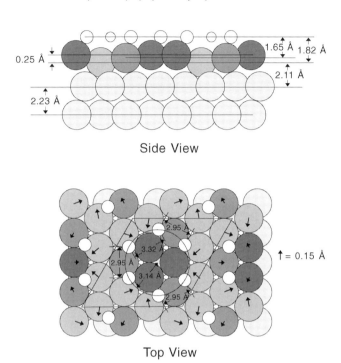

**Figure B1.21.6.** Side and top views of the best-fit structure of the Re(0001)–$(2\sqrt{3} \times 2\sqrt{3})R30°$–6S surface structure (with a half-monolayer coverage of sulfur), as determined by LEED [31]. A $(2\sqrt{3} \times 2\sqrt{3})R30°$ unit cell is outlined in the top view. Sulfur atoms are drawn as small open circles, Re atoms as large grey circles. Sulfur–sulfur distances in a ring of six alternate between 2.95 and 3.32 Å, expanded from the unrelaxed distance between hollow sites of 2.75 Å. Arrows represent lateral relaxations in the topmost metal layer, with the scale of displacements indicated by the lone arrow on the right. The bulk interlayer spacing in Re(0001) is 2.23 Å. Shades of grey identify atoms that are equivalent by symmetry in the sulfur and outermost rhenium layers. The darkest-grey rhenium atoms forming a triangle within a sulfur ring are pulled out of the surface by the adsorbed sulfur, relative to the lighter-grey rhenium atoms in the same layer.

Islands occur particularly with adsorbates that aggregate into two-dimensional assemblies on a substrate, leaving bare substrate patches exposed between these islands. Diffraction spots, especially fractional-order spots if the adsorbate forms a superlattice within these islands, acquire a width that depends inversely on the average island diameter. If the islands are systematically anisotropic in size, with a long dimension primarily in one surface direction, the diffraction spots are also anisotropic, with a small width in that direction. Knowing the island size and shape gives valuable information regarding the mechanisms of phase transitions, which in turn permit one to learn about the adsorbate–adsorbate interactions.

Lattice-gas disorder, in which adatoms occupy a periodic lattice of equivalent sites with a random occupation probability, produces diffuse intensity distributions between diffraction spots. For complete disorder, one observes such diffuse intensity throughout the diffraction pattern. If there is order in one surface direction, but disorder in the other, one observes streaking in the diffraction pattern: the direction of the streaks corresponds to the direction in which disorder occurs. In principle, the diffuse intensity distribution can be converted into a direct-space distribution, including a pair-correlation function between occupied sites,

# Mo(100)-c(4x2)-3S

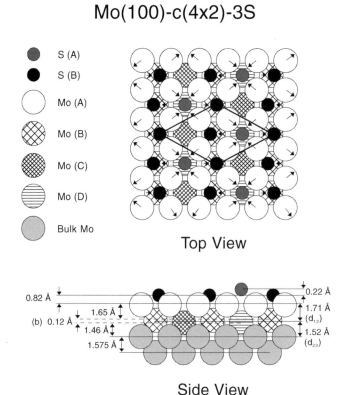

S (A)

S (B)

Mo (A)

Mo (B)

Mo (C)

Mo (D)

Bulk Mo

Top View

Side View

**Figure B1.21.7.** Top and side views of the best-fit structure of the Mo(100)–c(4 × 2)–3S surface structure (with a 3/4-monolayer coverage of sulfur), as determined by LEED [32]. A c(4 × 2) unit cell is outlined in the top view. The sulfur sizes (small black and dark grey circles) have been reduced from covalent for clarity, while the molybdenum atoms (large circles) are drawn with touching radii. The same cross-hatching has been assigned to molybdenum atoms that are equivalent by symmetry in the topmost two metal layers. Two-thirds of the sulfur atoms are displaced away from the centre of the hollow sites in which they are bonded: these displacements by 0.13 Å are drawn exaggerated. Arrows in the top view also indicate the directions and relative magnitudes of molybdenum atom displacements (these substrate atoms are drawn in their undisplaced positions, except for the buckling seen in the second molybdenum layer in the side view). The bulk interlayer spacing in Mo(100) is 1.575 Å.

e.g. by Fourier transformation. However, the diffuse intensity is too much affected by other diffraction effects (like multiple scattering) to be very useful in this manner. It nonetheless can be interpreted in terms of local structure, i.e. bond lengths and angles, by a procedure that is very similar to the multiple-scattering modelling for solving structures in full detail [20].

LEED has found a strong competitor for studying surface disorder: scanning tunnelling microscopy, STM (see chapter B1.20). Indeed, STM is the ideal tool for investigating irregularities in periodic surface structures. LEED (as any other diffraction method) averages its information content over macroscopic parts of the surface, giving only statistical information about disorder. By contrast, STM can provide a direct image of individual atoms or defects, enabling the observation of individual atomic behaviour. By observing a sufficiently large area, STM can also provide statistical information, if desired.

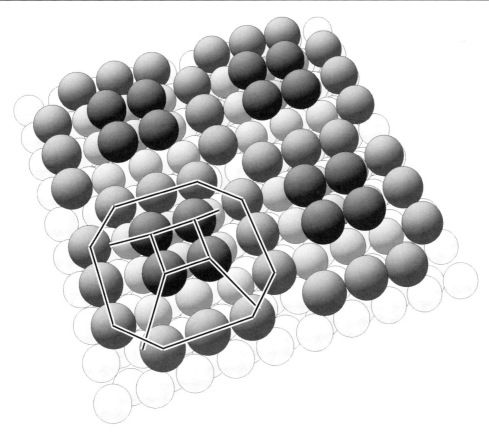

Cu(100)-(4x4)-10Li

**Figure B1.21.8.** Perspective view of the structure of the Cu(100)–(4 × 4)–10Li surface structure (with a 10/16-monolayer coverage of lithium), as determined by LEED [33]. The atoms are drawn with radii that are reduced by about 15% from covalent radii. The surface fragment shown includes four (4 × 4) unit cells. Lithium atoms are shown as larger spheres. In each unit cell, four lithium atoms (dark grey) form a flat-topped pyramid (as outlined): the lithium atoms rest in hollow sites on a 3 × 3 base of nine Cu atoms (lighter grey). Around each pyramid 12 lithium atoms occupy substitutional sites, i.e. have taken the place of Cu atoms: these lithium atoms are shown linked by an octagon. Since the lithium atoms are about 15% larger than the copper atoms that they replace, fewer lithium atoms can fit in the troughs evacuated by the copper atoms; thus, they do not fill the troughs completely, and leave a hole at each intersection between troughs (e.g. at the exact centre of the fragment). The lightest-grey atoms underneath are the bulk Cu(100) termination: some small local distortions in the atomic positions are also detected by LEED there. This and the following figure were produced with the SARCH/LATUSE/PLOT3D/BALSAC software, available from the author.

## B1.21.7   Full structural determination

In the previous sections we have emphasized the two-dimensional information available through the diffraction pattern observed in LEED. But, as mentioned before, one can extract the detailed atomic positions as well, including interlayer spacings, bond lengths, bond angles, etc. Here we sketch how this more complete structural determination is accomplished. We focus on the case of LEED, since this method has produced by

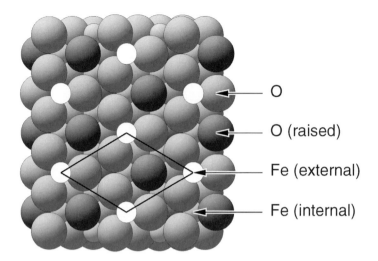

Top view

Fe$_3$O$_4$(111)-(1x1) (magnetite)

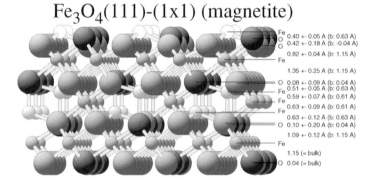

Side view

Fe$_3$O$_4$(111)-(1x1) (magnetite)

**Figure B1.21.9.** Perspective side and top view of the best-fit structure of Fe$_3$O$_4$(111), as determined by LEED [34]. A unit cell is outlined in the top view, in which all atoms are drawn with nearly touching radii, while smaller radii are used in the side view. This iron oxide was grown as an ultrathin film on a Pt(111) substrate, in order to prevent electrical charging of the surface. The free surface is at the top end of the side view, exposing 1/4 monolayer of 'external' iron ions (shown as small light-grey circles in both views). Large circles represent oxygen ions, forming hexagonally close-packed layers. In each such layer, one-fourth of the oxygen ions (drawn in darkest grey) is not coplanar with the others: in particular, in the outermost oxygen layer, these ions are raised outward by a large amount (0.42 Å, compared to 0.04 Å in the opposite direction in the bulk). Small circles below the surface represent iron ions in tetrahedral or octahedral interstitial positions between the O layers: the lightest-grey of these are in tetrahedral positions. Interlayer spacings as determined by LEED are given at the right with error bars and with corresponding bulk values in parentheses.

far the most structural determinations [5, 17, 18, 21]. The procedures employed to analyse PD data are in fact very similar to those for LEED, in many details. With XRD, the kinematic (single-scattering) nature of the problem makes the analysis simpler, but still considerable for complex structures: there also, a trial-and-error search for the solution is common.

To obtain spacings between atomic layers and bond lengths or angles between atoms, it is necessary to measure and analyse the *intensity* of diffraction spots. This is analogous to measuring the intensity of XRD reflections.

The measurement of LEED spot intensities is nowadays mostly accomplished by digitizing the image recorded by a video camera that observes the diffraction pattern, which is visibly displayed on a fluorescent screen within an ultra-high vacuum system [22]. The digitized image is then processed by computer to give the integrated spot intensity, after removal of the background. This is repeated for different incident electron energies. Thereby, the intensity of each spot is obtained as a function of the incident electron energy, resulting in an *IV curve* (intensity–voltage curve) for each spot. Computer codes for this purpose are available, and are normally packaged together with the required hardware [23]. The resulting IV curves form the experimental database to which theory can fit the atomic structure. It typically takes between minutes and an hour to accumulate such a database, once the sample has been prepared.

Since ED by a surface is a complicated process, there is no routine method available to *directly* and accurately extract atomic positions from the experimental data. Direct holographic methods have been proposed [24], but have not yet become routine methods, and in any case they yield only approximate atomic positions (with uncertainties on the scale of 0.2–0.5 Å) and work only for relatively simple structures; when they do work they have to be followed up by refinement using the same trial-and-error approach that we discuss next.

A detailed structural determination proceeds by modelling the full multiple scattering of the electrons that are diffracted through the surface structure. The multiple scattering means that an electron can bounce off a succession of atoms in an erratic path before emerging from the surface. Various theoretical and computational methods are available to treat this problem to any degree of precision: a compromise between precision and computing expense must be struck, with progress moving toward higher precision, even for more complex structures.

The modelling of the multiple scattering requires input of all atomic positions, so that the *trial-and-error approach* must be followed: one guesses reasonable models for the surface structure, and tests them one by one until satisfactory agreement with experiment is obtained. For simple structures and in cases where structural information is already known from other sources, this process is usually quite quick: only a few basic models may have to be checked, e.g. adsorption of an atomic layer in hollow, bridge or top sites at positions consistent with reasonable bond lengths. It is then relatively easy to refine the atomic positions within the best-fit model, resulting in a complete structural determination. The refinement is normally performed by some form of *automated steepest-descent optimization*, which allows many atomic positions to be adjusted simultaneously [21] Computer codes are also available to accomplish this part of the analysis [25]. The trial-and-error search with refinement may take minutes to hours on current workstations or personal computers.

In more complex cases, and when little additional information is available, one must test a larger number of possible structural models. The computational time grows rapidly with complexity, so that it may take hours to check a single model. More time-consuming, however, is often the human factor in guessing what are reasonable models to test. This is a much more difficult problem, which is the issue of finding the 'global optimum', not just a 'local optimum'. At present, several approaches to *global optimization* are being examined, such as *simulated annealing* [26] and *genetic algorithms* [27]. In any event, these will require larger amounts of computer time, since a wide variety of surface models must be tested in such a global search.

## B1.21.8 Present capabilities and outlook

Surface crystallography started in the late 1960s, with the simplest possible structures being solved by LEED [14]. Such structures were the clean Ni(111), Cu(111) and Al(111) surfaces, which are unreconstructed and essentially unrelaxed, i.e. very close to the ideal termination of the bulk shown in figure B1.21.1(a): typically, only one unknown structural parameter was fitted to experiment, namely the spacing between the two outermost atomic layers.

Progress in experiment, theory, computational methods and computer power has contributed to the capability to solve increasingly complex structures [28, 29]. Figure B1.21.5 quantifies this progress with three measures of complexity, plotted logarithmically: the achievable two-dimensional unit cell size, the achievable number of fit parameters and the achievable number of atoms per unit cell per layer: all of these measures have grown from 1 for simple clean metal surfaces, like Ni(111) (see figure B1.21.1(a)), to about 50–100 in the case of the reconstructed Si(111)–(7 × 7) surface, the most complicated structure examined to date [30] (note that the basic model which solved the Si(111)–(7 × 7) surface was mainly derived from another diffraction study, using TED [9]). All these measures thus exhibit a progression by about two orders of magnitude over less than 25 years.

Figures B1.21.6–B1.21.9 show several of the more complex structures solved by LEED in recent years. They exhibit various effects observed at surfaces:

- clustering of adatoms in Re(0001)–$(2\sqrt{3} \times 2\sqrt{3})R30°$–6S [31], see figure B1.21.6;

- hollow-site adsorption and adsorbate-induced relaxations of substrate atoms both in Re(0001)–$(2\sqrt{3} \times 2\sqrt{3})R30°$–6S [31] and in Mo(100)–c(4 × 2)–3S [32], see figure B1.21.7;

- adsorbate-induced reconstruction as well as substitutional adsorption in Cu(100)–(4 × 4)–10Li [33], see figure B1.21.8; note that this is the most complex surface structural determination by LEED to date, involving far more adjustable structural parameters than were fitted in the Si(111)–(7 × 7) structure [30];

- compound ionic surface with a large bulk unit cell and very large surface relaxations in $Fe_3O_4$(111) [34], see figure B1.21.9.

Further progress towards solving more complex surface structures is possible. The biggest challenge on the computational and theoretical side is the identification of the globally optimum structure. Holographic and other methods have not yet provided a convenient way to accomplish this, and would actually fail with structures that have the complexity of Cu(100)–(4 × 4)–10Li and Si(111)–(7 × 7). Global-search algorithms, like simulated annealing and genetic algorithms, may provide workable, if perhaps not cheap, solutions.

On the experimental side, a larger measured database is required than is commonly available to determine the large number of structural parameters to be fitted. For instance, LEED calculations for the Si(111)–(7 × 7) surface have been attempted to fit the many tens of unknown structural parameters; however, the amount of experimental data was insufficient for the task, resulting in a multitude of locally-optimum structures, without the ability to discriminate between them. Increasing the database size can be achieved by extending the energy range to higher energies, or by acquiring data at a number of different incidence directions: either way, the calculations become disproportionately more time-consuming, because the computing effort rises quickly with energy and non-symmetrical off-normal incidence directions.

## Acknowledgments

Professor K Hermann is gratefully thanked for the images in figure B1.21.1, which were produced with the program BALSAC [35].

# References

[1] Henzler M 1997 Capabilities of LEED for defect analysis *Surf. Rev. Lett.* **4** 489–500

[2] Rehr J J 1995 Multiple-scattering approach to surface EXAFS—theory versus experiment *Surf. Rev. Lett.* **2** 63–9

[3] Stöhr J 1992 *NEXAFS Spectroscopy* (Heidelberg: Springer)

[4] Feidenhans'l R 1989 Surface structure determination by x-ray diffraction *Surf. Sci. Rep.* **10** 105–88

[5] Van Hove M A, Weinberg W H and Chan C-M 1986 *LEED Experiment, Theory and Structural Determination* (Heidelberg: Springer)

[6] Ichimiya A, Ohno Y and Horio Y 1997 Structural analysis of crystal surfaces by reflection high energy electron diffraction *Surf. Rev. Lett.* **4** 501–11

[7] Kjems J K, Passell L, Taub H, Dash J G and Novaco A D 1976 Neutron scattering study of nitrogen adsorbed on basal plane-oriented graphite *Phys. Rev.* B **13** 1446–62

[8] McTague J P, Nielsen M and Passell L 1979 Neutron scattering by adsorbed monolayers *Crit. Rev. Solid State Sci.* **8** 125–56

[9] Takayanagi K 1990 Surface structure analysis by transmission electron diffraction—effects of the phases of structure factors *Acta Crystallogr.* A **46** 83–6

[10] Fadley C S *et al* 1997 Photoelectron diffraction: space, time and spin dependence of surface structures *Surf. Rev. Lett.* **4** 421–40

[11] Lee P A, Citrin P H, Eisenberger P and Kincaid B M 1981 Extended x-ray absorption fine structure—its strengths and limitations as a structural tool *Rev. Mod. Phys.* **53** 769–806

[12] Cowan P L, Golovchenko J L and Robbins M F 1980 X-ray standing waves at crystal surfaces *Phys. Rev. Lett.* **44** 1680–3

[13] Woodruff D P, Cowie B C C and Ettema A R H F 1994 Surface structure determination using x-ray standing waves: a simple view *J. Phys.: Condens. Matter* **6** 10 633–45

[14] Watson P R, Van Hove M A and Hermann K 1999 *NIST Surface Structure Database Ver. 3.0* (Gaithersburg, MD: NIST Standard Reference Data Program)

[15] Ohtani H, Kao C-T, Van Hove M A and Somorjai G A 1986 A tabulation and classification of clean solid surfaces and of adsorbed atomic and molecular monolayers as determined from low-energy electron diffraction *Prog. Surf. Sci.* **23** 155–316

[16] Lang B, Joyner R W and Somorjai G A 1972 LEED studies of high index crystal surfaces of platinum *Surf. Sci.* **30** 440–53

[17] Pendry J B 1974 *Low-Energy Electron Diffraction* (London: Academic)

[18] Clarke L J 1985 *Surface Crystallography, An Introduction to LEED* (Chichester: Wiley)

[19] Wood E A 1964 Vocabulary of surface crystallography *J. Appl. Phys.* **35** 1306–12

[20] Heinz K 1994 Diffuse LEED and local surface structure *Phys. Status Solidi* A **146** 195–204

[21] Van Hove M A, Moritz W, Over H, Rous P J, Wander A, Barbieri A, Materer N, Starke U and Somorjai G A 1993 Automated determination of complex surface structures by LEED *Surf. Sci. Rep.* **19** 191–229

[22] Heinz K and Müller K 1982 LEED intensities—experimental progress and new possibilities of surface structure determination *Springer Tracts in Modern Physics* **91** 1–54 (Heidelberg: Springer)

[23] e.g., SPECS GmbH, Voltastrasse 5, D-13355 Berlin, Tel. +49-30-46 78 24-0, Fax. +49-30-46 42 083, email: support@specs.de, http://www.specs.de

[24] Saldin D K, Chen X, Vamvakas J A, Ott M, Wedler H, Reuter K, Heinz K and De Andres P L 1997 Holographic LEED: a review of recent progress *Surf. Rev. Lett.* **4** 991–1001

[25] Van Hove M A http://electron.lbl.gov/software/software.html
Heinz K TLEED@fkp.physik.uni-erlangen.de

[26] Rous P J 1993 A global approach to the search problem in surface crystallography by low-energy electron diffraction *Surf. Sci.* **296** 358–73

[27] Döll R and Van Hove M A 1996 Global optimization in LEED structure determination using genetic algorithms *Surf. Sci.* **355** L393–8

[28] Van Hove M A 1996 Complex surface structures from LEED *Surf. Rev. Lett.* **3** 1271–84

[29] Van Hove M A 1997 Determination of complex surface structures with LEED *Surf. Rev. Lett.* **4** 479–87

[30] Tong S Y, Huang H, Wei C M, Packard W E, Men F K, Glander G and Webb M B 1988 Low-energy electron diffraction analysis of the Si(111)-(7 × 7) structure *J. Vac. Sci. Technol.* A **6** 615–24

[31] Barbieri A, Jentz D, Materer N, Held G, Dunphy J, Ogletree D F, Sautet P, Salmeron M, Van Hove M A and Somorjai G A 1994 Surface crystallography of Re(0001)–(2 × 2)–S and Re(0001)–(2$\sqrt{3}$ × 2$\sqrt{3}$)r30°–6S: a combined LEED and STM study *Surf. Sci.* **312** 10–20

[32] Jentz D, Rizzi S, Barbieri A, Kelly D, Van Hove M A and Somorjai G A 1995 Surface structures of sulfur and carbon overlayers on Mo(100): a detailed analysis by automated tensor LEED *Surf. Sci.* **329** 14–31

[33] Mizuno S, Tochihara H, Barbieri A and Van Hove M A 1995 Completion of the structural determination of and rationalizing of the surface-structure sequence (2 × 1) → (3 × 3) → (4 × 4) formed on Cu(001) with increasing Li coverage *Phys. Rev.* B **52** 11 658–61

[34] Barbieri A, Weiss W, Van Hove M A and Somorjai G A 1994 Magnetite $Fe_3O_4$(111): surface structure by LEED crystallography and energetics *Surf. Sci.* **302** 259–79

[35] Hermann K http://www.fhi-berlin.mpg.de/th/personal/hermann/balpam.html

## Further Reading

Pendry J B 1974 *Low-energy Electron Diffraction* (London: Academic)

A full description of the principles of low-energy electron diffraction (LEED) as of 1974, containing all the basic physics still in use today.

Van Hove M A, Weinberg W H and Chan C-M 1986 *LEED Experiment, Theory and Structural Determination* (Heidelberg: Springer)

Covers in great detail the practical application of low-energy electron diffraction (LEED) for structural studies, excepting more recent techniques like tensor LEED and holography.

Van Hove M A 1997 Determination of complex surface structures with LEED *Surf. Rev. Lett.* **4** 479–88

Describes the state of the art in surface structural determination by low-energy electron diffraction (LEED), focusing on complex structures.

Fadley C S *et al* 1997 Photoelectron diffraction: space, time and spin dependence of surface structures *Surf. Rev. Lett.* **4** 421–40

Summarizes the state of the art of photoelectron diffraction as a structural tool.

Rehr J J 1995 Multiple-scattering approach to surface EXAFS—theory versus experiment *Surf. Rev. Lett.* **2** 63–9

Addresses the need for advanced methods in surface extended x-ray absorption fine structure (SEXAFS) for accurate structural determination.

Feidenhans'l R 1989 Surface structure determination by x-ray diffraction *Surf. Sci. Rep.* **10** 105-188

Covers the experiment and analysis of surface x-ray diffraction, excepting the x-ray standing wave method.

# B1.22
# Surface characterization and structural determination: optical methods

## Francisco Zaera

### B1.22.1   Introduction

As discussed in more detail elsewhere in this encyclopaedia, many optical spectroscopic methods have been developed over the last century for the characterization of bulk materials. In general, optical spectroscopies make use of the interaction of electromagnetic radiation with matter to extract molecular parameters from the substances being studied. The methods employed usually rely on the examination of the radiation absorbed, emitted or scattered by a system, and may be based on simple linear optical processes, resonance transitions and/or nonlinear processes. Molecular spectroscopy probes energy transitions at all levels, from the excitation of spins in the radiofrequency range (NMR), to rotational (microwave), vibrational (infrared) and electronic valence (visible–UV) and core (x-rays) excitations. Additional diffraction- and polarization-based techniques provide structural information and laser-based pump–probe methods allow for the study of molecular dynamics down to the femtosecond time scale.

In spite of the wide range of applications of optical techniques for the study of bulk samples, however, they have so far found only limited use in the characterization of surfaces. One of the main reasons for this is the fact that it is quite difficult to discriminate optical signals originating from the surface from those arriving from the bulk of a given material. To illustrate this problem, imagine a typical metal sample consisting of a cube one centimetre long on each side. At a density of approximately $10$ g cm$^{-3}$, this represents about $0.1$ moles (for molybdenum, to pick an example), or approximately $6 \times 10^{22}$ atoms, and of those about $1 \times 10^{16}$, that is, only one in six million atoms, are on the surface of the cube. This means that if one wants to selectively characterize a surface phenomenon, one would need to develop a technique with a large dynamic range (of at least seven orders of magnitude) and/or the ability to discriminate between signals from surface and bulk elements.

Moreover, with the advent of relatively cheap vacuum technologies over the past decades, physicists have been able to develop a large number of alternative particle-based (electrons, ions, atoms) techniques capable of selectively probing solid surfaces. Most particles interact strongly with matter, and therefore cannot penetrate deeply into the substance being probed. Consequently, whatever information can be obtained from the interactions of those particles with solid samples, it must be related to the properties of the surface. The same argument does not work as well with optical techniques, because photons penetrate through most substances to depths comparable to their wavelength, microns in the case of IR radiation. In order to overcome this difficulty, alternative ways have been devised to gain surface sensitivity with optical spectroscopies. Among them are the following:

(1)   Increasing the surface-to-bulk ratio of the sample to be studied. This is easily done in the case of highly porous materials, and has been exploited for the characterization of supported catalysts, zeolites, sol-gels and porous silicon, to mention a few.

(2) Taking advantage of the intrinsic physical and chemical differences of surfaces introduced by the discontinuity of the bulk environment. Specifically, most solids display specific structural relaxations and reconstructions, surface phonons and surface electronic states easy to discriminate from those of the bulk. A clearer surface specificity is introduced in the study of adsorbates by the uniqueness of the molecules present at the interface.

(3) Taking advantage of the symmetry changes induced by the presence of a surface. Many nonlinear techniques rely on the fact that the surface breaks the centrosymmetrical nature of the bulk. The use of polarized light can also discriminate among dipole moments in different orientations.

(4) Illuminating the sample at grazing angles. The penetration depth of photons depends on the cosine of the incidence angle and, therefore, can be reduced by this procedure. Although such an approach has limited use, it has been successfully employed in a few instances, such as for x-ray diffraction experiments.

The power of optical spectroscopies is that they are often much better developed than their electron-, ion- and atom-based counterparts, and therefore provide results that are easier to interpret. Furthermore, photon-based techniques are uniquely poised to help in the characterization of liquid–liquid, liquid–solid and even solid–solid interfaces generally inaccessible by other means. There has certainly been a renewed interest in the use of optical spectroscopies for the study of more 'realistic' systems such as catalysts, adsorbates, emulsions, surfactants, self-assembled layers, etc.

In this chapter we review some of the most important developments in recent years in connection with the use of optical techniques for the characterization of surfaces. We start with an overview of the different approaches available to the use of IR spectroscopy. Next, we briefly introduce some new optical characterization methods that rely on the use of lasers, including nonlinear spectroscopies. The following section addresses the use of x-rays for diffraction studies aimed at structural determinations. Lastly, passing reference is made to other optical techniques such as ellipsometry and NMR, and to spectroscopies that only partly depend on photons.

## B1.22.2   IR spectroscopy

Perhaps the optical technique most used for surface characterization has been infrared (IR) spectroscopy. The reason for this may very well be because the vibrational modes identified by the interaction of IR radiation with matter are among the most specific and thus the most informative for chemical characterization. Not only can vibrational frequencies be easily identified with specific localized vibrational groups within a molecule (metal–adsorbate vibrations, O–H stretches, C–C–C deformation modes, etc), but they also depend strongly on the local environment in which the probed moiety is placed [1, 2]. The use of IR spectroscopy was greatly enhanced by the development of Fourier-transform (FTIR) spectrometers in the early 1970s, an event that brought about an enormous improvement in performance in terms of sensitivity, acquisition time, dynamic range and ease of data processing (spectra ratioing in particular) over the conventional scanning apparatus; this made the extension of IR spectroscopy to difficult systems quite feasible. The several experimental approaches pursued for the implementation of IR spectroscopy in surface studies include straight transmission, diffuse reflectance, reflection–absorption, attenuated total reflectance and emission. Each of these is discussed in some detail below.

### B1.22.2.1   Transmission IR spectroscopy

The most common use of spectroscopy in general is in its transmission mode, and this was the first method employed for surface characterization as well. The pioneering work of Terenin *et al* on porous glasses [3] and of Eischens and others on chemisorption over supported metals [4] has already been reviewed in the past [5, 6]. Extensive studies have been carried out since on the characterization of catalysts upon chemisorption of many reactants, from simple molecules such as carbon and nitrogen oxides to hydrocarbons and other complex species [7, 8]. In a recent use of transmission IR absorption to surface problems, the reactivity of silicon towards water and other gases was addressed by first creating highly reproducible porous surfaces by the controlled etching of silicon single crystals [9]. Unfortunately, the general application of transmission IR spectroscopy to surface studies faces some significant limitations, in particular the need for high-surface-area solids (which usually have quite heterogeneous and ill characterized surfaces) and the restricted range of frequencies available away from the regions where the solid absorbs (above 1300 cm$^{-1}$ for silica, 1050 cm$^{-1}$ for alumina, 1200 cm$^{-1}$ for titania, 800 cm$^{-1}$ for magnesia).

### B1.22.2.2   Diffuse-reflectance IR spectroscopy

Another useful technique for the IR characterization of surfaces in powders is diffuse-reflectance IR spectroscopy (DRIFTS) [10]. In the past, the challenge in using this approach has been in the development of efficient optics to collect the diffuse reflected radiation from the sample once illuminated with a focused IR beam, but nowadays this problem has been solved, and several cell designs are available commercially for this endeavour [11]. DRIFTS has, in theory, several advantages over conventional transmission arrangements. First, loose powders can be used without the need to press them into pellets, thus avoiding any sample distortions due to severe physical treatments, allowing for better exposures of the surface to adsorbates, and avoiding losses in the high-frequency range due to light scattering. Second, band intensities in the DRIFTS mode can be as much as four times more intense than in the transmission mode, possibly because of the potential multiple internal reflection of the light in the vicinity of the surface before its emergence towards the detector. Lastly, DRIFTS is better for opaque samples than transmission IR spectroscopy, although the diffuse reflectance may still be low in spectral regions where the absorptivity of the substrate is high. On the negative side, there is a potential lack of reproducibility in the intensities of the DRIFTS bands because of variations in scattering coefficients with cell geometry and sample-loading procedure. Furthermore, diffuse-reflectance spectroscopy suffers from the same key limitation in transmission IR spectroscopy, namely, it requires high-surface-area samples, and therefore provides average spectra only from many types of surface local ensembles and adsorption sites.

The use of DRIFTS for the characterization of surfaces has to date been limited, but has recently been used for applications in fields as diverse as sensors development [12], soils science [13], forensic chemistry [14], corrosion [15], wood science [16] and art [17]. Given that there is in general no reason for preferring transmission over diffuse reflectance in the study of high-area powder systems, DRIFTS is likely to become much more popular in the near future.

### B1.22.2.3   Reflection–absorption IR spectroscopy

The best way to perform IR spectroscopy studies on small samples is in the reflection–absorption (RAIRS) or attenuated total-reflectance (ATR) modes, which work best for opaque and transparent substrates, respectively. RAIRS has in fact become the method of choice for the study of adsorbates on well characterized metal samples, including single crystals. The first attempt to obtain spectra from adlayers on bulk metal samples was that of Pickering and Eckstrom, who in 1959 looked at the adsorption of carbon monoxide and hydrogen on metal films by using a multiple reflection technique with an incoming beam at close to normal incidence to the surface [18]. It soon became clear that better spectra could be obtained by using glancing incidence angles instead [19], and that the gain from using multiple reflection was not worth the complications connected with the required experimental set-up (the optimum number of reflections usually varies between 3 and 10, and results in signal intensity increases of only about 30–50% compared to those from single reflection) [20]. The theory for IR radiation reflection at metal surfaces was later developed by Greenler, who proved that only the p-polarized component of the incident beam is capable of strong interaction with adsorbates on metals, and that interference between that component of the incident and reflected rays sets an intense standing field at the surface which can yield an intensity enhancement of a factor of up to 25 compared to that from the perpendicularly polarized photons [21]. Many surface scientists have since taken advantage of these properties to perform reflection–absorption measurements of monolayers on solid metals [22–24]. Even though the initial RAIRS experiments were carried out with molecules with large dynamic moments such as CO (in order to take advantage of their large absorption cross sections), recent FTIR developments have led to the possibility of detecting submonolayer quantities of species like hydrocarbons with much weaker signals on single crystals of less than 1 $cm^2$ area [25].

A recent example of the usefulness of RAIRS for the characterization of supported catalyst surfaces is given in figure B1.22.1, which displays spectra obtained for a mixture of carbon and nitrogen monoxides coadsorbed on different palladium surfaces [26]. Both CO and NO stretching frequencies are quite sensitive to their adsorption sites, so they can be used to probe local surface sites by determining adsorption geometries in an analogous way as in organometallic discrete complexes. In this example, signals can be easily seen for the two-fold coordination of both CO and NO on Pd(100) surfaces and for three-fold, bridge and atop coordination of CO on Pd(111). The peaks from adsorption on single crystals are used as signatures for the different planes in palladium particles, so an estimate can be obtained on the relative abundance of (100) *versus* (111) sites available on the supported-metal system.

On metals in particular, the dependence of the radiation absorption by surface species on the orientation of the electrical vector can be fully exploited by using one of the several polarization techniques developed over the past few decades [27–30]. The idea behind all those approaches is to acquire the p-to-s polarized light intensity ratio during each single IR interferometer scan; since the adsorbate only absorbs the p-polarized component, that spectral ratio provides absorbance information for the surface species exclusively. Polarization-modulation methods provide the added advantage of being able to discriminate between the signals due to adsorbates and those from gas or liquid molecules. Thanks to this, RAIRS data on species chemisorbed on metals have been successfully acquired *in situ* under catalytic conditions [31], and even in electrochemical cells [32].

The polarization dependence of the photon absorbance in metal surface systems also brings about the so-called surface selection rule, which states that only vibrational modes with dynamic moments having components perpendicular to the surface plane can be detected by RAIRS [22–24]. This rule may in some instances limit the usefulness of the reflection technique for adsorbate identification because of the reduction in the number of modes visible in the IR spectra, but more often becomes an advantage thanks to the simplification of the data. Furthermore, the relative intensities of different vibrational modes can be used to estimate the orientation of the surface moieties. This has been particularly useful in the study of self-assembled and Langmuir–Blodgett monolayers, where RAIRS data have been unique in providing information on the orientation of the hydrocarbon chains [33]. Figure B1.22.2 shows an example in which RAIRS was used to determine a collective change in adsorption geometry for alkyl halides on metal single crystals as the surface coverage is increased past the half-monolayer [34].

The use of RAIRS has recently been extended from its regular mid-IR characterization of adsorbates on metals into other exciting and promising directions. For one, changes in optics and detectors have allowed for an extension of the spectral range towards the far-IR region in order to probe substrate–adsorbate vibrations [35]. The use of intense synchrotron sources in particular looks quite promising for the detection of such weak modes [36]. Thanks to the speed with which Fourier-transform spectrometers can acquire complete IR spectra, kinetic studies of surface reactions can be carried out as well. To date this has only been done in a few cases, usually for reactions that take seconds or more to occur [37], but the advent of step scanners promises the availability of time resolutions of $10^{-8}$ s or better in the near future [38]. In terms of the lifetime of the vibrational excitations themselves, this can in some instances be estimated from IR absorption line shapes. Because of the efficient coupling between the vibrations of adsorbate and phonons and other electronic surface states, the former are generally short-lived, and therefore yield IR absorption bands several wavenumbers

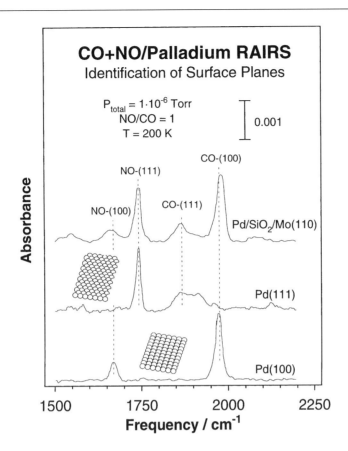

**Figure B1.22.1.** Reflection–absorption IR spectra (RAIRS) from palladium flat surfaces in the presence of a $1 \times 10^{-6}$ Torr 1:1 NO:CO mixture at 200 K. Data are shown here for three different surfaces, namely, for Pd(100) (bottom) and Pd(111) (middle) single crystals and for palladium particles (about 500 Å in diameter) deposited on a 100 Å thick $SiO_2$ film grown on top of a Mo(110) single crystal. These experiments illustrate how RAIRS titration experiments can be used for the identification of specific surface sites in supported catalysts. On Pd(100) CO and NO each adsorbs on twofold sites, as indicated by their stretching bands at about 1970 and 1670 cm$^{-1}$, respectively. On Pd(111), on the other hand, the main IR peaks are seen around 1745 cm$^{-1}$ for NO (on-top adsorption) and about 1915 cm$^{-1}$ for CO (threefold coordination). Using those two spectra as references, the data from the supported Pd system can be analysed to obtain estimates of the relative fractions of (100) and (111) planes exposed in the metal particles [26].

wide. Nevertheless, bands as narrow as 0.7 cm$^{-1}$ have been observed in some cases [39]. Finally, the use of RAIRS is not limited to metal surfaces. Although the surface selection rules change significantly for non-metal surfaces, they can still be used to obtain orientational information for adsorbates on transparent substrates, as recently demonstrated in the elegant study by Hoffmann *et al* on the adsorption of long-chain hydrocarbons on silicon (figure B1.22.3) [40], and even for the analysis of air–liquid interfaces [41]. There are many clear new directions still unexplored for the use of RAIRS in surface characterization studies.

### B1.22.2.4  *Attenuated total reflectance IR spectroscopy*

In 1960, Harrick demonstrated that, for transparent substrates, absorption spectra of adsorbed layers could be obtained using internal reflection [42]. By cutting the sample in a specific trapezoidal shape, the IR beam

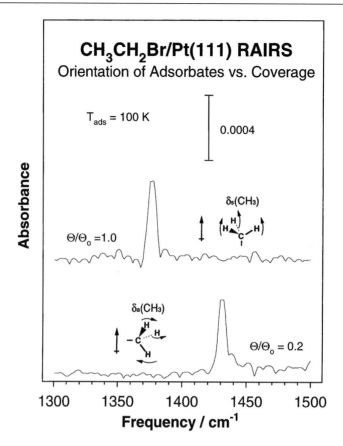

**Figure B1.22.2.** RAIRS data from molecular ethyl bromide adsorbed on a Pt(111) surface at 100 K. The two traces shown, which correspond to coverages of 20% and 100% saturation, illustrate the use of the RAIRS surface selection rule for the determination of adsorption geometries. Only one peak, but a different one, is observed in each case: while the signal detected at low coverages is due to the asymmetric deformation of the terminal methyl group (1431 cm$^{-1}$), the feature corresponding to the symmetric deformation (1375 cm$^{-1}$) is the one seen at high coverages instead. Given that only vibrations with dynamic dipole moments perpendicular to the surface are visible with this technique, it is concluded that a flat adsorption geometry prevails at low coverages but that a collective rearrangement of the adsorbates to a standing-up configuration takes place at about half-saturation [34].

can be made to enter through one end, bounce internally a number of times from the flat parallel edges, and exit the other end without any losses, leading to high adsorption coefficients for the species adsorbed on the external surfaces of the plate (higher than in the case of external reflection) [24]. This is the basis for the ATR technique.

In recent years, ATR has been used primarily in connection with the characterization of semiconductor surfaces. For instance, ATR studies have led to the detailed mapping of the complex series of reconstructions that silicon surfaces follow upon thermal treatment and/or hydrogen exposures [43]. Surface electronic excitations have been studied with this technique as well; see, for instance, the pioneering work of Mc-Combe *et al* on the characterization of inter-subband optical transitions in silicon MOS field-effect transistors (figure B1.22.4) [44]. One interesting additional extension of the use of multiple internal reflection to the characterization of non-transparent samples was discussed by Bermudez, who suggested that the sensitivity

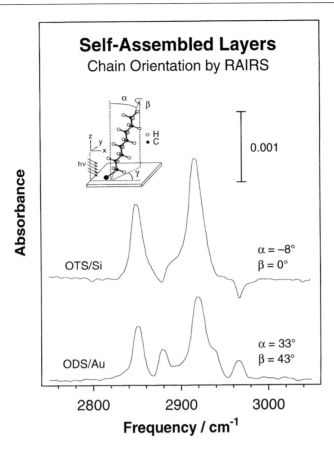

**Figure B1.22.3.** RAIRS data in the C–H stretching region from two different self-assembled monolayers, namely, from a monolayer of dioctadecyldisulfide (ODS) on gold (bottom), and from a monolayer of octadecyltrichlorosilane (OTS) on silicon (top). Although the RAIRS surface selection rules for non-metallic substrates are more complex than those which apply to metals, they can still be used to determine adsorption geometries. The spectra shown here were, in fact, analysed to yield the tilt ($\alpha$) and twist ($\beta$) angles of the molecular chains in each case with respect to the surface plane (the resulting values are also given in the figure) [40].

to adsorbates in IR-reflection spectroscopy can be enhanced by burying a metal layer beneath the surface of a dielectric material [45].

*B1.22.2.5  Other surface IR spectroscopy arrangements*

There have been a few other experimental set-ups developed for the IR characterization of surfaces. Photoacoustic (PAS), or, more generally, photothermal IR spectroscopy relies on temperature fluctuations caused by irradiating the sample with a modulated monochromatic beam: the acoustic pressure wave created in the gas layer adjacent to the solid by the adsorption of light is measured as a function of photon wavelength in order to determine the absorption spectra [11]. It has sometimes been thought that PAS is more surface sensitive than DRIFTS, but in fact that depends on the specific optical and thermal properties of the material being studied. In emission spectrometry (EMS), the IR radiation emitted by the sample is directly collected and analysed. The detection of the (non-monochromatized) IR radiation from thin films has recently being combined with

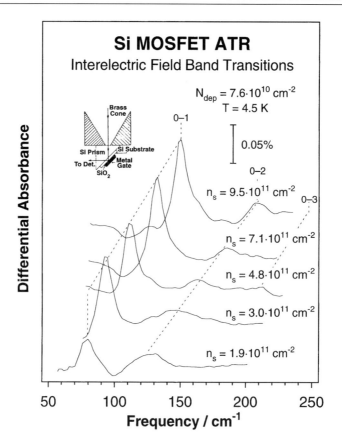

**Figure B1.22.4.** Differential IR absorption spectra from a metal–oxide silicon field-effect transistor (MOSFET) as a function of gate voltage (or inversion layer density, $n_s$, which is the parameter reported in the figure). Clear peaks are seen in these spectra for the 0–1, 0–2 and 0–3 inter-electric-field subband transitions that develop for charge carriers when confined to a narrow ($<100$ Å) region near the oxide–semiconductor interface. The inset shows a schematic representation of the attenuated total reflection (ATR) arrangement used in these experiments. These data provide an example of the use of ATR IR spectroscopy for the probing of electronic states in semiconductor surfaces [44].

molecular beam techniques in order to perform differential microcalorimetric measurements on adsorption processes [46]. Finally, the sample itself can be used as the detector of IR radiation. None of these techniques have found much use in surface studies to date.

## B1.22.3   Laser-based spectroscopies

Although the development of a large variety of lasers with different spectral ranges, intensities and temporal resolutions has led to the surge of many new optical characterization techniques, most of those have yet to make a large impact in surface science. As discussed above, the signals from surfaces are often weak and hard to differentiate from those from the bulk, and this is particularly troublesome in nonlinear techniques which rely on the absorption of more than one photon. Furthermore, the increase of the laser power to levels where signal intensities are no longer an issue may lead to damaging of the substrate. In spite of these limitations, some laser-based methods have already been developed for surface-characterization studies.

### B1.22.3.1 Raman spectroscopy

Perhaps the best known and most used optical spectroscopy which relies on the use of lasers is Raman spectroscopy. Because Raman spectroscopy is based on the inelastic scattering of photons, the signals are usually weak, and are often masked by fluorescence and/or Rayleigh scattering processes. The interest in using Raman for the vibrational characterization of surfaces arises from the fact that the technique can be used *in situ* under non-vacuum environments, and also because it follows selection rules that complement those of IR spectroscopy.

Regular Raman has been employed mainly for the characterization of high-surface-area solids [47]. Specifically, a good methodology has been developed for the determination of bond orders, bond lengths and local geometries in many metal oxides used for catalysis. Figure B1.22.5 illustrates this point by displaying some examples where the Raman vibrational signals for metal–oxygen single and double bonds as well as for oxygen–metal–oxygen deformations were used to determine the structure of a number of supported and highly dispersed transition-metal oxides [48].

In an interesting development in Raman spectroscopy, Fleischmann *et al* noticed in 1974 that there is a significant enhancement in the Raman signal intensities from solid surfaces if the substrate is comprised of small silver particles [49]. The same phenomenon has since been observed with copper, silver, gold, lithium, sodium, potassium, indium, platinum and rhodium, and has become the basis for surface-enhanced Raman spectroscopy (SERS) [50, 51]. The reasons for this enhancement are still not completely clear, but have been recognized to be the result of a combination of effects, including a surface electromagnetic field enhancement (in particular when illuminating rough samples with photons of energies near those of localized plasmons) and a chemical enhancement due to the change of polarizability in molecules when interacting with surfaces [52].

Since its initial development, SERS has been used for the surface characterization of a good number of systems. One important extension to the use of SERS has been in the determination of surface geometries. Figure B1.22.6 shows an example of the SERS C–H stretching frequency data used to determine the different chemisorption geometries of 2-butanol and 2-butanethiol on silver electrodes [53]. Notice that, being an optical technique, SERS works quite well in solid–liquid interfaces. On the other hand, the need for signal enhancement normally limits the use of SERS to a handful of metals and/or to samples with rough surfaces.

Another recent development in the use of Raman spectroscopy for the characterization of surfaces has been the employment of UV light for the initial excitation of the sample [54]. The advantage of UV over conventional Raman spectroscopy is twofold: (1) since the normal Raman scattering cross sections are proportional to the fourth power of the scattered light frequency, the use of higher-energy photons significantly increases the signal intensity and (2) by using UV light the spectral range is moved away from that where fluorescence de-excitation is observed. This allows for the Raman characterization of virtually any high-surface-area sample, including opaque solids such as black carbon. Unfortunately, UV-Raman spectroscopy is still not commercially available.

### B1.22.3.2 Other nonlinear optical techniques

Other nonlinear optical spectroscopies have gained much prominence in recent years. Two techniques in particular have become quite popular among surface scientists, namely, second harmonic (SHG) [55] and sum-frequency (SFG) [56] generation. The reason why both SHG and SFG can probe interfaces selectively without being overwhelmed by the signal from the bulk is that they rely on second-order processes that are electric-dipole forbidden in centrosymmetric media; by breaking the bulk symmetry, the surface places the molecular species in an environment where their second-order nonlinear susceptibility, the term responsible for the absorption of SHG and SFG signals, becomes non-zero.

In SHG the sample is illuminated with light of a single colour and the component at twice the initial frequency is filtered from the emitted light and analysed. Because these experiments usually involve near-IR

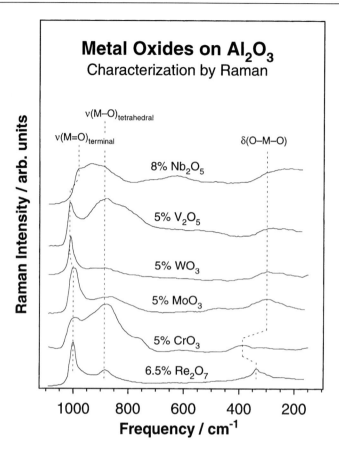

**Figure B1.22.5.** *In situ* Raman spectra from a family of transition-metal oxides dispersed on high-surface-area alumina substrates. Three distinct regions can be differentiated in these spectra, namely, the peaks around 1000 cm$^{-1}$, which are assigned to the stretching frequency of terminal metal–oxygen double bonds, the features about 900 cm$^{-1}$, corresponding to metal–oxygen stretches in tetrahedral coordination sites, and the low-frequency ($<400$ cm$^{-1}$) range associated with oxygen–metal–oxygen deformation modes. Data such as these can be used to determine the nature and geometry of supported oxides as a function of metal loading and subsequent treatment [48].

or visible–UV photons, SHG most often probes electronic transitions. In fact, SHG has often been used as a way to measure changes in work function or localized electrostatic surface potentials. When using polarized light, SHG can also be used to determine the geometrical alignment of polar molecules at interfaces and, by sweeping the incident photon energy, spectroscopic information can be obtained on molecular orbital energies as well. Figure B1.22.7 shows an example of the latter for the case of rhodamine 6G [55]. This figure also shows a clever extension of the technique as a microscope to provide spatial information on adsorbates.

By combining two beams on the surface, one of visible or IR fixed frequency and a second, of variable energy in the IR region, resonance absorption can be measured by detecting the intensity of the outgoing light resulting from the addition of the two incident beams as a function of the photon energy of the variable laser. The net effect of this SFG is the acquisition of vibrational absorption spectra for surface species where signals are seen only for the modes active in both IR and Raman. Vibrational information can be obtained with SFG for almost any interface as long as lasers are available to cover the frequency range of interest and the bulk materials are transparent to the laser light. Also, as with many of the other techniques described above,

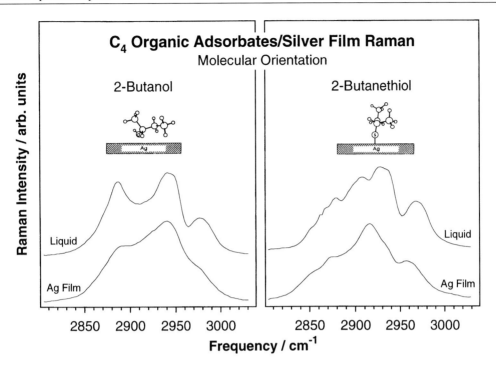

**Figure B1.22.6.** Raman spectra in the C–H stretching region from 2-butanol (left frame) and 2-butanethiol (right), each either as bulk liquid (top traces) or adsorbed on a rough silver electrode surface (bottom). An analysis of the relative intensities of the different vibrational modes led to the proposed adsorption structures depicted in the corresponding panels [53]. This example illustrates the usefulness of Raman spectroscopy for the determination of adsorption geometries, but also points to its main limitation, namely the need to use rough silver surfaces to achieve adequate signal-to-noise levels.

orientational information can be obtained with SFG as well. Figure B1.22.8 displays data demonstrating that an increase in the concentration of acetonitrile dissolved in water leads to a collective molecular orientation change at the air/water interface [57].

There are a few other surface-sensitive characterization techniques that also rely on the use of lasers. For instance surface-plasmon resonance (SPR) measurements have been used to follow changes in surface optical properties as a function of time as the sample is modified by, for instance, adsorption processes [58]. SPR has proven useful to image adsorption patterns on surfaces as well [59].

*B1.22.3.3    Time-resolved pump–probe experiments*

The dynamics of fast processes such as electron and energy transfers and vibrational and electronic de-excitations can be probed by using short-pulsed lasers. The experimental developments that have made possible the direct probing of molecular dissociation steps and other ultrafast processes in real time (in the femtosecond time range) have, in a few cases, been extended to the study of surface phenomena. For instance, two-photon photoemission has been used to study the dynamics of electrons at interfaces [60]. Vibrational relaxation times have also been measured for a number of modes such as the O–H stretching in silica and the C–O stretching in carbon monoxide adsorbed on transition metals [61]. Pump-probe laser experiments such as these are difficult, but the field is still in its infancy, and much is expected in this direction in the near future.

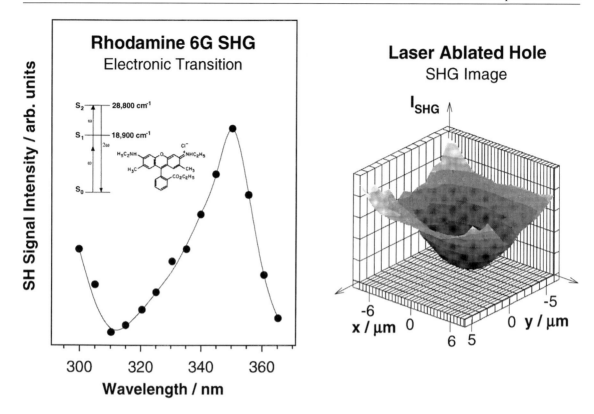

**Figure B1.22.7.** Left: resonant second-harmonic generation (SHG) spectrum from rhodamine 6G. The inset displays the resonant electronic transition induced by the two-photon absorption process at a wavelength of approximately 350 nm. Right: spatially resolved image of a laser-ablated hole in a rhodamine 6G dye monolayer on fused quartz, mapped by recording the SHG signal as a function of position in the film [55]. SHG can be used not only for the characterization of electronic transitions within a given substance, but also as a microscopy tool.

## B1.22.4   X-ray diffraction and x-ray absorption

Because x-rays are particularly penetrating, they are very useful in probing solids, but are not as well suited for the analysis of surfaces. X-ray diffraction (XRD) methods are nevertheless used routinely in the characterization of powders and of supported catalysts to extract information about the degree of crystallinity and the nature and crystallographic phases of oxides, nitrides and carbides [62, 63]. Particle size and dispersion data are often acquired with XRD as well.

One way to obtain surface sensitivity in XRD experiments with crystalline samples is to illuminate the substrate at glancing angles. Under normal conditions, x-rays projected onto the sample at incident angles of less than $10°$ still penetrate to a depth of $10$ $\mu$m or more but, beyond a critical angle, x-ray photons are completely reflected from the surface and light propagation into the solid is *via* a rapidly attenuating evanescent wave, and this renders the x-ray probe quite surface sensitive [64]. There are several inherent difficulties in implementing grazing-angle XRD experiments, which require focusing high-intensity x-rays onto surfaces at angles of the order of $0.1°$, but recent experiments have proved the usefulness of this technique in providing interesting information on the structure of robust surfaces such as oxides, nitrides, silicides and other thin films.

A related technique that also relies on the interference of x-rays for solid characterization is extended x-ray absorption fine structure (EXAFS) [65, 66]. Because the basis for EXAFS is the interference of outgoing

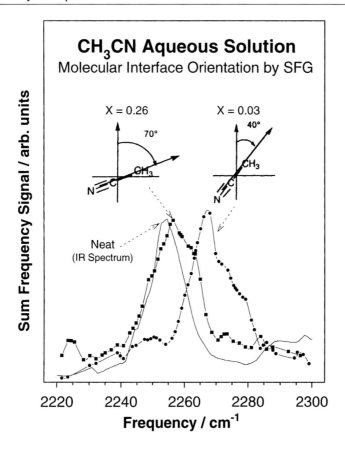

**Figure B1.22.8.** Sum-frequency generation (SFG) spectra in the C≡N stretching region from the air/aqueous acetonitrile interfaces of two solutions with different concentrations. The solid curve is the IR transmission spectrum of neat bulk $CH_3CN$, provided here for reference. The polar acetonitrile molecules adopt a specific orientation in the air/water interface with a tilt angle that changes with changing concentration, from 40° from the surface normal in dilute solutions (molar fractions less than 0.07) to 70° at higher concentrations. This change is manifested here by the shift in the C≡N stretching frequency seen by SFG [57]. SFG is one of the very few techniques capable of probing liquid/gas, liquid/liquid, and even liquid/solid interfaces.

photoelectrons with their scattered waves from nearby atoms, it does not require long-range order to work (as opposed to diffraction techniques), and provides information about the local geometry around specific atomic centres. Unfortunately, EXAFS requires the high-intensity and tunable photon sources typically available only at synchrotron facilities. Further limitations to the development of surface-sensitive EXAFS (SEXAFS) have come from the fact that it requires technology entirely different from that of regular EXAFS, involving in many cases ultrahigh-vacuum environments and/or photoelectron detection. One interesting advance in SEXAFS came with the design by Stöhr *et al* of fluorescence detectors for the x-rays absorbed by the surface species of small samples; that allows for the characterization of well defined systems such as single crystals under non-vacuum conditions [67]. Figure B1.22.9 shows the S K-edge x-ray absorption data obtained for a c($2 \times 2$)S–Ni(100) overlayer using their original experimental set-up. This approach has since been extended to the analysis of lighter atoms (C, O, F) on many different substrates and under atmospheric pressures [68].

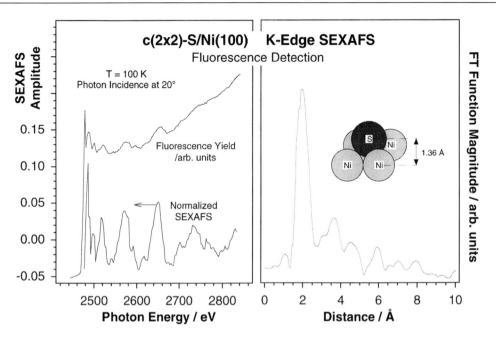

**Figure B1.22.9.** Fluorescence-yield, surface-extended x-ray absorption fine structure (FY-SEXAFS) spectra for the S K-edge of a c(2 × 2) ordered monolayer of sulfur atoms adsorbed on a Ni(100) surface. The upper trace in the left panel corresponds to the raw data obtained at an incident photon angle of 20° from the surface plane, while the bottom trace displays the background-subtracted and normalized SEXAFS oscillations calculated from the original spectrum. The right panel, which corresponds to the Fourier-transformed SEXAFS data, provides information on both the Ni–S bond length (2.22 Å) and the Ni near-neighbour coordination number around each sulfur atom (4) [67]. Because EXAFS relies on the interference of an outgoing photoelectron with its own scattering from nearby atoms, it provides local geometry information without requiring long-range order. This example also illustrates the high sensitivity of the technique (these experiments were carried out with a 1 cm² area single crystal). The use of fluorescence detection allows for the extension of this type of study to samples in non-vacuum environments [68].

Soon after the development of EXAFS it was recognized that the signal near the x-ray absorption edge is quite complex and provides information on electronic transitions from atomic core levels to valence bands and/or molecular orbitals [69]. The analysis of that signal constitutes the basis for a technique named near-edge x-ray absorption fine structure (NEXAFS, or XANES). The shape of the x-ray absorption spectra near the absorption edge has long been used as an empirical fingerprint for the local chemical environment of oxides and other supported catalysts, but newer developments allow for the extraction of a more detailed picture of the nature and geometrical arrangement of adsorbates from those data. This is possible thanks in great part to the combination of the polarized nature of synchrotron radiation and the simplicity of the electronic transition dipoles for absorption from core levels [70]. Figure B1.22.10 displays an example where the geometry of vinyl moieties adsorbed on Ni(100) surfaces was determined by using NEXAFS [71].

## B1.22.5   Other optical techniques

A few additional optical techniques need to be mentioned in this review. As discussed above, these are by and large well known spectroscopies for the study of bulk samples; it is only their extension to the study of surfaces what has not been realized to its fullest potential yet.

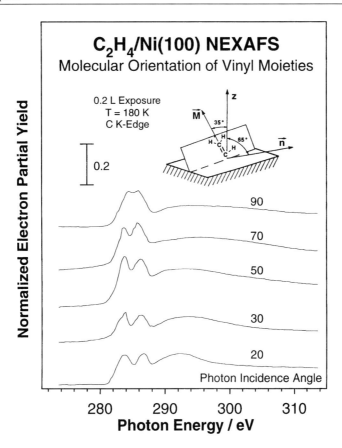

**Figure B1.22.10.** Carbon K-edge near-edge x-ray absorption (NEXAFS) spectra as a function of photon incidence angle from a submonolayer of vinyl moieties adsorbed on Ni(100) (prepared by dosing 0.2 l of ethylene on that surface at 180 K). Several electronic transitions are identified in these spectra, to both the pi (284 and 286 eV) and the sigma (>292 eV) unoccupied levels of the molecule. The relative variations in the intensities of those peaks with incidence angle can be easily converted into adsorption geometry data; the vinyl plane was found in this case to be at a tilt angle of about 65° from the surface [71]. Similar geometrical determinations using NEXAFS have been carried out for a number of simple adsorbate systems over the past few decades.

*B1.22.5.1    Other UV–visible optical techniques*

Spectroscopies such as UV–visible absorption and phosphorescence and fluorescence detection are routinely used to probe electronic transitions in bulk materials, but they are seldom used to look at the properties of surfaces [72]. As with other optical techniques, one of the main problems here is the lack of surface discrimination, a problem that has sometime been bypassed by either using thin films of the materials of interest [73, 74], or by using a reflection detection scheme. Modulation of a parameter, such as electric or magnetic fields, stress, or temperature, which affects the optical properties of the sample and detection of the AC component of the signal induced by such periodic changes, can also be used to achieve good surface sensitivity [75]. This latter approach is the basis for techniques such as surface reflectance spectroscopy, reflectance difference spectroscopy/reflectance anisotropy spectroscopy, surface photoadsorption spectroscopy and surface differential reflectivity [76–78]. Early optical characterization studies of solid surfaces were instrumental in the detection

# Oxazine/PMMA NSOM Image
## Single Molecule Detection

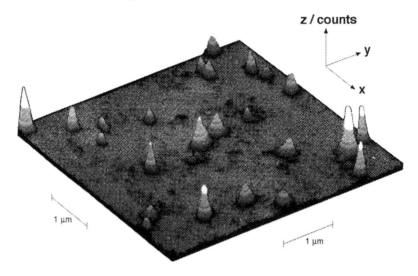

**Figure B1.22.11.** Near-field scanning optical microscopy fluorescence image of oxazine molecules dispersed on a PMMA film surface. Each protuberance in this three-dimensional plot corresponds to the detection of a single molecule, the different intensities of those features being due to different orientations of the molecules. Sub-diffraction resolution, in this case on the order of a fraction of a micron, can be achieved by the near-field scanning arrangement. Spectroscopic characterization of each molecule is also possible. (Reprinted with permission from [82]. Copyright 1996 American Chemical Society.)

and characterization of intrinsic surface electronic states due to the uniqueness of the interface environments. Ellipsometry is also a mature technique often used to obtain film thickness and other optical properties [79].

One interesting new field in the area of optical spectroscopy is near-field scanning optical microscopy, a technique that allows for the imaging of surfaces down to sub-micron resolution and for the detection and characterization of single molecules [80, 81]. When applied to the study of surfaces, this approach is capable of identifying individual adsorbates, as in the case of oxazine molecules dispersed on a polymer film, illustrated in figure B1.22.11 [82]. Absorption and emission spectra of individual molecules can be obtained with this technique as well, and time-dependent measurements can be used to follow the dynamics of surface processes.

### B1.22.5.2  Magnetic resonance

NMR has developed into a powerful analytical technique in the past decades, and has been used extensively in the characterization of a great number of chemical systems. Its extension to the study of surfaces, however, has been hampered by the need of large samples because of its poor sensitivity. On the other hand, the development of magic-angle-spinning NMR (MAS-NMR) and the extension of NMR to many nuclei besides hydrogen have opened the doors for the use of that technique to many solids. For instance, MAS $^1$H NMR has been quite useful in the research on Brønsted acidity in oxides [83]. Also, the study of zeolites with NMR is now practically routine: the chemical shifts in $^{29}$Si NMR data are easily interpreted in terms of the number of aluminium atoms next to a given silicon centre, and the position of the $^{27}$Al peaks provides information on the coordination number and geometry of the aluminium atoms [84]. $^{129}$Xe NMR has been used to probe both the local environment inside porous materials [83] and heterogeneities in adsorbates [85]. Dynamic studies on the thermal conversion of adsorbates on transition metal catalysts have been performed as well [86].

Other magnetic measurements of catalysts include electron paramagnetic resonance and magnetic susceptibility. Although those are not as common as NMR, they can be used to look at the properties of paramagnetic and ferromagnetic samples. Examples of these applications can be found in the literature [87, 88].

### B1.22.5.3 *Spectroscopies which rely only in part on photons*

A number of surface-sensitive spectroscopies rely only in part on photons. On the one hand, there are techniques where the sample is excited by electromagnetic radiation but where other particles ejected from the sample are used for the characterization of the surface (photons in; electrons, ions or neutral atoms or moieties out). These include photoelectron spectroscopies (both x-ray- and UV-based) [89–91], photon stimulated desorption [92], and others. At the other end, a number of methods are based on a particles-in/photons-out set-up. These include inverse photoemission and ion- and electron-stimulated fluorescence [93, 94]. All these techniques are discussed elsewhere in this encyclopaedia.

### References

[1] Nakamoto K 1978 *Infrared and Raman Spectra of Inorganic and Coordination Compounds* 3rd edn (New York: Wiley-Interscience)
[2] Socrates G 1994 *Infrared Characteristic Group Frequencies: Tables and Charts* 2nd edn (Chichester: Wiley)
[3] Yaroslavskii N G and Terenin A N 1949 Infrared absorption spectra of adsorbed molecules *Dokl. Akad. Nauk* **66** 885–8
[4] Eischens R P, Pliskin W A and Francis S A 1954 Infrared spectra of chemisorbed CO *J. Chem. Phys.* **22** 1986–7
[5] Little L H 1966 *Infrared Spectra of Adsorbed Species* (New York: Academic)
[6] Hair M L 1967 *Infrared Spectroscopy in Surface Chemistry* (New York: Marcel Dekker)
[7] Sheppard N and De La Cruz C 1996 Vibrational spectra of hydrocarbons adsorbed on metals. Part I. Introductory principles, ethylene, and higher acyclic alkenes *Adv. Catal.* **41** 1–112
[8] Sheppard N and De La Cruz C 1998 Vibrational spectra of hydrocarbons adsorbed on metals. Part II. Adsorbed acyclic alkynes and alkanes, cyclic hydrocarbons including aromatics and surface hydrocarbon groups derived from the decomposition of alkyl halides, etc *Adv. Catal.* **42** 181–313
[9] Gupta P, Dillon A C, Bracker A S and George S M 1991 FTIR studies of $H_2O$ and $D_2O$ decomposition on porous silicon surfaces *Surf. Sci.* **245** 360–72
[10] Willey R R 1976 Fourier transform infrared spectrophotometer for transmittance and diffuse reflectance measurements *Appl. Spectrosc.* **30** 593–601
[11] Griffiths P R and de Haseth J A 1986 *Fourier Transform Infrared Spectrometry* (New York: Wiley)
[12] Benitez J J, Centeno M A, Merdrignac O M, Guyader J, Laurent Y and Odriozola J A 1995 DRIFTS chamber for *in situ* and simultaneous study of infrared and electrical response of sensors *Appl. Spectrosc.* **49** 1094–6
[13] Vreugdenhil A J and Butler I S 1998 Investigation of MMT adsorption on soils by diffuse reflectance infrared spectroscopy DRIFTS and headspace analysis gas-phase infrared spectroscopy HAGIS *Appl. Organomet. Chem.* **2** 121–8
[14] Lennard C J, Mazzella W D and Margot P A 1993 Some applications of diffuse reflectance infrared Fourier transform spectroscopy DRIFTS in forensic science *Analysis* **21** M34–7
[15] Iob A, Somuah S K, Siddiqui A H and Abbas N M 1990 Diffuse reflectance infrared DRIFTS studies of corrosion products *Anal. Lett.* **23** 1537–52
[16] Kazayawoko M, Balatinecz J J and Woodhams R T 1997 Diffuse reflectance Fourier transform infrared spectra of wood fibers treated with maleated polypropylenes *J. Appl. Polymer Sci.* **66** 1163–73
[17] Zeine C and Grobe J 1997 Diffuse reflectance infrared Fourier transform DRIFT spectroscopy in the preservation of historical monuments: studies on salt migration *Mikrochim. Acta* **125** 279–82
[18] Pickering H L and Eckstrom H C 1959 Studies by infrared absorption *J. Phys. Chem.* **63** 512–17
[19] Francis S A and Ellison A H 1959 Infrared spectra of monolayers on metal mirrors *J. Opt. Soc. Am.* **49** 131–8
[20] Hollins P and Pritchard J 1985 Infrared studies of chemisorbed layers on single crystals *Prog. Surf. Sci.* **19** 275–350
[21] Greenler R G 1966 Infrared study of adsorbed molecules on metal surfaces by reflection techniques *J. Chem. Phys.* **44** 310–15
[22] Bradshaw A M 1982 Vibrational spectroscopy of adsorbed atoms and molecules *Appl. Surf. Sci.* **11/12** 712–29

[23] Hoffmann F M 1983 Infrared reflection–absorption spectroscopy of adsorbed molecules *Surf. Sci. Rep.* **3** 107–92

[24] Chabal Y J 1988 Surface infrared spectroscopy *Surf. Sci. Rep.* **8** 211–357

[25] Zaera F and Hoffmann H 1991 Detection of chemisorbed methyl and methylene groups: surface chemistry of methyl iodide on Pt(111) *J. Phys. Chem.* **95** 6297–303

[26] Xu X, Chen P and Goodman D W 1994 A comparative study of the coadsorption of CO and NO on Pd(100), Pd(111), and silica-supported palladium particles with infrared reflection–absorption spectroscopy *J. Phys. Chem.* **98** 9242–6

[27] Dowrey A E and Marcott C 1982 A double-modulation Fourier transform infrared approach to studying adsorbates on metal surfaces *Appl. Spectrosc.* **36** 414–16

[28] Ishida H, Ishino Y, Buijs H, Tripp C and Dignam M J 1987 Polarization-modulation FT-IR reflection spectroscopy using a polarizing Michelson interferometer *Appl. Spectrosc.* **41** 1288–94

[29] Hoffmann H, Wright N A, Zaera F and Griffiths P R 1989 Differential-polarization dual-beam FT-IR spectrometer for surface analysis *Talanta* **36** 125–31

[30] Barner B J, Green M J, Sáez E I and Corn R M 1991 Polarization modulation Fourier transform infrared reflectance measurements of thin films and monolayers at metal surfaces utilizing real-time sampling electronics *Anal. Chem.* **63** 55–60

[31] Hoffmann F M and Weisel M D 1993 Fourier transform infrared reflection absorption spectroscopy studies of adsorbates and surface reactions—bridging the pressure gap between surface science and catalysis *J. Vac. Sci. Technol.* A **11** 1957–63

[32] Bewick A and Pons S 1985 Infrared spectroscopy of the electrode–electrolyte solution interface *Advances in Infrared and Raman Spectroscopy* ed R J H Clark and R E Hester (New York: Wiley Heyden) **12** 1–63

[33] Parikh A N and Allara D L 1992 Quantitative determination of molecular structure in multilayered thin films of biaxial and lower symmetry from photon spectroscopies. I. Reflection infrared vibrational spectroscopy *J. Chem. Phys.* **96** 927–45

[34] Zaera F, Hoffmann H and Griffiths P R 1990 Determination of molecular chemisorption geometries using reflection–absorption infrared spectroscopy: alkyl halides on Pt(111) *J. Electron. Spectrosc. Relat. Phenom.* **54/55** 705–15

[35] Malik I J and Trenary M 1989 Infrared reflection–absorption study of the adsorbate–substrate stretch of CO on Pt(111) *Surf. Sci.* **214** L237–45

[36] Dumas P, Suhren M, Chabal Y J, Hirschmugl C J and Williams G P 1997 Adsorption and reactivity of NO on Cu(111): a synchrotron infrared reflection absorption spectroscopic study *Surf. Sci.* **371** 200–12

[37] Janssens T V W, Stone D, Hemminger J C and Zaera F 1998 Kinetics and mechanism for the H/D exchange between ethylene and deuterium over Pt(111) *J. Catal.* **177** 284–95

[38] Johnson T J, Simon A, Weil J M and Harris G W 1993 Applications of time-resolved step-scan and rapid-scan FT-IR spectroscopy: dynamics from ten seconds to ten nanoseconds *Appl. Spectrosc.* **47** 1376–81

[39] Agrawal V K and Trenary M 1989 Infrared spectrum from 400 to 1000 cm$^{-1}$ of PF$_3$ chemisorbed on the Pt(111) surface *J. Vac. Sci. Technol.* A **7** 2235–7

[40] Hoffmann H, Mayer U and Krischanitz A 1995 Structure of alkylsiloxane monolayers on silicon surfaces investigated by external reflection infrared spectroscopy *Langmuir* **11** 1304–12

[41] Mendelsohn R, Brauner J W and Gericke A 1995 External infrared reflection absorption spectrometry of monolayer films at the air–water interface *Ann. Rev. Phys. Chem.* **46** 305–34

[42] Harrick N J 1960 Physics and chemistry of surfaces from frustrated total internal reflections *Phys. Rev. Lett.* **4** 224–6

[43] Chabal Y J 1986 High-resolution infrared spectroscopy of adsorbates on semiconductor surfaces: hydrogen on silicon(100) and germanium(100) *Surf. Sci.* **168** 594–608

[44] McCombe B D, Holm R T and Schafer D E 1979 Frequency domain studies of intersubband optical transitions in Si inversion layers *Solid State Commun.* **32** 603–8

[45] Bermudez V M 1992 Infrared optical properties of dielectric/metal layer structures of relevance to reflection absorption spectroscopy *J. Vac. Sci. Technol.* A **10** 152–7

[46] Borroni-Bird C E and King D A 1991 An ultrahigh vacuum single crystal adsorption microcalorimeter *Rev. Sci. Instrum.* **62** 2177–85

[47] Stencel J M 1990 *Raman Spectroscopy for Catalysis* (New York: Van Nostrand Reinhold)

[48] Vuurman M A and Wachs I E 1992 *In situ* Raman spectroscopy of alumina-supported metal oxide catalysts *J. Phys. Chem.* **96** 5008–16

[49] Fleischmann M, Hendra P J and McQuillan A J 1974 Raman spectra of pyridine adsorbed at a silver electrode *Chem. Phys. Lett.* **26** 163–6

[50] Moskovits M 1985 Surface-enhanced spectroscopy *Rev. Mod. Phys.* **57** 783–826

[51] Creighton J A 1988 The selection rules for surface-enhanced Raman spectroscopy *Spectroscopy of Surfaces* ed R J H Clark and R E Hester (Chichester: Wiley) pp 37–89

[52] Kambhampati P, Child C M, Foster M C and Campion A 1998 On the chemical mechanism of surface enhanced Raman scattering: experiment and theory *J. Chem. Phys.* **108** 5013–26

[53] Pemberton J E, Bryant M A, Sobocinski R L and Joa S L 1992 A simple method for determination of orientation of adsorbed organics of low symmetry using surface-enhanced Raman scattering *J. Phys. Chem.* **96** 3776–82

[54] Stair P C and Li C 1997 Ultraviolet Raman spectroscopy of catalysts and other solids *J. Vac. Sci. Technol.* A **15** 1679–84

[55] Shen Y R 1989 Optical second harmonic generation at interfaces *Ann. Rev. Phys. Chem.* **40** 327–50

[56] Eisenthal K B 1996 Liquid interfaces probed by second-harmonic and sum-frequency spectroscopy *Chem. Rev.* **96** 1343–60

[57] Wang H, Borguet E, Yan E C Y, Zhang D, Gutow J and Eisenthal K B 1998 Molecules at liquid and solid surfaces *Langmuir* **14** 1472–7

[58] Caruso F, Jory M J, Bradberry G W, Sambles J R and Furlong D N 1998 Acousto-optic surface-plasmon resonance measurements of thin films on gold *J. Appl. Phys.* **83** 1023–8

[59] Jordan C E and Corn R M 1997 Surface plasmon resonance imaging measurements of electrostatic biopolymer adsorption onto chemically modified gold surfaces *Anal. Chem.* **69** 1449–56

[60] Harris C B, Ge N-H, Lingle R L Jr, McNeill J D and Wong C M 1997 Femtosecond dynamics of electrons on surfaces and at interfaces *Ann. Rev. Phys. Chem.* **48** 711–44

[61] Heilweil E J, Casassa M P, Cavanagh R R and Stephenson J C 1989 Picosecond vibrational energy transfer studies of surface adsorbates *Ann. Rev. Phys. Chem.* **40** 143–71

[62] Peiser H S 1960 *X-Ray Diffraction by Polycrystalline Materials* (London: Chapman and Hall)

[63] Cohen J B and Schwartz L H 1977 *Diffraction from Materials* (New York: Academic)

[64] Brunel M 1995 Glancing angle x-ray diffraction *Encyclopedia of Analytical Science* vol 8, ed A Townshend (London: Academic) **8** 4922–30

[65] Kongingsberger D C and Prins R (ed) 1988 *X-Ray Absorption: Principles, Applications, Techniques of EXAFS, SEXAFS and XANES* (New York: Wiley)

[66] Iwasawa Y (ed) 1996 *X-Ray Absorption Fine Structure for Catalysis and Surfaces* (Singapore: World Scientific)

[67] Stöhr J, Kollin E B, Fischer D A, Hastings J B, Zaera F and Sette F 1985 Surface extended x-ray-absorption fine structure of low-Z adsorbates studied with fluorescence detection *Phys. Rev. Lett.* **55** 1468–71

[68] Zaera F, Fischer D A, Shen S and Gland J L 1998 Fluorescence yield near-edge X-ray absorption spectroscopy under atmospheric conditions: CO and $H_2$ coadsorption on Ni(100) at pressures between $10^{-9}$ and 0.1 Torr *Surf. Sci.* **194** 205–16

[69] Stöhr J 1992 *NEXAFS Spectroscopy* (Berlin: Springer)

[70] Stöhr J and Jaeger R 1982 Adsorption-edge resonances, core-hole screening and orientation of chemisorbed molecules: CO, NO and $N_2$ on Ni(100) *Phys. Rev.* B **26** 4111–31

[71] Zaera F, Fischer D A, Carr R G and Gland J L 1988 Determination of chemisorption geometries for complex molecules by using near-edge X-ray absorption fine structure: ethylene on Ni(100) *J. Chem. Phys.* **89** 5335–41

[72] Sharma A and Khatri R K 1995 Surface analysis: optical spectroscopy *Encyclopedia of Analytical Science* ed A Townshend (London: Academic) **8** 4958–65

[73] Shanthi E, Dutta V, Banerjee A and Chopra K L 1980 Electrical and optical properties of undoped and antimony-doped tin oxide films *J. Appl. Phys.* **51** 6243–51

[74] Wruck D A and Rubin M 1993 Structure and electronic properties of electrochromic NiO films *J. Electrochem. Soc.* **140** 1097–104

[75] Chiarotti G 1994 Electronic surface states investigated by optical spectroscopy *Surf. Sci.* **299/300** 541–50

[76] McGlip J F 1990 Epioptics: linear and non-linear optical spectroscopy of surfaces and interfaces *J. Phys.: Condens. Matter* **2** 7985–8006

[77] Aspens D E and Dietz N 1998 Optical approaches for controlling epitaxial growth *Appl. Surf. Sci.* **130–132** 367–76

[78] Frederick B G, Power J R, Cole R J, Perry C C, Chen Q, Haq S, Bertrams T, Richardson N V and Weightman P 1998 Adsorbate azimuthal orientation from reflectance anisotropy spectroscopy *Phys. Rev. Lett.* **80** 4490–3

[79] Tompkins H G 1993 *A User's Guide to Ellipsometry* (Boston: Academic)

[80] Empedocles S A, Norris D J and Bawendi M G 1996 Photoluminescence spectroscopy of single CdSe nanocrystallite quantum dots *Phys. Rev. Lett.* **77** 3873–6

[81] Moerner W E 1996 High-resolution optical spectroscopy of single molecules in solids *Acc. Chem. Res.* **29** 563–71

[82] Xie X S 1996 Single-molecule spectroscopy and dynamics at room temperature *Acc. Chem. Res.* **29** 598–606

[83] Haw J F 1992 Nuclear magnetic resonance spectroscopy *Anal. Chem.* **64** R243–54

[84] Engelhardt G and Günter 1987 *High-Resolution Solid-State NMR of Silicates and Zeolites* (Chichester: Wiley)

[85] Chmelka B F, Pearson J G, Liu S B, Ryoo R, de Menorval L C and Pines A 1991 NMR study of the distribution of aromatic molecules in NaY zeolite *J. Phys. Chem.* **95** 303–10

[86] Wang P-K, Slichter C P and Sinfelt J H 1984 NMR study of the structure of simple molecules adsorbed on metal surfaces: $C_2H_2$ on Pt *Phys. Rev. Lett.* **53** 82–5

[87] Brey W S 1983 Applications of magnetic resonance in catalytic research *Heterogeneous Catalysis: Selected American Stories* ed B H Davis and W P Hettinger Jr (Washington: American Chemical Society)

[88] Deviney M L and Gland J L (eds) 1985 *Catalyst Characterization Science: Surface and Solid State Chemistry* (Washington, DC: American Chemical Society)

[89] Briggs D (ed) 1978 *Handbook of X-ray and Ultraviolet Photoelectron Spectroscopy* (London: Heyden)

[90] Ertl G and Küppers J 1985 *Low Energy Electrons and Surface Chemistry* (Weinheim: VCH)

[91] Woodruff D P and Delchar T A 1988 *Modern Techniques of Surface Science* (Cambridge: Cambridge University Press)

[92] Avouris P, Bozso F and Walkup R E 1987 Desorption via electronic transitions: fundamental mechanisms and applications *Nucl. Instrum. Methods Phys. Res.* B **27** 136–46

[93] Czanderna A W and Hercules D M (ed) 1991 *Ion Spectroscopies for Surface Analysis* (New York: Plenum)

[94] Zangwill A 1988 *Physics at Surfaces* (Cambridge: Cambridge University Press)

# B1.23
# Surface structural determination: particle scattering methods

*J Wayne Rabalais*

## B1.23.1  Introduction

The origin of scattering experiments has its roots in the development of modern atomic theory at the beginning of this century. As a result of both the Rutherford experiment on the scattering of alpha particles (He nuclei) by thin metallic foils and the Bohr theory of atomic structure, a consistent model of the atom as a small massive nucleus surrounded by a large swarm of light electrons was confirmed. Following these developments, it was realized that the inverse process, namely, analysis of the scattering pattern of ions from crystals, could provide information on composition and structure. This analysis is straightforward because the kinematics of energetic atomic collisions is accurately described by classical mechanics. Such scattering occurs as a result of the mutual Coulomb repulsion between the colliding atomic cores, that is, the nucleus plus core electrons. The scattered primary atom loses some of its energy to the target atom. The latter, in turn, recoils into a forward direction. The final energies of the scattered and recoiled atoms and the directions of their trajectories are determined by the masses of the pair of atoms involved and the closeness of the collision. By analysis of these final energies and angular distributions of the scattered and recoiled atoms, the elemental composition and structure of the surface can be deciphered.

Low-energy (1–10 keV) ion scattering spectrometry (ISS) had its beginning as a modern surface analysis technique with the 1967 work of Smith [1], which demonstrated both surface elemental and structural analysis. Over the next twenty years, it was clearly demonstrated [2–6] that direct surface structural information could be obtained from ISS. Most of the early workers used electrostatic analysers to measure the kinetic energies of the scattered ions. There are two problems with this technique. (i) It analyses only the scattered ions; these are typically only a very small fraction ($<5\%$) of the total scattered flux. Thus, high primary ion doses are required for spectral acquisition which are potentially damaging to the surface and adsorbate structures. (ii) Neutralization probabilities are a function of the ion beam incidence angle $\alpha$ with the surface and the azimuthal angle $\delta$ along which the ion beam is directed. This is not a simple behaviour since the probabilities depend on the distances of the ion to specific atoms. As a result, it is difficult to separate scattering intensity changes due to neutralization effects from those due to structural effects. The use of alkali primary ions [7] which have low neutralization probabilities leads to higher scattering intensities and pronounced focusing effects, however, the contamination of the sample surface by the reactive alkali ions is a potential problem with this method. Buck and co-workers [8], who had been developing time-of-flight (TOF) methods for ion scattering since the mid 1970s, used TOF methods for surface structure analysis in 1984 and demonstrated the capabilities and high sensitivity of the technique when both neutrals and ions are detected simultaneously. A TOF spectrometer system with a long flight path for separation of the scattered and recoiled particles and continuous variation of the scattering $\theta$ and recoiling $\phi$ angles was developed in 1990 [9]. This coupling of

TOF methods with detection of both scattered and recoiled particles led to the development of TOF scattering and recoiling spectrometry (TOF-SARS) as a tool for structural analysis [10]. A large, time-gated, position-sensitive microchannel plate detector was used in 1997 to obtain images of the scattered and recoiled particles, leading to the development of scattering and recoiling imaging spectrometry (SARIS) [11]. Several research groups [12–24] throughout the world are now engaged in surface structure determinations using some form of low-keV ISS.

This section will concentrate on TOF-SARS and SARIS, for it is felt that these are the techniques that will be most important in future applications. Also, emphasis will be placed on surface structure determinations rather than surface elemental analysis, because TOF-SARS and SARIS are capable of making unique contributions in the area of structure determination. The use of TOF methods and large-area position-sensitive detectors has led to a surface crystallography that is sensitive to all elements, including the ability to directly detect hydrogen adsorption sites. TOF detection of both neutrals and ions provides the high sensitivity necessary for non-destructive analysis. Detection of atoms scattered and recoiled from surfaces in simple collision sequences, together with calculations of shadowing and blocking cones, can now be used to make direct measurements of interatomic spacings and adsorption sites within an accuracy of less than 0.1 Å.

## B1.23.2   Basic physics underlying keV ion scattering and recoiling

There are two basic physical phenomena which govern atomic collisions in the keV range. First, repulsive interatomic interactions, described by the laws of classical mechanics, control the scattering and recoiling trajectories. Second, electronic transition probabilities, described by the laws of quantum mechanics, control the ion–surface charge exchange process.

### B1.23.2.1   *Kinematics of ion–surface collisions and elemental analysis*

The dynamics of ion surface scattering at energies exceeding several hundred electronvolts can be described by a series of binary collision approximations (BCAs) in which only the interaction of one energetic particle with a solid atom is considered at a time [25]. This model is reasonable because the interaction time for the collision is short compared with the period of phonon frequencies in solids, and the interaction distance is shorter than the interatomic distances in solids. The BCA simplifies the many-body interactions between a projectile and solid atoms to a series of two-body collisions of the projectile and individual solid atoms. This can be described with results from the well known two-body central force problem [26].

Within the BCA, the trajectories of energetic particles on the surface become a series of linear motion segments between neighbouring atoms. Both the scattered and recoiled atoms have high, discrete kinetic energy distributions. The simplest case of ion–surface scattering phenomena is quasi-single scattering (QSS), which represents the case of one large-angle deflection that is preceded and/or followed by a few small deflections. Figure B1.23.1 shows an example of QSS. This typically produces a sharp scattering peak whose energy is near that of the theoretical single-collision energy. The energies of scattered and recoiled particles in single scattering (SS) can be derived from the laws of conservation of energy and momentum. The energy $E_s$ of a projectile scattered from a stationary target is given as

$$E_s = E_0\{(\cos\theta \pm (A^2 - \sin^2\theta)^{1/2}\}^2/(1 + A)^2 \qquad (B1.23.1)$$

where $A = M_t/M_p$, and $E_0$, $M_t$, and $M_p$ are the initial energy of the projectile and the mass of the target and the projectile, respectively. If the mass of the impinging particle is less than or equal to that of the target atom then $A \geq 1$, and only the positive sign is used. If the scattering angle is chosen as $90°$, equation (B1.23.1) can be simplified to

$$\frac{E_s}{E_0} = \frac{A - 1}{A + 1}. \qquad (B1.23.2)$$

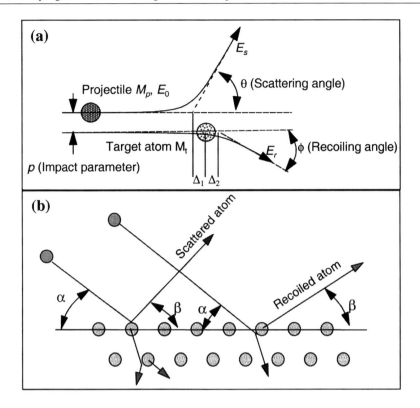

**Figure B1.23.1.** (a) Two-body collision of a projectile of mass $M_p$ and kinetic energy $E_0$ approaching a stationary target of mass $M_t$ with an impact parameter $p$. (b) Quasi-single scattering and direct recoiling with incident angle $\alpha$ and exit angle $\beta$, based on the BCA.

Solving for $M_t$ yields

$$M_t = M_p \frac{1 + B}{1 - B} \tag{B1.23.3}$$

where $B = E_s/E_0$. If the mass of the impinging particle is greater than that of the target atom, $M_p > M_t$, and both signs are used in the equation. The energy of the scattered particle is then found to be a double-valued function of the scattering angle $\theta$, i.e. there are two $E_s$ for each $\theta$. For the case of $A < 1$, the maximum SS scattering angle is

$$\theta_{max} = \sin^{-1} A. \tag{B1.23.4}$$

For angles greater than $\theta_{max}$, only multiple scattering can occur.

The energy of scattered or recoiled ions can be measured directly by means of an electrostatic energy analyser. If the TOF method is used, the relation between scattering energy $E_s$ and TOF $t_s$ is expressed as

$$E_s = \frac{1}{2} M_s v_s^2 = \frac{1}{2} M_s \left( \frac{d_{tof}}{t_s} \right)^2 \tag{B1.23.5}$$

and

$$t_s = d_{tof} \sqrt{\frac{M_s}{2 E_s}} \tag{B1.23.6}$$

where $d_{tof}$ is the flight distance of the scattered atom.

The intensity of SS $I_i$ from an element $i$ in the solid angle $\Delta\Omega$ is proportional to the initial beam intensity $I_0$, the concentration of the scattering element $N_i$, the neutralization probability $P_i$, the differential scattering cross section $d\sigma(\theta)/d\Omega$, the shadowing coefficient $f_{si}(\alpha, \delta_{in})$ and the blocking coefficient $f_{bi}(\alpha, \delta_{out})$ for the $i$th component on the surface:

$$I_i \approx I_0 N_i P_i f_{si} f_{bi} \frac{d\sigma}{d\Omega} \Delta\Omega. \qquad (B1.23.7)$$

Similar to QSS, direct recoil (DR) of surface atoms produces energetic atoms that have a relatively narrow velocity distribution. DR particles are those species which are recoiled from the surface layers as a result of a direct collision of the primary ion. They escape from the surface with little energy loss through collisions with neighbouring atoms. The energy $E_r$ of a DR surface atom can be expressed as

$$E_r = E_0\{4A\cos^2\phi\}/(1 + A)^2. \qquad (B1.23.8)$$

From geometry considerations, DR is observed only in the forward-scattering direction for which $\phi < 90°$. A similar expression as equation (B1.23.7) is applicable for recoiling intensity evaluation. All elements, including hydrogen, can be analysed by either scattering, recoiling, or both techniques. TOF peak identification of QSS and DR is straightforward using the equations above.

### B1.23.2.2  *Shadow cones, blocking cones, and structural analysis*

A simple interpretation based on the BCA yields some important concepts for ion scattering and recoiling: shadow cones and blocking cones. As shown in figure B1.23.2(a), scattering of ions by a target atom produces a region in which no ion can penetrate behind the target atom. This region is called a shadow cone. The cone dimensions can be evaluated from known interatomic distances in experiments. Figure B1.23.2(b) shows the normalized ion flux density across the shadow cone. There is zero flux density inside of the shadow cone and unit flux density far outside of the cone. Highly focused ion flux density appears at the boundary of the cone. This anisotropic distribution of ion flux after interaction with a target atom is the basis of ISS structural determinations. If a neighbouring atom lies inside the shadow cone (A in (a)), it cannot be scattered or recoiled. If it is well away from the cone (C), the cone has no effect on the intensity. When it lies in the focusing region (B), enhanced intensity scattering and/or recoiling intensity is observed. TOF-SARS measures the intensity change due to the shadow cone effect on neighbouring atoms as a function of incident beam direction.

Trajectories of ions interacting with an additional target atom (blocking atom) after scattering from the initial target atom (scattering atom) produce a hollow region behind the blocking atom called a blocking cone (figure B1.23.2(c)). It can be regarded as an interaction of a target atom (blocking atom) with ions emitted from an adjacent point ion source (scattering atom). A shadow cone is different because it originates from the interaction with ions from an infinitely distant point ion source (ion source in the ion beam line). Unlike a shadow cone, a blocking cone diverges with a measurable blocking cone angle $\xi$. The closer the blocking atom is to the scattering atom, the larger is the angle of divergence. In traditional ISS, the variation of the interactions of both shadowing and blocking cones are measured. It is possible to minimize the effect of blocking cones by a judicious choice of scattering geometry. Blocking cones are, however, inevitable, especially for scattering trajectories from deep subsurface layers.

Considering a large number of ions with parallel trajectories impinging on a target atom, the ion trajectories are bent by the repulsive potential such that there is an excluded volume, called the shadow cone, in the shape of a paraboloid formed behind the target atom as shown in figure B1.23.3(a). Ion trajectories do not penetrate into the shadow cone, but instead are concentrated at its edges much as rain pours off an umbrella. Atoms located inside the cone behind the target atom are shielded from the impinging ions. Similarly, if the scattered ion or recoiling atom trajectory is directed towards a neighbouring atom, that trajectory will be

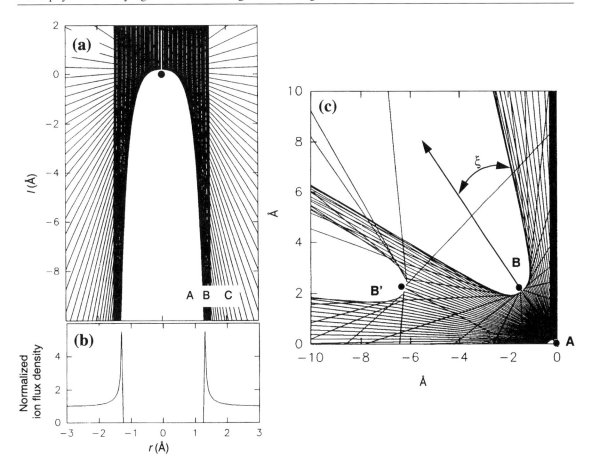

**Figure B1.23.2.** (a) Shadow cone of a stationary Pt atom in a 4 keV Ne$^+$ ion beam, appearing with the overlapping of ion trajectories as a function of the impact parameter. The initial position of the target atom that recoils in the collision is indicated by a solid circle. (b) Plot of the normalized ion flux distribution density across the shadow cone in (a). The flux density changes from 0 inside the shadow cone, to much greater than 1 in the focusing region, converging to 1 away from the shadow cone edge. (c) Blocking cones of Pt atoms (B, B′) cast by 4 keV Ne$^+$ ions scattered from another Pt atom (A). Note the different blocking angles of the two blocking atoms, which is due to the differences in the interatomic spacings between the scattering and blocking atoms.

blocked. For a large number of scattering or recoiling trajectories, a blocking cone will be formed behind the neighbouring atom into which no particles can penetrate, as shown in figure B1.23.3(b). The dimensions of the shadowing and blocking cones can be determined experimentally from scattering measurements along crystal azimuths for which the interatomic spacings are accurately known.

*B1.23.2.3  Scattering and recoiling anisotropy caused by shadowing and blocking cones*

When an isotropic ion fluence impinges on a crystal surface at a specific incident angle $\alpha$, the scattered and recoiled atom flux is anisotropic. This anisotropy is a result of the incoming ion's eye view of the surface, which depends on the specific arrangement of atoms and the shadowing and blocking cones. The arrangement of atoms controls the atomic density along the azimuths and the ability of ions to channel, that is, to penetrate

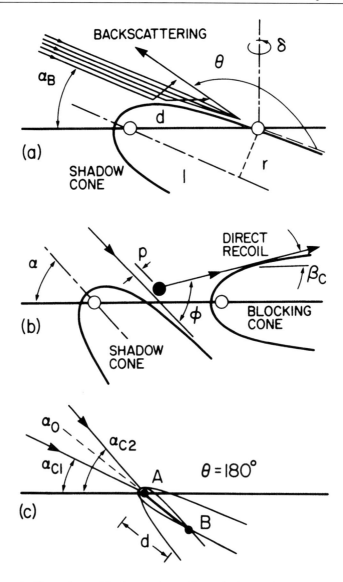

**Figure B1.23.3.** Schematic illustrations of backscattering with shadowing and direct recoiling with shadowing and blocking.

into empty spaces between atomic rows. The cones determine which nuclei are screened from the impinging ion flux and which exit trajectories are blocked, as depicted in figure B1.23.3. By measuring the ion and atom flux at specific scattering and recoiling angles as a function of ion beam incident $\alpha$ and azimuthal $\delta$ angles to the surface, structures are observed which can be interpreted in terms of the interatomic spacings and shadow cones from the ion's eye view.

*(a) Time of flight scattering and recoiling spectrometry (TOF-SARS)—shadow cone based experiment*

In TOF-SARS [9], a low-keV, monoenergetic, mass-selected, pulsed noble gas ion beam is focused onto a sample surface. The velocity distributions of scattered and recoiled particles are measured by standard

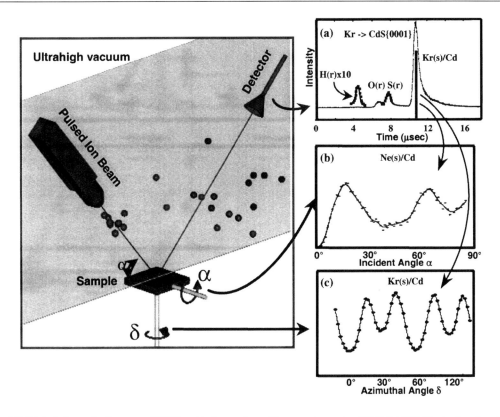

**Figure B1.23.4.** Schematic diagram of TOF scattering and recoiling spectrometry (TOF-SARS) illustrating the plane of scattering formed by the ion beam, sample and detector. TOF spectra (a) are collected with fixed positions of the ion beam, sample and detector. In order to measure the incident angle ($\alpha$) or azimuthal angle ($\delta$) intensity variation of a peak in a TOF spectrum, the sample is rotated about an axis that goes through its normal or through its plane, respectively. Such $\alpha$ and $\delta$ intensity variations of a peak are shown in (b) and (c).

TOF methods. A channel electron multiplier is used to detect fast ($>800$ eV) neutrals and ions. This type of detector has a small acceptance solid angle. A fixed angle is used between the pulsed ion beam and detector directions with respect to the sample as shown in figure B1.23.4. The sample has to be rotated to measure ion scattering and recoiling anisotropy as a function of the incident angle ($\alpha$) and azimuthal angle ($\delta$) of the incident beam. Since the sample rotation changes both the incident beam direction and the detector direction, the spectra are affected by both shadow cones and blocking cones. In order to reduce blocking cone effects for simpler interpretations, high-angle scattering is preferred. Elemental analyses are achieved by converting the velocity distributions into energy distributions and relating those to the masses of the target atoms through the kinematic relationship that describes classical scattering and recoiling (equations (B1.23.1) and (B1.23.8)). Structural analyses are achieved by monitoring the scattered and recoiled particles as a function of both beam incident angle $\alpha$ and crystal azimuthal angle $\delta$. The anisotropic features in these $\alpha$- and $\delta$-scans are interpreted by means of shadow cone and blocking cone analyses. It requires several hours to collect data needed to construct a contour map of intensities as a function of both $\alpha$ and $\delta$. Moreover, the experimental geometry is restricted to fixed scattering angles and in-plane scattering and recoiling trajectories.

*(b) Scattering and recoiling imaging spectrometry (SARIS)—blocking cone based experiment*

In an ideal SARIS system [11], it would be desirable to measure the velocity distributions of all the energetic particles scattered and recoiled from a sample surface in a short time period. This concept requires a hemispherical, time-resolving, position-sensitive detector which covers all of the solid angle space above the sample surface. Implementation of this concept is not currently feasible. If data collection time is unimportant, a point detector as described above, such as a channel electron multiplier mounted on a flexible goniometer which allows movement of the detector over a large solid angle, can be used to collect TOF spectra. In order to compromise the size of detector with data collection time, a hybrid configuration can be used. This is a large, time-resolving, position-sensitive microchannel plate detector mounted on a triple-axis UHV goniometer. This instrument makes it possible to capture scattering and recoiling intensity distributions without changing the incident beam direction. It gives a great advantage in comparison of experimental results with those of computer simulations. The large-area detector provides the intensity distribution over a limited solid angle. Since the optimum detector position and flight distance for a specific experiment are variable, the detector has to be moved around to cover a large solid angle, and to compromise TOF resolution with the detector acceptance solid angle.

### B1.23.3   Instrumentation

The basic requirements [9] for low-energy ion scattering are an ion source, a sample mounted on a precision manipulator, an energy or velocity analyser and a detector as shown in figure B1.23.5. The sample is housed in an ultra-high vacuum (UHV) chamber in order to prepare and maintain well defined clean surfaces. The UHV prerequisite necessitates the use of differentially pumped ion sources. Ion scattering is typically done in a UHV chamber which houses other surface analysis techniques such as low-energy electron diffraction (LEED), x-ray photoelectron spectroscopy (XPS) and Auger electron spectroscopy (AES). The design of an instrument for ion scattering is based on the type of analyser to be used. An electrostatic analyser (ESA) measures the kinetic energies of ions while a TOF analyser measures the velocities of both ions and fast neutrals.

*B1.23.3.1   Ion source and beam line*

The critical requirements for the ion source are that the ions have a small energy spread, there are no fast neutrals in the beam and the available energy is 1–10 keV. Both noble gas and alkali ion sources are common. For TOF experiments, it is necessary to pulse the ion beam by deflecting it past an aperture. A beam line for such experiments is shown in figure B1.23.5; it is capable of producing ion pulse widths of $\approx 15$ ns.

*B1.23.3.2   Analysers*

An ESA provides energy analysis of the ions with high resolution. A TOF analyser provides velocity analysis of both fast neutrals and ions with moderate resolution. In an ESA the energy separation is made by spatial dispersion of the charged particle trajectories in a known electrical field. ESAs were the first analysers used for ISS; their advantage is high-energy resolution and their disadvantages are that they analyse only ions and have poor collection efficiency due to the necessity for scanning the analyser. A TOF analyser is simply a long field-free drift region. It has the advantage of high efficiency since it collects both ions and fast neutrals simultaneously in a multichannel mode; its disadvantage is only moderate resolution.

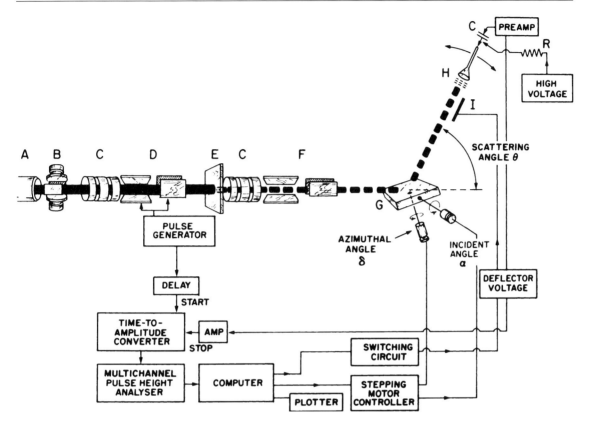

**Figure B1.23.5.** Schematic illustration of the TOF-SARS spectrometer system. A = ion gun, B = Wien filter, C = Einzel lens, D = pulsing plates, E = pulsing aperture, F = deflector plates, G = sample, H = electron multiplier detector with energy prefilter grid and I = electrostatic deflector.

### B1.23.3.3  Detectors

The most common detectors used for TOF-SARS are continuous dynode channel electron multipliers which are capable of multiplying the signal pulses by $10^6$–$10^7$. They are sensitive to both ions and fast neutrals. Neutrals with velocities $\gtrsim 10^6$ cm s$^{-1}$ are detected with the same efficiency as ions. Since the cones of these detectors are usually less than 1 cm$^2$ and the TOF flight paths are of the order of $\sim$1 m, the acceptance solid angles of such detectors are very small. Incident and azimuthal angle scans are made by rotating the sample with the detector at a fixed position.

SARIS overcomes the limitations of small-area detectors by using a large, time-resolving, position-sensitive microchannel plate (MCP) detector and TOF methods to capture images of both ions and fast neutrals that are scattered and recoiled from a surface. Due to the large solid angle subtended by the MCP, atoms that are scattered and recoiled in both planar and non-planar directions are detected simultaneously. For example, with a 75 × 95 mm MCP situated at a distance of 16 cm from the sample, it spans a solid angle of $\sim$0.3 sr corresponding to an azimuthal range of $\sim$26°. Using a beam current of $\sim$0.1 nA cm$^{-2}$, the four images required to make up a 90° azimuthal range can be collected in $\sim$2 min with a total ion dose of $\sim$10$^{11}$ ions cm$^{-2}$. The time gating of the MCP provides resolution of the scattered and recoiled atoms into time frames as short as 10 ns, thereby providing element-specific spatial-distribution images. These SARIS images contain features that are sharply focused into well defined patterns as a function of both space and time by the crystal structure of the target sample. If the MCP is mounted on a goniometer that provides both horizontal and vertical rotation and translation away from the sample, it is possible to change the solid angle of collection and the flight path length.

### B1.23.4  Computer simulation methods

It is extremely helpful to use classical ion trajectory simulations in order to visualize the ion trajectories to improve the understanding of ion behaviour in the surface region and to provide a systematic method for surface structure determination. Such simulations are based on the BCA in which the trajectories of the energetic particles are assumed to be a series of straight lines corresponding to the asymptotes of the scattering trajectories due to sequential binary collisions. The BCA has been proven to be valid for the keV range of energies.

#### B1.23.4.1  Binary collision approximation

Atom–surface interactions are intrinsically many-body problems which are known to have no analytical solutions. Due to the shorter de Broglie wavelength of an energetic ion than solid interatomic spacings, the energetic atom–surface interaction problem can be treated by classical mechanics. In the classical mechanical framework, the problem becomes a set of Newtonian equations of motion [26] for $i$th particle in an $N$-body problem.

$$F_i = m_i \frac{\mathrm{d}^2 r_i}{\mathrm{d}t^2} = -\nabla \Phi(r_1, r_2, \dots, r_N). \tag{B1.23.9}$$

The summation of pair-wise potentials is a good approximation for molecular dynamics calculations for simple classical many-body problems [27]. It has been widely used to simulate hyperthermal energy ($>1$ eV) atom–surface scattering:

$$F_i \approx -\sum_{j}^{N, i \neq j} \nabla \Phi(r_i, r_j). \tag{B1.23.10}$$

As the kinetic energy involved in the system goes higher, the interaction of energetic particles is more and more localized near the nuclei. When the interaction distance is much smaller than interatomic distances in the system, the BCA is valid:

$$F_i \approx -\nabla \Phi(r_i, r_j) \qquad j = \text{the nearest neighbour of } i. \tag{B1.23.11}$$

In the BCA, each collision process is regarded as an isolated event. Ion–solid interactions are approximated by a series of two-body interactions which are reduced to one-body problems in the centre-of-mass (CM) coordinates. The projectile is assumed to converge to the asymptote after a collision before interacting with the next collision partner. In the central force one-body problem, evaluation of scattering integrals and time integrals replaces the time-consuming numerical integration of a set of differential equations.

#### B1.23.4.2  Scattering integral and time integral

Once atom–surface scattering is reduced to the BCA, one can calculate the energy relationship between two particles involved in scattering with a known scattering angle and the laws of conservation of energy and momentum as shown in section B1.23.2.1. The scattering angle as a function of impact parameter necessitates evaluation of the scattering integrals. The scattering angle, $\chi$ in the CM coordinate system, and the CM energy $E$, are given by

$$\chi = \pi - 2 \int_{r_0}^{\infty} \frac{p\,dr}{r^2} \left( 1 - \frac{V(r)}{E} - \frac{p^2}{r^2} \right)^{-1/2} \quad (B1.23.12)$$

where

$$E = \frac{A}{1+A} E_0. \quad (B1.23.13)$$

$A = M_t/M_p$, $p$ is impact parameter and $r_0$ is the distance of closest approach (apsis) of the collision pair. The transformations from the CM coordinates (scattering angle $\chi$) to the laboratory coordinates with the scattering angle $\theta$ for the primary particle and $\phi$ for the recoiled surface atoms is given by

$$\tan \theta = \frac{A \sin \chi}{1 + A \cos \chi} \quad (B1.23.14)$$

and

$$\tan \phi = \frac{\sin \theta}{1 - \cos \chi}. \quad (B1.23.15)$$

For accurate ion trajectory calculation in the solid, it is necessary to evaluate the exact positions of the intersections of the asymptotes ($\Delta_1$, $\Delta_2$) of the incoming trajectory and that of the outgoing trajectories of both the scattered and recoiled particles in a collision. The evaluation of these values requires time integrals and the following transformation equations:

$$t = \sqrt{r_0^2 - p^2} - \int_{r_0}^{\infty} \left\{ \left( 1 - \frac{V(r)}{E_r} - \frac{p^2}{r^2} \right)^{-1/2} \left( 1 - \left( \frac{p^2}{r^2} \right) \right)^{-1/2} \right\} dr \quad (B1.23.16)$$

$$\Delta_1 = \frac{t + (A-1)p \tan \frac{1}{2}\chi}{1+A} \quad (B1.23.17)$$

$$\Delta_2 = \frac{p}{\tan \phi} - \Delta_1. \quad (B1.23.18)$$

Numerical integration methods are widely used to solve these integrals. The Gauss–Mühler method [28] is employed in all of the calculations used here. This method is a Gaussian quadrature [29] which gives exact answers for Coulomb scattering.

### B1.23.4.3 Potential function

One of the most important issues in simulation of energetic atom–surface scattering is the determination of the interaction potential between the colliding atoms. In the low-keV energy region, electrons have a screening effect on the Coulomb interaction of nuclei so that the actual nuclear charges affecting the trajectories are less than the atomic numbers ($Z$) of the atoms involved. This screening effect decreases the potential $V(r)$ by an amount which is expressed as a screening function $\Phi(r)$. The form of the potential function is

$$V(r) = \frac{Z_1 Z_2 e^2}{r} \Phi(r). \quad (B1.23.19)$$

The $\Phi(r)$ can be expressed in various forms [30], e.g. the Bohr, Born–Mayer, Thomas–Fermi–Firsov and Moliere models, as well as the 'universal potential' of Ziegler, Biersack and Littmark known as the ZBL potential [31]. The ZBL potential function is expressed as

$$\Phi(r) = 0.1818 \, e^{-3.2/ar} + 0.5099 \, e^{0.9423/ar} + 0.2802 \, e^{0.4029/ar} + 0.028\,17 \, e^{0.2016/ar} \quad (B1.23.20)$$

where the screening length $a = (0.8853C_{F}a_0/(Z_1^{0.23} + Z_2^{0.23}))$, $a_0$ is Bohr radius (0.53 Å), $Z_1$, $Z_2$ are the atomic numbers of the atoms involved and $C_F$ is a screening constant for adjusting the screening length to calibrate the potential to experimental scattering data. The ZBL potential provides good agreement between simulated and experimental results.

### B1.23.4.4  *General description of simulation program*

Classical ion trajectory computer simulations based on the BCA are a series of evaluations of two-body collisions. The parameters involved in each collision are the type of atoms of the projectile and the target atom, the kinetic energy of the projectile and the impact parameter. The general procedure for implementation of such computer simulations is as follows. All of the parameters involved in the calculation are defined: the surface structure in terms of the types of the constituent atoms, their positions in the surface and their thermal vibration amplitude; the projectile in terms of the type of ion to be used, the incident beam direction and the initial kinetic energy; the detector in terms of the position, size and detection efficiency; the type of potential functions for possible collision pairs.

    After defining the input parameters, the calculation of the trajectories of an incident ion begins with a randomly chosen initial entrance point on the surface. The next step is to find the first collision partner. Taking advantage of the symmetry of the crystal structure, one can list the positions of surface atoms within a certain distance from the projectile. The atoms are sorted in ascending order of the scalar product of the interatomic vector from the atom to the projectile with the unit velocity vector of the projectile. If the collision partner has larger impact parameter than a predefined maximum impact parameter ($p_{max}$), it is discarded. If a partner has a shorter impact parameter than $p_{max}$, the evaluation of the collision is initiated by converting three-dimensional information, such as the positions of the projectile and the target atom and the velocity of the projectile, into the parameters necessary to calculate the expressions for the impact parameter and the relative energy. After the equations are solved with these parameters, the values are converted back to three-dimensional information to search for a new collision partner with a new set of parameters calculated in the previous collision. This procedure is repeated until the kinetic energy of the projectile falls below a predefined cutoff energy or it ejects from the surface. If it is necessary to follow recoiled particles, they are regarded as new projectiles in subsequent collisions. After finishing a trajectory calculation, a new calculation starts with a new randomly chosen entrance point. Millions of trajectory calculations with different initial impact parameters are carried out in order to compare the results with those of experiments. In order to increase the speed of the calculation, a precalculated table of scattering angle and scattered energy as a function of impact parameter and kinetic energy is used. A two-dimensional spline method is used to interpolate a scattering angle and energy from the table.

### (a) Crystal models with thermal vibration inclusion

The lattice atoms in the simulation are assumed to vibrate independently of one another. The displacements from the equilibrium positions of the lattice atoms are taken as a Gaussian distribution, such as

$$\rho(x) = \frac{1}{\sqrt{2\pi}\sigma} e^{-\Delta x^2/2\sigma^2} \qquad (B1.23.21)$$

where $\sigma^2$ and $\Delta x$ are the variance and the displacement from the lattice equilibrium position, respectively. The variance of the distribution can be expressed as [32]

$$\sigma^2 = \overline{\Delta x^2} = \frac{2h^2 T}{4\pi^2 mk\Theta_D^2} \qquad (B1.23.22)$$

where $\Theta_D$ is the Debye temperature. The Box–Muller method [29] is used for generating random deviates with a Gaussian distribution.

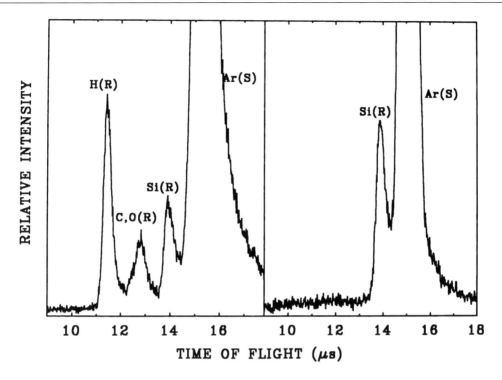

**Figure B1.23.6.** TOF spectra of a Si{100} surface with chemisorbed $H_2O$ (left) and clean Si (right). Peaks due to scattered Ar and recoiled H, O, and Si are observed. Conditions: 4 keV $Ar^+$, scattering angle $\theta = 28°$, incident angle $\alpha = 8°$.

*(b) Comparison of simulated and experimental data*

A systematic comparison of two sets of data requires a numerical evaluation of their likeliness. TOF-SARS and SARIS produce one- and two-dimensional data plots, respectively. Comparison of simulated and experimental data is accomplished by calculating a one- or two-dimensional reliability ($R$) factor [33], respectively, based on the $R$-factors developed for LEED [34]. The $R$-factor between the experimental and simulated data is minimized by means of a multiparameter simplex method [33].

### B1.23.5 Elemental analysis from scattering and recoiling

TOF-SARS and SARIS are capable of detecting all elements by either scattering, recoiling or both techniques. TOF peak identification is straightforward by converting equations (B1.23.1) and (B1.23.8) to the flight times of the scattered $t_s$ and recoiled $t_r$ particles as

$$t_s = L(M_1 + M_2)/(2M_1 E_0)^{1/2}\{\cos +[(M_2/M_1)^2 - \sin^2\theta]^{1/2}\} \qquad \text{(B1.23.23)}$$

and

$$t_r = L(M_1 + M_2)/(8M_1 E_0)^{1/2}\cos\phi \qquad \text{(B1.23.24)}$$

where $L$ is the flight distance, that is, the distance from target to detector. Collection of neutrals plus ions results in scattering and recoiling intensities that are determined by elemental concentrations, shadowing and blocking effects and classical cross sections. The main advantage of TOF-SARS for surface compositional

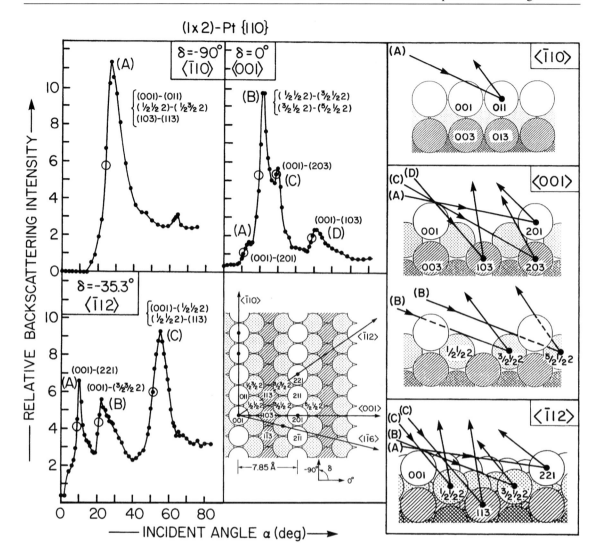

**Figure B1.23.7.** Scattering intensity versus incident angle $\alpha$ scans for 2 keV Ne$^+$ incident on $(1 \times 2)$-Pt{110} at $\theta = 149°$ along the $\langle \bar{1}10 \rangle$, $\langle 001 \rangle$ and $\langle \bar{1}12 \rangle$ azimuths. A top view of the $(1 \times 2)$ missing-row Pt{110} surface along with atomic labels is shown. Cross-section diagrams along the three azimuths illustrating scattering trajectories for the peaks observed in the scans are shown on the right.

analyses is its extreme surface sensitivity as compared to the other surface spectrometries, i.e. mainly XPS and AES. Indeed, with a correct orientation and aperture of the shadow cone, the first monolayer can be probed selectively. At selected incident angles, it is possible to delineate signals from specific subsurface layers. Detection of the particles independently of their charge state eliminates ion neutralization effects. Also, the multichannel detection requires primary ion doses of only $\approx 10^{11}$ ions cm$^{-2}$ or $\approx 10^{-4}$ ions/surface atom for spectral acquisition; this ensures true static conditions during analyses.

Examples of typical TOF spectra obtained from 4 keV Ar$^+$ impinging on a Si{100} surface with chemisorbed H$_2$O and H$_2$ are shown in figure B1.23.6 [35]. Peaks due to Ar scattering from Si and recoiled H, O and Si are observed. The intensities necessary for structural analysis are obtained by integrating the areas of fixed time windows under these peaks.

While qualitative identification of scattering and recoiling peaks is straightforward, quantitative analysis requires relating the scattered or recoiled flux to the surface atom concentration. The flux of scattered or recoiled atoms is dependent on several parameters as detailed in equation (B1.23.7). Compositional analyses by TOF-SARS and ISS have been applied in different areas of surface science, mainly in situations where the knowledge of the uppermost surface composition (first monolayer) is crucial. Some of these areas are as follows: gas adsorption, surface segregation, compounds and polymer blends, surface composition of real supported catalysts, surface modifications due to preferential sputtering by ion beams, diffusion, thin film growth and adhesion.

## B1.23.6   Structural analysis from TOF-SARS

The atomic structure of a surface is usually not a simple termination of the bulk structure. A classification exists based on the relation of surface to bulk structure. A *bulk truncated* surface has a structure identical to that of the bulk. A *relaxed* surface has the symmetry of the bulk structure but different interatomic spacings. With respect to the first and second layers, *lateral relaxation* refers to shifts in layer registry and *vertical relaxation* refers to shifts in layer spacings. A *reconstructed* surface has a symmetry different from that of the bulk symmetry. The methods of structural analysis will be delineated below.

### B1.23.6.1   *Scattering versus incident angle scans*

When an ion beam is incident on an atomically flat surface at grazing angles, each surface atom is shadowed by its neighbouring atom such that only forwardscattering (FS) is possible; these are large impact parameter ($p$) collisions. As $\alpha$ increases, a critical value $\alpha_{c,sh}^i$ is reached each time the $i$th layer of target atoms moves out of the shadow cone allowing for large-angle backscattering (BS) or small-$p$ collisions as shown in figure B1.23.3. If the BS intensity $I_{BS}$ is monitored as a function of $\alpha$, steep rises [36] with well defined maxima are observed when the focused trajectories at the edge of the shadow cone pass close to the centre of neighbouring atoms. This is illustrated for scattering of Ne$^+$ from a Pt(110) surface in figure B1.23.7. From the shape of the shadow cone, i.e. the radius ($r$) as a function of distance ($l$) behind the target atom (figure B1.23.3), the interatomic spacing ($d$) can be directly determined from the $I_{BS}$ versus $\alpha$ plots. For example, by measuring $\alpha_{c,sh}^1$ along directions for which specific crystal azimuths are aligned with the projectile direction and using $d = r/\sin \alpha_{c,sh}^1$, one can determine interatomic spacings in the first atomic layer. The first–second layer spacing can be obtained in a similar manner from $\alpha_{c,sh}^2$ measured along directions for which the first- and second-layer atoms are aligned, providing a measure of the vertical relaxation in the outermost layers.

### B1.23.6.2   *Scattering versus azimuthal angle $\delta$ scans*

Fixing the incident beam angle and rotating the crystal about the surface normal while monitoring the backscattering intensity provides a scan of the crystal azimuthal angles $\delta$ [37]. Such scans reveal the periodicity of the crystal structure. For example, one can obtain the azimuthal alignment and symmetry of the outermost layer by using a low $\alpha$ value such that scattering occurs from only the first atomic layer. With higher $\alpha$ values, similar information can be obtained for the second atomic layer. Shifts in the first–second layer registry can be detected by carefully monitoring the $\alpha_{c,sh}^2$ values for second-layer scattering along directions near those azimuths for which the second-layer atoms are expected, from the bulk structure, to be directly aligned with the first-layer atoms. The $\alpha_{c,sh}^2$ values will be maximum for those $\delta$ values where the first- and second-layer neighbouring atoms are aligned.

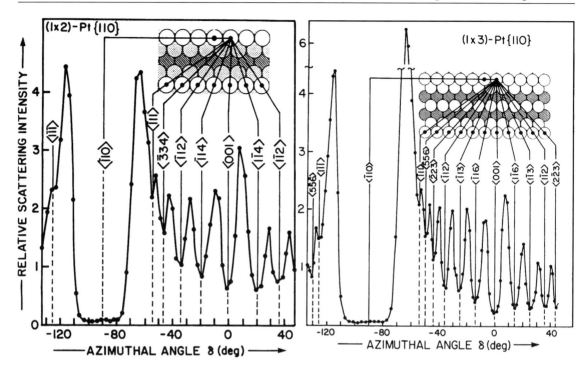

**Figure B1.23.8.** Scattering intensity of 2 keV Ne$^+$ *versus* azimuthal angle $\delta$ scans for Pt{110} in the (1 × 2) and (1 × 3) reconstructed phases. Scattering angle $\theta = 28°$ and incident angle $\alpha = 6°$.

When the scattering angle $\theta$ is decreased to a forward angle ($<90°$), both shadowing effects along the incoming trajectory and blocking effects along the outgoing trajectory contribute to the patterns. The blocking effects arise because the exit angle $\beta = \theta - \alpha$ is small at high $\alpha$ values. Surface periodicity can be read directly from these features [37], as shown in figure B1.23.8 for Pt{110}. Minima are observed at the $\delta$ positions corresponding to alignment of the beam along specific azimuths. These minima are a result of shadowing and blocking along the close-packed directions, thus providing a direct reading of the surface periodicity.

Azimuthal scans obtained for three surface phases of Ni{110} are shown in figure B1.23.9 [38]. The minima observed for the clean and hydrogen-covered surfaces are due only to Ni atoms shadowing neighbouring Ni atoms, whereas for the oxygen-covered surface minima are observed due to both O and Ni atoms shadowing neighbouring Ni atoms. Shadowing by H atoms is not observed because the maximum deflection in the Ne$^+$ trajectories caused by H atoms is less than 2.8°.

## B1.23.7    Structural analysis from SARIS

An example of the SARIS experimental arrangement is shown in figure B1.23.10 [39–41]. The velocity distributions of scattered and recoiled ions plus fast neutrals are measured by analysing the positions of the particles on the detector along with their correlated TOF from sample to detector. The detector is gated so that it can be activated in windows of several microseconds duration, which are appropriate for TOF collection of specific scattered or recoiled particles. These windows are divided into 255 time frames with the time duration of 16.7 ns for each frame. Good statistics are obtained in a total acquisition time of ∼1 min. The image

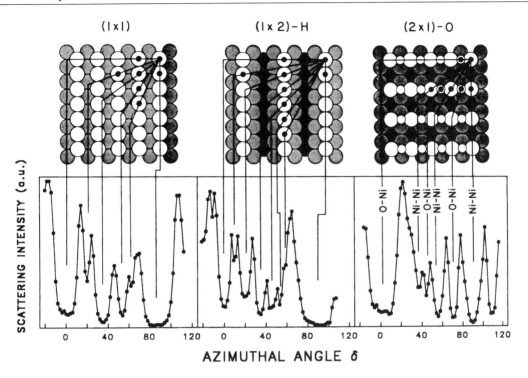

**Figure B1.23.9.** Scattering intensity of 4 keV Ne$^+$ *versus* azimuthal angle $\delta$ for a Ni{110} surface in the clean $(1 \times 1)$, $(1 \times 2)$-H missing row, and $(2 \times 1)$-O missing row phases. The hydrogen atoms are not shown. The oxygen atoms are shown as small open circles. O–Ni and Ni–Ni denote the directions along which O and Ni atoms, respectively, shadow the Ni scattering centre.

ordinate represents particle exit angles ($\beta$) and the abscissa represents the crystal azimuthal angles ($\delta$), i.e. an image in ($\beta, \delta$)-space.

*B1.23.7.1    Interaction of 4 keV Ar with Pt{111}*

*(a) Ar scattering*

The time-resolved images of Ar scattering [33] from Pt{111} of figure B1.23.11 correspond to selected frames of scattered Ar atoms with the azimuthal angle of the incident beam aligned along $\langle \bar{1}\bar{1}2 \rangle$. The overall scattering intensity is maximal at 1.17 $\mu$s (3.54 keV) for the scattered Ar atoms, corresponding to SS as predicted by the BCA. The two intense spots at 1.17 $\mu$s result from the scattering from a first-layer Pt atom and focusing of the scattered beam by an 'atomic lens' formed by neighbouring first-layer Pt atoms (2, 3, 4 in figure B1.23.11). The intense spots are at small $\beta$ since most of the Ar atoms are scattered and focused by first-layer Pt atoms. Focused high-$\beta$ scattering usually arises from subsurface collisions.

*(b) Pt recoiling*

The images of recoiled Pt atoms [33] by 4 keV Ar$^+$ are shown in figure B1.23.12. With increasing TOF, the recoil Pt images change from diffuse, to a focused recoil spot at $\beta \sim 25°$ and, finally, to movement of this spot to a higher $\beta$ that is partially off the MCP. This focused recoil is observed along the 0° $\langle \bar{2}11 \rangle$ and 60°

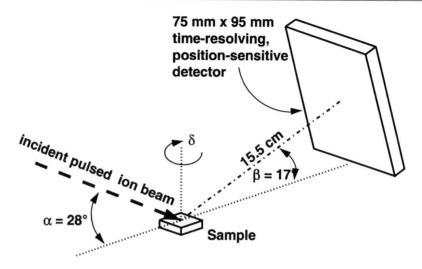

**Figure B1.23.10.** Schematic diagram of a scattering and recoiling imaging spectrometer (SARIS). A large-area ($95 \times 75$ mm$^2$), time-resolving, position-sensitive microchannel plate (MCP) detector captures a large solid angle of the scattering and recoiling particles. A triple-axis UHV goniometer moves the MCP inside the vacuum chamber in order to vary the scattering angle, the distance from detector to sample, the TOF resolution and the acceptance solid angle of the detector.

$\langle \bar{1}\bar{1}2 \rangle$ azimuths but not along the 30° $\langle \bar{1}01 \rangle$ and 90° $\langle 0\bar{1}1 \rangle$ azimuths. The diffuse images at short TOF, e.g. 3.77 $\mu$s, correspond to recoil of Pt atoms from the first layer. These recoils have more isotropic distributions and higher energies because there are no atoms above them by which they can be blocked. At longer TOF, e.g. 4.57 $\mu$s, the focused recoil is due to Pt atoms from the second layer which have been focused by an 'atomic lens' created by first-layer atoms (figure B1.23.12). The second-layer Pt atom can either be recoiled directly into the atomic lens or it can scatter from the neighbouring aligned third-layer atom into the atomic lens.

### B1.23.7.2   Interaction of 4 keV He with Pt{111}

A series of time-resolved He scattering images [33] taken as a function of azimuthal angle is shown in figure B1.23.13. The crystal was rotated about its surface normal by 3° for each image. Each image is taken from a 16.7 ns frame corresponding to the QSS TOF. The same intensity scale was used for all of the frames. The observed images are rich in features which change in position and intensity as a function of azimuthal angle. The regions of low intensity correspond to the positions of the centres of the blocking cones; these regions have mainly circular or oval shapes with distortions caused by other overlapping blocking cones. The regions of high intensity correspond to the positions of intersection or near-overlap of blocking cones; atom trajectories are highly focused along the edges of the cones.

The images at $\delta = 0°$ and 60° along the $\langle \bar{2}11 \rangle$ and $\langle \bar{1}\bar{1}2 \rangle$ azimuths, respectively, are symmetrical about a vertical line through the centre of the frame, as is the crystal structure along these azimuths as shown in figure B1.23.14. The shifts in the positions and sizes of the blocking cones can be monitored as the azimuthal angle $\delta$ is rotated away from the symmetrical 0° or 60° directions. There are large variations in the intensities as a function of $\delta$, with the highest intensities being observed along the directions $\delta = 22$–32° and 56–60°. These high-intensity features result from focusing of ions onto second-layer atoms by the shadow cones of first-layer atoms. The first-layer atoms are symmetrical; however, the second-layer atoms are in sites which

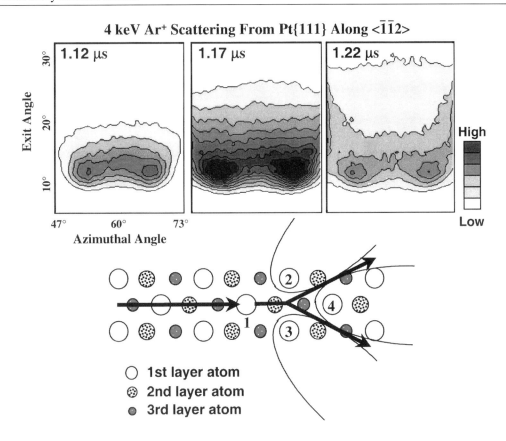

**Figure B1.23.11.** Above: selected time-resolved SARIS images of 4 keV Ar⁺ scattering from Pt{111} along ⟨1̄1̄2⟩. Below: view of Pt{111} surface along ⟨1̄1̄2⟩ showing Ar⁺ scattering from a first-layer Pt atom (1) and splitting into two focused beams by an 'atomic lens' formed by neighbouring first-layer Pt atoms (2, 3, 4).

are asymmetrical with respect to the first layer, resulting in non-planar scattering trajectories. Very intense features in asymmetrical positions are observed at higher exit angles. These intense features correspond to semichanneling in asymmetrical channels. Semichannels are 'valleys' in surfaces through which scattered ions are guided. Along ⟨1̄01⟩ the first-layer atoms form the 'walls' and the second-layer atoms form the 'floor' of the semichannel. However, the second-layer rows are not centred in the bottom of the channel, resulting in an asymmetrical channel. As a result, the scattered atom trajectories are bent and focused along directions determined by the asymmetry of the channel.

The frames along the 0° ⟨2̄11⟩ and 60° ⟨1̄1̄2⟩ azimuths in figure B1.23.13 were selected to compare with those of blocking cone analyses and classical ion trajectory simulations. The arrangement of the first-layer atoms is identical along both of these azimuths; however, the second- and third-layer atoms have a different arrangement with respect to the first-layer atoms. He atoms scattered from second- and third-layer atoms experience a different arrangement of blocking cones on their exit from the surface. The positions of the blocking cones were calculated [39] from the interatomic vectors of figure B1.23.14 and the critical blocking angles or sizes of the cones were calculated with the method described in section B1.23.4. The results are shown in figure B1.23.15. The blocking of scattering trajectories from $n$th-layer atoms by their neighbouring $n$th-layer atoms are observed at low $\beta$ since these atoms are all in the same plane. This first/first-layer atom scattering contributes most of the intensity at low $\beta$. The arcs corresponding to the edges of the blocking

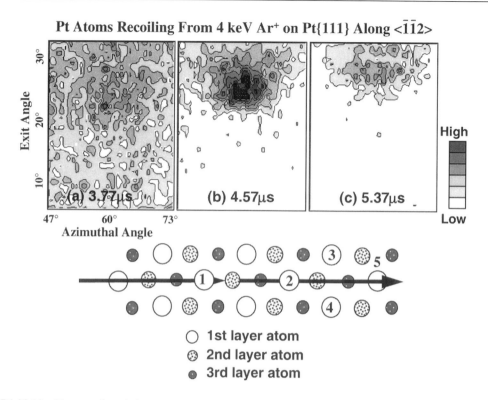

**Figure B1.23.12.** Above: selected time-resolved SARIS images of 4 keV Ar⁺ recoiling Pt atoms from Pt{111} along $\langle\bar{1}\bar{1}2\rangle$. Below: view of Pt{111} surface along $\langle\bar{1}\bar{1}2\rangle$ showing a focused second-layer Pt recoil trajectory (atoms 1–4 form a focusing 'atomic lens').

cones (figure B1.23.15) resulting from the vectors $A$, $B$, $D$ and $E$ in figure B1.23.14 occur at $\beta \sim 10°$. The features at higher $\beta$ correspond to scattering trajectories from second- and third-layer atoms that are blocked and focused by first-layer atoms. The cones resulting from the vectors $F$, $G$ and $O$ along $\langle\bar{2}11\rangle$ and $M$ and $N$ along $\langle\bar{1}\bar{1}2\rangle$ are due to scattering trajectories from second- and third-layer atoms that are blocked by first-layer atoms along these symmetrical directions. These are centred along the azimuths and are directed to higher $\beta$ values for shorter interatomic spacings. Blocking cones due to the vectors $H$, $I$, $K$ and $L$ result from second- and third-layer scattering and are observed at $\delta$ values off of the 0° and 60° directions due to non-planar scattering trajectories.

*(a) Quantitative analysis*

Quantitative analyses can be achieved by using the scattering and recoiling imaging code (SARIC) simulation and minimization of the $R$-factor [33] (section B1.23.4.4) between the experimental and simulated images as a function of the structural parameters. The SARIC was used to generate simulated images of 4 keV He⁺ scattering from bulk-terminated Pt{111} as a function of the first–second interlayer spacing $d$. Anisotropic thermal vibrations with an amplitude of 0.1 Å were included in the model. A two-dimensional reliability, or $R$, factor, based on the differences between the experimental and simulated patterns, was calculated as a function of the deviation $d$ of the first–second interlayer spacing from the bulk value. The plots shown in figure B1.23.16 exhibit minima at $d_{\min} = -0.005$ and $+0.005$ Å for the $\langle\bar{1}\bar{2}2\rangle$ and $\langle\bar{2}11\rangle$ azimuths, respectively. The optimized simulated images corresponding to $d_{\min}$ are shown in figure B1.23.16, rightmost frames; there

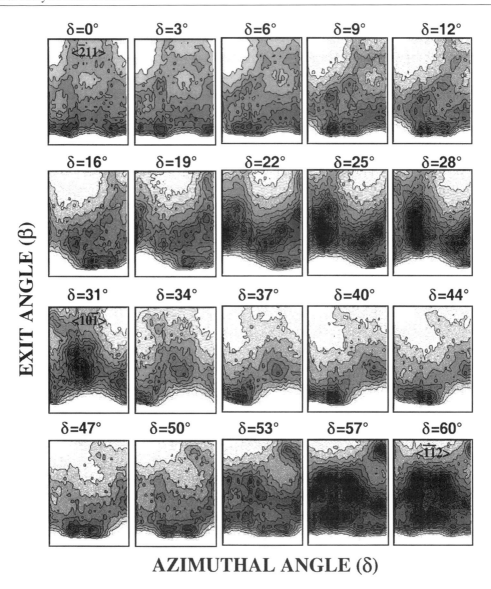

**EXIT ANGLE ($\beta$)**

**AZIMUTHAL ANGLE ($\delta$)**

**Figure B1.23.13.** A series of 20 time-resolved SARIS frames for 4 keV He$^+$ scattering from Pt{111}-(1 × 2) taken every 3° of rotation about the azimuthal angle $\delta$, starting with $\delta = 0°$ as the $\langle \bar{2}11 \rangle$ azimuth and 60° as the $\langle \bar{1}\bar{1}2 \rangle$ azimuth. Each frame represents a 16.7 ns window centred at the TOF corresponding to QSS as predicted by the BCA. The abscissa is the crystal azimuthal angle ($\delta$) and the ordinate is the particle exit angle ($\beta$).

is good agreement between these simulated and experimental images. The $R$-factors are sensitive to changes in the interlayer spacing at the level of 0.01 Å. Based on these data, we conclude that the Pt{111} surface is bulk terminated with the first–second layer spacing within +0.01 Å, or 0.4%, of the 2.265 Å bulk spacing. This sensitivity is less than the uncertainty due to the thermal vibrations because SARIS samples the average positions of lattice atoms.

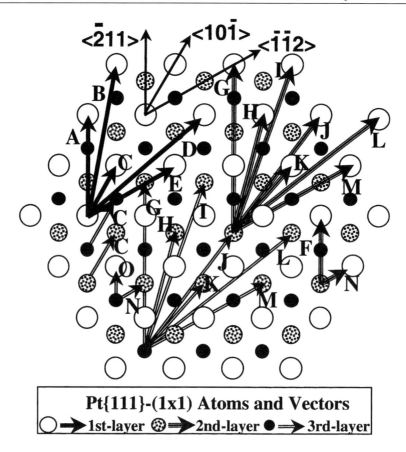

**Figure B1.23.14.** Schematic illustration of the Pt{111}-(1×1) surface. Arrows are drawn to indicate the nearest-neighbour first–first-, second–first-, and third–first-layer interatomic vectors.

### B1.23.8    Ion–surface electron exchange

One of the unsolved problems in the interaction of low-energy ions with surfaces is the mechanism of charge transfer and prediction of the charge composition of the flux of scattered, recoiled and sputtered atoms. The ability to collect spectra of neutrals plus ions and only neutrals provides a direct measure of scattered and recoiled ion fractions. SARIS images can provide electronic transition probability contour maps which are related to surface electron density and reactivity along the various azimuths.

Ion–surface electron transition probabilities are determined by electron tunnelling between the valence bands of the surface and the atomic orbitals of the ion [42]. Such transition probabilities are highest for close distances of approach. Since TOF-SARS is capable of directly measuring the scattered and recoiled ion fractions, it provides an excellent method for studying ion–surface charge exchange. For simplicity, electron exchange [43] between ions or atoms and surfaces can be discussed in terms of two regions: (i) along the incoming and outgoing trajectories where the particle is within Ångstroms of the surface and (ii) in the close atomic encounter where the core electron orbitals of the collision partners overlap. In region (i), the dominating processes are resonant and Auger electron tunnelling transitions, both of which are fast ($\tau < 10^{-15}$ s). Since the work functions of most solids are lower than the ionization potentials of most gaseous atoms, keV scattered and recoiled species are predominately neutrals as a result of electron capture from the solid. In region (ii),

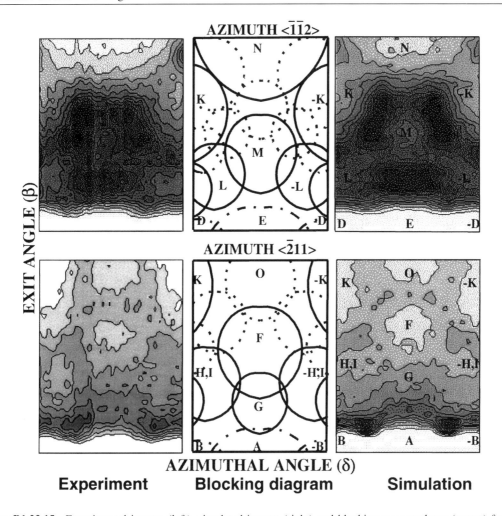

**Figure B1.23.15.** Experimental images (left), simulated images (right) and blocking cone analyses (centre) for He$^+$ scattering along the $\langle \bar{2}11 \rangle$ and $\langle \bar{1}\bar{1}2 \rangle$ azimuths. For the calculated blocking cones, first–first, first–second, and first–third layer interactions are identified by dash-dot, solid, and dotted lines, respectively. The scattering parameters are: scattering angle (with respect to incident beam direction) $\theta = 45°$; beam incident angle (with respect to surface plane) $\alpha = 28°$; exit angle of scattered particles along detector normal (relative to surface plane) $\beta = 17°$; flight path to detector (along the detector normal)$= 15.5$ cm. Angular space subtended by MCP is 27° of crystal azimuthal angle $\delta$ and 33° of particle exit angle $\beta$.

as the interatomic distance $R$ decreases, the atomic orbitals (AOs) of the separate atoms of atomic number $Z_1$ and $Z_2$ evolve into molecular orbitals (MOs) of a quasimolecule and finally into the AO of the 'united' atom of atomic number $(Z_1 + Z_2)$. As $R$ decreases, a critical distance is reached where electrons are promoted into higher-energy MOs because of electronic repulsion and the Pauli exclusion principle. This can result in collisional reionization of neutral species. The fraction of species scattered and recoiled as ions is sensitive to atomic structure through changes in electron density along the trajectories. A direct method for measuring the spatial dependence of charge transfer probabilities with atomic-scale resolution has been developed using the method of DR ion fractions [43]. The data demonstrate the need for an improved understanding of how

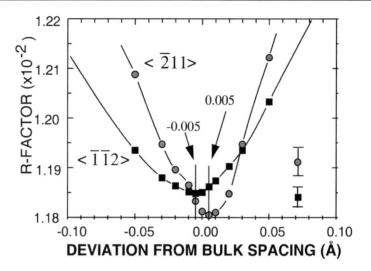

**Figure B1.23.16.** Plots of the two-dimensional $R$-factors as a function of the deviation ($d$) of the first–second interlayer spacing from the bulk value. The experimental and simulated images along the $\langle\bar{2}11\rangle$ and $\langle\bar{1}\bar{1}2\rangle$ azimuths of figure B1.23.15 were used in the comparison.

atomic energy levels shift and broaden near surfaces. These types of measurements, combined with theoretical modelling, can provide a detailed microscopic map of the local reactivity of the surface as well as electron tunnelling rates within the surface unit cell. This information is of crucial importance for the understanding of various impurity-induced promotion and poisoning phenomena in catalysis and electron-density maps from scanning tunnelling microscopy.

### B1.23.9    Role of scattering and recoiling among surface science techniques

Scattering and recoiling contribute to our knowledge of surface science through (i) elemental analysis, (ii) structural analysis and (iii) analysis of electron exchange probabilities. We will consider the merits of each of these three areas.

#### B1.23.9.1    Elemental analysis

There are two unique features of scattering and recoiling spectrometry: (1) sensitivity to the outermost atomic layer of a surface and (2) sensitivity to surface hydrogen. Using an ESA, it is possible to resolve ions scattered from all elements of mass greater than carbon. The TOF technique is sensitive to all elements, including hydrogen, although it has limited mass resolution. For general qualitative and quantitative surface elemental analyses, XPS and AES remain the techniques of choice.

#### B1.23.9.2    Structural analysis

The major role of TOF-SARS and SARIS is as surface structure analysis techniques which are capable of probing the positions of all elements with an accuracy of $\lesssim 0.1$ Å. They are sensitive to short-range order, i.e. individual interatomic spacings that are $<10$ Å. They provide a direct measure of the interatomic distances in the first and subsurface layers and a measure of surface periodicity in real space. *One of its most important applications is the direct determination of hydrogen adsorption sites by recoiling spectrometry* [12, 41]. Most other surface structure techniques do not detect hydrogen, with the possible exception of He atom scattering and vibrational spectroscopy.

TOF-SARS and SARIS are complementary to LEED, which probes long-range order, minimum domain size of 100–200 Å and provides a measure of surface and adsorbate symmetry in reciprocal space. Coupling TOF-SARS, SARIS and LEED provides a powerful combination for surface structure investigations. The techniques of medium- and high- (Rutherford backscattering) energy ion scattering only sample subsurface and bulk structure and are not as surface sensitive as TOF-SARS.

## B1.23.10    Low-energy scattering of light atoms

Atomic and molecular beams of light atoms such as He, H and $H_2$ formed from supersonic nozzle beam sources typically have kinetic energies of 20 to 100 meV [44]. Scattering of such low-energy light atoms from surfaces is predominantly elastic. Coherently scattered waves from regularly spaced surface atoms can interfere with each other, giving rise to well known diffraction phenomena. Such hyperthermal atoms have classical turning points that are typically about 3 Å above the centres of the outermost atomic layer of the surface. The diffraction data probe the outer regions of the atom–surface potential. This information is usually expressed as a corrugation function. Structural information such as bond angles and lengths is obtained by calculating the potential or corrugation function from assumed geometries [44].

The basic components of an apparatus for such atom scattering consist of a UHV scattering chamber equipped with a supersonic atomic and molecular beam source, a sample manipulator and a rotating mass spectrometer. Cryogenic sample temperatures are usually used in order to reduce the vibrational amplitudes of the surface atoms. Data are obtained from the in-plane and out-of-plane diffraction intensity distributions as a function of scattering angle. A set of corrugation parameters is derived from this data. These parameters can be calculated from a first principles approach based on a proportionality relation between the atom–surface interaction potential and the surface charge density. Such diffraction experiments of thermal He atoms coupled with theoretical simulations of the data have been shown to be a very useful structural tool for studying adsorption on surfaces. The method is extremely surface-sensitive, is capable of providing adsorption site information and is one of the few techniques that can detect surface hydrogen.

## B1.23.11    Summary

Emphasis in this chapter has been placed on the physical concepts and structural applications of TOF-SARS and SARIS. These techniques are now established as surface structural analysis methods that will have a significant impact in areas as diverse as thin-film growth, catalysis, hydrogen embrittlement and penetration of materials, surface reaction dynamics and analysis of interfaces. Surface crystallography is evolving from the classical concept of a static surface and the question of 'Where do atoms sit?' to the concept of a dynamically changing surface. The development of large-area detectors with rapid acquisition of scattering and recoiling structural images, as described in B1.23.7, will provide a technique for capturing time-resolved snapshots of such dynamically changing surfaces.

## References

[1]   Smith D P 1967 Scattering of low-energy noble gas ions from metal surfaces *J. Appl. Phys.* **38** 340–7
[2]   Heiland W and Taglauer E 1977 The backscattering of low energy ions and surface structure *Surf. Sci.* **68** 96–107
       Heiland W and Taglauer E 1975 Low energy ion scattering and Auger electron spectroscopy studies of clean nickel surfaces and adsorbed layers *Surf. Sci.* **47** 234–43
       Heiland W and Taglauer E 1973 Bombardment induced surface damage in a nickel single crystal observed by ion scattering and LEED *Rad. Effects.* **19** 1–6
       Heiland W, Iberl F, Taglauer E and Menzel D 1975 Oxygen adsorption on (110) silver *Surf. Sci.* **53** 383–92

[3]   Brongersma H H and Theeten J B 1976 The structure of oxygen adsorbed on Ni(0001) as determined by ion scattering spectroscopy
      *Surf. Sci.* **54** 519–24
      Brongersma H H and Mul P 1973 Analysis of the outermost atomic layer of a surface by low-energy ion scattering *Surf. Sci.* **35**
      393–412
[4]   Suurmijer E P Th M and Boers A L 1973 Low-energy ion reflection from metal surfaces *Surf. Sci.* **43** 309–52
[5]   DeWit A G J, Bronckers R P N and Fluit J M 1979 Oxygen adsorption on Cu(110): determination of atom positions with low
      energy ion scattering *Surf. Sci.* **82** 177–94
[6]   Aono M, Hou Y, Souda R, Oshima C, Otani S, Ishizawa Y, Matsuda K and Shimizu R 1982 Interaction potential between He$^+$ and
      Ti in a keV range as revealed by a specialized technique in ion scattering spectroscopy *Japan. J. Appl. Phys. Lett.* **21** L670–2
      Aono M, Hou Y, Oshima C and Ishizawa Y 1982 Low-energy ion scattering from the Si(001) surface *Phys. Rev. Lett.* **49** 567–70
      Aono M and Souda R 1985 Quantitative surface atomic structure analysis by low energy ion scattering spectroscopy *Japan. J. Appl.
      Phys. Part 1* **24** 1249–62
[7]   Niehus H 1984 Analysis of the Pt(110) × (1 × 2) surface reconstruction *Surf. Sci.* **145** 407–18
      Niehus H and Comsa G 1984 Determination of surface reconstruction with impact-collision alkali ion scattering *Surf. Sci.* **140**
      18–30
[8]   Marchut L, Buck T M, Wheatley G H and McMahon C J Jr 1984 Surface structure analysis using low energy ion scattering *Surf.
      Sci.* **141** 549–66
[9]   Grizzi O, Shi M, Bu H, and Rabalais J W 1990 Time-of-flight scattering and recoiling spectrometer (TOF-SARS) for surface
      analysis *Rev. Sci. Instrum.* **61** 740–52
[10]  Rabalais J W 1990 Scattering and recoiling spectrometry: an ion's eye view *Science* **250** 521–7
[11]  Kim C, Höfner C, Al-Bayati A and Rabalais J W 1998 Scattering and recoiling imaging spectrometer (SARIS) *Rev. Sci. Instrum.*
      **69** 1676–84
[12]  Aono M, Katayama M and Nomura E 1992 Exploring surface structures by coaxial impact collision ion scattering spectroscopy
      (CAICISS) *Nucl. Instrum. Methods* B **64** 29–37
[13]  Ghrayeb R, Purushotham M, Hou M and Bauer E 1987 Estimate of repulsive interatomic pair potentials by low-energy alkali-
      metal-ion scattering and computer simulation *Phys. Rev.* B **36** 7364–70
[14]  Chester M and Gustafsson T 1991 Geometric structure of the Si(111)-($\sqrt{3} \times \sqrt{3}$)R30°-Au surface *Surf. Sci.* **256** 135–46
[15]  Hetterich W, Höfner C and Heiland W 1991 An ion scattering study of the surface structure and thermal vibrations on Ir(110) *Surf.
      Sci.* **251/252** 731–6
[16]  O'Connor D J, King B V, MacDonald R J, Shen Y G and Chen X 1990 The study of surfaces using ion beams *Aust. J. Phys.* **43** 601
[17]  Dodonoy A I, Mashkova E S and Molchanov V A 1989 Medium-energy ion scattering by solid surfaces. III: ejection of fast recoil
      atoms from solids under ion bombardment *Rad. Eff. Def. Sol.* **110** 227–341
[18]  Mintz M H, Atzmony U and Shamir N 1987 Initial adsorption kinetics of oxygen on polycrystalline copper *Surf. Sci.* **185** 413–30
[19]  van de Riet E and Niehus A 1991 Application of low energy neutral ionizaton spectroscopy for surface structure analysis *Surf. Sci.*
      **243** 43–8
[20]  Niehus H, Spitzl R, Besocke K and Comsa G 1991 N-induced (2 × 3) reconstruction of Cu(110): evidence for long-range, highly
      directional interaction between Cu–N–Cu bonds *Phys. Rev.* B **43** 12 619–25
[21]  Shoji F, Kashihara K, Sumitomo K and Oura K 1991 Low-energy recoil-ion spectroscopy studies of hydrogen adsorption on
      Si(100)-2 × 1 surfaces *Surf. Sci.* **242** 422–7
[22]  Overbury S H, Mullins D R, Paffett M T and Koel B E 1991 Surface structure determination of Sn deposited on Pt(111) by low
      energy alkali ion scattering *Surf. Sci.* **254** 45–57
[23]  Taglauer E, Beckschulte M, Margraf R and Mehl D 1988 Recent developments in the applications of ion scattering spectroscopy
      *Nucl. Instrum. Methods* B **35** 404–9
[24]  Bracco G, Canepa M, Catini P, Fossa F, Mattera L, Terreni S and Truffelli D 1992 Impact-collision ion scattering study of Ag(110)
      *Surf. Sci.* **269/270** 61–7
[25]  Mashkova E S and Molchanov V A 1985 *Medium-Energy Ion Reflection From Solids* (Amsterdam: North-Holland)
[26]  Goldstein H 1980 *Classical Mechanics* 2nd edn (Reading, MA: Addison-Wesley)
[27]  Gibson J B, Goland A N, Milgram M and Vineyard G H 1960 Dynamics of radiation damage *Phys. Rev.* Series 2 **120** 1229–53
[28]  Johnson L W and Reiss R D 1982 *Numerical Analysis* (Berlin: Springer)
[29]  Press W H, Flannery B P, Teukolsky S A and Vetterling W T 1988 *Numerical Recipe in C* (Cambridge: Cambridge University
      Press) p 131
[30]  Parilis E S *et al* 1993 *Atomic Collisions on Solids* (New York: North-Holland)
[31]  Zeigler J F, Biersack J P and Littmark U 1985 *The Stopping and Range of Ions in Solids* (New York: Pergamon)
[32]  Zangwill A 1989 *Physics at Surfaces* (Cambridge: Cambridge University Press) p 117
[33]  Kim C, Höfner C and Rabalais J W 1997 Surface structure determination from ion scattering images *Surf. Sci.* **388** L1085–91
[34]  Xu M L and Tong S Y 1985 Multilayer relaxation for the clean Ni(110) surface *Phys. Rev.* B **31** 6332–6
[35]  Shi M, Wang Y and Rabalais J W 1993 Structure of the Si{100} surface in the clean (2 × 1), (2 × 1)-H monohydride, (1 × 1)-H
      dihydride, and c(4 × 4)-H phases *Phys. Rev.* B **48** 1678–88
[36]  Masson F and Rabalais J W 1991 Time-of-flight scattering and recoiling spectrometry (TOF-SARS) analysis of Pt{110}. I. Quan-
      titative structure study of the clean (1 × 2) surface *Surf. Sci.* **253** 245–57

Masson F and Rabalais J W 1991 Time-of-flight scattering and recoiling spectrometry (TOF-SARS) analysis of Pt{110}. II. The (1 × 2)-to-(1 × 3) interconversion and characterization of the (1 × 3) phase **253** 258–69

[37] Masson F and Rabalais J W 1991 Surface periodicity exposed through shadowing and blocking effects *Chem. Phys. Lett.* **179** 63–7

[38] Roux C D, Bu H and Rabalais J W 1991 Structure of the hydrogen induced Ni{110}-p(1 × 2)-H reconstructed surface *Surf. Sci.* **259** 253–65

Bu H, Roux C D and Rabalais J W 1992 Hydrogen adsorption site on the Ni{110}-p(1 × 2)-H surface from time-of-flight scattering and recoiling spectrometry (TOF-SARS) *Surf. Sci.* **271** 68–80

[39] Kim C and Rabalais J W 1997 Projections of atoms in terms of interatomic vectors *Surf. Sci.* **385** L938–44

[40] Kim C, Höfner C, Bykov V and Rabalais J W 1997 Element-, time-, and spatially-resolved images of scattered and recoiled atoms *Nucl. Instrum. Methods* B **125** 315–22

Kim C and Rabalais J W 1998 Focusing of He$^+$ ions on semichannel planes in the Pt{111} surface *Surf. Sci.* **395** 239–47

[41] Höfner C, Bykov V and Rabalais J W 1997 Three-dimensional focusing patterns of He$^+$ ions scattering from a Au{110} surface *Surf. Sci.* **393** 184–93

Kim C, Ahn J, Bykov V and Rabalais J W 1998 Element-, velocity-, and spatially-resolved images of Kr$^+$ scattering and recoiling from a CdS surface *Int. J. Mass Spectrom. Ion Phys.* **174** 305–15

[42] Hsu C C and Rabalais J W 1991 Structure sensitivity of scattered Ne$^+$ ion fractions from a Ni{100} surface *Surf. Sci.* **256** 77–86

[43] Hsu C C, Bu H, Bousetta A, Rabalais J W and Nordlander P 1992 Angular dependence of charge transfer probabilities between O- and a Ni{100}-c(2 × 2)-O surface *Phys. Rev. Lett.* **69** 188–91

[44] Engel T and Rieder K H 1982 Structural studies of surfaces with atomic and molecular beam diffraction *Structural Studies of Surfaces With Atomic and Molecular Beam Scattering (Springer Tracts in Modern Physics vol 91)* (Berlin: Springer) pp 55–180

## Further Reading

Rabalais J W 1994 Low energy ion scattering and recoiling *Surf. Sci.* **299/300** 219–32

A review of the ion scattering literature starting in the 1960s.

Rabalais J W 1992 Surface crystallography from ion scattering *Chemistry in Britain* **28** 37–71

A simple survey of ion scattering for students.

Niehus H, Heiland W and Taglauer E 1993 Low-energy ion scattering at surfaces *Surf. Sci. Rep.* **17** 213–304

An excellent review of ion scattering.

Rabalais J W (ed) 1994 *Low Energy Ion–Surface Interactions* (Chichester: Wiley)

A volume with contributions from several authors that treats ion–surface interactions at different energies.

Taglauer E 1997 Low-energy ion scattering and Rutherford backscattering *Surface Analysis; The Principal Techniques* ed J C Vickerman (Chichester: Wiley) pp 215–66

A textbook that treats the principal techniques of surface science.

# B1.24
# Rutherford backscattering, resonance scattering, PIXE and forward (recoil) scattering

*C C Theron, V M Prozesky and J W Mayer*

### B1.24.1   Introduction

The use of million electron volt (MeV) ion beams for materials analysis was instigated by the revolution in integrated circuit technology. Thin planar structures were formed in silicon by energetic ion implantation of dopants to create electrical active regions and thin metal films were deposited to make interconnections between the active regions. Ion implantation was a new technique in the early 1960s and interactions between metal films and silicon required analysis. For example, the number of ions implanted per square centimetre (ion dose) and thicknesses of metal layers required careful control to meet the specifications of integrated circuit technology. Rutherford backscattering spectrometry (RBS) and MeV ion beam analysis were developed in response to the needs of the integrated circuit technology. In turn integrated circuit technology provided the electronic sophistication used in the instrumentation in ion beam analysis. It was a synergistic development of analytical tools and the fabrication of integrated circuits.

Rutherford backscattering spectrometry is the measurement of the energies of ions scattered back from the surface and the outer microns (1 micron = 1 $\mu$m) of a sample. Typically, helium ions with energies around 2 MeV are used and the sample is a metal coated silicon wafer that has been ion implanted with about a monolayer ($10^{15}$ ions cm$^{-2}$) of electrically active dopants such as arsenic. Only moderate vacuum levels (about $10^{-4}$ Pa) are required so that sample exchange is rapid allowing the analysis of the number of implanted atoms per square centimetre and their distribution in depth to be carried out in periods of about 15 minutes or less. The sample is not damaged structurally during the analysis and therefore Rutherford backscattering spectrometry is considered non-destructive. This is in contrast to surface sensitive techniques such as Auger electron spectroscopy where surface erosion by sputtering is required for depth analysis. One of the strong features of Rutherford backscattering spectrometry is that the scattering cross sections are well known so that the analysis is quantitative. In other analytical techniques, such as secondary ion mass spectrometry (SIMS), the cross sections are not well defined. The relative ion yields can vary over three orders of magnitude depending on the nature of the surface. Rutherford backscattering has been a convenient way to calibrate secondary ion mass spectrometry, which in turn is more sensitive to the detection of trace elements than Rutherford backscattering. The two techniques are thus complementary.

Ion beam analysis grew out of nuclear physics research on cross sections and reaction products involved in atomic collisions. In this work million volt accelerators were developed and used extensively. As the energies of the incident particles increased, the lower energy accelerators became available for use in solid state applications. The early nuclear physics research used magnetic spectrometers to measure the energies of the particles. This analytical procedure was time consuming and the advent of the semiconductor nuclear particle detector allowed simultaneous detection of all particle energies. It was an energy dispersive spectrometer. The semiconductor detector is a Schottky barrier (typically a gold film on silicon) or shallow diffused p–n junction with the active region defined by the high electric field in the depletion layer. The active region extends tens of microns below the surface of the detector so that in almost every application the penetration of the energetic particles is confined within the active region. The response of the detector is linear with the energy of the particles providing a true particle energy spectrometer.

Analysis of ion implanted layers and metal–silicon interactions was carried out with Rutherford backscattering at 2.0 MeV energies and with semiconductor nuclear particle detectors for several years. Rutherford backscattering became well established and was utilized in materials analysis in industrial and university laboratories across the world. The importance of hydrogen and its influence in solid state chemistry led to the development of forward scattering in which one measures the energy of the recoiling hydrogen atom. The helium ion is heavier than that of hydrogen so that by tilting the sample it is possible to measure the recoil energy of the emerging hydrogen, again with a nuclear particle detector. In other words, the modification to the Rutherford backscattering spectrometry target chamber geometry was only to tilt the target and to move the detector. These forward recoil techniques have of course become more sophisticated with use of heavy incident ions and detectors which measure both the energy and the mass of the recoiling particles ($\Delta E$–$E$ or time of flight detector).

Analysis of silicon is an almost ideal experimental situation because the masses of most implanted atoms and metal layers exceed that of silicon. In Rutherford backscattering the mass of the atom must be greater than that of the silicon target to separate the energy signals of the target atom from those of the silicon spectrum. Oxygen is an exception. It is lighter than silicon and also is ubiquitous in surface and interface layers. The analysis of oxygen, and also carbon and nitrogen, are carried out in the same experimental chamber as used in Rutherford backscattering, but the energy of the incident helium ions is increased to energies where there are resonances in the backscattering cross sections. These resonances increase the yield of the scattered particle by nearly two orders of magnitude and provide high sensitivity to the analysis of oxygen and carbon in silicon. The use of these high energies, 3.04 MeV for the helium–oxygen resonance, is called resonance scattering and the word Rutherford is inappropriate for a descriptor.

By inserting a semiconductor x-ray detector into the analysis chamber, one can measure particle induced x-rays. The cross section for particle induced x-ray emission (PIXE) is much greater than that for Rutherford backscattering and PIXE is a fast and convenient method for measuring the identity of atomic species within the outer microns of the sample surface. The energy resolution in helium ion Rutherford backscattering spectrometry does not allow discrimination between the signals from high atomic number (high $Z$) elements close to each other in the periodic table. With conventional semiconductor detectors one cannot distinguish between gold and tungsten, for example, whereas the ion induced x-ray energies are easily distinguished for the two high $Z$ elements. PIXE, then, becomes another tool in the MeV ion analysis chamber and only requires the addition of a x-ray detector system.

The dimensions of the incident ion beam are typically 1 mm across the width of the incident beam impinging on the sample surface. This dimension can be easily obtained using slits in the beam handling system. The beam diameter can be reduced by orders of magnitude by using quadrupole or electrostatic lenses to focus the ion beam to diameters of about one micron on the sample surface. The beam is then rastered across the surface to provide a visual image of the surface with micron resolution. In this work the large cross sections for PIXE are important, because sample analysis can be performed without sample damage caused by the high current density of incident ions.

This overview covers the major techniques used in materials analysis with MeV ion beams: Rutherford backscattering, channelling, resonance scattering, forward recoil scattering, PIXE and microbeams. We have not covered nuclear reaction analysis (NRA), because it applies to special incident-ion–target-atom combinations and is a topic of its own [1, 2].

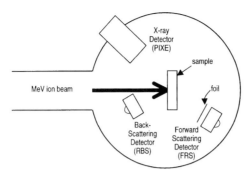

**Figure B1.24.1.** Schematic diagram of the target chamber and detectors used in ion beam analysis. The backscattering detector is mounted close to the incident beam and the forward scattering detector is mounted so that, when the target is tilted, hydrogen recoils can be detected at angles of about 30° from the beam direction. The x-ray detector faces the sample and receives x-rays emitted from the sample.

## B1.24.2 Rutherford backscattering spectrometry (RBS)

The discussion of Rutherford backscattering spectrometry starts with an overview of the experimental target chamber, proceeds to the particle kinematics that determine mass identification and depth resolution, and then provides an example of the analysis of a silicide.

### B1.24.2.1 Target chamber

Figure B1.24.1 shows the placement of the sample and detectors in the target chamber. The sample is located so that its surface is on an axis of rotation of a goniometer so that the beam position does not shift across the sample as the sample is tilted with respect to the incident ion beam. The backscattering detector is mounted as close to the incident beam as possible so that the average backscattering angle, $\theta$, is close to 180°, typically 170°, with a detector solid angle of about 3–5 millisteradians (msr). In some cases annular detectors are used with the incident beam passing through the centre of the detector aperture in order to provide larger analysis solid angles. The sample is rotated to glancing angle geometries when the forward scattering detector is used. Typically a thin foil is placed in front of the detector to block the helium ions while allowing the hydrogen ions to pass through with only minimal energy loss. The stopping power (energy loss) of MeV helium ions is ten times that of the recoiling hydrogen ions. As shown in section B1.24.8 below, the forward scattering detector system can be augmented to include a $\Delta E$–$E$ detector to allow identification of the ion mass as well as energy. The x-ray detector is placed so that the active region is in full view of the sample surface bombarded with the incident ions.

For conventional backscattering spectrometry with helium ions, the energy resolution of the semiconductor particle detector is typically 15 kiloelectron volts (keV). This resolution can be improved to 10 keV with special detectors and detector cooling. The output signal, which is typically millivolts in pulse height, is processed by silicon integrated circuit electronics and provides an energy spectrum in terms of number of particles versus energy. It is often displayed as particles versus channel number as the energy scale is divided into channels which must be calibrated to give the energy scale. The calibration between the measured particle energy and the channel number is independent of the ion energy and sample analysed and only depends on the semiconductor detector and associated electronics response to the energy of the ion beams.

The vacuum requirements in the target chamber are relatively modest ($10^{-4}$ Pa) and are comparable to those in the accelerator beam lines. All that is required is that the ion beam does not lose energy on its path to the sample and that there is minimal deposition of contaminants and hydrocarbons on the surface during analysis.

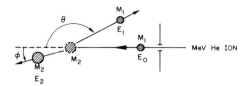

**Figure B1.24.2.** A schematic representation of an elastic collision between a particle of mass $M_1$ and energy $E_0$ and a target atom of mass $M_2$. After the collision the projectile and target atoms have energies of $E_1$ and $E_2$ respectively. The angles $\theta$ and $\phi$ are positive as shown. All quantities refer to the laboratory frame of reference.

### B1.24.2.2 Kinematics

In ion beam analysis the incident particle penetrates into the silicon undergoing inelastic collisions, predominantly with target electrons, and loses energy as it penetrates to the end of its range. The range of 2.5 MeV helium ions is about 10 microns in silicon; the range of comparable energy protons is about ten times that of the helium ions (the range of 3 MeV hydrogen is about 100 microns in silicon). During the penetration of the helium ions, a small fraction undergo elastic collisions with the target atom to give the backscattering signal.

Figure B1.24.2 is a schematic representation of the geometry of an elastic collision between a projectile of mass $M_1$ and energy $E_0$ with a target atom of mass $M_2$ initially at rest. After collision the incident ion is scattered back through an angle $\theta$ and emerges from the sample with an energy $E_1$. The target atom after collision has a recoil energy $E_2$. There is no change in target mass, because nuclear reactions are not involved and energies are non-relativistic.

The ratio of the projectile energies for $M_1 < M_2$ is given by

$$K = \frac{E_1}{E_2} = \left[ \frac{(M_2^2 - M_1^2 \sin^2 \theta)^{1/2} + M_1 \cos \theta}{M_2 + M_1} \right]^2. \tag{B1.24.1}$$

The energy ratio, called the kinematic factor $K = E_1/E_0$, shows that the energy after scattering is determined by the masses of the incident particle and target atoms and the scattering angle. For a direct backscattering through $180°$ the energy ratio has its lowest value given by

$$\frac{E_1}{E_0} = \left( \frac{M_2 - M_1}{M_2 + M_1} \right)^2. \tag{B1.24.2}$$

For incident helium ions ($M_1 = 4$) at $E_0 = 2.0$ MeV the energy $E_1$ of the backscattered particle for silicon ($M_2 = 28$) is 1.12 MeV and for palladium ($M_2 = 106$) the energy is 1.72 MeV.

The energy $E_2$ transferred to the target atom has a general relation given by

$$\frac{E_2}{E_0} = \frac{4 M_1 M_2}{(M_1 + M_2)^2} \cos^2 \phi \tag{B1.24.3}$$

and at $\theta = 180°$ the energy $E_2$ transferred to the target atom has its maximum value given by

$$\frac{E_2}{E_0} = \frac{4 M_1 M_2}{(M_1 + M_2)^2}. \tag{B1.24.4}$$

In collisions where $M_1 = M_2$ at $\theta = 180°$ the incident particle is at rest after the collision, with all the energy transferred to the target atom. For 2.0 MeV helium ions colliding with silicon the recoil energy $E_2$ is 0.88 MeV and from palladium is 0.28 MeV.

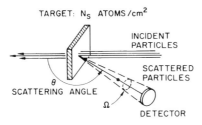

TARGET: $N_S$ ATOMS/cm$^2$

INCIDENT PARTICLES

SCATTERED PARTICLES

$\theta$

SCATTERING ANGLE

$\Omega$

DETECTOR

**Figure B1.24.3.** Layout of a scattering experiment. Only primary particles that are scattered within the solid angle $\Omega$ spanned by the solid state detector are counted.

The ability to identify different mass species depends on the energy resolution of the detector which is typically 15 keV full width at half maximum (FWHM). For example, silver has a mass $M_2 = 108$ and tin has a mass $M_2 = 119$. The difference between $K_{Ag} = 0.862$ and $K_{Sn} = 0.874$ is 0.012. For 2 MeV helium ions the difference in backscattering energy is 24 keV which is well outside the range of the detector resolution, indicating that signals from Ag and Sn on the surface can be resolved. The difference between gold and tungsten $K$ values is 0.005, and at 2 MeV energies one would not resolve the signals between gold and tungsten. With Rutherford backscattering and conventional detectors with energy resolution of 15 keV one can resolve the signals from and identify the elements of masses up to 100. One can resolve isotopes up to a mass of around 60. For example, all the silicon isotopes can be identified.

### B1.24.2.3   Scattering cross section

The identity of target elements is established by the energy of the scattered particles after an elastic collision. The number of atoms per unit area, $N_S$, is found from the number $Q_D$ of detected particles (called the yield, $Y$) for a given number $Q$ of particles incident on the target. The connection is given by the scattering cross section $\sigma(\theta)$ by

$$Y = Q_D = \sigma(\theta)\Omega Q N_S. \tag{B1.24.5}$$

This is shown schematically in figure B1.24.3. In the simplest approximation the scattering cross section $\sigma$ is given by

$$\sigma(\theta) = \left(\frac{Z_1 Z_2 e^2}{4E}\right)^2 \frac{1}{\sin^4 \theta/2}, \tag{B1.24.6}$$

the scattering cross section originally derived by Rutherford. For 2 MeV helium ions incident on silver, $Z_2 = 47$ at 180°, the cross section is $2.89 \times 10^{-24}$ cm$^2$ or 2.89 barns where the barn $= 10^{-24}$cm$^2$. The distance of closest approach is about $7 \times 10^{-4}$ Å which is smaller than the K-shell radius of silver ($10^{-2}$ Å). This means that the incident helium ion penetrates well within the innermost radius of the electrons so that one can use an unscreened Coulomb potential for the scattering. The distance of closest approach is sufficiently large that penetration into the nuclear core does not occur and one neglects nuclear reactions.

The cross sections are sufficiently large that one can detect sub-monolayers of most heavy mass elements on silicon. For example, the yield of 2.0 MeV helium ions from $10^{14}$ cm$^{-2}$ silver atoms (approximately one-tenth of a monolayer) is 800 counts for a current of 100 nanoamperes for 15 minutes and detector area of 5 msr. This represents a large signal for a small number of atoms on the surface. With care, $10^{-4}$ monolayers of gold on silicon can be detected.

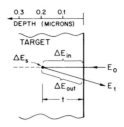

**Figure B1.24.4.** Energy loss components for a projectile that scatters from depth $t$. The particle loses energy $\Delta E_{in}$ via inelastic collisions with electrons along the inward path. There is energy loss $\Delta E_S$ in the elastic scattering process at depth $t$. There is energy lost to inelastic collisions $\Delta E_{out}$ along the outward path. For an incident energy $E_0$ the energy of the exiting particle is $E_1 = E_0 - \Delta E_{in} - \Delta E_S - \Delta E_{out}$.

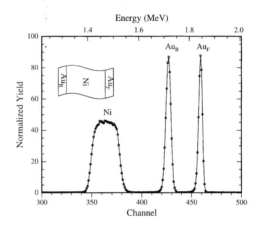

**Figure B1.24.5.** Backscattering spectrum of a thin Ni film (950 Å) with near monolayers ($\approx 30 \times 10^{15}$ at cm$^{-2}$) of Au on the front and back surfaces of the Ni film. The signals from the front and back layers of Au are shown and are separated in energy from each other by nearly the same energy width as the Ni signal.

### B1.24.2.4  Depth scale

Light ions such as helium lose energy through inelastic collision with atomic electrons. In backscattering spectrometry, where the elastic collision takes place at depth $t$ below the surface, one considers the energy loss along the inward path and on the outward path as shown in figure B1.24.4. The energy loss on the way in is weighted by the kinematic factor and the total is

$$\Delta E = \Delta t \left( K \frac{dE}{dx}\bigg|_{in} + \frac{1}{|\cos\theta|} \frac{dE}{dx}\bigg|_{out} \right) = \Delta t[S] \qquad (B1.24.7)$$

where $dE/dx$ is the rate of energy loss with distance and $[S]$ is the energy loss factor. An example illustrating the influence of depth on analysis is given in figure B1.24.5, which shows two thin gold layers on the front and back of a nickel film. The scattering from gold at the surface is clearly separated from gold at the back layer. The energy width between the gold signals is closely equal to that of the energy width of the nickel signal. This signal is nearly square shaped because nickel exists from the front to the back surface. The depth scales are determined from energy loss values, which are given in tabular form as a function of energy [1, 2]. It is often expressed as a stopping cross section in terms of $(1/N)\,dE/dx$, which gives values in eV cm$^2$. The depth resolution is given by dividing the detector resolution by the energy loss factor. For 2 MeV helium in

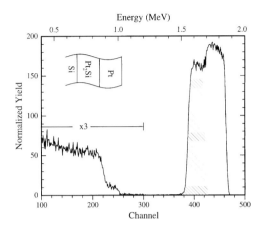

**Figure B1.24.6.** Backscattering spectrum of a layer of Pt on Si that has been thermally heated so that approximately half the Pt has been consumed in the formation of $Pt_2Si$, the first stage in the reaction of Pt with Si. In the spectrum the signals from Si have been multiplied by a factor of three for visibility, because the atomic number of Si (14) is much less than that of Pt (78). In the spectrum the signals from Pt are in the region of 1.6–1.8 MeV. The higher energy corresponds to scattering from unreacted Pt and the step around 1.7 MeV corresponds to the transition from Pt to $Pt_2Si$. The signal at 1.6 MeV corresponds to the back interface of the $Pt_2Si$ in contact with silicon. The silicon signal in the energy range from about 0.9–1.0 MeV corresponds to the Si in the $Pt_2Si$. At lower energies the spectrum represents signals from the Si substrate.

silicon one might expect a depth resolution of about 200 Å for 180° scattering geometries. This can be reduced to values of about 50 Å for glancing incident and exit angles. These values of depth resolution degrade as the particle penetrates into the sample and energy straggling becomes a factor.

### B1.24.2.5 Simulation

Rutherford backscattering spectra can be analysed by use of some of the available analysis programs. Programs such as RUMP and GISA provide a layer-by-layer signal for multielement targets [3, 4, 5]. These programs include detector resolution, energy straggling and individual isotopes, and can also be applied to forward recoil spectrometry for detection of light elements. These programs also include provisions for enhanced cross section for light elements such as carbon and oxygen.

### B1.24.2.6 Silicide formation

An example of Rutherford backscattering spectrometry of the formation of PtSi is shown in figure B1.24.6 for the case where the original Pt layer has reacted to form $Pt_2Si$. The backscattering signals at the high energy end near 1.8 MeV represent Pt at the surface of the sample. The plateau extends downward in energy to 1.7 MeV where there is a step down to the signals from Pt in $Pt_2Si$. In the Pt signal the contribution from $Pt_2Si$ is shown shaded. In the Si portion of the spectrum the signal steps upward around 1.0 MeV and represents the silicon in the $Pt_2Si$. The second step represents the Si signal from the Si substrate. In this case, the signals from the unreacted Pt, the Pt and Si in $Pt_2Si$ and the Si in the substrate are clearly identified and can be used to specify the thickness and composition of the silicide layer.

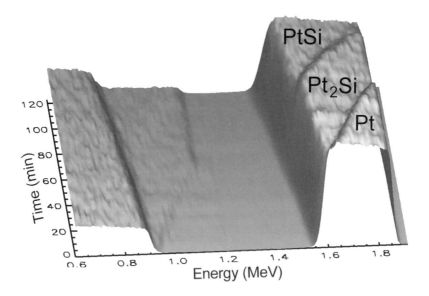

**Figure B1.24.7.** A three-dimensional plot of backscattering signal versus time for a Pt film deposited onto Si and heated at a rate of 1 °C min⁻¹. At time zero, the Pt signal shows a square-topped energy distribution. As time progresses and the sample is heated, a step appears in the Pt signal, indicating the formation of the first phase $Pt_2Si$. At longer times a second step appears, indicating the formation of the second phase PtSi after all the Pt has been consumed. The energy widths of the Pt signals give the thickness of the formed phases. The heights of the Pt signals, relative to those from Si, give the composition of the phases.

### B1.24.3  *In situ* real-time RBS

The essentially non-destructive nature of Rutherford backscattering spectrometry, combined with the its ability to provide both compositional and depth information, makes it an ideal analysis tool to study thin-film, solid-state reactions. In particular, the non-destructive nature allows one to perform *in situ* RBS, thereby characterizing both the composition and thickness of formed layers, without damaging the sample. Since only about two minutes of irradiation is needed to acquire a Rutherford backscattering spectrum, this may be done continuously to provide a real-time analysis of the reaction [6].

There are two main applications for such real-time analysis. The first is the determination of the chemical reaction kinetics. When the sample temperature is ramped linearly with time, the data of thickness of formed phase together with ramped temperature allows calculation of the complete reaction kinetics (that is, both the activation energy and the pre-exponential factor) from a single sample [6], instead of having to perform many different temperature ramps as is the usual case in differential thermal analysis [7–11]. The second application is in determining the contribution that each of the elements in the reaction couple makes to the overall atomic transport across the forming layer. For this purpose, thin, inert markers (analogous to the thin wires used by Kirkendall [12, 13]) are inserted into the layers to establish a reference frame within which to measure the contribution of each element's flux to the overall growth. Without the use of a real-time analysis technique, one must rely on the use of different samples, which, although nominally identical, do not necessarily behave identically since many of these reactions depend critically on the exact conditions at the interfaces between the layers. On the other hand, if analysis can be performed on a single sample, small changes in the position of the marker can then confidently be interpreted. Examples of these two applications are presented below.

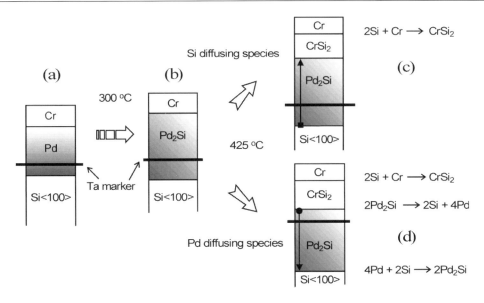

**Figure B1.24.8.** Schematic diagram of the reaction of Pd/Cr layers on (100) Si with a Ta marker placed inside the Pd layer. When the sample (a) is heated to 300 °C, the Pd reacts with the Si to form $Pd_2Si$ (b). Upon further heating Cr reacts with Si to form $CrSi_2$ on top of $Pd_2Si$. The required Si can either be supplied directly by the diffusion of Si atoms from the crystalline substrate (c) or by $Pd_2Si$ dissociation followed by Pd diffusion (d). The motion of a thin Ta marker, embedded in the $Pd_2Si$ layer, is used to distinguish between these two mechanisms. In (c) there is no movement of the marker relative to the $Pd_2Si$ layer, while in (d) the marker moves towards the $Pd_2Si/CrSi_2$ interface.

### B1.24.3.1  Pt–Si

When a thin (about 3000 Å) layer of Pt is deposited onto a Si wafer and then heated, the first phase that forms at the interface between Pt and Si is $Pt_2Si$. After all the Pt has been consumed, the newly formed $Pt_2Si$ layer reacts with the Si to form PtSi, which is stable in contact with excess Si. No further reaction is observed. Figure B1.24.7 shows the progress of this reaction as the temperature is ramped linearly at a rate of 1 °C min$^{-1}$. At time zero, the signal between 1.5 and 1.9 MeV is from the unreacted Pt layer, whereas the signal from the Si wafer appears below about 0.9 MeV. The signal from the Si has been magnified by a factor of three to compensate for the differences in cross sections between Pt ($Z = 78$) and Si ($Z = 14$).

After 60 minutes of annealing, all the Pt has reacted to form $Pt_2Si$. Almost immediately thereafter the reaction between $Pt_2Si$ and Si to form PtSi starts and after a further 60 minutes all the $Pt_2Si$ has reacted, resulting in a stable PtSi film on Si. The data of silicide thickness versus ramped temperature can be plotted in reduced form in an Arrhenius-like plot to give the activation energy [6, 14].

### B1.24.3.2  $Pd_2Si$ on $CrSi_2$

When a thin film structure of Si(100)/Pd/Cr (see figure B1.24.8(a)) is heated to 300 °C, the Pd quickly reacts with the Si to form $Pd_2Si$ (b). Upon further heating the Cr reacts with the Si to form $CrSi_2$ on top of the $Pd_2Si$. The required silicon can either be supplied directly by the diffusion of Si atoms from the crystalline substrate (c) or by $Pd_2Si$ dissociation followed by Pd diffusion (d). The motion of a thin Ta marker embedded in the $Pd_2Si$ layer is used to distinguish between these two mechanisms. In (c) there is no movement of the marker relative to the $Pd_2Si$ layer, while in (d) the marker moves towards the $Pd_2Si/CrSi_2$ interface.

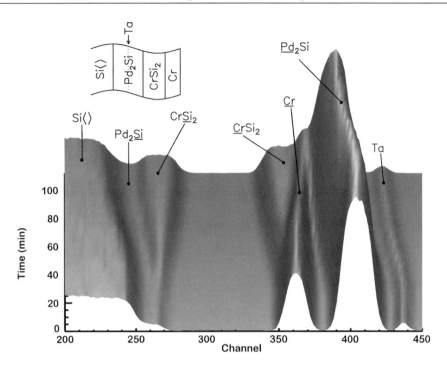

**Figure B1.24.9.** *In situ*, real-time, backscattering spectrum of the formation of CrSi$_2$ on Pd$_2$Si at 425 °C. The structure is shown in the inset. The elements underlined represent the origin of the signal; the Si signal in CrSi$_2$ is around channel 260 and the Cr signal from CrSi$_2$ appears around channel 350. The Ta marker embedded in the Pd$_2$Si layer shifts to lower energy during CrSi$_2$ formation indicating that Si diffusion through the Pd$_2$Si layer has occurred as indicated in diagram (c) of figure B1.24.8.

In figure B1.24.9 the *in situ*, real-time, RBS spectrum of the formation of CrSi$_2$ on Pd$_2$Si at 425 °C is shown. The Ta marker embedded in the Pd$_2$Si layer shifts to lower energies during CrSi$_2$ formation in agreement with the prediction for the case of Si diffusion (c). In the figure, the element from which backscattering has taken place has been underlined [14].

## B1.24.4   Channelling

If the sample is mounted on a goniometer so that the crystal axis or planes of a single crystal sample, such as silicon, are aligned within about 0.1 or 0.5 degrees, the crystal lattice can steer the trajectories of incident ions penetrating the crystal [15, 16]. This steering of the incident energetic beam is known as 'channelling' as the atomic rows and planes are guides that steer the energetic ions along the channels between rows and planes. The channelled ions do not closely approach the lattice atoms with the result that the backscattering yield can be reduced 100-fold (an aligned spectrum compared to that when the incident ions are misaligned from the lattice atoms gives a random spectrum). Channelling measurements can determine the amount of lattice disorder in which displaced atoms are located within the channels and hence accessible to backscattering collisions with the channelled ions. Channelling can also be used to measure the number of impurity atoms located sufficiently far from substitutional lattice sites that they are accessible to backscattering from the channelled ions.

Channelling phenomena were studied before Rutherford backscattering was developed as a routine analytical tool. Channelling phenomena are also important in ion implantation, where the incident ions can

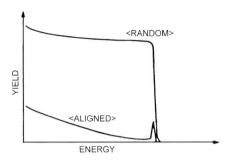

**Figure B1.24.10.** Random and aligned (channelled) backscattering spectrum from a single crystal sample of silicon. The aligned spectrum has a peak at the high energy end of the Si signal. This peak represents helium ions scattered from the outer layers of Si that are exposed to the incident beam. The yield behind the peak is 1/40th of the random yield because the Si atoms are shielded from close encounter elastic collisions from the ion beam that is channelled along the axial rows of the Si crystal.

be steered along the lattice planes and rows. Channelling leads to a deep penetration of the incident ions to depths below that found in the normal, near Gaussian, depth distributions characterized by non-channelled energetic ions. Even today, implanted channelled ions are of concern when one attempts to form shallow junctions in ion implantation of integrated circuit structures. Channelling effects can be overcome if the silicon crystal is amorphized by a prior implantation of silicon or germanium atoms.

Figure B1.24.10 shows schematically a random and aligned spectrum for MeV helium ions incident on silicon. The aligned spectrum is characterized by a peak at the high energy end of the spectrum. The peak represents ions scattered from the outermost layer of atoms directly exposed to the incident beam. This peak is called the 'surface peak'. Behind the surface peak, at lower energies, the aligned spectrum drops to a value of 1/40th of the silicon random spectrum indicating that nearly 98% of the incident ions are channelled and do not make close impact collisions with the lattice atoms. The rise in the aligned spectrum at lower energies represent the ions that are deflected from the steering by the lattice atoms and can then collide in close impact collisions with the lattice atoms and hence directly contribute to the backscattering spectra.

The application of channelling to Rutherford backscattering spectrometry is used to determine the amount of damage in ion-implanted crystal and the lattice location of ion-implanted dopant atoms. One of the important contributions of channelling to integrated circuit technology is the analysis of amorphous layer formation during ion implantation and its subsequent reanneal at temperatures near 550 °C, approximately half the melting temperature of silicon (1414 °C). Figure B1.24.11 shows a channelling spectrum in a silicon sample, where the outer 4200 Å of the silicon were converted into an amorphous layer by implantation of silicon atoms at liquid nitrogen temperatures [17]. In the spectrum of the as-implanted sample, marked '0 minutes', the yield of the silicon spectra matches that of the random spectra at energies of around 1 to 1.1 MeV. This shows that the implanted amorphous layer has atoms that are displaced from the underlying single crystal silicon. The silicon signal at 0 minutes shows a decrease at around 0.9 MeV. This decrease represents the fraction of channelled ions in the silicon lattice. The yield does not drop to the non-implanted level because the incident helium atoms suffer multiple collisions penetrating through the amorphous layer and their angular distribution is broadened well beyond the critical angle for channelling. The critical angle for channelling of 1 MeV helium ions along the $\langle 100 \rangle$ axis of silicon at room temperature is 0.63 degrees. As the sample is thermally annealed at temperatures above 500 °C, the amorphous layer reorders epitaxially on the silicon substrate. The rear edge of the amorphous spectrum moves towards the surface such that after 30 minutes half of the layer has recrystallized. The yield from the single-crystal silicon behind the implanted layer decreases since fewer of the incident ions suffer multiple collisions sufficient to make their angular distribution exceed

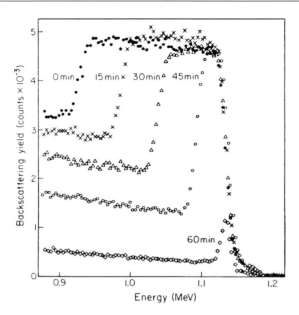

**Figure B1.24.11.** The backscattering yield from an Si sample that has been implanted with Si atoms to form an amorphous layer. Upon annealing this amorphous layer recrystallizes epitaxially leading to a shift in the amorphous/single-crystal interface towards the surface. The aligned spectra have a step between the amorphous and crystal substrate which shifts towards the surface as the amorphous layer epitaxially recrystallizes on the Si.

that of the critical angle. Finally, after 60 minutes annealing, almost all the implanted layer is recrystallized and one is left with a surface peak slightly greater than that in the non-implanted case.

Channelling only requires a goniometer to include the effect in the battery of MeV ion beam analysis techniques. It is not as commonly used as the conventional backscattering measurements because the lattice location of implanted atoms and the annealing characteristics of ion implanted materials is now reasonably well established [18]. Channelling is used to analyse epitaxial layers, but even then transmission electron microscopy is used to characterize the defects.

## B1.24.5    Resonances

At 2 MeV energies the incident helium ion does not penetrate through the barrier around the nucleus. At higher energies and for lighter target atoms such as carbon, nitrogen and oxygen, the helium ion can penetrate and resonances in the cross sections lead to enhanced backscattering yields. This allows one to investigate these target atoms within silicon and even higher mass substrates.

An example of the oxygen resonance cross section is shown in figure B1.24.12, which displays the cross section *versus* energy [19]. The resonance that occurs at 3.04 MeV shows a strong peak. This results in a peak in the backscattering spectra as shown in figure B1.24.13 for 3.05 MeV $He^4$ incident on an $LaCaMnO_3$ film on an $LaAlO_3$ substrate. In the analysis one increases the energy of the beam to move the resonance to increasing depths.

Carbon also has a resonance in its cross section leading to a 100-fold increase in the backscattering signal. This resonance has been very convenient for analysing 1% carbon in silicon–germanium films. The resonance for nitrogen is not as pronounced and has not been used extensively.

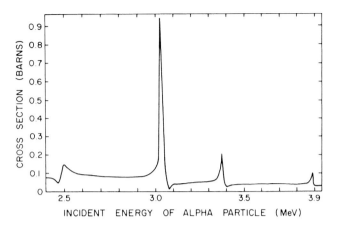

**Figure B1.24.12.** Elastic cross section of helium ions scattered from oxygen atoms. The pronounced peak in the spectrum around 3.04 MeV represents the resonance scattering cross section that is often used in detection of oxygen.

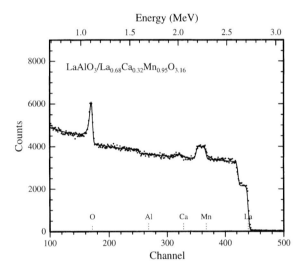

**Figure B1.24.13.** A thin film of $LaCaMnO_3$ on an $LaAlO_3$ substrate is characterized for oxygen content with 3.05 MeV helium ions. The sharp peak in the backscattering signal at channel 160 is due to the resonance in the scattering cross section for oxygen. The solid line is a simulation that includes the resonance scattering cross section and was obtained with RUMP [3]. Data from E B Nyeanchi, National Accelerator Centre, Faure, South Africa.

## B1.24.6   Particle-induced x-ray emission (PIXE)

The PIXE method [20] is based on the spectrometry of the radiation released during the filling of vacancies of inner atomic levels. These vacancies are produced by bombarding a sample with energetic (a few MeV) ions that are normally derived from a high-voltage accelerator. The binding energies of the electrons in the outer layers of the electron shell of an atom are of the order of eV and radiation produced from the rearrangement of electrons in these levels will be in the region of visible wavelength. On the other hand, the binding energies of the inner levels are of the order of keV and radiation produced from processes involving these levels will be in the x-ray region. More importantly, as the electron energy levels of each element are quantized and unique,

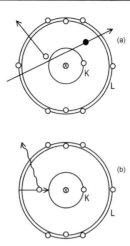

**Figure B1.24.14.** A schematic diagram of x-ray generation by energetic particle excitation. (a) A beam of energetic ions is used to eject inner-shell electrons from atoms in a sample. (b) These vacancies are filled by outer-shell electrons and the electrons make a transition in energy in moving from one level to another; this energy is released in the form of characteristic x-rays, the energy of which identifies that particular atom. The x-rays that are emitted from the sample are measured with an energy dispersive detector.

the measurement of the x-ray energy offers the possibility of determining the presence of a specific element in the sample. Furthermore, the x-ray intensity of a specific energy is proportional to the concentration of the corresponding element in the sample.

The relative simplicity of the method and the penetrative nature of the x-rays, yield a technique that is sensitive to elements with $Z > 10$ down to a few parts per million (ppm) and can be performed quantitatively from first principles. The databases for PIXE analysis programs [21–23] are typically so well developed as to include accurate fundamental parameters, allowing the absolute precision of the technique to be around 3% for major elements and 10–20% for trace elements. A major factor in applying the PIXE technique is that the bombarding energy of the projectiles is a few orders of magnitude more that that of the binding energies of the electrons in the atom and, as the x-rays are produced from the innermost levels, no chemical information is obtained in the process. The advantage of this is that the technique is also not matrix dependent and offers quantitative information regardless of the chemical states of the atoms in the sample. The major application of the technique is the determination of trace element concentrations and, due to the accuracy and non-destructive nature of the technique [24], there are few other techniques that can compete.

The PIXE technique is described schematically in figure B1.24.14. A beam of energetic ions (normally protons of around 3 MeV) is used to eject inner-shell electrons from atoms in a sample. This unstable condition of the atom cannot be maintained and these vacancies are filled by outer-shell electrons. This means that the electrons make a transition in energy in moving from one level to another, and this energy can be released in the form of characteristic x-rays, the energies of which identify the atom. In a competing process, called Auger electron spectroscopy, this energy can also be transferred to another electron that is ejected from the atom and can be detected by an electron detector. Therefore, the step from ionization to x-ray production is not 100% efficient. The x-ray production efficiency is called the fluorescence yield and must be included in the database for quantitative measurements. The x-rays that are emitted from the sample are measured using an energy dispersive detector that has a typical energy resolution around 2.5% (150 eV at 6 keV).

By convention, the transitions filling vacancies in the innermost shell are called K x-rays, those filling the next shell are L x-rays, etc. The energies of L x-rays are normally much lower than those of K x-rays,

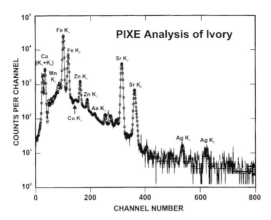

**Figure B1.24.15.** An example of a PIXE spectrum. This spectrum was obtained from the analysis of a piece of ivory to establish whether the origin of the ivory could be determined from trace element concentrations. The spectrum shows the contribution from the different elements, also showing the Ca yield originating from the Ca-rich matrix of the ivory. In this case an 80 $\mu$m Al filter was used to filter most of the x-rays from Ca, as they tend to dominate the spectrum. As interest was focused on the higher-energy part of the spectrum (the higher-energy x-rays are typically not absorbed as much as those of low energy through the same filter), this enabled better sensitivities for the heavier elements to be obtained. The x-ray peaks are situated on a continuous background of *bremsstrahlung* from the projectiles and secondary electrons and, typically, PIXE software programs perform non-linear iterative procedures to obtain accurate information on peaks and this background.

and similarly M x-rays have much lower energies than L x-rays. Due to the structure of the electron shells, there are naturally more possible transitions yielding L x-rays and even more possibilities of yielding M x-rays; therefore it becomes more complex to measure the higher-order x-rays. Typically, the analytical method is limited to K, L and M x-rays. The limitation of detecting elements with $Z > 10$ is due to the low energies of x-rays from the light elements that are absorbed before reaching the detector. The high yield of low-energy x-rays that originate from the major elements of a sample can be eliminated by a filter in front of the detector. Although the stopping of the bombarding ion is depth dependent, the measured x-ray energy gives no direct indication of the depth at which it was produced, and therefore the technique does not provide depth distribution information.

Typically, PIXE measurements are performed in a vacuum of around $10^{-4}$ Pa, although they can be performed in air with some limitations. Ion currents needed are typically a few nanoamperes and current is normally not a limiting factor in applying the technique with a particle accelerator. This beam current also normally leads to no significant damage to samples in the process of analysis, offering a non-destructive analytical method sensitive to trace element concentration levels.

An example of a PIXE spectrum is shown in figure B1.24.15: this spectrum was obtained from the analysis of a piece of ivory to establish whether its source could be determined from trace element concentrations [25]. The spectrum shows the contribution from the different elements, also showing the high Ca yield originating from the Ca-rich matrix of the ivory. In this case an 80 $\mu$m Al filter was used to filter most of the x-rays from Ca, as they tend to dominate the spectrum. As most interest was focused on the higher-energy part of the spectrum (the higher-energy x-rays are typically not absorbed as much as those of low energy through the same filter) this enabled better sensitivities for the heavier elements to be obtained. To maximize the sensitivity and statistical accuracy, the yields from all the K- or L-shell x-rays from an element are used together to determine the concentration for each element [21]. As can be seen in the figure, the x-ray peaks are situated on a continuous background due to *bremsstrahlung* of the projectiles and secondary electrons and, typically, PIXE software programs perform non-linear iterative procedures to obtain accurate information on peaks and this background [26].

### B1.24.7   Nuclear microprobe (NMP)

A microbe employs a focused beams of energetic ions, to provide information on the spatial distribution of elements at concentration levels that range from major elements to a few parts per million [27]. The range of techniques available that allowed depth information plus elemental composition to be obtained could all be used in exactly the same way; it simply became possible to obtain lateral information simultaneously.

The basis of the nuclear microprobe (NMP) is a source of energetic ions from a particle accelerator. These ions are then focused using either magnetic or electrostatic lenses to a minimum spot size less than 1 $\mu m^2$. The technology of focusing ions is still at the stage where the resolution of the NMP does not compete with the electron microprobe. A set of deflection plates allows movement or scanning of the ion beam and, using computer control, the position of the bombarding beam is known. At each point of irradiation, analytical data are acquired. In this way, the analytical information obtained can be presented as an image. Naturally, the imaging capabilities of the NMP is limited by the sensitivity of each technique used since an entire spectrum must be collected at each 'pixel' in the image. For example, a $128 \times 128$ pixel image is equivalent to 16 384 single-point analyses. For this reason the analytical techniques used in imaging are mostly limited to PIXE in the case of trace elements, RBS and forward recoil spectrometry (FRS) for depth information and light elements and nuclear reaction analysis (NRA) to the detection of elements at high concentrations. Naturally the NMP can also be used in a point analysis mode, as for example in the case of geological applications [28]. Here the grain sizes that need to be analysed are often of the order of a few microns and a reasonably small bombarding beam is necessary to limit the analysis to a specific grain.

An example of the application of the PIXE technique using the NMP in the imaging mode [29] is shown in figure B1.24.16. The figures show images of the cross section through a root of the *Phaseolus vulgaris L.* plant. In this case the material was sectioned, freeze-dried and mounted in vacuum for analysis. The scales on the right hand sides of the figures indicate the concentrations of the elements presented in ppm by weight. From the figures it is clear that the transports of the elements through the root are very different not only in the cases of the major elements Ca and K, but also in the case of the trace element Zn. These observations offer a wealth of information that is useful to a botanist studying dynamic processes in plant material.

The quantitative imaging capability of the NMP is one of the major strengths of the technique. The advanced state of the databases available for PIXE [21–23] allows also for the analysis of layered samples as, for example, in studying non-destructively the elemental composition of fluid inclusions in geological samples.

The application of RBS is mostly limited to materials applications, where concentrations of elements are fairly high. RBS is specifically well suited to the study of thin film structures. The NMP is useful in studying lateral inhomogeneities in these layers [30] as, for example, in cases where the solid state reaction of elements in the surface layers occur at specific locations on the surfaces. Other aspects, such as lateral diffusion, can also be studied in three-dimensions.

There are some special techniques that can be used with the NMP, specifically scanning transmission ion microscopy (STIM). In this case the bombarding ion beam is allowed to penetrate through a thin sample and the energies of the transmitted ions are measured. As the energy loss of the ions through the sample is directly proportional to the amount of material traversed, the sample 'thickness' can be imaged with very high efficiency. The technique is so efficient because every ion is counted. Beam currents of only a few fA are needed, thus permitting an imaging resolution of about 100 nm. The technique is well suited for the study of biological material where cell structure is easily identifiable due to the thickness differences in different parts of cells. An example is shown of STIM measurements of human oral cancer cells in figure B1.24.17.

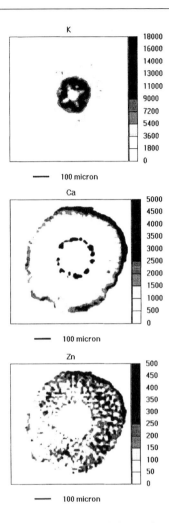

**Figure B1.24.16.** An example of the application of the PIXE technique using the NMP in the imaging mode. The figures show images of the cross section through a root of the *Phaseolus vulgaris L.* plant. In this case the material was sectioned, freeze-dried and mounted in vacuum for analysis. The scales on the right of the figures indicate the concentrations of the elements in ppm by weight. It is clear that the transports of the elements through the root are very different, not only in the cases of the major elements Ca and K, but also in the case of the trace element Zn.

The different images indicate areas of different thickness, starting from thin to thick areas. The technique offers a 'thickness' scan through the sample and, in this case, the cell walls of one specific cell can be seen in the areas dominated by thicker structures. There is relatively little material in the inner areas of the cell.

Another special application of the NMP is the measurement of single event upset (SEU) in memory structures of computer chips [31]. In this technique, a single ion is directed onto a part of the memory structure, with a subsequent measurement of whether the memory bit was changed by the ion impact. In this way, the radiation hardness of different parts of the memory can be imaged. This information is valuable for production of components for space applications, where devices are subjected to high fluxes of ionizing radiation. A modern trend is also to study SEUs in living biological material to detect structures susceptible to radiation damage. A similar technique is the study of ion-beam-induced charge (IBIC) collection [32]

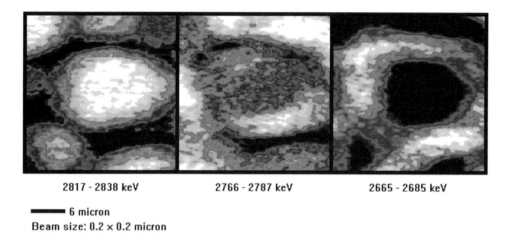

**2817 - 2838 keV**        **2766 - 2787 keV**        **2665 - 2685 keV**

▬▬ **6 micron**
**Beam size: 0.2 × 0.2 micron**

**Figure B1.24.17.** An example of scanning transmission ion microscopy (STIM) measurements of a human oral cancer cell. The different images indicate different windows in the energy of transmitted helium ions as indicated in the figure. White indicate areas of high counts. The technique offers a 'thickness' scan through the sample, and, in this case, the cell walls of one specific cell can be seen in the areas dominated by thicker structures (data from C A Pineda, National Accelerator Centre, Faure, South Africa).

from p–n junctions in semiconductor material. In this case, the ion beam is directed onto a p–n junction and the current flowing through the junction is measured. By rastering the beam an image can be obtained of the quality of these junctions.

### B1.24.8    Forward recoil spectrometry (FRS)

Forward recoil spectrometry (FRS) [33], also known as elastic recoil detection analysis (ERDA), is fundamentally the same as RBS with the incident ion hitting the nucleus of one of the atoms in the sample in an elastic collision. In this case, however, the recoiling nucleus is detected, not the scattered incident ion. RBS and FRS are near-perfect complementary techniques, with RBS sensitive to high-$Z$ elements, especially in the presence of low-$Z$ elements. In contrast, FRS is sensitive to light elements and is used routinely in the detection of H at sensitivities not attainable with other techniques [34]. As the technique is also based on an incoming ion that is slowed down on its inward path and an outgoing nucleus that is slowed down in a similar fashion, depth information is obtained for the elements detected.

The analytically important parameters in FRS are exactly those of RBS. Naturally, the target nucleus can only recoil in the forward direction and, therefore, thick targets must be bombarded at an oblique angle to allow detection of the recoil. Thin targets allow the recoiling nucleus to be transmitted through the target. In the case of thick targets, the incident ion also has a high probability of scattering from the target into the detector. It is common that a filter is applied in front of the detector to remove scattered projectiles. This is possible because the projectile has a higher $Z$ and lower energy and can be stopped while allowing the recoils through to the detector. Because of the kinematics as well as straggling, the depth resolution is somewhat worse than that obtainable with RBS. A simple schematic of the experimental setup for FRS is shown in figure B1.24.18.

The most common use of FRS is the detection of H using $^4$He of a few MeV, with the recoils being detected at 30° with respect to the beam direction and a stopper foil to keep $^4$He from hitting the detector. This set-up can be generalized to include an energy loss ($\Delta E$) detector in front of the detector, thus allowing the

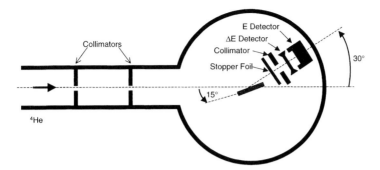

**Figure B1.24.18.** A simple schematic diagram of the experimental setup for FRS. The most common use of the technique is that of hydrogen detection using $^4$He of a few MeV with the recoils being detected at 30° with respect to the beam direction and using a stopper foil to keep $^4$He from hitting the detector. This set-up can be generalized to include an energy loss ($\Delta E$) detector in front of the detector thus allowing, in one experiment, the separation of signals due to hydrogen, deuterium and tritium.

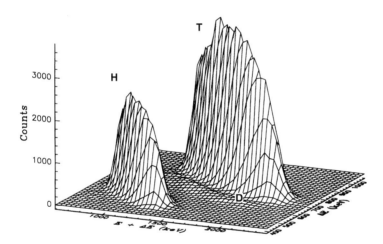

**Figure B1.24.19.** The FRS result of an experiment where a sample was analysed for hydrogen, deuterium and tritium content using a 3.8 MeV $^4$He beam, detecting the recoils at 30° with a 13.6 $\mu$m $\Delta E$ detector. The three-dimensional graph shows a plot of the counts obtained versus $\Delta E$ on the one axis and the total energy ($E + \Delta E$) on the other axis. The traces due to the three isotopes of hydrogen can clearly be seen, with the edges at high $E$ corresponding to the surface of the sample. The shape of the traces from the surface to lower energies are indicative of the depth distribution of these isotopes in the sample.

separation of signals from H, D and T in one experiment. The result of such an experiment [35] is shown in figure B1.24.19, where a sample was analysed for hydrogen, deuterium and tritium content using a 3.8 MeV $^4$He beam, and detecting the recoils at 30° with a 13.6 $\mu$m $\Delta E$ detector. The three-dimensional graph shows a plot of the counts obtained versus $\Delta E$ on the one axis and the energy measured (with a surface barrier detector) after passing through the $\Delta E$ detector on the other axis. The traces due to the three isotopes of hydrogen can clearly be seen, with the edges at high $E$ corresponding to the surface of the sample. The shape of the traces from the surface to lower energies are indicative of the depth distribution of isotopes in the sample.

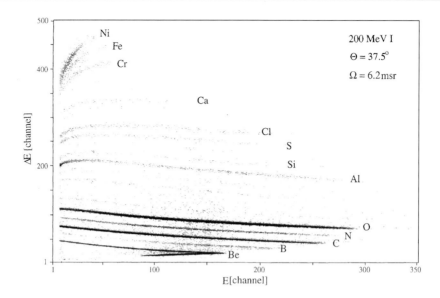

**Figure B1.24.20.** An example of a heavy-ion (iodine) FRS measurement where a large $\Delta E$–$E$ detector telescope was used to enable the discrimination of different elements. The plot shows counts (as intensity plot) versus $\Delta E$ and $E$. The sample analysed was graphite introduced in an experimental nuclear fusion device (tokamak). In this device, a plasma causes different elements to be deposited on the surface of the sample. These elements were determined using a 200 MeV I beam with the detector telescope at 37.5° with respect to the incident beam. It is clear that all the elements from Ni down to Be can be detected in one experiment. The starting points of the traces at high energies indicate concentration of the elements at or near the surface. The trend of the lines as a function of $E$ indicates the concentrations of the elements as a function of depth.

The most advanced applications of the FRS technique employ high-energy (some tens of MeV) heavy ions, such as Cl and I [36]. In this case a number of nuclei lighter than the projectile can be detected. A detector that can separate different nuclei is required. A two-stage detector is used in which either the time of flight or the energy loss $\Delta E$ is determined together with the energy, thus allowing the separation of nuclei with different mass (and $Z$ in the case of $\Delta E$).

An example of such a measurement is shown in figure B1.24.20, where a $\Delta E$–$E$ detector telescope was used to discriminate between different elements. When using heavy ions as incident particles in the analysis of surface layers, care must be taken not to damage the surface.

# References

[1]   Feldman L C and Mayer J W 1986 *Fundamentals of Surface and Thin Film Analysis* (Amsterdam: Elsevier)
[2]   Tesmer J R and Nastasi M (eds) 1995 *Handbook of Modern Ion Beam Materials Analysis* (Pittsburgh, PA: Materials Research Society)
[3]   Doolittle L R 1986 A semiautomatic algorithm for Rutherford backscattering analysis *Nucl. Instrum. Methods* B **15** 227
[4]   Saarilahti J and Rauhala E 1992 Interactive personal-computer data analysis of ion backscattering spectra *Nucl. Instrum. Methods* B **64** 734
[5]   Leavitt J A, McIntyre L C Jr and Weller M R 1995 Backscattering spectrometry *Handbook of Modern Ion Beam Materials Analysis* ed J R Tesmer and M Nastasi (Pittsburgh, PA: Materials Research Society) ch 4
[6]   Theron C C 1997 In situ, real-time characterization of solid-state reaction in thin films *PhD Thesis* University of Stellenbosch
[7]   Kissinger H E 1957 Reaction kinetics in differential thermal analysis *Anal. Chem.* **29** 1702
[8]   Mittemeijer E J, Cheng L, der Schaaf P J V, Brakman C M and Korevaar B M 1988 Analysis of nonisothermal transformation kinetics; tempering of iron–carbon and iron–nitrogen martensites *Metall. Trans.* A **19** 925

[9] Mittemeijer E J, Gent A V and der Schaaf P J V 1986 Analysis of transformation kinetics by nonisothermal dilatometry *Metall. Trans.* A **17** 1441

[10] Colgan E G 1995 Activation energy for $Pt_2Si$ and PtSi formation measured over a wide range of ramp rates *J. Mater. Res.* **10** 1953

[11] Colgan E G 1996 Activation energy for $Ni_2Si$ and NiSi formation measured over a wide range of ramp rates *Thin Solid Films* **279** 193

[12] Kirkendall E O 1942 Diffusion of zinc in $\alpha$-brass *Trans. Metall. Soc. AIME* **147** 104

[13] Smigelskas A D and Kirkendall E O 1947 Zn diffusion in $\alpha$-brass *Trans. Metall. Soc. AIME* **171** 130

[14] Theron C C, Mars J A, Churms C L, Farmer J and Pretorius R 1998 *In situ*, real-time RBS measurement of solid state reaction in thin films *Nucl. Instrum. Methods* B **139** 213

[15] Feldman L C, Mayer J W and Picraux S T 1982 *Materials Analysis by Ion Channelling* (New York: Academic)

[16] Swanson M L 1995 Channelling *Handbook of Modern Ion Beam Materials Analysis* ed J R Tesmer and M Nastasi (Pittsburgh, PA: Materials Research Society) ch 10, p 231

[17] Csepregi L, Kennedy E F, Gallagher T J, Mayer J W and Sigmon T W 1978 Substrate orientation dependence of the epitaxial regrowth rate from Si-implanted amorphous Si *J. Appl. Phys.* **49** 3906

[18] Rimini E 1995 *Ion Implantation: Basics to Device Fabrication* (Boston, MA: Kluwer)

[19] Cox R P, Leavitt J A and McIntyre L C Jr 1995 Non-Rutherford elastic backscattering cross sections *Handbook of Modern Ion Beam Materials Analysis* ed J R Tesmer and M Nastasi (Pittsburgh, PA: Materials Research Society) ch A7, p 481

[20] Johannson T B, Akselsson K R and Johannson S A E 1970 X-ray analysis: elemental trace analysis at the $10^{-12}$ g level *Nucl. Instrum. Methods* **84** 141

[21] Ryan C G, Cousins D R, Sie S H, Griffin W L, Suter G F and Clayton E 1990 Quantitative PIXE microanalysis of geological material using the CSIRO proton microprobe *Nucl. Instrum. Methods* B **47** 55

[22] Ryan C G and Jamieson D N 1993 Dynamic analysis—online quantitative PIXE microanalysis and its use in overlap-resolved elemental mapping *Nucl. Instrum. Methods* B **77** 203

[23] Maxwell J A, Campbell J L and Teesdale W J 1989 The Guelph PIXE software package *Nucl. Instrum. Methods* B **43** 218

[24] Campbell J L, Russell S B, Faiq S, Sculte C W, Ollerhead R W and Gingerich R R 1981 Optimization of PIXE sensitivity for biomedical applications *Nucl. Instrum. Methods* **181** 285

[25] Prozesky V M, Raubenheimer E R, van Heerden W F P, Grotepass W P, Przybyłowicz W J, Pineda C A and Swart R 1995 Trace element concentration and distribution in ivory *Nucl. Instrum. Methods* B **104** 638

[26] Vekemans B, Jannsens K, Vincze L, Adams F and van Espen F 1995 Comparison of several background compensation methods useful for evaluation of energy-dispersive x-ray fluorescence spectra *Spectrochim. Acta* B **50** 149

[27] Watt F, Grime G W and Hilger A *Principles and Applications of High-Energy Ion Microbeams* (Bristol: Institute of Physics)

[28] Ryan C G 1995 The nuclear microprobe as a probe of earth structure and geological processes *Nucl. Instrum. Methods* B **104** 69

[29] van As J A, Jooste J H, Mesjasz-Przybyłowicz J and Przybyłowicz W J 1995 Nuclear microprobe studies of the mechanisms of Zn uptake *Phaseolus vulgaris L., Microsc. Soc. of Southern Africa—Proceedings* **25** 34

[30] de Waal H S, Pretorius R, Prozesky V M and Churms C L 1997 The study of voids in the Au–Al thin-film system using the nuclear microprobe *Nucl. Instrum. Methods* B **130** 722

[31] Breese M B H, Jamieson D N and King P J C 1996 *Materials Analysis with a Nuclear Microprobe* (New York: Wiley)

[32] Jamieson D N 1998 Structural and electrical characterization of semiconductor materials using a nuclear microprobe *Nucl. Instrum. Methods* B **136–138** 1

[33] Tirira J, Serruys Y and Trocellier P 1996 *Forward Recoil Spectrometry —Applications to Hydrogen Determination in Solids* (New York: Plenum)

[34] Sweeney R J, Prozesky V M, Churms C L, Padayachee J and Springhorn K 1998 Application of a $\Delta E{-}E$ telescope for sensitive ERDA measurement of hydrogen *Nucl. Instrum. Methods* B **136–138** 685

[35] Prozesky V M, Churms C L, Pilcher J V and Springhorn K A 1994 ERDA measurement of hydrogen isotopes with a $\Delta E{-}E$ telescope *Nucl. Instrum. Methods* B **84** 373

[36] Behrisch R, Prozesky V M, Huber H and Assmann W 1996 Hydrogen desorption induced by heavy-ions during surface analysis with ERDA *Nucl. Instrum. Methods* B **118** 262

## Further Reading

Chu W K, Mayer J W and Nicolet M-A 1978 *Backscattering Spectrometry* (New York: Academic)

> Comprehensive and detailed coverage of Rutherford backscattering and channelling.

Feldman L C, Mayer J W and Picraux S T 1982 *Materials Analysis by Ion Channelling* (New York: Academic)

> Fundamental treatment suitable for both graduate students and researchers.

Feldman L C and Mayer J W 1986 *Fundamentals of Surface and Thin Film Analysis* (New York: Elsevier)

> General coverage of analytical techniques suitable for undergraduates, graduate students and researchers.

Tesmer J R and Nastasi M (ed) 1995 *Handbook of Modern Ion Beam Materials Analysis* (Pittsburgh, PA: Materials Research Society)

This comprehensive handbook covers all aspects of ion beam analysis from energy loss to radiological safety. It is valuable for the researcher.

Johanssen S A E and Campbell J L 1988 *PIXE: A Novel Technique for Elemental Analysis* (Chichester: Wiley)

Covers PIXE in detail and is a good reference for graduate students and researchers.

Tirira J, Serruys Y and Trocellier P 1996 *Forward Recoil Spectrometry* (New York: Plenum)

A clear description of applications to hydrogen determination in solids for students and researchers.

Watt F, Grime G W and Hilger A (ed) 1987 *Principles and Applications of High-Energy Ion Microbeams* (Bristol: Institute of Physics)

Fundamental description of all aspects of nuclear microprobes for students and researchers.

# B1.25
# Surface chemical characterization

*L Coulier and J W Niemantsverdriet*

## B1.25.1 Introduction

Chemical characterization of surfaces plays an important role in various fields of physics and chemistry, e.g. catalysis, polymers, metallurgy and organic chemistry. This section briefly describes the concepts and a few examples of the techniques that are most frequently used for chemical surface characterization, which are x-ray photoelectron spectroscopy (XPS), Auger electron spectroscopy (AES), ultraviolet photoelectron spectroscopy (UPS), secondary ion mass spectrometry (SIMS), temperature programmed desorption (TPD) and electron energy loss spectroscopy (EELS), respectively. We have tried to give examples in a broad range of fields. References to more extensive treatments of these techniques and others are given at the end of the section, see 'Further reading'.

Although the techniques described undoubtedly provide valuable results on various materials, the most useful information almost always comes from a combination of several (chemical and physical) surface characterization techniques. Table B1.25.1 gives a short overview of the techniques described in this chapter.

## B1.25.2 Electron spectroscopy (XPS, AES, UPS)

### B1.25.2.1 X-ray photoelectron spectroscopy (XPS)

X-ray photoelectron spectroscopy (XPS) is among the most frequently used surface chemical characterization techniques. Several excellent books on XPS are available [1–7]. XPS is based on the photoelectric effect: an atom absorbs a photon of energy $h\nu$ from an x-ray source; next, a core or valence electron with binding energy $E_b$ is ejected with kinetic energy (figure B1.25.1):

$$E_k = h\nu - E_b - \varphi \tag{B1.25.1}$$

where $E_k$ is the kinetic energy of the photoelectron, $h$ is Planck's constant, $\nu$ is the frequency of the exciting radiation, $E_b$ is the binding energy of the photoelectron with respect to the Fermi level of the sample and $\varphi$ is the work function of the spectrometer. Routinely used x-ray sources are Mg K$\alpha$ ($h\nu = 1253.6$ eV) and Al K$\alpha$ ($h\nu = 1486.3$ eV). In XPS one measures the intensity of photoelectrons $N(E)$ as a function of their kinetic energy $E_k$. The XPS spectrum is a plot of $N(E)$ versus $E_b$ ($= h\nu - E_k - \varphi$).

Photoelectron peaks are labelled according to the quantum numbers of the level from which the electron originates. An electron coming from an orbital with main quantum number $n$, orbital momentum $l$ (0, 1, 2, 3, ... indicated as s, p, d, f, ...) and spin momentum $s$ (+1/2 or −1/2) is indicated as $nl_{l+s}$. For every orbital momentum $l > 0$ there are two values of the total momentum: $j = l + 1/2$ and $j = l - 1/2$, each state filled with $2j + 1$ electrons. Hence, most XPS peaks come in doublets and the intensity ratio of the components is $(l+1)/l$. When the doublet splitting is too small to be observed, the subscript $l + s$ is omitted.

**Table B1.25.1.** Overview of the surface characterization techniques described in this chapter.

| Acronym | Full name | Principle of measurement | Key information |
|---|---|---|---|
| XPS | X-ray photoelectron spectroscopy | Absorption of a photon by an atom, followed by the ejection of a core or valence electron with a characteristic binding energy. | Composition, oxidation state, dispersion |
| AES | Auger electron spectroscopy | After the ejection of an electron by absorption of a photon, an atom stays behind as an unstable ion, which relaxes by filling the hole with an electron from a higher shell. The energy released by this transition is taken up by another electron, the Auger electron, which leaves the sample with an element-specific kinetic energy. | Surface composition, depth profiles |
| UPS | UV photoelectron spectroscopy | Absorption of UV light by an atom, after which a valence electron is ejected. | Chemical bonding, work function |
| SIMS | Secondary ion mass spectroscopy | A beam of low-energy ions impinges on a surface, penetrates the sample and loses energy in a series of inelastic collisions with the target atoms leading to emission of secondary ions. | Surface composition, reaction mechanism, depth profiles |
| TPD | Temperature programmed desorption | After pre-adsorption of gases on a surface, the desorption and/or reaction products are measured while the temperature increases linearly with time. | Coverages, kinetic parameters, reaction mechanism |
| EELS | Electron energy loss spectroscopy | The loss of energy of low-energy electrons due to excitation of lattice vibrations. | Molecular vibrations, reaction mechanism |

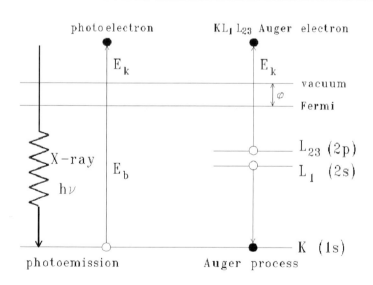

**Figure B1.25.1.** Photoemission and Auger decay: an atom absorbs an incident x-ray photon with energy $h\nu$ and emits a photoelectron with kinetic energy $E_k = h\nu - E_b$. The excited ion decays either by the indicated Auger process or by x-ray fluorescence.

Figure B1.25.2 shows the XPS spectra of two organoplatinum complexes which contain different amounts of chlorine. The spectrum shows the peaks of all elements expected from the compounds, the Pt 4f and 4d

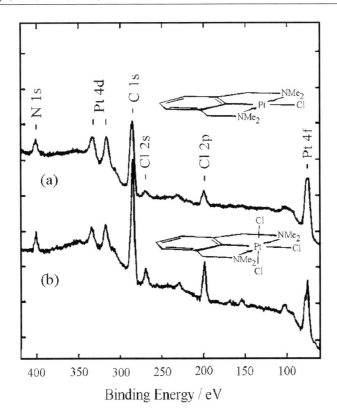

**Figure B1.25.2.** XPS scans between 0 and 450 eV of two organoplatinum complexes showing peaks due to Pt, Cl, N and C. The C 1s signal represents not only carbon in the compound but also contaminant hydrocarbon fragments, as in any sample. The abbreviation 'Me' in the structures stands for $CH_3$ (courtesy of J C Muijsers, Eindhoven).

doublets (the 4f doublet is unresolved due to the low energy resolution employed for broad energy range scans), Cl 2p and Cl 2s, N 1s and C 1s. However, the C 1s cannot be taken as characteristic for the complex only. All surfaces that have not been cleaned by sputtering or oxidation in the XPS spectrometer contain carbon. The reason is that adsorbed hydrocarbons from the atmosphere give the optimum lowering of the surface free energy and hence, all surfaces are covered by hydrocarbon fragments [9].

Because a set of binding energies is characteristic for an element, XPS can analyse chemical composition. Almost all photoelectrons used in laboratory XPS have kinetic energies in the range of 0.2 to 1.5 keV, and probe the outer layers of the sample. The mean free path of electrons in elemental solids depends on the kinetic energy. Optimum surface sensitivity is achieved with electrons at kinetic energies of 50–250 eV, where about 50% of the electrons come from the outermost layer.

Binding energies are not only element specific but contain chemical information as well: the energy levels of core electrons depend on the chemical state of the atom. Chemical shifts are typically in the range 0–3 eV. In general, the binding energy increases with increasing oxidation state and, for a fixed oxidation state, with the electronegativity of the ligands. Figure B1.25.3 illustrates the sensitivity of XPS binding energy to oxidation states for platinum in metal, and in the two organoplatinum complexes of figure B1.25.2. The Pt $4f_{7/2}$ peak of the metal comes at 71.0 eV, that of the complex where Pt has one Cl ligand at 72.0 eV, characteristic of $Pt^{2+}$, while the binding energy of the $Pt^{4+}$ in the complex with three Cl ligands on platinum is again 2 eV higher, 74.4 eV [9]. The binding energy goes up with the oxidation state of the platinum.

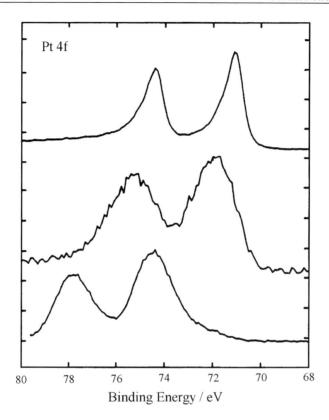

**Figure B1.25.3.** Pt 4f XPS spectra of platinum metal (top) and of the two organoplatinum compounds (a) and (b), middle and bottom respectively, shown in figure B1.25.2, illustrating that the Pt 4f binding energy reflects the oxidation state of platinum (from [9]).

The reason is that the 74 electrons in the $Pt^{4+}$ ion (lower curve) feel a higher attractive force from the nucleus with a positive charge of $78^{+}$ than the 76 electrons in $Pt^{2+}$ (middle curve) or the 78 in the neutral Pt atom (upper curve).

Note that XPS measures binding energies. These are not necessarily equal to the energy of the orbitals from which the photoelectron is emitted. The difference is caused by reorganization of the remaining electrons when an electron is removed from an inner shell. Thus, the binding energy of the photoelectron contains both information on the state of the atom before photoionization (the initial state) and on the core-ionized atom left behind after the emission of an electron (the final state) [6]. Fortunately, it is often correct to interpret binding energy shifts in terms of initial state effects.

Determining compositions is possible if the distribution of elements over the outer layers of the sample and the surface morphology is known. Two limiting cases are considered, namely a homogeneous composition throughout the outer layers and an arrangement in which one element covers the other.

For homogeneous mixed samples it is relatively easy to determine the relative concentrations of the various constituents. For two elements one has approximately:

$$n_1/n_2 = (I_1/S_1)/(I_2/S_2) \qquad (B1.25.2)$$

where $n_1/n_2$ is the ratio of elements 1 and 2, $I_1$, $I_2$ are the intensities of the peaks of elements 1 and 2 (i.e. the area of the peaks) and $S_1$, $S_2$ are atomic sensitivity factors which are tabulated [8].

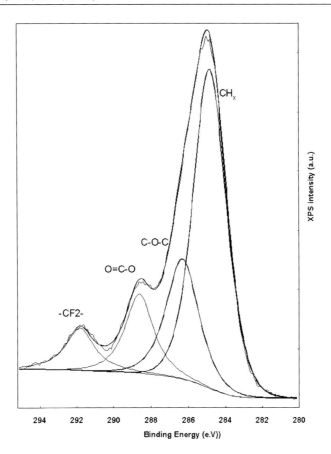

**Figure B1.25.4.** C 1s XPS spectrum of a polymer, illustrating that the C 1s binding energy is influenced by the chemical environment of the carbon. The spectrum clearly shows four different kinds of carbon, which corresponds well with the structure of the polymer (courtesy of M W G M Verhoeven, Eindhoven).

A more accurate calculation will account for differences in the energy dependent mean free paths of the elements and for the transmission characteristics of the electron analyser (see [7]).

An example in which XPS is used for studying surface compositions and oxidation states is illustrated in figures B1.25.2 and B1.25.3 for two organoplatinum complexes. The samples were prepared for XPS by letting a solution of the complexes in dichloromethane dry on a stainless steel sample stub. The sample should thus be homogeneous and the use of (B1.25.2) permitted. Figure B1.25.2 shows the wide scan up to a binding energy of 450 eV [9]. The figure shows immediately that the Cl peaks in the spectrum of the trichloride complex are about three times as intense as in the spectrum of the compound with one Cl. If we apply (B1.25.2) for the elements Pt, N and Cl, we obtain Pt:N:Cl = 1:1.9:4 for the trichloride complex, close to the true stoichiometry of 1:2:3.

In the case of metal particles distributed on a support material (e.g. supported catalysts), XPS yields information on the dispersion. A higher metal/support intensity ratio (at the same metal content) indicates a better dispersion [3].

Another good example of the application of XPS in a different field of chemistry is shown in figure B1.25.4. This figure shows the C 1s spectrum of a polymer [11]. Four different carbon species can be distinguished. Reference tables indicate that the highest binding energy peak is due to carbon–fluorine species [7, 8]. The other

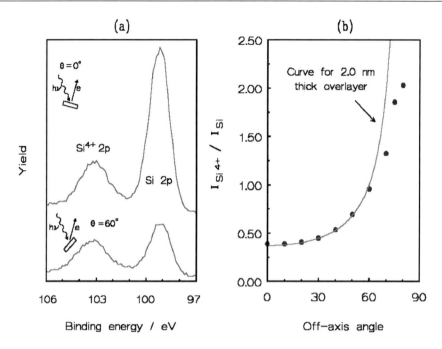

**Figure B1.25.5.** (a) XPS spectra at take-off angles of 0° and 60° as measured from the surface normal from a silicon crystal with a thin layer of SiO₂ on top. The relative intensity of the oxide signal increases significantly at higher take-off angles, illustrating that the surface sensitivity of XPS increases. (b) Plot of Si⁴⁺/Si 2p peak areas as a function of take-off angle. The solid line is a fit which corresponds to an oxide thickness of 2.0 nm (from [12]).

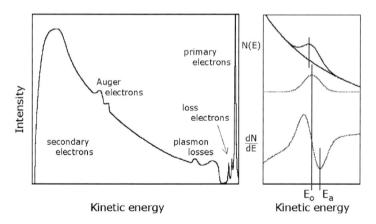

**Figure B1.25.6.** Energy spectrum of electrons coming off a surface irradiated with a primary electron beam. Electrons have lost energy to vibrations and electronic transitions (loss electrons), to collective excitations of the electron sea (plasmons) and to all kinds of inelastic process (secondary electrons). The element-specific Auger electrons appear as small peaks on an intense background and are more visible in a derivative spectrum.

three peaks are attributed (from high to low binding energy) to an ester species, an ether species and hydro-carbon/benzene fragments, respectively [8]. Hence, the carbon XPS spectrum nicely reflects the structure of the polymer. This example is also a nice illustration of the influence of the electronegativity of the ligands

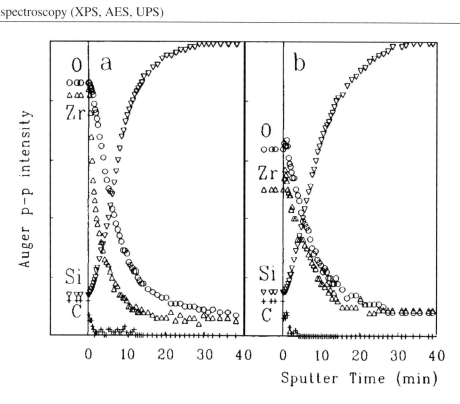

**Figure B1.25.7.** Auger sputter depth profile of a layered $ZrO_2/SiO_2/Si$ model catalyst. While the sample is continuously bombarded with argon ions that remove the outer layers of the sample, the Auger signals of Zr, O, Si and C are measured as a function of time. The depth profile is a plot of Auger peak intensities against sputter time. The profile indicates that the outer layer of the model catalyst contains carbon. Next Zr and O are sputtered away, but note that oxygen is also present in deeper layers where Zr is absent. The left-hand pattern is characteristic for a layered structure, and confirms that the zirconium is present in a well dispersed layer over the silicon oxide. The right-hand pattern is consistent with the presence of zirconium oxide in larger particles (from [27]).

on the binding energy of carbon: the binding energy of carbon increases with the electronegativity of the ligands.

Owing to the limited escape depth of photoelectrons, the surface sensitivity of XPS can be enhanced by placing the analyser at an angle to the surface normal (the so-called take-off angle of the photoelectrons). This can be used to determine the thickness of homogeneous overlayers on a substrate.

This is demonstrated by the XPS spectra in figure B1.25.5(a), which show the Si 2p spectra of a silicon crystal with a thin (native) oxide layer, measured under take-off angles of 0° and 60° [12]. When the take-off angle is high, relatively more photoelectrons from the oxide surface region reach the analyser and the $Si^{4+}/Si$ intensity ratio increases with increasing angle. Figure B1.25.5(b) shows the intensity ratio as a function of take-off angle, the line being a fit corresponding to a flat, homogeneous oxide layer with a uniform thickness of 2 nm.

An experimental problem in XPS is that electrically insulating samples may charge up during measurement, due to photoelectrons leaving the sample. Since the sample thereby acquires a positive charge, all XPS peaks in the spectrum shift by the same amount to higher binding energies. More serious than the shift itself is that different parts of the sample may acquire slightly different amounts of charge. This phenomenon, called differential charging, gives rise to broadening of the peaks and degrades the resolution. Correction for charging-induced shifts is made by using the binding energy of a known compound (in most cases one uses

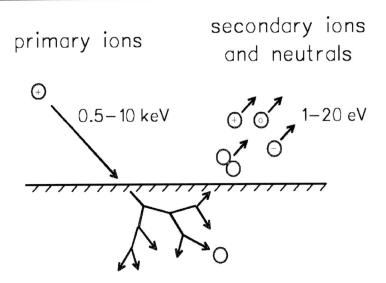

**Figure B1.25.8.** The principle of SIMS: Primary ions with an energy between 0.5 and 10 keV cause a collisional cascade below the surface of the sample. Some of the branches end at the surface and stimulate the emission of neutrals and ions. In SIMS, the secondary ions are detected directly with a mass spectrometer.

the C 1s binding energy of 284.6 eV). Alternatively, in certain circumstances, a low energy electron beam can be used to neutralize the charged surface and eliminate the shift.

Sensitive materials, such as metal salts or organometallic compounds, may decompose during XPS analysis, particularly when a standard x-ray source is used. Apart from the x-rays themselves, heat and electrons from the source may cause damage to the samples. In such cases, a monochromated x-ray source can offer a solution [9]. Damage is in particular an issue in imaging XPS, where the x-ray intensity is focused in a narrow spot. In this mode, a small hole in front of the analyser entrance enables one to select electrons from an area of a few micrometres, such that an image of the surface composition can be made.

### B1.25.2.2  Auger electron spectroscopy (AES)

Auger electron spectroscopy is a powerful technique in the fields of materials and surface science [2–4, 7]. In AES, core holes are created by exciting the sample with a beam of electrons. The excited ion relaxes by filling the core hole with an electron from a higher shell. The energy released by this transition is taken up by another electron, the Auger electron, which leaves the sample with an element-specific kinetic energy (figure B1.25.1). The Auger electrons appear as small peaks on a high background of secondary electrons, scattered by the sample from the incident beam. To enhance the visibility of the Auger peaks, spectra are usually presented in the derivative ($dN/dE$) mode, see figure B1.25.6.

Auger peaks are labelled according to the x-ray level nomenclature. For example, $KL_1L_2$ stands for a transition in which the initial core hole in the K shell is filled from the $L_1$ shell, while the Auger electron is emitted from the $L_2$ shell. Valence levels are indicated by 'V' as in the KVV transitions of carbon or oxygen. The energy of an Auger electron formed in a KLM transition is to a good approximation given by

$$E_{KLM} = E_K - E_L - E_M - \delta E - \varphi \tag{B1.25.3}$$

where $E_{KLM}$ is the kinetic energy of the Auger electron, $E_i$ is the binding energy of an electron in the $i$ shell ($i = K, L, M, \ldots$), $\delta E$ is the energy shift caused by relaxation effects and $\varphi$ is the work function of the

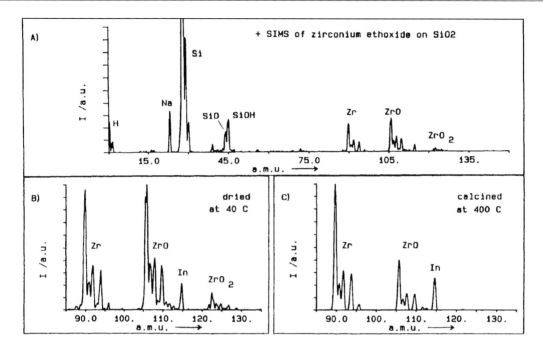

**Figure B1.25.9.** Positive SIMS spectra of a $ZrO_2/SiO_2$ catalyst, (a) after preparation from $Zr(OC_2H_5)$, (b) after drying at $40\,°C$ and (c) after calcination in air at $400\,°C$ (from [17]).

spectrometer. The $\delta E$ term accounts for the relaxation effect involved in the decay process, which leads to a final state consisting of a heavily excited, doubly ionized atom.

Auger peaks also appear in XPS spectra. In this case, the x-ray ionized atom relaxes by emitting an electron with a specific kinetic energy $E_k$. One should bear in mind that in XPS the intensity is plotted against the binding energy, so one uses (B1.25.1) to convert to kinetic energy.

The strong point of AES is that it provides a quick measurement of elements in the surface region of conducting samples. For elements having Auger electrons with energies in the range of 100–300 eV where the mean free path of the electrons is close to its minimum, AES is considerably more surface sensitive than XPS.

Auger electron spectroscopy allows for three types of measurement. First, it provides the elemental surface composition of a sample. If the Auger decay process involves valence electrons, one often obtains information on the oxidation state as well, although XPS is certainly the better technique for this purpose. Second, owing to the short data collection times, AES can be combined with sputtering to measure concentrations as a function of depth, see figure B1.25.7. Third, as electron beams are easily collimated and deflected electrostatically, AES can be used to image the composition into a chemical map of the surface (scanning Auger spectroscopy). The best obtained resolution is now around 25 nm [7].

A disadvantage of AES is that the intense electron beam easily causes damage to sensitive materials (polymers, insulators, adsorbate layers). Charging of insulating samples also causes serious problems.

### B1.25.2.3 Ultraviolet photoelectron spectroscopy (UPS)

Ultraviolet photoelectron spectroscopy (UPS) [2–4, 6] differs from XPS in that UV light (He I, 21.2 eV; He II, 40.8 eV) is used instead of x-rays. At these low exciting energies, photoemission is limited to valence electrons.

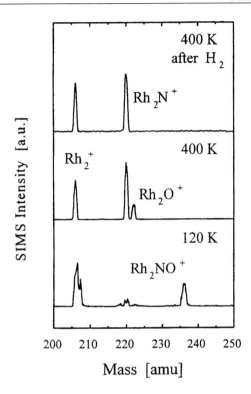

**Figure B1.25.10.** SIMS spectra of the Rh(111) surface after adsorption of 0.12 ML NO at 120 K (bottom), after heating to 400 K (middle) and after reaction with $H_2$ at 400 K (top) (from [19]).

Hence, UPS spectra contain important chemical information. At low binding energies, UPS probes the density of states (DOS) of the valence band (but images it in a distorted way in a convolution with the unoccupied states). At slightly higher binding energies (5–15 eV), occupied molecular orbitals of adsorbed gases may become detectable. UPS also provides a quick measure of the macroscopic work function, $\phi$, the energy separation between the Fermi and the vacuum level: $\phi = h\nu - W$, where $W$ is the width of the spectrum. UPS is a surface science technique typically applied to single crystals, the main reason being that all elements contribute peaks to the valence band region. As a result, the UPS spectra of compounds which contain more than two elements are rather complicated.

### B1.25.3 Secondary ion mass spectrometry (SIMS)

Secondary ion mass spectrometry (SIMS) is by far the most sensitive surface technique, but also the most difficult one to quantify. SIMS is very popular in materials research for making concentration depth profiles and chemical maps of the surface. For a more extensive treatment of SIMS the reader is referred to [3] and [14–16]. The principle of SIMS is conceptually simple: When a surface is exposed to a beam of ions ($Ar^+$, $Cs^+$, $Ga^+$ or other elements with energies between 0.5 and 10 keV), energy is deposited in the surface region of the sample by a collisional cascade. Some of the energy will return to the surface and stimulate the ejection of atoms, ions and multi-atomic clusters (figure B1.25.8). In SIMS, secondary ions (positive or negative) are detected directly with a mass spectrometer.

SIMS is, strictly speaking, a destructive technique, but not necessarily a damaging one. In the dynamic mode, used for making concentration depth profiles, several tens of monolayers are removed per minute.

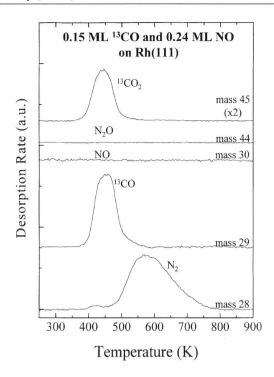

**Figure B1.25.11.** Temperature programmed reactions of 0.15 ML of $^{13}CO$ coadsorbed with 0.24 ML of NO. Adsorption was done at 150 K and the heating rate was 5 K s$^{-1}$ (from [23]).

In static SIMS, however, the rate of removal corresponds to one monolayer per several hours, implying that the surface structure does not change during the measurement (between seconds and minutes). In this case one can be sure that the molecular ion fragments are truly indicative of the chemical structure on the surface.

The advantages of SIMS are its high sensitivity (ppm detection limit for certain elements), its ability to detect hydrogen and the emission of molecular fragments which often bear tractable relationships with the parent structure on the surface. A disadvantage is that secondary ion formation is a poorly understood phenomenon and that quantitation is usually difficult. A major drawback is the matrix effect: Secondary ion yields of one element can vary tremendously with its chemical environment. This matrix effect and the elemental sensitivity variation of five orders of magnitude across the periodic table make quantitative interpretation of SIMS spectra of many compounds extremely difficult.

Figure B1.25.9(a) shows the positive SIMS spectrum of a silica-supported zirconium oxide catalyst precursor, freshly prepared by a condensation reaction between zirconium ethoxide and the hydroxyl groups of the support [17]. Note the simultaneous occurrence of single ions ($H^+$, $Si^{4+}$, $Zr^+$) and molecular ions ($SiO^+$, $SiOH^+$, $ZrO^+$, $ZrOH^+$, $ZrO_2^+$). Also, the isotope pattern of zirconium is clearly visible. Isotopes are important in the identification of peaks, because all peak intensity ratios must agree with the natural abundance. In addition to the peaks expected from zirconia on silica mounted on an indium foil, the spectrum in figure B1.25.9(a) also contains peaks from the contaminants, $Na^+$, $K^+$ and $Ca^+$. This is typical for SIMS: sensitivities vary over several orders of magnitude and elements such as the alkalis are already detected when present in trace amounts.

The most useful information is in the relative intensities of the $Zr^+$, $ZrO^+$, $ZrOH^+$ and $ZrO_2^+$ ions. This is illustrated in figures B1.25.9(b) and (c) which show the isotope patterns of these ions of a freshly dried and a calcined catalyst, respectively [17]. Note that the SIMS spectrum of the fresh catalyst contains small but significant contributions from ZrOH$^+$ ions (107 amu, $^{90}$ZrOH$^+$, and 111 amu, $^{94}$ZrOH$^+$). ZrOH$^+$ is most probably a fragment ion from zirconium ethoxide. In the spectrum of the catalyst which was oxidized at 400 °C, the isotope pattern in the ZrO range resembles that of Zr, indicating that ZrOH species are absent. Spectrum B1.25.9(b) of the zirconium ethoxide (O:Zr = 4:1) shows higher intensities of the ZrO$_2^+$ and ZrO$^+$ signals than the calcined ZrO$_2$ (O:Zr = 2:1) does. The way to interpret this information is to compare the spectra of the catalysts with reference spectra of ZrO$_2$ and zirconium ethoxide [17].

For single crystals, matrix effects are largely ruled out and excellent quantization has been achieved by calibrating SIMS yields by means of other techniques such as EELS and TPD (see further) [18]. Here SIMS offers the challenging perspective to monitor reactants, intermediates and products of catalytic reactions in real time while the reaction is in progress.

A good example of monitoring adsorbed species on surfaces with SIMS is shown in figure B1.25.10. This figure shows positive SIMS spectra of the interaction of NO with the Rh(111) surface [19]. The lower curve shows the adsorption of molecular NO (peak at 236 amu) on Rh(111) at 120 K. The middle curve shows the situation after heating the sample to 400 K. The presence of the peaks at 220 amu (Rh$_2$N$^+$) and 222 amu (Rh$_2$O$^+$) and the absence of the Rh$_2$NO$^+$ (236 amu) indicate that NO has dissociated. Heating the sample at 400 K in H$_2$ causes the removal of atomic oxygen and, thus, the disappearance of the Rh$_2$O$^+$ at 222 amu, as can be seen in the upper curve.

As in Auger spectroscopy, SIMS can be used to make concentration depth profiles and, by rastering the ion beam over the surface, to make chemical maps of certain elements. More recently, SIMS has become very popular in the characterization of polymer surfaces [14–16].

### B1.25.4   Temperature programmed desorption (TPD)

Thermal desorption spectroscopy (TDS) or temperature programmed desorption (TPD), as it is also called, is a simple and very popular technique in surface science. A sample covered with one or more adsorbate(s) is heated at a constant rate and the desorbing gases are detected with a mass spectrometer. If a reaction takes place during the temperature ramp, one speaks of temperature programmed reaction spectroscopy (TPRS).

TPD is frequently used to determine (relative) surface coverages. The area below a TPD spectrum of a certain species is proportional to the total amount that desorbs. In this way one can determine uptake curves that correlate gas exposure to surface coverage. If the pumping rate of the UHV system is sufficiently high, the mass spectrometer signal for a particular desorption product is linearly proportional to the desorption rate of the adsorbate [20, 21]:

$$r = -\mathrm{d}\theta/\mathrm{d}t = k_{\mathrm{des}}\theta^n = \nu(\theta)\theta^n \exp(-E_{\mathrm{des}}(\theta)/RT) \qquad (B1.25.4)$$

$$T = T_0 + \beta t$$

where $r$ is the rate of desorption, $E_{\mathrm{des}}$ is the activation energy of desorption, $\theta$ is the coverage in monolayers, $R$ is the gas constant, $t$ is the time, $T$ is the temperature, $k_{\mathrm{des}}$ is the reaction rate constant for desorption, $T_0$ is the temperature at the start, $n$ is the order of desorption, $\beta$ is the heating rate, equal to $\mathrm{d}T/\mathrm{d}t$ and $\nu$ is the preexponential factor of desorption.

With the aid of (B1.25.4), it is possible to determine the activation energy of desorption (usually equal to the adsorption energy) and the preexponential factor of desorption [21, 24]. Attractive or repulsive interactions between the adsorbate molecules make the desorption parameters $E_{\mathrm{des}}$ and $\nu$ dependent on coverage [22]. In the case of TPRS one obtains information on surface reactions if the latter is rate determining for the desorption.

Figure B1.25.11 shows the temperature programmed reaction between 0.15 ML of CO and 0.24 ML of NO adsorbed at 150 K on a Rh(111) single crystal [23]. The spectra show the desorption of species with masses 28, 29, 30, 44 and 45, corresponding to N$_2$, $^{13}$CO, NO, N$_2$O and $^{13}$CO$_2$ respectively, as functions of the

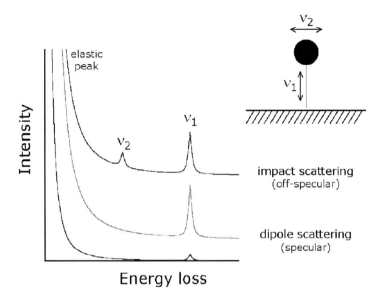

**Figure B1.25.12.** Excitation mechanisms in electron energy loss spectroscopy for a simple adsorbate system: Dipole scattering excites only the vibration perpendicular to the surface ($v_1$) in which a dipole moment normal to the surface changes; the electron wave is reflected by the surface into the specular direction. Impact scattering excites also the bending mode $v_2$ in which the atom moves parallel to the surface; electrons are scattered over a wide range of angles. The EELS spectra show the highly intense elastic peak and the relatively weak loss peaks. Off-specular loss peaks are in general one to two orders of magnitude weaker than specular loss peaks.

temperature. $N_2$, $^{13}CO$ and $^{13}CO_2$ are the only desorption products, indicating that NO is totally dissociated and all N atoms are converted to $N_2$. CO is not decomposed at all; part of it desorbs as CO and part of it reacts with atomic oxygen to $CO_2$. At higher NO coverages (not shown) the TPD spectrum has also a mass 30 signal, which is due to desorption of NO [28]. In this case, the coverage of adsorbed CO is too low to convert all NO to $N_2$ and $CO_2$.

The disadvantage of TPD is that, in order to derive the kinetic parameters, rather involved computations are necessary [21, 24]. As an alternative to the complete desorption analysis, many authors rely on simplified methods. The analysis of spectra using simplified analysis should be made with care, as simplified analysis methods may easily give erroneous results [21].

### B1.25.5  Electron energy loss spectroscopy (EELS)

Molecules possess discrete levels of vibrational energy. Vibrations in molecules can be excited by interaction with waves and with particles. In electron-energy loss spectroscopy (EELS, sometimes HREELS for high resolution EELS), a beam of monochromatic, low energy electrons falls on the surface, where it excites lattice vibrations of the substrate, molecular vibrations of adsorbed species and even electronic transitions. An energy spectrum of the scattered electrons reveals how much energy the electrons have lost to vibrations, according to the formula

$$E = E_0 - h\nu \qquad \qquad (B1.25.5)$$

where $E$ is the energy of the scattered electron, $E_0$ is the energy of the incident electrons, $h$ is Planck's constant and $\nu$ is the frequency of the excited vibration. The use of electrons requires that experiments are

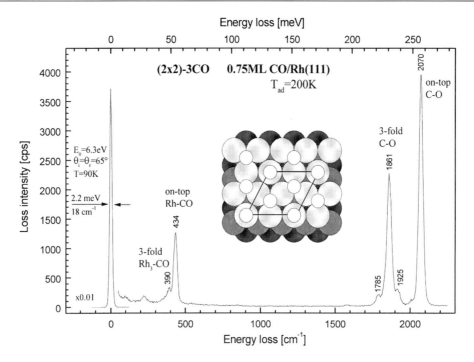

**Figure B1.25.13.** HREELS spectrum of CO adsorbed on Rh(111) at $T = 200$ K. Visible are the C–O vibration peaks at energy losses around 1800–2100 cm$^{-1}$ and the Rh–CO signals at energy losses 300–500 cm$^{-1}$ (courtesy of R Linke [26]).

done in high vacuum and preferably on the flat surfaces of single crystals or foils (making ultrahigh vacuum conditions desirable).

While infrared and Raman spectroscopy are limited to vibrations in which a dipole moment or the molecular polarizability changes, EELS detects all vibrations. Two excitation mechanisms play a role in EELS: dipole and impact scattering [4].

In dipole scattering we are dealing with the wave character of the electron. Close to the surface, the electron sets up an electric field with its image charge in the metal. This oscillating field is perpendicular to the surface and excites only those vibrations in which a dipole moment changes in a direction normal to the surface, similarly as in reflection absorption infrared spectroscopy. The outgoing electron wave has lost an amount of energy equal to $h\nu$, see (B1.25.5), and travels mainly in the specular direction.

The second excitation mechanism, impact scattering, involves a short range interaction between the electron and the molecule (put simply, a collision) which scatters the electrons over a wide range of angles. The useful feature of impact scattering is that all vibrations may be excited and not only the dipole active ones. As in Raman spectroscopy, the electron may also take an amount of energy $h\nu$ away from excited molecules and leave the surface with an energy equal to $E_0 + h\nu$.

Figure B1.25.12 illustrates the two scattering modes for a hypothetical adsorption system consisting of an atom on a metal [3]. The stretch vibration of the atom perpendicular to the surface is accompanied by a change in dipole moment; the bending mode parallel to the surface is not. As explained above, the EELS spectrum of electrons scattered in the specular direction detects only the dipole-active vibration. The more isotropically scattered electrons, however, undergo impact scattering and excite both vibrational modes. Note that the comparison of EELS spectra recorded in specular and off-specular direction yields information about the orientation of an adsorbed molecule.

A strong point of EELS is that it detects losses in a very broad energy range, which comprises the entire infrared regime and extends even to electronic transitions at several electron volts. EELS spectrometers have to satisfy a number of stringent requirements. First, the primary electrons should be monochromatic. Second, the energy of the scattered electrons should be measured with a high accuracy. Third, the low energy electrons must effectively be shielded from magnetic fields [25].

Figure B1.25.13 shows an HREELS spectrum of CO adsorption on a Rh(111) surface [26]. In the experiment 3 L of CO was adsorbed at a pressure of $1 \times 10^{-8}$ mbar and $T = 200$ K. At zero energy loss one observes the highly intense elastic peak. The other peaks in the spectrum are loss peaks. At high energy, loss peaks due to dipole scattering are visible. In this case they are caused by CO vibration perpendicular to the surface. The peak at 2070 cm$^{-1}$ is attributed to on-top adsorption of CO on Rh, while the peak at 1861 cm$^{-1}$ corresponds to CO adsorption on a threefold Rh site. The loss peaks at low energy loss are due to metal–adsorbate vibrations. In this case it is the Rh–CO bond. The peak at 434 cm$^{-1}$ is due to the Rh–CO vibration of the CO adsorbed on top, that at 390 cm$^{-1}$ to the Rh–CO vibration in threefold CO. The CO molecules order in a $(2 \times 2)$–3CO structure, with one linear and two threefold CO molecules per $(2 \times 2)$ unit cell.

## References

[1] Delgass W N, Haller G L, Kellerman R and Lunsford J H 1979 *Spectroscopy in Heterogeneous Catalysis* (New York: Academic)
[2] Feldman L C and Mayer J W 1986 *Fundamentals of Surface and Thin Film Analysis* (Amsterdam: North-Holland)
[3] Niemantsverdriet J W 1993 *Spectroscopy in Catalysis, an Introduction* (Weinheim: VCH)
[4] Ertl G and Kuppers J 1985 *Low Energy Electrons and Surface Chemistry* (Weinheim: VCH)
[5] Ghosh P K 1983 *Introduction to Photoelectron Spectroscopy* (New York: Wiley)
[6] Feuerbacher B, Fitton B and Willis R F (eds) 1978 *Photoemission and the Electronic Properties of Surfaces* (New York: Wiley)
[7] Briggs D and Seah M P (eds) 1983 *Practical Surface Analysis by Auger and X-ray Photoelectron Spectroscopy* (New York: Wiley)
[8] Wagner C D, Riggs W M, Davis L E, Moulder J F and Muilenburg G E 1979 *Handbook of X-ray Photoelectron Spectroscopy* (Eden Prairie, MN: Perkin Elmer)
[9] Muijsers J C, Niemantsverdriet J W, Wehman-Ooyevaar I C M, Grove D M and van Koten G 1992 *Inorg. Chem.* **31** 2655
[10] Somorjai G A 1981 *Chemistry in Two Dimensions, Surfaces* (Ithaca, NY: Cornell University Press)
[11] Verhoeven M W G M and Niemantsverdriet J W, unpublished results
[12] Gunter P L J, de Jong A M, Niemantsverdriet J W and Rheiter H J H 1992 *Surf. Interface Anal.* **19** 161
[13] Haas T W, Grant J T and Dooley G J 1972 *J. Appl. Phys.* **43** 1853
[14] Benninghoven A, Rudenauer F G and Werner H W 1987 *Secondary Ion Mass Spectrometry, Basic Concepts, Instrumental Aspects, Applications and Trends* (New York: Wiley)
[15] Vickerman J C, Brown A and Reed N M (eds) 1989 *Secondary Ion Mass Spectrometry, Principles and Applications* (Oxford: Clarendon)
[16] Briggs D, Brown A and Vickerman J C 1989 *Handbook of Static Secondary Ion Mass Spectrometry* (Chichester: Wiley)
[17] Meijers A C Q M, de Jong A M, van Gruijthuijsen L M P and Niemantsverdriet J W 1991 *Appl. Catal.* **70** 53
[18] Borg H J and Niemantsverdriet J W 1994 *Catalysis, Specialist Periodical Report* vol 11, ed J J Spivey and S K Agarwal (Cambridge: Royal Society of Chemistry) ch 11
[19] van Hardeveld R M, Borg H J and Niemantsverdriet J W 1998 *J. Mol. Catal.* A **131** 199
[20] King D A 1975 *Surf. Sci.* **47** 384
[21] de Jong A M and Niemantsverdriet J W 1990 *Surf. Sci.* **233** 355
[22] Cassuto A and King D A 1981 *Surf. Sci.* **102** 388
[23] Hopstaken M J P, van Gennip W J H and Niemantsverdriet J W 1999 *Surf. Sci.* 433–435
[24] Falconer J L and Schwarz J A 1983 *Catal. Rev. Sci. Eng.* **25** 141
[25] Ibach H 1990 *Electron Energy Loss Spectrometers: the Technology of High Performance* (Berlin: Springer)
[26] Linke R and Niemantsverdriet J W, to be published
[27] Eshelman L M, de Jong A M and Niemantsverdriet J W 1991 *Catal. Lett.* **10** 201

## Further Reading

Ertl G and Kuppers J 1985 *Low Energy Electrons and Surface Chemistry* (Weinheim: VCH)

Feldman L C and Mayer J W 1986 *Fundamentals of Surface and Thin Film Analysis* (Amsterdam: North-Holland)

Woodruff D P and Delchar T A 1986 *Modern Techniques of Surface Science* (Cambridge: Cambridge University Press)

Niemantsverdriet J W 1993 *Spectroscopy in Catalysis, an Introduction* (Weinheim: VCH)

# B1.26
# Surface physical characterization

*W T Tysoe and Gefei Wu*

### B1.26.1 Introduction

The physical structure of a surface, its area, morphology and texture and the sizes of orifices and pores are often crucial determinants of its properties. For example, catalytic reactions take place at surfaces. Simple statistical mechanical estimates suggest that a surface-mediated reaction should proceed about $10^{12}$ times faster than the corresponding gas-phase reaction for identical activation energies [1]. The catalyst operates by lowering the activation energy of the reaction to accelerate the rate. The reaction rate, however, also increases in proportion to the exposed surface area of the active component of the catalyst so that maximizing its area also strongly affects its activity. Catalysts often have complicated morphologies, consisting of exposed regions, and small micro- and meso-pores. A traditional method for measuring these areas, which is still the workhorse for the catalytic chemists, is to titrate the surface with molecules of known 'areas' and to measure the amount that just covers it. This is done by pressurizing the sample using probe gases and gauging when a single layer of adsorbate forms. This relies on developing robust theoretical methods for determining the equilibrium between the gas phase and the surface. This was done in 1938 by Brunauer, Emmett and Teller. Brunauer and Emmett were catalytic chemists and Teller a theoretical physicist who was persuaded to undertake the theoretical task of developing an adsorption isotherm [2]. This he apparently did in one day and the Brunauer–Emmett–Teller isotherm was born. This, with minor modifications, is the isotherm still used today.

On planar systems, morphologies can be generally measured by directly imaging them using optical (sections B1.19 and B1.21) or electron (B1.18) microscopies or using scanning probes (see section B1.20). Coarse morphologies can be measured using crude probes such as profilometers [3] and, more recently, at the atomic level, using atomic force microscopy (section B1.20). The measurement of the thickness and properties of thin films deposited onto planar surfaces is more of a challenge. Electron-based spectroscopic probes can measure the nature of the outer selvedge of planar films (see sections B1.6, B1.7, B1.21). For example, x-ray photoelectron spectroscopy is useful for measuring films of few Ångströms thick using the electron escape depth. The film itself can be probed using optical spectroscopic techniques such as infrared (B1.2) and Raman (B1.3) spectroscopies. Ellipsometry, the change in polarization of linearly polarized light as it passes through the film, is used to measure film thickness non-destructively over a wide range. It is particularly useful for probing surface coatings such as anti-reflection and protective films and has been used very effectively to probe overlayers in ultrahigh vacuum and the formation kinetics of self-assembled monolayers on gold.

The final technique addressed in this chapter is the measurement of the surface work function, the energy required to remove an electron from a solid. This is one of the oldest surface characterization methods, and certainly the oldest carried out *in vacuo* since it was first measured by Millikan using the photoelectric effect [4]. The observation of this effect led to the proposal of the Einstein equation:

$$E_k = h\nu - e\Phi \tag{B1.26.1}$$

where $\nu$ is the light frequency, $h$ is Planck's constant, $E_k$ the kinetic energy of the emitted electron, $e$ the charge on an electron and $\Phi$ the material work function. The resulting notion of wave–particle duality led directly to the development of quantum mechanics. This is not strictly a physical probe since the work function of a clean sample depends on its electronic structure. This is strongly affected by the presence of adsorbates, electronegative adsorbates leading to an increase in work function, and electropositive adsorbates to a decrease. The observations have technological implications, so that filaments used today as electron sources in cathode ray (television) and vacuum tubes (valves) are coated with electropositive alkaline earth compounds that lower the work function and enhance the thermionically emitted current. This allows the filaments to operate effectively at lower temperatures and thereby increases their lifetimes. The main experimental utility of this method is to measure, in a simple and direct way, the coverage of an adsorbate on a surface (see section A1.7).

## B1.26.2 The Brunauer–Emmett–Teller (BET) method

### B1.26.2.1 Principles

#### (a) Measurements of surface area by gas adsorption

The central idea underlying measurements of the area of powders with high surface areas is relatively simple. Adsorb a close-packed monolayer on the surface and measure the number $N$ of these molecules adsorbed per unit mass of the material (usually per gram). If the specific area occupied by each molecule is $A_m$, then the total surface area $S_A$ of the sample is simply given by:

$$S_A = N A_m. \tag{B1.26.2}$$

The saturation coverage during chemisorption on a clean transition-metal surface is controlled by the formation of a chemical bond at a specific site [5] and not necessarily by the area of the molecule. In addition, in this case, the heat of chemisorption of the first monolayer is substantially higher than for the second and subsequent layers where adsorption is *via* weaker van der Waals interactions. Chemisorption is often useful for measuring the area of a specific component of a multi-component surface, for example, the area of small metal particles adsorbed onto a high-surface-area support [6], but not for measuring the *total* area of the sample. Surface areas measured using this method are specific to the molecule that chemisorbs on the surface. Carbon monoxide titration is therefore often used to define the number of 'sites' available on a supported metal catalyst. In order to measure the total surface area, adsorbates must be selected that interact relatively weakly with the substrate so that the area occupied by each adsorbent is dominated by intermolecular interactions and the area occupied by each molecule is approximately defined by van der Waals radii. This generally necessitates experiments being carried out at low temperatures such that $kT \ll \Delta H_{(ads)}$, the heat of adsorption. Since now both the interaction of the first and subsequent absorbate layers is dominated by van der Waals forces, this leads not simply to the formation of a single monolayer, but also to the growth of second, third and subsequent layers. This distinction is shown in figure B1.26.1 which plots coverage versus pressure at constant temperature (an isotherm) for a molecule (hydrogen) which chemisorbs at the surface where the saturation of the monolayer is clearly evident from the appearance of a plateau [7]. In the case of physisorption, as demonstrated in figure B1.26.2, subsequent layers can grow so that the number of molecules adsorbed in the first monolayer is much more difficult to identify [8]. It is clear, in this case, that the first monolayer forms somewhere near the first 'knee' of the isotherm, labelled point 'B'. The importance of this point was first emphasized by Emmett [9]. In order to usefully measure the total surface area, the shape of the adsorption isotherm must be analysed to more clearly distinguish between monolayer (point B) and multilayer adsorption. In 1985,

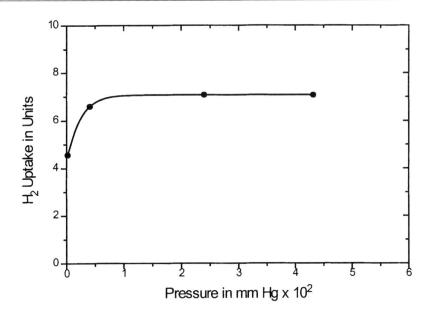

**Figure B1.26.1.** Sorption isotherm for chemisorption of hydrogen on palladium film at 273 K (Stephens S J 1959 *J. Phys. Chem.* **63** 188–94).

IUPAC introduced a classification of six different types of adsorption isotherm [11] (figure B1.26.3) exhibited by real surfaces. Types I–V were originally classified by Brunauer, Denning, Denning and Teller (BDDT) [10]. Type I represents the Langmuir isotherm [12] for monolayer coverage and is most often exhibited for chemisorption where the heat of adsorption in the first layer is much greater than that in subsequent layers, but also corresponds to physisorption by microporous absorbents (pore width <2 nm) within the solid. Type II are monolayer–multilayer isotherms and represent non-porous or macroporous absorbents (pore width >50 nm). Industrial absorbents and catalysts which possess mesoporous (pore width 2–50 nm) structures often exhibit type IV behaviour. The shapes of types III and V isotherms are analogous to type II and IV respectively, but with weak gas–solid interactions. The stepwise type VI isotherms can be obtained with well defined, uniform solids. The origin of the shapes of some of these isotherms will be discussed in greater detail below.

*(b) Typical isotherms for the physical adsorption of gases on surfaces*

As noted above, an isotherm plots the number of molecules adsorbed on the surface at some temperature in equilibrium with the gas at some pressure. Adsorption gives rise to a change in the free energy which, of course, depends on the number of molecules already adsorbed on the surface (defined as the coverage, $\theta$, see section A1.7). If it is assumed for simplicity that the structure of the adsorbed phase is similar to that of a solid (so that, at equilibrium, the chemical potential of the bulk phase equals the chemical potential for the gas phase in equilibrium with it) then:

$$\mu_{\text{bulk}} = \mu^0 + RT \ln P_0 \tag{B1.26.3}$$

where $P_0$ is the equilibrium vapour pressure. The chemical potential of the adsorbed phase in equilibrium with the gas at some temperature $T$ can similarly be written as:

$$\mu_{\text{ads}} = \mu^0 + RT \ln P. \tag{B1.26.4}$$

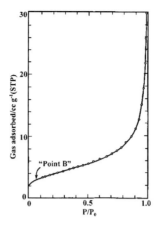

**Figure B1.26.2.** Adsorption isotherm for nitrogen on anatase at 77 K showing 'point B' (Harkins W D 1952 *The Physical Chemistry of Surface Films* (New York: Reinhold)).

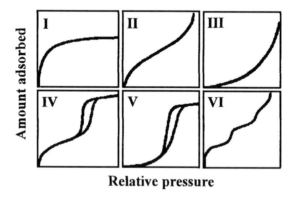

**Figure B1.26.3.** The IUPAC classification of adsorption isotherms for gas–solid equilibria (Sing K S W, Everett D H, Haul R A W, Mosoul L, Pierotti R A, Rouguerol J and Siemieniewska T 1985 *Pure. Appl. Chem.* **57** 603–19).

We assume for simplicity that the adsorbed phase has the same entropy as the solid so that only an energy change is associated with the transfer of material from the bulk to the adsorbed phase, then:

$$\Delta F = \mu_{ads} - \mu_{bulk} = N\Delta E \qquad (B1.26.5)$$

where $\Delta E$ is the change in energy per adsorbed atom or molecule in going from the solid to the adsorbed phase. Combining equations (B1.26.3)–(B1.26.5) yields:

$$\ln \frac{P}{P_0} = \frac{\Delta E}{kT}. \qquad (B1.26.6)$$

If we knew the variation in $\Delta E$ as a function of coverage $\theta$, this would be the equation for the isotherm. Typically the energy for physical adsorption in the first layer, $-\Delta E_1$, when adsorption is predominantly through van der Waals interactions, is of the order of $10kT$ where $T$ is the temperature and $k$ the Boltzmann constant, so that, according to equation (B1.26.6), the first layer condenses at a pressure given by $P/P_0 \sim 10^{-3}$. This accounts for the rapid initial rise in the isotherm for low values of $P/P_0$ shown in figure B1.26.2. In the

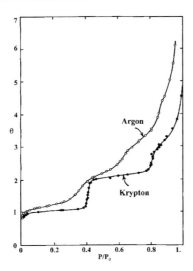

**Figure B1.26.4.** The adsorption of argon and krypton on graphitized carbon black at 77 K (Eggers D F Jr, Gregory N W, Halsey G D Jr and Rabinovitch B S 1964 *Physical Chemistry* (New York: Wiley) ch 18).

case of chemisorption, where the interaction is even stronger, the first monolayer saturates at even lower values of $P/P_0$. It is initially assumed that adsorption into the second layer is dominated by van der Waals interactions with the sample. The attractive van der Waals energy $\epsilon_a$ between two molecules in the gas-phase separated by a distance $r$ is $\epsilon_a \sim -K_a/r^6$ where $K_a$ is a constant. When combined with an $r^{-12}$ repulsive potential, this yields the Lennard-Jones 6–12 equation. An adsorbed species interacts with atoms in the truncated bulk of the sample, and the number of these increases as $\sim r^3$ so that the net van der Waals interaction with the surface varies as $r^{-3}$. This indicates that the energy of adsorption in the second layer $\Delta E_2$, assuming this to be dominated by van der Waals interactions with the surface, is given by: $\Delta E_2 = \Delta E_1/2^3$. From equation (B1.26.6), this yields $\ln(P/P_0) \sim -1.3$ so that the pressure at which this layer is complete should be $P/P_0 \sim 0.3$, a value significantly higher than that required to saturate the first layer. Similarly, the energy required to saturate the third layer will be reduced by a factor $3^3$, the fourth layer $4^3$ and so on. Therefore, if a value of $\Delta E$ is assumed for the first layer, the pressures at which the second and subsequent layer saturate can be calculated from this value. Writing this in terms of the coverage $\theta$ yields a simple form for the variation of $\Delta E$ as a function of coverage as $\Delta E_1/\theta^3$. This yields an isotherm for physisorption dominated by van der Waals interactions with the surface as:

$$\ln \frac{P}{P_0} = \frac{\Delta E_1}{kT\theta^3}. \tag{B1.26.7}$$

Such isotherms are shown in figure B1.26.4 for the physical adsorption of krypton and argon on graphitized carbon black at 77 K [13] and are examples of type VI isotherms (figure B1.26.3). Equation (B1.26.7) further predicts that a plot of $\ln(P/P_0)$ versus $1/\theta^3$ should be linear: such a plot is displayed in figure B1.26.5 [13] for the adsorption of argon on graphitized carbon black at 77 K and yields a good straight line.

It is clear, however, that not all isotherms display the stepwise behaviour shown in figure B1.26.4. For example, the isotherm for the adsorption of nitrogen on anatase [8] (figure B1.26.2) has a rapid increase in coverage at low pressures, corresponding to the saturation of the first monolayer, but varies much more smoothly with coverage at higher pressures. This effect is also seen in the data for argon on carbon black [13] (figure B1.26.4) where the steps become much less pronounced for multilayer adsorption. Part of the reason for this discrepancy is that, as the layer becomes thicker, intermolecular van der Waals interactions

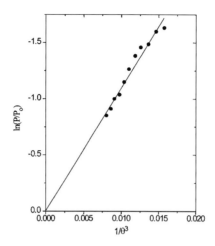

**Figure B1.26.5.** Plot of $\ln(P/P_0)$ versus $1/\theta^3$ for argon on graphitized carbon black at 77 K (from the argon data in figure B1.26.4) (Eggers D F Jr, Gregory N W, Halsey G D Jr and Rabinovitch B S 1964 *Physical Chemistry* (New York: Wiley) ch 18).

within the layer become more important. This has two effects. First, the difference in energy between layers becomes less pronounced, leading to a smoothing out of the curve. In addition, this decrease in energy difference for each layer means that subsequent layers start to grow even before previous monolayers have saturated. Finally, in the case of high-surface-area samples, the surface tends to become more heterogeneous, leading to a further smoothing out of the steps. The limiting case is an isotherm calculated on a different basis to that used for equation (B1.26.8). Here it is assumed that adsorption in the first layer is dominated by surface van der Waals interactions, but that adsorption into second and subsequent layers is dominated by intermolecular interactions between adsorbates. This clearly no longer results in a strong variation in $\Delta E$ with each layer and allows multilayer films to be formed in which another layer can start before the previous layer has been completed. This smooths out the isotherm resulting in a variation of coverage with $P$ that more closely resembles that shown in figure B1.26.2. Such an isotherm has the advantage that it often more closely mimics the behaviour of nitrogen physisorbed on high-surface-area materials and, as such, is more useful in reproducibly identifying 'point B'. This forms the basis of a reproducible method for measuring surface areas using the BET isotherm. The calculation of the BET isotherm assumes that:

1.  There are $B$ equivalent sites available for adsorption in the first layer.

2.  Each molecule adsorbed in the first layer is considered to be a possible adsorption site for molecules adsorbing into a second layer, and each molecule adsorbed in the second layer is considered to be a 'site' for adsorption into the third layer, and so on.

3.  All molecules in the second and subsequent layers are assumed to behave similarly to a liquid, in particular to have the same partition function. This is assumed to be different to the partition function (A2.2) of molecules adsorbed into the first layer.

4.  Intermolecular interactions are ignored for all layers.

Detailed derivations of the isotherm can be found in many textbooks and exploit either statistical thermo-dynamic methods [1] or independently consider the kinetics of adsorption and desorption in each layer and set these equal to define the equilibrium coverage as a function of pressure [14]. The most common form of

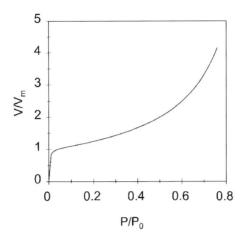

**Figure B1.26.6.** BET isotherm plotted from equation (B1.26.8) using a value of $C = 500$.

BET isotherm is written as a linear equation and given by:

$$\frac{P}{V(P - P_0)} = \frac{1}{V_m C} + \frac{P(C - 1)}{V_m C P_0}.$$ (B1.26.8)

Here $V_m$ is the volume of gas required to saturate the monolayer, $V$ the total volume of gas adsorbed, $P$ the sample pressure, $P_0$ the saturation vapour pressure and $C$ a constant related to the enthalpy of adsorption. The resulting shape of the isotherm is shown plotted in figure B1.26.6 for $C = 500$. A plot of $P/V(P - P_0)$ against $P/P_0$ should give a straight line having a slope $(C - 1)/V_m C$ and an intercept $1/V_m C$. The BET surface area is then calculated using the following equation:

$$S_A = V_m N A_m$$ (B1.26.9)

where $S_A$ is the required surface area of the sample, $V_m$ the volume of the adsorbed monolayer, $N$ Avogadro's number and $A_m$ the cross-sectional area of the adsorbed molecule. In the BET method, where nitrogen is generally used, the value of $A_m$ is taken to be $16.2$ Å$^2$ per nitrogen molecule. Classically, this BET equation is used for only for systems that exhibit type II and IV isotherms.

*(c) Measurement of BET isotherms*

Practically, using the BET method to measure surface area involves three steps: (1) obtaining a full adsorption isotherm, (2) evaluating the monolayer capacity and (3) the calculation of surface area using equation (B1.26.9). It should be emphasized that the BET surface area represents a 'standard' method for measuring the area of a high-surface-area material that allows samples from different laboratories to be compared. Note that, due to the simplifying assumptions of the derivation, the BET method does not work well for type III and V isotherms where the weak interaction between gas and solid makes it hard to discern the formation of the first layer. Attempts to modify the BET equation to take account of these situations have proven unreliable and impractical. Beyond the BET method, approaches such as the Gibbs adsorption isotherm [15], immersion calorimetry [16] or adsorption from solution [17] have been used but the BET method continues as a standard procedure for the determination of surface areas [18]. Generally the measured value of BET-area can be regarded as an effective area unless the material is ultramicroporous. In the case of porous materials, it is important to know the pore sizes and their distributions. This can be calculated for type IV absorbents where the mesopore size distribution can be obtained using the Kelvin equation:

$$\ln\left(\frac{P}{P_0}\right) = -\frac{2\gamma V}{rRT}.$$                                                                                      (B1.26.10)

This equation describes the additional amount of gas adsorbed into the pores due to capillary action. In this case, $V$ is the molar volume of the gas, $\gamma$ its surface tension, $R$ the gas constant, $T$ absolute temperature and $r$ the Kelvin radius. The distribution in the sizes of micropores may be determinated using the Horvath–Kawazoe method [19]. If the sample has both micropores and mesopores, then the $T$-plot calculation may be used [20]. The $T$-plot is obtained by plotting the volume adsorbed against the statistical thickness of adsorbate. This thickness is derived from the surface area of a non-porous sample, and the volume of the liquified gas.

### B1.26.2.2   Instrumentation

Two parameters must be measured to apply the BET equation, the pressure at the sample and the amount adsorbed at this pressure. There are three common methods for measuring the amount of gas adsorbed, called the volumetric method, the gravimetric method and the dynamic method, of which the volumetric method is the commonest [21].

#### (a) Volumetric method

This method essentially consists of admitting successive charges of gas (generally nitrogen) to the adsorbent using some form of volumetric measuring device such as a gas burette or pipette with the sample held at liquid nitrogen temperature (77 K). Nowadays, the amount of gas admitted to the sample can most conveniently be measured using a mass-flow controller. When equilibrium has been attained, the gas pressure in the dead space surrounding the sample is read using a manometer, and the quantity of gas remaining unadsorbed is then calculated with the aid of the gas law, assuming a perfect gas. The volume of the dead space of the apparatus must, of course, be accurately calibrated. A precision manometer should be employed. The quantity of gas adsorbed onto the surface can be calculated by subtracting the amount remaining unadsorbed from the total amount which has been admitted. This type of apparatus can be simply constructed from glass. Alternatively, commercial BET measuring apparatuses are also available using computers to collect the data and to calculate the resulting surface area [22]. Shown in figure B1.26.7 is a schematic diagram of a typical apparatus for measuring BET isotherms [23]. Helium, which does not adsorb at liquid nitrogen temperatures, is used to calibrate the volume of the dead space and the furnace is used to outgas the sample prior to gas adsorption. A mass-flow controller is often used instead of a burette or pipette to measure total amount of nitrogen that has been admitted.

#### (b) Gravimetric method

This method is simple but experimentally more cumbersome than the volumetric method and involves the use of a vacuum microbalance or beam balance [22]. The solid is suspended from one arm of a balance and its increase in weight when adsorption occurs is measured directly. The 'dead space' calculation is thereby avoided entirely but a buoyancy correction is required to obtain accurate data. Nowadays this method is rarely used.

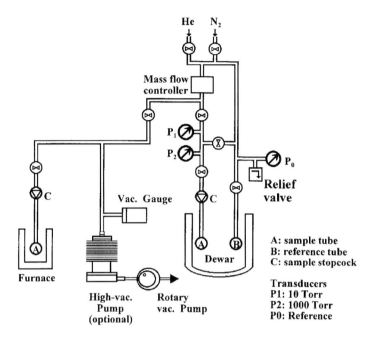

**Figure B1.26.7.** Diagram of a BET apparatus (representing an OMNISORB 100) (Beckman Coulter 1991 *OMNISORP Manual*).

*(c) Dynamic method*

This method has been developed using gas chromatographic techniques. The most popular way of implementing this method is by using a continuous nitrogen flow as first described by Nelsen and Eggertsen [24]. A known mixture of nitrogen and helium is passed through a bed of solid sample at ambient temperature (~300 K) where the exit gas is measured using a gas chromatographic detector. When the gas composition equilibrates as indicated by a constant base line on the GC recorder chart, the sample tube is immersed in a liquid nitrogen bath. The adsorption of nitrogen by the solid is then indicated by a negative excursion on the recorder chart corresponding to a loss of nitrogen from the system due to adsorption. After equilibrium is established at the particular partial pressure with the sample held at 77 K, the baseline attains its original level. The sample tube is then allowed to warm up to room temperature (300 K) by removal of the liquid nitrogen bath so that a positive excursion appears due to nitrogen desorption (see figure B1.26.8) [24]. The areas under these two curves should be equal and constitute a measure of the amount of nitrogen adsorbed. This method has drawn considerable interest due to its simplicity and speed and since it does not require a vacuum system. One of the main problems is that of deciding on the most appropriate conditions for 'outgassing' which can considerably affect the precision.

*B1.26.2.3   Experimental notes*

Nitrogen is the most widely used absorbent (at 77 K) for the BET method and has been employed almost universally. Argon is more suited to the measurement of microporous zeolites. Krypton may be used for the measurement of very low-surface-area (less than 3 m$^2$ g$^{-1}$) samples because it has a very low saturation vapour pressure (~2.45 Torr) at liquid nitrogen temperature so that the absolute pressure range is from 0 to approximately 0.75 Torr for a ratio of $P/P_0$ from 0 to 0.3. The absolute pressure range used for the

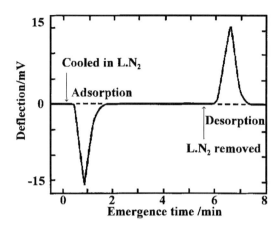

**Figure B1.26.8.** Adsorption/desorption peaks for nitrogen obtained with the continuous flow method (Nelsen F M and Eggertsen F T 1958 *Anal. Chem.* **30** 1387–90).

measurement depends upon the type of data reduction required and the absorbent's properties. The optimum pressure range needed for BET surface area determinations may be taken to be for $P/P_0$ up to 0.3 [23].

### B1.26.3   Ellipsometry

*B1.26.3.1   Principles*

The term ellipsometry was first coined by Rothen in 1945 [25] to refer to the measurement of thin films of materials by monitoring the light reflected from them at some incident angle $\theta$. The method is illustrated in figure B1.26.9. The technique was used extensively before this [26]. In the simplest case, the film is transparent with refractive index $n$, such that light can be reflected or transmitted at the first interface. The transmitted beam propagates through the material and is reflected from the substrate. The reflected portion of the beam can either subsequently reflect from the film/air interface or reflect once again into the film and undergo further reflections. The multiple beams are eventually emitted together so that, in general they are attenuated and one of the parameters that can most simply be measured is the reflectivity of the film. In addition, because of the phase shifts that occur because of path length differences as the beam passes through the film, there is also a change of the phase of this beam which depends on the film thickness $d$ and the wavelength of the light, $\lambda$. The way in which this phase change can be measured will be described below. Each of these values, that is, the reflectivity and the phase shift, depend on the polarization of the radiation (see below). When the light is polarized parallel to the surface it is said to be s polarized and when it is polarized perpendicularly to the surface, p polarized. The corresponding reflectivities are denoted $R^s$ and $R^p$ which, since there is a phase change on reflection, are generally complex numbers. The ratio of these values is defined as $\rho$ which is given by:

$$\rho = \frac{R^p}{R^s}. \tag{B1.26.11}$$

Since this is a complex number, it can be separated into an amplitude and a phase and written as:

$$\rho = \frac{|R^p|}{|R^s|} \exp(i\Delta) \tag{B1.26.12}$$

where $\Delta$ is the difference between the phase shifts for p- and s-polarized light: $\Delta = \delta_s - \delta_p$. By convention, the ratio of the moduli of the reflectivities of the p- and s-polarized light $|R^p|/|R^s|$ is written in terms of

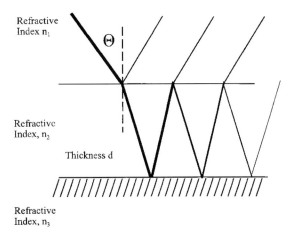

**Figure B1.26.9.** Schematic diagram showing the reflection of light incident at an angle $\Theta$ from a medium with refractive index $n_1$ through a film of thickness d with refractive index $n_2$.

another parameter $\Psi$ where $\tan \Psi = |R^p|/|R^s|$. Thus, the parameters that are measured in ellipsometry are $\Psi$ and $\Delta$ which can be related to the refractive index and thickness of the film.

*(a) The nature of electromagnetic radiation*

As shown by Maxwell, light consists of an oscillating electromagnetic field propagating in vacuum at the speed of light [27–33]. Both the electric and magnetic fields are oriented perpendicularly to the direction of propagation of the light and perpendicularly to each other. By convention, the direction in which the electric field points defines the polarization direction so that plane-polarized light has an electric field component only in one direction. A common way to obtain plane-polarized light is to use a polarizer which only transmits light of one polarization and absorbs the other. Polarized spectacles have lenses made from such polarizing material. Maxwell was also able to demonstrate that the velocity $v$ with which light propagates in a material is given by $v = 1/\sqrt{(\mu \epsilon \mu_0 \epsilon_0)}$ where $\epsilon$ is the permittivity of the material, $\epsilon_0$ that of vacuum, $\mu$ the permeability of the material and $\mu_0$ the value in vacuum. When light propagates in vacuum, this reduces to $v = 1/\sqrt{(\mu_0 \epsilon_0)}$ and yields the speed of light, $c$. This correspondence provided confirmation that light and electromagnetic radiation were one and the same. The refractive index of a material, $n$, is defined as $n = c/v$, and is therefore given by $n = \sqrt{(\mu \epsilon)}$. Since most materials under investigation are not magnetic, $\mu = 1$, and the equation for refractive index simplifies to $n = \sqrt{\epsilon}$.

As electromagnetic radiation propagates through space, the electric field converts into a magnetic field during the oscillation cycle. Thus, the energy present in the electric field converts to energy in a magnetic field. The magnetic field oscillates at the same frequency but $90°$ out of phase with the electric field so that, when the electric field is a maximum, the magnetic field is a minimum, and *vice versa*. This idea allows us to calculate the relative electric and magnetic field amplitudes in an electromagnetic wave. Let the electric field amplitude be $E_0$ and the corresponding magnetic field amplitude $H_0$. The energy in a magnetic field is proportional to $\mu H_0^2$ and that in an electric field is proportional to $\epsilon E_0^2$. Since the electric and magnetic fields interconvert in an electromagnetic wave, conservation of energy requires that these be equal, so that $H_0 = \sqrt{(\epsilon/\mu)}E_0$. Again, assuming that $\mu = 1$ for a non-magnetic material and using the equation for refractive index above yields: $H_0 = nE_0$.

Non-polarized electromagnetic radiation, of course, comprises two perpendicular polarizations, which can change both in amplitude and in phase with respect to each other. If the two polarizations are in phase

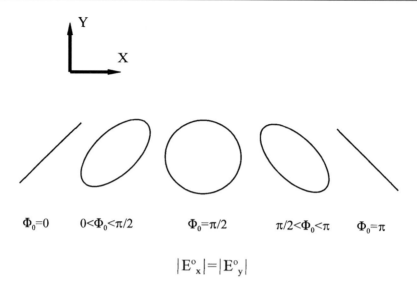

$$|E^o{}_x| = |E^o{}_y|$$

**Figure B1.26.10.** Various polarization configurations corresponding to different values of the phase shift, $\Phi$.

with each other, the resultant is just another linearly polarized beam, with the resultant polarization direction given by a simple vector addition of the two electric field components. When the electric fields of the two polarizations are out of phase with each other, the resulting electric field precesses as it propagates through space. For example, if the two electric fields are of equal magnitude but 90° out of phase with each other the electric field spirals as it moves through space. Depending on the relative phases, this rotation can be either clockwise or anti-clockwise. In this case, if we were able to look end on at the electric field vector, this would describe a circle and is therefore referred to as circularly polarized light. This particular combination of electric fields also carries with it angular momentum, and is responsible for the angular momentum selection rules in spectroscopy; this is really just the law of conservation of angular momentum. Different relative phases or amplitudes generally lead to elliptically polarized light, so that the phase shift $\Delta$ between the reflected and transmitted light measured in ellipsometry is manifested as elliptically polarized light. The different types of polarized light found as the phase shifts between 0 and 360° are shown in figure B1.26.10.

Light can also be absorbed by a material through which it passes. This leads to an attenuation in intensity of the light as it passes through the material, which decays exponentially as a function of distance through the material and is described mathematically by the Beer–Lambert law [34]:

$$I = I_0 \exp(-\epsilon cl) \qquad (B1.26.13)$$

where $c$ is the concentration of the absorbant, $l$ the path length and $\epsilon$ the extinction coefficient. This is represented in the mathematical description of the propagation of an electromagnetic wave by modifying the dielectric constant to add an imaginary part, $k$, which is generally written as: $n - ik$ where $i = \sqrt{-1}$.

### (b) Reflection of light from a dielectric surface

Before discussing multiple reflections at a thin film, we will first examine the reflection of light from one material (with real refractive index $n_1$) into another (with refractive index $n_2$) [27, 32, 33]. It is assumed that the material does not absorb light and this situation is depicted in figure B1.26.11. Since electromagnetic waves propagate in both materials, the behaviour at the interface is dictated by the boundary conditions for electric and magnetic fields. The components of $E$ and $H$ parallel to the surface (in the $x$ direction in figure B1.26.11) and the components of $D$ and $B$ perpendicular to the surface (in the $z$ direction in figure B1.26.11) are continuous across the boundary. Thus the reflected and transmitted wave amplitudes can be calculated by simply applying these equations. We will consider the reflection of p-polarized radiation from the surface. The calculation of the equations for s-polarized radiation is essentially identical. The beam is incident at an angle $\theta_i$, and a portion is reflected at an angle $\theta_r$. Of course, $\theta_i$ and $\theta_r$ are equal. The remaining light is transmitted at an angle $\theta_t$ which will be different to $\theta_i$, due to refraction at the surface. The electric field amplitudes are taken to be $E_i$ in the incident beam, $E_r$ in the reflected beam and $E_t$ in the transmitted beam. The corresponding values for magnetic field amplitudes are $H_i$, $H_r$ and $H_t$ respectively. Application of the above boundary conditions gives:

$$E_i^0 \cos \theta_i - E_r^0 \cos \theta_r = E_t^0 \cos \theta_t \tag{B1.26.14}$$

$$H_i^0 + H_r^0 = H_t^0. \tag{B1.26.15}$$

Using the above relationship between the electric and magnetic field amplitudes from equation (B1.26.15):

$$n_1 E_i^0 + n_1 E_r^0 = n_2 E_t^0. \tag{B1.26.16}$$

In order to calculate the reflected amplitude, $E_t^0$ can be eliminated from equations (B1.26.15) and (B1.26.16) to yield:

$$\frac{(E_i^0 - E_r^0) \cos \theta_i}{\cos \theta_t} = \frac{n_1 (E_i^0 - E_r^0)}{n_2} \tag{B1.26.17}$$

where we have used $\theta_i = \theta_r$. Writing the reflection coefficient for p-polarized radiation as $r^p$ (which equals $E_r^0 / E_i^0$) yields:

$$r^p = \frac{n_2 \cos \theta_i - n_1 \cos \theta_t}{n_2 \cos \theta_i + n_1 \cos \theta_t} \tag{B1.26.18}$$

which is the Fresnel equation for p-polarized radiation. A similar analysis can be carried out for s-polarized radiation using the boundary conditions in a similar way to yield:

$$r^s = \frac{n_1 \cos \theta_i - n_2 \cos \theta_t}{n_1 \cos \theta_i + n_2 \cos \theta_t}. \tag{B1.26.19}$$

Corresponding equations can also be written for the transmitted portion of the beam. For a non-absorbing sample, the refractive index is real, so that $\exp(i\phi)$ is real where $\phi$ is the phase shift on reflection. This means that $\phi$ is either 0 or 180°. If, however, the sample absorbs light, the refractive index can become complex, resulting in a phase change of the light. This, for example, must be taken into account when reflecting light from a metallic (mirror) surface.

The reflection coefficients $r^p$ and $r^s$ give the electric field in the reflected beam for each polarization. Since the intensity of light is proportional to the square of the electric field, the reflectances for s- and p-polarized light can be written as $R^p = |r^p|^2$ and $R^s = |r^s|^2$, respectively. These are plotted in figure B1.26.12 for a light beam incident from air ($n = 1$) onto a material with refractive index $n = 3$. It is evident that p-polarized radiation is reflected to a much lesser extent than s-polarized radiation, and exclusively s-polarized radiation is reflected at the polarizing angle. This effect is exploited in polarized sunglasses (as mentioned above) to minimize the reflective glare from surfaces by only allowing p-polarized light to be transmitted. At normal incidence, these equations reduce to:

$$R^s = R^p = \left( \frac{n_2 - 1}{n_2 + 1} \right)^2$$

which for glass with $n \sim 1.5$ gives about 4% reflectivity. The polarizing angle shown in figure B1.26.12 is given by:

$$\tan \theta_i = \frac{n_2}{n_1}$$

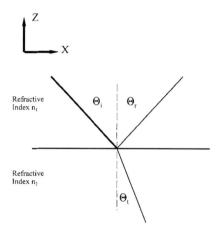

**Figure B1.26.11.** Diagram showing light impinging from a material of refractive index $n_1$ at an angle $\Theta_i$ onto a material with refractive index $n_2$ and reflected at an angle $\Theta_r$ and transmitted at an angle $\Theta_t$.

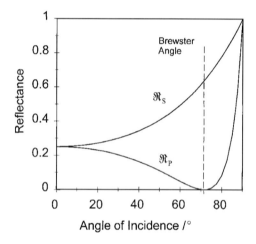

**Figure B1.26.12.** Plot of the reflectivity of s- and p-polarized light from a material with refractive index $n = 3$.

and is known as the Brewster angle. p-polarized radiation is perfectly transmitted at this angle and Brewster windows (oriented at this angle) are used in lasers to minimize the loss of radiation in the laser cavity. This often results in laser light being polarized. This effect is exploited in polarimeters (see below). Note finally that the phase shift on reflection from a dielectric is $0°$ below the Brewster angle and $180°$ above.

*(c) Reflection at multiple interfaces*

We are now in a position to calculate the reflections from multiple interfaces using the simple example of a thin film of material of thickness $d$ with refractive index $n_2$ sandwiched between a material of refractive index $n_1$ (where this is generally air with $n = 1$) deposited onto a substrate of refractive index $n_3$ [35, 36]. This is depicted in figure B1.26.9. The resulting reflectivities for p- and s-polarized light respectively are given by:

$$R^p = \frac{r_{12}^p + r_{23}^p \exp(-i2\beta)}{1 + r_{12}^p r_{23}^p \exp(-i2\beta)} \tag{B1.26.20}$$

and

$$R^{\text{s}} = \frac{r_{12}^{\text{s}} + r_{23}^{\text{s}} \exp(-\text{i}2\beta)}{1 + r_{12}^{\text{s}} r_{23}^{\text{s}} \exp(-\text{i}2\beta)} \qquad (B1.26.21)$$

where $r_{ij}^{\text{s}}$ and $r_{ij}^{\text{p}}$ are the reflection coefficients for s- and p-polarized radiation, respectively, at the interface between material $i$ and $j$ (equations (B1.26.18) and (B1.26.19)). The path length difference due to the film results in phase differences between different emerging beams giving rise to complex reflection coefficients and hence phase shifts. As noted above, this produces elliptically polarized light which can be analysed to yield the amplitude and phase shift and ultimately $\beta$. This parameter depends on the film thickness and the wavelength of light and is given by:

$$\beta = 2\pi \left(\frac{d}{\lambda}\right) n_2 \cos\theta_2. \qquad (B1.26.22)$$

These equations are generally too complex to be solved analytically even for relatively simpler systems and are therefore solved numerically. The way in which this is done will be described below. The measurement of $\Delta$ and $\Psi$ clearly depends on the wavelength of light directly through this equation. However, both the real and imaginary parts of the refractive indices also depend on the wavelength of light. Now the reflectivity of the surface for s- and p-polarized light is $\mathfrak{R}^{\text{p}} = |r^{\text{p}}|^2$ and $\mathfrak{R}^{\text{s}} = |R^{\text{p}}|^2$ respectively. This dependence is also exploited in ellipsometry by measuring $\Delta$ and $\Psi$ and a function of light wavelength in a technique known as spectroscopic ellipsometry [37].

### B1.26.3.2 Applications

*(a) Measurement of the optical constants of materials using ellipsometry*

In this case, the Fresnel equations (equations (B1.26.18) and (B1.26.19)) for reflection at a single interface are used. The phase shift is zero or 180° for a dielectric with a real refractive index, which can be measured directly from the reflectivities (figure B1.26.12). Intermediate phase shifts are found for absorbant materials with complex refractive indices which can also be measured from $\Delta$ and $\Psi$. This is generally done by numerically solving the Fresnel equations (B1.26.18) and (B1.26.19). Many ellipsometers include software to calculate these values directly and Fortran programs are also available to calculate these values. The variation in $\Psi$ plotted as a function of the imaginary part of the refractive index is shown in figure B1.26.13. This varies between 0 and 45° over the range of $K$ values. The variation in $\Delta$ as a function of $K$ with $n = 2$ and an incidence angle of 70° is displayed in figure B1.26.14. Again, the value of $\Delta$ varies as the imaginary part of the refractive index changes and can vary between 0 and 180°. It is important to emphasize that, unless a material is extremely pure, its optical constants can vary so that it is generally important to measure these parameters for a particular sample.

*(b) Measurement of film thickness and optical properties*

The way in which ellipsometry can be used to measure film thickness will be illustrated for the simple case of a material with thickness $d$ and complex refractive index $n_2$ deposited onto a substrate with complex refractive index $n_3$ with light incident from a material of refractive index $n_1$. Since light is generally incident from air, $n_1$ is usually taken to be unity. This is done by measuring $\Delta$ and $\Psi$ and using equations (B1.26.20) and (B1.26.21) to calculate the optical properties of the film and its thickness. Since, in addition to the film thickness, there are potentially six other variables in the system (the real and imaginary parts of each of the refractive indices) whereas only two parameters $\Delta$ and $\Psi$ are measured, several of these must be determined independently. The substrate parameters can, for example, be measured prior to film deposition as described in the previous section and the refractive index of air is known. The variation in $\Delta$ and $\Psi$ with the thickness $d$ of a film of silicon dioxide (with refractive index 1.46) deposited onto a silicon substrate (with refractive

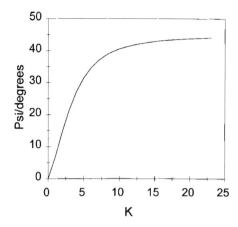

**Figure B1.26.13.** Plot of $\Psi$ *versus* $K$, the imaginary part of the refractive index.

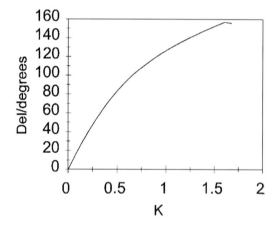

**Figure B1.26.14.** Plot of $\Delta$ *versus* $K$, the imaginary part of the refractive index.

index $3.872 - i0.037$) is shown plotted in figure B1.26.15 [37]. This yields a trajectory of the allowed values of $\Delta$ and $\Psi$ as the film thickness. Since the film thickness is measured from the interference between the light reflected from the boundary between air and the film and that reflected from the interface between the film and the substrate, the maximum film thickness $d_{max}$ that can be measured before the trajectory shown in figure B1.26.15 retraces itself is given by:

$$2d_{max} \cos \theta_t = \frac{\lambda}{n_2} \qquad (B1.26.23)$$

where $n_2$ is included to take account of the change in wavelength as the light passes through the film with this refractive index. The angle of the transmitted light ($\theta_t$) and the incidence angle ($\theta_i$) are related through Snell's law so that:

$$\cos \theta_t = \sqrt{1 - \frac{\sin^2 \theta_i}{n_2}} \qquad (B1.26.24)$$

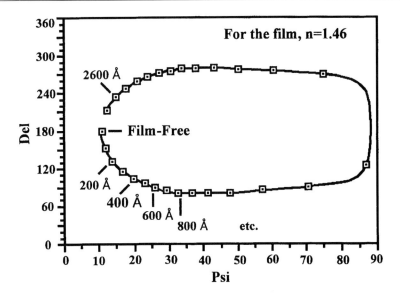

**Figure B1.26.15.** The Del/Psi trajectory for silicon dioxide on silicon with angle of incidence $\phi_1 = 70°$ and wavelength $\lambda = 6328$ Å (Tompkins H G 1993 *A Users Guide to Ellipsometry* (San Diego, CA: Academic)).

assuming that the light is incident from air ($n = 1$). Substitution into equation (B1.26.23) yields a value of $d_{max}$ as:

$$d_{max} = \frac{\lambda}{2\sqrt{n_2^2 - \sin^2 \theta_2}}. \tag{B1.26.25}$$

This shows that the values of $\Delta$ and $\Psi$ are identical for films of thickness $nd_{max}$ where $n$ is an integer. Thus the trajectory shown in figure B1.26.15 retraces itself for multiples of this maximum thickness. For the silicon dioxide film deposited onto silicon, this yields a value of $d_{max} = 2832$ Å. Thus, in figure B1.26.15, if values $\Delta$ and $\Psi$ were measured to be 90 and 25° respectively, this would correspond to a film of $600 + n \times 2832$ Å thick where $n = 0, 1, 2$ etc.

### B1.26.3.3   Instrumentation

The following components make up an ellipsometer:

(1) *Monochromatic light source.* This is generally a small laser, usually a helium–neon laser emitting red light at 6328 Å. The plasma tube of these lasers is usually terminated by a Brewster window (see section B1.26.3.1) to minimize losses in the laser cavity so that the light emitted by the laser is linearly polarized. If an unpolarized source is used, a polarizer is placed after the light source to produce linearly polarized light.

(2) *An element that converts linearly polarized light into elliptically polarized light.* This component is made of a birefringent material where the refractive index of light that passes through the material depends on the light polarization with respect to its crystal lattice [33] where there are generally two perpendicular axes with different refractive indices for light polarized along each of these directions. Since, as shown above, the velocity of light through a medium depends on the refractive index, where the velocity of light $v$ in the medium is given by $v = c/n$, this implies that light travels at different velocities depending on the polarization with respect to each of these directions. When the light is polarized along the direction

of the smallest refractive index, the light travels the fastest and this is known as the 'fast' axis. Since light polarized along the other, higher-refractive-index axis travels more slowly, this is known as the 'slow' axis. The refractive indices of the fast and slow axis are designated $n_f$ and $n_s$, respectively, where it follows from the definitions that $n_s > n_f$. If we now imagine linearly polarized light incident on this material with the electric field vector oriented at 45° to each of these axes, the electric field of the incident light along both the slow and fast axes will be equal at $E_0 \cos 45°$, where $E_0$ is the electric field amplitude of the polarized light. They will also be in phase. The frequency of the light, $\nu_0$, will be the same for both components within the material, but they will travel with different velocities. This means that after traversing a plate of this material of thickness $d$, the two components will no longer be in phase with each other so that, according to the discussion in section B1.26.3.1, the light will, in general, be elliptically polarized. In order to calculate this, we first calculate the time required for light polarized along the fast and slow axes to traverse a disc of the birefringent material of thickness $d$. This is given by $t_s = d/v_s$ and $t_f = d/v_f$ for the slow and fast axes respectively. The difference in transit time for the two beams, $\Delta t = t_s - t_f$. If the period of oscillation of the light is $\tau$ ($=1/\nu_0$), then if $\Delta t = \tau$, the slow beam is one period behind the fast beam and the phase difference is $2\pi$ radians. The phase difference for intermediate values of $\Delta t$ designated $\Delta\phi$ is given by ($2\pi \Delta t/\tau$). Combining these equations yields:

$$\Delta\phi = \frac{2\pi \nu_0 d}{c}(n_s - n_f). \tag{B1.26.26}$$

If the wavelength of the incident radiation *in vacuo* is $\lambda_0$, equation (B1.26.26) becomes:

$$\Delta\phi = \frac{2\pi d}{\lambda_0}(n_s - n_f) \tag{B1.26.27}$$

so the value of $\Delta\phi$ can be selected to be any desired value merely by varying the value of $d$ for a material with particular values of $n_s$, $n_f$ and $\lambda$. It is common to select $\Delta\phi = 90° = (2\pi)/4$ radians, and this is known as a quarter-wave plate for a particular wavelength and produces circularly polarized light from linearly polarized light if the polarization direction of the incident beam is oriented at 45° to the fast and slow axes. Orientating the incident polarization to intermediate angles with respect to the fast and slow axes yields elliptically polarized light. This effect is exploited in the ellipsometer.

(3) A polarizer which transmits only one polarization of radiation is required to define the state of polarization of the reflected beam.

(4) The intensity of the reflected light must also be measured. Historically, this was done using the eye. Since, in general, a null (a measurement of the point at which the light decreases to zero) is required, this can be relatively sensitive. However, nowadays, the light intensity is generally measured using a photomultiplier tube.

(5) These components must be mounted so that the incident and detection angles can be varied and kept equal and a place must be provided to mount the sample.

There are several possible configurations used to construct an ellipsometer [44]. We will describe one example of the one of the most common arrangements shown in figure B1.26.16. The linearly polarized light emerging from the laser passes through the quarter-wave plate, which can be rotated to yield elliptically polarized light. When this reflects from the sample this also produces elliptically polarized light. The quarter-wave plate is rotated to find the condition such that, when the elliptically polarized light interacts with the sample, the phase change produced on reflection exactly compensates for the elliptical polarization of the incident light to produce linearly polarized light. The angle of the resulting linearly polarized light can be accurately determined using the polarizer placed before the detector (known as the analyser) by rotating it so that no light reaches the detector (the null condition). Being able to achieve this depends, of course, on the

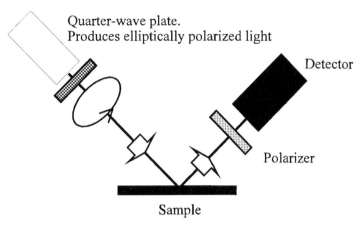

**Figure B1.26.16.** Schematic diagram of an ellipsometer.

quarter-wave plate being correctly oriented so as to exactly compensate for the effect of the sample and the analyser being oriented at exactly $90°$ to the direction of the resulting linearly polarized light. The experiment then consists of rotating both the quarter-wave plate and the analyser so that no light reaches the detector. This is often done automatically, and these resulting quarter-wave plate and analyser angles can be simply converted into the parameters $\Delta$ and $\Psi$ [37, 44].

### B1.26.3.4 Examples

#### (a) Basic studies

Dielectric constants of metals, semiconductors and insulators can be determined from ellipsometry measurements [38, 39]. Since the dielectric constant can vary depending on the way in which a film is grown, the measurement of accurate film thicknesses relies on having accurate values of the dielectric constant. One common procedure for determining dielectric constants is by using a Kramers–Kronig analysis of spectroscopic reflectance data [39]. This method suffers from the series-termination error as well as the difficulty of making corrections for the presence of overlayer contaminants. The ellipsometry method is for the most part free of both these sources of error and thus yields the most accurate values to date [39].

#### (b) Characterization of thin films and multilayer structures

Ellipsometry measurements can provide information about the thickness, microroughness and dielectric function of thin films. It can also provide information on the depth profile of multilayer structures non-destructively, including the thickness, the composition and the degree of crystallinity of each layer [39]. The measurement of the various components of a complex multilayered film is illustrated in figure B1.26.17 [40]. This also illustrates the use of different wavelengths of light to obtain much more information on the nature of the film. Here $\Delta$ and $\Psi$ are plotted versus the wavelength of light ($\bullet$) and the line drawn through these data represents a fit calculated for the various films of yttrium oxide deposited on silica as shown at the bottom of the figure [40].

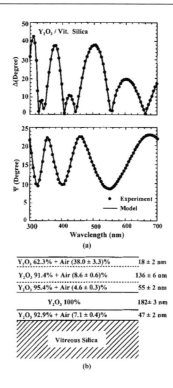

**Figure B1.26.17.** (a) Observed and calculated ellipsometric $[\Delta(\lambda), \Psi(\lambda)]$ spectra for the $Y_2O_3$ film on vitreous silica. Angle of incidence 75°. (b) Best-fit model of the $Y_2O_3$ film on vitreous silica (Chindaudom P and Vedam K 1994 *Physics of Thin Films* vol 19, ed K Vedam (New York: Academic) p 191).

*(c) Real-time studies*

With the development of multichannel spectroscopic ellipsometry, it is possible now to use real-time spectroscopic ellipsometers, for example, to establish the optimum substrate temperature in a film growth process [41, 42].

## B1.26.4   Work-function measurements

### B1.26.4.1   Principles

The work function ($\Phi$) is defined as the minimum work that has to be done to remove an electron from the bulk of the material to a sufficient distance outside the surface such that it no longer experiences an interaction with the surface electrostatic field [43–45]. In other words, it is the minimum energy required to remove an electron from the highest occupied level (the 'Fermi level') of a solid, through the surface, to the so-called vacuum reference level (figure B1.26.18). Thus it is influenced by two factors. The first is associated with the bulk electronic properties of the solid: work function increases with increasing binding energy. The second is associated with penetrating the surface dipole layer: work function changes with surface contamination and structure.

In order to understand the tendency to form a dipole layer at the surface, imagine a solid that has been cleaved to expose a surface. If the truncated electron distribution originally present within the sample does not relax, this produces a steplike change in the electron density at the newly created surface (figure B1.26.19(A)).

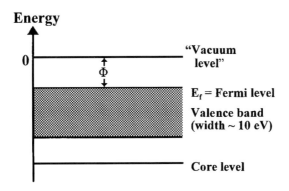

**Figure B1.26.18.** Schematic diagram of the energy levels in a solid.

Since the electron density $\rho(x) \propto |\psi(x)|^2$, where $\psi(x)$ is the electron wavefunction, this implies that the electron wavefunction varies in a similarly step-wise fashion at the interface. This indicates that $d^2\psi/dx^2|_s$, where s indicates that the derivative is evaluated at the surface, becomes infinite. Since the electron kinetic energy $E_K = (-\hbar^2/2m)\,(d^2\psi/dx^2)$, this creates an infinite-energy surface. This energy can be decreased by reducing $d^2\psi/dx^2$ and by allowing the wavefunction to become 'smoother' at the interface as shown in figure B1.26.19(B). This means that electron density previously within the sample extends outside the sample, producing a negative charge. Since the sample was originally electrically neutral, the excess charge outside the sample is balanced by a corresponding positive charge within it, resulting in an electric dipole moment at the surface. The work required to separate these charges increases the potential energy at the same time as the kinetic energy decreases. The equilibrium surface dipole moment corresponds to the minimum in this energy and this has been discussed in detail by Smoluchowski [46, 47]. In general, the greater the electron density of the sample, the larger will be the surface dipole. Thus, for the same metal, close-packed surfaces generally have the highest work functions; for example, in the case of copper, the work functions of the various surfaces are Cu(111): $\Phi = 4.94$ eV, Cu(100): $\Phi = 4.59$ eV, Cu(110): $\Phi = 4.48$ eV [45]. It is the presence of this surface dipole that renders the work function sensitive to changes in surface properties. For example, the surface dipole layer may change as a result of adsorption. Adsorbed species can be viewed as having a discrete dipole moment that tends to modify the total dipole layer at the surface by charge transfer and consequently change the work function. Thus, measurement of the work function change, $\Delta\Phi = \Phi_{adsorbate\ covered} - \Phi_{clean}$ yields important information on the degree of charge reorganization upon adsorption and the surface coverage of the adsorbate. For example, adsorption of an electronegative species (for example, chlorine, hydrogen etc) will tend to increase the surface dipole moment causing an increase in work function ($\Delta\Phi$ positive). Correspondingly, an electropositive adsorbate (for example, an alkali metal) decreases the surface dipole moment causing a decrease in the work function ($\Delta\Phi$ negative). These changes can be used to measure the adsorbate coverage as illustrated in figure B1.26.20. Here it is assumed that there are $N$ adsorbates per unit area each having an effective charge $q$. This is balanced by an equal and opposite 'image charge' in the substrate a distance $d$ away so that each adsorbate possesses a dipole moment $\mu = qd$. The separated layers of positive and negative charge can be thought of as forming a parallel-plate capacitor, where the potential difference between the capacitor plates corresponds to the change in work function of the sample. If the total charge per unit area on each of the plates is $Q$, then:

$$Q = C\Delta\Phi \qquad\qquad (B1.26.28)$$

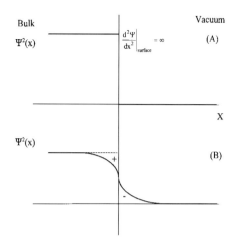

**Figure B1.26.19.** The variation of the electron density (A) from an unrelaxed surface and (B) showing the smoothing of the electron density to lower the kinetic density.

where $C$ is the capacitance per unit area and is $\varepsilon_0/d$. Since $Q = Nq$, equation (B1.26.28) becomes:

$$Nq = \frac{\Delta\phi\,\varepsilon_0}{d} \tag{B1.26.29}$$

which, remembering that $\mu = qd$, yields the Helmholz equation:

$$\Delta\Phi = \frac{N\mu}{\varepsilon_0} \tag{B1.26.30}$$

and shows that, for this simple case, the change in work function varies linearly with the adsorbate coverage $N$ and the surface dipole moment of the adsorbate $\mu$. A simplifying assumption in these equations is that either the coverage or the surface dipole moment or both are sufficiently small that the dipoles do not interact. If the distance between the dipoles decreases (as the coverage becomes larger) and/or if the dipole moment is large, the electric field created by one dipole can polarize adjacent dipoles to reduce their dipole moments. This effect is known as depolarization. This situation has been described by Topping [48] and Miller [49] and the resulting change in work function taking these effects into account is given by:

$$\Delta\Phi = \frac{N\mu(1 + 9\alpha N^{3/2})}{\varepsilon_0} \tag{B1.26.31}$$

where $\alpha$ is the polarizibility of the adsorbate. This reduces to equation (B1.26.30) for low coverages.

### B1.26.4.2  Instrumentation

There are many ways to measure the work function or work function change which can be generally classified as electron emission methods (thermionic emission, field emission and photoelectron emission), low-energy electron beam (retarding-potential) methods and capacitance methods [50–52]. Absolute work function values can be measured using emission methods while the other techniques measure only work-function changes.

The probes for measuring surface work functions are generally incorporated into an ultrahigh vacuum apparatus and supplement the existing vacuum-compatible, surface-sensitive probes (see for example sections B1.7, B1.9, B1.20, B1.21 and B1.25). Rather than measuring the absolute value of the work function,

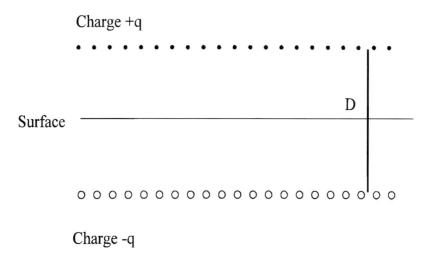

**Figure B1.26.20.** Diagram showing the dipole layer created on a surface by an electropositive adsorbate.

it is often more interesting to measure the change in work function caused by some change to the surface. It can either be measured by modifying equipment already present in the vacuum system, for example using the electron gun of low-energy electron diffraction optics for the electron beam method or from the high-binding-energy cut-off in ultraviolet photoelectron spectroscopy. Specific probes for rapidly and conveniently measuring work-function changes can also be introduced separately into the chamber. The most common of these is the Kelvin probe or vibrating capacitor.

*(a) Thermionic emission method*

When a metal sample is heated, electrons are emitted from the surface when the thermal energy of the electrons, $kT$, becomes sufficient to overcome the work function $\Phi$ [53]. The probability of this electron emission depends on work function $\Phi$ and temperature $T$ as expressed in the Richardson–Dushman equation:

$$J = A(1 - r)T^2 \exp(-e\Phi/kT) \tag{B1.26.32}$$

where $J$ is the thermionic emission current density, $A = 120$ A cm$^{-2}$ deg$^{-2}$ and $r$ is the reflection coefficient for electrons arriving at the work function barrier. Thus, plotting $\ln(J/T^2)$ against $1/kT$ yields a straight line with slope equal to $e\Phi$. The method is not generally suitable for monitoring adsorbates since the sample has to be heated to emit electrons.

*(b) Field emission method*

The barrier to electron removal from a surface can be reduced substantially by the presence of a strong electric field. The physical situation involved in this process is shown in figure B1.26.21 where, in the presence of an applied field, the work function is less than that at zero field. This increases the probability for electrons to 'tunnel' out of the sample [54]. This is a quantum mechanical phenomenon which can be formulated mathematically by considering a Fermi sea of electrons within the metal impinging on a potential barrier at the surface. The result is given by the Fowler–Nordheim equation as:

$$J = 6 \times 10^6 \left(\frac{\mu}{\Phi(\mu + \Phi)}\right)\left(\frac{E}{\alpha}\right)^2 \exp\left(\frac{-(6.8 \times 10^7 \alpha \Phi^{3/2})}{E}\right) \tag{B1.26.33}$$

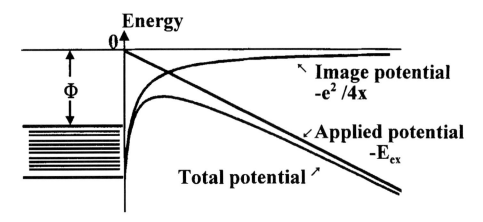

**Figure B1.26.21.** Potential energy curves for an electron near a metal surface. 'Image potential' curve: no applied field. 'Total potential' curve: applied external field $= -E_{ex}$.

where $J$ is the current arising from field emission, $E$ is the electric field strength and $\alpha$ is a tabulated function equal to $0.95 \pm 0.009$ over the range of current densities normally encountered. A plot of $\ln(J/E^2)$ against $(1/E)$ can be used to determine $\Phi$.

This method needs a highly specialized sample configuration and the sample has to be a very sharp point (radius $\sim 10^{-5}$ cm) so that sufficiently high fields can be maintained for emission measurements. This requirement often precludes the use of other surface analysis techniques. However, for specialized applications, the field emission method may be the only one or the most convenient one available. It has been shown to be able to directly measure the work function change induced by the adsorption of a single atom on a tungsten plane [55]. Combined with field emission microscopy, the work function of different crystal planes can be detected. Much of the early understanding of adsorption was obtained with this device and a large number of data on single-crystal work functions have been produced by this technique [52].

*(c) Photoelectron emission method*

When photons of sufficiently high frequency $v$ are directed onto a metal surface, electrons are emitted in a process known as photoelectron emission [56]. The threshold frequency $v_0$ is related to the work function by the expression

$$\Phi = hv_0/e. \tag{B1.26.34}$$

The total electron current generated in this process is given by the Fowler equation:

$$J = BT^2 f \left[ \frac{h(v - v_0)}{kT} \right] \tag{B1.26.35}$$

where $B$ is a parameter that depends on the material involved. The photoemitted current is measured as a function of photon energy; extrapolation allows the determination of $v_0$, and thus of $\Phi$, to be made.

This technique requires a photon source (a light source with monochromator or filters) of calibrated spectral intensity and variable energy in the range around $v_0$, and an electron collector. Both the work function and the work-function change may be determined conveniently from the cut-off in inelastic electrons in a photoelectron spectrum [47]. As demonstrated by figure B1.26.18, electrons with the minimum kinetic energy barely surmount the work function, while electrons from the Fermi level ($E_F$, the highest occupied level) will

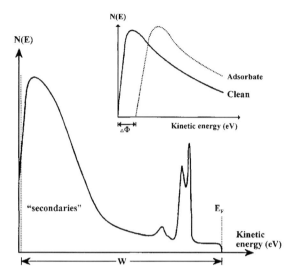

**Figure B1.26.22.** The energy width $W$ of an ultraviolet photoelectron spectrum from a solid may be used to determine the work function. Changes in work function may be obtained from changes in the 'cut-off' of the secondary electron peak (inset) (Attard G and Barnes C 1988 *Surfaces* (Oxford: Oxford University Press)).

have the maximum kinetic energy, $h\nu - \Phi$. The work function can be calculated from the relationship

$$\Phi = h\nu - W \tag{B1.26.36}$$

where $W$ is the energy width of the whole photoelectron spectrum. Adsorption changes the work function which in turn changes the width ($W$) of the UP spectrum. $\Phi$ and $W$ are inversely related. The work-function change upon adsorption manifests itself in a shift in the low-kinetic-energy 'secondary tail', as shown in the inset in figure B1.26.22 [47].

*(d) Low-energy electron beam (retarding-potential) methods*

In this method, the sample is designed as an anode. Electrons emitted from the cathode by thermionic emission normally impinge on the anode sample on which a retarding potential is applied. Basically this can be arranged as a diode (known as the diode method) [57] or a triode (called the Shelton triode method) [58]. Different circuit arrangements give rise to different $I$–$V$ relationships and the difference in the work functions of the two electrode surfaces is measured. When a low-energy electron diffraction (LEED) apparatus is available (see section B1.9), this method can be implemented using the low-energy electrons from the LEED electron gun. In this case, electrons of known, low energy are incident on the sample. As the potential across the sample is slowly made more negative, the current is measured and the relationship between sample and retarding voltage is shown in figure B1.26.23 [50]. At low retarding voltage, most of the impinging electrons are collected. At some larger retarding voltage, the impinging electrons do not posses sufficient energy to reach the sample and so are reflected. This voltage is determined by the work functions of the sample and the electron gun filament, and by the accelerating voltage of the electron gun. Changes in the work function of the sample, when the filament work function and the electron gun accelerating voltage are held constant, are manifested by a change in the cut-off retarding voltage (see figure B1.26.23).

A low-energy electron beam can also be obtained using a field emission tip and used in the field emission retarding-potential method. This combination provides an absolute measure of the sample work function and the resolution is excellent [52].

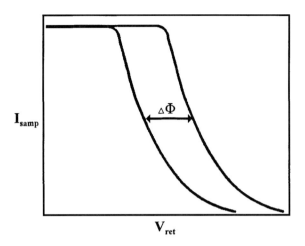

**Figure B1.26.23.** Current–voltage curves observed in the retarding potential difference method of work-function measurement (Hudson J B 1992 *Surface Science* (Stoneham, MA: Butterworth–Heinemann)).

*(e) Capacitance methods*

When an electrical connection is made between two metal surfaces, a contact potential difference arises from the transfer of electrons from the metal of lower work function to the second metal until their Fermi levels line up. The difference in contact potential between the two metals is just equal to the difference in their respective work functions. In the absence of an applied emf, there is electric field between two parallel metal plates arranged as a capacitor. If a potential is applied, the field can be eliminated and at this point the potential equals the contact potential difference of the two metal plates. If one plate of known work function is used as a reference electrode, the work function of the second plate can be determined by measuring this applied potential between the plates [52]. One can determine the zero-electric-field condition between the two parallel plates by measuring directly the tendency for charge to flow through the external circuit. This is called the static capacitor method [59].

Historically, the first and most important capacitance method is the vibrating capacitor approach implemented by Lord Kelvin in 1897. In this technique (now called the Kelvin probe), the reference plate moves relative to the sample surface at some constant frequency and the capacitance changes as the interelectrode separation changes. An AC current thus flows in the external circuit. Upon reduction of the electric field to zero, the AC current is also reduced to zero. Originally, Kelvin detected the zero point manually using his quadrant electrometer. Nowadays, there are many elegant and sensitive versions of this technique. A piezoceramic foil can be used to vibrate the reference plate. To minimize noise and maximize sensitivity, a phase-locked detection circuit is used and a feedback loop may automatically null the electric field. The whole process can be carried out electronically to provide automatic recording of the contact potential and any changes which occur.

This technique does not involve heating the sample to high temperature or exposing it to high electric fields. Nor is there a need for a hot filament which cannot be used at high pressures. When mounted on a linear motion manipulator, the reference plate assembly can be moved away from the sample leaving no interference with other techniques. One major shortcoming of this technique is that the work function of the reference plate, or electrode, must be precisely known for an absolute determination of the sample work function. Moreover, when relative changes are examined, the work function of the reference electrode must be stable, either unaffected by adsorption or cleanable before each experiment. This disadvantage may be

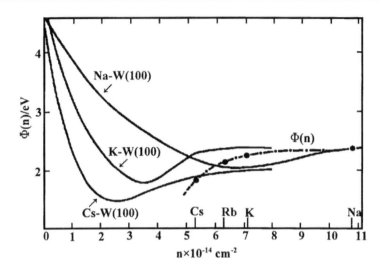

**Figure B1.26.24.** The change of work function of the (100) plane of tungsten covered by Na, K and Cs, and work function of alkali metals (dashed–dotted line) *versus* adatom concentration *n* (Kiejna A and Wojciechowski 1981 *Prog. Surf. Sci.* **11** 293–338).

compensated for by using an inert metal such as gold. In addition, the reference electrode must be well shielded to reduce the effects of external electric and magnetic fields on the experiment [52].

### B1.26.4.3   Applications

#### (a) Surface characterization

Measurement of the work function of a surface is an important part of overall surface characterization. Surface electron charge density can be described in terms of the work function and the surface dipole moment can be calculated from it (equations (B1.26.30) and (B1.26.31)). Likewise, changes in the chemical or physical state of the surface, such as adsorption or geometric reconstruction, can be observed through a work-function modification. For studies related to cathodes, the work function may be the most important surface parameter to be determined [52].

#### (b) Measurement of adsorption isotherms

Almost all adsorbates cause work function changes ($\Delta\Phi$). Plotting $\Delta\Phi$ *versus* pressure at various temperatures produces an adsorption isotherm which can be used to determine heats of adsorption and the surface coverage [60]. Much early understanding of adsorption was gained by this method. In particular alkali and alkali earth adsorption has been rather extensively studied in this way. This is illustrated in figure B1.26.24 for the adsorption of alkali metals on W(100) [61]. The change in work function is depicted as solid lines which initially decrease with increasing coverage as expected for an electropositive adsorbate. The dots connected by the dashed line represent the work functions of each of the bulk alkali metals and each of the work-function curves tends asymptotically to these values for large coverages. These curves are not linear as a function of coverage because of depolarization effects (equation (B1.26.31)) and so reach a minimum before attaining their bulk values.

# References

[1]   Clarke A 1970 *The Theory of Adsorption and Catalysis* (London: Academic)
[2]   Brunauer S, Emmett P H and Teller E 1938 Adsorption of gases in multimolecular layers *J. Am. Chem. Soc.* **60** 309–19
[3]   Halling J 1975 *Principles of Tribology* (New York: Macmillan) ch 1
[4]   Millikan R A 1916 A direct photoelectric determination of Planck's '*h*' *Phys. Rev.* **7** 355–88
[5]   Yates J T and Garland C 1961 Infrared studies of carbon monoxide chemisorbed on nickel surfaces *J. Catal.* **65** 617–24
[6]   Scholten J J and van Montfoort A 1962 The determination of the free-metal surface area of palladium catalysts *J. Catal.* **1** 85–92
[7]   Stephens S J 1959 Surface reactions on evaporated palladium films *J. Phys. Chem.* **63** 188–94
[8]   Harkins W D 1952 *The Physical Chemistry of Surface Films* (New York: Reinhold)
[9]   Emmett P H and Brunauer S 1937 The use of low temperature van der Waals adsorption isotherms in determining the surface area of iron synthetic ammonia catalysts *J. Am. Chem. Soc.* **59** 1553–64
[10]  Brunauer S, Deming L S, Deming W S and Teller E A 1940 Theory of the van der Waals adsorption of gases *J. Am. Chem. Soc.* **62** 1723–32
[11]  Sing K S W, Everett D H, Haul R A W, Mosoul L, Pierotti R A, Rouguerol J and Siemieniewska T 1985 Reporting physisorption data to the determination of surface area and porosity *Pure Appl. Chem.* **57** 603–19
[12]  Langmuir I 1916 The constitution and fundamental properties of solids and liquids *J. Am. Chem. Soc.* **38** 2221–95
[13]  Eggers D F Jr, Gregory N W, Halsey G D Jr and Rabinovitch B S 1964 *Physical Chemistry* (New York: Wiley) ch 18
[14]  Castellan G W 1983 *Physical Chemistry* (Reading, MA: Addison-Wesley) ch 18
[15]  Donohue M D and Aranovich G L 1998 Classification of Gibbs adsorption isotherms *Adv. Colloid Interface Sci.* **76/77** 137–52
[16]  Harkins W D and Jura G 1944 An absolute method for the determination of the area of a finely divided crystalline solid *J. Am. Chem. Soc.* **66** 1362–6
[17]  Giles C H, Smith D and Huitson A 1974 General treatment and classification of the solute adsorption isotherm *J. Colloid Interface Sci.* **47** 755–65
[18]  Sing K S W 1998 Adsorption methods for the characterization of porous materials *Adv. Colloid Interface Sci.* **76/77** 3–11
[19]  Horvath G and Kawazoe K 1983 Method for calculation of effective pore size distribution in molecular sieve carbon *J. Chem. Eng. Japan* **16** 470–5
[20]  Lippens B C and deBoer J H 1965 Studies on pore systems in catalysts V. The *T* method *J. Catal.* **4** 319–23
[21]  Gregg S J and Sing K S W 1967 *Adsorption, Surface Area and Porosity* (London, UK: Academic) ch 8; 1982 2nd edn, ch 6
[22]  'OMNISORP' Series, Beckman Coulter, USA; 'ASAP 2400', 'FlowSorb 2300', Micromeritics; 'Autosorb-1', 'nova 1000', Quantachrome, USA; 'Grimm 100', Labortechnik, Germany; 'Sorpty 1750', 'Sorptomatic', Carlo Erba, Italy
[23]  Beckman Coulter 1991 *OMNISORP Manual*
[24]  Nelsen F M and Eggertsen F T 1958 Determination of surface area: adsorption measurements by a continuous flow method *Anal. Chem.* **30** 1387–90
[25]  Rothen A 1945 The ellipsometer, an apparatus to measure thicknesses of thin surface films *Rev. Sci. Instrum.* **16** 26–30
[26]  Drude P 1901 *Theory of Optics* (New York: Longmans Green)
[27]  Jenkins F A and White H E 1957 *Fundamentals of Optics* (New York: McGraw-Hill)
[28]  Longhurst R S 1967 *Geometrical and Physical Optics* (New York: Wiley)
[29]  Klein M V 1970 *Optics* (New York: Wiley)
[30]  Welford W T 1988 *Optics* (Oxford: Oxford University Press)
[31]  Fincham W H A and Freeman M H 1980 *Optics* (London: Butterworths)
[32]  Born M and Wolf E 1969 *Principles of Optics* (New York: Pergamon)
[33]  Banerjee P P and Poon T C 1991 *Principles of Applied Optics* (Boston, MA: Asken)
[34]  Atkins P W 1998 *Physical Chemistry* (New York: Freeman) ch 19
[35]  Azzam R M A and Bashara N M 1977 *Ellipsometry and Polarized Light* (Amsterdam: North-Holland)
[36]  Hevens O S 1965 *Optical Properties of Thin Solid Films* (New York: Dover)
[37]  Tompkins H G 1993 *A Users Guide to Ellipsometry* (San Diego, CA: Academic)
[38]  Aspnes D E and Studna A A 1983 Dielectric functions and optical parameters of silicon and germanium *Phys. Rev.* B **27** 985–1009
[39]  Vedam K 1998 Spectroscopic ellipsometry: a historical overview *Thin Solid Films* **313/314** 1–9
[40]  Chindaudom P and Vedam K 1994 *Physics of Thin Films* vol 19, ed K Vedam (New York: Academic) p 191
[41]  Muller R H and Farmer J C 1984 Fast self-compensating spectral-scanning ellipsometer *Rev. Sci. Instrum.* **55** 371–4
[42]  Collins R W, An I, Fujiwara H, Lee J, Lu Y, Kol J and Rovira P I 1998 Advances in multichannel spectroscopic *Thin Solid Films* **313/314** 18–32
[43]  Riviere J C 1969 *Solid State Surface Science* vol I, ed M Green (New York: Dekker)
[44]  Wedler G 1976 *Chemisorption: an Experimental Approach* (London: Butterworth)
[45]  Smoluchowski R 1941 Volume magnetostriction of nickel *Phys. Rev.* **60** 249–51
[46]  Smoluchowski R 1941 The theory of volume magnetostriction *Phys. Rev.* **59** 309–17
[47]  Attard G and Barnes C 1998 *Surfaces* (Oxford: Oxford University Press)
[48]  Topping J 1927 Form and energy potential of atoms on surfaces *Proc. R. Soc.* A **114** 67–76

[49] Miller A R 1946 The variation of the dipole moment of adsorbed particles with the fraction of the surface covered *Proc. Camb. Phil. Soc.* **42** 292–303

[50] Hudson J B 1992 *Surface Science* (Stoneham, MA: Butterworth–Heinemann)

[51] Woodruff D P and Delchar T A 1994 *Modern Technologies of Surface Science* 2nd edn (Cambridge: Cambridge University Press)

[52] Swanson L W and Davis P R 1985 Work function measurements *Solid State Physics: Surfaces (Methods of Experimental Physics 22)* ed R L Park and M G Lagally (New York: Academic) ch 1

[53] Reimann A L 1934 *Thermionic Emission* (London: Chapman and Hall)

[54] Gomer R 1994 Field emission, field ionization, and field desorption *Surf. Sci.* **299/300** 129–52

[55] Todd C J and Rhodin T N 1974 Adsorption of single alkali atoms on tungsten using field emission and field desorption *Surf. Sci.* **42** 109–21

[56] Fowler R H 1931 The analysis of photoelectric sensitivity curves for clean metals at various temperatures *Phys. Rev.* **38** 45–56

[57] Haas G A and Thomas R E 1972 Thermionic emission and work function *Tech. Met. Res.* **6** 91–262

[58] Shelton H 1957 Thermionic emission from a planar tantalum crystal *Phys. Rev.* **107** 1553–7

[59] Delchar T, Eberhagen A and Tompkins F C 1963 A static capacitor method for the measurement of the surface potential of gases on evaporated metal films *J. Sci. Instrum.* **40** 105–7

[60] Conrad H, Ertl G and Latta E E 1974 Adsorption of hydrogen on palladium single crystal surfaces *Surf. Sci.* **41** 435–46

[61] Kiejna A and Wojciechowski 1981 Work function of metals: relation between theory and experiment *Prog. Surf. Sci.* **11** 293–338

# B1.27
# Calorimetry

*Kenneth N Marsh*

## B1.27.1  Introduction

Calorimetry is the basic experimental method employed in thermochemistry and thermal physics which enables the measurement of the difference in the energy $U$ or enthalpy $H$ of a system as a result of some process being done on the system. The instrument that is used to measure this energy or enthalpy difference ($\Delta U$ or $\Delta H$) is called a calorimeter. In the first section the relationships between the thermodynamic functions and calorimetry are established. The second section gives a general classification of calorimeters in terms of the principle of operation. The third section describes selected calorimeters used to measure thermodynamic properties such as heat capacity, enthalpies of phase change, reaction, solution and adsorption.

## B1.27.2  Relationship between thermodynamic functions and calorimetry

The first law of thermodynamics relates the energy change in a system at constant volume to the work done on the system $w$ and the heat added to the system $q$,

$$\Delta U = w + q. \tag{B1.27.1}$$

Both heat and work are a flow of energy, heat being a flow resulting from a difference in temperature between the system and the surroundings and work being an energy flow caused by a difference in pressure or the application of other electromechanical forces such as electrical energy. For the case where the system is thermally isolated from its surrounding, termed *adiabatically enclosed*, there is no heat flow to or from the surrounds, i.e. $q = 0$ so that

$$\Delta U = w_{\text{adiabatic}}. \tag{B1.27.2}$$

A calorimeter is a device used to measure the work $w$ that would have to be done under adiabatic conditions to bring about a change from state 1 to state 2 for which we wish to measure $\Delta U = U_2 - U_1$. This work $w$ is generally done by passing a known constant electric current $\Im$ for a known time $t$ through a known resistance $R$ embedded in the calorimeter, and is denoted by $w_{\text{elec}}$ where

$$w_{\text{elec}} = \Im^2 Rt. \tag{B1.27.3}$$

In general it is difficult to construct a calorimeter that is truly adiabatic so there will be unavoidable heat leaks $q$. It is also possible that non-deliberate work is done on the calorimeter such as that resulting from a change in volume against a non-zero external pressure $p_{\text{ext}}(-\int p_{\text{ext}}\,dV)$, often called $pV$ work. Additional work $w'$ may be done on the system by energy introduced from stirring or from energy dissipated due to self-heating of the device used to measure the temperature. The basic equation for the energy change in a calorimeter is

$$\Delta U = w_{\text{elec}} - \int p_{\text{ext}}\, dV + w' + q. \tag{B1.27.4}$$

The $pV$ work term is not normally measured. It can be eliminated by suspending the calorimeter in an evacuated space ($p = 0$) or by holding the volume of the calorimeter constant ($dV = 0$) to give

$$\Delta U = w_{\text{elec}} + w' + q. \tag{B1.27.5}$$

This is the working equation for a constant volume calorimeter. Alternatively, a calorimeter can be maintained at constant pressure $p$ equal to the external pressure $p_{\text{ext}}$ in which case

$$-\int_{V_1}^{V_2} p_{\text{ext}}\, dV = p(V_2 - V_1) \tag{B1.27.6}$$

and

$$\Delta U = U_2 - U_1 = w_{\text{elec}} - pV_1 - pV_2 + w' + q \tag{B1.27.7}$$

hence

$$\Delta(U + pV) = (U + pV)_2 - (U + pV)_1 = w_{\text{elec}} + w' + q. \tag{B1.27.8}$$

The quantity $U + PV$ is termed the enthalpy $H$, hence

$$\Delta H = H_2 - H_1 = w_{\text{elec}} + w' + q. \tag{B1.27.9}$$

is the working equation for a constant pressure calorimeter.

The heat capacity at constant volume $C_V$ is defined from the relations

$$\Delta U = \int_{T_1}^{T_2} C_V\, dT \tag{B1.27.10}$$

and

$$C_V \overset{\text{def}}{=} (\partial U / \partial T)_V \overset{\text{def}}{=} \lim_{T_2 \to T_1} [\{U(T_2, V_2) - U(T_1, V_1)\} / (T_2 - T_1)]. \tag{B1.27.11}$$

Values of $C_V(T)$ can be derived from a constant volume calorimeter by measuring $\Delta U$ for small values of $(T_2 - T_1)$ and evaluating $\Delta U / (T_2 - T_1)$ as a function of temperature. The energy change $\Delta U$ can be derived from a knowledge of the amount of electrical energy required to change the temperature of the sample + container from $T_1$ to $T_2$, $w_{\text{elec}}$ (sample + container), and the energy required to change the temperature of the container only from $T_1$ to $T_2$, $w_{\text{elec}}$ (container). If the volume of the sample is kept constant, and the calorimeter is adiabatic ($q = 0$) or the heat leak is independent of the amount of sample and no other work is done then:

$$\Delta U = U(T_2, V) - U(T_1, V) = w_{\text{elec}}(\text{sample} + \text{container}) - w_{\text{elec}}(\text{container}). \tag{B1.27.12}$$

Except for gases, it is very difficult to determine $C_V$. For a solid or liquid the pressure developed in keeping the volume constant when the temperature is changed by a significant amount would require a vessel so massive that most of the total heat capacity would be that of the container. It is much easier to measure the difference

$$\Delta H = H(T_2, p) - H(T_1, p) = w_{\text{elec}}(\text{sample} + \text{container}) - w_{\text{elec}}(\text{container}) \tag{B1.27.13}$$

between the enthalpies of the initial and final states when the pressure is kept constant and derive the heat capacity at constant pressure $C_p$ defined by

$$C_p \overset{\text{def}}{=} (\partial H / \partial T)_p \overset{\text{def}}{=} \lim_{T_2 \to T_1} [\{H(T_2, V_2) - H(T_1, V_1)\} / (T_2 - T_1)]. \qquad \text{(B1.27.14)}$$

The enthalpy change $\Delta H$ for a temperature change from $T_1$ to $T_2$ can be obtained by integration of the constant pressure heat capacity

$$\Delta H = \int_{T_1}^{T_2} C_p \, dT. \qquad \text{(B1.27.15)}$$

The entropy change $\Delta S$ for a temperature change from $T_1$ to $T_2$ can be obtained from the following integration

$$\Delta S = \int_{T_1}^{T_2} (C_p / T) \, dT. \qquad \text{(B1.27.16)}$$

### B1.27.3 Operating principle of a calorimeter

All calorimeters consist of the calorimeter proper and its surround. This surround, which may be a jacket or a bath, is used to control the temperature of the calorimeter and the rate of heat leak to the environment. For temperatures not too far removed from room temperature, the jacket or bath usually contains a stirred liquid at a controlled temperature. For measurements at extreme temperatures, the jacket usually consists of a metal block containing a heater to control the temperature. With non-isothermal calorimeters (calorimeters where the temperature either increases or decreases as the reaction proceeds), if the jacket is kept at a constant temperature there will be some heat leak to the jacket when the temperature of the calorimeter changes. Hence, it is necessary to correct the temperature change observed to the value it would have been if there was no leak. This is achieved by measuring the temperature of the calorimeter for a time period both before and after the process and applying Newton's law of cooling. This correction can be reduced by using the technique of adiabatic calorimetry, where the temperature of the jacket is kept at the same temperature as the calorimeter as a temperature change occurs. This technique requires more elaborate temperature control and it is primarily used in accurate heat capacity measurements at low temperatures.

With most non-isothermal calorimeters, it is necessary to relate the temperature rise to the quantity of energy released in the process by determining the calorimeter constant, which is the amount of energy required to increase the temperature of the calorimeter by one degree. This value can be determined by electrical calibration using a resistance heater or by measurements on well-defined reference materials [1]. For example, in bomb calorimetry, the calorimeter constant is often determined from the temperature rise that occurs when a known mass of a highly pure standard sample of, for example, benzoic acid is burnt in oxygen.

### B1.27.4 Classification of calorimeters

#### B1.27.4.1 Classification by principle of operation

*Isothermal calorimeters (more precisely, quasi-isothermal)*

These include calorimeters referred to as calorimeters with phase transitions. The temperatures of the calorimeter (i.e. the vessel) $T(c)$ and the jacket $T(s)$ in such calorimeters remain constant throughout the experiment. For calorimeters with phase transition, the calorimetric medium is usually a pure solid (its stable modification) in equilibrium with the liquid phase of the same substance, for example, ice and water. The reaction chamber is placed inside a vessel inside the layer of this substance. The jacket also contains an equilibrium mixture of two phases of the same substance. For an exothermic process, part of the solid substance melts in the vessel, and the volume change of the liquid is precisely measured. Another calorimeter of this type is an isothermal titration calorimeter. One of the reactants is added at such a rate that, for an endothermic system, the enthalpy or energy change is balanced by the simultaneous addition of electrical energy so the calorimeter remains isothermal. The energy added is a direct measure of the energy or enthalpy change. For an exothermic system electrical energy that is added at a constant rate is counterbalanced by the removal of energy at a constant rate (by, for example, a thermoelectric cooling device) to maintain the calorimeter isothermal. The reactant is then added at such a rate that the calorimeter remains isothermal when the addition of electrical energy is discontinued.

*Adiabatically-jacketed calorimeters*

The energy released when the process under study takes place makes the calorimeter temperature $T(c)$ change. In an adiabatically jacketed calorimeter, $T(s)$ is also changed so that the difference between $T(c)$ and $T(s)$ remains minimal during the course of the experiment; that is, in the best case, no energy exchange occurs between the calorimeter (unit) and the jacket. The thermal conductivity of the space between the calorimeter and jacket must be as small as possible, which can be achieved by evacuation or by the addition of a gas of low thermal conductivity, such as argon.

*Heat-flow calorimeters*

These calorimeters are enclosed in a thermostat ('heat sink') which has a much greater heat capacity than that of the calorimeter vessel proper. The energy released in the calorimeter is negligibly small compared with the heat capacity of the thermostat, and hence the thermostat temperature $T(s)$ does not change. The outer surface of the calorimeter (vessel) is in direct thermal contact with the inner surface of the thermostat, and the energy flow occurs through a series of thermopiles, which consist of a large number of thermocouples connected in series. The flow of energy through the thermocouples gives rise to a voltage. The thermopiles are designed so that the majority of energy that flows from the calorimeter to the thermostat flows through them as rapidly as possible and the area under the curve of the voltage produced against time is a measure of the overall quantity of the energy released (or taken up) in the process occurring in the calorimeter.

*Isoperibole calorimeters*

This type of calorimeter is normally enclosed in a thermostatted-jacket having a constant temperature $T(s)$. and the calorimeter (vessel) temperature $T(c)$ changes through the energy released as the process under study proceeds. The thermal conductivity of the intermediate space must be as small as possible. Most combustion calorimeters fall into this group.

*B1.27.4.2   Classification by design*

*Liquid calorimeters*

A liquid serves as the calorimetric medium in which the reaction vessel is placed and facilitates the transfer of energy from the reaction. The liquid is part of the calorimeter (vessel) proper. The vessel may be isolated from the jacket (isoperibole or adiabatic), or may be in good thermal contact (heat-flow type) depending upon the principle of operation used in the calorimeter design.

*Aneroid (liquidless) calorimeters*

The reaction vessel is situated inside a metal of high thermal conductivity having a cylindrical, spherical, or other shape which serves as the calorimetric medium. Silver is the most suitable material because of its high thermal conductivity, but copper is most frequently used.

*Combined calorimeters*

These are a combination of the liquid and aneroid types.

### B1.27.4.3 Selection of method of measurement

In designing a calorimeter, consideration must be given to the combination of the principle of its operation with the type of design and this depends on the ultimate goal. Thus, isoperibole calorimeters may be a liquid, aneroid, or a combined calorimeter type with a static or rotating vessel (dynamic). For adiabatically-jacketed calorimeters, one normally uses a combined or aneroid design that enables the creation of a vacuum between the unit and the jacket, which is essential for proper thermal isolation of the unit. Isothermal calorimeters require special consideration depending on their application. To characterize a calorimeter design, it is necessary to specify its type by all the classifications, for example, adiabatic aneroid static or isoperibole liquid dynamic calorimeter.

The selection of the operating principle and the design of the calorimeter depends upon the nature of the process to be studied and on the experimental procedures required. However, the type of calorimeter necessary to study a particular process is not unique and can depend upon subjective factors such as technical restrictions, resources, traditions of the laboratory and the inclinations of the researcher.

## B1.27.5 Calorimeters for specific applications

Various books and chapters in books are devoted to calorimeter design and specific applications of calorimetry. For several decades the Commission on Thermodynamics of the International Union of Pure and Applied Chemistry (IUPAC) has been responsible for a series of volumes on experimental thermodynamics and thermochemistry. *Experimental Thermochemistry*, volume I, published in 1956, edited by F D Rossini [2], dealt primarily with combustion calorimetry. Volume II published in 1962, edited by H A Skinner [3], primarily documented advances in combustion calorimetry since the first volume. In 1979 an update of much of the material covered in *Experimental Thermodynamics*, volumes I and II, was published under the title *Combustion Calorimetry* with editors S Sunner and M Månsson [4]. The first volume in the series *Experimental Thermodynamics, Calorimetry of Non-Reacting Systems*, edited by J P McCullough and D W Scott [5] was published in 1968. This volume covered the general principle of calorimeter design for non-reacting systems. It included a detailed discussion of adiabatic and drop calorimeters for the measurements of heat capacity, calorimeters for measurement of enthalpies of fusion and vaporisation, and calorimeters for the measurement of heat capacities of liquids and solutions close to room temperature. The second volume, *Experimental Thermodynamics of Non-Reacting Systems*, edited by B LeNeindre and B Vodar [6], published in 1975, was concerned with the measurement of a broader class of thermodynamic and transport properties over a wide range of temperature and pressure. A number of the techniques covered, such as density of a fluid as a function of temperature and pressure and speed of sound, allow the calculation of energy differences by non-calorimetric methods. Volume III, *Measurement of Transport Properties of Fluids*, edited by W A Wakeham, A Nagashima and J V Sengers [7], published in 1991, was concerned primarily with the measurement of the transport properties of fluids. Volume IV, *Solution Calorimetry*, edited by K N Marsh and P A G O'Hare [8] was published in 1994. This book covered calorimetric techniques for the measurement of enthalpies of reaction of organic substances, heat capacity and excess enthalpy of mixtures of organic compounds in both the liquid and

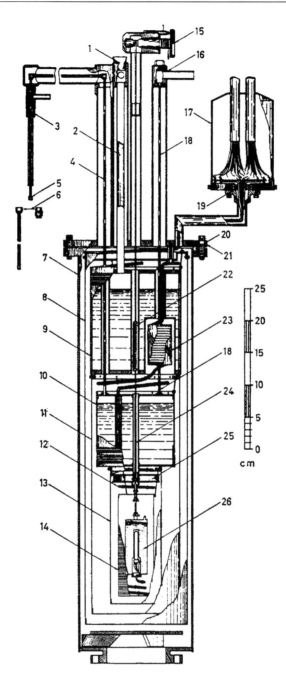

**Figure B1.27.1.** Aneroid-type cryostat for low-temperature adiabatic calorimeter: 1, 2, liquid nitrogen transfer; 3, 4, 5, 6, liquid helium transfer parts; 7, brass vacuum jacket; 8, outer floating radiation shield; 9, liquid nitrogen tank; 10, liquid helium tank; 11, nitrogen radiation shield; 12, lead wire; 13, helium radiation shield; 14, adiabatic shield; 15, windlass; 16, helium exit connector; 17, copper shield for terminal block; 18, helium exit tube; 19, vacuum seal; 20, O-ring gasket; 21, cover plate; 22, coil spring; 23, helium vapour exchanger; 24, supporting braided silk line; 25, floating ring; 26, calorimeter assembly. (Reprinted with permission from 1968 *Experimental Thermodynamics* vol I (Butterworth).)

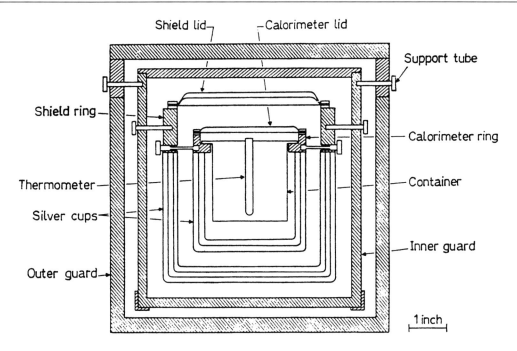

**Figure B1.27.2.** Schematic vertical section of a high-temperature adiabatic calorimeter and associated thermostat (Reprinted with permission from 1968 *Experimental Thermodynamics* vol I (Butterworth).)

gas phase, calorimetry of electrolyte solutions at high temperature and pressure, microcalorimetric application in biological systems, titration calorimetry, and the calorimetric determination of pressure effects. *IUPAC Chemical Data Series No 32, Enthalpies of Vaporization of Organic Compounds* by V Majer and V Svoboda [9], contains a detailed review of calorimeters used to measure enthalpy of vaporization. Other monographs dealing extensively with calorimetric techniques have been published. These include *Specialist Periodical Reports, Chemical Thermodynamics*, volume 1 [10], which covered combustion and reaction calorimetry, heat capacity of organic compounds, vapour-flow calorimetry and calorimetric methods at high temperature. *Physical Methods of Chemistry*, Volume VI, *Determination of Thermodynamic Properties* [11, 12] contains a chapter on calorimetry and a chapter devoted to differential thermal methods including differential thermal calorimetry.

### B1.27.5.1 Measurement of heat capacity

The most important thermodynamic property of a substance is the standard Gibbs energy of formation as a function of temperature as this information allows equilibrium constants for chemical reactions to be calculated. The standard Gibbs energy of formation $\Delta_f G^\circ$ at 298.15 K can be derived from the enthalpy of formation $\Delta_f H^\circ$ at 298.15 K and the standard entropy $\Delta S^\circ$ at 298.15 K from

$$\Delta_f G^\circ = \Delta_f H^\circ - T \Delta S^\circ. \tag{B1.27.17}$$

The enthalpy of formation is obtained from enthalpies of combustion, usually made at 298.15 K while the standard entropy at 298.15 K is derived by integration of the heat capacity as a function of temperature from $T = 0$ K to 298.15 K according to equation (B1.27.16). The Gibbs–Helmholtz relation gives the variation of the Gibbs energy with temperature

$$\{\partial(G/T)/\partial T\}_p = -H/T^2. \tag{B1.27.18}$$

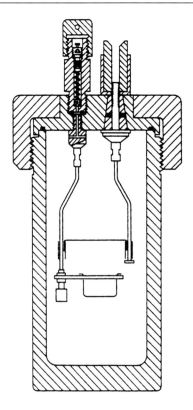

**Figure B1.27.3.** Typical static combustion bomb. (Reproduced with permission from A Gallencamp & Co. Ltd.)

Hence it is necessary to measure the heat capacity of a substance from near 0 K to the temperature required for equilibrium calculations to derive the enthalpy as a function of temperature according to equation (B1.27.15).

*Low temperature heat capacity*

For solids and non-volatile liquids accurate heat capacity measurements are generally made in an adiabatic calorimeter. A typical low temperature aneroid-type adiabatic calorimeter used to make measurements between 4 K and about 300 K is shown in figure B1.27.1. The primary function of the complex assembly is to maintain the calorimeter proper at any desired temperature between 4 K and 300 K. The only energy gain should be from the addition of electrical energy during a measurement. The upper part of the calorimeter contains vessels for holding liquid nitrogen and helium that provide low temperature heat sinks. Construction materials are generally those having high thermal conductivity (e.g. copper) plated with reflectant material (e.g. chromium) to reduce radiant energy transfer. The calorimeter proper and its surrounding adiabatic shield are suspended by silk lines and can be raised to bring them into good thermal contact with the lower tank, thereby cooling the calorimeter. When the calorimeter proper has reached its desired temperature, thermal contact is broken by lowering the calorimeter and the adiabatic shield. Adiabatic conditions are maintained by keeping the temperature of the adiabatic shield at the temperature of the calorimeter and heat conduction is minimized by maintaining a high vacuum ($10^{-3}$ Pa) inside the cryostat. The temperature is normally measured with high precision using a calibrated platinum resistance thermometer. A major source of heat leak is through the electrical leads. This can be minimized by tempering the leads as they pass through the nitrogen and helium tanks and then bringing them to the calorimeter temperature with an electrical heater on the floating ring. In operation, a known amount of electrical energy is added through the heater and the temperature rise (usually of the order of 5 K) is measured to within $10^{-3}$ K. The temperature of the adiabatic shield is automatically controlled to follow the temperature of the calorimeter.

*High temperature heat capacity*

Adiabatic and drop calorimetry are the primary methods used to make measurements of heat capacity above room temperature. In drop calorimetry, a known mass of a sample at a known high temperature is dropped into a calorimeter vessel, usually close to room temperature and its temperature rise is measured. This method gives enthalpy differences, which are usually represented as a power series in the temperature. The equation can be differentiated to give the heat capacity. This method can be used to very high temperatures with moderate accuracy but it gives poor results when the sample undergoes phase transitions during the cooling process, since there may not be a complete transformation in the calorimeter. For systems with known phase transitions adiabatic calorimetry is widely used. This technique is similar to that used in low temperature calorimetry except that no cooling is required. An example high temperature adiabatic calorimeter is shown in figure B1.27.2. An adiabatic shield that in the figure is the outer silver cup surrounds the calorimeter proper. Its temperature is controlled to be as close as possible to that of the calorimeter proper. The shield is surrounded by an inner and outer guard, which consist of multiple layers of thin aluminium. The inner guard is usually heated to a temperature close to the calorimeter temperature. Since the main heat loss mechanism at high temperature is radiation, the volume surrounding the calorimeter does not need to be evacuated.

Two methods are generally used when operating an adiabatic calorimeter. In the continuous-heating method, the calorimeter and shield are heated at a constant rate such that the temperature difference between the calorimeter and shield are minimal. At predetermined temperatures the power and time are recorded allowing the average heat capacity to be determined. This method allows for rapid measurement and the control of the shield temperature is less demanding. In the intermittent-heating method a known amount of power is added for a known time and the temperature change measured. The control of the adiabatic shield is more difficult because of the sudden changes in the rate of change of the temperature of the calorimeter at the beginning and end of the heating period. However, it is essential to use the intermittent method when a transition such as a solid–solid, liquid–solid, or an annealing process takes place.

*Heat capacity of gases*

The heat capacity of a gas at constant pressure $C_p$ is normally determined in a flow calorimeter. The temperature rise is determined for a known power supplied to a gas flowing at a known rate. For gases at pressures greater than about 5 MPa Magee *et al* [13] have recently described a twin-bomb adiabatic calorimeter to measure $C_V$.

*B1.27.5.2 Combustion calorimetry*

Combustion or bomb calorimetry is used primary to derive enthalpy of formation values and measurements are usually made at 298.15 K. Bomb calorimeters can be subdivided into three types: (1) static, where the bomb or entire calorimeter (together with the bomb) remains motionless during the experiment; (2) rotating-bomb calorimeters, where provision is made to rotate the bomb in the calorimetric media and (3) entirely rotating calorimeters, called dynamic. It is not necessary to use a rotating-bomb calorimeter for burning conventional organic compounds (containing only C, H, O and N). A stainless steel bomb without a corrosion-proof metal lining is suitable.

For burning organic substances containing heteroatoms of non-metals and metals, dynamic calorimeters of the combined or aneroid types are used. Liquid rotating-bomb calorimeters can also be used. For burning compounds containing halogens and sulphur, a bomb made of corrosion-resistant metal or lined with such a metal is generally used. The most resistant metal for the protection of the inner surface of a bomb used in combustion of chlorine-, sulphur- or bromine-containing organic compounds is tantalum, since it is very little affected by the products of combustion of these substances. Platinum can also be used as a protective layer, even though it is prone to react with the reaction products (e.g. $Cl_2 + HCl + H_2O$); the correction required to account for the enthalpy of such a reaction can be made by the analysis of the quantity of platinum dissolved. To study comparatively slow bomb processes, an adiabatically jacketed calorimeter designed as a combined or aneroid type or an isothermal calorimeter can be used only if the reaction can be conducted under static conditions.

*Static bomb calorimeter*

An example of a static bomb calorimeter used to measure energies of combustion in oxygen is shown in figure B1.27.3. The bomb is typically a heavy walled vessel capable of withstanding pressures of 20 MPa. A precisely known mass of the material to be burnt is held in a small platinum cup and oxygen is added to a pressure of about 3 MPa. The bomb is either immersed in a known mass of water or suspended in an evacuated vessel (aneroid type). The material is ignited by passing a current through a thin platinum wire stretched between the two metal posts, which causes an attached cotton or polythene fuse to burn. A small amount of water is added to the calorimeter to ensure that any solution that is formed is sufficiently dilute to allow the small corrections associated with the various solution processes to be calculated. The amount of material in the vessel is chosen to give a temperature rise of from 1 K to 3 K, for a typical bomb immersed in about 3 kg $H_2O$. For an organic compound this corresponds to a mass of from 0.5 g to 1.6 g. A static bomb calorimeter is used for substances containing only carbon, hydrogen, oxygen and nitrogen giving only carbon dioxide, water and $N_2$ and possibly small amounts of $HNO_2$ and $HNO_3$. The temperature rise is typically measured to between $10^{-4}$ K and $10^{-5}$ K and is measured with either a platinum resistance or a quartz thermometer. In order to relate the temperature rise to the energy of combustion, the calorimeter constant (the amount of energy required to increase the temperature of the calorimeter by 1 K) must be known. This can be obtained either by direct electrical calibration or by burning a certified reference material whose energy of combustion has been determined in specifically designed calorimeters. Direct electrical calibration is not simple as it involves the installation of a high power electrical heater within the bomb. Measurements on reference materials are usually made at a National Standards Laboratory. Reference materials suitable for the calibration of various calorimeters have been recommended by the International Union of Pure and Applied Chemistry (IUPAC) [1]. Benzoic acid is the most used reference material and its energy of combustion is known to about 1 part in 15 000. Involatile liquids or solids can usually be burnt directly. Volatile materials must be encapsulated in either plastic bags or glass ampoules. Combustion of these samples requires the addition of a known mass of an auxiliary material, which is usually an involatile oil whose energy of combustion is known.

*Rotating bomb calorimeter*

For substances containing elements additional to C, H, O and N a rotating bomb calorimeter is generally used. A typical rotating bomb calorimeter system is shown in figure B1.27.4. With this calorimeter considerably more water is added to the combustion bomb and the continuous rotation of the bomb both about the cylindrical axis and end over end ensures that the final solution is homogeneous and in equilibrium with the gaseous products. At the completion of the experiment this solution is withdrawn and analysed. Substances containing S, Si, P and the halogens can be studied in such calorimeters. To ensure that the products are in a known oxidation state it is generally necessary to add small quantities of reducing agents or other materials.

    Precision combustion measurements are primarily made to determine enthalpies of formation. Since the combustion occurs at constant volume, the value determined is the energy change $\Delta_c U$. The enthalpy of

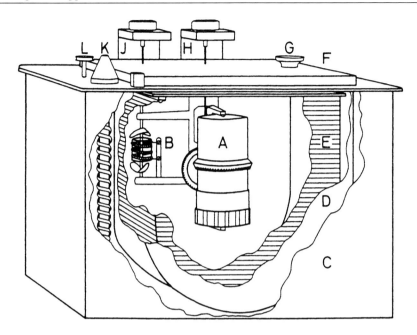

**Figure B1.27.4.** Rotating bomb isoperibole calorimeter. A, stainless steel bomb, platinum lined; B, heater; C, thermostat can; D, thermostat inner wall; E, thermostat water; G, sleeve for temperature sensor; H, motor for bomb rotation; J, motor for calorimeter stirrer; K, connection to cooling or heating unit for thermostat; L, circulation pump.

combustion $\Delta_c H$ can be calculated from $\Delta_c U$, provided that the change in the pressure within the calorimeter is known. This change can be calculated from the change in the number of moles in the gas phase and assuming ideal gas behaviour.

Enthalpies of formation of compounds that do not readily burn in oxygen can often be determined by combusting in fluorine and the enthalpy of formation of volatile substances can be determined using flame calorimetry. For compounds that only combust at an appreciable rate at high temperature, such as zirconium in chlorine, the technique of hot-zone calorimetry is used. In this method one heats the sample only very rapidly with a known amount of energy until it reaches a temperature where combustion will occur. Alternatively, a well characterized material such as benzoic acid can be used as an auxiliary material which, when it burns, raises the temperature sufficiently for the material to combust. These methods have been discussed in detail [2–4].

### B1.27.5.3    Enthalpies of phase change

Accurate enthalpies of solid–solid transitions and solid–liquid transitions (fusion) are usually determined in an adiabatic heat capacity calorimeter. Measurements of lower precision can be made with a differential scanning calorimeter (see later). Enthalpies of vaporization are usually determined by the measurement of the amount of energy required to vaporize a known mass of sample. The various measurement methods have been critically reviewed by Majer and Svoboda [9]. The actual technique used depends on the vapour pressure of the material. Methods based on vaporization into a vacuum are best suited for pressures from about 25 kPa down to $10^{-4}$ Pa. This method has been extensively developed by Sunner's group in Lund [14]. The most recent design allows measurement on samples down to 5 mg over a temperature range from 300 K to 423 K with an accuracy of about 1%. Methods based on vaporization into a steady stream of carrier gas, useful in

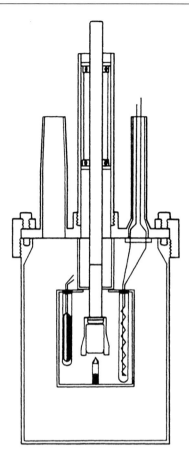

**Figure B1.27.5.** A typical solution calorimeter with thermometer, heater and an ampoule on the base of the stirrer which is broken by depressing it against the ampoule breaker. (Reproduced with permission from Sunner S and Wadsö I 1959 *Acta. Chem. Scand.* **13** 97.)

the range 0.05 Pa to 25 kPa, have also been developed in Lund under Wadsö [15] and gas flow cells based on their designs are commercially available. The method is accurate to between 0.2 and 0.5%. Methods based on vaporization into a closed system, useful in the range 5 kPa to 3 MPa fall into two types; recycle and controlled withdrawal. Both types can give an accuracy approaching 0.1%. Methods based on recycle often contain a second calorimeter to determine the heat capacity of the flowing gas.

### B1.27.5.4  Solution calorimetry

Solution calorimetry covers the measurement of the energy changes that occur when a compound or a mixture (solid, liquid or gas) is mixed, dissolved or adsorbed in a solvent or a solution. In addition it includes the measurement of the heat capacity of the resultant solution. Solution calorimeters are usually subdivided by the method in which the components are mixed, namely, batch, titration and flow.

*Batch calorimeters*

Batch calorimeters are instruments where there is no flow of matter in or out of the calorimeter during the time the energy change is being measured. Batch calorimeters differ in the way the reactants are mixed and in the method used to determine the enthalpy change. Enthalpy changes can be measured by the various methods outlined above; isothermal, adiabatic, heat flow or isoperibole. It is necessary to have the reactants separated in the calorimeter. The most common method is to maintain one of the reactants in an ampoule that is broken to release its contents, which initiates the reaction. Initially, thin walled glass ampoules were used but these usually required the narrow neck to be flame-sealed after the contents were added. In recent years there have been significant improvements in ampoule design. An ampoule particularly suited for solids consists of a stainless steel cylinder with replaceable thin glass windows at each end. The cylinder can be taken apart with one half forming a cup into which the solid can be added and weighed. Wadsö and coworkers have developed a variety of ampoules that attach to the stirrer. The glass window is broken by depressing the stirrer so as to impinge against an ampoule-breaking pin. This technique ensures good mixing of the reactants. A typical solution calorimeter is shown in figure B1.27.5

Ampoules are satisfactory when the presence of a vapour space is not important. When volatile organic compounds are mixed in the presence of a vapour space there can be a considerable contribution to the measured heat effect from the vaporization. This results from the change in the vapour composition that occurs so as to maintain vapour–liquid equilibrium with the liquid mixture. For a small enthalpy of mixing, this correction can be greater than the enthalpy of mixing itself. A batch method suitable for the measurement of enthalpies of mixing in the absence of a vapour space is shown in figure B1.27.6. Known masses of liquids A and B are separately confined over mercury and are mixed by rotation of the entire calorimeter. When liquids are mixed at constant pressure there is a volume change on mixing. The side arm C, which is partially filled with mercury, allows for the expansion or contraction of the mixture against the air space D. The calorimeter operates in the isoperibole mode.

Reviews of batch calorimeters for a variety of applications are published in the volume on *Solution Calorimetry* [8]: cryogenic conditions by Zollweg [22], high temperature molten metals and alloys by Colinet and Pasturel [19], enthalpies of reaction of inorganic substances by Cordfunke and Ouweltjes [16], electrolyte solutions by Simonson and Mesmer [24], and aqueous and biological systems by Wadsö [25].

*Titration or dilution calorimeters*

In a titration or dilution calorimetry one fluid is added from a burette, usually at a well-defined rate, into a calorimeter containing the second fluid. One titration is equivalent to many batch experiments. Calorimeters are typically operated in either the isoperibole or isothermal mode. The method has been extensively developed by Izatt, Christensen and co-workers at Brigham Young University for the measurement of formation constants and enthalpies of reaction for a variety of organic and inorganic compounds. This technique is described in detail by Oscarson *et al* [23]. A titration calorimeter typically has a vapour space, which for large enthalpies of reaction or solution in a solvent with a low vapour pressure does not give rise to significant errors. The vapour space can be eliminated by filling the cell completely and, as the titration proceeds, the excess liquid flows out of the calorimeter into a reservoir. The calculations of the enthalpy change for such a procedure is complex. To measure enthalpies of mixing of organic liquids, Stokes and Marsh [20] have used an alternative method that eliminates both the vapour space and the effect of volume changes on mixing. Their isothermal dilution calorimeter is shown in figure B1.27.7. The calorimeter proper, made from either stainless steel or glass, contains a stirrer, a sealed heater, a thermistor to measure the temperature, and a silver rod connected to a Peltier device. A known volume of mercury is added from a pipette and one component is added to completely fill the calorimeter. The calorimeter is then brought to isothermal conditions within $10^{-3}$ K. For endothermic reactions the Peltier device removes energy at a rate sufficient to counterbalance the power

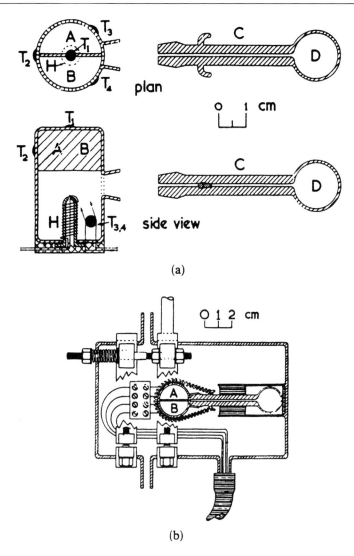

**Figure B1.27.6.** A calorimeter for enthalpies of mixing in the absence of a vapour space. (Reproduced with permission from Larkin J A and McGlashan M L 1961 *J. Chem. Soc.* 3245.)

introduced from the stirrer. The second component is then injected into the calorimeter from a burette, displacing the mercury. Electrical power is added to maintain the calorimeter approximately isothermal. Usually the rate of addition of the second component is adjusted so that the calorimeter remains isothermal to within $10^{-3}$ K during the addition. The injection is stopped at selected intervals and the calorimeter brought back to isothermal conditions by either the addition of additional electrical power or additional liquid from the burette. The volume of the mercury pipette is such that one run, which comprises about 20 individual measurement, covers over half the composition range. The components are then interchanged to determine the second half of the curve. The two runs should give results overlapping to within 1%. For exothermic systems the Peltier device is run at high power and the energy removed is counterbalanced by the addition of electrical energy to maintain the calorimeter isothermal. When the second component is added, the power is

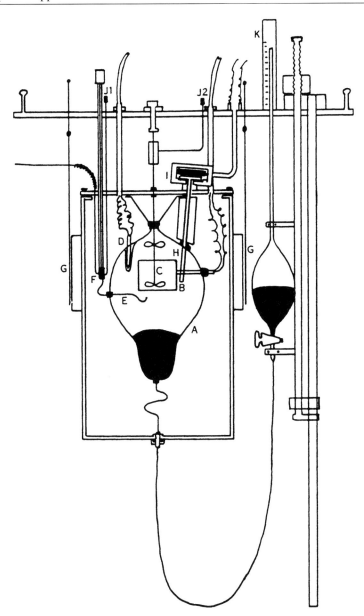

**Figure B1.27.7.** Schematic diagram of isothermal displacement calorimeter: A, glass calorimeter cell; B, sealed heater; C, stainless steel stirrer; D, thermistor; E, inlet tube; F, valve; G, window shutters; H, silver rod; I, thermoelectric cooler; J, small ball valves; K, levelling device. (reproduced by permission from Costigan M J, Hodges L J, Marsh K N, Stokes R H and Tuxford C W 1980 *Aust. J. Chem.* **33** 2103.)

turned off for known periods of time. This calorimeter has been modified to measure enthalpies of solution of gases in liquids and Stokes [20] has described a version of this calorimeter that uses the overflow technique. Titration and dilution calorimeters have the disadvantage that they are difficult to operate at high pressures or at temperatures considerably removed from ambient. Flow calorimetry does not suffer these disadvantages.

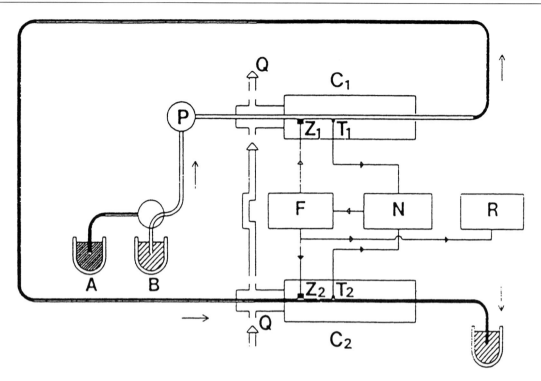

**Figure B1.27.8.** Schematic view of Picker's flow microcalorimeter. A, reference liquid; B, liquid under study; P, constant flow circulating pump; $Z_1$ and $Z_2$, Zener diodes acting as heaters; $T_1$ and $T_2$, thermistors acting as temperature sensing devices; F, feedback control; N, null detector; R, recorder; Q, thermostat. In the above A is the reference liquid and $C_2$ is the reference cell. When B circulates in cell $C_1$ this cell is the working cell. (Reproduced by permission from Picker P, Leduc P-A, Philip P R and Desnoyers J E 1971 *J. Chem. Thermo.* B **41**.)

*Flow calorimeters*

In a flow calorimeter one or more streams flow in and out of the calorimeter. Flow calorimeters are suited for measurements over a wide range of temperatures and pressures on both liquids and gases. A wide pressure range is possible because measurement are usually made in a small bore tube, and a wide temperature range is feasible because the in-flowing material can be readily brought to the calorimeter temperature by heat exchange with the out-flowing fluid. In a flow experiment there is generally no vapour space and changes in volume on mixing are inconsequential. Flow methods are not suitable for measurement involving solids, and usually large volumes of materials are required. Various flow calorimeters are available commercially.

Flow calorimeters have been used to measure heat capacities, enthalpies of mixing of liquids, enthalpy of solution of gases in liquids and reaction enthalpies. Detailed descriptions of a variety of flow calorimeters are given in *Solution Calorimetry* by Grolier [17], by Albert and Archer [18], by Ott and Wormald [21], by Simonson and Mesmer [24] and by Wadsö [25].

A flow calorimeter developed by Picker suitable for the measurement of heat capacity of a liquid is shown in figure B1.27.8. The method measures the difference in heat capacity between the fluid under study and some reference fluid. The apparatus contains two thermistors $T_1$ and $T_2$ used to measure the temperature change that occurs when the flowing fluid is heated by two identical heaters $Z_1$ and $Z_2$. The standard procedure is to flow the reference material from A through both cells. With the same fluid, same power and same flow rate the temperature change $\Delta T$ should be the same. The temperature difference observed on flowing the sample

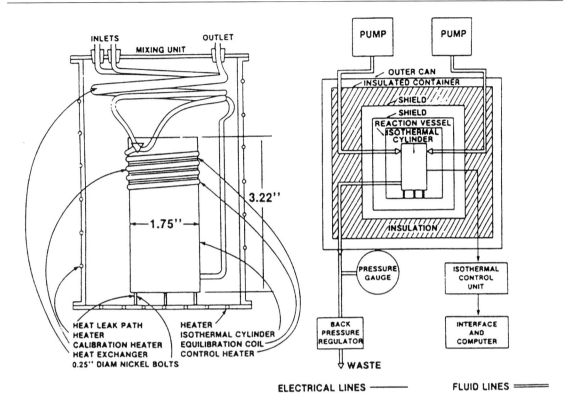

**Figure B1.27.9.** High-temperature heat-leak calorimeter. (Reproduced by permission from Christensen J J and Izatt R M 1984 An isothermal flow calorimeter designed for high-temperature, high-pressure operation *Thermochim. Acta* **73** 117–29.)

material from B through cell $C_1$ while the reference is still flowing through $C_2$ is a measure of the heat capacity difference between the two liquids. The flow method has been extensively developed for measurement on biological systems and on liquid mixtures at high temperatures and pressures. The apparatus constructed by Christensen and Izatt, shown in figure B1.27.9, can be used to measure positive and negative enthalpy changes at pressures up to 40.5 MPa and temperatures up to 673 K. Two high pressure pumps were used for the fluid flow. Mixing occurs in the top half of the isothermal cylinder where the fluid from the two pumps meet. A control heater encircles the cylinder, which is attached by three heat-leak rods to a base plate maintained 1 K below the cylinder temperature. A pulsed electrical current is passed through the control heater to maintain the temperature of the cylinder the same as that of the walls of the oven. During mixing the frequency of the pulses are either increased or decreased depending on the size of the enthalpy of mixing. Commercial calorimeters are available based on both the Picker and Christensen *et al* designs.

## B1.27.6 Differential scanning calorimetry

Boerio-Goats and Callanan [12] have recently reviewed different thermal methods, including differential scanning calorimetry. A differential scanning calorimeter (DSC) consists of two similar cells containing a sample and a reference material. In one type of DSC both cells are subjected to a controlled temperature change by applying power to separate heaters and the temperature difference between the sample and reference cells is observed. When an endothermic change occurs in the sample, for example, melting, the sample temperature

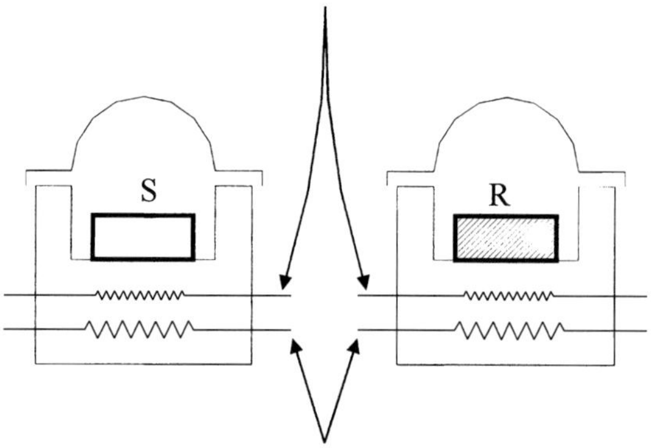

Temperature sensors

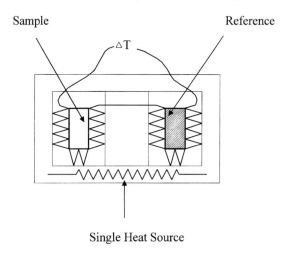

**Figure B1.27.10.** Schematic diagram of a power-compensated DSC.

**Figure B1.27.11.** Schematic diagram of a Tian–Calvet heat-flux or heat-conduction calorimeter.

lags behind that of the reference. In a power compensated DSC the power to that cell is increased to keep the heating rate of the sample and reference cells the same. A schematic diagram of a typical DSC is shown in figure B1.27.10. Another type of calorimeter, developed initially by Tian and Calvet, is also considered a differential scanning calorimeter but is called a heat-flux or heat-conduction calorimeter. A schematic diagram of this type of calorimeter is shown in figure B1.27.11. In this calorimeter two sets of thermopiles, consisting of multiple junction thermocouples, connect both cells to a large block enclosing the sample. The output of the two thermopiles, when connected in opposition gives a measure of the difference in energy flows between the sample and reference when both are heated at the same rate. Both types of DSC can be used to measure heat capacities, enthalpies of phase change, adsorption, dehydration, reaction and polymerization. The major advantage of DSC is the rapidity of the measurements, the small sample requirement, and the ready availability of commercially available equipment capable of operating from liquid nitrogen temperatures to well above 1000 K by unskilled personnel. With the majority of DSC instruments it is possible to obtain heat capacities with an accuracy approaching 2–3%, provided one uses the optimum sample size and scan rate

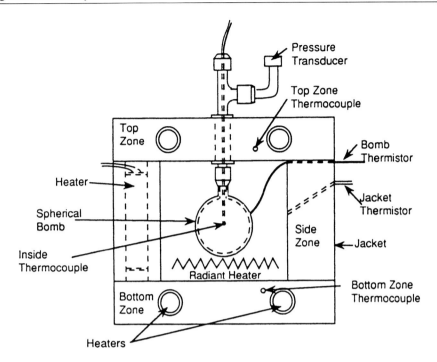

**Figure B1.27.12.** Schematic diagram of an accelerating rate calorimeter (ARC).

along with careful calibration of the temperature scale and the calorimetric response. Reference materials are used to calibrate a DSC and to check the correct operating conditions. Differential scanning calorimeters have been developed for specific applications. A very precise calorimeter, developed by Privilov and coworkers [17] and now available commercially, has been used to measure the heat capacity of very small amounts of biological materials in aqueous solutions.

### B1.27.7   Accelerating rate calorimetry

Special calorimeters have been developed to make thermal hazard evaluations. In an exothermic chemical reaction there is the possibility of a runaway reaction occurring where the energy released from the reaction increases the temperature with a consequent increase in the reaction rate, thus increasing the release of energy. If there are insufficient resources to remove the generated energy, hazardous temperature and pressure regimes can be encountered. An accelerating rate calorimeter or ARC, initially developed at Dow Chemical, is available commercially to study such thermal hazards. A schematic diagram is shown in figure B1.27.12. The reaction vessel consists of a spherical bomb that can withstand pressures greater than 20 MPa and temperatures to 770 K. The calorimeter operates in an adiabatic mode under computer control in a heat-wait-seek mode. After the reaction comes to thermal equilibrium the rate of temperature rise due to the reaction is determined. If this is less than a preset value the calorimeter temperature is increased in steps and the process repeated until the reaction rate is sufficient to give the preset temperature rise. The chemical reaction then proceeds at its own rate and the temperature and pressure recorded. From these measurements the kinetic parameters are determined and used to establish the conditions that could lead to a runaway reaction. A problem with this calorimeter is that the massive vessel required to withstand the pressure has a heat capacity well in excess of the heat capacity of the reactants. This problem can be overcome by having a thin-walled vessel within the bomb and the pressure in the space between the reaction vessel and the bomb is automatically controlled to the pressure in the calorimeter.

## B1.27.8 Specialized calorimeters

Ultra sensitive Calvet type microcalorimeters are available commercially to measure the deterioration of materials over relatively long periods. For example, the lifetime of a battery can be estimated from the energy released when it is placed in such a calorimeter on open circuit [27]. Similarly the shelf life of drugs and other digestible products can be evaluated from the very small calorimetric response that results from decomposition reactions [28]. Calorimeters have also been developed to measure the heat effects associated with the uptake of oxygen and carbon dioxide in living plants. Such measurements have been used to identify species that exhibit high rates of metabolism [29]. Calorimeters are also used to measure the rate of enzyme reaction and as a clinical tool to identify micro-organisms and test the effect of drugs in inhibiting the growth of such micro-organisms.

## B1.27.9 Recent developments

Recent developments in calorimetry have focused primarily on the calorimetry of biochemical systems, with the study of complex systems such as micelles, proteins and lipids using microcalorimeters. Over the last 20 years microcalorimeters of various types including flow, titration, dilution, perfusion calorimeters and calorimeters used for the study of the dissolution of gases, liquids and solids have been developed. A more recent development is pressure-controlled scanning calorimetry [26] where the thermal effects resulting from varying the pressure on a system either step-wise or continuously is studied.

## References

[1] Head A J and Sabbah R 1987 Enthalpy *Recommended Reference Materials for the Realization of Physicochemical Properties* ed K N Marsh (Oxford: Blackwell)
[2] Rossini F D (ed) 1956 *Experimental Thermochemistry* vol I (New York: Interscience)
[3] Skinner H A (ed) 1962 *Experimental Thermochemistry* vol II (New York: Interscience)
[4] Sunner S and Månsson M (eds) 1979 *Combustion Calorimetry* (Oxford: Pergamon)
[5] McCullough J P and Scott D W (eds) 1968 *Experimental Thermodynamics Calorimetry of Non-Reacting Systems* vol I (London: Butterworths)
[6] LeNeindre B and Vodar B (eds) 1975 *Experimental Thermodynamics Experimental Thermodynamics of Non-Reacting Systems* vol II (London: Butterworths)
[7] Wakeham W A, Nagashima A and Sengers J V (eds) 1991 *Experimental Thermodynamics Measurement of Transport Properties of Fluids* vol III (Oxford: Blackwell)
[8] Marsh K N and O'Hare P A G (eds) 1994 *Solution Calorimetry, Experimental Thermodynamics* vol IV (Oxford: Blackwell)
[9] Majer V and Svoboda V 1985 (IUPAC Chemical Data Series No 32) *Enthalpies of Vaporization of Organic Compounds* (Oxford: Blackwell)
[10] McGlashan M L (ed) 1973 *Specialist Periodical Reports, Chemical Thermodynamics* vol 1 (London: The Chemical Society)
[11] Oscarson J L and Izatt R M 1992 Calorimetry *Physical Methods of Chemistry Determination of Thermodynamic Properties* 2nd edn, vol VI, ed B W Rossiter and R C Baetzold (New York: Wiley)
[12] Boerio-Goates J and Callanan J E 1992 Differential thermal methods *Physical Methods of Chemistry Determination of Thermodynamic Properties* 2nd edn, vol VI, ed B W Rossiter and R C Baetzold (New York: Wiley)
[13] Magee J W, Blanco J C and Deal R J 1998 High-temperature adiabatic calorimeter for constant-volume heat capacity of compressed gases and liquids *J. Res. Natl Inst. Stand. Technol.* **103** 63
[14] Morawetz E 1972 Enthalpies of vaporization of *n*-alkanes from $C_{12}$ to $C_{20}$ *J. Chem. Thermodyn.* **4** 139
[15] Wadsö I 1968 Heats of vaporization of organic compounds: II. Chlorides, bromides and iodides *Acta Chem. Scand.* **22** 2438
[16] Cordfunke E H P and Ouweltjes W 1994 Solution calorimetry for the determination of enthalpies of reaction of inorganic substances at 298.15 K *Solution Calorimetry, Experimental Thermodynamics* vol IV, ed K N Marsh and P A G O'Hare (Oxford: Blackwell)
[17] Grolier J-P E 1994 Heat capacity of organic liquids *Solution Calorimetry, Experimental Thermodynamics* vol IV, ed K N Marsh and P A G O'Hare (Oxford: Blackwell)
[18] Albert H J and Archer D G 1994 Mass-flow isoperibole calorimeters *Solution Calorimetry, Experimental Thermodynamics* vol IV, ed K N Marsh and P A G O'Hare (Oxford: Blackwell)

[19] Colinet C and Pasturel A 1994 High temperature solution calorimetry *Solution Calorimetry, Experimental Thermodynamics* vol IV, ed K N Marsh and P A G O'Hare (Oxford: Blackwell)

[20] Stokes R H 1994 Isothermal displacement calorimeters *Solution Calorimetry, Experimental Thermodynamics* vol IV, ed K N Marsh and P A G O'Hare (Oxford: Blackwell)

[21] Ott J B and Wormald C J 1994 Excess enthalpy by flow calorimetry *Solution Calorimetry, Experimental Thermodynamics* vol IV, ed K N Marsh and P A G O'Hare (Oxford: Blackwell)

[22] Zollweg J A 1994 Mixing calorimetry at cryogenic conditions *Solution Calorimetry, Experimental Thermodynamics* vol IV, ed K N Marsh and P A G O'Hare (Oxford: Blackwell)

[23] Oscarson J L, Izatt R M, Hill J O and Brown P R 1994 Continuous titration calorimetry *Solution Calorimetry, Experimental Thermodynamics* vol IV, ed K N Marsh and P A G O'Hare (Oxford: Blackwell)

[24] Simonson J M and Mesmer R E 1994 Electrolyte solutions at high temperatures and pressures *Solution Calorimetry, Experimental Thermodynamics* vol IV, ed K N Marsh and P A G O'Hare (Oxford: Blackwell)

[25] Wadsö I 1994 Microcalorimetry of aqueous and biological systems *Solution Calorimetry, Experimental Thermodynamics* vol IV, ed K N Marsh and P A G O'Hare (Oxford: Blackwell)

[26] Randzio S L 1994 Calorimetric determination of pressure effects *Solution Calorimetry, Experimental Thermodynamics* vol IV, ed K N Marsh and P A G O'Hare (Oxford: Blackwell)

[27] Hansen L D and Hart R M 1978 Shelf-life prediction from induction period *J. Electrochem. Soc.* **125** 842

[28] Hansen L D, Eatough D J, Lewis E A and Bergstrom R G 1990 Calorimetric measurements on materials undergoing autocatalytic decomposition *Can. J. Chem.* **68** 2111–14

[29] Hansen L D, Hopkins M S and Criddle R S 1997 Plant calorimetry: a window to plant physiology and ecology *Thermochim. Acta* **300** 183–97

## Further Reading

Oscarson J L and Izatt R M 1992 Calorimetry *Determination of Thermodynamic Properties, Physical Methods of Chemistry* 2nd edn vol VI ed B W Rossiter and R C Baetzold (New York: Wiley)

Boerio-Goates J and Callanan J E 1992 Differential thermal methods *Determination of Thermodynamic Properties, Physical Methods of Chemistry* 2nd edn vol VI ed B W Rossiter and R C Baetzold (New York: Wiley)

Rossini F D (ed) 1956 *Experimental Thermochemistry* vol I (New York: Interscience)

Skinner H A (ed) 1962 *Experimental Thermochemistry* vol II (New York: Interscience)

Sunner S and Månsson M (eds) 1979 *Combustion Calorimetry* (Oxford: Pergamon)

McCollough J P and Scott D W (eds) 1968 *Calorimetry of Non-Reacting Systems, Experimental Thermodynamics* vol I (London: Butterworths)

Marsh K N and O'Hare P A G (eds) 1994 *Solution Calorimetry, Experimental Thermodynamics* vol IV (Oxford: Blackwell)

Head A J and Sabbah R 1987 Enthalpy *Recommended Reference Materials for the Realization of Physicochemical Properties* ed K N Marsh (Oxford: Blackwell)

Majer V and Svoboda V 1985 *Enthalpies of Vaporization of Organic Compounds* (Oxford: Blackwell)

# B1.28
# Electrochemical methods

*Alexia W E Hodgson*

### B1.28.1   Introduction

Electrochemical methods may be classified into two broad classes, namely potentiometric methods and voltammetric methods. The former involves the measurement of the potential of a working electrode immersed in a solution containing a redox species of interest with respect to a reference electrode. These are equilibrium experiments involving no current flow and provide thermodynamic information only. The potential of the working electrode responds in a Nernstian manner to the activity of the redox species, whilst that of the reference electrode remains constant. In contrast, in voltammetric methods the system is perturbed and involves the control of the electrode potential or the current as the independent variable, and measurement of the resulting current or potential.

The latter may be further subdivided into *transient experiments*, in which the current and potential vary with time in a non-repetitive fashion; *steady-state experiments*, in which a unique interrelation between current and potential is generated, a relation that does not involve time or frequency and in which the steady-state current achieved is independent of the method adopted and *periodic experiments*, in which current and potential vary periodically with time at some imposed frequency.

In this chapter, transient techniques, steady-state techniques, electrochemical impedance, photoelectro-chemistry and spectroelectrochemistry are discussed.

### B1.28.2   Introduction to electrode reactions

Electrode processes are a class of heterogeneous chemical reaction that involves the transfer of charge across the interface between a solid and an adjacent solution phase, either in equilibrium or under partial or total kinetic control. A simple type of electrode reaction involves electron transfer between an inert metal electrode and an ion or molecule in solution. Oxidation of an electroactive species corresponds to the transfer of electrons from the solution phase to the electrode (anodic), whereas electron transfer in the opposite direction results in the reduction of the species (cathodic). Electron transfer is only possible when the electroactive material is within molecular distances of the electrode surface; thus for a simple electrode reaction involving solution species of the form

$$O + n\,e^- \rightarrow R$$

in which species $O$ is reduced at the electrode surface to species $R$ by the transfer of $n$ electrons, the overall conversion may be divided into three steps [1, 3]:

$$O_{\text{bulk}} \xrightarrow{\text{mass transport}} O_{\text{electrode}}$$

$$O_{\text{electrode}} \xrightarrow{\text{electron transfer}} R_{\text{electrode}}$$

$$R_{\text{electrode}} \xrightarrow{\text{mass transport}} R_{\text{bulk}}.$$

The scheme involves the transport of the electroactive species from the bulk solution to the electrode surface, where it can undergo electron transfer, thus forming the reduced species $R$ at the electrode surface. Finally, the reduced species is transported from the electrode surface back to the bulk solution. The overall reaction rate will be limited by the slowest step, therefore a particular reaction might be controlled by either the kinetics of electron transfer or by the rate at which material is brought to or from the electrode surface. The rate of electron transfer can be experimentally controlled through the electrode potential imposed and can vary by several orders of magnitude in a small potential interval. For the steps involving the transport of species to and from the electrode surface there are three distinct modes of mass transport regime which can occur: diffusion, migration and convection.

The nature of electrode processes can, of course, be more complex and also involve phase formation, homogeneous chemical reactions, adsorption or multiple electron transfer [1–4].

### B1.28.2.1 Electron transfer

For a simple electron transfer reaction containing low concentrations of a redox couple in an excess of electrolyte, the potential established at an inert electrode under equilibrium conditions will be governed by the Nernst equation and the electrode will take up the equilibrium potential $E_e$ for the couple $O/R$. In terms of current density, the dynamic situation at the electrode surface is expressed by $j = \vec{j} + \overleftarrow{j} = 0$, the sum of the partial cathodic and partial anodic current densities, which have opposite signs and the magnitude of which at equilibrium potential is defined as $j_0 = -\vec{j} = \overleftarrow{j}$. The exchange current density, $j_0$, is a measure of the amount of electron transfer activity at the equilibrium potential. On applying a potential to the electrode, the system will seek to move towards a new equilibrium where the concentrations of the electroactive species are those demanded by the Nernst equation for the applied potential, and an associated current of reduction or oxidation will flow. The rate of electron transfer can be described by classical kinetics, and hence expressed by the product of a rate constant with the concentration of the reactant at the electrode surface. The rate of the heterogeneous electron transfer will depend on the potential gradient at the interface driving the transfer of electrons between the electrode and the solution phases and in general will take the form of $\vec{k} = k_0 \exp((-\alpha_c n F/RT)E)$ and $\overleftarrow{k} = k_0 \exp((-\alpha_A n F/RT)E)$ for a reduction and oxidation process respectively. $\alpha_C$ and $\alpha_A$ are the cathodic and anodic transfer coefficients, $F$ is the Faraday constant and $k_0$ the standard rate constant [1–4]. By substituting and defining the overpotential, $\eta = E - E_e$, as the deviation of the potential from the equilibrium value, the Butler–Volmer equation for current density may be derived:

$$j = j_0 \left\{ \exp\left( \frac{\alpha_A n F}{RT} \eta \right) - \exp\left( \frac{-\alpha_C n F}{RT} \eta \right) \right\}$$

which represents the fundamental equation of electrode kinetics. The equation may be simplified for the limiting cases in which very high positive or very high negative overpotentials are applied, leading to the Tafel equations, which provide a simple method for determining exchange current density and transfer coefficient (figure B1.28.1). For very low overpotentials the equation simplifies to $j = j_0(nF/RT)\eta$, indicating that very close to the equilibrium potential, the current density varies linearly with overpotential.

### B1.28.2.2 Mass transport

Diffusion, convection and migration are the forms of mass transport that contribute to the essential supply and removal of material to and from the electrode surface [1–4].

*Diffusion* may be defined as the movement of a species due to a concentration gradient, which seeks to maximize entropy by overcoming inhomogeneities within a system. The rate of diffusion of a species, the

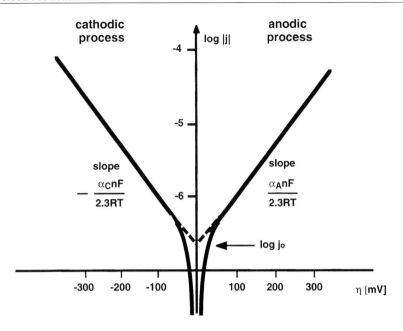

**Figure B1.28.1.** Schematic Tafel plot for the experimental determination of $j_0$ and $\alpha$.

flux, at a given point in solution is dependent upon the concentration gradient at that particular point and was first described by Fick in 1855, who considered the simple case of linear diffusion to a planar surface:

$$\text{Flux} = -D\frac{dc_i(x)}{dx}$$

where $dc_i/dx$ is the concentration gradient and $D$ is the diffusion coefficient (figure B1.28.2). The flux of species to and from the electrode surface must also be accompanied by the conversion of reactant to product and by the flux of electrons. The flux of material crossing the electrode boundary can therefore be converted to current density by equating the two fluxes:

$$\frac{j}{nF} = -D\frac{dc_i(x)}{dx}.$$

The second of Fick's laws expresses the change in concentration of a species at a point as a function of time due to diffusion (figure B1.28.2). Hence, the one-dimensional variation in concentration of material within a volume element bounded by two planes $x$ and $x + dx$ during a time interval $dt$ is expressed by $(\partial c_i(x, t)/\partial t) = D(\partial^2 c_i(x, t)/\partial x^2)$. Fick's second law of diffusion enables predictions of concentration changes of electroactive material close to the electrode surface and solutions, with initial and boundary conditions appropriate to a particular experiment, provide the basis of the theory of instrumental methods such as, for example, potential-step and cyclic voltammetry.

*Convection* is the movement of a species due to external mechanical forces. This can be of two types: natural convection, which arises from thermal gradients or density differences within the solution, and forced convection, which can take the form of gas bubbling, pumping or stirring. The former is undesirable and can occur in any solution at normal sized electrodes on the time scale of ten or more seconds. In contrast, the latter has the function of overcoming contributions from natural convection and of increasing the rate of mass transport and hence facilitates the study of the kinetics of electrode reactions. Forced convection usually

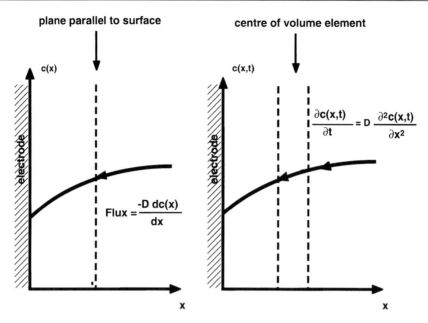

**Figure B1.28.2.** Fick's laws of diffusion. (a) Fick's first law, (b) Fick's second law.

possesses well defined hydrodynamic behaviour, thus enabling the quantitative description of the flow in the solution and the prediction of the pattern of mass transport to the electrode.

*Migration* is the movement of ions due to a potential gradient. In an electrochemical cell the external electric field at the electrode/solution interface due to the drop in electrical potential between the two phases exerts an electrostatic force on the charged species present in the interfacial region, thus inducing movement of ions to or from the electrode. The magnitude is proportional to the concentration of the ion, the electric field and the ionic mobility.

Most electrochemical experiments are designed so that one of the mass transport regimes dominates over the others, thus simplifying the theoretical treatment, and allowing experimental responses to be compared with theoretical predictions. Normally, specific conditions are selected where the mass-transport regime results only from diffusion or convection. Such regimes allow mass transport to be described by a set of mathematical equations, which have analytical solutions. A common experimental practice to render the migration of reactants and products negligible is to add an excess of inert supporting electrolyte, thus ensuring that any migration is dominated by the ions of the electrolyte. Electro-neutrality is also thus maintained, ensuring that electric fields do not build up in the solution. Furthermore, the addition of a high concentration of electrolyte increases the solution conductivity, compresses the double-layer region to dimensions of 10–20 Å, and ensures a constant ionic strength during the electrochemical experiment. As a consequence, the activities of the electroactive species and thus the applied potentials, as predicted by the Nernst equation and by the rate of electron transfer, remain constant throughout the experiment.

### B1.28.3   Transient techniques

Voltammetry relies on the registering of current–potential profiles, whether by controlling the potential of the working electrode and recording the resulting current or by measuring the potential response as a function of an applied current. The electrochemical cell, as well as a conducting medium, must also contain at least one other electrode. In a two-electrode configuration, the second electrode is a reference electrode that serves both as a standard against which the working potential is measured and as the necessary current-carrying electrode where the rate of charge transfer must be equal and opposite to that of the working electrode. Commonly, these two functions are separated in a three-electrode configuration, in which a secondary or counter electrode is employed as the current-carrying electrode and a separate reference electrode reports the potential of the working electrode. This prevents any undesirable polarization of the reference electrode, since only small currents flow in the reference electrode loop. Placement of the reference electrode close to the working electrode enables the exclusion of the majority of the solution $IR_u$ drop, which is often achieved by the use of a Luggin capillary [1, 2].

The measurement of the current for a redox process as a function of an applied potential yields a voltammogram characteristic of the analyte of interest. The particular features, such as peak potentials, half-wave potentials, relative peak/wave height of a voltammogram give qualitative information about the analyte electrochemistry within the sample being studied, whilst quantitative data can also be determined. There is a wealth of voltammetric techniques, which are linked to the form of potential program and mode of current measurement adopted. Potential-step and potential-sweep techniques are carried out under conditions where diffusion is the only mode of mass transport and the experiment is designed such that diffusion may be described by linear diffusion to a plane electrode and changes in concentration occur perpendicular to the surface [1–5].

### B1.28.3.1 Linear-sweep and cyclic voltammetry

Linear-sweep and cyclic voltammetry were first reported in 1938 and described theoretically in 1948 by Randles and Sevčik [1–6]. The techniques consist of scanning the potential between two chosen limits at a known sweep rate, $v$, and measuring the current response arising from any electron transfer process. In linear-sweep voltammetry the scan terminates at the chosen end potential, $E_f$, whereas in cyclic voltammetry, the potential is reversed back at $E_f$ toward the starting potential $E_i$, or another chosen potential limit. The potential limits define the electrode reactions that take place so that the potential scan is normally chosen to start at a potential value where no electrode reaction occurs and swept towards positive or negative potentials to investigate oxidation or reduction processes, respectively. The current–potential curves for a simple reversible electrode reaction are characterized by unsymmetrical peaks with the current density increasing as the sweep rate is raised. On the forward sweep, the current begins to rise as the potential reaches the vicinity of the reversible formal potential $E_e^0$, then passes through a maximum before decreasing again as the potential is sufficiently driven to produce a diffusion-limited current. The surface concentration of an electroactive species, $R$, decreases as the potential is made more positive and the rate of oxidation increases, until it becomes effectively zero, at which point the reaction is diffusion controlled. In terms of concentration profiles, the flux of species to the electrode surface increases with potential (hence time) and continues to increase until the surface concentration reaches zero, at which point the flux to the surface starts to decrease, since the surface concentration remains at zero, yielding the peak-shaped response (figure B1.28.3). On the reverse sweep, anodic current continues to flow until the potential is still sufficiently negative to cause the oxidation of $R$. When the potential, however, reaches the vicinity of $E_e^0$, the oxidized species produced can diffuse back to the electrode surface to be reduced, the current becomes cathodic and a similar peak-shaped response is obtained as the reaction becomes diffusion controlled.

The scan rate, $v = |dE/dt|$, plays a very important role in sweep voltammetry as it defines the time scale of the experiment and is typically in the range 5 mV s$^{-1}$ to 100 V s$^{-1}$ for normal macroelectrodes, although sweep rates of $10^6$ V s$^{-1}$ are possible with microelectrodes (see later). The short time scales in which the experiments are carried out are the cause for the prevalence of non-steady-state diffusion and the peak-shaped response. When the scan rate is slow enough to maintain steady-state diffusion, the concentration profiles with time are linear within the Nernst diffusion layer which is fixed by natural convection, and the current–potential response reaches a plateau steady-state current. On reducing the time scale, the diffusion layer cannot relax

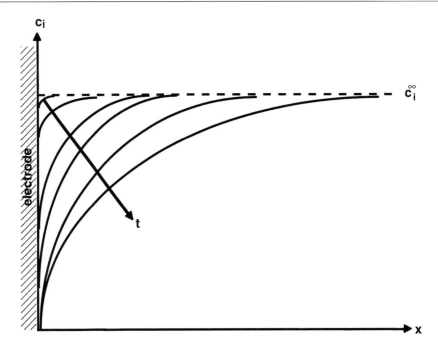

**Figure B1.28.3.** Concentration profiles of an electroactive species with distance from the electrode surface during a linear sweep voltammogram.

**Table B1.28.1.** Diagnostic tests for reversibility of electrode processes in cyclic voltammetry at 293 K.

| Reversible process | Irreversible process | Quasi-reversible process |
|---|---|---|
| $\Delta E_p = E_p^A - E_p^C = 59/n$ (mV) | Absence of reverse peak | $\Delta E_p > 59/n$ (mV), increases with $v$ |
| $|E_p - E_{p/2}| = 59/n$ (mV) | $|E_p - E_{p/2}| = 48/\alpha_C n_\alpha$ (mV) | $|I_p^A/I_p^C| = 1$, if $\alpha_C = \alpha_A = 0.5$ |
| $|I_p^A/I_p^C| = 1$ | $I_p^C \propto v^{1/2}$ | $I_p$ increases with $v^{1/2}$; $I_p$ not $\propto v^{1/2}$ |
| $I_p \propto v^{1/2}$ | $E_p^C$ dependent of $v$ | $E_p^C$ shifts with increasing $v$ |
| $E_p$ independent of $v$ | | |
| $E > E_p, I \propto t^{-1/2}$ | | |

to its equilibrium state, the diffusion layer is thinner and hence the currents in the non-steady-state will be higher.

Cyclic voltammetry provides a simple method for investigating the reversibility of an electrode reaction (table B1.28.1). The reversibility of a reaction closely depends upon the rate of electron transfer being sufficiently high to maintain the surface concentrations close to those demanded by the electrode potential through the Nernst equation. Therefore, when the scan rate is increased, a reversible reaction may be transformed to an irreversible one if the rate of electron transfer is slow. For a reversible reaction at a planar electrode, the peak current density, $j_p$, is given by

$$j_p = 2.69 \times 10^5 \, n^{3/2} D^{1/2} c_i^\infty v^{1/2}$$

where $n$ is the number of electrons, $D$ is the diffusion coefficient, $c_i^\infty$ the concentration of the electroactive species in the bulk and $v$ the sweep rate. Of particular importance is the proportionality of the peak current to

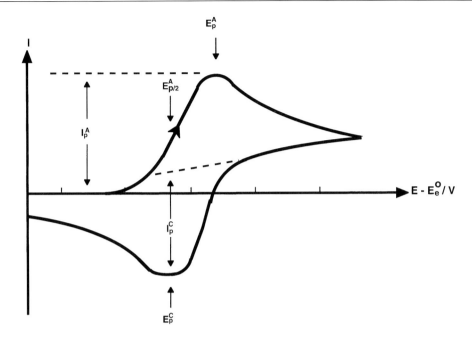

**Figure B1.28.4.** Cyclic voltammogram for a simple reversible electrode reaction in a solution containing only oxidized species.

the square root of the scan rate. In addition, for a reversible couple, the cathodic and anodic peak potentials are separated by $59/n$ mV, the reversible half-wave potential is situated midway between the peaks, the peak potential is independent of scan rate and the peak current ratio equals 1 (figure B1.28.4). As the response becomes less reversible, the separation between the peaks increases, as an overpotential is necessary to drive reduction and oxidation reactions, and the shape of the peaks will become more drawn out. Beyond the peak, however, the electrode reaction remains diffusion controlled. For totally irreversible systems the reverse peak disappears completely and the peak current density is expressed by

$$j_p = 2.99 \times 10^5 \, n (\alpha \, n_\alpha)^{1/2} c_i^\infty D^{1/2} v^{1/2}$$

where $\alpha$ is the transfer coefficient and $n_\alpha$ the number of electrons transferred up to and including the rate-determining step. The majority of redox couples fall between the two extremes and exhibit quasi-reversible behaviour. When investigating an electrode reaction for reversibility it is essential to obtain results over a sweep-rate range of at least two orders of magnitude, in order not to reach erroneous conclusions. In addition, although subsequent cyclic voltammograms enable valuable mechanistic information to be deduced, the first sweep cycle only should be considered for accurate analysis of kinetic data.

If adsorbed electroactive species are present on the electrode surface, the shape of the cyclic voltammogram changes, since the species do not need to diffuse to the electrode surface. In this case the peaks are symmetrical with coincident peak potentials provided the kinetics are fast.

On investigating a new system, cyclic voltammetry is often the technique of choice, since a number of qualitative experiments can be carried out in a short space of time to gain a feeling for the processes involved. It essentially permits an electrochemical spectrum, indicating potentials at which processes occur. In particular, it is a powerful method for the investigation of coupled chemical reactions in the initial identification of mechanisms and of intermediates formed. Theoretical treatment for the application of this technique extends to many types of coupled mechanisms.

*B1.28.3.2   Potential-step techniques*

In a potential-step experiment, the potential of the working electrode is instantaneously stepped from a value where no reaction occurs to a value where the electrode reaction under investigation takes place and the current *versus* time (chronoamperometry) or the charge *versus* time (chronocoulometry) response is recorded. The transient obtained depends upon the potential applied and whether it is stepped into a diffusion control, in an electron transfer control or in a mixed control region. Under diffusion control the transient may be described by the Cottrell equation obtained by solving Fick's second law with the appropriate initial and boundary conditions [1–6]:

$$|j| = \frac{nFD^{1/2}c_i^{\infty}}{\pi^{1/2}t^{1/2}}.$$

Immediately after the imposition of a large negative overpotential in a solution containing oxidized species, $O$, a large current is detected, which decays steadily with time. The change in potential from $E_e$ will initiate the very rapid reduction of all the oxidized species at the electrode surface and consequently of all the electroactive species diffusing to the surface. It is effectively an instruction to the electrode to instantaneously change the concentration of $O$ at its surface from the bulk value to zero. The chemical change will lead to concentration gradients, which will decrease with time, ultimately to zero, as the diffusion-layer thickness increases. At time $t = 0$, on the other hand, $(\partial c_i/\partial x)_{x=0}$ will tend to infinity. The linearity of a plot of $j$ *versus* $t^{-1/2}$ confirms whether the reaction is under diffusion control and can be used to estimate values for the diffusion coefficient. It is a good technique for determining exact kinetic parameters when a mechanism is fully understood. Under mixed control, where the rates of diffusion and electron transfer are comparable, the current decays less steeply: at short times it will be controlled by electron transfer but, as the surface concentration is depleted, mass transport will become the rate-limiting step.

When analysing the data, it is important to consider a wide time range to ensure the reliability of the data, since at short times, $<1$ ms, it will be determined by the charging time of the double layer, and at longer times, $>10$ s, by the effects of natural convection.

Potential-step techniques can be used to study a variety of types of coupled chemical reactions. In these cases the experiment is performed under diffusion control, and each system is solved with the appropriate initial and boundary conditions.

Double potential steps are useful to investigate the kinetics of homogeneous chemical reactions following electron transfer. In this case, after the first step—raising to a potential where the reduction of $O$ to $R$ occurs under diffusion control—the potential is stepped back after a period $\tau$, to a value where the reduction of $O$ is mass-transport controlled. The two transients can then be compared and the kinetic information obtained by looking at the ratio of the currents, which are a function of both $\tau$ and the homogeneous rate constant, $k$. This is a good method for obtaining exact information, provided that the mechanism is already understood.

*B1.28.3.3   Pulse voltammetry*

Pulse techniques were originally devised to provide enhanced sensitivity in classical polarography for analytical applications [7–9]. Sub-nanomolar detection limits can, in fact, be achieved with mercury electrodes where charging and background faradic currents are minimal. At solid electrodes (C, Pt, Au), charging currents and background currents arising from electrode surface reactions limit the level of analyte detection. However, pulse techniques remain particularly useful when looking at analyte concentrations of $10^{-5}$ M and lower, where voltammetric techniques such as linear-sweep voltammetry and cyclic voltammetry become limited by the difficulty of measuring faradic currents in the presence of background currents. Many step techniques have been devised, based on the succession of potential steps of varying height and in forward and reverse directions [2, 6, 7]. They find wide application in digitally based potentiostats, the electronics of which are suited to their exploitation. The current is normally sampled toward the end of the potential pulse,

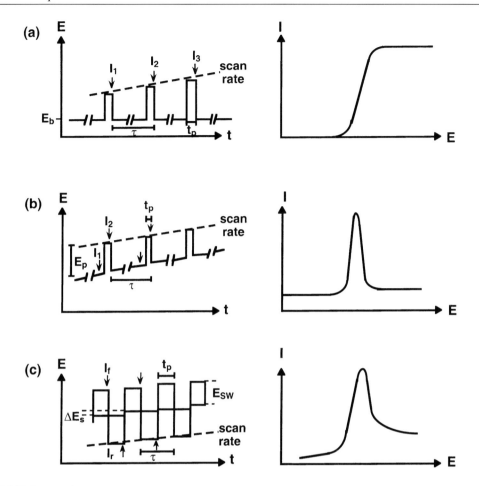

**Figure B1.28.5.** Applied potential–time waveforms for (a) normal pulse voltammetry (NPV), (b) differential pulse voltammetry (DPV), and (c) square-wave voltammetry (SWV), along with typical voltammograms obtained for each method.

after the capacitative current has decayed, and the pulse widths are adjusted to fit between this limit and the onset of natural convection. Normal-pulse voltammetry (NPV), differential-pulse voltammetry (DPV) and square-wave voltammetry (SWV) are perhaps the most widely used of a variety of pulse techniques that have been developed.

In NPV [2, 7, 10, 11], short pulses of increasing height are superimposed on a constant base potential, $E_b$, where no reaction occurs (figure B1.28.5(a)). At the end of the pulse of width, $t_P$ (typically 50–60 ms), the potential is returned to $E_b$, where it is held for another fixed period of a few seconds, before being pulsed again with a height increase determined by the scan rate. The current is sampled at the end of each pulse and the values are plotted against the potential to give a voltammetric profile similar to a steady-state voltammogram. The maximum current is given by the Cottrell equation, $j = nFDc_i^\infty/\pi^{1/2}t_S^{1/2}$, where $t_S$ is the time at which the current is measured. NPV is therefore a good technique for determining diffusion coefficients.

In DPV [2, 7, 11] the pulse height is kept constant and the base potential is either swept constantly or is incremented in a staircase (figure B1.28.5(b)). The current is sampled just before the end of the pulse and just before pulse application, and the difference between the two measurements is plotted as a function of the potential. The resulting voltammogram is peak-shaped since it essentially is the differential of a steady-state shaped response, and for this feature the technique particularly lends itself to analytical purposes, enabling lower detection limits to be achieved. For a reversible system, the peak is symmetric with $E_P = E_{1/2} - \Delta E/2$, where $\Delta E$ is the pulse amplitude. In general, DPV is better at eliminating capacitive contributions and the peak-shaped response is useful for distinguishing two waves with close half-wave potentials.

SWV is an alternative voltammetric technique, first reported in 1952 by Barker and Jenkins [12] and subsequently developed into the form known today by Osteryoung *et al* [13–16]. The potential–time wave-form is composed of a sequence of symmetrical square-wave pulses superimposed on an underlying ramp (figure B1.28.5(c)). The critical parameters are the step height of the underlying potential scan, $\Delta E_S$; the height of the square-wave pulse, $E_{SW}$; the pulse width, $t_P$ and the time at which the current is sampled on the forward and reverse pulses, $t_S$. Current measurements are made near the end of the pulse in each square-wave cycle: once at the end of the forward pulse and once at the end of the reverse pulse. However, capacitive contributions can be discriminated against before they decay, since over a small potential range between for-ward and reverse pulses, the capacity is constant and is thus annulled by subtraction. Consequently, shorter pulses than in DPV and NPV can be applied, enabling higher frequencies to be employed and much faster analysis to be carried out. The difference between the two currents, the net current, is plotted *versus* the base staircase potential, yielding a peak-shaped response. Since the square-wave modulation amplitude is large, the reverse pulses cause the reverse reaction to occur and, thus, the net current is larger than either the forward or the reverse components. This, coupled with the effective discrimination against charging currents, enables a more sensitive analysis. The resulting peak-shaped voltammograms are symmetrical with characteristic position, width and height: the peak potential, $E_P$, coincides with the half-wave potential of a redox couple, the peak width indicates the effective number of electrons transferred and the peak current is proportional to the analyte concentration. In addition, the peak shape and position have been found to be largely independent of the size and geometry of the electrode. The net current is generally compared with theoretical predictions of a dimensionless current $\Psi$, which are related by the Cottrell equation for the characteristic time:

$$ j = \left( nFc \left( \frac{D}{\pi t_P} \right)^{1/2} \right) \psi $$

where $t_P$ is the pulse width. As well as for analysis, SWV has been found to be well suited to kinetic investigations.

### B1.28.3.4   Stripping voltammetry

Stripping voltammetry involves the pre-concentration of the analyte species at the electrode surface prior to the voltammetric scan. The pre-concentration step is carried out under fixed potential control for a predetermined time, where the species of interest is accumulated at the surface of the working electrode at a rate dependent on the applied potential. The determination step leads to a current peak, the height and area of which is proportional to the concentration of the accumulated species and hence to the concentration in the bulk solution. The stripping step can involve a variety of potential waveforms, from linear-potential scan to differential pulse or square-wave scan. Different types of stripping voltammetries exist, all of which commonly use mercury electrodes (dropping mercury electrodes (DMEs) or mercury film electrodes) [7, 17].

Anodic-stripping voltammetry (ASV) is used for the analysis of cations in solution, particularly to determine trace heavy metals. It involves pre-concentrating the metals at the electrode surface by reducing the dissolved metal species in the sample to the zero oxidation state, where they tend to form amalgams with Hg. Subsequently, the potential is swept anodically resulting in the dissolution of the metal species back into solution at their respective formal potential values. The determination step often utilizes a square-wave scan (SWASV), since it increases the rapidity of the analysis, avoiding interference from oxygen in solution, and improves the sensitivity. This technique has been shown to enable the simultaneous determination of four to six trace metals at concentrations down to fractional parts per billion and has found widespread use in seawater analysis.

Cathodic stripping voltammetry follows a similar sequence of events, except that trace anionic species are reduced in the form of insoluble salts with metal constituents on the electrode surface, e.g. Ag and Hg, during application of a short, relatively positive deposition potential. The applied potential is then swept linearly or pulsed from the deposition potential in the cathodic direction resulting in the selective desorption of the anionic species according to the respective formal potential values. Cathodic stripping voltammetry can be used to determine organic and inorganic compounds that form an insoluble film at the electrode surface. Various inorganic analytes such as halide ions, sulphide ions and oxo-anions are capable of forming insoluble Hg salts which can be pre-concentrated on the Hg electrode surface and be measured.

Adsorptive stripping analysis involves pre-concentration of the analyte, or a derivative of it, by adsorption onto the working electrode, followed by voltammetric measurement of the surface species. Many species with surface-active properties are measurable at Hg electrodes down to nanomolar levels and below, with detection limits comparable to those for trace metal determination with ASV.

Improved sensitivities can be attained by the use of longer collection times, more efficient mass transport or pulsed waveforms to eliminate charging currents from the small faradic currents. Major problems with these methods are the toxicity of mercury, which makes the analysis less attractive from an environmental point of view, and surface fouling, which commonly occurs during the analysis of a complex solution matrix. Several methods have been reported for the improvement of the pre-concentration step [17, 18]. The latter is, in fact, strongly influenced by the choice of solvent, electrode material, pH, electrode potential and temperature. A constant mass-transport rate leads to better reproducibility and hence stirring is often used with static mercury drop electrodes and stationary electrodes. Hydrodynamic electrodes are also employed in order to increase the sensitivity and decrease the detection limits.

Recent years have witnessed the exploitation of stripping voltammetry in chemical sensors. Complex, fixed-site ASV analysers are used to determine a wide range of metals, such as Cr, Ni, Cu, V, Sn, As and Cd, in the effluents from mining, mineral processing, metal-finishing and related industries. The portable instrumentation and low power demands of stripping analysis satisfy many of the requirements for on-site *in situ* measurements. The development of remotely deployed submersible stripping probes, easy-to-use microfabricated metal sensor strips and micromachined, hand-held total stripping analysers have been reported to move the measurement of trace metals to the field and to perform them more rapidly, reliably and inexpensively [17–20].

## B1.28.4  Steady-state techniques

In the study of electrode reactions, the rates of electron transfer are very often high compared to mass transport, rendering the extrapolation of mechanistic and kinetic data unfeasible. It is therefore essential for the study of electrode reactions and the extrapolation of kinetic information to disrupt the equilibrium by increasing the rate of mass transport and forcing the process into a mixed-control region where the rate of electron transfer is comparable to that of mass transport. There are several methods available for increasing and varying the rate of mass transport in a controlled way, amongst which are hydrodynamic electrodes and microelectrodes [1–4]. In both cases, the regime may be described by solvable systems that may be used to predict the rate of mass transport and in the interpretation of experimental data. In hydrodynamic electrodes, the increased rate of mass transport of species is brought about by external mechanical forces, which can arise from the movement of the electrode, agitation of the solution or flowing of the solution past the electrode surface. The resulting forced convection leads to the thinning of the Nernst diffusion layer with a consequent increase in the linear concentration gradient that exists across it and hence to current densities as large as 100 times greater than the steady-state diffusion-limited value. By measuring the current–potential response as a function of mass transport it is thus possible to extrapolate kinetic information regarding an electrode reaction, provided it is under mixed control. There are a number of electrode designs that fall into the category of hydrodynamic electrodes, which include the rotating-disc electrode (RDE), the rotating ring–disc electrode (RRDE), the wall-jet electrode, the wall-pipe electrode, the tube electrode and the channel electrode [21–27]. The RDE and RRDE are perhaps the most commonly employed in kinetic and mechanistic studies, and these will be further discussed together with the channel electrode. Microelectrodes, scanning electrochemical microscopy (SECM) and sonoelectrochemistry are also discussed.

### B1.28.4.1   Rotating-disc electrodes

A rotating-disc electrode (RDE) consists of a disc of electrode material embedded into a larger insulating sheath, and attached to the rotor spindle *via* a suitable electrical contact. The disc and sheath are rotated about a vertical axis. Upon rotation, a pump-action flow is initiated, which brings solution perpendicularly to the electrode surface and throws it out in a radial direction on meeting disc and sheath (figure B1.28.6). A more quantitative description of the flow patterns can be made by the use of cylindrical polar coordinates, by looking at the variation of the solution-flow velocity components $V_x$, $V_r$ and $V_\theta$ as a function of $x$, the distance perpendicular to the surface of the electrode. The change in concentration of an electroactive species with time due to convection and diffusion may be written as [1, 2, 4, 5]

$$\frac{\partial c_i}{\partial t} = D\underbrace{\left[\frac{\partial^2 c_i}{\partial x^2} + \frac{\partial^2 c_i}{\partial r^2} + \frac{1}{r}\frac{\partial c_i}{\partial r} + \frac{1}{r^2}\frac{\partial^2 c_i}{\partial \theta^2}\right]}_{\text{diffusion}} - \underbrace{\left[V_x\frac{\partial c_i}{\partial x} + V_r\frac{\partial c_i}{\partial r} + \frac{V_\theta}{r}\frac{\partial c_i}{\partial \theta}\right]}_{\text{convection}}.$$

However, the equation can be simplified, since the system is symmetrical and the radius of the disc is normally small compared to the insulating sheath. The access of the solution to the electrode surface may be regarded as uniform and the flux may be described as a one-dimensional system, where the movement of species to the electrode surface occurs in one direction only, namely that perpendicular to the electrode surface:

$$\frac{\partial c_i}{\partial t} = D\frac{\partial^2 c_i}{\partial x^2} - V_x\frac{\partial c_i}{\partial x} \qquad \text{where} \quad V_x = -0.51\omega^{3/2}\nu^{-1/2}x^2.$$

The importance of convection in the system increases as the square of the distance from the electrode surface, and close to the surface it is not a dominant form of mass transport. Hence concentration changes will arise due to both diffusion and convection. In the Nernst diffusion model, this trend is exaggerated, and for the mass transport behaviour at an RDE, a plot of the concentration of electroactive species, $c_i$, *versus* the distance from the electrode surface, $x$, is divided into two distinct zones (figure B1.28.6). At the electrode surface, i.e. $x = 0$, the concentration of the electroactive species will be $c_i^\sigma$ and up to a distance $\delta$ away from the electrode, there is a stagnant layer, in which diffusion is the only form of mass transport (the Nernst diffusion layer). Outside this layer, mass transport is dominated by strong convection and the concentration is maintained at the bulk value, $c_i^\infty$. The diffusion-layer thickness is determined by the rotation rate of the disc, the layer becoming thinner with increasing rotation rate. In this model, the values of $c_i^\infty$ and $\delta$ will depend on the applied potential and the electrode rotation rate, respectively. In a linear-sweep experiment at a given rotation rate, the concentration profiles, $(\mathrm{d}c_i/\mathrm{d}x)$, within the diffusion layer will vary linearly as the applied potential at the RDE is swept from a value where no electron transfer occurs towards values positive to $E_e^0$. As the experiment is driven further, the surface concentration of the electroactive species eventually reaches zero, at which point the current response reaches its limiting plateau value. The limiting current density is expressed by [1, 2, 4, 5]

$$j_L = \frac{nFDc_i^\infty}{\delta} = nFk_m c_i^\infty$$

where $k_m$ is the mass-transport coefficient and the diffusion-layer thickness is $\delta = 1.61\nu^{1/6}c^\infty\omega^{-1/2}$, where $\omega$ is the rotational speed in rad s$^{-1}$, and $\nu$ is the kinematic viscosity of the solution. Substituting for $\delta$ leads

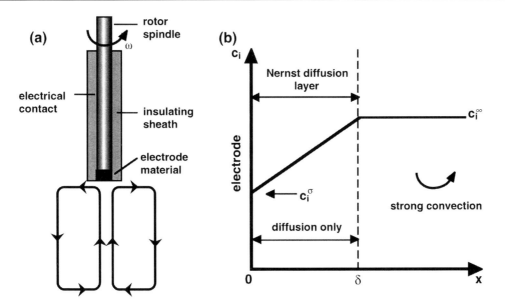

**Figure B1.28.6.** (a) Convection within the electrolyte solution, due to rotation of the electrode; (b) Nernst diffusion model for steady state.

to the Levich equation

$$j_L = 0.621 n F D^{2/3} v^{-1/6} c_i^\infty \omega^{1/2}.$$

The expression for the mass-transport-limiting current density may be employed together with the Nernst equation to deduce the complete current–potential response in a solution containing only oxidized or reduced species

$$E = E_e^0 + \frac{2.3RT}{nF} \log \frac{I_L - I}{I}$$

in which $I$ are current values from the rising portion of the curve (under mixed control). This equation, of course, only holds for fast electron-transfer reactions, where the surface concentrations are related through the Nernst equation. For a reversible electrode reaction, therefore, a plot of potential *versus* the logarithmic quotient should have a slope of $(59/n)$ mV (at 298 K) and the formal potential for the couple will coincide with the half-wave potential, $E_{1/2}$, of the curve (potential corresponding to $1/2\ I_L$) (figure B1.28.7). On reversing the scan, the current–potential curve will exactly retrace the forward scan, as the electroactive species will continue to be reduced or oxidized at the same potentials as in the forward sweep. The product formed at the electrode surface during both scans quickly disappears into the bulk solution through convection and is not available for the reverse electron transfer during the back-sweep. In addition to analysing the shape of a current–potential curve at a single rotation rate, the relationship between limiting current densities and mass transport can also be investigated. A common treatment for RDE data is to plot the limiting current density *versus* the square root of rotation speed, with a linear plot confirming conditions of mass-transport control, and the slope being used to determine the other parameters.

In the case of an irreversible electrode reaction, the current-potential curve will display a similar shape, with $j_L$ still proportional to $\omega^{1/2}$, but the curve is drawn out along the potential axis. The current-potential curve may be described by [1, 2, 4, 5]

$$E = E_{1/2} + \frac{2.3RT}{\alpha n F} \log \frac{I_L - I}{I}$$

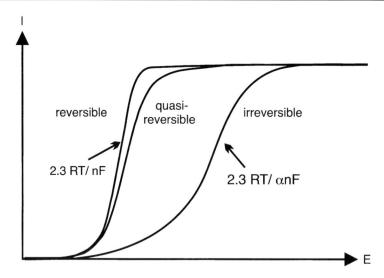

**Figure B1.28.7.** Schematic shape of steady-state voltammograms for reversible, quasi-reversible and irreversible electrode reactions.

in which the log plot remains linear but with a slope of $2.3RT/\alpha nF$ (figure B1.28.7). The most obvious feature of the irreversible voltammogram is that the half-wave potential no longer falls near the reversible formal potential, reflecting the sluggish electron transfer kinetics. An activation overpotential is required to drive the reaction.

The rotating-disc is also well suited to the study of coupled chemical reactions [2, 4].

It is essential for the rotating-disc that the flow remain laminar and, hence, the upper rotational speed of the disc will depend on the Reynolds number and experimental design, which typically is $1000$ s$^{-1}$ or $10\,000$ rpm. On the lower limit, $10$ s$^{-1}$ or $100$ rpm must be applied in order for the thickness of the boundary layer to be comparable to that of the radius of the disc.

The great advantage of the RDE over other techniques, such as cyclic voltammetry or potential-step, is the possibility of varying the rate of mass transport to the electrode surface over a large range and in a controlled way, without the need for rapid changes in electrode potential, which lead to double-layer charging current contributions.

### B1.28.4.2 Rotating ring–disc electrodes

The rotating ring-disc electrode (RRDE) consists of a central disc separated from a concentric ring electrode by a thin, non-conducting gap. It was first developed by Frumkin and Nekrasov to detect unstable intermediates in electrochemical reactions [1, 2, 22]. As with the RDE, on rotation of the disc, solution is pulled towards the centre of the disc and then thrown out radially across the surface of the structure. The ring is effectively situated downstream to the disc. This permits the intermediates formed on the disc, as the result of an oxidation or reduction process, to be detected at the ring following their mass transport across the insulating gap between the electrodes. Hence, information on intermediates can be obtained before they reach the bulk solution or react further with the electrolyte solution. The ring and the disc are independent from one another and can hence be potentiostatted independently.

In order to employ the RRDE for quantitative studies, it is necessary to describe the transport of species from disc to ring. In the absence of homogeneous chemical reactions, the electrogenerated species at the disc reaction is transported to the ring by diffusion across the stagnant layer at the electrode surface, by convection across the gap and diffusion across the stagnant layer at the ring electrode. The collection efficiency, $N_0$, is defined as the ratio of the mass-transport-controlled current for the electrode reactions at ring and disc, $N_0 = -i_{\text{ring}}/i_{\text{disc}}$, where the minus sign arises because the reactions at the ring and at the disc occur in the opposite direction. The collection efficiency thus represents the fraction of material produced at the disc that is detected at the ring. Analytical solutions of the convective–diffusion transport at the ring–disc enables the collection efficiency for specific disc and ring dimensions to be calculated:

$$N_0 = 1 - F\left(\frac{\alpha}{\beta}\right) + \beta^{2/3}[1 - F(\alpha)] - (1 + \alpha + \beta)^{2/3}\left\{1 - F\left[\left(\frac{\alpha}{\beta}\right)(1 + \alpha + \beta)\right]\right\}$$

where $\alpha$, $\beta$ and $F$ are defined as

$$\alpha = \left(\frac{r_2}{r_1}\right)^3 - 1$$

$$\beta = \left(\frac{r_3}{r_1}\right)^3 - \left(\frac{r_2}{r_1}\right)^3$$

$$F(\theta) = \frac{3^{1/2}}{4\pi}\ln\left\{\frac{(1 + \theta^{1/3})^3}{1 + \theta}\right\} + \frac{3}{2\pi}\arctan\left(\frac{2\theta^{1/3} - 1}{3^{1/2}}\right) + \frac{1}{4}$$

and $r_1$, $r_2$ and $r_3$ are the radius of the disc, the radius of the disc surrounded by the insulating sheath and the radius of the disc surrounded by sheath and ring, respectively. The collection efficiency is a function of $r_1$, $r_2$ and $r_3$ and does not depend on the rotation speed or the nature of the redox species. Since access of material to the ring is not uniform, some of the material will be transported back into solution and collection efficiencies are typically around 0.2–0.3 and strongly depend on the geometry of the electrodes and the distance between them. The rotation rate will affect the time taken for the intermediates to be transported from the disc to the ring, short-lived intermediates requiring higher rotation rates and the construction of RRDEs with thin inter-electrode gaps. The rotation rate will not, however, affect the efficiency, since the currents at both the generator and collector electrodes will be enhanced.

A number of different types of experiment can be designed, in which disc and ring can either be swept to investigate the potential region at which the electron transfer reactions occur, or held at constant potential (under mass-transport control), depending on the information sought.

The RRDE is very useful for the detection of short-lived intermediates, in the investigation of reaction mechanisms, but also in the distinction of free and adsorbed intermediates, as the latter are not transported to the ring.

### B1.28.4.3   Channel-flow electrodes

Forced convection can also arise from the movement of electrolyte solution over a stationary working electrode. In a channel electrode, the electrode is embedded smoothly in one wall of a thin, rectangular duct through which electrolyte is mechanically pumped [3, 6, 26, 27]. The design of the flow cell consists of two plates sealed together, with typical dimensions of 30–50 mm in length, <10 mm in width and a distance between the plates of less than 1 mm (the cell height). The electrode is embedded either at the centre of the base plate or attached to the centre of the cover plate by means of an adhesive.

The solution flow is normally maintained under laminar conditions and the velocity profile across the channel is therefore parabolic with a maximum velocity occurring at the channel centre. Thanks to the well defined hydrodynamic flow regime and to the accurately determinable dimensions of the cell, the system lends itself well to theoretical modelling. The convective–diffusion equation for mass transport within the rectangular duct may be described by

$$\frac{\partial c_i}{\partial t} = D\left(\frac{\partial^2 c_i}{\partial x^2} + \frac{\partial^2 c_i}{\partial y^2} + \frac{\partial^2 c_i}{\partial z^2}\right)c_i - \left(v_x\frac{\partial c_i}{\partial x} + v_y\frac{\partial c_i}{\partial y} + v_z\frac{\partial c_i}{\partial z}\right)$$

where $v_x$, $v_y$, $v_z$ are the solution-velocity profiles in the directions $x$, $y$, $z$. By convention, the direction of flow is designated as the $x$-direction and the $y$-direction is that normal to the electrode. The equation may be considerably simplified by removal of the time dependence under steady-state conditions and by neglecting axial diffusion to the macroelectrode, since convection is considerably faster. The diffusion layer is situated very close to the electrode surface and is small compared to the cell depth, decreasing as the flow rate is increased. The Lévêque approximation further simplifies the system by approximating the parabolic flow to a linear flow near the electrode surface, provided that the electrode is less wide than the channel (for edge effects to be neglected) and the height, $h$, of the channel is much greater than the width, $d$. In an analogous fashion as for the RDE, solution for a simple mass-transport-limited electrode reaction leads to the Levich equation

$$I_{lim} = 0.925 n F c_i^\infty D^{2/3} v_f^{1/3} (h^2 d)^{-1/3} w x_e^{2/3}$$

where $v_f$ is the solution volume flow rate, $x_e$ the length of the electrode, and $d$ and $w$ the height and width of the cell, respectively. Analytical solutions for the channel electrode also extend to more complicated electrode reactions involving coupled homogeneous reactions.

Amongst the greatest advantages of channel-flow electrodes is the possibility of controlling the rate of mass transport over a range of three orders of magnitude, from $10^{-4}$ to $10^{-1}$ cm$^3$ s$^{-1}$, and of varying the mass-transport coefficient by altering the cell depth and the electrode length. These rates are, in fact, not attainable at other hydrodynamic electrodes, such as the RDE and the wall-jet electrode. In addition, there is no risk of a build-up of a stagnant zone since the spent solution flows to waste.

The channel-flow electrode has often been employed for analytical or detection purposes as it can easily be inserted in a flow cell, but it has also found use in the investigation of the kinetics of complex electrode reactions. In addition, channel-flow cells are immediately compatible with spectroelectrochemical methods, such as UV/VIS and ESR spectroscopy, permitting detection of intermediates and products of electrolytic reactions. UV–VIS and infrared measurements have, for example, been made possible by constructing the cell from optically transparent materials.

### B1.28.4.4  *Microelectrodes*

A microelectrode is an electrode with at least one dimension small enough that its properties are a function of size, typically with at least one dimension smaller than 50 $\mu$m [28–33]. If compared with electrodes employed in industrial-scale electrosynthesis or in laboratory-scale synthesis, where the characteristic dimensions can be of the order of metres and centimetres, respectively, or electrodes for voltammetry with millimetre dimension, it is clear that the size of the electrodes can vary dramatically. This enormous difference in size gives microelectrodes their unique properties of increased rate of mass transport, faster response and decreased reliance on the presence of a conducting medium. Over the past 15 years, microelectrodes have made a tremendous impact in electrochemistry. They have, for example, been used to improve the sensitivity of ASV in environmental analysis, to investigate rapid electron transfer and coupled chemical reactions, and to study electrode reactions in low resistive media.

The *increased rate of mass transport* is one of the most attractive and advantageous properties of microelectrodes over conventional electrodes, as the increased transport of the reactant to the electrode surface allows it to reach steady-state regimes rapidly. The diffusion rates increase with decreasing electrode size beyond that obtained with other steady-state techniques. For example, with a 10 $\mu$m diameter disc, the steady-state, mass-transfer coefficient, $k_m$, is comparable to that of a rotating disc revolving at an experimentally impossible 250 000 rpm. The *discrimination against charging currents* is another very important property. In fact, the magnitude of the charging current depends on the area of the capacitor, and for a microelectrode it decreases with electrode area. Thus, a microelectrode has a very reduced interface capacitance, and the charging current decays much more quickly than with conventional electrodes and faster response times may be achieved. Another property of microelectrodes is the *decreased distortion from $IR_u$*, the potential drop between working and reference electrodes generated by the passage of current through a solution and expressed in terms of the product of the solution resistance and the current flowing in the circuit. With conventional electrodes, it is usual to add supporting electrolyte to minimize the solution resistance, but with microelectrodes, the current passing through the cell is low, often of the order of $10^{-9}$ A, and hence problems with $IR_u$ drop are greatly reduced. This proves to be an advantage to experiments with either a large current, $I$, or a large resistance, $R$, as for example in experiments with solvents of very low dielectric constant, in media with very low ionic strength, or in studies of solutions with high concentrations of electroactive species. Electrochemical measurements can be therefore made in new and unique chemical environments, which are not amenable at larger electrodes, and experiments have been reported in frozen acetonitrile, low-temperature glasses, ionically conductive polymers, oil-based lubricants and milk. In addition, the use of electrolyte-free organic media can greatly extend the electrochemical potential window, thus allowing studies of species with high redox potentials. Furthermore, such dimensions offer obvious analytical advantages, including the exploration of microscopic domains, measurement of local concentration profiles, detection in micro-flow systems and analysis of very small sample volumes [28].

Microelectrodes with several geometries are reported in the literature, from spherical to disc to line electrodes; each geometry has its own critical characteristic dimension and diffusion field in the steady state. The diffusional flux to a spherical microelectrode surface may be regarded as planar at short times, therefore displaying a transient behaviour, but spherical at long times, displaying a steady-state behaviour [28, 34]. If a potential is applied so that the reaction $O + ne^- \rightarrow R$, becomes diffusion controlled, the current density at a microsphere electrode can be expressed by

$$j = \frac{nFD^{1/2}c_i}{\pi^{1/2}t^{1/2}} + \frac{nFDc_i}{r}.$$

transient term steady-state term

This expression is the sum of a transient term and a steady-state term, where $r$ is the radius of the sphere. At short times after the application of the potential step, the transient term dominates over the steady-state term, and the electrode is analogous to a plane, as the depletion layer is thin compared with the disc radius, and the current varies with time according to the Cottrell equation. At long times, the transient current will decrease to a negligible value, the depletion layer is comparable to the electrode radius, spherical diffusion controls the transport of reactant, and the current density reaches a steady-state value. At times intermediate to the limiting conditions of Cottrell behaviour or diffusion control, both transient and steady-state terms need to be considered and thus the full expression must be used. However, many experiments involving microelectrodes are designed such that one of the simpler current expressions is valid.

Of course, in order to vary the mass transport of the reactant to the electrode surface, the radius of the electrode must be varied, and this implies the need for microelectrodes of different sizes. Spherical electrodes are difficult to construct, and therefore other geometries are often employed. Microdiscs are commonly used in the laboratory, as they are easily constructed by sealing very fine wires into glass epoxy resins, cutting perpendicular to the axis of the wire and polishing the front face of the disc that is created [30]. Because of its planar geometry, the diffusion field over the surface of a microdisc is non-uniform and the flux only approximates that of a hemisphere. The rate of diffusion to the edge of the disc will be higher than to the centre. Therefore, the rates of diffusion to the disc are estimated as space-averaged quantities and a factor of $4/\pi$ is required to adjust the equation for a spherical microelectrode to describe the diffusion of reactant to the surface of a microdisc electrode, which becomes $j = 4nFDc_i^\infty/\pi r$, where $r$ is now the radius of the disc.

Similarly to the response at hydrodynamic electrodes, linear and cyclic potential sweeps for simple electrode reactions will yield steady-state voltammograms with forward and reverse scans retracing one another, provided the scan rate is slow enough to maintain the steady state [28, 35–38]. The limiting current will be determined by the slowest step in the overall process, but if the kinetics are fast, then the current will be under diffusion control and hence obey the above equation for a disc. The slope of the wave in the absence of $IR_u$ drop will, once again, depend on the degree of reversibility of the electrode process.

All types of voltammetry may be applied to microelectrodes, including normal, reverse pulse and square-wave voltammetry. Pulse voltammetry and potential-step program at microelectrodes discriminate against charging currents and the boundary conditions are set much faster between pulses, since the electrode responds in a much more rapid fashion to a potential change [11, 15, 16]. Cyclic voltammetry measurements can be made on a much more rapid time scale than with electrodes of conventional size and be operated in the range of tens of nanoseconds, without important distortion by $IR_u$ drop and concerns regarding charging currents. This renders the characterization of rates and mechanisms of very fast chemical reactions as well as determination of trace quantities of transient species possible. At high sweep rates, however, only linear diffusion needs to be considered [28].

The advantages of microelectrodes for low-volume detection and spatially and temporally resolved measurements have been largely exploited in biology and medicine [39–41]. One of the most active and longer-standing fields is neuroscience, where the development of the electroanalysis of brain extracellular fluid has been remarkable, since it can be relatively non-invasive due to the small size and low currents flowing. An example is the monitoring of the release of neurotransmitters with carbon microelectrodes in either amperometric mode or using fast cyclic voltammetry [42, 43]. Carbon materials are the most common starting materials and have also been applied to other electroactive compounds such as histamine, anticancer drugs and ascorbic acid. Extension of the investigation of cellular systems to non-electroactive neurochemicals has led to the development of enzyme-modified microelectrodes for measurements of glutamate, glucose, and choline and acetylcholine. Voltammetric measurements have also been reported in single cells, although the living cells are separated from the parent organism.

Microelectrodes have also found widespread use in sensor technology and environmental analysis. Due to the high rate of steady-state diffusion at a microelectrode, their response is independent of convection, thus enabling their use for the analysis of flowing systems. In order to enhance the current response at microelectrodes, a number of approaches have been described. Amongst these are random arrays of microdisc electrodes [44–46] and interdigitated arrays of microband electrodes [47, 48]. Arrays of microelectrodes enable the enhancement of the current response, whilst retaining the properties of a single microelectrode, and have been used as highly sensitive detectors in flow-injection analysis and in liquid chromatography.

*B1.28.4.5   Scanning electrochemical microscopy*

SECM is a scanning-probe technique introduced by Bard *et al* in 1989 [49–51] based on previous studies by the same group on *in situ* STM [52] and simultaneous work by Engstrom *et al* [53–54], who were the first to show that an amperometric microelectrode could be used as a local probe to map the concentration profile of a larger active electrode. SECM may be envisaged as a 'chemical' microscope based on faradic current changes as a microelectrode is moved across a surface of a sample. It has proved useful for obtaining topographical and chemical information on a wide range of sample surfaces, including electrodes, minerals, polymers and biological materials.

The apparatus consists of a tip-position controller, an electrochemical cell with tip, substrate, counter and reference electrodes, a bipotentiostat and a data-acquisition system. The microelectrode tip is held on a piezoelectric pusher, which is mounted on an inchworm-translator-driven $x–y–z$ three-axis stage. This assembly enables the positioning of the tip electrode above the substrate by movement of the inchworm translator or by application of a high voltage to the pusher *via* an amplifier. The substrate is attached to the bottom of the electrochemical cell, which is mounted on a vibration-free table [55–58]. A number of different size and shape tips have been reported. The most common are disc shaped with diameters of 0.6–25 $\mu$m formed by sealing a Pt, Au wire or carbon fibre of the required radius in a glass capillary and polishing the sealed end. The glass wall surrounding the disc is sharpened to a conical shape to decrease the possibility of contact between glass and substrate as the tip is moved close to the latter. For most studies, the ratio of the diameter of the entire tip end, including the insulator, to that of the electrode itself should typically be $\approx$10. Metal electrodes down to the nanometre scale have also been fabricated by sealing an etched Pt or Pt–Ir wire in a suitable insulating material, leaving the etched end exposed. Commercial SECM instruments have only recently appeared on the market.

With SECM, almost any kind of electrochemical measurement may be carried out, whether voltammetric or potentiometric, and the addition of spatial resolution greatly increases the possibilities for the characterization of interfaces and kinetic measurements [55–59]. It may be employed as an electrochemical tool for the investigation of heterogeneous and homogeneous reactions, as an imaging device, or for microfabrication, making use of different modes of operation. In amperometric *feedback mode* a three- or four-electrode configuration is employed, in which a microelectrode tip serves as the working electrode, the potential is controlled *versus* the reference electrode and the current flows between tip and counter-electrodes. The potential of the sample may also be controlled and it may thus serve as a second working electrode. The electrolyte solution contains a redox mediator, e.g. a reducible species $O$, such that when a suitably negative potential is applied to the tip, its reduction takes place at a rate governed by diffusion of the electroactive species to the electrode. If the tip is more than several tip diameters away from the surface, the steady-state current is given by $I_{T_\infty} = 4nFDc_i^\infty r$, for a disc-shaped tip, where $r$ is the radius of the tip. However, when the tip is brought within a few tip radii to a conductive substrate, the reduced species formed at the tip diffuses to the substrate where it is re-oxidized. As a consequence, an additional flux of $O$ to the tip is produced which leads to an increase in the tip current, known as positive feedback. The smaller the tip–substrate distance the larger is the effect. In contrast, if the substrate is an electrical insulator, the reducible species cannot be regenerated and, since the diffusion of $O$ from the bulk is hindered at small distances to the substrate, the tip current will be smaller than $I_{T_\infty}$, i.e. negative feedback. Therefore, by scanning over the surface of a substrate, the variation in current can be related to changes in the distance and hence to the topography of the substrate. Besides feedback mode, several other modes exist, such as *generation/collection mode*, where species generated at one working electrode are detected at the second, *penetration mode*, in which a small tip is used to penetrate a microstructure and extract spatially resolved information about concentrations, kinetic and mass-transport parameters, and *ion-transfer feedback mode*, recently developed and useful for studies of ion-transfer reactions at liquid/liquid and liquid/membrane interfaces. The SECM methodologies are based on quantitative theory, which has been developed for a variety of systems involving heterogeneous and homogeneous processes and different tip and substrate geometries. In many cases, analytical approximations allow the generation of theoretical dependences and an analysis of experimental data [60–64].

The high rate of mass transfer in SECM enables the study of fast reactions under steady-state conditions and allows the mechanism and physical localization of the interfacial reaction to be probed. It combines the useful features of microelectrodes and thin-layer cells in dimensions not easily attainable in larger electrochemical cells. The mass-transfer rate in SECM is a function of the tip–substrate distance. At large distances, $d$, $k_m \approx D/r$, whereas for small distances ($d < r$), $k_m \approx D/d$. The large effective $k_m$ obtainable enables fast heterogeneous reaction rates to be measured under steady-state conditions. Zhou and Bard measured a rate constant of $6 \times 10^7$ Ms for the electro-hydrodimerization of acrylonitrile (AN) and observed the short-lived intermediate $AN^-$ for this process [65].

Heterogeneous reactions at a substrate can also be probed without the need for an external voltage, if for example the mediator regeneration is chemical in nature rather than electrochemical. This has opened the possibility of studying dissolution of ionic single crystals and locating individual sites of reactivity in multiphase systems. It can be used to map the local surface reactivity in either feedback or collection mode.

Feedback mode has been employed, for example, to probe the surface reactivity of a titanium substrate covered with $TiO_2$ and to individuate precursor sites for pitting corrosion, whereas collection mode has been used to image fluxes of species produced or consumed at the substrate, such as iontophorectic fluxes of electroactive species through porous membranes [56–58].

Among the systems most studied by SECM are heterogeneous electron transfer reactions at the metal/ solution interface. Nonetheless, the diversity of interfaces and processes that can be studied with SECM has grown to include liquid/liquid and liquid/gas interfaces [66], and materials of biological significance [67]. It is a promising technique for the mapping of biochemical activity, for example transport in tissues and immobilized enzyme kinetics. Detection of single molecules has also recently been reported [68].

### B1.28.4.6   Sonoelectrochemistry

The technique of applying ultrasound during electrochemical measurements and reactions is known as sonoelectrochemistry or sonovoltammetry and is a field that has grown rapidly in recent years [69–72]. The dominant ultrasonic effects are the enhanced mass transport of electroactive substrate to the electrode and the activation of the electrode surface through a cavitational cleaning action. The latter are violent collapses of oscillating bubbles, which can cause effects such as depassivation and erosion. The huge effects on the rate of mass transport to the electrode surface may be envisaged as an extremely thinned diffusion layer of uniform accessibility, partly induced by acoustic streaming, when ultrasonic horn transducers are employed, and by cavitational collapse of micro-bubbles at the solid–liquid interface.

Two major sources of ultrasound are employed, namely ultrasonic baths and ultrasonic immersion horn probes [70, 71]. The former consists of fixed-frequency transducers beneath the exterior of the bath unit filled with water in which the electrochemical cell is then fixed. Alternatively, the metal bath is coated and directly employed as electrochemical cell, but in both cases the results strongly depend on the position and design of the set-up. The ultrasonic horn transducer, on the other hand, is a transducer provided with an electrically conducting tip (often Ti6Al4V), which is immersed in a three-electrode thermostatted cell to a depth of 1–2 cm directly facing the electrode surface. The ultrasound intensity and the distance between the horn and the electrode may be varied at a fixed frequency, typically of 20 kHz. This cell set-up enables reproducible results to be obtained due to the formation of a macroscopic jet of liquid, known as acoustic streaming, which is the main physical factor in determining the magnitude of the observed current.

The effects of ultrasound-enhanced mass transport have been investigated by several authors [73–76]. Empirically, it was found that, in the presence of ultrasound, the limiting current for a simple reversible electrode reaction exhibits quasi-steady-state characteristics with intensities considerably higher in magnitude compared to the peak current of the response obtained under silent conditions. The current density can be described by $j_{lim} = nFDc_i^{\infty}/\delta$, where the diffusion layer $\delta$ depends on the distance between the horn and the electrode and on the ultrasound intensity. Superimposed on the faradic current a fluctuation or noise is also detected consistent with the turbulent nature of the macroscopic jet of liquid and the presence of oscillating and cavitating bubbles.

In an alternative design, the actual tip of the ultrasonic horn may be used as the working electrode after insertion of an isolated metal disc [77–79]. With this electrode, known as the sonotrode, very high limiting currents are obtained at comparatively low ultrasound intensities, and diffusion layers of less than 1 $\mu$m have been reported. Furthermore, the magnitude of the limiting currents has been found to be proportional to $D^{2/3}$, enabling a parallel to be drawn with hydrodynamic electrodes.

The cleaning or depassivation effect is of great importance in sonoelectrochemistry, as it can be employed to wash off surface-adsorbed species and reduce blocking of the electrode by adsorption of reaction products. This effect has been reported, for example, for the depassivation of iron electrodes and for the removal of deposits and in the presence of polymer films on the electrode surface. However, damage of the electrode surface, especially for materials of low hardness such as lead or copper, can also occur under harsh experimental conditions and applied intensities [70, 71, 80].

Sonoelectrochemistry has been employed in a number of fields such as in electroplating for the achievement of deposits and films of higher density and superior quality, in the deposition of conducting polymers, in the generation of highly active metal particles and in electroanalysis. Furthermore, the sonolysis of water to produce hydroxyl radicals can be exploited to initiate radical reactions in aqueous solutions coupled to electrode reactions.

### B1.28.5 Electrochemical impedance spectroscopy

In contrast to transient techniques, which involve the perturbation of a system and studying its relaxation with time, when the perturbation is sinusoidal, the analysis is performed in the frequency domain, which is obtained by applying Laplace transforms to the time-domain information. Alternating-current impedance techniques employ the ratio of an imposed sinusoidal voltage and the resulting sinusoidal current to define the impedance, which is a function of the frequency of the signal. When a steady-state system is perturbed by an applied AC voltage, it relaxes to a new steady state and the time taken for this relaxation is known as $\tau$. $\tau = RC$, where $R$ is the resistance and $C$ the capacitance of the system. Analysis of this relaxation process provides information about the system. In the frequency domain, fast processes, low $\tau$, occur at high frequencies, while slow processes, with high $\tau$, occur at low frequencies. Thus, dipolar properties may be studied at high frequencies, bulk properties at intermediate frequencies and surface properties at low frequencies.

Methods for measuring the impedance can be divided into controlled current and controlled potential [2, 4, 81]. Under controlled potential conditions, the potential of the electrode is sinusoidal at a given frequency with the amplitude being chosen to be sufficiently small to assure that the response of the system can be considered linear. The ratio of the response to the perturbation is the transfer function, or impedance, $Z$, when considering the response of an AC current to an AC voltage imposition and is defined as $E = IZ$, where $E$ and $I$ are the waveform amplitudes for the potential and the current respectively. Impedance may also be envisaged as the resistance to the flow of an alternating current.

Two different components contribute to impedance: the resistive or real component due to resistors and the reactive or imaginary component from AC circuitry elements, such as capacitors, inductors, etc. Unlike the resistive component, the reactive impedance affects not only the magnitude of the AC wave but also its time-dependent characteristic, the phase. For example, when an alternating voltage wave is applied to a capacitor, the resulting current waveform will lead the applied voltage by $90°$. Due to this reason the introduction of complex notation is convenient. Thus, when a system is perturbed by a sinusoidal potential, varying with time according to

$$E(t) = E_0 \exp(i\omega t)$$

the response can be expressed in terms of

$$I(t) = I_0 \exp(i\omega t - \varphi)$$

where i is the complex number, $E(t)$ and $I(t)$ are the instantaneous values, $E_0$ and $I_0$ the peak amplitude of the potential and the current respectively, $\varphi$ the phase angle difference and $\omega$ the angular frequency in radians ($\omega = 2\pi f$).

Introducing the complex notation enables the impedance relationships to be presented as Argand diagrams in both Cartesian and polar co-ordinates ($r, \varphi$). The former leads to the Nyquist impedance spectrum, where the real impedance is plotted against the imaginary and the latter to the Bode spectrum, where both the modulus of impedance, $r$, and the phase angle are plotted as a function of the frequency.

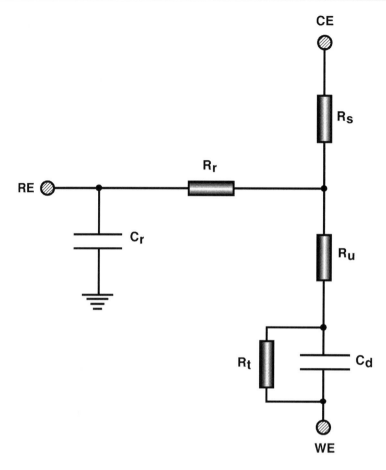

**Figure B1.28.8.** Equivalent circuit for a three-electrode electrochemical cell. WE, CE and RE represent the working, counter and reference electrodes; $R_s$ is the solution resistance, $R_u$ the uncompensated resistance, $R_t$ the charge-transfer resistance, $R_r$ the resistance of the reference electrode, $C_d$ the double-layer capacitance and $C_r$ the parasitic loss to the ground.

In AC impedance the cell is essentially replaced by a suitable model system in which the properties of the interface and the electrolyte are represented by appropriate electrical analogues and the impedance of the cell is then measured over a wide frequency range, usually between $10^4$ and $10^{-3}$ Hz. By comparing the measured results with values calculated from the model system, the suitability of the model and the values of the parameters can be evaluated. In fact, one of the advantages of EIS is that impedance functions frequently display many of the features exhibited by passive electrical circuits. The most important elements employed in equivalent circuits are the resistor $R$, which represents the resistance that charge carriers encounter in a specific medium, the capacitor $C$, which represents the accumulation of charged species, and the inductance $L$, which represents the deposition of surface layers (figure B1.28.8). An analogy, however, is not always feasible, due to the active nature of electrochemical interfaces and the chemical nature of charge transfer processes, as well as the non-ideal electric behaviour of real electrochemical systems. Furthermore, problems can arise in selecting a correct equivalent circuit out of a large number of possibilities, because of the uncertainty connected with the impedance at low frequency and because of the large number of possible combinations of mechanistic reactions that can produce the same impedance shape within error limits. Various methods have been reported to discriminate for the correct equivalent circuit and to obtain values for the elements in the circuit [81–83]. Spectra displaying one time constant are simple to interpret and may be easily resolved graphically directly from the spectra; however, more complex methods such as deconvolution and complex nonlinear least-square methods, or a combination of these are required for more complicated spectra. A number of software packages are based on these different types of methods.

AC impedance spectroscopy is widely employed for the investigation of both solid- and liquid-phase phenomena. In particular, it has developed into a powerful tool in corrosion technology and in the study of porous electrodes for batteries [84–87]. Its usage has grown to include applications ranging from fundamental studies of corrosion mechanisms and material properties to very applied studies of quality control and routine corrosion engineering. In corrosion, EIS enables one to obtain instantaneous corrosion-rate information, polarization resistance and information on the kinetics and mechanisms of charge transfer processes such as oxide growth and metal dissolution. The technique is frequently employed in the monitoring of polymer-coated metals to investigate the corrosion protection, the dielectric properties, the onset of defect formation and the processes of coating degradation [84].

## B1.28.6  Photoelectrochemistry

The combination of electrochemistry and photochemistry is a form of dual-activation process. Evidence for a photochemical effect in addition to an electrochemical one is normally seen in the form of photocurrent, which is extra current that flows in the presence of light [88–90]. In photoelectrochemistry, light is absorbed into the electrode (typically a semiconductor) and this can induce changes in the electrode's conduction properties, thus altering its electrochemical activity. Alternatively, the light is absorbed in solution by electroactive molecules or their reduced/oxidized products inducing photochemical reactions or modifications of the electrode reaction. In the latter case electrochemical cells (RDE or channel-flow cells) are constructed to allow irradiation of the electrode area with UV/VIS light to excite species involved in electrochemical processes and thus promote further reactions.

Conduction in semiconductors requires that electrons in the valence band be excited into the conduction band either by thermal or photochemical excitation. Upon excitation, an unoccupied vacancy (a hole) is left in the valence band. The hole and the excited electrons can move in response to an applied electric field and so permit the passage of current. Semiconduction can be controlled via doping of small quantities of material, which can be either electron donating or electron accepting, leading to n-type and p-type semiconductors [91–94]. In a solution containing a redox couple, electron transfer will occur until the electrochemical potentials of the semiconductor and the solution are equal (figure B1.28.9). The semiconductor will have a net positive or negative charge, which is situated near the surface of the solid, known as the space-charge layer (2–500 nm thickness). The bands may be envisaged as bent: band-bending downwards indicates excess of negative charge at the surface, whereas band-bending upwards indicates an excess of positive holes. Potentiostatic control of a semiconductor can change the energy of the conduction and valence bands with consequent changes in the band bending, leading to the supply and removal of charge carriers within the space layer and enabling electrolysis to occur at the solid/liquid interface. For the n-type semiconductor $TiO_2$, for example, the conduction and valence bands are bent upward at the surface, provided that a very negative potential is not applied to the electrode. An applied potential leading to no band-bending—i.e. to an absence of the space-charge layer—is called the flat-band potential of the semiconductor. Upon irradiation with light of an equal or greater energy than the band gap, the photochemically promoted electrons will be swept into the bulk of the material by the electric field present in the space-charge layer, whilst the holes in the valence band will migrate to the surface of the solid (figure B1.28.10). As a result, photo-oxidation processes will be promoted to occur at the solid–liquid interface.

In the last 30 or more years, research in the field of photoelectrochemistry and photocatalysis has greatly expanded, with advances being made in the fundamental understanding of the faradic processes that control

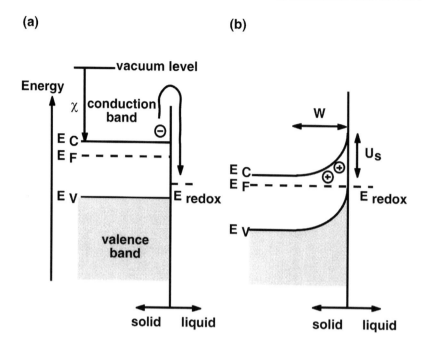

**Figure B1.28.9.** Energetic situation for an n-type semiconductor (a) before and (b) after contact with an electrolyte solution. The electrochemical potentials of the two systems reach equilibrium by electron exchange at the interface. Transfer of electrons from the semiconductor to the electrolyte leads to a positive space charge layer, $W$. $U_s$ is the potential drop in the space-charge layer.

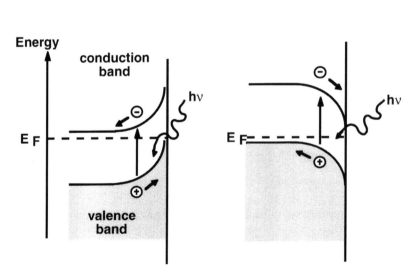

**Figure B1.28.10.** Schematic representation of an illuminated (a) n-type and (b) p-type semiconductor in the presence of a depletion layer formed at the semiconductor–electrolyte interface.

charge transfer at semiconductor/liquid junctions, and with the development of stable efficient and inexpensive photoelectrochemical cells [95–97]. Semiconductor photoelectrochemistry has had an impact in a number of fields. Photocorrosion has been exploited to prepare technologically useful structures such as lenses that are integrated with light-emitting diodes and mated optical fibres. It has also led to the formation of porous Si electrodes, which have received much attention due to their interesting optoelectronic properties. Photoelectrochemical surface preparation of semiconductor materials has been used to clean solids and to evaluate etch pit densities.

Photoelectrochemistry may be used as an *in situ* technique for the characterization of surface films formed on metal electrodes during corrosion. Analysis of the spectra allows the identification of semiconductor surface phases and the characterization of their thickness and electronic properties.

Furthermore, semiconductor powders can be employed for the catalytic generation of useful products such as $H_2$ and $O_2$ that can be used for the destruction of pollutants [98–100].

## B1.28.7 Spectroelectrochemistry

In addition to the mechanism of electrode reactions, readily deduced using voltammetric techniques, the electrochemist seeks knowledge of the chemical composition and properties of electro-generated intermediates and films formed on electrode surfaces. Spectroelectrochemistry allows the simultaneous acquisition of electrochemical and spectroscopic data, which offer additional information to the investigation of a wide range of complex surface and homogeneous processes occurring in electrochemical systems [101–104]. Advances made in instrumentation over the past three decades have enabled the adaptation of spectroscopic methods to *in situ* application in an electrochemical cell and the development of new techniques, which have found widespread use in structure characterization of electrode surfaces, in identification of homogeneous phase molecules, and in studies of species adsorbed at the electrode/electrolyte interface.

One of the first *in situ* combined electrochemical/spectroscopic techniques to be investigated employed UV/VIS detection, where solution-phase spectra of organic radicals generated at an electrode could be recorded. Typically, organic intermediates or products possess additional absorption bands not observed in the parent molecule, which can be used to fingerprint the electro-generated species. In addition to spectra, useful information may be obtained by monitoring the absorbance as a function of time during a potential-step experiment. When the potential is stepped from a value where no electrode transfer takes place to one where an electro-generated species is formed, the spectroscopic intensity–time response of the products may be analysed and, from the shape and size, estimates of the lifetime of the electrode intermediates can be extrapolated.

In UV–VIS spectroelectrochemistry, optically transparent electrodes (OTEs) are utilized. A beam of monochromatic UV–VIS light is directed perpendicularly through the OTE, then through the diffusion layer next to the electrode and the bulk solution, before passing out of the electrochemical cell through an exit window and being detected. The beam is attenuated by the presence of absorbing species in the solution, therefore enabling spectral and temporal information about the concentrations of such species in the diffusion layer at the electrode to be obtained. An alternative design is the optically transparent thin layer electrode, in which a minigrid electrode is employed [2, 103, 105].

Infrared spectroscopy has also been widely employed in electrochemistry [105–107]. Spectra aid the identification of reactants, of products and of long-lived intermediates and allow changes in the interfacial solvent to be tracked. A variety of spectral sampling and data acquisition methods have been developed to approach *in situ* detection of species. In external reflection sampling methods, the infrared beam is directed through a polarizer onto the front surface of a highly polished disc-shaped working electrode with high reflectivity in the infrared spectral region, such as Pt, Au or Ag. A special, thin-layer electrochemical cell is used that permits the infrared beam to enter and strike the disc, where it is reflected out of the cell and detected. In contrast, in attenuated total internal reflection sampling methods, the working electrode is a thin film of metal deposited on one surface of an ATR crystal. The metal film must be sufficiently thin to allow penetration of the IR evanescent wave beyond the metal solution interface. The ATR crystal forms the bottom of a chamber that holds the electrolyte solution and the counter and reference electrodes and the crystal is positioned so that the metal film is inside the chamber. This method has not been widely used in electrochemistry partly due to the difficulty in the preparation of the thin metal film working electrodes. Nonetheless, the latter design overcomes molecular transport limitations imposed by external reflection methods, where a thin solution layer of the order 1–5 $\mu$m between the front face of the working electrode and the infrared transparent window is required to minimize absorption of infrared radiation by the solvent. In fact, diffusion of species into and out of the thin-layer region is restricted and can lead to reactant depletion or product accumulation.

Luminescence has been used in conjunction with flow cells to detect electro-generated intermediates downstream of the electrode. The technique lends itself especially to the investigation of photoelectrochemical processes, since it can yield information about excited states of reactive species and their lifetimes. It has become an attractive detection method for various organic and inorganic compounds, and highly sensitive assays for several clinically important analytes such as oxalate, NADH, amino acids and various aliphatic and cyclic amines have been developed. It has also found use in microelectrode fundamental studies in low-dielectric-constant organic solvents.

One of the most important advances in electrochemistry in the last decade was the application of STM and AFM to structural problems at the electrified solid/liquid interface [108, 109]. Sonnenfield and Hansma [110] were the first to use STM to study a surface immersed in a liquid, thus extending STM beyond the gas/solid interfaces without a significant loss in resolution. *In situ* local-probe investigations at solid/liquid interfaces can be performed under electrochemical conditions if both phases are electronic and ionic conducting and this offers a great advantage since the Fermi levels of both substrate and tip can be precisely adjusted independently of each other. This opens the possibility of correlating structural to physical properties, since charge transfer central to electrochemical reactivity occurs within a few atomic diameters of the electrode surface, in the inner Helmholtz plane, and the detailed arrangement of atoms and molecules at this interface strongly controls the corresponding electrochemical reactivity. Since its introduction in electrochemistry, STM investigations have focused on studies of metal electrodes, i.e. Au, Pt, Pd and Rh, their surface charges in the double-layer potential region, and the surface changes caused by the formation of surface oxides. In addition, it has been employed in reconstruction and restructuring studies of metal surfaces, in studies of underpotential deposition of metals, in the investigation of adsorption/desorption processes, as well as in the understanding of processes controlling deposition and corrosion at semiconductor electrodes.

Amongst other spectroscopic techniques which have successfully been employed *in situ* in electrochemical investigations are ESR, which is used to investigate electrochemical processes involving paramagnetic molecules, Raman spectroscopy and ellipsometry.

## References

[1]   Pletcher D 1991 *A First Course in Electrode Processes* (Romsey: The Electrochemical Consultancy)
[2]   Greef R, Peat R, Peter L M, Pletcher D and Robinson J 1993 *Instrumental Methods in Electrochemistry* (Chichester: Southampton Electrochemistry Group/Ellis Horwood)
[3]   Fisher A C 1996 *Electrode Dynamics (Oxford Chemistry Primers)* (New York: Oxford University Press)
[4]   Bard A J and Falkner L R 1980 *Electrochemical Methods—Fundamentals and Applications* (New York: Wiley)
[5]   Evans D H 1991 Review of voltammetric methods for the study of electrode reactions *Microelectrodes: Theory and Applications (Nato ASI Series E vol 197)* ed M I Montenegro, M A Queirós and J L Daschbach (Dordrecht: Kluwer)
[6]   Brett C M A and Brett A M O 1998 *Electroanalysis (Oxford Chemistry Primers)* (New York: Oxford University Press)
[7]   Wang J 1994 *Analytical Electrochemistry* (Weinheim: VCH)
[8]   Bond A 1980 *Modern Polarographic Methods in Analytical Chemistry* (New York: Dekker)
[9]   Galus Z 1994 *Fundamentals of Electrochemical Analysis* (Chichester: Ellis Horwood/Polish Scientific Publishers PWN)
[10]  Osteryoung J 1983 Pulse voltammetry *J. Chem. Educ.* **60** 296
[11]  Osteryoung J and Murphy M M 1991 Normal and reverse pulse voltammetry at small electrodes *Microelectrodes: Theory and Applications (Nato ASI Series E vol 197)* ed M I Montenegro, M A Queirós and J L Daschbach (Dordrecht: Kluwer)

[12] Barker G C and Jenkins I L 1952 *Analyst* **77** 685

[13] Osteryoung J and O'Dea J J 1986 Square wave voltammetry *Electroanalytical Chemistry* ed A J Bard (New York: Dekker)

[14] O'Dea J J, Osteryoung J and Osteryoung R A 1981 *Anal. Chem.* **53** 695

[15] O'Dea J, Wojciechowski M and Osteryoung J 1985 Square wave voltammetry at electrodes having a small dimension *Anal. Chem.* **57** 954

[16] Osteryoung J 1991 Square-wave and staircase voltammetry at small electrodes *Microelectrodes: Theory and Applications (Nato ASI Series)* ed M I Montenegro, M A Queirós and J L Daschbach (Dordrecht: Kluwer)

[17] Wang J 1985 *Stripping Analysis: Principles, Instrumentation and Applications* (Weinheim: VCH)

[18] Wang J 1987 *Anal. Proc.* **24** 325

[19] Wang J, Tian B, Wang J, Lu J, Olsen C, Yarnitzky C, Olsen K, Hammerstrom D and Bennett W 1999 Stripping analysis into the 21st century: faster, smaller, cheaper, simpler and better *Anal. Chim. Acta* **385** 429

[20] Economu A, Fielden P R and Packham A J 1994 *Analyst* **119** 279

[21] Albery W J and Bruckenstein S 1983 Uniformly accessible electrodes *J. Electroanal. Chem.* **144** 105

[22] Albery W J and Hitchman M L 1971 *Ring-Disc Electrodes* (Oxford: Clarendon)

[23] Albery W J and Brett C M A 1983 The wall-jet ring disc electrode. 1. Theory *J. Electroanal. Chem.* **148** 201

[24] Albery W J and Brett C M A 1983 The wall-jet ring disc electrode. 2. Collection efficiency, titration curves and anodic stripping voltammetry *J. Electroanal. Chem.* **148** 201

[25] Albery W J 1985 The current distribution on a wall-jet electrode *J. Electroanal. Chem.* **191** 1

[26] Unwin P R and Compton R G 1989 *Comprehensive Chemical Kinetics* vol 29, ed R G Compton (Lausanne: Elsevier)

[27] Cooper J A and Compton R G 1998 Channel electrodes—a review *Electroanalysis* **10** 141

[28] Montenegro M I, Queirós M A and Daschbach J L (eds) 1991 *Microelectrodes: Theory and Applications (Nato ASI Series E vol 197)* (Dordrecht: Kluwer)

[29] Fleischmann M *et al* (eds) 1987 *Ultramicroelectrodes* (Morganton: Datatech Systems Inc. Science Publishing)

[30] Denuault G 1996 Microelectrodes *Chemistry and Industry* **18** 678

[31] Pons S and Fleischmann M 1987 The behavior of microelectrodes *Anal. Chem.* **59** 1391A

[32] Cassidy J F and Foley M B 1993 Microelectrodes—potential invaders *Chem. Br.* **29** 764

[33] Forster R J 1994 Microelectrodes—new dimensions in electrochemistry *Chem. Soc. Rev.* **23** 289

[34] Aoki K 1993 Theory of ultramicroelectrodes *Electroanalysis* **5** 627

[35] Wightman R M and Wipf D O 1989 Voltammetry at ultramicroelectrodes *Electroanal. Chem.* **15** 267

[36] Montenegro M I 1994 *Research in Chemical Kinetics* vol 2, ed R G Compton and G Hancock (Amsterdam: Elsevier)

[37] Oldham K B 1991 Steady-state microelectrode voltammetry as a route to homogeneous kinetics *J. Electroanal. Chem.* **313** 3

[38] de Carvalho R M, Kubota L T and Rohwedder J J 1999 *Quim. Nova* **22** 591

[39] Koudelka-Hep M and Van der Wal P D 2000 Microelectrode sensors for biomedical and environmental applications *Electrochim. Acta* **45** 2437

[40] Armstrong F A and Wilson G S 2000 Recent developments in faradaic bioelectrochemistry *Electrochim. Acta* **45** 2623

[41] Tanaka K and Tokuda K 1996 *In vivo* electrochemistry with microelectrodes *Experimental Techniques in Bioelectrochemistry* ed V Brabec, D Walz and G Milazzo (Basel: Birkhäuser)

[42] Stamford J A and Justice J B Jr 1996 Probing brain chemistry *Anal. Chem.* **68** 359A

[43] Stamford J A, Palij P, Davidson C and Trout J 1995 Fast cyclic voltammetry: neurotransmitter measurement in 'real time' and 'real space' *Bioelectrochem. Bioenerg.* **38** 289

[44] Fletcher S 1991 Random assemblies of microdisk electrodes (RAM electrodes) for nucleation studies—a tutorial review *Microelectrodes: Theory and Applications (Nato ASI Series)* ed M I Montenegro, M A Queirós and J L Daschbach (Dordrecht: Kluwer)

[45] Fletcher S and Horne M D 1999 Random assemblies of microelectrodes (RAM$^{TM}$ electrodes) for electrochemical studies *Electrochem. Commun.* **1** 502

[46] Fungaro D A and Brett C M A 1999 Microelectrode arrays: application in batch-injection analysis *Anal. Chim. Acta* **385** 257

[47] Schwarz J, Kaden H and Enseleit U 2000 Voltammetric examinations of ferrocene on microelectrodes and microarrayelectrodes *Electrochem. Commun.* **2** 606

[48] Morita M, Niwa O and Horiuchi T 1997 Interdigitated array microelectrodes as electrochemical sensors *Electrochim. Acta* **42** 3177–83

[49] Bard A J, Fan F-R F, Kwak J and Lev O 1989 Scanning electrochemical microscopy—introduction and principles *Anal. Chem.* **61** 132

[50] Kwak J and Bard A J 1989 Scanning electrochemical microscopy—theory of the feedback mode *Anal. Chem.* **61** 1221

[51] Kwak J and Bard A J 1989 Scanning electrochemical microscopy—apparatus and two-dimensional scans of conductive and insulating substrates *Anal. Chem.* **61** 1794

[52] Liu H Y, Fan F-R F, Lin C W and Bard A J 1986 Scanning electrochemical and tunnelling ultramicroelectrode microscope for high-resolution examination of electrode surfaces in solution *J. Am. Chem. Soc.* **108** 3838

[53] Engstrom R C, Webber M, Wunder D J, Burgess R and Winquist S 1986 Measurements within the diffusion layer using a microelectrode probe *Anal. Chem.* **58** 844

[54]  Winquist, Engstrom R C, Meaney T, Tople R and Wightman R M 1987 Spatiotemporal description of the diffusion layer with a microelectrode probe *Anal. Chem.* **59** 2005

[55]  Bard A J, Fan F-R F and Mirkin M V 1994 Scanning electrochemical microscopy *Electroanalytical Chemistry* vol 18, ed A J Bard (New York: Dekker)

[56]  Mirkin M V 1996 Recent advances in scanning electrochemical microscopy, analytical chemistry *Anal. Chem.* **68** 177A

[57]  Mirkin M V 1999 High resolution studies of heterogeneous processes with the scanning electrochemical microscope *Mikrochim. Acta* **30** 127

[58]  Mirkin M V and Horrocks B R 2000 Electroanalytical measurements using the scanning electrochemical microscope *Anal. Chim. Acta* **406** 119

[59]  Bard A J, Fan F-R F, Pierce D T, Unwin P R, Wipf D O and Zhou F 1991 Chemical imaging of surfaces with the scanning electrochemical microscope *Science* **254** 68

[60]  Unwin P R and Bard A J 1991 Scanning electrochemical microscopy—theory and application of the feedback mode to the measurement of following chemical-reaction rates in electrode processes *J. Phys. Chem.* **95** 7814

[61]  Unwin P R 1998 Dynamic electrochemistry as a quantitative probe of interfacial physicochemical processes *J. Chem. Soc., Faraday Trans.* **94** 3183

[62]  Fulian Q, Fisher A C and Denuault G 1999 Applications of the boundary element method in electrochemistry: scanning electrochemical microscopy *J. Phys. Chem.* B **103** 4387

[63]  Fulian Q, Fisher A C and Denuault G 1999 Applications of the boundary element method in electrochemistry: scanning electrochemical microscopy, part 2 *J. Phys. Chem.* B **103** 4393

[64]  Selzer Y and Manler D 2000 Scanning electrochemical microscopy. Theory of the feedback mode for hemispherical ultramicroelectrodes: steady-state and transient behavior *Anal. Chem.* **72** 2383

[65]  Zhou F and Bard A J 1994 Detection of the electrohydrodimerization intermediate acrylonitrile radical-anion by scanning electrochemical microscopy *J. Am. Chem. Soc.* **116** 393

[66]  Barker A L, Gonsalves M, Macpherson J V, Slevin C J and Unwin P R 1999 Scanning electrochemical microscopy: beyond the solid/liquid interface *Anal. Chim. Acta* **385** 223

[67]  Kranz C, Wittstock G, Wohlschläger H and Schumann W 1997 Imaging of microstructured biochemically active surfaces by means of scanning electrochemical microscope *Electrochim. Acta* **42** 3105

[68]  Bard A J and Fan F-R F 1996 Electrochemical detection of single molecules *Acc. Chem. Res.* **29** 572

[69]  Mason T J and Lorimer J P 1998 *Sonochemistry: Theory, Applications and Uses of Ultrasound in Chemistry* (Chichester: Ellis Horwood)

[70]  Compton R G, Eklund J C and Marken F 1997 Sonoelectrochemical processes: a review *Electroanalysis* **9** 509

[71]  Compton R G, Eklund J C, Marken F, Rebbitt T O, Akkermans R P and Waller D N 1997 Dual activation, coupling ultrasound to electrochemistry—an overview *Electrochim. Acta* **42** 2912

[72]  Akkermans R P, Wu M, Bain C D, Fidel-Suérez M and Compton R G 1998 Electroanalysis of ascorbic acid: a comparative study of laser ablation voltemmetry and sonovoltammetry *Electroanalysis* **10** 613

[73]  Compton R G, Eklund J C, Page S D, Mason T J and Walton D J 1996 Voltammetry in the presence of ultrasound: mass transport effects *J. Appl. Electrochem.* **26** 775

[74]  Walton D J, Phull S S, Chyla A, Lorimer J P, Mason T J, Burke L D, Murphy M, Compton R G, Eklund J C and Page S D 1995 Sonovoltammetry at platinum electrodes: surface phenomena and mass transport processes *J. Appl. Electrochem.* **25** 1083

[75]  Birkin P R and SilvaMartinez S 1995 The effect of ultrasound on mass-transport to a microelectrode *J. Chem. Soc., Chem. Commun.* **17** 1807

[76]  Birkin P R and SilvaMartinez S 1997 A study on the effects of ultrasound on electrochemical phenomena *Ultrasonics Sonochemistry* **4** 121

[77]  Reisse J, Francois H, Vandercammen J, Fabre O, Kirschdemesmäker A, Märschalk C and Delplancke J L 1994 *Electrochim. Acta* **39** 37

[78]  Eklund J C, Markem F, Waller D N and Compton R G 1996 Voltammetry in the presence of ultrasound, a novel sono-electrode geometry *Electrochim. Acta* **41** 1541

[79]  Compton R G, Eklund J C, Marken F and Waller D N 1996 Electrode processes at the surfaces of sonotrodes *Electrochim. Acta* **41** 315

[80]  Birkin P R, O'Connor R, Rapple C and SilvaMartinez S 1998 Electrochemical measurement of erosion from individual cavitation events generated from continuous ultrasound *J. Chem. Soc., Faraday Trans.* **94** 3365

[81]  MacDonald J R 1987 *Impedance Spectroscopy* (New York: Wiley)

[82]  Scully J R, Silverman D C and Kendig M W (eds) 1993 *Electrochemical Impedance—Analysis and Interpretation* (Philadelphia: ASTM)

[83]  Urquindi-Macdonald M and Egan P C 1997 Validation and extrapolation of electrochemical impedance spectroscopy data *Corr. Rev.* **15** 169

[84]  Amirudin A and Thierry D 1995 Application of electrochemical impedance spectroscopy to the study and degradation of polymer-coated metals *Prog. Org. Coat.* **26** 1

[85]  Grundmeier G, Schmidt W and Stratmann M 2000 Corrosion protection by organic coatings: electrochemical mechanism and novel methods of investigation *Electrochim. Acta* **45** 2515

[86] Gomes W P and VanMaelkelbergh D 1996 Impedance spectroscopy at semiconductor electrodes: review and recent developments *Electrochim. Acta* **41** 967

[87] Swarup J and Sharma P C 1996 Electrochemical techniques for the monitoring of corrosion of reinforcement in concrete structures *Bull. Electrochem.* **12** 103

[88] Gerischer H 1970 *Physical Chemistry* vol 9, ed H Eyring, D Henderson and W Jost (New York: Academic)

[89] Morrison S R 1977 *The Chemical Physics of Surfaces* (New York: Plenum)

[90] Pleskov Y V and Gurevich Y Y 1986 *Semiconductor Photoelectrochemistry* (New York: Plenum)

[91] Stimming U 1986 Photoelectrochemical studies of passive films *Electrochim. Acta* **31** 415

[92] Gerischer H 1990 On the interpretation of photoelectrochemical experiments with passive layers on metals *Corr. Sci.* **31** 81

[93] Kamat P V 1993 Photochemistry on non-reactive and reactive (semiconductor) surfaces *Chem. Rev.* **93** 267

[94] Gerischer H 1990 The impact of semiconductors on the concepts of electrochemistry *Electrochim. Acta* **35** 1677

[95] Chandra S 1985 *Photoelectrochemical Solar Cells* (New York: Gordon and Breach)

[96] Tryk D A, Fujishima A and Honda K 2000 Recent topics in photoelectrochemistry: achievements and future prospects *Electrochim. Acta* **45** 2363

[97] Lewis N S 1996 Photoelectrochemistry—energy conversion using semiconductor electrodes *ECS Inferface* Autumn, 28

[98] Rajeshwar K 1995 Photoelectrochemistry and the environment *J. Appl. Electrochem.* **25** 1067

[99] Pleskov Y V 1994 Semiconductor photoelectrochemistry for a cleaner environment: utilisation of solar energy *Environmental Oriented Electrochemistry (Studies in Environmental Science 59)* ed C A C Sequeira (Amsterdam: Elsevier)

[100] Haram S K and Santhanam K S V 1994 Prospective usage of photoelectrochemistry for environmental control *Environmental Oriented Electrochemistry (Studies in Environmental Science 59)* ed C A C Sequeira (Amsterdam: Elsevier)

[101] Gale R G (ed) 1988 *Spectroelectrochemistry, Theory and Practice* (New York: Plenum)

[102] Gutiérrez C and Melendres C 1990 *Spectroscopic and Diffraction Techniques in Interfacial Electrochemistry (NATO ASI Series C vol 320)* (Dordrecht: Kluwer)

[103] Christensen P A and Hamnett A 1994 *Techniques and Mechanisms in Electrochemistry* (Glasgow: Blackie) (an imprint of Chapman and Hall)

[104] Plieth W, Wilson G S and de la Fe C 1998 Spectroelectrochemistry: a survey of *in situ* spectroscopic techniques *Pure Appl. Chem.* **70** 1395

[105] Beden B 1995 On the use of '*in situ*' UV-visible and infrared spectroscopic techniques for studying corrosion products and corrosion inhibitors *Mater. Sci. Forum* **192–4** 277

[106] Korzeniewski C 1997 Infrared spectroscopy in electrochemistry: new methods and connections to UhV surface science *Crit. Rev. Anal. Chem.* **27** 81

[107] Christensen P and Hamnett A 2000 *In situ* techniques in electrochemistry—ellipsometry and FTIR *Electrochim. Acta* **45** 2443

[108] Gewirth A A and Niece B K 1997 Electrochemical applications of *in situ* scanning probe microscopy *Chem. Rev.* **97** 1129

[109] Lillehei P T and Bottomley L A 2000 Scanning probe microscopy *Anal. Chem.* **72** 189R

[110] Sonnenfield R and Hansma P K 1986 Atomic-resolution microscopy in water *Science* **232** 211

## Further Reading

Pletcher D 1991 *A First Course in Electrode Processes* (Romsey: The Electrochemical Consultancy)

Fisher A C 1996 *Electrode Dynamics (Oxford Chemistry Primers)* (New York: Oxford University Press)

Brett C M A and Brett A M O 1998 *Electroanalysis (Oxford Chemistry Primers)* (New York: Oxford University Press)

These books provide an excellent introduction to the subject.

Bard A J and Falkner L R 1980 *Electrochemical Methods-Fundamentals and Applications* (New York: Wiley)

For in-depth coverage of electrochemical methods including mathematical derivations.

Montenegro M I, Queirós M A and Daschbach J L 1991 *Microelectrodes: Theory and Applications (Nato ASI Series)* (Dordrecht: Kluwer)

An essential introduction to the field of microelectrodes.

MacDonald J R 1987 *Impedance Spectroscopy* (New York: Wiley)

For in-depth theory of impedance.

Sawyer D T, Sobkowiak A and Roberts J L Jr 1995 *Electrochemistry for Chemists* (New York: Wiley)

Brabec V, Walz D and Milazzo G (eds) 1996 *Experimental Techniques in Bioelectrochemistry* (Basel: Birkhäuser)

These are a good introduction of specific electrochemical techniques for organic chemists and biologists.

Christensen P A and Hamnett A 1994 *Techniques and Mechanisms in Electrochemistry* (Glasgow: Blackie) (an imprint of Chapman and Hall)

This contains a good overview of spectroelectrochemical techniques.

# B1.29
# High-pressure studies

*Malcolm F Nicol*

## B1.29.1   Introduction

This chapter introduces the physical chemistry of materials under high pressures. Space limitations permit only a broad-brush introductory survey. High-pressure studies range from designing equipment to generate, to confine and to measure high pressures to spectroscopic studies from $10^5$ Hz to beyond $10^{19}$ Hz at temperatures from below 1 K to $10^5$ K and beyond for all sorts of elements, compounds, solutions and mixtures. To say that these are extreme ranges of conditions is an understatement.

To gain a sense of the range of behaviours, consider what happens to one element familiar to every chemist and physicist: oxygen. At ambient temperature, oxygen, $O_2$, exists as the canonical odourless, colourless gas of elementary school and as a purple, orange, red, blue or black solid depending on the pressure and the direction from which you look at the crystals. It becomes a metal and, at low temperatures, an antiferromagnet or a superconductor. In the solid phase stable above 10 GPa, $O_2$ has a strong infrared vibrational absorption band in the stretching region. Then, of course, there is the $O_3$ isomer which has not been studied at high pressures.

Similarly 'strange' things happen to other materials. Above 5 GPa, CO spontaneously polymerizes; the structure of the product is $\left( O = C \diagup^{>C=O}_{\diagdown} \right)_n$. Indeed, almost every carbon compound with unsaturated bonds becomes unstable with respect to reactions that produce saturated compounds at this or slightly higher pressures. Somewhat above 100 GPa, CsI and Xe also are metals. By 100 TPa and $10^5$ K—yes, experiments have been done to these conditions—the density of Al exceeds 12 g cm$^{-3}$, or about five times greater than at ambient pressure. All but the 1s and possibly 2s electrons remain localized on an Al nucleus.

Books are available on many of these subjects. The objective here, therefore, is to introduce several fundamental issues and point to additional information by citing key references and suggesting further reading. We begin by briefly delimiting what we mean by *high pressure*. Then, we discuss how high pressures are achieved and measured before describing the behaviours of a few familiar materials at high pressures.

## B1.29.2   What is pressure?

Almost everyone has a concept of 'pressure' from weather reports of the pressure of the atmosphere around us. In this context, 'high pressure' is a sign of good weather while very low pressures occur at the 'eyes' of cyclones and hurricanes. In elementary discussions of mechanics, hydrostatics of fluids and the gas laws, most scientists learn to compute pressures in static systems as force per unit area, often treated as a scalar quantity. They also learn that unbalanced pressures cause fluids to flow. Winds are the flow of the atmosphere from regions of high to low pressures. However, high and low pressures in the atmosphere rarely deviate by as much as 10% from the local mean pressure, about 0.1 megapascal at 'sea level'. The pascal (Pa) is the SI unit of pressure, 1 Pa= 1 N m$^{-2}$ = $10^{-5}$ bar. One standard atmosphere is about $1.013 \times 10^{5}$ Pa. Local fluctuations in the pressure of the atmosphere are, however, much smaller than the difference between the average pressure at 'sea level' and at the peaks of high mountains. The average pressure near the top of Mount Everest is less than one-quarter of atmospheric pressure at 'sea level'.

This example of high and low pressure also shows the ambiguities of these terms in science. All these pressures are essentially *constant* in terms of the range of pressures encountered in nature. From negative pressures in solids under tension (e.g., on the wall of flask confining a fluid), pressure in nature increases through the very low-pressure vacuum of interplanetary space (less than $10^{-13}$ Pa) to well in excess of $10^{20}$ Pa at the centres of neutron stars! In these terms, *high pressure* and *low pressure* are relative terms with different meanings in different areas of chemistry and physics. Test this by searching an electronic database for 'high pressure'. A discussion of high-pressure studies, therefore, must decide what pressure is and what high means; just how high is high.

Relationships from thermodynamics provide other views of pressure as a macroscopic state variable. Pressure, temperature, volume and/or composition often are the controllable independent variables used to constrain equilibrium states of chemical or physical systems. For fluids that do not support shears, the pressure, $P$, at any point in the system is the same in all directions and, when gravity or other accelerations can be neglected, is constant throughout the system. That is, the equilibrium state of the system is subject to a hydrostatic pressure. The fundamental differential equations of thermodynamics:

$$dU = -PdV + TdS$$

$$dA = -PdV - SdT$$

identify $P$ through the Maxwell relations:

$$P = -(\partial U / \partial V)_S = -(\partial A / \partial V)_T.$$

Two other Maxwell relations define the direction systems change to achieve equilibrium:

$$V = \rho^{-1} = (\partial H / \partial P)_S = (\partial G / \partial P)_T.$$

In both mechanical (constant $S$, minimize $H$) and thermal (constant $T$, minimize $G$) contexts, pressure drives a system to become smaller or denser.

The situation is more complex for rigid media (solids and glasses) and more complex fluids: that is, for most materials. These materials have finite yield strengths, support shears and may be anisotropic. As samples, they usually do not relax to hydrostatic equilibrium during an experiment, even when surrounded by a hydrostatic pressure medium. For these materials, $P$ should be replaced by a stress tensor, $\sigma_{ij}$, and the appropriate thermodynamic equations are more complex.

The take-home lesson is that the vast majority of high-pressure studies are on solids or other rigid media and are not done under hydrostatic conditions. The stresses and stress-related properties may vary throughout the sample. Unless the probes are very local and focus on a small region of the sample, measurements are averages over a range of, often uncharacterized, conditions.

As well as macroscopic equations of state relating free energies, enthalpy, entropy, density, composition, temperatures and pressure, high-pressure science also concerns how changes of pressure and other macroscopic constraints affect the microscopic molecular and electronic structures of matter. At low pressures, the chemistry of most materials is described in terms of electrons tightly bound to specific atoms, molecules or ions and relatively weaker intermolecular van der Waals or ionic forces. The itinerant conduction electrons of metals are an exception; they are delocalized throughout the solid. The highly local nature of most electrons reflects the drive to minimize their potential energies.

To a rough approximation, the kinetic and potential energies of electrons in simple systems vary with density as $\rho^{2/3}$ and $-\rho^{1/3}$, respectively. This means that kinetic energy considerations should dominate at very high densities. Localized electrons should, therefore, eventually delocalize at very high pressures, converting ionic and molecular materials to more closely packed extended network structures and to metals at high pressures. Again, of course, the question occurs: how high is very high? Experiments provide the answer, or at least a lower limit. Where the answer is missing, experimenters are driven to try to attain even higher pressures.

Many experiments support the pressure-delocalization principle. Most unsaturated organic molecules including CO and $C_2N_2$ polymerize at pressures of the order of 10 GPa [1–3]. Layered covalent solids like graphite and hexagonal boron nitride (h-BN) transform to dense, three-dimensional network materials, diamond and cubic boron nitride (c-BN) [4–6]. Solid oxygen, solid and fluid iodine, fluid hydrogen and nitrogen, xenon, and cesium iodide are examples of materials developing metallic behaviour at pressures of the order of 100 GPa [7–22]. Later sections provide more details about some of these transformations. At intermediate pressures, the energies of atomic and molecular orbitals change with pressure. By measuring electronic spectra at high pressures, differences of these energy changes can be determined for various different orbitals. In some cases, spectral features change by as much as 0.1 eV GPa$^{-1}$. Drickamer named this phenomenon pressure-tuning spectroscopy and has written extensively about observations for many systems [23–25].

## B1.29.3    What pressures are high?

What then are high pressures? The answer to this question involves the bias of personal experience. I often remark in an off-hand manner that 'In my laboratory, we consider 5 kbar a low pressure'. We have several reasons for setting the low–high boundary around 1 GPa. This and slightly higher pressures can conveniently be achieved by use of commercial autoclaves several litres in size, mechanical compressors and fluid or even compressed-gas pressure media. Many commercial processes run at these and lower pressures. These include the Haber synthesis of ammonia, a method for producing high-density polyethylene, and recently developed methods for producing vaccines by denaturing viruses or for sterilizing (pascalizing) strawberry jam and other foods at ambient temperature which preserve their flavours better than pasteurizing at higher temperatures. The rates of several chemical reactions accelerate by factors of $10^4$ or more between 0.1 and 100 MPa at ambient temperature, so much interesting chemistry occurs at these lower pressures. At such 'low' pressures, Bridgman [26] even showed how to cook eggs at 'room' temperature.

At ambient temperature, however, few materials remain fluid at pressures much higher than 1 GPa. Fluids also are much more difficult to confine at higher pressures. Absolute pressures have been measured with dead-weight testers only to about 2.5 GPa, that is to the pressure of a solid–solid phase transition in elemental bismuth at 298 K [27]. The importance of non-hydrostatic stresses and changes to the technology of high-pressure studies above 1 GPa suggest the rough dividing line which I have adopted for this essay.

The energies of chemical changes provide a third criterion for defining high pressures. Many unsaturated organic compounds dimerize or polymerize at high pressures because the products are denser by about $10^{-5}$ m$^3$ mol$^{-1}$ (10 cm$^3$ mol$^{-1}$). At a pressure of 1 GPa, the corresponding decrease of the energy, enthalpy and free energy is 10 kJ mol$^{-1}$ or relatively modest compared with chemical bond energies. At 10 GPa, for the same difference of molar volume, the energies decrease by 100 kJ mol$^{-1}$, an amount comparable to bond energies. That is, chemical change can be anticipated at pressures somewhat above 1 GPa.

## B1.29.4    How are high pressures achieved?

Laboratory high-pressure studies follow many approaches. The pressure may remain constant (so-called static experiments) or be transient (so-called dynamic experiments where shock waves generated by an explosion, impact or laser ablation compress a sample for a few microseconds or shorter times). Many devices have been used for static experiments to about 20 GPa; Jayaraman described many of these in three articles [28–30]. Many static high-pressure cells are variants of the piston–cylinder apparatus frequently used to illustrate compression in elementary discussions of the thermodynamics of gases. An external force on a piston free to move within a cylinder applies pressure to a sample that is confined as long as the seal between the piston and cylinder remains leaktight. Bridgman's anvils [31] represent a different concept. The concept confines the sample between two pistons made of a hard material shaped as truncated cones and a crushable cylindrical gasket. The classical Bridgman design used cemented tungsten carbide anvils and a lava (pyrophyllite) gasket. External force applied to the anvils crush the cylinder, preventing the sample from 'blowing out', while applying pressure to the confined sample. Many dual-piston, tetrahedral and cubic cells elaborate on one or both of these concepts of compressing a sample within a confined, sealed volume.

All static studies at pressures beyond 25 GPa are done with diamond-anvil cells conceived independently by Jamieson [32] and by Weir *et al* [33]. In these variants of Bridgman's design, the anvils are single-crystal gem-quality diamonds, the hardest known material, truncated with small flat faces (culets) usually less than 0.5 mm in diameter. Diamond anvils with 50 $\mu$m diameter or smaller culets can generate pressures to about 500 GPa, the highest static laboratory pressures equivalent to the pressure at the centre of the Earth.

Dynamic experiments with conventional (chemical) explosives or projectiles accelerated in gas guns have achieved 1 TPa in favourable cases. Laser-driven shocks have produced higher shock pressures [34], and measurements to 75 TPa have been reported for shock waves generated during underground tests of nuclear explosives (for a recent discussion see [35]). Sample volumes in static experiments range from litres at pressures up to 10 GPa to 0.1 nl at 500 GPa. Samples for commercial dynamic high-pressure production of diamond powder was done on the 100 kl scale. Most samples for shock wave studies are smaller; laser-driven shock wave experiments often use microlitre samples.

In static experiments, the temperatures of samples can be controlled from less than 1 to more than 5000 K and can be measured with reasonable accuracy. For low temperatures, entire pressure vessels are thermostatted by mounting them in cryostats or surrounding them with heaters or furnaces. With these techniques, the temperatures are uniform throughout the sample. The strengths of the materials used to construct the vessel, however, limit the temperatures and pressures that can be achieved by such external heating methods. For higher temperatures, internal heating is used: that is, the sample is heated while the confining pressure vessel is kept at a lower temperature to maintain its mechanical strength. This can be done by surrounding the sample with heating elements mounted inside the pressure vessel around the sample or by irradiating the sample with an intense infrared or visible laser. Internal-heating methods may involve very large temperature gradients, up to $10^8$ K m$^{-1}$ (100 K $\mu$m$^{-1}$), and challenge thermometry. The irreversible nature of the work done when shock waves compress a sample necessarily increases the sample's temperature; however, the temperature of the shocked state is often impossible to characterize. Indeed, measuring temperatures achieved during dynamic experiments is one of the biggest unsolved problems of high-pressure research.

### B1.29.5   How are high pressures measured?

Absolute pressure measurements by dead-weight piston–cylinder methods have been made only to 2.5 GPa, although Getting recently developed a cell which may extend absolute measurements to 5 GPa [27]. Pressures achieved during shock experiments are computed with the Rankine–Hugoniot equations which assume that the shocked, high-temperature state is one of thermodynamic equilibrium and mass, momentum, and energy are conserved [35]. Several procedures have been developed to relate densities and other properties of states achieved during shock experiments to values of the same properties on ambient or zero-Kelvin isotherms. Pressures in static experiments above 2.5 GPa are often determined by measuring the density of a convenient material like NaCl or Au confined with the sample being studied by x-ray diffraction or by a secondary probe that has been calibrated in terms of x-ray densities. Typical secondary probes include luminescence spectra of ruby (dilute $Cr^{3+}$ in $Al_2O_3$) [36] or Sm:YAG [37] or Raman spectra of nitrogen [38] or diamond [39]. In each secondary probe, a spectral feature—a narrow emission line or vibrational band—whose energy can be measured precisely changes energy with pressure in an established, ideally simple manner.

### B1.29.6 High-pressure forms of familiar or useful materials: diamond, fluid metallic hydrogen, metallic oxygen, ionic carbon dioxide, gallium nitride

The most important commercial products of high-pressure science are the extremely hard materials, synthetic diamond [40–42] and (c-BN) [43]. At ambient pressure, diamond is less stable than the less dense allotrope, graphite. c-BN also may be less stable than the less dense h-BN isomer with its graphitic structure of rings of alternating B and N atoms. Diamond and c-BN with four equivalent $sp^3$ bonds per atom are denser than graphite and h-BN with three $sp^2$ bonds, each shorter than an $sp^3$ bond and one very long intermolecular van der Waals bond per atom. The volume change, $\Delta V$, for the transformations from graphite to diamond is negative. Thus, the $\Delta(PV) = P\Delta V$ contribution to the change of enthalpy or Gibbs' free energy is negative and becomes even more negative at higher pressures, so the denser forms of each material become more stable at higher pressures. Rearranging the bonds around each atom involves high-energy barriers that separate the low and high density forms (for a historical review of these processes see [44]). Although Yagi and Utsumi [45] showed that the graphite converted to the hexagonal form of diamond at ambient temperature above 10 GPa, complete conversion and recovery to ambient pressure was not possible unless the product was heated under pressure to more than 1100 K. That is, high temperatures and high pressures are used to overcome the thermodynamic and kinetic barriers to making the desirable dense hard materials that can be recovered metastably. The high barrier also impedes reversion of the recovered diamond and c-BN to the stable graphitic forms at low temperatures. Besides the high-pressure route, diamond can be synthesized at low pressures, even less than 0.1 MPa, by kinetically controlled gas–surface reactions.

Large quantities of both diamond and c-BN are produced by static or shock methods for industrial cutting applications. Most of the synthetic material is finely powdered and can be bound or compacted to make tools. Manufacturing costs for large crystals are too high for the commercial gem market; however, large diamonds of exceptionally high quality are made for special applications: e.g. x-ray monochromators for high-intensity synchrotrons. For cutting steels, c-BN is particularly valuable, because diamond tools tend to react with the steel, forming iron carbide.

Metallic hydrogen has been a holy grail of high-pressure research since Wigner and Huntington suggested that it might be stable above about 10 GPa [46]. Many claims to the contrary notwithstanding, metallic solid hydrogen has not been found at ambient temperatures to 342 GPa [47]. Weir *et al*, however, found that fluid hydrogen becomes highly conductive under shock compression at lower pressures, 140 GPa, and higher temperatures [11]. The fact that homonuclear diatomic molecular fluids become conductive at lower pressures—and necessarily higher temperatures—than solid phases of the same systems is evident in nitrogen and iodine, and may be a general phenomenon. Further careful experiments must be done to confirm this conjecture. Detailed studies of the conductivities of supercritical Cs and Hg show that the transition from low to high electrical conductivity has neither the characteristics of a thermodynamic phase transition nor a general relationship to the vapour–liquid critical point [48–51].

Oxygen is the low-$Z$ diatomic which is known to transform to a metal and, at about 1 K, a superconductor at high pressures. The transition pressure is slightly greater than 100 GPa [7]. The conductive phase consists of $O_2$ molecules; that is, it is not an atomic phase. Optical, infrared and visual spectral, and x-ray diffraction data show that the relevant $\varepsilon$ phase of oxygen is very anisotropic, and it is reasonable to conjecture that the electrical conductivity also depends upon crystallographic orientation. The other group VI elements also have metallic, superconductive phases at high pressures and low temperatures.

Recent work on the carbon dioxide system shows another unusual high-pressure behaviour. Raman spectra of carbon dioxide show that $CO_2$ molecules remain the basis of the phases to more than 40 GPa at temperatures below a few hundred Kelvin [52]. These results, however, do not mean that the molecular crystals are the stable phases; indeed, recent studies of the combustion of carbon at high pressures by Yoo *et al* [53] reach another conclusion. They initiated combustion of a mixture of carbon and oxygen at pressures between 7 and 13 GPa by heating the carbon with a Nd:YAG laser, quenching the products to ambient temperature under pressure and recording their Raman spectra. As well as features of unreacted $O_2$ and $CO_2$ in some samples, they found vibron bands characteristic of the carbonate ion near 734 and 1079 cm$^{-1}$, a band assigned to $CO^{++}$ near 2243 cm$^{-1}$, and several lattice modes between 100 and 350 cm$^{-1}$. These features, including the shape of the lattice-mode spectrum, closely match the spectra of $NO^+NO_3^-$, the ionic dimer of $NO_2$, reported by Agnew *et al* [54], apart from minor shifts because of different pressures, force constants and reduced masses. At higher pressures, heating carbon to ignition was more difficult because diamond formed, which greatly reduced the absorption of the Nd:YAG radiation. The $CO^{++}CO_3^{--}$ could not be quenched to ambient pressure at ambient temperature; it transformed to $CO_2$ below 2 GPa. When the $CO^{++}CO_3^{--}$ was compressed above 15 GPa at ambient temperature, the sharp $CO^{++}CO_3^{--}$ band near 1100 cm$^{-1}$ either disappeared or broadened above. This change was reversible with pressure and was attributed to either amorphorization of $CO^{++}CO_3^{--}$ or transitions to larger multimers.

An ionic dimer of carbon dioxide has a critical implication for understanding detonation chemistry of energetic organic molecules. This dimerization provides a reasonable explanation for the kink in shock compression Hugoniot of $CO_2$ [55], which, if correct, implies that the dimer forms on the time scale of shock loading and detonation. Because water and some nitrogen oxides also become more ionic at these high pressures, interactions and chemical reactivities of these ionic H–C–N–O species will differ from those of the neutral species assumed, in most models, to be the major detonation products over a wide range of high pressures and temperatures [56]. Furthermore, because ionic species are implicated in planetary 'ices' like $H_2O$, $CH_4$ and $NH_3$ [57–60], the ionic dimer should have some bearing in understanding the internal structure and magnetism of the Jovian planets [61].

Interest in AlN, GaN, InN and their alloys for device applications as blue light-emitting diodes and blue lasers has recently opened up new areas of high-pressure synthesis. Near atmospheric pressure, GaN and InN are unstable with respect to decomposition to the elements far below the temperatures where they might melt. Thus, large boules of these materials typically used to make semiconductor devices cannot be grown from the melt or annealed at temperatures approaching their melting points. Devices have been grown by heteroepitaxial methods, depositing GaN on $Al_2O_3$ or SiC substrates, although high defect concentrations because of mismatched lattice constants and thermal expansivities are a serious problem. Grzegory *et al* [62] showed how to overcome this limitation for GaN by growing large crystals from $N_2$ dissolved in liquid Ga at pressures up to 2 GPa. AlN but not InN also have been grown this way [63]. These relatively slow processes produce excellent materials.

Wallace *et al* explored another approach, metathesis reactions. By igniting a mixture of $GaI_3$ and $Li_3N$ confined at pressures of the order of 4 GPa, they produced fine-crystalline GaN [63]. The thermodynamic driving force for the process is the very negative enthalpy of forming LiI. With appropriate mixtures and pressures, they produced CrN, $Cr_2N$, TaN and other nitrides by this metathesis route. The metathesis method, like the direct reactions between elements that Yoo *et al* used to make c-BN, $\beta$-$Si_3N_4$, $B_2O_3$ and other materials [64, 65], yields more crystalline products at higher confining pressure. Recently, Wallace [66] devised combinations of reagents, chemical diluents and confinement so that GaN and InN crystals of similar quality can be made at as low as ambient pressure.

## B1.29.7  Spectroscopy at high pressures

Almost every modern spectroscopic approach can be used to study matter at high pressures. Early experiments include NMR [67], ESR [68]; vibrational infrared [33] and Raman [69]; electronic absorption, reflection and emission [23–25, 70]; x-ray absorption [71] and scattering [72], Mössbauer [73] and gc-ms analysis of products recovered from high-pressure photochemical reactions [74]. The literature contains too many studies to do justice to these fields by describing particular examples in detail, and only some general rules, appropriate to many situations, are given.

The frequencies of vibrational modes usually increase with increasing pressure because the corresponding potential wells become narrower and the force constants increase. In wavenumber terms, these increases range up to the order of $10\,\mathrm{cm}^{-1}\,\mathrm{GPa}^{-1}$. A notable exception is the stretching mode of an O-H$\cdots$O hydrogen bond. Other instances of modes whose frequencies decrease with increasing pressure suggest molecular or lattice instabilities that lead to phase transitions at higher pressures. Usually, the transition occurs before the frequency of the mode reaches zero.

Most electronic valence transitions shift to longer wavelengths at higher pressures: that is, the gap between the highest occupied orbital and lowest unoccupied orbital tends to decrease upon compression. The rates of shift usually are larger (1) for pure materials than for solutes in a solvent and (2) for stronger (more allowed) transitions. However, these correlations are not quantitative, and many transitions shift in the opposite direction. The largest shifts are of the magnitude $0.1\,\mathrm{eV}\,\mathrm{GPa}^{-1}$. Many d–d bands of transition element compounds vary linearly with the fifth power of the metal–ligand distance.

New methods appear regularly. The principal challenges to the ingenuity of the spectroscopist are availability of appropriate radiation sources, absorption or distortion of the radiation by the windows and other components of the high-pressure cells, and small samples. Lasers and synchrotron radiation sources are especially valuable, and use of beryllium gaskets for diamond-anvil cells will open new applications. Impulse-stimulated Brillouin [75], coherent anti-Stokes Raman [76, 77], picosecond kinetics of shocked materials [78], visible circular and x-ray magnetic circular dichroism [79, 80] and x-ray emission [72] are but a few recent spectroscopic developments in static and dynamic high-pressure research.

An especially interesting recent example is Benedetti *et al*'s use of circular dichroism (CD) spectroscopy to detect a pressure-induced change of the configuration at the metal centre of the octahedral chiral $\Delta$- and $\Lambda$-tris[cyclo O,O' 1(R),2(R)-dimethylethylene dithophosphato] chromium(III) [79]. The pressure medium was Nujol®. To measure the CD spectrum, they had to overcome the birefringence of the strained diamond windows of the high-pressure cell. They did this by recording and averaging spectra of the sample—and of a blank cell filled with Nujol®—for each of four $90°$ rotations of the cell around the axis normal to the windows. The measurements for the blank showed that the baseline obtained by this averaging procedure was close to ideal, although a small further correction was required.

# References

[1] Yoo C S and Nicol M F 1986 Chemical reactions and new phases of solid $C_2N_2$ at high pressure *J. Phys. Chem.* **90** 6726

[2] Yoo C S and Nicol M F 1986 Kinetics of the polymerization of solid $C_2N_2$ near 10 GPa *J. Phys. Chem.* **90** 6732

[3] Katz A I, Schiferl D and Mills R L 1981 New phases and chemical reactions in solid carbon monoxide under pressure *J. Phys. Chem.* **88** 3176

[4] Johnson Q and Mitchell A C 1972 First x-ray diffraction evidence for a phase transition during shock-wave compression *Phys. Rev. Lett.* **29** 1369

[5] Riter J R Jr 1973 Shock-induced graphite.far.wurtzite phase transformation in boron nitride and implications for stacking graphitic boron nitride *J. Chem. Phys.* **59** 1538

[6] Bundy F P 1980 The P, T phase and reaction diagram for elemental carbon, 1979 *J. Geophys. Res.* **85** 6930

[7] Shimizu K, Eremets M I, Suhara K and Amaya K 1998 Oxygen under high pressure—temperature dependence of electrical resistivity *Koatsuryoku no Kagaku to Gijutsu (Proc. Int. Conf. AIRAPT-16 and HPCJ-38 on High Pressure Science and Technology, 1997)* vol 7, p 1040

[8] Reichlin R, Schiferl D, Martin S, Vanderborgh C and Mills R L 1985 Optical studies of nitrogen to 130 GPa *Phys. Rev. Lett.* **55** 1464

[9] Goettel K A, Eggert J H and Silvera I F 1989 Optical evidence for the metallization of xenon at 132(5) GPa *Phys. Rev. Lett.* **62** 665

[10] Reichlin R, Ross M, Martin S and Goettel K A 1986 Metallization of cesium iodide *Phys. Rev. Lett.* **56** 2858

[11] Weir S T, Mitchell A C and Nellis W J 1996 Metallization of fluid molecular hydrogen at 140 GPa (1.4 Mbar) *Phys. Rev. Lett.* **76** 1860

[12]  Hemley R J and Mao H K 1990 Critical behavior in the hydrogen insulator–metal transition *Science* **249** 391
[13]  Lorenzana H E, Silvera I F and Goettel K A 1990 Order parameter and a critical point on the megabar-pressure hydrogen-A phase line *Phys. Rev. Lett.* **65** 1901
[14]  Takemura K, Minomura S, Shimomura O, Fujii Y and Axe J D 1982 Structural aspects of solid iodine associated with metallization and molecular dissociation under high pressure *Phys. Rev.* B **26** 998
[15]  Buontempo U, Degiorgi E and Postorino P 1998 Towards the metallization transition in liquid $I_2$: a spectroscopic study *Nuovo Cimento* D **20** 573
[16]  Ross M 1968 Shock compression of argon and xenon. IV. Conversion of xenon to a metal-like state *Phys. Rev.* **171** 777
[17]  Goettel K A, Eggert J H, Silvera I F and Moss W C 1989 Optical evidence for the metallization of xenon at 132(5) GPa *Phys. Rev. Lett.* **62** 665
[18]  Reichlin R, Brister K E, McMahan A K, Ross M, Martin S, Vohra Y K and Ruoff A L 1989 Evidence for the insulator–metal transition in xenon from optical, x-ray, and band-structure studies to 170 GPa *Phys. Rev. Lett.* **26** 669
[19]  Nellis W J, Holmes N C, Mitchell A C and Van Thiel M 1984 Phase transition in fluid nitrogen at high densities and temperatures *Phys. Rev. Lett.* **53** 1661
[20]  Hamilton D C, Mitchell A C and Nellis W J 1986 Electrical conductivity measurements in shock compressed liquid nitrogen *Shock Waves in Condensed Matter (Proc. 4th Am. Phys. Soc. Top. Conf.)* p 473
[21]  Eremets M I, Shimizu K, Kobayashi T and Amaya K 1998 Metallic CsI at pressures of up to 220 gigapascals *Science* **281** 1333
[22]  Shimizu K, Suhara K, Ikumo M, Eremets M I and Amaya K 1998 Superconductivity in oxygen *Nature* **393** 767
[23]  Drickamer H G and Frank C W 1973 *Electronic Transitions and the High-Pressure Chemistry and Physics of Solids* (London: Chapman and Hall)
[24]  Drickamer H G 1986 Pressure tuning spectroscopy *Acc. Chem. Res.* **19** 329
[25]  Drickamer H G 1990 Forty years of pressure tuning spectroscopy *Ann. Rev. Mater. Sci.* **20** 1
[26]  Bridgman P W 1914 The coagulation of albumin by pressure *Proc. Am. Acad. Arts Sci.* **49** 627
[27]  Heydemann P L M 1997 The Bi I–II transition pressure measured with a dead-weight piston gauge *J. Appl. Phys.* **38** 2640
       Getting I 1998 New determination of the bismuth I-II equilibrium pressure—a proposed modification to the practical pressure scale *Metrologica* **35** 119
[28]  Jayaraman A 1983 Diamond anvil cell and high-pressure physical investigations *Rev. Mod. Phys.* **55** 65
[29]  Jayaraman A 1984 The diamond-anvil high-pressure cell *Sci. Am.* **250** 54
[30]  Jayaraman A 1986 Ultrahigh pressures *Rev. Sci. Instrum.* **57** 1013
[31]  Bridgman P W 1941 Explorations towards the limit of utilizable pressures *J. Appl. Phys.* **12** 461
[32]  Jamieson J C, Lawson A W and Nachtreib N D 1959 New device for obtaining X-ray diffraction patterns from substances exposed to high pressures *Rev. Sci. Instrum.* **30** 1016
[33]  Weir C E, Lippencott E R, Van Valkenberg A and Bunting E N 1959 Infrared studies in the 1–15 micron region to 30,000 atmospheres *J. Res. Natl Bur. Stds* A **63** 55
[34]  Da Silva L B *et al* 1997 Absolute equation of state measurements on shocked liquid deuterium up to 200 GPa (2 Mbar) *Phys. Rev. Lett.* **78** 483
[35]  Trunin R F 1998 *Shock Compression of Condensed Matter* (Cambridge: Cambridge University Press)
[36]  Mao H K, Bell P M, Shaner J W and Steinberg D J 1978 Specific volume measurements of Cu, Mo, Pd, and Ag, and calibration of the ruby $R_1$ fluorescence pressure gauge from 0.06 to 1 Mbar *J. Appl. Phys.* **49** 3276
[37]  Zhao Y, Barvosa-Carter W, Theiss S D, Mitha S, Aziz M J and Schiferl D 1998 Pressure measurement at high temperature using ten Sm:YAG fluorescence peaks *J. Appl. Phys.* **84** 4049
[38]  Schmidt S C, Schiferl D, Zinn A S, Ragan D D and Moore D S 1991 Calibration of the nitrogen vibron pressure scale for use at high temperatures and pressures *J. Appl. Phys.* **69** 2793
[39]  Schiferl D, Nicol M, Zaug J M, Sharma S K, Cooney T F, Wang S-Y, Anthony T R and Fleischer J F 1997 The diamond $^{13}C/^{12}C$ isotope Raman pressure sensor system for high-temperature/pressure diamond-anvil cells with aqueous and other chemically reactive samples *J. Appl. Phys.* **82** 3256
[40]  Bundy F P, Hall H T, Strong H M and Wentorf R H 1955 Man-made diamonds *Nature* **176** 51
[41]  DeCarli P S and Jamieson J C 1961 Formation of diamond by explosive shock *Science* **133** 1821
[42]  DuPont 1965 Synthetic diamonds *UK Patent* 1115648
[43]  Wentorf R H 1957 Cubic form of boron nitride *J. Chem. Phys.* **26** 956
[44]  Hazen R M 1993 *The New Alchemists. Breaking Through The Barriers of High Pressure* (New York: Times Books)
[45]  Yagi T and Utsumi W 1993 Direct conversion of graphite into hexagonal diamond under high pressure *New Funct. Mater.* C 99
[46]  Wigner E and Huntington H B 1965 *J. Chem. Phys.* **3** 764
[47]  Narayana C, Luo H, Orloff J and Ruoff A L 1998 Solid hydrogen at 342 GPa: no evidence for an alkali metal *Nature* **393** 46
[48]  Hensel F and Franck E U 1966 Electric conductivity and density of supercritical, gaseous mercury at high pressures *Ber. Bunsenges. Phys. Chem.* **70** 1154
[49]  Hensel F and Franck E U 1968 Metal–nonmetal transition in dense mercury vapor *Rev. Mod. Phys.* **40** 697
[50]  Renkert H, Hensel F and Franck E U 1969 Metal–nonmetal transition in dense cesium vapor *Phys. Lett.* A **30** 494
[51]  Renkert H, Hensel F and Franck E U 1971 Electrical conductivity of liquid and gaseous cesium to 2000.deg. and 1000 bars *Ber. Bunsenges. Phys. Chem.* **75** 507

[52] Olijnyk H and Jephcoat A P 1998 Vibrational studies on $CO_2$ up to 40 GPa by Raman spectroscopy at room temperature *Phys. Rev.* B **57** 879

[53] Yoo C S, Cynn H and Nicol M, Carbon combustion at high pressures and temperatures: evidence for $CO^{++}CO_3^{--}$, an ionic dimer of carbon dioxide, in preparation

[54] Agnew S F, Swanson B I, Jones L H, Mills R L and Schiferl D 1983 Chemistry of nitrogen oxide ($N_2O_4$) at high pressure: observation of a reversible transformation between molecular and ionic crystalline forms *J. Phys. Chem.* **87** 5065

[55] Mitchell A C and Nellis W J 1982 Equation of state and electrical conductivity of water and ammonia shocked to the 100 GPa (1 Mbar) pressure range *J. Chem. Phys.* **76** 6273

[56] See, Ree F and Vanthiel M 1986 Modeling explosive behavior *Energy Technol. Rev.* UCRL-52000-86-3, 41

[57] Holzapfel W B and Franck E U 1966 Conductance and ion dissociation of water up to 1000° and 100 kilobars *Ber. Bunsenges. Phys. Chem.* **70** 1105

[58] Nellis W J, Hamilton D C, Holmes N C, Radousky H B, Ree F H, Mitchell A C and Nicol M 1988 The nature of the interior of Uranus based on studies of planetary ices at high dynamic pressure *Science* **240** 779

[59] Ancilotto F, Chiarotti G L, Scandolo S and Tosatti E 1997 Dissociation of methane into hydrocarbons at extreme (planetary) pressure and temperature *Science* **275** 1288

[60] Ross M 1981 The ice layer in Uranus and Neptune. Diamonds in the sky? *Nature* **292** 435

[61] Hubbard W B 1984 *Planetary Interiors* (New York: Van Nostrand-Reinhold)

[62] Grzegory I, Jun J, Bockowski M, Krukowski S, Wroblewski M, Lucznik B and Porowski S 1995 III–V nitrides—thermodynamics and crystal growth at high $N_2$ pressure *J. Phys. Chem. Solids* **56** 639

[63] Wallace C H, Rao L, Kim S-H, Heath J R, Nicol M and Kaner R B 1998 Solid-state metathesis reactions under pressure: a rapid route to crystalline gallium nitride *Appl. Phys. Lett.* **72** 596

[64] Yoo C S, Akella J and Nicol M 1996 Chemistry at high pressures and temperatures; *in-situ* synthesis and characterization of $\beta$-$Si_3N_4$ by DAC x-ray/laser-heating studies *Advanced Materials '96* ed M Akaishi *et al* (Tsukuba: National Institute for Research in Inorganic Materials) p 175

[65] Yoo C S, Akella J, Nicol M and Cynn H 1997 Direct elementary synthesis of hexagonal and cubic boron nitrides at high pressures and temperatures *Phys. Rev.* B **56** 140

[66] Wallace C H 1998 The rapid solid-state synthesis of group III and transition metal nitrides at ambient and high pressures *PhD Dissertation* University of California, Los Angeles

[67] Doverspike M A, Liu S B, Ennis P, Johnson T, Conradi M S, Luszczynski K and Norberg R E 1986 NMR in high-pressure phases of solid ammonia and ammonia-$d_3$ *Phys. Rev.* B **33** 14

[68] Johansen C R, Nelson H M and Gardner J H 1968 Method for measuring magnetization to high pressures *J. Appl. Phys.* **39** 2152

[69] Asell J F and Nicol M 1968 Raman spectrum of $\alpha$-quartz at high pressures *J. Chem. Phys.* **49** 5395

[70] Sonnenschein R, Syassen K and Otto A 1981 Effect of pressure on the first singlet exciton in crystalline anthracene *J. Chem. Phys.* **74** 4315

[71] Ingalls R, Crozier E D, Whitmore J E, Seary A J and Tranquada J M 1980 Extended x-ray absorption fine structure of sodium bromide and germanium at high pressure *J. Appl. Phys.* **51** 3158

[72] Kao C-C, Rueff J P, Strushkin V V, Shu J, Hemley R and Mao H-K 1999 Private communication

[73] Ingalls R, Drickamer H G and De Pasquali G 1967 Isomer shift of iron-57 in transition metals under pressure *Phys. Rev.* **155** 165

[74] Yin G Z and Nicol M 1985 Photochemistry of naphthalene in alcohol or alkane solutions at high pressures *J. Phys. Chem.* **89** 1171

[75] Brown J M, Slutsky L J, Nelson K A and Cheng L T 1988 Velocity of sound and equations of state for methanol and ethanol in a diamond-anvil cell *Science* **241** 65

[76] Schmidt S C, Moore D S, Schiferl D, Chatelet M, Turner T P, Shaner J W, Shampine D L and Holt W T 1986 Coherent and spontaneous Raman spectroscopy in shocked and unshocked liquids *NATO ASI Series* C **184** 425

[77] Hare D E, Franken J and Dlott D D 1995 A new method for studying picosecond dynamics of shocked solids: application to crystalline energetic materials *Chem. Phys. Lett.* **244** 224

[78] Chronister E L and Crowell R A 1991 Time-resolved coherent Raman spectroscopy of low-temperature molecular solids in a high-pressure diamond anvil cell *Chem. Phys. Lett.* **182** 27

[79] Benedetti M, Biscarini P and Brillante A, The effect of pressure on circular dichroism spectra of chiral transition metal complexes *Physica* B **265** 1

[80] Baudelet F, Odin S, Giorgetti C, Dartyge E, Itie J P, Polian A, Pizzini S, Fontaine A and Kappler J P 1997 $PtFe_3$ Invar studied by high pressure magnetic circular dichroism *J. Physique* IV **C7** 441

## Further Reading

This brief essay could neither deal with all fields of high-pressure science nor any one field in depth. The literature of articles, reviews, books, and conference proceedings for this field is extensive and only a few are cited in the references. Here, we suggest a few books and conference proceedings for the reader interested in exploring the field further. This somewhat arbitrary selection is the author's; it emphasizes recent books and a few classic works.

*Overviews and experimental methods*

Asay J R and Shahinpour M (eds) 1993 *High-Pressure Shock Compression of Solids* (New York: Springer)
Bridgman P W 1958 *The Physics of High Pressure* (London: G Bell and Sons)
Brooks H, Birch F, Holton G and Paul W (eds) 1964 *Collected Experimental Papers of P W Bridgman* vol I–VII (Cambridge: Harvard University Press)
Chéret R 1992 *Detonation of Condensed Explosives* (New York: Springer)
Drickamer H G and Frank C W 1973 *Electronic Transitions and the High-Pressure Chemistry and Physics of Solids* (London: Chapman and Hall)
Eremets M I 1996 *High Pressure Experimental Methods* (New York: Oxford University Press)
Graham R A 1993 *Solids under High-Pressure Shock Compression* (New York: Springer)
Horie Y and Sawaoka A B 1993 *Shock Compression Chemistry of Materials* (Tokyo: KTK Scientific)
Sawaoka A B (ed) 1993 *Shock Waves in Materials Science* (New York: Springer)
Trunin R F 1998 *Shock Compression of Condensed Materials* (Cambridge: Cambridge University Press)


*Phase diagrams*

Liu L-G and Bassett W A 1986 *Elements, Oxides, Silicates* (New York: Oxford University Press)
Young D A 1991 *Phase Diagrams of the Elements* (Los Angeles: University of California Press)


*Conference proceedings*

Hocheimer H D and Etters R D 1991 *Frontiers of High-Pressure Research* (New York: Plenum)
Polian A, Loubeyre P and Boccara N (eds) 1989 *Simple Molecular Systems at Very High Density* (New York: Plenum)
Winter R and Jonas J (eds) 1993 *High Pressure Chemistry, Biochemistry and Materials Science* (Dordrecht: Kluwer)
Proceedings of the annual conference of the European High Pressure Research Group, the most recent of which is:
Isaacs N S (ed) 1998 *High Pressure Food Science, Bioscience and Chemistry (Spec. Publ. vol 222)* (Cambridge, UK: Royal Society of Chemistry)
Proceedings of the biannual conference of AIRAPT, the International Association for the Advancement of High Pressure Research and Technology, the most recent of which is:
Nakahara M (ed) 1998 *Koatsuryoku no Kagaku to Gijutsu (Proc. Int. Conf. AIRAPT-16 and HPCJ-38 on High Pressure Science and Technology, 1997)* vol 7 (Kyoto: Japan Society of High Pressure Science and Technology)
Proceedings of the biannual Conference of the American Physical Society Topical Group on Shock Compression Science, the most recent of which is:
Schmidt S C, Dandekar D P and Forbes J W (eds) 1998 *Shock Compression of Condensed Matter, 1997 (AIP Conf. Proc. vol 429)* (College Park, MD: American Institute of Physics)
Proceedings of the US–Japan Seminars on High Pressure-Temperature Research, the most recent (fifth) of which is:
Manghnani M H and Yagi T (eds) 1998 *Properties of Earth and Planetary Materials* (Washington, DC: American Geophysical Union)

# PART B2

# DYNAMIC MEASUREMENTS

# B2.1
# Ultrafast spectroscopy

*Warren F Beck*

## B2.1.1   Introduction

The development of the millisecond and microsecond flash photolysis experiments by George Porter and co-workers [1, 2] in the 1950s marks the true birth of time-resolved spectroscopy. Porter's work, which provided for the first time a way to capture the absorption spectrum of a short-lived kinetic intermediate in a photochemical reaction, helped to start a new era in physical chemistry, one that was focused on the mechanism and dynamics of chemical reactions. Owing to the subsequent development of mode-locked laser sources, beginning with the picosecond ruby and neodymium–glass lasers in the 1960s, the sub-picosecond passively mode-locked dye laser in the late 1970s and, most recently, the femtosecond self-mode-locked Ti–sapphire laser in the early part of this decade, the time resolution for spectroscopic measurements has advanced three orders of magnitude, from the 10 ps to the 10 fs regime [3]. It is now possible to conduct a wide variety of spectroscopies with ultrashort laser pulses of photons selectable over the entire spectral range from the x-ray region [4, 5] to the terahertz or far-infrared (IR) region [6–10]. A variety of robust methods have been developed to probe the time evolution of populations and coherences. The shortest time scale that is now routinely accessible is comparable to or shorter than the period of molecular vibrations, the fundamental time scale of chemistry.

This chapter focuses on the primary experimental methods of ultrafast spectroscopy, as discussed in terms of studies on intramolecular dynamics in the condensed phase or in proteins. Ultrafast spectroscopy generally denotes spectroscopy that exploits the time resolution obtainable with mode-locked laser sources. The ultrafast regime encompasses electronic and vibrational energy transfer, charge transfer and structural dynamics involving isomerization and the breaking of bonds. In many cases, these processes can be optically triggered so that the time course can be studied with a delayed probing or gating pulse. The initial and delayed pulses are derived from a pulse train emitted by a single mode-locked light source; the ultrashort timing of the experiment is usually derived from the distance of flight of the optical pulses using a technique that is reminiscent of interferometry. After a discussion of the current ideas in producing and characterizing tunable ultrashort optical pulses for spectroscopy, the chapter discusses methods for time-resolved fluorescence spectroscopy, pump–probe methods for time-resolved absorption spectroscopy and multipulse photon-echo techniques for the measurement of coherence. The chapter closes with a brief discussion of the use of phase-controlled, multiple-pulse sequences for advanced, highly selective spectroscopies and for the control of chemical dynamics.

## B2.1.2   Femtosecond light sources

The development of ultrafast spectroscopy has paralleled progress in the technical aspects of pulse formation [11]. Because mode-locked laser sources are tunable only with difficulty, until recently the most heavily studied physical and chemical systems were those that had strong electronic absorption spectra in the neighbourhood of conveniently produced wavelengths.

As one important example, the introduction of the prism-controlled, colliding-pulse, mode-locked (CPM) dye laser [12, 13] led almost immediately to developments in measurement technique with pulses of less than 100 fs duration; Shank and co-workers used an amplified CPM laser [14] in their work with 6 fs pulses in 1987 [15]. Until recently, the pulses used in those experiments were the shortest optical pulses characterized. The transition-state spectroscopy of Zewail and Bernstein [16–22] exploited an amplified CPM laser after frequency doubling and/or continuum generation. The chemical systems that were most easily studied, however, were those that could be stimulated either by the 620 nm output of the CPM directly or after frequency doubling to 310 nm. In addition, the CPM laser and its contemporary, more tunable alternative, the pulse-compressed, synchronously pumped dye laser [11], were tools that could be effectively used only by researchers with extensive backgrounds in lasers and optics.

These limitations have recently been eliminated using *solid-state* sources of femtosecond pulses. Most of the femtosecond dye laser technology that was in wide use in the late 1980s [11] has been rendered obsolete by three technical developments: the self-mode-locked Ti–sapphire oscillator [23–27], the chirped-pulse, solid-state amplifier (CPA) [28–31], and the non-collinearly pumped optical parametric amplifier (OPA) [32–34]. Moreover, although a number of investigators still construct home-built systems with narrowly chosen capabilities, it is now possible to obtain versatile, nearly state-of-the-art apparatus of the type described below from commercial sources. Just as home-built NMR spectrometers capable of multidimensional or solid-state spectroscopies were still being home built in the late 1970s and now are almost exclusively based on commercially prepared apparatus, it is reasonable to expect that ultrafast spectroscopy in the next decade will be conducted almost exclusively with apparatus from commercial sources based around entirely solid-state systems.

Figure B2.1.1 depicts an instrument that takes advantage of many of the most recent technical developments. The best strategy for generating wavelength-tunable ultrashort laser pulses for time-resolved spectroscopy involves use of an OPA as the only wavelength-tunable element. This approach is organized around the principle that extremely stable, fixed wavelength, high-energy pulse trains can now be generated using an amplified Ti–sapphire-based system. The chief advantage of instruments like the one shown in figure B2.1.1 is that experimental demands for specific operating wavelengths are met by adjustment of the *last* device in the pulse-forming chain, the OPA. One can expect such a design to be considerably more robust and user-friendly than systems based on wavelength tunable oscillators, which demand manipulation of *every* device in the instrument in response to tuning to a new wavelength.

### B2.1.2.1  Oscillators

The most commonly used femtosecond oscillator at this point is the self-mode-locked Ti–sapphire laser [23–27], shown in figure B2.1.1, which can *routinely* produce pulses of light with durations adjustable over the 10–150 fs range. The wavelength of the Ti–sapphire oscillator can be tuned over the 700–1100 nm range using an intracavity slit or birefringent filter, providing pulse durations that are essentially limited by the bandwidth of the filtering element. (It should be emphasized, however, that tuning an oscillator of this type is not as routinely done as is tuning an OPA.) The pulse energy that can be directly obtained from the oscillator is typically limited to the 2 nJ/pulse regime, but the oscillator emits pulses at a high repetition rate, typically 75–100 MHz, depending on the cavity dimensions.

The Ti–sapphire oscillator is extremely useful as a stand-alone source of femtosecond pulses in the near-IR region of the spectrum. Some ultrafast experiments, especially of the pump–probe variety (see below), can be conducted with pulses obtained directly from the oscillator or after pulse selection at a lower repetition rate. Far-IR (terahertz) radiation is usually generated using a semiconductor (usually GaAs) substrate and focused Ti–sapphire oscillator pulses [7]. If somewhat higher-energy pulses are required for an experiment,

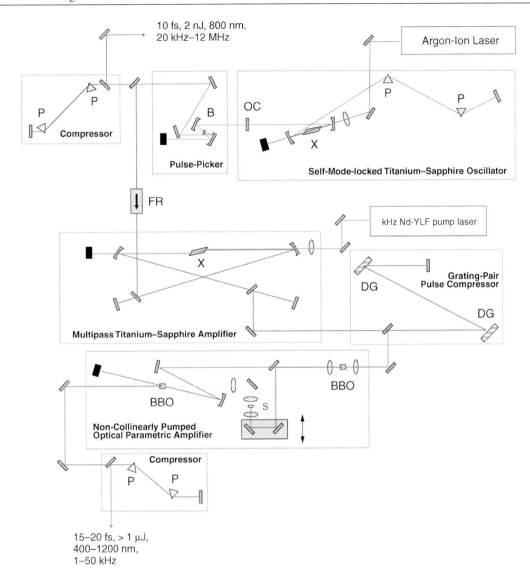

**Figure B2.1.1.** Femtosecond light source based on an amplified titanium–sapphire laser and an optical parametric amplifier. Symbols used: P, Brewster dispersing prism; X, titanium–sapphire crystal; OC, output coupler; B, acousto-optic pulse selector (Bragg cell); FR, Faraday rotator and polarizer assembly; DG, diffraction grating; BBO, $\beta$-barium borate nonlinear crystal.

the Ti–sapphire oscillator can be *cavity dumped* by an intracavity acousto-optical device known as a *Bragg cell*, producing perhaps 50 nJ pulses but at a reduced repetition rate, typically between 200 kHz and 1 MHz [35]. The energy available from a cavity-dumped Ti–sapphire laser is intense enough to generate a continuum-like source in a single-mode optical fibre. Wiersma and co-workers [36] generated 5 fs pulses by compressing these continuum pulses with a sequence of gratings and prisms.

Comparable oscillators have been described that exploit other types of solid-state gain media. Although Ti–sapphire crystals are widely used because the absorption spectrum overlaps favourably with the output spectra of argon-ion and frequency-doubled Nd-YVO$_4$ continuous-wave lasers, Cr-LISAF may be favoured in the future as a gain medium for femtosecond oscillators because it can be pumped by continuous-wave GaAs diode lasers [37, 38].

### B2.1.2.2  Amplification

Although many useful femtosecond spectroscopic experiments on condensed-phase targets can be easily performed with low-energy pulses, in the 100 pJ to 1 nJ regime, higher-energy pulses are required if wavelength tunability is desired. A femtosecond continuum [39] can be generated in water or sapphire if the pulse energy is higher than 200 nJ; OPA sources require even higher energies, in excess of 1 $\mu$J/pulse. Amplification of Ti–sapphire oscillators is at this point routinely performed, with excellent commercial systems readily available of the regenerative amplifier type [28, 30, 31, 40, 41], and there are simple multipass amplifier designs [42, 43] that are easily constructed in the laboratory.

Pulses are selected for amplification from the oscillator's 75–100 MHz pulse train at a much lower repetition rate, ranging in published designs from 10 Hz to 250 kHz, either by a Pockels cell or a Bragg cell (as shown in figure B2.1.1). The selected pulse train is amplified in Ti–sapphire gain media using a method known as *chirped-pulse amplification* [28–31]. In this scheme, oscillator pulses are stretched temporally well into the picosecond regime prior to amplification so that the damage threshold for the gain crystal is not exceeded [44]. If the amplifier is designed to operate in the $>10$ kHz regime, like the one depicted in figure B2.1.1, a stretcher may not be required. A grating-pair pulse compressor [45] is used to compress the pulse back nearly to its original duration after it emerges from the amplifier. Regenerative amplifiers capable of producing 75–150 fs pulses are the most common systems in use [30, 40], but recently a multipass ring amplifier has been described that produces 20 fs pulses [43]. The multipass amplifier depicted in figure B2.1.1 is a non-ring design that permits a more facile input and extraction of the amplified pulse.

The most common commercially prepared amplifier systems are pumped by frequency-doubled Nd-YAG or Nd-YLF lasers at a 1–5 kHz repetition rate; a continuously pumped amplifier that operates typically in the 250 kHz regime has been described and implemented commercially [40]. The average power of all of the commonly used types of Ti–sapphire amplifier systems approaches 1 W, so the energy per pulse required for an experiment effectively determines the repetition rate.

### B2.1.2.3  Optical parametric amplification

Perhaps the ultimate femtosecond light source, the OPA exploits a nonlinear parametric process to amplify a portion of a femtosecond continuum [32–34, 46–48]. In most designs, a portion of the output of a regenerative or multipass Ti–sapphire amplifier is frequency doubled in a nonlinear crystal to prepare an intense *pump* pulse. Less than 1 $\mu$J/pulse of the amplifier's output is reserved to seed a single-filament continuum [47] in a thin sapphire crystal. A second nonlinear crystal is used as the gain medium for the parametric process, which splits an input pump photon at frequency $\omega_3$ into two output photons, a signal photon at frequency $\omega_1$ and an idler photon at frequency $\omega_2$, with energy conserved ($\omega_3 = \omega_1 + \omega_2$). The parametric process is greatly enhanced by the presence of *seed* light at either $\omega_1$ or $\omega_2$, which is supplied by the continuum [46]. The apparatus can be adjusted to select a certain range of frequencies from the continuum for amplification in the nonlinear crystal, allowing the production of wavelength-tunable output pulses derived either from the signal or idler with adjustable pulse durations. At this point, $\beta$-barium borate (BBO) is the material of choice for the nonlinear crystal.

The OPA should not be confused with an optical parametric oscillator (OPO), a resonant-cavity parametric device that is synchronously pumped by a femtosecond, mode-locked oscillator. 14 fs pulses, tunable over much of the visible regime, have been obtained by Hache and co-workers [49, 50] with a BBO OPO pumped by a self-mode-locked Ti–sapphire oscillator.

Shortly after the development of high-energy/pulse Ti–sapphire regenerative amplifier systems, a number of investigators reported progress in using OPAs in producing tunable sources of very short pulses. Wilson and co-workers [46] showed early on that an *experimentally* useful source for femtosecond spectroscopy with <50 fs pulses was obtained through the use of continuum seeding of a type I nonlinear OPA crystal, which was pumped by the *fundamental* output of an amplified Ti–sapphire laser. The main problem with the early systems was inherent to the physics of *collinearly pumped* parametric amplification: the signal, idler and pump frequencies have different group velocities (see below) in the nonlinear crystal, which limits the amount and frequency bandwidth of the parametric gain. In other language, the phase-matching condition for the collinearly pumped OPA works only over a small bandwidth, which tends to limit the pulse duration to fairly long (100 fs) pulses, and tuning of the OPA to different signal wavelengths requires reoptimization of the crystal's orientation. The design advanced by Wilson's group takes advantage of the smaller mismatch in group velocities in the near-IR part of the spectrum; other designs employing pumping with the second harmonic of the Ti–sapphire laser provide direct access to visible signal pulses but with significantly longer durations (150 fs) [48].

Very recently, Hache and co-workers [49] found that a *non-collinear* pumping of an OPO crystal produces a phase-matching condition that is *independent of signal wavelength* over a very broad bandwidth. This discovery makes it possible to obtain very high parametric gain in an OPA with a single pass through the crystal and adjustable signal bandwidths. The result is a source of light with wide tunability and adjustable pulse durations, the ultimate femtosecond light source. The design for the OPA depicted in figure B2.1.1 is an adaptation of that recently described by Riedle and co-workers [32]. 400 nm light obtained by frequency doubling the output of a regenerative Ti–sapphire amplifier is overlapped at an angle of 3.7° with the femtosecond continuum light. This angle produces in BBO the wavelength-independent phase-matching condition noted by Hache and co-workers. Riedle and co-workers demonstrated that signal pulses of 15–20 fs duration could be obtained from this system, with tunability over most of the 400–800 nm visible range. Using a similar approach, but with some changes in the details of producing the continuum seed light that were intended to produce as broad a signal bandwidth as possible, De Silvestri and co-workers [34] subsequently showed that signal pulses as short as 7.2 fs in the visible regime could be produced. In contemporaneous work, Kobayashi and co-workers [33] obtained sub-10 fs pulses from the visible signal *and* near-IR idler from a comparable apparatus. These latter two results have nearly matched the legendary performance of the pulse-compressed CPM dye laser of Shank and co-workers [15].

A surprising aspect of the recent work on non-collinearly pumped OPA systems is that it appears that fairly long pump pulses, in the 150 fs regime, are to be preferred over shorter pulses if sub-10 fs signal pulses are desired. Mature commercial regenerative amplifier designs operating over the 1–250 kHz repetition-rate range are already capable of producing these pulses. Thus, the experimental motivation for developing amplifier systems capable of producing very short pulses *directly* has vanished for *spectroscopic* applications; however, there are a number of important applications for short, high-energy pulses, such as driving the formation of femtosecond pulses of x-rays [5]. High-energy physics applications for short, amplified pulses have been reviewed recently by Mourou and co-workers [51].

### B2.1.2.4    *Pulse compression*

Owing to its wide spectral bandwidth, a short optical pulse is distorted temporally by passing through the beamsplitters, lenses, filters, etc, that are required for an ultrafast spectroscopic experiment. The distortion arises from the dispersion of the speed of light in a medium as a function of wavelength, or group-velocity dispersion (GVD). The sweep of frequencies observed at a given spatial position is known as chirp (in analogy to the sound of a frequency-swept audio pulse) or group-delay dispersion (GDD). This phenomenon is usually described in terms of a Taylor series expansion of the phase $\phi$ of the optical pulse around the centre frequency, $\omega_0$ [11, 52]. The first derivative term represents the group delay, $t(\omega) = \mathrm{d}\phi/\mathrm{d}\omega$. The quadratic term, corresponding to the *linear* sweep in the group delay with respect to wavelength, can be corrected by an optical delay line with *negative* GDD (alternatively speaking, *anomalous dispersion*) [53], constructed from a pair of diffraction gratings [45], a pair of dispersing prisms with Brewster angled surfaces [54] or, less often, by a Gires–Tournois interferometer (GTI) [55, 56]. The amount of negative GDD is determined by the distance of separation between the two prisms or gratings or between the two reflective films in the interferometer.

In most femtosecond experiments, a double-passed pair of prisms is inserted into the beam prior to its reaching the measurement apparatus. In figure B2.1.1, a pair of prisms is depicted after the OPA and after the pulse-picker (for use in oscillator-only experiments). This practice allows one to *precompensate* for the GDD imparted to the beam by the optics in the measurement apparatus so that the pulses are as short as possible when they arrive at the sample's position. The prisms are typically manipulated by translating one or both of the prisms normal to the base so as to increase or decrease the amount of prism glass traversed; this permits a small amount of positive GDD to be added or subtracted. The pulse duration at the sample's position is usually determined with an intensity autocorrelation measurement, performed by replacing the sample with a nonlinear crystal, such as potassium dihydrogen phosphate (KDP) or BBO.

An even more significant problem associated with GDD is that the formation of a short pulse in an oscillator requires many round-trips through the gain medium, so it is routine to place a pair of prisms or a GTI in the oscillator's cavity to repair the damage suffered by the pulse on each trip. The design of Murnane and co-workers [25] depicted for the oscillator in figure B2.1.1 exemplifies this practice, which was first established in the final design for the CPM dye laser by Valdmanis and Fork [13]. In the future, oscillators may not require prism pairs; it is now possible to fabricate *chirped mirrors* that employ multiple layers of dielectric coatings to provide for a precise compensation of quadratic and cubic phase distortions that arise from the gain medium of a femtosecond oscillator [26]. In fact, a commercial design employing a chirped-mirror, Ti–sapphire oscillator and a solid-state continuous-wave Nd-YVO$_4$ pump laser in a single, compact, sealed (no user controls!) enclosure has just become available. Such a source would be expected to be unusually robust.

The shortest optical pulses actually used so far (1998) in ultrafast spectroscopic experiments were obtained by Shank and co-workers from an amplified CPM laser [57]. In these extraordinary experiments, a sequence of a pair of prisms and a pair of gratings was employed. The reason for this additional complexity was that the cubic-term phase distortion imparted by the prism pair is of opposite sign of that imparted by the grating pair. As a result, it was possible to null simultaneously the quadratic and cubic dispersion terms at the position of the sample [15].

Most optical materials exhibit *normal* dispersion; the index of diffraction increases monotonically with respect to frequency. Some materials, however, exhibit an anomalous dispersion regime in the near-IR part of the spectrum. Hochstrasser and co-workers exploited the finding that the mid-IR substrates CaF$_2$ and BaF$_2$, used, for instance, for beamsplitters and windows, imparts GDD to a transmitted beam that is of the opposite sign of that imparted by Ge, which was used as a long-pass filter [58]. Fused silica optical fibres exhibit anomalous dispersion at wavelengths above 1.3 $\mu$m [59, 60]. The implication of this phenomenon is that positively chirped IR pulses can be compressed by being propagated in a fibre. Further, short IR pulses can be propagated in a fibre for arbitrarily long distances without suffering a change in pulse duration [61, 62].

### B2.1.3  Femtosecond time-resolved spectroscopy

In *lieu* of electronic timing mechanisms, a form of interferometry is usually employed in ultrafast spectroscopy to obtain time resolution on the picosecond or shorter time scales. The simplest ultrafast measurement apparatus is that shown in figure B2.1.2, an *autocorrelator*. This instrument is used in various forms to measure the duration of an ultrashort laser pulse [63]. A pulse of light is first split into equal portions by a beamsplitter; the two parts are then recombined after they travel in the two arms of an interferometer

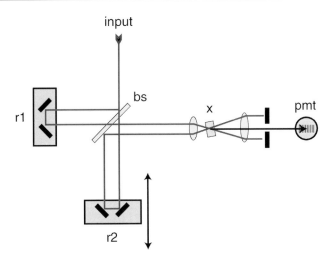

**Figure B2.1.2.** Modified Michelson interferometer for non-collinear intensity autocorrelation. Symbols used: r1, r2, retroreflecting mirror pair mounted on a translation stage; bs, beamsplitter; x, nonlinear crystal; pmt, photomultiplier tube.

(here, derived from the Michelson design) so that they are focused onto a thin (100 $\mu$m or less in thickness) nonlinear crystal. When the two pulses overlap *temporally and spatially*, the second harmonic is emitted along the direction that conserves momentum, between the direction of the two emerging fundamental pulses [64]. By moving one of the retroreflectors in the interferometer along the beam path towards or away from the beamsplitter, one of the pulses is made to scan *temporally* through the other pulse in the nonlinear crystal. Thus, the ultrafast timing of the experiment is provided by a simple measurement of displacement and knowledge of the speed of light. Reproducible displacements of the order of 1 $\mu$m, corresponding to a time-of-flight displacement of 6.67 fs in the interferometer, are routinely performed with computer-controlled linear actuators.

The intensity of this *upconverted* light detected by a photodetector is described by the equation $I(\tau) = \int_{-\infty}^{+\infty} I(t-\tau)I(t)\,\mathrm{d}t$, where $I(t)$ is the intensity of the pulse as a function of time $t$ at a given point in the nonlinear crystal, and $\tau$ is the shift in time of the variably delayed or *gating* pulse, as controlled by one of the retroreflectors in the interferometer. The photodetector integrates the upconverted intensity for a given delay $\tau$; an example of the signal obtained by scanning $\tau$ is shown in figure B2.1.3(a), with the input pulse train provided by a self-mode-locked Ti–sapphire laser. The symmetrical shape of the autocorrelation trace arises from the convolution of the input $I(t)$ pulse shapes. The true shape of $I(t)$ and the dependence of the phase on wavelength can be obtained in a slightly more elaborate experiment called frequency-resolved optical gating (FROG) [65–68], a form of which can be performed by recording the autocorrelation traces as a function of wavelength by dispersing the upconverted light in a spectrometer placed before the photodetector [69].

The spectrum of the femtosecond pulse provides some information on whether the input pulse is chirped, however, causing the temporal width of $I(t)$ to be broader than expected from the Heisenberg indeterminancy relationship. The full width at half maximum of the autocorrelation signal, 21 fs, corresponds to a pulse width of 13.5 fs if a sech$^2$ shape for the $I(t)$ function is assumed. The corresponding output spectrum shown in figure B2.1.3(b) exhibits a width at half maximum of approximately 700 cm$^{-1}$. The time–bandwidth product $\Delta\tau\,\Delta\nu$ is close to 0.3. This result implies that the pulse was compressed nearly to the Heisenberg indeterminacy (or Fourier transform) limit [53] by the double-passed prism pair placed in the beam path prior to the autocorrelator.

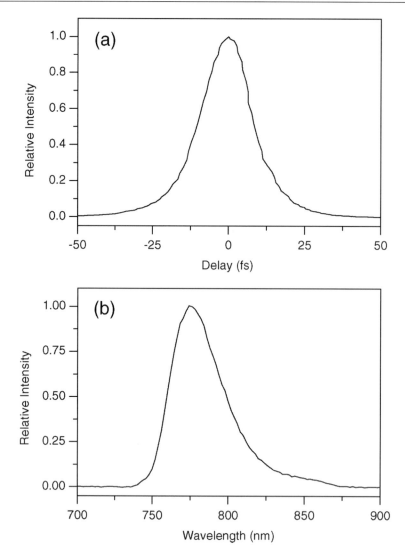

**Figure B2.1.3.** Output of a self-mode-locked titanium–sapphire oscillator: (a) non-collinear intensity autocorrelation signal, obtained with a 100 $\mu$m $\beta$-barium borate nonlinear crystal; (b) intensity spectrum.

The intensity autocorrelation measurement is comparable to all of the *spectroscopic* experiments discussed in the sections that follow because it exploits the use of a variably delayed, *gating* pulse in the measurement. In the autocorrelation experiment, the gating pulse is just a replica of the time-fixed pulse. In the spectroscopic experiments, the gating pulse is used to interrogate the populations and coherences established by the time-fixed pulse.

### B2.1.3.1 Fluorescence upconversion spectroscopy

Time-resolved fluorescence is perhaps the most direct experiment in the ultrafast spectroscopist's palette. Because only one laser pulse interacts with the sample, the method is essentially free of the problems with field-matter time orderings that arise in all of the subsequently discussed multipulse methods. The signal detected is usually directly proportional to the population in the resonantly prepared excited state alone. In systems that exhibit photochemistry, for instance, the time evolution of the fluorescence provides a direct view of the decay of the photochemically active state.

Fluorescence spectroscopy can be performed in the ultrafast regime with a nonlinear crystal and a short gating pulse in an experiment known as *fluorescence upconversion* [70, 71]. The interferometer shown in figure B2.1.2 is modified so that an incident pulse is split into two pulses: a weaker pulse that is used to excite a sample and a stronger, gating pulse that is overlapped spatially in a nonlinear crystal with the fluorescence that is collected from the sample. Sum-frequency generation in the nonlinear crystal produces output photons with frequency $\omega_3$ from the input gate photons of frequency $\omega_1$ and fluorescence photons of frequency $\omega_2$; since $\omega_3 = \omega_1 + \omega_2$, the output photons are *upconverted*, or transferred to a higher frequency by the gate pulse. The intensity of the beam of output photons is proportional to the product of the intensity of the gating and fluorescence photons; the gate photons slice out only those fluorescence photons that are temporally overlapped with the short gating pulse in the nonlinear crystal. This permits the time course of the fluorescence intensity at a particular frequency $\omega_2$ to be mapped out by scanning the time delay for the gate pulse.

The frequency $\omega_2$ that is detected in the fluorescence can be largely selected by adjusting the phase-matching condition for the nonlinear crystal. A double monochromator (or, alternatively, a prism and a single-stage monochromator) is used to discriminate between the upconverted fluorescence and the background interference from the second harmonic of the gating light. Even though the nonlinear crystal is angle-tuned to optimize the intensity of the upconverted fluorescence at the wavelength chosen by the monochromator, the strong gate pulse generates a significant second-harmonic signal that can often be as strong or stronger than the gated fluorescence. If the gate pulse is short, with a concomitantly large spectral bandwidth, the second-harmonic background may significantly overlap the fluorescence spectrum of the sample under study. In many respects this problem is comparable to that encountered in discriminating Rayleigh scattering from Raman scattering in the low Raman frequency regime. In Raman spectroscopy, however, the vibrational line shapes are usually much narrower than any fluorescence background; in fluorescence upconversion spectroscopy, the second-harmonic background from the gate pulse is often as broad as the fluorescence signal, making things comparably more difficult [71].

In figure B2.1.4 a design for a fluorescence upconversion spectrometer is depicted that is based on one discussed by Jimenez and Fleming [71]. In the most versatile instrument, an OPA would be used to generate the wavelength-tunable excitation pulse, while a portion of the direct output of the amplified Ti–sapphire laser that pumps the OPA would be reserved for use as a strong gate pulse. This practice has several advantages: the gating and excitation pulses are implicitly time-synchronized and the excitation pulse can be well removed in wavelength from the region of the fluorescence photon, in order to minimize the second-harmonic gate background mentioned above. The experimental set-up depicted in figure B2.1.4 employs two off-axis parabolic reflectors to collect and collimate the fluorescence emitted by the sample and then to focus it onto the nonlinear crystal. Other designs employ a single elliptical reflector to perform both tasks. The design shown here may permit low-temperature fluorescence studies to be executed, however, since the crystal and sample can be well separated on the optical table. Jimenez and Fleming [71] note that very few femtosecond fluorescence upconversion experiments have so far been attempted at low temperature.

Over the last few years, the time resolution attainable in fluorescence upconversion has reached the sub-100 fs regime. The main problems associated with pushing the time resolution down further are sensitivity and GDD. The problem of collecting enough fluorescence photons and imaging them onto the nonlinear crystal, where detection occurs, is at odds with obtaining pulse-width-limited time resolution. Large elliptical or parabolic reflectors enhance the number of photons collected, but the time-of-flight dispersion increases with the surface area used on the reflector, so in many implementations the reflectors are masked. Lenses were used for collection and focusing in early experiments, as in the design discussed by Barbara and co-workers

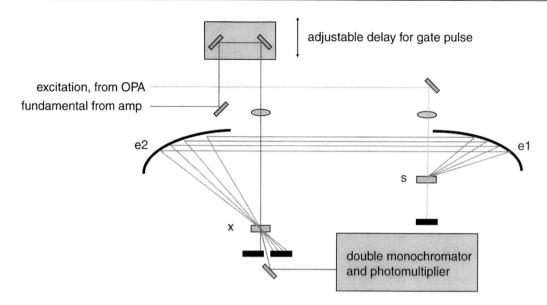

**Figure B2.1.4.** Fluorescence upconversion spectrometer based on the use of off-axis elliptical reflectors for the collection and focusing of fluorescence. Symbols used: e1, e2, off-axis elliptical reflectors; s, sample; x, nonlinear crystal. (After Jimenez and Fleming [71].)

[70], but the time resolution obtainable is typically limited to the 250 fs regime owing to the GDD suffered by the fluorescence photons in being transmitted through the material used to make the lens. In principle, spherical confocal reflective optics might be used without time-of-flight dispersion. Even so, the nonlinear crystal itself imparts GDD and distorts the time response.

The instrument response function (IRF) for the fluorescence upconversion experiment, then, cannot be shorter than the intensity *cross-correlation* function, which can be obtained using an instrument like that shown in figure B2.1.4 in which the excitation and gate pulses used in the upconversion experiment are injected separately into the two arms of the interferometer. In the simplest case, the actual time response of the fluorescence resembles the *integral* of the IRF, $S(\tau) = \int_{-\infty}^{+\infty} I(t - \tau)F(t)\,dt$. The fluorescence signal $F(\tau)$ at the detection frequency $\omega_2$ is, in the absence of other dynamics and rapid relaxation, a step function; the excitation pulse transfers a fraction of the ground-state population to the resonant excited state, which then emits fluorescence. As indicated in the above equation, the upconversion signal $S(\tau)$ responds as the gate pulse $I(t)$ is integrated by the step-response of $F(\tau)$. Of course, if ground-state recovery or energy-transfer processes deplete the resonant excited-state population, then $F(\tau)$ decays accordingly. The upconversion signal $S(\tau)$ then exhibits a shape that is effectively a *convolution* of the IRF and the excited-state population response $F(\tau)$. Since many experiments are intended to *determine* $F(\tau)$ via measurement of $S(\tau)$, one is often faced with *deconvolution* of the IRF from the measured signal in the course of data analysis. This problem has been extensively discussed in the literature [72]; the best solution is, is, if possible, to make the IRF much shorter than the dynamics exhibited by $F(\tau)$ so that the measured signal is not significantly distorted.

An important extension to the simplest upconversion experiment at a single detection frequency $\omega_2$ is the practice of measuring *time-resolved fluorescence spectra*, that is, the shape of the fluorescence spectrum emitted by the sample at a given gate delay $\tau$, $S(\omega_2, \tau)$. In most reported work, the $S(\omega_2, \tau)$ spectrum is built up by obtaining a family of single-wavelength transients, scanned at a given $\omega_2$ as a function of $\tau$. If the instrument provides for computer control of the angle tuning of the nonlinear crystal and the detection monochromator, it would be more efficient, with respect to experimental time, to directly scan $\omega_2$ at a given time delay $\tau$.

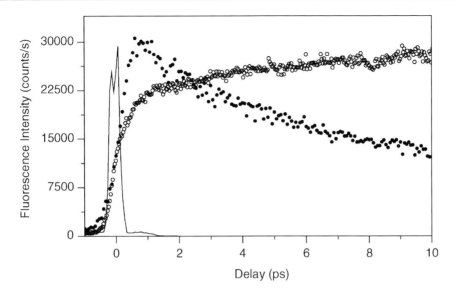

**Figure B2.1.5.** Fluorescence upconversion traces obtained at two observation wavelengths (full circles, 570 nm; open circles, 650 nm) at room temperature with an oxazine dye, phenoxazone, in methanol solvent. Figure courtesy of Professor S Rosenthal (Vanderbilt University).

Perhaps the best example of a situation requiring knowledge of the time evolution of the fluorescence spectrum is that associated with *dynamic solvation*, the time-dependent reorganization of solvent dipoles in response to a light-induced change in the dipole moment of a dissolved chromophore [73–78]. As an example, figure B2.1.5 shows two upconversion traces obtained with the dye phenoxazone dissolved in methanol at room temperature. When the observation wavelength is tuned to the blue edge of the emission spectrum, at 570 nm, the fluorescence is observed to decay with a time constant of several picoseconds. If the observation wavelength is tuned to the red edge of the absorption spectrum at 650 nm, the fluorescence transient exhibits a rise with a similar time constant. These two upconversion transients evidence a blue-to-red shift of the time-resolved fluorescence spectrum owing to dynamic solvation on the picosecond time scale.

The upconversion method can also be applied to the measurement of *anisotropy* [79], which returns information on the time dependence of the orientation of the excited-state transition-dipole moment with respect to time following excitation with a linearly polarized pulse of light. Anisotropy information can be used to study rotational diffusion [80, 81] in liquids, energy transfer between chromophores in proteins [79, 82] and excited-state isomerization [83, 84], to name just three common applications. The method takes advantage of photoselection of those molecules whose transition-dipole moments are aligned with the excitation pulse's plane of polarization to prepare an essentially polarized excited-state orientational distribution initially. The fluorescence that is emitted initially by this distribution is highly polarized; as time progresses, any mechanism that causes rotation of the excited-state transition-dipole moment, such as rotational diffusion, energy transfer and isomerization, will cause depolarization of the fluorescence.

The anisotropy function $r(t) = (I_{\parallel}(t) - I_{\perp}(t))/(I_{\parallel}(t) + 2I_{\perp}(t))$ is determined by two polarized fluorescence transients $I_{\parallel}(t)$ and $I_{\perp}(t)$ observed parallel and perpendicular, respectively, to the plane of polarization of the excitation pulse. In the upconversion experiment, the two measurements are most conveniently made by rotating the plane of polarization of the excitation pulse with respect to the fixed orientation of the input plane of the nonlinear crystal. Because the photoselected angular distribution for a set of isolated transition dipoles exhibits a $\cos(\theta)$ dependence with respect to the angle $\theta$ between the plane of polarization of the excitation pulse and the observation plane, the anisotropy $r(t)$ decays from an initial value of 0.4 as the depolarization proceeds. The time course of the anisotropy is essentially a measure of the time-correlation function $\langle P_2[\hat{\mu}_1(0) \cdot \hat{\mu}_2(t)]\rangle$ that describes the memory of the initial, photoselected dipole orientation $\hat{\mu}_1(0)$ as correlated with the probed dipole direction $\hat{\mu}_2(t)$ as a function of time [79, 84].

In certain situations involving coherently interacting pairs of transition dipoles, the initial fluorescence anisotropy value is expected to be larger than 0.4. As indicated by the theory described by Wynne and Hochstrasser [85, 86] and by Knox and Gülen [87, 88], the initial anisotropy expected for a pair of coupled dipoles oriented 90° apart, as an example, is 0.7, and the decay of the anisotropy from 0.7 to 0.4 is a measure of the time scale for the decay of the electronic coherence between the two states. This theory has been applied to the interpretation of the decay of anisotropy in the analogous anisotropy measurement obtained using pump–probe (see below) stimulated-emission measurements in magnesium tetraphenylporphyrin [89] and in exciton-coupled chromophore pairs in cyanobacterial light-harvesting proteins [90, 91].

### B2.1.3.2    Pump–probe spectroscopy

An interferometric method was first used by Porter and Topp [1, 92] to perform a *time-resolved absorption* experiment with a $Q$-switched ruby laser in the 1960s. The nonlinear crystal in the autocorrelation apparatus shown in figure B2.1.2 is replaced by an absorbing sample, and then the transmission of the variably delayed pulse of light is measured as a function of the delay $\tau$. This approach is known today as a *pump–probe* experiment; the first pulse to arrive at the sample transfers (*pumps*) molecules to an excited energy level and the delayed pulse *probes* the population (and, possibly, the coherence) so prepared as a function of time.

The pump–probe concept can be extended, of course, to other methods for detection. Zewail and co-workers [16, 18–20, 93] have used the probe pulse to drive population from a reactive state to a state that emits fluorescence [94–98] or photodissociates, the latter situation allowing the use of mass spectrometry as a sensitive and selective detection method [99, 100].

Pump–probe absorption experiments on the femtosecond time scale generally fall into two effective types, depending on the duration and spectral width of the pump pulse. If the pump spectrum is significantly narrower in width than the electronic absorption line shape, *transient hole-burning spectroscopy* [101–113] can be performed. The second type of experiment, *dynamic absorption spectroscopy* [57, 114–122], can be performed if the pump and probe pulses are short compared to the period of the vibrational modes that are coupled to the electronic transition.

Figure B2.1.6 depicts a standard type of apparatus used for the hole-burning type of time-resolved absorption experiment [112, 113, 123]. A pulse train from an amplified laser is split into two portions. The minor portion is used directly as a source of pump photons; the major portion is used to generate a broad-band probe source derived from a femtosecond continuum. In this application, the continuum is typically generated by focusing a $>1\ \mu J$ pulse of light into flowing water or ethylene glycol in a cuvette [39]; a continuum with particularly good optical properties can be generated in a thin sapphire crystal [47]. After the pump and probe pulses are overlapped in the sample, the transmitted probe light is dispersed in a monochromator and then detected either by a photodiode or by a multichannel detector, such as a charge-coupled device (CCD). The most common detection scheme involves using a mechanical chopper to modulate the intensity of the pump beam; the pump-induced changes in the transmission of the probe beam are then detected by using a lock-in amplifier.

As an example, a series of transient hole-burning spectra obtained with a chirp-compensated continuum probe with a light-harvesting protein is shown in figure B2.1.7 [112]. As the probe delay increases, the initially narrow transmission hole increases in width owing to vibrational redistribution and shifts about 500 cm$^{-1}$ to the red owing to dynamic solvation. Analysis of this series of spectra was made in terms of overlapping spectral contributions from ground-state depletion, stimulated emission, and excited-state absorption. The time resolution of the experiment is limited by the width of the pump–probe cross-correlation function,

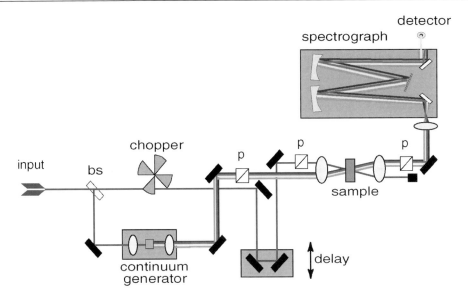

**Figure B2.1.6.** Femtosecond spectrometer for transient hole-burning spectroscopy with a continuum probe. Symbols used: bs, 10% reflecting beamsplitter; p, polarizer. The continuum generator consists of a focusing lens, a cell containing flowing water or ethylene glycol or, alternatively, a sapphire crystal and a recollimating lens.

which can be conveniently determined in this case using either FROG or a wavelength-resolved optical Kerr measurement [124].

The main cost of this enhanced time resolution compared to fluorescence upconversion, however, is the aforementioned problem of time ordering of the photons that arrive from the pump and probe pulses. When the probe pulse either precedes or trails the arrival of the pump pulse by a time interval that is significantly longer than the pulse duration, the action of the probe and pump pulses on the populations resident in the various resonant states is unambiguous. When the pump and probe pulses temporally overlap in the sample, however, all possible time orderings of field–molecule interactions contribute to the response and complicate the interpretation. Double-sided Feynman diagrams, which provide a pictorial view of the density matrix's time evolution under the action of the laser pulses, can be used to determine the various contributions to the sample response [125].

The part of the response arising from a coherent interaction between the temporally overlapped probe and pump pulses is the so-called *coherence spike* [126, 127], which makes its appearance in the zero-delay region in figure B2.1.7 essentially confined to the spectral region coinciding with the pump-pulse spectrum. Accordingly, the overall pump–probe signal's temporal shape when viewed at a single-probe wavelength is not just the integral of the convolution of the pump and probe pulse temporal shapes. The intensity of the coherence spike is strongly dependent on the duration of the laser pulses employed in the experiment and the time scale of dephasing [127, 128].

The first *dynamic absorption* studies that afforded a view of a molecular response stimulated by *impulsive* excitation with femtosecond laser pulses were performed by Shank and co-workers in 1988 with 6 fs pulses from a fibre-grating pulse-compressed CPM laser [57]. This experiment might be called a degenerate pump–probe experiment since the 6 fs pulses were used both for pumping and probing; the broad spectral bandwidth had been used previously as a short, chirp-free, continuum-like probe in time-resolved hole-burning spectroscopy [101]. Under the conditions of impulsive excitation, where the pulse duration is short relative to the period of the coupled vibrations, vibrational wavepackets [129, 130] are created on structurally displaced

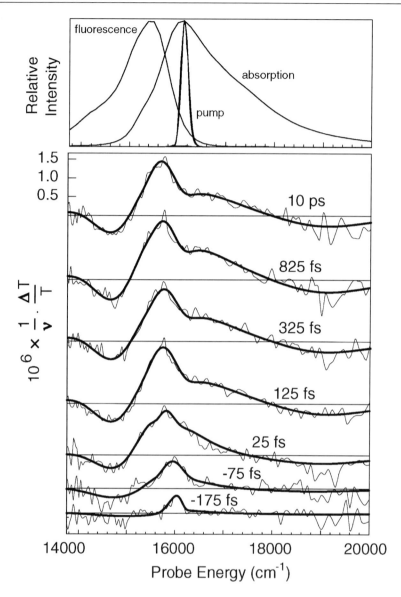

**Figure B2.1.7.** Transient hole-burned spectra obtained at room temperature with a tetrapyrrole-containing light-harvesting protein subunit, the $\alpha$ subunit of $C$-phycocyanin. Top: fluorescence and absorption spectra of the sample superimposed with the spectrum of the 80 fs pump pulses used in the experiment, which were obtained from an amplified CPM dye laser operating at 620 nm. Bottom: absorption-difference spectra obtained at a series of probe time delays.

excited-state potential energy surfaces. The wavepackets move back and forth under the forces of the excited-state potential and cause modulation of the stimulated-emission contribution to the pump–probe signal. A second pump-field–molecular interaction causes a stimulated transition of some of the moving excited-state wavepacket back down to the ground state at a position that is displaced from the equilibrium molecular geometry, causing a ground-state wavepacket motion; that corresponds to the signal detected in conventional resonance Raman spectroscopy [130, 131]. Accordingly, the modulation of the ground-state depletion signal

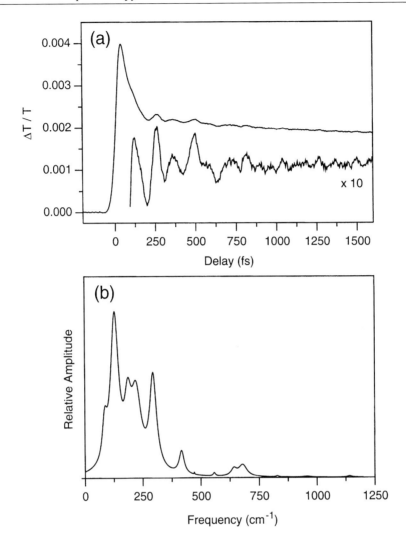

**Figure B2.1.8.** Dynamic absorption trace obtained with the dye IR144 in methanol, showing oscillations arising from coherent wavepacket motion: (a) transient observed at 775 nm; (b) frequency analysis of the oscillations obtained using a linear prediction, singular-value-decomposition method.

owing to the wavepacket motion on the ground-state surface is termed resonant-impulsive stimulated-Raman scattering (RISRS) [132]. Champion and co-workers [133–135] have made extensive use of RISRS in their studies of heam motions in response to ligand photodissociation. Nelson and co-workers [132] have used RISRS to look at collective motions in molecular crystals.

Figure B2.1.8 shows dynamic absorption results obtained with an IR dye in solution. The experiment was conducted with 13 fs pulses from a pulse-picked Ti–sapphire laser and a rapid-scanning pump–probe interferometer [136]. The single-wavelength transient was obtained by dispersing the transmitted probe light in a monochromator and monitoring at a single narrow range of wavelengths. The transient exhibits a modulation signal arising from excited-state vibrational wavepacket motions that is sustained for at least a picosecond. Modulations of this type can be frequency analysed using either Fourier transformation or a linear-prediction, singular-value decomposition (LPSVD) method, as was done in figure B2.1.8(b). The LPSVD method [137] fits the modulation pattern to a series of damped cosinusoids. The representation shown in figure B2.1.8(b) uses the frequencies and damping (dephasing) times to construct a spectral representation that resembles a conventional Raman spectrum. The modulation spectrum shown in figure B2.1.8(b) evidences contributions from several vibrational modes over the $100$–$1200$ cm$^{-1}$ range. The intensity of modulation for a given frequency is observed to depend strongly on the pulse duration; as the pulse duration is made shorter, the frequency window that provides impulsive excitation broadens. This windowing has been described by Lotshaw and McMorrow in their discussion of non-resonant optical Kerr effect studies, where impulsive excitation of intramolecular and intermolecular vibrational modes in neat liquids can be studied through the use of orthogonally polarized pump and probe beams and optically heterodyned detection methods [138–142].

Excited-state vibrational coherence of the type observed in figure B2.1.8 is potentially a very important tool for the elucidation of excited-state reaction dynamics. The most well known example of this type of work is that of Mathies, Shank and co-workers on the dynamics of rhodopsin [118, 143]. Elsaesser and co-workers [144] used a two-colour dynamic absorption technique to study ultrafast intramolecular proton transfer in a benzotriazole dye in solution. Wynne and Hochstrasser [145] observed vibrational coherence associated with charge transfer in a contact ion-pair in solution. Diffey *et al* [122] used a comparison of excited-state and RISRS wavepacket modulation patterns to study ultrafast charge transfer in a bacteriochlorophyll dimer system isolated from a purple-bacterial light-harvesting chromophore. Vos and co-workers [146–148] have observed excited-state vibrational coherence in purple-bacterial reaction centres; Stanley and Boxer observed analogous signals using fluorescence upconversion [149].

So far we have exclusively discussed time-resolved absorption spectroscopy with *visible* femtosecond pulses. It has become recently feasible to perform time-resolved spectroscopy with femtosecond IR pulses. Hochstrasser and co-workers [58, 150–157] have worked out methods to employ IR pulses to monitor chemical reactions following electronic excitation by visible pump pulses; these methods were applied in work on the light-initiated charge-transfer reactions that occur in the photosynthetic reaction centre [156, 157] and on the excited-state isomerization of the retinal pigment in bacteriorhodopsin [155]. Walker and co-workers [158] have recently used femtosecond IR spectroscopy to study vibrational dynamics associated with intramolecular charge transfer; these studies are complementary to those performed by Barbara and co-workers [159, 160], in which ground-state RISRS wavepackets were monitored using a dynamic-absorption technique with visible pulses.

In some extremely innovative recent experiments, Hochstrasser and co-workers [58] have described IR transient hole-burning experiments focused on characterizing inhomogeneous broadening in the amide I transition in several small polypeptides. 180 fs IR pulses centred at $1650$ cm$^{-1}$ were generated from a BBO OPA source through the use of difference-frequency mixing of the signal and idler pulses in a AgGaS$_2$ crystal. Narrower, 1 ps duration IR pulses were produced from the spectrally broad femtosecond IR pulses using a mechanically scannable Fabry–Perot etalon. The results were presented using a novel two-dimensional representation as shown in figure B2.1.9, in which the time-resolved hole-burned spectra were plotted against the centre frequency of the pump spectrum that burned the hole. The results provide information on the extent of delocalization of the amide vibrational wavefunction along the peptide backbone.

### B2.1.3.3  *Photon-echo and transient-grating spectroscopy*

The methods discussed so far, fluorescence upconversion, the various pump–probe spectroscopies, and the polarized variations for the measurement of anisotropy, are essentially conventional spectroscopies adapted to the femtosecond regime. At the simplest level of interpretation, the information content of these conventional time-resolved methods pertains to *populations* in resonantly prepared or probed states. As applied to chemical kinetics, for most *slow* reactions (on the ten picosecond and longer time scales), populations adequately specify the position of the reaction coordinate; intermediates and products show up as time-delayed spectral entities,

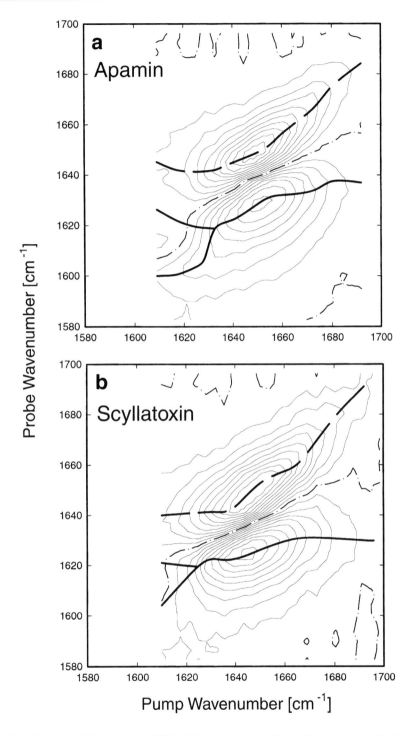

**Figure B2.1.9.** Two-dimensional time-resolved IR holeburning spectra obtained with two small polypeptides, apamin and scyllatoxin, by Hochstrasser and co-workers [58]. Figure courtesy Professor R M Hochstrasser (University of Pennsylvania).

and assignment of the transient spectra to chemical structures follows, in most cases, the same principles used in spectroscopic experiments performed with continuous wave or nanosecond pulsed lasers.

The multiple-pulse methods discussed in this section, in contrast, can be used to obtain information on the time evolution of electronic or vibrational coherence, the correlation of *phase* between two states. In *fast*, sub-picosecond chemical reactions and in energy transfer, as examples, knowledge of the time evolution of coherence *and* population is essential if a correct physical model is to be established. For example, in the purple-bacterial photosynthetic reaction centre, the 3 ps time scale for the charge transfer from the bacteriochlorophyll dimer that serves as the primary electron donor to the pheophytin that serves as the electron acceptor is comparable to that of vibrational dephasing and just longer than the fast part of electronic dephasing [146, 149, 156, 161–167]. It is certainly inadequate to describe this situation just in terms of populations and the states of the *individual* macrocycles that are involved in the charge-transfer reaction. A similar problem applies to the photophysics of retinal in the visual and proton-pumping proteins rhodopsin and bacteriorhodopsin. The light-initiated isomerization of retinal in these proteins occurs on the sub-picosecond time scale; it involves a time evolution from an excited electronic state to a isomerized ground state on a time scale that is shorter than that involved in vibrational dephasing [114, 117, 118, 120, 143].

Photon-echo and transient-grating spectroscopy exploit, in general, time-ordered interactions between three or more optical pulses [125, 168]. The principles involved are analogous to those that are well established for pulsed experiments in nuclear and electron magnetic resonance [169]. Although the methods discussed below were applied first for the study of electronic coherence, as picosecond and femtosecond IR sources were developed a set of corresponding methods were applied to the *direct* study of vibrational coherence. Simple multiple-pulse sequences, with control over relative delays between pulses, can be formed using interferometers with three (or more) arms. Importantly, the pulses can be aimed spatially so that the phase-matched (momentum-conserving) outgoing directions for the signal (echo or diffracted) pulses are spatially resolved from the transmitted excitation pulses, affording zero-background detection [170].

Transient-grating spectroscopy is performed using two pump pulses that are arranged to arrive at the sample at the same time but aimed to overlap at an angle [171]. A *transmission diffraction grating* is formed owing to the spatial interference between the two pump pulses. The grating consists of alternating regions of excited-state molecules, formed in the bright regions where the two incoming beams interfere constructively, and of ground-state molecules, left undisturbed in the dark regions where the two beams interfere destructively. A third, probing laser pulse is diffracted by the population grating into a direction that is resolved from the pump directions, allowing the intensity of the population grating to be measured as a function of time. Any physical process that causes the spatial pattern of ground- and excited-state molecules to fade away or become less distinct will be detected in terms of a decrease in the diffracted beam's intensity. Thus, in addition to being sensitive to population decay, the transient-grating experiment can be used to detect spatial motion of molecules owing to transport or diffusion. For example, Miller and co-workers [172–174] have employed transient-grating spectroscopy to study protein motions in heam proteins. Further, since the diffraction efficiency is sensitive to the orientation of the transition-dipole moments of the excited-state molecules, the transient-grating method can be used to characterize rotational diffusion. Fayer and co-workers have also exploited polarization gratings, formed by making the two coincident incoming beams be orthogonally polarized [171].

The simplest echo experiment, the two-pulse photon echo, reports information on the decay of coherence in terms of echo intensity. As in the corresponding spin-echo experiment in magnetic resonance, the first incident laser pulse rotates the Bloch vector [168, 175] from pure population in the ground state to an orientation corresponding to a coherent superposition state; the density matrix now exhibits off-diagonal elements as well as diagonal elements. During the waiting period $t$ between the first and second pulses, the off-diagonal (coherence) elements in the density matrix decay according to the dephasing time, $T_2$, while the on-diagonal (population) elements decay according to the lifetime of the resonantly prepared state, $T_1$. The second laser pulse rotates the Bloch vectors again so that the phase of rotation of the Bloch vectors is inverted, which leads to refocusing. The vectors are maximally refocused at an interval $t$ after the second pulse, so that the *spontaneous* emission arising from the ensemble of molecules is radiated with spatial coherence, forming an *echo* pulse. Thus, the intensity of the echo as a function of the waiting time $t$ can be described by an exponential decay with time constant $T_2$ [176]. If the two input pulses are directed along the directions $\vec{k}_1$ and $\vec{k}_2$, echo signal beams are emitted along the $2\vec{k}_1 - \vec{k}_2$ and $2\vec{k}_2 - \vec{k}_1$ directions. In practice, the input and echo beams are recollimated by a single lens after they emerge from the sample; iris apertures are used to spatially isolate the direction of either of the echo beams so that the emerging input beams are blocked.

In liquids, the Bloch equations (single-dephasing time scale) picture [175] used above to describe the formation of echo signals is apparently inadequate. It is now known that electronic dephasing occurs over a distribution of time scales, so a single time constant $T_2$ is insufficient to describe all of the line-broadening dynamics [177]. The two-pulse echo method described above only is sensitive to the fastest of processes; in organic molecules in solution, the two-pulse echoes typically decay on the 20 fs or shorter time scale [176]. This is the time scale usually assigned to homogeneous line broadening. The slower electronic dephasing processes that contribute to inhomogeneous line broadening, involving solvent-induced fluctuations or radiationless decay between uncorrelated states, extend over the 10 fs to 100 ps (or longer) time scales in liquids and proteins [77, 78, 177].

A three-pulse or stimulated photon-echo experiment can be employed to characterize dephasing on a much longer time scale than is accessible to the two-pulse photon echo experiment. Figure B2.1.10 shows a three-pulse interferometer and beam-input geometry employed by Fleming and co-workers [170, 178–181]. The modified forward-box beam-input geometry allows three-pulse echoes (or grating signals) to be detected in the phase-matched $\vec{k}_1 - \vec{k}_2 + \vec{k}_3$ and $-\vec{k}_1 + \vec{k}_2 + \vec{k}_3$ directions. As in the Hahn stimulated spin-echo sequence [169], for a given waiting time $t$ between the first two pulses, a plot of the echo intensity as a function of the time period $T$ between the second and third pulses returns just the lifetime $T_1$ of the resonantly prepared state. At a given delay $T$, a plot of the intensity as a function $t$ returns a decay related to the dephasing time $T_2$, but there is a subtle change in the shape of the intensity envelope that is discernible as $T$ is varied [170, 179, 182–185]. Figure B2.1.11 shows the results obtained from a three-pulse photon-echo experiment on a small protein subunit that binds an extended tetrapyrrole chromophore [186]. At early delays $T$, the shape is asymmetrical, with the maximum intensity shifted away from $t = 0$. As $T$ is increased, so that the ensemble evolves for longer time periods prior to rephasing by the third pulse, the envelope becomes more symmetrical, and the maximum shifts back to near $t = 0$. The asymmetrical shape observed at early delays $T$ reports the presence of an echo, but the symmetrical signal observed at longer delays $T$ arises from a free-induction decay only.

This experiment is now known as a three-pulse stimulated photon-echo peak-shift (3PEPS) experiment. Weiner and Ippen [182] were the first to describe the echo-envelope-shifting phenomenon and its relationship to inhomogeneous line broadening; application of the 3PEPS method to problems of dynamic solvation has been popularized especially by the groups of Fleming [177–179, 184, 185] and of Wiersma, who has advanced gated versions of the experiment that actually time-resolves the echo in order to obtain additional information [187–189]. The 3PEPS method returns, in general, superior information on solvation dynamics as compared to that returned by dynamic Stokes shift measurements by fluorescence upconversion or transient hole-burning spectroscopy because no line shape assumptions have to be made; in fact, a full analysis of the time-correlation and line-broadening functions obtained from the 3PEPS experiment can be used to obtain all pertinent spectroscopic observables, including the absorption and fluorescence line shapes. Owing to the mapping of coherence into population by the second pulse in the sequence, dephasing can be studied using the 3PEPS method over an enormous time scale, generally as long as the lifetime of the resonant state [177]. Figure B2.1.12 shows the entire 3PEPS profile obtained from a series of experiments on the sample used for figure B2.1.11, conducted with a series of $T$ delays. Several different time scales that contribute to electronic dephasing are notable, corresponding to a very fast decay on the <20 fs time scale, a roughly exponential decay on the 100 fs time scale, and a slower decay to a long-lived offset over the 200 fs to 1 ps time scale [186]. The magnitude of the peak shift for each component is proportional to the strength of coupling of

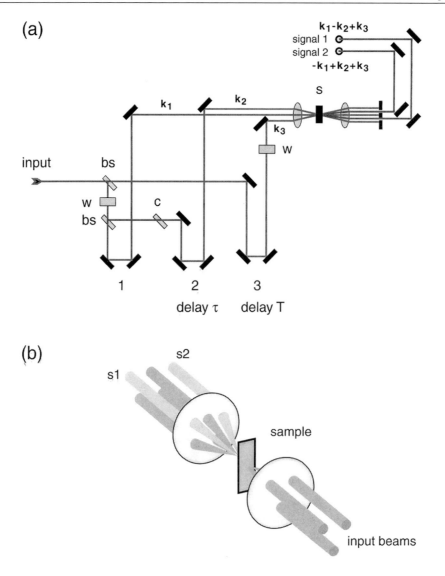

**Figure B2.1.10.** Three-pulse femtosecond spectrometer for use in stimulated-photon-echo and transient-grating measurements. Top: Mach–Zehnder interferometer. Symbols used: bs, beamsplitter; w, $\lambda/2$ plate and polarizer, in sequence; c, compensator plate; s, sample; $k_i$, direction vector for incoming beams $i = 1\ldots3$; s1, s2, signal beams. Bottom: modified forward-box beam geometry used to spatially resolve the signal beams s1 and s2 from the three input beams.

solvent fluctuations on a given time scale to the electronic dipole of the resonant electronic state that is used as a probe [177].

As implied above, comparable two- and three-pulse echo experiments can be conducted with femtosecond IR pulses in order to study *vibrational* dephasing. Notable work in this area has been conducted by Fayer and co-workers with picosecond IR pulses obtained from a free-electron laser at Stanford University [190–195]. Vibrational states can also be prepared with visible femtosecond pulses through the use of stimulated Raman *coherences*. In this family of methods, two laser pulses with frequencies $\omega_1$ and $\omega_2$ act simultaneously to

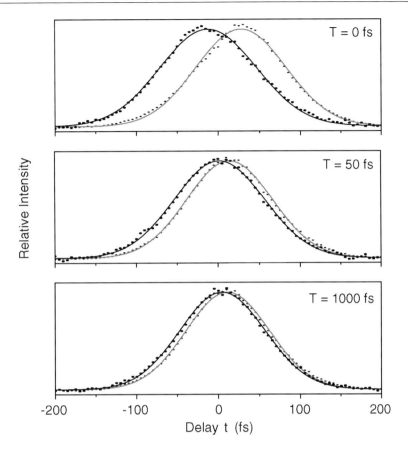

**Figure B2.1.11.** Stimulated photon-echo peak-shift (3PEPS) signals. Top: pulse sequence and interpulse delays $t$ and $T$. Bottom: echo signals scanned as a function of delay $t$ at three different population periods $T$, obtained with samples of a tetrapyrrole-containing light-harvesting protein subunit, the $\alpha$ subunit of $C$-phycocyanin.

transfer population to a vibrational level of frequency $\omega_1 - \omega_2$. The chief advantage of using pairs of visible pulses is that a wider frequency range becomes accessible owing to the availability of very short visible pulses; experiments conducted with IR pulses are limited by time–bandwidth considerations to lower frequencies. At present, however, the stimulated Raman methods have only been used *non-resonantly*, so very intense femtosecond laser pulses are required, typically in the $\mu J$/pulse regime. In contrast, photon-echo experiments conducted with resonant electronic states are generally conducted with pulse energies in the low nJ regime.

One important stimulated Raman method, known as the Raman echo experiment, is analogous to the three-pulse or stimulated photon-echo discussed above. Two pairs of visible pulses prepare and map vibrational coherences into population, respectively, and after a waiting period a fifth pulse is used to stimulate rephasing and echo formation. Berg and co-workers have used the Raman echo to probe the vibrational dephasing of rotating methyl groups in different solvent environments [196, 197]. Tokmakoff, Fleming, and their co-workers [198–201] have exploited the intrinsic two-dimensionality of the Raman echo experiment to explore anharmonic coupling between intermolecular modes in liquid $CS_2$; each axis of the experiment returns information that is analogous to that available from conventional stimulated coherent Raman spectroscopy, but in the two-dimensional representation cross-peaks appear that directly report coupling between two vibrational modes.

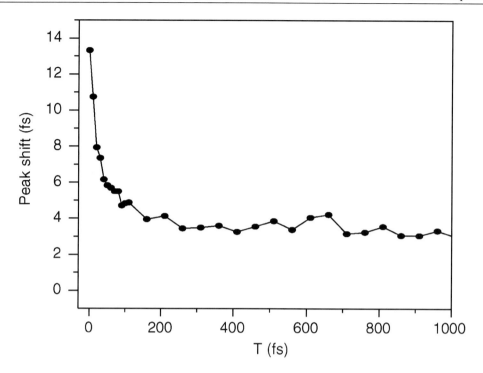

**Figure B2.1.12.** 3PEPS profile obtained at room temperature with samples of a tetrapyrrole-containing light-harvesting protein subunit, the $\alpha$ subunit of $C$-phycocyanin, as used in the previous figure.

*B2.1.3.4   Coherent control and future ultrafast spectroscopies*

A number of investigators are now developing pulse-shaping and modulation techniques that are useful with ultrashort laser pulses. These methods will permit preparation of precisely timed *and phased* multipulse sequences of arbitrary complexity for use in nonlinear spectroscopy. In addition, rather than just exploiting pulse sequences to project coherences into echo intensities and time shifts for spectroscopic purposes, as in the methods discussed above, several investigators are devising pulse sequences to focus wavefunctions onto potential surfaces in a non-statistical manner. This concept, known generally as *coherent control* [202–204], refers to attempts to control chemical reactions with specially constructed sequences of ultrashort laser pulses of known phase evolution and duration.

Scherer *et al* [205, 206] showed how to prepare, using interferometric methods, pairs of laser pulses with known relative phasing. These pulses were employed in experiments on vapour phase $I_2$, in which wavepacket motion was detected in terms of fluorescence emission. A more general approach, which can be used in principle to generate pulse sequences of any type, is to transform a single input pulse into a *shaped* output profile, with the intensity and phase of the output under control throughout. The idea being exploited by a number of investigators, notably Warren and Nelson, is to use a programmable dispersive delay line constructed from a pair of diffraction gratings spaced by an *active* device that is used either to absorb or phase shift selectively the frequency-dispersed wavefront. The approach favoured by Warren and co-workers exploits a Bragg cell driven by a radio-frequency signal obtained from a frequency synthesizer and a computer-controlled arbitrary waveform generator [207]. Nelson and co-workers use a computer-controlled liquid-crystal pixel array as a mask [208]. In the future, it is likely that one or both of these approaches will allow execution of currently impossible nonlinear spectroscopies with highly selective information content. One can take inspiration from the complex pulse sequences used in modern multiple-dimension NMR spectroscopy to suppress unwanted interfering resonances and to enhance selectively the resonances from targeted nuclei.

A simple example of what is possible now, even with two-pulse sequences, is the work by Shank and co-workers on focused RISRS wavepackets. Bardeen *et al* [209] used pump pulses prepared on purpose with a linear, negative chirp to enhance the magnitude of wavepackets driven to the ground state by impulsive stimulated Raman scattering. In the simplest application, this kind of approach might be used to make it easier to detect weakly displaced normal modes in dynamic absorption spectroscopy. In a mode-selective chemistry example, focusing of wavepackets [204, 210] might be used to prepare a certain vibrational superposition state, which could be subsequently excited by another laser pulse to produce an enhanced product yield [211]. The goal of this kind of work is to drive chemical reactions in directions that are normally not possible along normal kinetically and energetically controlled routes [204].

# References

[1] Porter G 1992 Chemistry in microtime *The Chemical Bond: Structure and Dynamics* ed A Zewail (Boston: Academic) pp 113–48

[2] Porter G 1995 Flash photolysis into the femtosecond—a race against time *Femtosecond Chemistry* ed J Manz and L Wöste (New York: VCH) pp 3–13

[3] Shank C V 1986 Investigation of ultrafast phenomena in the femtosecond domain *Science* **233** 1276–80

[4] Raksi F, Wilson K R, Jiang Z, Iklef A, Côté C Y and Kieffer J-C 1996 Ultrafast x-ray absorption probing of a chemical reaction *J. Chem. Phys.* **104** 6066–99

[5] Schoenlein R W, Leeman W P, Chin A H, Volfbein P, Glover T E, Balling P, Zolotorev M, Kim K-J, Chattopadhayay S and Shank C V 1996 Femtosecond x-ray pulses at 0.4 Å generated by 90° Thomson scattering: a tool for probing the structural dynamics of materials *Science* **274** 236–8

[6] Fattinger C and Grischkowsky D 1989 Terahertz beams *Appl. Phys. Lett.* **54** 490–2

[7] Katzenellenbogen N and Grischkowsky D 1991 Efficient generation of 380 fs pulses of THz radiation by ultrafast laser pulse excitation of a biased metal–semiconductor interface *Appl. Phys. Lett.* **58** 222–4

[8] Pedersen J E and Keiding S R 1992 THz time-domain spectroscopy of non-polar liquids *IEEE J. Quantum. Electron.* **28** 2518–22

[9] Harde H, Katzenellenbogen N and Grischkowsky D 1995 Line-shape transition of collision broadened lines *Phys. Rev. Lett.* **74** 1307–10

[10] Flanders B N, Cheville R A, Grischkowsky D and Scherer N F 1996 Pulsed terahertz transmission spectroscopy of liquid $CHCl_3$, $CCl_4$, and their mixtures *J. Phys. Chem.* **100** 11 824–35

[11] Shank C V 1988 Generation of ultrashort optical pulses *Ultrashort Laser Pulses and Applications* ed W Kaiser (Berlin: Springer) pp 5–34

[12] Valdmanis J A, Fork R L and Gordon J P 1985 Generation of optical pulses as short as 27 femtoseconds directly from a laser balancing self phase modulation, group velocity dispersion, saturable absorption and saturable gain *Opt. Lett.* **10** 131–3

[13] Valdmanis J A and Fork R L 1986 Design considerations for a femtosecond pulse laser balancing self phase modulation, group velocity dispersion, saturable absorption, and saturable gain *IEEE J. Quantum. Electron.* **22** 112–18

[14] Knox W H, Downer M C, Fork R L and Shank C V 1984 Amplified femtosecond optical pulses and continuum generation at 5 kHz repetition rate *Opt. Lett.* **9** 552–4

[15] Fork R L, Brito Cruz C H, Becker P C and Shank C V 1987 Compression of optical pulses to six femtoseconds by using cubic phase compensation *Opt. Lett.* **12** 483–5

[16] Zewail A H 1988 Laser femtochemistry *Nature* **328** 760–1

[17] Khundkar L R and Zewail A H 1990 Ultrafast molecular reaction dynamics in real-time: progress over a decade *Annu. Rev. Phys. Chem.* **41** 15–60

[18] Zewail A H 1990 The birth of molecules *Sci. Am.* **263** 76–82

[19] Gruebele M and Zewail A H 1990 Ultrafast reaction dynamics *Phys. Today* **43** 24–33

[20] Zewail A H 1991 Femtosecond transition-state dynamics *Faraday Discuss. Chem. Soc.* **91** 207–37

[21] Polanyi J C and Zewail A C 1995 Direct observation of the transition state *Acc. Chem. Res.* **28** 119–32

[22] Zewail A H 1995 Femtochemistry: concepts and applications *Femtosecond Chemistry* ed J Manz and L Wöste (New York: VCH) pp 15–128

[23] Spence D E, Kean P N and Sibbett W 1991 60 fs pulse generation from a self-mode-locked Ti:sapphire laser *Opt. Lett.* **16** 42–4

[24] Huang C-P, Asaki M T, Backus S, Murnane M M and Kapteyn H C 1992 17 fs pulses from a self-mode-locked Ti:sapphire laser *Opt. Lett.* **17** 1289–91

[25] Asaki M T, Huang C-P, Garvey D, Zhou J, Kapteyn H C and Murnane M M 1993 Generation of 11 fs pulses from a self-mode-locked Ti:sapphire laser *Opt. Lett.* **977** 977–9

[26] Stingl A, Lenzner M, Spielmann C, Krausz F and Szipšcs R 1996 Sub-10 fs mirror-dispersion-controlled Ti:sapphire laser *Opt. Lett.* **20** 602–4

[27] Jung I D, Kärtner F X, Matuschek N, Sutter D H, Morier-Genoud F, Zhang G, Keller U, Scheuer V, Tilsch M and Tschudi T 1997 Self-starting 6.5 fs pulses from a Ti:sapphire laser *Opt. Lett.* **22** 1009–11

[28] Bado P, Bourvier M and Coe J S 1987 Nd:YLF mode-locked oscillator and regenerative amplifier *Opt. Lett.* **12** 319–21

[29] Pessot M, Squier J, Mourou G and Harter D 1989 Chirped-pulse amplification of 100 fsec pulses *Opt. Lett.* **14** 797–9

[30] Vaillancourt G, Norris T B, Coe J S, Bado P and Mourou G 1990 Operation of a 1 kHz pulse-pumped Ti:sapphire regenerative amplifier *Opt. Lett.* **15** 317–19

[31] Rudd J V, Korn G, Kane S, Squier J, Mourou G and Bado P 1993 Chirped-pulse amplification of 55 fs pulses at a 1 kHz repetition rate in a TiAl$_2$O$_3$ regenerative amplifier *Opt. Lett.* **18** 2044–6

[32] Wilhelm T, Piel J and Riedle E 1997 Sub-20 fs pulses tunable across the visible from a blue-pumped single-pass noncollinear parametric converter *Opt. Lett.* **22** 1494–6

[33] Shirakawa A, Sakane I and Kobayasi T 1998 Pulse-front-matched optical parametric amplification for sub-10 fs pulse generation tunable in the visible and near infrared *Opt. Lett.* **23** 1292–4

[34] Cerullo G, Nisoli M, Stagira S and De Silvestri S 1998 Sub-8 fs pulses from an ultrabroadband optical parametric amplifier in the visible *Opt. Lett.* **23** 1283–5

[35] Pshenichnikov M S, de Boeij W P and Wiersma D A 1994 Generation of 13 fs, 5 MW pulses from a cavity-dumped Ti:sapphire laser *Opt. Lett.* **19** 572–4

[36] Baltuska A, Wei Z, Pshenichnikov M S and Wiersma D A 1997 Optical pulse compression to 5 fs at a 1 MHz repetition rate *Opt. Lett.* **22** 102–4

[37] Evans J M, Spence D E, Sibbett W, Chai B H T and Miller A 1992 50 fs pulse generation from a self-mode-locked Cr:LiSrAlF$_6$ laser *Opt. Lett.* **17** 1447–9

[38] Valentine G J, Hopkins J-M, Loza-Alvarez P, Kennedy G T, Sibbett W, Burns D and Valster A 1997 Ultralow-pump-threshold, femtosecond Cr$^{3+}$:LiSrAlF$_6$ laser pumped by a single narrow-stripe AlGaInP laser diode *Opt. Lett.* **22** 1639–41

[39] Fork R L, Shank C V, Hirlimann C, Yen R and Tomlinson W J 1983 Femtosecond white-light continuum pulses *Opt. Lett.* **8** 1–3

[40] Norris T B 1992 Femtosecond pulse amplification at 250 kHz with a Ti:sapphire regenerative amplifier and application to continuum generation *Opt. Lett.* **17** 1009–11

[41] Joo T, Jia Y and Fleming G R 1995 Ti:sapphire regenerative amplifier for ultrashort high-power multikilohertz pulses without an external stretcher *Opt. Lett.* **20** 389–91

[42] Le Blanc C, Grillon G, Chambaret J P, Migus A and Antonetti A 1993 Compact and efficient multipass Ti:sapphire system for femtosecond chirped-pulse amplification at the terawatt level *Opt. Lett.* **18** 140–2

[43] Backus S, Peatross J, Huang C P, Murnane M M and Kapteyn H C 1995 Ti:sapphire amplifier producing millijoule-level, 21 fs pulses at 1 kHz *Opt. Lett.* **20** 2000–2

[44] Kane S, Squier J, Rudd J V and Mourou G 1994 Hybrid grating-prism stretcher-compressor system with cubic phase and wavelength tunability and decreased alignment sensitivity *Opt. Lett.* **19** 1876–8

[45] Treacy E B 1969 Optical pulse compression with diffraction gratings *IEEE J. Quantum. Electron.* **5** 454–8

[46] Yakovlev V V, Kohler B and Wilson K R 1994 Broadly tunable 30 fs pulses produced by optical parametric generation *Opt. Lett.* **19** 2000–2

[47] Reed M K, Steiner-Shepard M K and Negus D K 1994 Widely tunable femtosecond optical parametric amplifier at 250 kHz with a Ti:sapphire regenerative amplifier *Opt. Lett.* **19** 1855–7

[48] Greenfield S R and Wasielewski M R 1995 Near-transform-limited visible and near-IR femtosecond pulses from optical parametric amplification using Type II $\beta$-barium borate *Opt. Lett.* **20** 1394–6

[49] Gale G M, Cavallari M, Driscoll T J and Hache F 1995 Sub-20 fs tunable pulses in the visible from an 82 MHz optical parametric oscillator *Opt. Lett.* **20** 1562–4

[50] Hache F, Zéboulon A, Gallot G and Gale G M 1995 Cascaded second-order effects in the femtosecond regime in $\beta$-barium borate: self-compression in a visible femtosecond optical parametric oscillator *Opt. Lett.* **20** 1556–8

[51] Umstadter D P, Barty C, Perry M and Mourou G A 1998 Tabletop, ultrahigh intensity lasers: dawn of nonlinear relativistic optics *Opt. Photon. News* **9** 41

[52] Johnson A M and Shank C V 1989 Pulse compression in single-mode fibres—picoseconds to femtoseconds *The Supercontinuum Laser Source* ed R R Alfano (New York: Springer) pp 399–449

[53] Ippen E 1997 Characterizing optical components for ultrafast laser applications *Optics 1997/98 Catalog* (Irvine, CA: Newport Corp.) pp 8-2–8-3

[54] Fork R L, Martinez O E and Gordon J P 1984 Negative dispersion using pairs of prisms *Opt. Lett.* **9** 150–2

[55] Gires F and Tournois P 1964 Interféromètre utilisable pour la compression d'impulsions lumineuses modulées en fréquence *Compte Rendue Acad. Sci. Paris* **258** 6112–15

[56] Kuhl J and Heppner J 1986 Compression of femtosecond optical pulses with dielectric multilayer interferometers *IEEE J. Quantum. Electron.* **22** 182–5

[57] Fragnito H L, Bigot J-Y, Becker P C and Shank C V 1989 Evolution of the vibronic absorption spectrum in a molecule following impulsive excitation with a 6 fs optical pulse *Chem. Phys. Lett.* **160** 101–4

[58] Hamm P, Lim M and Hochstrasser R M 1998 Structure of the amide I band of peptides measured by femtosecond nonlinear-infrared spectroscopy *J. Phys. Chem.* B **102** 6123–38

[59] Mollenauer L F, Stolen R H and Gordon J P 1980 Experimental observation of picosecond pulse narrowing and solitons in optical fibres *Phys. Rev. Lett.* **45** 1095–7

[60] Agrawal G P 1989 Ultrashort pulse propagation in nonlinear dispersive fibers *The Supercontinuum Laser Source* ed R R Alfano (New York: Springer) pp 91–116

[61] Hasegawa A 1983 Amplification and reshaping of optical solitons in glass fibre–IV *Appl. Phys. Lett.* **8** 650–2

[62] Mollenauer L F, Gordon J P and Islam M N 1986 Soliton propagation in long fibers with periodically compensated loss *IEEE J. Quantum. Electron.* **22** 157–73

[63] Fleming G R 1986 *Chemical Applications of Ultrafast Spectroscopy* (New York: Oxford University Press)

[64] Maznev A A, Crimmins T F and Nelson K A 1998 How to make femtosecond pulses overlap *Opt. Lett.* **23** 1378–80

[65] Trebino R and Kane D J 1993 Using phase retrieval to measure the intensity and phase of ultrafast pulses: frequency-resolved optical gating *J. Opt. Soc. Am.* A **10** 1101–11

[66] Kane D J and Terbino R 1993 Characterization of arbitrary femtosecond pulses using frequency-resolved optical gating *IEEE J. Quantum. Electron.* **29** 571–9

[67] Kane D J and Trebino R 1993 Single-shot measurement of the intensity and phase of an arbitrary ultrashort pulse by using frequency-resolved optical gating *Opt. Lett.* **18** 823–5

[68] Kane D J, Taylor A J, Trebino R and DeLong K W 1994 Single-shot measurement of the intensity and phase of a femtosecond UV laser pulse with frequency-resolved optical gating *Opt. Lett.* **19** 1061–3

[69] DeLong K W, Trebino R, Hunter J and White W E 1994 Frequency-resolved optical gating with the use of second-harmonic generation *J. Opt. Soc. Am.* B **11** 2206–15

[70] Walker G C, Jarzeba W, Kang T J, Johnson A E and Barbara P F 1990 Ultraviolet femtosecond fluorescence spectroscopy: techniques and applications *J. Opt. Soc. Am.* B **7** 1521–7

[71] Jimenez R and Fleming G R 1996 Ultrafast spectroscopy of photosynthetic systems *Biophysical Techniques in Photosynthesis* ed J Amesz and A J Hoff (Dordrecht: Kluwer) pp 63–73

[72] Cross A J and Fleming G R 1984 Analysis of time-resolved fluorescence anisotropy decays *Biophys. J.* **46** 45–56

[73] Simon J D 1988 Time-resolved studies of solvation in polar media *Acc. Chem. Res.* **21** 128–34

[74] Jarzeba W, Walker G C, Johnson A E and Barbara P F 1991 Nonexponential solvation dynamics of simple liquids and mixtures *Chem. Phys.* **152** 57–68

[75] Maroncelli M 1993 The dynamics of solvation in polar solvents *J. Mol. Liq.* **57** 1–37

[76] Jimenez R, Fleming G R, Kumar P V and Maroncelli M 1994 Femtosecond solvation dynamics of water *Nature* **369** 471–3

[77] Stratt R M and Cho M 1994 The short-time dynamics of solvation *J. Chem. Phys.* **100** 6700–8

[78] Stratt R M and Maroncelli M 1996 Nonreactive dynamics in solution: the emerging molecular view of solvation dynamics and vibrational relaxation *J. Phys. Chem.* **100** 12 981–96

[79] van Amerongen H and Struve W S 1995 Polarized optical spectroscopy of chromoproteins *Methods Enzymol.* **246** 259–83

[80] Tao T 1969 Time-dependent fluorescence depolarization and Brownian rotational diffusion coefficients of macromolecules *Biopolymers* **8** 609–32

[81] Cross A J, Waldeck D H and Fleming G R 1983 Time resolved polarization spectroscopy: level kinetics and rotational diffusion *J. Chem. Phys.* **78** 6455–67

[82] Matro A and Cina J A 1995 Theoretical study of time-resolved fluorescence anisotropy from coupled chromophore pairs *J. Phys. Chem.* **99** 2568–82

[83] Abrash S, Repinec S and Hochstrasser R M 1990 The viscosity dependence and reaction coordinate for isomerization of *cis*-stilbene *J. Chem. Phys.* **93** 1041–53

[84] Hochstrasser R M *et al* 1991 Anisotropy studies of ultrafast dipole reorientations *Proc. Indian Acad. Sci. (Chem. Sci.)* **103** 351–62

[85] Wynne K and Hochstrasser R M 1993 Coherence effects in the anisotropy of optical experiments *Chem. Phys.* **171** 179–88

[86] Wynne K and Hochstrasser R M 1995 Anisotropy as an ultrafast probe of electronic coherence in degenerate systems exhibiting Raman scattering, fluorescence, transient absorption and chemical reactions *J. Raman Spectrosc.* **26** 561–9

[87] Rahman T S, Knox R S and Kenkre V M 1979 Theory of depolarization of fluorescence in molecular pairs *Chem. Phys.* **44** 197–211

[88] Knox R S and Gülen D 1993 Theory of polarized fluorescence from molecular pairs *Photochem. Photobiol.* **57** 40–3

[89] Galli C, Wynne K, LeCours S, Therien M J and Hochstrasser R M 1993 Direct measurement of electronic dephasing using anisotropy *Chem. Phys. Lett.* **206** 493–9

[90] Edington M D, Riter R E and Beck W F 1995 Evidence for coherent energy transfer in allophycocyanin trimers *J. Phys. Chem.* **99** 15 699–704

[91] Riter R E, Edington M D and Beck W F 1997 Isolated-chromophore and exciton-state photophysics in C-phycocyanin trimers *J. Phys. Chem.* B **101** 2366–71

[92] Porter G and Topp M R 1968 Nanosecond flash photolysis and the absorption spectra of excited singlet states *Nature* **220** 1228-9

[93] Zewail A and Bernstein R 1992 Real-time laser femtochemistry: viewing the transition from reagents to products *The Chemical Bond: Structure and Dynamics* ed A Zewail (San Diego, CA: Academic) pp 223–79

[94] Scherer N F, Khundkar L R, Bernstein R B and Zewail A H 1987 Real-time picosecond clocking of the collision complex in a bimolecular reaction: the birth of OH from H + $CO_2$ *J. Chem. Phys.* **87** 1451–3

[95] Dantus M, Rosker M J and Zewail A H 1987 Real-time femtosecond probing of 'transition states' in chemical reactions *J. Chem. Phys.* **87** 2395–7

[96] Dantus M, Rosker M J and Zewail A H 1987 Femtosecond real-time probing of reactions. II. The dissociation reaction of ICN *J. Chem. Phys.* **89** 6128–40

[97] Rosker M J, Dantus M and Zewail A H 1988 Femtosecond real-time probing of reactions. I. The technique *J. Chem. Phys.* **89** 6113–27

[98] Rosker M J, Dantus M and Zewail A H 1988 Femtosecond clocking of the chemical bond *Science* **241** 1200–2

[99] Dantus M, Janssen M H M and Zewail A H 1991 Femtosecond probing of molecular dynamics by mass-spectrometry in a molecular beam *Chem. Phys. Lett.* **181** 281–7

[100] Pedersen S, Herek J L and Zewail A H 1994 The validity of the 'Diradical' hypothesis: direct femtosecond studies of the transition-state structures *Science* **266** 1359–64

[101] Brito Cruz C H, Fork R L, Knox W H and Shank C V 1986 Spectral hole burning in large molecules probed with 10 fs optical pulses *Chem. Phys. Lett.* **132** 341–5

[102] Loring R F, Yan Y J and Mukamel S 1987 Time-resolved fluorescence and hole-burning line shapes of solvated molecules: longitudinal dielectric relaxation and vibrational dynamics *J. Chem. Phys.* **87** 5840–57

[103] Vogel W, Welsch D-G and Wilhelmi B 1988 Time-resolved spectral hole burning *Chem. Phys. Lett.* **153** 376–8

[104] Brito Cruz C H, Gordon J P, Becker P C, Fork R L and Shank C V 1988 Dynamics of spectral hole burning *IEEE J. Quantum. Electron.* **24** 261–6

[105] Kinoshita S 1989 Theory of transient hole-burning spectrum for molecules in solution *J. Chem. Phys.* **91** 5175–84

[106] Kang T J, Yu J and Berg M 1990 Rapid solvation of a nonpolar solute measured by ultrafast transient hole burning *Chem. Phys. Lett.* **174** 476–80

[107] Kang T J, Yu J and Berg M 1991 Limitations on measuring solvent motion with ultrafast transient hole burning *J. Chem. Phys.* **94** 2413–24

[108] Yu J and Berg M 1992 Solvent-electronic state interactions measured from the glassy to the liquid state. I. Ultrafast transient and permanent hole burning in glycerol *J. Chem. Phys.* **96** 8741–9

[109] Murakami H, Kinoshita S, Hirata Y, Okada T and Mataga N 1992 Transient hole-burning and time-resolved fluorescence spectra of dye molecules in solution: evidence for ground-state relaxation and hole-filling effect *J. Chem. Phys.* **97** 7881–8

[110] Ma J, Bout D V and Berg M 1995 Solvation dynamics studied by ultrafast transient hole burning *J. Mol. Liq.* **65/66** 301–4

[111] Kovalenko S A, Ernsting N P and Ruthmann J 1996 Femtosecond hole-burning spectroscopy of the dye DCM in solution: the transition from the locally excited to a charge-transfer state *Chem. Phys. Lett.* **258** 445–54

[112] Riter R E, Edington M D and Beck W F 1996 Protein-matrix solvation dynamics in a subunit of C-phycocyanin *J. Phys. Chem.* **100** 14 198–205

[113] Edington M D, Riter R E and Beck W F 1997 Femtosecond transient hole-burning detection of interexciton-state radiationless decay in allophycocyanin trimers *J. Phys. Chem. B* **101** 4473–7

[114] Dexheimer S L, Wang Q, Peteanu L A, Pollard W T, Mathies R A and Shank C V 1992 Femtosecond impulsive excitation of nonstationary vibrational states in bacteriorhodopsin *Chem. Phys. Lett.* **188** 61–6

[115] Pollard W T, Dexheimer S L, Wang Q, Peteanu L A, Shank C V and Mathies R A 1992 Theory of dynamic absorption spectroscopy of nonstationary states. 4. Application to 12 fs resonant Raman spectroscopy of bacteriorhodopsin *J. Phys. Chem.* **96** 6147–58

[116] Bardeen C J and Shank C V 1993 Femtosecond electronic dephasing in large molecules in solution using mode suppression *Chem. Phys. Lett.* **203** 535–9

[117] Schoenlein R W, Peteanu L A, Wang Q, Mathies R A and Shank C V 1993 Femtosecond dynamics of cis-trans isomerization in a visual pigment analog: isorhodopsin *J. Phys. Chem.* **97** 12 087–92

[118] Peteanu L A, Schoenlein R W, Wang Q, Mathies R A and Shank C V 1993 The first step in vision occurs in femtoseconds: complete blue and red spectral studies *Proc. Natl Acad. Sci. USA* **90** 11 762–6

[119] Bardeen C J and Shank C V 1994 Ultrafast dynamics of the solvent-solute interaction measured by femtosecond four-wave mixing: LD690 in n-alcohols *Chem. Phys. Lett.* **226** 310–16

[120] Wang Q, Schoenlein R W, Peteanu L A, Mathies R A and Shank C V 1994 Vibrationally coherent photochemistry in the femtosecond primary event of vision *Science* **266** 422–4

[121] Wang Q, Kochendoerfer G G, Schoenlein R W, Verdegem P J E, Lugtenburg J, Mathies R A and Shank C V 1996 Femtosecond spectroscopy of a 13-demethylrhodopsin visual pigment analogue: the role of nonbonded interactions in the isomerization process *J. Phys. Chem.* **100** 17 388–94

[122] Diffey W M, Homoelle B J, Edington M D and Beck W F 1998 Excited-state vibrational coherence and anisotropy decay in the bacteriochlorophyll a dimer protein B820 *J. Phys. Chem. B* **102** 2776–86

[123] Edington M D, Riter R E and Beck W F 1996 Interexciton-state relaxation and exciton localization in allophycocyanin trimers *J. Phys. Chem.* **100** 14 206–17

[124] Yamaguchi S and Hamaguchi H 1995 Convenient method of measuring the chirp structure of femtosecond white-light continuum pulses *Appl. Spectrosc.* **49** 1513–15

[125] Mukamel S 1995 *Principles of Nonlinear Optical Spectroscopy* (New York: Oxford University Press)

[126] Balk M W and Fleming G R 1985 Dependence of the coherence spike on the material dephasing time in pump-probe experiments *J. Chem. Phys.* **83** 4300–7

[127] Cong P, Deuhl H P and Simon J D 1993 Using optical coherence to measure the ultrafast electronic dephasing of large molecules in room-temperature liquids *Chem. Phys. Lett.* **211** 367–73

[128] Cong P, Simon J D and Yan Y 1995 Probing the molecular dynamics of liquids and solutions *Ultrafast Processes in Chemistry and Photobiology* ed M A El-Sayed, I Tanaka and Y Molin (Oxford: Blackwell) pp 53–82

[129] Lee S-Y and Heller E J 1979 Time-dependent theory of Raman scattering *J. Chem. Phys.* **71** 4777–88

[130] Heller E J, Sundberg R L and Tannor D 1982 Simple aspects of Raman scattering *J. Phys. Chem.* **86** 1822–33

[131] Myers A B and Mathies R A 1987 Resonance Raman intensities: a probe of excited-state structure and dynamics *Biological Applications of Raman Spectroscopy* vol 2, ed T G Spiro (New York: Wiley-Interscience) pp 1–58

[132] Yan Y-X, Cheng L-T and Nelson K A 1988 Impulsive stimulated light scattering *Advances in Non-linear Spectroscopy* ed R J H Clark and R E Hester (Chichester: Wiley) pp 299–355

[133] Zhu L, Li P, Huang M, Sage J T and Champion P M 1994 Real time observation of low frequency heme protein vibrations using femtosecond coherence spectroscopy *Phys. Rev. Lett.* **72** 301–4

[134] Zhu L, Sage J T and Champion P M 1994 Observation of coherent reaction dynamics in heme proteins *Science* **266** 629–32

[135] Zhu L, Wang W, Sage J T and Champion P M 1995 Femtosecond time-resolved vibrational spectroscopy of heme proteins *J. Raman Spectrosc.* **26** 527–34

[136] Diffey W M and Beck W F 1997 Rapid-scanning interferometer for ultrafast pump–probe spectroscopy with phase-sensitive detection *Rev. Sci. Instrum.* 3296–300

[137] Johnson A E and Myers A B A 1996 A comparison of time- and frequency-domain resonance Raman spectroscopy in triiodide *J. Chem. Phys.* **104** 2497–507

[138] McMorrow D and Lotshaw W T 1990 The frequency response of condensed-phase media to femtosecond optical pulses: spectral-filter effects *Chem. Phys. Lett.* **174** 85–94

[139] McMorrow D and Lotshaw W T 1991 Intermolecular dynamics in acetonitrile probed with femtosecond Fourier transform Raman spectroscopy *J. Phys. Chem.* **95** 10 395–406

[140] McMorrow D and Lotshaw W T 1991 Dephasing and relaxation in coherently excited ensembles of intermolecular oscillators *Chem. Phys. Lett.* **178** 69–74

[141] McMorrow D and Lotshaw W T 1993 Evidence for low-frequency ($\approx 15$ cm$^{-1}$) collective modes in benzene and pyridine liquids *Chem. Phys. Lett.* **201** 369–76

[142] Lotshaw W T, McMorrow D, Thantu N, Melinger J S and Kitchenbaum R 1995 Intermolecular vibrational coherence in molecular liquids *J. Raman Spectrosc.* **26** 571–83

[143] Schoenlein R W, Peteanu L A, Mathies R A and Shank C V 1991 The first step in vision: femtosecond isomerization of rhodopsin *Science* **254** 412–15

[144] Chudoba C, Riedle E, Pfeiffer M and Elsaesser T 1996 Vibrational coherence in ultrafast excited-state proton transfer *Chem. Phys. Lett.* **263** 622–8

[145] Wynne K, Galli C and Hochstrasser R M 1994 Ultrafast charge transfer in an electron donor-acceptor complex *J. Chem. Phys.* **100** 4796–810

[146] Vos M H, Rappaport F, Lambry J-C, Breton J and Martin J-L 1993 Visualization of the coherent nuclear motion in a membrane protein by femtosecond spectroscopy *Nature* **363** 320–5

[147] Vos M H, Jones M R, Hunter C N, Breton J and Martin J-L 1994 Coherent nuclear dynamics at room temperature in bacterial reaction centers *Proc. Natl Acad. Sci. USA* **91** 12 701–5

[148] Vos M H, Jones M R, Breton J, Lambry J-C and Martin J-L 1996 Vibrational dephasing of long- and short-lived primary donor states in mutant reaction centers of *Rhodobacter sphaeroides Biochemistry* **35** 2687–92

[149] Stanley R J and Boxer S G 1995 Oscillations in the spontaneous fluorescence from photosynthetic reaction centers *J. Phys. Chem.* **99** 859–63

[150] Walker G C, Maiti S, Cowen B R, Moser C C, Dutton P L and Hochstrasser R M 1994 Time resolution of electronic transitions of photosynthetic reaction centers in the infrared *J. Phys. Chem.* **98** 5778–83

[151] Owrutsky J C, Raftery D and Hochstrasser R M 1994 Vibrational relaxation dynamics in solution *Annu. Rev. Phys. Chem.* **45** 519–55

[152] Lian T, Locke B, Kholodenko Y and Hochstrasser R M 1994 Energy flow from solute to solvent probed by femtosecond IR spectroscopy: malachite green and heme protein solutions *J. Phys. Chem.* **98** 11 648–56

[153] Owrutsky J C, Li M, Locke B and Hochstrasser R M 1995 Vibrational relaxation of the CO stretch vibration in hemoglobin-CO, myoglobin-CO, and protoheme-CO *J. Phys. Chem.* **99** 4842–6

[154] Lian T, Kholodenko Y and Hochstrasser R M 1995 Infrared probe of the solvent response to ultrafast solvation processes *J. Phys. Chem.* **99** 2546–51

[155] Diller R, Maiti S, Walker G C, Cowen B R, Pippenger R, Bogomolni R A and Hochstrasser R M 1995 Femtosecond time-resolved infrared laser study of the J–K transition of bacteriorhodopsin *Chem. Phys. Lett.* **241** 109–15

[156] Haran G, Wynne K, Moser C C, Dutton P L and Hochstrasser R M 1996 Level mixing and energy redistribution in bacterial photosynthetic reaction centers *J. Phys. Chem.* **100** 5562–9

[157] Wynne K, Haran G, Reid G D, Moser C C, Dutton P L and Hochstrasser R M 1996 Femtosecond infrared spectroscopy of low-lying excited states in reaction centers of *Rhodobacter sphaeroides J. Phys. Chem.* **100** 5140–8

[158] Wang C, Akhremitchev B and Walker G C 1997 Femtosecond infrared and visible spectroscopy of photoinduced intermolecular electron transfer dynamics and solvent–solute reaction geometries: Coumarin 337 in dimethylaniline *J. Phys. Chem.* A **101** 2735–8

[159] Walker G C, Barbara P F, Doorn S K, Dong Y and Hupp J T 1991 Ultrafast measurements on direct photoinduced electron transfer in a mixed-valence complex *J. Phys. Chem.* **95** 5712–15

[160] Tominaga K, Kliner D A V, Johnson A E, Levinger N E and Barbara P F 1993 Femtosecond experiments and absolute rate calculations on intervalence electron transfer of mixed-valence compounds *J. Chem. Phys.* **98** 1228–43

[161] Boxer S G, Goldstein R A, Lockhart D J, Middendorf T R and Takiff L 1989 Excited states, electron-transfer reactions, and intermediates in bacterial photosynthetic reaction centers *J. Phys. Chem.* **93** 8280–94

[162] Kirmaier C and Holten D 1988 Subpicosecond spectroscopy of charge separation in *Rhodobacter capsulatus* reaction centers *Isr. J. Chem.* **28** 79–85

[163] Jean J M, Chan C-K and Fleming G R 1988 Electronic energy transfer in photosynthetic bacterial reaction centers *Isr. J. Chem.* **28** 169–75

[164] Breton J, Martin J-L, Fleming G R and Lambry J-C 1988 Low-temperature femtosecond spectroscopy of the initial step of electron transfer in reaction centers from photosynthetic purple bacteria *Biochemistry* **27** 8276

[165] Holzapfel W, Finkele U, Kaiser W, Oesterhelt D, Scheer H, Stilz H U and Zinth W 1989 Observation of a bacteriochlorophyll anion radical during the primary charge separation in a reaction center *Chem. Phys. Lett.* **160** 1–7

[166] Holzapfel W, Finkele U, Kaiser W, Oesterhelt D, Scheer H, Stilz H U and Zinth W 1990 Initial electron-transfer in the reaction center from Rhodobacter sphaeroides *Proc. Natl Acad. Sci. USA* **87** 5168–72

[167] Stanley R J, King B and Boxer S G 1996 Excited state energy transfer pathways in photosynthetic reaction centers. 1. Structural symmetry effects *J. Phys. Chem.* **100** 12 052–9

[168] Levenson M D and Kano S S 1988 *Introduction to Nonlinear Laser Spectroscopy* (San Diego, CA: Academic)

[169] Slichter C P 1980 *Principles of Magnetic Resonance* (Berlin: Springer)

[170] Joo T and Albrecht A C 1993 Electronic dephasing studies of molecules in solution at room temperature by femtosecond degenerate four wave mixing *Chem. Phys.* **176** 233–47

[171] Fourkas J T and Fayer M D 1992 The transient grating: a holographic window to dynamic processes *Acc. Chem. Res.* **25** 227–33

[172] Genberg L, Richard L, McLendon G and Miller R J D 1991 Direct observation of global protein motion in hemoglobin and myoglobin on picosecond time scales *Science* **251** 1051–6

[173] Miller R J D 1994 Energetics and dynamics of deterministic protein motion *Acc. Chem. Res.* **27** 145–50

[174] Deak J, Richard L, Pereira M, Chui H-L and Miller R J D 1994 Picosecond phase grating spectroscopy: applications to bioenergetics and protein dynamics *Meth. Enzymol.* **232** 322–60

[175] Allen L and Eberly J H 1975 *Optical Resonance and Two-Level Atoms* (New York: Wiley)

[176] Becker P C, Fragnito H L, Bigot J-Y, Brito Cruz C H, Fork R L and Shank C V 1989 Femtosecond photon echoes from molecules in solution *Phys. Rev. Lett.* **63** 505–7

[177] Fleming G R and Cho M 1996 Chromophore-solvent dynamics *Annu. Rev. Phys. Chem.* **47** 109–34

[178] Joo T, Jia Y, Yu J-Y, Jonas D M and Fleming G R 1996 Dynamics in isolated bacterial light-harvesting antenna (LH2) of *Rhodobacter sphaeroides* at room temperature *J. Phys. Chem.* **100** 2399–409

[179] Joo T, Jia Y, Yu J-Y, Lang M J and Fleming G R 1996 Third-order nonlinear time domain probes of solvation dynamics *J. Chem. Phys.* **104** 6089–108

[180] Passino S A, Nagasawa Y, Joo T and Fleming G R 1997 Three-pulse echo peak shift studies of polar solvation dynamics *J. Phys. Chem.* A **101** 725–31

[181] Nagasawa Y, Passino S A, Joo T and Fleming G R 1997 Temperature dependence of optical dephasing in an organic polymer glass *J. Chem. Phys.* **106** 4840–52

[182] Weiner A M and Ippen E P 1985 Femtosecond excited state relaxation of dye molecules in solution *Chem. Phys. Lett.* **114** 456–60

[183] Bigot J-Y, Portella M T, Schoenlein R W, Bardeen C J, Migus A and Shank C V 1991 Non-Markovian dephasing of molecules in solution measured with three-pulse femtosecond photon echoes *Phys. Rev. Lett.* **66** 1138–41

[184] Cho M, Yu J-Y, Joo T, Nagasawa Y, Passino S A and Fleming G R 1996 The integrated photon echo and solvation dynamics *J. Phys. Chem.* **100** 11 944–53

[185] Passino S A, Nagasawa Y, Joo T and Fleming G R 1996 Photon echo measurements in liquids using pulses longer than the electronic dephasing time *Ultrafast Phenomena X* ed P Barbara, W Knox, W Zinth and J Fujimoto (Berlin: Springer) pp 199–200

[186] Homoelle B J, Edington M D, Diffey W M and Beck W F 1998 Stimulated photon-echo and transient-grating studies of protein-matrix solvation dynamics and interexciton-state radiationless decay in a phycocyanin and allophycocyanin *J. Phys. Chem.* B **102** 3044–52

[187] de Boeij W P, Pshenichnichov M S and Wiersma D A 1995 Phase-locked heterodyne-detected stimulated photon echo. A unique tool to study solute-solvent interactions *Chem. Phys. Lett.* **238** 1

[188] Pshenichnikov M S, Duppen K and Wiersma D A 1995 Time-resolved femtosecond photon echo probes bimodal solvent dynamics *Phys. Rev. Lett.* **74** 674–7

[189] de Boeij W P, Pshenichnikov M S and Wiersma D A 1996 Mode suppression in the non-Markovian limit by time-gated stimulated photon echo *J. Chem. Phys.* **105** 2953–60

[190] Tokmakoff A, Zimdars D, Urdahl R S, Francis R S, Kwok A S and Fayer M D 1995 Infrared vibrational photon echo experiments in liquids and glasses *J. Phys. Chem.* **99** 13 310–20

[191] Tokmakoff A and Fayer M D 1995 Homogeneous vibrational dynamics and inhomogeneous broadening in glass-forming liquids: infrared photon echo experiments from room temperature to 10 K *J. Chem. Phys.* **103** 2810–26

[192] Tokmakoff A and Fayer M D 1995 Infrared photon echo experiments: exploring vibrational dynamics in liquids and glasses *Acc. Chem. Res.* **28** 439–45

[193] Tokmakoff A and Fayer M D 1995 Infrared photon echo experiments: exploring vibrational dynamics in liquids and glasses *Acc. Chem. Res.* **28** 437–45

[194] Rella C W, Rector K D, Kwok A, Hill J R, Schwettman H A, Dlott D D and Fayer M D 1996 Vibrational echo studies of myoglobin–CO *J. Phys. Chem.* **100** 15 620–29

[195] Rector K D, Rella C W, Hill J R, Kwok A S, Sligar S G, Chien E Y T, Dlott D D and Fayer M D 1997 Mutant and wild-type myoglobin-CO protein dynamics: vibrational echo experiments *J. Phys. Chem.* B **101** 1468–75

[196] Vanden Bout D and Berg M 1995 Ultrafast Raman echo experiments in liquids *J. Raman Spectrosc.* **26** 503—11

[197] Berg M and Vanden Bout D A 1997 Ultrafast Raman echo measurements of vibrational dephasing and the nature of solvent–solute interactions *Acc. Chem. Res.* **30** 65–71

[198] Tokmakoff A, Lang M J, Larsen D S and Fleming G R 1997 Intrinsic optical heterodyne detection of a two-dimensional fifth order Raman response *Chem. Phys. Lett.* **272** 48–54

[199] Tokmakoff A and Fleming G R 1997 Two-dimensional Raman spectroscopy of the intermolecular modes of liquid $CS_2$ *J. Chem. Phys.* **106** 2569–82

[200] Tokmakoff A, Lang M J, Larsen D S, Fleming G R, Chernyak V and Mukamel S 1997 Two-dimensional Raman spectroscopy of vibrational interactions in liquids *Phys. Rev. Lett.* **79** 2702–5

[201] Tokmakoff A, Lang M J, Jordanides X J and Fleming G R 1998 The intermolecular interaction mechanisms in liquid $CS_2$ at 295 and 165 K probed with two-dimensional Raman spectroscopy *Chem. Phys.* **233** 231–42

[202] Warren W S 1993 Coherent control: the dream is alive *Science* **259** 1581

[203] Nelson K A 1994 Coherent control: optics, molecules, and materials *Ultrafast Phenomena IX* ed P F Barbara, W H Knox, G A Mourou and A H Zewail (Berlin: Springer) pp 47

[204] Kohler B, Krause J L, Raksi F, Wilson K R, Yakovlev V V, Whitnell R M and Yan Y 1995 Controlling the future of matter *Acc. Chem. Res.* **28** 133–40

[205] Scherer N F, Carlson R J, Matro A, Du M, Ruggiero A J, Romero-Rochin V, Cina J A, Fleming G R and Rice S A 1991 Fluorescence-detected wave packet interferometry: time resolved molecular spectroscopy with sequences of femtosecond phase-locked pulses *J. Chem. Phys.* **95** 1487–511

[206] Scherer N F, Matro A, Ziegler L D, Du M, Carlson R J, Cina J A and Fleming G R 1992 Fluorescence-detected wave packet interferometry. II. Role of rotations and determination of the susceptibility *J. Chem. Phys.* **96** 4180

[207] Dugan M A, Tull J X and Warren W S 1997 High-resolution acousto-optic shaping of unamplified and amplified femtosecond laser pulses *J. Opt. Soc. Am.* B **14** 2348–58

[208] Wefers M M and Nelson K A 1995 Analysis of programmable ultrashort waveform generation using liquid-crystal spatial light modulators *J. Opt. Soc. Am.* B **12** 1343–62

[209] Bardeen C J, Wang Q and Shank C V 1998 Femtosecond chirped pulse excitation of vibrational wave packets in bacteriorhodopsin *J. Phys. Chem.* A **102** 2759–66

[210] Krause J L, Whitnell R M, Wilson K R, Yan Y and Mukamel S 1993 Optical control of molecular dynamics: molecular cannons, reflectrons, and wave-packet focusers *J. Chem. Phys.* **99** 6562–78

[211] Bardeen C J, Che J, Wilson K R, Yakovlev V V, Cong P, Kohler B, Krause J L and Messina M 1997 Quantum control of NaI photodissociation reaction product states by ultrafast tailored light pulses *J. Phys. Chem.* A **101** 3815–22

## Further Reading

Manz J and Wöste L (eds) 1995 *Femtosecond Chemistry* (New York, VCH)

Theory and experimental techniques for study of chemical reaction dynamics with ultrafast spectroscopic methods.

El-Sayed M A, Tanaka I and Molin Y (eds) 1995 *Ultrafast Processes in Chemistry and Photobiology* (Oxford: Blackwell)

Applications of ultrafast spectroscopy to chemical dynamics, especially in the condensed phase and in proteins.

Mukamel S 1995 *Principles of Nonlinear Optical Spectroscopy* (New York: Oxford University Press)

A comprehensive theoretical treatment of nonlinear spectroscopy, with an emphasis on theory applicable to ultrafast nonlinear spectroscopy.

Fleming G R 1986 *Chemical Applications of Ultrafast Spectroscopy* (New York: Oxford University Press)

Fundamentals of technique and theory for ultrafast experiments in chemistry, written before the titanium–sapphire revolution but still indispensable.

Kaiser W (ed) 1988 *Ultrashort Laser Pulses and Applications* (Berlin: Springer)

Applications of ultrafast laser techniques for studies in solids, optoelectronics, condensed phase, and in biological systems.

Rullière C (ed) 1998 *Femtosecond Laser Pulses* (Berlin: Springer)

A current description of femtosecond laser technology, with a discussion of ultrafast spectroscopic applications.

# B2.2
# Electron, ion and atom scattering

## M R Flannery

### B2.2.1  Introduction

This chapter deals with quantal and semiclassical theory of heavy-particle and electron–atom collisions. Basic and useful formulae for cross sections, rates and associated quantities are presented. A consistent description of the mathematics and vocabulary of scattering is provided. Topics covered include collisions, rate coefficients, quantal transition rates and cross sections, Born cross sections, quantal potential scattering, collisions between identical particles, quantal inelastic heavy-particle collisions, electron–atom inelastic collisions, semiclassical inelastic scattering and long-range interactions.

### B2.2.2  Collisions

#### B2.2.2.1  Differential and integral cross sections

A uniform monoenergetic beam of *test* or *projectile* particles A with number density $N_A$ and velocity $v_A$ is incident on a single *field* or *target* particle B of velocity $v_B$. The direction of the relative velocity $v = v_A - v_B$ is along the $Z$-axis of a Cartesian $XYZ$ frame of reference. The incident current (or intensity) is then $j_i = N_A v$, which is the number of test particles crossing unit area normal to the beam in unit time. The differential cross section for scattering of the test particles into unit solid angle $d\Omega = d(\cos \psi) \, d\phi$ about the direction $\hat{v}'(\psi, \phi)$ of the final relative motion is

$$\frac{d\sigma(v; \psi, \phi)}{d\Omega} = \frac{\text{Number of test particles scattered by one field particle into unit solid angle per unit time}}{\text{Current } j_i \text{ of incident beam}}.$$

The number of particles scattered per unit time by the field particle and detected per unit time is then

$$\frac{dN_d}{dt} = j_i \frac{d\sigma}{d\Omega} \, d\Omega = N_A v \frac{d\sigma}{d\Omega} \, d\Omega = N_A v \frac{d\sigma}{d\Omega} \frac{dA}{r^2}$$

where the detector, located along the scattered direction $\hat{v}'(\psi, \phi)$, subtends an angle $d\Omega = dA/r^2$ at the scattering centre and projects an area $dA = r^2 d(\cos \theta) \, d\phi$ normal to the scattered beam. Thus $[d\sigma/d\Omega] \, d\Omega$ is the cross-sectional area of the beam that is intercepted by one target particle and scattered into the solid angle $dA/r^2$ of a cone with axis along $\hat{v}'(\psi, \phi)$ and vertical angle $d\psi$. In classical terms (figure B2.2.1), the number of particles detected per second about direction $(\psi, \phi)$ is the number $N_A v(b \, db \, d\phi)$ of incident particles crossing the initial areal element $b \, db \, d\phi$ per second. Hence

$$\frac{d\sigma}{d\Omega} = \frac{b \, db}{d(\cos \psi)}.$$

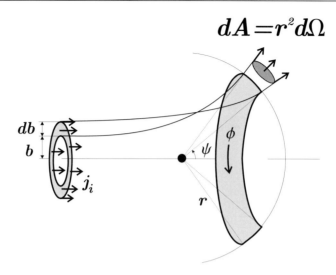

**Figure B2.2.1.** Scattering of a beam with current $j_i = N_A v$ particles per unit area incident between two cylinders of radii $b$ and $b + db$ by one particle at rest in the laboratory.

For an incident current flowing between two cylinders of radii $b$ and $b + db$, then $j_i 2\pi [d\sigma/d\Omega] d(\cos \psi)$ is the number of particles scattered per second between the two cones of semivertical angles $\psi, \psi + d\psi$ (figure B2.2.1).

The integral cross section for scattering over all directions is

$$\sigma(v) = \int_{-1}^{+1} d(\cos \psi) \int_0^{2\pi} \left[ \frac{d\sigma}{d\Omega}(v; \psi, \phi) \right] d\phi.$$

The integral cross section is therefore the effective area presented by each field particle B for scattering of the test particles A into all directions. The *probability* that the test particles are scattered into a given direction $\hat{v}'(\psi, \phi)$ is the ratio

$$\mathcal{P}(v; \psi, \phi) = \frac{d\sigma(v; \psi, \phi)}{d\Omega} / \sigma(v)$$

of the differential-to-integral cross sections.

### B2.2.2.2    *Collision rates, collision frequency and path length*

An electron or atomic beam of (projectile or test) particles A with density $N_A$ of particles per cm$^3$ travels with speed $v$ and energy $E$ through an infinitesimal thickness $dx$ of (target or field) gas particles B at rest with density $N_B$ particles per cm$^3$. The particles are scattered out of the beam by A–B collisions with integral cross section $\sigma(E)$ at a rate (cm$^{-3}$ s$^{-1}$) given by the total number of collisions between A and B particles

$$\frac{dN_A(E)}{dt} = -[N_A(E)v\sigma(E)]N_B$$
$$= -k(E)N_A(E)N_B$$
$$= -\nu_B(E)N_A(E)$$

in unit time and unit volume. The *microscopic rate coefficient* (cm$^3$ s$^{-1}$) for the scattering of one test particle by one field particle is $k(E) = v\sigma(E)$. The frequency (s$^{-1}$) of collision between one test particle and $N_B$ field particles (cm$^{-3}$) is $\nu_B = k(E)N_B$. Since $v = dx/dt$, the variation with $x$ of intensity $j_i$ of the attenuated beam is governed by

$$\frac{d j_i}{dx} = -[N_B \sigma(E)] j_i(x).$$

For constant density $N_B$ and speed $v$, the solution is

$$j_i(E, x) = j_i(E, 0) \exp[-N_B \sigma(E)x]$$
$$= j_i(E, 0) \exp(-x/\lambda)$$

where $\lambda \equiv 1/N_B \sigma(E) = v/\nu_B$ is the path length between collisions. Since $j_i = N_A v$, the density $N_A(E, x)$ obeys a similar equation. These equations describe the attenuation of a particle beam A travelling through a target gas B. For target gas particles with a distribution $f_B(v_B) dv_B$ in velocities $v_A$, the microscopic rate then becomes

$$k(E) = \int |v_A - v_B| \sigma(|v_A - v_B|) f_B(v_B) dv_B$$

where $E = \frac{1}{2} M_A v_A^2$ is the kinetic energy of the projectile beam. For an isothermal beam with an energy distribution $f_A(E) dE$ at temperature $T$, the macroscopic rate coefficient (cm$^3$ s$^{-1}$) or thermal rate constant is

$$k(T) = \int_0^\infty k(E) f_A(E) dE.$$

### B2.2.2.3  *Energy and angular momentum: centre of mass and relative velocity*

The velocity of the centre of mass (CM) of the projectile and target particles of respective masses $M_A$ and $M_B$ is

$$V = (M_A v_A + M_B v_B)/(M_A + M_B).$$

The relative velocity is

$$v = v_A - v_B.$$

The velocities of A and B in terms of $V$ and $v$ are

$$v_A = V + \frac{M_B}{M_A + M_B} v$$

$$v_B = V - \frac{M_A}{M_A + M_B} v.$$

The total kinetic energy then decomposes into the sum

$$E = \frac{1}{2} M_A v_A^2 + \frac{1}{2} M_B v_B^2 = \frac{1}{2} M V^2 + \frac{1}{2} M_{AB} v^2$$
$$= E_{CM}(A + B) + E_{rel}(AB)$$

of the energy $E_{CM} = \frac{1}{2} M V^2$ of the CM with mass $M = (M_A + M_B)$, and the energy $E_{rel} = \frac{1}{2} M_{AB} v^2$ of relative motion, where the reduced mass $M_{AB}$ is $M_A M_B/(M_A + M_B)$. Let $R$ be the position of the $CM$ relative to a fixed origin O and $r$ be the inter-particle separation. The total angular momentum about O similarly decomposes into the sum

$$L = R \times M V + r \times M_{AB} v$$
$$= L_{CM}(A + B) + L_{rel}(AB)$$

of angular momenta of the CM and of relative motion. For any collision in the absence of any external field, the energy $E_{CM}$ and angular momentum $L_{CM}$ of the CM are always conserved for all types of collision. The two species, A and B, may be electrons, ions, atoms or molecules, with or without any internal structure and may therefore possess internal energy and angular momentum which must be taken into account. For structured particles $E_{rel}$ and $L_{rel}$ can change in a collision.

### B2.2.2.4 Elastic scattering

$$A(\alpha) + B(\beta) \rightarrow A(\alpha) + B(\beta).$$

Elastic scattering involves no permanent changes in the internal structures (states $\alpha$ and $\beta$) of A and B. Both the energy $E_{rel}$ and angular momentum $L_{rel}(AB)$ of relative motion are therefore all conserved.

### B2.2.2.5 Inelastic scattering

$$A(\alpha) + B(\beta) \rightarrow A(\alpha') + B(\beta')$$

Inelastic scattering produces a permanent change in the internal energy and angular momentum state of one or both structured collision partners A and B, which retain their original identity after the collision. For inelastic $i \equiv (\alpha, \beta) \rightarrow f \equiv (\alpha', \beta')$ collisional transitions, the energy $E_{i,f} = \frac{1}{2} M_{AB} v_{i,f}^2$ of relative motion, before ($i$) and after ($f$) the collision satisfies the energy conservation condition,

$$E_i + \epsilon_\alpha(A) + \epsilon_\beta(B) = E_f + \epsilon_{\alpha'}(A) + \epsilon_{\beta'}(B)$$

where $\epsilon_{A,B}$ are the internal energies of A and B. The maximum amount of kinetic energy that can be transferred to internal energy is limited to the initial kinetic energy of relative motion, $E_{rel}(AB) = \frac{1}{2} M_{AB} v_i^2$. Excitation implies $\epsilon_i \equiv \epsilon_\alpha(A) + \epsilon_\beta(B) < \epsilon_{\alpha'}(A) + \epsilon_{\beta'}(B) \equiv \epsilon_f$, de-excitation (or superelastic) implies $\epsilon_f < \epsilon_i$ and energy resonance or excitation transfer implies $\epsilon_i = \epsilon_f$. Changes in angular momentum are limited by the conservation requirement that

$$L_{rel}(i) + L_\alpha(A) + L_\beta(B) = L_{rel}(f) + L_{\alpha'}(A) + L_{\beta'}(B)$$

where $L_{\alpha,\beta}$ denotes the internal angular momentum of each isolated species. Collisions, in which only angular momentum is transferred without any energy change, are called *quasi-elastic* collisions.

### B2.2.2.6 Reactive scattering

$$A + B \rightarrow C + D$$

Reactive scattering or a chemical reaction is characterized by a rearrangement of the component particles within the collision system, thereby resulting in a change of the physical and chemical identity of the original collision reactants A + B into different collision products C + D. Total mass is conserved. The reaction is exothermic when $E_{rel}(CD) > E_{rel}(AB)$ and is endothermic when $E_{rel}(CD) < E_{rel}(AB)$. A threshold energy is required for the endothermic reaction.

### B2.2.2.7 Centre-of-mass to laboratory cross section conversion

Theorists calculate cross sections in the CM frame while experimentalists usually measure cross sections in the laboratory frame of reference. The laboratory (Lab) system is the coordinate frame in which the target particle B is at rest before the collision i.e. $v_B = 0$. The centre of mass (CM) system (or barycentric system) is the coordinate frame in which the CM is at rest, i.e. $V = 0$. Since each scattering of projectile A into

$(\psi, \phi)$ is accompanied by a recoil of target B into $(\pi - \psi, \phi + \pi)$ in the CM frame, the cross sections for scattering of A and B are related by

$$\left\{\frac{d\sigma(\psi, \phi)}{d\Omega}\right\}_{CM} \equiv \left\{\frac{d\sigma_A(\psi, \phi)}{d\Omega}\right\}_{CM} = \left\{\frac{d\sigma_B(\pi - \psi, \phi + \pi)}{d\Omega}\right\}_{CM}$$

$$\left\{\frac{d\sigma_B(\psi, \phi)}{d\Omega}\right\}_{CM} = \left\{\frac{d\sigma(\pi - \psi, \phi - \pi)}{d\Omega}\right\}_{CM}.$$

In the *Lab* frame, the projectile is scattered by $\theta_A$ and the target, originally at rest, recoils through angle $\theta_B$. The number of particles scattered into each solid angle in each frame remains the same, the relative speed $v$ is now $v_A$ and $j_i = N_A v$ in each frame. Hence

$$\left\{\frac{d\sigma_A(\theta_A, \phi)}{d\Omega_A}\right\}_{Lab} = \left\{\frac{d\Omega}{d\Omega_A}\right\}\left[\frac{d\sigma(\psi, \phi)}{d\Omega}\right]_{CM}$$

$$\left\{\frac{d\sigma_B(\theta_B, \phi)}{d\Omega_B}\right\}_{Lab} = \left\{\frac{d\Omega}{d\Omega_B}\right\}\left[\frac{d\sigma(\psi, \phi)}{d\Omega}\right]_{CM}.$$

*(a) Two-body elastic scattering*

$$A(\alpha) + B(\beta) \rightarrow A(\alpha) + B(\beta).$$

The scattering and recoil angles $\theta_A$ and $\theta_B$ in the Lab frame are related to the CM scattering angle $\psi$ by

$$\tan \theta_A = \frac{\sin \psi}{1 + \gamma \cos \psi} \qquad \gamma = M_A/M_B$$

$$\theta_B = \tfrac{1}{2}(\pi - \psi) \qquad 0 \le \theta_B \le \tfrac{1}{2}\pi.$$

The elastic cross sections for scattering and recoil in the Lab-frame are related to the cross section in the CM-frame by

$$\left\{\frac{d\sigma_A(\theta_A, \phi)}{d\Omega_A}\right\}_{Lab} = \frac{(1 + \gamma^2 + 2\gamma \cos \psi)^{3/2}}{|1 + \gamma \cos \psi|}\left[\frac{d\sigma(\psi, \phi)}{d\Omega}\right]_{CM}$$

$$\left\{\frac{d\sigma_B(\theta_B, \phi)}{d\Omega_B}\right\}_{Lab} = |4 \sin \tfrac{1}{2}\psi|\left[\frac{d\sigma(\psi, \phi)}{d\Omega}\right]_{CM}.$$

*(b) Two-body inelastic or reactive scattering process $A + B \rightarrow C + D$*

The energies $E_i$ and $E_f$ of relative motion of A and B and of C and D, respectively satisfy $E_f/E_i = 1 - \epsilon_{fi}/E_i$, where $\epsilon_{fi} = \epsilon_f - \epsilon_i$ is the increase in internal energy. The scattering and recoil angles are

$$\tan \theta_C = \frac{\sin \psi}{(\gamma_C + \cos \psi)} \qquad \gamma_C = \left[\frac{M_A M_C}{M_B M_D}\right]^{1/2}\left(\frac{E_i}{E_f}\right)^{1/2}$$

$$\tan \theta_D = \frac{\sin \psi}{(|\gamma_D| - \cos \psi)} \qquad \gamma_D = -\left[\frac{M_A M_D}{M_B M_C}\right]^{1/2}\left(\frac{E_i}{E_f}\right)^{1/2}.$$

The Lab and CM cross sections are then related by

$$\left\{\frac{d\sigma_j(\theta_j, \phi)}{d\Omega_j}\right\}_{Lab} = \frac{[1 + 2\gamma_j \cos \psi + \gamma_j^2]^{3/2}}{|1 + \gamma_j \cos \psi|}\left[\frac{d\sigma(\psi, \phi)}{d\Omega}\right]_{CM}$$

where $j$ denotes C or D. The scattering of a beam from a stationary target is governed by these equations. A crossed beam experiment in which two beams intersect at an angle is not in the Lab-frame. In this case the measured quantities can be similarly transformed [1] to CM for comparison with theoretical calculations.

### B2.2.3    Macroscopic rate coefficients

*B2.2.3.1    Scattering rate*

A distribution $f_A(v_A)$ of $N_A(t)$ test particles (cm$^{-3}$) of species A in a beam collisionally interacts with a distribution $f_B(v_B)$ of $N_B(t)$ field particles of species B. Collisions with B will scatter A out of the beam at the loss rate (cm$^{-3}$ s$^{-1}$)

$$\frac{dN_A}{dt} = -kN_A(t)N_B(t) = -\nu_B N_A(t).$$

The macroscopic rate coefficient $k$ (cm$^3$ s$^{-1}$) for elastic collisions between the ensembles A and B is

$$k \ (\text{cm}^3 \ \text{s}^{-1}) = \int f_A(v_A)\,dv_A \int [v\sigma(v)]f_B(v_B)\,dv_B$$

in terms of the integral cross section $\sigma(v)$ for A–B elastic scattering at relative speed $v = |v_A - v_B|$. The microscopic rate coefficient is $v\sigma(v)$. The frequency $\nu_B$ (s$^{-1}$) of collision between one test particle A with $N_B$ field particles is $kN_B$.

   The rate coefficient for elastic scattering between two species with non-isothermal Maxwellian distributions is then

$$k \ (\text{cm}^3 \ \text{s}^{-1}) = \tilde{v}_{AB} \int_0^\infty \sigma(\tilde{\epsilon}_{rel})\tilde{\epsilon}_{rel}\exp(-\tilde{\epsilon}_{rel})\,d\tilde{\epsilon}_{rel}$$

where

$$\tilde{v}_{AB} = \left[\frac{8k_B}{\pi}\left(\frac{T_A}{M_A} + \frac{T_B}{M_B}\right)\right]^{1/2}$$

and

$$\tilde{\epsilon}_{rel} = \frac{1}{2}\frac{M_A M_B v^2}{k_B(M_A T_B + M_B T_A)}.$$

For isothermal distributions $T_A = T_B = T$, the rate is

$$k(T) = \langle v_{AB}\rangle \int_0^\infty \sigma(\epsilon)\epsilon\exp(-\epsilon)\,d\epsilon \qquad (\text{cm}^3 \ \text{s}^{-1})$$

where $\epsilon = \frac{1}{2}M_{AB}v^2/k_B T$ and $\langle v_{AB}\rangle = (8k_B T/\pi M_{AB})^{1/2}$. The rate of collisions of electrons A at temperature $T_e$ with a gas of heavy-particles B at temperature $T_B$ is

$$k(T_e) = \langle v_e\rangle \int_0^\infty \sigma(\epsilon_e)\epsilon_e\exp(-\epsilon_e)\,d\epsilon_e \qquad (\text{cm}^3 \ \text{s}^{-1})$$

where $\epsilon_e = \frac{1}{2}m_e v^2/k_B T_e$ and $\langle v_e\rangle = (8k_B T_e/\pi m_e)^{1/2}$.

*B2.2.3.2    Energy transfer rate*

Each of the species A transfers energy $\mathcal{E}_{AB}$ to each species B. The amount of energy transferred per unit volume in unit time from ensemble A to ensemble B is

$$\frac{d}{dt}[N_A\langle\mathcal{E}_{AB}\rangle] = -k_E N_A(t)N_B(t) = -\nu_{EB} N_A(t)$$

where the macroscopic rate coefficient $k_E$ (energy cm$^3$ s$^{-1}$) for the averaged energy loss $\langle\mathcal{E}_{AB}\rangle$ is

$$k_E = \int f_A(v_A)\,dv_A \int f_B(v_B)\,dv_B \int \mathcal{E}_{AB}(v_A, v_B; \psi, \phi)v\left(\frac{d\sigma}{d\Omega}\right)d\Omega.$$

The amount of energy lost in unit time, the energy-loss frequency, is $\nu_{EB} = k_E N_B(t)$. The energy-loss rate coefficient for two-temperature Maxwellian distributions is

$$k_E(T_A, T_B) = \frac{2M_A M_B}{(M_A + M_B)^2} k_B(T_A - T_B)\tilde{v}_{AB} \int_0^\infty \sigma_D(\tilde{\epsilon}_{rel})(\tilde{\epsilon}_{rel})^2 \exp(-\tilde{\epsilon}_{rel}) \, d\tilde{\epsilon}_{rel}$$

where $\sigma_D(\tilde{\epsilon}_{rel})$ is the momentum transfer cross section at reduced energy $\tilde{\epsilon}_{rel}$. For isothermal distributions, $T_A = T_B$ and the energy rate coefficient $k_E$ of course then vanishes.

### B2.2.3.3 Transport cross sections and collision integrals

Transport cross sections are defined for integer $n = 1, 2, 3\ldots$, as

$$\sigma^{(n)}(E) = 2\pi \left[1 - \frac{1 + (-1)^n}{2(n+1)}\right]^{-1} \int_{-1}^{+1} [1 - \cos^n \theta] \frac{d\sigma}{d\Omega}(\cos\theta).$$

The diffusion and viscosity cross sections are given by the transport cross sections $\sigma^{(1)}$ and $\frac{2}{3}\sigma^{(2)}$, respectively. Collision integrals are defined for integer $s = 0, 1, 2, \ldots$, as

$$\Omega^{(n,s)}(T) = [(s+1)!(k_B T)^{s+2}]^{-1} \int_0^\infty \sigma^{(n)}(E) E^{s+1} \exp(-E/k_B T) \, dE$$

$$= [(s+1)!]^{-1} \int_0^\infty \sigma^{(n)}(\epsilon)\epsilon^{s+1} \exp -\epsilon \, d\epsilon$$

where $\epsilon = \frac{1}{2}M_{AB}v^2/k_B T$. The external factors are chosen so that these expressions for $\sigma^{(n)}$ and $\Omega^{(n,s)}$ reduce to $\pi d^2$ for classical rigid spheres of diameter $d$. The rate coefficient $k$ (cm$^3$ s$^{-1}$) for scattering can then be expressed, in terms of the collision integral, as equal to $\tilde{v}_{AB}\Omega^{(0,0)}$. The amount of energy lost per cm$^3$ per second by collision can be expressed in terms of $\Omega^{(1,1)}$. Tables of transport cross sections and collision integrals for $(n, 6, 4)$ ion–neutral interactions are available [2, 3].

### (c) Chapman–Enskog mobility formula

When ions move under equilibrium conditions in a gas and an external electric field, the energy gained from the electric field $E$ between collisions is lost to the gas upon collision so that the ions move with a constant drift speed $v_d = KE$. The mobility $K$ of ions of charge $e$ in a gas of density $N$ is given in terms of the collision integral by the Chapman–Enskog formula [2]

$$K = \frac{3e}{16N}\left(\frac{\pi}{2Mk_B T}\right)^{1/2} [\Omega^{(1,1)}(T)]^{-1}.$$

## B2.2.4 Quantal transition rates and cross sections

### B2.2.4.1 Microscopic rate of transitions

In the general elastic/inelastic collision process

$$A(\alpha) + B(\beta) \rightarrow A(\alpha') + B(\beta')$$

the external scattering or deflection of a beam of projectile particles A (electrons, ions, atoms) by target particles B (atoms, molecules) is accompanied by transitions (electronic, vibrational, rotational) within the internal structure of either or both collision partners. For a beam with incident momentum $p_i = \hbar k_i$ in the range $(p_i, p_i + dp_i)$ or directed energy $E_i \equiv (E_i, \hat{p}_i)$ in the range $(E_i, E_i + dE_i)$, the translational states representing the A–B relative or external motion undergo free–free transitions $(E_i, E_i + dE_i) \rightarrow (E_f, E_f + dE_f)$ within the translational continuum, while the structured particles undergo bound–bound (excitation, de-excitation, excitation transfer) or bound–free (ionization, dissociation) transitions $i \equiv (\alpha, \beta) \rightarrow f \equiv (\alpha', \beta')$ in their internal electronic, vibrational or rotational structure. The transition frequency $(s^{-1})$ for this collision is

$$\frac{dW_{if}}{d\hat{p}_f}(s^{-1}) = \frac{2\pi}{\hbar} \frac{1}{g_i} \sum_{i,f} |V_{fi}|^2 \rho_f(E_f)$$

which is an average over the $g_i$ initial degenerate internal states $i$ and a sum over all $g_f$ final degenerate internal states $f$ of the isolated systems A and B. It is therefore the probability per unit time for scattering from a specified $E_i$—(external) continuum state into unit solid angle $d\hat{p}_f$ accompanied by a transition from any one of the $g_i$ initial states $(\alpha, \beta)$ to all final internal states $(\alpha', \beta')$ of degeneracy $g_f$ and to all final translational states $\rho_f(E_f) dE_f$ of relative motion consistent with energy conservation. The double summation $\sum_{i,f}$ is over the $g_i$ initial and $g_f$ final internal states of A and B with total energy $\epsilon_i$ and $\epsilon_f$ respectively.

 *Check*: The dimension of $[|V_{ij}|^2 \rho]$ is $E$, $[\hbar] = Et$ so that $dW_{if}/d\hat{p}_f$ indeed has the correct dimension of $t^{-1}$.

*(a) Interaction matrix element*

The matrix element

$$V_{fi} = \langle N_f \Phi_f | V(r_A, r_B, R) | N_i \Psi_i^+ \rangle_{r,R} = V_{if}^*$$

is an integration over the internal coordinates $r \equiv r_A, r_B$ of the electrons of A and B and over the channel vector $R$ for A–B relative motion. The matrix element of the mutual electrostatic interaction $V(r_A, r_B, R)$ couples the eigenfunction $N_i \Psi_i^+(R, r_A, r_B)$ of $[\hat{H}_{rel} + \hat{H}_{int} + V]$ for the complete collision system for all $R$ to the final $R \rightarrow \infty$ asymptotic state $N_f \Phi_f(R, r_A, r_B)$, which is an eigenfunction only of the unperturbed Hamiltonian $[\hat{H}_{rel} + \hat{H}_{int}]$. The wavefunction

$$\Psi_i^+(R, r_A, r_B) = \sum_j \Phi_j^+(R) \psi_\alpha(r_A) \phi_\beta(r_B)$$

$$\equiv \sum_j \Phi_j^+(R) \psi_j^{int}(r)$$

for the full collision system with Hamiltonian $\hat{H}_{rel} + \hat{H}_{int} + V$ tends at asymptotic $R$ to

$$\Psi_i^+ \sim \sum_j \left[ e^{ik_j \cdot R} \delta_{ij} + f_{ij}(\theta, \phi) \frac{e^{ik_j R}}{R} \right] \psi_j^{int}(r)$$

which represents an incoming plane wave of unit amplitude in the incident elastic channel $i$ and an outgoing spherical waves of amplitude $f_{ij}$ in all channels $j$, including $i$. The Kronecker symbol means $\delta_{ij} = 1, i = j$ and $\delta_{ij} = 0, i \neq j$. The final state at infinite separation $R$ is

$$\Phi_f(R, r_A, r_B) = e^{ik_f \cdot R} \psi_{\alpha'}(r_A) \phi_{\beta'}(r_B) \equiv e^{ik_f \cdot R} \psi_f^{int}(r)$$

which is an eigenfunction only of $\hat{H}_{rel} + \hat{H}_{int}$. The plane wave of unit amplitude describes the external relative motion with Hamiltonian $\hat{H}_{rel}$ and $\phi_{\alpha'}(r_A) \psi_{\beta'}(r_B)$ describes the internal, isolated, normalized atomic eigenstates of A and B with internal Hamiltonian $\hat{H}_{int}$. The factors $N_{i,f}$ provide the possibility of having translational (scattering) states with arbitrary amplitudes which are not necessarily unity.

*(b) Transition operator*

The interaction matrix element can also be written as

$$V_{fi} = \langle N_f \Phi_f | \hat{T} | N_i \Phi_i \rangle$$

where the transition operator, $\hat{T}$, is defined by $\hat{T}\Phi = V\Psi$. The transition operator $\hat{T}$ therefore couples states which are eigenfunctions of the same unperturbed Hamiltonian $\hat{H}_{rel} + \hat{H}_{int}$, in contrast to $V$ which couples states $\Psi_i^+$ and $\Phi_f$ belonging to different Hamiltonians.

### B2.2.4.2  Detailed balance between rates

The frequency (number per second) of $i \rightarrow f$ transitions from all $g_i$ degenerate initial internal states and from the $\rho_i \, dE_i$ initial external translational states is equal to the reverse frequency from the $g_f$ degenerate final internal states and the $\rho_f \, dE_f$ final external translational states. The *detailed balance relation* between the forward and reverse frequencies is therefore

$$[g_i \rho_i \, dE_i \, d\hat{\boldsymbol{p}}_i] \left( \frac{dW_{if}}{d\hat{\boldsymbol{p}}_f} \right) d\hat{\boldsymbol{p}}_f = [g_f \rho_f \, dE_f \, d\hat{\boldsymbol{p}}_f] \left( \frac{dW_{fi}}{d\hat{\boldsymbol{p}}_i} \right) d\hat{\boldsymbol{p}}_i$$

since $V_{if} = V_{fi}^*$. From energy conservation $\epsilon_i + E_i = \epsilon_f + E_f$, then $dE_i = dE_f$. The differential frequencies

$$\frac{dR_{if}}{d\hat{\boldsymbol{p}}_f} \equiv g_i \rho_i \frac{dW_{if}}{d\hat{\boldsymbol{p}}_f} = g_f \rho_f \frac{dW_{fi}}{d\hat{\boldsymbol{p}}_i} \equiv \frac{dR_{fi}}{d\hat{\boldsymbol{p}}_i}$$

for the forward and reverse transitions, $i \rightleftharpoons f$, are therefore equal.

### B2.2.4.3  Energy density of continuum states

The continuum wavefunctions $\phi_p(\boldsymbol{R})$ for the states of the $A$–$B$ relative motion satisfy the orthonormality condition

$$\int \rho(E) \, dE \int \phi_p(\boldsymbol{R}) \phi_{p'}^*(\boldsymbol{R}) \, d\boldsymbol{R} = 1.$$

The number of translational states per unit volume $d\boldsymbol{R}$ with directed energies $\boldsymbol{E} \equiv (E, \hat{\boldsymbol{p}})$ in the range $[\boldsymbol{E}, \boldsymbol{E} + d\boldsymbol{E}]$ is $\rho(E) \, dE$. This orthonormality condition for continuum states is analogous to the condition $\sum_j |\langle \phi_j | \phi_i \rangle|^2 = 1$ for bound states. For plane waves, $\phi_p(\boldsymbol{R}) = N \exp(i\boldsymbol{p} \cdot \boldsymbol{R}/\hbar)$, then

$$\langle \phi_{p'} | \phi_p \rangle = |N|^2 (2\pi\hbar)^3 \delta(\boldsymbol{p} - \boldsymbol{p}') = |N|^2 (2\pi)^3 \delta(\boldsymbol{k} - \boldsymbol{k}')$$

$$= |N|^2 \frac{(2\pi\hbar)^3}{mp} \delta(E - E') = \frac{1}{\rho(E)} \delta(E - E').$$

Note, irrespective of the method chosen to normalize the wavefunctions, that

$$|N|^2 \rho(E) = \frac{mp}{(2\pi\hbar)^3}$$

always. The amplitude $|N|$ does, however, depend on the choice of normalization.

(1)  For momentum normalized states, $\langle \phi_{p'} | \phi_p \rangle = \delta(\boldsymbol{p} - \boldsymbol{p}')$, $|N| = (2\pi\hbar)^{-3/2}$ and the density of states $\rho(E) = mp$.

(2)  For wavevector normalized states, $\langle \phi_{p'} | \phi_p \rangle = \delta(\boldsymbol{k} - \boldsymbol{k}')$, $|N| = (2\pi)^{-3/2}$ and the density of states $\rho(E) = (mp/\hbar^3)$.

(3)   For energy-normalized states, $\langle \phi_{p'} | \phi_p \rangle = \delta(E - E')$, $\rho(E) = 1$ and $|N| = (mp/h^3)^{1/2}$.

(4)   For waves with unit amplitude, $|N| = 1$ and $\rho(E) = (mp/h^3)$.

Note that $\langle \phi_{p'} | \phi_p \rangle \rho(E) \, dE$ is dimensionless for all cases and yields unity for a single particle when integrated over all $E$. The number of states in the phase-space element $dE \, dR$ is

$$dn = |N|^2 \rho(E) \, dE \, dR = dp \, dR/(2\pi\hbar)^3$$

i.e. each translational state occupies a cell of phase volume $(2\pi\hbar)^3$. The density of states in the interval $[E, E + dE]$ is $\rho(E) = 4\pi \rho(E)$. The number of translational states per unit volume with energy in the scalar range $[E, E + dE]$ is

$$|N|^2 \rho(E) \, dE = \frac{2}{\sqrt{\pi}} \frac{(2\pi m)^{3/2}}{h^3} E^{1/2} \, dE.$$

*Check.* The number of free particles with all momenta $p$ in equilibrium with a gas bath of volume $\mathcal{V}$ at temperature $T$ is the translational partition function $\mathcal{Z}_t$. Since the fraction of particles with energy $E$ is $\exp(-E/k_B T)/\mathcal{Z}_t$, the *Maxwell distribution*

$$\begin{aligned}
f_M(E) \, dE &= \frac{|N|^2 \mathcal{V} \rho(E) \, dE}{\mathcal{Z}_t} \exp -(E/k_B T) \\
&= \frac{2}{\sqrt{\pi}} (E/k_B T)^{1/2} \exp(-E/k_B T) \, d(E/k_B T)
\end{aligned}$$

is then recovered.

### Current

Current is the number of particles crossing unit area in unit time. The current in a beam with directed energy $E$ within the range $(E, E + dE)$ is

$$j \, dE = v|N|^2 \rho(E) \, dE = (p^2/h^3) \, dE.$$

The current per unit $dE$ is the *current density* $j = (p^2/h^3)$. The *quantal expression for current*

$$\boldsymbol{J} \equiv \frac{\hbar}{2mi} [\phi_p^* \nabla \phi_p - \phi_p \nabla \phi_p^*]$$

when applied to the plane wave $\phi_p = N \exp(i \boldsymbol{p} \cdot \boldsymbol{R}/\hbar)$, gives $\boldsymbol{J} = |N|^2 \boldsymbol{v}$. The current in a $(E, E + dE)$-beam of plane waves is then $J[\rho(E) \, dE]$ so that the current density is $j(E) = J\rho(E) = |N|^2 v \rho(E)$, as before.

### B2.2.4.4   *Inelastic cross sections*

The differential cross section $d\sigma_{if}/d\hat{\boldsymbol{p}}_f$ for $i \rightarrow f$ transitions from any one of the $g_i$ initial states is defined as $[dR_{if}/d\hat{\boldsymbol{p}}_f]/g_i j_i$, the transition frequency per unit incident current. Since current is the number of particles crossing unit area in unit time, the cross section is therefore the effective area presented by the target towards $i \rightarrow f$ internal transitions in the internal structures of the collision partners which are scattered into unit solid angle $d\hat{\boldsymbol{p}}_f$ about direction $\hat{\boldsymbol{p}}_f$ in the CM-frame.

**Table B2.2.1.** Continuum wavefunction normalization, density of states and cross section factors.

| Type | $\langle \Phi_{k'}|\Phi_k\rangle$ | $N$ | $N_\ell$ | $\rho(E)$ | $\gamma_{if}$ |
|---|---|---|---|---|---|
| Unit amplitude | $\begin{cases} (2\pi)^3\delta(k-k') \\ (2\pi\hbar)^3\delta(p-p') \end{cases}$ | $1$ | $\dfrac{4\pi}{k}$ | $mp/h^3$ | $\dfrac{v_f}{v_i}\left(\dfrac{1}{4\pi}\right)^2\left(\dfrac{2m_f}{\hbar^2}\right)^2$ |
| Wavenumber | $\delta(k-k')$ | $(2\pi)^{-3/2}$ | $\left(\dfrac{2}{\pi}\right)^{1/2}\dfrac{1}{k}$ | $mp/\hbar^3$ | $\dfrac{v_f}{v_i}(2\pi^2)^2\left(\dfrac{2m_f}{\hbar^2}\right)^2$ |
| Momentum | $\delta(p-p')$ | $(2\pi\hbar)^{-3/2}$ | $\left(\dfrac{2}{\pi\hbar}\right)^{1/2}\dfrac{1}{k}$ | $mp$ | $\dfrac{v_f}{v_i}(2\pi^2\hbar^3)^2\left(\dfrac{2m_f}{\hbar^2}\right)^2$ |
| Directed energy | $\delta(E-E')$ | $(mp/h^3)^{1/2}$ | $\left(\dfrac{2m}{\hbar^2}\dfrac{1}{\pi k}\right)^{1/2}$ | $1$ | $\dfrac{(2\pi)^4}{k_i^2}$ |

*(a) Basic expression for cross section*

The differential cross section for

$$A + B \rightarrow C + D$$

collisions is therefore defined as

$$\frac{d\sigma_{if}}{d\hat{p}_f} = \frac{1}{g_i j_i}\frac{dR_{if}}{d\hat{p}_f} = \frac{2\pi}{\hbar}\left(\frac{\rho_i\rho_f}{j_i}\right)\frac{1}{g_i}\sum_{i,f}|\langle N_f\Phi_f|V(r_A, r_B, R)|N_i\Psi_i^+\rangle|^2$$

which is an average over the $g_i$ initial internal degenerate states and a sum over the $g_f$ final degenerate states. Since $j_i = |N_i|^2 v_i\rho_i = p_i^2/h^3$, an alternative form [4] for the cross section is

$$\frac{d\sigma_{if}}{d\hat{p}_f} = \frac{2\pi}{\hbar v_i}\left(\frac{\rho_f}{g_i}\right)\sum_{i,f}|\langle N_f\Phi_f|V(r_A, r_B, R)|\Psi_i^+\rangle|^2.$$

*B2.2.4.5 Detailed balance between cross sections*

When cast in terms of cross sections, the detailed balance relation in section B2.2.4.2 is

$$g_i j_i(E_i)\frac{d\sigma_{if}}{d\hat{p}_f} = g_f j_f(E_f)\frac{d\sigma_{fi}}{d\hat{p}_i}.$$

The basic relationship satisfied by the differential cross sections for the forward and reverse $i \rightleftharpoons f$ transitions is

$$g_i p_i^2\frac{d\sigma_{if}(E_i)}{d\hat{p}_f} = g_f p_f^2\frac{d\sigma_{fi}(E_f)}{d\hat{p}_i}.$$

*(a) Collision strengths*

Collision strengths $\Omega_{if}$ exploit this detailed balance relation by being defined as

$$\Omega_{if} = g_i p_i^2\sigma_{if}(E_i) = g_f p_f^2\sigma_{fi}(E_f) = \Omega_{fi}.$$

They are therefore symmetrical in $i$ and $f$.

*(b) Reactive processes*

For any reactive process

$$A + B \rightleftharpoons C + D$$

the detailed balance relations involving differential/integral cross sections are

$$g_A g_B p_{AB}^2 \left[ \frac{d\sigma_{if}(E_{AB})}{d\hat{p}_{CD}} \right] = g_C g_D p_{CD}^2 \left[ \frac{d\sigma_{fi}(E_{CD})}{d\hat{p}_{AB}} \right]$$

$$g_A g_B p_{AB}^2 \sigma_{if}(E_{AB}) = g_C g_D p_{CD}^2 \sigma_{fi}(E_{CD})$$

where $p_{JK}^2 = 2M_{JK} E_{JK}$, in terms of the reduced mass $M_{JK}$ and relative energy $E_{JK}$ of species J and K.

### B2.2.4.6  *Examples of detailed balance*

*(a) Excitation–de-excitation*

$$e^-(E_i) + A_i \rightleftharpoons e^-(E_f) + A_f$$

$$\sigma_{if}(E_i) = \left( \frac{g_f}{g_i} \right) \left( \frac{E_f}{E_i} \right) \sigma_{fi}(E_f).$$

With energy conservation, $E_i = E_f + (\epsilon_f - \epsilon_i) \equiv E_f + \epsilon_{fi}$ the cross section for superelastic collisions $(E_f > E_i)$ can be obtained from $\sigma_{if}$ at energy $E_i$ *via* the relation

$$\sigma_{fi}(E_i - \epsilon_{fi}) = \left( 1 - \frac{\epsilon_{fi}}{E_i} \right)^{-1} \left( \frac{g_i}{g_f} \right) \sigma_{if}(E_i).$$

*(b) Dissociative recombination/associative ionization*

Dissociative recombination and associative ionization are represented by the forward and backward directions of

$$e^- + AB^+ \rightleftharpoons A + B^*.$$

The respective cross sections $\sigma_{DR}$ and $\sigma_{AI}$ are related by

$$\sigma_{DR}(E_e) = \left( \frac{g_A g_B}{g_e g_{AB^+}} \right) \left( \frac{k_{AB}^2}{k_e^2} \right) \sigma_{AI}(E_{AB})$$

where the statistical weight of each species $j$ involved is denoted by $g_j$.

*(c) Radiative recombination/photoionization*

Similarly, the cross sections for radiative recombination (RR) and for photoionization (PI), the forward and reverse directions of

$$A^+ + e^- \rightleftharpoons h\nu + A(n\ell)$$

are related by

$$\sigma_{RR}(E_e) = \left( \frac{g_A g_\nu}{g_e g_{A^+}} \right) \left( \frac{p_\nu^2}{p_e^2} \right) \sigma_{PI}(h\nu).$$

The photon statistical weight is $g_\nu = 2$, corresponding to the two directions of polarization of the photon. The photon energy $E$ is related to its momentum $p_\nu$ and wavenumber $k_\nu$ and to the ionization energy $I_{n\ell}$ of the atom $A(n\ell)$ by

$$E = h\nu = p_\nu c = \hbar k_\nu c = I_{n\ell} + E_e$$

where $c$ is the speed of light. This ratio is

$$\frac{p_\nu^2}{p_e^2} = \frac{(h\nu)^2}{(2E_e m_e c^2)} = \frac{\alpha^2 (h\nu)^2}{2 E_e \varepsilon_0}.$$

### B2.2.4.7 Four useful expressions for the cross section

The final expressions to be used for the calculation of cross sections depend on the particular choice of normalization of the continuum wavefunction for relative motion. Since it is often a vexing problem and is a continued source of confusion and error in the literature, these final expressions are worked out below. The external relative-motion part of the system wavefunction $N\Psi^+ \equiv \Psi_p(R, r_A, r_B)$ is $N\phi_p(R)$. Since $|N|^2 \rho = mp/h^3$, the density $\rho(E)$ of continuum states therefore depends on the choice of normalization factor $N$ adopted for the continuum wave. For future reference, the amplitude $N$ and the energy densities $\rho(E)$ associated with four common methods adopted for normalization of continuum waves are summarized in table B2.2.1. Also included is the amplitude $N_\ell$ of the corresponding radial partial wave

$$R_{\epsilon\ell}(r) \sim \frac{N_\ell}{r} \sin\left(kr - \frac{1}{2}\ell\pi + \eta_\ell\right)$$

of section B2.2.6.1. The external multiplicative factors $\gamma_{if} = (2\pi/\hbar)\left(\rho_i \rho_j / j_i\right)$ in the basic formula in section B2.2.4.4 for the cross section are also summarized in table B2.2.1 for the various normalization schemes. The reduced masses before and after the collision are $m_i = M_A M_B/(M_A + M_B)$ and $m_f = M_C M_D/(M_C + M_D)$, respectively.

### (a) Energy-normalized initial and final states

The wavefunctions

$$\chi_p = \rho^{1/2} N\Psi = (mp/2\pi\hbar^3)^{1/2} \Psi_p(R, r_A, r_B)$$

are energy-normalized according to

$$\langle \chi_p | \chi_{p'} \rangle = \delta(E - E').$$

The basic formula in section B2.2.4.4 with $j_i = p_i^2/(2\pi\hbar)^3$ yields

$$\frac{d\sigma_{if}}{d\hat{p}_f} = \frac{\pi}{k_i^2}\frac{1}{g_i}|T_{if}|^2 = \left(\frac{h^2}{8\pi m_i E_i}\right)\frac{1}{g_i}|T_{if}|^2.$$

The transition probability is

$$P_{if} = |T_{if}|^2 = \sum_{i,f} \int |2\pi \langle \tilde{\chi}_f | V | \chi_i^+ \rangle|^2 \, d\hat{k}_i$$

the magnitude squared of the element $T_{if}$ of the transition matrix $T$ between $\chi_i^+$ and $\tilde{\chi}_f$, the two energy-normalized eigenfunctions of $\hat{H}_{\text{rel}} + \hat{H}_{\text{int}} + V$ and $\hat{H}_{\text{rel}} + \hat{H}_{\text{int}}$, respectively. The detailed balance relation in this case is simply

$$|T_{if}|^2 = |T_{fi}|^2$$

thereby verifying that $|T_{if}|^2$ is indeed the $i \to f$ transition probability for transitions between all $g_i$ initial and $g_f$ final states. This type of normalization is convenient for *rearrangement collisions* such as *dissociative*, *radiative* and *dielectronic recombination*.

*(b) Unit amplitude initial and final states*

Here the initial and final wavefunctions with unit amplitude are $\Psi_i^+$ and $\Phi_f$. They are each normalized according to

$$\langle \Psi_{p'} | \Psi_p \rangle = (2\pi)^3 \delta(p - p').$$

The basic expression in section B2.2.4.4 with $j_i = |N_i|^2 v_i \rho_i$ and $|N_f|^2 \rho_f = m_f p_f / h^3$ reduces to

$$\frac{d\sigma_{if}}{d\hat{p}_f} = \frac{v_f}{v_i} |f_{if}(\theta, \varphi)|^2$$

where the scattering amplitude is

$$f_{if} = -\frac{1}{4\pi}(2m_f/\hbar^2)\langle \Phi_f | V | \Psi_i^+ \rangle$$

which couples scattering states $\Psi_i^+$ and $\Phi_f$ of unit amplitude. This expression is also applicable for rearrangement collisions $A + B \rightarrow C + D$ by including the reduced mass $m_f = M_C M_D / (M_C + M_D)$ of the reacted species after the collision. The integral cross section consistent with the above scattering amplitude is,

$$\sigma_{if}(E) = \frac{v_f}{v_i} \int_0^\pi d(\cos\theta) \int_0^{2\pi} |f_{if}(\theta, \varphi)|^2 \, d\varphi$$

at relative energy $E = k_i^2 \hbar^2 / 2M_{AB}$. The scattering amplitude consistent with the common use of

$$\sigma_{if}(E) = \frac{k_f}{k_i} \int_0^\pi d(\cos\theta) \int_0^{2\pi} |\tilde{f}_{if}(\theta, \varphi)|^2 \, d\varphi$$

for rearrangement collisions is

$$\tilde{f}_{if} = -\frac{1}{4\pi}(2\sqrt{m_i m_f}/\hbar^2)\langle \Phi_f | V | \Psi_i^+ \rangle.$$

Both conventions are identical only for direct collisions $A(\alpha) + B(\beta) \rightarrow A(\alpha') + B(\beta')$. This normalization is customary [5] for *elastic* and *inelastic* scattering processes.

For symmetrical potentials $V(r)$ scattering is confined to a plane and $f_{ij}$ depends only on scattering angle $\theta = \hat{k}_i \cdot \hat{k}_f$.

*(c) Momentum-normalized initial and final states*

Here the initial and final wavefunctions $\xi_p = (2\pi\hbar)^{-3/2}\Psi_i^+$ and $\tilde{\xi}_{p'} = (2\pi\hbar)^{-3/2}\Phi_f$ are normalized according to

$$\langle \xi_{p'} | \xi_p \rangle = \delta(p - p').$$

The cross section B2.2.4.4 is then

$$\frac{d\sigma_{if}}{d\hat{p}_f} = \frac{v_f}{v_i} |f_{if}(\theta, \varphi)|^2$$

where the scattering amplitude [6–8] is now

$$f_{if} = -(2\pi^2\hbar^3)(2m_f/\hbar^2)|\langle \tilde{\xi}_f | V | \xi_i^+ \rangle|.$$

*(d) Energy-normalized final and unit amplitude initial states*

Here the basic formula B2.2.4.4 yields

$$\frac{d\sigma_{if}}{d\hat{p}_f} = \frac{2\pi}{\hbar v_i} |\langle \tilde{\chi}_f | V | \Phi_i^+ \rangle|^2$$

which couples the initial scattering state $\Psi_i^+$ of unit amplitude with the energy-normalized final state $\tilde{\chi}_f = \rho_f^{1/2} N_f \Phi_f$. This normalization is customary for photoionization problems.

## B2.2.5   Born cross sections

Here an (undistorted) plane wave of unit amplitude is adopted for the channel wavefunction $\Phi_j(R)$ in the wavefunction $\Psi_i^+ = \Sigma_j \Phi_j(R) \psi_j^{\text{int}}(r)$ for the complete system. The differential cross section for elastic $(i = f)$ or inelastic scattering $(i \neq f)$ into $\hat{k}_f(\theta, \phi)$ is then

$$\frac{d\sigma_{if}}{d\Omega} = \frac{v_f}{v_i} |f_{if}(\theta)|^2.$$

The Born scattering amplitude for A–B collisions is

$$f_{if}^{(B)}(K) = -\frac{1}{4\pi} \frac{2M_{AB}}{\hbar^2} \int V_{fi}(R) \exp(iK \cdot R) \, dR$$

which is the Fourier transform of the interaction potential

$$V_{fi}(R) = \langle \psi_f^{\text{int}}(r) | V(r, R) | \psi_i^{\text{int}}(r) \rangle$$

which couples the initial and final isolated states $\psi_j^{\text{int}}(r) = \phi_j(r_A) \psi_j(r_B)$ of the atoms. The diagonal potential $V_{ii}(R)$ is the static interaction for elastic scattering. The Born scattering amplitude is a pure function only of the collisional momentum change

$$q = \hbar K = M_{AB}(v_i - v_f) = \hbar(k_i - k_f)$$

where $v$ is the A–B relative velocity. Since $K^2 = k_i^2 + k_f^2 - 2k_i k_f \cos\theta$, the *Born integral cross section* is

$$\sigma_{if}^B(k_i) = \frac{2\pi}{M_{AB}^2 v_i^2} \int_{q_-}^{q_+} |f_{if}^{(B)}(q)|^2 q \, dq$$

$$= \frac{2\pi}{(k_i a_0)^2} \int_{K_- a_0}^{K_+ a_0} |f_{if}^{(B)}(K)|^2 (K a_0) \, d(K a_0)$$

where $q_\pm = \hbar K_\pm = \hbar |k_i \pm k_f|$ are the maximum and minimum momentum changes consistent with energy conservation. For symmetric interactions $V_{fi}(R)$, then

$$f_{if}^B(K) = -\frac{2M_{AB}}{\hbar^2} \int V_{fi}(R) \frac{\sin KR}{KR} R^2 \, dR.$$

*B2.2.5.1   Fermi golden rules*

*Rule A.* The transition rate (probability per unit time) for a transition from state $\Phi_i$ of a quantum system to a number $\rho(E)\,\mathrm{d}E$ of continuum states $\Phi_E$ by an *external* perturbation $V$ is

$$w_{if} = \frac{2\pi}{\hbar}|\langle\Phi_E|V|\Phi_i\rangle|^2\rho_f(E) \equiv \frac{2\pi}{\hbar}|V_{i\epsilon}|^2\rho_f(E)$$

to first order in $V$. Since $|V_{i\epsilon}|^2\rho_f$ has the dimension of energy, $w_{if}$ has the dimension $\mathrm{t}^{-1}$.

   *Rule B.* When the direct coupling $V_{i\epsilon}$ from only the initial state to the continuum vanishes, but the coupling $V_{n\epsilon} \neq 0$ for $n \neq i$, the transition can then occur via the intermediate states $n$ at the rate

$$w_{if} = \frac{2\pi}{\hbar}\sum_n\left|\frac{V_{in}V_{n\epsilon}}{E-E_n}\right|^2\rho_f(E).$$

These rules, A and B (which are not exact) are useful for both scattering and radiative processes and are often referenced as Fermi's Rules 2 and 1, respectively.

*Scattering example*

The cross section for inelastic scattering of beam of particles by potential $V(r, R)$ is

$$\frac{\mathrm{d}\sigma_{if}}{\mathrm{d}\hat{p}_f} = \frac{w_{if}}{J_i}.$$

A plane-wave monoenergic beam, $\Phi_{i,f} = N_i, f\ \exp(\mathrm{i}p_{i,f}\cdot R/\hbar)\phi_{i,f}(r)$, has current $J_i = |N_i|^2 v_i$ and density determined from $|N_f|^2\rho_f(E) = M_{\mathrm{AB}}p_f/(2\pi\hbar)^3$. Hence

$$\frac{\mathrm{d}\sigma}{\mathrm{d}\hat{p}_f} = \frac{v_f}{v_i}\left|\frac{1}{4\pi}\frac{2M_{\mathrm{AB}}}{\hbar^2}\int V_{fi}(r)\,\mathrm{e}^{\mathrm{i}(p_i-p_f)\cdot r/\hbar}\,\mathrm{d}r\right|^2.$$

Since this agrees with the first Born differential cross section for (in)elastic scattering, Fermi's Rule 2 is therefore valid to first order in the interaction $V$.

*B2.2.5.2   Ion (electron)–atom collisions*

The electrostatic interaction between a structureless projectile ion P of charge $Z_\mathrm{P}e$ and an atom A with nuclear charge $Z_\mathrm{A}e$ is

$$V(r, R) = \frac{Z_\mathrm{A}Z_\mathrm{P}e^2}{R} - \sum_{j=1}^{N_\mathrm{A}}\frac{Z_\mathrm{P}e^2}{|R-r_j|}.$$

With the use of Bethe's integral

$$\int\frac{\mathrm{e}^{\mathrm{i}K\cdot R}}{|R-r_j|}\,\mathrm{d}R = \frac{4\pi}{K^2}\,\mathrm{e}^{\mathrm{i}K\cdot r_j}$$

the Born scattering amplitude (see B2.2.5) reduces to

$$|f_{if}^{(\mathrm{B})}(q)|^2 = \frac{4M_{\mathrm{PA}}^2 Z_\mathrm{P}^2 e^4}{q^4}|Z_\mathrm{A}\delta_{if} - F_{if}^\mathrm{A}(q)|^2$$

which is a function only of momentum transfer $q = \hbar K$. The dimensionless inelastic *form factor* for $i \to f$ inelastic transitions between states $\phi_{i,f}$ of atom A with $Z_A$ electrons is defined as

$$F_{fi}^A(q) = \langle \phi_f(r) | \sum_{j=1}^{Z_A} e^{iq \cdot r_j/\hbar} | \phi_i(r) \rangle$$

where the integration is over all electron positions denoted collectively by $r \equiv \{r_j\}$. The integrated cross section is

$$\sigma_{if}(v_{PA}) = \frac{8\pi Z_P^2 e^4}{v_{PA}^2} \int_{q_-}^{q_+} |Z_A \delta_{if} - F_{if}^A(q)|^2 \frac{dq}{q^3}$$

$$= \frac{8\pi Z_P^2 a_0^2}{(v_{PA}/v_0)^2} \int_{K-a_0}^{K+a_0} |Z_A \delta_{if} - F_{if}^A(K)|^2 \frac{d(Ka_0)}{(Ka_0)^3}$$

where $v_0 = e^2/\hbar$ is the atomic unit (au) of velocity. The dimensionless momentum change $q/m_e v_0$ is $Ka_0$. In the heavy-particle or high-energy limit, $q_+ \to \infty$ and

$$q_- \approx \frac{|\Delta E_{fi}|}{v_{PA}} \left[ 1 + \frac{\Delta E_{fi}}{2M_{PA} v_{PA}^2} \right]$$

where $\Delta E_{fi} = E_f - E_i$ is the energy lost by the projectile. Since

$$f_{if}^B(q) = f_C^{Z_P Z_A}(q)\delta_{if} + f_C^{eZ_P}(q)F_{if}(q)$$

can be expressed in terms of the individual two-body amplitudes $f_C^{z_1 z_2}$ for Coulomb elastic scattering between particles of charges $z_1$ and $z_2$, the Born cross section for inelastic collisions can be written [9, 11, 28] in the useful form

$$\sigma_{if}^B(v_i) = \frac{2\pi}{M_{eP}^2 v_i^2} \int_{q_-}^{q_+} P_{fi}(q) \left( \frac{d\sigma}{d\Omega} \right)_{el} q \, dq$$

where $P_{fi}(q) = |F_{fi}^A(q)|^2$ is the transition probability for which the impulsive transfer of momentum $q$ to atom A and where

$$\left( \frac{d\sigma}{d\Omega} \right)_{el} = \frac{4 M_{eP}^2 Z_P^2 e^4}{q^4}$$

is the differential cross section for elastic (Coulomb) scattering with momentum $q$ transferred from the projectile of charge $Z_P e$ to one electron of atom A.

*B2.2.5.3   Atom–atom collisions*

The Born integral cross section for specific $(\alpha\beta) \to (\alpha'\beta')$ transitions in the collision

$$A(\alpha) + B(\beta) \to A(\alpha') + B(\beta')$$

in terms of the atomic form factors is

$$\sigma_{\alpha\alpha'}^{\beta\beta'}(v_i) = \frac{8\pi a_0^2}{(v_i/v_0)^2} \int_{K-a_0}^{K+a_0} |Z_A \delta_{\alpha\alpha'} - F_{\alpha\alpha'}^A(K)|^2 |Z_B \delta_{\beta\beta'} - F_{\beta\beta'}^B(K)|^2 \frac{d(Ka_0)}{(Ka_0)^3}.$$

### B2.2.5.4   Quantal and classical impulse cross sections

In the impulse approximation [6, 9], the integral cross section for $(\alpha, \beta) \rightarrow (\alpha', \beta')$ transitions in A is

$$\sigma_{if}(v_i) = \frac{2\pi}{M_{\text{eB}}^2 v_i^2} \int_{q_-}^{q_+} |f_{\text{eB}}^{\beta\beta'}(q)|^2 |F_{\alpha\alpha'}^{\text{A}}(q)|^2 q \, dq$$

where $f_{\text{eB}}^{\beta\beta'}$ is the scattering amplitude for elastic $\beta = \beta'$ or inelastic $\beta \neq \beta'$ collisions between projectile B and an orbital electron of A. For structureless ions B, the Coulomb $f_{\text{eB}}^{\beta\beta}(q)$ for elastic electron–ion collisions reproduces the Born approximation for B–A collisions. When Born amplitudes $f_{\text{eB}}^{\beta\beta}(q)$ are used for fast atom B–e collisions, then the Born approximation for atom–atom collisions is also recovered for general scattering amplitudes $f_{\text{eB}}^{\beta\beta'}$. For slow atoms B, $f_{\text{eB}}$ is dominated by s-wave elastic scattering so that $f_{\text{eB}} = -a$ and $\sigma_{\text{eB}} = 4\pi a^2$ where $a$ is the scattering length. Then

$$\sigma_{if}(v_i) = \frac{2\pi a^2}{(v_i/v_0)^2} \int_{K-a_0}^{K+a_0} |F_{if}^{\text{A}}(K)|^2 (K a_0) \, d(K a_0)$$

which is a good approximation for collisional transitions $nl \rightarrow n'l'$ in Rydberg atoms A. The full quantal impulse cross section [6, 9] for general $f_{\text{eB}}^{\beta\beta'}$ has recently been presented in a valuable new form [28] which is the appropriate representation for direct classical correspondence. The classical impulse cross section was then defined [28] to yield the first general expression for the classical impulse cross section for $n\ell - n'\ell'$ and $n\ell - \epsilon\ell'$ electronic transitions. The cross section satisfies the optical theorem and detailed balance. Direct connection with the classical binary encounter approximation (BEA) was established and the derived $n\ell - n'$ and $n\ell - \epsilon$ cross sections reproduce the standard BEA cross sections.

### B2.2.5.5   Atomic form factor and generalized oscillator strength

In terms of the form factor $F_{if}(K)$, the generalized oscillator strength is defined as

$$f_{if}(K) = \left( \frac{2m_e E_{fi}}{q^2} \right) |F_{if}(q)|^2 = \frac{2 E_{fi}^{\text{a.u.}}}{(K a_0)^2} |F_{if}(K)|^2$$

which tends to the dipole oscillator strength in the $K \rightarrow 0$ limit.

#### (a) Sum rules

$$\sum_f f_{if}(K) = \sum_f f_{if}(K) + \int_0^\infty \frac{d f_{iE}}{dE} \, dE = N$$

$$\sum_f |F_{if}(K)|^2 = \sum_f F_{if}(K) + \int_0^\infty \frac{d F_{iE}}{dE} \, dE$$

$$= N + \sum_{j<k}^N |\langle \Psi_i | \exp(i\vec{K} \cdot (\vec{r}_j - \vec{r}_k)) | \Psi_i \rangle|^2$$

where $N$ is the number of electrons. The summation $\sum_f$ extends over all discrete and continuum states.

*(b) Energy-change moments*

The energy-change moments are defined as

$$S(\alpha, K) = \sum_{f \neq i} (2\Delta E_{fi}^{\text{a.u.}})^\alpha f_{if}(K)$$

$$= \sum_{f \neq i} (2\Delta E_{fi}^{\text{a.u.}})^{\alpha+1} |F_{if}(K)|^2 (Ka_0)^{-2}.$$

The exact energy-change moments for H(1s) are

$$S(-1, K) = \{1 - [1 + \tfrac{1}{4}(Ka_0)^2]^{-4}\}(Ka_0)^{-2}$$

$$S(0, K) = 1$$

$$S(1, K) = (Ka_0)^2 + \tfrac{4}{3}$$

$$S(2, K) = (Ka_0)^4 + 4(Ka_0)^2 + \tfrac{16}{3}.$$

### B2.2.5.6  Form factors for atomic hydrogen

The probability of a transition $i \to f$ resulting from any external perturbation which impulsively transfers momentum $q$ to the internal momenta of the electrons of the target system is

$$P_{if}(q) = |F_{fi}(q)|^2.$$

The impulse can be due to sudden collision with particles or to exposure to electromagnetic radiation. The physical significance of the form factor is that $P_{if}$ is the impulsive transition probability for any atom. For $nl \to n'l'$ transitions in atomic hydrogen,

$$P_{nl,n'l'}(q) = \sum_{m,m'} |\langle \Psi_{nlm}(r)|e^{iq\cdot r/\hbar}|\Psi_{n'l'm'}(r)\rangle|^2,$$

with $\Psi(r) = R_{nl}(r)Y_{lm}(\hat{r})$, can be decomposed as

$$P_{nl,n'l'}(q) = (2l + 1)(2l' + 1) \sum_{L=|l-l'|}^{l+l'} (2L + 1) \begin{pmatrix} L & l & l' \\ 0 & 0 & 0 \end{pmatrix}^2 [f_{nl,n'l'}^{(L)}(q)]^2$$

where $(\ldots)$ is the Wigner's $3j$-symbol and $f_{nl,n'l'}^L(q)$ is the radial integral

$$f_{nl,n'l'}^{(L)}(q) = \int_0^\infty R_{nl}(r)R_{n'l'}(r)j_L(qr)r^2 \, dr$$

where $j_L$ is the modified Bessel function. For $nlm \to n'l'm'$ subshell transitions, the amplitude decomposes as

$$F_{nlm,n'l'm'}(q) = 4\pi \sum_{L=|l-l'|}^{l+l'} i^L w_{lm'l'm'}^{(L)} f_{nl,n'l'}^{(L)}(q) Y_{L,M}(\hat{q})$$

where $M = m - m'$ and where the coefficients

$$w_{lm'l'm'}^{(L)} = \left[ \frac{(2l + 1)(2l' + 1)(2L + 1)}{4\pi} \right]^{\frac{1}{2}} \begin{Bmatrix} L & l & l' \\ M & -m & m' \end{Bmatrix} \begin{Bmatrix} L & l & l' \\ 0 & 0 & 0 \end{Bmatrix}.$$

Exact algebraic expressions for the probability

$$P_{n,n'}(q) = \sum_{l'l'} \sum_{m,m'} |\langle n'l'm'|e^{iq\cdot r/\hbar}|nlm\rangle|^2$$

of $n \to n'$ transitions in atomic hydrogen, have been recently derived [11] as analytical functions of $n$ and $n'$.

### B2.2.5.7 Rotational excitation

For ion-point dipole D interactions, only $\Delta J = \pm 1$ transitions are allowed. For ion-point quadrupole Q interactions only $\Delta J = 0, \pm 2$ transitions are allowed. The Born differential cross sections for $J \to J'$ transitions are

$$\frac{d\sigma^{(d)}}{d\hat{k}_f}(J \to J+1) = \frac{4}{3}\frac{k_f}{k_i}\left(\frac{J+1}{2J+1}\right)\frac{D^2}{K^2}$$

$$\frac{d\sigma^{(q)}}{d\hat{k}_f}(J \to J) = \frac{4}{45}\frac{J(J+1)}{(2J-1)(2J+3)}Q^2$$

$$\frac{d\sigma^{(q)}}{d\hat{k}_f}(J \to J+2) = \frac{2}{15}\frac{k_f}{k_i}\frac{(J+1)(J+2)}{(2J+1)(2J+3)}Q^2$$

which are all spherical symmetrical. The sum

$$\sum \frac{d\sigma^{(q)}}{d\hat{k}_f}(J \to J, J \pm 2) = \frac{4}{45}Q^2$$

is independent of the initial value of $J$. The integral cross sections

$$\sigma^{(d)}(J \to J+1) = \frac{8\pi}{3k_i^2}\left(\frac{J+1}{2J+1}\right)\ln\left(\frac{k_i+k_f}{k_i-k_f}\right)D^2$$

$$\sigma^{(q)}(J \to J, J+2) = \frac{8\pi}{15}\frac{k_f}{k_i}\frac{(J+1)(J+2)}{(2J+1)(2J+3)}Q^2$$

all satisfy the detailed balance relation

$$k_i^2(2J_i+1)\sigma(J_i \to J_f) = k_f^2(2J_f+1)\sigma(J_f \to J_i).$$

The summed diffusion cross sections are

$$\sigma^{(d)} = \int\left[\sum_{J\pm 1}\frac{d\sigma}{d\hat{k}_f}(J \to J')\right](1-\cos\theta)\,d\hat{k}_f$$

$$= \left(\frac{8\pi}{3k_i^2}\right)D^2$$

$$\sigma^{(q)} = (16\pi/45)Q^2.$$

### B2.2.5.8 List of Born cross sections for model potentials

$$k^2 = (2M_{AB}/\hbar^2)E \qquad K = 2k\sin\tfrac{1}{2}\theta$$
$$U = (2M_{AB}/\hbar^2)V \qquad U/k^2 = V/E.$$

For a symmetric potential, the scattering amplitude is

$$f_B(K) = -\int U(R)\frac{\sin KR}{KR}R^2\,dR.$$

The Born integral cross section is

$$\sigma_{if}^B(E) = \frac{2\pi}{k^2}\int_{K_-}^{K_+}|f_{if}^{(B)}(K)|^2K\,dK$$

which is independent of the sign of the potential $V$.

*(a) Exponential*

$V(R) = V_0 \exp(-\alpha R)$

$$f_B(K) = -\frac{2\alpha U_0}{(\alpha^2 + K^2)^2}$$

$$\sigma_B(E) = \frac{16}{3}\pi U_0^2 \left[\frac{3\alpha^4 + 12\alpha^2 k^2 + 16k^4}{\alpha^4(\alpha^2 + 4k^2)^3}\right] \xrightarrow{E \to \infty} \frac{4}{3}\pi\left(\frac{V_0}{E}\right)\left(\frac{U_0}{\alpha^4}\right).$$

*(b) Gaussian*

$V(R) = V_0 \exp(-\alpha^2 R^2)$

$$f_B(K) = -\left(\frac{\pi^{1/2} U_0}{4\alpha^2}\right)\exp(-K^2/4\alpha^2)$$

$$\sigma_B(E) = \left(\frac{\pi^2 U_0}{8\alpha^4}\right)\left(\frac{V_0}{E}\right)[1 - \exp(-2k^2/\alpha^2)].$$

*(c) Spherical well/barrier*

$V(R) = V_0$ for $R < a$, $V(R) = 0$ for $R > a$, $U_0 = (2M_{AB}/\hbar^2)V_0$

$$f_B(K) = -\frac{U_0}{K^3}[\sin Ka - Ka \cos Ka],$$

$$\sigma_B(E) = \frac{\pi}{2}\frac{V_0}{E}(U_0 a^4)[1 - (ka)^{-2} + (ka)^{-3}\sin 2ka - (ka)^{-4}\sin^2 2ka].$$

At low energies, $f_B \to (2M_{AB}/\hbar^2)V_0 a^3/3$ and the scattering is isotropic. At high energies, $\sigma_B(E) \sim E^{-1}$.

*(d) Screened Coulomb interaction*

$V(R) = V_0 \exp(-\alpha R)/R.$

$$f_B(K) = -\frac{U_0}{\alpha^2 + K^2}$$

$$\sigma_B(E) = \frac{4\pi U_0^2}{\alpha^2(\alpha^2 + 4k^2)}$$

where $U_0 = 2Z/a_0$. At low energies, $f_B = -U_0/\alpha^2$ is isotropic. At high energies, $\sigma_B \to \pi(V_0/E)(U_0\alpha^2)$.

*(e) Electron-atom model static interaction*

$V(R) = -N(e^2/a_0)[Z + a_0/R]\exp(-2ZR/a_0).$

$$f_B(\theta) = \frac{2N}{a_0}\left[\frac{2\alpha^2 + K^2}{(\alpha^2 + K^2)^2}\right] \quad \alpha = 2Z/a_0$$

$$\sigma_B(E) = \frac{\pi a_0^2 N^2[12Z^4 + 18Z^2 k^2 a_0^2 + 7k^4 a_0^2]}{3Z^2(Z^2 + k^2 a_0^2)^3}.$$

For atomic H (1s), $N = 1$ and $Z = 1$. For He ($1s^2$), the approximate parameters are $N = 2$ and $Z = 27/16$.

*(f) Polarization potential*

$$V(R) = V_0/(R^2 + R_0^2)^2$$

$$f_B(K) = -\frac{1}{4}\pi \left(\frac{U_0}{R_0}\right) \exp(-K R_0)$$

$$\sigma_B(E) = \left(\frac{\pi^3 U_0}{32 R_0^4}\right)\left(\frac{V_0}{E}\right)[1 - (1 + 4k R_0)\exp(-4k R_0)].$$

### B2.2.6  Quantal potential scattering

The *Schrödinger equation*

$$\left(-\frac{\hbar^2}{2M_{AB}}\nabla_r^2 + V(r)\right)\Psi_k^+(r) = E\Psi_k^+(r)$$

solved subject to the asymptotic condition

$$\Psi_k^+(r) \sim \exp(i\mathbf{k}\cdot\mathbf{r}) + \frac{1}{r}f(\theta,\phi)\exp(ikr)$$

for outgoing spherical waves is equivalent to the solution of the Lippman–Schwinger integral equation

$$\Psi_k^+(r) = \Phi_k^+(r) + \int G(r, r')U(r')\Psi_k^+(r')\,dr'$$

where the outgoing Green's function for a free particle is

$$G(r, r') = -\frac{1}{4\pi}\frac{\exp\{ik|r - r'|\}}{|r - r'|}.$$

Solution of the scattering amplitude may then be determined from the asymptotic form of $\Psi_k^+(r)$ directly or from the integral representation

$$f(\theta,\phi) = -\frac{1}{4\pi}(2M_{AB}/\hbar^2)\langle\exp(i\mathbf{k}_f\cdot\mathbf{r})|V(r)|\Psi_i^+\rangle.$$

The differential cross section for elastic scattering is

$$\frac{d\sigma}{d\Omega} = |f(\theta,\phi)|^2.$$

#### B2.2.6.1  *Partial wave expansion*

A plane wave of unit amplitude can be decomposed according to

$$\Phi_k(r) = \exp(i\mathbf{k}\cdot\mathbf{r}) = 4\pi \sum_{\ell,m} i^\ell j_\ell(kr)Y_{\ell m}^*(\hat{k})Y_{\ell m}(\hat{r})$$

where $j_\ell$ is the spherical Bessel function which varies asymptotically as

$$j_\ell(kr) \sim \frac{1}{kr}\sin\left(kr - \frac{1}{2}\ell\pi\right).$$

The addition theorem for spherical harmonics is

$$4\pi \sum_m Y_{\ell m}^*(\hat{k}) Y_{\ell m}(\hat{r}) = (2\ell + 1) P_\ell(\hat{k} \cdot \hat{r}).$$

Another useful identity is,

$$\frac{\sin Kr}{Kr} = 4\pi \sum_{\ell,m} j_\ell(k_i r) j_\ell(k_f r) Y_{\ell m}(\hat{k}_i) Y_{\ell m}^*(\hat{k}_f)$$

where $K^2 = k_i^2 + k_f^2 - 2k_i k_f \cos\theta$. The system wavefunction $\Psi_k^+(r) \sim N\Phi_k$ with amplitude $N$ is expanded according to

$$\Psi_k^+(r) = \sum_{\ell,m} \iota^\ell e^{\iota\eta_\ell} R_{\epsilon\ell}(r) Y_{\ell m}^*(\hat{k}) Y_{\ell m}(\hat{r})$$

$$= \frac{4\pi N}{kr} \sum_{\ell,m} \iota^\ell F_{\epsilon\ell}(r) Y_{\ell m}^*(\hat{k}) Y_{\ell m}(\hat{r}).$$

The radial wave $F_\ell$ is the solution of the radial Schrödinger equation

$$\frac{d^2 F_\ell}{dr^2} + \left[ k^2 - \left\{ U(r) + \frac{\ell(\ell+1)}{r^2} \right\} \right] F_\ell(r) = 0.$$

The reduced potential and energy are $U(r) = (2M_{AB}/\hbar^2)V(R)$ and $k^2 = (2M_{AB}/\hbar^2)E$, respectively. They both have dimensions of $[a_0^{-2}]$. Also $(ka_0)^2 = (2E/\varepsilon_0)(M_{AB}/m_e)$. Each $\ell$-partial wave is separately scattered since the angular momentum of relative motion is conserved for central forces. The radial waves $R_{\epsilon\ell}(r)$ and $F_{\epsilon\ell}(r)$ vary asymptotically as

$$R_{\epsilon\ell}(r) \sim \frac{N_\ell}{r} \sin(kr - \tfrac{1}{2}\ell\pi + \eta_\ell)$$

$$F_{\epsilon\ell}(r) \sim e^{\iota\eta_\ell} \sin(kr - \tfrac{1}{2}\ell\pi + \eta_\ell).$$

The amplitude of the partial radial wave $R_{\epsilon\ell}(r)$ is $N_\ell = 4\pi N/k$. In table B2.2.1 are displayed the amplitudes $N$ and $N_\ell$ appropriate to various choices for normalization of the continuum wavefunctions $\Psi_k(r)$.

### B2.2.6.2  Scattering amplitudes

For symmetric interactions $V = V(r)$, the wavefunctions $\Psi_i^+$ and $\exp(ik_f \cdot r)$ are decomposed into partial waves. From their asymptotic forms, the following partial wave expansions for the scattering amplitude

$$f(\theta) = \frac{1}{2ik} \sum_{\ell=0}^\infty (2\ell + 1)[\exp(2i\eta_\ell) - 1] P_\ell(\cos\theta)$$

$$f(\theta) = \frac{1}{2ik} \sum_{\ell=0}^\infty (2\ell + 1)[S_\ell(k) - 1] P_\ell(\cos\theta)$$

$$f(\theta) = \frac{1}{2ik} \sum_{\ell=0}^\infty (2\ell + 1) T_\ell(k) P_\ell(\cos\theta)$$

can be deduced. The *scattering*, *transition* and *reactance* matrix elements are defined, in terms of the *phase shift* $\eta_\ell$ suffered by each partial wave, as

$$S_\ell(k) = \exp(2i\eta_\ell)$$
$$T_\ell(k) = 2i \sin \eta_\ell \exp(i\eta_\ell)$$
$$K_\ell(k) = \tan \eta_\ell.$$

The asymptotic $(kr \to \infty)$ form of $F_\ell$ may then be written in terms of the following linear combinations:

$$F_\ell(kr) \sim \sin(kr - \ell\pi/2) + \left(\frac{T_\ell}{2i}\right) e^{i(kr-\ell\pi/2)}$$

$$= -\frac{1}{2i}[e^{-i(kr-\ell\pi/2)} - S_\ell e^{i(kr-\ell\pi/2)}]$$

$$= e^{i\eta_\ell} \cos \eta_\ell [\sin(kr - \ell\pi/2) + K_\ell \cos(kr - \ell\pi/2)]$$

expressed as a combinations of standing waves (trigonometric functions), of incoming ($-$) and outgoing ($+$) spherical waves (exponential functions) and of a standing wave and an outgoing spherical wave. The physical significance of the admixture coefficients $S_\ell$, $T_\ell$ and $K_\ell$ is then transparent. The elements are connected by

$$S_\ell = 1 + T_\ell = (1 + iK_\ell)/(1 - iK_\ell)$$

$K_\ell$ is real while both $S_\ell$ and $T_\ell$ are complex. In term of the full solutions $F_\ell$ of the radial *Schrödinger* equation, the $T$-matrix element for elastic scattering is

$$T_\ell = -\frac{2i}{k} \int_0^\infty F_\ell^{(0)}(r)U(r)F_\ell(r) \, dr$$

where $F_\ell^{(0)} = (kr)j_\ell(kr)$ is the radial component of the final plane wave. The Born approximation to $T_\ell$ is obtained upon the substitution $F_\ell^{(0)} = F_\ell$.

### B2.2.6.3  Integral cross sections

$$\sigma(E) = \frac{4\pi}{k^2} \sum_{\ell=0}^\infty (2\ell + 1) \sin^2 \eta_\ell$$

$$= \frac{\pi}{k^2} \sum_{\ell=0}^\infty (2\ell + 1)|T_\ell|^2$$

$$= \frac{2\pi}{k} \sum_{\ell=0}^\infty (2\ell + 1)[1 - \mathrm{Re}S_\ell].$$

The semiclassical version is obtained by the substitution $mvb = (\ell + \frac{1}{2})\hbar$ so that $k^2b^2 = (\ell + \frac{1}{2})^2$ in terms of the impact parameter $b$. Regarding $\ell$ as a continuous variable,

$$\sigma(E) = \frac{\pi}{k^2} \int_0^\infty (2\ell + 1)|T_\ell|^2 \, d\ell = 2\pi \int_0^\infty |T(b)|^2 b \, db.$$

The transition matrix $|T(b)|^2$ is therefore the probability of scattering particles with impact parameter $b$.

### B2.2.6.4  *Differential cross sections*

The differential cross section for elastic scattering is

$$\frac{d\sigma}{d\Omega} = |f(\theta)|^2 = A(\theta)^2 + B(\theta)^2$$

where the real and imaginary parts of $f(\theta)$ are, respectively,

$$A(\theta) = \frac{1}{2k} \sum_{\ell=0}^{\infty} (2\ell+1) \sin 2\eta_\ell \, P_\ell(\cos\theta)$$

$$B(\theta) = \frac{1}{2k} \sum_{\ell=0}^{\infty} (2\ell+1)[1 - \cos 2\eta_\ell] P_\ell(\cos\theta).$$

Their individual contributions to the integral cross sections are

$$\int A(\theta)^2 \, d\Omega = \frac{4\pi}{k^2} \sum_{\ell=0}^{\infty} (2\ell+1) \sin^2 \eta_\ell \cos^2 \eta_\ell$$

$$\int B(\theta)^2 \, d\Omega = \frac{4\pi}{k^2} \sum_{\ell=0}^{\infty} (2\ell+1) \sin^4 \eta_\ell.$$

### (a) Expansion in Legendre polynomials

When expanded as a series of Legendre polynomials $P_{\mathrm{L}}(\cos\theta)$, the differential cross section has the following form

$$\frac{d\sigma(E,\theta)}{d\Omega} = \frac{1}{k^2} \sum_{L=0}^{\infty} a_{\mathrm{L}}(E) P_{\mathrm{L}}(\cos\theta)$$

where the coefficients

$$a_{\mathrm{L}} = \sum_{\ell=0}^{\infty} \sum_{\ell'=|\ell-L|}^{\ell+L} (2\ell+1)(2\ell'+1)(\ell\ell'00 \mid \ell\ell'L0)^2 \sin\eta_\ell \sin\eta_{\ell'} \cos(\eta_\ell - \eta_{\ell'})$$

are determined by the phase shifts $\eta_\ell$ and the Clebsch–Gordon coefficients $(\ell\ell'mm' \mid \ell\ell'LM)$.

### (b) Example: three-term expansion in $\cos\theta$

The differential cross section can be expanded as

$$\frac{d\sigma(E,\theta)}{d\Omega} = \frac{1}{k^2}[(a_0 - \tfrac{1}{2}a_2) + a_1 \cos\theta + \tfrac{3}{2}a_2 \cos^2\theta].$$

The coefficients are

$$a_0 = \sum_{\ell=0}^{\infty} (2\ell+1) \sin^2 \eta_\ell$$

$$a_1 = 6 \sum_{\ell=0}^{\infty} (\ell+1) \sin\eta_\ell \sin\eta_{\ell+1} \cos(\eta_{\ell+1} - \eta_\ell)$$

$$a_2 = 5 \sum_{\ell=0}^{\infty} [b_\ell \sin^2 \eta_\ell + c_\ell \sin\eta_\ell \sin\eta_{\ell+2} \cos(\eta_{\ell+2} - \eta_\ell)]$$

where

$$b_\ell = \frac{\ell(\ell+1)(2\ell+1)}{(2\ell+1)(2\ell+3)}$$

$$c_\ell = \frac{3(\ell+1)(\ell+2)}{2\ell+3}.$$

*(c) Example: S- and P-wave contributions*

The combined S-, P-wave ($\ell = 0, 1$) contributions to the differential and integral cross sections are

$$\frac{d\sigma}{d\Omega} = \frac{1}{k^2}[\sin^2 \eta_0 + [6 \sin \eta_0 \sin \eta_1 \cos(\eta_1 - \eta_0)] \cos \theta + 9 \sin^2 \eta_1 \cos^2 \theta]$$

$$\sigma(E) = \frac{4\pi}{k^2}[\sin^2 \eta_0 + 3 \sin^2 \eta_1].$$

For pure S-wave scattering, the differential cross section (DCS) is isotropic. For pure P-wave scattering, the DCS is symmetric about $\theta = \pi/2$, where it vanishes; the DCS rises to equal maxima at $\theta = 0, \pi$. For combined S- and P-wave scattering, the DCS is asymmetric with forward–backward asymmetry.

### B2.2.6.5   Optical theorem

The optical theorem relates the integral cross section to the imaginary part of the forward scattering amplitude by

$$\sigma(E) = (4\pi/k)\,\text{Im}\,f(0).$$

This relation is a direct consequence of the conservation of flux. The target casts a shadow in the forward direction where the intensity of the incident beam becomes reduced by just that amount which appears in the scattered wave. This decrease in intensity or shadow results from interference between the incident wave and the scattered wave in the forward direction. Figure B2.2.2 for the density $|\Psi_k^+(r)|$ of section B2.2.6 illustrates how this interference tends to illuminate the shadow region at the right-hand side of the target. Flux conservation also implies that the phase shifts $\eta_\ell$ are always real. Thus

$$|S_\ell|^2 = 1 \qquad |T_\ell|^2 = \text{Im}\,T_\ell.$$

### B2.2.6.6   Levinson's theorem

For a local potential $V(r)$ which supports $n_\ell$ bound states of angular momentum $\ell$ and energy $E_n < 0$, the phase shift $\lim_{k\to 0} \eta_\ell(k))$ tends in the limit of zero collision energy to $n_\ell \pi$. When the well becomes deep enough so as to introduce an additional bound level $E_{n+1} = 0$ at zero energy, then $\lim_{k\to 0} \eta_0(k) = (n_0 + \frac{1}{2})\pi$.

### B2.2.6.7   Partial wave expansion for transport cross sections

The transport cross sections

$$\sigma^{(n)}(E) = 2\pi \left[ 1 - \frac{1 + (-1)^n}{2(n+1)} \right]^{-1} \int_{-1}^{+1} [1 - \cos^n \theta] \frac{d\sigma}{d\Omega}(\cos \theta)$$

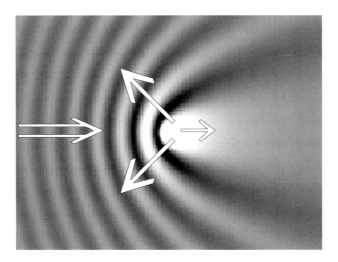

**Figure B2.2.2.** Scattering of an incident plane wave.

for $n = 1$–$4$ have the following phase shift expansions

$$\sigma^{(1)}(E) = \frac{4\pi}{k^2} \sum_{\ell=0}^{\infty} (\ell + 1) \sin^2(\eta_\ell - \eta_{\ell+1})$$

$$\sigma^{(2)}(E) = \frac{4\pi}{k^2} \left(\frac{3}{2}\right) \sum_{\ell=0}^{\infty} \frac{(\ell + 1)(\ell + 2)}{(2\ell + 3)} \sin^2(\eta_\ell - \eta_{\ell+2})$$

$$\sigma^{(3)}(E) = \frac{4\pi}{k^2} \sum_{\ell=0}^{\infty} \frac{(\ell + 1)}{(2\ell + 5)} \left[ \frac{(\ell + 2)(\ell + 3)}{(2\ell + 3)} \sin^2(\eta_\ell - \eta_{\ell+3}) + \frac{3(\ell^2 + 2\ell - 1)}{(2\ell - 1)} \sin^2(\eta_\ell - \eta_{\ell+1}) \right]$$

$$\sigma^{(4)}(E) = \frac{4\pi}{k^2} \left(\frac{5}{4}\right) \sum_{\ell=0}^{\infty} \frac{(\ell + 1)(\ell + 2)}{(2\ell + 3)(2\ell + 7)} \left[ \frac{(\ell + 3)(\ell + 4)}{(2\ell + 5)} \sin^2(\eta_\ell - \eta_{\ell+4}) + \frac{2(2\ell^2 + 6\ell - 3)}{(2\ell - 1)} \sin^2(\eta_\ell - \eta_{\ell+2}) \right].$$

The momentum-transfer or diffusion cross section is $\sigma^{(1)}$ and the viscosity cross section is $\frac{2}{3}\sigma^{(2)}$.

*B2.2.6.8   Born phase shifts*

For a symmetric interaction, the Born amplitude is

$$f_B(K) = -\int U(R) \frac{\sin KR}{KR} R^2 \, dR$$

where $U(r) = (2M_{AB}/\hbar^2)V(R)$. Comparison with the partial wave expansion for $f_B(K)$ and

$$\frac{\sin KR}{KR} = \sum_{\ell=0}^{\infty} (2\ell + 1)[j_\ell(kR)]^2 P_\ell(\cos \theta)$$

provides the Born phase shift

$$\tan \eta_\ell^B(k) = -k \int_0^\infty U(R)[j_\ell(kR)]^2 R^2 \, dR.$$

*(a) Examples of the Born S-wave phase shift*

$$\tan \eta_0^{\mathrm{B}}(k) = -\frac{1}{k} \int_0^\infty U(R) \sin^2(kR) \, \mathrm{d}R.$$

For the potential $U = U_0 \dfrac{\mathrm{e}^{-\alpha R}}{R}$,

$$\tan \eta_0^{\mathrm{B}} = -\frac{U_0}{4k} \ln[1 + 4k^2/\alpha^2].$$

For the potential $U = \dfrac{U_0}{(R^2 + R_0^2)^2}$,

$$\tan \eta_0^{\mathrm{B}} = -\frac{\pi U_0}{4k R_0^3} [1 - (1 + 2k R_0) \, \mathrm{e}^{-2k R_0}].$$

*(b) Born phase shifts (large $\ell$)*

For $\ell \gg ka$,

$$\tan \eta_\ell^{\mathrm{B}} = -\frac{k^{2\ell+1}}{[(2\ell + 1)!!]^2} \int_0^\infty U(R) R^{2\ell+2} \, \mathrm{d}R$$

valid only for finite-range interactions $U(R > a) = 0$. If $U = -U_0$, $R \leq a$ and $U = 0$, $R > a$, then

$$\tan \eta_\ell^{\mathrm{B}}(\ell \gg ka) = U_0 a^2 \frac{(ka)^{2\ell+1}}{[(2\ell + 1)!!]^2 (2\ell + 3)}.$$

The ratio $\eta_{\ell+1}/\eta_\ell \sim (ka/2\ell)^2$.

### B2.2.6.9 Coulomb scattering

For elastic scattering by the interaction $V(r) = Z_A Z_B e^2/r$, the Coulomb wave can be decomposed as

$$\Psi_k^{(C)}(r) = \frac{4\pi}{kr} \sum_{\ell,m} \imath^\ell \, \mathrm{e}^{\imath \eta_\ell} F_{\epsilon \ell}(r) Y_{\ell m}^*(\hat{k}) Y_{\ell m}(\hat{r})$$

where the radial wave varies asymptotically as

$$F_\ell \sim \sin(kR - \tfrac{1}{2}\ell\pi + \eta_\ell^{(C)} - \beta \ln 2kR)$$

where the parameter $\beta$ is $Z_A Z_B e^2/\hbar v$. The Coulomb phase shift is

$$\eta_\ell^{(C)} = arg\,\Gamma(\ell + 1 + i\beta) = \mathrm{Im} \ln \Gamma(\ell + 1 + i\beta)$$

to give the Coulomb $S$-matrix element

$$S_\ell^{(C)} = \exp[2i\eta_\ell^{(C)}] = \frac{\Gamma(\ell + 1 + i\beta)}{\Gamma(\ell + 1 - i\beta)}.$$

The Coulomb scattering amplitude is

$$f_C(\theta) = -\frac{\beta \exp[2i\eta_\ell^{(C)} - i\alpha \ln(\sin^2 \tfrac{1}{2}\theta)]}{2k \sin^2 \tfrac{1}{2}\theta}.$$

The Coulomb differential cross section $|f_C|^2$ is

$$\left(\frac{d\sigma}{d\Omega}\right)_{\text{Coul}} = \frac{Z_A^2 Z_B^2 e^4}{16E^2} \text{cosec}^4 \tfrac{1}{2}\theta.$$

This is the Rutherford scattering cross section. It is interesting to note that Born and classical theory also reproduce this cross section. Moreover,

$$\left(\frac{d\sigma}{d\Omega}\right)_{\text{Coul}}(q) = \frac{4M_{AB}^2 Z_A^2 Z_B^2}{q^4} = 4a_0^2 (M_{AB}/m_e)^2 \left[\frac{Z_A^2 Z_B^2}{(Ka_0)^4}\right]$$

is a function only of the momentum transferred $q = \hbar K = 2\hbar k \sin \tfrac{1}{2}\theta$ in the collision. Note that $q^2 = 8M_{AB}E \sin^2 \tfrac{1}{2}\theta$.

## B2.2.7   Collisions between identical particles

The identical colliding particles, each with spin $s$, are in a resolved state with total spin $S_t$ in the range $(0 \to 2s)$. The spatial wavefunction with respect to particle interchange satisfies $\Psi(R) = (-1)^{S_t} \Psi(-R)$. Wavefunctions for identical particles with even or odd total spin $S_t$ are therefore symmetric (S) or antisymmetric (A) with respect to particle interchange. The appropriate combinations are $\Psi_{S,A}(R) = \Psi(R) \pm \Psi(-R)$, where the positive sign (symmetric wavefunction S) and the negative sign (antisymmetric wavefunction A) are associated with even and odd values of the total spin $S_t$, respectively. The scattering wavefunction for a pair of identical particles in spatially symmetric (+) or antisymmetric (−) states behaves asymptotically as

$$\Psi_{S,A}(R) \to [\exp(i\mathbf{k} \cdot \mathbf{R}) \pm \exp(-i\mathbf{k} \cdot \mathbf{R})] + [f(\theta, \phi) \pm f(\pi - \theta, \phi + \pi)]\frac{\exp(ikR)}{R}.$$

The differential cross section for scattering of *both* the projectile and target particles into direction $\theta$ is

$$\left(\frac{d\sigma}{d\Omega}\right)_{S,A} = |f(\theta, \phi) \pm f(\pi - \theta, \phi + \pi)|^2$$

in the CM-frame where scattering of the projectile into polar direction $(\pi - \theta, \phi + \pi)$ is accompanied by scattering of the identical target particle into direction $(\theta, \phi)$. This is related to the probability that both identical particles are scattered into $\theta$. In the classical limit, where the particles are distinguishable, the classical cross section is

$$\left(\frac{d\sigma}{d\Omega}\right)_C = |f(\theta, \phi)|^2 + |f(\pi - \theta, \phi + \pi)|^2$$

the sum of the cross sections for observation of the projectile and target particles in the direction $(\theta, \phi)$. Since $P_\ell[\cos(\pi - \theta)] = (-1)^\ell P_\ell(\cos\theta)$, the differential cross section for $\phi$-independent amplitudes $f$ is then

$$\left(\frac{d\sigma}{d\Omega}\right)_{S,A} = \frac{1}{4k^2}\left|\sum_{\ell=0}^{\infty} \omega_\ell (2\ell + 1)[\exp 2i\eta_\ell - 1]P_\ell(\cos\theta)\right|^2.$$

For scattering in the symmetric (S) channel where $S_t$ is even, $\omega_\ell = 2$ for $\ell$ even and $\omega_\ell = 0$ for $\ell$ odd. For scattering in the antisymmetric channel where $S_t$ is odd, $\omega_\ell = 0$ for $\ell$ even and $\omega_\ell = 2$ for $\ell$ odd. The integral cross section is

$$\sigma_{S,A}(E) = \frac{8\pi}{k^2}\sum_{\ell=0}^{\infty} \omega_\ell (2\ell + 1) \sin^2 \eta_\ell.$$

Let $g_A$ and $g_S$ be the fractions of states with odd and even total spins $S_t = 0, 1, 2, \ldots, 2s$. When the $2s+1$ spin-states $S_t$ are unresolved, the appropriate combination of symmetric and antisymmetric cross sections is the weighted mean

$$\frac{d\sigma}{d\Omega} = g_S \left(\frac{d\sigma}{d\Omega}\right)_S + g_A \left(\frac{d\sigma}{d\Omega}\right)_A$$

$$\sigma(E) = g_S \sigma_S(E) + g_A \sigma_A(E).$$

### B2.2.7.1 Fermion and boson scattering

#### (a) Fermions

For fermions with half-integral spin $s$, the statistical weights are $g_S = s/(2s + 1)$ and $g_A = (s + 1)/(2s + 1)$. The differential cross section for fermion-fermion scattering is then

$$\frac{d\sigma_F}{d\Omega} = |f(\theta)|^2 + |f(\pi - \theta)|^2 - \left(\frac{2}{2s + 1}\right) \mathrm{Re}[f(\theta) f^*(\pi - \theta)].$$

The integral cross section fermion–fermion collisions is

$$\sigma_F = \tfrac{1}{2}[\sigma_S + \sigma_A] - \tfrac{1}{2}[\sigma_S - \sigma_A]/(2s + 1)$$

which reduces, for fermions with spin-$\tfrac{1}{2}$, to

$$\sigma_F(E) = \frac{2\pi}{k^2} \left[ \sum_{\ell = \text{even}}^{\infty} (2\ell + 1) \sin^2 \eta_\ell + 3 \sum_{\ell = \text{odd}}^{\infty} (2\ell + 1) \sin^2 \eta_\ell \right].$$

#### (b) Bosons

The statistical weights for bosons with integral spin $s$, are $g_S = (s + 1)/(2s + 1)$ and $g_A = s/(2s + 1)$. The differential cross section for boson–boson scattering is

$$\frac{d\sigma_B}{d\Omega} = |f(\theta)|^2 + |f(\pi - \theta)|^2 + \left(\frac{2}{2s + 1}\right) \mathrm{Re}[f(\theta) f^*(\pi - \theta)].$$

The integral cross section boson–boson collisions is

$$\sigma_B = \tfrac{1}{2}[\sigma_S + \sigma_A] + \tfrac{1}{2}[\sigma_S - \sigma_A]/(2s + 1)$$

which reduces, for bosons with zero spin, to

$$\sigma_B(E) = \frac{8\pi}{k^2} \left[ \sum_{\ell = \text{even}}^{\infty} (2\ell + 1) \sin^2 \eta_\ell \right].$$

Symmetry oscillations therefore appear in the differential cross sections for fermion–fermion and boson–boson scattering. They originate from the interference between unscattered incident particles in the forward ($\theta = 0$) direction and backward scattered particles ($\theta = \pi$, $\ell = 0$). A general differential cross section for scattering of spin-$s$ particles is

$$\frac{d\sigma}{d\Omega} = |f(\theta)|^2 + |f(\pi - \theta)|^2 + \frac{(-1)^{2s}}{2s + 1} 2 \, \mathrm{Re}[f(\theta) f^*(\pi - \theta)].$$

*B2.2.7.2    Coulomb scattering of two identical particles*

*(a) Two spin-zero bosons*

Two spin-zero bosons (e.g. $^4$He–$^4$He)

$$\frac{d\sigma}{d\Omega} = \frac{\beta^2}{4k^2}[\mathrm{cosec}^4 \tfrac{1}{2}\theta + \sec^4 \tfrac{1}{2}\theta + 2\mathrm{cosec}^2 \tfrac{1}{2}\theta \sec^2 \tfrac{1}{2}\theta \cos \gamma].$$

*(b) Two spin-$\frac{1}{2}$ fermions*

Two spin-$\frac{1}{2}$ fermions (e.g. H$^+$–H$^+$, e$^\pm$–e$^\pm$)

$$\frac{d\sigma}{d\Omega} = \frac{\beta^2}{4k^2}[\mathrm{cosec}^4 \tfrac{1}{2}\theta + \sec^4 \tfrac{1}{2}\theta - \mathrm{cosec}^2 \tfrac{1}{2}\theta \sec^2 \tfrac{1}{2}\theta \cos \gamma].$$

*(c) Two spin-1 bosons*

Two spin-1 bosons (e.g. deuteron–deuteron)

$$\frac{d\sigma}{d\Omega} = \frac{\beta^2}{4k^2}[\mathrm{cosec}^4 \tfrac{1}{2}\theta + \sec^4 \tfrac{1}{2}\theta + \tfrac{2}{3}\mathrm{cosec}^2 \tfrac{1}{2}\theta \sec^2 \tfrac{1}{2}\theta \cos \gamma].$$

(a)–(c) are the Mott formulae, where $\beta = (Ze)^2/\hbar v$ and $\gamma = 2\beta \ln(\tan \tfrac{1}{2}\theta)$.

*B2.2.7.3    Scattering of identical atoms*

Two ground-state hydrogen atoms, for example, interact via the $X\,^1\Sigma_g^+$ and $b\,^3\Sigma_u^+$ electronic states of H$_2$. The nuclei are interchanged by rotating the atom pair by $\pi$, then by reflecting the electrons first through the midpoint of $R$ and then through a plane perpendicular to the original axis of rotation. The mid-point reflection changes the sign only of the ungerade state wavefunction and both $\Sigma^+$ states are symmetric with respect to the plane reflection.

The cross section for scattering by the gerade potential is then the combination

$$\left(\frac{d\sigma}{d\Omega}\right)_g = \frac{1}{4}\left(\frac{d\sigma}{d\Omega}\right)_S + \frac{3}{4}\left(\frac{d\sigma}{d\Omega}\right)_A$$

of S and A cross sections which involve the phase shifts $\eta_\ell^S$ calculated under the singlet interaction. For scattering by the ungerade triplet interaction

$$\left(\frac{d\sigma}{d\Omega}\right)_u = \frac{1}{4}\left(\frac{d\sigma}{d\Omega}\right)_A + \frac{3}{4}\left(\frac{d\sigma}{d\Omega}\right)_S$$

where the S and A cross sections involve the phase shifts $\eta_\ell^T$ calculated under the triplet interaction. Since the electrons have statistical weights $\frac{1}{4}$ and $\frac{3}{4}$ for the $^1\Sigma_g^+$ and $^3\Sigma_u^+$ states, the differential cross section for H(1s) $-$ H(1s) scattering by both potentials is

$$\frac{d\sigma}{d\Omega} = \frac{1}{4}\left(\frac{d\sigma}{d\Omega}\right)_g + \frac{3}{4}\left(\frac{d\sigma}{d\Omega}\right)_u.$$

These combinations also hold for the integral cross sections.

*(c) Scattering of incident beam alone*

Since the current of incident particles $j_i = 2v$, the cross sections presented by the target (i.e. the number of incident particles removed from the beam in unit time per unit incident current) are $1/2$ of all those above. For example,

$$\left(\frac{d\sigma}{d\Omega}\right)^{\mathrm{I}}_{\mathrm{S,A}} = \frac{1}{2}|f(\theta) \pm f(\pi - \theta)|^2 = \frac{1}{2}\left(\frac{d\sigma}{d\Omega}\right)_{\mathrm{S,A}}$$

and

$$\sigma^{\mathrm{I}}_{\mathrm{S,A}}(E) = \tfrac{1}{2}\sigma_{\mathrm{S,A}}(E).$$

## B2.2.8   Quantal inelastic heavy-particle collisions

The wavefunction for the complete A–B collision system satisfies the Schrödinger equation

$$\mathcal{H}(r, R)\Psi(r, R) = \left[\hat{H}_{\mathrm{int}}(r) - \frac{\hbar^2}{2M_{\mathrm{AB}}}\nabla^2_R + V(r, R)\right]\Psi(r, R)$$

$$= E\Psi(r, R)$$

where the internal Hamiltonian is the sum $\hat{H}_{\mathrm{int}}(r) = H_{\mathrm{A}}(r_{\mathrm{A}}) + H_{\mathrm{B}}(r_{\mathrm{B}})$ of individual Hamiltonians $H_{A,B}$ for each isolated atomic or molecular species. The total energy (internal plus relative)

$$E = \frac{\hbar^2 k_i^2}{2M_{\mathrm{AB}}} + \epsilon_i = \frac{\hbar^2 k_f^2}{2M_{\mathrm{AB}}} + \epsilon_f$$

remains constant for all channels $f$ throughout the collision. The combined internal energy $\epsilon_j$ of A and B at infinite separation $R$ is $\epsilon_j(A) + \epsilon_j(B)$ which are the eigenvalues of the internal Hamiltonian $\hat{H}_{\mathrm{int}}$ corresponding to the combined eigenstates $\Phi_{\mathrm{A}}(r_{\mathrm{A}})\Phi_{\mathrm{B}}(r_{\mathrm{B}})$. There are two limiting formulations (*diabatic* and *adiabatic*) for describing the relative motion. These depend on whether the mutual electrostatic interaction $V(r, R)$ between A and B at nuclear separation $R$, or the variation in the kinetic energy of relative motion, is considered to be a perturbation to the system, i.e. on whether the incident speed $v_i$ is fast or slow in comparison with the internal motions, e.g. with the electronic speed of the electrons bound to A and B.

### B2.2.8.1   *Adiabatic formulation (kinetic coupling scheme)*

When relaxation of the internal motion during the collision is fast compared with the slow collision speed $v_i$, or when the relaxation time is short compared with the collision time, the kinetic energy operator $(2M_{\mathrm{AB}}/\hbar^2)\nabla^2_R$ is then considered as a small perturbation to the quasi-molecular A–B system at fixed $R$. The system wavefunction $\Psi(r, R) = \sum_n F_n(R)\Phi_n(r, R)$ can therefore be expanded in terms of the known 'adiabatic' molecular wavefunctions $\Phi_n(r, R)$ for the quasi-molecule AB at fixed nuclear separation $R$. This set of orthonormal eigenfunctions satisfies

$$[\hat{H}_{\mathrm{int}}(r) + V(r, R)]\Phi_n(r, R) = E_n(R)\Phi_n(r, R).$$

As $R \to \infty$, both $\Phi_n(r, R)$ and the eigenenergies $E_n(R)$ tend, in the limit of infinite nuclear separation $R$, to the (diabatic) eigenfunctions $\Phi_n(r_{\mathrm{A}}, r_{\mathrm{B}}) = \psi_i(r_{\mathrm{A}})\phi_j(r_{\mathrm{B}})$, of $\hat{H}_{\mathrm{int}}$ with eigenenergies $\epsilon_n$, respectively. The substitution $\Psi(r, R) = \sum_n F_n(R)\Phi_n(r, R)$ into the Schrödinger equation results in the following set

$$[\nabla^2_R + \mathcal{K}^2_n(R)]F_n(R) = \sum_j [X_{nj} \cdot \nabla_R + T_{nj}(r)]F_j(R)$$

of coupled equations for the relative motion functions $F_n$. The local momentum $\mathcal{K}_n$ is determined from $\mathcal{K}_n^2 = 2M_{AB}[E - E_n(R)]/\hbar^2$ and the coupling matrix elements are

$$X_{nj}(\mathbf{R}) = -2\langle \Phi_n(\mathbf{r}, \mathbf{R})|\nabla_R|\Phi_j(\mathbf{r}, \mathbf{R})\rangle_r$$

and

$$T_{nj}(\mathbf{R}) = -\langle \Phi_n(\mathbf{r}, \mathbf{R})|\nabla_R^2|\Phi_j(\mathbf{r}, \mathbf{R})\rangle_r.$$

Solution of this set for $F_n(\mathbf{R})$ represents the *adiabatic close-coupling method*. The adiabatic states are normally determined (via standard computational techniques of quantum chemistry) relative to a set of axes $(X', Y', Z')$ with the $Z'$-axis directed along the nuclear separation $\mathbf{R}$. On transforming to this set which rotates during the collision, then $\Psi(\mathbf{r}', \mathbf{R}')$, for the diatomic A–B case, satisfies

$$\left[\hat{H}_0(\mathbf{r}') + V(\mathbf{r}', \mathbf{R}') - \frac{\hbar^2}{2M_{AB}R'^2}\hat{K}\right]\Psi(\mathbf{r}', \mathbf{R}') = E\Psi(\mathbf{r}', \mathbf{R}')$$

where the perturbation operator to the molecular wavefunctions in the rotating frame is

$$\hat{K} = \frac{\partial}{\partial R'}\left(R'^2\frac{\partial}{\partial R'}\right) - (\hat{L}_{X'} - \hat{J}_{X'})^2 - (\hat{L}_{Y'} - \hat{J}_{Y'})^2$$

in terms of the operators $\hat{L}$ and $\hat{J}$ for the total and internal angular momentum $\mathbf{L}$ and $\mathbf{J}$ respectively of the collision system. Note $L_{Z'} = J_{Z'}$, for diatoms. An advantage of using this rotating system in the adiabatic treatment is that radial perturbations, which cause vibrational $v \to v'$ and electronic $nl \to n'l$ transitions, originate from the first term (radial) of $\hat{K}$ while angular perturbations (torques) which causes rotational $J \to J'$ and electronic $nl \to nl'$ transitions originate from the angular momentum operator products $[\hat{L}_{X'}\hat{J}_{X'} + \hat{L}_{Y'}\hat{J}_{X'}]$. The use of a rotating frame causes some complication, however, to the direct use of the asymptotic boundary condition for $\Psi(\mathbf{r}', \mathbf{R}')$.

### B2.2.8.2  *Diabatic formulation (potential coupling scheme)*

When relaxation of the internal motion is slow compared with the fast relative speed $v_i$, then $\Psi$ is expanded in terms of the known unperturbed (diabatic) orthonormal eigenstates $\Phi_j(\mathbf{r}_A, \mathbf{r}_B) = \psi_i(\mathbf{r}_A)\phi_k(\mathbf{r}_B)$ of $\hat{H}_{int}$ according to

$$\Psi(\mathbf{r}, \mathbf{R}) = \sum_j F_j(\mathbf{R})\Phi_j(\mathbf{r}).$$

Substituting into the Schrödinger equation, multiplying by $\Phi_n^*(\mathbf{r})$ and integrating over $\mathbf{r}$, shows that the unknown functions $F_n(\mathbf{R})$ for the relative motion in channel $n$ satisfy the infinite set of coupled equations

$$[\nabla_R^2 + \mathcal{K}_n^2(\mathbf{R})]F_n(\mathbf{R}) = \sum_{j \neq i} U_{nj}(\mathbf{R})F_j(\mathbf{R}).$$

The reduced potential matrix elements which couple the internal states $n$ and $j$ are

$$U_{nj}(\mathbf{R}) = \frac{2M_{AB}}{\hbar^2}V_{nj}(\mathbf{R}) = U_{jn}^*(\mathbf{R})$$

where the electrostatic interaction averaged over states $n$ and $j$ is

$$V_{nj}(\mathbf{R}) = \int \Phi_n^*(\mathbf{r})V(\mathbf{r}, \mathbf{R})\Phi_j(\mathbf{r})\,d\mathbf{r}.$$

The local wavenumber $\mathcal{K}_n$ of relative motion under the static interaction $V_{nn}$ is given by

$$\mathcal{K}_n^2(\boldsymbol{R}) = k_n^2 - U_{nn}(\boldsymbol{R}).$$

The diagonal elements $U_{nn}$ are the *distortion* matrix elements which distort the relative motion from plane waves in elastic scattering, while the off-diagonal matrix elements, $U_{if}$ and $U_{ij}$, $U_{jf}$ which couple states $i$ and $f$ either directly or via intermediate channels $j$ cause inelastic scattering and polarization contributions to elastic scattering. In contrast to the *adiabatic formulation*, radial and angular transitions originate in the *diabatic formulation* from the radial and angular components to the potential coupling elements $V_{nj}(\boldsymbol{R})$. The set of coupled are solved subject to the usual asymptotic $(R \to \infty)$ requirement that

$$F_j(\boldsymbol{R}) \sim \exp(\mathrm{i}k_i Z)\delta_{ij} + f_{ij}\exp(\mathrm{i}k_j R)/R$$

for the elastic $i = j$ and inelastic $i \neq j$ scattered waves. In terms of the amplitude $f_{ij}$ for scattering into direction $(\theta, \phi)$, the differential and integral cross sections for $i \to j$ transitions are

$$\frac{\mathrm{d}\sigma_{ij}}{\mathrm{d}\Omega} = \frac{v_j}{v_i}|f_{ij}(\theta, \phi)|^2$$

and

$$\sigma_{ij} = \frac{v_j}{v_i}\int_0^\pi \mathrm{d}(\cos\theta)\int_0^{2\pi}|f_{ij}(\theta, \phi)|^2\,\mathrm{d}\phi.$$

As well as obtaining the scattering amplitude from the above asymptotic boundary conditions, $f_{if}$ can also be obtained from the *integral representation* for the scattering amplitude is

$$f_{if}(\theta) = \langle \Phi_f(\boldsymbol{r})\exp(\mathrm{i}\boldsymbol{k}_f \cdot \boldsymbol{R})|V(\boldsymbol{r}, \boldsymbol{R})|\Psi(\boldsymbol{r}, \boldsymbol{R})\rangle_{r,R}.$$

### B2.2.8.3   Inelastic scattering by a central field

When the atom–atom or atom–molecule interaction is spherically symmetric in the channel vector $\boldsymbol{R}$, i.e. $V(\boldsymbol{r}, \boldsymbol{R}) = V(r, R)$, then the orbital $l$ and rotational $j$ angular momenta are each conserved throughout the collision so that an $\ell$-partial wave decomposition of the translational wavefunctions for each value of $j$ is possible. The translational wave is decomposed according to

$$F_j(\boldsymbol{R}) = \frac{4\pi N}{k_i R}\sum_{\ell, m} \imath^\ell F_{j\ell}(R) Y_{\ell m}^*(\hat{\boldsymbol{k}}) Y_{\ell m}(\hat{\boldsymbol{R}})$$

and inserted into the *diabatic* set of coupled equations (of section B2.2.8.2). The radial wavefunction $F_{j\ell}$ is then the solution of

$$\frac{\mathrm{d}^2 F_{j\ell}}{\mathrm{d}R^2} + \left[k_i^2 - \left\{U_{ii}(R) + \frac{\ell(\ell+1)}{R^2}\right\}\right]F_{j\ell}(R) = \sum_{j \neq i} U_{ij}(R) F_{j\ell}(R)$$

which is the direct generalization of the quantal radial equation for potential scattering to directly include other channels $j \neq i$. The coupled equations are now solved subject to the requirements that

$$F_{i\ell}(k_i R) \sim \sin(k_i R - \ell\pi/2) + \left\{\frac{T_{ii}^\ell}{2\mathrm{i}}\right\}\mathrm{e}^{\mathrm{i}(k_i R - \ell\pi/2)}$$

for the elastic scattered wave and

$$F_{j\ell}(k_j R) \sim \left(\frac{k_i}{k_j}\right)^{\frac{1}{2}}\left\{\frac{T_{ij}^\ell}{2\mathrm{i}}\right\}\mathrm{e}^{\mathrm{i}(k_j R - \ell\pi/2)}$$

for the inelastic wave. The transition-matrix elements for elastic and inelastic scattering are

$$T_{ij}^{\ell} = -\frac{2i}{(k_i k_j)^{\frac{1}{2}}} \int_0^{\infty} F_{j\ell}^{(0)}(r) U_{ji}(r) F_{i\ell}(r) \, dr$$

where $F_{j\ell}^{(0)} = (k_f r) j_{\ell}(k_f r)$ and $F_{i\ell}(r)$ are the solutions of these coupled radial equations. The differential cross section for inelastic scattering is

$$\frac{d\sigma_{ij}}{d\Omega} = (1/4k_i^2) \left| \sum_{\ell=0}^{\infty} (2\ell+1) T_{ij}^{\ell} P_{\ell}(\cos\theta) \right|^2.$$

The integral inelastic cross section is

$$\sigma_{ij}(E) = \frac{\pi}{k_i^2} \sum_{\ell=0}^{\infty} (2\ell+1) |T_{ij}^{\ell}|^2.$$

The transition matrix $\boldsymbol{T}^{\ell} = \{T_{ij}^{\ell}\}$ is symmetrical, $T_{ij}^{\ell} = T_{ji}^{\ell}$, and the cross sections satisfy detailed balance. Each transition matrix element $|T_{ij}^{\ell}|^2$ is the probability of an $i \to f$ transition in the target for each value $\ell$ of the (orbital) angular momentum of relative motion.

### B2.2.8.4 *Two-state treatment*

Here all couplings are ignored except the direct couplings between the initial and final states as in a two-level atom. The coupled equations to be solved are

$$[\nabla^2 + k_i^2 - U_{ii}(\boldsymbol{R})]\psi_i(\boldsymbol{R}) = U_{if}(\boldsymbol{R})\psi_f(\boldsymbol{R})$$

$$[\nabla^2 + k_f^2 - U_{ff}(\boldsymbol{R})]\psi_f(\boldsymbol{R}) = U_{fi}(\boldsymbol{R})\psi_i(\boldsymbol{R}).$$

*(a) Distorted-wave approximation*

Here all matrix elements in the two-level equations (section B2.2.8.4) are included, except the back coupling $V_{if}\Psi_f$ term which provides the influence of the inelastic channel on the elastic channel and is required to conserve probability. Distortion of the elastic and outgoing inelastic waves by the averaged (static) interactions $V_{ii}$ and $V_{ff}$ respectively is therefore included. The two-state equations can then be decoupled and effectively reduced to one-channel problems. An analogous static-exchange distortion approximation, where exchange between the incident and one of the target particles also follows from the two-level treatment.

*(b) Born approximation*

Here the distortion (diagonal) and back coupling matrix elements in the two-level equations (section B2.2.8.4) are ignored so that $\psi_i(\boldsymbol{R}) = \exp(\mathrm{i}\boldsymbol{k}_i \cdot \boldsymbol{R})$ remains an undistorted plane wave. The asymptotic solution for $\psi_f$ when compared with the asymptotic boundary condition then provides the Born elastic ($i = f$) or inelastic scattering amplitudes

$$f_{if}^{\mathrm{B}}(\theta, \phi) = -\frac{1}{4\pi}\frac{2M_{\mathrm{AB}}}{\hbar^2} \int V_{fi}(\boldsymbol{R}) \mathrm{e}^{\mathrm{i}\boldsymbol{k}\cdot\boldsymbol{R}} \, d\boldsymbol{R}.$$

The momentum change resulting from the collision is $\boldsymbol{Q} = \hbar\boldsymbol{K}$ where $\boldsymbol{k} = \boldsymbol{k}_i - \boldsymbol{k}_f$. The Born amplitude also follows by inserting $\Psi(\boldsymbol{r}, \boldsymbol{R}) = \Phi_i(\boldsymbol{r}) \exp \mathrm{i}(\boldsymbol{k}_i \cdot \boldsymbol{R})$ in the integral representation. Comparison with potential scattering shows that the elastic scattering of structured particles occurs in the Born approximation via the averaged electrostatic interaction $V_{ii}(\boldsymbol{R})$.

For electron–ion or ion–ion collisions, the plane waves $\exp(\mathrm{i}\boldsymbol{k}_{i,f} \cdot \boldsymbol{R})$ are simply replaced by Coulomb waves to provide the Coulomb–Born approximation.

*B2.2.8.5    Exact resonance*

The two-state equations of section B2.2.8.4 cannot, in general, be solved analytically except for the specific case of *exact resonance* when $k_i = k_f = k$ and $U_{ii} = U_{ff} = U$, $U_{if} = U_{fi}$. Then the equations can be decoupled by introducing the linear combinations $\psi^\pm(\boldsymbol{R}) = \frac{1}{\sqrt{2}}[\psi_i(\boldsymbol{R}) \pm \psi_f(\boldsymbol{R})]$, so the two-state set can be converted to two one-channel decoupled equations

$$[\nabla^2 + k^2 - (U \pm U_{if})]\psi^\pm(\boldsymbol{R}) = 0.$$

The problem has therefore been reduced to potential scattering by the interactions $U_\pm = (U \pm U_{if})$ associated with elastic scattering amplitudes $f^\pm$. Hence the elastic ($i = f$) and 'inelastic' ($i \neq f$) amplitudes are

$$f_{ii} = (f^+ + f^-)/2 \quad f_{if} = (f^+ - f^-)/2.$$

In terms of the phase shifts $\eta_l^\pm$ associated with potential scattering by $U_\pm$, the amplitudes for elastic and inelastic scattering are then

$$f_{in}(\theta) = \frac{1}{2ik} \sum_{l=0}^{\infty} (2l+1)[(e^{2i\eta_l^+} + e^{2i\eta_l^-})/2 - 1]P_l(\cos\theta)$$

and

$$f_{if}(\theta) = \frac{1}{2ik} \sum_{l=0}^{\infty} (2l+1)[(e^{2i\eta_l^+} - e^{2i\eta_l^-})/2 - 1]P_l(\cos\theta).$$

The corresponding differential cross sections $|f_{if}|^2$ will therefore exhibit interference oscillations. The integral cross sections are

$$\sigma_{ii} = \frac{4\pi}{k^2} \sum_{l=0}^{\infty} (2l+1)\left[\left(\frac{1}{2}\sin^2\eta_l^+ + \sin^2\eta_l^-\right) \Big/ - \frac{1}{4}\sin^2(\eta_l^+ - \eta_l^-)\Big/4\right]$$

and

$$\sigma_{if} = \frac{\pi}{k^2} \sum_{l=0}^{\infty} (2l+1)\sin^2(\eta_l^+ - \eta_l^-)$$

respectively.

*(a) Examples: atomic collisions with identical nuclei*

Important cases of *exact resonance* are the *symmetrical resonance charge transfer* collision

$$\text{He}_f^+(1s) + \text{He}_s(1s^2) \rightarrow \text{He}_f(1s^2) + \text{He}_s^+(1s)$$

which converts a fast ion beam $f$ to a fast neutral beam and the excitation transfer collision

$$\text{He}(1s2s^3S) + \text{He}(1s^2\,{}^1S) \rightarrow \text{He}(1s^2\,{}^1S) + \text{He}(1s2s^3S)$$

which transfers the internal excitation in the projectile beam fully to the target atom. The electronic molecular wavefunctions divide into even (gerade) or odd (ungerade) classes upon reflection about the mid-point of the internuclear line ($\boldsymbol{R} \rightarrow -\boldsymbol{R}$). In the separated atom limit, $\Psi_{g,u} \sim \phi(r_A) \pm \phi(r_B)$. The potentials $U_\pm$ in the former case are the gerade and ungerade interactions $V_{g,u}$. The phase shifts for elastic scattering by the resulting gerade ($g$) and ungerade ($u$) molecular potentials of $A_2^+$ are, respectively, $\eta_\ell^g$ and $\eta_\ell^u$. The charge transfer (X) and transport cross sections are then

$$\sigma_X(E) = \frac{\pi}{k^2} \sum_{\ell=0}^{\infty} (2\ell + 1) \sin^2(\eta_\ell^g - \eta_\ell^u)$$

$$\sigma_{u,g}^{(1)}(E) = \frac{4\pi}{k^2} \sum_{\ell=0}^{\infty} (\ell + 1) \sin^2(\beta_\ell - \beta_{\ell+1})$$

$$\sigma_{u,g}^{(2)}(E) = \frac{4\pi}{k^2} \left(\frac{3}{2}\right) \sum_{\ell=0}^{\infty} \frac{(\ell+1)(\ell+2)}{(2\ell+3)} \sin^2(\beta_\ell - \beta_{\ell+2}).$$

For ungerade potentials, $\beta_\ell = \eta_\ell^g$ for $\ell$ even and $\eta_\ell^u$ for $\ell$ odd. For gerade potentials, $\beta_\ell = \eta_\ell^u$ for $\ell$ even and $\eta_\ell^g$ for $\ell$ odd. The diffusion cross section $\sigma_{u,g}^{(1)}$ contains ($g/u$) interference. The viscosity cross section $\sigma_{u,g}^{(2)}$ does not. For charge transfer between the heavier rare gas ions Rg$^+$ with their parent atoms Rg, the degenerate states at large internuclear separations are not s states but p states. The states are then $\Sigma_{g,u}$ which arise from the p state with $m = 0$ and $\Pi_{g,u}$ which arises from $m = \pm 1$ with space quantization along the molecular axis. Since there is no coupling between molecular states of different electronic angular momentum, the scattering by the $^2\Sigma_{g,u}$ pair and the $^2\Pi_{g,u}$ pair of Ne$_2^+$ potentials (for example) is independent. The cross section is therefore the combination

$$\sigma_{el,X}(E) = \tfrac{1}{3}\sigma_\Sigma(E) + \tfrac{2}{3}\sigma_\Pi(E)$$

of cross sections $\sigma_\Sigma$ and $\sigma_\Pi$ for the individual contributions arising from the isolated $^2\Sigma_{g,u}$ and $^2\Pi_{g,u}$ states to elastic *el* or charge-transfer X scattering. See [12, 13] for further details on excitation-transfer and charge-transfer collisions.

*(b) Singlet–triplet spin-flip cross section*

This cross section is

$$\sigma_{ST}(E) = \frac{\pi}{k^2} \sum_{\ell=0}^{\infty} (2\ell + 1) \sin^2(\ell^s - \ell^t)$$

where $\eta_\ell^{s,t}$ are the phase shifts for individual potential scattering by the singlet and triplet potentials, respectively.

*B2.2.8.6 Partial wave analysis*

In order to reduce the three-dimensional *diabatic* or *adiabatic* set of coupled equations for atom–atom and atom–molecule scattering to a corresponding working set of coupled radial equations, analogous to those in section B2.2.8.3, the orbital angular momentum $l$ of relative motion must be distinguished from the combined internal angular momentum $j$ associated with the internal (rotational and electronic) degrees of freedom of the partners A and B at rest at infinite separation $R$. Both the orbital angular momentum $l$ of relative motion and the internal angular momentum $j$ of the atomic electrons or of molecular rotation are in general coupled. The total angular momentum $J = l + j$ and its component $J_z$ along some fixed direction (of incidence) are each conserved. Angular momentum may therefore be exchanged between the internal (rotational) and translational (orbital) degrees of freedom via the couplings $V_{nm}(R)$ or $\hat{K}$. Partial wave analysis is an exercise in angular momentum coupling and is well-established (e.g. [14]) for both the diabatic and adiabatic treatments of heavy-particle collisions.

### B2.2.9    Electron–atom inelastic collisions

*B2.2.9.1    Close-coupling equations for electron–atom (ion) collisions*

A partial wave decomposition provides the full close-coupling quantal method for treating A–B collisions, electron–atom, electron–ion or atom–molecule collisions. The method [15] is summarized here for the inelastic processes

$$\mathrm{e}^- + \mathrm{A}_i \rightarrow \mathrm{e}^- + \mathrm{A}_f$$

at collision speeds less or comparable with those target electrons actively involved in the transition. It is based upon an expansion of the total wavefunction $\Psi$ for the (e$^-$–A) - multi-electron system in terms of a sum of products of the known atomic target state wavefunctions $|\Phi_i\rangle$ and the unknown functions $F_i(r)$ for the relative motion. Here

$$\Psi^\Gamma(\mathbf{r}_1, \ldots, \mathbf{r}_N; \mathbf{r}) = \mathcal{A} \sum_i \Phi_i^\Gamma(\mathbf{r}_1, \mathbf{r}_2, \ldots, \mathbf{r}_N; \hat{\mathbf{r}}) \frac{1}{r} F_i^\Gamma(r)$$

involves a sum over all discrete and an integral over the continuum states of the target. The operator $\mathcal{A}$ antisymmetrizes the summation with respect to exchange of all pairs of electrons in accordance with the Pauli exclusion principle. The angular and spin momenta (denoted collectively by $\hat{\mathbf{r}}$) of the projectile electron have been coupled with the orbital and spin angular momenta of the target states $|\Phi_i\rangle$ to produce the 'channel functions' $\Phi_i^{LS\pi}(\mathbf{r}_1, \mathbf{r}_2, \ldots, \mathbf{r}_i; \hat{\mathbf{r}})$ which are eigenstates of the total orbital $L$, total spin $S$ angular momentum, their $Z$-components $M_L, M_S$ and parity $\pi$. The set $\Gamma \equiv LSM_LM_S\pi$ of quantum numbers are therefore conserved throughout the collision. By substituting the expansion for $\Psi^\Gamma$ into the Schrödinger equation,

$$H_{N+1}\Psi^\Gamma = \left[ \sum_{i=1}^{N+1} \left( -\frac{1}{2}\nabla_i^2 - \frac{Z}{r_i} \right) + \sum_{i>j=1}^{N+1} \frac{1}{r_{ij}} \right]\Psi^\Gamma$$

expressed in *atomic units*, the radial functions for the motion of the scattered electron satisfy the infinite set of coupled *integro-differential* equations

$$\left[ \frac{\mathrm{d}^2}{\mathrm{d}r^2} + k_i^2 - \frac{\ell_i(\ell_i+1)}{r^2} + \frac{2(Z-N)}{r} \right] F_i^\Gamma(r) = 2 \sum_j [V_{ij}^\Gamma(r) + W_{ij}^\Gamma(r)]F_j^\Gamma(r).$$

The direct potential couplings are represented by

$$V_{ij}^\Gamma(r) = Z_P \left[ \frac{Z_t}{r}\delta_{ij} + \sum_{k=1}^N \left\langle \Phi_i^\Gamma \left| \frac{1}{|\mathbf{r}_k - \mathbf{r}|} \right| \Phi_j^\Gamma \right\rangle \right].$$

The non-local exchange couplings are represented by

$$W_{ij}F_j^\Gamma(r) = \sum_{k=1}^N \left\langle \Phi_i^\Gamma \left| \frac{1}{|\mathbf{r}_k - \mathbf{r}|} \right| (\mathcal{A}-1)\Phi_j^\Gamma F_j^\Gamma \right\rangle.$$

The direct potential gives rise to the long-range polarization attraction which is very important for low-energy scattering. The exchange potentials are short range and are extremely complicated. Additional non-local potentials that arise from various correlations (which cannot be included directly but which can be constructed from pseudostates) can also be added to the right-hand side of the equations.

Numerical solution of this set of *close-coupled equations* is feasible only for a limited number of close target states. For each $N$, several sets of independent solutions $F_{ij}$ of the resulting close-coupled equations are determined subject to $F_{ij} = 0$ at $r = 0$ and to the reactance $\boldsymbol{K}$-matrix asymptotic boundary conditions,

$$F_{ij}^{\Gamma} \sim \sin\theta_i \delta_{ij} + K_{ij}^{\Gamma} \cos\theta_i$$

for $n$ open channels characterized by $k_i^2 = 2(E - E_i) > 0$. The argument is

$$\theta_i = k_i r - \frac{1}{2}\ell_i \pi + \frac{Z - N}{k_i}\ln(2k_i r) + \sigma_i$$

where $\ell_i$ is the orbital angular momentum of the scattered electron and where $\sigma_i = \arg\Gamma[\ell_i + 1 - \mathrm{i}(Z - N)k_i]$ is the Coulomb phase. For closed channels, $k_i^2 < 0$ and $F_{ij} \sim C_{ij}\exp(-|k_i|r)$ as $r \to \infty$. The scattering amplitude can then be expressed in terms of the elements $T_{ij}$ of the $(n \times n)$ $\boldsymbol{T}$-matrix which is related to the $\boldsymbol{K}$ and $\boldsymbol{S}$ matrices by,

$$\boldsymbol{T}^{\Gamma} = \frac{2\mathrm{i}\boldsymbol{k}^{\Gamma}}{\boldsymbol{I} - \mathrm{i}\boldsymbol{K}^{\Gamma}} = \boldsymbol{S}^{\Gamma} - \boldsymbol{I}.$$

The integral cross section for the transition $i \equiv \alpha_a L_i S_i \to f \equiv \alpha_f L_f S_f$ in the target atom, where $\alpha$ denotes the additional quantum numbers required to completely specify the state, is then

$$\sigma_{if}(k_i^2) = \frac{\pi}{k_i^2}\sum_{LS\pi,\ell_i\ell_j}\frac{(2L + 1)(2S + 1)}{2(2L_i + 1)(2S_i + 1)}|T_{ij}^{\Gamma}|^2.$$

According to detailed balance, the collision strength

$$\Omega_{if} = k_i^2(2L_i + 1)(2S_i + 1)\sigma_{if}(k_i^2)$$

is therefore dimensionless and is symmetric with respect to $i \to f$ interchange. Further extensions, simplifications and calculational schemes of the basic close-coupling and related methods are found in [15–17].

With modern high-speed computers, it is feasible to solve the coupled set of radial equations only for a restricted basis set of unperturbed states $\Phi_n(\boldsymbol{r})$ regarded as being closely and strongly coupled. For electron–atom (molecule) collisions at low energies $E$, the full quantal close-coupling method is extremely successful in predicting the cross sections and shapes and widths of resonances which appear at energies $E$ just below the various thresholds for excitation of the various excited levels. As $E$ increases past the threshold for ionization, it becomes less successful, and is plagued by problems with convergence both in the number of the basis states and in the number of partial waves used in the expansion for $\Psi^{\Gamma}$. Other methods for intermediate and high energies are therefore preferable. For heavy-particle collisions and for electron collisions at high energies, semiclassical versions (in section B2.2.10) of the close-coupling equations can be derived.

### B2.2.9.2  *Close coupling with pseudostates and correlation*

#### (a) Pseudostates

A partial acknowledgment of the influence of higher discrete and continuum states, not included within the wavefunction expansion, is to add, to the truncated set of basis states, functions of the form $\Psi_P(\boldsymbol{r})\Phi_P(\boldsymbol{r})$ where $\Phi_P$ is not an eigenfunction of the internal Hamiltonian $\hat{H}_{\mathrm{int}}$ but is chosen so as to represent some appropriate average of bound and continuum states. These pseudostates can provide full polarization distortion to the target by incident electrons and allows flux to be transferred from the the open channels included in the truncated set.

*(b) Correlation*

When the initial and final internal states of the system are not well-separated in energy from other states then the closed-coupling calculation converges very slowly. An effective strategy is to add a series of correlation terms involving powers of the distance $r_{ij}$ between internal particles of projectile and target to the truncated close-coupling expansion which already includes the important states.

### B2.2.9.3 *The R-matrix method*

This method, introduced originally in an analysis of nuclear resonance reactions, has been extensively developed [15–17] over the past 20 years as a powerful *ab initio* calculational tool. It partitions configuration space into two regions by a sphere of radius $r = a$, where $r$ is the scattered electron coordinate. In the internal region $r < a$, the electron–atom complex behaves almost as a bound state so that a configuration interaction expansion of the total wavefunction $\Psi$, as in atomic structure calculations, is appropriate. In the external region the scattered electron moves in the long-range multipole potential contained in the direct electrostatic interaction, and can be accurately represented by a perturbation approach. See [15–17] for further details, for other modern quantal approximations and for various computational methods useful for electron–atom collisions over a wide energy range.

### B2.2.9.4 *Electron–molecule collisions*

The close-coupling equations are also applicable to electron–molecule collision but severe computational difficulties arise due to the large number of rotational and vibrational channels that must be retained in the expansion for the system wavefunction. In the fixed nuclei approximation, the Born–Oppenheimer separation of electronic and nuclear motion permits electronic motion and scattering amplitudes $f_{nn'}(\mathbf{R})$ to be determined at fixed internuclear separations $R$. Then in the adiabatic nuclear approximation the scattering amplitude for $i \equiv n, v, J \rightarrow n', v', J' \equiv f$ transitions is

$$f_{if}() = \langle \chi_{n'v'}(R) Y_{J'M_{J}'}(\hat{\mathbf{R}}) | f_{nn'}(r) | \chi_{nv}(R) Y_{JM_{J}}(\hat{\mathbf{R}}) \rangle$$

and cross sections can be obtained. See [15] for further details.

## B2.2.10 Semiclassical inelastic scattering

The term *semiclassical* is used in scattering theory to denote many different situations.

(a) The use of some time-dependent classical path $\mathbf{R}(t)$ within a time-dependent quantal treatment of the response of the internal degrees of freedom of A and B to the time-varying field $V(\mathbf{R}(t))$ created by the approach of A towards B along the classical trajectory $\mathbf{R}(t)$. This procedure generalizes classical theory for potential scattering to structured collision partners and inelastic transitions.

(b) The use of the three-dimensional *eikonal-phase* $S(\mathbf{R})$, which is the solution of the *Hamilton–Jacobi equation*, for the channel wavefunction $\Psi^{+}(\mathbf{R})$, within the full quantal expression for the cross section.

(c) The use of JWKB approximate solutions of the radial *Schrödinger equation* for the radial wavefunction $R_{\epsilon,\ell}$ for A–B relative motion within the full quantum treatment of the A–B collision.

### B2.2.10.1 Classical path theory

The basic assumption here is the existence over the inelastic scattering region of a common classical trajectory $R(t)$ for the relative motion under an appropriately averaged *central* potential $\bar{V}[R(t)]$. The interaction $V[r, R(t)]$ between A and B may then be considered as time-dependent. The system wavefunction therefore satisfies

$$i\hbar \frac{\partial \Psi(r, t)}{\partial t} = [\hat{H}_{int}(r) + V(r, R(t))]\Psi(r, t)$$

and can be expanded in terms of the eigenfunctions $\Phi_n$ of $\hat{H}_{int}$ as

$$\Psi(r, t) = \sum_n A_n(t)\Phi_n(r)\exp(-iE_n t).$$

The transition amplitudes $A_n$ then satisfy the set

$$i\hbar \frac{\partial A_n(b, t)}{\partial t} = \sum_n A_j(b, t)V_{nj}(R(t))\exp(i\omega_{nj}t)$$

of first-order equations coupled by the matrix elements $V_{nj}(R)$ between states $n$ and $j$ with energy separation $\hbar\omega_{nj} = E_n - E_j$. Once the classical trajectory $R \equiv (R(t), \theta(t), \phi = \text{constant})$ is determined from the classical equations

$$\frac{dR}{dt} = \pm v[1 - b^2/R^2 - \bar{V}(R)/E]^{\frac{1}{2}}$$

$$\frac{db}{dt} = \frac{vb}{R^2}$$

of motion for impact parameter $b$ and kinetic energy $E = \frac{1}{2}M_{AB}v^2$, the coupled equations are solved subject to the requirement $A_n(b, t \to -\infty) = \delta_{ni}$. Since the probability for an $i \to f$ transition is $P_{if} = |A_f(b, t \to \infty)|^2$, the differential cross section for inelastic scattering is

$$\frac{d\sigma_{if}}{d\Omega} = \sum_n P_{if}(b_n, \phi)\left\{\frac{d\sigma_{el}}{d\Omega}\right\}$$

where $d\sigma_{el}/d\Omega$ is the differential cross section $|bdb/d(\cos\theta)|$ for elastic scattering by $\bar{V}(R)$ and where the summation is over all trajectories $b_n$ which pass through $(\theta, \phi)$. The integral cross section is

$$\sigma_{if} = 2\pi \int_0^\infty |A_f(b, \infty)|^2 b\,db.$$

### Impact parameter method

This normally refers to the use of the straight-line trajectory $R(t) = (b^2 + v^2 t^2)^{1/2}$, $\theta(t) = \arctan(b/vt)$ within the classical path treatment. See Bates [18, 19] for examples and further discussion.

### B2.2.10.2 Landau–Zener cross section

The Landau–Zener transition probability is derived from an approximation to the full two-state impact-parameter treatment of the collision. The single passage probability for a transition between the diabatic surfaces $H_{11}(R)$ and $H_{22}(R)$ which cross at $R_X$ is the Landau–Zener transition probability

$$P_{12}(R_X; b) = 1 - \exp(-2\pi|H_{12}(R_X)|^2/\hbar v_X|H'_{11} - H'_{22}|)$$

where $H_{12}$ is the interaction coupling states 1 and 2. The diabatic curves are assumed to have linear shapes in the vicinity of the crossing at $R_X$, i.e. $(H'_{11} - H'_{22} = \Delta F)$ and $H_{12}$ is assumed constant. The adiabatic surfaces

$$W^{\pm} = \tfrac{1}{2}(H_{11} + H_{22}) \pm \tfrac{1}{2}[(H_{11} - H_{22})^2 + 4H_{12}]$$

do not cross (avoided crossing). They are separated at $R_X$ by $W^+ - W^- = 2H_{12}(R_X)$. The probability for remaining on the adiabatic surface is $P_{12}(R_X)$. The probability for remaining on the diabatic surface or for pseudocrossing between the adiabatic curves is $1 - P_{12}(R_X)$. The overall transition probability for both the incoming and outgoing legs of the trajectory $R(t)$ is then

$$\mathcal{P}_{12} = 2P_{12}(1 - P_{12}).$$

The Landau–Zener cross section is

$$\sigma_{12} = 4\pi \int_0^{R_X} P_{12}(1 - P_{12})b\,\mathrm{d}b$$

where the variation of $P_{12}$ on impact parameter $b$ arises from the speed $v_X \approx v_0(1 - b^2/R_X^2)^{1/2}$ at the crossing point $R_X$. For rectilinear trajectories $R = b + v_0 t$,

$$\sigma_{12}(v_0) = 4\pi R_X^2[E_3(\alpha) - E_3(2\alpha)]$$

where $E_n(\alpha) = \int_1^{\infty} y^{-n} \exp(-\alpha y)\,\mathrm{d}y$ is the exponential integral with argument $\alpha = 2\pi|H_{12}|^2/\hbar v_0 \Delta F$. See Nikitin [20, 21] for more elaborate models which include interference effects arising from the phases or eikonals associated with the incoming and outgoing legs of the trajectory.

### B2.2.10.3  Eikonal theories

Here the relative motion wavefunction $F_n(R)$ is decomposed as [22]

$$F_n(R) = A_n(R) \exp \mathrm{i}S_n(R) \exp(-\chi_n(R)).$$

The classical action, or solution of the Hamilton–Jacobi equation $\nabla S_n(R) = \kappa_n(R)$, for relative motion under the channel interaction $V_{nn}(R)$, is

$$S_n(R) = k_n \cdot R + \int_{R_0}^{R} (\kappa_n - k_n) \cdot \mathrm{d}R_n$$

where $R_0$ is the initial point on the associated trajectory $R_n(t)$ where $\kappa_n = k_n$. The current $J_n$ in channel n, assumed elastic, satisfies the conservation condition $\nabla J_n = 0$, so that $\chi_n$ is the solution of

$$\nabla_R^2 S_n - 2(\nabla_R S_n) \cdot (\nabla_R \chi_n) = 0.$$

Flux in channel $n$ is therefore lost only via transition to another state $f$ with a probability controlled solely by $A_f$. When many wavelengths of relative motion can be accommodated within the range of $V_{nn}$, as at the higher energies favoured by the diabatic scheme, the fast $R$-variation of $F_n$ is mainly controlled by $S_n$, and the original diabatic set of coupled equations then reduce to the simpler set

$$\mathrm{i}\hbar \frac{\partial A_f(t)}{\partial t} = \sum_{n \neq f} A_n(t) V_{fn}(R_f(t)) \exp \mathrm{i}(S_n - S_f) \exp -(\chi_n - \chi_f)$$

of first-order coupled equations. When a common trajectory $R_n(t) = R(t)$ under some averaged interaction $\bar{V}(R)$ can be assumed for all channels $n$ then

$$S_n(R) - S_f(R) = \omega_{fn}t + \hbar^{-1} \int_{t_0}^{t} [V_{ff}(R(t)) - V_{nn}(R(t))]\,\mathrm{d}t$$

and the classical path equations are recovered [22].

*(a) Averaged potential*

The orbit common to all channels is found by choosing the potential governing the relative motion as the average [22]

$$\mathcal{V}(\boldsymbol{R}) = \langle \Psi(r, \boldsymbol{R}) | \hat{H}_{\text{int}}(r) + V(r, \boldsymbol{R}) | \Psi(r, \boldsymbol{R}) \rangle_r.$$

Hamilton's equations of motion for this interaction

$$\mathcal{V}(\boldsymbol{R}) = \sum_n \left[ |A_n|^2 + \sum_f A_f^* A_n V_{fn} \exp \mathrm{i}(S_n - S_f) \right]$$

are therefore coupled to the set of first-order equations for the transition amplitudes $A_f(\boldsymbol{R})$. An essential feature is that total energy is always conserved, being continually redistributed between the relative motion and internal degrees of freedom, as motion along the trajectory proceeds. In terms of the solutions $A_f(b_j, t)$ and the differential cross section $\mathrm{d}\sigma_{\text{el}}^{(j)}/\mathrm{d}\Omega$ for elastic scattering of particles with impact parameter $b_j(\theta)$ through $\theta$ by $\bar{V}(R)$, the semiclassical scattering amplitude is

$$f_{if}^j(\theta) = A_f[b_j(\theta), \infty] \exp \mathrm{i} S_f'(b_j(\theta), \infty) \{ \mathrm{d}\sigma_{\text{el}}^{(j)}/\mathrm{d}\Omega \}^{1/2}.$$

The accumulated classical action for orbit $b_j(\theta)$ is

$$S_f'(b_j, t) = \int_{R_0}^{R(t)} [\kappa_f(\boldsymbol{R}) - k_n(\boldsymbol{R})] \cdot \mathrm{d}\boldsymbol{R}.$$

When the same scattering angle originates from more than one impact parameter $b_j$, then interference effects originate from the different actions associated with the different orbits $b_j(\theta)$. The contributions arising from $N$-orbits which are well-separated combine according to

$$f_{if}(\theta) = -\mathrm{i} \sum_{j=1}^N \alpha_j \beta_j f_{if}^j(\theta).$$

The coefficients $\alpha_j = \exp \pm \pi/4$ depends on whether the scattered particle emerges on the same side (+) of the axis as it entered, as in a collision overall repulsive, or on the opposite side (−), as in an overall attractive collision. The coefficients $\beta_j$ is $\exp \pm \pi/4$ according to whether the sign of $\mathrm{d}b/\mathrm{d}\theta$ is (+) or (−). The differential cross section will therefore exhibit characteristic oscillations, directly attributable to interference between the action phases $S_f'(b_j)$ associated with each contributing classical path $b_j(\theta)$. The analysis can be extended, as in the uniform Airy function approximation to cover orbits which are not widely separated, as for the case of rainbow scattering or of caustics, in general, where the density of paths become infinite. This theory provides the basis of the multistate orbital treatment [22] which is successful for rotational and vibrational excitation in atom–molecule and ion–molecule collisions at higher energies $E_{\text{AB}} \geq 11$ eV. Other semiclassical treatments based on the JWKB approximation to the corresponding set of coupled equations for the radial wavefunction for relative motion can be found in [23–25].

In figure B2.2.3 cross sections for the quenching process

$$\text{He} + \text{H}_2(v = 1, J = 0) \rightarrow \text{He} + \text{H}_2(v = 0, J = 0)$$

for collision energies $E$ ranging from the ultracold to 1 keV are displayed. The full quantal results [26] are shown together with those calculated [27] from the semiclassical multistate orbital method [22]. It is seen that results from both methods complement and connect with each other very well, in that the quantal treatment is calculationally feasible up to $E - 1$ eV while semiclassical procedures are feasible at the higher collision energies.

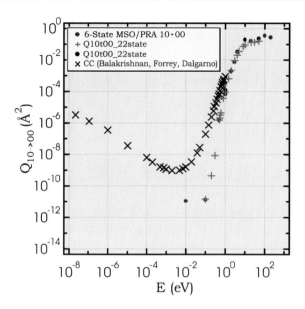

**Figure B2.2.3.** Vibrational relaxation cross sections (quantal and semiclassical) as a function of collision energy $E$.

### (b) Multichannel eikonal method

For electronic transitions in electron–atom and heavy-particle collisions at high impact energies, the major contribution to inelastic cross sections arises from scattering in the forward direction. The trajectories implicit in the action phases and set of coupled equations can be taken as rectilinear. The integral representation

$$f_{if}(\theta) = \langle \phi_f(r) \exp(\mathrm{i}k_f \cdot R) | V(r, R) | \Psi(r, R) \rangle_{r,R}$$

for the scattering amplitude, where

$$\Psi(r, R) = \sum_n A_n(R) \phi_n(r) \exp \mathrm{i} S_n(R)$$

then provides the basis of the multichannel eikonal treatment [28] valuable, in particular, for heavy-particle collisions and for electron (ion)–excited atom collisions where, due to the large effect of atomic polarization (charge-induced dipole), the collision is dominated by scattering in the forward direction.

### B2.2.11  Long-range interactions

#### B2.2.11.1  Polarization, electrostatic and dispersion interactions

The long-range interaction $V(R)$ between two atomic/molecular species can be decomposed into

$$V_{\mathrm{polarization}}(R) + V_{\mathrm{electrostatic}}(R) + V_{\mathrm{dispersion}}(R).$$

The polarization interaction arises from the interaction between the ion of charge $Ze$ and the multipole moments it induces in the atom or molecule AB. The dominant polarization interaction is the ion-induced dipole interaction

$$V_{\mathrm{pol}}(Ze; \mathrm{ind}D) = -\frac{\alpha_{\mathrm{d}} Z e^2}{2R^4} [1 + (\alpha'_{\mathrm{d}}/\alpha_{\mathrm{d}}) P_2(\hat{s} \cdot \hat{R})]$$

where the averaged dipole polarizability is $\alpha_d = (\alpha_\| + 2\alpha_\perp)/3$ and $\alpha_\|$ and $\alpha_\perp$ are the polarizabilities of AB in the directions parallel and perpendicular to the molecular axis $\hat{s}$ of AB. The anisotropic polarizability is $\alpha_d' = 2(\alpha_\| - \alpha_\perp)/3$. The next polarization interaction is the charge-induced quadrupole interaction, averaged over all molecular orientations

$$V_{\mathrm{pol}}(Ze; \mathrm{ind}Q) = -\bar{\alpha}_q Ze^2/2R^6$$

where $\bar{\alpha}_q$ is the averaged quadrupole polarizability. Additional polarization terms arise from permanent multipole moments of one partner and the dipole (or multipole) it induces in the other, averaged over all directions. The leading term is

$$V_{\mathrm{pol}}(\boldsymbol{D}; \mathrm{ind}\boldsymbol{D}) = -\frac{1}{R^6}(D_n^2 \bar{\alpha}_{di} + D_i^2 \bar{\alpha}_{dn})$$

where the subscripts i and n label the permanent dipole moments $D$ and the dipole-polarizabilities $\alpha_d$ of the ion and neutral, respectively. The variation $R^{-6}$ is similar to that for the charge-induced quadrupole interaction.

The electrostatic interaction results from the interaction of the ion with the permanent multipole moments of the neutral. For cylindrically symmetric neutrals or linear molecules, the ion–neutral multipole interaction is

$$V_{el}(Ze; \boldsymbol{D}, \boldsymbol{Q}) = -\frac{(Ze)D_n}{R^2} P_1(\hat{s} \cdot \hat{\boldsymbol{R}}) + \frac{(Ze)Q_n}{R^3} P_2(\hat{s} \cdot \hat{\boldsymbol{R}})$$

where $D_n = \int r\rho(r)\,dr$ and $Q_n = \int (3z^2 - r^2)\rho(r)\,dr$ are the permanent dipole and quadrupole moments of the neutral. The ion dipole–neutral dipole interaction is

$$V_{el}(\boldsymbol{D}_i; \boldsymbol{D}_n) = -\frac{D_i D_n}{R^3}[2\cos\theta_i \cos\theta_n - \sin\theta_i \sin\theta_n \cos(\phi_i - \phi_n)]$$

where $\theta_i$ and $\theta_n$ are the angles made by the ionic and molecular dipoles $\boldsymbol{D}_i$ and $\boldsymbol{D}_n$ and the line $\boldsymbol{R}$ of centres and $\phi_i$ and $\phi_n$ are the azimuthal angles of rotation about the line of centres. The dipole–molecular quadrupole interaction is

$$V_{el}(\boldsymbol{D}_i, \boldsymbol{Q}_n) = \frac{3D_i Q_n}{2R^4}[(3\cos^2\theta_n - 1)\cos\theta_i - 2\sin\theta_i \sin\theta_n \cos\theta_n \cos(\phi_i - \phi_n)].$$

The dispersion interaction arises between the fluctuating multipoles and the moments they induce and can occur even between spherically symmetric ions and neutrals. Thus,

$$V_{\mathrm{dispersion}} \sim -\frac{C_6}{R^6} - \frac{C_8}{R^8} - \frac{C_{10}}{R^{10}} \cdots$$

represents the interaction of the fluctuating dipole interacting with the induced dipole $C_6$ term and quadrupole $C_8$ term, respectively. The leading $R^{-6}$ term represents the van der Waal's attraction.

## References

[1] Catchen G L, Husain J and Zare R N 1978 *J. Chem. Phys.* **69** 1737
[2] Mason E A and McDaniel E W 1988 *Transport Properties of Ions in Gases* (New York: Wiley)
[3] Viehland L A, Mason E A, Morrison W F and Flannery M R 1975 *At. Data Nucl. Data Tables* **16** 495
[4] Rodberg L S and Thaler R M 1967 *Introduction to the Quantum Theory of Scattering* (New York: Academic) p 226
[5] Mott N F and Massey H S W 1965 *The Theory of Atomic Collisions* 3rd edn (Oxford: Clarendon)
[6] Goldberger M L and Watson K M 1964 *Collision Theory* (New York: Wiley)
[7] Newton R G (ed) 1966 *Scattering Theory of Waves and Particles* (New York: McGraw-Hill)
[8] Joachain C J 1975 *Quantum Collision Theory* (Amsterdam: North-Holland) p 383
[9] Flannery M R 1980 *Phys. Rev.* A **22** 2408
[10] Flannery M R and Vrinceanu D 2000 *Phys. Rev. Lett.* **85**

[11]  Vrinceanu D and Flannery M R 1999 *Phys. Rev.* A **6** 1053
[12]  Bransden B H and McDowell M R C 1992 *Charge Exchange and the Theory of Ion–Atom Collisions* (Oxford: Clarendon)
[13]  Bransden B H 1983 *Atomic Collision Theory* 2nd edn (Menlo Park, CA: Benjamin-Cummings)
[14]  Child M S 1996 *Molecular Collision Theory* (New York: Dover)
[15]  Burke P G 1996 *Atomic, Molecular and Optical Physics Handbook* ed G W Drake (New York: American Institute of Physics Press) ch 45
[16]  Bartschat K (ed) 1996 *Computational Atomic Physics* (New York: Springer)
[17]  McCarthy I E and Weigold E 1995 *Electron–Atom Collisions* (Cambridge: Cambridge University Press)
[18]  Bates D R 1961 *Quantum Theory I. Elements* ed D R Bates (New York: Academic) ch 8
[19]  Bates D R (ed) 1962 *Atomic and Molecular Processes* (New York: Academic) ch 14
[20]  Nikitin E E 1996 *Atomic, Molecular and Optical Physics Handbook* ed G W Drake (American Institute of Physics Press) ch 47
[21]  Nikitin E E and Umanskiĭ S Ya 1984 *Theory of Slow Atomic Collisions* (Berlin: Springer)
[22]  McCann K J and Flannery M R 1978 *J. Chem. Phys.* **12** 5275
      McCann K J and Flannery M R 1975 *J. Chem. Phys.* **63** 4695
[23]  Bates D R and Crothers D S F 1970 *Proc. R. Soc.* A **315** 465
[24]  Crothers D S F 1971 *Adv. Phys.* **20** 405
[25]  Child M S 1991 *Semiclassical Mechanics with Molecular Applications* (Oxford: Clarendon)
[26]  Balakrishnan N, Forrey R C and Dalgarno A 1998 *Phys. Rev. Lett.* **80** 3224
[27]  Flannery M R and McCann K J, in preparation (unpublished)
[28]  Flannery M R and McCann K J 1974 *Phys. Rev.* **9** 1947

## Further Reading

Drake G W (ed) 1996 *Atomic, Molecular and Optical Physics Handbook* (American Institute of Physics Press)
McDaniel E W and Mansky E J 1994 Guide to bibliographies, books, reviews and compendia of data on atomic collisions *Advances in Atomic and Molecular Physics* vol 33, ed B Bederson and H Walther p 389
Bates D R and Estermann I (eds) 1965 *Advances in Atomic and Molecular Physics* vol 1
Bates D R and Estermann I (eds) 1973 *Advances in Atomic and Molecular Physics* vol 1
Bates D R and Bederson B (eds) 1974 *Advances in Atomic and Molecular Physics* vol 10
Bates D R and Bederson B (eds) 1988 *Advances in Atomic and Molecular Physics* vol 25
Bates D R and Bederson B (eds) 1989 *Advances in Atomic Molecular and Optical Physics* vol 26
Bates D R and Bederson B (eds) 1993 *Advances in Atomic Molecular and Optical Physics* vol 31
Bederson B and Dalgarno A (eds) 1994 *Advances in Atomic Molecular and Optical Physics* vol 32
Bederson B and Walther H (eds) 1994 *Advances in Atomic Molecular and Optical Physics* vol 33
Bederson B and Walther H (eds) 1998 *Advances in Atomic Molecular and Optical Physics* vol 38
Scoles G (ed) 1988 *Atomic and Molecular Beam Methods* (New York: Oxford University Press)
Baer M (ed) 1985 *Theory of Chemical Reaction Dynamics* (Boca Raton, FL: CRC Press) vols 1–4
Bernstein R B (ed) 1979 *Atom–Molecule Collision Theory: A Guide for the Experimentalist* (New York: Plenum)
Miller W H (ed) 1976 *Dynamics of Molecular Collisions, Parts A and B* (New York: Plenum)
McDaniel E W and McDowell M R C (eds) 1969 *Case Studies in Atomic Collision Physics* (Amsterdam: North-Holland) vol 1
McDaniel E W and McDowell M R C (eds) 1972 *Case Studies in Atomic Collision Physics* (Amsterdam: North-Holland) vol 2
McDaniel E W, Čermák V, Dalgarno A, Ferguson E E and Friedman L (eds) 1970 *Ion–Molecule Reactions* (New York: Wiley)
Bates D R (ed) 1962 *Atomic and Molecular Processes* (New York: Academic)
Bates D R (ed) 1961 *Quantum Theory I. Elements* (New York: Academic)
Massey H S W, Burhop E H S and Gilbody H B (eds) 1969–74 *Electronic and Ionic Impact Phenomena* (Oxford: Clarendon) vols 1–5
McDaniel E W 1989 *Atomic Collisions: Electron and Photon Projectiles* (New York: Wiley)
McDaniel E W, Mitchell J B A and Rudd M E 1993 *Atomic Collisions: Heavy Particle Projectiles* (New York: Wiley)

# B2.3
# Reactive scattering

*Paul J Dagdigian*

## B2.3.1  Introduction

Reactive scattering is one of a number of gas-phase phenomena included in the field of molecular collision dynamics, which is the study of the molecular mechanism of elementary physical and chemical rate processes. Other such dynamical processes include photodissociation, vibrational and rotational energy transfer, electronic quenching, unimolecular decay, reactions within weakly bound complexes and gas–surface interactions. The object of studying the dynamics of these processes is to gain an understanding of the behaviour of a system at the molecular level. We would like to unravel the forces exerted on the nuclei, as described by the potential energy surface (PES) of interaction, during the collisional encounter. We also wish to learn whether the system has jumped to another PES through an electronically non-adiabatic transition. In this section, techniques appropriate to the study of the dynamics of chemical reactions are emphasized. However, these techniques are generally applicable to the study of a variety of gas-phase collisional processes.

The implementation of molecular beam techniques and introduction of laser-based detection methods has allowed chemical reaction dynamics to be elucidated in far greater detail than is possible from inferences based on the temperature dependence of reaction rate constants. In an ideal crossed-beam reactive scattering experiment, which is illustrated schematically in figure B2.3.1, two collimated beams of the reagents, of well defined velocities, are crossed in a collision centre, and the flux of reaction products in specified internal vibration–rotation quantum states scattered into particular solid angles is determined. In this way, the differential cross section for scattering of the reaction product in a given internal quantum state into a given solid angle element can be measured. There have been relatively few studies of reaction dynamics which have closely approximated this ideal, nor is the ideal experiment usually required in order to infer what one would like to understand about the dynamics of a particular reaction. While this section describes experimental techniques as applied to the study of chemically reactive collisions, these methods have also been applied to the study of a wide range of collisional phenomena, including non-reactive energy transfer collisions, photodissociation, and gas–surface scattering.

Two radically different approaches have been taken for the study of reactive scattering. In the first, which approximates the ideal reactive scattering experiment, collimated beams are crossed in a high-vacuum chamber, and the products are detected with a rotatable detector. In most scattering experiments, a mass spectrometer, with electron bombardment ionization and mass-resolved detection of the ions transmitted through a radio-frequency (RF) electric quadrupole [1], is employed as the detector, and the products are identified from the mass-to-charge ratio of the detected ion, either the parent or a fragment ion. This detection method is 'universal' in that every atom or molecule can, in principle, be detected mass spectrometrically. An excellent description of such a molecular beam apparatus is given by Lee *et al* [2].

In order to determine the partitioning of the energy available to the products into internal (vibrational and rotational) excitation and relative translational recoil energy of the products, the velocity of the detected

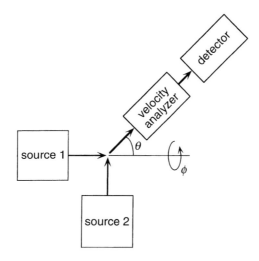

**Figure B2.3.1.** Schematic diagram of an idealized molecular beam scattering experiment.

products is determined, usually by a time-of-flight method [3]. In this way, the translational energy of the products can be determined. The mass spectrometer is essentially insensitive to the degree of internal excitation of the product, however, and the internal excitation of the products can be only determined indirectly, through energy conservation with the knowledge of the total energy available to the products (reaction exoergicity + translational and internal energy of the reagents).

The second approach to the study of reactive scattering involves the use of some spectroscopic method for the detection of the products in specified internal quantum states. Molecular spectroscopy is well suited to the determination of the relative populations in individual states since the quantum numbers of the upper and lower states of a molecular line in an assigned transition are known. Moreover, the intensities may be directly related to concentrations of specific internal states. The original implementation of this approach for the study of reactive scattering involved observation of spontaneous infrared emission from the radiative decay of vibrationally excited products [4, 5]. This approach is still being employed, however now usually with detection of the emission with Fourier transform [6], rather than grating-tuned spectrometers. In some cases, emission from electronically excited products can be observed for highly exothermic reactions.

For many reaction products and for the detection of molecules in their ground vibrational level, some laser-based spectroscopic method must be employed, rather than observation of spontaneous emission. The simplest spectroscopic method for determining concentrations of specified product internal states would involve the application of the Beer–Lambert law on resolved molecular lines in direct absorption. However, the optical density of the product will be very small and limited by the requirement that the nascent reaction products do not undergo any secondary, relaxing collisions before being detected. In the gas phase, the collision frequency can be conveniently reduced by changing the density. The average time between collisions is increased by reducing the total pressure, and hence the concentration of the products. Very recently [7], an ultrasensitive absorption method has been developed and applied for the detection of reaction products.

A more sensitive method of detecting absorption, through observation of a so-called 'action' or 'excitation' spectrum, has been mainly employed for the detection of the reaction products. In most such experiments, a wavelength-tunable laser, usually a dye laser, is scanned over an electronic band system of the reaction product in question, and a signal indicative of molecular absorption is recorded. The relative intensities of spectral lines or bands are then converted into relative populations of the reaction product in specified internal quantum states. In this way the disposal of the available reaction energy into the internal degrees of freedom of this product is directly determined. If the accompanying product is an atom or a molecule with little internal excitation, the relative translational energy of the products can be obtained from energy conservation and knowledge of the total available energy. In this second approach, the angular distribution of the products is usually not determined. However, recent experiments employing Doppler resolution of isolated spectral lines have allowed determination of a low-resolution angular distribution of the product.

Most often, fluorescence excitation to an excited electronic state with the fundamental or frequency-doubled output of a wavelength-tunable dye laser has been employed for laser-based detection of the reaction products. In this method, the total, spectrally unresolved, photon emission from the detection zone is monitored as the laser wavelength is scanned over a molecular transition. Such an excitation spectrum provides the same information as would be available from an absorption spectrum, but with much higher detection sensitivity. This increased sensitivity arises from two factors. Fluorescence detection is a 'zero-background' technique and is limited only by background due to scattered light and quantum counting statistics, while absorption requires the measurement of ratios of signals. In addition, the sensitivity of fluorescence detection is greatly enhanced by the high spectral intensity of laser radiation, as opposed to incoherent radiation from lamps. Product internal state distributions have been determined with laser fluorescence detection in both beam- and bulb-type experiments.

The first half of this section discusses the use of the crossed beams method for the study of reactive scattering, while the second half describes the application of laser-based spectroscopic methods, including laser-induced fluorescence and several other laser-based optical detection techniques. Further discussion of both non-optical and optical methods for the study of chemical reaction dynamics can be found in articles by Lee [8] and Dagdigian [9].

## B2.3.2    Crossed-beams method

### B2.3.2.1    The basic scattering experiment and signal intensity

An ideal scattering experiment requires that the velocity spread of the reagent beams be narrow, so that the relative translational energy of the reagents is well defined. Effusive beams have a very broad velocity distribution, and their use in a scattering experiment usually required the insertion of a slotted-disk velocity selector [10] to reduce the velocity spread to a reasonable width. The flux in an effusive beam is not large, and the insertion of a velocity selector reduces the reagent beam flux significantly. The introduction of supersonic beam sources, with a dramatic narrowing of the velocity spread [11], radically increased the flux from beam sources and made the modern era of crossed-beam reactions possible. The term 'supersonic' refers to the fact that the molecular velocities in such a source are greater than the local speed of sound. In such a source and in contrast to an effusive source, the backing pressure is high enough that the mean free path within the source is much smaller than the orifice diameter so that the gas behaves as a hydrodynamic fluid. Under ideal conditions the enthalpy is converted to net motion during the expansion of the gas into vacuum, and the local temperature becomes very low, leading to a very small spread of velocities about the mean. In addition to obviating the need of velocity selection, the increased backing pressure over that attainable with an effusive source leads to significantly higher downstream beam densities.

A molecular beam scattering experiment usually involves the detection of low signal levels. Thus, one of the most important considerations is whether a sufficient flux of product molecules can be generated to allow a precise measurement of the angular and velocity distributions. The rate of formation of product molecules, $dN/dt$, can be expressed as

$$dN/dt = n_1 n_2 v_{rel} \sigma V_{coll} \qquad (B2.3.1)$$

where the number densities of the reagent beams in the collision volume are given by $n_1$ and $n_2$; and $v_{rel}$, $\sigma$, $V_{coll}$ are the relative velocity between the reagents, the integral reaction cross section, and the scattering volume, respectively. (Equation B2.3.1 is just a re-expression of the law of mass action since the product

$v_{rel}\, \sigma$ is the microcanonical rate constant.) In an experiment with one supersonic beam and a velocity-selected effusive beam, typical values for the reagent beam densities are $10^{12}$ and $10^{10}$ molecules cm$^{-3}$, respectively, in a collision volume of $10^{-2}$ cm$^3$. The relative velocity will be approximately $10^5$ cm s$^{-1}$, and reaction cross section $10^{-15}$ cm$^2$. This leads to an estimate of $10^{10}$ molecules s$^{-1}$ for the total rate of production formation $dN/dt$.

The product molecules will scatter into a range of laboratory angles, depending upon the exoergicity of the reaction, the reaction dynamics, and kinematics, which is a function of the masses of the reagents and products. If we assume that the scattering is confined to 1 sr of solid angle (out of the total $4\pi$), then the detector will receive $\sim 3 \times 10^6$ molecules s$^{-1}$ if it subtends $1°$ in both directions (i.e. an angular acceptance of 1/3000 sr). If, instead, the molecules were isotropically scattered, the detector would see a considerably smaller flux of $\sim 2 \times 10^5$ molecules s$^{-1}$. Of course, we desire not only the angular distribution of the products, but also the velocity distribution at each scattering angle, for a fuller understanding of the reaction dynamics.

If the molecules could be detected with 100% efficiency, the fluxes quoted above would lead to impressive detected signal levels. The first generation of reactive scattering experiments concentrated on reactions of alkali atoms, since surface ionization on a hot-wire detector is extremely efficient. Such detectors have been superseded by the 'universal' mass spectrometer detector. For electron-bombardment ionization, the rate of formation of the molecular ions can be written as

$$d[M^+]/dt = I_e \sigma [M] \qquad (B2.3.2)$$

where $I_e$ is the intensity of the electron beam (typically 10 mA cm$^{-2}$, or $6 \times 10^{16}$ electrons cm$^2$ s$^{-1}$) and $\sigma$ is the ionization cross section (typically $10^{-16}$ cm$^2$ for electrons of 150 eV energy). This leads to an estimated ionization rate of $k_i = I_e \sigma = 6$ s$^{-1}$. The molecules are, of course, not stationary in the ionization region but are travelling with a typical velocity of $\sim 5 \times 10^4$ cm s$^{-1}$. With an ionization region of length $\sim 1$ cm, the probability of ionization is thus estimated to be $\sim 10^{-4}$, which is a low detection efficiency. For example, the above quoted product flux of $3 \times 10^6$ molecules s$^{-1}$ leads to only 360 detection ions s$^{-1}$. This ion count rate, or rates as low as even 1 ion s$^{-1}$, can be measured with good statistics in minutes if the background signal is not much larger than this. Hence, such beam-scattering experiments have been successful mainly through careful reduction of the background ion count rate. This discussion of expected product signal levels follows a discussion of intensities in crossed-beam reactive scattering experiments by Lee [8].

The background ion signal arises from two sources of molecules, namely the inherent background of molecules in a vacuum chamber and molecules effusing from the collision chamber into the detector chamber while the beams are on. The former arises from outgassing from the materials employed in the construction of the apparatus and the limitations in the pumps. The latter requires the careful design of differential pumping of the sources and the detector [2, 12].

### B2.3.2.2    Laboratory to centre-of-mass transformation

In a crossed-beam experiment the angular and velocity distributions are measured in the laboratory coordinate system, while scattering events are most conveniently described in a reference frame moving with the velocity of the centre-of-mass of the system. It is thus necessary to transform the measured velocity flux contour maps into the center-of-mass coordinate (CM) system [13]. Figure B2.3.2 illustrates the reagent and product velocities in the laboratory and CM coordinate systems. The CM coordinate system is travelling at the velocity $c$ of the centre of mass

$$c = [m_1 v_1 + m_2 v_2]/(m_1 + m_2) \qquad (B2.3.3)$$

where the velocities of the reagents are $v_1$ and $v_2$, with masses $m_1$ and $m_2$, respectively. Thus, the velocities in the two coordinate systems are related by

$$v_i = c + u_i \qquad \text{for } i = 1, 2. \qquad (B2.3.4)$$

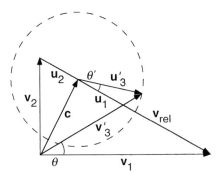

**Figure B2.3.2.** Velocity vector diagram for a crossed-beam experiment, with a beam intersection angle of 90°. The laboratory velocities of the two reagent beams are $v_1$ and $v_2$, while the corresponding velocities in the centre-of-mass coordinate system are $u_1$ and $u_2$, respectively. The laboratory and CM velocities for one of the products (assumed here to be in the plane of the reagent velocities) are denoted $v'_3$ and $u'_3$, respectively. The dashed circle denotes the possible laboratory velocities $v'_3$ for the full range of CM scattering angles $\theta'$.

The CM velocities are given by

$$u_1 = m_2 v_{\text{rel}}/(m_1 + m_2) \tag{B2.3.5a}$$

$$u_2 = -m_1 v_{\text{rel}}/(m_1 + m_2) \tag{B2.3.5b}$$

where $v_{\text{rel}} = v_1 - v_2$ is the relative velocity.

The relative translational energy of the reagents is given by

$$E_{\text{trans}} = \tfrac{1}{2}\mu v_{\text{rel}}^2. \tag{B2.3.6}$$

The energy available to the product equals

$$E'_{\text{tot}} = E_{\text{trans}} + E_{\text{int}} + \Delta E \tag{B2.3.7}$$

where $E_{\text{int}}$ is the internal excitation energy of the reagents and $\Delta E$ is the reaction exoergicity. The energy $E'_{\text{tot}}$ can be partitioned between translational and internal excitation of the products. The CM speed of one of the products can be expressed as

$$u'_3 = \frac{m_4}{m_3 + m_4}\left(\frac{2}{\mu'}(E'_{\text{tot}} - E'_{\text{int}})\right)^{1/2}. \tag{B2.3.8}$$

where $m_3$ is the mass of this product, $m_4$ is the mass of the other product, $\mu' = m_3 m_4/(m_3 + m_4)$ is the reduced mass of the products, and $E'_{\text{int}}$ is the internal excitation energy of the products.

It can be seen from figure B2.3.2 that scattering angles (relative to the direction of one of the reagent beams) are different in the laboratory and CM coordinate systems. If the detected product is very heavy compared with its partner, or if its translational energy is very small, then its speed will be small compared with the speed of the centre of mass of the system. In this case, the product is scattered into a small range of scattering angles about $c$, and determination of the CM angular distribution will be difficult. Moreover, the scattered intensity at one laboratory scattering angle can come from two CM scattering angles, as can be seen in figure B2.3.2. From the intensity estimates presented in section B2.3.2.1, this concentration of the scattered product into a small laboratory angular range will facilitate detection of the product molecules. By contrast, if the product CM speed is large, then the product can be scattered into all laboratory angles.

In addition to transforming the velocities and scattering angles between the laboratory and CM frames, we must also consider the transformation of the cross sections

$$\left(\frac{d^2\sigma}{dv'd\Omega}\right)_{\text{lab}} = \left(\frac{d^2\sigma}{du'd\omega}\right)_{\text{CM}} \frac{du'\,d\omega}{dv'\,d\Omega} \tag{B2.3.9}$$

where the last term on the right-hand side is the Jacobian of the transformation. Because the internal energies of the products are quantized, the velocity of a product scattered into a specific direction can have only discrete values. Equation (B2.3.9) is written with the velocities as continuous variables since the experimental resolution in reactive scattering experiments is not sufficient to resolve the discrete product velocities, due to the spread in the reagent beam velocities and angles. A detailed derivation of the Jacobian has been presented [13], and we obtain

$$\frac{du'\,d\omega}{dv'\,d\Omega} = \frac{v'^2}{u^2}. \tag{B2.3.10}$$

Equation (B2.3.10) shows that the scattered intensity observed in the laboratory is distorted from that in the CM coordinate system. Those products which have a larger laboratory velocity or a smaller CM velocity will be observed in the laboratory with a greater intensity.

The detection technique can also have an effect upon the angle- and velocity-dependent intensities. Cross sections refer to fluxes of molecules into a given range of velocities and angles. The commonly employed technique of mass spectrometric detection provides a measure of the density in the ionization region. Since density and flux are related by the velocity, we must include a factor of $1/v'$ in making the transformation indicated in equation (B2.3.10) from the CM cross sections to the measured laboratory intensities.

If the reagent velocity and angular spreads are sufficiently small, one can infer the CM angle-velocity distributions, i.e. the CM differential cross section on the right-hand side of equation (B2.3.10), directly from the measured laboratory intensities, by simply transforming the velocities to the CM frame and removing the transformation of the Jacobian, with inclusion of the velocity-dependent detection efficiency. On the other hand, as the scattering experiments are often carried out with limited resolution, it is usually necessary to deconvolute the results over the experimental spread [14]. More commonly, a forward convolution technique is employed, in which the CM angle-velocity distribution is adjusted until the laboratory distribution calculated by transformation of the coordinate system and convoluted over the experimental spreads agrees with the measured laboratory distribution.

### B2.3.2.3 Beam sources

Many reactive scattering experiments involve the reaction of an atomic species, such as hydrogen, oxygen, a halogen, or a metal atom, with a stable molecular reagent. A variety of techniques have been employed for the generation of the reagent atomic beam. Beams of halogen atoms have been prepared by thermal dissociation. At room temperature, these elements exist as diatomic molecules, while the equilibrium is shifted toward the monatomic species at sufficiently high temperatures. A detailed description of such a source for the production of Cl, Br, and I beams is given by Valentini et al [15]. The atomic beam is prepared by heating the halogen molecule, diluted in a rare gas, to 2000 °C in a graphite tube. At this temperature, dissociation to atoms is essentially complete. In order to reduce the spread in velocities, the gas mixture is expanded supersonically into vacuum. Problems with materials corrosion have, until recently, limited the intensities of atomic fluorine beams. Use of a nickel tube limits the temperature to 700 °C, for which dissociation yields of <15% are obtained. Recently a F atom source employing a tube made of single-crystal $MgF_2$ has been constructed, and dissociation fractions of ~80% have been achieved at tube temperatures near 1000 °C [16, 17].

Thermal dissociation is not suitable for the generation of beams of oxygen atoms, and RF [18] and microwave [19] discharges have been employed in this case. The first excited electronic state, $O(^1D)$, has a different spin multiplicity than the ground $O(^3P)$ state and is electronically metastable. The collision dynamics of this very reactive state have also been studied in crossed-beam reactions with a RF discharge source which has been optimized for production of $O(^1D)$ [20].

Beams of metal atoms have been prepared by many researchers through thermal vaporization from a heated crucible. An example of such a source, employed for the generation of beams of alkaline earth atoms, is described by Irvin and Dagdigian [21]. By striking an electrical discharge within this source, beams containing electronically excited metastable atoms could be prepared. For Ca, the conversion efficiency to the $3s3p\,^3P$ and $3s4d\,^1D$ metastable excited states was of the order of 80%. Laser ablation from a solid has been widely used to generate atomic atoms of refractory elements [22]. A detailed description of such a source for the production of a beam of atomic carbon has been given [23]. This source has been employed in crossed-beam studies of reactions of carbon atoms.

Laser photolysis of a precursor may also be used to generate a reagent. In a crossed-beam study of the $D + H_2$ reaction [24], a hyperthermal beam of deuterium atoms (0.5 to 1 eV translational energy) was prepared by 248 nm photolysis of DI. This preparation method has been widely used for the preparation of molecular free radicals, both in beams and in experiments in a cell, with laser detection of the products. Laser photolysis as a method to prepare reagents in experiments in which the products are optically detected is further discussed below.

In most reactive scattering experiments, the reagent beam sources, which are housed in differentially pumped enclosures, are fixed and cross at a $90°$ intersection angle, while the detector is rotated about the scattering centre. The stable molecular co-reagent is usually produced in an effusive or supersonic source of the pure reagent. Care must be taken to ensure that no clusters are formed in the beam source, for example by heating the source or by limiting the total pressure behind the source orifice.

### B2.3.2.4 An example reaction: $F + H_2 \rightarrow HF + H$

This reaction has been intensively studied because of its accessibility to both experimental and theoretical treatments. This reaction is also important because it is the pumping mechanism for the hydrogen fluoride infrared chemical laser. We present some data from the extensive study in 1985 by Lee and co-workers (Neumark *et al* [25, 26]) and recent, higher resolution experiments by Faubel *et al* [27, 28]. Figure B2.3.3 presents a schematic diagram of the apparatus in which Lee and co-workers carried out their experiments [29]. An effusive beam of fluorine atoms was prepared by thermal dissociation of $F_2$ in a nickel tube at $650°C$. The velocity spread was reduced to 11% by passage through a slotted-disk chopper. The fluorine atom beam was crossed with a supersonic molecular hydrogen beam, and the incident relative translational energy was varied by changing the temperature of the molecular beam source. The products were detected in a triply differentially pumped mass spectrometer employing electron-impact ionization. Cryogenically cooled surfaces also provided additional pumping to reduce the background signal, primarily from diffuse scattering from surfaces. The laboratory velocity distributions of the product at various laboratory scattering angles were measured by a time-of-flight method with a mechanical chopper. Crossed-beam scattering of a normal hydrogen beam, a *para*-hydrogen beam (all molecules in the $j = 0$ rotational level) [25], and beams of $D_2$ and HD [26] was studied.

We present results on the $F + D_2$ reaction. Study of the reaction of the $D_2$ isotopic reagent was easier because the background in the mass spectrometer was smaller at mass 21 (DF) than for mass 20 (HF). Moreover, the masses of the DF and D are less dissimilar, so that the DF product is less kinematically constrained in the range of accessible laboratory scattering angles. Figure B2.3.4 presents the laboratory angular distribution and a velocity vector diagram for the reaction, showing the accessible angular ranges for the product vibrational levels. From this plot, it already appears that the bulk of the DF products is made in the $v = 3$ and 4 vibrational levels that are scattered backward in the CM frame with respect to the incident F atom beam.

The above conjectures need to be verified by measurement of the doubly differential cross sections in angle and velocity. Typical time-of-flight distributions at several laboratory scattering angles from the more

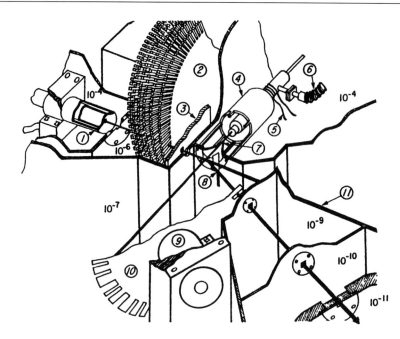

**Figure B2.3.3.** Crossed-molecular beam apparatus employed for the study of the $F + D_2 \rightarrow DF + D$ reaction. Indicated in the figure are: (1) the effusive F atom source; (2) slotted-disk velocity selector; (3) liquid-nitrogen-cooled trap; (4) $D_2$ beam source; (7) skimmer; (8) chopper; (9) cross-correlation chopper for product velocity analysis; and (11) rotatable, ultrahigh-vacuum, triply differentially pumped, mass spectrometer detector chamber. Reprinted with permission from Lee [29]. Copyright 1987 American Association for the Advancement of Science.

recent, higher resolution experiments of Faubel *et al* [27, 28] are presented in figure B2.3.5. We see that products formed in the different vibrational levels appear at distinct time intervals, corresponding to different laboratory velocities. From data such as those presented in figure B2.3.5 and after transformation from the laboratory to the CM frame, the CM velocity flux contour map is obtained. Figure B2.3.6 displays such a contour plot derived by Lee and co-workers (Neumark *et al* [26]) for the $F + D_2$ reaction at one collision energy. It can be seen that all DF vibrational levels are predominantly scattered into the backward hemisphere. The CM angular spread is seen to be larger for the highest energetically accessible level, $v = 4$.

Keil and co-workers (Dharmasena *et al* [16]) have combined the crossed-beam technique with a state-selective detection technique to measure the angular distribution of HF products, in specific vibration–rotation states, from the $F + H_2$ reaction. Individual states are detected by vibrational excitation with an infrared laser and detection of the deposited energy with a bolometer [30].

### B2.3.2.5   *Problems with product identification*

It is well known that the electron-impact ionization mass spectrum contains both the parent and fragment ions. The observed fragmentation pattern can be useful in identifying the parent molecule. This ion fragmentation also occurs with mass spectrometric detection of reaction products and can cause problems with identification of the products. This problem can be exacerbated in the mass spectrometric detection of reaction products because these internally excited molecules can have very different fragmentation patterns than thermal molecules. The parent molecules associated with the various fragment ions can usually be sorted out by comparison of the angular distributions of the detected ions [8].

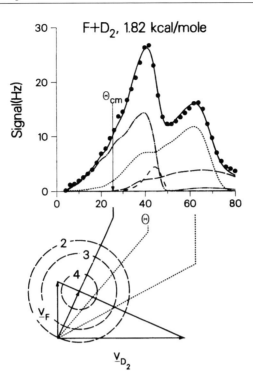

**Figure B2.3.4.** Laboratory angular distribution of DF products from the F + $D_2$ reaction at an incident relative translational energy of 1.82 kcal mol$^{-1}$ [26]. The full curve shows the fit with the derived CM angle–velocity contour. The angular distributions for the $v = 1, 2, 3$, and 4 vibrational levels are indicated by (— —), (———), (· · · · · ·), and (— · —), respectively. (By permission from AIP.)

Many of the problems associated with electron impact ionization, such as the formation of fragment ions, can be alleviated by the use of photoionization detection. When the photon wavelength is tuned below the dissociative ionization threshold, it is possible to ionize the molecule 'softly', with the formation of parent ions only. This advantage for photoionization arises because the cross sections generally rise rapidly from the energetic threshold, namely the ionization potential. Recently, a molecular beam scattering apparatus using photoionization mass spectrometric detection of the products has been constructed [12]. This apparatus, shown schematically in figure B2.3.7, takes advantage of the intense vacuum ultraviolet (VUV) radiation available from a third-generation synchrotron radiation source at a national facility, the advanced light source, at the Lawrence Berkeley National Laboratory. In most scattering experiments, the detector is rotated about the crossing point of the fixed reagent beams. In this newly constructed apparatus, the detector includes an electron storage ring and cannot be rotated; in this case, it is the sources that are rotated about the scattering centre. This apparatus makes liberal use of turbomolecular pumps, which can be positioned in any orientation and are convenient for vacuum pumps on rotating assemblies.

### B2.3.3 Optical detection of the reaction products

Optical methods, in both bulb and beam experiments, have been employed to determine the relative populations of individual internal quantum states of products of chemical reactions. Most commonly, such methods employ a transition to an excited electronic, rather than vibrational, level of the molecule. Molecular electronic

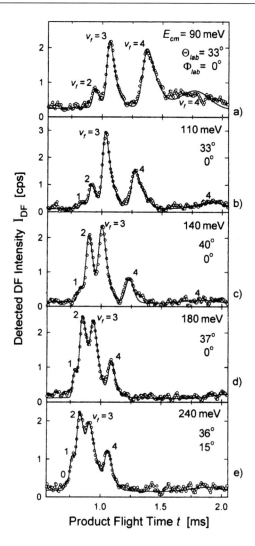

**Figure B2.3.5.** Typical time-of-flight spectra of DF products from the F + D$_2$ reaction [28]. The collision energies and in-plane ($\Theta_{lab}$) and out-of-plane ($\Phi_{lab}$) laboratory scattered angles are given in each panel. The DF product vibrational quantum number $v_f$ associated with each peak is indicated. Reprinted with permission from Faubel *et al* [28]. Copyright 1997 American Chemical Society.

transitions occur in the visible and ultraviolet, and detection of emission in these spectral regions can be accomplished much more sensitively than in the infrared, where vibrational transitions occur. In addition to their use in the study of collisional reaction dynamics, laser spectroscopic methods have been widely applied for the measurement of temperature and species concentrations in many different kinds of reaction media, including combustion media [31] and atmospheric chemistry [32].

### B2.3.3.1    Laser-induced fluorescence detection

The most widely employed optical method for the study of chemical reaction dynamics has been laser-induced fluorescence. This detection scheme is schematically illustrated in the left-hand side of figure B2.3.8.

## F+D$_2$→DF+D, 1.82 kcal/mole

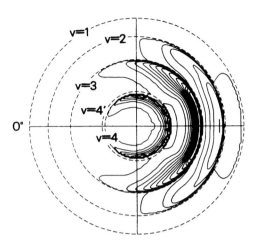

**Figure B2.3.6.** CM angle–velocity contour plot for the F + D$_2$ reaction at an incident relative translational energy of 1.82 kcal mol [26]. Contours are given at equally spaced intensity intervals. This CM differential cross section was used to generate the calculated laboratory angular distributions given in figure B2.3.4. (By permission from AIP.)

A tunable laser is scanned through an electronic band system of the molecule, while the fluorescence emission is detected. This maps out an 'action spectrum' that can be used to determine the relative concentrations of the various vibration–rotation levels of the molecule.

There are several requirements for this to be a suitable detection method for a given molecule. Obviously, the molecule must have a transition to a bound, excited electronic state whose wavelength can be reached with tunable laser radiation, and the band system must have been previously spectroscopically assigned. If the molecules are formed with considerable vibrational excitation, the available spectroscopic data may not extend up to these vibrational levels. Transitions in the visible can be accessed directly by the output of a tunable dye laser, while transitions in the ultraviolet can be reached by frequency-doubled radiation. The excited state must also have a reasonably short radiative lifetime (say $<10^{-5}$ s) with a near 100% fluorescence quantum yield (preferably independent of internal state). Finally, assignments of the individual rotational lines and the vibrational bands within the electronic transition must be available. These restrictions place considerable limits on the molecules which can be detected in this way—mainly diatomics and some triatomics. For incisive interpretation of the experimental observations, it is precisely those reactions involving small molecules whose collision dynamics can be treated theoretically with modern quantum mechanical methods.

Figure B2.3.9 presents a schematic diagram of a typical laser fluorescence experiment. In this apparatus, one of the reagents is prepared by photolysis of a suitable precursor [33] using radiation from an excimer laser (usually 248 nm from a KrF laser or 193 nm from a ArF laser). The tunable laser employed for fluorescence excitation counter-propagates along the beam of the excimer laser. Fluorescence of the product molecules is collected with a telescope and is imaged onto a photomultiplier. Because of their greater coverage of wavelengths, pulsed, rather than continuous (cw), lasers are almost universally employed. Thus, the photomultiplier output signal will typically appear at the 10–50 Hz repetition rate of the lasers and is usually sampled with a gated integrator, whose output is recorded with a laboratory computer. Analogue, rather than digital, electronics is usually employed because of pile-up of the detected photon counts in an experiment with reasonable product intensities.

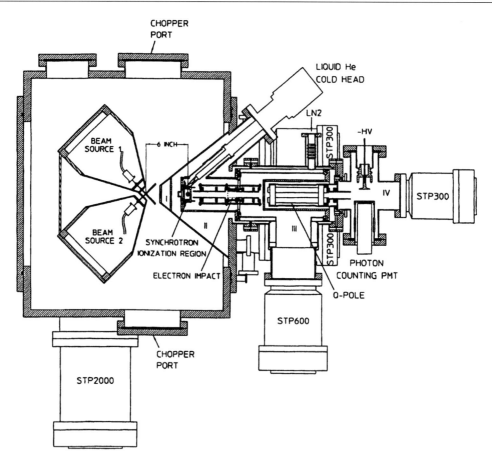

**Figure B2.3.7.** Schematic apparatus of crossed molecular beam apparatus with synchrotron photoionization mass spectrometric detection of the products [12]. To vary the scattering angle, the beam source assembly is rotated in the plane of the detector. (By permission from AIP.)

The principal source of background in laser fluorescence detection is laser light scattered diffusely from optical elements such as windows, and from surfaces such as the walls of the apparatus. The windows for entry and exit of the laser beam, which are significant sources of scattered light, are usually mounted on long (0.4–0.8 m) sidearms. Baffles are installed in the sidearms to prevent light from scattering off the inside of the sidearms. Further reduction of scattered light can be achieved through the use of imaging optics (as illustrated in figure B2.3.9) to relay fluorescence from the excitation zone to the photomultiplier detector, so that emission from only a well defined volume is detected. If the fluorescence is mainly at wavelengths greatly different from the excitation wavelength, as would be the case, for example, for a molecule with significantly different equilibrium internuclear separations in the ground and excited electronic state, and hence a very non-diagonal Franck–Cordon array, then spectral filtering can provide further reduction in the background signal.

### B2.3.3.2    Determination of product internal state distributions

Considerable spectroscopic data are required for the determination of the relative populations in the various internal quantum levels of the product from the relative intensities of various lines, or bands, in a spectrum.

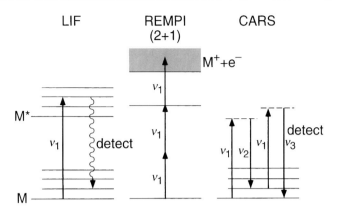

**Figure B2.3.8.** Energy-level schemes describing various optical methods for state-selectively detecting chemical reaction products: left-hand side, laser-induced fluorescence (LIF); centre, resonance-enhanced multiphoton ionization (REMPI); and right-hand side, coherent anti-Stokes Raman spectroscopy (CARS). The ionization continuum is denoted by a shaded area. The dashed lines indicate virtual electronic states. Straight arrows indicate coherent radiation, while a wavy arrow denotes spontaneous emission.

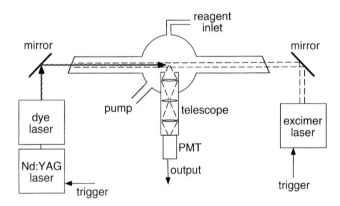

**Figure B2.3.9.** Schematic diagram of an apparatus for laser fluorescence detection of reaction products. The dye laser is synchronized to fire a short delay after the excimer laser pulse, which is used to generate one of the reagents photolytically.

As discussed above, the spectrum must be assigned, i.e. the quantum numbers of the upper and lower levels of the spectral lines must be available. In addition to the line positions, intensity information is also required.

To compare the relative populations of vibrational levels, the intensities of vibrational transitions out of these levels are compared. Figure B2.3.10 displays typical potential energy curves of the ground and an excited electronic state of a diatomic molecule. The intensity of a $(v', v'')$ vibrational transition can be written as

$$I(v', v'') = C N_{v''} p_{v',v''} \qquad (B2.3.11)$$

where $N_{v''}$ is the desired density of product molecules in the vibrational level $v''$, $p_{v',v''}$ is the vibrational band strength [34, 35], and $C$ is a proportionality constant. If the electronic transition moment [36] is constant as a function of the internuclear separation, then $p_{v',v''}$ is proportional to the Franck–Condon factor $q_{v',v''}$, i.e. the square of the overlap integral of the upper and lower vibrational wavefunctions (see figure B2.3.10). The band strengths are usually determined experimentally from measurement of the decay lifetime of the

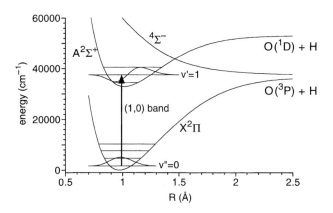

**Figure B2.3.10.** Potential energy curves [42] of the ground $X^2\Pi$ and excited $A^2\Sigma^+$ electronic states of the hydroxyl radical. Several vibrational levels are explicitly drawn in each electronic state. One vibrational transition is explicitly indicated, and the upper and lower vibrational wavefunctions are plotted. The upper and lower state vibrational quantum numbers are denoted $v'$ and $v''$, respectively. Also shown is one of the three repulsive potential energy curves which correlate with the ground $O(^3P) + H$ dissociation asymptote. These cause predissociation of the higher rotational and vibrational levels of the $A^2\Sigma^+$ state.

excited vibrational level and the branching of emission into the various ground state vibrational levels $v''$, as illustrated for the A–X transition of the hydroxyl radical [37]. Alternatively, if the potential energy curves of the two electronic states can be calculated from spectroscopic data, e.g. by the RKR method [38], then Franck–Condon factors can be computed [35] and used to estimate the relative band strengths.

The $H + NO_2 \rightarrow OH + NO$ reaction provides an excellent example of the use of laser fluorescence detection for the elucidation of the dynamics of a chemical reaction. This reaction is a prototype example of a radical–radical reaction in that the reagents and products are all open-shell free radical species. Both the hydroxyl and nitric oxide products can be conveniently detected by electronic excitation in the UV at wavelengths near 226 and 308 nm, respectively. Atlases of rotational line positions for the lowest electronic band systems of these molecules ($A^2\Sigma^+$–$X^2\Pi$ for both) are available [39, 40], and accurate band strengths for transition between various vibrational levels in the ground and excited electronic states have been reported [37, 41]. Because it is crossed by repulsive electronic states correlating with the ground state atoms $O(^3P) + H$ (see figure B2.3.10), the $OH(A^2\Sigma^+)$ state has low fluorescence quantum yields for rotational levels $N' > 25$, 17, and 4 for vibrational levels $v' = 0, 1,$ and 2, respectively [42]. This causes some problems for the detection of higher vibrational levels of the OH product since the intensities of the vibrational bands are strongest for the so-called diagonal bands, i.e. those for which $\Delta v = v' - v'' = 0$ [37]. Because of the excited-state predissociation, the higher levels must be detected through the weaker off-diagonal bands ($\Delta v < 0$).

The dynamics of the $H + NO_2$ reaction have been studied by several different techniques including laser fluorescence excitation of the products, infrared chemiluminescence, crossed-beam scattering, and electron paramagnetic resonance. We highlight the laser fluorescence studies by Sauder and Dagdigian [43] and by Irvine *et al* [44]. In the former experiment, the NO products, in vibrational levels $v \leq 2$, were detected at the intersection of an atomic hydrogen beam, generated by a microwave discharge source, with a pulsed beam of $NO_2$ diluted in Ar. Figure B2.3.11 illustrates an excitation spectrum for the detection of NO products in the ground ($v = 0$) vibrational level. The structure in the spectrum results from the energy differences between the rotational/fine-structure levels in the upper and lower electronic states and the degree of internal excitation of the products. Figure B2.3.12 illustrates the rotational transitions allowed in a $^2\Sigma^+$–$^2\Pi$ electronic transition. With such a diagram it is possible to determine the rotational/fine-structure level being detected through the excitation of a specific rotational line in the spectrum.

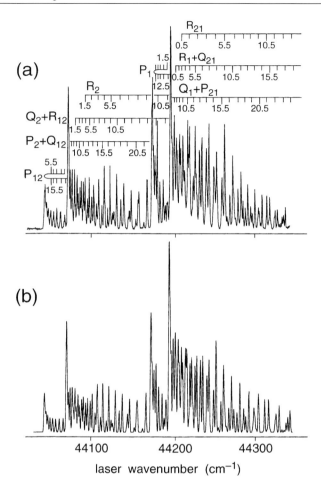

**Figure B2.3.11.** (a) Experimental laser fluorescence excitation spectrum of the $A\,^2\Sigma^+$–$X\,^2\Pi$ (0,0) band for the NO product from the $H + NO_2$ reaction [43]. Individual lines in the various rotational branches are denoted by the total angular momentum $J$ of the lower state. (b) Simulated spectrum with the NO rotational state populations adjusted to reproduce the spectrum in (a). (By permission from AIP.)

Both OH and NO are open-shell free radicals, with doublet electron spin multiplicity. Consequently, the coupling of the angular momentum of the unpaired electron with the angular momentum $N$ of nuclear rotation leads to a more complicated rotational energy level pattern than for a closed-shell molecule ($^1\Sigma^+$ electronic state) [45]. For the upper, $^2\Sigma^+$ electronic state, the electron spin $S = \frac{1}{2}$ can couple with the rotational angular momentum to yield two fine-structure levels, with total angular momenta $J = N + \frac{1}{2}$ and $N - \frac{1}{2}$. These are conventionally [34] denoted $F_1$ and $F_2$, respectively, in order of increasing energy for a given value of $J$. The rotational energy is given approximately by $BN(N + 1)$, where $B$ is the rotational constant, and the splitting of the fine-structure levels is usually much smaller and grows with increasing $N$. This pattern of rotational/fine-structure levels is illustrated in the upper portion of figure B2.3.12.

For the $^2\Pi$ state, the projection $\Lambda = 1$ of the electron orbital angular momentum along the internuclear axis can couple with the projection $\Sigma = \pm\frac{1}{2}$ to yield two spin–orbit levels, $^2\Pi_\Omega$, with $\Omega = \frac{1}{2}$ and $\frac{3}{2}$. The NO($X\,^2\Pi$) state follows so-called Hund's case (a) coupling [34], for which the spin–orbit splitting is much

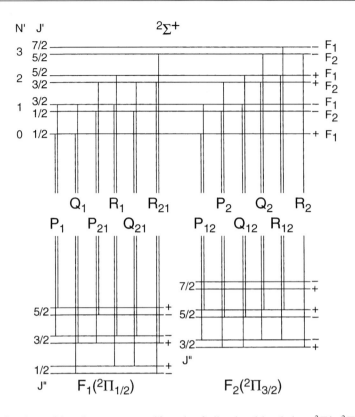

**Figure B2.3.12.** Rotational transitions between a specific pair of vibrational levels in a $^2\Sigma^+$–$^2\Pi$ electronic transition. The total angular momentum $J$ and the parity of the lowest rotational levels in each state are given. Hund's case (a) coupling is assumed for the $^2\Pi$ state. Conventional spectroscopic designations [34] are given for the allowed rotational transitions.

larger than the rotational energy. In this case, the rotational energy within each spin–orbit level is given approximately by $B[J(J + 1) - \Omega^2]$, with half-integral $J \geq \Omega$. It should be noted that a $\Pi$ state is orbitally degenerate, i.e. it has two components of the same energy. In Cartesian notation, these are often indicated as $\Pi_x$ and $\Pi_y$. As a result, the rotational/fine-structure levels appear as nearly degenerate pairs, with opposite parity, or symmetry with respect to reflection of all the coordinates through the space-fixed origin. These pairs of levels are called $\Lambda$-doublets.

For high rotational levels, or for a molecule like OH, for which the spin–orbit splitting is small, even for low $J$, the pattern of rotational/fine-structure levels approaches the Hund's case (b) limit. In this situation, it is not meaningful to speak of the projection quantum number $\Omega$. Rather, we first consider the rotational angular momentum $N$ exclusive of the electron spin. This is then coupled with the spin to yield levels with total angular momentum $J = N + \frac{1}{2}$ and $N - \frac{1}{2}$. As before, there are two nearly degenerate pairs of levels associated with each value of $J$.

The rotational/fine-structure levels of the lower, $^2\Pi$ electronic state in figure B2.3.12 are drawn for a molecule near the case (a) limit since NO falls into this coupling scheme. Also indicated in the figure are the electric-dipole allowed rotational lines, indicated with conventional spectroscopic notation [34]. In the spectrum displayed in figure B2.3.11, the individual rotational lines appear to pile up into so-called 'heads' [34] at four distinct wavenumbers. The splitting between the two pairs of heads can be roughly identified with the NO(X $^2\Pi$) spin–orbit splitting.

In a conventional spectroscopic experiment, the intensity of a rotational transition within a given vibrational band can be written as

$$I(J', J'') = C'[N_{J''}(2J'' + 1)^{-1}]S_{J',J''} \tag{B2.3.12}$$

where $N_{J''}$ is the density of the rotational/fine-structure level, $J''$ is its total angular momentum, $S_{J',J''}$ is the rotational line strength factor [34, 45], and $C'$ is a proportionality constant. The relative intensities of the rotational lines can be used with equation (B2.3.12) to derive the rotational/fine-structure state distribution associated with a given vibrational level. Zare [45] presents a detailed discussion of the calculation of rotational line strength factors for diatomic electronic transitions.

Strictly speaking, equation (B2.3.12) does not apply to a measurement of the concentration through laser-induced fluorescence detection, as would be observed in the apparatus schematically illustrated in figure B2.3.9. The rotational line strength factors $S_{J',J''}$ apply to the situation of isotropic irradiation and detection, which is clearly not the case for irradiation with a unidirectional polarized laser and detection of fluorescence emitted into a specific solid angle. Greene and Zare [46] have considered in detail the correct relationship between the molecular density and the intensity for arbitrary fluorescence excitation and detection geometries. In practice, because of the large angular momentum $J$ often found for reaction products, the factors $S_{J',J''}$ follow fairly closely the $J$-dependence of the correct line strength factors, as long as lines in the same rotational branch are employed.

An additional inadequacy of equation (B2.3.12) is the assumption of an isotropic $M_J$ distribution of product molecules in the detected rotational level. The product could be aligned because of the dynamics of the reaction. An extreme case is that of alignment imposed by kinematics for the mass combination $H + HL \rightarrow HH + L$, where $H$ and $L$ represent heavy and light atoms, respectively. From angular momentum conservation, we have $J_{\text{tot}} = L_i + J_i = L_f + J_f$, where $J_{\text{tot}}$ is the total angular momentum of the system. Here, the vectors $L$ and $J$ are the orbital and rotational angular momenta, respectively, and the subscripts $i$ and $f$ denote the reagents and products, respectively. Because of the small moment of inertia of the $HL$ reagent, $J_i$ will be small. Moreover, $L_f$ will also be small because of the small reduced mass of the HH–L combination of the products. Thus, we have $L_i = J_f$. This then implies that the product rotational angular momentum $J_f$ must be strongly polarized, with an anisotropic $M_J$ distribution, since $L_i$ is perpendicular to the initial relative velocity.

The anisotropy of the product rotational state distribution, or the polarization of the rotational angular momentum, is most conveniently parametrized through multipole moments of the $M_J$ distribution [45]. Odd multipoles, such as the dipole, describe the orientation of the angular momentum $J$, i.e. which way the tips of the $J$ vectors preferentially point. Even multipoles, such as the quadrupole, describe the alignment of $J$, i.e. the spatial distribution of the $J$ vectors, regarded as a collection of double-headed arrows. Orr-Ewing and Zare [47] have discussed in detail the measurement of orientation and alignment in products of chemical reactions and what can be learned about the reaction dynamics from these measurements.

At low laser powers, the fluorescence signal is linearly proportional to the power. However, the power available from most tunable laser systems is sufficient to cause partial saturation of the transition, with the result that the fluorescence intensity is no longer linearly proportional to the probe laser power. While more elaborate treatments have been given [48, 49], saturation can be simply described by a rate-equation model of radiative transitions with the help of the 2-level diagram in figure B2.3.13. If the laser pulse can be approximated as a rectangular pulse of length $T$, then the fraction $f$ of molecules originally in the lower state which are excited is

$$f = [W_{12}/(W_{12} + W_{21} + A_{21})][1 - \exp(-\{W_{12} + W_{21} + A_{21}\}T)]. \tag{B2.3.13}$$

The fluorescence signal is linearly proportional to the fraction $f$ of molecules excited. The absorption rate $W_{12}$ and the stimulated emission rate $W_{21}$ are proportional to the laser power. In the limit of low laser power, $f$ is proportional to the laser power, while this is no longer true at high powers ($W_{12}, W_{21} \gg A_{21}$). Care must

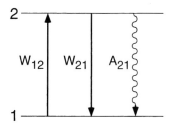

**Figure B2.3.13.** Model 2-level system describing molecular optical excitation, with first-order excitation rate constant $W_{12}$ proportional to the laser power, and spontaneous (first-order rate constant $A_{21}$) and stimulated (first-order rate constant $W_{21}$ proportional to the laser power) emission pathways.

thus be taken in a laser fluorescence experiment to be sure that one is operating in the linear regime, or that proper account of saturation effects is taken, since transitions with different strengths reach saturation at different laser powers.

Following the procedures outlined above, internal state distributions for the products of the $H + NO_2$ reaction have been determined [43, 44, 50]. Comparison of the intensities of various bands of the NO $A\,^2\Sigma^+$–$X\,^2\Pi$ electronic transitions, through equation (B2.3.11), allows determination of the ratio of the populations of the vibrational levels of the NO product, From spectra such as that in figure B2.3.11, the rotational/fine-structure state distribution of the NO product in a particular vibrational level can be deduced. Figure B2.3.14 presents the vibration–rotation state distribution derived for the NO product [43]. The vibrational state populations monotonically decrease with increasing $v$, up to $v = 3$, the highest detected level [50].

In the work of Irvine *et al* [44], the OH product was detected, as illustrated by the fluorescence excitation spectrum in figure B2.3.15. Since the rotational constant of OH is much larger than that of NO, the spectrum is much less congested. Since $OH(X\,^2\Pi)$ follows Hund's case (b) coupling, the spin–orbit splitting is not directly reflected in any separations between rotational lines. The distribution in the product OH rotational/fine-structure levels was determined by the same methods as employed for the analysis of the NO spectrum. The degree of product OH vibrational excitation was found to be significantly greater than for the NO product. The $H + NO_2$ reaction proceeds on the ground state HONO PES, which has a fairly deep well corresponding to the stable nitrous acid molecule. Because of this well, the collision complex has a transient existence. However, its lifetime is not sufficiently long that the available energy is randomized through all the degrees of freedom. The 'new' OH bond is found to have more vibrational energy than the 'old' NO bond.

### B2.3.3.3 *Preparation of reagents*

In most reactive scattering experiments in which the products are detected optically, a thermodynamically labile species is allowed to react with a stable molecular reagent. In many experiments, this involves allowing a beam of the unstable species, prepared in a separately pumped vacuum chamber, to impinge upon the scattering partner in a so-called beam–gas scattering arrangement. The beam is usually prepared by one of the methods described in section B2.3.2.3, e.g. a high-temperature source of a beam of metal atoms or a microwave discharge source for a beam of hydrogen atoms. In some cases, two beams are crossed, and the products are detected in the collision zone [43, 51, 52]. In a few cases, the product of a reaction of two labile reagents have been studied, e.g. the OD product from the $O(^3P) + ND_2$ reaction [53]. In this study, the oxygen atoms were prepared in a microwave discharge source while the $ND_2$ reagent was prepared by laser photolysis of $ND_3$.

Many optical studies have employed a quasi-static cell, through which the photolytic precursor of one of the reagents and the stable molecular reagent are slowly flowed. The reaction is then initiated by laser

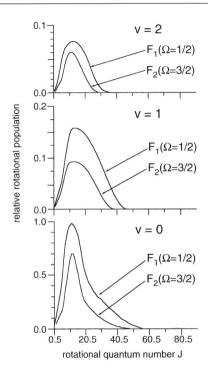

**Figure B2.3.14.** Experimentally derived vibration–rotation populations for the NO product from the H + NO$_2$ reaction [43]. The fine-structure labels $F_1$ and $F_2$ refer to the two ways that the projections $\Sigma$ and $\Lambda$ of the electron spin and orbital angular momenta along the internuclear axis of this open-shell can be coupled ($\Omega = \Lambda + \Sigma$). (By permission from AIP.)

photolysis of the precursor, and the products are detected a short time after the photolysis event. To avoid collisional relaxation of the internal degrees of freedom of the product, the products must be detected in a shorter time when compared to the time between gas-kinetic collisions, that depends inversely upon the total pressure in the cell. In some cases, for example in case of the stable NO product from the H + NO$_2$ reaction discussed in section B2.3.3.2, the products are not removed by collisions with the walls and may have long residence times in the apparatus. Study of such reactions are better carried out with pulsed introduction of the reagents into the cell or under crossed-beam conditions.

*B2.3.3.4    Extraction of angular information (correlations)*

With spectroscopic detection of the products, the angular distribution of the products is usually not measured. In principle, spectroscopic detection of the products can be incorporated into a crossed-beam scattering experiment of the type described in section B2.3.2. There have been relatively few examples of such studies because of the great demands on detection sensitivity. The recent work of Keil and co-workers (Dharmasena *et al* [16]) on the F + H$_2$ reaction, mentioned in section B2.3.3, is an excellent example of the implementation of state-selective optical detection in the measurement of the angular distribution of a reaction product.

The use of photolytically generated reagents in a cell, combined with sub-Doppler detection, has allowed the extraction of information on the angular distribution and also the alignment of the products in experiments carried out in a cell [54]. The theoretical treatment of Shafer *et al* [55] shows how, in principle, the reaction product angular distribution can be extracted from measurement of its laboratory velocity distribution when

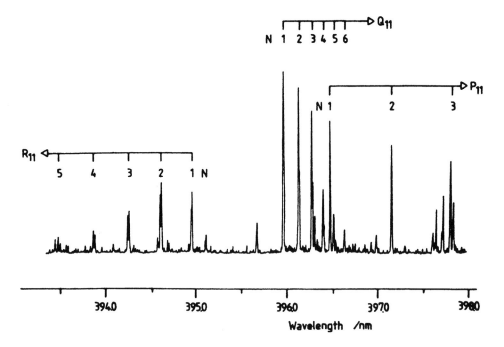

**Figure B2.3.15.** Laser fluorescence excitation spectrum of the A $^2\Sigma^+$–X $^2\Pi$ (1,3) band for the OH product, in the $v = 3$ vibrational level, from the H + NO$_2$ reaction [44]. (By permission from AIP.)

one of the reagents is prepared by photolysis. It is well known that the angular distribution of photolytically formed fragments can be expressed as [45]

$$P(\theta_{\text{phot}}) = [1 + \beta P_2(\cos \theta_{\text{phot}})]/(4\pi) \tag{B2.3.14}$$

where $\theta_{\text{phot}}$ is the angle between the $E$ vector of the photolysis laser and the fragment recoil direction, $P_2(x)$ is the second-order Legendre polynomial, and $\beta$ is the recoil anisotropy parameter. The parameter $\beta$ can vary from $-1$ (perpendicular-type transition) to 2 (parallel-type transition), where the type of transition refers to the direction of fragment recoil relative to the electronic transition moment of the dissociation transition [45].

In the bimolecular collision of the photolytically generated reagent, assumed to have a mass $m_1$ and laboratory speed $v_1$, the centre-of-mass speed will be

$$c = m_1 v_1/(m_1 + m_2) \tag{B2.3.15}$$

if the velocity of the co-reagent (mass $m_2$) can be neglected. In this case, the relative velocity vector $\boldsymbol{v}_{\text{rel}}$ and $\boldsymbol{c}$ are parallel to the laboratory velocity $\boldsymbol{v}_1$ of the photolytically generated reagent. The CM speed $u'_3$ of the detected product can be computed with equation (B2.3.8). Figure B2.3.16 illustrates how the laboratory velocity $\boldsymbol{v}'_3$ of this product is related to the CM scattering angle, $\theta'$. Shafer *et al* [55] show that the laboratory velocity distribution $f(v'_3)$ of the product is related to the CM differential cross section by

$$f(v'_3) = (2\,v'_3\,c\,u'_3)^{-1} \left(\frac{\mathrm{d}\sigma}{\mathrm{d}\omega}\right)_{\text{CM}} [1 + \beta P_2(\cos \alpha) P_2(\cos \theta'_{\text{phot}})] \qquad \text{for } |c - u'_3| < v'_3 < (c + u'_3)$$

$$f(v'_3) = 0 \qquad \text{for } v'_3 < |c - u'_3| \text{ or } v'_3 > (c + u'_3) \tag{B2.3.16}$$

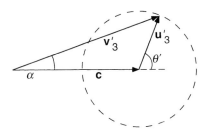

**Figure B2.3.16.** Velocity diagram for the reaction of a photolytically generated reagent with an assumed stationary co-reagent. In this case, the relative velocity $v_{rel}$ of the reagents is parallel to the velocity $c$ of the centre of mass.

where

$$\cos \alpha = [v_3'^2 + c^2 - u_3'^2]/(2\, v_3' c) \tag{B2.3.17}$$

and $\theta_{phot}'$ is the angle between $v_3'$ and the $\mathbf{E}$ vector of the photolysis laser.

There are several practical limitations to the use of equation (B2.3.16) for the determination of CM angular distributions. The optimum kinematics for the use of this equation is the case where the speed $c$ of the centre of mass is approximately equal to the product CM speed $u_3'$. In the limiting case where the latter is small, the product laboratory distribution is dominated by the angular distribution of the velocity $c$ of the centre of mass and nothing can be learned about the product CM angular distribution. In the opposite limiting case where $u_3'$ is much larger than $c$, the angular distribution of $v_3'$ is limited by the angular distribution of the photolytically-prepared reagent. Equation (B2.3.16) assumes that the velocity of the co-reagent can be neglected. This applies to the situation where the reagents are pre-cooled in a supersonic beam expansion [56]. The effects of thermal averaging have been considered to describe photo-initiated reactors in room-temperature cells [54]. Information on the anisotropy of the rotational angular momentum can also be determined through the study of photo-initiated reactions by variation of the direction of the $\mathbf{E}$ vector of the probe laser [54, 57].

*B2.3.3.5   Other spectroscopic techniques*

In addition to laser fluorescence excitation, several other laser spectroscopic methods have been found to be useful for the state-selective and sensitive detection of products of reactive collisions: resonance-enhanced multiphoton ionization [58], coherent anti-Stokes Raman scattering [59], bolometric detection with laser excitation [30], and direct infrared absorption [7]. Several additional laser techniques have been developed for use in spectroscopic studies or for diagnostics in reacting systems. Of these, four-wave mixing [60] is applicable to studies of reaction dynamics although it does have a somewhat lower sensitivity than the techniques mentioned above.

The most widely used of these techniques is resonance-enhanced multiphoton ionization (REMPI) [58]. A schematic energy-level diagram of the most commonly employed variant (2 + 1) of this detection scheme is illustrated in the centre of figure B2.3.8. The molecules are irradiated with the focused output of a tunable laser. As the wavelength of the laser is tuned through a 2-photon transition in the molecule, there will be some electronic excitation. If the photon energy is sufficiently large that the ionization continuum can be reached by absorption of an additional photon by the molecule, then molecular ions can be efficiently produced. While non-resonant laser ionization is possible, the efficiency of ionization will be strongly enhanced at a 2-photon resonance in the molecule. This particular resonant ionization scheme is called 2 + 1 REMPI. Other ionization schemes are possible, with different numbers of photons required for electronic excitation and ionization of the molecule.

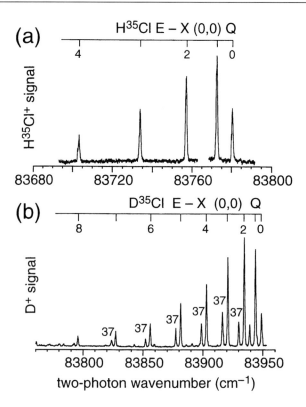

**Figure B2.3.17.** REMPI spectra of the HCl and DCl products from the reaction of Cl atoms with $(CH_3)_3CD$ [63]. The mass 36 and 2 ion signals are plotted as a function of the 2-photon wavenumber. Assignments of the $Q$-branch lines ($\Delta J = 0$) of the $E\,^1\Sigma^+$–$X\,^1\Sigma^+$ (0,0) bands of $H^{35}Cl$ and $D^{35}Cl$ are given. In (b), both the $D^{35}Cl$ and $D^{37}Cl$ isotopomers are observed since $D^+$ ions are monitored.

A REMPI spectrum is usually recorded by monitoring the molecular ion signal as the laser wavelength is scanned, although it is also possible to record the spectrum by monitoring the photoelectron signal. In most applications of REMPI, the laser-produced ions are mass analysed in a time-of-flight mass spectrometer (TOFMS) [61] in order to detect the desired molecular ion in the presence of background ions, for example from non-resonant ionization of other species in the reaction chamber. This mass discrimination is particularly important in probing chemical reaction products since the product ion signal could be obscured by a small degree of non-resonant ionization of reagent molecules present at much higher concentrations than the product.

This technique can be used both to permit the spectroscopic detection of molecules, such as $H_2$ and HCl, whose first electronic transition lies in the vacuum ultraviolet spectral region, for which laser excitation is possible but inconvenient [62], or molecules such as $CH_3$ that do not fluoresce. With 2-photon excitation, the required wavelengths are in the ultraviolet, conveniently generated by frequency-doubled dye lasers, rather than 1-photon excitation in the vacuum ultraviolet. Figure B2.3.17 displays $2 + 1$ REMPI spectra of the HCl and DCl products, both in their $v = 0$ vibrational levels, from the Cl + $(CH_3)_3CD$ reaction [63]. For some electronic states of HCl/DCl, both parent and fragment ions are produced, and the spectrum in figure B2.3.17 for the DCl product was recorded by monitoring mass 2 ($D^+$) ions. In this case, both isotopomers ($D^{35}Cl$ and $D^{37}Cl$) are detected.

In the ideal case for REMPI, the efficiency of ion production is proportional to the line strength factors for 2-photon excitation [64], since the ionization step can be taken to have a wavelength- and state-independent efficiency. In actual practice, fragment ions can be produced upon absorption of a fourth photon, or the ionization efficiency can be reduced through predissociation of the electronically excited state. It is advisable to employ experimentally measured ionization efficiency line strength factors to calibrate the detection sensitivity. With sufficient knowledge of the excited molecular electronic states, it is possible to understand the state dependence of these intensity factors [65].

Product angular and velocity distributions can be measured with REMPI detection, similar to Doppler probing in a laser-induced fluorescence experiment discussed in section B2.3.3.4. With appropriate time- and space-resolved ion detection, it is possible, in principle, to determine the three-dimensional velocity distribution of a product (see equation (B2.3.16)). The time-of-arrival of a particular mass in the TOFMS will be broadened by the velocity of the neutral molecule being detected. In some modes of operation of a TOFMS, e.g. space-focusing conditions [61], the shift of the arrival time from the centre of a mass peak is proportional to the projection of the molecular velocity along the TOFMS axis. In addition, Doppler tuning of the probe laser allows one component of the velocity perpendicular to the TOFMS axis to be determined. A more general approach for the two-dimensional velocity distribution in the plane perpendicular to the TOFMS direction involves the use of imaging detectors [66].

Welge and co-workers (Schnieder *et al* [67]) have developed a resonant ionization technique for hydrogen atoms which allows the determination of the velocity to $\sim$0.3% by a time-of-flight method. The hydrogen atoms are sequentially irradiated in the detection zone with a 121.6 nm laser, which is resonant with the $n = 2 \leftarrow n = 1$ transition, and $\sim$365 nm light to produce high-$n$ Rydberg atoms:

$$H(n = 1) + 121.6\,\text{nm} \rightarrow H^*(n = 2) + \sim 365\,\text{nm} \rightarrow H^{**}(n \approx 40). \tag{B2.3.18}$$

The Rydberg atoms are allowed to drift through a $\sim$1 m flight path, after which they are field ionized with a strong electric field, and the resulting ions collected with a particle detector. The key to the high-velocity resolution of this technique is to ionize the atoms far from the laser interaction region. By contrast, when the ions are produced in this region, space-charge effects can lead to a significant velocity spread. This detection technique has been applied to the study of the H+D$_2$ reaction through measurement of the velocity distribution of the D atom products [68]. The laboratory velocities of D atoms formed in coincidence with specific HD vibration–rotation states were resolved, and angularly resolved differential cross sections for the formation of HD products in specific states were determined. This H atom detection technique has also been extensively employed for the study of the dynamics of the photodissociation of hydride molecules [69].

Coherent anti-Stokes Raman spectroscopy (CARS) [59] has also found utility in the determination of the internal state distributions of products of chemical reactions. This is one of several coherent Raman spectroscopies based on the existence of vibrational resonances in four-wave mixing. As illustrated in the schematic energy level diagram in the right-hand side of figure B2.3.8, electric fields at three frequencies are mixed to produce a fourth field. In most CARS experiments, $\nu_1$ is held fixed, usually at 532 nm, the second harmonic of a Nd:YAG laser output, while $\nu_2$ is scanned. The intensity of the output field at $\nu_3$ is enhanced whenever the difference $\nu_1 - \nu_2$ equals the energy difference between two molecular levels connected by a Raman transition. Unlike the normal, spontaneous Raman process, this mixing is coherent, and the output light is a coherent beam propagating in a particular direction. The high intensity and directionality provides a great increase in detection sensitivity. In reactive scattering experiments, the CARS technique has been employed for the determination of the vibration–rotation state distributions of H$_2$ and HD products in reactions yielding hydrogen molecular products, e.g. the H + D$_2$ [70] and H + HX (X = halogen) [71] reactions.

Recently, the state-selective detection of reaction products through infrared absorption on vibrational transitions has been achieved and applied to the study of HF products from the F + H$_2$ reaction by Nesbitt and co-workers (Chapman *et al* [7]). The relatively low sensitivity for direct absorption has been circumvented by the use of a multi-pass absorption arrangement with a narrow-band tunable infrared laser and dual beam differential detection of the incident and transmission beams on matched detectors. A particular advantage of probing the products through absorption is that the absolute concentration of the product molecules in a given vibration–rotation state can be determined.

## B2.3.4   Conclusion

The molecular beam and laser techniques described in this section, especially in combination with theoretical treatments using accurate PESs and a quantum mechanical description of the collisional event, have revealed considerable detail about the dynamics of chemical reactions. Several aspects of reactive scattering are currently drawing special attention. The measurement of vector correlations, for example as described in section B2.3.3.4, continue to be of particular interest, especially the interplay between the product angular distribution and rotational polarization.

In most theoretical treatments of the collision dynamics, the reaction is assumed to proceed on a single PES. However, reactions involving open-shell reagents of products will involve several PESs. For example, in the $F + H_2$ reaction, discussed in section B2.3.2.4, three PESs emanate from the separated reagents, of which only one leads to the $H + HF$ products. The F atom ground $^2P_{3/2}$ spin–orbit state is connected to the products by this reactive PES, while the excited $^2P_{1/2}$ state is not. Nevertheless, the $^2P_{1/2}$ state does react with $F_2$ to a small extent because of non-adiabatic transitions between the PESs. There is considerable current interest in elucidating the role of such non-adiabatic transitions in collision dynamics.

The reaction of an atom with a diatomic molecule is the prototype of a chemical reaction. As the dynamics of a number of atom–diatom reactions are being understood in detail, attention is now being turned to the study of the dynamics of reactions involving larger molecules. The reaction of Cl atoms with small aliphatic hydrocarbons is an example of the type of polyatomic reactions which are now being studied [56, 63, 72, 73]. The idea of controlling the outcome of a chemical reaction by exciting a particular bond in a reagent has long held considerable appeal. Such bond-selected chemistry has been achieved with simple triatomic reagents such as partially deuterated water, HOD, by preparation of the reagent in a suitable vibrational level [74]. Current interest is focused on the extension to larger reagents. This is more difficult than in triatomics because of intramolecular redistribution of the initial excitation [75], which becomes more rapid in larger molecules.

## References

[1]   Dawson P H 1976 *Quadrupole Mass Spectrometry and Its Applications* (Amsterdam: Elsevier)
[2]   Lee Y T, McDonald J D, LeBreton P R and Herschbach D R 1969 Molecular beam reactive scattering apparatus with electron bombardment detector *Rev. Sci. Instrum.* **40** 1402–8
[3]   Auerbach D J 1988 Velocity measurements by time-of-flight methods *Atomic and Molecular Beam Methods* vol 1, ed G Scoles *et al* (New York: Oxford University Press) pp 362–79
[4]   Carrington T and Polanyi J C 1972 Chemiluminescent reactions *Chemical Kinetics, Int. Rev. Sci. Physical Chemistry* series 1, vol 9, ed J C Polanyi (London: Butterworths) pp 135–71
[5]   Leone S R 1983 Infrared fluorescence: a versatile probe of state-selected chemical dynamics *Acc. Chem. Res.* **16** 88–95
[6]   Sloan J J 1992 Fourier-transform methods: infrared *Atomic and Molecular Beam Methods* vol 2, ed G Scoles, D Lainé and U Valbusa (New York: Oxford University Press) pp 309–23
[7]   Chapman W B, Blackman B W, Nizkorodov S and Nesbitt D J 1998 Quantum-state resolved reactive scattering of $F + H_2$ in supersonic jets: Nascent HF($v, J$) rovibrational distributions via IR laser direct absorption methods *J. Chem. Phys.* **109** 9306–17
[8]   Lee Y T 1988 Reactive scattering I: nonoptical methods *Atomic and Molecular Beam Methods* vol 1, ed G Scoles *et al* (New York: Oxford University Press) pp 553–68
[9]   Dagdigian P J 1988 Reactive scattering II: optical methods *Atomic and Molecular Beam Methods* vol 1, ed G Scoles *et al* (New York: Oxford University Press) pp 596–629
[10]  van den Meijdenberg C J N 1988 Velocity selection by mechanical methods *Atomic and Molecular Beam Methods* vol 1, ed G Scoles *et al* (New York: Oxford University Press) pp 345–61

[11] Miller D R 1988 Free jet sources *Atomic and Molecular Beam Methods* vol 1, ed G Scoles *et al* (New York: Oxford University Press) pp 14–53

[12] Yang X, Lin J, Lee Y T, Blank D A, Suits A G and Wodtke A M 1997 Universal crossed molecular beams apparatus with synchrotron photoionization mass spectrometric product detection *Rev. Sci. Instrum.* **68** 3317–26

[13] Catchen G L, Husain J and Zare R N 1978 Scattering kinematics: transformation of differential cross sections between two moving frames *J. Chem. Phys.* **69** 1737–41

[14] Siska P E 1973 Iterative unfolding of intensity data, with application to molecular beam scattering *J. Chem. Phys.* **59** 6052–60

[15] Valentini J J, Coggiola M J and Lee Y T 1977 Supersonic atomic and molecular halogen nozzle beam source *Rev. Sci. Instrum.* **48** 58–63

[16] Dharmasena G, Copeland K, Young J H, Lasell R A, Phillips T R, Parker G A and Keil M 1997 Angular dependence for $v'$, $j'$-resolved states in $F + H_2 \rightarrow HF(v', j') + H$ reactive scattering using a new atomic beam source *J. Phys. Chem.* A **101** 6429–40

[17] Faubel M, Martinez-Haya B, Rusin L Y, Tappe U and Toennies J P 1996 An intense fluorine atom beam source *J. Phys. D: Appl. Phys.* **29** 1885–93

[18] Sibener S J, Buss R J, Ng C Y and Lee Y T 1980 Development of a supersonic $O(^3P_J)$, $O(^1D_2)$ atomic oxygen nozzle beam source *Rev. Sci. Instrum.* **51** 167–82

[19] Gorry P A and Grice R 1979 Microwave discharge source for the production of supersonic atom and free radical beams *J. Phys. E: Sci. Instrum.* **12** 857–60

[20] Casavecchia P, Balucani N and Volpi G G 1993 Reaction dynamics of $O(^3P)$, $O(^1D)$ and $OH(X\,^2\Pi)$ with simple molecules *Research in Chemical Kinetics* vol 1, ed R G Compton and G Hancock (Amsterdam: Elsevier) pp 1–63

[21] Irvin J A and Dagdigian P J 1980 Chemiluminescence from the $Ca(4s3d\,^1D) + O_2$ reaction: absolute cross sections, photon yield, and CaO dissociation energy *J. Chem. Phys.* **73** 176–82

[22] Dietz T G, Duncan M A, Powers D E and Smalley R E 1981 Laser production of supersonic metal cluster beams *J. Chem. Phys.* **74** 6511–12

[23] Kaiser R I and Suits A G 1995 A high-intensity, pulsed supersonic carbon course with $C(^3P_j)$ kinetic energies of 0.08–0.7 eV for crossed beam experiments *Rev. Sci. Instrum.* **66** 5405–11

[24] Continetti R E, Balko B A and Lee Y T 1990 Crossed molecular beams study of the reaction $D + H_2 \rightarrow DH + H$ at collision energies of 0.53 and 1.01 eV *J. Chem. Phys.* **93** 5719–40

[25] Neumark D M, Wodtke A M, Robinson G N, Hayden C C and Lee Y T 1985 Molecular beam studies of the $H + H_2$ reaction *J. Chem. Phys.* **82** 3045–66

[26] Neumark D M, Wodtke A M, Robinson G N, Hayden C C, Shobatake K, Sparks R K, Schafer T P and Lee Y T 1985 Molecular beam studies of the $F + D_2$ and $F + HD$ reactions *J. Chem. Phys.* **82** 3067–77

[27] Faubel M, Rusin L, Schlemmer S, Sondermann F, Tappe U and Toennies J P 1994 A high resolution crossed molecular beam investigation of the absolute cross sections and product rotational states for the reaction $F + D_2(v_i = 0, j_i = 0, 1) \rightarrow DF(v_f, j_i) + D$ *J. Chem. Phys.* **101** 2106–25

[28] Faubel M, Martinez-Haya R, Rusin L Y, Tappe U and Toennies J P 1997 Experimental absolute cross sections for the reaction $F + D_2$ at collision energies 90–240 meV *J. Phys. Chem.* A **101** 6415–28

[29] Lee Y T 1987 Molecular beam studies of elementary chemical processes *Science* **236** 793–8

[30] Miller R E 1995 Near-infrared laser optothermal techniques *Laser Techniques in Chemistry* vol 23, ed A B Myers and T R Rizzo (New York: Wiley) pp 43–69

[31] Kohse-Höinghaus K 1994 Laser techniques for the quantitative detection of reactive intermediates in combustion systems *Proc. Energy Combust. Sci.* **20** 203–79

[32] Crosley D R 1995 The measurement of OH and $HO_2$ in the atmosphere *J. Am. Sci.* **52** 3299–314

[33] Okabe H 1978 *Photochemistry of Small Molecules* (New York: Wiley)

[34] Herzberg G 1950 *Molecular Spectra and Molecular Structure I. Spectra of Diatomic Molecules* 2nd edn (Princeton: Van Nostrand)

[35] Zare R N 1964 Calculation of intensity distribution in the vibrational structure of electronic transitions: the $B\,^3\Pi_{0+u}$–$X\,^1\Sigma_{0+g}$ resonance series of molecular iodine *J. Chem. Phys.* **40** 1934–44

[36] Whiting E E, Schadee A, Tatum J B, Hougen J T and Nicholls R W 1980 Recommended conventions for defining transition moments and intensity factors in diatomic molecular spectra *J. Molec. Spectrosc.* **80** 249–56

[37] Chidsey I L and Crosley D R 1980 Calculated rotational transition probabilities for the A–X system of OH *J. Quant. Spectrosc. Radiat. Transfer* **23** 187–99

[38] Tellinghuisen J A 1974 A fast quadrature method for computing diatomic RKR potential energy curves *Comput. Phys. Commun.* **6** 221–8

[39] Dieke G H and Crosswhite H M 1963 The ultraviolet bands of OH: fundamental data *J. Quant. Spectrosc. Radiat. Transfer* **2** 97–199

[40] Engleman R Jr, Rouse P E, Peek H M and Biamonte V D 1970 Beta and gamma band systems of nitric oxide *Los Alamos Scientific Laboratory Report* no LA-4364

[41] Piper L G and Cowles L M 1986 Einstein coefficients and transition moment variation for the $NO(A\,^2\Sigma^+$–$X\,^2\Pi)$ transition *J. Chem. Phys.* **85** 2419–22

[42] Yarkony D R 1992 A theoretical treatment of the predissociation of the individual rovibronic levels of $OH/OD(A\,^2\Sigma^+)$ *J. Chem. Phys.* **97** 1838–49

[43] Sauder D G and Dagdigian P J 1990 Determination of the internal state distribution of NO produced from the H + NO$_2$ reaction *J. Chem. Phys.* **92** 2389–96

[44] Irvine A M L, Smith I W M, Tuckett R P and Yang X-F 1990 A laser-induced fluorescence determination of the complete internal state distribution of OH produced in the reaction: H + NO$_2$ → OH + NO *J. Chem. Phys.* **93** 3177–86

[45] Zare R N 1988 *Angular Momentum* (New York: Wiley)

[46] Greene C H and Zare R N 1983 Determination of product population and alignment using laser-induced fluorescence *J. Chem. Phys.* **78** 6741–53

[47] Orr-Ewing A J and Zare R N 1995 Orientation and alignment of the products of bimolecular reactions *The Chemical Dynamics and Kinetics of Small Radicals* vol 2, ed K Liu and A Wagner (Singapore: World Scientific) pp 936–1063

[48] Altkorn R and Zare R N 1984 Effects of saturation on laser-induced fluorescence measurements of population and polarization *Ann. Rev. Phys. Chem.* **35** 265–89

[49] Hefter U and Bergmann K 1988 Spectroscopic detection methods *Atomic and Molecular Beam Methods* vol 1, ed G Scoles *et al* (New York: Oxford University Press) pp 193–253

[50] Irvine A M L, Smith I W M and Tuckett R P 1990 A laser-induced fluorescence determination of the internal state distribution of NO produced in the reaction H + NO$_2$ → OH + NO *J. Chem. Phys.* **93** 3187–95

[51] Liu K, Macdonald R G and Wagner A F 1990 Crossed-beam investigations of state-resolved collision dynamics of simple radicals *Int. Rev. Phys. Chem.* **9** 187–225

[52] Scott D C, Winterbottom F, Scholfield M R, Goyal S and Reisler H 1994 Kinetic energy effects on product state distributions in the C($^3$P) + N$_2$O(X $^1\Sigma^+$) reaction: energy partitioning between the NO(X $^2\Pi$) and CN(X $^2\Sigma^+$) products *Chem. Phys. Lett.* **222** 471–80

[53] Patel-Misra D, Sauder D G and Dagdigian P J 1991 Internal state distribution of OD produced from the O($^3$P) + ND$_2$ reaction *J. Chem. Phys.* **95** 955–62

[54] Alexander A J, Brouard M, Kalogerakis K S and Simons J P 1998 Chemistry with a sense of direction—the stereodynamics of bimolecular reactions *Chem. Soc. Rev.* **27** 405–15

[55] Shafer N E, Orr-Ewing A J, Simpson W R, Xu H and Zare R N 1993 State-to-state differential cross sections from photoinitiated bulk reactions *Chem. Phys. Lett.* **212** 155–162

[56] Simpson W R, Orr-Ewing A J and Zare R N 1993 State-to-state differential cross sections for the reaction Cl($^2$P$_{3/2}$) + CH$_4$($v_3 = 1$, $J = 1$) → HCl($v' = 1$, $J'$) + CH$_3$ *Chem. Phys. Lett.* **212** 163–71

[57] Rakitzis T P, Kandel S A and Zare R N 1997 Determination of differential-cross-section moments from polarization-dependent product velocity distributions of photoinitiated bimolecular reactions *J. Chem. Phys.* **107** 9382–91

[58] Ashfold M N R and Howe J D 1994 Multiphoton spectroscopy of molecular species 1994 *Ann. Rev. Phys. Chem.* **45** 57–82

[59] Valentini J J 1985 Coherent anti-Stokes spectroscopy *Spectrometric Techniques* vol 4, ed G A Vanasse (New York: Academic) pp 1–62

[60] Vaccaro P H 1995 Resonant four-wave mixing spectroscopy: a new probe for vibrationally-excited species *Molecular Dynamics and Spectroscopy by Stimulated Emission Pumping (Advances in Chemistry Series)* vol 7, ed H-L Dai and R W Field (Singapore: World Scientific) p 1

[61] Wiley W C and McLaren I H 1955 Time-of-flight mass spectrometer with improved resolution *Rev. Sci. Instrum.* **26** 1150–7

[62] Hepburn J W 1995 Generation of coherent vacuum ultraviolet radiation: applications to high-resolution photoionization and photoelectron spectroscopy *Laser Techniques in Chemistry* vol 23, ed A B Myers and T R Rizzo (New York: Wiley) pp 149–83

[63] Varley D F and Dagdigian P J 1996 Product state resolved study of the Cl + (CH$_3$)$_3$CD reaction: comparison of the dynamics of abstraction of primary vs tertiary hydrogens *J. Phys. Chem.* **100** 4365–74

[64] Kummel A C, Sitz G O and Zare R N 1986 Determination of population and alignment of the ground state using two-photon nonresonant excitation *J. Chem. Phys.* **85** 6874–97

[65] Dagdigian P J, Varley D F, Liyanage R, Gordon R J and Field R W 1996 Detection of DCl by multiphoton ionization and determination of DCl and HCl internal state distributions *J. Chem. Phys.* **106** 10 251–62

[66] Heck A J R and Chandler D W 1995 Imaging techniques for the study of chemical reaction dynamics *Ann. Rev. Phys. Chem.* **46** 335–72

[67] Schnieder L, Maier W, Welge K H, Ashfold M N R and Western C M 1990 Photodissociation dynamics of H$_2$S at 121.6 nm and a determination of the potential energy function of SH(A $^2\Sigma^+$) *J. Chem. Phys.* **92** 7027–37

[68] Schnieder L, Seekamp-Rahn K, Wiede E and Welge K H 1997 Experimental determination of quantum state resolved differential cross sections for the hydrogen exchange reaction H + D$_2$ → HD + D *J. Chem. Phys.* **107** 6175–95

[69] Ashfold M N R, Mordaunt D H and Wilson S H S 1996 Photodissociation dynamics of hydride molecules: H atom photofragment translational spectroscopy *Adv. Photochem.* **21** 217–95

[70] Gerrity D P and Valentini J J 1984 Experimental study of the dynamics of the H + D$_2$ → HD + D reaction at collision energies of 0.55 and 1.30 eV *J. Chem. Phys.* **81** 1298–313

[71] Aker P M, Germann G J and Valentini J J 1989 State-to-state dynamics of H + HX collisions. I. The H + HX → H$_2$ + X (X = Cl, Br, I) abstraction reactions at 1.6 eV collision energy *J. Chem. Phys.* **90** 4795–808

[72] Hemmi N and Suits A G 1998 The dynamics of hydrogen abstraction reactions: crossed-beam reaction Cl + $n$-C$_5$H$_{12}$ → C$_5$H$_{11}$ + HCl *J. Chem. Phys.* **109** 5338–43

[73]  Kandel S A and Zare R N 1998 Reaction dynamics of atomic chlorine with methane: importance of methane bending and tortional excitation in controlling reactivity *J. Chem. Phys.* **109** 9719–27

[74]  Crim F F 1996 Bond-selected chemistry: vibrational state control of photodissociation and bimolecular reaction *J. Phys. Chem.* **100** 12 725–34

[75]  Nesbitt D J and Field R W 1998 Vibrational energy flow in highly excited molecules: role of intramolecular vibrational redistribution *J. Chem. Phys.* **100** 12 735–56

## Further Reading

Scoles G, Bassi D, Buck U and Lainé D (eds) 1988 *Atomic and Molecular Beam Methods* vol 1 (New York: Oxford University Press)

This book presents an extensive and detailed description of basic techniques for the generation and detection of atomic and molecular beams, as well as beam techniques for the study of molecular scattering processes.

Zare R N 1988 *Angular Momentum* (New York: Wiley)

This book presents a detailed exposition of angular momentum theory in quantum mechanics, with numerous applications and problems in chemical physics. Of particular relevance to the present section is an elegant and clear discussion of molecular wavefunctions and the determination of populations and moments of the rotational state distributions from polarized laser fluorescence excitation experiments.

Herzberg G 1989 *Molecular Spectra and Molecular Structure. I. Spectra of Diatomic Molecules* reprint (Malabar, FL: Krieger)

This book, originally published in 1950, is the first of a classic three-volume set on molecular spectroscopy. A rather complete discussion of diatomic electronic spectroscopy is presented. Volumes II (1945) and III (1967) discuss infrared and Raman spectroscopy and polyatomic electronic spectroscopy, respectively.

# B2.4
# NMR methods for studying exchanging systems

*Alex D Bain*

### B2.4.1   Introduction

No molecule is completely rigid and fixed. Molecules vibrate, parts of a molecule may rotate internally, weak bonds break and re-form. Nuclear magnetic resonance spectroscopy (NMR) is particularly well suited to observe an important class of these motions and rearrangements. An example is the restricted rotation about bonds, which can cause dramatic effects in the NMR spectrum (figure B2.4.1).

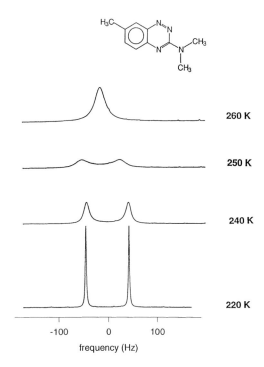

**Figure B2.4.1.** Proton NMR spectra of the N,N′-dimethyl groups in 3-dimethylamino-7-methyl-1,2,4-benzotriazine, as a function of temperature. Because of partial double-bond character, there is restricted rotation about the bond between the dimethylamino group and the ring. As the temperature is raised, the rate of rotation around the bond increases and the NMR signals of the two methyl groups broaden and coalesce.

These exchanges often occur while the system is in macroscopic equilibrium—the sample itself remains the same and the dynamics may be invisible to other techniques. It is merely the environment of a given nucleus that changes. Since NMR follows an individual nucleus, it can easily follow these dynamic processes. This is just one of several reasons that the study of chemical exchange by NMR is important.

First is the observation of the phenomenon itself—the exchange of axial and equatorial ligands in trigonal bipyramidal species, the scrambling of carbonyl ligands in metal complexes and dynamic behaviour of rings was mainly revealed and studied in detail by NMR methods. Not only does NMR give a detailed picture of the mechanism of the exchange, but it also provides excellent ways of measuring the reaction rate.

Secondly, NMR is a good example of spectroscopy in general. The spectroscopic transition probability can be shown to have a simple physical interpretation in NMR: the total magnetization is divided amongst individual observable transitions and the intensity of a transition is related to its share of the total. This can be further generalized to exchanging systems, in which the transition probability now becomes a complex number. The exchange lineshapes can be decomposed into a sum of transitions, whose phase, intensity, position and linewidths are governed by the real and imaginary parts of the transition probability.

Finally, exchange is a kinetic process and governed by absolute rate theory. Therefore, study of the rate as a function of temperature can provide thermodynamic data on the transition state, according to equation (B2.4.1). This equation, in which $k$ is Boltzmann's constant and $h$ is Planck's constant, relates the observed rate to the Gibbs free energy of activation, $\Delta G^{\dagger}$.

$$\text{Rate} = \frac{kT}{h}\, e^{-\Delta G^{\dagger}/RT} = \frac{kT}{h}\, e^{-\Delta H^{\dagger}/RT}\, e^{\Delta S^{\dagger}/R}. \tag{B2.4.1}$$

In order to separate the enthalpy and the entropy of activation, the rate is measured as a function of temperature. These data should give a straight line on an Eyring plot of $\log(\text{rate}/T)$ against $(1/T)$ (figure B2.4.2). The slope of the line gives $\Delta H^{\dagger}$, and the intercept at $1/T = 0$ is related to $\Delta S^{\dagger}$. A unimolecular reaction, such as many cases of exchange, might be expected to have a very small entropy change on going to the transition state. However, several systems have shown significant entropy contributions—entropy can make up more than 10% of the barrier. It is therefore important to measure the rates over as wide a range of temperatures as possible to obtain reliable thermodynamic data on the transition state.

There are several ways of measuring exchange rates with NMR, each with its own optimum range. Figure B2.4.2 illustrates this. A combination of spin–spin relaxation time ($T_2$) measurements at high temperature, bandshape analysis in the intermediate regime and selective inversions at low temperature, gives rates over a range of about five orders of magnitude. This provides a very well defined Eyring plot.

The first fully analysed example of chemical exchange in NMR was the proton spectrum of dimethylamides [1], observed at about the same time (and in the same laboratory) as the phenomenon of scalar coupling between nuclei. Figure B2.4.1 illustrates this type of behaviour. If there is no rotation about the bond joining the N, N'-dimethyl group to the ring, the proton NMR signals of the two methyl groups will have different chemical shifts. If the rotation were very fast, then the two methyl environments would be exchanged very quickly and only a single, average, methyl peak would appear in the proton NMR spectrum. Between these two extremes, spectra like those in figure B2.4.1 are observed. At low temperature, when the rate is slow, two sharp lines are seen. As the temperature is increased, the exchange becomes faster, the lines broaden, coalesce into a single line and finally become a single sharp line at the average chemical shift. This is perhaps the most familiar manifestation of chemical exchange in NMR.

The different types of chemical exchange in NMR are classified according to the rate relative to some NMR timescale. The example in figure B2.4.1 is called intermediate exchange, in which the exchange rate is comparable to the chemical shift differences and coupling constants. Intermediate exchange gives an array

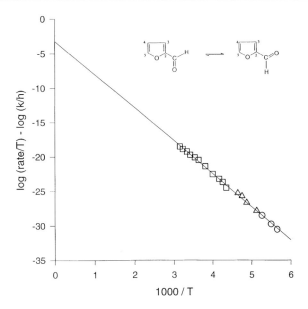

**Figure B2.4.2.** Eyring plot of $\log(\text{rate}/T)$ versus $(1/T)$, where $T$ is absolute temperature, for the *cis–trans* isomerism of the aldehyde group in furfural. Rates were obtained from three different experiments: $T_2$ measurements (squares), bandshapes (triangles) and selective inversions (circles). The line is a linear regression to the data. The slope of the line is $\Delta H^{\dagger}/R$, and the intercept at $1/T = 0$ is $\Delta S^{\dagger}/R$, where $R$ is the gas constant. $\Delta H^{\dagger}$ and $\Delta S^{\dagger}$ are the enthalpy and entropy of activation, according to equation (B2.4.1).

of unusual and characteristic lineshapes in the spectrum, which can be explained quite neatly in terms of a generalized transition probability. Fast exchange is the regime well after coalescence, when only a single line is observed. There is still observable broadening due to exchange, and rates are measured from the linewidth, or equivalently, the spin–spin relaxation time, $T_2$. In slow exchange, no dramatic line broadening is observed, but the exchange rate is comparable to the reciprocal of the spin–lattice relaxation time, $T_1$. In this regime, modifications of the inversion-recovery experiment, or techniques related to the nuclear Overhauser effect (NOE), are used to measure rates.

The timescale is just one sub-classification of chemical exchange. It can be further divided into coupled *versus* uncoupled systems, mutual or non-mutual exchange, inter- or intra-molecular processes and solids *versus* liquids. However, all of these can be treated in a consistent and clear fashion.

The NMR experimental methods for studying chemical exchange are all fairly routine experiments, used in many other NMR contexts. To interpret these results, a numerical model of the exchange, as a function of rate, is fitted to the experimental data. It is therefore necessary to look at the theory behind the effects of chemical exchange. Much of the theory is developed for intermediate exchange, and this is the most complex case. However, with this theory, all of the rest of chemical exchange can be understood.

## B2.4.2   Intermediate exchange

### B2.4.2.1   Introduction

Figure B2.4.1 shows the lineshape for intermediate chemical exchange between two equally populated sites without scalar coupling. For more complicated spin systems, the lineshapes are more complicated

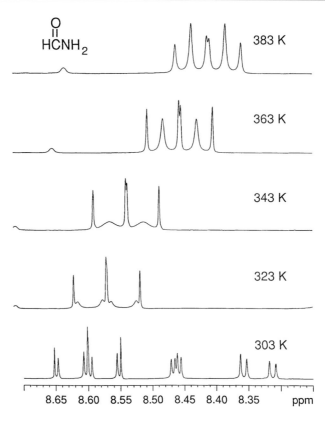

**Figure B2.4.3.** Proton NMR spectrum of the aldehyde proton in $^{15}$N-labelled formamide. This proton has couplings of 1.76 Hz and 13.55 Hz to the two amino protons, and a coupling of 15.0 Hz to the $^{15}$N nucleus. The outer lines in the spectrum remain sharp, since they represent the sum of the couplings, which is unaffected by the exchange. The inner lines of the multiplet broaden and coalesce, as in figure B2.4.1. The other peaks in the 303 K spectrum are due to the NH$_2$ protons, whose chemical shifts are even more temperature dependent than that of the aldehyde proton.

as well, since a spin may retain its coupling information even though its chemical shift changes in the exchange. Figure B2.4.3 shows an example of this in the aldehyde proton spectrum of $^{15}$N-labelled formamide. Some lines in the spectrum remain sharp, while others broaden and coalesce. There is no fundamental difference between the lineshapes in figures B2.4.1 and B2.4.3—only a difference in the size of the matrices involved. First, the uncoupled case will be discussed, then the extension to coupled spin systems.

The original analysis of the spectra in figure B2.4.1 was done by the groups of Gutowsky [1] and McConnell [2], both of whom treated the spectrum as a whole in the frequency domain. Reeves showed [3] somewhat later that the two-site exchange lineshape (figure B2.4.4(a) and (b)) can be deconstructed into two transitions. More recently, it was demonstrated [4] that the these transitions are defined by a generalized transition probability. This transition probability (now a complex number) which is just the product of how much coherence a transition receives at the start of an experiment, times how much the transition contributes to the total signal. This leads naturally into a discussion of the $xy$ magnetizations in the time domain. As with most NMR, the choice of the time domain or the frequency domain depends on the problem.

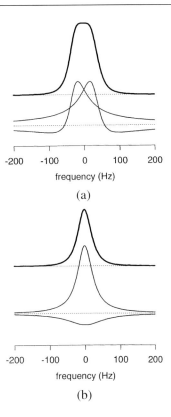

**Figure B2.4.4.** The two-site equally populated exchange lineshape (figure B2.4.1) decomposed into two individual transitions. The bottom spectrum (a) is the situation before coalescence: two symmetrically out-of-phase lines. In slow exchange, these become the signals of the two sites. The top spectrum (b) is after coalescence: the lineshape is made up of two central lines, one positive and one negative. In fast exchange, the negative line broadens and loses intensity, to leave a single positive line at the average chemical shift.

### B2.4.2.2  The Bloch equations approach

The Bloch equations for the motion of the $x$ and $y$ magnetizations (usually called the $u$- and $v$-mode signals), in the presence of a weak radiofrequency (RF) field, $B_1$, are given in equation (B2.4.2).

$$\frac{\mathrm{d}u}{\mathrm{d}t} + \frac{u}{T_2} - (\omega_0 - \omega)v = 0$$

$$\frac{\mathrm{d}v}{\mathrm{d}t} + \frac{v}{T_2} + (\omega_0 - \omega)u = \gamma B_1 M_z. \tag{B2.4.2}$$

In this equation, $\omega$ is the frequency of the RF irradiation, $\omega_0$ is the Larmor frequency of the spin, $T_2$ is the spin–spin relaxation time and $M_z$ is the $z$ magnetization of the spin system. The notation can be simplified somewhat by defining a complex magnetization, $M$, as in equation (B2.4.3).

$$M = u + \mathrm{i}v. \tag{B2.4.3}$$

With this definition, the Bloch equations can be written as in equation (B2.4.4).

$$\frac{\mathrm{d}M}{\mathrm{d}t} + \mathrm{i}(\omega_0 - \omega)M + \frac{1}{T_2}M = \mathrm{i}\gamma B_1 M_z. \tag{B2.4.4}$$

In chemical exchange, the two exchanging sites, A and B, will have different Larmor frequencies, $\omega_A$ and $\omega_B$. Assuming equal populations in the two sites, and the rate of exchange to be $k$, the two coupled Bloch equations for the two sites are given by equation (B2.4.5).

$$\frac{dM_A}{dt} + i(\omega_A - \omega)M_A + \frac{1}{T_2}M_A - kM_B + kM_A = i\gamma B_1 M_{zA}$$

$$\frac{dM_B}{dt} + i(\omega_B - \omega)M_B + \frac{1}{T_2}M_B - kM_A + kM_B = i\gamma B_1 M_{zB}. \tag{B2.4.5}$$

The observable NMR signal is the imaginary part of the sum of the two steady-state magnetizations, $M_A$ and $M_B$. The steady state implies that the time derivatives are zero and a little further calculation (and neglect of $T_2$ terms) gives the NMR spectrum of an exchanging system as equation (B2.4.6).

$$v = \frac{1}{2}\gamma B_1 M_z \frac{k(\omega_A - \omega_B)^2}{(\omega_A - \omega)^2(\omega_B - \omega)^2 + 4k^2(\omega - (\omega_A + \omega_B)/2)^2}. \tag{B2.4.6}$$

### B2.4.2.3  Matrix formulation of chemical exchange

Equation (B2.4.5) can be re-written in a matrix form [5] as equation (B2.4.7).

$$\frac{d}{dt}\begin{pmatrix} M_A \\ M_B \end{pmatrix} + i\mathbf{L}\begin{pmatrix} M_A \\ M_B \end{pmatrix} + \mathbf{R}\begin{pmatrix} M_A \\ M_B \end{pmatrix} + \mathbf{K}\begin{pmatrix} M_A \\ M_B \end{pmatrix} = i\gamma B_1 \begin{pmatrix} M_{zA} \\ M_{zB} \end{pmatrix}. \tag{B2.4.7}$$

In this equation, the matrices $\mathbf{L}$, $\mathbf{R}$ and $\mathbf{K}$ are given by equations (B2.4.8), (B2.4.9) and (B2.4.10).

$$\mathbf{L} = \begin{pmatrix} \omega_A - \omega & 0 \\ 0 & \omega_B - \omega \end{pmatrix} \tag{B2.4.8}$$

$$\mathbf{R} = \begin{pmatrix} \frac{1}{T_2} & 0 \\ 0 & \frac{1}{T_2} \end{pmatrix} \tag{B2.4.9}$$

$$\mathbf{K} = \begin{pmatrix} k & -k \\ -k & k \end{pmatrix}. \tag{B2.4.10}$$

The steady-state solution without saturation to this equation is obtained by setting the time derivatives to zero and taking the terms linear in $B_1$, as in equation (B2.4.11).

$$\begin{pmatrix} M_A \\ M_B \end{pmatrix} = (i\mathbf{L} + \mathbf{R} + \mathbf{K})^{-1} \begin{pmatrix} M_{zA} \\ M_{zB} \end{pmatrix}. \tag{B2.4.11}$$

Recall that $\mathbf{L}$ contains the frequency $\omega$ (equation (B2.4.8)). To trace out a spectrum, equation (B2.4.11) is solved for each frequency. In order to obtain the observed signal $v$, the sum of the two individual magnetizations can be written as the dot product of two vectors, equation (B2.4.12).

$$v = (1\ 1)\begin{pmatrix} M_A \\ M_B \end{pmatrix}. \tag{B2.4.12}$$

This apparently artificial way of re-writing the Bloch equations is important, since this form applies to all exchanging systems—coupled or uncoupled—in the frequency domain. The description starts with the equilibrium $z$ magnetizations. These are affected by all the NMR interactions: chemical shifts, relaxation and exchange. Finally, the observed signal is detected. This is the standard preparation–evolution–detection paradigm used in multi-dimensional NMR. There may be algebraic and numerical complications in setting up and solving the equations for different systems, but the form remains the same for all frequency-domain calculations.

### B2.4.2.4    Chemical exchange in the time domain

If the magnetizations, $M_A$ and $M_B$, are created (by a pulse) at time zero, and then the $B_1$ magnetic field is turned off, equation (B2.4.7) can be simplified to equation (B2.4.13). Note that $\omega$ in the matrix $\mathbf{L}$ (equation (B2.4.8)) is also zero.

$$\frac{\mathrm{d}}{\mathrm{d}t}\begin{pmatrix} M_A \\ M_B \end{pmatrix} = -(i\mathbf{L} + \mathbf{R} + \mathbf{K})\begin{pmatrix} M_A \\ M_B \end{pmatrix}. \tag{B2.4.13}$$

Equation (B2.4.13) is a pair of first-order differential equations, so its formal solution is given by equation (B2.4.14), in which $\exp()$ means the exponential of a matrix.

$$\begin{pmatrix} M_A(t) \\ M_B(t) \end{pmatrix} = \exp(-[i\mathbf{L} + \mathbf{R} + \mathbf{K}]t)\begin{pmatrix} M_A(0) \\ M_B(0) \end{pmatrix}. \tag{B2.4.14}$$

This is the description of NMR chemical exchange in the time domain. Note that this equation and equation (B2.4.11) are Fourier transforms of each other. The time-domain and frequency-domain pictures are always related in this way.

In practice, the matrix $(i\mathbf{L} + \mathbf{R} + \mathbf{K})$ is diagonalized first, with a matrix of eigenvectors, $\mathbf{U}$, as in equation (B2.4.15), to give a diagonal matrix, $\mathbf{\Lambda}$, with the eigenvalues, $\lambda_i$, of $\mathbf{L}$ down the diagonal.

$$\mathbf{\Lambda} = \mathbf{U}^{-1}(i\mathbf{L} + \mathbf{R} + \mathbf{K})\mathbf{U}. \tag{B2.4.15}$$

Equation (B2.4.14) becomes equation (B2.4.16).

$$\begin{pmatrix} M_A(t) \\ M_B(t) \end{pmatrix} = \mathbf{U}\exp(-\mathbf{\Lambda}t)\mathbf{U}^{-1}\begin{pmatrix} M_A(0) \\ M_B(0) \end{pmatrix}. \tag{B2.4.16}$$

The exponential of a diagonal matrix is again a diagonal matrix with exponentials of the diagonal elements, equation (B2.4.17).

$$\begin{pmatrix} M_A(t) \\ M_B(t) \end{pmatrix} = \mathbf{U}\begin{pmatrix} e^{-\lambda_1 t} & 0 \\ 0 & e^{-\lambda_2 t} \end{pmatrix}\mathbf{U}^{-1}\begin{pmatrix} M_A(0) \\ M_B(0) \end{pmatrix}. \tag{B2.4.17}$$

As was mentioned above, the observed signal is the imaginary part of the sum of $M_A$ and $M_B$, so equation (B2.4.17) predicts that the observed signal will be the sum of two exponentials, evolving at the complex frequencies $\lambda_1$ and $\lambda_2$. This is the free induction decay (FID). In the limit of no exchange, the two frequencies are simply $i\omega_A$ and $i\omega_B$, as expected. When $k$ is non-zero, the situation is more complex.

Without relaxation and exchange, $\mathbf{L}$ is a Hermitian matrix with real eigenvalues and eigenvectors. However, when the exchange contributes significantly, the Hermitian character is lost and the eigenvalues and eigenvectors have both real and imaginary parts. The eigenvalues are given by the roots of the characteristic equation, (B2.4.18), in which $\delta$ is $(\omega_A - \omega_B)/2$.

$$\begin{vmatrix} i\delta + \frac{1}{T_2} + k - \lambda & -k \\ -k & -i\delta + \frac{1}{T_2} + k - \lambda \end{vmatrix} = 0. \tag{B2.4.18}$$

The eigenvalues of equation (B2.4.16) are given in equation (B2.4.19).

$$\lambda = \left(\frac{1}{T_2} + k\right) \pm \sqrt{k^2 - \delta^2}. \tag{B2.4.19}$$

These eigenvalues are the (complex) frequencies of the lines in the spectrum: the imaginary part gives the oscillation frequency and the real part gives the rate of decay. If $k < \delta$ (slow exchange) then there are two different imaginary frequencies, which become $\pm\delta$ in the limit of small $k$. Figure B2.4.4(a) shows this decomposition. In fast exchange, when $k$ exceeds the shift difference, $\delta$, the quantity in the square root in equation (B2.4.19) becomes positive, so the roots are pure real. This means that the spectrum is still two lines, but they are both at the average chemical shift (offset of zero) and have different widths (figure B2.4.4(b)).

It is convenient, for simple systems, to have explicit expressions for equation (B2.4.17). Since the original matrix is non-Hermitian, the matrix formed by the eigenvectors will not be unitary, and will have four independent complex elements. Let them be $a$, $b$, $c$ and $d$, so that $\mathbf{U}$ is given by equation (B2.4.20).

$$\mathbf{U} = \begin{pmatrix} a & b \\ c & d \end{pmatrix}. \tag{B2.4.20}$$

Regardless of whether $\mathbf{U}$ is unitary, its inverse is given by equation (B2.4.21), where $\Delta$ is the determinant of the matrix.

$$\mathbf{U}^{-1} = \frac{1}{\Delta} \begin{pmatrix} d & -b \\ -c & a \end{pmatrix}. \tag{B2.4.21}$$

Equation (B2.4.16) then says that the signal is given by equation (B2.4.22), regardless of slow or fast exchange.

$$\text{Signal} = \frac{(a+c)(d-b)}{\Delta} e^{\lambda_1 t} + \frac{(b+d)(-c+a)}{\Delta} e^{\lambda_2 t}. \tag{B2.4.22}$$

For slow exchange, a convenient matrix of eigenvectors is given by equation (B2.4.23).

$$\begin{pmatrix} k & i\left(\sqrt{\delta^2 - k^2} + \delta\right) \\ -i\left(\sqrt{\delta^2 - k^2} + \delta\right) & k \end{pmatrix}. \tag{B2.4.23}$$

After coalescence, a possible set of eigenvectors is given in equation (B2.4.24). If these are substituted into (B2.4.22), the results are pure real, reflecting the fact that $k^2 - \delta^2$ is now positive.

$$\begin{pmatrix} \sqrt{k^2 - \delta^2} - i\delta & -\sqrt{k^2 - \delta^2} - i\delta \\ k & k \end{pmatrix}. \tag{B2.4.24}$$

Because of the role of the eigenvectors in equation (B2.4.16), the factor (amplitude) multiplying the complex exponential is itself complex. The magnitude of the complex amplitude gives the intensity of the line and its phase gives the phase of the line (the mixture of absorption and dispersion). In slow exchange, the two lines have the same real part, but the imaginary parts have opposite signs, so the phase distortion is opposite, as in figure B2.4.4(a). The sum of these distorted lineshapes gives the familiar coalescence spectrum. In fast exchange, the two lines are both in phase, but one line is negative (figure B2.4.4(b)). This negative line is very broad, and decreases in absolute intensity as the rate increases, leaving only the single, positive, in-phase line for fast exchange.

### B2.4.2.5   *Chemical exchange in coupled spin systems*

The development given in the previous section is simply a special case of the general density matrix treatment of chemical exchange. In an uncoupled system, the whole of the coherence from one site is transferred to the other, since the signal is directly associated with a given nucleus. This simplifies the calculation. In a coupled spin system, particularly a strongly coupled system, this is no longer true. The relation between the lines in the spectrum and individual nuclei can be much more complicated. Furthermore, the amount of 'mixing' of nuclei in a given spectrum depends on the chemical shifts and couplings. Therefore, when a nucleus exchanges in a coupled system, coherence that was associated with a single line in one site may be distributed amongst several lines in the other site. In dealing with chemical exchange in coupled systems, it is necessary to keep track of the details of each of the lines in the spectrum, but the fundamental approach is the same.

There is an important special case, called mutual exchange. In all exchange phenomena, a specific nucleus experiences a different magnetic environment when it moves from one site to the other. However, in many cases, the new arrangement is simply a permutation of the old, as in the case of formamide in figure B2.4.3. The two amide protons have switched places, but the chemical shifts and couplings are the same. All that has changed is the nuclei associated with each of them. This can be treated in the same way as all other exchanges, as two different sites. However, the permutation symmetry of the problem means that this is equivalent to copies of a single, mutual, exchange. The matrices are then reduced in size. It is possible to have quite complex permutations, so the analysis must be done carefully and systematically. It is therefore important to identify mutual exchange and treat it appropriately.

Binsch [6] provided the standard way of calculating these lineshapes in the frequency domain, and implemented it in the program DNMR3 [7]. Formally, it is the same as the matrix description given in section B2.4.2.3. The calculation of the matrices $\mathbf{L}$, $\mathbf{R}$ and $\mathbf{K}$ is more complex for a coupled spin system, but that should not interfere with the understanding of how the method works. This work will be discussed later, but first the time-domain approach will be developed.

The basic equation [8] is the equation of motion for the density matrix, $\rho$, given in equation (B2.4.25), in which $H$ is the Hamiltonian.

$$i\frac{h}{2\pi}\frac{\partial}{\partial t}\rho = [H, \rho].\qquad(B2.4.25)$$

It is more convenient to re-express this equation in Liouville space [8–10], in which the density matrix becomes a vector, and the commutator with the Hamiltonian becomes the Liouville superoperator. In this formulation, the lines in the spectrum are some of the elements of the density 'matrix' vector, and what happens to them is described by the superoperator matrix. Equation (B2.4.25) becomes (B2.4.26).

$$i\frac{h}{2\pi}\frac{\partial}{\partial t}\rho = \mathbf{L}\rho.\qquad(B2.4.26)$$

This Liouville-space equation of motion is exactly the time-domain Bloch equations approach used in equation (B2.4.13). The magnetizations are arrayed in a vector, and anything that happens to them is represented by a matrix. In frequency units ($h/2\pi = 1$), the formal solution to equation (B2.4.26) is given by equation (B2.4.27) (compare equation (B2.4.14)).

$$\rho(t) = \exp(-i\mathbf{L}t)\rho(0).\qquad(B2.4.27)$$

For a coupled spin system, the matrix of the Liouvillian must be calculated in the basis set for the spin system. Usually this is a simple product basis, often called product operators, since the vectors in Liouville space are spin operators. The matrix elements can be calculated in various ways. The Liouvillian is the commutator with the Hamiltonian, so matrix elements can be calculated from the commutation rules of spin operators. Alternatively, the angular momentum properties of Liouville space can be used. In either case, the chemical shift terms are easily calculated, but the coupling terms (since they are products of operators) are more complex. In section B2.4.2.7, the Liouville matrix for the single-quantum transitions for an AB spin system is presented.

Relaxation or chemical exchange can be easily added in Liouville space, by including a Redfield matrix, $\mathbf{R}$, for relaxation, or a kinetic matrix, $\mathbf{K}$, to describe exchange. The equation of motion for a general spin system becomes equation (B2.4.28).

$$\rho(t) = \exp(-i\mathbf{L} - \mathbf{R} - \mathbf{K})t\ \rho(0).\qquad(B2.4.28)$$

In NMR, the magnetization in the $xy$ plane is detected, so it is the expectation value of the $I_x$ operator that is measured. This is just the unweighted sum of all the $I_{xi}$ operators for the individual spins $i$. It may be a function of several time variables (multi-dimensional experiments), including the time during the acquisition, but it is always given by equation (B2.4.29).

$$\langle I_x(t) \rangle = \mathrm{trace}(I_x \rho(t)). \tag{B2.4.29}$$

In Liouville space, both the density matrix and the $I_x$ operator are vectors. The dot product of these Liouville space vectors is the trace of their product as operators. Therefore, the NMR signal, $S$, as a function of a single time variable, $t$, is given by equation (B2.4.30), in which the parentheses denote a Liouville space scalar product (compare equation (B2.4.12)).

$$S(t) = (I_x \mid \rho(t)). \tag{B2.4.30}$$

The experiment starts at equilibrium. In the high-temperature approximation, the equilibrium density operator is proportional to the sum of the $I_z$ operators, which will be called $F_z$. If there are multiple exchanging sites with unequal populations, $p_i$, the sum is a weighted one, as in equation (B2.4.31).

$$F_x = \sum_{i=1}^{n} p_i I_{xi}. \tag{B2.4.31}$$

A simple, non-selective pulse starts the experiment. This rotates the equilibrium $z$ magnetization onto the $x$ axis. Note that neither the equilibrium state nor the effect of the pulse depend on the dynamics or the details of the spin Hamiltonian (chemical shifts and coupling constants). The equilibrium density matrix is proportional to $F_z$. After the pulse the density matrix is therefore given by $F_x$ and it will evolve as in equation (B2.4.27). If (B2.4.28) is substituted into (B2.4.30), the NMR signal as a function of time $t$, is given by (B2.4.32). In this equation there is a distinction between the sum of the operators weighted by the equilibrium populations, $F_x$, from the unweighted sum, $I_x$. The detector sees each spin (but not each coherence!) equally well.

$$S(t) = (I_x \mid \exp([-\mathrm{i}\mathbf{L} - \mathbf{R} - \mathbf{K}]t)F_x). \tag{B2.4.32}$$

As with the uncoupled case, one solution involves diagonalizing the Liouville matrix, $\mathrm{i}\mathbf{L} + \mathbf{R} + \mathbf{K}$. If $\mathbf{U}$ is the matrix with the eigenvectors as columns, and $\mathbf{\Lambda}$ is the diagonal matrix with the eigenvalues down the diagonal, then (B2.4.32) can be written as (B2.4.33). This is similar to other eigenvalue problems in quantum mechanics, such as the transformation to normal co-ordinates in vibrational spectroscopy.

$$S(t) = (I_x \mid \mathbf{U} \exp(-\mathrm{i}\mathbf{\Lambda}t)\mathbf{U}^{-1} \mid F_x). \tag{B2.4.33}$$

Note that the Liouville matrix, $\mathrm{i}\mathbf{L} + \mathbf{R} + \mathbf{K}$ may not be Hermitian, but it can still be diagonalized. Its eigenvalues and eigenvectors are not necessarily real, however, and the inverse of $\mathbf{U}$ may not be its complex-conjugate transpose. If complex numbers are allowed in it, equation (B2.4.33) is a general result. Since $\mathbf{\Lambda}$ is a diagonal matrix it can be expanded in terms of the individual eigenvalues, $\lambda_j$. The inverse matrix $\mathbf{U}^{-1}$ can be applied ('backwards') to $I_x$, and we obtain equation (B2.4.34).

$$S(t) = \sum_j (\mathbf{U}^{-1}I_x)_j^* (\mathbf{U}F_x)_j \, \mathrm{e}^{\mathrm{i}\lambda_j t}. \tag{B2.4.34}$$

In this equation, the index $j$ runs over all the transitions and the exponents have both real and imaginary parts, which give the linewidth and position of the lines. The terms before the exponential are also complex, giving the intensity and phase.

For a general system, these sets of equations are huge, as written. For $n$ spins $-1/2$, the density matrix has $2^{2n}$ elements, so the Liouville matrix has $2^{4n}$ elements. However, the density matrix elements can be sorted according to coherence level—the number of quanta associated with the transition. In this case, the matrices block, and the largest single block is the one corresponding to the single-quantum transitions. Its size is the binomial coefficient $(2n)!/(n+1)!(n-1)!$. This can be further divided into blocks based on the factoring from spectral analysis. A transition can change the $z$ quantum number of the spin wavefunction by only $\pm 1$. For three spins $-1/2$, the eight wave functions are divided as follows: one with $z$ quantum number $+3/2$, three with $+1/2$, three with $-1/2$ and one with $-3/2$. Therefore, the 15 possible transitions are divided into two groups of three ($+3/2 \rightarrow +1/2$, and $-1/2 \rightarrow -3/2$), and a group of nine ($+1/2 \rightarrow -1/2$). Further reductions can be achieved using weak coupling approximations and magnetic equivalence. In practice, system of five or six spins can be treated with modern computers.

### B2.4.2.6 Generalized transition probabilities

The quantities $(\mathbf{U}^{-1}I_x)_j$ and $(\mathbf{U}F_x)_j$ in B2.4.34 are projections of the eigenvector $j$ along $I_x$. From the above equations, this can be interpreted as follows. The term $(\mathbf{U}F_x)_j$ is the amount that the transition $j$ received from the total $x$ magnetization created from the equilibrium state and $(\mathbf{U}^{-1}I_x)_j$ is how much that transition contributes to the observed signal. These two terms may not be equal, as will be seen in exchanging systems. An informal way of thinking about these terms is to consider the transition moment, $\langle \phi_f | I_x | \phi_i \rangle$. If the ket–bra operator $|\phi_f\rangle\langle\phi_i|$ represents the transition, then the transition moment is the projection of the transition operator along the $I_x$ operator.

In the usual preparation–evolution–detection paradigm, neither the preparation nor the detection depend on the details of the Hamiltonian, except in special cases. Starting from equilibrium, a hard pulse gives a density matrix that is just proportional to $F_z$. The detector picks up only the unweighted sum of the spin operators, $I_x$. It is only during an evolution (perhaps between sampling points in an FID) that these totals need be divided amongst the various lines in the spectrum. Therefore, one of the factors in the transition probability represents the conversion from preparation to evolution; the other factor represents the conversion back from evolution to detection.

Equations (B2.4.33) and (B2.4.34) are the basic equations for a time-domain description. For instance, they say that any time-domain NMR signal is the sum of decaying oscillations. This is obvious from the fact that it is described by a first-order differential equation, but (B2.4.34) gives a way of calculating the values of these exponentials for any system, static or dynamic. The distinctions amongst different types of spectrum lie in the eigenvalues and eigenvectors of the Liouville matrix $i\mathbf{L} + \mathbf{R} + \mathbf{K}$. Equation (B2.4.34) describes static spectra, spin relaxation and spectra showing the effects of chemical exchange or $T_2$ relaxation, in a single, unified picture.

### B2.4.2.7 Example of the AB spin system

For example, the observed transitions of an AB spin system have a Liouville matrix given in equation (B2.4.35). The coupling constant is $J$, and it is assumed that $\omega_B = -\omega_A = -\delta/2$, so that $\delta$ is the frequency difference between the two sites. The angle, $\theta$, is defined for the AB system by the equation $\tan(\theta) = J/2\delta$. The Liouville space basis used here is the superspin equivalent of the four product operators ($I_x^A$, $I_x^A I_z^B$, $I_x^B$, $I_x^B I_z^A$), and a set of rules for calculating these elements is given elsewhere [12].

$$
\begin{pmatrix}
i\omega_A & iJ/2 & 0 & -iJ/2 \\
iJ/2 & i\omega_A & -iJ/2 & 0 \\
0 & -iJ/2 & i\omega_B & iJ/2 \\
-iJ/2 & 0 & iJ/2 & i\omega_B
\end{pmatrix}.
\tag{B2.4.35}
$$

The four eigenvalues, which give the positions of the lines, are $\pm J/2 \pm ((J/2)^2 + (\delta/2)^2)^{1/2}$, as expected for an AB system. The matrix of eigenvectors as columns is given in equation (B2.4.36), in which $c = \cos(\theta)$,

$s = \sin(\theta)$ and $\delta$ is defined above.

$$\text{Eigenvectors} = \begin{pmatrix} c & s & c & s \\ c & s & -c & -s \\ -s & c & s & -c \\ -s & c & -s & c \end{pmatrix}. \tag{B2.4.36}$$

In the basis used in (B2.4.35), the total $x$ magnetization is proportional to the vector $(1, 0, 1, 0)$. Taking the dot product with the eigenvectors shows that the outer lines receive $(\cos\theta - \sin\theta)$ from the total, whereas the inner lines receive $(\cos\theta + \sin\theta)$. The squares of these terms give the familiar AB system intensities: $(1 - \sin 2\theta)$ and $(1 + \sin 2\theta)$.

Once the AB spin system is defined, the effects of chemical exchange can be calculated. This can be either non-mutual or mutual exchange. In the case of non-mutual exchange, there are two blocks, one for each site, and the exchange connecting them, as in equation (B2.4.37). For a simple product basis, the exchange always has this form: the off-diagonal blocks are themselves diagonal and the sum of the exchange contributions in any column must be zero, to preserve the number of spins. In this equation, zeros have been replaced by dots to emphasize the form of the matrix.

$$\begin{pmatrix}
i\omega_A - k' & iJ/2 & \cdot & -iJ/2 & k & \cdot & \cdot & \cdot \\
iJ/2 & i\omega_A - k' & -iJ/2 & \cdot & \cdot & k & \cdot & \cdot \\
\cdot & -iJ/2 & i\omega_B - k' & iJ/2 & \cdot & \cdot & k & \cdot \\
-iJ/2 & \cdot & iJ/2 & i\omega_B - k' & \cdot & \cdot & \cdot & k \\
k' & \cdot & \cdot & \cdot & i\omega'_A - k & iJ'/2 & \cdot & -iJ'/2 \\
\cdot & k' & \cdot & \cdot & iJ'/2 & i\omega'_A - k & -iJ'/2 & \cdot \\
\cdot & \cdot & k' & \cdot & \cdot & -iJ'/2 & i\omega'_B - k & iJ'/2 \\
\cdot & \cdot & \cdot & k' & -iJ'/2 & \cdot & iJ/2 & i\omega'_B - k
\end{pmatrix}. \tag{B2.4.37}$$

In this equation, the primes on the imaginary parts indicate that the Larmor frequencies and coupling constants will be different. Also, if the equilibrium constant for the exchange is not 1, then the forward and reverse rates will not be equal. Note that the 1, 2 block, in the top right, represents the rate from site 2 into site 1.

### B2.4.2.8   Mutual exchange in the AB system

In the case of mutual AB exchange this matrix can be simplified. The equilibrium constant must be 1, so $k = k'$. Also, $\omega'_A$ is equal to $\omega_B$ and vice versa, and the coupling constant is the same. For instance, if $\mathbf{L}$ is the Liouville matrix for one site, then the Liouville matrix for the other site is $\mathbf{P}^{-1}\mathbf{LP}$, where $\mathbf{P}$ is the matrix describing the permutation. The exchange matrix, $\mathbf{K}$, is just the rate, $k$, times the unit matrix. In block form, the full matrix for two sites is given in the eigenvalue equation, (B2.4.38).

$$\begin{pmatrix} i\mathbf{L} - \mathbf{K} & \mathbf{K} \\ \mathbf{K} & i\mathbf{P}^{-1}\mathbf{LP} - \mathbf{K} \end{pmatrix} \begin{pmatrix} a \\ \mathbf{P}^{-1}a \end{pmatrix} = \lambda \begin{pmatrix} a \\ \mathbf{P}^{-1}a \end{pmatrix}. \tag{B2.4.38}$$

This equation is equivalent to the pair of equations in (B2.4.39).

$$i\mathbf{L}a - \mathbf{K}(1 - \mathbf{P}^{-1})a = \lambda a$$

$$i\mathbf{P}^{-1}\mathbf{L}a - \mathbf{K}(\mathbf{P}^{-1} - 1)a = \lambda \mathbf{P}^{-1}a. \tag{B2.4.39}$$

Since $\mathbf{K}$ is a multiple of the unit matrix and the permutation is its own inverse, the two equations in (B2.4.39) are the same. The Liouvillian for a single site is set up and the exchange is described by $\mathbf{K}(1 - \mathbf{P}^{-1})$.

Application of this approach to equation (B2.4.37) gives equation (B2.4.40). If $\omega_B = -\omega_A = -\delta/2$, the symmetry of the matrix and one additional transformation means that it can be broken into two $2 \times 2$ complex matrices, which can be diagonalized analytically. The resulting lineshapes match the published solutions [13].

$$\begin{pmatrix} i\omega_A - k & iJ/2 & k & -iJ/2 \\ iJ/2 & i\omega_A - k & -iJ/2 & k \\ k & -iJ/2 & i\omega_B - k & iJ/2 \\ -iJ/2 & k & iJ/2 & i\omega_B - k \end{pmatrix}. \tag{B2.4.40}$$

### B2.4.2.9 Intermolecular exchange

The phenomenon of intermolecular exchange is very common. The loss of couplings to hydroxyl protons in all but the very purest ethanol samples was observed at a very early stage. Proton transfer reactions are still probably the most carefully studied [14] class of intermolecular exchange.

In classical kinetics, intermolecular exchange processes are quite different from the unimolecular, first-order kinetics associated with intramolecular exchange. However, the NMR of chemical exchange can still be treated as pseudo-first-order kinetics, and all the previous results apply. One way of rationalizing this is as follows. NMR follows a particular nucleus, but typically only 1 in $10^5$ nuclei is 'visible' (due to the small Boltzmann population difference). When a visible nucleus exchanges with another nucleus on another molecule, the probability is that the other nucleus is invisible. The exchange partners vastly overwhelm the visible nuclei.

However, all the nuclei have spin. An example of this occurs in the intermolecular exchange of an AB spin system, as in equation (B2.4.41).

$$AB \rightleftharpoons A + B. \tag{B2.4.41}$$

In this case, a spin A that was coupled to the $\alpha$ orientation of the B spin may end up, after the exchange, coupled to either $\alpha$ or $\beta$. Because of the Boltzmann distribution, the amounts of $\alpha$ and $\beta$ orientation are each half of the sample. The first exchange is degenerate, but the second is a change of the B spin. This can be treated as exchange with a site in which the shifts are the same, but the coupling constant is of the opposite sign. If these spin parameters are used in equation (B2.4.37), then the lineshapes in figure B2.4.5 are obtained.

### B2.4.2.10 Calculation of the spectrum

Once the basic work has been done, the observed spectrum can be calculated in several different ways. If the problem is solved in the time domain, then the solution provides a list of transitions. Each transition is defined by four quantities: the integrated intensity, the frequency at which it appears, the linewidth (or decay rate in the time domain) and the phase. From this list of parameters, either a spectrum or a time-domain FID can be calculated easily. The spectrum has the advantage that it can be directly compared to the experimental result. An FID can be subjected to some sort of apodization before Fourier transformation to the spectrum; this allows additional line broadening to be added to the spectrum independent of the simulation.

The Bloch equation approach (equation (B2.4.6)) calculates the spectrum directly, as the portion of the spectrum that is linear in a $B_1$ observing field. Binsch generalized this for a fully coupled system, using an exact density-matrix approach in Liouville space. His expression for the spectrum is given by equation (B2.4.42). Note that this is formally the Fourier transform of equation (B2.4.32), so the time domain and frequency domain are connected as usual.

$$S(\omega) = \text{Re}[F_-(i\mathbf{L} - \mathbf{R} - \mathbf{K})^{-1} M_0]. \tag{B2.4.42}$$

In practice, the spectrum is usually calculated by diagonalizing the matrix first [15], as was done in the time domain. This means that the large matrix does not have to be inverted for each point in the spectrum.

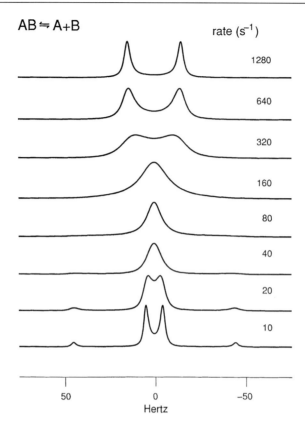

**Figure B2.4.5.** Simulated lineshapes for an intermolecular exchange reaction in which the bond joining two strongly coupled nuclei breaks and re-forms at a series of rates, given beside the lineshape. In slow exchange, the typical spectrum of an AB spin system is shown. In the limit of fast exchange, the spectrum consists of two lines at the two chemical shifts and all the coupling has disappeared.

However, for very large matrices, it may be numerically more efficient not to diagonalize, but rather to invert at each data point. For a six-spin system, the full matrix is $792 \times 792$, and each additional spin multiplies each dimension by roughly a factor of 4. Since the time for a diagonalization scales roughly as the cube of the dimension of the matrix, larger spin systems become impractical. However, modern sparse-matrix methods for matrix inversion do not suffer from the same dramatic scaling, and so will become more efficient for larger spin systems.

The method for studying intermediate exchange in NMR is to obtain an excellent equilibrium spectrum of the system as a function of temperature. Then the theoretical apparatus developed above can be used to simulate and to fit the experimental data, in order to obtain the rate data.

### B2.4.3   Fast exchange

*B2.4.3.1   Introduction*

In the limit of fast exchange, the lineshape of chemical exchange quickly becomes a single Lorentzian line. In figure B2.4.4(b), the negative line broadens directly as the rate, and loses absolute intensity as well. This combination of the increasing width and decreasing integral means that the negative line quickly becomes irrelevant to the experimental lineshape. This becomes a pure Lorentzian, whose width is proportional to $(\Delta\omega)^2/k$, where $\Delta\omega$ is the difference in Larmor frequency of the two sites and $k$ is the exchange rate. Measuring the rate is then equivalent to measuring the spin–spin relaxation time, $T_2$. The problem is that unless there is an estimate of $\Delta\omega$ (from a spectrum that is 'frozen out') an absolute value of the rate cannot be measured. However, $T_2$ measurements themselves have an associated timescale. If that can match the exchange rate, then an absolute rate can be measured.

### B2.4.3.2  $T_2$ measurements

In principle, $T_2$ can be measured directly from the linewidth of the spectrum. However, since experimental linewidths are also governed by inhomogeneous broadening (magnetic field inhomogeneities etc), careful $T_2$ determinations require methods that cancel out inhomogeneous effects [16]. Three techniques are commonly available: the Carr–Purcell–Meiboom–Gill (CPMG) spin echo experiment, the $T_{1\rho}$ experiment and the offset-saturation method [17]. All three can be implemented easily on modern spectrometers.

The Hahn spin echo, with a single refocusing pulse, eliminates effects due to magnetic field inhomogeneities, so a study of the echo intensity as a function of the echo time should yield $T_2$. However, if there is a field gradient present, diffusion in the sample can also attenuate the echo. Even with modern well shimmed magnets, the gradients are still large enough to affect a $T_2$ measurement in water. In the CPMG experiment, a series of closely spaced refocusing pulses suppresses diffusion effects. In this case, the echo intensity is measured as a function of echo number, and reliable values of $T_2$ can be obtained.

The parameter $T_{1\rho}$ is the longitudinal relaxation time of a magnetization which is spin locked along a radiofrequency magnetic field. The magnetization is flipped onto the $x$ axis (for instance) by an RF pulse along $y$. The RF phase is changed by $90°$, and the magnetization is then spin locked by the RF field, now along $x$. When the RF is shut off, the remaining $xy$ magnetization can be detected directly. An analysis of the signal as a function of the spin-locking time yields $T_2$.

The offset-saturation experiment [18] consists of irradiating the spin system with an known RF field at some offset from resonance until a steady state is achieved. The $z$ magnetization is then measured by a non-selective observe pulse. A plot (figure B2.4.6) of the partially saturated $z$ magnetization against offset from resonance will show a dip at resonance. The width of this dip is given by equation (B2.4.43).

$$\text{Dip width} = (\gamma B_2)\sqrt{T_1/T_2}. \tag{B2.4.43}$$

If the strength of the saturating RF, $\Delta B_2$, and the spin–lattice relaxation time, $T_1$, are known, then $T_2$ can be measured, again free of magnetic field inhomogeneities.

These $T_2$ measurements can allow reaction rates of $10^5$ s$^{-1}$ or more to be measured. In combination with slow- and intermediate-exchange methods, this means that rates can be measured over a range of more than five orders of magnitude. This means that excellent thermodynamic parameters can be obtained. Figure B2.4.2 shows some results on furfural, a system which has unequally populated sites. For this case, the range of rates over which lineshape methods are useful is quite small. In the case of two-site, unequally populated exchange, the minor peak broadens faster than the major peak (the relative rate of broadening is the ratio of the major population to the minor). The minor peak disappears into the baseline quickly, but $T_2$ measurements can still provide good data.

These experiments yield $T_2$ which, in the case of fast exchange, gives the ratio $(\Delta\omega)^2/k$. However, since the experiments themselves have an implicit timescale, absolute rates can be obtained in favourable circumstances. For the CPMG experiment, the timescale is the repetition time of the refocusing pulse; for the $T_{1\rho}$ experiment, it is the rate of precession around the effective RF field. If this timescale is fast with respect to the exchange rate, then the experiment effectively measures $T_2$ in the absence of exchange. If the timescale is slow, the apparent $T_2$ contains the effects of exchange. Therefore, the apparent $T_2$ shows a dispersion as the timescale of the measurement method is changed [19]. Practical spectrometer considerations of RF heating

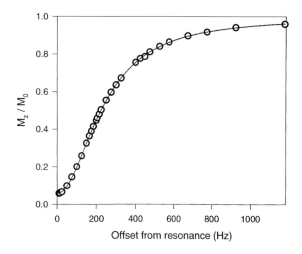

**Figure B2.4.6.** Results of an offset-saturation experiment for measuring the spin–spin relaxation time, $T_2$. In this experiment, the signal is irradiated at some offset from resonance until a steady state is achieved. The partially saturated $z$ magnetization is then measured with a $\pi/2$ pulse. This figure shows a plot of the $z$ magnetization as a function of the offset of the saturating field from resonance. Circles represent measured data; the line is a non-linear least-squares fit. The signal is normal when the saturation is far away, and dips to a minimum on resonance. The width of this dip gives $T_2$, independent of magnetic field inhomogeneity.

and duty cycle usually limit these timescales to tens of kilohertz. However, if the conditions are appropriate, this dispersion curve yields an absolute rate.

## B2.4.4   Slow exchange

### B2.4.4.1   Introduction

The term 'slow' in this case means that the exchange rate is much smaller than the frequency differences in the spectrum, so the lines in the spectrum are not significantly broadened. However, the exchange rate is still comparable with the spin–lattice relaxation times in the system. Exchange, which has many mathematical similarities to dipolar relaxation, can be observed in a NOESY-type experiment (sometimes called EXSY). The rates are measured from a series of EXSY spectra, or by performing modified spin–lattice relaxation experiments, such as those pioneered by Hoffman and Forsen [20].

In the absence of exchange (and ignoring dipolar relaxation), each $z$ magnetization will relax back to equilibrium at a rate governed by its own $T_1$, as in (B2.4.44).

$$\frac{\mathrm{d}}{\mathrm{d}t}[M(t) - M(\infty)] = -\frac{1}{T_1}[M(t) - M(\infty)]. \tag{B2.4.44}$$

If there are two sites, A and B, then an analogous equation can be written, as in (B2.4.45).

$$\frac{\mathrm{d}}{\mathrm{d}t}\begin{bmatrix} M_A(t) - M_A(\infty) \\ M_B(t) - M_B(\infty) \end{bmatrix} = \begin{pmatrix} \frac{1}{T_1^A} & 0 \\ 0 & \frac{1}{T_1^B} \end{pmatrix}\begin{bmatrix} M_A(t) - M_A(\infty) \\ M_B(t) - M_B(\infty) \end{bmatrix}. \tag{B2.4.45}$$

If the two sites exchange with rate $k$ during the relaxation, then a spin can relax either through normal spin–lattice relaxation processes, or by exchanging with the other site. Equation (B2.4.45) becomes (B2.4.46).

$$\frac{d}{dt}\left[\begin{array}{c} M_A(t) - M_A(\infty) \\ M_B(t) - M_B(\infty) \end{array}\right] = \left(\begin{array}{cc} -\frac{1}{T_1^A} - k & k \\ k & -\frac{1}{T_1^B} - k \end{array}\right)\left[\begin{array}{c} M_A(t) - M_A(\infty) \\ M_B(t) - M_B(\infty) \end{array}\right]. \qquad \text{(B2.4.46)}$$

This equation is very similar to (B2.4.13). The basic situation is just as in intermediate exchange, except that it describes $z$ magnetizations rather than $xy$. The frequencies are zero, and the matrix now has pure real eigenvalues, but the approach is the same. The time domain is a natural one for slow exchange, since a relaxation experiment follows the $z$ magnetizations as a function of time. As before, the time dependence is obtained by diagonalizing the relaxation/exchange matrix and calculating the magnetizations for each time at which they are sampled. In this case, the solution is given by equation (B2.4.47), the same as equation (B2.4.14), except there are no imaginary terms.

$$\left[\begin{array}{c} M_A(t) - M_A(\infty) \\ M_B(t) - M_B(\infty) \end{array}\right] = \exp(-[R + K]t)\left(\begin{array}{c} M_A(0) - M_A(\infty) \\ M_B(0) - M_B(\infty) \end{array}\right). \qquad \text{(B2.4.47)}$$

### B2.4.4.2   Two-dimensional methods

There are two main applications of slow chemical exchange: one is to determine the qualitative mechanism, and the other is to measure the rates of the processes as accurately as possible. For the first case, in which we have a spectrum in slow exchange, we need to establish the mechanism: which site is exchanging with which. For this purpose, the homonuclear two-dimensional experiment EXSY (the same pulse sequence as NOESY, but involving exchange) is by far the best technique to use. Exchange between sites leads to a pair of symmetrical cross-peaks joining the diagonal peaks of the same site, so the mechanism is very obvious.

The EXSY pulse sequence starts with two $\pi/2$ pulses separated by the incrementable delay, $t_1$. This modulates the $z$ magnetizations, so that the relaxation that occurs during the mixing time which follows, $t_m$, is frequency labelled. Finally, the $z$ magnetizations are sampled with a third $\pi/2$ pulse. Magnetization from a different site that enters via exchange will have a different frequency label. A two-dimensional Fourier transform then produces the spectrum. The initial rate of increase of the cross-peak gives the rate of exchange. A series of EXSY experiments as a function of mixing times will define the mechanism and give an estimate of the rates.

However, care must be taken in choosing the mixing time if there are multiple exchange processes. If the mixing time is too long, there is a substantial probability that a spin may have exchanged twice in that time, leading to spurious cross-peaks. Orrell and his group [21] have solved this problem by treating the effect of exchange on the $z$ magnetizations correctly, and have written a program which simulates the two-dimensional spectrum as a function of the mixing time.

Since exchange and coupled relaxation have the same mathematical form, both may contribute to a NOESY/EXSY spectrum, as in figure B2.4.7. For small molecules, since the NOE is positive and exchange creates saturation transfer (like a negative NOE), the NOESY and EXSY cross peaks have opposite signs. For macromolecules in the spin-diffusion limit, the peaks have the same sign, but exchange cross-peaks can usually be distinguished by their much stronger temperature dependence.

### B2.4.4.3   One-dimensional selective-inversion methods

For careful rate measurements, once the mechanism is established, it is our opinion that one-dimensional methods are superior to quantitative 2D ones. Apart from the fact that 1D spectra can be integrated more easily, there is also more control over the experiment. Modern spectrometers can create almost any type of selective excitation, so that there is control of the conditions at the start of the relaxation. For two sites, a

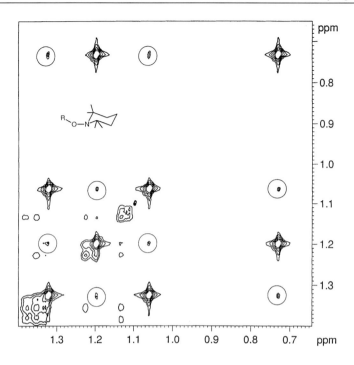

**Figure B2.4.7.** Contour plot of a phase-sensitive NOESY/EXSY spectrum of the derivative of TEMPO. The peaks are positive, with the exception of the circled peaks, which are negative. The spectrum shows the exchange of the four methyl groups in this molecule. A combination of a ring-flip and inversion at nitrogen means that the axial methyl on one side of the ring exchanges with the equatorial methyl on the other side. The positive cross-peaks show the exchange of the two sets of methyls. However, there are also NOE cross-peaks between methyl groups on the same side of the ring. Since this is a small molecule, the NOE peaks are of opposite sign to the exchange peaks. There are two NOE cross-peaks for each methyl, since the exchange process is relatively fast, and distributes the NOE between the two exchange partners.

non-selective inversion that inverts both sites equally will mask most of the exchange effects and the relaxation will be dominated by $T_1$. However, if one site is inverted selectively, then that site can regain equilibrium by either $T_1$ processes or by exchanging with the other site that was left at equilibrium [20]. The inverted signal will relax at roughly the sum of the exchange and spin–lattice relaxation rate, while the signal that was unperturbed at the start of the experiment shows a characteristic transient, as in figure B2.4.8. These one-dimensional selective-inversion experiments have been widely used in systems without scalar coupling, such as methyl groups or $^{13}$C spectra.

For multiple sites, a wide range of initial conditions is available. For instance, in a three-site exchange amongst A, B and C, the signal due to A can be inverted selectively. This will provide rate information on the A–B exchange and the A–C exchange, but relatively little about B–C. This is one example of using the initial conditions to suppress or enhance the observation of particular processes. The definition of how selective a selective inversion may be gives an added degree of control over the experiment.

The selective-inversion experiment gives excellent rate data. The description of the time evolution of the $z$ magnetizations in equation (B2.4.47) is exact. There are no assumptions about short mixing times or initial rates. Standard non-linear least-squares methods allow fitting these curves to the measured data and deriving values for the rates involved. Believing in the error estimates of these multi-parameter fitting procedures can be dangerous, however. A more reliable error estimate can be obtained by 'profiling'. In this procedure, a

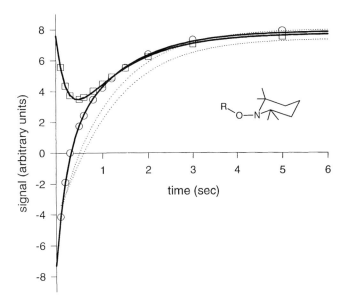

**Figure B2.4.8.** Relaxation of two of the exchanging methyl groups in the TEMPO derivative in figure B2.4.7. The dotted lines show the relaxation of the two methyl signals after a non-selective inversion pulse (a typical $T_1$ experiment). The heavy solid line shows the recovery after the selective inversion of one of the methyl signals. The inverted signal (circles) recovers more quickly, under the combined influence of relaxation and exchange with the non-inverted peak. The signal that was not inverted (squares) shows a characteristic transient. The lines represent a non-linear least-squares fit to the data.

global fit to all parameters is done first. Then the rate (or any other parameter of interest) is fixed at a different value, and the data are re-fitted using all the other parameters. As the rate is moved from the optimum value, the fit will become worse, as measured by the sum of the squares of the deviations between real data and the model. When the 'badness of fit' exceeds a critical level of the $F$ statistic, the value of the rate is at the end of a confidence interval. This confidence interval can be several times larger than the one calculated from the usual standard deviation in a nonlinear least-squares fit. Even with this error estimate, it is possible to measure rates with errors of less than 10%. Figure B2.4.8 shows the quality of result that is possible.

### B2.4.4.4    Selective inversions on coupled systems

In a selective-inversion experiment, it is the relaxation of the $z$ magnetizations that is being studied. For a system without scalar coupling, this is straightforward: a simple pulse will convert the $z$ magnetizations directly into observable signals. For a coupled spin system, this relation between the $z$ magnetizations and the observable transitions is much more complex [22].

In a coupled spin system, the number of observed lines in a spectrum does not match the number of independent $z$ magnetizations and, furthermore, the spectra depend on the flip angle of the pulse used to observe them. Because of the complicated spectroscopy of homonuclear coupled spins, it is only recently that selective inversions in simple coupled spin systems [23] have been studied. This means that slow chemical exchange can be studied using proton spectra without the requirement of single characteristic peaks, such as methyl groups.

The $z$ magnetizations of the spin system are key to the problem, since the exchange is measured in competition with their relaxation processes. However, for a coupled spin system, the lines in the spectrum do not directly reflect the $z$ magnetizations. Even for two weakly coupled spins, there are four lines in the spectrum (two doublets), but there are only three independent $z$ magnetizations. There are indeed four energy levels, but the sum of their populations must be constant. There are three independent quantities: the total magnetization of $A$, the total $X$ magnetization, and a shared $I_z I_z$ magnetization. For coupled systems, especially those with strong coupling, the relation of the $z$ magnetizations to the observed spectrum can be quite complex.

There are several complications. One is the flip angle dependence of the spectra [22, 24]. For a non-equilibrium state of a coupled spin system, the observed intensities of the lines depend on the flip angle used to observe them. In particular, spectra are only 'true' reflections of the $z$ magnetizations in the limit of small flip angles. A further complication arises because the $z$ magnetizations are part of a larger manifold of coherences that also includes the zero-quantum transitions. Both the $z$ magnetizations and the zero-quantum transitions have coherence level zero and they cannot be separated by pulses or phase cycling [11]. This is the problem with the zero-quantum coherences in NOESY, for instance. For instance, for three spins there are eight $z$ magnetizations, one of which is fixed as the total number of spins. However, there are 20 coherence level zero density matrix elements, leaving six pairs of zero quantum transitions. The 15 observable $xy$ magnetizations for a three-spin system cannot correspond directly to the eight $z$ magnetizations. Provided that these complications are recognized, they can be treated easily with standard spin-dynamics techniques.

The $xy$ magnetizations can also be complicated. For $n$ weakly coupled spins, there can be $n*2^{n-1}$ lines in the spectrum and a strongly coupled spin system can have up to $(2n!)/((n-1)!(n+1)!)$ transitions. Because of small couplings, and because some lines are weak combination lines, it is rare to be able to observe all possible lines. It is important to maintain the distinction between mathematical and practical relationships for the density matrix elements.

These complications require some careful analysis of the spin systems, but fundamentally the coupled spin systems are treated in the same way as uncoupled ones. Measuring the $z$ magnetizations from the spectra is more complicated, but the analysis of how they relax is essentially the same.

## B2.4.5   Exchange in solids

Exchange in the solid state follows the same basic principles as in liquids. The classic Cope re-arrangement of bullvalene occurs in both the liquid and solid state [25], and the lineshapes in the spectra are similar. However, because of chemical shielding anisotropy (CSA) and quadrupolar and dipolar effects, the Larmor frequency of a given spin depends on the orientation of the individual molecule. In a liquid, where there is isotropic tumbling, these effects average out and exchange is only evident if there is a change in isotropic chemical shift. In a solid, almost any type of molecular motion can cause lineshape and other effects. Furthermore, many NMR spectra of solids are run under the conditions of magic-angle spinning. This introduces a further timescale into the spectroscopy: the spinning rate, which can now go up to 25 kHz or more. The basic principles are the same, but the systems studied and the observed phenomena can be quite different.

Intermediate exchange is the regime in which lineshape changes are the most obvious manifestation. In $^{13}$C magic-angle spinning (MAS) spectra, there can be all the liquid-like coalescence phenomena [26]. These can be analysed just as before. Another type occurs when a molecule re-orients in the crystal lattice. Since the magnetic environment of the nucleus is anisotropic, the Larmor frequency of the spin changes. One of the most familiar examples is the effect of dynamics on powder patterns in deuterium spectra [27]. In a typical carbon–deuterium bond, the quadrupole coupling of the deuterium nucleus is about 160 kHz. This defines a timescale that is very useful for polymers and biological membranes, and deuterium spectra are widely used. Quite detailed information is available: lineshapes are significantly different for twofold jumps, threefold jumps or continuous diffusion. For instance, in a *tert*-butyl group, overall rotation can be distinguished from rotations of individual methyl groups.

Magic-angle spinning of solid samples provides an experimental parameter not available in liquids. A full analysis of the combined effects of dynamics and MAS for all relative timescales is a very complex problem, but for slow exchange there are techniques that are intuitive. The first is EXSY, which works well in MAS spectra, although the interpretation is confused by the phenomenon of spin diffusion. Cross-peaks can be due to the exchange of spin polarization (rather than the nuclei themselves) via the dipolar interaction. One-dimensional methods that use MAS are also available. The TOSS pulse sequence will eliminate spinning sidebands, provided there is no internal dynamics in the sample. If there is exchange on the timescale of a rotor period, the careful cancellation will no longer work, and sidebands will reappear [28]. Another use of spinning is the ODESSA [29] pulse sequence, which selectively inverts some of the sidebands, which then relax back due to chemical exchange.

## B2.4.6 Conclusions

In order to study chemical exchange in NMR, it is necessary to have a scale against which to measure it. In fast and intermediate exchange, the timescale is the difference in Larmor frequency between the two sites. As the exchange rate approaches this timescale, the lines in the one-dimensional spectrum broaden, coalesce and sharpen into single lines. For intermediate exchange, the lineshape provides the best information. In fast exchange, the rate is starting to dominate the timescale, but useful information can still be extracted from $T_2$ measurements. In slow exchange, the spin–lattice relaxation provides a timescale. In this regime, modifications of methods for measuring $T_1$ and the NOE are used. In MAS spectra of solids, the spinning rate provides another timescale. When the timescales match, dramatic effects can often be observed.

Once the exchange has been established, it is necessary to measure its rate. This is usually done by simulating the NMR experiment with a mathematical model and adjusting the rate in the model until it matches experiment. This means simulating the lineshape in intermediate exchange, or simulating the coupled relaxation of the $z$ magnetizations in slow exchange. In both these cases, the model is similar. There is a matrix which describes each site in the exchange. These matrices form the diagonal of a larger block-diagonal matrix. The blocks are then connected by the exchange process, to form one large matrix. The exchange is described by the eigenvalues and eigenvectors of the single large matrix.

The exact form of the matrices depend on the situation. In slow exchange, the matrices are real and they model the multi-exponential relaxation of the $z$ magnetizations. In intermediate exchange, the matrix has both real and imaginary parts, as do the eigenvalues and eigenvectors. The model in this case produces a series of transitions, whose intensity, phase, position and width are given by a complex-valued transition probability. The intermediate-exchange spectrum is just the sum of these transitions.

The NMR methods for studying chemical exchange are fundamentally no different from standard NMR methods. Chemical exchange effects appear in the spectrum and in measurements of the relaxation times, so careful measurement of these will provide good exchange data. Perhaps this is the single conclusion: apart from some algebraic and numerical details, chemical exchange is identical to 'normal' NMR.

## References

[1]  Gutowsky H S and Holm C H 1956 Rate processes and nuclear magnetic resonance spectra. II. Hindered internal rotation of amides *J. Chem. Phys.* **25** 1228–34
[2]  McConnell H M 1958 Reaction rates by nuclear magnetic resonance *J. Chem. Phys.* **28** 430–1
[3]  Reeves L W and Shaw K N 1970 Nuclear magnetic resonance studies of multi-site chemical exchange. I. Matrix formulation of the Bloch equations *Can. J. Chem.* **48** 3641–53
[4]  Bain A D and Duns G J 1996 A unified approach to dynamic NMR based on a physical interpretation of the transition probability *Can. J. Chem.* **74** 819–24
[5]  Sack R A 1958 A contribution to the theory of the exchange narrowing of spectral lines *Mol. Phys.* **1** 163–7
[6]  Binsch G 1969 A unified theory of exchange effects on nuclear magnetic resonance lineshapes *J. Am. Chem. Soc.* **91** 1304–9

[7] Kleier D A and Binsch G 1970 General theory of exchange-broadened NMR line shapes. II. Exploitation of invariance properties *J. Magn. Reson.* **3** 146–60

[8] Ernst R R, Bodenhausen G and Wokaun A 1987 *Principles of Nuclear Magnetic Resonance in One and Two Dimensions* (Oxford: Clarendon)

[9] Fano U 1964 Liouville representation of quantum mechanics with application to relaxation processes *Lectures on the Many Body Problem* vol 2, ed E R Caianiello (New York: Academic) pp 217–39

[10] Bain A D 1988 The superspin formalism for pulse NMR *Prog. Nucl. Magn. Reson. Spectrosc.* **20** 295–315

[11] Bain A D 1984 Coherence levels and coherence pathways in NMR. A simple way to design phase cycling procedures *J. Magn. Reson.* **56** 418–27

[12] Banwell C N and Primas H 1963 On the analysis of high-resolution nuclear magnetic resonance spectra. I. Methods of calculating NMR spectra *Mol. Phys.* **6** 225–56

[13] Alexander S 1962 Exchange of interacting nuclear spin in nuclear magnetic resonance. I. Intramolecular exchange *J. Chem. Phys.* **37** 967–74

[14] Limbach H H 1991 Dynamic NMR spectroscopy in the presence of kinetic hydrogen/deuterium isotope effects *NMR Basic Principles and Progress* vol 23, ed P Diehl, E Fluck, H Günther, R Kosfeld and J Seelig (Berlin: Springer) p 63–164

[15] Gordon R G and McGinnis R P 1968 Lineshapes in molecular spectra *J. Chem. Phys.* **49** 2455–6

[16] Freeman R and Hill H D W 1975 Determination of spin–spin relaxation time in high-resolution NMR *Dynamic Nuclear Magnetic Resonance Spectroscopy* ed L M Jackman and F A Cotton (New York: Academic) p 131–62

[17] Bain A D, Duns G J, Ternieden S, Ma J and Werstiuk N H 1994 The barrier to internal rotation and chemical exchange in N-acetylpyrrole. A study based on novel NMR methods and molecular modelling *J. Phys. Chem.* **98** 7458–63

[18] Bain A D and Duns G J 1994 Simultaneous determination of spin–lattice (T1) and spin–spin (T2) relaxation times in NMR; a robust and facile method for measuring T2. Optimization and data analysis of the offset-saturation experiment *J. Magn. Reson.* A **109** 56–64

[19] Deverell C, Morgan R E and Strange J H 1970 Studies of chemical exchange by nuclear magnetic relaxation in the rotating frame *Mol. Phys.* **18** 553–9

[20] Hoffman R A and Forsen S 1966 Transient and steady-state Overhauser experiments in the investigation of relaxation processes. Analogies between chemical exchange and relaxation *J. Chem. Phys.* **45** 2049–60

[21] Abel E W, Coston T P J, Orrell K G, Sik V and Stephenson D 1986 Two-dimensional NMR exchange spectroscopy. Quantitative treatment of multisite exchanging systems *J. Magn. Reson.* **70** 34–53

[22] Schäublin S, Höhener A and Ernst R R 1974 Fourier spectroscopy of non-equilibrium states. Application to CIDNP, Overhauser experiments and relaxation time measurements *J. Magn. Reson.* **13** 196–216

[23] McClung R E D and Aarts G H M 1995 Multisite magnetization transfer in strongly coupled spin systems *J. Magn. Reson.* A **115** 145–54

[24] Bain A D and Martin J S 1978 FT NMR of non-equilibrium states of complex spin systems: I. A Liouville space description *J. Magn. Reson.* **29** 125–35

[25] Olivier L, Poupko R, Zimmermann H and Luz Z 1996 Bond shift tautomerism of bibullvalenyl in solution and in the solid-state—a C-13 NMR-study *J. Phys. Chem.* **100** 17 995–18 003

[26] Lyerla J R, Yannoni C S and Fyfe C A 1982 Chemical applications of variable-temperature CPMAS NMR spectroscopy in solids *Accounts Chem. Res.* **15** 208–16

[27] Vold R R and Vold R L 1991 Deuterium relaxation in molecular solids *Adv. Magn. Opt. Reson.* **16** 85–171

[28] Yang Y, Schuster M, Blümich B and Spiess H W 1987 Dynamic magic-angle spinning NMR spectroscopy: exchange-induced sidebands *Chem. Phys. Lett.* **139** 239–43

[29] Gerardy-Montouillout V, Malveau C, Tekely P, Olender Z and Luz Z 1996 Odessa, a new 1D NMR exchange experiment for chemically equivalent nuclei in rotating solids *J. Magn. Reson.* A **123** 7–15

## Further Reading

Sandstrom J 1982 *Dynamic NMR Spectroscopy* (London: Academic)

An excellent introductory text.

Jackman L M and Cotton F A 1975 *Dynamic Nuclear Magnetic Resonance Spectroscopy* (New York: Academic)

A collection of reviews, covering essentially all of dynamic NMR up to about 1974.

Johnson C S 1965 Chemical rate processes and magnetic resonance *Adv. Magn. Reson.* **1** 33–102

One of the first complete reviews of the subject, and still very useful.

Orrell K G, Sik V and Stephenson D 1990 Quantitative investigations of molecular stereodynamics by 1D and 2D NMR methods *Prog. Nucl. Magn. Reson. Spectrosc.* **22** 141–208

Perrin C L and Dwyer T 1990 Application of two-dimensional NMR to kinetics of chemical exchange *Chem. Rev.* **90** 935–67

These two reviews (Orrell *et al* and Perrin and Dwyer) cover the modern pulse and two-dimensional NMR techniques for studying exchange.

# B2.5
# Gas-phase kinetics studies

*David Luckhaus and Martin Quack*

### B2.5.1   Introduction

The key to experimental gas-phase kinetics arises from the measurement of time, concentration, and temperature. Chemical kinetics is closely linked to time-dependent observation of concentration or amount of substance. Temperature is the most important single statistical parameter influencing the rates of chemical reactions (see chapter A3.4 for definitions and fundamentals).

The rich history of experimental chemical kinetics can be broadly classified according to various conceptual phases. The starting point of quantitative chemical kinetics was the formulation, in 1850 by Wilhelmy [1], of the time dependence of concentrations by a differential equation corresponding to a pseudo-first-order rate law for the hydrolysis ('inversion') of cane sugar. The observation of the concentration of cane sugar was carried out spectroscopically in the early experiments by following the time-dependent rotation of the plane of polarized light by the reaction mixture after mixing the reactant and the catalyst (the acid). During the following half-century, until about 1900, the nature of such phenomenological rate laws was clarified, as was the role of the rate constant and its temperature dependence. This epoch is characterized by the concepts introduced by van't Hoff [2] and Arrhenius [3]. It became clear that a distinction must be made between *phenomenological rate laws* for reactions following from a compound mechanism, such as the inversion of cane sugar, and rate laws and rate constants for *elementary reactions*, which can be combined into a compound mechanism. These new concepts characterize the second phase of chemical kinetics studies. For about half a century from 1900 to 1950, experimental investigations concentrated on elementary reactions and the mechanisms in which they are combined. The fathers of gas-phase kinetics, such as Bodenstein, Lindemann, and Hinshelwood, may be named as the representatives of this period.

During the course of these studies the necessity arose to study ever-faster reactions in order to ascertain their elementary nature. It became clear that the mixing of reactants was a major limitation in the study of fast elementary reactions. Fast mixing had reached its high point with the development of the accelerated and stopped-flow techniques [4, 5], reaching effective time resolutions in the millisecond range. Faster reactions were then frequently called 'immeasurably fast reactions' [6].

The new concept overcoming this limitation in the third phase of experimental kinetics started around 1950, and consisted of initiating a chemical reaction by a very fast physical perturbation and measuring the subsequent relaxation kinetics without a mixing step in the experiment. Various schools developed techniques along these lines, such as Norrish and Porter with flash photolysis [7, 8], and Eigen's school with $T$-jump and other relaxation techniques [6]. Weller's kinetics of fluorescence change can be classified as an indirect technique along these lines [9–11], and the Davidson and Jost schools developed shock-wave methods [12]. While the ideas for such techniques can be traced to earlier theoretical papers by Nernst and Einstein, the actual experimental developments started around 1950, initiating what has been called 'a race against time' [7, 8]. In particular, in relation to the laser-flash photolysis and pump–probe techniques developed after 1960, the essence of this race is to generate ever shorter laser pulses for 'pumping' a sample and well-controlled pulse delays for probing the sample's time evolution, where the lengths of the pulses define the limits of the time resolution of the techniques. Nanosecond resolution was available by about 1966, picosecond resolution around 1970, and by about 1985 the domain around 10 fs had been reached. Since that time progress by these techniques towards shorter times has slowed down somewhat, a typical value being about 5 fs today (see chapter B2.1). One might mention, however, a paper on the '0 fs' pulse [13] (dated 1 April 1990).

In parallel with this race against time, new experimental concepts were introduced, which escape the race by switching to a conceptually new approach. Among these are the molecular-beam-scattering techniques developed by Datz, Taylor, Martin, Herschbach, Lee, and others [14, 15], measuring reaction cross sections without time resolution instead of time evolution and reaction rates. NMR line-shape methods proposed by Gutowsky and Holm [16, 17] are further examples for alternative techniques, where time-dependent rates are calculated from time independent information. Neither of these techniques, however, is very well suited to study very fast intramolecular primary processes, which mark the starting point of all chemical reactions and may be considered to characterize the present fourth phase of experimental chemical kinetics. The new concept making these processes accessible to experimental investigation starts from high-resolution molecular spectra in the frequency domain (stationary or at least without short-time resolution) to derive ultimately the full molecular quantum-chemical kinetics in the time ranges from nanoseconds to attoseconds [18]. This experimental approach was, in fact, historically the first one to provide non-trivial three-dimensional molecular quantum wave packet dynamics [18–20], some time even before the first one-dimensional molecular quantum wave packet kinetics became available by short-pulse techniques (see [7, 21–23] and references cited therein).

The scope of the present chapter is to cover the most important experimental techniques in current use, including some of the well-established but also some of the most recent ones, together with current developments and to illustrate them with a few typical examples of results. The presentation by necessity is exemplary and not exhaustive.

On a modest level of detail, kinetic studies aim at determining overall phenomenological rate laws. These may serve to discriminate between different mechanistic models. However, to *prove* a compound reaction mechanism, it is necessary to determine the rate constant of each elementary step individually. Many kinetic experiments are devoted to the investigations of the temperature dependence of reaction rates. In addition to the obvious practical aspects, the temperature dependence of rate constants is also of great theoretical importance. Many statistical theories of chemical reactions are based on thermal equilibrium assumptions. Non-equilibrium effects are not only important for theories going beyond the classical transition-state picture. Eventually they might even be exploited to control chemical reactions [24]. This has led to the increased importance of energy or even quantum-state-resolved kinetic studies, which can be directly compared with detailed quantum-mechanical models of chemical reaction dynamics [25, 26].

Many experimental methods may be distinguished by whether and how they achieve time resolution—directly or indirectly. Indirect methods avoid the requirement for fast detection methods, either by determining relative rates from product yields or by transforming from the time axis to another coordinate, for example the distance or flow rate in flow tubes. Direct methods include (laser-) flash photolysis [27], pulse radiolysis [28] (see also chapter A3.5 on ion reactions), and the important relaxation techniques, which study the relaxation of a reacting system back into equilibrium after a perturbation [29]. Here one distinguishes two types of perturbation methods: (i) small perturbations from equilibrium leading to relaxation kinetics in the narrower sense with generalized first-order kinetics [6] and (ii) large perturbations from equilibrium leading to generally nonlinear rate laws. Typical examples for large perturbations are temperature jumps achieved through laser heating or in shock-tube experiments [30].

The time resolution of these methods is determined by the time it takes to initiate the reaction, for example the mixing time in flow tubes or the laser pulse width in flash photolysis, and by the time resolution of the detection. Relatively slow reactions can be monitored by taking samples and quickly quenching the reaction by cooling to low temperature or by dilution. The samples can then be analysed using conventional analytical techniques [31]. For time-dependent monitoring, any physical–chemical property changing during the course of a reaction can be used. Perhaps most important are spectroscopic detection techniques, particularly IR (infrared) and UV–VIS (ultraviolet–visible) absorption and fluorescence techniques. Directly recording spectroscopic signals as a function of time can achieve a time resolution of about 0.1 ns in favourable cases. Even higher time resolution of a few femtoseconds can be realised in pump–probe, laser flash-photolysis experiments [22, 32] (see also chapter B2.1 on ultrafast spectroscopy).

A completely different approach, in particular for fast unimolecular processes, extracts state-resolved kinetic information from molecular spectra without using any form of time-dependent observation. This includes conventional line-shape methods, as well as the quantum-dynamical analysis of rovibrational overtone spectra [18, 33–35].

At this point, we only mention the very important molecular-beam techniques [15, 27, 36–38] that allow the study of isolated molecules, largely without thermal congestion. They are ideally suited to the investigation of unimolecular processes, in particular dissociation reactions and energy redistribution processes (see chapter A3.13 on energy redistribution in reacting systems). The determination of state-resolved cross sections for bimolecular reactions in crossed molecular beams has paved the way for mechanistic investigations of elementary processes in the greatest possible detail (see chapter B2.3 *Reactive scattering*).

## B2.5.2 Flow tubes

Figure B2.5.1 schematically illustrates a typical flow-tube set-up. In gas-phase studies, it serves mainly two purposes. On the one hand it allows highly reactive shortlived reactant species, such as radicals or atoms, to be prepared at well-defined concentrations in an inert buffer gas. On the other hand, the flow replaces the time dependence, $t$, of a reaction by the dependence on the distance $x$ from the point where the reactants are mixed by the simple transformation with the flow velocity $v_f$:

$$t - t_0 = \frac{x - x_0}{v_f}. \tag{B2.5.1}$$

Instead of shifting the detector position, as indicated in figure B2.5.1, one often varies the location of the reactant mixing region using moveable injectors. This allows complex, possibly slow, but powerful, analytical techniques to be used for monitoring gas-phase reactions. In combination with mass-spectrometric detection, both reactants and products can be monitored quantitatively [39–41]. A further possibility consists of keeping the position $(x - x_0)$ in equation (B2.5.1) constant and varying the flow velocity $v_f$, thereby varying $(t - t_0)$. This technique is called 'accelerated flow'.

The time-to-distance transformation requires fast mixing and a known flow profile, ideally a turbulent flow with a well-defined homogeneous composition perpendicular to the direction of flow ('plug-flow'), as indicated by the shaded area in figure B2.5.1. More complicated profiles may require numerical transformations.

One of the major limiting factors for the time resolution of flow-tube experiments is the time required for mixing reactants and—to a lesser extent—the resolution of distance. With typical fast flow rates of more than 25 m s$^{-1}$ [42, 43] the time resolution lies between milliseconds and microseconds.

Modern applications of the technique include kinetic studies of post-combustion processes and their complex reaction systems. The influence of traces of $NO_x$ on the reaction kinetics of the $H_2/O_2$ and $CO/H_2O/O_2$ systems has recently been investigated in a high-pressure, turbulent-flow reactor at pressures up to 14 atm between 750 K and 1100 K [31, 44]. The reaction was monitored by taking samples at a fixed position and varying the location where the fuel ($CO/H_2O/NO_x$ or $H_2/NO_x$) is injected into a hot stream of $O_2$. The samples were instantly quenched in the hot-water-cooled sampling probe and analysed with a variety of analytical techniques including Fourier-transform infrared spectroscopy. The results were interpreted in terms of a reaction mechanism including 52 elementary reactions, assuming instant mixing and homogeneous composition perpendicular to the flow direction.

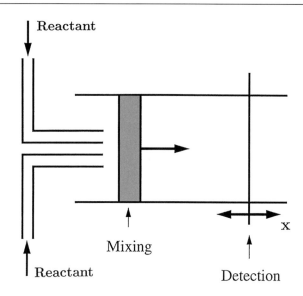

**Figure B2.5.1.** Schematic representation of a typical flow tube set-up with moveable detection. Adapted from [110].

NO generally catalyses 'fuel consumption' by transforming hydroperoxyl radicals into highly-reactive hydroxyl radicals:

$$HO_2 + NO \rightarrow NO_2 + OH \tag{B2.5.2}$$

$$H_2 + OH \rightarrow H_2O + H \tag{B2.5.3}$$

$$CO + OH \rightarrow CO_2 + H \tag{B2.5.4}$$

$$NO_2 + H \rightarrow NO + OH. \tag{B2.5.5}$$

As a stable radical, however, NO can also catalyze the recombination of radicals (X, Y) at higher concentrations, eventually inhibiting overall oxidation [45]:

$$X + NO \rightarrow XNO \tag{B2.5.6}$$

$$Y + XNO \rightarrow XY + NO. \tag{B2.5.7}$$

The balance of these two effects was found to depend delicately on the stoichiometry, pressure, and temperature. The results were used to develop a more comprehensive $CO/H_2O/O_2/NO_x$ reaction mechanism, incorporating the explicit fall-off behaviour of recombination reactions [46, 47].

### B2.5.3  Relaxation methods

Two types of relaxation techniques are distinguished, depending on whether the perturbation applied is small or large.

#### B2.5.3.1  *Relaxation after a small perturbation from equilibrium*

Perturbation or relaxation techniques are applied to chemical reaction systems with a well-defined equilibrium. An 'instantaneous' change of one or several state functions causes the system to relax into its new equilibrium [29]. In gas-phase kinetics, the perturbations typically exploit the temperature ($T$-jump) and pressure

($P$-jump) dependence of chemical equilibria [6]. The relaxation kinetics are monitored by spectroscopic methods.

$T$-jump techniques can achieve fast heating of the reaction system by pulsed radiation, for example with a microwave source or an IR laser. In the latter case one often adds an efficient inert absorber, such as $SF_6$. The heating of the reaction system then results from fast collisional relaxation of the initially-excited absorber molecules [48, 49].

When the perturbation is small, the reaction system is always close to equilibrium. Therefore, the relaxation follows generalized first-order kinetics, even if bi- or trimolecular steps are involved (see chapter A3.4). Take, for example, the reversible bimolecular step

$$A + B \overset{k_2}{\underset{k_{-2}}{\rightleftharpoons}} C + D. \tag{B2.5.8}$$

With equilibrium concentrations $c^{eq}$, the (small) deviation from equilibrium is given by

$$\Delta_c = c_A - c_A^{eq} = c_B - c_B^{eq} = c_C^{eq} - c_C = c_D^{eq} - c_D. \tag{B2.5.9}$$

Exploiting microscopic reversibility

$$k_2 c_A^{eq} c_B^{eq} = k_{-2} c_C^{eq} c_D^{eq} \tag{B2.5.10}$$

and neglecting terms quadratics in $\Delta_c$ leads to the approximate first-order rate law for this elementary step:

$$-\frac{dc_A}{dt} = -\frac{d\Delta_c}{dt} = k_{eff} \Delta_c \tag{B2.5.11}$$

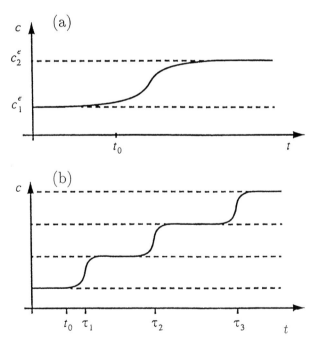

**Figure B2.5.2.** Schematic relaxation kinetics in a $T$-jump experiment. $c$ measures the progress of the reaction, for example the concentration of a reaction product as a function of time $t$ (abscissa with a logarithmic time scale). The reaction starts at $t_0$. (a) Simple relaxation kinetics with a single relaxation time. (b) Complex reaction mechanism with several relaxation times $\tau_i$. The different relaxation times $\tau_i$ are given by the turning points of $c$ as a function of $\ln(t)$. Adapted from [110].

$$k_{\text{eff}} = \{k_2(c_A^{eq} + c_B^{eq}) + k_{-2}(c_C^{eq} + c_D^{eq})\}. \tag{B2.5.12}$$

For this reaction alone, one would thus obtain a simple exponential relaxation with relaxation time

$$\tau_R = \frac{1}{k_{\text{eff}}}. \tag{B2.5.13}$$

More generally, the relaxation follows generalized first-order kinetics with several relaxation times $\tau_i$, as depicted schematically in figure B2.5.2 for the case of three well-separated time scales. The various relaxation times determine the turning points of the product concentration on a logarithmic time scale. These relaxation times are obtained from the eigenvalues of the appropriate rate coefficient matrix (chapter A3.4). The time resolution of $T$-jump relaxation techniques is often limited by the rate at which the system can be heated. With typical $T$-jumps of several Kelvin, the time resolution lies in the microsecond range.

$T$-jump experiments are particularly well-suited to the study of the dissociation kinetics of weakly-bound molecules or molecular complexes. Markwalder *et al* [49] used the laser-induced $T$-jump method to investigate the temperature and pressure dependence of $NO_2$ recombination kinetics:

$$N_2O_4 + M \underset{k_{\text{rec}}}{\overset{k_{\text{diss}}}{\rightleftharpoons}} 2NO_2 + M. \tag{B2.5.14}$$

With M = He, experiments were carried out between 255 K and 273 K with a few millibar $NO_2$ at total pressures between 300 mbar and 200 bar. Temperature jumps on the order of 1 K were effected by pulsed irradiation ($\ll 1$ $\mu$s) with a $CO_2$ laser at 9.2–9.6 $\mu$m and with $SiF_4$ or perfluorocyclobutane as primary IR absorbers ($\ll 1$ mbar). Under these conditions, the dissociation of $N_2O_4$ occurs within the irradiated volume on a time scale of a few hundred microseconds. $NO_2$ and $N_2O_4$ were monitored simultaneously by recording the time-dependent UV absorption signal at 420 nm and 253 nm, respectively. The recombination rate constant $k_{\text{rec}}$ can be obtained from the effective first-order relaxation time, $\tau_R$. A derivation analogous to (equations (B2.5.9)–(B2.5.12)) yields

$$k_{\text{rec}} = \frac{\tau_R^{-1}}{K_c + 4[NO_2]} \tag{B2.5.15}$$

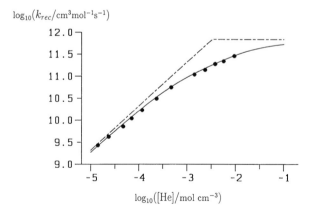

**Figure B2.5.3.** The fall-off curve of reaction (B2.5.14) with M = He between 0.3 bar and 200 bar. The dashed lines represent the extrapolated low- and high-pressure limits. $k_{\text{rec},0} = (2.1 \pm 0.2) \times 10^{14} \times$ [He] cm$^6$ mol$^{-2}$ s$^{-1}$ and $k_{\text{rec},\infty} = (7.0 \pm 0.7) \times 10^{11}$ cm$^3$ mol$^{-1}$ s$^{-1}$ yield the best fit (full curve) to the experimental data (full circles). Adapted from [49].

where $K_c$ is the equilibrium constant of equation (B2.5.14)). $k_{rec}$, $K_c$, and [NO$_2$] all refer to the final temperature. At 255 K, the authors obtained the typical fall-off curve depicted in figure B2.5.3. Even at 200 bar, the effective rate constant is still less than half the extrapolated high-pressure limit. The final results of the high- ($k_{rec,\infty}$) and low-pressure ($k_{rec,0}$) limiting rate constants (see chapter A3.4) were [49]

$$k_{rec,\infty} = (2.2 \pm 0.2) \times 10^6 (T/K)^{(2.3\pm0.2)} \text{ cm}^3 \text{ mol}^{-1} \text{ s}^{-1} \tag{B2.5.16}$$

$$k_{rec,0} = (7.5 \pm 0.8) \times 10^{35} (T/K)^{(-9.0\pm0.9)} \times [\text{He}] \text{ cm}^6 \text{ mol}^{-2} \text{ s}^{-1} \tag{B2.5.17}$$

where the temperature is given in Kelvin.

### B2.5.3.2  *Periodic small perturbation from equilibrium and ultrasound absorption*

The previous subsection described single-experiment perturbations by $T$-jumps or $P$-jumps. By contrast, sound and ultrasound may be used to induce small periodic perturbations of an equilibrium system that are equivalent to periodic pressure and temperature changes. A temperature amplitude $\delta T \approx 0.002$ K and a pressure amplitude $\delta P \approx 30$ mbar are typical in experiments with high-frequency ultrasound. Figure B2.5.4 illustrates the situation for different rates of chemical relaxation with the angular frequency of the sound wave $\omega$ and the relaxation time $\tau_R$:

$\omega\tau_R \ll 1$. The sample relaxes fast with the displacement from equilibrium synchronous with the sound wave.

$\omega\tau_R \simeq 1$. Compared with the sound wave, the system relaxes slowly. It lags behind, the phase is shifted, and amplitudes are reduced by damping.

$\omega\tau_R \gg 1$. Very slow relaxation.

As an example for the mathematical treatment, we take the bimolecular reaction

$$A + B \underset{k_b}{\overset{k_a}{\rightleftharpoons}} P + M. \tag{B2.5.18}$$

The turnover variable

$$x = c_A(t = 0) - c_A = c_B(t = 0) - c_B \tag{B2.5.19}$$

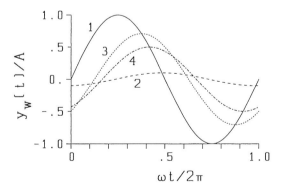

**Figure B2.5.4.** Periodic displacement from equilibrium through a sound wave. The full curve represents the temporal behaviour of pressure, temperature, and concentrations in the case of a very fast relaxation. The other lines illustrate various situations, with $\omega\tau_R$ according to table B2.5.1. $\omega$ is the angular frequency of the sound wave and $\tau_R$ is the chemical relaxation time. Adapted from [110].

**Table B2.5.1.** Form of the 'chemical wave' $y_w(t)$ (equation (B2.5.24)) for the various cases depicted in figure B2.5.4.

| Case | | Amplitude | Phase shift | $y_w(t)$ |
|------|------|-----------|-------------|----------|
| 1 | $\omega\tau_R \ll 1$ | $\approx a$ | $\approx 0$ | $a\sin(\omega t)$ |
| 2 | $\omega\tau_R = 10 \gg 1$ | $\approx a/(\omega\tau_R)$ | $\approx \pi/2$ | $-[a/(\omega\tau_R)]\cos(\omega t)$ |
| 3 | $\omega\tau_R = 1$ | $a/\sqrt{2}$ | $\pi/4$ | $(a/\sqrt{2})\sin(\omega t - \pi/4)$ |
| 4 | $\omega\tau_R = \sqrt{3}$ | $a/2$ | $\pi/3$ | $(a/2)\sin(\omega t - \pi/3)$ |

obeys a first-order rate law near the equilibrium value, $x_{eq}$, which, in turn, depends on the temperature change $\Delta T$ induced by the sound wave:

$$\frac{dx}{dt} = k_{eff}(x_{eq}(T + \Delta T) - x). \tag{B2.5.20}$$

For small $\Delta T$, the temperature dependence of the effective first-order rate constant, $k_{eff}$, can be neglected. With $y = x_{eq}(T) - x$ and with the shift of the equilibrium, $\Delta x_{eq} = x_{eq}(T) - x_{eq}(T + \Delta T)$, one obtains

$$-\frac{d\Delta y}{dt} = (y - \Delta x_{eq})\{k_b + k_a(c_A^{eq} + c_B^{eq})\} = \frac{y - \Delta x_{eq}}{\tau_R}. \tag{B2.5.21}$$

As long as $\Delta T$, $\Delta x_e$, and $\Delta x$ remain small, they will be proportional to the sinusoidal pressure wave. In particular

$$\Delta x_{eq} = a\sin(\omega t). \tag{B2.5.22}$$

This leads to

$$y(t) + \tau_R\frac{dy(t)}{dt} = a\sin(\omega t) \tag{B2.5.23}$$

with the general solution

$$y(t) = \left[y(0) + \frac{a\omega\tau_R}{1 + \omega^2\tau_R^2}\right]\exp\left(\frac{-t}{\tau_R}\right) + \frac{a}{1 + \omega^2\tau_R^2}\sin(\omega t) - \frac{a\omega\tau_R}{1 + \omega^2\tau_R^2}\cos(\omega t). \tag{B2.5.24}$$

For sufficiently long times (index $w$), the exponential can be neglected, leaving an oscillation of the turnover variable phase shifted with respect to the sound wave and with its amplitude reduced by the finite relaxation time $\tau_R$:

$$y_w(t) = \frac{a}{\sqrt{1 + \omega^2\tau_R^2}}\sin(\omega t - \arctan(\omega\tau_R)). \tag{B2.5.25}$$

The easily accessible frequency range of sound and ultrasound waves confines the range of applicability of this technique to relaxation times, $\tau_R$, between $10^{-4}$ s and $10^{-9}$ s. The derivation given here is, of course, independent of the underlying chemical process, as long as it is characterized by a single relaxation time. In general a complex relaxation spectrum is possible, so that the method reaches its limit for complex reactions as the interpretation of the results may become ambiguous. Extensive descriptions of relaxation experiments with small perturbations—both single and periodic—can be found in [29, 50]. Here one can also find numerous relaxation-time expressions for various equilibrium systems (uni-, bi-, trimolecular and reverse).

### B2.5.3.3 *Relaxation after large perturbation: shock-wave experiments*

A general limitation of the relaxation techniques with small perturbations from equilibrium discussed in the previous section arises from the restriction to systems starting at or near equilibrium under the conditions used. This limitation is overcome by techniques with large perturbations. The most important representative of this class of relaxation techniques in gas-phase kinetics is the shock-tube method, which achieves $T$-jumps of some 1000 K (accompanied by corresponding $P$-jumps) [30, 51–53]. Shock tubes are particularly useful for measuring the temperature dependence of reaction rates up to high temperatures. Figure B2.5.5 shows a schematic representation of the experimental set-up. The shock tube consists of a high- and a low-pressure part (R), separated by a diaphragm (d). The latter is filled with the reaction mixture and is operated either as a static cell or as a low-pressure flow tube. The high-pressure part is filled with a light, inert gas, such as $H_2$ or He, whose pressure is increased until the diaphragm breaks. This creates a shock wave travelling through the reaction mixture with supersonic speed. Behind the shock front the temperature can jump by more than 1000 K within 1 $\mu$s or less. After passing the detection zone, where the relaxation is followed spectroscopically, the wave front is reflected back at the end of the tube. The resulting change of the temperature as a function of time is depicted in figure B2.5.6. The temperature $T_2$ after the wave front is determined indirectly from the speed, $u$, at which the wave front travels through the tube. For a highly-dilute reaction mixture in a monoatomic gas with atomic mass $M$ one obtains

$$T_2 = T_1 \left( \frac{Mu^2}{5RT_1} - \frac{1}{3} \right) \left( \frac{Mu^2 + 1}{5RT_1} \right) \left( \frac{16Mu^2}{5RT_1} \right)^{-1} \tag{B2.5.26}$$

where $T_1$ is the temperature before the wave front has passed.

A classic shock-tube study concerned the high-temperature recombination rate and equilibrium for methyl radical recombination [54, 55]. Methyl radicals were first produced in a fast decomposition of diazomethane at high temperatures ($T > 1000$ K)

$$CH_3NNCH_3 \xrightarrow{[M]} 2CH_3 + N_2. \tag{B2.5.27}$$

Subsequently, the recombination of methyl radicals was studied by the high-temperature UV absorption of

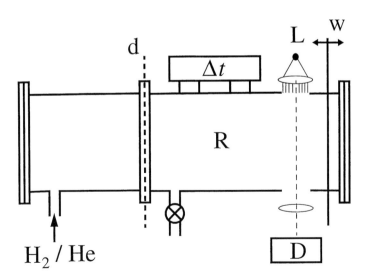

**Figure B2.5.5.** Schematic representation of a shock-tube apparatus. The diaphragm d separates the high-pressure part from the low-pressure reaction chamber R. The speed of the shock wave is determined by the time $\Delta t$ it takes to pass two observation points. The reaction itself is monitored spectroscopically using the light source L and a detector D close to the reflection wall W. Adapted from [110].

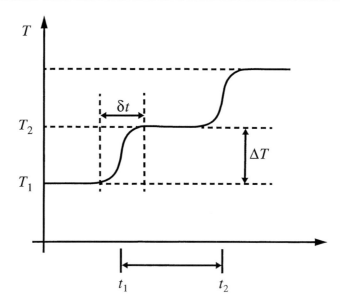

**Figure B2.5.6.** Temperature as a function of time in a shock-tube experiment. The first $T$-jump results from the incoming shock wave. The second is caused by the reflection of the shock wave at the wall of the tube. The rise time $\delta t$ typically is less than 1 $\mu$s, whereas the time delay between the incoming and reflected shock wave is on the order of several hundred microseconds. Adapted from [110].

the methyl radicals near 216 nm [54–56].

$$CH_3 + CH_3 \xrightarrow{[M]} C_2H_6. \tag{B2.5.28}$$

Figure B2.5.7 shows the absorption traces of the methyl radical absorption as a function of time. At the time resolution considered, the appearance of $CH_3$ is practically instantaneous. Subsequently, $CH_3$ disappears by recombination (equation (B2.5.28)). At temperatures below 1500 K, the equilibrium concentration of $CH_3$ is negligible compared with $C_2H_6$ (left-hand trace): the recombination is complete. At temperatures above 1500 K (right-hand trace) the equilibrium concentration of $CH_3$ is appreciable, and thus the technique allows the determination of both the equilibrium constant and the recombination rate [54, 55]. This experiment resolved a famous controversy on the temperature dependence of the recombination rate of methyl radicals. While standard RRKM theories [57, 58] predicted an increase of the high-pressure recombination rate coefficient $k_{rec,\infty}(T)$ by a factor of 10–30 between 300 K and 1400 K, the statistical-adiabatic-channel model predicts a slight decrease of $k_{rec,\infty}(T)$ with increasing temperature [59, 60], in agreement with experiment [54, 55]. This temperature dependence of the high-pressure recombination rate coefficient for radical–radical association is now a generally accepted feature, frequently reconfirmed for other examples in this class of reaction. The secondary isotope effect for the recombination of $CD_3$ has also been studied for this reaction [54, 55].

In a more recent example, a shock-tube experiment was used to study the thermal decomposition of methylamine between 1500 K and 2000 K [61, 62]:

$$CH_3NH_2 \xrightarrow{[M]} CH_3 + NH_2 \tag{B2.5.29}$$

$$k(T) = 8.17 \times 10^{16} \exp(-30\,710\ \text{K}/T)\ \text{cm}^3\ \text{mol}^{-1}\ \text{s}^{-1}(\pm 20\%). \tag{B2.5.30}$$

The pyrolysis of $CH_3NH_2$ (<1 mbar) was performed at 1.3 atm in Ar, spectroscopically monitoring the

a b

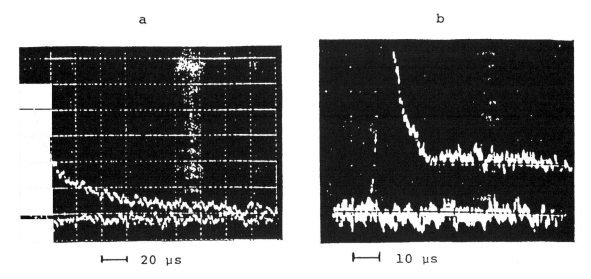

├──┤ 20 μs ├──┤ 10 μs

**Figure B2.5.7.** Oscilloscope trace of the UV absorption of methyl radical at 216 nm produced by decomposition of azomethane after a shock wave (after [54]): at (a) 1280 K and (b) 1575 K.

concentration of $NH_2$ radicals behind the reflected shock wave as a function of time. The interesting aspect of this experiment was the combination of a shock-tube experiment with the particularly sensitive detection of the $NH_2$ radicals by frequency-modulated, laser-absorption spectroscopy [61]. Compared with 'conventional' narrow-bandwidth laser-absorption detection the signal-to-noise ratio could be increased by a factor of 20, with correspondingly more accurate values for the rate constant $k(T)$.

## B2.5.4 Flash photolysis with flash lamps and lasers

One of the most important techniques for the study of gas-phase reactions is flash photolysis [8, 63]. A reaction is initiated by absorption of an intense light pulse, originally generated from flash lamps (duration $\approx 1$ $\mu$s). Nowadays these have frequently been replaced by pulsed laser sources, with the shortest pulses of the order of a few femtoseconds [22, 64].

### B2.5.4.1 Flash photolysis with flash lamps

The absorption of a light pulse 'instantaneously' generates reactive species in high concentrations, either through the formation of excited species or through photodissociation of suitable precursors. The reaction can then be followed spectroscopically by monitoring reactant and product concentrations. Among the classic studies using this technique, one may mention methyl radical spectroscopy used to study recombination at room temperature [56, 65–68].

A recent example of laser flash-lamp photolysis is given by Hippler *et al* [69], who investigated the temperature and pressure dependence of the thermal recombination rate constant $k_{rec}$ for the reaction

$$O + NO \xrightarrow{[M]} NO_2. \tag{B2.5.31}$$

The experiments were performed in a static reaction cell in a large excess of $N_2$ (2–200 bar). An UV laser pulse (193 nm, 20 ns) started the reaction by the photodissociation of $N_2O$ to form O atoms in the presence of NO. The reaction was monitored via the $NO_2$ absorption at 405 nm using a Hg–Xe high-pressure arc lamp, together with direct time-dependent detection. With a 20–200-fold excess of NO, the formation of $NO_2$ followed a pseudo-first-order rate law:

$$[NO_2] = [NO_2]_{t=\infty}(1 - \exp\{-k_{rec}[NO]t\}). \qquad (B2.5.32)$$

Direct time-dependent detection is limited by the response time of detectors, which depends on the frequency range, and the electronics used for data acquisition. In the most favourable cases, modern detector/oscilloscope combinations achieve a time resolution of up to 100 ps, but 1 ns is more typical. Again, this reaction has been of fundamental theoretical interest for a long time [59, 60].

### B2.5.4.2  Laser flash photolysis and pump–probe techniques

The so-called pump–probe technique uses a first photolysis pump pulse to generate reactive species and a second 'probe' pulse to detect reactant and product species. Figure B2.5.8 illustrates the experimental set-up. The time resolution is achieved through varying the delay between the pump pulse, which initiates the reaction, and the probe pulse, which monitors the reaction. Variable, short time delays in the picosecond range are conveniently realized through geometrical variation of the optical path length that the probe pulse travels. The probe pulse monitors reactant or product species either through direct absorption or by fluorescence excitation (laser-induced fluorescence, LIF), which generally is much more sensitive [70].

With the short pulses available from modern lasers, femtosecond time resolution has become possible [7, 71–73]. Producing accurate time delays between pump and probe pulses on this time scale represents a major challenge for the experimentalist, since light only travels 0.3 $\mu$m fs$^{-1}$. Table B2.5.2 summarizes typical laser pulses and characteristic times that are now available.

One of the early examples for kinetic studies on the femtosecond time scale is the photochemical pre-dissociation of NaI [74]:

$$NaI \xrightarrow{h\nu} NaI^* \longrightarrow Na + I. \qquad (B2.5.33)$$

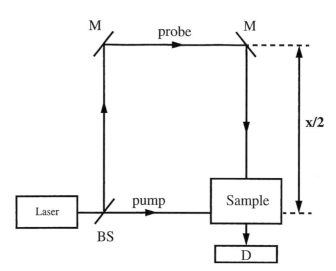

**Figure B2.5.8.** Schematic representation of laser-flash photolysis using the pump–probe technique. The beam splitter BS splits the pulse coming from the laser into a pump and a probe pulse. The pump pulse initiates a reaction in the sample, while the probe beam is diverted by several mirrors M through a variable delay line. The detector D monitors the absorption of the probe beam as a function of the delay between the pulses given by $x/2c$, where $c$ is the speed of light and $x$ is the difference between the optical path travelled by the probe and by the pump pulse. Adapted from [110].

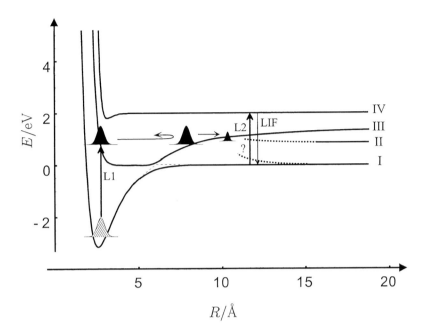

**Figure B2.5.9.** Schematic representation of the potential curves for the photodissociation of NaI as a function of the interatomic distance $R$. L1 and L2 are the pump and probe laser pulses, respectively. The dissociation limits are (I) Na $(^2S_{1/2})$ + I $(^2P_{3/2})$, (II) Na $(^2S_{1/2})$ + I $(^2P_{1/2})$, (III) Na$^+$ $(^1S_0)$ + I$^-$ $(^1S_0)$, and (IV) Na $(^2P)$ + I $(^2P_{3/2})$. $E$ is the excitation energy relative to the lowest dissociation limit (I). Adapted from Rosker *et al* [74].

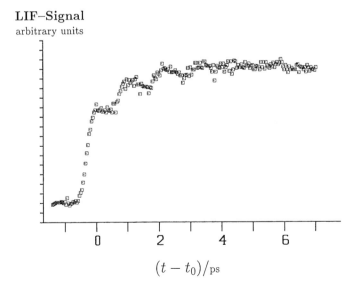

**Figure B2.5.10.** LIF signal of free Na atoms produced in the photodissociation of NaI. $t - t_0$ is the delay between the photolysis pulse (at $t_0$) and the probe pulse. Adapted from [111].

**Table B2.5.2.** Examples for pulsed lasers with different pulse durations and corresponding path lengths. For comparison the last column also gives the distance travelled by atoms with a velocity of 1000 ms$^{-1}$ (in parentheses) [81].

| Pulse duration | Laser | Availability | Optical path |
|---|---|---|---|
| 100–200 ns | Atmospheric CO$_2$ laser | Commercial | 30–60 m (0.1–0.2 mm) |
| 1–2 ns | Atmospheric CO$_2$ laser, mode coupled with saturable absorber | Available | 30–60 cm (1–2 $\mu$m) |
| 100 fs–1 ps | Solid-state laser (e.g. Ti:sapphire), dye laser | Commercial | 0.03–0.3 mm (100 pm–1 nm) |
| 8 fs | Laser with subsequent pulse compression | World record [108] | 2.4 $\mu$m (8.4 pm) |
| 6.5 fs | Ti:sapphire, mode coupling, with saturable absorber (semiconductor) | World record [64] | 2 $\mu$m (6.5 pm) |

The experiment is illustrated in figure B2.5.9. The initial pump pulse generates a localized wavepacket in the first excited $S_1$ state of NaI, which evolves with time. The potential well in the $S_1$ state is the result of an avoided crossing with the ground state. Every time the wavepacket passes this region, part of it crosses to the lower surface before the remainder is reflected at the outer wall of the $S_1$ potential. The crossing leads to ground-state dissociation products Na ($^2S_{1/2}$) + I ($^2P_{3/2}$). The crossing is monitored with the time-delayed probe pulse, which excites ground-state Na ($^2P^o \leftarrow {}^2S_{1/2}$). A photomultiplier detects the fluorescence back to the ground state. Figure B2.5.10 shows the resulting LIF signal as a function of the probe delay. One clearly recognizes the signature of the oscillatory wavepacket motion with a relatively long oscillation period of 1 ps, resulting from the flat potential and the heavy masses.

### B2.5.4.3  *The principle of continuous detection with uncertainty-limited time and frequency resolution*

In this approach one uses narrow-band continuous wave (cw) lasers for continuous spectroscopic detection of reactant and product species with high time and frequency resolution. Figure B2.5.11 shows an experimental scheme using detection lasers with a 1 MHz bandwidth. Thus, one can measure the energy spectrum of reaction products with very high energy resolution. In practice, today one can achieve an uncertainty-limited resolution given by

$$\Delta v \Delta t \geq \frac{1}{4\pi}. \tag{B2.5.34}$$

This technique with very high frequency resolution was used to study the population of different hyperfine structure levels of the iodine atom produced by the IR-laser-flash photolysis of organic iodides through multiphoton excitation:

$$CF_3I \xrightarrow{h\nu} CF_3 + I\,({}^2P_{3/2}, F = 1, 2, 3, 4). \tag{B2.5.35}$$

Figure B2.5.12 shows the energy-level scheme of the fine structure and hyperfine structure levels of iodine. The corresponding absorption spectrum shows six sharp hyperfine structure transitions. The experimental resolution is sufficient to determine the Doppler line shape associated with the velocity distribution of the I atoms produced in the reaction. In this way, one can determine either the temperature in an oven—as shown in figure B2.5.12—or the primary translational energy distribution of I atoms produced in photolysis, equation (B2.5.35).

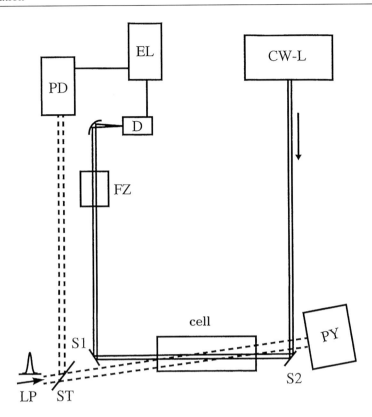

**Figure B2.5.11.** Schematic set-up of laser-flash photolysis for detecting reaction products with uncertainty-limited energy and time resolution. The excitation $CO_2$ laser pulse LP (broken line) enters the cell from the left, the tunable cw laser beam CW–L (full line) from the right. A filter cell FZ protects the detector D, which determines the time-dependent absorbance, from scattered $CO_2$ laser light. The pyroelectric detector PY measures the energy of the $CO_2$ laser pulse and the photon drag detector PD its temporal profile. A complete description can be found in [109].

## B2.5.5  Multiphoton excitation

### B2.5.5.1  Mechanisms of multiphoton excitation

The common flash-lamp photolysis and often also laser-flash photolysis are based on photochemical processes that are initiated by the absorption of a photon, $h\nu$. The intensity of laser pulses can reach GW cm$^{-2}$ or even TW cm$^{-2}$, where multiphoton processes become important. Figure B2.5.13 summarizes the different mechanisms of multiphoton excitation [75, 76, 112]. The direct multiphoton absorption of mechanism (i) requires an odd number of photons to reach an excited atomic or molecular level in the case of strict electric dipole and parity selection rules [117]. The Goeppert–Mayer two- (or multi-) photon absorption, mechanism (ii), may look similar, but it involves intermediate levels far from resonance with one-photon absorption. A third, quasi-resonant stepwise mechanism (iii), proceeds via single-photon excitation steps involving near-resonant intermediate levels. Finally, in mechanism (iv), there is the stepwise multiphoton absorption of incoherent radiation from thermal light sources or broad-band statistical multimode lasers. In principle, all of these processes and their combinations play a role in the multiphoton excitation of atoms and molecules, but one can broadly distinguish two situations.

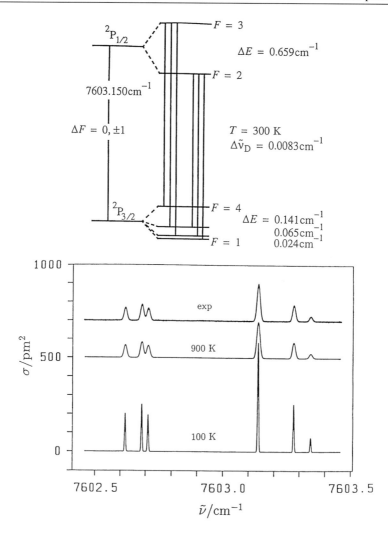

**Figure B2.5.12.** Hyperfine structure energy level scheme and spectrum for the I $(^2P_{3/2}) \leftrightarrow$ I $(^2P_{1/2})$ fine structure transition [109].

(A) During the multiphoton excitation of molecular vibrations with IR lasers, many (typically 10–50) photons are absorbed in a quasi-resonant stepwise process until the absorbed energy is sufficient to initiate a unimolecular reaction, dissociation, or isomerization, usually in the electronic ground state. The record in the number of absorbed photons (about 500 photons of a $CO_2$ laser) was reached with the $C_{60}$ molecule [77]. This case proved an exception in that the primary reaction was ionization. The IR multiphoton excitation is the starting point for a new gas-phase photochemistry, IR laser chemistry, which encompasses numerous chemical processes.

(B) The multiphoton excitation of electronic levels of atoms and molecules with visible or UV radiation generally leads to ionization. The mechanism is generally a combination of direct, Goeppert–Mayer, and quasi-resonant stepwise processes. Since ionization often requires only two or three photons, this type of multiphoton excitation is used for spectroscopic purposes in combination with mass-spectrometric detection of ions.

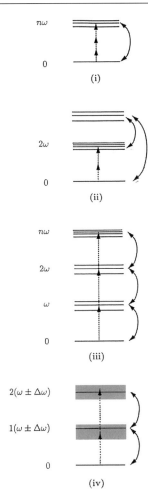

**Figure B2.5.13.** Schematic representation of the four different mechanisms of multiphoton excitation: (i) direct, (ii) Goeppert–Mayer, (iii) quasi-resonant stepwise and (iv) incoherent stepwise. Full lines (right) represent the coupling path between the energy levels and broken arrows the photon energies with angular frequency $\omega$ ($\Delta\omega$ is the frequency width of the excitation light in the case of incoherent excitation), see also [112].

### B2.5.5.2  *IR multiphoton excitation and IR laser chemistry*

The most commonly used laser-light source in IR laser chemistry is the atmospheric $CO_2$ laser, with IR emission lines between $900 \, \text{cm}^{-1}$ and $1100 \, \text{cm}^{-1}$, in the fingerprint range of the IR spectrum, where characteristic molecular vibrations can be excited. With a photon energy of about $12 \, \text{kJ} \, \text{mol}^{-1}$, on the order of 10–40 photons are needed to initiate a chemical reaction in the energy range of $100–500 \, \text{kJ} \, \text{mol}^{-1}$. The laser pulses are 100 ns–1 $\mu$s long, but a series of 1–2 ns pulses can be generated by mode coupling. Typical intensities are $100 \, \text{MW} \, \text{cm}^{-2}$. Figure B2.5.14 schematically illustrates the photodissociation of $CF_3I$ (equation (B2.5.35)) after multiphoton excitation via the CF stretching vibration at $1070 \, \text{cm}^{-1}$. More than 17 photons are needed to break the C–I bond, a typical value in IR laser chemistry. Contributions from direct absorption (i) are insignificant, so that the process almost exclusively follows the quasi-resonant mechanism (iii), which can be treated by generalized first-order kinetics. As an example, figure B2.5.15 illustrates the formation of I

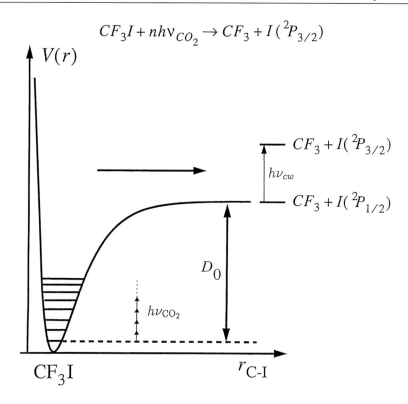

**Figure B2.5.14.** The IR laser chemistry of $CF_3I$ excited up to the dissociation energy $D_0$ with about 17 quanta of a $CO_2$ laser, $h\nu_{CO_2}$. The dissociation is detected by uncertainty limited cw absorption ($h\nu_{cw}$), see figures B2.5.11 and B2.5.12. The energy levels of the C–I stretching vibration are not drawn to scale. In reality their separation is much smaller. Adapted from [109].

atoms (upper trace) during excitation with the pulse sequence of a mode-coupled $CO_2$ laser (lower trace). In addition to the intensity, $I$, the fluence, $F$, of radiation is a very important parameter in IR laser chemistry (and more generally in multiphoton excitation):

$$F(t) = \int_0^t I(t')\,dt'.$$

(B2.5.36)

Consequently, the reaction yield $F_p$ in figure B2.5.15 is shown as a function of the fluence, $F$. At the end of a laser-pulse sequence with a typical fluence $F \simeq 3$ J cm$^{-2}$, practically 100% of the $CF_3I$ is photolysed. As described in section B2.5.4.3, the product-level distribution of the iodine atoms formed in this type of reaction can be determined spectroscopically. Table B2.5.3 shows the results of such an analysis of the population of hyperfine structure levels and of the translational energy distribution for the IR multiphoton dissociation of different organic iodides. The average product translational energy (in the centre-of-mass system) does not change much from the small $CF_3I$ to the much larger $C_6F_5I$ molecule. Its relative share of the total energy, however, decreases: much more energy appears as the internal energy of the $C_6F_5$ fragment. This can be readily understood assuming a roughly statistical distribution over the large number of internal degrees of freedom. Such results are crucial for a more accurate dynamical understanding of the processes taking place during a chemical reaction.

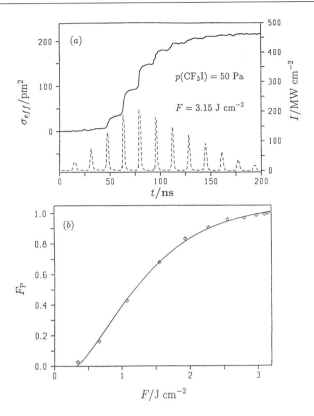

**Figure B2.5.15.** Iodine atom formation in the IR laser chemistry of $CF_3I$ (excitation at 1074.65 $cm^{-1}$, probe on the $F = 4 \rightarrow F = 3$ hyperfine structure transition, see figure B2.5.12.) (*a*) The absorbance as a function of time (effective absorption cross section $\sigma_{eff}$, full curve, left ordinate) shows clear steps at each maximum of the mode locked $CO_2$ laser pulse sequence (intensity, broken curve, right ordinate). (*b*) The fraction $F_P$ of dissociating molecules as a function of fluence $F$.

**Table B2.5.3.** Product energy distribution for some IR laser chemical reactions. $\langle E_t \rangle$ is the average relative translational energy of fragments, $\langle E_{int} \rangle$ is the average vibrational and rotational energy of polyatomic fragments, and $f_t$ is the fraction of the total product energy appearing as translational energy [109].

| Reaction | $\langle E_t \rangle/(kJ\ mol^{-1})$ | $\langle E_{int} \rangle/(kJ\ mol^{-1})$ | $f_t$ |
|---|---|---|---|
| $CF_3I \rightarrow CF_3 + I$ | 9.9 | 19.8 | 0.33 |
| $CF_3CHFI \rightarrow CF_3CHF + I$ | 10.9 | 100.9 | 0.097 |
| $C_6F_5I \rightarrow C_6F_5 + I$ | 13.5 | 233.0 | 0.055 |

In exceptional cases, the IR laser excitation can lead to ionization. An interesting example is the $CO_2$-laser-induced ionization of $C_{60}$, where $n \geq 500$ photons are absorbed and vibrations are excited far beyond the ionization threshold of the molecule

$$C_{60} \xrightarrow{nh\nu} C_{60}^+ + e^- \tag{B2.5.37}$$

$$C_{60}^+ \xrightarrow{mh\nu} C_{58}^+ + C_2. \tag{B2.5.38}$$

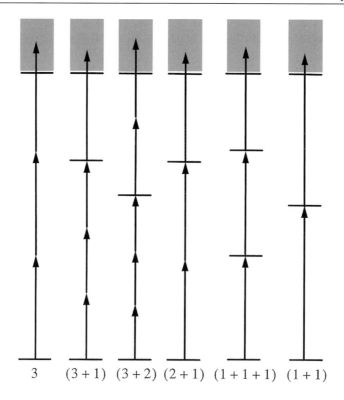

$$3 \quad (3+1) \quad (3+2) \quad (2+1) \quad (1+1+1) \quad (1+1)$$

**Figure B2.5.16.** Different multiphoton ionization schemes. Each scheme is classified according to the number of photons that lead to resonant intermediate levels and to the ionization continuum (hatched area). Adapted from [110].

The $C_{60}^+$ is excited further and decomposes stepwise into $C_{58}^+$, $C_{56}^+$, etc with the formation of $C_2$ units [77].

### B2.5.5.3 Multiphoton ionization

In contrast to the ionization of $C_{60}$ after vibrational excitation, typical multiphoton ionization proceeds via the excitation of higher electronic levels. In principle, multiphoton ionization can either be used to generate ions and to study their reactions, or as a sensitive detection technique for atoms, molecules, and radicals in reaction kinetics. The second application is more common. In most cases of excitation with visible or UV laser radiation, a few photons are enough to reach or exceed the ionization limit. A particularly important technique is resonantly enhanced multiphoton ionization (REMPI), which exploits the resonance of monochromatic laser radiation with one or several intermediate levels (in one-photon or in multiphoton processes). The mechanisms are distinguished according to the number of photons leading to the resonant intermediate levels and to the final level, as illustrated in figure B2.5.16. Several lasers of different frequencies may be combined.

As an example, we mention the detection of iodine atoms in their $^2P_{3/2}$ ground state with a $3 + 2$ multiphoton ionization process at a laser wavelength of 474.3 nm. Excited iodine atoms ($^2P_{1/2}$) can also be detected selectively as the resonance condition is reached at a different laser wavelength of 477.7 nm. As an example, figure B2.5.17 shows REMPI iodine atom detection after IR laser photolysis of $CF_3I$. This 'pump–probe' experiment involves two, delayed, laser pulses, with a 200 ns IR photolysis pulse and a 10 ns probe pulse, which detects iodine atoms at different times during and after the photolysis pulse. This experiment illustrates a fundamental problem of product detection by multiphoton ionization: with its high intensity,

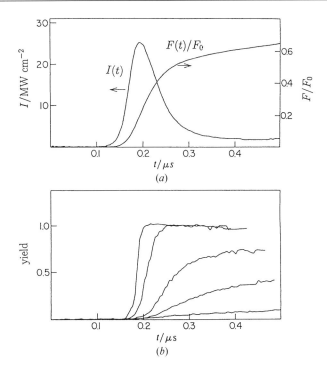

**Figure B2.5.17.** (*a*) Time-dependent intensity $I$ and reduced fluence $F/F_0$ for a single-mode $CO_2$ laser pulse used in the IR laser photolysis of $CF_3I$. $F_0$ is the total fluence of the laser pulse. (*b*) VIS–REMPI iodine atom signals obtained with $CO_2$ laser pulses of different fluence (after [113]).

the short-wavelength probe laser radiation alone can photolyse the reactant $CF_3I$ molecules. One cannot distinguish between iodine atoms produced by the photolysis pulse and those produced by the probe pulse. In the present example the problem is solved by the well-founded assumption that the photolysis of $CF_3I$ by a visible probe pulse produces excited iodine atoms ($^2P_{1/2}$), whereas the IR photolysis pulse leads to ground-state iodine atoms ($^2P_{3/2}$). In general, however, significant perturbations of the reaction system are to be expected from the REMPI spectroscopic detection of products.

### B2.5.5.4   *Laser isotope separation and mode-selective reactions*

Apart from the obvious property of defining pulses within short time intervals, the pulsed laser radiation used in reaction kinetics studies can have additional particular properties: (i) high intensity, (ii) high monochromaticity, and (iii) coherence. Depending on the type of laser, these properties may be more or less pronounced. For instance, the pulsed $CO_2$ lasers used in IR laser chemistry easily reach intensities between $MW\,cm^{-2}$ and $GW\,cm^{-2}$. Special lasers used in nuclear fusion experiments may even reach $10^{21}\,W\,cm^{-2}$ [78, 79]. Ideally the monochromaticity, $\Delta \nu$, is related to the pulse length, $\Delta t$, through

$$\Delta \nu \Delta t \simeq 1. \tag{B2.5.39}$$

Although this limit is not always reached. The same is true for the coherence of the radiation. Each of these properties can be exploited for particular chemical applications. The monochromaticity can be used to initiate a chemical reaction of particular molecules in a mixture. The laser isotope separation of $^{12}C$ and $^{13}C$ in natural abundance exploits the isotope shift of molecular vibrational frequencies. At $10–50\,cm^{-1}$, the

Intermolecular Selectivity

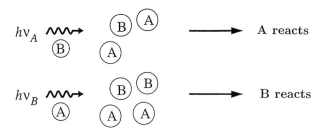

Intramolecular Selectivity

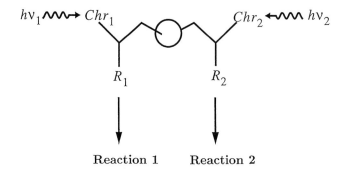

**Figure B2.5.18.** General scheme for *inter-* and *intra*molecular selectivity in laser chemistry. Intermolecular selectivity: a laser with frequency $\nu_A$ selectively excites molecules A, which subsequently react, in a mixture of A and B molecules. *Intra*molecular selectivity: a laser with frequency $\nu_1$ ($\nu_2$) selectively excites the chromophore Chr$_1$ (Chr$_2$) of a molecule which preferentially follows reaction 1 (2) at this position (after [75]).

corresponding shift of IR absorption wavenumbers is large compared to the spectral width of the $CO_2$ laser pulse ($\leq 0.1$ cm$^{-1}$), which makes the $^{13}$C isotope separation relatively easy. Table B2.5.4 summarizes this and other similar applications [75, 80, 81]. The intermolecular selectivity of IR-multiphoton excitation can be greatly increased by two-frequency–two-step schemes such as in the new spectroscopic technique of IRLAPS (InfraRed Laser Assisted Photofragment Spectroscopy [115]).

Figure B2.5.18 compares this *inter*molecular selectivity with *intra*molecular or mode selectivity. In an IR plus UV, two-photon process, it is possible to break either of the two bonds selectively in the same HOD molecule. Depending on whether the OH or the OD stretching vibration is excited, the products are either H + OD or HO + D [24]. In large molecules, *intra*molecular selectivity competes with fast *intra*molecular (i.e. unimolecular) vibrational energy redistribution (IVR) processes, which destroy the selectivity. In laser experiments with D-difluorobutane [82], it was estimated that, in spite of frequency selective excitation of the CHDF end group, no selective reaction would occur on time scales above $10^{-11}$ s, figure B2.5.18. In contrast to IVR processes, which can be very fast, the *inter*molecular energy transfer processes, which may reduce intermolecular selectivity, are generally much slower, since they proceed via bimolecular energy exchange, which is limited by the collision frequency (see chapter A3.13).

Strategies for achieving intra- and intermolecular selectivity are the subject of a very active field of current research with many open questions. Under the label 'coherent control' it includes approaches that

**Table B2.5.4.** Laser isotope separation (see also [75]).

| Isotope | Source | Comments |
|---|---|---|
| $^2$H | $CHF_2Cl$ | High selectivity at room temperature |
| $^{10}$B | $BCl_3$ | Early laser isotope separation after IR multiphoton excitation high selectivity at room temperature |
| $^{13}$C | $CHF_2Cl$ | Two-step separation scheme (220 mg $^{13}$C h$^{-1}$) |
| $^{14}$N,$^{15}$N | $CH_3NO_2$ | Selectivity through two absorption bands |
| $^{16}$O,$^{17}$O | OCS | IR–UV double resonance; also selective for S and C |
| $^{29}$Si,$^{30}$Si | $Si_2F_6$ | Reaction of both isotopes with high selectivity (high fluence) |
| $^{34}$S | $SF_6$ | Early report of laser isotope separation |
| $^{35}$Cl,$^{37}$Cl | $CF_2Cl_2$ | Also selective with respect to C |
| Mo | $MoF_6$ | Applied to several isotopes; low selectivity and yield |
| $^{235}$U | $UF_6$ | Dissociation with two lasers at different wavelengths (two-colour dissociation) |

exploit the coherence properties of laser radiation to control chemical reactions. Figure B2.5.18 summarizes the different schemes of intra- and intermolecular selectivity.

## B2.5.6 Chemical activation

The formation of reactive species by photodissociation of a precursor through flash photolysis can be regarded as a special case of chemical activation. More generally, this technique exploits the enthalpy of a chemical reaction to generate species with a non-equilibrium energy distribution (relative to the ambient temperature). Using different reactions to produce the same reactive species allows one to study the energy dependence of the ensuing reaction kinetics (or collisional deactivation). Historically, the method has played a central role in the experimental study of collisional energy-transfer processes and non-equilibrium effects on chemical reaction rates [83–87]. Although modern laser techniques can in principle achieve much narrower energy distributions, optical excitation is frequently not a viable method for the preparation of excited reactive species. Therefore chemical activation—often combined with (laser-) flash photolysis—still plays an important role in gas-phase kinetics, in particular of unstable species such as radicals [88]. Chemical activation also plays an important role in energy-transfer studies (see chapter A3.13).

A recent study of the vibrational-to-vibrational (V–V) energy transfer between highly-excited oxygen molecules and ozone combines laser-flash photolysis and chemical activation with detection by time-resolved LIF [89]. Partial laser-flash photolysis at 532 nm of pure ozone in the Chappuis band produces translationally-hot oxygen atoms $O(^3P)$. In the chemical-activation step they react with ozone to form an electronic ground-state $O_2(\tilde{X}^3\Sigma_g^-, v'')$ with up to $v'' = 27$ quanta of vibrational excitation in an excess of thermally populated ozone:

$$O(^3P) + O_3 \longrightarrow O_2 + O_2(\tilde{X}^3\Sigma_g^-, v \gg 1) \tag{B2.5.40}$$

$$k = 3.9 \times 10^{-11} \text{ cm}^3 \text{ s}^{-1}. \tag{B2.5.41}$$

The chemical-activation step is between one and two orders of magnitude faster than the subsequent collisional deactivation of vibrationally excited $O_2$. Finally, the population of individual vibrational levels $v''$ of $O_2$ is probed through LIF in the Schumann–Runge band ($B^3\Sigma_u^- \leftarrow X^3\Sigma_g^-$): after exciting the oxygen molecules to the vibrational ground state of their first electronically excited state ($v' = 0$), the ensuing fluorescence back to the electronic ground state is detected by a photomultiplier tube and recorded as a function of time. The

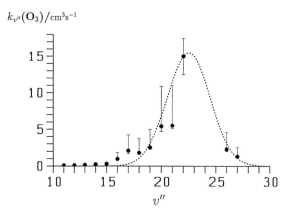

**Figure B2.5.19.** The collisional deactivation rate constant $k_{v''}(O_3)$ (equation (B2.5.42)) as a function of the vibrational level $v''$. Adapted from [89]. Experimental data are represented by full circles with error bars. The broken curve is to serve as a guide to the eye.

resulting collisional relaxation rate constants as a function of the vibrational excitation of $O_2$

$$\frac{d[O_2(v'')]}{dt} = k_{v''}(O_3)[O_2(v'')][O_3] \tag{B2.5.42}$$

show a pronounced maximum near $v'' = 23$, as illustrated in figure B2.5.19. At this value, the $v'' \rightarrow v'' - 1$ transition happens to be in almost perfect resonance with the symmetric stretch fundamental of $O_3$. The resonance enhancement by one to two orders of magnitude is typical for collisional V–V energy transfer of highly-excited molecules (see chapter A1.13).

## B2.5.7   Line-shape methods

Energy (or frequency) spectra are fundamentally related to the underlying time-dependent processes through the Fourier transformation. In practice, however, the relation between spectroscopically-observed line shapes and kinetic (reaction) processes is neither simple nor unambiguous [18]. There are many contributions to observed line shapes [33]. Apart from finite instrumental resolution, spectra may be inhomogeneously broadened through thermal congestion. A simple example is the Doppler broadening as a result of the Maxwell–Boltzmann velocity distribution leading to a Gaussian line shape.

Even if the homogeneous line shape can be extracted, many other processes can contribute. Every decay process contributes to the finite lifetime of an excited species, $A^*$, with an individual decay constant $k_j^*$

$$A^* \xrightarrow{k_j^*} A_j \tag{B2.5.43}$$

$$-\frac{d[A^*]}{dt} = \sum_j k_j^*[A^*] \tag{B2.5.44}$$

$$k_{\text{eff}} = \sum_j k_j^*. \tag{B2.5.45}$$

The exponential decay of the $A^*$ population corresponds to a Lorentzian line shape for the absorption (or emission) cross section, $\sigma$, as a function of energy $E$. The lineshape is centred around its maximum at $E_0$.

**Table B2.5.5.** The photochemical decomposition of methyl radicals (UV excitation at 216 nm). $\tilde{\Gamma}$ is the wavenumber linewidth of the methyl radical absorption and $k$ is the effective first-order decay constant [54].

| Decay process | $\tilde{\Gamma}$ (cm$^{-1}$) | $\Delta\nu = hc\tilde{\Gamma}$ (s$^{-1}$) | $k = 2\pi\Delta\nu$ (s$^{-1}$) | $\tau = 1/k$ (fs) |
|---|---|---|---|---|
| $CH_3^* \to CH_2 + H$ | 60 | $1.8 \times 10^{12}$ | $1.13 \times 10^{13}$ | 88 |
| $CD_3^* \to CD_2 + D$ | 8 | $2.4 \times 10^{11}$ | $1.51 \times 10^{12}$ | 663 |

The full-width at half-maximum ($\Gamma$) is proportional to $k_{\text{eff}}$:

$$\sigma(E) = \sigma(E_0)\frac{(\Gamma/2)^2}{(E - E_0)^2 + (\Gamma/2)^2} \tag{B2.5.46}$$

$$\Gamma = k_{\text{eff}}h/(2\pi). \tag{B2.5.47}$$

Apart from the *natural lifetime* due to spontaneous emission, both uni- and bimolecular processes can contribute to the observed value of $\Gamma$. One important contribution $k_{\text{col}}$ comes from *collisional broadening*, which can be distinguished by its pressure dependence (or dependence upon concentration [M] of the collision partner):

$$k_{\text{col}} = \left(\frac{8k_{B}T}{\pi\mu}\right)^{1/2}\langle\sigma_{\text{col}}\rangle[M]. \tag{B2.5.48}$$

Equation (B2.5.48) introduces the effective average collision cross section $\langle\sigma_{\text{col}}\rangle$. Here, the lifetime broadening results from the (collisional) perturbation of $A^*$ by collisions with M.

Lifetimes of 1 ps translate into linewidths of about 5 cm$^{-1}$. Thus, line-shape methods are ideally suited to measure very fast decay processes, in particular predissociation of excited species. An example is the predissociation of $O_2$ molecules excited above 50 000 cm$^{-1}$, which gives rise to the broadening of the Schumann–Runge bands

$$O_2^* \longrightarrow O + O. \tag{B2.5.49}$$

This is the source of ozone, through the reaction $O_2 + O \xrightarrow{[M]} O_3$. One obtains a pronounced dependence of the decay rate on the vibrational level of $O_2^*$ and to a lesser extent on its rotational state [90, 91]. Typical decay rate constants for this reaction range from $1.5 \times 10^{11}$ s$^{-1}$ to $7.5 \times 10^{11}$ s$^{-1}$. Another important example is the predissociation of methyl radicals [54]

$$CH_3^* \longrightarrow \text{Products}. \tag{B2.5.50}$$

The results are summarized in table B2.5.5. The rate constants $k_j$ of individual decay channels may be obtained from the relative yields of all primary reaction products, which can be determined in stationary experiments.

Similar considerations have been exploited for the systematic analysis of room-temperature and molecular-beam IR spectra in terms of intramolecular vibrational relaxation rates [33, 34, 92, 94] (see also chapter A3.13).

## B2.5.8 Intramolecular kinetics from high-resolution spectroscopy

Molecular spectroscopy offers a fundamental approach to intramolecular processes [18, 94]. The spectral analysis in terms of detailed quantum mechanical models in principle provides the complete information about the wave-packet dynamics on a level of detail not easily accessible by time-resolved techniques.

The approach is ideally suited to the study of IVR on fast timescales, which is the most important primary process in unimolecular reactions. The application of high-resolution rovibrational overtone spectroscopy to this problem has been extensively demonstrated. Effective Hamiltonian analyses alone are insufficient, as has been demonstrated by explicit quantum dynamical models based on *ab initio* theory [95]. The fast IVR characteristic of the CH chromophore in various molecular environments is probably the most comprehensively studied example of the kind [96] (see chapter A3.13). The importance of this question to chemical kinetics can perhaps best be illustrated with the following examples. The atom recombination reaction

$$H + H + H \rightarrow H_2 + H \qquad (B2.5.51)$$

is well known to occur as a very slow trimolecular process. By contrast, the polyatomic recombination

$$H \cdot + \cdot CR_1R_2R_3 \rightarrow CHR_1R_2R_3^* \qquad (B2.5.52)$$
$$CHR_1R_2R_3^* + M \rightarrow CHR_1R_2R_3 + M \qquad (B2.5.53)$$

happens quickly as a sequence of bimolecular recombination and collisional energy-transfer steps, with a relatively long-lived intermediate, $CHR_1R_2R_3^*$. The reason is the possibility of transferring energy intramolecularly from the initially excited C–H bond to other parts of the polyatomic molecule, according to the scheme

$$H \cdot + \cdot CR_1R_2R_3 \rightleftharpoons H \overset{***}{-} CR_1R_2R_3 \qquad (B2.5.54)$$
$$H \overset{***}{-} CR_1R_2R_3 \rightleftharpoons H \overset{*}{-} CR_1R_2^*R_3^* \qquad (B2.5.55)$$

$$H \overset{*}{-} CR_1R_2^*R_3^* \rightleftharpoons H–CR_1R_2^*R_3^{**} \qquad (B2.5.56)$$
$$H–CR_1R_2^*R_3^{**} \rightleftharpoons HCR_1R_2 + R_3. \qquad (B2.5.57)$$

This illustrates the steps of energy transfer from the initially highly-excited C–H bond to other parts of the molecule, subsequent concentration of energy in one part of the molecule ($CR_3^{**}$), and finally rupture of the corresponding bond. A typical example of this kind is the chemical activation reaction (abbreviated)

$$H + C_2H_5 \rightarrow C_2H_6^* \qquad (B2.5.58)$$
$$C_2H_6^* \rightarrow 2CH_3. \qquad (B2.5.59)$$

It is the first IVR step of (B2.5.59) that is investigated by high-resolution spectroscopy. The analysis, outlined in some detail in [18], follows the scheme in figure B2.5.20. This kind of analysis has been applied to the evolution of entropy in the single, isolated molecule $CHD_2F$, as shown in figure B2.5.21. In this case, entropy is investigated as a relevant time-dependent observable of kinetics (see chapter A3.4). In the example, the question of time-reversal symmetry on the femtosecond timescale has been studied [18, 114, 116], but many other applications can be thought of.

This kind of 'dynamical spectroscopic analysis' is not restricted to fast primary IVR processes. It would apply just as well to the study of completely unimolecular reactions, viz isomerizations such as H-atom transfer reactions, for example $CH_2O \rightleftharpoons HCOH$ [97], $HCN \rightleftharpoons HNC$ ([98] and references cited therein), and $HCCH \rightleftharpoons H_2CC$ ([99] and references cited therein) (although the spectroscopic aspects have not been fully exploited in these cases), as well as the carefully studied $NH_2OH \rightleftharpoons NH_3O$ [100, 101]. Recent studies on the tunnelling dynamics of hydrogen peroxide and aniline have actually carried through the method to a model for one of chemistry's most fundamental processes: the stereomutation of chiral molecular structures [102–106], see also [107]. Figure B2.5.22 illustrates the minimum energy path for the interconversion of the left- and right-handed forms of hydrogen peroxide, roughly corresponding to the

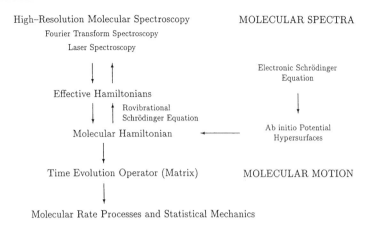

**Figure B2.5.20.** The combined experimental and theoretical approach 'Molecular spectra and motion' (after [18]).

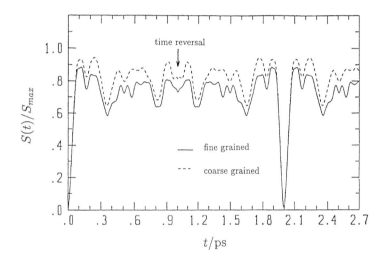

**Figure B2.5.21.** Time-dependent entropy $S(t)/S_{max}$ of $CHD_2F$ starting from a pure CH stretching excitation with six quanta ($v_s = 6$) at $t = 0$ fs. Time evolution with time reversal at $t = 1$ ps (after [114]).

torsion about the O–O bond. The quantum dynamics are governed by tunnelling through the low barrier in the *trans* configuration, even at very high energies, a phenomenon readily understood in terms of an adiabatic picture of the stereomutation kinetics. The detailed model extracted from experimental spectra with the support of quantum-chemical calculations allows one to describe the observed mode specificity of the stereomutation in terms of the full six-dimensional quantum wavepacket dynamics. The time-dependent probability density in the reaction coordinate (figure B2.5.23) illustrates the acceleration of the stereomutation by IR excitation of the antisymmetric bend vibration ($v_6$). An approximately Gaussian wavepacket initially localized on one side of the *trans* barrier (figure B2.5.22) moves periodically between the two potential wells. In the vibrational ground state ($v = 0$), this corresponds to the stereomutation of a chiral equilibrium structure through tunnelling to its enantiomer within 1.5 ps. Exciting the antisymmetric bend vibration ($v_6 = 1$) roughly halves the time required for stereomutation, an effect that could be considered as catalysis by IR (vibrational) excitation. Figure B2.5.23 also illustrates the high degree of adiabaticity of this process: the initial form of the

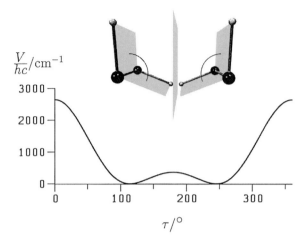

**Figure B2.5.22.** Potential $V$ along the minimum energy path for the stereomutation of hydrogen peroxide. Adapted from [103].

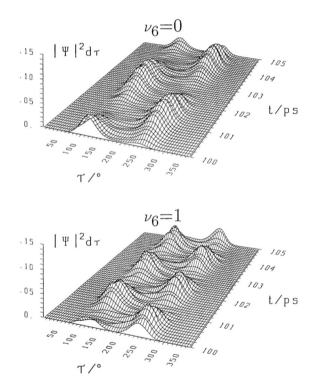

**Figure B2.5.23.** Mode-specific stereomutation tunnelling in hydrogen peroxide: time-dependent probability density $|\Psi|^2$ in the reaction coordinate $\tau$ (see figure B2.5.22). The probability density was integrated over the remaining coordinates (OH stretches, OH bends, OO stretch). The initial wavepacket at $t = 0$ was strictly localized on one side of the torsional barrier. $v_6 = 0$ refers to the vibrational ground state and $v_6 = 1$ to an initial state with one quantum of antisymmetric OOH bend excitation.

wavepacket in the reaction coordinate is found to be approximately conserved, even after about $10^2$ tunnelling periods, although the spectroscopic result (as analysed by theory) is exact in full six-dimensional dynamics. While $\nu_6$ can thus be considered to be a promoting mode for stereomutation, other vibrations of $H_2O_2$ (except torsion) have been shown to be inhibiting modes, slowing down the stereomutation process. Thus, one has cases of inhibition of a reaction by vibrational excitation. A certain degree of thermal averaging allows one to evaluate a relaxation time corresponding to rate constants more characteristic for ordinary racemization kinetics [103, 107] in contrast to the strictly periodic process shown in figure B2.5.23.

Related results of promotion (catalysis) and inhibition of stereomutation by vibrational excitation have also been obtained for the much larger molecule, aniline–NHD ($C_6H_5NHD$), which shows short-time chirality and stereomutation [104, 105]. This kind of study opens the way to a new look at kinetics, which shows 'coherent' and mode-selective dynamics, even in the absence of coherent external fields. The possibility of enforcing coherent dynamics by fields ('coherent control') is discussed in chapter A3.13.

### B2.5.9 Summarizing overview on gas-phase kinetics studies

Gas-phase kinetics studies are ideally concerned with the most fundamental events of chemical reactions related to 'isolated, single molecules' either as elementary unimolecular reactions, isolated bimolecular collisions, or trimolecular reactions. The experimental study of such fast elementary processes has progressed to a point where it is possible to 'prove a reaction mechanism' by identifying each elementary reaction contributing to the total reactive flux and by demonstrating that any conceivable additional contribution to the total reactive flux must be negligible. In fact gas-phase kinetics studies have even gone beyond this fundamental goal of reaction kinetics. By using the techniques of femtosecond spectroscopy and quantum-chemical kinetics from high-resolution spectroscopy it is possible to look into the very details of the primary processes that initiate chemical reactions. These fields are still in active development and most of the fruits from these fields still remain to be harvested.

### References

[1] Wilhelmy L 1850 über das Gesetz nach welchem die Einwirkung der Säuren auf Rohrzucker Stattfiudet *Ann. Physik* **81** 413–29

[2] van't Hoff J H 1884 *Études de Dynamique Chimique* (Amsterdam: Müller)

[3] Arrhenius S 1899 Zur Theorie der chemischen Reaktionsgeschwindigkeiten *Z. Physik. Chem.* **28** 318–35

[4] Chance B 1949 The reaction of catalase and cyanide *J. Biol. Chem.* **179** 1299–341

[5] Chance B 1951 Rapid and sensitive spectrophotometry. I. The accelerated and stopped-flow methods for the measurement of the reaction kinetics and spectra of unstable compounds in the visible region of the spectrum *Rev. Sci. Instrum.* **22** 619–27

[6] Eigen M 1996 *Die unmessbar schnellen Reaktionen (Ostwalds Klassiker der exakten Naturwissenschaften)* vol 281 (Thun und Frankfurt: Harri Deutsch)

[7] Manz J and Woeste L (eds) 1995 *Femtosecond Chemistry Proc. Berlin Conf. Femtosecond Chemistry (Berlin, March 1993)* vol 1 (Weinheim: Verlag Chemie)

[8] Porter G 1995 Flash photolysis into the femtosecond—a race against time *Femtosecond Chemistry Proc. Berlin Conf. Femtosecond Chemistry (Berlin, March 1993)* ed J Manz and L Woeste (Weinheim: Verlag Chemie) ch 1, pp 3–13

[9] Weller A 1952 Quantitative Untersuchungen der Fluoreszenzumwandlung bei Naphtholen *Ber. Bunsenges. Phys. Chem.* **56** 662–8

[10] Weller A 1957 Eine Verallgemeinerte Theorie diffusionsbestimmter Reaktionen und ihre Anwendung auf die Fluoreszenzlöschung *Z. Phys. Chem.* **13** 335–52

[11] Weller A 1961 Fast reactions of excited molecules *Progress in Reaction Kinetics* (Oxford: Pergamon) pp 187–214

[12] Jost W 1939 *Explosionen und Verbrennungsvorgänge in Gasen* (Berlin: Springer)

[13] Knox W H, Knox R S, Hoose J F and Zare R N 1990 Observation of the 0 fs pulse *Opt. Photon. News* **1** 44–5

[14] Herschbach D R 1987 Molecular dynamics of elementary chemical reactions *Angew. Chem.* **26** 1221–43

[15] Lee Y T 1987 Molecular beam studies of elementary chemical processes *Angew. Chem.* **26** 939–51

[16] Gutowsky H S and Holm C H 1956 Rate processes and nuclear magnetic resonance spectra. II. Hindered internal rotation of amides *J. Chem. Phys.* **25** 1228–34

[17] Gutowsky H S and Holm C H 1975 Time-dependent magnetic perturbations *Dynamic Nuclear Magnetic Resonance Spectroscopy* ed L M Jackman and F A Cotton (New York: Academic) pp 1–21

[18] Quack M 1995 Molecular femtosecond quantum dynamics between less than yoctoseconds and more than days: experiment and theory *Femtosecond Chemistry Proc. Berlin Conf. Femtosecond Chemistry (Berlin, March 1993)* ed J Manz and L Woeste (Weinheim: Verlag Chemie) ch 27, pp 781–818

[19] Marquardt R, Quack M, Stohner J and Sutcliffe E 1986 Quantum-mechanical wavepacket dynamics of the CH group in the symmetric top $X_3CH$ compounds using effective Hamiltonians from high-resolution spectroscopy *J. Chem. Soc. Faraday Trans. Series 2* **82** 1173–87

[20] Marquardt R and Quack M 1991 The wavepacket motion and intramolecular vibrational redistribution in $CHX_3$ molecules under infrared multiphoton excitation *J. Chem. Phys.* **95** 4854–67

[21] Zewail A H 1995 Femto chemistry: concepts and applications *Femtosecond Chemistry Proc. Berlin Conf. Femtosecond Chemistry (Berlin, March 1993)* ed J Manz and L Woeste (Weinheim: Verlag Chemie) ch 2, pp 15–128

[22] Zewail A H 2000 Femtochemistry: atomic-scale dynamics of the chemical bond using ultrafast lasers (Nobel lecture) *Angew. Chem.* **39** 2586–631

[23] Gerber R B, McCoy A B and Garcia-Vela A 1995 Dynamics of photoinduced reactions in the van der Waals and in the hydrogen–bonded clusters *Femtosecond Chemistry Proc. Berlin Conf. Femtosecond Chemistry (Berlin, March 1993)* ed J Manz and L Woeste (Weinheim: Verlag Chemie) pp 499–531

[24] Crim F F 1996 Bond-selected chemistry: vibrational state control of photodissociation and bimolecular reaction *J. Phys. Chem.* **100** 12 725

[25] Fernández-Alonzo F, Bean B D, Ayers J D, Pomerantz A E, Zare R N, Bañares L and Aoiz F J 2000 Evidence for scattering resonances in the $H + D_2$ reaction *Angew. Chem. Int. Ed. (Eng.)* **39** 2748–52

[26] Moore C B and Smith I W M 1996 State-resolved studies of reactions in the gas-phase *J. Phys. Chem.* **100** 12 848

[27] Bernstein R B (ed) 1982 Chemical dynamics via molecular beam and laser techniques *The Hinshelwood Lectures (Oxford, 1980)* (Oxford: Oxford University Press)

[28] Pagsberg P, Jodkowski J T, Ratajczak E and Sillesen A 1998 Experimental and theoretical studies of the reaction between $CF_3$ and $NO_2$ at 298 K *Chem. Phys. Lett.* **286** 138–44

[29] Bernasconi C F (ed) 1976 *Relaxation Kinetics* (New York: Academic)

[30] Sturtevant B, Shephard J E and Hornung H G (eds) 1996 *Proc. 20th Int. Symp. on Shock Waves* (Singapore: World Scientific)

[31] Mueller M A, Yetter R A and Dryer F L 1999 Flow reactor studies and kinetic modelling of the $H_2/O_2/NO_x$ reaction *Int. J. Chem. Kinetics* **31** 113–25

[32] Fleming G R 1986 *Chemical Applications of Ultrafast Spectroscopy* (Oxford: Oxford University Press)

[33] Quack M 1990 Spectra and dynamics of coupled vibrations in polyatomic molecules *Ann. Rev. Phys. Chem.* **41** 839–74

[34] Lehmann K, Scoles G and Pate B H 1994 Intramolecular dynamics from Eigenstate-resolved infrared spectra *Ann. Rev. Phys. Chem.* **45** 241–74

[35] Quack M 1995 Molecular infrared spectra and molecular motion *J. Mol. Struct.* **347** 245–66

[36] Scoles G (ed) 1988 *Atomic and Molecular Beam Methods* (Oxford: Oxford University Press)

[37] Faubel M and Toennies J P 1978 *Adv. Atom. Mol. Phys.* **13** 229

[38] Levine R D and Bernstein R B (eds) 1989 *Molecular Reaction Dynamics and Chemical Reactivity* (Oxford: Oxford University Press)

[39] Gehring M M, Hoyermann K, Schacke H and Wolfrum J 1973 *Proc. 14th Int. Symp. on Combustion* p 99

[40] Bedjanian Y, Le Bras G and Poulet G 1999 Kinetic study of the reactions of $Br_2$ with OH and OD *Int. J. Chem. Kinetics* **31** 698–704

[41] Lipson J B, Beiderhase T W, Molina L T and Molina M J 1999 Production of HCl in the OH + ClO reaction: laboratory measurements and statistical rate theory calculations *J. Phys. Chem. A* **103** 6540–51

[42] Daugey N, Bergeat A, Schuck A, Caubet P and Dorthe G 1997 Vibrational distribution in $CN(X\,^2\Sigma^+)$ from the $N + C_2 \rightarrow CN + C$ reaction *Chem. Phys.* **222** 87–103

[43] Bergeat A, Calvo T, Dorthe G and Loison J-C 1999 Fast-flow study of the CH + CH reaction products *J. Phys. Chem. A* **103** 6360–5

[44] Mueller M A, Yetter R A and Dryer F L 1999 Flow reactor studies and kinetic modelling of the $H_2/O_2/NO_x$ and $CO/H_2O/O_2/NO_x$ reactions *Int. J. Chem. Kinetics* **31** 705–24

[45] van den Bergh H and Troe J 1975 NO-catalyzed recombination of iodine atoms. Elementary steps of the complex mechanism *Chem. Phys. Lett.* **31** 351–4

[46] Gilbert R G, Luther K and Troe J 1983 Theory of thermal unimolecular reactions in the fall-off range. II. Weak collision rate constants *Ber. Bunsenges. Phys. Chem.* **87** 169–77

[47] Cobos C J, Hippler H and Troe J 1985 High-pressure falloff curves and specific rate constants for the reactions $H + O_2 \rightleftharpoons HO_2 \rightleftharpoons HO + O$ *J. Phys. Chem.* **89** 342–9

[48] Quack M 1984 On the mechanism of reversible unimolecular reactions and the canonical ('high pressure') limit of the rate coefficient at low pressures *Ber. Bunsenges. Phys. Chem.* **88** 94–100

[49] Markwalder B, Gozel P and van den Berg H 1992 Temperature-jump measurements on the kinetics of association and dissociation in weakly bound systems: $N_2O_4 + M = NO_2 + NO_2 + M$ *J. Chem. Phys.* **97** 5472–9

[50] Eigen M and de Maeyer L 1963 Relaxation methods *Technique of Organic Chemistry* vol 8, ed S L Friess, E S Lewis and A Weissberger (New York: Wiley) pp 895–1054

[51] Troe J 1975 Shock wave studies of elementary chemical processes *Modern Developments in Shock Tube Research* ed G Kamimoto (Japan: Shock Tube Research Society) pp 29–54

[52] Greene C H and Toennies J P 1964 *Chemical Reactions in Shock Waves* (London: Arnold)

[53] Jaumotte A L (ed) 1971 *Chocs et Ondes de Choc* (Paris: Masson)

[54] Glänzer K, Quack M and Troe J 1977 High temperature UV absorption and recombination of methyl radicals in shock waves *Proc. 16th Int. Symp. on Combustion* (Pittsburg, PA: The Combustion Institute) pp 949–60

[55] Glänzer K, Quack M and Troe J 1976 A spectroscopic determination of the methyl radical recombination rate constant in shockwaves *Chem. Phys. Lett.* **39** 304–9

[56] Herzberg G and Shoosmith 1956 Absorption spectrum of free $CH_3$ and $CD_3$ radicals *Can. J. Phys.* **34** 523–5

[57] Burcat A, Skinner G B, Crossley R W and Scheller K 1973 High temperature decomposition of ethane *Int. J. Chem. Kinetics* **5** 345–52

[58] Waage E V and Rabinovitch B S 1971 Some aspects of theory and experiment in the ethane–methyl radical system *Int. J. Chem. Kinetics* **3** 105–25

[59] Quack M and Troe J 1974 Specific rate constants of unimolecular processes ii. Adiabatic channel model *Ber. Bunsenges. Phys. Chem.* **78** 240–52

[60] Quack M and Troe J 1998 Statistical adiabatic channel models *Encyclopedia of Computational Chemistry* vol 4, ed P v R Schleyer *et al* (New York: Wiley) pp 2708–26

[61] Votsmeier M, Song S, Davidson D F and Hanson R K 1999 Shock tube study of monomethylamine thermal decomposition and $NH_2$ high temperature absorption coefficients *Int. J. Chem. Kinetics* **31** 323–30

[62] Votsmeier M, Song S, Davidson D F and Hanson R K 1999 Sensitive detection of $NH_2$ in shock tube experiments using frequency modulation spectroscopy *Int. J. Chem. Kinetics* **31** 445–53

[63] Porter G 1950 The absorption spectroscopy of substances of short life *Discuss. Faraday Soc.* **9** 60–9

[64] Jung D, Kärtner F X, Matuschek N, Suther D H, Morier-Genoud F, Zhang G, Keller U, Scheurer V, Tilsch M and Tschudi T 1997 Self-starting 6.5-fs pulses from a Ti:sapphire laser *Opt. Lett.* **22** 1009–11

[65] Callear A B and Metcalfe M P 1976 *Chem. Phys.* **14** 275

[66] van den Bergh H E, Callear A B and Norström R J 1969 An experimental determination of the oscillator strength of the 2160 Å band of the free methyl radical and a spectroscopic measurement of the combination rate *Chem. Phys. Lett.* **4** 101–2

[67] Herzberg G 1961 The spectra and structures of free methyl and free methylene *Proc. R. Soc.* A **262** 291–317

[68] Herzberg G 1971 *The Spectra and Structures of Simple Free Radicals. An Introduction to Molecular Spectroscopy* (Ithaca, NY: Cornell University Press)

[69] Hippler H, Siefke M, Staerk H and Troe J 1999 New studies of the unimolecular reaction $NO_2 \rightleftharpoons O + NO$. Part 1. High pressure range of the O + NO recombination between 200 and 400 K *Phys. Chem. Chem. Phys.* **1** 57–61

[70] Sinha M P, Schulz A and Zare R N 1973 Internal state distribution of alkali dimers in supersonic nozzle beams *J. Chem. Phys.* **58** 549–56

[71] Zewail A H 1993 Femtochemistry *J. Phys. Chem.* **97** 12 427–46

[72] Zewail A H 1994 *Femtochemistry. Ultrafast Dynamics of the chemical Bond (World Scientific Series in 20th Century Chemistry, vol 3)* (Singapore: World Scientific)

[73] Zewail A H 1995 Femtosecond dynamics of reactions: elementary processes of controlled solvation *Ber. Bunsenges. Phys. Chem.* **99** 474–7

[74] Rosker M J, Rose T S and Zewail A 1988 Femtosecond real-time dynamics of photofragment-trapping resonances on dissociative potential-energy surfaces *Chem. Phys. Lett.* **146** 175–9

[75] Lupo D W and Quack M 1987 IR-laser photochemistry *Chem. Rev.* **87** 181–216

[76] Quack M 1982 Reaction dynamics and statistical mechanics of the preparation of highly excited states by intense infrared radiation *Adv. Chem. Phys.* **50** 395–473

[77] Hippler M, Quack M, Schwarz R, Seyfang G, Matt S and Märk T 1997 Infrared multiphoton excitation, dissociation, and ionization of $C_{60}$ *Chem. Phys. Lett.* **278** 111–20

[78] Ditmire T, Zweiback J, Yanovsky V P, Cowan T E, Hays G and Wharton K B 1999 Nuclear fusion from explosions of femtosecond laser-heated deuterium clusters *Nature* **389** 489–92

[79] Pretzler G *et al* 1998 Neutron production by 200 mJ ultrashort laser pulses *Phys. Rev.* E **58** 1165–8

[80] Quack M 1989 Infrared laser chemistry and the dynamics of molecular multiphoton excitation *Infrared Phys.* **29** 441–66

[81] Quack M 1995 IR laser chemistry *Infrared Phys. Technol.* **36** 365–80

[82] Quack M and Thöne H J 1987 Absolute and relative rate coefficients in the IR-laser chemistry of bichromophoric fluorobutanes: tests for inter- and intra-molecular selectivity *Chem. Phys.* **135** 487–94

[83] Rabinovitch B S and Flowers M C 1964 Chemical activation *Q. Rev. Chem. Soc.* **18** 122–67

[84] Oref I and Rabinovitch B S 1979 Do highly excited polyatomic molecules behave ergodically? *Acc. Chem. Res.* **12** 166–75

[85] Flowers M C and Rabinovitch B S 1985 Localization of excitation energy in chemically activated systems. 3-ethyl-2-methyl-2-pentyl radicals *J. Phys. Chem.* **89** 563–5

[86]  von E Doering W, Gilbert J C and Leermakes P A 1968 Symmetrical distribution of energy in initially unsymmetrically excited products. Reaction of dideuteriodiazomethane with allene, methylenecyclopropane, and vinylcyclopropane *Tetrahedron* **29** 6863–72

[87]  Setser D W 1972 *International Review of Science. Physical Chemistry* ed J C Polyanyi (London: Butterworths)

[88]  Sang Kyu Kim, Ju Guo, Baskin J S and Zewail A H 1996 Femtosecond chemically activated reactions: concept of nonstatistical activation at high thermal energies *J. Phys. Chem.* **100** 9202–5

[89]  Mack J A, Mikulecky K and Wodtke A M 1997 Resonant vibration–vibration energy transfer between highly vibrationally excited $O_2(X\,^3\Sigma_g^-, v = 15{-}20)$ and $CO_2$, $N_2O$, $N_2$, and $O_3$ *J. Chem. Phys.* **105** 4105–16

[90]  Ackermann M and Biaume F 1970 Structure of the Schumann–Runge bands from the 0–0 to the 13–0 band *J. Mol. Spectrosc.* **35** 73–82

[91]  Cheung A S C, Yoshino K, Freeman D E, Friedman R S, Dalgarno A and Parkinson W H 1989 The Shumann–Runge absorption-bands of $^{16}O^{18}O$ in the wavelength region 175–205 nm and spectroscopic constants of isotopic oxygen molecules *J. Mol. Spectrosc.* **134** 362–89

[92]  von Puttkamer K, Dübal H-R and Quack M 1983 Time-dependent processes in polyatomic molecules during and after intense infrared irradiation *Faraday Discuss. Chem. Soc.* **75** 197–210

[93]  Quack M and Suhm M A 1991 Potential energy surfaces, quasiadiabatic channels, rovibrational spectra, and intramolecular dynamics of $(HF)_2$ and its isotopomers from quantum Monte Carlo calculations *J. Chem. Phys.* **95** 28–59

[94]  Quack M and Kutzelnigg W 1995 Molecular spectroscopy and molecular dynamics: theory and experiment *Ber. Bunsenges. Phys. Chem.* **99** 231–45

[95]  Beil A, Luckhaus D, Quack M and Stohner J 1997 Intramolecular vibrational redistribution and unimolecular reaction: concepts and new results on the femtosecond dynamics and statistics in CHFClBr *Ber. Bunsenges. Phys. Chem.* **101** 311–28

[96]  Quack M 1993 Molecular quantum dynamics from high resolution spectroscopy and laser chemistry *J. Mol. Struct.* **292** 171–96

[97]  Moore C B and Weisshaar J C 1983 Formaldehyde photochemistry *Ann. Rev. Phys. Chem.* **34** 525

[98]  Bowman J M and Gazdy B 1997 A new perspective on isomerization dynamics illustrated by HCN $\rightarrow$ HNC *J. Phys. Chem. A* **101** 6384–8

[99]  Kiefer J H, Mudipalli P S, Wagner A F and Harding L 1996 Importance of hindered rotations in the thermal dissociation of small unsaturated molecules: classical formulation and application to hcn and hcch *J. Chem. Phys.* **105** 1–22

[100] Luckhaus D 1997 The rovibrational spectrum of hydroxylamine: a combined high resolution experimental and theoretical study *J. Chem. Phys.* **106** 8409–26

[101] Luckhaus D 1997 The rovibrational dynamics of hydroxylamine *Ber. Bunsenges. Phys. Chem.* **101** 346–55

[102] Kuhn B, Rizzo T R, Luckhaus D, Quack M and Suhm M A 1999 A new six–dimensional analytical potential up to chemically significant energies for the electronic ground state of hydrogen peroxide *J. Chem. Phys.* **111** 2565–87

[103] Fehrensen B, Luckhaus D and Quack M 1999 Mode selective stereomutation tunnelling in hydrogen peroxide isotopomers *Chem. Phys. Lett.* **300** 312–20
       Fehrensen B, Luckhaus D and Quack M 2001 to be published

[104] Fehrensen B, Luckhaus D and Quack M 1999 Inversion tunneling in aniline from high resolution infrared spectroscopy and an adiabatic reaction path hamiltonian approach *Z. Phys. Chem.* **209** 1–19

[105] Fehrensen B, Hippler M and Quack M 1998 Isotopomer selective overtone spectroscopy by ionization detected IR + UV double resonance jet-cooled aniline *Chem. Phys. Lett.* **298** 320–8

[106] Luckhaus D 2000 6D vibrational quantum dynamics: generalized coordinate discrete variable representation and (a)diabatic contraction *J. Chem. Phys.* **113** 1329–47

[107] Quack M 1989 Structure and dynamics of chiral molecules *Angew. Chem.* **28** 571–86

[108] Shank C 1985 *Laser Focus* March

[109] He Y, Pochert J, Quack M, Ranz R and Seyfang G 1995 Dynamics of unimolecular reactions induced by monochromatic infrared radiation: experiment and theory for $C_nF_mXI \rightarrow C_nF_mX + I$ probed with hyperfine-, Doppler- and uncertainty limited time resolution of iodine atom infrared absorption *J. Chem. Soc. Faraday Discuss.* **102** 275–300

[110] Quack M and Jans-Bürli S 1986 *Molekulare Thermodynamik und Kinetik. Teil 1: Chemische Reaktionskinetik* (Zürich: Verlag der Fachvereine) (New English edition in preparation)

[111] Zewail A H 1988 Laser femtochemistry *Science* **242** 1645–53

[112] Quack M 1998 Multiphoton excitation *Encyclopedia of Computational Chemistry* vol 3, ed P v R Schleyer *et al* (New York: Wiley) pp 1775–91

[113] Quack M, Sutcliffe E, Hackett P A and Rayner D M 1986 Molecular photofragmentation with many infrared photons. Absolute rate parameters from quantum dynamics, statistical mechanics, and direct measurement *Faraday Discuss. Chem. Soc.* **82** 229–40

[114] Quack M and Stohner J 1993 Femtosecond quantum dynamics of functional groups under coherent infrared multiphoton excitation as derived from the analysis of high-resolution spectra *J. Phys. Chem.* **97** 12 574–90

[115] Settle R D F and Rizzo T R 1992 *J. Chem. Phys.* **97** 2823
       Boyarkine O V, Settle R D F and Rizzo T R 1995 Vibrational overtone spectra of jet-cooled CF$_3$H by infrared laser assisted photofragment spectroscopy *Ber. Bunsenges. Phys. Chem.* **99** 504–13

[116] Quack M 1999 Intramolekulare Dynamik, Irreversibilität, Zeitumkehrsymmetrie und eine absolute Moleküluhr *Nova Acta Leopoldina NF* **81** 137–73

[117] Douley E A, Marquardt R, Quack M, Stohner J, Thanopulos I and Wallenborn E-U 2001 *Mol. Phys.* to be published

## Further Reading

Bernasconi C F 1976 *Relaxation Kinetics* (New York: Academic)

Faraday Discussions of the Chemical Society **112** *Unimolecular Dynamics*

Fleming G R 1986 *Chemical Applications of Ultrafast Spectroscopy* (Oxford: Oxford University Press)

Johnston H S 1966 *Gas Phase Reaction Rate Theory* (Ronald)

Levine R D and Bernstein R B 1989 *Molecular Reaction Dynamics and Chemical Reactivity* (Oxford: Oxford University Press)

Lupo D W and Quack M 1987 IR-laser photochemistry *Chem. Rev.* **87** 181–216

Manz J and Woeste L (eds) 1995 *Femtosecond Chemistry Proc. Berlin Conf. Femtosecond Chemistry (Berlin, March 1993)* (Weinheim: Verlag Chemie)

Pilling M J and Smith I W M (eds) 1987 *Modern Gas Kinetics. Theory, Experiment and Application* (Oxford: Blackwell)

Quack M 1982 Reaction dynamics and statistical mechanics of the preparation of highly excited states by intense infrared radiation *Adv. Chem. Phys.* **50** 395–473

Quack M 1995 IR laser chemistry *Infrared Phys. Technol.* **36** 365–80

Quack M and Jans–Bürli S 1986 *Molekulare Thermodynamik und Kinetik. Teil 1. Chemische Reaktionskinetik* (Zürich: Verlag der Fachvereine)

Sandström J 1981 *Dynamic NMR Spectroscopy* (New York: Academic)

Steinfeld J I, Francisco S and Hase W L 1998 *Chemical Kinetics and Dynamics* 2nd edn (Englewood Cliffs, NJ: Prentice-Hall)

# PART B3

---

# TECHNIQUES FOR APPLYING THEORY

# B3.1
# Quantum structural methods for atoms and molecules

*Jack Simons*

### B3.1.1   What does quantum chemistry try to do?

Electronic structure theory describes the motions of the electrons and produces energy surfaces and wave-functions. The shapes and geometries of molecules, their electronic, vibrational and rotational energy levels, as well as the interactions of these states with electromagnetic fields lie within the realm of quantum structure theory.

#### B3.1.1.1   The underlying theoretical basis—the Born–Oppenheimer model

In the Born–Oppenheimer [1] model, it is assumed that the electrons move so quickly that they can adjust their motions essentially instantaneously with respect to any movements of the heavier and slower atomic nuclei. In typical molecules, the valence electrons orbit about the nuclei about once every $10^{-15}$ s (the inner-shell electrons move even faster), while the bonds vibrate every $10^{-14}$ s, and the molecule rotates approximately every $10^{-12}$ s. So, for typical molecules, the fundamental assumption of the Born–Oppenheimer model is valid, but for loosely held (e.g. Rydberg) electrons and in cases where nuclear motion is strongly coupled to electronic motions (e.g. when Jahn–Teller effects are present) it is expected to break down.

This separation-of-time-scales assumption allows the electrons to be described by electronic wavefunc-tions that smoothly 'ride' the molecule's atomic framework. These electronic functions are found by solving a Schrödinger equation whose Hamiltonian $\hat{H}_e$ contains the kinetic energy $T_e$ of the electrons, the Coulomb repulsions among all the molecule's electrons $V_{ee}$, the Coulomb attractions $V_{en}$ among the electrons and all of the molecule's nuclei, treated with these nuclei held clamped, and the Coulomb repulsions $V_{nn}$ among all of these nuclei, but it does not contain the kinetic energy $T_N$ of all the nuclei. That is, this Hamiltonian keeps the nuclei held fixed in space. The electronic wavefunctions $\psi_k$ and energies $E_k$ that result

$$\hat{H}_e \psi_k = E_k \psi_k$$

thus depend on the locations $\{Q_i\}$ at which the nuclei are sitting. That is, the $E_k$ and $\psi_k$ are parametric functions of the coordinates of the nuclei, and, of course, the wavefunctions $\psi_k$ depend on the coordinates of all of the electrons.

These electronic energies' dependence on the positions of the atomic centres cause them to be referred to as electronic energy surfaces such as that depicted below in figure B3.1.1 for a diatomic molecule. For nonlinear polyatomic molecules having $N$ atoms, the energy surfaces depend on $3N-6$ internal coordinates and thus can be very difficult to visualize. In figure B3.1.2, a 'slice' through such a surface is shown as a function of two of the $3N-6$ internal coordinates.

The Born–Oppenheimer theory is soundly based in that it can be derived from a Schrödinger equation describing the kinetic energies of all electrons and of all $N$ nuclei plus the Coulomb potential energies of

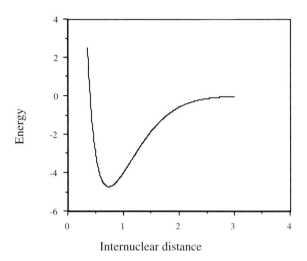

**Figure B3.1.1.** Energy as a function of internuclear distance for a typical bound diatomic molecule or ion.

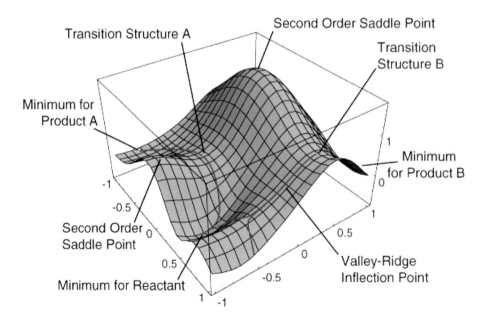

**Figure B3.1.2.** Two-dimensional slice through a $(3N - 6)$-dimensional energy surface of a polyatomic molecule or ion. After [2].

interaction among all electrons and nuclei. By expanding the wavefunction $\Psi$ that is an eigenfunction of this full Schrödinger equation in the complete set of functions $\{\psi_k\}$ and then neglecting all terms that involve derivatives of any $\psi_k$ with respect to the nuclear positions $\{Q_i\}$, one can separate variables such that:

(1) the electronic wavefunctions and energies obey

$$\hat{H}_e \psi_k = E_k \psi_k$$

(2) the nuclear motion (i.e. vibration/rotation) wavefunctions obey

$$(\hat{T}_N + E_k)\chi_{k,L} = E_{k,L}\chi_{k,L}$$

where $T_N$ is the kinetic energy operator for movement of all nuclei.

Each and every electronic energy state, labelled $k$, has a set, labelled $L$, of vibration/rotation energy levels $E_{k,L}$ and wavefunctions $\chi_{k,L}$.

### B3.1.1.2 Non-Born–Oppenheimer corrections—radiationless transitions

Because the Born–Oppenheimer model is obtained from the full Schrödinger equation by making approximations, it is not exact. Thus, in certain circumstances it becomes necessary to correct the predictions of the Born–Oppenheimer theory (i.e. by including the effects of the neglected coupling terms using perturbation theory). For example, when developing a theoretical model to interpret the rate at which electrons are ejected from rotationally/vibrationally hot $NH^-$ ions, we had to consider [3] coupling between:

(1) $^2\Pi$ $NH^-$ in its $v = 1$ vibrational level and in a high rotational level (e.g. $J > 30$) prepared by laser excitation of vibrationally 'cold' $NH^-$ in $v = 0$ having high $J$ (due to natural Boltzmann populations), see figure B3.1.3; and

(2) $^3\Sigma^-$ NH neutral plus an ejected electron in which the NH is in its $v = 0$ vibrational level (no higher level is energetically accessible) and in various rotational levels (labelled $N$).

Because NH has an electron affinity of 0.4 eV, the total energies of the above two states can be equal only if the kinetic energy $KE$ carried away by the ejected electron obeys

$$KE = E_{\text{vib/rot}}(NH^- (v = 1, J)) - E_{\text{vib/rot}}(NH (v = 0, N)) - 0.4 \text{ eV}.$$

In the absence of any coupling terms, no electron detachment would occur. It is only by the anion converting some of its vibration/rotation energy and angular momentum into electronic energy that the electron that occupies a bound $N_{2p}$ orbital in $NH^-$ can gain enough energy to be ejected.

My own research efforts [4] have, for many years, involved taking into account such non-Born–Oppenheimer couplings, especially in cases where vibration/rotation energy transferred to electronic motions causes electron detachment, as in the $NH^-$ case detailed above. Professor Yngve Öhrn has been active [5] in attempting to avoid using the Born–Oppenheimer approximation and, instead, treating the dynamical motions of the nuclei and electrons simultaneously. Professor David Yarkony has contributed much [6] to the recent treatment of non-Born–Oppenheimer effects and to the inclusion of spin–orbit coupling in such studies.

### B3.1.1.3 What is learned from an electronic structure calculation?

The knowledge gained via structure theory is great. The electronic energies $E_k(Q)$ allow one to determine [7] the geometries and relative energies of various isomers that a molecule can assume by finding those geometries $\{Q_i\}$ at which the energy surface $E_k$ has minima $\partial E_k / \partial Q_i = 0$, with all directions having positive curvature (this is monitored by considering the so-called Hessian matrix $H_{i,j} = \partial^2 E_k / \partial Q_i \partial Q_j$: if none of its

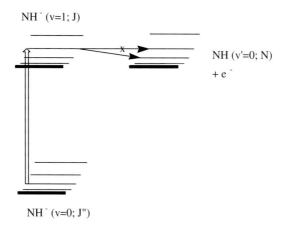

**Figure B3.1.3.** Energies of $NH^-$ and of NH pertinent to the autodetachment of $v = 1$, $J$ levels of $NH^-$ formed by laser excitation of $v = 0$, $J''NH^-$.

eigenvalues are negative, all directions have positive curvature). Such geometries describe stable isomers, and the energy at each such isomer geometry gives the relative energy of that isomer. Professor Berny Schlegel [8] has been one of the leading figures in using gradient and Hessian information to locate stable structures and transition states. Professor Peter Pulay [9] has done as much as anyone to develop the theory that allows us to compute gradients and Hessians for most commonly used electronic structure methods.

There may be other geometries on the $E_k$ energy surface at which all 'slopes' vanish $\partial E_k/\partial Q_i = 0$, but at which not all directions possess positive curvature. If the Hessian matrix has only one negative eigenvalue, there is only one direction leading downhill away from the point $\{Q_i\}$ of zero force; all the remaining directions lead uphill from this point. Such a geometry describes that of a *transition state*, and its energy plays a central role in determining the rates of reactions which pass through this transition state. The energy surface shown in figure B3.1.2 displays such transition states, and it also shows a second-order saddle point (i.e. a point where the gradient vanishes and the Hessian has two directions of negative curvature).

At any geometry $\{Q_i\}$, the gradient vector having components $\partial E_k/\partial Q_i$ provides the forces ($F_i = -\partial E_k/\partial Q_i$) along each of the coordinates $Q_i$. These forces are used in molecular dynamics simulations which solve the Newton $\boldsymbol{F} = m\boldsymbol{a}$ equations and in molecular mechanics studies which are aimed at locating those geometries where the $\boldsymbol{F}$ vector vanishes (i.e. the stable isomers and transition states discussed above).

Also produced in electronic structure simulations are the electronic wavefunctions $\{\psi_k\}$ and energies $\{E_k\}$ of each of the electronic states. The separation in energies can be used to make predictions on the spectroscopy of the system. The wavefunctions can be used to evaluate the properties of the system that depend on the spatial distribution of the electrons. For example, the $z$ component of the dipole moment [10] of a molecule $\mu_z$ can be computed by integrating the probability density for finding an electron at position $r$ multiplied by the $z$ coordinate of the electron and the electron's charge $e$: $\mu_z = \int e\psi_k^*\psi_k z \, dr$. The average kinetic energy of an electron can also be computed by carrying out such an average-value integral: $\int \psi_k^*(-\hbar^2/2m_e\nabla^2)\psi_k \, dr$. The rules for computing the average value of any physical observable are developed and illustrated in popular undergraduate text books on physical chemistry [11] and in graduate-level texts [12].

Not only can electronic wavefunctions tell us about the average values of all the physical properties for any particular state (i.e. $\psi_k$ above), but they also allow us to tell us how a specific 'perturbation' (e.g. an electric field in the Stark effect, a magnetic field in the Zeeman effect and light's electromagnetic fields in spectroscopy) can alter the specific state of interest. For example, the perturbation arising from the electric field of a photon interacting with the electrons in a molecule is given within the so-called electric dipole approximation [12] by:

$$\hat{H}_{\text{pert}} = \sum_j e^2 r_j \cdot E(t)$$

where $E$ is the electric field vector of the light, which depends on time $t$ in an oscillatory manner, and $r_j$ gives the spatial coordinates of the $j$th electron. This perturbation, $\hat{H}_{\text{pert}}$ can induce transitions to other states $\psi_{k'}$, with probabilities that are proportional to the square of the integral:

$$\int \psi_{k'}^* \hat{H}_{\text{pert}} \psi_k \, dr.$$

So, if this integral were to vanish, transitions between $\psi_k$ and $\psi_{k'}$ would not occur, and would be referred to as 'forbidden'. Whether such integrals vanish or not often is determined by symmetry. For example, if $\psi_k$ were of odd symmetry under a plane of symmetry $\sigma_v$ of the molecule, while $\psi_{k'}$ were even under $\sigma_v$, then the integral would vanish unless one or more of the three Cartesian components of the dot product $r_j \cdot E$ were odd under $\sigma_v$. The general idea is that for the integral not to vanish, the direct product of the symmetries of $\psi_k$ and of $\psi_{k'}$ must match the symmetry of at least one of the symmetry components present in $\hat{H}_{\text{pert}}$. Professor Poul Jørgensen [13] has been involved in developing such so-called response theories for perturbations that may be time dependent (e.g. as in the interaction of light's electromagnetic radiation).

### B3.1.1.4   Summary

In summary, computational *ab initio* quantum chemistry attempts to solve the electronic Schrödinger equation for the $E_k(R)$ energy surfaces and wavefunctions $\psi_k(r; R)$ on a 'grid' of values for the 'clamped' nuclear positions. Because the Schrödinger equation produces wavefunctions, it has a great deal of predictive power. Wavefunctions contain all the information needed to compute dipole moments, polarizability, etc and transition properties such as the electric dipole transition strengths among states. They also permit the evaluation of system responses with respect to external perturbations such as geometrical distortions [9], which provides information on vibrational frequencies and reaction paths.

## B3.1.2   Why is it so difficult to calculate electronic energies and wavefunctions with reasonable accuracy?

As a scientific tool, *ab initio* quantum chemistry is not yet as accurate as modern laser spectroscopic measurements, for example. Moreover, it is difficult to estimate the accuracies with which various methods predict bond energies and lengths, excitation energies and the like. In the opinion of the author, chemists who rely on the results of quantum chemistry calculations must better understand what underlies the concepts and methods of this field. Only by so doing will they be able to judge for themselves the value of given quantum chemistry data to their own research. There exist a variety of sources of further information on the 'jargon', underlying theory, methodologies, and current strengths and weaknesses of *ab initio* quantum chemistry. In 1996, Head-Gordon [14] produced a nice overview entitled 'Quantum chemistry and molecular processes', Schaefer *et al* [15] offered a very good discussion in 1995; Simons [16] offered a somewhat earlier perspective in 1991. The present chapter includes many of the ideas contained in these and other earlier descriptions of this field's impacts, but also attempts to extend the perspective to include more recent developments.

Returning now to the issue of the accuracy of various electronic structure predictions, it is natural to ask why it is so difficult to achieve reasonable accuracy (i.e. ca. 1 kcal mol$^{-1}$ in computed bond energies or activation energies) even with the most sophisticated and computer-resource-intensive quantum chemistry calculations. The reasons include the following.

(A)  *Many-body problems with $R^{-1}$ potentials are notoriously difficult.* It is well known that the Coulomb potential falls off so slowly with distance that mathematical difficulties can arise. The $4\pi R^2$ dependence of the integration volume element, combined with the $R^{-1}$ dependence of the potential, produce ill-defined interaction integrals unless attractive and repulsive interactions are properly combined. The classical or quantum treatment of ionic melts [17], many-body gravitational dynamics [18] and Madelung sums [19] for ionic crystals are all plagued by such difficulties.

(B) *The electrons require quantal treatment and they are indistinguishable.* The electron's small mass produces local de Broglie wavelengths that are long compared to atomic 'sizes', thus necessitating quantum treatment. Their indistinguishability requires that permutational symmetry be imposed on solutions of the Schrödinger equation.

(C) *All mean-field models of electronic structure require large corrections.* Essentially all *ab initio* quantum chemistry approaches introduce a 'mean field' potential $V_{mf}$ that embodies the average interactions among the $N$ electrons. The difference between the mean-field potential and the true Coulombic potential is termed [20] the *'fluctuation potential'*. The solutions $\{\Psi_k, E_k\}$ to the true electronic Schrödinger equation are then approximated in terms of solutions $\{\Psi_k^0, E_k^0\}$ to the model Schrödinger equation in which $V_{mf}$ is used. Improvements to the solutions of the model problem are made using perturbation theory or the variational method. Such approaches are expected to work when the difference between the starting model and the final goal is small in some sense.

The most elementary mean-field models of electronic structure introduce a potential that an electron at $r_1$ would experience if it were interacting with a *spatially averaged* electrostatic charge density arising from the $N - 1$ remaining electrons:

$$V_{mf}(r_1) = \int \rho_{N-1}(r') \frac{e^2}{|r_1 - r'|} \, dr'.$$

Here $\rho_{N-1}(r')$ represents the probability density for finding the $N - 1$ electrons at $r'$, and $e^2/|r_1 - r'|$ is the mutual Coulomb repulsion between electron density at $r_1$ and $r'$.

The magnitude and 'shape' of such a mean-field potential is shown below [21] in figure B3.1.4 for the two 1s electrons of a beryllium atom. The Be nucleus is at the origin, and one electron is held fixed 0.13 Å from the nucleus, the maximum of the 1s orbital's radial probability density. The Coulomb potential experienced by the second electron is then a function of the second electron's position along the $x$-axis (connecting the Be nucleus and the first electron) and its distance perpendicular to the $x$-axis. For simplicity, this second electron is arbitrarily constrained to lie on the $x$-axis. Along this direction, the Coulomb potential is singular, and hence the overall interactions are very large.

On the ordinate, two quantities are plotted: (i) the mean-field potential between the second electron and the other 1s electron computed, via the self-consistent field (SCF) process (described later), as the interaction of the second electron with a spherical $|1s|^2$ charge density centred on the Be nucleus; and (ii) the fluctuation potential ($F$) of this average (mean-field) interaction.

As a function of the inter-electron distance, the fluctuation potential decays to zero more rapidly than does the mean-field potential. However, the magnitude of $F$ is quite large and remains so over an appreciable range of inter-electron distances. The corrections to the mean-field picture are therefore quite large when measured in kcal mol$^{-1}$. For example, the differences (called pair correlation energies) $\Delta E$ between the true (state-of-the-art quantum chemical calculation as discussed later) energies of the interaction among the four electrons in the Be atom and the mean-field estimates of these interactions are given in table B3.1.1 in electronvolts (1 eV = 23.06 kcal mol$^{-1}$).

Another example of the difficulty is offered in figure B3.1.5. Here we display on the ordinate, for helium's $^1S$ ($1s^2$) state, the probability of finding an electron whose distance from the He nucleus is 0.13 Å (the peak of the 1s orbital's density) and whose angular coordinate relative to that of the other electron is plotted on the abscissa. The He nucleus is at the origin and the second electron also has a radial coordinate of 0.13 Å. As the relative angular coordinate varies away from $0°$, the electrons move apart; near $0°$, the electrons approach

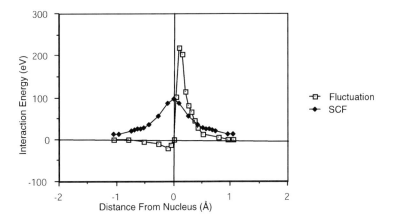

**Figure B3.1.4.** Fluctuation and mean-field SCF potentials for a 2s electron in Be.

**Table B3.1.1.** Pair correlation energies for the four electrons in Be.

| Orbital pair | $1s\alpha 1s\beta$ | $1s\alpha 2s\alpha$ | $1s\alpha 2s\beta$ | $1s\beta 2s\alpha$ | $1s\beta 2s\beta$ | $2s\alpha 2s\beta$ |
|---|---|---|---|---|---|---|
| $\Delta E$ (eV) | 1.126 | 0.022 | 0.058 | 0.058 | 0.022 | 1.234 |

one another. Since both electrons have opposite spin in this state, their mutual Coulomb repulsion alone acts to keep them apart.

What figure B3.1.5 shows is that, for a highly accurate wavefunction (one constructed using so-called Hylleraas functions [23] that depend explicitly on the coordinates of the two electrons as well as on their interparticle distance coordinate), one finds a 'cusp' in the probability density for finding one electron in the neighbourhood of another electron with the same spin. The probability plot for the Hylleraas function is the lower bold curve in figure B3.1.5. The line above the Hylleraas plot was extracted from a configuration interaction wavefunction for He obtained using a rather large atomic orbital (AO) basis set [22]. Even for such a sophisticated wavefunction (of the type used in many state-of-the-art *ab initio* calculations), the cusp in the relative probability distribution is, clearly, not well represented. Finally, the Hartree–Fock (HF) probability, which is not even displayed above, would, if plotted, be flat as a function of the angle shown above and thus clearly very much in error.

### B3.1.2.1    Summary

The above evidence shows why an *ab initio* solution of the Schrödinger equation is a very demanding task if high accuracy is desired. The HF potential takes care of 'most' of the interactions among the $N$ electrons (which interact via long-range Coulomb forces and whose dynamics requires the application of quantum physics and permutational symmetry). However, the residual fluctuation potential is large enough to cause significant corrections to the HF picture. The reality is that electrons in atoms and molecules undergo dynamical motions in which their Coulomb repulsions cause them to 'avoid' one another at every instant of time, not only in the average-repulsion manner that the mean-field models embody. The inclusion of instantaneous spatial correlations (usually called *dynamical correlations*) among electrons is necessary to achieve a more accurate description of the atomic and molecular electronic structure.

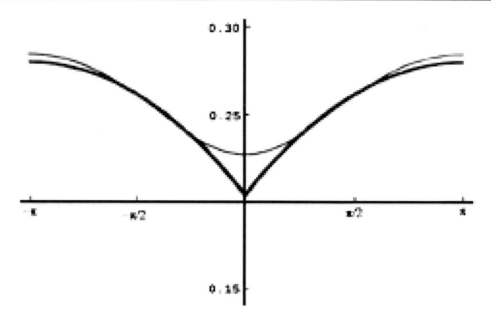

**Figure B3.1.5.** Probability (as a function of angle) for finding the second electron in He when both electrons are located at the maximum in the 1s orbital's probability density. The bottom line is that obtained using a Hylleraas-type function, and the other related to a highly-correlated multiconfigurational wavefunction. After [22].

### B3.1.3    What are the essential concepts of *ab initio* quantum chemistry?

The mean-field potential and the need to improve it to achieve reasonably accurate solutions to the true electronic Schrödinger equation introduce three constructs that characterize essentially all *ab initio* quantum chemical methods: *orbitals, configurations* and *electron correlation*.

#### B3.1.3.1    *Orbitals and configurations—what are they (really)?*

*(a) How the mean-field model leads to orbitals and configurations*

The mean-field potentials that have proven most useful are all one-electron additive: $V_{mf}(r) = \sum_j V_{mf}(r_j)$. Since the electronic kinetic energy $\hat{T} = \sum_j \hat{T}_j$ operator is also one-electron additive, so is the mean-field Hamiltonian $\hat{H}^0 = \hat{T} + \hat{V}_{mf}$. The additivity of $\hat{H}^0$ implies that the mean-field energies $\{E_k^0\}$ are additive and the wavefunctions $\{\Psi_k^0\}$ can be formed in terms of products of functions $\{\phi_k\}$ of the coordinates of the individual electrons.

   Thus, it is the *ansatz* that $V_{mf}$ is separable that leads to the concept of *orbitals*, which are the one-electron functions $\{\phi_j\}$ found by solving the one-electron Schrödinger equations: $(\hat{T}_1 + \hat{V}_{mf}(r_1))\phi_j(r_1) = \varepsilon_j\phi_j(r_1)$; the eigenvalues $\{\varepsilon_j\}$ are called *orbital energies*.

   Given the complete set of solutions to this one-electron equation, a complete set of $N$-electron mean-field wavefunctions can be written. Each $\Psi_k^0$ is constructed by forming a product of $N$ orbitals chosen from the set of $\{\phi_j\}$, allowing each orbital in the list to be a function of the coordinates of one of the $N$ electrons (e.g. $\Psi_k^0 = |\phi_{k1}(r_1)\phi_{k2}(r_2)\phi_{k3}(r_3)\ldots\phi_{kN-1}(r_{N-1})\phi_{kN}(r_N)|$, as above). The corresponding mean-field energy is evaluated as the sum over those orbitals that appear in $\Psi_k^0$: $E_k^0 = \sum_{j=1,N} \varepsilon_{kj}$.

Because of the indistinguishability of the $N$ electrons, the antisymmetric component of any such orbital product must be formed to obtain the proper mean-field wavefunction. To do so, one applies the so-called antisymmetrizer operator [24] $\hat{A} = \sum_P (-1)^P \hat{P}$, where the permutation operator $\hat{P}$ runs over all $N!$ permutations of the $N$ electrons. Application of $\hat{A}$ to a product function does not alter the occupancy of the functions $\{\phi_{kj}\}$ in $\Psi_k^0$, it simply scrambles the order which the electrons occupy the $\{\phi_{kj}\}$ and it causes the resultant function (which is often denoted $|\phi_{k1}(r_1)\phi_{k2}(r_2)\phi_{k3}(r_3)\dots\phi_{kN-1}(r_{N-1})\phi_{kN}(r_N)|$ and called a Slater determinant) to obey the Pauli exclusion principle.

Because the electrons also possess intrinsic spin, the one-electron functions $\{\phi_j\}$ used in this construction are taken to be eigenfunctions of $(\hat{T}_1 + \hat{V}_{mf}(r_1))$ multiplied by either an $\alpha$ or $\beta$ spin function. This set of functions is called the set of mean-field *spin orbitals*.

By choosing to place $N$ electrons into $N$ specific spin orbitals, one specifies a *configuration*. By making other choices of which $N\phi_j$ to occupy, one describes other configurations. Just as the one-electron mean-field Schrödinger equation has a complete set of spin–orbital solutions $\{\phi_j$ and $\varepsilon_j\}$, the $N$-electron mean-field Schrödinger equation has a complete set of antisymmetric $N$-electron Slater determinants. When these determinants are combined to generate functions that are eigenfunctions of the total $S^2$ and $S_z$ and eigenfunctions of the molecule's point group symmetry (or $\hat{L}^2$ and $\hat{L}_z$ for atoms), one has what are called *configuration state functions* (CSFs) $\Psi_k^0$ whose mean-field energies are also given by $E_k^0 = \sum_{j=1,N} \varepsilon_{kj}$.

## (b) The self-consistent mean-field (SCF) potential

The one-electron additivity of the mean-field Hamiltonian $\hat{H}^0$ gives rise to the concept of spin orbitals for *any* additive $\hat{V}_{mf}(r)$. In fact, there is no *single* mean-field potential; different scientists have put forth different suggestions for $\hat{V}_{mf}$ over the years. Each gives rise to spin orbitals and configurations that are specific to the particular $\hat{V}_{mf}$. However, if the difference between any particular mean-field model and the full electronic Hamiltonian is fully treated, corrections to all mean-field results should converge to the same set of exact states. In practice, one is never able to treat *all* corrections to any mean-field model. Thus, it is important to seek particular mean-field potentials for which the corrections are as small and straightforward to treat as possible.

In the most commonly employed mean-field models [25] of electronic structure theory, the configuration specified for study plays a central role in defining the mean-field potential. For example, the mean-field Coulomb potential felt by a $2p_x$ orbital's electron at a point $r$ in the $1s^2 2s^2 2p_x 2p_y$ configuration description of the carbon atom is:

$$\hat{V}_{mf}(r) = 2 \int |1s(r')|^2 e^2/|r - r'| \, dr' + 2 \int |2s(r')|^2 e^2/|r - r'| \, dr' + \int |2p_y(r')|^2 e^2/|r - r'| \, dr'.$$

The above mean-field potential is used to find the $2p_x$ orbital of the carbon atom, which is then used to define the mean-field potential experienced by, for example, an electron in the 2s orbital:

$$\hat{V}_{mf}(r) = 2 \int |1s(r')|^2 e^2/|r - r'| \, dr' + \int |2s(r')|^2 e^2/|r - r'| \, dr' + \int |2p_y(r')|^2 e^2/|r - r'| \, dr'$$
$$+ \int |2p_x(r')|^2 e^2/|r - r'| \, dr'.$$

Notice that the orbitals occupied in the configuration under study appear in the mean-field potential. However, it is $\hat{V}_{mf}$ that, through the one-electron Schrödinger equation, determines the orbitals. For these reasons, the solution of these equations must be carried out in a so-called SCF manner. One begins with an approximate description of the orbitals in $\Psi_k^0$. These orbitals then define $\hat{V}_{mf}$, and the equations $(\hat{T}_1 + \hat{V}_{mf}(r_1))\phi_j(r_1) = \varepsilon_j \phi_j(r_1)$ are solved for 'new' spin orbitals. These orbitals are then be used to define an improved $\hat{V}_{mf}$, which gives another set of solutions to $(\hat{T}_1 + \hat{V}_{mf}(r_1))\phi_j(r_1) = \varepsilon_j\phi_j(r_1)$. This iterative process is continued until the orbitals used to define $\hat{V}_{mf}$ are identical to those that result as solutions of $(\hat{T}_1 + \hat{V}_{mf}(r_1))\phi_j(r_1) = \varepsilon_j\phi_j(r_1)$. When this condition is reached, one has achieved 'self-consistency'.

### B3.1.3.2  What is electron correlation?

By expressing the mean-field interaction of an electron at $r$ with the $N - 1$ other electrons in terms of a probability density $\rho_{N-1}(r')$ that is independent of the fact that another electron resides at $r$, the mean-field models ignore spatial *correlations* among the electrons. In reality, as shown in figure B3.1.5, the conditional probability density for finding one of $N - 1$ electrons at $r'$, given that one electron is at $r$ depends on $r'$. The absence of a spatial correlation is a direct consequence of the spin–orbital *product nature* of the mean-field wavefunctions $\{\Psi_k^0\}$.

To improve upon the mean-field picture of electronic structure, one must move beyond the single-configuration approximation. It is essential to do so to achieve higher accuracy, but it is also important to do so to achieve a *conceptually* correct view of the chemical electronic structure. Although the picture of configurations in which $N$ electrons occupy $N$ spin orbitals may be familiar and useful for systematizing the electronic states of atoms and molecules, these constructs are approximations to the true states of the system. They were introduced when the mean-field approximation was made, and neither orbitals nor configurations can be claimed to describe the proper eigenstates $\{\Psi_k, E_k\}$. It is thus inconsistent to insist that the carbon atom be thought of as $1s^2 2s^2 2p^2$ while insisting on a description of this atom accurate to $\pm 1$ kcal mol$^{-1}$.

### B3.1.3.3  Summary

The SCF mean-field potential takes care of 'most' of the interactions among the $N$ electrons. However, for all mean-field potentials proposed to date, the residual or fluctuation potential is large enough to require significant corrections to the mean-field picture. This, in turn, necessitates the use of more sophisticated and computationally taxing techniques (e.g., high-order perturbation theory or large variational expansion spaces) to reach the desired chemical accuracy.

For electronic structures of atoms and molecules, the SCF model requires quite substantial corrections to bring its predictions in line with experimental fact. Electrons in atoms and molecules undergo dynamical motions in which their Coulomb repulsions cause them to 'avoid' one another at every instant of time, not only in the average-repulsion manner of mean-field models. The inclusion of *dynamical correlations* among electrons is necessary to achieve a more accurate description of atomic and molecular electronic structure. *No single spin–orbital product wavefunction is capable of treating electron correlation to any extent*; its product nature renders it incapable of doing so.

## B3.1.4  How to introduce electron correlation via configuration mixing

### B3.1.4.1  The multi-configuration wavefunction

In most of the commonly used *ab initio* quantum chemical methods [26], one forms a set of configurations by placing $N$ electrons into spin orbitals in a manner that produces the spatial, spin and angular momentum symmetry of the electronic state of interest. The correct wavefunction $\Psi$ is then written as a linear combination of the mean-field configuration functions $\{\Psi_k\}$: $\Psi = \sum_k C_k \Psi_k^0$. For example, to describe the ground $^1S$ state of the Be atom, the $1s^2 2s^2$ configuration is augmented by including other configurations such as $1s^2 3s^2$, $1s^2 2p^2$, $1s^2 3p^2$, $1s^2 2s3s$, $3s^2 2s^2$, $2p^2 2s^2$, etc, all of which have overall $^1S$ spin and angular momentum symmetry. The various methods of electronic structure theory differ primarily in how they determine the $\{C_k\}$ expansion coefficients and how they extract the energy $E$ corresponding to this $\Psi$.

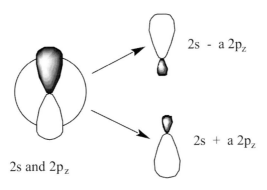

$2s - a \, 2p_z$

$2s + a \, 2p_z$

2s and $2p_z$

**Figure B3.1.6.** Polarized orbital pairs involving 2s and $2p_z$ orbitals.

### B3.1.4.2 The physical meaning of mixing in 'excited' configurations

When considering the ground $^1S$ state of the Be atom, the following four antisymmetrized spin–orbital products are found to have the largest $C_k$ amplitudes:

$$\Psi \cong C_1|1s^2 2s^2| - C_2[|1s^2 2p_x^2| + |1s^2 2p_y^2| + |1s^2 2p_z^2|].$$

The fact that the latter three terms possess the same amplitude $C_2$ is a result of the requirement that a state of $^1S$ symmetry is desired. It can be shown [27] that this function is equivalent to

$$\Psi \cong \tfrac{1}{6} C_1 |1s\alpha \, 1s\beta \{ [(2s - a2p_x)\alpha(2s + a2p_x)\beta - (2s - a2p_x)\beta(2s + a2p_x)\alpha]$$
$$+ [(2s - a2p_y)\alpha(2s + a2p_y)\beta - (2s - a2p_y)\beta(2s + a2p_y)\alpha]$$
$$+ [(2s - a2p_z)\alpha(2s + a2p_z)\beta - (2s - a2p_z)\beta(2s + a2p_z)\alpha]\}|$$

where $a = \sqrt{3C_2/C_1}$.

Here two electrons occupy the 1s orbital (with opposite, $\alpha$ and $\beta$ spins) while the other electron pair resides in $2s - 2p$ polarized orbitals in a manner that instantaneously correlates their motions. These *polarized orbital pairs* $(2s \pm a2p_{x, y \, or \, z})$ are formed by combining the 2s orbital with the $2p_{x, y \, or \, z}$ orbital in a ratio determined by $C_2/C_1$. This way of viewing an electron pair correlation forms the basis of the generalized valence bond (GVB) method that Professor Bill Goddard [28] pioneered.

This ratio $C_2/C_1$ can be shown to be proportional to the magnitude of the coupling $\langle 1s^2 2s^2|\hat{H}|1s^2 2p^2 \rangle$ between the two configurations involved and inversely proportional to the energy difference $(\langle 1s^2 2s^2 \hat{H}|1s^2 2s^2 \rangle - \langle 1s^2 2p^2|\hat{H}|1s^2 2p^2 \rangle)$ between these configurations. In general, configurations that have similar Hamiltonian expectation values and that are coupled strongly give rise to strongly mixed (i.e. with large $|C_2/C_1|$ ratios) polarized orbital pairs.

A set of polarized orbital pairs is described pictorially in figure B3.1.6. In each of the three equivalent terms in the above wavefunction, one of the valence electrons moves in a $2s + a2p$ orbital polarized in one direction while the other valence electron moves in the $2s - a2p$ orbital polarized in the opposite direction. For example, the first term $(2s - a2p_x)\alpha(2s + a2p_x)\beta - (2s - a2p_x)\beta(2s + a2p_x)\alpha$ describes one electron occupying a $2s - a2p_x$ polarized orbital while the other electron occupies the $2s + a2p_x$ orbital. The electrons thus reduce their Coulomb repulsion by occupying *different* regions of space; in the SCF picture $1s^2 2s^2$, both electrons reside in the same 2s region of space. In this particular example, the electrons undergo *angular correlation* to 'avoid' one another.

Let us consider another example. In describing the $\pi^2$ electron pair of an olefin, it is important to mix in 'doubly excited' configurations of the form $(\pi^*)^2$. The physical importance of such configurations can again

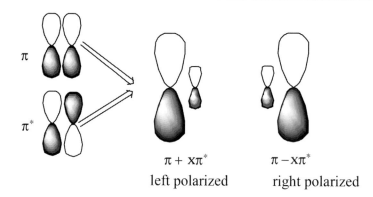

**Figure B3.1.7.** Left- and right-polarized orbital pairs involving $\pi$ and $\pi^*$ orbitals.

be made clear by using the identity

$$C_1|\ldots\phi\alpha\phi\beta\ldots| - C_2|\ldots\phi'\alpha\phi'\beta\ldots| = C_1/2\{|\ldots(\phi - x\phi')\alpha(\phi + x\phi')\beta\ldots| - |\ldots(\phi - x\phi')\beta(\phi + x\phi')\alpha\ldots|\}$$

where $x = (C_2/C_1)^{1/2}$.

In this example, the two non-orthogonal 'polarized orbital pairs' involve mixing the $\pi$ and $\pi^*$ orbitals to produce two left–right polarized orbitals as depicted in figure B3.1.7. Here one says that the $\pi^2$ electron pair undergoes left–right correlation when the $(\pi^*)^2$ configuration is introduced.

### B3.1.4.3   Are polarized orbital pairs hybrid orbitals?

It should be stressed that these polarized orbital pairs are *not* the same as hybrid orbitals. The latter are used to describe directed bonding, but polarized orbital *pairs* are each a 'mixture' of two mean-field orbitals with amplitude $x = (C_2/C_1)^{1/2}$ and with a *single electron* in each, thereby allowing the electrons to be spatially correlated and to 'avoid' one another. In addition, polarized orbital pairs are *not generally orthogonal* to one another; hybrid orbital sets are.

### B3.1.4.4   Relationship to the generalized valence bond picture

In these examples, the analysis allows one to *interpret* the combination of pairs of configurations that differ from one another by a 'double excitation' from one orbital ($\phi$) to another ($\phi'$) as equivalent to a singlet coupling of two polarized orbitals ($\phi - a\phi'$) and ($\phi + a\phi'$). As mentioned earlier, this picture is closely related to the GVB model that Goddard [28] and Goddard and Harding [29] developed. In the simplest embodiment of the GVB model, each electron pair in the atom or molecule is correlated by mixing in a configuration in which that pair is 'doubly excited' to a correlating orbital. The direct product of all such pair correlations generates the simplest GVB-type wavefunction.

In most *ab initio* quantum chemical methods, the correlation calculation is actually carried out by forming a linear combination of the mean-field configuration state functions and determining the $\{C_k\}$ amplitudes by some procedure. The identities discussed in some detail above are then introduced merely to permit one to interpret the presence of configurations that are 'doubly excited' relative to the dominant mean-field configuration in terms of polarized orbital pairs.

*B3.1.4.5   Summary*

The dynamical interactions among electrons give rise to instantaneous spatial correlations that must be handled to arrive at an accurate picture of the atomic and molecular structure. The single-configuration picture provided by the mean-field model is a useful starting point, but it is *incapable* of describing electron correlations. Therefore, improvements are needed. The use of doubly-excited configurations is a mechanism by which $\Psi$ can place electron *pairs*, which in the mean-field picture occupy the same orbital, into different regions of space thereby lowering their mutual Coulombic repulsions. Such electron correlation effects are referred to as *dynamical electron correlation*; they are extremely important to include if one expects to achieve chemically meaningful accuracy.

## B3.1.5   The single-configuration picture and the HF approximation

Given a set of $N$-electron space- and spin-symmetry-adapted configuration state functions $\{\Phi_J\}$ in terms of which $\Psi$ is to be expanded as $\Psi = \sum_J C_J \Phi_J$, two primary questions arise: (1) how to determine the $\{C_J\}$ coefficients and the energy $E$ and (2) how to find the 'best' spin orbitals $\{\phi_j\}$? Let us first consider the case where a single configuration is used so only the question of determining the spin orbitals exists.

*B3.1.5.1   The single-determinant wavefunction*

*(a) The canonical SCF equations*

The simplest trial function employed in *ab initio* quantum chemistry is the single Slater determinant function in which $N$ spin orbitals are occupied by $N$ electrons:

$$\Psi = |\phi_1 \phi_2 \phi_3 \dots \phi_N|.$$

For such a function, variational optimization of the spin orbitals to make the expectation value $\langle \Psi | \hat{H} | \Psi \rangle$ stationary produces [30] the canonical HF equations

$$\hat{F}\phi_i = \varepsilon_i \phi_i$$

where the so-called Fock operator $\hat{F}$ is given by

$$\hat{F}\phi_i = \hat{h}\phi_i + \sum_{j(\text{occupied})} [\hat{J}_j - \hat{K}_j]\phi_i.$$

The Coulomb ($\hat{J}_j$) and exchange ($\hat{K}_j$) operators are defined by the relations

$$\hat{J}_j \phi_i = \int \phi_j^*(r')\phi_j(r')/|r - r'| \, d\tau' \, \phi_i(r)$$

and

$$\hat{K}_j \phi_i = \int \phi_j^*(r')\phi_i(r')/|r - r'| \, d\tau' \, \phi_j(r)$$

the symbol $\hat{h}$ denotes the sum of the electronic kinetic energy, and electron–nuclear Coulomb attraction operators. The $d\tau$ implies integration over the spin variables associated with the $\phi_j$ (and, for the exchange operator, $\phi_i$), as a result of which the exchange integral vanishes unless the spin function of $\phi_j$ is the same as that of $\phi_i$; the Coulomb integral is non-vanishing no matter what the spin functions of $\phi_j$ and $\phi_i$.

*(b) The equations have orbital solutions for occupied and unoccupied orbitals*

The HF [31] equations $\hat{F}\phi_i = \varepsilon_i\phi_i$ possess solutions for the spin orbitals in $\Psi$ (the *occupied* spin orbitals) as well as for orbitals not occupied in $\Psi$ (the *virtual* spin orbitals) because the $\hat{F}$ operator is Hermitian. Only the $\phi_i$ occupied in $\Psi$ appear in the Coulomb and exchange potentials of the Fock operator.

*(c) The spin-impurity problem*

As formulated above, the HF equations yield orbitals that do not guarantee that $\Psi$ has proper spin symmetry. To illustrate, consider an open-shell system such as the lithium atom. If $1s\alpha$, $1s\beta$, and $2s\alpha$ spin orbitals are chosen to appear in $\Psi$, the Fock operator will be

$$\hat{F} = \hat{h} + \hat{J}_{1s\alpha} + \hat{J}_{1s\beta} + \hat{J}_{2s\alpha} - [\hat{K}_{1s\alpha} + \hat{K}_{1s\beta} + \hat{K}_{2s\alpha}].$$

Acting on an $\alpha$ spin orbital $\phi_{k\alpha}$ with $F$ and carrying out the spin integrations, one obtains

$$\hat{F}\phi_{k\alpha} = \hat{h}\phi_{k\alpha} + (2\hat{J}_{1s} + \hat{J}_{2s})\phi_{k\alpha} - (\hat{K}_{1s} + \hat{K}_{2s})\phi_{k\alpha}.$$

In contrast, when acting on a $\beta$ spin orbital, one obtains

$$\hat{F}\phi_{k\beta} = \hat{h}\phi_{k\beta} + (2\hat{J}_{1s} + \hat{J}_{2s})\phi_{k\beta} - (\hat{K}_{1s})\phi_{k\beta}.$$

Spin orbitals of $\alpha$ and $\beta$ type do *not* experience the same exchange potential in this model because $\Psi$ contains two $\alpha$ spin orbitals and only one $\beta$ spin orbital. A consequence is that the optimal $1s\alpha$ and $1s\beta$ spin orbitals, which are themselves solutions of $\hat{F}\phi_i = \varepsilon_i\phi_i$, do not have identical orbital energies (i.e. $\varepsilon_{1s\alpha} \neq \varepsilon_{1s\beta}$) and are not spatially identical. This resultant spin polarization of the orbitals gives rise to *spin impurities* in $\Psi$. The determinant $|1s\alpha 1s'\beta 2s\alpha|$ is not a pure doublet spin eigenfunction, although it is an $S_z$ eigenfunction with $M_s = 1/2$; it contains both $S = 1/2$ and $S = 3/2$ components. If the $1s\alpha$ and $1s'\beta$ spin orbitals were spatially identical, then $|1s\alpha 1s'\beta 2s\alpha|$ would be a pure spin eigenfunction with $S = 1/2$.

The above single-determinant wavefunction is referred to as being of the unrestricted Hartree–Fock (UHF) type because no restrictions are placed on the spatial nature of the orbitals in $\Psi$. In general, UHF wavefunctions are not of pure spin symmetry for any open-shell system or for closed-shell systems far from their equilibrium geometries (e.g. for $H_2$ or $N_2$ at long bond lengths) These are significant drawbacks of methods based on a UHF starting point. Such a UHF treatment forms the basis of the widely used and highly successful Gaussian 70 through Gaussian-9X series of electronic structure computer codes [32] which derive from Pople [32] and co-workers.

To overcome some of the problems inherent in the UHF method, it is possible to derive SCF equations based on minimizing the energy of a wavefunction formed by spin projecting a single Slater determinant starting function (e.g. using $\{|1s\alpha 2s\beta| - |1s\beta 2s\alpha|\}/2^{1/2}$ for the singlet excited state of He rather than $|1s\alpha 2s\beta|$). It is also possible for a trial wavefunction of the form $|1s\alpha 1s\beta 2s\alpha|$ to constrain the $1s\alpha$ and $1s\beta$ orbitals to have exactly the same spatial form. In both cases, one then is able to carry out what are called restricted Hartree–Fock (RHF) calculations.

*B3.1.5.2    The linear combinations of atomic orbitals to form molecular orbitals expansion of the spin orbitals*

The HF equations must be solved iteratively because the $J_i$ and $K_i$ operators in $F$ depend on the orbitals $\phi_i$ for which solutions are sought. Typical iterative schemes begin with a 'guess' for those $\phi_i$ that appear in $\Psi$, which then allows $\hat{F}$ to be formed. Solutions to $\hat{F}\phi_i = \varepsilon_i\phi_i$ are then found, and those $\phi_i$ which possess the space and spin symmetry of the occupied orbitals of $\Psi$ and which have the proper energies and nodal character are used to generate a new $\hat{F}$ operator (i.e. new $\hat{J}_i$ and $\hat{K}_i$ operators). This iterative HF SCF process is continued until the $\phi_i$ and $\varepsilon_i$ do not vary significantly from one iteration to the next, at which time one says that the process has converged.

In practice, solution of $\hat{F}\phi_i = \varepsilon_i\phi_i$ as an integro-differential equation can be carried out only for atoms [34] and linear molecules [35] for which the angular parts of the $\phi_i$ can be exactly separated from the radial because of axial- or full-rotation group symmetry (e.g. $\phi_i = Y_{l,m}(\theta, \phi) R_{n,l}(r)$ for an atom and $\phi_i = \exp(im\phi) R_{n,l,m}(\rho, z)$ for a linear molecule).

In the procedures most commonly applied to nonlinear molecules, the $\phi_i$ are expanded in a *basis* $\chi_\mu$ according to the linear combinations of AOs to form molecular orbitals (LCAO–MO) [36] procedure:

$$\phi_i = \sum_\mu C_{\mu,i}\chi_\mu.$$

This reduces $\hat{F}\phi_i = \varepsilon_i\phi_i$ to a matrix eigenvalue-type equation:

$$\sum_\nu F_{\mu,\nu}C_{\nu,i} = \varepsilon_i \sum_\nu S_{\mu,\nu}C_{\nu,i}$$

where $S_{\mu,\nu} = \langle\chi_\mu|\chi_\nu\rangle$ is the overlap matrix among the AOs and

$$F_{\mu,\nu} = \langle\chi_\mu|\hat{h}|\chi_\nu\rangle + \sum_{\delta,\kappa}[\gamma_{\delta,\kappa}\langle\chi_\mu\chi_\delta|\hat{g}|\chi_\nu\chi_\kappa\rangle - \gamma_{\delta,\kappa}^{\text{ex}}\langle\chi_\mu\chi_\delta|\hat{g}|\chi_\kappa\chi_\nu\rangle]$$

is the matrix representation of the Fock operator in the AO basis. Here and elsewhere, the symbol $\hat{g}$ is used to represent the electron–electron Coulomb potential $e^2/|r - r'|$.

The charge- and exchange-density matrix elements in the AO basis are:

$$\gamma_{\delta,\kappa} = \sum_{i(\text{occupied})} C_{\delta,i}C_{\kappa,i}$$

and

$$\gamma_{\delta,\kappa}^{\text{ex}} = \sum_{i(\text{occupied and same spin})} C_{\delta,i}C_{\kappa,i}$$

where the sum in $\gamma_{\delta,\kappa}^{\text{ex}}$ runs over those occupied spin orbitals whose $m_s$ value is equal to that for which the Fock matrix is being formed (for a closed-shell species, $\gamma_{\delta,\kappa}^{\text{ex}} = 1/2\gamma_{\delta,\kappa}$).

It should be noted that by moving to a matrix problem, one does not remove the need for an iterative solution; the $F_{\mu,\nu}$ matrix elements depend on the $C_{\nu,i}$ LCAO–MO coefficients which are, in turn, solutions of the so-called Roothaan [30] matrix HF equations: $\sum_\nu F_{\mu,\nu}C_{\nu,i} = \varepsilon_i \sum_\nu S_{\mu,\nu}C_{\nu,i}$. One should also note that, just as $\hat{F}\phi_i = \varepsilon_i\phi_i$ possesses a complete set of eigenfunctions, the matrix $F_{\mu,\nu}$, whose dimension $M$ is equal to the number of atomic basis orbitals, has $M$ eigenvalues $\varepsilon_i$ and $M$ eigenvectors whose elements are the $C_{\nu,i}$. Thus, there are *occupied and virtual* MOs each of which is described in the LCAO–MO form with the $C_{\nu,i}$ coefficients obtained via solution of $\sum_\nu F_{\mu,\nu}C_{\nu,i} = \varepsilon_i \sum_\nu S_{\mu,\nu}C_{\nu,i}$.

*B3.1.5.3   AO basis sets*

*(a) Slater-type orbitals and Gaussian-type orbitals*

The basis orbitals commonly used in the LCAO–MO process fall into two primary classes:

(1)   Slater-type orbitals (STOs) $\chi_{n,l,m}(r, \theta, \phi) = N_{n,l,m,\zeta} Y_{l,m}(\theta, \phi) r^{n-1} \exp(-\zeta r)$, are characterized by the quantum numbers $n$, $l$ and $m$ and the exponent (which characterizes the 'size') $\zeta$. The symbol $N_{n,l,m,\zeta}$ denotes the normalization constant.

(2)   Cartesian Gaussian-type orbitals (GTOs) $\chi_{a,b,c}(r, \theta, \phi) = N'_{a,b,c,\alpha} x^a y^b z^c \exp(-\alpha r^2)$, are characterized by the quantum numbers $a$, $b$ and $c$, which detail the angular shape and direction of the orbital, and the exponent $\alpha$ which governs the radial 'size'.

For both types of orbitals, the coordinates $r$, $\theta$ and $\phi$ refer to the position of the electron relative to a set of axes attached to the centre on which the basis orbital is located. Although STOs have the proper 'cusp' behaviour near the nuclei, they are used primarily for atomic- and linear-molecule calculations because the multi-centre integrals which arise in polyatomic-molecule calculations cannot efficiently be performed when STOs are employed. In contrast, such integrals can routinely be done when GTOs are used. This fundamental advantage of GTOs has led to the dominance of these functions in molecular quantum chemistry.

To overcome the primary weakness of GTO functions (i.e. their radial derivatives vanish at the nucleus whereas the derivatives of STOs are non-zero), it is common to combine two, three, or more GTOs, with combination coefficients which are fixed and *not* treated as LCAO–MO parameters, into new functions called *contracted* GTOs or CGTOs. Typically, a series of tight, medium, and loose GTOs are multiplied by *contraction coefficients* and summed to produce a CGTO, which approximates the proper 'cusp' at the nuclear centre.

Although most calculations on molecules are now performed using Gaussian orbitals (STOs are still commonly employed in atomic calculations), it should be noted that other basis sets can be used as long as they span enough of the region of space (radial and angular) where significant electron density resides. In fact, it is possible to use plane wave orbitals [37] of the form $\chi(r, \theta, \phi) = N \exp[i(k_x r \sin\theta \cos\phi + k_y r \sin\theta \sin\phi + k_z r \cos\theta)]$, where $N$ is a normalization constant and $k_x$, $k_y$ and $k_z$ are the quantum numbers detailing the momenta of the orbital along the $x$, $y$ and $z$ Cartesian directions. The advantage to using such 'simple' orbitals is that the integrals one must perform are much easier to handle with such functions; the disadvantage is that one must use many such functions to accurately describe sharply peaked charge distributions of, for example, inner-shell core orbitals.

### (b) Basis set libraries

Much effort has been devoted to developing sets of STO or GTO basis orbitals for main-group elements and the lighter transition metals. This ongoing effort is aimed at providing standard basis set libraries which:

(1)   yield predictable chemical accuracy in the resultant energies;

(2)   are cost effective to use in practical calculations;

(3)   are relatively transferable so that a given atom's basis is flexible enough to be used for that atom in various bonding environments.

*The fundamental core and valence basis.*      In constructing an AO basis, one can choose from among several classes of functions. First, the size and nature of the primary core and valence basis must be specified. Within this category, the following choices are common.

(1)   A *minimal basis* in which the number of STO or CGTO orbitals is equal to the number of core and valence AOs in the atom.

(2)   A *double-zeta* (DZ) basis in which twice as many STOs or CGTOs are used as there are core and valence AOs. The use of more basis functions is motivated by a desire to provide additional variational flexibility so the LCAO–MO process can generate MOs of variable diffuseness as the local electronegativity of the atom varies.

(3)   A *triple-zeta* (TZ) basis in which three times as many STOs or CGTOs are used as the number of core and valence AOs (and, yes, there now are quadruple-zeta (QZ) and higher-zeta basis sets appearing in the literature).

(4)  Dunning and Dunning and Hay [38] developed CGTO bases which range from approximately DZ to substantially beyond QZ quality. These bases involve contractions of primitive uncontracted GTO bases which Huzinaga [39] had earlier optimized. These Dunning bases are commonly denoted as follows for first-row atoms: (10s,6p/5s,4p), which means that 10 s-type primitive GTOs have been contracted to produce five separate s-type CGTOs and that six primitive p-type GTOs were contracted into four separate p-type CGTOs in each of the $x$, $y$ and $z$ directions.

(5)  Even-tempered basis sets [40] consist of GTOs in which the orbital exponents $\alpha_k$ belonging to series of orbitals consist of geometrical progressions: $\alpha_k = a\beta^k$, where $a$ and $\beta$ characterize the particular set of GTOs.

(6)  STO-3G bases [41] were employed some years ago, but have recently become less popular. These bases are constructed by least-squares fitting GTOs to STOs which have been optimized for various electronic states of the atom. When three GTOs are employed to fit each STO, a STO-3G basis is formed.

(7)  4-31G, 5-31G and 6-31G bases [42] employ a single CGTO of contraction length 4, 5, or 6 to describe the core orbital. The valence space is described at the DZ level with the first CGTO constructed from three primitive GTOs and the second CGTO built from a single primitive GTO.

(8)  More recently, the Dunning group has focused on developing basis sets that are optimal not for use in SCF-level calculations on atoms and molecules, but that have been optimized for use in correlated calculations. These so-called correlation-consistent bases [43] are now widely used because more and more *ab initio* calculations are being performed at a correlated level.

(9)  Atomic natural orbital (ANO) basis sets [44] are formed by contracting Gaussian functions so as to reproduce the natural orbitals obtained from correlated (usually using a configuration interaction with single and double excitation (CISD) level wavefunction) calculations on atoms.

Optimization of the *orbital exponents* ($\zeta$s or $\alpha$s) and the GTO-to-CGTO *contraction coefficients* for the kind of bases described above have undergone explosive growth in recent years. As a result, it is not possible to provide a single or even a few literature references from which one can obtain the most up-to-date bases. However, the theory group at the Pacific Northwest National Laboratories (PNNL) offer a webpage [45] from which one can find (and even download in a form prepared for input to any of several commonly used electronic structure codes) a wide variety of Gaussian atomic basis sets.

*Polarization functions.*    One usually enhances any core and valence functions with a set of so-called polarization functions. They are functions of one higher angular momentum than appears in the atom's valence orbital space (e.g. d-functions for C, N and O and p-functions for H), and they have exponents ($\zeta$ or $\alpha$) which cause their radial sizes to be similar to the sizes of the valence orbitals (i.e. the polarization p orbitals of the H atom are similar in size to the 1s orbital). Thus, they are *not* orbitals which describe the atom's valence orbital with one higher $l$ value; such higher-$l$ valence orbitals would be radially more diffuse.

The primary purpose of the polarization functions is to give additional angular flexibility to the LCAO–MO process in forming the valence MOs. This is illustrated below in figure B3.1.8 where polarization $d_\pi$ orbitals are seen to contribute to formation of the bonding $\pi$ orbital of a carbonyl group by allowing polarization of the carbon atom's $p_\pi$ orbital toward the right and of the oxygen atom's $p_\pi$ orbital toward the left.

The polarization functions are essential in strained ring compounds because they provide the angular flexibility needed to direct the electron density into the regions between the bonded atoms.

Functions with higher $l$ values and with 'sizes' like those of lower-$l$ valence orbitals are also used to introduce additional angular correlation by permitting angularly polarized orbital pairs to be formed. Optimal polarization functions for first- and second-row atoms have been tabulated and are included in the PNNL Gaussian orbital web site data base [45].

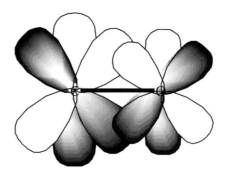

**Figure B3.1.8.** The role of d-polarization functions in the $\pi$ bond between C and O.

*Diffuse functions.*    When dealing with anions or Rydberg states, one must further *augment* the basis set by adding so-called diffuse basis orbitals. The valence and polarization functions described above do not provide enough radial flexibility to adequately describe either of these cases. Once again, the PNNL web site data base [45] offers a good source for obtaining diffuse functions appropriate to a variety of atoms.

    Once one has specified an AO basis for each atom in the molecule, the LCAO–MO procedure can be used to determine the $C_{v,i}$ coefficients that describe the occupied and virtual orbitals. It is important to keep in mind that the basis orbitals are *not* themselves the SCF orbitals of the isolated atoms; even the proper AOs are combinations (with atomic values for the $C_{v,i}$ coefficients) of the basis functions. The LCAO–MO–SCF process itself determines the magnitudes and signs of the $C_{v,i}$; alternations in the signs of these coefficients allow radial nodes to form.

### B3.1.5.4   *The physical meaning of orbital energies*

The HF–SCF equations $\hat{F}\phi_i = \varepsilon_i \phi_i$ imply that $\varepsilon_i$ can be written as

$$\varepsilon_i = \langle \phi_i | \hat{F} | \phi_i \rangle = \langle \phi_i | \hat{h} | \phi_i \rangle + \sum_{j(\text{occupied})} \langle \phi_i | \hat{J}_j - \hat{K}_j | \phi_i \rangle = \langle \phi_i | \hat{h} | \phi_i \rangle + \sum_{j(\text{occupied})} [J_{i,j} - K_{i,j}].$$

Thus $\varepsilon_i$ is the average value of the kinetic energy plus the Coulombic attraction to the nuclei for an electron in $\phi_i$ plus the sum over all of the spin orbitals occupied in $\Psi$ of the Coulomb minus exchange interactions.

    If $\phi_i$ is an occupied spin orbital, the term $[J_{i,i} - K_{i,i}]$ disappears and the latter sum represents the Coulomb minus exchange interaction of $\phi_i$ with all of the $N - 1$ *other* occupied spin orbitals. If $\phi_i$ is a virtual spin orbital, this cancellation does not occur, and one obtains the Coulomb minus exchange interaction of $\phi_i$ with all $N$ of the occupied spin orbitals.

    Hence the orbital energies of *occupied* orbitals pertain to interactions appropriate to a total of $N$ electrons, while the orbital energies of *virtual* orbitals pertain to a system with $N + 1$ electrons. This usually makes SCF virtual orbitals not very good for use in subsequent correlation calculations or for use in interpreting electronic excitation processes. To correlate a pair of electrons that occupy a valence orbital requires double excitations into a virtual orbital of similar size; the SCF virtual orbitals are too diffuse. For this reason, significant effort has been devoted to developing methods that produce so-called 'improved virtual orbitals' (IVOs) [46] that are of more utility in performing correlated calculations.

*(a) Koopmans' theorem*

Let us consider a model of the vertical (i.e. at fixed molecular geometry) detachment or attachment of an electron to an $N$-electron molecule.

(1)   In this model, *both* the parent molecule and the species generated by adding or removing an electron are treated at the single-determinant level.

(2)   The HF orbitals of the parent molecule are used to describe both species. It is said that such a model neglects '*orbital relaxation*' (i.e. the reoptimization of the spin orbitals to allow them to become appropriate to the daughter species).

Within this model, the energy difference between the daughter and the parent can be written as follows ($\phi_k$ represents the particular spin orbital that is added or removed):

(1)   For electron detachment

$$E^{N-1} - E^N = -\varepsilon_k.$$

(2)   For electron attachment

$$E^N - E^{N+1} = -\varepsilon_k.$$

So, within the limitations of the single-determinant, frozen-orbital model, the ionization potentials (IPs) and electron affinities (EAs) are given as the negative of the occupied and virtual spin–orbital energies, respectively. This statement is referred to as *Koopmans' theorem* [47]; it is used extensively in quantum chemical calculations as a means for estimating IPs and EAs and often yields results that are qualitatively correct (i.e., ±0.5 eV).

*(b) Orbital energies and the total energy*

The total SCF electronic energy can be written as

$$E = \sum_{i(\text{occupied})} \langle \phi_i | \hat{h} | \phi_i \rangle + \sum_{i>j(\text{occupied})} [J_{i,j} - K_{i,j}]$$

and the sum of the orbital energies of the occupied spin orbitals is given by

$$\sum_{i(\text{occupied})} \varepsilon_i = \sum_{i(\text{occupied})} \langle \phi_i | \hat{h} | \phi_i \rangle + \sum_{i,j(\text{occupied})} [J_{i,j} - K_{i,j}].$$

These two expressions differ in a very important way; the sum of occupied orbital energies double counts the Coulomb minus exchange interaction energies. Thus, within the HF approximation, the sum of the occupied orbital energies is *not* equal to the total energy.

*B3.1.5.5    Solving the Roothaan SCF equations*

Before moving on to discuss methods that go beyond the single-configuration mean-field model, it is important to examine some of the computational effort that goes into carrying out an SCF calculation.

Once atomic basis sets have been chosen for each atom, the *one- and two-electron integrals* appearing in $F_{\mu,\nu}$ must be evaluated. There are numerous, highly-efficient computer codes [48] which allow such integrals to be computed for s, p, d, f and even g, h and i basis functions. After executing one of these '*integral packages*' for a basis with a total of $P$ functions, one has available (usually on the computer's hard disk) of the order of $P^2/2$ one-electron ($\langle \chi_\mu | \hat{h} | \chi_\nu \rangle$ and $\langle \chi_\mu | \chi_\nu \rangle$) and $P^4/8$ two-electron ($\langle \chi_\mu \chi_\delta | \hat{g} | \chi_\nu \chi_\kappa \rangle$) integrals. When treating extremely large AO basis sets (e.g. 1000 or more basis functions), modern computer programs [49] calculate the requisite integrals, but never store them on the disk. Instead, their contributions to $F_{\mu,\nu}$ are accumulated 'on the fly' after which the integrals are discarded. Recently, much progress has been made towards achieving an evaluation of the non-vanishing (i.e. numerically significant) integrals [48] as well as solving the subsequent SCF equations in a manner whose effort *scales linearly* [50] with the number of basis functions for large $P$.

After the requisite integrals are available or are being computed on the fly, to begin the SCF process one must input into the computer routine which computes $F_{\mu,\nu}$ the *initial 'guesses'* for the $C_{\nu,i}$ values corresponding to the occupied orbitals. These initial guesses are typically made as follows.

(1)  If one has available the $C_{\nu,i}$ values for the system from a calculation performed at a nearby geometry, one can use these $C_{\nu,i}$ values.

(2)  If one has $C_{\nu,i}$ values appropriate to fragments of the system (e.g. for C and O atoms if the CO molecule is under study or for $CH_2$ and O if $H_2CO$ is being studied), one can use these.

(3)  If one has no other information available, one can carry out one iteration of the SCF process in which the two-electron contributions to $F_{\mu,\nu}$ are ignored (i.e. take $F_{\mu,\nu} = \langle \chi_\mu | h | \chi_\nu \rangle$) and use the resultant solutions to $\sum_\nu F_{\mu,\nu} C_{\nu,i} = \varepsilon_i \sum_\nu S_{\mu,\nu} C_{\nu,i}$ as initial guesses.

Once the initial guesses have been made for the $C_{\nu,i}$ of the occupied orbitals, the full $F_{\mu,\nu}$ matrix is formed and new $\varepsilon_i$ and $C_{\nu,i}$ values are obtained by solving $\sum_\nu F_{\mu,\nu} C_{\nu,i} = \varepsilon_i \sum_\nu S_{\mu,\nu} C_{\nu,i}$. These new orbitals are then used to form a new $F_{\mu,\nu}$ matrix from which new $\varepsilon_i$ and $C_{\nu,i}$ are obtained. This iterative process is carried on until the $\varepsilon_i$ and $C_{\nu,i}$ do not vary (within specified tolerances) from iteration to iteration, at which time the SCF process has reached self-consistency.

### B3.1.6   Methods for treating electron correlation

*B3.1.6.1   An overview of various approaches*

There are numerous procedures currently in use for determining the 'best' wavefunction of the form

$$\Psi = \sum_I C_I \Phi_I$$

where $\Phi_I$ is a spin- and space-symmetry-adapted CSF consisting of determinants $|\phi_{I1}\phi_{I2}\phi_{I3}\ldots\phi_{IN}|$ (see [14, 16, 26]). In all such wavefunctions there are two kinds of parameters that need to be determined—the $C_I$ and the LCAO–MO coefficients describing the $\phi_{Ik}$. The most commonly employed methods used to determine these parameters include the following.

*(a) The multiconfigurational self-consistent field method*

In this approach [51], the expectation value $\langle \Psi | \hat{H} | \Psi \rangle / \langle \Psi | \Psi \rangle$ is treated variationally and made stationary with respect to variations in the $C_I$ and $C_{\nu,i}$ coefficients. The energy functional is a quadratic function of the $C_I$ coefficients, and so one can express the stationary conditions for these variables in the secular form

$$\sum_J H_{I,J} C_J = E C_I.$$

However, $E$ is a quartic function of the $C_{\nu,i}$s because $H_{I,J}$ involves two-electron integrals $\langle \phi_i \phi_j | \hat{g} | \phi_k \phi_l \rangle$ that depend quartically on these coefficients.

It is well known that minimization of the function ($E$) of several nonlinear parameters (the $C_{\nu,i}$) is a difficult task that can suffer from poor convergence and may locate local rather than global minima. In a multiconfigurational self-consistent field (MCSCF) wavefunction containing many CSFs, the energy is only weakly dependent on the orbitals that appear in CSFs with small $C_I$ values; in contrast, $E$ is strongly dependent on those orbitals that appear in the CSFs with larger $C_I$ values. One is therefore faced with minimizing a function of many variables that depends strongly on several of the variables and weakly on many others.

For these reasons, in the MCSCF method the number of CSFs is usually kept to a small to moderate number (e.g. a few to several thousand) chosen to describe *essential correlations* (i.e. configuration crossings, near degeneracies, proper dissociation, etc, all of which are often termed *non-dynamical correlations*) and important dynamical correlations (those electron-pair correlations of angular, radial, left–right, etc nature that are important when low-lying 'virtual' orbitals are present).

### (b) The configuration interaction method

In this approach [52], the LCAO–MO coefficients are determined first via a single-configuration SCF calculation or an MCSCF calculation using a small number of CSFs. The $C_I$ coefficients are subsequently determined by making the expectation value $\langle \Psi | \hat{H} | \Psi \rangle / \langle \Psi | \Psi \rangle$ stationary.

The CI wavefunction is most commonly constructed from CSFs of $\Phi_J$ that include:

(1)  all of the CSFs in the SCF or MCSCF wavefunction used to generate the molecular orbitals $\phi_i$. These are referred to as the '*reference*' CSFs;

(2)  CSFs generated by carrying out single-, double-, triple-, etc, level 'excitations' (i.e. orbital replacements) relative to reference CSFs. CI wavefunctions limited to include contributions through various levels of excitation are denoted S (singly), D (doubly), SD (singly and doubly), SDT (singly, doubly, and triply) excited.

The orbitals from which electrons are removed can be restricted to focus attention on the correlations among certain orbitals. For example, if the excitations from the core electrons are excluded, one computes the total energy that contains no core correlation energy. The number of CSFs included in the CI calculation can be far in excess of the number considered in typical MCSCF calculations. CI wavefunctions including 5000 to 50 000 CSFs are routine, and functions with one to several billion CSFs are within the realm of practicality [53].

The need for such large CSF expansions should not be surprising considering (i) that each electron pair requires *at least* two CSFs to form polarized orbital pairs, (ii) there are of the order of $N(N-1)/2 = X$ electron pairs for $N$ electrons, hence (iii) the number of terms in the CI wavefunction scales as $2^X$. For a molecule containing ten electrons, there could be $2^{45} = 3.5 \times 10^{13}$ terms in the CI expansion. This may be an overestimate of the number of CSFs needed, but it demonstrates how rapidly the number of CSFs can grow with the number of electrons.

The $H_{I,J}$ matrices are, in practice, evaluated in terms of one- and two-electron integrals over the MOs using the Slater–Condon rules [54] or their equivalent. Prior to forming the $H_{I,J}$ matrix elements, the one- and two-electron integrals, which can be computed only for the atomic (e.g. STO or GTO) basis, must be transformed [55] to the MO basis. This transformation step requires computer resources proportional to the fifth power of the number of basis functions, and thus is one of the more troublesome steps in most configuration interaction calculations.

For large CI calculations, the full $H_{I,J}$ matrix is *not* formed and stored in the computer's memory or on disk; rather, 'direct CI' methods [56] identify and compute non-zero $H_{I,J}$ and immediately add up contributions to the sum $\sum_J H_{I,J} C_J$. Iterative methods [57], in which approximate values for the $C_J$ coefficients are refined through sequential application of $\sum_J H_{I,J}$ to the preceding estimate of the $C_J$ vector, are employed to solve these large eigenvalue problems.

### (c) The Møller–Plesset perturbation method

This method [58] uses the single-configuration SCF process to determine a set of orbitals $\{\phi_i\}$. Then, using an unperturbed Hamiltonian equal to the sum of the $N$ electrons' Fock operators $\hat{H}^0 = \sum_{i=1,N} \hat{F}(i)$, perturbation theory is used to determine the $C_I$ amplitudes for the CSFs. The MPPT procedure [59] is a special case of many-body perturbation theory (MBPT) in which the UHF Fock operator is used to define $\hat{H}^0$. The amplitude for the *reference* CSF is taken as unity and the other CSFs' amplitudes are determined by the Rayleigh–Schrödinger perturbation using $\hat{H} - \hat{H}^0$ as the perturbation.

In the MPPT/MBPT method, once the reference CSF is chosen and the SCF orbitals belonging to this CSF are determined, the wavefunction $\Psi$ and energy $E$ are determined in an order-by-order manner. The perturbation equations *determine* what CSFs to include and their particular order. This is one of the primary strengths of this technique; it does not require one to make further choices, in contrast to the MCSCF and *CI* treatments where one needs to choose which CSFs to include.

For example, the first-order wavefunction correction $\Psi^1$ is

$$\Psi^1 = -\sum_{i<j,m<n} [\langle i, j|\hat{g}|m, n\rangle - \langle i, j|\hat{g}|n, m\rangle][\varepsilon_m - \varepsilon_i + \varepsilon_n - \varepsilon_j]^{-1} \mid \Phi^{m,n}_{i,j}\rangle$$

where the SCF orbital energies are denoted $\varepsilon_k$ and $\Phi^{m,n}_{i,j}$ represents a CSF that is *doubly excited* ($\phi_i$ and $\phi_j$ are replaced by $\phi_m$ and $\phi_n$) relative to $\Phi$. Only doubly-excited CSFs contribute to the *first-order wavefunction*; the fact that the contributions from singly-excited configurations vanish in $\Phi^1$ is known as the *Brillouin theorem* [60].

The energy $E$ is given through second order as

$$E = E_{SCF} - \sum_{i<j,m<n} |\langle i, j|\hat{g}|m, n\rangle - \langle i, j|\hat{g}|n, m\rangle|^2/[\varepsilon_m - \varepsilon_i + \varepsilon_n - \varepsilon_j].$$

Both $\Psi$ and $E$ are expressed in terms of two-electron integrals $\langle i, j|\hat{g}|m, n\rangle$ coupling the virtual spin orbitals $\phi_m$ and $\phi_n$ to the spin orbitals from which the electrons were excited $\phi_i$ and $\phi_j$ as well as the orbital energy differences $[\varepsilon_m - \varepsilon_i + \varepsilon_n - \varepsilon_j]$ accompanying such excitations. Clearly, the major contributions to the correlation energy are made by double excitations into virtual orbitals $\phi_m\phi_n$ with large $\langle i, j|\hat{g}|m, n\rangle$ integrals and small orbital energy gaps $[\varepsilon_m - \varepsilon_i + \varepsilon_n - \varepsilon_j]$. In higher-order corrections, contributions from CSFs that are singly, triply, etc excited relative to $\Phi$ appear, and additional contributions from the doubly-excited CSFs also enter.

### (d) The coupled-cluster method

In the coupled-cluster (CC) method [61], one expresses the wavefunction in a somewhat different manner:

$$\Psi = \exp(T)\Phi$$

where $\Phi$ is a single CSF (usually the UHF determinant) used in the SCF process to generate a set of spin orbitals. The operator $\hat{T}$ is expressed in terms of operators that achieve spin–orbital excitations as follows:

$$T = \sum_{i,m} t^m_i \hat{m}^+\hat{i} + \sum_{i,j}^{m,n} t^{m,n}_{i,j} \hat{m}^+\hat{n}^+\hat{j}\hat{i} + \cdots$$

where the combination of operators $\hat{m}^+\hat{i}$ denotes the *creation* of an electron in the virtual spin orbital $\phi_m$ and the *removal* of an electron from the occupied spin orbital $\phi_i$ to generate a single excitation. The operation $\hat{m}^+\hat{n}^+\hat{j}\hat{i}$ therefore represents a double excitation from $\phi_i\phi_j$ to $\phi_m\phi_n$.

The amplitudes $t^m_i$, $t^{m,n}_{i,j}$, etc, which play the role of the $C_I$ coefficients in CC theory, are determined through the set of equations generated by projecting the Schrödinger equation in the form

$$\exp(-T)\hat{H}\exp(T)\Phi = E\Phi$$

against CSFs which are single, double, etc, excitations relative to $\Phi$:

$$\langle\Phi^m_i|\hat{H} + [\widehat{H, T}] + \tfrac{1}{2}[[\widehat{H, T}], T] + \tfrac{1}{6}[[[\widehat{H, T}], T], T] + \tfrac{1}{24}[[[[\widehat{H, T}], T], T], T]|\Phi\rangle = 0$$

$$\langle\Phi^{m,n}_{i,j}|\hat{H} + [\widehat{H, T}] + \tfrac{1}{2}[[\widehat{H, T}], T] + \tfrac{1}{6}[[[\widehat{H, T}], T], T] + \tfrac{1}{24}[[[[\widehat{H, T}], T], T], T]|\Phi\rangle = 0$$

$$\langle\Phi^{m,n,p}_{i,j,k}|\hat{H} + [\widehat{H, T}] + \tfrac{1}{2}[[\widehat{H, T}], T] + \tfrac{1}{6}[[[\widehat{H, T}], T], T] + \tfrac{1}{24}[[[[\widehat{H, T}], T], T], T]|\Phi\rangle = 0$$

and so on for higher-order excited CSFs.

It can be shown [62] that the expansion of the exponential operators truncates exactly at the fourth power in $T$. As a result, the exact CC equations are *quartic equations* for the $t_i^m$, $t_{i,j}^{m,n}$, etc amplitudes. The matrix elements appearing in the CC equations can be expressed in terms of one- and two-electron integrals over the spin orbitals including those occupied in $\Phi$ and the virtual orbitals not in $\Phi$.

These quartic equations are solved in an iterative manner and, as such, are susceptible to convergence difficulties. In any such iterative process, it is important to start with an approximation reasonably close to the final result. In CC theory, this is often achieved by neglecting all of the terms that are nonlinear in the $t$ amplitudes (because the $t$s are assumed to be less than unity in magnitude) and ignoring factors that couple different doubly-excited CSFs (i.e. the sum over $i'$, $j'$, $m'$ and $n'$). This gives $t$ amplitudes that are equal to the amplitudes of the first-order MPPT/MBPT wavefunction:

$$t_{i,j}^{m,n} = -\langle i, j|\hat{g}|m, n\rangle'/[\varepsilon_m - \varepsilon_i + \varepsilon_n - \varepsilon_j].$$

As Bartlett [63] and Pople have both demonstrated [64], there is a close relationship between the MPPT/MBPT and CC methods when the CC equations are solved iteratively starting with such an MPPT/MBPT-like initial 'guess' for these double-excitation amplitudes.

*(e) Density functional theories*

These approaches provide alternatives to the conventional tools of quantum chemistry. The CI, MCSCF, MPPT/MBPT, and CC methods move beyond the single-configuration picture by adding to the wavefunction more configurations whose amplitudes they each determine in their own way. This can lead to a very large number of CSFs in the correlated wavefunction and, as a result, a need for extraordinary computer resources.

The density functional approaches are different [65]. Here one solves a set of orbital-level equations

$$\left[-\hbar^2/2m_e\nabla^2 - \sum_A Z_A e^2/|\mathbf{r} - \mathbf{R}_A| + \int \rho(\mathbf{r}')e^2/|\mathbf{r} - \mathbf{r}'|\,d\mathbf{r}' + U(\mathbf{r})\right]\phi_i = \varepsilon_i\phi_i$$

in which the orbitals $\{\phi_i\}$ 'feel' potentials due to the nuclear centres (having charges $Z_A$), Coulombic interaction with the *total* electron density $\rho(\mathbf{r}')$ and a so-called *exchange-correlation* potential denoted $U(\mathbf{r}')$. The particular electronic state for which the calculation is being performed is specified by forming a corresponding density $\rho(\mathbf{r}')$. Before going further in describing how density functional theory (DFT) calculations are carried out, let us examine the origins underlying this theory.

The so-called Hohenberg–Kohn [66] theorem states that the *ground-state* electron density $\rho(\mathbf{r})$ describing an $N$-electron system uniquely determines the potential $V(\mathbf{r})$ in the Hamiltonian

$$\hat{H} = \sum_j \left\{-\hbar^2/2m_e\nabla_j^2 + V(\mathbf{r}_j) + \frac{1}{2}\sum_{k\neq j} e^2/r_{j,k}\right\}$$

and, because $\hat{H}$ determines the ground-state energy and wavefunction of the system, the ground-state density $\rho(\mathbf{r})$ determines the ground-state properties of the system. The proof of this theorem proceeds as follows.

(a) $\rho(\mathbf{r})$ determines $N$ because $\int \rho(\mathbf{r})\,d^3r = N$.

(b) Assume that there are two distinct potentials (aside from an additive constant that simply shifts the zero of total energy) $V(\mathbf{r})$ and $V'(\mathbf{r})$ which, when used in $\hat{H}$ and $\hat{H}'$, respectively, to solve for a ground state produce $E_0$, $\Psi(\mathbf{r})$ and $E_0'$, $\Psi'(\mathbf{r})$ that have the same one-electron density: $\int |\Psi|^2\,d\mathbf{r}_2\,d\mathbf{r}_3\ldots d\mathbf{r}_N = \rho(\mathbf{r}) = \int |\Psi'|^2\,d\mathbf{r}_2\,d\mathbf{r}_3\ldots d\mathbf{r}_N$.

(c) If we think of $\Psi'$ as trial variational wavefunction for the Hamiltonian $\hat{H}$, we know that

$$E_0 < \langle\Psi'|\hat{H}|\Psi'\rangle = \langle\Psi'|\hat{H}'|\Psi'\rangle + \int \rho(\mathbf{r})[V(\mathbf{r}) - V'(\mathbf{r})]\,d^3r = E_0' + \int \rho(\mathbf{r})[V(\mathbf{r}) - V'(\mathbf{r})]\,d^3r.$$

(d)  Similarly, taking $\Psi$ as a trial function for the $H'$ Hamiltonian, one finds that

$$E'_0 < E_0 + \int \rho(r)[V'(r) - V(r)]\, d^3r.$$

(e)  Adding the equations in (c) and (d) gives

$$E_0 + E'_0 < E_0 + E'_0.$$

A clear contradiction.

Hence, there cannot be two distinct potentials $V$ and $V'$ that give the same ground-state $\rho(r)$. So, the ground-state density $\rho(r)$ uniquely determines $N$ and $V$, and thus $\hat{H}$, and therefore $\Psi$ and $E_0$. Furthermore, because $\Psi$ determines all the properties of the ground state, then $\rho(r)$, in principle, determines all such properties. This means that even the kinetic energy and the electron–electron interaction energy of the ground state are determined by $\rho(r)$. It is easy to see that $\int \rho(r)V(r)\, d^3r = V[\rho]$ gives the average value of the electron–nuclear (plus any additional one-electron additive potential) interaction in terms of the ground-state density $\rho(r)$, but how are the kinetic energy $T[\rho]$ and the electron–electron interaction $V_{ee}[\rho]$ energy expressed in terms of $\rho$?

The main difficulty with DFTs is that the Hohenberg–Kohn theorem shows that the *ground-state* values of $T$, $V_{ee}$, $V$, etc are all unique functionals of the *ground-state* $\rho$ (i.e. that they can, in principle, be determined once $\rho$ is given), but it does not tell us what these functional relations are.

To see how it might make sense that a property such as the kinetic energy, whose operator $(-\hbar^2/2m_e)\nabla^2$ involves derivatives, can be related to the electron density, consider a simple system of $N$ non-interacting electrons moving in a three-dimensional cubic 'box' potential. The energy states of such electrons are known to be

$$E = (h^2/8m_eL^2)(n_x^2 + n_y^2 + n_z^2)$$

where $L$ is the length of the box along the three axes and $n_x$, $n_y$ and $n_z$ are the quantum numbers describing the state. We can view $n_x^2 + n_y^2 + n_z^2 = R^2$ as defining the squared radius of a sphere in three dimensions, and we realize that the density of quantum states in this space is one state per unit volume in the $n_x$, $n_y$ and $n_z$ space. Because $n_x$, $n_y$ and $n_z$ must be positive integers, the volume covering all states with energy less than or equal to a specified energy $E = (h^2/2m_eL^2)R^2$ is one-eighth the volume of the sphere of radius $R$:

$$\Phi(E) = \tfrac{1}{8}(4\pi/3)R^3 = (\pi/6)(8m_eL^2E/h^2)^{3/2}.$$

Since there is one state per unit of such volume, $\Phi(E)$ is also the number of states with energy less than or equal to $E$, and is called the *integrated density of states*. The number of states $g(E)\, dE$ with energy between $E$ and $E + dE$, the *density of states*, is the derivative of $\Phi$:

$$g(E) = d\Phi/dE = (\pi/4)(8m_eL^2/h^2)^{3/2}E^{1/2}.$$

If we calculate the total energy for $N$ electrons, with the states having energies up to the so-called *Fermi energy* $(E_F)$ (i.e. the energy of the highest occupied molecular orbital HOMO) doubly occupied, we obtain the ground-state energy:

$$E_0 = 2\int_0^{E_F} g(E)E\, dE = (8\pi/5)(2m_e/h^2)^{3/2}L^3E_F^{5/2}.$$

The total number of electrons $N$ can be expressed as

$$N = 2\int_0^{E_F} g(E)\, dE = (8\pi/3)(2m_e/h^2)^{3/2}L^3E_F^{3/2}$$

which can be solved for $E_F$ in terms of $N$ to then express $E_0$ in terms of $N$ instead of $E_F$:

$$E_0 = (3h^2/10m_e)(3/8\pi)^{2/3}L^3(N/L^3)^{5/3}.$$

This gives the total energy, which is also the kinetic energy in this case because the potential energy is zero within the 'box', in terms of the electron density $\rho(x, y, z) = (N/L^3)$. It therefore may be plausible to express kinetic energies in terms of electron densities $\rho(r)$, but it is by no means clear how to do so for 'real' atoms and molecules with electron–nuclear and electron–electron interactions operative.

In one of the earliest DFT models, the *Thomas–Fermi* theory, the kinetic energy of an atom or a molecule is approximated using the above type of treatment on a 'local' level. That is, for each volume element in $r$ space, one assumes the expression given above to be valid, and then one integrates over all $r$ to compute the total kinetic energy:

$$T_{TF}[\rho] = \int (3h^2/10m_e)(3/8\pi)^{2/3}[\rho(r)]^{5/3}\,\mathrm{d}^3r = C_F \int [\rho(r)]^{5/3}\,\mathrm{d}^3r$$

where the last equality simply defines the $C_F$ constant (which is 2.8712 in atomic units). Ignoring the correlation and exchange contributions to the total energy, this $T$ is combined with the electron–nuclear $V$ and Coulombic electron–electron potential energies to give the Thomas–Fermi total energy:

$$E_{0,TF}[\rho] = C_F \int [\rho(r)]^{5/3}\,\mathrm{d}^3r + \int V(r)\rho(r)\,\mathrm{d}^3r + e^2/2 \int \rho(r)\rho(r')/|r - r'|\,\mathrm{d}^3r\,\mathrm{d}^3r'.$$

This expression is an example of how $E_0$ is given as a *local density functional approximation* (LDA). The term local means that the energy is given as a functional (i.e. a function of $\rho$) which depends only on $\rho(r)$ at the points in space, but not on $\rho(r)$ at more than one point in space.

Unfortunately, the Thomas–Fermi energy functional does not produce results that are of sufficiently high accuracy to be of great use in chemistry. What is missing in this theory are the exchange energy and the correlation energy; moreover, the kinetic energy is treated only in the approximate manner described.

In the book by Parr and Yang [67], it is shown how Dirac was able to address the exchange energy for the 'uniform electron gas' ($N$ Coulomb *interacting* electrons moving in a uniform positive background charge whose magnitude balances the charge of the $N$ electrons). If the exact expression for the exchange energy of the uniform electron gas is applied on a local level, one obtains the commonly used Dirac *local density approximation to the exchange energy*:

$$E_{ex,Dirac}[\rho] = -C_x \int [\rho(r)]^{4/3}\,\mathrm{d}^3r$$

with $C_x = (3/4)(3/\pi)^{1/3} = 0.7386$ in atomic units. Adding this exchange energy to the Thomas–Fermi total energy $E_{0,TF}[\rho]$ gives the so-called Thomas–Fermi–Dirac (TFD) energy functional.

Because electron densities vary rather strongly spatially near the nuclei, corrections to the above approximations to $T[\rho]$ and $E_{ex,Dirac}$ are needed. One of the more commonly used so-called *gradient-corrected* approximations is that invented by Becke [68], and referred to as the Becke88 exchange functional:

$$E_{ex}(\text{Becke88}) = E_{ex,Dirac}[\rho] - \gamma \int x^2\rho^{4/3}(1 + 6\gamma x \sinh^{-1}(x))^{-1}\,\mathrm{d}r$$

where $x = \rho^{-4/3}|\nabla\rho|$, and $\gamma$ is a parameter chosen so that the above exchange energy can best reproduce the known exchange energies of specific electronic states of the inert gas atoms (Becke finds $\gamma$ to equal 0.0042). A common gradient correction to the earlier $T[\rho]$ is called the Weizsacker correction and is given by

$$\delta T_{\text{Weizsacker}} = (1/72)(\hbar/m_e)\int |\nabla\rho(r)|^2/\rho(r)\,\mathrm{d}r.$$

Although the above discussion suggests how one might compute the ground-state energy once the ground-state density $\rho(r)$ is given, one still needs to know how to obtain $\rho$. Kohn and Sham [69] (KS) introduced a set of so-called KS orbitals obeying the following equation:

$$\left\{ -(\hbar^2/2m^e)\nabla^2 + V(r) + e^2/2 \int \rho(r')/|r - r'| \, dr' + U_{xc}(r) \right\} \phi_j = \varepsilon_j \phi_j$$

where the so-called exchange-correlation potential $U_{xc}(r) = \delta E_{xc}[\rho]/\delta\rho(r)$ could be obtained by functional differentiation if the exchange-correlation energy functional $E_{xc}[\rho]$ were known. KS also showed that the KS orbitals $\{\phi_j\}$ could be used to compute the density $\rho$ by simply adding up the orbital densities multiplied by orbital occupancies $n_j$:

$$\rho(r) = \sum_j n_j |\phi_j(r)|^2.$$

Here $n_j = 0$, 1 or 2 is the occupation number of the orbital $\phi_j$ in the state being studied. The kinetic energy should be calculated as

$$T = \sum_j n_j \langle \phi_j(r)| -(\hbar^2/2m_e)\nabla^2 |\phi_j(r)\rangle.$$

The same investigations of the idealized 'uniform electron gas' that identified the Dirac exchange functional, found that the correlation energy (per electron) could also be written exactly as a *function* of the electron density $\rho$ of the system, but only in two limiting cases—the high-density limit (large $\rho$) and the low-density limit. There still exists no exact expression for the correlation energy even for the uniform electron gas that is valid at arbitrary values of $\rho$. Therefore, much work has been devoted to creating efficient and accurate interpolation formulae connecting the low- and high-density uniform electron gas expressions (see appendix E in [67] for further details). One such expression is

$$E_C[\rho] = \int \rho(r)\varepsilon_c(\rho) \, dr$$

where

$$\varepsilon_c(\rho) = A/2\{\ln(x/X) + 2b/Q \tan^{-1}(Q/(2x+b)) - bx_0/X_0[\ln((x-x_0)^2/X) + 2(b+2x_0)/Q \tan^{-1}(Q/(2x+b))]\}$$

is the correlation energy per electron. Here $x = r_s^{1/2}$, $X = x^2 + bx + c$, $X_0 = x_0^2 + bx_0 + c$ and $Q = (4c - b^2)^{1/2}$, $A = 0.062\,1814$, $x_0 = -0.409\,286$, $b = 13.0720$, and $c = 42.7198$. The parameter $r_s$ is how the density $\rho$ enters since $\frac{4}{3}\pi r_s^3$ is equal to $1/\rho$; that is, $r_s$ is the radius of a sphere whose volume is the effective volume occupied by one electron. A reasonable approximation to the full $E_{xc}[\rho]$ would contain the Dirac (and perhaps gradient corrected) exchange functional plus the above $E_C[\rho]$, but there are many alternative approximations to the exchange-correlation energy functional [68]. Currently, many workers are doing their best to 'cook up' functionals for the correlation and exchange energies, but no one has yet invented functionals that are so reliable that most workers agree to use them.

To summarize, in implementing any DFT, one usually proceeds as follows.

(1)  An AO basis is chosen in terms of which the KS orbitals are to be expanded.

(2)  Some initial guess is made for the LCAO–KS expansion coefficients $C_{j,a}$: $\phi_j = \sum_a C_{j,a} \chi_a$.

(3)  The density is computed as $\rho(r) = \sum_j n_j |\phi_j(r)|^2$. Often, $\rho(r)$ is expanded in an AO basis, which need not be the same as the basis used for the $\phi_j$, and the expansion coefficients of $\rho$ are computed in terms of those of the $\phi_j$. It is also common to use an AO basis to expand $\rho^{1/3}(r)$ which, together with $\rho$, is needed to evaluate the exchange-correlation functional's contribution to $E_0$.

(4)  The current iteration's density is used in the KS equations to determine the Hamiltonian $\{-1/2\nabla^2 + V(r) + e^2/2 \int \rho(r')/|r - r'| \, dr' + U_{xc}(r)\}$ whose 'new' eigenfunctions $\{\phi_j\}$ and eigenvalues $\{\varepsilon_j\}$ are found by solving the KS equations.

(5)  These new $\phi_j$ are used to compute a new density, which, in turn, is used to solve a new set of KS equations. This process is continued until convergence is reached (i.e. until the $\phi_j$ used to determine the current iteration's $\rho$ are the same $\phi_j$ that arise as solutions on the next iteration).

(6)  Once the converged $\rho(r)$ is determined, the energy can be computed using the earlier expression

$$E[\rho] = \sum_j n_j \langle \phi_j(r)| -(\hbar^2/2m_e)\nabla^2|\phi_j(r)\rangle + \int V(r)\rho(r)\,dr + e^2/2 \int \rho(r)\rho(r')/|r-r'|\,dr\,dr' + E_{xc}[\rho].$$

In closing this section, it should once again be emphasized that this area is currently undergoing explosive growth and much scrutiny [70]. As a result, it is nearly certain that many of the specific functionals discussed above will be replaced in the near future by improved and more rigorously justified versions. It is also likely that extensions of DFTs to excited states (many workers are actively pursuing this) will be placed on more solid ground and made applicable to molecular systems. Because the computational effort involved in these approaches scales much less strongly [71] with the basis set size than for conventional (MCSCF, CI, etc) methods, density functional methods offer great promise and are likely to contribute much to quantum chemistry in the next decade.

*(f) Efficient and widely distributed computer programs exist for carrying out electronic structure calculations*

The development of electronic structure theory has been ongoing since the 1940s. At first, only a few scientists had access to computers, and they began to develop numerical methods for solving the requisite equations (e.g. the HF equations for orbitals and orbital energies, the configuration interaction equations for electronic state energies and wavefunctions). By the late 1960s, several research groups had developed reasonably efficient computer codes (written primarily in Fortran with selected subroutines that needed to be written especially efficiently in machine language), and the explosive expansion of this discipline was underway. By the 1980s and through the 1990s, these electronic structure programs began to be used by practicing 'bench chemists' both because they became easier to use and because their efficiency and the computers' speed grew to the point where modest to large molecules could be studied.

Web page links [72] to many of the more widely used programs offer convenient access. At present, more electronic structure calculations are performed by non-theorists than by practicing theoretical chemists, largely because of the proliferation of such programs. This does not mean that all that needs to be done in electronic structure theory is done. The rates at which improvements are being made in the numerical algorithms used to solve the problems as well as at which new models are being created remain as high as ever. For example, Professor Rich Friesner [73] has developed and Professor Emily Carter [74] has implemented, for correlated methods, a highly efficient way to replace the list of two-electron integrals $(\phi_i\phi_j|1/r_{1,2}|\phi_k\phi_l)$, which number $N^4$, where $N$ is the number of AO basis functions, by a much smaller list $(\phi_i\phi_j|g)$ from which the original integrals can be rewritten as

$$(\phi_i\phi_j|1/r_{1,2}|\phi_k\phi_l) = \sum_g (\phi_i(g)\phi_j(g)) \int dr\, \phi_k(r)\phi_l(r)/|r - g|.$$

This tool, which they call *pseudospectral methods*, promises to reduce the CPU, memory and disk storage requirements for many electronic structure calculations, thus permitting their application to much larger molecular systems. In addition to ongoing developments in the underlying theory and computer implementation, the range of phenomena and the kinds of physical properties that one needs electronic structure theory to address is growing rapidly. There is every reason to believe that this sub-discipline of theoretical chemistry is continuing to blossom.

*B3.1.6.2    Computational requirements, strengths and weaknesses of various methods*

*(a) Computational steps*

Essentially all of the techniques discussed above require the evaluation of one- and two-electron integrals over the $N$ AO basis functions: $\langle \chi_a | \hat{f} | \chi_b \rangle$ and $\langle \chi_a \chi_b | \hat{g} | \chi_c \chi_d \rangle$. As mentioned earlier, there are of the order of $N^4/8$ such two-electron integrals that must be computed (and perhaps stored on disk); their computation and storage is a major consideration in performing conventional *ab initio* calculations. Much current research is being devoted to reducing the number of such integrals that must be evaluated using either the pseudo-spectral methods discussed earlier or methods that approximate integrals between product distributions (one such distribution is $\chi_a \chi_c$ and another is $\chi_b \chi_d$ when the integral $\langle \chi_a \chi_b | \hat{g} | \chi_c \chi_d \rangle$ is treated) whenever the distributions involve orbitals on sites that are distant from one another.

Another step that is common to most, if not all, approaches that compute orbitals of one form or another is the solution of matrix eigenvalue problems of the form

$$\sum_{\nu} F_{\mu,\nu} C_{\nu,i} = \varepsilon_i \sum_{\nu} S_{\mu,\nu} C_{\nu,i}.$$

The solution of any such eigenvalue problem requires a number of computer operations that scales as the dimension of the $F_{\mu,n}$ matrix to the third power. Since the indices on the $F_{\mu,n}$ matrix label AOs, this means that the task of finding all eigenvalues and eigenvectors scales as the cube of the number of AOs ($N^3$).

The DFT approaches involve basis expansions of orbitals $\phi_i = \sum_{\nu} C_{i,\nu} \chi_{\nu}$ and of the density $\rho$ (or various fractional powers of $\rho$), which is a quadratic function of the orbitals ($\rho = \sum_i n_i |\phi_i|^2$). These steps require computational effort scaling only as $N^2$, which is one of the most important advantages of these schemes. No cumbersome large CSF expansion and associated large secular eigenvalue problem arise, which is another advantage.

The more conventional quantum chemistry methods provide their working equations and energy expressions in terms of one- and two-electron integrals over the final MOs: $\langle \phi_i | \hat{f} | \phi_j \rangle$ and $\langle \phi_i \phi_j | \hat{g} | \phi_k \phi_l \rangle$. The MO-based integrals can only be evaluated by *transforming* the AO-based integrals [55]. Clearly, the $N^5$ scaling of the integral transformation process makes it an even more time-consuming step than the ($N^4$) atomic integral evaluation and a severe *bottleneck* to applying *ab initio* methods to larger systems. Much effort has been devoted to expressing the working equations of various correlated methods in a manner that does not involve the fully-transformed MO-based integrals.

Once the requisite one- and two-electron integrals are available in the MO basis, the multiconfigurational wavefunction and energy calculation can begin. Each of these methods has its own approach to describing the configurations $\{\Phi_J\}$ included in the calculation and how the $\{C_J\}$ amplitudes and the total energy $E$ are to be determined.

The *number of configurations* ($N_C$) varies greatly among the methods and is an important factor to keep in mind. Under certain circumstances (e.g. when studying reactions where an avoided crossing of two configurations produces an activation barrier), it may be *essential* to use more than one electronic configuration. Sometimes, one configuration (e.g. the SCF model) is adequate to capture the qualitative essence of the electronic structure. In all cases, many configurations will be needed if a highly accurate treatment of electron–electron correlations are desired.

The value of $N_C$ determines how much computer time and memory is needed to solve the $N_C$-dimensional $\sum_J H_{I,J} C_J = E C_I$ secular problem in the CI and MCSCF methods. Solution of these matrix eigenvalue equations requires computer time that scales as $N_C^2$ (if few eigenvalues are computed) to $N_C^3$ (if most eigenvalues are obtained).

So-called *complete active space* (CAS) methods form *all* CSFs that can be created by distributing $N$ valence electrons among $P$ valence orbitals. For example, the eight non-core electrons of $H_2O$ might be distributed, in a manner that gives $M_S = 0$, among six valence orbitals (e.g. two lone-pair orbitals, two OH

$\sigma$-bonding orbitals and two OH $\sigma^*$-antibonding orbitals). The number of configurations thereby created is 225. If the same eight electrons were distributed among ten valence orbitals 44 100 configurations result; for 20 and 30 valence orbitals, 23 474 025 and 751 034 025 configurations arise, respectively. Clearly, practical considerations dictate that CAS-based approaches be limited to situations in which a few electrons are to be correlated using a few valence orbitals.

*(b) Variational methods provide upper bounds to energies*

Methods that are based on making the functional $\langle\Psi|\hat{H}|\Psi\rangle/\langle\Psi|\Psi\rangle$ stationary yield *upper bounds* to the lowest energy state having the symmetry of the CSFs in $\Psi$. The CI and MCSCF methods are of this type. They also provide approximate excited-state energies and wavefunctions in the form of other solutions of the secular equation [75] $\sum_J H_{I,J}C_J = EC_I$. Excited-state energies obtained in this manner obey the so-called *bracketing theorem*; that is, between any two approximate energies obtained in the variational calculation, there exists at least one true eigenvalue. These are strong attributes of the variational methods, as is the long and rich history of developments of analytical and computational tools for efficiently implementing such methods.

*(c) Variational methods are not size-extensive*

Variational techniques suffer from a serious drawback, however: they are not necessarily *size extensive* [76]. The energy computed using these tools cannot be trusted to scale with the size of the system. For example, a calculation performed on two $CH_3$ species at large separation may not yield an energy equal to twice the energy obtained by performing the *same* kind of calculation on a single $CH_3$ species. Lack of size extensivity precludes these methods from use in extended systems (e.g. polymers and solids) where errors due to improper size scaling of the energy produce nonsensical results.

By carefully adjusting the variational wavefunction used, it is *possible* to circumvent size-extensivity problems for selected species. For example, the CI calculation on $Be_2$ using all $^1\sum_g$ CSFs formed by placing the four valence electrons into the $2\sigma_g$, $2\sigma_u$, $3\sigma_g$, $3\sigma_u$, $1\pi_u$, and $1\pi_g$ orbitals can yield an energy equal to twice that of the Be atom described by CSFs in which the two valence electrons of the Be atom are placed into the 2s and 2p orbitals in all ways consistent with a $^1S$ symmetry. Such CAS-space MCSCF or CI calculations [77] are size extensive, but it is impractical to extend such an approach to larger systems.

*(d) Most perturbation and CC methods are size-extensive, but do not provide upper bounds and they assume that one CSF dominates*

In contrast to variational methods, perturbation theory and CC methods achieve their energies by projecting the Schrödinger equation against a reference function $\langle\Phi|$ to obtain [78] a *transition formula* $\langle\Phi|\hat{H}|\Psi\rangle$, rather than from an expectation value $\langle\Psi|\hat{H}|\Psi\rangle$. It can be shown that this difference allows non-variational techniques to yield size-extensive energies.

This can be seen by considering the second-order MPPT energy of two non-interacting Be atoms. The reference CSF is $\Phi = |1s_a^2 2s_a^2 1s_b^2 2s_b^2|$; as discussed earlier, only doubly-excited CSFs contribute to the correlation energy through second order. These 'excitations' can involve atom a, atom b, or both atoms. However, CSFs that involve excitations on both atoms (e.g. $|1s_a^2 2s_a 2p_a 1s_b^2 2s_b 2p_b|$) give rise to one- and two-electron integrals over orbitals on both atoms (e.g. $\langle 2s_a 2p_a|\hat{g}|2s_b 2p_b\rangle$) that vanish if the atoms are far apart, so contributions due to such CSFs vanish. Hence, only CSFs that are excited on one or the other atom contribute to the energy. This, in turn, results in a second-order energy that is additive as required by any size-extensive method. In general, a method will be size extensive *if* its energy formula is additive *and* the equations that determine the $C_J$ amplitudes are themselves separable. The MPPT/MBPT and CC methods possess these characteristics.

However, size-extensive methods have two serious weaknesses. Their energies do *not* provide upper bounds to the true energies of the system (because their energy functional is not of the expectation-value form for which the upper bound property has been proven). Moreover, they express the correct wavefunction in terms of corrections to a (presumed dominant) reference function which is usually taken to be a single CSF (although efforts have been made to extend the MPPT/MBPT and CC methods to allow for multiconfigurational reference functions, this is not yet standard practice). For situations in which two CSFs 'cross' along a reaction path, the single-dominant-CSF assumption breaks down, and these methods can have difficulty.

### B3.1.7   There are methods that calculate energy differences rather than energies

In addition to the myriad of methods discussed above for treating the energies and wavefunctions as solutions to the electronic Schrödinger equation, there exists a family of tools that allow one to compute energy differences 'directly' rather than by first finding the energies of pairs of states and subsequently subtracting them. Various energy differences can be so computed: differences between two electronic states of the same molecule (i.e. electronic excitation energies $\Delta E$), differences between energy states of a molecule and the cation or anion formed by removing or adding an electron (i.e. IPs and EAs).

Because of space limitations, we will not be able to elaborate much further on these methods. However, it is important to stress that:

(1)   these so-called *Greens function* or *propagator* methods [71] utilize essentially the same input information (e.g. AO basis sets) and perform many of the same computational steps (e.g. evaluation of one- and two-electron integrals, formation of a set of mean-field MOs, transformation of integrals to the MO basis, etc) as do the other techniques discussed earlier;

(2)   these methods are now rather routinely used when $\Delta E$, IP, or EA information is sought. In fact, the 1998 version of the Gaussian program includes an electron propagator option.

The basic ideas underlying most, if not all, of the energy-difference methods follow

(1)   One forms a *reference wavefunction* $\Psi$ (this can be of the SCF, MPn, CC, etc variety); the energy differences are computed relative to the energy of this function.

(2)   One expresses the *final-state wavefunction* $\Psi'$ (i.e. describing the excited, cation, or anion state) in terms of an operator $\Omega$ acting on the reference $\Psi$: $\Psi' = \Omega\Psi$ Clearly, the $\Omega$ operator must be one that removes or adds an electron when one is attempting to compute IPs or EAs, respectively.

(3)   One writes equations which $\Psi$ and $\Psi'$ are expected to obey. For example, in the early development of these methods [80], the Schrödinger equation itself was assumed to be obeyed, so $\hat{H}\Psi = E\Psi$ and $\hat{H}'\Psi' = E'\Psi'$ are the two equations (note that, in the IP and EA cases, the latter equation, and the associated Hamiltonian $\hat{H}'$, refer to one fewer and one more electrons than does the reference equation $\hat{H}\Psi = E\Psi$).

(4)   One combines $\Omega\Psi = \Psi'$ with the equations that $\Psi$ and $\Psi'$ obey to obtain an equation that $\Omega$ must obey. In the above example, one: (a) uses $\Omega\Psi = \Psi'$ in the Schrödinger equation for $\Psi'$, (b) allows $\Omega$ to act from the left on the Schrödinger equation for $\Psi$ and (c) subtracts the resulting two equations to achieve $(\hat{H}'\hat{\Omega} - \hat{\Omega}\hat{H})\Psi = (E' - E)\Omega\Psi$ or, in commutator form $[\hat{H}, \hat{\Omega}]\Psi = \Delta E\hat{\Omega}\Psi$. By expressing the Hamiltonian in the second-quantization form, only one $\hat{H}$ appears in this final so-called *equation of motion* (EOM) $[\hat{H}, \hat{\Omega}]\Psi = \Delta E\hat{\Omega}\Psi$ (i.e. in the second-quantized form, $\hat{H}'$ and $\hat{H}$ are one and the same).

(5)   One can, for example, express $\Psi$ in terms of a superposition of configurations $\Psi = \sum_J C_J \Phi_J$ whose amplitudes $C_J$ have been determined from an MCSCF, CI or MPn calculation and express $\Omega$ in terms of second-quantization operators $\{O_K\}$ that cause single-, double-, etc, level excitations (for the IP (EA) cases, $\Omega$ is given in terms of operators that remove (add), remove and singly excite (add and singly excite) electrons): $\Omega = \sum_K D_K \hat{O}_K$.

(6)  Substituting the expansions for $\Psi$ and for $\hat{\Omega}$ into the EOM $[H, \hat{\Omega}]\Psi = \Delta E \hat{\Omega} \Psi$, and then projecting the resulting equation on the left against a set of functions (e.g. $\{\hat{O}_{K'}|\Psi\rangle\}$ or $\{\hat{O}_{K'}|\Phi_0\rangle$, where $\Phi_0$ is the dominant component of $\Psi$), gives a matrix eigenvalue–eigenvector equation:

$$\sum_K \langle \hat{O}_{K'} \Psi | [\hat{H}, \hat{O}_K] \Psi \rangle D_K = \Delta E \sum_K \langle \hat{O}_{K'} \Psi | \hat{O}_K \Psi \rangle \hat{D}_K$$

to be solved for the $\hat{D}_K$ operator coefficients and the excitation energies $\Delta E$. Such are the working equations of the EOM (or Greens function or propagator) methods.

In recent years, these methods have been greatly expanded and have reached a degree of reliability where they now offer some of the most accurate tools for studying excited and ionized states. In particular, the use of time-dependent variational principles have allowed the much more rigorous development of equations for energy differences and nonlinear response properties [81]. In addition, the extension of the EOM theory to include coupled-cluster reference functions [82] now allows one to compute excitation and ionization energies using some of the most accurate *ab initio* tools.

## B3.1.8   Summary of *ab initio* methods

At this time, it may not be possible to say which method is preferred for applications where all are practical. Nor is it possible to assess, in a way that is applicable to most chemical species, the accuracies with which various methods predict bond lengths and energies or other properties. However, there are reasons to recommend some methods over others in specific cases. For example, certain applications require a size-extensive energy (e.g. extended systems that consist of a large or macroscopic number of units or studies of weak intermolecular interactions), so MBPT/MPPT-, CC- or CAS-based MCSCF are preferred. Moreover, many chemical reactions and bond-breaking events require two or more 'essential' electronic configurations. For them, single-configuration-based methods such as conventional CC and MBTP/MPPT should be used only with caution; MCSCF or CI calculations are preferred. Very large molecules, in which thousands of AO basis functions are required, may be impossible to treat by methods whose effort scales as $N^4$ or higher; density functional methods would be the only choice then.

For all calculations, the choice of AO basis set must be made carefully, keeping in mind the $N^4$ scaling of the two-electron integral evaluation step and the $N^5$ scaling of the two-electron integral transformation step. Of course, basis functions that describe the essence of the states to be studied are essential (e.g. Rydberg or anion states require diffuse functions and strained rings require polarization functions).

As larger atomic basis sets are employed, the size of the CSF list used to treat a dynamic correlation increases rapidly. For example, many of the above methods use singly- and doubly-excited CSFs for this purpose. For large basis sets, the number of such CSFs ($N_C$) scales as the number of electrons squared $n_e^2$ times the number of basis functions squared $N^2$. Since the effort needed to solve the CI secular problem varies as $N_C^2$ or $N_C^3$ (the latter being to find all eigenvalues and vectors), a dependence as strong as $n_e^6 N^6$ can result. To handle such large CSF spaces, all of the multiconfigurational techniques mentioned in this paper have been developed to the extent that calculations involving of the order of 100–5000 CSFs are routinely performed and calculations using even several billion CSFs are possible [53].

Some of the most significant advances that have been made recently in expanding the applicability of the *ab initio* methods to larger systems are based on recognizing that many of the two-electron integrals and one- and two-electron density matrix elements arising in the pertinent working equations vanish if expressed in terms of localized (atomic or molecular) orbitals. For example, in a polymer consisting of $P$ monomer units (or a crystal composed of $P$ unit cells), the integrals and density matrix elements indexed by monomer units far distant from one another are negligible. Thus, if a method whose effort scales as the $k$th power of the number of AOs ($N$) per monomer (or unit cell) is applied to a system having $P$ units, the effort should

**Table B3.1.2.** Properties of commonly used methods.

| Method | Variational/size extensive | Computational scaling |
|---|---|---|
| HF | Yes/Yes | $N^4$ integrals; $N^3$ eigenvalues; $P^1$ |
| GVB | Yes/Yes | $N^4$ integrals |
|  |  | $N^4$ (per electron pair) GVB equations |
| DFT | No/Yes | $N^3$ eigenvalues; $N^2$ integrals; $P^1$ |
|  |  | $N^3$ orbital orthogonalization; $P^1$ |
| MP2 | No/Yes | $N^5$; $P^2$ |
| CI | Yes/No | $N^5$ transformed integrals; |
|  |  | $N_C^2$ to solve for *one* CI energy and eigenvector |
| CISD | Yes/No | $N^5$ transformed integrals; |
|  |  | $n^2 N^4$ to solve for one CI energy and eigenvector |
| CAS-MCSCF | Yes/Yes | $N^5$ transformed integrals; |
|  |  | $N_C^2$ to solve for CI energy; many iterations also needed |
| CCS | No/Yes | $N^4$ |
| CCSD | No/Yes | $N^6$ |
| CCSDT | No/Yes | $N^8$ |
| CCSD(T) | No/Yes | $N^7$ |

$N$ is the number of atomic basis functions, which usually is proportional to the number of electrons $n$. $N_C$ is the number of configurations; in CI calculations, $N_C$ is usually at least as large as the number of electrons squared times the number of orbitals squared $n_e^2 N^2$; in MCSCF calculations, $N_C$ is usually much smaller than $n_e^2 N^2$. $P$ is explained in section B3.1.8.

*not* scale as $(PN)^k$ but, hopefully, as $PN^k$. Indeed, for the DFT ($k = 3$), SCF ($k = 4$) and MP2 ($k = 5$) methods, specialized techniques [50] have allowed for the implementation of codes scaling linearly (or nearly so for MP2) with the system 'size' $P$ (i.e. the number of units).

Other methods, most of which can be viewed as derivatives of the techniques introduced above, have been and are still being developed; stimulated by the explosive growth in computer power and changes in computer architecture realized in recent years. All indications are that this growth pattern will continue; so *ab initio* quantum chemistry is likely to have an even larger impact on future chemistry research and education (through new insights and concepts). For many of the most commonly employed *ab initio* quantum chemistry tools, the computational efforts, as characterized by how they scale with the system size $P$ (i.e. the number of units), with basis set size $N$ and with the number of electronic configurations $N_C$, as well their variational nature and size extensivity are summarized in table B3.1.2.

Figure B3.1.9 [83] displays the errors (in picometres compared to experimental findings) in the equilibrium bond lengths for a series of 28 molecules obtained at the HF, MP2-4, CCSD, CCSD(T), and CISD levels of theory using three polarized correlation-consistent basis sets (valence DZ through to QZ).

Clearly, the HF method, independent of basis, systematically underestimates the bond lengths over a broad percentage range. The CISD method is neither systematic nor narrowly distributed in its errors, but the MP2 and MP4 (but not MP3) methods are reasonably accurate and have narrow error distributions if valence TZ or QZ bases are used. The CCSD(T), but not the CCSD, method can be quite reliable if valence TZ or QZ bases are used.

In closing this section and this chapter, I wish to remind the reader that my discussion has been limited to *ab initio* techniques; that is, to methods that begin with the electronic Schrödinger equation attempt to solve it without explicitly introducing any experimental data or any numerical results from another calculation. There exists a whole family of alternative approaches called *semi-empirical methods* [84] in which (a) overlaps between pairs of orbitals distant from one another are neglected, (b) many of the two-electron

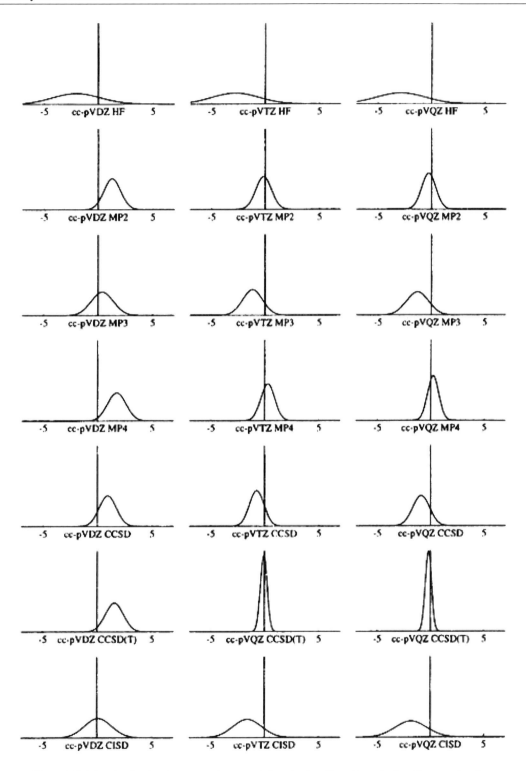

**Figure B3.1.9.** Distribution in errors (picometres) in calculated bond lengths for 28 test molecules.

integrals appearing in *ab initio* methods are neglected (because they are 'small' in some sense) and (c) certain combinations of one- and two-electron integrals that can be (approximately) related to orbital energies of a constituent atom are not computed explicitly but are replaced by experimental data (or data from an *ab initio* calculation) on that atom. Interested readers in these approaches to electronic structure are referred to the articles given in [84].

# References

[1] Born M and Oppenheimer J R 1927 *Ann. Phys., Lpz* **84** 457
It was then used within early quantum chemistry in the following references:
Kolos W and Wolniewicz L 1968 Improved theoretical ground-state energy of the hydrogen molecule *J. Chem. Phys.* **49** 404–10
Pack R T and Hirschfelder J O 1968 Separation of rotational coordinates from the $N$-electron diatomic Schrödinger equation *J. Chem. Phys.* **49** 4009
Pack R T and Hirschfelder J O 1970 Energy corrections to the Born–Oppenheimer approximation. The best adiabatic approximation *J. Chem. Phys.* **52** 521–34
[2] Schlegel H B 1995 *Modern Electronic Structure Theory* ed D R Yarkony (Singapore: World Scientific) ch 8
[3] Chalasinski G, Kendall R A, Taylor H and Simons J 1988 Propensity rules for vibration–rotation induced electron detachment of diatomic anions: application to $NH^- \rightarrow NH + e^-$ *J. Phys. Chem.* **92** 3086–91
[4] Early treatments of molecules in which non-Born–Oppenheimer terms were included were made in:
Kolos W and Wolniewicz L 1963 Nonadiabatic theory for diatomic molecules and its application to the hydrogen molecule *Rev. Mod. Phys.* **35** 473–83
Kolos W and Wolniewicz L 1964 *J. Chem. Phys.* **41** 3663
Kolos W and Wolniewicz L 1964 *J. Chem. Phys.* **41** 3674
Kolos W and Wolniewicz L 1965 Potential energy curves for the $X\,^1\Sigma_g^+$, $b\,^3\Sigma_u^+$, and $C\,^1\Pi_u$ states of the hydrogen molecule *J. Chem. Phys.* **43** 2429–41
Some of the work my students and I have done in this area can be found in:
Simons J 1981 Propensity rules for vibration-induced electron detachment of anions *J. Am. Chem. Soc.* **103** 3971–6
Acharya P K, Kendall R A and Simons J 1984 Vibration-induced electron detachment in molecular anions *J. Am. Chem. Soc.* **106** 3402–7
Acharya P K, Kendall R A and Simons J 1985 Associative electron detachment: $O^- + H \rightarrow OH + e^-$ *J. Chem. Phys.* **83** 3888–93
O'Neal D and Simons J 1989 Vibration-induced electron detachment in acetaldehydeenolate anion *J. Phys. Chem.* **93** 58–61
Simons J 1989 Modified rotationally adiabatic model for rotational autoionization of dipole-bound molecular anion *J. Chem. Phys.* **91** 6858–68
[5] Ohrn Y 2000 The Quantum Theory Project at the University of Florida webpage http://www.qtp.ufl.edu/˜ohrn/
[6] Yarkony D 2000 webpage http://jhuniverse.jhu.edu/˜chem/yarkony.html
For a good recent overview of some of his work, see:
Yarkony D R 1995 Electronic structure aspects of nonadiabatic processes in polyatomic systems *Modern Electronic Structure Theory* vol 2, ed D R Yarkony (Singapore: World Scientific) pp 642–721
[7] Simons J 1983 *Energetic Principles of Chemical Reactions* (Boston, MA: Jones and Bartlett)
[8] Schlegel B 2000 webpage http://www.science.wayne.edu/˜chem/schlegel.html
For a good recent overview see:
Schlegel H B 1995 Geometry optimization on potential energy surfaces *Modern Electronic Structure Theory* vol 2, ed D R Yarkony (Singapore: World Scientific) pp 459–500
Strategies for 'walking' on potential energy surfaces are overviewed by
Schlegel H B 1987 Optimization of equilibrium geometries and transition structures *Adv. Chem. Phys.* **67** 249–86
My own coworkers and I have also contributed to finding transition states, in particular. See, for example:
Simons J, Jørgensen P, Taylor H and Ozment J 1983 Walking on potential energy surfaces *J. Phys. Chem.* **87** 2745–53
Nichols J A, Taylor H, Schmidt P P and Simons J 1990 Walking on potential energy surfaces *J. Chem. Phys.* **92** 340–6
Simons J and Nichols J 1990 Strategies for walking on potential energy surfaces using local quadratic approximations *Int. J. Quant. Chem.* S **24** 263–76
[9] Pulay P 1995 Analytical derivative techniques and the calculation of vibrational spectra *Modern Electronic Structure Theory* vol 2, ed D R Yarkony (Singapore: World Scientific) pp 1191–240
Much of the early work is described in:
Pulay P 1977 Direct use of the gradient for investigating molecular energy surfaces *Modern Theoretical Chemistry* vol 4, ed H F III Schaefer (New York: Plenum) pp 53–185
One of the earliest applications to molecular structure is given in:

Thomsen K and Swanstrøm P 1973 Calculation of molecular one-electron properties using coupled Hartree–Fock methods I. Computational scheme *Mol. Phys.* **26** 735–50

More recent contributions are summarized in:

Jørgensen P and Simons J (eds) 1986 *Geometrical Derivatives of Energy Surfaces and Molecular Properties* (Boston, MA: Reidel)

Pulay P 1987 Analytical derivative methods in quantum chemistry *Advances in Chemical Physics* vol LXIX, ed K P Lawley (New York: Wiley–Interscience) pp 241–86

Helgaker T and Jørgensen P 1988 Analytical calculation of geometrical derivatives in molecular electronic structure theory *Adv. Quantum Chem.* **19** 183–245

[10]   The dipole moment is discussed in:

Karplus M and Porter R N 1970 *Atoms and Molecules* (New York: Benjamin)

[11]   Atkins P W 1992 *The Elements of Physical Chemistry* (New York: Freeman)

Berry R S, Rice S A and Ross J 1980 *Physical Chemistry* (New York: Wiley)

and also webpage www.whfreeman.com\echem/index.html

[12]   Levine I N 1991 *Quantum Chemistry* 4th edn (Englewood Cliffs, NJ: Prentice-Hall)

McQuarrie D A 1983 *Quantum Chemistry* (Mill Valley, CA: University Science)

Simons J and Nichols J 1997 *Quantum Mechanics in Chemistry* (New York: Oxford University Press)

Atkins P W and Friedman R S 1997 *Molecular Quantum Mechanics* 3rd edn (Oxford: Oxford University Press)

Szabo A and Ostlund N S 1989 *Modern Quantum Chemistry* 1st edn (revised) (New York: McGraw-Hill)

and also webpage http://www.emsl.pnl.gov:2080/docs/tms/quantummechanics/

[13]   For a good recent overview, see:

Olsen J and Jørgensen P 1995 Time-dependent response theory with applications to self-consistent field and multiconfigurational self-consistent field wave functions *Modern Electronic Structure Theory* vol 2, ed D R Yarkony (Singapore: World Scientific) pp 857–990

An earlier introduction of molecular properties in terms of wavefunctions and energies and their responses to externally applied fields is given in:

Jørgensen P and Simons J 1981 *Second Quantization-Based Methods in Quantum Chemistry* (New York: Academic)

Jørgensen P and Simons J (eds) 1986 *Geometrical Derivatives of Energy Surfaces and Molecular Properties* (Boston, MA: Reidel)

Amos R D 1987 Molecular property derivatives *Advances in Chemical Physics* vol LXVII, ed K P Lawley, pp 99–153

[14]   Head-Gordon M 1996 Quantum chemistry and molecular processes *J. Phys. Chem.* **100** 13 213–25

[15]   Schaefer H F III, Thomas J R, Yamaguchi Y, DeLeeuw B J and Vacek G 1995 The chemical applicability of standard methods in *ab initio* molecular quantum mechanics *Modern Electronic Structure Theory* vol 2, ed D R Yarkony (Singapore: World Scientific) pp 3–54

Schaefer F 2000 webpage http://zopyros.ccqc.uga.edu/

[16]   Simons J 1991 An experimental chemist's guide to *ab initio* quantum chemistry *J. Phys. Chem.* **95** 1017–29

[17]   Levy H A and Danford M D 1964 *Molten Salt Chemistry* ed M Blander (New York: Interscience)

Bredig M A 1969 *Molten Salts* ed G M Mamantov (New York: Dekker)

[18]   See, for example:

Synge J L and Griffith B A 1949 *Principles of Mechanics* (New York: McGraw-Hill)

[19]   Berry R S, Rice S A and Ross J 1980 *Physical Chemistry* (New York: Wiley) section 11.9

[20]   One of the earliest uses of this term was offered by:

Sinanoglu O 1961 Many-electron theory of atoms and molecules *Proc. US Natl Acad. Sci.* **47** 1217–26

[21]   Sinanoglu O 1962 Many-electron theory of atoms and molecules I. Shells, electron pairs vs many-electron correlations *J. Chem. Phys.* **36** 706–17

[22]   Helgaker T, Jørgensen P and Olsen J 1999 *Electronic Structure Theory* (New York: Wiley)

[23]   Hylleraas E A 1963 Reminiscences from early quantum mechanics of two-electron atoms *Rev. Mod. Phys.* **35** 421–31

[24]   Pilar F L 1968 *Elementary Quantum Chemistry* (New York: McGraw-Hill) section 11.5

[25]   Löwdin P O 1959 Correlation problem in many-electron quantum mechanics *Adv. Chem. Phys.* **2** 207–32

[26]   Excellent early overviews of many of these methods are included in:

Schaefer H F III (ed) 1977 *Modern Theoretical Chemistry* vol 3 (New York: Plenum)

Schaefer H F III (ed) 1977 *Modern Theoretical Chemistry* vol 4 (New York: Plenum)

Lawley K P (ed) 1987 *Advances in Chemical Physics* vol LXVII (New York: Wiley–Interscience)

Lawley K P (ed) 1987 *Advances in Chemical Physics* vol LXIX (New York: Wiley–Interscience)

Yarkony D R (ed) 1995 *Modern Electronic Structure Theory* (Singapore: World Scientific)

[27]   Simons J 1983 *Energetic Principles of Chemical Reactions* (Boston, MA: Jones and Bartlett) p 129

[28]   Goddard W A III 2000 webpage http://www.caltech.edu/~chemistry/Faculties/Goddard.html

When most quantum chemists were pursuing improvements in the molecular orbital method, he returned to the valence bond theory and developed the so-called GVB methods that allow electron correlation to be included within a valence bond framework

[29]   See, for example:

Goddard W A III and Harding L B 1978 The description of chemical bonding from *ab initio* calculation *Ann. Rev. Phys. Chem.* **29** 363–96

[30] The classic papers in which the SCF equations for closed- and open-shell systems are treated are:
Roothaan C C J 1951 New developments in molecular orbital theory *Rev. Mod. Phys.* **23** 69–89
Roothaan C C J 1960 Self-consistent field theory for open shells of electronic systems *Rev. Mod. Phys.* **32** 179–85

[31] The original works by the Hartree father and son team appear in:
Hartree D R 1928 The wave mechanics of an atom with a non-Coulomb central field. Part III. Term values and intensities in series in optical spectra *Proc. Camb. Phil. Soc.* **24** 426–37
Hartree D R, Hartree W and Swirles B 1940 Self-consistent field including exchange and superposition of configurations with some results for oxygen *Phil. Trans. R. Soc.* A **238** 229–47
The work of Fock is given in:
Fock V 1930 *Z. Phys.* **61** 126

[32] Frisch M J *et al* 1995 *Gaussian 94* Revision A.1 (Pittsburgh, PA: Gaussian)

[33] Pople J 2000 webpage http://www.chem.nwu.edu/brochure/pople.html
Pople made many developments leading to the suite of Gaussian computer codes that now constitute the most widely used electronic structure computer programs

[34] Froese-Fischer C 1970 A multi-configuration Hartree–Fock program *Comput. Phys. Commun.* **1** 151–66

[35] McCullough E A Jr 1975 The partial-wave self-consistent method for diatomic molecules computational formalism and results for small molecules *J. Chem. Phys.* **62** 3991–9
Christiansen P A and McCullough E A Jr 1977 Numerical Hartree–Fock calculations for $N_2$, FH, and CO comparison with optimized LCAO results *J. Chem. Phys.* **67** 1877–82

[36] Mulliken Prof. R S University of Chicago. He is the person who came up with the phrase 'molecular orbital' and the concepts behind it

[37] Jordan Prof. K 2000 webpage http://www.chem.pitt.edu/~jordan/index.html
Jordan compared the use of plane wave and conventional Gaussian basis orbitals within density functional calculations in:
Nachtigall P, Jordan K D, Smith A and Jønsson H 1996 Investigation of the reliability of density functional methods reaction and activation energies for Si–Si bond cleavage and $H_2$ elimination from silanes *J. Chem. Phys.* **104** 148–58

[38] Dunning T H Jr 1970 Gaussian basis functions for use in molecular calculations I. Contraction of (9s 5p) atomic basis sets for the first-row atoms *J. Chem. Phys.* **53** 2823–33
Dunning T H Jr and Hay P J 1977 Gaussian basis sets for molecular calculations *Methods of Electronic Structure Theory* vol 3, ed H F III Schaefer (New York: Plenum) pp 1–27

[39] Huzinaga S 1965 Gaussian-type functions for polyatomic system I. *J. Chem. Phys.* **42** 1293–1302

[40] Schmidt M W and Ruedenberg K 1979 Effective convergence to complete orbital bases and to the atomic Hartree–Fock limit through systematic sequences of Gaussian primitives *J. Chem. Phys.* **71** 3951–62

[41] Hehre W J, Stewart R F and Pople J A 1969 Self-consistent molecular-orbital method I. Use of Gaussian expansions of Slater-type atomic orbitals *J. Chem. Phys.* **51** 2657–64

[42] Ditchfield R, Hehre W J and Pople J A 1971 Self-consistent molecular-orbital methods IX. An extended Gaussian-type basis for molecular-orbital studies of organic molecules *J. Chem. Phys.* **54** 724–8
Hehre W J, Ditchfield R and Pople J A 1972 Self-consistent molecular-orbital methods XII. Further extension of Gaussian-type basis sets for use in molecular orbital studies of organic molecules *J. Chem. Phys.* **56** 2257–61
Hariharan P C and Pople J A 1973 The influence of polarization functions on molecular orbital hydrogenation energies *Theoret. Chim. Acta* **28** 213–22
Krishnan R, Binkley J S, Seeger R and Pople J A 1980 Self-consistent molecular orbital methods XX. A basis set for correlated wave functions *J. Chem. Phys.* **72** 650–4

[43] Helgaker T and Taylor P R 1995 Gaussian basis sets and molecular integrals *Modern Electronic Structure Theory* vol 2, ed D R Yarkony (Singapore: World Scientific) section 5.4, pp 725–856

[44] Helgaker T and Taylor P R 1995 Gaussian basis sets and molecular integrals *Modern Electronic Structure Theory* vol 2, ed D R Yarkony (Singapore: World Scientific) section 5.3, pp 725–856

[45] Pacific Northwest National Laboratories 2000 webpage www.emsl.pnl.gov:2080/forms/basisform.html

[46] Hunt W J and Goddard W A III 1969 Excited states of $H_2O$ using improved virtual orbitals *Chem. Phys. Lett.* **3** 414–18

[47] von Koopmans T 1934 Über die zuordnung von wellenfunktionen und eigenwerten zu den einzelnen elektronen eines atoms *Physica* **1** 104–13

[48] Some of the integral packages and the techniques used to evaluate the integrals are described in:
Csizmadia I G, Harrison M C, Moscowitz J W and Sutcliffe B T 1966 Commentationes. Non-empirical LCAO–MO–SCF–CI calculations on organic molecules with Gaussian type functions. Part I. Introductory review and mathematical formalism *Theoret. Chim. Acta* **6** 191–216
Clementi E and Davis D R 1967 Electronic structure of large molecular systems *J. Comp. Phys.* **1** 223–44
Rothenberg S, Kollman P, Schwartz M E, Hays E F and Allen L C 1970 Mole. A system for quantum chemistry I. General description *Int. J. Quantum Chem.* S **3** 715–25
Hehre W J, Lathan W A, Ditchfield R, Newton M D and Pople J A 1971 *Program* No 236 (Bloomington, IN: Quantum Chemistry Program Exchange)
Dupuis M, Rys J and King H F 1976 Evaluation of molecular integrals over Gaussian basis functions *J. Chem. Phys.* **65** 111–16

McMurchie L E and Davidson E R 1978 One- and two-electron integrals over Cartesian Gaussian functions *J. Comp. Phys.* **26** 218–31

Gill P M W 1994 Molecular integrals over Gaussian basis functions *Adv. Quantum Chem.* **25** 141–205

[49]  This concept of 'direct' calculations in which integrals are not stored but used 'on the fly' is discussed in:

Almlöf J, Faegri K and Korsell K 1982 Principles for a direct SCF approach to LCAO–MO *ab initio* calculations *J. Comput. Chem.* **3** 385–99

A good recent overview of direct methods in given by:

Almlöf J 1995 Direct methods in electronic structure theory *Modern Electronic Structure Theory* vol 2, ed D R Yarkony (Singapore: World Scientific) pp 110–51

[50]  Strain M C, Scuseria G E and Frisch M J 1996 Linear scaling for the electronic quantum coulomb problem *Science* **271** 51–3

White C A, Johnson B G, Gill P M W and Head-Gordon M 1994 The fast multipole method *Chem. Phys. Lett.* **230** 8–16

White C A, Johnson B G, Gill P M W and Head-Gordon M 1996 Linear scaling density functional calculations via the continuous fast multipole method *Chem. Phys. Lett.* **253** 268–78

Saebo S and Pulay P 1993 Local treatment of electron correlation *Ann. Rev. Phys. Chem.* **44** 213–36

[51]  Werner H-J 1987 Matrix-formulated direct multiconfiguration self-consistent field and multiconfiguration reference configuration-interaction methods *Advances in Chemical Physics* vol LXIX, ed K P Lawley (New York: Wiley–Interscience) pp 1–62

Shepard R 1987 The multiconfiguration self-consistent field method *Advances in Chemical Physics* vol LXIX, ed K P Lawley (New York: Wiley–Interscience) pp 63–200

describe several of the advances that have been made in the MCSCF method, especially with respect to enhancing its rate and range of convergence

Wahl A C and Das G 1977 The multiconfiguration self-consistent field method *Modern Theoretical Chemistry* vol 3, ed H F III Schaefer (New York: Plenum) pp 51–78

covers the 'earlier' history on this topic

Bobrowicz F W and Goddard W A III 1977 The self-consistent field equations for generalized valence bond and open-shell Hartree–Fock wave functions *Modern Theoretical Chemistry* vol 3, ed H F III Schaefer (New York: Plenum) pp 97–127

provide, in *Modern Theoretical Chemistry* vol 3, an overview of the GVB approach, which can be viewed as a specific kind of MCSCF calculation

See also:

Dalgaard E and Jørgensen P 1978 Optimization of orbitals for multiconfigurational reference states *J. Chem. Phys.* **69** 3833–44

Jensen H J Aa, Jørgensen P and Ågren H 1987 Efficient optimization of large scale MCSCF wave functions with a restricted step algorithm *J. Chem. Phys.* **87** 451–66

Lengsfield B H III and Liu B 1981 A second order MCSCF method for large CI expansions *J. Chem. Phys.* **75** 478–80

[52]  An early article on this method is:

Boys S F 1950 Electronic wave functions II. A calculation for the ground state of the beryllium atom *Proc. R. Soc.* A **201** 125–37

Shavitt I 1977 The method of configuration interaction *Modern Theoretical Chemistry* vol 3, ed H F III Schaefer (New York: Plenum) pp 189–275

Ross B O and Siegbahn P E M 1977 The direct configuration interaction method from molecular integrals *Modern Theoretical Chemistry* vol 3, ed H F III Schaefer (New York: Plenum) pp 277–318

give excellent overviews of the CI method

For a nice overview of recent work, see:

Schaefer H F III, Thomas J R, Yamaguchi Y, Deleeuw B J and Vacek G 1995 The chemical applicability of standard methods in *ab initio* molecular quantum mechanics *Modern Electronic Structure Theory* vol 2, ed D R Yarkony (Singapore: World Scientific) pp 3–54

[53]  Olsen J, Roos B, Jørgensen P and Jensen H J Aa 1988 Determinant based configuration interaction algorithms for complete and restricted configuration interaction spaces *J. Chem. Phys.* **89** 2185–92

Olsen J, Jørgensen P and Simons J 1990 Passing the one-billion limit in full configuration-interaction (Fci) calculations *Chem. Phys. Lett.* **169** 463–72

[54]  The so-called Slater–Condon rules express the matrix elements of any one-electron ($F$) plus two-electron ($G$) additive operator between pairs of antisymmetrized spin–orbital products that have been arranged (by permuting spin–orbital ordering) to be in so-called maximal coincidence. Once in this order, the matrix elements between two such Slater determinants (labelled $|\rangle$ and $|'\rangle$) are summarized as follows:

(i) if $|\rangle$ and $|'\rangle$ are identical, then

$$\langle |F+G| \rangle = \sum_i \langle \phi_i | f | \phi_i \rangle + \sum_{i>j} [\langle \phi_i \phi_j | g | \phi_i \phi_j \rangle - \langle \phi_i \phi_j | g | \phi_j \phi_i \rangle]$$

where the sums over $i$ and $j$ run over all spin orbitals in $|\rangle$;

(ii) if $|\rangle$ and $|'\rangle$ differ by a single spin–orbital mismatch ($\phi_p \neq \phi'_p$),

$$\langle |F+G| \rangle = \langle \phi_p | f | \phi'_p \rangle + \sum_j [\langle \phi_p \phi_j | g | \phi'_p \phi_j \rangle - \langle \phi_p \phi_j | g | \phi_j \phi'_p \rangle]$$

where the sum over $j$ runs over all spin orbitals in $|\rangle$ except $\phi_p$;

(iii) if $|\rangle$ and $|'\rangle$ differ by two spin orbitals ($\phi_p \neq \phi'_p$ and $\phi_q \neq \phi'_q$),

$$\langle|F+G|\rangle = \langle\phi_p\phi_q|g|\phi'_p\phi'_q\rangle - \langle\phi_p\phi_q|g|\phi'_q\phi'_p\rangle$$

(note that the $F$ contribution vanishes in this case);

(iv) if $|\rangle$ and $|'\rangle$ differ by three or more spin orbitals, then

$$\langle|F+G|\rangle = 0$$

[55]  Nesbet R K 1963 Computer programs for electronic wave-function calculations *Rev. Mod. Phys.* **35** 552–7

It would seem that the process of evaluating all $N^4$ of the $\langle\phi_i\phi_j|g|\phi_k\phi_l\rangle$, each of which requires $N^4$ additions and multiplications, would require computer time proportional to $N^8$. However, it is possible to perform the full transformation of the two-electron integral list in a time that scales as $N^5$ by first performing a transformation of the $\langle\chi_a\chi_b|g|\chi_c\chi_d\rangle$ to an intermediate array $\langle\chi_a\chi_b|g|\chi_c\phi_l\rangle = \sum_d C_{d,l}\langle\chi_a\chi_b|g|\chi_c\chi_d\rangle$ which requires $N^5$ multiplications and additions. The list $\langle\chi_a\chi_b|g|\chi_c\phi_l\rangle$ is then transformed to a second-level transformed array $\langle\chi_a\chi_b|g|\phi_k\phi_l\rangle = \sum_c C_{c,k}\langle\chi_a\chi_b|g|\chi_c\phi_l\rangle$, which requires another $N^5$ operations. This sequential transformation is repeated four times until the final $\langle\phi_i\phi_j|g|\phi_k\phi_l\rangle$ array is in hand

[56]  For early perspectives, see, for example:

Nesbet, R K 1965 Algorithm for diagonalization of large matrices *J. Chem. Phys.* **43** 311–12

Davidson E R 1976 The iterative calculation of a few of the lowest eigenvalues and corresponding eigenvectors of large real-symmetric matrices *J. Comput. Phys.* **17** 87–94

Roos B O and Siegbahn P E M 1977 The direct configuration interaction method from molecular integrals *Modern Theoretical Chemistry* vol 3, ed H F III Schaefer (New York: Plenum) pp 277–318

Roos B 1972 A new method for large-scale CI calculations *Chem. Phys. Lett.* **15** 153–9

For a good review, see:

Saunders V R and Van Lenthe J H 1983 The direct CI method a detailed analysis *Mol. Phys.* **48** 923–54

[57]  Davidson E 2000 webpage http://php.indiana.edu/~davidson/

Professor Davidson has contributed as much as anyone both to the development of the fundamentals of electronic structure theory and its applications to many perplexing problems in molecular structure and spectroscopy

[58]  The essential features of the MPPT/MBPT approach are described in the following articles:

Pople J A, Krishnan R, Schlegel H B and Binkley J S 1978 Electron correlation theories and their application to the study of simple reaction potential surface *Int. J. Quantum Chem.* **14** 545–60

Bartlett R J and Silver D M 1975 Many-body perturbation theory applied to electron pair correlation energies I. Closed-shell first-row diatomic hydrides *J. Chem. Phys.* **62** 3258–68

Krishnan R and Pople J A 1978 Approximate fourth-order perturbation theory of the electron correlation energy *Int. J. Quantum Chem.* **14** 91–100

[59]  Kelly H P 1963 Correlation effects in atoms *Phys. Rev.* **131** 684–99

Møller C and Plesset M S 1934 Note on an approximation treatment for many electron systems *Phys. Rev.* **46** 618–22

[60]  Szabo A and Ostlund N S 1989 *Modern Quantum Chemistry* 1st edn (revised) (New York: McGraw-Hill) p 128

[61]  The early work in chemistry on this method is described in:

Cizek J 1966 On the correlation problem in atomic and molecular systems. Calculation of wave function components in Ursell-type expansion using quantum-field theoretical methods *J. Chem. Phys.* **45** 4256–66

Paldus J, Cizek J and Shavitt I 1972 Correlation problems in atomic and molecular systems IV. Extended coupled-pair many-electron theory and its application to the BH$_3$ molecule *Phys. Rev.* A **5** 50–67

Bartlett R J and Purvis G D 1978 Many-body perturbation theory coupled-pair many-electron theory and the importance of quadruple excitations for the correlation problem *Int. J. Quantum Chem.* **14** 561–81

Purvis G D III and Bartlett R J 1982 A full coupled-cluster singles and doubles model. The inclusion of disconnected triples *J. Chem. Phys.* **76** 1910

[62]  Jørgensen P and Simons J 1981 *Second Quantization Based Methods in Quantum Chemistry* (New York: Academic) ch 4

[63]  Bartlett R 2000 webpage http://www.qtp.ufl.edu/~bartlett

Professor Bartlett brought the CC method, developed earlier by others, into the mainstream of electronic structure theory. For a nice overview of his work on the CC method see:

Bartlett R J 1995 Coupled-cluster theory: an overview of recent developments *Modern Electronic Structure Theory* vol 2, ed D R Yarkony (Singapore: World Scientific) pp 1047–131

[64]  Bartlett R J and Purvis G D 1978 Many-body perturbation theory coupled-pair many-electron theory and the importance of quadruple excitations for the correlation problem *Int. J. Quantum Chem.* **14** 561–81

Pople J A, Krishnan R, Schlegel H B and Binkley J S 1978 Electron correlation theories and their application to the study of simple reaction potential surfaces *Int. J. Quantum Chem.* **14** 545–60

[65]  Parr B 2000 webpage http://net.chem.unc.edu/faculty/rgp/cfrgp01.html

Professor Parr was among the first to push the density functional theory of Hohenberg and Kohn to bring it into the mainstream of electronic structure theory. For a good overview, see the book:

Parr R G and Yang W 1989 *Density Functional Theory of Atoms and Molecules* (New York: Oxford University Press)

[66]  Hohenberg P and Kohn W 1964 Inhomogeneous electron gas *Phys. Rev.* B **136** 864–72

[67] The Hohenberg–Kohn theorem and the basis of much of density functional theory are treated:
Parr R G and Yang W 1989 *Density-Functional Theory of Atoms and Molecules* (New York: Oxford University Press)
The original paper relating to this theory is [66]

[68] Professor Axel Becke of Queens University, Belfast has been very actively involved in developing and improving exchange-correlation energy functionals. For a good recent overview, see:
Becke A D 1995 Exchange-correlation approximations in density-functional theory *Modern Electronic Structure Theory* vol 2, ed D R Yarkony (Singapore: World Scientific) pp 1022–46
Becke A D 1983 Numerical Hartree–Fock–Slater calculations on diatomic molecules *J. Chem. Phys.* **76** 6037–45

[69] Kohn W and Sham L J 1965 Self-consistent equations including exchange and correlation effects *Phys. Rev.* A **140** 1133–8

[70] Many of the various density functional approaches that are under active development can be found in:
Jones R O 1987 Molecular calculations with the density functional formalism *Advances in Chemical Physics* vol LXVII, ed K P Lawley (New York: Wiley–Interscience) pp 413–37
Dunlap B I 1987 Symmetry and degeneracy in Xα and density functional theory *Advances in Chemical Physics* vol LXIX, ed K P Lawley (New York: Wiley–Interscience) pp 287–318
Dahl J P and Avery J (eds) 1984 *Local Density Approximations in Quantum Chemistry Solid State Physics* (New York: Plenum)
Parr R G 1983 Density functional theory *Ann. Rev. Phys. Chem.* **34** 631–56
Salahub D R, Lampson S H and Messmer R P 1982 Is there correlation in Xα analysis of Hartree–Fock and LCAO Xα calculations for $O_3$ *Chem. Phys. Lett.* **85** 430–3
Ziegler T, Rauk A and Baerends E J 1977 On the calculation of multiplet energies by the Hartree–Fock–Slater method *Theor. Chim. Acta* **43** 261–71
Becke A D 1983 Numerical Hartree–Fock–Slater calculations on diatomic molecules *J. Chem. Phys.* **76** 6037–45
Case D A 1982 Electronic structure calculation using the Xα method *Ann. Rev. Phys. Chem.* **33** 151–71
Labanowski J K and Andzelm J W (eds) 1991 *Density Functional Methods in Chemistry* (New York: Springer)
For a recent critical evaluation of situations where current DFT approaches experience difficulties, see:
Davidson E R 1998 How robust is present-day DFT? *Int. J. Quantum Chem.* **69** 241–5

[71] This is because no four-indexed two-electron integral like expressions enter into the integrals needed to compute the energy. All such integrals involve $\rho(r)$ or the product $\rho(r)\rho(r)$; because $\rho$ is itself expanded in a basis (say of $M$ functions), even the term $\rho(r)\rho(r)$ scales no worse than $M^2$. The solution of the KS equations for the KS orbitals $\phi_i$ involves solving a matrix eigenvalue problem; this is expected to scale as $M^3$. However, as discussed in section B3.1.8, the scalings of the DFT, SCF, and MP2 methods have been reduced even further

[72] Pacific Northwest National Laboratories is developing a suite of programs called NWChem
Pacific Northwest National Laboratories 2000 webpage http://www.emsl.pnl.gov:2080/
The MacroModel program of Professor C Still, Columbia University
webpage http://www.cc.columbia.edu/~chempub/mmod/mmod.html
The Gaussian suite of programs webpage http://www.gaussian.com
The GAMESS program webpage http://www.msg.ameslab.gov/GAMESS/GAMESS.html
The HyperChem programs of Hypercube, Inc webpage http://www.hyper.com
The CAChe software packages from Oxford Molecular webpage http://www.oxmol.com/getinfo/eduf
The MOPAC program of CambridgeSoft webpage http://www.camsoft.com
The Amber program of Professor Peter Kollman, University of California, San Francisco
webpage http://www.amber.ucsf.edu/amber/amber.html
The CHARMm program webpage charmm-bbs-request@emperor.harvard.edu
The programs of MSI, Inc webpage http://www.msi.com/info/index.html
The COLUMBUS program webpage shavitt@mps.ohio-state.edu
The CADPAC program of Dr Roger Amos webpage http://www.cray.com/PUBLIC/DAS/files/CHEMISTRY/CADPAC.txt
The programs of Wavefunction, Inc webpage http://wavefun.com/
The ACES II program of Professor Rod Bartlett webpage http://www.qtp.ufl.edu/Aces2/
The MOLCAS program of Professor Bjorn Roos webpage teobor@garm.teokem.lu.se
A nice compendium of various softwares is given in the appendix of reviews in:
Lipkowitz K B and Boyd D B (eds) 1996 *Computational Chemistry* vol 7 (New York: VCH)

[73] Friesner R 2000 webpage http://www.columbia.edu/cu/chemistry/faculty/raf.html
Professor Friesner built on earlier developments of:
Beebe N H F and Linderberg J 1977 Simplifications in the generation and transformation of two-electron integrals in molecular calculations *Int. J. Quantum Chem.* **12** 683–705
Feyereisen M, Fitzgerald G and Komornicki A 1993 Use of approximate integrals in *ab initio* theory an application in MP2 energy calculations *Chem. Phys. Lett.* **208** 359–63
to develop the pseudospectral methods that he and others now widely use. See:
Friesner R A 1987 Solution of the Hartree–Fock equations for polyatomic molecules by a pseudospectral method *J. Chem. Phys.* **86** 3522–31

[74] Carter E 2000 webpage http://www.chem.ucla.edu/dept/Faculty/carter.html

For an overview of Professor Carter's group's work using pseudospectral methods, see:

Martinez T J and Carter E A 1995 Pseudospectral methods applied to the electron correlation problem *Modern Electronic Structure Theory* vol 2, ed D R Yarkony (Singapore: World Scientific) pp 1132–65

[75] Hylleraas E A and Undheim B 1930 *Z. Phys.* **65** 759

MacDonald J K L 1933 Successive approximations by the Rayleigh–Ritz variation method *Phys. Rev.* **43** 830–3

[76] Pople J A 1973 Theoretical models for chemistry *Energy, Structure, and Reactivity* ed D W Smith and W B McRae (New York: Wiley) p 51–67

[77] Roos B O, Taylor P R and Siegbahn P E M 1980 A complete active space SCF method (CASSCF) using a density matrix formulated super-CI approach *Chem. Phys.* **48** 157–73

Roos B O 1987 The complete active space self-consistent field method and its applications in electronic structure calculations *Adv. Chem. Phys.* **69** 399–445

[78] Kelly H P 1963 Correlation effects in atoms *Phys. Rev.* **131** 684–99

[79] Good early overviews of the electron propagator (that is used to obtain IP and EA data) and of the polarization propagator are given in:

Jørgensen P and Simons J 1981 *Second Quantization Based Methods in Quantum Chemistry* (New York: Academic)

The very early efforts on these methods are introduced in:

Linderberg J and Öhrn Y 1973 *Propagator Methods in Quantum Chemistry* (New York: Academic)

More recent summaries include:

Cederbaum L S and Domcke W 1977 Theoretical aspects of ionization potentials and photoelectron spectroscopy a Green's function approach *Adv. Chem. Phys.* **36** 205–344

Oddershede J 1987 Propagator methods *Adv. Chem. Phys.* **69** 201–39

Ortiz J V 1997 The electron propagator picture of molecular electronic structure *Computational Chemistry: Reviews of Current Trends* vol 2, ed J Leszczynski (Singapore: World Scientific) pp 1–61

[80] The introduction of EOMs for energy differences and for operators that connect two states appears first in the nuclear physics literature; see for example:

Rowe D J 1968 Equation-of-motion method and the extended shell model *Rev. Mod. Phys.* **40** 153–66

I applied these ideas to excitation energies in atoms and molecules in 1971; see equation (2.1)–(2.6) in:

Simons J 1971 Direct calculation of first- and second-order density matrices. The higher RPA method *J. Chem. Phys.* **55** 1218–30

In 1973, the EOM method was then extended to treat IP and EA cases:

Simons J 1973 Theory of electron affinities of small molecules *J. Chem. Phys.* **58** 4899–907

In a subsequent treatment from the time-dependent response point of view, connection with the Greens function methods was made:

Simons J 1972 Energy-shift theory of low-lying excited electronic states of molecules *J. Chem. Phys.* **57** 3787–92

A more recent overview of much of the EOM, Greens function, and propagator field is given in:

Oddershede J 1987 Propagator methods *Adv. Chem. Phys.* **69** 201–39

[81] Olsen J and Jørgensen P 1995 Time-dependent response theory with applications to self-consistent field and multiconfigurational self-consistent field wave functions *Modern Electronic Structure Theory* vol 2, ed D R Yarkony (Singapore: World Scientific) pp 857–990

[82] A good overview of the recent status is given in:

Bartlett R J 1995 Coupled-cluster theory: an overview of recent developments *Modern Electronic Structure Theory* vol 2, ed D R Yarkony (Singapore: World Scientific) pp 1047–131

[83] Helgaker T, Gauss J, Jørgensen P and Olsen J 1997 The prediction of molecular equilibrium structures by the standard electronic wave functions *J. Chem. Phys.* **106** 6430–40

for a listing and for further details on this study

[84] Two review papers that introduce and compare the myriad of semi-empirical methods:

Stewart J J P 1991 Semiempirical molecular orbital methods *Reviews in Computational Chemistry* vol 1, ed K B Lipkowitz and D B Boyd (New York: VCH) pp 45–81

Zerner M C 1991 Semiempirical molecular orbital methods *Reviews in Computational Chemistry* vol 2, ed K B Lipkowitz and D B Boyd (New York: VCH) 313–65

A very recent overview, including efforts to interface semi-empirical electronic structure with molecular mechanics treatments of some degrees of freedom is given by:

Thiel W 1996 Perspectives on semiempirical molecular orbital theory *New Methods in Computational Quantum Mechanics (Adv. Chem. Phys. XCIII)* ed I Prigogine I and S A Rice (New York: Wiley) pp 703–57

Earlier texts dealing with semi-empirical methods include:

Pople J A and Beveridge D L 1970 *Approximate Molecular Orbital Theory* (New York: McGraw-Hill)

Murrell J N, Kettle S F A and Tedder J M 1965 *Valence Theory* 2nd edn (London: Wiley)

# B3.2
# Quantum structural methods for the solid state and surfaces

*Frank Starrost and Emily A Carter*

### B3.2.1 Introduction

We are entering an era when condensed matter chemistry and physics can be predicted from theory with increasing realism and accuracy. This is particularly important in cases where experiments lead to ambiguous conclusions, for regimes in which there still exists no experimental probe and for predictions of the properties of modern materials in order to select the most promising ones for synthesis and experimental testing. For example, continuing miniaturization in microelectronics heightens the importance of understanding of quantum effects, which computational materials theory is poised to provide, based to some degree on the methods presented here.

Our intention is to give a brief survey of advanced theoretical methods used to determine the electronic and geometric structure of solids and surfaces. The electronic structure encompasses the energies and wavefunctions (and other properties derived from them) of the electronic states in solids, while the geometric structure refers to the equilibrium atomic positions. Quantities that can be derived from the electronic structure calculations include the electronic (electron energies, charge densities), vibrational (phonon spectra), structural (lattice constants, equilibrium structures), mechanical (bulk moduli, elastic constants) and optical (absorption, transmission) properties of crystals. We will also report on techniques used to study solid surfaces, with particular examples drawn from chemisorption on transition metal surfaces.

In his chapter on the fundamentals of quantum mechanics of condensed phases (A1.3), James R Chelikowsky introduces the plane wave pseudopotential method. Here, we will complement his chapter by introducing in some detail tight-binding methods as the simplest pedagogical illustration of how one can construct crystal wavefunctions from atomic-like orbitals. These techniques are very fast but generally not very accurate. After reviewing some of the efforts made to improve upon the local density approximation (LDA, explained in A1.3), we will discuss general features of the technically more complex all-electron band structure methods, focusing on the highly accurate but not very fast linear augmented plane wave (LAPW) technique as an example. We will introduce the idea of orbital-free electronic structure methods based directly on density functional theory (DFT), the computational effort of which scales linearly with size, allowing very large systems to be studied. The periodic Hartree–Fock (HF) method and the promising quantum Monte Carlo (QMC) techniques will be briefly sketched, representing many-particle approaches to the condensed phase electronic structure problem.

In the final section, we will survey the different theoretical approaches for the treatment of adsorbed molecules on surfaces, taking the chemisorption on transition metal surfaces, a particularly difficult to treat yet extremely relevant surface problem [1], as an example. While solid state approaches such as DFT are often used, hybrid methods are also advantageous. Of particular importance in this area is the idea of embedding, where a small cluster of surface atoms around the adsorbate is treated with more care than the surrounding region. The advantages and disadvantages of the approaches are discussed.

### B3.2.2  Tight-binding methods

#### B3.2.2.1  Tight binding: from empirical to self-consistent

The wavefunction in a solid can be thought to originate from two different limiting cases. One extreme is the nearly free electron (NFE) approach. The idea here is that the valence electrons are hardly affected by the periodic potential of the atomic cores. Their wavefunctions can then be assumed to be easily described as linear combinations of the solutions for free electrons: the plane waves, $\exp(i\mathbf{k} \cdot \mathbf{r})$. The NFE approximation is particularly useful for so-called NFE metals, such as the alkali metals. At the other extreme, the solid can be viewed as constructed from individual atoms. The valence wavefunctions of the solid are then approximated as linear combinations of the wavefunctions of the valence electrons of the atoms (see also section A1.3.5.6). In this case, the electrons are considered to be 'tightly bound' to the atoms. This is a physically reasonable view of covalently bound solids and molecules, where localized chemical bonds are the norm (bulk silicon, organic or biomolecules etc). Methods which employ this view of the electrons in the solid are called tight-binding (TB) methods. The wavefunctions are generally expanded in atomic orbitals (in a linear combination of atomic orbitals (LCAO) formalism) or similarly localized functions.

An advantage of TB is that generally the number of basis functions linearly combined to give the wavefunctions is rather small. The solution of the Schrödinger equation in these bases is then fast because the matrices representing the operators are small. Also, the construction of the Hamiltonian matrix elements is fast, since generally a number of, sometimes drastic, approximations are made. At the same time, however, the small basis set generally limits the quality of the TB results, since the variational freedom for the solution of the Schrödinger equation is not as high as in other methods. The approximations of Hamiltonian matrix elements often further reduce the quality of the results.

Today, the term TB method is generally understood to refer to a technique using TB basis functions in which the Hamiltonian matrix elements are adjusted to reproduce results from experiments and/or from more sophisticated electronic structure methods [2]. Depending on the degree of dependence on external parameters, the methods are called empirical or semi-empirical TB. A number of approaches are used for the fitting of the TB parameters, generally a tough minimization task with many minima (using genetic algorithms has proved quite efficient [3, 4]). It has been noted that 'great care is needed to test the resulting model for reasonable behavior outside the range of the fit' [5, 6]. A disadvantage of the empirical methods is that it is difficult to distinguish to what extent the parametrization or the method itself is responsible for errors in the results.

Frequent approximations made in TB techniques in the name of achieving a fast method are the use of a minimal basis set, the lack of a self-consistent charge density, the fitting of matrix elements of the potential, the assumption of an orthogonal overlap matrix, a cut-off radius used in the integration to determine matrix elements, and the neglect of matrix elements that require three-centre integrals and crystal-field terms. We will now provide more details on these approximations.

Generally, the following *ansatz* for the wavefunction is made:

$$\psi_i(\mathbf{r}) = \sum_{\alpha l} c_{\alpha l}^i \varphi_{\alpha l}(\mathbf{r}),$$

where $\varphi_{\alpha l}(\mathbf{r}) = \langle r|\varphi_{\alpha l}\rangle$ represents an atomic orbital of symmetry $\alpha$ (such as s, $p_x$, $p_y$, $p_z$) at atom $l$.

This yields the generalized eigenvalue problem

$$\underline{H}c^i = \epsilon_i \underline{S}c^i, \tag{B3.2.1}$$

with the elements of the Hamiltonian matrix $H_{\alpha l\beta m} \equiv \langle \varphi_{\alpha l} | H | \varphi_{\beta m} \rangle$ and the overlap matrix $S_{\alpha l\beta m} \equiv \langle \varphi_{\alpha l} | \varphi_{\beta m} \rangle$. In the TB approximation, the basis functions are thought to be sufficiently localized such that contributions to the Hamiltonian matrix usually are accounted for only up to at most the third or fourth neighbour. Frequently a minimal basis set is used, i.e. a single orbital $\varphi_{\alpha l}$ is used per atom and per orbital symmetry to expand the wavefunction.

In orthogonal TB methods, the overlap matrix is assumed to be diagonal, even though the basis functions of adjacent sites ordinarily are not orthogonal [6]. Harrison has shown that this approximation can be compensated for by adjustments to the Hamiltonian matrix elements (these adjustments are arrived at automatically in methods depending on fitting, for example, a DFT band structure) [7]. However, this approach reduces the transferability of the TB parameters to other structures [8]. Including the overlap matrix brings with it the additional cost of its calculation and solving the generalized eigenvalue problem, see equation (B3.2.1), rather than an ordinary eigenvalue problem.

One can construct an effective potential, written here in the DFT language (see, for example, equation A1.3.38 of A1.3) as

$$v_{\text{eff}}(\boldsymbol{r}) = v_{\text{ext}}(\boldsymbol{r}) + v_{\text{H}}[\rho(\boldsymbol{r})] + v_{\text{xc}}[\rho(\boldsymbol{r})]. \qquad (B3.2.2)$$

To rationalize the 'two-centre approximation', the effective potential is written as

$$v_{\text{eff}}(\boldsymbol{r}) = \sum_l v_{\text{eff},l}(|\boldsymbol{r} - \boldsymbol{R}_l|),$$

where $v_{\text{eff},l}$ is centred on the atom $l$ and vanishes away from the atom, which need not involve any approximation.

In the calculation of the elements

$$H_{\alpha l\beta m} = \left\langle \varphi_{\alpha l} \middle| T + \sum_n v_{\text{eff},n} \middle| \varphi_{\beta m} \right\rangle$$

with $T = -\frac{1}{2}\nabla^2$ the kinetic energy operator, several types of potential matrix elements can be distinguished [6]:

(1) Three-centre terms, i.e. $l \neq m \neq n$. These are frequently neglected, in what is called the two-centre approximation, based on the assumed strong localization of the orbitals $\varphi_{\alpha l}(\boldsymbol{r})$.

(2) Inter-atomic two-centre matrix elements $\langle \varphi_{\alpha l} | v_{\text{eff},l} + v_{\text{eff},m} | \varphi_{\beta m} \rangle$. These matrix elements represent the hopping of electrons from one site to another. They can be described [7] as linear combinations of so-called Slater–Koster elements [9]. The coefficients depend only on the orientation of the atoms $l$ and $m$ in the crystal. For elementary metals described with s, p, and d basis functions there are ten independent Slater–Koster elements. In the traditional formulation, the orientation is neglected and the two-centre elements depend only on the distance between the atoms [6]. (In several models [6, 10], they have been made dependent on the environment of the atoms $l$ and $m$.) These elements are generally fitted to reproduce DFT results such as the band structure or the values of DFT matrix elements in diatomics.

(3) Intra-atomic matrix elements, or on-site terms, with $l = m$. Traditionally, the potential contributions from other atomic sites, $v_{\text{eff},n\neq l=m}$, so-called crystal-field terms, are neglected [10]. In this case, then the only non-zero on-site terms have $\alpha = \beta$, since basis functions on the same site are orthogonal atomic orbitals. There are methods which include these crystal-field terms [11, 12]. Physically, these diagonal elements represent the energy required to place an electron in a specific orbital. In some implementations, they are set to the orbital energy values of the neutral free atom [13], guaranteeing the correct limit for isolated atoms. However, this approach ignores the potential contributions to the diagonal elements due to different environments in a molecule or crystal; these are taken into account in other variants of the method [6, 10, 11].

Most TB approaches are not charge self-consistent. This means that they do not ensure that the charge derived from the wavefunctions yields the effective potential $v_{eff}$ assumed in their calculation. Some methods have been developed which yield charge densities consistent with the electronic potential [14–16].

The localized nature of the atomic basis set makes it possible to implement a linear-scaling TB algorithm, i.e. a TB method that scales linearly with the number of electrons simulated [17]. (For more information on linear scaling methods, see section B3.2.3.3.)

The accuracy of most TB schemes is rather low, although some implementations may reach the accuracy of more advanced self-consistent LCAO methods (for examples of the latter see [18–20]). However, the advantages of TB are that it is fast, provides at least approximate electronic properties and can be used for quite large systems (e.g., thousands of atoms), unlike some of the more accurate condensed matter methods. TB results can also be used as input to determine other properties (e.g., photoemission spectra) for which high accuracy is not essential.

### B3.2.2.2   Applications of tight-binding methods

TB methods have been widely used to study properties of simple semiconductors such as Si [11] and GaN [16]. In the latter study, the effect of dislocations on the electronic structure of GaN was investigated with a view toward understanding how dislocations affect the material's optical properties. The large supercell of 224 atoms led to TB as the method of choice. This particular variant of TB fits TB matrix elements to DFT-LDA results and solves self-consistently for atomic charges. It has also been used to predict reaction energetics of organic molecules, the structure of large biomolecules and the surface geometry and band structure of III–V semiconductors [15]. The TB method is expected to provide qualitatively reasonable results for systems where localized atomic charges make sense and hence is not expected to perform as well for metallic systems. Despite potential problems of TB for metals, the TB approach has also been used to study the phonon spectrum of the transition metal molybdenum [6], the elastic constants, vacancies and surfaces of monatomic transition and noble metals and the Hall coefficient of complex perovskite crystals [10]. As an example of data available from a TB calculation, a TB variant of extended Hückel theory [21, 22] was used to describe the initial states in photoemission from GaN [23]. The parameters were fitted to the bulk band structure $E_n(k)$ (for a definition, see section A1.3.6). As displayed in figure B3.2.1, good agreement is found for the occupied states (negative energies), while larger differences for the conduction bands (positive energies) reveal a typical problem of the TB methods: they are far less capable of describing the delocalized conduction band states (the same is true for delocalized valence states in a metal, as mentioned above). In figure B3.2.2, we show a series of calculated photoemission spectra compared to experimental results [23]. The dispersion of the main peaks as a function of emission angle and photon energy agrees reasonably well in theory and experiment.

### B3.2.3   First-principles electronic structure methods

In this section, we briefly review the basic elements of DFT and the LDA. We then focus on improvements suggested to remedy some of the shortcomings of the LDA (see section B3.2.3.1). A wide variety of techniques based on DFT have been developed to calculate the electron density. Many approaches do not calculate the density directly but rather solve for either a set of single-electron orbitals, or the Green's function, from which the density is derived.

In section B3.2.3.2, we introduce a number of techniques commonly referred to as *ab initio* all-electron electronic structure methods. *Ab initio* methods, in particular, aim at calculating the energies of electrons and their wavefunctions as accurately as possible, introducing as few adjustable parameters as possible. (Empirical or semi-empirical methods include the empirical pseudopotential approach (see section A1.3.5.5) and many TB techniques (see section B3.2.2).) Within the *ab initio* band structure approach, two communities exist that differ in their treatment of the singular nature of realistic, Coulomb-like crystal potentials. In the

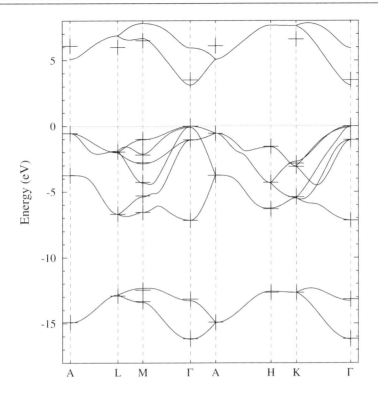

**Figure B3.2.1.** The band structure of hexagonal GaN, calculated using EHT-TB parameters determined by a genetic algorithm [23]. The target energies are indicated by crosses. The target band structure has been calculated with an *ab initio* pseudopotential method using a quasiparticle approach to include many-particle corrections [194].

pseudopotential approach discussed by Chelikowsky in chapter A1.3, the Coulomb singularity $(-Z/r)$ of the crystal potential is replaced by a smoother function, whereas in the so-called 'all-electron' approach, the Coulomb singularity is retained. The pseudopotential transformation limits the range of electron energies which can be accessed. However, since the pseudo-wavefunction is much smoother than the all-electron wavefunction (which has large oscillations near the nucleus), the pseudopotential allows the use of a plane wave basis set, which is comparatively easy to handle. In principle, the all-electron methods have no limitation on the energy range of calculations. This is achieved by a sophisticated representation of the wavefunction.

The so-called orbital-free DFT technique, which aims to directly calculate the electron density for which the total energy is minimal, is presented as an example of methods whose computational effort scales linearly with system size (see section B3.2.3.3). In section B3.2.3.4, we discuss the periodic HF method, an alternative approach to DFT that offers a well defined starting point for many-particle corrections. Finally, the two most frequently used QMC techniques are described in section B3.2.3.5.

*B3.2.3.1   The local density approximation and beyond*

In DFT, the electronic density rather than the wavefunction is the basic variable. Hohenberg and Kohn showed [24] that all the observable ground-state properties of a system of interacting electrons moving in an external potential $v_{ext}(r)$ are uniquely dependent on the charge density $\rho(r)$ that minimizes the system's total energy. However, there is no known formula to calculate from the density the total energy of many electrons moving in a general potential. Hohenberg and Kohn proved that there exists a universal functional of the density,

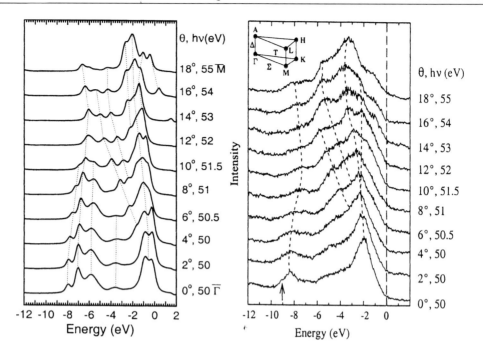

**Figure B3.2.2.** A series of photoemission spectra. The angles give the polar angle of electron emission at the stated photon energy scanning the surface Brillouin zone from $\bar{\Gamma}$ to $\bar{M}$. Left: A calculation using the tight-binding parametrization (given the band structure in figure B3.2.1) for the initial states [23]. Right: Experimental spectra by Dhesi *et al* [195]. The difference in binding energies is due to the experimental difficulty in determining the Fermi energy [23]. (Experimental figure by Professor K E Smith.)

called $G[\rho]$, such that the expression

$$E[\rho] = \int v_{\text{ext}}(r)\rho(r)\,\mathrm{d}^3r + \frac{1}{2}\int \frac{\rho(r)\rho(r')}{|r - r'|}\mathrm{d}^3r\,\mathrm{d}^3r' + G[\rho] \qquad \text{(B3.2.3)}$$

has as its minimum value the correct ground-state energy associated with $v_{\text{ext}}(r)$. Here, the first term on the right-hand side represents the energy due to an external potential, including the electron–nuclear potential, while the second term is the classical Coulomb energy of the electronic system. The functional $G[\rho]$ is valid for any number of electrons and any external potential, but it is unknown and further steps are necessary to approximate it.

Kohn and Sham [25] decompose $G[\rho]$ into the kinetic energy of an analogous set of non-interacting electrons with the same density $\rho(r)$ as the interacting system,

$$T_s[\rho] = \sum_i \left\langle \psi_i \left| -\frac{1}{2}\nabla^2 \right| \psi_i \right\rangle$$

(where $\psi_i(r) = \langle r|\psi_i\rangle$ is the wavefunction of electron $i$), and the exchange and correlation energy of an interacting system with density $\rho(r)$, $E_{\text{xc}}[\rho]$. The functional $E_{\text{xc}}[\rho]$ is not known exactly. Physically, it represents all the energy corrections beyond the Hartree term to the independent-particle model, i.e. the non-classical many-body effects of exchange and correlation (xc) and the difference between the kinetic energy of the interacting electron system $T[\rho]$ and the analogous non-interacting system $T_s[\rho]$.

In the LDA, the exchange and correlation energy is approximated using the exchange and correlation energy of the homogeneous electron gas at the same density (see section A1.3.3.3). The crystal density is obtained by solving the single-particle Kohn–Sham equation

$$\left(-\frac{1}{2}\nabla^2 + v_{\text{eff}}(\boldsymbol{r})\right)\psi_i(\boldsymbol{r}) = E_i\psi_i(\boldsymbol{r}), \tag{B3.2.4}$$

for a self-consistent potential $v_{\text{eff}}$, i.e. a potential which is produced by the density $\rho$. In bulk crystal calculations, the index $i$ runs over both the Bloch vector $\boldsymbol{k}$ (see section A1.3.4) and the band index $n$ (in a simple crystal, this band could be derived, for example, entirely from s states). The solutions to equation (B3.2.4) are often called Kohn–Sham orbitals. The crystal density is then

$$\rho(\boldsymbol{r}) = \sum_i \psi_i^*(\boldsymbol{r})\psi_i(\boldsymbol{r}).$$

The eigenenergy $E_i$ can be defined as the derivative of the total energy of the many-electron system with respect to the occupation number of a specific orbital [26]. In HF theory (where equation (B3.2.4) applies and the $v_{\text{eff}}$ contains a non-local exchange operator, see section A1.3.1.2 and chapter B3.1), Koopmans' theorem states that the single-particle eigenvalue is the negative of the ionization energy (neglecting the relaxation of the electronic system). In contrast, the identification of the highest occupied Kohn–Sham eigenvalue with the negative of the ionization energy is a controversial subject [27]. While there is no rigorous connection between eigenvalue differences and excitation energies in either HF or DF theory, comparisons of these values are common practice (see below for more appropriate methods). Relative differences among occupied single-particle energies often agree well with the experiment. Even though DFT only provides a solution for the ground state of the electronic system, the energy differences in the lower conduction bands, i.e. low-energy excited states, often are represented surprisingly well, too. However, in LDA calculations of semiconductors and insulators, almost always the size of the gap between the valence band maximum and the conduction band minimum is underestimated, since many-particle effects are incorrectly represented by the parametrized exchange-correlation energy (see, for example, [28]). One *ad hoc* remedy, which works well for many systems and which is employed in the examples presented here, is to use what is amusingly referred to as a *scissor operator*, i.e. a rigid shift, to correct the gap size [29, 30]. Typically the shift is determined by knowing, for example, the DFT error in predicting the measured optical band gap. The entire conduction band is shifted rigidly upward by the amount to match the experimental band gap.

More advanced techniques take into account quasiparticle corrections to the DFT-LDA eigenvalues. Quasiparticles are a way of conceptualizing the elementary excitations in electronic systems. They can be determined in band structure calculations that properly include the effects of exchange and correlation. In the LDA, these effects are modelled by the exchange–correlation potential $v_{\text{xc}}^{\text{LDA}}$. In order to more accurately account for the interaction between a particle and the rest of the system, the notion of a local potential has to be generalized and a non-local, complex and energy-dependent exchange–correlation potential has to be introduced, referred to as the self-energy operator $\Sigma(\boldsymbol{r}, \boldsymbol{r}'; E)$. The self-energy can be expanded in terms of the screened Coulomb potential $W$, where $W = \epsilon^{-1}v$ is the Coulomb interaction $v$ screened by the inverse dielectric function $\epsilon^{-1}$. In a lowest order expansion in $W$, the self-energy can be approximated as $\Sigma = GW$, giving the GW approximation [31]. Here $G$ is the one-electron Green's function describing the propagation of an additional electron injected into a system of other electrons (it can also describe the extraction of an electron).

To be a bit more explicit (following [32, 33]), the quasiparticle energies and wavefunctions are given by

$$(T + v_{\text{ext}} + v_{\text{H}})\psi_{nk}(\boldsymbol{r}) + \int \mathrm{d}\boldsymbol{r}'\Sigma(\boldsymbol{r}, \boldsymbol{r}'; E_{nk})\psi_{nk}(\boldsymbol{r}') = E_{nk}\psi_{nk}(\boldsymbol{r}),$$

where $T$ is the kinetic energy operator, $v_{\text{ext}}$ is the external potential due to the ions, and $v_H$ is the Hartree Coulomb interaction. Since the self-energy operator in general is non-Hermitian, the quasiparticle energies $E_{nk}$ are complex in general, and the imaginary part gives the lifetime of the quasiparticle. To first order in $W$, the self-energy is then given by

$$\Sigma(r, r'; E) = \frac{i}{2\pi} \int d\omega e^{-i\delta\omega} G(r, r'; E - \omega) W(r, r'; \omega)$$

where $\delta$ is a positive infinitesimal and $\omega$ corresponds to an excitation frequency. The inputs are the full interacting Green's function,

$$G(r, r'; E) = \sum_{nk} \frac{\psi_{nk}(r)\psi_{nk}^*(r')}{E - E_{nk} - i\delta_{nk}},$$

where $\delta_{nk}$ is an infinitesimal and the dynamically screened Coulomb interaction,

$$W(r, r'; \omega) = \Omega^{-1} \int dr'' \epsilon^{-1}(r, r''; \omega) v(r'' - r'),$$

where $\epsilon^{-1}$ is the inverse dielectric matrix, $v(r) = 1/|r|$ and $\Omega$ is the volume of the system. Usually the calculations start with the construction of the Green's function and the screened Coulomb potential from self-consistent LDA results. The self-energy $\Sigma$ then has to be obtained together with $G$ in a self-consistent procedure. However, due to the severe computational cost of this procedure, it is usually not carried out (see, for example, [34]). Instead, it is common practice to construct the self-energy operator non-self-consistently using the self-consistent LDA results to determine quasiparticle corrections to the LDA energies, resulting in the quasiparticle band structure. The GW approximation has been applied to a wide range of metals, semiconductors and insulators, where it has been found to lead to striking improvements in the agreement of optical excitation spectra with the experiment (see, for example [32, 35–37]). Recent studies also found that the GW charge density is close to the experiment for diamond structure semiconductors [38], and lifetimes of low-energy electrons in metals have been calculated [39].

Another disadvantage of the LDA is that the Hartree Coulomb potential includes interactions of each electron with itself, and the spurious term is not cancelled exactly by the LDA self-exchange energy, in contrast to the HF method (see A1.3), where the self-interaction is cancelled exactly. Perdew and Zunger proposed methods to evaluate the self-interaction correction (SIC) for any energy density functional [40]. However, full SIC calculations for solids are extremely complicated (see, for example [41–43]). As an alternative to the very expensive GW calculations, Pollmann *et al* have developed a pseudopotential built with self-interaction and relaxation corrections (SIRC) [44]. The pseudopotential is derived from an all-electron SIC-LDA atomic potential. The relaxation correction takes into account the relaxation of the electronic system upon the excitation of an electron [44]. The authors speculate that '. . .the ability of the SIRC potential to produce considerably better band structures than DFT-LDA may reflect an extra nonlocality in the SIRC pseudopotential, related to the nonlocality or orbital dependence in the SIC all-electron potential. In addition, it may mimic some of the energy and the non-local space dependence of the self-energy operator occurring in the GW approximation of the electronic many body problem' [45].

The LDA also fails for strongly correlated electronic systems. Examples of such systems are the late 3d transition-metal mono-oxides MnO, FeO, CoO, and NiO. Within the local spin density approximation (LSDA), the energy gaps calculated for MnO and NiO are too small [46] and, even worse, FeO and CoO are predicted to be metallic, whereas experimentally they have been found to be large-gap insulators. While the GW approximation yields an energy gap of NiO in reasonable agreement with experiment [47], the computational cost of this procedure is very high. The SIC-LDA method reproduces quite well the strong localization of the d electrons in transition metal compounds, but the orbital energies obtained by SIC are usually in strong disagreement with experimental results (for transition metal oxides, for example, occupied d bands are approximately $\frac{1}{2}$ Hartree below the oxygen valence band—a separation not seen in spectroscopic data: see, for example, the experimental results in [48]) [49]. An alternative solution to this problem is offered by the LDA+$U$ method [49, 50], where LDA encompasses the LSDA. In the LDA+$U$ technique, the electrons are divided into two subsystems which are treated separately: the strongly localized (d or f) electrons and the delocalized s and p electrons. The latter are treated by standard LDA. The on-site interactions among the strongly localized electrons on each atom, however, are taken into account by a term $\frac{1}{2}U\sum_{i\neq j}n_in_j$, where $n_i$ are the occupation numbers of the strongly localized orbitals and $U$ is the Coulomb interaction parameter (for details on the first-principles calculation of $U$, see [51]). At least for localized d or f states, the LDA+$U$ technique may be viewed as an approximation to the GW approximation [49]. Band gaps, valence band widths and magnetic moments have been calculated with LDA+$U$ that agree with experiment for a variety of transition metal compounds [49, 52], among other applications.

### B3.2.3.2   All-electron DFT methods

#### (a) Introduction

When the highest accuracy is sought for the electronic and geometric properties of crystals, all the electrons of the atoms in the crystal and the full Coulomb singularity of the nuclear potential must be accounted for. All-electron approaches, which do just that, generally cannot compete with pseudopotential techniques in speed and simplicity of algorithm. However, the latter suffer from severe drawbacks when it comes to the construction of the very pseudopotentials these methods depend upon: even for so-called *ab initio* potentials, the pseudopotentials are far from uniquely determined. Additionally, problems with transferability and the construction of potentials for such elements as the transition metals remain. All-electron techniques can deal with any element and there are no worries about transferability of the potential. However, the accuracy comes at a price: due to the Coulomb singularity of the potential at the nuclear positions, the wavefunctions are highly oscillatory close to the nucleus. For those all-electron methods that use wavefunctions to represent the electrons (a Green's function method, for example, does not), this means that a simple plane wave basis set cannot be used for the expansion of the wavefunctions. To reach convergence of a plane wave $\exp(i\mathbf{k}\cdot\mathbf{r})$ expansion would require a prohibitive number of basis functions. Thus, specialized basis sets have been invented for all-electron calculations.

We now discuss the most important theoretical methods developed thus far: the augmented plane wave (APW) and the Korringa–Kohn–Rostoker (KKR) methods, as well as the linear methods (linear APW (LAPW), the linear muffin-tin orbital [LMTO] and the projector-augmented wave [PAW]) methods.

In the early all-electron techniques, the crystal was separated into spheres around the atoms, so-called 'muffin-tin' spheres, and the interstitial region in between. Inside the spheres, the potential was approximated as spherically symmetric, while in the interstitial region it was assumed to be constant. This shape approximation of the potential is reasonable for close-packed crystals such as hexagonal close-packed metals, where the spheres cover a large fraction of the crystal volume. However, in less densely arranged crystals, such as diamond structure semiconductors (see figure A1.3.4), the muffin-tin approximation leads to large errors. In the diamond and the related zincblende structures, only 34% of the volume is covered by touching muffin-tin spheres (figure B3.2.3). For all of the all-electron methods, versions have been developed that are not restricted to shape approximations of the potential. These techniques are referred to as general, or full, potential methods.

#### (b) The augmented plane wave method

The APW technique was proposed by Slater in 1937 [53, 54]. It remains the most accurate of the band structure methods for the muffin-tin approximation of the potential. The wavefunction is expanded in basis functions $\varphi_i(\mathbf{k}+\mathbf{G}_i, E, \mathbf{r})$, the APWs, each of which is identical to the plane wave $\exp(i(\mathbf{k}+\mathbf{G}_i)\cdot\mathbf{r})$ in the interstitial

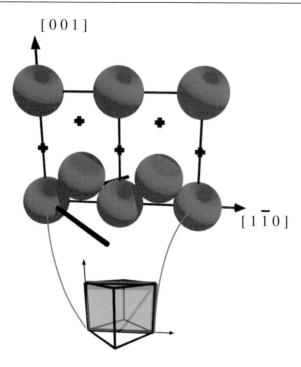

**Figure B3.2.3.** The muffin-tin spheres in the (110) plane of a zincblende crystal. The nuclei are surrounded by spheres of equal size, covering about 34% of the crystal volume. Unoccupied tetrahedral positions are indicated by crosses. The conventional unit cell is shown at the bottom; the crystal directions are noted.

region, where $G_i$ are the reciprocal lattice vectors (see section A1.3.4). The plane waves are augmented, i.e. they are joined continuously at the surface of the spheres by solutions of the radial Schrödinger equation. This means that in the spherical harmonic expansion of a plane wave around the centre of a muffin-tin sphere, the respective Bessel function inside the sphere is replaced by a solution $\phi_{li}(r, E)$ of the radial Schrödinger equation for a given energy. The radial function matches the Bessel function, $j_l(|k + G_i|r)$, value at the sphere boundary and must be regular (non-singular) at the origin. With the basis functions $\varphi_i(k + G_i, E, r)$, a variational solution is sought to the Kohn–Sham equation, equation (B3.2.4). Since the Hamiltonian matrix elements now depend nonlinearly upon the energy due to the energy-dependent basis functions, the resulting secular equation is solved by finding the roots of the determinant of the $\underline{H}(E) - E\underline{S}(E)$ matrix. (The problem cannot be treated by the eigenvalue routines of linear algebra.)

Numerically, the determination of the roots can be difficult because the determinant's value may change by several orders of magnitude when the energy $E$ is changed by only a few meV. Another difficulty can result at degenerate roots where the value of the determinant does not change sign. Additionally, the secular equation becomes singular when a node of the radial solution falls at the muffin-tin sphere boundary (the so-called 'asymptote problem').

Physically, the APW basis functions are problematic as they are not smooth at the sphere boundary, i.e., they have discontinuous slope. While in a fully converged solution of the secular equation, this discontinuity should disappear, alternative methods have been sought instead. Following a suggestion by Marcus [55] in 1967, the LAPW provided a way to avoid the above-mentioned drawbacks of the APW technique, as we now discuss.

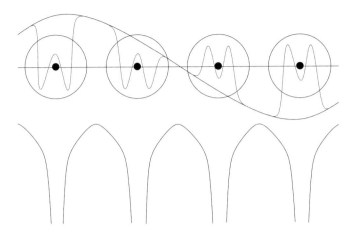

**Figure B3.2.4.** A schematic illustration of an energy-independent augmented plane wave basis function used in the LAPW method. The black sine function represents the plane wave, the localized oscillations represent the augmentation of the function inside the atomic spheres used for the solution of the Schrödinger equation. The nuclei are represented by filled black circles. In the lower part of the picture, the crystal potential is sketched.

*(c) The linear augmented plane wave method*

The main disadvantage of the APW technique is that it leads to a nonlinear secular problem because the basis functions depend on the energy. A number of attempts have been made to construct linear versions of the APW approach by introducing energy-independent basis functions in different ways. In 1970, Koelling invented the *alternative* APW [56] and Bross the *modified* APW [57]. In 1975, Andersen constructed the LAPW [58] formalism, which today is the most popular APW-like band structure method. Further extensions of the linear methods appeared in the early 1990s: Singh developed the LAPW *plus localized orbitals* (LAPW+LO [59]) in 1991 and Krasovskii the *extended* LAPW (ELAPW [60]) in 1994. Recently the APW+LO technique has been implemented by Sjöstedt and Nordström [61] according to an idea by Singh. While the LAPW technique is generally used in combination with DFT approaches, it has also been applied based on the LDA+$U$ [62] and HF theories [63].

The LAPW method, as suggested in 1975 [58, 64], avoids the problem of the energy dependence of the Hamiltonian matrix by introducing energy-independent APW basis functions. Here, too, the APWs are derived from plane waves by augmentation: Bessel functions $j_l(|\boldsymbol{k} + \boldsymbol{G}_i|r)$ in the Rayleigh decomposition inside the muffin-tin sphere are replaced by functions $u_{li}(r)$ derived from the spherical potential, which are *independent* of the energy of the state that is sought and that match the Bessel functions at the sphere radius in value and in slope (see figure B3.2.4). The plane wave part of the basis remains the same but the energy-independent APWs allow the energies and the wavefunctions to be determined by solving a standard generalized eigenvalue problem.

In linearizing the APW problem as it is done in the LAPW method, the variational freedom of the APW basis set is reduced. The reason is that the wavefunction inside the spheres is rigidly coupled to its plane wave expansion in the interstitial region [65]. This means that the method cannot yield an accurate wavefunction even if the eigenvalue is within a few eV of the chosen energy parameters [66]. Flexibility is defined in this context as the possibility to change the wavefunction inside the spheres independently from the wavefunction in the interstitial region. Flexibility can be achieved in the linear band structure methods by adding basis functions localized inside the spheres whose value and slope vanish at the sphere boundary [54, 67, 68]. A 'flexible' basis set extending the LAPW with localized functions is preferable to the one used in the pure

LAPW technique. Flexible linear methods are the MAPW, the LAPW+LO and the ELAPW, the latter of which provides a necessary degree of flexibility with a minimal number of basis functions [65].

The additional functions increase the matrix dimension slightly and thus the computational effort. However, the increased flexibility of the basis set makes possible a number of extensions of the LAPW method. One is a $k \cdot p$ formulation of the ELAPW method [68], which would lead to large errors in the regular LAPW due to its lesser flexibility. The augmented Fourier components (AFC) technique [69] for treating a general potential is based on this. The AFC method is an alternative to the full-potential LAPW (FLAPW) method [70, 71]. (Recently progress has been made in increasing the computational efficiency of the FLAPW method [72].) The AFC method does not have the same demanding convergence criteria as the FLAPW method but yields physically equivalent results [69].

The general potential LAPW techniques are generally acknowledged to represent the state of the art with respect to accuracy in condensed matter electronic-structure calculations (see, for example, [62, 73]). These methods can provide the best possible answer within DFT with regard to energies and wavefunctions.

### (d) The Korringa–Kohn–Rostoker technique

The KKR method uses multiple-scattering theory to solve the Kohn–Sham equations [74, 75]. Rather than calculate the wavefunction, modern incarnations calculate the Green's function $G$. The Green's function is the solution to the equation schematically given by $(H - E)G(E) = -\delta$, where $H$ is the Hamiltonian, $E$ the single-electron energy and $\delta$ the delta function $\delta(r - r')$. The properties of the system, such as the electron density, the density of states and the total energy can be derived from the Green's function [73]. The crystal is represented as a sum of non-overlapping potentials; in the modern version, there are no shape approximations, i.e. the potentials are space-filling [76]. Within the multiple-scattering formalism, the wavefunction is built up by taking into account the scattering and rescattering of a free-electron wavefunction by scatterers. The scatterers are (generally) the atoms of the crystal and the single-scattering properties (the properties of the isolated scatterer) are derived from the effective, singular potentials of the atoms (given in equation (B3.2.2)). The Green's matrix is then constructed from the knowledge of the scattering properties of the single scatterers and the analytically known Green's function of the free electron. The full-potential KKR method has been shown to have the same level of accuracy as the full-potential LAPW method [73]. The Green's function formulation offers the advantage of easy inclusion of defects in the bulk or clean surfaces. Such calculations start with the Green's function of the periodic crystal and include the perturbation through a Dyson equation [77]. Yussouff states that the difference in speeds between the linear methods and his 'fast' KKR technique is at most a factor of ten, in favour of the former [78]. While the KKR technique has an accuracy comparable to the APW method, it has the disadvantage of not being a linear approach, limiting speed and simplicity.

### (e) The linear muffin-tin orbital method

The LMTO method [58, 79] can be considered to be the linear version of the KKR technique. According to official LMTO historians, the method has now reached its 'third generation' [79]: the first starting with Andersen in 1975 [58], the second commonly known as TB-LMTO. In the LMTO approach, the wavefunction is expanded in a basis of so-called muffin-tin orbitals. These orbitals are adapted to the potential by constructing them from solutions of the radial Schrödinger equation so as to form a minimal basis set. Interstitial properties are represented by Hankel functions, which means that, in contrast to the LAPW technique, the orbitals are localized in real space. The small basis set makes the method fast computationally, yet at the same time it restricts the accuracy. The localization of the basis functions diminishes the quality of the description of the wavefunction in the interstitial region.

In the commonly used atomic sphere approximation (ASA) [79], the density and the potential of the crystal are approximated as spherically symmetric within overlapping muffin-tin spheres. Additionally, all integrals, such as for the Coulomb potential, are performed only over the spheres. The limits on the accuracy of the method imposed by the ASA can be overcome with the full-potential version of the LMTO (FP-LMTO) which gives highly accurate total energies [79, 80]. It was found that the FP-LMTO is 'at least as accurate as, and much faster than,' pseudopotential plane wave calculations in the determination of structural and dynamic properties of silicon [80]. The FP-LMTO is considerably slower than LMTO-ASA, however, and it has been found that ASA calculations can yield accurate results if the full expansion, rather than only the spherical part, of the charge is used in what is called a full-charge (rather than a full-potential) method and the integrals are performed exactly [73, 79].

The LMTO method is the fastest among the all-electron methods mentioned here due to the small basis size. The accuracy of the general potential technique can be high, but LAPW results remain the 'gold standard'.

## (f) The projector augmented wave technique

The projector augmented-wave (PAW) DFT method was invented by Blöchl to generalize both the pseudopotential and the LAPW DFT techniques [81]. PAW, however, provides all-electron one-particle wavefunctions not accessible with the pseudopotential approach. The central idea of the PAW is to express the all-electron quantities in terms of a pseudo-wavefunction (easily expanded in plane waves) term that describes interstitial contributions well, and one-centre corrections expanded in terms of atom-centred functions, that allow for the recovery of the all-electron quantities. The LAPW method is a special case of the PAW method and the pseudopotential formalism is obtained by an approximation. Comparisons of the PAW method to other all-electron methods show an accuracy similar to the FLAPW results and an efficiency comparable to plane wave pseudopotential calculations [82, 83]. PAW is also formulated to carry out DFT dynamics, where the forces on nuclei and wavefunctions are calculated from the PAW wavefunctions. (Another all-electron DFT molecular dynamics technique using a mixed-basis approach is applied in [84].)

PAW is a recent addition to the all-electron electronic structure methods whose accuracy appears to be similar to that of the general potential LAPW approach. The implementation of the molecular dynamics formalism enables easy structure optimization in this method.

## (g) Illustrative examples of the electronic and optical properties of modern materials

As an indication of the types of information gleaned from all-electron methods, we focus on one recent approach, the ELAPW method. It has been used to determine the band structure and optical properties over a wide energy range for a variety of crystal structures and chemical compositions ranging from elementary metals [60] to complex oxides [85], layered dichalcogenides [86, 87] and nanoporous semiconductors [88]. The $k \cdot p$ formulation has also enabled calculation of the complex band structure of the Al (100) surface [89].

As an illustration of the accuracy of the AFC ELAPW-$k \cdot p$ method, we present the dielectric function of GaAs. The dielectric function is a good gauge of the quality of a method, since not only do the energies enter the calculation, but also the wavefunctions *via* the matrix elements of the momentum operator $-i\nabla$. For the calculation of the dielectric function (equation (A1.3.87)) of GaAs, the conduction bands were rigidly shifted so that the highest peak agreed in both experiment and theory, a shift of 0.75 eV. The imaginary part of the dielectric function is shown in figure B3.2.5. Comparing the energy differences between the three peaks, we find that they agree to within 2 meV. For a wider comparison, we plot the results of two more experiments (which only have measured the two peaks at lower photon energy) and several all-electron calculations of the dielectric function of GaAs in figure B3.2.6. The FLAPW results agree almost exactly with the AFC ELAPW values. The discrepancies compared to the experimental results found for the other methods are considerably larger than for the general potential LAPW results, particularly for $E_1$.

A recent study of a class of nanoporous materials, the cetineites [88], offers further illustration of the possibilities offered by the modern band structure methods. The crystal is constructed of tubes of 0.7 nm

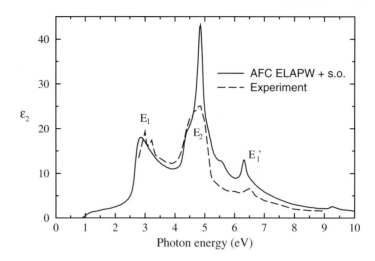

**Figure B3.2.5.** The imaginary part of the dielectric function of GaAs, according to the AFC ELAPW-$k \cdot p$ method (solid curve) [195] and the experiment (dashed curve) [196]. To correct for the band gap underestimated by the local density approximation, the conduction bands have been shifted so that the $E_2$ peaks agree in theory and experiment.

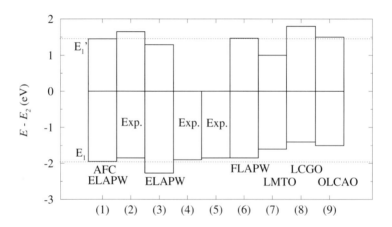

**Figure B3.2.6.** The energies of the $E_1$ and $E_1'$ peaks relative to the $E_2$ peak of the imaginary part of the dielectric function of GaAs, calculated by self-consistent DFT all-electron methods. These energies do not depend on the gap size. The theoretical methods are noted, as are experimental results obtained by ellipsometry (see chapter B1.26). The lower (upper) histogram gives the energy of peak $E_1$ ($E_1'$) relative to $E_2$. LCGO designates a linear-combination-of-Gaussian-orbitals method, OLCAO an orthogonalized linear-combination-of-atomic-orbitals approach. Sources: (1) [195], (2) [196], (3) [197], (4) [198], (5) [199], (6) [200], (7) [199], (8) [201], (9) [202].

diameter arranged in a two-dimensional hexagonal structure with 'flattened' $SbSe_3$ pyramids arranged between the tubes (see figure B3.2.7). Cetineites are of potential technological interest because, singularly among nanoporous materials, they are semiconductors rather than insulators. In figure B3.2.8, we show the comparison of the predicted density of states to the ultraviolet photoemission spectrum (PES, see chapter B1.1). The DOS can explain the two main structures in the PES at about $-3$ and $-12$ eV. Their relative intensities agree with those suggested by the DOS curve. Three structures in the DOS at $-1$, $-6$ and $-9$ eV

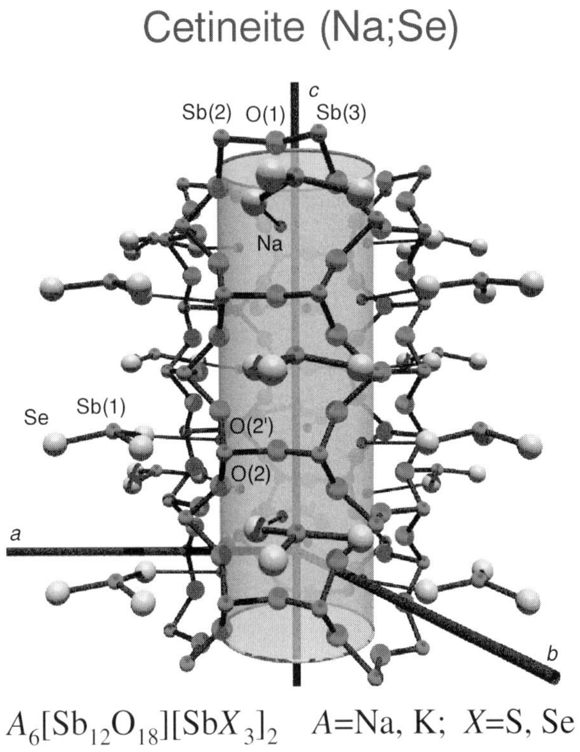

## Cetineite (Na;Se)

$$A_6[Sb_{12}O_{18}][SbX_3]_2 \quad A=Na, K; \ X=S, Se$$

**Figure B3.2.7.** A perspective view of the cetineite (Na;Se). The height of the figure is three lattice constants $c$. The shaded tube is included only as a guide to the eye. (From [88].)

are not resolved in the PES. This may be due to the selection rules of the photoemission process, not accounted for in the theory, or perhaps due to incomplete angle integration experimentally. The experimental results confirm, in particular, that the number of states is very high close to the valence band maximum. An orbital analysis shows that these states are derived mainly from the p states of the O and Se constituents of the crystal, with the chalcogen dominating near the top of the valence band. Electrons in the Se p states are thus most easily excited into the conduction band. This, together with their high DOS, makes the Se p states located on the pyramids the prime candidates for the initial states of the photoconductivity observed in the cetineites.

As another example of properties extracted from all-electron methods, figure B3.2.9 shows the results of a PAW simulation of benzene molecules on a graphite surface. The study aimed to show the extent to which the electronic structure of the molecule is modified by interaction with the surface, and why the images do not reflect the molecular structure. The PAW method was used to determine the structure of the molecule at the surface, the strength of the interaction between the surface and the molecules, and to predict and explain scanning tunnelling microscope (STM) images of the molecule on the surface [90] (the STM is described in section B1.19).

### B3.2.3.3 Linear-scaling electronic structure methods

DFT calculations such as the ones mentioned in chapter A1.3 and section B3.2.3.2 become computationally very expensive when the unit cell of the interesting system becomes large and complex, with certain parts of the computational algorithm typically scaling cubically with system size. A recent objective for treating large

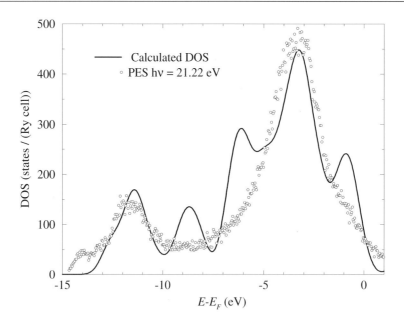

**Figure B3.2.8.** Comparison of the photoemission spectrum for the cetineite (Na;Se) and the density of states calculated by the AFC ELAPW-$k \cdot p$ method [88].

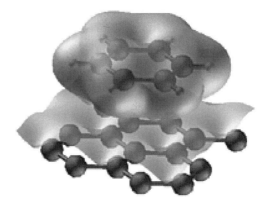

**Figure B3.2.9.** A benzene molecule on a graphite surface [90]. The geometry and the charge density (indicated by the surfaces of constant density) have been obtained using the PAW method. (Figure by Professor P E Blöchl.)

systems is to have the computational burden scale no more than linearly with system size. Methods achieving this are called linear-scaling or $O(N)$ (order $N$) methods, most of which are based on the Kohn–Sham equation (see equation (B3.2.4)), aiming to calculate single-electron wavefunctions, the Kohn–Sham orbitals. These methods tend to be faster than the conventional Kohn–Sham approach above a few hundred atoms [20, 91–93]. Another class of methods is based directly on the DFT of Hohenberg and Kohn [24]. With these techniques one seeks to determine directly the density that minimizes the total energy; they are often referred to as orbital-free methods [94–97]. Such orbital-free calculations do not have the bottlenecks present in orbital-based $O(N)$ DFT calculations, such as the need to localize orbitals to achieve linear scaling, orbital orthonormalization, or Brillouin zone sampling. Without such bottlenecks, the calculations become very inexpensive.

Equation (B3.2.3) lists the terms comprising the calculation of the total energy. The term due to the external potential and the Hartree term describing the Coulomb repulsion energy among the electrons already explicitly depend on the density instead of on orbitals. More difficult to evaluate is $G[\rho] = T_s[\rho] + E_{xc}[\rho]$, a functional which is not known exactly. However, over the years a number of high-quality exchange–correlation functionals have been developed for all kinds of systems. Only quite recently have more accurate kinetic energy density functionals (KEDFs) become available [97–99] that afford linear-scaling computations.

One current limitation of orbital-free DFT is that since only the total density is calculated, there is no way to identify contributions from electronic states of a certain angular momentum character $l$. This identification is exploited in non-local pseudopotentials so that electrons of different $l$ character 'see' different potentials, considerably improving the quality of these pseudopotentials. The orbital-free methods thus are limited to local pseudopotentials, connecting the quality of their results to the quality of the available local potentials. Good local pseudopotentials are available for the alkali metals, the alkaline earth metals and aluminium [100, 101] and methods exist for obtaining them for other atoms (see section VI.2 of [97]).

The orbital-free method has been used for molecular-dynamics studies of the formation of the self-interstitial defect in Al [102], pressure-induced glass-to-crystal transitions in sodium [103] and ion–electron correlations in liquid metals [101]. Calculations of densities for various Al surfaces have shown excellent agreement between the charge densities as calculated by Kohn–Sham DFT and an orbital-free method using a KEDF with a density-dependent response kernel [99]. The method was used recently to examine the metal–insulator transition in a two-dimensional array of metal quantum dots [104], where the theory showed that minute overlap of the nanoparticle's wavefunctions is enough to transform the array from an insulator to a metal. As an example of the ease with which large simulations can be performed, figure B3.2.10 shows a plot of the charge density from an orbital-free calculation of a vacancy among 255 Al atoms [98], carried out on a workstation.

### B3.2.3.4   The Hartree–Fock method in crystals

The HF method (discussed in section A1.3.1.2) is an alternative to DFT approaches. It does not include electron correlation effects, i.e. non-classical electron–electron interactions beyond the Coulomb and exchange interactions. The neglect of these terms means that the Coulomb interaction is unscreened, and hence the electron repulsion energy is too large, overestimating ionic character, which leads to band gaps that are too large by a factor of two or more and valence band widths that are too wide by 30–40% [63]. However, the HF results can be used as a well defined starting point for the inclusion of many-particle corrections such as the GW approximation [31, 32] or, with considerably less computational effort, the results can be improved considerably by accounting for the Coulomb hole and screening the exchange interaction using the dielectric function [63, 105].

*Ab initio* HF programs for crystals have been developed [106, 107] and have been applied to a wide variety of bulk and surface systems [108, 109]. As an example, a periodic HF calculation using pseudopotentials and an LCAO basis predicted binding energies, lattice parameters, bulk moduli and central-zone phonon frequencies of 17 III–V and IV–IV semiconductors. The authors find that '. . .[o]n the whole, the HF LCAO data appear no worse than other *ab initio* results obtained with DF-based Hamiltonians' [110]. They suggest that the largest part of the errors with respect to experiment is due to correlation effects and to a lesser extent due to the imperfections of the pseudopotentials [110]. More recently, the electronic and magnetic properties of transition metal oxides and halides such as perovskites, which had been a problem earlier, have been investigated with spin-unrestricted HF [111]. In general, the periodic HF method is best suited for the study of highly ionic, large band gap crystals because such systems are the least sensitive to the lack of electron correlation.

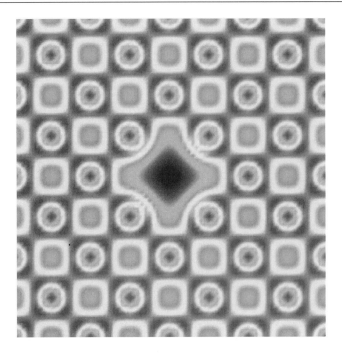

**Figure B3.2.10.** Contour plot of the electron density obtained by an orbital-free Hohenberg–Kohn technique [98]. The figure shows a vacancy in bulk aluminium in a 256-site cell containing 255 Al atoms and one empty site, the vacancy. Dark areas represent low electron density and light areas represent high electron density. A Kohn–Sham calculation for a cell of this size would be prohibitively expensive. Calculations on smaller cell sizes using both techniques yielded densities that were practically identical.

### B3.2.3.5  Quantum Monte Carlo

QMC techniques provide highly accurate calculations of many-electron systems. In variational QMC (VMC) [112–114], the total energy of the many-electron system is calculated as the expectation value of the Hamiltonian. Parameters in a trial wavefunction are optimized so as to find the lowest-energy state (modern methods instead minimize the variance of the local energy $\frac{H\Psi}{\Psi}$ [115]). A Monte Carlo (MC) method is used to perform the multi-dimensional integrations necessary to determine the expectation value,

$$E = \frac{\int |\Psi|^2 \frac{H\Psi}{\Psi} d\tau}{\int |\Psi|^2 d\tau}$$

where $\Psi$ is the trial wavefunction and $|\Psi|^2 / \int |\Psi|^2 d\tau$ is a normalized probability distribution. The integration is performed by summing up the local energy at points, corresponding to electron configurations, given by the probability distribution. A random walk algorithm, such as the Metropolis algorithm [116], is used to sample those regions of configuration space more heavily where the probability density is high. The standard Slater–Jastrow trial wavefunction is the product of a Slater determinant of single-electron orbitals and a Jastrow factor, a function which includes the description of two-electron correlation. As an example, the trial wavefunction used for a silicon crystal contained 32 variational parameters whose optimization required the calculation of the local energy for 10 000–20 000 statistically independent electron configurations [117]. In contrast to the DMC technique described below, the accuracy of a VMC calculation depends on the quality of the many-particle wavefunction used [114]. In figure B3.2.11, we show the determination of the lattice

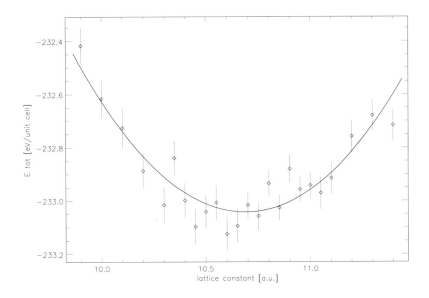

**Figure B3.2.11.** Total energy *versus* lattice constant of gallium arsenide from a VMC calculation including 256 valence electrons [118]; the curve is a quadratic fit. The error bars reflect the uncertainties of individual values. The experimental lattice constant is 10.68 au, the QMC result is 10.69 ($\pm$0.1) au (Figure by Professor W Schattke).

constant of GaAs by VMC by minimization of the total energy [118]. This figure illustrates the roughness of the potential energy surface due to statistical errors, which poses a challenge then for the calculation of forces with QMC.

In the diffusion QMC (DMC) method [114, 119], the evolution of a trial wavefunction (typically wavefunctions of the Slater–Jastrow type, for example, obtained by VMC) proceeds in imaginary time, $\tau = it$, according to the time-dependent Schrödinger equation, which then becomes a diffusion equation. All components of the wavefunction except for the ground-state wavefunction are damped by the time evolution operator $\exp(-iHt) = \exp(-H\tau)$. The DMC was developed as a simplification of the Green's function MC technique [113]. A particularly well known use of the Green's function MC technique was the determination by Ceperley and Alder of the energy of the uniform electron gas as a function of its density [120]. This $E(\rho)$ was subsequently parametrized by Perdew and Zunger for the commonly used LDA exchange–correlation potential [40]. Usually two approximations are made to make DMC calculations tractable: the fixed-node approximation, in which the nodes, the places where the trial function changes sign, are kept fixed for the solution to enforce the fermion symmetry of the wavefunction and the so-called short-time approximation, whose effect can be made very small [114]. Excited states have been calculated by replacing an orbital in the Slater determinant of the trial wavefunction by a conduction-band orbital [121].

Recently, a method has been proposed to overcome the problems associated with calculating forces in both VMC and DMC [122]. It has been suggested that the use of QMC in the near future to tackle the energetics of systems as challenging as liquid binary iron alloys is not unthinkable [123].

### B3.2.3.6 *Summary and comparisons*

As we have outlined, a very wide variety of methods are available to calculate the electronic structure of solids. Empirical TB methods (such as discussed in section B3.2.2) are the least expensive, affording the calculation of unit cells with large numbers (e.g. $10^3$) of atoms, or to provide cheap input to subsequent methods, at the price of quantitative accuracy. DFT methods (sections A1.3.5.4 and B3.2.3.3), on the other hand, are responsible for many of the impressive results obtained in computational materials theory in recent years. The tradeoff for DFT is the opposite: its expense, except in the not-yet-general linear scaling methods, limits it typically to systems with at most a few hundred atoms. Once the O(N) DFT methods become more general (for example, when orbital-free DFT can treat non-metallic systems), then the DFT method will be able routinely to treat systems as large as those treated now with TB.

The diversity of approaches based on HF (section B3.2.3.4) is small at present compared to the diversity found for DFT. For solids, HF appears to yield results inferior to DFT due to the neglect of electron correlation, but being a genuine many-particle theory it offers the possibility for consistent corrections, in contrast to DFT. Finally, the QMC techniques (section B3.2.3.4) hold promise for genuine many-particle calculations, yet they are still far from able to offer the same quantities for the same range of materials and geometries as the theories mentioned before. With this wide range of methods now introduced, we will look at their application to chemisorption on solid surfaces.

## B3.2.4    Quantum structural methods for solid surfaces

### B3.2.4.1    Introduction

First-principles models of solid surfaces and adsorption and reaction of atoms and molecules on those surfaces range from *ab initio* quantum chemistry (HF; configuration interaction (CI), perturbation theory (PT), etc: for details see chapter B3.1) on small, finite clusters of atoms to HF or DFT on two-dimensionally infinite slabs. In between these two extremes lie embedded cluster models, which recognize and attempt to correct the drastic approximation made by using a finite cluster to describe, for example, a metallic conductor whose electronic structure is inherently delocalized or an ionic crystal with long-range Coulomb interactions. Upon chemisorption, the binding of an atom or a molecule to a surface involves significant sharing of electrons in the bond between the adsorbate and surface atoms and this breaking of the crystal symmetry will induce localization of the electrons. The attractive feature of the embedded cluster idea is that it preserves the strengths of the cluster approach, namely it allows one to describe the very local process of chemisorption to a high degree of accuracy by, for example, quantum chemical methods, while at the same time attempting to account for the presence of the rest of the surface and bulk. Surface reconstruction and molecular adsorption have been studied on a variety of surfaces, including insulators, semiconductors and metals. To illustrate these methods, we will focus on those used to examine adsorption of atoms and molecules on transition metal surfaces. This is not a comprehensive review of each approach; rather, we provide selected examples that demonstrate the range of techniques and applications, and some of the lessons learned.

### B3.2.4.2    The finite cluster model

The most straightforward molecular quantum mechanical approach is to treat adsorption on a small, finite cluster of transition metal atoms, ranging from as small as four atoms up to ~40 atoms. Though all-electron calculations can be performed, typically the core electrons of transition metal atoms are replaced by an effective core potential (ECP, the quantum chemistry version of a pseudopotential that accounts approximately for the core–valence electron interaction), while the valence electrons of each metal atom are treated explicitly within a HF, CI, PT, or DFT formalism. Typically, a few atoms in the chemisorption region contain the valence (or all) electrons explicitly, while surrounding atoms tend to be described more crudely with, for example, a one-electron ECP representation, model pseudopotentials or, in the case of ionic crystals, a finite array of point charges. Generally, the structure of the cluster is chosen to be a fixed fragment of the bulk. Examples of this type of approach include the early work of Upton and Goddard [124], who examined adsorption of electronegative and electropositive atoms on a $Ni_{20}$ cluster designed to mimic various low-index faces of Ni. In this model, only the 4s electrons on each Ni atom were treated explicitly, while the 3d electrons were subsumed into an ECP. They made predictions concerning preferred binding sites, geometries, vibrational frequencies and binding energies. Bagus *et al* [125] published an important comparison study showing that it is more accurate to treat metal atoms directly interacting with an adsorbate at an all-electron level, while it is sufficient to describe the surrounding metal atoms with ECPs. Panas *et al* [126] proposed the idea that a cluster should be 'bond-prepared', namely that one should study an electronic state of the finite cluster that has enough singly-occupied orbitals of the correct symmetry to interact with the incoming admolecule to form the necessary covalent bonds between the adsorbate and the metal. In one of the first studies of a metal surface reaction, Panas *et al* [127] examined dissociative chemisorption pathways at the multi-reference CI level for $O_2$ on a $Ni_{13}$ cluster, generally using ECPs for all but the 4s electrons. Salahub and co-workers [128] used DFT-LDA with a Gaussian basis to examine chemisorption of C, O, H, CO and HCOO on Ni clusters containing up to 16 atoms meant to represent various low-index faces of Ni. Gradient corrections to the LDA scheme improved dramatically the binding energies for hydrogen bound to small Ni clusters, when compared to experimental results for Ni(111) and Ni(100) [129]. Multiple adsorbates were also studied by DFT-LDA: for example, in the case of hydrogen on Pd clusters modelling Pd(110) [130]. Diffusion barriers were also calculated by DFT-LDA for clusters containing up to 13 metal atoms of Pd, Rh, Sn, and Zn [131]. Other examples include HF calculations of K adsorbed on Cu clusters [132], HF and Møller–Plesset second-order PT (MP2) calculations of acetylene on Cu and Pd clusters [133], modified coupled pair functional (CPF) calculations for CO on Cu clusters [134], averaged CPF calculations of hydrogen adsorption on relaxed Cu clusters [135], HF, CASSCF (complete active space self-consistent field) and multireference CI and PT calculations for CO [136] and O [137] on Pt clusters, and spin-polarized DFT of c-$CH_2N_2$ on Pd and Cu tetramers [138] and of K and CO on $Pd_{8,14}$ [139]. The advantage of the finite-cluster model is that one can systematically include high levels of electron correlation; this is to be balanced against the lack of a proper band structure, the presence of edge effects and the fact that it is generally limited to modelling low coverages. Next we outline current strategies for ameliorating some of these difficulties.

*B3.2.4.3    Finite-cluster model in contact with a classical background*

Several modifications of the finite-cluster model meant to account for the background Fermi sea of electrons and to compensate for the lack of a proper band structure have been developed. They rely on simple approximations of the surface/bulk, usually involving classical electrostatic interactions and usually applied to ionic crystals (see, for example, [140]). Of these, the model invented by Nakatsuji is the primary one that has considered adsorption on metal surfaces [141]. The so-called 'dipped adcluster model' [142] considers a small cluster plus an adsorbate as the 'adcluster' that is 'dipped' onto the Fermi sea of electrons of the bulk metal. A normal HF calculation on the small system is performed, in which electrons are added to or removed from the cluster in each calculation. By comparing the variation in the total energy with respect to the fractional electron transfer, $dE/dn$, to the work function of the metal, $\mu$, the extent of electron transfer between the adcluster and the bulk metal can be established. Thus, charges on a small cluster are optimized and an image charge correction is also accounted for. In certain cases, integral charges are transferred between the cluster and the 'surroundings'; then electron correlation calculations, for example CI, can be carried out. This is a purely classical electrostatic approach to accounting for the background electrons in an implicit, rather than explicit, manner. Nakatsuji has used this to study adsorption of ionic adsorbates on metals, and finds that one can describe the polarization of the metal reasonably well. We have worked briefly with this approach [143], but found that there is a problem with extending the method beyond two-dimensional clusters, because of an ambiguity of where to place the image plane. Indeed, Nakatsuji's examples are always small one- or two-dimensional clusters. It is also likely that the wavefunction for such small clusters (typically ≤4 metal atoms) would not adequately represent a true metal surface wavefunction.

A simple, implicit means of describing the metallic band structure [144] was introduced by Rösch, using a Gaussian broadening of the cluster energy levels in order to determine a cluster Fermi level within DFT,

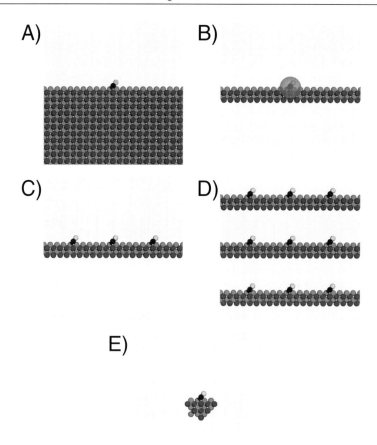

**Figure B3.2.12.** Schematic illustration of geometries used in the simulation of the chemisorption of a diatomic molecule on a surface (the third dimension is suppressed). The molecule is shown on a surface simulated by (A) a semi-infinite crystal, (B) a slab and an embedding region, (C) a slab with two-dimensional periodicity, (D) a slab in a supercell geometry and (E) a cluster.

originally by the $X_\alpha$ method (a simplified version of DFT-LDA; see section A1.3.3.3). Recent applications of this method have utilized more accurate forms of gradient-corrected spin-polarized DFT to look at adsorption of, for example, acetylene on $Ni_{14,20}$ clusters [145], CO adsorption on Ni, Pd and Pt clusters of eight or nine atoms [146] and NO adsorption on Ru [147].

*B3.2.4.4   Slab calculations*

The other extreme of modelling chemisorption is to use a slab described by DFT or HF. The slab is typically taken to be periodic in the directions parallel to the surface and contains a few atomic layers in the direction normal to the surface. For the adatoms not to influence each other, unless that is intended, the unit cell needs to be sufficiently large, parallel to the surface. For computational reasons, it is advantageous in some methods, namely plane wave techniques, to have periodicity in three dimensions. In the supercell geometry, this periodicity is gained by considering slabs which are periodic in the direction perpendicular to the surface but separated from each other by vacuum regions. The vacuum region has to be thick enough so that there is no influence between the surfaces facing each other (the same is true for the slab thickness). For a schematic description of several simulation model geometries, see figure B3.2.12.

Freeman and co-workers developed the FLAPW method (see B3.2.2) during the early 1980s [70, 148]. This was a major advance, because the conventional 'muffin-tin' potential was eliminated from their calculation allowing general-shape potentials to be evaluated instead. Freeman's group first developed this for thin films and then for bulk metals. As mentioned before, the LAPW basis, along with the elimination of any shape approximations in the potential, allows for highly accurate calculations on transition metal surfaces, within the DFT-LDA and the generalized gradient approximation, GGA (see section A1.3.3.3). For the 'stand-alone' slab geometry, figure B3.2.12(C), the LAPW basis functions decay exponentially into the vacuum. The numerous interfacial systems examined by Freeman's group include, for example, CO with K or S coadsorption on Ni(001) [149], adsorption of sulfur alone on Ni(001) [150], Fe monolayers on Ni(111) [151], Ag monolayers on MgO(001) [152], Au-capped Fe monolayers on MgO(001) [153], NO adsorption on Rh, Pd and Pt [154], and Li on Ru(001) [155]. Typical properties predicted are the equilibrium positions, magnetic moments, charge densities and surface densities of states.

More recently, other groups—primarily in Europe—have begun doing pseudopotential plane wave (often gradient-corrected) DFT supercell slab calculations (figure B3.2.12(D)) for chemisorption on metals. The groups of Nørskov [156], Scheffler [157], Baerends [158, 159] and Hafner and Kresse [160, 161] have been the most active. Adsorbate–metal surface systems examined include: alkalis and $N_2$ on Ru [156], NO on Pd [156], $H_2$ on Al [156], Cu [156, 158], Pd [157, 158] and sulfur-covered Pd [157], CO oxidation on Ru [157], CO on Ni, Pd and Pt [158], O on Pt [160], and $H_2$ on Rh, Pd and Ag [161].

An interesting study by te Velde and Baerends [159] compared slab- and cluster-DFT results for CO absorption on Cu(100). They found large oscillations in the chemisorption binding energy of CO to finite copper clusters as a function of cluster size. This suggests that the finite-cluster model (figure B3.2.12(E)) is likely to be inadequate, at least for modelling metal surfaces. By contrast, the slab calculations converge quickly with the number of Cu layers for the CO heat of adsorption and CO–CO distances.

The supercell plane wave DFT approach is periodic in three dimensions, which has some disadvantages: (i) thick vacuum layers are required so the slab does not interact with its images, (ii) for a tractably sized unit cell, only high adsorbate coverages are modelled readily and (iii) one is limited in accuracy by the form of the exchange–correlation functional chosen. In particular, while DFT, especially using gradient-corrected forms of the exchange–correlation functional (GGA), has proven to be remarkably reliable in many instances, there are a number of examples for chemisorption in which the commonly used GGAs have been shown to fail dramatically (errors in binding energies of $\sim 1$ eV or greater) [162, 163]. This naturally motivates the next set of approaches, namely the embedded cluster strategy.

### B3.2.4.5 Embedded-cluster schemes: cluster in cluster

Whitten and co-workers developed a metal cluster embedding scheme appropriate for CI calculations during the 1980s [164]. In essence, the method consists of: (i) solving for a HF minimum basis set (one 4s orbital/atom) description of a large cluster (e.g., $\sim 30$–90 atoms); (ii) localizing the orbitals *via* exchange energy maximization with atomic basis functions on the periphery; (iii) using these localized orbitals to set up effective Coulomb and exchange operators for the electrons within the cluster to be embedded; (iv) improving the basis set on the atoms comprising the embedded cluster and (v) performing a small CI calculation ($O(10^3)$ configurations) within orbitals localized on the embedded cluster. This strategy provides an approximate way of accounting for nearby electrons outside the embedded cluster itself. Whitten and co-workers have applied it to a variety of adsorbates (H, N, O, C—containing small molecules) on, primarily, Ni surfaces. Duarte and Salahub recently reported a DFT-cluster-in-DFT-cluster variant of Whitten's embedding, with a couple of twists on the original approach (for example, fractional orbital occupancies and charges, and an extra buffer region) [165]. Earlier, Sellers developed a related scheme for embedding a MP2 cluster within another cluster, where the background was modelled with screened ECPs [166]. Also, Ravenek and Geurts [167] and Fukunishi and Nakatsuji [168] extended the Green's matrix method of Pisani [169] (who developed it mainly for ionic crystals) to again embed a cluster within a cluster by introducing a semiorthogonal basis and renormalizing the charge on the cluster. It was implemented within the $X_\alpha$ method in the former case [167], and by broadening each discrete energy level to mimic the bulk band structure within HF theory for the cluster in the latter case [168].

Pisani [169] has used the density of states from periodic HF (see B3.2.2.4) slab calculations to describe the host in which the cluster is embedded, where the applications have been primarily to ionic crystals such as LiF. The original calculation to derive the external Coulomb and exchange fields is usually done on a finite cluster and at a low level of *ab initio* theory (typically minimum basis set HF, one electron only per atom treated explicitly).

The main drawback of the cluster-in-cluster methods is that the embedding operators are derived from a wavefunction that does not reflect the proper periodicity of the crystal: a two-dimensionally infinite wavefunction/density with a proper band structure would be preferable. Indeed, Rösch and co-workers pointed out recently a series of problems with such cluster-in-cluster embedding approaches. These include the lack of marked improvement of the results over finite clusters of the same size, problems with the orbital space partitioning such that charge conservation is violated, spurious mixing of virtual orbitals into the density matrix [170], the inherent delocalized nature of metallic orbitals [171], etc.

### B3.2.4.6    *Embedding of clusters in periodic background*

One of the first cluster embedding schemes was put forth by Ellis and co-workers [172]. They were interested in studying transition metal impurities in NiAl alloys, so they considered a $TMAl_xNi_y$ cluster embedded in a periodic self-consistent crystal field appropriate for bulk $\beta'$-NiAl. The field was calculated via $X_\alpha$ calculations, as was the cluster itself. The idea was to provide a relatively inexpensive alternative to supercell DFT calculations.

Perhaps the most sophisticated embedding scheme for describing metal surfaces to date is the LDA-based self-consistent Green's function method for semi-infinite crystals. Inglesfield, Benesh, and co-workers embed the near-surface layers using an embedding potential constructed from the bulk Green's function within an all-electron approach, using an LAPW basis [173]. Scheffler and co-workers developed a similar approach using a Gaussian basis for the valence electrons and pseudopotentials [174]. The formulation of the latter method is somewhat different from Inglesfield's and Benesh's, in that a reference system is chosen for which the Green's function and density are known (typically the bulk metal), and a $\Delta$(Green's function) is solved for in order to get a $\Delta$(embedding potential) and hence a $\Delta$(density). This allows one to solve for the embedding potential locally in a small region around the adsorbate. These methods allow for an economical yet accurate calculation of the embedding density, which yields a trustworthy description of charge transfer and other equilibrium properties, though subject to the accuracy limitations inherent in DFT-LDA.

In the late 1980s, Feibelman developed his Green's function scattering method using LDA with pseudopotentials to describe adsorption on two-dimensionally infinite metal slabs [175], based on earlier work by Williams *et al* [176]. The physical basis for the technique is that the adsorbate may be considered a defect off which the Bloch waves of the perfect substrate scatter. The interaction region is short-range because of screening by the electron gas of the metal. Feibelman has used this technique to study, for example, the chemisorption of an $H_2$ molecule on Rh(001) [177], S adatoms on Al(331) [178] and Ag adatoms on Pt(111) [179]. Charge densities, relative energies for various adsites and diffusion barriers (the latter in good agreement with experiment) were the typical quantities predicted.

Krüger and Rösch implemented within DFT the Green's matrix approach of Pisani within an approximate periodic slab environment [180]. They were able to successfully extend Pisani's embedding approach to metal surfaces by smoothing out the step function that determines the occupation numbers near the Fermi level. Keys to the numerical success of their method included: (i) symmetric orthogonalization of the Bloch basis to produce a localized set of functions that yielded a balanced distribution of charge in the system and (ii) self-consistent evaluation of the Fermi energy by fixing the charge on the cluster to be neutral. The slab was described with a Slater basis at the DFT-LDA level, while the embedded cluster orbitals were expanded in terms of Gaussian functions at the DFT-LDA level. While some properties exhibited non-monotonic behaviour with increasing cluster size, the charge transfer between the metal surface and the adsorbate seemed to be well described. They concluded that properties are not well converged in this method if the cluster does not contain shells of metal atoms that are at least next-nearest-neighbours to the adsite metal atoms.

Head and Silva used occupation numbers obtained from a periodic HF density matrix for the substrate to define localized orbitals in the chemisorption region, which then defines a cluster subspace on which to carry out HF calculations [181]. Contributions from the surroundings also only come from the bare slab, as in the Green's matrix approach. Increases in computational power and improvements in minimization techniques have made it easier to obtain the electronic properties of adsorbates by supercell slab techniques, leading to the Green's function methods becoming less popular [182].

Cortona embedded a DFT calculation in an orbital-free DFT background for ionic crystals [183], which necessitates evaluation of kinetic energy density functionals (KEDFs). Wesolowski and Warshel [184] had similar ideas to Cortona, except they used a frozen density background to examine a solute in solution and examined the effect of varying the KEDF. Stefanovich and Truong also implemented Cortona's method with a frozen density background and applied it to, for example, water adsorption on NaCl(001) [185].

### B3.2.4.7 *Embedding explicit correlation methods in a DFT background*

In principle, DFT calculations with an ideal exchange–correlation functional should provide consistently accurate energetics. The catch is, of course, that the exact exchange–correlation functional is not known. While various GGAs have been remarkably successful, there are notable exceptions [186, 187], including ones specific to surface adsorption mentioned earlier, where the binding-energy errors can be more than an eV [162, 163]. As another example, Louie and Cohen and co-workers found no systematic improvement over the LDA when gradient corrections were included in calculations of Al, Nb and Pd bulk properties, including the cohesive energy [186]. Indeed, the design of exchange–correlation functionals constitutes an active field of research (see, for example, [188]). The lack of completely systematic means to improve these functionals is an unappealing aspect of these calculations.

A first step towards a systematic improvement over DFT in a local region is the method of Aberenkov *et al* [189], who calculated a correlated wavefunction embedded in a DFT host. However, this is achieved using an analytic embedding potential function fitted to DFT results on an indented crystal. One must be cautious using a bare indented crystal to represent the surroundings, since the density at the surface of the indented crystal will have inappropriate Friedel oscillations inside and decay behaviour at the indented surface not present in the real crystal.

We have developed a different first-principles embedding theory that combines DFT with explicit correlation methods. We sought to develop a method for treating bulk or surface phases that is more accurate than current implementations of DFT. The idea is to provide more accurate predictions for local energetics, such as chemisorption binding energies and adsorbate electronic excitation energies. To achieve this, our theory improves upon the DFT description of electron correlation in a local region. This is accomplished by an embedding theory that treats a small region within an accurate quantum chemistry approach [190, 191], which interacts with its surroundings *via* an embedding potential, $v_{embed}(r)$. This $v_{embed}(r)$ is derived from a periodic DFT calculation on the total system. It is expressed purely in terms of orbital-free DFT (kinetic and potential energy) interaction terms between the embedded region and its surroundings *à la* Cortona and, in particular, purely in terms of functionals of the total density, $\rho_{tot}$, and the density of the embedded region,

$\rho_{\mathrm{I}}$. We thus avoid construction of localized orbitals to describe the electrons in the surrounding environment. This is especially important for metal surfaces, where the extensive $k$-point sampling required to get a well converged density makes localization impractical (very expensive). This way of expressing the embedding operator also eliminates problems that occur in other forms of embedding, such as those of matching conditions at the embedding boundary, or spurious charge transfer, since the electrostatic potential and the density are continuous by construction. Its only real disadvantage is that there is an arbitrariness associated with the choice of $T_{\mathrm{s}}$. Development of optimal $T_{\mathrm{s}}$ functionals is an active area of research in our group [97–99].

The self-consistent embedding cycle proceeds as follows. First, a well converged density, $\rho_{\mathrm{tot}}$, is calculated for the extended metal surface in the presence of an adsorbate. This is accomplished within a standard pseudopotential plane wave DFT calculation (see chapter A1.3). Second, we partition the system into the region of interest (typically the adsorbate and neighbouring metal atoms at or near the surface) and its surroundings (all the other atoms in the periodic unit cell). The embedded region is defined by the integral number of electrons and nuclei within that region but not by a particular physical, fixed boundary. This allows for the electron density from the embedded region to expand or contract variationally into the surroundings, thus affording some effective charge polarization to occur as needed.

The electron density, $\rho_{\mathrm{I}}$, of the embedded cluster/adsorbate atoms is calculated using quantum chemistry methods (HF, PT, multireference SCF, or CI). The initial step in this iterative procedure sets $v_{\mathrm{embed}}(r)$ to zero, since $\rho_{\mathrm{I}}$ is needed in order to calculate it. On subsequent iterations, the third step is to use $\rho_{\mathrm{I}}$ and $\rho_{\mathrm{tot}}$ to calculate $v_{\mathrm{embed}}(r)$, then insert it, as a one-electron operator expressed in matrix form in the atomic orbital basis of the adsorbate/cluster, into the quantum chemistry calculation of step two, and then $\rho_{\mathrm{I}}$ is updated (*via* the wavefunction). We repeatedly update $v_{\mathrm{embed}}(r)$ and then $\rho_{\mathrm{I}}$ until full self-consistency is achieved, with fixed $\rho_{\mathrm{tot}}$. In this way, we variationally optimize both the quantum chemistry wavefunction and, implicitly, the density of the surroundings, subject to fixed $\rho_{\mathrm{tot}}$. We tacitly assume that the DFT-slab density for the total system, $\rho_{\mathrm{tot}}$, is in fact a good representation and does not need to be adjusted.

We have shown that our embedding total energies may be written in terms of the total energy obtained in step one (the DFT total energy for the entire system), plus a correction term, that subtracts out the DFT energy in the local region I and adds back in an *ab initio* total energy for that same region,

$$E_{\mathrm{tot}}^{\mathrm{embed}} = E_{\mathrm{tot}}^{\mathrm{DFT}} + (E_{\mathrm{I}}^{\mathrm{ab\,initio}} - E_{\mathrm{I}}^{\mathrm{DFT}}).$$

Thus, another way to think of the embedding is that the *ab initio* treatment of region I is *correcting* the DFT results in the same region, for the same self-consistent density. We expect, then, that such a treatment should reduce, for example, the famous LDA overbinding problem (LDA bond energies are generally significantly overestimated). We have indeed seen a smooth decrease in the LDA overbinding as a function of increasing electron correlation. We benchmarked the method against nearly exact calculations on a small system and then further corroborated it on experimentally well studied chemisorption systems: CO on transition metal surfaces. Our binding energies are in good agreement with nearly full configuration interaction in the former and experimental adsorbate binding energies in the latter. Very recently, we have demonstrated that excitation energies for adsorbed CO are dramatically improved compared to experiment upon inclusion of the embedding potential [192]. In the future, we hope this method will provide a general means for accurate predictions of the local electronic structure of condensed matter.

## B3.2.5  Outlook

Computational solid-state physics and chemistry are vibrant areas of research. The all-electron methods for high-accuracy electronic structure calculations mentioned in section B3.2.3.2 are in active development, and with PAW, an efficient new all-electron method has recently been introduced. Ever more powerful computers enable more detailed predictions on systems of increasing size. At the same time, new, more complex materials require methods that are able to describe their large unit cells and diverse atomic make-up. Here, the new orbital-free DFT method may lead the way. More powerful techniques are also necessary for the accurate treatment of surfaces and their interaction with atoms and, possibly complex, molecules. Combined with recent progress in embedding theory, these developments make possible increasingly sophisticated predictions of the quantum structural properties of solids and solid surfaces.

## Acknowledgments

The authors would like to thank Professor P E Blöchl, Dr H Eckstein, Professor W Schattke, Professor K E Smith and T Strasser for making figures available for this publication. FS thanks Dr E E Krasovskii for introducing him to the LAPW method.

## References

[1] Whitten J L and Yang H 1996 Theory of chemisorption and reactions on metal surfaces *Surf. Sci. Rep.* **24** 59–124

[2] Mehl M J and Papaconstantopoulos D A 1998 Tight-binding parametrization of first-principles results *Topics in Computational Materials Science* ed C Y Fong (Singapore: World Scientific);
URL http://cst-www.nrl.navy.mil/~mehl/review/rev4.html

[3] Starrost F, Bornholdt S, Solterbeck C and Schattke W 1996 Band-structure parameters by genetic algorithm *Phys. Rev. B* **53** 12 549; *Phys. Rev. B* **54** 17 226E
Strasser T, Starrost F, Solterbeck C and Schattke W 1997 Valence-band photoemission from GaN(001) and GaAs: GaN surfaces *Phys. Rev. B* **56** 13 326

[4] Klimeck G, Brown R C, Boykin T B, Salazar-Lazaro C, Cwik T A and Stoica A 2000 Si tight-binding parameters from genetic algorithm fitting *Superlattices Microstruct.* **27** 10
Klimeck G, Brown R C, Boykin T B, Salazar-Lazaro C, Cwik T A and Stoica A 1999 *Preprint* 1006/spmi.1999.0797

[5] Cohen R E, Mehl M J and Papaconstantopoulos D A 1994 Tight-binding total-energy method for transition and noble metals *Phys. Rev. B* **50** 14 694–7

[6] Haas H, Wang C Z, Fähnle M, Elsässer C and Ho K M 1998 Environment-dependent tight-binding model for molybdenum *Phys. Rev. B* **57** 1461

[7] Harrison W A 1989 *Electronic Structure and the Properties of Solids* (New York: Dover)

[8] Menon M and Subbaswamy K R 1994 Transferable nonorthogonal tight-binding scheme for silicon *Phys. Rev. B* **50** 11 577

[9] Slater P C and Koster G F 1954 Simplified LCAO method for the periodic potential problem *Phys. Rev.* **94** 1498–524

[10] Mehl M J and Papaconstantopoulos D A 1996 Applications of a tight-binding total-energy method for transition and noble metals: Elastic constants, vacancies and surfaces of monatomic metals *Phys. Rev. B* **54** 4519
Mazin I I, Papaconstantopoulos D A and Singh D J 2000 Tight-binding Hamiltonians for Sr-filled ruthenates: Application to the gap anisotropy and Hall coefficient in $Sr_2RuO_4$ *Phys. Rev. B* **61** 5223

[11] Mercer J L Jr and Chou M Y 1994 Tight-binding model with intra-atomic matrix elements *Phys. Rev. B* **49** 8506

[12] Watson S C, Carter E A, Walters M K and Madden P A (unpublished)

[13] Seifert G, Eschrig H and Bieger W 1986 An approximate variation of the LCAO-Xα method *Z. Phys. Chem.* **267** 529

[14] Horsfield A P 1997 Efficient *ab initio* tight binding *Phys. Rev. B* **56** 6594–602

[15] Elstner M, Porezag D, Jungnickel G, Elsner J, Haugk M, Frauenheim Th, Suhai S and Seifert G 1998 Self-consistent-charge density-functional tight-binding method for simulations of complex materials properties *Phys. Rev. B* **58** 7260

[16] Lee S M, Belkhir M A, Zhu X Y, Lee Y H, Huang Y G and Frauenheim Th 2000 Electronic structure of GaN edge dislocations *Phys. Rev. B* **61** 16 033

[17] Bowler D R, Aoki M, Goringe C M, Horsfield A P and Pettifor D G 1997 A comparison of linear scaling tight-binding methods *Modelling Simulation Mater. Sci. Eng.* **5** 199

[18] Wang C S and Callaway J 1978 BNDPKG. A package of programs for the calculation of electronic energy bands by the LCGO method *Comput. Phys. Commun.* **14** 327

[19] Voß D, Krüger P, Mazur A and Pollmann J 1999 Atomic and electronic structure of WSe₂ from *ab initio* theory: bulk crystal and thin film systems *Phys. Rev. B* **60** 14 311

[20] Artacho E, Sánchez-Portal D, Ordejón P, García A and Soler J M 1999 Linear-scaling *ab initio* calculations for large and complex systems *Phys. Status Solidi* B **215** 809

[21] Hoffmann R 1963 An extended Hückel theory. I. Hydrocarbons *J. Chem. Phys.* **39** 1397

[22] Henk J, Schattke W, Carstensen H, Manzke R and Skibowski M 1993 Surface-barrier and polarization effects in the photoemission from GaAs(110) *Phys. Rev. B* **47** 2251

[23] Strasser T, Solterbeck C, Starrost F and Schattke W 1999 Valence-band photoemission from the GaN(0001) surface *Phys. Rev. B* **60** 11 577

[24] Hohenberg P and Kohn W 1964 Inhomogeneous electron gas *Phys. Rev.* **136** B864

[25] Kohn W and Sham L J 1965 Self-consistent equations including exchange and correlation effects *Phys. Rev.* **140** A1133

[26]  Janak J F 1978 Proof that $\partial E/\partial n_i = \varepsilon_i$ in density-functional theory *Phys. Rev.* B **18** 7165–8
[27]  Perdew J P, Parr R G, Levy M and Balduz J L Jr 1982 Density-functional theory for fractional particle number: derivative
        discontinuities of the energy *Phys. Rev. Lett.* **49** 1691–4
        Kleinman L 1997 Significance of the highest occupied Kohn–Sham eigenvalue *Phys. Rev.* B **56** 12 042–5
        Perdew J P and Levy M 1997 Comment on 'Significance of the highest occupied Kohn–Sham eigenvalue' *Phys. Rev.* B **56** 16 021–8
        Kleinman L 1997 Reply to 'Comment on 'Significance of the highest occupied Kohn–Sham eigenvalue'' *Phys. Rev.* B **56** 16 029–30
[28]  Bechstedt F 1992 Quasiparticle corrections for energy gaps in semiconductors *Adv. Solid State Phys.* **32** 161
[29]  Fiorentini V and Baldereschi A 1995 Dielectric scaling of the self-energy scissor operator in semiconductors and insulators *Phys.
        Rev.* B **51** 17 196
[30]  Pulci O, Onida G, Shkrebtii A I, Del Sole R and Adolph B 1997 Plane-wave pseudopotential calculation of the optical properties
        of GaAs *Phys. Rev.* B **55** 6685
[31]  Hedin L and Lundqvist S 1969 Effects of electron–electron and electron–phonon interactions on the one-electron states of solids
        *Solid State Phys.* **23** 1
[32]  Hybertsen M S and Louie S G 1985 First-principles theory of quasiparticles: Calculation of band gaps in semiconductors and
        insulators *Phys. Rev. Lett.* **55** 1418
[33]  Louie S G 1987 Theory of quasiparticle energies and excitation spectra of semiconductors and insulators *Electronic Band Structure
        and Its Applications (Lecture Notes in Physics vol 283)* ed M Youssouf (Berlin: Springer)
[34]  Rohlfing M, Krüger P and Pollmann J 1997 Quasiparticle calculations of semicore states in Si, Ge, and CdS *Phys. Rev.* B **56**
        R7065–8
[35]  Godby R W, Schlüter M and Sham L J 1988 Self-energy operators and exchange–correlation potentials in semiconductors *Phys.
        Rev.* B **37** 10159–75
[36]  Massidda S, Continenza A, Posternak M and Baldereschi A 1997 Quasiparticle energy bands of transition-metal oxides within a
        model GW scheme *Phys. Rev.* B **55** 13 494–502
[37]  Shirley E L 1998 Many-body effects on bandwidths in ionic, noble gas, and molecular solids *Phys. Rev.* B **58** 9579–83
[38]  Rieger M M and Godby R W 1998 Charge density of semiconductors in the GW approximation *Phys. Rev.* B **58** 1343
[39]  Campillo I, Silkin V M, Pitarke J M, Chulkov E V, Rubio A and Echenique P M 2000 First-principles calculations of hot-electron
        lifetimes in metals *Phys. Rev.* B **61** 13 484–92
[40]  Perdew J P and Zunger A 1981 Self-interaction correction to density-functional approximations for many-electron systems *Phys.
        Rev.* B **23** 5048
[41]  Svane A and Gunnarsson O 1990 Transition-metal oxides in the self-interaction-corrected density-functional formalism *Phys.
        Rev. Lett.* **65** 1148
[42]  Szotek Z, Temmerman W M and Winter H 1993 Application of the self-interaction correction to transition-metal oxides *Phys.
        Rev.* B **47** 4029
[43]  Svane A, Temmerman W and Szotek Z 1999 Theory of pressure-induced phase transitions in cerium chalcogenides *Phys. Rev.* B
        **59** 7888
[44]  Vogel D, Krüger P and Pollmann J 1997 Structural and electronic properties of group-III nitrides *Phys. Rev.* B **55** 12 836, and
        references therein
[45]  Stampfl C, van de Walle C G, Vogel D, Krüger P and Pollmann J 2000 Native defects and impurities in InN: First-principles
        studies using the local-density approximation and self-interaction and relaxation-corrected pseudopotentials *Phys. Rev.* B **61**
        R7846–9
[46]  Terakura K, Williams A R, Oguchi T and Kübler J 1984 Transition-metal monoxides: Band or Mott insulators *Phys. Rev. Lett.*
        **52** 1830
        Terakura K, Oguchi T, Williams A R and Kübler J 1984 Band theory of insulating transition-metal monoxides: Band-structure
        calculations *Phys. Rev.* B **30** 4734
[47]  Aryasetiawan F and Gunnarsson O 1995 Electronic structure of NiO in the GW approximation *Phys. Rev. Lett.* **74** 3221
[48]  Anisimov V I, Kuiper P and Nordgren J 1994 First-principles calculation of NiO valence spectra in the impurity-Anderson-model
        approximation *Phys. Rev.* B **50** 8257–65
[49]  Anisimov V I, Aryasetiawan F and Liechtenstein A I 1997 First-principles calculations of the electronic structure and spectra of
        strongly correlated systems: The LDA+U method *J. Phys.: Condens. Matter* **9** 767
[50]  Anisimov V I, Zaanen J and Andersen O K 1991 Band theory and Mott insulators: Hubbard $U$ instead of Stoner $I$ *Phys. Rev.* B
        **44** 943
[51]  Gunnarsson O, Andersen O K, Jepsen O and Zaanen J 1989 Density-functional calculation of the parameters in the Anderson
        model: Application to Mn in CdTe *Phys. Rev.* B **39** 1708–22
        Anisimov V I and Gunnarsson O 1991 Density-functional calculation of effective Coulomb interactions in metals *Phys. Rev.* B
        **43** 7570–4
[52]  Kwon S K and Min B I 2000 Unquenched large orbital magnetic moment in NiO *Phys. Rev.* B **62** 73
[53]  Slater J C 1937 Wave functions in a periodic potential *Phys. Rev.* **51** 846
[54]  Singh D J 1994 *Planewaves, Pseudopotentials and the LAPW Method* (Norwell, MA: Kluwer)
[55]  Marcus P M 1967 Variational methods in the computation of energy bands *Int. J. Quantum Chem.* **1** S 567
[56]  Koelling D D 1970 Alternative augmented-plane-wave technique: theory and application to copper *Phys. Rev.* B **2** 290–8

[57] Bross H, Bohn G, Meister G, Schubö W and Stöhr H 1970 New version of the modified augmented-plane wave method *Phys. Rev.* B **2** 3098–103

[58] Andersen O K 1975 Linear methods in band theory *Phys. Rev.* B **12** 3060

[59] Singh D and Krakauer H 1991 H-point phonon in molybdenum: Superlinearized augmented-plane-wave calculations *Phys. Rev.* B **43** 1441–5

[60] Krasovskii E E, Yaresko A N and Antonov V N 1994 Theoretical study of ultraviolet photoemission spectra of noble metals *J. Electron Spectrosc. Relat. Phenom.* **68** 157

[61] Sjöstedt E, Nordström L and Singh D J 2000 An alternative way of linearizing the augmented plane-wave method *Solid State Commun.* **114** 15

[62] Shick A B, Liechtenstein A I and Pickett W E 1999 Implementation of the LDA+U method using the full-potential linearized augmented plane-wave basis *Phys. Rev.* B **60** 10 763

[63] Massidda S, Posternak M and Baldereschi A 1993 Hartree–Fock LAPW approach to the electronic properties of periodic systems *Phys. Rev.* B **48** 5058

[64] Koelling D D and Arbman G O 1975 Use of energy derivative of the radial solution in an augmented plane wave method: application to copper *J. Phys. F: Met. Phys.* **5** 2041

[65] Krasovskii E E 1997 Accuracy and convergence properties of the extended linear augmented-plane-wave method *Phys. Rev.* B **56** 12 866

[66] Krasovskii E E, Nemoshkalenko V V and Antonov V N 1993 On the accuracy of the wavefunctions calculated by LAPW method *Z. Phys.* B **91** 463

[67] Singh D 1991 Ground-state properties of lanthanum: treatment of extended-core states *Phys. Rev.* B **43** 6388

[68] Krasovskii E E and Schattke W 1995 The extended-LAPW-based $k*p$ method for complex bandstructure calculations *Solid State Commun.* **93** 775

[69] Krasovskii E E, Starrost F and Schattke W 1999 Augmented Fourier components method for constructing the crystal potential in self-consistent band-structure calculations *Phys. Rev.* B **59** 10 504

[70] Wimmer E, Krakauer H, Weinert M and Freeman A J 1981 Full-potential self-consistent linearized-augmented-plane-wave method for calculating the electronic structure of molecules and surfaces: $O_2$ molecule *Phys. Rev.* B **24** 864

[71] Weinert M 1981 Solution of Poisson's equation: beyond Ewald-type methods *J. Math. Phys.* **22** 2433

[72] Petersen M, Wagner F, Hufnagel L, Scheffler M, Blaha P and Schwarz K 2000 Improving the efficiency of FP-LAPW calculations *Comp. Phys. Commun.* **126** 294–309

[73] Asato M, Settels A, Hoshino T, Asada T, Blügel S, Zeller R and Dederichs P H 1999 Full-potential KKR calculations for metals and semiconductors *Phys. Rev.* B **60** 5202

[74] Korringa J 1947 On the calculation of the energy of a Bloch wave in a metal *Physica (Amsterdam)* **13** 392–400

[75] Kohn W and Rostoker N 1954 Solution of the Schrödinger equation in periodic lattices with an application to metallic lithium *Phys. Rev.* **94** 1111–20

[76] Drittler B, Weinert M, Zeller R and Dederichs P H 1991 Vacancy formation energies of fcc transition metals calculated by a full potential Green's function method *Solid State Commun.* **79** 31

[77] Podloucky R, Zeller R and Dederichs P H 1980 Electronic structure of magnetic impurities calculated from first principles *Phys. Rev.* B **22** 5777

[78] Yussouff M 1987 Fast self-consistent KKR method *Electronic Band Structure and Its Applications (Lecture Notes in Physics vol 283)* ed M Yussouff (Berlin: Springer) pp 58–76

[79] Tank R W and Arcangeli C 2000 An introduction to the third-generation LMTO method *Phys. Status Solidi* B **217** 89

[80] Methfessel M, Rodriguez C O and Andersen O K 1989 Fast full-potential calculations with a converged basis of atom-centered linear muffin-tin orbitals: structural and dynamic properties of silicon *Phys. Rev.* B **40** 2009–12

[81] Blöchl P E 1994 Projector augmented-wave method *Phys. Rev.* B **50** 17 953

[82] Holzwarth N A W, Matthews G E, Dunning R B, Tackett A R and Zeng Y 1997 Comparison of the projector augmented-wave, pseudopotential and linearized augmented-plane-wave formalisms for density-functional calculations of solids *Phys. Rev.* B **55** 2005

[83] Alfè D, Kresse G and Gillan M J 2000 Structure and dynamics of liquid iron under Earth's core conditions *Phys. Rev.* B **61** 132

[84] Ohtsuki T, Ohno K, Shiga K, Kawazoe Y, Maruyama Y and Masumoto K 1998 Insertion of Xe and Kr atoms into C60 and C70 fullerenes and the formation of dimers *Phys. Rev. Lett.* **81** 967–70

[85] Krasovska O V, Krasovskii E E and Antonov V N 1995 *Ab initio* calculation of the optical and photoelectron properties of $RuO_2$ *Phys. Rev.* B **52** 11 825

[86] Leventi-Peetz A, Krasovskii E E and Schattke W 1995 Dielectric function and local field effects of TiSe2 *Phys. Rev.* B **51** 17 965

[87] Traving M, Boehme M, Kipp L, Skibowski M, Starrost F, Krasovskii E E, Perlov A and Schattke W 1997 Electronic structure of WSe2: a combined photoemission and inverse photoemission study *Phys. Rev.* B **55** 10 392–9

[88] Starrost F, Krasovskii E E, Schattke W, Jockel J, Simon U, Adelung R and Kipp L 2000 Cetineites: electronic, optical, and conduction properties of nanoporous chalcogenoantimonates *Phys. Rev.* B **61** 15 697

[89] Krasovskii E E and Schattke W 1997 Surface electronic structure with the linear methods of band theory *Phys. Rev.* B **56** 12 874

[90] Fisher A J and Blöchl P E 1993 Adsorption and scanning-tunneling-microscope imaging of benzene on graphite and MoS2 *Phys. Rev. Lett.* **70** 3263–6

[91]   Goedecker S 1999 Linear scaling electronic structure methods *Rev. Mod. Phys.* **71** 1085
[92]   Galli G 2000 Large-scale electronic structure calculations using linear scaling methods *Phys. Status Solidi* B **217** 231
[93]   Fattebert J-L and Bernholc J 2000 Towards grid-based O(N) density-functional theory methods: optimized nonorthogonal orbitals and multigrid acceleration *Phys. Rev.* B **62** 1713–22
[94]   Wang L-W and Teter M P 1992 Kinetic-energy functional of the electron density *Phys. Rev.* B **45** 13 196–220
[95]   Perrot F 1994 Hydrogen–hydrogen interaction in an electron gas *J. Phys.: Condens. Matter* **6** 431–46
[96]   Smargiassi E and Madden P A 1994 Orbital-free kinetic-energy functionals for first-principles molecular dynamics *Phys. Rev.* B **49** 5220–6
[97]   Wang Y A and Carter E A 2000 Orbital-free kinetic-energy density functional theory *Theoretical Methods in Condensed Phase Chemistry (Progress in Theoretical Chemistry and Physics Series)* ed S D Schwartz (Boston: Kluwer) pp 117–84
[98]   Wang Y A, Govind N and Carter E A 1998 Orbital-free kinetic energy functionals for the nearly-free electron gas *Phys. Rev.* B **58** 13 465
       Wang Y A, Govind N and Carter E A 1999 *Phys. Rev.* B **60** 17 162E
[99]   Wang Y A, Govind N and Carter E A 1999 Orbital-free kinetic-energy density functionals with a density-dependent kernel *Phys. Rev.* B **60** 16 350
[100]  Watson S, Jesson B J, Carter E A and Madden P A 1998 *Ab initio* pseudopotentials for orbital-free density functional *Europhys. Lett.* **41** 37–42
[101]  Anta J A, Jesson B J and Madden P A 1998 Ion–electron correlations in liquid metals from orbital-free *ab initio* molecular dynamics *Phys. Rev.* B **58** 6124–32
[102]  Jesson B J, Foley M and Madden P A 1997 Thermal properties of the self-interstitial in aluminum: an *ab initio* molecular-dynamics study *Phys. Rev.* B **55** 4941–6
[103]  Aoki M I and Tsumuraya K 1997 *Ab initio* molecular-dynamics study of pressure-induced glass-to-crystal transitions in the sodium system *Phys. Rev.* B **56** 2962–8
[104]  Watson S C and Carter E A 2000 Linear-scaling parallel algorithms for the first principles treatment of metals *Comp. Phys. Commun.* **128** 67–92
[105]  Hedin L 1965 New method for calculating the one-particle Green's function with application to the electron–gas problem *Phys. Rev.* **139** A796
[106]  Pisani C, Dovesi R and Roetti C 1988 *Hartree–Fock Ab Initio Treatment of Crystalline Systems (Lecture Notes in Chemistry, vol 48)* (Berlin: Springer)
[107]  CRYSTAL98 is the current version of the commercial HF program developed at the University of Torino and at Daresbury Laboratory (http://www.dl.ac.uk/TCS/Software/CRYSTAL/)
[108]  Su Y-S, Kaplan T A, Mahanti S D and Harrison J F 1999 Crystal Hartree–Fock calculations for $La_2NiO_4$ and $La_2CuO_4$ *Phys. Rev.* B **59** 10 521–9
[109]  Fu L, Yaschenko E, Resca L and Resta R 1999 Hartree–Fock studies of surface properties of $BaTiO_3$ *Phys. Rev.* B **60** 2697–703
[110]  Causà M, Dovesi R and Roetti C 1991 Pseudopotential Hartree–Fock study of seventeen III-V and IV-IV semiconductors *Phys. Rev.* B **43** 11 937–43
[111]  Chartier A, D'Arco P, Dovesi R and Saunders V R 1999 *Ab initio* Hartree–Fock investigation of the structural, electronic, and magnetic properties of $Mn_3O_4$ *Phys. Rev.* B **60** 14 042–8, and references therein
[112]  McMillan W L 1965 Ground state of liquid $^4$He *Phys. Rev.* **138** A442
[113]  Ceperly D M and Kalos M H 1986 Quantum many-body problems, *Monte Carlo Methods in Statistical Physics (Topics in Current Physics, vol 7)* 2nd edn, ed K Binder (Berlin: Springer) pp 145–94
[114]  Rajagopal G, Needs R J, James A, Kenney S D and Foulkes W M C 1995 Variational and diffusion quantum Monte Carlo calculations at nonzero wave vectors: theory and application to diamond-structure germanium *Phys. Rev.* B **51** 10 591–600
[115]  Umrigar C J, Wilson K G and Wilkins J W 1988 Optimized trial wavefunctions for quantum Monte Carlo calculations *Phys. Rev. Lett.* **60** 1719–22
[116]  Metropolis N, Rosenbluth A W, Rosenbluth M N, Teller A H and Teller E 1953 Equation of state calculations by fast computing machines *J. Chem. Phys.* **21** 1087
[117]  Kent P R C, Hood R Q, Williamson A J, Needs R J, Foulkes W M C and Rajagopal G 1999 Finite-size errors in quantum many-body simulations of extended systems *Phys. Rev.* B **59** 1917–29
[118]  Eckstein H, Schattke W, Reigrotzki M and Redmer R 1996 Variational quantum Monte Carlo ground state of GaAs *Phys. Rev.* B **54** 5512–15
[119]  Hammond B L, Lester W A and Reynolds P J 1994 *Monte Carlo Methods in Ab Initio Quantum Chemistry* (Singapore: World Scientific)
[120]  Ceperly D M and Alder B J 1980 Ground state of the electron gas by a stochastic method *Phys. Rev. Lett.* **45** 566–9
[121]  Towler M D, Hood R Q and Needs R J 2000 Minimum principles and level splitting in quantum Monte Carlo excitation energies: application to diamond *Phys. Rev.* B **62** 2330–7
[122]  Filippi C and Umrigar C J 2000 Correlated sampling in quantum Monte Carlo: a route to forces *Phys. Rev.* B **61** R16 291
[123]  Alfè D, Gillan M J and Price G D 2000 Constraints on the composition of the Earth's core from *ab initio* calculations *Nature* **405** 172–5

[124] Upton T H and Goddard W A III 1981 Chemisorption of H, Cl, Na, O, and S atoms on Ni(100) surfaces: a theoretical study using Ni$_{20}$ clusters *Crit. Rev. Solid State Mater. Sci.* **10** 261–96

[125] Bagus P S, Bauschlicher C W Jr, Nelin C J, Laskowski B C and Seel M 1984 A proposal for the proper use of pseudopotentials in molecular orbital cluster model studies of chemisorption *J. Chem. Phys.* **81** 3594–602

[126] Panas I, Schüle J, Siegbahn P and Wahlgren U 1988 On the cluster convergence of chemisorption energies *Chem. Phys. Lett.* **149** 265–72

Siegbahn P E M, Nygren M A and Wahlgren U 1992 *Cluster Models for Surface and Bulk Phenomena* ed G Pacchioni, P S Bagus and F Parmigiani *(NATO ASI Series B: Physics vol 283)* (New York: Plenum) p 267

[127] Panas I, Siegbahn P and Wahlgren U 1989 The mechanism for the O$_2$ dissociation on Ni(100) *J. Chem. Phys.* **90** 6791–801

[128] Fournier R and Salahub D R 1990 Chemisorption and magnetization: A bond order-rigid band model *Surf. Sci.* **238** 330–40

Ushio J, Papai I, St-Amant A and Salahub D R 1992 Vibrational analysis of formate adsorbed on Ni(110): LCGTO-MCP-LSD study *Surf. Sci.* **262** L134–8

[129] Mlynarski P and Salahub D R 1991 Local and nonlocal density functional study of Ni$_4$ and Ni$_5$ clusters. Models for the chemisorption of hydrogen on (111) and (100) nickel surfaces *J. Chem. Phys.* **95** 6050–6

[130] Papai I, Salahub D R and Mijoule D 1990 An LCGTO=MCP-LSD study of the (2 × 1) H-covered Pd(110) surface *Surf. Sci.* **236** 241–9

[131] Rochefort A, Andzelm J, Russo N and Salahub D R 1990 Chemisorption and diffusion of atomic hydrogen in and on cluster models of Pd, Rh, and bimetallic PdSn, RhSn, and RhZn catalysts *J. Am. Chem. Soc.* **112** 8239–47

[132] Bagus P S and Pacchioni G 1995 Ionic and covalent electronic states for K adsorbed on Cu$_5$ and Cu$_{25}$ cluster models of the Cu(100) surface *J. Chem. Phys.* **102** 879

[133] Clotet A and Pacchioni G 1996 Acetylene on Cu and Pd(111) surfaces: A comparative theoretical study of bonding mechanism, adsorption sites, and vibrational spectra *Surf. Sci.* **346** 91

[134] Bauschlicher C W Jr 1994 A theoretical study of CO/Cu(100) *J. Chem. Phys.* **101** 3250

[135] Triguero L, Wahlgren U, Boussard P and Siegbahn P 1995 Calculations of hydrogen chemisorption energies on optimized Cu clusters *Chem. Phys. Lett.* **237** 550

[136] Illas F, Zurita S, Marquez A M and Rubio J 1997 On the bonding mechanism of CO to Pt(111) and its effect on the vibrational frequency of chemisorbed CO *Surf. Sci.* **376** 279

[137] Illas F, Rubio J, Ricart J M and Pacchioni G 1996 The importance of correlation effects on the bonding of atomic oxygen on Pt(111) *J. Chem. Phys.* **105** 7192

[138] Rochefort A, McBreen P and Salahub D R 1996 Bond selectivity in the dissociative adsorption of c-CH$_2$N$_2$ on single crystals: a comparative DFT-LSD investigation for Pd(110) and Cu (110) *Surf. Sci.* **347** 11

[139] Filali Baba M, Mijoule C, Godbout N and Salahub D R 1994 Coadsorption of K and CO on Pd clusters: a density functional study *Surf. Sci.* **316** 349

[140] Kantorovich L N 1988 An embedded-molecular-cluster method for calculating the electronic structure of point defects in non-metallic crystals. I. General theory *J. Phys. C: Solid State Phys.* **21** 5041

Meng J, Pandey R, Vail J M and Kunz A B 1989 Impurity potentials derived from embedded quantum clusters: Ag$^+$ and Cu$^+$ transport in alkali halides *J. Phys.: Condens. Matter* **1** 6049–58

Grimes R W, Catlow C R A and Stoneham A M 1989 A comparison of defect energies in MgO using Mott–Littleton and quantum mechanical procedures *J. Phys.: Condens. Matter* **1** 7367–84

Zuo J, Pandey R and Kunz A B 1991 Embedded-cluster study of the lithium trapped-hole center in magnesium oxide *Phys. Rev.* B **44** 7187–91

Zuo J, Pandey R and Kunz A B 1992 Embedded-cluster study of Cu$^+$-induced lattice relaxation in alkali halides *Phys. Rev.* B **45** 2709–11

Visser O, Visscher L, Aerts P J C and Nieuwpoort W C 1992 Molecular open shell configuration interaction calculations using the Dirac–Coulomb Hamiltonian: the f$^6$-manifold of an embedded EuO$_6^{9-}$ cluster *J. Chem. Phys.* **96** 2910–19

Pisani C, Orlando R and Cora F 1992 On the problem of a suitable definition of the cluster in embedded-cluster treatments of defects in crystals *J. Chem. Phys.* **97** 4195–204

Martin R L, Pacchioni G and Bagus P S 1992 *Cluster Models for Surface and Bulk Phenomena* ed G Pacchioni *et al (NATO ASI Series B: Physics vol 283)* (New York: Plenum) p 485

Martin R L, Pacchioni G and Bagus P S 1992 *Cluster Models for Surface and Bulk Phenomena* ed G Pacchioni *et al (NATO ASI Series B: Physics vol 283)* (New York: Plenum) p 305

Pisani C 1993 Embedded-cluster techniques for the quantum-mechanical study of surface reactivity *J. Mol. Catal.* **82** 229

Hermann K 1992 *Cluster Models for Surface and Bulk Phenomena* ed G Pacchioni *et al (NATO ASI Series B: Physics vol 283)* (New York: Plenum) p 209

[141] Nakatsuji H 1987 Dipped adcluster model for chemisorptions and catalytic reactions on metal surface *J. Chem. Phys.* **87** 4995–5001

Nakatsuji H and Nakai H 1990 Theoretical study on molecular and dissociative chemisorptions of an O$_2$ molecule on an Ag surface: dipped adcluster model combined with symmetry-adapted cluster-configuration interaction method *Chem. Phys. Lett.* **174** 283–6

Nakatsuji H, Nakai H and Fukunishi Y 1991 Dipped adcluster model for chemisorptions and catalytic reactions on a metal surface: Image force correction and applications to Pd-O$_2$ adclusters *J. Chem. Phys.* **95** 640–7

Nakatsuji H and Nakai H 1992 Dipped adcluster model study for the end-on chemisorption of $O_2$ on an Ag surface *Can. J. Chem.* **70** 404–8

Nakatsuji H, Kuwano R, Morita H and Nakai H 1993 Dipped adcluster model and SAC-CI method applied to harpooning, chemical luminescence and electron emission in halogen chemisorption on alkali metal surface *J. Mol. Catal.* **82** 211–28

Zhen-Ming Hu and Nakatsuji H 1999 Adsorption and disproportionation reaction of OH on Ag surfaces: dipped adcluster model study *Surf. Sci.* **425** 296–312

[142] Nakatsuji H 1997 Dipped adcluster model for chemisorption and catalytic reactions *Prog. Surf. Sci.* **54** 1

[143] Chang T-M, Martinez T J and Carter E A 1994 unpublished results

[144] Rösch N, Sandl P, Gorling A and Knappe P 1988 Toward a chemisorption cluster model using the LCGTO-X$\alpha$ method: application to Ni(100)/Na *Int. J. Quantum Chem. Symp.* **22** 275

[145] Weinelt M, Huber W, Zebisch P, Steinrück H-P, Ulbricht P, Birkenheuer U, Boettger J C and Rösch N 1995 The adsorption of acetylene on Ni(110): an experimental and theoretical study *J. Chem. Phys.* **102** 9709

[146] Pacchioni G, Chung S-C, Krüger S and Rösch N 1997 Is CO chemisorbed on Pt anomalous compared with Ni and Pd? An example of surface chemistry dominated by relativistic effects *Surf. Sci.* **392** 173

[147] Staufer M *et al* 1999 Interpretation of x-ray emission spectra: NO adsorbed on Ru(001) *J. Chem. Phys.* **111** 4704–13

[148] Weinert M, Wimmer E and Freeman A J 1982 Total-energy all-electron density functional method for bulk solids and surfaces *Phys. Rev.* B **26** 4571–8

Jansen H J F and Freeman A J 1984 Total-energy full-potential linearized augmented plane-wave method for bulk solids: electronic and structural properties of tungsten *Phys. Rev.* B **30** 561–9

[149] Wimmer E, Fu C L and Freeman A J 1985 Catalytic promotion and poisoning: all-electron local-density-functional theory of CO on Ni(001) surfaces coadsorbed with K or S *Phys. Rev. Lett.* **55** 2618–21

[150] Fu C L and Freeman A J 1989 Covalent bonding of sulfur on Ni(001): S as a prototypical adsorbate catalytic poisoner *Phys. Rev.* B **40** 5359

[151] Wu R and Freeman A J 1992 Structural and magnetic properties of Fe/Ni(111) *Phys. Rev.* B **45** 7205

[152] Li C, Wu R, Freeman A J and Fu C L 1993 Energetics, bonding mechanism, and electronic structure of metal–ceramic interfaces: Ag/MgO(001) *Phys. Rev.* B **48** 8317–22

[153] Wu R and Freeman A J 1994 Magnetism at metal–ceramic interfaces: effects of a Au overlayer on the magnetic properties of Fe/MgO(001) *J. Magn. Magn. Mater.* **137** 127–33

[154] Mannstadt W and Freeman A J 1997 Dynamical and geometrical aspects of NO chemisorption on transition metals: Rh, Pd, and Pt *Phys. Rev.* B **55** 13 298

[155] Mannstadt W and Freeman A J 1998 LDA theory of the coverage dependence of the local density of states: Li adsorbed on Ru(001) *Phys. Rev.* B **57** 13 289

[156] Mortensen J J, Hammer B and Norskov J K 1998 Alkali promotion of $N_2$ dissociation over Ru(0001) *Phys. Rev. Lett.* **80** 4333

Hammer B and Norskov J K 1997 Adsorbate reorganization at steps: NO on Pd(211) *Phys. Rev. Lett.* **79** 4441

Hammer B, Scheffler M, Jacobsen K W and Norskov J K 1994 Multidimensional potential energy surface for $H_2$ dissociation over Cu(111) *Phys. Rev. Lett.* **73** 1400

Gundersen K, Jacobsen K W, Norskov J K and Hammer B 1994 The energetics and dynamics of $H_2$ dissociation on Al(110) *Surf. Sci.* **304** 131

[157] Wei C M, Gross A and Scheffler M 1998 *Ab initio* calculation of the potential energy surface for the dissociation of $H_2$ on the sulfur-covered Pd(100) surface *Phys. Rev.* B **57** 15 572

Tomanek D, Wilke S and Scheffler M 1997 Hydrogen-induced polymorphism of the Pd(110) surface *Phys. Rev. Lett.* **79** 1329

Stampfl C and Scheffler M 1997 Mechanism of efficient carbon monoxide oxidation at Ru(0001) *J. Vac. Sci. Technol.* A **15** 1635

Stampfl C and Scheffler M 1997 Anomalous behavior of Ru for catalytic oxidation: a theoretical study of the catalytic reaction CO+1/2 $O_2$ to $CO_2$ *Phys. Rev. Lett.* **78** 1500

Stampfl C and Scheffler M 1996 Theoretical study of O adlayers on Ru(0001) *Phys. Rev.* B **54** 2868

[158] Philipsen P H T, van Lenthe E, Snijders J G and Baerends E J 1997 Relativistic calculations on the adsorption of CO on the (111) surfaces of Ni, Pd and Pt within the zeroth-order regular approximation *Phys. Rev.* B **56** 13 556

Olsen R A, Philipsen P H T, Baerends E J, Kroes G J and Louvik O M 1997 Direct subsurface adsorption of hydrogen on Pd(111): quantum mechanical calculations on a new two-dimensional potential energy surface *J. Chem. Phys.* **106** 9286

Wiesenekker G, Kroes G J and Baerends E J 1996 An analytical six-dimensional potential energy surface for dissociation of molecular hydrogen on Cu(100) *J. Chem. Phys.* **104** 7344

Philipsen P H T, te Velde G and Baerends E J 1994 The effect of density-gradient corrections for a molecule-surface potential energy surface. Slab calculations on Cu(100)c(2x2)-CO *Chem. Phys. Lett.* **226** 583

[159] te Velde G and Baerends E J 1993 Slab versus cluster approach for chemisorption studies, CO on Cu(100) *Chem. Phys.* **177** 399

[160] Feibelman P J, Hafner J and Kresse G 1998 Vibrations of O on stepped Pt(111) *Phys. Rev.* B **58** 2179–84

[161] Eichler A, Kresse G and Hafner J 1998 Ab-initio calculations of the 6D potential energy surfaces for the dissociative adsorption of $H_2$ on the (100) surfaces of Rh, Pd and Ag *Surf. Sci.* **397** 116–36

[162] Rösch N 1998 *Lecture Given at the 7th International Symposium on Theoretical Aspects of Heterogeneous Catalysis, Cambridge, 25-28 August*

[163] Hammer B, Hansen L B and Nørskov J K 1999 Improved adsorption energetics within density functional theory using revised Perdew–Burke–Enerhof functionals *Phys. Rev.* B **59** 7413–21

[164] Whitten J L and Pakkanen T A 1980 Chemisorption theory for metallic surfaces: Electron localization and the description of surface interactions *Phys. Rev.* B **21** 4357–67

Madhavan P and Whitten J L 1982 Theoretical studies of the chemisorption of hydrogen on copper *J. Chem. Phys.* **77** 2673–83

Cremaschi P and Whitten J L 1987 The effect of hydrogen chemisorption on titanium surface bonding *Theor. Chim. Acta.* **72** 485–96

Whitten J L 1992 *Cluster Models for Surface and Bulk Phenomena* ed G Pacchioni *et al* (*NATO ASI Series B: Physics vol 283*) (New York: Plenum) p 375

Whitten J L 1993 Theoretical studies of surface reactions: embedded cluster theory *Chem. Phys.* **177** 387–97

[165] Duarte H A and Salahub D R 1998 Embedded cluster model for chemisorption using density functional calculations: oxygen adsorption on the Al(100) surface *J. Chem. Phys.* **108** 743

[166] Sellers H 1991 On modeling chemisorption processes with metal cluster systems. II. Model atomic potentials and site specificity of N atom chemisorption on Pd(111) *Chem. Phys. Lett.* **178** 351–7

[167] Ravenek W and Geurts F M M 1986 Hartree–Fock–Slater–LCAO implementation of the moderately large-embedded-cluster approach to chemisorption. Calculations for hydrogen on lithium (100) *J. Chem. Phys.* **84** 1613–23

[168] Fukunishi Y and Nakatsuji H 1992 Modifications for *ab initio* calculations of the moderately-large-embedded-cluster model. Hydrogen adsorption on a lithium surface *J. Chem. Phys.* **97** 6535–43

[169] Pisani C 1978 Approach to the embedding problem in chemisorption in a self-consistent-field-molecular-orbital formalism *Phys. Rev.* B **17** 3143

Pisani C, Dovesi R and Nada R 1990 *Ab initio* Hartree–Fock perturbed-cluster treatment of local defects in crystals *J. Chem. Phys.* **92** 7448

Pisani C 1993 Embedded-cluster techniques for the quantum-mechanical strudy of surface reactivity *J. Mol. Catal.* **82** 229

Casassa S and Pisani C 1995 Atomic-hydrogen interaction with metallic lithium: an *ab initio* embedded-cluster study *Phys. Rev.* B **51** 7805

[170] Gutdeutsch U, Birkenheuer U, Krüger S and Rösch N 1997 On cluster embedding schemes based on orbital space partitioning *J. Chem. Phys.* **106** 6020

[171] Gutdeutsch U, Birkenheuer U and Rösch N 1998 A strictly variational procedure for cluster embedding based on the extended subspace approach *J. Chem. Phys.* **109** 2056

[172] Ellis D E, Benesh G A and Byrom E 1978 Self-consistent embedded-cluster model for magnetic impurities: $\beta'$-NiAl *J. Appl. Phys.* **49** 1543

Ellis D E, Benesh G A and Byrom E 1979 Self-consistent embedded-cluster model for magnetic impurities: Fe, Co, and Ni in $\beta'$-NiAl *Phys. Rev.* B **20** 1198

[173] Benesh G A and Inglesfield J E 1984 An embedding approach for surface calculations *J. Phys. C: Solid State Phys.* **17** 1595

Inglesfield J E and Benesh G A 1988 Surface electronic structure: embedded self-consistent calculations *Phys. Rev.* B **37** 6682

Aers G C and Inglesfield J E 1989 Electric field and Ag(001) surface electronic structure *Surf. Sci.* **217** 367

Colbourn E A and Inglesfield J E 1991 Effective charges and surface stability of O on Cu(001) *Phys. Rev. Lett.* **66** 2006

Crampin S, van Hoof J B A N, Nekovee M and Inglesfield J E 1992 Full-potential embedding for surfaces and interfaces *J. Phys.: Condens. Matter* **4** 1475

Benesh G A and Liyanage L S G 1994 Surface-embedded Green-function method for general surfaces: application to Al(111) *Phys. Rev.* B **49** 17 264

Trioni M I, Brivio G P, Crampin S and Inglesfield J E 1996 Embedding approach to the isolated adsorbate *Phys. Rev.* B **53** 8052–64

[174] Scheffler M, Droste Ch, Fleszar A, Maca F, Wachutka G and Barzel G 1991 A self-consistent surface-Green-function (SSGF) method *Physica* B **172** 143

Wachutka G, Fleszar A, Maca F and Scheffler M 1992 Self-consistent Green-function method for the calculation of electronic properties of localized defects at surfaces and in the bulk *J. Phys.: Condens. Matter* **4** 2831

Bormet J, Neugebauer J and Scheffler M 1994 Chemical trends and bonding mechanisms for isolated adsorbates on Al(111) *Phys. Rev.* B **49** 17 242

Wenzien B, Bormet J and Scheffler M 1995 Green function for crystal surfaces I *Comp. Phys. Commun.* **88** 230

[175] Feibelman P J 1987 Force and total-energy calculations for a spatially compact adsorbate on an extended, metallic crystal surface *Phys. Rev.* B **35** 2626

[176] Williams A R, Feibelman P J and Lang N D 1982 Green's-function methods for electronic-structure calculations *Phys. Rev.* B **26** 5433

[177] Feibelman P J 1991 Orientation dependence of the hydrogen molecule's interaction with Rh(001) *Phys. Rev. Lett.* **67** 461

[178] Feibelman P J 1994 Sulfur adsorption near a step on Al *Phys. Rev.* B **49** 14 632

[179] Feibelman P J 1994 Diffusion barrier for a Ag adatom on Pt(111) *Surf. Sci.* **313** L801

[180] Krüger S and Rösch N 1994 The moderately-large-embedded-cluster method for metal surfaces; a density-functional study of atomic adsorption *J. Phys.: Condens. Matter* **6** 8149

Krüger S, Birkenheuer U and Rösch N 1994 Density functional approach to moderately large cluster embedding for infinite metal substrates *J. Electron Spectrosc. Relat. Phenom.* **69** 31

[181]  Head J D and Silva S J 1996 A localized orbitals based embedded cluster procedure for modeling chemisorption on large finite
       clusters and infinitely extended surfaces *J. Chem. Phys.* **104** 3244
[182]  Brivio G P and Trioni M I 1999 The adiabatic molecule–metal surface interaction: theoretical approaches *Rev. Mod. Phys.* **71**
       231–65
[183]  Cortona P 1991 Self-consistently determined properties of solids without band structure calculations *Phys. Rev. B* **44** 8454
       Cortona P 1992 Direct determination of self-consistent total energies and charge densities of solids: A study of the cohesive
       properties of the alkali halides *Phys. Rev. B* **46** 2008
[184]  Wesolowski T A and Warshel A 1993 Frozen density functional approach to *ab initio* calculations of solvated molecules *J. Phys.
       Chem.* **97** 8050
       Wesolowski T A and Warshel A 1994 *Ab initio* free energy perturbation calculations of solvation free energy using the frozen
       density functional approach *J. Phys. Chem.* **98** 5183
[185]  Stefanovich E V and Truong T N 1996 Embedded density functional approach for calculations of adsorption on ionic crystals
       *J. Chem. Phys.* **104** 2946
[186]  Garcia A, Elsässer C, Zhu J, Louie S G and Cohen M L 1992 Use of gradient-corrected functionals in total-energy calculations
       for solids *Phys. Rev. B* **46** 9829
[187]  Nachtigall P, Jordan K D, Smith A and Jónsson H 1996 Investigation of the reliability of density functional methods: reaction
       and activation energies for Si-Si bond cleavage and $H_2$ elimination from silanes *J. Chem. Phys.* **104** 148
[188]  Perdew J P, Burke K and Ernzerhof M 1996 Generalized gradient approximation made simple *Phys. Rev. Lett.* **77** 3865
       Proynov E I, Sirois S and Salahub D R 1997 Extension of the LAP functional to include parallel spin correlation *Int. J. Quantum
       Chem.* **64** 427
       Tozer D J, Handy N C and Green W H 1997 Exchange-correlation functionals from *ab initio* electron densities *Chem. Phys. Lett.*
       **273** 183
       Filatov M and Thiel W 1997 A new gradient-corrected exchange-correlation density functional *Mol. Phys.* **91** 847
       van Voorhis T and Scuseria G E 1998 A novel form for the exchange-correlation energy functional *J. Chem. Phys.* **109** 400
       Zhang Y and Yang W 1998 *Phys. Rev. Lett.* **80** 890
       Tozer D J and Handy N C 1998 The development of new exchange-correlation functionals *J. Chem. Phys.* **108** 2545
[189]  Abarenkov I V, Bulatov V L, Godby R, Heine V, Payne M C, Souchko P V, Titov A V and Tupitsyn I I 1997 Electronic-structure
       multiconfiguration calculation of a small cluster embedded in a local-density approximation host *Phys. Rev. B* **56** 1743
[190]  Govind N, Wang Y A, da Silva A J R and Carter E A 1998 Accurate *ab initio* energetics of extended systems via explicit correlation
       embedded in a density functional environment *Chem. Phys. Lett.* **295** 129
[191]  Govind N, Wang Y A and Carter E A 1999 Electronic structure calculations by first principles density-based embedding of
       explicitly correlated systems *J. Chem. Phys.* **110** 7677
[192]  Kluener T, Wang Y A, Govind N and Carter E A 2000 in preparation
[193]  Rubio A, Corkill J L, Cohen M L, Shirley E L and Louie S G 1993 Quasiparticle band structure of AlN and GaN *Phys. Rev. B* **48**
       11 810–16
[194]  Dhesi S S, Stagarescu C B, Smith K E, Doppalapudi D, Singh R and Moustakas T D 1997 Surface and bulk electronic structure
       of thin-film wurtzite GaN *Phys. Rev. B* **56** 10 271–5
[195]  Starrost F 1999 *PhD Thesis* Christian-Albrechts-Universität Kiel
       Starrost F, Krasovskii E E and Schattke W 1999 unpublished
[196]  Günther O, Janowitz C, Jungk G, Jenichen B, Hey R, Däweritz L and Ploog K 1995 Comparison between the electronic dielectric
       functions of a GaAs/AlAs superlattice and its bulk components by spectroscopic ellipsometry using core levels *Phys. Rev. B*
       **52** 2599–609
[197]  Starrost F, Krasovskii E E and Schattke W 1998 An alternative full-potential ELAPW method *Verhandl. DPG (VI)* **33** 741
[198]  Aspnes D E and Studna A A 1983 Dielectric functions and optical parameters of Si, Ge, GaP, GaAs, GaSb, InP, InAs, and InSb
       from 1.5 to 6.0 eV *Phys. Rev. B* **27** 985–1009
[199]  Logothetidis S, Alouani M, Garriga M and Cardona M 1990 $E_2$ interband transitions in $Al_xGa_{1-x}As$ alloys *Phys. Rev. B* **41**
       2959–65
[200]  Hughes J L P and Sipe J E 1996 Calculation of second-order optical response in semiconductors *Phys. Rev. B* **53** 10 751–63
[201]  Wang C S and Klein B M 1981 First-principles electronic structure of Si, Ge, GaP, GaAs, ZnS and ZnSe. II. Optical properties
       *Phys. Rev. B* **24** 3417–29
[202]  Huang Ming-Zhu and Ching W Y 1993 Calculation of optical excitations in cubic semiconductors. I. Electronic structure and
       linear response *Phys. Rev. B* **47** 9449–63

## Further Reading

Pisani C (ed) 1996 *Quantum-Mechanical Ab-initio Calculation of the Properties of Crystalline Materials (Lecture Notes
in Chemistry vol 67)* (Berlin: Springer)

A general introduction.

Dreizler R M and Gross E K U 1990 *Density Functional Theory: an Approach to the Quantum Many-body Problem* (Berlin: Springer)

A monograph on the foundations of density functional theory.

Pisani C, Doves R and Roetti C 1988 *Hartree–Fock Ab Initio Treatment of Crystalline Systems (Lecture Notes in Chemistry vol 48)* (Berlin: Springer)

An introduction to periodic Hartree–Fock.

Nemoshkalenko V V and Antonov V N 1998 *Computational Methods in Solid State Physics* (Amsterdam: Gordon and Breach)

An explicit introduction to the all-electron methods.

Singh D J 1994 *Planewaves, Pseudopotentials and the LAPW Method* (Norwell, MA: Kluwer)

A textbook on plane-wave and LAPW methods.

Whitten J L and Yang H 1996 Theory of chemisorption and reactions on metal surfaces *Surf. Sci. Rep.* **24** 59–124

# B3.3
# Statistical mechanical simulations

*Michael P Allen*

## B3.3.1 Introduction

Computer simulation, at the molecular level, has grown enormously in importance over the last 50 years. Affordable computer chips have historically doubled in power every 18 months, so the computer simulator, regarded as an experimentalist, has the unique advantage of rapidly improving apparatus. With the recent explosion in personal computing, there seems every prospect that this situation will continue, allowing computer simulation to become of even more practical value in fields such as the design of drugs and molecular materials. This provides a stimulus to develop simulation methods, and an industry has grown up marketing the necessary software.

This chapter concentrates on describing molecular simulation methods which have a connection with the statistical mechanical description of condensed matter, and hence relate to theoretical approaches to understanding phenomena such as phase equilibria, rare events, and quantum mechanical effects.

### B3.3.1.1 The aims of simulation

We carry out computer simulations in the hope of understanding bulk, macroscopic properties in terms of the microscopic details of molecular structure and interactions. This serves as a complement to conventional experiments, enabling us to learn something new; something that cannot be found out in other ways.

Computer simulations act as a bridge between microscopic length and time scales and the macroscopic world of the laboratory (see figure B3.3.1). We provide a guess at the interactions between molecules, and obtain 'exact' predictions of bulk properties. The predictions are 'exact' in the sense that they can be made as accurate as we like, subject to the limitations imposed by our computer budget. At the same time, the hidden detail behind bulk measurements can be revealed. Examples are the link between the diffusion coefficient and velocity autocorrelation function (the former easy to measure experimentally, the latter much harder); and the connection between equations of state and structural correlation functions.

Simulations act as a bridge in another sense: between theory and experiment (see figure B3.3.2). We can test a theory using idealized models, conduct 'thought experiments', and clarify what we measure in the laboratory. We may also carry out simulations on the computer that are difficult or impossible in the laboratory (for example, working at extremes of temperature or pressure).

Ultimately we may want to make direct comparisons with experimental measurements made on specific materials, in which case a good model of molecular interactions is essential. The aim of so-called *ab initio* molecular dynamics is to reduce the amount of fitting and guesswork in this process to a minimum. On the other hand, we may be interested in phenomena of a rather generic nature, or we may simply want to discriminate between good and bad theories. When it comes to aims of this kind, it is not necessary to have a perfectly realistic molecular model; one that contains the essential physics may be quite suitable.

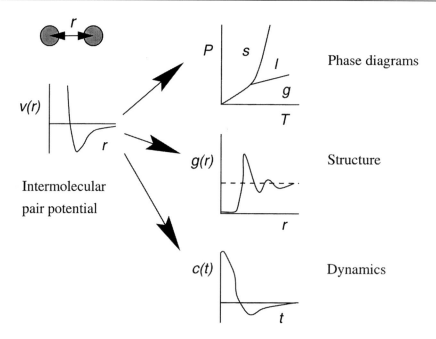

**Figure B3.3.1.** Simulations as a bridge between the microscopic and the macroscopic. We input details of molecular structure and interactions; we obtain predictions of phase behaviour, structural and time-dependent properties.

The two main families of simulation technique are molecular dynamics (MD) and Monte Carlo (MC). Additionally, there is a whole range of hybrid techniques which combine features from both MC and MD.

### B3.3.1.2   The techniques of simulation

Molecular dynamics consists of the brute-force solution of Newton's equations of motion. It is necessary to encode in the program the potential energy and force law of interaction between molecules; the equations of motion are solved numerically, by finite difference techniques. The system evolution corresponds closely to what happens in 'real life' and allows us to calculate dynamical properties, as well as thermodynamic and structural functions. For a range of molecular models, packaged routines are available, either commercially or through the academic community.

Monte Carlo can be thought of as a prescription for sampling configurations from a statistical ensemble. The interaction potential energy is coded into the program, and a random walk procedure adopted to go from one state of the system to the next. MC programs can be relatively easy to program; they allow us to calculate thermodynamic and structural properties, but not exact dynamics. It is relatively simple to specify external conditions (constant temperature, pressure etc.) and many tricks may be devised to improve the efficiency of the sampling.

Both MD and MC techniques evolve a finite-sized molecular configuration forward in time, in a step-by-step fashion. (In this context, MC simulation 'time' has to be interpreted liberally, but there is a broad connection between real time and simulation time (see [1, chapter 2]).) Common features of MD and MC simulation techniques are that there are limits on the typical timescales and length scales that can be investigated. The consequences of finite size must be considered both in specifying the molecular interactions, and in analysing the results.

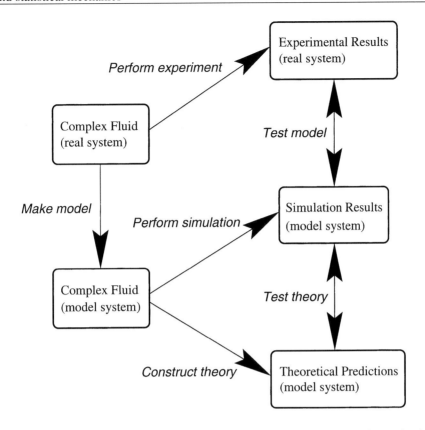

**Figure B3.3.2.** Simulation as a bridge between theory and experiment. We may test a theory by conducting a simulation using the same model. We may test the model by comparing with experimental results.

### B3.3.2  Simulation and statistical mechanics

Here we consider various aspects of statistical mechanics (see also chapter A2.3 and [2, 3]) that have a direct bearing on computer simulation methodology.

#### B3.3.2.1  *Simulation time and length scales*

Simulation runs are typically short ($t \sim 10^3$–$10^6$ MD or MC steps, corresponding to perhaps a few nanoseconds of real time) compared with the time allowed in laboratory experiments. This means that we need to test whether or not a simulation has reached equilibrium before we can trust the averages calculated in it. Moreover, there is a clear need to subject the simulation averages to a statistical analysis, to make a realistic estimate of the errors.

How long should we run? This depends on the system and the physical properties of interest. Suppose that we are interested in a variable $\mathcal{X}$, defined such that its ensemble average $X = \langle \mathcal{X} \rangle = 0$. (Here and throughout we use script letters for instantaneous dynamical variables, i.e., functions of coordinates and momenta, to distinguish them from averages and thermodynamic quantities.) A characteristic time, $\tau$, may be defined, over which the correlations $\langle \mathcal{X}(0)\mathcal{X}(t) \rangle$ decay towards zero. The simulation run time $t_{\text{run}}$ should be significantly longer than $\tau$. The time scales of properties of interest will vary from one system to another; they may not be predictable in advance, and this will have a bearing on the length of simulation required.

Similar considerations apply to the size of system simulated. The samples involved are typically quite small on the laboratory scale. Most fall in the range $N \sim 10^3$–$10^6$ particles, thus imposing a restriction on the length scales of the phenomena that may be investigated, in the nanometre–submicron range. Indeed, in many cases, there is an overriding need to do a system-size analysis of simulation results, to quantify these effects.

How large a simulation do we need? Once more this depends on the system and properties of interest. From a spatial correlation function $\langle \mathcal{X}(0)\mathcal{X}(r) \rangle$ relating values computed at different points $r$ apart, we may define a characteristic distance $\xi$ over which the correlation decays. The simulation box size $L$ should be significantly larger than $\xi$ in order not to influence the results.

The ratios, $t_{\mathrm{run}}/\tau$ and $L/\xi$, appear in expressions for estimating the errors on simulation-averaged quantities. Roughly speaking, a simulation sample can be regarded as a collection of $\sim(L/\xi)^3$ sub-samples, each making a statistically independent contribution to the average properties. Also, a simulation run may be regarded as a succession of $\sim t_{\mathrm{run}}/\tau$ statistically independent sub-runs. Then, the usual rules for combining independent samples apply, and estimated error bars are inversely proportional to the square root of the run time. For further information see [4–7].

Near critical points, special care must be taken, because the inequality $L \gg \xi$ will almost certainly not be satisfied; also, critical slowing down will be observed. In these circumstances a quantitative investigation of finite size effects and correlation times, with some consideration of the appropriate scaling laws, must be undertaken. Examples of this will be seen later; one of the most encouraging developments of recent years has been the establishment of reliable and systematic methods of studying critical phenomena by simulation.

### B3.3.2.2    *Periodic boundary conditions*

Small sample size means that, unless surface effects are of particular interest, periodic boundary conditions need to be used. Consider 1000 atoms arranged in a $10 \times 10 \times 10$ cube. Nearly half the atoms are on the outer faces, and these will have a large effect on the measured properties. Surrounding the cube with replicas of itself takes care of this problem. Provided the potential range is not too long, we can adopt the *minimum image convention* that each atom interacts with the nearest atom or image in the periodic array. In the course of the simulation, if an atom leaves the basic simulation box, attention can be switched to the incoming image. This is shown in figure B3.3.3. Of course, it is important to bear in mind the imposed artificial periodicity when considering properties which are influenced by long-range correlations. Special attention must be paid to the case where the potential range is not short: for example, for charged and dipolar systems. Methods for handling this are discussed later.

### B3.3.2.3    *Molecular interactions*

Let us denote a 'state of the system' by $\Gamma$. For the purposes of discussion, we shall concentrate on a system composed of atoms, and for this $\Gamma$ represents the complete set of coordinates $r^{(N)} = (r_1, r_2, \ldots r_N)$ and conjugate momenta $p^{(N)} = (p_1, p_2, \ldots p_N)$. Then the energy, or Hamiltonian, may be written as a sum of kinetic and potential terms $\mathcal{H} = \mathcal{K} + \mathcal{V}$. For atomic systems, $\mathcal{V}$ is a function of coordinates only and $\mathcal{K}$ may be written as a function of momenta; in molecular systems represented as rigid bodies, or in terms of generalized coordinates, the kinetic energy may also depend on the coordinates [8].

Sticking, for simplicity, with a simple atomic system, the kinetic energy may be written

$$\mathcal{K}(p_1, p_2, \ldots, p_N) = \sum_{i=1}^{N} \sum_{\alpha=x,y,z} p_{i\alpha}^2/2m_i.$$

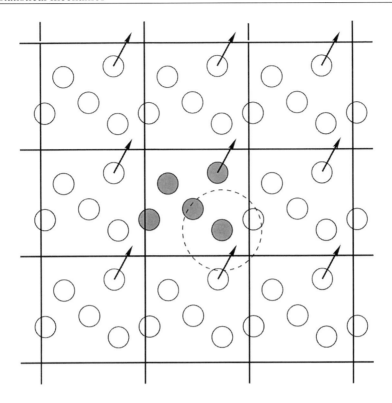

**Figure B3.3.3.** Periodic boundary conditions. As a particle moves out of the simulation box, an image particle moves in to replace it. In calculating particle interactions within the cutoff range, both real and image neighbours are included.

The potential energy $\mathcal{V}$ is traditionally split into one-body, two-body, three-body . . . terms:

$$\mathcal{V}(\boldsymbol{r}_1, \boldsymbol{r}_2, \ldots, \boldsymbol{r}_N) = \sum_i v^{(1)}(\boldsymbol{r}_i) + \sum_i \sum_{j>i} v^{(2)}(\boldsymbol{r}_i, \boldsymbol{r}_j) + \sum_i \sum_{j>i} \sum_{k>j} v^{(3)}(\boldsymbol{r}_i, \boldsymbol{r}_j, \boldsymbol{r}_k) + \cdots.$$

The $v^{(1)}$ term represents an externally applied potential field or the effects of the container walls; it is usually dropped for fully periodic simulations of bulk systems. Also, it is usual to neglect $v^{(3)}$ and higher terms (which in reality might be of order 10% of the total energy in condensed phases) and concentrate on $v^{(2)}$. For brevity henceforth we will just call this $v(r)$. There is an extensive literature on the way these potentials are determined experimentally, or modelled theoretically (see, e.g., [9–11]). In simulations, it is common to use the simplest models that faithfully represent the essential physics: the hard-sphere, square-well, and Lennard-Jones potentials have the longest history. The latter has the functional form

$$v^{\mathrm{LJ}}(r) = 4\varepsilon \left\{ \left(\frac{d}{r}\right)^{12} - \left(\frac{d}{r}\right)^6 \right\}$$

with two parameters: $d$, the diameter, and $\varepsilon$, the well depth. This potential was used, for instance, in the earliest studies of the properties of liquid argon [12, 13]. For molecular systems, we simply build the molecules out of site–site potentials of this, or similar, form (figure B3.3.4). If electrostatic charges are present, we add the appropriate Coulomb potentials

$$v^{\mathrm{Coulomb}}(r) = \frac{Q_1 Q_2}{4\pi \epsilon_0 r}$$

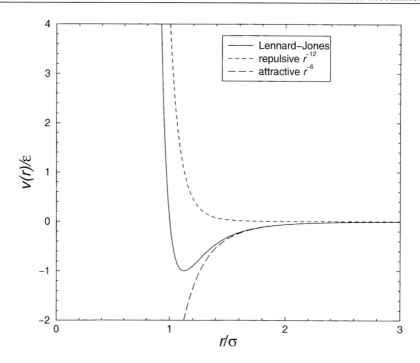

**Figure B3.3.4.** Lennard-Jones pair potential showing the $r^{-12}$ and $r^{-6}$ contributions.

where $Q_1$, $Q_2$ are the charges. We may also use rigid-body potentials which depend on centre of mass positions and orientations. An example is the Gay–Berne potential [14]

$$v^{\text{GB}}(r, \hat{u}_1, \hat{u}_2) = 4\varepsilon(\hat{r}, \hat{u}_1, \hat{u}_2)[\varrho^{-12} - \varrho^{-6}]$$

with

$$\varrho = \frac{r - d(\hat{r}, \hat{u}_1, \hat{u}_2) + d_0}{d_0}$$

which depends upon the molecular axis vectors $\hat{u}_1$ and $\hat{u}_2$, and on the direction $\hat{r}$ and magnitude $r$ of the centre–centre vector $r = r_1 - r_2$. The parameter $d_0$ determines the smallest molecular diameter and there are two orientation-dependent quantities in the above shifted Lennard-Jones form: a diameter $d(\hat{r}, \hat{u}_1, \hat{u}_2)$ and an energy $\varepsilon(\hat{r}, \hat{u}_1, \hat{u}_2)$. Each quantity depends in a complicated way (not given here) on parameters characterizing molecular shape and structure. This potential has been extensively used in the study of molecular liquids and liquid crystals [15–20].

It is common practice in classical computer simulations *not* to attempt to represent intramolecular bonds by terms in the potential energy function, because these bonds have very high vibration frequencies and should really be treated in a quantum mechanical way rather than in the classical approximation. Instead, the bonds are treated as being constrained to have fixed length, and some straightforward ways have been devised to incorporate these constraints into the dynamics (see later).

For a wide range of physical problems, a lattice spin system provides a useful, if very coarse-grained, description. The great advantage of such an approach is the speed with which such systems may be simulated, especially when a single spin may be taken to represent not just one molecule but a larger region of the physical system. The state $\Gamma$ of such systems may be specified by a set of discrete or continuous spin values $\sigma^{(N)} = (\sigma_1, \sigma_2, \ldots, \sigma_N)$, where $\sigma_i = \pm 1$ for the archetypal Ising model, but takes other values for other models. For Ising-like systems, the energy may be written $\mathcal{H} = -J \sum_{\langle ij \rangle} \sigma_i \sigma_j$, where $\sum_{\langle ij \rangle}$ indicates a sum over nearest neighbour spins $i$, $j$ and $J$ is a coupling constant. Again, there is much flexibility in the nature of the energy function. There is an extensive literature on spin simulations [6, 21–23] especially in relation to the theory of critical phenomena [24]. Spin models are not restricted to the obvious area of magnetic solids; it has proved possible to include, for instance, polymer liquids in this class [25], allowing the study of otherwise inaccessible behaviour.

### B3.3.2.4 Simulations and ensembles

One of the flexibilities of computer simulation is that it is possible to define the thermodynamic conditions corresponding to one of many statistical ensembles, each of which may be most suitable for the purpose of the study. A knowledge of the underlying statistical mechanics is essential in the design of correct simulation methods, and in the analysis of simulation results. Here we describe two of the most common statistical ensembles, but examples of the use of other ensembles will appear later in the chapter.

The microcanonical ensemble corresponds to an isolated system, with specified number of particles $N$, volume $V$, and energy $E$. The fundamental thermodynamic potential is the entropy, and it is related to statistical mechanical quantities as follows:

$$S = k \ln \Omega_{NVE}$$
$$\Omega_{NVE} = \int d\Gamma \, \delta[\mathcal{H}(\Gamma) - E]$$
$$\varrho_{NVE}(\Gamma) = \Omega_{NVE}^{-1} \delta[\mathcal{H}(\Gamma) - E]$$
$$\langle \mathcal{X} \rangle_{NVE} = \int d\Gamma \, \varrho_{NVE}(\Gamma) \, \mathcal{X}(\Gamma). \tag{B3.3.1}$$

Here, $\Omega_{NVE}$ is the number of states available to the system at given $NVE$, written as an integral over a thin energy shell; $\delta(\ldots)$ is the Dirac delta function. The ensemble average $\langle \mathcal{X} \rangle_{NVE}$ is defined in terms of the ensemble probability density function $\varrho_{NVE}(\Gamma)$.

The canonical ensemble corresponds to a system of fixed $N$ and $V$, able to exchange energy with a thermal bath at temperature $T$, which represents the effects of the surroundings. The thermodynamic potential is the Helmholtz free energy, and it is related to the partition function $Q_{NVT}$ as follows:

$$A = E - TS = -kT \ln Q_{NVT}$$
$$Q_{NVT} = \int d\Gamma \, e^{-\beta \mathcal{H}(\Gamma)} = \int dE \, \Omega_{NVE} \, e^{-\beta E}$$
$$\varrho_{NVT}(\Gamma) = Q_{NVT}^{-1} \, e^{-\beta \mathcal{H}(\Gamma)}$$
$$\langle \mathcal{X} \rangle_{NVT} = \int d\Gamma \, \varrho_{NVT}(\Gamma) \, \mathcal{X}(\Gamma). \tag{B3.3.2}$$

Here $\beta = 1/kT$. In a real system the thermal coupling with surroundings would happen at the surface; in simulations we avoid surface effects by allowing this to occur homogeneously. The state of the surroundings *defines* the temperature $T$ of the ensemble.

Since $\mathcal{H} = \mathcal{K} + \mathcal{V}$, the canonical ensemble partition function factorizes into ideal gas and excess parts, and as a consequence most averages of interest may be split into corresponding ideal and excess components, which sum to give the total. In MC simulations, we frequently calculate just the excess or configurational parts: in this case, $\Gamma$ consists just of the atomic coordinates, not the momenta, and the appropriate expressions are obtained from equation B3.3.2 by replacing $\mathcal{H}$ by the potential energy $\mathcal{V}$. The ideal gas contributions are usually easily calculated from exact expressions, in which the integrations over atomic momenta have been carried out analytically.

*B3.3.2.5   Averages and distributions*

It is generally well known that, for most averages, differences between ensembles disappear in the thermodynamic limit. However, for finite-sized systems of the kind studied in simulations, it is necessary to consider the differences between ensembles, which will be significant for mean-squared values (fluctuations) and, more generally, for the probability distributions of measured quantities. For example, energy fluctuations in the constant-$NVE$ ensemble are (by definition) zero, whereas in the constant-$NVT$ ensemble they are not. Since these points have a bearing on various aspects of simulation methodology, we expand on them a little here.

It is a standard result in the canonical ensemble that energy fluctuations are related to the heat capacity $C_V = (\partial E/\partial T)_V$:

$$kT^2 C_V = \langle \mathcal{H}^2 \rangle - \langle \mathcal{H} \rangle^2 = \langle \delta \mathcal{H}^2 \rangle.$$

Since $C_V$ and $E$ are both extensive properties ($\propto N$), the root-mean-square energy fluctuations are smaller, by a factor $1/\sqrt{N}$, than typical average energies $E$. As the system size increases, the relative magnitude of fluctuations decreases, and the thermodynamic limit is achieved.

It is instructive to see this in terms of the canonical ensemble probability distribution function for the energy, $\mathcal{P}_{NVT}(E)$. Referring to equations (B3.3.1) and (B3.3.2), it is relatively easy to see that

$$\mathcal{P}_{NVT}(E) = \langle \delta(\mathcal{H}(\Gamma) - E) \rangle_{NVT} = \frac{\Omega_{NVE}\, e^{-E/kT}}{Q_{NVT}}.$$

The product of a rapidly increasing function of energy, $\Omega_{NVE}$, and a rapidly decreasing function $e^{-E/kT}$, gives a distribution of energies which is very sharply peaked about the most likely value, as shown in figure B3.3.5. A reasonable first approximation to $\mathcal{P}_{NVT}(E)$ is a Gaussian function, centred on this most probable value, with a width determined by $C_V$.

In principle, these formulae may be used to convert results obtained at one state point into averages appropriate to a neighbouring state point. For any canonical ensemble average

$$\langle \mathcal{X} \rangle_{NVT_1} = \frac{\langle \mathcal{X}\, e^{-\delta\beta\mathcal{H}} \rangle_{NVT_0}}{\langle e^{-\delta\beta\mathcal{H}} \rangle_{NVT_0}} \tag{B3.3.3}$$

where $\delta\beta = \beta_1 - \beta_0$. Choosing $\mathcal{X} = \delta(\mathcal{H} - E)$ gives a way of re-weighting the energy distribution

$$\mathcal{P}_{NVT_1}(E) \equiv \langle \delta(\mathcal{H} - E) \rangle_{NVT_1} = \mathcal{P}_{NVT_0}(E)\, \frac{e^{-\delta\beta E}}{\langle e^{-\delta\beta\mathcal{H}} \rangle_{NVT_0}}. \tag{B3.3.4}$$

Such histogram re-weighting techniques have a long history [26–29]. The usefulness of this equation depends sensitively on accurate sampling of energies in the region of interest, at $T_1$, which may be far away from the maximum in $\mathcal{P}_{NVT_0}(E)$. We have seen how the mean-squared fluctuations in $E$ are related to the heat capacity; higher-order, non-Gaussian terms in $\mathcal{P}_{NVT}(E)$ are also related to thermodynamic derivatives [30, 31], but they are smaller, and so are hard to measure accurately. This limits the extension to nearby state points; a common theme of computer simulation is the devising of techniques to get around this problem, and we shall return to this later. Similar considerations apply to volume distributions in constant-pressure ensembles, and indeed to other cases of thermodynamically conjugated pairs of variables.

Statistical mechanics may be used to derive practical microscopic formulae for thermodynamic quantities. A well-known example is the virial expression for the pressure, easily derived by scaling the atomic coordinates in the canonical ensemble partition function

$$PV = NkT - \frac{1}{3}\left\langle \sum_i \sum_{j \neq i} w_{ij} \right\rangle.$$

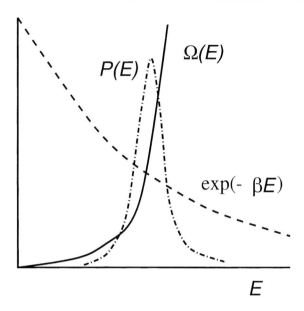

**Figure B3.3.5.** Energy distributions. The probability density is proportional to the product of the density of states and the Boltzmann factor.

Here we assumed pairwise additivity $\mathcal{V} = \sum_i \sum_{j \neq i} v_{ij}$, and defined $w(r) = r(dv(r)/dr)$. Also easily derived in the canonical ensemble is the general virial-like form, where $q$ may be any coordinate or momentum,

$$\left\langle \mathcal{X} \frac{\partial \mathcal{H}}{\partial q} \right\rangle = kT \left\langle \frac{\partial \mathcal{X}}{\partial q} \right\rangle.$$

A well known example of this is obtained by setting $\mathcal{X} = p_{i\alpha}$, $\alpha = x, y, z$, any component of momentum, giving the equipartition-of-energy relation

$$\left\langle \sum_{i\alpha} p_{i\alpha}^2 / 2m_i \right\rangle = \frac{3}{2} NkT.$$

This is commonly used to measure the temperature in a MD simulation. Less well known is the hypervirial relation obtained by setting $\mathcal{X} = -(\partial \mathcal{V}/\partial r_{i\alpha}) = f_{i\alpha}$, a component of the force:

$$\left\langle \sum_{i\alpha} \left( \frac{\partial \mathcal{V}}{\partial r_{i\alpha}} \right)^2 \right\rangle = kT \left\langle \sum_{i\alpha} \left( \frac{\partial^2 \mathcal{V}}{\partial r_{i\alpha}^2} \right) \right\rangle$$

which relates the Laplacian of the potential with the mean-squared force. Butler *et al* [32] have suggested using this expression to measure the 'configurational' temperature (as a check) in MC simulations; they also provide a derivation of the corresponding expression in the microcanonical ensemble (see also [33]); it is surprising that this useful proposal has only been made so recently.

  Finally, by considering increasing the number of particles by one in the canonical ensemble (looking at the excess, non-ideal, part), it is easy to derive the Widom [34] test-particle formula

$$\mu^{\text{ex}} \approx A_{N+1}^{\text{ex}} - A_N^{\text{ex}} = -kT \ln\langle e^{-\beta v_{\text{test}}} \rangle. \tag{B3.3.5}$$

Here we have separated terms in the potential energy which involve the extra 'test' particle, $\mathcal{V}_{N+1} = \mathcal{V}_N + v_{\text{test}}$. The ensemble average here includes an unweighted average over inserted particle coordinates. In practice this means randomly inserting a test particle, many times, and averaging the Boltzmann factor of the associated energy change. More details of free energy calculations will be given later.

### B3.3.2.6  Time dependence

A knowledge of time-dependent statistical mechanics is important in three general areas of simulation. First, in recent years there have been significant advances in the understanding of MD algorithms, which have arisen out of an appreciation of the formal operator approach to classical mechanics. Second, an understanding of equilibrium time correlation functions, their link with dynamical properties and especially their connection with transport coefficients, is essential in making contact with experiment. Third, the last decade has seen a rapid development of the use of nonequilibrium MD, with a better understanding of the formal aspects, particularly the link between the dynamical algorithm, dissipation, chaos and fractal geometry. Space does not permit a full description of this here: the interested reader should consult [35, 36] and references therein.

The Liouville equation dictates how the classical statistical mechanical distribution function $\varrho(r^{(N)}, p^{(N)}, t)$ evolves in time. (Also, quantum dynamics may be expressed in a formally equivalent way, but we shall concentrate exclusively on classical systems here.) From considerations of standard, Hamiltonian, mechanics [8] and the flow of representative systems in an ensemble through a particular region of phase space, it is easy to derive the Liouville equation

$$\frac{\partial \varrho}{\partial t} = -\left\{ \sum_{i\alpha} \dot{r}_{i\alpha} \frac{\partial}{\partial r_{i\alpha}} + \dot{p}_{i\alpha} \frac{\partial}{\partial p_{i\alpha}} \right\} \varrho \equiv -i\hat{L}\varrho$$

defining the Liouville operator $\hat{L}$. Compare this equation for $\varrho$ with the time evolution equation for a dynamical variable $\mathcal{X}(r^{(N)}, p^{(N)})$, which comes directly from the chain rule applied to Hamilton's equations

$$\dot{\mathcal{X}} = \sum_{i\alpha} \dot{r}_{i\alpha} \frac{\partial \mathcal{X}}{\partial r_{i\alpha}} + \dot{p}_{i\alpha} \frac{\partial \mathcal{X}}{\partial p_{i\alpha}} \equiv i\hat{L}\mathcal{X}.$$

The formal solutions of the time evolution equations are

$$\varrho(t) = e^{-i\hat{L}t}\varrho(0) \qquad \text{and} \qquad \mathcal{X}(t) = e^{i\hat{L}t}\mathcal{X}(0). \qquad (B3.3.6)$$

A number of manipulations are possible, once this formalism has been established. There are useful analogies both with the Eulerian and Lagrangian pictures of incompressible fluid flow, and with the Heisenberg and Schrödinger pictures of quantum mechanics [37, chapter 7], [38, chapter 11]. These analogies are particularly useful in formulating the equations of classical response theory [39], linking transport coefficients with both equilibrium and nonequilibrium simulations [35].

The Liouville equation applies to any ensemble, equilibrium or not. Equilibrium means that $\varrho$ should be *stationary*, i.e., that

$$\partial \varrho / \partial t = 0.$$

In other words, if we look at any phase-space volume element, the rate of incoming state points should equal the rate of outflow. This requires that $\varrho$ be a function of the constants of the motion, and especially $\varrho = \varrho(\mathcal{H})$. Equilibrium also implies $d\langle \mathcal{X} \rangle / dt = 0$ for any $\mathcal{X}$. The extension of the above equations to nonequilibrium ensembles requires a consideration of entropy production, the method of controlling energy dissipation (thermostatting) and the consequent non-Liouville nature of the time evolution [35].

### B3.3.3 Molecular dynamics

The solution of Newton's or Hamilton's equations on the computer

$$\dot{r}_i = p_i/m_i \qquad \text{and} \qquad \dot{p}_i = f_i$$

where $m_i$ is the mass of atom $i$, and $f_i$ is the total force acting on it, is intrinsically a simple task. Many methods exist to perform step-by-step numerical integration of systems of coupled ordinary differential equations. Characteristics of these equations are: (a) they are 'stiff', i.e., there may be short and long time scales, and the algorithm must cope with both; (b) calculating the forces is expensive, typically involving a sum over pairs of atoms, and should be performed as infrequently as possible.

Also we must bear in mind that the advancement of the coordinates fulfils two functions: (i) accurate calculation of dynamical properties, especially over times as long as typical correlation times $\tau$; (ii) accurately staying on the constant-energy hypersurface, for much longer times $t_{run}$. Exact time reversibility is highly desirable (since the original equations are exactly reversible). To ensure rapid sampling of phase space, we wish to make the time step as large as possible, consistent with these requirements. For these reasons, simulation algorithms have tended to be of *low order* (i.e., they do not involve storing high derivatives of positions, velocities etc): this allows the time step to be increased as much as possible without jeopardizing energy conservation. It is unrealistic to expect the numerical method to accurately follow the true trajectory for very long times $t_{run}$. The 'ergodic' and 'mixing' properties of classical trajectories, i.e., the fact that nearby trajectories diverge from each other exponentially quickly, make this impossible to achieve.

All these observations tend to favour the Verlet algorithm in one form or another, and we look closely at this in the following sections. For historical reasons only, we mention the more general class of predictor–corrector methods which have been optimized for classical mechanics simulations, [40,41]; further details are available elsewhere [7,42,43].

### B3.3.3.1 The Verlet algorithm

There are various, essentially equivalent, versions of the Verlet algorithm, including the original method employed by Verlet [13,44] in his investigations of the properties of the Lennard-Jones fluid, and a 'leapfrog' form [45]. Here we concentrate on the 'velocity Verlet' algorithm [46], which may be written

$$r_i(t+\delta t) = r_i(t) + \delta t\, p_i(t)/m_i + \tfrac{1}{2}\delta t^2 f_i(t)/m_i \qquad p_i(t+\delta t) = p_i(t) + \tfrac{1}{2}\delta t[f_i(t) + f_i(t+\delta t)].$$

This advances the coordinates and momenta over a small time step $\delta t$. A piece of pseudo-code illustrates how this works:

```
call force(r,f)
do step = 1, nstep
    r = r + dt*p/m + (0.5*dt**2)*f/m
    p = p + 0.5*dt*f
    call force(r,f)
    p = p + 0.5*dt*f
enddo
```

The forces are calculated from the positions at the start of a simulation. They are used to advance the positions, and 'half-advance' the velocities or momenta. The new forces $f(t+\delta t)$ are calculated, and these are used to complete the momentum update. At the end of the step, positions, momenta, and forces all conveniently refer to the same time point. Moreover, as we shall see shortly there is an interesting theoretical derivation of this version of the algorithm.

Important features of the Verlet algorithm are: (a) it is *exactly* time reversible; (b) it is *low* order in time, hence permitting long time steps; (c) it is easy to program.

*B3.3.3.2   Propagators and the Verlet algorithm*

The velocity Verlet algorithm may be derived by considering a standard approximate decomposition of the Liouville operator which preserves reversibility and is *symplectic* (which implies that volume in phase space is conserved). This approach [47] has had several beneficial consequences.

The Liouville operator of equation B3.3.6 may be written [48]

$$e^{i\hat{L}t} = (e^{i\hat{L}\delta t})^P_{\text{approx}} + \mathcal{O}(P\delta t^3)$$

where $\delta t = t/P$ and an approximate propagator, correct at short time steps $\delta t \to 0$, appears in the parentheses. This is a formal way of stating what we do in MD, when we split a long time period $t$ into a large number $P$ of small time steps $\delta t$, using an *approximation* to the true equations of motion over each time step. It turns out that useful approximations arise from splitting $\hat{L}$ into two parts

$$\hat{L} = \hat{L}_p + \hat{L}_r.$$

The following approximation

$$e^{i\hat{L}\delta t} = e^{(i\hat{L}_p + i\hat{L}_r)\delta t} \approx e^{i\hat{L}_p\delta t/2} e^{i\hat{L}_r\delta t} e^{i\hat{L}_p\delta t/2} \qquad (B3.3.7)$$

is asymptotically exact in the limit $\delta t \to 0$. For nonzero $\delta t$ this is an approximation to $e^{i\hat{L}\delta t}$ because in general $\hat{L}_p$ and $\hat{L}_r$ do not commute, but it is still exactly time reversible. Tuckerman *et al* [47] set

$$i\hat{L}_p = \sum_{i\alpha} \dot{p}_{i\alpha} \frac{\partial}{\partial p_{i\alpha}} = \sum_{i\alpha} f_{i\alpha} \frac{\partial}{\partial p_{i\alpha}} \qquad i\hat{L}_r = \sum_{i\alpha} \dot{r}_{i\alpha} \frac{\partial}{\partial r_{i\alpha}} = \sum_{i\alpha} (p_{i\alpha}/m) \frac{\partial}{\partial r_{i\alpha}}.$$

A straightforward derivation (not reproduced here) shows that the effect of the three successive steps embodied in equation (B3.3.7), with the above choice of operators, is precisely the velocity Verlet algorithm. This approach is particularly useful for generating multiple time-step methods.

*B3.3.3.3   Multiple time steps*

An important extension of the MD method allows it to tackle systems with multiple time scales: for example, molecules which have very strong internal springs representing the bonds, while interacting externally through softer potentials, or molecules consisting of both heavy and light atoms. A simple MD algorithm will have to adopt a time step short enough to handle the fast-varying internal motions.

Tuckerman *et al* [47] set out methods for generating time-reversible Verlet-like algorithms using the Liouville operator formalism described above. Here we suppose that there are two types of force in the system: slow-moving external forces $\boldsymbol{F}_i$ and fast-moving internal forces $\boldsymbol{f}_i$. The momentum satisfies $\dot{\boldsymbol{p}}_i = \boldsymbol{f}_i + \boldsymbol{F}_i$. Then we break up the Liouville operator $i\hat{L} = i\hat{L}_p + i\hat{\ell}$:

$$i\hat{L}_p = \sum_{i\alpha} F_{i\alpha} \frac{\partial}{\partial p_{i\alpha}} \qquad i\hat{\ell} = \sum_{i\alpha} \dot{r}_{i\alpha} \frac{\partial}{\partial r_{i\alpha}} + f_{i\alpha} \frac{\partial}{\partial p_{i\alpha}} \equiv i\hat{\ell}_r + i\hat{\ell}_p.$$

The propagator approximately factorizes

$$e^{i\hat{L}\Delta t} \approx e^{i\hat{L}_p\Delta t/2} e^{i\hat{\ell}\Delta t} e^{i\hat{L}_p\Delta t/2}$$

where $\Delta t$ represents a long time step. The middle part is then split again, using the conventional separation, and iterating over small time steps $\delta t = \Delta t/P$:

$$e^{i\hat{\ell}\Delta t} \approx [e^{i\frac{1}{2}\delta t\hat{\ell}_p} e^{i\delta t\hat{\ell}_r} e^{i\frac{1}{2}\delta t\hat{\ell}_p}]^P.$$

So the fast-varying forces must be computed many times at short intervals; the slow-varying forces are used just before and just after this stage, and they only need be calculated once per long time step.

This actually translates into a fairly simple algorithm, based closely on the standard velocity Verlet method. Written in a Fortran-like pseudo-code, it is as follows. At the start of the run we calculate both rapidly-varying (f) and slowly-varying (F) forces, then, in the main loop:

```
do STEP = 1, NSTEP
   p = p + 0.5*DT*F
   do step = 1, nstep
              r = r + dt*p/m + (0.5*dt**2)*f/m
              p = p + 0.5*dt*f
              call force(r,f)
              p = p + 0.5*dt*f

   enddo
   call FORCE(r,F)
   p = p + 0.5*DT*F
enddo
```

The entire simulation run consists of NSTEP long steps; each step consists of nstep shorter sub-steps. DT and dt are the corresponding time steps, DT = nstep*dt.

A particularly fruitful application, which has been incorporated into the computer code ORAC [49], is to split the interatomic force law into a succession of components covering different ranges: the short-range forces change rapidly with time and require a short time step, but advantage can be taken of the much slower time variation of the long-range forces, by using a longer time step and less frequent evaluation for these. Having said this, multiple time step algorithms are still under active study [50], and there is some concern that resonances may occur between the natural frequencies of the system under study, and the various time steps used in schemes of this kind [51].

### B3.3.3.4   *Constraints*

Although, in principle, multiple time steps provide a method of integrating stiff degrees of freedom, such as intramolecular bonds, the alternative of rigidly constraining bond lengths is still very popular. In classical mechanics, constraints are introduced through the Lagrangian [8] or Hamiltonian [52] formalisms. Given a set of algebraic relations between atomic coordinates, for example, a fixed bond length $b$ between atoms 1 and 2

$$\sigma(r_1, r_2) = (r_1 - r_2) \cdot (r_1 - r_2) - b^2 = 0$$

the constraint force between the atoms will have the following form

$$g_1 = \lambda \frac{\partial \sigma}{\partial r_1} \qquad \text{and} \qquad g_2 = \lambda \frac{\partial \sigma}{\partial r_2}$$

and it will appear in the equations of motion along with the normal forces. It is easy to derive an exact expression for the multiplier $\lambda$; for many constraints, a system of equations (one per constraint) is obtained. In practice, since the equations of motion are only solved approximately, the constraints will be increasingly violated as the simulation proceeds. The breakthrough in this area came with the proposal of a scheme, SHAKE, to solve the equations for the constraint forces *approximately* (i.e., to the same level of approximation as the dynamical algorithm) in such a way that the constraints are satisfied exactly at the end of each time step [53, 54]; for a review see [55]. The appropriate version of this scheme for the velocity Verlet algorithm is called RATTLE [56].

It is important to realize that a simulation of a system with rigidly constrained bond lengths is not equivalent to a simulation with, for example, harmonic springs representing the bonds, even within the limit of very strong springs. One obvious point is that the momenta conjugated to the bond coordinates are nonzero and store some kinetic energy in the spring case, while they are zero by definition in the constrained case. A subtle, but crucial, consequence of this is that it has an effect on the distribution function for the other coordinates. If we obtain the configurational distribution function by integrating over the momenta, the difference arises because in one case a set of momenta is set to zero, and not integrated, while in the other an integration is performed, which may lead to an extra term depending on particle coordinates. This is frequently called the 'metric tensor problem'; it is explained in more detail in [7, 57], and there are well-established ways of determining when the difference is likely to be significant [58] and how to handle it, if necessary [59].

Some people prefer to use the multiple time step approach to handle fast degrees of freedom, while others prefer to use constraints, and there are situations in which both techniques are applicable. Constraints also find an application in the study of rare events, where a system may be studied at the top of a free energy barrier (see later), or for convenience when it is desired to fix a thermodynamic order parameter or ordering direction [17].

### B3.3.3.5  Neighbour lists

In the inner loops of MD and MC programs, we consider an atom $i$ and loop over all atoms $j$ to calculate the minimum image separations. If $r_{ij} > r_c$, the potential cutoff, the program skips to the end of the inner loop, avoiding expensive calculations, and considers the next neighbour. In this method, the time to examine all pair separations is proportional to $N^2$; for every pair, one must compute at least $r_{ij}^2$; this still consumes a lot of time. Some economies result from the use of lists of nearby pairs of atoms.

Verlet [13] suggested a technique for improving the speed of a program by maintaining a list of neighbours. The potential cutoff sphere, of radius $r_c$, around a particular atom is surrounded by a 'skin', to give a larger sphere of radius $r_{list}$, as shown in figure B3.3.6. At the first step in a simulation, a list is constructed of all the neighbours of each atom, for which the pair separation is within $r_{list}$. Over the next few MD time steps, only pairs appearing in the list are checked in the force routine. From time to time the list is reconstructed: it is important to do this before any unlisted pairs have crossed the safety zone and come within interaction range. It is possible to trigger the list reconstruction automatically, if a record is kept of the distance travelled by each atom since the last update. The choice of list cutoff distance $r_{list}$ is a compromise: larger lists will need to be reconstructed less frequently, but will not give as much of a saving on cpu time as smaller lists. This choice can easily be made by experimentation.

For larger systems ($N \geq 1000$ or so, depending on the potential range) another technique becomes preferable. The cubic simulation box (extension to noncubic cases is possible) is divided into a regular lattice of $n_c \times n_c \times n_c$ cells; see figure B3.3.7. These cells are chosen so that the side of the cell $r_{cell} = L/n_c$ is greater than the potential cutoff distance $r_c$. If there is a separate list of atoms in each of those cells, then searching through the neighbours is a rapid process: it is only necessary to look at atoms in the same cell as the atom of interest, and in nearest neighbour cells. The cell structure may be set up and used by the method of linked lists [45, 60]. The first part of the method involves sorting all the atoms into their appropriate cells. This sorting is rapid, and may be performed at every step. Then, within the force routine, pointers are used to scan through the contents of cells, and calculate pair forces. This approach is very efficient for large systems with short-range forces. A certain amount of unnecessary work is done because the search region is cubic, not (as for the Verlet list) spherical.

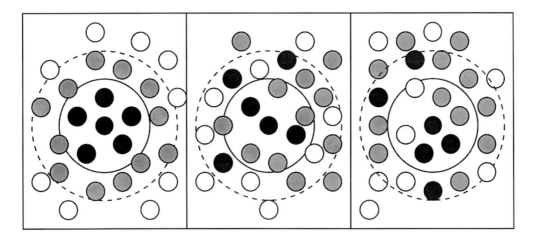

**Figure B3.3.6.** The Verlet list on its construction, later, and too late. The potential cutoff range, and the list range, are indicated. The list must be reconstructed before particles originally outside the list range have penetrated the potential cutoff sphere.

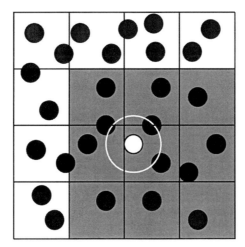

**Figure B3.3.7.** The cell structure. The potential cutoff range is indicated. In searching for neighbours of an atom, it is only necessary to examine the atom's own cell, and its nearest-neighbour cells.

### B3.3.3.6  Long-range forces

Many realistic simulations will involve the Coulomb interaction between charges, which decreases with separation as $r^{-1}$, and the dipole–dipole interaction, which decreases as $r^{-3}$. These cannot be treated simply by applying a spherical cutoff: it is essential to consider the effects of the surrounding medium. Two somewhat different techniques have been used in the majority of computer simulations to handle long-range forces: the reaction field method and the Ewald sum.

In the reaction field method, the space surrounding a dipolar molecule is divided into two regions: (i) a cavity, within which electrostatic interactions are summed explicitly, and (ii) a surrounding medium, which is assumed to act like a smooth continuum, and is assigned a dielectric constant $\varepsilon_\ell$. Ideally, this quantity will be equal to the dielectric constant $\varepsilon$ of the liquid itself, but calculating this, of course, is frequently one of the goals of the simulation, not one of the input parameters. The essence of the reaction field method is to calculate the total dipole moment of the molecules in the cavity, hence obtaining the polarization of the surrounding continuum, and to use this to work out the reaction field on the molecule at the centre. This supplements the direct electrostatic interaction with molecules in the cavity, in the calculation of the total energy. The reaction field method was used by Barker and Watts [61] in early simulations of water and has been discussed by Neumann and Steinhauser [62] and Patey *et al* [63].

In the Ewald method, a lattice sum is performed over charges within the periodically repeating simulation box. This is a subtle matter, since the sum, for Coulomb potentials, is only conditionally convergent: the result depends on the order of terms. Nonetheless, the procedure has been carefully analysed by de Leeuw *et al* [64] and Felderhof [65]. To make the summation a practical proposition, a trick is used: each point charge is screened by a surrounding, Gaussian, charge distribution, which makes the interactions short-ranged: these interactions are tackled in real space in the usual way. The contribution of an equal and opposite set of Gaussians is tackled in reciprocal space, using Fourier transforms. The choice of the width of the Gaussians is a parameter which may be varied to optimize the speed of calculation (for a given accuracy): Perram *et al* [66] have shown that the optimal choice leads to an algorithm whose expense grows as $N^{3/2}$.

When carried out properly, the results of the reaction field method and the Ewald sum are consistent [67]. Recently, the reaction field method has been recommended on grounds of efficiency and ease of programming [68, 69]. The expense of the Ewald method, particularly as the system size grows, has led to the search for alternative formulations [see, e.g., 70]. Recently, the practical implementation of the Ewald method has been significantly improved. The smooth particle mesh Ewald method, inspired by the approach of [45], employs a mesh and an interpolation scheme to allow evaluation of the reciprocal space sums using fast Fourier transforms [71, 72]. This approach has been incorporated into a standard code [49] and seems very promising for large biomolecular systems. It has to be said that there are still some subtleties involved in the handling of long-range forces [see, e.g., 73] and the reader should consult carefully the references if approaching the simulation of such systems from the beginning.

### B3.3.4   Monte Carlo

It is important to realize that MC simulation does not provide a way of calculating the statistical mechanical partition function: instead, it is a method of sampling configurations from a given statistical ensemble and hence of calculating ensemble averages. A complete sum over states would be impossibly time consuming for systems consisting of more than a few atoms. Applying the trapezoidal rule, for instance, to the configurational part of $Q_{NVT}$, entails discretizing each atomic coordinate on a fine grid; then the dimensionality of the integral is extremely high, since there are $3N$ such coordinates, so the total number of grid points is astronomically high.

The MC integration method is sometimes used to estimate multidimensional integrals by randomly sampling points. This is not feasible here, since a very small proportion of all points would be sampled in a reasonable time, and very few, if any, of these would have a large enough Boltzmann factor to contribute significantly to the partition function. MC simulation differs from such methods by sampling points in a nonuniform way, chosen to favour the important contributions.

### B3.3.4.1   *Importance sampling*

MC simulation is a method of concentrating the sampled points in the important regions, namely the regions with high Boltzmann factor $e^{-\beta\mathcal{H}}$: a random walk is devised, moving from one point to the next, with a biasing probability chosen to generate the desired distribution. Unfortunately, a consequence of this approach is that it is no longer possible to estimate the partition function itself, merely *ratios* of sums over states, that is, ensemble averages. Suppose that we have succeeded in selecting states $\Gamma$ with probability proportional to $\varrho(\Gamma) = \exp\{-\beta\mathcal{H}(\Gamma)\}$. Then, if we have conducted $N_t$ 'observations' or 'steps' in the process, the ensemble average becomes an average over steps

$$\langle\mathcal{X}\rangle_{NVT} = \sum_\Gamma \varrho(\Gamma)\mathcal{X}(\Gamma) = \frac{1}{N_t}\sum_{t=1}^{N_t} \mathcal{X}_t.$$

The Boltzmann weight appears *implicitly* in the way the states are chosen. The form of the above equation is like a time average as calculated in MD. The MC method involves designing a stochastic algorithm for stepping from one state of the system to the next, generating a trajectory. This will take the form of a *Markov chain*, specified by transition probabilities which are independent of the prior history of the system.

Write $\varrho(\Gamma) \equiv \varrho_\Gamma$ treating it as a component of a (very large) column vector. Consider an *ensemble* of systems all evolving at once. Specify a matrix whose elements $\pi_{\Gamma'\leftarrow\Gamma}$ give the probability of going to state $\Gamma'$ from state $\Gamma$, for every pair of states. The matrix must satisfy $\sum_{\Gamma'} \pi_{\Gamma'\leftarrow\Gamma} = 1$, to conserve probability. At each step, implement jumps with this transition matrix. This generates a *Markov chain* of states, i.e., one in which the transition probabilities do not depend on the history. Feller's theorem [74] tells us that, subject to some reasonable conditions, there exists a limiting (equilibrium) distribution of states and the system will tend towards this limiting distribution. (Recently [75], it has been shown that the Markov condition can be relaxed, and the system will still behave in this way.) A little thought shows that the limiting distribution will satisfy

$$\varrho_{\Gamma'} = \sum_\Gamma \pi_{\Gamma'\leftarrow\Gamma}\varrho_\Gamma$$

which is a matrix eigenvalue equation. The eigenvector is already known: it is the Boltzmann distribution. The MC method is specified by choosing a transition matrix which satisfies this equation. One way of guaranteeing this is to ensure that

$$\pi_{\Gamma\leftarrow\Gamma'}\varrho_{\Gamma'} = \pi_{\Gamma'\leftarrow\Gamma}\varrho_\Gamma$$

which is usually termed the *microscopic reversibility* condition. An immediate consequence of this is that the ratio of probabilities $\varrho_{\Gamma'}/\varrho_\Gamma$ is equal to the ratio of transition matrix elements $\pi_{\Gamma'\leftarrow\Gamma}/\pi_{\Gamma\leftarrow\Gamma'}$. This relationship is analogous to that relating the equilibrium constant for a chemical reaction to the ratio of forward and backward rate constants.

The most commonly used prescription [76] is

$$\pi_{\Gamma'\leftarrow\Gamma} = \alpha_{\Gamma'\leftarrow\Gamma}\min(1,\varrho_{\Gamma'}/\varrho_\Gamma) \qquad \Gamma \neq \Gamma'$$
$$\pi_{\Gamma\leftarrow\Gamma} = 1 - \sum_{\Gamma'\neq\Gamma}\pi_{\Gamma'\leftarrow\Gamma} \qquad \text{otherwise.}$$

Here, an underlying matrix, with elements $\alpha_{\Gamma'\leftarrow\Gamma}$, sets the probability of *attempting* a move like $\Gamma' \leftarrow \Gamma$, and the other factor gives the probability of *accepting* such a move. This scheme only requires a knowledge of the ratio $\varrho_{\Gamma'}/\varrho_\Gamma$:

$$\min(1,\varrho_{\Gamma'}/\varrho_\Gamma) = \min(1, e^{-\beta(\mathcal{H}(\Gamma')-\mathcal{H}(\Gamma))}) = \min(1, e^{-\beta\delta\mathcal{H}}).$$

It does not require knowledge of the factor normalizing the $\varrho$, i.e., the partition function.

For atomic and molecular systems, the partition function is split into a product of 'ideal' (exactly calculable) and 'excess' terms: the position and momentum distributions also factorize, and we wish to sample

$$\varrho_{NVT}(r) \propto \exp\{-\beta\mathcal{V}(r)\}.$$

The prescription for accepting or rejecting moves is exactly as written before, but with $\mathcal{V}$ replacing $\mathcal{H}$. Assuming that the interaction potential is short-ranged, it is not necessary to perform a complete recalculation of $\mathcal{V}$ every time an atom is moved: just the part involving that atom. For a given trial move, this is done twice: once before the attempted move and once after. Some improvement in efficiency may be obtained by using neighbour lists, as described earlier for MD.

Selecting trial moves in an unbiased way typically means (a) choose an atom 'randomly', with equal probability from the complete set; (b) displace it by random amounts in the $x$, $y$ and $z$ directions, chosen independently and uniformly from a predefined range (symmetric about the origin). These choices use a random number generator; the quality of the random numbers may be an issue. A key aspect of move selection is that the probabilities for *attempting* forward and reverse moves must be equal, so $\alpha_{\Gamma' \leftarrow \Gamma} = \alpha_{\Gamma \leftarrow \Gamma'}$. For the above prescription, it should be evident that this is true, by considering the number of ways of selecting an atom, and the number of positions it might be moved to (assuming a fine discretization of space). In the case of a rigid molecule, move selection will include a procedure for randomly rotating a molecule in an unbiased way. The magnitudes of trial moves are parameters of the method, chosen to give a reasonable acceptance rate, traditionally 50% or so. There is no special reason for this value. Ideally, for every study, one would investigate which choice gives the most efficient sampling of phase space.

The analysis of Manousiouthakis and Deem [75] mentioned above has demonstrated that it is also correct to choose atoms sequentially rather than randomly: it has been tacitly assumed for many years that this violation of the Markovian restriction is acceptable, so a proof of this kind is very welcome.

### B3.3.4.2 Weighted and biased sampling

It is useful to write down here the basic formulae for sampling with an additional weight function applied, sometimes called non-Boltzmann or umbrella sampling, and for sampling when the selection of trial moves is done in a biased way, i.e., the $\alpha$ matrix is not symmetrical.

A weight factor $\mathcal{W}(\Gamma)$ may be introduced in the MC sampling algorithm, to generate a modified, or 'weighted', distribution,

$$\varrho_W \propto \mathcal{W}(\Gamma) \exp\{-\beta \mathcal{H}(\Gamma)\}.$$

The usual MC procedure is adopted, with trial moves $\Gamma' \leftarrow \Gamma$ selected as usual, but now accepted with probability

$$\min\left(1, \frac{\mathcal{W}(\Gamma')}{\mathcal{W}(\Gamma)} e^{-\beta \delta \mathcal{H}}\right).$$

In the calculation of ensemble averages, we correct for the weighting as follows

$$\langle \mathcal{X} \rangle = \frac{\langle \mathcal{X}/\mathcal{W} \rangle_W}{\langle 1/\mathcal{W} \rangle_W}$$

where $\langle \ldots \rangle_W$ represents the weighted simulation averages. This kind of sampling may be useful when the most important states for our purposes are not those which have the highest weights in the canonical ensemble: for example, when we wish to compute a free energy difference between two states. We will return to this later.

Biased move selection means that $\alpha_{\Gamma' \leftarrow \Gamma} \neq \alpha_{\Gamma \leftarrow \Gamma'}$. Suppose that we wish, nonetheless, to sample the canonical distribution. To do this, we need to calculate the ratio $\alpha_{\Gamma' \leftarrow \Gamma}/\alpha_{\Gamma \leftarrow \Gamma'}$ as well as $\varrho_{\Gamma'}/\varrho_{\Gamma} = e^{-\beta \delta \mathcal{H}}$. Then we accept the move with probability

$$\min\left(1, \frac{\alpha_{\Gamma \leftarrow \Gamma'}}{\alpha_{\Gamma' \leftarrow \Gamma}} e^{-\beta \delta \mathcal{H}}\right).$$

A consideration of the transition probabilities allows us to prove that microscopic reversibility holds, and that canonical ensemble averages are generated. This approach has greatly extended the range of simulations that can be performed. An early example was the preferential sampling of molecules near solutes [77], but more recently, as we shall see, polymer simulations have been greatly accelerated by this method.

## B3.3.5   Simulation in different ensembles

It is very convenient to be able to choose a nonstandard ensemble for a simulation. Generally, it is more straightforward to do this in MC, but MD techniques for various ensembles have been developed. We consider MC implementations first.

### B3.3.5.1   *MC in different ensembles*

The isothermal–isobaric ensemble corresponds to a system whose volume and energy can fluctuate, in exchange with its surroundings at specified $NPT$. The thermodynamic driving force is the Gibbs free energy $G = A + PV$. The configurational distribution function may be written

$$\varrho_{NPT}(s^{(N)}, V) \propto V^N \, e^{-\beta PV} \, e^{-\beta \mathcal{V}}.$$

Here we have introduced scaled coordinates $s^{(N)} = L^{-1} r^{(N)}$ where $L$ is the box length (assumed cubic).

This ensemble is a weighted superposition of $NVT$ ensembles for different volumes. A typical MC sweep consists of $N$ attempted single-particle moves, exactly as for constant-$NVT$ MC, followed by one attempt to scale, homogeneously, the volume of the simulation box, together with the coordinates of all the particles in it. This is accepted or rejected so as to generate the above distribution. One prescription for selecting the volume move is to attempt to change $V \rightarrow V' = V + \delta V$ where $\delta V$ is uniformly sampled from an interval $[-\delta V_{\max} \ldots \delta V_{\max}]$. The new box length is computed, and all the particle coordinates scaled by an appropriate factor; then the new potential energy is computed. Assuming that the selection of $\delta V$ is unbiased, the probability ratio to use in the Metropolis prescription is just the ratio of the two ensemble densities, $\varrho_{NPT}(V')/\varrho_{NPT}(V)$ and the move is accepted with probability

$$\min(1, e^{-\beta \delta \mathcal{W}}) \qquad \text{where} \quad \delta \mathcal{W} = \delta \mathcal{V} + P \delta V - NkT \ln(V'/V).$$

Here $\delta \mathcal{V} = \mathcal{V}' - \mathcal{V}$ and $\delta V = V' - V$. The maximum attempted volume change is chosen to give a reasonable acceptance rate, traditionally 35–50% or so; there is no firm reason for this choice.

The above prescription for selecting volume changes is not unique. It may seem more natural to make random, uniform, changes in the box length $L$; some people prefer to sample $\ln V$ uniformly [78]. These choices are *not* precisely consistent with the accept/reject procedure described above. They can be regarded as unbiased sampling of a variable different from $V$ (in which case a simple transformation of variables is needed to convert $\varrho_{NPT}$ into the new form, and additional powers of $V$ will appear in it) or as biased sampling in $V$-space, in which case a small correction factor (the same extra powers of $V$) will appear in the accept/reject procedure. An analysis of the forward and backward transition probabilities will give the appropriate acceptance/rejection criterion in each case.

The grand canonical ensemble corresponds to a system whose number of particles and energy can fluctuate, in exchange with its surroundings at specified $\mu VT$. The relevant thermodynamic quantity is the grand potential $\Omega = A - \mu N$. The configurational distribution is conveniently written

$$\varrho_{\mu VT}(s^{(N)}, N) \propto (N!)^{-1} V^N z^N \exp\{-\beta \mathcal{V}\}.$$

Here again we have introduced scaled coordinates $s^{(N)} = L^{-1} r^{(N)}$ where $L$ is the box length (assumed cubic), $z = \exp\{\beta \mu\}/\Lambda^3$ is the activity, and $\Lambda = h/\sqrt{2\pi mkT}$ is the thermal de Broglie wavelength.

This ensemble is a weighted superposition of $NVT$ ensembles with different values, of $N$. As a rule of thumb, a typical MC sweep consists of $N$ attempted moves, each of which is chosen randomly to be (i) a displacement (handled exactly as in constant-$NVT$ MC); (ii) the creation of a new particle at a randomly selected position; (iii) the destruction of a randomly selected particle from the system. The probabilities for attempting creation and destruction must be equal (for consistency with what follows), but they need not be equal to the probability for attempting displacement (although they often are).

For a creation attempt, a position is chosen uniformly at random within the box, and an attempt made to create a new particle there. The probability ratio for creation is:

$$\frac{\varrho_{\mu VT}(N+1)}{\varrho_{\mu VT}(N)} = \frac{zV}{N+1} \exp\{-\beta\delta\mathcal{V}\} \equiv \exp\{-\beta\delta\mathcal{Z}^{\text{create}}\}$$

where $\delta\mathcal{V} = \mathcal{V}' - \mathcal{V}$ is the potential energy change associated with inserting the new particle. In a Metropolis scheme, the creation attempt is accepted with probability $\min(1, \exp\{-\beta\delta\mathcal{Z}^{\text{create}}\})$.

For a destruction attempt, one of the existing $N$ particles is selected at random, and an attempt made to destroy it. The probability ratio to use is

$$\frac{\varrho_{\mu VT}(N-1)}{\varrho_{\mu VT}(N)} = \frac{N}{zV} \exp\{-\beta\delta\mathcal{V}\} \equiv \exp\{-\beta\delta\mathcal{Z}^{\text{destroy}}\}$$

where $\delta\mathcal{V}$ is the potential energy change associated with removing the particle. In a Metropolis scheme, the destruction attempt is accepted with probability $\min(1, \exp\{-\beta\delta\mathcal{Z}^{\text{destroy}}\})$. These expressions can be shown to satisfy microscopic reversibility [57].

In a dense system, the acceptance rate of particle creation and deletion moves will decrease, and the number of attempts must be correspondingly increased: eventually, there will come a point at which grand canonical simulations are not practicable, without some tricks to enhance the sampling.

### B3.3.5.2  MD in different ensembles

In this section we discuss MD methods in the constant-$NVT$ ensemble, and the constant-$NPT$ ensemble.

There are three general approaches to conducting MD at constant temperature rather than constant energy. One method, simple to implement and reliable, is to periodically reselect atomic velocities at random from the Maxwell–Boltzmann distribution [79]. This is rather like an occasional random coupling with a thermal bath. The resampling may be done to individual atoms, or to the entire system; some guidance on the reselection frequency may be found in [79].

A second approach, due originally to Nosé [80] and reformulated in a useful way by Hoover [81], is to introduce an extra 'thermal reservoir' variable into the dynamical equations:

$$\dot{r}_i = p_i/m \qquad \dot{p}_i = f_i - \zeta p_i$$
$$\dot{\zeta} = \frac{\sum_{i\alpha} p_{i\alpha}^2/m - gkT}{Q} \equiv v_{\text{T}}^2\left[\frac{\sum_{i\alpha} p_{i\alpha}^2/m}{gkT} - 1\right] = v_{\text{T}}^2\left[\frac{\mathcal{T}}{T} - 1\right].$$

Here $\zeta$ is a friction coefficient which is allowed to vary in time; $Q$ is a thermal inertia parameter, which may be replaced by $v_{\text{T}}$, a relaxation rate for thermal fluctuations; $g \approx 3N$ is the number of degrees of freedom. $\mathcal{T}$ stands for the instantaneous 'mechanical' temperature. It may be shown that the distribution function for the ensemble is proportional to $\exp\{-\beta\mathcal{H}'\}$ where $\mathcal{H}' = \mathcal{H} + \frac{1}{2}3NkT\zeta^2/v_{\text{T}}^2$. These equations lead to the following time variation of the system energy $\mathcal{H} = \sum_{i\alpha} p_{i\alpha}^2/2m + \mathcal{V}$, and for the variable $\mathcal{H}'$:

$$\dot{\mathcal{H}} = \sum_{i\alpha} p_{i\alpha}\dot{p}_{i\alpha}/m - \sum f_{i\alpha}\dot{r}_{i\alpha} = -\zeta\sum_{i\alpha} p_{i\alpha}^2/m \qquad \dot{\mathcal{H}}' = -3NkT\zeta.$$

If $\mathcal{T} > T$, i.e., the system is too hot, then the 'friction coefficient' $\zeta$ will tend to increase; when it is positive the system will begin to cool down. If the system is too cold, the reverse happens, and the friction coefficient may become negative, tending to heat the system up again. In some circumstances, this approach generates non-ergodic behaviour, but this may be ameliorated by the use of chains of thermostat variables [82]. Tobias *et al* [83] give an example of the use of this scheme in a biomolecular simulation.

As an alternative to sampling the canonical distribution, it is possible to devise equations of motion for which the 'mechanical' temperature is constrained to a constant value [84–86]. The equations of motion are

$$\dot{r}_i = p_i/m \qquad \dot{p}_i = f_i - \zeta p_i \qquad \zeta = \sum_{i\alpha} f_{i\alpha} p_{i\alpha} / \sum_{i\alpha} p_{i\alpha}^2.$$

Here the friction coefficient $\zeta$ is completely determined by the instantaneous values of the coordinates and momenta. It is easy to see that the kinetic energy $\mathcal{K} = \sum_{i\alpha} p_{i\alpha}^2/2m$ is now a constant of the motion:

$$\dot{\mathcal{K}} = \sum_{i\alpha} p_{i\alpha} \dot{p}_{i\alpha}/m_i = \sum_{i\alpha} p_{i\alpha} f_{i\alpha}/m_i - \zeta \sum_{i\alpha} p_{i\alpha}^2/m_i = 0.$$

It is possible to devise extended-system methods [79, 87] and constrained-system methods [88] to simulate the constant-$NPT$ ensemble using MD. The general methodology is similar to that employed for constant-$NVT$, and in the course of the simulation the volume $V$ of the simulation box is allowed to vary, according to the new equations of motion. A useful variant allows the simulation box to change shape as well as size [89, 90]. It is also possible to extend the Liouville operator-splitting approach to generate algorithms for MD in these ensembles; examples of explicit, reversible, integrators are given by Martyna *et al* [91].

### B3.3.6 Free energies, chemical potentials and weighted sampling

A major drawback of MD and MC techniques is that they calculate *average* properties. The free energy and entropy functions cannot be expressed as simple averages of functions of the state point $\Gamma$. They are directly connected to the logarithm of the partition function, and our methods do not give us the partition function itself. Nonetheless, calculating free energies is important, especially when we wish to determine the relative thermodynamic stability of different phases. How can we approach this problem?

*B3.3.6.1    Free energy differences*

It is possible to calculate *derivatives* of the free energy directly in a simulation, and thereby determine free energy differences by thermodynamic integration over a range of state points between the state of interest and one for which we know $A$ exactly (the ideal gas, or harmonic crystal for example):

$$(\beta A)_2 - (\beta A)_1 = \int_{\beta_1}^{\beta_2} E \, d\beta \qquad \text{or} \qquad A_2 - A_1 = -\int_{V_1}^{V_2} P \, dV.$$

This is reliable and fairly accurate, if tedious. It was used, for example, by Hoover [92] to locate the melting parameters for soft-sphere systems. The only point to watch out for is that one should not cross any phase transitions in taking the path from 1 to 2: it must be *reversible*.

A free energy difference between two systems with energy functions $\mathcal{H}_0$ and $\mathcal{H}_1$, respectively, may be written in a way analogous to equation B3.3.3

$$A_1 = A_0 - kT \ln\langle \exp\{-\beta\Delta\mathcal{H}\}\rangle_0$$

where $\langle\ldots\rangle_0$ is an ensemble average for system 0 and $\Delta\mathcal{H} = \mathcal{H}_1 - \mathcal{H}_0$. In the extreme case $\mathcal{H}_0 \equiv 0$, this would give an unweighted estimate of the partition function of system 1, but would be extremely poorly sampled for the reasons discussed in section B3.3.2. Reasonable estimates will result when the two systems do not differ 'too much', so that contributing values of the Boltzmann factor are given a significant weight by the sampling over states in system 0. One famous example of this is the test-particle insertion formula, equation (B3.3.5), for estimating the chemical potential, where system 1 contains an additional particle. Another example is where a molecule in the system can be mutated into another. The efficiency of the sampling may depend critically on the direction of the perturbation change: estimating $\mu^{\text{ex}}$ by particle *removal*, for instance, is formally possible but usually less accurate than particle insertion. Kofke and Cummings [93] have reviewed various approaches in this field and make the general recommendation that the change should take place in the direction of *decreasing entropy*.

### B3.3.6.2  *Histogram re-weighting*

A way of looking at the points raised in the previous section is to compare energy *distributions* in two systems whose free energies we wish to relate. In particular, consider measuring, in a simulation of system 0, the function $\mathcal{P}_0(\Delta E)$, i.e., the probability density per unit $\Delta E$ of configurations for which $\mathcal{H}_0$ and $\mathcal{H}_1$ differ by the prescribed amount $\Delta E$. The distribution $\mathcal{P}_1(\Delta E)$ may be similarly calculated by simulating system 1. These two functions may be straightforwardly related [94]:

$$\ln \mathcal{P}_1(\Delta E) = \ln \mathcal{P}_0(\Delta E) + \beta \Delta A - \beta \Delta E.$$

Therefore, apart from an unknown constant $\beta \Delta A$, and a known linear term $\beta \Delta E$, these are the same function. Bennett [94] suggested two graphical methods for determining $\beta \Delta A$ from $\mathcal{P}_0(\Delta E)$ and $\mathcal{P}_1(\Delta E)$, which rely on the two distributions, at worst, nearly overlapping (i.e., being measurable, with good statistics, for the same or similar values of $\Delta E$). To broaden the sampling into the wings of the distribution, thereby improving statistics and extending the overlap region, we may use weighted sampling as described in section B3.3.4.2. There are many related approaches, variously called umbrella [95], multicanonical [96] and entropic [97] sampling, simulated tempering [98] and expanded ensembles [99].

Windowing is a special case of umbrella sampling: the weight function is a constant inside a specified region of configuration space, and zero outside. In MC we simply reject moves which would take the system outside the window, and otherwise proceed as usual. This allows us to examine a distribution function, and hence a free energy curve, piece by piece, matching up the resulting curves afterwards. The way to do this combination of histograms has been discussed by Ferrenberg and Swendsen [100], and the statistical errors in histogram re-weighting have been discussed by Swendsen [101] and Ferrenberg *et al* [102]. Ultimately, this approach leads back to the idea of performing simulations in a (nearly)-*microcanonical* ensemble and relating the results at nearby energies, as we do in thermodynamic integration; as emphasized in the review of Dünweg [103], 'for any thermodynamic integration procedure there is an equivalent multistage sampling or histogram procedure, and *vice versa*'.

### B3.3.6.3  *The chemical potential*

The ensemble average in the Widom formula, $\langle\langle \exp\{-\beta v_{\text{test}}\} \rangle\rangle$, is sometimes loosely referred to as the 'insertion probability'. It becomes very low for dense fluids. For example, for hard spheres, we can use the scaled-particle theory [104] or the Carnahan–Starling equation of state [105] to estimate it (see figure B3.3.8). The insertion probability falls below $10^{-4}$, well before the freezing transition at $\eta \approx 0.49$. Similar estimates can be made for the Lennard-Jones fluid. The lower this factor becomes, the poorer the statistics, and the more unreliable will be the estimate of $\mu^{\text{ex}}$. The problem is particularly acute for dense molecular fluids where, as a first guess, one could take the overall Boltzmann factor to be the *product* of the individual atomic values.

A simple method of improving the efficiency of test particle insertion [106–109] involves dividing the simulation box into small cubic regions, and identifying those which would make a negligible contribution to the Widom formula, due to overlap with one or more atoms. These cubes are excluded from the sampling, and a correction applied afterwards for the consequent bias.

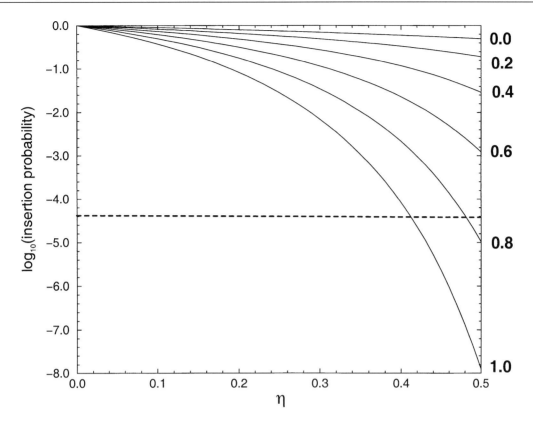

**Figure B3.3.8.** Insertion probability for hard spheres of various diameters (indicated on the right) in the hard sphere fluid, as a function of packing fraction $\eta$, predicted using scaled particle theory. The dashed line is a guide to the lowest acceptable value for chemical potential estimation by the simple Widom method.

Another trick is applicable to, say, a two-component mixture, in which one of the species, A, is smaller than the other, B. From figure B3.3.8, for hard spheres, we can see that A need not be *particularly* small in order for the test particle insertion probability to climb to acceptable levels, even when insertion of B would almost always fail. In these circumstances, the chemical potential of A may be determined directly, while that of B is evaluated indirectly, relative to that of A. The related 'semi-grand' ensemble has been discussed in some detail by Kofke and Glandt [110].

This naturally leads to the idea of estimating $\mu^{ex}$ by gradual insertion [111–113]. This can be thought of as a thermodynamic integration pathway, connecting the states with $N$ and $N+1$ particles via a set of intermediate points, characterized by a parameter $\lambda$, $0 \leq \lambda \leq 1$ which determines the degree to which the extra '$\lambda$-particle' is 'switched on'. A MC scheme is constructed, which (in addition to the usual moves) allows $\lambda$ to vary either continuously or in predefined discrete jumps. Then the chemical potential is expressed $\mu_{ex} = \Delta A_{ex} = A_{ex}(\lambda = 1) - A_{ex}(\lambda = 0)$. It is advantageous to apply a weighting function (see section B3.3.4.2) [112, 113], to ensure more or less uniform sampling of the different $\lambda$ states. Consider the probability histogram $\mathcal{P}(\lambda)$ of the sampled values of $\lambda$ during the runs. Without an external biasing potential this will be directly related to a Landau free energy

$$\mathcal{P}(\lambda) \propto \exp\{-\beta \mathcal{F}(\lambda)\}$$

where $\mathcal{F}(1) - \mathcal{F}(0) = \Delta A_{ex}$ is the desired free energy difference; to obtain a uniform distribution, a weight function $\mathcal{W}(\Gamma) \propto \exp\{-\beta\Psi(\lambda)\}$ with a biasing potential $\Psi(\lambda) = -\mathcal{F}(\lambda)$ would be used. This ideal weighting function is not known at the start of the simulation, but an initial guess may be iteratively refined from the measured $\mathcal{P}(\lambda)$ in a series of runs.

It is also advantageous to ensure that the $\lambda$-particle samples a wide range of positions in the fluid: this is achieved by attempting large-scale moves to new, randomly-selected positions from time to time, and also frequently attempting exchanges of position with a randomly-selected full-size particle. The former moves will have a high probability of success when $\lambda$ is small, and the latter when $\lambda$ is large. Camp *et al* [114] provide an example of this method in action, for a model of liquid crystals.

### B3.3.6.4  *Free energy of solids*

Early attempts to calculate solid-state free energies, and hence locate the melting transition, introduced the idea of conducting a thermodynamic integration along an artificial pathway [115, 116]. Each atom is artificially restricted to a single cell in space so that, as the density is lowered, the system converts more or less smoothly into a kind of 'lattice gas'. More recently, Frenkel *et al* [117] proposed a method in which $\lambda$ corresponds to switching between the true potential and a harmonic spring, which couples each atom at instantaneous position $r_i$ to its ideal lattice site $r_i^{(0)}$:

$$\mathcal{V}(\lambda) = \lambda\mathcal{V}_1 + (1 - \lambda)\mathcal{V}_0.$$

Here $\mathcal{V}_1$ is the original, many-body potential energy function, while $\mathcal{V}_0$ is a sum of single-particle spring potentials proportional to $|r_i - r_i^{(0)}|^2$. As $\lambda \to 0$ the system becomes a perfect Einstein crystal, whose free energy is exactly calculable. Recently, a combination of approaches has been used [118]: first the solid is subjected to a set of one-particle spring potentials, and then the influence of the interparticle forces is reduced to zero by expanding the crystal. This method was used to locate the melting transition for a model of nitrogen at $T = 300$ K.

Density functional theory arguments [119, 120] suggest that the springs should be of the correct strength to produce the same mean-squared displacement in the Einstein limit as in the original crystal. More precisely, it is best to switch over to a one-body potential in such a way that the one-body density $\rho(r)$ is the same at all points along the integration path. Such a path is guaranteed not to traverse a first-order phase transition.

## B3.3.7  Configuration-biased MC

The biased-sampling approach may be considerably generalized, to allow the construction of MC moves step-by-step, with each step depending on the success or failure of the last. Such a procedure is biased, but it is then possible to correct for the bias (by considering the possible reverse moves). The technique has dramatically speeded up polymer simulations, and is capable of wider application.

The idea may be illustrated by considering first a method for increasing the acceptance rate of moves (but at the expense of trying, and discarding, several other possible moves). Having picked an atom to move, calculate the new trial interaction energy $v_t$ for a range of trial positions $t = 1 \ldots k$. Pick the actual attempted move from this set, with a probability proportional to the Boltzmann factor. This biases the move selection, towards high-probability states, but we can calculate the contribution of the bias to $\alpha_{\Gamma'\leftarrow\Gamma}$. Then we must calculate $\alpha_{\Gamma\leftarrow\Gamma'}$ for the hypothetical reverse move: we do this by selecting $k - 1$ possible trial positions around the new position of the atom, plus the place it originally came from, making $k$ in all. The ratio $\alpha_{\Gamma\leftarrow\Gamma'}/\alpha_{\Gamma'\leftarrow\Gamma}$ is used in the accept/reject decision, along with the relevant Boltzmann factors (see section B3.3.4.2). For $k = 1$ this gives the usual Metropolis prescription; for $k \to \infty$ it is easy to show that the acceptance rate tends to unity, but the method becomes very expensive, since all the work has gone into calculating the biasing factors.

The expense is justified, however, when tackling polymer chains, where reconstruction of an entire chain is expressed as a succession of atomic moves of this kind [121]. The first atom is placed at random; the second selected nearby (one bond length away), the third placed near the second, and so on. Each placement of an atom is given a greater chance of success by selecting from multiple locations, as just described. Biasing factors are calculated for the whole multi-atom move, forward and reverse, and used as before in the Metropolis prescription. For further details see [122–125]. A nice example of this technique is the study [126, 127] of the distribution of linear and branched chain alkanes in zeolites.

## B3.3.8   Phase transitions

Here we discuss the exploration of phase diagrams, and the location of phase transitions. See also [128–131] and [22, chapters 8–14]. Very roughly we classify phase transitions into two types: first-order and continuous. The fact that we are dealing with a finite-sized system must be borne in mind, in either case.

### B3.3.8.1   First-order and continuous transitions

At a continuous phase transition, a correlation length $\xi$ (see section B3.3.2.1) diverges and an order parameter, typically the ensemble average of the corresponding dynamical variable, becomes macroscopically large. The divergence heralding the transition is describable in terms of universal exponent relations. Effects of finite size close to continuous phase transitions are well studied [24, 132]. By contrast, a first-order phase transition is abrupt, as one phase becomes thermodynamically more stable than another; there are no transition precursors. In the thermodynamic limit, there is a step-function discontinuity in most properties, including thermodynamic derivatives of the free energy. Again it is possible to describe the effects of finite size [132, 133].

For both first-order and continuous phase transitions, finite size *shifts* the transition and *rounds* it in some way. The shift for first-order transitions arises, crudely, because the chemical potential, like most other properties, has a finite-size correction $\mu(N) - \mu(\infty) \sim \mathcal{O}(1/N)$. An approximate expression for this was derived by Siepmann *et al* [134]. Therefore, the line of intersection of two chemical potential surfaces $\mu_{\mathrm{I}}(T, P)$ and $\mu_{\mathrm{II}}(T, P)$ will shift, in general, by an amount $\mathcal{O}(1/N)$. The rounding is expected because the partition function only has singularities (and hence produces discontinuous or divergent properties) in the limit $L \to \infty$; otherwise, it is analytic, so for finite $N$ the discontinuities must be smoothed out in some way. The shift for continuous transitions arises because the transition happens when $\xi \to L$ for the finite system, but when $\xi \to \infty$ in the infinite system. The rounding happens for the same reason as it does for first-order phase transitions: whatever the nature of the divergence in thermodynamic properties (described, typically, by critical exponents) it will be limited by the finite size of the system.

In either case, first-order or continuous, it is useful to consider the probability distribution function for variables averaged over a spatial block of side $L$; this may be the complete simulation box (in which case we must specify the ensemble and boundary conditions) or it may be a sub-system. For purposes of illustration we shall not distinguish these possibilities.

### B3.3.8.2   Continuous phase transitions

Here we discuss only briefly the simulation of continuous transitions (see [132, 135] and references therein). Suppose that the transition is characterized by a non-vanishing order parameter $X$ and a corresponding divergent correlation length $\xi$. We shall be interested in the block average value $X_L \equiv \langle \mathcal{X} \rangle_L$, where the $L$ reminds us of the system size. In a magnetic system, $X$ is the magnetization; in a fluid it might be the density. The basic idea of finite size scaling analysis is that the values of properties of the system are dictated by the ratio $\xi/L$, and that no other length scales enter the problem, near a critical point. Any property can be written

$$X_L = X \times \Phi(\xi/L)$$

where $X$ is the average value in the infinite system limit and $\Phi$ is some scaling function. There will be exponent laws dictating the behaviour of $X$ in the vicinity of the phase transition, and more scaling laws stating how $\xi$ behaves inside the function $\Phi$. We can apply a scaling analysis to the distribution function $\mathcal{P}(X_L)$ [136, 137]. Actually at the critical point, the distribution can be calculated by simulation, or predicted by renormalization group theory [136–139]; different universal forms will be seen for different universality classes. Examination of these functions is a powerful way of locating and characterizing critical points, and in the critical region the histogram reweighting method is a particularly useful way of maximizing the information obtained from individual simulations. For example, the prewetting critical point has been shown to lie in the $d = 2$ Ising universality class in this way [139]. A further example is the study of the critical point of the $d = 2$ Lennard-Jones fluid [140, 141]. For this, long runs of order $10^6$–$10^8$ sweeps were needed, but the system sizes were relatively small: $N \approx 100$ and 400.

### B3.3.8.3  *First-order phase transitions*

Consider simulating a system in the canonical ensemble, close to a first-order phase transition. In one phase, $\mathcal{P}_{NVT}(E)$ is essentially a Gaussian centred around a value $E_I$, while in the other phase the peak is around $E_{II}$. Far from the transition, one or other of these will apply. Close to the phase transition we will see contributions from both Gaussians, and a double-peaked distribution. The weight of each Gaussian changes as the temperature is varied. Thus, a smooth crossover occurs from one branch of the equation of state $E(T) = \langle \mathcal{H} \rangle_{NVT}$ to the other. In the transition region we may expect to see anomalies such as an increased specific heat: the double-peaked distribution is wider than its constituent single-peaked ones, and recall that $C_V$ is linked to $\langle \delta \mathcal{H}^2 \rangle$. The corresponding Landau free energy

$$\mathcal{F}_{NVT}(E) = -kT \ln \mathcal{P}_{NVT}(E)$$

has two minima separated by a barrier. The high-probability, low free energy values correspond to the single phase configurations; the intermediate values are for mixtures of the two phases, with an interfacial free energy penalty.

In the microcanonical ensemble, the signature of a first-order phase transition is the appearance of a 'van der Waals loop' in the equation of state, now written as $T(E)$ or $\beta(E)$. The $\beta(E)$ curve switches over from one branch, phase I, of the equation of state to the other, phase II, tracing out a loop in the transition region. This loop is a finite-size effect, due to the interfacial free energy contributions in the transition region, just mentioned. For a larger system size the loop will flatten out, becoming a horizontal line in the thermodynamic limit, joining the two coexisting energies at the transition temperature; for $N \to \infty$ the interfacial properties contribute a negligible amount to the total free energy. (Calling it a 'van der Waals loop' is therefore misleading: it has no connection with the loop in the approximate van der Waals equation of state for fluids, which in any case is independent of system size.) It is possible to inter-relate the form of this loop, and the double-peaked structure of the energy distribution, with the thermodynamic coexistence conditions [1, 132, 142–145].

The previous discussions translate directly over into pressure–volume variables, if we compare the constant-$NVT$ and constant-$NPT$ ensembles. Double-peaked distributions of volumes are seen near a transition at constant pressure.

Direct coexistence of solid and fluid phases of hard spheres and disks was observed in the early simulations of Wood and Jacobson [146] and Alder and Wainwright [147, 148]; the appearance of a 'van der Waals loop' in the equation of state was explained in some detail shortly afterwards [142, 149, 150]. Very detailed analyses of this situation, especially in relation to spin systems, have appeared in recent years [143–145, 151, 152]. Histogram reweighting can be useful here [153, 154] and measuring the height of the interfacial free energy barrier as a function of system size has been recommended as a test for first-order behaviour [155]. A nice example of this approach for a spin model of a liquid crystal, which exhibits a tricky *weak* first-order transition, is the work of Zhang *et al* [156, 157], following on from earlier work by Fabbri and Zannoni [158]. An example for a *strong* first-order transition is the study of melting and nucleation barriers in the Lennard-Jones system [159] and models of metallic systems [160]. This approach used windowing and biased sampling techniques.

Recently, Orkoulas and Panagiotopoulos [161] have shown that it is possible to use histogram reweighting and multicanonical simulations, starting with individual simulations near the critical point, to map out the liquid–vapour coexistence curve in a very efficient way.

### B3.3.8.4    *Gibbs ensemble*

Simulations in the Gibbs ensemble attempt to combine features of Widom's test particle method with the direct simulation of two-phase coexistence in a box. The method of Panagiotopoulos *et al* [162, 163] uses two fully-periodic boxes, I and II.

In the simplest version, a one-component system is simulated at a given temperature $T$ in both boxes; particles in different boxes do not interact directly with each other; however, volume moves and particle creation and deletion moves are coupled such that the total volume $V$ and the total number of particles $N$ are conserved. With appropriate acceptance/rejection probabilities for the volume exchange and particle exchange moves, together with the usual MC procedure for moving particles around within the two boxes, the thermodynamic conditions for mechanical and chemical equilibrium between the boxes are ensured. A typical MC cycle would consist of: one attempted move per particle in each box; one attempt to exchange volumes between boxes; a predetermined number of attempts to exchange particles. The technique has been reviewed by Panagiotopoulos [131, 164, 165] and Smit [129]. The partition function for the two-box system is simply the usual canonical sum over all possible states, including a sum over all the distributions of particles between the boxes such that $N_I + N_{II} = N$, and an integral over all box volumes such that $V_I + V_{II} = V$. The probability distribution function for the ensemble, and the acceptance and rejection rules for particle and volume exchanges, are easily derived.

The characteristic feature of the technique is the behaviour of the system if the overall density $N/V$ lies in a two-phase region. For a single simulation box, both phases would appear, with an interface between them; in the Gibbs ensemble, the interface free energy penalty can be avoided by the system arranging to have each phase entirely in its own box. This phase separation happens automatically during the equilibration stage of the simulation.

The great advantages of the technique are its avoidance of interfacial properties, and the semi-automatic way that it converges on the coexisting densities without the need to input chemical potentials or guess equations of state. Unavoidably, it suffers from the same problems as the Widom test-particle method: at high density the particle exchange moves are accepted with very low probability, and special techniques are required to overcome this. It is essential to monitor the success rate of exchanges, and carry out enough of them to ensure that a few percent of molecules are exchanged at each step.

The Gibbs ensemble method has been outstandingly successful in simulating complex fluids and mixtures. For a multicomponent system, it is possible to simulate at constant pressure rather than constant volume, as separation into phases of different compositions is still allowed. The method allows one to study straightforwardly phase equilibria in confined systems such as pores [166]. Configuration-biased MC methods can be used in combination with the Gibbs ensemble. An impressive demonstration of this has been the determination by Siepmann *et al* [167] and Smit *et al* [168] of liquid–vapour coexistence curves for *n*-alkane chain molecules as long as 48 atoms.

As we have seen, insertion of small molecules can be dramatically easier than large ones; this leads to a 'semi-grand' version of the Gibbs ensemble [131, 169, 170]: the smaller particles are exchanged between the boxes, while moves that interconvert particle species are carried out within the boxes. The ideas seen before for computing the chemical potential by gradual insertion [111–113] can be naturally generalized to the Gibbs ensemble: at each stage a given molecule may be in an intermediate state of transfer between one box and another. As an example, Escobedo and de Pablo [171] use an expanded Gibbs ensemble for polymers, and Nath *et al* [172] have computed the liquid–vapour envelopes for long-chain alkanes, using different potential models and comparing with previous work [167, 168].

### B3.3.8.5  *Thermodynamic methods*

The alternative to direct simulation of two-phase coexistence is the calculation of free energies or chemical potentials together with solution of the thermodynamic coexistence conditions. Thus, we must solve (say) $\mu_I(P) = \mu_{II}(P)$ at constant $T$. A reasonable approach [173–176] is to conduct constant-$NPT$ simulations, measure $\mu$ by test-particle insertion, and also to note that the simulations give the derivative $\partial\mu/\partial P = \langle V \rangle/N$ directly. Thus, conducting one or two simulations may be enough for a preliminary fit to the equations of state $\mu_I(P)$, $\mu_{II}(P)$ allowing one to home in on the intersection point quite quickly.

Once a point on the coexistence line has been found, one can trace out more of it using the approach of Kofke [177, 178] to numerically integrate the Clapeyron equation

$$\left(\frac{\mathrm{d}P}{\mathrm{d}\beta}\right)_\sigma = -\frac{\Delta h}{\beta\Delta v}.$$

Here, $\Delta h = h_\beta - h_\alpha$ is the difference in molar enthalpies of the coexisting phases, and $\Delta v$ is the difference in molar volumes; the suffix $\sigma$ indicates that the derivative is to be evaluated along the coexistence line.

The method consists of solving the above equation in a standard step-by-step manner, for example using a predictor–corrector algorithm. The right-hand side is calculated by simulating both phases at constant $T$ and $P$ in separate, uncoupled boxes. At intervals, a small change in $T$ (the independent variable) is made in both boxes, and this is accompanied by a change in $P$ (the dependent variable) as dictated by the differential equation solver. The approach relies on a starting point at which the two phases are at thermodynamic equilibrium, $\Delta\mu = 0$; thereafter the Clapeyron equation, if solved accurately, should guarantee that equilibrium is maintained. The method has been applied to the liquid–vapour coexistence curve [177, 178] and to the melting and sublimation curves [179] for the Lennard-Jones system; it was also extended by Agrawal and Kofke [180] to study the melting transition of a large family of soft-sphere systems, showing the emergence of the bcc phase as being stable relative to fcc for high enough softness parameters. Various technical details of this approach have been discussed [178, 179] and possible sources of inaccuracy considered.

An example of the use of this method in a complex situation is the study by Bolhuis and Kofke [181, 182] of the freezing of polydisperse hard spheres (a system in which there is a distribution of atomic diameters). A semi-grand ensemble imposes a distribution of chemical potential differences (for different hard-sphere diameters) on the system: the width of this distribution controls the polydispersity. The same chemical potentials (for all the different species) apply in two coexisting phases. The thermodynamic integration technique may then be used to map out the freezing–melting line in the pressure–polydispersity plane, starting from the monodisperse limit (simple hard spheres). The resulting phase diagram, in volume fraction–polydispersity variables, is shown in figure B3.3.9. An important result is that fractionation between the two phases allows a highly polydisperse fluid to precipitate a solid which is only slightly polydisperse—never higher than 5.7% in terms of the average sphere diameter.

### B3.3.8.6  *Studies of interfaces*

Simulation of both bulk phases in a single box, separated by an interface, is closest to what we do in real life. It is necessary to establish a well defined interface, most often a planar one between two phases in 'slab' geometry. A large system is required, so that one can characterize the two phases far from the interface, and read off the corresponding bulk properties. Naturally, this is the approach of choice if the interfacial properties (for example, the surface tension) are themselves of interest. The first stage in such a simulation

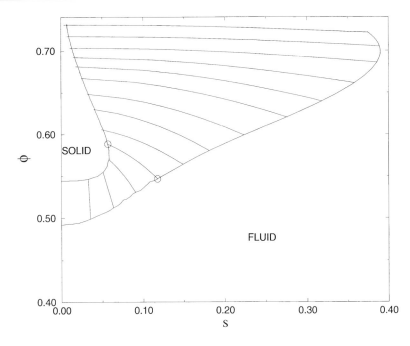

**Figure B3.3.9.** Phase diagram for polydisperse hard spheres, in the volume fraction ($\phi$)–polydispersity ($s$) plane. Some tie-lines are shown connecting coexisting fluid and solid phases. Thanks are due to D A Kofke and P G Bolhuis for this figure. For further details see [181, 182].

is to prepare bulk samples of each phase, as close to the coexisting densities as possible, in cuboidal periodic boundaries, using boxes whose cross sections match. The two boxes are brought together, to make a single longer box, giving the desired slab arrangement with two planar interfaces. There must then follow a period of equilibration, with mass transfer between the phases if the initial densities were not quite right.

Equilibration of the interface, and the establishment of equilibrium between the two phases, may be very slow. Holcomb *et al* [183] found that the density profile $\rho(z)$ equilibrated much more quickly than the profiles of normal and transverse pressure, $P_N(z)$ and $P_T(z)$, respectively. The surface tension is proportional to the $z$-integral of $P_N(z) - P_T(z)$. The bulk liquid in the slab may continue to contribute to this integral, indicating lack of equilibrium, for very long times if the initial liquid density is chosen a little too high or too low. A recent example of this kind of study, is the MD simulation of the liquid–vapour surface of water at temperatures between 316 and 573 K by Alejandre *et al* [184].

## B3.3.9    Rare events

By definition, rare events happen infrequently, but this does not mean that they happen slowly. Accordingly, molecular simulations may contribute greatly to our understanding of such events, but some special sampling tricks will be needed since the 'natural' timespans of simulations are already very short. The simplest example is where the system crosses a free energy barrier from one region to another in phase space, and a single 'reaction coordinate' can be identified to characterize the two stable regions and the transition state. The statistical mechanical background to barrier crossing rates, in terms of linear response theory, has been given by Chandler [2, 185]. The transition rate is typically a product of two factors: the equilibrium probability density for finding the system at the top of the barrier, and a dynamical quantity, essentially the inverse of a relaxation time for the system to settle from the barrier into one or other stable region.

We have already discussed weighted sampling methods for exploring regions of high free energy, so the first part of this problem is tractable. The calculation of time-dependent functions, which start from the barrier top, is facilitated by the so-called 'blue-moon' ensemble [186] in which a constraint is applied to keep the system exactly on the desired hypersurface. This allows sampling of the starting conditions with good statistics; then the constraint may be released and subsequent dynamics accumulated. (Metric tensor factors associated with the constraint are discussed elsewhere [187, 188].) To compute the time-dependent part of the barrier-crossing rate, special approaches have been developed to suppress transient behaviour and statistical noise [187].

In many cases, it may not be possible to identify a single reaction coordinate. A good example of a free-energy surface depending on two variables is found in the study by ten Wolde and Frenkel [189] of the mechanism of protein crystal nucleation, and the possible influence of fluctuations induced by a nearby metastable critical point. Here, the relevant variables are 'density' and 'crystallinity', as illustrated in figure B3.3.10. At high levels of supercooling, well away from the hidden critical point, nucleation follows a route whereby a crystalline nucleus forms from the start: this involves a high free energy barrier. Close to the critical point, a different mechanism operates: critical fluctuations encourage formation of a liquid-like nucleus, and only after this has grown to a certain size does a crystal start to form. This pathway involves a much lower free energy barrier. This work is an example of the insight obtainable from simulations into a previously poorly understood area, namely the art of obtaining good protein crystals for structure determination, as well as illustrating the qualitatively different mechanisms that may operate under different conditions.

The more general problem of finding a transition pathway when the relevant reaction coordinates are not obvious, has recently been tackled [190, 191]. The basic idea [192] is to generate chains of states linking the two stable states, through a weighted sampling procedure which makes no assumptions about the mechanism. The method is very general, but inevitably expensive.

### B3.3.10 Quantum simulation using path integrals

In this section we look briefly at the problem of including quantum mechanical effects in computer simulations. We shall only examine the simplest technique, which exploits an isomorphism between a quantum system of atoms and a classical system of ring polymers, each of which represents a path integral of the kind discussed in [193]. For more details on work in this area, see [22, 194] and particularly [195–197].

The coordinate representation of the density matrix, in the canonical ensemble, may be written

$$\varrho(r^{(N)}, r^{(N)'}) = Q_{NVT}^{-1}(r^{(N)}|e^{-\beta\hat{\mathcal{H}}}|r^{(N)'})$$

and correspondingly for the partition function

$$Q_{NVT} = \int dr^{(N)}(r^{(N)}|e^{-\beta\hat{\mathcal{H}}}|r^{(N)}).$$

Here we have adopted a Dirac bracket notation $(\ldots|\ldots|\ldots)$ which should be distinguished from the ensemble average $\langle\ldots\rangle$. Actually evaluating this is tricky, because the Hamiltonian is the sum of the kinetic and potential energy operators, $\hat{\mathcal{H}} = \hat{\mathcal{K}} + \hat{\mathcal{V}}$, which do not commute. Hence

$$e^{-\beta(\hat{\mathcal{K}}+\hat{\mathcal{V}})} \neq e^{-\beta\hat{\mathcal{K}}}e^{-\beta\hat{\mathcal{V}}}.$$

When the exponent is small (e.g., at high temperature), reasonable approximations exist. This problem is attacked in a manner similar to that used to derive expressions for the propagator $e^{i\hat{L}t}$, as a succession of small time step propagators, in section B3.3.3.2: we split the exponential up into smaller pieces. So, we write

$$e^{-\beta\hat{\mathcal{H}}} = [e^{-\beta\hat{\mathcal{H}}/P}]^P$$

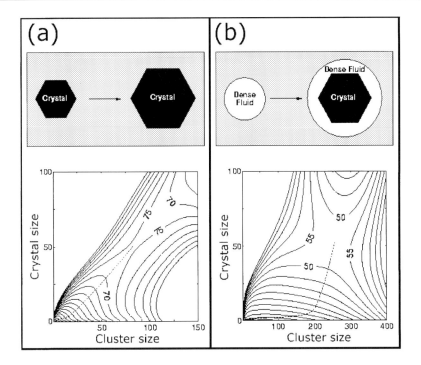

**Figure B3.3.10.** Contour plots of the free energy landscape associated with crystal nucleation for spherical particles with short-range attractions. The axes represent the number of atoms identifiable as belonging to a high-density cluster, and as being in a crystalline environment, respectively. (a) State point significantly below the metastable critical temperature. The nucleation pathway involves simple growth of a crystalline nucleus. (b) State point at the metastable critical temperature. The nucleation pathway is significantly curved, and the initial nucleus is liquidlike rather than crystalline. Thanks are due to D Frenkel and P R ten Wolde for this figure. For further details see [189].

and insert this into the expression for the partition function

$$Q_{NVT} = \int \mathrm{d}r^{(N)} (r^{(N)} | \mathrm{e}^{-\beta \hat{\mathcal{H}}/P} \, \mathrm{e}^{-\beta \hat{\mathcal{H}}/P} \ldots \mathrm{e}^{-\beta \hat{\mathcal{H}}/P} | r^{(N)}).$$

Now we do one of the standard quantum mechanical tricks, inserting the identity operator as a complete sum of states in the coordinate representation:

$$\hat{1} = \int \mathrm{d}r^{(N)'} | r^{(N)'} )(r^{(N)'} |$$

in between each exponential. This will introduce $P - 1$ additional integrations over coordinates. Each of the contributions $(r^{(N)} | \mathrm{e}^{-\beta \hat{\mathcal{H}}/P} | r^{(N)'})$ is an un-normalized, off-diagonal, density matrix $\varrho(r^{(N)}, r^{(N)'})$ evaluated at a temperature a factor $P$ higher than the temperature of the real system. For more background on this approach to quantum mechanics, see [193].

As in the case of the propagator, we shall be applying a symmetrical version of the Trotter formula [48] to the high-temperature density matrix

$$(r^{(N)} | \mathrm{e}^{-\beta \hat{\mathcal{H}}/P} | r^{(N)'}) \approx (r^{(N)} | \mathrm{e}^{-\frac{1}{2}\beta \hat{\mathcal{V}}/P} \, \mathrm{e}^{-\beta \hat{\mathcal{K}}/P} \, \mathrm{e}^{-\frac{1}{2}\beta \hat{\mathcal{V}}/P} | r^{(N)'}).$$

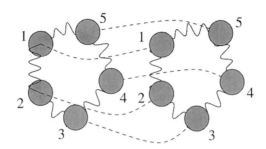

**Figure B3.3.11.** The classical ring polymer isomorphism, for $N = 2$ atoms, using $P = 5$ beads. The wavy lines represent quantum 'spring bonds' between different imaginary-time representations of the same atom. The dashed lines represent real pair–potential interactions, each diminished by a factor $P$, between the atoms, linking corresponding imaginary times.

The potential energy part is diagonal in the coordinate representation, and we drop the hat indicating an operator henceforth. The kinetic energy part may be evaluated by transforming to the momentum representation and carrying out a Fourier transform. The result is

$$Q_{NVT} \approx \left(\frac{Pm}{2\pi\beta\hbar^2}\right)^{dP/2} \int \cdots \int d\boldsymbol{r}_1^{(N)} \ldots d\boldsymbol{r}_P^{(N)} \, e^{-\beta \mathcal{V}_{qu}} \, e^{-\beta \mathcal{V}_{cl}}$$

$$\mathcal{V}_{qu} = -\frac{Pm}{2\beta^2\hbar^2}(|\boldsymbol{r}_1^{(N)} - \boldsymbol{r}_2^{(N)}|^2 + |\boldsymbol{r}_2^{(N)} - \boldsymbol{r}_3^{(N)}|^2 + \cdots + |\boldsymbol{r}_P^{(N)} - \boldsymbol{r}_1^{(N)}|^2)$$

$$\mathcal{V}_{cl} = \frac{1}{P}(\mathcal{V}(\boldsymbol{r}_1^{(N)}) + \mathcal{V}(\boldsymbol{r}_2^{(N)}) + \cdots + \mathcal{V}(\boldsymbol{r}_P^{(N)}))$$

This is better understood with a picture: see figure B3.3.11. The discretized path-integral is isomorphic to the classical partition function of a system of $N$ ring polymers each having $P$ atoms. Each atom in a given ring corresponds to a different 'imaginary time' point $p = 1 \ldots P$. $\mathcal{V}(\boldsymbol{r}^{(N)})$ represents the interatomic interactions (for example, Lennard-Jones) between the atoms of the real system. This couples together only *correspondingly labelled* atoms, i.e., atoms with the same index $p$. So, between each pair of the original atoms, there are $P$ such interactions, each one weaker than the true potential by a factor $1/P$. In addition, harmonic quantum 'springs' couple together successively indexed atoms within a ring polymer. We may simulate this classical ring polymer system by conventional MC or MD.

Temperature appears in the partition function in an unusual way. The average energy takes the form

$$E = \tfrac{3}{2}NPkT + \langle \mathcal{V}_{cl} \rangle - \langle \mathcal{V}_{qu} \rangle.$$

As $P$ is increased, the partial cancellation between the kinetic part and the spring part may worsen the statistics on $E$. This has led to suggestions of alternative ways of estimating $E$ [198]. As $P$ goes up, the springs become stronger, the interactions in $\mathcal{V}_{cl}$ become (individually) weaker, and this leads to sampling problems. In MD, one needs to use multiple time step methods to ensure proper handling of the spring vibrations, and there is a possible physical bottleneck in the transfer of energy between the spring system and the other degrees of freedom which must be handled properly [199]. In MC, one needs to use special methods to sample configuration space efficiently [200, 201].

For fermions (especially) and bosons there are additional problems. Let $\hat{P}$ be one of the $N!$ permutations of particle labels. Then the fermion density matrix $\varrho_F$ has the symmetry

$$\varrho_F(\boldsymbol{r}^{(N)}, \boldsymbol{r}^{(N)'}) = (-1)^{\hat{P}} \varrho_F(\hat{P}\boldsymbol{r}^{(N)}, \boldsymbol{r}^{(N)'}) = (-1)^{\hat{P}} \varrho_F(\boldsymbol{r}^{(N)}, \hat{P}\boldsymbol{r}^{(N)'}).$$

It is possible to relate this to the 'Boltzmann' (i.e., distinguishable particle) density matrix $\varrho(r^{(N)}, r^{(N)'})$ by

$$\varrho_F(r^{(N)}, r^{(N)'}) = \frac{1}{N!} \sum_{\hat{P}} (-1)^{\hat{P}} \varrho(\hat{P}r^{(N)}, r^{(N)'}).$$

It is necessary to sum over these permutations in a path integral simulation. (The same sum is needed for bosons, without the sign factor.) For fermions, odd permutations contribute with negative weight. Near-cancelling positive and negative permutations constitute a major practical problem [196].

## B3.3.11    Car–Parrinello simulations

Car and Parrinello [202] proposed a technique for efficiently solving the Schrödinger equation which has had an enormous impact on materials simulation (for reviews, see [203–206]). The technique is an *ab initio* one, i.e., free of empirical parameters, and is based on the use of a quantum mechanical orthonormal basis set $\psi^{(n)} \equiv \{\psi_i(r)\}$ to describe the electronic degrees of freedom. Specifically, the aim is to obtain the electron density $\rho(r)$; the total energy of the system may then be written as a functional of this density, whose minimization yields the ground state energy [207]. Pseudopotentials [208] represent the effects of the atomic cores on the valence electrons, allowing some economies. The energy functional is written

$$E[\psi^{(n)}, R^{(N)}] = \sum_i \langle \psi_i | \hat{\mathcal{K}} + \hat{\mathcal{V}}_{\text{ps}} | \psi_i \rangle + \frac{1}{2} \int dr\, dr' \frac{\rho(r)\rho(r')}{|r - r'|} \tag{B3.3.8}$$

$$+ E_{\text{xc}}[\rho] + \frac{1}{2} \sum_{I \neq J} \frac{Q_I Q_J}{|R_I - R_J|}. \tag{B3.3.9}$$

Here we distinguish between nuclear coordinates $R$ and electronic coordinates $r$; $\hat{\mathcal{K}}$ is the single-particle kinetic energy operator, and $\hat{\mathcal{V}}_{\text{ps}}$ is the total pseudopotential operator for the interaction between the valence electrons and the combined nucleus + frozen core electrons. The electron–electron and nucleus–nucleus Coulomb interactions are easily recognized, and the remaining term $E_{\text{xc}}[\rho]$ is the electronic exchange and correlation energy functional. This is usually treated in the local density approximation, using ground-state data for the homogeneous electron gas [209]; the most promising improvements seem to be based on the addition of gradient corrections [205, 206].

For each configuration of the nuclei, minimization of the total energy with respect to the electron density yields the instantaneous value of a potential energy function $\mathcal{V}(R^{(N)})$, and the corresponding forces on the nuclei. In principle, assuming an adiabatic separation between nuclear and electronic motion, Newton's equations for the nuclei may be solved in the usual way, while the electrons are allowed to evolve according to Schrödinger's equations, remaining on the instantaneous ground-state surface. This turns out to be very inefficient in practice, and the breakthrough came with the suggestion [202] that a classical dynamical evolution of the electronic configuration could be used to stay in the ground state. A Lagrangian, involving both nuclear and electronic degrees of freedom, is written down; the electrons are given fictitious masses, and minimization of the electronic energy may be performed by introducing a friction coefficient.

The Car–Parrinello method has found wide applicability, especially for studying systems in which structure and bonding are inseparable, or for materials under extreme conditions for which empirical potential would be unreliable. Examples are the studies by Alfe and Gillan [210] and de Wijs *et al* [211] of iron in the Earth's core, at temperatures of several thousand Kelvin and pressures sufficient to compress the metal to about half its normal volume. It was concluded that the liquid iron in the core is not exceptionally viscous (as has been suggested by some seismic measurements) and that dissolved sulphur atoms show no tendency to form clusters or chains (which might have a large effect on viscosity). This is shown in figure B3.3.12.

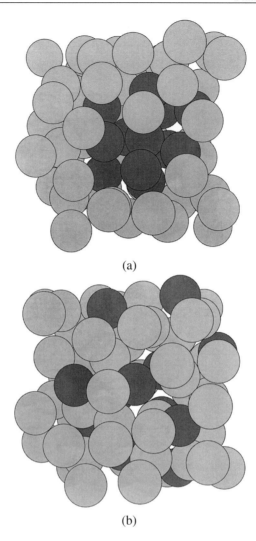

(a)

(b)

**Figure B3.3.12.** Sulphur atoms in liquid iron at the Earth's core conditions, simulated by first-principle Car–Parrinello molecular dynamics. (a) Initial conditions, showing a manually-prepared initial cluster of sulphur atoms. (b) A short time later, indicating spontaneous dispersal of the sulphur atoms, which mingle with the surrounding iron atoms. Thanks are due to D Alfe and M J Gillan for this figure. For further details see [210, 211].

Additionally, the simulations suggest that the solid part of the core has the hcp crystal structure, contrary to that inferred from experiments at lower pressure and temperature.

A similar approach, in spirit, has been proposed [212] for the study of two-component classical systems, for example polyelectrolytes, which consist of mesoscopic, highly-charged, polyions, and microscopic, oppositely-charged, counterions. This time, an 'effective free energy', depending parametrically on the polyion coordinates $R$, arises by integration over the counterion coordinates $r$. A unified dynamical scheme, in which the counterions are given fictitious masses, allows the counterion density to adjust adiabatically to the slower motion of the polyions, and hence permits the free energy to be minimized as the system evolves.

**Figure B3.3.13.** Intersecting stacking faults in a fcc crystal at the impact plane induced by collision with a momentum mirror for a square cross section of side 100 unit cells. The shock wave has advanced half way to the rear (~250 planes). Atom shading indicates potential energy. Thanks are due to B Holian for this figure. For further details see [219].

### B3.3.12    Parallel simulations

MD programs may be efficiently parallelized, that is, the computational work divided between many processors to result in faster execution. Well established message-passing standards and software make it relatively easy to write portable and efficient codes. The algorithm to advance positions and momenta is trivially handled, with each processor being responsible for a subset of the atoms. The critical considerations are (i) the parallelization of the time-consuming force calculation, and (ii) the overheads associated with communicating information between processors. Two general methodologies seem to be most promising. In the *replicated data* method, all the processors hold copies of all the atomic coordinates; however, in the double loop over pair interactions, each processor deals with a subset of pairs. Some care needs to be taken to balance the load between processors, and the results of the force calculations must be broadcast to all other processors, which may be time-consuming, but perhaps the biggest drawback is the memory requirement of holding copies of all data on all nodes. Nonetheless, this method is easy to program and reasonably efficient for many purposes [213–215]. In the *domain decomposition* method, the simulation box is split into (usually) cubic regions, and each processor is responsible only for the atoms in a given region; there is some communication of information from neighbouring domains before the force calculation, and also some redistribution of atoms as they move around the system. This approach may be integrated with the link-cell approach of section B3.3.3.5, and is especially efficient for systems with short-range forces [213, 215–218]. An example of the capabilities of such an approach on a massively parallel supercomputer is the study by Holian and Lomdahl [219] of shock waves in a fcc crystal of 10 million atoms, as illustrated in figure B3.3.13. In this case, a shock is generated by reflecting atoms at a piston face, i.e., imposing a momentum mirror. A system of this size was essential to ensure that the periodic boundaries do not limit the plastic flow induced by the shock: this can be seen in the randomly-spaced plaid pattern of the figure. The wave propagates back about 60 lattice spacings, generates a large number of stacking faults distributed randomly on the four {111} slip systems, and eventually produces a nonplanar propagation front.

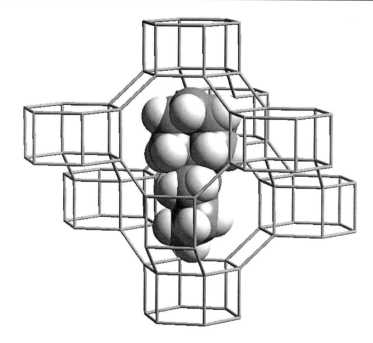

**Figure B3.3.14.** Template molecule in a zeolite cage. The CHA structure (periodic in the calculation but only a fragment shown here) is drawn by omitting the oxygens which are positioned approximately halfway along the lines shown connecting the tetrahedral silicon atoms. The molecule shown is 4-piperidinopiperidine, which was generated from the dicyclohexane motif suggested by computer. Thanks are due to D W Lewis and C R A Catlow for this figure. For further details see [225].

Although this section has concentrated on MD, it should not be forgotten that lattice-based MC codes may be parallelized very efficiently; for more information on parallel simulation methods see [220–223] and references therein.

### B3.3.13 Outlook

With the rapid development of computer power, and the continual innovation of simulation methods, it is impossible to predict what may be achieved over the next few years, except to say that the outlook is very promising. The areas of rare events, phase equilibria, and quantum simulation continue to be active.

An easily recognizable trend is the increasing application of simulation methods to problems of direct practical benefit to industry [224]. A pointer to this kind of use is provided by the synthesis of a small-pore microporous material, using a structure-directing template molecule designed by computer [225]. The aim is to promote formation of a desired material (here a cobalt aluminophosphate catalyst in the so-called CHA structure) without generating competing microporous phases. The simulation procedure allows molecular entities to be grown from a seed molecule by adding standard fragments, under the control of a cost function to minimize non-bonded overlaps with the surrounding CHA framework. Likely templates are ranked by binding energy, which measures how well they fit in the pore. The result is illustrated in figure B3.3.14. This led to a successful synthetic route: the suggested template molecule forms the desired pure mesoporous material in 4 h at 180°C, without the formation of competing structures which are found with other templates.

A further theme is the development of techniques to bridge the length and time scales between truly molecular-scale simulations and more coarse-grained descriptions. Typical examples are dissipative particle dynamics [226] and the lattice-Boltzmann method [227]. Part of the motivation for this is the recognition that brute-force molecular simulation will always be limited in time scale by achievable chip speeds, even if increased use of parallel computers allows one to tackle larger length scales. Nonetheless, there will always be the need to relate such work to underlying molecular parameters, through statistical mechanics. A more detailed discussion of these techniques would take us beyond the scope of this chapter.

# References

[1] Binder K and Heermann D W 1997 *Monte Carlo Simulation in Statistical Physics* 3rd edn, vol 80 *Solid State Sciences* (Berlin: Springer)
[2] Chandler D 1987 *Introduction to Modern Statistical Mechanics* (New York: Oxford University Press)
[3] Hansen J-P and McDonald I R 1986 *Theory of Simple Liquids* 2nd edn (London: Academic Press)
[4] Müller-Krumbhaar H and Binder K 1973 Dynamic properties of the Monte-Carlo method in statistical mechanics *J. Stat. Phys.* **8** 1–24
[5] Ferrenberg A M, Landau D P and Binder K 1991 Statistical and systematic errors in Monte-Carlo sampling *J. Stat. Phys.* **63** 867–82
[6] Binder K (ed) 1995 *The Monte Carlo Method in Condensed Matter Physics* vol 71 *Topics in Applied Physics* 2nd edn (Berlin: Springer)
[7] Allen M P and Tildesley D J 1987 *Computer Simulation of Liquids* (Oxford: Clarendon)
[8] Goldstein H 1980 *Classical Mechanics* 2nd edn (Reading, MA: Addison-Wesley)
[9] Maitland G C, Rigby M, Smith E B and Wakeham W A 1981 *Intermolecular Forces: Their Origin and Determination* (Oxford: Clarendon)
[10] Gray C and Gubbins K E 1984 *Theory of Molecular Fluids* (Oxford: Clarendon)
[11] Sprik M 1993 Effective pair potentials and beyond *Computer Simulation in Chemical Physics* vol 397 *NATO ASI Series C* ed M P Allen and D J Tildesley (Dordrecht: Kluwer) pp 211–59
[12] Rahman A 1964 Correlations in the motion of liquid argon *Phys. Rev.* A **136** 405–11
[13] Verlet L 1967 Computer experiments on classical fluids. I. Thermodynamical properties of Lennard-Jones molecules *Phys. Rev.* **159** 98–103
[14] Gay J G and Berne B J 1981 Modification of the overlap potential to mimic a linear site–site potential *J. Chem. Phys.* **74** 3316–19
[15] de Miguel E, Rull L F, Chalam M K and Gubbins K E 1991 Liquid crystal phase diagram of the Gay–Berne fluid *Mol. Phys.* **74** 405–24
[16] Berardi R, Emerson A P J, Smith W and Zannoni C 1993 Monte Carlo investigations of a Gay–Berne liquid crystal *J. Chem. Soc. Faraday Trans.* **89** 4069–78
[17] Allen M P, Warren M A, Wilson M R, Sauron A and William S 1996 Molecular dynamics calculation of elastic constants in Gay–Berne nematic liquid crystals *J. Chem. Phys.* **105** 2850–8
[18] Bates M A and Luckhurst G R 1996 Computer simulation studies of anisotropic systems. 26. Monte Carlo investigations of a Gay–Berne discotic at constant pressure *J. Chem. Phys.* **104** 6696–709
[19] Wilson M R 1997 Molecular dynamics simulations of flexible liquid crystal molecules using a Gay–Berne/Lennard-Jones model *J. Chem. Phys.* **107** 8654–63
[20] Billeter J and Pelcovits R 1998 Simulations of liquid crystals *Comput. Phys.* **12** 440–8
[21] Binder K 1992 *The Monte Carlo Method in Condensed Matter Physics* (Berlin: Springer)
[22] Binder K and Ciccotti G (ed) 1996 *Monte Carlo and Molecular Dynamics of Condensed Matter Systems* vol 49 (Bologna: Italian Physical Society)
[23] Newman M E J and Barkema G T 1999 *Monte Carlo Methods in Statistical Physics* (Oxford: Clarendon)
[24] Binney J J, Dowrick N J, Fisher A J and Newman M E J 1992 *The Theory of Critical Phenomena* (Oxford: Oxford University Press)
[25] Kremer K 1996 Computer simulation methods for polymer physics *Monte Carlo and Molecular Dynamics of Condensed Matter Systems* vol 49, ed K Binder and G Ciccotti (Bologna: Italian Physical Society) pp 669–723
[26] Salsburg Z W, Jacobson J D, Fickett W and Wood W W 1959 Application of the Monte Carlo method to the lattice gas model. Two dimensional triangular lattice *J. Chem. Phys.* **30** 65–72
[27] Chesnut D A and Salsburg Z W 1963 Monte Carlo procedure for statistical mechanical calculation in a grand canonical ensemble of lattice systems *J. Chem. Phys.* **38** 2861–75
[28] McDonald I R and Singer K 1967 Calculation of thermodynamic properties of liquid argon from Lennard-Jones parameters by a Monte Carlo method *Discuss. Faraday Soc.* **43** 40–9
[29] Valleau J P and Card D N 1972 Monte Carlo estimation of the free energy by multistage sampling *J. Chem. Phys.* **57** 5457–62
[30] Rickman J M and Phillpot S R 1991 Temperature dependence of thermodynamic quantities from simulations at a single temperature *Phys. Rev. Lett.* **66** 349–52
[31] Allen M P 1993 Back to basics *Computer Simulation in Chemical Physics* vol 397 *NATO ASI Series C* ed M P Allen and D J Tildesley (Dordrecht: Kluwer) pp 49–92

[32] Butler B D, Ayton O, Jepps O G and Evans D J 1998 Configurational temperature: verification of Monte Carlo simulations *J. Chem. Phys.* **109** 6519–22

[33] Rugh H H 1997 Dynamical approach to temperature *Phys. Rev. Lett.* **78** 772–4

[34] Widom B 1963 Some topics in the theory of fluids *J. Chem. Phys.* **39** 2808–12

[35] Evans D J and Morriss G P 1990 *Statistical Mechanics of Nonequilibrium Liquids* (London: Academic)

[36] Holian B L 1996 The character of the nonequilibrium steady state: beautiful formalism meets ugly reality *Monte Carlo and Molecular Dynamics of Condensed Matter Systems*, vol 49, ed K Binder and G Ciccotti (Bologna: Italian Physical Society) pp 791–822

[37] Reichl L E 1980 *A Modern Course in Statistical Physics* (Austin, TX: University of Texas Press)

[38] Friedman H L 1985 *A Course in Statistical Mechanics* (Englewood Cliffs, NJ: Prentice-Hall)

[39] Holian B L and Evans D J 1985 Classical response theory in the Heisenberg picture *J. Chem. Phys.* **83** 3560–6

[40] Gear C W 1966 The numerical integration of ordinary differential equations of various orders *ANL 7126* Argonne National Laboratory

[41] Gear C W 1971 *Numerical Initial Value Problems in Ordinary Differential Equations* (Englewood Cliffs, NJ: Prentice-Hall)

[42] Haile J M 1992 *Molecular Dynamics Simulation: Elementary Methods* (New York: Wiley)

[43] Rapaport D C 1995 *The Art of Molecular Dynamics Simulation* (Cambridge: Cambridge University Press)

[44] Verlet L 1968 Computer experiments on classical fluids. II. Equilibrium correlation functions *Phys. Rev.* **165** 201–14

[45] Hockney R W and Eastwood J W 1988 *Computer Simulations Using Particles* (Bristol: Adam Hilger)

[46] Swope W C, Andersen H C, Berens P H and Wilson K R 1982 A computer simulation method for the calculation of equilibrium constants for the formation of physical clusters of molecules: application to small water clusters *J. Chem. Phys.* **76** 637–49

[47] Tuckerman M, Berne B J and Martyna G J 1992 Reversible multiple time scale molecular-dynamics *J. Chem. Phys.* **97** 1990–2001

[48] Trotter H F 1959 On the product of semi-groups of operators *Proc. Am. Math. Soc.* **10** 545–51

[49] Procacci P, Darden T A, Paci E and Marchi M 1997 ORAC: a molecular dynamics program to simulate complex molecular systems with realistic electrostatic interactions *J. Comput. Chem.* **18** 1848–62

[50] Deuflhard P, Hermans J, Leimkuhler B, Mark A E, Reich S and Skeel R D (ed) 1998 *Computational Molecular Dynamics: Challenges, Methods, Ideas* vol 4 *Lecture Notes in Computational Science and Engineering* (Berlin: Springer)

[51] Schlick T, Mandziuk M, Skeel R D and Srinivas K 1998 Nonlinear resonance artifacts in molecular dynamics simulations *J. Comput. Phys.* **140** 1–29

[52] de Leeuw S W, Perram J W and Petersen H G 1990 Hamilton equations for constrained dynamic systems *J. Stat. Phys.* **61** 1203–22

[53] Ryckaert J-P, Ciccotti G and Berendsen H J C 1977 Numerical integration of the Cartesian equations of motion of a system with constraints: molecular dynamics of $n$-alkanes *J. Comput. Phys.* **23** 327–41

[54] Ciccotti G, Ferrario M and Ryckaert J-P 1982 Molecular dynamics of rigid systems in cartesian coordinates. A general formulation *Mol. Phys.* **47** 1253–64

[55] Ciccotti G and Ryckaert J P 1986 Molecular dynamics simulation of rigid molecules *Comput. Phys. Rep.* **4** 345–92

[56] Andersen H C 1983 RATTLE: a 'velocity' version of the SHAKE algorithm for molecular dynamics calculations *J. Comput. Phys.* **52** 24–34

[57] Frenkel D and Smit B 1996 *Understanding Molecular Simulation: From Algorithms to Applications* (San Diego: Academic)

[58] van Gunsteren W F 1980 Constrained dynamics of flexible molecules *Mol. Phys.* **40** 1015–19

[59] Fixman M 1974 Classical statistical mechanics of constraints: a theorem and application to polymers *Proc. Natl Acad. Sci.* **71** 3050–3

[60] Knuth D 1973 *The Art of Computer Programming* 2nd edn (Reading, MA: Addison-Wesley)

[61] Barker J A and Watts R O 1973 Monte Carlo studies of the dielectric properties of water-like models *Mol. Phys.* **26** 789–92

[62] Neumann M and Steinhauser O 1980 The influence of boundary conditions used in machine simulations on the structure of polar systems *Mol. Phys.* **39** 437–54

[63] Patey G N, Levesque D and Weis J J 1982 On the theory and computer simulation of dipolar fluids *Mol. Phys.* **45** 733–46

[64] de Leeuw S W, Perram J W and Smith E R 1980 Simulation of electrostatic systems in periodic boundary conditions. I. Lattice sums and dielectric constant *Proc. R. Soc.* A **373** 27–56

[65] Felderhof B U 1980 Fluctuation theorems for dielectrics with periodic boundary conditions *Physica* A **101** 275–82

[66] Perram J W, Petersen H G and DeLeeuw S W 1988 An algorithm for the simulation of condensed matter which grows as the 3/2 power of the number of particles *Mol. Phys.* **65** 875–93

[67] Neumann M, Steinhauser O and Pawley G S 1984 Consistent calculation of the static and frequency-dependent dielectric constant in computer simulations *Mol. Phys.* **52** 97–113

[68] Gray C G, Sainger Y S, Joslin C G, Cummings P T and Goldman S 1986 Computer simulation of dipolar fluids. Dependence of the dielectric constant on system size: a comparative study of Ewald sum and reaction field approaches *J. Chem. Phys.* **85** 1502–4

[69] Gil-Villegas A, McGrother S C and Jackson G 1997 Reaction-field and Ewald summation methods in Monte Carlo simulations of dipolar liquid crystals *Mol. Phys.* **92** 723–34

[70] Greengard L and Rokhlin V 1987 A fast algorithm for particle simulations *J. Comput. Phys.* **73** 325–48

[71] Darden T, York D and Pedersen L 1993 Particle mesh Ewald—an N.log(N) method for Ewald sums in large systems *J. Chem. Phys.* **98** 10 089–92

[72] Essmann U, Perera L, Berkowitz M L, Darden T, Lee H and Pedersen L G 1995 A smooth particle mesh Ewald method *J. Chem. Phys.* **103** 8577–93

[73] Procacci P, Marchi M and Martyna G J 1998 Electrostatic calculations and multiple time scales in molecular dynamics simulation of flexible molecular systems *J. Chem. Phys.* **108** 8799–803

[74] Feller W 1957 *An Introduction to Probability Theory and its Applications* 2nd edn, vol 1 (New York: Wiley)

[75] Manousiouthakis V I and Deem M W 1999 Strict detailed balance is unnecessary in Monte Carlo simulation *J. Chem. Phys.* **110** 2753–6

[76] Metropolis N, Rosenbluth A W, Rosenbluth M N, Teller A H and Teller E 1953 Equation of state calculations by fast computing machines *J. Chem. Phys.* **21** 1087–92

[77] Owicki J C and Scheraga H A 1977 Preferential sampling near solutes in Monte Carlo calculations on dilute solutions *Chem. Phys. Lett.* **47** 600–2

[78] Eppenga R and Frenkel D 1984 Monte Carlo study of the isotropic and nematic phases of infinitely thin hard platelets *Mol. Phys.* **52** 1303–34

[79] Andersen H C 1980 Molecular dynamics simulations at constant pressure and/or temperature *J. Chem. Phys.* **72** 2384–93

[80] Nosé S 1984 A molecular dynamics method for simulations in the canonical ensemble *Mol. Phys.* **52** 255–68

[81] Hoover W G 1985 Canonical dynamics: equilibrium phase-space distributions *Phys. Rev.* A **31** 1695–7

[82] Martyna G J, Klein M L and Tuckerman M 1992 Nosé–Hoover chains: the canonical ensemble via continuous dynamics *J. Chem. Phys.* **97** 2635–43

[83] Tobias D J, Martyna G J and Klein M L 1993 Molecular dynamics simulations of a protein in the canonical ensemble *J. Phys. Chem.* **97** 12 959–66

[84] Hoover W G, Ladd A J C and Moran B 1982 High strain rate plastic flow studied via nonequilibrium molecular dynamics *Phys. Rev. Lett.* **48** 1818–20

[85] Ladd A J C and Hoover W G 1983 Plastic-flow in close-packed crystals via non-equilibrium molecular-dynamics *Phys. Rev.* B **28** 1756–62

[86] Evans D J 1983 Computer experiment for nonlinear thermodynamics of Couette flow *J. Chem. Phys.* **78** 3297–302

[87] Nosé S 1984 A unified formulation of the constant-temperature molecular dynamics methods *J. Chem. Phys.* **81** 511–19

[88] Evans D J and Morriss G P 1983 The isothermal isobaric molecular dynamics ensemble *Phys. Lett.* A **98** 433–6

[89] Parrinello M and Rahman A 1980 Crystal structure and pair potentials: a molecular dynamics study *Phys. Rev. Lett.* **45** 1196–9

[90] Parrinello M and Rahman A 1981 Polymorphic transitions in single crystals: a new molecular dynamics method *J. Appl. Phys.* **52** 7182–90

[91] Martyna G J, Tuckerman M, Tobias D J and Klein M L 1996 Explicit reversible integrators for extended systems dynamics *Mol. Phys.* **87** 1117–57

[92] Hoover W G, Ross M, Johnson K W, Henderson D, Barker J A and Brown B C 1970 Soft sphere equation of state *J. Chem. Phys.* **52** 4931–41

[93] Kofke D A and Cummings P T 1997 Quantitative comparison and optimization of methods for evaluating the chemical potential by molecular simulation *Mol. Phys.* **92** 973–96

[94] Bennett C H 1976 Efficient estimation of free energy differences from Monte Carlo data *J. Comput. Phys.* **22** 245–68

[95] Torrie G M and Valleau J P 1977 Nonphysical sampling distributions in Monte Carlo free energy estimation: umbrella sampling *J. Comput. Phys.* **23** 187–99

[96] Berg B A and Neuhaus T 1992 Multicanonical ensemble—a new approach to simulate 1st-order phase transitions *Phys. Rev. Lett.* **68** 9–12

[97] Lee J 1993 New Monte Carlo algorithm—entropic sampling *Phys. Rev. Lett.* **71** 211–14

[98] Marinari E and Parisi G 1992 Simulated tempering: a new Monte Carlo scheme *Europhys. Lett.* **19** 451–8

[99] Lyubartsev A P, Martsinovski A A, Shevkunov S V and Vorontsov-Velyaminov P N 1992 New approach to Monte Carlo calculation of the free-energy—method of expanded ensembles *J. Chem. Phys.* **96** 1776–83

[100] Ferrenberg A M and Swendsen R H 1989 Optimized Monte Carlo data analysis *Phys. Rev. Lett.* **63** 1195–8

[101] Swendsen R H 1993 Modern methods of analyzing Monte Carlo computer simulations *Physica* A **194** 53–62

[102] Ferrenberg A M, Landau D P and Swendsen R H 1995 Statistical errors in histogram reweighting *Phys. Rev.* E **51** 5092–100

[103] Dünweg B 1996 Simulation of phase transitions: critical phenomena *Monte Carlo and Molecular Dynamics of Condensed Matter Systems* vol 49, ed K Binder and G Ciccotti (Bologna: Italian Physical Society) pp 215–54

[104] Reiss H, Frisch H L and Lebowitz J L 1959 Statistical mechanics of rigid spheres *J. Chem. Phys.* **31** 369–80

[105] Carnahan N F and Starling K E 1969 Equation of state for nonattracting rigid spheres *J. Chem. Phys.* **51** 635–6

[106] Deitrick G L, Scriven L E and Davis H T 1989 Efficient molecular simulation of chemical potentials *J. Chem. Phys.* **90** 2370–85

[107] Yoon K, Chae D G, Ree T and Ree F H 1981 Computer simulation of a grand canonical ensemble of rodlike molecules *J. Chem. Phys.* **74** 1412–23

[108] Lee Y S, Chae D G, Ree T and Ree F H 1981 Computer simulations of a continuum system of molecules with a hard-core interaction in the grand canonical ensemble *J. Chem. Phys.* **74** 6881–7

[109] Swope W C and Andersen H C 1995 A computer simulation method for the calculation of chemical potentials of liquids and solids using the bicanonical ensemble *J. Chem. Phys.* **102** 2851–63

[110] Kofke D A and Glandt E D 1988 Monte Carlo simulation of multicomponent equilibria in a semigrand canonical ensemble *Mol. Phys.* **64** 1105–31

[111] Mon K K and Griffiths R B 1985 Chemical potential by gradual insertion of a particle in Monte Carlo simulation *Phys. Rev.* A **31** 956–9

[112] Nezbeda I and Kolafa J 1991 A new version of the insertion particle method for determining the chemical potential by Monte Carlo simulation *Mol. Simul.* **5** 391–403

[113] Attard P 1993 Simulation of the chemical potential and the cavity free energy of dense hard-sphere fluids *J. Chem. Phys.* **98** 2225–31

[114] Camp P J, Mason C P, Allen M P, Khare A A and Kofke D A 1996 The isotropic–nematic transition in uniaxial hard ellipsoid fluids: coexistence data and the approach to the Onsager limit *J. Chem. Phys.* **105** 2837–49

[115] Hoover W G and Ree F H 1967 Use of computer experiments to locate the melting transition and calculate the entropy in the solid phase *J. Chem. Phys.* **47** 4873–8

[116] Hoover W G and Ree F H 1968 Melting transition and communal entropy for hard spheres *J. Chem. Phys.* **49** 3609–17

[117] Frenkel D, Mulder B M and McTague J P 1984 Phase-diagram of a system of hard ellipsoids *Phys. Rev. Lett.* **52** 287–90

[118] Meijer E J, Frenkel D, LeSar R A and Ladd A J C 1990 Location of melting point at 300 K of nitrogen by Monte Carlo simulation *J. Chem. Phys.* **92** 7570–5

[119] Lovett R 1995 Can a solid be turned into a gas without passing through a first order phase transition? *Observation, Prediction and Simulation of Phase Transitions in Complex Fluids* vol 460 *NATO ASI Series C* ed M Baus, L F Rull and J-P Ryckaert (Dordrecht: Kluwer) pp 641–54

[120] Sheu S-Y, Mou C-Y and Lovett R 1995 How a solid can be turned into a gas without passing through a first-order phase transformation *Phys. Rev.* E **51** R3795–8

[121] Rosenbluth M N and Rosenbluth A W 1995 Monte Carlo calculation of the average extension of molecular chains *J. Chem. Phys.* **23** 356–9

[122] Harris J and Rice S A 1988 A lattice model of a supported monolayer of amphiphile molecules—Monte Carlo simulations *J. Chem. Phys.* **88** 1298–306

[123] Siepmann J I and Frenkel D 1992 Configurational bias Monte Carlo—a new sampling scheme for flexible chains *Mol. Phys.* **75** 59–70

[124] Frenkel D, Mooij G A M and Smit B 1992 Novel scheme to study structural and thermal properties of continuously deformable molecules *J. Phys.: Condens. Matter* **4** 3053–76

[125] de Pablo J J, Laso M and Suter U W 1992 Simulation of polyethylene above and below the melting point *J. Chem. Phys.* **96** 2394–403

[126] Smit B, Loyens L D J C and Verbist G L M M 1997 Simulation of adsorption and diffusion of hydrocarbons in zeolites *Faraday Disc. Chem. Soc.* **106** 93–104

[127] Vlugt T J H, Krishna R and Smit B 1999 Molecular simulations of adsorption isotherms for linear and branched alkanes and their mixtures in silicalite *J. Phys. Chem.* B **103** 1102–18

[128] Frenkel D 1986 Free-energy computation and first-order phase transitions *Molecular Dynamics Simulation of Statistical Mechanical Systems* ed G Ciccotti and W G Hoover (Amsterdam: North-Holland) pp 151–88

[129] Smit B 1993 Computer simulations in the Gibbs ensemble *Computer Simulation in Chemical Physics* vol 397 *NATO ASI Series C* ed M P Allen and D J Tildesley (Dordrecht: Kluwer) pp 173–209

[130] Frenkel D 1995 Numerical techniques to study complex liquids *Observation, Prediction and Simulation of Phase Transitions in Complex Fluids* vol 460 *NATO ASI Series C* ed M Baus, L F Rull and J-P Ryckaert (Dordrecht: Kluwer) pp 357–419

[131] Panagiotopoulos A Z 1995 Gibbs ensemble techniques *Observation, Prediction and Simulation of Phase Transitions in Complex Fluids* ed M Baus, L F Rull and J-P Ryckaert, vol 460 *NATO ASI Series C* (Dordrecht: Kluwer) pp 463–501

[132] Privman V (ed) 1990 *Finite Size Scaling and Numerical Simulation of Statistical Systems* (Singapore: World Scientific)

[133] Binder K 1995 Introduction *The Monte Carlo Method in Condensed Matter Physics* vol 71 *Topics in Applied Physics* ed K Binder (Berlin: Springer) pp 1–22

[134] Siepmann J I, McDonald I R and Frenkel D 1992 Finite-size corrections to the chemical potential *J. Phys.: Condens. Matter* **4** 679–91

[135] Cardy J L (ed) *Finite-Size Scaling* vol 2 *Current Physics—Sources and Comments* (Amsterdam: North-Holland)

[136] Binder K 1981 Finite size scaling analysis of Ising-model block distribution-functions *Z. Phys.* B. *Condens. Matter.* **43** 119–40

[137] Bruce A D 1981 Probability density functions for collective coordinates in Ising-like systems *J. Phys.* C: *Solid State Phys.* **14** 3667–88

[138] Nicolaides D and Bruce A D 1988 Universal configurational structure in two-dimensional scalar models *J. Phys.* A: *Math. Gen.* **21** 233–44

[139] Nicolaides D and Evans R 1989 Nature of the prewetting critical-point *Phys. Rev. Lett.* **63** 778–81

[140] Bruce A D and Wilding N B 1992 Scaling fields and universality of the liquid-gas critical point *Phys. Rev. Lett.* **68** 193–6

[141] Wilding N B and Bruce A D 1992 Density fluctuations and field mixing in the critical fluid *J. Phys.: Condens. Matter* **4** 3087–108

[142] Wood W W 1968 Monte Carlo studies of simple liquid models *Physics of Simple Liquids* ed H N V Temperley, J S Rowlinson and G S Rushbrooke (Amsterdam: North Holland) chapter 5, pp 115–230

[143] Milchev A, Binder K and Heermann D W 1986 Fluctuations and lack of self-averaging in the kinetics of domain growth *Z. Phys. B. Condens. Matter.* **63** 521–35

[144] Challa M S S, Landau D P and Binder K 1986 Finite-size effects at temperature-driven 1st-order transitions *Phys. Rev. B* **34** 1841–52

[145] Brown F R and Yegulalp A 1991 Microcanonical simulation of 1st-order phase transitions in finite volumes *Phys. Lett. A* **155** 252–6

[146] Wood W W and Jacobson J D 1957 Preliminary results from a recalculation of the Monte Carlo equation of state of hard spheres *J. Chem. Phys.* **27** 1207–8

[147] Alder B J and Wainwright T E 1957 Phase transition for a hard sphere system *J. Chem. Phys.* **27** 1208–9

[148] Alder B J and Wainwright T E 1962 Phase transition in elastic disks *Phys. Rev.* **127** 359–61

[149] Mayer J E and Wood W W 1965 Interfacial tension effects in finite periodic two-dimensional systems *J. Chem. Phys.* **42** 4268–74

[150] Wood W W 1968 Monte Carlo calculations for hard disks in the isothermal–isobaric ensemble *J. Chem. Phys.* **48** 415–34

[151] Binder K and Landau D P 1984 Finite size scaling at 1st-order phase transitions *Phys. Rev. B* **30** 1477–85

[152] Binder K and Heermann D W 1988 *Monte Carlo Simulation in Statistical Physics* vol 80 *Solid State Sciences* (Berlin: Springer)

[153] Ferrenberg A M and Swendsen R H 1988 New Monte-Carlo technique for studying phase-transitions *Phys. Rev. Lett.* **61** 2635–8

[154] Ferrenberg A M 1989 Addition *Phys. Rev. Lett.* **63** 1658

[155] Lee J and Kosterlitz J M 1990 New numerical method to study phase transitions *Phys. Rev. Lett.* **65** 137–40

[156] Zhang Z, Mouritsen O G and Zuckermann M J 1992 Weak first-order orientational transition in the Lebwohl–Lasher model of liquid crystals *Phys. Rev. Lett.* **69** 2803–6

[157] Zhang Z, Zuckermann M J and Mouritsen O G 1993 Phase transition and director fluctuations in the 3-dimensional Lebwohl–Lasher model of liquid crystals *Mol. Phys.* **80** 1195–221

[158] Fabbri U and Zannoni C 1986 A Monte Carlo investigation of the Lebwohl–Lasher lattice model in the vicinity of its orientational phase transition *Mol. Phys.* **58** 763–88

[159] van Duijneveldt J S and Frenkel D 1992 Computer simulation study of free-energy barriers in crystal nucleation *J. Chem. Phys.* **96** 4655–68

[160] Lynden-Bell R M, van Duijneveldt J S and Frenkel D 1993 Free-energy changes on freezing and melting ductile metals *Mol. Phys.* **80** 801–14

[161] Orkoulas G and Panagiotopoulos A Z 1999 Phase behaviour of the restricted primitive model and square-well fluids from Monte Carlo simulations in the grand canonical ensemble *J. Chem. Phys.* **110** 1581–90

[162] Panagiotopoulos A Z 1987 Direct determination of phase coexistence properties of fluids by Monte Carlo simulation in a new ensemble *Mol. Phys.* **61** 813–26

[163] Panagiotopoulos A Z, Quirke N, Stapleton M and Tildesley D J 1988 Phase equilibria by simulation in the Gibbs ensemble. Alternative derivation, generalization and application to mixture and membrane equilibria *Mol. Phys.* **63** 527–45

[164] Panagiotopoulos A Z 1992 Direct determination of fluid phase equilibria by simulation in the Gibbs ensemble: a review *Mol. Simul.* **9** 1–23

[165] Panagiotopoulos A Z 1994 Molecular simulation of phase equilibria *Supercritical Fluids—Fundamentals for Application NATO ASI Series E* ed E Kiran and J M H Levelt Sengers (Dordrecht: Kluwer)

[166] Panagiotopoulos A Z 1987 Adsorption and capillary condensation of fluids in cylindrical pores by Monte Carlo simulation in the Gibbs ensemble *Mol. Phys.* **62** 701–19

[167] Siepmann J I, Karaborni S and Smit B 1993 Simulating the critical behaviour of complex fluids *Nature* **365** 330–2

[168] Smit B, Karaborni S and Siepmann J I 1995 Computer simulations of vapor–liquid phase equilibria of *n*-alkanes *J. Chem. Phys.* **102** 2126–40

[169] Panagiotopoulos A Z 1989 Exact calculations of fluid-phase equilibria by Monte Carlo simulation in a new statistical ensemble *Int. J. Thermophys.* **10** 447–57

[170] de Pablo J J and Prausnitz J M 1989 Phase equilibria for fluid mixtures from Monte Carlo simulation *Fluid Phase Equilibria* **53** 177–89

[171] Escobedo F A and de Pablo J J 1996 Expanded grand canonical and Gibbs ensemble Monte Carlo simulation of polymers *J. Chem. Phys.* **105** 4391–4

[172] Nath S K, Escobedo F A and de Pablo J J 1998 On the simulation of vapor–liquid equilibria for alkanes *J. Chem. Phys.* **108** 9905–11

[173] Möller D and Fischer J 1990 Vapour liquid equilibrium of a pure fluid from test particle method in combination with $NpT$ molecular dynamics simulations *Mol. Phys.* **69** 463–73

[174] Möller D 1992 Correction *Mol. Phys.* **75** 1461–2

[175] Lotfi A, Vrabec J and Fischer J 1992 Vapour liquid equilibria of the Lennard-Jones fluid from the $NpT$ plus test particle method *Mol. Phys.* **76** 1319–33

[176] Boda D, Liszi J and Szalai I 1995 An extension of the NpT plus test particle method for the determination of the vapour-liquid equilibria of pure fluids *Chem. Phys. Lett.* **235** 140–5

[177] Kofke D A 1993 Gibbs–Duhem integration: a new method for direct evaluation of phase coexistence by molecular simulation *Mol. Phys.* **78** 1331–6

[178] Kofke D A 1993 Direct evaluation of phase coexistence by molecular simulation via integration along the saturation line *J. Chem. Phys.* **98** 4149–62

[179] Agrawal R and Kofke D A 1995 Thermodynamic and structural properties of model systems at solid–fluid coexistence. II. Melting and sublimation of the Lennard-Jones system *Mol. Phys.* **85** 43–59

[180] Agrawal R and Kofke D A 1995 Thermodynamic and structural properties of model systems at solid–fluid coexistence. I. Fcc and bcc soft spheres *Mol. Phys.* **85** 23–42

[181] Bolhuis P G and Kofke D A 1996 Monte Carlo study of freezing of polydisperse hard spheres *Phys. Rev.* E **54** 634–43

[182] Kofke D A and Bolhuis P G 1999 Freezing of polydisperse hard spheres *Phys. Rev.* E **59** 618–22

[183] Holcomb C D, Clancy P and Zollweg J A 1993 A critical study of the simulation of the liquid–vapour interface of a Lennard-Jones fluid *Mol. Phys.* **78** 437–59

[184] Alejandre J, Tildesley D J and Chapela G A 1995 Molecular dynamics simulation of the orthobaric densities and surface tension of water *J. Chem. Phys.* **102** 4574–83

[185] Chandler D 1978 Statistical mechanics of isomerisation dynamics in liquids and the transition state approximation *J. Chem. Phys.* **68** 2959–70

[186] Carter E A, Ciccotti G, Hynes J T and Kapral R 1989 Constrained reaction coordinate dynamics for the simulation of rare events *Chem. Phys. Lett.* **156** 472–7

[187] Ruiz-Montero M J, Frenkel D and Brey J J 1997 Efficient schemes to compute diffusive barrier crossing rates *Mol. Phys.* **90** 925–41

[188] Ciccotti G and Ferrario M 1998 Constrained and nonequilibrium molecular dynamics *Classical and Quantum Dynamics in Condensed Phase Simulations* ed B J Berne, G Ciccotti and D F Coker (Singapore: World Scientific) pp 157–77

[189] ten Wolde P R and Frenkel D 1997 Enhancement of protein crystal nucleation by critical density fluctuations *Science* **277** 1975–8

[190] Chandler D 1998 Finding transition pathways: throwing ropes over rough mountain passes, in the dark *Classical and Quantum Dynamics in Condensed Phase Simulations* (Singapore: World Scientific) pp 51–66

[191] Dellago C, Bolhuis P G, Csajka F S and Chandler D 1998 Transition path sampling and the calculation of rate constants *J. Chem. Phys.* **108** 1964–77

[192] Pratt L R 1986 A statistical method for identifying transition states in high dimensional problems *J. Chem. Phys.* **85** 5045–8

[193] Feynman R P and Hibbs A R 1965 *Quantum Mechanics and Path Integrals* (New York: McGraw-Hill)

[194] Berne B J, Ciccotti G and Coker D F (ed) 1998 *Classical and Quantum Dynamics in Condensed Phase Simulations* (Singapore: World Scientific)

[195] De Raedt H 1996 Quantum theory *Monte Carlo and Molecular Dynamics of Condensed Matter Systems* ed K Binder and G Ciccotti (Bologna: Italian Physical Society) pp 401–42

[196] Ceperley D M 1996 Path integral Monte Carlo for fermions *Monte Carlo and Molecular Dynamics of Condensed Matter Systems* vol 49, ed K Binder and E G Ciccotti (Bologna: Italian Physical Society) pp 443–82

[197] Tuckerman M E and Hughes A 1998 Path integral molecular dynamics: a computational approach to quantum statistical mechanics *Classical and Quantum Dynamics in Condensed Phase Simulations* ed B J Berne, G Ciccotti and D F Coker (Singapore: World Scientific) pp 311–57

[198] Herman M F, Bruskin E J and Berne B J 1982 On path integral Monte Carlo simulations *J. Chem. Phys.* **76** 5150–5

[199] Tuckerman M, Berne B J, Martyna G J and Klein M L 1993 Efficient molecular dynamics and hybrid Monte Carlo algorithms for path integrals *J. Chem. Phys.* **99** 2796–808

[200] Sprik M, Klein M L and Chandler D 1985 Staging—a sampling technique for the Monte-Carlo evaluation of path-integrals *Phys. Rev.* B **31** 4234–44

[201] Berne B J and Thirumalai D 1986 On the simulation of quantum systems—path integral methods *Ann. Rev. Phys. Chem.* **37** 401–24

[202] Car R and Parrinello M 1985 Unified approach for molecular dynamics and density-functional theory *Phys. Rev. Lett.* **55** 2471–4

[203] Remler D K and Madden P A 1990 Molecular dynamics without effective potentials via the Car–Parrinello approach *Mol. Phys.* **70** 921–66

[204] Galli G and Pasquarello A 1993 First-principles molecular dynamics *Computer Simulation in Chemical Physics* vol 397 *NATO ASI Series C* ed M P Allen and D J Tildesley (Dordrecht: Kluwer) pp 261–313

[205] Car R 1996 Molecular dynamics from first principles *Monte Carlo and Molecular Dynamics of Condensed Matter Systems* vol 49 ed K Binder and G Ciccotti (Bologna: Italian Physical Society) pp 601–34

[206] Sprik M 1998 Density functional techniques for simulation of chemical reactions *Classical and Quantum Dynamics in Condensed Phase Simulations* ed B J Berne, G Ciccotti and D F Coker (Singapore: World Scientific) pp 285–309

[207] Hohenberg P C and Kohn W 1964 Inhomogeneous electron gas *Phys. Rev.* B **136** 864–71

[208] Vanderbilt D 1990 Soft self-consistent pseudopotentials in a generalized eigenvalue formalism *Phys. Rev.* B **41** 7892–5

[209] Ceperley D M and Alder B J 1980 Ground state of the electron gas by a stochastic method *Phys. Rev. Lett.* **45** 566–9

[210] Alfe D and Gillan M J 1998 First-principles simulations of liquid Fe–S under Earth's core conditions *Phys. Rev.* B **58** 8248–56

[211] de Wijs G A, Kresse G, Vočadlo L, Dobson D, Alfe D, Gillan M J and Price G D 1998 The viscosity of liquid iron at the physical conditions of the Earth's core *Nature* **392** 805–7

[212] Löwen H, Hansen J-P and Madden P A 1993 Nonlinear counterion screening in colloidal suspensions *J. Chem. Phys.* **98** 3275–89

[213] Smith W 1991 Molecular dynamics on hypercube parallel computers *Comput. Phys. Commun.* **62** 229–48

[214] Smith W 1992 A replicated data molecular dynamics strategy for the parallel Ewald sum *Comput. Phys. Commun.* **67** 392–406

[215] Wilson M R, Allen M P, Warren M A, Sauron A and Smith W 1997 Replicated data and domain decomposition molecular dynamics techniques for the simulation of anisotropic potentials *J. Comput. Chem.* **18** 478–88

[216] Rapaport D C 1991 Multi-million particle molecular dynamics II. Design considerations for distributed processing *Comput. Phys. Commun.* **62** 217–28

[217] Esselink K, Smit B and Hilbers P A J 1993 Efficient parallel implementation of molecular dynamics on a toroidal network. I. Parallelizing strategy *J. Comput. Phys.* **106** 101–7

[218] Beazley D M and Lomdahl P S 1993 Message-passing multi-cell molecular dynamics on the Connection Machine 5 *Parallel Comput.* **20** 173–95

[219] Holian B L and Lomdahl P S 1998 Plasticity induced by shock waves in nonequilibrium molecular-dynamics simulations *Science* **280** 2085–8

[220] Hilbers P A J and Esselink K 1992 Parallel molecular dynamics *Parallel Computing: From Theory to Sound Practice* ed W Joosen and E Milgrom (Amsterdam: IOS Press) pp 288–99

[221] Hilbers P A J and Esselink K 1993 Parallel computing and molecular dynamics simulations *Computer Simulation in Chemical Physics* vol 397 *NATO ASI Series C* ed M P Allen and D J Tildesley (Dordrecht: Kluwer) pp 473–95

[222] Heermann D W and Burkitt A N 1995 Parallel algorithms for statistical physics problems *The Monte Carlo Method in Condensed Matter Physics* vol 71 *Topics in Applied Physics* ed K Binder (Berlin: Springer) pp 53–74

[223] Heermann D W 1996 Parallelization of computational physics problems *Monte Carlo and Molecular Dynamics of Condensed Matter Systems* vol 49, ed K Binder and G Ciccotti (Bologna: Italian Physical Society) pp 887–906

[224] Gubbins K E and Quirke N 1996 *Molecular Simulation and Industrial Applications* (Reading: Gordon and Breach)

[225] Lewis D W, Sankar G, Wyles J K, Thomas J M, Catlow C R A and Willock D J 1997 Synthesis of a small-pore microporous material using a computationally designed template *Angew. Chem. Int. Ed. Engl.* **36** 2675–7

[226] Groot R D and Warren P B 1997 Dissipative particle dynamics: bridging the gap between atomistic and mesoscopic simulation *J. Chem. Phys.* **107** 4423–35

[227] Chen S and Doolen G D 1998 Lattice Boltzmann method for fluid flows *Ann. Rev. Fluid Mech.* **30** 329–64

## Further Reading

Allen M P and Tildesley D J 1989 *Computer Simulation of Liquids* (Oxford: Clarendon)

A comprehensive introduction to the field, covering statistical mechanics, basic Monte Carlo, and molecular dynamics methods, plus some advanced techniques, including computer code.

Frenkel D and Smit B 1996 *Understanding Molecular Simulation: From Algorithms to Applications* (San Diego: Academic)

A comprehensive and up-to-date introduction to the ideas of molecular dynamics and Monte Carlo, with statistical mechanical background, advanced techniques and case studies, supported by a Web page for software download.

Binder K and Ciccotti G (ed) 1995 *Monte Carlo and Molecular Dynamics of Condensed Matter Systems: Proc. Euroconference (Como, Italy, 3–28 July 1995)* vol 49 (Bologna: Italian Physical Society)

One of the most comprehensive and up-to-date summer school proceedings, with contributions from many of the world's leading experts. Almost every aspect of molecular simulation is covered.

Berne B J, Ciccotti G and Coker D F (ed) 1998 *Classical and Quantum Dynamics in Condensed Phase Simulations: Proc. Euroconference (Lerici, Italy, 7–18 July, 1997)* (Singapore: World Scientific)

A substantial summer school proceedings, concentrating on modern techniques for studying rare events and quantum mechanical phenomena.

Rapaport D C 1995 *The Art of Molecular Dynamics Simulation* (Cambridge: Cambridge University Press)

A detailed and comprehensive book on molecular dynamics, with a tutorial approach plus many examples of computer code, supported by a Web page for software download.

Binder K and Heermann D W 1997 *Monte Carlo Simulation in Statistical Physics* vol 80 *Solid State Sciences* 3rd edn (Berlin: Springer)

A compact and readable introduction to Monte Carlo, with examples and exercises, plus useful pointers to the literature on lattice models.

# B3.4
# Quantum dynamics and spectroscopy

*Sybil M Anderson, Rovshan G Sadygov and Daniel Neuhauser*

### B3.4.1 Introduction

The study of quantum effects associated with nuclear motion is a distinct field of chemistry, known as quantum molecular dynamics. This section gives an overview of the methodology of the field; for further reading, consult [1–5].

The importance of non-classical behaviour in molecular dynamics has its origins in the inherently quantal nature of atomic motion. The de Broglie wavelengths of atoms are small but non-vanishing. For example, hydrogen has a de Broglie wavelength that can be as high as $\approx 1.0$ Å at room temperature. Specific quantum effects in atomic and molecular motion include: zero-point motion, most notably for hydrogen which has a zero-point energy of $\sim 10$ kcal mol$^{-1}$ in many of its covalent interactions; interference resonances, which are spikes in reaction or pre-dissociation probabilities associated with quasi-bound states of molecules [6–9]; and tunnelling of nuclei which is important in many catalytic reactions [10–14].

In its most fundamental form, quantum molecular dynamics is associated with solving the Schrödinger equation for molecular motion, whether using a single electronic surface (as in the Born–Oppenheimer approximation—section B3.4.2) or with the inclusion of multiple electronic states, which is important when discussing non-adiabatic effects, in which the electronic state is changed [15–19].

Sections B3.4.3–B3.5 describe methods of solving the Schrödinger equation for scattering events. Sections B3.4.6 and B3.4.7 proceed to discuss photo-dissociation and bound states.

As these methods are explored, it is quickly realized that the numerical effort in the theoretical description grows prohibitively large with the number of atoms in a molecule. The difficulty lies in precisely what makes molecular motion fundamentally quasi-classical, i.e. the large molecular masses (relative to the mass of the electron). Consequently, a molecular wavefunction has many oscillations and is difficult to model numerically. There have been many attempts at developing alternate approaches for representing quantum wavefunctions and observables without the use of large grids or basis sets, ranging from approximations to path-integral descriptions. The basics of these approaches are described in section B3.4.8. Later, section B3.4.9 describes the issues involved in the study of non-adiabatic phenomena.

Finally, section B3.4.10 touches on the application of quantum molecular dynamics to a very exciting field: laser interactions with molecules. This field presents, in principle, the opportunity to influence chemistry by lasers rather than to simply observe it.

The scope of this section restricts the discussion. One omitted topic is the collision and interaction of molecules with surfaces (see [20, 21] and section A3.9). This topic connects quantum molecular dynamics in gas and condensed phases. Depending on the time scales of the interaction of a molecule with a surface, the reactions are similar to those in one phase or the other. If the collision is fast, so that one may neglect the motion of the surface molecules and treat them as frozen, it is effectively a gas-phase reaction. On the other extreme, if the motion of the adsorbate is slow or comparable in time to the motion of surface and subsurface molecules, then the collision problem becomes very similar to the interaction of molecules in condensed phases. The latter is a subject of a separate section C3.5.

Another modern and highly exciting topic, omitted here due to lack of space, is the motion of very cold molecules [22–24], which can have de Broglie wavelengths that are as large or larger than the distances between the molecules. The simplest examples are essentially extensions of floppy van der Waals structures, but at the extreme, when the wavelength is extremely large and there are many molecules per molecular wavelength, one ends up with Bose–Einstein condensates (where the wavefunctions of the molecules coalesce to form one giant coherent molecular function) and even molecular lasers (i.e. lasers where the fundamental particles are atoms or molecules rather than photons [25]) can be made. Section C1.4 provides an overview of this new field.

As in any field, it is useful to clarify terminology. Throughout this section an 'atom' more specifically refers to its nuclear centre. Also, for most of the section the $\hbar = 1$ convention is used. Finally, it should be noted that in the literature the label 'quantum molecular dynamics' is also sometimes used for a purely classical description of atomic motion under the potential created by the electronic distribution.

Finally, this section is related to several others, especially section A3.11 on formal scattering, which should be carefully consulted.

### B3.4.2   Quantum motion on a single electronic surface

A corner-stone of a large portion of quantum molecular dynamics is the use of a single electronic surface. Since electrons are much lighter than nuclei, they typically adjust their wavefunction to follow the nuclei [26]. Specifically, if a collision is started in which the electrons are in their ground state, they typically remain in the ground state. An exception is non-adiabatic processes, which are discussed later in this section.

The single-surface assumption, known also as the Born–Oppenheimer approximation, implies that the nuclei are described by a single wavefunction ($\psi(x, t)$ where $x$ is a multi-dimensional vector describing the nuclear position). The time-dependent equation for the evolution of the wavefunction is simply

$$i\frac{\partial \psi}{\partial t} = \hat{H}\psi \equiv [\hat{K} + V(x)]\psi \tag{B3.4.1}$$

where $\hat{H}$ is the Hamiltonian governing the motion of the nuclei (or the atomic motion, as typically denoted), $\hat{K}$ is the kinetic term (a sum of terms of the form $-(1/2m_j)(\partial^2/\partial x_j^2)$ for each atom $j$) and $V$ is the Born–Oppenheimer potential which is defined as the electronic ground-state energy for nuclear configuration $x$, including the nuclear–nuclear repulsion.

As a word of caution, the Born–Oppenheimer assumption is not universally valid. There are many reactions in which, for example, non-adiabatic curve-crossing processes occur. In these cases, two electronic potentials—e.g., the ground state and an excited state—are locally equal to one another at some configuration. Curve-crossing effects are highly non-trivial and can affect the nuclear dynamics even at energies which are way below the curve crossing. These points are briefly discussed in section B3.4.9.

### B3.4.3   Scattering

#### B3.4.3.1   Collinear motion

Equation (B3.4.1) is general and applies to both scattering and bound state spectroscopy. Scattering will be considered first. For simplicity, the discussion uses the collinear model for the $A + BC \rightarrow AB + C$ reaction (i.e. assuming all particles lie on a line). This model is easy to visualize and embodies most elements of three-dimensional (3D) scattering of larger molecules.

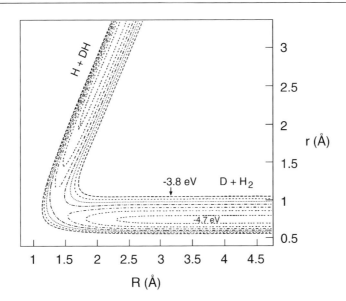

**Figure B3.4.1.** The potential surface for the collinear $D + H_2 \rightarrow DH + H$ reaction (this potential is the same as for $H + H_2 \rightarrow H_2 + H$, but to make the products and reactants identification clearer the isotopically substituted reaction is used). The $D + H_2$ reactant arrangement and the $DH + H$ product arrangement are denoted. The coordinates are $r$, the $H_2$ distance, and $R$, the distance between the D and the $H_2$ centre of mass. Distances are measured in angströms; the potential contours shown are 4.7 eV, $-4.55$ eV,..., $-3.8$ eV. (The potential energy is zero when the particles are far from each other. Only the first few contours are shown.) For reference, the zero-point energy for $H_2$ is $-4.47$ eV, i.e. 0.27 eV above the $H_2$ potential minimum ($-4.74$ eV); the room-temperature thermal kinetic energy is approximately 0.03 eV. The graph uses the accurate Liu–Seigbahn–Truhlar–Horowitz (LSTH) potential surface [195].

After removal of centre-of-mass motion, there are two independent distances which need to be considered for a collinear problem, $r_{BA}$ ($\equiv r_B - r_A$, where $r_B$ and $r_A$ denote the positions of A and B on the line) and $r_{CB}$, which is similarly defined. Unfortunately, the kinetic energy is not conveniently described with these coordinates; therefore, alternate systems are used. The most convenient one is reactant Jacobi coordinates $(r, R)$, where $r = r_{CB}$, and $R$ is the distance between A and the centre of mass of B and C [27]. In these coordinates, the kinetic energy gets a simple and separable form so that the Schrödinger equation is

$$i\frac{\partial \psi}{\partial t} = \left[ -\frac{1}{2M}\frac{\partial^2}{\partial R^2} - \frac{1}{2\mu}\frac{\partial^2}{\partial r^2} + V(R, r) \right]\psi \qquad (B3.4.2)$$

where $\mu$ is the reduced mass associated with the BC vibration, and $M$ is the mass associated with the $A - BC$ motion.

Figure B3.4.1 shows the potential surface for a simple collinear reaction, $D + H_2 \rightarrow HD + H$. The most notable aspect of this potential is the angle between the products and reactant arrangement. Below breakup ($D + H_2 \rightarrow D + H + H$, which can occur only at several electronvolts above the reaction threshold), the potential surface has three relevant regions: the reactants ($D + H_2$) asymptote (arrangement) at large $R$ and small $r$; the products ($H + DH$) asymptote; and a strong interaction region where all three atoms are closely spaced. The strong-interaction region is extended over a region of $\approx 1$ Å $\times$ 1 Å, containing several oscillations of the full ($DH_2$) wavefunction. Since the potential is not separable in $R$ and $r$, upon reaction the particles would exchange vibrational and translational energy (and in the three-dimensional case, also rotational energy).

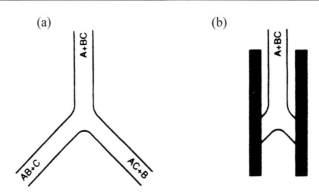

**Figure B3.4.2.** (a) A schematic potential surface showing bifurcation for a triatomic reactive system. (b) By blocking the products' arrangement with an absorbing potential (shaded area) the reactive system is reduced to one arrangement; this scheme enables calculation of both total reactivities and state-to-state information. Reprinted from [46] with permission.

The collinear model does not include bifurcation, i.e. the possibility of *several* product channels which the system can access. A model potential surface for an A + BC → AB + C, AC + B reaction is shown in figure B3.4.2. Both of these examples will be used in the discussion below.

### B3.4.3.2   *Boundary conditions*

The eventual goal in scattering calculations is essentially to obtain the scattering matrix, $S$ (see section A3.11 and equation (B3.4.4) below). The scattering matrix can be obtained by reference to the solution of the time-independent Schrödinger equation, fulfilling

$$(\hat{H} - E)\psi = 0 \tag{B3.4.3}$$

which is the Fourier transform of equation (B3.4.1). However, this wavefunction must obey the appropriate boundary conditions. It should have components associated with a single 'incoming' channel (an incoming wave associated with the translational A+BC motion, $e^{-ikR}$, multiplied by a wavefunction for BC at a specific initial target state, $\phi_{n_0}(r)$). In addition, there are components associated with all 'outgoing' channels. (See section A3.11 and [27].) For example, for the bifurcating potential case where three asymptotic arrangements are formally possible, the wavefunction (with an index $n_0$ attached) is

$$\psi_{n_0}(x, E)' = {}'\phi_{n_0}(r) \frac{e^{-ik_{n_0}R}}{\sqrt{k_{n_0}/M}} - \sum_n \phi_n(r) \frac{e^{ik_n R}}{\sqrt{k_n/M}} S_{nn_0}$$

$$- \sum_{\bar{n}} \phi_{\bar{n}}(\bar{r}) \frac{e^{ik_{\bar{n}}\bar{R}}}{\sqrt{k_{\bar{n}}/\bar{M}}} S_{\bar{n}n_0} - \sum_{\bar{\bar{n}}} \phi_{\bar{\bar{n}}}(\bar{\bar{r}}) \frac{e^{ik_{\bar{\bar{n}}}}}{\sqrt{k_{\bar{\bar{n}}}/\bar{\bar{M}}}} S_{\bar{\bar{n}}n_0} \tag{B3.4.4}$$

where $\phi_n$ is the $n$th vibrational state of the BC diatomic and $k_n$ is the translational momentum of A when BC is in the $n$th state ($k_n^2/2M = E - \epsilon_n$). Quantities with a bar ($\bar{r}$, $\bar{R}$, etc) refer to the product channel AB+C, so $\bar{R}$ is the distance between AB+C, etc. In addition, for cases in which the AC+B channel is open, a double-bar notation is used ($\bar{\bar{n}}$, $\bar{\bar{r}}$, etc). $S_{\bar{n}n_0}$ is the scattering matrix associated with the amplitude of the system to emerge at product channel $\bar{n}$ when it is initially at reactant channel $n_0$. The equality sign is in quotes to denote that this relation is only valid in the asymptote where the system is separated into an atom and a diatom.

### B3.4.3.3   Scattering techniques

The presence of the multiple arrangements make molecular scattering very challenging theoretically. After much trial and error, several techniques have been developed. These techniques generally fall into two broad categories:

- methods which aim at treating all possible arrangements simultaneously;
- arrangement-decoupling approaches [28–31] where an absorbing potential is used to convert multiple-arrangement problems to inelastic-scattering (or even bound-state-like) problems. In recent years, these approaches have become very powerful for large-scale applications.

### B3.4.3.4   All-arrangement methods

#### (a) Wavefunction expansion

The conceptually simplest approach to solve for the $S$-matrix elements is to require the wavefunction to have the form of equation (B3.4.4), supplemented by a bound function which vanishes in the asymptote [32–35]. This approach is analogous to the full configuration-interaction (CI) expansion in electronic structure calculations, except that now one is expanding the nuclear wavefunction. While successful for intermediate size problems, the resulting matrices are not very sparse because of the use of multiple coordinate systems, so that this type of method is prohibitively expensive for diatom–diatom reactions at high energies.

#### (b) Close coupling

Alternatively, one can use close-coupling methods. These methods are easiest to understand for single arrangement problems (i.e. when both the AB + C and AC + B product arrangements are very high in energy so that only the A + BC reactant arrangement can be accessed). Then one writes

$$\psi_{n_0}(R, r, E) = \sum_n a_{nn_0}(R)\phi_n(r) \tag{B3.4.5}$$

and it is readily shown that the Schrödinger equation can be written as

$$\frac{\partial^2}{\partial R^2} a_{nn_0} = 2M \sum_m (U_{nm}(R) - E\delta_{mn}) a_{mn_0} \tag{B3.4.6}$$

where

$$U_{nm}(R) = \int \phi_n(r)\left( -\frac{1}{2\mu}\frac{\partial^2}{\partial r^2} + V(R, r) \right)\phi_m(r)\,dr. \tag{B3.4.7}$$

Equation (B3.4.6) is solved by starting at a small value of $R$, denoted by $R_{start}$, where the potential is high and the wavefunction is exponentially vanishing, and picking random values for $a_{nn_0}$ and $\partial a_{nn_0}/\partial R$. Then, the equations are propagated towards larger $R$. Eventually, $\psi(R, r)$ is resolved at a large value of $R$ to yield the $S$-matrix [36]. In practice, one has to avoid linear dependence between the solutions associated with different initial conditions. This is achieved by simple stabilization approaches.

   The close-coupling approach works readily and simply if the reaction is purely 'inelastic'. The method can also be made to work very simply for a single product arrangement (as in collinear reactions), by using a 'twisted' coordinate system, most conveniently reaction path coordinates [37–39] as shown in figure B3.4.3.

   The complications which occur with bifurcation, i.e. when more than one product arrangement is accessible, can be solved by various methods. Historically, the first close-coupling approaches for multiple product channels employed fitting procedures [40], where the close-coupling equations are simultaneously propagated

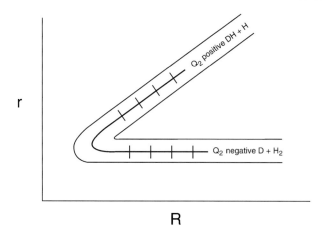

**Figure B3.4.3.** A schematic figure showing, for the DH$_2$ collinear system, a reaction-path coordinate $Q$ connecting continuously the reactants and the single products asymptote. Also shown are the cuts denoting the coordinate perpendicular to $Q$.

from each of the asymptotes inwards and then are fitted together at a dividing surface. This approach has been replaced in recent calculations by two methods. One is based on using absorbing potentials to turn the reactive problem into an inelastic one, as explained later. The other is to use hyperspherical coordinates for carrying out the close-coupling propagation [41–45]. The hyperspherical coordinates consist of a single radius $\rho$, which is zero at the origin (when all nuclei are stuck together) and increases outwards, and a set of angles. For the collinear problem as well as the atom–diatom problem (involving three independent distances) the hyperspherical coordinates are typically just the regular spherical coordinates. Close-coupling propagation starts at $\rho = 0$ and moves outward until a large value of $\rho$ is reached. When the asymptote are reached one fits the wavefunction to have the form of equation (B3.4.4) and thus obtains the scattering matrix.

### B3.4.4   Arrangement decoupling by absorbing potentials

A simplifying approach to scattering is to eliminate all the product asymptote. This can be done efficiently and rigorously [28, 30, 31, 46] by inserting in the Hamiltonian a negative-imaginary potential (or more generally a complex potential with a negative imaginary term) [30, 47–49]. This potential, denoted as $-iV_I(R, r)$, acts to 'chop' away the product arrangements, while retaining the correct form of the wavefunction (see figure B3.4.4).

To understand this unique feature of the negative imaginary potential it is easiest to refer to the time-dependent language, discussed later in the section. Heuristically, note that the time-dependent propagator, $e^{-i(\hat{H}-iV_I)t}$ essentially contains a $e^{-V_I t}$ term which decays the wavefunction in regions where $V_I$ is positive. The negative imaginary potential therefore prevents reflection and thus imposes 'outgoing boundary conditions' [31] on the wavefunctions to which they are applied.

There is considerable freedom in the choice of absorbing potentials; they are simply required [30] to be sufficiently extended to absorb any wavefunction which impinges on them, while not rising too sharply to avoid reflection from their rising slopes before the wave gets absorbed. This implies that they typically need to extend only over approximately one to two de Broglie wavelengths, which is usually short enough to add only a negligible overhead to the size of the required grids.

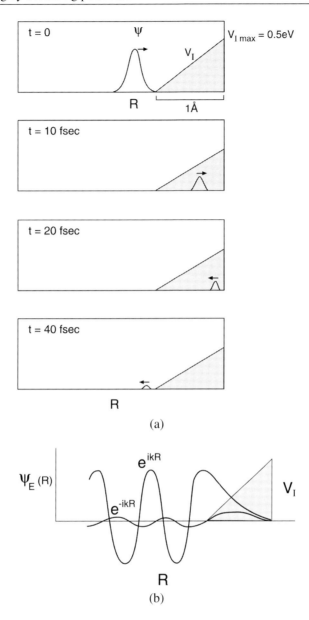

**Figure B3.4.4.** (a) Schematic evolution in a 1D problem of a wavepacket impinging on an absorbing potential with typical parameters (shaded). The width and magnitude of the absorbing potential must be sufficiently large so that a wavepacket impinging on it would eventually be completely absorbed, with very little reflected. (b) In time-independent language, the absorbing potential forces the wavefunction in the region preceding it to have an outgoing ($e^{ikR}$) form, with very little reflected ($e^{-ikR}$) component.

The negative imaginary potentials can be applied in any scattering formalism. In close coupling, they can be implemented to block any product arrangement [31] (see figure B3.4.2) and this thereby converts the reactive problem to an inelastic one; the only cost is the propagation of a complex matrix, $a_{nn_0}$, rather than a real one.

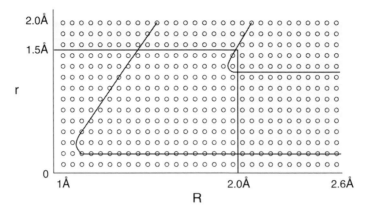

**Figure B3.4.5.** Schematic plot of a two-dimensional potential surface for D + $H_2$ restricted by an absorbing potential (shaded area). The absorbing potential $iV_I(R, r)$ (shaded region) rises gently outward towards the edges of the grid. In practice, the grid needs to be extended only by $\approx 0.5$–1 Å, or less for heavier mass systems. The absorbing potential imposes the correct boundary condition on the wavefunction in the inner region. This basic paradigm of a small grid (denoted by dots), used in the strong-interaction region to describe the main part of the wavefunction, applies in several different formulations: time-independent arrangement-decoupling scattering, where the time-independent wavefunction $\chi_{n_0}(R, r, E)$ is placed on this grid (supplemented by a function describing the initial wavefunction); time-dependent scattering (where it is used to describe the non-elastic part of the wavefunction, and the elastic part is represented on a separate grid); flux–flux studies, and photodissociation. All desired scattering information can be obtained from the information on the wavefunction in the strong-interaction region.

Alternately, absorbing potentials can also be applied to convert scattering to a bound-state-like problem. One method is to write the Schrödinger wavefunction as a sum of two terms $\psi_{n_0}(E) = \chi_{n_0}(E) + \zeta_{n_0}(E)$, where $\zeta_{n_0}$ includes the known incoming wave term, and $\chi$ includes the unknown outgoing wave part (see figure B3.4.5). The final equation is then [28]

$$(E - (\hat{H} - iV_I))\chi_{n_0} = (\hat{H} - E)\zeta_{n_0} \tag{B3.4.8}$$

where the absorbing potentials are inserted to impose the correct boundary conditions on $\chi_{n_0}$.

This approach has one key advantage [30]. Although when solving for $\chi$ one needs to invert $(E - \hat{H} + iV_I)$ (more precisely calculate the action of the so-called Green's function $(E - \hat{H} + iV_I)^{-1}$ on the initial state), the operator is defined in a finite region so that the wavefunction can be written on a single small grid covering the small-interaction region (as no asymptotic regions are involved). This makes the operation by $\hat{H}$ very rapid (formally, $\hat{H}$ is then very sparse) so that efficient iterative methods can be used [50–54]. It is then possible to handle grid sizes of more than a million points. Thus systems like AB + CD rearrangement scattering, with six or more floppy distances (and $\approx 10$ points per degree of freedom) are now routinely done with arrangement decoupling approaches based on absorbing potentials. The most widely applied iterative method with absorbing potentials and arrangement decoupling was developed within a time-dependent formulation, and is discussed below.

### B3.4.4.1   The time-dependent method

The approaches discussed so far are generally called time-independent methods, since they start from the time-independent Schrödinger equation, $(\hat{H} - E)\psi_{n_0} = 0$. An alternative is to use the time-dependent Schrödinger equation [28, 29, 50, 55–64]. Conceptually, the time-dependent approach is very simple: prepare an initial wavepacket on an appropriate grid; propagate the initial wavepacket for a sufficiently long time; and analyse the results. The approach is efficient, since propagation of a wavepacket is relatively cheap and since results on scattering at many energies are extracted at once. Correct boundary conditions have a very simple meaning in time-dependent approaches: the scattered component of the wavepacket is not returned from the edges of the grid. This is done by adding an absorbing potential to the Hamiltonian, which absorbs any component of the wavefunction that reaches the edge of the grid.

In more detail, in time-dependent approaches an initial wavepacket associated with the separate parts of the colliding system is prepared. The most efficient approach for a grid construction is to use two grids. Thus, the total wavefunction is divided into two parts [30, 65], a simple one which defines the initial wavefunction and a more complicated part, represented (for collinear scattering) on a two-dimensional grid of $R, r$ values, which is used to carry most of the wavefunction (essentially the scattered part) and which is padded with absorbing potentials (see figure B3.4.5). With this approach and with methods which reduce the number of grid points (in a given region) to at most two per oscillation [50, 66, 67], the total number of grid points can be reduced to 150–800 for collinear scattering involving hydrogen with energies of up to 2 eV.

Once the grid (or two grids) are prepared, there are two similar types of approaches to propagate the initial wavefunction forward with time. One approach is split-operator methods, [59] where the short-time propagator is divided into a kinetic and potential parts so that

$$|\psi_{n_0}(t+dt)\rangle \approx \exp\left[-i\frac{(V-iV_I)\,dt}{2}\right]\exp\left[-i\left(\frac{P^2}{2M}+\frac{p^2}{2\mu}\right)dt\right]\exp\left[-\frac{i(V-iV_I)\,dt}{2}\right]|\psi_{n_0}(t)\rangle \quad \text{(B3.4.9)}$$

where a bra–ket notation is used. The action by the potential is trivial, since it is local in coordinate space, and amounts to multiplication of $\psi_{n_0}(R, r, t)$ by $\exp(-iV(R, r)\,dt/2)$, and by a damping term, $\exp(-V_I(R, r)\,dt/2)$.

The action by the kinetic term is only slightly more complicated: the coordinate grid wavefunction, $\psi_{n_0}(R, r)$, undergoes a fast Fourier transform (FFT) to convert it to momentum space:

$$\psi_{n_0}(P, p, t) = \sum_{R,r} e^{-i(pr+PR)}\psi_{n_0}(R, r, t). \quad \text{(B3.4.10)}$$

The function is then multiplied by $\exp(-i((P^2/2M)+(p^2/2\mu))dt)$ and then returned to coordinate space by the inverse of equation (B3.4.10).

The key to this method is thus to act with each operator (exponential of the potential or kinetic term) in the representation (coordinate or momentum grid) in which it is local [50, 66, 67].

An alternative to split operator methods is to use iterative approaches. In these methods, one notes that the wavefunction is formally $|\psi_{n_0}(t)\rangle = \exp(-i\hat{H}t)|\psi_{n_0}\rangle$, and the action of the exponential operator is obtained by repetitive application of $H$ on a function (i.e. on the computer, by repetitive applications of the sparse matrix $H$ on wavefunction vectors). The simplest iterative method is the Taylor expansion of $e^{-i\hat{H}t}|\psi_{n_0}\rangle$ as $\sum_n((-i)^n t^n/n!)\hat{H}^n|\psi_{n_0}\rangle$. On the computer, this expansion would be performed by acting with $\hat{H}$ on $\psi_{n_0}$, then acting with $\hat{H}$ on the resulting vector, etc, and adding the contribution to the sum at each stage. The action by $\hat{H}$ on a vector is straightforward, as in the split operator approach: the potential is local, and the kinetic energy is evaluated by Fourier transforming back and forth onto the momentum grid.

The Taylor series by itself is not numerically stable, since the individual terms can be very large even if the result is small, but other polynomials which are highly convergent can be found, e.g. Chebyshev [50, 62–64] or Lancosz polynomials [51, 68].

The wavepacket is propagated until a time where it is all scattered and is away from the interaction region. This time is short (typically 10–100 fs) for a direct reaction. However, for some types of systems, e.g. for reactions with wells, the system can be trapped in resonances which are quasi-bound states (see section B3.4.7).

There are efficient ways to handle time-dependent scattering even with resonances, by propagating for a short time and then extracting the resonances and adding their contribution [69].

The last stage is the extraction of energy-resolved information, obtained automatically and simultaneously at many energies, by Fourier transforming the wavefunction to produce an energy-resolved state:

$$\psi_{n_0}(R, r, E) = \frac{1}{a_E} \int_{-\infty}^{\infty} \psi_{n_0}(R, r, t)e^{iEt}\, dt \qquad (B3.4.11)$$

where $a_E$ is related to the energy content of the initial wavepacket. It is easy to show that $\psi(R, r, E)$ fulfills the time-independent Schrödinger equation. (In practice, $\psi$ is known analytically prior to $t = 0$, so that the wavefunction only needs to be propagated forward in time.) The scattering matrix is then obtained from either of several formulae, all of the form [46, 65, 70–74]

$$S_{\bar{n}n_0} = \langle \xi_{\bar{n}} | \psi_{n_0}(E) \rangle \qquad (B3.4.12)$$

where $\xi_{\bar{n}}$ is a simple function which is associated with the final state $\bar{n}$. These formulae extract long-range scattering information from the wavefunction values in the strong-interaction region. Scattering information can therefore be extracted even when absorbing potentials are used to remove the asymptotic regions.

An interesting side point is that it is possible to recast the time-dependent approach, as described here, in a purely time-independent fashion, since from the equations above it follows that [74]

$$\psi_{n_0}(E) = \text{constant} \frac{1}{E - \hat{H} + iV_I} \psi_{n_0}.$$

The time-dependent approach is thus just one technique for evaluating the action of the Green's function on the initial wavepacket.

### B3.4.4.2  *Large-scale applications*

Both close-coupling approaches (hyperspherical or with absorbing potentials) and iterative/time-dependent absorbing-potential arrangement-decoupling approaches are readily extended to three-dimensional atom–molecule and molecule–molecule scattering. The wavefunction representation becomes more complicated and includes rotational matrices, but the essence and application of the method remains analogous [58, 65, 75, 76].

Iterative approaches, including time-dependent methods, are especially successful for very large-scale calculations because they generally involve the action of a very localized operator (the Hamiltonian) on a function defined on a grid. The effort increases relatively mildly with the problem size, since it is proportional to the number of points used to describe the wavefunction (and not to the cube of the number of basis sets, as is the case for methods involving matrix diagonalization). Present computational power allows calculations with optimized grids with sizes of $10^5$–$10^7$ points or more. This enables efficient simulations of four-body reactions involving six independent distances and up to two overall rotational coordinates. Thus far there have been several four-body reactions reported using this method, including $H_2 + OH \rightleftharpoons H_2O + H$ [75–78] and $CO + HO \rightleftharpoons CO_2 + H$ [79, 80], as well as surface reactions [58, 81] (see figure B3.4.6 for an example).

### B3.4.5  Coarse information

The methodology presented so far allows the calculations of state-to-state $S$-matrix elements. However, often one is not interested in this high-level of detail but prefers instead to find more average information, such as the initial-state selected reaction probability, i.e. the probability of rearrangement given an initial state $n_0$. In general, this probability is

$$P_{n_0}(E) = \sum_{\bar{n}} |S_{\bar{n}n_0}(E)|^2. \qquad (B3.4.13)$$

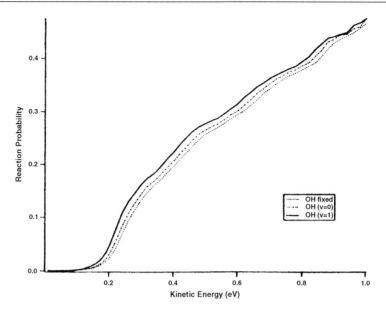

**Figure B3.4.6.** Reaction probabilities for the initial-state-selected process $H_2(v = 0, j = 0)+OH(v, j = 0) \rightarrow H_2O+H$, for zero total angular momentum. Taken from [75] with permission.

For example, for the collinear reaction A + BC this would be the probability that if initially the diatom BC is in a vibrational state $\phi_{n_0}(r)$, then after the reaction a diatom AB is formed (in *any* product vibrational state). In practice, the initial-state selected probability is easily calculated from the flux of the wavefunction $\psi_{n_0}(R, r, E)$ calculated at the product arrangement (e.g., at a large value of $r$, the B–C separation).

At times, however, even the information presented by $P_{n_0}(E)$ is too detailed. If one wants to rigorously calculate the thermal rate of rearrangement reactions, the initial vibrational state is not important. The relevant quantity is the sum of the initial-state-selected probabilities

$$N(E) = \sum_{n_0} P_{n_0}(E) = \sum_{n_0 \bar{n}} |S_{\bar{n}n_0}(E)|^2. \tag{B3.4.14}$$

$N(E)$ is called the cumulative reaction probability. It is directly related to the thermal reaction rate $k(T)$ by

$$k(T) = \frac{\int N(E)e^{-E/kT} \, dE}{Q} \tag{B3.4.15}$$

where $T$ is the temperature and $Q$ is the reactants' partition function.

A major achievement [71, 82–88] was the development of a simple quantum ('flux–flux') expression for the cumulative reaction probability, $N(E)$, with the final result [88]

$$N(E) \propto \text{Im} \, \text{Tr}(FGFG^*) \tag{B3.4.16}$$

where the Green's function is, as mentioned earlier,

$$G(E) = \frac{1}{E - (\hat{H} + iV_I)} \tag{B3.4.17}$$

and $F$ is the flux operator. In this expression, the trace is evaluated over a small grid region. In principle, the grid has to contain only a small-interaction region, in which the system 'decides' its final arrangement

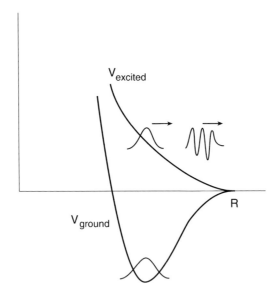

**Figure B3.4.7.** Schematic example of potential energy curves for photo-absorption for a 1D problem (i.e. for diatomics). On the lower surface the nuclear wavepacket is in the ground state. Once this wavepacket has been excited to the upper surface, which has a different shape, it will propagate. The photoabsorption cross section is obtained by the Fourier transform of the correlation function of the initial wavefunction on the excited surface with the propagated wavepacket.

(i.e. with what probability to react). This expression does not refer to the scattering matrix and therefore the asymptotic region does not have to be included in the grid.

The flux–flux expression and its extensions have been used to calculate reaction probabilities for several important reactions, including $H_2 + O_2 \rightarrow H + H_2O$, by explicit calculation of the action of $G$ in a grid representation with absorbing potentials. The main power of the flux–flux formula over the long run will be the natural way in which approximations and semi-classical expressions can be inserted into it to treat larger systems.

## B3.4.6 Photo-dissociation

The time-dependent approach has the advantage that it is easy to visualize the propagation of a simple wavepacket and make intuitive sense of a large body of chemical phenomena. This is especially powerful in photo-initiated processes. As a result of a photon absorption, the ground-state wavefunction is 'jumped' to a higher potential energy surface of a different electronic state and propagates on this new surface. The initial excitation is, by the Frank–Condon principle, essentially vertical (i.e. the nuclear position and momentum do not change, only the electronic state). The subsequent process (see figure B3.4.7) is the response of the nuclear coordinates to the change in the electronic state.

For two Born–Oppenheimer surfaces (the ground state and a single electronic excited state), the total photo-dissociation cross section for the system to absorb a photon of energy $\omega$, given that it is initially at a state $|\chi\rangle$ with energy $E_0$ can be shown, by simple application of second-order perturbation theory, to be [89]

$$\sigma(\omega) = \text{constant} \cdot \int e^{i(\omega + E_0)t} c(t) \, dt \qquad \text{(B3.4.18)}$$

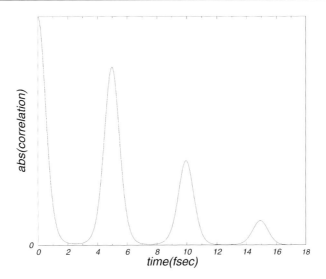

**Figure B3.4.8.** The correlation function $c(t) = \langle\psi_0|\psi(t)\rangle$ as a function of time for photodissociation in a collinear (or three-dimensional) polyatomic case. There are three relevant time scales; $T_1$, which measures how rapidly the initial wavefunction dephases; $T_2$, which measures how long it takes this initial wavefunction to regroup; and $T_3$ which measures how long the wavefunction takes to 'leak' to other degrees of freedom. In practice, photodissociation experiments may yield spectra which are more blurred, if $T_1$, $T_2$ and/or $T_3$ are not well separated.

where the correlation function is defined as

$$c(t) = \langle\Phi|e^{-i\hat{H}_{\text{exc}}t}|\Phi\rangle \tag{B3.4.19}$$

$\Phi = \mu\chi$, $\mu$ is the dipole moment and $\chi$ is the initial vibrational state on the ground surface (with energy $E_0$). $\hat{H}_{\text{exc}}$ is the excited-state potential energy. This expression has a clear physical meaning. Take an initial wavepacket, $\chi$, multiply it by the dipole moment, and use the resulting packet ($\Phi \equiv \mu\chi$) as an initial function so that it is propagated under the excited-state potential ($\Phi(t) \equiv e^{-i\hat{H}_{\text{exc}}t}\Phi$).

Equation (B3.4.18) makes a very powerful statement: absorption is only related to the Fourier transform of the correlation function. All that is needed is to know how the wavepacket propagates in time on the upper surface. Keeping in mind the time–energy uncertainty principle, one can then qualitatively understand the spectral features in photo-absorption [89]. Referring to figure B3.4.8, the correlation function starts at 1 at $t = 0$, then undergoes a period of decay over a time scale denoted by $T_1$. This initial decay is more rapid if the excited potential surface is steep, slower if it is shallow. This is the shortest time scale in the correlation function, so it corresponds to the broadest feature in the energy spectrum, i.e. the envelope in figure B3.4.9. In figure B3.4.8, the correlation peaks every time the wavepacket returns to the initial placement (denoted by $T_2$). In the energy picture, the peak spacing is $2\pi/T_2$. The correlation peaks are decreasing in magnitude over time $T_3$ as the wavepacket either decays to other modes or moves to another region of the potential energy surface. $1/T_3$ is therefore the peak's width in the energy spectrum.

This time-dependent method allows one to nicely connect the theoretical and experimental observations. As mentioned earlier, the correlation function and its generalizations yield the spectra for a large number of other photospectroscopy processes, such as Raman processes [90], as well as molecular scattering [73, 74].

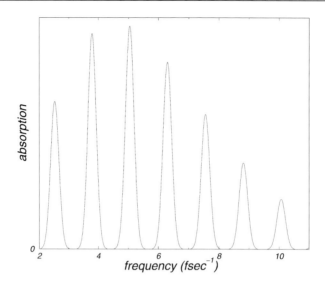

**Figure B3.4.9.** The Fourier transform of the correlation function, from figure B3.4.8, which gives the absorption spectrum as a function of frequency.

### B3.4.7   Bound states and resonances–extraction

*B3.4.7.1   Resonances—formalism*

The quantum dynamics of bound and scattered systems is closely correlated through the concept of resonances which are, heuristically, quasi-bound states in which the system can spend time [6–9, 91–94]. More formally, resonances are poles of the $S$-matrix. (See section A3.11.) In a scattering process, the cross section typically exhibits peaks as a function of the scattering energy, exactly at (or near) the energy of the resonances. For example, in a one-dimensional scattering off a double well, the scattering probabilities exhibit sharp peaks when the collision energy matches the energy of the quasi-bound states in the well (figure B3.4.10). (See figure B3.4.11 for a realistic example.)

The classical counterpart of resonances is periodic orbits [91, 95–98]. For example, a purely classical study of the $H + H_2$ collinear potential surface reveals that near the transition state for the $H + H_2 \rightarrow H_2 + H$ reaction there are several trajectories (in $R$ and $r$) that are periodic. These trajectories are not stable but they nevertheless affect strongly the quantum dynamics. A study of the resonances in $H + H_2$ scattering as well as many other triatomic systems (see, e.g., [99]) reveals that the scattering peaks are closely related to the frequencies of the periodic orbits and resonance wavefunctions are large in the regions of space where the periodic orbits reside.

Theoretically, resonances are essentially solutions of the Schrödinger equation at complex energies. These specific solutions have the property that they are mainly concentrated in the strong-interaction region and at the asymptote are outgoing waves. For one-dimensional predissociation (figure B3.4.12) where the coordinate is labelled $R$, the resonance wavefunction is asymptotically (i.e. for large positive $R$) the following outgoing wave:

$$\psi_{\text{Res}}(R) \approx a\, e^{ikR} \tag{B3.4.20}$$

where $a$ is a constant here and $k$ is the energy-dependent wavevector, $k^2/2m = E$. The existence of such a resonance function seems to be baffling, since graduate level quantum mechanical texts [100] prove that the only bound solutions to the Schrödinger equation are those with real energies and zero flux. Heuristically,

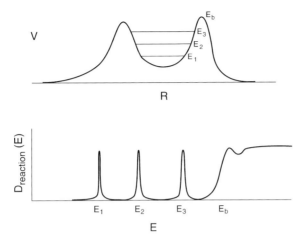

**Figure B3.4.10.** Schematic figure of a 1D double-well potential surface. The reaction probabilities exhibit peaks whenever the collision energy matches the energy of the resonances, which are here the quasi-bound states in the well (with their energy indicated). Note that the peaks become wider for the higher energy resonances—the high-energy resonance here is less bound and 'leaks' more toward the asymptote than do the low-energy ones.

the solution to this difficulty is that the formal resonance energies are complex. Thus, $k$ is complex and $e^{ikR}$ blows up when $R$ is very large. Therefore, resonance functions are not bound functions and the regular proofs do not apply to them.

The key to practical calculations of resonances is to limit the extent of of the grids used for describing the wavefunctions. In the original approach, called 'stabilization' [92–94], a finite grid or basis set is used and the Hamiltonian is diagonalized on that grid (or for that basis set). Those few eigenvalues which change very little when the grid size is modified are associated with the wavefunction of the resonances. (The resonances are concentrated in the interaction region, so that they are not sensitive to the details of the grid end points.) The main difficulty with this approach is that it necessitates many diagonalizations of the Hamiltonian matrix, one for each grid size.

An alternate and formally very powerful approach to resonance extraction is complex scaling [7, 101–107] whereby a new Hamiltonian is solved. In this Hamiltonian, the grid's multi-dimensional coordinate (e.g., $x$) is multiplied by a complex constant $\alpha$. The kinetic energy gains a constant complex factor $(\partial^2/\partial x^2 \to (1/\alpha^2)(\partial^2/\partial x^2))$, while the potential needs to be evaluated at points with a complex argument $V(\alpha x)$. In a typical calculation, one diagonalizes the resulting complex Hamiltonian for several complex values of $\alpha$, and the complex resonances are those which do not change appreciably with respect to $\alpha$ (analogous to the stabilization approach). Complex scaling has been applied mostly to analytical potentials [7, 101–104]; however, it could also be used for numerically-derived potentials [105–107].

Finally, the simplest approach to extract resonances is to add to the Hamiltonian an absorbing potential [8, 48, 108, 109], and then look for the complex eigenvalues of the Hamiltonian $\hat{H} - iV_I$. The absorbing potential ensures that the resonance wavefunction has the correct form (is outgoing) in the asymptotic region immediately preceding $V_I$ (see figure B3.4.4). Again, resonance functions are found by varying the parameters (the length and magnitude of $V_I$).

### B3.4.7.2   *Numerically extracting bound states and resonance functions*

As explained above, the practical extraction of resonance eigenfunctions and eigenvalues (in complex scaling or with absorbing potentials) amounts to extraction of eigenvalues of a complex Hamiltonian and is thus

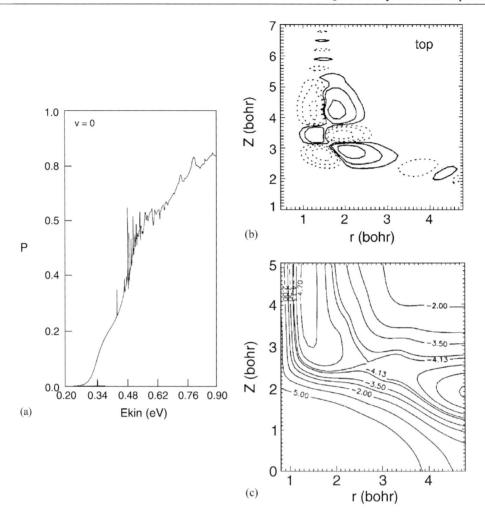

**Figure B3.4.11.** (a) Reaction probability for a 4D study of the dissociation of incident $H_2$ on CO. The probability exhibits sharp peaks whenever the energy matches that of a resonance wavefunction. (b) Plot of the resonance wavefunction associated with one of the peaks, as well as (c) a 2D cut of the potential surface. Note that the resonance wavefunction decays near the end of the grid, due to the use of an absorbing potential, which localizes its effects to the strong-interaction region. Taken from [196], with permission.

completely equivalent to extraction of bound states. One method for extracting the complex eigenstates of the Hamiltonian is simply to expand it in terms of a fixed basis set, and diagonalize directly the resulting matrix. This approach works for small- and intermediate-scale problems and/or low-energy eigenstates.

An alternative is to use iterative methods. The simplest iterative technique for calculating bound state or resonances is to pick a random initial wavefunction $\psi_0(x)$ and propagate it forward in time, producing a wavepacket:

$$\psi(x, t) = e^{-i\hat{H}t}\psi_0(x).\qquad\text{(B3.4.21)}$$

$\hat{H}$ refers here to the complex Hamiltonian, i.e. after complex scaling or inclusion of an absorbing potential;

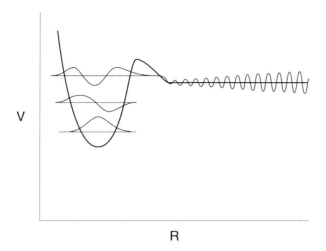

**Figure B3.4.12.** A schematic 1D vibrational pre-dissociation potential curve (wide full line) with a super-imposed plot of the two bound functions and the resonance function. Note that the resonance wavefunction is associated with a complex wavevector and is slowly increasing at very large values of $R$. In practice this increase is avoided by using absorbing potentials, complex scaling, or stabilization.

$x$ is the grid (or basis set) used to represent $\psi$. Fourier transform $\psi$ with respect to $E$

$$\psi(x, E) = \int_0^T e^{iEt} \psi(x, t)\, dt \qquad (B3.4.22)$$

where $T$ is a large time. It is clear that the squared norm of $\psi(x, E)$ (i.e. $\int |\psi(x, E)|^2\, dx$) has a peak whenever $E$ is near a resonance or bound-state energy, $\epsilon_n$, since $\psi(x, t)$ has contributions varying as $e^{-i\epsilon_n t}$ from each eigenenergy $\epsilon_n$.

This 'direct filter' technique is very powerful [56, 59] in extracting highly excited states, since only the propagation of a wavepacket is required. However, it is inefficient when there are closely-lying eigenvalues ($T$ needs to be larger than the inverse level spacing, $|\epsilon_{n+1} - \epsilon_n|^{-1}$, as a manifestation of Heisenberg's uncertainty relation) or $\epsilon_n$ is a wide resonance (i.e. its imaginary eigenvalue is large in absolute magnitude compared with the level spacing, so that its contribution to the wavepacket, $e^{-i\epsilon_n t}$, is washed out rapidly as a function of time).

To avoid these difficulties an alternate approach, labelled filter diagonalization, was developed [110–117]. The approach is powerful for extracting highly excited energies. Mechanistically, filter diagonalization is simple (see figure B3.4.13). The initial wavepacket is propagated for a short time $T$, and the filtered functions $\psi(x, E)$ are prepared as in equation (B3.4.22) but using a short time $T$. The filtering is carried out for several (typically ≈50) closely spaced energies which could be way above the ground-state energy. The key is to note that even after a short propagation time, $T$, the $\psi(x, E)$ functions would only contain contributions from eigenstates within a narrow strip of sampled energies. Thus, the filtered functions $\psi(x, E)$ are an excellent basis for eigenstates and eigenvalues within the filtered energy range. The true eigenvalues within this range are then found by diagonalizing the small (e.g., $50 \times 50$) matrix of the Hamiltonian operator within the $\psi(x, E)$ basis. The advantage of filter diagonalization is that it avoids the long propagation times of the pure filter approach as well as the large matrix diagonalizations associated with pure diagonalization.

As a side note, filter diagonalization is also useful in a more general context. It can be shown that it is an efficient approach for extracting frequencies from a short-time segment of a general signal [112–114, 118, 119], so that it is not even necessary to use a wavepacket! All one needs is a signal. This feature is very important

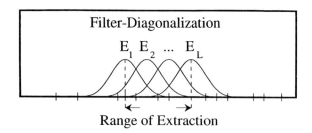

**Figure B3.4.13.** The basic premise in filter diagonalization is to filter a wavefunction for a short time at several energies, $E_1$, $E_2$, ..., so that, in energy space, the resulting set of several filtered functions (denoted by full bell-shapes) spans the eigenstates (short bars) in the energy range of interest. The short-time filtered wavefunctions can therefore be used to extract the eigenstates at the desired energy range, with a modest cost, since only short-time filter and small-matrix diagonalizations are used.

in semi-classical and path integral simulations discussed below, where all the information is extracted from a time-dependent correlation function, because the quality of the simulations degrades as a function of time (the number of trajectories is typically increased exponentially as the time is increased); therefore, information must be extracted from the shortest time possible.

## B3.4.8  Beyond grids

The formalism outlined in the previous sections is very useful for small systems, but is, as explained, impractical for more than six to ten strongly interacting degrees of freedom. Thus, alternate approaches are required to represent dynamics for large systems. Currently, there are many new approaches developed and tested for this purpose, and these approaches are broadly classified as follows:

- frozen and zero-point approximations;
- mean-field methods and their extensions;
- Gaussian-wavepacket-based techniques;
- path-integral and semi-classical approaches.

These approaches are generally interwoven, and some of the most exciting developments in chemical dynamics have been associated with their combinations. This section very briefly describes the motivation behind and the application of these techniques.

### B3.4.8.1  *Frozen and zero-point approximations*

Often a degree of freedom moves very slowly; for example, a heavy-atom coordinate. In that case, a plausible approach is to use a 'sudden' approximation, i.e. fix that coordinate and do reduced dimensionality quantum-dynamics simulations on the remaining coordinates. A common application of this technique, in a three-dimensional case, is to fix the angle of approach to the target [120, 121] (see figure B3.4.14).

The sudden approach has been applied widely, starting with atom–diatom calculations and continuing today for diatom–diatom calculations [79, 122, 123]. This approach and a related approximation (coupled states or CS, which involves the neglect of Coriolis coupling terms in three dimensions [120, 121, 124, 125]) are much more powerful when combined with the arrangement decoupling with absorbing potentials approach discussed earlier. The reason is that approximations are typically much easier to formulate and apply, and are more valid, for single-arrangement approaches. For an example of the successful merging of approximations and arrangement decoupling, see [126].

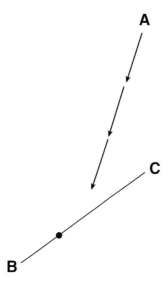

**Figure B3.4.14.** The infinite-order-sudden approximation for A+BC → AB+C. In this approximation, the BC molecule does not rotate until reaction occurs.

A related and particularly simple approximation is $J$-shifting [127, 128]. This method is a simple (and generally useful) trick for calculating reaction probabilities in three dimensions from a single calculation with zero total angular momentum ($J = 0$), by approximating the effect of the non-zero angular momentum as shifting a transition state:

$$P_J(E) = (2J + 1)P_{J=0}(E - C_J) \qquad (B3.4.23)$$

where $C_J$ is determined from the contribution of angular degrees of freedom to the transition state energy.

A related type of approximation is the reaction path method [122, 123, 129–132], in which the coordinates are divided into those which are relatively rigid during the reaction and are typically associated with harmonic oscillators, and the remaining coordinates (reaction coordinates) which undergo significant change during the reaction. The contribution of those degrees of freedom which are replaced by harmonic oscillators can be taken, for low-energy reactions, simply by the zero point energy of the harmonic oscillators. More sophisticated treatments, appropriate for higher temperatures, are actively developed now in the area of liquid reaction dynamics where they are used to describe effects of solvents (see section C3.5 for details and further references). In addition, proper inclusion of rotational states of the fragments was recently shown to yield accurate results in a molecular multi-dimensional reaction-path-like approach [133].

### B3.4.8.2 Mean-field methods and their extensions

The mean field technique is one of the most robust and simple methods used to handle larger molecules in gas and liquid environments [50, 134–136]. The basic premise of all mean-field methods is that the full wavefunction represents $N$ very weakly coupled modes ($Q_i$) and can be approximated as

$$\psi(\{Q_i\}; t) = \prod_{i=1}^{N} \chi(Q_i, t). \qquad (B3.4.24)$$

The result of this approximation is that each mode is subject to an effective average potential created by all the expectation values of the other modes. Usually the modes are propagated self-consistently. The effective potentials governing the evolution of the mean-field modes will change in time as the system evolves. The advantage of this method is that a multi-dimensional problem is reduced to several one-dimensional problems.

The fundamental disadvantage of the mean-field method is that it does not allow modes to respond in a correlated manner to each other. This problem can be somewhat alleviated by a good definition of the relevant coordinate system [134, 136]. (An extension of mean-field methods that does allow for coupling [137–139] will be discussed later.)

If one is interested in low-lying eigenvalues or low-energy scattering a CI-like approach can be applied, in which one uses zero-order eigenfunctions of a simple Hamiltonian to expand the wavefunction. Spectroscopic calculations including up to 20 to 30 degrees of freedom have been carried out using such an approach [140–143].

### B3.4.8.3  *Gaussian-wavepacket based techniques*

Gaussian wavepackets are very special functions which, in a sense, bridge the gap between classical and quantum descriptions [89, 144]. They are defined as

$$\psi(x, t) = \text{constant} \times \exp(-(x - \bar{x}(t))^2/2\sigma^2) \exp(i\bar{p}(t) \cdot x) \qquad (B3.4.25)$$

where $\bar{x}(t)$, $\sigma$ and $\bar{p}(t)$ are the average position, width and momentum of the packet, respectively. A Gaussian wavepacket represents the ground state of a harmonic oscillator, shifted in position and momentum. (Gaussian wavepackets are also known as Glauber or 'coherent' states—although the latter definition is sometimes applied to more general functions.)

The primary property of Gaussian wavepackets is that on a harmonic potential surface they are solutions of the time-dependent Schrödinger equation. Specifically, if one places a Gaussian wavepacket at $t = 0$ on a harmonic oscillator with an arbitrary average position and momentum, $\bar{x}(t = 0)$ and $\bar{p}(t = 0)$, the resulting wavefunction will remain a Gaussian. The average position and momentum change in time exactly like a classical particle would.

From this basic fact, several related approaches have emerged. First, a technique in which one propagates classical trajectories forward in time, and uses a single or multiple sets of Gaussian wavepackets (one for each classical trajectory) as an ansatz for the full wavefunction is introduced. For each Gaussian the position and momentum are specified by the classical trajectory [144]. This technique is already able to account for much of the zero-point energy effects. In addition, interference and tunnelling effects can be partially accounted for by adding several multi-dimensional Gaussians [144–147]. The method has been shown recently to be very powerful for non-adiabatic coupling problems (see section B3.4.9).

Gaussian wavepackets have also found use in a new approach that improves on mean-field techniques [137–139, 148]. In this method, the degrees of freedom are divided into those which change strongly during the reaction (the 'system' coordinates), which are treated by an explicit wavepacket; the remaining 'bath' coordinates are treated by Gaussians. The parameters for the bath Gaussians are dependent on the system state, and this introduces explicitly a correlation between the system and bath modes (the bath responds differently to different parts of the system). Use of this technique in multi-dimensional reactions involving tunnelling has shown it to be significantly superior to mean-field techniques, while requiring modest numerical effort even for multi-dimensional systems.

### B3.4.8.4  *Path integral and semi-classical dynamics*

In the 1940s, Feynmann realized that quantum mechanics can be recast in a simple form which is very reminiscent of classical mechanics [149, 150]. This approach, path integrals, has been heavily researched in chemical dynamics since, if properly convergent, it could allow calculations on very large systems. Even

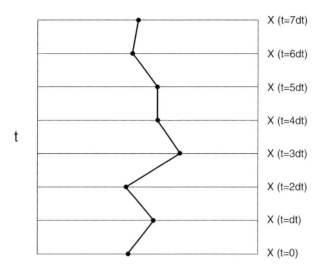

**Figure B3.4.15.** A possible Feynmann path trajectory for a 1D variable as a function of time. This trajectory carries an oscillating $e^{iS/\hbar}$ component with it, where $S$ is the action of the trajectory. The trajectory is highly fluctuating; its values at each time step ($x(dt)$, $x(2dt)$, ..., etc) are not correlated.

earlier, van Vleck [151] postulated a semi-classical approach, i.e. a method which (like the Wentzel–Kramers–Brillouin (WKB) approximation) captures quantum interference effects and is accurate when the masses or energies are large so that $\hbar$ can be considered small. This section briefly describes Feynmann's approach, semi-classical dynamics, and recent developments and improvements.

For concreteness, assume that we consider several particles described together by a multi-dimensional coordinate $x$ and assume that the kinetic energy is the usual $(-1/2M)(\partial^2/\partial x^2)$, where $M$ is a single relevant mass. (Hamiltonians in quantum dynamics can usually be brought to this form upon rescaling of the coordinates by constants.) The basic ingredient in Feynmann's approach is the correlation function for a general function $\psi$:

$$\langle \psi | \psi(\tau) \rangle = \langle \psi | e^{-i\hat{H}\tau} | \psi \rangle = \int \psi^*(x'') G(x'', x', \tau) \psi(x', \tau) \, dx' \, dx'' \qquad \text{(B3.4.26)}$$

where the time-dependent Green's function $G$ is very simple:

$$G(x'', x', \tau) = \text{constant} \cdot \sum_{\text{path}} e^{iS(\text{path})/\hbar}. \qquad \text{(B3.4.27)}$$

In this section, we insert $\hbar$ explicitly. The sum has the following meaning: for each pair of points $x'$ and $x''$ draw a path $x(t)$ that starts at $x'$ ($x(t=0)=x'$) and ends at $x''$ ($x(t=\tau)=x''$). This path need not be a classical path (see figure B3.4.15). Each such path contributes $e^{iS/\hbar}$, where $S$ is the action of the path. $S$ (unrelated to the scattering matrix) is calculated very simply as

$$S = \int \left[ \frac{M}{2} \left( \frac{dx}{dt} \right)^2 - V(x(t)) \right] dt \qquad \text{(B3.4.28)}$$

where $V$ is the potential. This, in principle, gives a very simple prescription. All one has to do in order to calculate quantum mechanical properties is to sum over 'many' quantum trajectories.

Unfortunately, this simple approach is not plausible numerically. The integral, as presented, will not converge, even for short times. The problem is that even trajectories which are 'wild', i.e. highly fluctuating, contribute, per path, the same as trajectories that are 'gentle'. The key to introducing convergence is to realize that highly fluctuating trajectories or trajectories that lie in regions of high potential energy give contributions that are cancelled by those of other nearby trajectories. Thus, a *bundle* of highly fluctuating trajectories gives only a very weak contribution. The only trajectories that give very strong contribution are *classical* trajectories (for example, if there is no potential, the classical trajectory would run with constant velocity from $x'$ to $x''$). This realization is the key to the development of semi-classical approaches, i.e. approaches in which the exact calculation of the time-dependent Green's function is replaced by the sum of contributions of classical trajectories. The basic semi-classical expression is [151–158]

$$\langle \psi | \psi(t) \rangle = \text{constant} \int \sum_{\text{path}} \frac{e^{iS_{\text{path}}/\hbar}}{\sqrt{|\partial p''/\partial x'|}} \psi^*(x'') \psi(x') \, dx'' \, dx'. \tag{B3.4.29}$$

Note the meaning of this expression: for each choice of the initial and final position $x'$ and $x''$, calculate the classical path that takes you from $x'$ to $x''$ in time $t$. Specifically, calculate the momentum along the path and the final momentum, $p''$, and find out how $p''$ varies with the initial position. This would give, for a multi-dimensional problem, a matrix $\partial p_i''/\partial x_j'$ whose absolute determinant needs to be inverted.

There is a simple physical explanation to the inverse determinant in equation (B3.4.29). Each classical trajectory has a 'volume' of quantum trajectories nearby which have similar phases to it. Beyond that volume, the phases are becoming random. Thus, the larger that volume, the greater contribution would that bundle give to the final semi-classical 'propagator'. If $p''$ varies slowly when $x'$ is changed (and $|\partial p''/\partial x'|$ is small), the action's phase varies slowly under a change in the end-point position, so the volume of the quantum trajectories that surround the classical trajectory is also large, and a large contribution is expected from the semi-classical propagator.

Expression (B3.4.29) is still not well suited for classical simulations due to several reasons. First, $|\partial p''/\partial x'|$ can vanish at specific times, which leads to infinities in the result. (In classical scattering this is related to the existence of 'scattering rainbows'.) This is easily circumvented by changing integration parameters, from $x''$ to $p'$ (i.e. from the final position to the initial momentum)

$$dx' \, dx'' = \left| \frac{\partial x''}{\partial p'} \right| dx' \, dp' \tag{B3.4.30}$$

leading to [156, 157]

$$\langle \psi | \psi(t) \rangle = \int \left| \frac{\partial p''}{\partial x'} \right|^{\frac{1}{2}} \psi^*(x'') \psi(x') e^{iS(x', p', t)/\hbar} \, dx' \, dp'. \tag{B3.4.31}$$

This transform also solves the boundary value problem, i.e. there is no need to find, for an initial position $x'$ and final position $x''$, the trajectory that connects the two points. Instead, one simply picks the initial momentum and position $p'$, $x'$ and calculates the classical trajectories resulting from them at all times. Such methods are generally referred to as initial variable representations (IVR).

Finally, one problem still remains. There are complex terms which need to be associated with the determinant. The complex terms (Maslov indices) have to do with the square root of the determinant, which may be negative, and also appear in the related WKB approximation. They can be calculated, albeit with difficulty [152–158].

Even expression (B3.4.31), although numerically preferable, is not the end of the story as it does not fully account for the fact that nearby classical trajectories (those with similar initial conditions) should be averaged over. One simple methodology for that averaging has been through the division of phase space into parts, each of which is 'covered' by a set of Gaussians [159, 160]. This is done by recasting the initial wavefunction as

$$\psi(x) = \int\int d\bar{x} \, d\bar{p} \langle g_{\bar{x}\bar{p}} | \psi \rangle g_{\bar{x}\bar{p}}(x) \tag{B3.4.32}$$

where $g_{\bar{x}\bar{p}}(x)$ is the Gaussian that is centred at point $\bar{x}$ with momentum $\bar{p}$

$$g_{\bar{x}\bar{p}}(x) = \text{constant}\, e^{-(x-\bar{x})^2/2\sigma^2}\, e^{i\bar{p}\cdot x/\hbar} \tag{B3.4.33}$$

and $\sigma$ is arbitrary. By applying these formulae two times, using the semi-classical approximation and eventually summing again over phases, the following expression results [161–163]:

$$\langle \psi | \psi(t) \rangle = \int dx'\, dp'\, e^{iS(x',p',t)/\hbar} F(x',p',t). \tag{B3.4.34}$$

where the exact form of $F$ is given clearly in [163]. The resulting expression is not difficult to numerically propagate.

Numerical applications of the formalism have been very successful lately [162–167]. This technique is very powerful for situations where short-time propagation is sufficient. For long-time processes, where the number of required trajectories is large, it is necessary to introduce other ingredients, such as methods for reducing the total propagation time—for example, the filter diagonalization method discussed above which was applied recently to the semi-classical approximation [113, 165], or backward-forward propagation schemes which tend to make the semiclassical integrand much smoother [168].

### B3.4.9 Non-adiabatic effects

*B3.4.9.1 Formalism*

The discussion in the previous sections assumed that the electron dynamics is adiabatic, i.e. the electronic wavefunction follows the nuclear dynamics and at every nuclear configuration only the lowest energy (or more generally, for excited states, a single) electronic wavefunction is relevant. This is the Born–Oppenheimer approximation which allows the separation of nuclear and electronic coordinates in the Schrödinger equation.

This assumption breaks down in many molecules, especially upon photo-excitation, since excited states are often close to each other or even cross one another (i.e. have the same electronic energy at a given nuclear position). Thus, the full Schrödinger wavefunction needs to be considered:

$$\Psi(x, r_e, t) = \sum_n \psi_n(x, t)\Phi_n(r_e; x) \tag{B3.4.35}$$

where $n$ is now the index of the electronic states and $r_e$ is the position of the electron. $\Phi_n$ are eigensolutions, for each position of $x$, of the electronic part of the Hamiltonian (every part of the electron–nuclear Hamiltonian except for the nuclear kinetic energy). The associated eigenvalues are labelled $u_n$, and are the adiabatic ground- and excited-state energies. From this expansion there follows an equation for the nuclear wavefunction [15–18, 169]. A complication is that the adiabatic electronic states, $\Phi_n(r_e; x)$, depend themselves on the nuclear coordinate $x$. Thus, the nuclear kinetic-energy terms in the Schrödinger equation, which have derivatives in them $(\partial^2/\partial x^2)$, also operate on $\Phi_n$. The resulting time-dependent Schrödinger equation is then straightforwardly shown to be

$$-\frac{1}{2M}\frac{\partial^2}{\partial x^2}\psi_n + \sum_m \frac{\tau_{nm}^1}{2M}\frac{\partial}{\partial x}\psi_m + \sum_m \frac{\tau_{nm}^2}{2M}\frac{\partial^2}{\partial x^2}\psi_m + u_n\psi_n = i\frac{\partial}{\partial t}\psi_n \tag{B3.4.36}$$

where the matrices $\tau^1$ and $\tau^2$ are

$$\tau_{nm}^1(x) = \int \Phi_n(r_e; x)\frac{\partial}{\partial x}\Phi_m(r_e; x)\, dr_e \tag{B3.4.37}$$

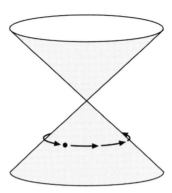

**Figure B3.4.16.** A generic example of crossing 2D potential surfaces. Note that, upon rotating around the conic intersection point, the phase of the wavefunction need not return to its original value.

$$\tau_{nm}^2(x) = \int \Phi_n(r_e; x) \frac{\partial^2}{\partial^2 x} \Phi_m(r_e; x) \, dr_e. \qquad (B3.4.38)$$

The effect of $\tau^2$ is usually negligible, due to the $1/M$ factor. However, $\tau^1$ is fundamentally important, yet mathematically difficult to treat. Specifically, it can be shown that $\tau_{nm}^1$ is proportional to the inverse of the energy difference between electronic states $n$ and $m$ so its value can become infinite. Several ways to simplify the equation have been developed, starting with the work of Baer [15–18, 26, 169, 170]. The mathematical theory is too intricate to discuss here in detail; it is sufficient to say that one does a 'gauge' transform, i.e. a transformation from adiabatic wavefunctions (the $\psi_n$) to a new set of 'diabatic functions' for which the derivative coupling (i.e. the $\tau^1$ coefficient of $\partial/\partial x$) is minimized or completely absent. Instead, a new part of the potential appears as a coupling of the electronic states (i.e. an off-diagonal potential). This coupling potential introduces a new 'phase' problem. The difficulty is that the diabatic functions (i.e. the functions which are obtained after the transformation) are defined in terms of a linear combination (sometimes complex) of the adiabatic functions. [169] These linear combinations are non-unique. Consider the generic example of two crossing potential surfaces in two (nuclear) dimensions (see figure B3.4.16). When the nuclei move around the potential contours, the linear combination changes. Upon return to the same starting point, the phase of the states need not be the same! Thus the nuclear diabatic functions are not uniquely defined. This is a very important effect (called the molecular phase or Berry phase [15–17, 19, 26, 171]), since it would appear even at low energies, way below the energies in which the conical interaction appears, i.e. way below the energies of the excited states! This phenomenon has been recently shown to be important in scattering of $H + H_2$ and its isotopic analogy [19]. (A simple interpretation of the molecular phase phenomenon in a $H + H_2$ type reaction system is that the full wavefunction is symmetric under exchange of any of the two hydrogen atoms and it is antisymmetric under electron–electron exchange, so that the nuclear part of it should be antisymmetric under a change of any pair of the three nuclei. This modifies the Schrödinger equation for the atomic motion.) In addition, it was shown in a model system that they can affect state-to-state transition probabilities in a reactive system [172].

The molecular phase effects are especially important when the system has some type of symmetry. Nevertheless, the typical treatment of non-adiabatic effects ignores the adiabatic phase, although, as cautioned, this is a problematic step.

In the remainder of this section, we will follow this simplifying (and problematic) assumption, and postulate that, upon the adiabatic to diabatic transformation, the Schrödinger equation has the form:

$$i\frac{\partial \psi_n}{\partial t} = -\frac{1}{2M}\frac{\partial^2}{\partial x^2}\psi_n + \sum_m V_{mn}\psi_m \qquad (B3.4.39)$$

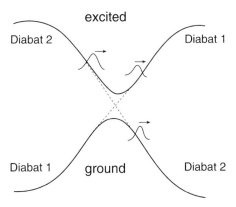

**Figure B3.4.17.** When a wavepacket comes to a crossing point, it will split into two parts (schematic Gaussians). One will remain on the same adiabat (different diabat) and the other will hop to the other adiabat (same diabat). The adiabatic curves are shown by full lines and denoted by 'ground' and 'excited'; the diabatic curves are shown by dashed lines and denoted 1, 2.

where $V_{nm}$ is called the diabatic potential matrix. (Note that if we were to use a finite set of molecular valence orbitals, $\Phi_n(r_e)$, that do not depend on the nuclear orbitals, and expand the Schrödinger wavefunction in terms of $\Phi_n$, we would automatically obtain this equation. Such a basis would be automatically diabatic.)

To see physically the problem of motion of wavepackets in a non-diagonal diabatic potential, we plot in figure B3.4.17 a set of two adiabatic potentials and their diabatic counterparts for a 1D problem, for example, vibrations in a diatom (as in metal–metal complexes). As figure B3.4.17 shows, if a wavepacket is started away from the 'crossing' point, it would slide towards this crossing point (where $V_{11} = V_{22}$) where it would branch; a part of it would continue on the same *adiabatic* state (i.e. shift to a different diabatic state) and the other part would 'jump' to a different adiabatic state.

The problem of branching of the wavepacket at crossing points is very old and has been treated separately by Landau and by Zener [15, 173, 174]. The model problem they considered has the following diabatic coupling matrix:

$$V = \begin{pmatrix} -F(R - R_0) & \Gamma \\ \Gamma & F(R - R_0) \end{pmatrix} \tag{B3.4.40}$$

where $2F$ is the difference in slope of the potentials (i.e. the difference in force felt in each state), $\Gamma$ is the coupling element and $R_0$ is the crossing point. Landau and Zener showed that, in such a case, the probability for the wavefunction to transfer from the higher adiabatic level to the lower one (i.e. to remain on the same diabat) is

$$\rho_{\text{non-adiabatic}} = \exp\left(\frac{-\pi |\Gamma|^2}{v|F|}\right) \tag{B3.4.41}$$

while the probability to remain on the same adiabatic level is

$$\rho_{\text{adiabatic}} = 1 - \rho_{\text{non-adiabatic}} \tag{B3.4.42}$$

where $v$ is the velocity of the wavepacket at the crossing point. Note two things about this formula: the steeper the difference is in the potentials, the higher the probability of a non-adiabatic transfer; in addition, if the mass is large (i.e. the velocity is low), then the motion is adiabatic ($\rho_{\text{adiabatic}} \approx 1$).

*B3.4.9.2    Numerical approaches for simulating non-adiabatic processes*

The simplest approach to simulating non-adiabatic dynamics is by surface hopping [175, 176]. In its simplest form, the approach is as follows. One carries out classical simulations of the nuclear motion on a specific adiabatic electronic state (ground or excited) and at any given instant checks whether the diabatic potential associated with that electronic state is intersecting the diabatic potential on another electronic state. If it is, then a decision is made as to whether a 'jump' to the other adiabatic electronic state should be performed, based on the values for $\rho_{\text{adiabatic}}$ and $\rho_{\text{non-adiabatic}}$ (when $\rho_{\text{non-adiabatic}}$ is close to 1, a jump to the other electronic state is made with a high probability so that the particle remains on the same *diabatic* potential). If a jump is made, the particle continues its motion along the new adiabatic potential surfaces, with the same instantaneous position and momentum.

This approach is very simple and powerful. It has been used in numerous studies (for references see [176, 177]) and generally captures the essentials of the adiabatic versus non-adiabatic branching. It is especially useful in circumstances where the nuclear motion is essentially classical (i.e. zero point motion and tunnelling can be ignored).

This basic hopping model has a major disadvantage, however, as it fails to say much about the phases of the wavefunction. Consider a case where the wavefunction visits a region where surface hopping occurs, so a part of it hops, and at some later time it re-visits this region and again a part of it undergoes hopping. These two parts would interfere together and the interference may be constructive or destructive, but the hopping model does not specify this information.

To remedy this difficulty, several approaches have been developed. In some methods, the phase of the wavefunction is specified after hopping [178]. In other approaches, one expands the nuclear wavefunction in terms of a limited number of basis-set functions and works out the quantum dynamical probability for jumping. For example, the quantum dynamical basis functions could be a set of Gaussian wavepackets which move forward in time [147]. This approach is very powerful for short and intermediate time processes, where the number of required Gaussians is not too large.

The ultimate approach to simulate non-adiabatic effects is through the use of a full Schrödinger wavefunction for both the nuclei and the electrons, using the adiabatic–diabatic transformation methods discussed above. The whole machinery of approaches to solving the Schrödinger wavefunction for adiabatic problems can be used, except that the size of the wavefunction is now essentially doubled (for problems involving two-electronic states, to account for both states). The first application of these methods for molecular dynamical problems was for the charge-transfer system

$$H_2(v=0) + H^+ \rightarrow \begin{array}{ll} H_2(v') + H^+ & \text{Energy transfer} \\ H_2^+(v') + H & \text{Charge transfer.} \end{array}$$

Here quantum-mechanical vibrational state-to-state differential cross sections were calculated for a translational energy of $E_{\text{tr}} = 20\,\text{eV}$ and compared with experiments, with very good agreement between experiment and theory. In another application of this approach, state-selected integral cross sections were calculated for the $(\text{Ar} + H_2)^+$ system. Reactive (exchange), charge-transfer and spin transitions processes were treated simultaneously in one single calculation, and they compared very well with experiments [179, 180].

Finally, semi-classical approaches to non-adiabatic dynamics have also been formulated and successfully applied [167, 181]. In an especially transparent version of these approaches [167], one employs a mathematical 'trick' which converts the non-adiabatic surfaces to a set of coupled oscillators; the number of oscillators is the same as the number of electronic states. This method is also quite accurate, except that the number of required trajectories grows with time, as in any semi-classical approach.

## B3.4.10    Controlling molecular motion

The preceding sections were concerned with the description of molecular motion. An ambitious goal is to proceed further and influence molecular motion. This lofty goal has been at the centrepiece of quantum dynamics in the past decade and is still under intense investigation [182–194]. Here we will only describe some general concepts and schemes.

The basic Hamiltonian describing the motion of atoms and molecules under a strong laser is simple in the dipole approximation,

$$\hat{H} = \hat{H}_0 - \mu \cdot E(t) \tag{B3.4.43}$$

where $E = E(t)$ is the time-dependent electric field at the molecule, while $\hat{H}_0$ is the Hamiltonian of the static system; $\mu$ is the transition dipole operator, which typically connects different electronic states.

There are several different possible goals in controlling molecular dynamics. One goal can be the localization of excitations to a specific bond in a molecule, and the molecule could be broken along that bond [188]. Alternatively, one can try to transfer the molecule completely to an excited electronic state [186]. Another is the control of alignment (so that a molecule would point in a certain direction) [189, 190]. Still another goal would be the control of branching ratios; for example, in a reaction of an atom with a diatom, A + BC, one may want to control the branching into products [182–184]:

$$A + BC \rightarrow \begin{array}{l} AB + C \\ AC + B \\ BC + A \\ A + B + C. \end{array}$$

Finally, one may want to control the emission of light from molecules.

The conceptually simplest approach towards controlling systems by laser field is by 'teaching' the field [188, 191–193]. Typically, the field is experimentally prepared as, for example, a sum of Gaussian pulses with variable height and positions. Each experiment gives an outcome which can be quantified. Consider, for example, an A + BC reaction where the possible products are AB + C and AC + B; if the AB + C product is preferred one would seek to optimize the branching ratio

$$P_{\text{branch}} \equiv \frac{P(A + BC \rightarrow AB + C)}{P(A + BC \rightarrow AB + C, AC + B)}. \tag{B3.4.44}$$

In a purely experimental (non-theory) approach [188, 191–193] the branching ratio can be controlled by repeating the experiment many times, each with a randomly chosen set of pulse magnitudes and start times. One can repeat the experiment, varying the electrical field somewhat each time until the best outcome is achieved. This approach maybe the most appropriate one for large systems where little is known about the underlying dynamics and it has recently been demonstrated to work very well on dissecting large molecules [188].

Closely related to these 'experimental' approaches are optimal control procedures, in which one simulates theoretically the effects of the electric field on the system, and then modifies the electric field to give the best objective, i.e. a desired output (in this case: a high branching ratio) [182, 183]. The optimal control algorithm can be recast in a very powerful mathematical form which makes the calculation converge rapidly to give an excellent field for any objective, if it is possible to simulate the system motion theoretically.

A different set of approaches uses simple physical properties to control the system [184]. To demonstrate this type of problem, consider an even simpler branching problem where, upon excitation, two possible degenerate products are simultaneously produced. An example would be to photo-dissociate a diatom AB and produce different states of the system: one state labelled A + B*, in which B is electronically excited and A is receding away slowly; and another state, labelled A + B, in which B is in the ground state and A

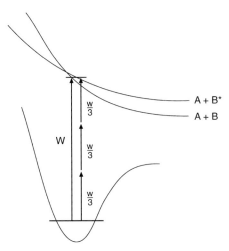

**Figure B3.4.18.** A schematic use of coherent control in $AB \rightarrow A+B$, $A+B^*$ dissociation: use of a single high-frequency photon ($\omega$) or three low-intensity ($\omega/3$) photons would lead to emerging wavefunctions in both arrangements. However, by properly combining the amplitudes and phases of the single- and three-photon paths, the wavefunction would emerge in a single channel.

is receding rapidly (so that the total energy is in both cases equal). The simplest method of controlling the $A+B^*$ *versus* $A+B$ production rate would be to mix two different pathways for obtaining $A+B$ and $A+B^*$; for example, mixing a field of frequency $\omega$ with a phase-lagged third harmonic of a field which is three times lower in frequency (see figure B3.4.18):

$$E_1 \cos \omega t + E_3 \cos \left( \frac{\omega t}{3} + \phi \right). \tag{B3.4.45}$$

Other possible choices are to use two pairs of frequencies which together have the same energies. The key point is that quantum interference between the two pathways can be used to control the branching ratio. This coherent-control approach is very general and can be used in virtually any branch of molecular dynamics, including scattering and photo-dissociation.

## References

[1]   Levine R D and Bernstein R B 1974 *Molecular Reaction Dynamics* (New York: Oxford University Press)
[2]   Zhang J Z H 1999 *Theory and Application of Quantum Molecular Dynamics* (River Edge, NJ: World Scientific)
[3]   Baer M (ed) 1985 *Theory of Chemical Reaction Dynamics* vols I–IV (Boca Raton, FL: CRC Press)
[4]   Wyatt R E and Zhang J Z H (eds) 1996 *Dynamics of Molecules and Chemical Reactions* (New York: Dekker)
[5]   Connor J N L (ed) 1999 Chemical reaction theory *Phys. Chem. Chem. Phys.* **1**(6) (special issue)
[6]   Truhlar D G (ed) 1984 *Resonances* (Washington, DC: ACS)
[7]   Moiseyev N 1998 Quantum theory of resonances: calculating energies, widths and cross-sections by complex scaling *Phys. Rep.* **302** 212
[8]   Bowman J M 1998 Resonances: bridge between spectroscopy and dynamics *J. Phys. Chem.* A **102** 3006
[9]   Levine R D and Wu S F 1971 Resonances in reactive collisions: computational study of the H + H$_2$ collision *Chem. Phys. Lett.* **11** 557
[10]  Truong T N 1997 Thermal rates of hydrogen exchange of methane with zeolite: a direct *ab initio* dynamics study on the importance of quantum tunneling effects *J. Phys. Chem.* B **101** 2750
[11]  Asscher M, Haase G and Kosloff R 1990 Tunneling mechanism for the dissociative chemisorption of N$_2$ on metal surfaces *Vacuum* **41** 269

[12] Antoniou D and Schwartz S D 1998 Activated chemistry in the presence of a strongly symmetrically coupled vibration *J. Chem. Phys.* **108** 3620

[13] Gamarnik A, Johnson B A and Garcia-Garibay M A 1998 Effect of solvents on the photoenolization of omicron-methylanthrone at low temperatures. Evidence for H-atom tunneling from nonequilibrating triplets *J. Phys. Chem.* A **102** 5491

[14] Jortner J and Pullman B (eds) 1986 *Tunneling* (The Netherlands: Reidel)

[15] Baer M 1985 The theory of electronic non-adiabatic transitions in chemical reactions *Theory of Chemical Reaction Dynamics* vol II, ed M Baer (Boca Raton, FL: CRC Press) p 281

[16] Baer M 1975 Adiabatic and diabatic representations for atom–molecule collisions: treatment of the collinear arrangement *Chem. Phys. Lett.* **35** 112

[17] Mead C A and Truhlar D G 1982 Conditions for the definition of a strictly diabatic electronic basis for molecular systems *J. Chem. Phys.* **77** 6090

[18] Sadygov R G and Yarkony D R 1998 On the adiabatic to diabatic states transformation in the presence of a conical intersection: a most diabatic basis from the solution to a Poisson's equation. I *J. Chem. Phys.* **109** 20

[19] Kuppermann A and Wu Y S M 1993 The geometric phase effect shows up in chemical reactions *Chem. Phys. Lett.* **205** 577

[20] Jackson B 1994 Quantum and semiclassical calculations of gas surface energy transfer and sticking *Comput. Phys. Commun.* **80** 119

[21] Baer R and Kosloff R 1997 Quantum dissipative dynamics of adsorbates near metal surfaces: a surrogate Hamiltonian theory applied to hydrogen on nickel *J. Chem. Phys.* **106** 8862

[22] Chu S 1998 The manipulation of neutral particles *Rev. Mod. Phys.* **70** 685

[23] Cohen-Tannoudji C N 1998 Manipulating atoms with photons *Rev. Mod. Phys.* **70** 707

[24] Phillips W D 1998 Laser cooling and trapping of neutral atoms *Rev. Mod. Phys.* **70** 721

[25] Hagley E W, Deng L, Kozuma M, Wen J, Helmerson K, Rolston S L and Phillips W D 1999 A well-collimated quasi-continuous atom laser *Science* **283** 1706

[26] Yarkony D R (ed) 1995 *Modern Electronic Structure Theory* (River Edge, NJ: World Scientific)

[27] Baer M 1985 The general theory of reactive scattering: the differential equations approach *Theory of Chemical Reaction Dynamics* vol I, ed M Baer (Boca Raton, FL: CRC Press) p 91

[28] Neuhauser D and Baer M 1989 The time dependent Schrödinger equation: application of absorbing boundary conditions *J. Chem. Phys.* **90** 4351

[29] Neuhauser D and Baer M 1989 The application of wavepackets to reactive atom–diatom systems: a new approach *J. Chem. Phys.* **91** 4651

[30] Neuhauser D and Baer M 1990 A new accurate (time independent) method for treating three-dimensional reactive collisions: the application of optical potentials and projection operators *J. Chem. Phys.* **92** 3419

[31] Neuhauser D, Baer M and Kouri D J 1990 The application of optical potentials for reactive scattering: a case study *J. Chem. Phys.* **93** 2499

[32] Miller W H 1969 Coupled equations and the minimum principle for collisions of an atom and a diatomic molecule, including rearrangements *J. Chem. Phys.* **50** 407

[33] Zhang J Z H and Miller W H 1989 Quantum reactive scattering via the S-matrix version of the Kohn variational principle—differential and integral cross sections for D + $H_2$ → HD + H *J. Chem. Phys.* **91** 1528

[34] Manolopoulos D E, Dmello M and Wyatt R E 1989 Quantum reactive scattering via the log derivative version of the Kohn variational principle—general theory for bimolecular chemical reactions *J. Chem. Phys.* **91** 6096

[35] Truhlar D G, Schwenke D W and Kouri D J 1990 Quantum dynamics of chemical reactions by converged algebraic variational calculations *J. Phys. Chem.* **94** 7346

[36] Stechel E B, Walker R B and Light J C 1978 R-matrix solution of coupled equations for inelastic scattering *J. Chem. Phys.* **69** 3518

[37] Marcus R A 1966 On the analytical mechanics of chemical reactions. Quantum mechanics of linear collisions *J. Chem. Phys.* **45** 4500

[38] Hofacker G L 1963 Quanten chemisher reaktionen *Natarforshung* **189** 607

[39] Walker R B, Stechel E B and Light J C 1978 Accurate $H_3$ dynamics on an accurate $H_3$ potential surface *J. Chem. Phys.* **69** 2922

[40] Schatz G C and Kuppermann A 1975 Quantum mechanical reactive scattering: an accurate three-dimensional calculation *J. Chem. Phys.* **62** 2502

[41] Aquilanti V and Cavalli S 1997 The quantum-mechanical Hamiltonian for tetraatomic systems in symmetric hyperspherical coordinates *J. Chem. Soc. Faraday Trans.* **93** 801

[42] Bacic Z, Kress J D, Parker G A and Pack R T 1990 Quantum reactive scattering in 3 dimensions using hyperspherical (APH) coordinates .4. discrete variable representation (DVR) basis functions and the analysis of accurate results for F + $H_2$ *J. Chem. Phys.* **92** 2344

[43] Pogrebnya S K, Echave J and Clary D C 1997 Quantum theory of four-atom reactions using arrangement channel hyperspherical coordinates: Formulation and application to OH + $H_2$ ⇔ $H_2$O + H *J. Chem. Phys.* **107** 8975

[44] Launay J M and Ledourneuf M 1990 Quantum-mechanical calculation of integral cross sections for the reaction F + $H_2(v = 0, j = 0)$ → FH$(v', j')$ + H by the hyperspherical method *Chem. Phys. Lett.* **169** 473

[45] Kuppermann A 1996 Reactive scattering with row-orthonormal hyperspherical coordinates. I. Transformation properties and Hamiltonian for triatomic systems *J. Phys. Chem.* **100** 2621

[46] Neuhauser D 1990 State-to-state reactive probabilities from single-arrangement propagation with absorbing potentials *J. Chem. Phys.* **93** 7836

[47] Kosloff R and Kosloff D 1986 Absorbing boundaries for wave propagation problems *J. Comput. Phys.* **63** 363

[48] Jolicard G, Leforestier C and Austin E J 1988 Resonance states using the optical potential model. Study of Feshbach resonances and broad shape resonances *J. Chem. Phys.* **88** 1026

[49] D'Mello M, Duneczky C and Wyatt R E 1988 Recursive generation of individual S-matrix elements: application to the collinear H + H$_2$ reaction *Chem. Phys. Lett.* **148** 169

[50] Kosloff R 1988 Time-dependent quantum-mechanical methods for molecular dynamics *J. Phys. Chem.* **92** 2087

[51] Moiseyev N, Friesner R A and Wyatt R E 1986 Natural expansion of vibrational wave functions: RRGM with residue algebra *J. Chem. Phys.* **85** 331

[52] Manthe U, Seideman T and Miller W H 1993 Full-dimensional quantum mechanical calculation of the rate constant for the H + H$_2$O $\rightarrow$ H$_2$ + OH reaction *J. Chem. Phys.* **99** 10 078

[53] Edlund A and Peskin U 1998 A parallel Green's operator for multidimensional quantum scattering calculations *Int. J. Quantum Chem.* **69** 167

[54] Peskin U, Miller W H and Edlund A 1995 Quantum time evolution in time-dependent fields and time-independent reactive-scattering calculations via an efficient Fourier grid preconditioner *J. Chem. Phys.* **103** 10 030

[55] McCullough E A and Wyatt R E 1971 Dynamics of the collinear H + H$_2$ reaction. I. Probability density and flux *J. Chem. Phys.* **54** 3578

[56] De-Leon N, Davis M J and Heller E J 1984 Quantum manifestations of classical resonance zones *J. Chem. Phys.* **80** 794

[57] Jackson B and Metiu H 1985 An examination of the use of wave packets for the calculation of atom diffraction by surfaces *J. Chem. Phys.* **83** 1952

[58] Mowrey R C and Kouri D J 1987 Application of the close coupling wave packet method to long lived resonance states in molecule–surface scattering *J. Chem. Phys.* **86** 6140

[59] Feit M D and Fleck J A 1983 Solution of the Schrödinger equation by a spectral method II. Vibrational energy levels of triatomic molecules *J. Chem. Phys.* **78** 301

[60] Neuhauser D, Judson R S, Kouri D J, Adelman D E, Shafer N S, Kliner D A and Zare R N 1992 State-to-state rates for the D + H$_2$($v = 1$, $j = 1$) $\rightarrow$ HD($v'$, $j'$) + H reaction: predictions and measurements *Science* **257** 522

[61] Neuhauser D, Baer M, Judson R S and Kouri D J 1989 Time-dependent three-dimensional body frame quantal wavepacket treatment of the atomic hydrogen + molecular hydrogen exchange reaction on the Liu–Siegbahn–Truhlar–Horowitz (LSTH) surface *J. Chem. Phys.* **90** 5882

[62] Gray S K and Balint-Kurti G G 1998 Quantum dynamics with real wave packets, including application to three-dimensional ($J = 0$)D + H$_2$ $\rightarrow$ HD + H reactive scattering *J. Chem. Phys.* **108** 950

[63] Kroes G J and Neuhauser D 1996 Performance of a time-independent scattering wave packet technique using real operators and wave functions *J. Chem. Phys.* **105** 8690

[64] Mandelshtam V A and Taylor H S 1995 A simple recursion polynomial expansion of the Green's function with absorbing boundary conditions. Application to the reactive scattering *J. Chem. Phys.* **102**

[65] Neuhauser D, Judson R S, Baer M and Kouri D J 1997 State-to-state time-dependent wavepacket approach to reactive scattering: State-resolved cross-sections for D + H$_2$($v = 1$, $j = 1$, $m$) $\rightarrow$ H + DH($\bar{v}$, $\bar{j}$), *J. Chem. Soc. Faraday Trans.* **93** 727

[66] Bacic Z and Light J C 1986 Highly excited vibrational levels of floppy triatomic molecules—a discrete variable representation—distributed Gaussian-basis approach *J. Chem. Phys.* **85** 4594

[67] Colbert D T and Miller W H 1992 A novel discrete variable representation for quantum mechanical reactive scattering via the S-matrix Kohn method *J. Chem. Phys.* **96** 1982

[68] Leforestier C *et al* 1991 A comparison of different propagation schemes for the time dependent Schrödinger equation *J. Comput. Phys.* **94** 59

[69] McCormack D A, Kroes G J and Neuhauser D 1998 Resonance affected scattering: Comparison of two hybrid methods involving filter diagonalization and the Lanczos method *J. Chem. Phys.* **109** 5177

[70] Neuhauser D 1992 Reactive scattering with absorbing potentials in general coordinate systems *Chem. Phys. Lett.* **200** 173

[71] Seideman T and Miller W H 1992 Quantum mechanical reaction probabilities via a discrete variable representation-absorbing boundary condition Green function *J. Chem. Phys.* **97** 2499

[72] Balint-Kurti G G, Dixon R N and Marston C C 1990 The Fourier grid Hamiltonian method for bound state eigenvalues and eigenfunctions *J. Chem. Soc. Faraday Trans.* **86** 1741

[73] Tannor D J and Weeks D E 1993 Wave packet correlation function formulation of scattering theory—the quantum analog of classical S-matrix theory *J. Chem. Phys.* **98** 3884

[74] Kouri D J, Huang Y, Zhu W and Hoffman D K 1994 Variational principles for the time-independent wave-packet–Schrödinger and wave-packet–Lippmann–Schwinger equations *J. Chem. Phys.* **100**

[75] Neuhauser D 1994 Fully quantal initial-state-selected reaction probabilities ($J = 0$) for a four-atom system—H$_2$($v = 0, 1$, $j = 0$) + OH($v = 0, 1$, $j = 0$) $\rightarrow$ H + H$_2$O *J. Chem. Phys.* **100** 9272

[76] Zhang D H and Zhang J Z H 1994 Full-dimensional time-dependent treatment for diatom–diatom reactions—the $H_2$ +OH reaction *J. Chem. Phys.* **101** 1146

[77] Zhu W, Dai J Q, Zhang J Z H and Zhang D H 1996 State-to-state time-dependent quantum calculation for reaction $H_2 + OH \rightarrow H + H_2O$ in six dimensions *J. Chem. Phys.* **105** 4881

[78] Zhang D H and Light J C 1996 Quantum state-to-state reaction probabilities for the $H+H_2O \rightarrow H_2$ +OH reaction in six dimensions *J. Chem. Phys.* **105** 1291

[79] Goldfield E M, Gray S K and Schatz G C 1995 Quantum dynamics of a planar model for the complex forming $OH+CO \rightarrow H+CO_2$ reaction *J. Chem. Phys.* **102** 8807

[80] Zhang D H and Zhang J Z H 1995 Quantum calculations of reaction probabilities for $HO + CO \rightarrow H + CO_2$ and bound states of HOCO *J. Chem. Phys.* **103** 6512

[81] McCormack D A, Kroes G J, Olsen R A, Baerends E J and Mowrey R C 1999 Rotational effects on vibrational excitation of $H_2$ on Cu(100) *Phys. Rev. Lett.* **82** 1410

[82] Toba M, Kubo R and Saito N 1992 *Statistical Physics I. Equilibrium Statistical Mechanics* (New York: Springer)

[83] Yamamoto T 1960 Quantum statistical mechanical theory of the rate of exchange chemical reactions in the gas phase *J. Chem. Phys.* **33** 281

[84] Miller W H 1974 Quantum mechanical transition state theory and a new semiclassical model for reaction rate constants *J. Chem. Phys.* **61** 1823

[85] Miller W H 1975 Semiclassical limit of quantum mechanical transition state theory for nonseparable systems *J. Chem. Phys.* **62** 1899

[86] Miller W H, Schwartz S D and Tromp J W 1983 Quantum mechanical rate constants for bimolecular reactions *J. Chem. Phys.* **79** 4889

[87] Pollak E and Liao J L 1998 A new quantum transition state theory *J. Chem. Phys.* **108** 2733

[88] Seideman T and Miller W H 1992 Calculation of the cumulative reaction probability via a discrete variable representation with absorbing boundary conditions *J. Chem. Phys.* **96** 4412

[89] Heller E J 1975 Time-dependent approach to semiclassical dynamics *J. Chem. Phys.* **62** 1544

[90] Lee S-Y and Heller E J 1979 Time-dependent theory of Raman scattering *J. Chem. Phys.* **71** 4777

[91] Main J, Mandelshtam V A, Wunner G and Taylor H S 1998 Harmonic inversion as a general method for periodic orbit quantization *Nonlinearity* **11** 1015

[92] Hazi A U and Taylor H S 1970 Stabilization method of calculating resonance energies: model problem *Phys. Rev.* A **1** 1109

[93] Mandelshtam V A, Ravuri T R and Taylor H S 1994 The stabilization theory of scattering *J. Chem. Phys.* **101** 8792

[94] Mandelshtam V A, Taylor H S, Jung C, Bowen H F and Kouri D J 1995 Extraction of dynamics from the resonance structure of $H + H_2$ spectra *J. Chem. Phys.* **102** 7988

[95] Pollak E 1985 Periodic orbits and the theory of reactive scattering *Theory of Chemical Reaction Dynamics* vol III, ed M Baer (Boca Raton, FL: CRC Press)

[96] Schinke R, Weide K, Heumann B and Engel V 1991 Diffuse structures and periodic orbits in the photodissociation of small polyatomic molecules *Faraday Discuss. Chem. Soc.* **91** 31

[97] Ezra G S 1996 Periodic orbit analysis of molecular vibrational spectra–spectral patterns and dynamical bifurcations in Fermi resonant systems *J. Chem. Phys.* **104** 26

[98] Kellman M E 1997 Nonrigid systems in chemistry: a unified view *Int. J. Quantum Chem.* **65** 399

[99] Sadeghi R and Skodje R T 1995 Barriers, thresholds and resonances—spectral quantization of the transition state for the collinear $D + H_2$ reaction *J. Chem. Phys.* **102** 193

[100] Messiah A 1961 *Quantum Mechanics* (Amsterdam: North-Holland)

[101] Reinhardt W P 1982 Complex coordinates in the theory of atomic and molecular structure and dynamics *Ann. Rev. Phys. Chem.* **35** 223

[102] Chu S I 1991 Complex quasivibrational energy formalism for intense-field multiphoton and above-threshold dissociation—complex-scaling Fourier-grid Hamiltonian method *J. Chem. Phys.* **94** 7901

[103] Lipkin N, Lefebvre R and Moiseyev N 1992 Resonances by complex nonsimilarity transformations of the Hamiltonian *Phys. Rev.* A **45** 4553

[104] Moiseyev N, Certain P R and Weinhold F 1978 Resonance properties of complex-rotated Hamiltonians *Molec. Phys.* **36** 1613

[105] Mandelshtam V A and Moiseyev N 1996 Complex scaling of *ab initio* molecular potential surfaces *J. Chem. Phys.* **104** 6192

[106] Leforestier C and Museth K 1998 Response to 'Comment on "On the direct complex scaling of matrix elements expressed in a discrete variable representation: application to molecular resonances"' *J. Chem. Phys.* **109** 1204

[107] Museth K and Leforestier C 1996 On the direct complex scaling of matrix elements expressed in a discrete variable representation—application to molecular resonances *J. Chem. Phys.* **104** 7008

[108] Leforestier C and Wyatt R E 1983 Optical potential for laser induced dissociation *J. Chem. Phys.* **78** 2334

[109] Riss U V and Meyer H D 1998 The transformative complex absorbing potential method: a bridge between complex absorbing potentials and smooth exterior scaling *J. Phys. B: At. Mol. Opt. Phys.* **31** 2279

[110] Neuhauser D 1990 Bound state eigenfunctions from wave packets—time $\rightarrow$ energy resolution *J. Chem. Phys.* **93** 2611

[111] Neuhauser D 1994 Circumventing the Heisenberg principle—a rigorous demonstration of filter-diagonalization on a LiCN model *J. Chem. Phys.* **100** 5076

[112] Wall M R and Neuhauser D 1995 Extraction, through filter-diagonalization, of general quantum eigenvalues or classical normal mode frequencies from a small number of residues or a short-time segment of a signal. I. Theory, and application to a quantum-dynamics model *J. Chem. Phys.* **102** 8011

[113] Pang J W, Dieckmann T, Feigon J and Neuhauser D 1998 Extraction of spectral information from a short-time signal using filter-diagonalization: recent developments and applications to semiclassical reaction dynamics and nuclear magnetic resonance signals *J. Chem. Phys.* **108** 8360

[114] Wall M R, Dieckmann T, Feigon J and Neuhauser D 1998 Two-dimensional filter-diagonalization: spectral inversion of 2D NMR time-correlation signals including degeneracies *Chem. Phys. Lett.* **291** 465

[115] Mandelshtam V A and Taylor H S 1997 Spectral analysis of time correlation function for a dissipative dynamical system using filter diagonalization: application to calculation of unimolecular decay rates *Phys. Rev. Lett.* **78** 3274

[116] Beck M H and Meyer H D 1998 Extracting accurate bound-state spectra from approximate wave packet propagation using the filter-diagonalization method *J. Chem. Phys.* **109** 3730

[117] Chen R Q and Guo H 1996 A general and efficient filter diagonalization method without time propagation *J. Chem. Phys.* **105** 1311

[118] Narevicius E, Neuhauser D, Korsch H J and Moiseyev M 1997 Resonances from short time complex-scaled cross-correlation probability amplitudes by the filter-diagonalization method *Chem. Phys. Lett.* **276** 250

[119] Main J, Mandelshtam V A and Taylor H S 1997 High resolution quantum recurrence spectra: beyond the uncertainty principle *Phys. Rev. Lett.* **78** 4351

[120] Parker G A and Pack R T 1978 Rotationally and vibrationally inelastic scattering in the rotational IOS approximation. Ultrasimple calculation of total (differential, integral and transport) cross sections for nonspherical molecules *J. Chem. Phys.* **68** 1585

[121] Kouri D J 1977 *Atom–Molecule Collision Theory: A Guide for the Experimentalist* ed R B Bernstein (New York: Plenum)

[122] Wang D and Bowman J M 1992 Reduced dimensionality quantum calculations of mode specificity in OH + $H_2$ → $H_2$O + H *J. Chem. Phys.* **96** 8906

[123] Clary D C 1994 Four-atom reaction dynamics *J. Phys. Chem.* **98** 10678

[124] Park T J and Light J C 1989 Accurate quantum thermal rate constants for the three-dimensional H + $H_2$ reaction *J. Chem. Phys.* **91** 974

[125] Baer M, Loesch H J, Werner H J and Last I 1994 Integral and differential cross sections for the Li + HF to LiF + H process. A comparison between $J_z$ quantum mechanical and experimental results *Chem. Phys. Lett.* **219** 372

[126] Baer M, Faubel M, Martinez-Haya B, Rusin L Y, Tappe U and Toennies J P 1998 A study of state-to-state differential state cross-sections for the: F + $D_2(v_i = 0, j_i)$ → $DF(v_f, j_f)$ + D reactions: a detailed comparison between experimental and three dimensional quantum mechanical results *J. Chem. Phys.* **108** 9694

[127] Cho S W, Wagner A F, Gazdy B and Bowman J M 1992 Theoretical studies of the reactivity and spectroscopy of H + CO to or from HCO. I. Stabilization and scattering studies of resonances for $J = 0$ on the Harding *ab initio* surface *J. Chem. Phys.* **96** 2812

[128] Zhang D H and Zhang J Z H 1994 Accurate quantum calculations for $H_2$ + OH → $H_2$O + H—reaction probabilities, cross sections and rate constants *J. Chem. Phys.* **100** 2697

[129] Heidrich D (ed) 1995 *The Reaction Path in Chemistry: Current Approaches and Perspectives* (Boston: Kluwer Academic)

[130] Miller W H, Handy N C and Adams J E 1980 Reaction path Hamiltonian for polyatomic molecules *J. Chem. Phys.* **72** 99

[131] Fast P L and Truhlar D G 1998 Variational reaction path algorithm *J. Chem. Phys.* **109** 3721

[132] Billing G D 1992 Quantum classical reaction-path model for chemical reactions *Chem. Phys.* **161** 245

[133] Zhang J Z H 1999 The semirigid vibrating rotor target model for quantum polyatomic reaction dynamics *J. Chem. Phys.* **111** 3929

[134] Gerber R B, Buch V and Ratner M A 1982 Simplified time-dependent self consistent field approximation for intramolecular dynamics *Chem. Phys. Lett.* **91** 173

[135] Peskin U and Steinberg M 1998 A temperature-dependent Schrödinger equation based on a time-dependent self consistent field approximation *J. Chem. Phys.* **109** 704

[136] Hammes-Schiffer S 1998 Quantum dynamics of multiple modes for reactions in complex systems *Faraday Discuss. Chem. Soc.* **110** 391

[137] Diz A, Deumens E and Ohrn Y 1990 Quantum electron-nuclear dynamics *Chem. Phys. Lett.* **166** 203

[138] Anderson S M, Zink J I and Neuhauser D 1998 A simple and accurate approximation for a coupled system-bath: locally propagating Gaussians *Chem. Phys. Lett.* **291** 387

[139] Anderson S M, Park T J and Neuhauser D 1999 Locally propagating Gaussians: flexible vs. frozen widths *Phys. Chem. Chem. Phys.* **1** 1343

[140] Carter S and Bowman J M 1998 The adiabatic rotation approximation of rovibrational energies of many-mode systems: description and tests of the method *J. Chem. Phys.* **108** 4397

[141] Kosloff R and Hammerich A D 1991 Nonadiabatic reactive routes and the applicability of multi configuration time dependent self consistent field approximations *Faraday Discuss. Chem. Soc.* **91** 239–47

[142] Manthe U 1996 A time-dependent discrete variable representation for mulitconfigurational Hartree methods *J. Chem. Phys.* **105** 2646

[143] McCoy A B and Siebert E L 1996 Canonical Van Vleck pertubation theory and its applications to studies of highly vibrationally excited states of polyatomic molecules *Dynamics of Molecules and Chemical Reactions* ed R E Wyatt and J Z H Zhang (New York: Dekker) p 151

[144] Heller E J 1981 Frozen Gaussians: a very simple semiclassical approximation *J. Chem. Phys.* **75** 2923

[145] Blake N P and Metiu H 1995 Efficient adsorption line shape calculations for an electron coupled to many quantum degrees of freedom. applications to an electron solvated in dry sodalites and halo-sodalites *J. Chem. Phys.* **103** 4455

[146] Markovic N and Billing G D 1997 Semi-classical treatment of chemical reactions: extension to 3D wave packets *Chem. Phys.* **224** 53

[147] Martinez T J, BenNun M and Levine R D 1996 Multi-electronic-state molecular dynamics—a wavefunction approach with applications *J. Phys. Chem.* **100** 7884

[148] Makri N 1990 Time-dependent self-consistent field approximation with explicit 2-body correlations *Chem. Phys. Lett.* **169** 541

[149] Feynman R P and Hibbs A R 1965 *Quantum Mechanics and Path Integrals* (New York: McGraw-Hill)

[150] Schulman L S 1981 *Techniques and Applications of Path Integration* (New York: Wiley)

[151] van Vleck J H 1928 The correspondence principle in statistical interpretation of quantum mechanics *Proc. Natl. Acad. Sci.* **14** 178

[152] Pechukas P 1969 Time-dependent semiclassical scattering theory. I. Potential scattering *Phys. Rev.* **181** 166

[153] Miller W H 1974 Classical-limit quantum mechanics and the theory of molecular collisions *Adv. Chem. Phys.* **25** 69

[154] Child M S 1991 *Semiclassical Mechanics with Molecular Applications* (Oxford: Clarendon)

[155] Campolieti G and Brumer P 1994 Semiclassical propagation: phase indices and the initial-value formalism *Phys. Rev.* A **50** 997

[156] Miller W H 1991 Comments on: Semiclassical time evolution without root searches *J. Chem. Phys.* **95** 9428

[157] Heller E J 1991 Reply to Comments on: Semiclassical time evolution without root searches *J. Chem. Phys.* **95** 9431

[158] Gutzwiller M C 1967 Phase-integral approximation in momentum space and the bound states of an atom *J. Math. Phys.* **8** 1979

[159] Huber D, Ling S, Imre D G and Heller E J 1989 Hybrid mechanics *J. Chem. Phys.* **90** 7317

[160] Herman M F and Kluk E 1984 A semiclassical justification for the use of non-spreading wavepackets in dynamics calculations *Chem. Phys.* **91** 27

[161] Herman M F, Kluk E and Davis H L 1986 Comparison of the propagation of semiclassical frozen Gaussian wave functions with quantum propagation for a highly excited anharmonic oscillator *J. Chem. Phys.* **84** 326

[162] Kay K G 1994 Semiclassical propagation for multidimensional systems by an initial value method *J. Chem. Phys.* **101** 2250

[163] Walton A R and Manolopoulos D E 1996 A new semiclassical initial value method for Franck-Condon spectra *Mol. Phys.* **87** 961

[164] Sun X, Wang H B and Miller W H 1998 Semiclassical theory of electronically nonadiabatic dynamics: Results of a linearized approximation to the initial value representation *J. Chem. Phys.* **109** 7064

[165] Grossmann F, Mandelshtam V A, Taylor H S and Briggs H S 1997 Harmonic inversion of semiclassical short time signals *Chem. Phys. Lett.* **279** 355

[166] Thompson K and Makri N 1999 Influence functionals with semiclassical propagators in combined forward–backward time *J. Chem. Phys.* **110** 1343

[167] Stock G and Thoss M 1997 Semiclassical description of nonadiabatic quantum dynamics *Phys. Rev. Lett.* **78** 578

[168] Thompson K and Makri N 1999 Rigorous forward-backward semiclassical formulation of many-body dynamics *Phys. Rev.* E **59** 4729

[169] Baer M, Yahalom A and Engelman R 1998 Time-dependent and time-independent approaches to study effects of degenerate electronic states *J. Chem. Phys.* **109** 6550

[170] Mead C A and Truhlar D G 1979 On the determination of Born–Oppenheimer nuclear motion wave functions including complications due to conical intersections and identical nuclei *J. Chem. Phys.* **70** 2284

[171] Berry M 1990 Anticipations of the geometric phase *Physics Today* **43** 34

[172] Baer R, Charutz D M, Kosloff R and Baer M 1996 A study of conical intersection effects on scattering processes—the validity of adiabatic single-surface approximations within a quasi-Jahn–Teller model *J. Chem. Phys.* **105** 9141

[173] Landau L D 1932 Zur theorie der Energieübertragung bei Stössen. I *Phys. Zts. Sowjet.* **2** 46

[174] Zener C 1932 Nonadiabatic crossing of energy levels *Proc. R. Soc.* A **137** 696

[175] Tully J C and Preston R K 1971 Trajectory surface hopping approach to nonadiabatic molecular collisions: the reaction of H$^+$ with D$_2$ *J. Chem. Phys.* **55** 562

[176] Hammes-Schiffer S and Tully J C 1995 Nonadiabatic transition state theory and multiple potential energy surfaces molecular dynamics of infrequent events *J. Chem. Phys.* **103** 8528

[177] Niv M Y, Krylov A I and Gerber R B 1997 Photodissociation, electronic relaxation and recombination of HCl in Ar-n(HCl) clusters—non-adiabatic molecular dynamics simulations *Faraday Discuss. Chem. Soc.* **108** 243–54

[178] Zhu C and Nakamura H 1994 Theory of nonadiabatic transition for general curved potentials I. *J. Chem. Phys.* **101** 10 630

[179] Baer M 1992 *State Selected and State-to-State Ion–Molecule Reaction Dynamics. Part II: Theory* ed M Baer and C Y Ng (New York: Wiley)

[180] Baer M, Niedner-Shcattenburg G and Toennies J P 1989 A 3-dimensional quantum mechanical study of vibrationally resolved charged transfer processes in H$^+$ + H$^2$ at $E_{CM} = 20$ eV *J. Chem. Phys.* **91** 4169

[181] Meyer H D and Miller W H 1979 A classical analog for electronic degrees of freedom in nonadiabatic collision processes *J. Chem. Phys.* **70** 3214

[182] Tannor D J, Kosloff R and Rice S A 1986 Coherent pulse sequence induced control of selectivity of reactions: exact quantum mechanical calculations *J. Chem. Phys.* **85** 5805

[183] Neuhauser D and Rabitz H 1993 Paradigms and algorithms for controlling molecular motion *Acc. Chem. Res.* **26** 496

[184] Seideman T, Shapiro M and Brumer P 1989 Coherence chemistry—controlling chemical reactions with lasers *Chem. Phys.* **90** 7132

[185] Smith T J and Cina J A 1996 Toward preresonant impulsive Raman preparation of large amplitude *J. Chem. Phys.* **104** 1272

[186] Krause J L, Messina M, Wilson K R and Yan Y J 1995 Quantum control of molecular dynamics—the strong response regime *J. Phys. Chem.* **99** 13 736

[187] Meyer S and Engel V 1997 Vibrational revivals and the control of photochemical reactions *J. Phys. Chem.* A **101** 7749

[188] Assion A, Baumert T, Bergt M, Brixner T, Kiefer B, Seyfried V, Strehle M and Gerber G 1998 Control of chemical reactions by feedback-optimized phase-shaped femtosecond laser pulses *Science* **282** 919

[189] Althorpe S C and Seideman T 1999 Molecular alignment from femtosecond time-resolved photoelectron angular distributions: nonperturbative calculations on NO *J. Chem. Phys.* **110** 147

[190] Friedrich B and Herschbach D 1996 Alignment enhanced spectra of molecules in intense non-resonant laser fields *Chem. Phys. Lett.* **262** 41

[191] Judson R S and Rabitz H 1992 Teaching lasers to control molecules *Phys. Rev. Lett.* **68** 1500

[192] Bardeen C J, Yakovlev V V, Wilson K R, Carpenter S D, Weber P M and Warren W S 1997 Feedback quantum control of molecular electronic population transfer *Chem. Phys. Lett.* **280**

[193] Zare R N 1998 Laser control of chemical reactions *Science* **279** 1875

[194] Bigwood R, Gruebele M, Leitner D M and Wolynes P G 1998 The vibrational energy flow transition in organic molecules: theory meets experiment *Proc. Natl Acad. Sci.* **95** 5960

[195] Truhlar D G and Horowitz C J 1978 Functional representation of Liu and Siegbahn's accurate *ab initio* potential energy calculations for H + $H_2$ *J. Chem. Phys.* **68** 2466

[196] Kroes G J, Wiesenekker G, Baerends E J, Mowrey R C and Neuhauser D 1996 Dissociative chemisorption of $H_2$ on Cu(100)—a four-dimensional study of the effect of parallel translational motion on the reaction dynamics *J. Chem. Phys.* **105** 5979

# B3.5
# Optimization and reaction path algorithms

*Peter Pulay and Jon Baker*

### B3.5.1  Introduction

A quantum mechanical treatment of molecular systems usually starts with the Born–Oppenheimer approximation, i.e., the separation of the electronic and nuclear degrees of freedom. This is a very good approximation
for well separated electronic states. The expectation value of the total energy in this case is a function of
the nuclear coordinates and the parameters in the electronic wavefunction, e.g., orbital coefficients. The
wavefunction parameters are most often determined by the variation theorem: the electronic energy is made
stationary (in the most important ground-state case it is minimized) with respect to them. The optimized
energy, calculated as a function of the nuclear coordinates, is known as the potential energy surface (PES).
Finding its (local) minimum gives the calculated equilibrium structure. The latter, although experimentally
not directly accessible, is perhaps the most satisfactory definition of molecular geometry, and serves as the
best starting point for a treatment of molecular vibrations. A large part of the total effort expended in quantum
chemical calculations is spent in optimizing either electronic parameters or molecular geometries; the latter
task is dominant in empirical force field calculations. The optimization of electronic wavefunctions is usually treated separately from the optimization of molecular geometries; however, there are enough similarities
between the two problems to discuss them together.

Both the electronic and the geometry optimization problem, particularly the latter, may have more than
one solution. For small, rigid molecules, the approximate molecular geometry is chemically obvious, and
the presence of multiple minima is not a serious concern. For large, flexible molecules, however, finding
the absolute minimum, or a complete set of low-lying equilibrium structures, is only a partially solved
problem. This topic will be discussed in the last section of this chapter. The rest of the article deals with
local optimization, i.e., finding a minimum from a reasonably close starting point. We will also discuss
the determination of other stationary points—most importantly saddle points-constrained optimization, and
reaction paths. Several reviews have been published on geometry optimization [1, 2]. The optimization of
SCF-type wavefunctions is often highly nonlinear, particularly for the multiconfigurational case, and this has
received most attention [3, 4].

The most important consideration affecting the choice of the method for locating minima, or stationary
points in general, is the availability of analytical derivatives of the object function, in our case the energy.
Zeroth-order (energy only) methods can be used for a few variables but are notoriously inefficient for a larger
number of degrees of freedom. First-order methods, which use both energy and gradient (first-derivative)
information, are particularly useful in quantum chemistry because the extra effort needed to evaluate all first
derivatives is usually comparable to the calculation of the energy itself and may be less, particularly for the
electronic degrees of freedom [5]. Second-order methods, which use second derivatives, further improve the
convergence of the optimization process. However, calculating second derivatives tends to be much more
expensive than calculating the gradient, and full second-order methods are usually cost efficient only when

first-order methods have severe convergence problems. Derivatives higher than the second have been used occasionally, but they are not generally available and are expensive to calculate. Consequently, this article will mainly concentrate on first- and second-order methods.

The electronic energy $W$ in the Born–Oppenheimer approximation can be written as $W = W(\mathbf{q}, \mathbf{p})$, where $\mathbf{q}$ is the vector of nuclear coordinates and the vector $\mathbf{p}$ contains the parameters of the electronic wavefunction. The latter are usually orbital coefficients, configuration amplitudes and occasionally nonlinear basis function parameters, e.g., atomic orbital positions and exponents. The electronic coordinates have been integrated out and do not appear in $W$. Optimizing the electronic parameters leaves a function depending on the nuclear coordinates only, $E = E(\mathbf{q})$. We will assume that both $W(\mathbf{q}, \mathbf{p})$ and $E(\mathbf{q})$ and their first derivatives are continuous functions of the variables $q_i$ and $p_j$.

## B3.5.2   Overview of techniques for local optimization

### B3.5.2.1   Characterization of stationary points

In this section, we will discuss general optimization methods. Our example is the geometry optimization problem, i.e., the minimization of $E(\mathbf{q})$. However, the results apply to electronic optimization as well. There are a number of useful monographs on the minimization of continuous, differentiable functions in many variables [6, 7].

For a point on the potential energy surface to be a stationary point, all its first derivatives, $\partial E/\partial q_i$, must vanish, and thus the whole gradient vector $\mathbf{g} = \{\partial E/\partial q_i, i = 1, n\}$ should be zero. The character of a stationary point, i.e., whether it is a local minimum, maximum or saddle point, can be determined by examining the second derivatives. Expanding the energy change in the neighbourhood of a stationary point $\mathbf{q}^0$ in a power series in terms of displacement coordinates from the stationary point, $\delta q_i = q_i - q_i^0$, gives

$$E(\mathbf{q}) - E(\mathbf{q}^0) = \sum_i (\partial E/\partial q_i)|_{\mathbf{q}^0} \delta q_i + 1/2 \sum_{i,j} (\partial^2 E/\partial q_i \partial q_j)|_{\mathbf{q}^0} \delta q_i \delta q_j + \text{ higher terms.} \qquad (B3.5.1)$$

As $q_0$ is a stationary point, the linear terms $(\partial E/\partial q_i)|_{\mathbf{q}^0} \delta q_i$ in equation (B3.5.1) vanish, and higher-order terms do not have to be considered for local characterization of stationary points, because lower-order terms (quadratics in this case) always dominate for sufficiently small displacements. Introducing the force constant, or Hessian matrix, i.e., the matrix of second derivatives $\mathbf{H}$, $H_{ij} = (\partial^2 E/\partial q_i \partial q_j)|_{\mathbf{q}^0}$, the above equation can be written in a convenient matrix notation as

$$\Delta E = E(\mathbf{q}) - E(\mathbf{q}^0) = \tfrac{1}{2} \delta \mathbf{q}^\dagger \mathbf{H} \delta \mathbf{q}. \qquad (B3.5.2)$$

Let us express the displacement coordinates as linear combinations of a set of new coordinates $\mathbf{y}$: $\delta \mathbf{q} = \mathbf{U} \mathbf{y}$; then $\Delta E = \mathbf{y}^\dagger \mathbf{U}^\dagger \mathbf{H} \mathbf{U} \mathbf{y}$. $\mathbf{U}$ can be an arbitrary non-singular matrix, and thus can be chosen to diagonalize the symmetric matrix $\mathbf{H}$: $\mathbf{U}^\dagger \mathbf{H} \mathbf{U} = \mathbf{\Lambda}$, where the diagonal matrix $\mathbf{\Lambda}$ contains the (real) eigenvalues of $\mathbf{H}$. In this form, the energy change from the stationary point is simply $\Delta E = \tfrac{1}{2} \sum_i \Lambda_i y_i^2$. It is clear now that a sufficient condition for a minimum is that all eigenvalues of $\mathbf{H}$ be positive, i.e., $\mathbf{H}$ must be a positive definite matrix. Otherwise choosing $y_i \neq 0$, all other $y_j = 0$, where $\Lambda_i$ is a negative eigenvalue, will decrease the energy, i.e., the stationary point cannot be a minimum. Zero eigenvalues of the Hessian (inflection points) need not be considered because their probability in the general case is vanishingly small. Stationary points with only one negative Hessian eigenvalue are called first-order saddle points. They have considerable importance as transition states in chemical reactions. The energy difference between a transition state and the reactant(s) is the barrier corresponding to the reaction path passing through that transition state. Stationary points with two or more negative eigenvalues are far less important, as in this case there is always a reaction path with lower barrier which determines the reaction probability (however, in symmetrical systems higher-order saddle points may be preferred reference geometries [8]).

The simplest smooth function which has a local minimum is a quadratic. Such a function has only one, easily determinable stationary point. It is thus not surprising that most optimization methods try to model the unknown function with a local quadratic approximation, in the form of equation (B3.5.1).

### B3.5.2.2 Energy-only methods

As noted in the introduction, energy-only methods are generally much less efficient than gradient-based techniques. The simplex method [9] (not identical with the similarly named method used in linear programming) was used quite widely before the introduction of analytical energy gradients. The intuitively most obvious method is a sequential optimization of the variables (sequential univariate search). As the optimization of one variable affects the minimum of the others, the whole cycle has to be repeated after all variables have been optimized. A one-dimensional minimization is usually carried out by finding the minimum of a parabola fitted to three points obtained by varying one of the variables (keeping the others constant), changing values so as to bracket the minimum, and zeroing in on the minimum by diminishing the step size [6]. Generalized to any vector direction on the surface, this is called a line search. The convergence rate of the sequential univariate search can be exceedingly slow if the variables are strongly coupled and, thus, this method is not recommended. A better alternative is to convert gradient methods, covered in the next section, to energy-only methods by calculating the gradients numerically. One of the most widely used energy-only methods is the modified Fletcher–Powell method described by Schlegel [1]; perhaps better is the numerical version of Baker's Eigenvector Following (EF) algorithm (see later) [10]. In spite of these ingenious algorithms, the general consensus among researchers in the field is that energy-only methods are simply not cost effective for systems with more than a few degrees of freedom.

### B3.5.2.3 Gradient methods

All efficient optimization methods require the gradient vector, i.e., the first derivatives of the function to be optimized. As the quantum mechanical energy as a function of nuclear coordinates is the result of an iterative procedure, and ordinary first-order perturbation theory is inapplicable in the usual case where the basis functions move with the nuclei, this is not a trivial problem. The introduction of analytical energy derivatives (forces on the atoms in the context of geometry optimization) of the SCF energy [11], and later generalizations to more complex wavefunctions (for reviews see, e.g., [5, 12]) improved the efficiency of geometry optimizations by one or two orders of magnitude, depending on the molecular size, making possible structure optimization for large polyatomic molecules.

### (a) Newton's method

Most gradient optimization methods rely on a quadratic model of the potential surface. The minimum condition for the quadratic energy expression

$$E(\mathbf{q}) = E(\mathbf{q}^0) + \delta\mathbf{q}^\dagger\mathbf{g} + \tfrac{1}{2}\delta\mathbf{q}^\dagger\mathbf{H}\delta\mathbf{q}$$

using the symmetry of $\mathbf{H}$, leads to

$$\mathbf{g} + \mathbf{H}\delta\mathbf{q} = \mathbf{0} \text{ and } \delta\mathbf{q} = -\mathbf{H}^{-1}\mathbf{g}. \tag{B3.5.3}$$

Here $\mathbf{g}$ is the gradient vector at $\mathbf{q}^0$, $\mathbf{g}_i = (\partial E/\partial q_i)|_{\mathbf{q}^0}$. The minimizer $\mathbf{q}^0 + \delta\mathbf{q}$ is exact on a quadratic surface and requires only the solution of a linear system of equations. For a nonlinear surface, this method has to be iterated. Near the solution, the iterative procedure is quadratically convergent. In practice, this means that three or four iterations usually suffice for locating the minimum to high accuracy. This is the basic Newton (or Newton–Raphson) method. Despite its rapid convergence for nearly quadratic surfaces, it suffers from a number of shortcomings and in its original form is seldom used for geometry optimization. It has some importance for difficult cases of wavefunction optimization. The principal defect of Newton's method is that it requires the Hessian (second-derivative) matrix at every iteration. Second derivatives are typically much more expensive than gradients in quantum chemistry applications. Another problem is that far from the minimum the Hessian may not be positive definite. The energy change in a Newton–Raphson cycle is $\delta \mathbf{q}^{\dagger} \mathbf{g} = -\frac{1}{2} \mathbf{g}^{\dagger} \mathbf{H}^{-1} \mathbf{g}$ in the quadratic approximation. Thus the energy does not necessarily decrease, even for small steps, unless $\mathbf{H}$ (and obviously its inverse) is positive definite.

### (b) Simple relaxation

Both defects of the Newton method can be eliminated by replacing the exact inverse Hessian $\mathbf{H}^{-1}$ by a (fixed) positive definite approximation to it, $\mathbf{F}$. This method is known as simple relaxation. In both geometry and wavefunction optimization, it is usually possible to construct a fairly good approximate Hessian. For geometry optimization, this can be based on the molecular connectivity and transferability of potential parameters, or on previous low-level calculations. For wavefunction optimization, a guess based on orbital energy differences is often reasonably accurate. Far from the minimum, approximate Hessian methods using positive definite matrices are preferable to the Newton method, as they have the descent property, i.e., the energy decreases for sufficiently small steps. However, they lack the quadratic terminal convergence rate of the Newton method. Instead, the residual error vector (the distance from the accurate minimum) is given by $\mathbf{r}^{(n)} = (\mathbf{I} - \mathbf{FH})^{n} \mathbf{r}^{(0)}$ on a quadratic surface. Here $\mathbf{I}$ is the unit matrix and $^{(n)}$ denotes the $n$th cycle. The ultimate convergence rate is governed by the magnitude of the largest eigenvalue of the matrix $(\mathbf{I} - \mathbf{FH})$. This will be small if $\mathbf{F}$ is a good approximation to $\mathbf{H}^{-1}$. To show this, we introduce new variables, through a linear transformation of the old ones, which diagonalize $\mathbf{FH}$. Using these coordinates, the $k$th component of the residue in step $n$ is $\lambda_{k}^{n} \mathbf{r}_{k}^{(0)}$ where $\lambda_{k}$ is the $k$th eigenvalue of $(\mathbf{I} - \mathbf{FH})$. This explains a common property of simple relaxation: it usually shows good initial convergence but slows down later as the surviving components of the residuum take on directions in which the Hessian is poorly estimated. If one of the eigenvalues of $(\mathbf{I} - \mathbf{F}^{-1}\mathbf{H})$ exceeds 1 in absolute magnitude then simple relaxation without a line search will ultimately diverge.

If there is no approximate Hessian available, then the unit matrix is frequently used, i.e., a step is made along the gradient. This is the steepest descent method. The unit matrix is arbitrary and has no invariance properties, and thus the resulting step may be made arbitrarily large or small by scaling the coordinates. Therefore, steepest descent methods require a line search for a minimum along the direction of the gradient vector. Line searches are often recommended in general optimization texts. However, they tend to be less efficient in quantum chemistry, as the evaluation of the gradient vector costs roughly the same as the calculation of the energy for a wide range of methods, and supplies much more information. Nevertheless, they may be necessary for strongly non-quadratic functions (or, what is essentially the same thing, at points far from the minimum). A good compromise which requires no additional energy evaluations was suggested by Schlegel [13]: a polynomial is fitted to the energies at two points, and to the gradients projected on the line connecting them, and its minimum is located. The polynomial can be cubic or, as recommended by Schlegel, a special quartic with only one minimum. If a line search is used, the energy of simple relaxation and steepest descent steps should always decrease, and they should ultimately converge. However, convergence may be very slow if $\mathbf{F}$ is a poor approximation to the inverse Hessian $\mathbf{H}^{-1}$, as is usually the situation in the steepest descent method. In this case, illustrated in figure B3.5.1, the optimization takes a zigzag path converging slowly to the minimum.

In simple relaxation (the fixed approximate Hessian method), the step does not depend on the iteration history. More sophisticated optimization techniques use information gathered during previous steps to improve the estimate of the minimizer, usually by invoking a quadratic model of the energy surface. These methods can be divided into two classes: variable metric methods and interpolation methods.

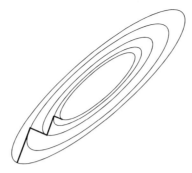

**Figure B3.5.1.** Contour line representation of a quadratic surface and part of a steepest descent path zigzagging toward the minimum.

*(c) Variable metric methods*

In these methods, also known as quasi-Newton methods, the approximate Hessian is improved (updated) based on the results in previous steps. For the exact Hessian and a quadratic surface, the quasi-Newton equation $\Delta \mathbf{g}^{(n)} = \mathbf{H} \Delta \mathbf{q}^{(n)}$ and its analogue $\mathbf{H}^{-1} \Delta \mathbf{g}^{(n)} = \Delta \mathbf{q}^{(n)}$ must hold (where $\Delta \mathbf{g}^{(n)} = \mathbf{g}^{(n+1)} - \mathbf{g}^{(n)}$, and similarly for $\Delta \mathbf{q}^{(n)}$). These equations, which have only $n$ components, are obviously insufficient to determine the $n(n+1)/2$ independent components of the Hessian or its inverse. Therefore, the updating is arbitrary to a certain extent. It is desirable to have an updating scheme that converges to the exact Hessian for a quadratic function, preserves the quasi-Newton conditions obtained in previous steps, and—for minimization—keeps the Hessian positive definite. Updating can be performed on either $\mathbf{F}$ or its inverse, the approximate Hessian. In the former case repeated matrix inversion can be avoided. All updates use dyadic products, usually built from $\Delta \mathbf{q}^{(n)}$ and $\mathbf{F} \Delta \mathbf{g}^{(n)}$. Fletcher [6] gives a detailed description of various update techniques. The most important update formulae are the Murtagh–Sargent update [14]:

$$\mathbf{F}^{(n+1)} = \mathbf{F}^{(n)} + (\Delta \mathbf{q}^{(n)} - \mathbf{F}^{(n)} \Delta \mathbf{g}^{(n)})(\Delta \mathbf{q}^{(n)} - \mathbf{F}^{(n)} \Delta \mathbf{g}^{(n)})^{\mathrm{T}} / \{(\Delta \mathbf{q}^{(n)} - \mathbf{F}^{(n)} \Delta \mathbf{g}^{(n)})^{\mathrm{T}} \Delta \mathbf{g}^{(n)}\}$$

the Davidon–Fletcher–Powell (DFP) update [15]:

$$\mathbf{F}^{(n+1)} = \mathbf{F}^{(n)} + s \Delta \mathbf{q}^{(n)} \Delta \mathbf{q}^{\mathrm{T}(n)} - s \mathbf{F}^{(n)} \Delta \mathbf{g}^{(n)} \Delta \mathbf{g}^{\mathrm{T}(n)} \mathbf{F}^{(n)}$$

and the Broyden–Fletcher–Goldfarb–Shanno (BFGS) update [6]:

$$\mathbf{F}^{(n+1)} = \mathbf{F}^{(n)} + s(1 + s \Delta \mathbf{g}^{\mathrm{T}(n)} \mathbf{F}^{(n)} \Delta \mathbf{g}^{(n)}) \Delta \mathbf{q}^{\mathrm{T}(n)} \Delta \mathbf{q}^{(n)} - s(\Delta \mathbf{q}^{(n)} \Delta \mathbf{g}^{\mathrm{T}(n)} \mathbf{F}^{(n)} + \mathbf{F}^{(n)} \Delta \mathbf{g}^{(n)} \Delta \mathbf{q}^{\mathrm{T}(n)})$$

where $s = 1/(\Delta \mathbf{q}^{\mathrm{T}(n)} \Delta \mathbf{g}^{(n)})$. A linear combination of these updates is also possible. Both the DFP and BFGS updates preserve positive definite $\mathbf{F}$ matrices provided $\Delta \mathbf{q}^{(n)\mathrm{T}} \Delta \mathbf{g}^{(n)} > 0$; current opinion is that the latter is the best update to use for general minimization.

For transition state searches, none of the above updates is particularly appropriate as a positive definite Hessian is not desired. A more useful update in this case is the Powell update [16]:

$$\mathbf{H}^{(n+1)} = \mathbf{H}^{(n)} + \{v \Delta \mathbf{q}^{(n)\mathrm{T}} + \Delta \mathbf{q}^{(n)} v^{\mathrm{T}} - (v^{\mathrm{T}} \Delta \mathbf{q}^{(n)\mathrm{T}})(\Delta \mathbf{q}^{(n)} \Delta \mathbf{q}^{(n)\mathrm{T}}) / t\} / t$$

where $v = \Delta \mathbf{g}^{(n)} - \mathbf{H}^{(n)} \Delta \mathbf{q}^{(n)}$ and $t = (\Delta \mathbf{q}^{(n)\mathrm{T}} \Delta \mathbf{q}^{(n)})$. The Powell update allows the signature of the Hessian, i.e., the number of negative eigenvalues, to change, which is necessary if the region of the potential energy surface is inappropriate for the stationary point being sought. Perhaps the best Hessian update for transition state searches is a linear combination of the Powell and Murtagh–Sargent updates proposed by Bofill [17, 18].

For a very large number of variables, the question of storing the approximate Hessian or inverse Hessian $\mathbf{F}$ becomes important. Wavefunction optimization problems can have a very large number of variables, a million or more. Geometry optimization at the force field level can also have thousands of degrees of freedom. In these cases, the initial inverse Hessian is always taken to be diagonal or sparse, and it is best to store the upgrade vectors and associated scalars and generate the inverse Hessian *in situ*, rather than store the full updated inverse Hessian itself.

A more general update method, widely used in the Gaussian suite of programs [19], is due to Schlegel [13]. In this method, the Hessian in the $n$-dimensional subspace spanned by taking differences between the current $\mathbf{q}^{(n)}$ and previous geometries $\mathbf{q}^{(n-1)}, \ldots, \mathbf{q}^{(0)}$ is calculated numerically. This is possible (although not terribly accurate), as the $n \, \Delta\mathbf{q}$ and $n \, \Delta\mathbf{g}$ values suffice for the calculation of an $n$-dimensional Hessian by forward differences. The Hessian in the small subspace is then projected back to the full space. A line search along the new correction vector is avoided by using the constrained quartic interpolation scheme described above.

*(d) Interpolation methods*

For a quadratic surface, the gradient vector is a linear function of the coordinates. An alternative way of using information gathered during the optimization is to interpolate among the coordinate vectors obtained in the preceding cycles. The basic interpolation method is the preconditioned conjugate gradient (CG) method [20]. Although usually formulated in a different way, it is equivalent to first making a simple relaxation step using an approximate inverse Hessian $\mathbf{F}$, called the preconditioner, and replacing the calculated displacement by a linear combination of the current and all previous coordinate displacement vectors $\Delta\mathbf{q}^{(i)}$:

$$\Delta\mathbf{q}^{(n+1)} = -\mathbf{F}\mathbf{g}^{(n+1)} + \sum_{i=1}^{n} \beta_i \Delta\mathbf{q}^{(i)}.$$

The coefficients $\beta_i$ are chosen so that, on a quadratic surface, the interpolated gradient becomes orthogonal to all $\Delta\mathbf{q}^{(i)}$. This condition is equivalent to minimizing the energy in the space spanned by the displacement vectors. In the quadratic case, a further simplification can be made as it can be shown that all $\beta_i$ with the exception of $\beta_n$ vanish. The latter is given by

$$\beta_n = \mathbf{g}^{(n+1)\mathrm{T}}\mathbf{F}\mathbf{g}^{(n+1)}/(\mathbf{g}^{(n)\mathrm{T}}\Delta\mathbf{q}^{(n)})$$

although there are several forms which are equivalent for a quadratic function but not in general [6], e.g., the Polak–Ribiere form [21]. The gradient at the (interpolated) new point is not recalculated but is itself interpolated. For very large problems, the conjugate gradient method has the advantage that it needs to store only a few vectors (the preconditioner is usually diagonal or sparse, and must be positive definite).

A similar method, direct inversion in the iterative subspace (DIIS) [22, 23] tries to minimize the norm of the error vector (in most cases the gradient) by interpolating in the subspace spanned by the previous vectors. Unlike the CG method, DIIS is able to converge to saddle points. DIIS is now the standard method for the SCF optimization problem. It is also useful for geometry optimization [24]. It does not have the conjugate property and therefore requires the storage of previous coordinate and gradient vectors (in practice, usually restricted to about 20 or fewer). However, not using the CG property, which is valid for quadratic surfaces only, probably adds to the stability of the method.

To derive the DIIS equations, let us consider a linear combination of coordinate vectors $\mathbf{q}^{(1)}, \ldots, \mathbf{q}^{(n)}$, $\mathbf{q} = \sum_i^n c_i \mathbf{q}^{(i)}$. On a quadratic surface, the gradient (or any linear function of the gradient) is an analogous linear combination if the coefficients sum to unity:

$$\mathbf{g} = \sum_i^n c_i \mathbf{g}^{(i)}. \tag{B3.5.4}$$

Minimizing the square of the gradient vector under the condition $\sum_i^n c_i = 1$ yields the following linear system of equations

$$
\begin{pmatrix}
B_{11} & \ldots & B_{1n} & -1 \\
\vdots & & \vdots & \vdots \\
B_{n1} & \ldots & B_{nn} & -1 \\
-1 & \ldots & -1 & 0
\end{pmatrix}
\begin{pmatrix}
c_1 \\
\vdots \\
c_n \\
\lambda
\end{pmatrix}
=
\begin{pmatrix}
0 \\
\vdots \\
0 \\
-1
\end{pmatrix}
$$

where $B_{ij} = \mathbf{g}^{(i)\mathrm{T}}\mathbf{g}^{(j)}$. Due to the wide dynamic range of the $B_{ij}$ coefficients, it is best to normalize the diagonal elements of this equation to 1. This procedure yields an extrapolated geometry and gradient (B3.5.4). The next step is calculated by relaxing the extrapolated gradient with the approximate inverse Hessian and adding it to the extrapolated geometry. Hessian updating can be combined with DIIS, but the method works well even with a static Hessian. Any linear function of the gradient can be used instead of the gradient. This changes the weighting of the error somewhat.

### B3.5.2.4   Second-order methods

As mentioned in the introduction, full second-order methods (e.g., the Newton method) are usually not cost efficient, particularly for geometry optimization, due to the high cost of Hessian evaluation. For wavefunction optimization, the explicit evaluation of the full Hessian is not practicable due to the large number of degrees of freedom. Second-order methods can still be utilized by using direct methods, i.e., finding the solution of the Newton–Raphson equation $\mathbf{H}\delta\mathbf{q} = -\mathbf{g}$ without explicitly constructing and inverting $\mathbf{H}$. In such a case, second-order methods are competitive, and in difficult cases superior to first-order methods in the quadratic region, i.e., close to the minimum. Optimization of transition states also frequently requires the explicit evaluation of the Hessian or a submatrix of it.

### B3.5.2.5   Damping methods

Particularly in the early stages of an optimization, when the gradient is large and Hessian information is inaccurate, the computed step size may be too great; using such large steps may lead to divergence or convergence to unwanted minima. Methods which incorporate a line search are usually immune to this problem; however, as discussed above, line searches are inefficient in quantum chemistry because the evaluation of the full gradient vector can often take *less* time than the evaluation of a single energy.

The simplest way to deal with large, potentially disastrous steps is to limit the step size, either its maximum component or its norm. For geometry optimization, 0.2 to 0.3 Å or rad appears to be a reasonable value for a maximum single component; 0.3 rad is also appropriate for the maximum orbital rotation component in wavefunction optimization. A better method than simply scaling the displacement is to use a trust radius. The idea behind the trust radius is to restrict the step taken so that it lies in the local region of the energy surface where the truncation of the original power series expansion (equation (B3.5.1)) to quadratic terms only is valid. The neighbourhood about the current point where quadratic behaviour holds is called the trust region.

If the computed step size exceeds the trust radius, $t$, its direction is reoptimized under the condition that $|\Delta\mathbf{q}| = t$, i.e., the Lagrangian

$$
E(\mathbf{q}, d) = E(\mathbf{q}^0) + \Delta\mathbf{q}^\mathrm{T}\mathbf{g} + \tfrac{1}{2}\Delta\mathbf{q}^\mathrm{T}\mathbf{H}\Delta\mathbf{q} + \tfrac{1}{2}d(|\Delta\mathbf{q}|^2 - t^2)
$$

is minimized. The solution is $\Delta\mathbf{q} = -(\mathbf{H} + d\mathbf{I})^{-1}\mathbf{g}$ where the positive denominator shift $d$ is a complex function of $t$ and is usually determined iteratively from the condition $|\Delta\mathbf{q}| = t$. The trust radius can be adjusted based on the accuracy of the energy difference predicted by the quadratic model, compared with the actual energy difference [2, 6]. One problem with the trust radius method, and other methods which limit $|\Delta\mathbf{q}|$, is that they do not scale properly with the system size: they are not 'size consistent'. For instance, if

$n$ identical, non-interacting molecules are optimized simultaneously, the maximum displacement norm for each decreases like $n^{-1/2}$. For this reason, it is perhaps best to limit the maximum component, rather than the norm of the displacement [6].

An alternative, and closely related, approach is the augmented Hessian method [25]. The basic idea is to interpolate between the steepest descent method far from the minimum, and the Newton–Raphson method close to the minimum. This is done by adding to the Hessian a constant shift matrix which depends on the magnitude of the gradient. Far from the solution the gradient is large and, consequently, so is the shift $d$. One can, e.g., choose $d$ to be proportional to the expected energy lowering, $d = -\alpha^2 \mathbf{g}^\mathrm{T} \Delta \mathbf{q}$ (note that $\mathbf{g}^\mathrm{T} \Delta \mathbf{q}$ is negative for minimization and thus $d$ is positive), and solve the damped equation

$$(\mathbf{H} + d\mathbf{I})\Delta \mathbf{q} = -\mathbf{g}.$$

This is equivalent to finding the lowest eigenvalue $\lambda$ (which is always negative and approaches zero at convergence) of the generalized eigenvalue equation

$$\begin{pmatrix} \mathbf{H} & \mathbf{g} \\ \mathbf{g}^\mathrm{T} & 0 \end{pmatrix} \begin{pmatrix} \Delta \mathbf{q} \\ 1 \end{pmatrix} = \lambda \begin{pmatrix} \alpha^2 \mathbf{I} \\ \mathbf{0}^\mathrm{T} \end{pmatrix} \begin{pmatrix} \mathbf{0} & \Delta \mathbf{q} \\ 1 & 1 \end{pmatrix}. \tag{B3.5.5}$$

Equation (B3.5.5) is, in turn, equivalent to the minimum condition on the rational function

$$E(\mathbf{q}) = E(\mathbf{q}^0) + (\mathbf{g}^\dagger \Delta \mathbf{q} + \tfrac{1}{2} \Delta \mathbf{q}^\mathrm{T} \mathbf{H} \Delta \mathbf{q}) / (1 + \alpha^2 \Delta \mathbf{q}^\mathrm{T} \Delta \mathbf{q}).$$

For $\alpha = 0$, minimization of this expression yields the Newton–Raphson formula for $\Delta \mathbf{q}$. For large values of $\alpha$, $\Delta \mathbf{q}$ becomes asymptotically $\alpha^{-1} \mathbf{g}/|\mathbf{g}|$, i.e., the steepest descent formula with a step length $1/\alpha$. The augmented Hessian method is closely related to eigenvector (mode) following, discussed in section B3.5.5.2. The main difference between rational function and trust radius optimizations is that, in the latter, the level shift is applied only if the calculated step exceeds a threshold, while in the former it is imposed smoothly and is automatically reduced to zero as convergence is approached.

### B3.5.3   The optimization of wavefunctions

The basic self-consistent field (SCF) procedure, i.e., repeated diagonalization of the Fock matrix [26], can be viewed, if sufficiently converged, as local optimization with a fixed, approximate Hessian, i.e., as simple relaxation. To show this, let us consider the closed-shell case and restrict ourselves to real orbitals. The SCF orbital coefficients are not the best set of coordinates to work with because, being constrained to be orthonormal, they are not independent. A better set of parameters can be chosen as the above-diagonal elements of an antisymmetric matrix $\mathbf{K}$, used to build an orthogonal matrix $\mathbf{U}$ by $\mathbf{U} = \exp(\mathbf{K})$ [27] or, alternatively $\mathbf{U} = 2(\mathbf{I} - \tfrac{1}{2}\mathbf{K})^{-1} - \mathbf{I}$ (see, e.g., [23]). $\mathbf{U}$ describes a generalized rotation between the orbitals. We start with an orthonormal set of orbitals $\mathbf{C}_0$, defined as $\varphi^\mathrm{T} = \chi^\mathrm{T} \mathbf{C}_0$, where $\varphi^\mathrm{T}$ and $\chi^\mathrm{T}$ are row vectors of molecular orbitals and atomic basis functions, respectively. All possible orthonormal orbital sets can be expressed as $\mathbf{C} = \mathbf{C}_0 \mathbf{U}$. Not all elements of $\mathbf{K}$ are relevant. Rotations between virtual orbitals can obviously be omitted, and rotations between two occupied orbitals have no effect on the energy because the determinantal wavefunction is invariant against such rotations. Therefore, only rotations between occupied and virtual orbitals, i.e., the elements $K_{ia}$, $i \leq n$, $a > n$ are needed if, as usual, the $n$ occupied orbitals $\varphi_i$ precede the virtual ones $\varphi_a$.

The gradient and second derivative components of the SCF energy can be expressed for both kinds of parametrization (see [28]) as

$$\tfrac{1}{4} \partial E / \partial K_{ia} = F_{ia} \tag{B3.5.6}$$

$$\tfrac{1}{4} \partial^2 E / \partial K_{ia} \partial K_{jb} = F_{ab} \delta_{ij} - F_{ij} \delta_{ab} + 4(ia|jb) - (ij|ab) - (ib|ja) \tag{B3.5.7}$$

where, e.g., $(ij|ab)$ is a two-electron integral in the usual Mulliken notation. In a typical SCF iteration near convergence, the Fock matrix is nearly diagonal, and the orbital rotation parameter corresponding to a small occupied-virtual (Brillouin-violating) element $F_{ia}$ is, from first-order perturbation theory, $K_{ia} = F_{ia}/(\varepsilon_a - \varepsilon_i)$. Comparing this with (B3.5.7) and noting that in a canonical orbital basis $F_{ii} = \varepsilon_i$, $F_{aa} = \varepsilon_a$ and the off-diagonal elements of **F** are zero, it is clear that the ordinary SCF iteration is equivalent to neglecting the two-electron integrals in the electronic Hessian, equation (B3.5.7). This explains the observation that straight SCF iteration frequently slows down as the SCF procedure progresses, cf. B3.5.2.3.

For ordinary SCF problems, interpolation methods are particularly suitable, as they require the storage of only a limited amount of information. The standard method for closed-shell or simple open-shell problems is DIIS (direct inversion in the iterative subspace) [22, 23]. Equation (B3.5.6) is not appropriate for the error vector because each error vector is expressed in a different basis. Transforming the occupied-virtual block of the orbital Fock matrix to a common basis, e.g., to the atomic orbital basis, yields the commutator **SDF** − **FDS**, arranged as a vector, for the gradient [22, 23]. DIIS usually converges well for closed shell systems from a reasonable starting wavefunction. Several modifications of this method have been proposed [29, 30]. For unrestricted (UHF) wavefunctions, DIIS is widely used but it is less appropriate. As it is a gradient norm minimization technique, it has a tendency to converge to the closest stationary point. For an even number of electrons, the closed-shell wavefunction is a formal solution of the UHF equations, and DIIS, unless started close to the expected minimum, may converge uphill to the closed-shell solution. The preconditioned conjugate gradient method (the preconditioner being the SCF approximation to the Hessian) is probably more appropriate in this case. Similarly, in density functional calculations with a plane wave basis set, the basis set is often huge, and the conjugate gradient method, with its limited storage requirement, is preferable [31, 32].

Due to the large number of variables in wavefunction optimization problems, it may appear that full second-order methods are impractical. For example, the storage of the Hessian for a modest closed-shell wavefunction with 500 basis functions and 200 electrons requires more than $(100 \times 400)^2/2 = 8 \times 10^8$ words. However, as shown by Bacskay [28] for closed- and open-shell SCF, and Lengsfield and Liu for MC-SCF [33], using techniques analogous to direct configuration interaction [34], the solution of the linear system of equations (B3.5.3) can be accomplished iteratively, each micro-iteration (to be distinguished from the SCF iterations, which are called macro-iterations) taking about the same effort as an SCF cycle. For closed- and open-shell SCF, the resulting doubly iterative algorithm is comparable in efficiency with DIIS [35] but it is more complex, and thus less widely used.

The situation is different for the multi-configurational SCF (MC-SCF) case. Although DIIS has been used successfully for simpler cases [36], the strong coupling between orbital rotations and configuration interaction (CI) coefficients mandates the use of second-order or approximate second-order methods (see the reviews [3, 4] and references therein). As the signature of the Hessian is frequently incorrect, the augmented Hessian (rational function) method, which forces a step in the right direction, is generally employed. Perhaps the most efficient method is that proposed by Werner and Meyer [37] and further expanded by Werner and Knowles [38]. In this method, an approximate MC-SCF energy expression is defined, which is accurate to second order in terms of the orbital coefficients **C** and not the unitary parameters **K**. Through this change of variables, the effect of orthonormality and the periodicity of the orbitals as functions of the orbital rotations are taken into account correctly, resulting in a large increase of the radius of convergence of the Newton–Raphson method.

## B3.5.4   Optimization of molecular geometries

This section discusses techniques specific to the optimization of molecular geometries.

There are four main factors that influence the rate of convergence of molecular structure optimizations: (1) the initial guess geometry; (2) the optimization algorithm; (3) the quality of the Hessian matrix and (4) the coordinate system. The first of these is obvious; the closer the starting geometry is to the final converged geometry, the fewer optimization cycles it should take to get there. Optimization algorithms will not be discussed here, as with a reasonable starting geometry and Hessian, most standard methods (see section B3.5.2) perform well. The choice of algorithm is, however, much more crucial for transition states, and one method, the Eigenvector Following algorithm, will be described in the section dealing with transition state optimization. The third point can also be dealt with briefly. Most current optimization algorithms use approximate second-derivative (Hessian) information with updating to help predict the next step. Assuming that the surface can be adequately modelled by a quadratic function, the more reliable the initial Hessian information and the updating is, the better will be the predicted step and the fewer cycles it should take to converge. Lower-level calculations, force fields and simple universal force constant [39–41] formulae can be employed to generate the initial Hessian. The fourth factor, the coordinates used to carry out the optimization, is now recognized as being vitally important and it is the choice of coordinates that is largely responsible for the efficiency of modern geometry optimization algorithms.

### B3.5.4.1   The coordinate system

As noted above, the coordinate system is now recognized as being of fundamental importance for efficient geometry optimization; indeed, most of the major advances in this area in the last ten years or so have been due to a better choice of coordinates. This topic is seldom discussed in the mathematical literature, as it is in general not possible to choose simple and efficient new coordinates for an abstract optimization problem. A nonlinear molecule with $N$ atoms and no symmetry needs $3N - 6$ internal coordinates to specify its geometry. Unless symmetry or other constraints fix the values of some coordinates, it is not possible to omit coordinates. However, it is possible, and sometimes useful to use more. The coordinates are then not independent, a situation called redundancy.

The two key factors that make a good set of coordinates are to minimize the degree of coupling between them, and make the potential energy surface more quadratic. In general, the less coupling the better, as variation of one particular coordinate will then have minimal impact on the other coordinates. Coupling manifests itself primarily as relatively large mixed partial derivative terms between different coordinates. For example, a strong harmonic coupling between two different coordinates, $i$ and $j$, results in a large off-diagonal element, $H_{ij}$, in the Hessian matrix. Cubic and higher-order couplings are even more deleterious for optimization, as they cannot be eliminated by a linear transformation.

Cartesian coordinates are an obvious choice as they can be defined for all systems, and gradients and second derivatives are calculated directly in Cartesians. Unfortunately, they normally make a poor coordinate set for optimization as they are fairly heavily coupled, their only advantage being their simplicity and completely general nature. If the quadratic model holds, i.e., for very small displacements, and if the gradient and Hessian are properly transformed, Cartesians are equivalent to any other coordinate set [42]. Of course, the source of good Hessian data is often a force field expressed in valence coordinates, so the latter are used implicitly. A further minor inconvenience of Cartesians is that the Hessian is singular, due to the presence of translational degrees of freedom (interestingly, rotations cause singularity of the force constant matrix—and zero vibrational frequencies—only at stationary points, a problem first discussed in [12] and rediscovered many times since). Cartesian optimization is used almost exclusively in molecular mechanics, despite its inefficiency, as it requires no transformation of the coordinates and derivatives. While the computational effort required by these transformations is negligible in *ab initio* work, it becomes significant in force field methods with energy and gradient computation being so rapid. Recent work promises to improve the efficiency of methods using valence internal coordinates [43–45].

Z-matrix coordinates are widely used to define molecular geometries. A Z matrix specifies the molecular geometry in a treelike manner, by connecting each new atom in the system to those that have been defined previously. The first three atoms in the Z matrix are unique, with the first atom at the origin, the second lying on the Z axis (connected to the first by a single stretch) and the third lying in the $XZ$ plane (connected to either

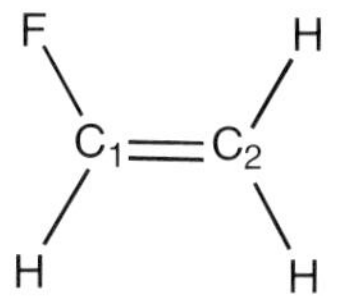

| C1 | | | | | | |
|----|----|----|----|----|----|----|
| C2 | C1 | L1 | | | | |
| F | C1 | L2 | C2 | A1 | | |
| H | C1 | L3 | C2 | A2 | F | 180.0 |
| H | C2 | L4 | C1 | A3 | F | 180.0 |
| H | C2 | L5 | C1 | A4 | F | 0.0 |
| L1 | | 1.3 | | | | |
| L2 | | 1.365 | | | | |
| L3 | | 1.08 | | | | |
| L4 | | 1.08 | | | | |
| L5 | | 1.08 | | | | |
| A1 | | 122.5 | | | | |
| A2 | | 118.0 | | | | |
| A3 | | 120.0 | | | | |
| A4 | | 120.0 | | | | |

**Figure B3.5.2.** Example $Z$ matrix for fluoroethylene. Notation: for example, line 4 of the $Z$ matrix means that a H atom is bonded to carbon atom C1 with bond length L3 (ångströms), making an angle with carbon atom C2 of A3 (degrees) and a dihedral angle with the fluorine atom of 180.0°. All parameters given lettered variable names (L1, A1 etc) will be optimized; the dihedral angles are given explicitly as these are fixed by symmetry (the molecule is planar). Simple constraints can be imposed by removing parameters from the optimization list.

the first or second atom via a stretch and defining a bend with the unconnected atom). Each new atom after the third is defined with respect to atoms previously defined in the $Z$ matrix using, for example, one stretch, one bend and one torsion. An example of a typical $Z$ matrix (for fluoroethylene) is shown in figure B3.5.2.

Initially, the $Z$ matrix was utilized simply as a means of geometry input. It was subsequently found that optimization was generally more efficient in $Z$-matrix coordinates than in Cartesians, especially for acyclic systems. This is not always the case, and care must be taken in constructing a suitable $Z$ matrix. A short discussion on good $Z$-matrix construction strategy is given by Schlegel [39].

The first *ab initio* gradient geometry optimizations were performed in what are now called natural internal coordinates [46], although they were formally defined only later [47]. These coordinates are derived from vibrational spectroscopy and are appropriate for covalent (mainly organic) molecules. They include all individual bond stretching coordinates, but only non-redundant linear combinations of bond angles and torsions as deformational coordinates. Suitable linear combinations of bends and torsions (the two are considered separately) are selected using group theoretical arguments based on approximate local symmetry. The major advantage of natural internal coordinates in geometry optimization is that they significantly reduce the coupling, both harmonic *and* anharmonic, between the various coordinates. Compared to natural internals, $Z$-matrix coordinates arbitrarily omit some angles and torsions—to prevent redundancy—and this can induce strong anharmonic coupling between the coordinates, especially with a poorly constructed $Z$ matrix. Successful minimizations can be carried out in natural internals with only an approximate (e.g., diagonal) Hessian provided at the starting geometry but a good starting Hessian is still needed for a transition state search. Using a suitable set of internal coordinates can reduce the number of cycles required to converge compared to the corresponding Cartesian optimization by an order of magnitude or more, depending on the system.

Despite their clear advantages, natural internals have only become popular relatively recently, principally because in early programs they had to be user defined, a tedious and error-prone procedure for large molecules. This situation changed with the development of algorithms capable of generating natural internals automatically from input Cartesians [48, 49]. For minimization, natural internals and their successors have become the coordinates of choice [50, 51].

However, there are some disadvantages to natural internal coordinates. Their automatic construction proceeds by an exhaustive topological analysis involving thousands of lines of code and, for molecules with a complex structure, e.g., multiply fused rings and cages, the algorithm may be unable to generate a suitable non-redundant set of coordinates. Additionally, *more* coordinates than the $3N - 6$ (where $N$ is the number of atoms) required may be generated. The redundancies can be removed by eliminating some coordinates, but this is arbitrary and may negatively influence convergence.

Various methods have been suggested for dealing with redundant coordinates. The normal coordinate optimization method [52] can use a force field defined in redundant coordinates, but is restricted to rectilinear coordinates. A general force field, expressed in redundant coordinates, can be transformed to a non-redundant set [13]. The current method of choice is to carry out the optimization directly in the redundant coordinate space [53] using the concept of a generalized inverse. If the total number of internal coordinates, including redundancies, is $n > 3N - 6$, then one constructs and diagonalizes the $n \times n$ matrix $\mathbf{G} = \mathbf{BB}^{\mathrm{T}}$ where $\mathbf{B}$ is the first-order transformation matrix from Cartesians to internal coordinates, $\Delta\mathbf{q} = \mathbf{B}\Delta\mathbf{x}$. Diagonalization of $\mathbf{G}$ results in two sets of eigenvectors; a set of $m = 3N - 6$ eigenvectors with eigenvalues $\lambda > 0$, and a set of $n - m$ eigenvectors with eigenvalues $\lambda = 0$ (to numerical precision). The eigenvalue equation for $\mathbf{G}$ can be written

$$\mathbf{G(UR)} = \mathbf{(UR)} \begin{pmatrix} \Lambda & \mathbf{0} \\ \mathbf{0} & \mathbf{0} \end{pmatrix} \qquad \text{(B3.5.8)}$$

and the generalized inverse of $\mathbf{G}$, $\mathbf{G}^-$, involves inverting the non-zero eigenvalues only and back-transforming

$$\mathbf{G}^- = \mathbf{U}\Lambda^-\mathbf{U}^{\mathrm{T}}.$$

In this way the optimization can be cast in terms of the original coordinate set, including the redundancies. Exactly the same transformations between Cartesian and internal coordinate quantities hold as for the non-redundant case (see the next section), but with the generalized inverse replacing the regular inverse.

The redundant optimization scheme [53] can be applied to natural internal coordinates, which are sometimes redundant for polycyclic and cage compounds. It can also be applied directly to the underlying primitives. This has the disadvantage that the coordinate space is larger, and contains many redundancies, but it is simpler to implement than a full natural internal coordinate scheme and can handle essentially any molecule, regardless of the topology, thus avoiding any failure in the generating algorithm. The well known Gaussian *ab initio* program [19] now uses, as a default, this type of algorithm [54].

As originally implemented, the redundant optimization scheme involved solution of equation (B3.5.8) at the beginning of *every* optimization cycle, which may be expensive in semiempirical or force field methods. A scheme which involves a *single* diagonalization was introduced by Baker *et al* [55]. Diagonalization of the $\mathbf{G}$ matrix partitions the original coordinate space into two subspaces, a redundant subspace spanned by the set of vectors in $\mathbf{R}$ and a non-redundant subspace spanned by $\mathbf{U}$. Since $\mathbf{R}$ is redundant, it can be discarded and the set of vectors in $\mathbf{U}$ defines a complete, non-redundant coordinate set which can be retained throughout the entire optimization. Unlike natural internals, which are linear combinations of just a few of the primitives localized in small regions of the molecule, each vector in $\mathbf{U}$ is potentially a linear combination of *all* of the primitives and is delocalized over the entire molecule; they are known as *delocalized internal coordinates* [55]. Despite their apparent complexity, delocalized internals perform in practice as well as natural internals.

We present in table B3.5.1 a comparison of Cartesian, Z-matrix and delocalized internal coordinate optimizations, using the semiempirical PM3 method [56] on ten typical medium-sized organic molecules.

**Table B3.5.1.** Number of cycles to converge for geometry optimizations of some typical organic molecules using Cartesian, Z-matrix and delocalized internal coordinates[a].

| Molecule | Formula | Symmetry | Cycles to converge | | | |
|---|---|---|---|---|---|---|
| | | | CART[b] | ZMAT[b] | INT[b] | INT[c] |
| azulene | $C_{10}H_8$ | $C_{2v}$ | 24 | 13 | 16 | 10 |
| lumazine | $C_6H_4N_4O_2$ | $C_s$ | 26 | 18 | 14 | 8 |
| dichloropropane (*gauche*) | $C_3H_6Cl_2$ | $C_1$ | 48 | 24 | 33 | 7 |
| 3,3,3-trifluoro-2-methyl propene | $C_4H_5F_3$ | $C_s$ | 24 | 10 | 13 | 7 |
| cyanomethyl methyl ether | $C_3H_5NO$ | $C_1$ | 45 | 25 | 33 | 9 |
| salicylic acid | $C_7H_6O_3$ | $C_s$ | 32 | 45[d] | 13 | 10 |
| isoxanthopterin | $C_6H_5N_5O_2$ | $C_s$ | 36 | 18 | 12 | 9 |
| pyrroloquinoline quinone anion (3−) | $C_{14}H_3N_2O_8$ | $C_1$ | 167[e] | F[f] | 109 | 36 |
| 2,5-bis-(4-aminophenyl)-1,3,4-oxadiazol | $C_{14}H_{12}N_4O$ | $C_{2v}$ | 27 | F[f] | 14 | 7 |
| permethyl-nonasilane (*gauche* conformer) | $Si_9(CH_3)_{20}$ | $C_1$ | 355[e] | 113[g] | 213 | 47 |

[a] Calculations using the semiempirical PM3 method with standard convergence criteria of 0.0003 au on the maximum component of the gradient vector and *either* an energy change from the previous cycle of $<10^{-6}$ hartree *or* a maximum predicted displacement for the next step of $<0.0003$ au.

[b] Started with a unit Hessian matrix.

[c] Started with a Hessian diagonal in the space of *primitive* internals using the recipe of Schlegel [39].

[d] Poor Z matrix.

[e] Converged prematurely with too high energy due to small energy changes between steps.

[f] Z matrix generated using Cartesian → Z-matrix conversion program. Severe converge problems with energy oscillation; halted after 90 cycles with energy higher than at starting geometry.

[g] Acyclic system; good Z matrix.

All optimizations were started with a unit Hessian and used the EF algorithm (see later) [57] with a BFGS Hessian update [6] to compute the optimization step; the only difference is in the coordinate system. The final column shows our best results using an initial non-unit Hessian matrix, diagonal in the space of *primitive* internals, with diagonal force constants estimated using the recipe of Schlegel [39]. The results, in terms of the number of cycles required for convergence, clearly show the advantages of using a good set of coordinates combined with a reliable estimate of the corresponding force constants.

### B3.5.4.2  Transformation between coordinate systems

This section deals with the transformation of coordinates and forces [11, 47] between different coordinate systems. In particular, we will consider the transformation between Cartesian coordinates, in which the geometry is ultimately specified and the forces are calculated, and internal coordinates which allow efficient optimization.

*(a) Transformation of first and second derivatives*

Let us consider the energy expanded through second order in two sets of displacement coordinates $\Delta\mathbf{x}$ and $\Delta\mathbf{q}$. The two coordinate systems are related by

$$\Delta q_i = \sum_a B_{ia}\Delta x_a + \tfrac{1}{2}\sum_{a,b} C^i_{ab}\Delta x_a \Delta x_b + \cdots. \tag{B3.5.9}$$

The potential energy is given as

$$E = E_0 + \sum_i \Delta q_i g_i + \tfrac{1}{2}\sum_{i,j} H_{ij}\Delta q_i \Delta q_j + \cdots = E_0 + \sum_a \Delta x_a f_a + \tfrac{1}{2}\sum_{a,b} K_{ab}\Delta x_a \Delta x_b + \cdots \tag{B3.5.10}$$

or, in matrix notation,

$$E = E_0 + \Delta\mathbf{q}^{\mathrm{T}}\mathbf{g} + \tfrac{1}{2}\Delta\mathbf{q}^{\mathrm{T}}\mathbf{H}\Delta\mathbf{q} + \cdots = E_0 + \Delta\mathbf{x}^{\mathrm{T}}\mathbf{f} + \tfrac{1}{2}\Delta\mathbf{x}^{\mathrm{T}}\mathbf{K}\Delta\mathbf{x} + \cdots$$

where $\mathbf{g}$ and $\mathbf{H}$ are the gradient and the force constant matrix, respectively, in internal coordinates, and $\mathbf{f}$ and $\mathbf{K}$ are the same in Cartesians. Substituting (B3.5.9) into equation (B3.5.10) and equating equal powers one obtains

$$\mathbf{f} = \mathbf{B}^{\mathrm{T}}\mathbf{g}$$

$$\mathbf{K} = \mathbf{B}^{\mathrm{T}}\mathbf{H}\mathbf{B} + \sum_i \mathbf{C}^i g_i.$$

If the two coordinate systems are connected by a non-singular transformation then, defining $\mathbf{A} = (\mathbf{B}^{\mathrm{T}})^{-1}$, the more important inverse transformations are given by

$$\mathbf{g} = \mathbf{A}\mathbf{f}$$

$$\mathbf{H} = \mathbf{A}\mathbf{K}\mathbf{A}^{\mathrm{T}} - \sum_i g_i \mathbf{A}\mathbf{C}^i \mathbf{A}^{\mathrm{T}}. \tag{B3.5.11}$$

Thus the transformation matrix for the gradient is the inverse transpose of that for the coordinates. In the case of transformation from Cartesian displacement coordinates ($\Delta\mathbf{x}$) to internal coordinates ($\Delta\mathbf{q}$), the transformation is singular because the internal coordinates do not specify the six translational and rotational degrees of freedom. One could augment the internal coordinate set by the latter but a simpler approach is to use the generalized inverse [58]

$$\mathbf{A} = (\mathbf{B}\mathbf{M}\mathbf{B}^{\mathrm{T}})^{-1}\mathbf{B}\mathbf{M}$$

where $\mathbf{M}$ is any non-singular $3N \times 3N$ matrix, in the simplest case the $3N$-dimensional unit matrix.

The second term in equation (B3.5.11) deserves comment. This term shows that Hessian (second-derivative) matrices in different coordinate systems are not related simply by a similarity transformation, except at stationary points. In particular, its signature (number of negative eigenvalues) does not have to be the same in different coordinate systems. Properly chosen internal coordinates tend to make the Hessian positive definite, and are probably one of the reasons why internal coordinates are preferable for molecular geometry optimization.

*(b) Transformation of molecular geometries*

When working with any coordinate system other than Cartesians, it is necessary to transform finite displacements between Cartesian and internal coordinates. Transformation from Cartesians to internals is seldom a problem as the latter are usually geometrically defined. However, to transform a geometry displacement from internal coordinates to Cartesians usually requires the solution of a system of coupled nonlinear equations. These can be solved by iterating the first-order step [47]

$$\Delta \mathbf{x} = \mathbf{A}^{\mathrm{T}} \Delta \mathbf{q}$$

where $\Delta \mathbf{q}$ is the difference between the current internal coordinates and their desired values, calculated to full accuracy from the Cartesians. If $\Delta \mathbf{q}$ is large, it may be better to proceed in stages, converging $\Delta \mathbf{x}$ roughly between each stage.

### B3.5.4.3  Constrained optimization

Constrained optimization refers to optimizations in which one or more variables (usually some internal parameter such as a bond distance or angle) are kept fixed. The best way to deal with constraints is by elimination, i.e., simply remove the constrained variable from the optimization space. Internal constraints have typically been handled in quantum chemistry by using $Z$ matrices; if a $Z$ matrix can be constructed which contains all the desired constraints as individual $Z$-matrix variables, then it is straightforward to carry out a constrained optimization by elimination.

The situation is more complicated in molecular mechanics optimizations, which use Cartesian coordinates. Internal constraints are now relatively complicated, nonlinear functions of the coordinates, e.g., a distance constraint between atoms $i$ and $j$ in the system is $R_{ij} = \sqrt{\{(x_i - x_j)^2 + (y_i - y_j)^2 + (z_i - z_j)^2\}} = R_0$, and this cannot be handled by simple elimination. There are two main approaches if elimination is not possible, penalty functions and Lagrange multipliers.

The general constrained optimization problem can be considered as minimizing a function of $n$ variables $F(\mathbf{x})$, subject to a series of $m$ constraints of the form $C_i(\mathbf{x}) = 0$. In the penalty function method, additional terms of the form $\frac{1}{2}\sigma_i C_i(\mathbf{x})^2$, $\sigma_i > 0$, are formally added to the original function, thus

$$F(\mathbf{x})^{\mathrm{pen}} = F(\mathbf{x}) + \sum \tfrac{1}{2}\sigma_i C_i(\mathbf{x})^2$$

with the summation over all $m$ constraints. If the constraint is satisfied, the additional term is zero, but if not then the value of the function increases in proportion to the square of the deviation, i.e., the additional term penalizes any geometries that do not satisfy the constraint. In practice, the value of the function is left unaltered and what is done is to modify the gradient according to

$$\partial F(\mathbf{x})^{\mathrm{pen}}/\partial x_j = \partial F(\mathbf{x})/\partial x_j + \sum \sigma_i \partial C_i(\mathbf{x})/\partial x_j.$$

Exactly the same types of step as for an unconstrained optimization can then be taken, using the modified as opposed to the regular gradient.

The performance of the penalty function algorithm is heavily influenced by the value chosen for $\sigma_i$. The larger the value of $\sigma_i$ the better the constraints are satisfied but the slower the rate of convergence. Optimizations with very high values of $\sigma_i$ encounter severe convergence problems. However, the method is very general and easy to apply.

A better approach is the method of Lagrange multipliers. This introduces the Lagrangian function [59]

$$L(\mathbf{x}, \lambda) = F(\mathbf{x}) - \sum \lambda_i C_i(\mathbf{x})$$

which replaces the function $F(\mathbf{x})$ in the unconstrained case. Here the $\lambda_i$ are the so-called Lagrange (or unknown) multipliers, one for each constraint. Differentiating with respect to $\mathbf{x}$ and $\lambda$ gives

$$\partial L(\mathbf{x}, \lambda)/\partial x_j = \partial F(\mathbf{x})/\partial x_j - \sum \lambda_i \partial C_i(\mathbf{x})/\partial x_j$$

and

$$\partial L(\mathbf{x}, \lambda)/\partial \lambda_i = -C_i(\mathbf{x}).$$

At a stationary point of the Lagrangian function, we have $\nabla \mathbf{L} = \mathbf{0}$, i.e., all $\partial L/\partial x_j = 0$ and all $\partial L/\partial \lambda_i = 0$. This latter condition means that all $C_i(\mathbf{x}) = 0$ and so all constraints are satisfied. Hence finding a set of values $(\mathbf{x}, \lambda)$ for which $\nabla \mathbf{L} = \mathbf{0}$ gives a solution to the constrained optimization problem in exactly the same way as finding an $\mathbf{x}$ for which $\nabla \mathbf{F} = \mathbf{0}$ gives a solution to the corresponding unconstrained problem.

A major difference between the penalty function and Lagrange multiplier methods is that in the latter the unknown multipliers are part of the optimization space and are treated essentially as additional variables. The Lagrange multiplier method usually converges significantly faster than the penalty function method and has the further advantage that constraints are satisfied essentially exactly. Note that, in both methods, constraints do not need to be satisfied in the starting geometry, but are instead satisfied at convergence. An efficient algorithm (within the context of Cartesian optimization) for imposing constraints in Cartesian coordinates, which incorporates both penalty functions and Lagrange multipliers, was presented by Baker in 1992 [60], with further improvements in the following year [61].

By combining the Lagrange multiplier method with the highly efficient delocalized internal coordinates, a very powerful algorithm for constrained optimization has been developed [62]. Given that delocalized internal coordinates are potentially linear combinations of *all* possible primitive stretches, bends and torsions in the system, cf. Z-matrix coordinates which are *individual* primitives, it would seem very difficult to impose any constraints at all; however, as shown in the original reference [55], any desired internal constraint can be imposed using a relatively simple Schmidt orthogonalization procedure. By projecting unit vectors (with unit components corresponding to particular primitives) onto the non-redundant subspace $\mathbf{U}$ (see equation (B3.5.8)) and Schmidt orthogonalizing the resultant vectors against all other vectors in $\mathbf{U}$, it is possible to isolate individual primitives in a consistent manner into *single* vectors. By removing these vectors from the optimization space, optimizations can be carried out in which the primitives involved retain their initial values throughout the optimization. The resulting algorithm has all the advantages of redundant internal coordinate optimizations in terms of efficiency, combined with the advantage of the Lagrange multiplier method that desired constraints do not have to be satisfied in the starting geometry. As constraints do become satisfied, they can simply be eliminated from the optimization space (this cannot be done with Cartesian optimizations due to the form of the constraints). By starting with an appropriate constraint vector, it is possible to impose constraints on linear combinations of variables rather than on individual primitives. It is also possible to perform constrained transition state searches. For more details see [55] and [62].

There are alternative methods for imposing constraints, but they are less satisfactory than those discussed above. One commonly used alternative is to use projection techniques [12, 63]. In this approach, components in directions that would result in motion that would violate the constraints are projected out of the gradient vector and Hessian matrix before calculating the next step. Unfortunately, while projection works fine for constraints that are linear in the coordinates (the standard method for imposing the Eckart conditions is by projection), nonlinear constraints have to be linearized, and consequently 'feasibility corrections' [63] must be applied to prevent deviations from the desired constraints increasing as the optimization progresses. The interesting method of Taylor and Simons [64] for combining linearized geometrical constraints with mode following also suffers from this drawback.

Another way of attempting to constrain variables without eliminating them from the optimization space is to set the appropriate force constants to very large values. This is what is currently done in Schlegel's redundant internal coordinate algorithm to prevent motion in the redundant subspace [54]. Perhaps a better method is to set the corresponding rows and columns in the inverse Hessian to zero. This method must begin with a geometry that satisfies the constraints.

## B3.5.5  Optimization of transition states

Searching for transition states poses additional difficulties compared to minimization. The first problem is the starting geometry. There is a host of structural information that can be called upon to provide a good estimate of the likely geometry of a local minimum. Far less knowledge is available about transition state geometries. Second is the structure of the Hessian. In the region of a transition structure, the Hessian must have one, and only one, negative eigenvalue. Unlike the situation for a minimum search, there is no simple and cheap method for guessing a reasonable starting Hessian with the appropriate eigenvalue structure. Even if you calculate an exact initial Hessian, if your starting geometry is poor, its eigenstructure will probably be inappropriate. One thing that is often done for transition states is to calculate a few rows and columns of the Hessian—those corresponding to variables in the 'active site' where most of the geometrical changes are expected—by finite difference on the gradient, and guess diagonal Hessian matrix elements for the rest of the system in the same way one would do for a minimum search. The starting Hessian will then hopefully have an appropriate negative eigenvalue.

In addition to the problem of generating a starting Hessian with the correct signature, there are problems in retaining it. The Hessian updates commonly used for minimization generally retain positive definiteness (see section B3.5.2.3); there are no such guarantees for retaining a negative eigenvalue during a transition state search. Once the desired region of the potential energy surface (PES) has been reached, quasi-Newton techniques can be used to refine the geometry; however they must be able to correct for the occasional bad update which may destroy the Hessian eigenstructure. Transition state searches can thus be separated into two parts; first find the correct region on the PES and then home in onto the transition state. Many of the methods described below for locating approximate transition states have the advantage that they require no second-derivative information.

### B3.5.5.1  Locating the correct neighbourhood

#### (a) Coordinate driving

A commonly used approach is coordinate driving. Here an appropriate internal coordinate, or a linear combination of coordinates, is chosen as a reaction coordinate. At various intervals along this coordinate, between its value in the reactants and in the products, all the other variables are optimized. This then defines a minimum energy path. The energy maximum on this path can be shown to be the transition state geometry. Usually, however, the maximum on the path is located only approximately. Coordinate driving involves several minimizations in $(n-1)$ variables; consequently it is quite expensive. Moreover, its success depends on a good definition of the reaction coordinate; it should be roughly parallel with the true reaction path. If, at any point along the path, the reaction coordinate becomes nearly perpendicular to the reaction path, the latter may become discontinuous. The minimum energy path defined in this way has little physical significance, as different choices of reaction coordinate can produce different pathways.

#### (b) Synchronous transit

Another approach requiring less intuition is the synchronous transit method [65]. Here the path between reactants and products is interpolated linearly between the reactant and the product. The interpolation can be carried out in Cartesians, internal coordinates or, perhaps best, in terms of distance coordinates [66]; the results depend somewhat on the interpolation method. A maximum is first found along this linear synchronous transit path. This is followed by alternate minimization along directions *orthogonal* to the original direction, combined with maximum searches along a parabolic path (the quadratic synchronous transit) joining the reactant, the product and the current estimate of the transition state. For very curved reaction paths the quadratic synchronous transit path may be a poor approximation, and the reaction path may have to be approximated piecewise. A similar algorithm, which involves minimization in a space *conjugate* to the maximum search direction, was developed by Bell and Crighton [67]. This last reference also contains a good discussion of various transition-state search strategies.

Both of the above methods can be considered as approximations to the Fukui reaction path, discussed later in this article. The maximum of the Fukui reaction path also yields the transition state, although usually at significantly more expense than coordinate driving or the synchronous transit method. A more modern development of these ideas has been given by Ionova and Carter [68].

### (c) Walking uphill

These algorithms try to walk up to transition states from minima, usually along the shallowest path, i.e., along the eigenvector of the Hessian which has the lowest eigenvalue [69–71]. They are more important when the transition state has already been located approximately and will be discussed in the next section.

### (d) Bracketing the transition state

These methods try to bracket the transition state from both the reactant and the product side [72, 73]. For example, in the method of Dewar *et al* [73], two structures, one in the reactant valley and one in the product valley, are optimized simultaneously. The lower-energy structure is moved to reduce the distance separating the two structures by a small amount, e.g. by 10%, and its structure is reoptimized under the constraint that the distance is fixed. This process is repeated until the distance between the two structures is sufficiently small.

### (e) Seam crossing

Here the transition state is approximated by the lowest crossing point on the seam intersecting the *diabatic* (non-interacting) potential energy surfaces of the reactant and product. The method was originally developed for excited state surfaces [74], and has subsequently been used to locate approximate transition states [75, 76].

### B3.5.5.2  Refining the transition state

It is usually not efficient to use the methods described above to refine the transition state to full accuracy. Starting from a qualitatively correct region on the potential surface, in particular one where the Hessian has the right signature, efficient gradient optimization techniques, with minor modifications, are usually able to zero in on the transition state quickly.

### (a) Direct inversion in the iterative subspace

One of the methods which is appropriate is DIIS applied to geometry optimization [24]; being a gradient norm minimization method, it will converge to any stationary point. This is, however, one of its problems as it may converge to the wrong point. The convergence radius of augmented Hessian type methods is larger. One of these methods, the eigenvector following (EF) method [57], is generally useful for both transition states and minima and will be described here as an example of a modern optimization algorithm.

### (b) Eigenvector following

The EF algorithm [57] is based on the work of Cerjan and Miller [69] and, in particular, Simons and coworkers [70, 71]. It is closely related to the augmented Hessian (rational function) approach [25]. We have seen in section B3.5.2.5 that this is equivalent to adding a constant level shift (damping factor) to the diagonal elements of the approximate Hessian $\mathbf{H}$. An appropriate level shift effectively makes the Hessian positive definite, suitable for minimization.

Although a single shift parameter can also be used to find a transition state, the eigenvector following algorithm utilizes *two* level shifts: one for the Hessian (transition state) mode along which the energy is to be maximized and the other for modes for which it is minimized. In terms of a diagonal Hessian representation, transforming the gradient appropriately, we have the two eigenvalue equations

$$
\begin{pmatrix} b_2 & \cdots & 0 & g_2' \\ & \ddots & & \vdots \\ 0 & & b_n & g_n' \\ g_2' & & g_n' & 0 \end{pmatrix} \begin{pmatrix} \Delta q_2 \\ \vdots \\ \Delta q_n \\ 1 \end{pmatrix} = \lambda_n \begin{pmatrix} \Delta q_2 \\ \vdots \\ \Delta q_n \\ 1 \end{pmatrix}
\tag{B3.5.12}
$$

and

$$
\begin{pmatrix} b_1 & g_1' \\ g_1' & 0 \end{pmatrix} \begin{pmatrix} \Delta q_1 \\ 1 \end{pmatrix} = \lambda_p \begin{pmatrix} \Delta q_1 \\ 1 \end{pmatrix}.
\tag{B3.5.13}
$$

In these two equations, the $b_i$ are the eigenvalues of $\mathbf{H}$($b_1 < b_2 < \cdots < b_n$), $\mathbf{g}'$ is the gradient vector transformed to the basis of the eigenvectors $\mathbf{U}$ of $\mathbf{H}$: $\mathbf{g}' = \mathbf{U}^T\mathbf{g}$, and we have (arbitrarily) set the factor $\alpha$ in equation (B3.5.5) to unity. Note that $\lambda_n$ is the lowest eigenvalue of equation (B3.5.12) (it is always negative and approaches zero at convergence), while $\lambda_p$ is the highest eigenvalue of equation (B3.5.13) (it is always positive and again approaches zero at convergence). Once suitable values of $\lambda_p$ and $\lambda_n$ have been determined, the final step is given by

$$
\Delta\mathbf{q} = -g_1'\mathbf{u}_1/(b_1 - \lambda_p) - \sum g_i'\mathbf{u}_i/(b_i - \lambda_n) \quad (i = 2, \ldots, n)
$$

where is it assumed that we are maximizing along the lowest Hessian mode $\mathbf{u}_1$, and minimizing along all the others. This holds *regardless* of the Hessian eigenvalue structure (unlike the Newton–Raphson step), and so the algorithm can handle Hessian matrices with the wrong signature.

It is also possible to maximize along modes other than the lowest and, in this way perhaps, locate transition states for alternative rearrangements/dissociations from the same initial starting point. For maximization along the $k$th mode (instead of the lowest), $b_1$ would be replaced by $b_k$, and the summation would now exclude the $k$th mode but include the lowest. Since what was originally the $k$th mode is the mode along which the negative eigenvalue is required, then this mode will eventually become the lowest mode at some stage of the optimization. To ensure that the original mode is being followed smoothly from one cycle to the next, the mode that is actually followed is the one with the greatest overlap with the mode followed on the previous cycle. This procedure is known as mode following. For more details and some examples, see [57]. Mode following can work well for small systems, but for larger, flexible molecules there are usually a number of soft modes which lead to transition states for conformational rearrangements and not to the more interesting reaction saddle points. Moreover, each eigenvector can be followed in two opposite directions and frequently only one leads to a reaction.

Although it was originally developed for locating transition states, the EF algorithm is also efficient for minimization and usually performs as well as or better than the standard quasi-Newton algorithm. In this case, a single shift parameter is used, and the method is essentially identical to the augmented Hessian method.

## B3.5.6 Simultaneous optimization of geometries and wavefunctions

So far, we have considered the optimization of wavefunction and geometry parameters separately. In view of the much shorter timescale and higher energy associated with the former, this is reasonable. However, additional savings can be potentially obtained by optimizing the wavefunction and the geometry simultaneously. This was first proposed for density functional methods [77] and later for traditional quantum chemistry techniques [78]. With the large increase of computing speed compared to disk input/output speed, direct techniques [79] were generally adopted. In direct methods, the large disparity between calculating the gradients of the molecular energy with respect to electronic parameters (the Fock matrix in SCF theory) and nuclear coordinates disappeared; gradients are now only a few times more expensive than a Fock matrix evaluation, making simultaneous wavefunction-geometry optimization much more attractive. In spite of this, such methods are not yet widely used, except in the crude form of relaxing the SCF convergence criteria if the geometry parameters are far from convergence.

The molecular dynamics method introduced by Car and Parrinello [80], though not strictly an optimization method, has many features in common with simultaneous optimization of the wavefunction and geometry. In this method, the electronic wavefunction and energy are close to, but not identical with, the Born–Oppenheimer energy. The basic idea is to consider the electronic degrees of freedom as dynamical variables, along with the nuclear coordinates. The Lagrangian contains the kinetic energy of the nuclei, the potential energy as a function of both the nuclear and electronic degrees of freedom and a fictitious kinetic energy term which is the square of the time derivative of the electronic wavefunction multiplied by a small mass. The inertia of this fictitious electronic mass causes the wavefunction to deviate slightly from the Born–Oppenheimer surface. The Car–Parrinello method is most efficient for plane wave basis sets, as the calculation of the nuclear gradient is very inexpensive in this method, but it has also been introduced into SCF theory [81, 82].

## B3.5.7   Reaction path algorithms

The reaction path is defined by Fukui [83] as the line $\mathbf{q}(s)$ leading down from a transition state along the steepest descent direction

$$\partial\mathbf{q}(s)/\partial s = -\mathbf{g}(s)/|\mathbf{g}(s)|. \tag{B3.5.14}$$

Here $s$ is the path length, $ds = (dq_1^2 + \cdots + dq_n^2)^{1/2}$. The reaction path is, unfortunately, dependent on the coordinate system. This should perhaps be emphasized more than is generally the case. Scaling one coordinate by a factor $\alpha > 1$ increases the coordinate value but decreases the corresponding gradient component, so that if the reaction path was antiparallel to the gradient before the scaling it will not be so after scaling. For qualitative studies of chemical reactions, there is little to recommend one particular reaction path over another. However, for dynamical studies, the intrinsic reaction coordinate (IRC) [83], defined as the path length along the reaction path in mass-weighted Cartesian coordinates, $\xi_i = m_i^{1/2}x_i$, has advantages over other definitions (for example the kinetic energy matrix is the unit matrix). Here $m_i$ is the atomic mass corresponding to the Cartesian coordinate $x_i$, making the reaction path isotope dependent. A major difficulty with reaction paths is that to decide whether a given point is on the path the whole path must be constructed; local information (energy, gradient, force constants etc.) is insufficient.

### B3.5.7.1   Following the reaction path downhill

The most widely used methods try to follow the gradient downhill starting from a transition state. At the transition state itself, the gradient vanishes and the first step must be made along the imaginary eigenvector of the Hessian in the proper coordinates, i.e., mass-weighted Cartesians for the IRC path. As pointed out by Schlegel [1, 2], (B3.5.14) is a stiff differential equation and its integration by simply making small downhill steps along the gradient, a method equivalent to Euler's method, requires very small steps and consequently much effort. Otherwise, the calculated reaction path diverges from the true one, at first slowly and then more rapidly. To deal with this problem requires either constrained minimization steps at each point on the path, or alternatively second-order (both gradient and Hessian) information. This increases the cost of the individual steps but allows much larger steps to be taken.

The method of Ishida *et al* [84] includes a minimization in the direction in which the path curves, i.e. along $(\mathbf{g}/|\mathbf{g}| - \mathbf{g}'/|\mathbf{g}'|)$, where $\mathbf{g}$ and $\mathbf{g}'$ are the gradient at the beginning and the end of an Euler step. This technique, called the stabilized Euler method, performs much better than the simple Euler method but may become numerically unstable for very small steps. Several other methods, based on higher-order integrators for differential equations, have been proposed [85, 86].

Page *et al* [87] use a local quadratic model for the surface. This requires the Hessian, but once it is available, the reaction path can be inexpensively determined for a quadratic (or even higher-order) analytical surface (see also [88]). Gonzales and Schlegel [89, 90] approximate the reaction path by an arc of a circle. They first make a step along the gradient of length half the current stepsize to an intermediate point. From this, they make another half step so that the energy is minimized, subject to the stepsize constraint. The wavefunction and the gradient need not be evaluated at the intermediate point. This method is implemented in the Gaussian series of programs [19] and is widely used. It does not need the exact Hessian, but a good estimate should be available so that the many local optimizations converge rapidly. An advantage of this method is that it yields the curvature of the reaction path at the transition state correctly.

### B3.5.7.2 *Approaching the reaction path from the side*

These methods, which probably deserve more attention than they have received to date, simultaneously optimize the positions of a number of points along the reaction path. The method of Elber and Karplus [91] was developed to find transition states. It furnishes, however, an approximation to the reaction path. In this method, a number (typically 10–20) equidistant points are chosen along an approximate reaction path connecting two stationary points $a$ and $b$, and the average of their energies is minimized under the constraint that their spacing remains equal. This is obviously a numerical quadrature of the integral $S^{-1} \int_a^b E(q(s)) \, ds$ where $S$ is the path length between the points $a$ and $b$. The Euler equation to this variation problem yields the condition for the reaction path, equation (B3.5.14). A similar method has been proposed by Stachó and Bán [92].

### B3.5.7.3 *Bifurcation of the reaction path and valley–ridge inflection points*

As shown by Valtazanos and Ruedenberg [93], steepest descent paths (e.g., the Fukui intrinsic reaction coordinate) can bifurcate, i.e., split in two, only at stationary points. Thus, the intuitive notion of a reaction path forking, e.g., upon ascent in a valley to two different transition states, or, starting down from a transition state to two different minima, is impossible. This should be regarded as an inherent limitation of the standard definition of a reaction path, not as a physical impossibility. Such cases, in which a valley floor is gradually transformed into a ridge are, in fact, quite common. Mathematically, they are characterized by fact that one of the eigenvalues of the Hessian in the subspace perpendicular to the path changes from positive to negative. The point at which the eigenvalue is zero is called the valley–ridge inflection point. The reaction path, started at a stationary point, will run directly along the ridge and thus becomes non-physical past a valley–ridge inflection point. The actual reaction, of course, will not follow the reaction path in this case, not even qualitatively. Steepest descent paths started a little away from the reaction path will veer away from the latter after passing the valley–ridge inflection point. Baker and Gill [94] have devised a method for locating valley–ridge inflection points (which they call branching points). The reader is reminded, however, that the signature of the Hessian at non-stationary points depends strongly on the coordinate system. Thus, the location of a valley–ridge inflection point may be quite different in Cartesians or mass-weighted Cartesians than in internal coordinates. In particular, the Hessian in Cartesian coordinates may have spurious negative eigenvalues corresponding to rotational coordinates.

### B3.5.8    Global optimization

For our purposes, global optimization refers to the location of the *lowest* minimum on a given potential energy surface. As mentioned in the introduction, this is currently only a partially solved problem. The number of conformational minima, e.g., for a large protein, increases enormously with the size of the system, and the only way to be *absolutely* sure that the lowest-energy structure has been found is to do an exhaustive search of the entire energy surface; for large molecules this is essentially impossible. Even if the lowest-energy structure were successfully located, this would likely have only limited chemical significance, as there would be many structures energetically close to the global minimum (within a kcal or so) which would need to be considered for an accurate treatment of the thermodynamics. It is almost a certainty (though the authors are unaware of a formal proof) that finding the global minimum on molecular potential energy surfaces is computationally NP complete, and thus scales factorially with the size of the problem. Such problems are generally regarded as insoluble (however, this does not exclude their solution in a given case).

With systematic PES searches excluded, random (stochastic) methods have become the most common techniques for global minimization. The two most popular methods are simulated annealing [95] and genetic algorithms [96]. The former method derives its name from the annealing process in condensed matter physics in which a solid is melted in a bath and the temperature is then slowly decreased; the particles are expected to settle into their lowest-energy states, provided the initial temperature is sufficiently high and the cooling rate is sufficiently low. In practical optimizations, cooling is represented by local minimizations and heating by random jumps, i.e., random displacements of some or all of the atoms. After a 'sufficient number' of local minimization/random jump cycles, the procedure is terminated with the lowest-energy structure found so far taken as the global minimum.

The genetic algorithm method takes its name from the trading of genetic information in chromosomes between parents to produce an offspring. A random population of individuals (geometrical structures for the system in question) is created, and local minimizations are performed on each individual. Selected structural components (genes) from mostly the lowest-energy individuals are allowed to exchange, producing a new set of individuals for the next round of local minimizations. After a sufficiently large number of rounds, the global minimum should be located.

Both of these global optimization methods require a very large number of essentially full local optimizations and, consequently, are normally restricted to moderate-sized systems described using mechanics force fields. A somewhat different approach has been developed by Piela and coworkers [97], utilizing the diffusion or heat conduction equation. In this method, a surface containing multiple local minima is smoothly deformed in such a way that wells on the surface gradually disappear, with shallower wells vanishing faster than deeper, lower-energy wells. Eventually a surface will be derived which has just one minimum, related to the lowest-energy, global minimum on the original surface. By carefully reversing the procedure, keeping track of the minimum as it evolves, one is (hopefully) led back to the global minimum as the original surface is reformed.

Other deterministic methods for global optimization have also been developed (see, e.g., [98]).

### References

[1]   Schlegel H B 1987 Optimization of equilibrium geometries and transition structures *Adv. Chem. Phys.* **67** 249
[2]   Schlegel H B 1995 Geometry optimization on potential energy surfaces *Modern Electronic Structure Theory* ed D Yarkony (Singapore: World Scientific) pp 459–500
[3]   Werner H-J 1987 Matrix-formulated direct multiconfigurational self-consistent field and multireference configuration interaction methods *Adv. Chem. Phys.* **69** 1
[4]   Shepard R 1987 The multiconfiguration self-consistent field method *Adv. Chem. Phys.* **69** 63
[5]   Pulay P 1987 Analytical derivative methods in quantum chemistry *Adv. Chem. Phys.* **69** 241
[6]   Fletcher R 1981 *Practical Methods of Optimization: Vol 1—Unconstrained Optimization* (New York: Wiley)

[7] Dennis J E and Schnabel R B 1983 *Numerical Methods for Unconstrained Optimization and Non-linear Equations* (Englewood Cliffs, NJ: Prentice-Hall)

[8] Miller W H 1983 Symmetry-adapted transition-state theory and a unified treatment of multiple transition states *J. Phys. Chem.* **87** 21

[9] Spendley W, Hext G R and Himsworth F R 1962 Sequential application of simplex designs in optimization and evolutionary operation *Technometrics* **4** 441

[10] Baker J 1987 An algorithm for geometry optimization without analytical gradients *J. Comput. Chem.* **8** 563

[11] Pulay P 1969 *Ab initio* calculation of force constants and equilibrium geometries in polyatomic molecules. I. Theory *Mol. Phys.* **17** 197

[12] Pulay P 1977 Direct use of the gradients for investigating molecular energy surfaces *Applications of Electronic Structure Theory* ed H F Schaefer III (New York: Plenum) p 153

[13] Schlegel H B 1982 Optimization of equilibrium geometries and transition states *J. Comput. Chem.* **3** 214

[14] Murtagh B A and Sargent R W 1970 Computational experience with quadratically convergent minimisation methods *Comput. J.* **13** 185

[15] Fletcher R and Powell M D 1963 A rapidly convergent descent method for minimization *Comput. J.* **6** 163

[16] Powell M J D 1971 Recent advances in unconstrained optimization *Math. Prog.* **1** 26

[17] Bofill J M 1994 Updated Hessian matrix and the restricted step method for locating transition structures *J. Comput. Chem.* **15** 1

[18] Baker J and Chan F 1996 The location of transition states: a comparison of Cartesian, Z-matrix, and natural internal coordinates *J. Comput. Chem.* **17** 888

[19] Frisch M J *et al* 1995 *Gaussian 94* revision C.3, Gaussian (Pittsburgh, PA)

[20] Fletcher R and Reeves C M 1964 Function minimization by conjugate gradients *Comput. J.* **7** 149

[21] Polak E 1971 *Computational Methods in Optimization: a Unified Approach* (New York: Academic)

[22] Pulay P 1980 Convergence acceleration in iterative sequences: the case of SCF iteration *Chem. Phys. Lett.* **73** 393

[23] Pulay P 1982 Improved SCF convergence acceleration *J. Comput. Chem.* **3** 556

[24] Császár P and Pulay P 1984 Geometry optimization by direct inversion in the iterative subspace *J. Mol. Struct. (Theochem)* **114** 31

[25] Lengsfield B H III 1980 General second-order MC-SCF theory: a density matrix directed algorithm *J. Chem. Phys.* **73** 382

[26] Szabo A and Ostlund N S 1982 *Modern Quantum Chemistry: Introduction to Advanced Electronic Structure Theory* (New York: Macmillan)

[27] Levy B 1969 Multi-configuration self-consistent wavefunctions for formaldehyde *Chem. Phys. Lett.* **4** 17

[28] Bacskay G B 1981 A quadratically convergent Hartree–Fock (QC-SCF) method. Applications to the closed-shell case *Chem. Phys.* **61** 385

[29] Sellers H L 1991 ADEM-DIOS, an SCF convergence algorithm for difficult cases *Chem. Phys. Lett.* **180** 461

[30] Ionova I V and Carter E A 1995 Orbital-based direct inversion in the iterative subspace for the generalized valence bond method *J. Chem. Phys.* **102** 1251

[31] Štich I, Car R, Parrinello M and Baroni S 1989 Conjugate gradient minimization of the energy functional: a new method for electronic structure calculation *Phys. Rev.* B **39** 4997

[32] Payne M C, Teter M P, Allan D C, Arias T A and Joanopoulos J D 1992 Iterative minimization techniques for *ab initio* total energy calculations: molecular dynamics and conjugate gradient *Rev. Mod. Phys.* **64** 1045

[33] Lengsfield B H III and Liu B 1981 A second-order MCSCF method for large CI expansions *J. Chem. Phys.* **75** 478

[34] Roos B 1972 A new method for large-scale CI calculations *Chem. Phys. Lett.* **15** 153

[35] Chaban G, Schmidt M W and Gordon M S 1997 Approximate second order methods for orbital optimization of SCF and MCSCF wavefunctions *Theor. Chim. Acta* **97** 88

[36] Hamilton T P and Pulay P 1986 Direct inversion in the iterative subspace (DIIS) optimization of open-shell, excited-state and small multiconfigurational SCF wavefunctions *J. Chem. Phys.* **84** 5728

[37] Werner H-J and Meyer W 1981 A quadratically convergent MCSCF method for the simultaneous optimization of several states *J. Chem. Phys.* **74** 5794

[38] Werner H-J and Knowles P 1985 A second order multiconfiguration SCF procedure with optimum convergence *J. Chem. Phys.* **82** 5053

[39] Schlegel H B 1984 Estimating the Hessian for gradient-type geometry optimizations *Theor. Chim. Acta* **66** 333

[40] Fischer T H and Almlöf J 1992 General methods for geometry and wavefunction optimization *J. Phys. Chem.* **96** 9768

[41] Lindh R, Bernhardsson A, Karlström G and Malmqvist P-Å 1995 On the use of a Hessian model function in molecular geometry optimizations *Chem. Phys. Lett.* **241** 423

[42] Baker J and Hehre W J 1991 Geometry optimization in Cartesian coordinates: The end of the Z-matrix? *J. Comput. Chem.* **12** 606

[43] Paizs B, Fogarasi G and Pulay P 1998 An efficient direct method for geometry optimization of large molecules *J. Chem. Phys.* **109** 6571

[44] Farkas Ö and Schlegel H B 1998 Methods for geometry optimization in large molecules. I. An $O(N^2)$ algorithm for solving systems of linear equations for the transformation of coordinates and forces *J. Chem. Phys.* **109** 7100

[45] Baker J, Kinghorn D and Pulay P 1999 Geometry optimization in delocalized internal coordinates: An efficient quadratically scaling algorithm for large molecules *J. Chem. Phys.* **110** 4986

[46] Pulay P and Meyer W 1971 *Ab initio* calculation of the force field of ethylene *J. Mol. Spectrosc.* **40** 59

[47] Pulay P, Fogarasi G, Pang F and Boggs J E 1979 Systematic *ab initio* gradient calculation of molecular geometries, force constants and dipole moment derivatives *J. Am. Chem. Soc.* **101** 2550

[48] Fogarasi G, Zhou X, Taylor P W and Pulay P 1992 The calculation of *ab initio* molecular geometries: efficient optimization by natural internal coordinates and empirical correction by offset forces *J. Am. Chem. Soc.* **114** 8191

[49] Pye C C and Poirier R A 1998 Graphical approach for defining natural internal coordinates *J. Comput. Chem.* **19** 504

[50] Baker J 1993 Techniques for geometry optimization: a comparison of Cartesian and natural internal coordinates *J. Comput. Chem.* **14** 1085

[51] Eckert F, Pulay P and Werner H-J 1997 *Ab initio* geometry optimization for large molecules *J. Comput. Chem.* **18** 1473

[52] Sellers H L, Klimkowski V J and Schäfer L 1978 Normal coordinate *ab initio* force relaxation *Chem. Phys. Lett.* **58** 541

[53] Pulay P and Fogarasi G 1992 Geometry optimization in redundant internal coordinates *J. Chem. Phys.* **96** 2856

[54] Peng C, Ayala P Y, Schlegel H B and Frisch M J 1996 Using redundant internal coordinates to optimize equilibrium geometries and transition states *J. Comput. Chem.* **17** 49

[55] Baker J, Kessi A and Delley B 1996 The generation and use of delocalized internal coordinates in geometry optimization *J. Chem. Phys.* **105** 192

[56] Stewart J J P 1989 Optimization of parameters for semiempirical wavefunctions *J. Comput. Chem.* **10** 209, 221

[57] Baker J 1986 An algorithm for the location of transition states *J. Comput. Chem.* **7** 385

[58] Crawford B Jr and Fletcher W H 1951 The determination of normal coordinates *J. Chem. Phys.* **19** 141

[59] Fletcher R 1981 *Practical Methods of Optimization: vol. 2—Constrained Optimization* (New York: Wiley)

[60] Baker J 1992 Geometry optimization in Cartesian coordinates: constrained optimization *J. Comput. Chem.* **13** 240

[61] Baker J and Bergeron D 1993 Constrained optimization in Cartesian coordinates *J. Comput. Chem.* **14** 1339

[62] Baker J 1997 Constrained optimization in delocalized internal coordinates *J. Comput. Chem.* **18** 1079

[63] Lu D-H, Zhao M and Truhlar D G 1991 Projection operator method for geometry optimization with constraints *J. Comput. Chem.* **12** 376

[64] Taylor H and Simons J 1985 Imposition of geometrical constraints on potential energy walking procedures *J. Phys. Chem.* **89** 684

[65] Halgren T A and Lipscomb W N 1977 The synchronous transit method for determining reaction pathways and locating molecular transition states *Chem. Phys. Lett.* **49** 225

[66] Ehrenson S 1974 Analysis of least motion paths for molecular deformations *J. Am. Chem. Soc.* **96** 3778

[67] Bell S and Crighton J 1984 Locating transition states *J. Chem. Phys.* **80** 2464

[68] Ionova I V and Carter E A 1993 Ridge method for finding saddle points on potential energy surfaces *J. Chem. Phys.* **98** 6377

[69] Cerjan C J and Miller W H 1981 On finding transition states *J. Chem. Phys.* **75** 2800

[70] Simons J, Jørgensen P, Taylor H and Ozment J 1983 Walking on potential energy surfaces *J. Phys. Chem.* **87** 2745

[71] Banerjee A, Adams N, Simons J and Shepard R 1985 Search for stationary points on surfaces *J. Phys. Chem.* **89** 52

[72] Müller K and Brown L D 1979 Location of saddle points and minimum energy paths by a constrained simplex optimization procedure *Theor. Chim. Acta* **53** 75

[73] Dewar M J S, Healy E F and Stewart J J P 1984 Location of transition states in reaction mechanisms *J. Chem. Soc. Faraday Trans. II* **80** 227

[74] Koga N and Morokuma K 1985 Determination of the lowest energy point on the crossing seam between two potential surfaces using the energy gradient *Chem. Phys. Lett.* **119** 371

[75] McDouall J J W, Robb M A and Bernardi F 1986 An efficient algorithm for the approximate location of transition structures in a diabatic surface formalism *Chem. Phys. Lett.* **129** 595

[76] Jensen F 1994 Transition structure modeling by intersecting potential energy surfaces *J. Comput. Chem.* **15** 1199

[77] Bendt P and Zunger A 1982 New approach for solving the density functional self-consistent field problem *Phys. Rev.* B **26** 3114

[78] Head-Gordon M and Pople J A 1988 Optimization of wavefunction and geometry in the finite basis Hartree–Fock method *J. Phys. Chem.* **92** 3063

[79] Almlöf J 1995 Direct methods in electronic structure theory *Modern Electronic Structure Theory* ed D Yarkony (Singapore: World Scientific) pp 110–51

[80] Car R and Parrinello M 1985 Unified approach for molecular dynamics and density functional theory *Phys. Rev. Lett.* **55** 2471

[81] Field M J 1991 Constrained optimization of *ab initio* and semiempirical Hartree–Fock wavefunctions using direct minimization or simulated annealing *J. Phys. Chem.* **95** 5104

[82] Hartke B and Carter E A 1992 Spin eigenstate-dependent Hartree–Fock molecular dynamics *Chem. Phys. Lett.* **189** 358

[83] Fukui K 1970 A formulation of the reaction coordinate *J. Phys. Chem.* **74** 4161

[84] Ishida K, Morokuma K and Komornicki A 1977 The intrinsic reaction coordinate. An *ab initio* calculation for HCN → HNC and H$^-$ + CH$_4$ → CH$_4$ + H$^-$ *J. Chem. Phys.* **66** 2153

[85] Baldridge K K, Gordon M S, Steckler R and Truhlar D G 1989 *Ab initio* reaction paths and direct dynamics calculations *J. Phys. Chem.* **93** 5107

[86] Melissas V S, Truhlar D G and Garrett B C 1992 Optimized calculations of reaction paths and reaction-path functions for chemical reactions *J. Chem. Phys.* **96** 5758

[87] Page M, Doubleday C and McIver J W Jr 1990 Following steepest descent reaction paths. The use of higher energy derivatives with *ab initio* electronic structure methods *J. Chem. Phys.* **93** 5634 and references therein

[88] Sun J-Q and Ruedenberg K 1993 Quadratic steepest descent on potential energy surfaces. I. Basic formalism and quantitative assessment *J. Chem. Phys.* **99** 5257

[89] Gonzales C and Schlegel H B 1991 Improved algorithms for reaction path following: higher-order implicit algorithms *J. Chem. Phys.* **95** 5853

[90] Schlegel H B 1994 Some thoughts on reaction-path following *J. Chem. Soc. Faraday Trans.* **90** 1569

[91] Elber R and Karplus M 1987 A method for determining reaction paths in large molecules: application to myoglobin *Chem. Phys. Lett.* **139** 375

[92] Stachó L L and Bán M I 1992 A global strategy for determining reaction paths *Theor. Chim. Acta* **83** 433

[93] Valtazanos P and Ruedenberg K 1986 Bifurcations and transition states *Theor. Chim. Acta* **69** 281

[94] Baker J and Gill P M W 1988 An algorithm for the location of branching points on reaction paths *J. Comput. Chem.* **9** 465

[95] Kirkpatrick S, Gelatt C D Jr and Vecchi M P 1983 Optimization by simulated annealing *Science* **220** 671

[96] Golding D E 1989 *Genetic Algorithms in Search, Optimization and Machine Learning* (Reading, MA: Addison Wesley)

[97] Piela L, Kostrowicki J and Scheraga H A 1989 The multiple-minima problem in the conformational analysis of molecules. Deformation of the potential energy hypersurface by the diffusion equation method *J. Phys. Chem.* **93** 3339

[98] Floudas C and Pardalos P M 1991 *Recent Advances in Global Optimization* (Princeton, NJ: Princeton University Press)

## Further Reading

Fletcher R 1981 *Practical Methods of Optimization: Vol 1—Unconstrained Optimization; Vol. 2—Constrained Optimization* (New York: Wiley)

A classic in the field, very readable, highly recommended, full of practical advice.

Polak E 1997 *Optimization: Algorithms and Consistent Approximations* (New York: Springer)

A complete and mathematically precise treatment of the subject. Includes topics which are not usually discussed in introductory texts. Complete with exercises. Best suited for the mathematically inclined reader.

Dennis J E and Schnabel R B 1983 *Numerical Methods for Unconstrained Optimization and Non-linear Equations* (Englewood Cliffs, NJ: Prentice-Hall)

A very pedagogical, highly readable introduction to quasi-Newton optimization methods. It includes a modular system of algorithms in pseudo-code which should be easy to translate to popular programming languages like C or Fortran.

Schlegel H B 1995 Geometry optimization on potential energy surfaces *Modern Electronic Structure Theory* ed D Yarkony (Singapore: World Scientific) pp 459–500

An excellent, up-to-date treatise on geometry optimization and reaction path algorithms for *ab initio* quantum chemical calculations, including practical aspects.

Werner H-J 1987 Matrix-formulated direct multiconfigurational self-consistent field and multireference configuration interaction methods *Adv. Chem. Phys.* **69** 1

A lucid and carefully written exposition of this difficult subject from one of the authors of the highly acclaimed MOLPRO suite of programs. It contains examples and plenty of physical insight.

Shepard R 1987 The multiconfiguration self-consistent field method *Adv. Chem. Phys.* **69** 63

A very detailed, pedagogical treatment of the subject, including much of the mathematical background and a nearly complete list of references prior to 1987.

Pulay P 1995 Analytical derivative techniques and the calculation of vibrational spectra *Modern Electronic Structure Theory* ed D Yarkony (Singapore: World Scientific) pp 1191–240

A concise introduction to the calculation of analytical derivatives in quantum chemistry, with applications to simulating vibrational spectra.

# B3.6
# Mesoscopic and continuum models

*Marcus Müller*

### B3.6.1 Introduction

Many systems in physical chemistry exhibit structure on length scales that greatly exceed the atomic dimensions. Systems containing surfactants—detergents or milk, for instance—often consist of droplets of one component dissolved in another phase. The size of these droplets exceeds the extension of the molecular constituents by far. Very generally, mesoscopic and continuum models describe the properties of materials on length scales larger than the atomic dimensions by incorporating the details of the underlying atomic structure only in terms of a reduced number of effective variables. In this very broad sense, the Navier–Stokes equation [1], which describes the motion of a fluid via a density, energy, and velocity field and elasticity theory [2], and which describes solids in terms of stress and displacement fields, also belongs to this class of model. In both approaches, the subject of the model is not the properties of individual atoms (e.g., their position or quantum state) but rather their average properties (like the density or velocity) in a small coarse-graining volume. Usually the coarse graining is not performed explicitly, but it is understood that the averaging volume is large enough to result in a continuous spatial variation of the variables of the mesoscopic model and yet still be smaller than the characteristic length scale of the phenomena under consideration.

In the following entry we shall restrict ourselves to discussing mesoscopic and continuum models for complex fluids in chemical physics. The wide span of time and length scales in these materials is illustrated in figure B3.6.1 for a blend of two polymers. On the atomistic scale each polymer consists of chemical repeat units joined together to form the chain molecule. The length scale is set by the distance between the atoms along the backbone of the polymer, typically in the range of 1–2 Å. The vibrations of the atoms occur on the timescale of picoseconds. In a dense melt, the flexible chain molecules adopt a random-walk-like conformation. The 'step length' of the random walk, or persistence length $b$, is typically of the order of a few nanometres. Since several thousands of repeat units form a polymer, the overall size of a single molecule, as specified by its radius of gyration, exceeds the persistence length by 1–3 orders of magnitude. On this range of length scales the structure of the polymer is self-similar. If the two components of the blend are not miscible, as it is generally the case, one species forms droplets that are dispersed in a matrix of the other species. The size of the droplets is in the micrometre range. On even larger length scales (say 1 mm) the material appears homogeneous. Clearly the properties on the mesoscopic length scale are important for application properties. A decrease of the droplet size or even the formation of a connected morphology (i.e. a microemulsion) improves the mechanical properties of the composite material. A similar span of time and length scales is encountered in many other systems (e.g., mixtures of oil, water and surfactant or glassy materials) and this behaviour is rather typical for complex fluids.

A unified model that describes the structure from the atomistic length scale up to macroscopic properties is not analytically tractable. Even state-of-the-art supercomputers cannot cope with such a broad spread of time and length scales in numerical simulations. Today's largest simulated systems in thermal equilibrium

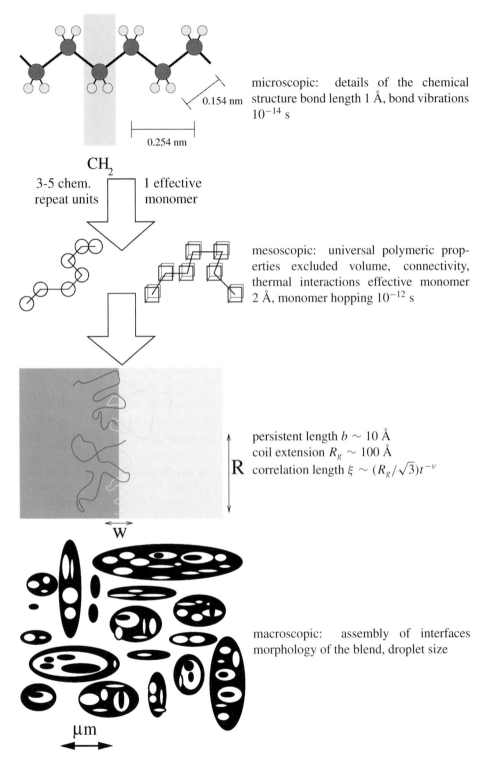

**Figure B3.6.1.** Illustration of the wide span of length scale in a binary polymer blend. (See the text for further explanation.)

comprise about $10^7$ particles and, hence, span about 2–3 decades in length scales. With the increase of computing power and progress in simulation methodology, simulating larger and larger system sizes will become feasible, but computer modelling from atomic to macroscopic scales in the framework of a single, unified model is not feasible at present or in the near future.

Another caveat for the modelling from the atomistic level up to the macroscopic level is the requirement of sufficiently accurate interaction potentials. Minor inaccuracies in calculations on small length scales can give rise to pronounced effects on the mesoscopic scale. Consider, for instance, the self-assembly of amphiphilic molecules (see section B3.6.3) into a spatially ordered structure. The free-energy difference between the different morphologies can be as small as $10^{-4}kT$ per molecule. The *ab initio* prediction of such a small free-energy difference is certainly a formidable task.

Mesoscopic and continuum models do not attempt to describe large-scale phenomena starting from the smallest atomic length scale, but rather incorporate the local structure via a small number of effective parameters. Mesoscopic models lump a small number of atoms into an effective particle. These particles interact via coarse-grained interactions. By this coarse-graining procedure much of the atomistic detail is lost, and only those interactions pertinent to the phenomena on the mesoscopic length scales are retained. Even if the interactions on the microscopic scale are extremely complex (e.g., hydrophobic interactions [3] in lipid water mixtures), they can often be captured by simple expressions on the mesoscopic length scale. Coarse-grained models thus yield valuable insights into the structure on large length scales. For specific examples the effective interactions are derived by eliminating the degrees of freedom on the smallest (atomistic) length scales, retaining only those on larger length scales; for some systems (e.g., polymer chains in the gas phase) this coarse-graining procedure has a formal justification due to the self-similar structure on a large range of length scales; for other systems the mapping between the atomistic/microscopic level and the mesoscopic description is rather a concept than a practicable procedure. In this latter case, the application of mesoscopic models rests on the observation that different systems (e.g., diblock copolymers and lipid water mixtures) share a common behaviour on mesoscopic scales. Universal mesoscopic behaviour that does not depend on the details on the atomistic level in a qualitative way is the subject of mesoscopic models.

Continuum models go one step further and drop the notion of particles altogether. Two classes of models shall be discussed: field theoretical models that describe the equilibrium properties in terms of spatially varying fields of mesoscopic quantities (e.g., density or composition of a mixture) and effective interface models that describe the state of the system only in terms of the position of interfaces. Sometimes these models can be derived from a mesoscopic model (e.g., the Edwards Hamiltonian for polymeric systems) but often the Hamiltonians are based on general symmetry considerations (e.g., Landau–Ginzburg models). These models are well suited to examine the generic, universal features of mesoscopic behaviour.

Mesoscopic and continuum models bridge the gap between atomistic realistic simulations and the description on the macroscopic level, (e.g., elasticity theory). The objectives of mesoscopic models are twofold. On the one hand, they help identify interactions that are necessary to bring about the phenomena on a mesoscopic scale (e.g., phase separation or self-assembly) and they aid in investigating the dependence of the mesoscopic behaviour on the effective interactions. This information also yields some qualitative insight into how the microscopic parameters influence mesoscopic behaviour (e.g., the dependence of the structure in a self-assembled system on the architecture/shape of the amphiphilic molecules). On the other hand, this class of models elucidates universal behaviour on the mesoscopic scale (e.g., identifying various morphologies into which systems can self-assemble, the relation between confinement and phase behaviour, or the consequences of fluctuations) and establishes a relation between behaviour on large length scales and experimentally accessible (mesoscopic) quantities (e.g., Flory–Huggins parameter, interfacial tension, or bending rigidity of membranes).

The hierarchy of models is complemented by a variety of methods and techniques. Mesoscopic models that incorporate some fluid-like packing (e.g., spring–bead models for polymer solutions) are investigated by Monte Carlo simulations, molecular dynamics or density functional techniques. Lattice models are studied by

Monte Carlo simulations. The larger the span of length scales considered, the larger the computational effort required. Models without pronounced packing effects (e.g., the Edwards Hamiltonian) are investigated by self-consistent field techniques. Continuum models are often analytically tractable, at least in the mean field approximation, and simple analytical expressions for various quantities (e.g., interfacial tension between two immiscible polymers) can be obtained in some limiting cases. The effect of fluctuations has been assessed by computer simulations, transfer matrix calculations and renormalization group techniques.

At the heart of mesoscopic and continuum models lies the question: Which degrees of freedom are to be retained as relevant and which can be ignored? The answer depends on the specific problem. By comparing different models, the degree of universality and the relevance of interactions can be gauged. This yields much insight into the mechanisms which underly the phenomena. Mesoscopic and continuum models make contact with chemical models on the atomistic level as well as with the macroscopic descriptions. Effort is being made to incorporate more chemical realism into the models as well as to extend them to larger length scales.

In the following we shall describe various applications of mesoscopic models to complex fluids. The examples extend from applications that are quite close to the atomistic level (e.g., coarse-grained polymer models) to highly idealized models (e.g., effective interface Hamiltonians or Ginzburg–Landau models). Moreover, we restrict ourselves mainly to the description of thermodynamic equilibrium. The remainder of this entry is organized as follows. In section B3.6.2 we discuss applications of coarse-grained models to systems involving homopolymers. Mesoscopic models for the description of self-repelling chains, polymer solutions, polymer melts and binary blends are introduced. From these models, more coarse-grained descriptions can be derived in terms of Ginzburg–Landau expansions or effective interface Hamiltonians. Section B3.6.3 then considers amphiphilic molecules. Their co-operative behaviour on the supramolecular level has been explored in the framework of models with various degrees of detail. Chain models retain the salient features of the amphiphile's architecture, while lattice models or continuum models yield a description in terms of a spatially varying concentration. On even larger scales, the statistical mechanics of interfaces has been investigated via random interface models. This article closes with a brief look at the application of mesoscopic and continuum models to dynamical phenomena.

## B3.6.2   Polymeric systems

### B3.6.2.1   Polymer solutions

Coarse-grained models have a longstanding history in polymer science. Long-chain molecules share many common mesoscopic characteristics which are independent of the atomistic structure of the chemical repeat units [4–6]. The self-similar structure [7–10] on large length scales is only characterized by a single length scale, the chain extension $R$.

The important interactions in polymer solutions are the connectivity of the segments along the chain molecules and interactions between segments. The solvent molecules are often not treated explicitly, but their effect is incorporated into the effective interactions between polymer segments. A good solvent corresponds to an effective repulsion between segments and the polymer chains adopt a swollen configuration. A bad solvent gives rise to an attraction between the polymer segments and leads to a collapse.

The observation of the universality and self-similarity of the large-length-scale properties has a theoretical basis. In 1972 de Gennes [11] related the structure of a polymer chain in a good solvent to a field theory of a $n$ component vector model in the limit $n \to 0$. This class of models (see the entry on phase transitions and critical phenomena; A2.5) exhibits a continuous phase transition and the properties close to this critical point have been investigated extensively with renormalization group calculations [8–10]. The inverse chain length plays the role of the distance from the critical point of the $n = 0$ component vector model. As in the theory of critical phenomena, the behaviour in the vicinity of this critical point (i.e. $1/N \ll 1$) is governed by a universal scaling behaviour that is brought about by only a few relevant interactions. The relation between the behaviour of polymer chains in the limit of $N \to \infty$ and the critical behaviour justifies the use of highly coarse-grained models that incorporate only two relevant interactions: connectivity along the chain and binary segmental interactions.

Lattice models of polymer solutions are a particularly simple and computationally efficient realization, and therefore have attracted abiding interest [12]. In simple lattice models, a small group of atomistic repeat units is represented by a site on a simple cubic lattice. Segments along a polymer occupy neighbouring lattice sites and multiple occupation of lattice sites is forbidden (excluded volume). The latter constraint corresponds to the repulsive binary interaction under good solvent conditions. Isolated chains on the lattice adopt configurations of self-avoiding walks. The polymer's end-to-end distance $R$ scales with the chain length like $R \sim N^{\nu}$. The exponent $\nu = 0.588$ has been calculated using renormalization group techniques [9, 10], enumeration techniques for short chain lengths and Monte Carlo simulations [13].

The application of lattice models to study the behaviour of multi-chain systems (i.e. dilute and semi-dilute solutions, and dense melts) is straightforward in principle. The equilibration of dense multi-chain systems is, however, a challenging problem for computer simulations, and simple lattice models have been a testing bed for many algorithms. Some methods are tailored to isolated chains or very dilute systems (e.g., the pivot algorithm [13] or the construction of a chain via the pruned-enriched Rosenbluth method [14]); other methods provide an effective relaxation of the overall chain dimensions in dense systems (e.g., configurational bias Monte Carlo [15, 16] or the recoil growth algorithm [17]).

Though these simple lattice models reproduce the universal features of polymer solutions, it is difficult to incorporate details of the chain architecture. The simple lattice model allows only for two bond angles which makes the investigation of orientational effects prone to lattice artefacts. Moreover, the particles in real fluids arrange to form neighbouring shells. This local packing structure of the fluid does not affect the universal scaling behaviour but it is pertinent to the relation between the coarse-grained effective interactions and the underlying microscopic potentials. Since the vacancies on the lattice and the polymer segments have the same size, packing effects in the density correlation function are largely absent. More sophisticated lattice models (e.g., the bond fluctuation model [18]), in which monomers are represented by extended objects (e.g., a whole unit cube) on the lattice, have been explored. These models exhibit packing effects and a large number of bond angles while still retaining the computational advantages of lattice models. They also allow for a diffusive dynamics of the polymers on the lattice which consists of random local displacements of the monomers. Moreover, the bond vectors can be chosen such that the excluded volume constraint prevents bonds from crossing through each other in the course of these local displacements. This non-crossability takes account of topological effects which are important for the dynamical properties of linear chains [19] and influence the conformational statistics of ring polymers [20–22] (e.g., in order to avoid topological interactions rings collapse in a concentrated solution).

Off-lattice models enjoy a growing popularity. Again, a particle corresponds to a small number of atomistic repeat units along the backbone of the polymer. Off-lattice models allow simulations at constant pressure or the calculation of the pressure via the virial expression. This yields direct access to the $pVT$ behaviour. By modelling polymers as a sequence of tangent hard spheres in continuous space, computer simulators have investigated the equation of state in polymer solutions and the detailed packing structure of polymer solutions in contact with a hard wall. This class of model is particularly suited for comparing the results to analytical theories (e.g., Wertheim's theory [23] or density functional approaches [24–26]) because of the existence of elaborated analytical descriptions for the corresponding hard-sphere monomer fluid.

Hard-sphere models lack a characteristic energy scale and, hence, only entropic packing effects can be investigated. A more realistic modelling has to take hard-core-like repulsion at small distances and an attractive interaction at intermediate distances into account. In non-polar liquids the attraction is of the van der Waals type and decays with the sixth power of the interparticle distance $r$. It can be modelled in the form of a Lennard-Jones potential $V_{LJ}(r)$ between segments

$$V_{LJ}(r) = 4\epsilon \left\{ \left( \frac{\sigma}{r} \right)^{12} - \left( \frac{\sigma}{r} \right)^{6} \right\} \tag{B3.6.1}$$

where the exponent of the first, repulsive term is chosen for computational convenience. The Lennard-Jones radius $\sigma$ sets the microscopic length scale and $\epsilon$ sets the energy scale. In many simulational applications the potential is truncated and shifted so as to yield a continuous, finite ranged potential. This does not alter the qualitative behaviour but shifts the temperature and density of the liquid–vapour critical point. The Lennard-Jones particles are tied together to form chain molecules. The constraint of fixed bond length or an harmonic bonding potential has been employed. Another popular choice is the FENE potential [27, 28]. It takes the form

$$V_{FENE}(r) = -\frac{k}{2} R_0^2 \ln \left( 1 - \frac{r^2}{R_0^2} \right) \qquad \text{with} \quad R_0 = 1.5\sigma. \tag{B3.6.2}$$

The parameter $k$ tunes the stiffness of the potential. It is chosen such that the repulsive part of the Lennard-Jones potential makes a crossing of bonds highly improbable (e.g., $k = 30$). This off-lattice model has a rather realistic equation of state and reproduces many experimental features of polymer solutions. Due to the attractive interactions the model exhibits a liquid–vapour coexistence, and an isolated chain undergoes a transition from a self-avoiding walk at high temperatures to a collapsed globule at low temperatures. Since all interactions are continuous, the model is tractable by Monte Carlo simulations as well as by molecular dynamics. Generalizations of the Lennard-Jones potential to anisotropic pair interactions are available: e.g., the Gay–Berne potential [29]. This latter potential has been employed to study non-spherical particles that possibly form liquid crystalline phases.

In the limit that the number of effective particles along the polymer diverges, but the contour length and chain dimensions are held constant, one obtains the Edwards model of a polymer solution [9, 30]. Polymers are represented by random walks that interact via zero-ranged binary interactions of strength $v$. The partition function of an isolated chain is given by

$$\mathcal{Z} = \int \mathcal{D}[r(t)] \exp \left( -\frac{3}{2b^2} \int_0^N dt \left( \frac{dr}{dt} \right)^2 \right) \exp \left( -\int dr \, dr' \hat{\rho}(r) v \delta(r - r') \hat{\rho}(r') \right) \tag{B3.6.3}$$

where the density field $\hat{\rho}$ is related to the configuration $\{r(t)\}$ of the polymer via

$$\hat{\rho}(r') = \int_0^N dt \, \delta(r' - r(t)). \tag{B3.6.4}$$

The path integral $\mathcal{D}$ sums over all polymer conformations $r(t)$, where $0 \leq t \leq N$ denotes the contour parameter along the polymer. The second term represents the connectivity along the molecule and $b$ denotes the persistence length (i.e. the 'step length' of the random walk). In the absence of the third term, the partition function describes a Gaussian chain with the end-to-end distance $R = b\sqrt{N}$ (Gaussian chain model). This is the only length scale in the problem. Very much like in quantum mechanics, the path integral results in a diffusion equation (i.e. the polymer analogue of Schrödinger's equation for the propagator) for the probability of finding a chain's segment after $t$ steps along the chain at position $r$ in space. The third term describes the interactions: if segments $t$ and $t'$ are located at the same position they interact with the strength $v$.

This model is very popular for analytical calculations and generalizations to multi-chain systems are straightforward. Properties of polymer solutions have been obtained via renormalization group techniques [8–10]. Similar to the simple lattice model, the Edwards model includes only the chain connectivity and binary segmental interactions: the detailed structure of the underlying fluid is omitted. For $v = 0$ the self-similar Gaussian statistics persist on all length scales and there is no rod-like behaviour on smaller scales. Generalizations of thread-like models to stiff polymers and orientational interactions, however, have been explored. The most popular one is the wormlike chain model, in which the second term in the Edwards Hamiltonian (B3.6.3) is replaced by [31]

$$\exp\left(-\frac{\eta}{2}\int_0^N dt \left(\frac{du}{dt}\right)^2\right) \tag{B3.6.5}$$

where $u(t) = dr/dt$ denotes the tangent vector along the path with unit norm. The parameter $\eta$ controls the local stiffness of the path. On small distances along the chain the tangent vectors $u$ are highly correlated and the stiffness parameter $\eta$ controls the decay of orientational correlations along the chain's contour: $\langle u(t)u(t')\rangle = \exp(-|t - t'|/\eta)$. On large length scales, however, the Gaussian behaviour is recovered and the end-to-end distance is given by $R^2 = 2\eta N$.

### B3.6.2.2  Polymer blends

The above models can be generalized to multicomponent systems by modifying the segmental interactions. Most applications deal with a binary blend in a common solvent. The excluded volume interaction between segments limits density fluctuations in a dense polymer liquid. Therefore many models of dense multi-component systems neglect the finite compressibility and model the excluded volume interactions by enforcing a uniform segment density. In 1941, Flory [32] and Huggins [33] employed a simple lattice model to calculate the phase diagram for a dense binary polymer blend in mean field approximation. The two polymer species—denoted A and B—are modelled as walks on a lattice, and the binary interactions of strengths $\epsilon_{AA}$, $\epsilon_{AB}$ and $\epsilon_{BB}$ act between neighbours on the lattice. Since all lattice sites are occupied, only the difference of the segmental interactions (the Flory–Huggins parameter)

$$\chi_{FH} = \frac{z}{kT}\left(\epsilon_{AB} - \frac{\epsilon_{AA} + \epsilon_{BB}}{2}\right) \tag{B3.6.6}$$

determines the phase diagram. Here, $z = 6$ denotes the coordination number of the simple cubic lattice. Typical experimental values of the Flory–Huggins parameter $\chi$ are in the range $10^{-2}$–$10^{-5}$ for partially compatible blends while the individual interactions between the segments $\epsilon_{ij}$ ($i, j = $ A, B) are of the order of $k_B T$. This illustrates that the phase behaviour is governed by a delicate cancellation of interactions. Starting from a model with atomistic details and performing an *ab initio* calculation of the packing structure and the effective segmental interactions in a binary blend would require extremely accurate interatomic potentials as input and a very high numerical quality of the calculation. Therefore, predicting the value of the Flory–Huggins parameter on an *ab initio* basis is virtually impossible. The concept of describing the effective incompatibility of two polymer species by a single mesoscopic parameter $\chi_{FH}$ has proven remarkably successful, however. When $\chi_{FH}$ is used as an adjustable parameter the mean field theory of Flory and Huggins is quite successful in describing many experimental observations. The values of the $\chi_{FH}$ parameter of various pairs of polymers have been extracted from a comparison between theory and experiment, and are compiled in, for example, [34]. In the framework of the mean field theory, the excess free energy of mixing per segment takes a particularly simple form:

$$\frac{\Delta F}{\rho V k T} = \frac{\phi_A}{N_A}\ln\phi_A + \frac{\phi_B}{N_B}\ln\phi_B + \chi_{FH}\phi_A\phi_B \tag{B3.6.7}$$

where $N_A$ and $N_B$ denote the chain length of the two polymer species and $\phi_A$ and $\phi_B$ denote the relative amount of A or B segments, respectively. $\phi_A + \phi_B = 1$. The first term represents the translational entropy of mixing. Due to the connectivity of the segments it is reduced by a factor $1/N_A$ or $1/N_B$, respectively. The second term describes the repulsion between unlike segments. The chain conformations are assumed to be independent of the composition. Therefore the conformational entropy does not give any contribution to the free energy of mixing in the Flory–Huggins treatment. Most notably, the theory rationalizes the fact that long macromolecules tend to demix, because a small repulsion is sufficient to far outweigh the entropy of mixing, which is reduced by the factor $1/N$. This expression for the excess free energy of mixing also forms the basis for self-consistent field models of spatially inhomogeneous systems.

In order to gain qualitative insight into how to relate the Flory–Huggins parameter to the architectural properties of the components, mesoscopic models with various degrees of structural detail have been investigated. Complex lattice models allow monomeric units to occupy more than one lattice site. In the lattice cluster model of Freed and co-workers [35] the effect of explicit monomer structure has been explored. The partition function of the model is expressed in a systematic double expansion with respect to the inverse temperature and the inverse coordination number of the underlying lattice. To zero order the approach recovers the results of the original Flory–Huggins theory. Higher-order terms account for geometric packing on the monomer scale and non-random mixing effects. This approach has been successful in predicting various subtle influences of the monomer architecture, including the occurrence of entropic contributions to the Flory–Huggins parameter.

Similar questions can be addressed by the P-RISM (polymer reference interaction site model) theory of Curro and Schweizer [36]. This integral equation theory generalizes the Ornstein–Zernike equation to polymeric systems in order to account for the fluid-like packing structure. Details of the molecular architecture enter via the single-chain structure factor. The P-RISM approach yields a detailed description of the phase behaviour and the local structure and has been applied to models with various degrees of structural detail. In the limit that the chains are modelled as infinitely thin Gaussian paths the results are very similar to the Flory–Huggins theory. The theory has been applied to fairly realistic chain models taking the experimentally measured single-chain structure factors as input. More recently, this approach has been applied self-consistently to calculate the change of the molecular conformation upon blending.

The bond fluctuation model [37] and off-lattice [38, 39] models have been used to investigate the binary polymer blends within Monte Carlo simulations. Attention has focused on rather different topics: (i) Monte Carlo simulations appropriately account for the effect of composition fluctuations. They are important in the vicinity of the critical temperature of the unmixing transition. When the chain length is increased, this fluctuation-dominated region shrinks and one observes a crossover between the 3D Ising universality class and mean field critical behaviour [40]. (ii) The relation between the polymer architecture and the Flory–Huggins parameter has been explored in simulations. Disparities in the architecture on the scale of the coarse-grained monomers (e.g., different local stiffness of the chains or different monomer shapes) alter the packing structure and give rise to enthalpic and entropic contributions to the Flory–Huggins parameter [37, 39]. When comparing experimental data to the predictions of the mean field theory, deviations from the simple proportionality $\chi_{FH} \sim 1/T$ of the Flory–Huggins parameter are rather the rule than an exception. (iii) Monte Carlo simulations reveal that the chains in the minority phase shrink. By reducing their size, they increase the local density of their own monomers and reduce the number of unfavourable contacts with the opposite species. The latter effect is, however, not captured in simple mean field theories. (iv) Off-lattice models have been employed to study binary blends at constant pressure and to explore the effect of compressibility on the miscibility behaviour [38, 39].

These coarse-grained approaches investigate the generic behaviour and the qualitative dependence on the chain architecture. Again it should be pointed out that these simulations and analytical methods cannot predict the absolute value of the Flory–Huggins parameter of a specific pair of polymers. However, by a careful choice of the coarse-grained model, they help in identifying relevant parameters for the miscibility on a coarse-grained scale.

### B3.6.2.3 Self-consistent field approach and Ginzburg–Landau models

Long polymers tend to demix and the properties of the interfaces between the coexisting phases have attracted longstanding interest. Using the Gaussian chain model, Helfand and Tagami [41] investigated the interfacial properties in the self-consistent field theory. Within the mean field approximation the problem of interacting polymers is formulated in terms of a single-chain problem in an effective, external field. This effective, external field replaces the interactions with the surrounding polymers in the binary A/B blend. The effective field

$$w_A(r) = \xi(r)\{\phi_A(r) + \phi_B(r) - 1\} + \chi\phi_B(r) \tag{B3.6.8}$$

acts on a monomer of species A at position $r$ with $\phi_A$ and $\phi_B$ denoting the local composition of the blend. A similar equation holds for $w_B$. The first term enforces the incompressibility; the factor $\xi$ is adjusted to comply with the constraint $\phi_A(r) + \phi_B(r) = 1$ everywhere. The second term describes the repulsion between different species parameterized by $\chi$. The local composition, in turn, depends on the fields and is obtained as the Boltzmann average of isolated A and B chains in the fields $w_A$ and $w_B$, respectively. For Gaussian chains, described by the first part of the Hamiltonian (B3.6.3), this leads to a diffusion equation in the external potential $w_A$ and $w_B$. Since the fields depend on the local composition the equations have to be solved self-consistently.

Helfand and Tagami calculated the composition profiles across the interface and determined the interfacial tension. In general, the self-consistent field equations have to be solved numerically. Different schemes in real space [42], on lattices [43] and in Fourier representation [44] have been devised. There are, however, two interesting limits in which simple analytical expressions for the interfacial width and the interfacial tension can be obtained. The limit in which the width of the interface is much smaller than the extension of the polymer and yet larger than the persistence length $b$ is called the strong segregation limit. It corresponds to the range $1 \gg \chi \gg 1/N$ of incompatibility. This strong segregation limit is only accessible for long chain lengths $N$ and corresponds to truly polymeric behaviour. The interfacial width $w$ and tension $\gamma$ are described by the simple forms

$$w = b/\sqrt{6\chi} \qquad \text{and} \qquad \gamma = \rho b\sqrt{\chi/6} \tag{B3.6.9}$$

where $\rho$ denotes the monomer density. The leading corrections to the strong segregation behaviour are of the order $1/\chi N$ and have been the subject of much investigation [45–47]. Of course, the Gaussian chain model cannot describe the structure on length scales smaller than or comparable to the persistence length $b$ of the polymer. This restricts the application of this mesoscopic model to the range $\chi \ll 1$.

The binary polymer blend exhibits a second-order unmixing transition. Close to the critical temperature the concentration of the two coexisting phases does not differ very much and the characteristic length scale of composition fluctuations $\xi$ or the interfacial width $w$ are large compared with the size $R$ of the polymer coil. In this weak segregation limit polymer blends behave very similarly to mixtures of small molecules in the vicinity of the critical point. The difference between the composition of the coexisting phases and the composition of the mixture at the critical point defines the order parameter $m$ of the unmixing transition (see the entry on phase transitions and critical phenomena; A2.5). It increases with a universal power law upon cooling the system below the critical temperature $T_c$:

$$m \sim t^\beta \qquad \text{with} \qquad t = \frac{T_c - T}{T_c} > 0. \tag{B3.6.10}$$

$\beta$ is the critical exponent and $t$ denotes the reduced distance from the critical temperature. In the vicinity of the critical point, the free energy can be expanded in terms of powers and gradients of the local order parameter $m(r) = \phi_A(r) - \phi_B(r)$:

$$\frac{F[m(r)]}{\rho kT} \sim \int d^3r \left\{ f(m) + \frac{l^2}{2}(\nabla m)^2 \right\} \qquad \text{with} \qquad f(m) = -\frac{t}{2}m^2 + \frac{1}{12}m^4. \tag{B3.6.11}$$

This form is called a Ginzburg–Landau expansion. The first term $f(m)$ corresponds to the free energy of a homogeneous (bulk-like) system and determines the phase behaviour. For $t > 0$ the function $f$ exhibits two minima at $m = \pm\sqrt{3t}$. This value corresponds to the composition difference of the two coexisting phases. The second contribution specifies the cost of an inhomogeneous order parameter profile. $l$ sets the typical length scale. The general form of the expansion is dictated by very general symmetry considerations; the specific coefficients for the example of a polymer blend can be derived from the self-consistent field theory. For a binary blend this yields $l^2 = b^2 N/18$. In mixtures of small molecules the coefficient is determined by the range of the interactions; in polymeric systems the coefficient is associated with the conformational entropy. It is the shape of the extended molecule and its deformation at a spatial inhomogeneity that gives rise to the free energy cost.

Ginzburg–Landau models constitute a widely used example of continuum models. This class of continuum models describes the generic behaviour of all binary mixtures close to the unmixing transition. The properties of the specific model enter only via the coefficients of the expansion which set the energy scale and length scale. Extensions to different transitions (e.g., first-order transitions or microemulsion) are available (see also section B3.6.3). This approach, however, does not incorporate any structural detail of the underlying systems and hence becomes quantitatively inaccurate at lower temperatures, where the coexisting phases differ more strongly in their composition or the characteristic length scale (i.e. the correlation length $\xi$) becomes comparable to the size of the molecules.

Within this continuum approach Cahn and Hilliard [48] have studied the universal properties of interfaces. While their elegant scheme is applicable to arbitrary free-energy functionals with a square gradient form we illustrate it here for the important special case of the Ginzburg–Landau form. For an ideally planar interface the profile depends only on the distance $z$ from the interfacial plane. In mean field approximation, the profile $m(z)$ minimizes the free-energy functional (B3.6.11). This yields the Euler–Lagrange equation

$$\frac{\delta \mathcal{F}}{\delta m} = 0 \qquad \Rightarrow \qquad -tm + \frac{1}{3}m^3 - l^2\frac{\mathrm{d}^2 m}{\mathrm{d}z^2} = 0 \qquad\qquad (\text{B3.6.12})$$

which, in turn, is solved by a simple function

$$m = \pm\sqrt{3}t^{1/2}\tanh\left(\frac{z}{w}\right) \qquad \text{with} \qquad w = \frac{\sqrt{2}l}{t^{1/2}}. \qquad\qquad (\text{B3.6.13})$$

In the vicinity of the critical point (i.e. $|t| \ll 1$) the interfacial width $w$ is much larger than the microscopic length scale $l$ and the Landau–Ginzburg expansion is applicable.

Both Monte Carlo simulations of lattice models [49, 50] and spring–bead models [51] have been employed to study interfaces in polymeric systems. The simulations yield insight into the local properties of the polymeric fluid. Unlike in the Landau–Ginzburg expansion, the notion of polymers is retained and the orientation of the extended molecules at the interface or the enrichment of end segments have been studied. Moreover, the simulations incorporate fluctuations, which are ignored in the mean field approximation. In the vicinity of the critical temperature composition fluctuations are important. The mean field treatment overestimates the critical point and the binodals are flatter in the simulations which exhibit 3D Ising critical behaviour ($\beta = 0.324$) than in the mean field case ($\beta = \frac{1}{2}$). The importance of composition fluctuations can be gauged by the Ginzburg criterion [52]: The neglect of fluctuations is justified when the order parameter fluctuations in one 'correlation volume' of size $\xi^3$ are small compared with the order parameter itself. In the case of a symmetric binary polymer blend this condition yields

$$\frac{\chi - \chi_c}{\chi_c} \gg \frac{N^2}{\rho^2 R^6} \sim \frac{1}{N}. \qquad\qquad (\text{B3.6.14})$$

Ultimately, in the vicinity of the critical point, composition fluctuations are important, but the region in which these fluctuations dominate the behaviour decreases with the chain length $N$. Qualitatively, the behaviour

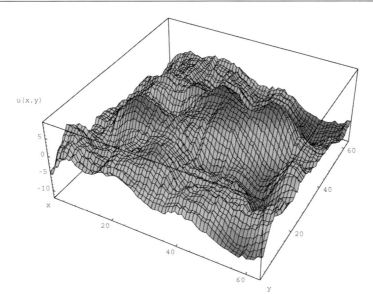

**Figure B3.6.2.** Local interface position in a binary polymer blend. After averaging the interfacial profile over small lateral patches, the interface can be described by a single-valued function $u(r_\parallel)$. (Monge representation). Thermal fluctuations of the local interface position are clearly visible. From Werner *et al* [49].

can be understood as follows: long-chain molecules do not fill space and strongly interdigitate; the number of other chains in the volume of a reference chain increases like $\sqrt{N}$ with chain length. This large number of interaction partners results in a strong suppression of fluctuations in the interactions on the level of a whole molecule and, hence, replacing the interactions by a non-fluctuating mean field is a good approximation.

Another important difference between the mean field treatment and the simulations or experiments are fluctuations of the local interfacial position. While the mean field treatment assumes a perfectly flat, planar interface right from the outset, the local interfacial position fluctuates in experiments and simulations. A typical snapshot of the local interface position, as obtained from a Monte Carlo simulation of a binary polymer blend, is depicted in figure B3.6.2. On not too small length scales the local position of the interface is smooth and without bubbles or overhangs. The system configuration can be described by two ingredients: the position $u(r_\parallel)$ of the centre of the interface as a function of the lateral coordinates $r_\parallel$ and the local structure described by profiles across the interface. The latter quantities depend only on the coordinate normal to the interface. In many applications the coupling between the long-wavelength fluctuations of the local interfacial position $u$ and the intrinsic profile is neglected. In this case the intrinsic profiles describe the variation of quantities across an ideally planar interface. The apparent interfacial profile $p_{app}(z)$, which is averaged over fluctuations of the local interfacial position in experiments or simulations, can be approximated by a convolution of the intrinsic profile $p_{int}(z)$ and the distribution $P(u)$ of the local interface position [53]

$$p_{app}(z) = \int du\, P(u)\, p_{int}(z-u) \tag{B3.6.15}$$

If one is only interested in the properties of the interface on scales much larger than the width of the intrinsic profiles, the interface can be approximated by an infinitely thin sheet and the properties of the intrinsic profiles can be cast into a few effective parameters. Using only the local position of the interface, effective interface Hamiltonians describe the statistical mechanics of fluctuating interfaces and membranes.

### B3.6.2.4  *Effective interface Hamiltonians*

The fluctuations of the local interfacial position increase the effective area. This increase in area is associated with an increase of free energy $\mathcal{H}$ which is proportional to the interfacial tension $\gamma$. The free energy of a specific interface configuration $u(r_\parallel)$ can be described by the capillary wave Hamiltonian:

$$\mathcal{H}[u(r_\parallel)] = \gamma \int dx\, dy \left\{ \sqrt{1 + \left(\frac{du}{dx}\right)^2} \sqrt{1 + \left(\frac{du}{dy}\right)^2} - 1 \right\} \approx \frac{\gamma}{2} \int d^2 r_\parallel (\nabla u)^2. \tag{B3.6.16}$$

The functional $\mathcal{H}[u]$ can be diagonalized via a Fourier transformation with respect to the lateral coordinates $r_\parallel$. This results in

$$\mathcal{H}[u_q] = \frac{\gamma}{2} \sum_q q^2 |u_q|^2. \tag{B3.6.17}$$

In this Fourier representation the Hamiltonian is quadratic and the equipartition theorem yields for the thermal fluctuations: $\langle u^2(q) \rangle = k_B T / \gamma q^2$. This spectrum corresponds to a Gaussian distribution of the local interface position:

$$P(u) = \frac{1}{\sqrt{2\pi s^2}} \exp\left(-\frac{u^2}{2s^2}\right) \quad \text{with} \quad s^2 = \frac{1}{4\pi^2} \int d^2 q_\parallel \langle u^2(q_\parallel) \rangle = \frac{kT}{2\pi\gamma} \ln\left(\frac{q_{max}}{q_{min}}\right) \tag{B3.6.18}$$

where a short and long wave length scale cut-off $q_{max}$ and $q_{min}$ have to be introduced to avoid the divergence at $q \to \infty$ and $q \to 0$.

The interfacial fluctuations broaden laterally averaged profiles. Within the convolution approximation (B3.6.15) one obtains a profile with the shape of the erfc function [49]:

$$w_{app}^2 = w_{int}^2 + \frac{kT}{4\gamma} \ln\left(\frac{q_{max}}{q_{min}}\right). \tag{B3.6.19}$$

Thus, the apparent interfacial width $w_{app}$, which is measured in simulations or experiments, is larger than the intrinsic width $w_{int}$ and depends via the wavevector cut-offs on the geometry considered. This can actually be used to measure the interfacial tension in computer simulations.

For a free interface the cut-off at large length scales is determined by the lateral patch size on which the interface is observed. In simulations this is set by the size of the simulation cell. In scattering experiments (e.g., neutron reflectivity) it is associated with the lateral coherence length of the beam. If the coexisting phases differ in density, gravitation will give rise to a large-scale cut-off for capillary waves [54]

$$q_{max} = \sqrt{\frac{g \Delta\rho}{\gamma}} \tag{B3.6.20}$$

where $\Delta\rho$ is the density difference and $g$ the gravitational constant. Similarly, interactions with boundaries (e.g. van der Waals forces) limit fluctuations and give rise to a large length scale cut-off. In this case the cut-off depends on the distance between the interface and the wall, and the cut-off imparts a dependence of the apparent interfacial width on the distance between the wall and the interface.

On short length scales the coarse-grained description breaks down, because the fluctuations which build up the (smooth) intrinsic profile and the fluctuations of the local interface position are strongly coupled and cannot be distinguished. The effective interface Hamiltonian can describe the properties only on length scales large compared with the width $1/q_{max} \sim w$ of the intrinsic profile. The absolute value of the cut-off is difficult to determine: the apparent profiles are experimentally accessible, but in order to use equation (B3.6.19) the width of an hypothetically flat interface without fluctuations has to be known. Polymer blends are suitable candidates for investigating this problem. Since the self-consistent field theory gives an accurate description of the interface profile except for fluctuations, it yields a quantitative description of the intrinsic, ideally flat, profile. The comparison with Monte Carlo simulations, which include fluctuations, then yields $q_{max}$. Simulations of a coarse-grained polymer blend by Werner *et al* find $q_{max} = 1.65/w_{int}$ [49] in the strong segregation limit, in rather good agreement with the value $q_{max} = 2/w_{int}$ suggested by analytical theory [55].

An important application of effective interface Hamiltonians are wetting phenomena. If a binary mixture is confined, the wall of the container will favour one component of the mixture, say A. This component forms an enrichment layer at the wall, while the B component is expelled from the wall region. Rather than describing the detailed composition profile at the wall, the effective interface Hamiltonian specifies the system configuration solely by the distance between the A-rich enrichment layer at the wall and the B component further away. This coarse graining concept is sketched in figure B3.6.3. The profile is distorted in the vicinity of the wall and this gives rise to a short-range effective interaction between the wall and the interface. The length scale of the interaction is set by the characteristic length scale of the (free) interface profile. Dispersion forces give rise to an additional long-ranged effective interaction, which decays like a power-law with the distance $l$. The effective interfacial Hamiltonian takes the form:

$$\mathcal{H}[l(\mathbf{r}_{\parallel})] = \int d^2 r_{\parallel} \left\{ \frac{\gamma(l)}{2} (\nabla l)^2 + g(l) \right\} \tag{B3.6.21}$$

where $g(l)$ denotes the effective interaction between the wall and a portion of the interface at a distance $l$. The first term corresponds to the capillary wave Hamiltonian. In general, the coefficient $\gamma(l)$ in front of the square gradient depends on the distance between the wall and the interface [56], because the intrinsic profile is distorted in the presence of the wall. Only for large distances $l \to \infty$ does the effective interfacial tension $\gamma(l)$ tend to its macroscopic value. The second term describes the effective interface potential between the wall and the interface. Depending on the shape of the interface potential $g(l)$ different situations are encountered [57]: if the effective potential exhibits a minimum in the vicinity of the wall, the interface is bound to the wall. This corresponds to a microscopically thin layer of the preferred component A at the wall. One says: A does not wet the wall. If $g(l)$ has a minimum at infinite distance $l \to \infty$ there is a macroscopically thick layer of A at the wall: component A wets the wall. The transition between both states is the wetting transition, which can be continuous (i.e. the thickness of the enrichment layer diverges upon approaching the transition temperature) or (most often) discontinuous (i.e. the thickness of the layer jumps from a microscopic value to a macroscopic one).

In the mean field considerations above, we have assumed a perfectly flat interface such that the first term in the Hamiltonian (B3.6.21) is ineffective. In fact, however, fluctuations of the local interface position are important, and its consequences have been studied extensively [57, 58].

### B3.6.3 Amphiphilic models

Another important class of materials which can be successfully described by mesoscopic and continuum models are amphiphilic systems. Amphiphilic molecules consist of two distinct entities that like different environments. Lipid molecules, for instance, comprise a polar head that likes an aqueous environment and one or two hydrocarbon tails that are strongly hydrophobic. Since the two entities are chemically joined together they cannot separate into macroscopically large phases. If these amphiphiles are added to a binary mixture (say, water and oil) they greatly promote the dispersion of one component into the other. At low amphiphile concentrations the molecules enrich at the interface so as to place their different ends into the corresponding phases. This displaces the water and the oil from the oil/water interface and greatly reduces the interfacial tension. At larger concentration of amphiphiles the molecules self-assemble into complex morphologies. These might either be isotropic (i.e. a microemulsion) or possess liquid crystalline order. The spatial structure is selected by a balance to minimize the contacts between the different entities and to fill space. Some of the

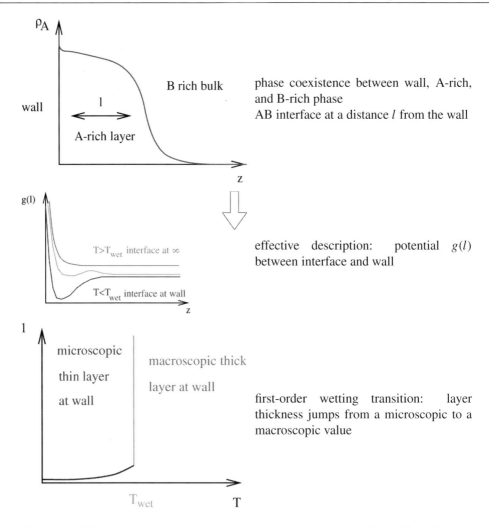

**Figure B3.6.3.** Sketch of the coarse-grained description of a binary blend in contact with a wall. (a) Composition profile at the wall. (b) Effective interaction $g(l)$ between the interface and the wall. The different potentials correspond to complete wetting, a first-order wetting transition and the non-wet state (from above to below). In case of a second-order transition there is no double-well structure close to the transition, but $g(l)$ exhibits a single minimum which moves to larger distances as the wetting transition temperature is approached from below. (c) Temperature dependence of the thickness $l$ of the enrichment layer at the wall. The jump of the layer thickness indicates a first-order wetting transition. In the case of a continuous transition the layer thickness would diverge continuously upon approaching $T_{wet}$ from below.

possible morphologies are displayed in figure B3.6.4. Analogous morphologies are encountered in polymeric systems involving block copolymers.

The relation between the architecture of the molecules and the spatial morphology into which they assemble has attracted longstanding interest, because of their importance in daily life. Lipid molecules are important constituents of the cell membrane. Amphiphilic molecules are of major importance for technological applications (e.g., in detergents and the food industry).

The large length scale on which the self-assembly occurs and the universality of the morphologies borne out in experiments on a large variety of different systems make mesoscopic and continuum models suitable

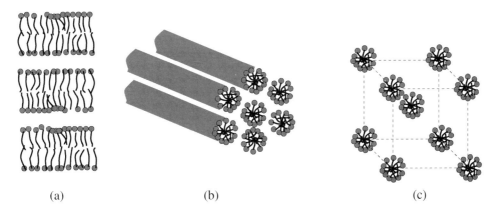

(a)                                          (b)                                          (c)

**Figure B3.6.4.** Illustration of three structured phases in a mixture of amphiphile and water. (a) Lamellar phase: the hydrophilic heads shield the hydrophobic tails from the water by forming a bilayer. The amphiphilic heads of different bilayers face each other and are separated by a thin water layer. (b) Hexagonal phase: the amphiphiles assemble into a rod–like structure, where the tails are shielded in the interior from the water and the heads are on the outside. The rods arrange on a hexagonal lattice. (c) Cubic phase: amphiphilic micelles with a hydrophobic centre order on a BCC lattice.

tools for investigating the underlying universal mechanism. Experiments suggest that many of the generic features can be captured by the amphiphilicity of the molecules. The models that have been employed can be broadly divided into models that aim at correlating the molecular architecture with the morphology and those models which investigate the generic phase behaviour and the influence of fluctuations.

*B3.6.3.1    Chain models*

The architecture of the lipid molecules or the diblock copolymers results in the typical amphiphilic properties, like surface activity and self-assembly. On the most qualitative level, understanding of the self-assembly in lipid systems [3] is provided by a characterization of the molecules as a simple geometrical object ('wedge') parameterized by its volume, the maximum chain length and the area per head group. The different phases result from simple geometric packing considerations. Similar arguments on the balance between chain stretching and interfacial tension yield the qualitative features of the phase diagrams in systems containing diblock copolymers [59].

Chain models capture the basic elements of the amphiphilic behaviour by retaining details of the molecular architecture. Ben-Shaul *et al* [60] and others [61] explored the organization of the hydrophobic portion in lipid micelles and bilayers by retaining the conformational statistics of the hydrocarbon tail within the RIS (rotational isomeric state) model [4, 5] while representing the hydrophilic/hydrophobic interface merely by an effective tension. By invoking a mean field approximation and calculating the properties of the tails by an enumeration of a large sample of conformations, they investigated the packing effects inside the hydrocarbon core for various detailed chain architectures. This mean field technique has been extended, for example, to include a modelling of the hydrophilic head and to study the self-assembly of lipids in aqueous solutions [62] or to investigate the absorption of proteins at surfaces covered with a polymer brush [63].

Many simulational approaches use a coarse-grained description of the amphiphiles by representing them via short-chain molecules on a lattice. The lattice is there only for computational convenience but is assumed to play no role otherwise. Typically, the number of lattice sites to model the amphiphiles is small and does not exceed 32. Each site is conceived as a small number of atomistic units along the amphiphilic molecule. A particularly popular model has been suggested by Larson [64]. There are two types of sites: hydrophilic and hydrophobic. Hydrophobic sites correspond to the oil or the hydrocarbon tail of the amphiphiles; hydrophilic sites represent the polar head of the amphiphiles or water. Oil and water are modelled as single-site entities. There is a short-range repulsion between unlike segments. The phase diagram of ternary oil/amphiphiles/water and binary amphiphile/water mixtures has been investigated by Monte Carlo simulations. Many phases observed in experiments (disordered, lamellar, hexagonal and even the gyroid phase) can be obtained as a function of temperature, composition and architecture of the amphiphile. Of course, special care has to be devoted to the study of finite-size effects. Typically, only a small number of unit cells of the spatially periodic structure fit into a simulation cell. If the size of the simulation box is close to a multiple of the unit cell size the stability of the phase might be greatly enhanced; if the size of the simulation box is incompatible with the spatially periodic structure the morphology is strongly distorted and its stability reduced. Very similar effects occur in nature if a spatially periodic structure is confined into a thin film.

A multitude of different variants of this model has been investigated using Monte Carlo simulations (see, for example [65]). The studies aim at correlating the phase behaviour with the molecular architecture and revealing the local structure of the aggregates. This type of model has also proven useful for studying rather complex structures (e.g., vesicles or pores in bilayers).

For structures with a high curvature (e.g., small micelles) or situations where orientational interactions become important (e.g., the gel phase of a membrane) lattice-based models might be inappropriate. Off-lattice models for amphiphiles, which are quite similar to their counterparts in polymeric systems, have been used to study the self-assembly into micelles [66], or to explore the phase behaviour of Langmuir monolayers [67] and bilayers. In those systems, various phases with a nematic ordering of the hydrophobic tails occur.

Since the amphiphilic nature is essential for the phase behaviour, systems of small molecules (e.g., lipid water mixtures) and polymeric systems (e.g., homopolymer copolymer blends) share many common features. Within the mean field approximation, the phase behaviour of block-copolymer models can conveniently be explored in the framework of the Gaussian chain model. The investigation of the self-assembly into various complex phases takes advantage of a Fourier decomposition of the spatially varying densities. The phase diagrams for pure diblock copolymers, binary blends of diblock copolymers and binary and ternary solutions have been investigated [44, 48]. These calculations reveal a rich variety of different morphologies as a function of the incompatibility, architecture and amount of homopolymer 'solvent'. In binary and ternary solutions, highly swollen phases are found in which the periodicity of the structure far exceeds the radius of gyration. An example of the possible phases in a ternary blend of two homopolymers and a symmetric diblock copolymer is presented in figure B3.6.5. At a fixed incompatibility one finds a complex phase diagram, including disordered homopolymer–rich phases, a symmetric lamellar phase $L$ and asymmetric swollen lamellar phases $L_A$ and $L_B$, which accommodate different amounts of the homopolymer components. Note that very similar phase diagrams are found in ternary oil water amphiphile mixtures. In the self-consistent field calculations not only the phase behaviour but also effective properties of the internal interfaces (e.g., the interfacial tension of bending moduli) are accessible [69]. The latter information might serve as input to effective interface Hamiltonians.

In general, the self-consistent field calculations are accurate for long polymers; for short chains however, fluctuations become important. Self-consistent field calculations and simulations of polymeric models show, for example, that the bending rigidity of copolymer-laden interfaces might be quite small for short chain lengths. In a region where the self-consistent field theory predicts a highly swollen lamellar phase, the lamellar order is unstable with respect to interfacial fluctuations and a microemulsion forms [70]. Polymeric microemulsions have been observed in simulations [71] and experiments [72], but they are much more common in small molecular amphiphilic systems. Similarly, fluctuations easily destroy the body-centred cubic arrangement of micelles found in the self-consistent field theory and lead to formation of a micellar solution.

These chain models are well suited to investigate the dependence of the phase behaviour on the molecular architecture and to explore the local properties (e.g., enrichment of amphiphiles at interfaces, molecular conformations at interfaces). In order to investigate the effect of fluctuations on large length scales or the shapes of vesicles, more coarse-grained descriptions have to be explored.

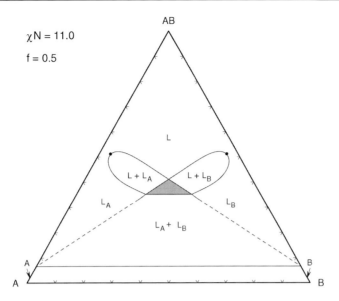

**Figure B3.6.5.** Phase diagram of a ternary polymer blend consisting of two homopolymers, A and B, and a symmetric AB diblock copolymer as calculated by self-consistent field theory. All species have the same chain length $N$ and the figure displays a cut through the phase prism at $\chi N = 11$ (which corresponds to weak segregation). The phase diagram contains two homopolymer-rich phases A and B, a symmetric lamellar phase $L$ and asymmetric lamellar phases, which are rich in the A component $L_A$ or rich in the B component $L_B$, respectively. From Janert and Schick [68].

### B3.6.3.2  Lattice models

A further step in coarse graining is accomplished by representing the amphiphiles not as chain molecules but as single site/bond entities on a lattice. The characteristic architecture of the amphiphile—the hydrophilic head and hydrophobic tail—is lost in this representation. Instead, the interaction between the different lattice sites, which represent the oil, the water and the amphiphile, have to be carefully constructed in order to bring about the amphiphilic behaviour.

As early as 1969, Wheeler and Widom [73] formulated a simple lattice model to describe ternary mixtures. The bonds between lattice sites are conceived as particles. A bond between two positive spins corresponds to water, a bond between two negative spins corresponds to oil and a bond connecting opposite spins is identified with an amphiphile. The contact between hydrophilic and hydrophobic units is made infinitely repulsive; hence each lattice site is occupied by either hydrophilic or hydrophobic units. These two states of a site are described by a spin variable $s_i$, which can take the values $+1$ and $-1$. Obviously, oil/water interfaces are always completely covered by amphiphilic molecules. The Hamiltonian of this Widom model takes the form

$$\mathcal{H} = -h \sum_i s_i - J \sum_{\langle ij \rangle} s_i s_j - 2M \sum_{\langle\langle ij \rangle\rangle} s_i s_j - M \sum_{\langle\langle\langle ij \rangle\rangle\rangle} s_i s_j \qquad (B3.6.22)$$

where $\langle ij \rangle$, $\langle\langle ij \rangle\rangle$, and $\langle\langle\langle ij \rangle\rangle\rangle$ denote nearest, next-nearest and fourth-nearest neighbours on the lattice, respectively. The first two terms correspond to the Ising Hamiltonian. $h$ acts as a chemical potential which favours positive spins, while $J$ controls the incompatibility between water and oil. If only these two terms were present the model would describe a simple binary mixture. The additional terms have to be incorporated to bring about the amphiphilic properties. The case of negative $M$ has been much investigated. In this case, the third term imparts some kind of bending rigidity to the oil/water interface, while the fourth term favours sequences of the form $(\cdots + + - - + + - - \cdots)$. This leads to the formation of lamellar phases which are not directly tied to the lattice spacing.

Slightly more complex models treat the water, the amphiphile and the oil as three distinct variables corresponding to the spin variables $S = +1, 0,$ and $-1$. The most general Hamiltonian with nearest-neighbour interactions has the form

$$\mathcal{H} = -\sum_{\langle ij \rangle} \{ J S_i S_j + K S_i^2 S_j^2 + C(S_i^2 S_j + S_j^2 S_i) \} - \sum_i \{ H S_i - \Delta S_i^2 \}. \qquad \text{(B3.6.23)}$$

This Blume–Emery–Griffiths (BEG) model [74] has been studied both by mean field calculations as well as by simulations. There is no pronounced difference between the amphiphile molecules $S = 0$, the oil or the water. Indeed, the model was first suggested in a quite different context. An extension of the model by Schick and Shih [75] includes an additional interaction of the form

$$\Delta \mathcal{H} = -L \sum_{(ijk)} S_i (1 - S_j^2) S_k \qquad \text{(B3.6.24)}$$

where $(ijk)$ denotes three sites in a line. For negative values of $L$ the term favours local conformations in which the amphiphile sits between the water and the oil. The model exhibits an oil-rich phase, a water-rich phase, a lamellar phase and a disordered phase, which exists between the lamellar phase and the oil–water coexistence. The disordered phase consists of water and oil domains separated by amphiphile sheets. It is homogeneous on large length scales, but shows—for certain parameter regions—oscillating structure on smaller length scales. These two length scales are associated with the structure of a microemulsion. The period of the oscillations characterize the local domain size in the microemulsion, while this nearly liquid-crystalline order 'dephases' on larger length scales. This defines the persistence length $\xi$. The latter length scale is mesoscopic (i.e. of the order of 100 Å), while the former is roughly the size of the molecules.

Lattice models have been studied in mean field approximation, by transfer matrix methods and Monte Carlo simulations. Much interest has focused on the occurrence of a microemulsion. Its location in the phase diagram between the oil-rich and the water-rich phases, its structure and its wetting properties have been explored [76]. Lattice models reproduce the reduction of the surface tension upon adsorption of the amphiphiles and the progression of phase equilibria upon increasing the amphiphile concentration. Spatially periodic (lamellar) phases are also describable by lattice models. However, the structure of the lattice can interfere with the properties of the periodic structures.

### B3.6.3.3  Continuum models

An even coarser description is attempted in Ginzburg–Landau-type models. These continuum models describe the system configuration in terms of one or several, continuous order parameter fields. These fields are thought to describe the spatial variation of the composition. Similar to spin models, the amphiphilic properties are incorporated into the Hamiltonian by construction. The Hamiltonians are motivated by fundamental symmetry and stability criteria and offer a unified view on the general features of self-assembly. The universal, generic behaviour—the possible morphologies and effects of fluctuations, for instance—rather than the description of a specific material is the subject of these models.

An important example is the one-order-parameter model invented by Gompper and Schick [77], which describes a ternary mixture in terms of the density difference $\phi$ between water and oil:

$$\mathcal{F}[\phi(r)] = \int dr \{ f(\phi) + g(\phi) |\nabla \phi|^2 + c |\triangle \phi|^2 \}. \qquad \text{(B3.6.25)}$$

The first two terms resemble the Ginzburg–Landau Hamiltonian for the polymeric systems. $f(\phi)$ describes the bulk free energy and there is a gradient square term to account for the free-energy costs of a spatially

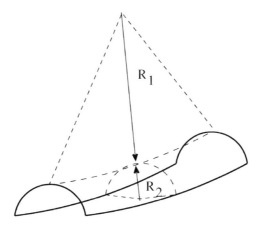

**Figure B3.6.6.** Illustration of the two principal radii of curvature for a membrane.

varying order parameter profile. In principle, the functions $f$, $g$ and $c$ of this Ginzburg–Landau expansion can be derived from a more microscopic model (e.g., the lattice models of the previous section). Such a derivation serves to relate the input parameters of the Ginzburg–Landau theories to microscopic parameters (e.g., length of the amphiphile) of the underlying model. The coefficients $f$, $g$ and $c$ are also related to the scattering intensity and, hence, some guidance from the experiment is available on how to choose them. Though the amphiphiles do not occur explicitly in the description, they determine the density dependence of the functions $f$ and $g$. In order to model three-phase coexistence between an oil-rich phase, a water-rich phase and a microemulsion with roughly equal amounts of oil and water, the function $f$ has to exhibit three minima. The amphiphiles decrease the free-energy cost of interfaces. This is modelled by a negative value of $g$ in some intermediate composition range. This favours the formation of interfaces and, therefore, the third term (with $c > 0$) is required to ensure thermodynamic stability.

By virtue of their simple structure, some properties of continuum models can be solved analytically in a mean field approximation. The phase behaviour, interfacial properties and the wetting properties have been explored. The effect of fluctuations is investigated in Monte Carlo simulations as well as non-equilibrium phenomena (e.g., phase separation kinetics). Extensions of this one-order-parameter model are described in the review by Gompper and Schick [76]. A very interesting feature of these models is that effective quantities of the interface—like the interfacial tension and the bending moduli—can be expressed as a functional of the order parameter profiles across an interface [78]. These quantities can then be used as input for an even more coarse-grained description.

### B3.6.3.4 Random interface models

Most characteristics of amphiphilic systems are associated with the alteration of the interfacial structure by the amphiphile. Addition of amphiphiles might reduce the free-energy costs by a dramatic factor (up to $10^{-2}$ dyn cm$^{-1}$ in the oil/water/amphiphile mixture). Adding amphiphiles to a solution or a mixture often leads to the formation of a microemulsion or spatially ordered phases. In many aspects these systems can be conceived as an assembly of internal interfaces. The interfaces might separate oil and water in a ternary mixture or they might be amphiphilic bilayers in binary solutions. Random interface models study the large-scale structure of amphiphilic systems by describing the configuration of the local, instantaneous interfacial position.

The effective free energy of the system of interfaces takes the general form [79–81]

$$\mathcal{H} = \int dS \{\gamma + \lambda_s H + 2\kappa H^2 + \bar{\kappa} K\} \tag{B3.6.26}$$

where $dS$ denotes the surface element, $H$ the local mean curvature and $K$ the local Gaussian curvature of the interface. The latter two quantities are related to the two principal radii of curvature via $2H = 1/R_1 + 1/R_2$ and $K = 1/R_1 R_2$. The interfacial tension $\gamma$ controls the area of the interface. $\lambda_s$ describes a spontaneous curvature of the interface, which is related to an asymmetry of the interface. This might occur even in a bilayer, when it is composed of an amphiphilic mixture and the two sheets have different compositions. The coefficients $\kappa$ and $\bar{\kappa}$ characterize the bending rigidity and the saddle-splay modulus, respectively. If the interface is closed, the Gauss–Bonnet theorem relates $\int dS\, K = 2\pi \chi_E$ to the Euler characteristic $\chi_E$. Since this quantity is a topological invariant, the last term in the Hamiltonian can be omitted if the topology of the interface does not change (e.g., in the case of a vesicle).

One can regard the Hamiltonian (B3.6.26) above as a phenomenological expansion in terms of the two invariants $K$ and $H$ of the surface. To establish the connection to the effective interface Hamiltonian (B3.6.16) it is instructive to consider the limit of an almost flat interface. Then, the local interface position $u$ can be expressed as a single-valued function of the two lateral parameters $u(r_\parallel)$. In this Monge representation the interface Hamiltonian can be written as

$$\mathcal{H} = \int d^2 r_\parallel \left\{ \frac{\gamma}{2} |\nabla u|^2 + \frac{\kappa}{2} |\triangle u|^2 \right\}. \tag{B3.6.27}$$

Among the different problems which have been tackled with random interface Hamiltonians are the following. (i) The phase diagram of the random interface Hamiltonian has been explored by Huse and Leibler [82]. The phase diagram comprises a droplet phase, in which the minority component is dissolved into the matrix of the majority component, disordered phases and lamellar phases. (ii) Much interest has focused on the role of fluctuations. In the presence of a wall or another interface, the fluctuations of the local interface position are restricted. This gives rise to an entropic repulsion between the fluctuating interface and confining boundaries (Helfrich interaction [83]). (iii) In order to avoid the free-energy cost of a rim, membranes close up to form vesicles. The shapes of vesicles as a function of the bending rigidity and the pressure difference between the vesicle's interior and the outside have been mapped out [84].

### B3.6.4   Applications to dynamic phenomena

Though this entry has focused on equilibrium properties, mesoscopic and continuum models in chemical physics can also describe non-equilibrium phenomena, and we shall mention some techniques briefly.

Mesoscopic models can often be treated by molecular dynamics simulations. This method generates a realistic (Hamiltonian) trajectory in the phase space of the model from which information about the equilibrium dynamics can readily be extracted. The application to non-equilibrium phenomena (e.g., the kinetics of phase separation) is, in principle, straightforward.

Exploring the hydrodynamic behaviour of complex fluids with conventional molecular dynamics models poses a challenge to computational resources, because the hydrodynamic behaviour appears only on large time and length scales. Coarse-grained models [85–87] have been explored in which a particle does not correspond to a molecule or a small number of atoms but rather to a fluid element. These 'fluid particles' [86] interact with an extremely soft potential, which does not diverge as two effective particles approach each other (see the Lennard-Jones potential (B3.6.1)) but increases only linearly with the interparticle distance. This very soft repulsive potential allows for very large time steps in a molecular dynamics simulation. As the particles correspond to coarse-grained fluid elements they do not conserve energy when they collide. This provides a motivation for a dissipative friction force and a random force. The strength of the friction and the noise are related by a fluctuation–dissipation theorem, which ensures that the equilibrium distribution corresponds to the canonical ensemble. Unlike the standard implementation of noise and friction forces in molecular dynamics schemes, noise and friction in the dissipative particle dynamics do not act on the velocity of a single particle but on pairs of particles. In this way momentum is conserved. The macroscopic behaviour is not diffusive but hydrodynamic (note, however, that the energy is not conserved and there is no transport equation of the energy). This method promises to be an efficient way to study dynamic effects on the mesoscopic scale of complex fluids. An application of dissipative particle dynamics to a binary homopolymer blend is described in [87].

Monte Carlo schemes generate a stochastic trajectory through phase space (see the entry about statistical mechanical simulations; B3.3). If the Monte Carlo moves resemble the configurational changes in a realistic dynamics (e.g., the conformations evolve via small displacements of particles) some dynamical information can be gained. Since there is no momentum in Monte Carlo simulations the dynamics is diffusive. However, many Monte Carlo algorithms employ moves that involve rather large changes in the system conformation (e.g., deletion of a molecule and subsequent insertion at a random position). These 'unphysical' moves are extremely efficient in propagating the system through configuration space, but they do not allow for a dynamic interpretation of the trajectory.

A lattice scheme which does capture hydrodynamic behaviour is the lattice Boltzmann method [88–91]. This method has been devised as an effective numerical technique of computational fluid dynamics. The basic variables are the time-dependent probability distributions $f_\alpha(x, t)$ of a velocity class $\alpha$ on a lattice site $x$. This probability distribution is then updated in discrete time steps using a deterministic local rule. A careful choice of the lattice and the set of velocity vectors minimizes the effects of lattice anisotropy. This scheme has recently been applied to study the formation of lamellar phases in amphiphilic systems [92, 93].

Analytic techniques often use a time-dependent generalization of Landau–Ginzburg free-energy functionals. The different universal dynamic behaviours have been classified by Hohenberg and Halperin [94]. In the simple example of a binary fluid (model B) the concentration difference can be used as an order parameter $m$. A gradient in the local chemical potential $\mu(r) = \delta F / \delta m(r)$ gives rise to a current $j$

$$j = -\Lambda \nabla \frac{\delta F}{\delta m(r)} \qquad (B3.6.28)$$

which strives to minimize the free energy. The kinetic coefficient $\Lambda$ denotes a phenomenological constant which sets the time scale. In complex fluids (e.g., polymer blends) the relation between the gradient of the chemical potential and the current is non-local and the kinetic coefficient has to be generalized [95]. If the order parameter is conserved (e.g., in the demixing of a binary mixture) the change of the order parameter and the current are related by the continuity equation

$$\frac{\partial m(r, t)}{\partial t} = \nabla j = -\nabla \Lambda \nabla \frac{\delta F}{\delta m(r)}. \qquad (B3.6.29)$$

This time development of the order parameter is completely deterministic; when the equilibrium $\mu(r) = \text{const}$ is reached the dynamics comes to rest. Noise can be added to capture the effect of thermal fluctuations. This leads to a Langevin dynamics for the order parameter.

Time-dependent Ginzburg–Landau models can be generalized to models with or without conserved order parameters. Also, the effect of additional conservation laws (for example, the inclusion hydrodynamic effects) has been explored. More complicated forms of the free-energy functional can be used to incorporate more details of the systems and alleviate the restriction to small order parameters inherent in the Ginzburg–Landau expansion. Shi and Noolandi [96] have used the free energy functional of the self-consistent field theory to explore fluctuations in spatially structured phases of diblock copolymers. A similar free-energy functional is employed by Fraaije and co-workers [97] to study the kinetics of self-assembly in amphiphilic systems. Extensions of the time-dependent Ginzburg–Landau equation to a formal scheme for the time evolution of non-equilibrium systems in terms of a set of coarse-grained variables have been explored [98].

## References

[1] Landau L D and Lifshitz E M 1959 *Fluid Mechanics (Course of Theoretical Physics vol 6)* (Oxford: Pergamon)
[2] Landau L D and Lifshitz E M 1970 *Theory of Elasticity (Course of Theoretical Physics vol 7)* (Oxford: Pergamon)
[3] Israelachvili J 1991 *Intermolecular and Surface Forces* 2nd edn (New York: Academic)
[4] Flory P J 1953 *Principles of Polymer Chemistry* (Ithaca, NY: Cornell University Press)
[5] Flory P J 1969 *Statistical Mechanics of Chain Molecules* (New York: Wiley–Interscience)
[6] Grosberg A Y and Khokhlov A R 1994 *Statistical Physics of Macromolecules (AIP Series in Polymers and Complex Materials)* (New York: AIP)
[7] de Gennes P G 1979 *Scaling Concepts in Polymer Physics* (Ithaca, NY: Cornell University Press)
[8] Freed K F 1987 *Renormalization Group Theory of Macromolecules* (New York: Wiley–Interscience)
[9] des Cloizeaux J and Jannink G 1990 *Polymers in Solution: Their Modelling and Structure* (Oxford: Oxford Science Publications)
[10] Schäfer L 1999 *Excluded Volume Effects in Polymer Solutions* (Berlin: Springer)
[11] de Gennes P G 1972 Exponents for the excluded volume problem as derived by the Wilson method *Phys. Lett.* A **38** 339
[12] Kremer K and Binder K 1988 Monte Carlo simulations of lattice models for macromolecules *Comp. Phys. Rep.* **7** 259
[13] Sokal A D 1995 *Monte Carlo and Molecular Dynamics Simulations in Polymer Science* ed K Binder (New York: Oxford University Press) ch 3
[14] Grassberger P 1997 Pruned-enriched Rosenbluth method: simulations of theta polymers of chain length up to 1,000,000 *Phys. Rev.* E **56** 3682
[15] Frenkel D, Mooij G C A M and Smit B 1992 Novel scheme to study structural and thermal properties of continuously deformable molecules *J. Phys.: Condens. Matter* **4** 3053
[16] Laso M, dePablo J J and Suter U W 1992 Simulation of phase equilibria for chain molecules *J. Chem. Phys.* **97** 2817
[17] Consta S, Wilding N B, Frenkel D and Alexandrowicz Z 1999 Recoil growth: an efficient simulation method for multi-polymer systems *J. Chem. Phys.* **110** 3220
[18] Carmesin I and Kremer K 1988 The bond fluctuation method—a new effective algorithm for the dynamics of polymers in all spatial dimensions *Macromolecules* **21** 2819
[19] Doi M and Edwards S F 1986 *The Theory of Polymer Dynamics* (Oxford: Clarendon)
[20] Deutsch J M and Cates M E 1986 Conjectures on the statistics of ring polymers *J. Physique* **47** 2121
[21] Khokhlov A R and Nechaev S K 1985 Polymer chain in an array of obstacles *Phys. Lett.* A **112** 156
[22] Müller M, Wittmer J P and Cates M E 1996 Topological effects in ring polymers: a computer simulation study *Phys. Rev.* E **53** 5063
[23] Wertheim M S 1987 Thermodynamic perturbation theory of polymerization *J. Chem. Phys.* **87** 7323
[24] Yethiraj A and Woodward C E 1995 Monte Carlo density functional theory of nonuniform polymer melts *J. Chem. Phys.* **102** 5499
[25] Kierlik E and Rosinberg M L 1993 Perturbation density functional theory for polyatomic fluids III: application to hard chain molecules in slitlike pores *J. Chem. Phys.* **100** 1716
[26] Sen S, Cohen J M, McCoy J D and Curro J G 1994 The structure of a rotational isomeric state alkane melt near a hard wall *J. Chem. Phys.* **101** 9010
[27] Kremer K and Grest G S 1990 Dynamics of entangled linear polymer melts: a molecular-dynamics simulation *J. Chem. Phys.* **92** 5057
[28] Bishop M, Ceperley D, Frisch H L and Kalos M H 1980 Investigation of static properties of model bulk polymer fluids *J. Chem. Phys.* **72** 3228
[29] Gay J G and Berne B J 1981 Modification of the overlap potential to mimic a linear site–site potential *J. Chem. Phys.* **74** 3316
[30] Edwards S F 1966 The theory of polymer solutions at intermediate concentration *Proc. Phys. Soc.* **88** 265
[31] Saito N, Takahashi K and Yunoli Y 1967 The statistical mechanical theory of stiff chains *J. Phys. Soc. Japan* **22** 219
[32] Flory P J 1941 *J. Chem. Phys.* **9** 660
[33] Huggins M L 1941 *J. Chem. Phys.* **9** 440
[34] Orwoll R A and Arnold P A 1996 Polymer–solvent interaction parameter χ *Physical Properties of Polymers, Handbook* ed J E Mark (Woodbury, NY: AIP) ch 14
[35] Foreman K W and Freed K F 1998 Lattice cluster theory of multicomponent polymer systems: chain semiflexibility and specific interactions *Adv. Chem. Phys.* **103** 335
[36] Schweizer K S and Curro J G 1997 Integral equation theories of the structure, thermodynamics and phase transitions of polymer fluids *Adv. Chem. Phys.* **98** 1
[37] Müller M 1999 Miscibility behavior and single chain properties in polymer blends: a bond fluctuation model study *Macromol. Theory Simul.* **8** 343

[38] Escobedo F A and de Pablo J J 1999 On the scaling of the critical solution temperature of binary polymer blends with chain length *Macromolecules* **32** 900

[39] Taylor-Maranas J K, Debenedetti P G, Graessley W W and Kumar S K 1997 Compressibility effects in neutron scattering by polymer blends *Macromolecules* **30** 6943

[40] Deutsch H-P and Binder K 1993 Mean-field to Ising crossover in the critical behavior of polymer mixtures—a finite size scaling analysis of Monte Carlo simulations *J. Physique* II **3** 1049

[41] Helfand E and Tagami Y 1972 Theory of the interface between immiscible polymers *J. Polym. Sci. Polym. Lett.* **9** 741
Helfand E and Tagami Y 1972 *J. Chem. Phys.* **56** 3592

[42] Hong K M and Noolandi J 1981 Theory of inhomogeneous multicomponent polymer systems *Macromolecules* **14** 727

[43] Fleer G J, Cohen Stuart M A, Scheutjens J M H M, Cosgrove T and Vincent B 1993 *Polymers at Interfaces* (London: Chapman and Hall)

[44] Matsen M W and Schick M 1994 Stable and unstable phases of a diblock copolymer melt *Phys. Rev. Lett.* **72** 2660

[45] Broseta D, Fredrickson G H, Helfand E and Leibler L 1990 Molecular-weight effects and polydispersity effects at polymer–polymer interfaces *Macromolecules* **23** 132

[46] Helfand E, Bhattacharjee S M and Fredrickson G H 1989 Molecular weight dependence of the polymer interfacial tension and concentration profile *J. Chem. Phys.* **91** 7200

[47] Semenov A N 1996 Theory of long-range interactions in polymer systems *J. Physique* II **6** 1759

[48] Cahn J W and Hilliard J E 1958 Free energy of a nonuniform system: I. Interfacial free energy *J. Chem. Phys.* **28** 258

[49] Werner A, Schmid F, Müller M and Binder K 1997 Anomalous size-dependence of interfacial profiles between coexisting phases of polymer mixtures in thin film geometry: a Monte-Carlo study *J. Chem. Phys.* **107** 8175

[50] Werner A, Schmid F, Müller M and Binder K 1999 Intrinsic profiles and capillary waves at homopolymer interfaces: a Monte Carlo Study *Phys. Rev. E* **59** 728

[51] Lacasse M D, Grest G S and Levine A J 1998 Capillary–wave and chain length effects at polymer/polymer interfaces *Phys. Rev. Lett.* **80** 309

[52] Ginzburg V L 1960 *Sov. Phys. Solid State* **1** 1824
deGennes P G 1977 Qualitative features of polymer demixtion *J. Physique Lett.* **38** 441

[53] Jasnow D 1984 Critical phenomena at interfaces *Rep. Prog. Phys.* **47** 1059

[54] Rowlinson J S and Widom B 1982 *Molecular Theory of Capillarity* (Oxford: Clarendon)

[55] Semenov A N 1994 Scattering of statistical structure of polymer–polymer interfaces *Macromolecules* **27** 2732

[56] Jin A J and Fisher M E 1993 Effective interface Hamiltonians for short-range critical wetting *Phys. Rev.* **47** 7365

[57] Schick M 1990 Introduction to wetting phenomena *Les Houches Lectures: Liquid at interfaces* ed J Charvolin, J F Joanny and J Zinn-Justin (Amsterdam: Elsevier)

[58] Dietrich S 1988 *Phase Transitions and Critical Phenomena* vol 12, ed C Domb and J Lebowitz (London: Academic)

[59] Semenov A N 1985 Contribution to the theory of microphase layering in block-copolymer melts *Sov. Phys.–JETP* **61** 733

[60] Ben-Shaul A, Szleifer I and Gelbart W M 1985 Chain organization and thermodynamics in micelles and bilayers: I. Theory *J. Chem. Phys.* **83** 3597

[61] Gruen D W R 1984 A model for the chains in amphiphilic aggregates: I. Comparison with a molecular dynamics simulation of a bilayer *J. Phys. Chem.* **89** 645

[62] Müller M and Schick M 1998 Calculation of the phase behavior of lipids *Phys. Rev. E* **57** 6973

[63] Szleifer I and Carignano M A 1996 Tethered polymer layers *Adv. Chem. Phys.* **94** 165

[64] Larson R G 1996 Monte Carlo simulations of the phase behavior of surfactant solutions *J. Physique* II **6** 1441

[65] Liverpool T B and Bernardes A T 1995 Monte Carlo simulation of the formation of layered structures and membranes by amphiphiles *J. Physique* II **5** 1003
Liverpool T B and Bernardes A T 1995 Monte Carlo simulation of the formation of layered structures and membranes by amphiphiles *J. Physique* II **5** 1457

[66] Smit B, Schlijper A G, Rupert L A M and van Os N M 1994 *J. Chem. Phys.* **94** 6933
Karaborni S, Esselink K, Hilbers P A J, Smit B, Karthäuser J, van Os N M and Zana R 1994 Simulating the self-assembly of Gemini (dimeric) surfactants *Science* **266** 254

[67] Stadler C and Schmid F 1999 Phase behavior of grafted chain molecules: influence of head size and chain length *J. Chem. Phys.* **110** 9697

[68] Janert P K and Schick M 1997 Phase behavior of ternary homopolymer/diblock blends: microphase unbinding in the symmetric system *Macromolecules* **30** 3916

[69] Matsen M W 1999 Elastic properties of a diblock copolymer monolayer and their relevance to bicontinuous microemulsion *J. Chem. Phys.* **110** 4658

[70] de Gennes P G and Taupin C 1982 Microemulsion and the flexibility of oil/water interfaces *J. Phys. Chem.* **86** 2294

[71] Müller M and Schick M 1996 Bulk and interfacial thermodynamics of a symmetric, ternary homopolymer–copolymer mixture: a Monte Carlo study *J. Chem. Phys.* **105** 8885

[72] Fredrickson G H and Bates F S 1997 Design of bicontinuous polymeric microemulsions *J. Polym. Sci. B* **35** 2775

[73] Wheeler J C and Widom B 1968 *J. Am. Chem. Soc.* **90** 3064

[74] Blume M, Emry V and Griffiths R B 1971 Ising model for the $\lambda$ transition and phase separation in $He^3$–$He^4$ mixtures *Phys. Rev. A* **4** 1071

[75] Schick M and Shih W-H 1987 Simple microscopic model of a microemulsion *Phys. Rev. Lett.* **59** 1205
[76] Gompper G and Schick M 1994 Self-assembling amphiphilic systems *Phase Transitions and Critical Phenomena* vol 16, ed C Domb and J Lebowitz (New York: Academic)
[77] Gompper G and Schick M 1990 Correlation between structural and interfacial properties in amphiphilic systems *Phys. Rev. Lett.* **65** 1116
[78] Gompper G and Zschocke S 1991 Elastic properties of interfaces in a Ginzburg–Landau theory of swollen micelles, droplet crystals and lamellar phases *Euro. Phys. Lett.* **16** 731
[79] Canham P B 1970 The minimum energy of bending as a possible explanation of the biconcave shape of the human red blood cell *J. Theoret. Biol.* **26** 61
[80] Helfrich W 1973 Elastic properties of lipid bilayers: theory and possible experiments *Z. Naturf.* c **28** 693
[81] Evans E A 1974 Bending resistance and chemically induced moments in membrane bilayers *Biophys. J.* **14** 923
[82] Huse D A and Leibler S 1988 Phase behavior of an ensemble of nonintersecting random fluid films *J. Physique* **49** 605
[83] Helfrich W 1977 Steric interaction of fluid membranes in multilayer systems *Z. Naturf.* a **33** 305
[84] Gompper G and Kroll D M 1995 Phase diagram and scaling behavior of fluid vesicles *Phys. Rev. E* **51** 514
[85] Hoogerbrugge P J and Koelman J M V A 1992 Simulating microscopic hydrodynamic phenomena with dissipative particle dynamics *Euro. Phys. Lett.* **19** 155
[86] Espanol P and Warren P 1995 Statistical mechanics of dissipative particles dynamics *Euro. Phys. Lett.* **30** 191
Espanol P 1996 Dissipative particle dynamics for a harmonic chain: a first-principles derivation *Phys. Rev. B* **53** 1572
[87] Groot R D and Warren P B 1997 Dissipative particles dynamics: bridging the gap between atomistic and mesoscopic simulation *J. Chem. Phys.* **107** 4423
[88] McNamara G R and Zanetti G 1998 Use of the Boltzmann equation to simulate lattice–gas automata *Phys. Rev. Lett.* **61** 2332
[89] Quian Y H, D'Humieres D and Lallemand P 1992 Lattice BGK models for Navier–Stokes equation *Euro. Phys. Lett.* **17** 479
[90] Doolen G D (ed.) 1990 *Lattice Gas Methods for Partial Differential Equations* (Redwood City, CA: Addison-Wesley)
[91] Doolen G D (ed.) 1991 *Lattice Gas Methods: Theory, Applications, and Hardware* (Cambridge, MA: MIT)
[92] Gonnella G, Orlandini E and Yeomans J M 1997 Spinodal decomposition to a lamellar phase: effect of hydrodynamic flow *Phys. Rev. Lett.* **78** 1695
[93] Theissen O, Gompper G and Kroll D M 1998 Lattice–Boltzmann model of amphiphilic systems *Euro. Phys. Lett.* **42** 419
[94] Hohenberg P C and Halperin B I 1977 Theory of dynamic critical phenomena *Rev. Mod. Phys.* **49** 435
[95] Binder K 1983 Collective diffusion, nucleation, and spinodal decomposition in polymer mixtures *J. Chem. Phys.* **79** 6387
[96] Shi A C, Noolandi J and Desai R C 1996 Theory of anisotropic fluctuations in ordered block copolymer phases *Macromolecules* **29** 6487
[97] Fraaije J G E M 1993 Dynamic density functional theory for micro-phase separation kinetics of block copolymer melts *J. Chem. Phys.* **99** 9202
Fraaije J G E M 1994 *J. Chem. Phys.* **100** 6984 (erratum)
[98] Öttinger H C 1997 General projection operator formalism for the dynamics and thermodynamics of complex fluids *Phys. Rev. E* **57** 1416

## Further Reading

Grosberg A Y and Khokhlov A R 1995 *Statistical Physics of Macromolecules (AIP Series in Polymers and Complex Materials)* (New York: AIP)
Schäfer L 1999 *Excluded Volume Effects in Polymer Solutions* (Berlin: Springer)
Fleer G J, Cohen Stuart M A, Scheutjens J M H M, Cosgrove T and Vincent B 1993 *Polymers at Interfaces* (London: Chapman and Hall)
Rowlinson J S and Widom B 1982 *Molecular Theory of Capillarity* (Oxford: Clarendon)
Dietrich S 1988 *Phase Transitions and Critical Phenomena* vol 12, ed C Domb and J Lebowitz (London: Academic)
Gompper G and Schick M 1994 Self-assembling amphiphilic systems *Phase Transitions and Critical Phenomena* vol 16, ed C Domb and J L Lebowitz (New York: Academic)